QL805 .F86 2001
Functional anatomy of the vertebrates : an evolutionary perspective

D0138643

is

A Harcourt Higher Learning Company

Now you will find Saunders College Publishing's distinguished innovation, leadership, and support under a different name . . . a new brand that continues our unsurpassed quality, service, and commitment to education.

We are combining the strengths of our college imprints into one worldwide brand: Harcourt

Our mission is to make learning accessible to anyone, anywhere, anytime—reinforcing our commitment to lifelong learning.

We are now Harcourt College Publishers. Ask for us by name.

One Company
"Where Learning Comes to Life."

www.harcourtcollege.com
www.harcourt.com

Publisher: Emily Barrosse
Acquisitions Editor: Nedah Rose
Marketing Strategist: Kathleen Sharp
Developmental Editor: Melanie Cann
Project Editor: Theodore Lewis
Production Manager: Alicia Jackson
Art Director: Caroline McGowan

FUNCTIONAL ANATOMY OF THE VERTEBRATES
An Evolutionary Perspective, third edition

ISBN: 0-03-022369-5
Library of Congress Card Number: 00-107968

Copyright © 2001, 1994, 1987 by Harcourt, Inc.

All rights reserved. No part of this publication may be reproduced or transmitted in any form or by any means, electronic or mechanical, including photocopy, recording, or any information storage and retrieval system, without permission in writing from the publisher.

Requests for permission to make copies of any part of the work should be mailed to: Permissions Department, Harcourt, Inc., 6277 Sea Harbor Drive, Orlando, Florida 32887-6777.

Portions of this work were previously published.

Address for domestic orders:
Harcourt College Publishers
6277 Sea Harbor Drive
Orlando, FL 32887-6777
1-800-782-4479
e-mail collegesales@harcourt.com

Address for international orders:
International Customer Service, Harcourt, Inc.
6277 Sea Harbor Drive, Orlando FL 32887-6777
(407) 345-3800
Fax (407) 345-4060
e-mail hbintl@harcourt.com

Address for editorial correspondence:
Harcourt College Publishers
Public Ledger Building, Suite 1250
150 S. Independence Mall West
Philadelphia, PA 19106-3412

Web Site Address
http://www.harcourtcollege.com

Printed in the United States of America
0123456789 048 10 987654321

Photo Credits

Cover: Humpback whales. © *Amos Nachoum*

Endsheets: Skeletons of various osteichthyan vertebrates. © *Lance Grande and John Weinstein*

Part I
Inset: Preening flamingo. © *Ron Sanford/Tony Stone*
Thumbstrip: Pink flamingo feathers. © *John Warden/Tony Stone*

Part II
Inset: Red-eyed tree frog. © *Tim Flach/Tony Stone*
Thumbstrip: Retina of a South American lungfish. *William E. Bemis*

Part III
Inset: Giraffe. *William E. Bemis*
Thumbstrip: Giraffe skin pattern. © *Siede Preis/Animal Patterns*

Part IV
Inset: Salmon egg embryos. © *Natalie Fobes/Tony Stone*
Thumbstrip: Yellowspot emperor fish. © *Stuart Westmorland/Tony Stone*

Part V
Inset: Varying hare. © *Tim Fitzharris*
Thumbstrip: Detail of the Weberian apparatus of a fallfish. *William E. Bemis*

Functional Anatomy *of the* Vertebrates

An Evolutionary Perspective

third edition

To the teachers, mentors, and researchers who have personally influenced our professional development:

Hobart Muir Smith
University of Illinois

Glenn Northcutt
University of California at San Diego

Alfred Sherwood Romer
Harvard University

Gareth Nelson
formerly of the American Museum of Natural History

Functional Anatomy *of the* Vertebrates

An Evolutionary Perspective

third edition

Karel F. Liem
Museum of Comparative Zoology
Harvard University

William E. Bemis
University of Massachusetts, Amherst

Warren F. Walker, Jr.
Oberlin College

Lance Grande
Field Museum of Natural History, Chicago

Art Development by
William E. Bemis and William B. Sillin

HARCOURT COLLEGE PUBLISHERS
Fort Worth Philadelphia San Diego New York Orlando Austin
San Antonio Toronto Montreal London Sydney Tokyo

Preface

Purpose

Increasingly, comparative anatomy is becoming one of the most integrative fields in biology. It now encompasses traditional descriptive anatomy, embryology, functional studies of structure, physiology, systematics, paleontology, behavior, and ecology all within a phylogenetic context so that we can better understand the history and diversity of vertebrate life. To better interpret and integrate these diverse fields, Drs. Liem and Walker are privileged to have two distinguished biologists join this edition of *Functional Anatomy of the Vertebrates* as coauthors: Dr. William E. Bemis, of the University of Massachusetts at Amherst, and Dr. Lance Grande, of the Field Museum of Natural History in Chicago.

Our purpose and basic approach remain the same as in the first two editions: to help students understand vertebrate form and function as well as the modes of vertebrate life that have occurred during 500 million years of evolution. Classical comparative anatomy emphasizes the morphological changes, but structures **do** things; they perform functions that must be integrated with the functions of other body parts, with the animal's lifestyle, and with the environment in which the animal lives. In a sense, form and function are two sides of the same coin; one does not exist without the other. Functional anatomy not only is the description of form and function but also asks how form–function changes have come about. By studying form and function together in a modern comparative anatomy course, students will see how the threads of form, function, and ecology weave together to form a coherent tapestry. This is one of today's most challenging and dynamic fields of research.

We assume that students taking a course in comparative functional anatomy have had a college course in biology or zoology and therefore are familiar with the basic principles of cell and developmental biology, physiology, evolutionary theory, and ecology. We have not developed these subjects in our text but rather summarized essential points and then built upon them. The physics necessary to understand functional anatomy is presented as we develop the subject.

Distinctive Features and Approach

The most distinctive feature of this book is the discussion of structures and their functional analysis together in a phylogenetic context. To do this, we have organized the book into five parts. In **Part I, Background for the Study of Vertebrate Anatomy,** we discuss topics that students need in order to understand the evolution of the ten organ systems. In Chapter 1, *Introduction,* we begin with examples of functional anatomy to illustrate our approach, go on to consider the relation between functional anatomy and development, lay the necessary groundwork to interpret modern methods of phylogenetic analysis, and end with a brief explanation of anatomical terminology. Chapter 2, *Phylogenetic Relationships of Chordates and Craniates,* places craniates (nearly all are also vertebrates) in a phylogenetic context and explains the origin of their distinctive characteristics. In Chapter 3, *Diversity and Phylogenetic History of Craniates,* we begin with an overview of the living clades of craniates and go on to discuss the major groups (including important fossil

Cleared and double-stained skeletons (cartilage, blue; bone, red) of various osteichthyans (i.e., actinopterygians and sarcopterygians).

taxa), their diagnostic characters, evolutionary relationships, diversity, and modes of life. Chapter 4, *Early Development and Comparative Embryology,* lays the essential foundations for understanding adult structure. We trace development through the organogenesis of the nervous system and sense organs and build on this foundation as we discuss the further development of the organ systems in later chapters. We consider the unique organization of the vertebrate head in a special section. In Chapter 5, *Form and Function,* we consider many of the mathematical and physical principles that link form to function, emphasizing those dealing with biomechanics. Others are introduced later. Our intention is that students use Part I as a frequent reference while reading Parts II through IV.

Parts II through IV deal with the structure and function of the organ systems in a phylogenetic context, but we have grouped the organ systems to discuss together those systems related to common functions. In **Part II, Protection, Support, and Movement,** we discuss those organ systems related to the protection of the body and its support and movement in different environments: Chapter 6, *The Integument;* Chapter 7, *The Cranial Skeleton;* Chapter 8, *The Postcranial Skeleton: The Axial Skeleton;* Chapter 9, *The Postcranial Skeleton: The Appendicular Skeleton;* Chapter 10, *The Muscular System;* and Chapter 11, *Functional Anatomy of Support and Locomotion.* **Part III, Integration,** deals with those systems that integrate the activity of the muscles, which were studied near the end of Part II and all other organ systems. Chapter 12 deals with *The Sense Organs;* Chapter 13, *The Nervous System I: Organization, Spinal Cord, and Peripheral Nerves;* Chapter 14, *The Nervous System II: The Brain;* and Chapter 15, *Endocrine Integration.* It is unusual to introduce these systems so early in a comparative anatomy text, but doing so allows consideration of the integration of the other organ systems as they are being studied. In the laboratory, in particular, parts of the nervous system and the endocrine glands are observed as other organ systems are dissected, so it is both sensible and convenient to describe them in the middle of the course. **Part IV, Metabolism and Reproduction,** deals with the systems through which the organism obtains and utilizes the energy needed to sustain life and with the continuity of life: Chapter 16, *The Digestive System: Oral Cavity and Feeding Mechanisms;* Chapter 17, *The Digestive System: Pharynx, Stomach, and Intestine;* Chapter 18, *The Respiratory System;* Chapter 19, *The Circulatory System;* Chapter 20, *The Excretory System and Osmoregulation;* and Chapter 21, *The Reproductive System and Reproduction.* In **Part V, Conclusion,** we illustrate the integrative aspect of modern comparative anatomy with a functional and phylogenetic analysis of two astounding methods of hearing that have evolved independently in fishes and mammals and the importance of this in the evolution of these groups.

The systems approach we use facilitates understanding evolutionary changes in organ systems, but it makes it easy to lose sight of the integration of systems within an entire organism. We make it clear how changes in organ systems relate to the animal as a whole, to the environment in which it lives, and to its mode of life.

Teaching Considerations

Comparative anatomy is taught in many ways and at many depths according to the goals of the instructor and the time available. Enough material is included in this text for an in-depth course, but we recognize that many courses cannot cover all of the material. We prepared this book for maximum flexibility so that instructors can select what they need for their own courses. The Contents, Chapter Outlines, and cross-references between chapters allow instructors to present material in a different sequence, to omit or abbreviate topics, to cover certain material by reading assignments, or to use certain sections for reference. We believe that the topics are presented clearly enough that students can study them with a minimum of instructor assistance.

Several pedagogical devices are included to help students study. A Precis introduces students to the purposes of each chapter, and a Summary at the end of each chapter highlights the major points developed. Some terms in the text are boldfaced near their first use to indicate where to look for basic information. We try to help students learn the terms for structures because effective communication necessitates such knowledge. Because most terms are a description in Latin or Greek of some aspect of the structure, some knowledge of the terms' derivation helps students learn and understand them. When introducing an important term that is likely to be unfamiliar, we have given its classical derivation so students can see that the term is descriptive of the structure. Because most of the classical roots are used repeatedly, students will soon recognize them, and learning new terms will become easier. Giving the derivations when terms are introduced is a natural way for students to learn anatomical terminology. To facilitate reading at any point in the text, a Glossary is included. Our Glossary is more complete than are those in most textbooks because we provide not only definitions but also guides to pronunciation and classical derivation.

New to the Third Edition

Advances in vertebrate phylogeny necessitated the revision, expansion, and reinterpretation of comparative and evolutionary perspectives to reflect the vitality of the increasingly integrative studies and approaches in the study of animal form. It is especially important to introduce and apply phylogenetic approaches to functional anatomy to discover patterns in the distribution of anatomical characters. This, in turn, is essential to any theories about processes that have influenced vertebrate evolution. We have made a special effort to interpret patterns in a modern phylogenetic context throughout the text, and these are presented in concluding sections of the chapters. However, in adding this phylogenetic component, we have not sacrificed functional anatomical aspects.

So much progress has been made in experimental zoology since the second edition that several chapters have been completely rewritten, every chapter has been thoroughly revised, and new Focus boxes have been added, without substantially lengthening the text. Completely rewritten chapters are those dealing with craniate origins, diversity and phylogenetic history of craniates, early development and comparative embryology, and form and function. The postcranial skeleton has been divided into two chapters, one on the axial and one on the appendicular skeleton. The three chapters on the sense organs and nervous system were revised to reflect the extraordinary strides in comparative vertebrate neurobiology. All material on endocrinology, except for that involving the reproductive hormones, has been consolidated into one chapter. The former edition's chapter on body cavities was consolidated into Chapter 4. A new Epilogue has been added. More citations to the original literature are given than is customary in a book at this level, but past users have found this to be helpful. Most importantly, all of the illustrations have been digitally redrawn and much revised, and many new figures have been added. Color has been used throughout the illustrations to enhance their didactic value, and we have tried to integrate better the illustrations with the text. The authors' tight control of the art program—unusual for most modern textbooks—has been possible because of the remarkable talents of William B. Sillin of the University of Massachusetts, Amherst, who invested nearly two years in the preparation of the new illustrations. We hope that readers will find them helpful in pursuing studies of vertebrate functional anatomy.

A Note to Potential Adopters

Harcourt College Publishers may provide complimentary instructional aids and supplements or supplement packages to those adopters qualified under our adoption policy. Please contact your sales representative for more information. If as an adopter or potential user you receive supplements you do not need, please return them to your sales representative or send them to

Attn: Returns Department
Troy Warehouse
465 South Lincoln Drive
Troy, MO 63379

Acknowledgments

We are grateful to many people for help in writing and production. The outstanding staff of Harcourt College Publishers, listed on the copyright page, guided and helped us in many ways. Colleagues whose intellectual, didactic, and other skills have contributed significantly to this text are: A. W. Crompton, K. Hartel, A. Summers, F. Galis, C. Wilga, G. V. Lauder, E. Brainerd, S. L. Sanderson, E. Hilton, A. Richmond, N. Kley, R. Lederman, D. Saulnier, D. Baker, W. Bassham, and K. Doyle.

Dr. Liem is especially grateful to C. Souza for her tireless efforts, technical skills, and dedication in producing the numerous versions of his parts of the manuscript. Dr. Walker has also benefited from her skills. He is also grateful to his wife, Tensy Walker, for her support and careful proofreading of manuscripts and figures.

Dr. Bemis enjoyed working with William B. Sillin, designing, revising, and fine-tuning illustrations; this has been an extremely pleasant professional relationship. The comparative anatomy class of 2000 at the University of Massachusetts, Amherst, was especially helpful in improving the illustrations for this book.

Finally, we would like to thank the following persons, who reviewed portions of the manuscript for the third edition. We are most grateful for their help and suggestions, but we are responsible for any errors that remain.

Jessica Bolker
University of New Hampshire
Durham, New Hampshire

Brooks M. Burr
Southern Illinois University at Carbondale
Carbondale, Illinois

Ann B. Butler
Krasnow Institute for Advanced Study
George Mason University
Fairfax, Virginia

George R. Cline
Jacksonville State University
Jacksonville, Alabama

Frank E. Fish
West Chester University
West Chester, Pennsylvania

Terry Grande
Loyola University
Chicago, Illinois

Brian Hall
Dalhousie University
Halifax, Nova Scotia

Chris Haynes
Shelton State Community College
Tuscaloosa, Alabama

John W. Hermanson
Cornell University
Ithaca, New York

Christine Janis
Brown University
Providence, Rhode Island

John Long
Vassar College
Poughkeepsie, New York

R. Glenn Northcutt
University of California, San Diego
San Diego, California

G. G. E. Scudder
University of British Columbia
Vancouver, British Columbia

Larry Thomas Spencer
Plymouth State College
Plymouth, New Hampshire

Renn Tumlison
Henderson State University
Arkadelphia, Arkansas

Karel F. Liem
William E. Bemis
Warren F. Walker, Jr.
Lance Grande

November 2000

Contents Overview

Contents

part **I**

Background for the Study of Vertebrate Anatomy

1

Introduction

PRECIS

Concepts that will help you begin your study of vertebrate anatomy, function, and evolution are introduced in this chapter.

OUTLINE

Anatomical and Functional Interpretation

Since the publication of Charles Darwin's *On the Origin of Species* in 1859 and general acceptance of the theory of organic evolution, biologists have studied broad patterns of evolution primarily by comparing the anatomy of different species. **Comparative anatomy** serves us well because it enables us to develop hypotheses tracing the evolution of organisms and organs and simultaneously allows us to see ourselves in a broader biological context. We have inherited a wealth of comparative anatomical information from previous generations, but an astonishing amount of basic comparative anatomical research remains to be done.

Using a variety of technologies, comparative anatomists can make detailed investigations about the functions of anatomical structures. **Functional anatomy** examines the performance of structures within organisms, such as cells, tissues, organs, organ systems, and other complex functional units. An organ's form is linked to its function just as the design of a musical instrument is linked to the sounds it can make. The link between form and function has been an organizing principle for many aspects of the sciences and humanities at least since the 18th century. Philosophically, form and function are two sides of the same coin: one cannot exist without the other. In studying the form of a particular anatomical structure, it is therefore natural to ask about its function. Biologists sometimes refer to the combination of an anatomical structure and its immediate or most essential function as a "form–function complex." Most importantly, however, **living animals** must be analyzed to discover how form–function complexes work in intact organisms, and for this reason, functional anatomy and physiology often interface with studies of animal behavior.

Functional anatomists not only describe form and function but also ask, "How and why did a particular form–function complex evolve?" To answer this question, we must not only study the anatomy and function of various organ systems but also learn how to study evolutionary history. Scientists derive hypotheses about patterns of evolutionary history using **phylogenetic analyses** and depict the resulting **phylogenetic hypotheses** using diagrams known as **cladograms.** Inclusion of fossil vertebrates in phylogenetic analyses can yield important additional details about evolutionary changes in anatomy. Then, by mapping functional anatomical features onto cladograms, we can hypothesize how particular form–function complexes evolved. We can see, for example, that flight in birds evolved in several steps rather than a single one, with the evolution of feathers as thermal insulation preceding the evolution of the wing downstroke and flight feathers that are used for powered flight.

In any study of evolutionary anatomy, it is essential to remember that new anatomical designs evolve only by changing preexisting structures. Change can be radical, and over evolutionary time, a structure may take on a new form while maintaining its original function, or it may lose its original function and acquire a new one, or it may acquire both a new form and a new function. Flight feathers are an excellent example of this last possibility, with their original function of insulation only offering the slightest hint of their potential to evolve into flight feathers. Unlike an engine, which can be turned off when mechanical adjustments are made, evolutionary changes in the machinery of a vertebrate's body must always take place while the engine is running (Frazzetta, 1975).

To further understand the comparative anatomy of vertebrates, it is important to understand changes in form and function that occur during an animal's individual development, or **ontogeny.** Also, ontogeny may offer clues about why certain forms and functions have changed in particular ways during the course of evolution. Some link appears to exist between ontogeny and phylogeny, although the nature of this linkage has proved elusive for more than 150 years and remains controversial. Still, there is no doubt that the types and extent of evolutionary modifications in form and function of an adult structure depend, at least in part, on the pattern and mechanism of its development in the embryo.

Our Approach to Teaching Comparative and Functional Anatomy

Because comparative anatomy is one of the oldest areas of the natural sciences, many different approaches to teaching students about the subject have been tried. Three basic approaches stand out, although, of necessity, each incorporates elements of the other two.

First, one can teach comparative anatomy using a **regional approach,** comparing the structure and function of different general body regions, such as the head, in different groups of vertebrates. A work that follows this approach is *Studies on the Structure and Development of Vertebrates* by E. S. Goodrich (1930), one of the most important texts ever written on the subject of vertebrate anatomy.

A second basic organization for teaching may be termed a **taxonomic approach,** in which the biology of a particular group of vertebrates is described in detail, followed by the next group, then the next group until the diversity of vertebrates has been treated in the appropriate depth. Several popular textbooks and courses are organized around such an approach.

We adopt the third basic organizational model, the **systems approach,** in our textbook. This approach allows us to emphasize matches between anatomical structure and function. We will note close matches at many levels of organization, from the exact match of certain DNA regulatory proteins with specific sites on DNA molecules to the lightweight bones and special arrangements of muscles correlated with flight in birds. Our systems approach also allows us to trace the evolution of organ systems using phylogenetic trees, a process termed **character analysis.** We compare aspects of each organ system across vertebrate diversity to illustrate anatomical and functional changes over the course of vertebrate evolution. A major advantage of the systems approach is that it highlights how different vertebrates have "solved" similar general problems or evolved specializations for living in unique habitats.

We provide background information in Part I of this book, in chapters intended to be read and consulted as reference material. We describe essential facts concerning the origin of chordates and vertebrates in Chapter 2, review vertebrate diversity in Chapter 3, summarize vertebrate development in Chapter 4, and outline general principles of structure and mechanics as applied to vertebrate bodies in Chapter 5.

The ten organ systems of vertebrates are listed in Table 1-1; detailed treatment of the anatomy and function of each of these organ systems is the purpose of Part II through Part IV. We present the **integumentary system** (skin) as a multifunctional system and

TABLE 1-1 The Ten Organ Systems of Vertebrates

1. Integumentary system	6. Digestive system
2. Skeletal system	7. Respiratory system
3. Muscular system	8. Circulatory system
4. Nervous system (including sense organs)	9. Excretory system
5. Endocrine system	10. Reproductive system

give several examples of its evolutionary diversifications. The **skeletal system** is central to analyses of many aspects of vertebrate anatomy, function, and evolution, and so we divide its treatment into three chapters: one focused on the cranial skeleton; one describing the axial skeleton; and one, the appendicular skeleton. After introducing general aspects of the **muscular system,** including the basic arrangement and structural modifications of the head muscles, we provide a chapter highlighting biomechanical aspects of the skeleton and muscles as related to locomotion. Our treatment of the **nervous system** and associated sense organs requires three chapters because the comparative anatomy of these structures is a fascinating and rapidly growing field. Like the nervous system, the **endocrine system** has important roles in integrating the functions of other organ systems; but in this book, we provide only a relatively brief overview of major aspects of endocrine glands because an in-depth treatment of this area now requires more advanced knowledge of molecular biology than can be included in a book this size. Our treatment of the **digestive system** illustrates how form and function respond in unison for vertebrates to exploit a great array of foods. Studies on both the **respiratory** and **circulatory systems** have yielded many fundamental theories and examples of vertebrate evolution, such as those related to the transition from aquatic to terrestrial life or the essential anatomical and physiological changes that occur at birth in a placental mammal. Our description of the **excretory** and **reproductive systems** includes examples of how developmental mechanisms may have governed the direction and extent of evolutionary change in these two closely linked organ systems. Functional adjustments in one or both of these systems often are necessary in order to cope with changes in the environment, such as metamorphic transitions in amphibians or migratory movements of bony fishes from salt to fresh water and vice versa. We conclude, in an Epilogue, by analyzing an example of one model system (hearing) as a case study of the integration of form and function and the significance of comparative studies.

Examples of Functional Anatomy

As a first example of how functional anatomy is practiced, we briefly summarize elegant research on starling flight by Dial et al. (1991). The major flight muscles and associated skeletal elements were dissected in dead starlings, and their anatomy was described and illustrated (Fig. 1-1). Thorough anatomical knowledge was needed in order to generate testable hypotheses of function, such as the prediction that a particular muscle performs a particular function during flight. Once these hypotheses were generated, functional analysis could begin. The starlings were trained either to fly down a 50-m-long hallway to a landing platform or to fly in a wind tunnel. Their major flight muscles were then surgically implanted with electrodes. Using a technique termed **electromyography,** the electrical activity of the muscles was recorded; simultaneously, a synchronized, high-speed motion picture of the wing actions of the flying bird was taken (Fig. 1-2*A*). The camera recorded 400 images per second, a rate sufficient to accurately study movements of the wings throughout complete wingbeat cycles. The average activity of the flight muscles was then studied in relation to wing movements (Fig. 1-2*B*). The results revealed which muscles were active during the downstroke of the wing (*black bars*) and which were active during the upstroke (*gray bar*). Some overlapping muscle action (*white bar*) occurred during the transitions between the downstroke and the upstroke. On the bases of these results, some hypotheses of musculoskeletal function could be rejected, that is, the muscles do not perform exactly as proposed on the basis of a static anatomical examination of starlings. We will return to more details from this study when we discuss flight in Chapter 11. With electromyography, high-speed motion analysis, and other powerful analytical tools, functional anatomists have made discoveries that are revolutionizing our concepts of vertebrate anatomy, function, and evolution.

Our second example of functional morphology illustrates the close match between structure and func-

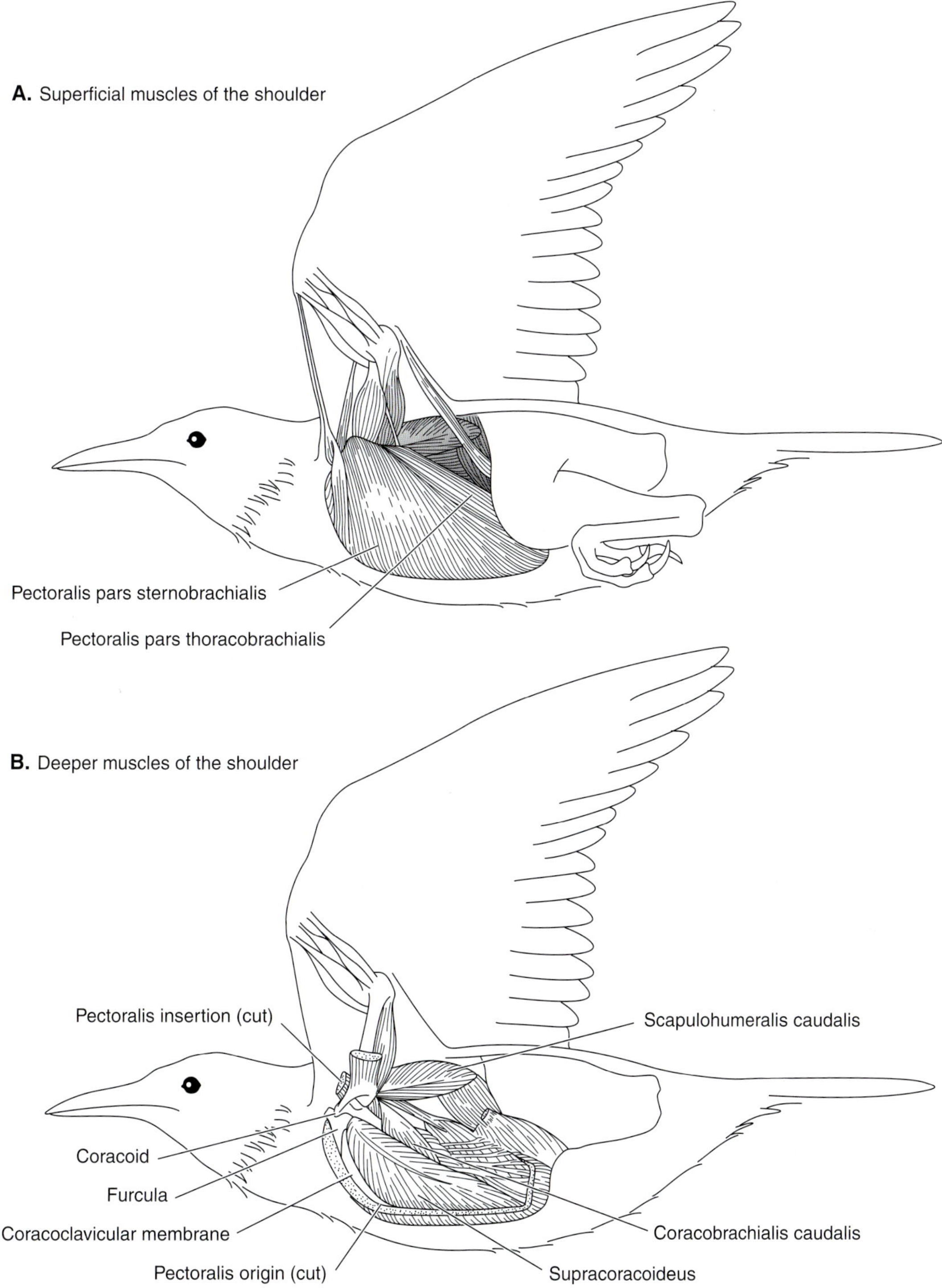

FIGURE 1-1
Lateral views of the muscles of the shoulder of the European starling, *Sturnus vulgaris*. *A,* Superficial muscles. *B,* Deeper muscles after resection of most of the pectoralis major. *(Modified from Dial et al.)*

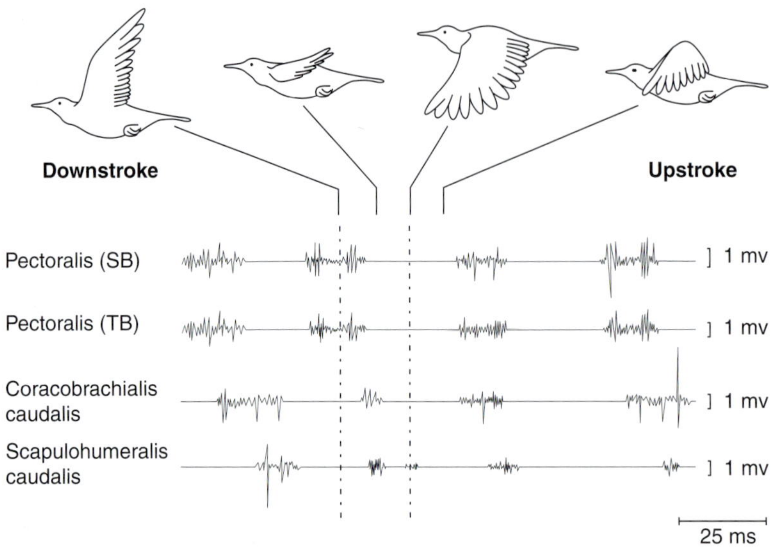

A. Electromyographic traces from a flying starling

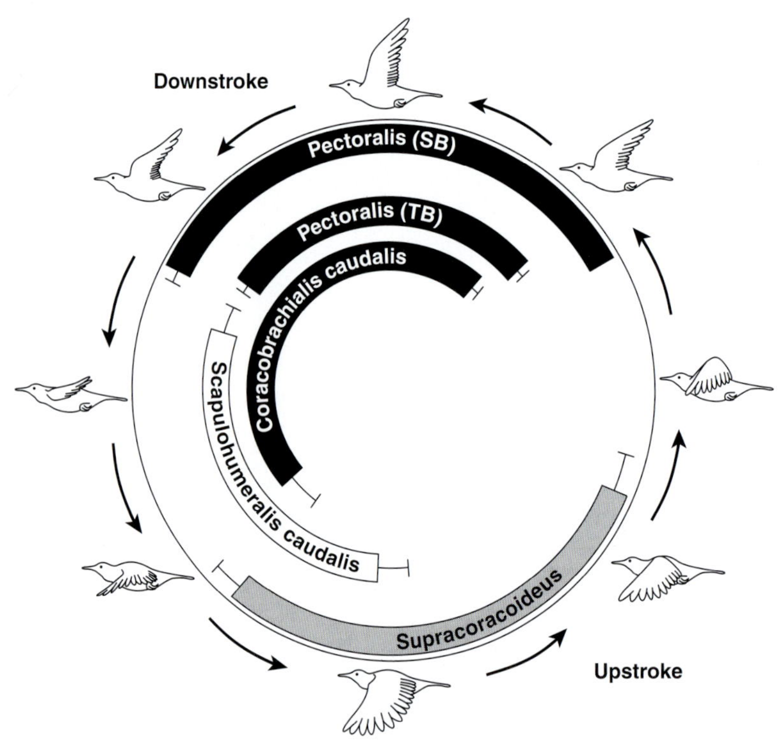

B. Average muscle activity during a wingbeat cycle

FIGURE 1-2

Functional morphology of starling flight. *A,* Representative traces of electromyographic signals recorded from shoulder muscles of the European starling during flight. The strength of the signals in millivolts (mv) is shown on the vertical axis and their duration in milliseconds (ms) on the horizontal axis. SB = sternobrachial part of the pectoralis; TB = thoracobrachial part. Sketches of flying birds above key the signals to phases of flight. The first sketch is the start of the downstroke, and the third sketch is the start of the upstroke. *B,* A summary of the mean onset and offset of the activity of selected shoulder muscles of the European starling during a wing cycle of 72 ms. Muscles indicated in black are used in the downstroke; those indicated in gray, during the upstroke; and those not shaded, during the transition between downstroke and upstroke. Thin lines extending beyond the range of mean muscle activity indicate the standard error. *(Modified from Dial et al.)*

tion at the cellular level. Humans and other vertebrates have approximately 200 named cell types. Four of these types are shown in Figure 1-3, and even a brief examination hints at the great range of possible cellular specializations for different functions. For example, the cytoplasm of the **skeletal muscle cell** shown in Figure 1-3*A* contains many **sarcomeres,** which are highly organized arrays of the contractile proteins **myosin** and **actin.** The contractile properties of the entire skeletal muscle cell, which are the basis for muscular movements, are a direct consequence of the types of actin and myosin and the arrangement of the sarcomeres. Note the peripheral location of the **nucleus.** Each skeletal muscle cell forms by the fusion of many separate cells during development and so is both long (so long that only part of it can be shown at this magnification) and multinucleate. Skeletal muscle cells in most types of muscles have many **mitochondria,** which are used to produce adenosine triphosphate (ATP) needed for muscle contraction by oxidation of glucose. An

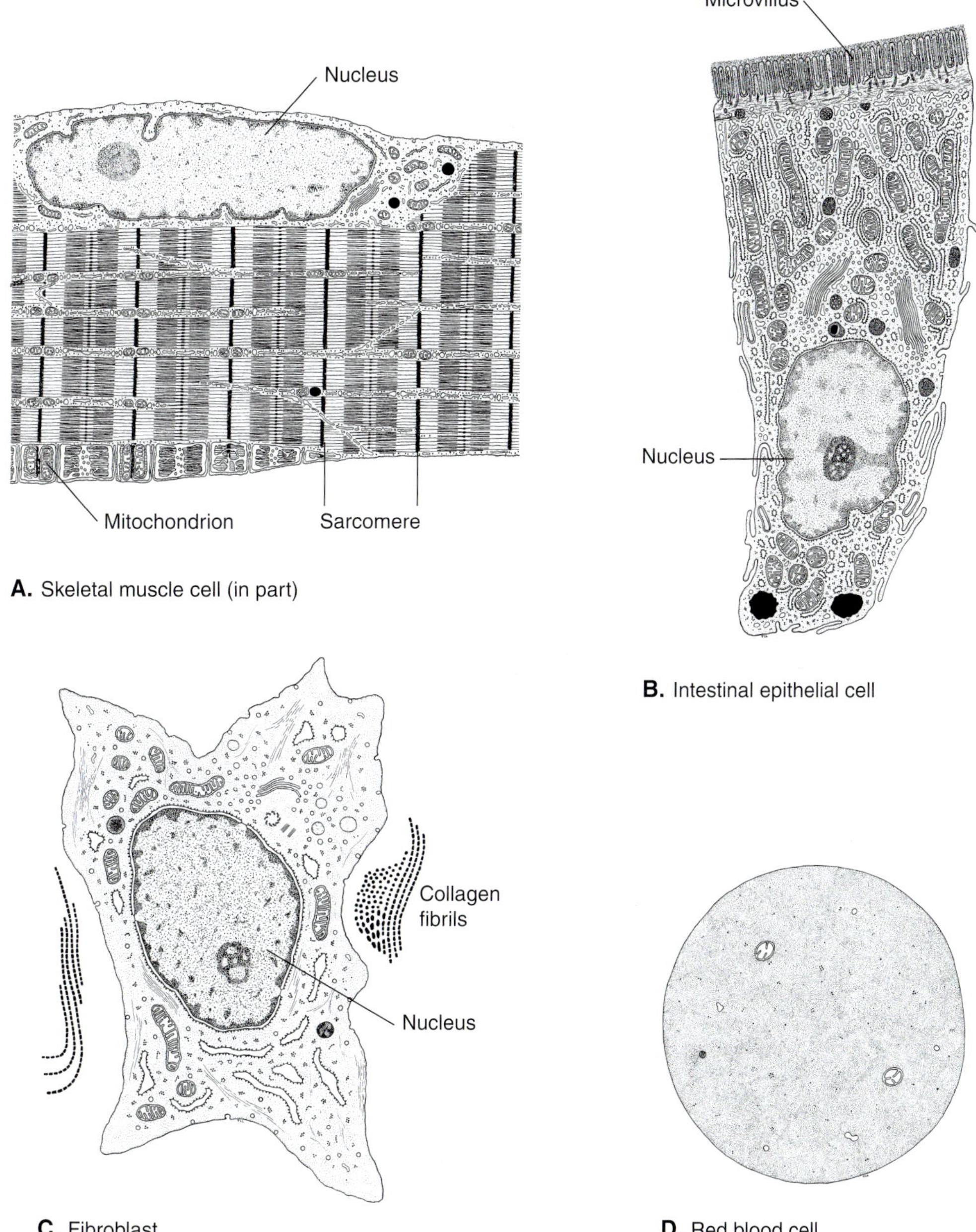

FIGURE 1-3
A selection of cell types found in humans. *A,* Skeletal muscle cell. *B,* Intestinal epithelial cell. *C,* Fibroblast. *D,* Red blood cell. *(Modified from Lentz.)*

intestinal epithelial cell (Fig. 1-3*B*) has many **microvilli** on its apical surface, which serve to greatly increase its surface area for absorption of nutrients from the intestine. Its nucleus is located nearer to the base of the cell. Organelles in the cytoplasm of a **fibroblast** (Fig. 1-3*C*) synthesize and secrete **collagen.** Collagen is the most abundant extracellular protein in the body, and it is essential to all biomechanical systems for it is the main structural material in tendons, ligaments, and other connective tissues. Finally, mature mammalian red blood cells (Fig. 1-3*D*) lack a nucleus.[1] The cytoplasm of these circulating elements of the blood is completely packed with the respiratory molecule **hemoglobin.** Dense packing with hemoglobin not only increases the cell's efficiency in transporting oxygen to the tissues of the body but also limits the potential life span of each individual red blood cell, necessitating that new ones constantly differentiate from stem cells in the bone marrow.

From these examples, it is clear that a comprehensive understanding of vertebrate structure and function ultimately draws from all levels of biological organization. As a unifying theme, we can predict that correlations exist between a structure and its function (or functions) across all levels of organization.

Functional Anatomy and Development

As we noted above, the course of an individual's development from a fertilized egg to death is called its **ontogeny.** Often, comparative and functional anatomists study ontogenetic changes in structure or function. We also are concerned with the conceptual relationship between development and evolution, particularly how differences in the timing of a structure's development might lead to a radically different structure in a descendant species.

Growth and Changing Proportions of Body Parts

Differential growth occurs during the development and growth of individual animals as it has in the course of evolution (Chapter 5). Body size increases rapidly during embryonic development. If all parts of an embryo increased in size at equal rates (the growth pattern known as **isometry**), then surface–volume relationships would soon result in the physiological exchange surfaces becoming too small to supply gas, nutrients, and other products to the increasing mass. How do vertebrates solve this fundamental problem?

As concisely stated by Gilbert (1991), "The most successful growth strategy to circumvent the surface–volume problem is invagination."[2] The digestive tract and lungs are examples of invaginated regions of the body that increase an animal's surface area relative to its mass. As size increases further, these gas and nutrient exchange surfaces become more distant from the cells to be supplied. This is solved by the development of the circulatory system, which shortens diffusion distances by bringing materials to be exchanged closer to the cells. Because an organism continues to function as it develops, vertebrates undergo continuous differential changes and remodeling of exchange surfaces, the circulatory system, the skeleton, and other organ systems.

Body proportions also change during embryonic development. Intrauterine development in marsupial mammals is very short (only about two weeks for the American opossum, *Didelphis*). A newborn marsupial is essentially an embryo, but its pectoral appendages, face, and jaws have grown at a much faster rate than the rest of the body. Thus, it can pull itself into the mother's pouch, or **marsupium,** where it attaches itself to a nipple. Pouch life is several times longer than intrauterine life, and "embryonic" development is completed there.

Heterochrony

Some evolutionary shape changes result from changes in the timing of embryonic development of a structure relative to the timing of its development in an ancestor. This general phenomenon is known as **heterochrony** (Gr., *heteros* = other + *chronos* = time). Heterochronic processes can produce two major evolutionary outcomes. The first of these is **paedomorphosis** (Gr., *paid-*, from *pais* = child + *morphe* = shape), which is a retention of embryonic or larval characteristics by adults. The second is **peramorphosis** (Gr., *pera-* = beyond + *morphe* = shape), which is development of entirely new shapes, beyond that of a typical adult form. Both paedomorphosis and peramorphosis can be associated with radical evolutionary transformations of parts of organisms or whole organisms (Reilly et al., 1997). For brevity, and because it is important for understanding one theory about the origin of vertebrates (Chapter 2), we emphasize paedomorphosis

[1]With a few interesting exceptions, such as camels and Dalmatian dogs.

[2]An invagination is an infolding of a group of cells to form a pocket.

here, but both paedomorphosis and peramorphosis are modeled in Figure 1-4.

Larvae and juveniles possess body shapes and other features that differ considerably from their adult counterparts. Retention of such larval or juvenile features in sexually mature individuals can happen if differential embryonic growth rates exist between reproductive organs and other parts of the body, so that certain larval or juvenile features are retained in sexually mature individuals. Paedomorphosis has been especially well studied in salamanders, but groups as distant as lungfishes (Bemis, 1984), teleost fishes (Grande, 1994), and hominids (Gould, 1977) appear to have been shaped, at least in part, by paedomorphosis. For example, the endoskeleton of the extant species of lungfishes is largely cartilaginous, although these same skeletal elements were fully ossified bones in Paleozoic lungfishes.

As an evolutionary outcome, paedomorphosis can be produced in theory by three different processes, which are most easily understood by studying the diagrams shown in Figure 1-4*A*. In these models, the independent variable is time, with **α** being the time at which development of a structure begins; **β,** the time at which development of the structure stops; and **k,** the rate of development of the structure. Changes in α, β, or k will influence the final form or **shape** of a structure. For example, the rate of development (k) of a structure could decrease, or **decelerate,** over evolu-

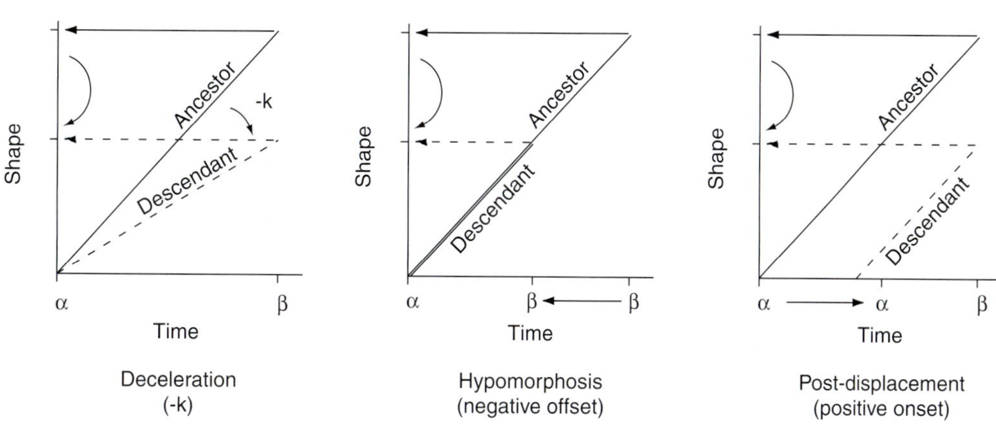

A. Paedomorphosis—development of shape is truncated.

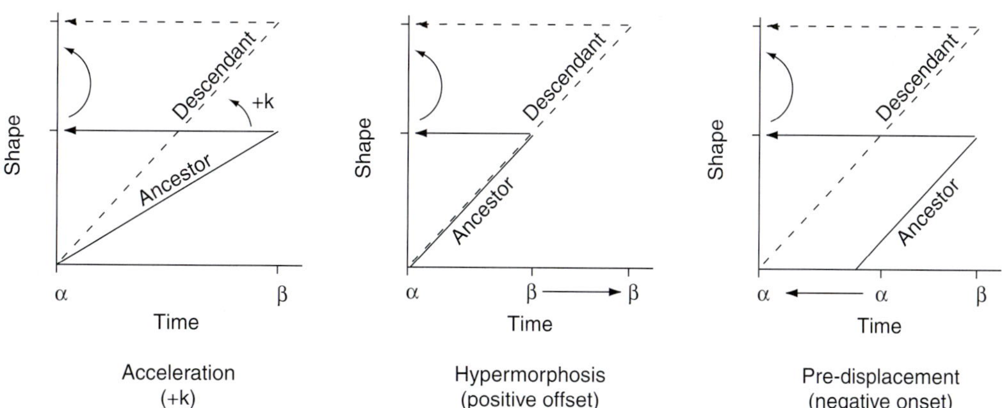

B. Peramorphosis—development of shape is extended.

FIGURE 1-4

Models of the six simple heterochronic processes. These are theoretical models and not necessarily distinguishable in nature. *A,* Paedomorphosis could theoretically be produced by three different processes: deceleration, hypomorphosis, and post-displacement. *B,* Peramorphosis could theoretically be produced by acceleration, hypermorphosis, and pre-displacement. See text for additional explanation. *(Modified from Reilly et al.)*

tionary history. If this occurs, then the shape found in the descendant will appear truncated relative to that found in the ancestor. Alternatively, a structure could simply stop developing at an earlier time in development, which is modeled in Figure 1-4*A* as a leftward shift of β. This phenomenon is termed **hypomorphosis;** note that, like deceleration, hypomorphosis also truncates the final shape of the structure. In the third mechanism, the start of development of a structure could be delayed, or **postdisplaced,** relative to its ancestor. In this case, α is shifted to the right, and truncation of shape results (Fig. 1-4*A*).

Rarely, if ever, can we distinguish which of these three processes operated to produce a particular paedomorphic result, but some generalizations can be made. For example, deceleration in the rate of development of structures tends to be correlated with evolutionary increases in size. Deceleration has operated in the case of many **perennibranchiate** (L., *perennis* = throughout the year + Gr., *branchia* = gills) salamanders, such as the mudpuppy, *Necturus.* Perennibranchiate salamanders never metamorphose to become terrestrial adults, as do other salamanders. They retain their gills, as the name suggests, and other juvenile features, such as a finned tail.

Paedomorphosis probably has been important in vertebrate evolution because it produces new morphological combinations of juvenile and adult characters that might not otherwise occur. The potential exists for a paedomorphic population to take a new evolutionary direction that would have been impossible for an adult population highly adapted to a restricted mode of life. Paedomorphosis also may explain the large morphological gaps we sometimes see between major groups of animals. For example, the theory for the origin of vertebrates that we discuss in this book invokes paedomorphosis (Chapter 2).

From this background, you should be able to work through the three hypothetical models of peramorphosis shown in Figure 1-4*B*. In this second set of models, the rate of development (k) is accelerated, or the endpoint of development (β) is shifted to the right, or the start of development (α) is shifted to the left.

Systematic Biology and Comparative Interpretation

Systematics deals with naming and classifying both species and higher groups and investigating their evolutionary relationships and history. As a science, systematics has old roots, many conventions, a formidable literature, and a reputation for complex philosophical arguments and deeply held convictions. Its products, however, are essential: biologists cannot operate without the evolutionary units that we familiarly know as "species," and they also need classification schemes and phylogenetic trees. This need for accurate, up-to-date interpretations and the often intense philosophical controversies make systematics an extremely exciting area for research. The work of systematists includes **taxonomy** and **nomenclature** (i.e., the naming of species and higher groups), **description** and **classification** (e.g., a more detailed comparative study than is usual for a publication that names a new species), and analyses of data for **reconstructing patterns** (e.g., trees of character data for examining patterns of interrelationships among species or trees indicating patterns of relationship among geographical areas used for historical biogeographical studies).

Contrary to what you may expect, we do not yet have anything approaching a single "consensus" evolutionary tree of vertebrates. In fact, we do not even have a complete inventory of all of the species of vertebrates alive today, nor are we ever likely to. Among other reasons, this is because some habitats, such as the deep sea, remain difficult to sample, and other habitats, such as portions of certain river systems and forests, will be destroyed long before a complete inventory of their vertebrate species can be made.

Phylogeny and Classification

Evolutionary relationships are graphically represented using a **phylogeny** (Gr., *phylon* = tribe + *genesis* = birth). As shown on the left side of Figure 1-5, each pair of divergent taxa in a phylogeny is represented by **dichotomous branches** (Gr., *dicheres* = dividing in two), with the two lines diverging from a **node** that represents the hypothetical ancestor of the two taxa. For example, Node *JK* is indicated at the divergence of *Jus jus* and *Jus kus.* The oldest divergence forms the base, or **root,** of the phylogenetic tree, while the **terminal branches** of the tree include the most recently evolved taxa. In such a phylogeny, it is the **pattern of branching,** not the lengths of the lines between taxa, that matters. This is because we cannot assume anything about the absolute times of divergence of the taxa based only on the pattern of branching. The **relative times of divergence** can be discerned from the branching pattern (e.g., in Fig. 1-5, the divergence at Node *IJ* to Genus *Ius* and Genus *Jus* must have preceded the divergence at Node *JK* into Species *Jus jus* and Species *Jus kus*).

The term **taxon** generally refers to any named group of organisms, such as a genus, family, order, and so on. In a carefully constructed phylogeny, each taxon

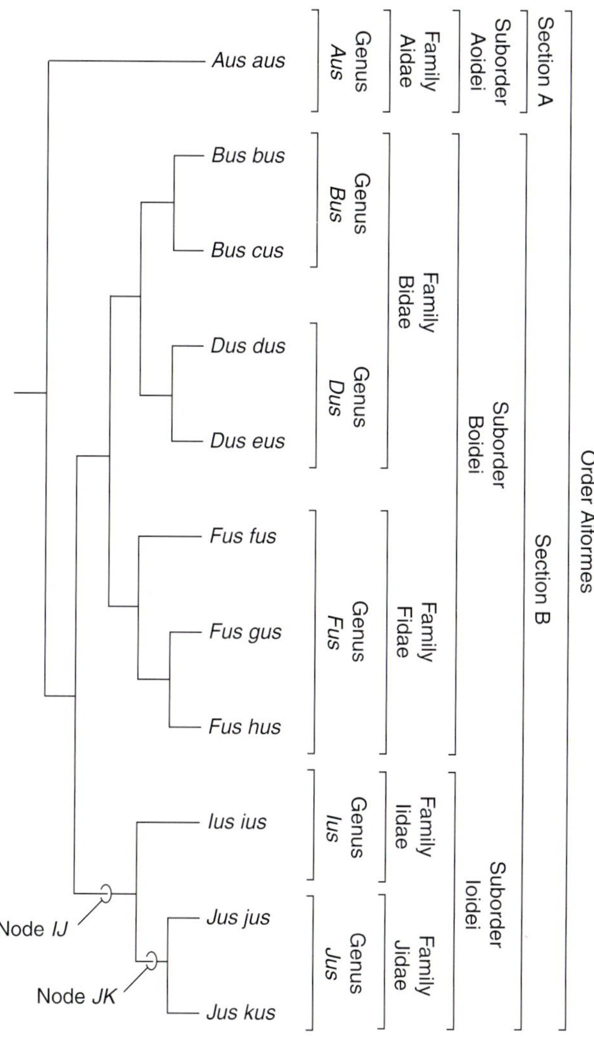

FIGURE 1-5
A phylogeny for 11 hypothetical species in the hypothetical order Aiformes. Note the dichotomous branching pattern of the taxa. The classification of these species into a nested hierarchy of progressively inclusive groups is shown on the right side. This same phylogenetic classification is presented in indented form in the text (pages 12–13).

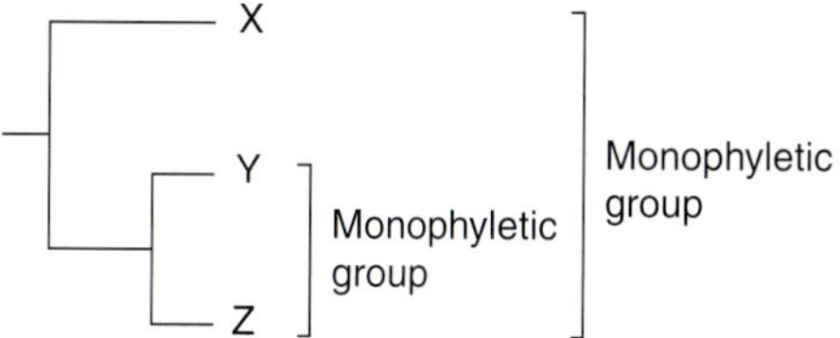

A. Groups containing only Y & Z or X, Y & Z are monophyletic.

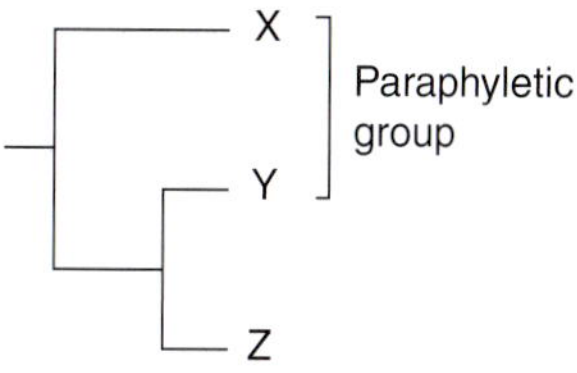

B. Group containing only X & Y is paraphyletic.

FIGURE 1-6
Composition of monophyletic and paraphyletic groups. *A,* Each monophyletic group includes a hypothetical ancestral species and all of its descendant species. *B,* A paraphyletic group does not contain all of the descendant species of a single ancestor.

contains all of the known species derived from a hypothetical common ancestor. A taxon that includes the hypothetical common ancestor and all of its descendant species is said to be a **monophyletic group** (Gr., *monos* = single + *phyle* = tribe) or **clade**. The concepts of **monophyly** and a contrasting method of grouping taxa, termed **paraphyly,** are illustrated in Figure 1-6. As a real-world example, we believe that all cats form a monophyletic group, and that cats, dogs, and bears, who share a more remote common ancestor, form a larger monophyletic group. In contrast, the animals that we commonly know as fishes do not constitute a monophyletic group unless land vertebrates are included because land vertebrates are descendants of fishes.[3]

Formally stated, a phylogeny is a hypothesis of evolutionary relationships among groups of organisms based on monophyly. Detection of phylogenetic patterns begins with an analysis of characters and construction of a **character matrix** (Focus 1-1). As scientists make continuing discoveries in paleontology, anatomy, molecular biology, behavior, and ecology, our information base changes. In particular, new discoveries of either taxa or characters test phylogenetic hypotheses and often modify previously published phylogenetic trees and classifications.

To give hierarchical order to classifications, biologists use a system of nomenclature developed by Karl von Linnaeus in the 18th century. Linnaeus' central concept was to recognize, diagnose, and name **species** as discrete units in nature and to classify them into nested, hierarchical groups. Each species receives a unique **scientific name** in order to avoid confusion with other taxa. To further promote stability and uni-

[3]It is common practice in systematic literature to put quotation marks around the name of a group that is not monophyletic. We have elected not to do this in this textbook because it is distracting to readers. Where we deem it important to know that a group is non-monophyletic, we specifically note it in the text or footnotes.

FOCUS 1-1 *Cladistics and Character Matrices*

Systematic biologists reconstruct phylogenies by using **character congruence** to determine relative relationships among taxa. The method known as **phylogenetic systematics** or **cladistics** offers clear and logical rules for making phylogenetic reconstructions.

At least three taxa (plus an outgroup) are needed for a phylogenetic analysis. Note in Figure A that the interrelationships of the three taxa can be resolved in three ways. Taxa Y and Z might be more closely related to each other than either is to Taxon Z; or Taxa X and Y might be more closely related to each other than either is to Taxon Z; or Taxa Z and X might be more closely related to each other than either is to Taxon Y. In each of these resolutions, cladists speak of the two more closely related taxa as **sister groups,** and the third taxon as **outgroup** to them. Any phylogenetic problem can be broken down into sets of hierarchical three-taxon statements, and this is the simplest and often most helpful way to think about phylogenetic questions.

If three possible phylogenetic resolutions exist for any three taxa, then how do we rule out two of them in favor of the third? Cladists do this by searching for character congruence through the principle of **parsimony**. When two groups appear to share a feature that is not present in any other group, then this feature may be evidence of a close phylogenetic relationship. In the language of phylogenetic systematics, such a special similarity is hypothesized to be a **shared derived character** or **synapomorphy** (Gr., *syn* = together + *apo* = away from + *morphe* = shape). Each hypothetical synapomorphy is subjected to

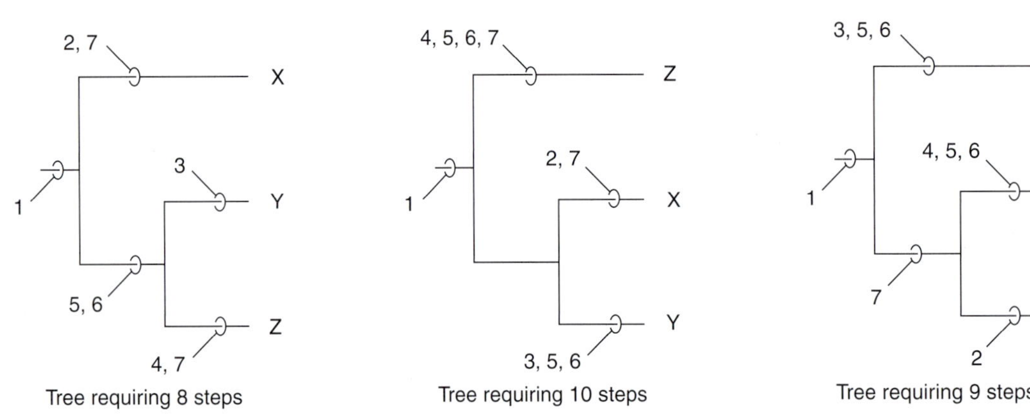

A. Three possible phylogenetic resolutions for Taxa X, Y, and Z, indicating the minimum number of steps needed to accommodate data from Table A. Outgroup is omitted from cladogram.

versality across different languages, the scientific name is given in Latinized form. Latin is a language that follows strict conventions and that is no longer spoken, and so it is unlikely to change appreciably over time. The scientific name of an animal is a **binomial** (L., *bi* = two + *nomen* = name), which consists of a **genus name,** the first letter of which is always capitalized, and a **species epithet,** which is always in lower case. Because species names are words in another language, it is customary to italicize them in a printed work.[4]

As shown on the right side of Figure 1-5, species can be arranged into a **nested hierarchy** of progressively inclusive groups or taxa based on a phylogeny. The phylogeny in Figure 1-5 leads to the following phylogenetic classification in indented form:

Order: Aiformes
- Section A: Unnamed
  - Suborder: Aoidei
    - Family: Aidae
      - Genus: *Aus*
        - Species: *Aus aus*
- Section B: Unnamed
  - Suborder: Boidei
    - Family: Bidae
      - Genus: *Bus*
        - Species: *Bus bus*, *Bus cus*
      - Genus: *Dus*
        - Species: *Dus dus*, *Dus eus*

[4]In the absence of an italic font, the conventional proofreader's mark of single underlining is used to signal a species or genus name (or other word in a foreign language) that should be printed in italics.

a **test of congruence,** that is, it is tested by comparing its pattern of distribution with that of other characters among the three taxa. The most efficient fit of the character data reveals the **most parsimonious** phylogenetic resolution. "Parsimony" means that, all other things being equal, we should accept the simplest possible explanation—the one that allows us to accept the tree topology with the smallest number of steps.

This process is best understood by examining a character matrix, in which the taxa are the rows and the characters are the columns (Table A). In this matrix, the presence of a character is coded as Character State 1; the absence of a feature is coded as Character State 0. In the example character matrix, Character 1 is present in all ingroup taxa; it is said to be **plesiomorphic** for each of these taxa individually, and it cannot offer any insight about their interrelationships. Character 2 occurs only in Taxon X, Character 3 occurs only in Taxon Y, and Character 4 occurs only in Taxon Z. Because of the restricted distribution of Characters 2, 3, and 4, none of them offers insight about interrelationships of Taxa X, Y, and Z.

When we plot the distribution of our characters from the data matrix on the three possible resolved trees, a different number of steps is required for each tree to accommodate the characters. Tree A requires at least eight steps, Tree B requires at least ten steps, and Tree C requires nine steps. Thus, Tree A requires the fewest evolutionary steps and is the most parsimonious resolution or tree. This gives us a **working phylogenetic hypothesis,** subject to improvement as additional characters or additional taxa are discovered and studied. Real-life examples are far more complex—and interesting—than this simple model. The principles are similar and often can be explored most easily using computer software, such as Phylogenetic Investigator (Brewer and Haffner, 1996), PAUP* (Swofford, 1999), MacClade (Maddison and Maddison, 1992), or Hennig86 (Farris, 1986) and CLADOS (Nixon, 1992).

TABLE A An Example of a Character Data Matrix

Taxon	Character 1	Character 2	Character 3	Character 4	Character 5	Character 6	Character 7
Outgroup	0	0	0	0	0	0	0
X	1	1	0	0	0	0	1
Y	1	0	1	0	1	1	0
Z	1	0	0	1	1	1	1

Family: Fidae
Genus: *Fus*
Species: *Fus fus*
Fus gus
Fus hus
Suborder: Ioidei
Family: Iidae
Genus: *Ius*
Species: *Ius ius*
Genus: *Jus*
Species: *Jus jus*
Jus kus

Genera and more inclusive groups, such as families, suborders, sections, and orders, are referred to as **higher taxa.** For instance, returning to our real-world example, lions, tigers, and closely related large cats are placed in the genus *Panthera;* the domestic cat and wild cat of the Old World are in the genus *Felis;* and the lynx and other species are in yet other genera. All cats are grouped into the cat family (Felidae), which, together with the dog family (Canidae), bear family (Ursidae), and other related families, constitute the order of carnivores (Carnivora). Zoologists have some helpful conventions concerning these names of higher taxa. For example, formal family-rank names end in *-idae,* subordinal ranks end in *-oidei,* and many ordinal-rank names end in *-iformes.*

Linnaeus and most pre-Darwinian zoologists assigned species to groups based on overall similarity (e.g., all backboned animals with gills were classified as fishes). Contemporary systematic biologists attempt to arrange and to name their groups on the basis of phylogenetic relationships, which are founded not on overall similarity but on specific, derived similarities known as **synapomorphies** (Focus 1-1). But what

FOCUS 1-2 *Reconstructing a Phylogeny of Amniotes Using Phylogenetic Systematics*

As you master the tools for making simple phylogenetic analyses, you will find yourself thinking about many biological problems from the perspective of a phylogeneticist. To demonstrate a simple but real phylogenetic analysis, and to introduce you to some living groups of terrestrial vertebrates, we analyze the distribution of homologous features in a series of taxa united by an embryonic specialization, known as the **amnion.** This group of vertebrates is known as the **amniotes.** For this analysis, we omit Linnean ranks and formal Latin names of the taxa. The taxa in our ingroup are turtles, lepidosaurs (tuataras, lizards, and snakes), crocodilians, birds, monotremes (egg-laying mammals), marsupials (pouched mammals), and eutherians (placental mammals). We summarize some of the available character information in a character matrix (Table A). If a feature is present,

TABLE A A Simple Character Data Matrix for Amniotes

Taxon	Character 1. Amniotic egg	Character 2. Hair	Character 3. Endothermy	Character 4. Single centrale bone in ankle	Character 5. Diapsid skull	Character 6. Trunk encased in shell	Character 7. Fenestra in lower jaw
Distant Outgroup	0	0	0	0	0	0	0
Near Outgroup	0	0	0	0	0	0	0
Turtles	1	0	0	0	0	1	0
Lepidosaurs	1	0	0	1	1	0	0
Crocodilians	1	0	0	1	1	0	1
Birds	1	0	1	1	1	0	1
Monotremes	1	1	1	1	0	0	0
Marsupials	1	1	1	0	0	0	0
Eutherians	1	1	1	0	0	0	0

Taxon	Character 8. Transverse cloacal slit	Character 9. Feathers	Character 10. Foramen of Panizza	Character 11. Viviparity	Character 12. Unique jaw opening muscle	Character 13. Chorio-allantoic placenta	Character 14. Marsupium
Distant Outgroup	0	0	0	0	0	0	0
Near Outgroup	0	0	0	0	0	0	0
Turtles	0	0	0	0	0	0	0
Lepidosaurs	1	0	0	P[1]	0	P[1]	0
Crocodilians	0	0	1	0	0	0	0
Birds	0	1	0	0	0	0	0
Monotremes	0	0	0	0	1	0	0
Marsupials	0	0	0	1	0	0	P[2]
Eutherians	0	0	0	1	0	1	0

[1]Some lepidosaurs are viviparous, and some are not. Those that are viviparous have chorio-allantoic placentas similar to those seen in eutherian mammals. Thus, this character is scored as polymorphic for lepidosaurs.

[2]Not all marsupials have a pouch. Thus, this character is scored as polymorphic for marsupials.

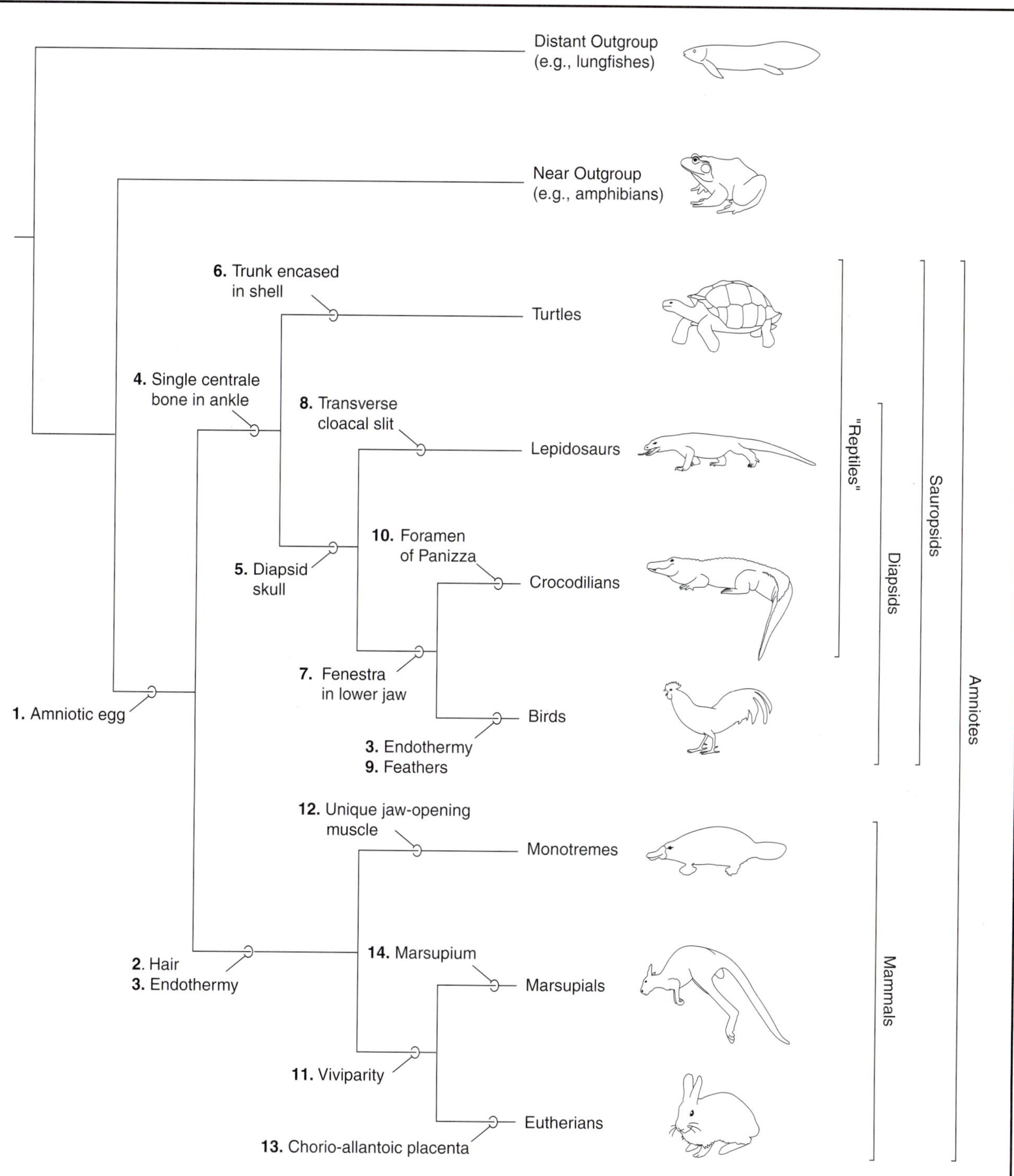

A. A cladogram of amniotes, keyed to the character data matrix shown in Table A.

(Continued)

FOCUS 1-2 *Reconstructing a Phylogeny of Amniotes Using Phylogenetic Systematics* (continued)

then it is scored as 1; if it is absent, then it is scored as 0; if it is present in only some of the group's taxa or individuals, then it is scored as P for polymorphic (Gr., *poly* = many + *morphe* = forms).

In the matrix (for Focus 1-2), note that all of the ingroup taxa have Character State "1," indicating the presence of an amniotic egg, and that both the sister group to the ingroup (which is the near ingroup) and the more distant outgroup have Character State "0," indicating the absence of an amniotic egg (Fig. A). The presence (State 1) of an **amniotic egg** is a synapomorphy for amniotes because it is shared across all of these taxa, not found in the outgroups, and congruent with the most parsimonious tree. Turtles, lepidosaurs, crocodilians, and birds have a unique type of ankle, so Character 4 is a synapomorphy for this entire group, which is known as Sauropsida. Now, trace for yourself the distributions of Characters 3 to 14 in the matrix and on the cladogram.

Some examples of convergent evolution emerge from the analysis. First, Character 3, **Endothermy** (having "warm blood"; Chapter 3) appears twice on the cladogram: once in the line leading to birds and once in the line leading to mammals. Based on the distribution of all other characters, endothermy appears to have evolved convergently in birds and mammals. Second, some lepidosaurs give birth to live young and thus can be described as **viviparous** (Character 11). These animals have evolved a specialized fetal–maternal exchange organ, or **placenta** (Character 13), that is comparable to those found in certain eutherian mammals but that must be regarded as convergently evolved based on the distribution of all other characters (see Mossman, 1987).

Note that the classification on the right side of the phylogeny names four monophyletic groups: Amniotes, Sauropsids, Diapsids (Gr., *di* = two + *apsis* = arch, referring to the presence of both a lower and an upper temporal fenestra), and Mammals. The classification also names one paraphyletic group, enclosed in quotation marks: "Reptiles." Animals traditionally grouped as "reptiles" include turtles, lepidosaurs, and crocodilians. But if these taxa are artificially forced into a group without including the birds, then the resulting group "Reptilia" will be paraphyletic. In this book, we try to minimize reference to paraphyletic groups as a way to clarify our communication about the characteristics of organisms.

For the cladograms presented throughout the rest of this book, we do not present character matrices, preferring instead to either chart the distribution of selected characters directly on the diagrams or to discuss them in the text. Bear in mind that any formal phylogenetic analysis, however, invariably requires collection and analysis of data for the preparation of a complete character matrix.

happens if a phylogenetic hypothesis changes because of new character information or discovery of new taxa? Must the classification also change? This has been a hotly contested issue in systematic biology. To allow efficient communication, stability in names and grouping arrangements of higher taxa is desirable. Still, most systematists agree that classifications ought to be maximally informative about evolutionary relationships and reflect new phylogenetic information when it becomes available. Therefore, names and grouping arrangements will continue to change as our phylogenetic knowledge increases. We need to keep in mind that each phylogenetic tree and each phylogenetic classification derived from a tree are only hypotheses, which in science always are subject to change.

A real-world example of a phylogenetic tree is given in Focus 1-2, which concerns the phylogeny of amniotes. Phylogenetic arrangements of taxa are extremely important to functional anatomists because they offer a way to select the most meaningful organisms for comparative interpretation. Consider this: about 50,000 species of craniates[5] are living today, and many thousands more are known only as fossils. Given such enormous diversity, we rely on phylogenetic trees to help us distill a manageable yet still meaningful sample of taxa for study.

Homology and Homoplasy

As a first step in building the character matrices needed for a phylogenetic analysis, biologists compare different taxa and search for similarities that appear to be uniquely shared among closely related taxa. Structural, functional, molecular, genetic, or other features may be compared. All contribute to our understanding of relationships among animals, but anatomists emphasize

[5]**Craniata** is the name for all animals with a braincase, such as hagfishes and vertebrates. **Vertebrata** includes all animals with both a braincase and a vertebral column. The overwhelming majority of living craniates are vertebrates; see Chapters 2 and 3.

structural ones. In making comparisons, an investigator must clearly understand the types of features being compared and how they are similar. The two types of similarities important to biologists are homology and homoplasy.

Several kinds of **homology** have been defined. In this book, we are concerned primarily with the types of homology compared among taxa, specifically phylogenetic homology and evolutionary homology. These two concepts are intuitively connected but operationally different. A **phylogenetic homology** is equivalent to a synapomorphy (Focus 1-1) and is the product of a rigorous phylogenetic analysis of empirical data. When structures in two taxa have a fundamental resemblance to each other not shared by other closely related taxa, then we can hypothesize that these structures are phylogenetic homologues and include them in a character matrix. In contrast, the term **evolutionary homology** relates to a specific evolutionary scenario in which we assume a particular phylogeny to be true and wish to propose an idea about evolutionary transformation.

Usually, similar topographical relationships and connections to surrounding parts are the first indication that features in different taxa may be homologues. Similarities in shape and other features may also be influential. As a real-world example, consider the bones in the forelimbs of a human, horse, and seal (Fig. 1-7). Because the limbs are used in different types of locomotion, we should not expect them to be identical, yet the bones have very similar topographical relationships. Sometimes the resemblance is less obvious, as in the case of the auditory ossicles of mammals and portions of the gills arch skeleton of early vertebrates, but the resemblance sometimes becomes evident when one examines a series of different organisms representing stages in the transformation (Chapter 7). Also, the auditory ossicles and gill arches arise from the same em-

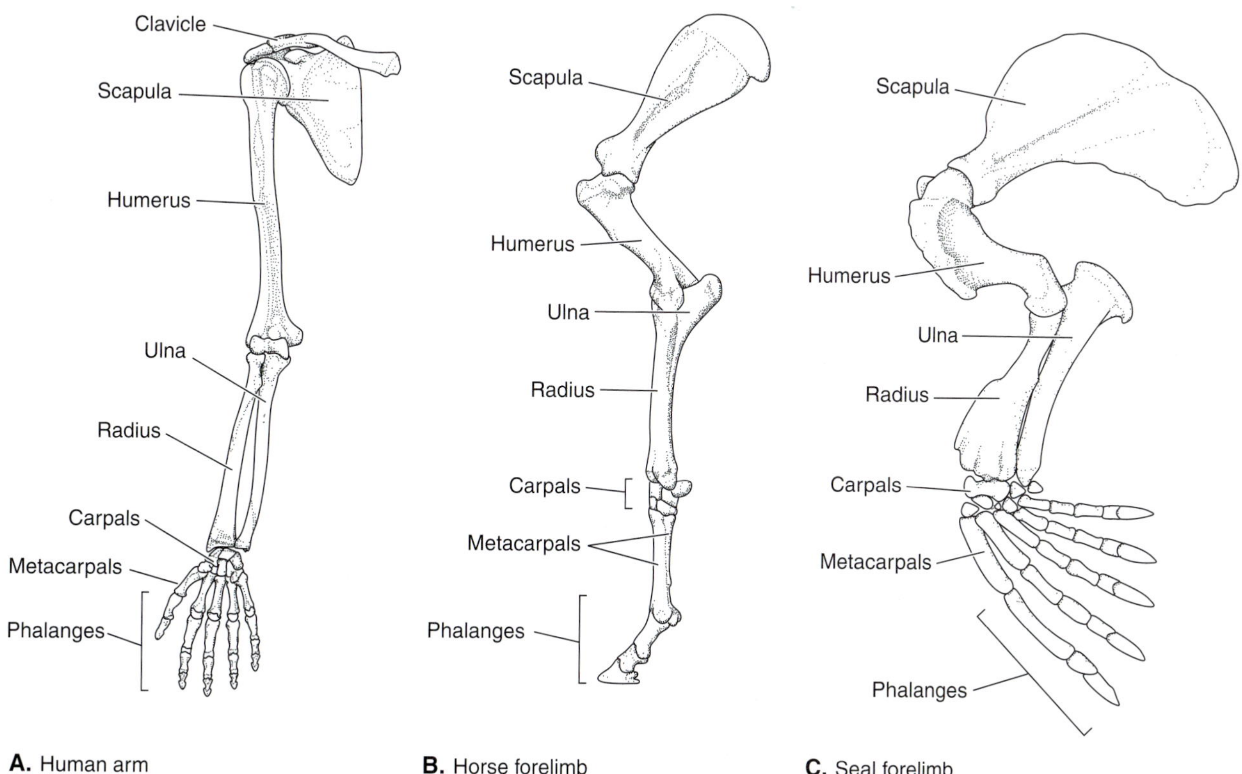

FIGURE 1-7
Homology demonstrated by the limb bones of three mammals. The bones of the left pectoral girdle and limb of a human (*A*), horse (*B*), and seal (*C*) illustrate similar topographical relationships of the major skeletal elements despite differences in limb function: grasping, running, and a steering tool, respectively. The human limb is viewed from the front; the others are from a lateral view. (A, *Modified from Grant;* B, *modified from Sisson;* C, *modified from Howell.)*

bryonic structures, and such ontogenetic information can offer important clues for detecting homology. Homologous structures may or may not have similar functions. Similar function is not necessarily a criterion for homology.

Serial homology refers to similar repeating structures within the same individual. For example, the individual vertebrae are serially homologous to each other, and each develops from the same part of a linear series of embryonic body segments. Serial homology is less relevant to evolutionary studies than is phylogenetic homology (i.e., a synapomorphy suggests more about phylogenetic relationships than does the knowledge that one vertebra is a serial homologue of the next vertebra). **Sexual homology** refers to entities in different sexes of the same species that can be traced back to the same embryonic source (e.g., the glans penis of the male and the glans clitoris of the female, which both derive from the phallus of an embryo).

Structurally similar organs are not necessarily homologous. If structural similarity evolved independently in two or more taxa, then the similarity is **homoplasic** (Gr., *homos* = like + *plastos* = molded). The difference between homology and homoplasy is illustrated in Figure 1-8. In a phylogenetic analysis, homoplasic characters are those that are incongruent with the most parsimonious distribution of all other characters. In an evolutionary scenario, we explain such incongruent characters as examples of convergent or parallel evolution. Familiar examples of homoplasy are the wings of birds, bats, and †pterosaurs[6] (Fig. 1-9). All three types of wings are specialized forelimbs. However, based on many synapomorphies, craniate phylogeny indicates that birds, bats, and †pterosaurs each evolved powered flight independently (i.e., each group had a different flightless ancestor). We say that birds, bats, and †pterosaurs **converged** as they adapted to a flying mode of life and evolved wings. Although the individual bones in the wings are homologues and the wings are all superficially similar airfoils, the three different types of wing did not evolve from each other. This phylogenetic interpretation is supported by differences in the airfoil surfaces. A bird's wing is composed of feathers that grow out from the caudal border of the limb, whereas a bat's wing is essentially an enlarged, webbed hand. †Pterosaurs had a single, elongated digit that supported a stiff, tear-resistant web of skin. The wings of these animals and other such homoplasic structures that have closely similar functions often are referred to as **analogies.**

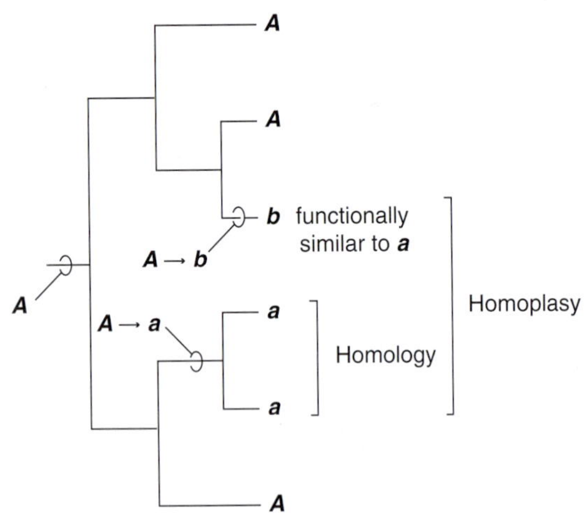

FIGURE 1-8
Contrast between homology and homoplasy. In the bottom lineage, character state ***A*** evolves to ***a***; the presence of character state ***a*** in the two species is an example of homology. In the top lineage, character state ***A*** evolves to ***b***, which is functionally similar to ***a***, which is an example of homoplasy. We may place secondary, process-oriented interpretations on this, such as that ***a*** and ***b*** are examples of convergent evolution.

Earth History and Vertebrate Evolution

It is customary to subdivide the geological record of Earth history into eons, eras, and periods (Fig. 1-10). On the basis of radioactive dating methods, the oldest rocks on Earth are estimated to be 4.6 billion years old, but the earliest known evidence of complex multicellular organisms is far younger (about 1 billion years old; Seilacher et al., 1998). The appearance in the fossil record of animals with hard skeletons occurred more than 600 million years ago (MY) and demarcates the start of the **Phanerozoic eon** (Gr., *phaneros* = visible + *zoos* = animals). Within the Phanerozoic eon, we recognize the **Paleozoic era** (Gr., *palaios* = old), **Mesozoic era** (Gr., *mesos* = middle), and **Cenozoic era** (Gr., *kainos* = recent).

Diverse invertebrate groups, including mollusks, arthropods, and echinoderms, are known from the **Cambrian period,** which is the earliest period of the Paleozoic era (Fig. 1-10). The oldest fossils that are unquestionably craniates are known from rocks of the **Ordovician period.**[7] By the early part of the **Devonian period,** fossils of all major groups of fishes can be found, and the Devonian period is sometimes popu-

[6]Throughout this book, a dagger symbol (†) preceding the name of a group indicates that the group is extinct.

[7]Some fragmentary Late Cambrian fossils have been questionably considered to be remains of craniates.

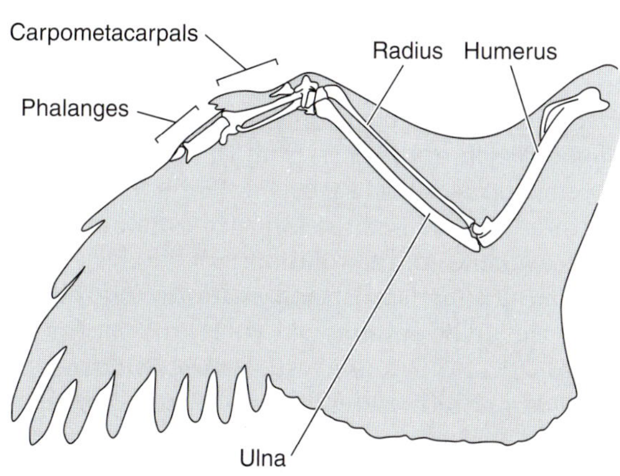

A. Bird wing with airfoil surface made of feathers

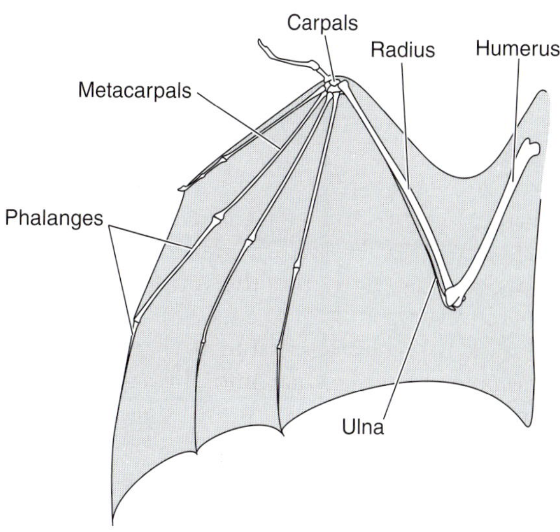

B. Bat wing with airfoil surface made of skin supported by four elongated digits

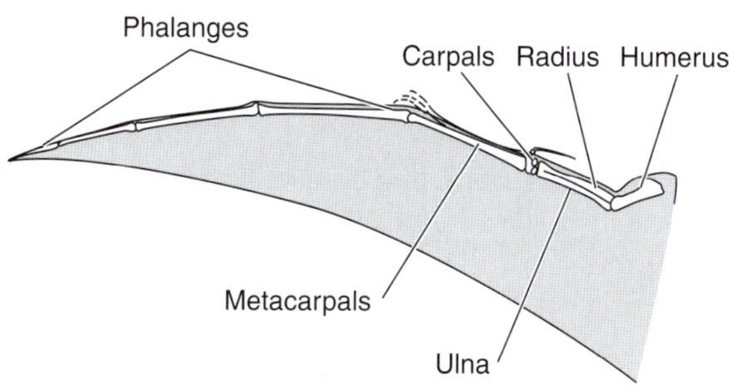

C. †Pterosaur wing with airfoil surface made of skin supported by a single elongated digit

FIGURE 1-9
Homoplasy demonstrated by three independent evolutionary origins of wings for powered flight. *A,* Skeleton and outline of the airfoil surface in a bird, based on Young. *B,* Skeleton and outline of the airfoil surface in a bat, based on Owen. *C,* Skeleton and outline of the airfoil surface in a †pterosaur, based on Gregory. As wings, all three examples are homoplasic because each type of wing evolved independently. However, at the level of comparison of appendages, all three wings are homologous because each represents a modification of a terrestrial vertebrate forelimb.

larly referred to as the "Age of Fishes." Already by the end of the Devonian period, land vertebrates (tetrapods) had evolved. Tetrapods diversified to occupy a range of terrestrial habitats during the **Carboniferous** and **Permian periods.** During the Mesozoic, one clade of tetrapods, the amniotes, diversified extensively. Rocks from the **Triassic period** offer the earliest known occurrences of mammals. Many clades of amniotes popularly known as dinosaurs also occur in the Mesozoic; Dinosauria is a group that, in phylogenetic terms, also includes the birds (Chapter 3). The Mesozoic era is often called the "Age of Reptiles," but the "Age of Diapsids" might be more phylogenetically informative. At the end of the Mesozoic, many groups of amniotes, including the clades of giant dinosaurs, became extinct. Mammals underwent a rapid evolutionary radiation during the Cenozoic era, which is sometimes called the "Age of Mammals." The Cenozoic era could as well be called the "Second Age of Fishes" for during this era, the higher bony fishes known as teleosts exhibited an unequaled radiation in the world's aquatic environments.

Continental drift, which is the incremental movement over the Earth's surface of the tectonic plates that bear the continents, was an important factor in vertebrate evolution. The simplified illustrations in Figure 1-10*A–D* show the position of the continents at several points in time. In our daily lives, we are conditioned to thinking of the continents as fixed on the Earth's surface and having the arrangement in Figure 1-10*A*. But for most of vertebrate history, this continental arrangement did not exist. For example, during the rapid diversification of fishes in the middle part of the Paleozoic era, the continents formed a single "supercontinent," named **Pangea** (Gr., *pantos* = all + *gaia* = Earth), as shown in Figure 1-10*D*. Thus, many

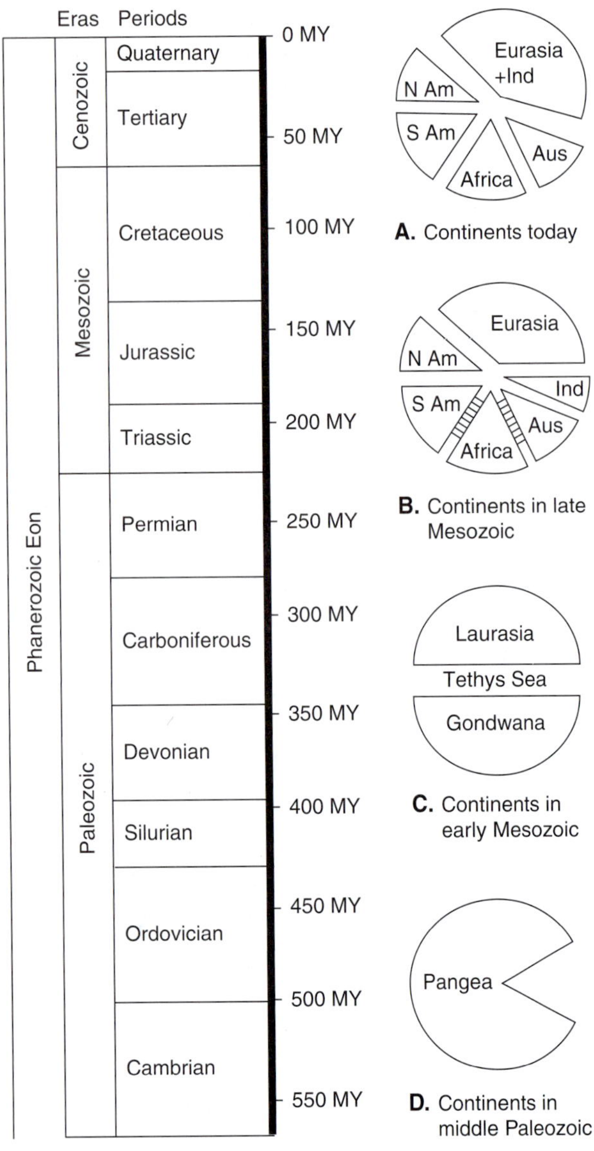

FIGURE 1-10
Geologic time scale and continental drift. *A,* Configuration of the continents (except Antarctica) in the present day. *B,* Configuration of the continents (except Antarctica) in the late Mesozoic. *C,* Configuration of the continents (except Antarctica) in the early Mesozoic. *D,* Configuration of the continents (except Antarctica) in the middle Paleozoic. Aus = Australia; Ind = India; MY = millions of years before present; N Am = North America; S Am = South America.

taxa of early vertebrates that evolved on Pangea had remarkably large distributions across the Earth's surface. Freshwater rhipidistian fishes in the extinct genus †*Eusthenopteron,* for example, occurred across much of the ancient continent of Pangea, for we find their fossils today in sites as far apart as Scotland, Canada, Nevada, and Australia. In the later periods of the Paleozoic era, Pangea split into two supercontinents: **Gondwana** in the south and **Laurasia** in the north (Fig. 1-10*C*). This split affected the global distribution of many groups of vertebrates well into the Mesozoic, when additional rounds of continental breakup divided Gondwana and Laurasia as diagrammed in Figure 1-10*B*. One of the more recent events was the rafting of a continental block northward from the southern hemisphere to Eurasia (Fig. 1-10*A*), where it formed peninsular India and uplifted the Himalayan mountain range as a consequence of the collision.

We model how such Earth processes as continental drift or the uplift of mountain ranges may be linked to phylogenetic history using **historical biogeography** (Fig. 1-11). The first step is a phylogenetic analysis of characters, which in this case yields a hypothesis about relationships among species A, B, and C (Fig. 1-11*A*). Next, we add the biogeographic data for each of these species to the tree (Fig. 1-11*B*), yielding a pattern of **area relationships** (Fig. 1-11*C*). The hypothesized area relationships predict the history of the geographic areas occupied by the three taxa (Fig. 1-11*D*). The prediction in this case is that at Time 1, Areas Z, X, and Y were contiguous. By Time 2, this area had divided to form Areas Z and Y + X, leading eventually to the present pattern in which Areas Z, X, and Y are separate from each other. We may test this prediction by discovering other taxa with the same pattern of area relationships or by recovering relevant Earth history data.

As a real-life example of this method and how continental drift impacted vertebrate history, consider Figure 1-12. These maps chart the arrangement of the continents today (Fig. 1-12*A*) and their appearance during the Early Cretaceous period, 118 MY (Fig. 1-12*B*). The geographic distributions of two tribes of †vidalamiine fishes, indicated by **H** and **I,** are plotted on the maps. Such distributions are referred to as **biogeographical patterns.** The biogeographical pattern of the two tribes does not fit well with present-day geography because deep ocean basins divide the localities where the fossils were found. But if we plot these patterns on a map of the Earth in Cretaceous times, when these †vidalamiine fishes lived, then they make much better geographic sense. For instance, note that the northern group, Group I, occurred along a contiguous continental margin stretching from western North America to the Middle East. The southern group, Group H, was restricted to a smaller portion of the joined continents of Africa and South America. This result suggests that the movement of continents played a role in the evolution of †vidalamiines and their allies; it also corroborates a result from phylogenetic analysis of a character matrix. Because the patterns recovered

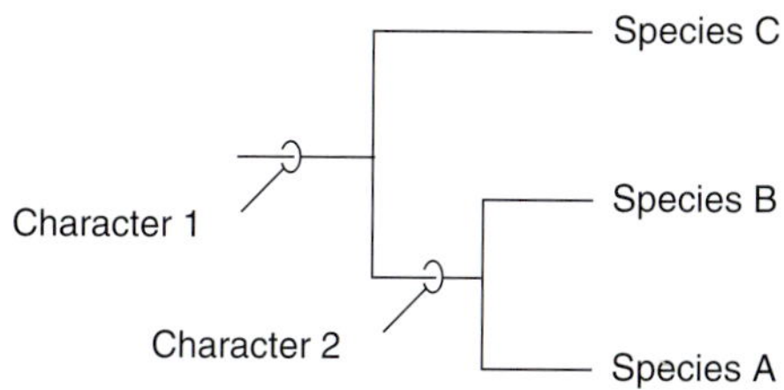

A. Phylogenetic analysis of Characters 1 & 2, indicating interrelationships of Species A, B & C (outgroup implied for character polarities)

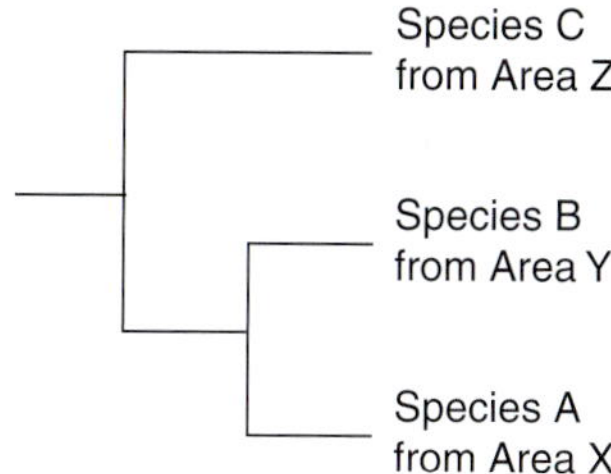

B. Addition of biogeographic information for Species A, B & C

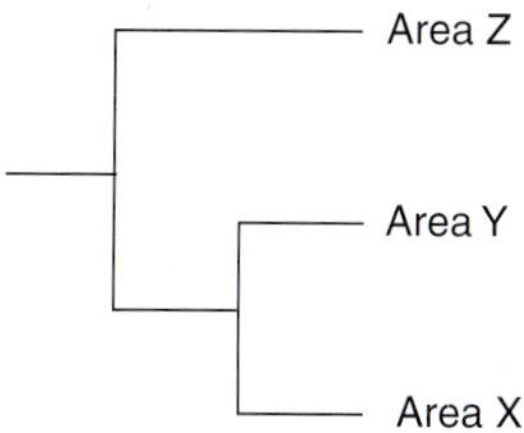

C. Hypothesized area relationships for Areas X, Y and Z

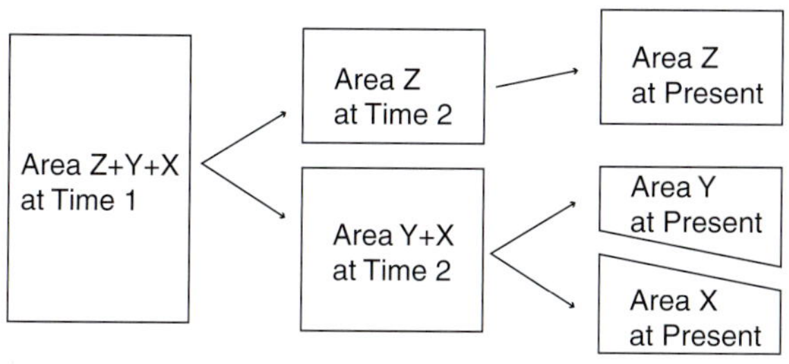

D. Geographic prediction

FIGURE 1-11
The conceptual link between phylogeny and Earth history. *A,* Phylogenetic analysis of Characters 1 and 2 in species A, B, and C. *B,* The areas inhabited by each species are then inserted onto the phylogeny. *C,* Hypothesized relationships of Areas X, Y, and Z based on the phylogeny. *D,* The geographic prediction from C expressed in the form of maps at three different times: Time 1, Time 2, and Present.

from Earth history and phylogenetic analysis corroborate each other, we say they are examples of **repeating patterns** in Earth and phylogenetic history.

Anatomical Terminology

Before continuing our structural and functional analysis of the evolution of vertebrate organ systems, you should understand the bases for anatomical terminology.

Over the years, many authors have used different terms for the same organs. For example, the vagus nerve that carries parasympathetic fibers to the thoracic and abdominal organs was at one time called the "pneumogastric nerve." Medical anatomists long ago recognized the confusion that could be caused by such duplicate terminology and agreed on a code of anatomical nomenclature known as the *Nomina Anatomica.* This code has been updated periodically and has been extended to include embryological and histological terms. In the meantime, veterinary anatomists agreed on a *Nomina Anatomica Veterinaria* that applies human terms to quadrupedal mammals insofar as possible. Major differences relate to differences in posture (e.g., "superior" in human beings becomes "cranial" in quadrupeds). Bird anatomists have agreed on a *Nomina Anatomica Avium.* These codes are not binding, but they are helpful, and we have used them as guides for the terminology in this book. Unfortunately, no guide, other than conventional usage, is available for structures in anamniotes (= fishes and amphibians) and reptiles.

Most anatomical terms derive from Latin or Greek, and they often are written in Latin in scientific papers, as are the names of animals. For example, our biceps muscle is technically known as *Musculus biceps brachii* (a Latin name, and thus italicized), but anatomical terms are given in the vernacular in most writing, and thus we will refer to it as the "biceps brachii muscle" (an Anglicized term, and thus not italicized). The derivation of many widely used terms from their Greek or Latin roots is indicated in this book. The roots are repeated in many combinations, and learning the vast anatomical terminology will become far easier for you to manage if you study the derivations of each new term as you come to it.

Because terms that define positions within the body are needed to describe the location of organs, they must be understood at the outset. Most of these should be clear from Figure 1-13. Notice the differences in terminology between quadrupeds and bipeds. The terms **dorsal** and **ventral** are not used in adult bipeds, and **anterior** and **posterior** are used differently in adult

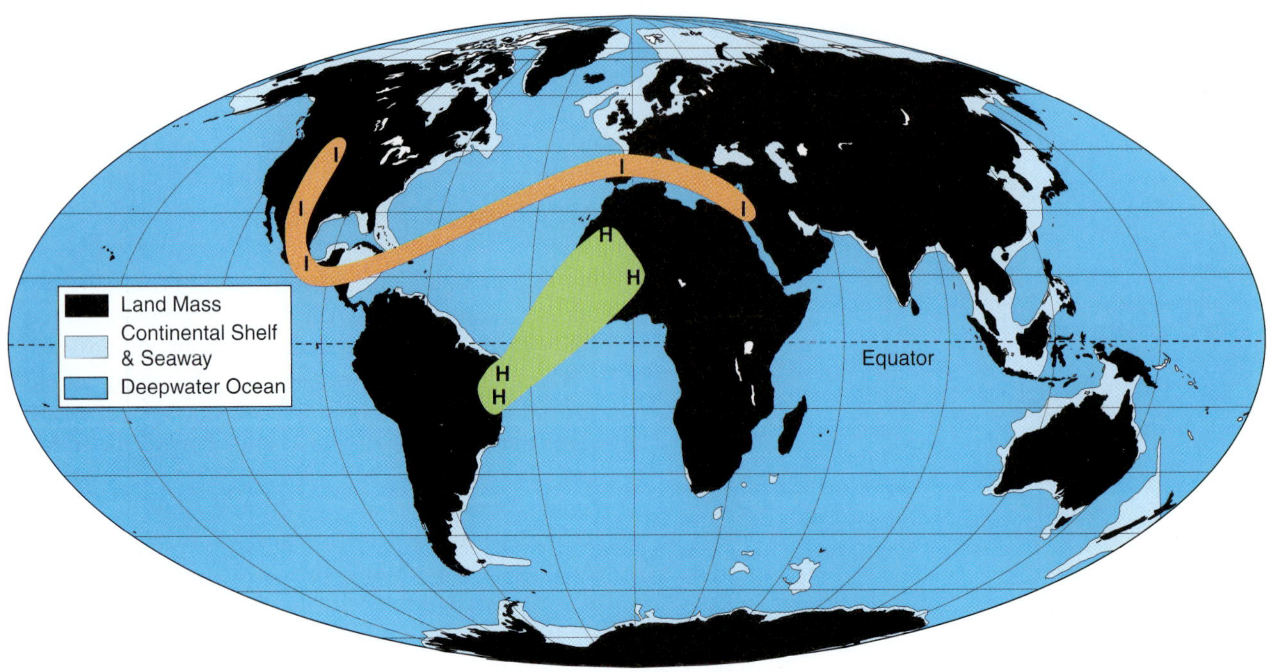

A. Present-day geography

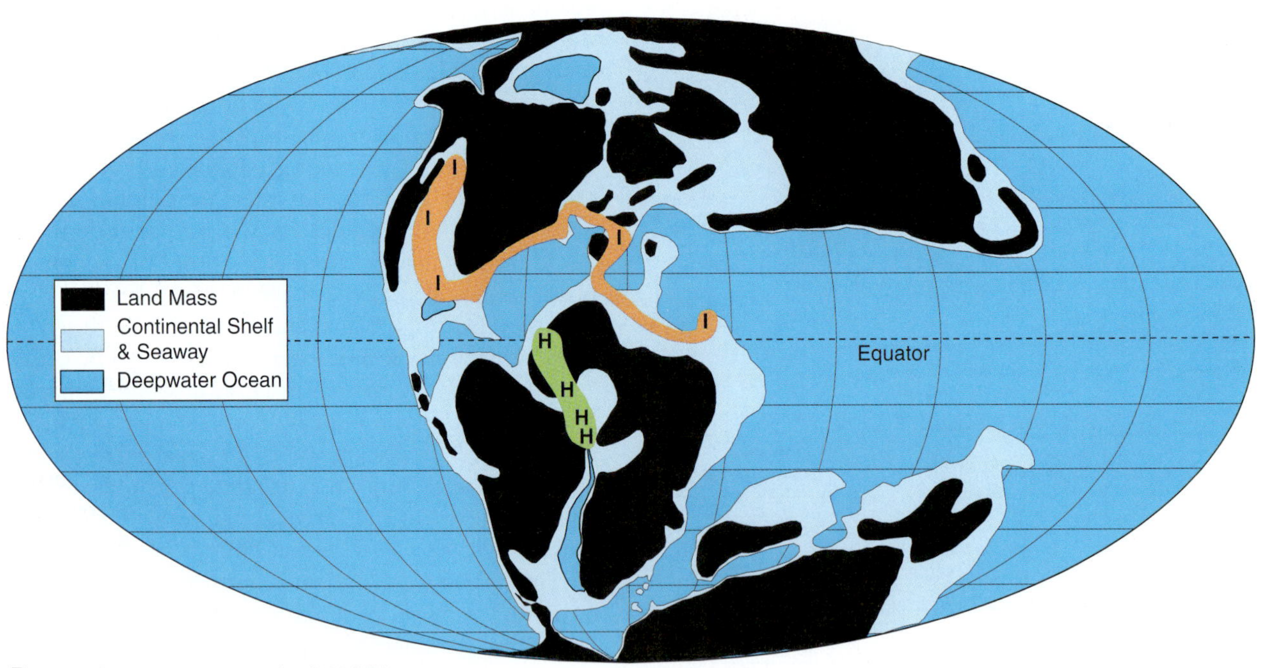

B. Early Cretaceous geography (118 MY)

FIGURE 1-12
Maps showing the distribution of two tribes (H and I) of actinopterygian fishes in the subfamily †Vidalamiinae. *A,* Distribution of fossil localities plotted on present-day map. *B,* Distribution of fossil localities plotted on a map of the continental configurations in the early Cretaceous when these taxa were alive. The biogeographical distribution of †vidalamiines is much better explained when the history of continental drift is considered. *(From Grande and Bemis.)*

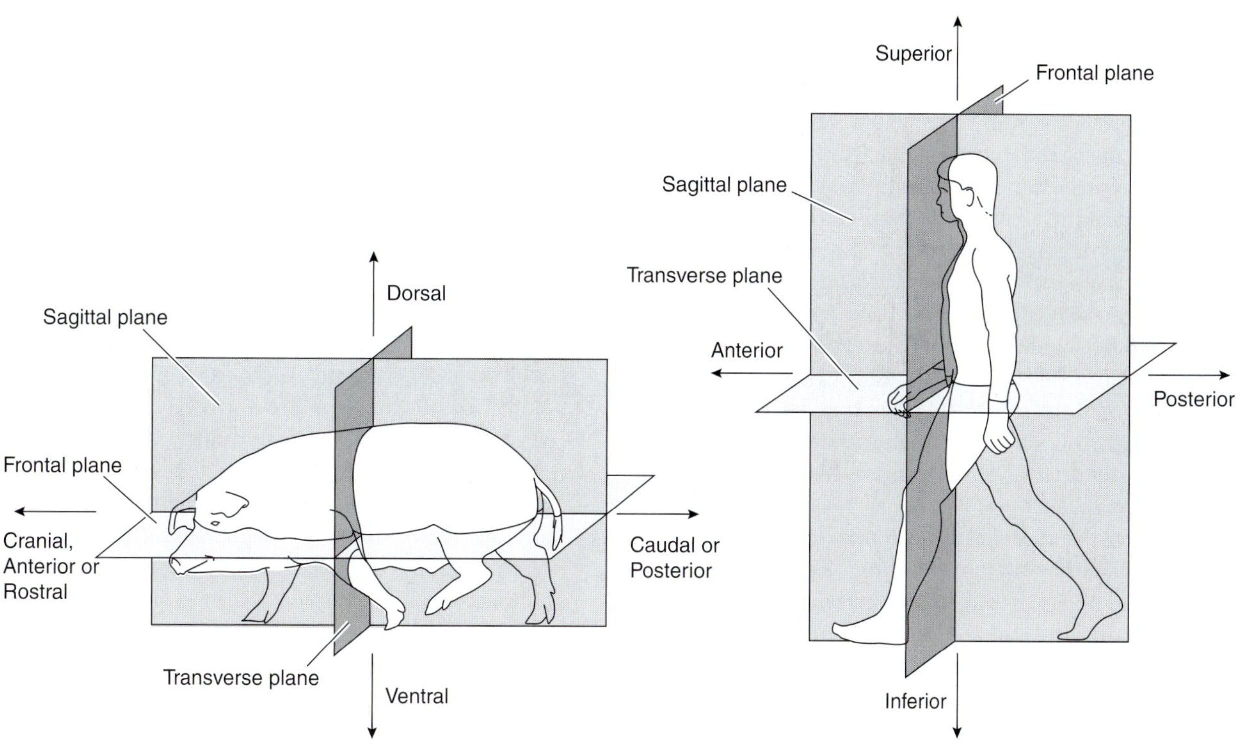

FIGURE 1-13
Terminology for directions and planes of vertebrate bodies in reference to the trunk. *A,* Terminology that applies to most vertebrates and that will be used in this book. *B,* Terminology used for humans, which reflects our bipedal stance.

bipeds and quadrupeds. Some workers avoid using the terms "anterior" and "posterior" for adult quadrupeds, preferring instead the terms **cranial** and **caudal. Rostral** may also be used for "cranial," especially for structures within the head. For elongated structures attached to the body, such as limbs, we call the part toward the base where it joins the body **proximal,** while the part toward the free end is named **distal.**

Some couplets of terms are used to describe movement of parts of a body relative to each other. In each case, the descriptors refer to antagonistic movements (= in opposite directions). Among the more useful couplets describing body movements are **flexion** (i.e., a bending of two parts of a body toward each other) versus **extension** (i.e., a bending of two parts of a body away from each other), **protraction** (i.e., a movement that moves a structure relatively forward) versus **retraction** (i.e., a movement that moves a structure relatively backward), and **abduction** (i.e., movement away from the midline of the body) versus **adduction** (movement of a structure toward the midline of the body). In a few cases, more specialized usages and definitions of these terms are appropriate, and these will be defined in the text as needed.

SUMMARY

1. Functional anatomy examines how the various components of organisms perform specific functions and the linkage between form and function. It examines why and how changes in form and function evolved.
2. Changes in the rate and timing of development over evolutionary time are termed "heterochrony." One example of a heterochronic phenomenon is paedomorphosis, which is the retention of some larval or juvenile features by otherwise sexually mature individuals. Theoretically, paedomorphosis may result from three different processes: deceleration, hypomorphosis, and delayed development (also known as "postdisplacement"). Paedomorphosis is implicated in adapting some living populations to different environmental conditions and may have

played a role in some major evolutionary transitions.

3. Evolutionary relationships among taxa can be depicted by a phylogeny, in which each pair of lineages is displayed by lines that diverge from their most recent common ancestor. Using methods of phylogenetic systematics (cladistics), we develop character matrices and search for shared derived characters (synapomorphies) that indicate monophyletic groups (clades).
4. Cladograms express hypotheses about the pattern of branching of taxa and thereby communicate information about the relative time of divergence. By themselves, however, cladograms do not say anything about the absolute time of divergence of any pair of taxa.
5. Our system of classification groups organisms into a nested hierarchy of progressively inclusive groups, or taxa. The most informative classifications accurately reflect careful phylogenetic analyses. As phylogenetic information increases and phylogenetic trees change, classifications also must be changed.
6. The underlying basis for constructing phylogenies is the search for special similarities, which are known as "synapomorphies" or "phylogenetic homologies." In searching for synapomorphies, investigators must distinguish between phylogenetic homology (features thought to be the result of inheritance from a common ancestor) and homoplasy (features thought to be the result of convergence that have evolved independently in different lineages).
7. Continental drift provides a source of historical patterns independent from the patterns generated by phylogenetic analysis. When a historical biogeographic pattern is congruent with patterns from a phylogenetic analysis, it suggests that the evolution of the Earth's surface and its biota are intimately connected.

GENERAL REFERENCES

Our references include books and articles cited in the text and others that will lead students to important additional sources of information. We emphasize material published in the past decade or two. Older works are included only if they are cited in the text, are classics, or are of particular interest. References listed in this section are works of general interest and works that apply to more than one organ system or group of organisms. Those related to the subject of particular chapters are placed at the ends of the chapters.

Alexander, R. McNeil, 1974: *Functional Design in Fishes,* 3rd edition. London, Hutchinson & Co.

Alexander, R. McNeil, 1981: *The Chordates,* 2nd edition. Cambridge, Cambridge University Press.

Alexander, R. McNeil, 1983: *Animal Mechanics,* 2nd edition. Oxford, Blackwell Scientific.

Bellairs, A. d'Arcy, 1969: *The Life of Reptiles.* London, Weidenfeld & Nicolson.

Bellairs, A. d'Arcy, and Cox, C. B., 1976: *Morphology and Biology of the Reptiles.* London, published for the Linnean Society by Academic Press.

Bemis, W. E., Burggren, W. W., and Kemp, N. E., 1986: *The Biology and Evolution of the Lungfishes.* New York, Alan R. Liss.

Bolk, L., Göppert, E., Kallius, E., and Lubosch, W., editors, 1931–1938: Handbuch der vergleichenden Anatomie der Wirbeltiere. 6 vols. Reprinted: Amsterdam, A. Asher & Co., 1967.

Bond, C. E., 1996: *Biology of Fishes,* 2nd edition. Philadelphia, W. B. Saunders.

Brewer, S. D., and Hafner, R., 1996. Phylogenetic Investigator 2.0: Software for teaching phylogenetic inference. 1996 BioQuest Library (4th edition), ePress Project. College Park, University of Maryland Press.

Carroll, R. L., 1988: *Vertebrate Paleontology and Evolution.* New York, W. H. Freeman & Co.

Cogger, H. G., and Zweifel, R. G., editors, 1998: *Encyclopedia of Reptiles and Amphibians,* 2nd edition. San Diego, Academic Press.

Corliss, C. E., 1976: *Patten's Human Embryology.* New York, McGraw-Hill.

Dawson, T. J., 1983: *Monotremes and Marsupials: The Other Mammals.* Southampton, Edward Arnold.

Dorst, J., 1974: *The Life of Birds.* Translated by J. C. J. Galbraith. New York, Columbia University Press.

Duellman, W. E., and Trueb, L., 1985: *Biology of the Amphibia.* New York, McGraw-Hill.

Dullemeijer, P., 1974: *Concepts and Approaches in Animal Morphology.* Assen, The Netherlands, Van Gorcum & Co.

Farris, J. S., 1988: Hennig86, version 1.5. Computer program and documentation. Port Jefferson Station, New York.

Fawcett, D. M., 1986: *Bloom and Fawcett, A Textbook of Histology,* 11th edition. Philadelphia, W. B. Saunders.

Feder, M. E., and Burggren, W. W., editors, 1992: *Environmental Physiology of the Amphibians.* Chicago, University of Chicago Press.

Feduccia, A., and McCrady, E., 1991: *Torrey's Morphogenesis of the Vertebrates,* 5th edition. New York, John Wiley & Sons.

Ferguson, M. W. J., editor, 1985: *Structure, Development and Evolution of Reptiles.* Symposia of the Zoological Society of London, no. 52. London, Academic Press.

Forey, P. L., 1998: *History of the Coelacanth Fishes.* London, Chapman and Hall.

Forshaw, J., editor, 1998: *Encyclopedia of Birds,* 2nd edition. San Diego, Academic Press.

Gans, C., 1974: *Biomechanics, An Approach to Vertebrate Biology.* Philadelphia, J. B. Lippincott. Reprinted 1980, Ann Arbor, Mich., University of Michigan Press.

Gans, C., et al., editors, 1969–1992: *Biology of the Reptilia.* New York, Academic Press.

Goodenough, J., McGuire, B., and Wallace, R., 1993: *Perspectives in Animal Behavior.* New York, John Wiley & Sons.

Goodrich, E. S., 1930: *Studies on the Structure and Development of Vertebrates.* London, Macmillan & Co.

Gould, E., and McKay, G., editors, 1998: *Encyclopedia of Mammals,* 2nd edition. San Diego, Academic Press.

Grande, L., and Bemis, W. E., 1998: *A comprehensive phylogenetic study of amiid fishes (Amiidae) based on comparative skeletal anatomy: An empirical search for interconnected patterns of natural history.* Society of Vertebrate Paleontology, Memoir 4, Supplement to *Journal of Vertebrate Paleontology* 18.

Grande, L., and Rieppel, O., editors, 1994: *Interpreting the Hierarchy of Nature: From Systematic Patterns of Evolutionary Process Theories.* San Diego, Academic Press.

Grassé, P. P., editor, 1948–1973: *Traité de Zoologie.* Volumes 11–17 deal with tunicates, cephalochordates, and vertebrates. Paris, Masson et Cie.

Gregory, W. K., 1951: *Evolution Emerging.* New York, Macmillan.

Hardisty, M. W., 1979: *Biology of Cyclostomes.* London, Chapman and Hall.

Harrison, R. J., editor, 1972: *Functional Anatomy of Marine Mammals.* New York, Academic Press.

Helfman, G. S., Collette, B. B., and Facey, D. E., 1997: *The Diversity of Fishes.* Malden, Mass., Blackwell Scientific.

Hildebrand, M., 1995: *Analysis of Vertebrate Structure,* 4th edition. New York, John Wiley & Sons.

Hildebrand, M., Bramble, D. M., Liem, K. F., and Wake, D. B., editors, 1985: *Functional Vertebrate Morphology.* Cambridge, Mass., Harvard University Press.

Hoar, W. S., 1975: *General and Comparative Physiology,* 2nd edition. Englewood Cliffs, N.J., Prentice-Hall.

Hoar, W. S., and Randall, D. J., editors, volumes 1–17, 1969–1998: *Fish Physiology.* New York, Academic Press.

Howell, A. B., 1930: *Aquatic Mammals: Their Adaptations to Life in the Water.* Springfield, Ill., Charles C Thomas.

Jaeger, E. C., 1978: *A Source-Book of Biological Names and Terms,* 3rd edition. Springfield, Ill., Charles C Thomas.

Janvier, P., 1996: *Early Vertebrates.* New York, Oxford University Press.

Jarvik, E., 1980–1981: *Basic Structure and Evolution of Vertebrates.* London, Academic Press.

Jollie, M., 1973: *Chordate Morphology.* Huntington, N. Y., Robert E. Krieger.

Kemp, T. S., 1982: *Mammal-Like Reptiles and the Origin of Mammals.* London, Academic Press.

King, A. S., and McLelland, J., 1979: *Form and Function in Birds.* London, Academic Press.

Kitching, I. J., Forey, P. L., and Humphries, C., 1998: *Cladistics: The Theory and Practice of Parsimony Analysis,* 2d edition. Oxford, Oxford University Press.

Kluge, A. G., Frye, B. E., Johansen, K., Liem, K. F., Noback, C. R., Olsen, I. D., and Waterman, A. J., 1977: *Chordate Structure and Function,* 2nd edition. New York, Macmillan.

Maddison, W. P., and Maddison, D. R., 1992: *MacClade, version 3: Analysis of Phylogeny and Character Evolution.* Sunderland, Mass., Sinauer Associates.

Moyle, P. B., and Cech, J. J., Jr., 1999: *Fishes: An Introduction to Ichthyology,* 4th edition. Englewood Cliffs, N. J., Prentice-Hall.

Moy-Thomas, J. A., and Miles, R. S., 1971: *Paleozoic Fishes.* Philadelphia, W. B. Saunders.

Neal, H. V., and Rand, H. W., 1936: *Comparative Anatomy.* Philadelphia, P. Blakiston's Son & Co.

Nelson, G., and Platnick, N., 1981: *Systematics and Biogeography: Cladistics and Vicariance.* New York, Columbia University Press.

Nelson, J. S., 1994: *Fishes of the World,* 3rd edition. New York, John Wiley & Sons.

Nixon, K. C., 1992: CLADOS 1.1 IBM PC-compatible character analysis program. Ithaca, New York.

Paxton, J. R., and Eschmeyer, W. N., editors, 1998: *Encyclopedia of Fishes,* 2nd edition. San Diego, Academic Press.

Piveteau, J., editor, 1952–1968: *Traité de Paleontologie.* Paris, Masson et Cie.

Portmann, A., 1948: *Einfuhrung in die Vergleichende Morphologie der Wirbeltiere.* Basel, Benno Schwabe & Co. Verlag.

Pough, F. H., Andrews, R. M., Cadle, J. E., Crump, M. L., Savitzky, A. H., and Wells, K. D., 1998: *Herpetology.* Englewood Cliffs, N. J., Prentice-Hall.

Pough, F. H., Janis, C. M., and Heiser, J. B., 1998: *Vertebrate Life,* 5th edition. Upper Saddle River, N. J., Prentice-Hall.

Prosser, C. L., editor, 1991: *Environmental and Metabolic Animal Physiology, Comparative Animal Physiology,* 4th edition. New York, Wiley-Liss.

Prosser, C. L., editor, 1991: *Neural and Integrative Animal Physiology, Comparative Animal Physiology,* 4th edition. New York, Wiley-Liss.

Romer, A. S., and Parsons, T. S., 1986: *The Vertebrate Body,* 6th edition. Philadelphia, Saunders College Publishing.

Schmidt-Nielsen, K., 1983: *Animal Physiology: Adaptation and Environment,* 3rd edition. Cambridge, Mass., Cambridge University Press.

Schmidt-Nielsen, K., 1984: *Scaling, Why Is Animal Shape So Important?* Cambridge, Mass., Cambridge University Press.

Schmidt-Nielsen, K., Bolis, L., Taylor, C. R., Bentley, P. J., and Stevens, C. E., editors, 1980: *Comparative Physiology: Primitive Mammals.* Cambridge, Mass., Cambridge University Press.

Schultze, H. P., and Trueb, L., editors, 1991: *Origins of the Higher Groups of Tetrapods.* Ithaca, N. Y., Cornell University Press.

Shuttleworth, T. V., editor, 1988: *Physiology of Elasmobranch Fishes.* Berlin, Springer-Verlag.

Sisson, S., 1930: *The Anatomy of the Domestic Animals.* Philadelphia, W. B. Saunders.

Smith, H. M., 1960: *Evolution of Chordate Structure.* New York, Holt.

Starck, D., 1978–1982: *Vergleichende Anatomie der Wirbeltiere.* Berlin, Springer-Verlag.

Stiassny, M. L. J., Parenti, L. R., and Johnson, G. D., editors, 1996: *Interrelationships of Fishes.* San Diego, Academic Press.

Sturkie, P. D., editor, 1986: *Avian Physiology,* 4th edition. New York, Springer-Verlag.

Swofford, D. L., 2000: PAUP* 4.0 BETA. Sunderland, Mass., Sinauer Associates.

Taylor, C. R., Johansen, K., and Bolis, L., editors, 1982: *A Companion to Animal Physiology.* Cambridge, Mass., Cambridge University Press.

Tyndale-Biscoe, H., 1973: *Life of Marsupials.* London, Edward Arnold, Australia.

Vaughan, T., Ryan, J. M., and Czaplewski N., 2000: *Mammalogy.* Philadelphia, Saunders College Publishing.

Wake, D. B., and Roth, G., 1989: *Complex Organismal Function: Integration and Evolution in Vertebrates.* New York, John Wiley & Sons.

Wake, M. H., editor, 1979: *Hyman's Comparative Anatomy,* 3rd edition. Chicago, University of Chicago Press.

Walker, W. F., Jr., and Homberger, D. G., 1992: *Vertebrate Dissection,* 8th edition. Philadelphia, Saunders College Publishing.

Weiner, J., 1994: *The Beak of the Finch: A Story of Evolution in Our Time.* New York, Knopf.

Welty, J. C., and Baptista, L., 1988: *The Life of Birds,* 4th edition. Philadelphia, Saunders College Publishing.

Williams, P. L., Berry, M. M., Collins, P., Dyson, M., Dussek, J. E., and Ferguson, M. W. T., editors, 1995: *Gray's Anatomy,* 38th British edition. Edinburgh, Churchill Livingstone.

Wolfe, R. G., 1991: *Functional Chordate Anatomy.* Lexington, Mass., D. C. Heath & Co.

Young, J. Z., 1981: *The Life of Vertebrates,* 3rd edition. Oxford, Clarendon Press.

Young, J. Z., and Hobbs, M. J., 1975: *The Life of Mammals,* 2nd edition. Oxford, Clarendon Press.

Zug, G. R., 1993: *Herpetology: An Introductory Biology of Amphibians and Reptiles.* San Diego, Academic Press.

REFERENCES FOR CHAPTER 1

Alberch, P., Gould, S. J., Oster, G. F., and Wake, D. B., 1979: Size and shape in ontogeny and phylogeny. *Paleobiology,* 5: 296–317.

Alexander, R. McNeil, 1985: Body support, scaling, and allometry. *In* Hildebrand, M., Bramble, D. M., Liem, K. F., and Wake, D. B., editors: *Functional Vertebrate Morphology.* Cambridge, Mass., Harvard University Press.

Baumel, J. J., King, A. S., Lucas, A. M., Breazile, J. E., and Evans, H. E., 1979: *Nomina Anatomica Avium.* London, Academic Press.

Bemis, W. E., 1984: Paedomorphosis and the evolution of the Dipnoi. *Paleobiology,* 10: 293–307.

Calder, W. A. III, 1984: *Size, Function, and Life History.* Cambridge, Mass., Harvard University Press.

Charig, A., 1982: Systematics in biology: A fundamental comparison of some major schools of thought. *In* Joysey, K. A., and Friday, A. E., editors: *Problems of Phylogenetic Reconstruction.* New York, Academic Press.

Darwin, C., 1859: *The Origin of Species by Means of Natural Selection or the Preservation of Favored Races in the Struggle for Life.* New York, Modern Library Edition.

DeBeer, G. R., 1958: *Embryos and Ancestors,* 3rd edition. London, Oxford University Press.

Dial, K. P., Goslow, G. E., Jr., and Jenkins, F. A., Jr., 1991: The functional anatomy of the shoulder of the European starling (*Sturnus vulgaris*). *Journal of Morphology,* 207: 327–344.

Eldredge, N., and Cracraft, J., 1980: *Phylogenetic Patterns and the Evolutionary Process.* New York, Columbia University Press.

Frazzetta, T. H., 1975: *Complex Adaptations in Evolving Populations.* Sunderland, Mass., Sinauer Associates.

Gilbert, S. F., 1991: *Developmental Biology,* 3rd edition. Sunderland, Mass., Sinauer Associates.

Gould, S. J., 1966: Allometry and size in ontogeny and phylogeny. *Biological Reviews,* 41: 587–640.

Gould S. J., 1977: *Ontogeny and Phylogeny.* Cambridge, Mass., Harvard University Press.

Grande, L., 1985: The use of paleontology in systematics and biogeography, and a time control refinement for historical biogeography. *Paleobiology,* 11: 234–243.

Grande, L., and Rieppel, O., 1994a: Introduction to pattern and process perspectives. *In* Grande, L., and Rieppel, O., editors: *Interpreting the Hierarchy of Nature: From Systematic Patterns to Evolutionary Process Theories.* San Diego, Academic Press.

Grande, L., and Rieppel, O., 1994b: Glossary. *In* Grande, L., and Rieppel, O., editors: *Interpreting the Hierarchy of Nature: From Systematic Patterns to Evolutionary Process Theories.* San Diego, Academic Press.

Grande, T., 1994: Phylogeny and paedomorphosis in an African family of freshwater fishes (Gonorynchiformes: Kneriidae). Fieldiana, *Zoology,* 1459: 1–20.

Grant, J. C. B., 1972: *An Atlas of Anatomy.* Baltimore, Williams & Wilkins.

Huxley, J., 1972: *Problems of Relative Growth.* Reprinted in 1932: New York, Dover Publications.

International Committee on Veterinary Anatomical Nomenclature, 1983: *Nomina Anatomica Veterinaria,* 3rd edition. Ithaca, N. Y., Department of Veterinary Anatomy, Cornell University.

Janvier, P., 1999: Catching the first fish. *Nature,* 402: 21–22.

Lauder, G. V., 1981: Form and function: Structural analysis in evolutionary morphology. *Paleobiology,* 7: 430–442.

Lentz, T., 1971: *Cell Fine Structure.* Philadelphia, W. B. Saunders.

Mayr, E., 1969: *Principles of Systematic Zoology.* New York, McGraw-Hill.

McMahon, T., 1973: Size and shape in biology. *Science,* 179: 1201–1204.

Mossman, H. W., 1987: *Vertebrate Fetal Membranes.* New Brunswick, N. J., Rutgers University Press.

Owen, R., 1866–1868: *On the Anatomy of Vertebrates, Volumes 1–3.* London, Longmans, Green and Company.

Reilly, S. M., Wiley, E. O., and Meinhardt, D. J., 1997: An integrative approach to heterochrony: The distinction between interspecific and intraspecific phenomena. *Zoological Journal of the Linnean Society,* 60: 119–143.

Rieppel, O., and Grande, L., 1994: Summary and comments on systematic pattern and evolutionary process. *In* Grande, L., and Rieppel, O., editors: *Interpreting the Hierarchy of Nature: From Systematic Patterns to Evolutionary Process Theories.* San Diego, Academic Press.

Seilacher A., Bose, P. K., and Pfluger, F., 1998: Triploblastic animals more than 1 billion years ago: Trace fossil evidence from India. *Science,* 282: 80–83.

Thompson, D'Arcy W., 1961: *On Growth and Form,* abridged edition. Edited by J. T. Bonner. Cambridge, Mass., Cambridge University Press.

Wiley, E. O., 1981: *Phylogenetics: The Theory and Practice of Phylogenetic Systematics.* New York, John Wiley & Sons.

2

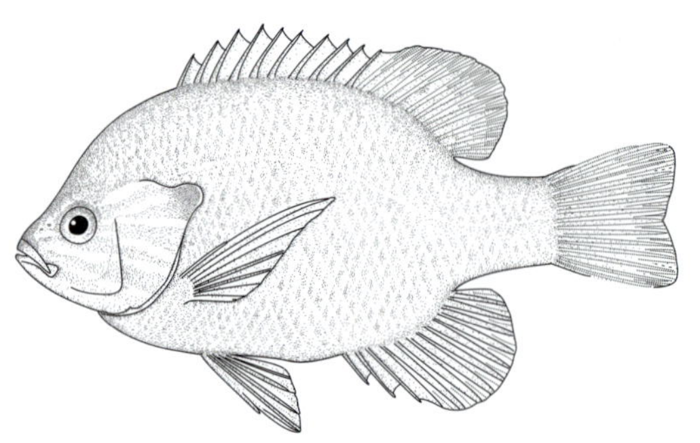

Phylogenetic Relationships of Chordates and Craniates

PRECIS

This chapter explores characters of chordates and other metazoans with a close phylogenetic relationship to craniates. This material enables us to better understand the distinctive derived characters of these groups and provides necessary background for analyses of their functional anatomy and evolution.

OUTLINE

As noted in Chapter 1, **Craniata** is the name of the group that includes the living hagfishes and vertebrates. Such animals are the main focus of this book, but to understand them we must first look outward to other groups of metazoans. Particularly important are two groups of small, soft-bodied, filter-feeding marine animals known as sea squirts (Tunicata) and amphioxus (Cephalochordata). In the scheme that we adopt, Tunicata + Cephalochordata + Craniata belong to Chordata (Fig. 2-1). Craniates are generally active animals with well-developed heads. In contrast, an adult sea squirt is sac-shaped and lives attached to the ocean floor. An adult amphioxus is superficially fish shaped, but it is a relatively inactive filter-feeder that lives in its shallow burrow in the sand. Although the three clades of chordates contain quite different kinds of animals, we regard Chordata as monophyletic because, at some stage of its development, every chordate exhibits five uniquely derived characters or synapomorphies of the group: (1) pharyngeal pouches that grow laterally from the pharynx and often open to the surface as pores or gill slits; (2) a groove in the pharyngeal floor known as the endostyle, or a thyroid gland derived from part of the endostyle; (3) a stiff, longitudinal rod of turgid cells along the dorsal part of the body that is called a notochord; (4) a single, tubular nerve cord that is located dorsal to the notochord; and (5) a larva or embryo with a postanal tail. To understand the significance of these five chordate synapomorphies, and to understand hypotheses concerning the evolutionary origin of chordates and craniates, we must look even further outward

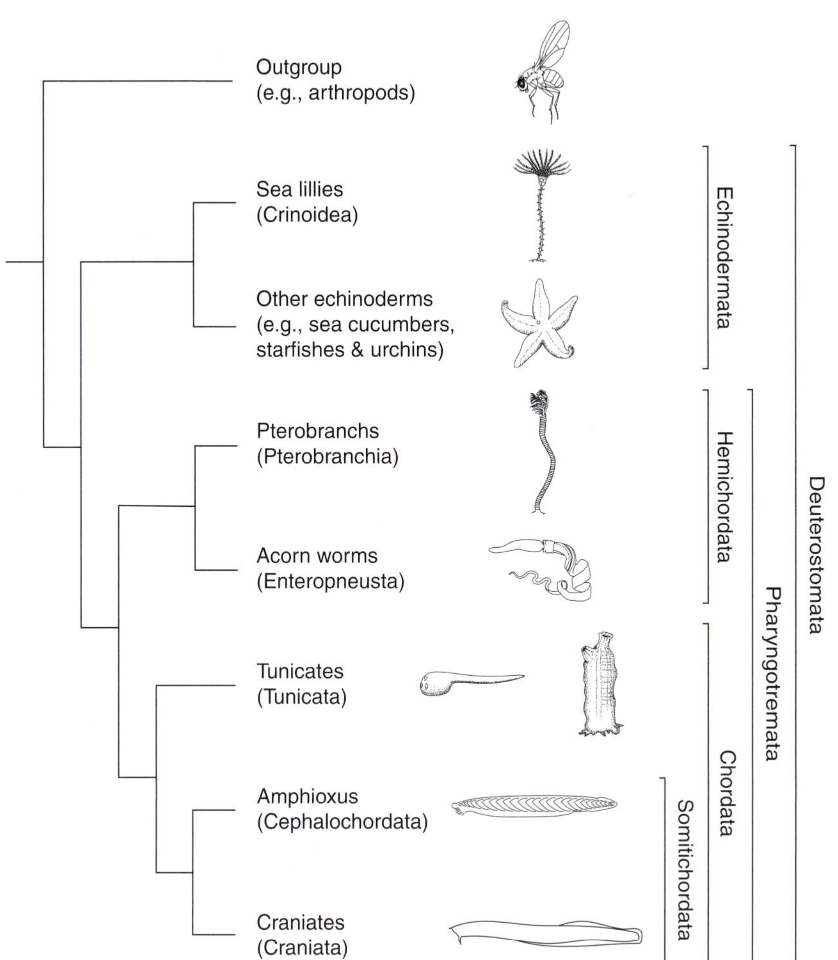

FIGURE 2-1
Cladogram showing phylogenetic relationships of major groups of living deuterostomes. Characters supporting this interpretation are discussed throughout this chapter and listed in the chapter summary.

to consider some other members of the larger group, Deuterostomata, to which Chordata belongs.

Characteristics and Phylogenetic Relationships of Deuterostomes

Echinoderms, hemichordates, and chordates belong to a major group within the animal kingdom known as **Deuterostomata** (Fig. 2-1). This section briefly reviews characteristics of echinoderms and hemichordates and describes evidence for deuterostome and pharyngotremate monophyly, including the derivation and meaning of these names.

Echinoderms and Hemichordates

Echinodermata (Gr., *echinos* = spiny + *derma* = skin) includes sea lilies, brittle and basket stars, sea cucumbers, sea urchins, and starfishes, as well as many other extinct clades. All known living and extinct echinoderms are marine. Echinoderms are considered to form a monophyletic group, based on synapomorphies such as their peculiar water vascular system and tube feet. As adults, echinoderms exhibit radial symmetry and body plans so different from those of craniates that you might never suspect that they could be closely related to them. In particular, though, echinoderms and vertebrates share some important characters of early embryonic development, which we discuss later. The sea lilies, or crinoids, have a stalk and a series of arms used for filter-feeding (Fig. 2-1)[1].

Hemichordata (L., *hemi* = half + *chorde* = cord) includes two clades: the pterobranchs and enteropneusts (Fig. 2-1). Like echinoderms, all known hemi-

[1]Most living crinoids belong to Comatulida (the feather stars), a group in which the stalk is absent in the adult stage. Feather stars have a stalk in their early post-larval stages.

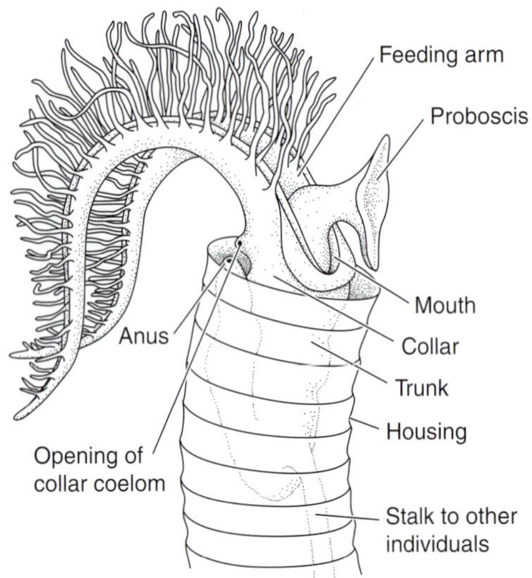

FIGURE 2-2
Anatomy of a representative pterobranch. The animal illustrated belongs to the genus *Rhabdopleura,* which forms colonies; only one individual of the colony is shown here. All hemichordates have a tripartite body consisting of a proboscis, collar, and trunk.

chordates are marine animals. We regard Hemichordata as monophyletic, based on the synapomorphy of a tripartite body consisting of a proboscis, collar, and trunk. Figure 2-2 shows a representative **pterobranch** (Gr., *pter* = wing + *branchia* = gill). These small, colonial animals superficially resemble plants. Each individual is lodged in a little cup of a branching, trunklike house that is secreted by members of the colony as they multiply by budding. Individuals may or may not remain connected to each other, but each has a distinct proboscis, collar, and trunk, each with its own body cavity, or **coelom** (Gr., *koiloma* = a hollow). The collar extends into one or more pairs of branching, ciliated feeding arms that collect food. Species in one genus, *Cephalodiscus,* have a single pair of pharyngeal slits; other pterobranchs have none. The intestine forms a U-shaped loop, so that the anus opens near the feeding arms.

More familiar hemichordates are the approximately 80 species of acorn worms, or **enteropneusts** (Gr., *enteros* = gut + *pneusta* = hole), such as *Saccoglossus* (Fig. 2-3). Some species of acorn worms reach more than 2 m in total length. Like a pterobranch, the body of an acorn worm is divided into a proboscis, collar, and trunk, each of which contains a coelomic cavity. The proboscis and collar coelom cavities connect by pores to the outside so that they can be inflated with seawater. Enteropneusts move through the sand by the ciliary action of epidermal cells assisted by actions of the proboscis and collar. The proboscis has longitudinal and circular muscles, which enable it to contract and lengthen. Once the inflated collar anchors the body, the proboscis can be pushed forward into the sand. Next, the proboscis can be inflated and anchored, and the rest of the body can be pulled up behind it. The proboscis is also a food-gathering organ. Food particles falling on it are trapped in mucus and carried by cilia into the mouth, which is located ventrally just in front of the collar (Fig. 2-3*B* and *C*). Excess water escapes from the pharynx through many paired **pharyngeal pores** located dorsolaterally along the anterior part of the trunk. These pores increase in number during development by the formation of **tongue bars** that subdivide them. In some species, folds of the body wall extend dorsally to form an atriumlike space that partly covers the pharyngeal slits. The intestine extends caudally from the pharynx to the terminal anus.

The nervous system of enteropneusts consists of a subepidermal network of neurons that becomes condensed dorsally and ventrally into solid nerve cords (Fig. 2-3*C*). The dorsal nerve strand may contain a few cavities in the collar region, but these are not regarded as homologous to the tubular nerve cord characteristic of chordates because of differences in the way that they form. The base of the proboscis contains a pharyngeal diverticulum known as the **stomochord** (Fig. 2-3*C*). Its wall is composed of vacuolated cells that resemble those that make up the distinctive notochord of chordates. Some investigators consider the stomochord and notochord to be homologous, but this is doubtful because their positional relationships and embryonic development are different.

Chordata is the third major group of Deuterostomata that we will discuss (Fig. 2-1), but we defer our treatment of chordates until after the next section.

Monophyly of Deuterostomes and Pharyngotremates

As shown in Figure 2-1, we consider that echinoderms, hemichordates, and chordates belong to a monophyletic group known as Deuterostomata. Evidence for the monophyly of deuterostomes comes from similarities in their early embryonic development. Patterns seen in deuterostome development differ from those seen in other groups of animals, such as annelids (i.e., earthworms, leeches, and polychaetes), arthropods (i.e., insects, crustaceans, arachnids, and others), and mollusks (i.e., snails, clams, cephalopods, and others). In annelids, arthropods, and mollusks, the division, or

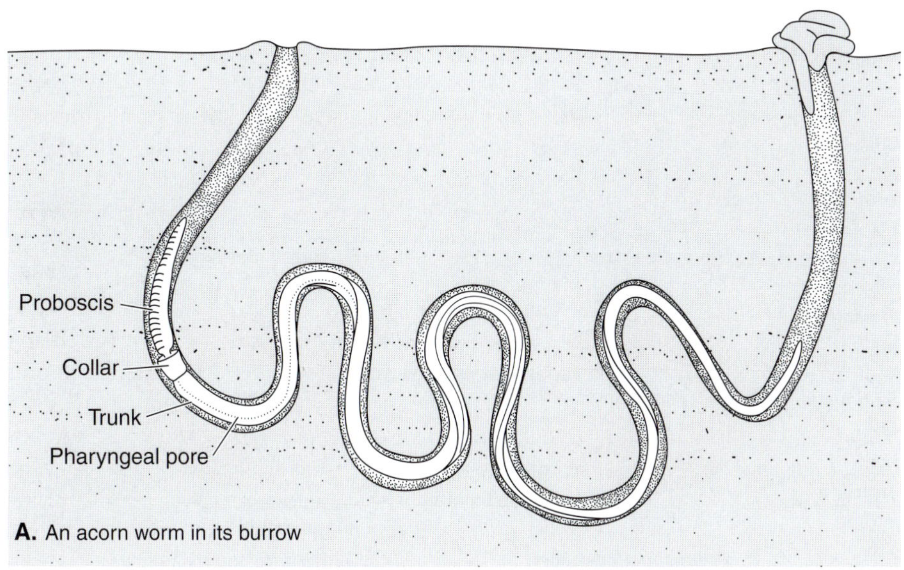

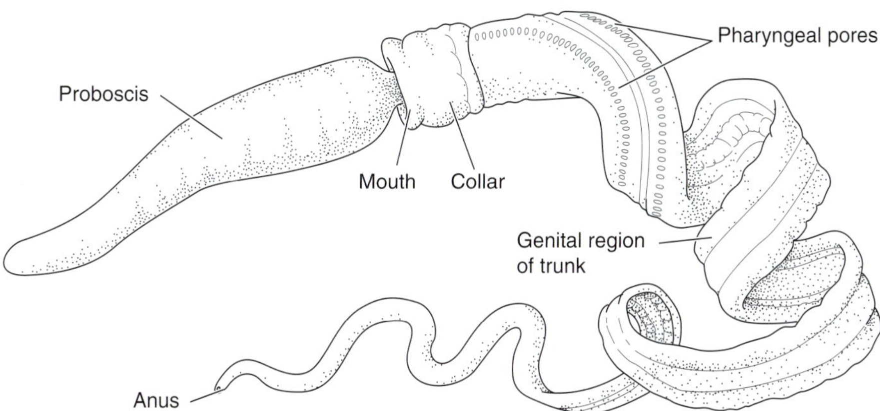

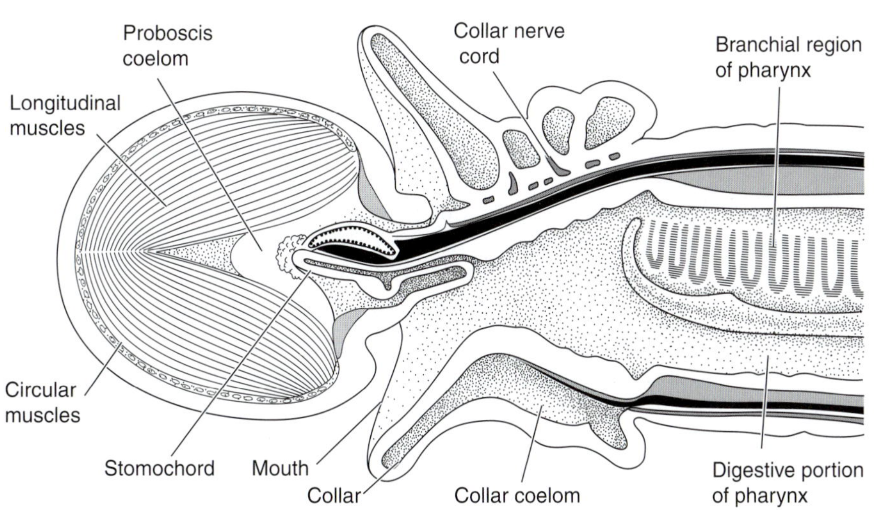

FIGURE 2-3
Anatomy of a representative enteropneust. *A,* An acorn worm in its burrow. *B,* External anatomy of *Saccoglossus;* note the proboscis, collar, trunk, pharyngeal pores, and terminal anus. The proboscis is shown in its extended state, which is achieved by filling it with sea water. *C,* Sagittal section through anterior part of the body with proboscis contracted. Note the mouth, pharynx, coelomic spaces, nerve cord, and stomochord, which some workers consider homologous to the notochord of chordates. *(B and C, modified from Sherman and Sherman.)*

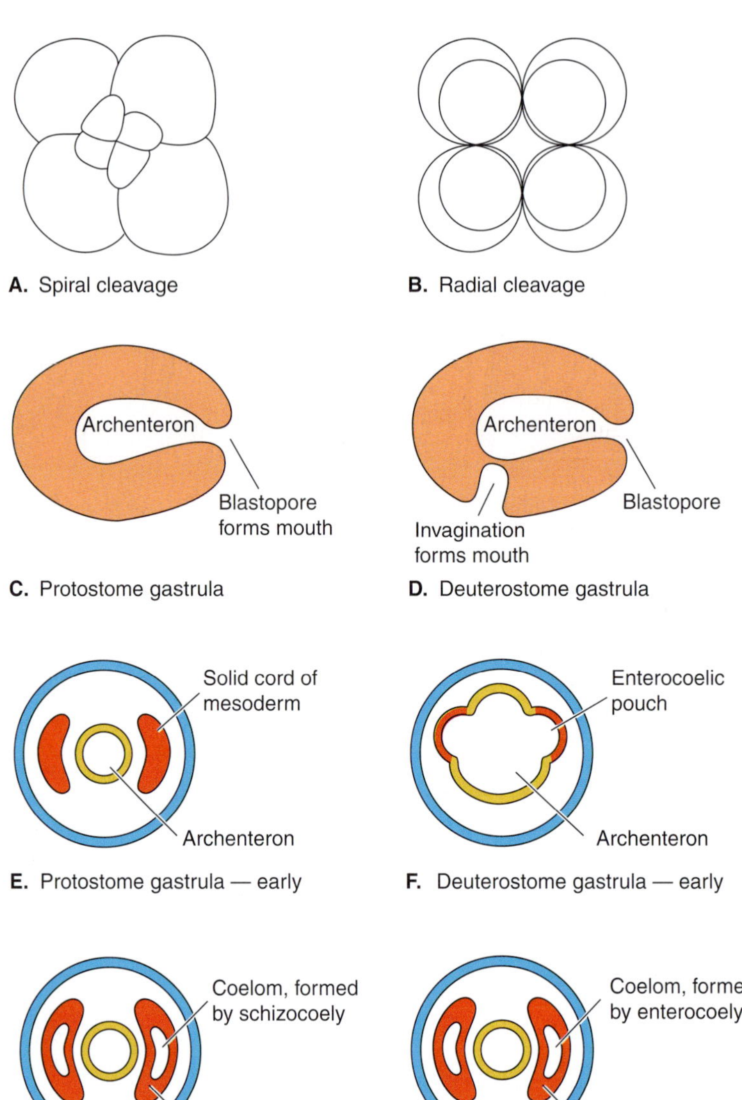

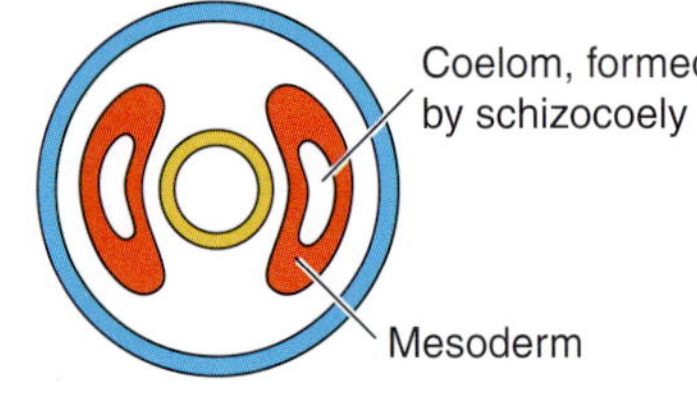

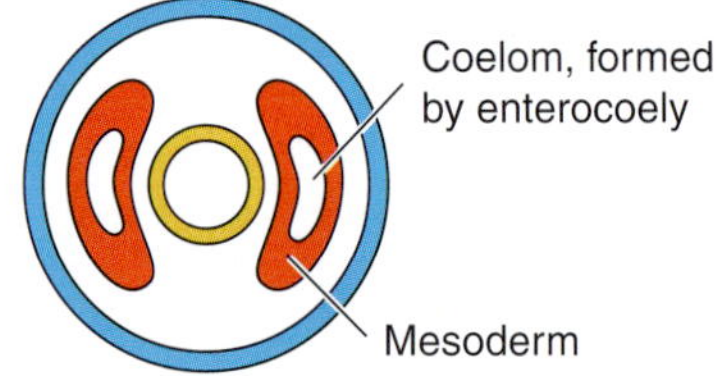

FIGURE 2-4
Schematic diagrams comparing early developmental features of protostomes and deuterostomes. *A,* Schematic diagram of spiral cleavage seen in early development of protostomes. *B,* Schematic diagram of radial cleavage seen in early development of deuterostomes. *C,* A lateral view of a protostome gastrula showing archenteron and blastopore. *D,* Lateral view of a deuterostome gastrula, showing archenteron and blastopore as well as mouth formation by means of a secondary invagination. *E,* Early gastrula of a protostome, showing solid cords of mesoderm lying between the ectoderm and endoderm. *F,* Early gastrula of a deuterostome, showing enterocoelic pouches bulging outward from the archenteron. *G,* Late gastrula of a protostome, showing coelom formation by schizocoely. *H,* Late gastrula of a deuterostome, showing coelom formation by enterocoely.

cleavage, of the fertilized egg into a sphere of cells has a precise, spiral pattern, and the part of the body to which cleavage cells give rise is determined very early (Fig. 2-4*A*). In echinoderms, hemichordates, and chordates, however, cleavage is neither spiral nor determinate (Fig. 2-4*B*). As development continues, the sphere of cells known as a **blastula** becomes converted into a **gastrula** that contains a simple gut cavity, or **archenteron** (Gr., *arche* = origin + *enteron* = gut), that opens to the surface through a **blastopore** (Fig. 2-4*C* and *D*). In annelids, arthropods, and mollusks, the embryonic blastopore becomes, or at least contributes to, the adult mouth (Fig. 2-4*C*), which is why these animals are called **protostomes** (Gr., *protos* = first + *stoma* = mouth). Echinoderms, hemichordates, and chordates are called **deuterostomes** (Gr., *deuteros* = second) because the mouth arises not from the blastopore but from a second invagination at the anterior end of the larva that pushes in to connect with the archenteron (Fig. 2-4*D*). The blastopore becomes the anus, or is located near the future site of the anus, in deuterostomes.

Deuterostomes also share a unique pattern of formation of the **mesoderm** (Gr., *mesos* = middle + *derma* = skin), the embryonic tissue layer that gives rise to "middle" tissues of the body, such as the muscles and circulatory system (Chapter 4). Lying within the mesoderm is the coelomic cavity. In basal members of all deuterostome groups, mesoderm develops from paired enterocoelic pouches that pinch off from the wall of the embryonic gut tube or **archenteron** (Fig. 2-4*F* and *H*). This pattern of development, in which the coelom develops from an open connection with the gut cavity, becomes modified in advanced craniates, but the coelom in all deuterostomes still is considered to be an **enterocoele.** In contrast, the mesoderm of a protostome arises by an earlier segregation of cells, and the coelom is never continuous with the gut cavity. The coelom of a protostome develops by cavity formation within originally solid bands of mesoderm (Fig. 2-4*E* and *G*) and is thus referred to as a **schizocoele** (Gr., *schizein* = to cleave).

Basal deuterostomes also share a type of larva in which a ciliated band loops around the body surface (Fig. 2-5). Figure 2-5*A* shows a **bipinnaria larva** typical of starfishes and some other groups of echinoderms; Figure 2-5*B* illustrates a **tornaria larva** of a hemichordate. Such larvae use their ciliated bands for locomotion, and they are thought by some researchers to be precursors to tissues that become incorporated into the nervous system of chordates.

Patterns of early development appear to be excellent indicators of phylogeny, for it is unlikely that animals with radically different patterns can be related closely. Such embryological evidence suggests a major—and very ancient—branching point in the evolution of Metazoa. Our brief review of these features indicates four synapomorphies of deuterostomes: (1) cleavage is nonspiral and indeterminate; (2) the blastopore becomes, or is near the site of formation of, the anus; (3) the coelom is an enterocoele; and (4) the larva has a loop-shaped ciliated band that is used for locomotion.

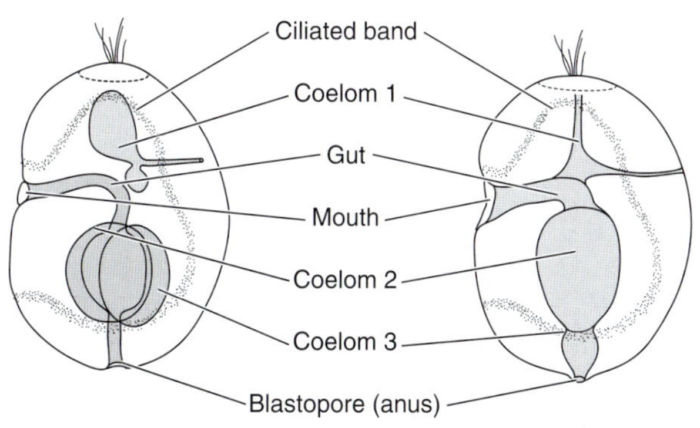

FIGURE 2-5
Comparison of echinoderm and enteropneust larvae. *A,* Bipinnaria larva of an echinoderm. *B,* Tornaria larva of a hemichordate. In addition to the ciliated bands that are used for locomotion, note the overall similarity in larval shape, the organization of the gut cavity, and the tripartite coelom. We regard this type of larva as synapomorphic for deuterostomes.

As shown in Figure 2-1, hemichordates and chordates belong to a group known as **Pharyngotremata** (Gr., *pharyngo* = pharynx + *tremata* = opening), with echinoderms as an outgroup. This interpretation is currently the most widely accepted view about relationships among deuterostomes. In support of this view, we note that pharyngotremates have (1) a pharyngeal skeleton and (2) ciliated pharyngeal slits, which occur as just a single pair in the pterobranch, *Cephalodiscus,* but which are numerous in acorn worms, such as *Saccoglossus* (Fig. 2-3*B*) and in all basal chordates. We also note in passing that some molecular evidence suggests that echinoderms and hemichordates are sister taxa and that some workers consider that chordates arose directly from echinoderms (see Jefferies, 1986; Jefferies et al., 1996; and Gee, 1996).

Characteristics and Phylogenetic Relationships of Chordates

As shown in Figure 2-1, the three clades of chordates are the tunicates, cephalochordates, and craniates.[2] In this section, we briefly review characteristics of tunicates and cephalochordates and describe evidence for chordate and somitichordate (= cephalochordates + craniates) monophyly.

Tunicates

Tunicata contains approximately 2000 species of small marine animals (Fig. 2-1). All but about 100 of these are sea squirts, or **Ascidiacea.**[3] Many sea squirts are colonial, and others are solitary. All adult sea squirts are **sessile** (L., *sessilus* = fit for sitting), a word that refers to their permanent attachment to the substrate. Some species have broad geographic distributions, at least partly because of introductions by humans. For example, some ascidians are fouling organisms that encrust the hulls and ballast tanks of ships. Such species have been introduced to ports around the world, from which they have spread.

An adult sea squirt (Fig. 2-6) is a small, sac-shaped animal encased in a protective and supportive **tunic** (Gr., *tunica* = coating). The tunic is composed of a

[2]Some texts refer to tunicates and amphioxus as "protochordates." Based on our phylogeny, "Protochordata" is not a monophyletic group, so we avoid using the term.

[3]The other 100 species of tunicates belong to the group **Thaliacea**; these peculiar pelagic animals will not be considered further here because they are not directly relevant to understanding the origin of craniates.

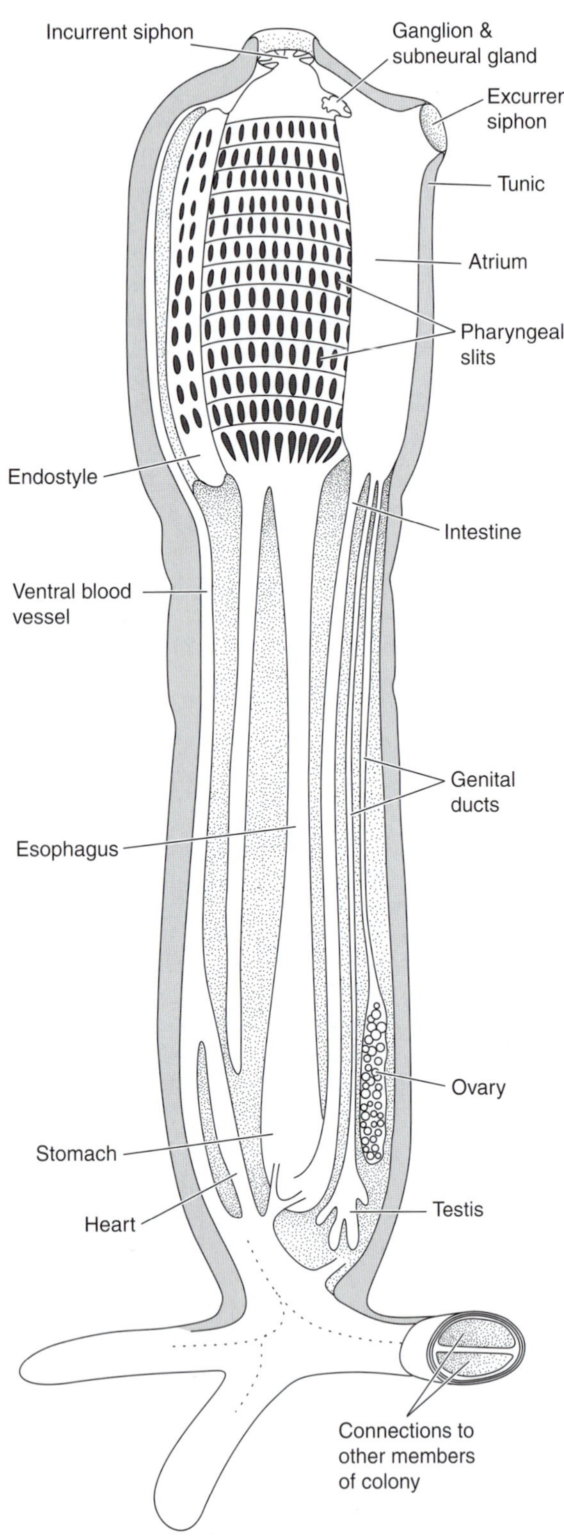

FIGURE 2-6
Anatomy of a colonial adult ascidian. *(Modified from Jollie.)*

mucopolysaccharide that contains fibers of **tunicin,** a carbohydrate similar to cellulose. It is secreted by the epidermis of the thin body wall, or mantle, and by amoeboid cells that move through the tunic. Water and food particles enter the large pharynx of a sea squirt through the **incurrent siphon.** Food is trapped in mucus secreted by a ciliated groove in the pharynx known as the **endostyle.** Excess water escapes through hundreds of minute **pharyngeal slits** into the **atrium,** a chamber that surrounds the pharynx laterally and ventrally. The atrium discharges through an **excurrent siphon** located near the incurrent one. Food passes through a simple digestive tract, which consists of an **esophagus, stomach,** and **intestine.** The stomach secretes digestive enzymes for extracellular digestion, and nutrient absorption occurs in the intestine. The intestine empties into the atrium, and fecal matter exits through the excurrent siphon.

The circulatory system of a sea squirt consists of a small **heart** and blood vessels that extend from it to the various organs and into the tunic. The vessels extending from the heart form a continuous closed system of channels, but they are not completely lined with endothelial cells (Chapter 19). Blood is pumped by peristaltic contractions of the heart in one direction for several minutes, and then the flow reverses. No specialized excretory organs are present, and so excretion occurs by diffusion. A **ganglion,** sometimes termed the "brain," lies in the mantle between the siphons. Nerves extend from it to the siphons, mantle muscles, and some visceral organs. A **subneural gland** lies ventral to the ganglion.

Sea squirts can reproduce asexually by budding, but sexual reproduction also occurs in this group. Individuals are **hermaphroditic,** and their **ovary** and **testis** connect via genital ducts to the atrium, so that the eggs and spermatozoa ultimately are discharged through the excurrent siphon. Nearby individuals simultaneously shed gametes into the water column, where external fertilization occurs. A fertilized egg develops quickly into a small, tadpole-shaped larva that does not feed. After a few days, the larva attaches to the bottom and metamorphoses into a sessile adult.

Pharyngeal slits and an endostyle are present in adult ascidians, but three other important characters, which relate to locomotion, can be seen only in their larvae. The tadpole-shaped larva has an expanded anterior end (Fig. 2-7*A*) and a slender tail (shown in cross-section in Fig. 2-7*B*). The tail is used for locomotion and contains a notochord of approximately 40 vacuolated and turgid cells. Unsegmented columns of longitudinal muscles flank the notochord. A **tubular nerve cord** lies dorsal to the notochord and integrates locomotor movements. It is, however, much simpler in organization than the neural tube of cephalochordates

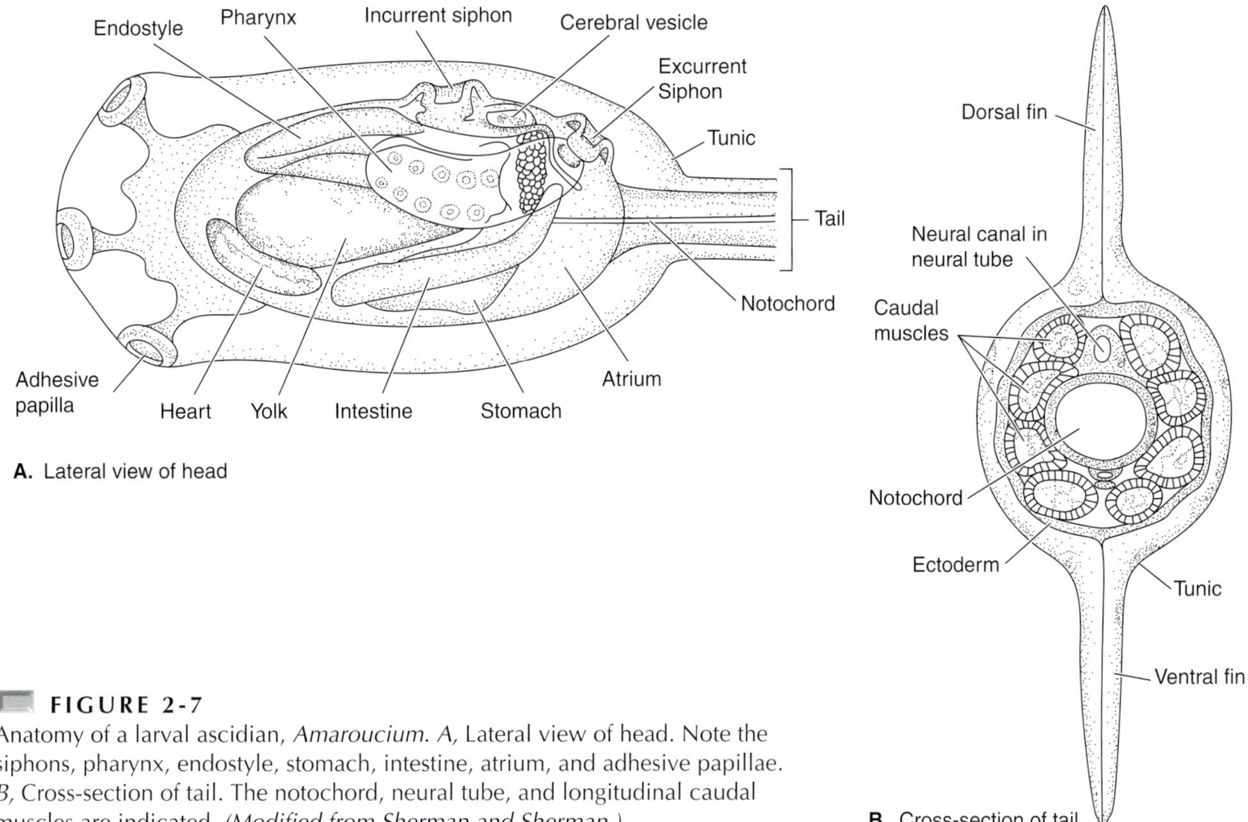

FIGURE 2-7
Anatomy of a larval ascidian, *Amaroucium. A,* Lateral view of head. Note the siphons, pharynx, endostyle, stomach, intestine, atrium, and adhesive papillae. *B,* Cross-section of tail. The notochord, neural tube, and longitudinal caudal muscles are indicated. *(Modified from Sherman and Sherman.)*

and craniates, consisting only of the lining layer of the tube, known as the ependyma, and axons coursing outward from the cerebral vessel. The anterior end of the neural tube expands into a **cerebral vesicle.** The cerebral vesicle is not readily comparable to the vertebrate brain, although some gene expression data suggest that regions homologous to portions of the brain of a craniate can be distinguished (see Wada et al., 1998). An ascidian larva has a simple eyespot and a balancing organ known as a statolith, which it uses to find a suitable place to settle. The pharynx, endostyle, digestive tract, and atrium begin to develop ventral to the cerebral vesicle. At metamorphosis, the larva attaches to the substrate using three **adhesive papillae** located at its anterior end. The tail is absorbed, and part of the cerebral vesicle becomes the ganglion.

Cephalochordates

Cephalochordata comprises approximately 45 species, most of which belong to the genus *Branchiostoma,* commonly called lancelets or amphioxus.[4] Cephalochordates are widely distributed along the ocean coasts of the world. In a few locations, they occur in astonishingly high densities: 5000 individuals per square meter of sandy bottom (Stokes and Holland, 1998). These small animals, about 3 cm to 5 cm long, are pointed at each end (hence the name amphioxus: Gr., *amphi* = both + *oxys* = sharp). They have short, postanal tails and distinct anterior ends but lack well-developed heads. An amphioxus is far more active than is a tunicate and resembles a craniate in some ways.

The notochord of an amphioxus extends from the anterior end of the body to nearly the tip of the tail (Fig. 2-8). Its extension to the anterior tip of the body is the basis for the name Cephalochordata (Gr., *kephale* = head + *chorde* = cord). The notochord stiffens each end of the animal and facilitates pushing into the sand in either direction. Each notochordal cell forms a highly vacuolated, disk-shaped plate, the flat surface of which lies in the transverse plane of the body so that the notochord as a whole resembles a stack of checkers. Electron microscopy reveals that the notochordal cells contain striated muscle fibrils and have slender cytoplasmic processes that extend into the neural tube, where they synapse with motor nerve cells. Stimulation of notochordal cells causes them to contract and increases their internal fluid pressure. Thus an amphioxus can regulate the turgidity of its notochord.

[4]The name *Branchiostoma* Costa, 1834, has priority over *Amphioxus* Yarrell, 1836, but *Amphioxus* had been so widely used before the synonomy was recognized that it is retained as the common name for the animal.

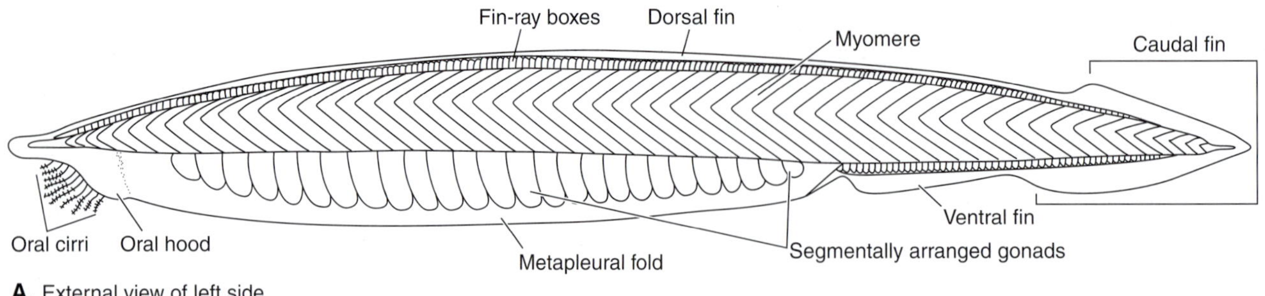

A. External view of left side

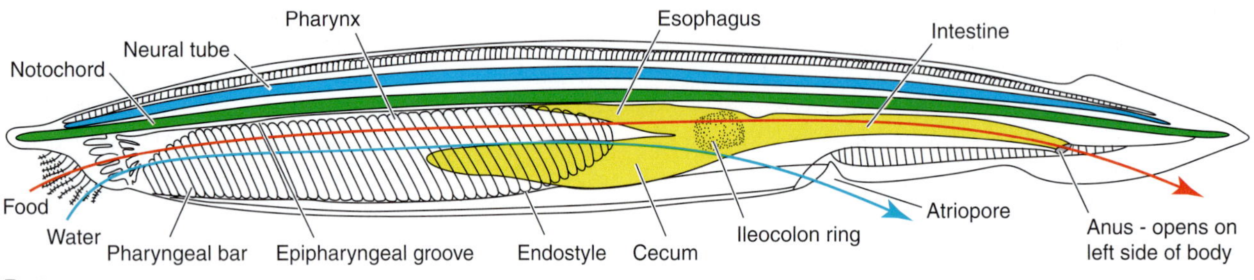

B. General internal anatomy

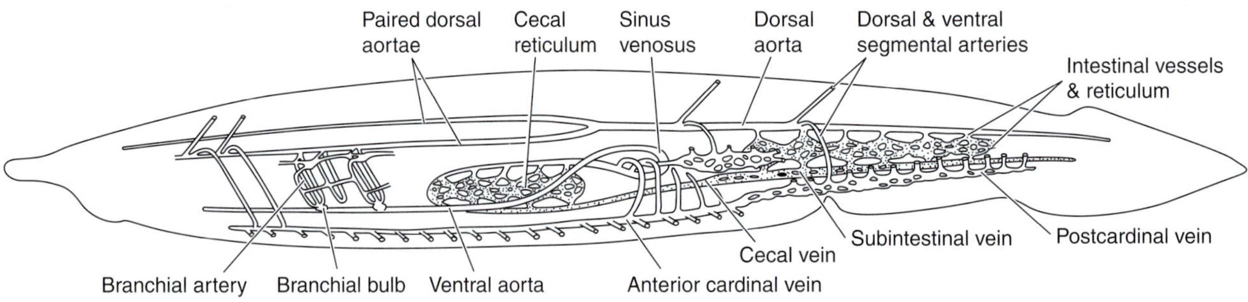

C. Circulatory system

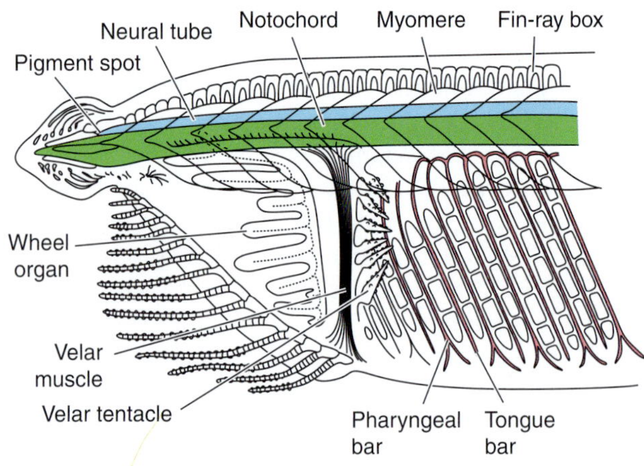

D. Anterior portion of body

FIGURE 2-8
Anatomy of amphioxus. *A,* External view of left side, showing arrangement of serial myomeres, fins and fin-ray boxes, and segmentally arranged gonads. *B,* General internal anatomy, as seen from the left side of a cleared and stained specimen. Note the relative positions of the neural tube, notochord, and pharynx; also note that the notochord extends to the anterior tip of the head. *C,* Schematic diagram of circulatory system. See text for description of blood flow. *D,* Anterior portion of the body of amphioxus as seen from the left side of a cleared and stained specimen. Structures of the feeding apparatus and pharyngeal region are emphasized. *(A and B, modified from Eddy et al.; C and D, modified from Jollie.)*

Although an amphioxus burrows in the sand most of the time, it also swims. Small **dorsal, caudal,** and **ventral fins** are present (Fig. 2-8*A*). These median fins are supported by structures known as **fin-ray boxes.** Paired ventrolateral folds of skin, known as **metapleural folds,** may provide additional stability for swimming (Figs. 2-8*A* and 2-9). The metapleural folds extend along the anterior two thirds of the body and unite posteriorly in the midline. The longitudinal muscle fibers in the body wall are not continuous but are grouped into a series of cone-shaped muscle segments known as **myomeres;** the myomeres develop from embryonic **somites,** which are segmented blocks of tissue that form lateral to the notochord (Chapter 4). Separate waves of contraction can begin at one end of the body (either anterior or posterior) and sweep to the other end, causing a series of **lateral undulations** (Chapter 11). Effective swimming requires that the functional posterior end of the body oscillate more than the other end. The great length of the notochord and the ability to control its rigidity differentially along its length are adaptations that allow an amphioxus to swim and burrow in either direction. Segmentation of the muscle into myomeres imposes a pattern on much of the rest of the body, as is evidenced by the segmentally arranged blood vessels that supply the myomeres and intersegmental nerves that pass between them.

The thin, unpigmented skin of an amphioxus consists of an **epidermis** of only a single layer of columnar epithelial cells resting on a collagenous basal lamina. An amphioxus has no organized respiratory structures, but its small size and compressed shape allow for gas exchange through the skin and pharyngeal wall.

When feeding, an amphioxus lies partly buried in the sand, with only its anterior end protruding (Fig. 2-10*H*). All filter-feeders must spend a great deal of time feeding, and the feeding and digestive organs make up much of the body. The mouth lies deep

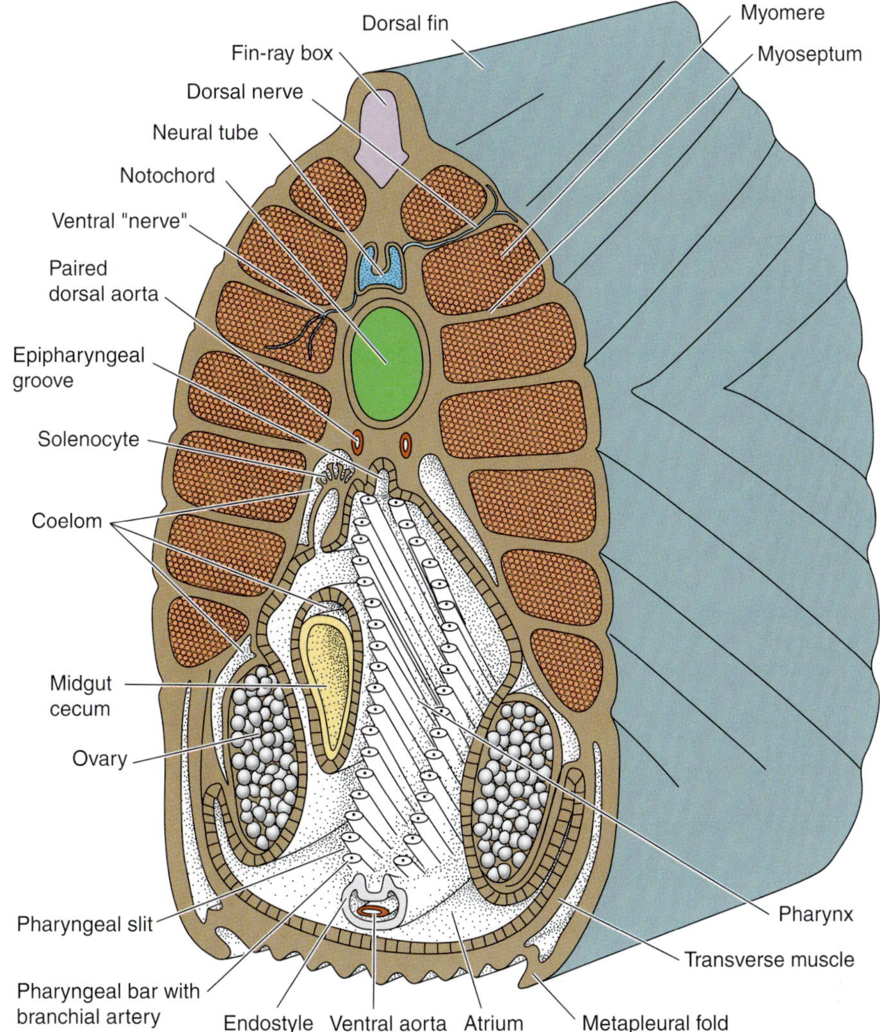

FIGURE 2-9
Schematic cross-section through the posterior portion of the pharynx of amphioxus. The section is viewed from the front, so that the right side of the drawing shows the left side of the animal (e.g., the midgut caecum lies on the right side of the pharynx).

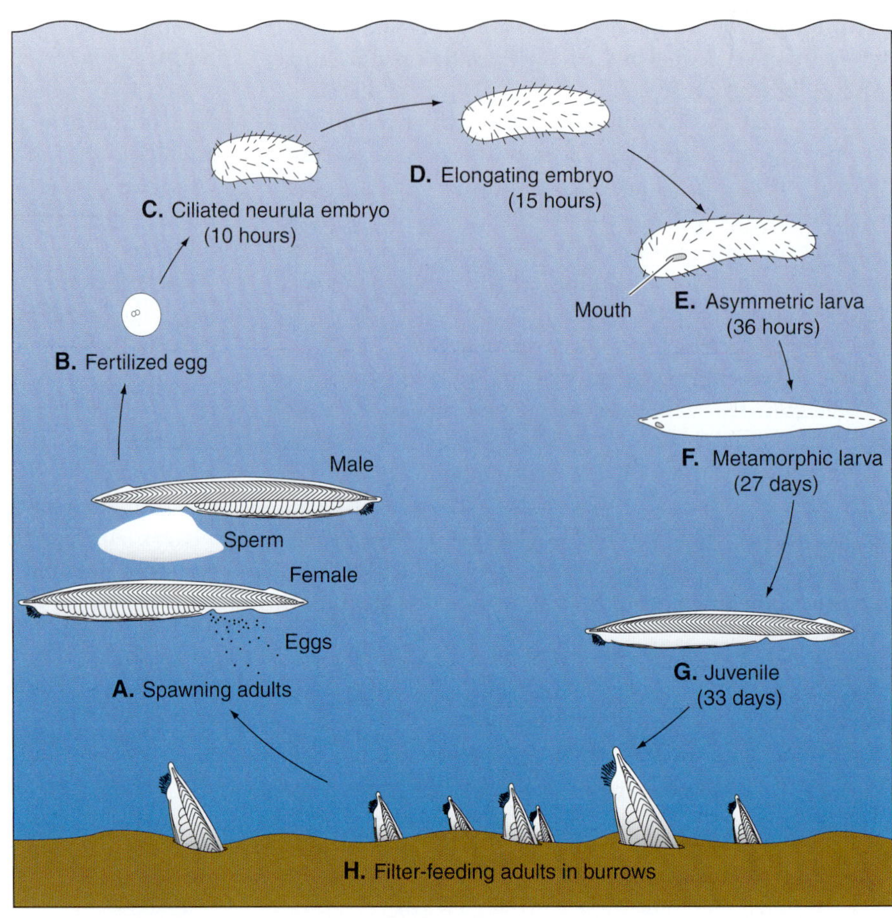

FIGURE 2-10 Life cycle of amphioxus, based on *Branchiostoma floridae.* All parts of this figure are not to the same scale. *A,* Adults spawn during warm summer months. *B,* Fertilized egg develops rapidly. *C,* Neurula stage embryo showing external cilia used for swimming in plankton. *D,* Neurula elongates and begins to use intermittently its trunk muscles for swimming. *E,* Mouth opening signals beginning of larval period. *F,* Four weeks later, larva metamorphoses to juvenile stage; cilia no longer important in locomotion. *G,* Juveniles (approximately 1 cm long) become oriented to bottom and begin to burrow. *H,* Adults live in burrows, exposing their preoral cirri and oral hood to allow continuous filter-feeding. *(Modified from Stokes and Holland.)*

within the **oral hood** and is surrounded by **velar tentacles** (Fig. 2-8*D*). The oral hood is fringed with **oral cirri,** which strain out coarse particles, and it is lined by a ciliated structure known as the **wheel organ.** Water is drawn into the pharynx by ciliary action, using cilia located on the wheel organ and in the large pharynx. The pharyngeal wall is perforated by many vertically elongated pharyngeal slits separated by fibrocartilaginous pharyngeal bars. The number of slits and bars increases as the animal grows because secondary tongue bars extend downward from the dorsal wall of the pharynx between the first-formed, or primary, pharyngeal bars (Fig. 2-8*D*). Cells in the endostyle in the pharyngeal floor secrete mucus that entraps minute food particles (Figs. 2-8*B* and 2-9). Some endostylar cells also secrete iodinated proteins. The mucus and food are carried by ciliary action dorsally along the pharyngeal bars to the **epipharyngeal groove.**

Excess water taken in with the food escapes through the pharyngeal slits into the atrium that surrounds the pharynx (Fig. 2-9). Contractions of muscles in the floor of the atrium expel the water through the **atriopore** (blue arrow, Fig. 2-8*B*). The atrium protects the delicate pharyngeal bars from abrasion during burrowing. Amphioxus also gains oxygen and discharges carbon dioxide from the water flowing through the pharynx even though specialized gills are not present.

The pharynx narrows caudally into an **esophagus** (Fig. 2-8*B*). The midgut and hindgut follow, and the **intestine** opens by an anus that is located on the left side of the caudal fin. No muscle fibers are present in the gut wall. Instead, a ring of cilia, known as the **ileocolon ring,** moves the cord of mucus and food caudally within the gut tube. A large **midgut cecum** extends from the floor of the midgut forward into the atrium along the right side of the pharynx (Figs. 2-8*B* and 2-9). Food particles break off from the food cord and are carried by ciliary action into the cecum, where both extracellular and intracellular digestion and some nutrient absorption occur. Absorbed food is stored as glycogen and lipid in some of the cecal cells. The cecum thus has functions resembling those of a craniate's liver (storage) and others associated with the pancreas (enzyme secretion).

The **ventral aorta** carries blood forward beneath the pharynx, and paired **branchial arteries** carry it dorsally through each pair of pharyngeal bars (Figs. 2-8*C* and 2-9). Blood is collected from the branchial

FOCUS 2-1 *The Excretory Organs of Amphioxus*

Excretion in amphioxus occurs via clusters of highly specialized cells called **solenocytes** that lie in the dorsal coelomic canals (Fig. A). The solenocytes of amphioxus are remarkable because they have both filtration and pumping effects. Each cell has a group of footlike processes that wrap around an arterial branch from the dorsal aorta. Each solenocyte also has a pump consisting of a slotted, tubelike structure that extends through the dorsal coelomic canal to enter a **nephridioduct.** The nephridioduct opens into the atrium. The slotted tube wall consists of ten long microvilli that extend from the cell body of the solenocyte. A long flagellum lies within the tube, and its movements pump and propel liquid. Pressure in the artery forces water and wastes through the arterial wall, between the footlike processes of the solenocyte, and into the dorsal coelomic canal. Once in the dorsal coelomic canal, water and waste products are drawn through the slots and into the cavity of the microvillar tube by the movements of the flagellum and then are pumped into the nephridioduct. Water and wastes are then driven into the atrium. The loss of water that this mechanism entails is not a problem because amphioxus has the same salt concentration in its body fluid as in sea water. Lost water readily diffuses back into the body.

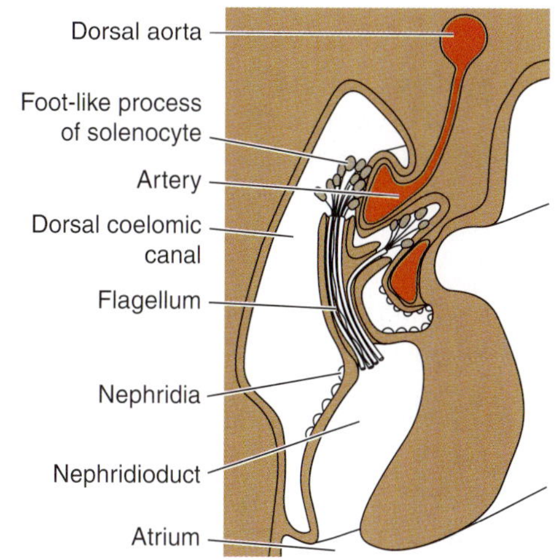

A. Cross-section through the pharyngeal region of amphioxus (*Branchiostoma*) showing details of the relationships of one solenocyte to surrounding structures. *(Modified from Brandenburg and Kummel.)*

arteries by **paired dorsal aortae**, which unite caudally to form a single **dorsal aorta,** which distributes blood to the tissues (Fig. 2-8*C*). This pattern of blood flow, anteriorly along the ventral side of the body and posteriorly along the dorsal side of the body, closely resembles that seen in a craniate. The flow pattern is opposite of that seen in other animals with well-organized circulatory systems, such as arthropods. Blood flows through tissue spaces because amphioxus has no true capillaries. Blood is collected by a system of body wall veins and a **subintestinal vein** that converge on a chamber known as the **sinus venosus,** located at the caudal end of the ventral aorta. No heart is present. Instead, blood is propelled by the slow contraction of many vessels, including muscular **branchial bulbs** at the bases of the branchial arteries. The blood contains amoebocytes but no other cells. Oxygen-binding blood pigments, such as hemoglobin, are lacking, so oxygen is carried in physical solution.

A pair of **dorsal coelomic canals** lies above the atrium on either side of the pharynx (Fig. 2-9). They contain the excretory organs, called **solenocytes,** which remove waste products from the blood and discharge them into the atrium (Focus 2-1).

Amphioxus lacks highly organized sense organs, such as a nose, eyes, and ears, but specialized sensory cells are present. Simple **tactile receptors** occur in the skin, atrium, and part of the gut wall. **Chemoreceptors** on the body surface and on the velar tentacles detect changes in the chemical environment and provide information on the type of particles entering the pharynx. Noxious particles are rejected. Pigmented **photoreceptors** lie in the floor of the central canal of the neural tube. Each consists of a single cell shielded ventrally by pigment, so it responds only to light coming from above through the thin integument. Several photoreceptors are aggregated to form a **pigment spot** at the front of the neural tube (Fig. 2-8*D*). Sensory neurons from the receptor cells in the skin pass between the myomeres to the neural tube. These fibers contribute to the **dorsal nerves** (Fig. 2-9), but the cell bodies of these nerve cells lie within the spinal cord and not in peripheral spinal ganglia, as they do in craniates (Chapter 12). Visceral motor neurons leave the neural tube and travel in the dorsal nerves to part of the gut and atrial walls. All neurons of amphioxus lack myelin sheaths. The separate **ventral "nerves"** (Fig 2-9) are composed of the processes of myomeric muscle cells, not neurons.

Although the central canal of the neural tube expands slightly at the anterior end of the neural tube, the cord itself does not. Groups of cells in the anterior end of the neural tube appear to be homologous to portions of the vertebrate brain (Chapter 14). The behavioral repertoire of an amphioxus consists of little more than feeding, escape movements, and reproductive behavior.

Unlike sea squirts, the sexes are separate in amphioxus, as in nearly all craniates. Testes or ovaries develop in a series of segmented coelomic sacs that bulge from the ventral part of the body wall into the atrium (Figs. 2-8*A* and 2-9). Mature gametes are discharged into the atrium, which they exit via the atriopore. Fertilization and development occur externally, and the gonads develop anew in the next reproductive season. The life cycle of *Branchiostoma floridae* is shown in Figure 2-10. Spawning occurs in the summer. Early development yields a ciliated, swimming neurula stage embryo (Fig. 2-10*C*), which elongates and begins to use muscles for locomotion before the mouth opens (Fig. 2-10*E*). Over the next month, the larva feeds and grows to about 1 cm (Fig. 2-10*F*). It then metamorphoses into a juvenile and takes up the adult behavior of living in a shallow burrow.

Craniata is the third major group of Chordata (Fig. 2-1), but we defer discussion of craniates until after the next section.

Monophyly of Chordates and Somitichordates

As described earlier, both adult and larval sea squirts have pharyngeal slits and an endostyle, and larvae also have a notochord, neural tube, and elongated postanal tail. In contrast, all five of these features can be seen and studied easily in an adult amphioxus. Pharyngeal slits and an endostyle are characters related to a unique method of filter-feeding, whereas the notochord, neural tube, and larva with an elongated postanal tail are characters that evolved in connection with a unique method of locomotion.

Pharyngeal slits open into the atrium of tunicates and cephalochordates or to the body surface in craniates. Specialized respiratory gills are not present in the pharyngeal slits of tunicates or cephalochordates, as they are in craniates. Instead, the pharyngeal slits of tunicates and cephalochordates are part of their filter-feeding mechanism. This was probably their original function in ancestral chordates. Tunicates and cephalochordates use cilia to draw a current of water in through the incurrent siphon or mouth and into the pharynx. Cells in a ciliated groove in the floor of the pharynx, the endostyle, secrete mucus that spreads over the inside of the pharyngeal wall. Minute food particles suspended in the water are trapped in this mucus, and excess water escapes through the pharyngeal slits. Although many animals filter-feed, the combined use of pharyngeal slits and an endostyle to do this is unique to chordates. Certain cells in the endostyle also bind proteins with iodine, and we consider these cells to be homologous to thyroid secretory cells of craniates, which also incorporate iodine into the synthesis of thyroid hormone.

The notochord, which all chordates have at some stage of their development, is a skeletal structure important in locomotion. It is an example of a hydrostatic skeleton, or **hydroskeleton**, that consists of a longitudinal column of cells, each of which contains a large, central vacuole full of liquid (see Chapter 5 for a general discussion of hydroskeletons). A sheath of dense connective tissue surrounds the vacuolated notochordal cells and maintains their turgidity. The primary function of the notochord is to resist compression and so prevent the body from shortening or telescoping (in the manner of an earthworm) when the longitudinal muscle fibers in the body wall contract. Because of the stiff notochord, muscle contraction causes instead a lateral bending from side to side, which propels the animal forward or backward.

All chordates have at some stage of life a single, dorsally located and tubular nerve cord of a unique type that integrates the contraction of the locomotor muscles in the body wall. When present in other animals, such as arthropods, the nerve cords are paired, ventrally located, and solid.

Tunicates, cephalochordates, and more than half of the living species of craniates have a unique swimming larval type, one with an elongate postanal tail. Tailed embryos are present in virtually all of the remaining species of craniates.

In summary, then, the five synapomorphies of chordates are (1) pharyngeal slits in the wall of the pharynx, which primitively serve for filter-feeding and only later evolved to serve a respiratory function; (2) an endostyle, or its derivative, the thyroid gland; (3) a notochord; (4) a single, dorsal, and tubular nerve cord; and (5) a tailed larva. These features are not always present (or easily observed) in adult stages, and some clades of craniates have secondarily lost the tailed larva.

As shown in Figure 2-1, cephalochordates and craniates belong to a group known as **Somitichordata.** Somitichordate synapomorphies include (1) somites and (2) the retention of larval features as adults, specifically, the notochord, neural tube, and tail. The name Somitichordata refers to the fact that these are chordates with somites.

Characteristics of Craniates

Craniates clearly exhibit all five chordate synapomorphies. As embryos, all craniates have pharyngeal pouches, and these perforate the body wall to form gill slits in fishes and larval amphibians. As adults, all craniates have a thyroid gland, which develops in lampreys (a member of an early radiation of craniates) from an endostyle-like subpharyngeal gland. All craniate embryos possess a notochord, and some retain it as adults. A single, dorsal, and tubular nerve cord develops in craniate embryos and persists in adults. More than half of the living species of craniates are aquatic animals and have larvae with elongate, post-anal tails. The remaining species have embryos with post-anal tails. Most adult craniates also retain elongate, post-anal tails.

The evolution of many characteristics of craniates is correlated with their relatively high level of activity and large sizes. In very general terms, a craniate's body consists of a head, trunk, and tail (Fig. 2-11*A*). We say that craniates are **cephalized,** because a craniate's head usually is very distinct from its trunk, particularly in comparison to an amphioxus. The head houses the mouth and gill slits that exit from the pharynx. An active animal also needs a concentration of sense organs and nerves at its anterior end because this is the part of the body that first encounters new environments. Therefore, the head of craniates also contains well-developed paired sense organs, including the **nose, lateral eyes,** and **ears** (Fig. 2-11). Many craniates also have a single **median eye** located on the dorsal surface of the head.

The nose and ears develop embryonically from **neurogenic placodes.**[5] Neurogenic placodes are unique to craniates and are only found in the head of an embryo, although structures derived from them may become located on the trunk or tail by migration of cells during development (Chapter 4). Neurogenic placodes invaginate or delaminate[6] to form (or induce to form, in some cases) sensory receptor cells and sensory neurons. The sensory neurons grow from the receptor epithelium back into the brain stem during later development and become important components of the **cranial nerves** (Chapter 12). Other neurogenic placodes form the **lateral line** and **electroreceptive system,** which ramifies over the head and extends caudally along the side of the trunk and tail. These systems provide fishes and larval amphibians with information about water movements and electric fields impinging on the body. Neurogenic placodes associated with the pharyngeal region induce the formation of the **gustatory system** (i.e., taste buds). Although its placode does not give rise to any neurons and thus cannot be described as neurogenic, even the lens of the eye develops from a placodal thickening in the ectoderm.

In craniates, the anterior end of the neural tube is enlarged to form a distinct **brain** (Fig. 2-11*B* and *C*) that receives and integrates information incoming from the sense organs and coordinates motor outflow to control locomotion and feeding. The brain has three parts. The **forebrain** integrates olfactory information from the nose, and the **midbrain** processes information from the eyes and auditory information from the ears. The **hindbrain** is a transitional region between the brain and spinal cord; it also receives sensory nerve input from the gustatory system and the ear, lateral line, and electroreceptive systems.

The sense organs and brain are encased and protected by a skeletal **cranium,** or **braincase,** formed by cartilage or bone. Bone is a dense connective tissue strengthened with a calcium phosphate compound; its complex histological organization adds greatly to its strength (Chapters 5 and 6). Bone occurs only in craniates. Calcified structures in the body wall of other animals, such as echinoderms, differ from bone in that they are composed of calcium carbonate.

An active animal needs an efficient locomotor system. Most craniates swim by lateral undulations of their trunks and tails. These movements are generated, as in amphioxus, by sequential contraction of a series of segmental myomeres. The action of the myomeres and the larger size of most craniates require a stronger strut than a notochord to resist compression as the animals push through the water. A **vertebral column** composed of many individual **vertebrae** does this. Vertebrae develop around the neural tube and notochord, and the vertebral bodies largely replace the notochord in most adult craniates.

Increased activity and size are impossible without increased metabolic activity. More oxygen and carbon dioxide must be exchanged with the environment, a larger volume of food must be digested and absorbed, and materials must be distributed efficiently throughout the body. Craniates have **gills** in the pharyngeal slits. The gills are supported partly by a series of cartilaginous or bony skeletal supports known as **gill**

[5]The generic term placode is used to refer to many different types of thickenings in the ectoderm that contribute to the formation of many different structures, which is why the modifier neurogenic is important.

[6]During invagination, a neurogenic placode "pockets" into deeper tissues of the body, just as it would if you had pushed your finger against its outer surface. During delamination, the epithelium breaks down, and the formerly epithelial cells of the placode take on a migratory, mesenchymal morphology. See Chapter 4 for more information.

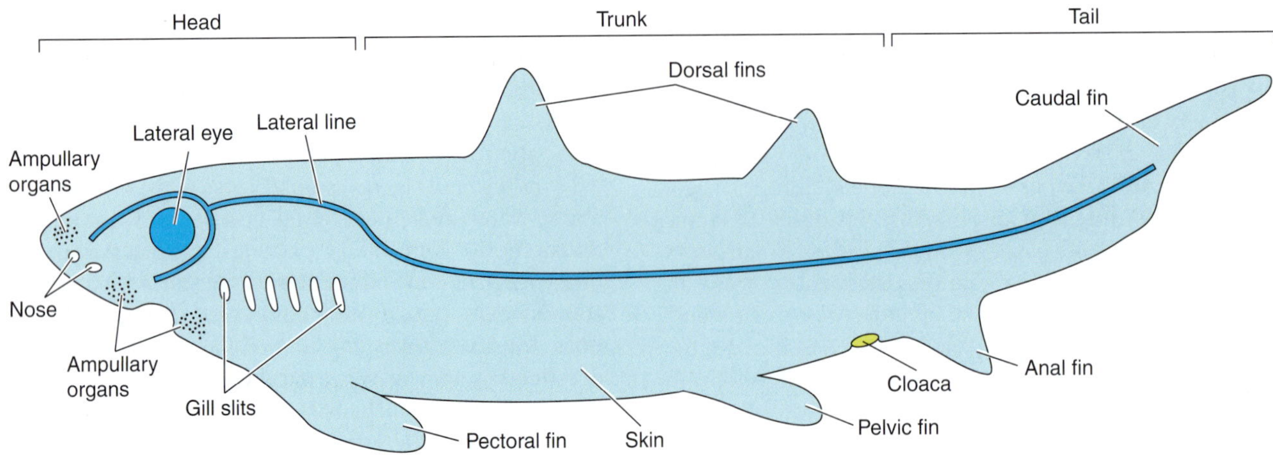

A. Lateral view

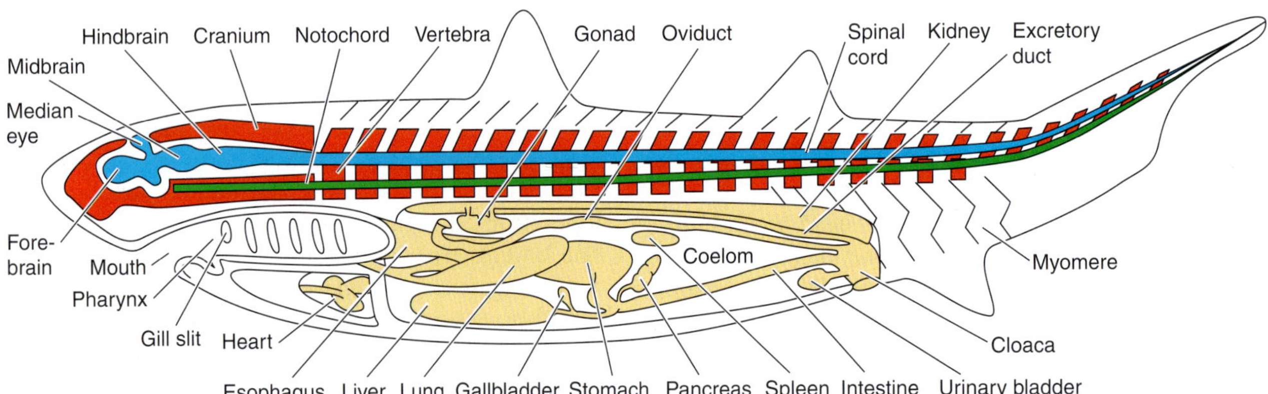

B. Mid-sagittal section

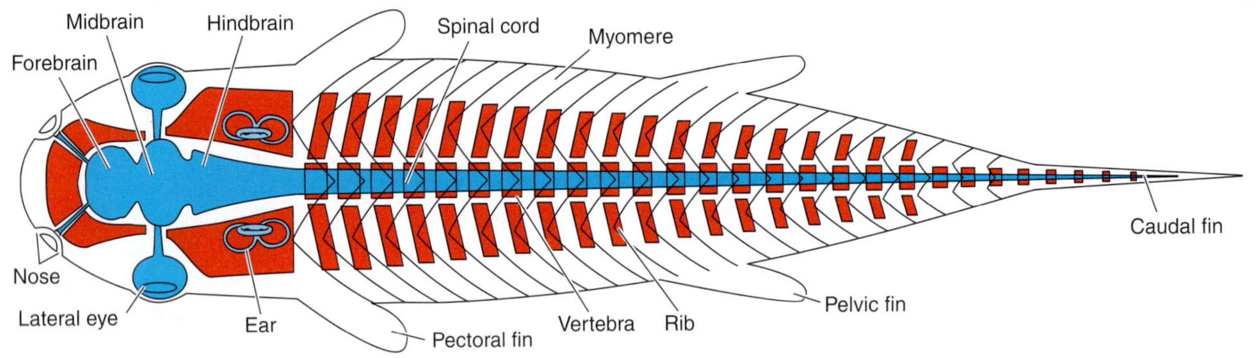

C. Frontal section

FIGURE 2-11

Diagram of major organ systems of craniates, based on a gnathostome. *A,* Lateral view. *B,* Mid-sagittal section. *C,* Frontal section. Many features are greatly simplified, and some (e.g., most of the cranial nerves) are not shown.

arches, which develop in the wall of the pharynx between the pharyngeal slits. The gut wall also becomes muscularized, which it is not in tunicates and cephalochordates. The gill arches and muscles in the wall of the pharynx are innervated by motor and sensory cranial nerves. Specialized **gill arch muscles** become associated with the gill arches, and their actions are to expand and compress the pharynx. These movements draw a respiratory current of water into the mouth and expel it across the gills. Biologists believe that the earliest craniates captured prey by such pumping of the pharynx, for such mechanisms are preserved today in larval lampreys. Later in craniate evolution, an anterior gill arch was transformed into jaws, and many new types of feeding subsequently evolved. Muscles in the wall of the gut efficiently move food through a series of successive chambers specialized for storage, digestion, and absorption. A **liver** evolved that, among its many functions, stores considerable energy as glycogen or lipid. The **pancreas** manufactures and secretes digestive enzymes into the **intestine.**

Efficiency of the circulatory system increased with the evolution of a muscularized, multichambered **heart.** Craniates also have a system of closed capillaries connecting the arteries and veins, and the respiratory pigment **hemoglobin** that binds reversibly with oxygen. Specialized **kidney tubules** allow for better elimination of nitrogenous wastes and control of water balance.

Movements of the gut, heart, and other visceral organs are facilitated by the coelom that surrounds them. The coelom of basal craniates is subdivided into a **pericardial cavity** around the heart and a **peritoneal cavity** around most of the digestive organs. These coelomic spaces separate the gut tube and heart from the body wall. The skeleton, muscles, and nerves in the body wall constitute the **somatic part** of the body. The somatic part of the body is associated with locomotion. The skeleton, muscles, and nerves in the gut tube and heart form the **visceral part** of the body. This part is associated primarily with metabolic functions.

Endocrine organs, sometimes called ductless glands, secrete hormones into the circulation to regulate processes such as metabolism, growth, and reproduction. Craniates have several major **endocrine organs,** including the **hypophysis, thyroid gland,** and **adrenal glands.** Other organs, such as the pancreas and gonads, also have important endocrine functions.

Many distinctive features are related to an embryonic tissue unique to craniates: the **neural crest.** Neural crest appears in early embryos as groups of cells that separate from the developing neural tube. The neural tube develops by the elevation and joining of a pair of longitudinal, ectodermal folds; as these meet in the dorsal midline, neural crest cells are "pinched off" to form ridges along the dorsolateral wall of the neural tube (Chapter 4). The neural tube forms in a similar way in cephalochordates, but neural crest does not develop. Cells from the neural crest migrate to many parts of the body, give rise to many important structures, and regulate aspects of development. Some neural crest cells contribute to the distinctive gill arch skeleton of craniates. Others form cranial motor neurons that innervate muscles in the wall of the visceral tube. Still others give rise to sensory neurons of craniates, particularly for the skin. In contrast, such sensory neurons arise in cephalochordates directly from the neural tube. Migratory neural crest cells form the dentine of teeth, which in turn induces the formation of enamel. Dentine and enamel are found not only in teeth but also in scales of early craniates. Virtually all pigment cells in the body derive from the neural crest, as well as portions of some endocrine organs. We will see many other derivatives of the neural crest as we consider the development and structure of the organ systems, for this important synapomorphy underlies many of the functions so characteristic of craniate life.

We conclude this section by listing some of the many synapomorphies of craniates, including (1) the neural crest, (2) neurogenic placodes, (3) the braincase, (4) complex sense organs, (5) cranial nerves, (6) tripartite brain, (7) a complex endocrine system, (8) muscularization of the wall of the gut tube, (9) differentiated digestive organs and extensive regionalization along the length of the gut tube, (10) gills, (11) a heart, and (12) hemoglobin. Each of these features is treated at length in later chapters of the book.

A Scenario About Craniate Origins

Evolutionary scenarios are inherently speculative because we do not have a time machine that might allow us to witness actual evolutionary events. Still, scenarios can be very useful learning devices when based on a well-supported phylogeny, such as that shown in Figure 2-1.

Because craniates are active animals, investigators in the 19th century sought to demonstrate evolutionary relationships between them and active invertebrates, such as mollusks (i.e., snails, clams, and especially cephalopods), annelids (i.e., earthworms, leeches, and polychaetes), and arthropods (i.e., insects, spiders, centipedes, millipedes, and crustaceans). The type of segmentation and the basic body organization of craniates differ from those seen in mollusks, annelids, and arthropods. For example, body segmentation in the annelid-arthropod evolutionary line affects not only the body wall muscles but also the coelom, which is divided into a series of compartments.[7] Contractions of the body wall of annelids can vary the fluid pressure in the coelomic compartments, so that the coelom acts as a flexible hydroskeleton. The nerve cord of annelids and arthropods is a solid, ganglionated, double strand near the ventral surface of the body. The direction of blood flow in annelids and arthropods is opposite of that of craniates, with blood moving posteriorly on the ventral surface of the body and anteriorly on the dorsal side. At the very least, it would be necessary to turn an annelid or arthropod upside down to make it resemble a craniate. Early investigators had good imaginations, however, and some ingenious hypotheses were proposed. All beg the question, however, of the origin of tunicates and cephalochordates, other than regarding them as degenerate craniates.

Later investigators sought the origin of craniates not among the highly specialized annelids and arthropods but among filter-feeding tunicates and cephalochordates and early, filter-feeding echinoderms. As we have seen, this line of inquiry led to the discovery of four synapomorphies of deuterostomes, which rule out the idea that craniates are related to the annelid–arthropod group of invertebrates. Based on this phylogenetic interpretation, it is reasonable to speculate that the common ancestor of deuterostomes was a filter-feeder that separated food particles from debris and water before the food entered the mouth because it lacked pharyngeal slits.

We may further speculate that the common ancestor of pharyngotremates was a pterobranch-like animal with feeding arms, a U-shaped gut, and a pair of pharyngeal slits. We see such a combination of features in one extant genus of pterobranchs, *Cephalodiscus.* Pharyngeal slits apparently were an effective way to eliminate excess water and to concentrate food. The pharyngeal slits increased in number, and feeding arms were reduced and eventually lost. Enteropneusts adapted to a burrowing mode of life. The body became longer, the U-shaped gut became straight, and a longer proboscis became part of both the feeding and the burrowing mechanisms.

Other pterobranch-like creatures with multiple pharyngeal slits and reduced feeding arms remained sessile. They lost the proboscis and collar, kept the U-shaped gut, improved the pharyngeal slit filtering mechanism, and developed a protective atrium around the slits. These animals would have been very much like the living sea squirts.

[7]These compartments are secondarily reduced in arthropods, such as insects.

From this point, most investigators favor a hypothesis originally suggested by Bateson (1886) and later developed and modified by Garstang (1928), Berrill (1955), Northcutt and Gans (1983), and many others. Echinoderms and hemichordates rely on small, ciliated larvae to disperse their populations. Chordates, however, have larger larvae. Because mass is a cubic function and surface area is a square function (see Chapter 5 for discussion of scaling), ciliated bands on the body surface are inadequate to propel a large larva. As larval size increased, muscles became more important in locomotion. The larval tail, with its notochord, neural tube, and longitudinal muscles, may have evolved in this locomotor context, along with a simple eye and a balancing organ at the anterior end. The resulting larva would have been better able to disperse and to find a suitable habitat in which to settle.

Somitichordates may have evolved from such an organism by paedomorphosis (Chapter 1), that is, by retaining the larval characteristics of a tail and notochord into the adult stage. The evolution of somites enhanced their locomotor capabilities. Cephalochordates became burrowing filter-feeders capable of relatively limited swimming movements. Craniates, in contrast, became more mobile, larger, and even more active animals that could exploit a wider range of habitats. Increased activity was made possible by the evolution of better sensory systems, specifically the nose, eyes, taste buds, ears, and lateral line and electroreceptive systems. These new sense organs in turn required a brain capable of integrating the information. Additional metabolic machinery was needed to support increased activity, including a gut with a muscular wall, gills, a heart, and a unique type of kidney tubule.

Northcutt and Gans (1983) were the first to emphasize that many structural and functional innovations of craniates are linked to organs that develop embryonically from either the neural crest or the neurogenic placodes. As we have already noted, these two embryonic precursor tissues are unique to craniates, and they will be discussed at length in Chapter 4. Northcutt and Gans postulated that the neural crest and neurogenic placodes evolved from the subepidermal nerve net found in other deuterostomes.

Ancestral craniates were probably marine or estuarine, and Berrill (1955) proposed that increased activity evolved as ancestral craniates invaded freshwater habitats. In another view, Northcutt and Gans (1983) proposed that increased activity evolved when ancestral craniates shifted from filter-feeding to predation upon small animals that could be scooped up and crushed in the muscularized pharynx. The hypothesis that predation drove the early evolution of craniates is very attractive, for it concisely explains the evolution of many features seen in basal living and early fossil craniates. Many ideas developed by Northcutt and Gans have been widely accepted by biologists.

In concluding this chapter, we note that the entire field of metazoan phylogeny is in an extremely exciting period. Like other aspects of systematics, the study of metazoan relationships has blossomed as a result of molecular phylogenetic techniques. New phylogenetic interpretations appear often, and some of these are radically different from the interpretation shown in Figure 2-1. For insight into the many strengths—and pitfalls—of these contemporary studies, we refer interested readers to McHugh and Halanych (1998) and Jenner and Schram (1999).

SUMMARY

1. Echinoderms, hemichordates, and chordates are deuterostomes. Deuterostomata is named for the pattern of formation of the mouth in embryos: the mouth arises not from the blastopore at the caudal end of the gastrula but from a second invagination at the anterior end of the larva, which pushes in to connect with the archenteron. Four synapomorphies provide evidence for deuterostome monophyly: (1) cleavage is nonspiral and indeterminate; (2) the blastopore becomes, or is near the site of formation of, the anus; (3) the coelom is an enterocoele; and (4) the larvae have a loop-shaped, ciliated band used for locomotion.
2. Hemichordata includes two clades: Pterobranchia and Enteropneusta (i.e., acorn worms). Monophyly of this group is supported by their tripartite body, which consists of a proboscis, collar, and trunk. Hemichordates have pharyngeal slits, but they lack an endostyle.
3. Pharyngotremata includes Hemichordata and Chordata. Two synapomorphies of this group are (1) gill skeleton and (2) ciliated pharyngeal slits.
4. At some stage of their development, chordates exhibit (1) pharyngeal pouches that grow laterally from the pharynx and often open to the surface as pores or slits; (2) a groove in the pharyngeal floor known as the endostyle, or a thyroid gland derived from part of the endostyle; (3) a stiff, longitudinal rod of turgid cells along the dorsal part of the body that is called a notochord; (4) a single, dorsal, and tubular nerve cord that is located dorsal to the notochord; and (5) a larva or embryo with an elongate, postanal tail. These five characteristics are synapomorphies for Chordata that evolved in connection with unique methods of feeding and locomotion.
5. Tunicata includes the sessile sea squirts (Ascidia). Adult ascidians have pharyngeal slits and an endostyle, but the notochord, neural tube, and elongate tail occur only in their nonfeeding but motile larvae.
6. Somitichordata includes Cephalochordata and Craniata. Two synapomorphies of this group are (1) the presence of somites and (2) the retention of larval characteristics such as the notochord, neural tube, and postanal tail as adults. This second character is usually attributed to paedomorphosis.
7. Cephalochordata is represented by amphioxus. Its anatomy and mode of life clearly illustrate the uses and significance of the derived chordate characters.
8. Members of Craniata exhibit many synapomorphies, such as (1) the neural crest, (2) neurogenic placodes, (3) a braincase, (4) complex sense organs, (5) cranial nerves, (6) tripartite brain, (7) a complex endocrine system, (8) muscularization of the wall of the gut tube, (9) differentiated digestive organs and extensive regionalization along the length of the gut tube, (10) gills, (11) a heart, and (12) hemoglobin. Most of these synapomorphies are linked to the evolution of larger bodies, more powerful locomotor systems, and the greater metabolic needs that large size imposes.
9. Among craniate synapomorphies, the presence of neural crest and neurogenic placodes stands out as particularly interesting because many other derived characters of craniates are developmentally linked to these two embryonic precursor tissues. Neurogenic placodes are restricted initially to the head of a craniate embryo and give rise, via invagination or delamination, to sense organs and sensory nerves. Neural crest cells separate from the neural tube early in embryonic development and form components of the skin, nervous system, skeleton, and endocrine organs.
10. Scenarios are speculative but useful devices. With this caveat, we may speculate that the common ancestor of deuterostomes was a filter-feeding animal. From such a form, we speculate that the common ancestor of pharyngotremates had at least one pair of gill slits as in an extant ptero-

branch, *Cephalodiscus.* We speculate that the common ancestor of chordates resembled extant sea squirts and possessed a relatively large larva with a motile tail containing a notochord and distinctive neural tube. The earliest somitichordates may have been derived from such an ancestor by paedomorphic retention of three larval features: the (1) notochord, (2) neural tube, and (3) elongate postanal tail. Greater activity and mobility were probably important in the early history of craniates, which could be correlated with the evolution of new sense organs, a large brain, a powerful locomotor apparatus, and the metabolic machinery needed to support increased activity.

REFERENCES

Akam, M., Holland, P., and Wray, G., 1994: The evolution of developmental mechanism: Preface. *Development,* (Supplement S):5–6.

Barrington, E. J. W., 1965: *The Biology of the Hemichordata and Protochordata.* Edinburgh, Oliver & Boyd.

Barrington, E. J. W., and Jefferies, R. P. S., editors, 1975: Protochordates. *Symposia of the Zoological Society of London,* no. 36. Published for the society by Academic Press.

Bateson W., 1886: The Ancestry of the Chordata. *Quarterly Journal of the Microscopical Society* 26:535–571.

Berrill, N. J., 1955: *The Origin of Vertebrates.* Oxford, Clarendon Press.

Bone, Q., 1972: *The Origin of Chordates.* Oxford Biology Readers. London, Oxford University Press.

Brandenburg, J., and Kummel, G., 1961: Die Feinstruktur der Solenocyten. *Journal of Ultrastructure Research,* 5:437–452.

Costa, O. G., 1834: Cenni Zoologici, ossia deszcrizione delle specie nuove di animali discoperti in diverse coutrade del regno nell'anno 1834. *Annuario del Museo Zoologico della Universita de Napoli,* 12.

Eddy, S., Oliver, C. P., and Turner, J. P., 1964: *Atlas of Drawings for Vertebrate Anatomy.* New York, John Wiley & Sons.

Flood, P. R., 1968: Structure of the segmental trunk muscles in amphioxus with notes on the course and "endings" of the so-called ventral root fibers. *Zeitschrift für Zellforschung und mikroskopische Anatomie,* 8:389–416.

Flood, P. R., 1975: Fine structure of the notochord of amphioxus. *In* Barrington and Jefferies, *q.v.*

Gans, C., and Northcutt, R. G., 1983: Neural crest and the origin of vertebrates: A new head. *Science,* 220:268–274.

Gans, C., Kemp, N., and Poss, S., 1996: The lancelets: A new look at some old beasts. *Israel Journal of Zoology,* 42 (supplement):1–446.

Garstang, W., 1928: The morphology of the Tunicata and its bearings on the phylogeny of the Chordata. *Quarterly Journal of Microscopical Science,* 72:51–185.

Gee, H., 1996: *Before the Backbone: Views on the Origin of Vertebrates.* London, Chapman & Hall.

Halanych, K. M., 1998: Considerations for reconstructing metazoan history: Signal, resolution, and hypothesis testing. *American Zoologist,* 38:929–941.

Halanych, K. M., Bacheller, J. D., Aguinaldo A. M. A., Liva, S. M., Hillis, D. M, and Lake, J. A., 1995: Evidence from 18s ribosomal DNA that the lophophorates are protostome animals inarticulate. *Science,* 267:1641–1643.

Holland, L. Z., and Holland, N. D., 1999: Chordate origins of the vertebrate central nervous system. *Current Opinion in Neurobiology,* 9:596–602.

Holland, L. Z., and Holland, N. D., 1998: Developmental gene expression in amphioxus: New insights into the evolutionary origin of vertebrate brain regions, neural crest, and rostrocaudal segmentation. *American Zoologist,* 38:647–658.

Holland, N. D., and Holland, L. Z., 1999: Amphioxus and the utility of molecular genetic data for hypothesizing body part homologies between distantly related animals. *American Zoologist,* 39:630–640.

Holland, P. W. H., 1999: Gene duplication: Past, present and future. *Seminars in Cell and Developmental Biology,* 10:541–547.

Holland, P. W. H., Garcia-Fernandez, J., Williams, N. A., and Sidow, A., 1994: Gene Duplications and the Origins of Vertebrate Development. *Development,* (Supplement S):125–133.

Jefferies, R. P. S., 1986. *The Ancestry of the Vertebrates.* London, British Museum (Natural History).

Jefferies, R. P. S., Brown, N., and Daly, P. E. J., 1996: The early phylogeny of chordates and echinoderms and the origin of chordate left-right asymmetry and bilateral symmetry. *Acta Zoologica (Stockholm),* 77:101–122.

Jenner, R. A., and Schram, F. R., 1999: The grand game of metazoan phylogeny: Rules and strategies. *Biological Reviews of the Cambridge Philosophical Society,* 74:121–142.

Jollie, M., 1973: *Chordate Morphology.* Huntington, N.Y., Robert E. Krieger.

Knoll, A. H., and Carroll, S. B., 1999: Early animal evolution: Emerging views of comparative biology and geology. *Science,* 284:2129–2137.

Lacalli, T. C., 1996: Landmarks and subdomains in the larval brain of *Branchiostoma:* Vertebrate homologies and invertebrate antecedents. *In* Gans et al., *q.v.*

Lacalli, T. C., 1999: Tunicate tails, stolons, and the origin of the vertebrate trunk. *Biological Reviews of the Cambridge Philosophical Society,* 74:177–198.

Lacalli, T. C., Holland, N. D., and West, J. E., 1994: Landmarks in the anterior central nervous system of amphioxus larvae. *Philosophical Transactions of the Royal Society of London B,* 344:165–185.

McHugh, D., and Halanych, K. M., 1998: Introduction to the symposium: Evolutionary relationships of metazoan phyla: Advances, problems, and approaches. *American Zoologist,* 38:813–817.

McHugh, D., 1998: Deciphering metazoan phylogeny: The need for additional molecular data. *American Zoologist,* 38:859–866.

Nielsen, C., Scharff, N., and Eibye-Jacobsen, D., 1996: Cladistic analyses of the animal kingdom. *Biological Journal of the Linnean Society,* 57:385–410.

Northcutt, R. G., 1996: The origin of craniates: Neural crest, neurogenic placodes, and homeobox genes. *In* Gans et al., *q.v.*

Northcutt, R. G., and Gans, C., 1983: The genesis of neural crest and epidermal placodes: A reinterpretation of vertebrate origins. *Quarterly Review of Biology,* 58:1–28.

Ruppert, E. E., and Smith, P. R., 1988: The functional organization of filtration nephridia. *Biological Reviews,* 63: 231–258.

Schaeffer, B., 1987: Deuterostome monophyly and phylogeny. *Evolutionary Biology,* 21:179–235.

Sherman, I. W., and Sherman, V. G., 1970: *The Invertebrates: Function and Form.* New York, Macmillan.

Stokes, M. D., and Holland, N. D., 1998: The lancelet. *American Scientist,* 86:552–560.

Stokes, M. D., and Holland, N. D., 1995: Embryos and larvae of a lancelet, *Branchiostoma floridae,* from hatching through metamorphosis: Growth in the laboratory and external morphology. *Acta Zoologica (Stockholm),* 76: 105–120.

Wada, H., Saiga, H., Satoh, N., and Holland, P. W. H., 1998: Tripartite organization of the ancestral chordate brain and the antiquity of placodes: Insights from ascidian *Pax-2/5/8, Hox* and *Otx* genes. *Development,* 125: 1113–1122.

Webb, J. E., 1973: The role of the notochord in forward and reverse swimming and burrowing in the amphioxus, *Branchiostoma lanceolatum. Journal of Zoology (London),* 170:325–338.

Welsch, U., 1975: The fine structure of the pharynx, cyrtopodocytes, and digestive caecum of amphioxus (*Branchiostoma lanceolatum*). *Symposium of the Zoological Society of London,* 36:17–41.

Wray, G. A., Levinton, J. S., and Shapiro, L. H., 1996: Molecular evidence for deep Precambrian divergences among metazoan phyla. *Science,* 274:568–573.

Yarrell, W., 1836: *A History of British Fishes,* 1st edition. London, J. van Voorst.

3

Diversity and Phylogenetic History of Craniates

PRECIS

This chapter begins with an overview of craniate phylogeny (evolutionary relationships), followed by brief discussions of major craniate groups, their diagnostic characters (synapomorphies), their species diversity, and their modes of life. We present and discuss cladograms and examine the role of fossils in reconstructing craniate phylogeny. We refer back to these groups and cladograms in subsequent chapters as we describe additional details about the function and evolution of craniate organ systems.

OUTLINE

An Overview of the Major Living Clades of Craniates
Craniates and the Origin of Gnathostomes
- *Hagfishes*
- *Vertebrates*
- *The Origin of Gnathostomes*

Elasmobranchiomorphs
- *Cartilaginous Fishes*

Bony Fishes and the Diversification of Teleosts
- *Actinopterygians*
- *Neopterygians*
- *Teleosts*
- *Euteleosts*

Sarcopterygians and the Origin of Tetrapods
- *Coelacanths*
- *Rhipidistians*
- *The Origin of Tetrapods*

Amphibians
- *Lissamphibians*
- *Batrachians*

Reptilomorphs and the Origin of Amniotes
- *The Origin of Amniotes*

Sauropsids and the Origin of Diapsids
- *The Origin and Early Diversification of Diapsids*

Lepidosaurs
Archosauromorphs
- *Saurischians*

Theropods and the Origin and Diversification of Birds
- *Origin and Diversification of Birds*
- *"Ratites"*
- *Modern Birds*
- *Perching Birds*

Synapsids
- *Therapsids*
- *Cynodonts*

Mammals
- *Early Mammals*
- *Monotremes*

Therians and the Diversification of Marsupials
- *Diversification of Marsupials*

Eutherians
- *Edentates*
- *Glires*
- *Insectivores*
- *Carnivores*
- *Archontans*
- *Primates*
- *Ungulates*

Focus 3-1 *Practical Considerations About Fossils and Anatomical Reconstructions*
Focus 3-2 *Adapting to a Phylogenetic Classification of Craniata*
Focus 3-3 *Thinking About the Origin of Tetrapods*
Focus 3-4 *Ectothermy and Endothermy*
Focus 3-5 *Early Mammals and the Origin of Mammalian Endothermy*
Appendix: Example Synamorphies for Clades Figured in Chapter 3

Craniata contains nearly 50,000 living species and many more species that have become extinct during the past 500 million years. We will consider phylogenetic relationships among craniates in order to establish an organizational framework for studies of comparative functional anatomy. This task is possible only because we can organize these species into nested sets of monophyletic groups, or **clades,** using techniques of phylogenetic analysis introduced in Chapter 1. By organizing your study of craniate diversity in such a phylogenetic framework, you can focus on hierarchical patterns of anatomical characters without having to learn attributes of tens of thousands of species. Conveniently, we can concisely express diagnostic features of each subgroup of craniates by summarizing its shared derived characters, or **synapomorphies** (Chapter 1). Remember that a cladogram identifies hypothesized branching points but can give no absolute indications about how long ago such branching occurred. We may establish some ideas about divergence times based on the fossil record and the relative pattern of branching on a cladogram.

Learning to think as a phylogeneticist will enable you to understand the latest research in functional anatomy, for phylogenetics is the language of modern comparative biology. This is an exciting time because researchers are applying phylogenetic techniques and adding new sources of character data, such as molecular sequence information, to the study of craniate evolution. Just as with other information-retrieval systems, such as library catalogs, retrievability is maximized when details are organized hierarchically. Hierarchical treatment of species and characters also allows us to reconstruct more easily the evolution of specific functional anatomical systems, such as the wings and flight systems independently evolved by birds, bats, and †pterosaurs,[1] and to clearly distinguish between hypotheses of **homology** and **homoplasy** (see Chapter 1 for more on homology and homoplasy). It is inevitable that clade branching patterns will change as our knowledge of morphology, species, and higher taxa increases. Still, the cladograms and characters presented in this chapter should serve you well in guiding your study of craniate anatomy and function.

Craniate skeletons are well suited to fossilization, and fossil craniates provide much information against which to test hypotheses derived from analyses of living species. Fossils alone cannot solve phylogenetic questions, but extinct species often demonstrate different mixtures of **plesiomorphic** (= primitive) and **apomorphic** (= derived) traits than we can observe in any living species. Unfortunately, however, we never can recover as much or as detailed character information from a fossil species as from a living species, which makes it challenging to make rigorous phylogenetic studies that integrate fossil and living species. Ultimately, structures in fossils can be reliably interpreted only with the help of comparative anatomical information about living species. Empirical information from fossils usually relates only to the skeletal system, and it has proven essential to limit extrapolation concerning other organ systems in extinct animals (Focus 3-1 [p. 52]). Surprising as it may seem, empirical knowledge about the anatomy of many living species is so incomplete that a study of fossils often begins with new research on living species. It is never easy to seamlessly integrate fossils into a phylogenetic analysis or classification, and inclusion of fossil taxa sometimes complicates phylogenetic understanding. This is partly because adding taxa to a data matrix increases the number of possible trees. Also, because many fossil taxa exhibit different mixtures of plesiomorphic and apomorphic characters than are observed in any living taxa, their inclusion in an analysis may decrease confidence about some branching points. Such challenges to conventional interpretations are a normal and good part of the research process, but they can be distracting to those just beginning a study of vertebrate phylogeny. In an abbreviated treatment, then, it is best to begin with living craniates and then to integrate fossil taxa opportunistically wherever they can contribute to the overall evolutionary picture.

In keeping with this philosophy, we begin with a cladogram of the major living groups of Craniata (Fig. 3-1). Clarifying the phylogenetic arrangement of the 14 terminal taxa in this cladogram is one goal of this chapter. We also present 16 other cladograms in this chapter, most of which include selected extinct taxa to help summarize current notions about phylogenetic relationships, character evolution, or major points in craniate history. It is important to remember that cladograms are only hypotheses about phylogenetic relationships and that the number of characters analyzed to generate such cladograms is large. As a general guide to character evolution in craniates, we include in the Appendix to Chapter 3 an example synapomorphy for each major group indicated on the cladograms for this chapter. In selecting information to present, we limited ourselves to a practical subset of well-accepted, widely cited characters.

As in Chapters 1 and 2, we show our phylogenetic classification by a **nested series of brackets** on the right side of phylogenetic diagrams. As explained in Focus 1-2, we do not recognize **paraphyletic groups,** which are groups that do not include all of the descen-

[1] As noted in Chapter 1, the dagger symbol (†) preceding a taxon indicates that all members of that taxon are extinct; our reasons for adopting this policy are discussed later in the text.

dent taxa derived from a single common ancestor. Recognition of paraphyletic groups inherently conflicts with our goal of teaching you useful generalizations about the evolutionary history of craniates. Unfortunately, many paraphyletic groups have been named in the past, and some of these names are entrenched deeply in scientific and popular works. At first, it may seem uncomfortable to you to find that we reject a familiar grouping even when it is demonstrably paraphyletic (e.g., the living animals commonly known as reptiles form a paraphyletic group if we exclude the birds; see Focus 1-2). This is, however, a necessary step if one is to approach systematics logically and with the goal of improving communication (Focus 3-2 [p. 52]). In this chapter, where reference to a paraphyletic group is essential to understand classification schemes used in older works, we enclose the name in quotation marks (e.g., "Agnatha," and "Reptilia"). Many vernacular or **common names** are given in the text, and generic names are given for those taxa that we either illustrate, deem especially important, or refer to elsewhere in this book. Throughout our text and figures, we use a **dagger symbol (†)** to indicate exclusively extinct taxa. This convention is intended to remind you that the information about that particular taxon is far less complete than for a living taxon.

Our phylogenetic arrangement and classification of Craniata is intended as a teaching tool. Although it benefits from the logic and clarity of communication made possible by phylogenetic systematics, it is not comprehensive for two reasons. First, comprehensive coverage to even the familial level would require an entire book. Second, many groups remain too poorly studied to allow for the development of well-corroborated phylogenies. In selecting taxa to discuss, we have been guided by our own interests as well as the relevance of the taxa to functional anatomical studies of craniates.

An Overview of the Major Living Clades of Craniates

The phylogenetic hypothesis and classification for living craniates shown in Figure 3-1 serve as our framework. Several clades in Figure 3-1 include aquatic animals with internal gills as adults, which are commonly known as "fishes" (the common name "fishes" and the formal name "Pisces," while perhaps readily understood, refer to a paraphyletic group without phylogenetic value because, as generally used, they exclude a major descendant clade, the Tetrapoda). Among these are hagfishes (**Myxiniformes**), lampreys (**Petromyzontiformes**), cartilaginous fishes (**Chondrichthyes**), and ray-finned fishes (**Actinopterygii**). Roughly half of all living species of craniates are actinopterygians.

The remaining species of living craniates are lobe-finned fishes (**Sarcopterygii**). Because we use a phylogenetic classification, Sarcopterygii includes two groups of living "fishes," **Actinistia** (coelacanths) and **Dipnoi** (lungfishes), as well as **Tetrapoda** (land vertebrates). We consider lungfishes to be the living sister group of tetrapods and recognize the clade **Rhipidistia** to contain lungfishes and tetrapods. As shown in Figure 3-1, **Lissamphibia** (salamanders, frogs, and caecilians) forms the living sister group to all other living tetrapods, which form the clade **Amniota.** Living amniotes belong to two clades: **Sauropsida** and **Mammalia.** Sauropsida includes **Testudines** (turtles), which is the living sister group of **Diapsida.** Living diapsids include **Lepidosauria** (tuatara, lizards, and snakes), **Crocodilia** (alligators and crocodiles), and **Aves** (birds). Mammalia includes **Monotremata** (platypus and echidnas), **Metatheria** (i.e., opossums and kangaroos, commonly called marsupials), and **Eutheria** (commonly called placental mammals).

If you study the phylogenetic classification on the right side of Figure 3-1, then you can follow its logic. Begin with the most inclusive group, clade **Craniata.** Then note that:

Vertebrata includes all living craniates except Myxiniformes.

Gnathostomata includes all living vertebrates except Petromyzontiformes.

Osteichthyes includes all living gnathostomes except Chondrichthyes.

Sarcopterygii includes all living osteichthyans except Actinopterygii.

Rhipidistia includes all living sarcopterygians except Actinistia.

Tetrapoda includes all living rhipidistians except Dipnoi.

Amniota includes all living tetrapods except Lissamphibia.

Sauropsida includes all living amniotes except Mammalia.

Diapsida includes all living sauropsids except Testudines.

Archosauria includes all living diapsids except Lepidosauria.

Aves includes all living archosaurs except Crocodilia.

Theria includes all living mammals except Monotremata.

Eutheria includes all living therians except Metatheria.

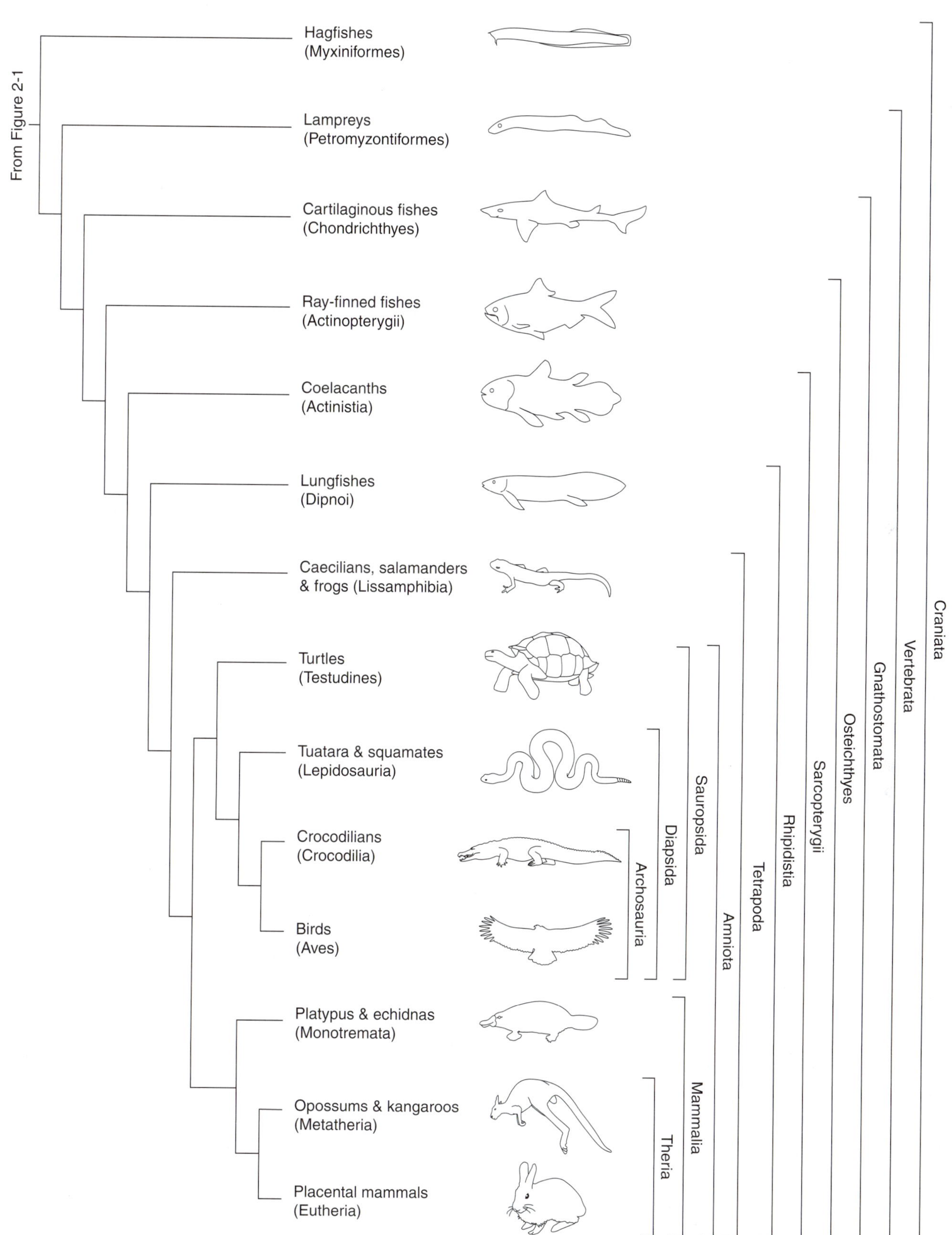

FIGURE 3-1
Phylogeny of the major extant (i.e., living) clades of craniates, highlighting major patterns in vertebrate evolution and the phylogenetic classification used in this book.

FOCUS 3-1 *Practical Considerations About Fossils and Anatomical Reconstructions*

One of the chief problems in studying vertebrate fossils is their lack of completeness. A fossil never reveals the same quality of data as an extant animal because it is missing soft tissues, such as muscles, nerves, and digestive organs. In fact, it is relatively rare to recover even a complete skeleton, and many extinct taxa that are judged to be of great phylogenetic significance are known only from very limited material.

Even with fossil taxa known from abundant, well-preserved specimens from the same locality, paleontologists rarely have a single "perfect" specimen for photography or illustration. Instead, they often employ artists' drawings, known as **reconstructions,** to illustrate their conception of the animal's appearance when it was alive. Some reconstructions are based on detailed study of very complete fossils, but others are not, and there is no easy way to tell without going back to the original description of the specimens. Often, reconstructions include features not known from actual specimens, and other features that have been influenced by dissimilar fossil or living taxa that were used as models. Once published, such a reconstruction may have an extremely long life. For example, in the 1950s, paleontologist Erik Jarvik published a famous reconstruction of an early tetrapod, †*Ichthyostega*. His reconstruction is misleading because he included the front leg, which was unknown in the fossils. Although Jarvik later discussed the possibility that †*Ichthyostega* had more than five digits (based on specimens with intact hindlimbs), his pentadactylous (five-toed) reconstruction lived on. It was not until specimens of †*Acanthostega* were described in the 1980s that researchers realized that the earliest tetrapods had more than five fingers and toes. Such is the power of reconstructions that Jarvik's mistake will probably live on; indeed, we used it as the basis for our illustration of †*Ichthyostega* (Fig. 3-17*C*)—although we increased the number of toes!

Given these complications, how much can you rely on reconstructions? Not much, and certainly never for scoring character states for phylogenetic analysis. However, given their ease of study, we use many reconstructions in this text. The warnings noted here offer yet another good reason for our convention of using a † symbol to indicate a fossil taxon.

FOCUS 3-2 *Adapting to a Phylogenetic Classification of Craniata*

Empirically based phylogenetic classifications are routinely included in research papers in systematics and evolutionary biology. Yet to date, few textbooks and general reference works on vertebrate anatomy have used phylogenetics as a basis for their organization. This focus offers some tips on using Chapter 3 and phylogenetic classifications in general.

It is essential to understand the logical connection between cladograms and the naming of groups. Phylogenetic classifications are built on the premise that we can recognize and diagnose monophyletic groups of organisms (for a definition and discussion of the concept of monophyly, see Chapter 1). If we lack character evidence to support a group's monophyly, then we normally do not name the group. On the other hand, just because a phylogenetic hypothesis indicates that a particular group is monophyletic does not mean that we must name it, especially because name proliferation can impede communication. Often in this book we do not name monophyletic groups that are of little significance to our purpose of providing a general background in craniate phylogeny.

The Appendix to Chapter 3 includes example characters for the named nodes on the cladograms in this chapter. Regard these as a starting point for understanding character evolution in craniates and not as the final word. Remember that cladograms presented in Chapter 3 may differ from those published elsewhere. This is because every phylogenetic pattern is only a hypothesis based on available character evidence, and different workers may interpret the evidence differently. As data on characters and species accumulate, the resolution and branching patterns of cladograms will change. This is progress. Also, it is certainly not our intent that all of the cladograms and characters we present in this chapter should be memorized. Rather, our goal is to include enough basic information so that our cladograms and phylogenetic classification can become reference tools for placing your study of comparative functional morphology in a phylogenetic context.

Craniates and the Origin of Gnathostomes

Craniata (Fig. 3-2) is named for one of its many synapomorphies, the **braincase,** which is a skeletal trough or box that protects the brain. Other craniate synapomorphies include neural crest, neurogenic placodes, complex sense organs, cranial nerves, a tripartite brain, a complex endocrine system, muscularization of the wall of the gut tube, differentiated digestive organs, respiratory gills, a heart, and hemoglobin (Chapter 2). Thus, much evidence shows that craniates are monophyletic. One functional anatomical synapomorphy of craniates is their use of muscles to draw in a feeding and respiratory current of water. Such a system is far more powerful than the cilia used for filter-feeding by tunicates and cephalochordates. Muscularization of the wall of the gut tube undoubtedly was a factor in the increase in size and level of activity of early craniates.

The earliest craniates lacked jaws. Such animals can be described accurately as **agnathous** (Gr., *a* = without + *gnathos* = jaw). Many authors refer to a group "Agnatha" containing these forms and excluding their descendants. The absence of jaws, however, is plesiomorphic for craniates (i.e., no outgroup taxa have true jaws). "Agnatha" is paraphyletic and is not a helpful group for our phylogenetic survey. Hagfishes and lampreys have been considered to belong to a monophyletic group known as "Cyclostomata" (Gr., *cyclos* = round + *stoma* = mouth). A few anatomical and molecular characters would support the monophyly of this group, but based on the congruence of other characters, we do not recognize "Cyclostomata" in this text.

Hagfishes

The 45 living species of hagfishes form a distinctive monophyletic group, known as **Myxiniformes,** which we regard as the sister group of all other craniates (Fig. 3-2). All hagfishes have many integumentary glands (Fig. 3-3*A*) that secrete great quantities of protective mucus, which is the source for one of their common names, the slime hags. The group has a poor fossil record, and the earliest well-preserved hagfish occurs relatively late in the fossil record of craniates (the late Paleozoic of Illinois). Hagfishes exhibit craniate synapomorphies, such as a braincase, a brain and cranial nerves, a nose with a single median nostril, two lateral eyes, a mid-dorsal pineal eye, a lateral line system, an inner ear with a semicircular duct, a ventral liver, a heart, renal tubules, and gills located in pouch-shaped gill chambers that open to the surface by ducts. Hagfishes also retain many plesiomorphic features. Like amphioxus, tunicates, and echinoderms, hagfishes lack jaws, paired appendages, and bone. Also, hagfishes lack any trace of vertebrae, and thus the notochord is their only axial support.

All extant hagfishes live in marine environments. Like marine invertebrates but unlike all other extant craniates, hagfishes have the same overall salt concentration in their body fluids as sea water. This **isosmotic** (Gr., *iso* = equal + *osmos* = pushing) condition suggests that hagfishes have been marine throughout their evolution and that they diverged from early marine craniates. The small eyes of hagfishes are considered to be degenerate modifications related to their habit of burrowing in the mud along continental shelves. They find food with three pairs of sensory tentacles around the mouth and with the single terminal nostril (Fig. 3-3*A*). Hagfishes have a unique mode of feeding, in which they use keratinized "teeth" and "jaws"[2] to tear off pieces of polychaete worms and dead fishes. Pacific hagfishes (*Bdellostoma*) have as many as 14 pairs of gill pouches, each of which opens by a pore on the body surface. Atlantic hagfish (*Myxine*) have fewer gill pouches, and branchial ducts lead from them to a common branchial aperture far caudally on the trunk. Hagfishes have a single gonad and appear to be **serially hermaphroditic,** meaning that they are one sex (in this case, male) when newly matured but eventually become the other sex (in this case, female; see Chapter 21 for more on sex reversal in craniates). They are oviparous, but few fertile eggs have been collected. Our knowledge of their embryology depends chiefly on specimens collected nearly a century ago.

Vertebrates

Vertebrata (Fig. 3-2) is characterized by three synapomorphies. First, vertebrates have a backbone composed of **vertebrae,** which are skeletal elements that surround the spinal cord and notochord and increase the stiffness of the body (Chapter 8). Vertebrates also have at least two **vertical semicircular ducts,** which are specializations of the inner ear that allow them to more accurately detect roll, pitch, and yaw as they swim (Chapter 12). **Radial fin muscles** associated with the bases of the median and other fins allow vertebrates to control the shape and movements of their fins and thus to enhance swimming ability (Chapter 10). Fossils of the earliest known vertebrates

[2]The "teeth" and "jaws" of hagfishes are not homologous to the teeth and jaws of gnathostomes (i.e., the "jaws" of hagfishes are not considered true jaws).

Text continues on page 56

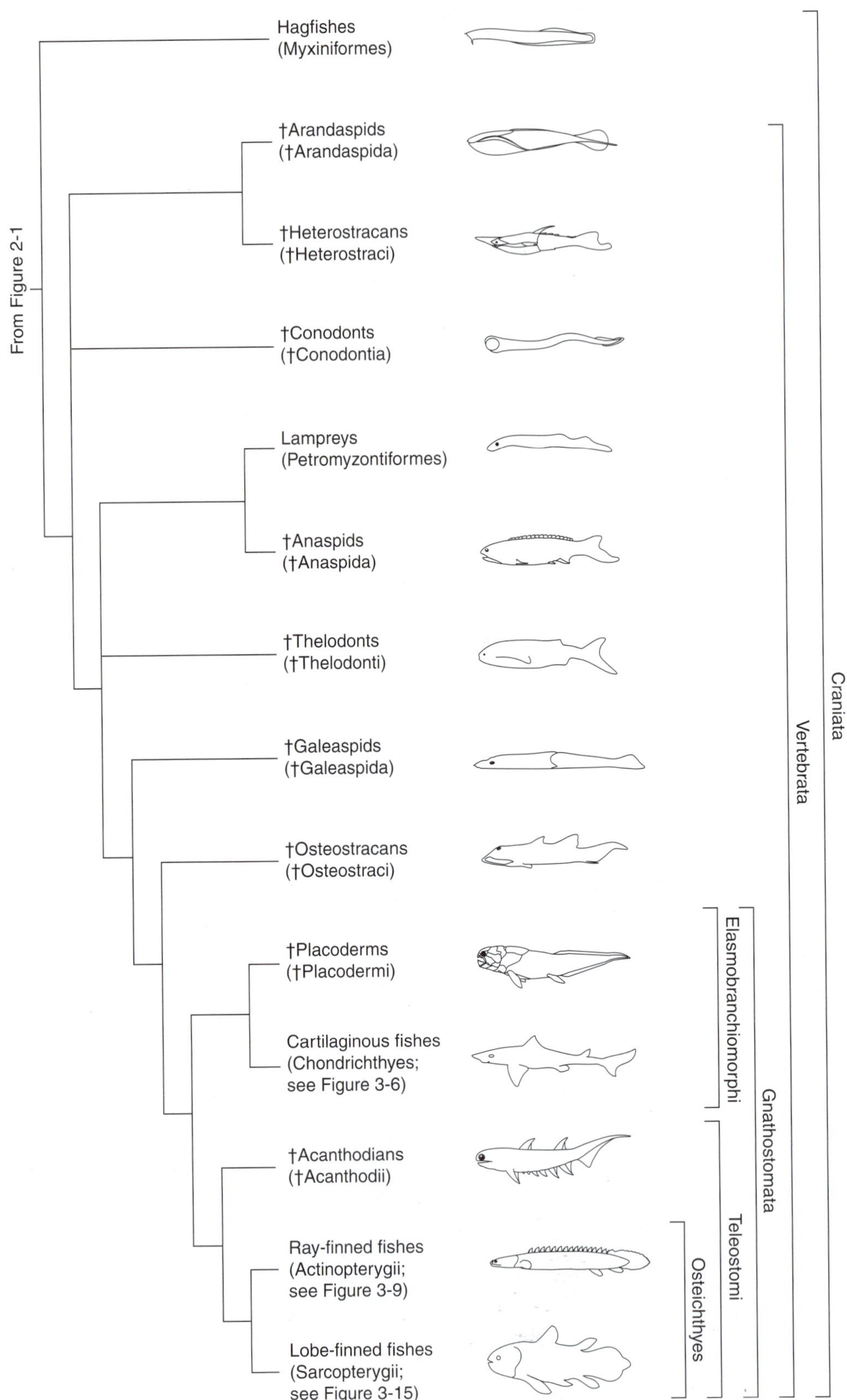

FIGURE 3-2
Phylogeny of selected living and fossil craniates (Craniata), highlighting the origin of gnathostomes (Gnathostomata), elasmobranchiomorphs (Elasmobranchiomorphi), teleostomes (Teleostomi), bony fishes (Osteichthyes), ray-finned fishes (Actinopterygii), and lobe-finned fishes (Sarcopterygii).

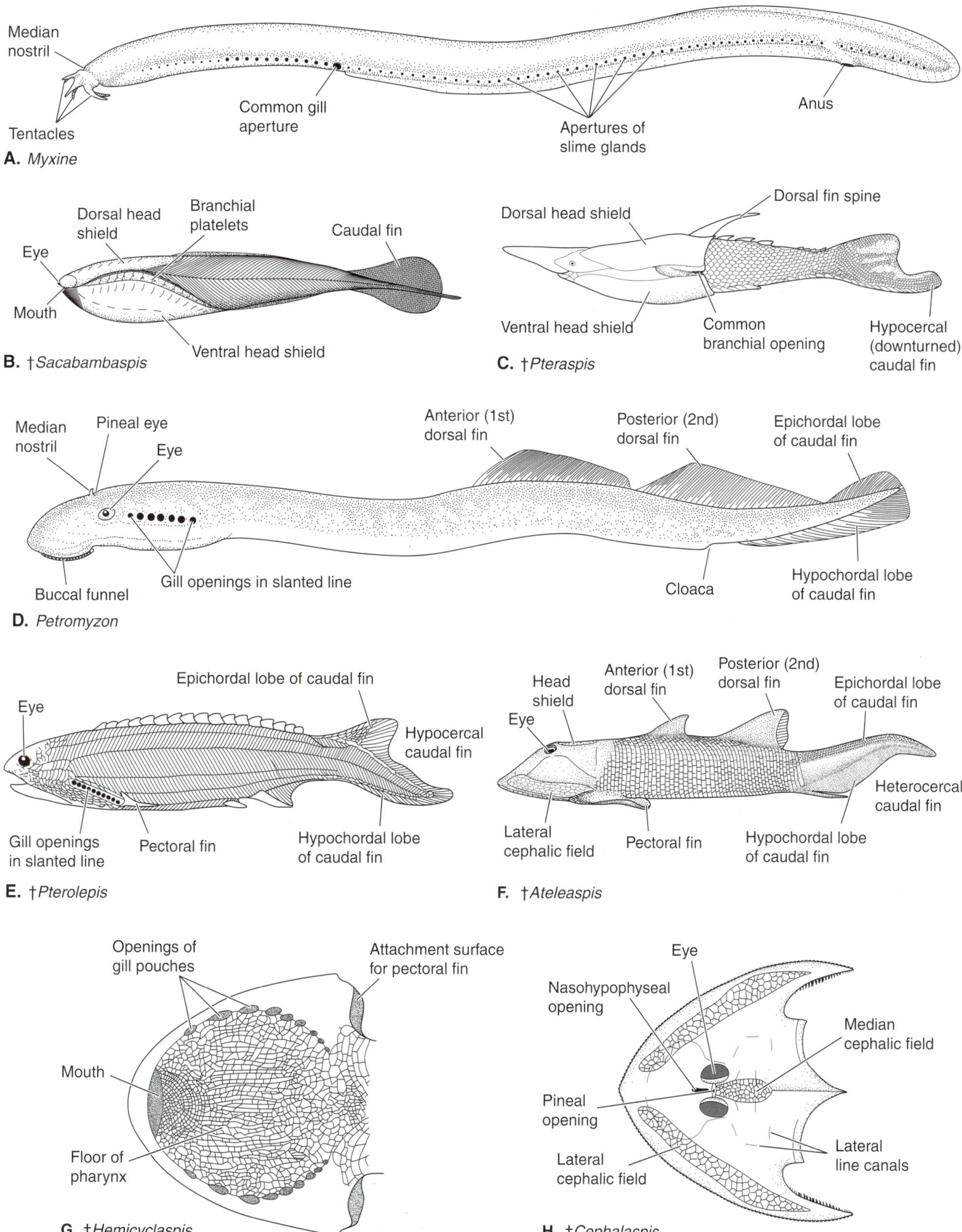

FIGURE 3-3

Representative craniates and vertebrates. *A,* Lateral view of the body of a hagfish, *Myxine,* showing features of the head, the single common gill aperture, and the openings of the many slime glands. *B,* Reconstruction of an †arandaspid, †*Sacabambaspis,* from the Ordovician of Bolivia. *C,* Reconstruction of a †heterostracan, the †pteraspidiform, †*Pteraspis. D,* Lateral view of an adult sea lamprey, *Petromyzon. E,* Reconstruction of an †anaspid, †*Pterolepis,* showing the downturned hypocercal tail and row of gill openings in slanted line. *F,* Reconstruction in lateral view of †*Ateleaspis* showing the two dorsal fins found in some †osteostracans. *G,* Ventral view of the head of an †osteostracan, †*Hemicyclaspis. H,* Dorsal view of head shield of an †osteostracan, †*Cephalaspis.*

occur in rocks that formed in the early to middle Paleozoic. Like hagfishes, these early vertebrates lacked jaws, but unlike hagfishes, some had pectoral fins, and many had mineralized exoskeletons. The name "ostracoderm" often has been applied to such animals because of their exoskeleton, which usually takes the form of shell-like bony plates in their skin (Gr., *ostrakon* = shell + *derma* = skin). Although it is a good descriptor, "ostracoderm" refers to a paraphyletic group, so it does not appear in our phylogenetic classification (Fig. 3-2).

Early Paleozoic vertebrates were small, aquatic animals, usually less than 20 cm in length. A few fragmented plates found in late Cambrian marine deposits 500 million years old are sometimes regarded as evidence of vertebrates. More complete fossils that can be confidently assigned to Vertebrata occur in Ordovician rocks that formed in marine or estuarine environments (scientists base the assessment of this paleoenvironment on the co-occurrence of fossil echinoderms and other exclusively marine groups). The earliest vertebrates, then, like living hagfishes and outgroup deuterostomes, were marine animals. In contrast, many remarkably complete specimens of early vertebrates come from younger Silurian and Devonian freshwater deposits. Fresh water appears to have hosted subsequent evolutionary radiations of vertebrates that led some groups back to the sea and some onto the land. Because many early vertebrate fossils are fragmentary, detailed studies of their mineralized tissues are a primary source of character information. For example, such studies show that bone in many early vertebrates differs from typical vertebrate bone (e.g., mammalian bone) in that it lacks enclosed bone cells. Such bone is referred to as **acellular bone,** or **aspidine.**

One of the earliest well-preserved vertebrates is †*Sacabambaspis,* known from Ordovician rocks of Bolivia deposited in near-shore marine environments (Fig. 3-3*B*). †*Sacabambaspis* is a member of a larger group, known as **†Arandaspida** (e.g., †*Pteraspis,* Fig. 3-2). It is torpedo shaped, with a paddle-shaped caudal fin but no other median or paired fins (Fig. 3-3*B*). Its jawless mouth and eyes are located at the anterior tip of the body. On the dorsal surface of the head shield are openings for light-receptive pineal and parapineal organs and for the canals of the mechanosensory lateral line system. Branchial platelets cover the gill region.

†Heterostraci includes †Pteraspida (e.g., †*Pteraspis;* Fig. 3-3*C*) and other groups. The head of a †heterostracan was armored primarily by two plates of bone, known as the **dorsal** and **ventral head shields** (or disks), and all †heterostracans had only a single pair of common branchial openings. The **hypocercal** tail had a large ventral lobe. Some authors consider †heterostracans to be the sister group of gnathostomes based on the presence of paired nostrils in both groups, a condition known as **diplorhiny.** Based on the congruence of other characters, we regard diplorhiny as convergently evolved in †heterostracans and gnathostomes.

For decades, geologists used tiny, cone-shaped and toothlike fossils known as **†conodont elements** to define and correlate Paleozoic and Mesozoic sedimentary rocks, but no one knew what they were. Partial specimens were discovered in the 1960s, and in the 1980s better-preserved fossils containing †conodont elements were discovered in England and South Africa. The †conodont elements turned out to be the teeth of the †conodont animals. These "teeth," however, are composed of a different mineral than are typical vertebrate teeth and bones and so are not homologous to the teeth found in jawed vertebrates. These fossils also reveal gill arches and traces of muscular segments that convince us to accept **†Conodontia** as vertebrates of uncertain phylogenetic affinities (Fig. 3-2).[3]

There are about 40 extant species of lampreys (**Petromyzontiformes**; Fig. 3-2), and a few fossil forms. Adults of most living species attach to prey (usually a bony fish) by a round, suctorial buccal funnel. Keratinized, tooth-like structures cover the tongue, which can be protruded by a piston cartilage and used to rasp a hole in the prey. Buccal glands secrete an anticoagulant that prevents the prey's blood from clotting while the lamprey feeds. The buccal funnel and piston cartilage are synapomorphic for lampreys. The sucking pharynx is divided longitudinally into a dorsal food passage (conventionally called the esophagus) and a ventral respiratory tube that ends blindly. Many other aspects of their internal anatomy, including a notochord that persists into the adult stage and neural arches surrounding the spinal cord, are thought to reflect the plesiomorphic condition for Vertebrata. The phylogeny shown in Figure 3-2 implies secondary losses of the mineralized exoskeleton and pectoral fins. Such losses are believable because of insights from convergent evolution: several other groups of eel-like vertebrates have secondarily lost their body armor (e.g., scales) and one or both pairs of paired fins.

Most populations of the sea lamprey (*Petromyzon*) are **anadromous,** meaning that they live in the ocean as adults and return to rivers to spawn; other lampreys live entirely in fresh water. Adult lampreys die after they have spawned. Their tiny eggs hatch within about

[3]†Conodontia is thought by some workers to belong to Craniata but not Vertebrata, but we adopt the latter view in this book; see Figure 3-2 and Donoghue, Forey, and Aldridge (2000).

two weeks into **ammocoetes larvae** that live and grow as filter feeders in the bottom of the stream for six or seven years. After metamorphosing into an adult, a sea lamprey normally migrates downstream to the ocean, where it lives for a year or two and grows dramatically before returning to fresh water to spawn. In contrast, some lampreys do not feed as adults but reproduce and die soon after metamorphosis. Superficially, an ammocoetes larva resembles an amphioxus. They have the same general shape, and the notochord and myomeres are obvious in both. The pharynx is not as large as that of amphioxus, nor is it surrounded by an atrium. Unlike the feeding and respiratory systems in amphioxus, a larval lamprey draws water into its pharynx by expanding its walls and by the action of a pair of muscular flaps, the velum. Food is entrapped in mucus that is secreted by an endostyle-like subpharyngeal gland.

A group thought to be related closely to lampreys is †**Anaspida** (Fig. 3-2). A reconstruction of an †anaspid known as †*Pterolepis* is shown in Figure 3-3*E*. These streamlined vertebrates have a **hypocercal** (= downturned) caudal fin and an anterior pair of fins that we regard as homologous to the pectoral fins of gnathostomes. They lack head shields, having instead a series of small scales or plates. As many as ten openings for gill pouches extended posteroventrally from the eye, forming a diagonal line that resembles the condition in living lampreys.

Some other groups of extinct jawless vertebrates are indicated in Figure 3-2 and include †**Thelodonti,** which had paired pectoral fins, a series of branchial openings, and small or micromeric (Gr., *micro* = small + *meros* = part or unit) body scales that resemble those of gnathostomes. Members of †**Galeaspida** (Fig. 3-2) occur in Devonian deposits of China and Vietnam. Significantly, †galeaspids share with †osteostracans (the next group on Fig. 3-2) and gnathostomes the presence of **perichondral bone.** Perichondral bone is one type of cartilage-replacement bone; it forms as a thin ring immediately surrounding a cartilage (Chapter 5).

†**Osteostraci** (Fig. 3-2) includes some well-preserved Silurian and early Devonian freshwater vertebrates. The large head shield and plates along the trunk were composed of bone overlain by dentine and enamel-like tissue (Fig. 3-3*F–H*). This surface armor may have protected them from predacious aquatic arthropods, such as giant sea scorpions (†eurypterids), or provided a metabolic reservoir for calcium and phosphate, or insulated electroreceptive organs (although this is unlikely; see Chapter 12). Thanks to extensive perichondral ossification of the braincase, investigators have been able to determine the shape of the brain, the inner ear, and the courses of the cranial nerves. Other parts of the internal skeleton, such as vertebrae, are not known. †Osteostracans and gnathostomes share the presence of **cellular bone.** This refers to the occurrence of cells within the bone matrix, and from a functional perspective, it means that the skeleton can be easily remodeled during growth (see Chapter 5 for a description of bone development and remodeling). By facilitating repair and ontogenetic change of the skeleton, cellular bone may be linked to the evolution of larger body sizes.

The flattened ventral surface and dorsally directed eyes of †osteostracans indicate that they had **benthic** (Gr., *benthos* = bottom) habits, meaning that they lived close to the bottom of their Paleozoic ponds and streams. The large, dorsal lobe of the tail was stiffened by an upturned extension of the notochord. Such a **heterocercal tail** (Gr., *heteros* = other + *kerkos* = tail) is synapomorphic for †osteostracans and gnathostomes. The heterocercal tail is linked to derived swimming methods and buoyancy control (Chapter 11). Two **dorsal fins** were present in †osteostracans such as †*Ateleaspis* (Fig. 3-3*F*), and we regard this as another synapomorphy of †osteostracans and gnathostomes.[4]

The structure of the mouth and pharynx suggests that †osteostracans were predators that ate small, soft-bodied animals. The jawless mouth was located at the front of the ventral surface of the head. The large pharynx was floored with many small plates that must have given it considerable mobility (Fig. 3-3*G*). We assume that the pharynx contracted and expanded by muscular action, as it does in living vertebrates, and that the suction this created was used to suck in water and food. Food was trapped and probably crushed in the pharynx, and the water escaped through many pairs of saccule-shaped gill pouches that opened to the surface through pore-shaped gill slits. Some †osteostracans had as many as 15 pairs of gill pouches.

The brain and cranial nerves were well developed in †osteostracans. Their major sensory systems consisted of a single median nostril, a pair of lateral eyes, a single pineal eye, and an inner ear containing two semicircular ducts. They also had three **cephalic fields** on the dorsal surface of the head, consisting of groups of small plates underlain by canals for branches of cranial nerves (Fig. 3-3*H*). The function of these cephalic fields is uncertain, for nothing strictly like them occurs in any living vertebrate. Because the fields appear to be richly innervated, many paleontologists and neurobiologists consider them to be related to the lateral line and electroreceptive systems.

[4]Although some †osteostracans have only one dorsal fin, primitively the group had two.

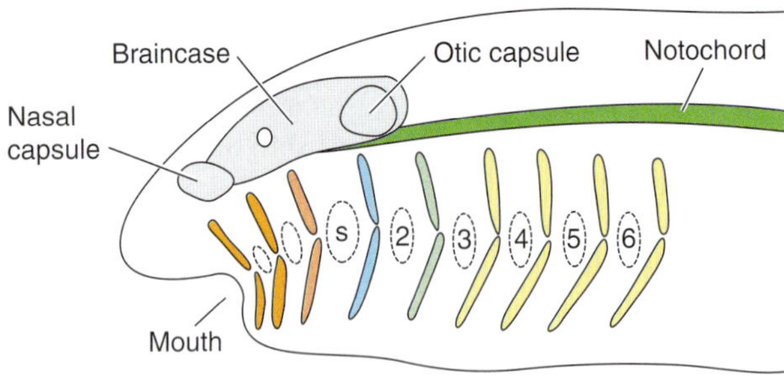

A. Hypothetical jawless condition

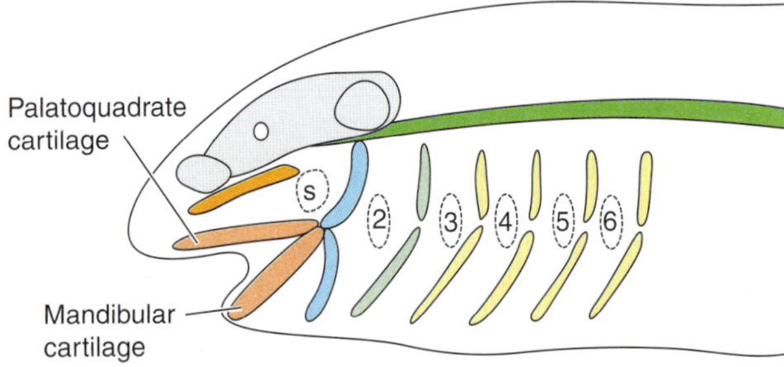

B. Mandibular arch functions as jaws

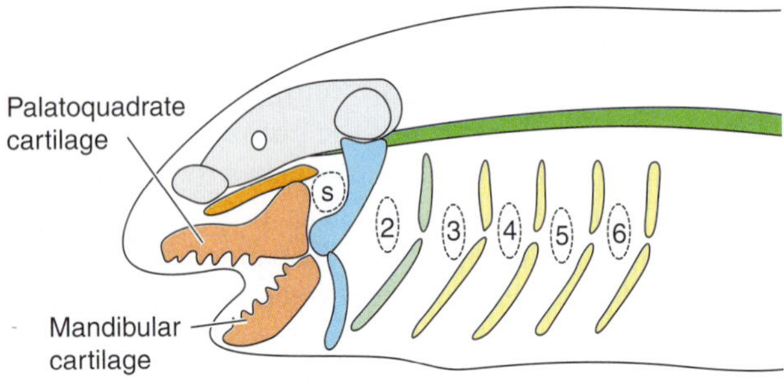

C. Jaws associated with braincase

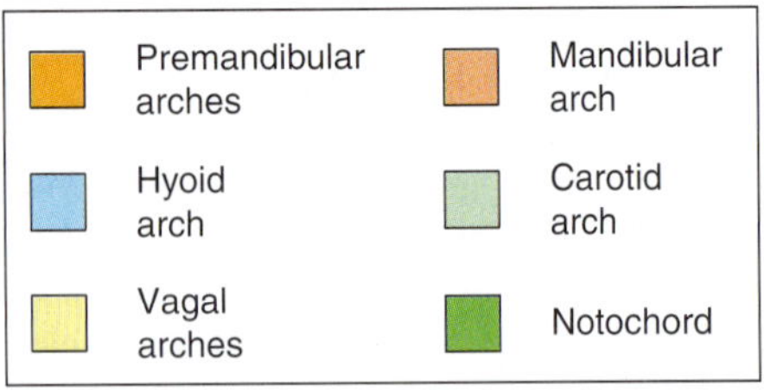

FIGURE 3-4
Origin of jaws. *A,* Hypothetical jawless condition in which the gill arches lie in a series beneath the braincase and notochord. *B,* The mandibular arch functions as jaws and is supported by the hyoid arch. *C,* Jaws associated with the braincase. *S* indicates the spiracle, which is the gill opening between the mandibular and hyoid arches. Numbers 2 through 6 indicate gill openings between posterior gill arches.

The Origin of Gnathostomes

†Osteostracans were freshwater animals. The first known vertebrates with jaws also come from late Silurian freshwater deposits. Because of their jaws, such vertebrates are known as **Gnathostomata** (Fig. 3-2; Gr., *gnathos* = jaw + *stoma* = mouth). Gnathostomes also have **paired nostrils,** five **gill slits,** and a series of visceral or **gill arches.** The gill arches of gnathostomes are deeply situated, lying next to the pharyngeal cavity, and they are articulated, which means that they consist of a jointed series of parts. These features are different from the gill arches of jawless vertebrates, which lie superficially just beneath the skin and which are not articulated. For these reasons, we cannot be sure that the gill arches of gnathostomes are homologous to those of jawless vertebrates.

The origin of the distinctive jaws is not certain (Chapter 7), but the conventional hypothesis is that jaws evolved from an anterior gill arch (not necessarily the first one) that lay close to the mouth opening (Fig. 3-4). Anatomists call this the **mandibular arch,** and it is considered by convention to be the first arch of a gnathostome. We think that its teeth are derived from modified dermal scales or plates. The mandibular arch is composed of cartilage or of cartilage replacement bone. Its dorsal part is the **palatoquadrate cartilage;** its ventral is the **mandibular (Meckel's) cartilage.** When these ossify, their caudal ends form the **quadrate bone** (from the palatoquadrate cartilage) and the **articular bone** (from the mandibular cartilage). The caudal ends of these cartilages, or the quadrate and articular bones, bear the jaw joint in all gnathostomes except mammals. Dermal bones, which are derived from superficial bones and plates found in groups such as †Osteostraci, overlie the mandibular arch in most gnathostomes and contribute much to the jaws (but not to the jaw joint itself; see Chapter 7 for more on the origin and structure of jaws). The second arch is known as the **hyoid arch,** and it often helps to suspend the mandibular arch from the cranium (Fig. 3-4*B*, *C*). Posterior to the hyoid arch are five **branchial,** or typical, **gill arches.** Gill pouches that open to the surface through gill slits lie between the branchial arches. The gill slit between the hyoid and mandibular arches is either reduced to a small **spiracle** or lost.

The **paired fins** of gnathostomes may have evolved from paired flaps of skin along the ventrolateral body wall. Like the origin of jaws, however, no clearly intermediate fossils are known (see Chapter 9 for more on the structure and origin of paired fins). Gnathostome ears have a **horizontal semicircular duct.** This brings their complement to three semicircular ducts, a pattern retained throughout gnathostome evolution.

Jaws, two sets of paired appendages, and three semicircular ducts in the ears gave gnathostomes special abilities. Jaws enable gnathostomes to seize and to eat a much wider range of food, and two sets of paired appendages provide increased stability and maneuverability and enable them to swim more efficiently. Gnathostome ears allow better three-dimensional orientation in the water column. A large and very rapid radiation followed the evolution of these structures, making Devonian waters the first environments in Earth history to be dominated by large, active, predatory vertebrates. The ecological ascendancy of gnathostomes may have contributed to the extinction of "ostracoderms" and restricted hagfishes and lampreys to specialized modes of life.

Elasmobranchiomorphs

The two major groups of gnathostomes are Elasmobranchiomorphi and Teleostomi (Fig. 3-2). We regard †placoderms (†Placodermi, a diverse extinct group) as the sister group of cartilaginous fishes (Chondrichthyes, which includes the sharks, rays, and chimaeras), and the clade consisting of †Placodermi and Chondrichthyes is called **Elasmobranchiomorphi** (Fig. 3-2). The presence of an **optic pedicel,** a small, cartilaginous prop that supports the back of the eye, is unique to elasmobranchiomorphs. Several groups of Silurian and Devonian gnathostomes form **†Placodermi** (Gr., *plax* = plate + *derma* = skin; Fig. 3-2). †Placoderms had a ring of armor on the anterior part of the trunk, which is synapomorphic for the group. The small scales on the rest of the trunk were, in some †placoderms, reduced to minute denticles. The internal skeleton was cartilaginous, and although **vertebral centra** (which are the spool-shaped vertebral bodies that surround the notochord) were absent or not ossified, neural and hemal arches were present. The tail was heterocercal. Many †placoderms appear to have been benthic predators that laid in wait for prey. Early species lived in fresh water, but later species were marine. †Placoderms are not known to have lived past the very early Carboniferous. Two groups within †Placodermi serve to illustrate the diversity of the group: the †Arthrodira and the †Antiarcha. †Arthrodires (Gr., *arthrodes* = well jointed) had a unique joint between the skull and the thoracic plates that allowed the skull to be raised to increase the gape (Fig. 3-5*A–C*). Most †arthrodires lacked teeth, but dermal bones overlying the mandibular arch had sharp, cutting edges resembling meat cleavers. The †arthrodire †*Dunkleosteus* from Devonian marine shale deposits in Cleveland, Ohio, attained a length of 6 m. †*Bothriolepis* from Devonian deposits at Miguasha, Quebec, is a well-studied †antiarch (Fig. 3-5*D*). Its long, jointed pectoral spines were covered with dermal plates that superficially resemble the limb of an arthropod. These spines may have elevated the body or held it stationary in flowing water. Based on internal molds of the body cavity, some workers have interpreted that †*Bothriolepis* had lungs; if so, then lungs evolved independently in †antiarchs and osteichthyans based on the distribution of other characters (Chapter 18). The pelvic fins of some †antiarchs are modified to form an intromittent organ for internal fertilization.

Cartilaginous Fishes

Approximately 800 living species of sharks, skates, rays, and chimaeras belong to **Chondrichthyes** (Gr., *chondros* = cartilage + *ichthyos* = fish; Fig. 3-6). Their internal skeleton is cartilaginous during early ontogeny, and none of this cartilage becomes replaced by bone. Chondrichthyans also lack the large bony dermal plates found in †osteostracans, †placoderms, and teleostomes. Based on the phylogenetic distribution of other characters, the absence of these bones must be a derived condition; chondrichthyans evidently evolved from an ancestor with bone in both its endoskeleton and exoskeleton.

Much of the cartilage of chondrichthyans is strengthened by calcification, but this tissue differs structurally from bone. Chondrichthyans have a unique type of calcified cartilage in which calcium salts form a superficial layer of prismatic plates, called tesserae (L., *tessera* = tile). Such **prismatic calcification of cartilage** is synapomorphic for chondrichthyans. Chondrichthyans also have unique scales, known as **placoid scales.** Each placoid scale consists of an acellular, bony base that supports a rose-thorn–shaped denticle. The exposed tips of the placoid denticles point caudally. Spines occur along the anterior border of the median fins of some species, and in some species of skates and rays, the placoid scales on the surfaces of the pectoral fins form enlarged spikes. Chondrichthyan fin rays are composed of giant bundles of collagen termed **ceratotrichia** (Gr., *keratos* = horn + *trichia* = hair). Because chondrichthyans lack dermal bones, the teeth are supported by jaws consisting solely of the palatoquadrate and mandibular cartilages.

Fertilization is internal in all chondrichthyans, and the pelvic fins of a male chondrichthyan are modified to form intromittent organs known as **claspers.** Internal fertilization in animals often correlates with the production by females of a small number of relatively large eggs, and chondrichthyans follow this model.

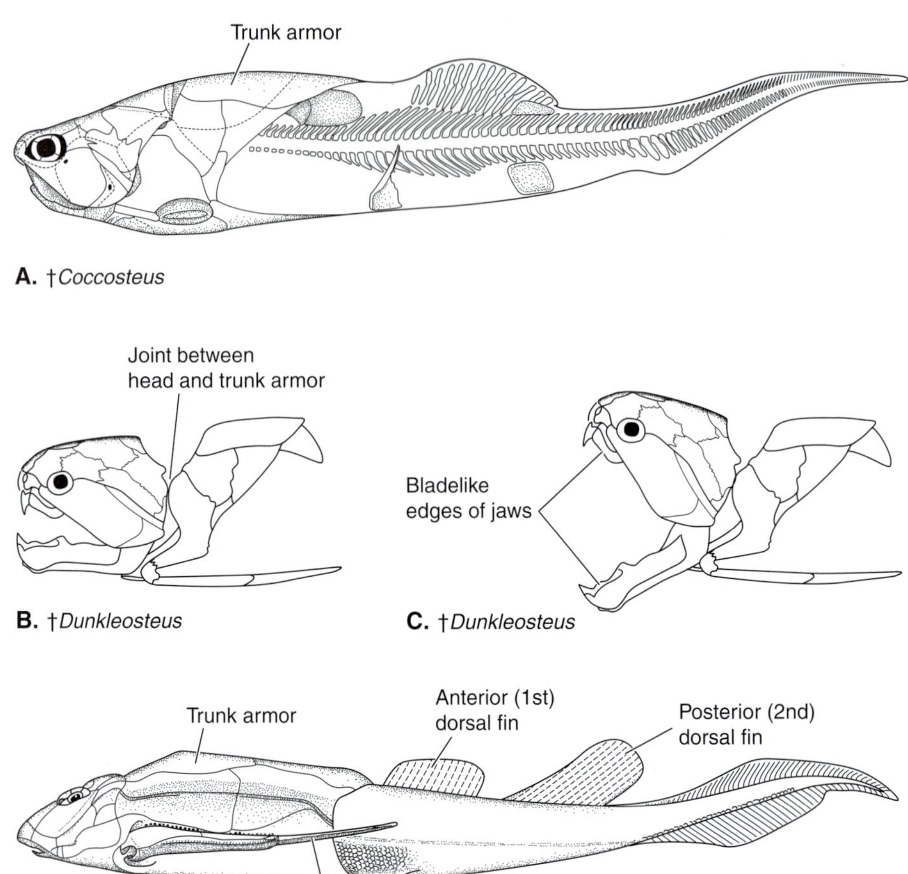

FIGURE 3-5 Representative †placoderms. *A,* Reconstruction of an early †arthrodire, †*Coccosteus*. *B,* Cephalothoracic armor of the giant arthrodire †*Dunkleosteus,* which attained a length of 6 m. *C,* Same as *B* but with mouth open to show the mechanism of the joint between the skull and thoracic armor. *D,* Reconstruction of a small antiarch, †*Bothriolepis*.

Oviparity (egg laying) is plesiomorphic for chondrichthyans, and many living species produce large, yolky eggs enclosed in egg cases. Some chondrichthyans are **ovoviviparous,** which means that the female retains fertilized eggs inside her reproductive tract until they hatch, and the young are born alive. Still others are **viviparous,** which means that the developing young remain inside the female's reproductive tract and receive nutrients via uterine secretions, a placenta, oophagy (i.e., eating eggs as they pass down the reproductive tract), uterine cannibalism (i.e., eating other developing siblings), or other means. Such nutrients augment those available from the egg yolk, and the large young are born alive.

Many chondrichthyans retain a heterocercal tail. Although chondrichthyans lack lungs or swim bladders (i.e., air-filled sacs within the body that increase buoyancy; see Chapter 18), lipids in the liver provide some buoyancy. A skeleton of calcified cartilage, which is not as dense as bone, as well as the reduced dermal armor in the form of placoid scales, also lighten the body and may facilitate swimming. Although our image is one of roving predators, many chondrichthyans are sluggish and rest on the bottom when not swimming. Chondrichthyan sense organs include arrays of electroreceptors known as **ampullary organs,** or ampullae of Lorenzini (Chapter 12). In conjunction with sophisticated sensory systems and advanced locomotory and feeding modes, some sharks and rays have very large brains. The fossil record of Chondrichthyes begins in the Devonian, but the clade may have originated earlier than this. Most fossils come from marine deposits, and a few extinct and living groups of chondrichthyans have adapted to live in fresh water.

Sharks, skates, and rays constitute **Elasmobranchii** (Figs. 3-6 and 3-7*A–C*). The name derives from the thin plates of connective tissue that bear the gills and separate the vertically elongated gill pouches (Gr., *elasmos* = thin plate + *branchia* = gills). This feature is synapomorphic for elasmobranchs. Paleozoic sharks include the Devonian †Cladoselachidae (Fig. 3-6), which had terminal mouths, long jaws, and a notochord that persisted in adults. We regard †cladoselachids as the

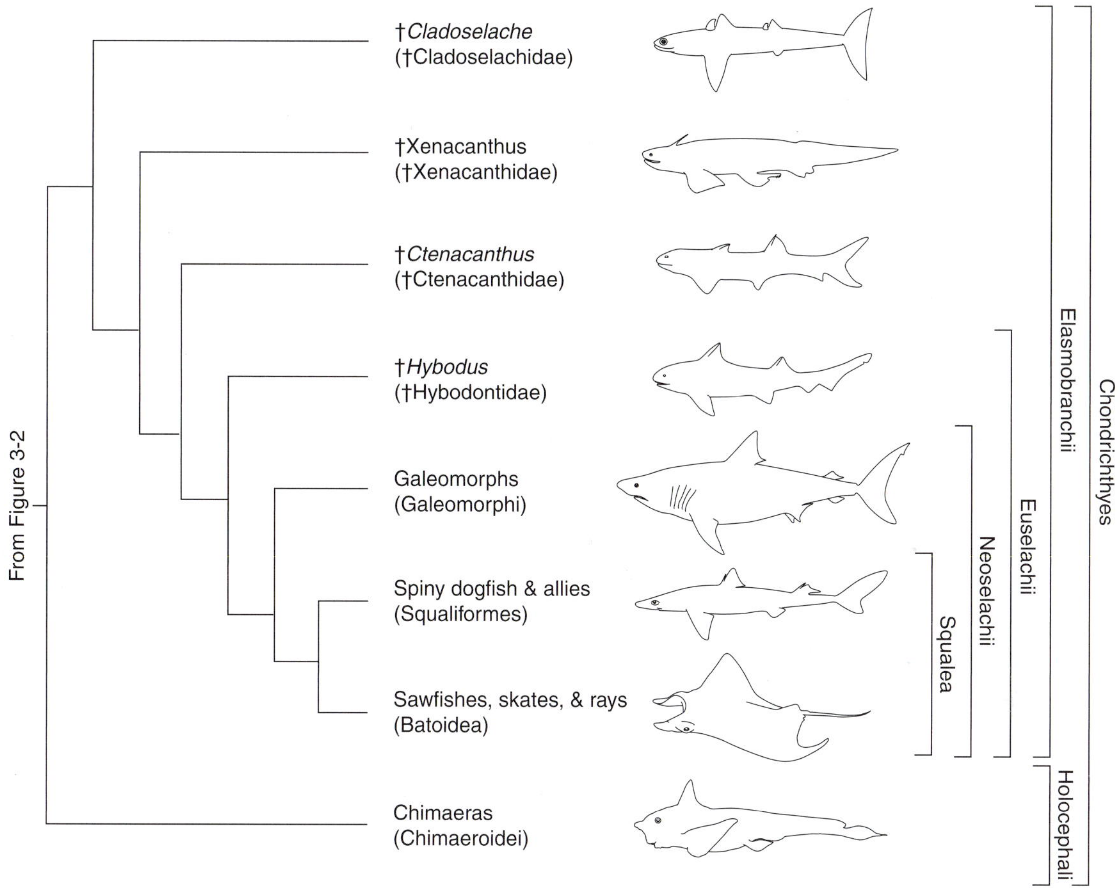

FIGURE 3-6
Phylogeny of selected fossil and living cartilaginous fishes (Chondrichthyes).

sister group of all other sharks. As Paleozoic sharks diversified, one line led to the freshwater †Xenacanthidae (Fig. 3-6). The dominant sharks of the late Paleozoic were †ctenacanthids, represented by forms such as †*Ctenacanthus* (up to 1.5 m in length).

During the Mesozoic, **euselachians** evolved (Fig. 3-6). Euselachians have several synapomorphies. For example, the endoskeleton of the fins is reduced, and the large fin webs become supported chiefly by exoskeletal ceratotrichia (i.e., collagenous fin rays). Euselachians also have a **subterminal mouth,** which means that the mouth is on the ventral surface of the head. Their mobile jaws allow them to gouge out large pieces from their prey. Jaw mobility also means that euselachians that eat small, whole prey can generate greater suction for prey capture. Finally, euselachians have **vertebral centra,** which largely replace the notochord and provide a stronger vertebral column. An early clade of euselachians is †Hybodontidae (Fig. 3-6). †Hybodontids had the characteristic subterminal mouth and short, powerful jaws that could be raised and lowered relative to the braincase.

All living species of euselachians belong to the clade **Neoselachii** (Fig. 3-6), which is characterized by skeletal features of the pectoral girdle. Living neoselachians belong to two large clades: **Galeomorphi** and **Squalea.** Galeomorphs typify the Hollywood ideal for a shark because many are large, active, open-water predators, such as the great white shark, *Carcharodon carcharias* (Fig. 3-7*A*). White sharks reach lengths of 8 m and often feed in shallow coastal waters, where they prey on marine mammals. Several species of galeomorphs are pelagic filter-feeders, such as the basking shark, *Cetorhinus maximus* (Fig. 3-7*B*). In contrast to the open-water predatory habits of most galeomorphs, most squalean sharks have a benthic orientation and lifestyle. **Squaliformes,** or dogfish sharks, is the group that includes the spiny dogfish, *Squalus acanthias,* which is a widespread species commonly studied in teaching laboratories. *Squalus* retains two dorsal fins with large fin spines but lacks an anal fin. Schools of spiny dogfish migrate north along the eastern coast of North America in the spring, reaching Nova Scotia in the summer, where the young are born after an 18-month gestation.

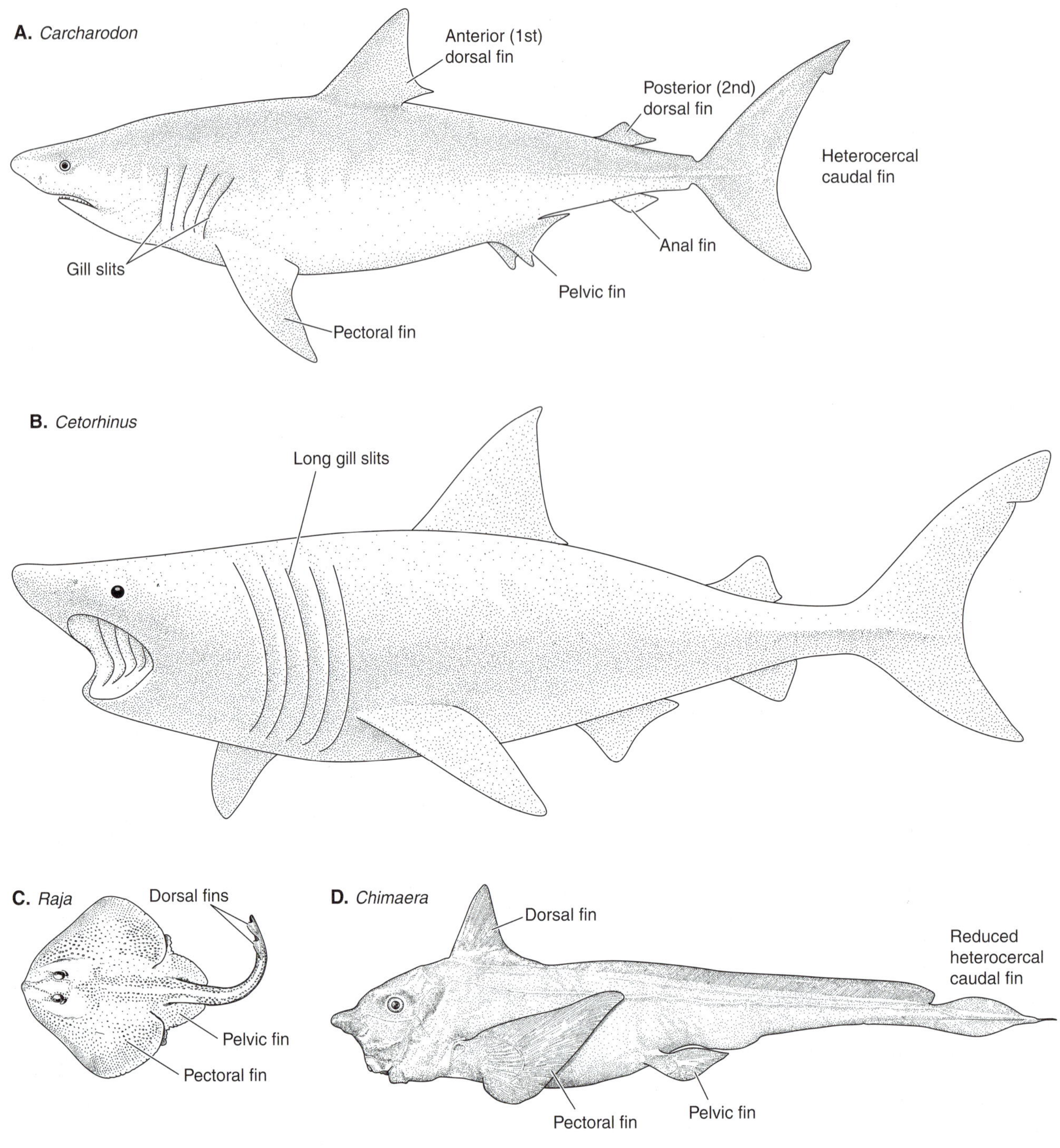

FIGURE 3-7
Representative chondrichthyans. *A,* A large predatory galeomorph, the great white shark, *Carcharodon. B,* A filter-feeding shark, the basking shark, *Cetorhinus. C,* The little skate, *Raja,* showing the enlarged pectoral fins found in batoids. *D,* A holocephalan, *Chimaera.*

Squalea also includes a radiation of more than 400 living species of sawfishes, skates, and rays, together known as **Batoidea** (Fig. 3-6). Batoids have flat bodies and dorsally positioned eyes that are adaptations for bottom-dwelling (Fig. 3-7*C*). During early ontogeny, the enlarged pectoral fins of a batoid grow forward and fuse to the sides of the head (a synapomorphy of the group). Batoids swim by flapping their pectoral fins or by passing undulatory waves along their margins. Respiratory water can be drawn into the pharynx through a pair of large spiracles located behind the eyes; water exits via gill slits on the ventral surface. Most batoids eat mollusks or crustaceans, which they crush using specialized flattened tooth plates. Such feeding on hard prey items is called **durophagy** (L., *durus* = hard + Gr., *phagy* = eat). Skates (Rajidae) are perhaps the most familiar batoids, and the genus *Raja* (Fig. 3-7*C*) is the most speciose among living elasmobranchs (about 150 species

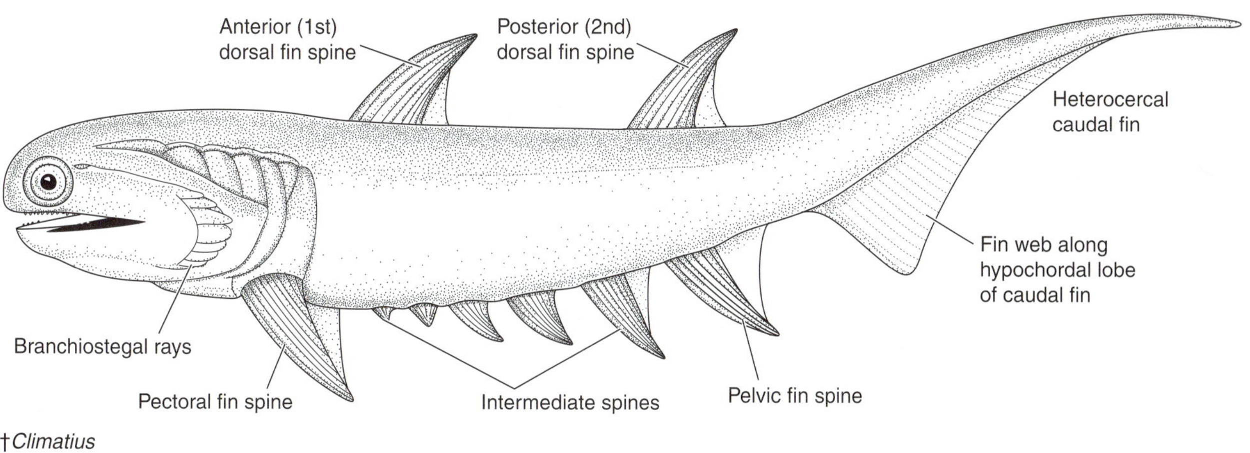

FIGURE 3-8
Reconstruction of an acanthodian, †*Climatius,* in lateral view showing arrangement of the fins and intermediate spines. Body scales are not shown.

are known). Skates are more dorsoventrally flattened than other batoids and have a strong benthic orientation. Sting rays (Dasyatidae) have serrated dorsal fin spines that can inflict a painful injury, particularly when venom from a gland at the base of the spine enters the wound. Sting rays generally live on the bottom but also swim in the water column. The most derived batoids are the cow-nose, eagle, devil, and manta rays. These pelagic batoids can be found swimming in open water far offshore, using pectoral fin flapping as their main method of locomotion.

Holocephali includes shark-like Paleozoic forms and 30 species of peculiar living marine chondrichthyans commonly known as **chimaeras** (Fig. 3-7*D*). Unlike elasmobranchs, the upper jaw of holocephalans is firmly united to the cranium, a condition termed **autostyly** (Chapter 7) and the source of the group's name (Gr., *holos* = whole + *kephale* = head). Holocephalans are durophagous and feed on mollusks and other shelled invertebrates, using specialized tooth plates that are synapomorphic for the group. The gills of chimaeroids are covered by an external fleshy flap, known as an operculum, that has a single external opening just anterior to the pectoral fin. The fleshy operculum of chimaeroids is not homologous to the bony operculum of osteichthyans. In addition to their pelvic claspers, male chimaeras have a spiny head clasper, which is used to grasp a female's pectoral fin or operculum during courtship and mating.

As shown in Figure 3-2, the clade containing †Acanthodii and Osteichthyes is known as the **Teleostomi** in recognition of their terminal mouths (Gr., *teleos* = terminal + *stome* = mouth). Teleostomes also have a deep and narrow braincase and a **bony operculum** with many associated branchiostegal rays. †Acanthodii (Gr., *akanthodes* = spiny) is an extinct group of teleostomes sometimes called spiny sharks because their heterocercal tail gives them a shark-like appearance and stout accessory paired spines precede their fins (Fig. 3-8). Unlike true sharks, however, †acanthodians had small, thick, bony scales and a partly ossified internal skeleton. As many as five additional pairs of spines lay between the pectoral and pelvic fins. These spines often are cited as indirect evidence for the lateral fin-fold hypothesis, but this has been called into question because little evidence shows that the spines supported fin webs. †Acanthodian fossils are known from the Silurian to the Permian. Early forms come from freshwater deposits, but some later ones occur in marine deposits.

Bony Fishes and the Diversification of Teleosts

So far in our survey, we have explored only a fraction of the diversity of living vertebrate species. Now, we come to the bony fishes, or **Osteichthyes** (Gr., *osteon* = bone + *ichthyes* = fishes; Fig. 3-2), a clade that includes about 49,000 living species. Some osteichthyans retain cartilage in their internal skeletons, but most have well-ossified skeletons. The heavy, bony surface armor found in †osteostracans is reduced, but most aquatic osteichthyans retain thin, bony scales embedded in the skin. The spiracle is retained in some osteichthyans, but it is lost in most.[5]

[5]Loss of the spiracle is one of several synapomorphies for Neopterygii, a group discussed later. The spiracle also was lost independently in Sarcopterygii.

A pair of **lungs,** or a swim bladder that evolved from the lungs, is present in all groups of osteichthyans except for some tetrapods and teleosts, in which they have been lost secondarily. Lungs develop as a ventral outgrowth of the floor of the pharynx and remain connected to it by an air duct. We believe that lungs evolved as accessory respiratory organs and postulate that the common ancestor of osteichthyans lived in warm, freshwater habitats where such an organ supplemented gill breathing (warm water holds less oxygen than cold water, and stagnant bodies of fresh water can become deoxygenated). In contrast, a swim bladder develops during ontogeny as a dorsal outgrowth of the pharynx and is usually a dedicated hydrostatic organ (see Chapter 18 for more on the evolution of lungs and swim bladders). By controlling the amount of gas in the swim bladder, buoyancy can be regulated to maintain depth in the water column with little muscular effort. Osteichthyans have bony fin rays, or **lepidotrichia** (Gr., *lepidos* = scale + *trichia* = hair). Lepidotrichia are formed by half-rings of bone that surround bundles of giant collagen fibers termed **actinotrichia** (Gr., *aktin* = ray + *trichia* = hair). Lepidotrichia strengthen the fin rays and give them a jointed appearance.

Actinopterygians

Except for six species of lungfishes and one species of coelacanth, all living fishlike osteichthyans belong to **Actinopterygii** (Gr., *aktin* = ray + *pteryg* = wing or fin), a group commonly known as the ray-finned fishes (Fig. 3-9). Actinopterygians have a single **dorsal fin** (e.g., Fig. 3-10) and rhombic trunk scales, with a unique peg-and-socket articulation between the scales. Many layers of an enamel-like surface tissue, **ganoine,** cover the scales and some dermal bones of early actinopterygians. The endoskeleton and fin muscles do not extend far into the paired fins of actinopterygians; instead, elongate, flexible fin rays support the fin web. Individual fin rays can be moved to spread or collapse the fin for use in different types of locomotion (Chapter 11).

Actinopterygians rapidly radiated in the Devonian, coming to ecologically dominate aquatic environments by the late Paleozoic. Early species inhabited both fresh and coastal marine waters. Many Paleozoic actinopterygians are sometimes loosely referred to as "†paleoniscoids." More than 100 extinct genera have been included in this group, but as commonly used, the term "†paleoniscoids" refers to a paraphyletic group. By the late Permian, a clade of actinopterygians with modified fins appeared, and these are known as **Neopterygii** (Gr., *neo* = new + *pteryg* = fin; Fig. 3-9). Neopterygians diversified in the Mesozoic and eventually ecologically replaced nearly all earlier clades of actinopterygians (although we can only speculate about what may have caused this replacement to occur). By the end of the Triassic, one neopterygian lineage gave rise to a great radiation, **Teleostei** (Gr., *teleos* = end + *osteon* = bone; Fig. 3-9). Teleosts radiated in the Mesozoic and throughout the Cenozoic and eventually ecologically replaced nearly all other neopterygian clades.

In Figure 3-9, a Devonian genus, †*Cheirolepis,* is shown as the sister group of all other actinopterygians. It exhibits the diagnostic single dorsal fin, but its tiny scales lacked peg-and-socket articulations. The next group shown in Figure 3-9 is Polypteriformes, which includes 13 extant species of freshwater bichirs (*Polypterus;* Fig. 3-10*A*) and reedfish (*Erpetoichthys*). Polypteriforms have a long dorsal fin divided into finlets, each bearing a sharp anterior spine. They have paired ventral lungs and will drown if prevented from breathing air. Although superficially similar, the fleshy pectoral fin of *Polypterus* is structurally and developmentally different from the lobed fins of sarcopterygians. The relatively poor fossil record of this group extends only from the Cretaceous to Recent.

Thanks to remarkable specimens from the Devonian Gogo Formation in Australia, we know much about forms such as †*Moythomasia* (Fig. 3-10*B*). For example, the maxillary bone that formed most of the margin of the upper jaw was firmly connected to adjacent bones, so it could not move. They had a simple, scissors-like jaw action that allowed for minimal lateral movement or flaring of the buccal cavity. We believe that such a jaw action typified the feeding systems of early actinopterygians (Chapter 16).

Chondrostei[6] (Fig. 3-9) includes three extinct families from the Mesozoic as well as the living paddlefishes (Polyodontidae) and sturgeons (Acipenseridae). All chondrosteans retain a heterocercal tail, spiracle, and some other plesiomorphic features but have lost the heavy ganoid scales (Fig. 3-10*C*). Members of this group have highly modified jaws and jaw-suspension systems compared with other actinopterygians. The 25 living species of sturgeons (e.g., *Acipenser* and *Scaphirhynchus*) and the living Chinese paddlefish (*Psephurus*) are suction-feeding predators, but the living paddlefish from eastern North America (*Polyodon*) is a pelagic filter-feeder that uses its long gill rakers to trap zooplankton.

[6]We restrict the name Chondrostei to this monophyletic group, but many authors have used this name in a paraphyletic sense to include members of several different clades of actinopterygians.

Text continues on page 67

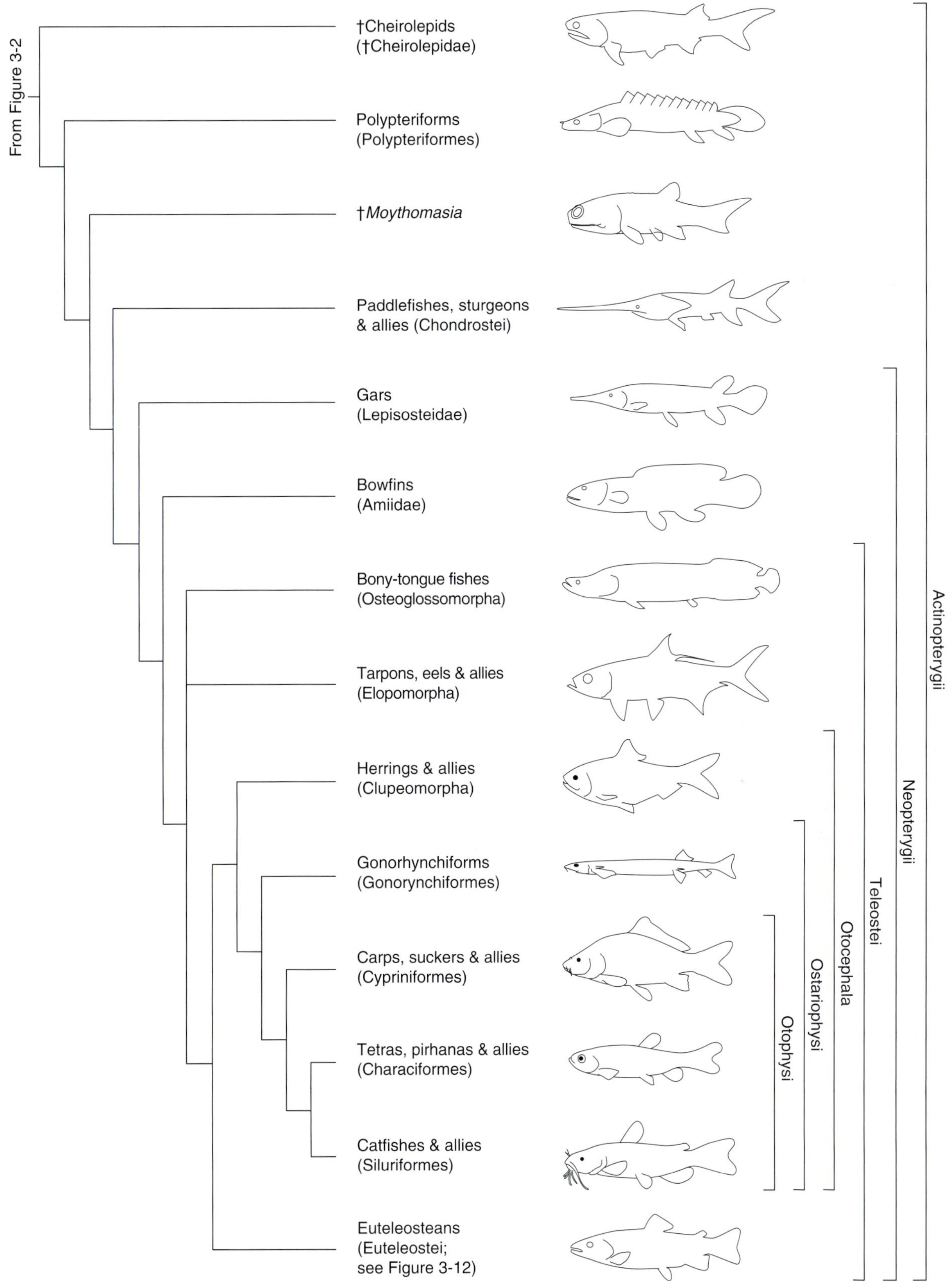

FIGURE 3-9

Phylogeny of selected living and fossil actinopterygians (Actinopterygii), highlighting the origin of neopterygians (Neopterygii), teleosts (Teleostei), and euteleosts (Euteleostei).

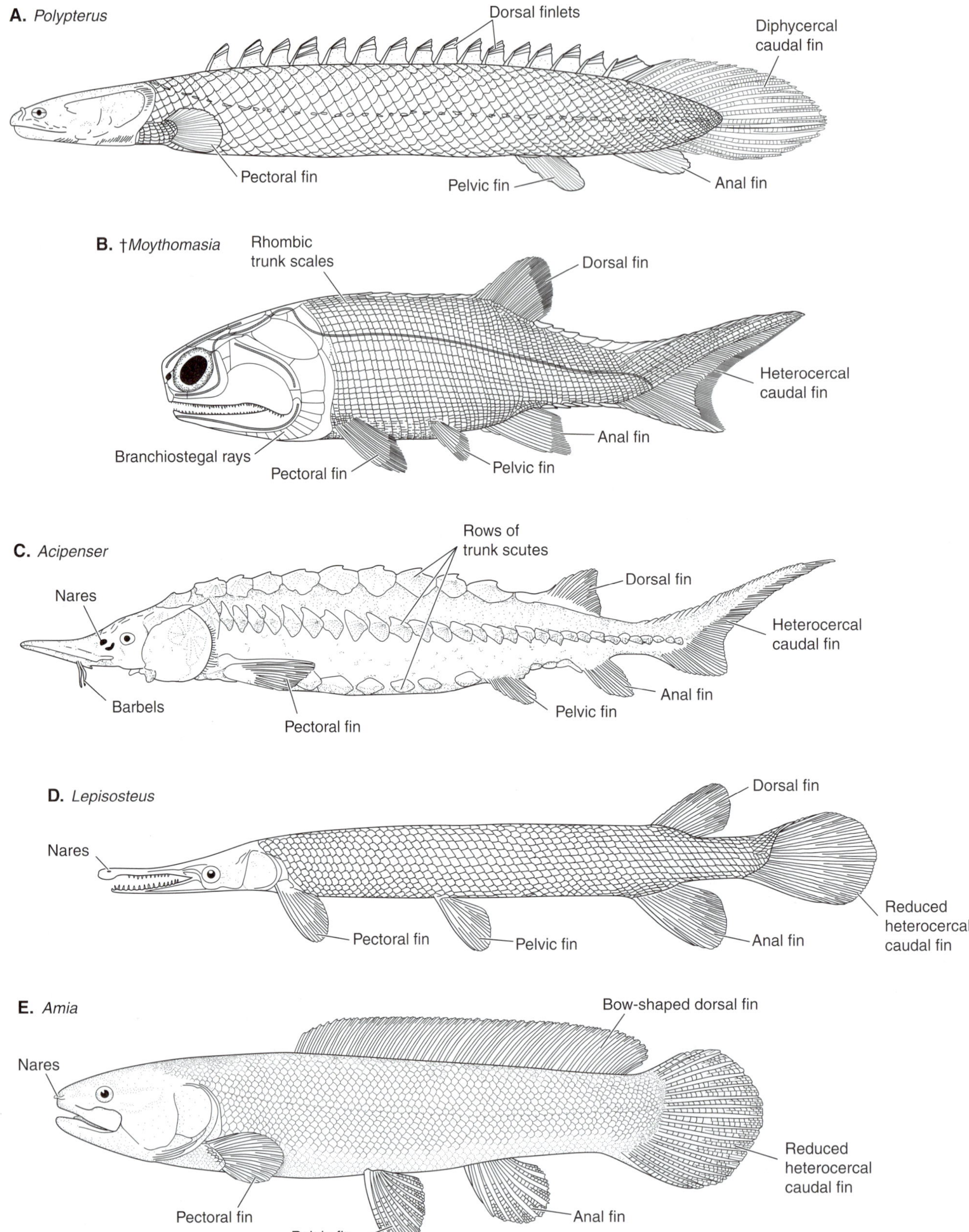

FIGURE 3-10

Representative nonteleostean actinopterygians. *A,* A bichir *Polypterus,* in which the single dorsal fin is divided into finlets. *B,* Reconstruction of a well-preserved Devonian actinopterygian, †*Moythomasia,* showing the single dorsal fin characteristic of actinopterygians. *C,* A sturgeon, *Acipenser,* with characteristic rows of scutes armoring the trunk. *D,* A gar, *Lepisosteus,* demonstrating the reduced heterocercal caudal fin found in neopterygians. *E,* The bowfin, *Amia,* is named for its elongate, single dorsal fin.

Neopterygians

Members of **Neopterygii** (Fig. 3-9) share many synapomorphies, such as a reduced number of endoskeletal elements supporting the fins. As a result of this feature, a neopterygian can more easily collapse or spread its fins than can a chondrichthyan, sarcopterygian, or chondrostean. The ability to change the shape and area of a fin has important consequences for locomotion (Chapter 11). Median fins of neopterygians can be moved from side to side by specialized fin muscles. The shape of the caudal fin in neopterygians is described as reduced heterocercal, in that its dorsal lobe is far less prominent than it is in outgroup taxa (Figs. 3-10 and 8-10). Neopterygians lack a spiracle and have more mobile jaws than do actinopterygians such as †*Moythomasia* and *Polypterus*, and they can move their jaws to form a round mouth opening. The jaw suspension and its musculature allow neopterygians to laterally flare the buccal cavity. Lateral flaring can generate a rapid increase in the volume of the oral cavity during feeding, which results in a powerful sucking action like that of a pipette. Through variations on these structures, neopterygians evolved diverse feeding modes. One of the least-understood characters of neopterygians is their loss of electroreceptive ampullary organs as well as the nerves and brain centers associated with processing electroreceptive information (Chapter 12). Electroreception independently evolved in at least two groups of teleosts, but always with a different arrangement than found in non-neopterygians.

Hundreds of extinct species of nonteleostean neopterygians are known, but only two living groups exist: **Lepisosteidae** (gars) and **Amiidae** (bowfins). Some workers place gars and bowfins as sister taxa in a group termed "Holostei." Some anatomical and molecular phylogenetic evidence supports such a group, but this is not the view we show in Figure 3-9. The seven living species of gars are restricted today to eastern North and Central America. They are instantly recognizable for their elongate jaws and teeth, thick rhombic scales, posterior placement of anal and dorsal fins, and lurking predatory habits (Fig. 3-10*D*). Only a single living species of Amiidae exists: the North American bowfin, *Amia calva* (Fig. 3-10*E*). Bowfins inhabit backwaters of rivers and ponds in the eastern United States. These stout-bodied predators have thin, round scales and a slightly heterocercal tail. The common name bowfin refers to the elongate, bow-shaped dorsal fin. They breathe air and can live for long periods out of the water. Amiidae has an extensive fossil record (Jurassic to Recent) and formerly ranged over much of the Northern Hemisphere. Bowfins have been intensively studied, at least in part because the living bowfin retains many features considered plesiomorphic for actinopterygians.

Bowfins share with teleosts a lever system of skull, opercular, and other bones and associated ligaments and muscles, which is called the **opercular chain mechanism.** This mechanism allows for more rapid jaw opening (Chapter 16). The evolution of this derived jaw-opening system may be linked to the tremendous diversification of feeding systems seen among Teleostei.

Teleosts

Because of its great diversity (more than 24,000 living species in more than 400 families), we restrict our treatment of **Teleostei** to its major living groups. The caudal fin of teleosts is more externally symmetrical than in early neopterygians and is known as a **homocercal tail** (Chapter 8). Internally, a teleost's tail has modified neural arches (known as uroneural bones) that ensure similar stiffnesses of the dorsal and ventral lobes of the caudal fin. The highly mobile maxilla of a teleost lacks bony articulations with cheek and infraorbital bones, and thus it can move in ways not possible for other actinopterygians. Most teleosts use their swim bladder as a dedicated hydrostatic organ for regulating buoyancy. Finally, teleostean scales are reduced to thin, round, flexible elements that enable the body to be bent easily. Our knowledge of teleostean phylogeny has improved over the past 25 years, but many groups of teleosts still are inadequately studied.

The bony-tongue fishes, or **Osteoglossomorpha** (Gr., *osteon* = bone + *glossa* = tongue; Fig. 3-9), have a unique feeding system. Teeth on gill-arch elements (= the "tongue") shear against teeth on the roof of the mouth. All living osteoglossomorphs live in fresh water, including the arowanas (e.g., *Osteoglossum;* Fig. 3-11*A*) and more than 200 species of mormyrids, commonly called elephant-nose fishes (e.g., *Gnathonemus;* Fig. 3-11*B*). Many mormyrids live in dark or turbid waters, and they use an electrosensory system (probably derived from modified lateral line organs) and electrogenic tissues (derived from muscles in the caudal peduncle) to detect prey and communicate with other individuals (Chapter 12). The enlarged cerebellum in the brain of mormyrids is correlated with processing of electroreceptive information.

Elopomorpha includes tarpons, eels, and allies (Fig. 3-9). Elopomorphs have unique, leaf-shaped larvae, known as a **leptocephalus larvae** (Gr., *lepto* = leaf + *cephalus* = head). Fishlike elopomorphs include tarpons (*Megalops;* Fig. 3-11*C*), a near-shore and estuarine group relatively unchanged in morphology from Mesozoic times. Elopomorpha also includes more than 750 species of eels, such as the marine moray eels (e.g., *Gymnothorax;* Fig. 3-11*D*) and the migratory freshwa-

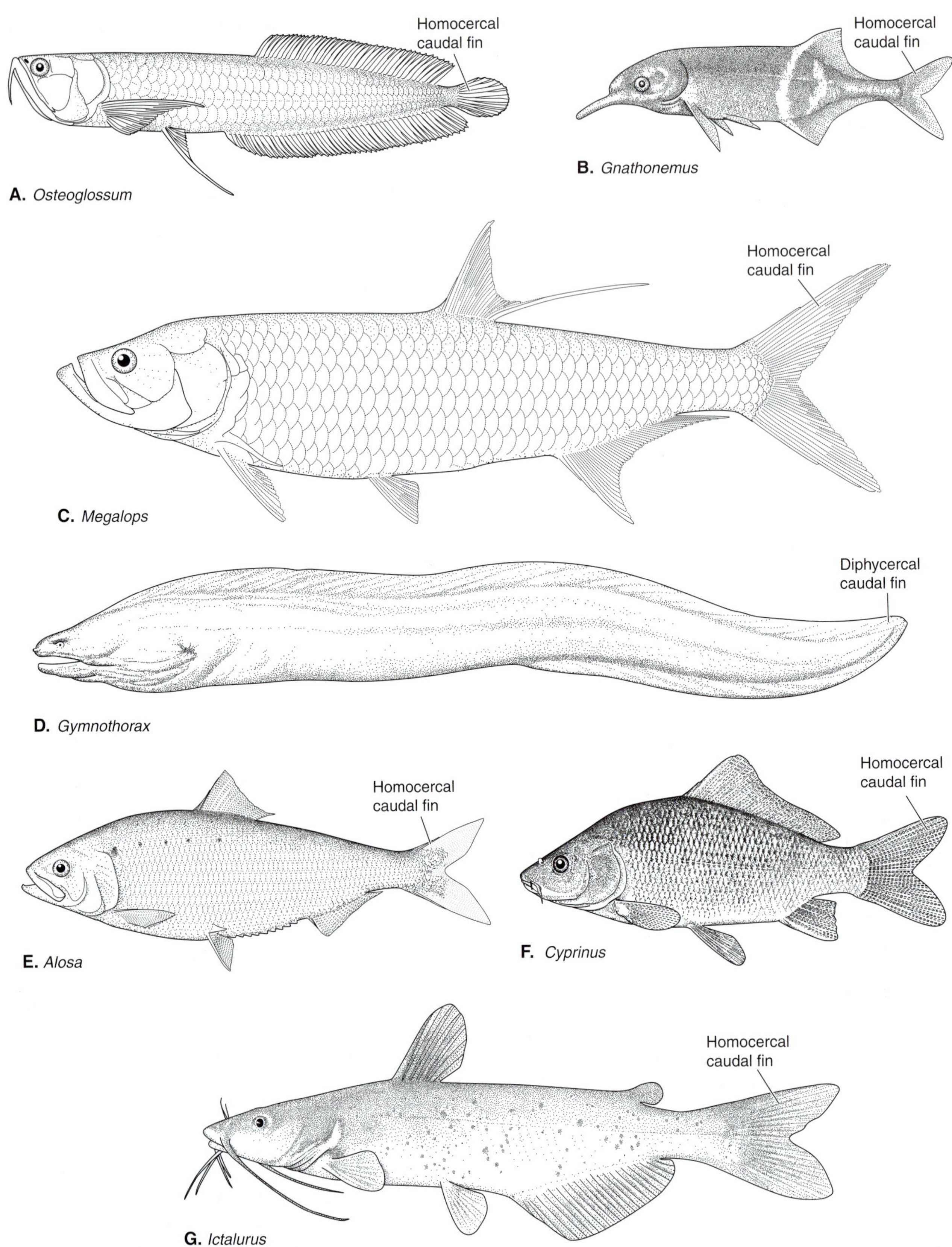

FIGURE 3-11
Representative teleosts. *A,* An arowana, *Osteoglossum. B,* An elephant-nose fish (Mormyridae), *Gnathonemus.* Mormyrids communicate using weak electrical discharges. *C,* An example of a fishlike elopomorph, the tarpon, *Megalops. D,* The moray eel, *Gymnothorax,* an example of an eel-like elopomorph. *E,* A clupeomorph, the American shad, *Alosa. F,* A cypriniform, the carp, *Cyprinus. G,* A siluriform, the channel catfish, *Ictalurus.*

ter eels (*Anguilla*). *Anguilla* is noteworthy for its **catadromous** life history pattern, in which spawning occurs in salt water but adults live in fresh water. The saltwater spawning grounds of North American and European anguillids lie in the Sargasso Sea, far out in the Atlantic Ocean, where spawning is believed to occur at great depth. Leptocephalus larvae of *Anguilla* are carried by oceanic currents to the coasts of North America or Europe, where they metamorphose and migrate up rivers to mature.

Otocephala (Fig. 3-9) includes about 7000 species of herrings (Clupeomorpha) and ostariophysans (Ostariophysi). **Clupeomorpha** includes several fossil groups, as well as herrings (*Clupea*), shads (*Alosa;* Fig. 3-11*E*), sardines, anchovies, and related forms. These silver, fast-swimming, schooling fishes can be extremely abundant: schools of Atlantic herring may exceed 1 billion individuals. Most species of clupeomorphs are marine. Anterior diverticula of the swim bladder extend into the otic region of the skull, which houses the inner ear, and allow for improved detection of high-frequency sounds.

Members of **Ostariophysi** (Fig. 3-9) have a divided swim bladder. The anterior chamber of the swim bladder is specialized for detection of low-amplitude, high-frequency sounds (Chapters 12 and 22), whereas the posterior chamber serves as a hydrostatic organ. Gonorhynchiformes (Fig. 3-9) includes a few species, but most species of ostariophysans belong to a derived group known as **Otophysi** (Fig. 3-9). Otophysans share a remarkable synapomorphy, known as the **Weberian apparatus.** This chain of four bones is derived from portions of ribs and vertebrae, and it transmits vibrations from the anterior chamber of the swim bladder to the inner ear. By means of the Weberian apparatus, otophysans can detect weak or high-frequency sounds (Chapter 12). Most of the 6500 species of otophysans are small to medium-sized and live in fresh water. Members of the otophysan group **Cypriniformes** lack teeth in their oral jaws but have a well-developed pharyngeal dentition. Most cypriniforms are small, stream-dwelling herbivores, but a few large, carnivorous species are native to rivers of western North America. Cypriniformes includes carp (*Cyprinus;* Fig. 3-11*F*), goldfish (*Carrasius*), minnows, loaches, zebrafish (*Danio*), suckers (*Catostomus*), and many others. Cypriniforms occur in fresh waters throughout much of the world except South America. Another large otophysan group is **Characiformes** (Fig. 3-9), which includes the carnivorous tetras and pirhanas from Africa and South America. **Siluriformes** (Fig. 3-9) includes knife fishes and catfishes. The most famous knife fish is the electric eel (*Electrophorus*), which can produce powerful electric fields to stun prey or to defend itself. Other knife fishes produce weak electric fields used for social communication and orientation. Ichthyologists recognize more than 2500 species of catfishes. Catfishes have up to three pairs of barbels on the head, and each may be covered with thousands of external tastebuds. Some can detect weak electric fields using organs that are components of the lateral line system (Chapter 12). These specializations allow catfishes to live and forage in turbid or dark waters and to be active at night. The North American family Ictaluridae includes the commercially cultured channel catfish (*Ictalurus;* Fig. 3-11*G*) and about 40 other species. Most catfishes live in fresh water but some, such as the sea catfishes from the Atlantic coast of North America (family Ariidae), are marine.

Euteleosts

Euteleostei includes all of the remaining groups of teleosts (Fig. 3-12). The first of these groups is **Protacanthopterygii** (Fig. 3-12), which includes the salmons and trouts (Salmonidae) native to cooler waters of the northern hemisphere and widely introduced by humans in the southern hemisphere. Streamlined and colorful, a salmon offers an excellent model of a generalized teleost. Many species are anadromous, but others never migrate and instead pass their lives within a single body of fresh water. Atlantic salmon and European brown trout belong to the genus *Salmo.* Pacific salmon as well as rainbow and cutthroat trouts of western North America belong to the genus *Oncorhynchus* (Fig. 3-13*A*). **Esociformes** (Fig. 3-12) includes the pickerels and pikes (*Esox*), which are widely distributed in cooler northern lakes and streams of the northern hemisphere. Pikes have elongate bodies with dorsal and anal fins placed far caudally (Fig. 3-13*B*). These features are specializations for rapidly starting from a resting position, which suits them for lurking predatory habits.

Bristlemouths and allies constitute the group **Stenopterygii** (Fig. 3-12). Many are specialized for life at great depths in the oceans. Deep oceanic realms are the largest habitats on Earth, so it is not surprising that some of these forms are extremely abundant. Many deep-sea teleosts have light-producing organs used for social communication, prey luring, or even crypsis (= hiding or camouflaging). Such **bioluminescence** usually results from symbiotic associations with light-emitting bacteria that live in specialized compartments within the skin. Radical specializations of sense organs occur in many stenopterygians, such as the telescope eyes that marine hatchetfishes use to detect faint light. Many stenopterygians have long, fanglike teeth (Fig. 3-13*C*) that allow them to seize relatively large

Text continues on page 72

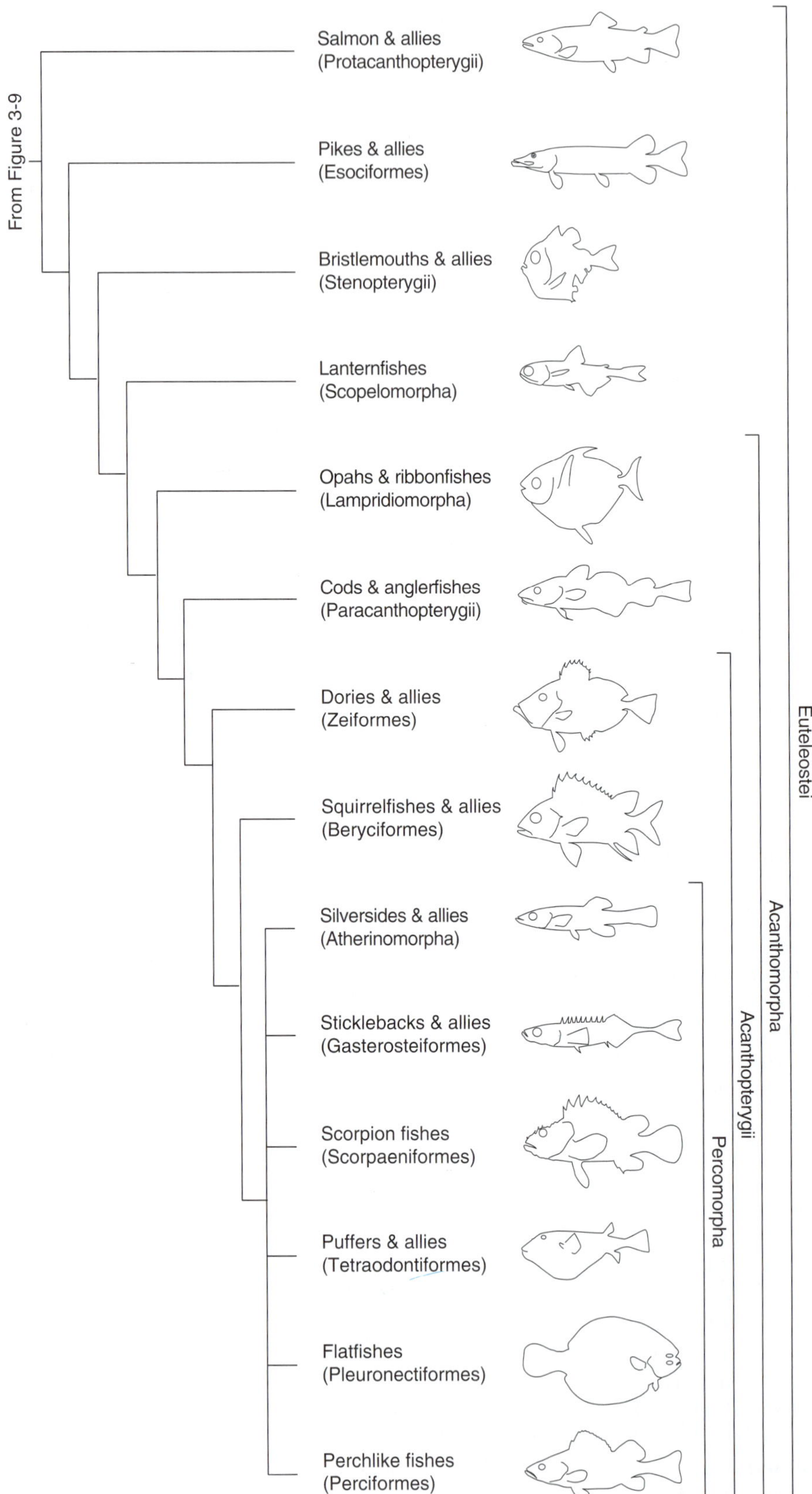

FIGURE 3-12
Phylogeny of living euteleosteans (Euteleostei). The flatfish shown is a righteye flounder, which accordingly is lying on its left side. The left side of all other specimens is shown.

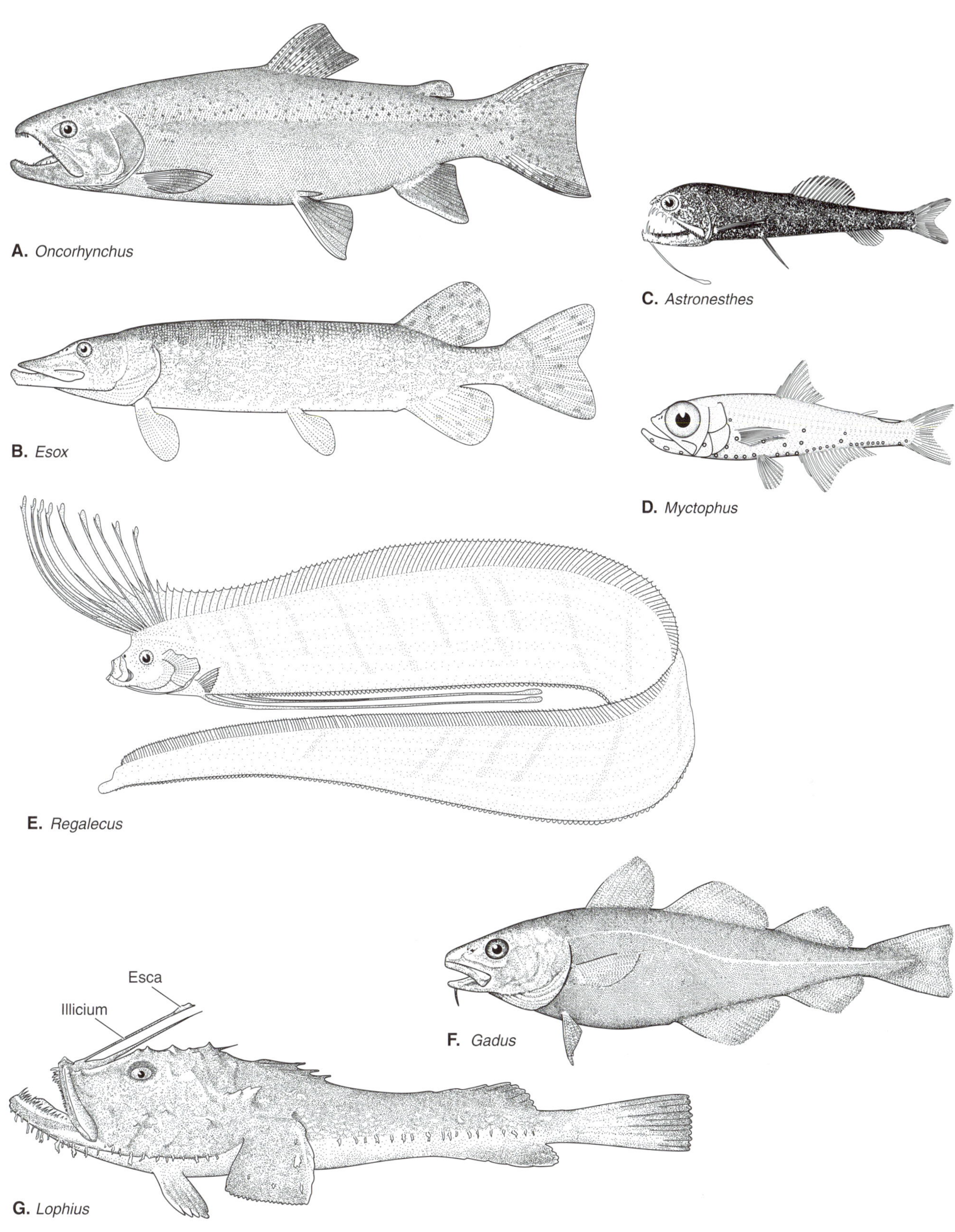

FIGURE 3-13

Representative euteleosts. *A,* A protacanthopterygian, the Pacific salmon *Oncorhynchus. B,* An esociform, the pike, *Esox. C,* A stenopterygian, the bristlemouth, *Astronesthes. D,* A myctophiform, *Myctophus. E,* A lampridiform, the giant oarfish, *Regalecus. F,* A paracanthopterygian, the cod, *Gadus. G,* A paracanthopterygian, the anglerfish, *Lophius.*

prey, which may be important in environments such as the deep sea, where food is scarce. **Scopelomorpha** (Fig. 3-12) is exemplified by members of the family Myctophidae (Fig. 3-13*D*), commonly known as lanternfishes. They can be so abundant that schools of them produce a "false bottom" (or deep scattering layer) when detected by sonar. Myctophids descend to depths during the day but return to surface waters to feed at night. Thus, lanternfishes quickly convey nutrients between depth zones of the ocean. The name lanternfish refers to their many light-producing photophore organs, which are located chiefly on the ventral surface of the body. **Lampridiomorpha** (Fig. 3-12) is exemplified by the 8-m-long oarfish (*Regalecus;* Fig. 3-13*E*), which is among the animals that inspired legends of sea serpents. **Paracanthopterygii** (Fig. 3-12) is a large and diverse group of benthically oriented marine teleosts. Their pelvic fins are located far anteriorly, sometimes anterior to the insertion of the pectoral fins. Paracanthopterygii contains several subgroups, but we note just the cods (Gadidae) and anglerfishes (Lophiiformes). The Atlantic cod, *Gadus* (Fig. 3-13*F*), long has been important to commercial fisheries, but in many areas it has been seriously depleted by overfishing. Anglerfishes have an **ilicium,** which serves as its "fishing pole," tipped with an **esca,** which serves as a lure when the fish is angling for prey. The ilicium and esca are derived from modified dorsal fin spines. Prey is attracted to the lure and then quickly sucked into the large mouth. The unusual appearance and cryptic coloration of many anglerfishes relate to their sit-and-wait style of predation. Examples include the goosefishes (*Lophius;* Fig. 3-13*G*).

The remaining groups of teleosts belong to a clade known as **Acanthopterygii** (Fig. 3-12). Acanthopterygians share synapomorphies related to their feeding systems, such as highly mobile tooth plates in the pharynx that are used for crushing food and highly protrusible jaws used for prey capture (Chapter 16). Examples of acanthopterygians are the dories (Zeiformes; Fig. 3-12) and the squirrelfishes, soldierfishes, and flashlight fishes (Beryciformes; Fig. 3-12). **Percomorpha** is a group within Acanthopterygii (Fig. 3-12) that contains more than 12,000 living species. Most species of percomorphs are marine, but they live nearly everywhere there is water: coral reefs; deep seas; open oceans; cold Antarctic waters[7]; saline desert pools; and freshwater rivers, lakes, and ponds. Some live in deoxygenated waters and even migrate over land. As they penetrated and exploited these environments, an amazing array of body and fin shapes evolved, including specialized jaws, unusual sense organs, and radically different respiratory systems. The phylogeny of Percomorpha is unresolved. Instead of attempting to force an artificial resolution about the interrelationships of the six percomorph subgroups, we instead show them in Figure 3-12 as having uncertain phylogenetic relationships. The first of these six subgroups is **Atherinomorpha** (Fig. 3-12), which includes familiar aquarium fishes, such as guppies (*Poecilia*) and mosquitofish (*Gambusia;* Fig. 3-14*A*). Male guppies and mosquitofish have a **gonopodium,** an organ composed of modified anal fin rays and used for internal fertilization; young guppies are born alive, which is why aquarists call them livebearers. The second percomorph subgroup includes the sticklebacks, seahorses, and their allies (**Gasterosteiformes;** Fig. 3-12); these percomorphs have bony abdominal plates (Gr., *gaster* = stomach + *osteon* = bone) that can give the body a bizarre external appearance (Fig. 3-14*B*). Most scorpionfishes (**Scorpaeniformes;** Fig. 3-12) have many spines on the head and strongly spinous fins, such as the slowly swimming, cryptically colored lionfishes and firefishes (e.g., *Dendrochirus;* Fig. 3-14*C*). Glands associated with the sharp dorsal fin spines of scorpion fishes can carry a potent neurotoxin, as in the stonefishes (*Synanceia*). Puffers and their allies (**Tetraodontiformes;** Fig. 3-12) exhibit unusual body shapes and swimming modes. For example, puffers and porcupine fish (Fig. 3-14*D*) can inflate the body with air or water by filling the stomach. Their short vertebral column makes the trunk relatively inflexible. As a result, they emphasize dorsal and anal fins for slow swimming. Familiar examples of flatfishes (**Pleuronectiformes;** Fig. 3-12) include the 2-m Atlantic halibut (*Hippoglossus;* Fig. 3-14*E*). Larval flatfishes are bilaterally symmetrical, but adults are asymmetrical and specialized for life on the bottom. Development of asymmetry happens at metamorphosis, when one eye migrates from the left side of the head to the right side (or vice versa, depending on the group of flatfishes).

In terms of diversity and trophic position, the percomorph subgroup **Perciformes** (Fig. 3-12) dominates oceanic environments and some tropical freshwater systems. In this abbreviated treatment, we discuss examples of only a few of the many subgroups of Perciformes. A representative member of the large group **Percoidei** is the yellow perch (Fig. 3-14*F*), a common freshwater fish that is often studied in teaching laboratories. Jacks (Carangidae), exemplified by the lookdown (*Selene;* Fig. 3-14*G*) are pelagic marine fishes. This group is noteworthy to functional anatomists because they swim using a locomotor mode known as **carangiform locomotion,** which emphasizes force application to the water by the narrow caudal peduncle and semilunate (= crescent-shaped) tail rather than the

[7]Which may dip below 0°C because the salt in the ocean depresses the freezing point.

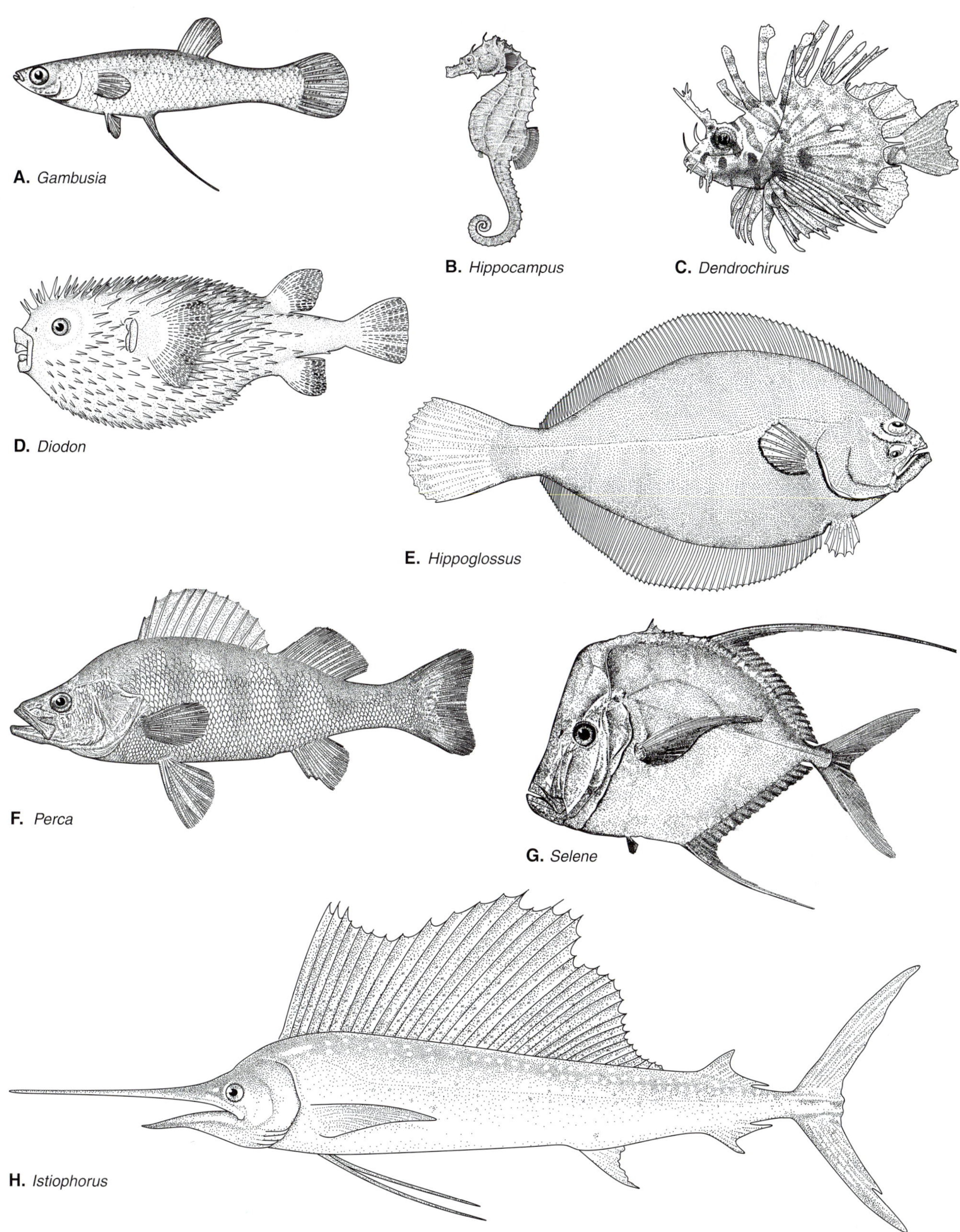

FIGURE 3-14
Representative percomorphs. *A,* A male mosquitofish, *Gambusia,* showing the long anal fin specialized as an intromittent organ used for internal fertilization. *B,* A gasterosteiform, the seahorse, *Hippocampus. C,* A scorpaeniform, the zebra firefish, *Dendrochirus. D,* A tetraodontiform, the porcupine fish, *Diodon. E,* A pleuronectiform, the halibut, *Hippoglossus,* in which both eyes are on the right side of the head. *F,* A percid, the yellow perch, *Perca. G,* A carangid, the lookdown, *Selene,* showing a tall, semilunate caudal fin. *H,* A scombroid, the sailfish, *Istiophorus.*

anterior part of the trunk (Chapter 11). Cichlids, damselfishes, wrasses, and allies (**Labroidei**) are small to medium-sized marine and freshwater perciforms. Many labroids employ **labriform locomotion,** a swimming mode in which the body is held straight and the pectoral fins are used to generate lift and thrust. Mackerels and allies (**Scombroidei**) are fast-swimming, oceanic predators, such as barracudas, mackerels, and billfishes such as the sailfishes (*Istiophorus;* Fig. 3-14*H*). These pelagic speed specialists have a tall, deeply forked tail and demonstrate **thunniform locomotion,** in which body undulations are minimal and the tail is swept from side to side to power swimming. The air-breathing labyrinth fishes (**Anabantoidei**) live in tropical fresh water. The cabbage-shaped **labyrinth organ** in the opercular chamber is their gas-exchange organ. Familiar examples include the Siamese fighting fish (*Betta*) and many species of gouramis.

Sarcopterygians and the Origin of Tetrapods

Sarcopterygii is the sister group of Actinopterygii (Fig. 3-2). It includes living and extinct fishlike forms as well as the terrestrial vertebrates (Fig. 3-15). The name Sarcopterygii refers to their **muscular fins** (or limbs in land vertebrates; Gr., *sarkodes* = fleshy +

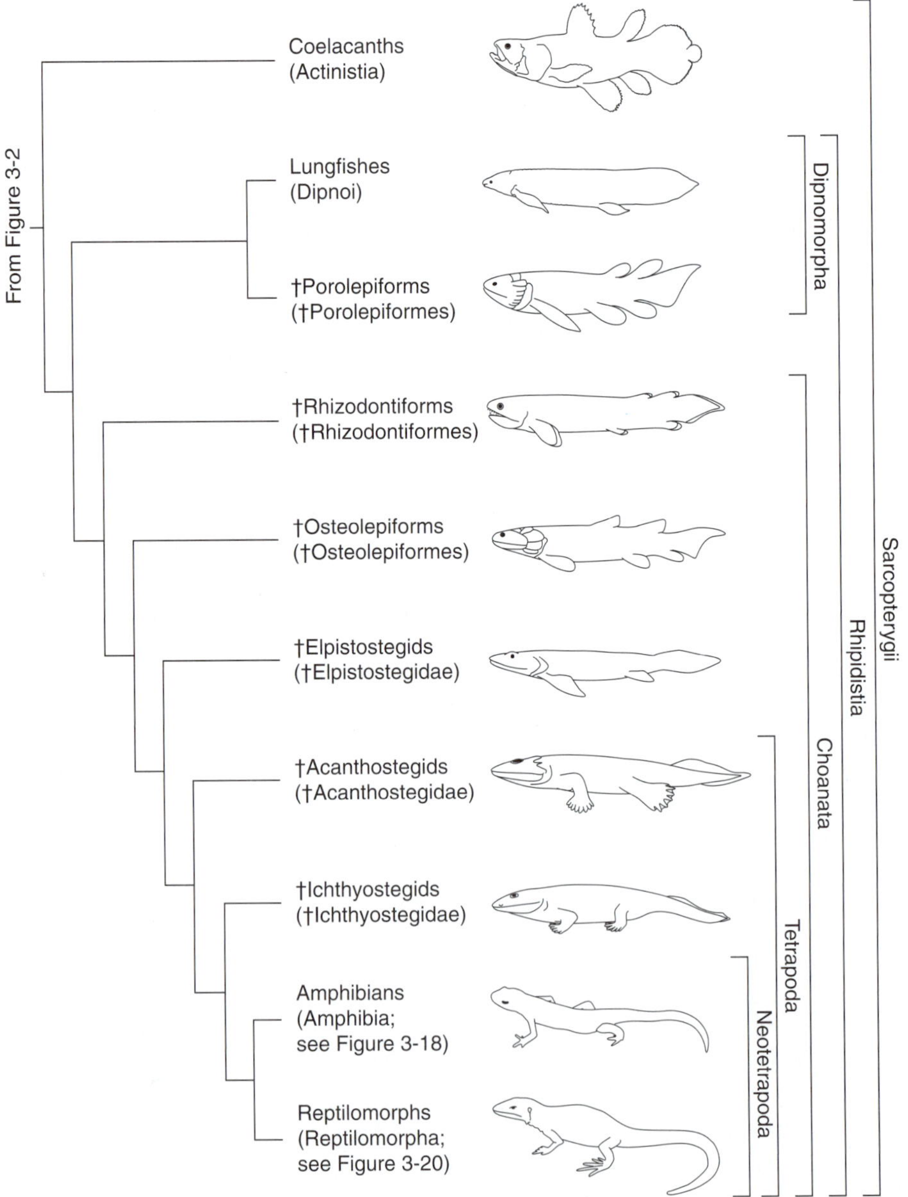

FIGURE 3-15
Phylogeny of living and fossil sarcopterygians (Sarcopterygii), highlighting the origin of choanates (Choanata), tetrapods (Tetrapoda), and neotetrapods (Neotetrapoda).

pteryg = wing or fin). This type of fin may have evolved as an adaptation for benthic life, where the fins could be used as props or to push against the substrate. The paired fins of sarcopterygians are **monobasic,** which is a formal way of saying that a humerus is present in each pectoral fin (or forelimb in land vertebrates) and a femur in each pelvic fin (or hindlimb in land vertebrates) (Chapter 9). Paleozoic sarcopterygians had a transverse joint in the braincase, known as the **intracranial joint,** which was lost in the lineages leading to lungfishes and tetrapods. Early sarcopterygians had cosmoid scales, which consisted of a thick layer of **cosmine** (dentine overlain by enamel) on top of the bony scales. Sarcopterygians retained the two dorsal fins and heterocercal caudal fin inherited from early gnathostomes. Independently in each major lineage of sarcopterygians, the ancestral heterocercal tail became less heterocercal and more symmetrical internally and externally. Such a condition is referred to as a **diphycercal tail** (Chapter 8). Throughout their long history, most fishlike sarcopterygians have inhabited fresh water, although some early fossil lungfishes are found in marine deposits and the living coelacanths inhabit marine environments. Fishlike sarcopterygians were diverse, abundant, and widespread during the Paleozoic, but the living coelacanths and lungfishes have restricted distributions. Polypteriforms (an actinopterygian group discussed earlier), coelacanths, and certain fossil sarcopterygians were known as "crossopterygians" in older literature, but because this is not a monophyletic group, we do not use the term.

Coelacanths

Although once more diverse, **Actinistia** (commonly known as coelacanths) is today represented by one genus, *Latimeria* (Fig. 3-16*A*). It lives at moderate depths (circa 200 m) off the southeastern coast of Africa, the Comores Islands, and Madagascar, and in 1998, it was first reported from Indonesia. (The species from the western Indian Ocean is known as *Latimeria chalumnae*. The form from Indonesian waters now appears to be a distinct species and is called *Latimeria menadoensis.*) *Latimeria* has fleshy paired fins, two dorsal fins, and cosmoid scales, and lacks internal nostrils. The vestigial lung is largely filled with fat. Studies and films of *Latimeria* made from submersibles show that it uses its muscular fins not to walk on the bottom but to slowly swim in midwater. Coelacanths are the only living vertebrates with an **intracranial joint.** The joint lies between the anterior and posterior halves of the braincase, and motion at the joint is powered by large, paired, basicranial muscles. The basicranial muscles are homologous to paired eye muscles (i.e., the bulbar retractor muscles) found in tetrapods (Chapter 13). An electrosensory organ in the snout, known as the rostral organ, is unique to coelacanths. It consists of three paired tubes, each about 3 mm to 5 mm in diameter in an adult, leading to a median cavity in the snout. Although this cavity is preserved in fossil coelacanths, the organ that occupied it was unknown until the discovery of *Latimeria,* and its electrosensory function remained unknown for another 40 years (Chapter 12).

Rhipidistians

Members of the group **Rhipidistia** (Fig. 3-15) have teeth in which the enamel folds inward, which accordingly are termed labyrinthodont teeth. †Porolepiformes and Dipnoi (lungfishes) make up the rhipidistian clade **Dipnomorpha** (Fig. 3-15). Dipnomorphs have long, paired fins with many central axial elements. Living lungfishes use these fins to "walk" on the bottom using the same gait found in tetrapods. Perhaps the best known †porolepiform is †*Holoptychius* (Fig. 3-16*B*). Large pores in the cosmine covering the scales and dermal bones of †porolepiforms are the source of the group's name (Gr., *poros* = hole + *lepi* = scale + *form* = structure).

Since the discovery of living lungfishes in the 1830s, dipnoans have been central to discussions about the classification and evolution of land vertebrates (Focus 3-3). The name Dipnoi (Gr., *di* = two + *pnoe* = breath) refers to their ability to breathe using either gills or lungs. Their remarkable fossil record makes them a favorite group for evolutionary studies. The internal skeleton of Devonian lungfishes, such as †*Dipterus* (Fig. 3-16*C*), was partly ossified, but extant lungfishes have poorly ossified internal skeletons. Reduced ossification and the confluence of the originally separate dorsal, caudal, and anal fins may be a result of paedomorphosis within the group (Chapter 1). Lungfishes have an autostylic jaw suspension, in which the palatoquadrate is fused to the neurocranium. All living and most fossil lungfishes have triangular tooth plates on the palate and lower jaw, with which they crush plants or the shells of invertebrates: the anterior nares of lungfishes are inside of the posterior edge of the upper lip, and the posterior nares open within the oral cavity. Only three genera of lungfishes exist today in tropical rivers and lakes. The Australian lungfish (*Neoceratodus forsteri;* Fig. 3-16*D*) occurs in a few rivers of Queensland. There is a single species from South America (*Lepidosiren paradoxa*) and four species from Africa (all in the genus *Protopterus;* Fig. 3-16*E*). The gills and circulatory systems of South American and African lungfishes are highly modified to allow for life

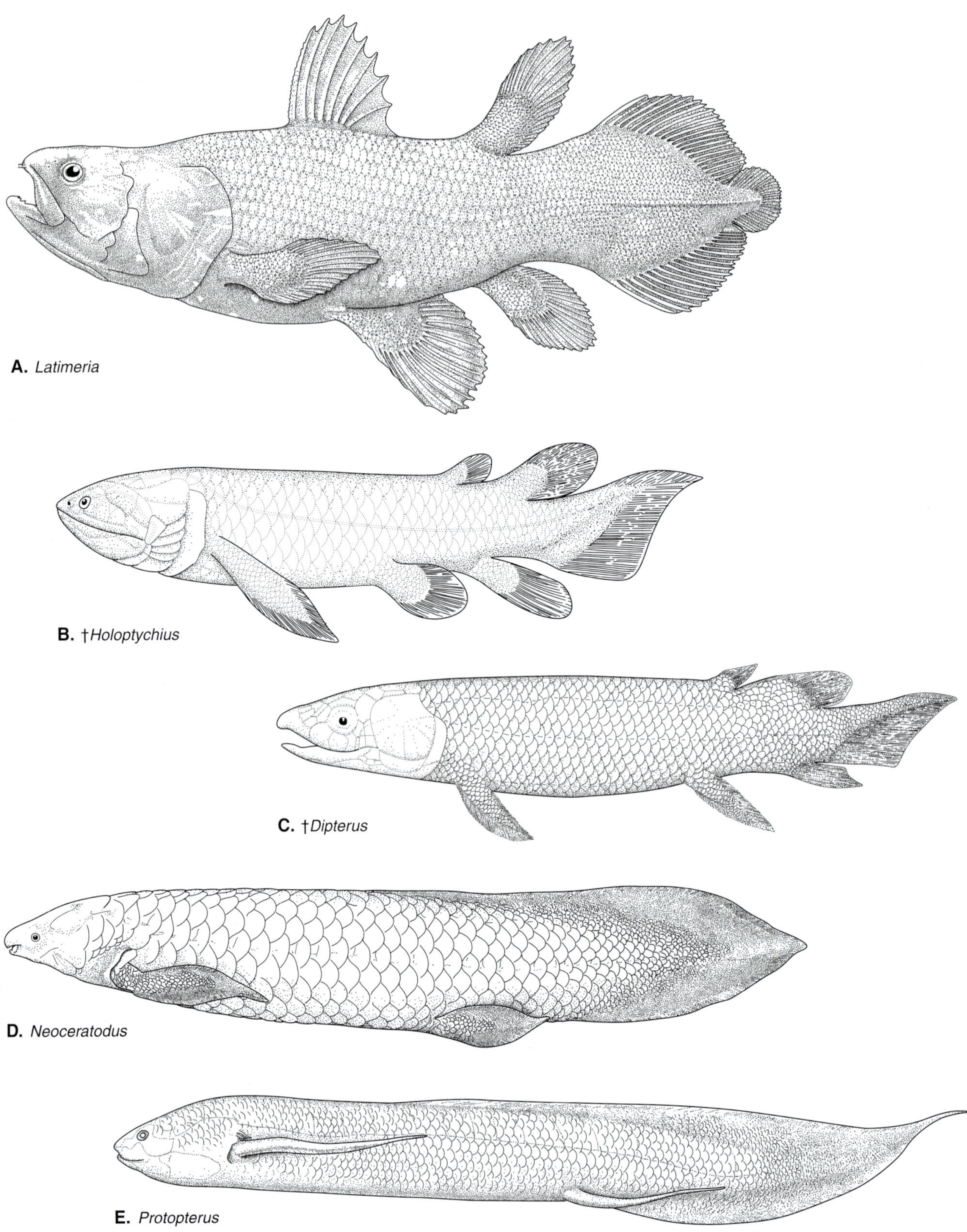

FIGURE 3-16
Representative sarcopterygians. *A, Latimeria,* the living coelacanth genus, showing the large muscular lobes of the paired fins, the anal fin, and the second dorsal fin. *B,* Reconstruction of a †porolepiform, †*Holoptychius. C,* Reconstruction of a Devonian dipnoan, †*Dipterus,* in lateral view; note that all four median fins are present. *D,* The living Australian lungfish, *Neoceratodus;* note that only a single median fin is present. *E,* One of four species of African lungfishes in the genus *Protopterus.* African lungfishes have elongate, tendril-like paired fins.

FOCUS 3-3 *Thinking About the Origin of Tetrapods*

Because of the different physical and chemical properties of water and air, terrestrial tetrapods experience different environments than did their aquatic ancestors, and many organ systems of tetrapods are modified for life on land. Air is less dense than water and does not provide the buoyancy of water, but it offers less resistance to movement. Accordingly, the musculoskeletal systems of tetrapods support the weight of the body from a strengthened vertebral column and power movement by walking on limbs. Sound waves travel more slowly in air than in water, but fewer wavelengths of light are absorbed by the atmosphere than by water, and these differences are reflected in the new or reconfigured sense organs of tetrapods. Oxygen is more abundant in air, and gills do not function well for gas exchange if they cannot be periodically moistened, so it is not surprising that terrestrial adult tetrapods lack gills. Water is in short supply in many terrestrial environments, and a permeable skin can be a liability. Thus, water-conserving specializations, including behaviors for avoiding desiccation and thicker, more cornified skins, evolved in tetrapods. Finally, air does not have the thermal stability of water, meaning that ambient temperatures on land can fluctuate rapidly and widely. Tetrapods must be able to retreat to more stable, sheltered environments or have systems for internal heat production and conservation, such as the types of endothermy found in birds and mammals.

The earliest known tetrapods occur in Devonian deposits. One way to approach questions such as, "Which ecological and behavioral factors triggered or allowed the initial invasion of land?" is to study living animals that engage in similar transitions from water to land. Reasoning by analogy, we can hypothesize that a shift from aquatic to terrestrial life was triggered by factors such as decreasing resources for food, decreasing opportunities for reproduction, and increasing intraspecific and interspecific competition in Devonian waters. Simultaneously, food and other terrestrial resources were becoming more abundant, for land plants were spreading across the Earth's surface by the Devonian period. Living air-breathing fishes, such as the bichir (*Polypterus*), the walking catfish (*Clarias*), and the climbing perch (*Anabas*), regularly engage in terrestrial excursions in response to naturally occurring or experimentally induced dwindling food resources or increasing population densities (Liem, 1987). Fishlike rhipidistians may have engaged in similar terrestrial sojourns to escape competition or to search for prey or breeding habitats. Superior air-breathing capacities with enlarged lungs, sturdy muscular fins, and a strong skeleton would have been advantageous for such terrestrial excursions. By the late Devonian, land was becoming a hospitable environment for vertebrates, with continuously warm and humid climates. A rich cover of primitive plants existed, and insects and other terrestrial invertebrates were abundant. Thus opportunities for the evolution of terrestriality must have increased during the Devonian.

The origin of tetrapods is a major evolutionary event in the history of vertebrates, yet the details of this transition continue to be debated. The debate traces from the 1830s, a period of tremendous exploration and discovery in zoology, but one that occurred two decades before Darwinian evolutionary thinking entered the general consciousness of zoologists. Then, as now, zoologists were striving to perfect classification schemes and to reach generalizations about groups of organisms. Working in England, Richard Owen (later, Sir Richard Owen, discoverer of dinosaurs and founder of the British Museum of Natural History) described the anatomy of the African lungfish (*Protopterus*). More or less simultaneously, Leopold Fitzinger in Germany described the anatomy of the South American lungfish (*Lepidosiren*). The two scientists reached quite different conclusions about the classification of these animals. Fitzinger considered his lungfish to be some type of amphibian, similar to the genus *Siren* from North America but with scales (thus the name *Lepidosiren;* Gr., *lepidos* = scale). Owen, on the other hand, argued that his lungfish was a fish, based on the structure of its olfactory system, among many other features. From these roots, the debate about lungfishes and tetrapods has continued, despite the introduction of major new paradigms, such as Darwinian evolution in the mid-19th century and phylogenetic systematics in the mid-20th century. The debate also survived the injection of many new fossil and living taxa, including the Australian lungfish, *Neoceratodus* (1871); a well-preserved †osteolepiform, †*Eusthenopteron* (1881); the first fragmentary fossil †elpistostegid, †*Elpistostega* (1938); and living coelacanths, *Latimeria* (1939). We merely rephrase an old debate when we ask questions such as, "Are coelacanths or lungfishes more closely related to tetrapods?" Everything old is new again.

in poorly oxygenated water. They are obligate air breathers and will drown if denied access to air (Chapters 18 and 19). Many investigators agree with the late Alfred S. Romer (a leading American paleontologist and anatomist of the 20th century) that the bodies of fresh water in which Devonian fishlike sarcopterygians lived were subject to stagnation and periodic drying up. An African lungfish survives the drying of its swamp by burrowing and forming a cocoon of mucus and dried mud about itself, lowering its metabolism, and breathing air through a small hole in the cocoon. Some Paleozoic lungfishes also made cocoons, as evidenced by fossilized burrows with the animals still inside.

Members of **Choanata** (Fig. 3-15) have a **choana,** a specialization that allows air to be drawn into the oral cavity through the nose. Choanates have only one pair of external narial openings, termed **incurrent nares,** and air passing into them has more or less direct access to the oral cavity and pharynx. There has been much controversy about the presence or absence of a choana among sarcopterygian groups. Here, we consider that coelacanths, lungfishes, and †porolepiforms do not have a choana but that a choana is present in †rhizodontiforms, †osteolepiforms, †elpistostegids, and tetrapods (see Chapters 12 and 18 for illustrations and additional discussion of the choana). Choanates also have a concave glenoid fossa (a depression on the side of the pectoral girdle) to receive the head of the humerus, and the humerus bears bony crests associated with the insertion of muscles used in locomotion.

The predatory **†Rhizodontiformes** (Fig. 3-15) reached lengths of up to 2 m. All **†Osteolepiformes** (Fig. 3-15) have enlarged axillary scutes, which can be easily seen at the bases of the two dorsal and anal fins as well as the paired fins of †*Eusthenopteron* (Fig. 3-17*A*). Many skeletal details observed in †osteolepiforms are retained by tetrapods, such as the basic patterns of bones in the skull roof and paired appendages. Another extinct choanate group is **†Elpistostegidae** (Figs. 3-15 and 3-17*B*). This group is interpreted in Figure 3-15 as the sister group of tetrapods because its members share several characters with early tetrapods, such as patterns of bones in the skull roof, an elongate humerus, loss of the dorsal fin, and loss of the anal fin.

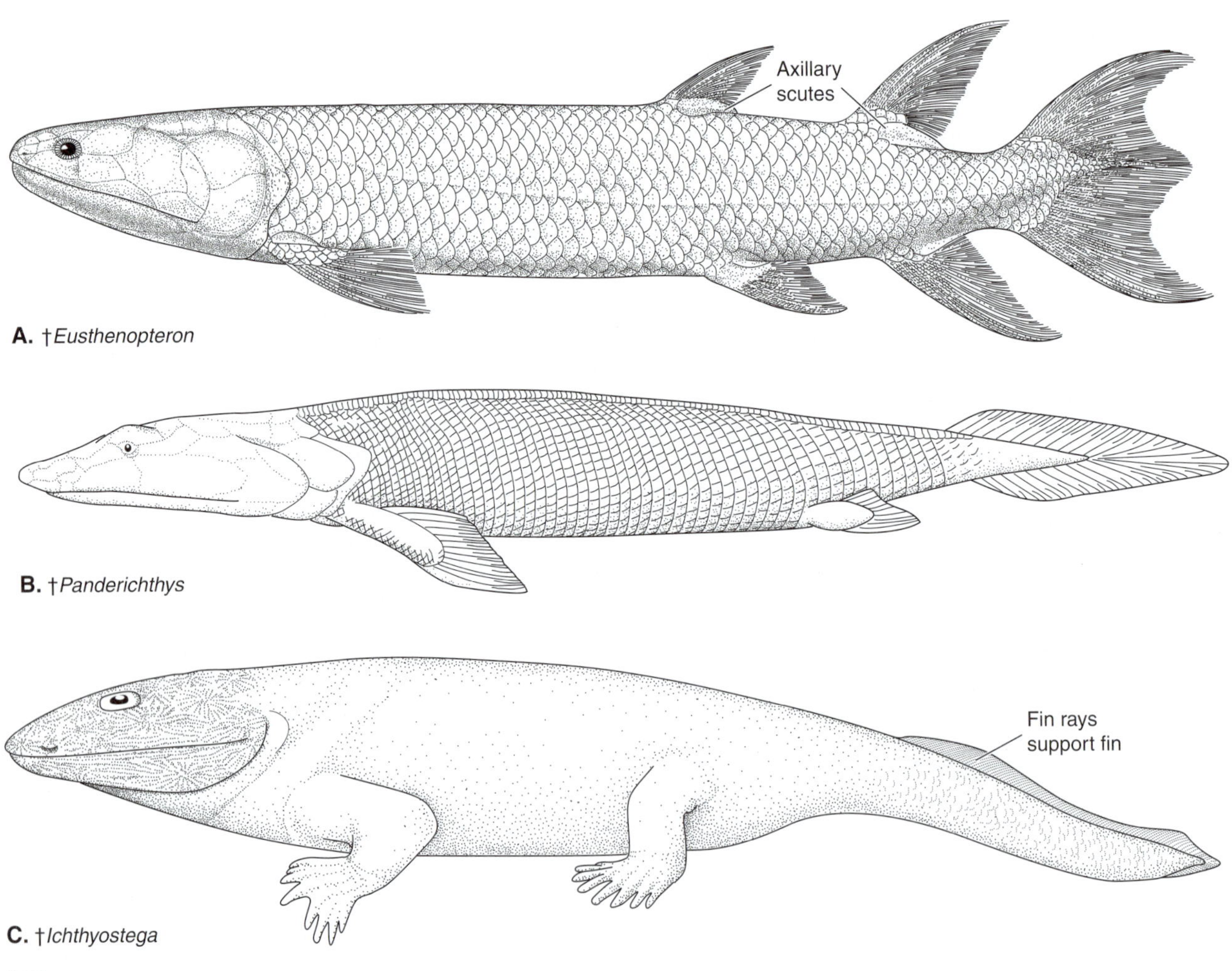

FIGURE 3-17
Representative choanates. *A*, Reconstruction of †*Eusthenopteron,* one of the best known Devonian osteolepiforms. *B*, Reconstruction of the †elpistostegid †*Panderichthys* in lateral view. Note the absence of dorsal and anal fins. *C*, Reconstruction of †*Ichthyostega*.

The Origin of Tetrapods

We turn now to the terrestrial vertebrates, or **Tetrapoda** (Gr., *tetra* = four + *podos* = foot; Fig. 3-15). Their paired appendages are not fins but **dactylous** limbs, meaning that they have digits. In a few groups of tetrapods, the limbs have been secondarily reduced or lost. Major changes also are present in the vertebral column and limb girdles. The vertebral column of a tetrapod is strengthened by the evolution of well-developed articulations, called **zygapophyses,**[8] between the dorsal parts of successive vertebrae (Chapter 8). This extra strengthening is needed because water no longer helps to bear the weight of the body. The weight of the body is transferred from the trunk to the pelvic girdle and hind legs by one or more specialized **sacral vertebrae** and **sacral ribs,** but tetrapods have lost all skeletal connection between the skull and pectoral girdle.

The earliest known tetrapods belong to †**Acanthostegidae** and †**Ichthyostegidae** from Upper Devonian rocks in Scandinavia and Greenland (Figs. 3-15 and 3-17*C*). These groups mark early stages in the transition from water to land. Such a fundamental shift in ecology affected the behavior of the animals and virtually all aspects of their anatomy and physiology (Focus 3-3). Early tetrapods displayed a mixture of aquatic and terrestrial features. We presume that they made at least brief ventures onto land because of their dactylous limbs and modified limb girdles. Interestingly, digit number in early tetrapods was not stabilized at what we consider the normal five but instead ranged from six to eight. Having inherited a choana and lungs from their ancestors, they must have been air breathers, but they retained components of the gill ventilatory system. They also retained deep canals in the skull for the lateral line sensory system, a sensory system that is useful only in water (Chapter 12). The tail of †*Ichthyostega* (Fig. 3-17*C*) even retained a fin supported by fin rays. Given this mixture of characters, we presume that early tetrapods were primarily aquatic animals that made forays onto land to escape from predators or to search for food.

Neotetrapoda (Fig. 3-15) includes all remaining tetrapods. Neotetrapods have five digits on the hands and feet, a condition known as **pentadactyly.** Digits were lost during the evolutionary history of many groups of neotetrapods, and polydactyly (= more than five digits) evolved secondarily in a few groups. Many extinct groups of neotetrapods are known (e.g., †Baphetidae, †Crassigyrinidae, and †Aïstopoda), but their interrelationships are problematic, and we will not discuss them here. The remaining neotetrapods belong to two clades: **Amphibia** is exemplified by the living salamanders, frogs, and caecilians; **Reptilomorpha** is exemplified by the living turtles, lepidosaurs (= tuatara + "lizards" + snakes), crocodilians, birds, and mammals.

Amphibians

A phylogeny and classification of major groups of Amphibia is shown in Figure 3-18. An easily observed synapomorphy for the entire group is the reduction in the number of digits in the hand to four or fewer. Most living amphibians exhibit a bimodal life history, in which eggs laid in fresh water hatch into larvae that metamorphose into terrestrial adults. This life history is the source of the name Amphibia (Gr., *amphi* = double + *bios* = life). We believe that Paleozoic amphibians had such a bimodal life history, although fossilized larvae have been discovered in only a few cases. We are uncertain about the affinities of many early amphibians (e.g., †nectrideans, †microsaurs, and †colosteids; Fig. 3-18), but the remaining amphibians belong to a well-characterized group known as **Temnospondyla** (Gr., *temnos* = to cut + *spondyla* = vertebra; Fig. 3-18). The ear of temnospondyls is specialized for the detection of airborne sounds. The **columella** has become a much more slender bone than in early tetrapods, such as †*Ichthyostega*. Also, the columella is directed dorsally, rather than ventrally, and it contacts a thin membrane of skin known as the **tympanum,** or eardrum. The tympanum is very evident on the heads of frogs (e.g., *Rana* in Fig. 3-19*E*). This mechanism is convergently similar to sound-detecting systems found in derived groups of reptilomorphs (discussed later and in Chapter 12). Temnospondyla includes extinct groups, such as †Eryopoidea (Figs. 3-18 and 3-19*A*). †Eryopoids ranged from aquatic to fully terrestrial forms, and some reached lengths of about 3 m. We may speculate that their biology was perhaps more similar to the (unrelated) crocodiles of today than to any of the living amphibians.[9] An extinct Permian group, †Dis-

[8]The zygapophyses of tetrapod vertebrae are not homologous to the complex intervertebral articulations found in some teleosts, such as marlins, which also have been called zygapophyses.

[9]Such speculation is based on the convergently evolved similarities of †eryopoids and crocodilians, such as a relatively flat head and large teeth.

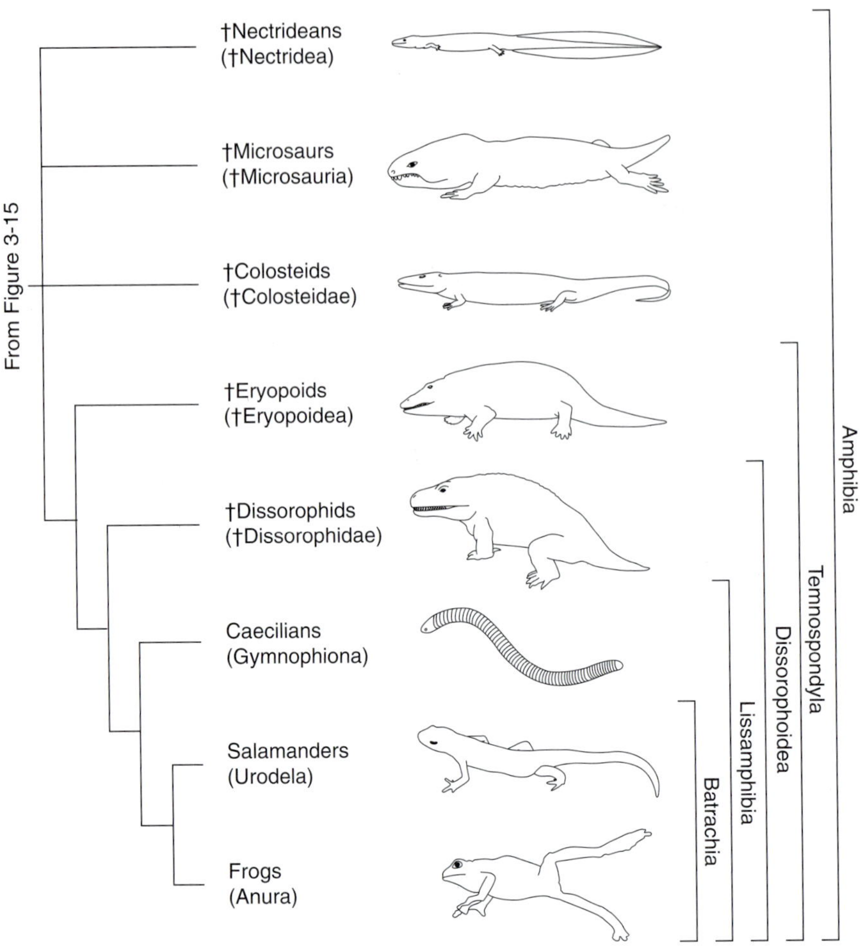

FIGURE 3-18
Phylogeny of living and fossil Amphibia highlighting the origin and diversity of lissamphibians (Lissamphibia).

sorophidae, shares unique tooth characteristics with living amphibians with which they are grouped as **Dissorophoidea** (Fig. 3-18). Dissorophoid teeth are bicuspid (i.e., each has two distinct high points, or cusps) and pedicellate (i.e., each tooth crown is separate from its supporting base, or pedicel).

Lissamphibians

Lissamphibia (Gr., *lissos* = smooth; Fig. 3-18) is the subgroup of temnospondyls that includes the three living groups of amphibians: caecilians, salamanders, and frogs. The skin of lissamphibians is thin, smooth, and glandular. In contrast to earlier groups, the ribs of lissamphibians are short and sometimes fused to the vertebrae. Their wrist and ankle bones are incompletely ossified, and they share a distinctive sensory structure, known as the **amphibian papilla,** in the inner ear (Chapter 12). Most extant lissamphibians exhibit structural, functional, and behavioral adaptations correlated with a bimodal life. During **metamorphosis,** aquatic larval features are rapidly lost, and those related to terrestrial life are attained. For example, gills, which are well suited for gas exchange in the water but poorly suited for life on land, are lost during metamorphosis. Other changes are linked to changes in diet and feeding mechanics: most frog tadpoles are herbivorous, whereas most adult frogs are carnivorous, and such changes require rapid remodeling of the mouth parts and gut. Also during metamorphosis, limbs emerge, and the median fins become reduced. The skin of adult lissamphibians, although slightly keratinized, is thin, moist, and vascular. It supplements, or in some groups replaces, the lungs as a respiratory membrane. Such a thin and moist skin allows for considerable evaporative loss of body water. Amphibians compensate for this by living in moist habitats in or beside ponds or streams, under rocks and logs, in damp forests, or in other humid habitats. Many tropical species are fully terrestrial in humid habitats. A few frogs live in deserts, where they survive by burrowing into the soil and becoming torpid (= estivation, a state in which metabolic activities are greatly reduced), except during rare and brief rainy periods.

Lissamphibians are ectothermic, and terrestrial ectotherms face more complex problems than do aquatic ones (Focus 3-4). Ambient temperatures in a terrestrial

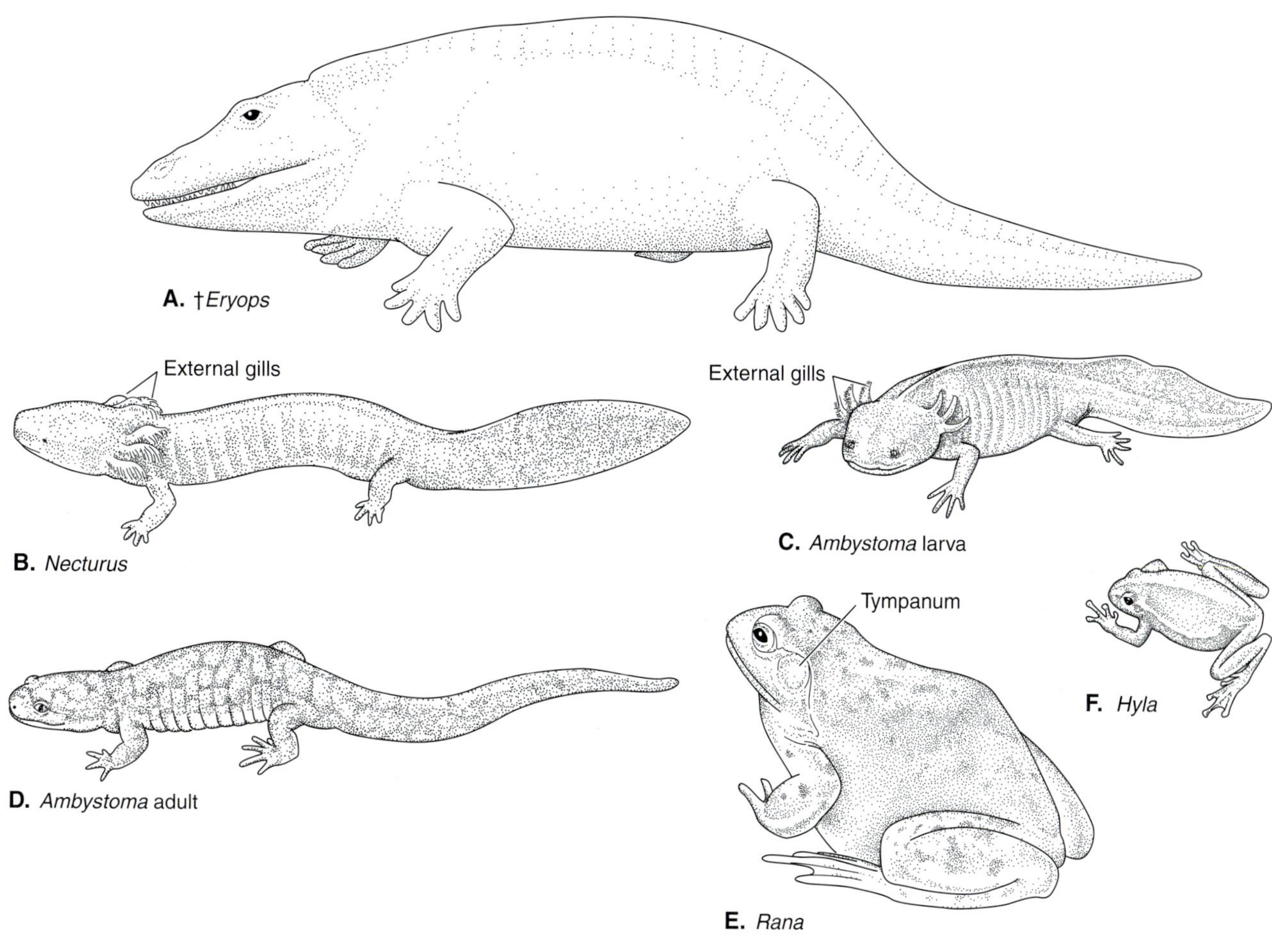

FIGURE 3-19
Representative amphibians. *A,* Reconstruction of an early temnospondyl, †*Eryops. B,* The mudpuppy, *Necturus,* lives in water and retains gills throughout life. *C,* Larva of a tiger salamander *Ambystoma,* showing bushy external gills. *D,* Metamorphosed adult of a tiger salamander *Ambystoma.* Gills and other features related to life in water are lost during metamorphosis. *E,* The bullfrog, *Rana.* Note the large tympanum. *F,* A tree frog, *Hyla,* has specialized toe tips for adhering to steep surfaces.

habitat may fluctuate rapidly and greatly. Electromagnetic heat waves, which are absorbed by water, reach a terrestrial animal directly and warm it. Also, the evaporative loss of body water through the respiratory passages and skin, which is not a problem for aquatic animals, involves the conversion of water to water vapor, which in turn causes a loss of heat. Because 584 calories are needed for the evaporative conversion of 1 g of water to water vapor, the amount of body heat lost in this way is considerable. (By comparison, it requires only one fifth as many calories to bring 1 g of water from 0°C to its boiling point at 100°C.) The terrestrial environment, however, is a mosaic of microclimates ranging from exposed areas where solar radiation and temperatures vary greatly to more shady, humid, and thermally stable habitats. Even so, ambient temperatures in the temperate regions of the world can fall in winter far below the level that amphibians can tolerate. They respond by burrowing into soft ground or leaf litter or by retreating to the bottoms of ponds, where their metabolic rate drops, and they live on food reserves stored in their bodies. Some ectotherms overwinter by being freeze tolerant.[10]

As shown in Figure 3-18, the tropical caecilians in the order **Gymnophiona** are the sister group to Batrachia, which is the group that includes the salamanders and frogs. Gymnophiona (Gr., *gymnos* = naked + *ophion* = serpent) includes an early Jurassic form with rudimentary legs, but all of the living species are limb-

[10]Freeze tolerance involves shifting water from inside cells to extracellular portions of the body, such as the connective tissue matrix, where the water can freeze without injuring cells. See Storey and Storey, 1996.

FOCUS 3-4 *Ectothermy and Endothermy*

Central to understanding vertebrate evolution is the realization that a vertebrate's metabolic rate, in common with all chemical processes, is affected greatly by temperature. An increase in body temperature of only 10°C can more than double an animal's metabolic rate and hence its potential for activity. Increases in metabolic rate are costly in that they ultimately require a greater rate of food intake to support them. Body temperature is determined by the rate at which heat is produced internally by metabolic processes and the rate at which heat is lost to, or gained from, the surrounding thermal environment.

The way in which a vertebrate relates to its thermal environment can restrict the range of environments and habitats in which it can live. For example, a trout lacks a specialized insulating layer at its body surface and is surrounded by water, which is an excellent conductor of heat. As a result, its body temperature is determined largely by the temperature of the water in which it lives. Such an animal is said to be **ectothermic** (L., *ecto* = outside + *thermos* = temperature; the term "cold blooded" is a loose synonym, but it is not an accurate descriptor because an ectotherm in a warm environment can have a high body temperature). Although water temperature changes with latitude and with depth, an individual trout typically lives in a relatively stable thermal environment. Water has a high thermal stability, which means that much energy is required to change its temperature. Diurnal temperature fluctuations in even small bodies of water are dampened by the thermal stability of water, and the more extensive seasonal changes in water temperature occur slowly. Thus, a trout can adjust its enzyme systems to maintain a rate of metabolism that is no higher than that needed to sustain activity. This minimizes the amount of food and other resources that it must obtain and allows it to be active at the lowest cost. A trout cannot tolerate rapid dramatic temperature changes, but its enzyme systems do adapt to slower seasonal changes in water temperature by the process termed **acclimatization.** An acclimatized trout may have the same rate of metabolism at 5°C in the winter as at 25°C in the summer. In contrast, terrestrial ectotherms, such as salamanders or snakes, often experience more variable environments because air is not as thermally stable as water. As a general rule, terrestrial ectotherms are less active at night, when solar energy is absent, and are less abundant and diverse in areas with cold climates. Ectothermy does have one major advantage, however: it is a "low-energy approach" to survival because relatively little food (= energy) is used to maintain a high metabolic rate and thus food intake can be correspondingly low.

A vertebrate that is **endothermic** (L., *endo* = within + *thermos* = heat) is one that has specializations for retaining heat generated by the body ("warm blooded" is a loose synonym). Examples of these specializations include insulating layers, such as hair or feathers, and specialized countercurrent exchange systems that help limit the loss of heat through the surface structures. Endothermy has evolved convergently several times among the vertebrates. For example, some sharks and tunas generate heat from their muscular activity and can conserve some of it by means of specialized networks of blood vessels that work as countercurrent exchange systems. Some billfishes even have specialized brain-heater organs, which allow them to keep the temperature of the brain higher than the rest of the body. Without activity, however, such animals cannot long maintain an overall body temperature that is different from the water—an essentially infinite heat sink—in which they live.

Most species of birds and mammals have metabolic rates three or four times greater than that of ectotherms of comparable size and activity. The dominant aspect of the heat balance of a bird or mammal is the internal production of heat by its high metabolic level. Although birds and mammals are endothermic, they descended from different groups of amniotes, and evolved endothermy independently as they adapted to active modes of life.

Birds and mammals can maintain a high and constant body temperature both day and night over a wide range of ambient temperatures. Body temperature varies with the species, but it is roughly 35°C to 40°C. Because of their high body temperatures, birds and mammals can be active over a far greater range of environmental conditions than can an ectotherm. Heat is produced by the activity of all parts of the body, but the visceral organs contribute a disproportionately large share. Usually, body core temperature is higher than environmental temperature, so that heat flows from the core to the body surface to the environment. Temperature regulation is accomplished primarily by controlling the amount of heat lost at the body surface. Birds and mammals evolved insulating layers of subcutaneous fat and surface feathers or hair that conserve heat. Both feathers and hair entrap a still layer of air next to the skin. Because the thermal conductivity coefficient of air is only

less, tropical forms specialized for burrowing or swimming. Caecilians have solidly ossified skulls; long, wormlike trunks; short tails; greatly reduced eyes; and a peculiar chemosensory organ in the head, known as a **tentacle,** which can be moved by means of an eye muscle no longer used to move the eyeball. Their unusually "loose" skin allows them to burrow using an internal concertina mechanism. To do this, the vertebral column is thrown into a series of curves (independent of the skin) and pressed against the walls of the burrow. The head then is forced into the soil, which it compacts and pushes aside. As the burrow is extended, the vertebral

0.00006 calories/second • cm • °C (that of water is 0.001 calories/second • cm • °C), heat loss to the environment is minimized. (Thermal conductivity coefficients give an estimate of heat flow; thus a material such as air, which has a low thermal conductivity coefficient, conducts heat poorly.) The degree to which the feathers or hairs are elevated above the surface can vary the thickness of the layer of still air and hence the rate of heat loss. Aquatic birds and mammals have very oily feathers or hair that prevent water from reaching the body surface, and they also have thick layers of subcutaneous fat. The amount and rate of blood flow through the skin also can be controlled by the degree of vasoconstriction of cutaneous vessels and by the heart rate. Flow is extensive and rapid when heat needs to be lost; opposite changes occur when heat needs to be conserved. Finally, birds and mammals can regulate heat loss by controlling the amount of water evaporation. Birds and many mammals pant; some mammals also can sweat. Control of heat loss at the skin surface does not require the expenditure of much additional energy. A high and constant body temperature can be maintained at rest over a fluctuation in environmental temperatures of 5°C to 10°C with little extra metabolic work. This range is called the **thermal neutral zone,** and most birds and mammals spend much of their time in environments that fall within it. The control of body temperature when environmental temperatures exceed this range requires significant increases in metabolism. More heat is produced internally in cold weather. Heat loss at the surface is increased in hot weather by rapid blood circulation through the skin and by rapid panting or sweating.

Turtles, lepidosaurs, and crocodilians have no body insulation and are ectothermic, but they have mechanisms that give them more control over their body temperature than amphibians. For example, when the sun rises, the body temperature of a turtle increases toward a species-specific optimum level, metabolic rate rises, and the animal becomes active. Features that maximize the rate of heat gain, minimize the rate of cooling, and enable maintenance of a relatively high and constant body temperature during their period of activity differentiate them from amphibians. As another example, a species of Andean iguanian can maintain a body temperature close to 35°C, about the same as in many mammals, even when the air temperature is 0°C to 10°C. In the cool nighttime, the animal burrows shallowly into the ground, which retains some heat. When the sun rises, it first warms up by heat conduction through the soil, and then it emerges and gains heat rapidly by solar radiation. The animal increases the radiant energy it receives by dispersing a dark pigment in the skin cells known as chromatophores. When it is warm enough, the pigment is retracted. Iguanians detect the amount of solar radiation received, often by the median parietal eye, and regulate their behavior accordingly. They move in or out of the shade as needed or change their orientation to the rays of the sun. They lie perpendicular to the rays of the sun to maximize heat absorption and parallel to them to minimize heat absorption. A keratinized skin reduces loss of heat through the evaporation of water better than an amphibian's thin skin. Some ability to control heat loss or gain is possible by varying heart rate and hence the amount of blood flowing through the skin. Early in the morning, when core body temperature is low, heart rate increases so that heat is transferred rapidly from the body surface being warmed by the sun to deeper body tissues. Heart rate slows when core body temperature reaches an optimum level. Animals that depend on solar radiation for thermoregulation are sometimes called **heliotherms** (Gr., *helios* = sun + *thermos* = heat).

These sorts of temperature-control mechanisms do not require the expenditures of much energy, and they allow an animal to maintain a high and relatively constant body temperature on sunny days at a relatively low metabolic cost. They can enter more stressful environments than amphibians and be active under a wider range of conditions. But terrestrial ectotherms are still limited in the ecological niches that they can exploit. Most are diurnal, and only a few tropical species are active at night. Ectotherms from the temperate zone must hibernate during the winter. Very large ectotherms, such as crocodilians and giant tortoises, have some thermal stability that derives from their body size because a large mass with a relatively small surface area changes temperature slowly.[a] Some workers believe that early (i.e., nonavian) dinosaurians were endothermic and that they maintained a high body temperature by an increased metabolic rate. Whether or not this hypothesis is correct, it is reasonable to regard the large Mesozoic dinosaurs as having relatively stable body temperatures because of their favorable surface-to-volume ratios.

[a]See Chapter 5 for discussion of scaling of body mass and surface area.

column straightens until it is necessary to reposition the body for the next burrowing push. The lower jaw of caecilians has a unique jaw-closing system that permits them to generate large bite forces without having a large head diameter and thus minimizes the cost of subterranean burrowing (a larger-diameter burrow requires more work to construct because more soil must be pushed away). Some caecilians retain tiny, ossified scales in the skin. Some species are viviparous, and the developing young have specialized teeth that they use to scrape away at and feed on the uterine lining during their long gestation.

Batrachians

Batrachia (Gr., *batrachos* = a frog; Fig. 3-18) includes the salamanders (**Urodela**) and frogs (**Anura**). Batrachians have an **opercular apparatus,** which is a unique sound-conducting system. We will discuss this system in Chapter 12, but briefly, it consists of a tiny bone known as the **operculum** (not homologous to the operculum of bony fishes) that sits in a fenestra in the otic region of the skull. The operculum is connected by a muscle to the pectoral girdle. The opercular apparatus permits detection of low-frequency, ground-borne vibrations and supplements the columella–tympanum system, which is attuned to higher frequency airborne sounds. Batrachians also share the loss of several skull bones that were present in outgroup amphibians. Also, batrachians lack any trace of scales. Because of these and other losses, it has been suggested that paedomorphosis played a role in batrachian evolution.

Salamanders retain the long tail and general body shape of Paleozoic temnospondyls, but most are small, secretive, terrestrial animals. Some salamanders never metamorphose and instead retain gills and remain aquatic throughout life. These are described as **perennibranchiate** (L., *perennis* = throughout the year + *branchia* = gills). Significantly, the perennibranchiate condition has evolved multiple times among salamanders, so this condition does not characterize any specific monophyletic group. *Necturus* (Fig. 3-19*B*) is a good example of a perennibranchiate salamander found in streams and lakes in northern North America that is commonly studied in the laboratory. Many populations and a few species of mole salamanders (Ambystomatidae; Fig. 3-19*C* and *D*) also never metamorphose. As adults, these nonmetamorphosing ambystomatids retain gills and other larval features and provide an outstanding example of paedomorphosis (Chapter 1). Plethodontidae is by far the largest extant family of salamanders. Conventionally, herpetologists consider that ancestral plethodontids evolved in mountain streams. The adaptive explanation presumes that lungs could have been a disadvantage for life in fast-flowing water because a buoyant animal might easily be swept downstream; also, a small salamander living in cool, well-oxygenated stream water can rely on the skin for cutaneous gas exchange. In addition to those species of plethodontids that live in the southern central Appalachian Mountains, many also live in the western United States and the neotropics (i.e., Central and South America); a few species occur in Europe. Many plethodontids are fully terrestrial and no longer are tied to water for reproduction or larval life. Some plethodontids have remarkable tongue-projection systems that enable them to capture insects at a distance (Chapter 15).

The fossil history of frogs and toads (Anura; Fig. 3-18) extends from the Triassic. This is the most speciose and structurally diverse clade of lissamphibians. Anurans are highly specialized for jumping, or **saltation,** and move on both land and water by powerful thrusts of their elongate hind legs. A short trunk, elongate pelvic girdle, and loss of the tail (L., *a* = without + *nura* = tail) are among anurans' many adaptations for saltation (Chapter 11). **Vocalization,** or calling, is very important in the biology of many anurans, and males often have anatomical specializations for sound production. The "tailed" frog (*Ascaphus*) from mountain streams of northern California and the Pacific Northwest is the sister group of all other extant frogs. Its "tail" is actually a diverticulum of the cloaca that is used by males as an intromittent organ. This condition is unique to its family among anurans. The peculiar pipoids are aquatic anurans from South American and sub-Saharan Africa exemplified by the African clawed frogs (*Xenopus*), which often are studied by embryologists. Bufonids are among the most terrestrial of anurans. Commonly known as true toads, *Bufo* and others have a dryer and more heavily keratinized skin than do other anurans. They also have large poison glands on the back of the head (**parotid glands**) that produce a secretion that discourages predators. The arrow poison frogs (Dendrobatidae) of Central America and northern South America produce highly toxic skin secretions used by indigenous peoples to poison the tips of arrows or darts. Arrow poison frogs walk slowly on the forest floor, their safety from predators ensured by their bright aposematic (= warning) coloration. Ranoidea includes the family Ranidae, or true frogs (Fig. 3-19*E*). Perhaps the most familiar ranid in North America is the largely aquatic bullfrog, *Rana catesbeiana*. Hylidae (Fig. 3-19*F*) includes many species of tree frogs that use specialized toe tips to adhere to slanted or even vertical surfaces.

Reptilomorphs and the Origin of Amniotes

Reptilomorpha (Figs. 3-15 and 3-20) includes the remaining groups of tetrapods. As in early tetrapods and the fishlike rhipidistians from which they arose, early reptilomorphs had a vertebral structure in which each vertebral segment consisted of two parts (pleurocentrum and intercentrum; see Chapter 8). Over the course of reptilomorph evolution, the pleurocentrum became larger, and some subgroups of reptilomorphs completely lost the intercentrum. Their feet have a reduced and standardized number of phalanges, which

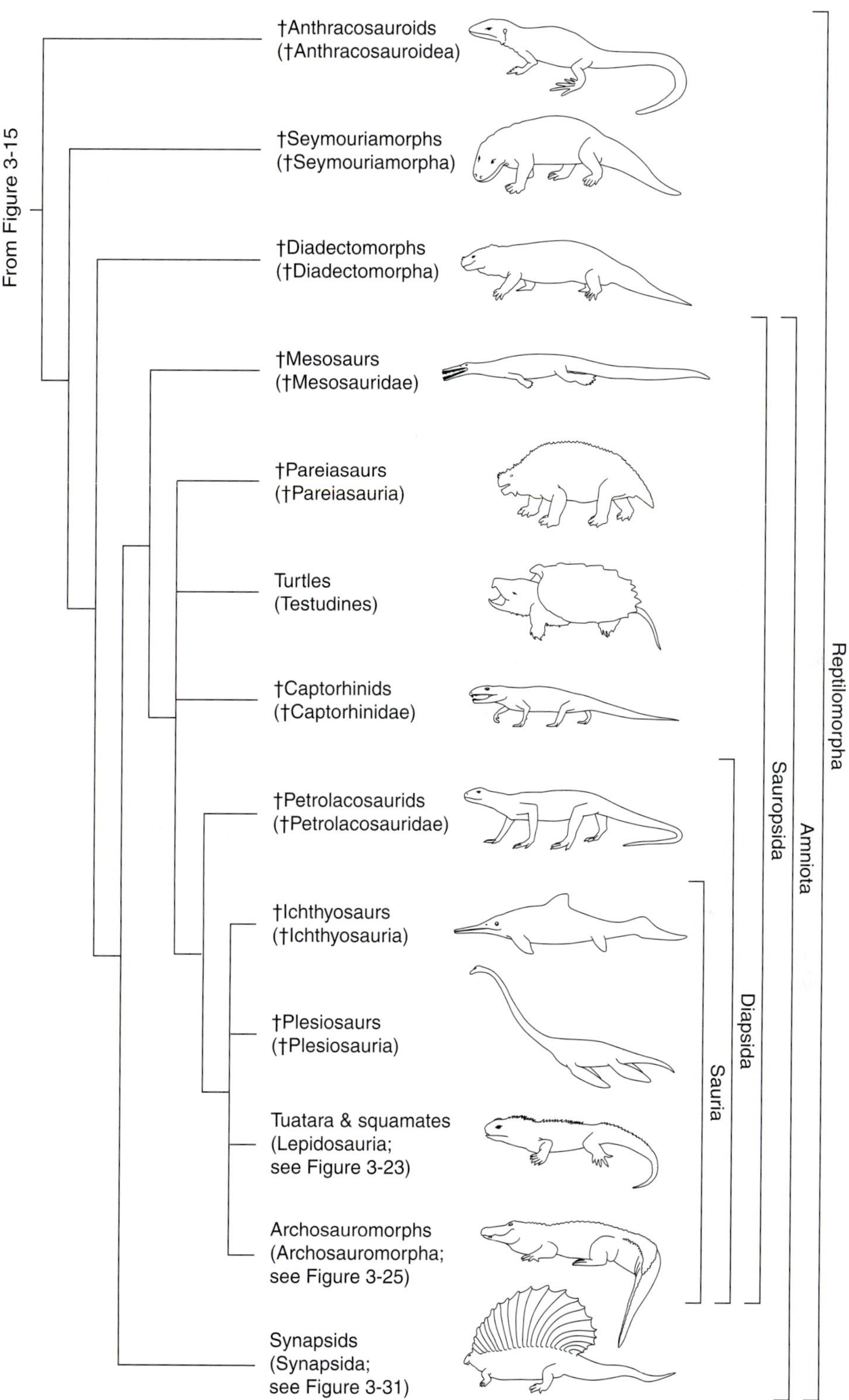

FIGURE 3-20
Phylogeny of living and fossil Reptilomorpha, highlighting the origin and diversity of diapsids (Diapsida).

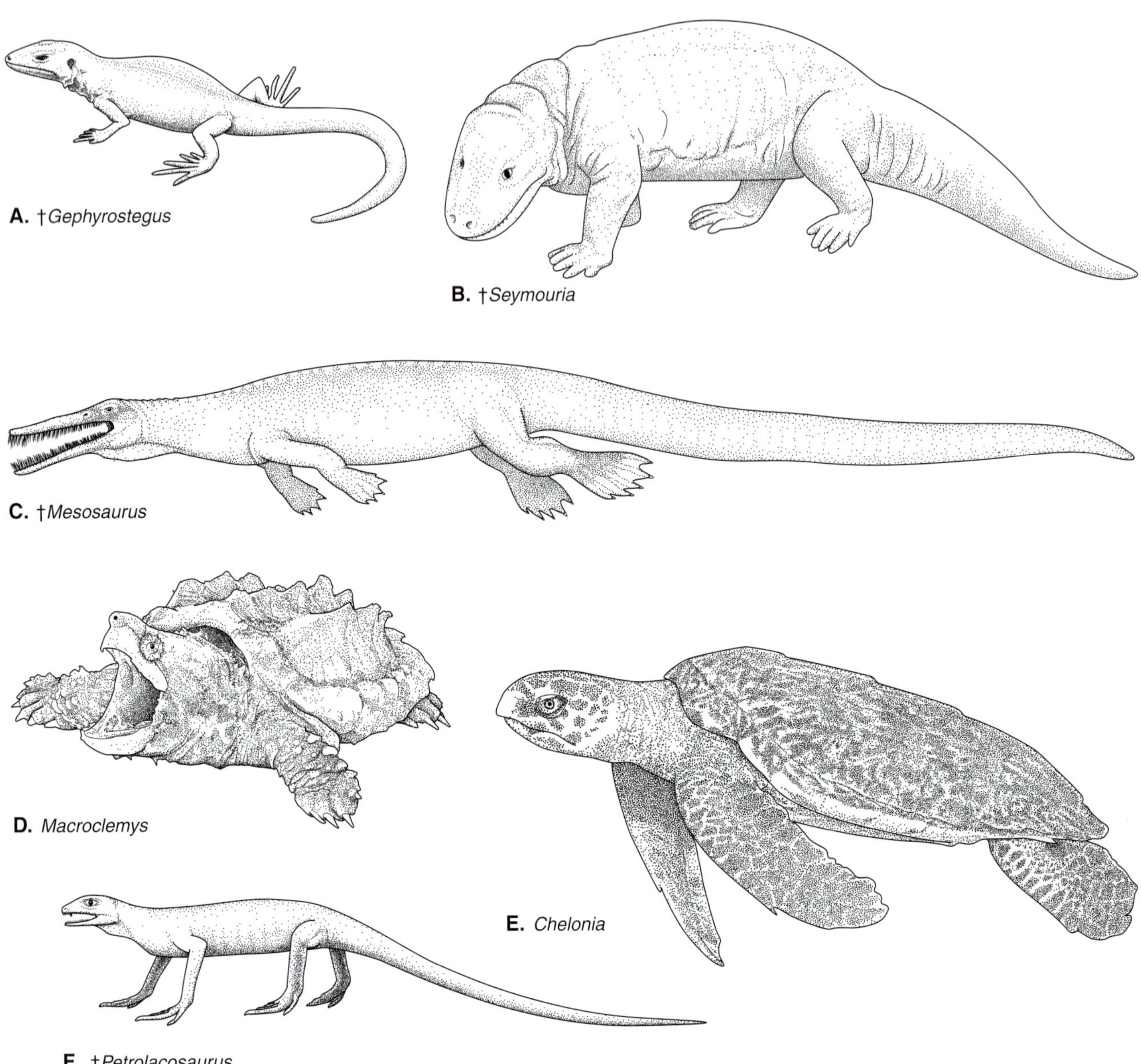

FIGURE 3-21
Representative reptilomorphs. *A,* Reconstruction of an †anthracosaur, †*Gephyrostegus.* *B,* Reconstruction of a †Seymouriamorph, †*Seymouria.* *C,* Reconstruction of †*Mesosaurus,* an early sauropsid. *D,* Alligator snapping turtle, *Macroclemys.* *E,* Green sea turtle, *Chelonia mydas,* showing front limbs modified as flippers. *F,* Reconstruction of an early diapsid, †*Petrolacosaurus.*

are the bones that support the fingers. We express this arrangement by counting the number of phalanges in each digit of the hand, beginning with digit one (the thumb). Reptilomorphs have a **phalangeal formula** of 2, 3, 4, 5, 4. (To interpret this formula, count the phalanges in your own hand to confirm that your phalangeal formula is reduced even further to 2, 3, 3, 3, 3.) The palate of reptilomorphs also has unique flanges on the pterygoid bone that serve as attachment sites for specialized jaw muscles (Chapters 7 and 10).

Extinct reptilomorphs known as †Anthracosauroidea (Fig. 3-20) inhabited aquatic and terrestrial environments in the Carboniferous (the name of the group refers to the discovery of fossils in coal beds, with anthracite being a type of coal). For example, the terrestrial †*Gephyrostegus* was a moderately sized, superficially lizard-like animal (Fig. 3-21*A*). †Seymouriamorpha (Fig. 3-20) is named for Seymour, Tex., the discovery site of †*Seymouria*; it was a terrestrial reptilomorph that reached about 1 m in total length (Fig.

3-21*B*). It had an erect posture, which means that the four legs were tucked under the trunk and not sprawling out to the sides, as was the case for most earlier tetrapods. †Diadectomorpha is a late Carboniferous to early Permian reptilomorph group exemplified by the large (3-m) terrestrial †*Diadectes.* Its broad, flat cheek teeth could have been used for grinding tough plant material, and its incisors project anteriorly like the teeth that mammalian herbivores use to crop plant stems.

The Origin of Amniotes

Amniota (Fig. 3-20) is a reptilomorph subgroup characterized by the presence of an **amniotic membrane** during development (Focus 1-1 and Chapter 4). The resulting **cleidoic egg** enables amniotes to bypass an aquatic larval stage. A cleidoic egg is self-contained (Gr., *kleid* = key), and in a figurative sense it has been the key to fully terrestrial life. Its large store of yolk is suspended from the embryo in a **yolk sac.** Fluid contained within the amniotic membrane forms a liquid cushion around the embryo, and other extraembryonic membranes further protect the embryo and provide for gas exchange and the elimination of nitrogenous wastes (Chapter 4). We speculate that the extraembryonic membranes of the earliest amniotes were surrounded by albumen and by a porous, parchment-like or calcareous shell secreted by the oviduct. Such eggs must be deposited on land or retained within the oviduct until they hatch; they cannot be laid in water, for the embryo depends on gas exchange with the air or the lining of its mother's reproductive tract. Amniotes thus hatch from eggs or are born as miniature adults. Such a shelled cleidoic egg is retained by many living amniotes (e.g., squamates and birds). Most living mammals do not lay a shelled egg but retain the amnion and other extraembryonic membranes.

Amniotes are well adapted to life on land in other ways. For example, the outer layer of the skin consists of dry, keratinized cells that offer protection against dehydration. Such characters cannot be seen in most fossil specimens, but amniotes share several osteological synapomorphies that help us to place fossil taxa, such as attachment of the pelvic girdle to the vertebral column by at least two sacral vertebrae, the presence of a distinctive ankle bone (the **astragalus**), and extra ossifications in the scapulocoracoid of the pectoral girdle. As amniotes exploited terrestrial resources newly available to them, they evolved diverse feeding systems, which are reflected by changes in skull morphology, and new locomotor patterns, which are reflected by changes in stance, posture, and limb and foot structure. Thus, as we trace the phylogeny of amniotes, many of the characters we cite relate to feeding or locomotion. Of the two large clades of Amniota, which are Sauropsida and Synapsida (Fig. 3-20), we turn first to Sauropsida.

Sauropsids and the Origin of Diapsids

Members of **Sauropsida** (Fig. 3-20) have a modified ankle that allows the foot to be rotated underneath the body, which in turn affects their stance and posture. Some bony elements are lost in the jaws of sauropsids. Living sauropsids have a hardened type of keratin in the skin (beta keratin) that is a structural component of scales and feathers. Many extinct sauropsids are known. For example, the small, aquatic sauropsids known as **†mesosaurs** (Figs. 3-20 and 3-21*C*) occur in early Permian freshwater deposits. Their long skulls had many long, fine teeth that look as though they functioned to strain food from water. Similar fossil †mesosaurs occur in Africa and South America. Because all †mesosaurs are restricted to freshwater deposits, biogeographers reasoned in the 1960s that such animals could not have crossed an open expanse of ocean and therefore that the South Atlantic did not exist during the Permian. This provided important—and early—support for the notion that continental drift is responsible for current continental configurations (see also analysis of †vidalamiine fishes in Fig. 1-12).

The phylogenetic relationships of turtles (**Testudines**) are problematic. An extinct group, the †Pareiasauria (Fig. 3-20), has been proposed by some workers as the sister group of turtles, although we do not accept this view. There are other theories about the higher relationships of turtles, including the hypothesis that turtles are closely related to †captorhinds, lepidosaurs, or crocodilians. Whatever their phylogenetic affinities, turtles form one of the most distinctive groups of craniates. They lack teeth, and a keratinized beak covers the jaws.[11] A turtle's short trunk is encased in a bony shell covered by keratinized plates or thick skin with embedded bony nodules called osteoderms. The **carapace** covers the trunk dorsally; its bones are partly modified vertebrae and greatly expanded ribs. No movement is possible between the adjacent trunk vertebrae that contribute to the shell. The bony component of the ventral **plastron** derives from gastralia (i.e., abdominal ribs of

[11]The beak of turtles is not homologous to the beak in birds.

dermal origin; Chapter 8) and portions of the pectoral girdle. As a result, the pectoral and pelvic girdles are enclosed within the rib cage, a condition unique among all craniates. In contrast to the immobile trunk vertebrae, the neck can be extremely mobile and even retracted for protection under the margins of the shell. Some groups, such as the North American box turtles, have hinged plastrons to provide additional protection for the retracted head and limbs. Most extant turtles live on land or in fresh water, but one lineage is marine. Today, the side-neck and snake-neck turtles (**Pleurodira**) occur only in the southern hemisphere, although their fossil record is geographically larger. Their common name refers to the lateral folding of the neck during retraction (Gr., *pleuro* = side + *deire* = neck). An example of a pleurodire is the aquatic matamata (*Chelus*) from South America, which is a specialized suction-feeder that engulfs prey, such as fishes, when they come into range; it relies on cryptic skin flaps and algal overgrowth of the shell to disguise it against the bottom of shallow bodies of water. Most extant turtles belong to **Cryptodira.** The neck of a cryptodire folds in the midsagittal plane during retraction and becomes "hidden" underneath the lip of the shell (Gr., *cryptos* = hidden + *deire* = neck). An example of a cryptodire is the snapping turtle (*Chelydra;* Fig. 3-21*D*). These large, aggressive, aquatic turtles have a deeply emarginated skull that accommodates their large jaw muscles. The two extant families of sea turtles are exemplified by the green sea turtle, *Chelonia mydas* (Fig. 3-21*E*). The forelimbs of sea turtles are modified into flippers, and they fly[12] through the water by moving the flippers up and down in unison. Female sea turtles must return to beaches to lay their eggs because the membranes that enclose the developing embryos are able to exchange gases only with air; were the eggs to be laid in water, the embryos would drown. The largest living turtle is the leatherback sea turtle, *Dermochelys,* which reaches a carapace length of 2 m or more. Its shell is composed of small dermal ossifications not fused to the vertebrae, as is more typical for turtles.

†**Captorhinidae** (Fig. 3-20) includes small, superficially lizard-like forms that retained a solid, bony roof over the jaw muscles in the temporal region of the skull (Chapter 7). The condition of a solid skull roof is described as **anapsid** (Gr., *a* = without + *apsid* = loop or bar; i.e., no arches of bone are present between temporal openings). Note that such a solid skull roof is a plesiomorphic character and does not diagnose any monophyletic group of sauropsids.

The Origin and Early Diversification of Diapsids

In the later part of the Carboniferous, sauropsids with two openings in the temporal region of the skull appeared. These are **diapsids** (Gr., *di* = two + *apsid* = loop or bar; Fig. 3-20), and their emergence and diversification resulted in dramatic changes in tetrapod faunas. In addition to their modified skulls, early diapsids had long, slender limbs and a ridge-and-groove ankle joint that presumably allowed them to run rapidly. †*Petrolacosaurus* is regarded as the earliest well-preserved diapsid. It was a relatively small animal with long limbs, and it generally is reconstructed with an erect posture (Fig. 3-21*F*).

Members of **Sauria** (Fig. 3-20) have a slender, rodlike columella that connects an eardrum (= tympanum) to the oval window in the otic capsule and serves to improve detection of airborne sounds.[13] This mechanism for carrying vibrations from an eardrum to the inner ear is convergently similar to one found in temnospondyls (discussed on page 79) and one found in synapsids (discussed on page 101). Thus, at least three separate origins of the eardrum seem to have happened in neotetrapods, and such a striking example of convergent evolution should serve as a reminder that even very complex and functionally similar morphological systems can evolve independently. Patterns of relationships among the four clades of saurians shown in Figure 3-20 (i.e., †ichthyosaurs, †plesiosaurs, lepidosaurs, and archosauromorphs) are unresolved. We briefly discuss two extinct groups within Sauria, †Ichthyosauria and †Plesiosauria, before turning to Lepidosauria and Archosauromorpha.

Mesozoic seas were home to remarkable saurians belonging to †**Ichthyosauria** (Gr., *ichthyos* = fish + *sauros* = lizard; Fig. 3-20). †Ichthyosaurs lack any articulation between the pelvic girdle and the vertebral column; instead, the pelvic girdle was suspended in muscles of the body wall, in much the same way as the pelvic girdle of fishes. Also diagnostic for †Ichthyosauria are the last few vertebrae in the tail, which were bent sharply downward to support the ventral lobe of the caudal fin. †Ichthyosaurs demonstrate some remarkable convergences with living aquatic

[12]Sea turtles can use their flippers for either lift-based or drag-based propulsion.

[13]Turtles also have a slender, rodlike columella; we interpret that it evolved independently from the slender columella of saurians.

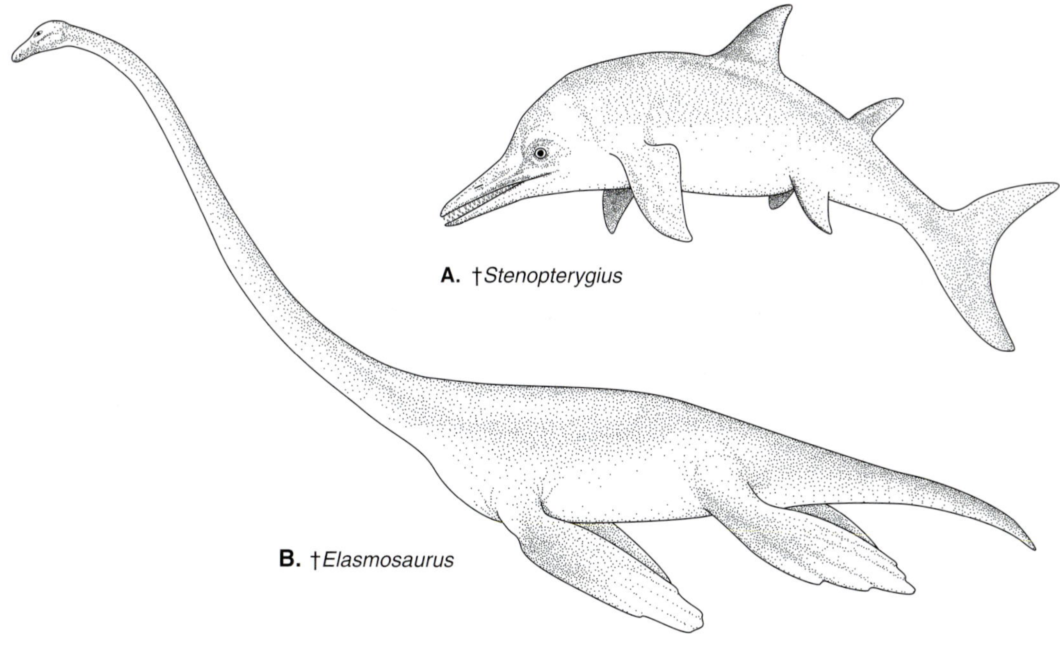

FIGURE 3-22
†Ichthyosaur and †plesiosaur. *A,* Reconstruction of an advanced †ichthyosaur, †*Stenopterygius. B,* Reconstruction of an advanced, long-necked †plesiosaur, †*Elasmosaurus.*

gnathostomes, and these analogous features allow us to build an image of the life of an animal such as †*Stenopterygius* (Fig. 3-22*A*). The tall tail was flat from side to side, like that of a pelagic shark or tuna. The paired limbs were paddle-shaped or fin-shaped, like those of sea turtles or cetaceans (dolphins and whales). Like dolphins and toothed whales, the elongate jaws of †*Stenopterygius* bore rows of similarly shaped and similarly sized teeth, a condition known as **homodonty.** The vertebral column in the trunk consisted of simple disklike vertebrae that resemble those of cetaceans.[14] Their large eyes suggest that †ichthyosaurs were diurnal sight hunters. Like cetaceans, †ichthyosaurs gave birth to live young. Thus, we imagine that †*Stenopterygius* and other forms were fast, pelagic carnivores that probably lived like some of the extant species of dolphins and porpoises.

†**Plesiosauria** (Gr., *plesios* = near + *sauros* = lizard; Fig. 3-20) is exemplified by †*Elasmosaurus* from the Upper Cretaceous of western North America (Fig. 3-22*B*). The forelimbs and hindlimbs were equally sized, elongated flippers that may have been used like sea-turtle flippers or penguin wings to generate lift for subaqueous flight. The greatly elongated neck seen in many species of †plesiosaurs supported a surprisingly small head; the jaws and dentition suggest that long-necked †plesiosaurs ate fish. Other †plesiosaurs had much shorter necks and larger skulls.

[14]Not all cetacean vertebrae are as simple as the simplest vertebrae of †ichthyosaurs.

Lepidosaurs

Members of **Lepidosauria** (Fig. 3-20) have a **transverse cloacal slit** rather than a longitudinal one. The name lepidosaur (Gr., *lepidos* = scale + *saurus* = lizard) refers to the imbricated epidermal scales found in members of this group (Chapter 6). Lepidosaurs also have many derived skeletal features, such as intravertebral breakage zones, termed autotomy planes, in the tail. Autotomy planes allow lepidosaurs to break off their tails to distract and escape from potential predators. Paleozoic lepidosaurs were quadrupedal, like their ancestors, and retained a typical diapsid skull, palatal teeth, and a median eye. The tuatara, *Sphenodon,* which today lives only on a few islands off the New Zealand coast, retains many plesiomorphic features and is interpreted as the living sister group of all other lepidosaurs (Fig. 3-23). In external appearance, *Sphenodon* resembles a "lizard" (compare Fig. 3-24*A* and *B*), but *Sphenodon* is the only living lepidosaur to

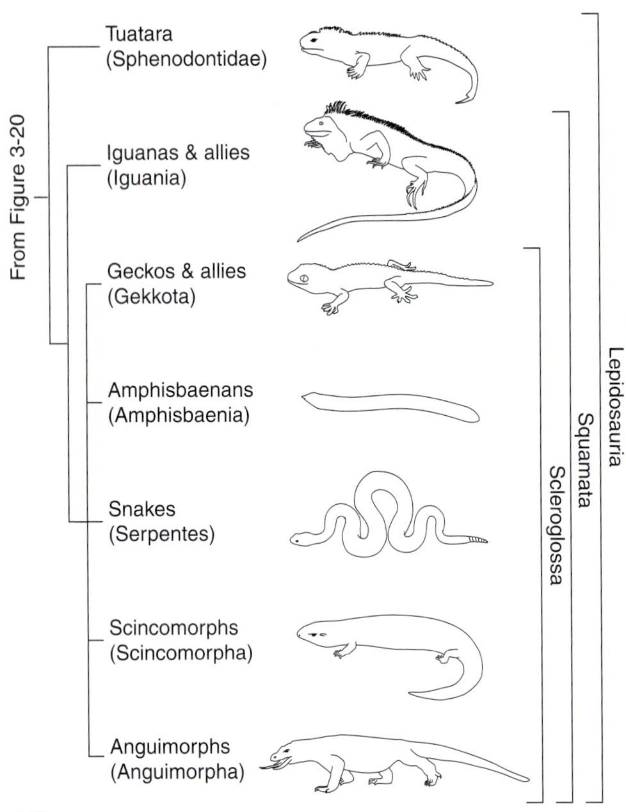

FIGURE 3-23
Phylogeny of living lepidosaurs (Lepidosauria).

retain both upper and lower temporal bars in its skull (Chapter 7).

Squamata (L., *squama* = scale; Fig. 3-23) includes almost 6000 species of "lizards," amphisbaenians, and snakes. Male squamates have unique paired copulatory organs, known as **hemipenes.** During mating, one is everted from the wall of the cloaca and inserted into the cloaca of a female, into which it transfers sperm. The bar of bone bordering the lower temporal fenestra is absent in squamates. This loss not only lightens the skull but also increases the potential for intracranial kinesis, which is well developed in many squamates (Chapter 16). Several derived groups of squamates have independently elongated the trunk and reduced or lost the legs, features that appear to have evolved at least initially as adaptations for burrowing.

Iguania (Fig. 3-23) is a group of squamates exemplified by the Central American green iguana (*Iguana iguana,* Fig. 3-24*B*). An iguana uses its fleshy, pad-like tongue to both bring food into the mouth and manipulate it during chewing. Chamaeleons also belong to Iguania; they have complex skin pigmentation systems, an elongate projectile tongue, highly specialized eyes, and climbing feet in which the toes are able to grip tightly on small branches (a condition known as **zygodactyly;** Gr., *zygon* = yoke + *dactyl* = finger). **Scleroglossa** (Fig. 3-23) is a group of squamates in which the anterior portion of the tongue (known as the foretongue) is flattened and the surface of the posterior portion of the tongue (the hindtongue) is heavily keratinized (Gr., *sclera* = horny + *glossa* = tongue). Scleroglossans use their modified tongues in many new ways, from cleaning the eyes to detecting prey. Generally, however, the food acquisition role of the tongue is reduced relative to other groups of squamates. Scleroglossa includes geckos, skinks, amphisbaenians, snakes, varanids, and allies. **Gekkota** (Fig. 3-23) is exemplified by members of the family Gekkonidae, such as the tokay gecko, *Gekko.* Most gekkonids are nocturnal, and some species make extensive use of vocalization. The toe tips of gekkonids have brushlike **digital setae** that enable them to climb vertical surfaces and even to hang suspended from overhanging structures. Loss of limbs and trunk elongation evolved independently several times in Scleroglossa (Fig. 3-23). Examples include the rarely seen, mostly tropical, burrowing amphisbaenids (**Amphisbaenia;** Fig. 3-23), certain skinks and anguimorphs (the glass lizards), and most notably the snakes (**Serpentes**). Loss of limbs in snakes also entails the complete loss of limb girdles and limb skeletal elements (only boas and pythons retain traces of the hindlimbs). Body elongation also may entail reduction of some internal organs: most snakes have only a single lung. Snakes have further simplified the diapsid skull by the loss of the upper post-temporal bar, which increases its potential for intracranial movement (Chapters 7 and 16). As a result, some snakes can subdue and swallow prey larger than the diameter of their heads. The foretongues of snakes are deeply notched, or forked, and used almost exclusively as sense organs for odor detection. Snakes include the heavy-bodied boas and pythons, the abundant, mostly nonvenomous colubrids (Colubridae, an assemblage of more than 1500 species that is probably not monophyletic), and several families of venomous species: cobras, coral snakes, and sea snakes (Elapidae); vipers (Viperidae); and pit vipers (Crotalidae; Fig. 3-24*C*). **Scincomorpha** (Fig. 3-23) includes the mostly secretive skinks (Scincidae), found in temperate and tropical environments ranging from forests to deserts. Skinks have glossy skin, and many have osteoderms (bony plates in the skin). Most skinks have limbs, but elongated trunks and reduced or even lost limbs evolved in several lineages of skinks. **Anguimorpha** (Fig. 3-23) includes glass and alligator lizards, the Gila monster and beaded lizard, and monitors. Monitors (Varanidae) are active, alert animals of arid and tropical regions. They walk with a more upright stance than do most squamates and are efficient killers of mammalian

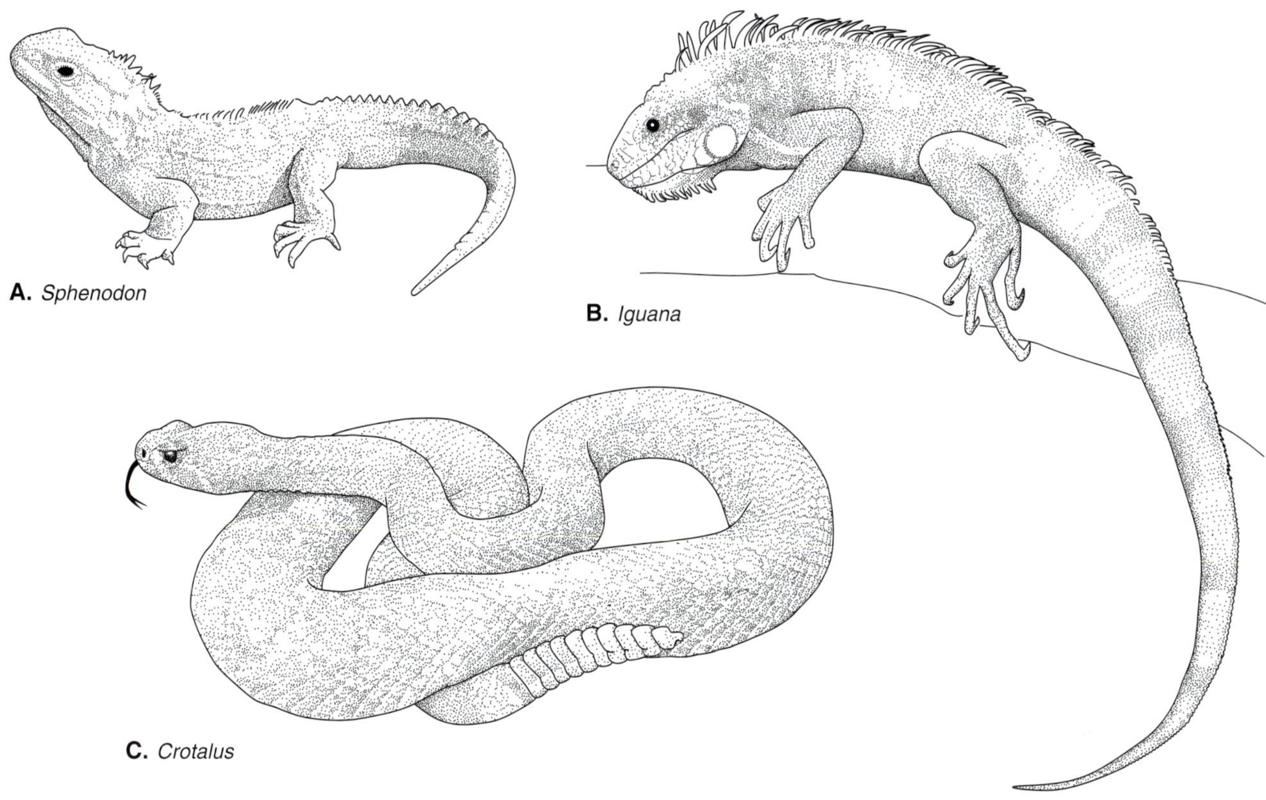

FIGURE 3-24
Representative lepidosaurs. *A,* The tuatara, *Sphenodon,* of New Zealand is the only living lepidosaur to retain both bony arcades found in the skulls of early fossil diapsids. *B,* A representative iguanian, *Iguana. C,* A rattlesnake, *Crotalus,* a venomous snake with infrared heat–sensory pits and elongate, erectile fangs.

prey. The Komodo dragon, *Varanus komodoensis,* is the largest quadrupedal squamate.

Archosauromorphs

Archosauromorphs (Figs. 3-20 and 3-25) rear up on their hind legs, at least when moving rapidly, and many are fully bipedal. Among their adaptations for bipedalism are elongations of the ankle and other parts of the hind leg, changes in the ankle that allow the foot to be directed forward, a reduction in the number of toes, and a heavy tail that acts as a counterweight to the front of the body when the animal stands on its hind legs. Many groups of Archosauromorpha are extinct (e.g., †Rhyncosauria, †Prolacertiformes, †Proterosuchidae, and †Erythrosuchidae; see Fig. 3-25). Members of a derived subgroup, **Archosauria** (Fig. 3-25), became the dominant amniotes during the Mesozoic. Archosauria includes the crocodilians, †pterosaurs, †ornithischians (= †bird-hipped dinosaurs), and saurischians (= †lizard-hipped dinosaurs and birds). **Crocodylotarsi** (Fig. 3-25) includes two extinct lineages (†Phytosauridae and †Pseudosuchia) as well as the living crocodilians and their extinct relatives. Crocodilians (Fig. 3-26*A*) are large to very large semiaquatic predators of tropical and subtropical zones. Their bony secondary palate allows them to breathe and to eat at the same time. Placement of the nostrils at the tip of the snout allows a crocodilian to breathe while it keeps most of its head and body submerged under water, which facilitates a stealthy approach to prey. Crocodilians may appear sedentary, but they can move very rapidly when necessary and even gallop using their powerful hind legs. The New World alligators (*Alligator*) and caimans are broad-snouted crocodilians that feed largely on turtles and other aquatic vertebrates.

The earliest known †**pterosaurs** (Fig. 3-25) are from the Triassic. We noted in Chapter 1 some of the many convergently evolved features of †pterosaurs, birds, and bats, such as a lightly built skeleton and large airfoil surfaces. In each group of vertebrates with powered flight, the forelimbs are modified (in different ways; see Fig. 1-9 and Chapter 11) to form wings that can be moved up and down to generate the necessary lift and power.

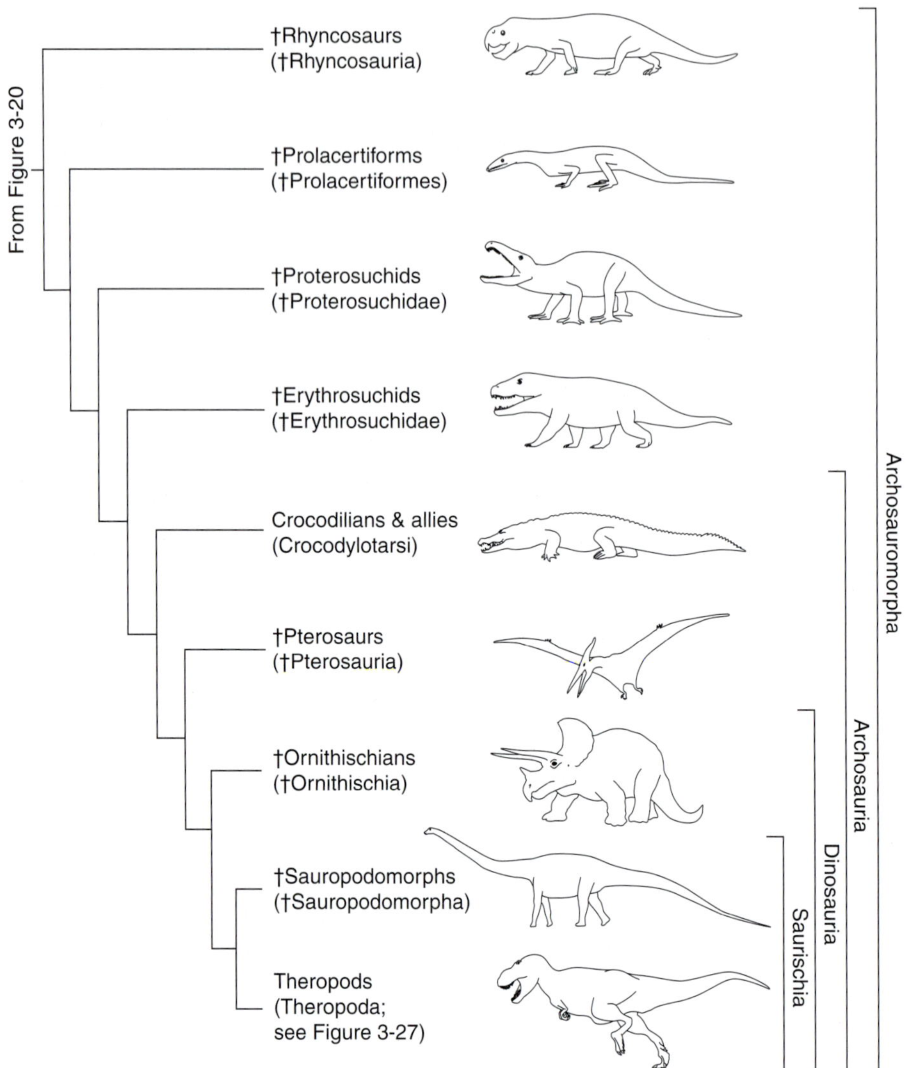

FIGURE 3-25
Phylogeny of living and fossil archosauromorphs, highlighting the origin of crocodiles and allies (Crocodylotarsi) and saurischians (Saurischia).

Wings of †pterosaurs were supported by the fourth finger, each phalanx of which was elongated. The wing membrane consisted of skin, but other aspects of wing structure remain controversial. Some workers believe that †pterosaurs had an insulating layer analogous to hair covering all or part of the body. Most of the 100 species of †pterosaurs known from the Mesozoic were small forms that have been found in shallow marine lagoonal deposits (e.g., †*Rhamphorynchus;* Fig. 3-26*B*), which suggests that most species of †pterosaurs inhabited coastal regions and fed on fishes or other marine prey. Some large late Cretaceous †pterosaurs existed, such as †*Pteranodon* (7-m wingspan), but †pterosaur diversity was already decreasing in the Cretaceous, and none are known to have survived into the Tertiary.

Dinosauria (Fig. 3-25) consists of †Ornithischia and Saurischia. Many early dinosaurs were the size of chickens, but the group is best known for the large size of later species. Synapomorphies of Dinosauria include the incorporation of additional vertebrae into the sacroiliac region, a feature that is linked to greater reliance on hindlimb locomotion. Some dinosaurs remained quadrupedal or reverted to a quadrupedal gait, but their enlarged hind legs and other features indicate bipedalism in their ancestry. Of the large and diverse clade Dinosauria, only the birds survive today. **†Ornithischia** (Fig. 3-25) is one of the largest completely extinct clades of vertebrates. Many synapomorphies of †Ornithischia are known, but the best known relate to a posteriorly directed pubis, which gives the pelvis an appearance superficially similar to the pelvis of extant birds. This is also the source of the group's name (Gr., *ornitho* = bird + *ischia* = hip). Most †ornithischians were herbivorous, and many evolved specialized jaws and dentition for efficiently grinding plant food, such as the †ceratopsians (e.g., †*Triceratops;* Fig. 3-26*C*). †Duck-billed dinosaurs (†Hadrosauridae) are thought to have lived on the

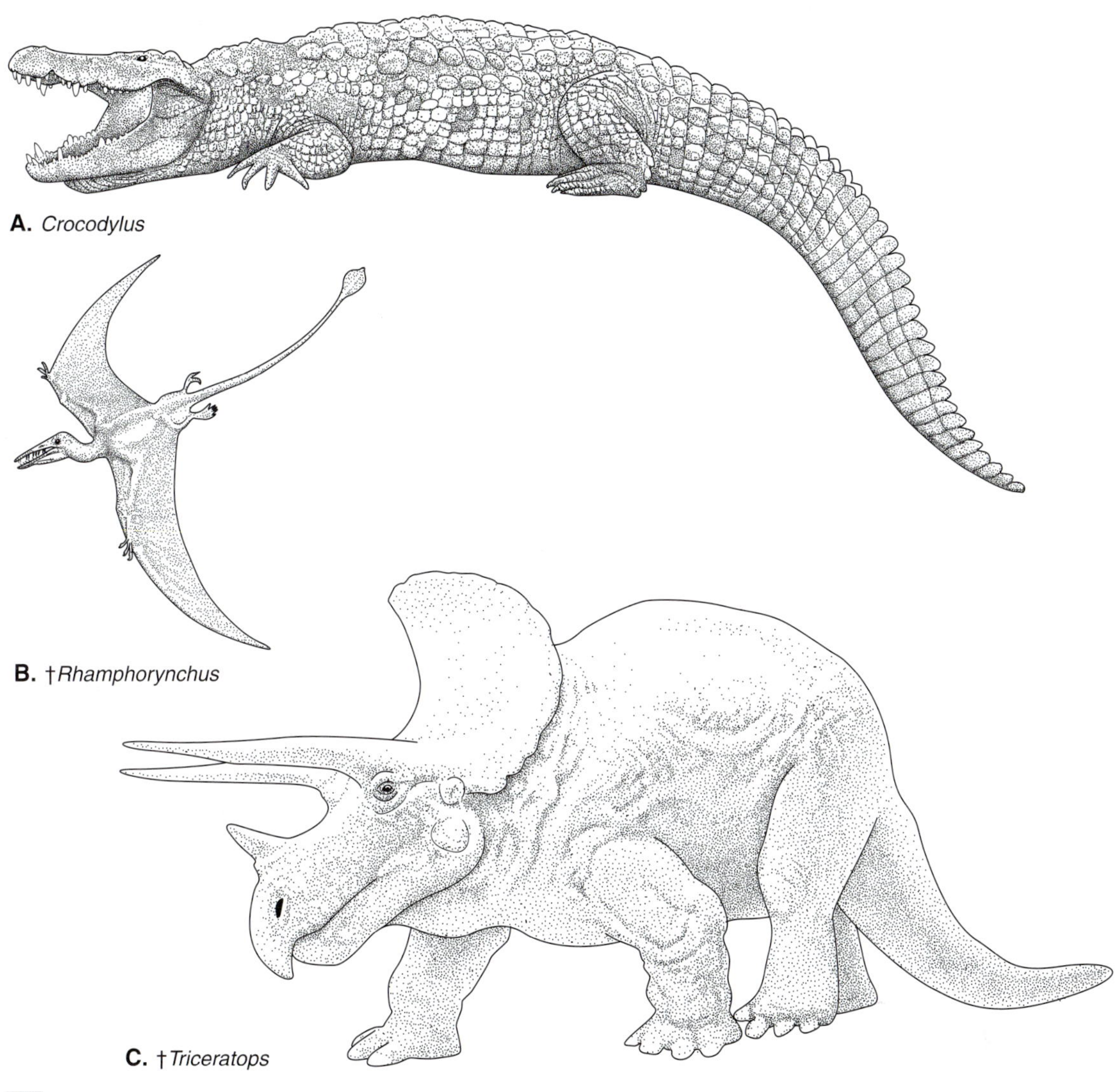

FIGURE 3-26
Representative archosaurs. *A,* A crocodile, *Crocodylus. B,* Reconstruction of a †pterosaur, †*Rhamphorynchus. C,* Reconstruction of an †ornithischian dinosaur, †*Triceratops.*

borders of swamps and to have consumed aquatic plants. Many †hadrosaur nests have been found, and it has been proposed that they may have given extensive parental care to their young.

Saurischians

Saurischia is the second clade of Dinosauria (Fig. 3-25). In contrast to †ornithischians, early saurischians were carnivorous, and carnivory remained a theme throughout their evolution. Their enlarged jaw muscles extend onto bones of the skull roof (a synapomorphy for the group). Saurischia includes two clades: the extinct †Sauropodomorpha, which includes the largest tetrapods that ever lived, and Theropoda, which includes the birds. †**Sauropodomorpa** (Fig. 3-25) includes giant herbivorous dinosaurs, such as †*Brachiosaurus,* which reached sizes of 30 m and 80,000 kg. †Sauropodomorphs had small heads and often very long necks (up to 19 cervical vertebrae), but most of their obvious anatomical specializations relate to mechanisms for supporting their great mass. For example, the legs of †sauropodomorphs formed stout vertical pillars directly under the body, which is known as a **graviportal posture.** Formerly, paleontologists thought that †sauropodomorphs needed to live in swamps in order to gain additional support for their bodies from the water, but now †sauropodomorphs often are reconstructed as rearing up on their hind legs. Because of their great mass, †sauropodomorphs must have achieved thermal inertia in body temperature that allowed them to be active in both day and night.

Theropods and the Origin and Diversification of Birds

Theropoda (Figs. 3-25 and 3-27) includes small, fleet-footed dinosaurs; giant, carnivorous forms; some remarkably specialized predators; and birds. Theropod bones are thin-walled and hollow, with an internal strutwork that gives them great strength (Chapter 5). We regard the late Triassic to late Jurassic †**ceratosaurs** as the sister group of all other theropods (Fig. 3-27), and †ceratosaurs often are reconstructed as colorful, alert, and agile predators. Giant †**carnosaurs,** such as †*Tyrannosaurus,* are also theropods (Fig. 3-27). †*Tyrannosaurus* was an obligate biped with a massive skull, dagger-like teeth deeply set into sockets, and tiny forelimbs that may have helped it to rise from a resting position on the ground (Fig. 3-28*A*). Its large tail is generally reconstructed as a counterbalance for the rest of the body. The more lightly built †**Ornithomimisaurs** and †**Troodonids** (Fig. 3-27) probably had a long stride and agile forelimbs (e.g.,

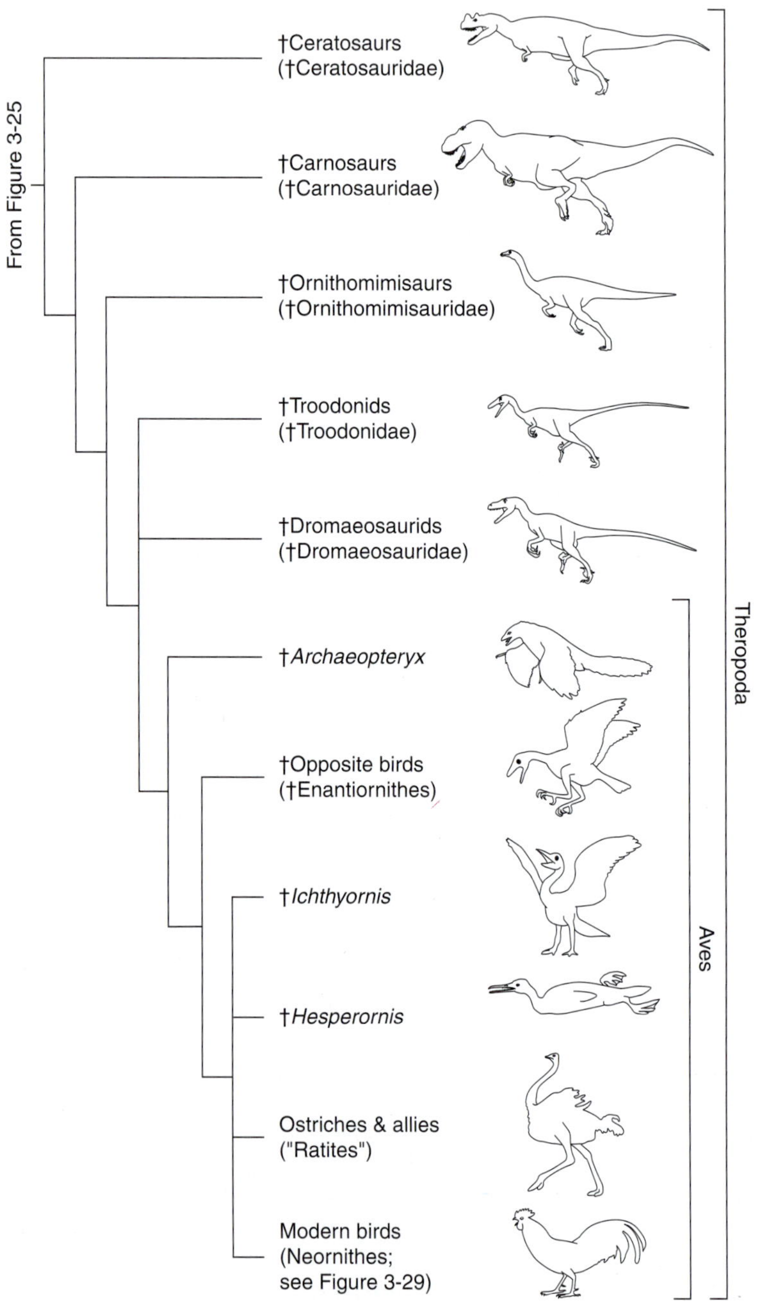

FIGURE 3-27
Phylogeny of living and fossil theropods, highlighting the origin of birds (Aves) and neognathous birds (Neornithes).

FIGURE 3-28
Representative theropods. *A,* The giant †carnosaur, †*Tyrannosaurus,* had tiny forelimbs. *B,* †*Struthiomimus,* an †ornithomimisaur with long legs specialized for running; the living ostriches (*Struthio*) convergently resemble it. *C,* A Jurassic bird, †*Archaeopteryx,* found in the fine-grained limestone of the Solnhofen Formation in southern Germany. *D,* The living ostrich, *Struthio.*

†*Struthiomimus;* Fig. 3-28*B*). Their large eyes and well-developed ears are suggestive of the improved sensory systems seen today in living birds. †**Dromaeosauridae** (Fig. 3-27) is exemplified by †*Velociraptor,* made famous as the "raptors" of the book and movie *Jurassic Park.* If you have seen that movie, then you have a very graphic idea about the biology of †dromaeosaurids.[15] These moderately sized but powerful carnivorous theropods had elongate teeth and an enlarged killing claw on the second toe of each foot. Because these claws are on the foot and not the hand, they could only have been used while the body was balanced on the other foot or when leaping at a prey item, with both feet off the ground. This suggests great agility and coordination, interpretations that helped to fuel the cursorial theory about the origin of flight (discussed later).

[15]With perhaps appropriate poetic license for a movie, the size of the "raptors" in *Jurassic Park* was exaggerated.

Origin and Diversification of Birds

Aves (L., *aves* = birds; Fig. 3-27) is a derived group of theropods characterized by **feathers** and unique adaptations for endothermy (Focus 3-4) and flight. Feathers and subcutaneous fat provide insulation, and excess heat is lost by panting. Flight requires wings, a lightweight body (or at least low wing-loading, a term that refers to the relationship between the size of the wing and the weight of the body), and a high energy output. The pectoral appendages are modified as wings, and their flying surfaces are composed chiefly of large, strong, and lightweight **primary flight feathers** (Chapter 11). In most birds, other flight feathers fan out from a group of fused caudal vertebrae known as the **pygostyle.** The broad, keeled sternum found in derived birds provides a large area for the origin of flight muscles. Bird bones are exceptionally strong and many contain air sacs that connect with the respiratory system. A bird's hind legs act as shock absorbers when a bird lands, and when not flying, birds are bipeds or swimmers. The pelvis of derived birds is rigid because of the fusion between the sacral and trunk vertebrae and pelvic bones to form a **synsacrum.** Bones of the lower leg and foot also are fused. The digestive, respiratory, and circulatory systems are adapted to sustain a high level of metabolism, and a keratinized bill, or beak, replaces teeth in all extant birds.

Two general models explain the evolutionary origin of avian flight. The **arboreal theory** postulates that the ancestors of birds lived in trees and that they glided within the forest canopy using skin membranes that were the precursors of wings. But in no other case do vertebrate gliders appear to have given rise to forms with powered flight. Anatomical specializations needed for a powerful downstroke appear to be incompatible with the rigid airfoil needed by a glider, and it is difficult to imagine how such a transition could have occurred. The **cursorial theory,** also known as the ground-up theory, is based on the interpretation that extinct theropods closely related to birds were probably agile runners. Wings might have evolved initially in such cursors to serve as balancers when leaping onto prey.

Whatever the origin of powered flight, many well-known fossil birds exist. Seven specimens of the crow-sized †***Archaeopteryx*** (Gr., *arche* = first + *pteryg* = wing; Fig. 3-28*C*) are known from Jurassic marine deposits of the Solnhofen Formation in southern Germany. These are near-shore, lagoonal deposits, and we presume that the specimens of †*Archaeopteryx* fell or were washed short distances into such lagoons because the fossils are remarkably complete. The rock matrix of the Solnhofen Formation is a very fine-grained limestone that faithfully preserves even small anatomical details, such as impressions of feathers. †*Archaeopteryx* retained a long tail, a typical theropod pelvis and hind legs, and teeth. Were it not for the impression of feathers in the fossil, the animal probably would have been regarded as just another small, flightless theropod. The discovery of †*Archaeopteryx* in the 1860s helped to promote Darwinian theories of evolutionary change because it presents such an interesting mixture of plesiomorphic (e.g., dinosaurian) and derived (e.g., avian) features and because it followed the publication of Darwin's *On the Origin of Species* by only a few years. Details of feather structure indicate that †*Archaeopteryx* was capable of flapping, powered flight, but because the keel on the sternum is not well developed, it usually has been assumed that †*Archaeopteryx* could not have been a strong flier. Other Mesozoic birds are being discovered and described at a great rate. For example, a 135-million-year-old, sparrow-sized bird known as †*Sinornis* from the early Cretaceous of China offers a further glimpse into the early evolution of avian flight and perching. †*Sinornis* had a broad, well-ossified sternum with a keel for the attachment of flight muscles, and its posterior vertebrae were fused into a pygostyle. These features suggest that it was capable of sustained flight around its inland lake habitats. †*Sinornis* and most other Cretaceous birds belong to †**Enantiornithes** (Fig. 3-27), commonly known as †opposite birds because the pattern of fusion between tarsal elements in the foot is opposite of that found in living birds. †Opposite birds also have a relatively short tail, partially united fingers in the wing, and more fusion between vertebrae, among the pelvic bones, and in foot bones than is seen in †*Archaeopteryx*. †**Ichthyornis** and †**Hesperornis** (Fig. 3-27) were Cretaceous birds that retained teeth. Some authors have placed them in a group called "†Odontognathae," but evidence for monophyly of this group is unclear. The much larger †*Hesperornis* had reduced wings and was secondarily flightless; it had broad, webbed feet for swimming.

"Ratites"

"Ratites" (Fig. 3-27) is an assemblage of mostly flightless, often large birds, such as ostriches (Fig. 3-28*D*). Its members are known to share only plesiomorphic features, such as a theropod-like bony palate, known as a paleognathous palate. The name ratite refers to their broad, keel-less sternum (L., *ratis* = raft), and all but one group of the living ratites are flightless. If ratites is not a monophyletic group, then this type of sternum evolved convergently in association with multiple independent derivations of flightlessness. The flightless ratites occur in areas where they are not subject to terrestrial predation. They use their large and powerful legs for running and have greatly reduced wings. The broad sternum and wing vestiges attest to their derivation from an ancestor that flew.

Modern Birds

Modern birds, or **Neornithes** (Fig. 3-29), have a **neognathous palate** in which the changes in the palatal bones allow for greater movement during feeding (Chapter 16). Ornithologists recognize 23 monophyletic groups within Neornithes and at least 9000 extant species. Their fossil record is incomplete, but this group, like percomorph fishes and eutherian mam-

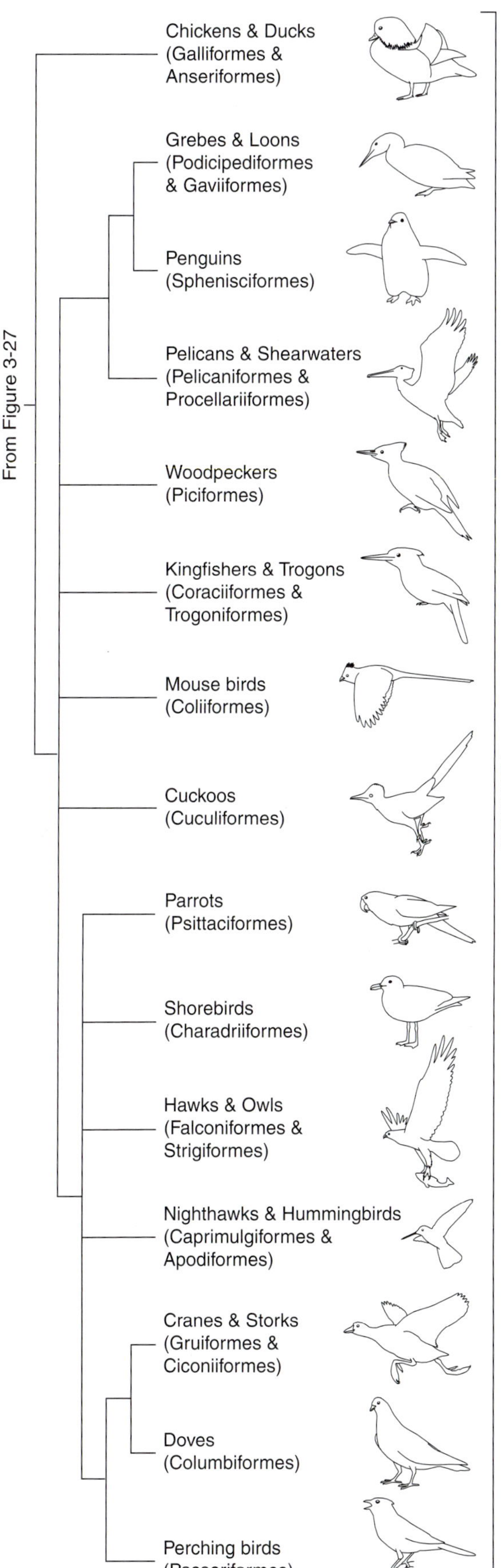

FIGURE 3-29
Phylogeny of living neognathous birds (Neornithes).

mals, radiated extensively in the Cenozoic. Even a casual comparison of ducks, penguins, gulls, eagles, hummingbirds, and finches demonstrates that modern birds evolved myriad specializations of diet, habitat, and methods of locomotion. As a result, the major groups of Neornithes differ in bill structure, wing and tail size and form, and morphology of their legs and feet. In a few cases, such as on small islands where they were not subject to terrestrial predation, some neornithans reduced the size of the wings and readapted to a completely terrestrial mode of life. Long-distance migrations are made by many species of Neornithes. Features related to migration include aspects of wing and muscle structure, reproductive biology, and nervous and sensory systems.

In our phylogeny, a clade composed of the chickens and ducks (**Galliformes** and **Anseriformes**) is the sister group of all other Neornithes (Fig. 3-29). Galliformes, such as the chicken (*Gallus*) and turkeys (Meleagridae), are ground-oriented birds that use their feet and short, strong bills to scratch in the soil. Anseriformes are aquatic vegetarians with broad, flat bills. Grebes and loons (Fig. 3-29) also are water birds, as are penguins (**Sphenisciformes;** Fig. 3-29), which are among the most highly specialized birds. A familiar example is the Adelie penguin *Pygoscelis adeliae* (Fig. 3-30*A*), which lives in Antarctica. Other penguins live in more temperate and even tropical portions of the southern hemisphere, but all living species lack primary and secondary feathers on the wings and cannot fly in air. Their short and bristly feathers are not restricted to the limited feather tracts typical of other birds (Chapter 6). Instead, penguin feathers are broadly distributed over the body surface to form a continuous insulating layer. Penguins use their short, solidly built wings for subaqueous "flight" and are capable of remarkable bursts of speed when pursuing prey. Albatrosses (**Procellariiformes;** Fig. 3-29) exhibit an almost complete contrast to this way of life. The narrow, elongate wings of the wandering albatross (*Diomeda exulans*) can span 3.5 m, and this striking specialization is ideal for dynamic soaring above the ocean's surface (Chapter 11). A woodpecker (**Piciformes;** Fig. 3-29) uses its tongue, supported by an elongate hyoid apparatus, to retrieve insects from holes that they punch into wood using their sharp bills. Their zygodactylous feet, in which two toes point forward and two toes point backward, allow them to cling to vertical tree trunks.[16] Colorful kingfishers,

[16]The zygodactylous feet of woodpeckers are convergently similar to the zygodactylous feet of chameleons.

rollers, hornbills, and hoopoes (**Coraciiformes;** Fig. 3-29) nest in cavities in tree trunks or in the ground. Most coraciiforms have elongate bills, and in some, such as the African hornbills, the bill is very large. Some are specialized carnivores, such as the kingfishers, which dive to catch fishes with their sharp beaks. Trogons (**Trogoniformes;** Fig. 3-29) are cavity nesters from the tropics exemplified by the quetzal from Mexico. The small group of South African mousebirds (**Coliiformes;** Fig. 3-29) is considered by some to be the sister group to the remaining Neornithes on the basis of anatomical studies of the hindlimb. **Cuculiformes** (Fig. 3-29) includes the cuckoos, the anis, and the specialized terrestrial roadrunners.

Parrots (**Psittaciformes;** Fig. 3-29) are widely distributed tropical to subtropical herbivores with large brains and large eyes; many parrots are brilliantly colored. A parrot's bill is sharply hooked and can be used to pry open hard foods as well as to tear open fruits. Parrots change the shape of the tongue using intrinsic vascular erectile tissue, which gives them fine control for manipulating objects and for vocalization. Terns, gulls, sandpipers, auks, murres, and puffins are shorebirds (**Charadriiformes;** Fig. 3-29). These water-oriented carnivores nest on the ground, whether on beaches, sea cliffs, or fields far inland. Subaqueous flight evolved in some shorebirds and was particularly well developed in the now extinct †great auk, which was exterminated by humans in the 1840s. Eagles, hawks, and vultures (**Falconiformes;** Fig. 3-29) are diurnal carnivores. Many species feed on carrion, but some, such as the peregrine falcon (*Falco peregrinus;* Fig. 3-30*B*) are

FIGURE 3-30
Representative neognathous birds. *A,* A penguin, *Pygoscelis. B,* A peregrine falcon, *Falco. C,* The extinct dodo, †*Raphus. D,* A crow, *Corvus.*

specialized for killing other birds on the wing. Owls (**Strigiformes;** Fig. 3-29) are nocturnal predators with specialized feathers that allow for nearly silent flight. They detect their small mammalian prey using keenly developed senses of hearing and sight. Goatsuckers and nighthawks (**Caprimulgiformes;** Fig. 3-29) forage for insects while flying at dusk; the bill is surrounded by a fan of fine, bristle-like feathers that help them trap prey. Legs are reduced in the small swifts and hummingbirds (**Apodiformes;** Fig. 3-29). The mostly tropical hummingbirds exhibit many remarkable anatomical adaptations for hovering flight. They use their long, narrow bills and brushy-tipped tongues to gather nectar from flowers.

Cranes and allies (**Gruiformes;** Fig. 3-29) are marsh birds with long legs for wading. Another group of long-legged wading birds is **Ciconiiformes** (Fig. 3-29), which includes the storks, herons, spoonbills, and flamingos. Flamingos filter-feed on small crustaceans while the head is upside-down under water, a feeding system that entails many anatomical specializations of the bill and tongue. Pigeons and doves (**Columbiformes;** Fig. 3-29) have short bills. This group includes large, flightless, ground-dwelling forms, such as the †dodo (†*Raphus;* Fig. 3-30*C*), formerly found on islands in the Indian Ocean. The †dodo became extinct because of direct predation by humans as well as domestic animals. The nearly 300 species of pigeons and doves are fast-flying tree or cliff nesters. Most species eat grains or seeds; the young are fed on crop milk produced by the crop (a region of the esophagus).

Perching Birds

The feet of perching birds (**Passeriformes;** Fig. 3-29) are specialized for perching on the branches of trees, and most passeriforms also nest in trees. Ornithologists recognize more than 5000 species of passeriforms. Phylogenetic relationships within this group are complicated to assess because of their recent divergence and many convergent specializations. In these respects, problems facing a passeriform systematist parallel those facing an ichthyologist attempting to examine relationships among perciforms. Not surprisingly, there have been many different phylogenetic interpretations for both of these large groups. An example of a basal passeriform group from North America is the tyrant flycatchers (Tyrannidae). Tyrannids have a distinctive hunting style: they sit on a high perch and watch for flying insects to pass by, which they then dart out to catch. The largest diagnosable monophyletic group within Passeriformes is **Oscini,** or songbirds. Oscines have a structurally modified **syrinx,** which allows them a great range of vocalization. Correlated with this improved singing ability are enhanced song centers in the brain for processing and interpreting songs. Most groups of oscines are diurnal, visually oriented animals, and many are strikingly colorful. Within Oscini, larks (Alaudidae) and swallows (Hirundinidae) retain several plesiomorphic anatomical features. More derived songbirds are exemplified by crows and jays (Corvidae; Fig. 3-30*D*); warblers (Parulidae); wrens (Troglodytes); the American robin and other thrushes (Turdidae); starlings (Sturnidae); blackbirds and orioles (Icteridae); and the finches, sparrows, and grosbeaks (Fringillidae). Many songbirds migrate seasonally between tropical and temperate zones, with the unfortunate consequence that they may be exposed to environmentally degraded habitats at both ends of their migration.

Synapsids

We now turn to **Synapsida,** noted in Figure 3-20 and shown in expanded form in Figure 3-31. Early synapsids have a single, laterally placed, **lower temporal fenestra** in the skull roof, and a similar, although more open, lower temporal fenestra is retained in derived synapsids (Gr., *syn* = with + *apse* = arch). Many extinct groups of synapsids are known from the late Paleozoic and Mesozoic. As a general observation, Mesozoic synapsids were less diverse than were Mesozoic sauropsids. But Synapsida includes the mammals, a group that came to dominate terrestrial ecosystems in the Cenozoic.[17] Synapsids have left an excellent fossil record, and the character transitions leading to the origin of mammals have been proposed in detail.

Some of the more familiar early synapsids are the "†sail-back synapsids," with large "sails" on the back, consisting of a web of skin supported by elongated neural spines. (Fig. 3-32*A*). Sail-backed synapsids long were known as "†pelycosaurs," but as traditionally defined, such a group is paraphyletic because sails appear to have evolved at least twice: once in the line leading to the herbivorous **†edaphosaurids** and once in the line leading to the carnivorous **†sphenacodontids** (Fig. 3-31). The neural spines that support the dorsal sail of

[17]The terrestrial dominance of mammals during the Cenozoic relates to several different measures, including (1) biomass of terrestrial animals exceeding 10 kg, (2) diversity of species exceeding 100 kg, and (3) diversity and prevalence of top predators.

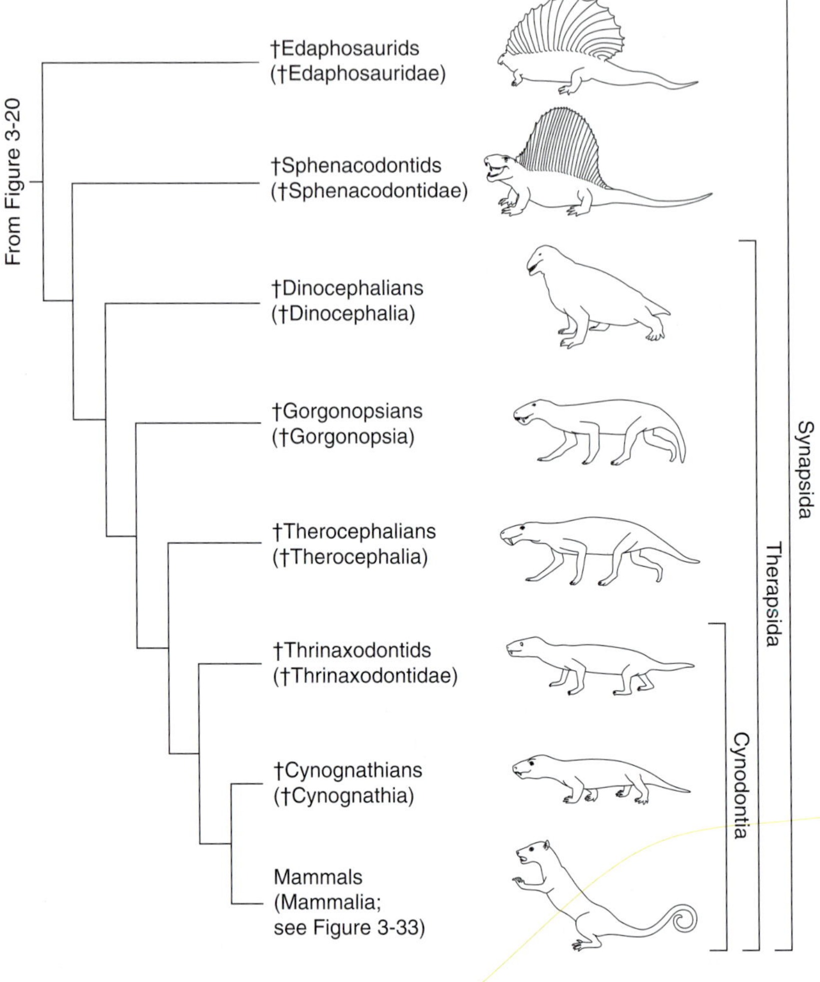

FIGURE 3-31
Phylogeny of living and fossil synapsids, highlighting the origin of mammals (Mammalia).

†edaphosaurids bear lateral projecting spines, whereas those of †sphenacodontids are smooth. †Sphenacodontidae includes such Carboniferous and Permian carnivores as †*Dimetrodon* (Fig. 3-32*A*), which had large, sharp teeth, including a **canine tooth** in the maxillary bone.

Therapsids

The lower temporal fenestra and the canine teeth are enlarged in members of **Therapsida** (Fig. 3-31). Therapsid limbs are long and slender and are located directly beneath the body. These and other features suggest that early therapsids were active and agile animals. Early therapsids also exhibit many osteological features that foreshadow the mammalian condition. Extinct groups of therapsids shown in Figure 3-31 are the **†Dinocephalia, †Gorgonopsia,** and **†Therocephalia**. Most †dinocephalians were large herbivores, whereas †gorgonopsians were small, terrestrial carnivores with well-developed and regionally differentiated dentition, including incisor, canine, and simple cheek teeth. We will trace the further evolution of these different types of teeth when we come to Mammalia (below). Most †therocephalians were large predators, but this group also includes small insectivorous and herbivorous species.

Cynodonts

Members of **Cynodontia** (Fig. 3-31) share several synapomorphies related to the feeding system. The cheek muscle (= masseter) of cynodonts facilitates grinding of food between the postcanine teeth. The masseter is hypothesized to have evolved from a portion of the temporalis muscle, and its evolution is coupled to the presence of a larger lower temporal fenestra. As the fenestra enlarged, the temporal muscle increased in size, and the portion that became the masseter muscle became associated with the outer surface of the zygomatic arch (see Chapter 7 for more on the evolutionary history of the skull and the temporalis and masseter muscles). The postcanine teeth of cynodonts have **cusps,** which are rounded or sharp points on a tooth. Because a cusp has a small surface

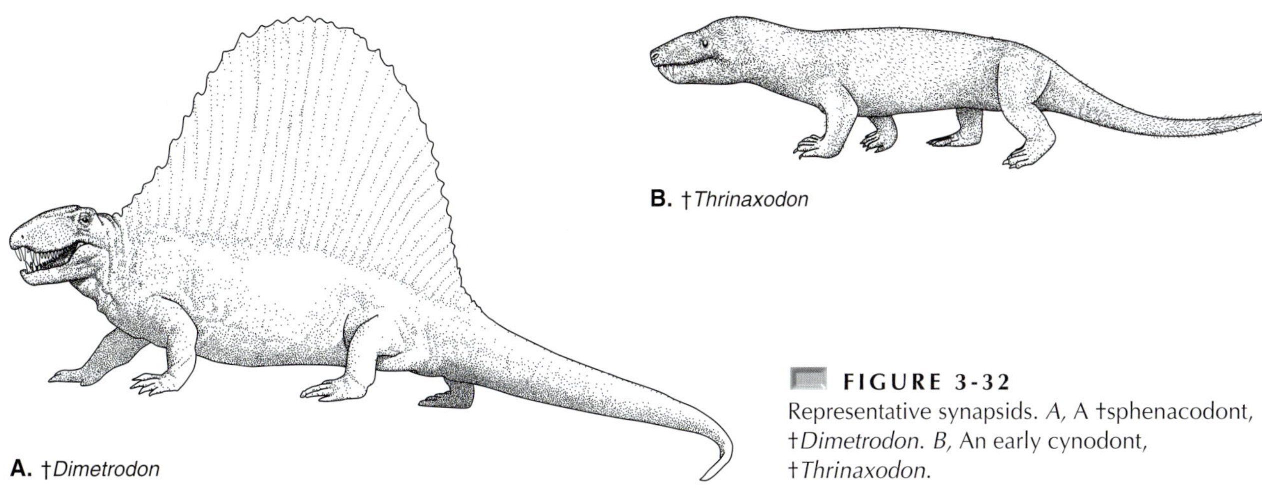

FIGURE 3-32
Representative synapsids. *A,* A †sphenacodont, †*Dimetrodon. B,* An early cynodont, †*Thrinaxodon.*

area, it can concentrate the bite force applied to food as it is chewed, which helps to break down food. Changes in tooth cusps and cusp patterns are a major theme in mammalian evolution. Cynodonts also have lost the ribs from the lumbar region of the vertebral column. This feature reflects the increasing regionalization and specialization of the vertebral column for locomotion. We think of early cynodonts as catlike, and leaping and pouncing on prey are facilitated by such a flexible vertebral column. Some workers postulate that early cynodonts were endothermic, based on skeletal features such as a well-developed secondary palate. Their reasoning is that the secondary palate allows cynodonts to separate the food from the respiratory passages in the mouth and much of the pharynx. Being able to chew food and breathe at the same time would have allowed early cynodonts to more efficiently consume the calories needed to support endothermy (Focus 3-5). Early cynodonts include †*Thrinaxodon,* which was a small, insectivorous animal from the Triassic that is known from very well-preserved specimens (Fig. 3-32*B*). A Triassic group known as †**Cynognathia** (Fig. 3-31) includes carnivorous forms with a well-developed dentition. †Cynognathians share with mammals an enlarged dentary bone in the lower jaw and a corresponding reduction of the other bones of the lower jaw. These features help us understand evolution of the mammalian jaw joint and middle ear bones (Chapters 7, 12, and 22).

Mammals

Mammals (Fig. 3-31) have two tiny middle ear bones, or ossicles, known as the **malleus** and **incus.** These two ossicles evolved from the articular (= malleus) and quadrate (= incus). These bones ossify in the caudal end of the mandibular cartilage (= articular) and palatoquadrate cartilage (= quadrate), which form the jaw joint in all other gnathostomes. The columella continues in mammals but it forms a small, stirrup-shaped bone called the **stapes** (L., stapes = stirrup). Together with the stapes, the malleus and incus form a three-element chain connecting the tympanum to the inner ear. The remarkable evolutionary conversion of these former jaw-joint bones to their new function as part of the vibration-conducting chain of middle ear bones is discussed in Chapters 7, 12, and 22. A correlate of the conversion of the articular and quadrate bones to the malleus and incus is that all adult mammals have a jaw joint that lies between the dentary of the lower jaw and the squamosal bone of the skull roof. Another correlate of this new connection between the tympanum and middle ear is an elongate **cochlea,** which is the sound-detecting part of the inner ear (Chapter 12). In living mammals, the cochlea is coiled into a compact, snail shape (L., *cochlea* = snail shell). Its extra length increases the potential for frequency discrimination and gives mammals a very keen sense of hearing.

Mammals are active, agile, endothermic animals with large brains and complex social, feeding, and reproductive behaviors. Most vertebrate systematists consider that mammals evolved endothermy independently from birds, for the synapsid line of evolution diverged from other amniotes many branches before the origin of archosauromorphs, which led eventually to theropods and to birds (Figs. 3-20, 3-25, and 3-27; see also Focus 3-4 and 3-5). A mammalian synapomorphy, **hair,** supplements subcutaneous fat in retaining heat produced internally by metabolism. Heat can be lost by sweating in many species as well as by panting. As in birds, mammalian digestive, respiratory, circulatory, and excretory systems are adapted to sustain

FOCUS 3-5 *Early Mammals and the Origin of Mammalian Endothermy*

Many, if not most, characteristics of mammals are linked to the evolution of endothermy. We cannot be certain when endothermy evolved. Some believe it evolved gradually among derived synapsids, as is suggested by the structure of their skeletons (text), but activity and a high metabolism are not necessarily coupled. Crompton et al. (1978) proposed an alternative hypothesis based on our knowledge of the metabolism of living mammals (e.g., insectivores) and early fossil mammals (e.g., †*Morganucodon*).

The earliest mammals were small creatures that probably fed on insects, worms, and other invertebrates. Their eyes were small, but their auditory and olfactory organs were extraordinarily well developed. This sensory apparatus suggests that they were active at night and that they located their prey by smell and hearing. This notion is also consistent with ecological theory. Instead of competing directly with other amniotes, early mammals probably occupied different ecological niches, and a nocturnal, insectivorous niche would not have been available to most ectothermic amniotes. But to be active at night, early mammals must have had some way of maintaining body temperature when the sun set. Probably they had an insulating layer of fat and fur. Early mammals may have lived like tenrecs found today in Madagascar. Tenrecs avoid the heat of the day by burrowing into the ground but emerge at night to forage on insects, worms, and grubs found by smell and hearing. Tenrecs are endothermic but maintain body temperatures of about 28°C to 30°C, just a few degrees above the ambient nocturnal temperature. This temperature is significantly lower than the body temperature of most mammals. Insulating layers of fat and fur enable tenrecs to maintain this body temperature at a relatively low energy cost. They have a relatively low level of metabolism; that is, they consume no more energy than an ectothermic amniote of the same size, body temperature, and level of activity. Setting a body temperature slightly higher than the prevalent ambient temperature reduces the need to cool off by evaporative loss of precious body water, which is particularly critical for a small animal, for when ambient temperatures exceed body temperature, its large surface-to-volume ratio would result in the rapid uptake of heat and the need to lose a great deal of water. Crompton et al. believe that tenrecs and the related hedgehogs reflect these adaptations of early mammals.

Many groups of amniotes became extinct at the end of the Cretaceous, and mammals began to exploit the previously monopolized diurnal resources. For a small endotherm to become active in the daytime, body temperature should be about 10°C higher than that of a nocturnal species, which also would reduce the need for evaporative cooling during the daytime. Setting the body temperature higher than ambient daytime temperatures—and maintaining that setting at night—requires a much greater energy expenditure. Crompton et al. propose that endothermy and a high metabolic rate initially evolved as adaptations for a nocturnal niche and only later increased as mammals shifted to diurnal niches. Some investigators question this reasoning by pointing out that many nocturnal mammals, including monotremes (e.g., echidnas) and metatherians (e.g., opossums), have body temperatures and metabolic rates similar to those of diurnal mammals. Crompton et al. counter this by suggesting that echidnas and opossums are only secondarily nocturnal.

a high level of metabolism, but many differences exist in the ways this is accomplished. Scrolls of bone known as **turbinate bones** increase the internal surface area of the nasal cavities. This extra surface area enables mammals to have a keen sense of smell and to help clean, moisten, and warm inspired air. Most mammals have flaps of skin that form an external ear, or **pinna.**

Mammalian teeth are specialized for many different aspects of food capture and processing. Because mammals extensively chew their food, precise **occlusion** between the upper and lower teeth is necessary. Tooth replacement is limited to a deciduous (or milk) dentition and an adult (or permanent) dentition. Mammals have four categories of teeth. In series from the front, these are the **incisors, canines, premolars,** and **molars.** Incisors are typically chisel-shaped and used for nipping or gnawing. Canine teeth have a single, round, stabbing cusp used to impale prey. Premolar and molar teeth are sometimes spoken of together as the **cheek teeth,** and all mammalian cheek teeth have at least two roots. Like the incisor and canine teeth, premolar teeth erupt as part of the milk dentition and are later replaced by the permanent set. Molar teeth, however, erupt later in development and are never replaced. Many mammals have secondarily lost or modified some teeth, and the number of teeth in different categories is an important phylogenetic tool for mammalogists.

An embryo of an endothermic vertebrate must develop in a warm environment while its own thermoregulatory mechanism is developing. Thus, the egg-laying monotremes brood their eggs, as do birds. The embryos of other mammals undergo at least part of their development within a uterus. Newly hatched or born mammals are fed on milk secreted by the **mammary glands.** The term mammal derives from this feature (L., *mamma* = breast).

Early Mammals

Many concepts about the biology of early mammals stem from studies of **†morganucodontians** (Fig. 3-33). †*Morganucodon* was a small mammal, only 10 cm to 15 cm in total length (Fig. 3-34*A*). Small size is the main evidence suggesting that early mammals were shrew-like. On the basis of its size and dentition, we think that †*Morganucodon* was insectivorous. Its claws would have allowed it to climb or scramble over the substrate. It is generally thought that early mammals were nocturnal and relied primarily on the sense of smell, although such interpretations are extrapolations based on what we know about features of living mammals rather than evidence directly from fossils. The premolar and molar teeth of †*Morganucodon* demonstrate the plesiomorphic pattern of occlusion found among mammals. In this pattern, occlusion occurs between the major cusps of the upper and lower cheek teeth. Another extinct group of early mammals is **†Multituberculata** (Fig. 3-33). †Multituberculates are known from late Mesozoic and early Tertiary deposits, and are the most abundant late Mesozoic mammals. These omnivorous, rodent-like forms are diagnosed by their unique cheek teeth.

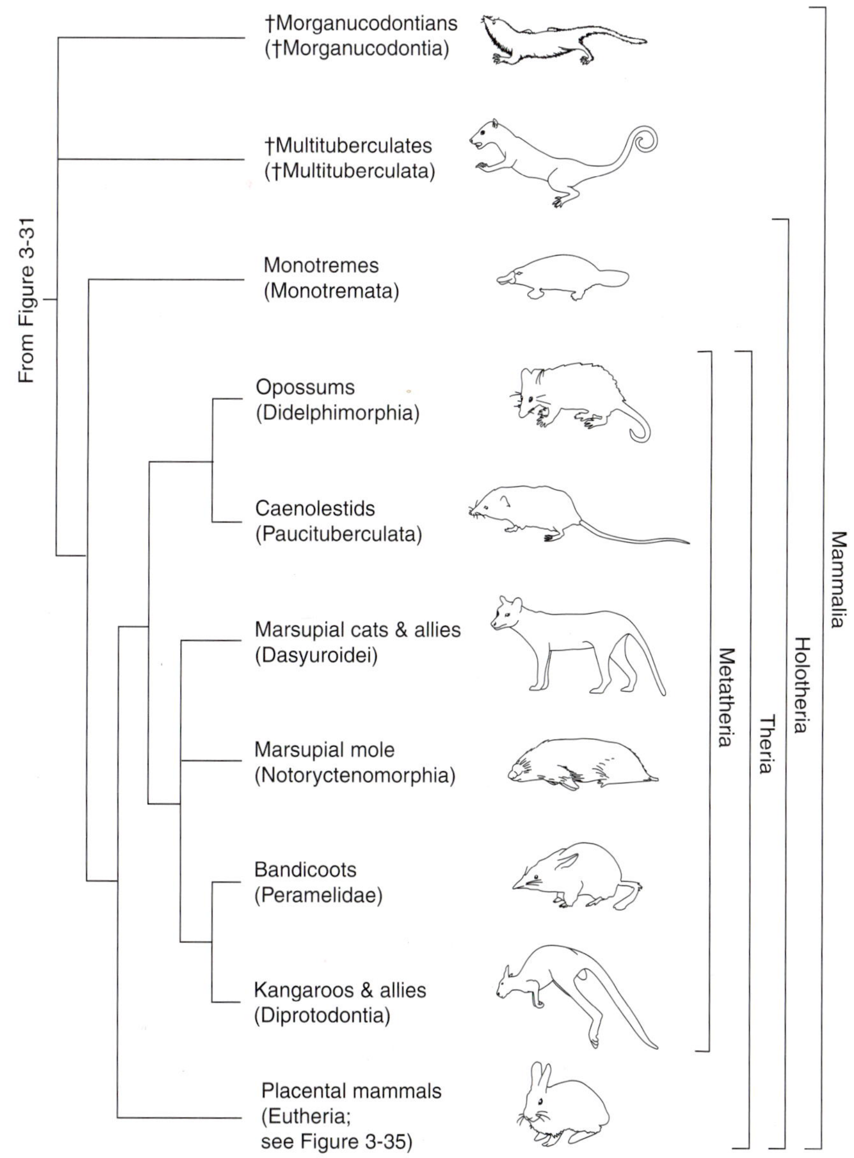

FIGURE 3-33
Phylogeny of living and fossil mammals highlighting the diversity of marsupials (Metatheria) and the origin of placental mammals (Eutheria).

FIGURE 3-34
Representative mammals. *A,* Reconstruction of a Mesozoic mammal, †*Morganucodon.* *B,* Platypus, *Ornithorhynchus.* *C,* Australian spiny anteater, *Tachyglossus.* *D,* Virginia opossum, *Didelphis.* *E,* Red kangaroo, *Macropus,* with young in pouch.

Monotremes

Holotheria (Gr., *holos* = all + *therion* = wild beast, in this case referring to mammals) includes the living monotremes and therian mammals (Fig. 3-33). Holotherians have a unique occlusal pattern in which cusps of the upper cheek teeth interlock between adjacent lower teeth to create what is known as a **reversed triangular pattern of occlusion** (Chapter 16). Many Mesozoic holotherians are known, but most are fragmentary remains of teeth and jaws, so we turn directly to **Monotremata** (Gr., *monos* = single + *trema* = hole, referring to a single opening for fecal and urogenital products; Fig. 3-33). Extant monotremes are the platypus (*Ornithorhynchus*) and echidnas (*Tachyglossus* and *Zaglossus*) of Australia and New Guinea (Fig. 3-34*B* and *C*). The semiaquatic platypus feeds in rivers using its ducklike bill. An unusual electrosensory system in the skin of the bill allows them to locate invertebrate prey, such as worms (Chapter 12). Echidnas gather termites and ants with the long beak and specialized tongue. It long was difficult to associate monotremes with other groups of mammals because adult platypus lack teeth, and teeth are absent in echidnas. Significantly, a fossil platypus that retains molar teeth with the reversed triangular occlusal pattern diagnostic for Holotheria has been described. Living monotremes are endothermic and have all diagnostic characters of mammals, including mammary glands. Yet they lay cleidoic eggs and retain many plesiomorphic skeletal and soft anatomical features. For example, the cloaca is not divided into separate digestive and urogenital passages, as in other mammals, so they retain a single opening for the discharge of fecal and urogenital products.

Therians and the Diversification of Marsupials

Theria (Gr., *therion* = wild beast) includes Metatheria, or marsupials, and Eutheria, or placental mammals (Fig. 3-33). Therians have diagnostic molar teeth known as **tribosphenic molars.** Teeth of this type have triangular crowns with three cusps next to a rounded, lingual border. The most familiar feature of living therians, however, is that they give birth to live young. Correlated with live birth, therian eggs are small and have little yolk.

Diversification of Marsupials

Members of **Metatheria** are commonly known as marsupials (Fig. 3-33). They have an inturned angular process on the lower jaw and limited replacement of teeth. Except in certain derived groups, marsupials retain the mammalian incisor pattern of five upper incisors; the term polyprotodont (Gr., *poly* = many + *proto* = front + *odont* = teeth) describes this plesiomorphic condition.[18] Marsupials exhibit a unique pattern of viviparous reproduction characterized by a very short intrauterine life. After birth, the embryos attach themselves to nipples, usually located in a skin pouch called the **marsupium,** where the young complete development.

Marsupials appear in the fossil record during the early Cretaceous, when the large southern continent of Gondwanaland had begun to break up by continental rifting. South America, Antarctica, and Australia were still connected, and a chain of islands lay between South and North America. Many investigators believe that marsupials evolved in North America, where the earliest fossils and the first radiation of marsupials have been found, and that they spread to Europe and South America by the late Cretaceous. Later, marsupial groups in Europe and North America became extinct. When South and North America again became connected in the early Tertiary, placental mammals moved into South America, where many marsupials subsequently became extinct. Some small, opossum-like marsupials survived in South and Central America, and the Virginia opossum, *Didelphis virginiana,* subsequently invaded North America, where it is well established.

Marsupials reached Australia at least as early as the Oligocene, presumably via a bridge of islands connecting Australia with South America by way of Antarctica (where marsupial fossils have been found). As Australia separated from Antarctica and drifted into lower latitudes, climatic conditions changed dramatically. Australia today has diverse climates, ranging from tropical rain forests to temperate regions to deserts. Because only a few bats, rodents, and marine eutherians also reached Australia, many ecological opportunities were available. Australian marsupials underwent an extensive radiation and occupied most of the insectivorous, carnivorous, and herbivorous niches occupied by eutherian mammals in other regions of the world. Most of the 240 extant species of marsupials live in Australia.

Didelphimorphia (Fig. 3-33) includes the omnivorous Virginia opossum, *Didelphis* (Fig. 3-34*D*), which has been intensively studied for clues about the feeding systems of early mammals (Chapter 16). *Didelphis* and some other didelphimorphs have **prehensile tails** that allow them to hang from branches. The rat opposums, or caenolestids (**Paucituberculata;** Fig. 3-33) are native to western South America.

Dasyuroidea (Fig. 3-33) includes the "native cats" of Australia, the ant-eating numbat, and the Tasmanian wolf, which is perhaps extinct. Most dasyurids are small scavengers or predators that feed on invertebrates and small rodents, but the sturdy skull of the Tasmanian devil (*Sacrophilus*) is remarkably convergent on the skull of placental carnivores, such as bears. Of uncertain affinities is the marsupial mole, **Notoryctes** (Fig. 3-33). It is a subterranean burrower convergently similar to the true moles (Talpidae), which are placental mammals. Like the true moles, *Notoryctes* uses its enlarged front feet as shovels to push soil aside as it burrows. Bandicoots (**Peramelidae;** Fig. 3-33) come from Australia and New Guinea. These hopping, rat-sized carnivorous marsupials forage in ground litter and dig among roots for food. Most of the living species of Australian marsupials are herbivores. They form the **Diprotodontia** (Fig. 3-33) and have three or fewer incisor teeth (Gr., *di* = two + *proto* = front + *odont* = teeth). The group includes animals commonly called possums, gliders, koalas, wombats, kan-

[18]Some classifications recognize a group called "Polyprotodontia"; it is paraphyletic and not recognized here.

garoos, and wallabies. Australian possums are chiefly nocturnal, arboreal herbivores. Gliders, such as the feather-tailed glider (*Acrobates*) have loose skin folds between their front and hind legs and glide gracefully from tree to tree as they forage. The diminutive honey possum (*Tarsipes*) has a long, tubelike mouth for feeding on nectar and pollen. The koala (*Phascolarctos*) feeds on the leaves of only a few species of eucalyptus trees. Wombats (Vombatidae) resemble large woodchucks or pigs. They shelter during the day in burrows and emerge in the evening to feed on plants. The dominant herbivores of Australia today are the saltatory wallabies of rocky and bushy terrain and the large kangaroos or macropodids (L., *macro* = big + *poda* = foot; Fig. 3-34*E*) of open country.

Eutherians

Members of **Eutheria** (Fig. 3-35) are commonly called placental mammals, although this term is not entirely appropriate because marsupials also have a placenta (Chapters 4 and 21). Eutherian mammals have a longer gestation period than do marsupials, and the young are born at a much more advanced stage of development. No evidence shows that eutherian reproduction is more efficient than is the marsupial pattern. Early eutherians, such as the Upper Cretaceous †*Zalambdestes,* were small, presumably insectivorous, nocturnal creatures, but eutherians rapidly diversified early in their history to occupy many different ecological niches. We trace the evolution of eutherians using the phylogeny shown in Figure 3-35.

Edentates

The name **Edentata** (Fig. 3-35) refers to the toothless jaws found in members of this group. We recognize two edentate subgroups: Xenarthra and Pholidota (Fig. 3-35). **Xenarthra** (Gr., *xenos* = different + *arthros* = articulation, referring to the unique articulations between vertebrae) includes the armadillos, anteaters, and sloths (e.g., the three-toed sloth, *Bradypus;* Fig. 3-36*A*). Teeth are retained in armadillos and sloths, although they are highly modified from those found in early eutherians. Armadillos have bony plates in the skin that form a carapace over the back; when threatened, they curl up to present this armor to potential predators. Sloths eat leaves while suspended from tree branches by their long arms and claws. Some other edentates remain generalized insectivores, but the anteaters became highly specialized for a diet of ants and termites, which they capture using an

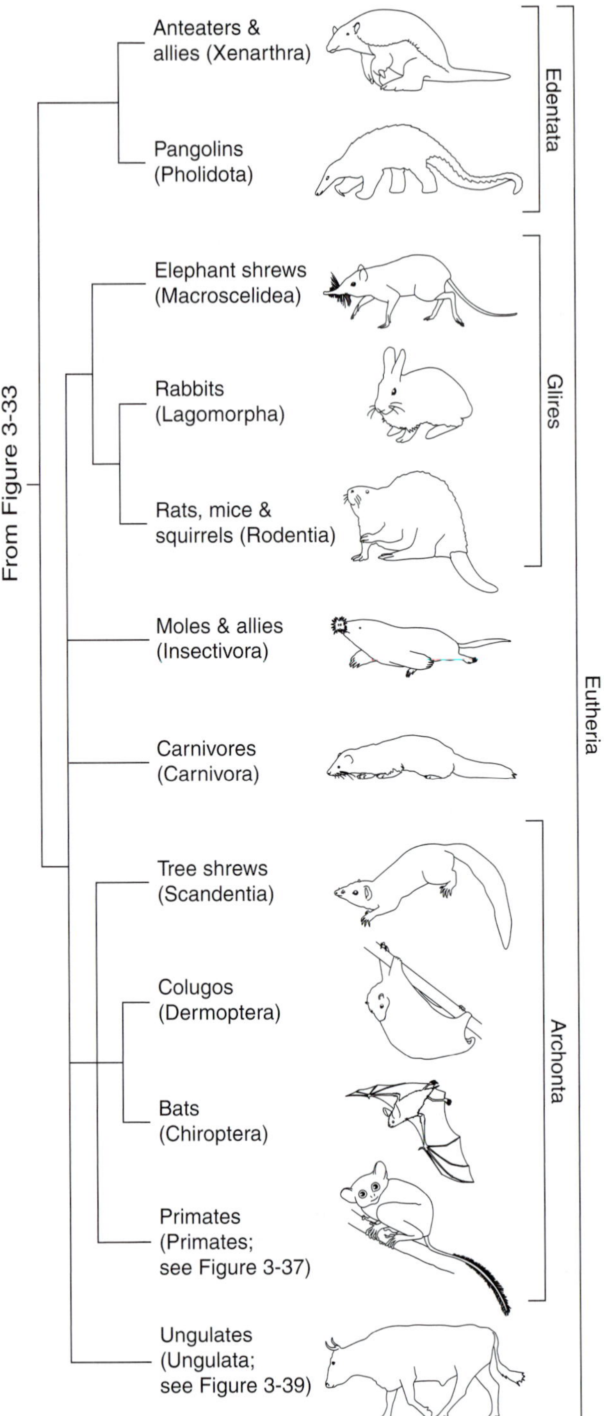

FIGURE 3-35
Phylogeny of living eutherian mammals (Eutheria).

elongate tongue. **Pholidota** (Gr., *pholidotos* = armed with scales) includes the scaly anteaters, or pangolins, that today are found only in the Old World tropics. This group independently evolved specializations for anteating, including long tongues, toothless jaws, and claws for breaking into termite mounds. Like armadil-

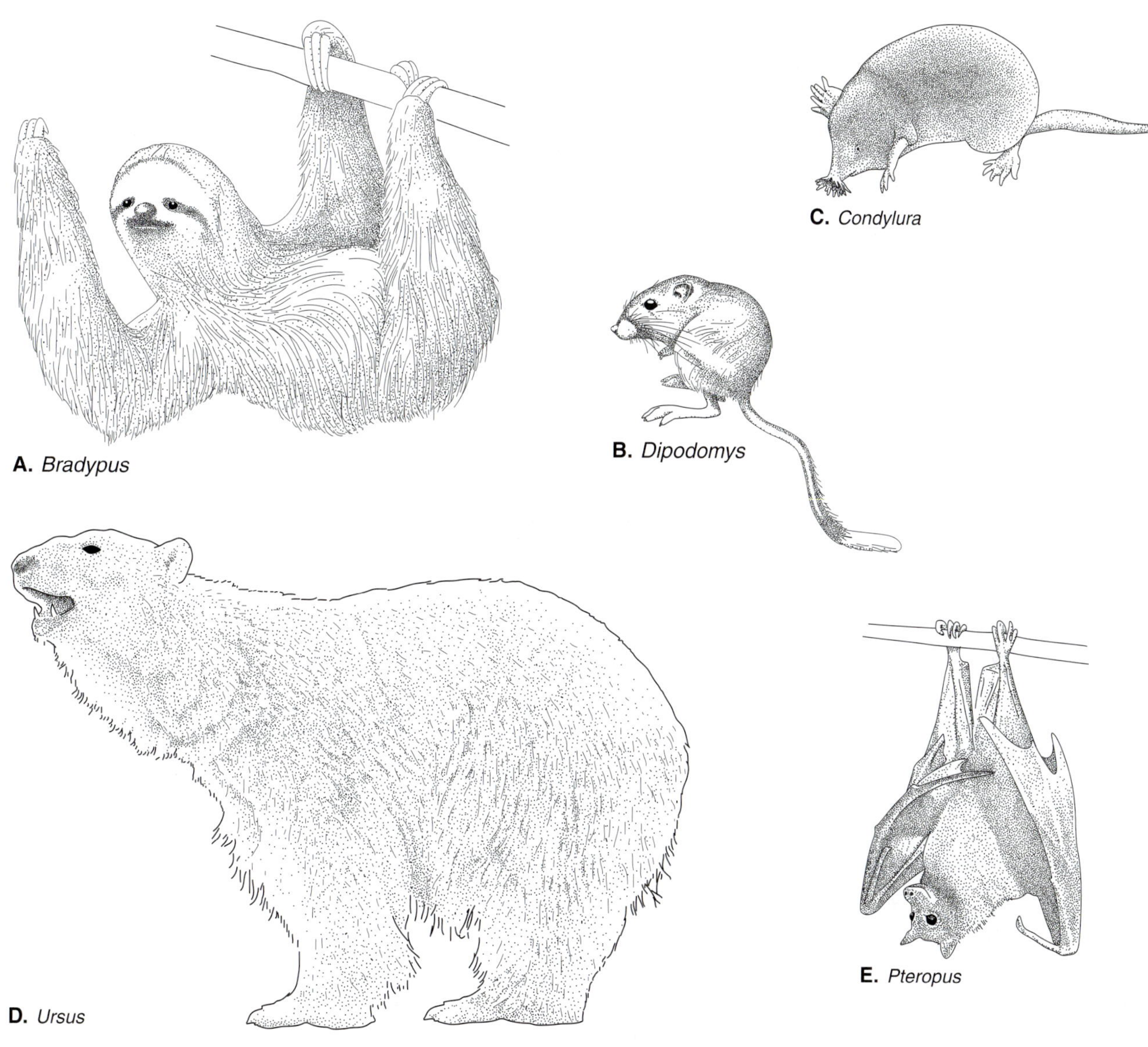

FIGURE 3-36
Representative eutherians. *A,* Three-toed sloth, *Bradypus. B,* Kangaroo rat, *Dipodomys. C,* Star-nosed mole, *Condylura.* The star has thousands of specialized touch receptors, known as Eimer's organs. *D,* Polar bear, *Ursus. E,* A megachiropteran bat, the flying fox, *Pteropus.*

los, the dorsal surface of a pangolin's body is armored, but pangolins accomplished this in an entirely different way by using keratinized plates derived from modified hair.

Glires

As shown in Figure 3-35, the eutherian group **Glires** (L., *gliris* = a doormouse) includes elephant shrews (**Macroscelidea**), rabbits (**Lagomorpha**), and rodents (**Rodentia**). Elephant shrews use their snouts as probes to locate invertebrate prey in leaf litter or soil and their elongated hindlimbs for saltation (= jumping), a feature convergently evolved in many other groups of small mammals. We regard the chiefly herbivorous rabbits (Lagomorpha) and rodents (Rodentia) as sister taxa (Fig. 3-35). Both lagomorphs and rodents lack canine teeth and have a **diastema** (a gap in the tooth row between the incisor teeth and the cheek teeth). Rabbits and their close relatives have a unique arrangement of the upper incisor teeth, in which a pair of peglike incisors is located immediately behind a pair of elongate, chisel-like incisors. This allows a rabbit to use its lower incisors to shear against the chisel-like upper incisors and pound against the peglike upper incisors. Most rabbits have long ears and short, tufted tails. Saltation is very well developed in some, such as the jackrabbits (*Lepus*). In contrast to rabbits, rodents have only a single pair of incisors in both the upper and lower jaws. Rodent incisors are rootless and ever-growing, and enamel is present only on their anterior surfaces, which makes the incisors self-sharpening and contributes to their function in gnawing. Rodentia includes an astonishing diversity of species, with more than 30 extant families, such as beavers (Castoridae), squirrels (Sciuridae), Norway rats and domestic mice (Muridae), hamsters (Cricetidae), New-World porcupines (Erethizontidae), gophers (Geomyidae), and kangaroo rats (Dipodidae; Fig. 3-36*B*). Some rodents exhibit striking functional anatomical specializations, such as legs, vertebrae, and tails modified for jumping; forelimbs modified for burrowing; and kidneys modified for water conservation.

Insectivores

Insectivora (Fig. 3-35) includes several families that retain many plesiomorphic features of eutherians. Most are small, nocturnal, insect-eating animals with a well-developed nose and ears but small eyes. Their full complement of teeth is well adapted for piercing, killing, cutting, and crushing their food. Five clawed toes are present, and the first toe can oppose the others to some extent. The foot is placed flat on the ground, a posture termed **plantigrade.** Some insectivorans, such as moles (Talpidae; Fig. 3-36*C*) burrow through soil using forefeet modified into spadelike structures. The shrew family (Soricidae) includes the smallest living eutherians, with some species maturing at a mass of just 2 g. The tenrecs of Madagascar (Tenrecidae) superficially resemble hedgehogs (Erinaceidae) in that both families include forms with spinelike fur that can be erected for defense.

Carnivores

Carnivora (Fig. 3-35) includes nine extant families of predominantly flesh-eating eutherians: the cats (Felidae), hyaenas (Hyaenidae), otters and weasels (Mustelidae), wolves and dogs (Canidae), bears (Ursidae; Fig. 3-36*D*), raccoons (Procyonidae), eared seals (Phocidae), sea lions (Otaridae), and walruses (Odobenidae). Cats, weasels, and dogs are predacious carnivores, but raccoons, bears, and a few other species of carnivorans have secondarily become omnivores. Eared seals, sea lions, and walruses are marine carnivores.

Archontans

Archonta (Fig. 3-35) includes several groups. The five extant genera of tree shrews (**Scandentia**) are native to forests of Southeast Asia. Because they resemble early Primates, some authors consider tree shrews to belong to Primates, although this is not the view depicted in Figure 3-35. Most mammalogists regard colugos (**Dermoptera**) and bats (**Chiroptera**) as sister taxa, and we adopt this view (Fig. 3-35). The two species of colugos (*Cynocephala*) are herbivorous forest dwellers from Southeast Asia. They sometimes are called "flying lemurs," but they are not lemurs, nor do they fly. Instead, they glide from tree to tree by extending loose skin folds between their front and hind legs. In contrast, bats engage in powered flight using wings composed of the fingers of the hand to support a web of elastic skin between them. Mammalogists recognize two groups of Chiroptera: **Megachiroptera** (Gr., *mega* = big + *cheiro* = hand + *ptera* = wing) and **Microchiroptera** (Gr., *micro* = small + *cheiro* = hand + *ptera* = wing). Flying foxes (*Pteropus;* Fig. 3-36*E*) and most other megachiropterans are frugivorous (fruit-eating) bats from the Old World tropics. They roost in large social groups, called camps, and fly out at night to forage. Vision is very important to megachiropterans, and they have large eyes.

Microchiropterans form one of the most diverse clades of mammals and have many fascinating anatomical and physiological specializations. Most microchiropterans are nocturnal insectivores, but others eat fruits, nectar, fishes, blood, or other foods. As their name suggests, most species of Microchiroptera are small. Unlike megachiropterans, the second digit of the hand of a microchiropteran is clawless and is enclosed fully in the wing membrane. Microchiropterans roost in social groups, sometimes reaching millions of individuals, as is the case for Mexican free-tail bats (*Molossus*) at Carlsbad Caverns in New Mexico. The external ears of microchiropterans often are large and complex, and the nose and face may have a peculiar appearance (Chapter 12). These features are related to **ultrasonic echolocation,** a sensory system used by microchiropterans to locate food and other objects. To do this, a bat uses its mouth to produce **ultrasonic vocalizations.**[19] The sound is reflected by objects, which the bat then detects and processes in its brain. All of this takes place so rapidly that the bat can adjust its flight to avoid collisions or to capture prey. Insectivorous bats catch prey on the wing directly in the

[19]By definition, an ultrasonic frequency is one that cannot be detected by human auditory systems (a young person can detect up to about 20,000 Hz [= 20,000 cycles/sec]). Echolocating bats typically use frequencies between 20,000 Hz and 100,000 Hz.

mouth or in the wing membranes. Among the more famous microchiropterans are the New-World vampire bats (*Desmodus*), which feed on vertebrate blood. Seasonal hibernation is very important for many bats that live in the temperate zone, such as the little brown bat (*Myotis lucifigus*) from North America.

Primates

We limit the order **Primates**[20] to lemurs, "monkeys," and apes (which includes humans; Fig. 3-37). The presence of fingernails is synapomorphic for primates. Early primates resembled insectivores, differing primarily in features related to foraging for insects, fruits, and soft plant food in trees. Most living primates are arboreal, but some, including humans, adapted to a terrestrial mode of life. **Lemuriformes** (Fig. 3-37) includes the lorises (Lorisidae), which are nocturnal, tree-dwelling lemuriforms of Southeast Asia and Africa, and lemurs (Lemuridae), which are restricted today to Madagascar, where they occupy many different ecological niches. The ring-tailed lemur, *Lemur catta,* is one of the most familiar (Fig. 3-38*A*). The three species of tarsiers (**Tarsiiformes;** Fig. 3-37) are specialized for a nocturnal life in the trees. They have very large eyes, and their elongated ankles help them leap through the forest canopy.

All remaining primates belong to **Anthropoidea** (Fig. 3-37). Anthropoids have an enlarged brain and

[20]Primates is an unusual case of a name that is spelled the same way in both formal and common use, the only difference being the capitalization of the formal name.

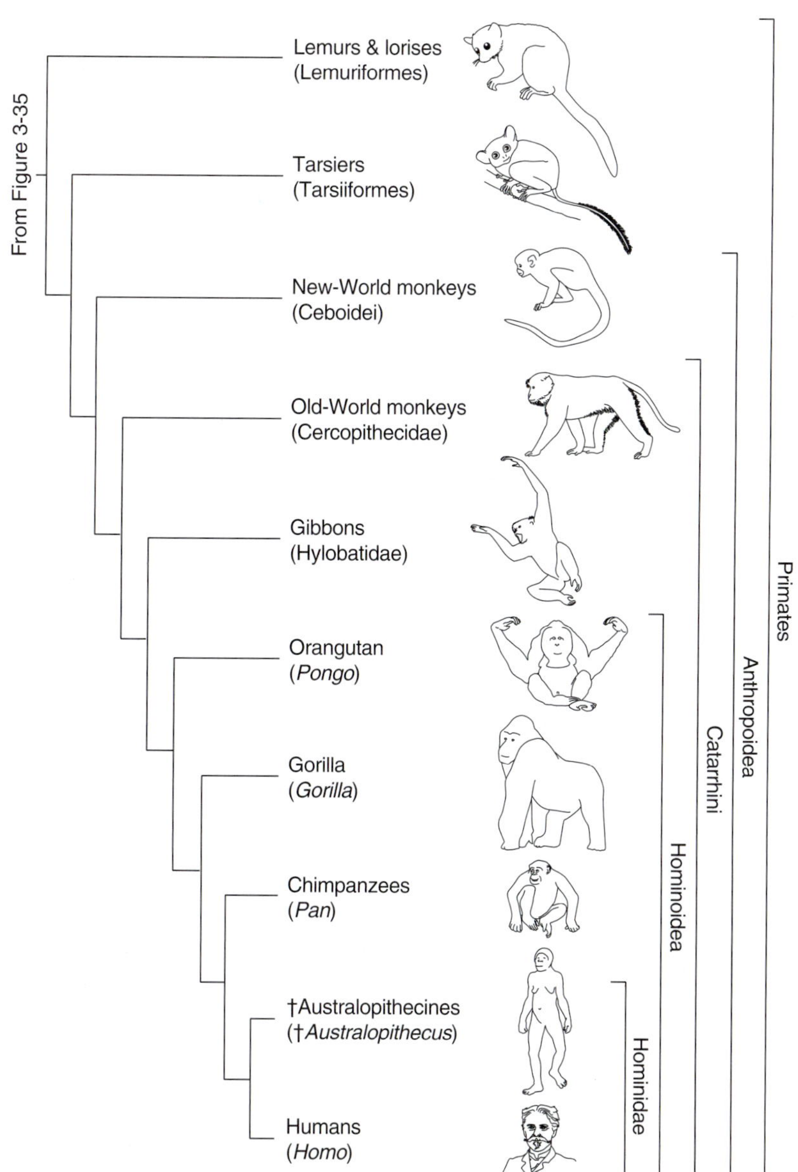

FIGURE 3-37
Phylogeny of living primates (Primates).

FIGURE 3-38
Representative Primates. *A,* Ring-tailed lemur, *Lemur. B,* Gorilla, *Gorilla.*

braincase, and a bony partition separates the orbit from the fossa behind it, which lodges the temporalis muscle. Also, the upper lip of anthropoids is not divided in the midline. The olfactory lobe of the brain is reduced, but the surface of the brain is more convoluted than in outgroup forms. Most anthropoids are social, diurnal animals. Many brachiate, that is, they use their long arms to swing suspended from the branches of trees. The opposable hallux (i.e., big toe) of anthropoids is used for grasping.

In Figure 3-37, note that primates commonly termed "monkeys" do not form a monophyletic group. The pollex (= thumb) of New-World monkeys (**Ceboidei;** Fig. 3-37) is only weakly capable of being opposed against other digits or the palm. Ceboidei includes howler monkeys, which have unique specializations of the larynx used to produce loud vocalizations; spider monkeys (*Ateles*), which use their long, prehensile tails while moving through the canopy; and the colorful, small marmosets and tamarins, which scramble along tree branches like squirrels.

The thumb can oppose the other digits of the palm in other anthropoids, which is synapomorphic for **Catarrhini** (Fig. 3-37). This feature is important in some types of locomotion and manipulation of objects. The nostrils of catarrhines are close together and ventrally directed, and the square quadritubercular molar teeth have cusps at each corner of the square. Old-World monkeys (**Cercopithecidae;** Fig. 3-37) occur today from southern Europe (Gibraltar); across Africa; to Arabia, Southeast Asia, and Japan. The rhesus monkey (*Macaca mulatta*) is a familiar, diurnal, omnivorous cercopithecid that is strongly terrestrial. Arboreal cercopithecids can brachiate, but their short tails are never prehensile.

Gibbons (**Hylobatidae;** Fig. 3-37) share the loss of the tail with their sister group, the Hominoidea. Gibbons are excellent brachiators. They do not grasp a limb but rather use their fingers as a U-shaped hook. Their thumb is relatively short. Living members of **Hominoidea** (Fig. 3-37) are the tree-dwelling orangutan (*Pongo*) from Southeast Asia, the terrestrial gorilla (*Gorilla;* Fig. 3-38*B*) from the mountains of central Africa, two species of chimpanzees from Africa (*Pan*), and humans (**Hominidae**). The fossil record of the human family includes genera such as †*Australopithecus,* which is known from several species found in the Pliocene and

Pleistocene of Africa, as well as the living genus *Homo*.

Ungulates

Ungulata consists of several familiar groups of large terrestrial and aquatic mammals (Fig. 3-39). The jaws and teeth of terrestrial ungulates are specialized for browsing or grazing on plant foods, and most ungulates have elongate limbs that allow them to outrun predators. Each toe bears a **hoof,** which is the source of the name Ungulata (L., *ungula* = hoof). All ungulates walk or stand on their digits, with the heel and ankle off the ground, which can be termed a **digitigrade posture;** many walk on just the toe tips and hooves, an additional modification that is known as an **unguligrade posture** (Chapters 9 and 11). Members of the order **Artiodactyla** (Gr., *artios* = even + *daktylos* = finger or toe; Figs. 3-39 and 3-40*A*) have an even number of toes or digits. They also share a unique modification of an ankle bone (the astragalus), and its double-pulley shape allows for great movement between the shank and the foot. The presence of four toes is the plesiomorphic condition for artiodactyls; this condition is retained in pigs (Suidae), peccaries (Tayassuidae), and hippopotamuses (Hippopotamidae). Camels, llama, and vicuña (Camelidae) are didactylous (= having two toes on each foot). Ruminantia is the group that contains deer (Cervidae), giraffes (Giraffidae), and antelopes (Bovidae). Bovids have horns, and this diverse family includes more than 100 species, including goats (*Capra*), sheep (*Ovis*), bison (*Bison*), cattle (*Bos*), and many African antelopes.

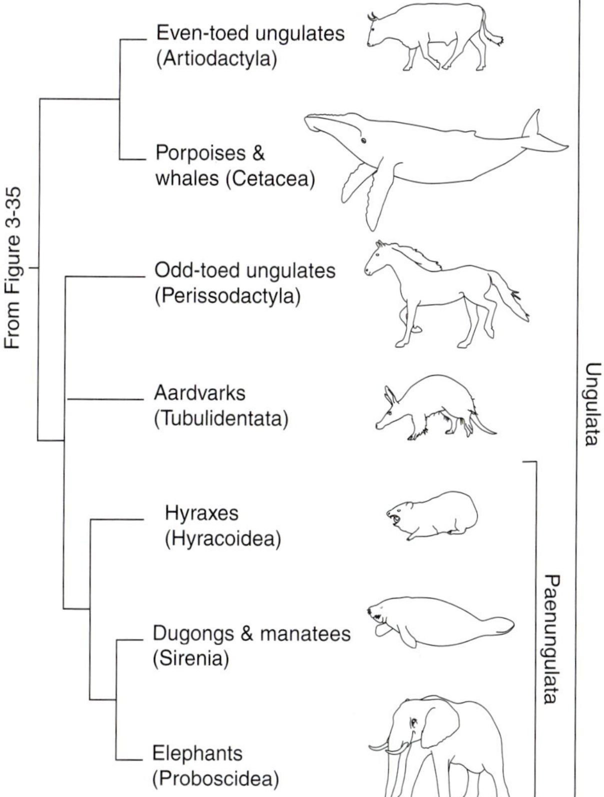

FIGURE 3-39
Phylogeny of living ungulates (Ungulata).

Living **Cetacea** (Fig. 3-39) are highly specialized aquatic ungulates with streamlined bodies, no hindlimbs, a horizontal tail with two flukes, and dorsally located nostrils known as blowholes. Most are marine; a few live in fresh water. The terrestrial ancestry of cetaceans is revealed by a series of remarkable fossil intermediates. Hair is reduced in cetaceans, and blubber (i.e., fat and other tissues in the dermis of the skin) provides thermal insulation. Cetacean skulls are highly modified and often have striking left–right asymmetries, a feature that is related to their sophisticated echolocation systems for detecting objects under water. Although most species of cetaceans retain teeth, the dentition is highly modified relative to other mammals in that it consists of rows of similarly shaped (homodont) teeth. Examples of toothed cetaceans include the dolphins (*Delphinus*), porpoises (*Phocoena*), pilot whales (*Globicephala*), killer whales (*Orcina*), and sperm whales (e.g., *Physeter catodon*). The most specialized cetaceans are the baleen whales, which lack teeth but have plates of baleen (commonly called whalebone, although it is actually a keratinized skin derivative similar to hair) hanging from the roof of the mouth. All baleen whales are filter-feeders. This group includes the humpback whale (*Megaptera novaeangliae;* see cover) and the largest known vertebrate, the blue whale (*Balaenoptera musculus;* Fig. 3-40*B*), which reaches a length of 31 m and a weight of 160,000 kg.

Perissodactyla (Gr., *perissos* = odd + *daktylos* = finger or toe; Fig. 3-39) is represented today by the tapirs (Tapiridae), rhinoceroses (Rhinocerotidae), and horses (Equidae; Fig. 3-40*C*). These animals have an odd number of toes, either three (plesiomorphic state) or one (derived state). They walk on the plantar surface of these digits, using an unguligrade posture. Perissodactyls have an extensive early Cenozoic fossil record, but their species diversity has declined over the past 30 million years as grasslands have expanded and artiodactyls have diversified. Another small group of ungulates is **Tubulidentata,** which contains a single living species, the African aardvark (*Orycteropus afer*). Aardvarks specialize in eating termites.

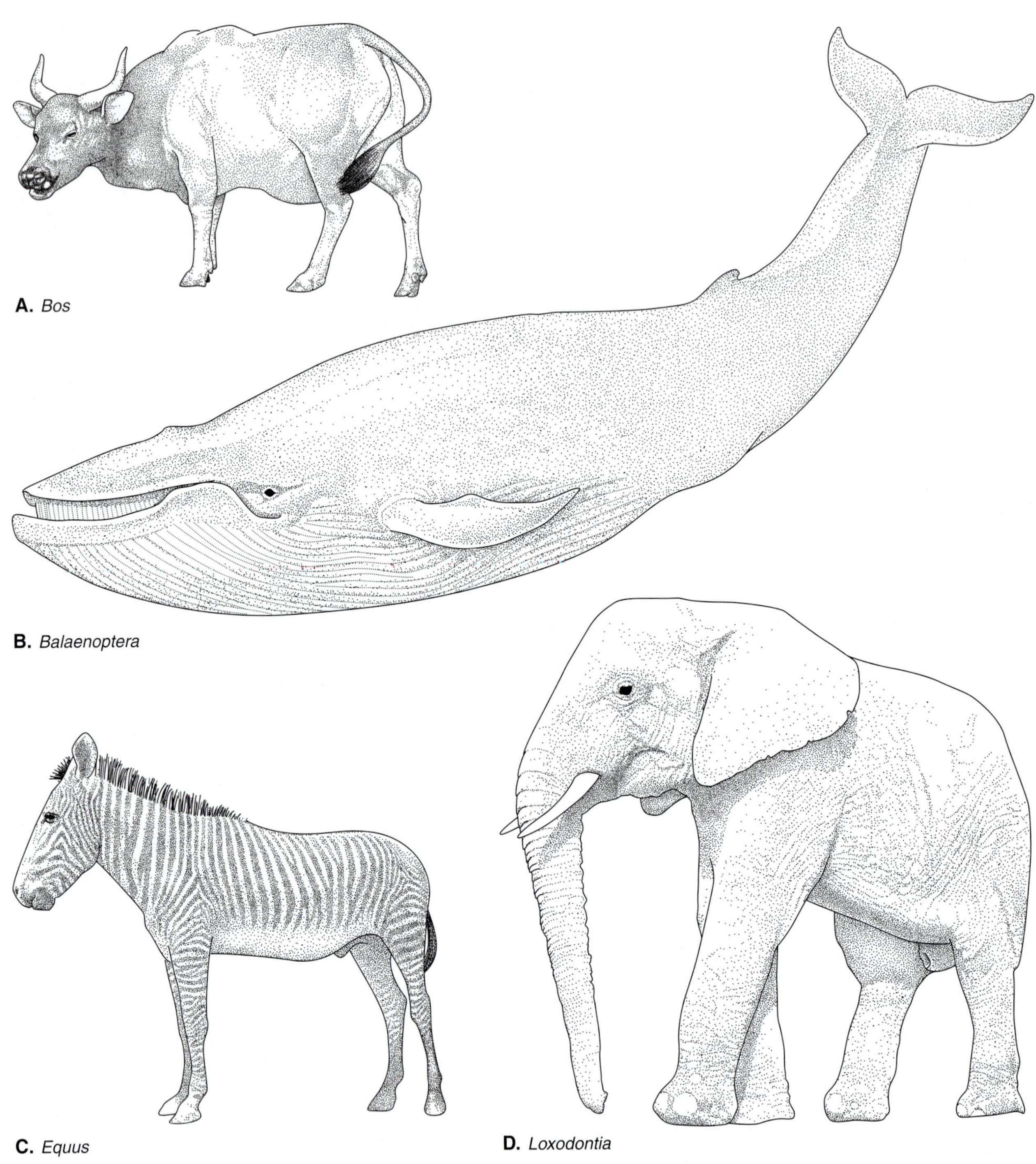

FIGURE 3-40
Representative ungulates. *A,* Cow, *Bos. B,* Blue whale, *Balaenoptera. C,* Zebra, *Equus. D,* African elephant, *Loxodontia.*

Hyraxes (**Hyracoidea**), dugongs and manatees (**Sirenia**), and elephants (**Proboscidea**) belong to the group **Paenungulata** (Fig. 3-39). The three genera of hyraxes are herbivorous, rabbit-like animals known today only from Africa and the Middle East. Like the elephants, hyraxes have ever-growing, tusk-like incisors composed primarily of dentine. Dugongs and manatees are large, aquatic herbivores that live in near-shore marine environments or in estuaries or rivers. They lack externally visible hindlimbs but have a single, broad, horizontal, caudal fluke. Because of our familiarity with the Florida manatee

(*Trichechus*), we often think of sirenians as subtropical to tropical forms. However, until exterminated by humans in the 18th or early 19th century, the †Steller sea cow (†*Hydrodamalis*) inhabited very cold waters of the Bering Sea. Two species of proboscideans—the Indian elephant (*Elephas maximus*) and the African elephant (*Loxodontia africana;* Fig. 3-40*D*)—are alive today. Elephants have unique cheek teeth, pillar-like limbs, large ear pinnae, tusklike incisors, and a trunk.

SUMMARY

1. Major patterns in craniate evolution can be efficiently summarized using cladograms and phylogenetic classifications. This approach differs from those that have been used traditionally and in some other textbooks, but it allows clearer reasoning about taxa and characters. In particular, it is important to minimize recognition of paraphyletic groups because it can confuse clear thinking about the distribution of characters, homology, and relationships.

2. Taxa known only from fossil remains have much to offer the study of craniate phylogeny, yet it is not always easy to integrate fossils into analyses. Fossil taxa are indicated with a dagger symbol (†) to easily distinguish them from extant taxa. The best approach for integrated studies of fossil and living organisms is to start with an understanding about the evolutionary relationships of living taxa and then to incorporate fossil taxa wherever they can contribute to the overall evolutionary picture.

3. Phylogenetic analysis allows us to "zoom in" or "zoom out" to whatever level of evolutionary history we are interested in studying. For example, we can focus on specific events in the evolutionary history of craniates and organize our understanding of them using phylogenies. Many different events in craniate history might be deemed important, depending on what you are interested in or who you ask, but some especially prominent events include the evolution of jaws and paired fins (Gnathostomata), the evolution of terrestrial vertebrates (Tetrapoda), the evolution of tetrapods with an amnion during development (Amniota), the evolution of amniotes with feathers (Aves), and the evolution of amniotes with three middle ear bones (Mammalia). You can begin a study of these and other events in the history of craniates using the tools provided in this chapter (which are the cladograms, characters listed in the Appendix to Chapter 3, and the short summaries of biodiversity).

4. An understanding of the diversity and phylogeny of Craniata will help you frame an understanding of the great diversity of functional anatomical specializations that they have evolved.

REFERENCES

Ahlberg, P. E., and Clack, J. A., 1998: Lower jaws, lower tetrapods: A review based on the Devonian genus *Acanthostega. Transactions of the Royal Society of Edinburgh-Earth Sciences,* 89:11–46.

Arratia, G., and Schultze, H.-P., 1999: *Mesozoic Fishes, Systematics and the Fossil Record.* München, F. Pfeil.

Arratia, G., and Viohl, G., 1996: *Mesozoic Fishes, Systematics and Paleoecology.* München, F. Pfeil.

Bartholomew, G. A., 1982: Physiological control of body temperature. *In* Gans, C., and Pough, F. H., editors: *Biology of the Reptilia,* vol. 12, New York, Academic Press.

Bemis, W. E., and Hetherington, T., 1982: The rostral organ of *Latimeria chalumnae:* Morphological evidence of an electroreceptive function. *Copeia,* 1982:467–471.

Bemis, W. E., and Grande, L., 1999: Development of the median fins of the paddlefish, *Polyodon spathula,* with comments on the lateral fin fold hypothesis. *In* Arratia and Schultze, *op cit.*

Bemis, W. E., and Northcutt, R. G., 1991: Innervation of the basicranial muscle of *Latimeria chalumnae. In* Musick, J. A., Bruton, M. N., and Balon, E. K., editors: *The Biology of* Latimeria chalumnae *and the Evolution of Coelacanths.* Dordrecht, Kluwer.

Bemis, W. E., and Northcutt, R. G., 1992: Skin and blood vessels of the snout of the Australian lungfish, *Neoceratodus forsteri,* and their significance for interpreting the cosmine of Devonian lungfishes. *Acta Zoologica (Stockholm),* 73:115–139.

Bemis, W. E., Burggren, W., and Kemp, N. E., 1986: *The Biology and Evolution of Lungfishes.* New York, Alan R. Liss.

Bemis, W. E., 1984: Paedomorphosis and the evolution of the Dipnoi. *Paleobiology,* 10:293–307.

Benton, M. J., and Harper, D. A. T., 1996: *Basic Paleontology.* New York, Addison-Wesley Longman.

Benton, M. J., 1993: *The Fossil Record 2.* London, Chapman & Hall.

Benton, M. J., 1988: *The Phylogeny and Classification of the Tetrapods,* volume 1: *Amphibians, Reptiles and Birds,* volume 2, *Mammals.* Oxford, Clarendon Press.

Birstein, V. J., Waldman, J., and Bemis, W. E., 1997: *Sturgeon Biodiversity and Conservation.* Dordrecht, Kluwer.

Block, B. A., Finnerty, S. A., Jr., and Kidd, J., 1993: Evolution of endothermy in fish: Mapping physiological traits on a molecular phylogeny. *Science,* 260:210–214.

Bolt, J. R., 1977: Dissorophoid relationships and ontogeny, and the origin of the Lissamphibia. *Journal of Paleontology,* 51:235–249.

Bond, C. E., 1996: *The Biology of Fishes.* Philadelphia, W. B. Saunders.

Carey, F. G., Teal, J. M., Kanwisher, J. W., Lawson, K. D., and Beckett, J. S., 1971: Warm-bodied fish. *American Zoologist,* 11:137–145.

Carroll, R. L., 1987: *Vertebrate Paleontology and Evolution.* New York, Freeman.

Clack, J. A., 1997. The earliest tetrapods lived in water. *Recherche,* 296:58–61.

Cloutier, R., and Ahlberg, P. E., 1995: Sarcopterygian interrelationships: How far are we from phylogenetic consensus? *In* Lelièvre et al., *op cit.*

Coates, M. I., 1999: Endocranial preservation of a Carboniferous actinopterygian from Lancashire, UK, and the interrelationships of primitive actinopterygians. *Philosophical Transactions of the Royal Society of London Series B: Biological Sciences,* 354:435–462.

Coates, M. I., 1996: The Devonian tetrapod *Acanthostega gunnari* Jarvik: Postcranial anatomy, basal tetrapod interrelationships and patterns of skeletal evolution. *Transactions of the Royal Society of Edinburgh: Earth Sciences,* 87:363–421.

Cogger, H., and Zweiffel, R., 1998: *Encyclopedia of Amphibians and Reptiles,* 2nd edition. San Diego, Academic Press.

Cracraft, J., 1988: The major clades of birds. *In* Benton, M. J., editor: *The Phylogeny and Classification of the Tetrapods,* volume 1. Oxford, Clarendon Press.

Crompton, A. W., Taylor, C. R., and Jagger, J. A., 1978: Evolution of homeothermy in mammals. *Nature,* 272:333–336.

Currie, P. J., and Padian, K., 1997: *Encyclopedia of Dinosaurs.* San Diego, Academic Press.

Dean, B., 1909: Studies on fossil fishes. *Memoirs of the American Museum of Natural History,* 9:211–287.

Donoghue, P. C. J., Forey, P. L., and Aldridge, R. J., 2000: Conodont affinity and chordate phylogeny. *Biology Review,* 75:191–251.

Duellman, W. E., and Trueb, L., 1986: *Biology of Amphibians.* New York, McGraw-Hill.

Erdmann, M. V., Caldwell, R. L., and Moosa, M. K., 1998: Indonesian 'king of the sea' discovered. *Nature,* 395:335.

Eschmeyer, W. N., Ferraris, C. J., Jr., and Hoang, M. D., 1998: *Catalog of Fishes.* San Francisco, California Academy of Sciences.

Fastovsky, D. E., and Weishampel, D. B., 1996: *The Evolution and Extinction of the Dinosaurs.* New York, Cambridge University Press.

Forey, P. L., 1998: *History of the Coelacanth Fishes.* London, Chapman & Hall.

Forey, P. L., and Janvier, P., 1993: Agnathans and the origin of jawed vertebrates. *Science,* 361:129–134.

Forshaw, J., 1998: *Encyclopedia of Birds,* 2nd edition. San Diego, Academic Press.

Gabbott, S. E., Aldridge, R. J., and Theron, J. N., 1995: A giant conodont with preserved muscle tissue from the Upper Ordovician of South Africa. *Nature,* 374:800–803.

Gagnier, P.Y., and Wilson, M. V. H., 1996: Early Devonian acanthodians from northern Canada. *Palaeontology,* 39: 241–258.

Gardiner, B. G., 1984: Devonian palaeoniscid fishes: New specimens of *Mimia* and *Moythomasia* from the Upper Devonian of Western Australia. London, British Museum (Natural History).

Gauthier, J. A., Kluge, A. G., and Rowe, T., 1988: Amniote phylogeny and the importance of fossils. *Cladistics,* 4:105–209.

Gauthier, J. A., Kluge, A. G., and Rowe, T., 1988: The early evolution of the amniotes. *In* Benton, M. J., editor: *The Phylogeny and Classification of the Tetrapods,* volume 1. Oxford, Clarendon Press.

Gingerich, P. D., Raza, S. M., Arif, M., Anwar, M., and Zhou, X. Y., 1994: New whale from the Eocene of Pakistan and the origin of cetacean swimming. *Nature,* 368:844–847.

Goujet, D., and Young, G. C., 1995: Interrelationships of placoderms revisited. *In* Lelièvre et al., *op cit.*

Gould, E., and McKay, G., 1998: *Encyclopedia of Mammals,* 2nd edition. San Diego, Academic Press.

Grande, L., and Bemis, W. E., 1991: Osteology and phylogenetic relationships of fossil and recent paddlefishes (Polyodontidae) with comments on the interrelationships of Acipenseriformes. Society of Vertebrate Paleontology Memoir 1. *Journal of Vertebrate Paleontology,* 11 (supplement 1):1–132.

Grande, L., and Bemis, W. E., 1998: A comprehensive phylogenetic study of amiid fishes (Amiidae) based on comparative skeletal anatomy: An empirical search for interconnected patterns of natural history. Society of Vertebrate Paleontology Memoir 4. *Journal of Vertebrate Paleontology,* 18 (supplement 1):1–690.

Greene, H. W., 1997: *Snakes: The Evolution of Mystery in Nature.* Berkeley, CA, University of California Press.

Griffiths, M., 1978: *The Biology of Monotremes.* New York, Academic Press.

Hardisty, M. W., and Potter, I. C., 1971–1982: *The Biology of Lampreys.* New York, Academic Press.

Hedges, S. B., and Poling, L. L., 1999: A molecular phylogeny of reptiles. *Science,* 283:998–1001.

Heilmann, G., 1927: *The Origin of Birds.* New York, D. Appleton & Co.

Holder, M. T., Erdmann, M. V., Wilcox, T. P., Caldwell, R. L., and Hillis, D. M., 1999: Two living species of coelacanths? *Proceedings of the National Academy of Sciences,* 96:12616–12620.

Honeycutt, R. L., and Adkins, R. M., 1993: Higher-level systematics of eutherian mammals: An assessment of molecular characters and phylogenetic hypotheses. *Annual Review of Ecology and Systematics,* 24:279–305.

Janis, C. M., and Farmer, C., 1999: Proposed habitats of early tetrapods: Gills, kidneys, and the water-land transition. *Zoological Journal of the Linnean Society,* 126:117–126.

Janvier, P., 1996: *Early Vertebrates.* Oxford, Clarendon Press.

Jarvik, E., 1980: *Basic Structure and Evolution of Vertebrates.* London, Academic Press.

Johanson, Z., and Ahlberg, P. E., 1998: A complete primitive rhizodont from Australia. *Nature,* 394:569–573.

Johnson, G. D., and Anderson, W. D., Jr., 1993: Proceedings of the symposium on phylogeny of the Percomorpha. *Bulletin of Marine Science,* 52:1–620.

Kemp, T. S., 1999: *Fossils and Evolution.* New York, Oxford University Press.

Kemp, T. S., 1982: *Mammal-like Reptiles and the Origin of Mammals.* New York, Academic Press.

Lauder, G. V., and Liem, K. F., 1983: The evolution and interrelationships of the actinopterygian fishes. *Bulletin of the Museum of Comparative Zoology,* 150:95–197.

Lelièvre, H., Wenz, S., Blieck, A., and Cloutier, R., 1995: Premiers Vertébrés et Vertébrés inférieurs. *Géobios,* Mémoire Spécial 19.

Liem, K. F., 1987: Functional design of the air ventilation apparatus and overland excursions by teleosts. *Fieldiana: Zoology* 1379:1–29.

Lombard, R. E., and Sumida, S. S., 1992: Recent progress in understanding early tetrapods. *American Zoologist,* 32:609–622.

Maddison, W. P., and Maddison, D. R., 1999: *MacClade: Analysis of Phylogeny and Character Evolution.* Sunderland, Mass., Sinauer Associates.

Maisey, J. G., 1986: Heads and tails: A chordate phylogeny. *Cladistics,* 2:201–256.

Maisey, J. G., 1996: *Discovering Fossil Fishes.* New York, Henry Holt.

McKenna, M. C., and Bell, S. K., 1997: *Classification of Mammals above the Species Level.* New York, Columbia University Press.

Meng, J., and McKenna, M. C., 1998: Faunal turnovers of Palaeogene mammals from the Mongolian plateau. *Nature,* 394:364–367.

Morell, V., 1993: *Archaeopteryx*: Early bird catches a can of worms. *Science,* 259:764–765.

Moyle, P. B., and Cech, J. J., Jr., 1995: *Fishes, An Introduction to Ichthyology,* 3rd edition. Englewood Cliffs, N.J., Prentice-Hall.

Moy-Thomas, J. A., and Miles, R. S., 1971: *Palaeozoic Fishes,* 2nd edition. Philadelphia, W. B. Saunders.

Musick, J. A., Bruton, M. N., and Balon, E. K., 1991: *The Biology of* Latimeria chalumnae *and the Evolution of Coelacanths.* Dordrecht, Kluwer.

Nelson, J. S., 1994: *Fishes of the World,* 3rd edition. New York, John Wiley & Sons.

Novacek, M. J., Wyss, A. R., and McKenna, M. C., 1988: The major groups of eutherian mammals. *In* Benton, M. J., editor: *The Phylogeny and Classification of the Tetrapods,* volume 2: *Mammals.* Oxford, Clarendon Press.

Padian, K., and Chiappe, L. M., 1998: The origin and early evolution of birds. *Biological Reviews of the Cambridge Philosophical Society,* 73:1–42.

Panchen, A. L., 1980: *The Terrestrial Environment and the Origin of Land Vertebrates.* London, Academic Press.

Patterson, C., Williams, D. M., and Humphries, C. J., 1993. Congruence between molecular and morphological phylogenies. *Annual Review of Ecology and Systematics,* 24:153–188.

Patterson, C., 1982: Morphology and interrelationships of primitive actinopterygian fishes. *American Zoologist,* 22: 241–259.

Paxton, J. R., and Eschmeyer, W. N., 1998: *Encyclopedia of Fishes,* 2nd edition. San Diego, Academic Press.

Pough, F. H., Janis, C. M., and Heiser, J. B., 1999: *Vertebrate Life,* 5th edition. Upper Saddle River, N.J., Prentice-Hall.

Pough, F. H., editor, 2001: *Herpetology,* 2nd edition. Upper Saddle River, N.J., Prentice-Hall.

Pouyaud, L., Wirjoatmodjo, S., Rachmatika, I., Tjakrawidjaja, A., Hadiaty, R., and Hadie, W., 1999: A new species of Coelacanth. *Comptes Rendus, Series III,* 322:261–267.

Prothero, D. R., and Schoch, R. M., 1994: *Major Features of Vertebrate Evolution.* Knoxville, The Paleontological Society.

Reynolds, J. E., and Rommel, S. A., 1999: *Biology of Marine Mammals.* Washington, D. C., Smithsonian Institute Press.

Rieppel, O. C., and deBraga, M., 1996: Turtles as diapsid reptiles. *Nature,* 384:453–455.

Rieppel, O. C., 1988: *Fundamentals of Comparative Biology.* Basel, Birkhauser Verlag.

Rosen, D. E., Forey, P. L., Gardiner, B. G., and Patterson, C., 1981: Lungfishes, tetrapods, paleontology and plesiomorphy. *Bulletin of the American Museum of Natural History,* 167:159–276.

Rougier, G. W., and Novacek, M. J., 1998: Early mammals: Teeth, jaws and finally . . . a skeleton! *Current Biology,* 8:R284–R287.

Schmidt-Nielsen, K., Bolis, L., Taylor, C. R., Bentley, P. J., and Stevens, C. E., 1980: *Comparative Physiology, Primitive Mammals.* New York, Cambridge University Press.

Schultze, H.-P., and Cloutier, R., 1996: *Devonian Fishes and Plants of Miguasha, Quebec, Canada.* München, F. Pfeil.

Schultze, H.-P., 1991: A comparison of controversial hypotheses on the origin of tetrapods. *In* Schultze, H.-P., and Trueb, L., editors: *Origins of the Higher Groups of Tetrapods: Controversy and Consensus.* Ithaca, N.Y., Comstock Publishing Associates.

Schultze, H.-P., and Trueb, L., 1991: *Origins of the Higher Groups of Tetrapods: Controversy and Consensus.* Ithaca, N. Y., Comstock Publishing Associates.

Sereno, P. C., 1999: The evolution of dinosaurs. *Science,* 284:2137–2147.

Sereno, P. C., and Chenggang, R., 1992: Early evolution of avian flight and perching: New evidence from the Lower Cretaceous of China. *Science,* 255:845–848.

Shaffer, H. B., Meylan, P., and McKnight, M. L., 1997: Tests of turtle phylogeny: Molecular, morphological, and paleontological approaches. *Systematic Biology,* 46:235–268.

Shipman, P., 1998: *Taking Wing: Archaeopteryx and the Evolution of Bird Flight.* New York, Simon & Schuster.

Shoshani, J., and McKenna, M. C., 1998: Higher taxonomic relationships among extant mammals based on morphology, with selected comparisons of results from molecular data. *Molecular Phylogenetics and Evolution,* 9:572–584.

Stiassny, M. L. J., Parenti, L., and Johnson, G. D., 1996: *Interrelationships of Fishes.* San Diego, Academic Press.

Storey, K. B., and Storey, J. M., 1996: Natural freezing survival in animals. *Annual Review of Ecology and Systematics,* 27:365–386.

Szalay, F. S., Novacek, M. J., and McKenna, M. C., 1993: *Mammal Phylogeny: Mesozoic Differentiation, Multituberculates, Monotremes, Early Therians, and Marsupials.* New York, Springer-Verlag.

Tarsitano, S., 1991: *Archeopteryx*: Quo Vadis? *In* Schultze, H.-P., and Trueb, L., editors: *Origins of Higher Groups of Tetrapods: Controversy and Consensus.* Ithaca, N.Y., Comstock Publishing Associates.

Thomson, K. S., 1999: The coelacanth: Act three. *American Scientist,* 87:213–215.

Trueb, L., 1991: *Origins of the Higher Groups of Tetrapods: Controversy and Consensus.* Ithaca, N.Y., Comstock Publishing Associates.

Trueb, L., and Cloutier, R., 1991: A phylogenetic investigation of the inter and intrarelationships of the Lissamphibia (Amphibia: Temnospondyli). *In* Schultze and Trueb, *q.v.*

Van Valkenburgh, B., 1999: Major patterns in the history of carnivorous mammals. *Annual Review of Earth and Planetary Sciences,* 27:463–493.

Vaughan, T. A., Ryan, J. M., and Czaplewski, N.J., 2000: *Mammalogy,* 4th edition. Philadelphia, W. B. Saunders.

Weishampel, D. B., Dodson, P., and Osmólska, H., 1992: *The Dinosauria.* Berkeley, University of California Press.

Welty, J. C., Baptista, L. F., and Welty, C., 1997: *The Life of Birds,* 4th edition. Philadelphia, W. B. Saunders.

Wilson, D. E., and Reeder, D. M., 1993: *Mammal Species of the World: A Taxonomic and Geographic Reference.* Washington, D. C., Smithsonian Institution Press.

Wilson, M. V. H., and Caldwell, M. W., 1998: The Furcacaudiformes: A new order of jawless vertebrates with thelodont scales, based on articulated Silurian and Devonian fossils from northern Canada. *Journal of Vertebrate Paleontology,* 18:10–29.

Wilson, M. V. H., and Caldwell M. W., 1993: New Silurian and Devonian fork-tailed thelodonts are jawless vertebrates with stomachs and deep bodies. *Nature,* 361: 442–444.

Zhu, M., Yu, X.-B., and Janvier, P., 1999: A primitive fossil fish sheds light on the origin of bony fishes. *Nature,* 397: 607–610.

APPENDIX *Example Synapomorphies for Clades Figured in Chapter 3*

Craniata—Chordates with a Braincase (Figs. 3-1 and 3-2)
Vertebrata—Craniates with a Backbone (Figs. 3-1 and 3-2)
Gnathostomata—Vertebrates with Jaws (Figs. 3-1 and 3-2)
Elasmobranchiomorphi—Gnathostomes with an Optic Pedicel (Fig. 3-2)
†Placodermi—Elasmobranchs with a Ring of Armor on the Trunk (Fig. 3-2)
Chondrichthyes—Elasmobranchiomorphs with Prismatic Calcified Cartilage (Figs. 3-1, 3-2 and 3-6)
Elasmobranchii—Chondrichthyans with Plate Gills (Fig. 3-6)
Euselachii—Elasmobranchs with a Subterminal Mouth (Fig. 3-6)
Neoselachii—Euselachians with a Fused Coracoid Bar (Fig. 3-6)
Galeomorphi—Neoselachians with an Anterior Location of the Hyomandibular Fossa (Fig. 3-6)
Squalea—Neoselachians with a New Origin for the Suborbital Muscle (Fig. 3-6)
Holocephali—Chondrichthyans with Autostyly (Fig. 3-6)
Teleostomi—Gnathostomes with Terminal Mouths (Fig. 3-2)
†Acanthodii—Teleostomes with Intermediate Fin Spines (Fig. 3-2)
Osteichthyes—Teleostomes with Lungs (Fig. 3-2)
Actinopterygii—Osteichthyans with a Single Dorsal Fin (Figs. 3-1, 3-2 and 3-9)
Neopterygii—Actinopterygians with Modernized Fins (Fig. 3-9)
Teleostei—Neopterygians with Homocercal Tails (Fig. 3-9)
Otocephala—Clupeocephalans with a Modified Caudal Skeleton (Fig. 3-9)
Ostariophysi—Otocephalans with a Divided Swim Bladder (Fig. 3-9)
Otophysi—Ostariophysans with a Weberian Apparatus (Fig. 3-9)
Euteleostei—Teleosts with a Membrane Bone Extension of Uroneural 1 (Figs. 3-9, 3-12)
Acanthomorpha—Euteleosts with Spiny Fins (Fig. 3-12)
Acanthopterygii—Acanthomorphs with Mobile Pharyngeal Jaws (Fig. 3-12)
Percomorpha—Acanthopterygians with Transforming Ctenoid Scales (Fig. 3-12)
Sarcopterygii—Osteichthyans with Muscular Lobe Fins (Figs. 3-1, 3-2, 3-15)
Rhipidistia—Sarcopterygians with Folded Teeth (Figs. 3-1, 3-15)
Dipnomorpha—Rhipidistians with Elongate Paired Fins (Fig. 3-15)
Choanata—Rhipidistians with a Choana (Fig. 3-15)
Tetrapoda—Rhipidistians with Dactylous Limbs (Figs. 3-1, 3-15)
Neotetrapoda—Tetrapods with Five Digits (Fig. 3-15)
Amphibia—Neotetrapods with Four Digits in the Hand (Fig. 3-15, 3-18)
Temnospondyla—Amphibians with a Sound-Conducting Stapes and a Tympanum (Fig. 3-18)

Dissorophoidea—Temnospondyls with Bicuspid, Pedicellate Teeth (Fig. 3-18)
Lissamphibia—Dissorophoids with Smooth Skin (Fig. 3-18)
Batrachia—Lissamphibians with an Opercular Apparatus (Fig. 3-18)
Reptilomorpha—Neotetrapods with Enlarged Pleurocentra in the Trunk Vertebrae (Figs. 3-15, 3-20)
Amniota—Reptilomorphs with Amniotic Eggs (Figs. 3-1, 3-20)
Sauropsida—Amniotes with a Single Centrale Bone in the Ankle (Figs. 3-1, 3-20)
Diapsida—Sauropsids with Two Temporal Fenestrae (Figs. 3-1, 3-20)
Sauria—Diapsids with a Slender, Rodlike Stapes (Fig. 3-20)
Lepidosauria—Saurians with a Transverse Cloacal Slit (Figs. 3-1, 3-20, 3-23)
Squamata—Lepidosaurs with Hemipenes (Fig. 3-23)
Scleroglossa—Squamates with Flattened, Keratinized Tongues (Fig. 3-23)
Archosauromorpha—Saurians with a Forwardly Directed Foot (Figs. 3-20, 3-25)
Archosauria—Archosauromorphs with Serrated Teeth (Figs. 3-1, 3-25)
Crocodylotarsi—Archosaurs with a Crocodile Type of Ankle (Fig. 3-25)
Dinosauria—Archosaurs with Three or More Sacral Vertebrae (Fig. 3-25)
Saurischia—Dinosaurs with Temporal Muscles Extending onto the Frontal Bones (Fig. 3-25)
Theropoda—Saurischians with Thin-Walled Hollow Long Bones (Figs. 3-25, 3-27)
Aves—Theropods with Feathers (Figs. 3-1, 3-27)
Neornithes—Aves (Birds) with a Neognathous Palate (Figs. 3-27, 3-29)
Synapsida—Reptilomorphs with a Lower Temporal Fenestra (Figs. 3-20, 3-31)
Therapsida—Synapsids with an Enlarged Lower Temporal Fenestra (Fig. 3-31)
Cynodontia—Therapsids with a Masseter Muscle (Fig. 3-31)
Mammalia—Cynodonts with a Malleus and Incus (Fig. 3-1, 3-31, 3-33)
Holotheria—Mammals with a Reversed Triangular Occlusal Pattern (Fig. 3-33)
Theria—Mammals that Give Birth to Live Young (Figs. 3-1, 3-33)
Metatheria—Therians with an Inturned Angular Process on the Lower Jaw (Figs. 3-1, 3-33)
Eutheria—Therians with a Chorioallantoic Placenta (Figs. 3-1, 3-33, 3-35)
Primates—Eutherians with Fingernails (Fig. 3-35, 3-37)
Anthropoidea—Primates with a Bony Postorbital Wall (Fig. 3-37)
Catarrhini—Anthropoids with Opposable Thumbs (Fig. 3-37)
Hominoidea—Catarrhines with Enlarged Brains (Fig. 3-37)
Hominidae—Hominoids with Upright Posture (Fig. 3-37)
Ungulata—Eutherians with Hooves (Figs. 3-35, 3-39)
Paenungulata—Ungulates with Ever-Growing Incisor Teeth (Fig. 3-39)

4

Early Development and Comparative Embryology

PRECIS

After a brief introduction to some concepts in developmental biology, we will examine the development of chordates from the structure of the gametes to the establishment of the embryonic axis and primordia of the organ systems. Our coverage emphasizes gastrulation; early development of the nervous system, neural crest, neurogenic placodes, and sense organs; mesoderm and coelom formation; and the basic structure and organization of the head and pharynx. These topics provide a foundation for understanding the subsequent development of the ten organ systems of vertebrates and introduce you to important developmental landmarks for the study of comparative vertebrate anatomy. We conclude with comments on the control of segmentation by Hox *genes and how duplication of these regulatory genes early in vertebrate history may have profoundly influenced the course of vertebrate evolution.*

OUTLINE

One of the greatest marvels of life is the transformation of one seemingly simple cell, a fertilized egg, into a structurally and functionally complex organism composed of multitudes of cells. Development is an exciting and dynamic process to observe, and its study has exploded in recent years as new techniques and approaches have been applied to classic questions. Development also offers many exquisite functional anatomical examples, from the match between the structure and function of the gametes to the patterns of interactions among cells and tissues during embryogenesis.

General Concepts in Developmental Biology

Beginning in the 19th century, comparative embryologists studied and named the succession of morphological changes during normal development (or **ontogeny**) of craniates. A summary of the succession of developmental stages during the life cycle of a typical vertebrate is shown in Figure 4-1. **Embryogenesis** is the series of stages during which a fertilized egg is converted into a self-sustaining individual organism. These stages are easily recognizable across the diversity of vertebrates and consist of fertilization, followed by cleavage, gastrulation, neurulation, organogenesis, and cytodifferentiation. Each of these six aspects of embryogenesis will be covered in this chapter, with specific examples from elasmobranchs, teleosts, amphibians, birds, and placental mammals. The end result of vertebrate embryogenesis is usually either a free-living larva, which will later metamorphose to achieve the general morphology of an adult (= **indirect development**), or a small "model" of an adult, which will change less dramatically during the course of its later development (= **direct development**). These aspects of later development (i.e., everything after embryogenesis) will not be treated here but will be covered in subsequent chapters.

The somatic cells of a vertebrate's body normally contain the same genes, but not all genes are active, or active to the same extent, during different periods of development or in the various cells and tissues of the adult. In contrast, a fertilized egg cell can be described as **totipotent,** in that it can develop the full variety of cell types present in an adult. As described in Chapter 1, the bodies of adult vertebrates are built from a great diversity of cell types. We are beginning to understand how gene activity is regulated during vertebrate development to produce cells with different **fates.** For example, differential distributions of certain key proteins and other materials in the cytoplasm of an egg cell can influence the way the genes contained in the nuclei of its daughter cells become expressed. Influencing the fate of a particular group of cells by means of such differential distributions of cytoplasmic materials is termed **cytoplasmic specification.** As cells with different properties begin to emerge during embryogenesis, their products, in turn, influence the way adjacent cells or tissues respond. This phenomenon is called **induction,** and it was discovered in 1901. In the 1920s, Hans Spemann, Hilde Mangold, and other experimental embryologists advanced our understanding of induction by observing the effects of transplanting bits of tissue from one amphibian embryo to another or from one part of an embryo to another. We now know that many of the substances responsible for inductions are proteins. Because these molecules can produce discernible morphological changes in the embryo, they often are known as **morphogens** (agents of morphogenesis). Cells that receive an inductive signal respond by regulating expression of their own genes. For example, a morphogen that stimulates cell division triggers a

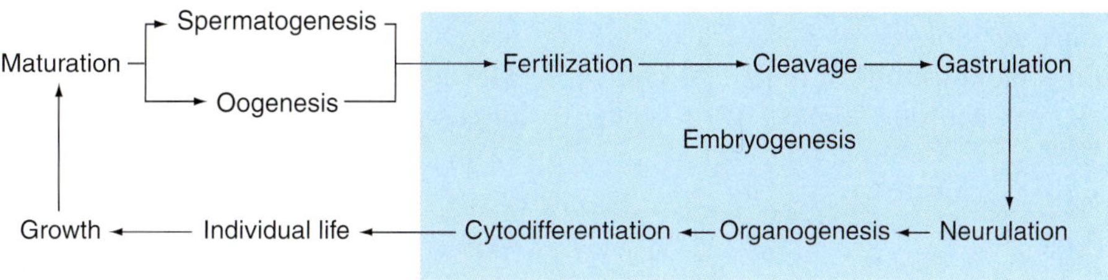

FIGURE 4-1
Life cycle of a vertebrate showing the context for embryogenesis. Events of embryogenesis are the primary subject of this chapter; some comments on gametogenesis and gametes are included to set the stage for fertilization. Individual life marks the endpoint of embryogenesis; this point can be hard to specify. For present purposes, embryogenesis ends when the yolk or any other maternal nutrient is exhausted, and the young is capable of ingesting food.

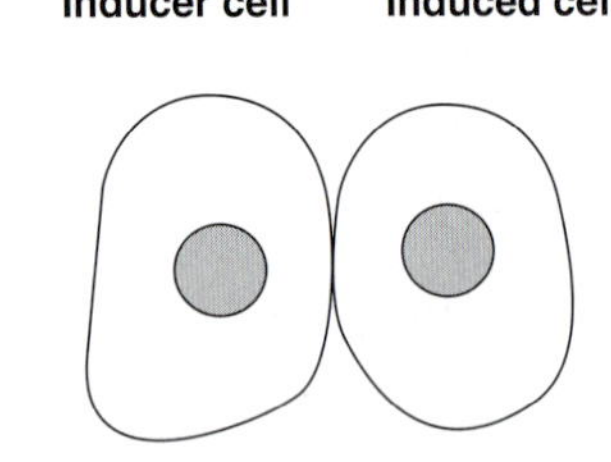

A. Cell–cell contact

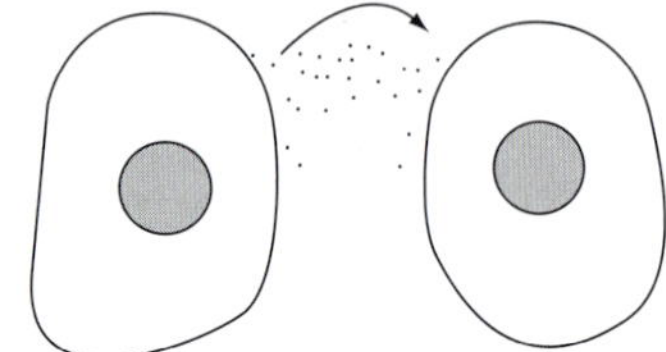

B. Morphogen

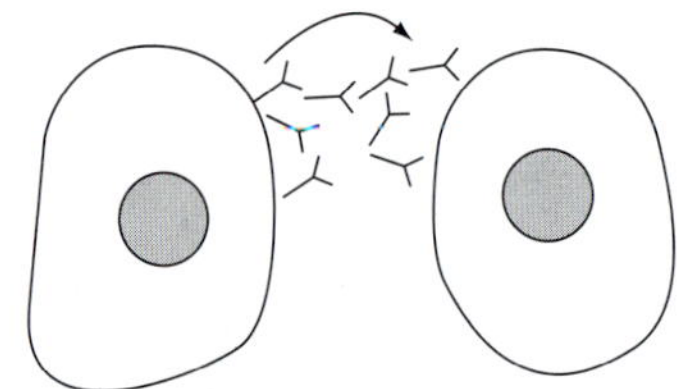

C. Extracellular matrix

FIGURE 4-2
Three modes of induction. *A,* Cell–cell contact, in which inductive signals pass directly from one cell to another. *B,* Morphogens diffuse down a gradient from one cell to target cells. *C,* Extracellular matrix molecules secreted by one cell may affect the movements of adjacent cells. These modes of embryonic induction are analogous to modes of cell communication in adults. See text for discussion. *(Based on a figure in Gilbert.)*

target cell to express genes needed for cell division to occur.

Figure 4-2 shows three basic ways that such inductions occur. The first method is by close cell-to-cell contact, in a fashion analogous to (but much less complex than) the way that neurons connect with other neurons or muscles (see Chapters 10 and 13). The second method is by diffusion of small, typically soluble morphogens, in a fashion analogous to the way that circulating hormones can influence cells at a distance (see Chapter 15). When such signaling molecules diffuse directly through intervening tissues or extracellular matrix before acting on the target cells, they are said to be a **paracrine secretion.** In contrast, if signaling molecules are delivered to their target cells via the circulatory system, they are spoken of as **endocrine secretions.** Most morphogens important in early development act as paracrine secretions. The third method is by local changes in the composition of the extracellular matrix molecules around a cell, which then influence the differentiation of adjacent cells. This latter method may remain important in the self-organization and repair of connective tissues throughout life.

Much of embryogenesis appears to be regulated by series, or **cascades,** of inductions. Cells and tissues respond to inductive influences only during a limited period, when they are said to be **competent** to respond. As tissues differentiate, they express some of the genes associated with their final fate (Fig. 4-3). As differentiation proceeds, cells lose their competence to respond to some inductive influences but not to others. Their potential fates are said to have been **restricted.** Eventually, the fate of most cells is **determined** by subsequent inductive interactions. This final restriction in fate is followed by expression of the many genes associated with that particular cell type and eventual achievement of the morphology typical for that cell type. Developmental regulation may ultimately be viewed as a special case of the broader field of homeostatic regulation, by which all parts of an organism can respond to changes while continuing to function as an integrated whole. Studies of developmental regulation will surely prove to be among the most important aspects of comparative developmental biology because changes in the regulation of developmental events likely provided mechanisms for morphological diversification during vertebrate evolution. An example of this is the role of *Hox* genes in axial patterning of vertebrates, which is discussed at the end of this chapter.

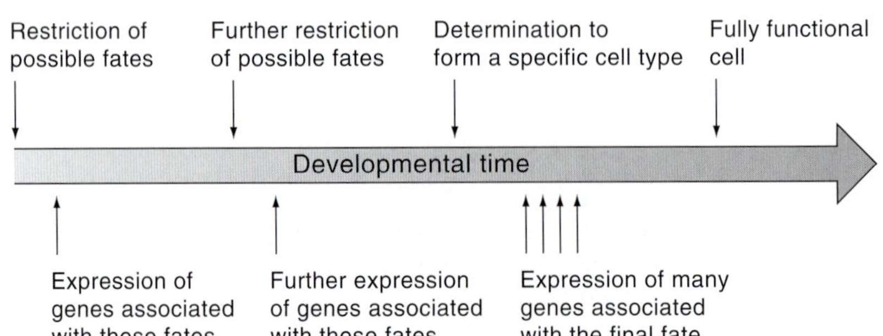

FIGURE 4-3
The roles of restriction, expression, and determination during differentiation of cell types. *(Based on a figure by Wessells.)*

To emphasize the importance of induction and developmental regulation, comparative embryologists often speak about **epigenetics** (Gr., *epi* = upon or on top of + *genetics* = in this case, the genome level). As a term, epigenetics refers to cell fates and the patterns of interactions and changes that occur at the cellular level during embryogenesis (Fig. 4-4). At the bottom of Figure 4-4 is the genetic level, which is a traditional focus of molecular and cellular biology. Genes code for proteins, such as structural proteins, enzymes, morphogens, or other regulatory molecules. Some of these proteins regulate the expression of other genes. Other proteins, chiefly structural proteins and enzymes, endow cells with their specific **cell properties.** The term cell properties covers an array of possible functions, from a cell's ability to adhere to other cells; to its ability to synthesize and secrete morphogens that alter the function or state of differentiation of adjacent cells; to cell death, which is a normal developmental fate for a surprisingly large percentage of embryonic cells.

Exceedingly complex cell–cell interactions and morphogenetic movements can occur at the epigenetic

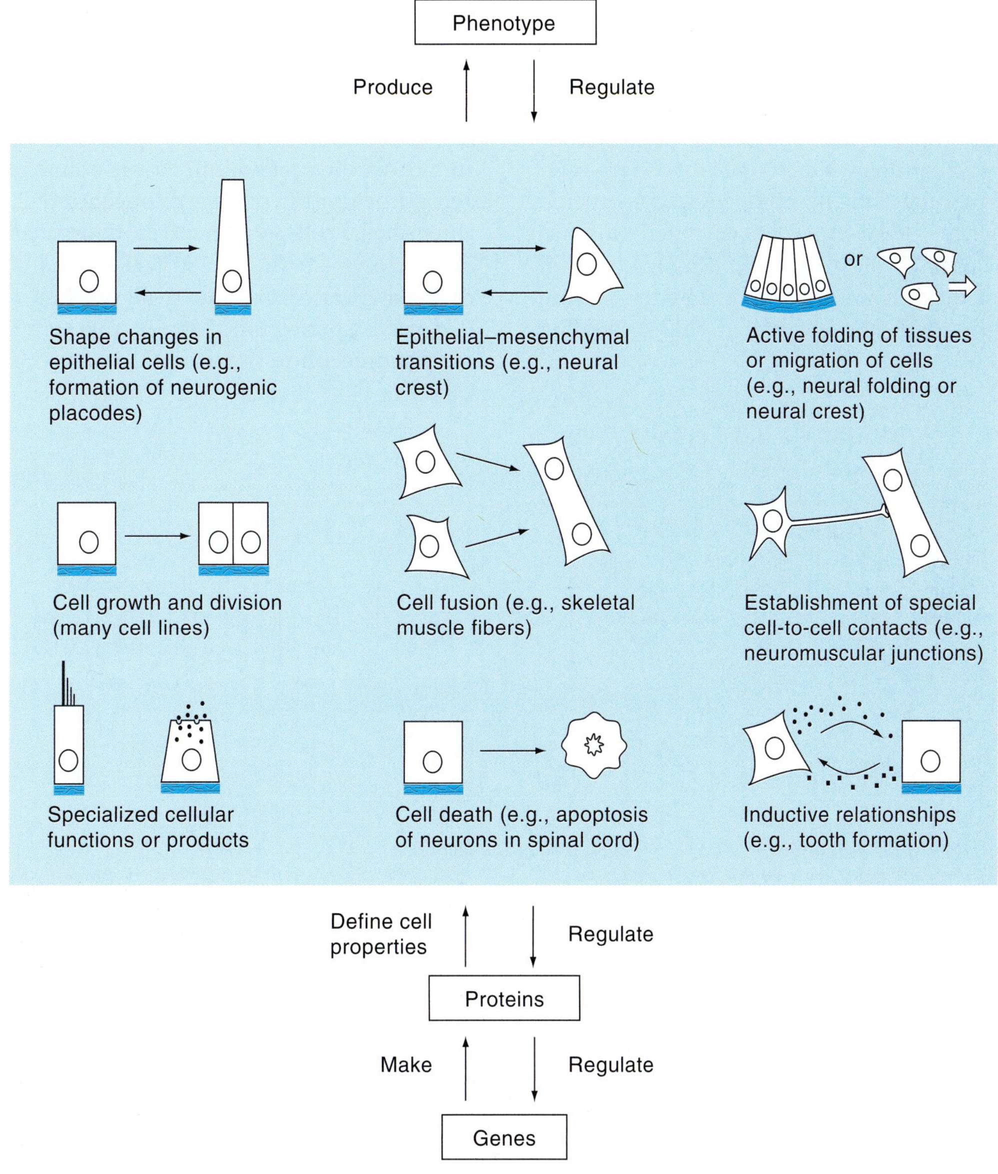

FIGURE 4-4
Examples of epigenetics. The shaded blue box represents epigenetic events, literally, on top of the genome. The small diagrams illustrate some of the many possible fates and interactions of cells and tissues during embryogenesis. These epigenetic phenomena act as a "filter" between the genes and the adult phenotype. *(Based on a figure by Alberch.)*

level. For example, if all of the cells in a flat sheet of epithelium change shape simultaneously, then the epithelium rolls up or folds into a tube or sphere that may serve as the precursor to an organ. The outcome of cell interactions at the epigenetic level also can regulate gene expression. In a mouse embryo, clusters of migrating neural crest cells (discussed later) come to lie beneath portions of the skin covering the jaw, where they participate in reciprocal inductive relationships with the overlying skin cells in the differentiation of a tooth. The skin cells determine what type of tooth will form (e.g., an incisor or a molar), and the neural crest cells then take over control of tooth morphogenesis (see Chapter 16). As another example, an alpha motor neuron that fails to establish contact with a muscle fiber will begin to express genes for cell death and die. As a concept, epigenetics is useful chiefly because it focuses our attention on the regulatory and cellular processes of development, which are a necessary filter between the genome and the completely constructed adult. In practice, however, we have much to learn about the specifics of most epigenetic interactions in most tissues of the vast majority of vertebrates, leaving plenty of room for basic research in this fascinating area.

Another important concept from research in cell and developmental biology concerns two basic ways in which cells can be organized into tissues in a vertebrate's body. We can contrast cells that are organized as an **epithelium** with those that are organized as a **mesenchyme** (Fig. 4-5). Cells in an epithelium form a closely packed array on top of a flexible foundation termed a **basal lamina** (Fig. 4-5*A*). A basal lamina is composed largely of extracellular materials secreted by the overlying epithelium, including proteins (particularly the mat-like form of collagen known as collagen type IV), glycoproteins (e.g., laminin and entactin), and proteoglycans rich in heparin sulfate. Together, these extracellular materials represent the **extracellular matrix** of the epithelium. Cells in an epithelium are usually **polarized,** which means that they exhibit basal specializations at the pole of the cell adjacent to the basal lamina and apical specializations at the free surface of the epithelium. As an example, the nucleus and mitochondria often are concentrated at the base of an epithelial cell, so that secretory organelles can be located closer to the free surface of the cell. Epithelial cells may bear cilia or microvilli on their apical surfaces that serve special functions of that cell type, such as mechanoreception or absorption of nutrients. Adjacent

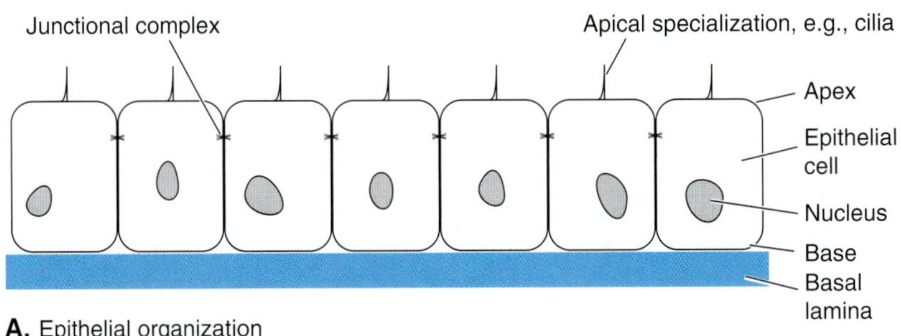

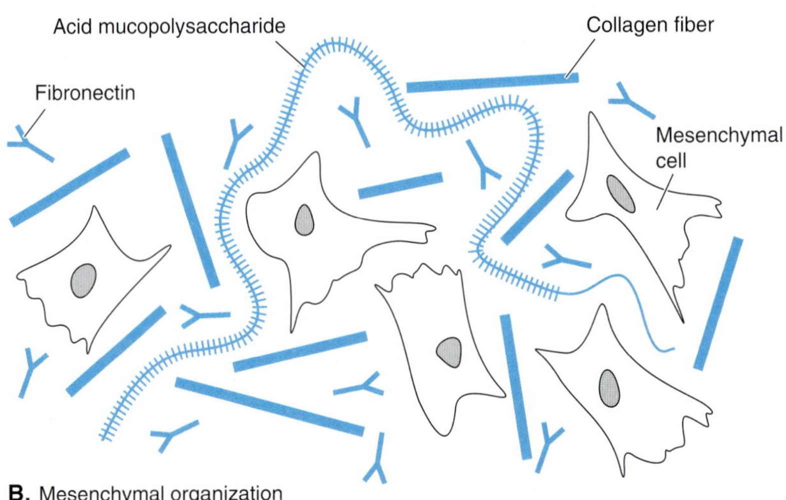

FIGURE 4-5
Models of epithelial and mesenchymal organization. Blue indicates extracellular matrices. *A,* Epithelial organization. In an epithelium, the cells are closely packed and arranged like bricks on a foundation. *B,* Mesenchymal organization. In a mesenchyme, the cells are loosely scattered in an extensive, gelatinous extracellular matrix. These two fundamental ways in which cells can be organized into tissues are important in embryogenesis because certain cell populations change from one type of organization to the other. Also, epithelial–mesenchymal interactions are important in the development of a great many organs.

cells in an epithelial sheet are linked by **junctional complexes** that can be thought of as fasteners that strengthen the continuity of the sheet and help limit passage of materials between them. In this way, epithelia often serve as barriers or filters. During embryogenesis, cells in an epithelium produce shape changes as a collective group. For example, if all of the cells in an epithelium are induced to narrow their apical surfaces, then the surface area of that epithelium will decrease, and folding may result (Fig. 4-4).

In contrast, a mesenchyme consists of a loosely packed array of stellate (= star-shaped) cells (Fig. 4-5*B*). The extracellular matrix of a mesenchyme has a very high water content, and contains fibrillar forms of collagen, such as collagen type I, and proteoglycans containing many chains of hyaluronic acid. These proteins give the extracellular matrix a gel-like consistency, so that the mesenchymal cells can be thought of as "raisins in Jell-O." Especially important is the protein **fibronectin,** which has attachment sites for cell surface proteins and other extracellular matrix molecules. By means of these attachment sites, fibronectin functions as a "glue" between cells and other molecules of the extracellular matrix. Mesenchymal cells generally lack cell surface specializations, such as cilia, and are not polarized. They have little or no physical contact with neighboring cells, so junctional complexes are absent. As a result of these characteristics, small molecules readily diffuse through the extracellular matrix of a mesenchyme. During morphogenesis, mesenchymal cells typically produce shape changes in an embryo by migrating as individual cells and then aggregating to form **organ rudiments** or **primordia** (or **anlagen,** from the German word for plan or design). Mesenchyme also is the source for various connective tissues and support structures of the body (see Chapter 5).

After drawing such a sharp distinction between epithelial and mesenchymal organization, it may come as a surprise to learn that many cell lines actually change from one mode of organization to the other and then back again during normal development. For example, as neural crest cells differentiate, they **delaminate** from the epithelium of the neural plate (discussed later) and then migrate as individual cells for some distance away from the parent epithelium before later reaggregating to form organ rudiments that may become reorganized as epithelia. The process of delamination involves breaking down the local epithelial organization, particularly the basal lamina and any cell junctions that held the cells in a sheet. Also, the organization of some tissues of adult vertebrates does not easily fit into either one of these two categories. Still, it is important to begin analysis of vertebrate embryogenesis with an understanding about these two extremes in tissue organization.

Having introduced some basic aspects of developmental biology, we now turn our attention to gametes and embryogenesis. Our coverage is designed to help you understand the later development, or **organogenesis,** of the organ systems. Knowledge of organogenesis helps us understand organs, their structural and functional complexity, and their homologies, which in turn positions us to interpret the evolution of organs and organ systems. Some of the variation in early development among different groups of vertebrates is considered in this chapter, but an extensive examination of this topic is beyond the scope of this book.[1] Thus, we focus on a few taxa for understanding vertebrate embryogenesis, including elasmobranchs (*Squalus* and *Raja*), frogs (*Rana*), salamanders (*Ambystoma*), and chickens (*Gallus*). (See Focus 4-1.)

Gametes and Fertilization

The reproductive cells, called **gametes,** develop in the gonads. More specifically, spermatozoa (or sperm) form in the seminiferous ampullae or seminiferous tubules of the testis, and ova (or eggs) form in the follicles of the ovary (for gonadal structure, see Chapter 21). The process of sperm formation is called **spermatogenesis,** the process of egg formation is called **oogenesis,** and together these events are termed **gametogenesis** (Fig. 4-6). Two major processes occur during gametogenesis: (1) reduction division, or **meiosis,** and (2) the acquisition of cellular specializations needed for fertilization and embryogenesis.

During gametogenesis, the gamete-producing cells undergo two meiotic divisions, so that mature gametes are haploid; that is, each has only a single set of chromosomes rather than the double set present in other body cells (Fig. 4-6). These two meiotic divisions are called **meiosis I** and **meiosis II.** Terms used to describe gametes during their differentiation are based on their meiotic state. Precursor cells of spermatozoa are **spermatogonia;** those for eggs are **oogonia.** When these cells enter into meiosis I, they are termed **primary spermatocytes** and **primary oocytes,** respectively. In males, completion of meiosis I yields two **secondary spermatocytes;** completion of meiosis II yields four **spermatids.** Spermatids are haploid (= 1N), which means that they have half the diploid complement of chromosomes. The spermatids continue to differentiate during the process known as **spermiogenesis**

[1]Some key references, such as Nelsen (1953), are listed at the end of this chapter for those interested in this topic.

FOCUS 4-1 *Model Species and Vertebrate Embryogenesis*

What species should we study to understand the diversity of developmental mechanisms among vertebrates? As pointed out in Chapter 3, about 50,000 species of vertebrates are alive today. Yet very few of these species have been the subject of detailed embryological investigation. In fact, our knowledge of vertebrate embryogenesis is based on remarkably few "model species" that exhibit characteristics making them particularly suitable for the experimental techniques required for research in modern developmental biology. These few species are by far the overwhelming focus of current research on vertebrate development. Among the vertebrate species commonly studied by developmental biologists, six stand out: (1) a small freshwater teleost fish from southern Asia, the zebrafish (*Danio rerio*); (2) an aquatic salamander from Mexico, the axolotl (*Ambystoma mexicanum*); (3) ranid frogs from the northern hemisphere (e.g., the grass frog, *Rana pipiens*); (4) various species of aquatic frogs, commonly called African clawed frogs (e.g., *Xenopus laevis*); (5) the domestic chicken (*Gallus gallus*); and (6) domesticated strains of the European house mouse (*Mus domesticus*). By almost any standard, these six species do not represent an ideal cross section of vertebrate taxa from the standpoint of phylogenetic breadth. Despite this, many people have generalized findings based on the embryology of these six model species to other taxa of vertebrates, usually by very informal phylogenetic analyses. This practice is now changing for two reasons. First, more researchers are studying and describing in detail the early development of species, such as various species of chondrichthyan fishes or metatherian mammals, that are not traditionally considered "model species" for developmental biology. Second, the types of comparisons being made have become more explicitly phylogenetic, in that researchers are treating embryological features as characters in phylogenetic data sets in exactly the same ways as any other characteristics of organisms. These two progressive changes seem certain to improve our insight into the evolution of development in vertebrates. Still, it is important to ask, why do we so restrict the taxa used for studies of vertebrate embryogenesis?

Primary considerations in selection of model species for embryological research have been convenience and practicality. For example, to simplify the husbandry of a captive breeding colony of vertebrates, researchers find it helpful to study small animals and especially helpful to study small, aquatic species that are broadly tolerant of varying environmental conditions. It is very useful to have a network of similarly minded colleagues and suppliers so that adults (or, even better, fertilized eggs) can be obtained throughout the year, a practice that tends to produce clusters of workers studying development of a particular model species. To help reduce the overall size of a laboratory colony, it is good to study a species that produces large numbers of eggs or young and that can do this frequently (or, even better, constantly under laboratory conditions). Genetic studies needed for the development of mutations and molecular tools for research now are essential components for the most powerful technologies in experimental developmental biology. To apply these technologies, however, the animals must mature rapidly and be easily bred in the laboratory so that the maximum number of generations can be produced in a given time.

Each of the six species listed earlier—with the exception of ranid frogs—meets most of these criteria for convenience and practicality. But can any other taxa be candidates for modern experimental research programs on the comparative embryogenesis of chordates?

The most important candidates from the perspective of improving our phylogenetic coverage of embryogenesis are nontetrapods, and virtually all of the most interesting taxa fail the tests of convenience and practicality. For example, the recent development of methods by Linda Holland and Nick Holland for breeding amphioxus (*Branchiostoma*) in captivity has yielded many new insights into their early development, but the adults still must be collected in the wild, so that developmental genetic studies are limited, at least for now. As another example, directly developing amphibians (i.e., species that bypass the larval stage during development and hatch as miniature adults) have yielded many insights into head formation (see papers by Hanken et al. 1997 and Jennings and Hanken 1998). Nevertheless, most are not good candidates for developmental genetic studies. For the foreseeable future, then, we seem likely to study phylogenetic aspects of embryogenesis using descriptive, rather than experimental, analyses of species not traditionally used as models in developmental biology. This is not all bad, for careful description is fundamental to comparative biology. By means of such descriptions, we seem certain to discover new variations in vertebrate embryogenesis and perhaps to answer some of the most intriguing questions about the origin of these patterns.

into four mature **spermatozoa.** In females, completion of meiosis I yields one **secondary oocyte** and the tiny **first polar body;** completion of meiosis II yields a single haploid **ovum** and the **second polar body** (which is also haploid). The first polar body may also undergo meiosis II to yield two **tertiary polar bodies** that are genetically equivalent to the mature ovum, although they lack any reproductive future because their tiny size limits the amount of yolk and other materials needed for normal development.

Gametes are highly specialized cells, the structures of which closely match their functions (Fig. 4-7). Spermatozoa are specialized for mobility, whereas ova are essentially packages of materials and information that

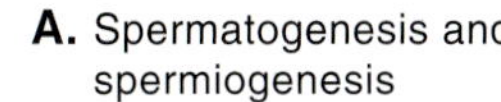

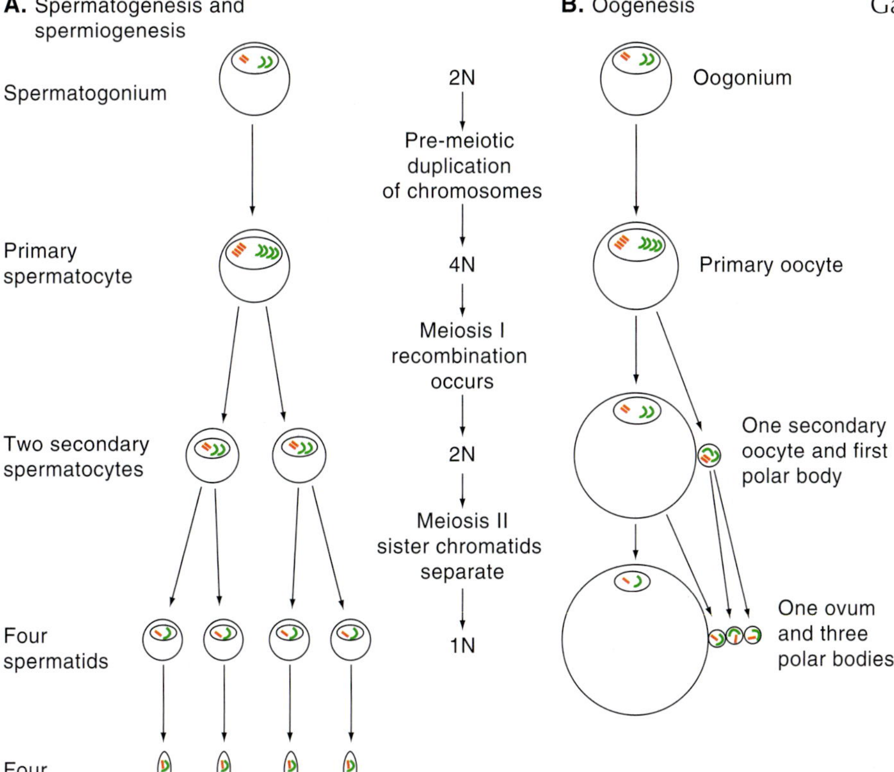

FIGURE 4-6
Changes in chromosome number and DNA content during gametogenesis. *A,* Spermatogenesis and spermiogenesis. *B,* Oogenesis. The terminology for spermatogenesis and oogenesis is explained in the figure.

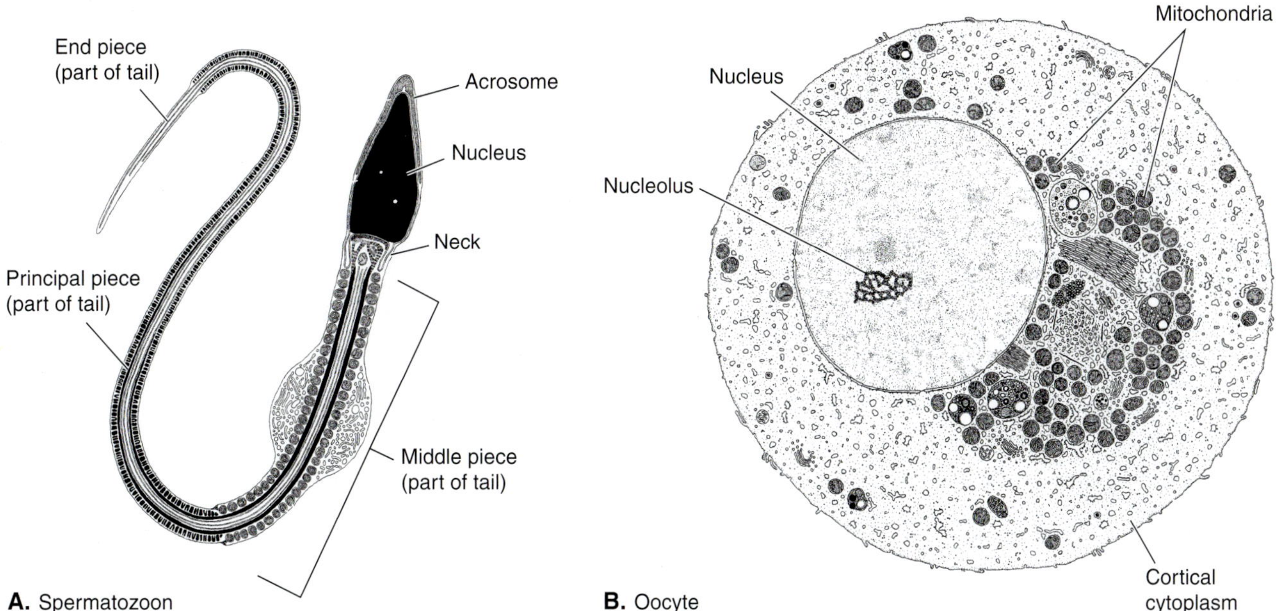

FIGURE 4-7
Human gametes. *A,* Spermatozoon. *B,* Oocyte. These are immature gametes. Before spermiogenesis is complete, spermatozoa lose most of the cytoplasmic droplet, visible as a bulge of cytoplasm in the middle piece. The oocyte shown is in an early stage of vitellogenesis, and it will accumulate additional materials before gametogenesis is complete. Major structures found in spermatozoa of other vertebrates generally resemble the condition in humans, but sperm shape varies greatly in different groups. Placental mammals have secondarily reduced the yolk content of oocytes, which thus are much smaller than eggs of other vertebrates. *(Modified from Lentz.)*

will be needed by the developing embryo. Thus, there is great disparity in the sizes of sperm and egg cells and a corresponding difference in the metabolic cost of producing them. Because of their smaller size, spermatozoa can be produced in extremely large numbers, and it is often the number of eggs that limits fecundity in species of vertebrates.

Spermatozoa

A mature mammalian sperm is a very small cell that has lost most of its cytoplasm during the process of spermiogenesis. It consists of a head, neck, and tail (Fig. 4-7*A*). The tail may be further divided into a principal piece, middle piece, and end piece. The **head** contains the greatly condensed nucleus and enzymes needed to penetrate the egg's surrounding membranes and any adhering cells. These enzymes are located in a membrane-bounded vesicle, the **acrosome,** which covers the front of the nucleus. In some species, the acrosome also contains molecular components of species-specific recognition systems that help to ensure that spermatozoa of one species will not accidentally fertilize an ovum of another species. The **middle piece** contains coiled mitochondria that have the enzymes needed for the release of energy to the flagellum-like axial filament that powers locomotion. Because a spermatozoon is small, it lacks extensive stores of energy, and once it begins to move, it soon exhausts most of its potential for locomotion. Thus, spermatozoa are maintained inside a male's reproductive tract in an immobile state and are activated to begin swimming only by changes encountered in a surrounding aqueous medium or by the fluids in the female's reproductive tract. Often, the trigger that activates spermatozoa to swim is a change in pH.

Spermatozoa of different species of vertebrates vary greatly in the sizes and in the shapes of their heads, and sperm shape may be phylogenetically informative. For example, spermatozoa of rodents typically have sickle-shaped heads, whereas those of most other species of mammals are elliptical. Spermatozoa of some groups of bony fishes normally have two tails and are referred to as biflagellate spermatozoa.

If fertilization occurs in the environment, outside of the female's reproductive tract, it is termed **external fertilization.** If fertilization occurs inside the reproductive tract of the female, it is termed **internal fertilization.** External fertilization is plesiomorphic for vertebrates and occurs in many extant species of aquatic vertebrates, such as lampreys, many species of bony fishes, and many species of frogs. In such species, a male releases spermatozoa directly from his reproductive tract near or onto eggs as they are extruded from the female's reproductive tract into the environment. Most species of vertebrates that have internal fertilization have evolved specialized mechanisms that allow a male to introduce spermatozoa directly into a female's reproductive tract by using an **intromittent organ** (see also Chapter 21). For example, many groups of fishes have specializations of the pelvic fins (e.g., claspers of chondrichthyans) or anal fins (e.g., gonopodia of many actinopterygians) that are used for intromission. In salamanders with internal fertilization, however, spermatozoa are packaged into **spermatophores** by the reproductive tract and cloacal glands of males. During courtship, a male salamander deposits spermatophores on the substrate, which may be either the bottom of a pond or leaf litter on land, depending on where that species breeds. The male then leads a female over the spermatophore, which she encloses with her cloacal lips to bring the packet into her reproductive tract. The package dissolves, liberating the spermatozoa to fertilize the eggs. Another interesting variant concerns **sperm storage,** which is remarkably common among vertebrates (the anatomy of reproductive tracts, including specializations for sperm storage, is described in Chapter 21). For example, in certain species of caecilians, the female reproductive tract has regions specialized for the storage and nourishment of sperm, so that sperm can be stored for extended periods before activation of their swimming mechanism—and fertilization—occur.

Egg Cells

Unlike spermatozoa, eggs are large, spherical, nonmotile cells. Within any given species, a mature egg, or ovum, is a larger cell than a spermatozoon chiefly because it contains the energy reserves and other materials in its cytoplasm to initiate embryonic development. Beyond that generalization, the sizes of vertebrate egg cells vary enormously: a human egg is about 0.15 mm in diameter, whereas the egg of a coelacanth (*Latimeria*) reaches 90 mm in diameter. This 600-fold difference in linear dimensions translates to a 200 million–fold increase in the total volume of materials inside the egg! Ova develop in **follicles** within an **ovary** (see Chapter 21). All egg cells have, in addition to their plasma-cell membrane, a **primary egg cell membrane** immediately surrounding them. This extracellular, largely proteinaceous structure is secreted by the egg cell or by its surrounding follicle cells during oogenesis. The primary egg cell membrane has different names in different groups of vertebrates. One frequently used term is **vitelline membrane.** In mammals, the primary egg cell membrane is known as the **zona pellucida.** In many actinopterygian fishes, it is toughened to withstand a harsh external environment and is known as a **chorion.**[2] The thick chorion of actinopterygian fishes has special openings, termed **micropyles,** to admit spermatozoa. In some vertebrates, layers of follicle cells adhere to and surround the primary egg cell membrane. In mammals these adhering follicle cells are known as the **corona radiata,** and they pose an additional structural barrier that a spermatozoon must penetrate before fertilizing the egg.

The cellular organization of an ovum can be exceedingly complex (Fig. 4-7*B*). The cytoplasm of egg cells has a thick, gelatinous **cortex,** which is the zone of cytoplasm lying immediately adjacent to the egg's plasma membrane. The cortex often contains pigment granules and **cortical granules,** which are important in fertilization, as described later. The interior cytoplasm of an egg cell may be more fluid than its cortex, but it is still highly structured and packed with materials needed for development. Ova contain variable amounts of yolk, which is composed primarily of protein, phospholipids, and neutral fats. The process of yolk deposition in the egg cell is termed **vitellogenesis,** and depending on the amount of yolk, vitellogenesis may require many months or even years. Yolk of vertebrate eggs is usually synthesized exogenously (i.e., not in the oocyte itself) in the mother's liver and delivered to the ovary via the circulatory system. Once inside an egg cell, much of the yolk is organized into organelles known as **yolk platelets.**

The amount of yolk in an ovum determines how long the embryo will be nourished by food stored in the egg. There is little yolk in the **microlecithal eggs** (Gr., *mikros* = small + *lekithos* = yolk) of amphioxus. These eggs hatch very soon into larvae that do not feed and that quickly metamorphose into feeding juveniles. Nonteleostean actinopterygian fishes, lungfishes, and amphibians have an intermediate amount of yolk in their **mesolecithal eggs.** Their eggs hatch into feeding larvae, which then metamorphose into juveniles. The eggs of most teleost fishes are physically small but have a very high proportion of yolk, which affects their cleavage (see Focus 4-2). Like other actinopterygians, teleosts follow the pattern of having feeding larvae. Much yolk is present in the large **macrolecithal eggs** of chondrichthyan fishes, coelacanths, reptiles, birds, and monotreme mammals, the embryos of which develop into miniature adults before hatching or birth. We also can describe the organization of the yolk itself. The term **plasmolecithal** (Gr., *plasmo* = fluid) describes an ovum in which the yolk is in suspension in the cytoplasm. In contrast is the descriptor **telolecithal** (Gr., *telos* = end), in which a great deal of yolk is packaged into yolk platelets.

Materials in an egg are not distributed randomly but establish and follow gradients. The small amount of yolk in the microlecithal egg of amphioxus is evenly distributed. In vertebrate eggs, yolk usually is most concentrated toward one end, the **vegetal pole,** and least concentrated at the other end, the **animal pole** (Fig. 4-8*A*). This gradient is particularly evident in macrolecithal eggs, where the egg nucleus and most of the cytoplasm are restricted to the animal pole. The future anteroposterior axis of the embryo is not the same as the egg axis but is related to it and is usually finally determined by the point of entry of sperm. Often the anterior end of the embryo is determined about midway between the animal pole and the equator of the egg (Fig. 4-8*B*).

In addition to yolk, vertebrate eggs are packaged with cytological necessities for early development of the zygote. Billions of **ribosomes** (related to protein synthesis) and thousands of **mitochondria** (related to ATP production) can be packed into a frog egg only a few millimeters in diameter. These organelles provide for future protein synthetic and energy needs of the embryo. Some **mRNA transcripts** of genes for later translation into proteins needed by the zygote are already prepositioned in the egg cell's cytoplasm. Still other information-containing molecules, such as the **germinal granules** of frogs, already are localized to parts of the oocyte, where they will help determine the differentiation of the germ cells in the adult.

At **ovulation,** an egg is discharged from its follicle and the ovary. A stimulus of some sort is needed to activate the egg and initiate further development. Normally, the stimulus is sperm penetration, but other chemical or physical stimuli are effective in some species, in which case the egg develops **parthenogenetically** (i.e., without the contribution of the DNA of the sperm). If an egg is not activated within a few hours of ovulation, its delicately balanced internal organization breaks down, and the egg degenerates.

[2]The "chorion" of an actinopterygian egg (a primary egg cell membrane) is not homologous to the chorion of amniotes (which is a secondary egg cell membrane; see section, "Secondary Egg Cell Membranes and Extraembryonic Structures").

Text continues on page 130

FOCUS 4-2 *Cleavage, Gastrulation, and Neurulation of Actinopterygians*

As discussed in Chapter 3, Actinopterygii is an enormous group, including more than 25,000 living species. The species diversity of actinopterygians is reflected by many differences in their patterns of development. This focus compares aspects of the development of nonteleostean actinopterygians (the North American paddlefish, *Polyodon spathula,* and closely related sturgeons in the genus *Acipenser*), to derived teleosts (exemplified by zebrafish, *Danio,* and killifish, *Fundulus*). These taxa differ greatly in adult size, mode of life, spawning biology, and many developmental characteristics.

In the figure, *A* outlines the main periods of early development in the paddlefish: the terminology summarized in this figure can be applied to most actinopterygians. The embryonic period is defined as the period up to hatching. Much of early organogenesis is completed in this period, but the yolk has only been partially used. During the yolk sac larval period (*B*), the hatchling rapidly uses its yolk reserves to complete later organogenesis. The yolk sac larval period is usually a very dynamic time during embryogenesis of actinopterygians: the mouth opens, gill ventilatory and locomotory movements start, and the sensory systems finish their development, until, by the end of the period, the free-living larva is ready to start feeding (*C*). The free-living larva feeds on small plankton and grows very rapidly, and soon thereafter takes on the appearance of a miniature adult, at which time it may be called a juvenile (*D;* see Bemis and Grande, 1992).

The eggs of most actinopterygian fishes are small, usually no more than a few millimeters in diameter. Typically, they are surrounded by a tough outer chorion (ho-

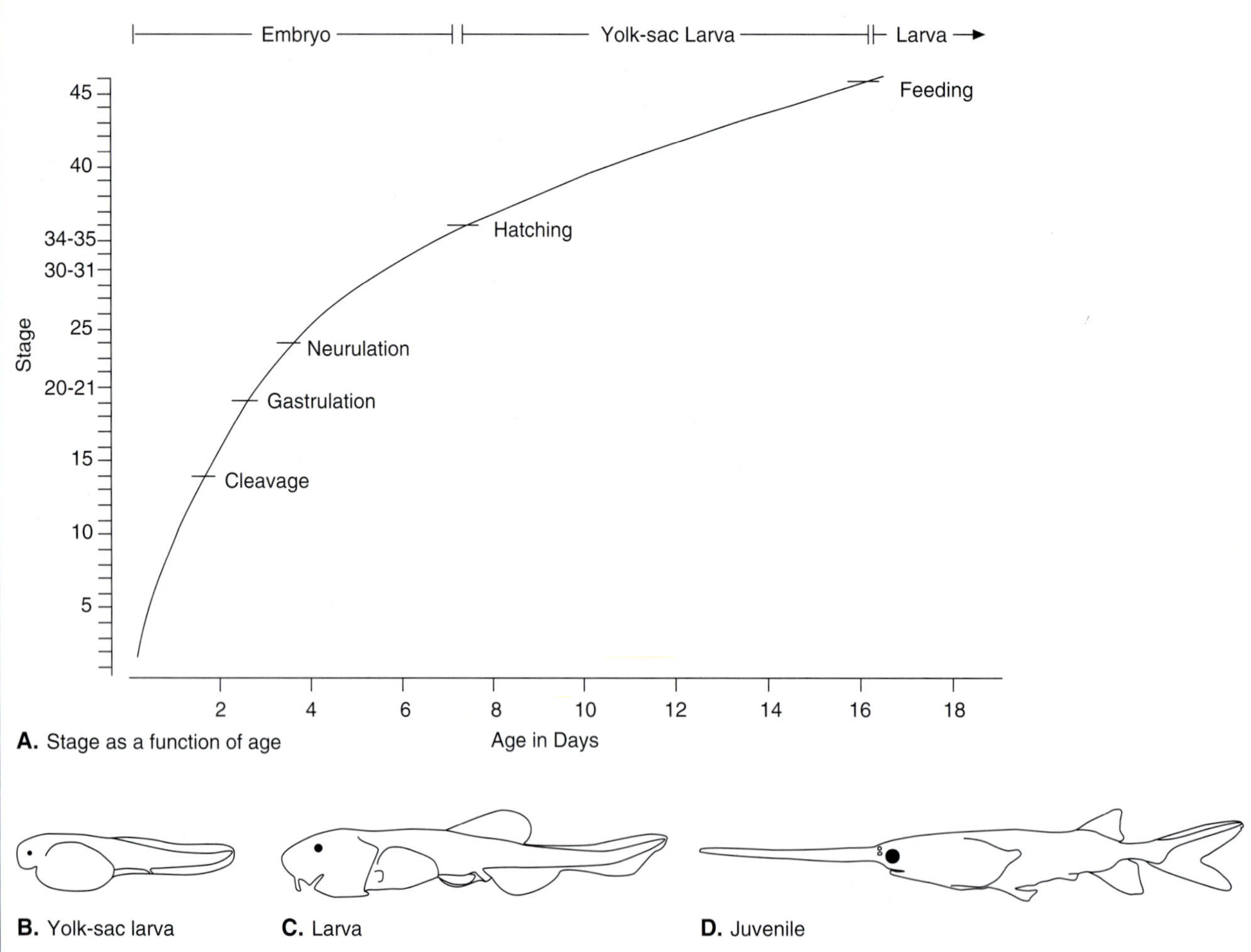

A. Stage as a function of age

B. Yolk-sac larva **C.** Larva **D.** Juvenile

mologous to the primary egg cell membrane), which has tiny openings, known as micropyles, to admit sperm. Teleostean eggs are usually yolky and often have a large oil droplet floating in the cytoplasm. The oil serves as nutrient for the embryo and as a buoyancy regulator in many taxa. In nonteleostean actinopterygians, such as paddlefishes or sturgeons, eggs are spawned directly onto the gravel bottom of a river. Eggs that develop in contact with the bottom are said to be demersal. The opposite condition is seen in many marine teleosts, in which the eggs are said to be pelagic because their enclosed oil droplet causes them to float at or near the water's surface. The eggs of other species of marine teleosts and virtually all freshwater teleosts are denser than the water, and, if spawned in midwater, sink to the bottom. This is the mode of spawning seen in zebrafish. Still other teleosts, such as killifish, attach individual eggs to chosen spots in the environment, such as submerged leaves, relying on the sticky jelly coat surrounding the egg to hold it in place throughout development.

Cleavage in paddlefishes or sturgeons is unequal and holoblastic and generally resembles patterns seen in amphibians. Gastrulation involves movements of cells at both the animal and vegetal poles. The resulting embryo neurulates by rolling up the neural tube in a pattern very similar to that seen in amphibians. This type of cleavage and gastrulation is presumably the plesiomorphic pattern for osteichthyans.

Teleosts exhibit many differences from these patterns. The egg cell nucleus in species such as the zebrafish (*Danio rerio*) undergoes repeated nuclear divisions, or karyokinetic events, yielding a yolky egg cell with many nuclei. Such yolk is termed syncytial yolk. In teleosts, cleavage is discoidal and meroblastic, resulting in a blastoderm that rises as a cap of cells at the animal pole on top of an uncleaved mass of yolk (*E*).

During gastrulation in teleosts, the blastoderm expands greatly, a movement known as epiboly. Epiboly sweeps the margins of the blastoderm (known as the germ ring) around the egg (*F–I*). Epiboly simultaneously encloses the yolk and generates the three primary germ layers as well as the embryo's axis. The rapid expansion of the blastoderm to cover the yolk is accomplished largely by the cells of the enveloping layer on the outer surface of the blastoderm. Note in Figure I that the embryo's nervous system, brain, optic vesicles, and somites rapidly form before gastrulation is even completed (gastrulation is not complete until the yolk plug is no longer visible). Interestingly, neurulation in teleosts occurs by the formation of a solid rod of cells that subsequently develops its central canal via a process known as cavitation (Fig. 4-19*B*). This differs from the pattern of neural tube development seen in most other vertebrates. At first, the embryo's head and tail are tightly bound to the underlying yolk. Soon, however, they will lift away during processes known as head formation and tail bud formation.

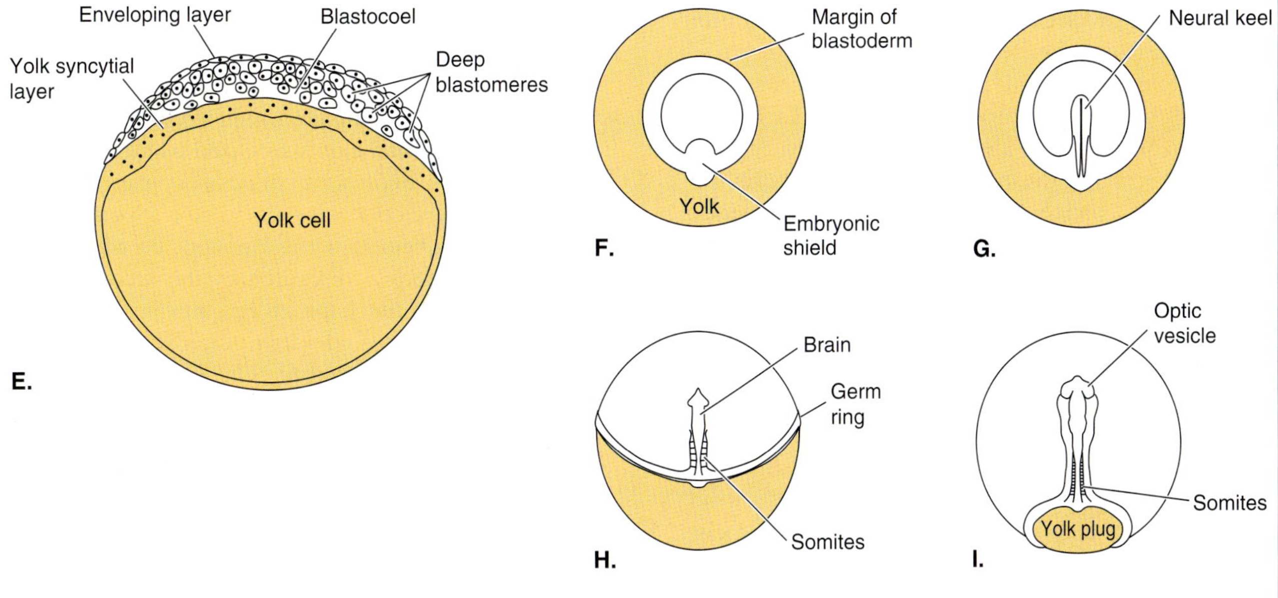

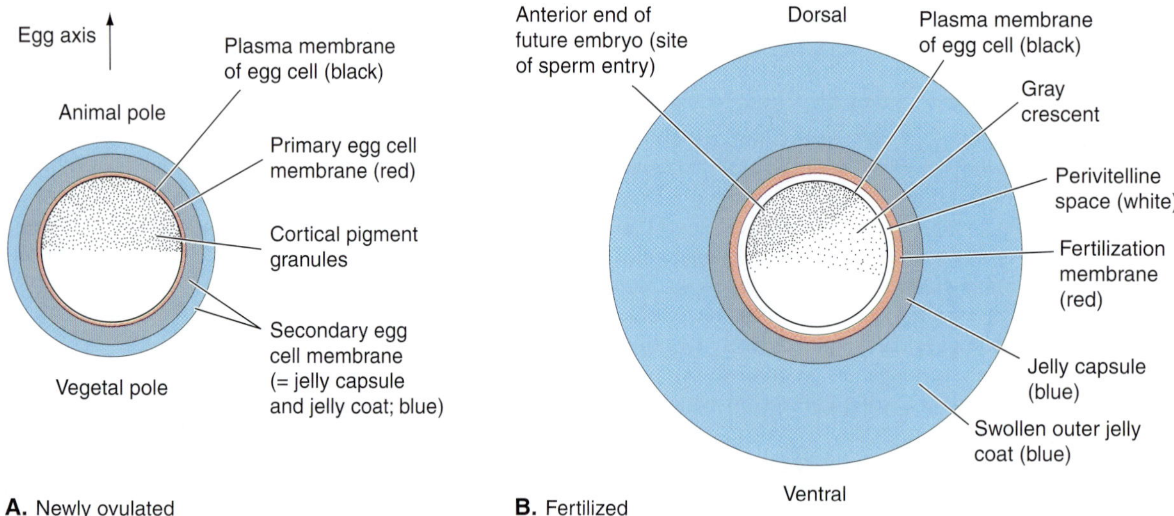

FIGURE 4-8
An amphibian egg, based on the frog, *Rana*. *A,* Newly ovulated. *B,* Some time after fertilization but before the first cleavage. The primary egg cell membrane (*pink*) lifts away from the egg cell's plasma membrane during the process of fertilization, which creates the perivitelline space. Once this has occurred, the primary egg cell membrane can be termed the fertilization membrane. The jelly capsule and outer jelly coat (*blue*) are secondary egg cell membranes that are secreted by the female's oviduct after the egg is ovulated. The outer jelly swells after contact with water. The point of fertilization is indicated; cortical cytoplasm rotates toward the site of sperm entry soon after fertilization has occurred. This leaves a few of the cortical pigment granules to form the gray crescent. The dark pigment absorbs solar radiation, which warms the egg and quickens development. Dark pigment also helps camouflage the egg. Jelly coats surrounding the egg insulate and protect it from predation and provide some mechanical protection.

Fertilization

Fertilization involves sperm penetration, the combination of male and female nuclear material, and egg activation. Sperm penetration can be complicated because each egg cell is surrounded not only by its own plasma membrane but also by the primary egg cell membrane secreted during oogenesis. The sperm undergoes many changes as it approaches a **conspecific** egg. (The term conspecific refers to another individual of the same species.) Membranes around its acrosome break down, releasing enzymes that will break down the primary egg cell membrane, and a stiff acrosomal filament may form. A mammalian sperm is said to be **capacitated** when it is ready to penetrate the primary egg cell membrane and any other materials, such as adhering follicular cells of the corona radiata, that may surround the egg.

Contact of the sperm head with the plasma membrane of the egg initiates a complex **cortical reaction** in the egg. The cortical reaction draws the sperm head into the egg and concurrently releases materials from the cortical granules that raise the primary egg cell membrane from the egg surface, preventing other sperm from entering. The primary egg cell membrane is now called the **fertilization membrane** (Fig. 4-8*B*).

At the time of ovulation, the eggs of many vertebrates have not yet completed their meiotic divisions. For example, a human egg is fertilized while it is still a secondary oocyte because its nucleus is arrested midway through the second meiotic division. Sperm entry triggers the completion of this division, by which one set of chromosomes is discarded in the second polar body. This leaves the haploid egg nucleus ready to combine with the haploid sperm nucleus to form the diploid nucleus of the **zygote** (Gr., *zygon* = yoke or union).

Finally, fertilization triggers a redistribution of materials within the egg's cytoplasm that activates the egg and establishes the plane of bilateral symmetry, if one was not already established in the unfertilized egg. Such redistribution is particularly evident in eggs of frogs in the genus *Rana,* in which the animal hemisphere is heavily pigmented. Cortical pigment granules shift after sperm entry and leave a **gray crescent** on one margin of the equator of the egg (Fig. 4-8*B*). The gray crescent marks the presumptive posterodorsal part of the embryo. The first cell division, or cleavage, of the zygote occurs soon after this redistribution of materials.

Cleavage

Cleavage is a period of rapid cell division, during which the unicellular zygote is converted into a multicellular embryo known as the **blastula** (Gr., *blastos* = germ or bud). Unlike typical mitotic divisions, no cytoplasmic growth occurs during cleavage, so the cells, which are called **blastomeres** (Gr., *meros* = part), become smaller and smaller. The rate of cell divisions during cleavage is higher than that at any other stage in development, and the number of cells in the zygote increases rapidly (Fig. 4-9).

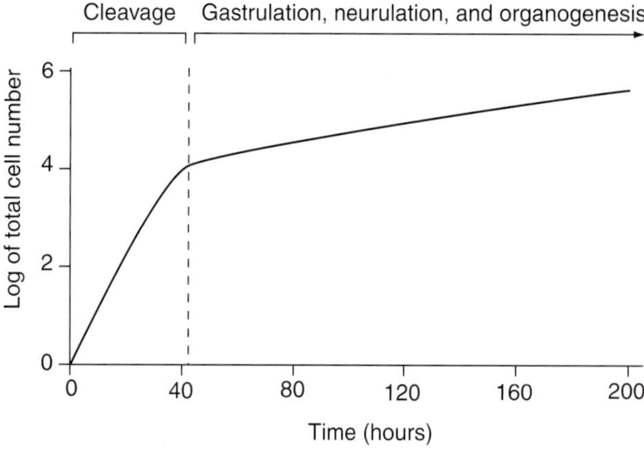

FIGURE 4-9
Rate of cell increase during cleavage. This diagram is based on a frog embryo; similar patterns occur in other vertebrates. *(Based on a figure in Balinsky.)*

The pattern of cleavage is correlated with the amount and distribution of yolk. In microlecithal and mesolecithal eggs, the entire zygote divides, so cleavage is described as complete, or **holoblastic. Cleavage furrows** mark the separation between daughter cells as the egg divides. The first cleavage lies in the vertical plane, extends from the animal to the vegetal pole, and divides the embryo into prospective left and right sides (Fig. 4-10*A* and *B*). This cleavage bisects the gray crescent in amphibian zygotes. The second cleavage, also in the vertical plane (except in mammals[3]), is at right angles to the first and results in the formation of four cells (Fig. 4-10*A* and *B*). The large amount of yolk in mesolecithal eggs slows the separation of the blastomeres, so the first cleavage furrow does not reach the vegetal pole before the second one begins at the animal pole (Fig. 4-10*B*).

The third cleavage usually lies in the horizontal plane, and it divides the embryo into eight cells. It lies near the equator in embryos that developed from microlecithal eggs (Fig. 4-10*A*), but it is displaced toward the animal pole in those that came from mesolecithal eggs (Fig. 4-10*B*). The resulting blastomeres in such embryos are unequal in size, with those near the vegetal pole being much larger than those near the animal pole. Cleavage continues in this fashion, tending to alternate between the vertical and horizontal planes. The cells that are formed remain close to the periphery, where gas and other exchanges with the environment occur. Because no growth occurs, and stored energy is used, the mass of the embryo decreases. A space termed the **blastocoele** appears within the embryo. The blastocoele is centrally located within blastulas that develop from microlecithal eggs, but it is displaced toward the animal pole in those that develop from mesolecithal eggs because of the large size of the yolk-filled cells in the vegetal hemisphere. The blastocoele does not contribute to any structures of the adult and is soon obliterated during the process of gastrulation (discussed later).

So much yolk is in macrolecithal eggs of turtles, squamates, crocodilians, and birds that cleavage is limited to a **cytoplasmic disk** at the animal pole (Fig. 4-10*C*). Cleavage in such eggs is described as incomplete, or **meroblastic,** because the cleavage furrows do not extend into the large yolk mass. As a consequence, cleavage results in the formation of a thin, skinlike disk of cells, the **blastoderm** (Gr., *derma* = skin) or **blastodisk** that lies on top of the yolk. A narrow cavity, termed the subgerminal space, lies between the blastoderm and yolk. Part of this cavity is considered homologous to the blastocoele.

The significance of cleavage is that it converts a unicellular zygote into a multicellular embryo. The ratio of the volume of the nucleus to the volume of the cytoplasm, which is low at the onset of cleavage, increases nearly to that of adult body cells. All nuclei of vertebrate blastomeres carry the same genetic information, but the nuclei lie in cytoplasm with different qualities because the cytoplasmic materials in the egg cell were unequally distributed from the start. Thus, individual blastomeres differ in the amount of yolk and

[3]Most mammals exhibit a variant form termed rotational cleavage, in which the second cleavage passes through the horizontal plane of one of the two blastomeres; see section, "Developmental Modifications of Eutherian Mammals" on page 144.

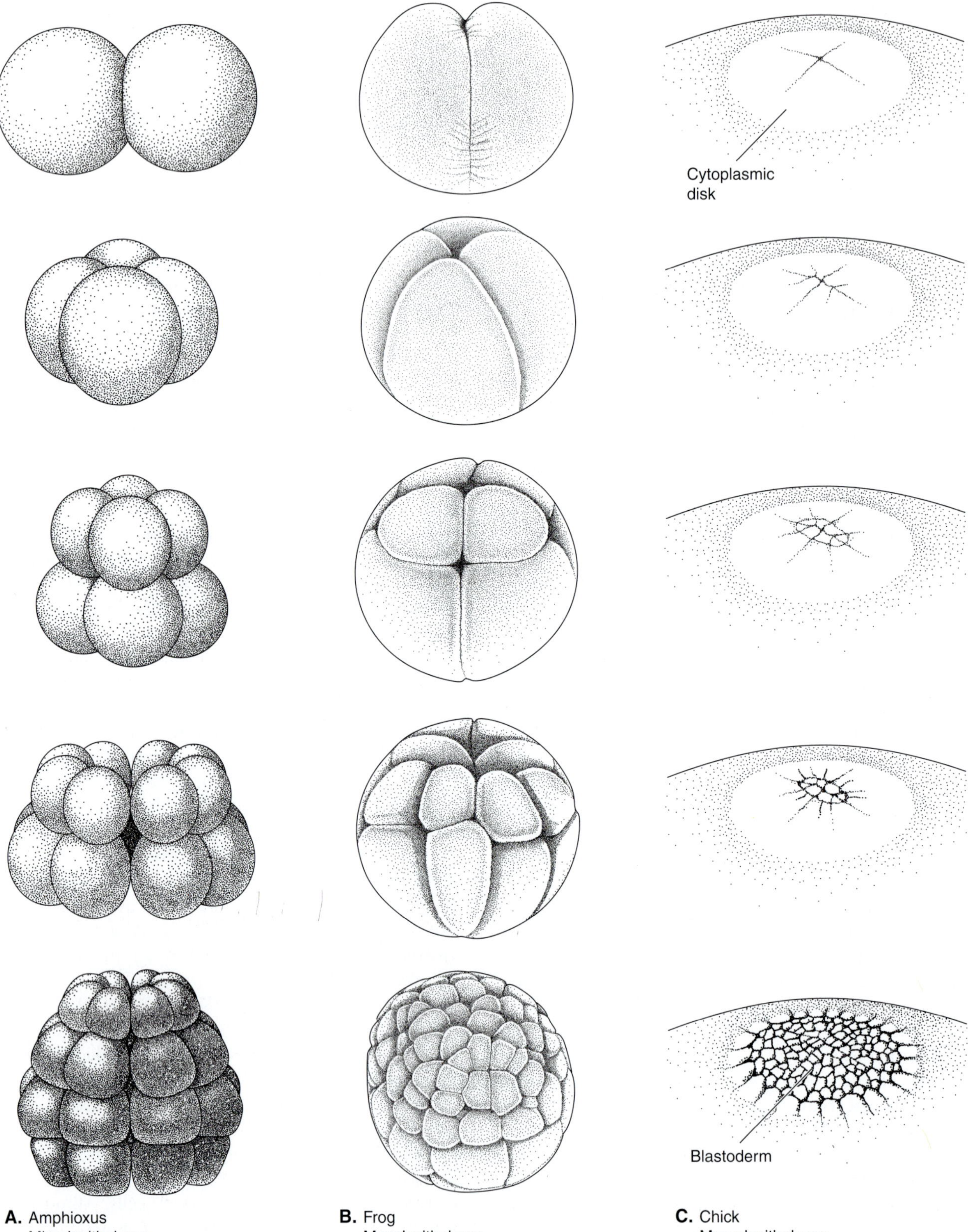

FIGURE 4-10
Three patterns of cleavage from the two-cell stage to the early blastula of amphioxus, an amphibian, and a bird. *A*, Amphioxus (*Branchiostoma*). *B*, Frog (*Rana*). *C*, Chicken (*Gallus*). *(Modified from figures in Balinsky.)*

in the types of nucleic acid transcripts, organelles, enzymes, morphogens, and other substances that they contain. These differences influence the selection and sequence of expression of nuclear genes during subsequent stages of development.

Fate Mapping

The general term **fate mapping** is used to describe methods for tracing the fates of cells, tissues, or organ rudiments during development. These techniques provide some of our most powerful insight into the development of patterns in embryos. Vertebrate embryologists adopted standard colors for fate maps, using yellow for endoderm, red for mesoderm, green for chordamesoderm, and blue for ectoderm. Fate maps often are constructed by marking a population of cells and then following their subsequent development or by transplanting a portion of one embryo with different characteristics (i.e., pigmentation) into another embryo. A third general method is to remove, or extirpate, a portion of an embryo and observe what deficiencies exist in later stages.

In its simplest form, fate mapping can be used to study the fates of the surface cells of a blastula. To do this, parts of the surface of the blastula are stained with vital dyes, and the color spots are then traced as development proceeds. By this method, the general fates of different regions of the blastula stage were demonstrated long ago. Some regions will become gut, others, neural tube, epidermis, notochord, head mesoderm, and so on. More recently developed methods that offer precision tracing include injecting single cells with fluorescent carbocyanin dyes, such as DiI. In a few species, such as chickens, retroviruses also can be injected to infect a population of cells, which can be later identified as constituents within a complex organ, such as a muscle. Sometimes cell movements can be traced by noting the types of organelles that they contain. For example, it is possible to trace migrating neural crest cells in amphibian embryos because they have smaller yolk platelets than the stationary cells that they are passing.

One of the most powerful of the transplantation methods has been the **quail-chick chimaera technique,** which allows for high-resolution study of the development of chick embryos. This has proved important for studies of organogenesis (especially of the nervous system and muscles) in amniotes. Another important transplantation method utilizes pigmented and albino axolotls, *Ambystoma mexicanum.* Transplantation of cells or organ rudiments from pigmented embryos into albino embryos allows for easy visualization of their fate. This has been important in understanding the genesis of taste buds and lateral line sense organs (see Chapter 12).

It is useful to compare simplified fate maps of amphibian and avian blastulas because this will help you understand basic differences in their modes of gastrulation. The choice of amphibian species for this comparison is important, for not all species of amphibians have comparable fate maps. The generalized example shown in Figure 4-11*A* is based on frogs of the genus *Rana;* it also approximates the condition in salamanders in the genus *Ambystoma.* Frogs of the genus *Xenopus,* however, are different from the species shown in that no cells fated to form mesoderm (red) are exposed on the surface of the blastula.[4] The significance of this phylogenetic difference is unclear, but *Xenopus* is one of the most widely studied model species in developmental biology, which makes this difference of more than just passing interest (see Focus 4-1). The generalized example shown in Figure 4-11*B* is based on the domestic chicken, *Gallus.* First, note that the blastula of the frog is spherical, whereas that of the chicken is flat. Next, notice that all three of the primary germ layers (i.e., ectoderm, mesoderm, and endoderm) are visible on the surface of both blastulas. In the frog, these primary germ layers form three horizontal bands around the egg. In the chick, the primary germ layers form three off-center targets, with cells fated to form endoderm in the middle, those fated to form mesoderm surrounding the endoderm, and those fated to form ectoderm forming the outermost circle.

Gastrulation, Mesoderm Formation, and Early Neurulation

Cleavage is followed by **gastrulation,** when cells with different fates move to appropriate parts of the embryo for further differentiation. Such **morphogenetic movements** result from active migration of cells, changes in the sizes and shapes of cells, and different rates of cell division in different parts of the embryo. The single-layered blastula is converted to a **gastrula** (L., *gastrula* = little stomach) with well-defined layers of tissue known as the **germ layers.** One germ layer, the endoderm, turns inward to form the gut, or **archenteron,** which at this stage opens to the surface

[4]Recent research indicates that a few mesodermal cells may be exposed on the surface of the blastula of *Xenopus.*

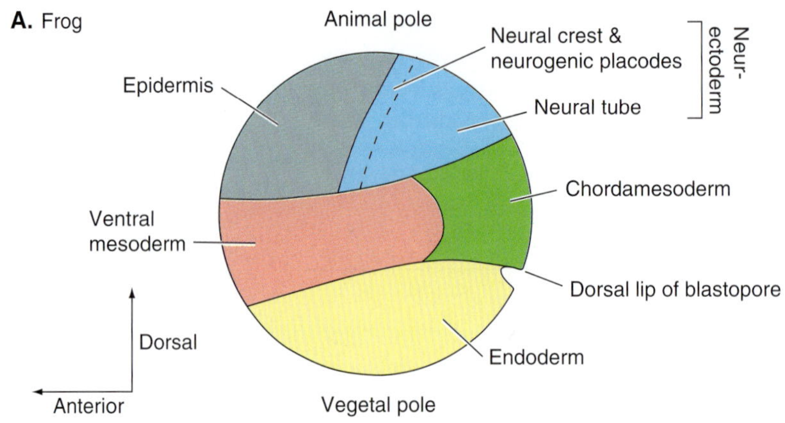

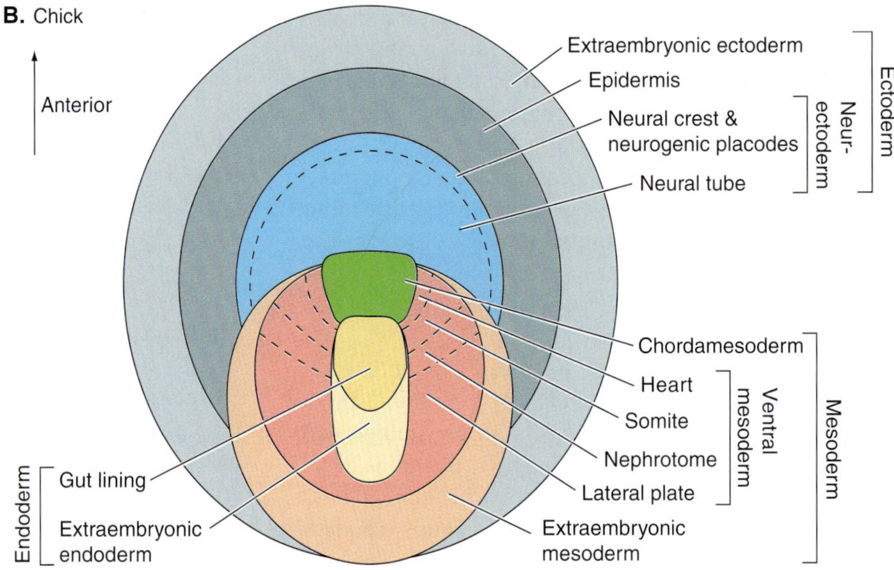

FIGURE 4-11
Simplified fate maps of the blastula of vertebrate embryos. *A,* View of the left side of the blastula of a frog (based on *Rana*). *B,* View looking down on the cytoplasmic disk of an amniote (based on the chicken, *Gallus*). Shades of yellow denote endoderm, shades of red denote mesoderm, green denotes chordamesoderm, and shades of blue denote ectoderm. Inset arrows indicate orientation. The dorsal lip of the blastopore shown in *A* marks the point of involution during gastrulation.

only by a **blastopore** at the posterior end of the gastrula (Fig. 4-12*D*). The blastopore is at or near the prospective anus of the embryo, a condition that characterizes all deuterostomes (e.g., echinoderms, chordates, and relatives; see Chapter 2). The mouth will later break through at the opposite end of the embryo. The **endoderm** will form the lining of most of the digestive tract and the glandular cells that develop from it (Table 4-1). A second germ layer, the **ectoderm,** covers the surface of the embryo and will form the epidermis of the skin; the nervous system; and major sense organs, such as the nose, eye, taste buds, ear, lateral line, and electrosensory systems (Table 4-1). The **neural crest** is a special population of cells derived from the ectoderm that will form components of the nervous system as well as many other structures, including pigment cells, cartilages, and bones (see the section, "Neural Crest and Neurogenic Placodes", page 149). **Neurogenic placodes** are another source of neurons for the developing nervous system (see later section, "Neural Crest and Neurogenic Placodes"). Cells that will form the third germ layer, the **mesoderm,** move into the roof of the archenteron (Table 4-1). The mesodermal cells that form the mid-dorsal part of the roof of the archenteron are referred to as the **chordamesoderm** (Fig. 4-12*F*), which will give rise to the notochord. Separation of the mesoderm from the other germ layers may occur during or after gastrulation. The mesoderm will form cartilages, bones, tendons, ligaments, muscles, the heart, blood vessels, the lining layers of the coelomic cavities, and many other tissues and organs (Table 4-1). As gastrulation ends, the ectoderm overlying the chordamesoderm is induced by the primordial notochord to thicken to become the **neural plate,** from which the brain, spinal cord, and neural crest will be derived (Fig. 4-12*E*). Gastrulation, neurulation, and mesoderm formation occur as three sequential and overlapping waves of development, generally in a cranial (anterior) to caudal (posterior) sequence.

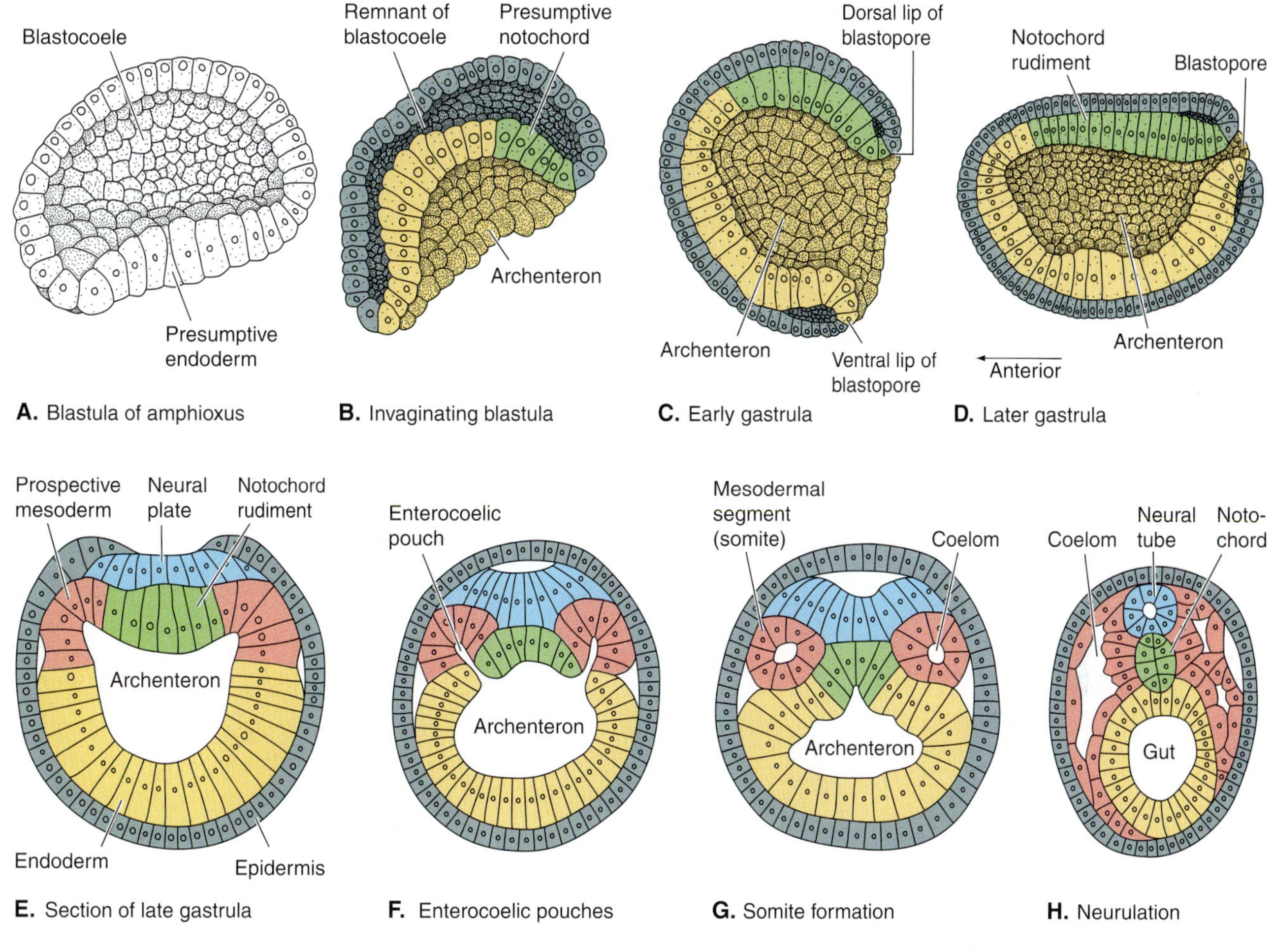

FIGURE 4-12
Gastrulation, mesoderm formation, and neurulation in amphioxus. *A–D,* Sagittal sections of the formation of the gastrula. *E–H,* Transverse sections through late gastrulas and neurula to show the formation of the mesoderm, notochord, and neural tube. Ectoderm is shown in blue, endoderm in yellow, presumptive notochord in green, and mesoderm in red. (*A–D, After Conklin; E–H, after Hatschek.*)

The organization of the neurula stage of the embryo foreshadows the structure of the adult. Table 4-1 summarizes the developmental pathways leading to formation of the various organs. Each pathway leads from one of the three germ layers. Each organ is identified with a particular germ layer, although in nearly all cases more than a single germ layer contributes to the final adult form of the organ. Consider, for example, the pancreas: pancreatic secretory cells develop from embryonic endoderm cells lining the gut tube, but its blood vessels and connective tissue develop from mesoderm, and its nerve supply develops from neural crest.

Chordates differ in the mechanisms by which they gastrulate. Such differences are described in the following brief comparative overview. Bear in mind, however, that the end products of gastrulation are very comparable among all vertebrates. When such developmental stages are recognizably similar across broad phylogenetic gaps, they are known as **phylotypic stages.**

Gastrulation by Invagination: Amphioxus

The pattern of gastrulation is relatively simple in amphioxus, and its serves as a good model for the major events seen in other types of gastrulation. Its microlecithal egg develops into a blastula composed of small cells; those of the vegetal hemisphere are not much larger than other ones. The vegetal pole flattens and then folds inward by a process called **invagination**

TABLE 4-1 Derivatives of the Three Primary Germ Layers

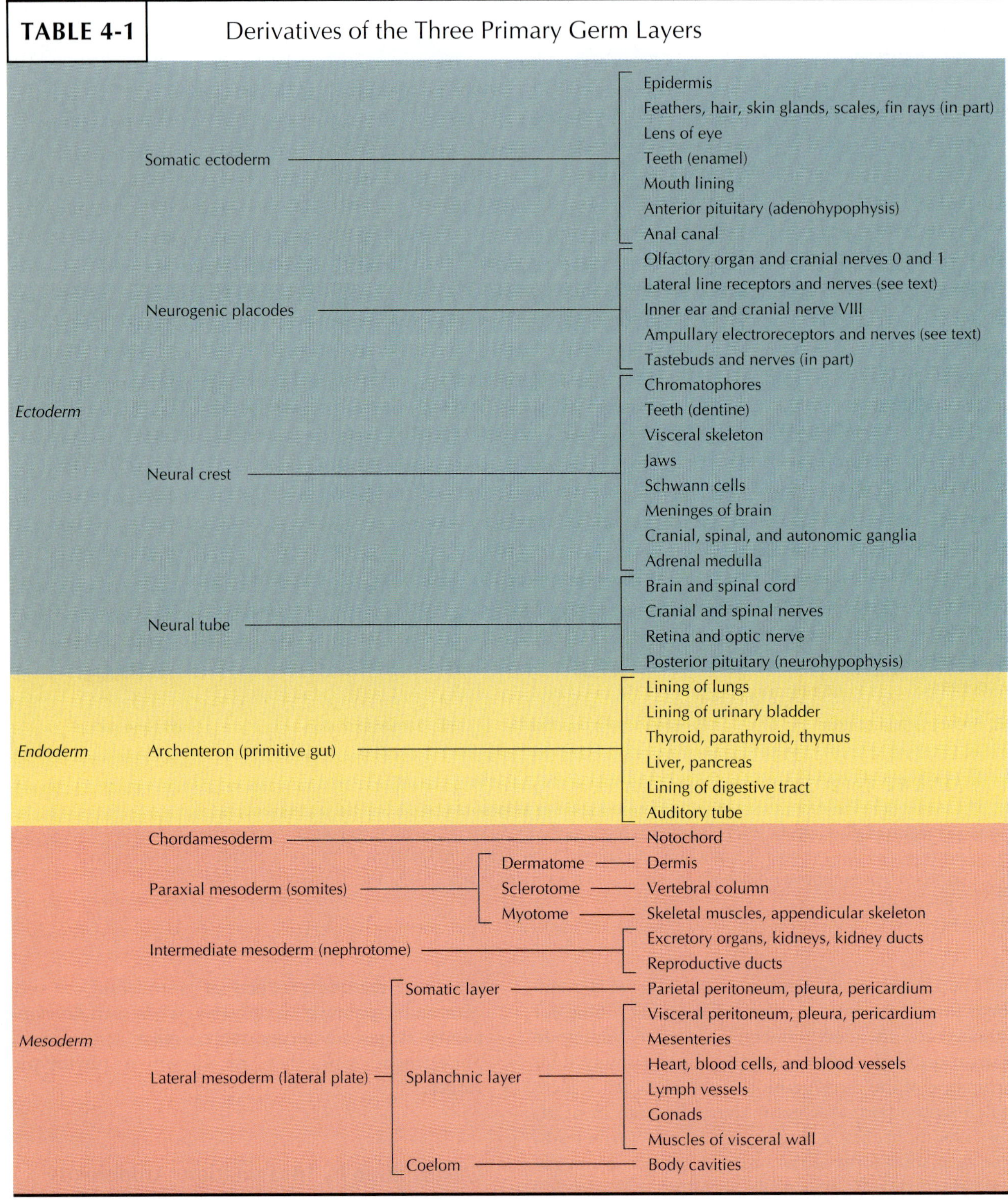

Germ layer	Primordium	Subdivision	Derivatives
Ectoderm	Somatic ectoderm		Epidermis
			Feathers, hair, skin glands, scales, fin rays (in part)
			Lens of eye
			Teeth (enamel)
			Mouth lining
			Anterior pituitary (adenohypophysis)
			Anal canal
	Neurogenic placodes		Olfactory organ and cranial nerves 0 and 1
			Lateral line receptors and nerves (see text)
			Inner ear and cranial nerve VIII
			Ampullary electroreceptors and nerves (see text)
			Tastebuds and nerves (in part)
	Neural crest		Chromatophores
			Teeth (dentine)
			Visceral skeleton
			Jaws
			Schwann cells
			Meninges of brain
			Cranial, spinal, and autonomic ganglia
			Adrenal medulla
	Neural tube		Brain and spinal cord
			Cranial and spinal nerves
			Retina and optic nerve
			Posterior pituitary (neurohypophysis)
Endoderm	Archenteron (primitive gut)		Lining of lungs
			Lining of urinary bladder
			Thyroid, parathyroid, thymus
			Liver, pancreas
			Lining of digestive tract
			Auditory tube
Mesoderm	Chordamesoderm		Notochord
	Paraxial mesoderm (somites)	Dermatome	Dermis
		Sclerotome	Vertebral column
		Myotome	Skeletal muscles, appendicular skeleton
	Intermediate mesoderm (nephrotome)		Excretory organs, kidneys, kidney ducts
			Reproductive ducts
	Lateral mesoderm (lateral plate)	Somatic layer	Parietal peritoneum, pleura, pericardium
		Splanchnic layer	Visceral peritoneum, pleura, pericardium
			Mesenteries
			Heart, blood cells, and blood vessels
			Lymph vessels
			Gonads
			Muscles of visceral wall
		Coelom	Body cavities

(Fig. 4-12*A–D*). The process of invagination in amphioxus resembles indenting the wall of a balloon or ball with your hand. The pocket that you create with your hand will be the embryo's gut, and the two "layers" created by the indentation correspond to the ectoderm (on the outside of the balloon) and the endoderm plus mesoderm (the lining of the pocket). As invagination continues, the inward folding layer forms the archenteron, and the blastocoele becomes obliterated. The opening of the archenteron created by this process is the **blastopore;** it marks the future posterior end of the animal. Immediately following gastrulation, folds begin to separate groups of epithelial cells as organ primordia (Fig. 4-12*E–H*). Mesoderm forms as a series of **enterocoelic pouches** that bud off the dorsolateral walls of the archenteron. These pouches form a series of mesodermal segments, or **somites.** In a few of the anterior somites, a portion of the archenteric cavity is pinched off as small cavities that will later coalesce to become the coelom. Such coelomic spaces, however, develop by **cavitation** (= formation of small spaces or cavities between cells that eventually fuse to form a large cavity) within more posterior somites, and the coelomic spaces that develop by cavitation are never connected with the lumen of the archenteron. The formation of mesoderm and coelom in amphioxus resembles their development from enterocoelic pouches in echinoderms. As the somites develop, the roof of the archenteron between them separates as the notochord and the neural tube forms from the ectoderm overlying the notochord.

Gastrulation by Involution: Amphibians

The large, yolk-filled cells that form in the vegetal hemisphere during the cleavage of a mesolecithal egg, as found in *Rana* or *Ambystoma,* cannot invaginate in the same way as in amphioxus. Instead, invagination of a few cells on the margin of the gray crescent of the embryo forms a cleft that is the beginning of the archenteron (Fig. 4-13*A*). The dorsal margin of the cleft is termed the **dorsal lip of the blastopore.** The cleft lengthens as time goes on, grows laterally and ventrally, and eventually forms a circular blastopore. As the blastopore develops, cells move from the dorsolateral surface of the embryo toward the blastopore, roll over its lip, and then continue to move forward beneath the ectoderm, thereby deepening and enlarging the archenteron (Fig. 4-13*B*). This process, in which a sheet of epithelium moves the reference point of the blastopore lips, is known as **involution.** Involution begins at the dorsal lip of the blastopore, but soon cells begin to involute at its lateral lips, and eventually a few cells involute at its ventral lip. The blastocoele becomes obliterated as the archenteron enlarges. Movement of surface cells toward the lips of the blastopore occurs faster than they can be involuted. As a result, the prospective ectodermal cells, which originally were limited to the animal hemisphere, overgrow the yolky cells of the vegetal hemisphere in a process called **epiboly.** Finally, only a small "plug" of yolk-filled cells can be seen through the blastopore (Fig. 4-13*C*). This stage is termed a yolk plug gastrula.

Cells that involute form endoderm, chordamesoderm, and mesoderm proper (Fig. 4-13*D*). Prospective chordamesoderm cells in the mid-dorsal part of the archenteron roof will form the notochord (Fig. 4-13*E*, *green*). Prospective mesoderm cells lie more laterally (Fig. 4-13*E*, *red*). Mesodermal cells move forward as a sheet between the endoderm and the ectoderm (Fig. 4-13*D*). The notochord separates from the laterally placed mesoderm, and endodermal folds meet beneath it to complete the roof of the archenteron (Fig. 4-13*D* and *E*). At the boundary between the notochord and endoderm is a small patch of cells, termed the **prechordal plate,** that extends anterior to the notochord and will play a role in head formation (not shown in Fig. 4-13). The cells of the ectoderm overlying the notochord thicken to form the **neural plate** (Fig. 4-13*D* and *E*). Some other vertebrates with mesolecithal eggs, such as lungfishes, bichirs, sturgeons, and paddlefishes, exhibit comparable methods of gastrulation by involution.

Gastrulation by Ingression: Birds

The process of gastrulation is different in macrolecithal eggs, such as those of chondrichthyan fishes, reptiles, and birds. Because large, yolky eggs evolved independently in chondrichthyan fishes and amniotes, it should not be surprising that the methods of gastrulation differ. Gastrulation movements are best known in chickens, which have become a standard model for gastrulation in amniotes. The central part of the blastoderm in chick embryos becomes clear and forms the **area pellucida** (clear area); the peripheral part remains opaque and is termed the **area opaca** (opaque area; Fig. 4-14*A*). Cells separate as a sheet, or **delaminate,** from the deep surface of the area pellucida at its prospective caudal end and migrate forward, forming a layer known as the **hypoblast** (Fig. 4-14*A*). Between the hypoblast and the mass of uncleaved yolk is a small space known as the **subgerminal cavity.** The overlying cells of the chick blastoderm now constitute the **epiblast,** and the space between these two layers corresponds to the blastocoele. Cells then begin to migrate from the periphery of the epiblast toward its center, where they form a thickened, longitudinal ridge known as the **primitive streak** (Fig. 4-14*B*). As more cells move toward the primitive streak, cells that are already there turn inward ventrally and spread out laterally and anteriorly beneath the epiblast. Many of the cells that move through the primitive streak delaminate and migrate as individual mesenchymal cells. The first cells to move inward along the primitive streak displace hypoblast cells and form endoderm. Cells that move inward later form mesoderm. When the process is complete, epiblast cells that remain on the surface are the ectoderm. All three germ layers thus are derived from the epiblast. By comparing the fate map of a chicken blastula in Figure 4-11*B* with

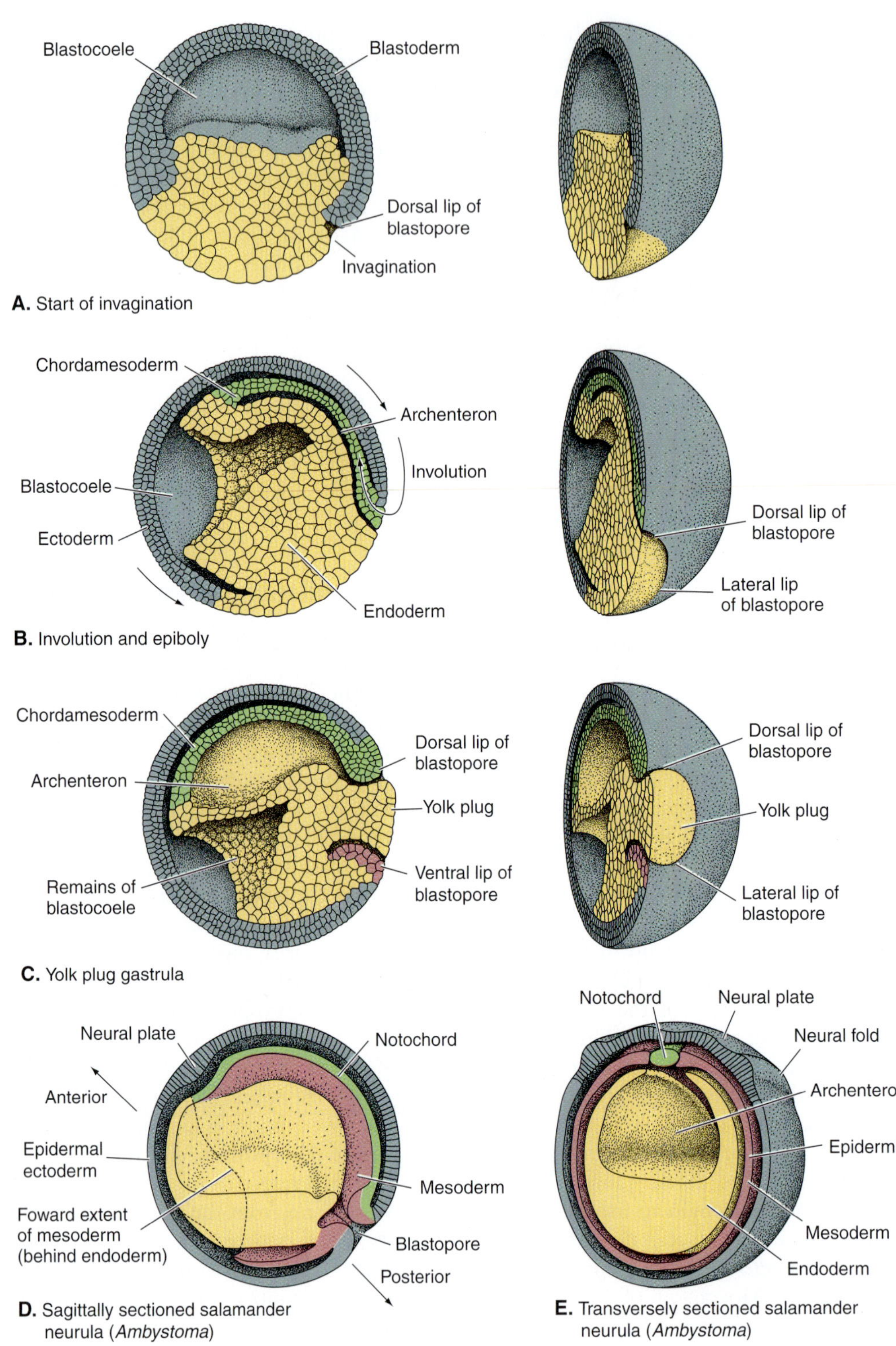

FIGURE 4-13
Gastrulation and mesoderm formation in amphibians. *A,* Sagittal section and posterior view of an early stage of a frog embryo, just as invagination is beginning. *B,* Similar views of a later stage, when involution and epiboly are occurring. *C,* The blastopore now is complete, and the yolk plug protrudes through it. *D,* A sagittal section of a later gastrula of a salamander, showing the spread of the mesoderm between the ectoderm and endoderm. *E,* A transverse section of a later stage, when the notochord has formed, endodermal folds are completing the roof of the archenteron, and the neural plate has been induced. *(A–C, Frog* [Rana], *from Balinsky and Fabian; D and E, salamander* [Ambystoma], *from Hamburger.)*

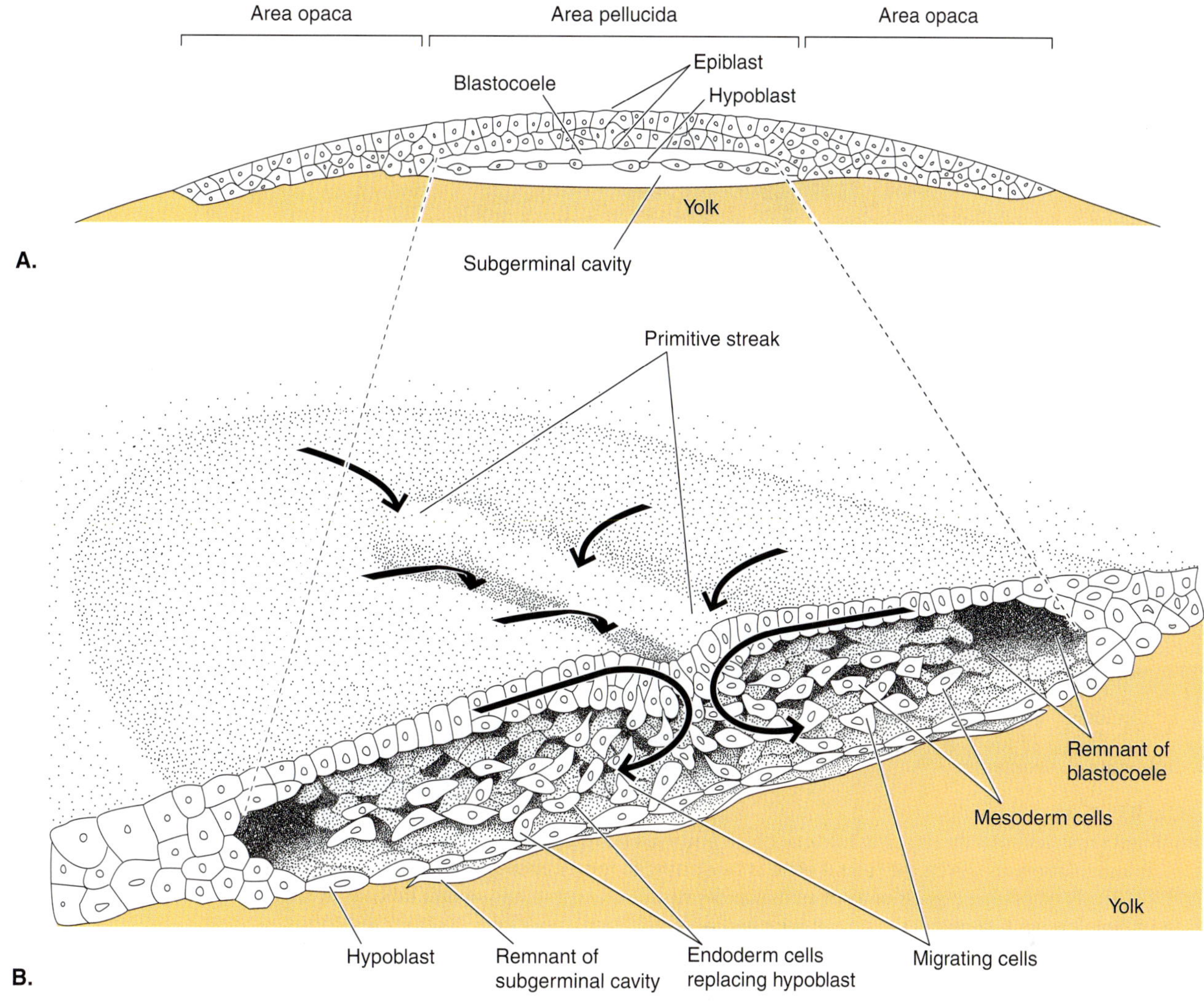

FIGURE 4-14
Blastula and gastrulation in an embryonic chicken (*Gallus*). *A,* Cross section through chicken blastoderm showing the area pellucida and area opaca. The hypoblast and epiblast have differentiated. The mass of uncleaved yolk is indicated in yellow. *B,* Stereodiagram of the anterior half of the area pellucida, showing the primitive streak. Dashed lines indicate the relationship of the diagram to earlier stage shown in *A*. Black arrows show ingression of prospective endodermal and mesodermal cells. *(Redrawn from Balinsky.)*

Figure 4-14, you should be able to visualize how gastrulation via a primitive streak rearranges the cells of the blastula.

The inward movement of cells along the primitive streak has been called **ingression,** but it is functionally comparable to the involution of cells in amphibian embryos. Indeed, the primitive streak has been considered the homologue of the blastopore since the 1870s. The process of ingression of cells along the primitive streak is completed first near the anterior end of the embryo, and the wave of completion spreads caudally. As ingression is completed, the primitive streak retreats caudally (from anterior to posterior), leaving the in-turned endoderm and mesoderm. Just as in amphibians, the notochord and mesoderm separate, and the ectoderm overlying the notochord thickens as the neural plate (Fig. 4-15*A*). Mesoderm immediately lateral to the notochord and developing neural tube, the **paraxial mesoderm,** differentiates into a series of somitomeres in the head (discussed in the section, "Basic Organization of the Vertebrate Head," page 164) and **somites** in the trunk (Fig. 4-15*B*). These processes, too, progress from anterior to posterior. The coelom forms, as it does in amphibians, by cavitation in the mesoderm lying lateral to the somites.

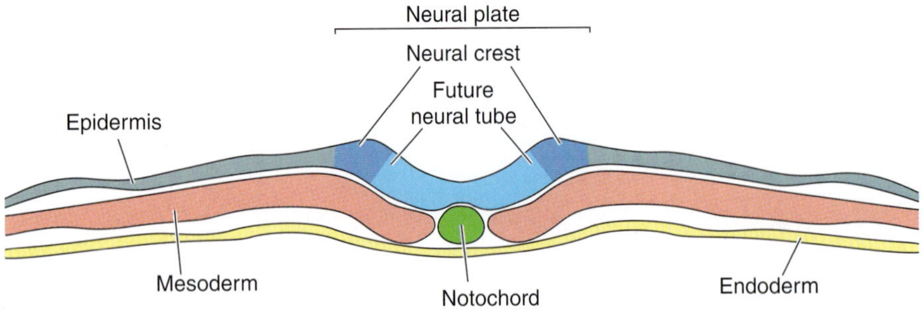

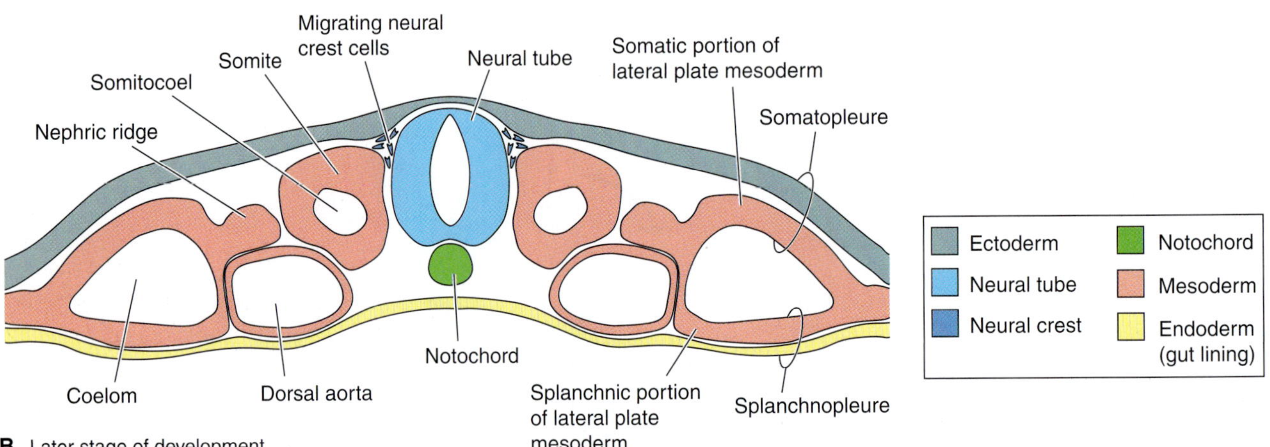

FIGURE 4-15
Simplified cross sections through the trunk of chick embryos (*Gallus*) at two times during neurulation. *A,* Early neurulation, in which the neural plate is beginning to form neural folds along its lateral margins. Trunk ectoderm consist of three parts: epidermis, neural crest, and neural tube. *B,* A later stage of development, showing the neural tube and differentiating mesoderm. Neural crest forms along the length of the head and trunk in vertebrates, but no neurogenic placodes form in the trunk (compare with Fig. 4-21*A*). *(Redrawn from Balinsky.)*

Mesoderm Differentiation

As gastrulation ends, several other processes go on more or less concurrently. These are: (1) mesoderm differentiates into parts that will form primordia of many organ systems; (2) extraembryonic membranes develop in amniotes, surround the developing embryo, protect it, and serve other functions; and (3) neurulation occurs, and the nervous system and sense organs begin to differentiate. We consider these topics in this sequence, but keep in mind that these processes are going on at the same time.

Mesoderm, somites, and the coelom usually begin to form near the end of gastrulation, and mesodermal development continues through neurulation. Idealized cross sections through amphibian embryos at two different stages during neural tube formation are shown in Figure 4-16. By the time the neural tube is complete, three bands of mesoderm can be recognized on each side of the embryo (Figs. 4-15*B* and 4-16). Thickened, block-like segmental **somites,** or **paraxial mesoderm,** lie lateral to the neural tube and notochord (Fig. 4-16). A broad, unsegmented lateral plate, or **lateral plate mesoderm,** extends laterally and ventrally between the archenteron and surface ectoderm. A nephric ridge, or **intermediate mesoderm,** which is segmented anteriorly, lies between the somites and lateral plate mesoderm. The definitive coelom develops by cavitation in the lateral plate. Portions of the coelom extend into the nephric ridge, and small, ephemeral spaces, known as **somitocoeles,** may occur within the somites.

As development proceeds, each somite differentiates into three regions (Fig. 4-16*B*). Initially, somites are epithelially organized, but their cells will undergo epithelial–mesenchymal transitions and migrate to achieve their final locations in the body. The dorsolateral group of somite cells constitutes the **dermatome.** Cells in the dermatome differentiate into mesenchymal cells that migrate out beneath the surface ectoderm to form most of the dermis of the skin. Cells

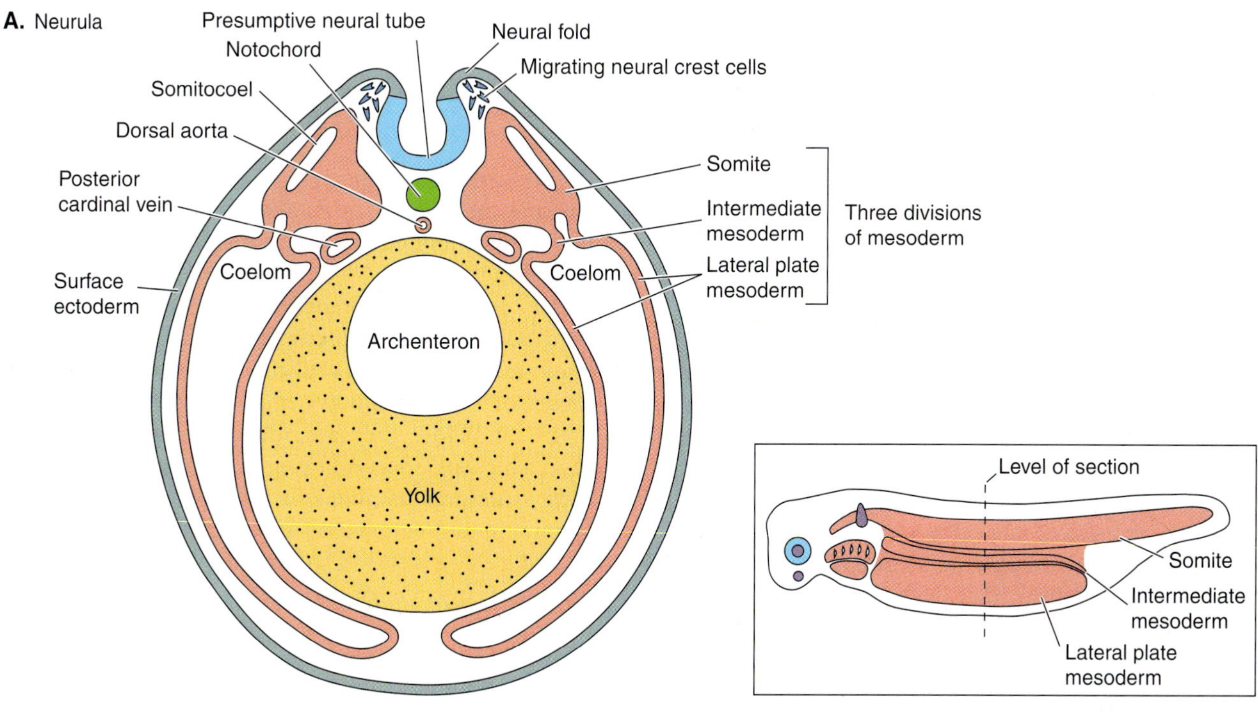

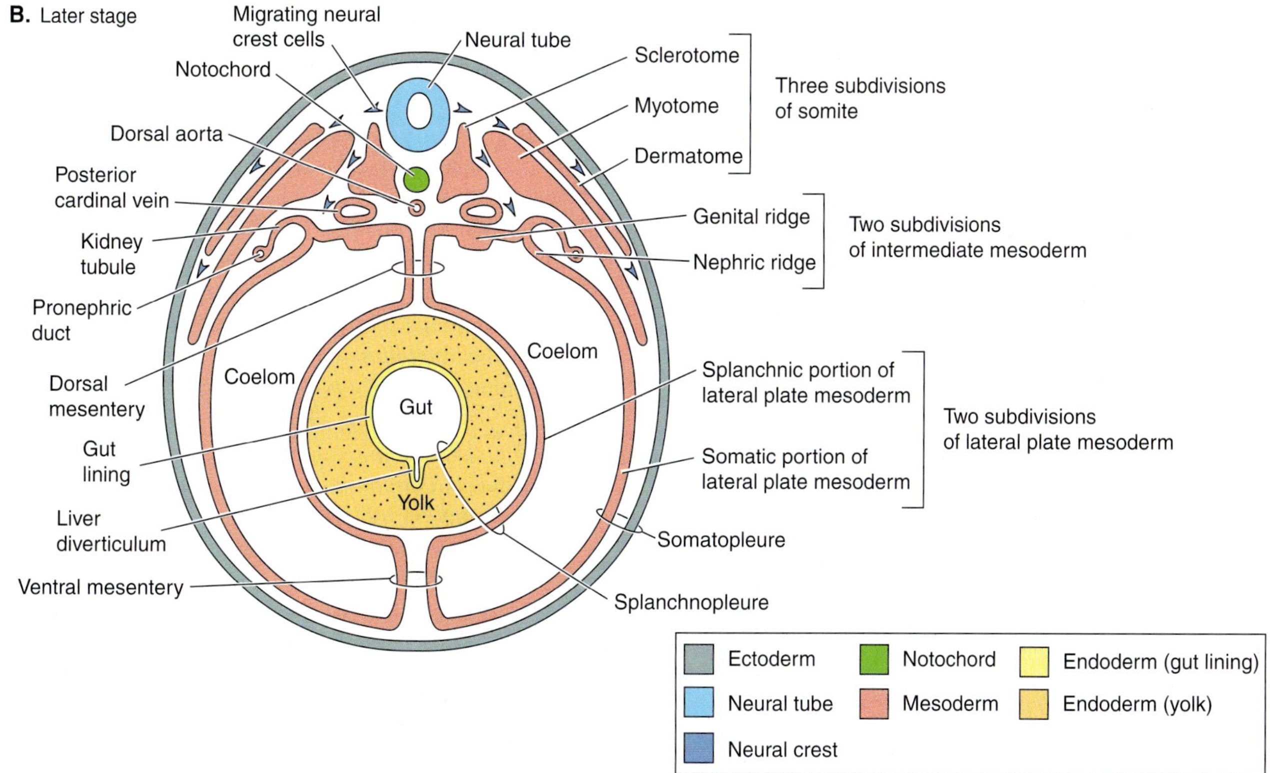

FIGURE 4-16

Differentiation of trunk mesoderm in an idealized amphibian embryo. *A,* Early neurula. The inset shows a lateral view and the plane of section for the figure in *A*. It also diagrams the basic organization of the three divisions of trunk mesoderm. In amphibian embryos (and other vertebrates with mesolecithal eggs) the moderate amount of yolk is incorporated into yolk-laden cells in the floor of the archenteron. *B,* Later neurula, showing the structures that develop from each somite and each germ layer. Note that each of the three divisions of trunk mesoderm diagrammed in *A* in turn gives rise to subdivisions. See text for additional descriptions.

deep to the dermatome form the **myotome,** or embryonic muscle segment. The segmental myotomes extend ventrally between the ectoderm and lateral plate mesoderm and differentiate into muscle cells that form all (in anamniotes) or most (in amniotes) of the somatic muscles of the body wall and appendages. The ventromedial part of the somite is the **sclerotome.** It differentiates into mesenchymal cells that migrate around the neural tube and notochord to form the vertebral column and occipital region of the skull. Other parts of the skull develop from mesenchymal cells derived from the cranial neural crest (= ectomesenchyme, described later) and from the dermatome. Other aspects of braincase and skull development will be treated in Chapter 7, and vertebral development will be described in Chapter 8.

The intermediate mesoderm develops two subdivisions, termed the **nephric** and **gonadal ridges** (Fig. 4-16*B*). The nephric ridge will differentiate into the kidney and its excretory ducts, as well as some portions of the reproductive ducts. The additional development of these organ systems will be treated in Chapters 20 and 21. The gonadal ridge forms medial to the nephric ridge. Gonadal development involves not only the mesoderm of the gonadal ridge but also migration of primary germ cells from endoderm, and is described in Chapter 21.

The lateral plate mesoderm is split into two layers, thereby creating the coelom (Fig. 4-16*B*). The medial **splanchnic layer** of the lateral plate mesoderm next to the endoderm will form the connective tissue and visceral muscles of the gut tube and heart walls. Splanchnic mesoderm forms the circulatory system, including the heart, arteries and veins, and circulating blood cells, with two important exceptions: (1) in the head region, neural crest cells contribute to blood vessel formation; and (2) the first generation of blood cells is made in the blood islands of the yolk sac. Part of the splanchnic mesoderm will form **mesenteries** and the **coelomic epithelium** (the **visceral peritoneum**) that covers the visceral organs. The lateral **somatic layer** of the lateral plate mesoderm forms the coelomic epithelium bounding the coelom laterally (the **parietal peritoneum**), and it also may contribute to the somatic musculature and other tissue of the lateroventral body wall. In contrast to the splanchnic layer, the somatic layer of lateral plate mesoderm does not form blood vessels or blood cells.

It is convenient to use the term **splanchnopleure** (Gr., *splanchnon* = gut + *pleure* = wall) to refer to the mesodermal and endodermal components that form the wall of the gut (Fig. 4-16*B*). Similarly, the term **somatopleure** (Gr., *somatikos* = body + *pleure* = wall) refers to the mesodermal and ectodermal components that make up the body wall (Fig. 4-16*B*).

Secondary Egg Cell Membranes and Extraembryonic Structures

We described the primary egg cell membrane, or vitelline membrane, in the earlier section on gametes and fertilization. **Secondary egg cell membranes** may surround the primary egg cell membrane. All secondary egg cell membranes of vertebrates are produced by the female's reproductive tract after an egg is ovulated from the ovary. Fertilization may occur after synthesis of the secondary egg cell membranes (e.g., frogs) or before they are added around the egg cell (e.g., chickens). Secondary egg cell membranes include such structures as adhesive jelly coats, egg capsules, egg cases, and egg shells. An example of the jelly coat of a frog's egg is shown in Figure 4-8. Because they are produced outside of the ovary, secondary egg cell membranes may contain one egg cell (as in chickens) or many individual egg cells (as in the egg capsules of some elasmobranchs). As a general rule, secondary egg cell membranes protect the egg and embryo. In those species with secondary egg cell membranes, the individual "hatches" twice: first from its vitelline membrane and, usually much later, from its secondary egg cell membranes.

In the case of a chicken egg, fertilization occurs soon after ovulation, near the ostium or infundibulum of the oviduct (see Fig. 21-17). The fertilized egg travels down the oviduct to the shell gland. During this passage, the egg accumulates **egg white,** which consists of water (about 80% by weight) and proteins (mostly albumens). Initially, the egg is rotated, and at each end becomes coated with spirally wrapped layers of albumen, which form the **chalaza.** More layers of albumen adhere to the chalaza as the egg continues down the oviduct. Next, the egg and its adhering albumen are enclosed in a tough **inner shell membrane** composed of the protein keratin. The surrounding **outer shell membrane** entraps a small pocket of air known as the **air space.** Finally, the slightly porous calcareous shell is added in the **shell gland.** Secretion of all of these secondary egg cell membranes requires about 22 hours, after which the egg is laid, and its incubation can begin. The shell and shell membranes are protective layers, whereas the albumen (together with the yolk) provides raw materials, water, and energy needed by the embryo during its development.

With the evolution of the cleidoic egg (see Chapters 1 and 3), amniotes were able to bypass an aquatic larval stage and reproduce on land. This is possible because the embryos of amniotes possess four **extraembryonic membranes** that protect the embryo and

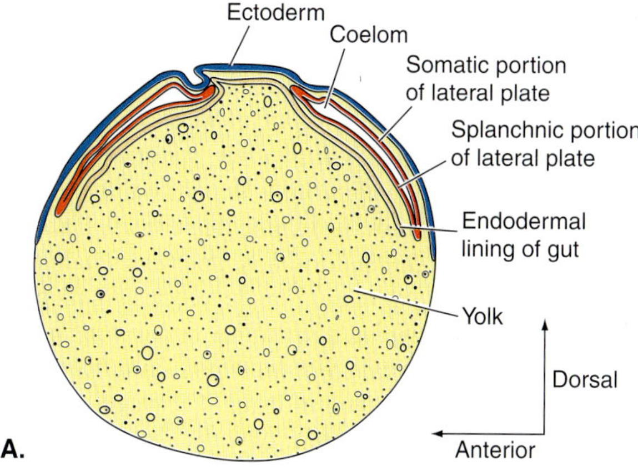

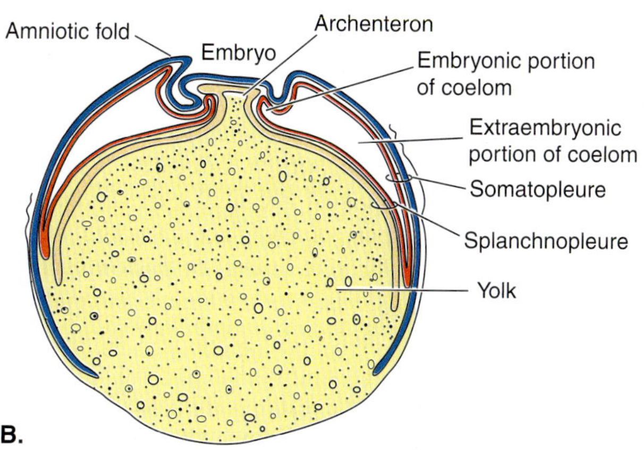

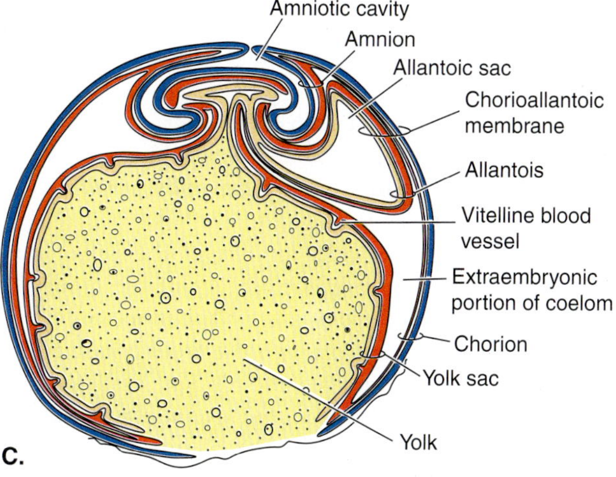

FIGURE 4-17
Schematic diagrams of the formation of extraembryonic membranes of a chick embryo. *A,* Early stage. *B,* Stage showing amniotic folds extending above the embryo's body. *C,* Later stage, in which all four extraembryonic membranes (i.e., amnion, chorion, allantois, and bilaminar yolk sac) can be discerned. See text for additional description. *(Modified from Patten.)*

sustain its metabolism. Our model for the extraembryonic membranes of amniotes is based on chickens because, with only a few modifications, this model can be applied to amniotes in general.

As a chick embryo develops in the blastodisk on top of its large yolk mass, tissue layers extend from the embryonic body to form extraembryonic membranes (Fig. 4-17). First to develop is the **yolk sac,** which forms by the spreading of tissue layers over the yolk. Endoderm and splanchnic mesoderm eventually surround the yolk. Thus, the yolk sac of amniotes is referred to as a bilaminar (two-layer) yolk sac. This is in contrast to the yolk sac membrane of actinopterygian fishes, which contains all three germ layers and thus is termed a trilaminar yolk sac.

An **extraembryonic coelom** extends from the embryo between the somatic and splanchnic layers of the part of the lateral plate mesoderm that contribute to the yolk sac. As the yolk sac of a chick embryo develops, the ectoderm and somatic layer of the lateral plate mesoderm become elevated as **amniotic folds** that arch over the embryo and meet above it (Fig. 4-17*B*). As the **head folds** and **tail folds** deepen and undercut the embryo, and as lateral body folds form, the embryo is raised off the yolk mass, but its archenteron remains connected to the yolk sac by a narrow yolk stalk.

Blood vessels and blood cells begin their development in the splanchnic mesoderm of the yolk sac and spread into the embryo. Embryonic circulation is soon established, and a rich network of blood vessels conveys materials from the yolk to the embryo. These vessels also can bring in materials from the egg white because the periphery of the yolk sac that is spreading over the yolk is in contact with the egg white for a long time.

Two additional extraembryonic membranes form when the amniotic folds meet and fuse above the embryo's body (Fig. 4-17*C*). The inner limbs of these folds form the **amniotic membrane** (or amnion), which surrounds the embryo's body. The outer limbs of the fold form the **chorionic membrane** (or chorion), which surrounds the amnion and, eventually, the entire yolk sac. **Amniotic fluid** accumulates in the **amniotic cavity** between the amnion and the embryo, so that the embryo continues its development in a liquid environment. It consists chiefly of water with salts and a few proteins, as well as living cells sloughed off from the embryo's ectoderm. Although a cleidoic egg must be laid on land, the embryo's immediate environment remains aquatic, like that of fish and amphibian larvae. Amniotic fluid also provides an important protective liquid cushion around the embryo that buffers it against physical bumps. Additional protection is afforded by the chorion itself, as well as by the egg white and shell.

Amniotic fluid is increasingly important in the analysis of human genetics. It can be sampled after about 18 weeks of gestation using a method called **amniocentesis.** In this procedure, a long hypodermic needle is inserted through the mother's abdominal and uterine walls and into the amniotic cavity itself. The positions of the fetus and the tip of the needle are monitored continuously using ultrasound imaging. A small sample of amniotic fluid is withdrawn, from which it is possible to culture living cells that can be used to study the baby's chromosomes.

The third extraembryonic membrane of amniotes, the **allantois,** forms as an evagination of the posterior part of the archenteron, so its wall is composed of endoderm and vascularized splanchnic mesoderm (Fig. 4-17*C*). The allantois enlarges, nearly fills the extraembryonic coelom, and eventually contacts the chorion. Frequently, these two membranes fuse to form a **chorioallantoic membrane.** The chorion and amnion are avascular membranes because they develop before any blood vessels enter the somatic mesoderm. In contrast, the allantois is richly vascular, and it **vascularizes** the chorion, bringing embryonic vessels close to the inside of the porous inner and outer egg membranes and shell. Embryonic gas exchange occurs by this route. Nitrogenous excretory products, in the form of inert crystals of uric acid, accumulate in the lumen of the allantois. After hatching or birth, the base of the allantois remains in many amniotes as the urinary bladder (Chapters 20 and 21).

Developmental Modifications of Eutherian Mammals

The ancestors of therian mammals laid macrolecithal cleidoic eggs, but therians evolved a pattern of reproduction in which embryos are retained within a female's uterus and born as miniature adults. This form of **viviparity** (live birth; see Chapter 21 for additional discussion) entailed many evolutionary modifications in the reproductive tract of females and in maternal immune systems as well as embryonic development. Here we will examine a condition typical for eutherians (placental mammals); reproductive aspects of metatherians (marsupial mammals) will be described in Chapter 21.

Implantation of an embryonic placental mammal establishes close contact between the embryo's extraembryonic membranes and the lining of the mother's uterus to form a **placenta.** Eutherians exhibit many variations in placental structure (see Fig. 21-21). After a eutherian embryo's body form becomes apparent, it can be termed a **fetus.** Placentation allows physiological exchange between an embryo and its mother. Because a eutherian mother will provide the embryo with nutrients and other materials via placental exchange, very little yolk is deposited in the egg cells of placental mammals during oogenesis. Eutherian eggs have become secondarily microlecithal, and this simplifies their cleavage. For development to proceed, it is essential that implantation occur very early, for without a placenta, the embryo's physiological needs cannot be met by the mother. Cleavage of a eutherian egg differs from that of other microlecithal eggs in that the second plane of cleavage is partially horizontal, a condition termed **rotational cleavage.** Cleavage continues, eventually yielding a ball of cells known as a **blastocyst** (Fig. 4-17*A*). The outermost cells of the blastocyst form a layer known as the **trophoblast** (Gr., *trophe* = nourishment + *blastos* = germ or bud), which is homologous to the chorionic ectoderm of the eggs of other amniotes. The trophoblast is the precursor of the fetal part of the placenta. The maternal part of the placenta is the vascularized and glandular uterine lining, or **endometrium** (Gr., *metra* = uterus). In the uterine environment, the trophoblast soon begins to grow and spread out on the endometrium. In human beings and other eutherians in which the fetal part of the placenta grows aggressively into the uterine lining, the trophoblast is said to be **invasive.** Contact or penetration of the endometrium by the trophoblast produces implantation of the embryo. In ways not fully understood, the trophoblast also provides an immunological barrier that prevents the mother from developing antibodies against the embryo, which otherwise would be rejected as foreign tissue because half of its genes are paternal. The trophoblast thus prevents immunological rejection of the embryo, a key innovation in the evolution of eutherian viviparity.

A small sphere of cells known as the **inner cell mass** lies within the trophoblast. All of the embryo's body will be derived from cells of the inner cell mass. The inner cell mass attaches to the trophoblast at the presumptive posterodorsal side of the embryo, but elsewhere it is separated from the trophoblast by a cavity, which is homologous to the blastocoele of other vertebrates. Details of gastrulation and differentiation of the embryo, as well as its extraembryonic membranes, differ considerably among eutherian taxa; we will describe the human condition. Endodermal cells differentiate on the underside of the inner cell mass and spread laterally and ventrally within the cavity of the blastocyst to form the endodermal part of the yolk sac (Fig. 4-18*B*). There is no yolk in the yolk sac of a human embryo (or other eutherian mammals), but its

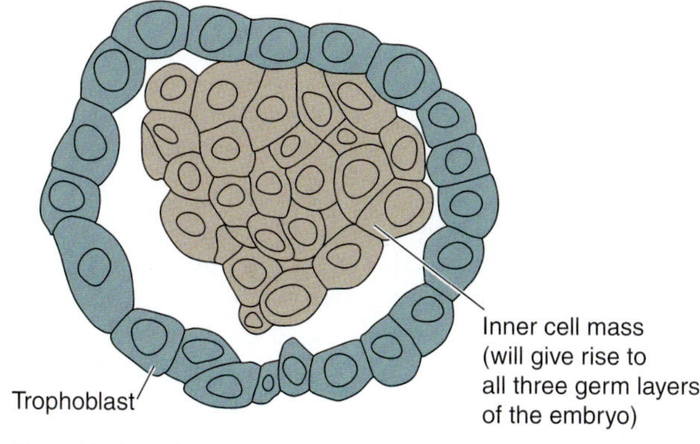

A. Early blastocyst

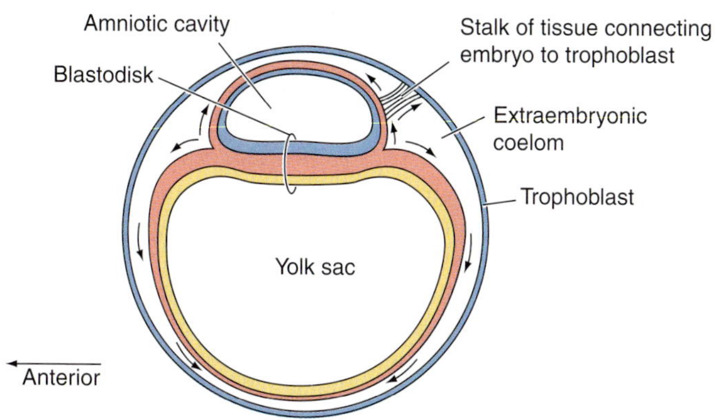

B. Inner cell mass differentiation

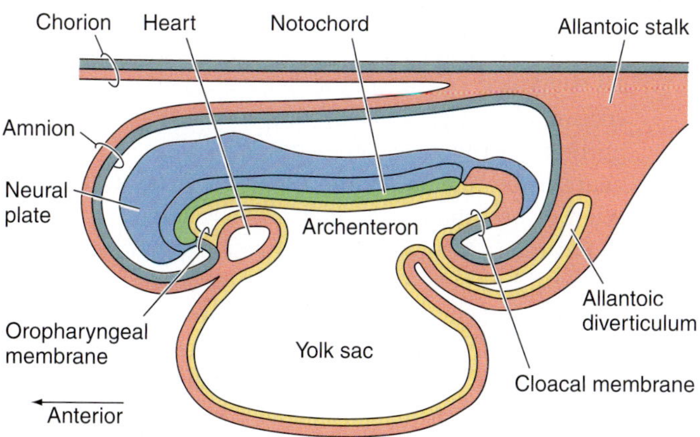

C. Embryo

FIGURE 4-18
Early development of a eutherian mammal. *A,* Early blastocyst. *B,* The inner cell mass has differentiated into the amnion, blastodisk, and yolk sac. Mesoderm ingresses along the primitive streak and spreads as indicated by the arrows. *C,* The blastodisk differentiates into the embryo. *(After Balinsky and Fabian.)*

dorsal part eventually will become separated from the rest of the yolk sac to form the archenteron. A space that forms by cavitation among the cells in the dorsal part of the inner cell mass is the beginning of the amniotic cavity. Cells lining this cavity are regarded as ectoderm. The two-layered disk of cells (ectoderm and endoderm) between the cavities of the yolk sac and the amnion is the blastodisk, from which the embryo's body will develop.

As in chick embryos (Fig. 4-14), gastrulation in a human embryo involves movements of prospective mesoderm cells toward a longitudinal primitive streak, through which they ingress and spread out between ectoderm and endoderm. Neurulation, mesoderm differentiation, and the separation of the embryo from the yolk sac by body folds resemble the corresponding processes in chick embryos. Mesoderm also spreads over the surfaces of the amnion and yolk sac, so that these become typical two-layered extraembryonic membranes as in other amniotes (Fig. 4-18*C*). A stalk of tissue persists near the posterior end of the embryo and connects it with the ectodermal trophoblast. Mesoderm spreads via this stalk over the underside of the trophoblast and converts it to a chorion that more closely resembles that of avian and reptilian embryos. An allantois grows from the posterior part of the archenteron into this stalk. The cavity inside the allantois never becomes large in human embryos, but its wall enables blood vessels from the embryo to reach the avascular chorion and to vascularize the fetal part of the placenta. This part of a eutherian placenta is homologous to the chorioallantoic membrane of reptiles and birds. As the embryo enlarges, it rises above the yolk sac and allantois. Thus, the embryo is connected to its placenta by a long, cordlike stalk that also contains the allantois and yolk sac. This cordlike structure is the **umbilical cord,** which must be broken or cut at birth.

In summary, the development of a mammalian embryo resembles that of reptiles and birds except for (1) the reduction of yolk, (2) simplification and modification of early cleavage patterns, (3) production of a blastocyst, and (4) the precocious development of extraembryonic parts that contribute to the placenta and the umbilical cord.

Organogenesis of the Nervous System and Sense Organs

Organ formation in the nervous and sensory systems provides several important examples of organogenesis and general concepts in embryology. The nervous system forms very early and is so pervasive in its structural and functional impact on other organ systems that an understanding of its early development will help you to understand many other aspects of the vertebrate body. Our first topic is neurulation, during which the neural tube forms and begins to differentiate into the brain and spinal cord. Next, we describe eye formation, which offers a particularly important demonstration of interactions of components of the central nervous system with other differentiating tissues to form a complex organ. In addition to the neural tube, vertebrates have two other neurogenic tissues: the neural crest and the neurogenic placodes. These embryonic precursor tissues are described, and some of their fates are discussed. Finally, we discuss the early development of other major cranial sense organs, including the nose, taste buds, ear, lateral line, and electroreceptive systems.

Neurulation

As gastrulation nears completion and somites form, the chordamesoderm induces the overlying neural plate to differentiate into the neural tube characteristic of chordates (Fig. 4-19). Several stages in this process, termed **neurulation,** are defined based on patterns of development observed in sharks, frogs, and amniotes. First, the ectoderm overlying the notochord thickens and flattens to form a **neural plate** along the dorsal midline of the embryo. In this process, cells of the neural plate are said to **palisade,** which describes a change in their shape from a short, cuboidal form to a tall, columnar form. Second, the margins of the neural plate elevate as a pair of **neural folds** with a **neural groove** between them (Fig. 4-19*A*). This is accomplished in part by a different type of shape change in the cells of the neural plate, in which the apices of the cells narrow or constrict to produce folding of the entire sheet. The neural folds are largest and most widely spaced at the anterior end of the embryo, in the region of the presumptive brain. Third, the neural folds meet dorsally, where the folds of the right and left sides fuse to form the **neural tube.** The neural groove becomes the **central canal** that characterizes the central nervous system of chordates. The epithelium immediately adjacent to the central canal is termed **ependymal epithelium.** The ependymal cells bear cilia that face the central canal. Initially, the walls of the neural tube may be one or two cells thick, but these cells rapidly divide to produce a wall several cell layers thick (see Fig. 13-9). Much of the central nervous system will differentiate from the walls of the neural tube, including portions of the eyes (see pages 146–149). Finally, the neural tube separates from overlying ectoderm, which is left to become the epidermis of the skin.

A variant form of neural tube formation occurs in teleost fishes (Fig. 4-19*B*), and it has been regarded as a synapomorphy of teleosts. This is important because about half of all living vertebrates are teleosts, including a currently popular model for vertebrate development, the zebrafish (*Danio rerio*). In teleosts, a **neural keel** forms as a solid rod dorsal to the notochord, instead of the flat neural plate found in embryos of outgroup taxa. This neural keel subsequently **cavitates** to form the central canal of the central nervous system (Fig. 4-19*B*).

Soon after the left and right neural folds close in the head region, the three **primary** (or embryonic) **brain regions** can be distinguished: the **prosencephalon, mesencephalon,** and **rhombencephalon** (Fig. 4-20). In turn, the prosencephalon will form two of the five brain regions found in adults, which are the **telencephalon** and the **diencephalon.** The embryonic rhombencephalon will form the **metencephalon** and

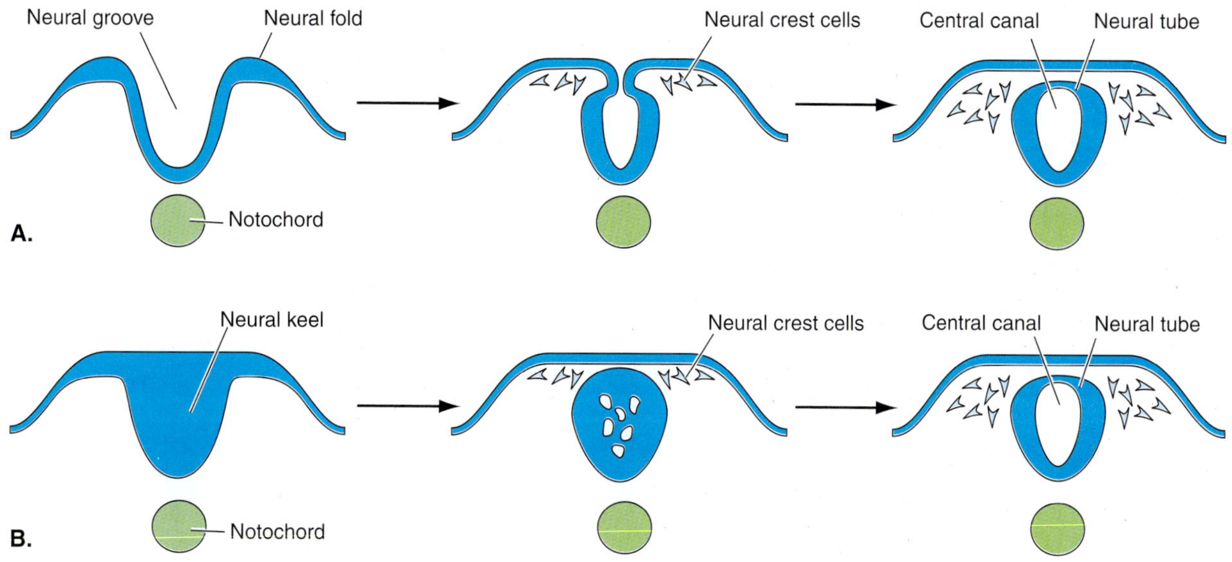

FIGURE 4-19
Formation of neural tube in the trunk via neural folds or cavitation. *A,* Neural folds meet and fuse in the midline, pinching off the hollow neural tube. *B,* A solid neural keel forms; cavities form within it and fuse to form the hollow neural tube. The scheme shown in *A* occurs in sharks, basal osteichthyan fishes, and tetrapods. The scheme shown in *B* occurs in lampreys and teleosts as well as in portions of the neural tube in some other groups.

myelencephalon of adults. The rhombencephalon shows evidence of segmental organization in its walls, in the form of a series of thickenings known as **rhombomeres.** The rhombomere segments are related to segments observed in the branchial region, and are discussed later in the section, "Organization of the Head in Amniote Embryos."

As the neural tube forms and its anterior part differentiates into the brain, the embryo continues to lengthen along its anteroposterior axis. **Head** and **tail folds** form in the embryos of most vertebrates, and the embryo begins to separate from the large mass of yolk beneath it.

Eye Formation as an Example of Induction

Vertebrate eye formation offers excellent examples of epigenetic interactions among tissues to yield a complex organ. An understanding of some basic aspects of eye development will help you appreciate the ways in which we think cells and tissues interact throughout embryogenesis to produce a complex organism.

An idealized schematic diagram of eye development (based on the chick) is shown in Figure 4-20*A–E.* The first indication of eye development is the evagination of paired **optic vesicles** from the lateral walls of the diencephalon (Fig. 4-20*A*). As each vesicle grows toward the body surface, its proximal part narrows as an **optic stalk,** and its distal part invaginates to form a two-layered **optic cup.** The outer layer of the optic cup becomes the pigment layer of the retina, and the inner layer differentiates into the photoreceptive cells (rods and cones) and neuronal layers of the retina. Because of the way it develops, the retina can be considered to be a part of the brain. The outer, receptive segments of the rods and cones develop from cilia of the ependymal epithelium lining the optic cup. Because of the way the optic cup forms, these cilia are still directed toward its lumen, which was part of the body surface earlier in development. After the lumen has narrowed later in development, the receptive segments of the rods and cones lie next to the pigment layer of the retina, so that light must pass through the nervous layers of the retina to reach them (see Chapter 12).

As the optic cup approaches the body surface, it induces the surface ectoderm first to thicken as a **lens placode** and then to invaginate and form a **lens vesicle** that will differentiate into the lens. Adjacent mesenchyme encapsulates the lens and optic cup to form the strong, vascularized wall of the eyeball. Tissue from the ectodermal optic cup contributes to the **iris,** which regulates the amount of light entering the eye, and the **ciliary body,** which forms the muscles controlling the iris.

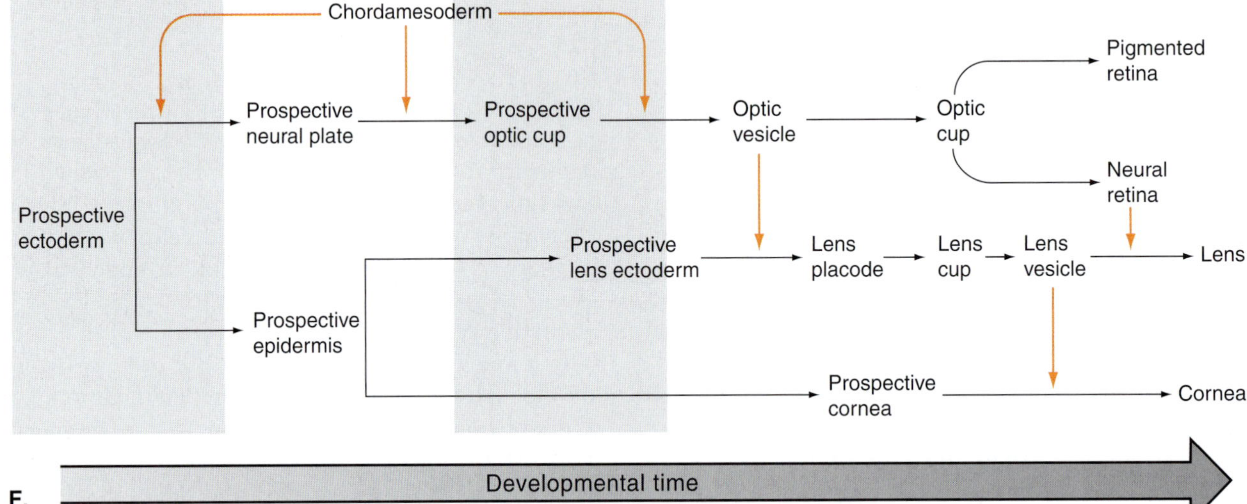

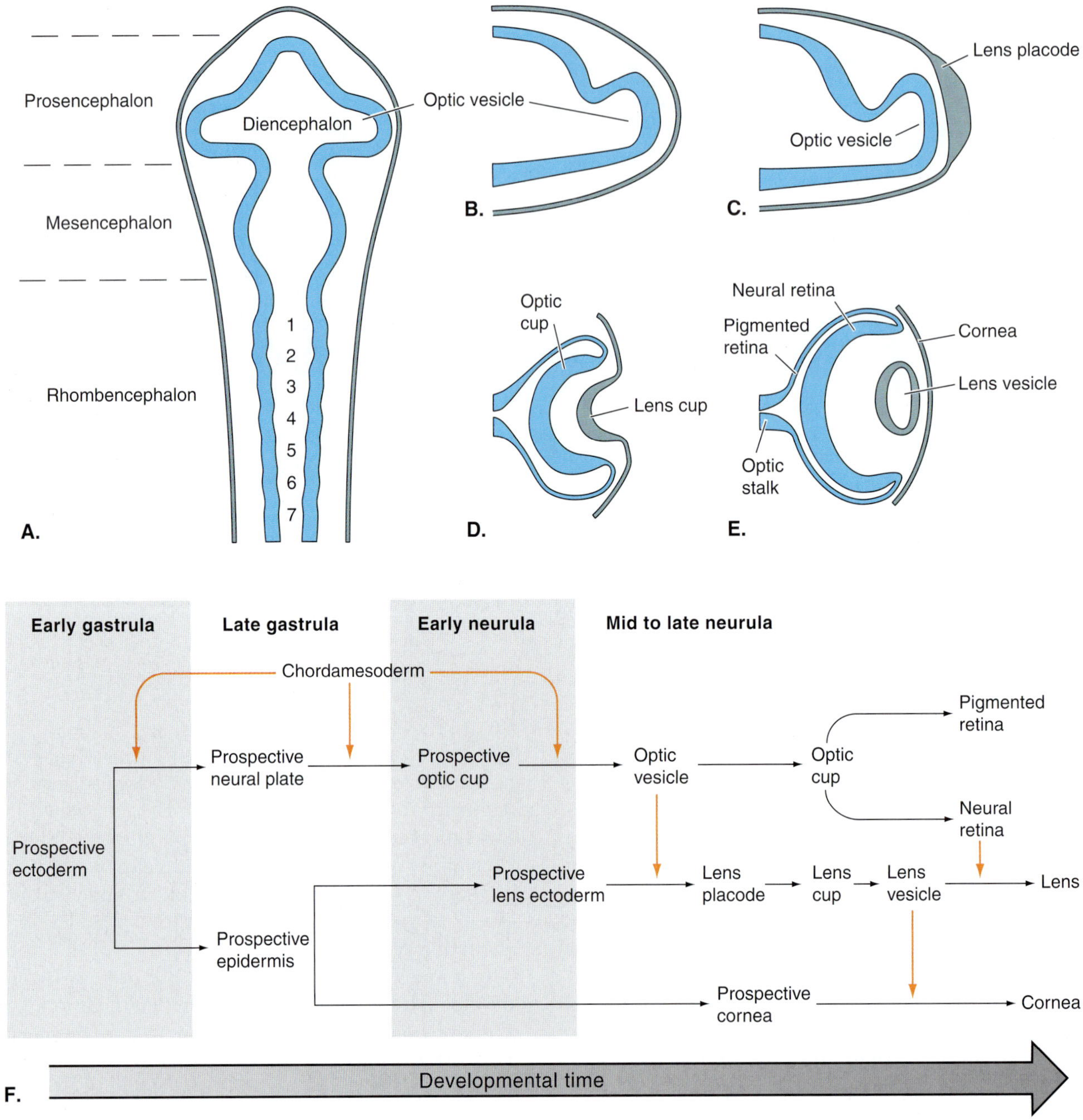

FIGURE 4-20
Eye development and the cascade of inductions that produces it. *A*, Schematic dorsal view of the head of an amniote embryo, showing primary brain regions and optic vesicles evaginating laterally from the walls of the diencephalon. Rhombomeres are numbered. *B*, Cross section through the optic vesicle in the stage shown in *A*. *C*, Cross section through the optic vesicle as it contacts the skin and induces formation of the lens placode. *D*, Invagination of the optic vesicle to form the optic cup, and the lens placode to form the lens cup. *E*, Lens vesicle stage, with neural and pigmented retina indicated. *F*, Cascades of inductions (*orange arrows*) during eye formation in an amphibian embryo. *(A–E, Based on a figure in Karp and Berrill; F, based on a figure in Gilbert.)*

Figure 4-20*F* summarizes the epigenetic interactions that take place during eye formation, based largely on studies of amphibian embryos. Inductive interactions are shown by the orange arrows. In the early gastrula stage, the prospective ectoderm in the animal pole is induced by involuting chordamesoderm to form the neural plate. Chordamesoderm again induces a population of cells within the neural plate to differentiate into the prospective optic cup. This population of cells is induced again by chordamesoderm to bulge outward from the diencephalon to the optic vesicle, and eventually to form the optic cup. We know that chordamesoderm is necessary in all three of these inductions because, if it is extirpated prior to each of these stages, then the subsequent steps in differentiation will not occur.

Meanwhile, most of the surface ectoderm remaining after neural induction has been fated to form epidermis of the skin. By the time of early neurulation, a subpopulation of these cells has been designated as prospective lens. As the optic vesicle grows laterally, it induces these cells to form the lens placode, which soon invaginates to form the lens cup and eventually the lens vesicle. The lens vesicle receives an inductive signal from the neural retina that causes it to differentiate into the definitive lens.

As the lens placode invaginates, it leaves behind a population of cells in the prospective epidermis that now is fated to form the cornea. A final inductive signal from the lens vesicle is needed for corneal development to occur. Together, these cellular-level and tissue-level interactions constitute an **inductive cascade** (by analogy to a waterfall or rapids), in which a step in the differentiation of a tissue triggers a step in the differentiation of another tissue.

Neural Crest and Neurogenic Placodes

Neurogenic precursor tissues are those embryonic tissues that give rise to nerve cells or **neurons** during development. Amphioxus has only a single type of neurogenic precursor tissue, the neural tube itself (Fig. 4-12*H*). Two general fates occur for neurons originating in the wall of the neural tube: (1) some will grow out from the neural tube to reach their target tissues, whereas (2) others will establish all of their connections to other neurons within the central nervous system. Unlike amphioxus, vertebrates have three types of neurogenic precursor tissues: (1) the neural tube, (2) the neural crest, and (3) neurogenic placodes (Fig. 4-21). Although the neural tube of vertebrates generates more neurons than the other two neurogenic precursor tissues, the other two sources of neurons are closely associated with many of the most remarkable structures—particularly the sense organs and cranial nerves—that are found in vertebrates.

The topography and developmental details of these three neurogenic precursor tissues have not been worked out for most species of vertebrates. During neurulation of an amphibian embryo, however, the neural folds of the head have distinct lateral and medial walls (Fig. 4-21*A*). Most of the medial wall will form the neural tube proper, which will develop into the brain and spinal cord. The crest of the fold will form the **neural crest** (Fig. 4-19), which will develop into

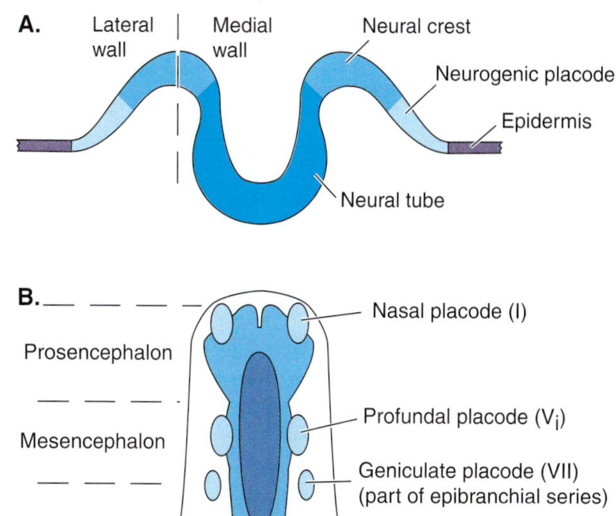

FIGURE 4-21
Positional relationships of neurogenic precursor tissues in the head of a neurulating embryo. *A,* Schematic diagram of neural folds based on an amphibian embryo. Note the relative positions of cells fated to form the brain, neural crest, and neurogenic placodes. All three of these precursor tissues will form neurons. *B,* Dorsal view showing idealized scheme of brain, neural crest, and neurogenic placodes in the head of an amniote embryo. See Table 4-3. *(A, Based on Northcutt; B, compiled from various sources, chiefly Le Douarin and Noden.)*

many different structures (Table 4-2 and the section, "Migration and Fates of Neural Crest Cells," page 151). Portions of the lateral wall of the neural fold will form **neurogenic placodes** (Fig. 4-21 and Table 4-3). The general definition of a placode is any ectodermal thickening caused by either increasing the basal–apical height of cells in that region or by increasing the number of cells. In addition to the lens of the eye (discussed earlier), many different tissues and organs of the integument originate at least partially as placodal thickenings, including scales, teeth, skin glands, and hair follicles (see Chapter 6). This is why the modifier "neurogenic" is used to distinguish those placodes that give rise to neurons.

The neural crest and the neurogenic placodes are among the most intriguing synapomorphies of craniates, for they form or participate in the formation of many structures not found in outgroups. In the 1980s, Glenn Northcutt and Carl Gans pointed out that neural

TABLE 4-2 Derivatives of Cranial and Trunk Neural Crest

	Cranial Neural Crest	Trunk Neural Crest
Pigment cells	Some	Melanocytes Xanthophores Iridophores
Sensory systems	Trigeminal nerve (V) Facial nerve (VII) Glossopharyngeal nerve (IX) Vagal nerve (X)	Spinal ganglia Vagal ganglia (X)
Autonomic nervous system	Parasympathetic ganglia	Sympathetic ganglia Adrenal medulla Parasympathetic ganglia
Skeletal and connective tissues	Gill bars Trabeculae Parachordals Odontoblasts Membrane bones	Walls of aortic arches
Endocrine organs		Adrenal medulla Calcitonin cells Carotid body cells (Type I) Parafollicular cells of thyroid

crest and neurogenic placodes are similar in that both (1) are derivatives of ectoderm, (2) migrate, and (3) form sensory neurons and special sense organs. Several key differences exist, however. Unlike neural crest, neurogenic placodes are initially restricted to the head, although some placodes may later migrate out onto the trunk (see section, "Sensory Systems and Nerves Derived from Neurogenic Placodes," page 154). Neural crest can form motor neurons, but neurogenic placodes do not. Cells from neurogenic placodes can form sensory receptor cells, such as the olfactory epithelium or hair cells of the ear, but neural crest cells do not.

TABLE 4-3 Major Derivatives of Neurogenic Placodes

Placode	Sensory Receptor Epithelium	Sensory Ganglion[a]
Nasal placode	Olfactory epithelium	
Profundal placode		Profundal ganglion
Geniculate placode	Induces taste buds	Geniculate ganglion
Anterodorsal lateral line placode	Supraorbital and infraorbital lateral lines and electroreceptor fields	Anterodorsal lateral line ganglion
Anteroventral lateral line placode	Mandibular lateral line and electroreceptors	Anteroventral lateral line ganglion
Otic lateral line placode	Otic lateral line and electroreceptors	Otic lateral line ganglion
Otic placode	Membranous labyrinth	Octaval ganglion[b]
Middle lateral line placode	Middle lateral line	Middle lateral line ganglion
Supratemporal lateral line placode	Supratemporal lateral line	Supratemporal lateral line ganglion
Trunk lateral line placode	Trunk line	Posterior lateral line ganglion
Petrosal placode	Induces taste buds	Petrosal ganglion
Nodose placode(s)	Induces taste buds	Nodose ganglion

[a]See Chapter 13 for additional details on these ganglia.

[b]This ganglion also is known as the vestibulocochlear ganglion; see Chapter 13.

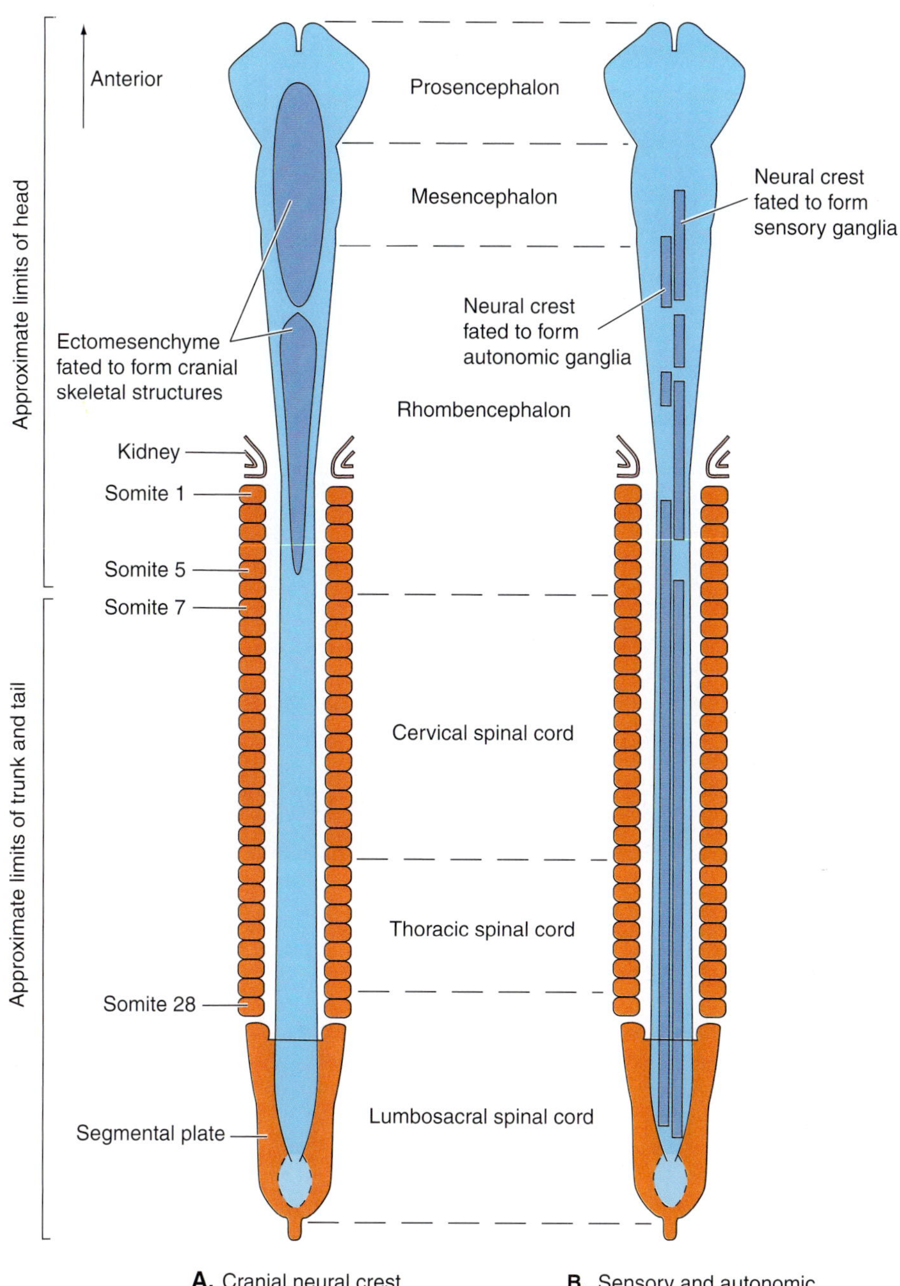

FIGURE 4-22
Simplified fate map of the neural crest in the head and trunk of an amniote (chicken, *Gallus*) embryo. Anterior is at the top. The neural tube is already divisible into the three embryonic brain regions (i.e., prosencephalon, mesencephalon, and rhombencephalon) and the spinal cord (i.e., cervical, thoracic, and lumbosacral regions). *A,* Cranial neural crest. In the head, much of the neural crest becomes ectomesenchyme, which will form portions of the cranial skeleton, teeth, aortic arches, branchiomeres, and other structures. *B,* Neural crest in the head and trunk is fated to form components of the sensory and autonomic nervous systems. Pigment cells (not indicated in this fate map) form from neural crest along its length. *(Based on illustrations by Le Douarin.)*

Fate maps of the neural crest and neurogenic placodes can be constructed using some of the methods described in the earlier section on fate mapping. For example, fate maps of neural crest cells in a chicken embryo are shown in Figure 4-22. Figure 4-22*A* shows the cranial neural crest that will form ectomesenchyme. In Figure 4-22*B*, the fate of the population of neural crest cells that will form sensory or autonomic ganglia is indicated. Pigment cells form from neural crest produced along the entire length of the neural tube; the fates of these pigment cells are not mapped in Figure 4-22.

Migration and Fates of Neural Crest Cells

Although they are derived from an epithelial layer, most neural crest cells transform into a loosely packed mesenchyme. Most mesenchymally organized cells in vertebrate embryos originate from mesoderm, but the neural crest is a primary source of mesenchymal cells in the head region. Mesenchyme of neural crest origin is called **ectomesenchyme** to reflect its derivation from the ectodermal neural crest. Ectomesenchyme participates in formation of many cranial structures (Table 4-2). Some neural crest cells remain close to their site

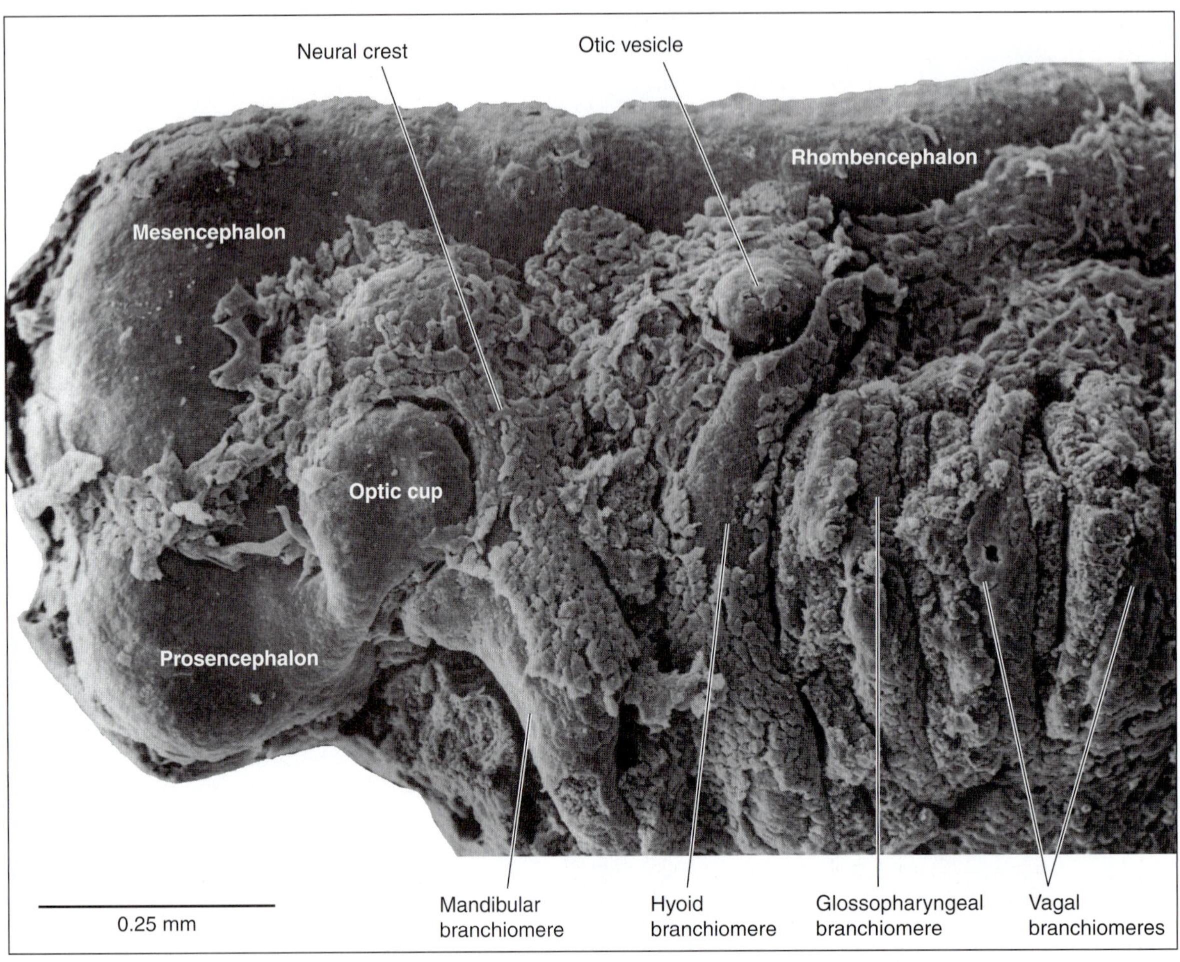

FIGURE 4-23
Scanning electron micrograph showing neural crest cell migration in the head of an axolotl embryo (*Ambystoma mexicanum*). The ectoderm has been dissected away to reveal the streams of stellate (= star-shaped) neural crest cells migrating between the pharyngeal pouches into the branchiomeres. *(From Northcutt and Brändle.)*

of origin, but most migrate. Movements of neural crest cells can be influenced by local environmental conditions within the embryo. In the head, the streams of neural crest cells migrate ventrally around the developing eye and into the pharyngeal region (Figs. 4-23 and 4-24). These streams of neural crest cells contribute substantially to the branchiomeres, the serial structures in the pharyngeal wall that will give rise to branchial arches and many other cranial structures (see "Branchiomeres and Pharyngeal Organization," page 165).

Pathways of neural crest cell migration in the trunk of a vertebrate embryo are diagrammed in Figure 4-25. Many neural crest cells spread out between the surface ectoderm and the segmented somatic mesoderm to form dermal pigment cells. Other populations migrate a short distance and transform into spinal ganglion cells, which will send a short process into the wall of the neural tube and a longer process out into peripheral tissues (see Chapter 13 and Fig. 13-6). Some populations of neural crest cells migrate ventrally to form the sympathetic chain ganglia, a major portion of the autonomic nervous system of the trunk. Still others migrate between the neural tube and developing somites to reach the gut wall, where they form other components of the autonomic nervous system.

How do neural crest cells actually migrate through an embryo? First, the extracellular matrix of a vertebrate neurula-stage embryo consists chiefly of water and proteoglycans rich in hyaluronic acid, giving it a gelatinous consistency. Very little collagen is present at this stage. Thus, it is possible for an amoeboid neural cell to migrate relatively easily through the extracellular spaces of the body. Second, it appears that at least some movements of neural crest cells can be guided by fibronectin "track ways" (Fig. 4-26). After neural crest cells adopt a mesenchymal organization, they decrease the number

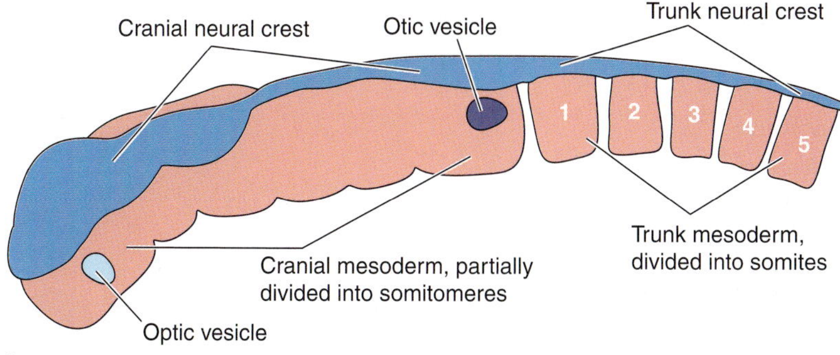

A. Early stage of head development in an amniote

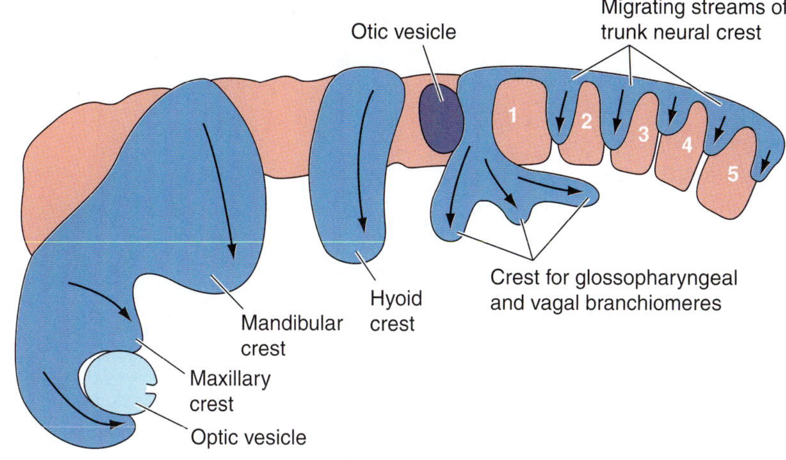

B. Later stage of head development in an amniote

FIGURE 4-24
Neural crest migration in the head of a generalized amniote (chick, *Gallus*) as seen in lateral view. *A,* An early stage of head development, in which the cranial neural crest (ectomesenchyme) still is located dorsally. *B,* A later stage of head development, when neural crest cells are migrating actively around the eye and down into the maxillary and mandibular branchiomeres. The stream of hyoid crest migrates anterior to the ear; glossopharyngeal and vagal crest migrates posterior to the ear. The relationships between the streams of neural crest cells and the mesodermal somitomeres in the head region and the somites in the body also are indicated. *(Based on Noden.)*

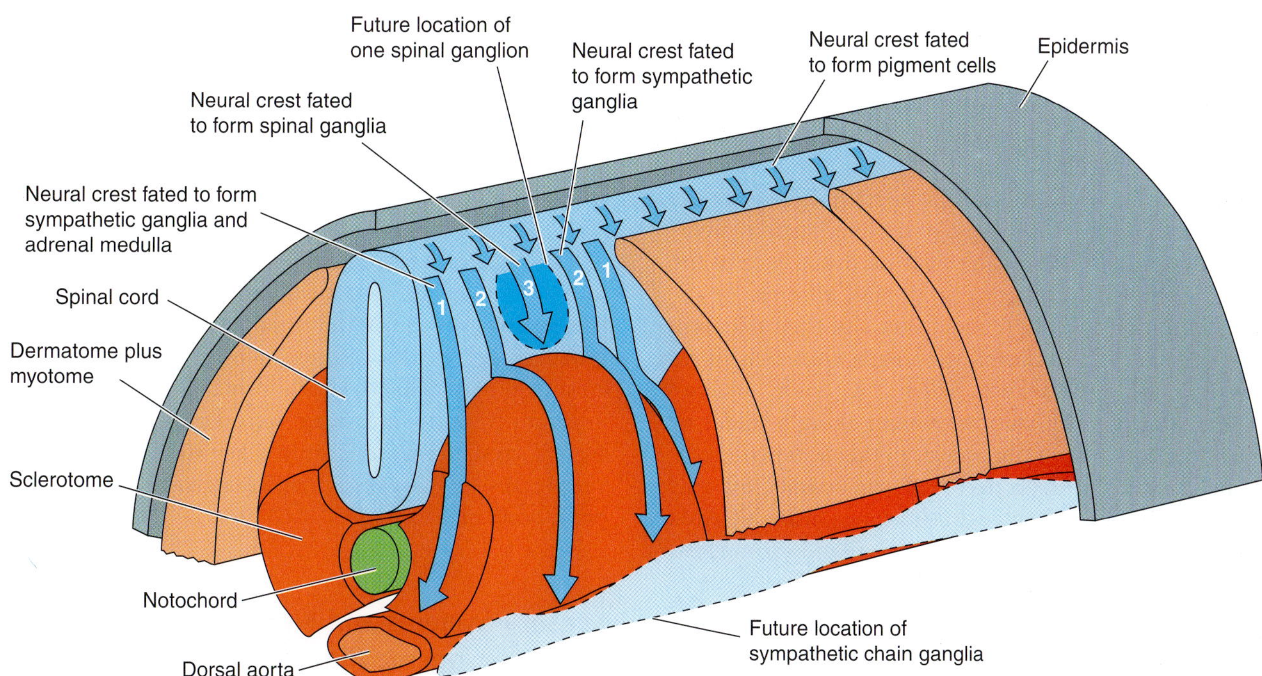

FIGURE 4-25
Routes of migration of neural crest cells in the trunk of amniotes. This stereodiagram shows an idealized view of the routes of neural crest cells relative to the somites. In relation to a single somite, neural crest cells fated to form portions of the peripheral nervous system follow one of three paths. Path 1 carries the cells deep between adjacent somites or through the rostral half of each somite to form sympathetic chain ganglia or the adrenal medulla. Cells following Path 2 migrate between the sclerotome and overlying dermatome and will form portions of the sympathetic chain ganglia. Cells following Path 3 will form spinal (dorsal root) ganglia and sensory nerves for the skin and other somatic structures of the trunk. Pigment cells (*short arrows*) migrate out along the entire length of the neural tube. *(Based on Le Douarin.)*

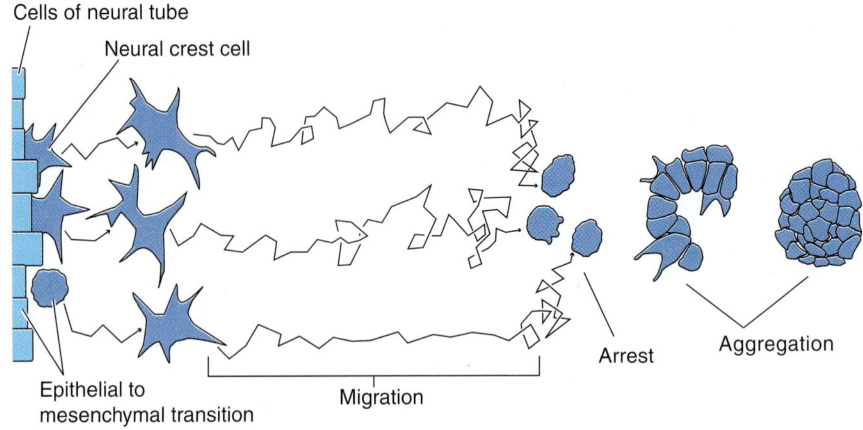

A. Delamination, migration and aggregation of neural crest cells

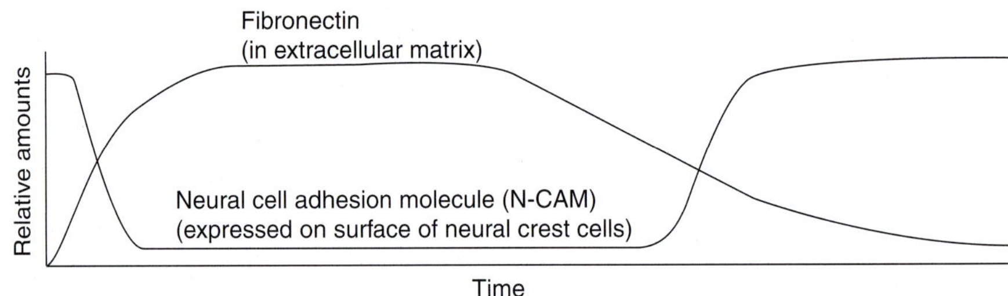

B. Corresponding levels of fibronectin and N-CAM along migration path

FIGURE 4-26
Functional morphology of neural crest cell migration. *A,* Diagram showing delamination and migration of neural crest cells from the neural tube (*left*) to their reaggregation as organ rudiments (*right*). *B,* Corresponding levels of fibronectin and neural cell adhesion molecule (N-CAM) during the course of the migration. As the neural crest cells delaminate from the epithelium, they initially round up before assuming a stellate, migratory morphology. As they migrate through the body, they pass through regions rich in fibronectin. When fibronectin concentrations decrease, they stop migrating and reaggregate to form rudiments of ganglia and other structures. *(Modified from Le Douarin.)*

of neural cell-adhesion molecules (N-CAMs) on their cell membranes. N-CAMs provide a molecular way for cells to attach at their surfaces, so a decrease in the number of N-CAMs decreases their attachments to each other. In this phase of their migration, they are exposed to high levels of the extracellular matrix molecule, fibronectin. Other molecules on the surface of the neural crest cells recognize and bind to sites on fibronectin molecules. By binding to, and then releasing attachments to, fibronectin, the neural crest cells move through the extracellular matrix. A crude analogy might be hand-over-hand movement along a ladder or a series of bars. When the bars stop, that is, when the fibronectin concentration decreases, the neural crest cells stop moving. They then begin to express more NCAMs on their surfaces, adhere to each other, and form organ rudiments that have an epithelial organization.

Neural crest cells give rise to many tissues and structures (Table 4-2). These include the cartilaginous visceral arches in the pharyngeal wall, including the upper and lower jaws; much of the braincase; all pigment cells (except those of the retina of the eye); parts of the teeth and bony scales; portions of the heart; smooth muscle cells; the sensory neurons of all of the spinal nerves and many of the cranial nerves; large portions of the autonomic nervous system; endocrine organs, such as the adrenal medulla; and the myelinated cellular sheaths (Schwann cells) that surround most peripheral neurons. They also contribute to the connective tissue layers (meninges) that surround the brain and spinal cord. The diversity of neural crest derivatives also helps us make sense of several otherwise inexplicable medical syndromes, such as the frequent association of albinism with deafness or craniofacial malformations with heart defects. Neural crest cells not only contribute directly to many important structures but also appear to regulate many aspects of development. For example, as noted earlier in the discussion of epigenetics, neural crest cells participate in tooth formation by a reciprocal interaction with overlying ectoderm of the jaws. Neural crest is thus not only a key derived feature of vertebrates but also essential in organizing the vertebrate body plan.

Sensory Systems and Nerves Derived from Neurogenic Placodes

To frame this discussion, Figure 4-27 compares the developing nervous system of the spiny dogfish, *Squalus,* and the domestic chicken, *Gallus.* By this point in embryogenesis, the neural folds have completely closed along the entire length of the embryo, and the three primary brain regions have differentiated.

The position of neurogenic placodes on each side of the head also is indicated. Most of the neurogenic

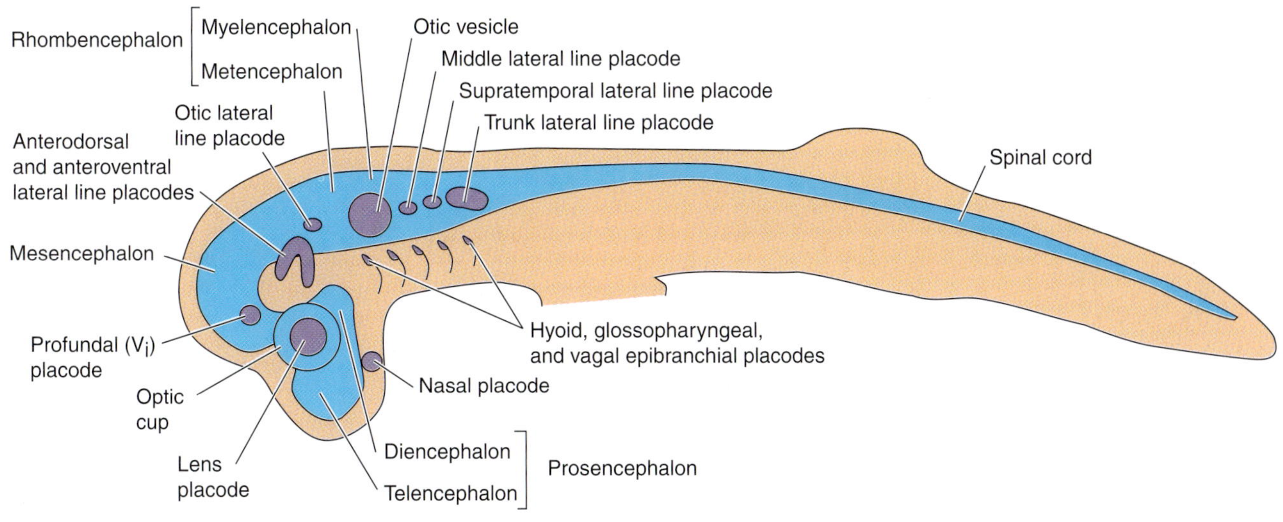

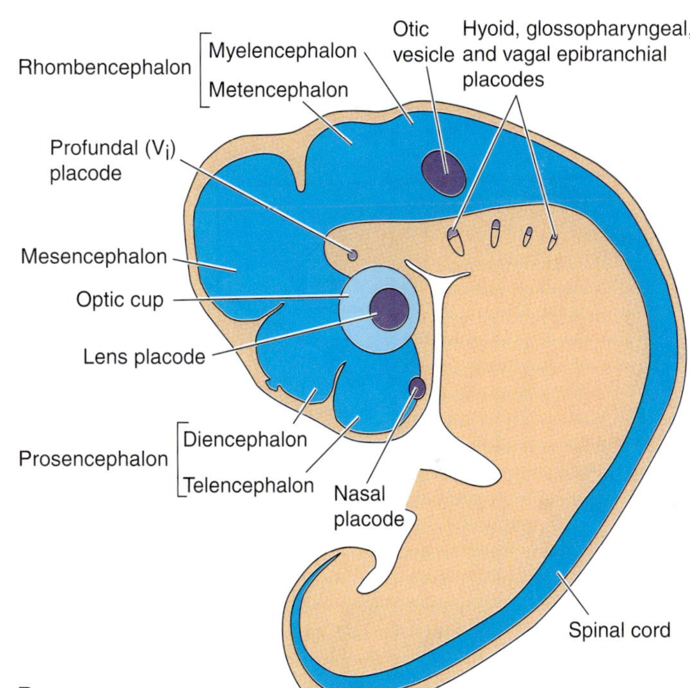

FIGURE 4-27
Comparison of the central nervous system and neurogenic placodes in elasmobranch and chick embryos. *A,* Elasmobranch embryo (based on *Squalus* and *Raja*). *B,* Amniote embryo (based on the chick, *Gallus*). See Table 4-3 for fates of neurogenic placodes. Note that amniotes lack both preotic and postotic lateral line placodes. *(Based on Ballard.)*

placodes of special interest here give rise to specialized sensory receptor cells as well as to the nerves. For example, the **nasal placode** (Fig. 4-28) gives rise to the olfactory epithelium, and the nerves that grow out from this epithelium constitute the first cranial nerve (I). The fibers find their way to connections in the forebrain (Chapters 12 through 14). The **terminal nerve** also develops from the nasal placode. What the placodally derived sensory systems have in common is a network of sensory cells distributed in the skin or structures (e.g., the **otic vesicle,** which forms the inner ear) derived embryonically from the skin. General similarities also exist in the cell types associated with these sensory systems. For instance, in most cases, the sensory receptor cells are ciliated. According to this model, the inner ear is simply an in-folded piece of embryonic skin, an interpretation that is supported by its development.

Cells derived from **dorsolateral neurogenic placodes** form the otic vesicle and inner ear in all vertebrates and other important sensory systems in anamniotes, specifically, the lateral line mechanoreceptive and electroreceptive systems. **Epibranchial neurogenic placodes** (epibranchial placodes, for short) give rise to sensory nerves and ganglia for the chemoreceptive taste buds.

The details by which neurogenic placodes contribute cells to sensory systems are beginning to be un-

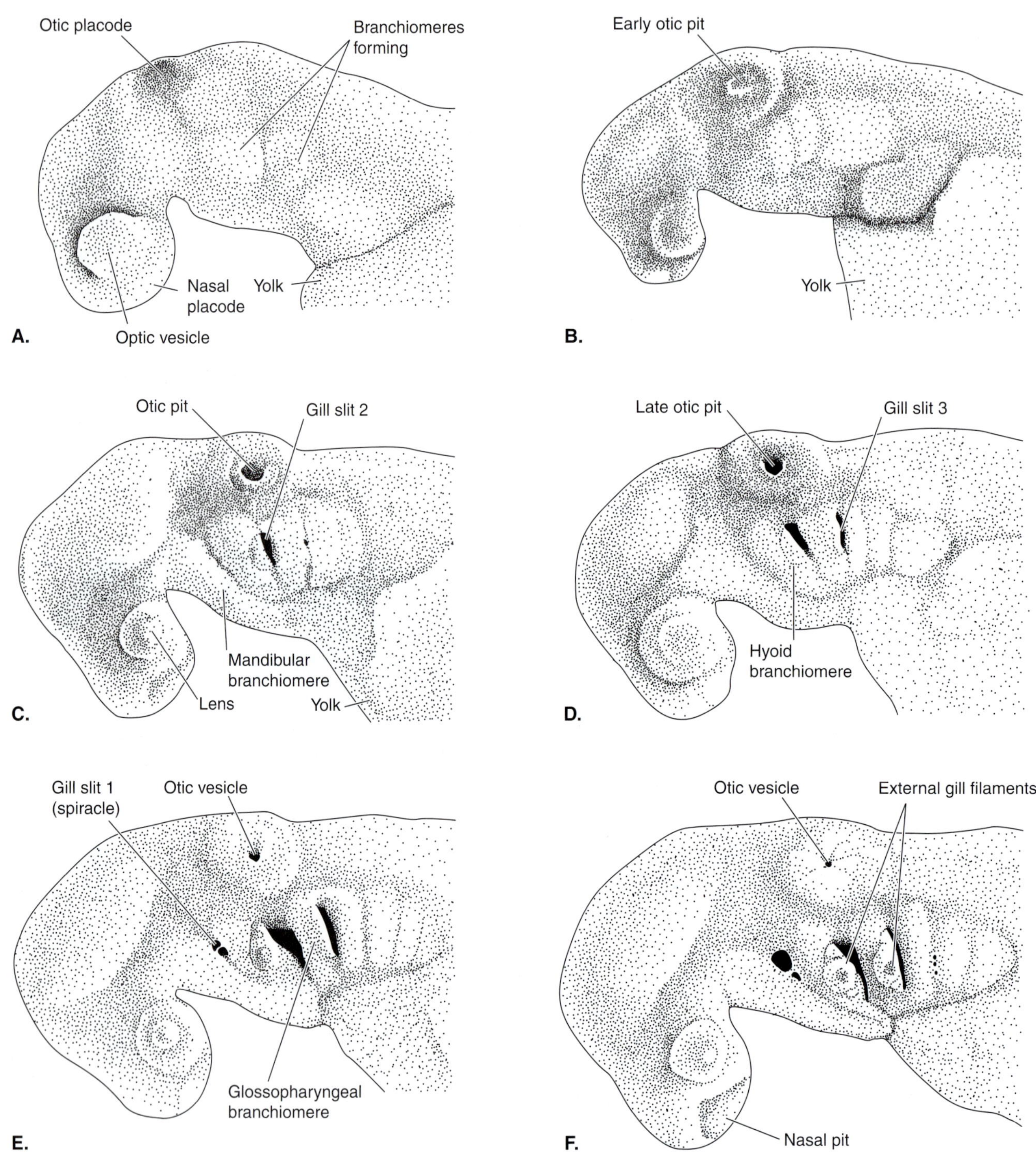

FIGURE 4-28
Ear formation and other aspects of head formation in a developmental series of skate embryos, *Raja erinacea*. *A,* Otic placode stage. *B,* Early otic pit stage. *C,* Otic pit stage. *D,* Late otic pit/early otic vesicle stage. *E,* Slightly later otic vesicle stage. *F,* Closing otic vesicle stage. Features of the branchiomeres and gill slits also are indicated.

derstood, particularly in axolotls (*Ambystoma mexicanum*). Essentially, subpopulations of placodal cells delaminate from the ectoderm and migrate as mesenchymal cells back toward the central nervous system. **Sensory nerves** and **sensory ganglia** associated with each of these sensory systems are formed from such placodally derived cells. Other subpopulations of placodal cells remain associated with the ectoderm, either staying where they formed or migrating before differentiation into sensory cells. Typically, sensory nerves derived from neurogenic placodes reach the brain stem in conjunction with cranial nerves derived from neural crest. For example, the **profundal placode** contributes sensory fibers to the profundal ramus of the trigeminal nerve (V_i), which enters the brain stem with the maxillary and mandibular rami of the trigeminal nerve ($V_{ii, iii}$; see Chapter 13).

By far the best studied of the dorsal series of neurogenic placodes is the **otic** (or octaval) placode, which gives rise to the otic vesicle and, eventually, the inner ear (Fig. 4-28). Like the other neurogenic placodes, the otic placode appears initially as a thickening of the ectoderm lateral to the rhombencephalon. The otic placode usually is apparent before the closure of the neural folds in the head region. The otic placode invaginates as an **otic pit** and eventually sinks beneath the overlying epithelium as the **otic vesicle.** Elasmobranchs retain a duct, known as the **endolymphatic duct,** connecting the vesicle to the skin's surface (see Chapter 12). Figure 4-29 shows the positions of the dorsolateral neurogenic placodes in a developmental series of a salamander (*Ambystoma*). Table 4-3 summarizes the different fates of the placodes diagrammed in Figure 4-29. In developing ambystomatid salamanders, it is easy to observe migration of the neurogenic placode that develops into the **main lateral line of the trunk.** This placode originates at the posterior end of the head and migrates as far back as the tail, "dropping off" small populations of cells en route. It is so large that this placode often can be seen even with a dissecting microscope as a small "bump" on the lateral surface of the trunk. Figure 4-30 shows the distribution of lateral line receptor organs, or **neuromasts,** in the skin of a juvenile axolotl (*Ambystoma mexicanum*). In addition to the array of neuromast organs that comprise the main lateral line, note the presence of lines of neuromasts on the head. Also note that fields of **ampullary organs,** specialized for electroreception, lie adjacent to the lines of neuromasts. Ampullary organs are restricted to the head in the axolotl and most other species of gnathostomes that have electroreception. On each side of the head, the cranial lines of neuromasts develop from five dorsolateral placodes. Three of these placodes originate anterior to the otic vesicle and elon-

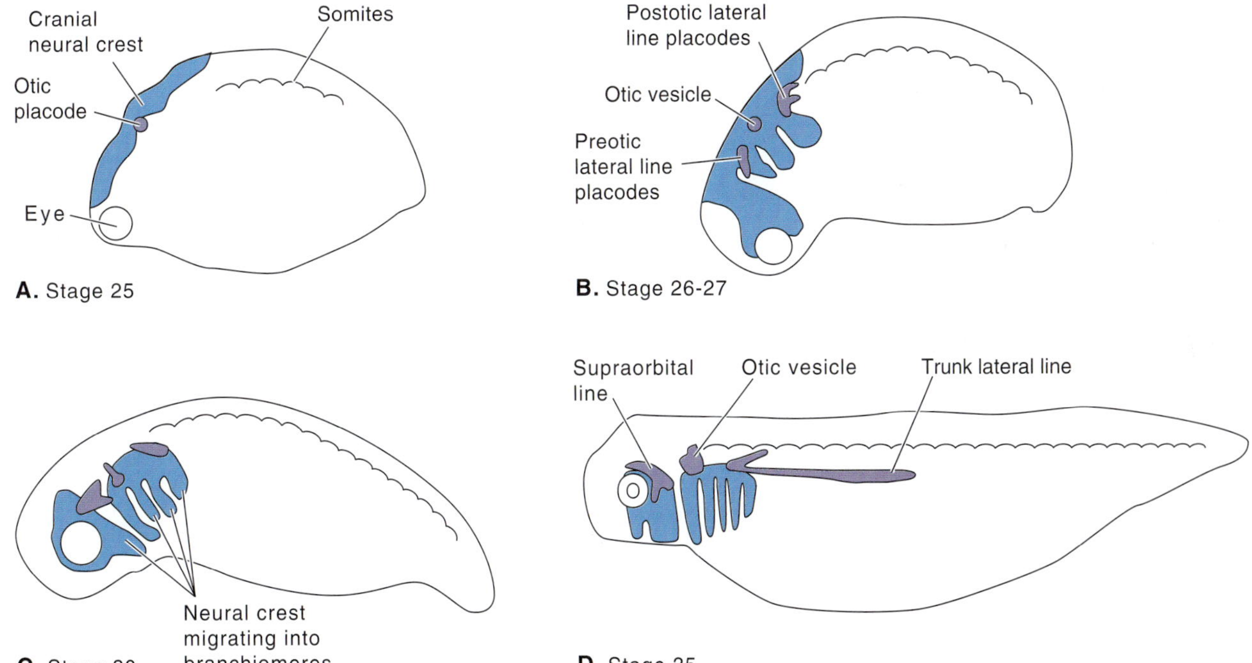

FIGURE 4-29
Simplified diagrams of neural crest and neurogenic placodes in lateral views of salamander embryos and larvae (*Ambystoma*) at different stages of development. The preotic lateral line placodes (indicated on the figure as a group to facilitate presentation at this scale) are anterodorsal lateral line placode, anteroventral lateral line placode, and otic lateral line placode (Table 4-3). The postotic lateral line placodes (indicated on the figure as a group to facilitate presentation at this scale) are the middle lateral line placode, supratemporal lateral line placode, and trunk lateral line placode (Table 4-3). The profundal and epibranchial placodes are not indicated. *A,* Stage 25. *B,* Stage 26–27. *C,* Stage 30. *D,* Stage 35. *(Redrawn and modified from Stone; for more detailed descriptions of the neural crest and neurogenic placodes of* Ambystoma, *see Northcutt and Brändle,* 1995.*)*

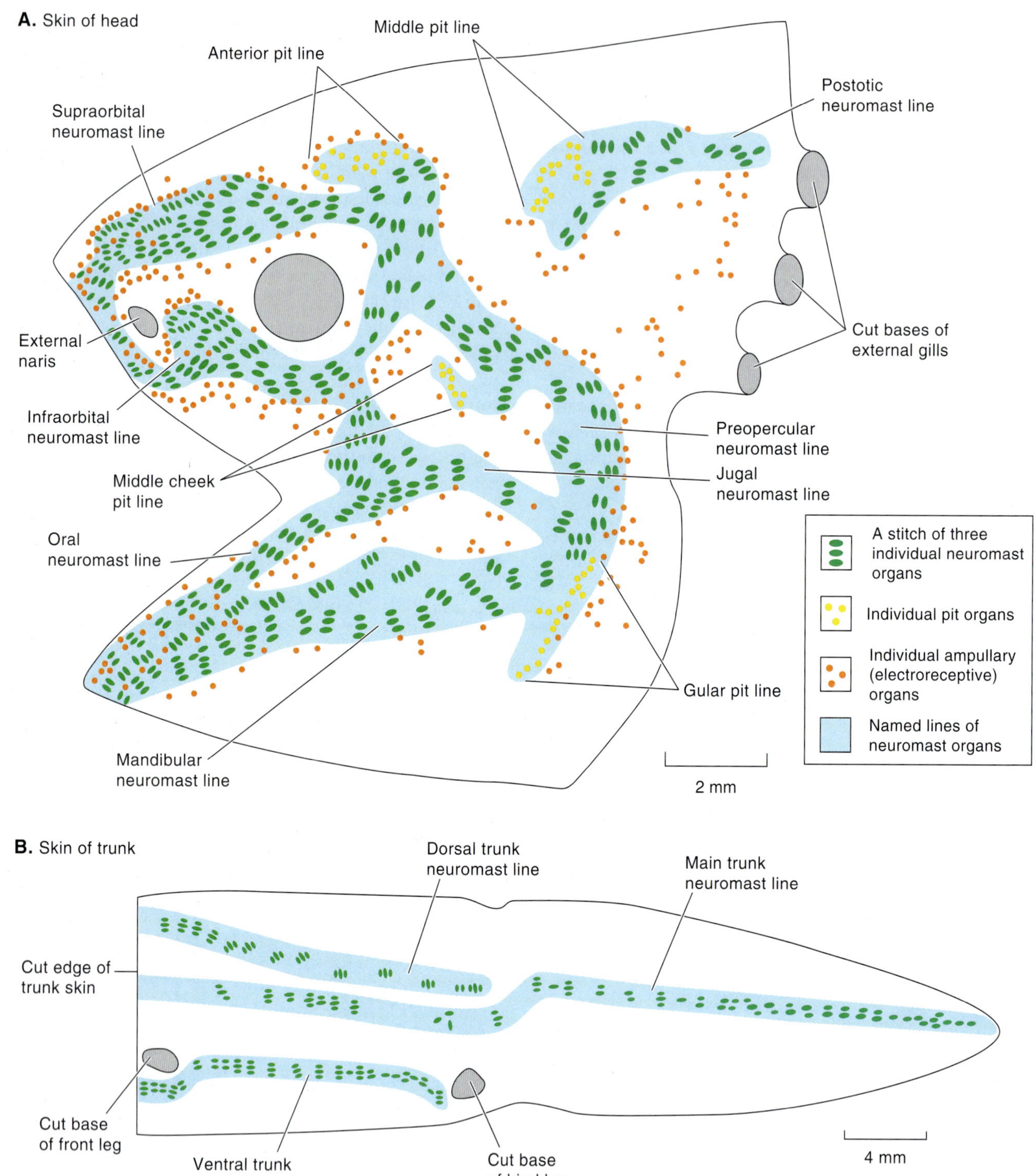

FIGURE 4-30
Distribution of neuromasts and electroreceptors in the skin of a juvenile axolotl. *A,* Skin of the head, cut in the dorsal and ventral midlines and flattened onto a glass slide. *B,* Skin of the trunk, prepared similarly. The lines of neuromast organs are shaded in blue. Note that the ampullary electroreceptors (*orange*) flank these lines. Pit organs shown in yellow are specialized mechanoreceptive components of the lateral line system. *(Redrawn from Northcutt.)*

gate (but do not migrate) to generate the supraorbital, infraorbital, and mandibular lines of neuromasts and their adjacent fields of electroreceptors. Behind the otic vesicle, two other dorsolateral placodes give rise to portions of the otic and supratemporal lines of neuromasts and adjacent fields of electroreceptors. Additional information concerning the structure, development, and function of these organs is included in Chapter 12.

Development of the Coelomic Cavity and Mesenteries

Early aspects of mesoderm differentiation were considered earlier (see page 140, "Mesoderm Differentiation"). This section traces the later development of the coelomic cavity that is defined by the wings of the lateral plate mesoderm. The definitive coelom is a space or network of spaces within the mesoderm that surrounds the visceral organs. It is lined by **serosa,** a thin membrane consisting of a simple squamous epithelium, which is called the coelomic epithelium, or **mesothelium,** and a thin layer of connective tissue beneath the mesothelium. The serosa secretes a small amount of watery, serous fluid into the cavity. Coelomic cavities allow room for the beating of the heart, changes in lung volume, the filling and emptying of the digestive tract, and other changes in the sizes and shapes of the organs. Thin membranes, the **mesenteries,** extend between the organs and from the organs to the body wall. Most mesenteries are identified by the prefix *meso-*, followed by the name of the organ to which they connect (e.g., the mesentery that extends between the dorsal body wall and the stomach is called the mesogaster; see later discussion for additional details). Some mesenteries are called ligaments (e.g., the falciform ligament extends between the ventral body wall and the liver). Although such ligaments are supporting structures, they differ both embryologically and structurally from the ligaments that link bones. Mesenteries consist of two layers of mesothelium with a thin layer of loose connective tissue between them. Mesenteries keep the organs in proper relationship to each other and provide passageways for ducts, blood vessels, and nerves going from one organ to another and between the organs and the body wall. Fat is stored in some mammalian mesenteries.

The coelom develops as a space within the part of the lateral plate mesoderm that extends from the level of the heart to the caudal end of the intestine (Fig. 4-31). As development continues, the splanchnic wall of the lateral plate mesoderm enfolds the gut and its derivatives (Fig. 4-31). These organs thus come to be surrounded by the coelom and are suspended by **dorsal** and **ventral mesenteries** formed by the coming together of lateral plate mesoderm dorsal and ventral to the gut tube. The liver diverticulum develops as a ventral outgrowth of the gut tube (Fig. 4-16). It grows into and expands within the ventral mesentery, dividing it into the **lesser omentum** that extends from the stomach and intestine to the liver, and the **falciform ligament** that extends from the liver to the ventral body wall (Fig. 4-31*B*). In a similar fashion, the dorsal pancreas grows into the dorsal mesentery. The heart develops by the fusion of a pair of blood vessels in the ventral mesentery just anterior to the liver (Fig. 4-31*A*; other aspects of heart development are traced in detail in Chapter 19). Continuous dorsal and ventral mesenteries are present early in development, but those supporting the heart, called **mesocardia,** as well as most of the ventral mesentery caudal to the liver are soon lost (Fig. 4-31*C*). Thus, the originally separate left and right halves of the coelom become continuous both anterior and posterior to the liver.

Most of the embryonic dorsal mesentery persists in adult vertebrates. The parts suspending different organs have different names. For example, the **mesogaster** supports the stomach. Parts of the dorsal mesentery may become complexly folded as the gut tube differentiates during later development. Additional mesenteries support the paired reproductive organs, which enlarge during development and "push" into the coelom from the dorsal body wall: a **mesorchium** goes to each testis; a **mesovarium,** to each ovary; and a **mesotubarium,** to each oviduct. The kidneys do not experience this same developmental "push" into the coelomic cavity but instead remain close to the body wall, plastered to it by the overlying peritoneal tissues. Thus, kidneys are described as **retroperitoneal** (= behind the peritoneal membrane).

In all vertebrates, a partition known as the **transverse septum** develops between the liver and the heart. It divides the coelom into an anterior **pericardial cavity** around the heart and a posterior **pleuroperitoneal cavity** around the abdominal viscera and lungs or swim bladder, if such organs are present (Fig. 4-32). Coelomic epithelium within the pericardial cavity is known as **pericardium.** The visceral pericardium covers the surface of the heart, whereas the parietal pericardium lies adjacent to the body wall (Fig. 4-32). The coelomic epithelium within the pleuroperitoneal cavity is called **peritoneum,** whereas that in the pleural cavities containing the lungs, when these cavities become distinct in amniotes, is known as the **pleura.**

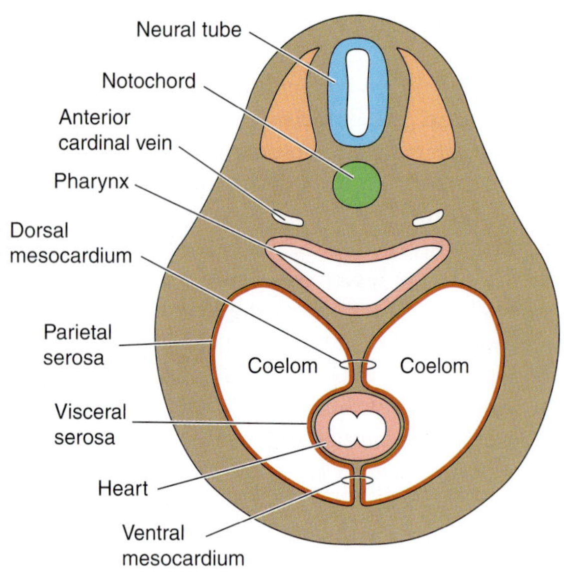

A. Transverse section at level of heart

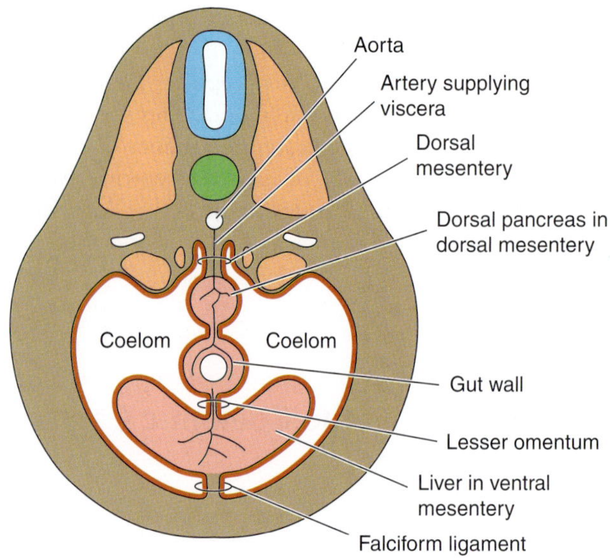

B. Transverse section at level of liver

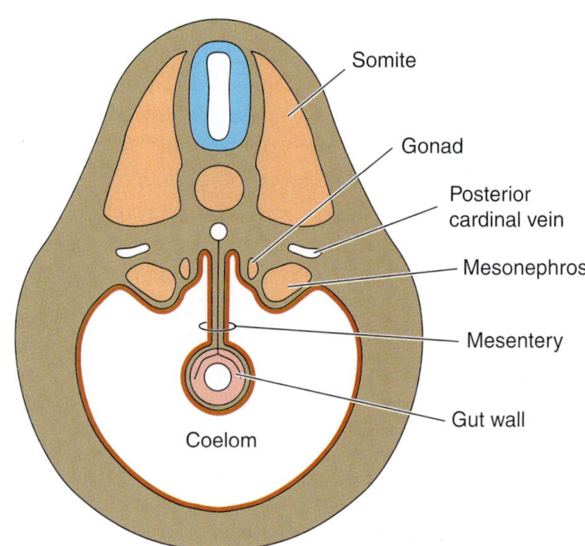

C. Transverse section at level of intestine

FIGURE 4-31
Structural relationships of coelom, mesenteries, and viscera in transverse sections of idealized vertebrate embryos. *A,* The heart develops by fusion of left and right precardiac vessels in the ventral mesentery (mesocardia) cranial to the liver. *B,* The liver grows into the ventral mesentery, and the dorsal pancreas grows into the dorsal mesentery. *C,* Much of the ventral mesentery disappears caudal to the liver, and the left and right coelomic spaces become confluent. *(Modified from Corliss.)*

Development of the Transverse Septum and Coelomic Divisions of Sharks

To visualize the development of the transverse septum, imagine the liver expanding laterally in the ventral mesentery (Fig. 4-33*A*) and carrying the coelomic epithelium that covers it laterally toward the body wall. When the liver touches the body wall (Fig. 4-33*B*), the visceral and parietal layers unite (Fig. 4-33*C*). Their unification forms a partition between the ventral parts of the pericardial and the pleuroperitoneal cavities. The liver grows caudally in the ventral mesentery during its subsequent enlargement, leaving the partition that it formed as the ventral part of the transverse septum (Fig. 4-33*C*). The ventral mesentery caudal to the liver disappears, but the liver remains connected to the transverse septum by a mesentery known as the **coronary ligament,** through which the large hepatic veins draining the liver enter the heart. Simultaneously, the paired common cardinal veins that carry blood from the dorsal part of the body wall to the heart push in medially from the dorsolateral parts of the body wall to the caudal end of the heart. They carry with them sheets of coelomic epithelium, which eventually form the dorsal part of the transverse septum. In sharks a small passage, the **pericardioperitoneal canal,** remains between the folds carrying the common cardinal veins. This passage interconnects the two parts of the coelom (Fig. 4-32*A*).

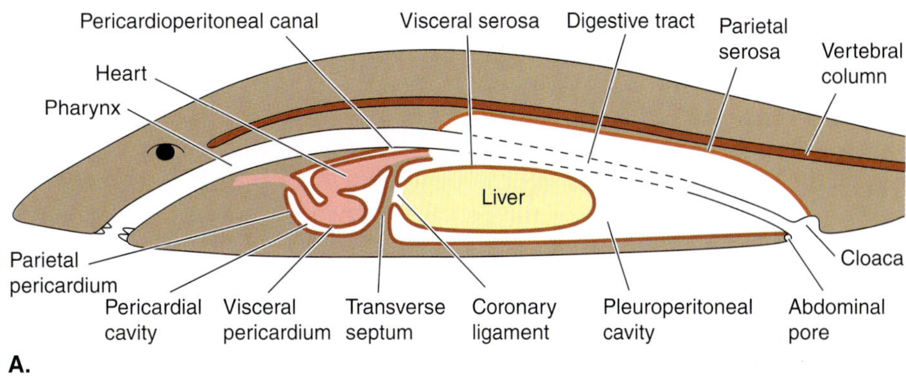

▲
FIGURE 4-32
The coelom and its divisions in an idealized gnathostome, based on a shark (*Squalus*). *A,* Lateral view. *B,* Transverse section through pharynx and pericardial cavity.

▲
FIGURE 4-33
Expansion of the liver and its relationship to coelom formation in *Squalus* shown in diagrammatic frontal sections. *A,* An early development stage. *B,* A later development stage. *C,* Adult condition.

The separation of the pericardial cavity from the pleuroperitoneal cavity is partly a by-product of the development of the transverse septum. However, the nearly complete separation of the two cavities also allows organs in one cavity to move independently of organs in the other cavity. For example, as we will discuss in Chapter 19, the presence of the pericardial cavity also allows the development of a reduced pressure around the fish heart, and this plays a role in the dynamics of blood circulation.

The heart and pericardial cavity of an adult shark are located far forward, ventral to the pharynx, and muscles of the body wall and the pharyngeal floor surround them (Fig. 4-32*B*). The heart thus is close to the gills, through which the heart must pump blood before it is distributed to the body. The transverse septum is located about the level of the pectoral girdle and lies in the transverse plane.

In some chondrichthyans and bony fishes a pair of small **abdominal pores** lead from the caudal end of the pleuroperitoneal cavity to the cloaca. Their significance is uncertain, but they may represent an ancestral passage for sperm and eggs from the coelom to the outside. Gametes are discharged in this way in lampreys.

Modifications of the Coelom and Mesenteries of Tetrapods

As in an adult shark, the heart and pericardial cavity of an embryonic tetrapod are located far forward in the body and ventral to the pharynx. During later development, however, the heart of tetrapods migrates caudally, closer to the lungs. The pericardial cavity thus comes to lie ventral to the anterior part of the pleuroperitoneal cavity, and the transverse septum assumes an oblique orientation, as seen in adult amphibians (Fig. 4-34*A*). The lungs are located ventrolateral to the digestive tract and its supporting mesenteries in the part of the pleuroperitoneal cavity overlying the pericardial cavity (Fig. 4-34*B*). The areas containing the lungs are called the **pleural recesses.** They frequently extend ventrally on each side of the pericardial cavity, thereby partly separating the pericardial cavity from the body wall (Fig. 4-34*B*). The membrane between pleuroperitoneal and pericardial cavities is homologous to the transverse septum of a shark, but the part of it separating the pleural recesses from the pericardial cavity often is called the **pleuropericardial membrane.**

In most living diapsids (some lizards, snakes, crocodiles, and birds) as well as all mammals, additional

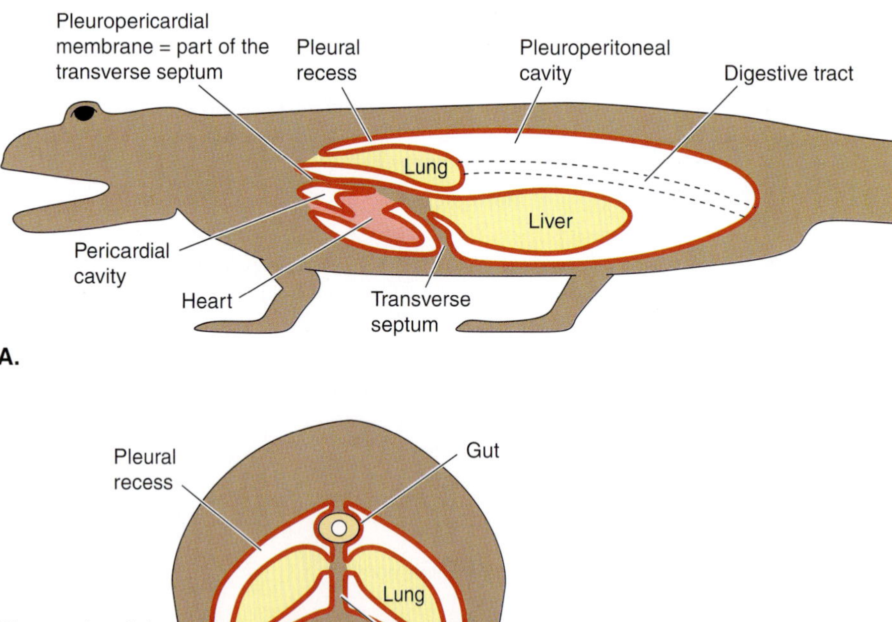

FIGURE 4-34
The coelom and its two divisions in a basal tetrapod, such as a salamander. *A,* Lateral view. *B,* Transverse section through pleural recesses and pericardial cavity.

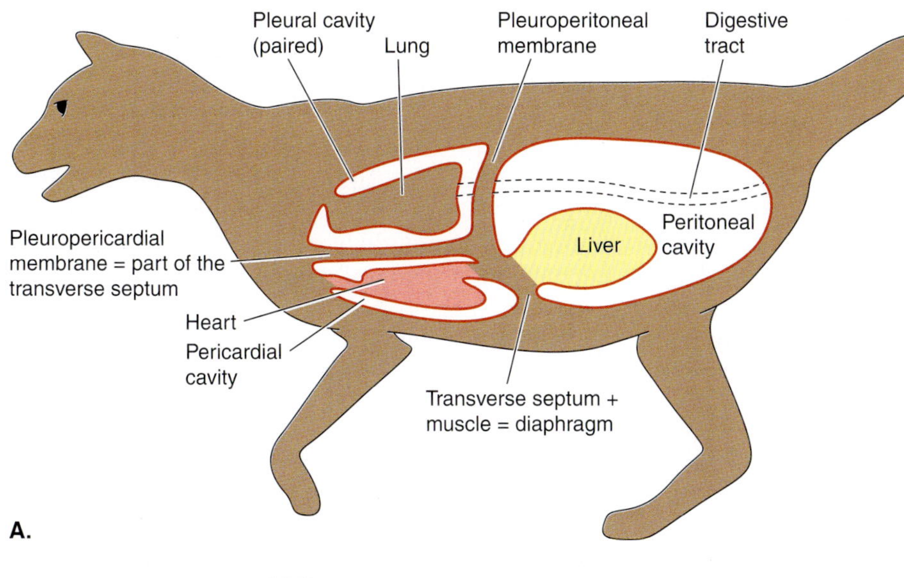

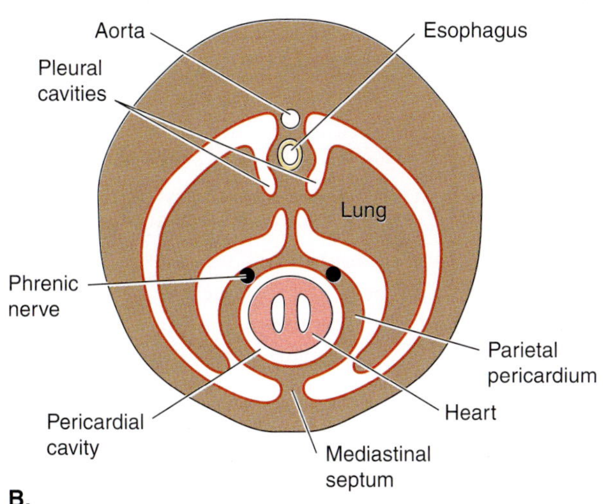

FIGURE 4-35
The coelom and its three divisions in a mammal, such as a cat. *A*, Lateral view. *B*, Transverse section through pleural and pericardial cavities.

folds of coelomic epithelium separate the paired pleural recesses from the rest of the pleuroperitoneal cavity. The coelom of these animals thus consists of four compartments: the pericardial cavity, two pleural cavities, and the peritoneal cavity (Fig. 4-35*A*). The location of the lungs in distinct pleural cavities allows them to expand and contract independently of other organs.

In reptiles and birds, the folds separating the pleural cavities from the peritoneal cavity form the oblique septum. In mammals, the separation between the two pleural cavities and the peritoneal cavity develops by the **pleuroperitoneal membranes,** which push in from the dorsolateral body wall, and by other folds that extend laterally from the mesenteries and medially from the body wall to meet the pleuroperitoneal membranes (Fig. 4-35*A*). Somatic muscles invade these membranes and part of the transverse septum separating the pericardial and peritoneal cavities to form the **diaphragm** (Fig. 4-35*A*). This is the primary respiratory muscle of a mammal. Because these developmental processes occur early in organogenesis, when the pericardial cavity and heart are located far forward, the somatic muscles entering the diaphragm develop from somites in the cervical (neck) region. When these structures shift caudally later in development, branches of cervical spinal nerves that innervate the diaphragmatic musculature become greatly elongated and pass through the anterior part of the trunk as the **phrenic nerves** (Fig. 4-35*B*).

The enlarging lungs and pleural cavities of mammals grow laterally and ventrally to surround the pericardial cavity and heart, often meeting ventral to the pericardial cavity (Fig. 4-35*B*). This effectively separates the pericardial cavity from contact with the body wall. In some mammalian species, the heart and pericardium also are separated from the diaphragm by a caudal and ventral growth of the lungs and pleural cavities. The wall of the pericardial cavity thus consists of the parietal pericardium and parietal pleura, with a thin layer of connective tissue sandwiched between them. This combined wall often is called the **pericardium** or **pericardial sac.**

Many organs lie between the pleural cavities of mammals: the pericardial cavity and heart; the esophagus; major arteries and veins; the phrenic and other nerves; and, in embryonic and young mammals, the thymus. The area between the two pleural cavities that contains these structures is called the **mediastinum.** The mesentery formed where the medial walls of the two pleural cavities meet above and below these structures is the mediastinal septum. Other aspects of mesentery formation and body cavities will be discussed in connection with the gut tube and its derivatives (Chapter 17).

Basic Organization of the Vertebrate Head

Embryogenesis of cranial structures is a major focus in evolutionary morphology because the brain, major sense organs, and many parts of the feeding and respiratory systems are located in the head. Also, the cranial skeleton develops from several different sources, which can be most easily understood from a developmental perspective (see Chapter 7). Elasmobranchs (sharks, skates, and rays) are convenient vertebrates in which to study head development. In contrast to the embryos of most bony fishes, which are typically small and thus strongly curved around their yolk, elasmobranch embryos are large and have a "straight" pharyngeal region that is easy to study (Fig. 4-36). In embryos of amniotes (e.g., the chick) the development of the pharyngeal region has been dramatically altered from the basic plan seen in early vertebrates. No amniotes retain gill slits as an adult, nor do they develop the full complement of aortic arches, nor do they have the lateral line mechanosensory and electrosensory systems that are found in early vertebrates. All three of these features are fundamental to understanding the plesiomorphic structure and organization of the head, thus making elasmobranch embryos the specimens of choice.

Figure 4-37 is a schematic lateral diagram of the head of an idealized shark embryo, based on a famous illustration by an English comparative anatomist and embryologist, Edwin S. Goodrich. Goodrich was interested in detecting the role of segmentation in patterning the structures of the vertebrate head, and his interpretive drawing has influenced generations of anatomists since its publication in the 1930s. The basic arrangement of the cranial nerves is particularly clear from his diagram, and it will be a useful reference throughout your study of vertebrate anatomy.

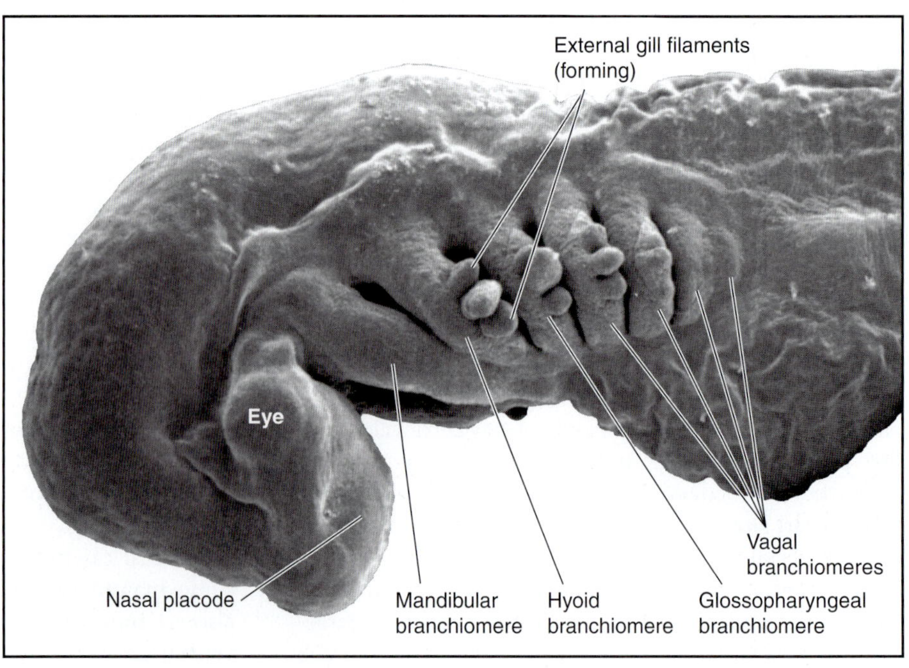

FIGURE 4-36
Scanning electron micrograph showing organization of the gill arches in a skate embryo (*Raja erinacea*). This specimen is at a later stage of development than the one shown in Figure 4-28*F*. Note the well-developed eye; the nearly complete series of branchiomeres and gill slits; and the still small external gill filaments forming on the hyoid, glossopharyngeal, and vagal branchiomeres. The nasal placode is beginning to invaginate to form the nasal pit.

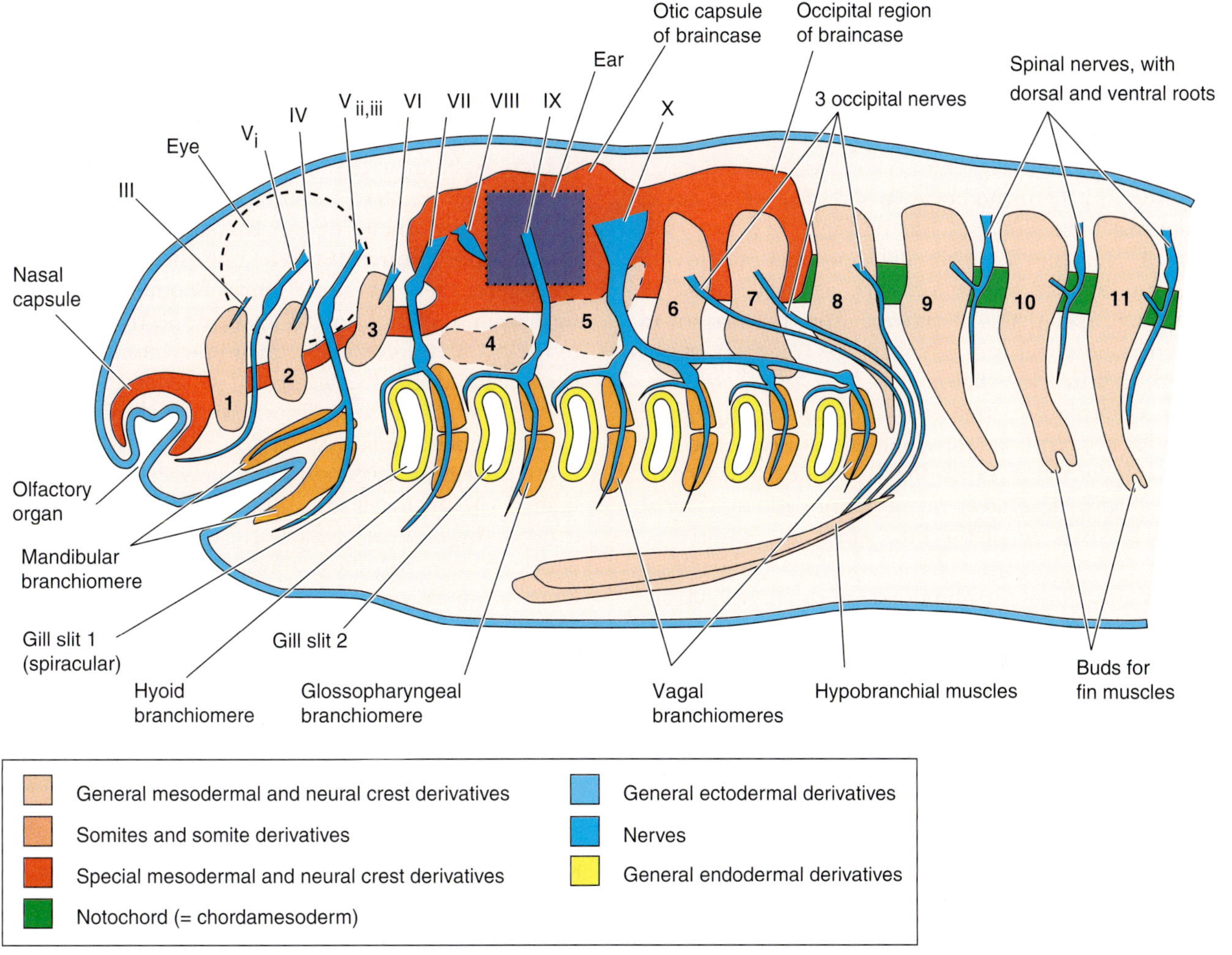

FIGURE 4-37
Semitransparent schematic view of vertebrate head development, based on a shark embryo. The brain is omitted. The diagram illustrates segmentation of the cranial somites and branchiomeres. Somites bear Arabic numerals and are innervated by cranial nerves III, IV, and VI. Branchiomeres located between the gill slits and including the jaws are innervated by cranial nerves V_{ii}, V_{iii}, VII, IX, and X. The braincase and associated nasal and otic sensory capsules are indicated. The myotomes and somatic motor nerves of segments 4 and 5 degenerate and are indicated with dashed lines. *(Based on a figure by Goodrich.)*

Branchiomeres and Pharyngeal Organization

In many textbooks and in the primary literature, the terms "pharyngeal arch," "visceral arch," and "branchial arch" are used inconsistently. Noden (1991) considered the term "branchial arch" least prone to the misinterpretation that all of the cells forming the arches develop from the visceral tube. However, using the word "arch" leaves open the possibility of confusion with the aortic arches or the skeletal arch that forms within a "branchial arch." Also, in adult fishes, "branchial arch" is used to refer to fully differentiated gill arches that carry gill filaments. In the absence of a single, wholly satisfactory term, we consider it best to refer to these embryonic structures collectively as the branchial segments, or **branchiomeres** (Bemis & Grande, 1992). This term branchiomere reflects the obvious segmentation of the gill region while avoiding confusion about either the embryonic sources of their cells or the fully differentiated adult condition. By tradition and because they are good descriptors, we use the terms **visceral pouch** or **pharyngeal pouch** to describe an outpocketing of the pharynx lying between two adjacent branchiomeres.

The basic pattern of the branchiomeres in a lateral view of an embryo of a skate (*Raja*) is apparent in Figure 4-36 (see also Fig. 4-28). Working from anterior to posterior, the mandibular, hyoid, glossopharyngeal, and vagal branchiomeres can be seen. Between each pair of branchiomeres are gill openings. These gill openings break through very early and never close in embryonic skates; this is different from the pattern in amniotes, in which gill openings are transitory structures if they develop at all.

An example of a transitory developmental feature in skates concerns the development and regression of external gills. Clear, functional reasons exist as to why external gills develop in many embryonic and larval fishes. In the case of the little skate, the embryonic period (from just after fertilization up to hatching from its vitelline membrane) and larval period are passed inside a proteinaceous capsule known as a mermaid's purse. The embryonic and larval periods can last more than 150 days, and at the end, a miniature adult breaks out of the mermaid's purse. How does efficient gas exchange occur within the confines of the mermaid's purse? In Figure 4-36, three small buds are visible on each of the postmandibular branchiomeres. Each of these buds will grow out to form a long, filamentous, capillary loop that serves to exchange respiratory gases. These external gills begin to develop as soon as the heart is sufficiently developed to circulate blood through the aortic arches. In order to have an efficient gas exchange surface, however, it is also necessary to move water over it. Larval skates generate a current by beating the tip of the tail to create a flow of aerated water through small openings in the egg case. This is functionally important because, in such a young larva, the development of the cranial muscles and skeleton is insufficient to allow for the generation of a respiratory current of water by the mechanism used in adults. About halfway through the developmental period, the mouth and relevant cranial skeletal muscular elements are fully functional, and it becomes possible to irrigate the gills by the normal buccal pumping mechanism used by adults (see Chapter 18). From this point on, the external gills regress until they are indistinguishable from the other gill filaments of the gill (i.e., the capillary loops that formed the external gills are retained as part of the regular series of gill filaments in an adult; see Pelster and Bemis, 1992).

A schematic diagram of a horizontal section through the pharynx of an elasmobranch embryo is shown in Figure 4-38. Note in Figure 4-38*A* that each branchiomere contains rami of the cranial nerve associated with that branchiomere,[5] as well as branchiomeric muscle, a skeletal rod, and an aortic arch. In each branchiomere, the lateral surface is covered with ectoderm, and the medial surface is covered with endoderm.

The nomenclature of the branchiomeres in elasmobranchs is summarized in Figure 4-38*B*. The first two branchiomeres are so important in craniofacial development that they are given special names, the **mandibular branchiomere** and the **hyoid branchiomere**; these are also commonly referred to as the **first arch** and the **second arch,** respectively. Caudal to the hyoid branchiomere, each branchiomere may be best denoted by the name of its associated cranial nerve. Thus, the next in series is the **glossopharyngeal branchiomere,** followed by four or even five **vagal branchiomeres** (depending on the species of elasmobranch chosen for study). Branchiomeres also can be denoted by numbers, which correspond to the numbering system for their aortic arches (Fig. 4-38*B*). Note, however, that the numbering system for the cranial nerves does not match that of the aortic arches and branchiomeres. The trigeminal (Vth) nerve supplies the mandibular branchiomere. The facial (VIIth) nerve supplies the hyoid branchiomere. Branchiomere III is supplied by the glossopharyngeal (IXth) nerve, and branchiomeres IV to VII are supplied by rami of the vagal (Xth) nerve (Chapter 13).

In elasmobranchs, **gill slits** develop between each branchiomere and the one anterior to it (Figs. 4-36 and 4-38*B*). Each gill slit or pouch thus has the same number as the branchiomere immediately anterior to it, although these numbers are conventionally written as Arabic numerals instead of Roman numerals. The development of the gill slits is shown in the developmental series of skate embryos illustrated in Figure 4-28. Gill slit 2 is the first to open, followed by gill slit 3, and next by gill slit 1 (which receives the special name of **spiracle**). All but the last gill slit are open in the older skate embryo shown in Figure 4-36. The lack of strict anterior-to-posterior sequence in the opening of the gill slits is undoubtedly related to differences in size and function of the openings in adult elasmobranchs. Before each gill slit "breaks through," we refer to the tissue that occupies the opening as the **branchial membrane.** The pocket on the medial side of each branchial membrane lined with endoderm is known as a **pharyngeal pouch.** Endodermal cells derived from these regions have many different fates, including formation of endocrine glands and organs of the immune system. These fates are traced in Chapter 15.

[5]The post-trematic ramus of the appropriate nerve supplies the skeletal muscles of a branchiomere, so it is usually the largest nerve bundle in a branchiomere; see Figure 13-18*C*.

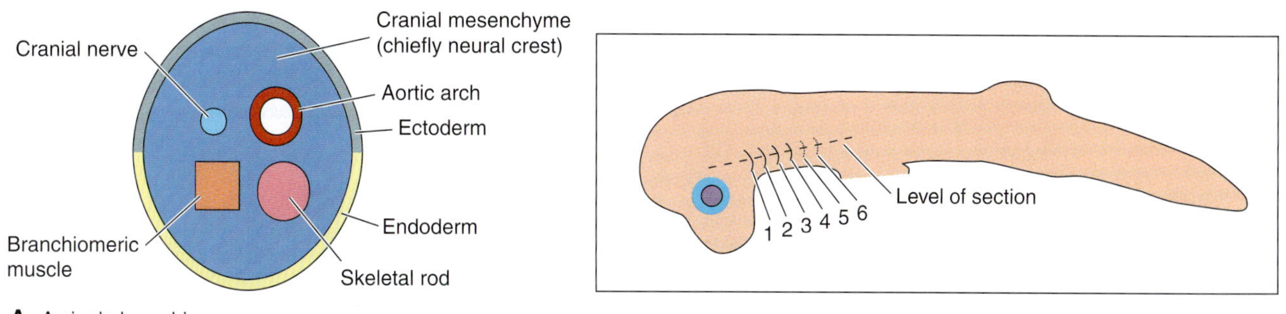

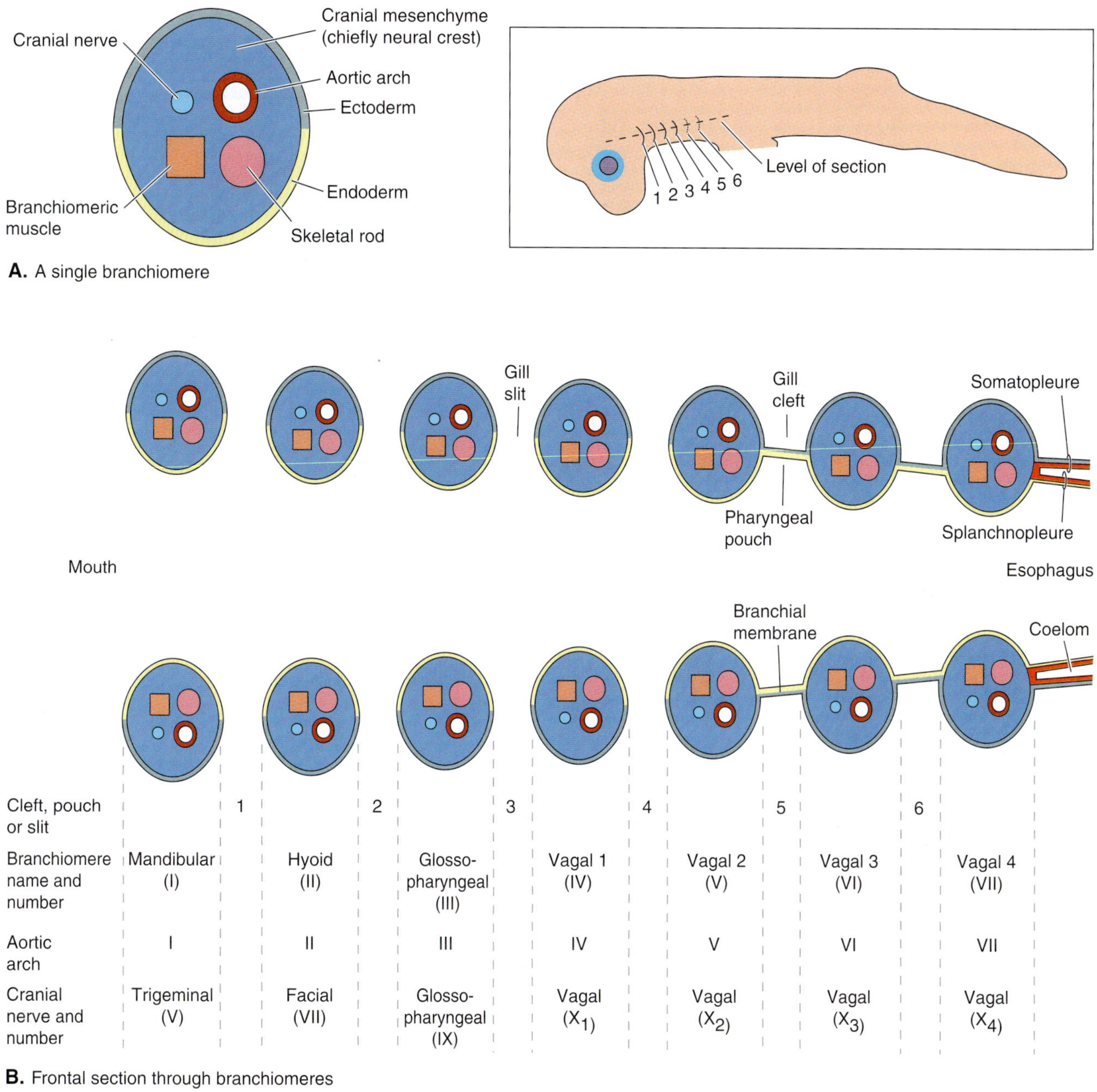

Cranial nerve
Cranial mesenchyme (chiefly neural crest)
Aortic arch
Ectoderm
Endoderm
Skeletal rod
Branchiomeric muscle
A. A single branchiomere
Level of section
1 2 3 4 5 6
Gill slit
Gill cleft
Somatopleure
Pharyngeal pouch
Splanchnopleure
Mouth
Esophagus
Branchial membrane
Coelom
Cleft, pouch or slit
1
2
3
4
5
6
Branchiomere name and number
Mandibular (I)
Hyoid (II)
Glosso-pharyngeal (III)
Vagal 1 (IV)
Vagal 2 (V)
Vagal 3 (VI)
Vagal 4 (VII)
Aortic arch
I
II
III
IV
V
VI
VII
Cranial nerve and number
Trigeminal (V)
Facial (VII)
Glosso-pharyngeal (IX)
Vagal (X_1)
Vagal (X_2)
Vagal (X_3)
Vagal (X_4)
B. Frontal section through branchiomeres

FIGURE 4-38
Idealized schematic frontal sections through branchiomeres of an elasmobranch embryo (e.g., *Raja* or *Squalus*). The inset shows a lateral view with the gill slits and clefts numbered and an approximate plane of section for the schematic figure in *B*. In the inset, gill clefts 5 and 6 have not yet broken through, so their location is indicated by dotted lines. *A,* A single branchiomere, showing its four major components: branchiomeric muscle, a cranial nerve, an aortic arch, and a skeletal rod. *B,* The complete series of branchiomeres, showing the terminology for the branchiomeres and intervening gill clefts, pouches, and slits.

Organization of the Head in Amniote Embryos

Head development in amniote embryos differs in some important ways from what we have seen in elasmobranchs, and as already noted, elasmobranchs are far better for understanding plesiomorphic features of cranial organization. Much more research, however, has been conducted on amniote embryos, particularly chickens (*Gallus*) and mice (*Mus*). Figure 4-39 presents an idealized schematic diagram of the cranial organ systems in an amniote embryo, redrawn from

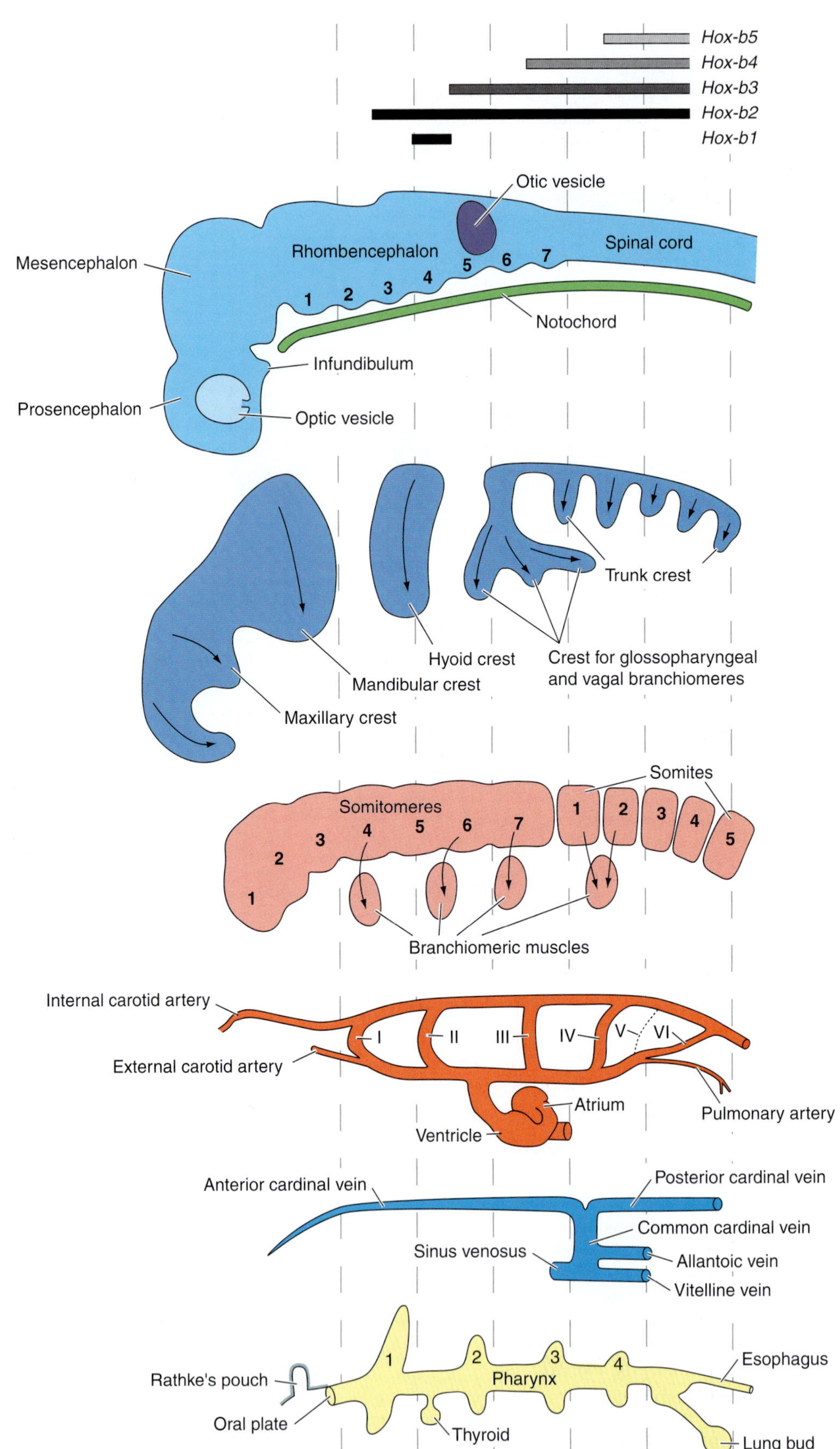
A. Expression domains of Hox-b genes
Hox-b5
Hox-b4
Hox-b3
Hox-b2
Hox-b1
B. Neural tube, notochord, optic vesicle and otic vesicle
Otic vesicle
Mesencephalon
Rhombencephalon
Spinal cord
1 2 3 4 5 6 7
Notochord
Infundibulum
Prosencephalon
Optic vesicle
C. Neural crest
Trunk crest
Hyoid crest
Crest for glossopharyngeal and vagal branchiomeres
Mandibular crest
Maxillary crest
D. Paraxial mesoderm
Somites
Somitomeres
1 2 3 4 5 6 7
1 2 3 4 5
Branchiomeric muscles
E. Heart and aortic arches
Internal carotid artery
External carotid artery
I II III IV V VI
Atrium
Ventricle
Pulmonary artery
F. Veins
Anterior cardinal vein
Posterior cardinal vein
Common cardinal vein
Sinus venosus
Allantoic vein
Vitelline vein
G. Pharyngeal pouches, thyroid gland and Rathke's pouch
Rathke's pouch
Oral plate
1 2 3 4
Pharynx
Esophagus
Thyroid
Lung bud
Branchiomere number
I II III IV V VI

the elegant synthetic work of Drew Noden (1991). Using gray tic lines to demarcate branchiomeric boundaries, the diagram presents major cranial tissues in registration with each other and with the expression domains of *Hox* genes known to play a role in cranial patterning (Fig. 4-39*A*).

Figure 4-39*B* shows dorsal axial structures, including the neural tube and notochord. The prosencephalon, mesencephalon, and rhombencephalon (the three primary brain regions) are labeled, and the seven rhombomeres in the wall of the hindbrain are numbered. The paired otic and optic vesicles also are shown; these are easily detected reference points when comparing embryonic vertebrates. The anterior tip of the notochord is another important reference point for comparing vertebrate embryos. It also marks the beginning of the "new head" postulated by Northcutt and Gans (1983; see Chapter 2). The **hypophysis** forms immediately anterior to the notochord; its location is demarcated in this diagram by the **infundibulum,** which contributes to formation of the hypophysis (see Fig. 15-4*A*).

Figure 4-39*C* diagrams the cranial neural crest migrating ventrally into the branchiomeres, in movements already introduced in Figure 4-24*B*. The many different fates of these cells were briefly discussed earlier and are summarized in Table 4-2.

Figure 4-39*D* shows the organization of paraxial mesoderm in the head. The anterior portion of the paraxial mesoderm is incompletely divided in amniote embryos. Instead, it consists of a series of seven **somitomeres,** which can be detected using scanning electron microscopy to study carefully dissected embryos. Somitomeres are transient, mesenchymally organized structures, which, unlike true somites, never become "epithelialized" and thus never contain somitocoele cavities. Note that the series of seven somitomeres in amniotes is numbered independently from the somites in the trunk. This convention differs from the standard nomenclature used for elasmobranch embryos, which is shown in Figure 4-37. This is because anterior portions of the paraxial mesoderm of elasmobranchs subdivide into well-developed, epithelially organized somites containing somitocoeles (somites 1–3 in Fig. 4-37). It has proved difficult for researchers to agree on a single alignment and consistent numbering system equating the anterior cranial somites of elasmobranchs with the somitomeres of amniotes. Despite this and the many differences in their organization, cells derived from the paraxial mesoderm are thought to have similar fates in elasmobranchs and amniotes. Much of the paraxial mesoderm forms skeletal muscles of the head, including the extrinsic eye muscles and muscles of the branchiomeres. As determined by Noden (1991) and shown in Figure 4-39*D,* amniote somitomeres 4, 6, and 7, along with several anterior somites, contribute to branchiomeric muscles in branchiomeres I (mandibular), II (hyoid), and III (glossopharyngeal). Somitomeres 1 to 3 and 5 contribute to the extrinsic ocular muscles (see Fig. 13-20).

The major cranial blood vessels are diagrammed in Figure 4-39*E* and *F.* The numbering system for aortic arches used for elasmobranchs (Fig. 4-38*B*) also applies to amniotes, but because amniotes develop fewer vagal branchiomeres, they consequently have fewer aortic arches. Also, the fifth aortic arch never forms in amniotes, and so it is indicated with a dashed line (Fig. 4-39*E*). Major cranial veins, such as the **anterior cardinal vein,** develop by fusions of a plexus of smaller vessels. As in the trunk, cranial blood vessels derive from mesoderm. The precursor cells, known as **angioblasts,** migrate early from their sources and move invasively throughout the head mesenchyme. Angioblasts differentiate into the lining cellular layer of the blood vessels, which is known as **endothelium.** In contrast to veins, arteries tend to develop by branching of existing vessels, a process known as **angiogenesis.** This difference in their mode of formation helps to explain why veins are anatomically more variable than arteries.

The paired pharyngeal pouches are shown in Figure 4-39*G*. As in the scheme shown for elasmobranchs in Figure 4-38, the pharyngeal pouches are numbered with Arabic numerals. The rudiment of the **thyroid**

◀**FIGURE 4-39**
Anatomical relationships among the components of the head in an amniote embryo in lateral view. The diagram is comparable to a stage 14 chick (45 hours, 22 somites). Light gray lines demarcate the approximate boundaries of branchiomeres I, II, III, IV, and V. *A,* Expression domains of genes of the *Hox-b* cluster. *B,* Axial structures including the neural tube and notochord. *C,* Neural crest migrating ventrally into the branchiomeres. *D,* Paraxial mesoderm, incompletely divided into somitomeres anteriorly and completely divided into somites in the trunk; somitomeric contributions to branchiomeres I, II, and III also are indicated. *E,* Heart and aortic arches. *F,* Major veins. *G,* Pharyngeal pouches and Rathke's pouch. Amniote embryos have four paired pharyngeal pouches lined with endoderm; in all gnathostomes, Rathke's pouch forms as a single median diverticulum of ectoderm just anterior to the oral plate. *(Modified from Noden.)*

gland originates as a median ventral diverticulum from the floor of the pharynx between the mandibular and hyoid branchiomeres. It extends caudally during later development and ceases to be connected to the pharynx. The developing lung also originates as a median ventral diverticulum; its retained connection to the gut tube will become the trachea. **Rathke's pouch** (Fig. 4-39*G*) is a median dorsal diverticulum of ectoderm that forms just anterior to the oral plate, the region known as the **stomodeum.** As it extends dorsally, Rathke's pouch comes to lie adjacent to the infundibulum, which is a ventral outpocketing of the neural tube (Fig. 4-39*B*). Rathke's pouch and the infundibulum together form the master endocrine gland, the **hypophysis.** The dual embryonic origins of the hypophysis continue to be reflected in the types and mechanisms of hormone production and secretion used by its component parts (see Chapter 15, especially Fig. 15-4).

Hox Genes, Segmentation, and the Evolution of the Vertebrate Body Plan

The connection between development and evolution has drawn the attention of biologists and philosophers for more than a century. The actively growing field of evolutionary developmental biology examines connections between embryology and evolution using tools from molecular biology, phylogenetics, and comparative anatomy. It has long been thought that detailed knowledge of DNA regulatory genes, which are those that control expression of other genes, will prove essential to understanding major features of vertebrate evolution. Perhaps 10% of the genome of a vertebrate consists of DNA regulatory genes, and we still have much to learn about all aspects of the story. To illustrate the type of synthesis that may emerge, we discuss *Hox* genes and their role in the development of the vertebrate body plan.

A **segment** is a set of body parts that is present in a repeated series in an embryo or adult. Segmental organization is fundamental to the vertebrate body plan, and all vertebrates are more or less obviously segmented during their development. Segmentation is most obvious in embryonic stages but often becomes masked in adults, particularly in the cranial region. Nevertheless, we have seen that the head displays several types of interrelated segmentation. Branchiomeres provide clear evidence of segmental organization, as do the rhombomeres of the hindbrain and other components of the head (Fig. 4-39*B*). Cranial neural crest cells migrate ventrally to contribute to the branchiomeres, following the same segmental pattern (Fig. 4-39*C*). The paraxial mesoderm that lies to each side of the developing brain is also segmentally organized, whether as true somites found in elasmobranch embryos (Fig. 4-37) or as the transient somitomeres of amniotes (Fig. 4-39*D*).

The connection between development and evolution can be demonstrated by considering the role of *Hox* genes in regulating the development of such segmented structures.[6] A series of these genes or their homologues is expressed differentially along the length of the body axis; this differential expression pattern helps establish the basic pattern of the segments and instruct the subsequent development of segment-specific features. In this way, *Hox* genes can be thought of as generating unique tags for specific locations in the body.

Hox genes code for **helix-turn-helix transcription factors,** which are DNA binding proteins (Fig. 4-40*A* and *B*). These proteins contain variable regions (green in Fig. 4-40) and highly conserved regions (red in Fig. 4-40). They fold in such a way that their tertiary structure consists of two helical regions separated by a short, variable "turn" region. The very highly conserved sequence of about 60 amino acids, known as the **homeodomain,** forms a recognition helix that can be presented to the major groove in the double-helical DNA molecule (Fig. 4-40*B*).

The portion of a *Hox* gene that codes for the homeodomain portion of the transcription factor is called a **homeobox.** Many families of genes contain homeoboxes. Thus, not all genes containing homeobox sequences are *Hox* genes, but these are the best known genes in this class. Wherever homeobox sequences occur in the genome, they code for remarkably similar amino acid sequences. These sequences also are preserved across remarkable phylogenetic gaps: 59 of the 60 amino acid residues in some homeodomains of vertebrates and fruit flies are identical. Such evolutionary conservation is an outstanding example of the tight link between structure and function at the molecular level because, if the homeodomain portion of the transcription factor does not fit precisely into the major groove of the DNA double helix, then it cannot function. The highly conserved homeodomain region, however, is just one part of the transcription factor. The remaining regions of the protein can be of variable length and differ from one Hox protein to the next. It is the regions outside the homeodomain that allow the transcription factor to bind with other specific factors in order to recognize specific genes along a DNA strand (Fig. 4-40*C*–*E*). By means

[6]By convention, the word *Hox* is italicized to indicate a DNA sequence or gene; when not italicized, the word Hox indicates an amino acid sequence or protein.

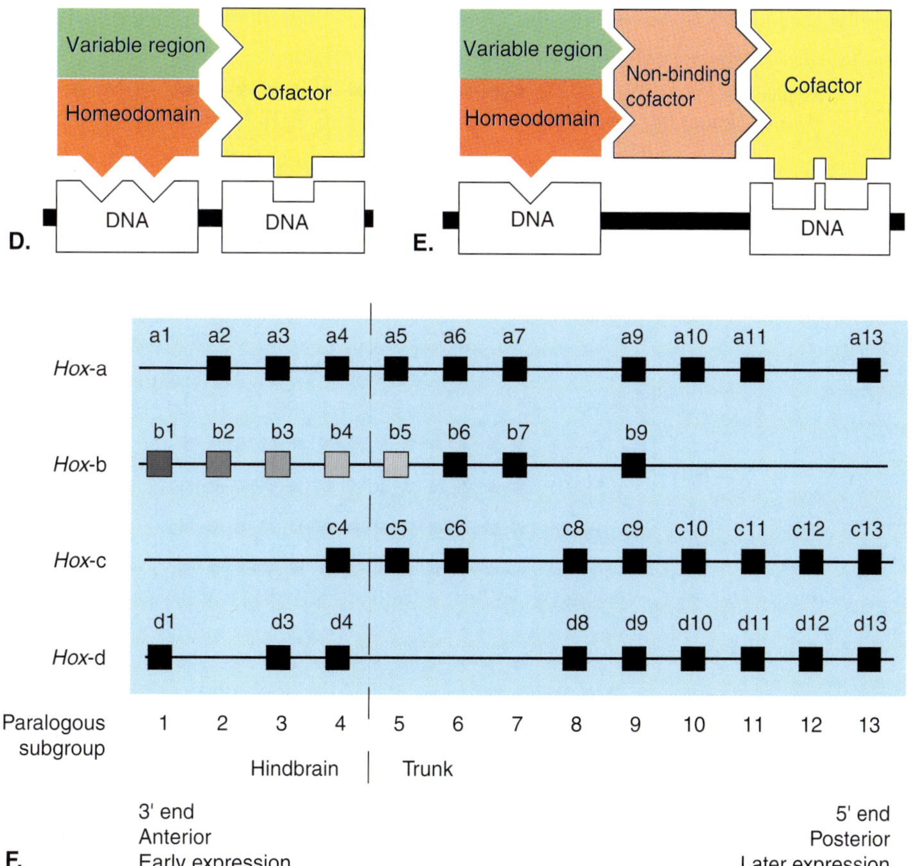

FIGURE 4-40
The organization of Hox proteins and *Hox* genes. *A,* Schematic diagram of a Hox protein, which is an example of a helix-turn-helix transcription factor, showing the relationship between highly conserved region (*red*) and adjacent variable regions (*green*). *B,* Schematic diagram showing how a folded Hox protein interacts with a DNA double helix. The 60 amino acid homeodomain folds up into an alpha-helix, termed a recognition helix, that fits within the major groove of the DNA. *C–E,* Models showing how Hox proteins may interact with other factors to recognize specific sites within the genome in order to regulate gene expression. *F,* The structure of *Hox* gene clusters in gnathostomes, based on the mouse (*Mus*). Genes of the *Hox-b* cluster are shaded to correspond with their expression domains indicated in Figure 4-39*A*. *(A and B, Modified from Latchman; D, modified from Krumlauf.)*

of their variable regions, different Hox proteins can regulate expression of different genes.

Proteins made by a particular *Hox* gene are found inside cells in a restricted and specific set of segments of the body. The places where a particular gene is expressed are known as that gene's **expression domain.** By regulating the expression of other genes, *Hox* genes determine the features that are characteristic of each body segment. This is the result of unique expression domains or unique combinations of overlapping expression domains. Perturbation experiments, known as "knockout" experiments, involve the deletion or inactivation of a particular gene, usually by mutation. A *Hox* gene that has been knocked out may result in a specific segmentation defect or defects; defects do not necessarily result because other genes can sometimes take over the role of the missing gene. In contrast, the insertion of an extra *Hox* gene can result in the formation of additional segmental structures.

The *Hox* genes of gnathostomes are organized in four clusters, a through d, each of which is located on a separate chromosome (Fig. 4-40*F*). Each cluster consists of as many as 13 **paralogous subgroups.**[7] The presence of these paralogous subgroups appears to be the result of tandem duplications of *Hox* genes that occurred prior to the common ancestry of arthropods and vertebrates. It is possible to align the *Hox* genes of one cluster with those of other clusters based on sequence similarity. We follow the nomenclature of Scott (1992), Krumlauf (1994), and others in naming the genes by their cluster and sequence order from the 3′ to 5′ end of the DNA.

[7]A paralogue is a copy of a gene within the same genome.

The opposite ends of a single strand of DNA are referred to as 3′ and 5′. At the 3′ end, a hydroxyl group is attached to the third carbon in the sugar (ribose) ring. At the opposite end of the DNA strand, the fifth carbon atom is joined to a phosphate group. This convention of naming genes in order from 3′ to 5′ ends is used because the normal direction of mRNA transcription proceeds from the 3′ end of a gene toward the 5′ end.

Some *Hox* genes have been lost in the evolutionary lines that lead to living amniotes, which is why genes such as *Hox-a8* or *Hox-d2* are missing from Figure 4-40*F*.

Within each cluster, the *Hox* genes occur in a highly conserved, linear order corresponding to their anterior-to-posterior expression domains along the body axis. This phenomenon is known as **colinearity.** Although each individual gene is transcribed from the 5′

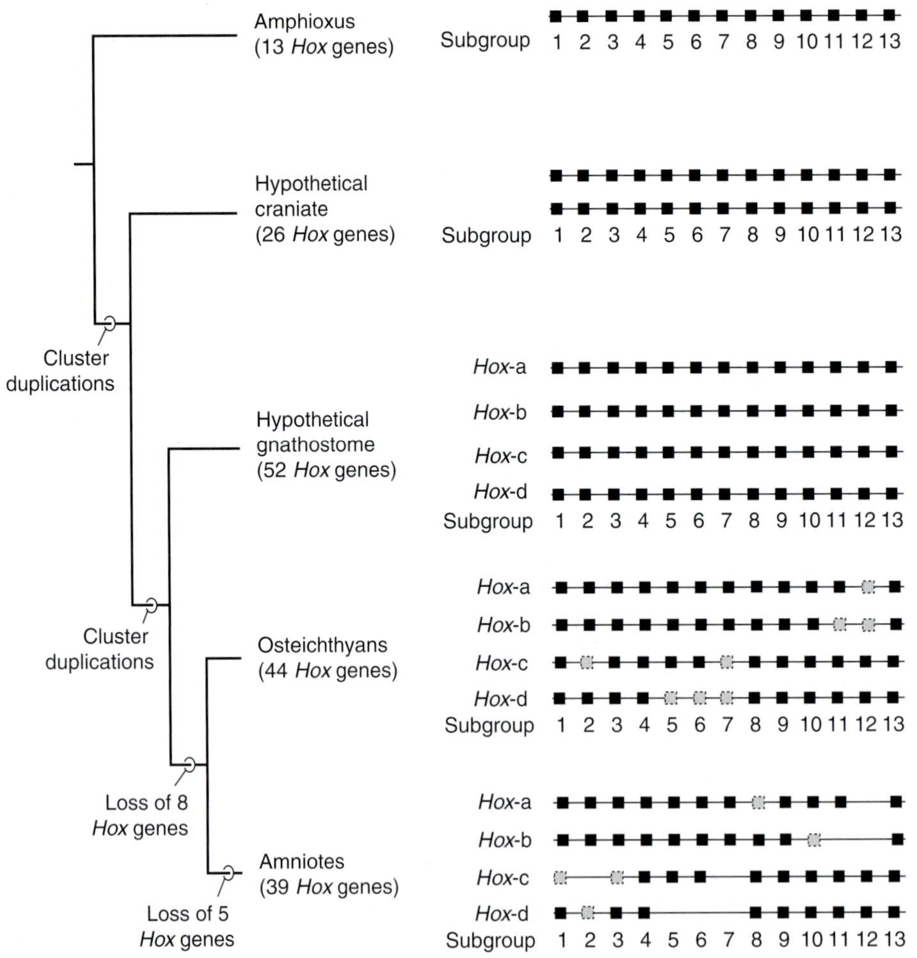

FIGURE 4-41
Phylogenetic tree of chordates showing position of *Hox* gene cluster duplications by Holland et al. (1994), and the subsequent losses of individual *Hox* genes proposed by Meyer (1998). Based on analyses of *Hox* genes in amphioxus, hagfish, lampreys, basal osteichthyans, and amniotes. Genes shown with dashed lines have been lost. *(Modified from Holland et al. and Meyer.)*

to 3′ direction, those that lie toward the 3′ end of each cluster are transcribed earlier in development than those nearer the 5′ end. Because the *Hox* genes are expressed sequentially in this linear order, they can establish the pattern of segmentation along the anteroposterior axis of the embryo. Thus, for example, we can recognize a boundary between *Hox* genes expressed in tissues of the hindbrain and those expressed in the trunk.

Colinearity between the DNA transcription level and the appearance of segmentation ranks as one of the most intriguing aspects of the story of the *Hox* genes. For example, *Hox-b1*, *Hox-b2*, *Hox-b3*, *Hox-b4*, and *Hox-b5* are arranged along the DNA molecule in an order that closely corresponds to their pattern of expression in tissues of the head, with the boundaries of their expression domains corresponding to boundaries between rhombomeres (Fig. 4-39*A*). *Hox-b2* is expressed from rhombomere 3 caudally into the trunk, *Hox-b1* is expressed in alignment with rhombomere 4, *Hox-b3* is expressed from rhombomere 5 caudally into the trunk, *Hox-b4* is expressed from rhombomere 7 caudally into the trunk, and *Hox-b5* is expressed posterior to all of the rhombomeres (Fig. 4-39*A*). Within this series, note that only *Hox-b1* fails to follow strictly the colinear pattern between cluster order and expression domain.

The occurrence of homologues of *Hox* genes in taxa such as the fruit fly, *Drosophila*, supports the notion that these genes first evolved at least 600 million years ago in flatworms or other early bilateral metazoans. Their truly remarkable constancy suggests that once the molecular system for determining the anteroposterior axis of animals was established, it was retained throughout evolutionary history and has served as the basis for the establishment of diverse body plans in animal phyla.

Comparative studies on *Hox* gene clusters suggest that two rounds of gene duplication occurred in the line leading to craniates, with the first round of duplications between amphioxus and craniates, and the second between hagfishes and gnathostomes (Fig. 4-41). *Hox* genes also were lost in some lineages leading to extant vertebrates. It is tempting to attribute the great diversification of craniates to these duplications, for gene duplication makes it possible for copies of particular genes to evolve rapidly without impairing any existing functions. Then, at a later time, some altered duplicate sequences might be "recaptured" for use in generating new phenotypes. The comparative analysis of *Hox* genes has become an extremely exciting area for research because it combines genetics, development, and systematics in the search for explanations in vertebrate evolution.

SUMMARY

1. Gametes are haploid cells that contain only a single set of chromosomes. Sperm are small, motile gametes that are capable of reaching and penetrating an egg. Eggs are much larger, nonmotile gametes that contain the metabolic reserves and information molecules needed to initiate development. Materials in an egg are not distributed randomly. The amount of yolk in the egg correlates with the future pattern of nutrition of the embryo or larva.
2. Fertilization is the union of a haploid sperm with a haploid egg to form a diploid zygote. It involves several events: sperm penetration of the egg, combination of male and female nuclear material, and activation of the egg.
3. During cleavage, the single-celled zygote is converted to a multicellular embryo called a blastula, but no growth occurs. Because the materials in the egg were not distributed evenly, the cells (blastomeres) in different parts of the blastula contain different enzymes, mRNAs, and other components. As a result, the blastomeres differ in their presumptive fates.
4. During gastrulation, the single-layered blastula is converted into a two-layered gastrula. The gastrula is covered by ectoderm and has a simple gut cavity, the archenteron, which is lined by endoderm and opens to the surface posteriorly at the blastopore. Mesoderm spreads between the ectoderm and endoderm. Patterns of gastrulation and mesoderm formation are greatly influenced by the amount of yolk and vary considerably in different groups of craniates.
5. Formation of the neural tube is induced by the underlying chordamesoderm. The central nervous system develops from the neural tube.
6. In addition to the neural tube, vertebrates have two other neurogenic tissues: the neural crest and neurogenic placodes.
7. Ectodermal neural crest cells separate during the formation of the neural tube and spread throughout the embryo. They give rise to many structures (i.e., branchial skeleton, other cranial bones, pigment cells, many peripheral neurons, parts of teeth, and bony scales) and also help regulate development and the emergence of the

vertebrate body plan by patterning mesoderm and ectoderm.

8. Neurogenic placodes give rise to sensory epithelia, nerves innervating sensory epithelia, and other components of the cranial nerves. Major derivatives include the nose, ear, and lateral line system (including electroreceptors in many groups of vertebrates).
9. Amniotes have lost the neurogenic placodes associated with the lateral line and electroreceptive systems and lack these sensory systems as adults.
10. The mesoderm differentiates into a series of segmental somites that lie lateral to the neural tube and notochord, a nephric ridge, and a lateral plate that spreads around the archenteron and yolk. The coelom develops by cavitation within the lateral plate, but it may enter the nephric ridge and somites.
11. Each somite breaks up into three regions. Laterally, the dermatome becomes reorganized as a mesenchyme, and its cells spread out beneath the ectoderm to form the dermis of the skin. Myotome gives rise to all or most of the somatic muscles of the body. Medially, the sclerotome reorganizes as a mesenchyme, and its cells migrate around the neural tube and notochord to form most of the axial skeleton.
12. The nephric ridge gives rise to the kidney tubules and the urinary and genital ducts.
13. The coelom divides the lateral plate mesoderm into an outer somatic and an inner splanchnic layer. The somatic layer forms the parietal peritoneum and contributes to the somatic muscles in many species. The splanchnic layer forms the connective tissue and visceral muscles of the gut and the heart walls and visceral peritoneum.
14. The yolk of macrolecithal eggs becomes suspended from the embryo in an extraembryonic membrane known as the yolk sac. The yolk sac is trilaminar in actinopterygian fishes, including all three germ layers, but is bilaminar in reptiles and birds, consisting only of the endoderm and splanchnic mesoderm.
15. The ectoderm and somatic mesoderm rise off the yolk in amniotes to form additional extraembryonic membranes. The amnion protects the embryo in a cushion of water, and a protective chorion surrounds the amnion, embryo, and yolk sac. Albumen, a shell membrane, and a shell, all of which are secreted by the oviduct, lie peripheral to the chorion.
16. The last extraembryonic membrane of amniotes, the allantois, extends into the extraembryonic coelom. It functions in gas exchange and serves as a site for the accumulation of excretory products. Its base may remain as the urinary bladder.
17. The development of eutherian mammals is similar to the pattern seen in birds except for the secondary reduction of yolk, the simplification of cleavage, and the precocious development of those extraembryonic membranes that contribute to the placenta. The placenta is composed of the embryonic chorioallantoic membrane and part of the maternal endometrium.
18. The definitive coelom is a space lined with epithelium in the lateral plate mesoderm. It surrounds the visceral organs and allows their functional movements.
19. Mesenteries consist of a double layer of coelomic epithelium. These membranes hold the visceral organs in place and provide routes for the passage of ducts, blood vessels, and nerves from the body wall to the organs and between the organs.
20. Early in development, a left and a right coelom are separated from each other by continuous dorsal and ventral mesenteries. Much of the ventral mesentery later disappears caudal and cranial to the liver.
21. The coelom of a fish becomes divided into an anterior pericardial cavity and a posterior pleuroperitoneal cavity by a transverse septum, through which blood vessels enter the heart.
22. The heart and pericardial cavity of tetrapods such as amphibians shift caudally, from a position beneath the gills closer to the lungs. As a consequence, the pericardial cavity underlies the cranial part of the pleuroperitoneal cavity, and the transverse septum assumes an oblique orientation.
23. In some reptiles and in birds and mammals, folds separate the portions of the pleuroperitoneal cavity containing the lungs from the part containing the other visceral organs. The coelom of these tetrapods is thus divided into two pleural cavities—a peritoneal cavity and a pericardial cavity.
24. In mammals, the pleuroperitoneal folds and the ventral part of the transverse septum are invaded by somatic musculature of cervical origin and form the diaphragm.
25. The pleural cavities of mammals extend ventrally during development, separating the pericardial cavity from the body wall. The area between the pleural cavities, which contains the pericardial cavity and other organs, is known as the mediastinum.
26. Many important landmarks in the head and pharynx can be recognized in embryonic vertebrates. Many of the patterns observed are conserved across vertebrate history, but some features, such as the patterning of cranial mesoderm or the occurrence of neurogenic placodes that contribute to the lateral line system, show marked differences between elasmobranchs and amniotes.

27. The expression of segmental structures in the hindbrain of vertebrates is regulated by clusters of *Hox* genes, which code for helix-turn-helix transcription factors. Homologous genes also occur in many groups of invertebrates. The homeobox region of a *Hox* gene codes for the large helix of the transcription factor, which fits into the major groove of a DNA double helix. Homeobox sequences are remarkably conserved across Metazoa. This transcription factor system must have evolved at least 600 million years ago, before arthropods and vertebrates diverged.
28. *Hox* genes of gnathostomes are basically organized as four clusters, a through d (variations exist that were not discussed in this chapter). The genes within a cluster follow a strictly conserved order along the DNA. The term colinearity describes the close association between a *Hox* gene's position within a cluster and its expression domain along the anterior-to-posterior axis of the embryo.
29. It has been proposed that two major rounds of gene duplication occurred early in vertebrate history. The duplications of regulatory genes, such as *Hox* genes, may have allowed rapid evolutionary change in vertebrates because the copies would be free to accumulate mutations silently before being "recaptured" for use by descendant taxa.

REFERENCES

Alberch, P., 1982: Developmental constraints in evolutionary processes. *In* Bonner J. T., editor: *Evolution and Development,* Dahlem Konferenzen. Berlin, Springer.

Alberts, B., Bray, D., Lewis, J., Raff, M., Roberts, K., and Watson, J. D., 1989: *Molecular Biology of the Cell,* 2nd edition. New York, Garland.

Arey, L. B., 1974: *Developmental Anatomy,* 7th edition. Philadelphia, W. B. Saunders.

Balinsky, B. I., assisted by Fabian, B. C., 1981: *An Introduction to Embryology,* 5th edition. Philadelphia, Saunders College Publishing.

Balinsky, B. I., 1970: *An Introduction to Embryology,* 3rd edition. Philadelphia, W. B. Saunders.

Ballard, W. W., 1964: *Comparative Anatomy and Embryology.* New York, Ronald Press.

Bemis, W. E., and Grande, L., 1999: Development of the median fins of the North American paddlefish (*Polyodon spathula*) with comments on the lateral fin-fold hypothesis. *In* Arratia, G., and Schultze, H.-P., editors: *Mesozoic Fishes II: Systematics and the Fossil Record.* Munich, Pfeil.

Bemis, W. E., and Grande, L., 1992: Early development of the actinopterygian head: I. External development and staging of the paddlefish *Polyodon spathula. Journal of Morphology,* 213:47–83.

Bolker, J., 1995: Model systems in developmental biology. *BioEssays,* 17:451–455.

Carlson, B. M., 1988: *Patten's Foundations of Embryology.* New York, McGraw-Hill.

Conklin, E. G., 1932: The embryology of amphioxus. *Journal of Morphology,* 54:69–118.

Corliss, C. E., 1976: *Patten's Human Embryology.* New York, McGraw-Hill.

De Robertis, E., Oliver, G., and Wright, C. V. E., 1990: Homeobox genes and the vertebrate body plan. *Scientific American,* 263:46–52.

Dean, B., 1895: *Fishes, Living and Fossil: An Outline of their Forms and Probable Relationships.* London, Macmillan.

Duellman, W. E., and Trueb, L., 1988: *Biology of Amphibians.* New York, McGraw-Hill Book Company.

Feduccia, A., and McCrady, E., 1990: *Torrey's Morphogenesis of the Vertebrates,* 3rd edition. New York, John Wiley.

Gans, C., and Northcutt, R. G., 1983: Neural crest and the origin of vertebrates: A new head. *Science,* 220:268–274.

Gee, H., 1996: *Before the Backbone: Views on the origin of the vertebrates.* London, Chapman & Hall.

Gilbert, S. F., 1997: *Developmental Biology,* 5th edition. Sunderland, Mass., Sinauer Associates.

Gilbert, S. F., 1991: *Developmental Biology,* 3rd edition. Sunderland, Mass., Sinauer Associates.

Goodrich, E. S., 1930: *Studies on the Structure and Development of Vertebrates.* London, Macmillan.

Hall, B. K., 1992: *Evolutionary Developmental Biology.* London, Chapman & Hall.

Hall, B. K., and Horstadius, S., 1988: *The Neural Crest.* Oxford, Oxford University Press.

Hamburger, V., 1960: *A Manual of Experimental Embryology,* revised edition. Chicago, University of Chicago Press.

Hamilton, W. J., Boyd, J. D., and Mossman, H. W., 1972: *Human Embryology,* 4th edition. Baltimore, Williams & Wilkins.

Hanken, J., Jennings, D. H., and Olsson, L., 1997: Mechanistic basis of life-history evolution in anuran amphibians: Direct development. *American Zoologist,* 37:160–171.

Hatschek, B., 1883: The amphioxus and its development. London, Sonnenschein.

Holland, P. W., 1997: Vertebrate evolution: something fishy about *Hox* genes. *Current Biology,* 7:R570–R572.

Holland, P. W., Garcia-Fernàndez, H. J., Williams, N. A., and Sidow, A., 1994: Gene duplication and the origin of vertebrate development. *Development,* (Supplement S): 125–133.

Janvier, P., 1996: *Early Vertebrates.* Oxford, Oxford Scientific.

Jennings, D. H., and Hanken, J., 1998: Mechanistic basis of life history evolution in anuran amphibians: Thyroid gland development in the direct-developing frog, *Eleutherodactylus coqui. General and Comparative Endocrinology,* 111:225–232.

Jollie, M., 1962: *Chordate Morphology*. New York, Reinhold.

Karp, G., Berrill, N. J., 1981: *Development*, 2nd edition. New York, McGraw-Hill.

Kerr, J. K., 1919: *Text-Book of Embryology: Vertebrata*, volume 2. London, Macmillan.

Kessel, M., Balling, R., and Grass, P., 1990: Variations of cervical vertebrae after expression of a *Hox-1.1* transgene in mice. *Cell*, 61:301–308.

Krumlauf, R., 1994: *Hox* genes in vertebrate development. *Cell*, 78:191–201.

Latchman, D., 1995: *Gene Regulation: A Eukaryotic Perspective*, 2nd edition. London, Chapman & Hall.

Le Douarin, N. M., 1986: Cell line segregation during peripheral nervous system ontogeny. *Science*, 231:1515–1522.

Le Douarin, N. M., 1982: *The Neural Crest*. Cambridge, Cambridge University Press.

Lentz, T. L., 1971: *Cell Fine Structure*. Philadelphia, W. B. Saunders.

Lillegraven, J. A., 1985: Use of the term "trophoblast" for tissues in therian mammals. *Journal of Morphology*, 183:293–299.

Lillegraven, J. A., 1975: Biological considerations of the marsupial-placental dichotomy. *Evolution*, 29:707–722.

Mathews, W. W., 1986: *Atlas of Descriptive Embryology*, 4th edition. New York, Macmillan.

Meyer, A., 1998: *Hox* gene variation and evolution. *Nature*, 225:227–228.

Nelsen, O. E., 1953: *Comparative Embryology of the Vertebrates*. New York, McGraw-Hill.

Noden, D. M., 1991: Vertebrate craniofacial development: The relation between ontogenetic process and morphological outcome. *Brain, Behavior and Evolution*, 38: 190–225.

Noden, D. M., and de Lahunta, D., 1985: *The Embryology of Domestic Animals*. Baltimore, Williams & Wilkins.

Northcutt, R. G., 1997: Evolution of gnathostome lateral line ontogenies. *Brain Behavior and Evolution*, 50:25–37.

Northcutt, R. G., 1996: The origin of craniates: Neural crest, neurogenic placodes, and homeobox genes. *Israel Journal of Zoology*, 42(supplement):273–313.

Northcutt, R. G., and Brändle, K., 1995; Development of branchiomeric and lateral line nerves in the axolotl. *The Journal of Comparative Neurology*, 355:427–454.

Northcutt, R. G., 1992: Distribution and innervation of lateral line organs in the axolotl. *The Journal of Comparative Neurology*, 325:95–123.

Northcutt, R. G., and Gans, C., 1983: The genesis of neural crest and epidermal placodes: A reinterpretation of vertebrate origins. *Quarterly Review of Biology*, 58:1–28.

Patten, B. M., 1951: *Early Embryology of the Chick*, 4th edition. Philadelphia, Blakiston.

Patten, B. M., 1948: *Embryology of the Pig*, 3rd edition. New York, McGraw-Hill.

Pelster, B., and Bemis, W. E., 1992: Structure and function of the external gill filaments of embryonic skates (*Raja erinacea*). *Respiration Physiology*, 89:1–13.

Raff, R. A., 1996: *The Shape of Life: Genes, Development, and the Evolution of Animal Form*. Chicago, University of Chicago Press.

Raff, R. A., and Kaufmann, P. C., 1983: *Embryos, Genes and Evolution*. New York, Macmillan.

Scott, M. P., 1992: Vertebrate homeobox gene nomenclature. *Cell*, 71:551–553.

Smith, K. K., 1997: Comparative patterns of craniofacial development in eutherian and metatherian mammals. *Evolution*, 51:1663–1678.

Stone, L. S., 1922: Experiments on the development of the cranial ganglia and the lateral-line sense organs in *Ambystoma punctatum*. *Journal of Experimental Zoology*, 35:421–495.

Stone, L. S., 1933. The development of lateral-line sense organs on amphibians observed in living and vital stained preparations. *Journal of Comparative Neurology*, 57:507–568.

Wessells, N. K., 1977: *Tissue Interactions and Development*. Philadelphia, W. A. Benjamin.

5

Form and Function

PRECIS

We conclude Part I, the Background for the Study of Vertebrate Anatomy, by examining the basic physical principles that link form and function. After discussing physical quantities, we consider forces and the effects they have on the vertebrate body. We describe the structural materials of the body and explain how they are adapted to these forces. The skeletal elements are linked at joints of various kinds and often form lever systems that transmit muscular forces to some point of application, such as the feet and teeth. We explain how forces on skeletal parts affect the distribution of materials within the parts. We conclude by looking at the profound effect that changes in body size have on the form and function of organs and organisms.

OUTLINE

A bridge, a piano, a tea kettle, and other structures constructed by human beings are designed to perform certain functions; a close linkage exists between the form of a structure and the function it performs. A bridge, for example, is designed to cross a river of a particular width, to resist storms of a certain magnitude, and to carry certain maximum loads. If the form did not match the function, the bridge might be much larger and more expensive than needed or it might be too weak and fail. So too in the living world, the evolution of the form of an organism and its parts goes hand and hand with the evolution of the functions they perform. A goal of functional anatomists is to find the connections between form and function. Sometimes the relationship is obvious, but sometimes the linkage is more elusive. This poses a challenge that enlivens research in this field.

We will be examining the evolution of form and function throughout this book, but certain mathematical and physical principles link form and function and must be understood. We will explore many of these in this chapter, particularly those that deal with **biomechanics,** that is, the science of applying the principles of the study of mechanics to living systems. These principles are especially important in analyzing the skeletomuscular systems and locomotion, topics that we treat in Part II. We will introduce other physical concepts as they become important for the subject at hand.

After reviewing physical quantities you need to be familiar with, we will concentrate on forces. Forces put a stress upon the body and its parts and determine the nature of the structural material of the body. Bones and other supporting elements are linked at joints and form lever systems that transmit forces. The way forces are transmitted determines the distribution of materials within bones. Finally, we will consider the importance of body size and the effect changes in size have on body form and function.

TABLE 5-1 Seven Basic Physical Quantities

Dimension	Symbol	SI Unit	Symbol
Length	L	Meter	m
Mass	M	Gram	g
Time	T	Second	s
Molecular amount of a substance	mol	Mole	mol
Temperature	*T*	Degree Celsius	°C
Electric current	A	Ampere	A
Luminous intensity	cd	Candela	cd

Physical Quantities

To analyze form and function, we must be able to measure physical quantities, such as length, mass, time, and force. Each **physical quantity** has a name (called a **dimension**) that describes its nature and a magnitude expressed as some **unit.** Length is a dimension the unit of which, in the metric system, is the meter. Time is a dimension the unit of which is the second. At a General Conference of Weights and Measures held in 1960, seven basic physical quantities were agreed on to measure phenomena. Their dimensions are length, mass, time, the molecular amount of a substance, temperature, electric current, and luminous intensity (Table 5-1). Their magnitudes are expressed in a system of standard units known as the SI units (from the French, *Système International d'Unités*). These dimensions and units have been adopted by scientists throughout the world and will be used throughout this book. The basic units can be reduced or amplified by factors of 10, as shown in Table 5-2.

Most of the dimensions can be understood from their names, but mass and mole may not be intuitively clear. Mass is often confused with weight. **Mass** in essence is the physical quantity of matter in a body. One cubic centimeter of water (under certain standard conditions) equals 1 g, and grams are the units of mass. **Weight** is mass × the acceleration of gravity. Because the phenomena we measure on Earth are subject to the same gravitational force, weight also is expressed in grams, but weight and mass are sometimes not the same. A person has the same mass whether he or she is on Earth or in a spacecraft, but this person's weight might vary from, say, 75 kg on Earth to near 0 kg in space. **Mole** is the amount of a chemical compound the mass (g) of which is equivalent to its molecular weight, that is, to the sum of the atomic weights of its constituent atoms.

Many other physical quantities can be derived from the seven basic ones, but we will encounter only a few in the study of functional anatomy. The dimensions and SI units for the following important quantities are given in Table 5-3.

Area = $(\text{length})^2$.

Volume = $(\text{length})^3$. Although not an SI unit, **liter** is a widely used measure for the volume of a liquid. One liter is the volume occupied by 1000 cm^3 of liquid.

TABLE 5-2 SI Unit Multiples

Factor	Prefix	Symbol	Example for Meters
1×10^9	giga-	G	1×10^9 m = 1 gigameter (Gm)
1×10^6	mega-	M	1×10^6 m = 1 megameter (Mm)
1×10^3	kilo-	k	1×10^3 m = 1 kilometer (km)
Basic unit	—	m	1 meter
1×10^{-1}	deci-	d	1×10^{-1} m = 1 decimeter (dm)
1×10^{-2}	centi-	c	1×10^{-2} m = 1 centimeter (cm)
1×10^{-3}	milli-	m	1×10^{-3} m = 1 millimeter (mm)
1×10^{-6}	micro-	μ	1×10^{-6} m = 1 micrometer (μm)
1×10^{-9}	nano-	n	1×10^{-9} m = 1 nanometer (nm)

Density = the mass of an object divided by its volume.

Velocity = the change in distance with time and usually is stated as meters per second (m/s).

Acceleration = the change in velocity with time.

Force (F) = mass (M) × acceleration (a). The SI unit of force is the **Newton** (N). 1 N gives a mass of 1 kg an acceleration of 1 m/s per second. That is, it increases its velocity by 1 m/s every second. This unit honors Sir Isaac Newton (1642–1727), a British physicist who formulated many basic laws of physics.

Pressure = force per unit area. Its SI unit is the **Pascal** (Pa). 1 Pa = 1 N per square meter (m^2). The Pascal honors Blaise Pascal (1623–1662), a French physicist, mathematician, and religious philosopher.

Energy and work have much in common. **Energy** is the capacity to do work, and **work,** in the physical sense, is done when energy generates a force and moves a mass a certain distance. Both are measured in **Joules** (J).1 J = a force of 1 N applied over 1 m. The Joule honors James Prescott Joule (1818–1889), a British physicist.

Many types of energy exist: potential, kinetic, chemical, electric, and thermal. Any type can do work. According to the principle of the **conservation of energy,** one form of energy can be changed to other forms of energy. No energy is lost or gained in the conversion. When an entire organism does work, the input energy usually is converted into motion of some sort (kinetic energy), but some overcomes friction and is lost as heat. The efficiency of the conversion of energy to useful work is the energy that appears as useful work divided by the energy put into the system. This ratio often is stated as a percentage.

Power is the rate of doing work. Its SI unit is the **Watt** (W), which honors James Watt (1736–1819), a Scottish engineer. 1 W = the use of 1 J in 1 s.

TABLE 5-3 Physical Quantities That Can Be Derived from the Seven Basic Physical Quantities

Quantity	Definition	Dimension	SI Units
Area (A)	Two-dimensional space	L^2	m^2
Volume (V)	Three-dimensional space	L^3	m^3
Density (p)	Mass/volume	M/L^3 or ML^{-3}*	kg m^{-3}
Velocity (u)	Change in distance/time	LT^{-1}	ms^{-1}
Acceleration (a)	Change in velocity/time	LT^{-2}	ms^{-2}
Force (F)	Mass × acceleration	MLT^{-2}	1 N = 1 kg ms^{-2}
Pressure (p), stress	Force/unit area	$ML^{-1}T^{-2}$	1 Pa = 1 kg $m^{-1}s^{-2}$
Work (W), energy, heat	Force × length	ML^2T^{-2}	1 J = 1 kg m^2s^{-2}
Power (P)	Work/time	ML^2T^{-3}	1 W = 1 kg m^2s^{-3}
Frequency (f)	Cycles/time	fT^{-1}	1 Hz = fs^{-1}

*Power notation is recommended.

Hz = Hertz; J = joule; L = Length; M = Mass; N = Newton; Pa = Pascal; T = Time; W = Watt.

Frequency measures periodic phenomena, such as sound waves, and its unit is the Hertz (Hz), named after Heinrich Rudolph Hertz (1857–1894), a German physicist. 1 Hz is the frequency of a periodic phenomenon that has a period of 1 s.

Although SI units are normally used in scientific writing, some other units are in common usage. Conversion factors between these other units and SI units are given in the appendix at the end of the chapter.

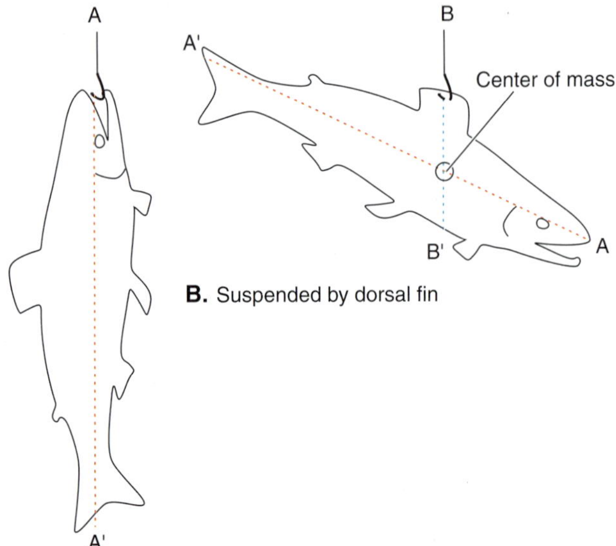

FIGURE 5-1
The center of mass of a body can be located by suspending it from two different points and locating the intersection of the supporting lines, A to A′ and B to B′. *A,* Suspended by jaw. *B,* Suspended by dorsal fin.

Forces

Newton's Laws of Motion

All structures are subject to gravity and often to many other forces as well. Their designs must meet the demands of the forces. The concept of force is basic to all functional anatomical analyses. Newton studied forces and motion and formulated three fundamental laws of motion:

1. First law of motion. A body retains its state of rest or motion unless acted on by an external force. This is often called the law of inertia.
2. Second law of motion. A force gives a body acceleration in the direction of the force. The acceleration is proportional to the force and inversely proportional to the mass of the body. This law is the basis for the definition of force, $F = Ma$.
3. Third law of motion. If one body (A) exerts a force on another (B), then B exerts an equal but opposite force on A. You are exerting a force on your chair as you read this, and the chair exerts an opposite but equal force on you. If not, you would fall to the floor.

Center of Mass

In analyzing a force acting on the entire body, we can think of the force as acting on a single point, the **center of mass,** also commonly called the center of gravity. The center of mass is the point about which a solid body is evenly balanced. One can locate this by suspending an animal, such as a fish, on a cord from any convenient point, such as its jaw (Fig. 5-1*A*). The body comes to rest when its center of mass is in line with the cord (A to A′). Next suspend the fish by another convenient point, its dorsal fin (Fig. 5-1*B*), and find another line (B to B′) that passes through the center of mass. The intersection of these two lines is the center of mass. If an animal has considerable mass in its limbs, the center of mass shifts according to the positions of the limbs and the overall shape they give to the body. The center of mass of a quadruped would be in one position if all limbs are extended and a somewhat different position if the front legs are extended and the hind limbs are folded up.

Vectors and Components of Force

Many quantities, such as length, mass, and temperature, have a magnitude and no direction. A structure may have a mass of 1 kg, but nothing is implied about a direction. Such quantities are described as **scalar quantities.** Force and velocity are described as **vector quantities** because they have both a magnitude and a direction. An arrow often represents a vector, such as a force. The arrow shows the direction of the force, and the length of the arrow is proportional on some convenient scale to the magnitude of the force. Consider a bipedal dinosaur walking (Fig. 5-2). The body moves forward as first one leg and then the other is placed on the ground. The center of mass projects to the ground between the alternately supporting feet. The leg supporting and pro-

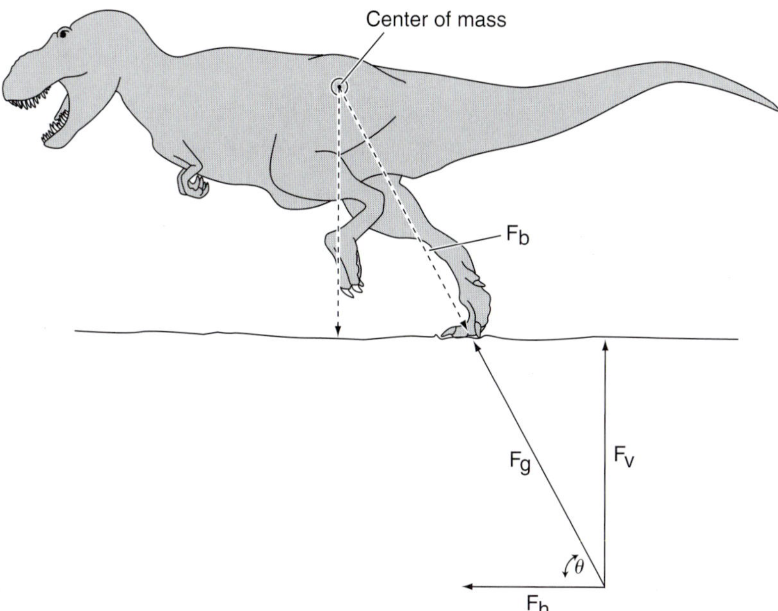

FIGURE 5-2
Forces acting on a walking bipedal dinosuar can be diagrammed as vectors and resolved into a supporting vertical force (F_v) and a propelling horizontal force (F_h).

pelling the body at any particular time exerts a downward and backward force on the ground (F_b), and, in conformity to Newton's third law of motion, the ground exerts an equal and opposite reaction force (F_g). We can ask how much of this force acts vertically to support the body (F_v) and how much acts horizontally to propel the body (F_h). These questions can be answered geometrically by drawing a vertical arrow from the beginning of the reaction force arrow to represent the force supporting the body, and a horizontal arrow to represent the force propelling the body. The length of these arrows, which is the magnitude of the forces, should be such that they form a parallelogram (in this case, a rectangle) the diagonal of which is the vector of the body or ground force. Because we are dealing with right-angled triangles, the same problem can be solved trigonometrically if you know the magnitude of any one of the forces and the angle it makes with any of the other forces. For example, if F_g and angle θ are known, then $F_v = F_g \text{ sine } \theta$; $F_h = F_g \text{ cosine } \theta$.

In Figure 5-2, the body force traveling down the leg and the ground reaction force make an angle with the ground of about 60°. This angle will obviously change during the course of a step, and so too will the vertical and horizontal components of the body or ground reaction forces. When the leg is nearly directly under the body, for example, the vertical force would reach a maximum value, and very little, if any, horizontal force will result. Many problems in vertebrate functional anatomy can be studied by vector analysis (Focus 5-1).

Stress and Strain

Many forces act on the body and its parts. Gravity acting on the center of mass of a terrestrial vertebrate exerts a strong downward force on the feet, supporting skeletal elements, and joint surfaces. The contraction of muscles pulls on their attachments. Biting on food exerts a force on the teeth and jaws. These forces, called **stresses,** can be measured as pressure (force per unit cross-sectional area, or Pascals) on the element subjected to the force. Stresses tend to cause a deformation of the material, which we call **strain.** Strain is very evident if you stress a rubber band by pulling on its ends, but even rigid material, such as bone, is subject to some degree of strain.

The major types of stresses and the strains they generate in bones or other structural materials of the body are shown in Figure 5-3. Two parallel forces moving directly toward each other subject the material to **compression** stress. The strain is a shortening and widening of the material. Two parallel forces pulling directly away from each other subject the material to **tension** stress. Strain resulting from tension is the lengthening and narrowing of the material. You can demonstrate this easily by stretching a thick rubber band and noting the change in its thickness. Two parallel forces moving toward each other, but not directly opposite each other, subject the material to **shear.** This causes one part of the material to slide relative to another part. Rotational forces applied in opposite directions induce **torsion** and cause the material to twist. Although one type of stress may predominate, many structural ele-

FOCUS 5-1 *Vector Analysis*

We have seen in Figure 5-2 how a force can be resolved into vertical and horizontal components. The converse is often done. The vertical and horizontal forces on the feet of a walking animal can be measured directly as the animal crosses a force platform. Given this information, the resultant force (F_r) can be determined in one of two ways. First, construct a parallelogram of force diagram the length of whose sides equal the vertical and horizontal forces. The resultant force, which summarizes the effect of the two forces, is the diagonal of the parallelogram. Second, construct a **triangle of force** diagram. Leave one vector (either the horizontal or vertical force) in its original position, and move the other without changing its length or angle, so that its head touches the tail of the first arrow. The line that completes the triangle is the resultant force (F_r; Fig. A).

In the examples we have considered, two of the vectors made a right angle, but this need not be the case. You can resolve a force into vectors going in directions of your choice by applying the same principles (Fig. B).

Sometimes it is necessary to find the resultant of more than two vectors acting on a single point. This can be done by vector addition using the procedures used in constructing a triangle of force diagram. Leave one vector connected to the point in question, and connect the others head to tail without changing their lengths or angles. If the final vector gets back to the starting point, then the resultant of all of the vectors is zero. The forces are said to be in balance. If another line is needed to complete the polygon, then this line is the resultant (Fig. C). All of these problems can be solved with precision trigonometrically, but vector analysis is a good way to visualize the situation and determine approximate values.

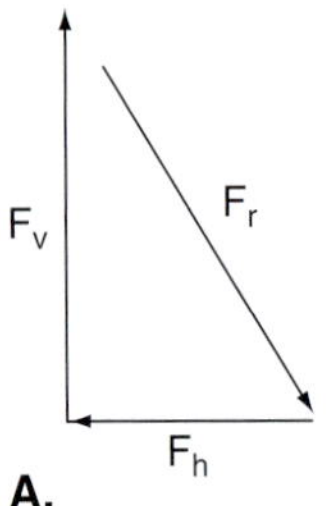

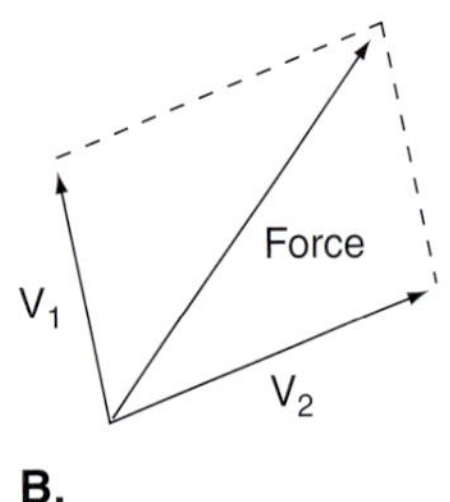

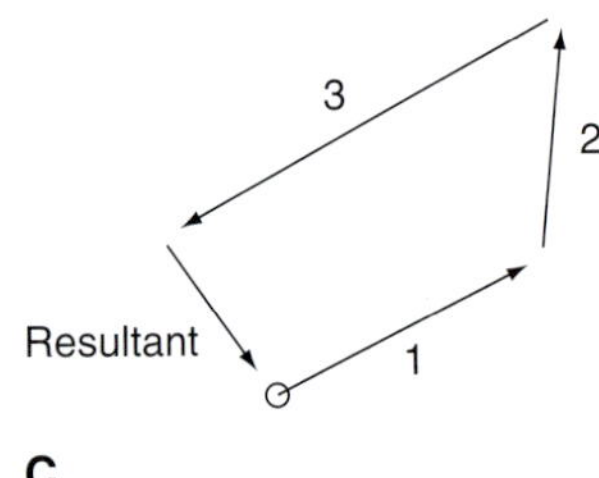

ments of the body are subjected to a complex combination of these forces as they support the weight of the body and resist the pull of muscles that attach to the bones. Bending, for example, results from the interaction of compression and tension. These forces, as well as genetic factors, affect the type of structural material utilized (e.g., ligament, cartilage, or bone), the shape of the structural element, and the distribution of materials within it.

When an elastic material, such as a rubber band, is pulled, it yields relatively easily to the stress, and it stretches. The energy used to stretch the rubber band is stored as **elastic energy.** You can sense the elastic energy by the force you must expend to hold the rubber band outstretched. When you release the rubber band, most of the elastic energy is recovered (a small amount is lost as heat). By contrast, the deformation in a rigid material, such as bone or steel, is less when it is stressed, but some always occurs.

The relationship between stress and the deformation produced can be expressed graphically in a **stress–strain curve** (Fig. 5-4). In a rigid material such as bone, strain, or deformation, at first increases only slightly as stress increases greatly. This portion of the curve is known as the **elastic region** because the bone will return to its original shape when the stress is released; that is, both stress and strain return to zero. The area under this portion of the curve represents the elastic energy that caused the deformation. When stress reaches the **yield point** of the material, a further increase in stress causes considerable deformation or strain, and the bone is said to behave **plastically.** The bone may bend or otherwise be deformed. When stress is released and falls to zero, the strain does not return to zero; that is, the bone is permanently deformed. Forces on bones normally lie within the elastic region of the curve, but bones can behave plastically over a considerable range of strain before they reach the **fracture point** and break.

The slope of the curve in the elastic region represents **Young's modulus of elasticity,** which, because it is linear, also can be defined as stress divided by strain. A rubber band, which deforms greatly under low stress, has a low modulus of elasticity, whereas

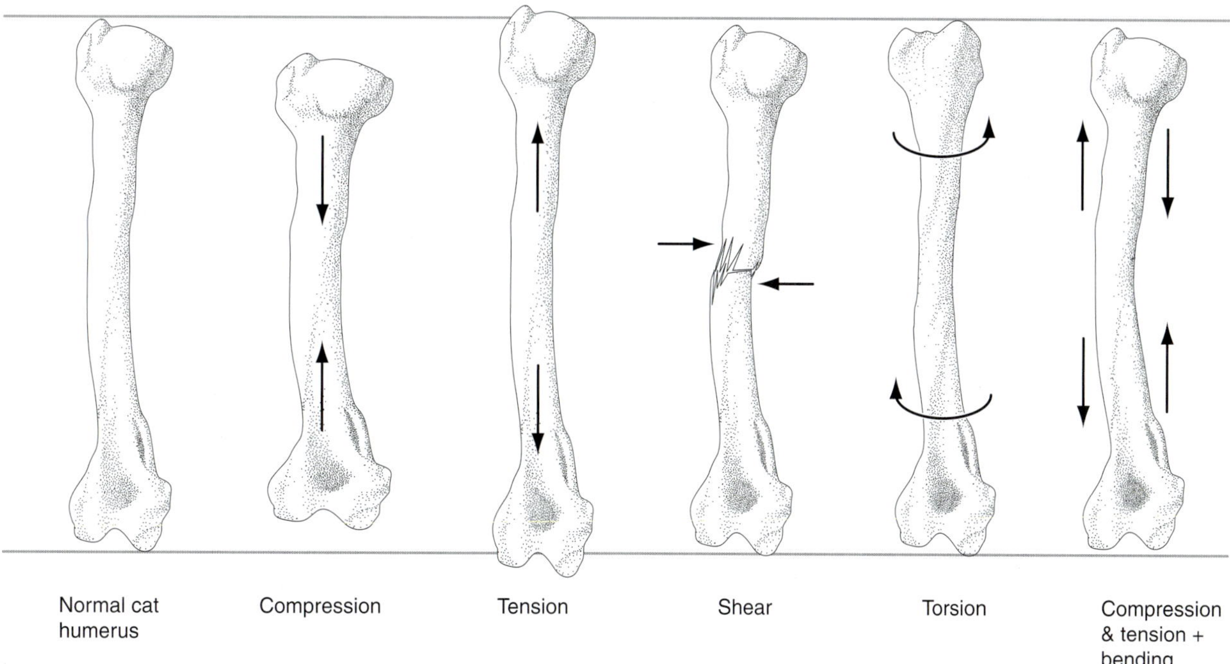

FIGURE 5-3
The major stresses to which materials can be subjected are illustrated on a humerus of the cat. The directions of the stresses are shown by arrows, and the resulting strains, by deformations in the humerus. A normal (unstressed) humerus is shown on the left for comparison. The deformations have been exaggerated for purposes of illustration.

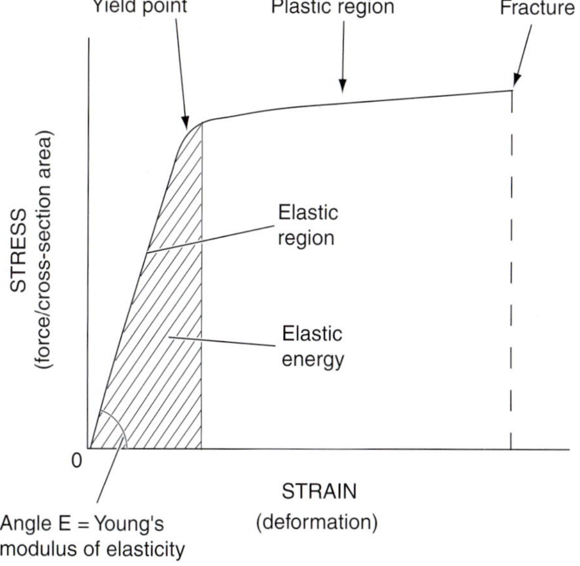

FIGURE 5-4
A stress–strain curve for a fairly rigid material, such as bone. The material at first deforms only slightly as stress increases greatly. *(After Currey.)*

bone, which deforms only slightly under high stress, has a high modulus of elasticity. The modulus of elasticity applies only to stresses up to the yield point of the material.

Properties and Growth of Structural Materials

Structural materials of the body differ with respect to the stress or combination of stresses they are adapted to meet, and this is an important factor that determines their distribution and use in the body.

Hydrostats

The pressure of the gas within them supports an inflated balloon or an inflated tire. The wall resists tension, and the gas within them resists compression. Many vertebrates have parts that are internally pressurized, but the parts are cylindrical and the internal fluid is water, which is incompressible. Such structures are called **hydrostats** (Gr., *hydro-* = water + *states* = to make stand) or **hydroskeletons.** The proboscis and colar coelom of enteropneust hemichordates can be filled with seawater and act as a hydrostat when these animals burrow. The notochord, which is the primary axial structural element of larval tunicates, amphioxus, and early craniates, is also a hydrostat. The structure of the liquid-filled cells of the notochord and the firm, fibrous sheath surrounding them combine to resist a change in the diameter and in the length of the notochord. Therefore, the notochord resists compression when the longitudinal muscles in the body wall contract but allows the body to bend from side to side during swimming movements.

The way a pressurized cylinder bends depends on the arrangement of the fibers in its wall. They could be arranged circularly and longitudinally, or they could take the form of left-hand and right-hand helices extending the length of the cylinder (Fig. 5-5). If the fibers are circular and longitudinal, the cylinder can easily kink when it is bent because the longitudinal fibers do not resist compression. A rope kinks when parts are pushed together. A helical pattern allows for smooth bending, and this is the pattern seen in the notochord. An analogy is an expensive garden hose, which is made of layers of rubber and a spiral wrapping of reinforcing fibers. Less expensive hoses made only of rubber or plastic are prone to kinking. A helical pattern of fibers does allow for shortening and lengthening of a structure, but the notochord does not change its length because it is composed of many turgid cells encased by a sheath that maintains internal pressure. It is not a hollow cylinder filled with water.

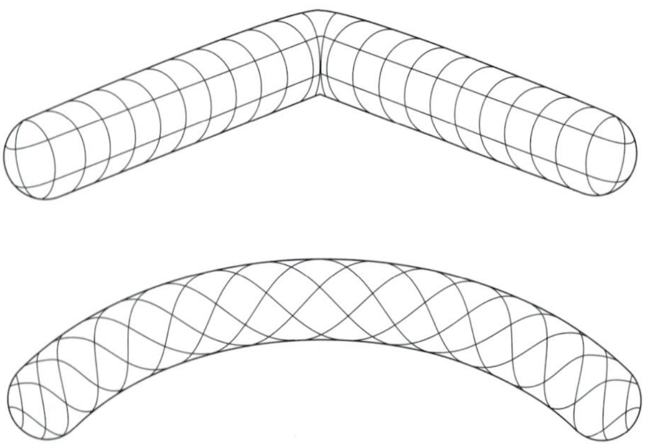

FIGURE 5-5
Two arrangements of reinforcing fibers in the wall of a pressurized cylinder. A helical arrangement permits the cylinder to bend without kinking. *(After Wainwright.)*

During embryonic development of most vertebrates, the centra of the vertebrae develop around the notochord, and the notochord is greatly reduced or disappears. The embryonic notochord enlarges and persists as a well-formed structure in the adults of hagfishes, lampreys, lungfishes, and a few other fishes and early tetrapods. In fossil lungfishes, choanate fishes, and ancestral tetrapods, for example, the large space extending through the centra of the vertebral column lodged the notochord.

The entire trunk and tail of a shark is another example of a hydrostat because the entire trunk and tail are encased by crossed helically arranged fibers in the skin, which therefore acts as a pressure-resistant membrane. Muscles attach to the inner surface of the skin, so the skin also acts as an extensive external tendon. We return to this point later.

Muscular hydrostats are organs that combine support and movement. They are made up of a mass of muscle fibers, which, as with all cells, are largely composed of water. The muscle fibers within the organ are arranged in antagonistic groups. Contraction of circular fibers, for example, squeezes on the mass of cells and lengthens the organ because the water-filled cells maintain their volume. If the cells become narrower, they must become longer. Contraction of longitudinal fibers shortens and widens the organ. Examples of muscular hydrostats are the tongues of many tetrapods and the elephant's trunk (Kier and Smith, 1985). The protrusion of the tongue of a lizard, or even our own, depends largely on the compression and lengthening of muscle fibers.

Connective Tissues

Connective tissues may be defined as those tissues in which extracellular materials, such as water, proteins, or carbohydrates, predominate over the cellular components. Blood can be considered a connective tissue because its cells are surrounded by an extensive matrix of water and proteins in solution. Most connective tissues have a fibrous matrix. **Loose fibrous connective tissue,** which is the most widely distributed connective tissue, is a loose web of cells, fibers, and other extracellular material. It permeates all of the organs of the body, binding other tissues together. We find it beneath the skin and in muscles and in the walls of blood vessels and digestive organs. Elongated cells known as **fibroblasts** (L., *fibra* = fiber + Gr., *blastos* = bud) produce loose connective tissue. Fibroblasts also branch, but their branches lie in the same plane, so they appear spindle shaped in side view. They develop embryonically from mesenchyme cells, and as with the mesenchyme cells, the fibroblasts are embedded in an **extracellular matrix** that they secrete (Fig. 5-6*A*).

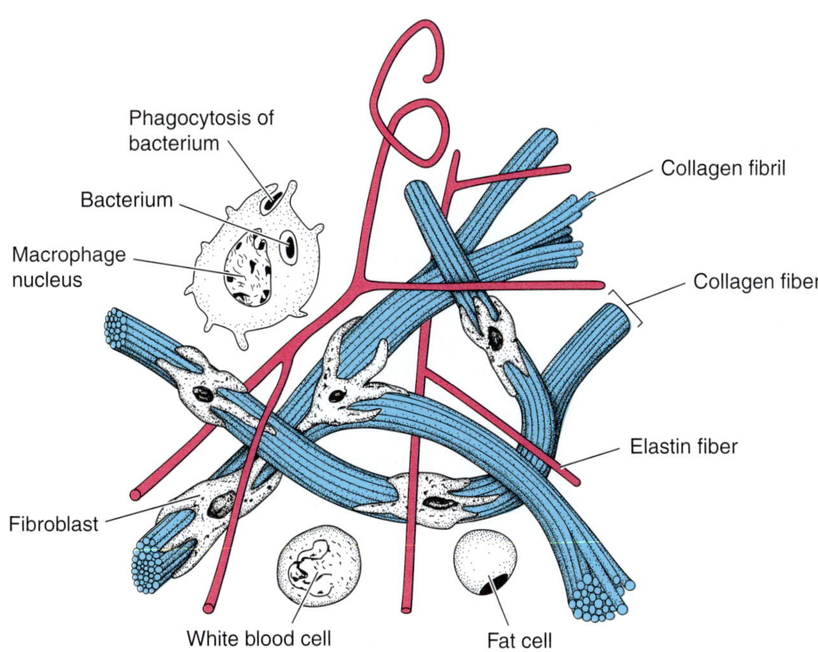

A. Components of loose connective tissue

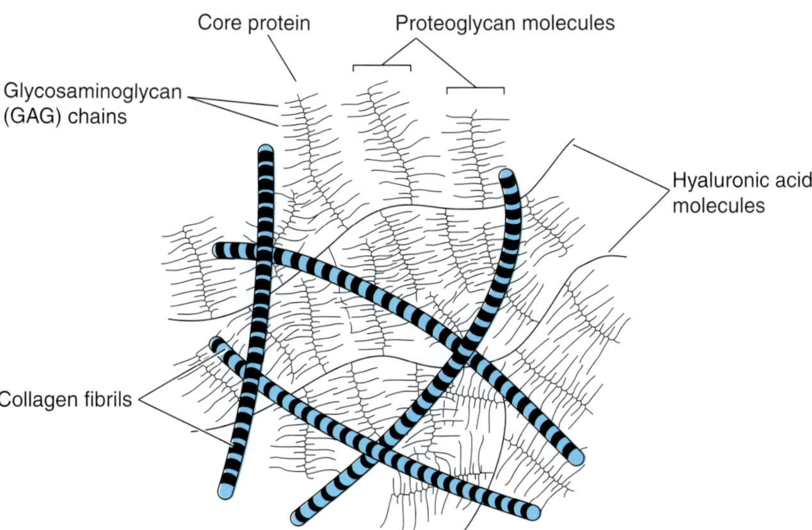

B. Ground substance of loose connective tissue

FIGURE 5-6
Loose fibrous connective tissue. *A,* A diagram as seen in high magnification showing the major components of the tissue. *B,* A molecular diagram of the structure of the ground substance at much greater magnification. Other constituents (e.g., fibronectin) are not shown.

The matrix consists of **collagen fibers** and **elastin fibers** embedded in a gel-like **ground substance.** Collagen is a protein consisting of filamentous molecules of **tropocollagen** that are organized first into **collagen fibrils,** which are visible only with the electron microscope. The collagen fibrils are assembled into the **collagen fibers,** which are visible by normal microscopy. Cross linkages between tropocollagen molecules give collagen a great tensile strength, but it is flexible and bends easily. Collagen itself has a relatively low elasticity, but elastin stretches easily and recoils to its normal length. Collagen is analogous to rope, and elastin, to a rubber band.

Collagen is not a single type of protein but a family of closely related proteins that differ to some extent in amino acid sequences in their chains and in the degree to which the chains organize into collagen fibrils and fibers. Collagen Types I, II, and III form microscopically visible fibrils or fibers of different sizes. Type I collagen forms the largest fibers and occurs in the dermis of the skin, bones, tendons, and the capsules of many organs. Type III collagen, which forms somewhat smaller fibers, occurs in the loose connective tissues. The fibrils of Type II collagen, which occurs in hyaline cartilage, do not aggregate into fibers. Other types of collagen can be detected only by biochemical means.

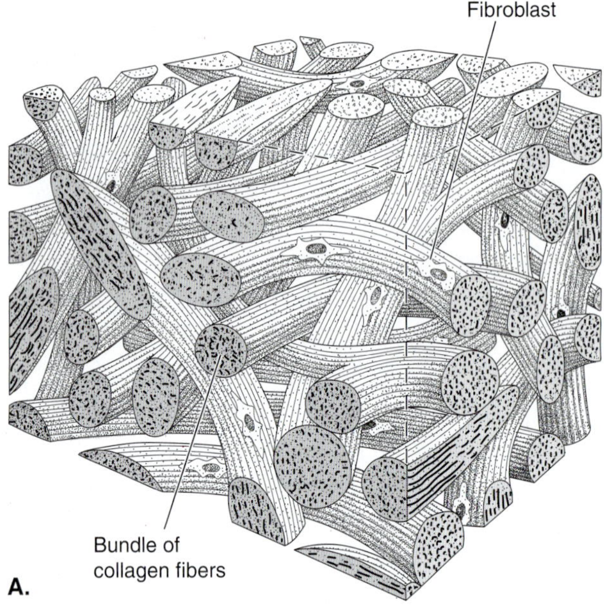

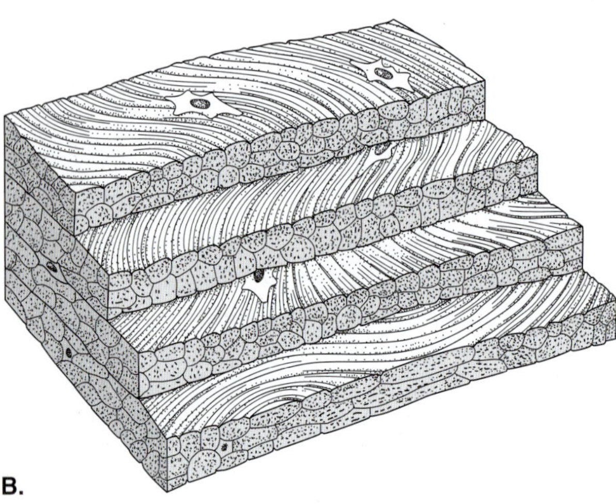

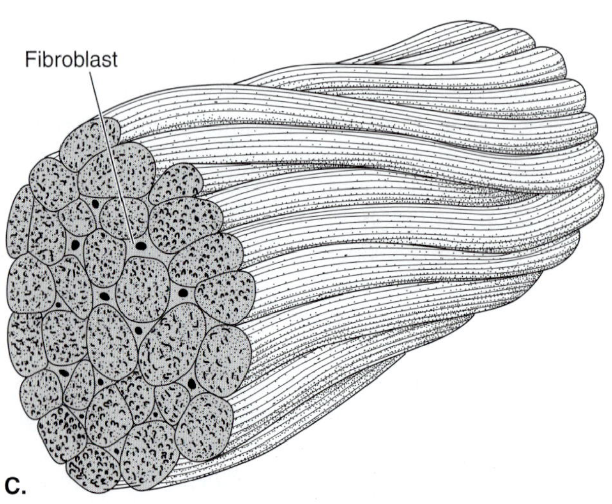

FIGURE 5-7
The arrangement of collagen fibers as seen at high magnification in some dense connective tissues. *A,* The numerous fibers are interwoven in dense irregular connective tissue as in the dermis of the skin. *B,* The fibers form layers in a ligament. Notice that the fiber direction is different in adjacent layers. *C,* The fibers have a cable-like arrangement in a tendon. *(After Williams et al.)*

The ground substance (Fig. 5-6*B*) consists of many **proteoglycan** molecules, which resemble little bottle-brushes. Each proteoglycan molecule consists of a core protein to which are attached many polysaccharide side chains known as **glycosaminoglycans** (GAGs). The proteoglycan molecules, in turn, are attached to very long molecules of another polysaccharide called **hyaluronic acid.** The whole complex is very viscous in aqueous solution, which gives the ground substance its gel-like quality.

Loose fibrous connective tissue may also include other cellular types, including fat cells, scattered white blood cells, and **macrophages.** Macrophages are large phagocytic cells derived from monocytes (a type of white blood cell) that help protect the body from microorganisms by engulfing and degrading them. In the fat depots of the body, fat cells are very abundant, and the extracellular matrix is reduced. This type of connective tissue is known as **adipose tissue.**

Many structural materials are specialized forms of fibrous connective tissue in which fibroblasts are less numerous than in loose fibrous connective tissues, and the collagen fibers form densely packed bundles (Fig. 5-7). **Dense irregular connective tissue,** which forms capsules around such organs as the kidneys and forms the dermis of the skin, consists of an irregular array of collagen bundles. Ligaments and tendons are composed of very tightly packed bundles of collagen fibers. Collagen fibers in **ligaments** often form layers, with fibers going in different directions. Ligaments unite bones at joints, holding them together but permitting movement. In **tendons,** the bundles of collagen fibers have a more cable-like orientation. Tendons link the muscles to bones and cartilages. Because of the high tensile strength of collagen, ligaments and tendons resist tension and torsion very well, but they bend easily and do not resist compression. In some cases, as along the vertebral axis in some large dinosaurs, tendons ossify and provide surfaces for more muscle attachments.

Cartilage

Both cartilage and bone are dense connective tissues the structure of which makes them far more rigid than ligaments and tendons. Their rigidity enables cartilage

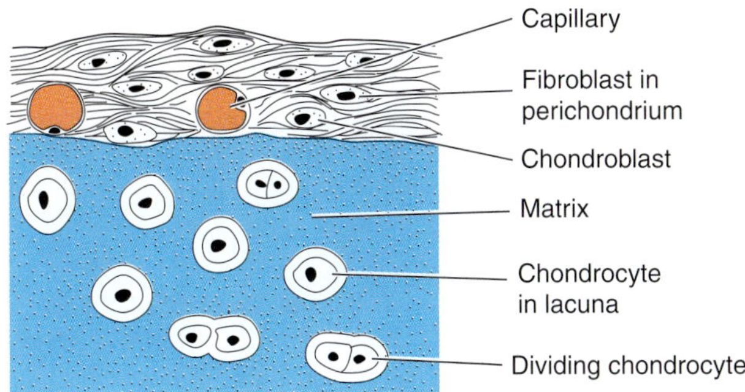

FIGURE 5-8
Hyaline cartilage.

and bone to resist compression and hence to provide support. Cartilage is nearly as strong as bone in resisting compression, but it does not resist tension and shear as well. It has a lower modulus of elasticity than does bone, and so cartilage is more flexible and elastic. Cartilage develops when mesenchyme cells transform into **chondroblasts** (Gr., *chondros* = cartilage + *blastos* = bud) and begin to secrete and deposit an extensive extracellular matrix. The mature condroblasts, which are called **chondrocytes** (Gr., *kytos* = cell), lie in small spaces, the **lacunae** (L., *lacuna* = small cavity), within the matrix (Fig. 5-8). The matrix contains Type II collagen fibrils embedded in a ground substance. Compared with those of loose connective tissue and bone, the collagen fibrils of cartilage do not assemble into coarser bundles, so they only can be seen with electron microscopy. The ground substance is similar to that of loose fibrous connective tissue (Fig. 5-6*B*), except that the proteoglycan molecules are in a much higher concentration, so the matrix is very firm. The principal side chains of the proteoglycan molecules are **chondroitin sulfate** and **keratin sulfate.** Negative charges on the ends of the side chains bind water molecules, and more water occurs in the interstices among the molecules. The ground substance of cartilage acts very much like a mop in its capacity to hold water. The resistance of cartilage to compression, its flexibility, and its smoothness derive from the water within it. When cartilage is under compression, some of the water is expelled and contributes to the lubrication at those joint surfaces covered by cartilage. The cartilage becomes thinner. When loads are released, the negative charges of the side chains draw water back into the ground substance, and the cartilage returns to its original thickness. Because of these changes in the cartilage, human beings who spend a great deal of time standing or walking are approximately 1 cm shorter at the end of the day than at the beginning. Unlike other connective tissues, cartilage seldom has a vascular or nerve supply. The metabolic rate of chondrocytes is low, and most cartilaginous skeletal elements are small enough to be supplied by diffusion from blood vessels in the fibrous **perichondrium** (Gr., *peri* = around), which covers the free surface of cartilages. Cartilage canals containing blood vessels may enter the larger cartilages of large vertebrates.

Because their matrix contains several types of materials and spaces that lodge the cellular elements, cartilage and bone are **composite materials.** Composite materials are much stronger than are homogeneous materials, such as glass. A crack can propagate rapidly through a homogeneous material, but the force at the apex of a crack spreading through a composite material becomes blunted and dissipates when it impinges on a different material or a space containing a cartilage or bone cell. Fiberglass, which consists of minute glass fibers embedded in resin, is a familiar example of a composite material.

Cartilages vary considerably in details of their structure. In adult mammals, for example, the most common type of cartilage, **hyaline cartilage** (Gr., *hyalos* = glass), occurs on the ends of the limb bones and on the ventral ends of the ribs and forms the cartilages of the larynx and tracheal rings. The extensive matrix is homogeneous and translucent, or glasslike (Fig. 5-8). The matrix of **elastic cartilage** of the mammalian external ear and epiglottis contains a particularly dense network of branching elastin fibers. The matrix of **fibrocartilage** contains many large collagen fibers and often grades into the dense connective tissues of tendons and ligaments. It often is found in the intervertebral disks as well as the mandibular and pubic symphyses.

Much embryonic cartilage contains many more cells relative to the matrix than other types of cartilage and is called **cellular cartilage.** The cartilage of hagfishes and lampreys is highly cellular, and the matrix is not based on collagen but on proteins with a different molecular structure. The protein is **myxin** in hagfishes and **lamprin** in lampreys (Wright and Youson, 1983; Robson et al., 1993).

Cartilage often is strengthened in adults by the deposition of calcium salts within it. **Calcified cartilage** often forms a prismatic layer around a core of hyaline cartilage. The prismatic layer is composed of tiny, mineralized blocks called **tessarae.** Prismatic cartilage of this type is synapomorphic for Chondrichthyes and can be seen in the vertebrae of many of these fishes. Although calcified cartilage has rigidity similar to that of bone, its histological structure is quite different.

Cartilage grows on its surface by the recruitment of chondroblasts, some of which remain in the inner part of the perichondrium, and interstitially by the mitotic division of chondrocytes. Daughter chondrocytes separate and synthesize more matrix. Because of the ease and speed with which cartilage grows without requiring complex remodeling, it is an excellent embryonic skeletal material. Most of the internal skeleton of a vertebrate embryo is cartilage. Later in the embryonic development of the majority of vertebrate species, nearly all of the cartilage is replaced by bone, but some cartilage persists where its ease of growth, smoothness, or elasticity are particularly important qualities. Such sites occur at or near the ends of limb bones, on the articular surfaces of bones at movable joints, and at the distal ends of the ribs. The internal skeleton of adult chondrichthyan fishes is entirely cartilaginous. Chondrichthyan fishes lack the hydrostatic swim bladder that allows most bony fishes to float at any depth in the water with little muscular effort. Consequently they are relatively dense fishes that tend to sink. A skeleton of cartilage, which is less dense than bone, makes them slightly more buoyant.

Much cartilage also occurs in the internal skeleton of early bony fishes, such as lungfishes, coelacanths, sturgeons, and paddlefishes, although whether this represents a plesiomorphic condition for bony fishes is not clear.

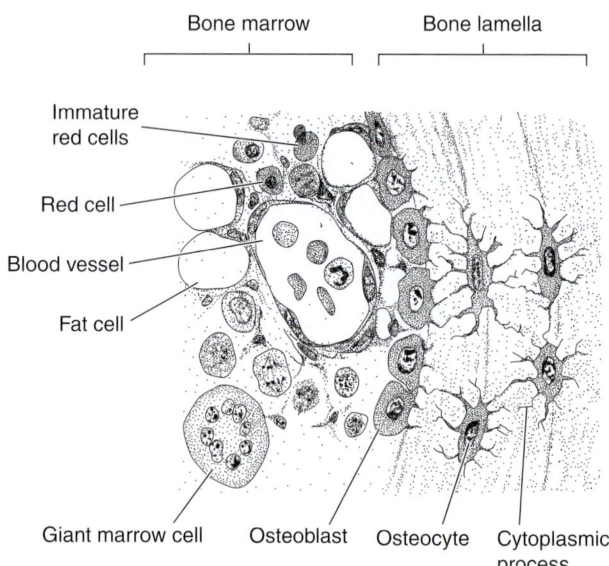

FIGURE 5-9
A microscopic section of cellular bone showing its formation and histological structure. This section of bone lies beside the marrow cavity. *(After Corliss.)*

Bone

Bone is a highly vascular, mineralized, dense connective tissue that is hard, resilient, and capable of slowly changing in structure as forces on the body change during an individual's life. Bone is particularly strong because its mineral component imposes rigidity and resists compression, and its fibrous component provides some flexibility and resists tension and torsion. Vertebrates exhibit tremendous variety in the structure of bone, and only a fraction of that variety will be considered here.

Bone is the primary skeletal tissue of most adult vertebrates (Fig. 5-9). Bone-forming cells, or **osteoblasts** (*osteon* = bone + *blastos* = bud), produce a matrix of polysaccharides and many collagen fibers. The bone becomes calcified by the binding of calcium phosphate, in the form of crystals of **hydroxyapatite,** to the collagen fibers. In many of the early jawless fishes, osteoblasts retreated peripherally as they laid down the bony matrix and did not become incorporated into the matrix. This type of bone is called **acellular bone,** or **aspidine** (Chapter 3). Acellular bone also occurs in the bony scales of many fishes (Chapter 6). In most vertebrates, however, the osteoblasts become trapped as **osteocytes** in small spaces, called lacunae, within the matrix they produce. This common type of bone is called **cellular bone.** Osteocytes are star-shaped cells the radiating processes of which lie within minute **canaliculi** that permeate the matrix. The relatively large Type I collagen fibers, which comprise about 40% of the dry weight of many bones, form randomly organized bundles in the **woven-fibered bone** of young individuals. As an individual matures, the collagen bundles become organized into parallel sheets, or **lamellae** (L., *lamella* = layer), with rows of osteocytes between the layers. Layers in **compact** or **lamellar bone** commonly have different fiber directions, which, as with the different layers of plywood, add to the strength of the bone.

It probably is significant that vertebrates use calcium phosphate in their skeletons whereas invertebrates deposit crystalline calcium carbonate, or calcite. Ruben and Bennett (1987) proposed that this correlates with the greater degree of activity that characterizes vertebrates in contrast to their early chordate ancestors. Increased activity causes a lowering of pH in body fluids, and this causes calcium compounds to dissolve. Some dissolution is normal and necessary, and calcium and phosphate ions in bones and body fluids constantly are interchanging. Calcium phosphate has an advantage in an active animal, however, for it is both more stable and less soluble than is calcium carbonate.

The degree of calcification of bones differs according to their functions. The mammalian femur, which is a load-bearing bone, has a high mineral content (67%) and is stiff. A deer's antler, which does not support weight but must resist impact, bending, and shear when males butt and lock heads, is only 59% mineral. Because so many variables are involved (e.g., location, function, sex, age, and species), it is difficult to generalize about the strength of bone, but it is clear that bone is a very strong material. Most bone can resist compression that is about four times the compressive strength of concrete. Compact bone resists tension about as well as tendons and ligaments. However, tendons and ligaments have an advantage when tension is the only stress because they can be made more easily, are more flexible, and weigh less.

Bone Growth

A bone can form embryonically directly within connective tissue or within and around a cartilage, which it gradually replaces. When bone forms in connective tissue, it is called **membrane bone.** Because most membrane bones develop near the body surface in or just beneath the dermis of the skin, membrane bone is often also called **dermal bone.** Many of the superficial skull bones and some of the pectoral girdle bones are membrane bone. The clavicle is a example. When forming a membrane bone, mesenchyme cells transform into bone-forming cells, or **osteoblasts.** The osteoblasts line up and begin to produce a matrix of fibers and ground substance called **osteoid** (Fig. 5-10*A*). The osteoid soon becomes calcified and forms a small bony rod called a **trabecula** (L., *trabecula* = small beam). Trabeculae coalesce when they meet (Fig. 5-10*B*). As embryonic development continues, layers of bone are deposited around the periphery of the latticework of trabeculae (Fig. 5-10*C*).

Many other bones, such as those of the limbs, form more deeply in the body in association with cartilaginous primordia. Such a bone appears in the developing embryo first as a condensation of mesenchyme. The mesenchyme then chondrifies to form a small, cartilaginous primordium of the adult bone (Fig. 5-11). As

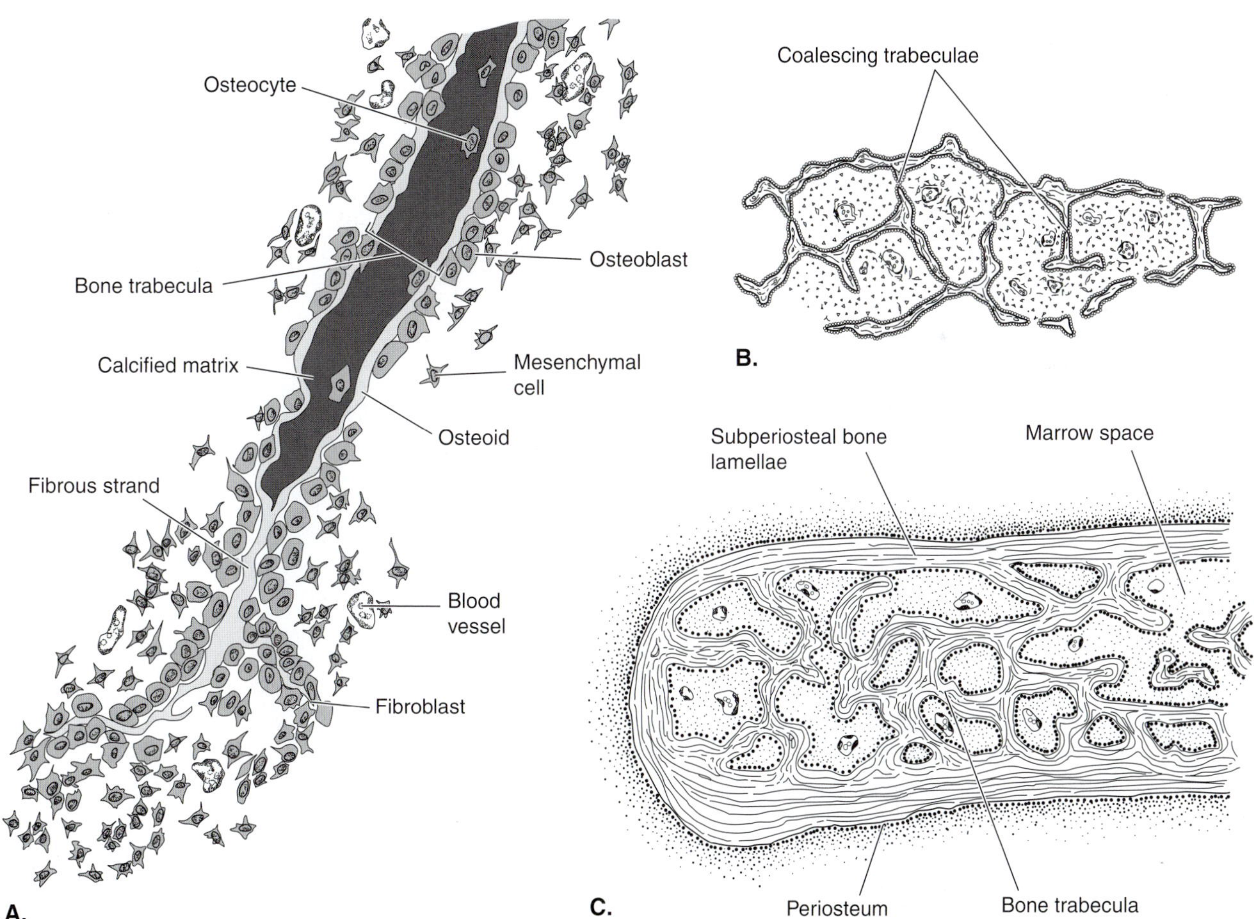

FIGURE 5-10
Stages in the formation of membrane or dermal bone. *A,* Osteoblasts, which are lined up on a group of collagen fibers, are producing more bone matrix and mineralizing it to form a bony trabecula. As the trabecula forms, osteoblasts become trapped as osteocytes in the matrix.*B,* Trabeculae coalesce. *C,* Layers, or lamellae, of bone develop around the periphery of a network of trabeculae to form a flat bone. (*After Corliss.*)

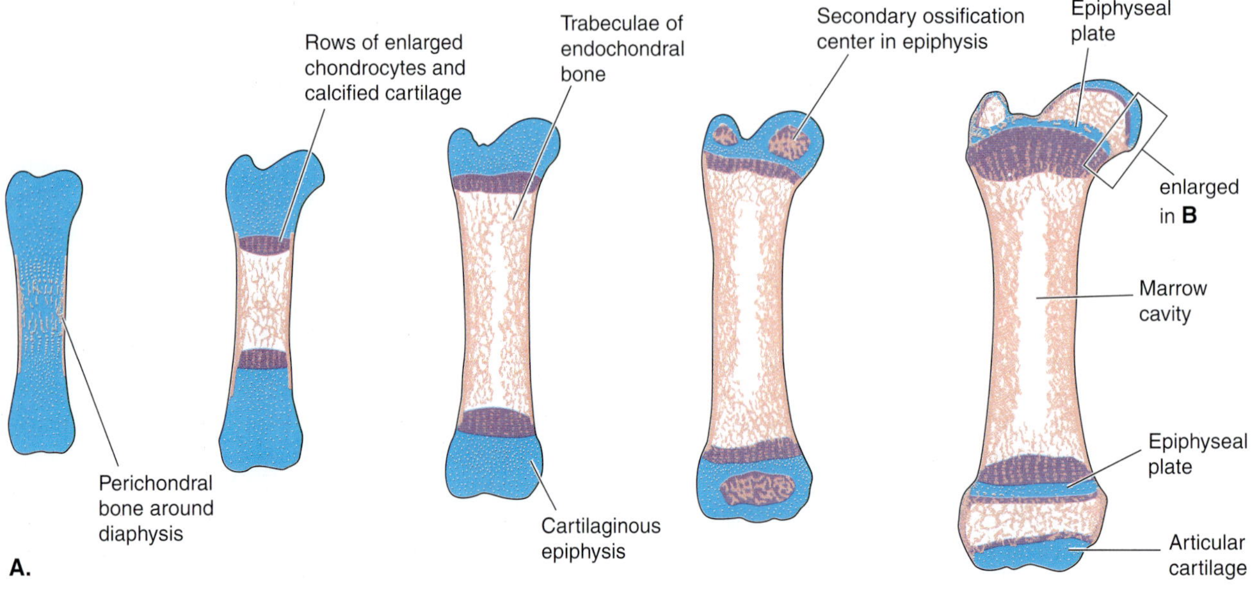

FIGURE 5-11
Development of the femur in a late embryo or fetus of a mammal. *A,* The original cartilage of the femur is replaced by bone that develops around the periphery of the femur (perichondral bone) and within it (endochondral bone). Together, perichondral bone and endochondral bone form the cartilage replacement bone of the mature femur. *B,* An enlarged detail of bone formation. See text for explanation. Blue = cartilage; purple = calcified cartilage; pink = bone. *(A, after Corliss; B, after Williams et al.)*

embryonic development continues, the cartilage grows, and bone begins to form both within the cartilage and around its periphery. Bone that forms within the cartilage, which it gradually replaces, is called **endochondral bone;** that which forms in connective tissue on the surface of the bone is called **perichondral bone.** Perichondral bone is very ancient and has been found in some of the earliest jawless vertebrates (†galeaspids and †osteostracans) as well as in all gnathostomes (Chapter 3). Technically, perichondral bone can be considered a type of membrane bone, but it usually is located more deeply in the body than other membrane bones. It is customary to call all of the bone that forms in association with cartilages (both endochondral and perichondral bone) **cartilage-replacement bone** or simply **replacement bone.** Membrane bones and cartilage-replacement bones are composed of the same types of cells and matrix, but the architecture of cartilage-replacement bones is often more complex, for they resist a more complex set of stresses.

Details of ossification patterns vary among vertebrates (Shapiro, 1992), but the major sequence of events for a mammal is evident in a late embryo (Fig. 5-11). The first ossification in the long bone of a mammal is the formation of a collar of perichondral bone around the middle of the shaft, or **diaphysis,** of the cartilaginous primordium. The bone-forming osteoblasts are derived from undifferentiated cells in the perichondrium, which now becomes the **periosteum.** Next, chondrocytes within the diaphysis enlarge and line up in rows. Some of the matrix is reabsorbed, but calcium is deposited in the remaining matrix adjacent to the enlarged chondrocytes. As the chondrocytes become cut off from nutritional and gas exchanges with blood vessels in the periosteum, they break down, and the cartilage becomes honeycombed. Vascular connective tissue invades the area from the periosteum. Some of the invading cells differentiate into osteoblasts, line up along the remnants of the matrix, and begin to form interlacing bony trabeculae. The spaces between the trabeculae become filled with **bone marrow.**

Endochondral ossification proceeds from the center of the diaphysis toward each end of the cartilaginous primordium. This process would soon replace the original cartilaginous primordium of the bone except that the cartilage at each end, which is called the **epiphysis,** continues to grow as bone forms in the diaphysis. In this way, the entire element increases in length. In many vertebrates, including many anamniotes, reptiles, and birds, the epiphyses remain cartilaginous throughout life, although the rate of growth slows down greatly as the animal matures. Secondary centers of ossification develop in the epiphyses of mammals. However, cartilaginous **epiphyseal plates** remain between the epiphyses and diaphysis, and cartilage continues to grow in these plates until the adult stage, when the ossification centers in the epiphyses and diaphysis unite. Union occurs at different ages in different bones, and these differences can help determine the age of a skeleton. The epiphyseal plates of mammals, together with the secondary ossification centers in the epiphyses, allow growth in length of the bone while maintaining strength at the ends of the bone next to the joint.

Growth in the diameter of a bone occurs by the continued formation of bone around its periphery. As a bone grows in length and diameter, it must be continually remodeled; the rigidity of mineralized bone precludes it from expanding interstitially as cartilage does. Bone is removed by the action of **osteoclasts** (Gr., *osteon* = bone + *klastos* = broken), large, multinucleated cells that differentiate by the fusion of precursor cells from the bone marrow. Osteoclasts establish a tight seal between themselves and a compartment next to the bone into which they secrete a strong acid that dissolves the mineral component of bone and enzymes that digest the collagen. The degraded products are transported across the cell to the free surface of the osteoclast (Mostov and Werb, 1997). Osteoclasts effectively remove old bone, and osteoblasts form new bone. The pattern of bone removal and formation varies among vertebrates. In some reptiles, birds, and mammals, the osteoclasts tunnel into old bone (Fig. 5-12). Blood vessels invade the tunnel and osteoblasts line up on the tunnel's inner wall. The osteoblasts deposit concentric layers of bone to form a columnar **osteon,** or **Haversian system.** The most recently formed layer of bone in an osteon is next to the central vascular space. As the bone continues to be remodeled, newly formed osteons replace the old ones, but fragments of former osteons often lie between the current ones (Fig. 5-13). The marrow cavity also enlarges as the bone grows, so most of the bone is located peripherally. Bone tissue on the surface is deposited in layers that go around the entire element (Fig. 5-13). More deeply, the bone tissue is deposited as osteons, most of which extend vertically through the bone. The collagen fibers in adjacent lamellae of an osteon spiral around the central vascular space at different pitches, and this, together with the columnar shape of an osteon, gives bone great strength. The dense, peripheral bone tissue is **compact bone.** Bone tissue near the ends of the bone and adjacent to the marrow cavity tends to retain its trabecular structure and is known as **cancellous bone.**

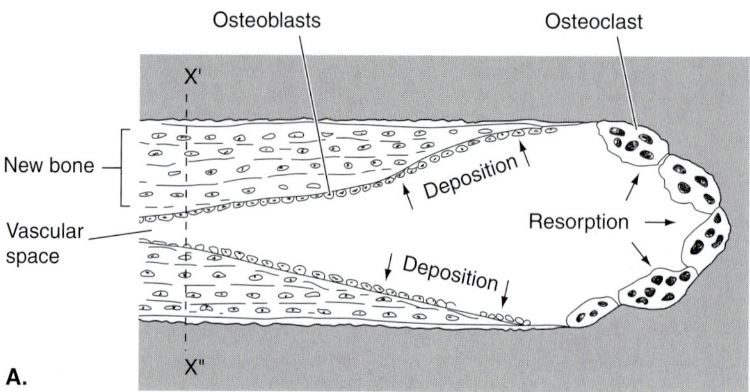

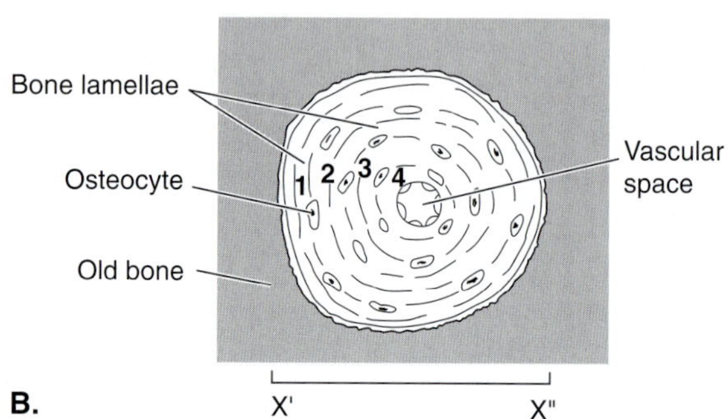

FIGURE 5-12
The formation of an osteon, or Haversian system, within old bone. *A,* A longitudinal section. As osteoclasts tunnel into old bone, osteoblasts produce layers, or lamellae, of new bone behind them. *B,* A transverse section of an osteon at the level X′ to X″ indicated in A. Numbers indicate the sequence in which the layers were formed. The vascular space remains and carries small blood vessels, which nourish the bone. *(After Lanyon and Rubin.)*

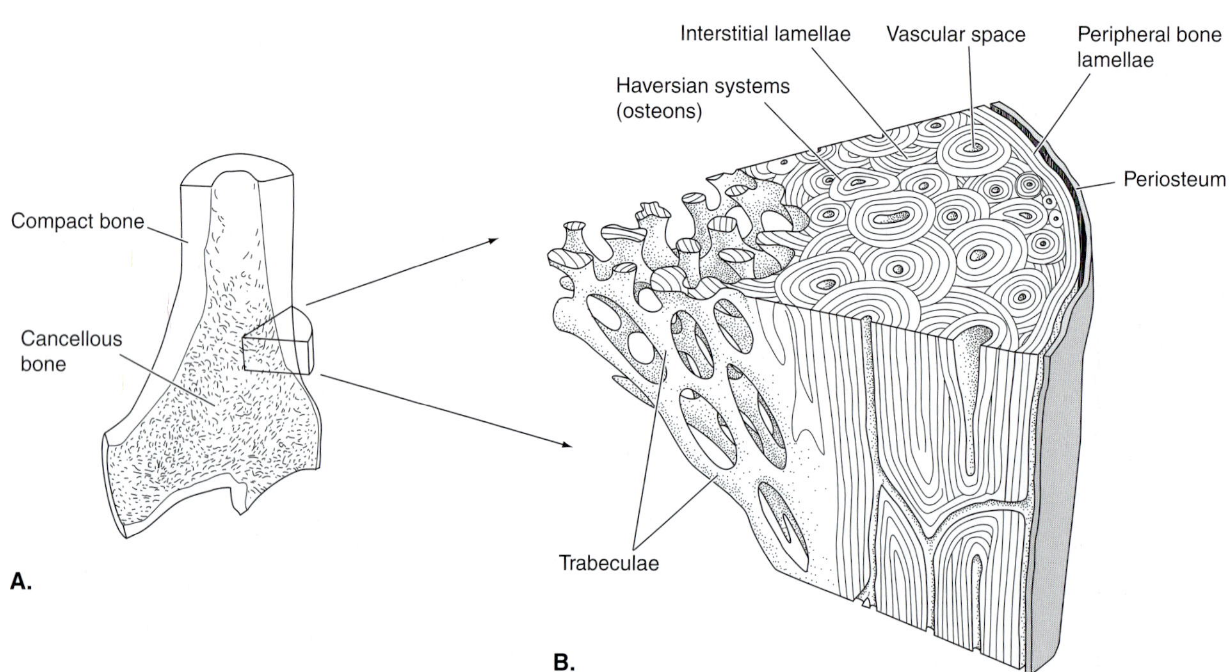

FIGURE 5-13
Mature bone structure. *A,* A longitudinal section near the end of a limb bone. *B,* An enlargement of a portion of it.

Other Functions of Bone

Bones are primarily structural elements that protect deeper tissue (e.g., the central nervous system), support the body, and form lever systems used in locomotion and feeding. Bones also have other vital functions. Bone tissue is an important reservoir of calcium and phosphate ions, which are essential in many biochemical processes, including muscle contraction and the storage and release of energy. Calcium and phosphate ions continually move between the bones and body fluids. The early evolution of bone among ancestral vertebrates may have made it possible for them to move from a marine environment into a freshwater one, where these ions are in short supply.

The cylindrical cavities that develop in the long bones of the appendages, as well as the spaces between the trabeculae of all bones, are filled with bone marrow. During fetal and much of juvenile life of a mammal, all of the marrow is **red bone marrow.** This highly vascular tissue contains **stem cells** that differentiate into both red blood cells and many types of white blood cells. Red blood cells play a crucial role in gas transport in the blood, and white blood cells are an essential part of the body's defense and immune systems. As a mammal ages, much of the red marrow in the long bones is replaced by fat cells, forming **yellow bone marrow.** Red bone marrow persists in many other bones, including the vertebrae and skull bones.

Joints and Kinematic Chains

The individual cartilages and bones of the skeleton are connected to one another at joints, or **articulations.** The structure of joints is determined by the degree and direction of movement needed, by the forces acting on the joint, and by the nature of the articulating elements. Cartilages may articulate with cartilages, bones with bones, or bones with cartilages. We normally think of joints as allowing free movement between skeletal elements, but movement is restricted at many joints. Joints where movement is restricted are known as **synarthroses** (Gr., *syn* = together, joined + *arthrosis* = joint). In addition to providing great strength, synarthroses allow for growth and often for limited movement. Common synarthroses are sutures, synchondroses, and symphyses. Most of the joints in the mammalian skull roof are **sutures.** Here two membrane bones are separated but united by a **sutural ligament** of connective tissue (Fig. 5-14*A*). The connective tissue periosteum,

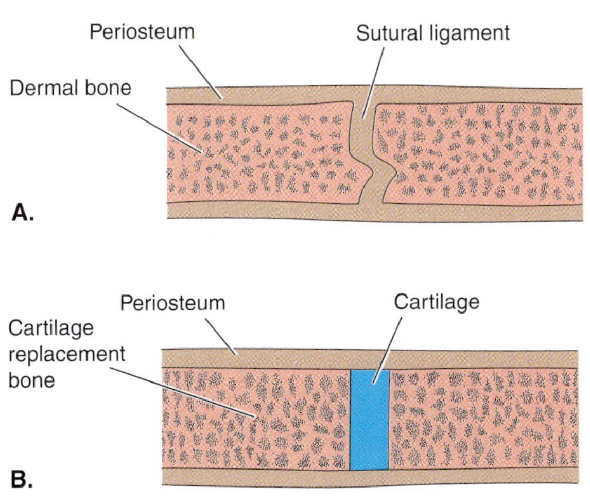

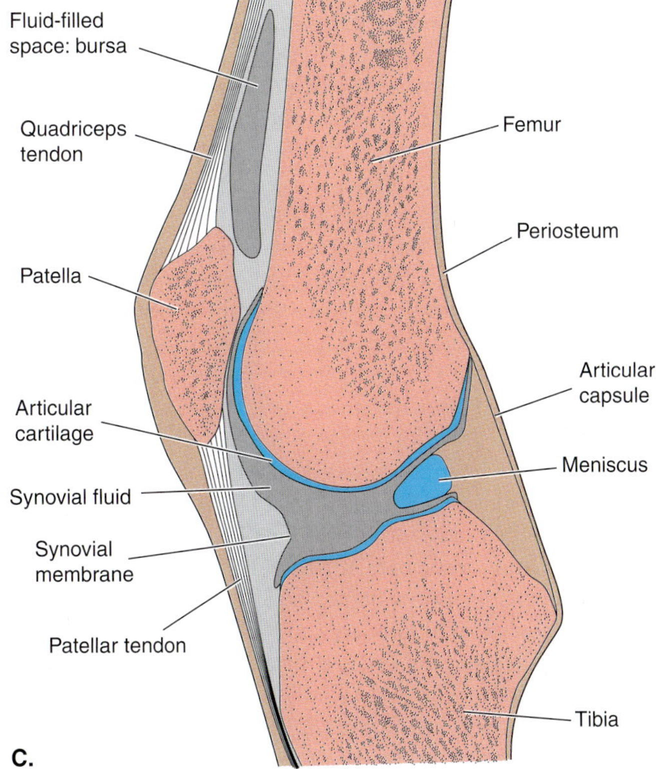

FIGURE 5-14
Diagrams of representative joints. *A,* Two bones are united by a thin layer of connective tissue (sutural ligament) in a suture. *B,* Two bones are connected by a thin plate of cartilage in a synchondrosis. *C,* The synovial knee joint of a mammal. Blue = cartilage; red = bone. *(After Williams et al.)*

which covers the bones, continues across the joint and helps bind the bones together. The sutural ligament is a remnant of the sheet of connective tissue in which the bones developed and in old age may become ossified. As we shall see (Chapter 7), many bones in the skull floor ossify in a continuous plate of cartilage. As the bones enlarge, they grow closer together but remain separated by a small plate of cartilage. These joints are examples of **synchondroses** (Fig. 5-14*B*). Synchondroses also occur during development of the long bones of the appendages, where the epiphysis and diaphysis are separated by an epiphyseal plate of cartilage. Examples of **symphyses** are the mental symphysis at the chin between the two halves of the lower jaw, the intervertebral symphyses between the centra of the vertebral column, and the pubic symphysis between the two halves of the pelvic girdle. All lie in the median plane of the body. Usually a deformable disk of fibrocartilage separates the bones and permits some motion. In some cases, as in the mental symphysis of a snake, considerable movement is allowed. Ligaments, in addition to the periosteum, cross the joint (see Focus 16-1).

Many other joints allow for considerable movement, and these are known as **diarthroses** (Gr., *di* = two + *arthrosis* = articulation), or **synovial joints.** Synovial joints often are classified by the type of motion they allow, such as a hinge action, a rotary action, a pivot action, a sliding action (called **translation**), or some combination of these. Often the shapes of the bones at a synovial joint are reciprocal, making for a very good fit, which increases joint strength but limits the amount of movement, called **degrees of freedom,** allowed. Bones at many other joints do not have such a good fit and permit more degrees of freedom. In a cat's jaw joint, for example, a cylindrical condyle on the lower jaw fits into a deep groove in the skull. Only two degrees of freedom are present; a hinge action and a sliding from side to side. Sliding the jaw from side to side brings the cutting teeth in the upper and lower jaws close together, which a cat needs in its carnivorous mode of life. The condyle of a rabbit's lower jaw is less cylindrical and fits into a flatter surface on the skull. Three degrees of freedom are present: a hinge action, sliding from side to side, and movement forward and backward. All these actions are used by an herbivore in grinding food. We will compare feeding in carnivores and herbivores further when we consider feeding mechanisms in Chapter 16.

Bones at a synovial joint are capped by smooth **articular cartilages** and are encased by a strong, fibrous **articular capsule** (Fig. 5-14*C*). A synovial membrane lines the enclosed articular cavity, but the membrane does not cover the cartilages. A small amount of a viscous **synovial fluid,** which is secreted by the synovial membrane, acts as a lubricant and helps nourish the cells in the articular cartilages. The coefficient of friction on the articular surfaces is very low, about comparable to that of ice on ice. Thickened bands of fibers within the articular capsule sometimes form **intrinsic ligaments.** Other **extrinsic ligaments** usually cross the joint external to the articular capsule, strengthening it further and sometimes restricting movement in certain directions. Muscles that span the joint also reinforce it.

Some synovial joints contain a complete **disk** of fibrocartilage, or a partial disk known as a **meniscus** (Fig. 5-14*C*). The disk or meniscus often improves the fit between the bones and may act as a shock absorber. Joints containing a disk or meniscus often have a translational movement between the bones in addition to hinge, rotary, or other motions.

Sometimes cartilages and bones are articulated in such a way that the movement of one requires the movement of the others. Such arrangements are known as **kinematic chains** (Gr., *kinema* = movement). A simple four-bar kinematic chain is shown in Figure 5-15*A*. If any one bar is held in position, a movement of any one of the others will cause the re-

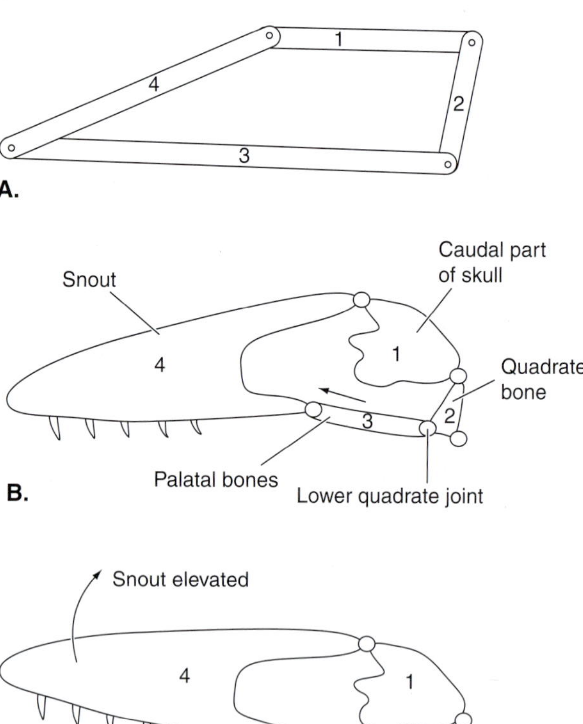

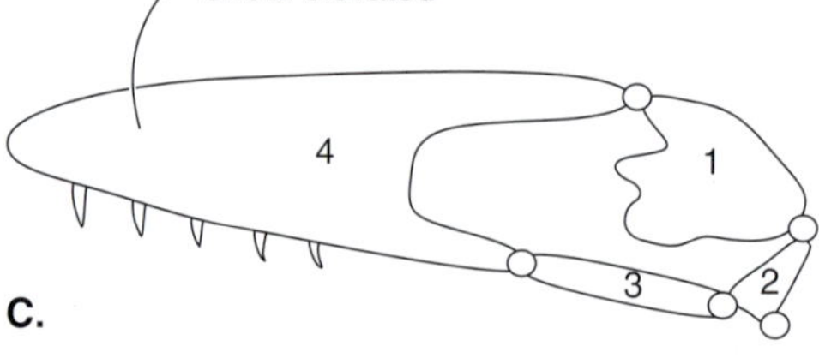

FIGURE 5-15
Kinematic chains in which the movement of one element requires the movement of others. *A*, A four-bar kinematic chain. *B* and *C*, The kinematic skull of a lizard in which the forward movement of the palatal bones in *B* causes an elevation of the snout in *C*.

maining two to move. Similar kinematic chains are found in the skulls of many fishes, lizards, snakes, and birds (Fig. 5-15*B* and *C*). The posterior part of the skull (1) is stable, but during feeding, muscles pull the palatal bones (3) upward and forward. This causes the quadrate bone (2), which bears the jaw joint, to move forward and the tip of the snout (4) to be elevated. (The lower jaw is lowered at the same time, but this is an independent action.) Many of the joints in this kinematic chain are synovial, but the joint on the skull roof is a suture, a synarthrosis. The connective tissue between the bones of the skull roof allows a hingelike movement of the snout. In some small skulls, the bones themselves are thin and flexible enough to allow for some bending. Skulls in which parts move relative to one another are described as **kinetic skulls** as opposed to the **akinetic skull** of a mammal, in which all parts are firmly united with each other. We will return to cranial kinesis during our discussion of feeding mechanisms in Chapter 16.

Lever Systems

Machines are devices that transmit force from one place to another, usually changing the magnitude and direction of the force in the process. Simple and basic machines include the inclined plane, lever, wedge, pulley, screw, and wheel. Levers are very common in a body composed of rigid skeletal parts connected to each other at joints. Many skeletal elements form rigid bars or **levers** that rotate or pivot about a fixed point, known as the **fulcrum.** A muscular force, called the **in-force** (F_i), is applied to one part of the lever, and the lever transmits this force to the part where the force is applied (e.g., another skeletal element, the jaws, or the feet). The applied force is called the **out-force** (F_o). The relative magnitudes of the in-force and out-force, and the distances and speeds with which the ends of the levers move, are determined by the points at which the forces are applied relative to the fulcrum.

A seesaw (Fig. 5-16*A*) is a familiar example of a lever system. A force applied to the end of the plank on which one sits (the in-force) will raise a load (the out-force) at the other end. We learned as children that if the plank were placed eccentrically on the fulcrum, then a small child at the long end of the plank could raise a heavy person at the other. This is because the turning forces, called **moments** or **torques,** are the products of the forces multiplied by the lengths of their lever arms (L_i and L_o). In this case,

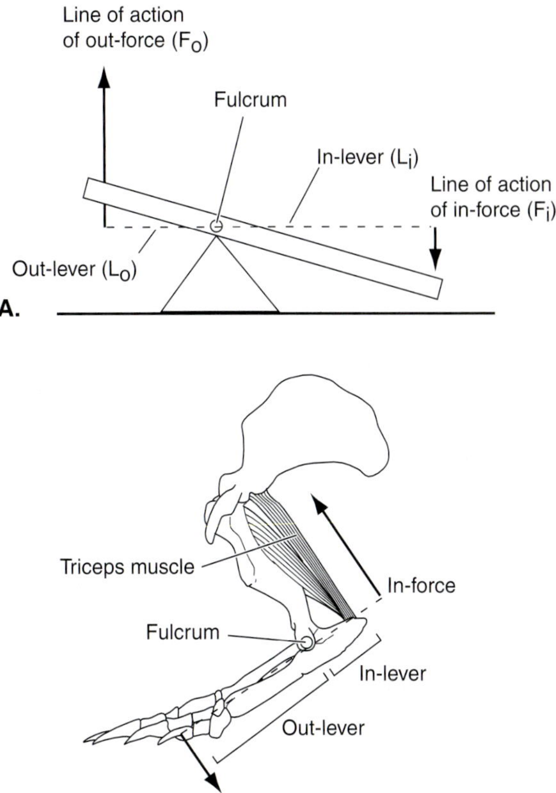

FIGURE 5-16
First-order lever systems. *A,* A seesaw. *B,* The extension of the forearm and hand in an armadillo. A short lever arm requires a stronger force (*long arrows*) to balance a long lever arm and weaker force (*short arrows*).

the in-torque has a long lever arm, so the in-torque can counter a heavy load at the out-torque end. When the seesaw is in balance, the in- and out-torques are in equilibrium:

$$F_i \times L_i = F_o \times L_o$$

It is important to realize in making the calculations that the effective lengths of the lever arms are the perpendicular distances from the line of action of the forces to the fulcrum (Fig. 5-16*A*). Because the forces (in this case, the pull of gravity) act vertically, the lengths of the lever arms would correspond to the length of the plank from the point of application of the forces to the fulcrum only when the plank is horizontal. Engineers call lever systems of the type we have described, in which the in-force is applied on one side of the fulcrum and the out-force is delivered on the other side of the fulcrum, **first-order levers.**

The extension of the forearm at the elbow is an anatomical example of a seesaw or first-order lever

(Fig. 5-16*B*); however, in the body, the in-lever arm is the shorter. The elbow joint is the fulcrum, the in-force is delivered by the triceps muscle to the proximal end of the ulna, and the in-lever arm is the perpendicular distance from the line of action of this force to the elbow joint. The out-force pushes down and backward on the ground, and its lever arm is the perpendicular distance from its line of action to the elbow joint.

In **second-order levers,** both forces are applied on the same side of the fulcrum, but the in-force is applied closer to the fulcrum than is the out-force. An example is the retraction of the entire forelimb at the shoulder joint (Fig. 5-17*A*). The fulcrum is the shoulder joint, the in-force is applied by retractor muscles that attach on the humerus, and the out-force is delivered by the distal end of the limb to the ground.

In **third-order levers,** both forces also lie on the same side of the fulcrum. The in-force is applied to the end of the lever farthest from the fulcrum, and the out-force is delivered between the in-force and the fulcrum. A wheelbarrow is a familiar example (Fig. 5-17*B*). An anatomical example of a third-order lever is the extension of the foot by the gastrocnemius and soleus muscles when one raises the body on the toes (Fig. 5-17*C*).

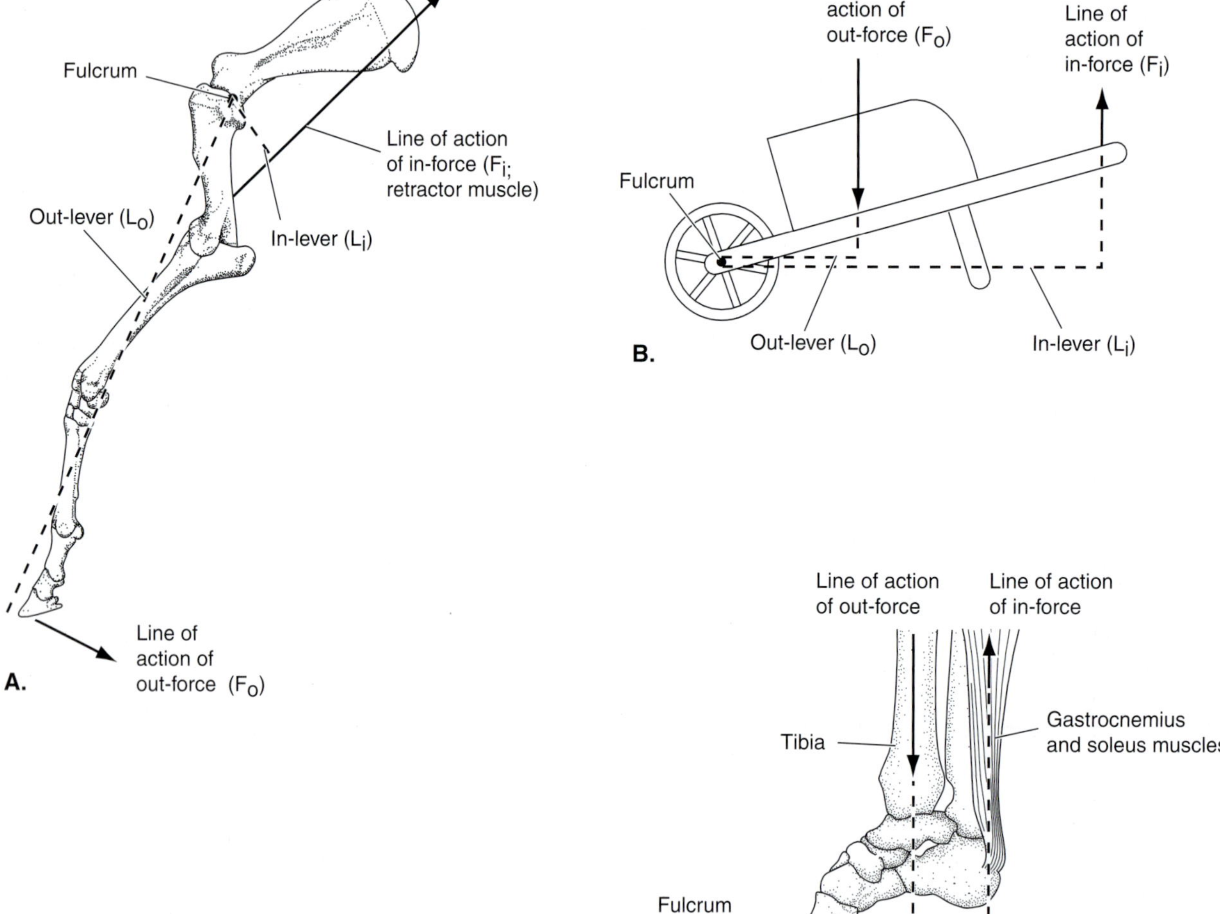

FIGURE 5-17
More lever systems. *A,* Retraction of the front leg of a horse at the shoulder joint, a second-order lever. *B,* A wheelbarrow, a third-order lever. *C,* Action of the gastrocnemius and soleus muscles in raising a human body upon the toes, a third-order lever.

In the third-order lever illustrated in Figure 5-16*C*, the contraction of the gastrocnemius and soleus muscles can lift the entire body weight on the toes because the in-lever is longer than the out-lever. This is a mechanically efficient system. Both first- and second-order levers, which are more common than third-order levers in musculoskeletal systems, are mechanically inefficient. Because the in-lever is shorter than the out-lever, the force delivered by the muscles to the in-lever must be greater than the force delivered by the out-lever. First- and second-order levers, however, place the muscles close to the fulcrum; the limbs are not as bulky as they might be. Also, a short contraction of the muscles can cause an extensive and rapid movement at the out-end of the levers. Power and speed of lever systems are inversely related. The designs of limbs reflect modifications for power or speed by slight shifts in the relative lengths of in-lever and out-lever arms. We will explore this further when we consider types of locomotion (Chapter 11).

Distribution of Materials

The properties of the structural materials, and the mechanical principles we have discussed, affect the distribution of these materials in the body. When an engineer designs a bridge, he or she carefully analyzes the nature and magnitude of the stresses that must be resisted and their distribution through the structure. Suitable materials are placed along the lines, or **trajectories,** of those stresses. It would be dangerous to use a material designed to meet tension, such as a cable, when compression must be resisted, and it would be uneconomical and inefficient to place materials where no stress trajectories are present. So, too, biological materials are not organized in a random way. Bone is present where compression, or a combination of compression, tension, and shear, must be met. Ligaments occur where only tension must be resisted. Moreover, stress lines in an animal change during growth and must meet the needs of a moving animal. These needs are far more complex and changeable than in a static bridge.

The size and shape of bones, and the distribution of materials within them, are determined by genetic factors and by the combination of forces that act on them. The basic shape of each bone is determined genetically and will develop even when a limb is paralyzed and not subject to any stress. But the distribution of materials within a bone, its cross-sectional shape and thickness, the size and shape of the processes to which muscles attach, and the size and shape of its articular surfaces require stressing to develop normally. The rodlike trabeculae within a developing bone have a random architecture at first. As stresses are applied, bone is remodeled, and the trabeculae form tracts that follow the stress lines. The tessarae in calcified cartilage (p. 187) sometimes also are arrayed in struts or trabeculae that follow stress lines (Summers et al., 1998). This is the case in the cartilaginous jaws of stingrays, which bear tooth plates adapted for crushing mollusks and other invertebrates with hard shells.

Many bones are beams that resist compression or a combination of compression and tension. Their length and cross-sectional size and shape are determined by the physics of beams. Some bones are subject only to compression stresses that run parallel to their long axes. The stresses are evenly distributed and are met by trabeculae that parallel the long axis of the bone. This is the case for the centrum of a human vertebra (Fig. 5-18). Because the capacity of a beam or bone to resist compression is inversely proportional to its length, a bone resisting only compression must be short and wide.

As a beam or bone becomes longer, a slight shift of the compressive force away from its longitudinal axis will cause bending. A meter stick that is in a vertical position, for example, bends if a load on top of it is not perfectly centered. Often bones are eccentrically loaded and tend to bend. This is the case in the mammalian femur in which body weight falls on the medi-

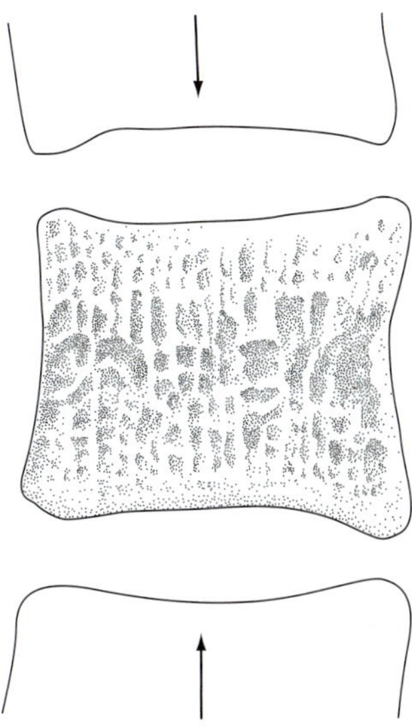

FIGURE 5-18
A vertebral centrum of a human being standing erect is subject to compression (*arrows*). As can be seen in this longitudinal section, the bony trabeculae tend to align with the compression forces.

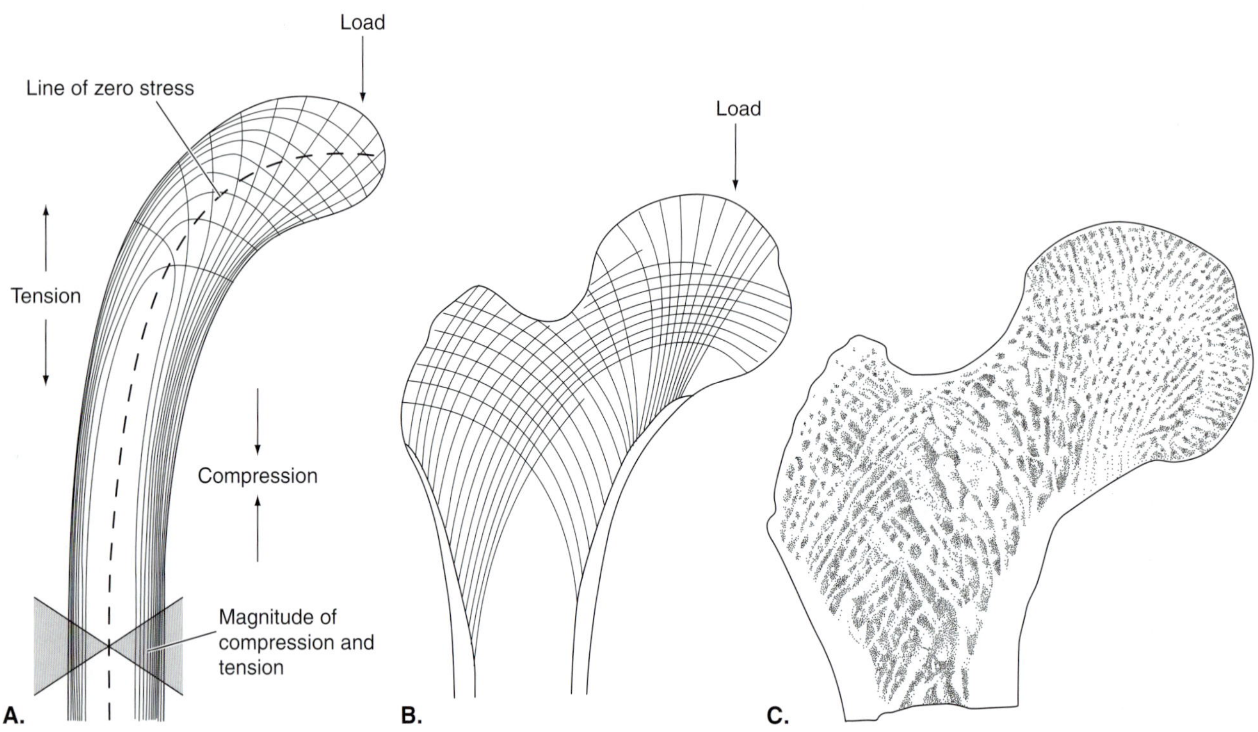

FIGURE 5-19
Stress trajectories in a simple Fairbairn crane (*A*) and the proximal end of a human femur (*B*) in response to a load applied to the medial side of the crane and to the head of the femur. Material on the load-bearing side is under compression; that on the opposite side is under tension. The magnitudes of both stresses decrease to zero at the neutral plane in the center. *C,* A drawing of an x-ray of the proximal end of the human femur showing that the orientation of the trabeculae approximates the expectation. *(A, after Murray.)*

ally projecting head of the femur. The femur is analogous to a simple Fairbairn crane (Fig. 5-19). Eccentric loading leads to a complex distribution of forces. Material on the concave side is under compression, whereas that on the convex side is under tension. The stresses are greatest at the surface and fall to zero in the middle of the crane or bone. The middle plane, where the stresses are zero, is described as the neutral plane. For this reason, a long limb bone can be a hollow cylinder with layers of compact **cortical bone** around the periphery or cortex and a marrow cavity in its center. Limb bones are cylindrical because forces may come from many directions, depending on the position of the limb.

The capacity of a beam or a limb bone to resist bending depends on its **flexural stiffness.** Flexural stiffness is determined by the intrinsic characteristics of the material (in this case, bone) and a quantity called the **second moment of inertia.** The second moment of inertia represents the contribution of each particle of material in each position of the beam or bone to the resistance to bending. Figure 5-19*A* clearly shows that the farther away the material is from the neutral plane, the more it contributes to the resistance to bending. Indeed, the effect of this distance (called y) is squared. If the beam or bone is a cylinder, as the femur is, an increase in radius greatly increases stiffness because the cortical bone is a greater distance from the neutral plane. Fibers and wires are very flexible because their radius is so small. The femur is much stiffer because it has a much larger radius.

At the top of the crane, or in the neck and head of the femur, compression and tension lines cross each other nearly at right angles. The stresses are met in the femur by crossing bone trabeculae. The situation is more complex in the femur than in the simple crane because of the attachment and pull of muscles on bone processes.

Muscle forces acting on a bone sometimes reduce the stresses in the bone that are generated by supporting body weight. As explained earlier, the lateral side of an eccentrically loaded femur is under tension, and the medial side is under compression. The contraction of certain muscles on the lateral surface of the thigh that extend from the hip to the distal end of the femur compress the lateral side of the femur, which reduces

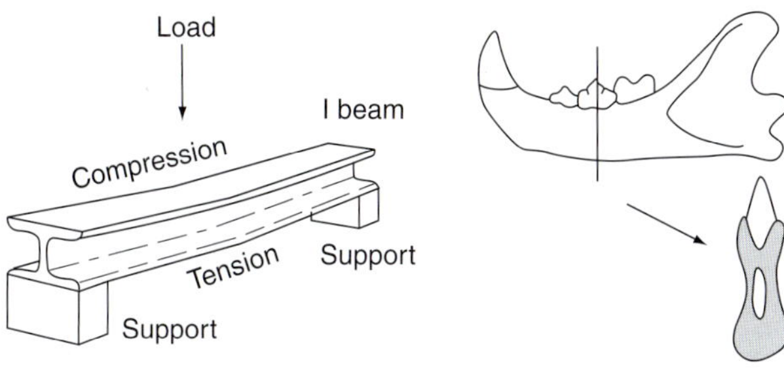

FIGURE 5-20
A, An I-beam supported at each end and loaded in the center tends to bend in the middle, so the top surface is under compression and the bottom surface is under tension. *B,* A biological example of an I-beam–like structure is the mandible of a mammal.

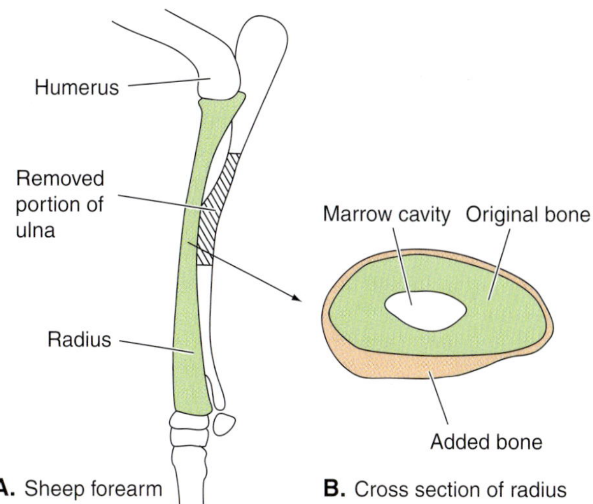

FIGURE 5-21
A, The forearm of a sheep showing the portion of the ulna that was experimentally removed. *B,* A cross section of the radius adjacent to the excised ulna after one year. *(After Lanyon et al.)*

the tension that supporting body weight generates on this side. The contraction of muscles on the lateral side of the thigh also subjects the medial side to tension and reduces the compression normally there.

If a bone need resist bending in only one direction, then the engineer's I-beam or the carpenter's joist illustrates a more economical use of materials than does a cylinder. A load on the middle of an I-beam, for example, causes it to bend slightly. The top comes under compression stress, and the bottom comes under tension (Fig. 5-20*A*). A neutral plane runs halfway between the top and bottom surfaces. The I-beam concentrates materials on the top and bottom, where the tension and compression stresses are the greatest. Just as a larger radius of a cylinder increases resistance to bending by increasing the distance (y) of the material from the neutral plane, so too separating the top and bottom surfaces of the I-beam by a vertical section is important. The formula for resisting bending in this situation is:

Resistance to bending = [(a constant determined by the nature of the material) × (the width of the beam)] × (the height of the vertical segment)^2

Biological examples of these principles can be seen in sections through many bones, such as the ilium of the pelvic girdle and the lower jaw. The height of the lower jaw, for example, is well adapted to meet the strong vertical forces generated in biting or chewing.

The pattern of bone deposition clearly shows that bone develops where stresses must be met. This fact also has been demonstrated experimentally. In one study, Lanyon et al. (1982) removed a portion of the ulna shaft from a sheep's limb (Fig. 5-21), thus increasing the load-bearing stresses on the radius. The animal was walked for 1 hour each day. After a year, the radius had increased in thickness over control animals by the equivalent of the amount of bone removed from the ulna. Strenuous exercise also can affect remodeling of mature bone. Loitz and Zernicke (1992) ran roosters on a treadmill for 1 hour a day, five days a week, for nine weeks at 70% to 75% of maximum aerobic capacity. At the end of the experiment, samples were taken from the proximal foot bone (the tarsometatarsus). The thickness of the wall of the bone in the experimental animals had increased significantly over that in the controls.

Much research is being done to understand the factors that control the remodeling of bones to meet the stresses upon them. As we have mentioned, an interchange of calcium and phosphate ions between the body fluids and bone continually occurs. Estimates in human beings are that nearly one quarter of the blood calcium interchanges with bone calcium every minute, but blood levels of calcium remain remarkably constant (Brookes, 1987). Interchanges are regulated by many hormones (including growth hormone, parathyroid hormone, and calcitonin), by certain vitamins (e.g., vitamin D); and by several factors derived from bone marrow cells. But it is difficult to see how this generalized chemical environment can target bone reabsorption and deposition in a way that leads to remodeling to meet stresses. Intermittent and variable stress appears to be necessary for bone deposition. Bone is reabsorbed when gravitational stresses are reduced for a prolonged time, as in bed-ridden patients and astronauts subject to zero gravity. A constant pressure also promotes bone reabsorption. For example, orthodontists move displaced teeth through the jawbone of young patients by using braces that apply a constant pressure.

The selective reabsorption and deposition of bone in response to changing stresses requires that a strain generates some signal that can selectively activate or inhibit osteoblasts and osteoclasts. The signal is not known. Possibly, a stressed bone generates an electric potential, perhaps by the piezoelectric effect of either the mineral or the collagen component of bone, but the origin of the electric potential is not certain. Some investigators are doubtful that this is the signal.

Scaling: Isometric and Allometric Growth

The anatomy and function of organs depend partly on body size. Because vertebrates vary greatly in size, biologists use **scaling analyses** to study the structural and functional consequences of changes in size among otherwise similar organisms. Scaling analyses can be illustrated by examining the effects of doubling the size of a cube (Fig. 5-22). If the linear dimension of the cube (i.e., the length of each side) is doubled, then the small and large cubes are said to be geometrically similar, or **isometrically similar** (Gr., *isos* = equal + *metron* = measure). The surface area and volume of a cube, however, do not change in proportion to its linear dimensions. If a linear dimension of a cube is doubled, then its surface area, which is a two-dimensional quantity, increases by the square of the linear increase, or four times ($2^2 = 4$), and its volume, which is a three-dimensional quantity, increases by the cube of the linear increase, or eight times ($2^3 = 8$). If the linear dimensions of the cube were quadrupled, then the surface area would increase 16 times (4^2), and the volume, 64 times (4^3). Stated in other terms:

$$\text{Surface area} = (\text{height})^2$$

$$\text{Volume} = (\text{height})^3\text{, or height} = (\text{volume})^{1/3}$$

$$\text{Surface area} = (\text{volume})^{2/3}$$

The last equation, which derives from the first two,[1] states that as geometrically similar objects increase in size, their surface area increases more slowly than does their volume (at only 2/3 the power of the volume). When surface area is plotted against volume on an arithmetic plot, the slope of the curve decreases as volume increases (Fig. 5-23*A*). When the same data are plotted on logarithmic coordinates, the curve is a straight line with a slope of 2/3, or 0.67 (Fig. 5-23*B*).

[1]By substituting height = $(\text{volume})^{1/3}$ into the first equation. Surface area = $[(\text{volume})^{1/3}]^2 = (\text{volume})^{2/3}$.

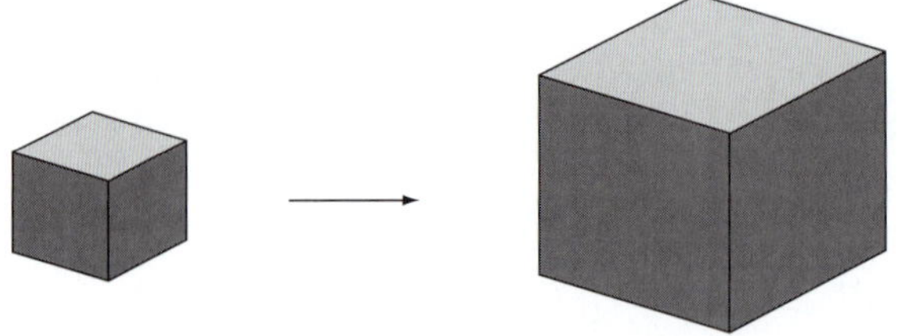

A. Effect of doubling a cube's linear dimensions

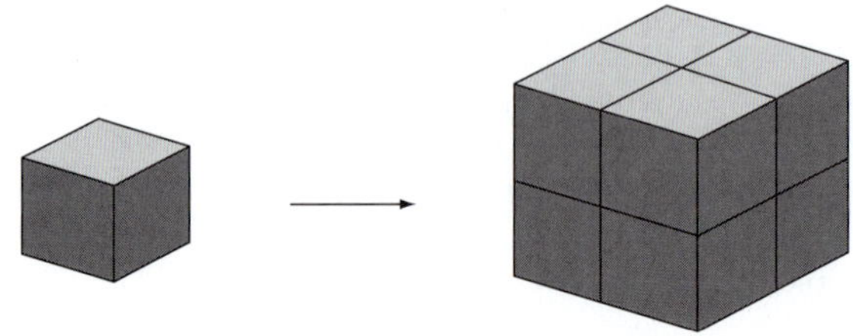

B. Area relationships increase 4 times

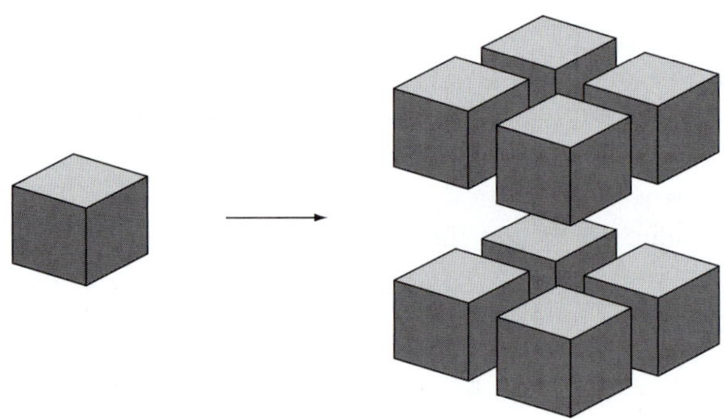

C. Volume relationships increase 8 times

FIGURE 5-22
A–C, Changes in the surface area and volume of a cube when its linear dimensions are doubled.

The same rules apply to animals that increase (or decrease) in size during evolutionary history or their own ontogeny. In the cube we discussed, each dimension was doubled (that is, changed in the same proportion, or isometrically). However, the linear dimensions of comparable parts of animals usually do not change in the same proportion. Instead, they tend to change in different proportions, or **allometrically** (Gr., *allos* = other + *metron* = measure). The range in size of adult vertebrates is enormous, going from some very small fishes that weigh 1 g to a blue whale, which can weigh 100 tons (100 million g, or 10^8 g). This is a range of eight orders of magnitude! Mass increases as the cube of the increase in linear dimensions, but the strength of

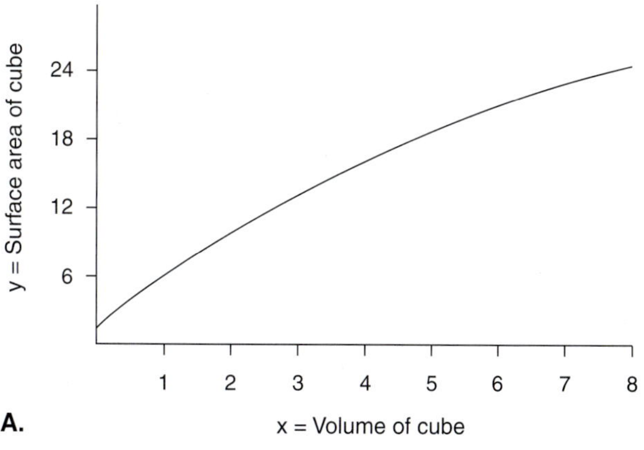

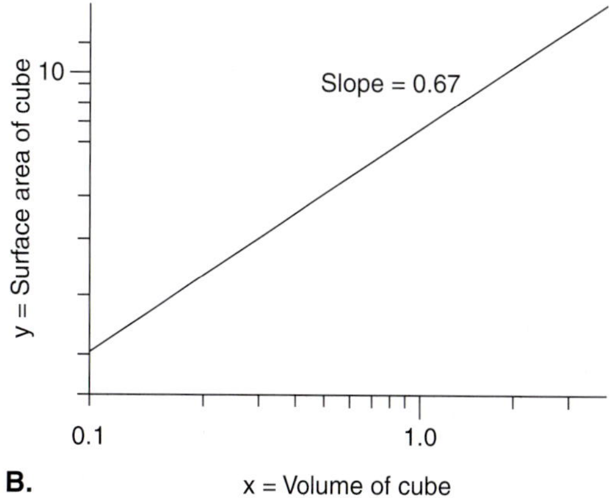

FIGURE 5-23
Graphs showing the relationship of surface area to volume as cube increases in size. *A*, Arithmetic coordinates. *B*, Logarithmic coordinates. *(After Schmidt-Nielsen.)*

supporting materials (bones) and the muscles to move them increases in proportion to their cross-sectional area, a square function. If isometry were maintained during change in size, then large animals could not support or move themselves. Moreover, as animals become larger, the surface area across which food is absorbed, gases are exchanged, or waste products are eliminated would become inadequate. Although surface area increases in absolute terms as animals become larger, the surface area decreases relative to the mass requiring these processes.

Allometric scaling is the analysis of the change in one body part (e.g., limb diameter), or rate of activity (e.g., metabolism), relative to another (usually body mass as measured by weight). The general formula of the allometric equation is:

$$y = a \times x^b$$

where y = the dependent variable (size or activity of some part); a = the proportionality coefficient, which is the intercept of the regression line on the y-axis when x = 0; x = the independent variable (usually body mass); and b = the body mass exponent, or slope of the regression line.

The exponent b (the slope) is not always 2/3 (which describes the relation of surface area to volume) but has different values for different structures and functions. A slope of 1 indicates that the size of the part or function in question increases at the same rate as the body mass, so no proportional change occurs (Fig. 5-24*A*, *dashed line*). An example is the relative

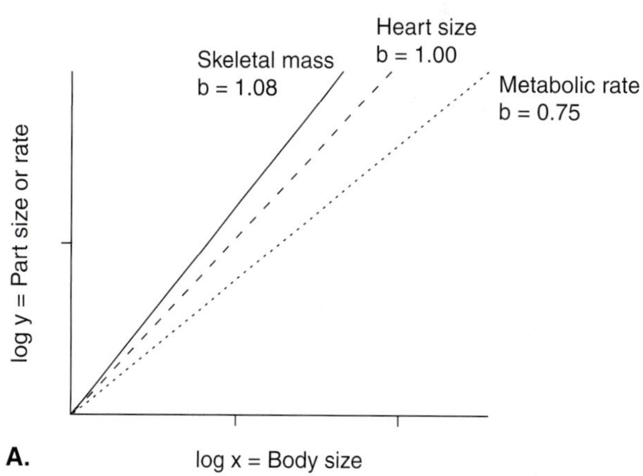

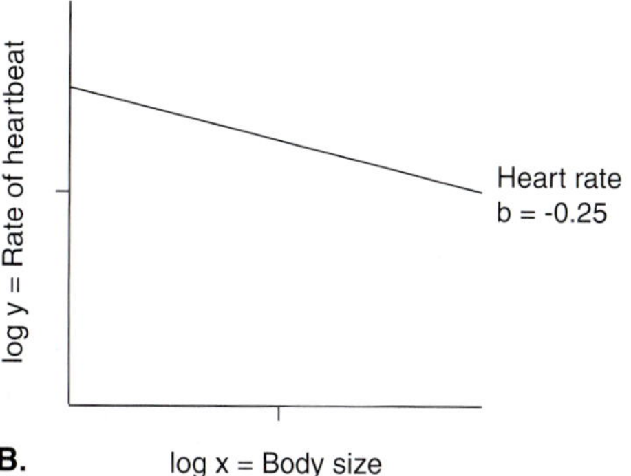

FIGURE 5-24
Graphs, drawn on logarithmic coordinates, of the relationship of the size of an organ, or rate of activity, to body size. *A*, Skeletal mass (*solid line*), heart size (*dashed line*), and metabolic rate (*dotted line*). *B*, Heart rate (b = the slope of the regression line). *(After Schmidt-Nielsen.)*

size of the heart in mammals. Larger mammals have larger hearts than do small animals, but in all cases, the heart is about 0.6% of body mass. A slope greater than 1 indicates that the part in question increases at a faster rate than does body mass, so its relative size becomes greater as the animal increases in size (Fig. 5-24*A*, *solid line*). An example is skeletal mass, which is relatively greater in large animals than in small ones. A slope less than 1 means that the part in question increases at a slower rate than does body mass as the animal increases in size, so it becomes relatively smaller (Fig. 5-24*A*, *dotted line*). This is the relationship of surface area relative to body mass. A negative slope means that the part or function in question decreases with increasing body mass (Fig 5-24*B*), which is the case for heart rate in mammals. An elephant's heart rate is only about 1/8 of that of a mouse, but, of course, the elephant's heart is much larger, so it pumps a larger volume of blood per stroke.

Allometric ratios describe such structural and functional correlations, but the challenge is to find the underlying reasons for these correlations. Often the reason for a relative increase in the size of a structure is found in the changing surface-to-volume ratio. As bridges, buildings, or animals become larger, more weight must be supported or moved, but weight is a function of volume and consequently increases as the cube of the increase in linear dimension. The strength of supporting materials (e.g., bones) and the strength of muscles, however, increase in proportion to their cross-sectional area, a square function. If all body parts grew isometrically, then the bones and muscles of large animals would not be strong enough to support and move the animal. Similarly, as animals become larger, the surface area across which food is absorbed, gases are exchanged, or wastes are eliminated decreases relative to the mass requiring these processes. To overcome problems of this sort, three parameters can change:

1. The relative dimensions of certain structures can change. For instance, large stone buildings have relatively thicker walls than do small ones, and the supporting bones of large animals have a relatively greater diameter than do those of small ones.
2. The nature of the structural materials may change. In a building, steel may replace stone; in an animal, bone may replace cartilage, or the bones may become denser.
3. A fundamental design change may be needed to accomplish the task. For instance, to cross a wide river, a suspension bridge with a more complex structure is used instead of a simple stone arch. The ancestors of craniates, which were relatively small animals, exchanged gases across their body surfaces by diffusion. Diffusion across the body surface is not adequate for most craniates because the surface area of a craniate's body is smaller relative to body mass in these larger animals. Thus, aquatic craniates have gills (a design innovation that greatly increases the surface area for gas exchange) and methods for pumping water across the external surfaces of the gills and blood through internal vessels of the gills. These methods are based on complex musculoskeletal systems and more complex hearts and circulatory systems than were present in the ancestors of craniates.

Comparing the skeletons of a cat and an elephant (Fig. 5-25) illustrates two of these ways around the

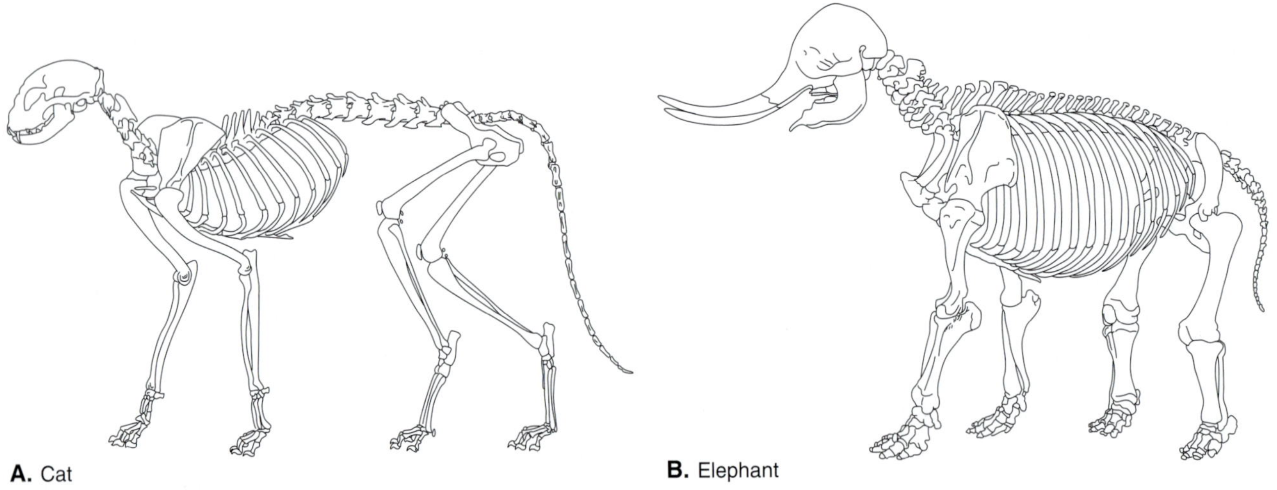

FIGURE 5-25
Skeletons of a cat (*A*) and an elephant (*B*) drawn to about the same size to facilitate comparison of relative limb bone diameters and proportions of the limb segments.

FOCUS 5-2 *Cartesian Transformations*

D'Arcy Thompson pointed out many years ago (originally in 1917) that changes in the growth rate of one body part relative to others can help describe morphological transformations. One can overlay, for example, the drawing of a skull of an Eocene rhinoceros (†*Hyrachyus*) with an equally spaced grid of Cartesian coordinates. One can then draw the skull of a latter species to the same size (†*Aceratherium* from the Miocene–Pliocene), identify the original grid intersection points on the new drawing, and connect them by lines (left and right). It becomes evident that during the evolution of rhinoceroses, some parts of the skull grew more rapidly than did other parts. This created a curvature of the skull and a vertical expansion of the caudal parts of the skull. Cartesian transformations help us visualize what has happened. Genetic factors that underlie changes in growth rates are now being studied.

Cartesian transformation of a late species of rhinoceros (†*Aceratherium*) from an earlier one (†*Hyrachyus*). (After D'Arcy Thompson.)

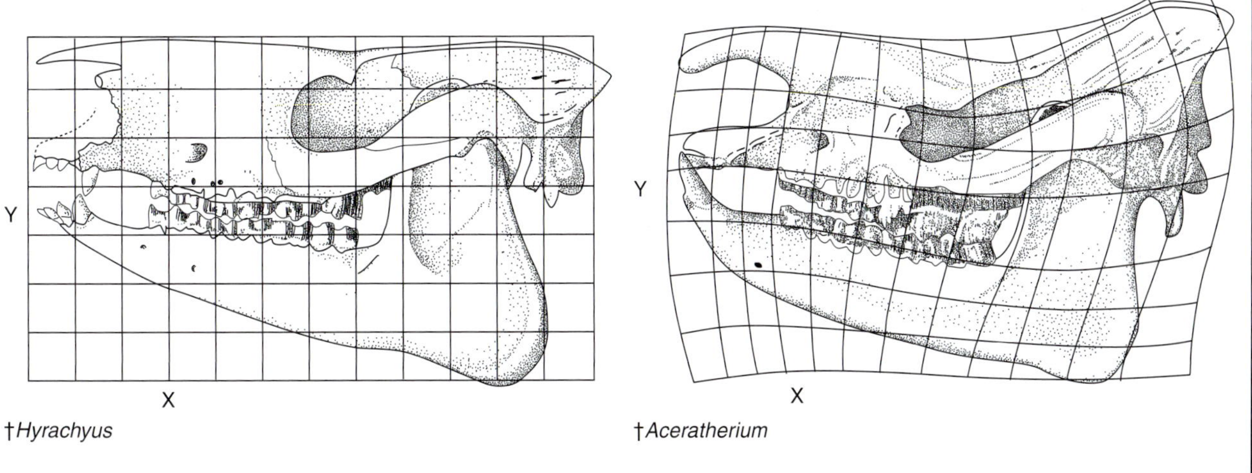

constraints of isometry. A cat cannot retain its slim limbs and grow to be the size of an elephant. Among other things, the bones must grow at a faster rate than mass as the animal increases in size, that is, allometrically. The diameter of the limb bones is relatively greater in an elephant, and the limb bones are relatively larger and more massive. Design changes also occur. The distal parts of the limbs, and especially the feet, are relatively shorter in an elephant than in a cat. The limb bones of elephants also assume a different orientation and form nearly vertical, supporting pillars under the body.

Although changes in surface–volume relationships help explain many phenomena, they do not explain others. For example, it is well known that small mammals have higher metabolic rates than large ones. The allometric formula for metabolic rate is:

$$\text{Metabolic rate} = 70 \times M^{0.75}$$

In this equation, M = body mass. Because (body mass)$^{0.75}$ approaches that of surface area, which scales to (body mass)$^{0.67}$, biologists first proposed that the higher metabolic rate of small mammals compensates for the heat lost through their relatively larger surface area. However, the ratio of metabolic rate to body mass in mammals is significantly higher than would be expected if heat loss through the surface were the explanation. Many alternative explanations have been explored. A model developed by McMahon (1973) is currently favored by many investigators. McMahon showed that the maximum power output (which is work per unit of time) of a muscle is proportional to its cross-sectional area and scales to body mass as follows:

$$P_{\text{max}} = M^{0.75}$$

If this formula applies to all of the metabolic variables involved in supplying muscles with energy and oxygen, as is likely, then metabolism would also scale to body mass to the power of 0.75. Thus, the power output of muscles relative to body mass appears to explain the allometry of metabolic rates better than do than surface–volume relationships.

Because the metabolic rate of mammals (*b* in the allometric equation) decreases as body mass increases, the relative need for oxygen decreases. Relative heart size remains the same (about 0.6% of body mass), so the relatively decreased oxygen need is met by a decrease in heart rate. Thus, heart rate scales as $(\text{body mass})^{-0.25}$.[2]

Scaling analyses help us understand many quantitative evolutionary changes in form and function, and we will demonstrate other examples as we consider the functional and structural evolution of vertebrates. However, not all factors are explained as the result of simple mathematical or geometrical relationships. The force of gravity on an animal, the laws of thermodynamics, and similar physical and chemical factors do not change with body size. Similarly, the properties of many biological materials are independent of body size. For example, because of the architecture and contractile properties of muscle as a tissue (Chapter 10), the force of muscle contraction per unit of cross-sectional area is the same for a mouse and an elephant.

We have been considering how a part of the body or some function changes with a change in body size. During an ontogenetic or phylogenetic series, changes also may occur in the proportions of the body as a whole or within some part of the body, such as the skull, whether or not the body or body part changes in size. Differential growth rates along the x- and y-axes can cause profound changes in proportions (Focus 5-2). Scaling changes and changes in relative growth rates focus our attention on the importance of allometry, but keep in mind that changes in rates are descriptive of the observed phenomena and do not explain underlying causes for the change.

SUMMARY

1. We define the physical quantities that will be used in this book and focus on force. Forces are central to all biomechanical functional analyses. They can be considered to act through the center of mass of a body and can be portrayed as vectors.
2. Forces exert stresses upon a body including compression, tension, shear, and torsion. Stresses cause a deformation or strain upon structures. The energy expended when a structure is strained is stored within the structure as elastic energy, and, if the yield point of the material has not been exceeded, most of this energy is recovered when the stress is released.
3. We examine the properties and growth of structural materials in the context of the stresses that they resist. Hydrostats of vertebrates are internally pressurized cylinders composed of liquid-filled cells and wrapped in a sheath of helical-arranged fibers. The notochord is an example of a hydrostat. It is well designed to bend but resists modest compression.
4. Connective tissues, in which extracellular material predominates over cellular components, often form supportive structures.
5. Loose fibrous connective tissue permeates all organs, binding other tissues together. More dense connective tissues form capsules around certain organs, ligaments, and tendons. These tissues are well suited to meet tension and torsion but not compression.
6. Cartilage and bone are dense connective tissues in which a firm matrix enables them also to resist compression.
7. Cartilage is firm but flexible, and its surface is very smooth. The properties of cartilage, and the ease with which it can grow interstitially by the division of chondrocytes, makes it an ideal supporting material for embryos and for adult structures, such as costal cartilages and the joint surfaces of limb bones, where flexibility, smoothness, or both are needed. Because cartilage is a nonvascular tissue, it cannot form large structures.
8. Bone is a highly vascular dense connective tissue in which crystals of calcium phosphate bind to the collagen fibers. Bone is an important reservoir for calcium and phosphate ions. The fibrous component of bone provides some flexibility and resistance to tension and to torsion, and its mineral component resists compression.
9. Bone has a complex histological structure, and it is remodeled throughout life as stresses change on the body and its parts. Osteoblasts produce new bone, and osteoclasts digest and remove old bone.
10. Two categories of bone are recognized on the basis of their embryonic development. Membrane, or dermal, bones develop directly within connective tissue. They form the superficial bony scales and plates of fishes, much of the skull, and parts of the pectoral girdle. Cartilage-replacement bones develop within and around deeper cartilaginous rudiments, which they largely replace.
11. The individual cartilages and bones of the body are connected to one another at joints. Movement is restricted at synarthroses and synchondroses, but considerable movement occurs in diarthroses or synovial joints. Ligaments and muscles may reinforce the joint.

[2]See West et al., 1997, for an analysis of scaling, with special reference to transport systems, such as the cardiovascular and respiratory systems.

12. In kinematic chains, cartilages and bones are connected in such a way that movement of one element causes the movement of the others.
13. Levers are a common type of machine in a body composed of rigid skeletal parts connected at joints. Levers transfer muscle forces from muscles to some point of application, such as the feet and jaws. An in-force applied by muscles at one part of a lever causes the lever to rotate about a fulcrum and exert an out-force at another part of the lever.
14. The relative magnitude of the in-force and out-force, and the distance and speed with which the parts of a lever move, are determined by the points at which the forces are applied relative to the fulcrum.
15. The properties of structural materials and the ways they are used affect the distribution of these materials in a bone. Structural materials are not randomly distributed but tend to lie along stress lines or trajectories.
16. Many bones, especially limb bones, such as the femur, are eccentrically loaded and tend to bend. Material on the concave side of the bone is under compression, whereas that on the convex side is under tension.
17. Stresses are greatest at the surface of the bone and drop to zero at the neutral plane near the center of the shaft of the bone. Material that is farthest from the neutral plane contributes the most to the resistance to bending. Bones such as the femur, therefore, are hollow cylinders with dense cortical bone on the surface and a marrow cavity in the center. Also, the greater the diameter of the bone, the greater its resistance to bending because material is farther from the neutral plane.
18. We conclude this chapter with an analysis of the effect on structures and functions of changes in body size. As an animal changes in size, its mass, which is a three-dimensional quantity, changes as the cube of the change in linear dimensions. Surface area, which is a two-dimensional quantity, changes as the square of the linear change. This has profound effects on animals because surface areas through which exchanges take place, or the strength of material (also a square function), do not keep pace with the increase in mass. Animals solve such problems by differential growth (allometry) of various parts, by changes in structural materials, or by design changes.
19. Allometric growth occurs both in evolutionary and ontogenetic sequences. The invagination of surfaces during development is one device that helps the surface area keep pace with the increase in mass.
20. Differential growth rates along the x- and y-axes in an ontogenetic or phylogenetic series also may occur whether or not a change in size occurs, which can have profound effects on the proportions and appearance of the body and its parts and doubtless has been a factor in evolutionary changes.

REFERENCES

Alexander, R. McNeil, 1983: *Animal Mechanics,* 2nd edition. Oxford, Blackwell Scientific Publications.

Brookes, M., 1987: Bone blood flow measurement: Part 2. *Bone Clinical Biochemistry News Review,* 4:33–36.

Calder, W. A. III, 1984: *Size, Function, and Life History.* Cambridge, Harvard University Press.

Corliss, C. E., 1976: *Patten's Human Embryology.* New York, McGraw-Hill.

Currey, J., 1984a: Comparative mechanical properties and histology of bone. *American Zoologist,* 24:5–12.

Currey, J., 1984b: *The Mechanical Adaptations of Bones.* Princeton, Princeton University Press.

Fawcett, D. D., 1994: *Bloom and Fawcett: A Textbook of Histology,* 12th edition. New York, Chapman & Hall.

Gans, C., 1974: *Biomechanics: An Approach to Vertebrate Biology,* 1980 Reprint. Ann Arbor, Mich., University of Michigan Press.

Halstead, L. B., 1974: *Vertebrate Hard Tissues.* London, Wykeham.

Huxley, J. S., 1972: *Problems of Relative Growth,* 2nd ed. New York, Dover.

Kier, W. M., and Smith, K. K., 1985: Tongues, tentacles and trunks: The biomechanics of movement in muscular hydrostats. *Zoological Journal of the Linnean Society of London,* 83:307–324.

Lanyon, L. E., Goodship, H. E., Pye, C., and McFie, H., 1982: Mechanically adaptive bone remodeling: A qualitative study on functional adaptation in the radius following ulna osteotomy in the sheep. *Journal of Biomechanics,* 15:141–154.

Lanyon, L. E., and Rubin, C. T., 1985: Functional adaptations in skeletal structures. *In* Hildebrand, M., Bramble, D. M., Liem, K. F., and Wake, D. B., editors: *Functional Vertebrate Morphology.* Cambridge, Harvard University Press.

Loitz, B. J., and Zernicke, R. F., 1992: Strenuous exercise-induced remodeling of mature bone: Relationships between *in vivo* strains and bone mechanics. *Journal of Experimental Biology,* 170:1–18.

Martin, R. B., Burr, D. B., and Sharkey, N. A., 1998: *Skeletal Tissue Mechanics.* New York, Springer-Verlag.

McLean, F. C., and Urist, M. R., 1968: *Bone.* Chicago, University of Chicago Press.

McMahon, T. A., 1984: *Muscles, Reflexes, and Locomotion.* Princeton, Princeton University Press.

Mostov, K., and Werb, Z., 1997: Journey across the osteoclast. *Science,* 276:219–220.

Murray, P. D. F., 1936: *Bones.* Cambridge, Cambridge University Press.

Prange, H. D., Anderson, J. F., and Rahn, H., 1979: Scaling of skeletal mass to body mass in birds and mammals. *American Naturalist,* 11:103–122.

Robson, P., et al., 1993: Characterization of lamprin, an unusual matrix protein from lamprey cartilage. *The Journal of Biochemistry,* 268:1440–1447.

Ruben, J. A., and Bennett, A. A., 1987: The evolution of bone. *Evolution,* 41:1187–1197.

Schmidt-Nielsen, K., 1984: *Scaling: Why Is Animal Size So Important?* Cambridge, Cambridge University Press.

Shapiro, F., 1992: Vertebral development in the chick embryo during days 3–9 of incubation. *Journal of Morphology,* 213:317–333.

Summers, A. P., Koob, T. J., and Brainerd, E. L., 1998: Stingray jaws strut their stuff. *Nature,* 395:450–451.

Thompson, D'Arcy, 1961: *On Growth and Form,* 2nd abridged edition, edited by Bonner, J. T. Cambridge, Cambridge University Press.

Vogel, S., 1988: *Life's Devices.* Princeton, Princeton University Press.

Wainwright, S. A., 1988: *Axis and Circumference, The Cylindrical Shape of Plants and Animals.* Cambridge, Harvard University Press.

West, G. B., Brown, J. H., and Enquist, B. J., 1997: A general model of the origin of allometric scaling laws in biology. *Science,* 276:122–126.

Williams, P. L., et al., 1995: *Gray's Anatomy,* 38th British edition. Edinburgh, Churchill Livingstone.

Wright, G. M., and Youson, J. H., 1983: Ultrastructure of cartilage from young adult sea lamprey, *Petromyzon marinus* L.: A new type of vertebrate cartilage. *The American Journal of Anatomy,* 167:59–70.

APPENDIX *Conversion Factors Among Units for Physical Quantities*

Sometimes it is helpful to be able to convert between SI units and others in common use. The following conversion factors can be used.

Length

1 m = 3.280 ft

1 ft = 0.305 m

Mass

1 g = 2.205×10^{-3} lb

1 lb = 453.592 g

Temperature

°F = 9/5 × °C + 32

°C = 5/9 × °F − 32

Velocity

1 m/s = 3.281 ft/s

1 ft/s = 0.305 m/s

Force

1 N = 10^5 dynes

1 dyne = 10^{-3} N

Pressure

1 N/m^2 = 1.50×10^{-4} lb/in^2

1 lb/in^2 = 6.895×10^{-3} N/m^2

1 atm = 1.013×10^5 N/m^2, or 14.701 lb

Work and Energy

1 J = 0.239 cal

1 cal = 4.187 J

part **II**

Protection, Support, and Movement

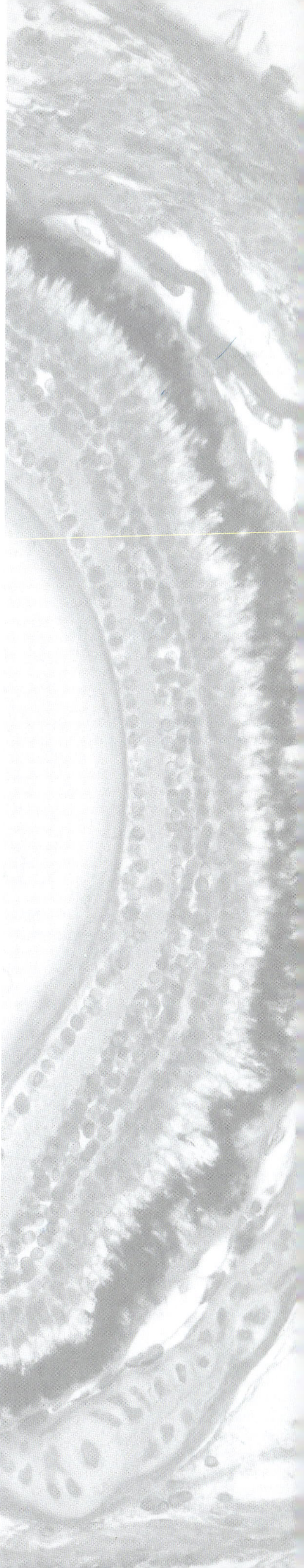

6

The Integument

PRECIS

The skin, or integument, is the interface between the external and internal environments and therefore plays a vital role in protecting the body and maintaining the integrity of the internal environment. Its structure and its derivatives (scales of many types, glands, feathers, hair) are adapted to the diverse characteristics of the vertebrate's environments and modes of life.

OUTLINE

We begin our consideration of Part II with the skin, or **integument** (L., *integumentum* = covering), which is one of the boundary layers of the body that forms the interface between the animal's internal and external environments. Other important boundary layers are the linings of the digestive and respiratory tracts. As the most exposed boundary layer, the skin is a very complex organ system and consists of up to 15% of the total body weight in humans. It is composed of many tissues, including epithelium, connective tissue, fat, and smooth muscle, and it contains blood vessels, glands, sensory receptors, nerves, and other structures. Because vertebrates maintain an internal environment distinctly different from their external environment, the skin has many important functions. It protects the body against abrasion, undue exchanges of water and salts, ultraviolet radiation, and other assaults of the external world, but it does not isolate the body. Many sensory signals are received through the skin, and exchanges of water, ions, and gases occur through it in some species. Secretions of its glands may be protective, serve as chemical signals in communication, or have nutritive functions. The skin is involved in vitamin D synthesis and the regulation of blood pressure and has an important immune function. Coloration of the skin frequently camouflages an animal; warns potential predators; and plays a role in many other aspects of behavior, such as aggression and courtship. In birds and mammals, and to a lesser extent in reptiles, the skin helps regulate body temperature. (In this chapter, the term reptiles is used in the classical paraphyletic sense to include turtles, lizards, the tuatara, and crocodilians.) Finally, the skin also is a complex ecosystem containing a characteristic flora and fauna of viruses, bacteria, fungi, yeasts, mites, and other arthropods. It is truly multifunctional.

General Structure and Development of the Skin

Although the integument varies considerably among species and even among parts of the body in an individual, its basic structure and development are much the same in all vertebrates (Fig. 6-1). An outer **epidermis** (Gr., *epi* = upon + *derma* = skin), which is composed of a stratified, squamous epithelium, develops embryonically from the ectoderm. It rests on a basal lamina of delicate fibrils and overlies a dermis of fibrous connective tissue. The underlying **dermis** develops from mesenchymal cells, most of which are derived from the mesodermal dermatomes of the somites and neural crest cells.

One or two cell layers of the epidermis, which are located just above the basal lamina, constitute the **stratum germinativum.** These cuboidal or columnar cells multiply mitotically. Cells that do not remain in the stratum germinativum move toward the body surface, differentiate, flatten, and eventually slough off either individually or in sheets. Epidermal cells have the capacity to synthesize **keratin** (Gr., *kerat* = horn), a water-insoluble, horny protein that may fill the cells and replace other organelles. This capacity is best developed in terrestrial vertebrates, where dead, keratin-filled cells form a **stratum corneum** (L., *cornu* = horn) on the skin surface. The epidermis is only a few cells deep in fishes and amphibians but relatively thick in reptiles and mammals.

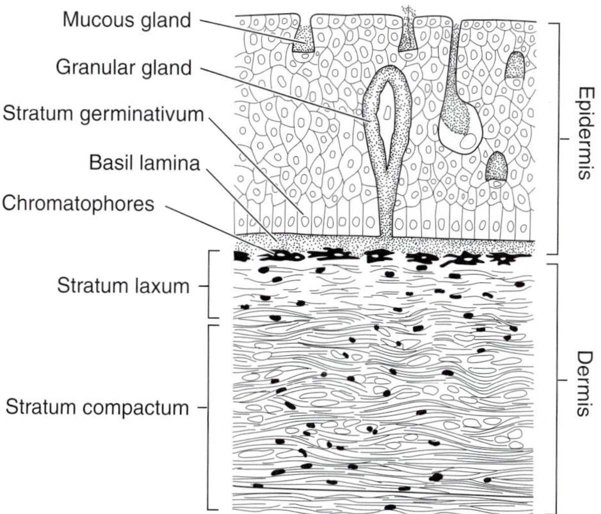

FIGURE 6-1
The structure of fish skin. This diagram is based on a species without bony scales, so the cellular layers of the skin are clear. *(After Portmann.)*

The dermis is the distinctive part of vertebrate skin; invertebrate integuments lack such a layer. The dermis is a fibrous connective tissue composed primarily of an extracellular matrix of **collagen** and **elastic fibers** embedded in a ground substance of **proteoglycans** and other macromolecules, many of which bind to water, giving the dermis a gelatinous consistency. The primary cellular elements of the dermis are elongated **fibroblasts,** which synthesize and extrude the fibers. Other cell types include pigment cells, or chromatophores, fat cells, smooth muscle fibers, scattered white blood cells, and macrophages. **Macrophages** are phagocytic cells that help protect the body against foreign microorganisms by engulfing and degrading them. Fibroblasts, collagen fibers, and other dermal structures differentiate embryonically just beneath the basal lamina of the epidermis and sink inward.

In mature skin, the dermis is thicker than the epidermis and typically consists of two layers. Collagen fibers are irregularly arranged in the superficial **stratum laxum** (L., *laxus* = loose) and more tightly packed in the deeper **stratum compactum.** Skins with a dermis with a very high collagen content are processed as leather. Nerves and sense organs occur in the dermis, and free nerve endings penetrate the epidermis in some species. Fat and muscle may be present in deeper parts. Many blood vessels travel through the dermis. Protrusions of the dermis into the epidermis are called **dermal papillae.** They bring capillaries close to the body surface, but capillaries normally do not enter the epidermis. Capillaries do enter the epidermis in some fishes and salamanders that depend on cutaneous respiration (Chapter 18).

Teeth, bony scales, horny scales, feathers, hair, and other skin derivatives develop as a result of **epithelial–mesenchymal interactions** between the dermis and the overlying epidermis. The cells that form these structures must be competent to respond to inductive influences of adjacent cells and to environmental influences. Neural crest cells that migrate between the epidermis and developing dermis are implicated in the formation of many of these structures. Feathers, hair, and other keratinized structures, as well as glands, are composed of epidermal cells. Parts of these structures grow down into the dermis during development. In contrast, superficial bones are dermal derivatives, whereas teeth and the bony scales of many fishes are composed of both epidermal and dermal products. Many of these structures are hard and perform the supportive and protective functions that we associate with skeletons, so they are appropriately called the dermal skeleton or **integumentary skeleton.** The integumentary skeleton develops within or just beneath the skin and should not be confused with the deeper cartilage and bones of the **endoskeleton** or with the **exoskele-**

ton of arthropods, mollusks, and many other invertebrates, which is a secretion on the body surface. Glands, bony scales, and keratinized structures differ greatly among the different groups of vertebrates, according to their specific environments.

Skin Coloration and Its Functions

The skin of all vertebrates, except albinos, contains pigments of various types within cells collectively known as **chromatophores** (Gr., *chroma* = color + *phoros* = bearing, from *pherein* = to bear). Chromatophores are of neural crest origin. They lie in the upper part of the dermis in fishes, amphibians, and reptiles, but they penetrate or are located in the epidermis in birds and mammals. When the pigment is a dark melanin, the cells are called **melanophores** (Gr., *melania* = blackness). The melanin may be brown or black or more yellow and reddish. Melanophores are star-shaped cells with long, branching processes (Fig. 6-2*A*). The pigment itself is synthesized and contained within cellular organelles, the **melanosomes.** In anamniotes and many reptiles, such as lizards, the melanosomes may migrate into the processes of the melanophores, which maximizes the color, or concentrate near the center of the cell. Pigment is synthesized within the melanophores of birds and mammals, but most of it is transferred to feather, hair, and other epidermal cells.

Most mammals only have melanophores, but other vertebrates also have brighter colored pigment cells. **Iridophores** (Gr., *iris* = rainbow) contain reflective platelets of purine, usually guanine, that may give them a silvery appearance. **Xanthophores** (Gr., *xanthos* = yellow) and **erythrophores** (Gr., *erythros* = red) contain yellowish or reddish pigments, respectively, that are composed of pteridines and carotenoids. Frequently, three or more chromatophores are organized

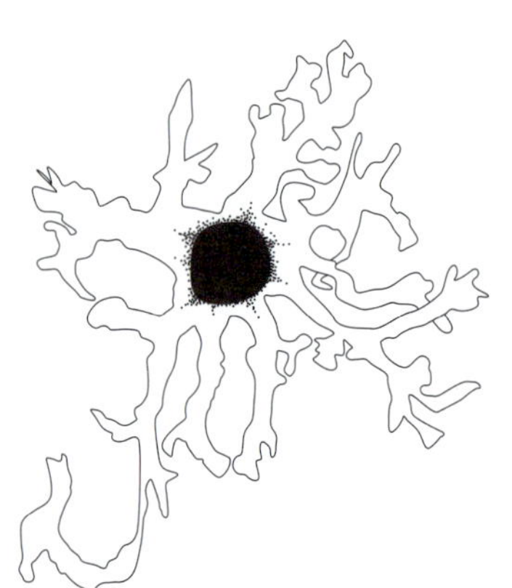

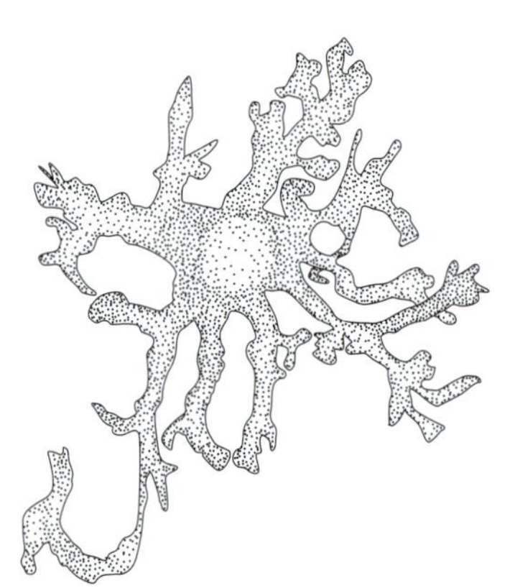

A. Surface view of a melanophore

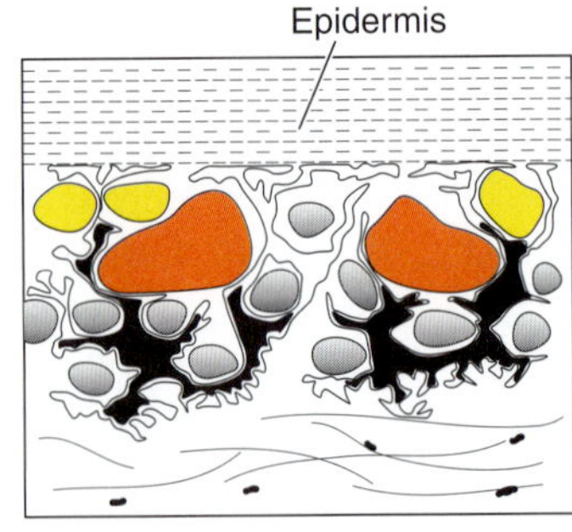

B. Skin of the rainbow lizard

FIGURE 6-2
Pigment cells. *A,* A surface view of a melanophore in which the pigment is concentrated in the center of the cell (left) and dispersed throughout the cell (right). *B,* Vertical sections through the skin of the rainbow lizard, showing two dermal chromatophore units. Pigment is withdrawn from the melanophore processes in the left figure and the lizard is orange; in the middle figure a partial dispersal of pigment produces a terracotta color; in the right figure full dispersal turns the animal a chocolate brown. *(After Harris.)*

into a **dermal chromatophore unit** (Fig. 6-2*B*). Iridophores in the center of the unit are surrounded by processes of a deeper melanophore and are overlain by xanthophores or erythrophores. Different colors result from different combinations of these cells and patterns of pigment dispersal within the melanophores. If melanin is dispersed in the processes of the melanophore overlying the iridophore, the body appears dark, or perhaps yellowish if pigment is well dispersed in the xanthophore. If pigment is withdrawn from the overlying melanophore processes, light reaches the iridophores and is dispersed. Short wavelengths (blue) are reflected through the skin, and longer ones are absorbed by the underlying melanophore. When the blue light comes back through a filter of yellow xanthophores, the color is green. An African bird, the touraco, is the only vertebrate in which a true green pigment is known.

In some species, minute ridges on surface cells refract light and produce iridescent colors that change with the angle of observation. This is known as **structural coloration** (p. 220). Hummingbirds have such colors. In a few vertebrates, the degree of vascularity of the skin produces color changes. Blushing is a familiar example.

Pigmentary colors are subject to change. More or less pigment may be synthesized as an animal slowly adjusts to its background, to different seasons, or to the degree of solar radiation (as in a suntan). Slow changes of these types are referred to as **morphological color changes.** Many fishes and amphibians and some reptiles can undergo rapid **physiological color change.** Rapid changes occur by a migration of pigment within the chromatophores. Migration of pigment within the melanophores of many fishes is controlled both by action of sympathetic nerves, which leads to an aggregation of melanosomes, and by melanophore-stimulating hormone produced by the pituitary gland, which causes a dispersal of melanosomes. Control in amphibians and reptiles appears to be primarily or exclusively hormonal. Melanophore-stimulating hormone promotes pigment dispersal; norepinephrine from the adrenal gland causes pigment aggregation.

Skin color plays an important role in the life of vertebrates. It may be concealing, or **cryptic** (Gr., *cryptos* = hidden) and help hide an organism from its predators or enable a predator to stalk its prey undetected (Focus 6-1). It may be *aposematic* (Gr., *apo* = away + *semat* = signal) and advertise the presence of dangerous, venomous, or distasteful species that profit from making known their presence. After a few unpleasant experiences, in which a few individuals may be injured or killed, predators learn to avoid members of these species. Skin coloration is also important in species recognition, establishment of territories, courtship, and other types of communication. Color changes help many reptiles thermoregulate (Chapter 3). Darker pigmentation aids in collecting solar radiation to warm the body faster than does light pigmentation.

Skin pigmentation also protects deeper body tissues from injury by ultraviolet radiation. In some small fishes and other vertebrates, especially those active in the daytime in brightly lit habitats, melanophores also occur in membranes around the nervous system and even in the peritoneal lining of the body cavity. Pigmentation of human skin also affords some protection against ultraviolet radiation.

The Skin of Fishes

Structure

Integumentary structure is correlated closely with a vertebrate's mode of life and environment. All fishes are aquatic and encounter similar environmental problems, and therefore the basic structure of the skin is similar in most groups, apart from the nature of their bony scales. The epidermis is relatively thin, and most of its cells are living (Fig. 6-1). The surface cells frequently are covered with complex patterns of microridges, visible only with electron microscopy. Microridges might increase the surface area available for exchanges between the animal and its environment, but their function is uncertain in most cases. Keratin may be deposited in limited areas, as in the horny teeth of lampreys and the nuptial tubercles that develop in many male teleosts during the breeding season, but most epidermal cells generally do not synthesize keratin.

Unicellular glands, most of which produce mucus, are abundant in the epidermis. Their secretions, together with secretions of the surface cells, form a mucous cuticle and a generally slimy surface. Surface mucus is believed to reduce water exchanges between the fish and its environment and so help the excretory organs maintain a stable internal environment (Chapter 20). Mucus also protects the body from bacterial invasion and the attachment of ectoparasites. It may also reduce friction and drag, especially during sudden bursts of speed, as when a fish tries to escape a predator. Granular unicellular glands are also present, but the functions of many are unknown. In minnows, catfishes, and other ostariophysans, some of these glandular cells contain an alarm substance that is released when the fish is injured and the skin is ruptured. This acts as a chemical messenger, or **pheromone** (Gr., *pherein* = to bear + *hormaein* = to excite) and triggers

FOCUS 6-1 *Adaptive Camouflage in Flatfishes*

The ability of organisms to change appearance according to their immediate local surroundings is called adaptive camouflage. This ability reaches a pinnacle in flatfishes (flounders, sole). This mechanism is adaptive, and concealed individuals survive predation significantly better than do less cryptic ones. The disguise evolved by flatfishes is so good that a viewer is unlikely to distinguish between the fish and its surroundings. The disguise functions like a malleable mirror having a reflection that is interpreted only in relation to the immediate local surroundings (see Figure).

Animals have only a limited number of designs to minimize their visibility. These designs are based on four principles: (1) color resemblance; (2) obliterative shading (a graded lightening and darkening on the surface that diminish the appearance of relief); (3) disruptive coloration (a pattern of colors and tones that serves to divide the continuous body outline and body surface into sections); and (4) shadow elimination. Camouflage in flatfishes is primarily based on disruptive coloration (Saidel, 1988) involving adaptive changes in skin reflectance and contrast between discrete skin areas (Figure). These changes occur on a mural provided by the distribution of different chromatophores and dermal reflecting cells (iridophores). By modulating the various dermal chromatophore units via sympathetic nerve stimulation, melanophore-stimulating hormone, and the melanophore-concentrating hormone (Kawauchi et al., 1983), the flatfish can make rapid changes of its skin appearance, blending it with its local surroundings. The entire control system is superimposed by the visual system. Blinded flatfishes darken and are unable to engage the adaptive camouflage mechanism when placed on light or contrasting backgrounds. At first it was thought that a flatfish can reproduce its background very faithfully; however, only a limited range of preset patterns of stimulation of chromatophore units, in turn, have fixed patterns of distribution. Thus the fish has only a restricted range of options (Figure) by which it can remain unseen by the viewer. The flatfish's camouflage strategy is to condition the viewer's eyes for certain characteristics of the local surroundings and to eliminate the sharp boundaries of its body by means of transparent fins along the entire perimeter of its body. With the transparency of the fins, the viewer sees a combination of the substratum and skin textures and no sharp boundary. At the same time, the fish adapts its skin to its local surroundings by producing an average reflectance of its background (Figure). In the natural habitat of the flatfish, this strategy will result in the fish being unseen not only by the human, but also by the predator.

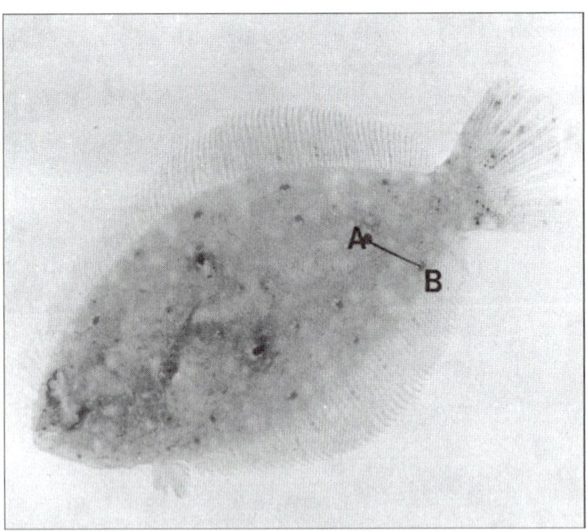

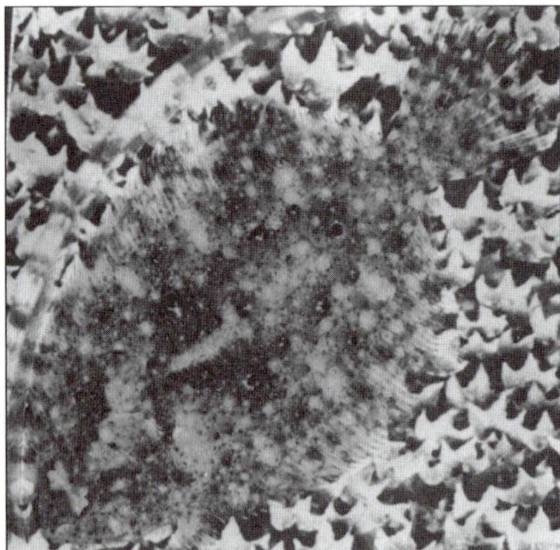

Adaptive changes in the flounder, *Paralichthys lethostima,* to a white (*left*) and patterned (*right*) background. The spots identified by A and B are maintained on both backgrounds. *(From Saidel.)*

a fright reaction in nearby members of the species (Chapter 13).

Multicellular glands that grow down into the dermis are relatively uncommon in fishes. Hagfishes have large slime glands; some teleosts have poison glands that produce toxic materials and often are associated with fin spines; some deep-sea species have light-generating glands, or **photophores.** In some cases, the secretions of the photophores and glandular cells nourish a bacterial colony that generates light; in others, chemical reactions within the gland produce light. In some species, the pattern of lights generated by the organ is

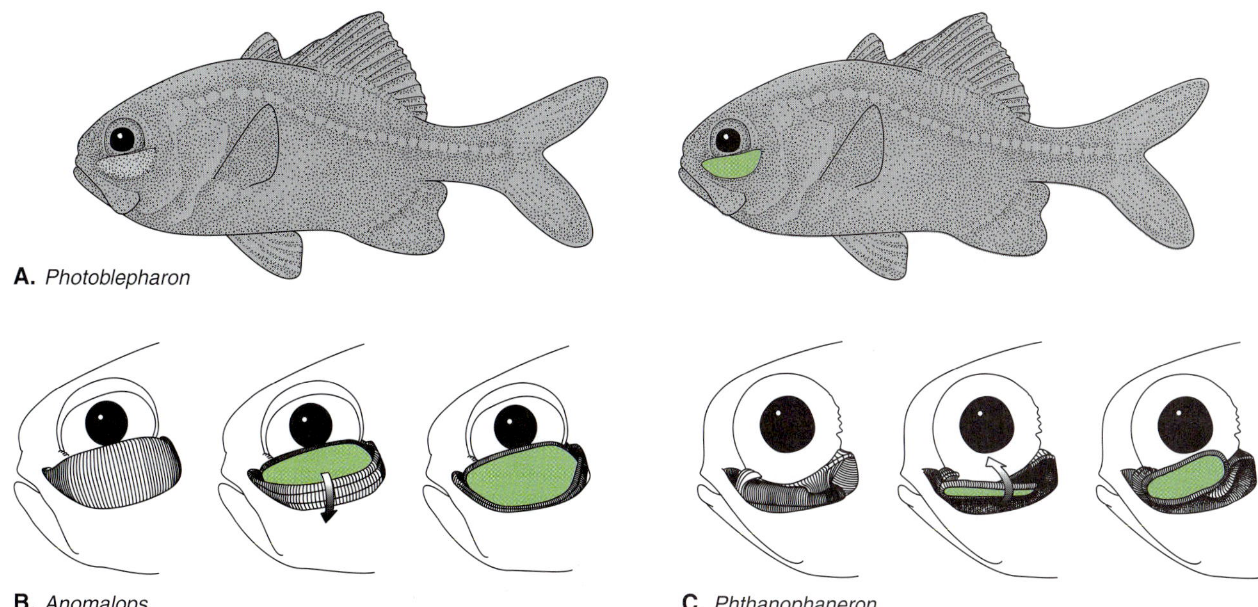

FIGURE 6-3
The photophores beneath the eye of flashlight fishes. In some species *(Photoblepharon, A)* a black shutter of elastic skin can be drawn over the organ (*left*) or lowered (*right*); in other species (*Anomalops, B*) the photophore is rotated; and in others (*Phthanophaneron, C*), a combination of these mechanisms occurs. *(B and C, After Johnson and Rosenblatt.)*

of value in species recognition; in some nocturnal species, it is used in feeding; and in the deep-sea angler fishes, a light organ dangled over the mouth on a modified dorsal fin spine acts as a lure. The photophore of the circumtropical flashlight fishes (Anomalopidae) contains symbiotic bacteria that emit light continuously, and shutter mechanisms have evolved that can cover or expose the organ (Fig. 6-3).

The collagen fibers of fish dermis frequently are arranged more regularly than they are in other vertebrates. They develop in layers that spiral around the body at an approximately 45° angle to the longitudinal axis. This is termed **helical cross fibering.** Adjacent layers of fibers are nearly perpendicular to each other. As in plywood, this arrangement strengthens the skin so that body shape is maintained during undulatory swimming, and the skin itself acts as an exotendon transmitting muscular forces (Chapter 11).

Bony Scales

Bony scales, more than any other feature, characterize the skin of most fishes. Tissues that may contribute to scales are bone, dentine, and enamel. **Bone** consists of an extracellular matrix of collagen fibers embedded in an organized ground substance of protein polysaccharides and calcium phosphate. This matrix is laid down by bone-forming cells, known as **osteoblasts** (Gr., *osteon* = bone + *blastos* = bud), which differentiate from mesenchymal cells of the dermis. Calcium phosphate crystals known as **hydroxyapatite,** which are synthesized by the osteoblasts, bind to the collagen fibers, and often constitute 70% of the bone by weight. As the bone develops, the bone-forming cells, now called **osteocytes,** usually become entrapped in the matrix. They lie in small cavities, the **lacunae** (Fig. 6-4), which are interconnected by minute canals, the **canaliculi.** Processes of the osteocytes extend through the canaliculi. This type of bone is called **cellular bone** (Chapter 5). In some cases, notably in the scales of teleost fishes, the bone is **acellular.** In such scales, the bone-forming cells are found on the periphery of the devel-

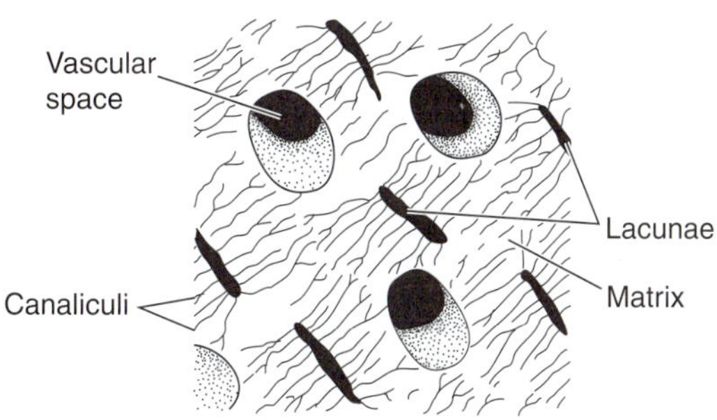

FIGURE 6-4
A microscopic section of bone.

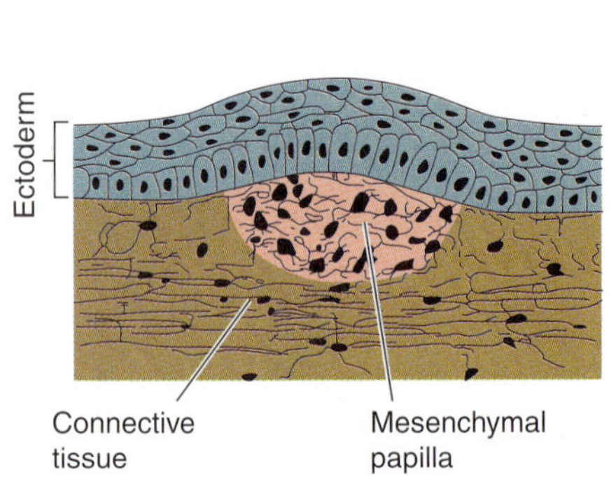

A. Development of mesenchymal papilla

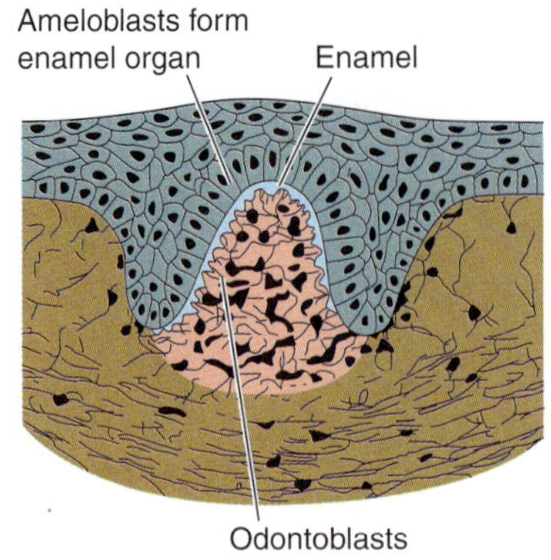

B. Enamel organ formation

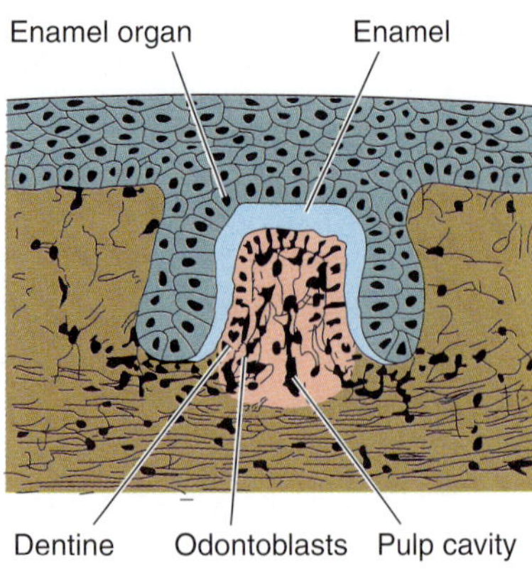

C. Dentine formation

FIGURE 6-5
A–C, Three vertical sections through the skin of a fish to show the interaction between a mesenchymal papilla in the dermis and the epidermis in forming enamel and dentine. Enamel and dentine are added to the surface of bony plates and scales in many fishes.

oping scale. As the bone matrix is formed, the cells move centrifugally, away from the center of the scale. No bone cells or processes are left behind.

Dentine and **enamel** (or enamel-like) layers may be deposited on top of the bone on the surface of the scales. The structure of scales and contributions of these layers vary considerably in different fish groups. Investigators often disagree about scale structure and terminology. Often, scales develop embryonically in a manner similar to teeth (Fig. 6-5). First, mesenchymal cells aggregate in small papillae just beneath the basal lamina of the epidermis. Although located in the dermis, many of these cells are believed to be of neural crest origin. The basal cells of the overlying epidermis respond by differentiating into **ameloblasts** (Middle English, *amel* = enamel + Gr., *blastos* = bud) that collectively form an **enamel organ.** Under the inductive influence of the enamel organ, underlying dermal cells differentiate into **odontoblasts** (Gr., *odont* = tooth), which begin to lay down dentine between themselves and the enamel organ. The enamel organ then produces enamel on top of the dentine. This process continues, with the enamel organ retreating in one direction and the odontoblasts in the other.

Enamel, which is of epidermal origin, is the hardest tissue in the body: approximately 96% by weight are crystals of hydroxyapatite. Its matrix includes distinctive proteins called **amelogenins.** As enamel is laid down and the enamel organ retreats, no cells or cell processes are left behind in the enamel. In some cases, the enamel is deposited in successive waves and has a lamellar appearance. This type of enamel, which occurs on many scales, was long thought to be different from the enamel on teeth and was called **ganoine** (Gr., *ganos* = sheen). It is now believed that the enamel of teeth differs from that of ganoine only in the pattern of deposition. They are the same material, and both are derivatives of the ectodermal epidermis.

Dentine, which is produced by cells of neural crest origin, resembles bone and differs primarily in the arrangement of the cells and matrix. As the odontoblasts form dentine and retreat, they leave long, cytoplasmic processes that become lodged in **dentine tubules. Cosmine** (Gr., *kosmios* = well ordered) is a variety of dentine in which the dentine tubules are grouped into radiating tufts.

The primary dermal armor of the ancestral jawless vertebrates consists of thick **cosmoid plates** containing all of these hard materials (Fig. 6-6*A*). The deepest part of the plate is made up of layers of bone, usually cellular. Spongy bone lies superficial to the lamellar bone. Many of the spaces in the scale accommodate blood vessels that nourish the developing scale (Bemis and Northcutt, 1992). Others contain mucous glands and parts of the aquatic lateral line and electroreceptive systems (Chapter 12) that enable these fishes to detect water movements of various kinds and weak electric currents. The surface of the plate is composed of a series of denticles. Each denticle contains dentine of the cosmoid type and a thin cap of enamel-like material. Because this enamel-like material may be hard cosmine that developed from the dermis (unlike enamel, which is always of epidermal origin), it is often called **enameloid.** Most of these early jawless vertebrates were small fishes that moved slowly as they fed along the bottoms of shallow seas or brackish waters. The heavy armor may have helped

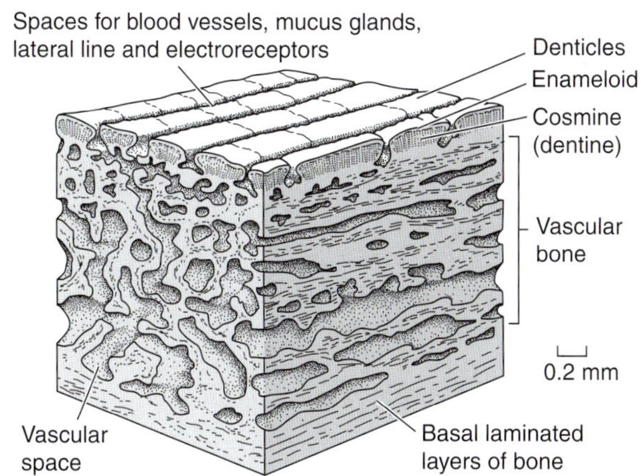

A. †Heterostracan

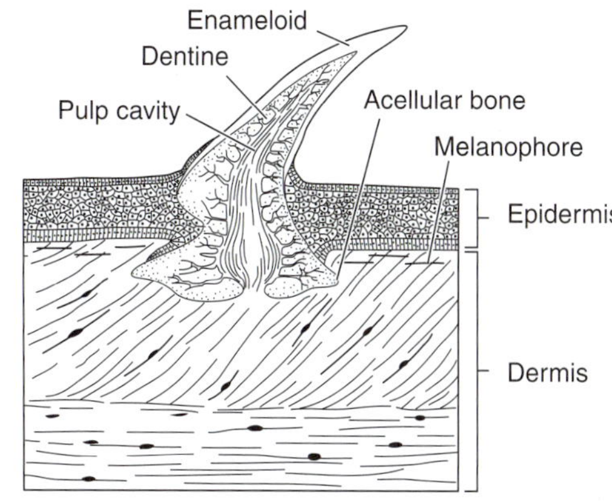

B. Shark skin

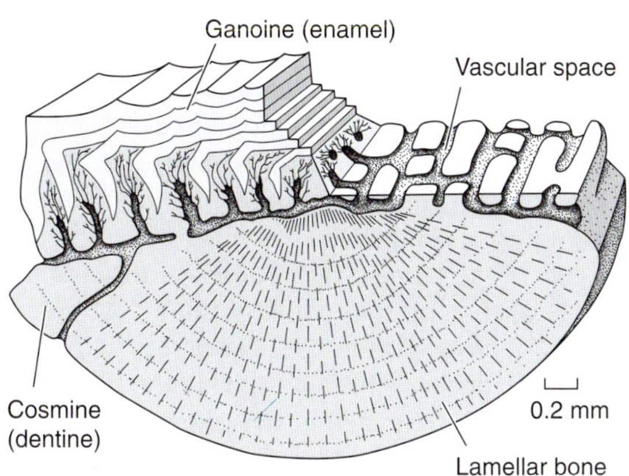

C. †Early actinopterygian ("†Paleoniscoid")

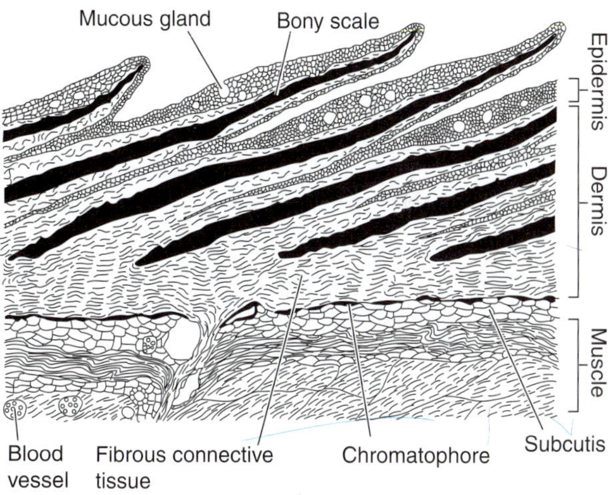

D. Teleost skin

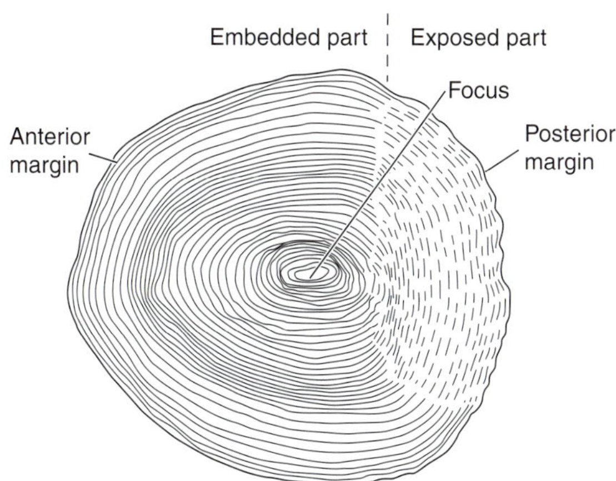

E. Cycloid scale

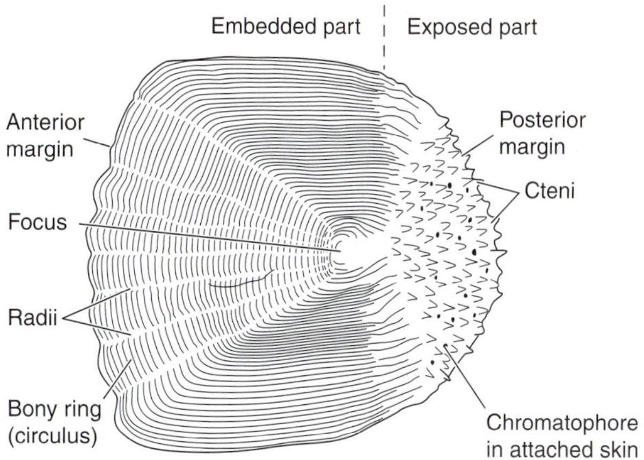

F. Ctenoid scale

FIGURE 6-6

Some bony scales of fishes. *A,* A vertical section through the cosmoid plate of an early jawless fish (a †heterostracan). *B,* The dermal denticle of a cartilaginous fish. *C,* A vertical section through the ganoid scale of a primitive actinopterygian (a "†palaeoniscoid"). *D,* A vertical section through the skin and cycloid scales of a teleost. *E,* Surface view of a cycloid scale. *F,* Surface view of a ctenoid scale. *(A, After Kiaer; C, after Ørvig.)*

keep them close to the bottom and protected them from predators, such as aquatic scorpions, or †eurypterids. The bone of the scales also may have provided a reservoir of needed calcium and phosphate ions. A source of calcium and phosphate would be a prerequisite for early vertebrates to penetrate fresh water, where these ions are in short supply. Finally, the scales may have helped a fish in fresh water solve its water-balance problem. Because the concentration of salts within the fish's body fluids was greater than that in the surrounding fresh water, large amounts of water would diffuse into the body by osmosis through gills and any other permeable surface. The enamel or enameloid on the scales would have reduced the surface area available for such an influx of water over the general body surface.

As fishes evolved and became more active creatures, a less cumbersome armor became advantageous. The thick, bony plates became reduced in different ways in different lineages. Living hagfishes and lampreys have no scales at all. Presumably, the ancestral armor of lampreys was lost, but hagfishes may have evolved from some unknown, unarmored species. †Placoderms lost the armor on most of the trunk and tail but retained heavy denticulate cosmoid scales and plates, similar to the ancestral armor, on the head and thorax. Thus a good part of the body was unarmored and therefore flexible enough for lateral undulations during swimming. Most of the bony material disappeared in cartilaginous fishes. The spiny **dermal denticles** or **placoid scales** of sharks and other cartilaginous fishes are composed of dentine surrounding a vascular pulp cavity and capped by a hard material (Fig. 6-6*B*). The nature of this material has long been a puzzle. Recent studies show that it contains the amelogenin proteins found in enamel, but it also contains some fibrous material of dermal origin. Many investigators now call this superficial layer enamel; others hedge and call it enameloid. A thin layer of what appears to be acellular bone underlies the dentine. In a sense, the dermal denticles represent isolated denticles of the ancestral armor.

Early bony fishes retained heavy, bony scales. Ancestral lungfishes and early rhipidistian fishes had rhomboid **cosmoid scales** without denticles and similar in composition to plates of early jawless fishes. Ancestral †acanthodians and actinopterygians had rhomboid **ganoid scales** (Fig. 6-6*C*) composed of lamellar bone overlain by many layers of enamel (ganoine). Some ganoid scales contain small amounts of vascular spongy bone and dentine; others do not. Ganoid scales occur today only in *Polypterus* and in the gars, *Lepisosteus*. Other actinopterygians have very thin scales that develop in overlapping skinfolds, so that the scales themselves overlap, a condition called **imbricating.** Such scales are composed primarily of what appears to be acellular bone underlain by dense, fibrous material (Fig. 6-6*D*). This gives the scale great flexibility. The surface of **cycloid scales** is sculptured by a pattern of growth rings (Fig. 6-6*E*). **Ctenoid scales** have, in addition, a series of comblike projections, **ctenii,** on the posterior part of the scale nearest the skin surface (Fig. 6-6*F*). The origin of the sculpturing on the most superficial part of the scale is uncertain, but some investigators believe that it may represent a thin layer of dentine or enamel. Contemporary lungfishes also have cycloid scales, which evolved independently.

Bony scales associated with the fins of osteichthyans line up in branching columns to form flexible, supporting **fin rays,** or **lepidotrichia** (Gr., *lepis* = scale + *trich* = hair; Fig. 6-7*A*). Because scales lie on both surfaces of fins, lepidotrichia are paired structures when seen in an end view. The fins of chondrichthyan fishes, in contrast, are supported by horny **ceratotrichia** (Gr., *kerat* = horn). One or more **spines** often lie in the leading edge of a fin, where they stiffen the fin and act as a cutwater. Usually each spine has evolved by the enlargement of a single scale so they are unitary structures and are not jointed, branching, or paired (Fig. 6-7*B*). Poison glands are associated with the spines in some species. The lionfishes (*Pterois*) of the Indo-Pacific Ocean have highly poisonous spines and may lunge with them at a potential predator or prey.

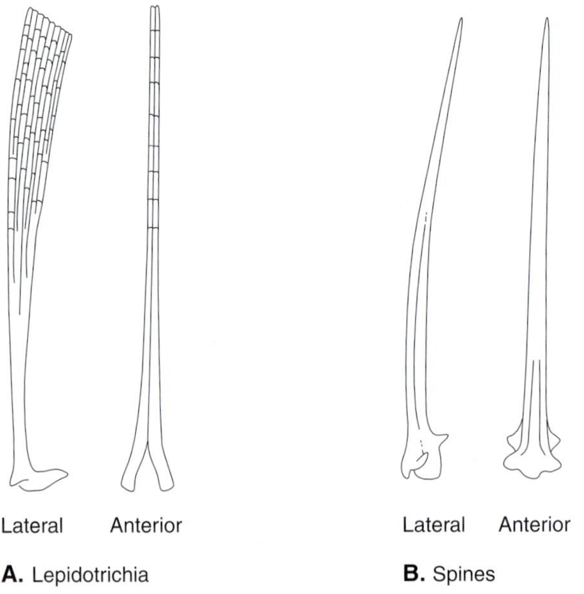

FIGURE 6-7
Lepidotrichia (*A*) and spines (*B*) of bony fishes as seen in lateral (*left*) and anterior (*right*) views.

The Skin of Amphibians

The aquatic larvae of amphibians have skin similar to that of fishes, although they lack scales. After metamorphosis, the epidermis remains relatively thin, but its cells synthesize keratin. As keratin accumulates, the cells die and form a horny layer, the **stratum corneum,** on the skin surface (Fig. 6-8). The stratum corneum is thin in amphibians, seldom more than several layers of cells thick, so cutaneous respiration is still possible while providing protection against desiccation and abrasion, which confront all land vertebrates. Its outer part is sloughed off (desquamation) periodically in large sheets only one or two cells thick. The discarded tissue usually is eaten. Desquamation is hormonally controlled. Evidence for this is that hypophysectomized toads do not slough their keratinized layer; instead, keratinized cells merely continue to pile up. During breeding seasons, nuptial pads may appear on digits or limbs of male frogs or salamanders. Such pads are calluses of keratinized epidermis and help the male hold the female during mating.

Unicellular glands, so prevalent in fishes, are absent, but multicellular, alveolus-shaped glands grow down into the dermis from the stratum germinativum. Most are **mucus glands,** the secretions of which protect the skin surface, reduce the loss of body water, and keep the surface moist. In recent amphibians, considerable gas and ionic exchanges occur across the skin—hence the prolific dermal vascularization. **Granular glands** secrete toxic substances. For example, the **parotid gland** or **parotoid gland** (Gr., *para* = beside + *ot* = ear) of toads that forms a bulge behind the ear is an aggregation of granular glands. Its irritating secretions discourage experienced snakes, raccoons, and other predators. Some tropical frogs have highly toxic skin glands. Their presence often is coupled with a brilliant red or yellow aposematic (warning) coloration. Some Amazonian Indians use the toxin on their arrow tips. Glands on the back of the Surinam toad, *Pipa pipa*, help nourish the tadpoles that develop in pits in the skin on the mother's back. The secretions of the cutaneous glands of amphibians probably also protect them from pathogenic bacteria and fungi, which often are abundant in the wetlands and damp groundcover where most species live. Research is underway to determine whether some of these secretions have medical or veterinary applications (Clarke, 1997).

Amphibian coloration is as highly developed as in fishes. Chromatophores include melanophores in both epidermis and dermis, and numerous xanthophores and iridophores in the dermis.

Small bony scales, or **osteoderms,** were present in the dermis of early tetrapods, but they were lost in most amphibians. Caecilians and some anurans retain a few bony osteoderms, which are homologues of dermal scales, deep in the grooves in their skin. In primitive tetrapods, one large osteoderm became incorporated into the pectoral girdle as the **interclavicle.**

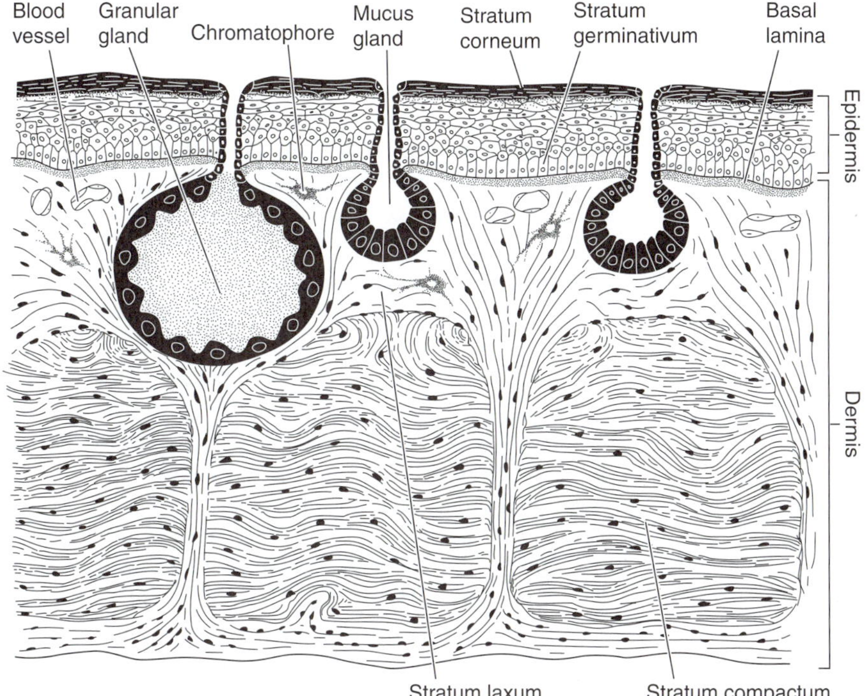

FIGURE 6-8
A vertical section through the skin of a frog. *(After Portmann.)*

The Skin of Reptiles

As vertebrates adapted more completely to the terrestrial environment, the skin became more important in protecting them against abrasion and loss of body water. The epidermis of reptiles (i.e., turtles, lepidosaurs, and crocodiles) has a thick stratum corneum composed of many layers of dead, keratin-filled cells. These are organized as **horny plates** in turtles and as **horny scales** in lizards and snakes (Fig. 6-9*A*). Reptiles lose less water by evaporation through the epidermis than do amphibians. The large amount of keratin is partly responsible, but phospholipids bound to the keratin are even more important in reducing water loss.

The epidermis of lizards and snakes is particularly complex. The functional scale consists of many layers of cells. The outer layers are heavily keratinized, but the deepest several layers are living. All these cells constitute a mature **outer epidermal generation** (Fig. 6-9*B*). An immature **inner epidermal generation** of cells, which has been produced more recently by the stratum germinativum, lies beneath. The same keratinized and living layers that characterize the outer generation begin to differentiate in the inner generation prior to skin shedding. A new layer of scales is ready to take over for the old layer before the old layer is lost. During shedding or molting, technically called **ecdysis,** a deep, unkeratinized layer of cells of the outer epithelial generation undergoes autolysis and breaks down. A **fission zone** develops, and the outer epidermal generation is shed as a unit. The horny plates of the shell of most turtles are not shed but rather wear away at the surface. As the turtle grows, newly formed horn protrudes beyond the margin of the old plates, resulting in a series of growth rings around the margin of each plate.

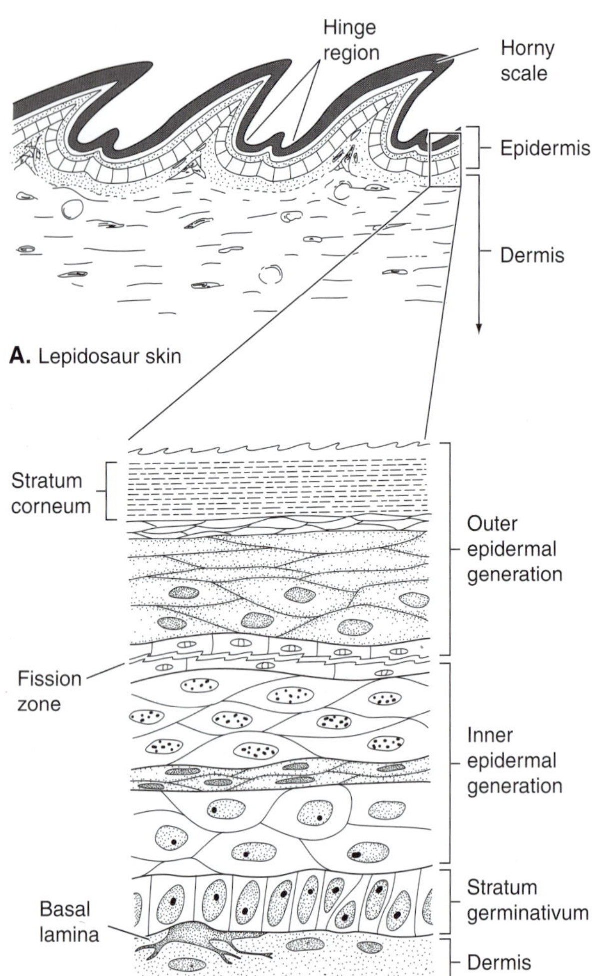

FIGURE 6-9
Reptile skin. *A,* A vertical section through the horny scales and the skin of squamate reptile. *B,* An enlargement of a portion of the skin. *(After Maderson.)*

The terminal phalanx of each digit is encased in a protective **claw.** Claws also help the animal grip the substratum during locomotion. Although claws are present in a few amphibians—the common African clawed frog (*Xenopus laevis*) is a familiar example—all reptiles with limbs have them. Claws, like horny scales, are derivatives of the stratum germinativum. Their hardness derives from the incorporation of calcium salts along with keratin.

Some epidermal scales are modified for particular functions. In geckos the scales over the tips of their digits have become small hairlike structures, called setae, which form adhesion forces with the surface area by means of intermolecular forces (Autumn et al., 2000), enabling geckos to crawl up vertical walls or move upside down on the ceiling. Modified ventral scales of snakes play an important role in locomotion. Except for **apical pits,** reptilian skin is generally devoid of sense organs. Apical pits are situated one to seven in a row, at or near the posterior margin of the epidermal scales. A tiny hairlike filament protrudes from each pit and has a tactile function.

Reptilian dermis contains no mucous glands. The few glands that are present are **scent glands,** producing musk and other secretions used in courtship and mating behavior. These occur in specific places in different species: under the thigh in some lizards, on the underside of the lower jaw in crocodiles, and near the cloaca in some turtles.

Some reptiles (e.g., the skinks among lizards) retain small, thin osteoderms in their dermis, which are thought to be modified remnants of the original dermal armor of primitive fishes. In crocodilians, some lizards, and *Sphenodon,* osteoderms form a series of riblike structures called **gastralia** in the ventrolateral abdominal wall. They stiffen the pleuroperitoneal cavity as part of the mechanism for ventilating the lungs.

Reptilian Color

Many reptiles, especially some chameleons, can undergo a kaleidoscopic color change, which is made possible by the presence of chromathophores in the dermis. Erythrophores and xanthophores produce red and yellow hues, respectively, wheras melanocytes contain black or dark brown pigment. Various colors are under the control of adrenaline and sympathetic nerves, which make the pigments expand within the chromatophores. Iridescent effects, on the other hand, are accomplished by iridocytes, which contain light-reflective crystals. Snakes that have subterranean ancestors are unable to change color appreciably.

The Skin of Birds

The epidermis of birds is relatively thin and is not heavily cornified over most of the body surface, but horny scales develop on the legs and feet of most birds. These scales are not shed. Claws are present, and the margins of the toothless jaws are covered by a horny beak.

The most conspicuous derivatives of the epidermis are the **feathers.** In addition to forming the flying surfaces of the wings and tail, feathers entrap air and reduce airflow next to the skin, which reduces evaporative loss of body water and forms an insulating layer that enables birds to maintain a high and constant body temperature (Chapter 3). These functions of feathers may have evolved before their role in flight. Those that cover the body surface are the **contour feathers.** The larger and stiffer contour feathers that form the flying surfaces are **flight feathers.** A typical contour feather (Fig. 6-10*A* and *B*) has a central axis, the base of which, called the quill, or **calamus** (L., *calamus* = reed), is lodged in a feather follicle in the dermis. The distal part, known as the shaft, or **rachis** (Gr., *rhachis* = spine), supports a **vane.** During development, dermal tissue and blood vessels enter the proximal end of the quill through an opening, the inferior umbilicus. The vane is composed of many **barbs** that branch obliquely from each side of the rachis, so the feather has a bipinnate structure. The barbs in turn bear many **barbules,** which have minute **hooklets** that interlock with the hooklets of adjacent barbules, so the vane is a coherent, strong, and flexible structure. If barbules become separated, the bird can interlock them again by preening its feathers with its bill. The mechanism of interlocking of hooklets is analogous to that of Velcro. Because hooklets are absent from the barbules of ostriches and many other flightless species, the feathers are fluffy. The contour feathers of some primitive birds are double, for a small second vane, known as the **after-feather,** arises from a superior umbilicus located at the distal end of the calamus. Traces of after-feathers are sometimes represented in more advanced species by a small tuft of barbules.

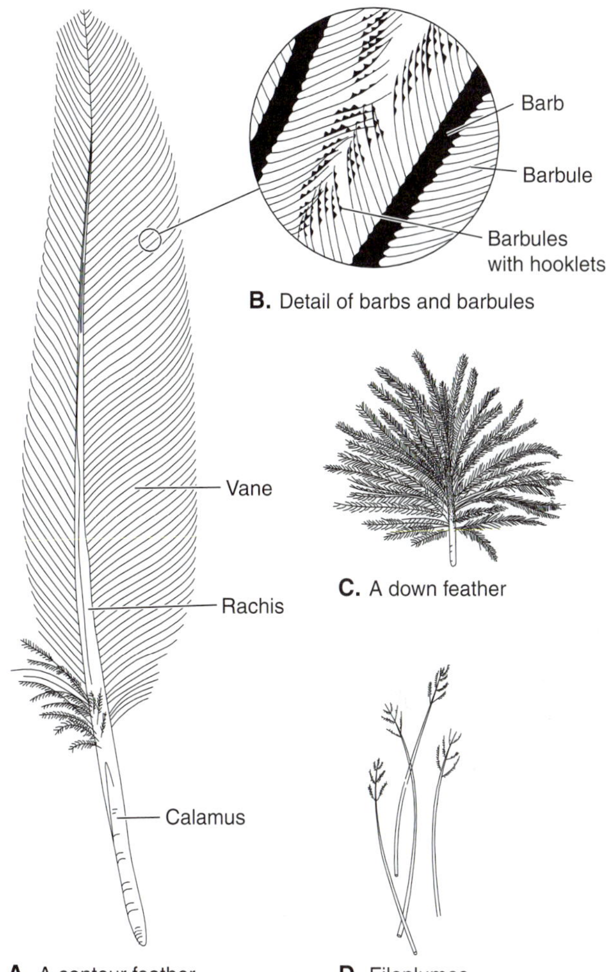

FIGURE 6-10
The structure of representative feathers of a bird. *A,* An entire feather. *B,* An enlargement of a small part of the vein. *C,* Down. *D,* Filoplumes. (Also see Fig. 11-29.)

Other types of feathers include down and bristles (Fig. 6-10*C*). **Down,** which is found beneath the contour feathers in some species, consists of very fluffy barbs that arise from the distal end of the quill. Down is excellent insulation. It is particularly important in young birds, the small body size of which gives them a large heat-losing surface relative to their heat-producing mass. Bristles, or **filoplumes** (L., *filum* = thread), are short, stiff feathers in which the barbs are reduced or lost (Fig. 6-10*D*). Those around the nostrils and eyes of some species keep out dirt. Longer ones around the mouths of nighthawks and flycatchers help them net insects.

Signaling systems in bird behavior have reached extreme levels, mainly because of the very colorful developments. Colors in feathers are the product of two mechanisms: physical structure and the presence of chemical pigments. Red, yellow, brown, and black are chemical pigments, whereas physical structure and the associated light scattering produce white, blue, and iridescent colors. Brown and black are caused by melanin taken up from the germinal region of the follicle. Red, orange, and the pink of the flamingo are caused by carotenoids in the keratinized cells of the feather and are of dietary origin. White is produced by air in cells and by the polygonal shape of the barbule cells, which breaks up the light and reflects and refracts all wavelengths. Blue is caused by the scattering action of particles in cells located beneath the outer layer of the feather barbs, resulting in the reflection of blue, whereas the longer wavelengths are transmitted. The combination of structural blue with yellow pigment results in green. This effect can be seen in hummingbirds.

The dermis from various regions in the body determines whether overlying epidermis remains flat and devoid of feathers, forms scales, or gives rise to different types of feathers. Feathers develop by an epithelial–mesenchymal interaction between the epidermis and the dermal mesenchyme that leads first to the formation of a cone-shaped dermal papilla that pushes up the overlying epidermis (Fig. 6-11). Mitotic divisions in a collar-like zone of the stratum germinativum near the base of the papilla form a crown of barbs. These are covered by a horny **feather sheath** composed of epidermis. As development proceeds, cells on one side of the collar divide more rapidly than do others to form a shaft that extends upward, carrying with it the barbs formed in the collar. Meanwhile, the base of the feather recedes into the skin, accompanied by layers of epithelial cells that form the feather follicle. Except in a few primitive species of birds, feathers do not develop uniformly over the body surface but fan out from distinct feather tracts called **pterylae.**

Feathers wear out and are replaced. Most birds have an annual molt in the summer after the breeding season, but much variation occurs among species. Many songbirds have two molts a year and distinct winter and breeding plumages. Usually, only a few feathers are lost at a time, so the molting period extends over several weeks and the bird remains covered and able to fly.

Avian dermis lacks ossifications. It is a thin, well-vascularized layer, with interlacing collagen fibers and several types of sensory receptors. Many smooth muscles are associated with the feathers. Feather position is crucial for flight, behavior, and thermoregulation. The position of a feather can be regulated by a group of muscles. **Erectors** lift feathers, **depressors** lower them, **retractors** pull them inward, and **rotators** turn them (Fig.

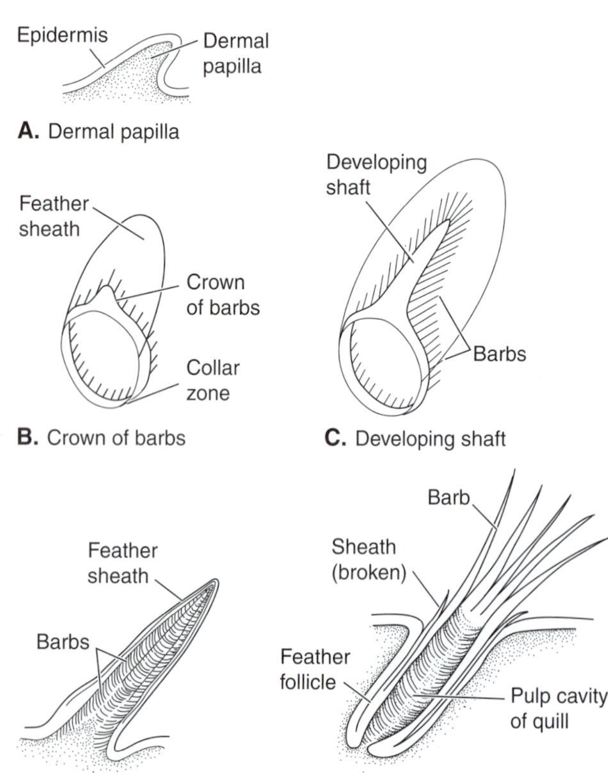

FIGURE 6-11
A–E, Five stages in the development of a feather. *(After Lillie.)*

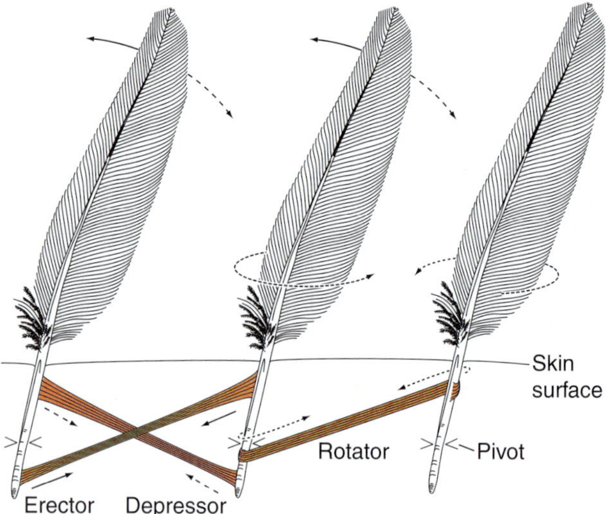

FIGURE 6-12
A simplified diagram of feather muscles showing the erector muscle running from the posterior margin of one feather, below its pivot point, to the anterior margin of an adjacent feather above its pivot point. The depressor muscle extends between the posterior margin of one feather, above its pivot point, to the posterior margin of the next feather below its pivot point. A rotator muscle wraps around the shaft of two successive feathers, attaching below the pivot point of one feather and above the pivot point of the adjacent feather. Combined actions of these muscles can position the feathers in many ways.

FOCUS 6-2 *The Evolution of Feathers*

Recently, Prum (1999) proposed a plausible scenario for feather evolution based on their embryonic development. The first feather-like structure (Stage I) was a simple epidermal cone or filament with a dermal core that protruded from a follicle. This is very similar to an early stage in feather embryonic development. This stage was followed by the development of a tuft of unbranched barbs on the cone so it resembles a down feather (Stage II). These early feathers might have had one or more functions, including communication (if they had a color pattern), thermal insulation, and water repellency. They could not have had an aerodynamic function until the later evolution of a bipinnate feather with a coherent vane (Stages III–V).

This hypothesis is supported by recent discoveries in China of well preserved theropod dinosaurs that were close to the ancestry of birds (Xu et al., 1999). These specimens and others described a year earlier have integumentary structures consisting of unbranched, and possibly branched, filaments ranging in length up to 70 mm. Some investigators believe that these filaments are an unusual array of collagen fibers within the skin, but this argument could not hold for the longer ones, which are much longer than the thickness of the skin. The filaments resemble closely Prum's Stages I and II of feather evolution. Other theropods that clearly have well-developed bipinnate feathers on their arms but ones too small for flight (Ji et al., 1998) have been discovered This is consistent with the cursorial theory of feather evolution (Chapter 3) in which it is proposed that feathers on the arm helped a bipedal, running cursor balance, maneuver, and catch prey with its arms and hands.

6-12). Because of the complex configuration, each of these muscles may affect adjacent feathers (e.g., one muscle may depress one feather and elevate an adjacent one).

Horny scales also begin their development when an underlying dermal papilla pushes up the overlying epidermal cells, which suggests that feathers and scales are homologous structures and that feathers may have evolved from the horny scales of reptiles. Feathers on the legs of owls and some other birds develop at the tips of horny scales, which reinforces the notion of a homology between these structures. Several investigators recently have studied the evolution of feathers (Focus 6-2).

The evolution of feathers and skin structure is closely correlated with the evolution of endothermy (see Focus 3-4). Air can be trapped by the feathers and held close to the skin as an insulating layer. The thickness of the layer of air is controlled by the dermal smooth muscles, which determine the positions of the feathers. The skin is also very vascular. Smooth muscles in the walls of the blood vessels regulate the blood flow through the skin and the degree to which body heat may be lost or conserved.

Although not organized as distinct glands, many epidermal cells of birds secrete lipids that help protect and waterproof the skin (Merton, 1984). The most conspicuous integumentary gland of birds is a single, branched, alveolar **uropygial gland** (Gr., *oura* = tail + *pyge* = rump), which is located above the base of the tail and produces fatty and waxy secretions that a bird spreads over its feathers as it preens. The gland is largest in aquatic species, such as penguins, which use its secretions to waterproof the feathers. It is absent from ostriches and a few other species. Beyond this, a few wax glands occur in the external ear canal.

The Skin of Mammals

The Epidermis and Its Derivatives

Mammals have a thick epidermis with a many-layered stratum corneum of dead and flattened cells filled with keratin bound to phospholipids (Fig. 6-13). It reduces water loss and other exchanges between the body and the external environment. Various transitional layers can be recognized between the stratum germinativum and stratum corneum as newly formed cells differentiate, accumulate keratin, and die. The stratum corneum is exceptionally thick and forms **footpads** on the soles of the feet and toes of many mammals.

Hair and Other Keratin Derivatives **Hair** is the distinctive keratinized derivative of mammalian skin. In furred species, it affords mechanical protection and entraps a layer of still air that provides excellent insulation. Human body hairs are much reduced and are primarily tactile. A representative hair consists of a **shaft** of dead, keratinized cells, the base of which is embedded in a **hair follicle** in the dermis (Fig. 6-13). Most of the follicle wall is composed of epidermal cells that grow down into the dermis with the developing hair. The epidermal part of the follicle is surrounded by fibrous dermal tissue containing nerve endings. The base of the follicle is enlarged, forming a **root,** into which a conical **hair papilla** protrudes. The papilla contains nerve endings and many capillaries that nourish the developing hair. As new cells are formed by mitosis just above the hair papilla, they elongate and are added to the base of the hair shaft. A hair shaft usually

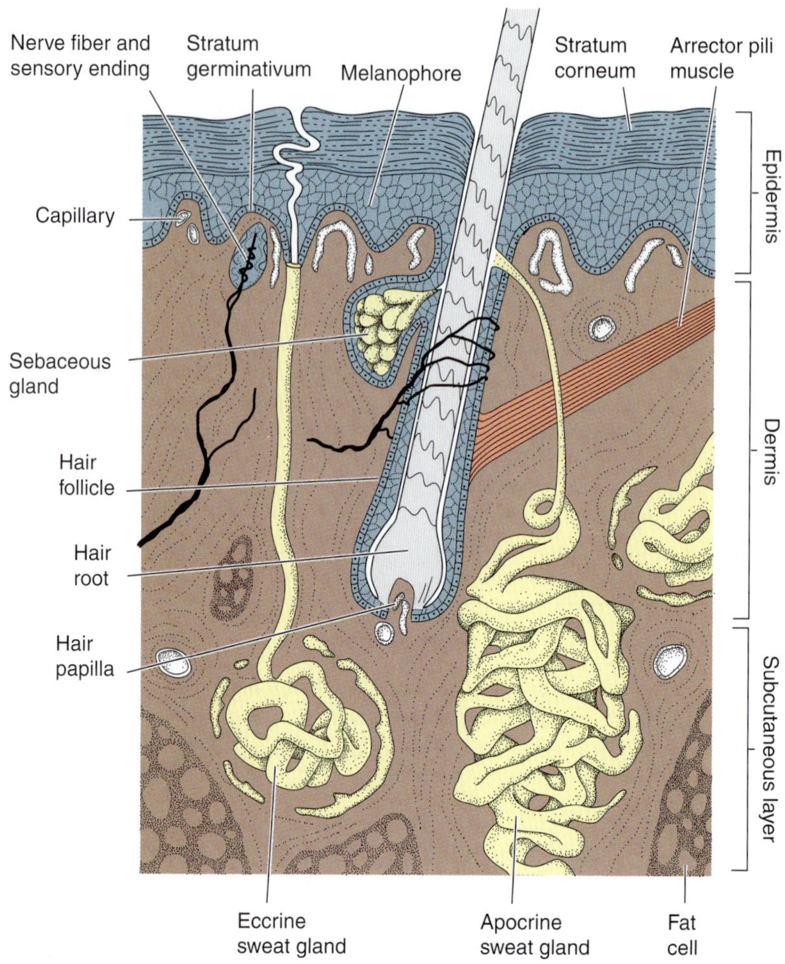

FIGURE 6-13
A vertical section through human skin.

has a **medulla** of shrunken, dead cells, a thickened **cortex** that imparts most of the strength, and an outer cuticle of overlapping **cuticular scales** (Fig. 6-14). The pattern of the cuticular scales and medulla differs enough among mammals that it can be used to distinguish many species.

Hair color results from pigment within the cells of the cortex. White is caused by the absence of pigment together with the refraction of light from air spaces between the cells of the medulla. The amount of pigment determines the shade of hair color.

Each follicle and hair shaft usually lies somewhat oblique to the skin surface. A group of smooth-muscle fibers of dermal origin form an **arrector pili muscle** (L., *arrectus* = set upright + *pilus* = hair), which originates from the upper layer of the dermis and attaches to one side of the follicle in such a way that its contraction can pull the hair upright. Marine mammals have no muscles associated with their hair. These muscles are under the control of the sympathetic nervous system and often contract under stress. Human goose bumps and the hackles that rise along the back of the neck when two hostile dogs confront each other are familiar examples. As temperature in the external environment drops, the change is detected by peripheral sensory receptors in the skin (see Chapter 12), and messages are relayed to the central centers in the spinal cord and brain. Responses are then sent via the sympathetic nerves to the arrector pili muscles, raising the hairs and thereby increasing the thickness of the layer of insulating air trapped near the skin surface. Regulating the thickness of the insulating air layer helps mammals maintain a relatively constant internal temperature through a wide range of temperature changes. The degree of blood flow through the skin helps regulate the exposure of blood to air temperature, thereby playing a role in thermoregulation (see Focus 3-4).

Hairs vary greatly on different parts of the body. Our fine body hairs lack a medulla and grow very little after they are formed, whereas scalp hair grows continuously. Hair life ranges from a few months for hair in our armpits to nearly four years for scalp hair. Growth stops when mitosis ceases in the root. After a period of quiescence, during which the base of the old hair shaft separates from the root, mitosis resumes and the newly forming shaft pushes out the old hair.

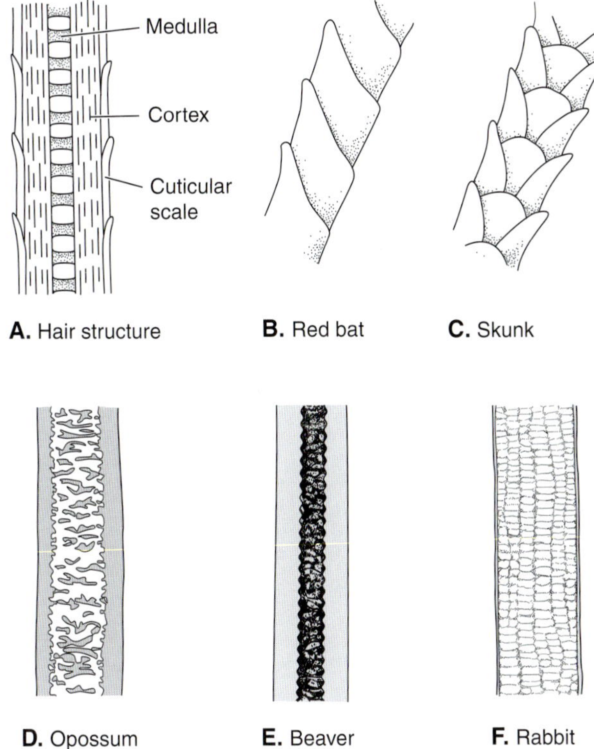

FIGURE 6-14
Structure of hair. *A,* Longitudinal section through a hair showing the medulla, cortex, and cuticular scales. *B,* Surface view of the cuticular scales of a red bat. *C,* Surface view of the cuticular scales of a skunk. *D,* Longitudinal section showing the medullary structure of opossum hair. *E,* Longitudinal section showing the medullary structure of beaver hair. *F,* Longitudinal section showing the medullary structure of rabbit hair. *(After Tumlison).*

Hairs also differ greatly among species. Most furred mammals have long **guard hairs** that protect a denser coat of soft **underhair.** A thick covering of underhair and guard hairs forms fur, or **pelage.** Guard hairs are modified as **quills** in echidnas, hedgehogs, and porcupines. Long, tactile whiskers, **vibrissae,** occur on the snouts of cats, dogs, rats, and many other mammals (Chapter 12). Patterns of hair replacement vary. Kangaroos, foxes, and some other species molt once a year. Arctic foxes, snowshoe hare, the ermine, and some weasels molt twice a year and have different-colored white winter and brown summer pelages. Hair is nearly absent in cetaceans, where its insulating function is performed by a layer of subcutaneous blubber (p. 228). Hair is also reduced in armadillos, elephants, rhinoceroses, hippopotamuses, and some other large mammals living in hot climates.

Zoologists believe that hair is a unique derivative of the epidermis and did not evolve from horny scales. Unlike horny scales and feathers, hair develops from an ingrowth of the epidermis into the dermis (Fig. 6-15). As further evidence that hair did not evolve from scales, both hairs and scales are present on the tails of some rodents. The hairs emerge in groups of three to five between the scales. This pattern of hair clustering can sometimes be seen on parts of the body where no scales are present. Maderson (1972) proposed that hairs evolved as tiny outgrowths between the scales that may at first have had a tactile function. As these outgrowths became more numerous, they began to serve as insulation.

At the end of each digit, the keratinizing system of the skin produces curved, laterally compressed, com-

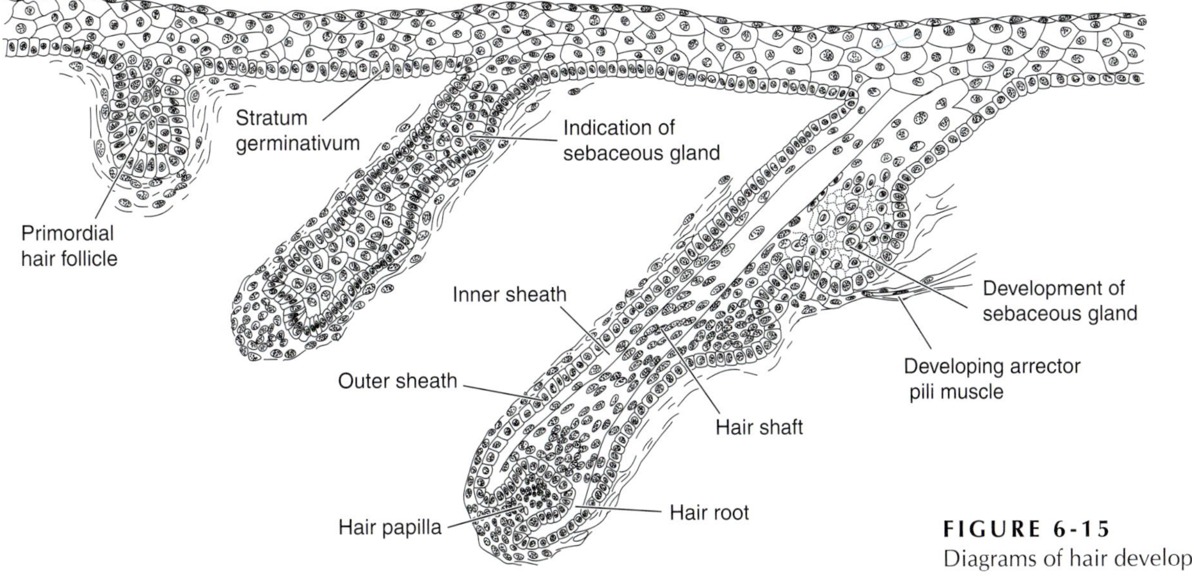

FIGURE 6-15
Diagrams of hair development.

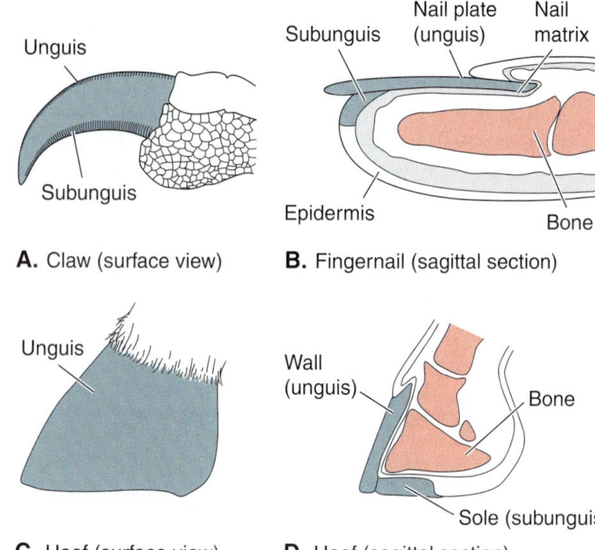

FIGURE 6-16
Surface view (*A*) of a claw, sagittal section (*B*) of a fingernail, surface view (*C*) and sagittal section (*D*) of a hoof. *(After Waterman.)*

pacted, cornified projections, which we call **claws** (Fig. 6-16), present in most mammals, most birds, reptiles, and even in some amphibians. However, in primates, claws are modified as **nails,** which are plates of tightly compacted, cornified epithelial cells on the surface of fingers and toes. New nails are formed from the nail matrix at the nail base by pushing the existing nail forward to replace the worn or broken free edge. Nails stabilize the skin at the tips of fingers and toes while establishing a secure friction grip during grasping. Some mammals have other cornified derivatives of the skin. The **horns** of sheep, cattle, and African antelopes, which are used in defense and courtship, consist of a core of dermal bone covered by a thick layer of keratinized cells that develop from the stratum germinativum (Fig. 6-17). Horns usually occur in both sexes, grow and wear away continually, and do not branch. They are not shed, unlike the true horn of bovids (cattle). The pronghorn of western North America has a unique type of a forked horn with one branch and a covering that is shed annually.

Horns should not be confused with **antlers,** which perform similar behavioral functions. The branched antlers of deer and related mammals are bony outgrowths of the skull that are covered by skin, the **velvet,** only during their growth. As antlers mature, the velvet is sloughed off. Antlers usually are found only on males; caribou are an exception. After the mating season, bone is reabsorbed at the base of the antlers, and they fall off.

The horn of the rhinoceros is yet another type. It consists of a solid mass of hairlike keratin fibers ce-

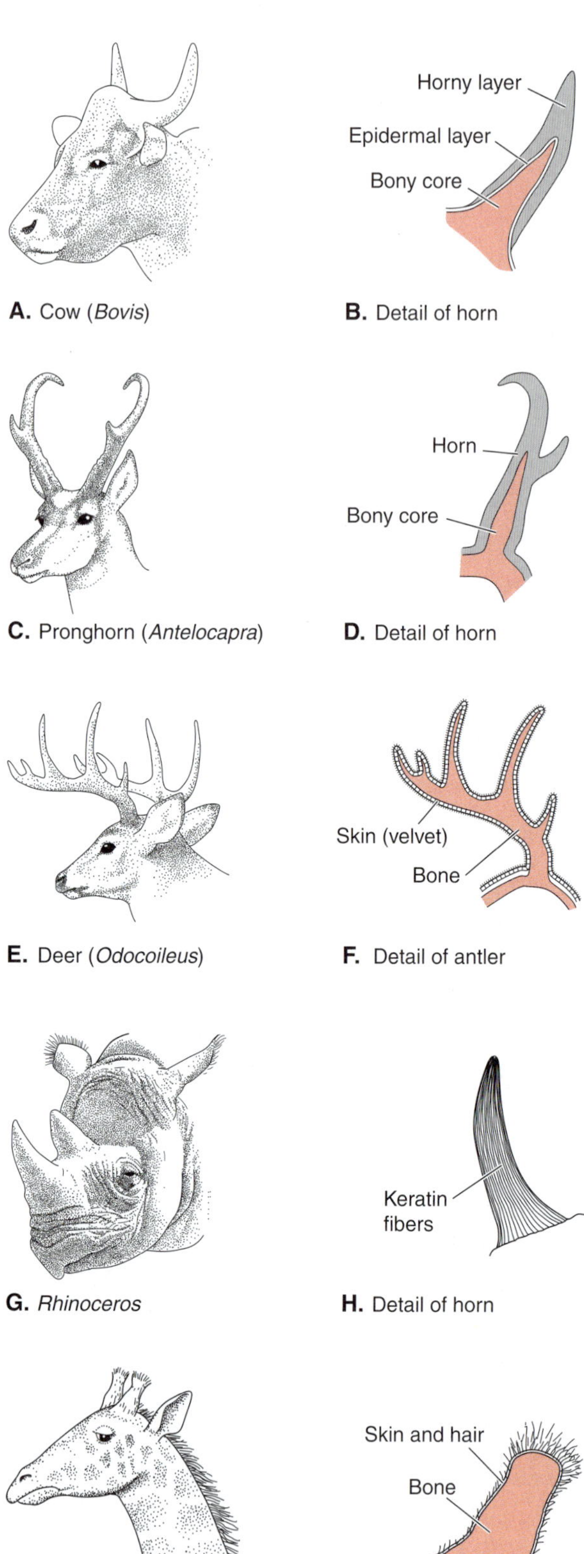

FIGURE 6-17
Types of horns and antlers as seen in surface views and sagittal sections. *A* and *B,* Cow. *C* and *D,* Pronghorn. *E* and *F,* Deer. *G* and *H,* Rhinoceros. *I* and *J,* Giraffe.

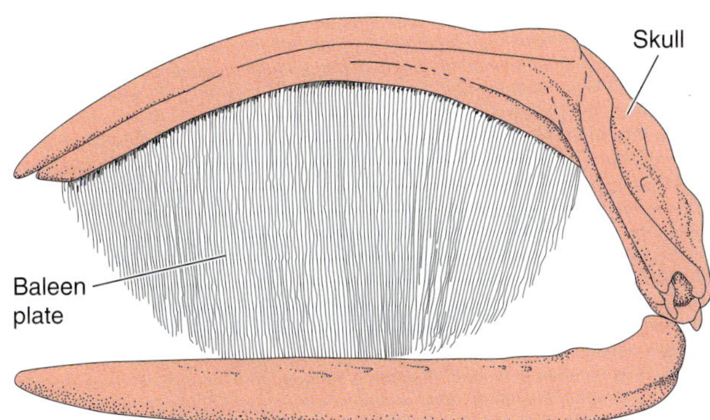

A. Skull and baleen of an Atlantic right whale (*Eubalaena*)

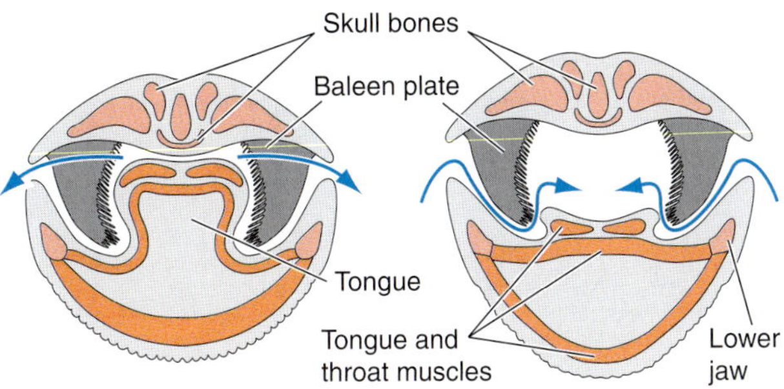

B. Water flow out of mouth **C.** Water flow into mouth

FIGURE 6-18
Baleen plates and feeding system of a mysticete whale. *A,* Lateral view of the skull and baleen of an Atlantic right whale (*Eubalaena*). *B,* Transverse section through the mouth and baleen plates, showing the tongue in an elevated position, with water flowing out of the mouth. *C,* Transverse section through the mouth and baleen plates, showing the tongue in a depressed position, with water flowing into the mouth. *(A, After Vaughan; B and C, after Portmann.)*

mented together. It is tragic that its alleged magical and aphrodisiac properties have encouraged poachers and placed rhinoceroses on the endangered species list. The giraffe's horns are small, ossified knobs covered by skin.

The large, toothless whales have plates of **baleen** that hang down from each side of the palate and filter plankton from mouthfuls of water (Fig. 6-18). Each plate is composed of hairlike keratin fibers that are cemented together at the base of the plate but free distally. In the 19th and early 20th centuries, baleen plates were used as corset stays and spokes in umbrellas.

Glands Several types of multicellular cutaneous glands evolved during mammalian evolution. Their characteristics have been reviewed by Blackburn (1991). **Sebaceous glands** (L., *sebum* = tallow) are branched alveolar glands (Fig. 6-13) that produce oily and waxy secretions, which are dis
cells producing the materials break d
of secretion is called **holocrine.** Seb
velop embryonically as buds from the
6-15) and usually discharge into the hair follicles, lubricating and waterproofing the hairs, but those on the nipples and some other parts of the body open directly to the surface. Long sebaceous glands, the **tarsal glands,** open on the edge of the eyelids. Their oily secretion helps protect the surface of the eyeball and coats the rim of the eyelid so that tears do not overflow.

Sweat glands occur in most mammals, but they usually are not as widely distributed on the skin surface as in humans. Cetaceans, sea cows, and a few other mammals have none. Sweat glands are coiled, tubular glands (Fig. 6-13), the secretions of which usually are released without cell destruction (**merocrine secretion**), or only the tip of the cells ruptures (**apocrine secretion**). The contraction of myoepithelial cells around the secretory cells helps discharge the sweat. Two types are recognized. **Eccrine sweat glands** develop as invaginations from the skin surface and secrete a watery solution that is discharged directly onto the skin surface. In most mammals, these surfaces are frictional and sensory surfaces, such as footpads and the snout. Sweat is mostly water, although it may contain some urea and salts, and the water appears to increase frictional adhesion and enhance tactile perception. Sweat plays an important role in the evaporative cooling of the body only in certain anthropoid primates, such as humans, in which the glands are abundant over the general body surface. Secretion is under nervous control.

Apocrine sweat glands resemble eccrine glands in most ways but, like sebaceous glands, develop as outgrowths from the hair follicles and usually discharge their secretion into the follicles and not directly onto the body surface. Most of these glands are limited to the armpits and genital areas in human beings, but they are more widespread in other mammals. Their secretions are complex, organic molecules often responsible for body odors. The **scent glands** of mammals, which may occur on any part of the body, are modified apocrine glands. Their secretions are pheromones used in marking trails and territories, in courtship behavior, and sometimes in defense. Some scent glands discharge into the anal or urinary areas, making the pheromones a component of the feces or urine. We have all seen dogs leaving identifying markers on hydrants and trees. The wax glands of our external ears also are modified apocrine glands.

The distinctive **mammary glands,** from which mammals get their name, display a mixture of the characteristics of sebaceous glands and apocrine sweat

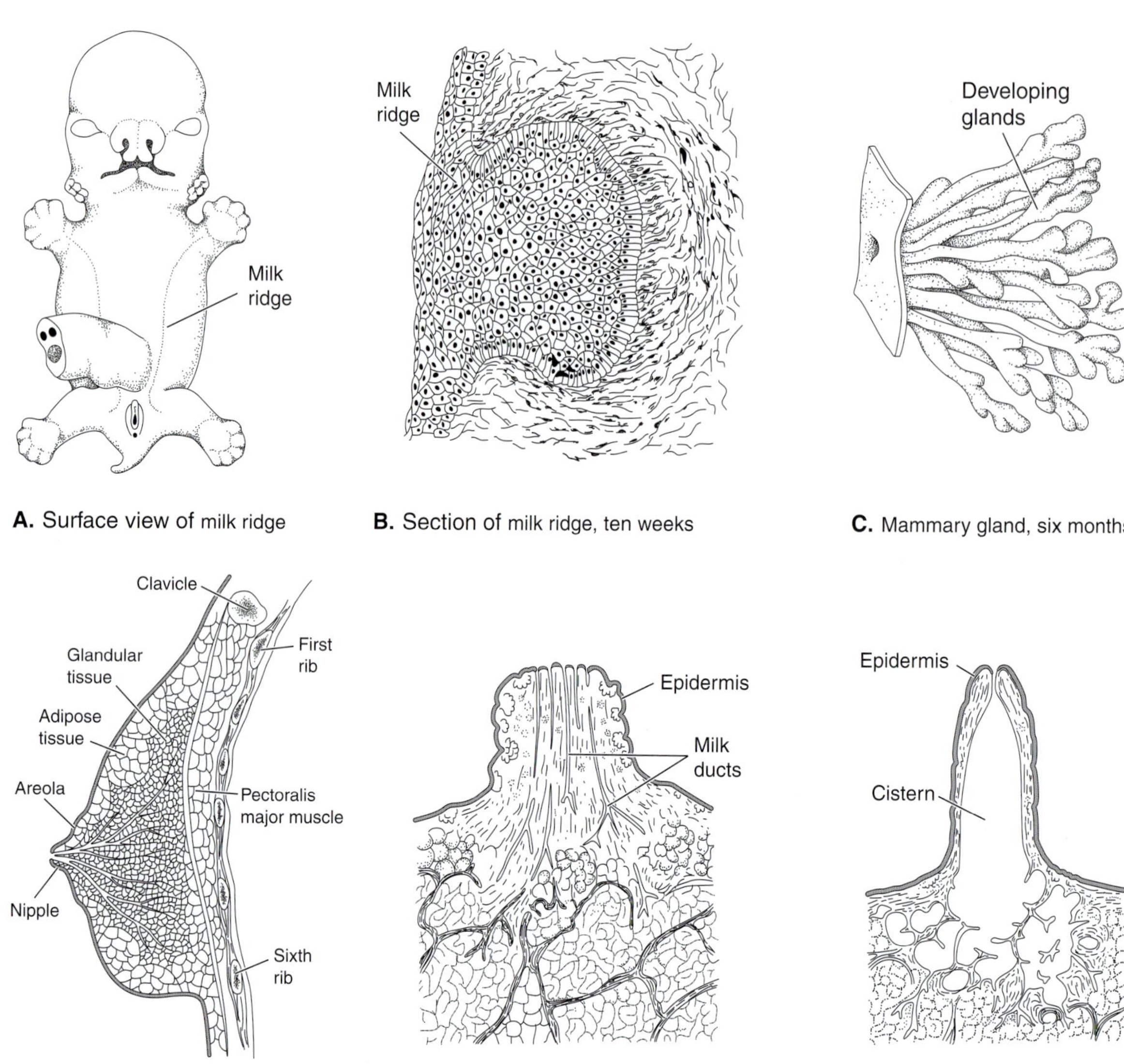

FIGURE 6-19
Mammary glands. *A–C,* The development of a mammary gland from the milk ridge. *D,* Section through a mature human mammary gland. *E,* A vertical section through a nipple. *F,* A vertical section through a teat. *(A–C, After Corliss; D, after Kluge; E and F, after Rand.)*

glands. They resemble sebaceous glands in having a branched alveolar structure, in synthesizing complex organic molecules, in beginning their maturation at the onset of puberty under the influence of steroid hormones, and in lacking a motor innervation to their secretory cells. Mammary glands resemble apocrine glands in having myoepithelial cells and a more copious secretion. Mammary glands develop embryologically from a pair of continuous epithelial milk ridges that grow down into the dermis and extend from the armpit to the groin (Fig. 6-19*A–C*). The definitive glands may develop along most of the ridge, as in the long series of mammary glands in pigs, or from limited parts of it. Glands develop near the groin in ungulates; primates have a single pair of mammary glands located in the pectoral region, whereas elephants and edentates have teats in the armpits. Extra patches of glandular material may develop abnormally along parts of the milk ridge and usually atrophy.

Mammary glands contain considerable adipose tissue along with secretory cells (Fig. 6-19*D*). They secrete milk under hormonal influence only during lactation (Chapter 15). Milk contains water, carbohydrates, fat, protein, various minerals, and antibodies. Parts of the duct system are enlarged and store milk until the young are fed. The ducts open to the surface in different ways. In monotremes, the ducts discharge into a small depression on the belly. The parent rolls over and, because monotremes have leathery bills rather than fleshy lips and cheeks and cannot suck efficiently, the young lap the milk from the hairs. Both male and female monotremes have functional glands. In other mammals, the functional glands are restricted to females, and the ducts open onto **nipples** or into **teats** (Fig. 6-19*E* and *F*). Many ducts reach the surface of nipples. A teat, as in a cow, is formed by the elevation of a collar of skin around a large cistern into which the ducts discharge. The teat itself has a single opening from the cistern. Mammary glands are a particularly important derived feature of mammals, for they necessitate a close association between the young and mother. This association provides young mammals more time to learn to cope with the rigors of their world, and it plays a key role in the evolution of social structure in mammals.

It has long been believed that mammary glands evolved as a modification of sebaceous or apocrine sweat glands, but the peculiar mixture of characters from both gland types made it uncertain from which one. Blackburn (1991) proposed that mammary glands did not evolve from either one of these glands but as a neomorphic mosaic. Mammals clearly have the genetic information needed to produce sebaceous and both types of sweat glands. Therefore, within certain developmental constraints, mammals should have the genetic potential to recombine certain characteristics of these glands into a new association, the mammary gland. If the organic compounds secreted by a protomammary gland of ancestral mammals, although possibly not copious, were beneficial in some way to the newborn, natural selection would favor further development of the gland and an increase in their secretion at the time of birth.

Vitamin D Synthesis Mammalian epidermis is also the site for the synthesis of vitamin D, the "sunshine" vitamin. Under the influence of ultraviolet rays, a precursor of vitamin D called the provitamin **7-dehydrocholesterol** in epidermal cells is converted to **vitamin D.** Vitamin D plays a key role in the absorption of calcium from the intestine and the deposition of calcium and other minerals in bone. The rate of vitamin D synthesis depends on the degree of skin pigmentation, because heavy pigmentation limits the amount of ultraviolet radiation that penetrates the skin. Many investigators believe that different degrees of pigmentation among people living in different parts of the world are adaptations to the amount of ultraviolet radiation, which helps explain the marked correlation between the heavy pigmentation of people living at equatorial latitudes and the lighter pigmentation of people living at higher latitudes. Normal tanning in the summer in lightly pigmented people may be a mechanism to maintain a constant rate of vitamin] the seasonal increase in ultraviolet ra

The Dermis and Subcutaneous Tissue

The dermis of mammals is similar to that of other vertebrates. Dermal bony elements of the skull and pectoral girdle develop in the deeper parts of the skin or just beneath it; otherwise, dermal bone is uncommon in mammalian skin. Bone is found under the keratinized epidermal plate of armadillos. In the related, extinct glyptodonts, most of the body was sheathed by dermal plates that formed a shield several centimeters thick.

Mammalian dermis has a particularly extensive vasculature (Fig. 6-20). Its networks of arterioles, capillaries, and venules play important roles in thermoregulation and blood pressure control. **Precapillary sphincters** in arterioles, and **postcapillary sphincters** in the venules, regulate capillary blood flow. A free flow of blood in the dermal capillaries, which possess a very large surface area, allows heat to radiate from the body, thereby cooling it. When heat needs to be conserved, the sphincters constrict, and blood flow through the capillaries is reduced greatly. The degree to which the hairs are elevated also affects heat loss from the surface.

The dermal vasculature also affects blood pressure because various shunts and ateriovenous anastomoses in the skin regulate blood volume in other parts of the body. If body blood pressure rises, arteriovenous shunts close and precapillary sphincters in the dermis open, allowing a greater volume of blood to flow through the extensive capillary beds, which reduces the

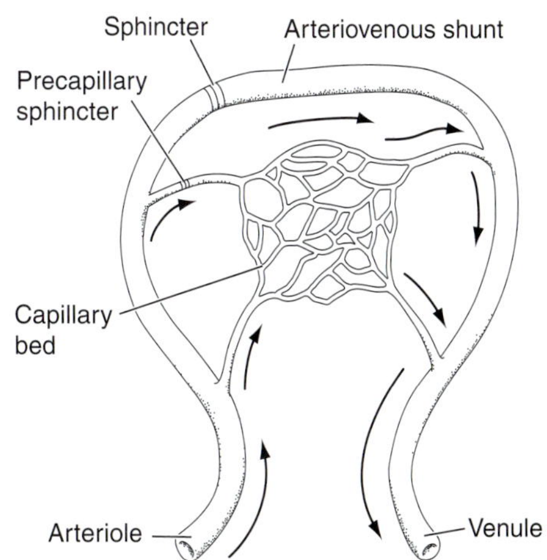

FIGURE 6-20
A simplified scheme of a dermal vascular arteriovenous shunt and capillary bed, with the location of various sphincters indicated.

volume of blood flowing through other parts of the body, resulting in a general lowering of blood pressure. When blood pressure in the body drops too low, the sphincters for the arteriovenous shunts in the dermis open. Blood is shunted away from the skin, increasing blood volume in the rest of the body, and blood pressure rises.

A layer of subcutaneous tissue, which is composed of bundles of connective tissue fibers interspersed with fat, lies between the dermis and the muscles of the body wall. This layer is far thicker in birds and mammals than in other vertebrates and helps insulate the body. It forms an exceptionally thick layer of **blubber** in cetaceans and other marine mammals. The blubber represents 40% or more of body weight in some of the larger whales, and its fat content is 40% to 60%. The blubber of whales helps streamline body shape, provides buoyancy, and is an important fuel reserve. Migrating whales do most of their feeding in feeding grounds close to the polar regions, where cold water and upwelling of minerals make plankton and other food abundant. Blubber thickness increases greatly during this period. Whales feed less frequently, or not at all, as they migrate to their temperate breeding grounds. They utilize the fuel in blubber, and the thickness of this layer decreases significantly. Blubber is important insulation for these animals in their cold feeding grounds. A thinning of the blubber when the whales enter warmer waters allows more body heat to be lost and prevents heatstroke. Heat loss is a problem for whales because their skin is far less vascular than is that of other mammals, and they do not sweat or pant. However, heat is lost in thermal windows, where the blubber is thin, that is, in flukes and flippers.

The Evolution of Vertebrate Skin

As we have seen, the earliest known jawless vertebrates, such as the †osteostracans, had a dermal armor of bony plates. This primary dermal armor underwent radical changes during the evolution of the vertebrates. The first transformation occurred in the †placoderms, in which the armor broke up into smaller units, the denticulated cosmoid scales (Fig. 6-21). With the evolution of teeth from scales in the mouth, the †placoderms could perform new predatory functions, occupy new feeding niches, and assume new habits. The selective advantage of teeth was so great that they were retained throughout the subsequent adaptive radiation of vertebrates, with the notable exception of turtles and modern birds. In the Chondrichthyes, the dermal denticles of the denticulate cosmoid scales became transformed into placoid scales.

The denticulate cosmoid scales lost the superficial dermal denticles and became adenticulate in ancestral bony fishes. The adenticulate cosmoid scales of early rhipidistian fishes gave rise to the dermal pectoral girdle and dermatocranium, both of which were retained but modified in the subsequent evolution of the vertebrates. Adenticulate cosmoid scales also gave rise to the osteoderms and gastralia in some reptiles. Within the actinopterygian fishes, the adenticulate cosmoid scales became transformed into ganoid scales, which, in turn, became modified successively into cycloid and ctenoid scales. The evolutionary history of the dermal skeleton is that of gradual morphological transformations accompanied by drastic functional changes. Thus the original protective function of the adenticulate cosmoid armor was changed into a locomotory function with the establishment of the dermal pectoral girdle, which resulted in a shift of the rhipidistians into an entirely new adaptive zone with an array of new selection pressures. Once the pectoral girdle had emerged, it led to an avalanche of evolutionary diversifications in the locomotory apparatus of the vertebrates.

Feathers and hair often are considered **key evolutionary** innovations, which can be defined as any changes in structure and function of an organ that permit organisms to assume radically new ways of life. Feathers and hair are essential to homeothermic endothermy of birds and mammals, respectively. Homeothermy is expressed in a high body temperature that is maintained at nearly a constant level (see Focus 3-4).This made possible a higher level of metabolic activity (endothermy), which is a prerequisite for flight in birds, and enabled mammals to be nocturnally active in a world dominated by diurnal reptiles. Constant high metabolic rates enable the nervous system to function more rapidly, enhancing the rate of information processing and, thereby, prey capture and predator avoidance. Endothermy conveys significantly greater stamina and aerobic capacity to produce continuous muscle contractions.

Feathers are unquestionably derived from reptilian scales. It is reasonable to accept the hypothesis that theropod reptiles ancestral to the birds already had feathers that functioned originally in a way not connected with flight, such as thermoregulation. These ancestral feathers represented one of the building blocks that led to the new function: flight. This example explains how an incipient structure could be favored by natural selection before reaching the elaboration for a radically new role. Similarly, it can be argued that mammalian hair originally functioned as tactile structures and that these tactile structures became converted into devices (hairs) for endothermy. Because we lack essential early fossil evidence, we can only speculate on early stages of hair evolution. Hair probably conveyed such a high selective advantage that it evolved very rapidly

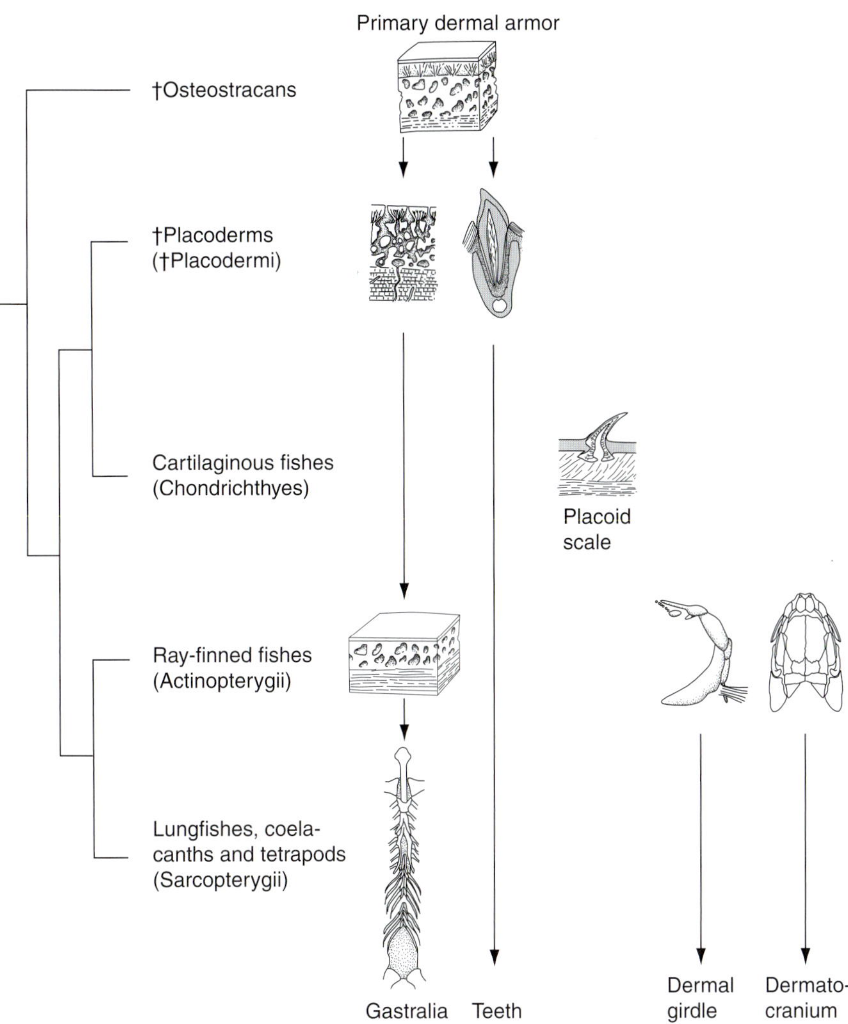

FIGURE 6-21 A cladogram shov of bony scales anc elements.

without intermediate stages. Thus feathers and hair are hypothesized to have functioned originally in a different way than their current primary function. Feathers, originally of value in thermoregulation, were also **preadaptations** for flight; and hair, originally tactile, was thought to be a preadaptation for thermoregulation. An organ is thought to be a preadaptation when an adaptive feature originated in a context different from its eventual function. The organ was then redeployed during evolution as a result of having been fortuitously suited to the new function (Skelton, 1985).

SUMMARY

1. The skin forms the interface between a vertebrate and its outside world. It protects the body in many ways; contains receptors; participates in some species in gas, water, and ion exchanges; sometimes helps nourish young; has a thermoregulatory role in birds and mammals; and participates in vitamin D synthesis and blood pressure regulation in mammals.
2. The skin consists of an epithelial epidermis of ectodermal origin and a connective tissue dermis derived from the dermatomes of the somites. Neural crest cells migrate between these layers, form chromatophores, and participate in many of the inductive interactions between the dermis and the epidermis that lead to the formation of skin derivatives: scales of many types, feathers, hair, and glands.
3. Skin color results primarily from the combination of chromatophore types and the degree of dispersal of pigment within them. Color can change morphologically, depending on the amount of pigment synthesized, and physiologically (in

anamniotes and reptiles) as pigment migrates within the melanophores.

4. Skin color is important in the lives of most vertebrates. Color may help conceal an individual, advertise an individual's presence, serve for species recognition, and send signals for courtship and establishing territories. Color also protects deeper body tissues from ultraviolet radiation, and it has a thermoregulatory role in some reptiles.
5. Fishes live in a more stable environment than do terrestrial vertebrates and have a simpler skin structure. Typically, no keratinization occurs, and most glands are unicellular mucous glands.
6. Most fishes have bony scales that develop in the skin from dermal–epidermal interactions. The scales of ancestral vertebrates were thick structures overlain by dentine and enamel-like materials. These scales may have offered mechanical protection and served as a reservoir for calcium and phosphate ions, and the enamel-like covering may have reduced water exchanges between the animal and its environment.
7. Scales became much thinner and lighter in later fishes and were lost in some groups. Cartilaginous fishes have only dermal denticles. Early actinopterygians had ganoid scales, and early sarcopterygian fishes had cosmoid scales, but these are reduced to thin disks of bone (cycloid or ctenoid scales) in recent species.
8. Some keratinization occurs in amphibian skin, and their skin glands are multicellular mucous or poison glands. Bony scales have been lost in most recent groups.
9. Heavily keratinized skin and horny scales evolved in reptiles. Most species shed their scales periodically. Keratin and phospholipids in the cells of the scales reduce evaporative water loss through the skin. Osteoderms, some of which form riblike gastralia, are present in many reptiles. Apart from scent glands in some species, reptile skin is aglandular.
10. Bird skin is characterized by the presence of feathers, which are homologous to horny scales. Feathers protect the body surface, reduce evaporative water loss, and insulate the body. The large flight feathers on the tail and wings form the flying surfaces. No dermal ossifications are present in bird skin. Distinct skin glands are generally absent except for a uropygial oil gland above the tail base and wax glands in the external ear canal.
11. Hair is the most obvious derived feature of mammalian skin. Hair protects the body surface and, with subcutaneous fat, is a thermal insulating layer. Hair distribution over the body and its modifications vary greatly among mammals.
12. Claws are retained in most mammals but are modified as nails in primates and as hooves in ungulates.
13. Other keratinized derivatives of the skin are horns of various types (not to be confused with bony antlers) and the baleen plates of the mysticetes.
14. Numerous glands evolved in mammalian skin, including sebaceous glands, sweat glands, and mammary glands, by which females nourish their infants. Nutrition of the young by mammary glands requires a longer association between mother and offspring than occurs in most other vertebrates.
15. Mammalian epidermis is the site for the ultraviolet-light–mediated conversion of a provitamin into vitamin D.
16. The dermis of mammals has an extensive and complex vasculature. The degree to which blood flows through the dermal capillaries, or bypasses them through arteriovenous shunts, affects thermoregulation and blood pressure.
17. During the evolution of vertebrate skin, the primitive dermal armor of ancestral jawless vertebrates broke up into smaller denticulated cosmoid scales in placoderms. Certain of these scales became teeth, and teeth enabled vertebrates to occupy new feeding niches and habitats. The superficial parts of the denticulated scales became the placoid scales of cartilaginous fishes. The ancestral, denticulated scales evolved into a wide variety of scales in bony fishes and formed the dermal parts of the pectoral girdle and skull and the osteoderms of early tetrapods.
18. Feathers and hair are key evolutionary innovations that enabled birds and mammals to become homeothermic. Homeothermy opened new niches to birds and mammals and made possible many changes in their modes of life.

REFERENCES

Autumn, K., Liang, Y. A., Hsieh, S. T., Zesch, W., Chan, W. I., Kenny, T. W., Fearing, R., and R. J. Full, 2000: Adhesive force of a single gecko foot-hair. *Nature,* 405:681–685.

Bagnara, J. T., and Hadley, M. E., 1973: *Chromatophores and Color Change.* Englewood Cliffs, N. J., Prentice-Hall.

Blackburn, D. G., 1991: Evolutionary origins of the mammary gland. *Mammals Reviews,* 21:81–96.

Bemis, W. E., and Northcutt, R. G., 1992: Skin and blood vessels of the snout of the Australian lungfish, *Neoceratodus forsteri,* and their significance for interpreting the cosmine of Devonian lungfishes. *Acta Zoologica (Stockholm),* 73:115–139.

Clarke, B. T., 1997: The natural history of amphibian skin secretions, their normal functions and potential medical applications. *Biological Reviews,* 72:365–379.

Corliss, C. E., 1976: *Patten's Human Embryology.* New York, McGraw-Hill.

Cott, H. B., 1957: *Adaptive Coloration in Animals,* 2nd edition. London, Methuen & Co.

Heatwolfe, H., Barthalmus, G. T., and Heatwolfe, A. Y., 1994: *Amphibian Biology,* volume 1, *The Integument.* New South Wales, Surrey Beatty, Chipping Norton.

Ji, Q., Currie, P. J., Norell, M. A., and Ji, S. A., 1998: Two feathered dinosaurs from northeastern China. *Nature,* 396:753–761.

Johnson, G. D., and Rosenblatt, R. H., 1988: Mechanisms of light organ occlusion in flashlight fishes, Anomalopidae (Teleostei: Beryciformes), and the evolution of the group. *Zoological Journal of the Linnean Society,* 94:65–96.

Kawauchi, H., Kawazoe, I., Tsubokawa, M., Kishida, M., and Baker, B. I., 1983: Characterization of melanin-concentrating hormone in chum salmon pituitaries. *Nature,* 305:321–323.

Kemp, N. E., 1984: Organic matrices and mineral crystallites in vertebrate scales, teeth and skeletons. *American Zoologist,* 24:965–976.

Kiaer, J., 1928: The structure of the mouth of the oldest known vertebrates, pteraspids and cephalaspids. *Paleobiologica,* 1:117–134.

Kluge, A., 1977: *Chordate Structure and Function,* 2nd edition. New York, Macmillan.

Lillie, F. R., 1942: On the development of feathers. *Biological Reviews of the Cambridge Philosophical Society,* 17:247–266.

Maderson, P. F. A., symposium organizer, 1972: The vertebrate integument. *American Zoologist,* 12:12–171.

Merton, G. K., 1984: Glandular functio
Overview. *Journal of the Yamashing I*
ogy, 16:1–12.

Montagna, W., and Parakkal, P. F., 1972
Function of the Skin, 3rd edition. New York, Academic Press.

Ørvig, T., 1957: Paleohistological notes: I. On the structure of the bone tissue in certain Palaeonisciformes. *Arkiv für Zoologie,* 10:481–490.

Parakkal, P. F., and Alexander, N. J., 1972: *Keratinization: A Survey of Vertebrate Epithelia.* New York, Academic Press.

Portmann, A., 1959: *Einführung in die vergleichende Morphologie der Wirbeltiere.* Basel, B. Schwabe.

Prum R. O., 1999: Development and the evolutionary origin of feathers. *Journal of Experimental Zoology,* 285: 291–306.

Rand, H. W., 1950: *The Chordates.* Philadelphia, Blakiston.

Rao, K. R., and Fingerman, M., symposium organizers, 1983: Chromatophores and color changes. *American Zoologist,* 23:461–592.

Saidel, W. M., 1988: How to be unseen: An essay in obscurity. *In* Atema, J., Fay, R. R., Popper, A. N., and Tavolga, W. N., editors: *Sensory Biology of Aquatic Animals.* New York, Springer-Verlag.

Skelton, P. W., 1985: Preadaptation and evolution in rudist bivalves. *Paleontology,* 33:159–173.

Sloan, C. P., and Mazzatena, O. L., 1999: Feathers for *T. rex*? *National Geographic,* 196:98–107.

Storer, T. I., 1943: *General Zoology.* New York, McGraw-Hill.

Tumlison, R., 1983: An annotated key to the dorsal guard hairs of Arkansas game mammals and furbearers. *Southwestern Naturalist,* 28:315–323.

Waterman, A. J., 1971: *Chordate Structure and Function.* New York, Macmillan.

Xu, X., Tang, Z. L., and Wang, X. L., 1999: A therinzinosauroid dinosaur with integumentary structures from China. *Nature,* 399:350–354.

7

The Cranial Skeleton

PRECIS

Skeletons form the framework of the body. We examine the cranial skeleton in this chapter and the postcranial in the following two. The cranial skeleton protects the brain, sense organs, and other soft tissues of the head and plays important roles in feeding and respiration. It is composed of an endoskeletal chondrocranium, which encases much of the brain and encapsulates the inner ear and nose; an endoskeletal splanchnocranium, which contributes to the jaws and forms arches supporting the pharynx wall and gills; and a dermal dermatocranium, which largely encases the previous two parts. We begin our consideration of the cranial skeleton by examining its condition in jawless craniates. Then we discuss the origin and suspension of jaws and go on to examine the cranial skeleton in the representative gnathostome groups. Emphasis is given to the evolution of the mammalian cranial skeleton.

OUTLINE

The Skeleton

The skeletons of animals form the framework of the body. They give the body its overall shape; provide support; protect many internal organs; and sometimes house blood-forming tissues, as in the marrow cavities of many vertebrate bones. Many bones form systems of levers that transfer muscle forces in support, locomotion, feeding, and respiration.

We often think of skeletons as dead and static, but they are dynamic, living tissues capable of growth, repair, and changing their composition and configuration as an animal grows and adapts to changing forces acting on its body. As we have seen (Chapter 5), bone is an important reservoir for calcium and phosphate ions, which are continuously interchanged with these ions in body fluids. We often take for granted the dynamic nature of the skeleton and its capacity for growth and repair, but these properties are unique to living organisms. They are not found in lumber, steel, concrete, and other structural materials used by builders.

So much can be learned from the skeleton that vertebrate morphologists often put more emphasis on this organ system than any other. The skeleton, including bony scales and teeth, fossilizes better than any other tissues of vertebrates; therefore, an analysis of these fossil skeletal remains contributes greatly to our understanding of vertebrate history. Because fishes live in water, their chance of fossilization in waterborne sediments is greater than for terrestrial vertebrates. Many entire fish skeletons have fossilized, but this is seldom the case for tetrapods. Paleontologists, however, often can hypothesize the entire skeleton by combining the known parts of several skeletons of the same species. Often, missing parts can be reconstructed from adjacent parts with which they articulate or by comparison with closely related contemporary species. Given a good reconstruction of the skeleton, much can be deduced about the soft parts and mode of life of the species. Muscles frequently attach on skeletal protuberances or leave attachment scars on the surface of bones. With this information and a knowledge of the muscular structure of related contemporary species, an investigator can hypothesize a reconstruction of the muscular system of an extinct species. The shapes of joint surfaces of limb bones, combined with an analysis of the proportion of limb segments, tell us a great deal about the posture of the animal and its mode of locomotion. Jaw proportions and the structures of the teeth provide information on feeding habits. Sensory canals on the head leave grooves or pits in the skulls that are clues to the nature of sensory systems. Cranial nerves and blood vessels pass through foramina in the skull. Casts of the inside of the skull often show the size and shape of parts of the brain. Canals in the skull that lodge the semicircular ducts of the inner ear tell us the position in which the head was held. This information is helpful, but as we cautioned in Chapter 3, reconstructions also can introduce errors.

Divisions of the Craniate Skeleton

For the purposes of discussion, we divide the complex skeletal system into smaller and more manageable parts. It should be recognized, however, that any subdivisions are abstractions that we make to suit various purposes. We can look at the craniate skeleton in at least three ways.

Dermal Skeleton and Endoskeleton

First of all, the craniate skeleton can be divided into a superficial **integumentary** or **dermal skeleton** and a deeper **endoskeleton.** Although separated by position within the body, the dermal and endoskeletons also differ in their mode of embryonic development. The dermal skeleton consists of bony scales and larger bony plates that develop as membrane bones in or just beneath the skin. The vertebrate dermal skeleton sometimes is called an exoskeleton because it lies so near the body surface, but we prefer to avoid this term for it leads to confusion with a true exoskeleton, that is, the hard secretion on the body surface found in insects, crustaceans, snails, clams, and many other invertebrates. Most of the craniate skeleton is a deeper endoskeleton composed of cartilage or cartilage-replacement bone. It is easy to recognize the distinction between dermal and endoskeleton in principle, but the reality is often more complex. Some dermal elements move inward and become intimately associated with endoskeletal elements in the formation of the skull and parts of the pectoral girdle.

Somatic Skeleton and Visceral Skeleton

Romer (1972) proposed that the vertebrate body is organized as a tube within a tube. The "outer" body wall tube is described as somatic, and the "inner" gut tube is described as visceral. Parts of the skeleton, muscles, nerves, and blood vessels associated with these two parts of the body can be called somatic and visceral. Thus we can divide the skeleton into a **somatic skeleton** and a **visceral skeleton** that are located, respectively, in the "outer" and "inner" tubes of the body. The somatic skeleton includes the dermal skeleton and most of the endoskeleton, but the part of the endoskeleton in the wall of the pharynx belongs to the visceral skeleton. The visceral skeleton forms arches of cartilage or bone that support the wall of the pharynx and the gills and help support the jaws in most vertebrates. The division of the body into somatic and visceral parts is helpful; however, the concept breaks down to some extent in the head, where neural crest cells of ectodermal origin migrate extensively and contribute to both the somatic and visceral skeletons.

Cranial Skeleton and Postcranial Skeleton

A particularly useful way of looking at the skeleton simply is to divide it regionally into a **cranial skeleton** in the head and a **postcranial skeleton** in the rest of the body. The cranial skeleton includes endoskeletal elements that encase and protect much of the brain, nose, and inner ear (**chondrocranium**); the endoskeletal visceral arches (**splanchnocranium**); and dermal elements that surround the other parts (**dermatocranium**). Except for the visceral arches, all parts are somatic.

The postcranial skeleton can be subdivided into an **axial skeleton** and an **appendicular skeleton** by focusing on the location of its parts within the body wall. The axial skeleton includes those parts of the skeleton that lie along the longitudinal axis of the body. Part of the chondrocranium may be considered to be a forward extension of the axial skeleton, but we limit the term to the postcranial axial structures: the notochord; the vertebral column, which develops embryonically around the notochord; the ribs; and, when present, the skeleton of the median fins and sternum. The appendicular skeleton lies in the body wall lateral to the axial elements, and it is composed of the skeletal elements of the paired appendages and their supporting girdles. The axial and appendicular skeletons are entirely somatic and, except for dermal plates associated with the pectoral girdle of most craniates, are composed of endoskeletal elements. A common method of dividing the skeleton, which we follow in this book, is listed here. All parts are derived from the endoskeleton except for the bones of the dermatocranium and dermal elements of the pectoral girdle. All are parts of the somatic skeleton except for the arches of the splanchnocranium.

- Cranial skeleton
  - Chondrocranium
    - Splanchnocranium (visceral skeleton)
    - Mandibular arch
    - Hyoid arch
    - Branchial arches
  - Dermatocranium
- Postcranial skeleton
  - Axial skeleton
    - Vertebral column (and notochord)
    - Ribs
    - Sternum
    - Median fins
  - Appendicular skeleton
    - Pectoral girdle and appendages
    - Pelvic girdle and appendages

We will examine the composition and evolution of the cranial skeleton in this chapter and the postcranial skeleton in the next two chapters.

Components of the Cranial Skeleton

As we have explained (Chapter 2), craniates are distinguished from tunicates and amphioxus by the presence of a well-developed head containing a concentration of sense organs, the brain, mouth, and (in fishes and larval amphibians) gills. All of these structures became protected and supported by a group of cartilages and bones that form the cranial skeleton. The cranial skeleton, therefore, has three primary functions: (1) protecting soft tissues, especially the brain and sense organs; (2) food gathering and sometimes food processing; (3) providing a passage for the respiratory flow of water or air and, in fishes and amphibians, providing the mechanism for generating this flow.

The three components of the cranial skeleton (chondrocranium, splanchnocranium, and dermatocranium) are quite distinct in their embryonic origin, and they can be distinguished easily in early fishes. During vertebrate evolution, the components become increasingly united and integrated, so it becomes difficult to distinguish them in the adults of advanced fishes and tetrapods. The cranial skeleton often is called the **skull,** but the term "skull" is most properly used for that part of the cranial skeleton that houses the brain and major sense organs and forms the upper jaws. The skull in this restricted sense forms nearly the entire cranial skeleton of mammals, but the cranial skeleton of fishes includes an extensive array of skeletal arches supporting the gills and jaws. Other vertebrates lie between these extremes. In order to understand their basic structure, we will examine each part separately in vertebrates in which the part is quite distinct and clear. After that, we will examine the evolution of the cranial skeleton.

The Chondrocranium

The function of the **chondrocranium** (Gr., *chondros* = cartilage + *kranion* = skull), sometimes called the **neurocranium,** is protection. The chondrocranium usually is a trough of cartilage and cartilage-replacement bone that covers the brain ventrally, caudally, and partly laterally. It also encapsulates the inner ear and nose.

Most of the chondrocranium arises from neural crest cells. Its structure can be understood by examining its embryonic development. A lizard (Fig. 7-1) is quite representative. The embryonic notochord extends forward beneath the brain, nearly as far as the pituitary gland. A pair of cartilaginous rods, called the **parachordals,** develops on either side of the rostral end of the notochord. As they enlarge, they unite and form a **basal plate.** A pair of rod-shaped **trabeculae** extends rostrally beside the pituitary gland and continues to the nasal capsules. The trabeculae form the floor of the rostral part of the braincase, leaving a space between them for the pituitary gland. Their rostral ends may unite to form an **ethmoid plate,** which lies between the pair of nasal capsules.

Caudally, several occipital elements surround the notochord and the back of the brain. They develop from mesodermal sclerotomes (Chapter 4) so are serially homologous to parts of vertebrae. Their union forms the **occipital arch.** The **foramen magnum** perforates the occipital arch, allowing for passage for the spinal cord. Depending on the species, one or a pair of **occipital condyles,** which articulate with the first vertebra, develop ventral or lateral to the foramen magnum. The dogfish has a pair of occipital condyles, but most fishes and tetrapods have a single one.

Nasal capsules develop around the nasal sacs and unite with the ethmoid plate. **Otic capsules** (Gr., *ot* = ear) develop around the parts of the ear that lie within the chondrocranium. This part of the ear, known as the **inner ear,** is composed of the semicircular ducts and associated sacs that contain the receptive cells for equilibrium and hearing. Most tetrapods also have external and middle ears that receive and transmit ground-borne or airborne sound waves to the inner ears; these are not present in fishes. **Optic capsules** (not shown in Fig. 7-1) begin to form around the eyeballs, but they do not unite with the rest of the chondrocranium because the eyeballs must be free to move. The optic capsules contribute to the fibrous tunic in the walls of the eyeballs (Chapter 12) and sometimes ossify as a ring of **sclerotic bones** (Gr., *skleros* = hard).

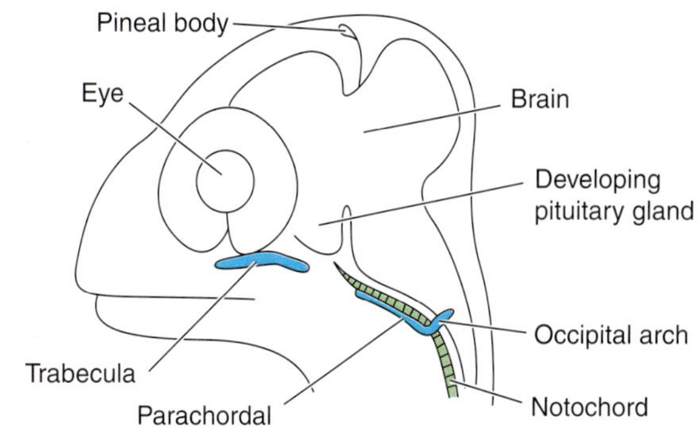

A. Embryo of *Lacerta* showing braincase *in situ*

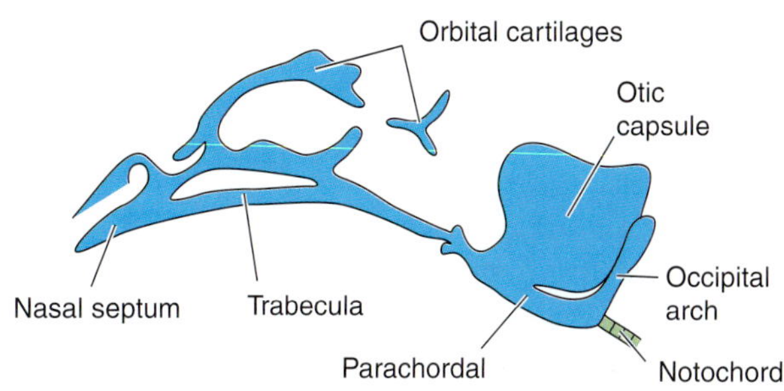

B. Braincase in an early embryo of *Lacerta*

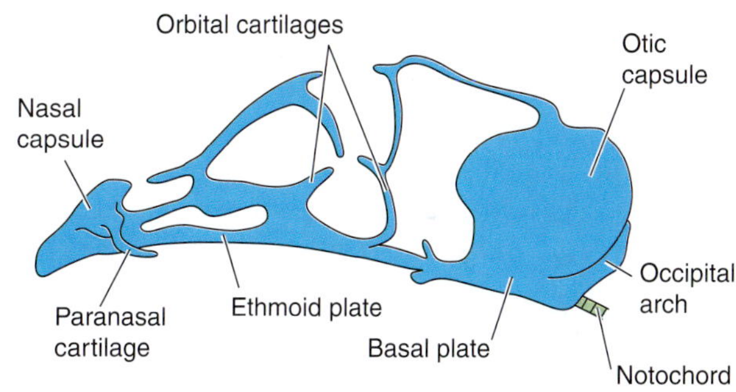

C. Braincase in an older embryo of *Lacerta*

FIGURE 7-1
Lateral view of three stages in the embryonic development of the chondrocranium of a lizard. *(After DeBeer.)*

The lateral walls of the chondrocranium between the eyes develop from a complex set of rods and pillars known as the **orbital cartilages.** These cartilages coalesce with each other and with adjacent parts of the chondrocranium, often leaving wide gaps through which the optic and other cranial nerves pass.

Usually, only one or two rods of cartilage cross the top of the brain. The brain is covered dorsally in most

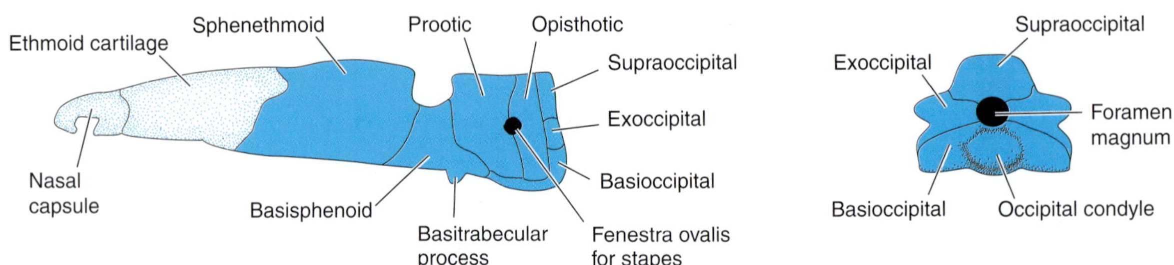

FIGURE 7-2
Structure of the adult chondrocranium based on that of an early tetrapod, †*Palaeoherpeton*. *A*, Lateral view. *B*, Posterior view. *(Modified after Romer and Parsons.)*

vertebrates by dermal bones. The term **braincase** or **cranium** is used for those elements, regardless of their origins, that encase the brain.

As development proceeds in most vertebrates, cartilage-replacement bones ossify within the chondrocranium and replace much of the cartilage. The basic pattern of these bones is clear in the ossified chondrocranium of †*Palaeoherpeton,* an early tetrapod (Fig. 7-2). Four occipital bones ossify in the occipital arch around the foramen magnum. Their names describe their position: a **basioccipital,** a pair of **exoccipitals,** and a **supraoccipital.** Two otic bones ossify in the otic capsule, a **prootic** rostrally and an **opisthotic** (Gr., *opisthen* = behind) caudally. A **basisphenoid** bone forms the floor and part of the lateral wall of the chondrocranium rostral to the basioccipital and otic bones. In **kinetic skulls,** in which other parts of the skull can move relative to the braincase, the basisphenoid may bear a basitrabecular process, which forms a movable joint with palatal bones. A trough-shaped **sphenethmoid** bone often lies between the orbits and rostral to the basisphenoid. The ethmoid region and nasal capsules seldom ossify.

The Splanchnocranium

The function of the **splanchnocranium** (Gr., *splanchnon* = gut), or **visceral skeleton,** of the earliest craniates may simply have been respiration, that is, supporting the gills and moving water across them, but it soon also became an integral part of the feeding mechanism. The splanchnocranium consists of a series of arches of cartilage or cartilage-replacement bone of neural crest origin that lie in the wall of the pharynx between the pharyngeal pouches. The splanchnocranium is better developed in fishes, in which the arches support the gills and add flexibility and elasticity to the pharynx, than it is in tetrapods. Muscle contraction compresses the arches, pharynx, and pouches, and the elastic recoil of the arches contributes to the expansion of these structures. The expansion and contraction of the pharynx are important movements in feeding and respiration of fishes and early tetrapods.

The composition of the splanchnocranium is shown in Figure 7-3, which is based on its condition in sharks. The first visceral arch, known as the **mandibular arch,** consists of a dorsal **palatoquadrate cartilage,** which forms the upper jaw of sharks, and a lower **mandibular cartilage** (also called **Meckel's cartilage**), which forms the lower jaw. The second visceral arch, known as the **hyoid arch,** lies close behind the mandibular arch. Its dorsal segment, the **hyomandibula,** extends from the otic capsule of the chondrocranium toward the caudal end of the palatoquadrate cartilage. It articulates with the palatoquadrate and helps to suspend the jaws. The hyoid arch continues ventrally into the floor of the mouth and pharynx. It is composed of the articulated segments shown in Figure 7-3. Space is insufficient for a complete gill pouch between the mandibular and hyoid arches. The pharyngeal pouch that one would expect to find between these arches is reduced to a **spiracle** or lost entirely.

The remaining visceral arches are associated with the gill apparatus and are known collectively as **branchial arches** (Gr., *branchia* = gill), or **gill arches.** Nearly all jawed fishes have five branchial arches (visceral arches 3–7 or branchial arches 1–5), but a few specialized sharks have secondarily acquired more. The first four branchial arches lie in the base of the gills. The last arch lies caudal to the last pharyngeal pouch, and sometimes it is very small. The branchial arches also may be given names according to some structure with which they are associated (Fig. 3-4). A representative branchial arch is a V-shaped, or sometimes a W-shaped, structure composed of the series of articulated cartilaginous rods shown in Figure 7-3. The **epibranchial cartilages** are serially homologous to the hyomandibular and palatoquadrate, and the **ceratobranchial cartilages** are serially homologous to the

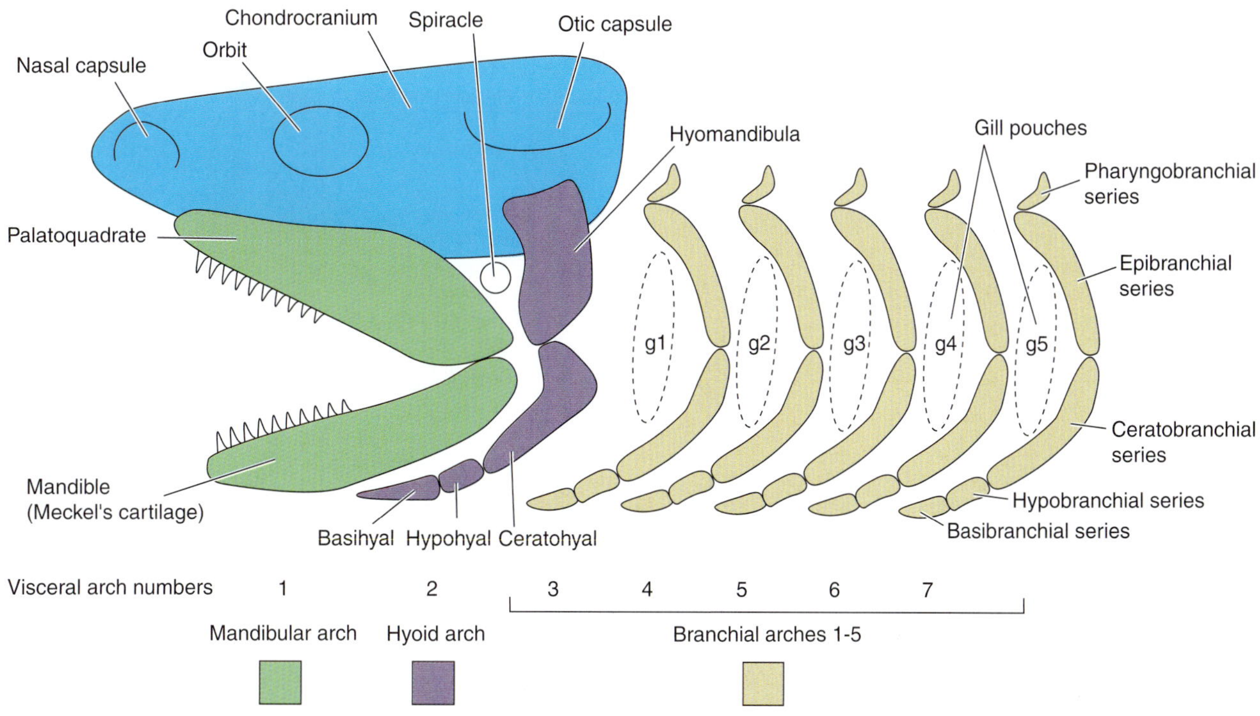

FIGURE 7-3
Lateral view of the structure of the splanchnocranium of an idealized early gnathostome. Loosely based on a shark to show terminology for the jaws, hyoid arch, and branchial arches. The chondrocranium is also included. Gill pouches are numbered g1 to g5. (The complete color coding that will be used throughout this chapter is shown in Figure 7-4.)

mandibular cartilage. The **basibranchials** of adjacent arches frequently fuse.

The Dermatocranium

Superficial dermal bones nearly completely cover the chondrocranium, splanchnocranium, muscles associated with the splanchnocranium and jaws, and eyeball (Fig. 7-4). Collectively, these bones form the **dermatocranium** (Gr., *derma* = skin). These bones arise embryonically largely from ectomesenchyme cells of neural crest origin, but some cells from the mesodermal dermatome contribute to them. In describing dermatocranial bones, it is convenient to group them into six series:

1. A **dermal roof** covers the top and sides of the head. The chondrocranium lies beneath it. Jaw muscles, the eyeball, and the palatoquadrate cartilage lie between it and the chondrocranium. The ventral border of the dermal roof contributes to the upper jaw. Caudally, the roof unites with the quadrate bone.
2. A **palatal series** of dermal bones develops in the roof of the mouth and covers most of the ventral surface of the palatoquadrate cartilage, leaving at least one large opening, the **subtemporal fenestra.** This fenestra allows for the passage of jaw muscles from their origin on the chondrocranium and underside of the temporal roof to their insertion on the lower jaw.
3. A slender **parasphenoid bone** lies on the ventral surface of the chondrocranium.
4. A **lower jaw series** of dermal bones nearly completely encases the mandibular cartilage and unites with the articular bone.
5. In bony fishes, an **opercular series** of dermal bones covers the branchial region laterally.
6. In bony fishes, a **gular series** of dermal bones covers the branchial region ventrally. The opercular and gular bones participate in feeding and respiratory movements.

Many jawed fishes have **teeth** attached to the jaws and, frequently, to the palatal bones and parasphenoid. Teeth may also attach to the branchial arches, especially in teleosts, many of which lack teeth on the jaws. Tooth structure and mode of attachment to skeletal elements vary considerably according to a vertebrate's diet and method of feeding. In the majority of verte-

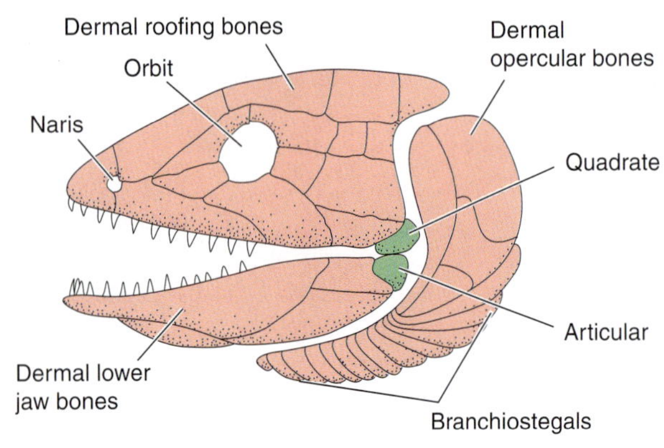

A. Dermatocranium in lateral view

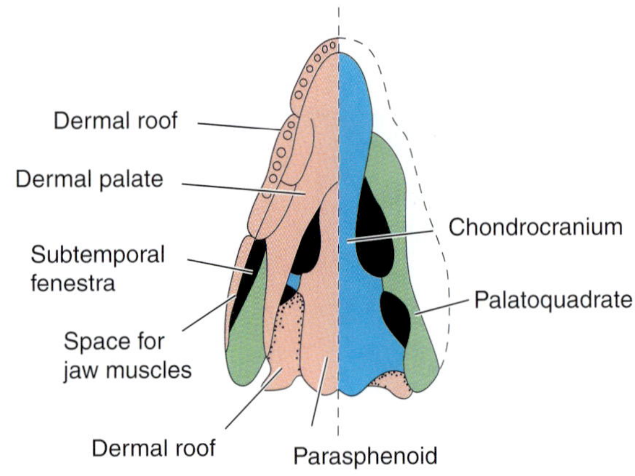

B. Ventral view of skull with dermatocranium removed from right side

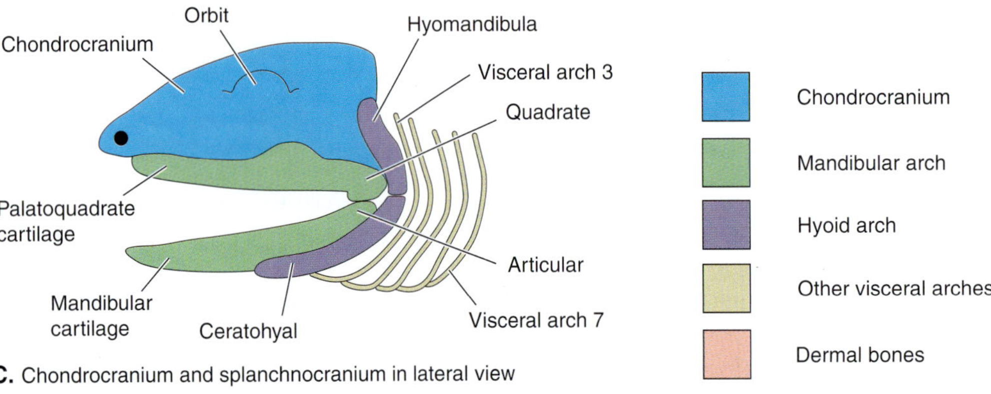

C. Chondrocranium and splanchnocranium in lateral view

FIGURE 7-4
Diagrams of the components of the cranial skeleton of a generalized early bony fish, loosely based on *Amia*. *A,* A lateral view to show the dermatocranial bones that cover most of the other components. *B,* A ventral view of the skull with dermatocranial bones removed from the right side of the drawing to expose other components. *C,* A lateral view after the removal of the dermatocranium, leaving the chondrocranium and splanchnocranium.

brate species, the teeth, although they may differ in size, are structurally alike, a condition called **homodont** (Gr., *homos* = the same + *odous, odont* = *tooth*). They usually are attached loosely to skeletal elements and not set in sockets. Teeth are functionally a part of an animal's food-gathering apparatus, and their distribution, structure, and pattern of replacement are related closely to feeding habits (Chapter 16).

Phylogeny of the Cranial Skeleton

Cranial Skeleton of Jawless Craniates

Hagfishes are the earliest surviving craniates (Fig. 3-2). Although they lack any trace of vertebrae (and thus are not vertebrates), they do have a cranial skeleton consisting of a small chondrocranium and cartilaginous visceral arches associated with the pharyngeal pouches. They are scavengers and, although they lack jaws, they can nibble off bits of their food with tooth plates that can be everted through their mouths.

Among jawless vertebrates, the cranial skeleton of the †osteostracans is particularly well known (Fig. 3-3*G* and *H*). Their large cephalothoracic bony shield, which covered the head dorsally and laterally, is certainly comparable with the dermatocranium. Their endoskeletal chondrocranium was ossified, and in some fossil †osteostracans it is complete enough to enable investigators to identify many details of brain, cranial nerve, and inner ear structure. Skeletal visceral arches lie between the pharyngeal pouches and were united with the cephalothoracic armor. Movements of a very flexible pharynx floor, which was covered by small, bony plates, mediated the movement of water and

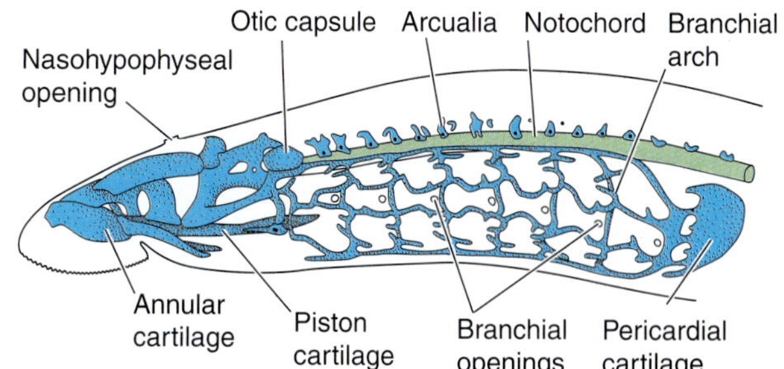

A. *Petromyzon*, lateral view of cranial skeleton

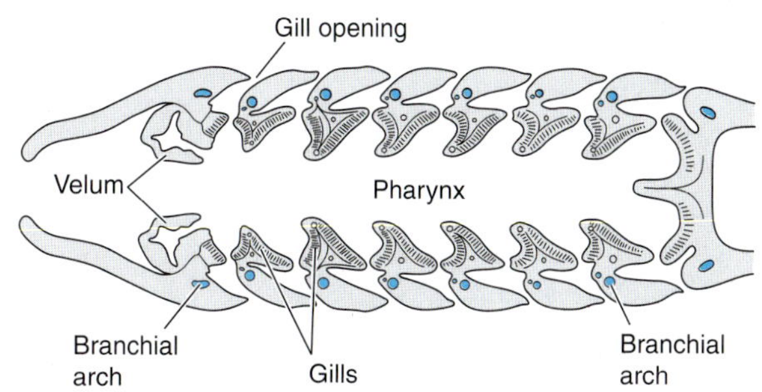

B. *Petromyzon*, frontal section through pharynx

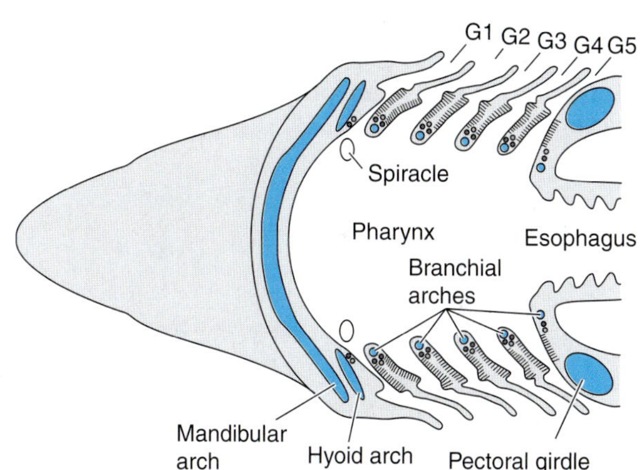

C. *Squalus*, frontal section through pharynx

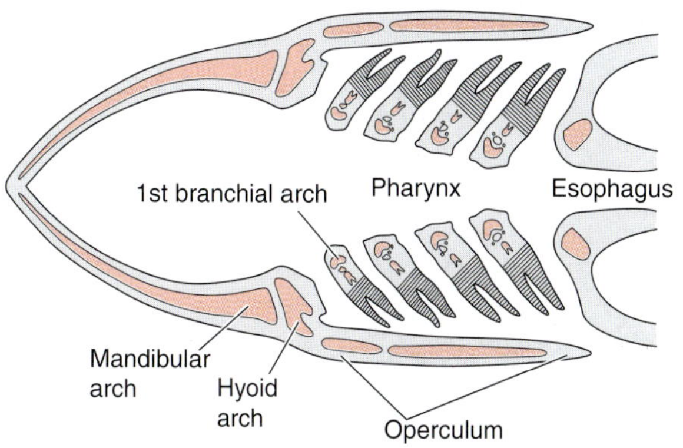

D. *Perca*, frontal section through pharynx

▶FIGURE 7-5
The splanchnocranium of representative craniates showing its relation to surrounding structures. *A,* Lateral view of the cranial skeleton of the lamprey. *B–D,* The location of the splanchnocranium as seen in frontal sections through the head. *B,* A lamprey. *C,* A shark. *D,* An actinopterygian fish. *(A and C, From Walker and Homberger, after Young; B and D, after Jarvik.)*

food into the mouth and through the pharynx. Water was expelled through many, ventrally placed gill slits.

Lampreys are the only surviving jawless vertebrates. The familiar sea lamprey, *Petromyzon marinus,* feeds by attaching to its prey, protruding its toothed tongue, rasping through the prey's skin, and sucking blood. Some brook lampreys do not feed as adults, for they reproduce quickly and die soon after metamorphosis from the feeding larval stage. Lampreys lack a dermal component to the cranial skeleton, which probably results from a secondary loss of dermal bones because most jawless vertebrates had dermal bones, and dermal bones also are present in the †anapsids, the sister group of lampreys (Fig. 3-2). The cranial skeleton of lampreys consists only of a small chondrocranium and a series of eight visceral arches, which are united to each other to form a **branchial basket** (Fig. 7-5*A*). As in hagfishes, †osteostracans, and other early jawless craniates, the visceral arches are rather superficial, lying peripheral to the gill pouches (Fig. 7-5*B*). They are composed of cartilages or bones that are fused together. In contrast, the visceral arches of jawed craniates lie deeply, next to the lumen of the pharynx (Fig. 7-5*C* and *D*), and they are composed of articulated segments. Because of these fundamental differences between the visceral arches of jawless and jawed craniates, we cannot be sure that the visceral arches are homologous in the two groups.

Gnathostomes: Jaws and their Suspension

Jawed vertebrates, or **gnathostomes,** have many features in common with the jawless †osteostracans and appear to be their sister group (Fig. 3-2). We are not certain how jaws evolved or how they were first used. The prevalent theory for the origin of jaws is based on the close resemblance between the mandibular arch and more caudal visceral arches in sharks (Figs. 7-3 and 7-4). It is argued that an anterior arch, which may originally have supported gills, enlarged and formed the jaws as the mouth opening moved caudally. A visceral arch near the mouth opening would be in a strategic position to help the fish open its mouth widely and suck in or seize food. Mallatt (1996) believes that the mandibular arch first had a respiratory function and only later was used in feeding. By opening the mouth, the mandibular arch would help expand the pharynx and draw in a current of water. By shutting the mouth, the arch would help expel the water across the gills and out the gill slits. Whatever its use, the present mandibular arch may not have been the first in the series. One or more **premandibular arches** that were later lost may have lain rostral to it. Supporting the notion of at least one lost premandibular arch is the resemblance in some ways of the paired trabeculae of the chondrocranium to the dorsal part of an arch. The trabeculae are rod shaped and arise from neural crest cells, as visceral arches do. Early craniates also have an "extra" cranial nerve (the profundus, or V_i, Chapter 14) that may have been associated with a premandibular arch. This nerve becomes a branch of cranial nerve V in most adult vertebrates.

Smith (1999) argues that the mandibular arch is different from other visceral arches and may have evolved from skeletal supports in a velum, a flaplike structure at the entrance to the pharynx in lampreys and probably extinct jawless craniates (Fig. 7-5*B*). No evidence shows that the mandibular arch ever supported gills, as most visceral arches do. The trigeminal (Vth cranial) nerve, which supplies structures associated with the mandibular arch, also supplies those associated with the velum. Some evidence indicates a difference in the genetic coding for the mandibular arch in comparison with other visceral arches, but more research is needed.

However the mandibular arch evolved, it is now an integral part of the jaw mechanism in fishes. Jaws were a major development because they freed craniates from a suspension-feeding and scavenging mode of life and allowed vertebrates to feed in many different ways and on many types of food. Many opportunities for diversification were opened, and fishes soon exploited them.

When the mandibular arch became, or contributed to, the jaws, the upper part of the mandibular arch, the palatoquadrate, needed to be anchored in some way. Its caudal end then could serve as a fulcrum for the movement of the lower jaw. The palatoquadrate is suspended in different ways among the many groups of jawed fishes. In †placoderms, the palatoquadrate was firmly attached to the underside of the chondrocranium. This is called a **primitive autostylic** mode of suspension (Gr., *auto* = self + *stylos* = pillar; Fig. 7-6).

In early chondrichthyan and in early bony fishes, the palatoquadrate had one or more movable articulations with the chondrocranium, and the hyomandibula extended as a prop from the otic capsule of the chondrocranium to the caudal end of the palatoquadrate. This type of jaw suspension is called **amphistylic** (Gr., *amphi* = both). An amphistylic suspension is also found in the choanate ancestors of terrestrial vertebrates.

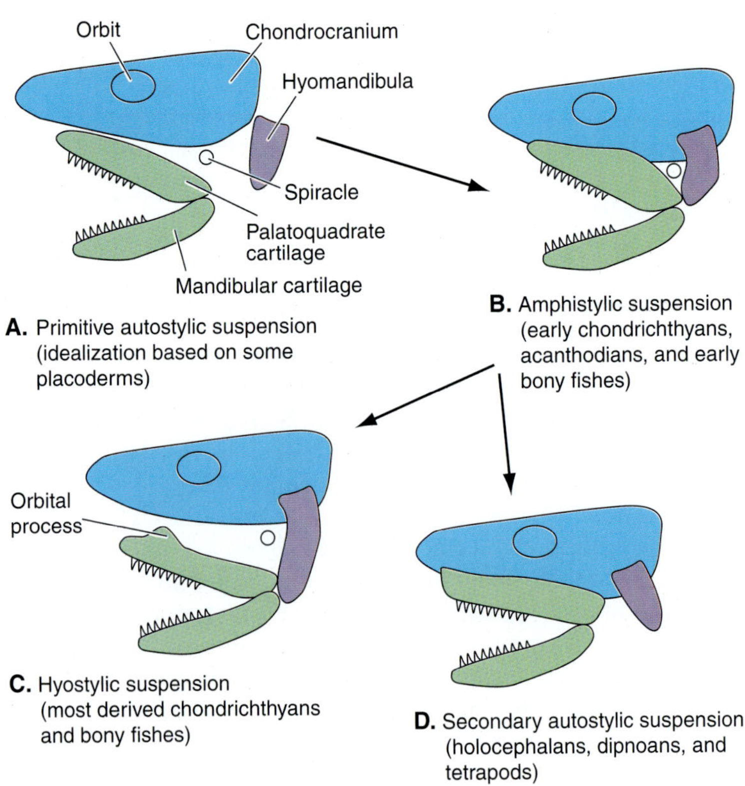

FIGURE 7-6
A–D, The probable evolution of the suspension of the palatoquadrate cartilage in fishes. The embryonic cartilaginous elements are shown, but many of these become ossified in bony fishes. See Figure 7-4 for color code.

Advanced chondrichthyan and bony fishes have a more flexible jaw mechanism that allows the jaws to be protruded downward, forward, or both during feeding, as we will discuss in Chapter 16. A more flexible jaw evolved when the palatoquadrate lost its primitive connection to the chondrocranium. The palatoquadrate is stabilized only by the hyomandibula. This type of suspension, called **hyostylic,** appears to be the most advanced type of suspension in fishes. It evolved independently, and in somewhat different ways, in the more advanced cartilaginous and bony fishes as their methods of feeding changed. Although the palatoquadrate of contemporary sharks is not articulated to the chondrocranium, the palatoquadrate does have an orbital process that extends dorsally beside the chondrocranium just in front of the basitrabecular processes. The orbital processes act as "guide rails," permitting the jaws to protrude but not to move from side to side.

Holocephalans, lungfishes, and other species that feed on shellfishes and other hard food have re-evolved an autostylic suspension in which the palatoquadrate is fused or firmly articulated to the chondrocranium, and the hyomandibula is not involved. This is called a **secondary autostylic** suspension. A secondary autostylic suspension also evolved independently in early tetrapods in which the hyomandibula assumed other functions.

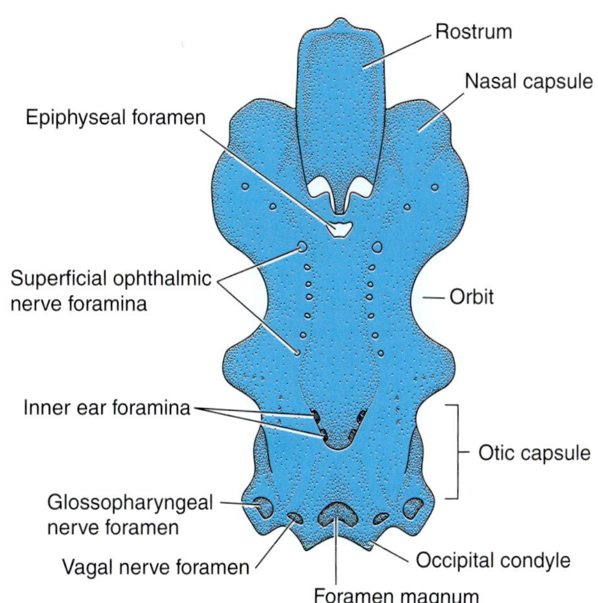

FIGURE 7-7
A dorsal view of the chondrocranium of a chondrichthyan fish, *Squalus. (From Walker and Homberger.)*

Cranial Skeleton of Jawed Fishes

As we have discussed (Chapter 3), more species of jawed fishes exist than all other vertebrates combined. They have specialized to every conceivable mode of life and method of feeding. Their range in cranial morphology is enormous and beyond the scope of this book. We summarize the differences in the three major components of the cranial skeleton in representative groups.

Chondrocranium The chondrocranium is a relatively conservative part of the cranial skeleton, differing primarily among groups of fishes in its degree of ossification. The chondrocranium is completely unossified in chondrichthyan fishes. It covers the dorsal as well as other surfaces of the brain because these fishes lack a dermatocranium. The chondrocranium is perforated by foramina for nerves and blood vessels and, dorsally, by an **epiphyseal foramen** and by foramina that connect with the inner ear (Fig. 7-7). The epiphysis is homologous to the pineal eye. Frequently, a rostrum extends forward from the nasal capsules into a long snout.

Usually the chondrocranium is well ossified in bony fishes except for its rostral end and sometimes the interorbital region, which remain cartilaginous. Only the exoccipitals ossify in lungfishes. The chondrocranium of the choanates ancestral to tetrapods formed a distinct rostral portion (ethmosphenoid) and a caudal portion (otic–occipital). An intracranial joint between them allowed for cranial kinesis, presumably coupled with feeding.

Splanchnocranium The cartilages of the mandibular arch form the jaws of chondrichthyan fishes. In other groups of jawed fishes, the mandibular arch and other visceral arches largely ossify and become covered by dermal bones. The caudal part of the palatoquadrate cartilage ossifies as a rectangular **quadrate bone,** which articulates with an **articular bone,** which ossifies in the caudal part of the mandibular cartilage. These are, of course, cartilage-replacement bones. We first see these bones in the fossil record in the †acanthodians (Fig. 7-8), a group of basal teleostomes (Fig. 3-2). The jaw joint of all jawed vertebrates, except for mammals, involves the quadrate and articular bones, or the posterior ends of the palatoquadrate and mandibular cartilages. The palatoquadrate rostral to the quadrate bone ossifies as one or two bones that, in †acanthodians, articulate with the chondrocranium. The hyomandibula of †acanthodians articulates with the quadrate, so these fishes have an amphistylic jaw suspension. Stiff **gill rays** extend outward from the branchial arches into the tissue supporting the gills, but they are not always seen in fossil specimens. **Gill rakers,** which prevent food from passing from the

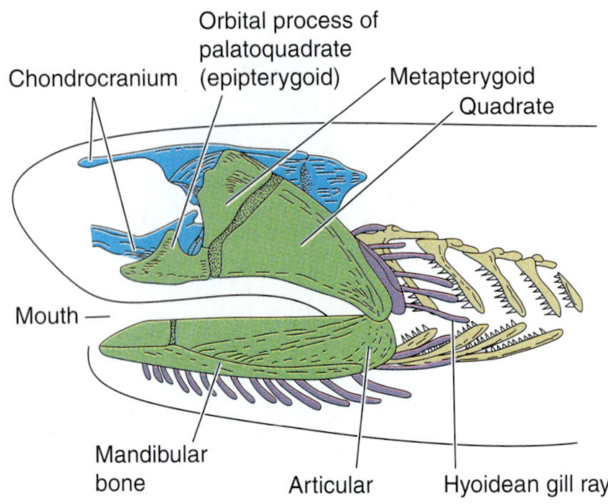

A. †*Acanthodes* cranial skeleton

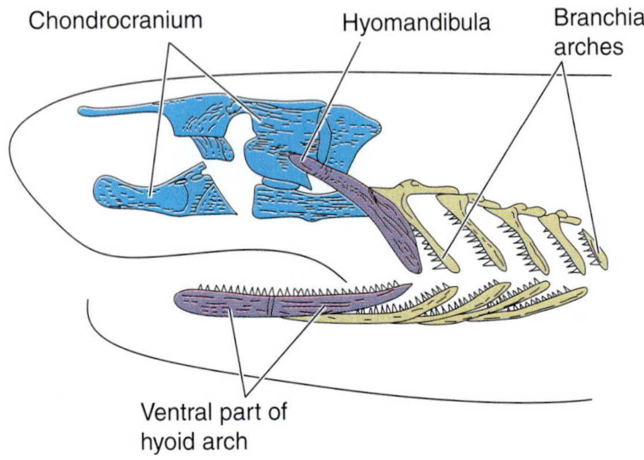

B. †*Acanthodes* cranial skeleton with mandibular arch removed

FIGURE 7-8
The cranial skeleton of an early teleostome, †*Acanthodes*. *A,* Dermal bones have been removed to expose the chondrocranium and splanchnocranium. *B,* The mandibular arch has been removed to show the hyoid arch. *(After Jarvik.)*

pharynx cavity into the gill pouches, often occur on the branchial arches.

Dermatocranium Hagfishes, lampreys, and chondrichthyans lack a dermatocranium. In most other fish groups, the dermatocranium is the most conspicuous part of the cranial skeleton because it largely covers the other two parts. It is perforated dorsally by a pineal foramen in many early bony fishes. We show (Fig. 7-9*A*) the groups of bones into which it can be divided. The pattern of individual bones in the dermatocranium of bony fishes is extremely diverse. Much of this diversity is related to different feeding mechanisms (Chapter 16). The pattern of the bones in the bowfin, *Amia,* is quite representative of early neopterygians (Fig. 7-9*B*). The dermal roof is not a solid shield, but gaps along its ventral and caudal borders expose palatal bones and the hyomandibula. These gaps allow for some cranial kinesis.

Cranial Skeleton of Early Tetrapods

Unfortunately, the cranial skeleton of living amphibians is highly specialized and is not at all representative of the cranial skeleton of ancestral tetrapods. The skull of an early reptilomorph, such as a lizard, is somewhat better, but its cranial skeleton also is specialized in some respects. In order to appreciate the evolution of the tetrapod cranial skeleton, we must first examine its condition in a basal tetrapod living during the Carboniferous period, shortly after tetrapods evolved. We base our description of the early tetrapod cranial skeleton on well-studied material from †*Palaeoherpeton,* an †anthracosauroid (Fig. 3-20).[1] The cranial skeleton of †*Palaeoherpeton* also retains many of the features found in the cranial skeletons of early choanate fishes that were close to the ancestry of tetrapods: the †osteolepiformes, as represented by †*Eusthenopteron,* and the †elpistostegids, as represented by †*Panderichthys* (Fig. 3-15).

General Features of the Early Tetrapod Skull The skull of ancestral tetrapods has the three basic components that we have been considering: (1) chondrocranium, (2) splanchnocranium, and (3) dermatocranium. But changes in feeding, gas exchange, and sensory systems that accompanied the transition of vertebrates from water to land did affect cranial morphology.

A major change occurred in skull proportions (compare Figs. 7-10*A* and 7-11*A*). The facial portion of the skull, from the eyes to the end of the snout, is short in choanate fishes (such as †*Eusthenopteron*) close to the ancestry of tetrapods but became much longer in the early tetrapod (†*Palaeoherpeton*). The resulting longer jaws are probably correlated with differences in feeding. Feeding changes appear to have taken place because the vertical intracranial joint near the center of the chondrocranium of choanate fishes (see Fig. 13-22*C*), which would have allowed the upper jaw and front of the skull to move up and down slightly relative to the back of the skull, is lost. A prominent suture remained at this site in early tetrapods.

Both choanate fishes and early tetrapods had a pair of internal nostrils, or **choanae** (Gr., *choane* = funnel), that opened near the front of the roof of the mouth (Figs. 7-10*B* and 7-11*B*). Choanae lead from the nasal

[1]†Anthracosaurs have been regarded as early amphibians lying close to the point of divergence of reptilomorphs. Paleontologists now regard †anthracosaurs as basal reptilomorphs. Regardless of just where †*Palaeoherpeton* lies in the scheme of classification, its cranial skeleton is a good example of that of an ancestral tetrapod.

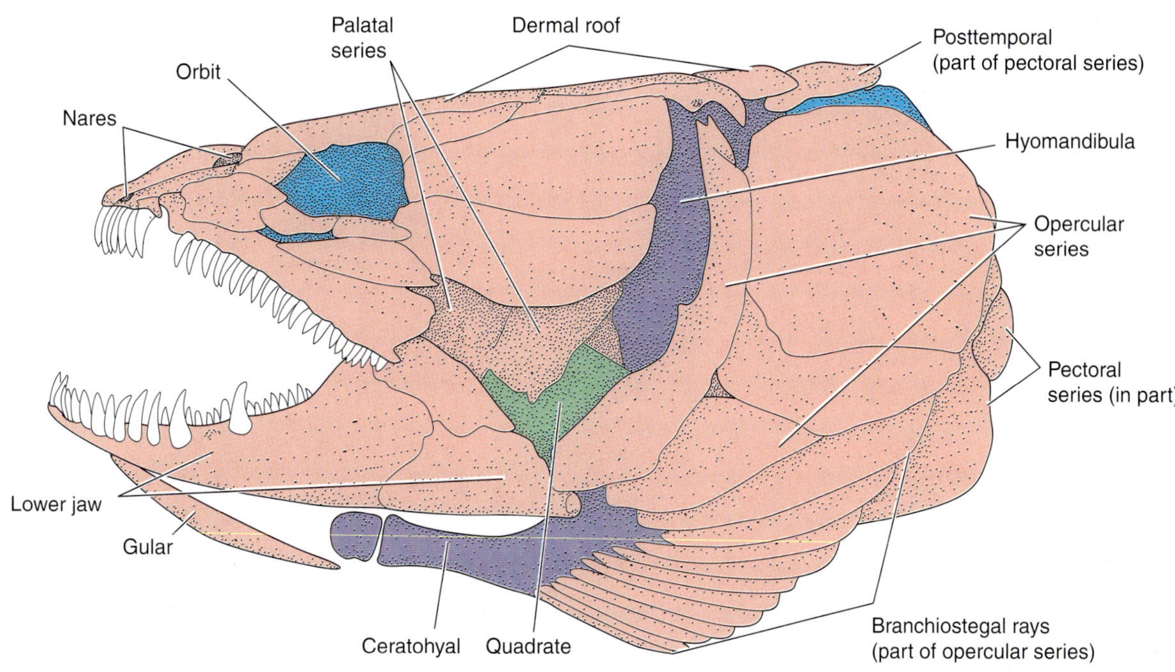

A. Components of the cranial skeleton of *Amia*

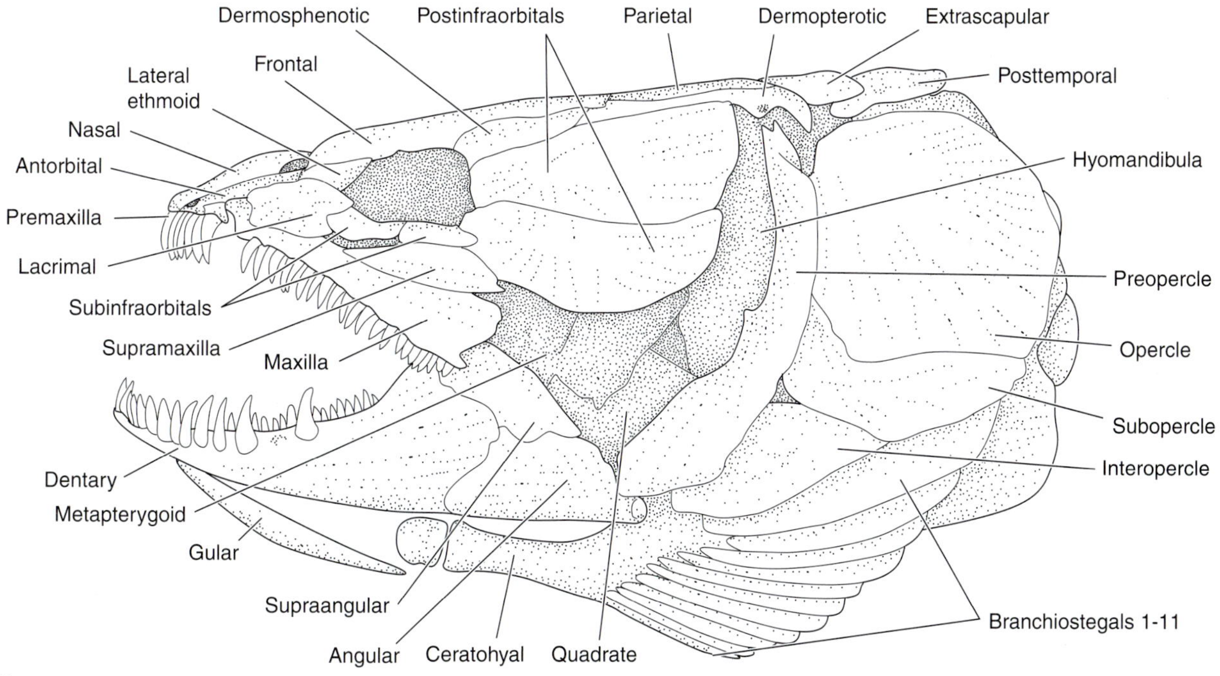

B. Bones of the cranial skeleton of *Amia*

FIGURE 7-9
Lateral views of the cranial skeleton of *Amia*. *A*, The components of the cranial skeleton. See Figure 7-4 for color code. *B*, Individual bones are identified. *(From Walker and Homberger; terminology after Grande and Bemis.)*

cavities to the mouth and are part of the air passages leading to lungs in living vertebrates. Their presence in these extinct species indicates that lungs were probably present. Choanate fishes also breathed with gills because they had a complete set of branchial arches. Opercular and gular bones covered them. Gills were lost in the transition from water to land, and with them, the opercular and gular bones.

The loss in ancestral tetrapods of much of the branchial apparatus, together with the loss of dermal

FIGURE 7-10
A, Dorsal view of the skull of an early choanate fish, the †osteolepiform †*Eusthenopteron. B,* Palatal view. Compare these views with the views in Figure 7-11 of an early tetrapod. Color code in Figure 7-4. *(After Jarvik.)*

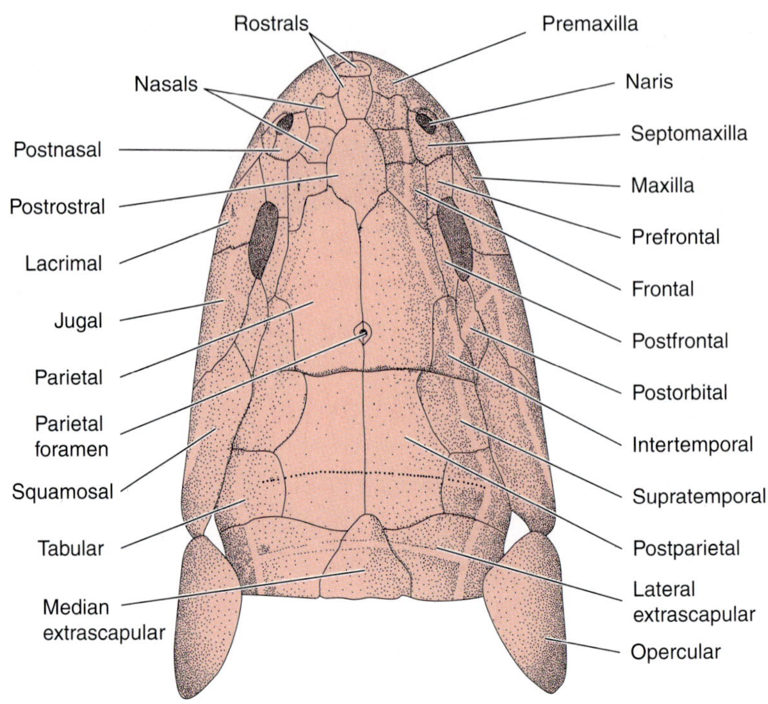

A. Dorsal view of skull of †*Eusthenopteron*

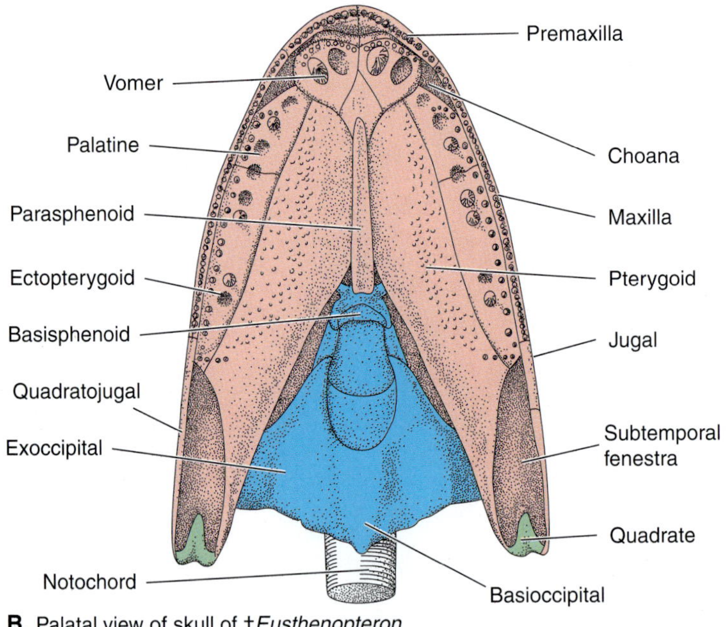

B. Palatal view of skull of †*Eusthenopteron*

bones that connected the pectoral girdle to the back of the skull in choanate fishes, enabled the head to move independently of the trunk. A short neck region began to develop.

In common with many jawless fishes and bony fishes, choanate fishes had a foramen on the top of the skull for a median eye. This foramen was retained in early tetrapods (Fig. 7-10*A* and 7-11*A*). The median eye in early tetrapods is usually a parietal, and not a pineal, eye (Chapter 12). The earliest known tetrapods (the anthostegids and ichthyostegids; Fig. 3-15) retained **lateral line canals** on the skull, but this sensory system, which detects water movements, was lost in most early tetrapods. Changes also occurred in other sense organs as vertebrates moved from water to land, but few affected the cranium.

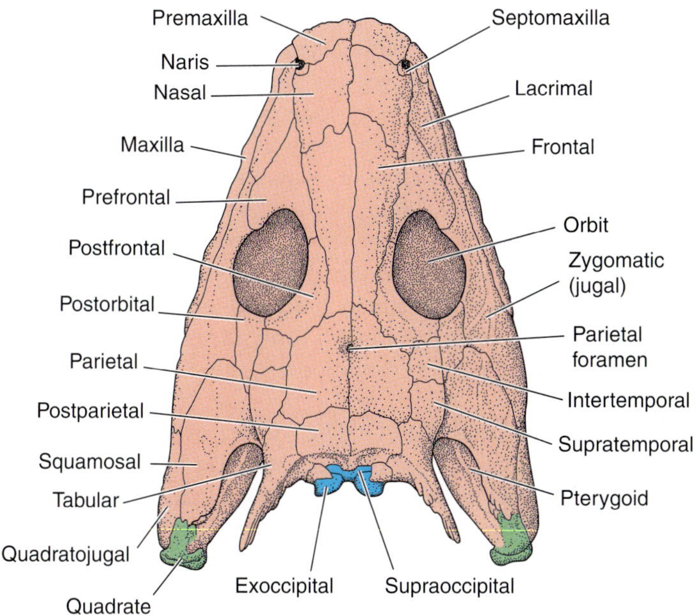

A. Dorsal view of skull of †*Palaeoherpeton*

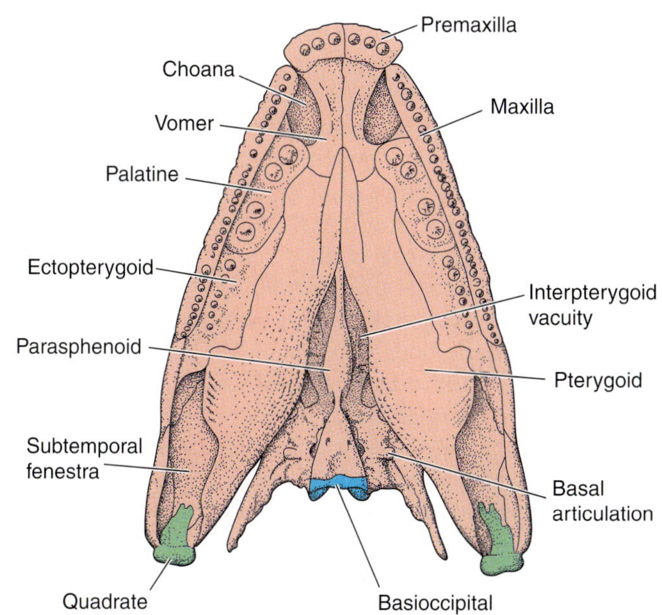

B. Palatal view of skull of †*Palaeoherpeton*

FIGURE 7-11
A, Dorsal view of the skull of an early tetrapod (an †anthracosaurid) based on †*Palaeoherpeton* and †*Protogyrinus. B,* Palatal view. Color code in Figure 7-4. *(From Romer and Parsons.)*

The Chondrocranium We already described the chondrocranium of an early tetrapod because we use an †anthracosauroid chondrocranium to exemplify an adult chondrocranium when we discussed the components of the cranial skeleton (Fig. 7-2).

The Splanchnocranium As in bony fishes, the caudal portions of the palatoquadrate and mandibular cartilages were ossified as the **quadrate** and **articular** bones, respectively. Rostral to the quadrate, the palatoquadrate ossified as an **epipterygoid,** which articulated with the chondrocranium and helped fill in a gap on its side (Fig. 7-12*A*). Other parts of the mandibular arch are not known in early tetrapods. They probably remained cartilaginous, as they do in contemporary amphibians.

The dorsal part of the hyoid arch of early tetrapods, the hyomandibula, formed an ossified rod called the columella (L., *columella* = small column),[2] which ex-

[2]The columella is homologous to the mammalian stapes, but the element usually is called the columella in nonmammalian tetrapods because of its resemblance to a little column.

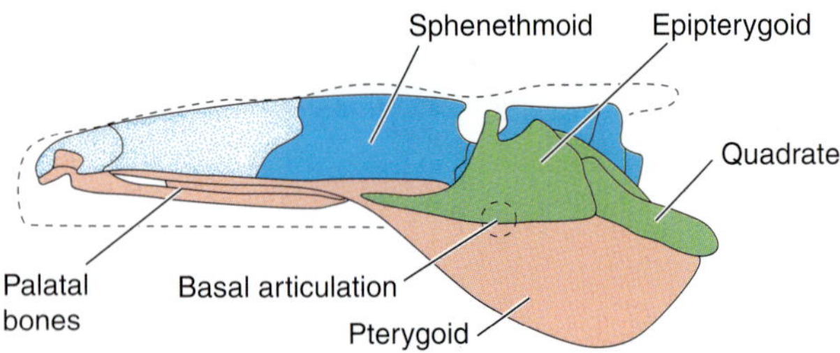

A. Lateral view with dermatocranium removed

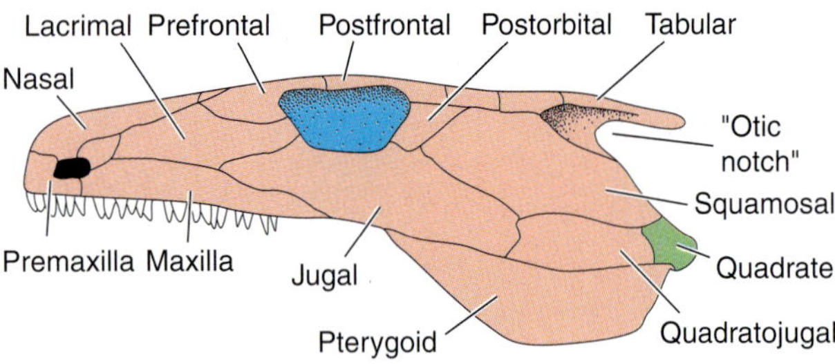

B. Lateral view of dermatocranium

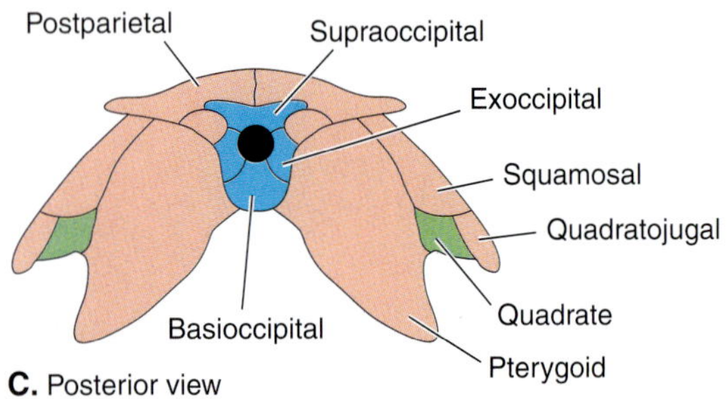

C. Posterior view

FIGURE 7-12
The skull of an early tetrapod based on †*Protogyrinus* and †*Palaeoherpeton. A,* Lateral view after removal of the dermal roof. *B,* Lateral view with the dermal roof in place. *C,* Posterior view. *(After Romer and Parsons.)*

tended between the otic region of the chondrocranium and the quadrate. It is not well known in †*Palaeoherpeton*, but is very evident in the skull of the †captorhinid, *Paleothyris* (see Fig. 7-17*B*). Suspension of the palatoquadrate became secondary autostylic in later amphibians and amniotes. The freed columella in frogs is a slender auditory ossicle that transmits high-frequency, airborne vibrations from the tympanic membrane to the fenestra ovalis on the side of the otic capsule. We are uncertain of the function of the columella in early tetrapods. An "otic notch" (Fig. 7-12*B*) high on the dermal roof may have lodged a tympanic membrane, but some investigators doubt this interpretation. Their columella was too large and massive a bone to have responded to the high-frequency vibrations of the sort detected by a tympanic ear. The columella may have transmitted lower-frequency, high-intensity vibrations by bone conduction from the jaw joint and cheek region to the otic capsule (Clack, 1989), or it simply may have braced the braincase against the cheek. It is clear from fundamental differences between the tympanic ear of frogs, many sauropsids, and mammals that a tympanic ear evolved independently among terrestrial vertebrates several times and was not inherited from a common type present in ancestral terrestrial vertebrates.

With the loss of gills in the transition from water to land, the branchial arches were reduced. The branchial skeleton of ancestral tetrapods is not well known. Presumably, the remaining branchial arches and the ventral part of the hyoid arch form a hyobranchial apparatus, as they do in contemporary amphibians (p. 248).

The Dermal Roof The dermatocranium of early tetrapods consisted of bones that formed a dermal roof and palate and encased the lower jaw. Many of the small bones in the snout of choanate fishes were lost during the evolutionary transition to tetrapods, but the pattern of the remaining roofing bones stayed much the same (Figs. 7-10*A* and 7-11*A*). Most of their names are descriptive of a bone's position, and for purposes of description, the roofing bones can be grouped into five easily remembered series (Fig. 7-13):

1. A lateral tooth-bearing series formed the margin of the upper jaw: **premaxilla** (sometimes called the **incisive** bone in mammals) and **maxilla.**
2. A median series lay along the top of the skull just lateral to the mid-dorsal line: **nasal, frontal, parietal,** and **postparietal.**
3. A circumorbital series surrounded the orbit: **lacrimal, prefrontal, postfrontal, postorbital,** and **jugal** (the **zygomatic** bone of a mammal). The lacrimal usually reached the external nostril, but in some species a small **septomaxilla** lay rostral to the lacrimal.
4. A temporal series was located lateral to the mid-dorsal bones and caudal to the circumorbital group: **intertemporal, supratemporal,** and **tabular.**
5. Finally, a cheek series occupied the cheek region: **squamosal** and **quadratojugal.** The last bone extended between the jugal and the quadrate along the margin of the skull. Some of these bones continue to mammals, some are lost, and some become parts of the mammalian occipital and temporal bones.

The area beneath the temporal and cheek bones and lateral to the chondrocranium is known as the **temporal fossa.** A large jaw-closing, adductor mandibulae muscle arose from the deep surface of the roofing bones and the side of the chondrocranium and passed through a large **subtemporal fenestra** to the lower jaw (Fig. 7-11*B*). The complete temporal roof characteristic of choanate fishes and early tetrapods is called an **anapsid roof** (Gr., *a* = without + *apsis* = loop or bar). (As we shall see, skulls with openings in the temporal roof have distinct bars of bone bordering the openings.)

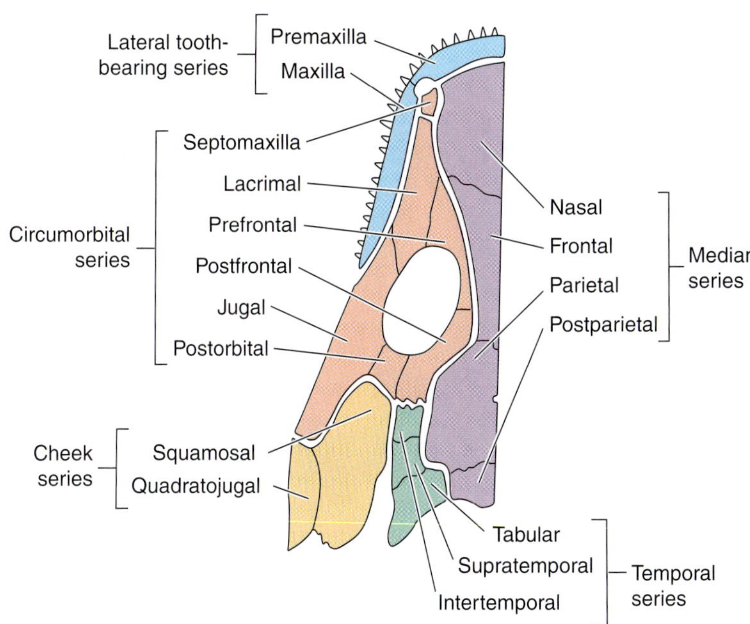

FIGURE 7-13
A diagram of the dermal roof of an early tetrapod with the bones grouped into regional series for ease in remembering them. Colors used here distinguish the groups of bones in the skull roof. All are dermal. *(From Romer and Parsons.)*

The Palatal Bones A group of paired dermal, palatal bones ossified in the roof of the mouth: **vomer, pterygoid, palatine,** and **ectopterygoid** (Figs. 7-10*B* and 7-11*B*). The **epipterygoid bone,** an ossification of the palatoquadrate cartilage, lay dorsal to the pterygoid and gave the palate a movable articulation with the braincase (Fig. 7-12*A*). A dermal **parasphenoid** lay along the ventral surface of the chondrocranium medial to the palatal bones. Spaces known as **interpterygoid vacuities,** which lay between the pterygoid bones and the chondrocranium, permitted the upper jaws and palate to move relative to the chondrocranium.

Dermal Bones of the Lower Jaw Dermal bones ensheathe the mandibular cartilage, leaving only its ossified caudal end, the articular, exposed. These bones are very similar in choanate fishes and early tetrapods (Fig. 7-14). A **dentary** formed much of the front half of the lower jaw. Two **splenials** lay ventral to the dentary; two or three **coronoids** were medial and posterior to

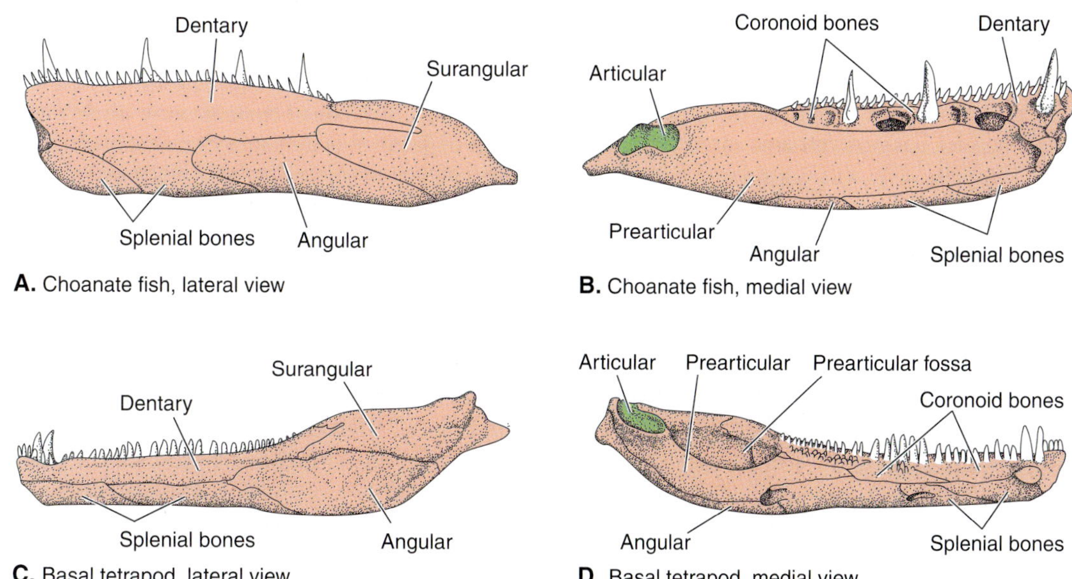

FIGURE 7-14
The lower jaw of a choanate fish (*A* and *B*) and early tetrapod (*C* and *D*) as seen in lateral view (*A* and *C*) and medial view (*B* and *D*). *(After Romer and Parsons.)*

the dentary. An **angular** and **surangular** lay posterior to the dentary on the lateral surface of the jaw, and a **prearticular** extended between the dentary and articular on the medial side. Of all of these bones, only the dentary remains in the lower jaw of a mammal. A **prearticular fossa,** located rostral to the articular, lay between the dermal bones on the medial surfaces of the lower jaw. The temporal muscle emerging through the subtemporal fenestra attached along the margins of the fossa, and blood vessels and nerves entered the lower jaw through it.

Teeth Early tetrapods retained homodont, conical teeth loosely attached to the jaws and not set in sockets (Figs. 7-12*B* and 7-14). The teeth of most fishes and early tetrapods are used simply to grasp food, not to masticate it. They occurred on the premaxilla and maxilla of the upper jaw, on many of the bones of the palate, and often on the parasphenoid. The dentary was the primary tooth-bearing bone of the lower jaw, but a few teeth occurred on the coronoids.

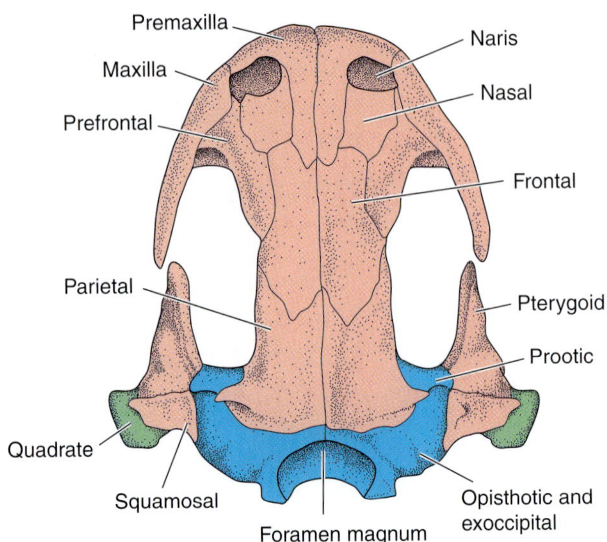

A. Dorsal view of skull of *Ambystoma*

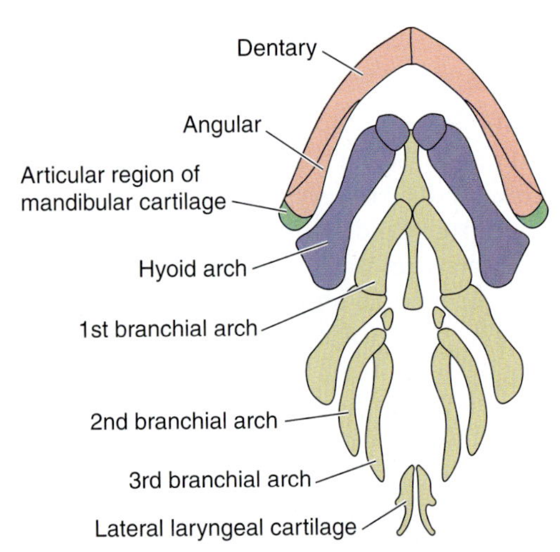

B. Lower jaw and hyobranchial apparatus of *Necturus*

FIGURE 7-15
The cranial skeleton of contemporary amphibians. *A,* Dorsal view of the skull of a salamander, *Ambystoma. B,* Ventral view of the lower jaw, hyobranchial apparatus, and larynx of the salamander, *Necturus.* See Figure 7-4 for color coding. *(A, after Romer and Parsons; B, after Walker and Homberger.)*

The Cranial Skeleton of Living Amphibians

In the evolution of contemporary amphibians, the skull became broad and flat (Figs. 7-15*A* and 7-16). A broad head allows for a broad sheet of muscles in the floor of the mouth and pharynx, which is important in feeding and breathing of amphibians. Most of the chondrocranium is unossified. The basioccipital region is absent, and paired occipital condyles are borne on the exoccipitals. Dermal bones in the circumorbital and temporal series of the roof have been lost. Most of the others remain. The frontal and parietal bones are fused in frogs but remain distinct in salamanders. A frog has a well-developed, rod-shaped columella that extends from the tympanic membrane to the otic capsule, but a salamander has no tympanic membrane and only a small columella. We will consider the unusual hearing mechanisms of contemporary amphibians in Chapter 12.

The mandibular cartilage usually remains unossified, but its caudal end (the articular region) articulates with the quadrate. Many of the dermal lower jaw bones are lost, but the dentary, angular, and sometimes the splenials remain (Fig. 7-15*B*). Salamanders retain teeth on the lower jaw, but frogs have lost them.

Larval amphibians and paedomorphic adults retain gills and a well-developed branchial skeleton. The branchial arches of adults are much reduced compared with those of fishes. The ventral parts of remaining arches unite with the ventral part of the hyoid arch to form the **hyobranchial apparatus** (Fig. 7-15*B*), much of which is not ossified. The hyobranchial apparatus is embedded in the base of the tongue. Ligaments extend from the ventral part of the hyoid arch to the skull base, attaching near the columella. The hyobranchial apparatus forms a sling for the support of the tongue and larynx and for the attachment of muscles. Hyobranchial movements are used in food capture, swallowing, and gas exchange. The ventral parts of the hyoid arch and of the first three branchial arches (visceral arches 3–5) contribute to the hyobranchial apparatus. The sixth visceral arch forms the **lateral laryn-**

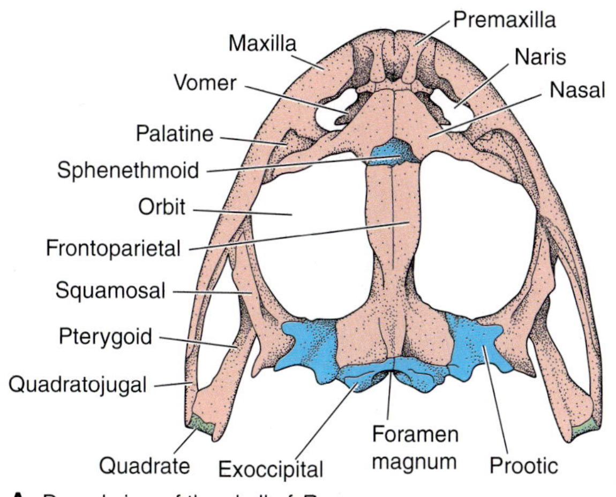

A. Dorsal view of the skull of *Rana*

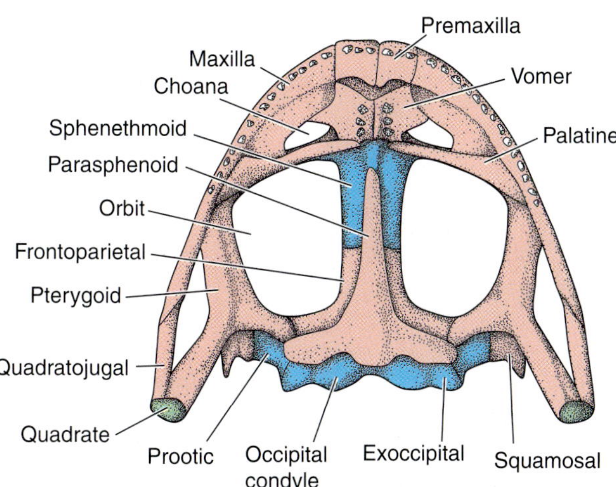

B. Palatal view of the skull of *Rana*

FIGURE 7-16
Dorsal and palatal views of the skull of a bullfrog, *Rana*.

geal cartilages of the larynx. The seventh visceral arch probably was lost completely in terrestrial vertebrates.

The Anapsid Skull of Early Reptilomorphs

The cranial skeleton of living amphibians is specialized in comparison with that of early tetrapods (†*Palaeoherpeton*), but that of reptilomorphs is much closer. We take as our example of a fairly early reptilomorph cranial skeleton that of the †captorhinids (Fig. 3-20) as represented by the Lower Permian †*Hylonomus* and †*Paleothyris*. These were small (200 mm long, including tail), superficially lizard-like insect eaters. The skull of †*Paleothyris* is the better known (Fig. 7-17*A* and *B*). Its skull was similar in many ways to that of ancestral tetrapods, but it was much smaller, higher, and somewhat narrower and lacked an otic notch. †*Paleothyris* retained the same complement of dermal bones as in †*Palaeoherpeton*, and the temporal region of the skull remained complete, or **anapsid.** The parietal foramen for a median eye was retained. The occipital and otic region of the chondrocranium was ossified, but the rostral part was largely cartilaginous. A single occipital condyle was present. A large columella buttressed the chondrocranium by extending from the otic capsule to the quadrate. The medial end of the columella fitted into a fenestra ovalis in the otic capsule, but hearing must have been limited to detecting low-frequency, high-intensity vibrations through bone conduction. None of the bones in the ear region was shaped in such a way that it could hold a tympanic membrane. Many simple, conical teeth were present.

The skull of turtles has an anapsid temporal roof (Fig. 7-17*C*). Investigators are not in agreement as to whether this is a retention of the primitive condition seen in †captorhinids or whether it secondarily evolved from a roof with temporal openings (Hedges and Poling, 1999). In any case, the nature of an anapsid roof is very evident in sea turtles, but in other species, the temporal roof has been partly lost by the development of a deep notch or emargination that extends rostrally from its posterior edge (Fig. 7-18*B*). The skull of turtles is highly specialized in other ways. Turtles have lost many of the dermal bones present in †captorhinids. The snout is very short. Teeth have been lost in contemporary species and are functionally replaced by horny sheaths over the jaw margins. Palatal teeth were retained in Triassic species of †*Triassochelys*. The pineal foramen also was lost. Turtles evolved a tympanic ear well adapted to detect airborne sound waves. The large quadrate bears a notch that supports a tympanic membrane, and the columella is a slender rod that extends from the membrane to the fenestra ovalis in the chondrocranium.

Temporal Fenestration in Reptilomorphs

Temporal fenestrae developed in the temporal region of the roof in other lines of reptilomorph evolution (Fig. 7-18). The temporal fenestrae appear to have evolved as early amniotes radiated and adapted to many different lifestyles and modes of feeding. Jaws and teeth changed in size and shape, as did the size and distribution of the adductor, or jaw-closing, muscles. Forces acting on the temporal roof from which parts of these muscles arose changed. As we have seen (Chapter 5), bone tends to develop and thicken in stressed areas and to thin out and disappear where stresses are absent. Sutures where several bones meet in the temporal region would be potential weak points, and a redistribution of forces in the temporal roof avoided stressing them. Bone appears

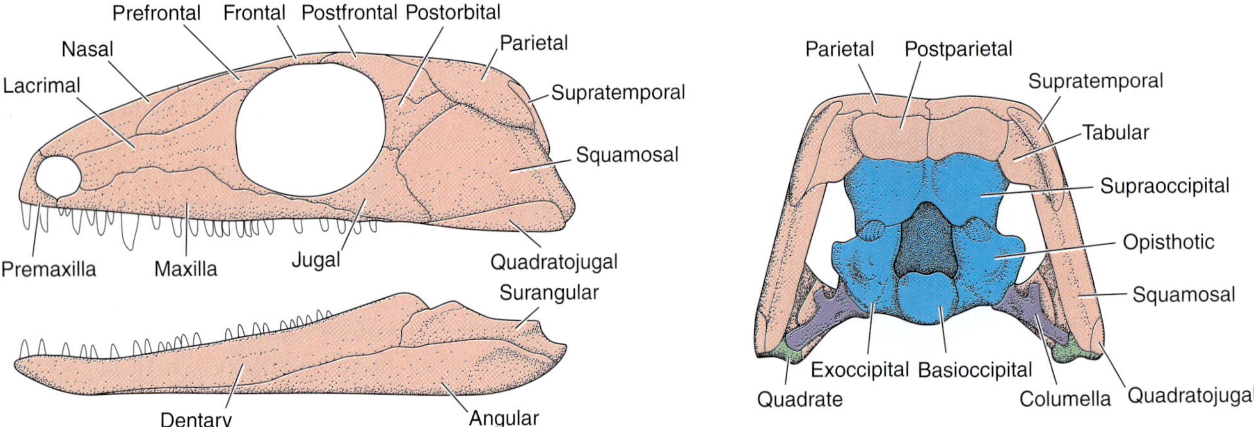

A. Lateral view of †*Paleothyris*

B. Posterior view of †*Paleothyris*

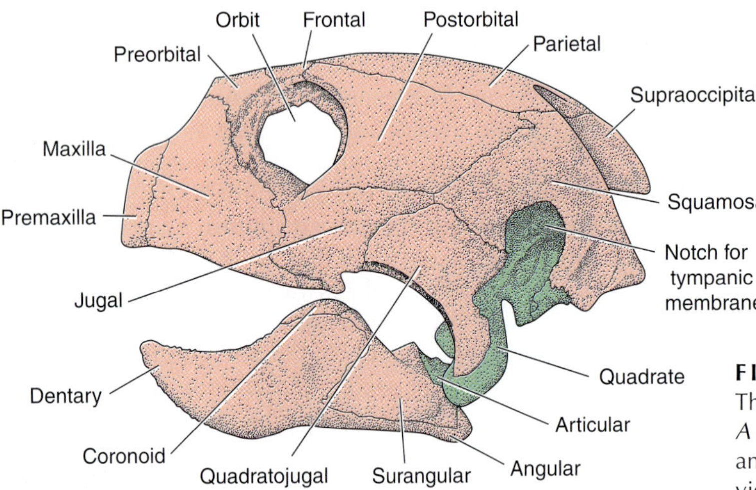

C. Lateral view of *Caretta*

FIGURE 7-17
The skull and lower jaw of anapsid reptilomorphs. *A* and *B*, lateral and posterior views of the skull of an early reptilomorph, †*Paleothyris; C,* A lateral view of the skull and lower jaw of the sea turtle, *Caretta. (A and B, after Carroll.)*

to have *initially* disappeared in such locations. The edges of the fenestrae provided extra points of attachment for jaw-closing muscles. As the fenestrae enlarged, they also provided an area into which underlying jaw muscles could bulge when they contracted and shortened. It is unlikely, however, that this bulging could have been the initial cause of fenestration, for the fenestrae would have been too small at first to serve this function.

Synapsids Different patterns of fenestration developed in different lines of reptilomorph evolution (Fig. 7-18). An early line to diverge was the synapsids. They are characterized by having a synapsid skull (Gr., *syn* = together) characterized by a single temporal fenestra located low on each side of the skull and initially bounded dorsally by the squamosal and postorbital bones (Fig. 7-18*G*). Mammals retain a modification of this type of skull, although the postorbital bone has been lost. The fenestra is much larger than in earlier synapsids and often merges with the orbit. We will return to the evolution of the mammal cranial skeleton presently.

Diapsids In other lines of reptilomorph evolution, an upper and a lower temporal opening developed below and above the postorbital and squamosal bones (Fig. 7-18*C*). This is the **diapsid** skull (Gr., *di* = two). It is retained in the tuatara, *Sphenodon,* of New Zealand (Fig. 7-19*A*), in crocodilians and other archosaurs (Fig. 7-20), and in modified form in squamates (Figs. 7-18*D* and *E* and 7-19*B*). Lizards have lost the arch of bone ventral to the lower opening, and snakes, in addition, have lost the squamosal–postorbital arch between the two openings. Most squamates have a kinetic skull, and the mouth gape is particularly large. The quadrate, and in snakes the squamosal as well, is free to participate in jaw move-

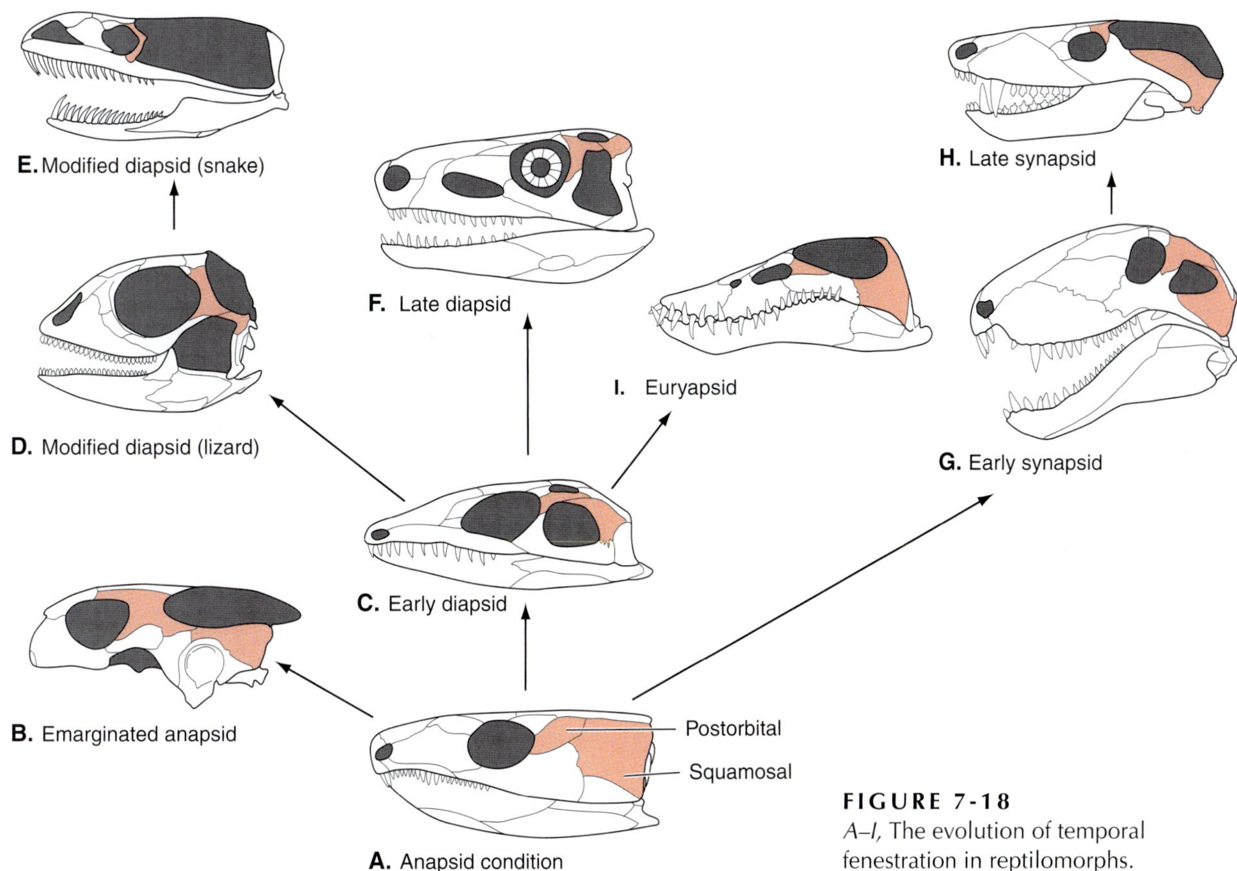

FIGURE 7-18
A–I, The evolution of temporal fenestration in reptilomorphs.

ments in addition to the usual movement between the quadrate and articular bones. This design enables snakes to swallow prey larger than the diameter of their bodies (Chapter 16).

Two groups of extinct marine reptiles, the †plesiosaurs and †ichthyosaurs, have a single temporal fenestra located high on the skull dorsal to the postorbital and squamosal (Fig. 7-18*I*). This is called an **euryapsid** skull (Gr., *eurys* = wide). Intermediate fossils suggest that the †plesiosaur skull evolved from a diapsid type by the loss of the lower temporal bar and the broadening of the remaining one. This may have been true for †ichthyosaurs too, but the evidence is less clear.

The Cranial Skeleton of Living Diapsid Reptiles

The tuatara (*Sphenodon*), squamates, and archosaurian reptiles, represented by alligators and crocodiles, have lost many of the small dermal bones that were present in the temporal region of the skulls of †captorhinids, but they retain most of the others (Figs. 7-19 and 7-20). *Sphenodon* and many lizards retain the foramen for a parietal eye, but this is lost in snakes and most archosauromorphs. The columella is a slender bone. A tympanic ear is well developed in many species but is modified in snakes. The part of the chondrocranium between the eyes is frequently unossified. The dentary bone expands in the lower jaw and bears all of the teeth. Splenials and one coronoid are lost, but the other lower jaw bones remain (Fig. 7-19*A*). The articular bone bears a prominent **retroarticular process,** to which a muscle that opens the jaws attaches.

Sphenodon and squamates have kinetic skulls with a palate capable of moving on the braincase, but in most other groups, the dermal palatal bones are united firmly to the braincase. In crocodilians, shelflike extensions of the premaxilla, maxilla, palatine, and pterygoid bones have grown ventrally and medially and unite ventral to the original roof of the mouth, forming a **secondary palate** (Fig. 7-20*B*)**.** The choanae now open far back in the mouth cavity. A flap of flesh completes the separation of the mouth cavity and respiratory passages. The secondary palate of crocodilians is an adaptation for aquatic life. A crocodile can seize and

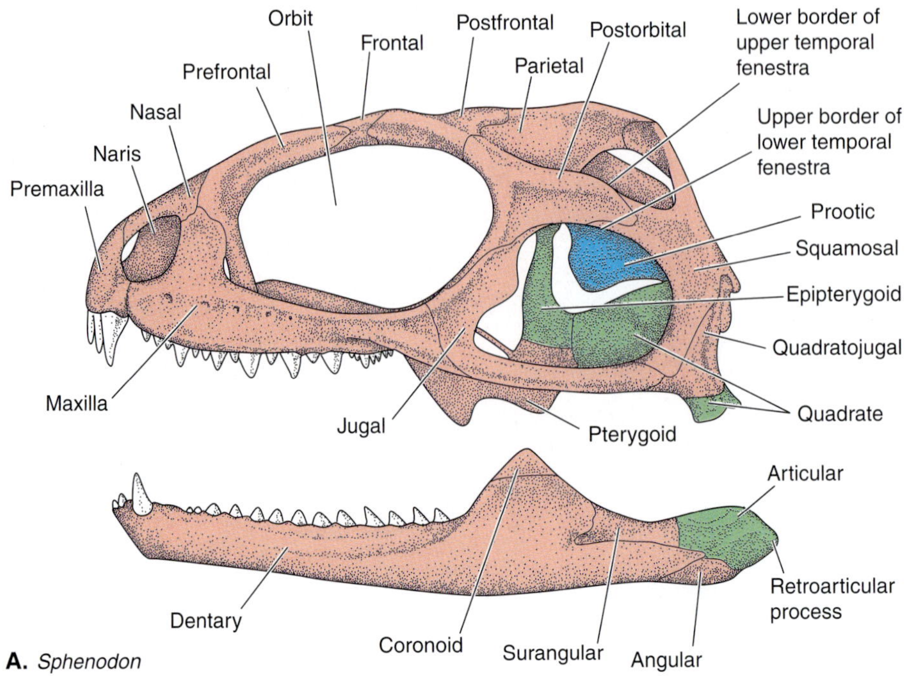

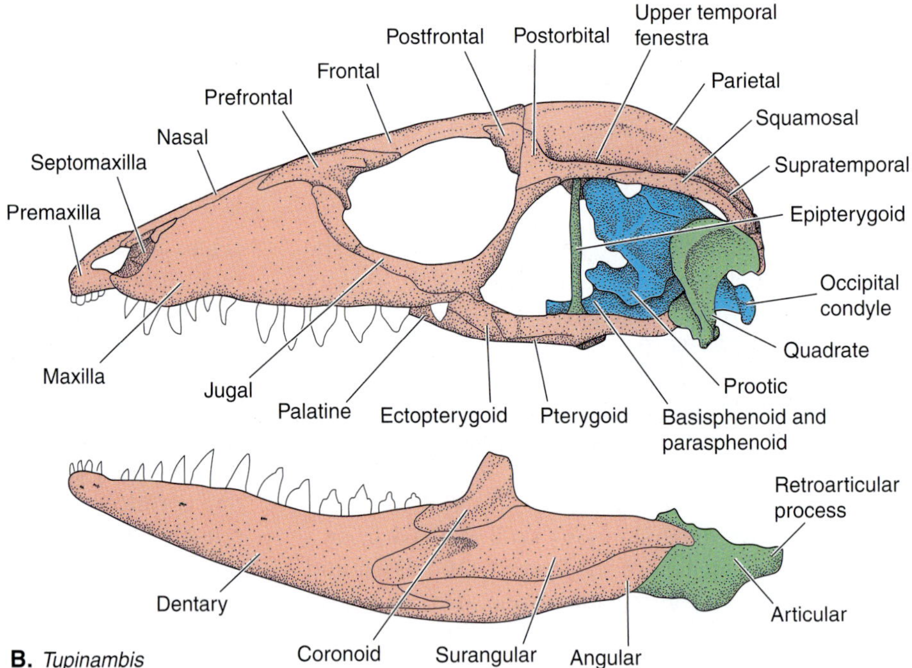

FIGURE 7-19
Lateral views of the skull and lower jaw of *Sphenodon* (*A*) and a tegu lizard (*B*). See Figure 7-4 for color coding. *(B, After Jollie.)*

manipulate prey beneath the water while continuing to breathe, provided the tip of its snout is out of the water.

Only the hyoid arch and the first two branchial arches contribute to the hyobranchial apparatus of contemporary reptiles (Fig. 7-21*A* and *C*). Their larynx includes a ring-shaped cricoid cartilage in addition to the lateral laryngeal cartilages present in amphibians (Fig. 7-15*B*). The third branchial arch, no longer a part of the hyobranchial apparatus, contributes to the cricoid cartilages.

The Cranial Skeleton of Birds

Birds belong to a group of dinosaurs known as theropods (Fig. 3-27). The skull in the best known fossils of †*Archaeopteryx* is badly crushed, but it is clear

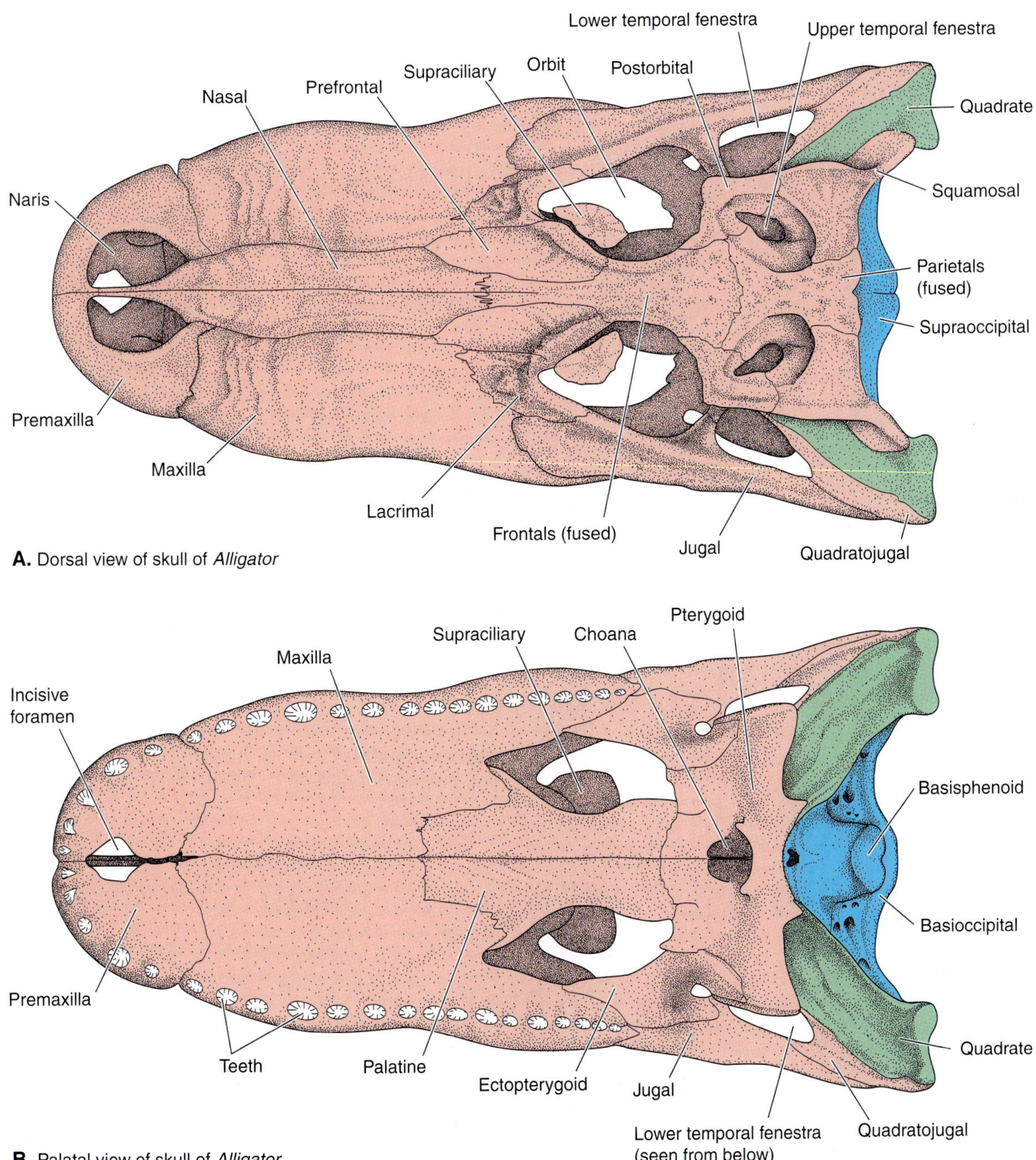

FIGURE 7-20
Dorsal (*A*) and palatal (*B*) views of the skull of an alligator.

that teeth were present and the braincase was modest in size in this earliest known bird. The temporal region of the skull appears to have been diapsid. Partly as an adaptation for flight and weight reduction, the skull of recent birds is very light but strong. Most of the bones are thin. The bar between the two temporal fenestrae has disappeared in modern birds, forming a single temporal fenestra, which merges with the orbit (Fig. 7-22). Only a slender bar of bone, composed of the jugal and quadratojugal, forms the boundary of the temporal fenestra ventrally. The brain enlarged considerably during the evolution of modern birds. The bird braincase, composed of the chondrocranium and dermal roofing elements, is much larger than in earlier

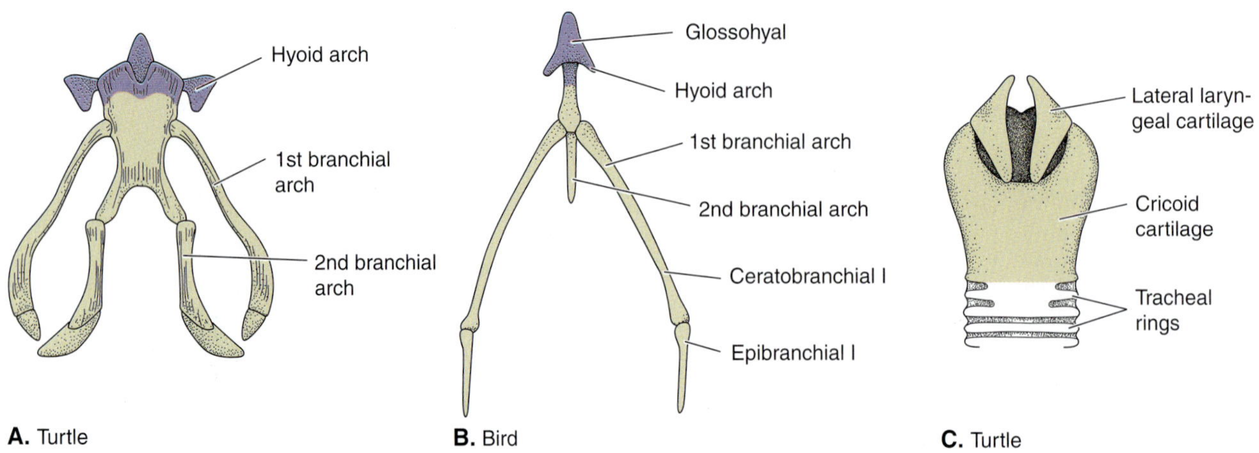

FIGURE 7-21
The hyobranchial apparatus and larynx of a turtle and a bird. *A,* A dorsal view of the hyobranchial apparatus of a turtle. *B,* A comparable view of the hyobranchial apparatus of a bird. *C,* A dorsal view of the larynx of a turtle. *(A and C, After Romer and Parsons.)*

reptilomorphs, globular in shape, and completely ossified. Sight is particularly important for birds, and the eyes and orbits are exceptionally large. A ring of **sclerotic bones** has developed in the wall of the eyeball in some species. The jaws form a long beak. Teeth have been lost in living species, and a horny sheath covers the beak.

Among extant birds, the flightless ratites have a **paleognathous palate** resembling their theropod ancestors (Fig. 7-23*A*). The vomer is a relatively large bone, and the palatine and pterygoid are firmly united with each other. Remaining recent species have a **neognathous palate** with a movable joint between the palatine and pterygoids. Birds have a highly kinetic

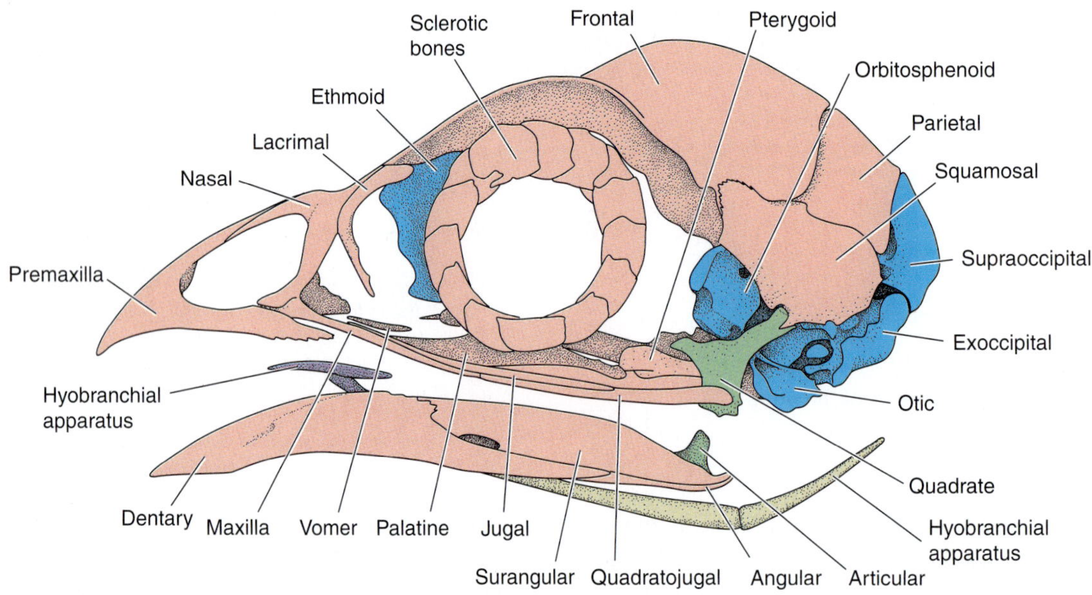

FIGURE 7-22
A lateral view of the skull and lower jaw of a gosling, *Anser ferus.* Most of the bones of the braincase are fused together in an adult goose. *(After Heilman.)*

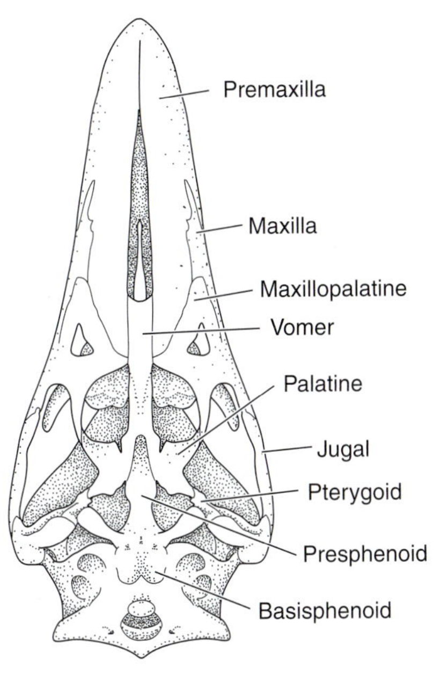

A. Paleognathous palate of *Rhea*

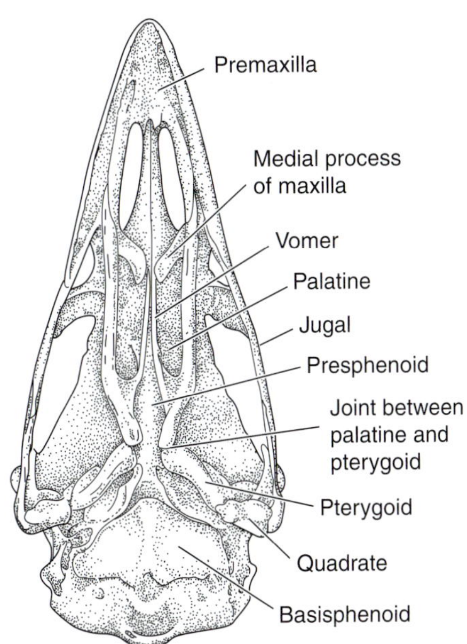

B. Neognathous palate of turkey (*Melagris*)

FIGURE 7-23
Ventral views of bird skulls to show (*A*) the paleognathous palate of a living ratite (the rhea) and (*B*) the neognathous palate of a modern bird (a turkey). *(After Feduccia.)*

skull as discussed in Chapter 16. The avian hyobranchial apparatus (Fig. 7-21*B*) is very similar to the reptilian one.

The Evolution of the Mammalian Cranial Skeleton

The line of synapsid evolution to mammals passed through the "sailed reptiles" (†edaphosaurids and †sphenacodontids, Fig. 3-31), which collectively often are called "†pelycosaurs," to the therapsids. Early synapsids very much resembled †captorhinids, except for a small synapsid-type temporal fenestra ventral to the postorbital and squamosal bones (Fig. 7-18*G*). They retained a heavy tail and a sprawled gait, with the humerus and femur moving close to the horizontal plane. Therapsids became more mammal-like in body proportions, limb posture, and cranial skeleton.

Mammals are endothermic animals with a high level of metabolism. They are active and inquisitive animals, very much alert to the world around them. Keen sense organs and an enlarged and complex brain support a lifestyle in which they avoid predators while searching for the increased food supply that endothermy requires. Obtaining more food and digesting it rapidly were accomplished partly by a change in feeding methods. Mammals differ from other reptilomorphs in that they masticate their food, cutting it up into smaller pieces. Smaller food particles also can be processed more rapidly in the digestive system because more surface area is available for the action of digestive enzymes. The mammalian feeding pattern requires a strong akinetic skull with powerful jaws and jaw muscles. A high level of metabolism requires not only more food but also increased gas exchange and a rapid distribution of materials between all parts of the body. Because the skull and lower jaw form an integrated unit, changes in one area affect stresses and the shape of elements in other areas. It is difficult, however, to discuss the skull as a whole. For purposes of analysis, we will break it up into smaller units.

Increasing Skull Strength As the brain increased in size and the jaws and jaw musculature became larger and more powerful, the skull as a whole enlarged and became stronger. Components of the skull united and became integrated into a solid unit. Kinesis between palatal bones and chondrocranium was lost. Some bones expanded, and as they did so, other bones became smaller and finally completely disappeared. Loss of some bones and the fusion of others reduced the numbers of sutures between them. Except for the jugal and lacrimal, the circumorbital bones of the dermal roof were lost, as were all of those in the temporal series. Only the jugal remains in the cheek series, and only the dentary remains in the lower jaw. The following bones present in ancestral tetrapods have been lost in mammals. The tabular may be incorporated in the interparietal.

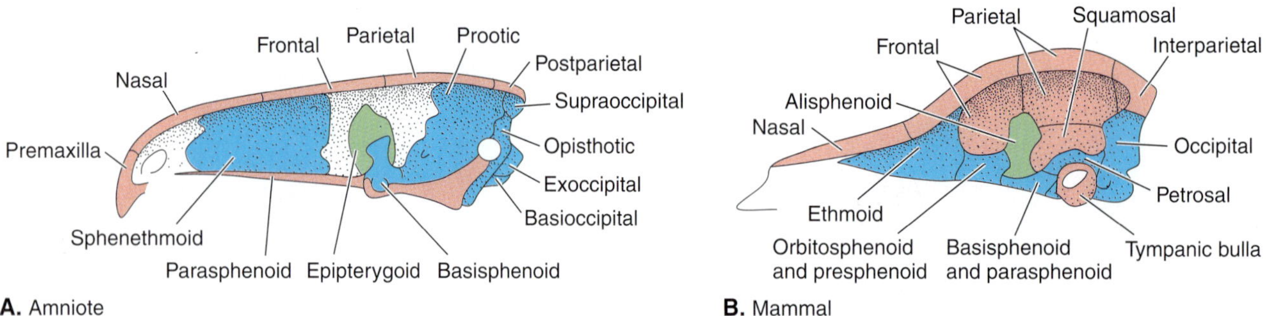

FIGURE 7-24
Sagittal sections through the skull of an ancestral amniote (*A*) and a mammal (*B*) to show the evolution of the bones that form the braincase.

Dermal Roof
- Circumorbital series
  - Prefrontal
  - Postfrontal
  - Postorbital
- Temporal series
  - Intertemporal
  - Supratemporal
  - Tabular ?
- Cheek series
  - Quadratojugal

Dermal Lower Jaw
  - Splenials
  - Surangular
  - Coronoids

Some bones also united with adjacent elements to form large, compound elements: the occipital, sphenoid, and temporal bones. We discuss these below.

Braincase Evolution An enlarging brain necessitated a remodeling of the braincase. Enlargement of the chondrocranium, which covers most of the lateral surface of the brain in ancestral amniotes, did not keep pace with the increase in brain size. Although the mammalian chondrocranium contains all of the parts originally present, it forms only the rear, floor, and front of the braincase (Figs. 7-24 to 7-26). Other bones encase the sides and top of the brain. The chondrocranium is completely ossified. The four occipital bones of ancestral terrestrial vertebrates unite in many adult mammals and often join with the dermal postparietals (now called an **interparietal**) to form a single

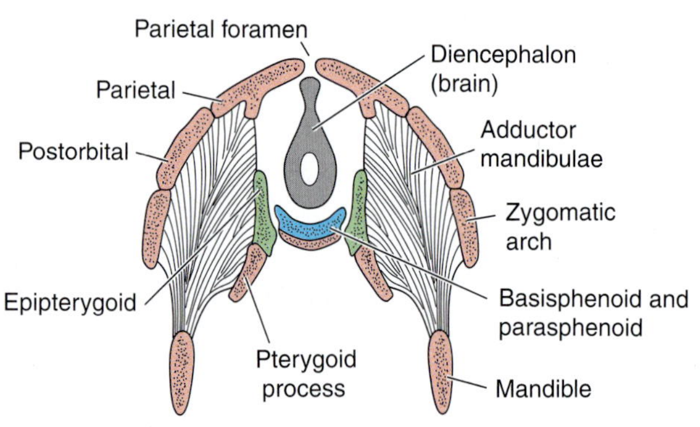

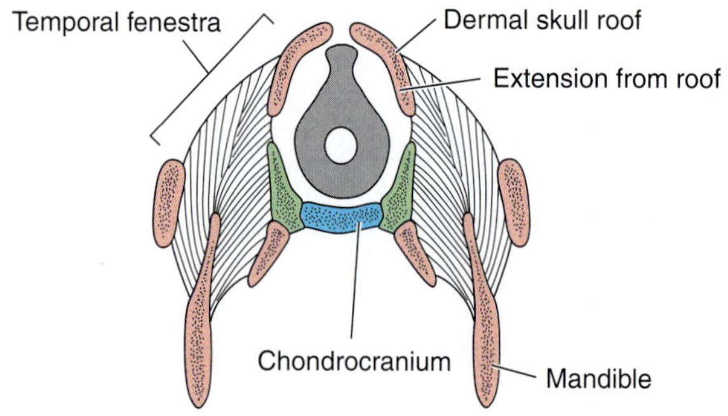

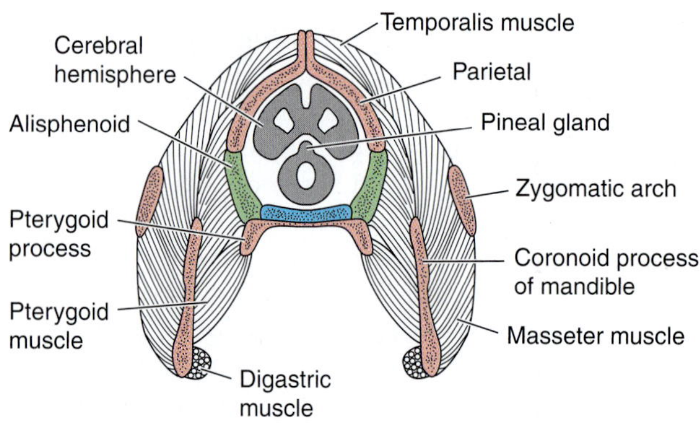

FIGURE 7-25▶
Evolution of the mammalian skull and lower jaw as seen in cross sections through the temporal roof. *A,* An ancestral amniote. *B,* An early synapsid. *C,* A mammal.

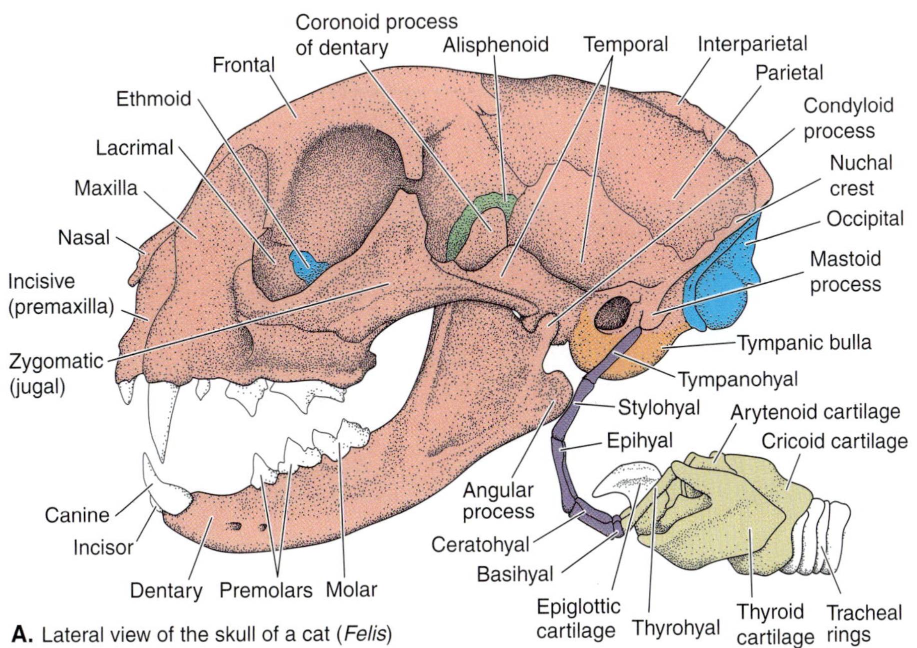

A. Lateral view of the skull of a cat (*Felis*)

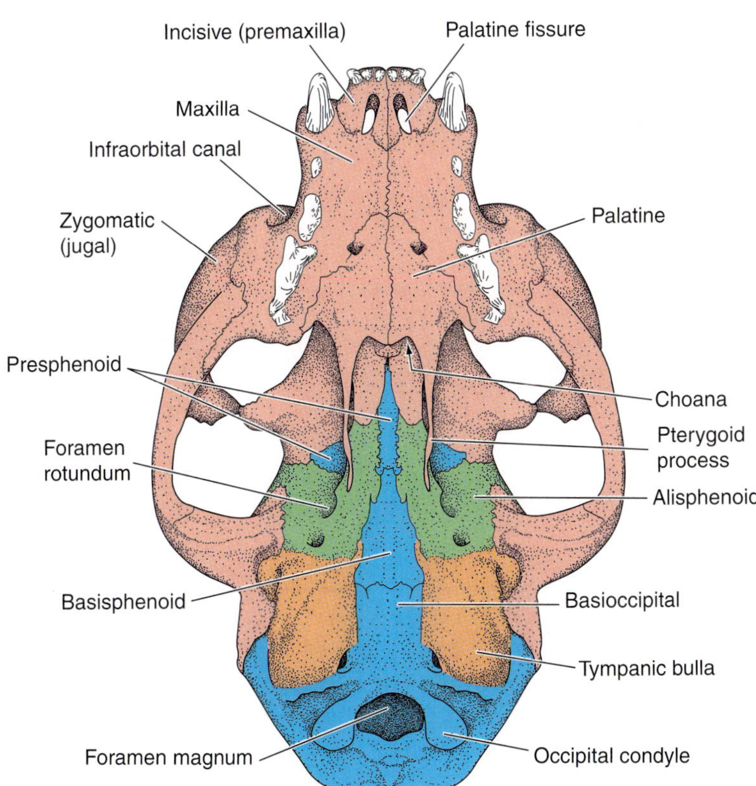

B. Palatal view of the skull of a cat (*Felis*)

FIGURE 7-26
Composition of the mammalian cranial skeleton as seen in a cat. See Figure 7-4 for color coding. *A,* A lateral view of the skull, lower jaw, hyoid apparatus, and larynx. *B,* A palatal view of the skull. The part of the tympanic bulla colored orange is a new cartilage-replacement bone without a homologue in reptiles. *(After Walker and Homberger.)*

occipital bone. The dermal tabulars may join the interparietal. Table 7-1 shows the parts of the compound skull bones and auditory ossicles of mammals and their homologues in an early tetrapod.

The originally single occipital condyle of †captorhinids and early synapsids, which was located ventral to the foramen magnum and borne primarily by the basioccipital bone (Fig. 7-17*B*), became double as it shifted dorsally and laterally to the exoccipital bones. Mammals have a pair of occipital condyles located lateral to the foramen magnum (Fig. 7-26*B*). Changes in the configuration of the occipital condyles, coupled

TABLE 7-1 Mammalian Compound Skull Bones and Auditory Ossicles

	Bone Name			
Early Tetrapod	**Occipital**	**Sphenoid**	**Temporal**	**Auditory Ossicles**
Chondrocranial elements				
Supraoccipital	Most of Bone			
Exoccipitals	Most of Bone			
Basioccipital	Most of Bone			
Opisthotic			Petrosal	
Prootic			Petrosal	
Basisphenoid		Basisphenoid		
Sphenethmoid (part)		Presphenoid		
		Orbitosphenoid		
Splanchnocranial elements				
Mandibular arch				
Epipterygoid		Alisphenoid		
Quadrate				Incus
Articular				Malleus
Hyoid arch				
Stapes				Stapes
Segment below stapes			Styloid process	
Dermal elements				
Roof and palate				
Postparietal	Interparietal			
Tabular ?	Interparietal			
Squamosal			Squamous part	
			Zygomatic part	
Parasphenoid		Part of basisphenoid		
Pterygoid		Pterygoid process		
Ectotympanic ?		Pterygoid process		
Lower jaw				
Angular			Ectotympanic	
Prearticular				Anterior process of malleus
No homologue			Endotympanic	

with changes in the first two cervical vertebrae (Chapter 8), allow the head to move in all directions as mammals explore their environment and feed.

The two otic bones on the lateral wall of the chondrocranium united to form a hard **petrosal bone** (Gr., *petros* = stone) that encases the inner ear (Fig. 7-24). The petrosal united with the squamosal of the dermal roof to form most of the mammalian **temporal bone** (Table 7-1).

The basisphenoid and the sphenethmoid (represented in embryonic mammals by a median presphenoid and a pair of lateral orbitosphenoids) form the floor and part of the lateral wall of the chondrocranium rostral to the occipital (Figs. 7-24 and 7-26*B*). Basisphenoid and presphenoid remain separate in some mammals, but in many species all unite to form the greater part of the wedge-shaped **sphenoid bone** (Gr., *sphen* = wedge + *eidos* = form; Table 7-1). The dermal parasphenoid united with them.

The epipterygoid, which formed a movable articulation between the palate and the chondrocranium in early tetrapods and ancestral amniotes, enlarged greatly and united firmly with the basisphenoid in the mammalian akinetic skull. It is known in mammals as the **alisphenoid** (L., *ala* = wing), for it forms the great wing of the sphenoid that contributes to the median wall of the orbit (Figs. 7-24 and 7-26*B*). The alisphenoid is a cartilage-replacement bone, but its phylogenetic origin is the palatoquadrate cartilage of the splanchnocranium and not the chondrocranium (Table 7-1).

The ethmoid region of the chondrocranium and the nasal capsules, which were largely unossified in early tetrapods and ancestral amniotes, are represented in mammals by the ethmoid and turbinate bones, respectively. The **ethmoid bone** (Gr., *ethmos* = sieve) forms the front of the braincase adjacent to the olfactory bulbs of the brain (Fig. 7-24). It is perforated by many sieve-like **cribriform foramina** (L., *cribrum* = sieve + *forma* = shape), through which processes of olfactory cells pass from the nasal epithelium to the olfactory bulbs. Part of the ethmoid bone extends rostrally to form most of the vertical nasal septum between the nasal cavities.

The paired frontals and parietals cover the dorsal and part of the lateral surfaces of the brain. The squamosals complete covering the brain laterally (Figs. 7-24 and 7-26*A*). All are dermal roofing bones. Sheetlike or platelike extensions of all of these bones extended from the dermal roof inward, passing medial to the temporal jaw muscles, and united with the alisphenoid and chondrocranial elements (Fig. 7-25). These dermal extensions form most of the dorsal and lateral walls of the mammalian braincase. Although derived from the dermal roof, the dermal extensions are not part of the original roof itself, which lay lateral to the jaw muscles. Only the portions of the frontals and parietals adjacent to the mid-dorsal line are part of the original roof. And only the zygomatic portion of the squamosal, which contributes to the zygomatic arch ventral to the orbit, is a part of the original dermal roof. Most of the original lateral wall of the dermal roof in the temporal region was lost as the synapsid temporal fenestra evolved and enlarged in early therapsids and mammals.

Skull Changes Correlated with Changes in Sense Organs Evolutionary changes in the ears, nose, and median eye affected several aspects of the skull. The median pineal–parietal eye complex of ancestral terrestrial vertebrates remained in basal synapsids and most therapsids. The parietal eye of contemporary lizards monitors solar radiation and affects the animal's behavior accordingly (Chapter 12). Presumably it had the same function in basal synapsids and early therapsids. It was lost in advanced therapsids (†cynodontids; Fig. 3-31), probably because of climatic changes (Roth et al., 1986). The median eye was transformed in mammals into the pineal gland. The pineal gland is an endocrine gland attached to the top of the diencephalon of the brain, and it usually is covered by the expanded cerebral hemispheres. The retreat of the pineal–parietal complex from the surface of the skull was accompanied by the loss of the parietal foramen (Fig. 7-25).

The senses of smell and hearing became particularly important in ancestral mammals. The otic capsule of the chondrocranium was large, and spaces within it indicate that a cochlea had evolved. The cochlea is the part of the inner ear that gives mammals their keen sense of hearing. The nasal cavities also enlarged, and the scroll-shaped turbinate bones, which were derived from the nasal capsules, greatly increased the surface area of the mucous membrane in the nasal cavity. The mucous membrane contains the olfactory receptors, and it cleans, moistens, and warms inspired air. Conditioning inspired air is important for an animal that is becoming more active and endothermic. Ancestral mammals probably had a keen sense of smell and hearing, as do contemporary nocturnal insectivores. Early mammals likely occupied a similar ecological niche. Their contemporary large, diurnal, and ectothermic †archosaurs did not exploit this niche.

Palatal Evolution Changes in feeding mechanisms enabled later therapsids and mammals to gather more food and to cut up and chew the food in the mouth before swallowing it. Jaw muscles became more powerful and complex, and the jaws and bite force became stronger. Correlated with these changes, the palate lost its primitive kinesis and became firmly united with the braincase. The dermal pterygoid bones were reduced to winglike, pterygoid processes on the ventral surface of the sphenoid (Figs. 7-25 and 7-26*B*). The names of these bones and processes come from their appearance in mammals (Gr., *pteryg* = wing or fin + *eidos* = form). The ectopterygoids may be incorporated in these processes. Pterygoid muscles extend from them to the medial side of the lower jaw.

The choanae of basal synapsids and early therapsids continued to open into the mouth cavity through the rostral part of the roof of the mouth (Fig. 7-27*A*). In †cynodontids close to the mammal transition, shelflike extensions of the premaxilla, maxilla, and palatine bones grew ventrally and medially and united to form a small hard palate that lay ventral to the original roof of the mouth, or primary palate (Fig. 7-27*B*). As the hard palate evolved and extended caudally, the rostral part of the primary palate regressed and provided space for the enlarging nasal cavities. In mammals the originally paired vomer bones became a single, median element that lies dorsal to the hard palate and forms the ventral part of the nasal septum (Fig. 7-27*C*). The choanae of mammals open far back in the mouth cavity near the caudal border of the hard palate. A fleshy soft palate continues caudally from the hard palate, attaching along the ventral bor-

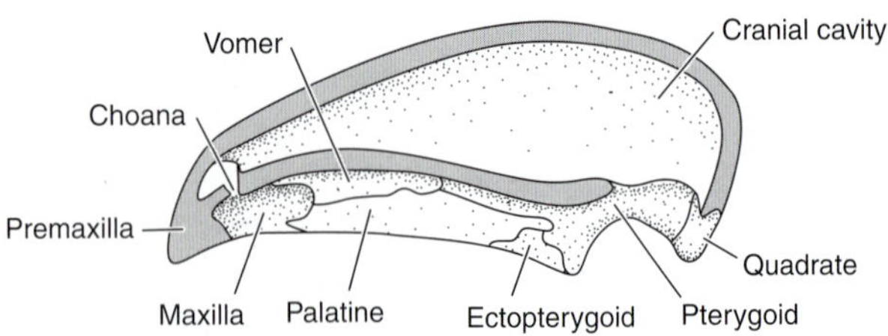

A. Early synapsid (†*Dimetrodon*)

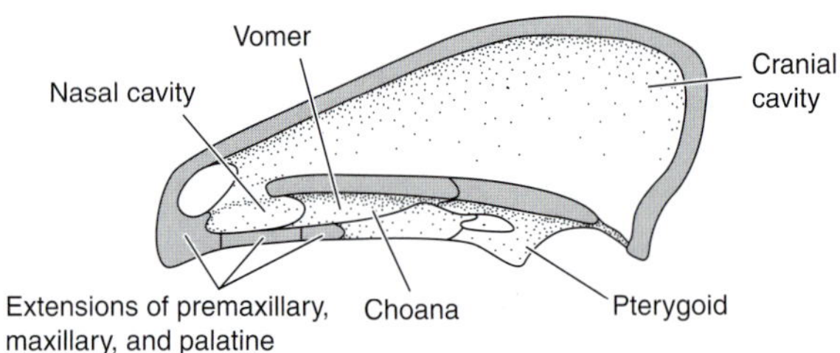

B. Therapsid (†*Probainognathus*)

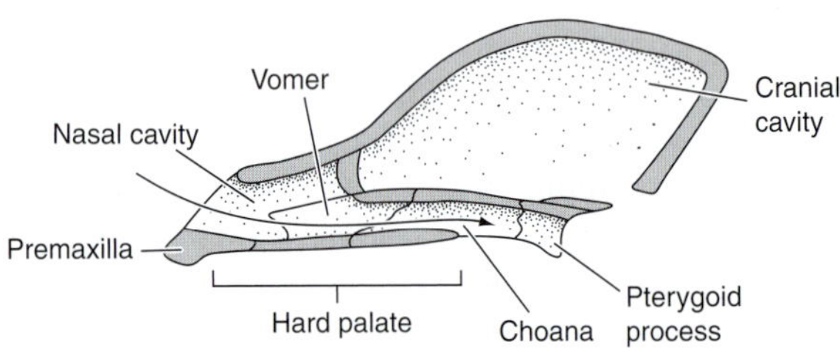

C. Mammal (*Canis*)

FIGURE 7-27
Sagittal sections through the skull of representative vertebrates to show the evolution of the mammalian hard palate. *A,* An early synapsid. *B,* A therapsid. *C,* A mammal. *(Modified after Romer and Parsons.)*

der of the pterygoid processes. The soft palate separates the nasal pharynx, into which the choanae lead, from the oral pharynx, into which the mouth cavity leads. Air and food passages are separated until they come together in the more caudal, laryngeal portion of the pharynx.

The evolution of the hard and soft palates, together known as the **secondary palate,** allows mammals to manipulate and chew food while still breathing. Breathing need be stopped only momentarily when food is swallowed and passes through the laryngeal pharynx to the esophagus. Separating food and breathing passages also allows neonatal mammals to suckle. In many neonatal mammals, humans included, the cranial end of the larynx extends into the nasal pharynx and completely separates food and air passages, which facilitates suckling. The infant can develop the reduced pressure in its mouth cavity needed to draw in milk from the mother. Later in life of most species, the larynx migrates caudally, which expands the human pharynx and makes it a better resonating chamber for speech but increases the likelihood of choking. An increased intake of food and a large volume of gas exchange are important for an endotherm, and the evolution of the secondary palate is correlated, at least partly, with the evolution of endothermy. But mammals also chew their food, and chewing requires strong upper and lower jaws. Thomason and Russell (1986) showed that the secondary palate greatly strengthens the upper jaw and snout of mammals, especially against torsion and lateromedial bending. This factor quite possibly contributed to the evolution of the hard palate, especially in early stages, when endothermy may not have been a factor.

Temporal Fenestration and Jaw Muscles A small, synapsid type of temporal fenestra appeared in basal synapsids (†*Dimetrodon*). The primary jaw-closing muscle, the **adductor mandibulae,** arose from the surface of the braincase, from the periphery of the fenestra, and from a tendinous sheet that covered the fenestra (Fig. 7-28*A*). The adductor mandibulae muscle inserted along a small, dorsally projecting **coronoid eminence** on the lower jaw. A **pterygoideus muscle** arose from the ventral side of the palate and inserted on the medial side of the lower jaw (Fig. 7-28*C*).

The adductor mandibulae muscle became much larger and more powerful in †cynodonts (†*Probainognathus*) and differentiated into a **temporalis muscle** and a **masseter muscle.** The masseter later subdivided into superficial and deep portions (Fig. 7-28*B*). The temporal fenestra enlarged considerably as jaw muscle mass increased and the brain and braincase became larger. As the fenestra enlarged, much of the dermal roof disappeared. The only parts of the original dermal roof remaining in this region are the postorbital bar, a bar of bone (now called the **zygomatic arch**) that bounded the fenestra and orbit ventrally, and a narrow, mid-dorsal strip of bone. The jugal bone (called zygomatic bone in mammals) and part of the squamosal (temporal) bone formed the zygomatic arch (Fig. 7-26*A*). The temporalis muscle continued to arise from the surface of the braincase and from the margins of the fenestra, but its insertion on the dentary was now on a large **coronoid process,** which evolved from the smaller coronoid eminence. The temporalis muscle attached on the dorsal border and medial side of this process (Fig. 7-28*C*). The masseter muscle took its origin from the zygomatic arch. An outward-bowing zygomatic arch allowed for increased muscle size. The deep masseter inserts on the lateral surface of

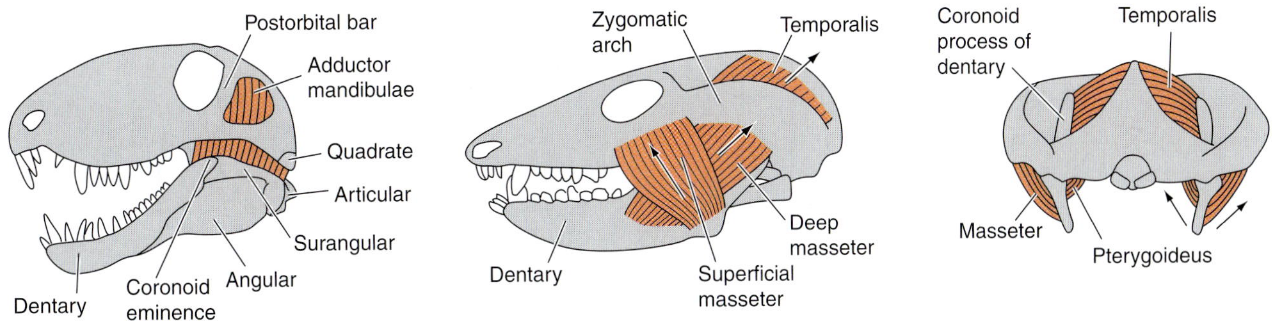

FIGURE 7-28
Jaw muscles in synapsids. *A,* A lateral view of the jaw muscles of a basal synapsid, based on †*Dimetrodon. B,* A similar view of the jaw muscles in a therapsid close to the mammal transition, †*Probainognathus. C,* A posterior view of therapsid close to the mammal transition. *(A, After Kemp; B and C, after Carroll.)*

the coronoid process and adjacent parts of the mandible. The superficial masseter inserted near the posterior angle of the lower jaw. The pterygoideus muscle remained essentially unchanged.

These muscle changes increased bite force and gave the control over the movements of the lower jaw needed to position the teeth for cutting and masticating food. Temporalis and deep masseter muscles pull upward and caudally; the anterior masseter, upward and rostrally (Fig. 7-28*B* and *C*). The masseter complex as a whole also pulls somewhat laterally, thereby balancing the medial pull of the pterygoideus.

Jaw and Middle Ear Changes Changes in the jaws, which culminated in a new jaw joint and two additional auditory ossicles in the middle ear, accompanied changes in jaw muscles. In basal synapsids, as in other reptilomorphs, the jaw joint lay between a large quadrate bone in the upper jaw and a large articular bone in the lower jaw (Fig. 7-29*A*). There is no indication from bone shapes that a tympanic membrane lay behind the quadrate, as one does in turtles and some lizards. As in ancestral amniotes (Fig. 7-17*B*), the columella was large and extended from the otic capsule to the quadrate. It was too massive a bone to have responded to high-frequency, airborne vibrations. Lower-frequency vibrations that were strong enough probably could be transmitted from the lower jaw through the jaw joint (articular and quadrate) to the columella and otic capsule. The dentary was the largest bone of the lower jaw and the primary tooth-bearing element. Postdentary bones (angular, surangular, and articular) still were relatively large and firmly united through sutures with each other and with the dentary.

In the evolution from early synapsids (Fig. 7-29*B*) to more advanced synapsids (the †cynodontids), the dentary became progressively larger as muscle forces acting on it increased in strength (Fig. 7-29*C*). Its caudal angle approached the squamosal and quadrate. At the same time, the postdentary bones became progressively small. They also became less firmly articulated with the dentary, and they lay partly in a groove on the medial surface of the dentary. The small prearticular (sometimes called the goniale) lay on the medial side of the articular. The quadrate and articular also became much smaller and less firmly articulated with their neighbors. The jaw joint, however, still lay between them, which poses a paradox: if the jaws and bite were becoming stronger, how was it possible for the bones bearing the jaw joint (articular and quadrate) to become smaller and loosely articulated with their neighbors? Part of the answer appears to be a redistribution of forces in the jaw such that, although the bite was stronger, there was little resultant force at the joint (Focus 7-1). Another part of the answer was the evolution at the same time of a tympanic ear sensitive to high-frequency, airborne vibrations. We will consider the details of ear evolution later (Chapter 12). Briefly, a reflected process or lamina of the angular bone in †cynodontids was shaped to hold a tympanic membrane, and the loosening of the joint-bearing bones (articular and quadrate) from the rest of the jaw made them more responsive to vibrations received by the tympanic membrane. They, in turn, transmitted vibrations to the columella, to which the quadrate has always been attached.

As these trends continued in very late †cynodontids, a part of the dentary bone (the condyloid process) made contact with the enlarging squamosal bone lateral

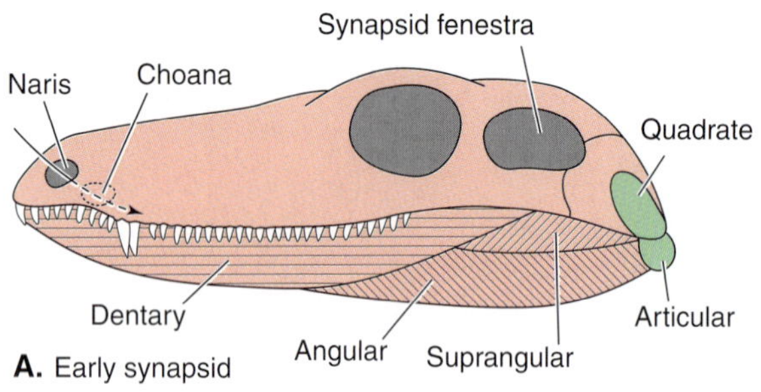

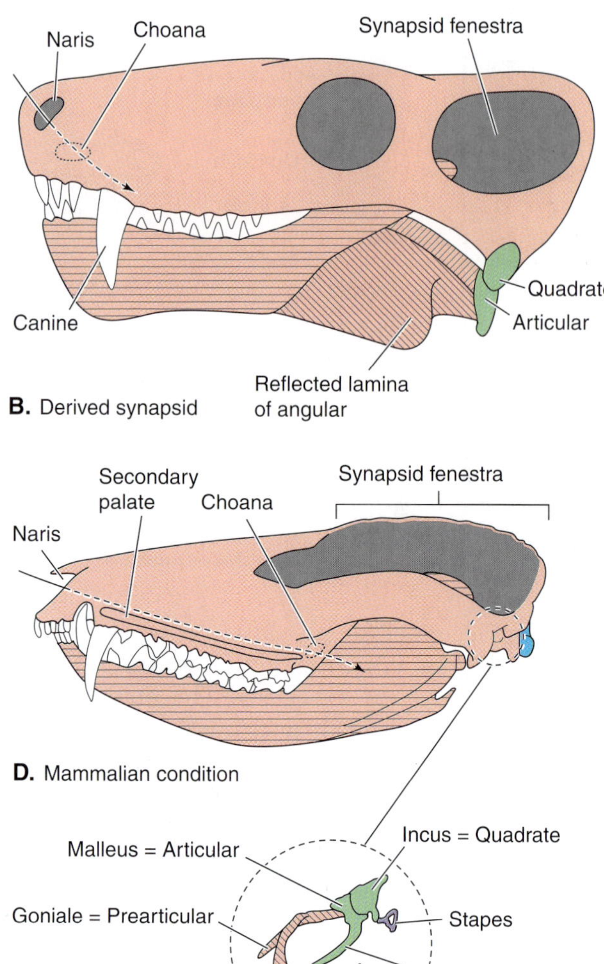

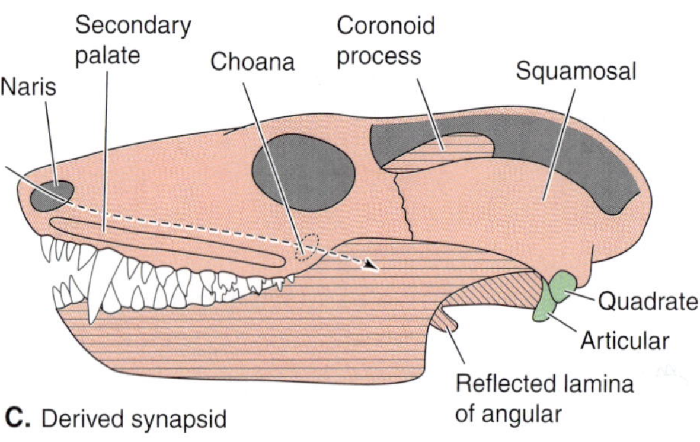

FIGURE 7-29
A–E, Evolution of the mammalian jaw joint and auditory ossicles.

to the articular and quadrate (Fig. 7-30). Four bones now participated in the jaw joint. The rounded articular condyle of the lower jaw was composed of the articular and dentary bones. The quadrate and squamosal bones formed the mandibular fossa in the skull that received the condyle. At the transition to mammals, the dentary–squamosal component of the joint enlarged. Soon it formed the entire joint. The articular and quadrate, already partly free from surrounding bones, became further reduced in size and detached from the joint. These bones then became incorporated into an expanding middle ear cavity as the additional auditory ossicles characteristic of mammals: articular = **malleus** and quadrate = **incus** (Fig. 7-29*E*). The quadrate, now the incus, continued to attach to the columella. The columella is now a very small, stirrup-shaped bone known in mammals as the **stapes** (L., *stapes* = stirrup). The small prearticular bone of the lower jaw formed the manubrium, a process on the malleus. The angular bone and its reflected lamina became the **ectotympanic bone,** which holds the tympanic membrane and partly encases the middle ear cavity. The encasement of this cavity is completed in eutherian mammals by the **endotympanic bone,** a new cartilage-replacement bone without a homologue in early amniotes. The ectotympanic and endotympanic bones unite with the squamosal and otic capsule to complete the mammalian temporal bone (Table 7-1). Other postdentary bones were lost, so the mammalian lower jaw consists of only the dentary bone. A single lower jaw bone, a dentary–squamosal jaw joint, and three auditory ossicles (malleus, incus, and stapes) are important derived features that define mammals.

Teeth

Changes in feeding mechanisms that affected the skull and lower jaw also affected the teeth. Early synapsids had conical teeth loosely attached to their palate and jaws and not set in sockets. These teeth were well adapted for capturing and holding small prey, which

FOCUS 7-1 *Forces Acting at the Jaw Joint*

Ancestral amniotes had a single adductor muscle (Fig. A), the contraction of which rotated the lower jaw from the jaw joint upward against the upper jaw. This generated a bite force. In order to press firmly against the upper jaw, however, the lower jaw must be in balance, or equilibrium. But the jaw cannot be in equilibrium through the action of these two forces alone because the forces are not in line. A strong downward reaction force is also needed at the jaw joint. A simple experiment illustrates this. Place the pointed end of a pencil (= the lower jaw) under the edge of a desk (= the upper jaw) and pull upward near the middle of the pencil (= adductor force), leaving the eraser end free to move. The part of the pencil under the desk rotates downward and does not exert a force (= bite force) on the desk unless you also press down on the eraser end of the pencil (= the reaction force at the jaw joint). Because there was a strong reaction force at the jaw joint, the bones at the joint of early amniotes (quadrate and articular) were large.

As we have seen (Fig. 7-28), the adductor musculature of late synapsids differentiated into pterygoideus, temporalis, and deep and superficial masseter muscles. The pterygoideus muscle on the medial side of the jaw and the deep masseter on its lateral surface had similar lines of action and balanced each other. We can focus then on the lines of action of the temporalis and superficial masseter and see the effects of their lines of action on forces at the jaw joint.

Early in synapsid evolution (Fig. B), the lines of action of the temporalis, superficial masseter, and bite forces did not line up and converge at a single point. A strong reaction force still was needed at the jaw joint.

During the evolution of synapsids, the dentary enlarged. The jaw musculature also became larger. This is indicated by the increased space available for the muscles. Larger jaw muscles imply a stronger bite force. Shifts in the lines of action of the temporalis and masseter brought these forces closer to the line of action of the bite force. The lines of action of these forces in a therapsid close to the mammal transition meet (or project) to a single point (Fig. D). This means that the forces could be in balance without a strong reaction force at the jaw joint. Whether or not they were in balance depends on the magnitude of the forces. We can only estimate their magnitude by extrapolating from contemporary, related species. Assuming the forces were in balance, the quadrate and articular and postdentary bones could be very small, as indeed they were.

Another way of looking at the forces acting on the jaw is to compare the resultant (Fig. D, R′) of the temporalis (T) and superficial masseter forces (M) with the bite force (Fig. D). To do this (Focus 5-1), place the head of vector T at the tail of vector M (but keep the orientation and length of vector T the same). The heavy dashed line (T′) shows this. The resultant of these forces is the vector R′. You will notice that the resultant is very close to the bite force and nearly parallel with it. Only a small reaction force, if any, would be needed at the jaw joint to balance the muscle and bite forces.

Figure C shows an intermediate condition, and many more are known from the fossil record. You can see the shift in forces that occurred during synapsid evolution by comparing Figures A, B, and C. Notice that the shift in the lines of action of the temporalis and superficial masseter brought their intersection increasingly close to the line of action of the bite force, until all forces converge on a single point. Notice too that the line of action of the temporalis shifts toward the horizontal, and that of the superficial masseter shifts toward the vertical. This brought their resultant closer to the bite force. The redistribution of forces appears to have gradually reduced the reaction force needed at the jaw joint until it finally became negligible. This enabled the jaw joint bones and the postdentary bones to become increasingly small.

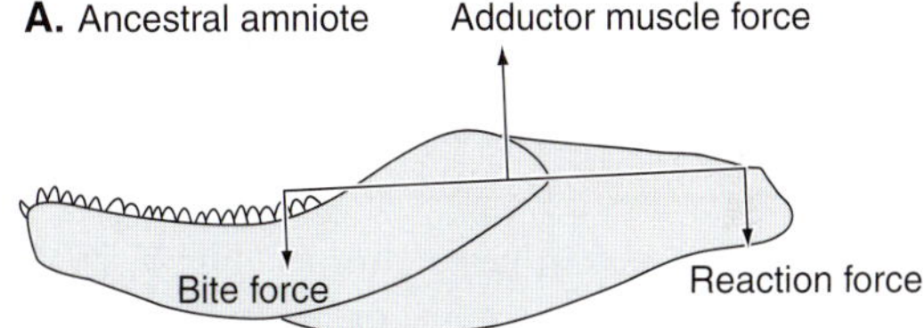

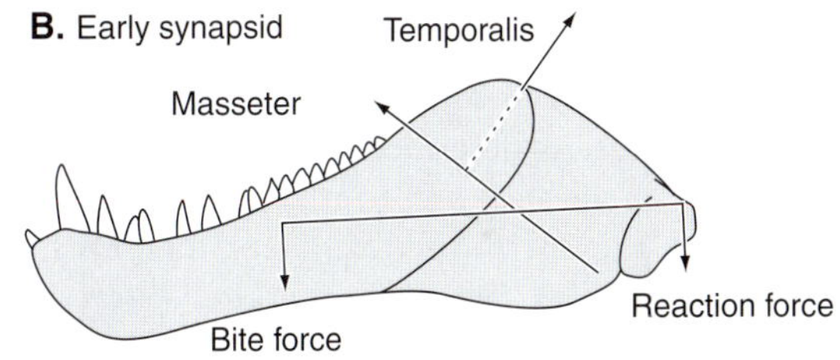

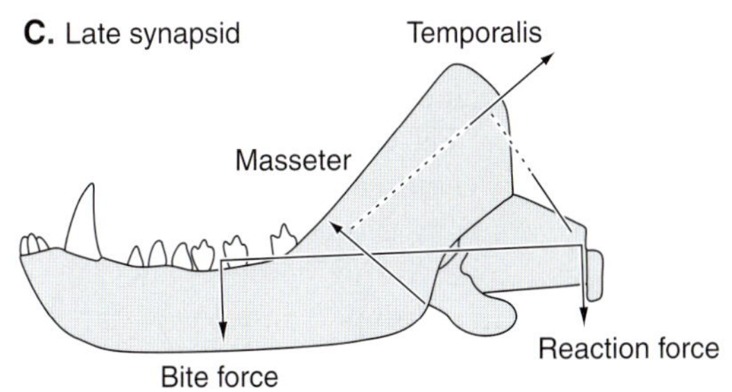

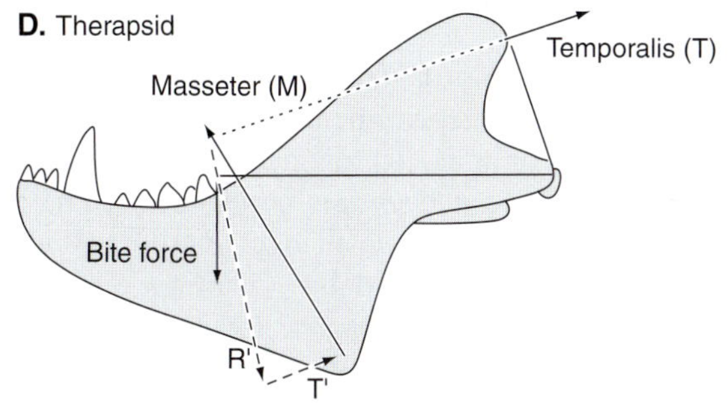

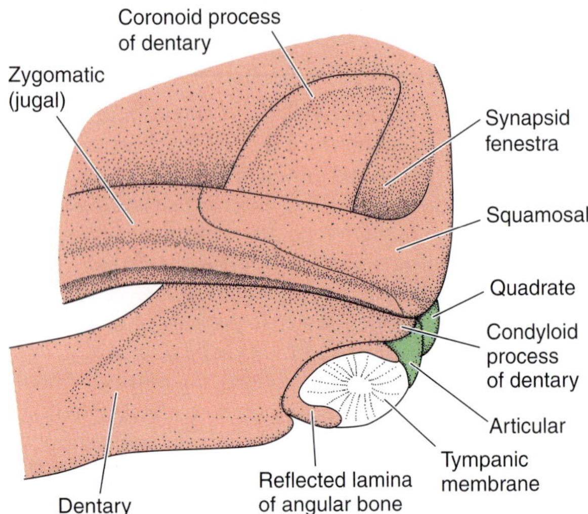

FIGURE 7-30
A lateral view of the jaw joint region of the skull and lower jaw in a †cynodont that was close to the mammal transition. Four bones form the jaw joint. *(After Allin.)*

they swallowed whole or in large chunks (Fig. 7-29*A*). With the evolution of a secondary palate in †cynodontids, palatal teeth were lost. Other teeth were limited to the bones forming the jaw margins, specifically to the premaxilla and maxilla in the upper jaw and to the dentary in the lower jaw. The teeth became firmly set in sockets, a condition described as **thecodont** (Gr., *theke* = case + *odous, odont* = tooth). The teeth also differentiated as they performed different functions, a condition described as **heterodont** (Gr., *heteros* = other, different). Teeth at the front of the jaws became small, nipping **incisors,** a pair of large **canine teeth** followed them, and the **cheek teeth** were adapted for cutting and tearing apart prey (Fig. 7-29*C* and *D*).

Cheek teeth differentiated in mammals into **premolars** and **molars** (see Fig. 16-6). Infant mammals suckle and lack teeth. When teeth do appear later during postnatal development, the control of jaw movements made possible by the more complex jaw musculature allows the teeth of upper and lower jaws to mesh, or occlude, very precisely. Specialized teeth, firmly set in sockets and with a good occlusion, enable mammals to cut and chew food and process the large amount of food needed to sustain an endothermic metabolism. Good occlusion is maintained during growth by a limited tooth replacement. As the jaw grows larger, the first, or milk, set of incisors, canines, and premolars are replaced by larger permanent teeth. In some species, morphological changes in the cusps of the permanent teeth are needed to maintain good occlusion (Tumlison and McDaniel, 1984). Molars are not replaced but appear sequentially as the jaw enlarges. Teeth are considered further in Chapter 16.

The Mammalian Hyoid Apparatus and Larynx

The hyobranchial apparatus of mammals, usually simply called the **hyoid apparatus**, is an integral part of the feeding mechanism, for it and its muscles participate in tongue movements, in opening and closing the jaws, and in swallowing (Chapter 16). The hyoid apparatus consists of a transverse body, or **basihyal**, a lesser horn, or **ceratohyal**, and a greater horn, or **thyrohyal** (Fig. 7-26*A*). All are embedded in the musculature in the base of the tongue. Either a ligament or a chain of small bones (**epihyal**, **stylohyal**, and **tympanohyal**) extends from each lesser horn to the tympanic region of the skull base. In some mammals, including humans, the dorsal part of this chain of bones has fused to the temporal bone, forming its **styloid process** (Gr., *stylos* = pillar + *eidos* = form). The thyrohyal attaches to the thyroid cartilage of the larynx. The thyrohyal is a derivative of the third visceral arch; the rest of the hyoid apparatus derives from the ventral half of the hyoid arch.

The fourth and fifth visceral arches, which are incorporated in the hyobranchial apparatus of amphibians (Fig. 7-15*B*), form the **thyroid cartilage** of the larynx of eutherian mammals (Gr., *thyreos* = door-shaped shield + *eidos* = form). The sixth arch, which forms small lateral laryngeal cartilages in amphibians, differentiates into the **cricoid cartilage** (Gr., *krikos* = ring) and a pair of **arytenoid cartilages** (Gr., *arytaina* = ladle). The arytenoids extend into the vocal cords and can move them together or apart. The evolution of a more complex larynx enables mammals to have a larger repertoire of sounds than recent reptiles, and the resulting improved communication has been an important feature of mammalian evolution. A newly evolved **epiglottic cartilage** supports the mammalian epiglottis, which helps deflect food around the entrance to the larynx and into the esophagus.

Conclusion: Changing Form and Function

The evolution of the cranial skeleton is an excellent case study of the way form changes as functions change. It shows the plasticity of the evolutionary process; that is, how structures that are used in one way can be remodeled to meet changing functional requirements. Seldom are completely new structures needed. Together, the three components of the cranial skeleton (chondrocranium, splanchnocranium, and dermatocranium) protect the brain and major sense organs and play important roles in feeding and breathing. However, the way in which the components participate

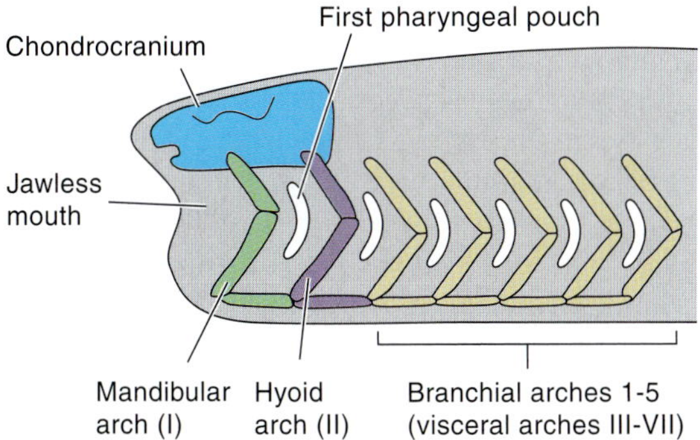

A. Hypothetical jawless condition

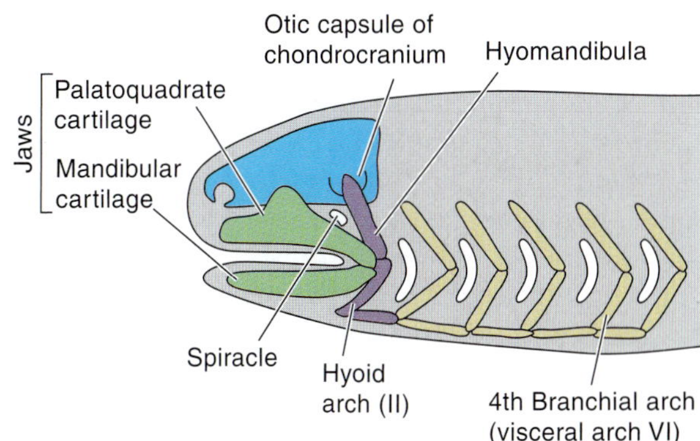

B. Gnathostome

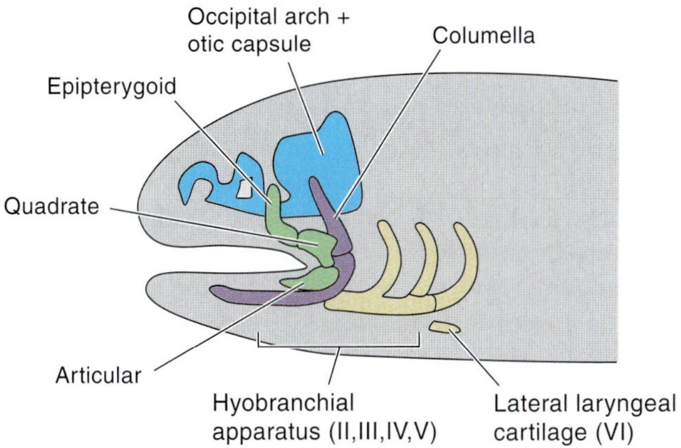

C. Hypothetical early tetrapod

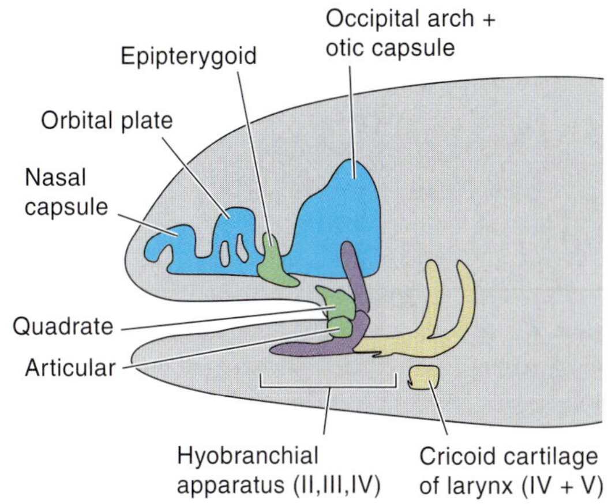

D. Hypothetical advanced therapsid

Alisphenoid
Stapes
Styloid process (humans)
Incus
Malleus
Stylohyoid ligament or chain of ossicles
Hyoid
Body and lesser horn (II)
Greater horn (III)
Cricoid cartilage (VI)
Arytenoid cartilage (VI)
Thyroid cartilage (VI + V)
Larynx

E. Placental mammal

FIGURE 7-31
Diagrams in lateral view of the evolution of the splanchnocranium and its relationship to the chondrocranium. *A,* Hypothetical ancestral jawless vertebrate. *B,* Jawed fish, based on a shark. *C,* An early terrestrial vertebrate based on an early tetrapod and a contemporary amphibian. *D,* An advanced therapsid in which the articular and quadrate bones are becoming quite small but still bear the jaw joint. *E,* A placental mammal.

in these functions changes greatly during the course of vertebrate evolution as the modes of life of the animals, and the environments in which they live, change.

Changing form and function are particularly well illustrated by summarizing the evolution of the splanchnocranium (Fig. 7-31). In ancestral jawless vertebrates the splanchnocranium probably consisted of a series of similar arches between the pharyngeal pouches, the sole function of which was supporting the gills and providing elasticity to the wall of the pharynx for breathing and feeding movements (Fig. 7-31*A*). In gnathostomes, an anterior arch became the mandibular arch, which forms or contributes to the jaws. The dorsal part of the hyoid arch, the hyomandibula, soon became involved in jaw suspension and movements. The remaining branchial arches retained their original form and function (Fig. 7-31*B*).

Profound changes occurred in the splanchnocranium during the transition of vertebrates from water to land as lungs replaced gills as the site for gas exchange. The caudal ends of the palatoquadrate and mandibular cartilages (quadrate and articular bones) continued to form the jaw joint. The part of the palatoquadrate rostral to the quadrate, the epipterygoid bone, remained, forming an articulation between the palate and the chondrocranium. This was originally a movable joint permitting cranial kinesis during feeding. Jaws became secondary autostylic, and the hyomandibula, which was no longer needed in jaw suspension, transformed into the columella, a sound-transmitting element (Fig. 7-31*C*). In the first terrestrial vertebrates, the columella was a large, stout element that probably transmitted strong, ground-borne vibrations from the jaws to the otic capsule and inner ear. In many groups of later tetrapods, the columella independently became a slender rod transmitting airborne vibrations from a tympanic membrane to the inner ear.

The branchial arches were greatly reduced with the loss of gills, but their ventral parts, along with the ventral part of the hyoid arch, formed a hyobranchial apparatus, which supported the floor of the mouth cavity and pharynx, and the newly evolved tongue. Movements of the hyobranchial apparatus continue to have a role in breathing in amphibians, but now they pump air into lungs and help expel it. Movements of the tongue are used in food capture and transport. A reduced caudal arch, the sixth visceral arch, formed cartilages located in the wall of a small larynx (Fig. 7-31*C*).

Ribs and, in mammals, diaphragm movements move air in out of the lungs in amniotes. The hyobranchial apparatus is further reduced and involved only with tongue and swallowing movements (Fig. 7-31*D*). The larynx enlarges and incorporates, in mammals, the fourth and fifth visceral arches as well as the sixth. The mammalian larynx contains the vocal cords. (Bird vocal cords lie more caudally in a syrinx.) With their keen sense of hearing, vocalization is an important part of the behavioral repertoire of birds and mammals.

The articular and quadrate bones remain as jaw joint bones in sauropsids and early therapsids (Fig. 7-31*D*). The quadrate and articular became very small in †cynodontids. When a new jaw joint evolved between the squamosal and dentary bones at the transition to mammals, the quadrate and articular became the two additional auditory ossicles (incus and malleus, respectively) characteristic of mammals. The small columella is called the stapes. The small dermal prearticular bone of the lower jaw of a therapsid forms a small process on the malleus (Fig. 7-31*E*).

The epipterygoid remains in sauropsids as an articulation between palate and chondrocranium. This frequently is a movable joint in their kinetic skulls. The epipterygoid becomes firmly united with the braincase (as the alisphenoid) in the akinetic skull of a mammal.

SUMMARY

1. The bones and cartilages of the craniate skeleton can be grouped in several ways: (1) by development (integumentary skeleton of dermal bones and an endoskeleton of cartilage or cartilage-replacement bone), (2) by body tube in which they lie (somatic skeleton and visceral skeleton), (3) by region of the body (cranial skeleton and postcranial skeleton).
2. The cranial skeleton consists of (1) a chondrocranium, surrounding all or much of the brain and major sense organs; (2) a splanchnocranium, consisting of a series of visceral arches between the pharyngeal pouches; and (3) a dermatocranium of dermal bones largely encasing the other parts.
3. The chondrocranium provides most of the protection for the brain and special sense organs in fishes, for it covers nearly all but the top of the brain, and it also encapsulates the inner ear and nose. Only the top of the brain and the eyeball are protected by the dermatocranium in most fishes, and the chondrocranium does this in chondrichthyan fishes and contemporary jawless craniates.
4. As the brain expanded in terrestrial vertebrates, and especially in the line of evolution toward birds and mammals, the dermatocranium became increasingly important in providing protection. Most of the dorsal and lateral walls of the mammalian braincase are dermal extensions from the roof that grew ventrally to unite with the chondrocranium. A derivative of the splanchnocranium (the alisphenoid), which originally articulated the palate with the chondrocranium, completes the mammalian braincase laterally.
5. The nasal capsule portion of the chondrocranium of mammals forms three scroll-shaped turbinate bones. These not only increase surface area for olfaction but also warm and otherwise condition in-

spired air. Conditioning the air is important for an endotherm.

6. The inner ear within the otic capsule of the chondrocranium becomes more elaborate in birds and mammals. This contributes to their keen sense of hearing and increased behavior repertoire.
7. Capturing or gathering food, transporting it through the mouth and pharynx, and swallowing always have been important functions of the splanchnocranium and dermatocranium. Ancestral, jawless craniates were suspension feeders and scavengers. Expansion and contraction of the pharynx appear to have been the engine for circulating a feeding and respiratory current of water through the pharynx.
8. Possibly an anterior gill arch became the jaws in gnathostomes. The palatoquadrate and mandibular cartilages of the mandibular arch form the jaws of chondrichthyans.
9. Dermal bones that ensheathe the mandibular arch form the main parts of the jaws in most vertebrates. The jaw joint, however, remains between the caudal ends of the palatoquadrate and mandibular cartilages, or their ossifications (quadrate and articular bones) until mammals.
10. The dorsal part of the hyoid arch, the hyomandibula, extends from the otic capsule containing the inner ear to the caudal end of the palatoquadrate in most fishes. It helps to suspend the jaws and often participates in their movements.
11. Some fishes use their jaws to seize and capture prey, but most fishes are suction feeders. The upper jaw is movably articulated with the braincase through palatal bones.
12. The splanchnocranium continues to have an important role in feeding in aquatic amphibians and some aquatic turtles and also in tongue protrusion in some terrestrial amphibians. The splanchnocranium becomes less important in prey capture in reptilomorphs but plays an important role in food transport and in swallowing.
13. Loss of gills also allows loss of the dermal gular and opercular bones that covered them, and loss of the dermal connection between the pectoral girdle and skull roof. A neck began to evolve. Terrestrial vertebrates can move their head independently of the trunk as they search for prey and capture prey with their jaws, or sometimes by a protrusive tongue.
14. Fenestration of the temporal part of the dermal roof in many lines of reptilomorph evolution correlates with increasing size and complexity of jaw muscles. A synapsid-type fenestra characterizes the line of evolution toward mammals. It enlarges greatly as the braincase expands and jaw muscles become larger and more powerful.
15. In the line of evolution to mammals, a redistribution of muscle forces on the lower jaw correlated with a great expansion of the dentary bone of the lower jaw and a reduction in size of postdentary dermal bones and joint-bearing bones (quadrate and articular).
16. Concurrent expansions of the squamosal bone in the dermal roof and a caudal extension of the dentary bone led to the formation of a second jaw joint just lateral to the original one.
17. The squamosal–dentary joint becomes the only one when the quadrate and articular leave the jaw to enter the middle ear as additional auditory ossicles. The small dermal angular bone of the lower jaw becomes the ectotympanic that surrounds the tympanic membrane. The middle ear is encased ventrally in eutherian mammals by an endotympanic, a new endochondral bone without a homologue in reptiles.
18. As changes occur in the lower jaw, a secondary palate evolves. This strengthens the upper jaw and largely separates the food and air passages. Breathing need not be stopped as the animal cuts up and masticates its food with teeth firmly set in sockets and differentiated into several types. This is an important step in the evolution of endothermy.

REFERENCES

Allin, E. F., 1975: Evolution of the mammalian ear. *Journal of Morphology,* 147:403–438.

Barghusen, H. R., 1972: The origin of the mammalian jaw apparatus. *In* Schumacher, G. H., editor: *Morphology of the Maxillo-Mandibular Apparatus.* Leipzig, Georg Thieme.

Bramble, D. M., 1978: Origin of the mammalian feeding complex, models and mechanisms. *Paleobiology,* 4:271–301.

Carroll, R. L., 1986: The skeletal anatomy and some aspects of the physiology of primitive reptiles. *In* Hotton III, N., MacLean, P. D., Roth, J. J., and Roth, E. C., editors: *The Ecology and Biology of Mammal-like Reptiles.* Washington, Smithsonian Institution Press.

Clack, J. A., 1989: Discovery of the earliest known tetrapod stapes. *Nature,* 342:425–427.

Crompton, A. W., and Hylander, W. L., 1986: Changes in mandibular function following the acquisition of a dentary–squamosal jaw articulation. *In* Hotton III, N., MacLean, P. D., Roth, J. J., and Roth, E. C., editors: *The Ecology and Biology of Mammal-like Reptiles.* Washington, Smithsonian Institution Press.

Crompton, A. W., and Parker, P., 1978: Evolution of the mammalian masticatory apparatus. *American Scientist,* 66:192–201.

Crompton, A. W., 1963: The evolution of the mammalian jaw. *Evolution,* 17:431–439.

DeBeer, G. R., 1937: *The Development of the Vertebrate Skull.* London, University of Oxford Press.

Feduccia, A., 1980: *The Age of Birds.* Cambridge, Harvard University Press.

Frazzetta, T. H., 1968: Adaptive problems and possibilities in the temporal fenestration of tetrapod skulls. *Journal of Morphology,* 125:145–158.

Grande, L., and Bemis, W. E., 1998: A comprehensive phylogenetic study of amiid fishes (Amiidae) based on comparative skeletal anatomy. An empirical search for interconnected patterns in natural history. Society of Vertebrate Paleontology, Memoir 4. *Journal of Vertebrate Paleontology,* 18:1690.

Hanken, J., and Hall, B., editors, 1993: *The Vertebrate Skull.* Chicago, University of Chicago Press.

Hedges, S. B., and Poling, L. L., 1999: A molecular phylogeny of reptiles. *Science,* 283:998–1001.

Heilman, G., 1927: *The Origin of Birds.* Reprinted 1972, New York, Dover Publications.

Hopson, J. A., 1966: The origin of the mammalian middle ear. *American Zoologist,* 6:437–450.

Jarvik, E., 1980: *Basic Structure and Evolution of Vertebrates,* volume 1. London, Academic Press.

Jollie, M., 1962: *Chordate Morphology.* Huntington, New York, Krieger.

Kemp, T. S., 1982: *Mammal-like Reptiles and the Origin of Mammals.* London, Academic Press.

Mallatt, J., 1996: Ventilation and the origin of jawed vertebrates: A new mouth. *Journal of the Linnean Society of London,* 117:329–404.

Moore, W. J., 1981: *The Mammalian Skull.* Cambridge, Cambridge University Press.

Nickel, R., Schummer, A., and Seiferle, E., 1979, 1981, 1986: *The Anatomy of Domestic Animals,* volumes 1–3. New York, Springer-Verlag.

Romer, A. S., 1972: The vertebrate as a dual animal: Somatic and visceral. *In* Dobzhansky, T., Hecht, M. K., and Steere, W. C., editors: *Evolutionary Biology,* 6:121–156. New York, Appleton-Century-Crofts.

Romer, A. S., and Parsons, T. S., 1977: *The Vertebrate Body.* Philadelphia, Saunders College Publishing.

Roth, J. J., Roth, E. C., and Hotton III, N., 1986: The parietal foramen and eye: Their functions and fate in therapsids. *In* Hotton III, N., MacLean, P. D., Roth, J. J., and Roth, E. C., editors: *The Ecology and Biology of Mammal-like Reptiles.* Washington, Smithsonian Institution Press.

Russell, A. P., and Thomason, J. J., 1993: Mechanical analysis of the mammalian head skeleton. *In* Hanken, J., and Hall, B. K., editors: *The Skull,* volume 3, *Functional and Evolutionary Mechanisms.* Chicago, University of Chicago Press.

Smith, M. M., 1999: Jaw evolution and function. *In* Teaford, M. F., Smith, M. M. and Ferguson, M. W. J., editors: *Teeth: Development, Evolution and Function.* London, Cambridge University Press.

Thomason, J. J., and Russell, A. P., 1986: Mechanical factors in the evolution of the mammalian secondary palate: A theoretical analysis. *Journal of Morphology,* 189:199–213.

Tumlison, R., and McDaniel, V. R., 1984: Morphology, replacement mechanisms, and functional conservation in dental replacement in the bobcat (*Felis rufus*). *Journal of Mammalogy,* 65:111–117.

Walker, W. F., and Homberger, D. G., 1992: *Vertebrate Dissection.* Philadelphia, Saunders College Publishing.

Young, J. Z., 1981: *The Life of Vertebrates.* Oxford, Clarendon Press.

8

The Postcranial Skeleton: The Axial Skeleton

PRECIS

The postcranial skeleton forms the framework of the body and plays an essential role in body support and movement, and parts are involved in breathing movements. It can be divided into the axial skeleton (notochord, vertebral column, median fins, ribs, and sternum), which we consider in this chapter, and the appendicular skeleton, which we will discuss in the next chapter. We correlate changes in structure with functional changes that occur as vertebrates move from water to land and become very active, agile animals, but many functional aspects of the axial skeleton will be deferred to Chapter 11, Support and Locomotion.

OUTLINE

Parts of the cranial skeleton lie in the longitudinal axis of the body, but we use the term "axial skeleton" for the postcranial axial elements (notochord, vertebral column, median fins, ribs, and sternum). These and the appendicular skeleton (paired fins and limbs, and the girdles to which they attach) constitute the postcranial skeleton. The axial and appendicular skeleton form the framework of the body and, together with associated muscles, support and move the body. Because the physical conditions on land are quite different from those encountered by aquatic vertebrates, profound evolutionary changes occurred in the axial skeleton when vertebrates moved from water to land and became fully adapted to terrestrial life and, in some cases, to a flying mode of life.

We examine the structure, embryonic development, and evolution of the axial skeleton in this chapter and the appendicular skeleton in the next one. Although we consider some aspects of support and locomotion in these chapters, we will analyze these subjects more thoroughly after we have considered the appendicular skeleton and muscles. This will allow us to integrate the structural and functional aspects of all the parts of the body involved in support and locomotion and correlate this with the animal's lifestyle.

Components of the Axial Skeleton

We begin our examination of the axial skeleton by looking at the adult structure and the embryonic development of its individual parts: the vertebrae, ribs, and sternum. Then we will discuss the evolution of the axial skeleton as a whole.

Vertebral Structure

The embryonic notochord persists in adult hagfishes as their only axial support. Vertebrae are not present in these most basal craniates, but at least traces of vertebrae are present in all adult vertebrates. Vertebral structure varies considerably among species and also among body regions of an individual, but certain common components are found in nearly all vertebrae. A representative vertebra has a **vertebral arch,** or **neural arch,** which extends dorsally around the spinal cord (Fig. 8-1*A*). **Intervertebral foramina,** through which spinal nerves pass from the spinal cord, usually lie between the bases of successive neural arches (Fig. 8-1*B*). In a few fishes, the foramina perforate the neural arch. Most vertebrae in the tail region of fishes and many tetrapods also have a **hemal arch,** which extends ventrally around the caudal artery and vein. Neural and hemal arches protect the structures they surround and also serve for the attachment of muscles used in body support and locomotion. The surface area of the arches available for muscle attachment, and the mechanical advantage of the arches as lever arms, are increased in many species by spinelike processes extending from the apices of the arches: **neural spines** and **hemal spines.**

A notochord lies ventral to the nerve cord and neural arches in the embryos of all vertebrates, and hemal arches lie ventral to the notochord in the tail region. The notochord expands during embryonic development and persists in the adults of early fishes and terrestrial vertebrates. Neural and hemal arches rest directly on it, or are anchored to ossifications (arch bases) that develop on the surface of the notochord. Most vertebrates have disklike **vertebral bodies,** or **centra,** which largely or completely replace the notochord during development, and the arches unite with them (Fig. 8-1*A* and *B*). Normally, only one centrum occurs per body segment; however, an increase in the number of centra, especially in the tail region, increases vertebral column flexibility during locomotion. Two centra in a body segment occurs in certain body regions of some early fishes and basal tetrapods and is referred to as **diplospondyly** (Gr., *diplos* = two + *spondylos* = vertebra; Fig. 8-1*C*).

Successive vertebrae are joined together to form the **vertebral column.** The shapes of the cranial and caudal surfaces of the centra differ among groups of vertebrates, and these differences have functional implications for the firmness of the union between vertebrae and the degrees of freedom of movement allowed. Both surfaces of a centrum of many fishes and early tetrapods are concave; a shape termed **amphicoelous** (Gr., *amphi* = both + *koilos* = hollow; Fig. 8-2*A*). Flexible **intervertebral pads,** which are derived from the notochord, are located in the concavities between

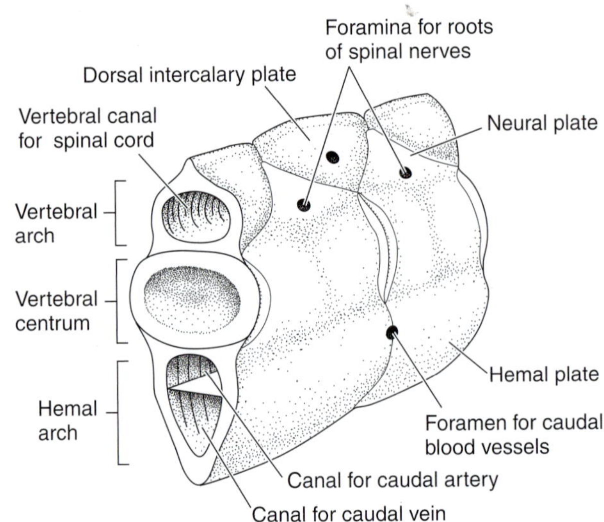

A. Chondrichthyan

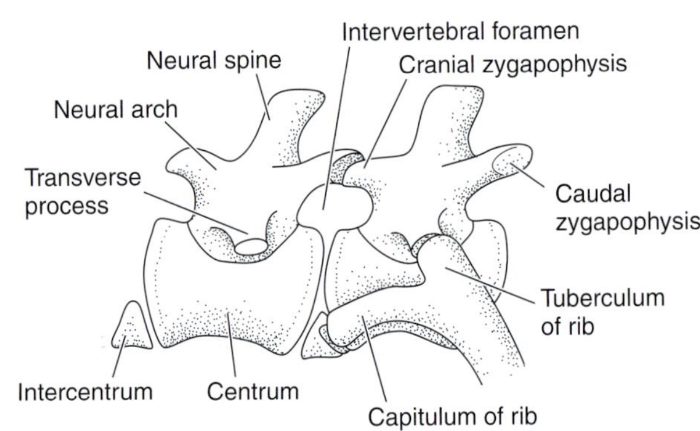

B. Reptilomorph

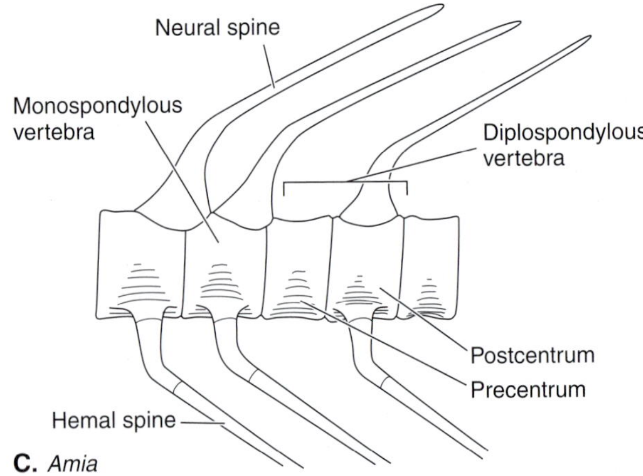

C. *Amia*

FIGURE 8-1
The structure of representative vertebrae; anterior is toward the left. *A,* The caudal vertebra of a chondrichthyan fish. *B,* Lateral view of two trunk vertebrae of an early sauropsid with an articulated rib. *C,* Lateral view of three caudal vertebrae of *Amia* showing the transition to diplospondyly. *(A, After Walker and Homberger; B, after Romer and Parsons; C, after Jollie.)*

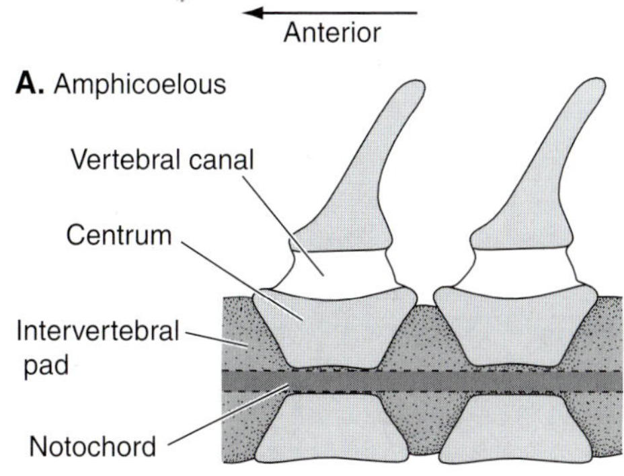

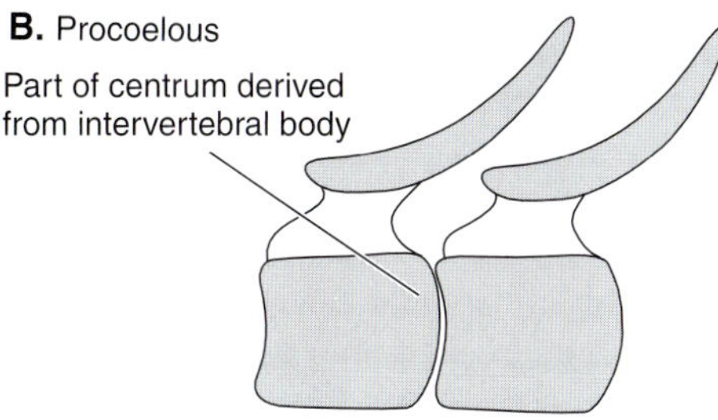

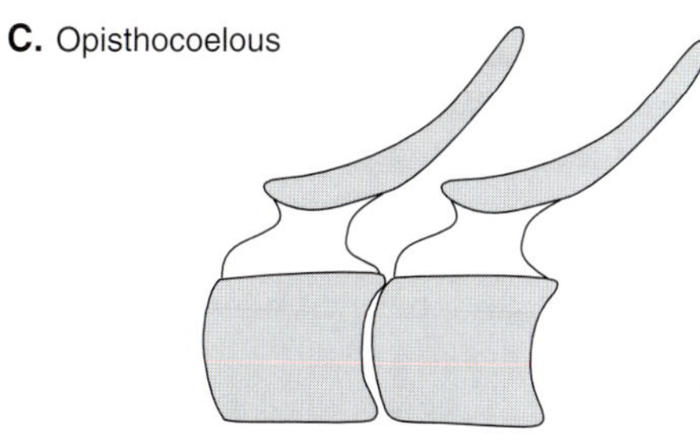

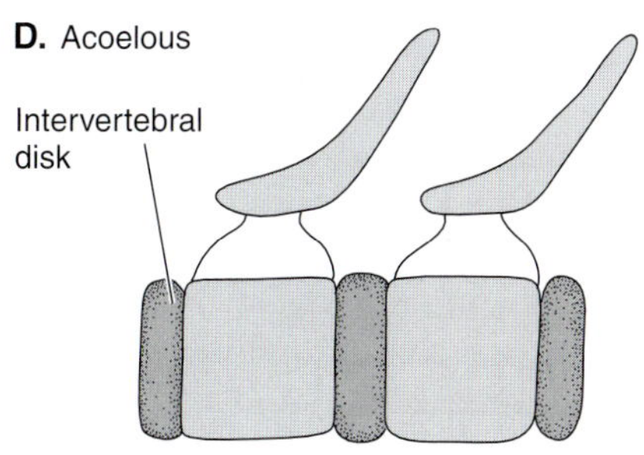

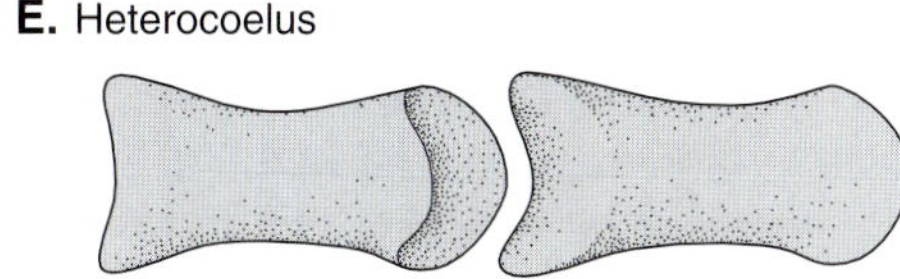

FIGURE 8-2
Shapes of vertebral centra; anterior is toward the left. *A,* Amphicoelous vertebrae of a fish. *B,* Procoelous vertebrae of an early sauropsid. *C,* Opisthocoelous vertebrae of an early sauropsid. *D,* Acoelous vertebrae of a mammal. *E,* Dorsal view of the centra of two heterocoelous cervical vertebrae of a bird. *(E, After Wake)*

amphicoelous centra and help to unite the vertebrae into a coherent vertebral column. A strand of notochord often runs through the centra. The intervertebral pads consist of a fibrous sheath surrounding vacuolated cells, similar to those in a notochord, and central extracellular fluid-filled spaces. An extensive filament network interlaces all parts. The whole pad is well suited to resist the compression to which fish vertebrae are subjected (Schmitz, 1995). They also probably help distribute forces when the vertebral column bends laterally; one side of the pad is compressed, and the other may bulge slightly.

Many vertebral shapes occur in terrestrial vertebrates because the degree and direction of vertebral movement and the amount of support needed vary among groups. Strength is increased considerably, and the possibility of dislocation is reduced, in a ball-and-socket arrangement in which one surface of the centrum is concave and the other forms a knoblike bump that fits into the concavity of the adjacent centrum. The knob is often formed by the ossification of the intervertebral pad (called an **intervertebral body** in tetrapods) and its fusion to one centrum. If the concavity is on the cranial surface of the centrum, the shape is **procoelous** (Gr., *pro-* = before; Fig. 8-2*B*); if on the caudal surface, **opisthocoelous** (Gr., *opisthen* = behind; Fig. 8-2*C*). Many amphibians and early reptilomorphs have vertebrae of one of these types. The trunk vertebrae of some early reptilomorphs, of birds, and of mammals have **acoelous** centra (Gr., *a-* = without), in which both surfaces are nearly flat (Fig. 8-2*D*). **Intervertebral disks** composed of connective tissue, remnants of the notochord, and often fibrocartilage lie between successive centra and help bind them together. The neck vertebrae of birds have **heterocoelous** centra (Gr., *hetero-* = different), in which one surface of the centrum is saddle-shaped in the horizontal plane and articulates with a reciprocal, vertical, saddle-shaped surface on the centrum of the adjacent vertebra (Fig. 8-2*E*). To visualize this, abduct your thumbs and forefingers, then bring your hands together at right angles to each other at the base of the thumbs. This vertebral arrangement is strong and prevents twisting but allows vertical and lateral movements at each joint. The neck as a whole is very flexible and allows birds to move their heads in many directions.

Strength of the vertebral column as a whole is increased, and its resistance to twisting and movements in certain directions is limited, in some teleosts and most terrestrial vertebrates by the presence of paired articular processes called **zygapophyses**[1] (Gr., *zygon* =

[1]Although called by the same name, the zygapophyses of certain teleosts are not homologous to those of tetrapods.

yoke + *apophysis* = offshoot, outgrowth). Zygapophyses extend forward and backward from each neural arch (Fig. 8-1*B*). In tetrapods, the articular surfaces of the caudal zygapophyses of one vertebra overlap those of the cranial zygapophyses of the next caudal vertebra. Paired **transverse processes** of several types extend laterally from the centrum, neural arch, or both and serve for the attachment of ribs and certain muscles. We will sort out the types of transverse processes when we consider the ribs.

Vertebral Development

The parts of vertebrae develop embryonically from mesenchyme cells that migrate from the sclerotome portion of the somites and gather around the developing spinal cord and notochord (Fig. 4-16*B*). These mesenchyme cells tend to aggregate where the transverse **myosepta,** which lie between embryonic myotomes, intersect with other connective tissue septa of the body (Fig. 8-3). These septa are called skeletogenous septa because skeletal tissues often form within them. A **dorsal skeletogenous septum** extends dorsally from the neural arches to the mid-dorsal line of the body, and, in the trunk, a **lateral skeletogenous septum** extends ventrally around each side of the coelom. The lateral skeletogenous septa may unite ventral to the coelom to form a **ventral skeletogenous septum,** which continues to the midventral line of the body. In the tail, where no coelom is present, the ventral skeletogenous septum extends ventrally from the notochord to the midventral line of the body. Jawed vertebrates also have a **horizontal skeletogenous septum,** which passes inward from the skin toward the spinal cord and notochord. The myotomes, which lie between the myosepta, define the body segments. Because mesenchyme cells that will form the vertebrae tend to aggregate at the intersection of a myoseptum with other septa, the vertebrae develop in an intersegmental position. The development of the vertebrae between the myotomes is functionally significant because it allows longitudinal muscle fibers from a single myotome to span the joint between vertebrae and attach onto adjacent vertebrae. Contraction of muscle fibers, therefore, can bend the vertebral column at the intervertebral joints.

Early in the embryonic development of the vertebrae, mesenchyme cells condense lateral to the notochord and spinal cord between the spinal nerves. These groups of cells form the neural plates, the primary component of the arches (Fig. 8-4*A*). Hemal plates and arches develop in the caudal region between branches of the caudal artery and vein. As the neural and hemal arches differentiate, they first form plates of cartilage, often collectively called **arcualia** in fishes. In most vertebrates, the cartilage is later replaced by bone. In some cases, the mesenchyme cells transform

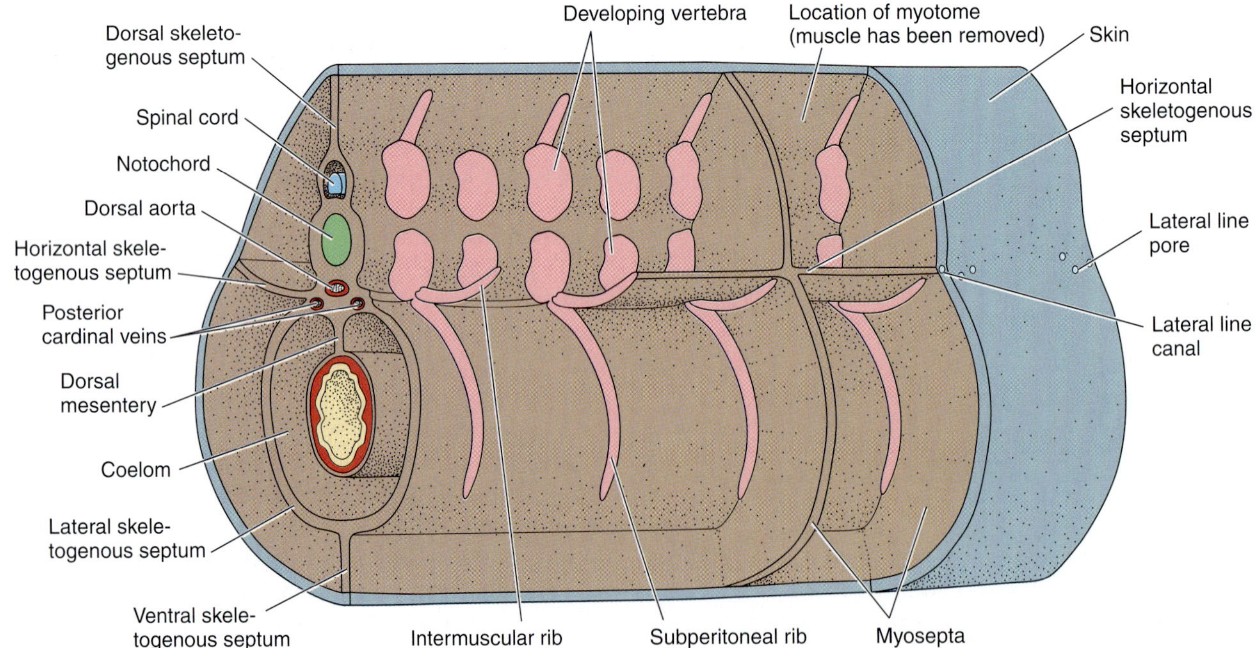

FIGURE 8-3
A stereodiagram of the vertebral axis, skeletogenous septa, and ribs of a jawed fish. *(After Goodrich.)*

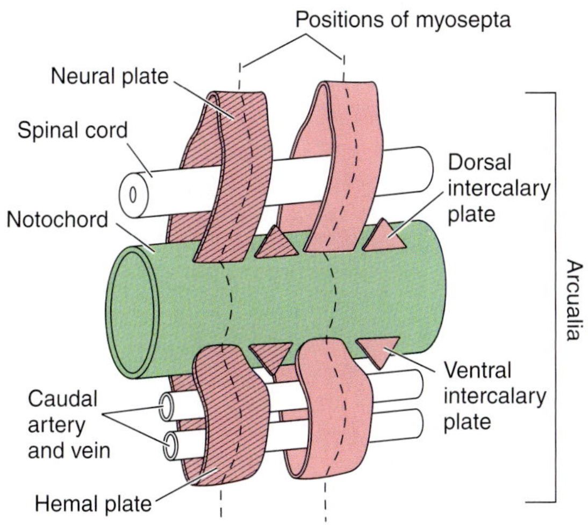

A. Neural and hemal arches

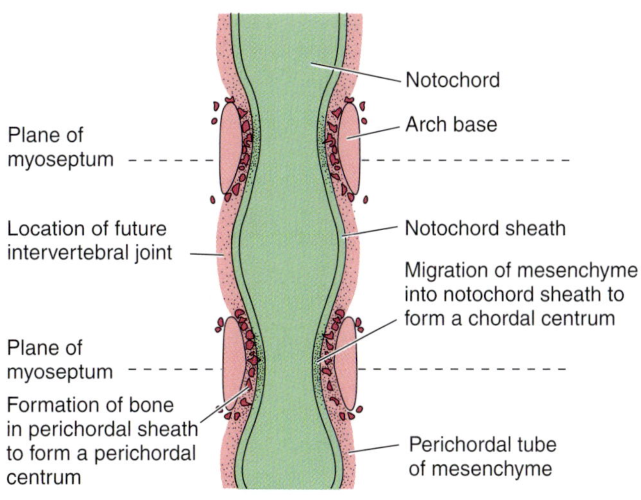

B. Formation of chordal or perichordal centra in an actinopterygian

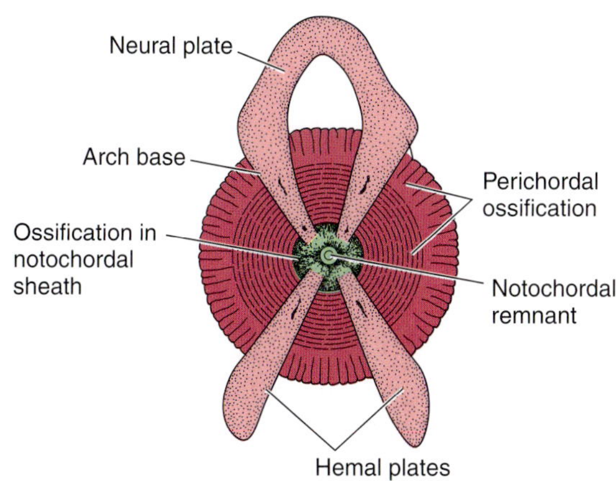

C. Transverse section through an actinopterygian vertebra

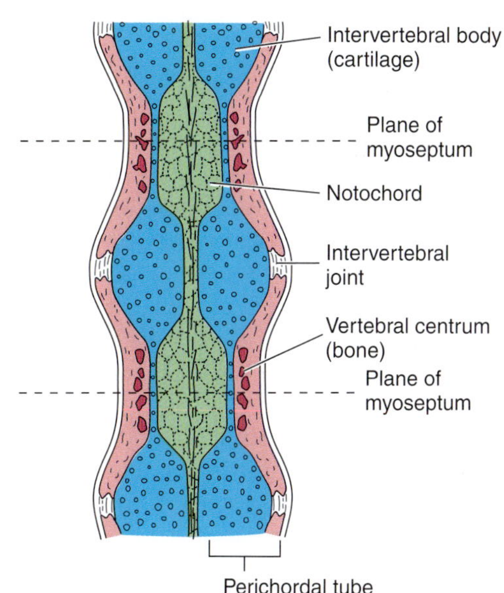

D. Formation of vertebrae in a lissamphibian

FIGURE 8-4

Embryonic development of vertebrae in anamniotes. *A,* A stereodiagram of the formation of neural and hemal arches in a fish. *B,* A frontal section through the notochord and perichordal tube to show differences in the formation of a chordal centrum and perichordal centrum in fishes. *C,* A transverse section through the vertebra of a bony fish in which most of the centrum forms as a perichordal ossification, but the arch bases and some ossification within the notochordal sheath make a contribution. *D,* A frontal section through the notochord and perichordal tube of a recent amphibian showing the development of the centra and intervertebral bodies. *Blue* = cartilage forming in perichordal tube; *green* = notochord and notochordal sheath; *pink* = arcualia; *red* = perichordal tube and bone forming in the perichordal tube. *(C, After Jollie; D, after Wake.)*

directly into bone-forming cells. Membrane bone may be added peripherally to cartilage-replacement bone in some species. The neural arches develop from a single pair of **neural plates** in most vertebrate species, but a second pair of **dorsal intercalary plates** develops between the neural plates in chondrichthyan and some early actinopterygian fishes and contributes to the neural arch. The hemal arches form from **hemal plates** that develop around the caudal artery and vein. Chondrichthyans and early actinopterygians may also have small **ventral intercalary plates** lying between the bases of the hemal plates. All of the hemal arches form

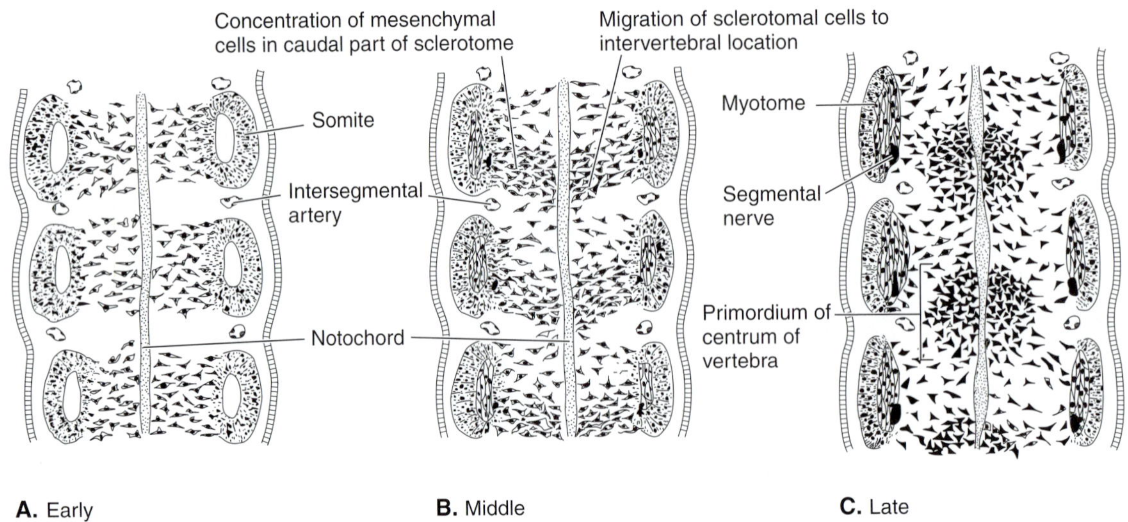

FIGURE 8-5
A–C, Three frontal sections through the notochord and somites of an amniote, showing the embryonic development of the vertebrae by condensation of sclerotomal mesenchyme. *(After Corliss.)*

a continuous sheath around the caudal artery and vein of fishes. Hemal arches are reduced in most tetrapods. Hemal arches are not found in the trunk region, but small, serial equivalents of hemal plates, called **ventral arch bases,** develop beneath the trunk vertebrae of some fishes. They extend a short distance into the lateral skeletogenous septa, forming basapophyses, onto which certain ribs attach (p. 275).

The way the centra develop, which occurs concurrently with arch formation, is quite variable. We can only consider the broad outlines of this complex topic. In most fishes, other mesenchyme cells of sclerotome origin condense to form a continuous **perichordal tube** around the notochord and its sheath (Fig. 8-4*B*). The centra of chondrichthyan fishes develop by the chondrification of mesenchyme cells that invade the notochordal sheath at intersegmental intervals. These are **chordal centra** (Fig. 8-4*B*). The centra of actinopterygian fishes do not pass through a cartilaginous stage but develop from the direct intersegmental deposition of bone by the mesenchyme cells around the perichordal tube (Fig. 8-4*B*). These are called **perichordal centra,** even though some mesenchyme cells may invade the notochordal sheath and ossify there. As the ossifying centra enlarge, they may spread over the bases of the neural and hemal arches and incorporate them into the centra (Fig. 8-4*C*).

Centra formation is somewhat similar in living amphibians (Fig. 8-4*D*). Cartilage and bone begin to form in a continuous perichordal tube, which gradually thickens. Cartilage becomes particularly thick in the intervertebral areas, where it forms a series of **intervertebral bodies.** In many adult salamanders, each intervertebral body remains as a distinct cartilaginous body between ossified amphicoelous centra. In adult frogs, each intervertebral body usually ossifies and fuses with either the front of a centrum to form an opisthocoelous vertebra or to its caudal surface to form a procoelous vertebra.

Centra develop differently in amniotes, for a continuous perichordal tube does not develop. Mesenchyme cells become more densely packed during early development in the caudal half of each sclerotome than in the cranial part (Fig. 8-5*A* and *B*). Mesenchyme cells of the caudal parts of the sclerotomes concentrate further around the notochord in an intersegmental position, where they form the vertebral centra (Fig. 8-5*C*). The fate of the cells from the cranial halves of the sclerotomes is less certain. Anatomists long believed that the cranial cells of a particular sclerotome migrated forward to join the caudal half of the next cranial sclerotome. According to this view, each vertebral centrum is derived from sclerotomal cells from two adjacent body segments. More recent investigators contend that the cranial sclerotomal cells simply dissipate and make no contribution to the centra.

Ribs and Transverse Processes

The term **rib** refers to hard, rod-shaped structures that develop in the myosepta lateral to the vertebrae. At first they are cartilaginous, but they ossify later in most vertebrate groups. Ribs strengthen the myosepta and body wall and provide attachments for many trunk and tail muscles. If ribs extend far ventrally, they also protect the visceral organs. Several types of ribs occur in actinopterygians (Fig. 8-3). **Intermuscular ribs,** or **dorsal ribs,** develop in the myosepta at their intersection with the horizontal skeletogenous septum. They extend laterally between the epaxial trunk muscles, lying dorsal to the horizontal skeletogenous septum, and the hypaxial muscles, lying ventral to it. Each attaches by a single head to the lateral surface of a centrum. **Subperitoneal ribs, or ventral ribs,** develop in the myosepta at their intersection with the lateral skeletogenous septum. Each attaches by a single head to a lateral extension of the ventral arch base, which is called a **basapophysis** (Gr., *basis* = base + *apophysis* = offshoot, outgrowth). Basapophyses are a type of transverse process that are serially homologous to the bases of the hemal arches. Subperitoneal ribs extend ventrally in the body wall next to the coelom.

Some species of fishes have intermuscular ribs, some have subperitoneal ribs, some have both types, and some have ribs that do not fit easily into either category. Chondrichthyan fishes, for example, have short ribs in an intermuscular position, but these attach to the centra far ventrally on basapophyses, an attachment occupied by subperitoneal ribs in actinopterygians. The potential for rib development exists in myosepta, where they intersect with other skeletogenous septa. Ribs in one group of vertebrates are homologous to ribs in another group in a general sense, but ribs seem to have evolved independently and in somewhat different sites among fishes, so precise homologies are not always possible.

Trunk and tail muscles also attach onto the relatively long ribs of tetrapods. Beyond this, tetrapod ribs usually help support the weight of the body against gravity when the animal is lying prone, and they prevent undue pressure on the lungs and other visceral organs. Movements of the ribs in most amniotes are also important for ventilating the lungs. Tetrapod ribs form differently than in fishes. They are subperitoneal in position, but they attach to the vertebrae farther dorsally than either the subperitoneal or intermuscular ribs of fishes. Typically, each tetrapod rib has two points of attachment to a vertebra (Figs. 8-1*B* and 8-6). The head, or **capitulum,** of a rib, which is located at its proximal end, attaches on the lateral surface of a centrum, and sometimes on a small transverse process known as a **parapophysis** (Gr., *para-* = along side of). The **tuberculum** of a rib, which is slightly distal to the capitulum, articulates on a transverse process that extends laterally from the base of the neural arch. This type of transverse process is called a **diapophysis** (Gr., *dia* = through, across). The tuberculum of a rib and diapophysis are not always present. A pair of embryonic ribs begins to develop on all tetrapod vertebrae except for the more distal caudal vertebrae. These rib primordia may form distinct ribs in adults, or in some regions of the vertebral column, they may become inseparably fused with the diapophysis and parapophysis to form enlarged transverse processes called **pleurapophyses** (Gr., *pleura* = side, rib).[2]

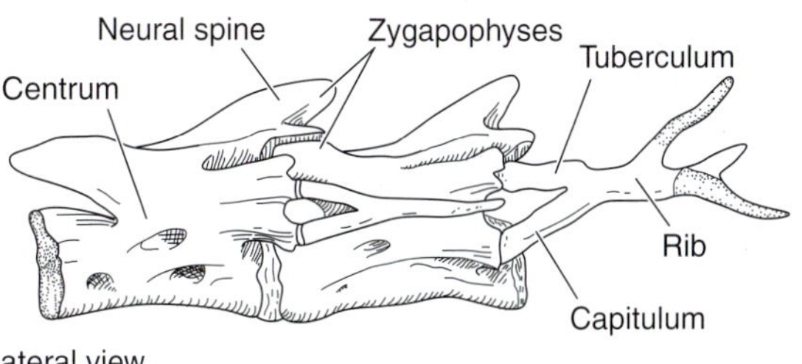

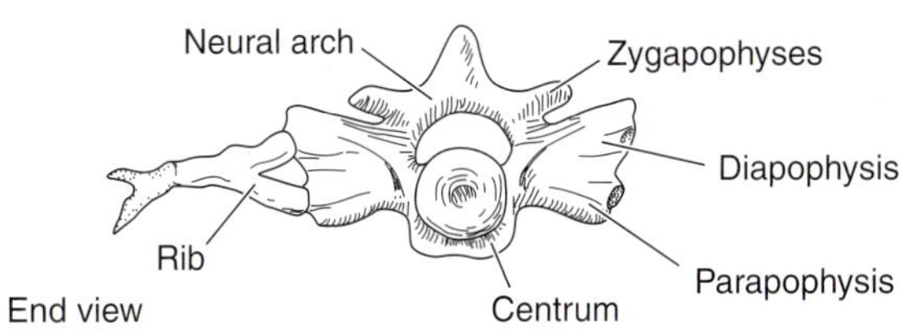

FIGURE 8-6
A lateral and an end view of the trunk vertebrae of the salamander, *Necturus,* showing rib attachments. *(After Goodrich.)*

Although ribs typically ossify, their distal ends usually remain cartilaginous, forming **costal cartilages** (L., *costa* = rib). Often ribs attach ventrally to a sternum (see subsequent discussion), and the flexibility of the costal cartilages permits breathing movements. In birds, the costal cartilages ossify as distinct **sternal ribs.** Caudally projecting **uncinate processes** (L., *uncinatus* = hooklike) develop on the dorsal portion of the ribs of birds (see Fig. 8-17*A*) and some early reptilomorphs. They serve for the attachment of certain trunk muscles and also strengthen the ribcage.

Sternum

Most living terrestrial vertebrates have a **sternum,** or breastbone, although one is not present in turtles and snakes. In salamanders, the sternum lies ventrally between the anterior portions of the left and right pectoral girdles (Fig. 8-7*A*); in frogs, it unites with the pectoral girdles (see Fig. 9-11). The short ribs of recent amphibians do not attach to the sternum. In amniotes, most of the trunk ribs attach to the sternum by

[2]Technically, the pleurapophysis is the rib component of the transverse process, but because it is indistinguishable from the other components in an adult, the term usually is applied to the entire transverse process.

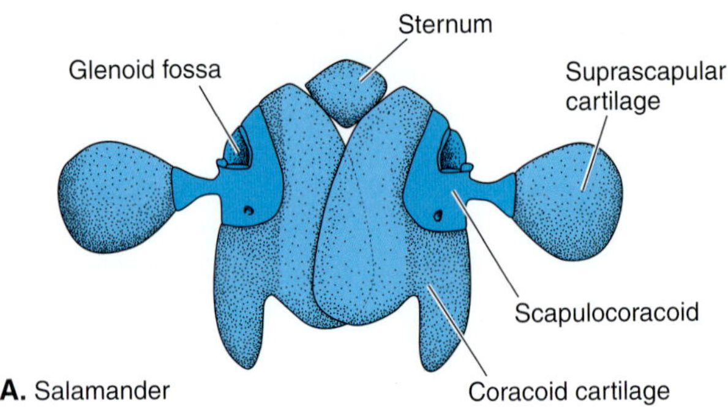

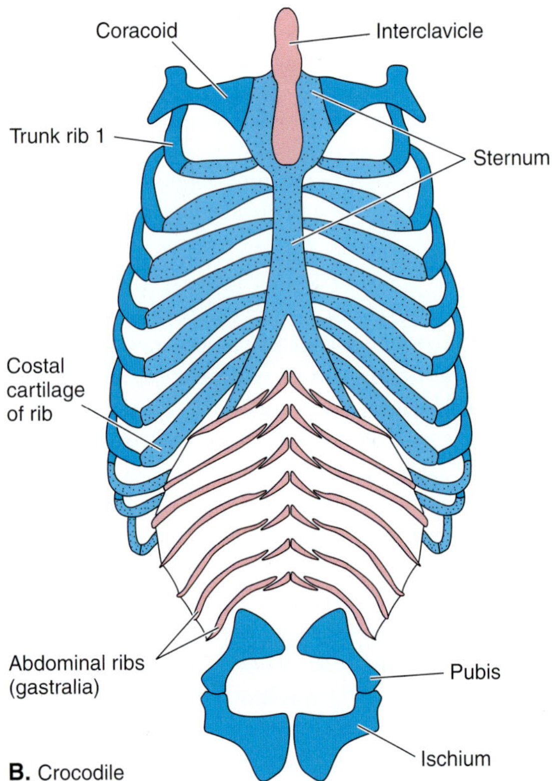

FIGURE 8-7
Ventral views of the sternum of a salamander and a crocodile. *A,* The small, unossified sternum of a salamander lies cranial to the coracoid cartilages of the pectoral girdle. *B,* The costal cartilages of ribs attach to the sternum of a crocodile. *Blue stippled* = cartilage; *blue* = cartilage replacement bone; *red* = dermal bone. *(A, After Noble; B, after Romer and Parsons.)*

way of their costal cartilages to form a solid ribcage that protects the viscera and participates in lung ventilation (Fig. 8-7*B*). Some pectoral muscles arise from the sternum.

The sternum develops embryonically from a single or paired cartilaginous elements in the midventral connective tissue septum and parts of the myosepta. It may remain cartilaginous or ossify in one or more pieces. The sternum is considered a new phylogenetic structure not derived either from ribs or the pectoral girdle. We know little about its evolutionary history, for it has not been found in basal tetrapods, fossil amphibians, or basal reptilomorphs. A cartilaginous one may have been present in these species but did not fossilize, or a sternum may have evolved more recently and independently in several early tetrapod lines.

The Evolution of the Axial Skeleton

Fishes

The Vertebral Axis The evolution of the vertebral column and ribs, their degree of ossification, and their degree of differentiation in body regions correlate with the environment in which a species lives (water or land), its method of locomotion, and the forces acting on its body. Because of the buoyancy of water, a fish's axial skeleton plays only a small role in support against gravity. A fish's notochord or vertebral column is not subjected to strong, vertical bending forces; rather, it must resist compression as a fish pushes through the relatively dense water, and it must prevent shortening when longitudinal muscle fibers in the body wall contract. Because telescoping cannot occur, and the notochord and vertebral column can bend from side to side, muscle contraction causes a series of lateral undulations that sweep down the body of a fish to provide the propulsive thrust. (We will examine swimming patterns and movements more thoroughly in Chapter 11.) Ligaments between vertebrae prevent excessive bending and dislocation of the vertebrae. In some cases, especially in fishes with asymmetrical tails, the vertebral axis also must resist some degree of twisting, or torsion.

An enlarged notochord encased in an elastic sheath persists in the adults of hagfishes and lampreys. As in amphioxus, the notochord resists compressive forces as these fishes swim and resists body shortening when myomeres contract. Hagfishes have no trace of vertebral elements, but a series of small, cartilaginous neural arches (often simply called arcualia) partially encase the spinal cord of lampreys (see Fig. 7-5*A*). No other vertebral elements exist. We know little about the vertebral axis of the extinct jawless vertebrates. Many species were encased in a heavy dermal armor that provided considerable support, but jawless vertebrates too probably had an enlarged and persistent notochord. Traces of small vertebral elements flanking the spinal cord and notochord have been found in only a few species (†heterostracans and †osteostracans; Fig. 3-2) in which some ossification of an internal skeleton occurred.

In other fishes, the vertebral column can be divided into two major regions: (1) **trunk vertebrae** occupy the trunk cranial to the cloaca and (2) **caudal vertebrae** lie in the tail. The first trunk vertebra is modified slightly and articulates with the skull. The remaining trunk vertebrae are essentially alike, differing slightly among species with respect to the parts of the trunk that undulate during swimming and the degree of undulation. Most of the trunk undulates in early species, but undulation becomes restricted to caudal regions in many derived species. Caudal vertebrae differ in having hemal arches encasing the caudal artery and vein. These prevent the caudal blood vessels from collapsing when waves of muscle contraction sweep down the tail during locomotion (Chapter 19). Such arches are unnecessary in the trunk, where the major blood vessels lie between the vertebral column and the body cavity. The posterior caudal vertebrae are modified to support the expanded caudal fin.

Neural arches located intersegmentally are present in most fishes, but they are not united by zygapophyses except in the more posterior trunk vertebrae of a few teleosts in which torsion must be resisted. The spinal cord of chondrichthyans is encased in a continuous series of cartilage plates rather than by separate neural arches (Fig. 8-1*A*). This continuity does not restrict lateral undulations because of the flexibility of cartilage.

Vertebral centra are absent and an enlarged notochord is present in early sharks, some †placoderms, lungfishes, the living coelacanths (*Latimeria*), and sturgeons and paddlefishes (Fig. 8-8*A*). A notochord adequately resists compressive forces in these species, most of which do not swim rapidly and powerfully. Centra develop and strengthen the vertebral axis in other groups of fishes (Fig. 8-8*B*). Chondrichthyans have chordal centra, which are strengthened by prismatic calcification deep within the cartilage next to a small, persistent notochord. Advanced actinopterygians have perichordal centra.

Basal choanate fishes (e.g., †*Eusthenopteron,* an †osteolepiform; Figs. 3-15 and 3-17*A*) lying close to the ancestry of tetrapods retained a large notochord, but the vertebral axis was strengthened by centra, which developed around the notochord. Each centrum was composed of several bony pieces that probably were united to each other and to the neural arch by fibrous tissue (Fig. 8-8*C* and *D*). A pair of small **pleurocentra** lay along the dorsolateral surface of the notochord just caudal to the base of the neural arch. A larger U-shaped **intercentrum** lay along the ventrolateral surface of the notochord just cranial to the pleurocentra. The intercentrum may have developed by the fusion of a pair of lateral halves. Each centrum thus is composed

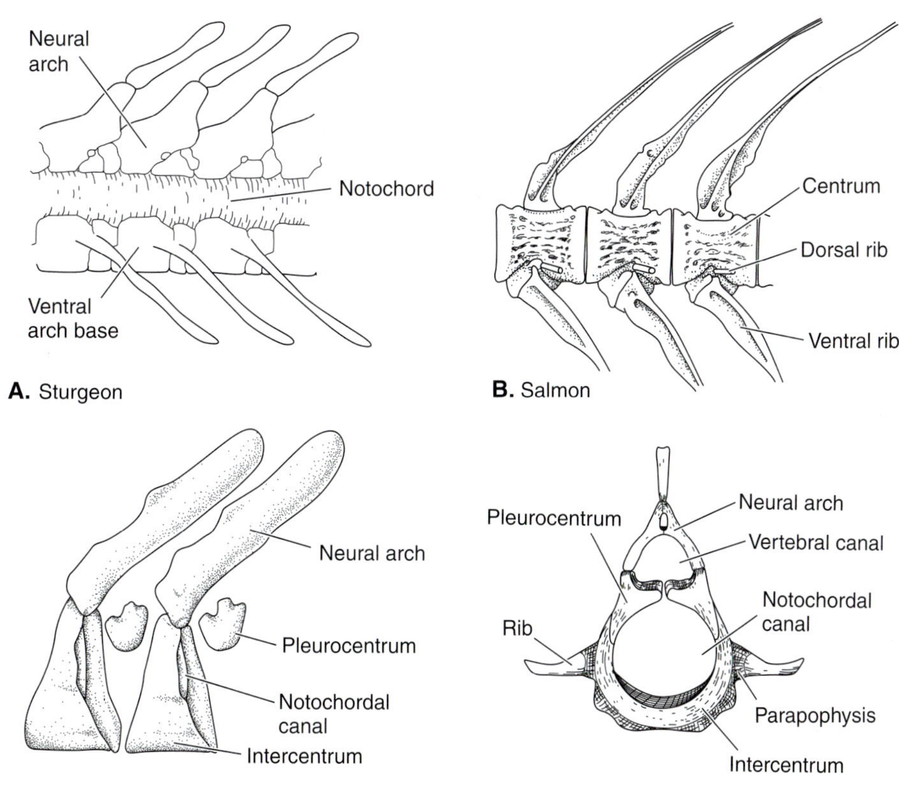

FIGURE 8-8
Some trunk vertebrae of fishes. *A,* A lateral view of the vertebral axis of a sturgeon in which the notochord persists in the adult. *B,* A lateral view of three vertebrae of a salmon. *C,* A lateral view of two vertebrae of an †osteolepiform (a choanate fish ancestral to tetrapods). *D,* A posterior view of the vertebra of an †osteolepiform. *(A, After Goodrich; B, after Jollie; C and D, after Jarvik.)*

of three blocks of bone—a pair of pleurocentra and an intercentrum.

Ribs are absent in hagfishes and lampreys and poorly developed in chondrichthyans. Many actinopterygians have both intermuscular and subperitoneal ribs, and some have additional accessory ribs that develop in other parts of the myosepta. Because fishes exchange gases through gills, their ribs have no function in respiratory movements; rather, they are part of the locomotor system. They strengthen the myosepta and help transfer the forces of muscular contraction to the vertebral axis.

Median and Caudal Fins The vertebral axis of fishes ends in an expanded **caudal fin** that delivers a strong propulsive thrust for swimming. Most fishes also have one or two **dorsal fins** along the midline of the back, and many species also have a median **anal fin** located on the tail just caudal to the cloacal aperture. Dorsal and anal fins help stabilize a fish as it moves through the water by reducing its tendency to roll from side to side and the tendency of its front end to yaw and move from left to right. Some fish species swim slowly and maneuver by undulating these fins rather than their trunk and tail.

All of these fins, and the paired fins that we will discuss with the appendicular skeleton, are supported distally by slender fin rays that develop in the skin on each surface of the fin (Fig. 8-9). The fin rays of chondrichthyans are horny **ceratotrichia**; those of actinopterygians and sarcopterygians are bony **lepidotrichia** (Chapter 6). The fin rays of caudal fins attach to vertebral elements. Those of dorsal and anal fins attach to deeper cartilages and bones collectively called **pterygiophores** (Gr., *pterygion* = fin or wing + *phoros* = bearing), which sometimes extend inward to neural or hemal spines.

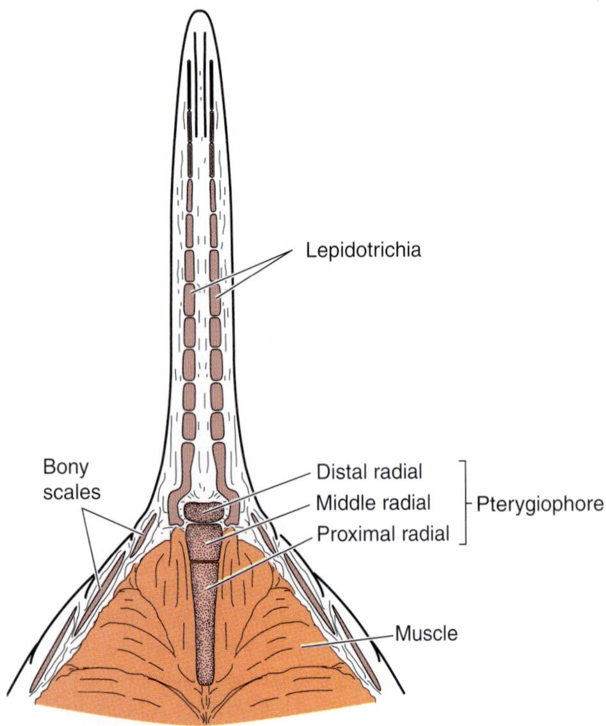

FIGURE 8-9
A transverse section through the dorsal fin of a bony fish. *(After Goodrich.)*

The shape of the caudal fin differs among fishes according to their methods of swimming and buoyancy control (Fig. 8-10). †Osteostracans (among jawless vertebrates), most chondrichthyans, and early actinopterygians and sarcopterygians have a **heterocercal tail** (Gr., *heteros* = other, different + *keros* = tail), in which the vertebral axis turns upward into an expanded dorsal lobe of the tail, which it stiffens. This type of tail is widespread among early fishes and is plesiomorphic for gnathostomes. As we will describe in the discussion on swimming (Chapter 11), it contributes to lift in these heavy fishes. An early variation was a reverse heterocercal, or **hypocercal tail** (Gr., *hypo* = under) found in some jawless vertebrates (†heterostracans and †anapsids; Fig. 3-3*C* and *E*) in which the vertebral axis turned downward into an expanded ventral lobe. This tail shape appears to be correlated with a unique pattern of gathering food from the bottom by swimming with the head pointed downward.

Osteichthyans have either lungs or a swim bladder. These contribute to buoyancy in fishlike osteichthyans, and the caudal fins became symmetrical in most derived fish groups. In the line of evolution toward teleosts, the caudal fin became symmetrical only externally. Internally, at the caudal end of the fin, the reduced centra of several terminal vertebrae have fused to form a spikelike **urostyle** (Gr., *oura* = tail + *stylos* = pillar), which tips sharply upward. Most of the caudal fin is comparable with the ventral portion of the heterocercal tail, for enlarged hemal spines known as **hypural bones** support it. Neural spines or **epural bones** support only a small part of the fin lying dorsal to the urostyle. This type of tail is known as a **homocercal tail** (Gr., *homos* = the same). Early neopterygians, such as *Amia* and *Lepisosteus,* have an **abbreviated heterocercal tail** that is intermediate between the heterocercal and homocercal types. Some sarcopterygians evolved a different type of symmetrical tail, called a **diphycercal tail** (Gr., *diphyes* = twofold) in which the vertebral axis straightened out and dorsal and ventral lobes became equal in size. This type of tail evolved independently in recent lungfishes and coelacanths, the ancestors of both of which had heterocercal tails.

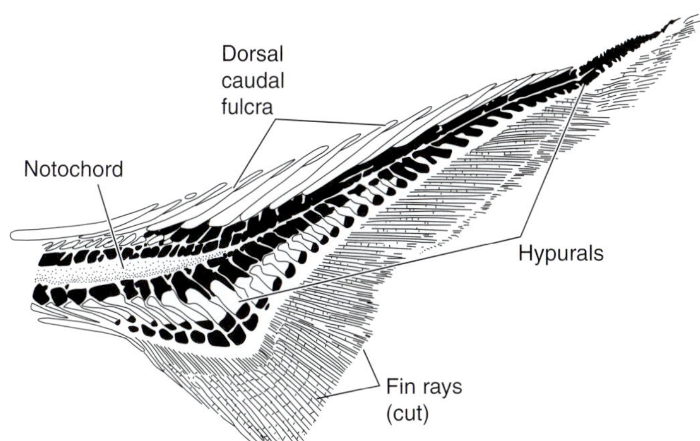

A. Caudal skeleton of *Polyodon* (heterocercal)

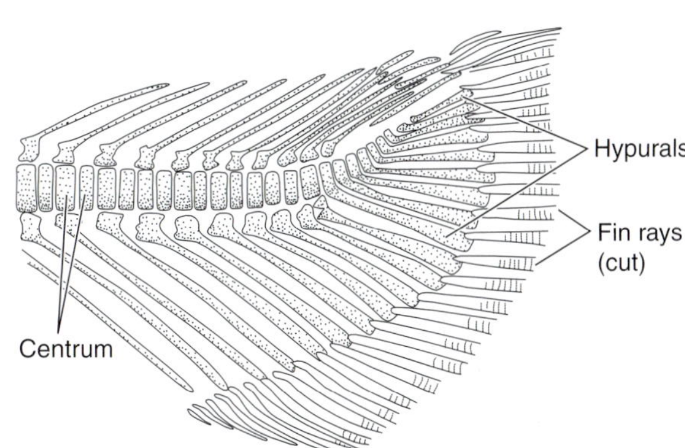

B. Caudal skeleton of *Amia* (abbreviated heterocercal)

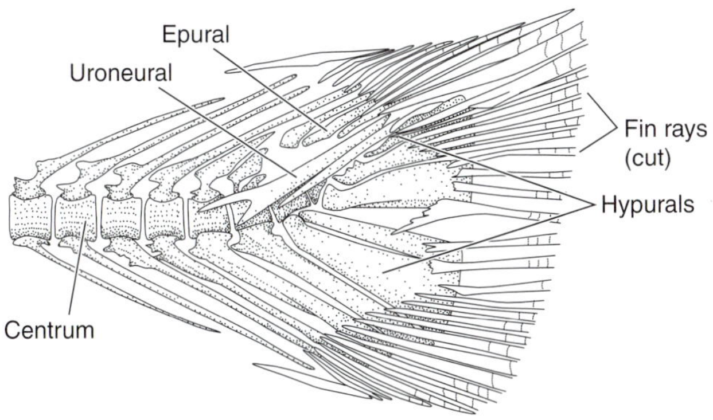

C. Caudal skeleton of *Elops* (homocercal)

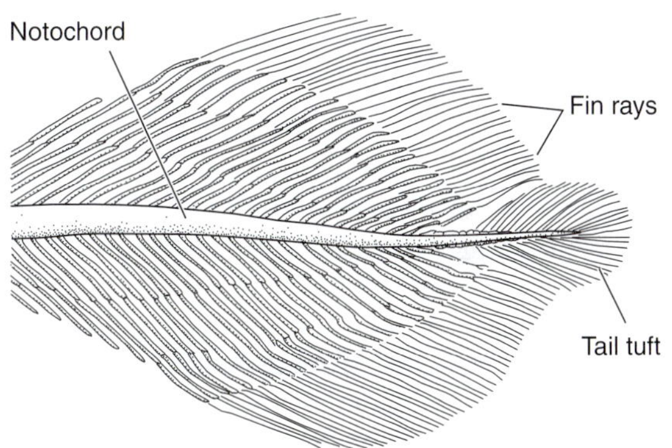

D. Caudal skeleton of *Latimeria* (diphycercal)

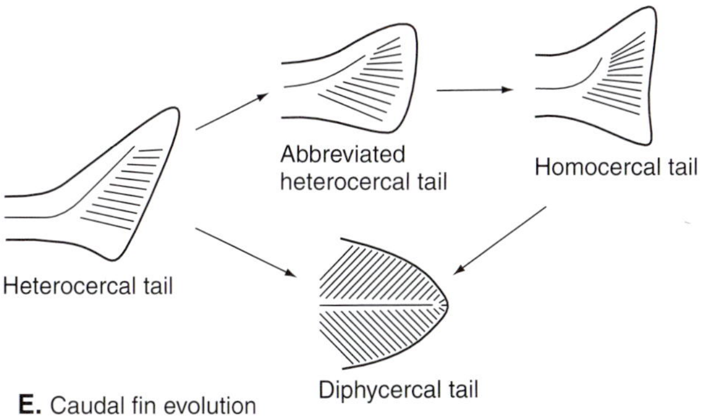

E. Caudal fin evolution

FIGURE 8-10
Major caudal fin and caudal skeleton types of bony fishes. *A,* Heterocercal caudal fin of a paddlefish, *Polyodon. B,* Abbreviated heterocercal caudal fin of a bowfin, *Amia. C,* Homocercal caudal fin of a ladyfish, *Elops. D,* Diphycercal caudal fin of a coelacanth, *Latimeria. E,* Caudal fin evolution in bony fishes; a heterocercal tail is plesiomorphic. Within actinopterygians, abbreviated heterocercal and homocercal tails evolved. Diphycercal types evolved independently in sarcopterygians (e.g., coelacanths and lungfishes) and in actinopterygians (e.g., cods and many eel-like teleosts). *(A, After Grande and Bermis; B and C, after Patterson; D, after Jarvik.)*

Basal Tetrapods, Amphibians

Air, which is less dense than water, provides no support for a terrestrial vertebrate against the pull of gravity. The weight of the body is in danger of collapsing the lungs and other internal organs. On the other hand, air offers little resistance to movement. Thrust of the limbs against the ground, rather than the trunk and tail against the water, becomes the main propulsive force. However, fishlike undulations of the trunk and tail continue in early amphibians and reptilomorphs and help advance and retract the limbs (Fig. 8-11*A*). But walking generates torsion forces in the trunk. For example, when a diagonally opposite front and a hind leg support the body, vertical forces extending through the supporting limbs tend to rotate the trunk in opposite directions (Fig. 8-11*B*).

All of these factors and others eventually had a profound effect on the evolution of the axial skeleton, but the transition from water to land was a gradual process. Although the adults of basal tetrapods such as the †ichthyostegids had legs and ventured onto the land, they spent a great deal of time in the water. Walking at first supplemented swimming and did not completely replace it. The vertebral column of early tetrapods gradually strengthened as it became a supporting beam

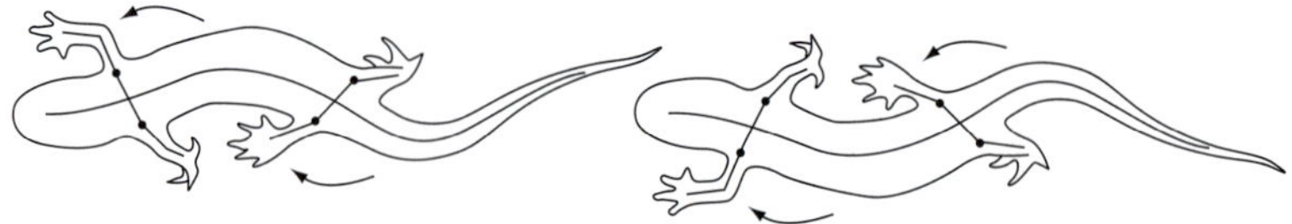

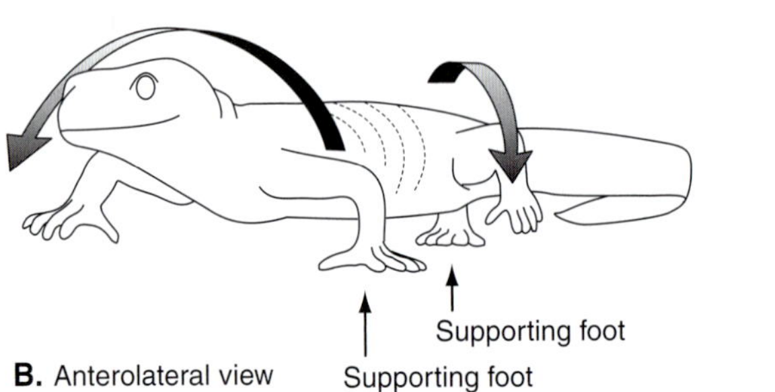

FIGURE 8-11
A, Fishlike lateral undulations of the trunk of a salamander help advance and retract the feet. *B,* The support of the body by a diagonal opposite front and hind legs at certain points during a stride subjects the trunk to torsion stresses. *(A, After Romer and Parsons; B, after Carrier.)*

that transferred body weight to the girdles and appendages. It must resist bending in the vertical plane, so these low-slung animals do not drag along the ground, and it must resist torsion. Finally, the head and trunk no longer need to move as a unit as they must in fishes pushing through the denser water. The head of terrestrial vertebrates loses its connection with the pectoral girdle. This, and the specialization of the cranial part of the vertebral column, allows the head to move independently of the trunk as the animal explores its environment and feeds.

A representative vertebra from the trunk of an †ichthyostegid (see Fig. 3-17*C*) was stronger than that of their †osteolepiform ancestors, but, apart from this, a remarkable resemblance in vertebral structure exists. †Ichthyostegids had **rhachitomous vertebrae** (Gr., *rhachis* = spine + *tomos* = cut), in which each centrum was "cut" into a large intercentrum and two smaller pleurocentra (Fig. 8-12*A*). This closely resembled the vertebrae of the †osteolepiforms (Fig. 8-8*C* and *D*). The notochord persisted but was more constricted than in †osteolepiforms. The intercentrum was a U-shaped or nearly circular piece perforated by a small space for the notochord. The pleurocentra were larger than in †osteolepiforms. The neural arch also was more massive and bore strong zygapophyses, which restricted vertical bending and torsion but allowed for lateral flexion. The neural arch articulated with the parts of the centrum in most basal tetrapods; in living amphibians and in amniotes, it is fused with the centrum.

Most basal tetrapod groups had a similar rhachitomous centrum in which the intercentrum was the larger component, but there were several departures from this type. One variant is seen in the aquatic, snakelike †aistopods (p. 79), in which the centrum is a single unit firmly united with the neural arch (Fig. 8-12*B*). We have no way of determining whether this **holospondylous vertebra** retained the intercentrum, pleurocentrum, or both fused together. Living lissamphibians have similar holospondylous vertebrae, but they probably evolved independently. An unusual **embolomerous vertebra** is found in a group of aquatic †anthracosauroids called the †embolomeres. In their case, both the intercentrum and pleurocentrum are equal in size and form disk-shaped units notched or perforated for the notochord (Fig. 8-12*C*). (This is another example of diplospondyly, also seen in some fishes.)

Most of the vertebrae of †ichthyostegids and most basal tetrapods bore ribs. Those on the trunk were strong and well developed, and they curved ventrally. Because the presence of a sternum is not known, we do not know whether ribs articulated with one. Each rib was two headed (Fig. 8-12*D*). The capitulum of a rib articulated on the intercentrum, and its tuberculum articulated on the diapophysis of the neural arch. Because some lateral undulations of the trunk and tail probably occurred during locomotion, the ribs continued to provide the attachment for locomotor muscles, but an important function probably was to strengthen the body wall of these relatively large animals. Ribs also helped to prevent the weight of the body from collapsing the lungs and abdominal viscera when these animals lay on the ground. We do not know whether rib movements were used in ventilating the lungs.

Lissamphibians are small animals that do not weigh much and which pump air into their lungs with movements of the floor of their mouth and pharynx. Ribs are not important for body support or lung ventilation. Adult salamanders and caecilians have short ribs (Fig. 8-13), but short embryonic ribs fuse to the sides of the

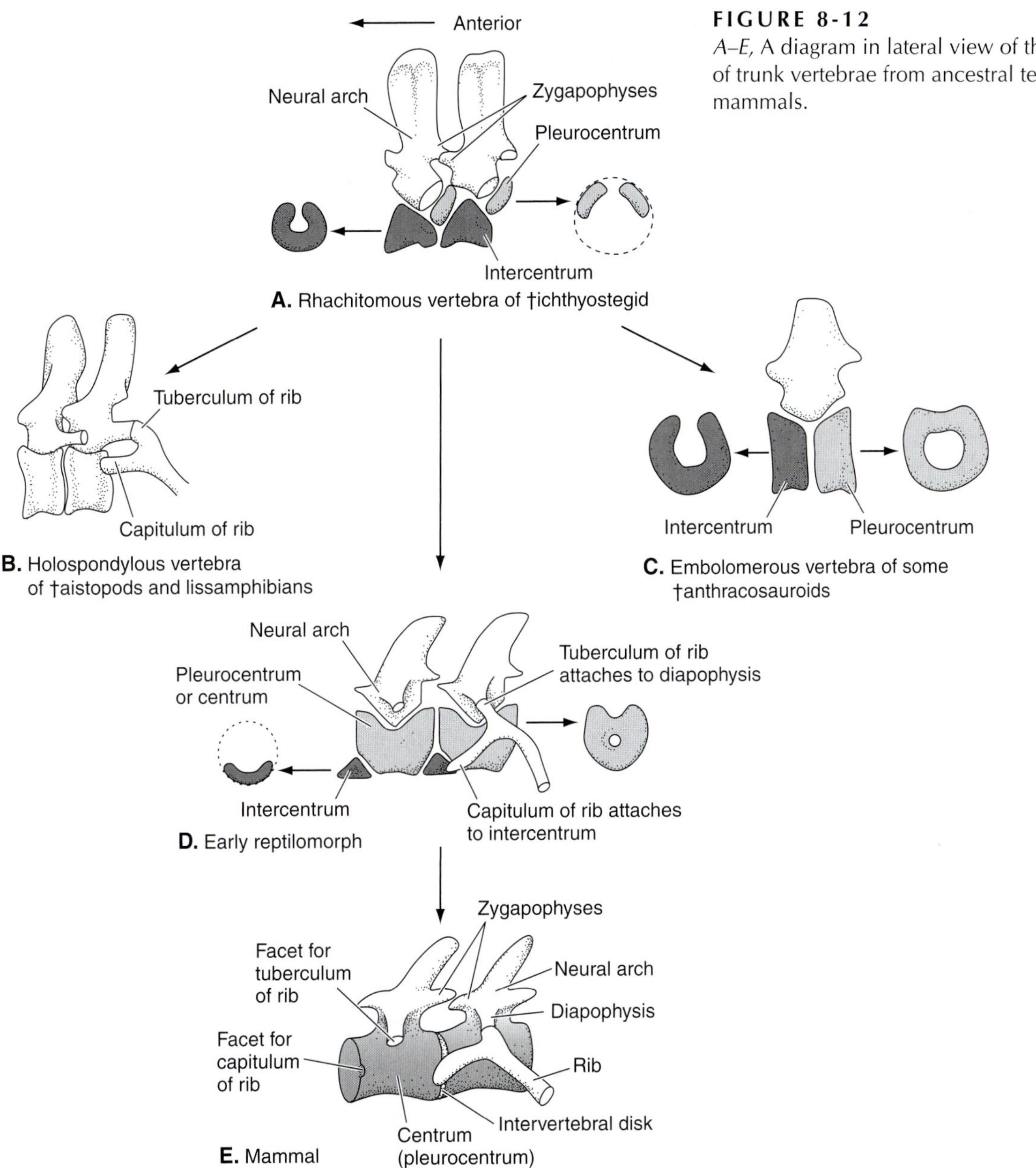

FIGURE 8-12
A–E, A diagram in lateral view of the evolution of trunk vertebrae from ancestral tetrapods to mammals.

vertebrae to form pleurapophyses to which muscles attach during the embryonic development of most adult frogs. Rib reduction is heterochronic, and traces of ribs are found not only in the adults of very primitive frogs but in larvae of different ages in several evolutionary lineages of frogs (Blanco and Schmitz, 2000). Frogs and most salamanders have a small sternum associated with the ventral part of the pectoral girdle (Fig. 8-7*A*).

The axial skeleton of early tetrapods has more regional differentiation than that of a fish because the functions it performs and the stresses it resists are more varied (Fig. 8-13). Although the skull is no longer united with the pectoral girdle, the "neck" of early tetrapods is barely distinguishable from the trunk. Amphibians have a single **cervical vertebra** (L., *cervix* = neck), called the **atlas,** which allows the head limited mobility (Table 8-1). The **trunk vertebrae** are essentially alike. A single **sacral vertebra** (so called because this part of the body of domestic animals was offered in sacrifices) and pair of ribs articulate with an expanded pelvic girdle and transfer body weight to the hind legs. Weight transfer to the pectoral girdle of nearly all tetrapods is not by bone but by muscular connections between the trunk skeleton and the scapula. **Caudal vertebrae** follow the sacral vertebra. Reduced hemal arches are present on the caudal vertebrae that lie behind the cloaca. Ribs and zygapophyses are usually absent in caudal vertebrae. The caudal vertebrae become

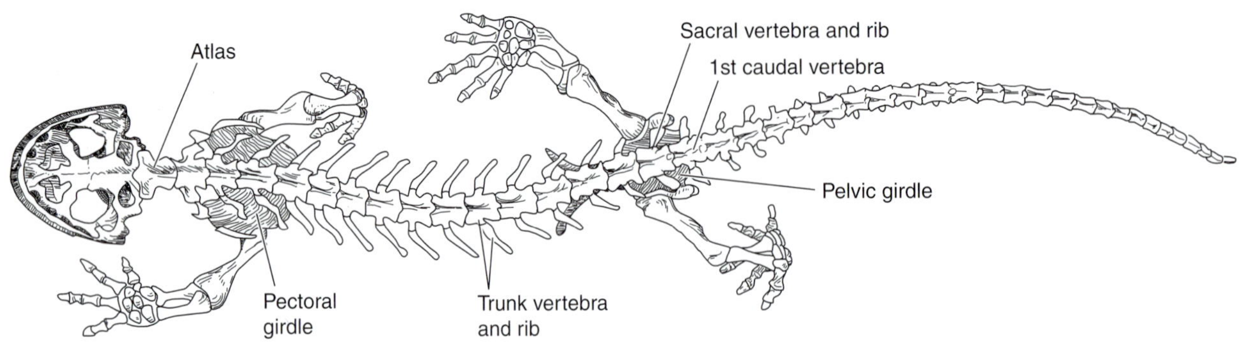

FIGURE 8-13
A dorsal view of the skeleton of a salamander, showing regional specialization of the vertebral column. *(From Romer and Parsons, after Schaeffer.)*

progressively smaller toward the tail tip, and the hemal and neural arches gradually disappear.

Trunk length varies greatly among amphibians. Some caecilians have 285 vertebrae; most frogs only have 8 plus a long, terminal urostyle that evolved from several fused caudal vertebrae. The short, strong, vertebral column of a frog is an adaptation for jumping and swimming by powerful thrusts of the hind legs (Chapter 11).

Reptiles

Correlated with their increased activity and greater penetration of the terrestrial environment, reptilomorphs evolved stronger axial skeletons than basal tetrapods and amphibians. A synapomorphy for the group is that the pleurocentra are always larger than the intercentrum. In both the sauropsid line of reptilomorph evolution culminating in birds and the synapsid line culminating in mammals (Fig. 3-20), the intercentrum first became reduced to a small piece located ventrally between the definitive centra, which are formed by the fused pleurocentra and finally are lost completely (Fig. 8-12*E*). (The term "intercentrum" derives from its position in early tetrapods.) The centrum becomes fused with the neural arch. Intercentra generally also are lost in recent sauropsids, although they persist in parts of the vertebral column in some groups. Intercentra may remain in the caudal vertebrae as points of attachment of the hemal arches, now called **chevron bones,** or they may become incorporated in the chevron bones. *Sphenodon* retains intercentra in the trunk vertebrae and many lizard-like squamates retain them in cervical vertebrae. Remnants of the notochord perforate the centra of *Sphenodon*, but otherwise notochord remnants are limited to the intervertebral disks.

Ribs are present on most of the vertebrae of reptiles extending from the atlas through the anterior caudal vertebrae. Those in the anterior part of the trunk are long and articulate by costal cartilages with a sternum (Fig. 8-7*B*). The sternum also unites with elements of the pectoral girdle (coracoid and interclavicle). With the loss of the intercentrum to which it originally attached, the capitulum of a rib usually has an intervertebral articulation, part of it attaching to the posterior margin of one centrum and part to the anterior margin of the next caudal centrum (Fig. 8-12*E*). Crocodiles

TABLE 8-1 Regional Differentiation of the Vertebral Column*

Vertebrate Group	Vertebral Region				
		(14–30 combined)			
Mammals	Cervical (7)	Thoracic	Lumbar	Sacral (3)	Caudal
Birds	Cervical (11–25)	Thoracic (3–19)	Lumbar +	Sacral (10–23)	Caudal (6–7 + pygostyle)
Reptiles	Cervical (6–10)	Trunk (10–22)		Sacral (2–3)	Caudal
Amphibians	Cervical (1)	Trunk (8–22)		Sacral (1)	Caudal
Fishes	—	Trunk		—	Caudal

*The number of vertebrae indicated represent a typical number or range; exceptions occur.

retain primitive, two-headed ribs, but the tuberculum tends to be lost from the ribs in other recent reptiles.

Regional differentiation of the vertebral column of early reptilomorphs is more extensive than in basal tetrapods and amphibians. Turtles, *Sphenodon,* squamates, and crocodiles have longer necks than do amphibians. More cervical vertebrae are present, which allows the head to move in many directions. The atlas and the second cervical vertebra, which is called the **axis,** begin to show some of the modifications for head movements that reach a higher degree of specialization in mammals (see Fig. 8-19). The remaining cervical vertebrae are characterized by having relatively short ribs, none of which reach the sternum. Turtles and many lizard-like squamates have eight cervical vertebrae; crocodilians, nine; and numerous extinct diapsids had more. Often, many of the cervical vertebrae and the anterior trunk vertebrae have a ventral projection from the centrum known as a **hypapophysis.** Hypapophyses are points of attachment for muscles and ligaments that help move the head (Fig. 8-14*A* and *B*).

The first trunk vertebra bears a pair of ribs that articulate by costal cartilages with the sternum (Fig. 8-7*B*). Only two additional ribs reach the sternum in *Sphenodon,* but eight or nine reach it in crocodilians. Trunk ribs help support the body and protect the lungs from excessive pressure when the animal is lying on the ground. Movements of the trunk ribs also are the major mechanism for ventilating the lungs in most reptiles (Chapter 18). Ribs on the remaining trunk vertebrae decrease in size toward the sacrum. The most posterior trunk ribs of lizard-like squamates fuse with the vertebrae to form pleurapophyses.

Reptiles have a stronger sacral region than do amphibians, for most species have two sacral vertebrae and ribs. Some species have more. The sacral ribs are distinct elements in most recent species, but they are represented by pleurapophyses in squamates.

Tail length and the number of caudal vertebrae vary considerably. Chevron bones are present on a few caudal vertebrae posterior to the cloaca. Distinct ribs are present on the proximal caudal vertebrae of many early reptiles but have become pleurapophyses in lizards.

Many lizard-like squamates can spontaneously drop off, or **autotomize,** much of the tail when attacked by a predator. The tail writhes on the ground and attracts the attention of the predator as the rest of the animal escapes and eventually regenerates a new tail. One or more caudal vertebrae have an **autotomy septum** that extends through the center of the vertebra, and cleavage of the tail occurs at these points. A few salamanders also can autotomize their tails, but separation occurs between vertebrae.

Many early reptiles and a few fossil amphibians have **abdominal ribs** embedded in the ventral abdominal wall (Fig. 8-7*B*). Presumably, they help support and protect this part of the body. Some abdominal ribs, known as **gastralia,** are dermal and may have evolved from rows of bony scales. Others, called **parasternalia,** are cartilage or cartilage-replacement bone. The sternum may have evolved from parasternalia.

Reptiles have undergone such an extensive adaptive radiation that we can consider only a few of the specializations of their trunk skeletons in a book of this scope. A good survey of this topic can be found in Romer's *The Osteology of the Reptiles* (1956). The trunk region of a turtle is short and includes only ten vertebrae and ribs. The trunk is encased in a bony **carapace** dorsally and a **plastron** ventrally (Fig. 8-15). These structures develop primarily from ossifications in the dermis of the skin, but the endoskeletal ribs become incorporated into the carapace, which places the ribs dorsal to the girdles, unlike their position in all other vertebrates, in which the ribs lie medial to the girdles (Fig. 8-16*C* and *D*). Burke (1989) has studied the embryonic development of the carapace. She concludes that interactions between the ectodermal epithelium and the underlying dermal mesenchyme in a **carapacial ridge** located at the margin of the developing carapace induce presumptive rib-

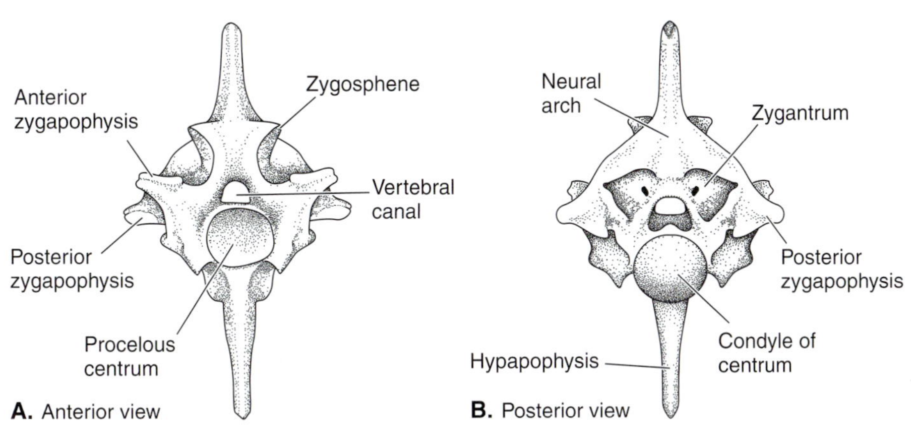

FIGURE 8-14
Anterior (*A*) and posterior (*B*) views of a cervical vertebra of a boa constrictor. *(After Bellairs.)*

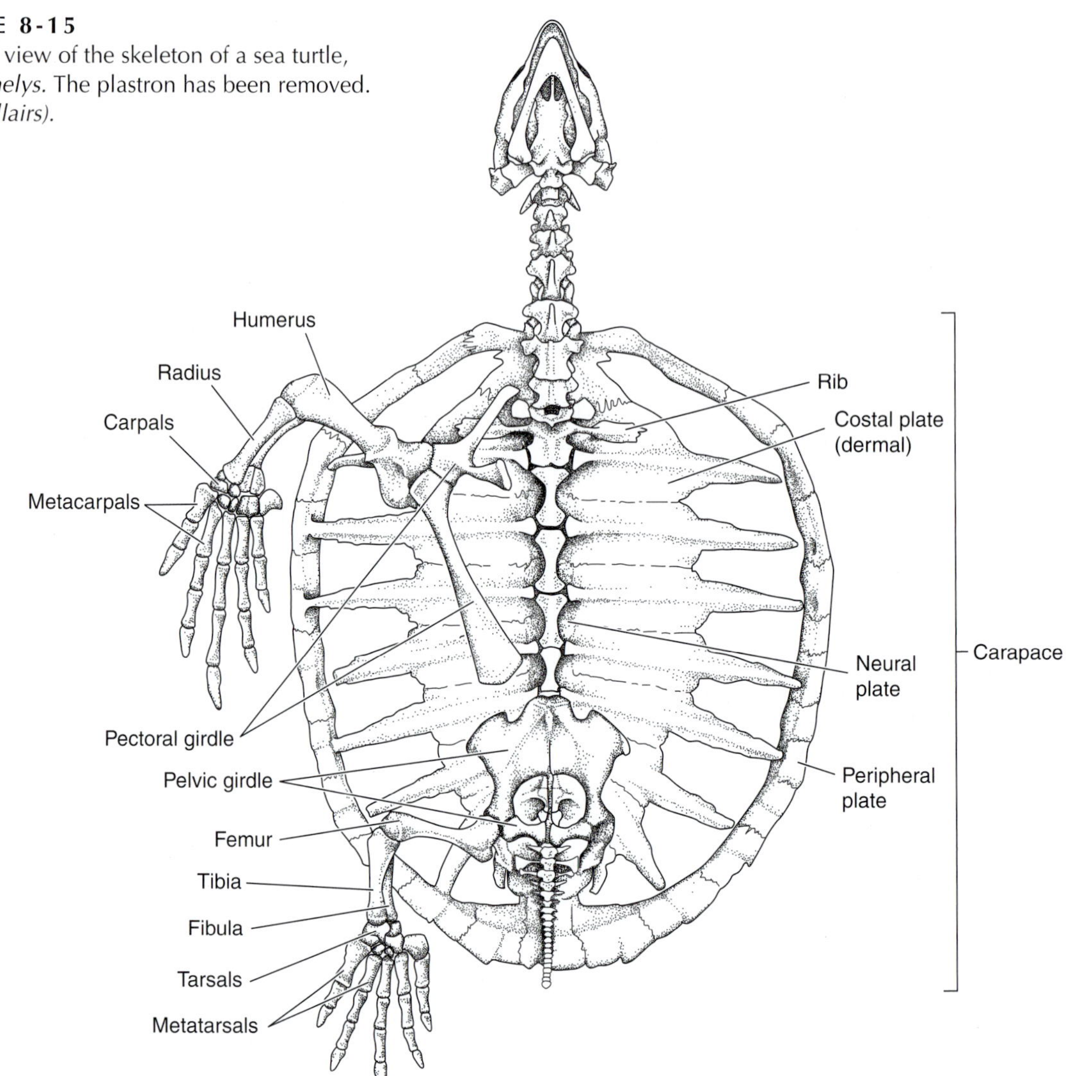

FIGURE 8-15
A ventral view of the skeleton of a sea turtle, *Eretmochelys*. The plastron has been removed. *(After Bellairs).*

forming, or costal, cells to migrate laterally into the carapace instead of taking their usual pathway ventrally (Fig. 8-16*A* and *B*). We have seen how other epithelium–mesenchyme interactions induce the formation of many dermal derivatives, such as bony scales, teeth, and feathers (Chapter 6).

Carapace and plastron provide turtles with a protective shell into which the limbs and head can be withdrawn when the animal is threatened. The head is withdrawn in recent turtles by forming an S-shaped loop in the cervical vertebrae. In our familiar cryptodire species of the northern hemisphere, the cervical loop lies in the vertical plane of the body, but it lies in the horizontal plane in the pleurodire, or side-necked turtles, of the southern hemisphere (Chapter 3).

The elongated body of a snake may contain 200 or more vertebrae. A snake moves primarily by lateral undulations of the trunk and tail and often twines itself around tree branches or prey. Not surprisingly, extra intervertebral joints help articulate and maintain the integrity of such a long and flexible body. A pair of processes called **zygosphenes** (Gr., *zygon* = yoke + *sphen* = wedge), which are located on the anterior surface of a neural arch dorsal to the zygapophyses, project into a pair of sockets, the **zygantra** (Gr., *antron* = cave) on the posterior surface of the next anterior neural arch (Fig. 8-14). Some other elongated sauropsids also have these extra articulations. Because a snake lacks legs and a sternum, distinct vertebral regions, apart from the atlas–axis complex and tail, cannot be recognized. All the vertebrae of snakes, except for the axis and atlas and caudal vertebrae, bear ribs that curve ventrally and may attach to the large ventral scales, or scutes. In addition to supporting the body and ventilating the lungs, rib and ventral scute movements are important in locomotion in some species of snakes.

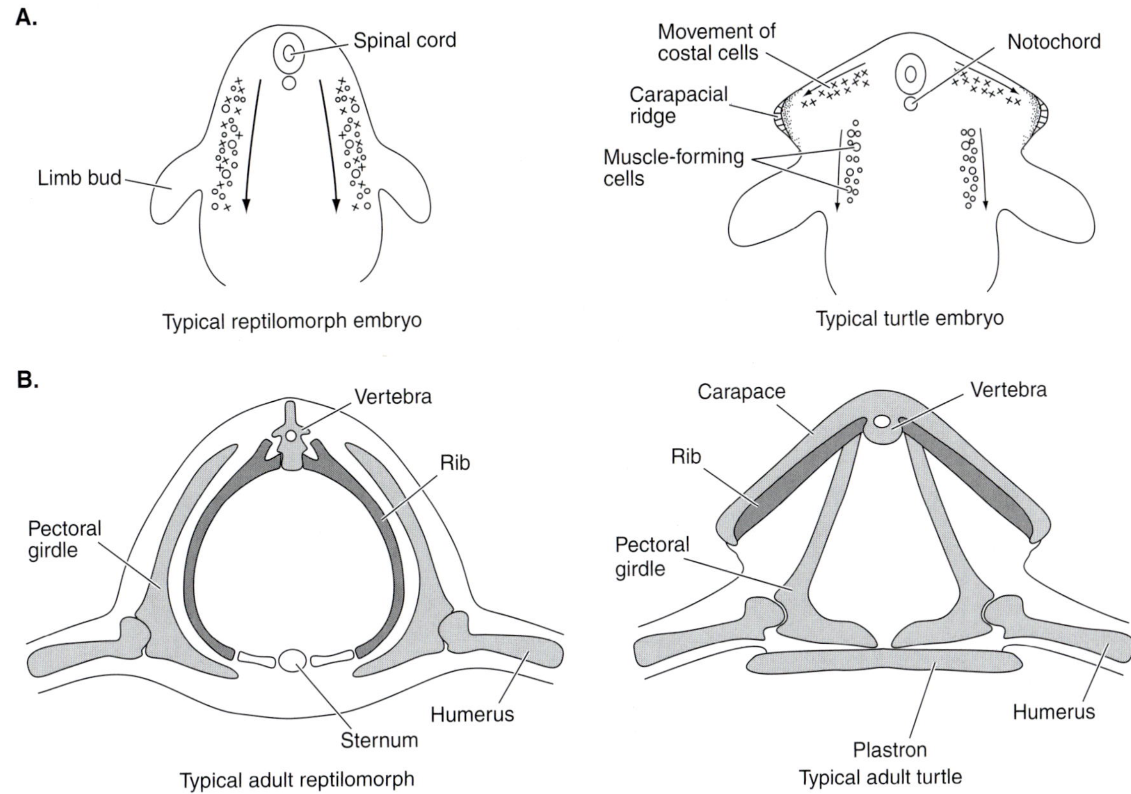

FIGURE 8-16
Diagrams in the transverse plane showing the embryonic development of ribs in relationship to the pectoral girdle in a typical reptilomorph (*left column*) and a turtle (*right column*). *A*, Embryonic stages; *B*, adults. *(After Burke.)*

Birds

Modifications of the axial skeleton of birds are correlated with their unique patterns of locomotion—flying and bipedal walking. Because the pectoral appendages are so highly specialized as wings, they cannot be used in other ways. The head and bill perform all feeding, nest building, and other manipulative functions often fulfilled by the pectoral appendages in other terrestrial vertebrates. An exceedingly long and flexible neck allows for the great mobility of the head (Fig. 8-17*A*). A bird can turn its head 180° to the left or right. Recent species have 11 to 25 cervical vertebrae with heterocoelous centra (Fig. 8-2*E*).

Trunk vertebrae are of two types: **thoracic vertebrae,** to which thoracic ribs attach, and **lumbar vertebrae,** which lack ribs. The back must be short and rigid because it forms the fulcrum on which the wings move up and down. Three to ten thoracic vertebrae are partially fused together. Bipedal locomotion and the action of the hind legs as shock absorbers when a bird alights necessitate a strong sacrum. The lumbar vertebrae fuse with the two sacral vertebrae of ancestral reptilomorphs, and often with several caudal vertebrae, to form a **synsacrum** (Gr., *syn* = with, together) that includes 10 to 23 vertebrae (Fig. 8-17*B*). The synsacrum, in turn, has fused with the pelvic girdle.

Ancestral birds had a long, reptilian tail, but in contemporary species, the terminal four to seven caudal vertebrae have united to form a large element, the **pygostyle** (Gr., *pyge* = rump + *stylos* = pillar), which supports the tail feathers. Six or seven more proximal caudal vertebrae remain independent and allow for the changes in tail position needed in flight and maneuvering. Movement within the vertebral column thus is limited to the cervical region, to the region between the trunk vertebrae and synsacrum, and to the proximal caudal vertebrae.

Short **cervical ribs** are present on most of the cervical vertebrae. The trunk vertebrae bear **trunk ribs,** which unite with an expanded and, in most species, a strongly keeled sternum. The sternum provides an origin for powerful flight muscles (Fig. 8-17*A*). The ventral, or sternal, portions of the ribs, which are represented by costal cartilages in most terrestrial vertebrates, are ossified. Movable joints unite dorsal and ventral parts of the ribs. The ribcage is strong but flexible enough to ventilate the lungs. Most of the dor-

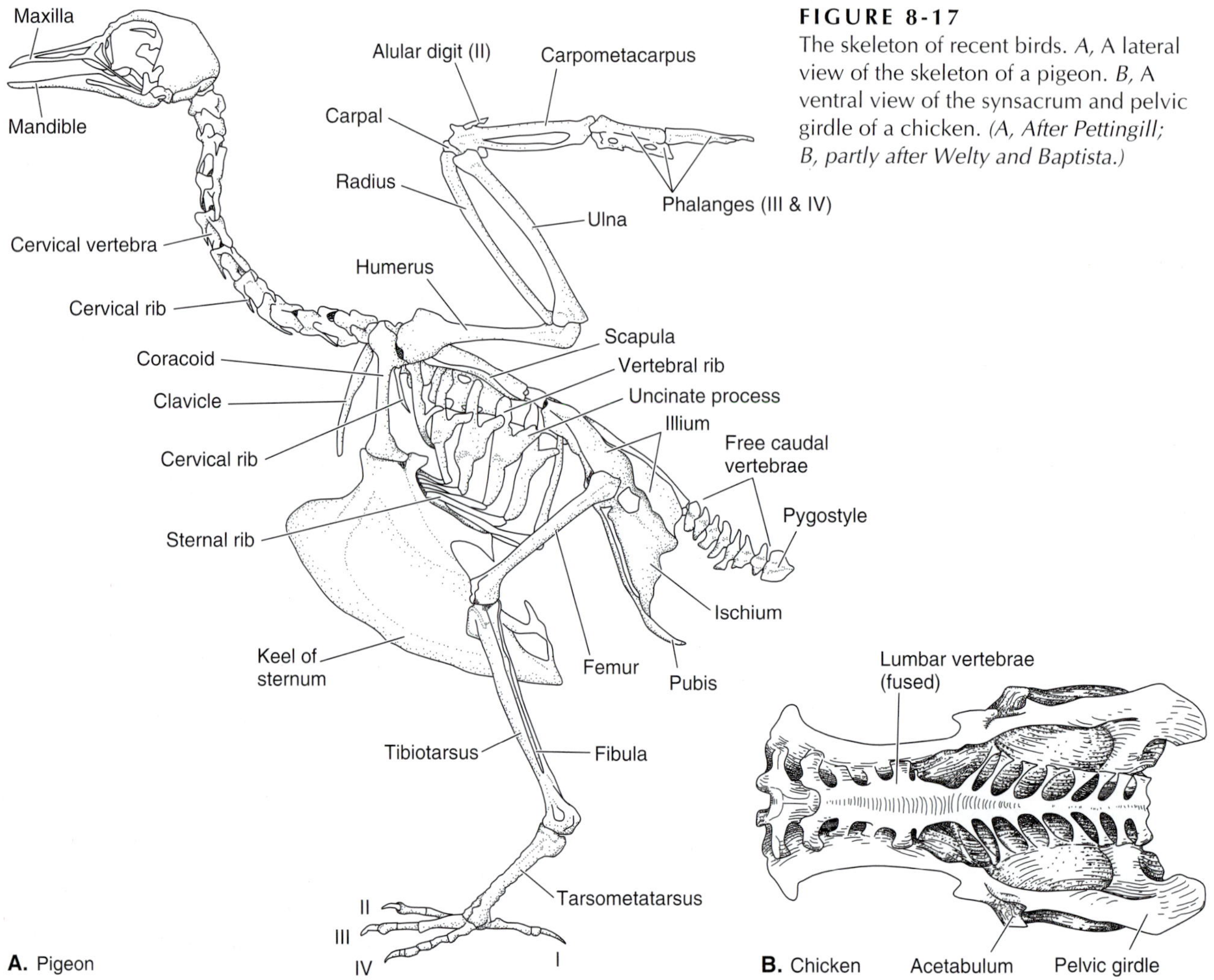

FIGURE 8-17
The skeleton of recent birds. *A,* A lateral view of the skeleton of a pigeon. *B,* A ventral view of the synsacrum and pelvic girdle of a chicken. *(A, After Pettingill; B, partly after Welty and Baptista.)*

sal rib segments bear uncinate processes that overlap the next posterior ribs. These processes further strengthen the ribcage and serve as lever arms onto which certain respiratory muscles attach.

Mammals

Most mammals are active, agile animals with strong, supportive but flexible axial skeletons that allow for free head movements, transfer body weight to the girdles and appendages, may participate in locomotor movements, and have an important role in respiration. A great deal of research has been done on the interesting biomechanics of the mammalian axial skeleton. We explore these topics in Chapter 11 after we have examined the muscles, which play an important role in the biomechanics of the axial skeleton. At this time, we will emphasize the regional differentiation of the vertebral column. Because functions and stresses vary all along the vertebral column, no two vertebrae are exactly alike. Nevertheless, mammalian vertebrae can be sorted into five groups (Fig. 8-18 and Table 8-1).

With few exceptions, all mammals have seven **cervical vertebrae** regardless of neck length (Focus 8-1). Even the very long neck of a giraffe has only seven cervical vertebrae, but, except for the atlas and axis, they are very long. The cervical vertebrae are very short and may fuse together in the entirely aquatic cetaceans (Focus 8-2) as an adaptation to their unique mode of life. Fusion of posterior cervical vertebrae also occurs in jumping rodents and kangaroos and prevents head bobbing during locomotion. Some fusion of cervical vertebrae stiffens the neck in burrowing species.

Because several vertebral elements usually contribute to the occipital region of the skull during embryonic development (Chapter 7), the craniovertebral joint of vertebrates is a modified intervertebral joint. In all terrestrial vertebrates, the first cervical vertebra, the

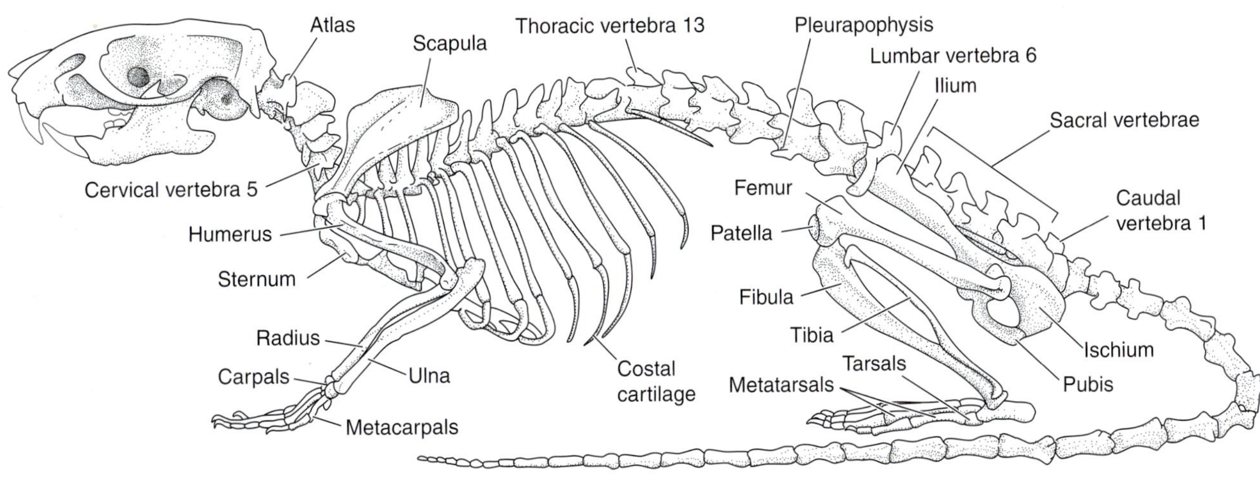

FIGURE 8-18
A lateral view of the skeleton of a rat. (*After Hebel and Stromberg.*)

atlas, is specialized to permit the head to move independently of the trunk. The second cervical vertebra, the axis, contributes to this movement in some reptilomorphs. The atlas and axis become highly specialized in the evolution from the therapsid ancestors of mammals to mammals and, together with changes in the occipital condyle on the back of the skull, form a universal joint that allows exceptional freedom of movement of the head. The occipital condyle of most reptilomorphs is a single hemispherical knob ventral to the foramen magnum that articulates with a concavity on the centrum of the atlas. Movement occurs here, but rotation and dorsoventral flexion of the head are limited by a **proatlas** that lies dorsally between the atlas and skull (Fig. 8-19*A*). The proatlas represents a neural arch of a vertebra, the rest of which probably is incorporated into the skull. In the therapsid ancestors of mammals, the single occipital condyle divided into a

Text continues on page 290.

FOCUS 8-1 *Why Seven Cervical Vertebrae?*

As we have seen, during the development of mammalian cervical vertebrae, embryonic ribs fuse onto the side of the vertebrae to form pleurapophyses. The ribs remain articulated to the vertebrae in the thoracic region, and the first rib typically articulates with the sternum. The absence or presence of a distinct rib is the main difference between cervical and thoracic vertebrae in mammals. A possible reason for the fairly consistent number of seven cervical vertebrae in mammals has been explored by Galis (1999). She points out that individual variations are known in humans and mice in which the seventh vertebra retains a rib and hence resembles an extra thoracic vertebra. In manatees and one species of sloth of the genus *Choloepus,* this normally happens so they have six cervical vertebrae. In sloths of the genus *Bradypus,* the two most anterior thoracic vertebrae become cervical vertebrae by the fusion of their ribs to their vertebra, so these mammals have nine cervical vertebrae.

Hox gene expression plays an important role in patterning the vertebral axis, and mutations in these genes can cause the retention of a cervical rib on the seventh vertebra. These mutations, which also lead to oxidative DNA damage, are coupled in humans and mice with severe congenital abnormalities, including several early childhood cancers. Galis proposes that strong stabilizing selection for seven cervical vertebrae among mammals has occurred because of the deleterious pleiotropic effects of the mutant for the retention of a cervical rib on the seventh vertebra. Reptiles display considerable variation in the number of cervical vertebrae. Presumably, they are not adversely affected because their low rate of metabolism makes them less susceptible to oxidative DNA damage. This also may be the case in manatees and sloths. Given the high rate of metabolism in birds, their low susceptibility to oxidative DNA damage is a puzzle but may result from a remarkably low level of free-radical production.

FOCUS 8-2 *Back to the Water: An Example of Convergent Evolution*

Many groups of terrestrial vertebrates have readapted in varying degrees to life in the water. The aquatic environment is rich in resources such as food and shelter, and it is more stable than are terrestrial environments subject to pronounced diurnal and seasonal changes in temperature, light, and other environmental parameters. Arguably, no recent group of vertebrates with a terrestrial ancestry has readapted more completely to an aquatic life than the porpoises, dolphins, and whales of the mammalian order Cetacea. Indeed, these species can no longer live on land.

Terrestrial vertebrates that readapt completely to life in the water must meet a set of common constraints. Because only a limited number of "solutions" to these constraints exist, the organisms come to resemble each other, at least superficially. That is, their adaptations converge as they come to resemble fishes in many ways. The increased density of water compared with air is a con-

A. Convergent evolution in body shape among a shark, †ichthyosaur, and porpoise.

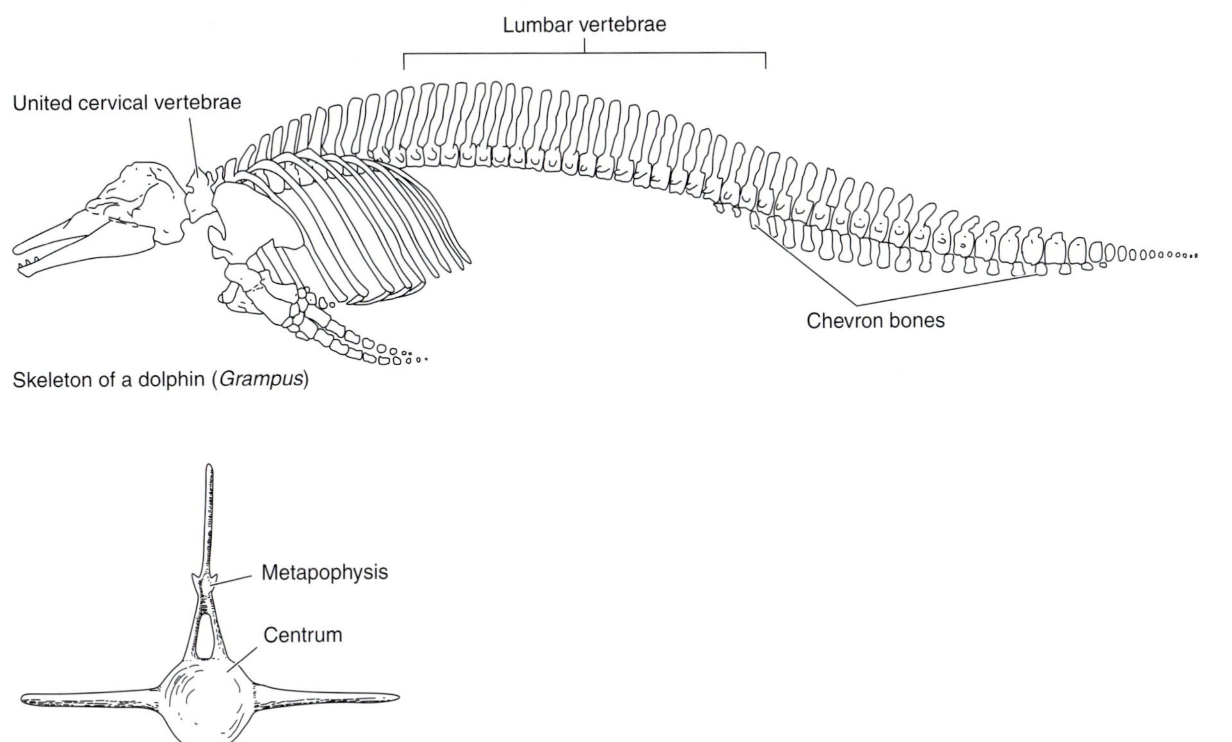

B. *Top,* The skeleton of the dolphin, *Grampus griseus. Bottom,* Anterior view of a lumbar vertebra of a porpoise. *(After Howell.)*

straint that has greatly affected body shape and the axial skeleton. All fast-moving aquatic vertebrates have a similar streamlined body shape. We see this in a shark, an ichthyosaur (an extinct group of marine reptiles), and a cetacean (Fig. A). The contours of the head merge with those of the trunk. A distinct neck and a freely movable head would be impediments. Cetaceans retain the seven cervical vertebrae characteristic of mammals, but the head merges with the trunk because the individual cervical vertebrae are very short and united in many species into a single unit (Fig. B, *top*). A shark, †ichthyosaur, and cetacean have a stabilizing dorsal fin and flaplike paired fins. The dorsal fin of sharks is supported by skeletal elements, but only a very dense connective tissue supports those of ichthyosaurs and cetaceans.

Because water provides considerable buoyancy, the axial skeleton of these aquatic vertebrates need not resist vertical bending as it must in a terrestrial species. The vertebral column, however, must resist strong compressive forces as these animals push themselves through relatively dense water, and it must be flexible enough to allow for propulsive undulations. The vertebral structure of a shark, ichthyosaur, and cetacean converge. All have large, disk-shaped vertebral centra well suited to resist compression (Fig. B, *bottom*). Zygapophyses on the neural arches are reduced or lost completely. They are present only in the more anterior thoracic vertebrae of cetaceans. The vertebral column need not resist vertical bending forces, and zygapophyses restrict undulation. Trunk and tail oscillate up and down in a vertical plane as a cetacean swims. Movement is most pronounced in the long lumbar region, which may contain 21 lumbar vertebrae compared with the 4 to 7 usually present in terrestrial mammals. The thoracic and lumbar vertebrae of cetaceans have **metapophyses** at the base of the neural spines that restrict lateral bending. Long neural spines on most of the vertebrae and processes of the caudal chevron bones act as lever arms for the locomotor muscles. Many cetaceans retain vestiges of the hindlimbs and pelvic girdle embedded in the body wall, but the girdle does not connect with the vertebral column, so no sacrum is present.

Ribs and a small sternum are present in the thoracic region of cetaceans, but they are not strong enough to support the body on land. Cetaceans stranded on beaches cannot ventilate their lungs adequately and eventually suffocate.

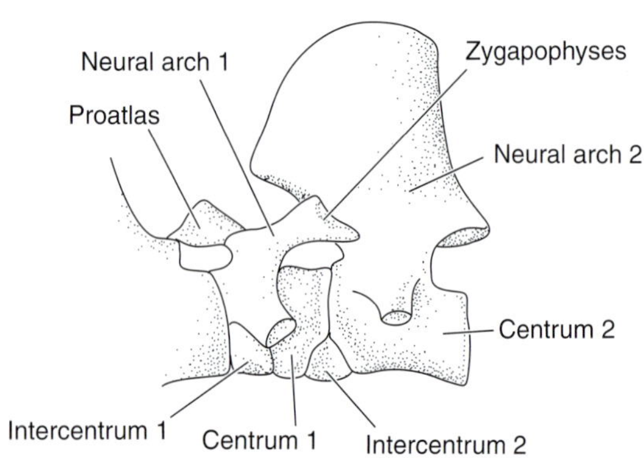

A. Atlas of †*Dimetrodon*

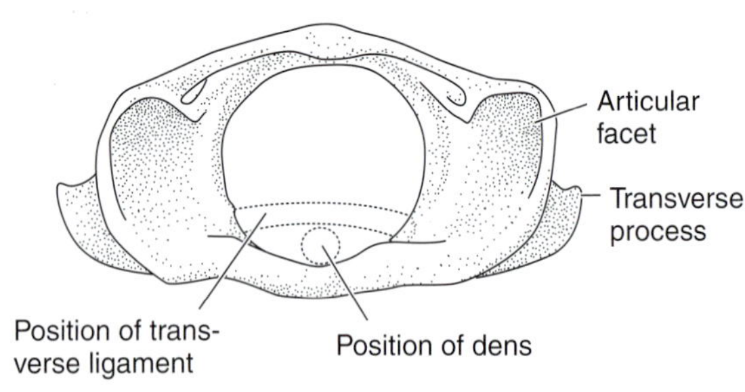

B. Atlas of a mammal, anterior view

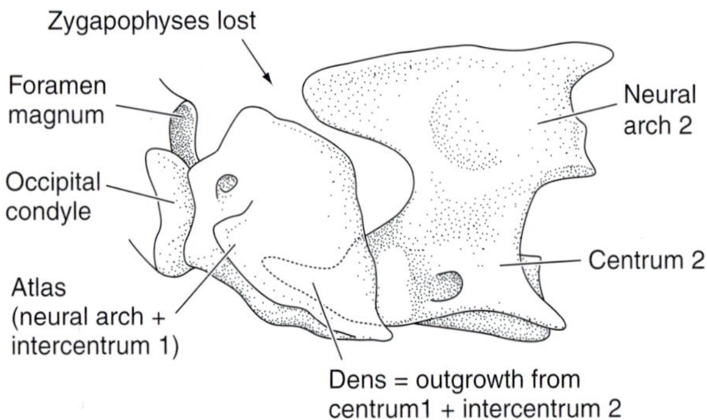

C. Atlas and axis complex of a mammal, lateral view

FIGURE 8-19
Evolution of the atlas and axis. *A,* A lateral view of the first two cervical vertebrae of an early synapsid, †*Dimetrodon*. *B,* An anterior view of the atlas of a mammal. *C,* A lateral view of the atlas and axis of a mammal. *(A, After Jollie; C, after Romer and Parsons.)*

pair that shifted dorsally and came to lie lateral to the foramen magnum. Complementary articular facets evolved on the neural arch of the atlas, and the centrum of the atlas is represented only by its small intercentrum (Fig. 8-19*A* and *B*). The proatlas was lost in mammals, as was the neural spine of the atlas. These changes allowed considerable up-and-down movement of the head, but excessive dorsoventral flexion, which would stress the spinal cord, was prevented by changes in the axis. The centrum of the axis evolved a long, toothlike process, the **dens,** which protrudes into the vertebral canal of the atlas (Fig. 8-19*C*). A **transverse ligament** within the vertebral canal of the atlas crosses the dorsal surface of the dens and prevents hyperflexion (Fig. 8-19*B*). The dens evolved as an outgrowth of the centrum of the axis, but it probably incorporates the intercentrum of the axis and the missing pleurocentrum of the atlas. Loss of the zygapophyses between the axis and atlas, and the evolution of new articular surfaces between the centrum of the axis and the neural arch of the atlas, allow rotational movements to occur at the axis–atlas joint.

The trunk of a mammal has differentiated into **thoracic** (Gr., *thorax* = chest) and **lumbar** (L., *lumbus* = loin) **regions** (Fig. 8-18). **Thoracic vertebrae** bear ribs, most of which connect by costal cartilages with the sternum. The short, embryonic ribs of the **lumbar vertebrae** fuse to the sides of the vertebrae and form conspicuous pleurapophyses. Differentiation of the trunk into the thoracic and lumbar regions is correlated with the evolution of a diaphragm that attaches to the distal ends of the caudal ribs and separates the pleural cavities, containing the lungs, from the peritoneal cavity. The ribs maintain the integrity of the pleural cavities as the diaphragm changes their volume and ventilates the lungs. Rib movements participate in strong ventilator movements. Ribs also transfer the weight of the anterior portion of the trunk to the pectoral girdle via a muscular sling that extends to the scapula (see Fig. 11-11).

Mammals have between 14 and 30 thoracic and lumbar vertebrae combined; usually more thoracic than lumbar vertebrae are present (Fig. 8-18). In many species, extra processes on the lateral surface of thoracic and lumbar vertebrae are points of attachment for muscles and ligaments and reinforce articulations between vertebrae. For example, edentates (sloths, anteaters, and armadillos) have **xenarthrous articulations** (Gr., *exenos* = strange + *arthron* = joint) on the lateral surface of their lumbar vertebrae ventral to the zygapophyses (Fig. 8-20*A*). These articulations help brace the trunk when armadillos burrow (Gaudin and Biewener, 1992) and when anteaters dig for ants and help arboreal sloths support a relatively heavy body as they reach out to grasp another tree limb.

Intervertebral articulations reach an extreme in the hero shrew (*Scutisorex somereni*) of eastern Africa (Fig. 8-20*B*). The intervertebral joint between the zygapophyses and centra (Fig. 8-20*B*) is greatly strengthened by the presence of many interdigitating tubercles. This is accompanied by a great increase of

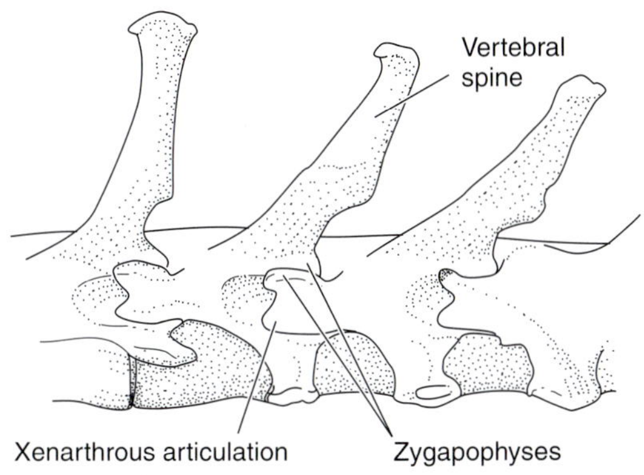

A. Lumbar vertebrae of armadillo (*Dasypus*) in lateral view

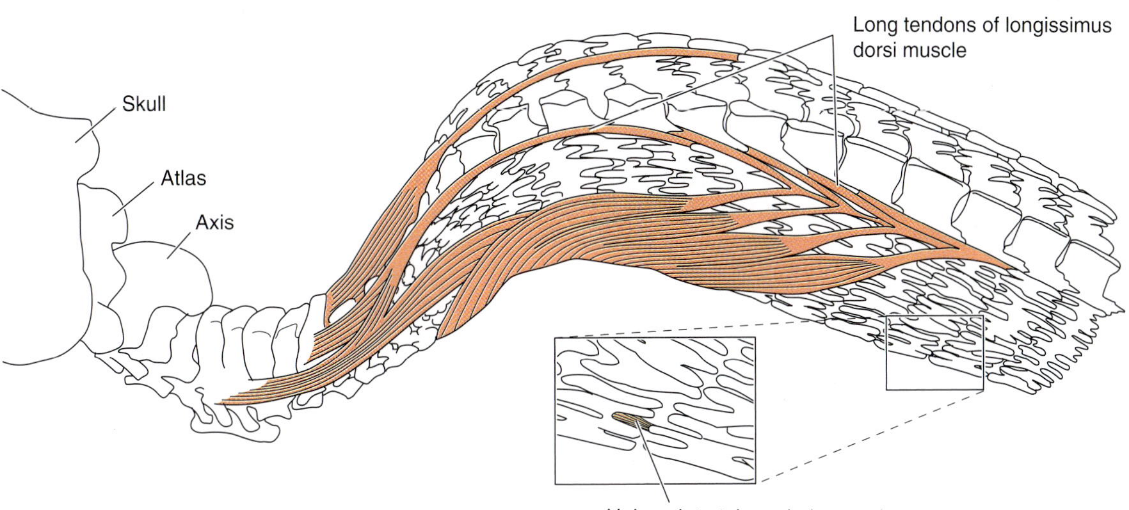

B. Vertebrae of hero shrew (*Scutisorex*) in lateral view

FIGURE 8-20
Some mammalian vertebral modifications. *A,* A lateral view of three lumbar vertebrae of an armadillo, showing the additional xenarthrous articulations. *B,* A lateral view of the cervical, thoracic, and many of the lumbar vertebrae of the hero shrew. *(A, After Vaughan; B, after Cullinane and Aleper.)*

bone mass of the vertebral column and ribs in comparison with other mammals of similar size (Cullinane et al., 1998) and by modifications of dorsal muscles associated with the spine (Cullinane and Aleper, 1998). All of these modifications suggest that the spine can resist very large compressive and torsional forces. However, we do not know the unique functional circumstances correlated with the evolution of this bizarre axial skeleton because little is known about the habits of this species.

The weight of the posterior part of the body of a terrestrial mammal is transferred to the pelvic girdle through a strong **sacrum** containing at least three sacral vertebrae and their ribs, all of which are usually fused together into a single unit. More sacral vertebrae are present in species in which stresses on the sacrum are greater. All the weight of the upper part of the body of a human, a biped, passes through the pelvic girdle, and five sacral vertebrae and ribs are incorporated in the sacrum.

Tail functions and length vary considerably among mammals. Cetaceans have powerful tails, the distal ends of which bear the horizontal flukes used in locomotion. The flukes are stiffened by dense connective tissue, not by skeletal elements. Only three to five reduced caudal vertebrae are present internally in human beings. They usually are fused to form a **coccyx** (Gr., *kokkyx* = cuckoo, because of its resemblance to a cuckoo's bill) to which certain anal and perineal muscles attach. Most mammals have modest-sized tails. They may be used in locomotion (the prehensile tails of New World monkeys), in balance (arboreal species, such as squirrels), helping to maneuver (kangaroo rats), or as part of a mammal's behavioral repertoire to express emotions (a dog's wagging tail) or warning (white-tailed deer). Proximal caudal vertebrae may bear small chevron bones, but these are lost distally, as are most other parts of the vertebrae. Terminal vertebrae consist only of reduced centra.

SUMMARY

1. The axial skeleton consists of the notochord, vertebral column, ribs, median fins (in fishes), and sternum (in terrestrial vertebrates). It forms the fundamental framework of the body and transfers body weight to the girdles and appendages in tetrapods. Rib movements help ventilate the lungs in most tetrapods.
2. Vertebrae develop in an intervertebral position by the condensation of mesenchyme of sclerotomal origin. The mesenchyme first chondrifies in the neural and hemal arches, and, in most species, the cartilage is later replaced by bone. The centra develop in different ways in different taxa.
3. Ribs develop in the myosepta at their intersection with other skeletogenous septa. Fishes have many rib types. Tetrapod ribs are subperitoneal in position and usually attach both on the side of the centrum (often on a parapophysis) and on a conspicuous transverse process known as the diapophysis. In some body regions of tetrapods, embryonic ribs fuse with the diapophysis, parapophysis, or both to form a compound transverse process known as a pleurapophysis.
4. The vertebral axis of a fish is an integral part of its locomotor apparatus. It resists compression of the body and converts the contraction of segmental longitudinal muscles into lateral undulations. Because all parts of the vertebral column have nearly the same function, there is little regional differentiation. The first vertebra articulates with the skull, and the caudal vertebrae bear hemal arches.
5. The notochord persists in the adult of living jawless craniates, lungfishes, sturgeons and paddlefishes, and some other species that do not move rapidly through the water. Contemporary sharks and their allies have cartilaginous centra that are partly calcified; advanced actinopterygians have well-ossified centra.
6. The vertebral axis of fishes ends in an expanded caudal fin that is supported internally by vertebral elements and distally by horny ceratotrichia (in chondrichthyans) or bony lepidotrichia (in fishlike osteichthyans). The types of caudal fins correlated with patterns of locomotion and buoyancy control.
7. Although the vertebral axis participates in lateral undulations in some amphibians and early reptilomorphs, it is primarily a supporting beam. It must resist torsion and bending in the vertical plane so that these low-slung animals do not drag on the ground when they walk.
8. The vertebrae of early basal tetrapods resembled those of their choanate fish ancestors. Each vertebral centrum consisted of a large intercentrum and a pair of smaller pleurocentra that flanked a notochord. Vertebrae became stronger in other species as the intercentrum, pleurocentra, or all elements enlarged.
9. Ribs were well developed in basal tetrapods and attached by two heads to most of the vertebrae. In contemporary amphibians they are very short and usually contribute to pleurapophyses.
10. The vertebral axis of an amphibian is stronger and more complex than in a fish. A well-developed cervical vertebra (the atlas) permits movements between the head and trunk, a sacral vertebra and pair of ribs articulate the vertebral column with the pelvic girdle, and caudal vertebrae lie in the tail.
11. Evolutionary trends that began in the axial skeleton of amphibians continue in reptiles. Intercentra are lost in most contemporary species, and the definitive centrum develops from the expanded pleurocentra. The neck is longer and more mobile, and there are more cervical vertebrae. At least two sacral vertebrae and their ribs articulate with the pelvic girdle. Tail length varies.
12. The anterior trunk ribs of living reptiles connect by flexible costal cartilages with the sternum in most species. Abdominal ribs may be present caudal to the sternum.
13. The structure of the axial skeleton of birds correlate with their specializations for flight and bipedal walking. The long and flexible neck allows the head and bill to be used in many manipulative functions that the specialized forelimbs can no longer do. Many trunk vertebrae are fused to form a firm fulcrum for wing action; others have united with the sacral vertebrae to form a solid synsacrum to which the pelvic girdle unites. The terminal caudal vertebrae have united to form a pygostyle, to which large tail feathers attach.
14. The sternum of birds is ossified and usually bears a large keel that increases the area for the attachment of flight muscles.
15. Mammals are active, agile vertebrates with strong but flexible axial skeletons. Most have seven cervical vertebrae, of which the first two (atlas and

axis) are specialized to allow for free movement of the head. The trunk region has differentiated into a thoracic region, which bears distinct ribs, and a lumbar region, in which the ribs have contributed to pleurapophyses. The thoracic region contains the lungs and transfers body weight to the pectoral girdle. Three or more sacral vertebrae usually unite, forming a strong sacrum. Tail length and functions vary among mammalian taxa.

REFERENCES

Bellaires, A., 1969: *The Life of Reptiles.* London, Weidenfeld and Nicolson.

Blanco, M. J., and Sanchiz, B., 2000: Evolutionary mechanisms of rib loss in anurans: A comparative developmental approach. *Journal of Morphology,* 244:57–67.

Burke, A. C., 1989: Development of the turtle carapace: Implications for the evolution of a novel Bauplan. *Journal of Morphology,* 199:363–378.

Carrier, D. R., 1993: Action of the hypaxial muscles during walking and swimming in the salamander *Dicamptodon. Journal of Experimental Biology,* 180:75–83.

Cave, A. J. E., 1975: The morphology of mammalian pleurapophyses. *Journal of Zoology (London),* 177:377–393.

Corliss, C. E., 1976: *Patten's Human Embryology.* New York, McGraw-Hill.

Cullinane, D. M., and Aleper, D., 1998: The functional and biomechanical modifications of the spine of *Scutisorex somereni,* the hero shrew: Spinal musculature. *Journal of Zoology (London),* 244:453–458.

Cullinane, D. M., Aleper, D., and Berytram, E. A., 1998: The functional and biomechanical modifications of the spine of *Scutisorex,* the hero shrew: Skeletal scaling relationships. *Journal of Zoology (London),* 244:447–452.

Flower, W. H., 1885: *An Introduction to the Osteology of the Mammalia.* Reprinted in Amsterdam, A. Asher & Co., 1966.

Gadow, H. F., 1933: *The Evolution of the Vertebral Column.* Cambridge, Cambridge University Press.

Galis, F., 1999: Why do most mammals have seven cervical vertebrae? Developmental constraints, *Hox* genes and cancer. *Journal of Experimental Zoology,* 285:19–26.

Gaudin, T. J., and Biewener, A. A., 1992: The functional morphology of xenarthrous vertebrae in the armadillo, *Dasypus novemcinctus* (Mammalia, Xenarthra). *Journal of Morphology,* 214:63–81.

Goodrich, E. S., 1930: *Studies on the Structure and Development of Vertebrates.* London, Macmillan.

Hebel, R., and Stromberg, M. W., 1976: *Anatomy of the Laboratory Rat.* Baltimore, Williams & Wilkins.

Howell, A. B., 1930: *Aquatic Mammals.* Reprinted in New York, Dover Publications, 1970.

Jarvik, E., 1980: *Basic Structure and Evolution of Vertebrates.* London, Academic Press.

Jenkins, F. A., Jr., 1969: The evolution and development of the dens of the mammalian axis. *Anatomical Record,* 164:173–184.

Jollie, M., 1973: *Chordate Morphology.* Huntington, New York, Krieger.

Kemp, T. S., 1969: The atlas–axis complex of the mammal-like reptiles. *Journal of Zoology (London),* 159:223–248.

Noble, G. K., 1931: *The Biology of the Amphibia.* New York, McGraw-Hill.

Panchen, A. L., 1977: The origin and early evolution of tetrapod vertebrae. *In* Andrews, S. M., Miles, R. S., and Walker, A. D., editors: *Problems in Vertebrate Evolution.* Linnean Society of London, Symposium no. 4. London, Academic Press.

Pettingill, O. S., Jr., 1985: *Ornithology in Laboratory and Field,* 5th edition. Orlando, Academic Press.

Romer, A. S., 1956: *The Osteology of the Reptiles.* Chicago, University of Chicago Press.

Romer, A. S., and Parsons, T. S., 1986: *The Vertebrate Body,* 6th edition. Philadelphia, Saunders College Publishing.

Schaeffer, B., 1967: Osteichthyan vertebrae. Linnean Society of London, *Zoology Journal,* 47:185–195.

Schmitz, R. J., 1995: Ultrastructure and function of cellular components of the intervertebral joint in the precoid vertebral column. *Journal of Morphology,* 226:124.

Slijper, E. J., 1962: *Whales.* London, Hutchinson.

Vaughan, T. A., 1978. *Mammalogy,* 2nd edition. Philadelphia, Saunders College Publishing.

Wake, D. B., and Lawson, R., 1973: Development and adult morphology of the vertebral column of the plethodontid salamander, *Eurycea bislineata,* with comments on vertebral evolution in the Amphibia. *Journal of Morphology,* 139:251–300.

Wake, M. H., editor, 1979: *Hyman's Comparative Vertebrate Anatomy,* 3rd edition. Chicago, University of Chicago Press.

Walker, W. F., and Homberger, D. G., 1992: *Vertebrate Dissection,* 8th edition. Fort Worth, Texas, Saunders College Publishing.

Welty, J. C., and Baptista, L., 1988: *The Life of Birds,* 4th edition. New York, Saunders College Publishing.

Williams, E. E., 1959: Gadow's arcualia and the development of tetrapod vertebrae. *Quarterly Review of Biology,* 34:1–32.

9

The Postcranial Skeleton: The Appendicular Skeleton

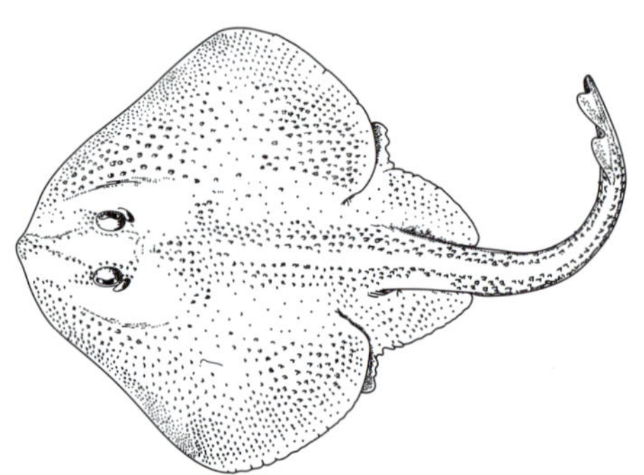

PRECIS

The appendicular skeleton consists of the paired appendages and their supporting girdles. Most fishes have paired fins, which help to stabilize the body and are used in maneuvering as the animal swims. Terrestrial vertebrates have much larger and stronger paired limbs, which play a major role in support and locomotion. We examine the structure and evolution of the appendicular skeleton in this chapter, considering how it changes as vertebrates move from water to land and in some cases become adapted to life in the air. Patterns of locomotion and biomechanical aspects of support and locomotion are deferred to Chapter 11 after we have described the muscular system.

OUTLINE

The appendicular skeleton consists of paired **pectoral appendages** and **pectoral girdles** in the shoulder region of the body and **pelvic appendages** and **pelvic girdles** in the hip region. Because fishes are supported by the water in which they live, the appendages of most species are small fins that assist the dorsal and anal fins to stabilize the body and reduce the tendency of the fish to roll, pitch, and yaw[1] as they move through the water. Paired fins also are used in turning and other maneuvers, and in some species their undulations help the animal to swim. The paired appendages become much larger **limbs** in terrestrial vertebrates, and they play a major role in supporting and moving the body on land. Pectoral or pelvic girdles in the body wall support paired fins or limbs. Girdles are small and weak in most fishes, but in tetrapods, they are large, sturdy structures that transfer body weight from the axial skeleton to the limbs.

Origin of the Appendicular Skeleton

Recent hagfishes and lampreys have no paired fins, which was also the case for most of the early jawless fishes. Early jawless fishes lay or swam slowly along the bottom and would not need the type of stabilizing mechanism provided by paired fins. Evidence for bottom dwelling is their heavy body armor, a somewhat flattened body shape, and dorsally directed

[1]Roll is the tendency of a moving body to rotate about its longitudinal axis; pitch is a vertical rotation about the transverse axis; yaw is the tendency for the head or tail to move from side to side in the horizontal plane.

eyes. Although most jawless fishes lack paired fins, several groups of early jawless fishes (Fig. 3-2) had some type of paired fins, at least in the pectoral region. Some †anaspids (†*Pharyngolepis;* Fig. 9-1*A*) had paired ventrolateral fins extending much of the length of the trunk. The †osteostracan †*Hemicyclaspis* (Fig. 9-1*B*) had a relatively large and probably muscular flap just behind its head shield. This has been interpreted as a pectoral fin. The †thelodont †*Phlebolepis* (Fig. 9-1*C*) had a pectoral fin with a broad base attaching to the body.

There has been much speculation as to how paired fins evolved. Some investigators believe that the long lateral fin of †*Pharyngolepis* and the broad base of the pectoral fin of †*Phlebolepis* are consistent with the **fin-fold hypothesis** proposed by Balfour (1876) and Thacher (1877) long before these fossils were well known. Balfour and Thacher based their theory on the very broad attachment to the body of the paired fins in elasmobranch embryos. Proponents of the fin-fold theory postulate that early vertebrates had a continuous fin fold on the lateral side of the body, somewhat like the long metapleural folds of amphioxus. Paired fins evolved by the loss of intermediate parts of the fin-fold (Fig. 9-2). The fin-fold hypothesis is doubtless an oversimplification, but the idea of a continuous lateral zone that gives rise to paired appendages gains some support from studies on fish and chick embryos. A continuous, longitudinal **limb-forming zone** of opaque cells exists at an early stage in the lateral plate mesoderm. The portions of this potential limb-forming zone that differentiate into pectoral and pelvic fins or limbs appear to be controlled by the location of the boundaries of *Hox* gene clusters (Coates and Cohn, 1999), but a lateral embryonic limb-forming zone is not a continuous lateral fin-fold. Bemis and Grande (1999) have reviewed the history of the fin-fold hypothesis and find no empirical paleontological or embryonic evidence to support it. They regard the fin-fold hypothesis as an example of late 19th and early 20th century idealistic morphology from which we should retreat.

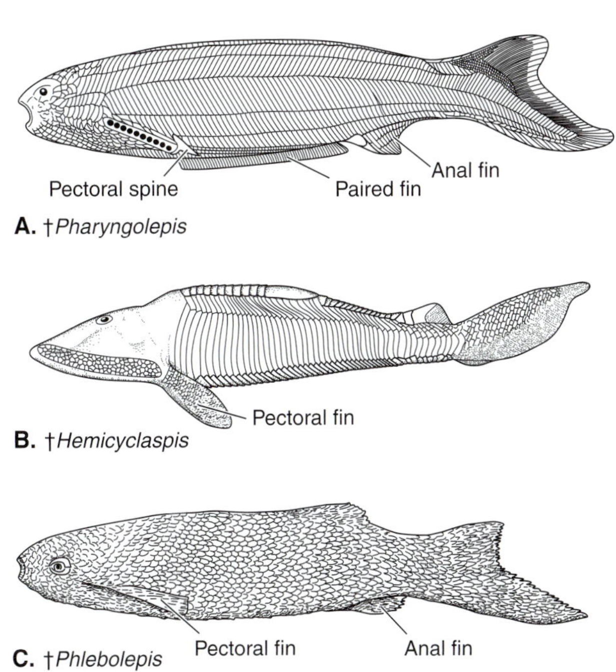

FIGURE 9-1
Paired fins in early jawless vertebrates. *A,* An †anaspid, †*Pharyngolepis. B,* An †osteostracan, †*Hemicyclaspis. C,* A †thelodont, †*Phlebolepis.* (*After Carroll.*)

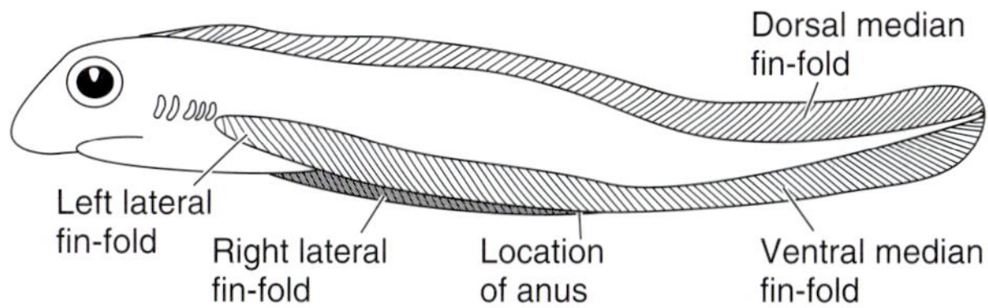

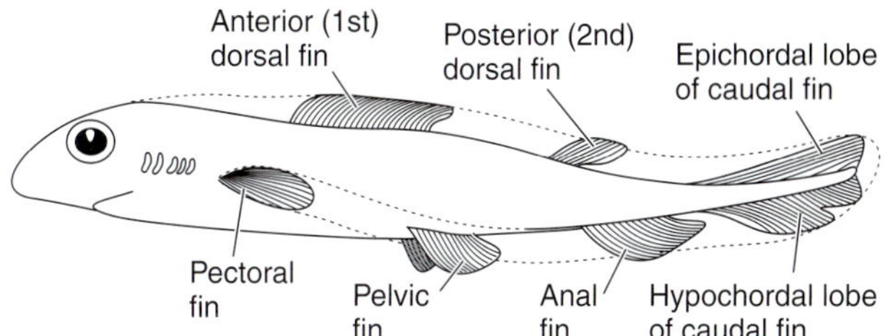

FIGURE 9-2
The fin-fold hypothesis of the origin of paired fins in an embryo (*A*) and an adult (*B*). Pectoral and pelvic fins evolved by the loss of parts of a continuous lateral fin-fold present in ancestral vertebrates. (*After Weidersheim.*)

Their streamlined body form and some reduction of the heavy, ancestral armor indicate that early jawed fishes became much more active animals than most jawless fishes. They probably swam by lateral undulations of the trunk and tail as recent fishes do. As fishes became more active and adapted to more diverse habitats and ecological niches, natural selection would favor the enlargement of any lateral protuberances or flaps that assisted the median fins in reducing roll and pitch. Large pectoral fins attached low near the front of the body also would act as hydroplanes and help to raise a heavy-bodied fish off of the bottom as it swam. Paired fins capable of some rotation and change of angle could help a fish maneuver.

Paired fins are supported by fin rays (ceratotrichia or lepidotrichia; Chapter 6) in the same way as median fins. The fin rays attach to deeper pterygiophores of cartilage or cartilage-replacement bone. As fins became larger, inward extensions of the pterygiophores probably coalesced to form the **endoskeletal part** of the pectoral and pelvic girdles, which is composed of cartilage or cartilage-replacement bone. Overlying dermal plates in the pectoral region formed the **dermal part** of the pectoral girdle. The dermal part of the pectoral girdle lies directly posterior to the gill region and serves as an anchorage for trunk muscles as well as helping to support the pectoral fin by its connection to the back of the skull. The pelvic girdle never has a dermal component. In whatever way formed, fins and girdles likely evolved independently in different lineages because distinct types of paired fins characterize early groups of jawed fishes.

The Appendicular Skeleton of Jawed Fishes

In one group of †placoderms, the †arthrodires, well-developed dermal plates covered the front of the trunk, and the skull had a distinctive and movable articulation with them (Fig. 3-5*A*). Many of these thoracic plates (the **cleithral elements** and the **clavicle**) appear comparable to the dermal part of a pectoral girdle (Fig. 9-3). A small **scapulocoracoid cartilage,** which formed the endoskeletal part of the pectoral girdle, lay beneath the ventral part of the thoracic armor and bore an articular surface for a pectoral fin, whose structure is not well known. Another group of †placoderms, the †antiarchs, had a jointed pectoral appendage that was covered with dermal plates and resembled an arthropod's limb (Fig. 3-5*D*). It may have been used to crawl or to elevate the front of the body. Small pelvic fins also were present.

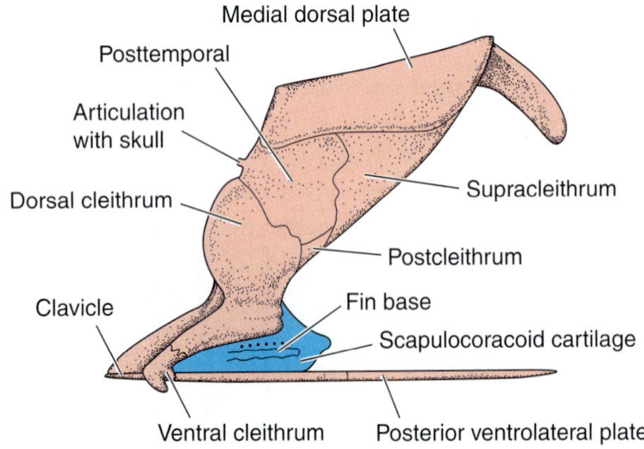

FIGURE 9-3
A lateral view of the thoracic plates of the †arthrodire, †*Dunkleosteus*. In this and other figures in this chapter, elements of the dermal girdle are shown in red; the cartilages and cartilage-replacement bones of the endoskeleton, in blue. (*After Westoll.*)

Early chondrichthyans, such as †*Cladoselache,* had large pectoral fins and smaller pelvic ones (Fig. 9-4*A*). Studies of †*Cladoselache* suggested that the fins had very broad bases, or attachments to the body wall, and some investigators believe that this reinforces the fin-fold hypothesis of fin origin. The pectoral fins also were large enough and attached low enough on the body to act as hydroplanes and help raise this heavy-bodied fish from the bottom. The fins were supported primarily by many cartilaginous **radial pterygiophores,** which attached to the girdles. In the absence of dermal bone in the pectoral girdle, the **scapulocoracoid** of a chondrichthyan is larger than that in other groups of fishes and provides a sufficient area for the attachment of appendicular muscles. Part of it extends across the midventral line and interconnects the left and right girdles. The pelvic girdle is a simple transverse rod of cartilage in the ventral body wall just anterior to the cloaca.

Recent chondrichthyans have narrow fin bases, which allows the fins to rotate and help in turning, braking, and other maneuvers, as well as acting as stabilizers. The pectoral fin has three **basal pterygiophores;** the pelvic fin has two (Fig. 9-4*C* and *D*). The pectoral fin is often described as a **tribasic fin**. The most posterior basal pterygiophore is always the largest and forms the fin axis, and most **radial pterygiophores** attach to it. The location of the fin axis along the posterior border of the fin also is seen in later groups of fishes, including those that gave rise to tetrapods.

Skates and rays among the chondrichthyan fishes have a dorsoventrally flattened body form that is adaptive to the bottom-dwelling mode of life of most species. Their pectoral fins are greatly enlarged and are broadly attached to the body from the head nearly to the middle of the trunk (Fig. 9-4*B*). Skates swim by wave-like undulations of the pectoral fins rather than lateral undulations of the trunk and tail. The tail usually is very slender. When on the bottom, skates move by synchronous thrusts of the cranial part of the pelvic fins. This part of the fin is separated by a deep notch from the rest of the fin, is strong and muscular, and has very limb-like actions (Koester and Spirito, 1999).

Among the early bony fishes, †acanthodians had conspicuous pectoral and pelvic spines on the leading edge of their pectoral and pelvic fins (Figs. 3-8 and 9-5*A*). These spines probably functioned as **cutwaters,** that is, stiff structures that broke the water on the leading edge of the fin and protected the rest of the fin. Traces of basal and radial pterygiophores, which at-

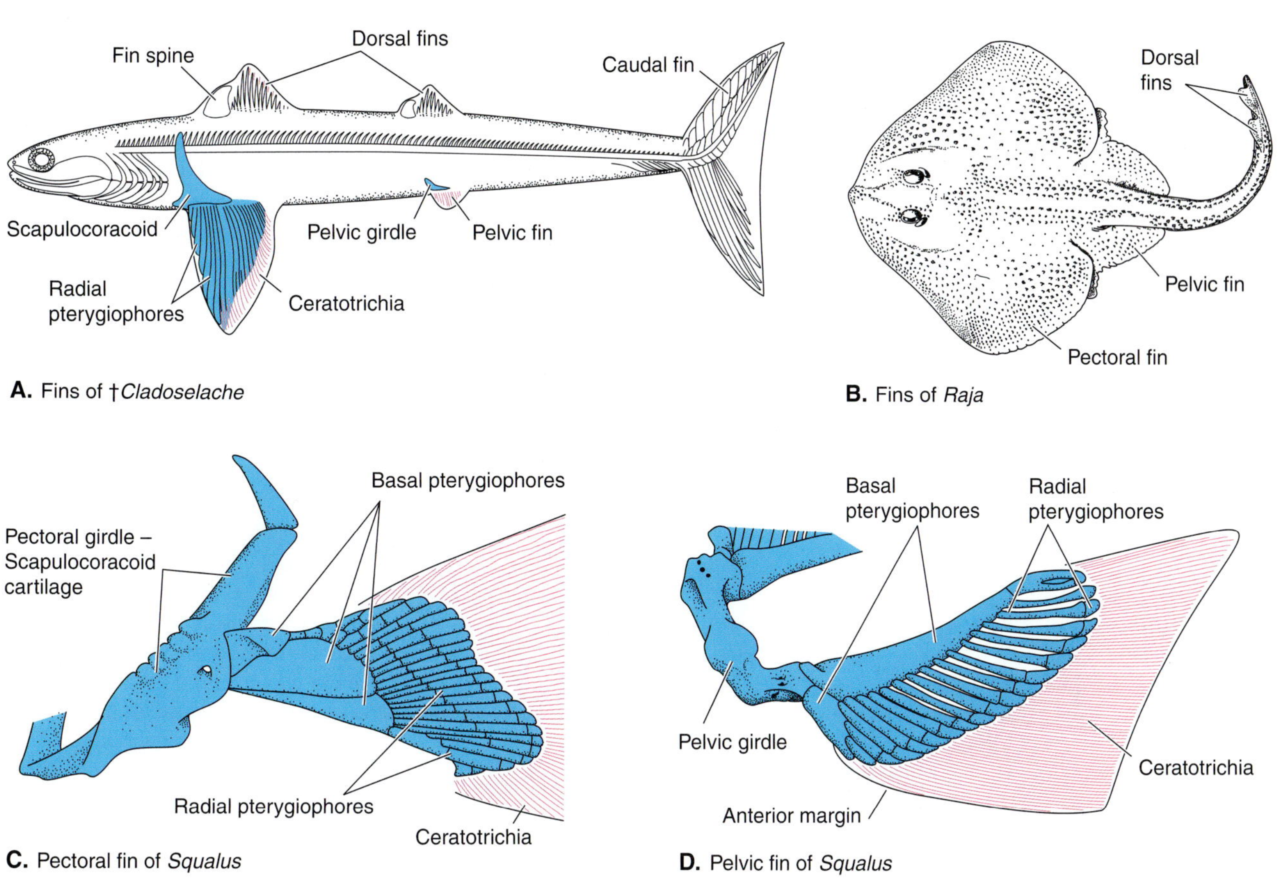

A. Fins of †*Cladoselache*

B. Fins of *Raja*

C. Pectoral fin of *Squalus*

D. Pelvic fin of *Squalus*

FIGURE 9-4
The appendicular skeleton of chondrichthyans. *A,* A lateral view of the skeleton of an early shark, †*Cladoselache. B,* A dorsal view of a female skate, *Raja. C* and *D,* Laterodorsal views of the pectoral (*C*) and pelvic (*D*) appendicular skeletons of a recent, small female shark, *Squalus.* (*A, After Zangerl; C and D, after Jollie.*)

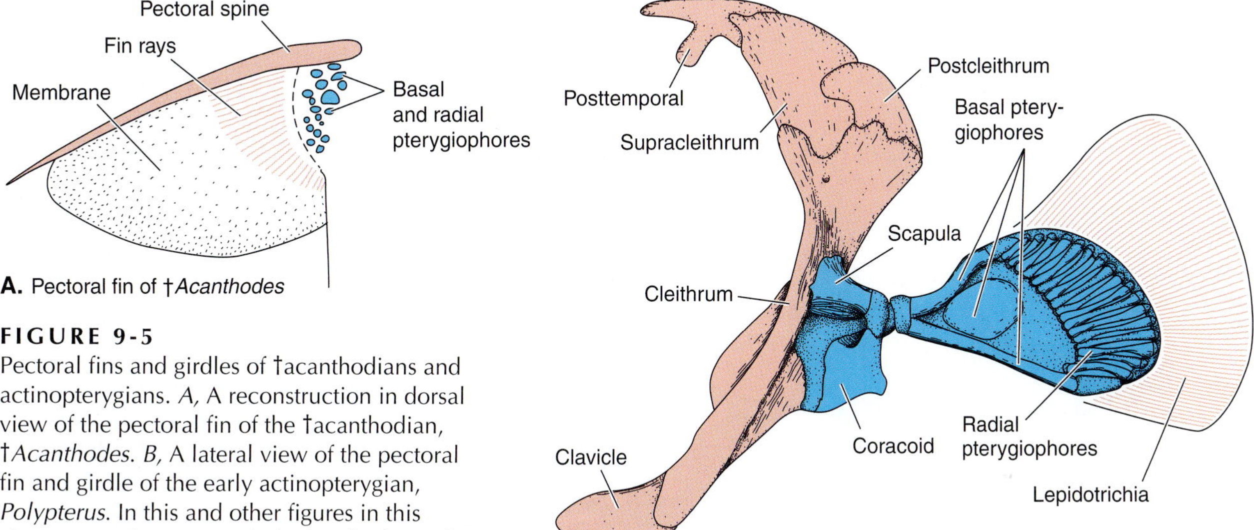

A. Pectoral fin of †*Acanthodes*

B. Pectoral girdle and fin of *Polypterus*

FIGURE 9-5
Pectoral fins and girdles of †acanthodians and actinopterygians. *A,* A reconstruction in dorsal view of the pectoral fin of the †acanthodian, †*Acanthodes. B,* A lateral view of the pectoral fin and girdle of the early actinopterygian, *Polypterus.* In this and other figures in this chapter, cartilage is stippled when both cartilage and cartilage-replacement bone are present in the girdle. (*A, After Stahl; B, partly after Jarvik.*)

tached to a scapulocoracoid, have been found. A series of paired spines lay between the pectoral and pelvic spines, but most did not support fins.

Throughout their evolution, their narrow-based, fan-shaped paired fins have characterized actinopterygians. The pectoral fin of the polypteriform *Polypterus* is representative (Figs. 3-9 and 3-10*A*). It is a tribasic fin composed of three basal pterygiophores, to which many radials attach (Fig. 9-5*B*). Most of the fin is supported by lepidotrichia, which develop in the skin. A small scapulocoracoid cartilage forms the endoskeletal part of the pectoral girdle. A distinct **scapula** and **coracoid** ossify within the scapulocoracoid in *Polypterus* and in many other actinopterygians. A much larger **cleithrum,** to which the scapulocoracoid attaches, forms the major part of the dermal portion of the pectoral girdle. A small **clavicle** extends ventrally from the cleithrum in early actinopterygians, but it is lost in more derived species. Many species have one or more **supracleithral** and **postcleithral** elements. A **posttemporal** bone anchors the pectoral girdle to the back of the skull. Head and trunk move as a unit in fishes. The structure of the pelvic fin is similar to that of the pectoral fin, but the pelvic girdle consists only of a pair of small ventral plates of cartilage or cartilage-replacement bone. The pelvic girdle does not articulate

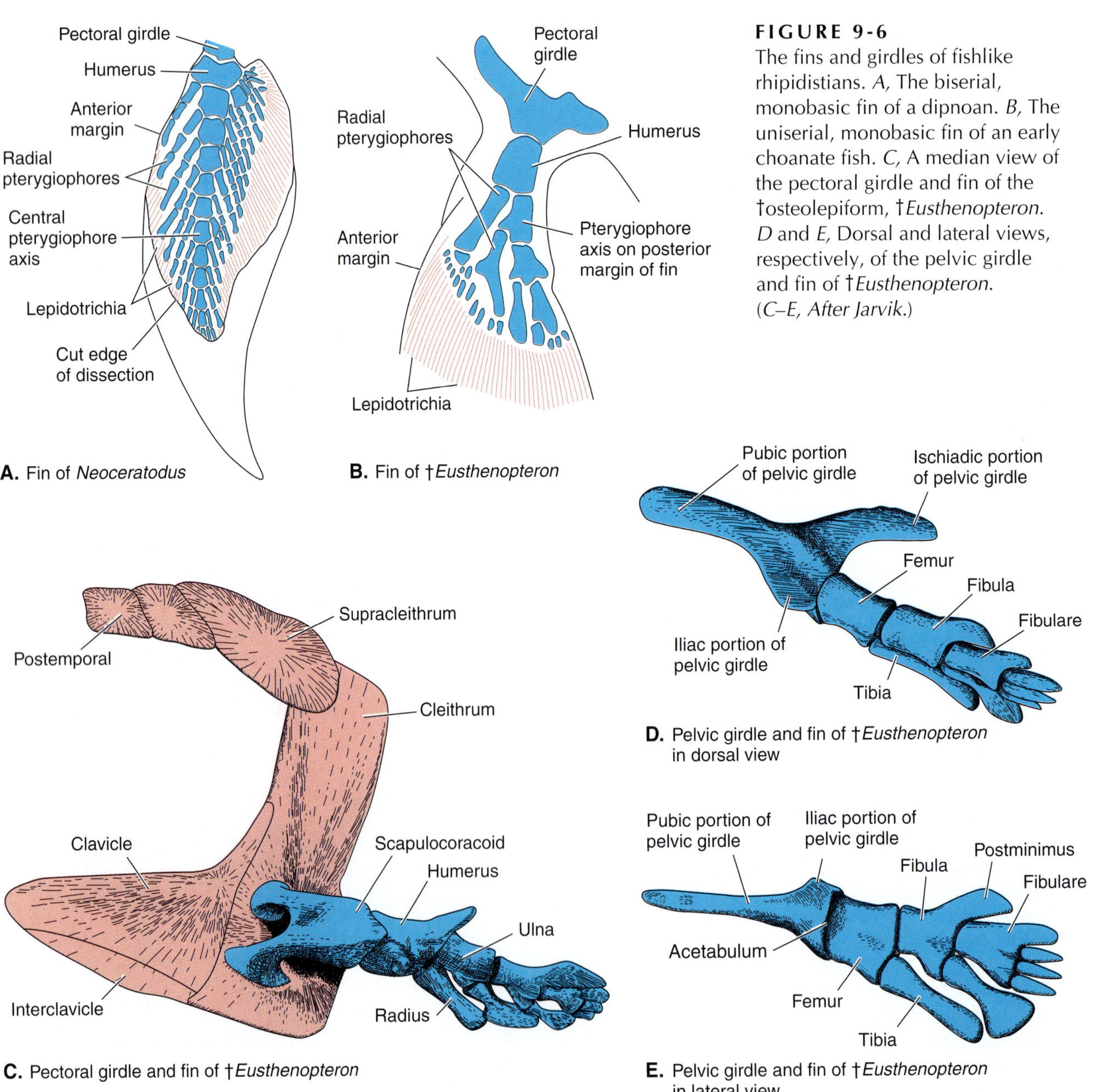

FIGURE 9-6
The fins and girdles of fishlike rhipidistians. *A,* The biserial, monobasic fin of a dipnoan. *B,* The uniserial, monobasic fin of an early choanate fish. *C,* A median view of the pectoral girdle and fin of the †osteolepiform, †*Eusthenopteron.* *D* and *E,* Dorsal and lateral views, respectively, of the pelvic girdle and fin of †*Eusthenopteron.* (*C–E, After Jarvik.*)

with the vertebral column in fishes because the body is supported by water, not by limbs.

Sarcopterygian fishes (Fig. 3-15) have only a single basal pterygiophore in their lobate fleshy fins. These fins are often called **monobasic fins** in contrast to the tribasic fins of most other recent fishes. The monobasic fin of early dipnoans also is a **biserial fin**, for its pterygiophores form a long, central axis with many short, radial elements extending from it to both borders of the fin (Fig. 9-6*A*). This type of fin was called an **archipterygium** (Gr., *arche* = origin + *pteryg-* = fin or wing) by the late 19th-century German anatomist Gegenbaur because he regarded it as ancestral. Gegenbaur believed that girdles were displaced visceral arches and that pterygiophores were enlarged branchial rays. This hypothesis became untenable when anatomists recognized the many fundamental differences between the visceral skeleton, to which visceral arches belong, and the somatic skeleton, to which the appendicular skeleton belongs. The recent Australian lungfish, *Neoceratodus*, retains an archipterygium (Fig. 3-16*D*), but the fins of recent African and South American lungfishes are greatly reduced and form little more than tendril-like structures.

Among the sarcopterygians, coelacanths (*Latimeria;* Fig. 3-16*A*) and early choanate fishes, such as †osteolepiforms (†*Eusthenopteron;* Fig. 3-17*A*) and †elpistostegids (†*Panderichthys;* Fig. 3-17*B*), have a **uniserial monobasic fin.** In this type of fin, the pterygiophore axis extends along the posterior border of the fin, and radial elements extend from the axis only to the anterior border of the fin (Fig. 9-6*B*). This type of monobasic fin is called a **crossopterygium** (Gr., *krossoi* = tassels). The crossopterygian fin resembles a tetrapod limb in many ways because it has a single basal element, which is homologous to the tetrapod humerus (in the pectoral appendage) or femur (in the pelvic appendage). The first radial pterygiophore and the second axial pterygiophore follow the single basal pterygiophore. These two elements are homologous to the two bones in the tetrapod forearm and shin. Distal pterygiophores of uncertain homology follow. The peripheral part of the archipterygium and crossopterygium are supported by dermal lepidotrichia, as in all osteichthyan fishes.

The pectoral girdle of early choanate fishes is similar to that of primitive actinopterygians except for the presence of an **interclavicle,** which connects the two sides ventrally (Fig. 9-6*C*). Some early dipnoans also had an interclavicle, but the pectoral girdle of recent species is much reduced. The partial support of the scapulocoracoid by an enlarged first rib is a unique feature of dipnoians. The pelvic girdle of early choanate fishes consists of a pair of ventral cartilages or bones (Fig. 9-6*D* and *E*). Each has a large, anteroventral **pubic portion,** a smaller, medioventral **ischiadic portion,** and a small, dorsally projecting **iliac portion.** The iliac portion does not extend far dorsally to reach the ribs and vertebral column, as it does in tetrapods.

The Appendicular Skeleton of Early Tetrapods

Limb Terminology

Before considering the origin of the tetrapod limb and its subsequent evolution, you should be familiar with the terminology of the typical five-digit limb, or **cheiropterygium** (Gr., *cheir* = hand), which occurs in most tetrapods. The cheiropterygium has three distinct segments in both the pectoral and the pelvic appendages: (1) the proximal **stylopodium** (Gr., *stylos* = pillar + *podion* = small foot), (2) a middle **zeugopodium** (Gr., *zeugma* = a yolking or joining together), and (3) a distal **autopodium** (Gr., *autos* = self; Fig. 9-7). The stylopodium consists of the upper arm, or **brachium,** in the front leg, and the **thigh** in the hind leg. The brachium contains a single element, the **humerus,** and the thigh contains the **femur.** The zeugopodium consists of the forearm, or **antebrachium,** which contains a **radius** and **ulna,** and the shin, or **crus,** which contains the **tibia** and **fibula.** The ulna and fibula lie on the posterior or lateral border of the limb, depending on limb position. The autopodium consists of the hand (**manus**) or foot (**pes**). The hand or foot is composed of a series of small **podial** elements, which are called **carpals** in the wrist and **tarsals** in the ankle, and the digits extending distally from the podials. The proximal part of each digit, which lies in the palm of the hand or the sole of the foot, is formed by elongated **metapodials** (**metacarpals** and **metatarsals**); the free part of each digit is formed by a series of **phalanges** (Gr., *phalanx* = soldiers in a battle line). The axis of the limb lies along its posterior or lateral border. It extends through the ulna or fibula in the zeugopodium and the fourth digit of the autopodium.

The numerous podial elements are arranged in a proximal and distal series with several central elements, known as **centralia,** between them. The three proximal tarsals and carpals are named according to their positions relative to the adjacent antebrachial or shin bones: **radiale, intermedium,** and **ulnare** in the manus, and **tibiale, intermedium**, and **fibulare** in the pes. Many tetrapods have five **distal carpals** and **distal tarsals,** one for each digit, but often the last two digits share one.

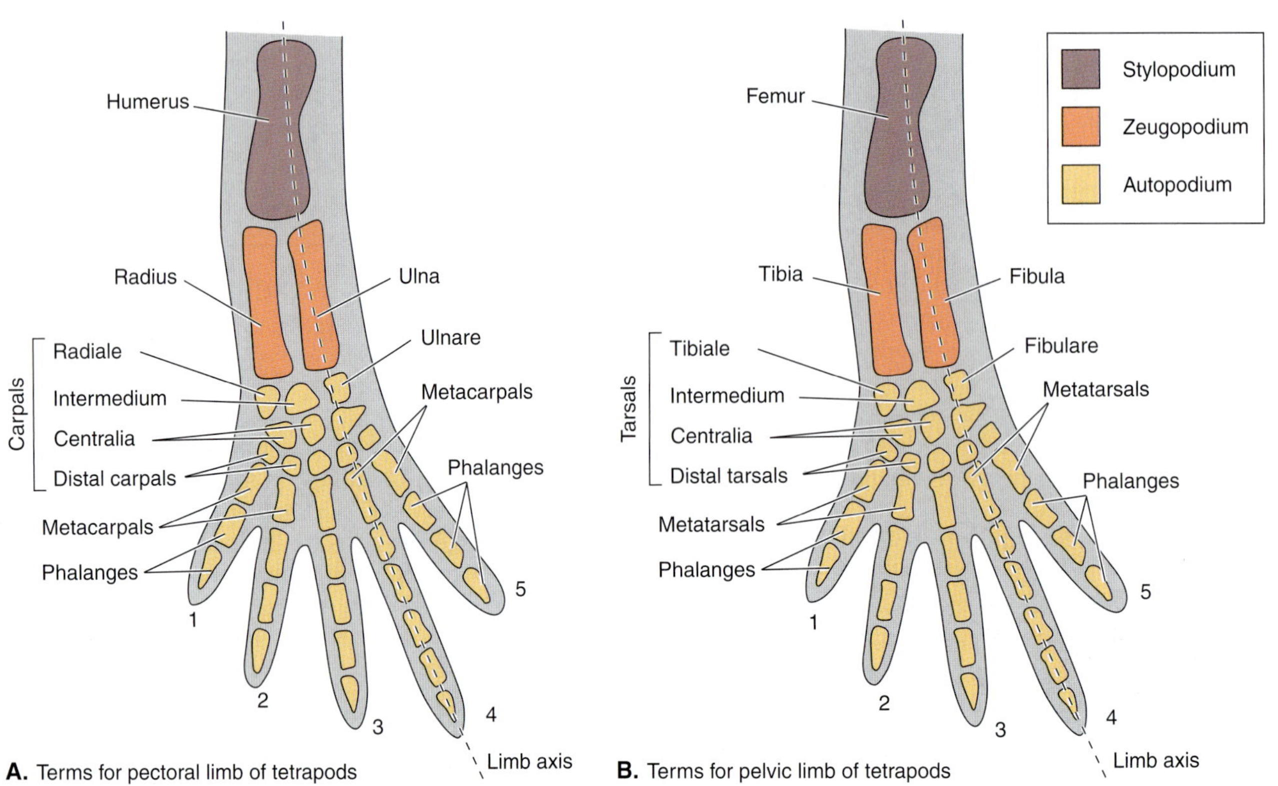

FIGURE 9-7
Diagrams of a generalized five-digit cheiropterygium of a tetrapod, showing terms used for elements in the pectoral (*A*) and pelvic (*B*) appendages. Note that the limb axis extends along the posterior border of the limb and into the fourth digit.

The typical cheiropterygium has five digits. The fifth digit lies on the posterior or lateral side of the hand or foot adjacent to the ulna or fibula. The number of phalanges varies with digit length. The pattern of locomotion of early tetrapods requires an increase in digit length from the medial toward the lateral side of the foot. The last digit decreases abruptly in length. Early reptilomorphs, such as †anthracosauroids (Fig. 3-20), and the ancestral tetrapods from whom they evolved, had two phalanges in the first digit, three in the second, four in the third, five in the fourth, and then a decrease to three or four in the fifth. This is expressed as a **phalangeal formula** of 2-3-4-5-3 or 4.

The Limbs and Girdles of Early Tetrapods

As we have pointed out, the crossopterygium resembles the tetrapod limb in many ways. It is easy to homologize the bones of the stylopodium and zeugopodium of early tetrapods, such as †acanthostegids and †ichthyostegids (Fig. 3-15), with those of an †osteolepiform (Fig. 9-8). The axis of the fin or limb runs along its posterior or lateral border. In the pectoral appendage the humerus and ulna, and possibly the ulnare, lie along this axis. In the pelvic appendage the axial elements are the femur, fibula, and possibly the fibulare. The radius or tibia evolves from the first radial element branching from the crossopterygium axis. But the numerous elements in the tetrapod autopodium, possibly excepting the ulnare and fibulare, defy easy comparison with elements in the crossopterygium. The structure of the manus and pes of early tetrapods is incompletely known, but it is clear that far more cartilages or cartilage-replacement bones exist in the autopodium than do cartilaginous elements in the crossopterygium, and autopodial elements could not have evolved from dermal lepidotrichia. The autopodium likely is a newly evolved segment of the appendage (Focus 9-1).

The numbers of toes in the earliest tetrapods were variable (Fig. 3-17*C*). Biologists have long assumed that the five digits of the cheiropterygium were a synapomorphy for tetrapods, that is, the primitive ancestral number. Apparently not so! †*Acanthostega* had eight fingers in the hand (its foot is unknown), and †*Ichthyostega* had seven toes in the foot (Coates and Clark, 1990). †*Tulerpeton* had six fingers and toes (Lebedev and Coates, 1995).

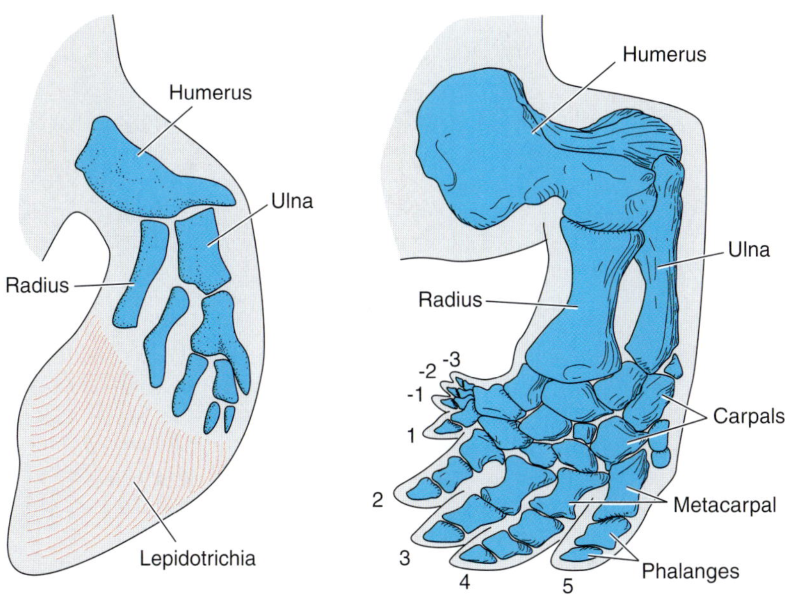

A. Pectoral fin of †*Eusthenopteron* **B.** Pectoral limb of †*Acanthostega*

FIGURE 9-8
Changes in the pectoral appendage during the transition from water to land. *A,* The crossopterygium fin of a choanate fish. *B,* The limb of an early tetrapod (an †acanthostegid).

The humerus and femur of †acanthostegids and †ichthyostegids were short and stocky bones because these animals did not raise themselves far from the ground when they walked, and their limbs were splayed (Figs. 9-8*B* and 9-9*A*). That is, when the animal was standing, its humerus and femur extended laterally from the girdles. The bones of the antebrachium and shin were flexed and extended vertically to the ground. A sharp bend or extension at the wrist and ankle placed the palm of the hand and sole of the foot nearly flat on the ground. Large processes for the attachment of muscles usually lay beside the heads of the humerus and femur on the proximal end of these bones. Prominent ridges for the attachment of other muscles often extended down the shaft of the humerus. The radius and ulna of the forearm, and the tibia and fibula of the shin, were also short and stocky, and they were nearly the same size and shared equally in the transfer of body weight.

Because the girdles of terrestrial vertebrates transfer body weight to the appendages and receive the thrusts of the legs, they must be strong. The endoskeletal part of the pectoral girdle of †acanthostegids and †ichthyostegids expanded greatly relative to the size of the girdle in fishes and had become the major part of the girdle. It usually ossified as a single **scapulocoracoid** element (Fig. 9-9*B*). As was the case in the choanate fishes that gave rise to tetrapods, the lateral surface of the scapulocoracoid bore a socket, the **glenoid fossa,** which received the humeral head. The scapular region extended dorsally from the glenoid fossa, and the broad, ventral coracoid region extended medially toward the midventral line. The coracoid was perforated by a **coracoid foramen** for a nerve. The dermal part of the pectoral girdle was much smaller than in early choanate fishes. The cleithrum remained, but the postcleithral and supracleithral elements and the posttemporal bone were lost. This freed the pectoral girdle from the skull, and head and trunk could now move independently. The clavicle remained. The interclavicle of early choanates also remained and connected the left and right sides of the girdle ventrally, thereby strengthening the pectoral arch. A terrestrial vertebrate does not have a direct skeletal attachment between the pectoral girdle and the vertebral column. Weight is transferred from the vertebral column to the scapula by a muscular sling (see Fig. 11-11). Ventrally, the girdle of recent amphibians connects with the sternum, but the sternum of early tetrapods is not known.

The pelvic girdle of an early tetrapod (Fig. 9-9*C*) had a prominent **ilium,** which extended dorsally to articulate directly with the one pair of **sacral ribs,** which in turn articulated with a **sacral vertebra.** Part of the ilium also extended caudally and served for the attachment of powerful tail muscles. The ventral part of the pelvic girdle was greatly expanded and ossified as an anterior **pubis** and posterior **ischium.** An **obturator foramen** for the obturator nerve perforated the pubis. The union of pubis and ischium of opposite sides at a midventral **pelvic symphysis** further strengthened the girdle. The head of the femur articulated with a socket known as the **acetabulum,** located at the area where the three pelvic bones united.

FOCUS 9-1 *A New Hand and Foot?*

A long-favored hypothesis for the origin of the autopodium is that a multiplication of radial elements in the crossopterygium gave rise to the podial elements and digits. The axis of the fish fin continued through the fourth digit (Fig. A, *left*). A postaxial radial gave rise to the fifth digit, and newly evolved preaxial radials gave rise to the other digits.

Recent studies of the embryonic development of the zebrafish (*Danio*) fin and the chick hind leg cast doubt on this hypothesis (Sordino et al., 1995; Nelsen et al., 1996). An inductive interaction between the apical ectoderm of the limb bud and the underlying mesenchyme causes the limb bud to enlarge (Fig. A, *right*). Within the limb bud, the gene *HoxD-11* is expressed along the limb axis (i.e., the protein it produces can be identified), so it is involved in the formation of elements along the axis. In the fish, the expression of this gene is confined to the posterior margin of the developing fin, but in the chick hind leg, after extending down the posterior border of the limb, the zone of gene expression turns sharply toward the preaxial border. Applying these findings to the early tetrapod limb (Fig. A, *center*), the limb axis crosses the distal part of the wrist or ankle. The digits develop on the postaxial side of the changed axis. The digits, therefore, cannot be radial elements, which develop on the preaxial side of the axis; rather, the digits and most of the podial elements are **neomorphs** (Gr., *neos* = new + *morphe* = form), that is, completely newly evolved structures. Some other *Hox* genes also have changed their expression and participate in this transformation.

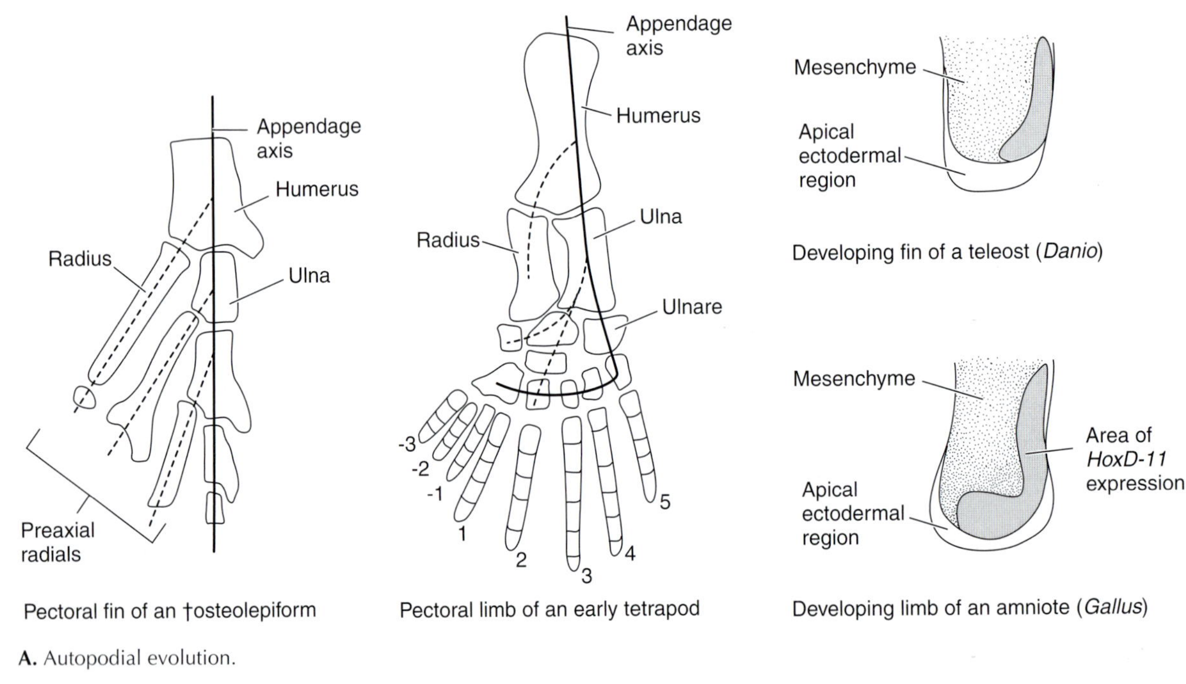

A. Autopodial evolution.

Origin of the Tetrapod Limb and Girdle

The similarities of the crossopterygium and girdles of choanate fishes to the limbs and girdles of early tetrapods are quite striking. Indeed, they contribute to the evidence that tetrapods evolved from choanate fishes rather than dipnoians. It is easy to visualize the morphological transition, but the transformation of a crossopterygium into a tetrapod limb did not come about abruptly. The choanates ancestral to terrestrial vertebrates probably lived in freshwater environments subject to stagnation and periodic drought (Chapter 3). Their sturdy, paired fins doubtless enabled them to push through shallow, vegetation-choked water; dig into mud at the bottoms of ponds; and probably briefly emerge on land to escape enemies or search for food. As these forays on land became more extensive, selection would have favored stronger appendages and girdles and their transformation into the ancestral tetrapod condition.

Although the early tetrapods had limbs and certainly could move out of the water onto the land, the presence of a caudal fin (Fig. 9-9*A*) indicates that they also spent a great deal of time in the water. Their broad hand and foot, containing six to eight digits, probably served as paddles when they were in the water. Reduc-

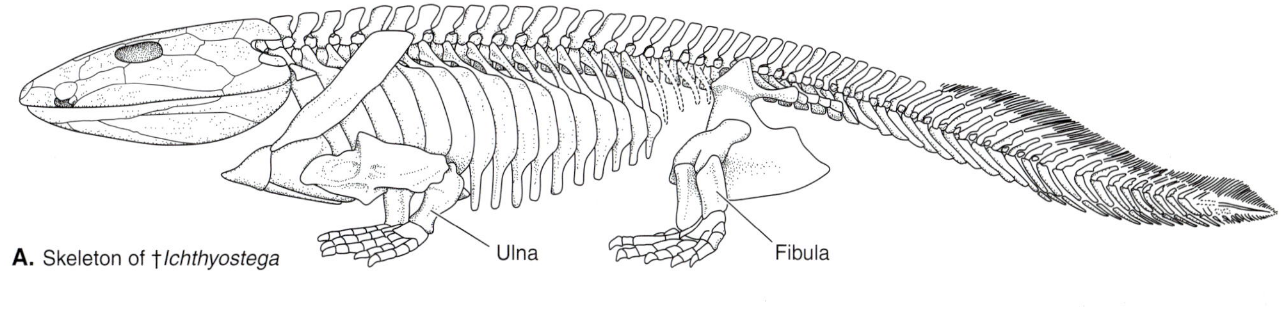

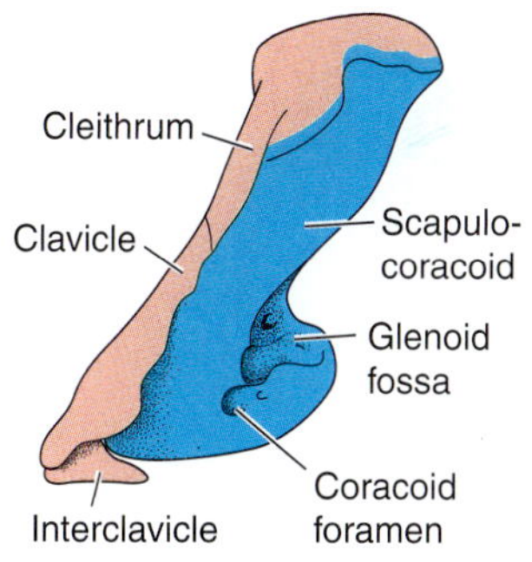

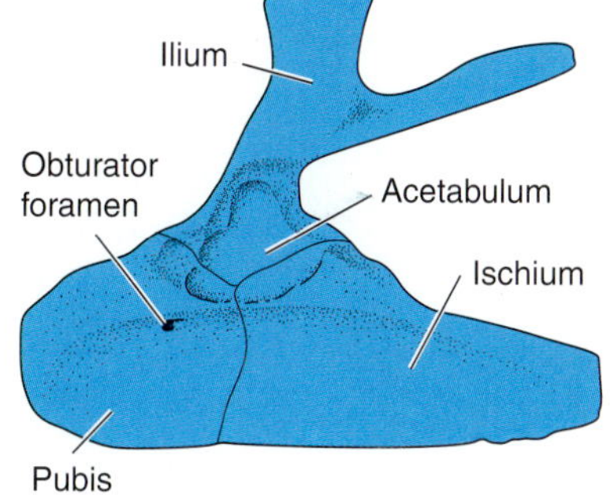

FIGURE 9-9
Lateral views of the girdles and limbs of early tetrapods. *A,* †*Ichthyostega,* one of the earliest known tetrapods; *B,* The pectoral girdle of †*Eryops*. *C,* The pelvic girdle of †*Archeria*. *Eryops* and *Archeria* are very early amphibians belonging to the Eryopoidea. (*A, After Jarvik; B and C, after Romer and Parsons.*)

tion in digit number occurred in later, more completely terrestrial neotetrapods (Fig. 3-15). The number of digits stabilized at five in those early tetrapods that gave rise to reptilomorphs. Recent amphibians have no more than four fingers, but most of them have five toes. The fossil record for the origin of recent amphibians is incomplete, so we cannot be certain whether they lost the first finger or evolved from a group of early tetrapods that had a different pattern of digit stabilization.

The Limbs and Girdles of Other Amphibians and Early Reptilomorphs

We will discuss locomotion thoroughly in Chapter 11, but you need to understand the basic pattern of walking to appreciate the structure of the tetrapod appendicular skeleton. The action of appendicular muscles elevates the body from the ground but only slightly so in most amphibians and early reptilomorphs[2] because their limbs are splayed and the trunk is carried close to the ground. Appendicular muscles, assisted by lateral undulations of the axial skeleton in many amphibians and early reptilomorphs (Fig. 8-11*A*), also help to advance an appendage in preparation for a step. When a foot is placed on the ground, retractor appendicular muscles, assisted by lateral trunk and tail undulations in some species, move the body forward relative to the foot placed on the ground. During slow walking, only one foot is removed from the ground and advanced at a time. The other limbs are in different stages of retraction. The feet of salamanders and lizards tend to point laterally and slightly anteriorly when first placed on the ground, but the feet rotate laterally during a step.

The structures of the bones of the limbs and girdles of amphibians and of early reptilomorphs are variations on the pattern established in early tetrapods. The long bones of the limbs become longer and more slender in recent amphibians and early reptilomorphs than they were in basal tetrapods (Fig. 9-10). The ulna of some early reptilomorphs begins to develop a process, the **olecranon,** which extends dorsally distal to the elbow joint. Powerful limb-extensor muscles attach onto it. The long bones of the antebrachium and crus sometimes fuse in species where extra rigidity is needed, as in the limbs of frogs (Fig. 11-27*B*).

The earliest amphibians, the †eryopids (Fig. 3-18), had a well-ossified carpus and tarsus containing the elements that we described earlier (Fig. 9-7). The number tends to be reduced in later amphibians and in reptilomorphs. The carpals and tarsals are unossified in many

[2]The early reptilomorphs we are considering are turtles and lepidosaurs in the sauropsid line of evolution, but early reptilomorphs in the synapsid line of evolution had very similar appendicular skeletons.

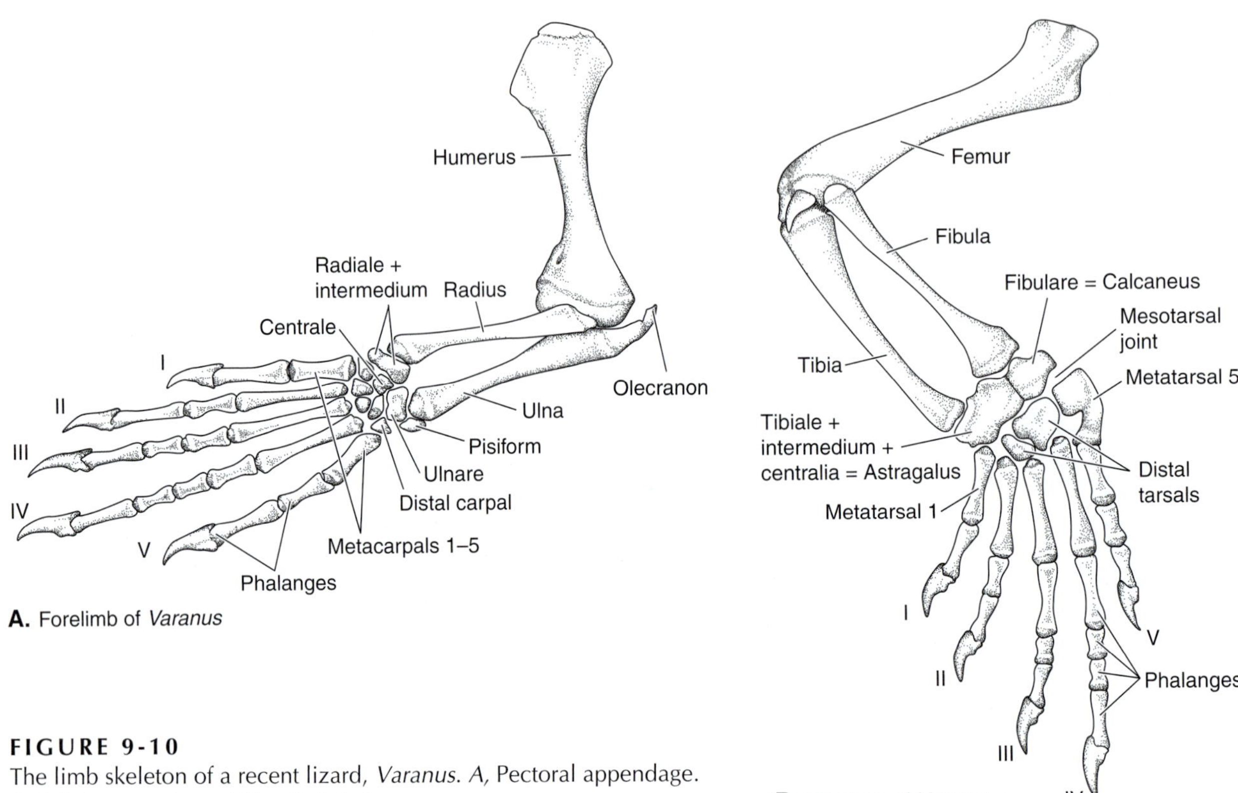

FIGURE 9-10
The limb skeleton of a recent lizard, *Varanus*. *A*, Pectoral appendage. *B*, Pelvic appendage. (*After Bellairs.*)

recent amphibians (lissamphibians), but the tibiale and fibulare in the foot of frogs are ossified and elongated, giving the foot more leverage in developing the powerful thrust of the hind legs in swimming and jumping (Fig. 11-27*B*).

A distinctive feature of the pes of reptilomorphs is the fusion of the tibiale, intermedium, and a centrale into a single element called the **astragalus**.[3] In the sauropsid line of reptilomorph evolution, the astragalus fuses or articulates very securely with the fibulare (often called the **calcaneus** in reptilomorphs) to form a single unit (Fig. 9-10*B*). The functional ankle joint lies between this unit and the distal tarsals. Because the joint lies near the center of the ankle, it is called a **mesotarsal joint.** A mesotarsal joint increases the capacity for foot rotation. Although the hind foot points nearly anteriorly when placed on the ground, it rotates laterally and eventually posteriorly during a step. The functional ankle joint lies between the zeugopodium and autopodium in amphibians and mammals. Reptilomorphs also usually have a small bone, called the **pisiform,** on the ulnar side of the wrist (Fig. 9-10*A*). The pisiform is not a supporting part of the skeleton but rather a **sesamoid bone** (Gr., *sesamon* = sesame seed + *eidos* = form). Sesamoid bones lie in the tendons of certain muscles and alter the direction of pull of the muscle, as the pisiform does, or facilitate the movement of a tendon across a joint.

As pointed out earlier, recent amphibians never have more than four digits in their hand, although their foot usually has five digits. They typically have only one to three phalanges in their digits. Most reptilomorphs retain the phalangeal formula of 2-3-4-5-3 or 4 seen in ancestral species. An increase in digit length from the medial to the lateral side of their autopodium helps to maintain digit contact with the ground during a step.

The endoskeletal part of the pectoral girdle ossifies as a distinct scapula and coracoid in recent amphibians and in sauropsids (Fig. 9-11*A*, *C–E*). The coracoid should be called an **anterior coracoid** to distinguish it from an additional **posterior coracoid,** which appears in the synapsid line of evolution leading to mammals. The anterior coracoid can be distinguished by a small **coracoid foramen** for a pectoral nerve. The dermal part of the pectoral girdle is reduced. The clavicle and interclavicle usually are retained, but the cleithrum is lost except in a few primitive frogs. The clavicles and interclavicles are incorporated into the anterior part of the plastron in turtles; and a few sauropsids, including crocodiles, lose the clavicle (Fig. 9-11*E*).

Most recent amphibians and sauropsids also have a midventral **sternum** attached to the caudal part of the

[3]Turtles have lost the tibiale, so their astragalus consists of the intermedium and a centrale (Burke and Alberch, 1985).

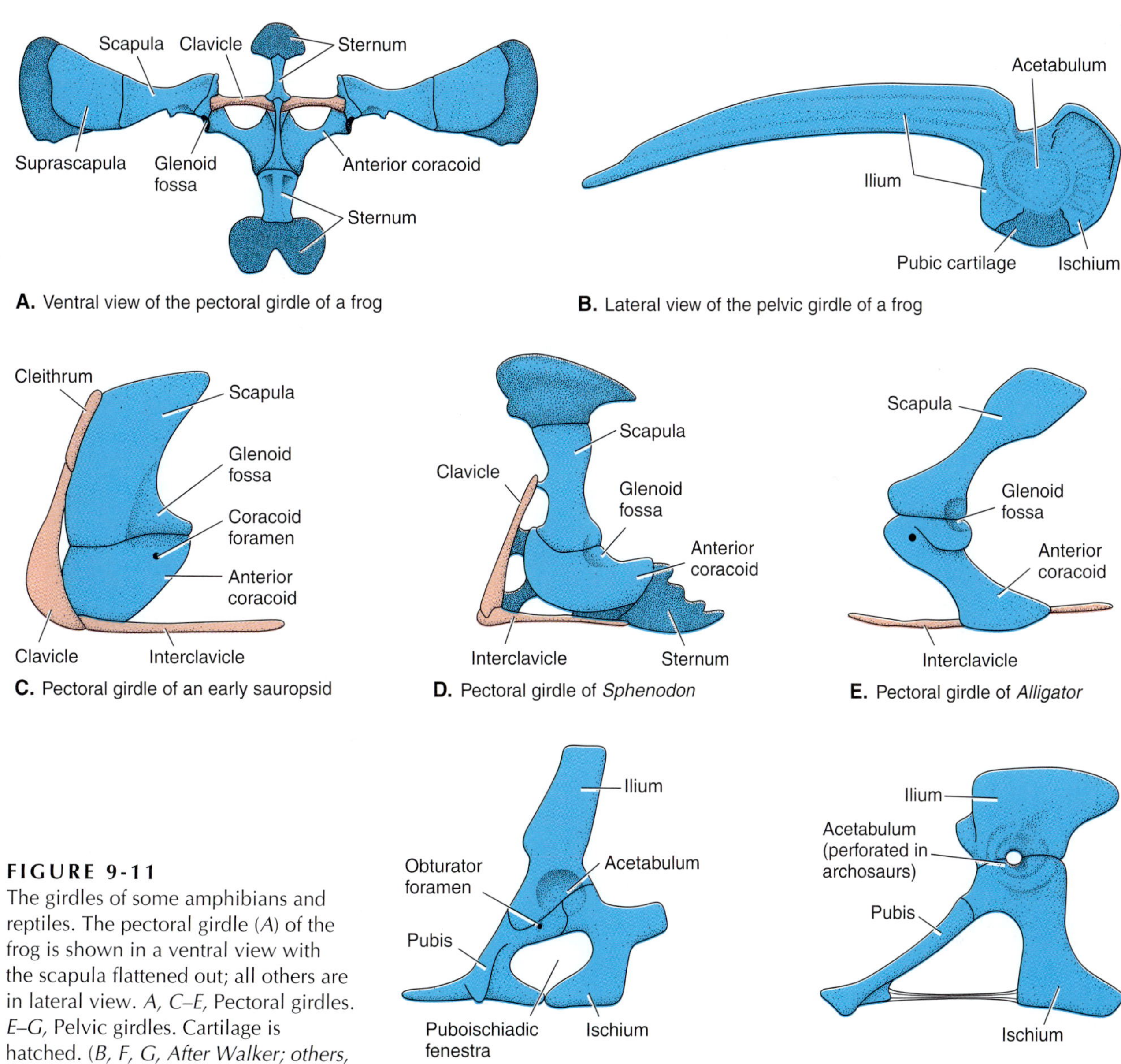

FIGURE 9-11
The girdles of some amphibians and reptiles. The pectoral girdle (*A*) of the frog is shown in a ventral view with the scapula flattened out; all others are in lateral view. *A, C–E,* Pectoral girdles. *E–G,* Pelvic girdles. Cartilage is hatched. (*B, F, G, After Walker; others, partly after Jollie.*)

coracoid plate. It is cartilaginous in most species but partly calcified in diapsids. As we have seen (Chapter 8), the anterior trunk ribs of sauropsids also attach to the sternum, but the very short ribs of amphibians do not. No fossil evidence suggests a sternum in ancestral amphibians and reptilomorphs, but a cartilaginous sternum, which might not have been fossilized, cannot be ruled out.

A peculiarity of the pelvic girdle of recent amphibians is the failure of the pubic region to ossify (Fig. 9-11*B*). Moreover, frogs have an exceptionally long ilium, which is related to the development of powerful thrusts by their hind legs during swimming and jumping (Chapter 11). A **puboischiadic fenestra** develops between the pubis and ischium in recent sauropsids (Fig. 9-11*F* and *G*). Certain pelvic muscles arise from its periphery. The sauropsid ilium articulates with at least two sacral ribs and vertebrae.

As sauropsids entered the many habitats and ecological niches available to them, their methods of locomotion and the structure of the appendicular skeletons changed considerably. For example, extinct †plesiosaurs (Fig. 3-22*B*) evolved large, oar-like appendages as they adapted to the marine environment, but the appendages of the marine †ichthyosaurs (Fig. 3-22*A*), which swam by fishlike undulations of the trunk and tail, became small, stabilizing keels (Focus 9-2). Snakes evolved long, slender, undulating trunks, and most species have lost their appendicular skeletons completely. Only traces of the femora remain in

FOCUS 9-2 A Limb Becomes a Fin

We have discussed how a fish paired fin became a limb. During the evolution of the extinct, fishlike †ichthyosaurs, evolution proceeded in the opposite direction: a limb became a paired fin used not for propulsion but, as in a fish, for stability and maneuvering. The fin moves as a unit, so joints within it are reduced. Stylopodial and zeugopodial bones become short and chunky, and the autopodium elongates (Fig. A). The resemblance of the pectoral fin to an arm is very striking in an early †ichthyosaur (†*Mixosaurus*, of the Middle Triassic). All of the individual elements, including the pisiform, can be recognized. Only five fingers are present, but the number of phalanges has increased in the anterior fingers, a condition called **polyphalangy**. The individual elements are shorter and more closely united in a late †ichthyosaur (†*Ichthyosaurus* from the Lower Jurassic). The numbers of digits has increased (**polydactyly**), as has the number of phalanges. Remarkably similar changes have occurred in the convergent evolution of a limb to a fin in cetaceans.

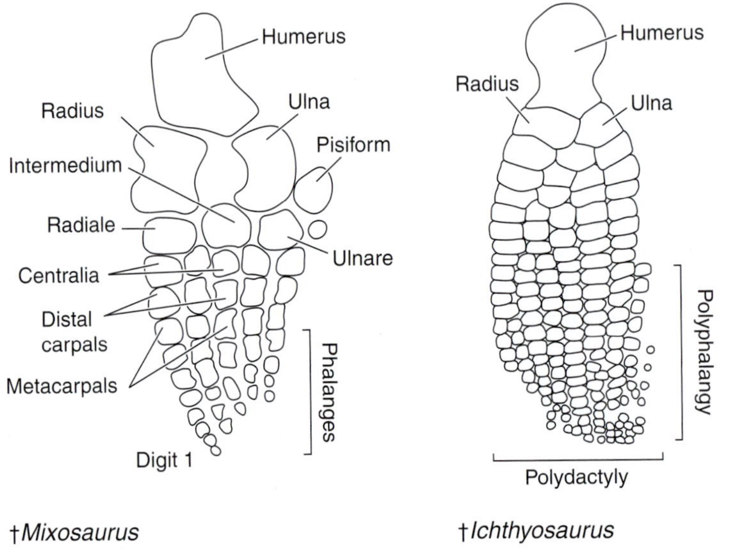

A. Transformation of a limb into a paired fin during the evolution of †ichthyosaurs.

pythons and boa constrictors in the form of **anal spurs.** The spurs lie beside the cloacal aperture, are larger in males than in females, and play a role in courtship behavior. Loss of the appendicular skeleton probably was originally an adaptation for burrowing, but most snakes now live above ground. The flying †pterosaurs evolved wings (see next section).

The Appendicular Skeletons of Birds and Other Flying Vertebrates

Wings and the Pectoral Girdle

†Pterosaurs (a group of extinct reptilomorphs; Fig. 3-25), birds, and bats adapted for powered flight and evolved wings. In each group the wing is a modified pectoral appendage, but differences in wing construction indicate that the wings evolved independently (Fig. 9-12). †Pterosaur wings are large membranes that attach to the side of the body and proximal part of the hind leg. The wings were supported and moved by the arm and a greatly elongated fourth finger. The fifth digit and the terminal, claw-bearing phalanx of the fourth digit are lost. The first three digits are free of the wing and bear grasping claws. A new **pteroid bone** in the wrist supports a forward extension of the wing membrane. Bat wings are also membranes, but the last four elongated fingers support it. The first digit bears a grasping hook and remains free of the wing. Bat wings attach along the side of the body and include the small hind legs and sometimes the tail.

The flying surface of birds' wings is not a membrane; rather, it is formed by long flight feathers that attach along the posterior border of the humerus, ulna, and hand (Figs. 8-17*A* and 9-12*B*). Tail feathers arising from the enlarged caudal pygostyle form another flying surface that is used primarily at low air speeds and in maneuvering. Feathers are modified horny scales (Chapter 6) and form a very lightweight and adjustable flying surface, which is easily repaired because worn feathers are replaced periodically. The bones of a

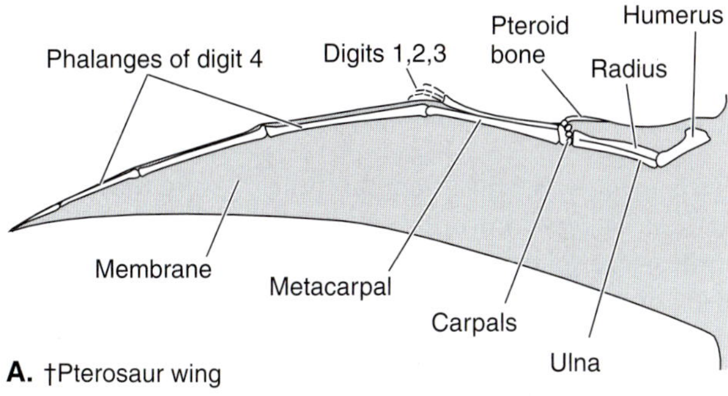

A. †Pterosaur wing

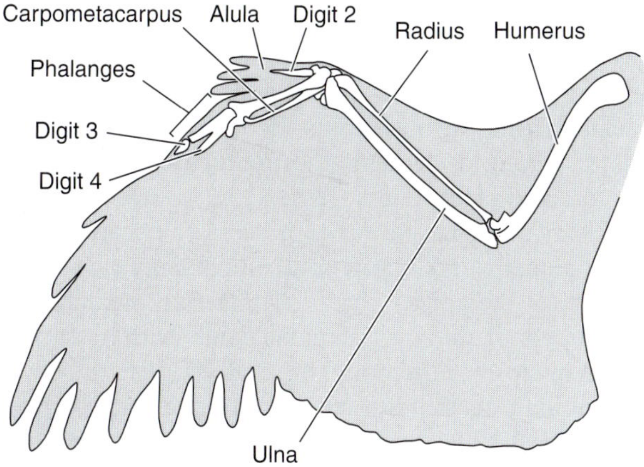

B. Bird wing

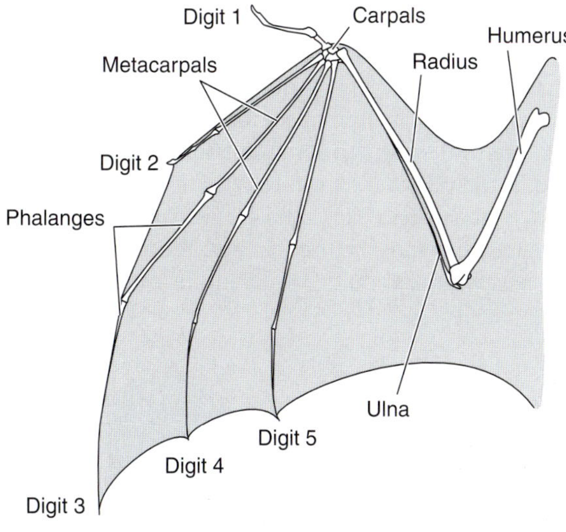

C. Bat wing

FIGURE 9-12
A–C, Dorsal views of extended vertebrate wings. (*After Norberg.*)

bird's wings, like the rest of the skeleton, are exceptionally lightweight but strong. Although the surface cortical bone is thin, internal struts and ridges along stress lines provide strength where needed. An air sac from the lungs extends into the humerus through a foramen at its proximal end. No rotation occurs at the elbow, and the radius and ulna extend parallel to each other to the wrist.

The bird hand is greatly modified. Only three fingers are present, and they are reduced. Which fingers remain is in question. The phalangeal formula in specimens of †*Archeopteryx,* the first known bird, is clearly 2-3-4. This phalangeal formula suggests that the fingers are the first, second, and third ones. But the pattern of cartilage formation in the developing hand of many amniotes, including a variety of birds, shows clearly that the arm axis runs through the humerus and into digit 4. (We noted this axis in discussing the origin of the tetrapod limb.) This axis extends through the last digit in the developing bird wing, which identifies the bird digits as 2, 3, and 4 (Burke and Feduccia, 1997). Phalangeal formulas apparently are not as definitive in establishing homologies as we have believed. Only two distinct carpals remain in the adult wrist, but these allow the hand and the feathers it bears to rotate relative to the rest of the wing. The other carpals and the metacarpals of the posterior two fingers have fused to form a long **carpometacarpus.** The most anterior digit (number two) can move independently of the other two and bears a small, distinct tuft of feathers, known as the **alula** (L., *alula* = little wing). The alula can be elevated from the rest of the wing to form an air slot that is important in increasing lift at certain stages of flight.

The pectoral girdle and thoracic skeleton of birds form a strong support for the action of the wings (Fig. 8-17*A*). The scapula is a narrow blade, the anterior coracoid is a strong strut that braces the shoulder joint, and the two clavicles have fused together ventrally to form the wishbone or **furcula** (L., *furcula* = small fork). Major flight muscles arise from the broad, keeled sternum. †Pterosaurs and bats also have a broad sternum but a much lower keel than seen in birds.

The Pelvic Girdle and Appendage

Because the pectoral appendages are wings, birds are bipeds. Indeed, bipedalism must have evolved before flight because the front legs could not have been modified as wings if they were needed in quadrupedal locomotion. Bipedal adaptations of birds include long and strong hind legs and an exceptionally strong pelvic girdle. Legs and girdle support all the weight of the bird when it is standing and the legs also act as shock absorbers when it lands. The pubis has turned caudally beside the ilium, and all the pelvic bones are firmly united (Fig. 8-17*A*). The long ilium is fused with the synsacrum. The midventral pelvic symphysis has been lost, so the pelvic canal between the two halves of the girdle is much larger than in other terrestrial vertebrates. This permits a caudal displacement of the viscera, which, together with a shortening of the trunk, brings the center of gravity of the bird over the hind legs. A large pelvic canal also allows for the passage of large eggs with fragile shells.

The hind legs of birds have rotated from the primitive splayed position to a position beneath the body. The head of the femur necessarily shifts to the medial side of the proximal end of the bone. Many birds, like human beings and many other bipeds, maintain balance on one leg when the other swings forward by adducting the leg. The supporting leg is pulled toward the midventral line, and body weight is shifted over this leg so that the foot is directly beneath the body's center of gravity. Many other birds hop on both legs.

The tibia is the primary supporting element of the shin, and the fibula is reduced to a thin splint needed for the attachment of certain muscles (Fig. 8-17*A*). The ankle joint, as in other sauropsids, is a mesotarsal joint. Because the legs are directly beneath the body, the only movement at the ankle is a hinge action. The joint is strengthened by the fusion of the proximal tarsals with the tibia. Accordingly, the tibia is most accurately described as a **tibiotarsus.** The fifth toe has been lost, but the metatarsals of the remaining toes and the distal tarsals are fused to form a long **tarsometatarsus.** The remaining toes have the characteristic early reptilomorph phalangeal formula of 2-3-4-5. The first toe usually is turned caudad; this increases stability.

The hind legs of bats are small and their feet form hooklike structures by which bats cling upside down when at rest. The hind legs of †pterosaurs are relatively long and slender. We are uncertain as to just how they were used, but recently discovered tracks of †pterosaurs suggest that they walked awkwardly with their hind limbs splayed (Unwin, 1996).

The Evolution of the Mammalian Appendicular Skeleton

Early synapsids, the †"sailed reptiles," were among the earliest of reptilomorphs (Figs. 3-20 and 3-31). Their limbs were splayed, and the humerus and femur moved back and forth close to the horizontal plane, as they did in †ichthyostegids and other early tetrapods. Their limb bones were also short and chunky and resembled those of †ichthyostegids (Figs. 9-8*B* and 9-9*A*). Their girdles were greatly expanded ventrally to accommodate the powerful limb adductor musculature, which was needed to raise them from the ground. A technical, but phylogenetically important, difference between synapsids and sauropsids was the ossification in synapsids of a second element, the **posterior coracoid,** in the coracoid plate (Fig. 9-13*A*).

Advanced synapsids, the therapsids, became more active animals and began to acquire many of the appendicular features that characterize generalized quadruped mammals. Mammals walk with their legs pulled in close to the body, or rotated under the body, and moving back and forth close to the vertical plane (Fig. 9-14). The elbow points posteriorly and the knee, anteriorly. We discuss the functional anatomy of locomotion later (Chapter 11), but this limb position evidently provides better mechanical support and enables the limbs to swing through longer arcs, thus increasing the efficiency of locomotion.

Changes in girdle and limb structure are correlated with the changes in limb position and movements. A strong, ventral adductor musculature is no longer needed in the pectoral region to raise the body from the ground and support it. Dorsal musculature, on the other hand, expands because it becomes more important in bracing the shoulder joint and in the fore and aft

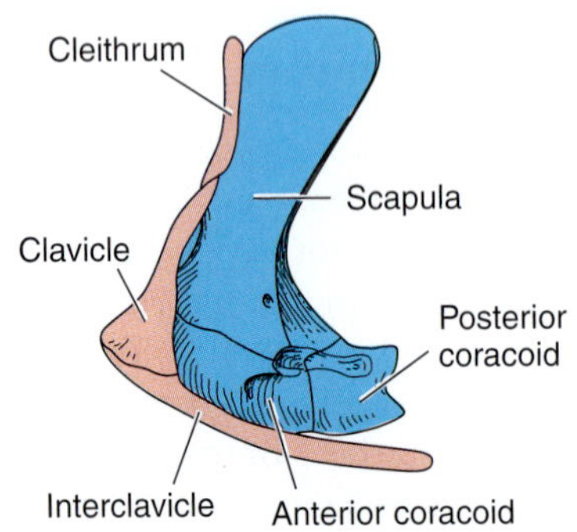

A. Pectoral girdle of †*Dimetrodon*

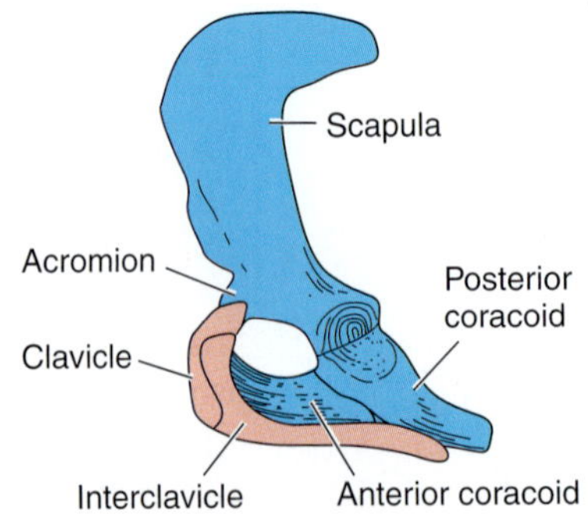

B. Pectoral girdle of *Ornithorhynchus*

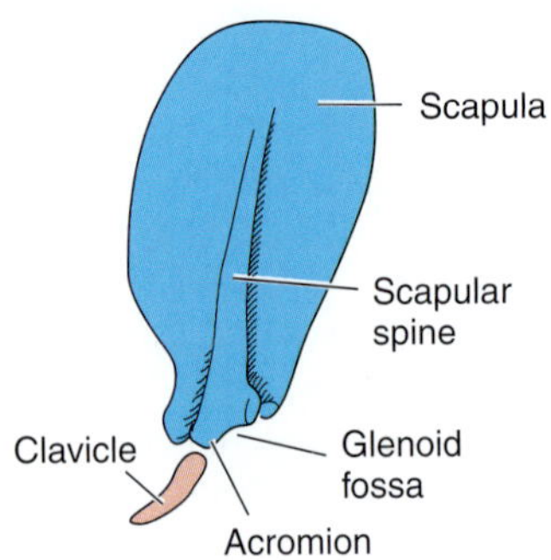

C. Pectoral girdle of *Didelphis*

FIGURE 9-13
A–C, Lateral views showing stages in the evolution of the mammalian pectoral girdle. (*A and B, After Jollie; C, after Romer and Parsons.*)

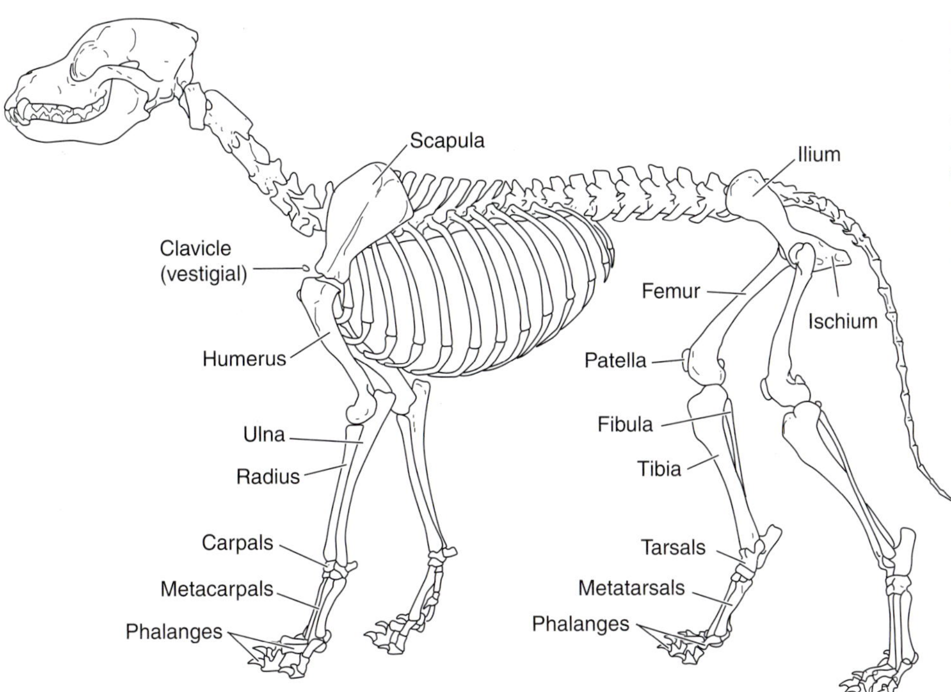

A. Skeleton of a dog (*Canis*) in standing posture

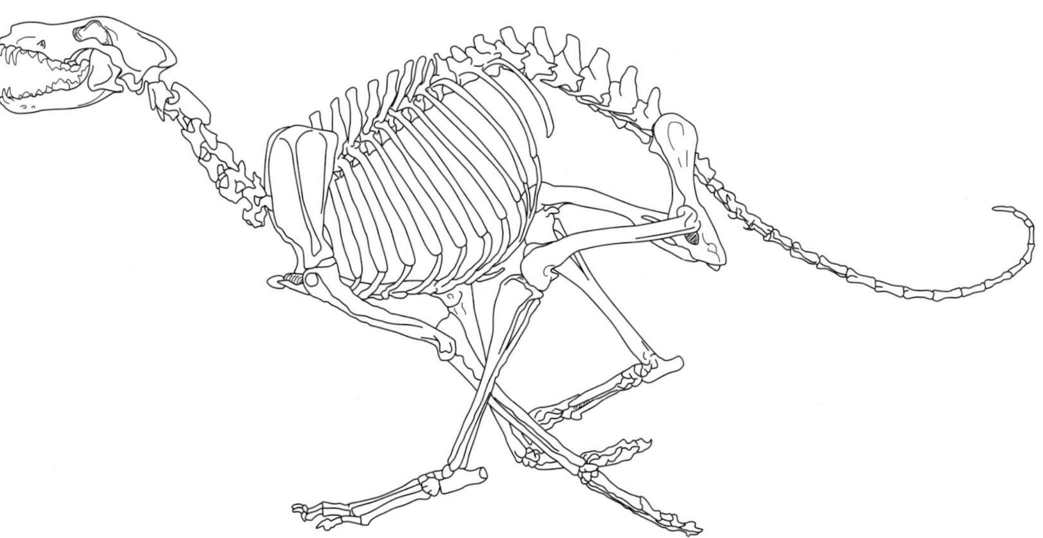

B. Skeleton of a dog (*Canis*) in flight stage of a full gallop

FIGURE 9-14
The skeleton of a dog standing (*A*) and in the flight stage of a gallop (*B*). The limbs are carried under the body, and the limb bones are long and slender. The galloping stage illustrates the great flexibility in the postcranial skeleton. (*A, After Miller et al.*)

swing of the foreleg at the shoulder. Indeed, some of the original ventral musculature migrates dorsally (Chapter 10). As ventral muscles become less important, the coracoid region becomes progressively smaller (Fig. 9-13*B* and *C*). Early therapsids and monotreme mammals, such as the platypus, *Ornithorhynchus,* retain two small coracoids, but the anterior one becomes excluded from the glenoid fossa. The anterior coracoid is lost in therian mammals, *Didelphis,* and the posterior coracoid is reduced to a small **coracoid process** on the scapula beside the glenoid fossa. Loss of the coracoids allows the glenoid fossa to face ventrally directly over the head of the humerus. The scapula expands to accommodate the increased dorsal musculature. The original cranial border of the scapula, which can be identified by a process, the **acromion,** to which the lateral end of the clavicle attaches, rolls outward to form a ridge known as the **scapular spine.** The supraspinous portion of the scapula, which is located cranial and dorsal to the spine, represents a newly evolved shelf of bone.

Reduction continues in the dermal portion of the pectoral girdle (Fig. 9-13*B* and *C*). The cleithrum is lost in all mammals. Monotremes retain an interclavicle, but this is lost in therians. The medial end of the clavicle, which articulates with the interclavicle in early therapsids, now articulates with the anterior end of the sternum. The clavicle persists in many mammals as a brace that helps stabilize the position of the shoulder, but it is reduced in running species (Fig. 9-14).

The ventral portion of the pelvic girdle also is reduced, but not to the extent as in the pectoral girdle. The large pubis and ischium in the ventral region of early synapsids (the †"sailed reptiles") become smaller in early therapsids and mammals, but both elements remain (Fig. 9-15). The acetabulum continues to face laterally. The obturator foramen, through which the obturator nerve passes in early synapsids (Fig. 9-15*A*), enlarges to form a much larger opening in late synapsids and mammals (Fig. 9-15*B–D*). Certain pelvic muscles arise from the margins of the larger obturator foramen, and the increased space also accommodates the bulging of certain pelvic muscles when they contract. A change in the size and positions of certain pelvic muscles also leads to an anterior expansion of the ilium and a reduction and eventual loss of its posteriorly directed portion. The ilium articulates with at least three sacral ribs and vertebrae, which usually are united to form a sacrum. The three bones that form the pelvic girdle are quite distinct in most synapsids and young mammals, but they fuse together to form a single element in adult mammals. This strengthens the girdle. Monotremes and marsupials of both sexes have a pair of rod-shaped **epipubic bones,** which extend forward from the pubis into the ventral abdominal wall. Their function is uncertain, but White (1989) has proposed that muscles associated with them help protract the pelvic limb during locomotion.

The humerus and femur and the bones of the forearm and shin become longer during synapsid evolution than in early tetrapods (Fig. 9-14), and this, too, increases step length. Processes for muscle attachment on the proximal end of the humerus and femur still are present, but they are relatively smaller than in early tetrapods, except in burrowing species of mammals, which have exceptionally powerful limbs. Because the glenoid fossa faces ventrally in mammals, the head of the humerus remains at the proximal end of the bone. The head of the femur shifts to the medial side of the bone in most mammals because the acetabulum with which it articulates continues to be directed laterally. As an adaptation to carrying great weight in elephants and other graviportal species, the acetabulum faces more ventrally, and the femoral head is near the proximal end of the bone.

Because the limbs are carried close to the body, the bones of the forearm and shin no longer share equally in weight transfer in mammals, as they did in early synapsids (Fig. 9-14). The radius transfers most body weight in the front leg, but the ulna contributes to the elbow joint, and it also is the axis about which the radius rotates. Because the elbow points caudally, a rotation of the radius across the anterior surface of the ulna is essential for the toes to point forward and for the sole of the front foot to be on the ground. You can demonstrate this rotation in your arm by first placing your arm in the

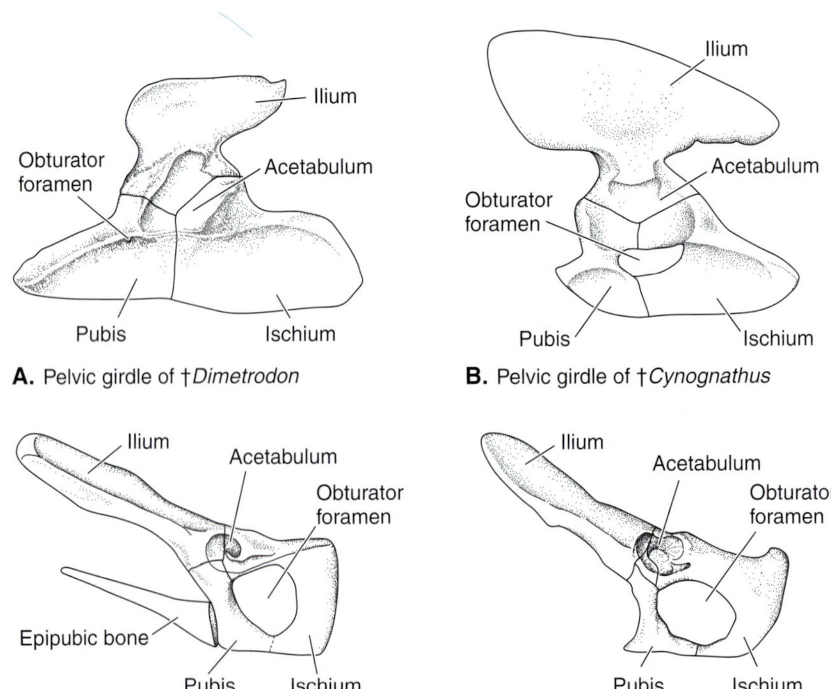

FIGURE 9-15
Lateral views showing stages in the evolution of the mammalian pelvic girdle. *A,* An early synapsid, †*Dimetrodon. B,* An advanced therapsid ancestral to mammals, †*Cynognathus. C,* The opossum, *Didelphis. D,* A eutherian mammal, *Felis.* (*After Romer and Parsons.*)

splayed early tetrapod position, with the hand directed laterally. The entire radius lies on the anterior border of the forearm. Now slowly move your arm into the quadruped mammalian position with the hand pointing forward. The distal end of the radius is now on the medial side of the forearm, but its proximal end is now lateral. The radius has rotated about the ulna. The capacity of the radius to rotate around the ulna persists in many mammals. We can direct the palm of the hand toward the ground in the **prone** position, or upward in the **supine** position. The capacity to rotate the forearm is lost in many running species, such as the horse, in which the limb is essentially a pendulum that swings fore and aft. The front foot remains prone at all times. The ulna is reduced. Only its proximal end, which helps form the elbow joint, remains (see Fig. 11-21*A*).

Because both the knee and the hind foot point forward (Fig. 9-14), the major movement at both the knee and ankle is a hinge action. Rotation is limited. A major muscle group (the quadriceps femoris), which extends the lower leg, crosses the front of the knee joint, and a sesamoid bone called the **patella** evolves in its tendon. The patella facilitates the movement of this tendon as it slides across the knee during the movements of the lower leg and also gives the quadriceps femoris additional leverage. Tibia and fibula lie parallel to each other. The tibia is the major weight transfer bone, and it is substantially larger than the fibula. The fibula is greatly reduced in some species.

The ability of the hand and foot to rotate to some extent at the wrist and ankle, which is present in early tetrapods, is reduced in mammals. A hinge action remains, and this is the primary movement at the wrist and ankle because the feet point forward at all times during a step cycle. A reduction in the number of carpals and tarsals accompanies the reduced degrees of freedom of movement in the wrist and ankle (Fig. 9-16). Only one distinct centrale remains in the carpus

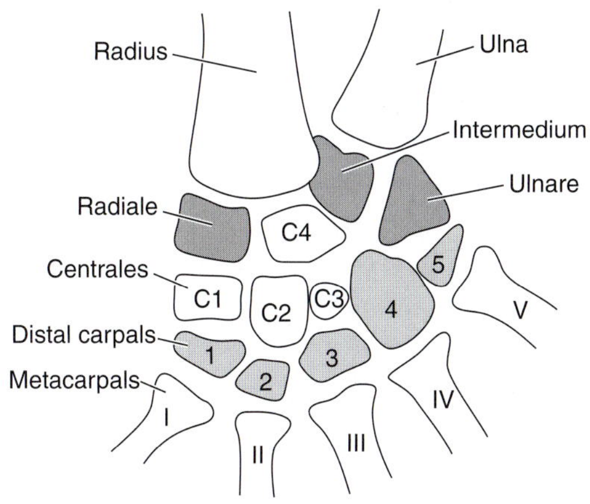

A. Wrist of an early tetrapod

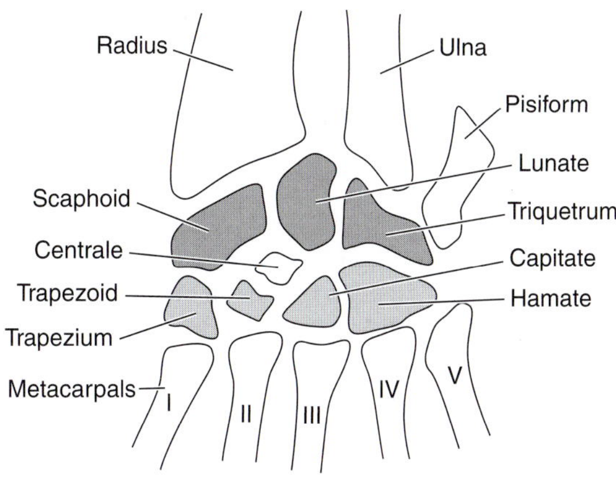

B. Wrist of a mammal

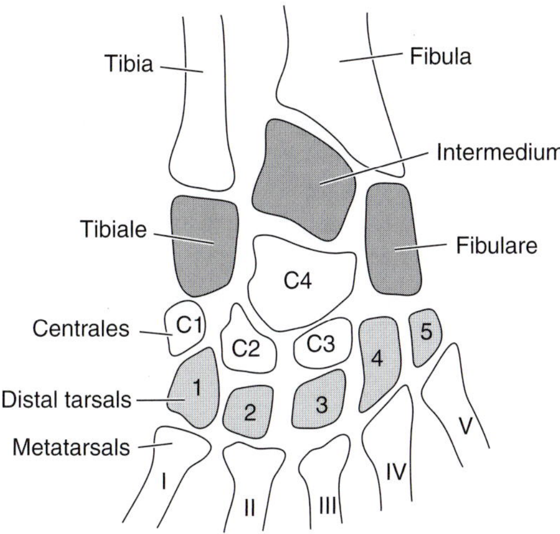

C. Ankle of an early tetrapod

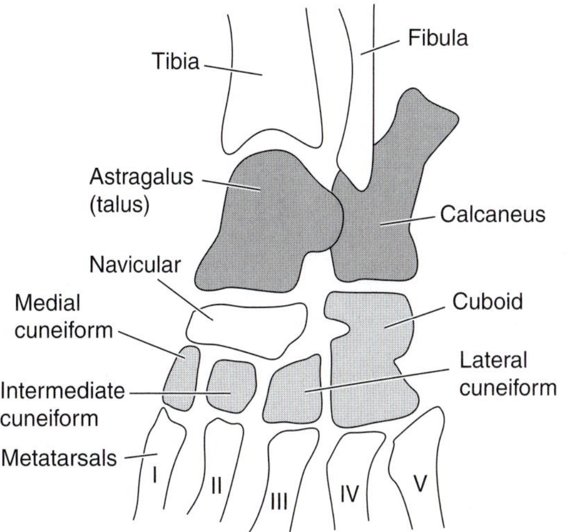

D. Ankle of a mammal

FIGURE 9-16
The evolution of the mammalian carpus (*A* and *B*) and tarsus (*C* and *D*). The wrist and ankle of an early tetrapod are shown in *A* and *C;* those of a mammal, in *B* and *D*. (*Modified from Romer and Parsons.*)

and tarsus of most species, and humans normally do not have the one in the wrist. As in sauropsids, the tibiale, intermedium, and probably another centrale unite in the tarsus to form the **astragalus.** The hinge action at the ankle occurs between the astragalus and tibia rather than at a mesotarsal joint as in sauropsids. The fibulare, now called the **calcaneus,** develops a caudally directed process that acts as a lever arm for the attachment of shank muscles that extend the foot. Alternative names are available for other carpals and tarsals, based on their shapes in humans (Fig. 9-16*B* and *D*).

Toes are more equal in length in the forward-directed feet of mammals than they were in early synapsids. The axis of the foot runs through the third digit in most species, and this one often is slightly longer than the others. The phalangeal formula of mammals is reduced to 2-3-3-3-3. Reduction begins in the therapsids ancestral to mammals, and stages are known in which the extra phalanges that were present in the last three digits of ancestral tetrapods are reduced to small nubbins of bone.

The appendicular skeleton of different groups of mammals has changed considerably from the generalized quadruped condition of early mammals during the extensive radiation of mammals and their adaptation to different patterns of locomotion. Some mammals remain generalized quadrupeds; others become adapted for running, climbing, flying, swimming, or burrowing; and humans are bipeds. We will explore many of these adaptations when we consider locomotion in Chapter 11.

SUMMARY

1. The earliest jawless vertebrates, most of whom were slow-moving, bottom-dwelling species, had little in the way of an appendicular skeleton. None is present in hagfishes and lampreys, and only a few extinct jawless fishes had pectoral fins.
2. Jawed fishes are streamlined and more active animals. Pectoral and pelvic fins help to stabilize them as they swim and are used in maneuvering. Paired fins probably evolved independently in each of the major groups of jawed fishes because chondrichthyans, actinopterygians, and sarcopterygians have quite different types of fins.
3. The paired fins are supported by small endoskeletal girdles of cartilage or cartilage-replacement bone. A larger dermal component of the pectoral girdle connects the girdle to the skull. The head and trunk move as a unit in fishes.
4. Terrestrial vertebrates are not supported by water and need a larger and much stronger appendicular skeleton. Although undulations of the trunk and tail help move early terrestrial vertebrates, thrust of the limbs upon the ground becomes increasingly important in locomotion in derived tetrapods.
5. Tetrapod paired limbs evolved from the uniserial monobasic fin of choanate fishes. The axis of a fin or limb extends down the posterior or lateral side of the fin or tetrapod limb. The tetrapod stylopodium contains only a single element, the humerus or femur. The axial element of the zeugopodium is the ulna or fibula, and then the axis continues into the fourth digit of the autopodium.
6. Elements in the tetrapod stylopodium and zeugopodium can easily be homologized with elements in the crossopterygian fin, but most of the elements of the tetrapod autopodium appear to be neomorphic.
7. The earliest tetrapods raised themselves only slightly from the ground and walked with their limbs in a splayed position. The long bones of the limbs were short and chunky and bore large processes for muscular attachment. The radius and ulna in the forearm and the tibia and fibula in the shin were about the same size and shared equally in transferring body weight to the ground.
8. The girdles were greatly enlarged, especially their ventral portions, which provided the origin for the powerful adductor musculature needed to raise the body. The dermal component of the pectoral girdle lost its connection to the skull. The head can now move independently of the trunk. Body weight is transferred directly to the pelvic girdle by a connection of the girdle to a single sacral rib and vertebra. Weight is transferred to the pectoral girdle by a muscular sling.
9. Although the earliest tetrapods had limbs and ventured onto the land, they also had caudal fins and spent a great deal of time in the water. Their hand and foot had six to eight toes so must have been quite broad and perhaps paddle-like. Only in later tetrapods, the digit number becomes stabilized at five in the foot of recent amphibians and in the hand and foot of reptilomorphs. Recent amphibians have only four digits in their hand.
10. Recent turtles and lepidosaurs are well adapted to the terrestrial environment. They are agile creatures with longer and more slender limbs than their early tetrapod ancestors. They can move very rapidly, but most still walk with a splayed gait.

11. Digits increase in length from the medial one to the fourth. This helps maintain digit contact with the ground as the foot rotates laterally and sometimes posteriorly during a step.
12. The girdles are strong. The endoskeletal part of the pectoral girdle ossifies as a dorsal scapula and a ventral anterior coracoid. The dermal part of the pectoral girdle is reduced to the clavicle and interclavicle.
13. The ventral part of the pelvic girdle develops a large puboischiadic fenestra, which accommodates certain pelvic muscles, between the pubis and ischium. The ilium has a firmer union with the axial skeleton because turtles and lepidosaurs have at least two sacral ribs and vertebrae.
14. The pectoral appendages are modified as wings capable of powered flight in three groups of tetrapods, but the wings evolved independently. Each group has a different wing structure. The extinct †pterosaurs had a membranous wing supported by the arm and elongated fourth finger. The first three digits were free of the wing and form small, grasping structures.
15. The bird wing is composed of feathers, which are modified horny scales that are supported by the arm and an elongated hand containing only three fingers.
16. The bat wing is also membranous, but is supported by the arm and the elongated second to fifth digits. The first digit is free of the wing and forms a small hook.
17. The pectoral girdle and thoracic skeleton of flying vertebrates are particularly strong because they serve as the fulcra for wing action. The sternum is broad and usually bears a keel onto which powerful flight muscles attach.
18. The bird pelvic appendages and girdle are particularly strong because birds are bipeds on the ground and the legs must support all of the body weight. The legs also act as shock absorbers when a bird alights. The legs of †pterosaurs and bats are much smaller. Bat hind legs form hooks with which a bat hangs upside down when at rest.
19. Mammals are very active, inquisitive animals that have a wide range of behaviors. In the line of evolution through early synapsids to mammals, the appendicular skeleton became modified to provide strong support in the terrestrial environment and to allow the animals to move rapidly. The limbs rotated close to or under the body, where they provide better mechanical support, and allow the limbs to swing through longer arcs, increasing step length and speed.
20. Limbs carried close to the body require a less powerful adductor musculature to raise and hold the body off of the ground. The ventral parts of the girdles become reduced, especially in the pectoral girdle. The dorsal parts expand to accommodate muscles that brace the legs at the shoulder and hip joints and swing the leg fore and aft.
21. The pectoral girdle of early synapsids had both an anterior coracoid and a posterior coracoid. The anterior coracoid becomes smaller and is eventually lost in therian mammals. The posterior coracoid is reduced to the coracoid process on the scapula, which lies beside the glenoid fossa. Only the clavicle remains in the dermal part of the girdle, and it often is quite small.
22. An expanded ilium in mammals articulates with at least three sacral ribs and vertebrae, which are usually fused together to form the sacrum.
23. The limbs lengthen, and this also increases speed. The radius and tibia become the major weight-transferring bones in the forearm and shin, respectively, and the ulna and fibula often are reduced.
24. Mammalian feet point forward throughout a step cycle. The capacity for foot rotation decreases, especially in the hind foot. The number of carpals and tarsals is reduced slightly, and the toes become more nearly equal in length. Often, the middle digit is the longest. The phalangeal formula is reduced to 2-3-3-3-3.

REFERENCES

Ashley-Ross, M. A., 1994: Hindlimb kinematics during terrestrial locomotion in a salamander (*Dicamptodon tenebrosus*). *Journal of Experimental Biology,* 193:255–283.

Balfour, F. M., 1876: The development of elasmobranch fishes. *Journal of Anatomy and Physiology,* 11:128–172.

Bellairs, A., 1969: *The Life of Reptiles.* London, Weidenfeld and Nicolson.

Bemis, W. E., and Grande, L., 1999: Development of the median fins of the North American paddle fish (*Polydon spathula*), and a reevaluation of the lateral fin-fold hypothesis. *In* Arratia, G., and Schultze, H.-P., 1999: *Mesozoic Fishes 2. Systematics and Fossil Record*:4168. München, Pfeil.

Bond, C. E., 1979: *The Biology of Fishes.* Philadelphia, Saunders College Publishing.

Burke, A. C., and Alberch, P., 1985: The development and homology of the chelonian carpus and tarsus. *Journal of Morphology,* 186:119–131.

Burke, A. C., and Feduccia, A., 1997: Developmental patterns and the identification of homologies in the avian hand. *Science,* 278:666–668.

Carroll, R. L., 1988: *Vertebrate Paleontology and Evolution.* New York, Freeman.

Coates, M. I., and Clark, J. A., 1990: Polydactyly in the earliest known tetrapod limbs. *Nature,* 347:66–69.

Coates, M. I., and Cohn, M. J., 1999: Vertebrate axial and appendicular patterning: The early development of the paired appendages. *American Zoologist,* 39:676–685.

Cracraft, J., 1971: The functional morphology of the hind limb of the domestic pigeon, *Columba livia. Bulletin of the American Museum of Natural History,* 144:173–267.

Edwards, J. L., 1989: Two perspectives on the evolution of the tetrapod limb. *American Zoologist,* 29:235–254.

Hebel, R., and Stromberg, M. W., 1976: *Anatomy of the Laboratory Rat.* Baltimore, Williams & Wilkins.

Jarvik, E., 1980: *Basic Structure and Evolution of Vertebrates.* London, Academic Press.

Jenkins, F. A., 1971: Limb posture and locomotion in the Virginia opossum (*Didelphis marsupialis*) and in other non-cursorial mammals. *Journal of Zoology (London),* 165:303–315.

Jollie, M., 1973: *Chordate Morphology.* Huntington, New York, Krieger.

Koester, D. M., and Spirito, C. P., 1999: Pelvic fin locomotion in the skate, *Leucoraja erinacea* [abstract 328]. *American Zoologist,* 39.

Kummer, B., 1959: *Bauprinzipien des Saugerskelets.* Stuttgart, Georg Thieme Verlag.

Lebedev, O. A., and Coates, M. I., 1995: The postcranial skeleton of the Devonian tetrapod *Tulerpeton curtum* Lebedev. *Zoological Journal of the Linnean Society,* 114:307–348.

Miller, E. M., Christensen, G. C., and Evans, H. E., 1964: *The Anatomy of the Dog.* Philadelphia, W. B. Saunders.

Nelson, C. E., et al., 1996: Analysis of *Hox* gene expression in the chick limb bud. *Development,* 122:1449–1466.

Norberg, U. M.: Flying, gliding, and soaring. *In* Hildebrand, M., Bramble, D. M., Liem, K. F., and Wake, D. B., editors, 1985: *Functional Vertebrate Morphology.* Cambridge, The Belknap Press of Harvard University Press.

Nursall, J. R., 1962: Swimming and the origin of paired appendages. *American Zoologist,* 2:127–141.

Ostrom, J. H., 1979: Bird flight: How did it begin? *American Scientist,* 67:46–56.

Padin, K., 1988: The flight of pterosaurs. *Natural History,* 97:58–66.

Pettingill, O. S., Jr., 1985: *Ornithology in Laboratory and Field,* 5th edition. Orlando, Academic Press.

Rackoff, J. S., 1980: The origin of the tetrapod limb and the ancestry of tetrapods. *In* Panchen, A. L., editor: *The Terrestrial Environment and the Origin of Land Vertebrates.* Systematics Association, Special Volume, no. 15. London, Academic Press.

Romer, A. S., 1956: *The Osteology of Reptiles.* Chicago, University of Chicago Press.

Romer, A. S., and Parsons, T. S., 1977: *The Vertebrate Body,* 5th edition. Philadelphia, Saunders College Publishing.

Schaeffer, B., 1967: Osteichthyan vertebrae. Linnean Society of London, *Zoology Journal,* 47:185–195.

Schaeffer, B., 1941: The morphological and functional evolution of the tarsus in amphibians and reptiles. *Bulletin of the American Museum of Natural History,* 78:395–472.

Sordino, P., Hoeven, F. van der, and Duboule, D. F., 1995: Hox gene expression in teleost fins and the origin of the vertebrate digits. *Nature,* 375:678–681.

Stahl, B. J., 1985: *Vertebrate History: Problems in Evolution.* New York, Dover.

Stephens, T. D., Sanders, D. D., and Yap, Y. F., 1992: Visual demonstration of the limb-forming zone in the chick embryo lateral plate. *Journal of Morphology,* 213:305–316.

Thacher, J. K., 1877: Median and paired fins, a contribution to the history of vertebrate limbs. *Transaction of the Connecticut Academy of Arts and Sciences,* 3:281–310.

Unwin, D. M., 1997: Pterosaur tracks and the terrestrial ability of pterosaurs. *Lethaia,* 29:373–386.

Walker, W. F., Jr., 1998: *Dissection of the Frog.* San Francisco, W. H. Freeman & Co.

Weidersheim, R., 1906: *Vergleichende Anatomie der Wirbeltierte.* Jena, Gustav Fischer.

Westoll, T. S., 1958: The lateral fin-fold theory and the pectoral fins of ostracoderms and early fishes. *In* Westoll, T. S., editor: *Studies of Fossil Vertebrates.* London, University of London Press.

White, T. D., 1989: An analysis of epipubic bone function in mammals using scaling theory. *Journal of Theoretical Biology,* 139:343–357.

Zangerl, R., 1981: Chondrichthyes: I. Paleozoic elasmobranchs, *In* Schultze, H. P., editor: *Handbook of Paleoichthyology,* volume 3. Stuttgart, Gustav Fischer.

10

The Muscular System

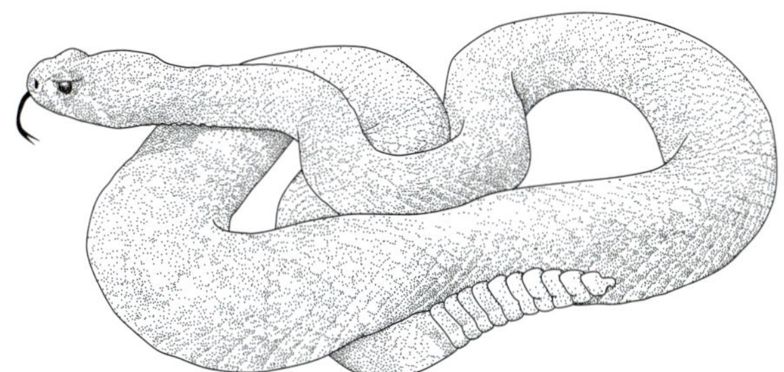

PRECIS

Muscles are the primary effector organs of vertebrates. Their actions bring about most of an animal's responses as well as help support the animal and generate body heat. We will examine the structure and actions of muscles and how the various types are used. After considering the development of muscles and their grouping, we will trace their major evolutionary changes.

OUTLINE

Muscles brace bones across joints and thereby help to support the body. They move the body and most materials through the digestive tract, the airways, the blood vessels, and most of the ducts and other passages of the body. Cilia assist the muscles in moving materials in some passages, including mucus in the respiratory ducts. Muscles are responsible for vocalization and other forms of sound production. Heat is generated as a by-product of muscle contraction. Heat is dissipated in most vertebrates, but much of it is utilized to maintain a high body temperature in endothermic birds and mammals and an elevated temperature in selected organs in some ectothermic vertebrates.

We begin this chapter by examining the types of muscle tissue. Then we will explore the structure of muscles and consider many aspects of muscle contraction. After this background, we will discuss the various muscle groups and their evolution. We conclude the chapter by considering how certain muscles are modified in some fishes as electric organs, which are used in electrolocation and defense.

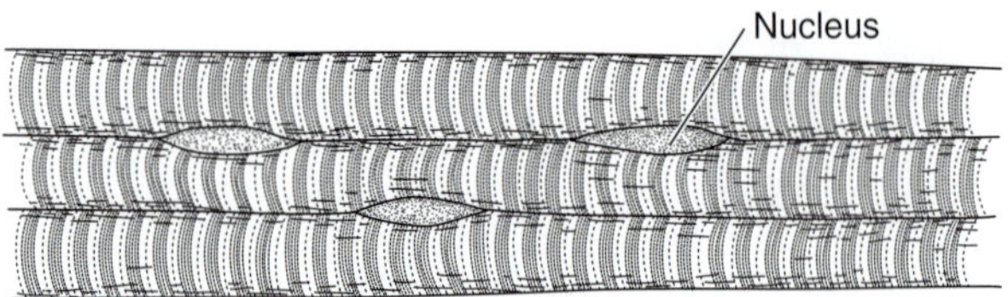

A. Skeletal muscle fibers

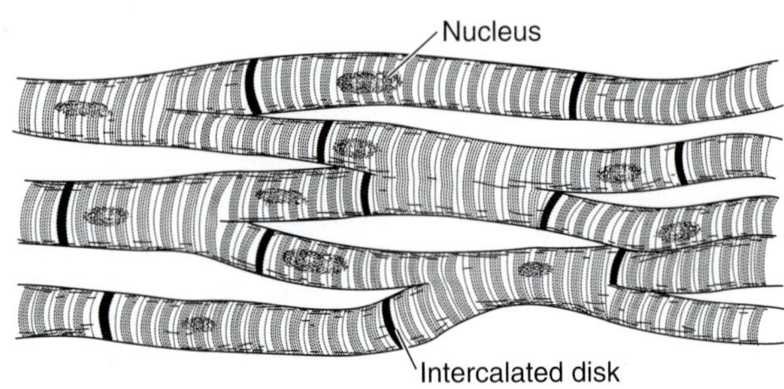

B. Cardiac muscle fibers

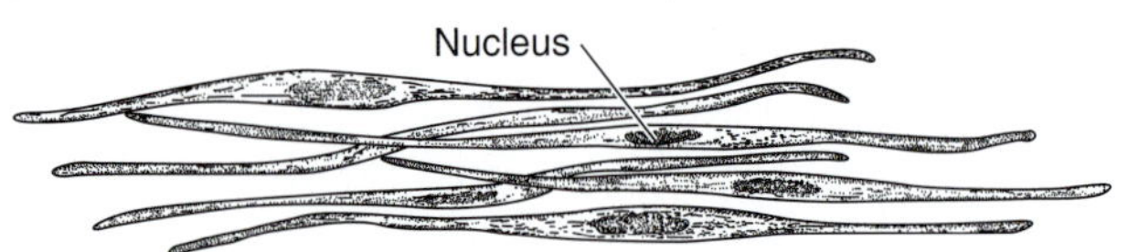

C. Smooth muscle fibers

FIGURE 10-1
A–C, The three types of muscle tissue. (*Modified from Romer and Parsons.*)

Types of Muscle Tissue

Muscle cells develop embryonically by the elongation of mesenchyme cells to form **myoblasts** (Gr., *myo-* = muscle + *blastos* = germ, bud). Myoblasts continue to divide, but at some point division stops. The myoblasts then elongate further as **actin** and **myosin**, and other muscle-specific proteins develop and become organized into **myofilaments**. The myoblasts become the adult muscle cells, which are known as **myocytes,** or **muscle fibers.** The cessation of myoblast division and their differentiation into muscle cells are regulated by a family of **myoblast regulatory genes** first discovered in 1987 by Davis and his coworkers. The development of cross bridges between myofilaments and a sliding action between the myofilaments leads to the development of tension along the longitudinal axis of the muscle fibers. This tension enables muscles to perform their functions.

Muscles can be classified in many ways: by their mode of embryonic development and location within the body (somatic or visceral muscles), by their method of nervous control (voluntary or involuntary muscles), and by their microscopic appearance. In this part, we examine their microscopic appearance and recognize three broad categories of muscle tissue (Fig. 10-1): (1) **skeletal muscles** are associated with the skeleton, (2) **cardiac muscles** form the musculature of the heart wall, and (3) **smooth muscles** contribute to the walls of blood vessels and many visceral organs. Each type is described here, and their characteristics are summarized in Table 10-1.

Skeletal Muscle

Skeletal muscle tissue is the most common type in the body because it forms the many individual muscles that attach to skeletal elements. The cells, or muscle fibers, of skeletal muscle are very large, multinucleated cells. Mammalian skeletal muscle fibers range in diameter from 0.1 mm to 0.5 mm and in length from 1 cm to 30 cm. Some muscle fibers extend the length of short muscles, but most end in the connective tissue within a muscle. Skeletal muscle fibers develop embryonically when the myoblasts cease to multiply, adhere to one another, line up longitudinally, and finally fuse. Each adult skeletal muscle fiber is thus a **syncytium** (Gr., *syn* = together + *kytos* = hollow vessel or cell), that is, a single cell that may contain several hundred nuclei. The nuclei are located peripherally in mammalian cells.

TABLE 10-1 Characteristics of Mammalian Muscle Types

Characteristic	Skeletal Muscle	Cardiac Muscle	Unitary Smooth Muscle	Multiunit Smooth Muscle
Location in body	Muscles attaching to skeleton	Heart wall	Walls of many visceral organs: digestive and urinary tracts	Iris, blood vessel walls
Cell size, shape	Long (≤30 cm), blunt-ending fibers	Moderately elongated, single, branching cells; united to others at intercalated disks	Moderately elongated (≤500 μm) single, spindle-shaped cells	Slightly elongated (≤15 μm), single, spindle-shaped cells
Nuclei	Many nuclei located peripherally	Single nucleus near cell center	Single nucleus at widest part of cell	Single nucleus at widest part of cell
Embryonic origin	Fusion of many myoblasts	Single myoblast	Single myoblast	Single myoblast
Myofilament arrangement	Partially overlap, fiber appears striated	Partially overlap, fiber appears striated	Scattered, fiber appears smooth	Scattered, fiber appears smooth
Force of contraction	Strong	Strong	Weaker	Weaker
Control	Voluntary, but muscle interactions controlled subconsciously	Involuntary	Involuntary	Involuntary
Number of neuron endings per cell	Single (twitch fiber) or multiple (tonic fiber)	Few neuron endings in the muscle	Few neuron endings in the muscle	Neuron ending on most cells
Stimulus for contraction	Neurogenic	Myogenic with some neuron modulation	Myogenic with some neuron modulation	Neurogenic
Rate of contraction	Varies with muscle, can be fast	Moderate speed	Slow, sustained	More rapid
Fatigue	Fatigues, but degree varies	Does not fatigue	Does not fatigue	Does not fatigue

The actin and myosin myofilaments of skeletal muscle fibers partially overlap, and they are arrayed in such a way that the muscle fibers appear striated (Figs. 10-1*A* and 10-2). The partial overlapping of actin and myosin myofilaments increases the ability of the myofilaments to interact with each other and, therefore, skeletal muscle can contract with considerable force. Contraction is initiated by nerve impulses and so is described as **neurogenic.** A single nerve cell or nerve fiber, which is called a **neuron,** branches and terminates at **motor end plates** on the muscle fibers within a muscle. When a muscle fiber is stimulated, a redistribution of ions across its plasma membrane leads to the development of an electrical **action potential,** which travels rapidly to all parts of the muscle fiber, and the fiber contracts. The activity of most skeletal muscles can be controlled voluntarily in the sense that an animal can decide whether to stand or move, but the complex sequence of contractions of different muscles needed for each activity is executed subconsciously. We need not consciously consider which muscles to contract when we walk.

Cardiac Muscle

Cardiac muscle of the heart wall is composed of moderately elongated cells—about 80 micrometers (μm) long—that frequently branch (Fig. 10-1*B*). Each cell originates from a single myoblast, so it contains a single nucleus, which is located near the center of the cell. As in skeletal muscle fibers, the myofilaments partially overlap and are arranged in such a way that the muscle cell appears striated. The individual cells are firmly united with each other, end to end, at specialized junctions known as **intercalated disks.** Unlike skeletal muscle, the contraction of cardiac muscle is involuntary. One cannot voluntarily control the heart rate. Contraction is **myogenic** (Gr., *genesis* = origin), that is, it originates within the muscle tissue itself. A contraction originates in a region of specialized muscle fibers in the heart wall known as the **pacemaker,** or **sinoatrial node** (Chapter 19), and the action potential spreads rapidly from cell to cell across the intercalated disks. Bundles of specialized cardiac muscles, the **Purkinje**

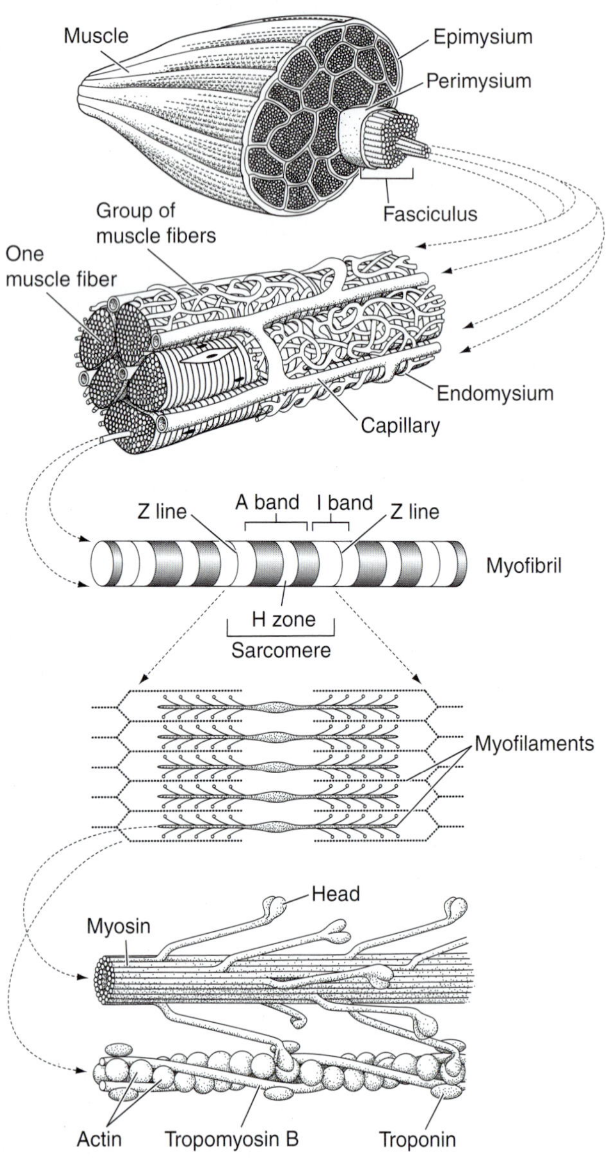

FIGURE 10-2
The structure of skeletal muscle. Successive stages of magnification have been used to show the structure from an entire muscle to the ultramicroscopic myofilaments of actin and myosin. (*Modified from Williams et al.*)

cells, carry the action potential rapidly between more distant parts of the heart. Nerve fibers terminate on the pacemaker and on many other cardiac muscle fibers, so the inherent **myogenic** rhythm of cardiac muscle is modulated by **neurogenic** control. Cardiac muscle does not fatigue, and the partial overlapping of its myofilaments enables cardiac muscle to contract with a force nearly as great as in some skeletal muscles.

Smooth Muscle

Smooth muscle fibers are elongated, spindle-shaped cells. Because each arises from a single myoblast, it has a single nucleus near the center of the cell (Fig. 10-1*C*). The cells range in length from about 15 μm in the walls of small arteries to more than 500 μm in the wall of the uterus. Actin and myosin myofilaments are present in the cytoplasm, but they are not lined up in a regular manner, so groups of them cannot be seen by light microscopy. Thus, smooth muscle, unlike other types of muscle tissue, appears to have a homogeneous texture.

Smooth muscle fibers are a part of the walls of blood vessels and many visceral organs, and they also attach to the hairs in mammalian skin. Their actions are involuntary, and their contractions tend to be slow and sustained. They do not fatigue. Two types of smooth muscle fibers, which differ in their modes of action and innervation, are recognized in amniotes.

Unitary smooth muscle fibers are found in the walls of the digestive tract, uterus, and urinary ducts. They have spontaneous, rhythmic contractions that are usually initiated by the stretching of the muscle fibers as the organs of which they are a part fill. When a fiber becomes active, an action potential spreads along its surface. The action potential of active fibers spreads slowly to others. Neurons terminate on some of the muscle fibers, and nerve impulses appear to modulate their rate and force of contraction, but the contraction itself is primarily myogenic. Unitary smooth muscle is well suited for the slow and sustained contractions needed to move food down the digestive tract, urine along the ureter, and so forth.

Multiunit smooth muscle fibers are found in the walls of many blood vessels, in the iris of the eye, and in the walls of the sperm ducts. Neurons terminate on most of the cells, and contraction is initiated by nerve impulses; that is, it is neurogenic. The degree of contraction of arteries and the pupil of the eye, and the ejaculation of sperm, are involuntary actions that must be more carefully regulated than the movements of the gut.

Muscle Structure

Muscle Organization and Connective Tissue

In all types of muscle tissue, the individual muscle fibers are enveloped by a thin layer of connective tissue, known as the **endomysium** (Gr., *endon* = within + *my-* = muscle), through which blood vessels and nerves that supply the fibers travel (Fig. 10-2). Smooth and cardiac muscle fibers are organized as sheets or layers within the walls of the visceral organs. Their fibers do not have well-defined points of attachment but pull on each other and the wall of the organs of which they are a part. They act on the contents of an organ by changing the organ's size and shape (as in the stomach or heart) or by changing the tension in its wall (as in some blood vessels).

Skeletal muscle tissue is more complexly organized than is smooth or cardiac muscles and forms distinct units. Groups of skeletal muscle fibers surrounded by their endomysium form small bundles, or **fasciculi** (L., *fasciculus* = small bundle), which are held together by a layer of connective tissue known as the **perimysium.** Many fasciculi, in turn, are surrounded by an **epimysium** and aggregated into units that we recognize as the individual **muscles.**

Muscle Attachments

Skeletal muscles have distinct attachments to skeletal elements or to well-defined connective tissue septa by **tendons** (Fig. 10-3). Tendons consist of an extension of the connective tissue within a muscle into the connective tissue **periosteum** that surrounds a bone. Frequently, the connective tissue of the tendon penetrates the bone. When the muscle and bone to which it attaches are close together, as in the humeral origin of the triceps muscle (Fig. 10-3), the attachments to the bone are inconspicuous. But tendons usually are conspicuous, cordlike bands. Some cordlike tendons are very long, extending, for example, from the muscle tissue in the proximal part of the antebrachium or crus to the digits. There are mechanical advantages to keeping bulky muscles proximally (Chapter 11). Sometimes tendons form broad, thin sheets, in which case they are called **aponeuroses** (Gr., *apo* = from + *neuron* = nerve, sinew). The tendons and connective tissue within muscles do play a role in their actions, as we shall see.

Skeletal muscles or their tendons extend across one or more joints and either move skeletal elements relative to one another, as when you bend your knee, or stabilize the skeletal elements at the joint so they do not move, as when you stand still. It is convenient to describe the opposite attachments of a skeletal muscle as its origin and insertion (Fig. 10-3). The **origin** is often thought of as the attachment that remains fixed in position when the muscle shortens, and the **insertion** as the end attached to the element that moves, but the end that moves may change with circumstances. By convention, the origin of a limb muscle is its proximal end, and the insertion is its distal end.

Motor Units

The muscle fibers within a skeletal muscle are organized into **motor units.** Each motor unit consists of a motor neuron and the muscle fibers it supplies. All muscle fibers in a motor unit are activated when the neuron supplying them is activated. Muscle fibers belonging to a single motor unit may be widely dispersed through a muscle, and many motor units, which are interspersed with each other, are present in a single muscle. Muscles differ with respect to the number of mus-

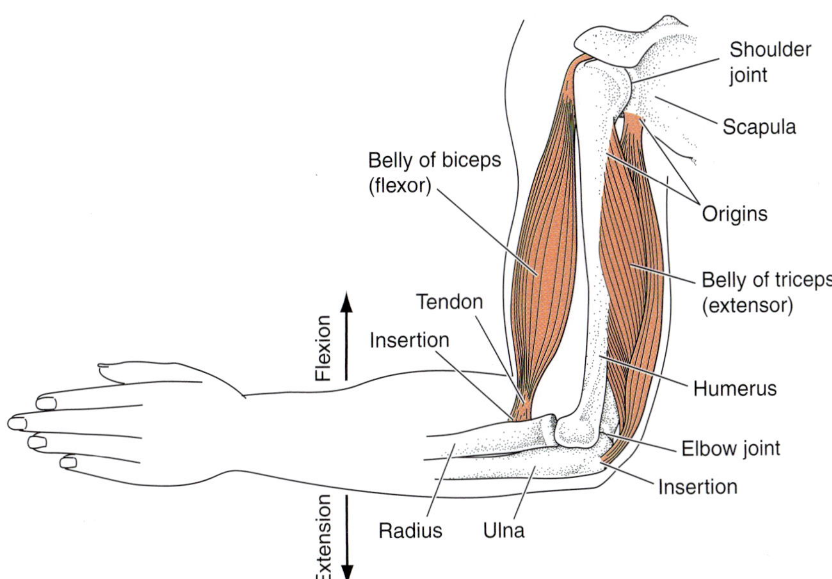

FIGURE 10-3
The attachments and actions of two antagonistic muscles of the upper arm, showing origins, insertions, and actions.

cle fibers in a motor unit. Where a very fine regulation of contraction is needed, as in the small muscles that control the movements of the eyeball, each motor unit contains only a dozen or fewer muscle fibers. Activating only one or a few motor units can cause a delicate movement. Where a strong force must be generated, as in the large leg muscles that maintain posture, each motor unit may contain 2000 muscle fibers or more. During normal muscle activity, an ever-changing rotation of active, relaxing, and quiescent motor units occurs. As functional demands on a muscle change, the proportion of active motor units increases or decreases.

Muscle Contraction

Skeletal Muscle Ultrastructure

Examination of skeletal muscle by electron microscopy at different stages of contraction, coupled with biochemical analysis, has revealed how muscle contracts. Although the biochemistry of muscle contraction is beyond the scope of this book, it is important to recognize that muscle contraction results from the interaction of actin and myosin myofilaments. Within a single skeletal muscle fiber, the myofilaments are arrayed in groups, called **myofibrils,** that can just be seen with the light microscope (Fig. 10-2). The myofibrils have a banded appearance with dark **A bands** alternating with light **I bands.** The striated appearance of a muscle fiber is caused by the A and I bands of adjacent myofibrils being in register, that is, each type of band of one myofibril lies in the same transverse plane within a muscle fiber as the corresponding band in adjacent myofibrils. The banding of a myofibril, in turn, results from the arrangement of the myofilaments within it. Thicker myosin myofilaments overlap thin actin myofilaments in the darker A bands, and only actin myofilaments occur in the lighter I bands. Another light area, the **H zone,** which is within an A band, is a region where only myosin myofilaments are present. A thin, dark **Z line,** to which actin myofilaments attach, crosses the I band. The Z lines demarcate functional units of the myofibril, known as the **sarcomeres** (Gr., *sark* = flesh + *meros* = part).

An action potential initiated on the membrane of a muscle fiber by a nerve impulse or by other means initiates a series of biochemical changes that lead to the formation of cross bridges between myosin and actin myofilaments. The heads on the myosin myofilaments bind to sites that have become active on the actin myofilaments (Fig. 10-2). The heads swivel, and the actin myofilaments are pulled into the array of myosin myofilaments. The heads let go of the actin myofilaments, reattach at new sites, and the process continues. This mechanism is known as the **sliding filament hypothesis** of muscle contraction. The attachment and detachment of each myosin head require one molecule of energy in the form of ATP (adenosine triphosphate). As the actin myofilaments slide between the myosin ones, the sarcomeres shorten (i.e., the Z lines come closer together) and the H zone becomes narrower or disappears. The amount of tension or force that a muscle fiber can generate is a function of the number of actin–myosin attachments that can be made at one time. A maximum is reached when a fiber is slightly stretched (Focus 10-1).

Modes of Muscle Contraction

The tension that results from the interaction of myosin and actin myofilaments within a muscle is called **muscle contraction.** Depending on circumstances, the muscle may or may not shorten. If the muscle shortens, it will cause bones or other structures to which it attaches to move, or the organ of which the muscle fibers are a part may change size or shape, as when a bladder contracts. A muscle contraction that induces a shortening of a muscle against a constant load or force is called **isotonic contraction** (Gr., *isos* = equal + *tonos* = strain). Movements of the body and its parts are caused by isotonic contractions. **Work,** as the term is used by physicists, is performed during isotonic contraction because:

$$\text{Work} = (\text{the force developed}) \times (\text{the distance through which the force works})$$

(Chapter 5). In **isometric contraction** (Gr., *metron* = measure), tension develops, but little if any shortening of the muscle occurs because the ends of the muscle are held in place by other forces. Muscles that support an animal or hold a part in a fixed position contract isometrically, but we usually are not aware of their activity. We can sense the tension that develops during isometric contraction by trying to lift an object that is firmly attached to the floor. Maximum muscle force is developed during isometric contraction, but no work in the physical sense is performed, for no movement occurs. Sometimes a muscle may increase in length as tension develops, which is called **negative work contraction.** An example is the action of the hamstring muscles (semimembranosus, semitendinosus, and biceps femoris) on the posterior surface of your thigh when you stand from a sitting position. Place your hand on these muscles or their tendons near your knee while sitting and then stand up. You can feel tension developing in these muscles as they help pull the thigh

FOCUS 10-1 *Tension–Length Curve for Muscle Fibers*

When the forces or tension that can be developed within muscle fibers is measured as the muscle fibers are stretched over a range of lengths, a **tension–length curve** can be drawn as shown in the figure. When a fiber is fully contracted, it can generate no tension. As the fiber is stretched, tension rises to a certain maximum and then begins to fall. A fully stretched fiber also generates no tension. These observations, which were made before the ultrastructure of muscle fibers was known, support the sliding filament hypothesis of muscle contraction. When a fiber is fully contracted, the actin myofilaments are pulled completely into the array of myosin myofilaments, so no further contraction is possible. Tension generation is thus zero. When a muscle fiber is fully stretched, actin and myosin filaments are completely separated and cannot make attachments with each other, and tension is again zero. Only when a muscle fiber is partly stretched can attachments be made. The maximum number of attachments is reached at an intermediate length, which is close to the length the fibers have in a body muscle at rest. A resting muscle is ready to generate its maximum force but will not do so unless all of its motor units are activated.

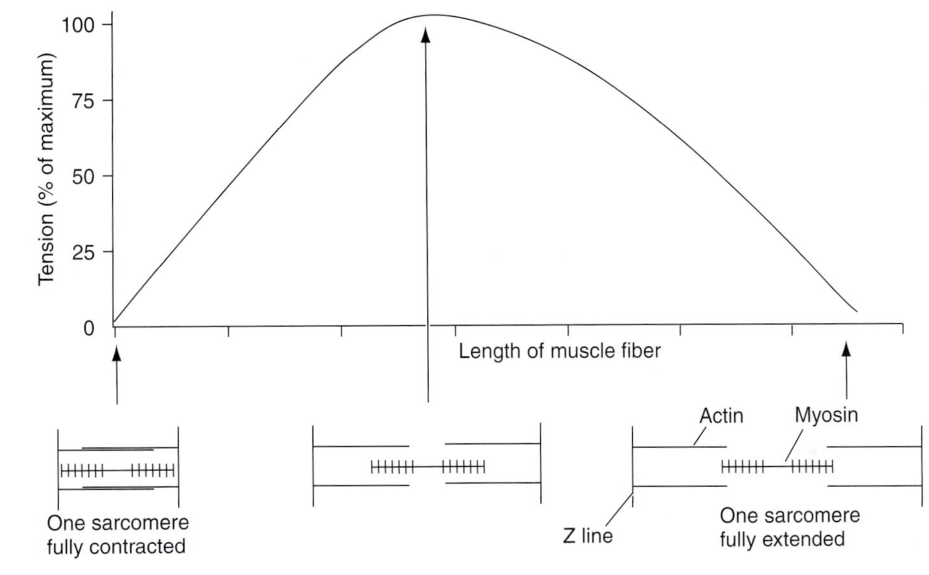

caudad under the body, but other muscles stretch them as the leg straightens at the knee, and the distance between their points of attachment increases.

Types of Muscle Action

Muscles perform their functions by developing tension and often shortening. They are restored to their resting length upon relaxation by an **antagonistic force** that operates in a direction opposite to the direction of contraction. Usually, muscles are arranged in antagonistic groups such that one muscle pulls a structure in one direction, and its antagonist pulls it in the opposite direction and restores the resting length. One example of an antagonistic set is the biceps and triceps muscles, which move the forearm in opposite directions (Fig. 10-3). The longitudinal and circular layers of muscle in the wall of the intestine are another example. The longitudinal muscles shorten the intestine and exert a force on the intestinal contents that causes the diameter of the intestine to increase; the contraction of circular muscles has the opposite effect, decreasing the diameter of the intestine and increasing its length. The antagonistic force need not be another muscle. The walls of blood vessels, for example, contain only circularly arranged muscle fibers. These work against and help control the hydrostatic pressure of the blood within the vessels. The hydrostatic force is the antagonistic force.

Sets of terms define many antagonistic actions. The movement of a distal limb segment toward a more proximal one, that is, bending the limb by decreasing the angle between the two segments, is called **flexion. Extension** is the opposite movement that straightens the limb. In general, flexion causes some type of bending, and extension causes a straightening. A ventral bending of the head or trunk also is called flexion; a dorsal movement of these parts is called extension. Ventral appendicular muscles, which lower the appendage and bend its distal segments, often are called flexor muscles; dorsal ones, which elevate an appendage and straighten its distal segments, are called extensor muscles. The forward movement of the brachium or thigh at the shoulder or hip joints is called **protraction. Retraction** is the opposite movement. **Adduction** describes the movement of a part, such as a leg, toward some point of reference—in this case, the midventral line of the body. **Abduction** is the opposite movement. **Rotation** is the movement of a bone around some axis. Pronation and supination of the hand are special cases of rotation. **Pronation** is the rotation of the radius around the ulna such that the palm of the hand faces the ground. **Supination** is a rotation that causes the palm to face upward. A simple sliding of one bone on another, called **translation,** occurs at some joints, including many joints between bones within the wrist and ankle. Some muscles are named according to the actions they cause. Most terms are self-evident: **levators, depressors, sphincters, dilators.**

These are useful descriptive terms, but muscle action is more complex than they imply. Electromyographic studies have shown that muscles seldom act individually but rather work together in **synergistic groups** (Gr., *syn* = together + *ergon* = work). One muscle stabilizes a joint so it maintains its position, another may initiate a movement at the joint, and others become active later in the movement as skeletal elements move apart or toward each other. One muscle may restrict certain movements at a joint and allow other muscles to cause a more limited movement. For example, contracting flexor muscles in your forearm and hand will cause your hand to form a fist. However, the simultaneous contracting of certain extensor muscles on the back of the hand will prevent the proximal phalanges from flexing, so you flex just the ends of your fingers. Movements also require that the antagonistic muscles relax in synchrony with the contracting muscles.

Tonic and Twitch (Phasic) Muscles

Skeletal muscles vary in their physiological contractile properties and in their biochemistry. Two broad categories are recognized: tonic and phasic. Each muscle fiber in a **tonic muscle** receives multiple motor end plates from the neuron that supplies it. The action potential resulting from a single nerve impulse does not propagate far, so only sarcomeres near the motor end plates become active. A rapid succession of nerve impulses causes a more extensive propagation of the action potential and activates more sarcomeres. In this case, the extent and force of contraction are graded by the frequency of nerve stimulation. The rate of contraction of tonic muscles also is rather slow, and tonic muscles do not fatigue easily. These muscles are well adapted for slow, sustained, and carefully graded contraction. They are often small muscles. The muscles that move the eyeball are among the few tonic muscles in mammals, but some of the postural muscles of early tetrapods are tonic.

Most vertebrate skeletal muscles are **twitch (phasic) muscles.** Each muscle fiber in a twitch muscle has a single motor end plate. One nerve impulse may not be adequate to initiate an action potential, but several in rapid succession will do so. The action potential propagates easily and rapidly along the muscle fiber, so the contraction of individual fibers is not graded, as in tonic muscle, but is **all or none.** If the stimulus is adequate to initiate contraction of a muscle, the entire muscle is activated after a brief latent period, and a brief contraction, or muscle **twitch,** results (Fig. 10-4). Muscle tension decreases during a relaxation period. Repetitive stimulation before the muscle has relaxed from the first twitch can cause a second or third twitch to be superimposed, or **summated,** on the first. Very rapid stimulation can lead to a maximum **tetanic contraction** of the muscle. The force of contraction can be graded in part by the rate of nerve stimulation. Because phasic muscles are usually larger than tonic ones and contain many motor units, the number of motor units active at a given time also grades force.

Twitch muscle fibers differ in their metabolism and many aspects of their contraction (Table 10-2). **Slow-twitch fibers,** also called **slow oxidative (SO) fibers,** derive most of their energy from oxidative metabolism. They have a high mitochondrial content because mitochondria contain the enzymes for oxidative metabolism. They are richly supplied with blood vessels that carry an abundant supply of oxygen to the muscles, and they contain a large amount of **myoglobin,** which is a hemoglobin-like molecule that facilitates the transfer of oxygen from the blood to the muscle fibers. Their myoglobin content gives them a reddish color, and sometimes they are called **red muscle.** These muscle fibers contract relatively slowly and no faster than oxygen can be delivered to them, so they do not fatigue easily. They are particularly well adapted for isometric contractions and for slow, repetitive, isotonic contractions.

Fast-twitch fibers, also called **fast glycolytic (FG) fibers,** are adapted for rapid movements of brief duration. They derive their energy primarily from glycolysis, the anaerobic metabolism of glycogen, so they can contract more rapidly than oxygen can be delivered to them. As would be expected, glycogen content

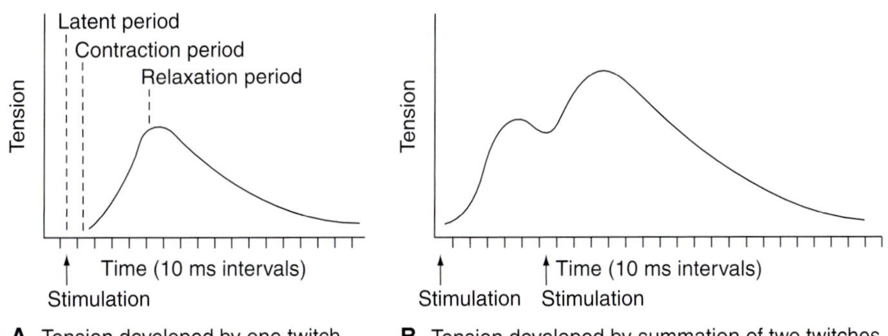

A. Tension developed by one twitch

B. Tension developed by summation of two twitches

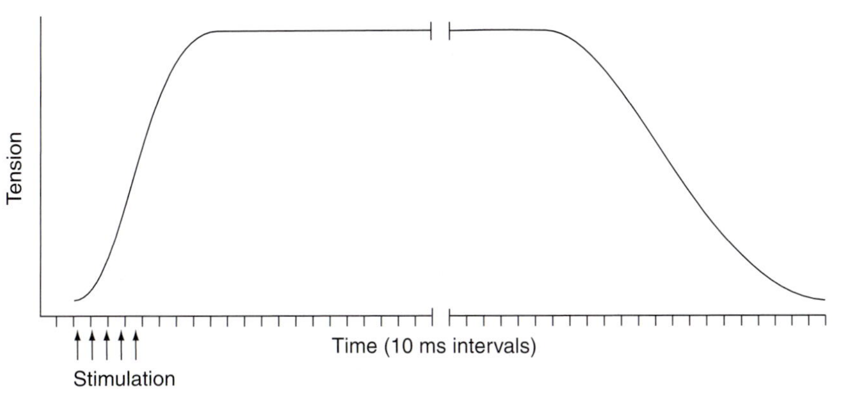

C. Tension developed by tetanic fusion of twitches

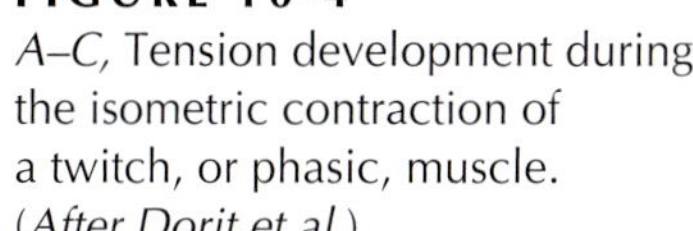

FIGURE 10-4
A–C, Tension development during the isometric contraction of a twitch, or phasic, muscle. (*After Dorit et al.*)

is higher than in the slow-twitch fibers, and mitochondrial content is lower. Fast-twitch fibers contain little myoglobin and appear white, so they are sometimes called **white muscle.** Because little of the pyruvic acid resulting from glycolysis is oxidized during the bursts of activity, it is transformed into lactic acid. Fast-twitch fibers fatigue quickly and go into an oxygen debt, which represents the amount of oxygen that must eventually be delivered to oxidize and remove the lactic acid. Other twitch fibers have intermediate properties. One intermediate type is the **fast oxidative glycolytic (FOG) fibers,** or **pink muscle.**

Many muscles contain populations of each fiber type. The slow-twitch fibers are recruited first and are

TABLE 10-2 Characteristics of Major Twitch Type Muscle Fiber Types*

Characteristic	Slow-Twitch Fibers (SO)	Fast-Twitch Fibers (FG)
Metabolism	Oxidative	Glycolytic
Mitochondria	Abundant	Fewer
Glycogen content	Low	High
Blood supply	Rich	Moderate
Oxygen supply	Rich	Moderate
Myoglobin content	High	Low or absent
Color of fibers	Red	White
Contraction rate	Slow	Fast
Fatigue	Do not fatigue	Fatigue quickly
Oxygen debt	Do no go into debt	Debt develops that is worked off later by oxidation of accumulated lactic acid
Use in body	Isometric contractions, slow isotonic contractions	Rapid isotonic contractions

*Some twitch fibers, which are called fast-oxidative glycolytic (FOG), or pink muscle, have intermediate characteristics.

used in maintaining posture and in slow movements. When rapid, brief contractions are needed, fast-twitch fibers or some intermediate fiber type (i.e., fast-oxidative glycolytic) become active. In a sense, the muscle has two or more gears: low, intermediate, and high. Other muscles are composed entirely of one fiber type or the other. Many fishes have a band of dark, slow-twitch fibers along each side of the flank that are used in slow swimming; white, fast-twitch fibers that make up the rest of the trunk and tail musculature are recruited when bursts of speed are needed. The dark meat of a chicken's leg is composed of slow-twitch fibers well adapted for sustaining posture and walking. The white meat of the breast is composed of fast-twitch fibers used in rapid wing movements of brief duration when chickens try to fly.

Whole Muscle Contraction

Role of Muscle Elasticity

In addition to a **contractile component** (the interaction of myofilaments), muscles contain elastic elements that have a profound effect on force generation. A **parallel elastic component** consists of elastic elements that parallel the myofibrils, that is, the plasma membranes of the muscle fibers and the connective tissue surrounding individual fibers and groups of fibers (Figs. 10-2 and 10-5). A **series elastic component** consists of elastic elements that lie in series with the myofilaments; among these are the tendons of the muscles. In many repetitive movements, such as the lateral undulations of a swimming fish or the oscillations of the limbs in a moving tetrapod, groups of muscle act in concert. Consider, for example, the movements of a mammal's limb. The first group of muscles to contract (the protagonists) lifts a limb from the ground and moves it forward. In doing this, they stretch their antagonistic muscles, and elastic energy is stored in their elastic components. When the antagonistic muscles become active, they place the foot on the ground and exert through the limb a force on the ground that moves the mammal forward. This force derives both from the contraction of the antagonistic muscles and from the release of their stored elastic energy. The cycle continues. The stored elastic energy contributes significantly to the total force that the muscles generate (Focus 10-2). The faster the mammal runs, the greater the extent of limb oscillation, and the more elastic energy is stored and released.

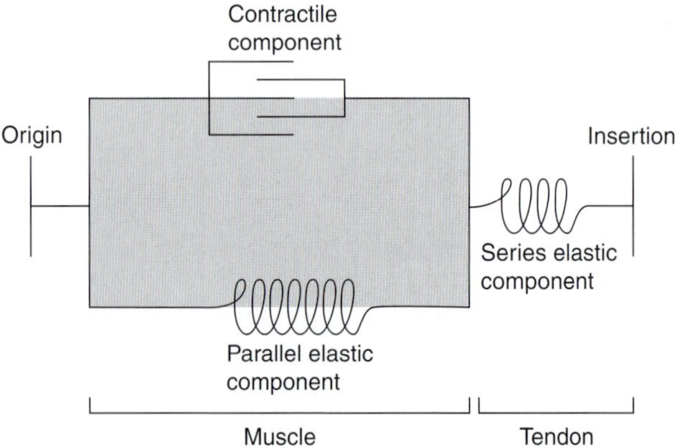

FIGURE 10-5
The contractile and elastic components of a muscle.

The protagonist and antagonist muscles can be thought of as a pair of springs linked though the common lever that they move back and forth. The contraction of the protagonistic spring stretches the antagonistic spring, the stored elastic energy in the antagonistic spring stretches the protagonistic spring, and the release of this energy again stretches the antagonistic spring. Muscle contraction starts the cycle of the storage and release of elastic energy, and a continuing input of energy from muscle contraction is needed to keep the mechanism going, but much of the total force and work generated comes from the storage and release of elastic energy. Lou et al. (1999) have found that all of the energy stored in the series elastic component in fish myomeres is recovered in useful work.

Importance of Muscle Architecture

Muscles also differ in the length and arrangement of the muscle fibers within them—that is, in their architecture (Fig. 10-6). **Strap-shaped muscles** contain long, parallel fibers and have relatively broad attachments. **Fusiform muscles** are similar, except that their fibers lead into narrow tendons at the ends of a muscle, so the force of contraction is concentrated on a smaller area. **Pennate muscles** contain short, diagonally arranged fibers that insert on a tendon on one side of the muscle (unipennate muscles) or on one or more tendons more centrally located (bipennate or multipennate muscles). Examples of these types of muscles can be seen in the shoulder and arm of a cat (see Fig. 10-21*C:* teres major [strap], biceps [fusiform], and subscapularis [multipennate]).

These variables in muscle architecture affect the degree a muscle can contract, the velocity of contraction, the force of contraction, and the power a muscle can develop. The degree to which a muscle can contract depends on the length of its muscle fibers. Most muscle fibers can contract about one third of their resting

FOCUS 10-2 *Tension–Length Curve for Whole Muscle*

In Focus 10-1, we discussed the tension–length curve for individual muscle fibers. A similar analysis can be done for a whole muscle by stretching the muscle to a series of different lengths, holding the ends in a fixed position at each length so contraction will be isometric, and maximally stimulating the muscle (see the figure). As with individual muscle fibers, **total tension** increases as muscle length increases up to an intermediate peak. This phase of tension development results from the contraction of the muscle fibers. The peak coincides with the length that the muscle has within the body when at rest (i.e., its resting length). Total tension levels off or falls slightly after this peak, as it does for the contraction of individual muscle fibers. But as length continues to increase, tension again rises and continues to increase until the muscle cannot be stretched further. The difference between this curve and the tension–length curve for muscle fibers results from the elastic components in whole muscle.

The effect of the elastic components, called the **passive tension,** can be determined separately by stretching a muscle (without stimulation), and measuring its tension at each length. The same could be done with a spring. No passive tension occurs until the muscle is at its resting length, and then passive tension increases along a curve, becoming greater and greater. (In a living vertebrate, the development of the passive tension derives partly from the pull of the muscle against its attachments and partly from the action of antagonistic muscles.)

The difference between the total tension and passive tension curves can be called the **developed tension.** The developed tension is the same as the total tension up to the intermediate peak. Thereafter, it falls, as shown by the dashed line in Figure A. The developed tension is the tension developed by the contractile component of the muscle, and not surprisingly, it has the same shape as the tension–length curve for muscle fibers. Stated in other terms:

Total tension = (developed tension) + (passive tension)

As the developed tension begins to fall, passive tension begins to increase, and it increases at a faster rate than the fall in developed tension, so the total tension rises sharply. It is clear from such an analysis that the elastic components of a muscle add significantly to the total tension that a muscle can develop.

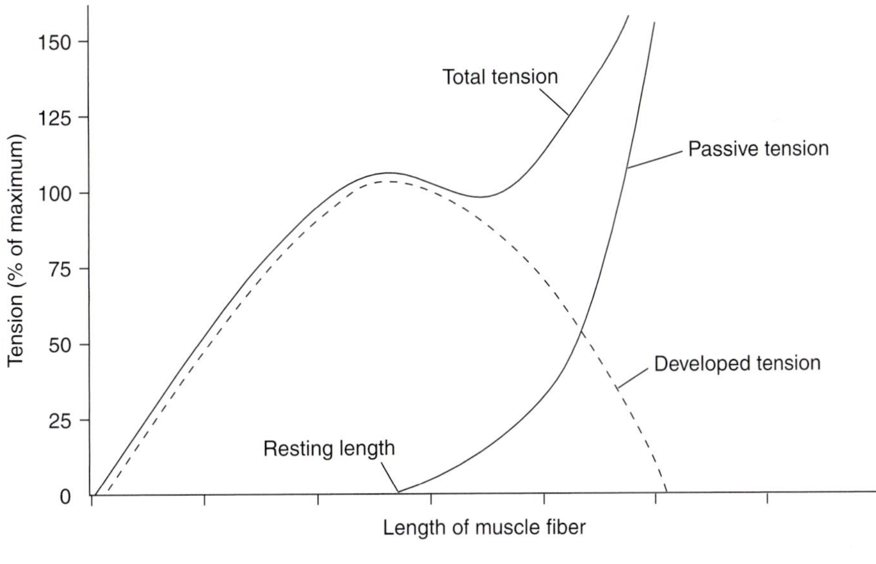

length before the actin myofilaments are completely pulled into the array of myosin ones. Strap and fusiform muscles contain longer fibers than pennate ones of the same mass, so they can contract a longer distance and can induce a more extensive movement. They also contract faster because the rate of contraction or shortening is a function of the number of sarcomeres in series. Because each sarcomere contracts at the same rate, the more sarcomeres in series, the faster the muscle as a whole can contract. Long fibers have more sarcomeres than do short ones. A strap or fusiform muscle is used when movements are extensive or rapid.

Pennate muscles, on the other hand, can develop a greater force than strap or fusiform muscles of the same mass. This is partly because they contain a great

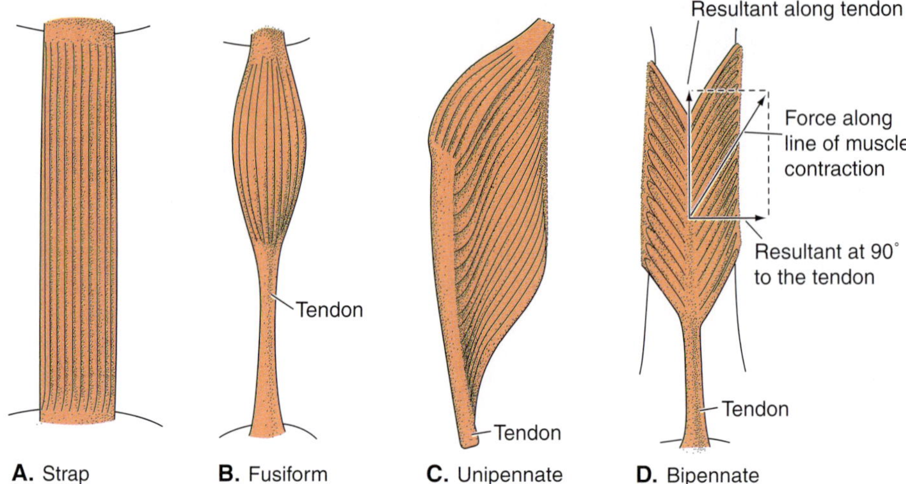

FIGURE 10-6
A–D, The architecture, or arrangement, of muscle fibers within a muscle. (*After Williams et al.*)

deal of elastic material (the connective tissue), but primarily because the force a muscle develops is a function of the number of myosin–actin connections, or cross bridges, made at one time. Thus, force depends to a great extent on the number of myofilaments in a fiber and the number of muscle fibers in a muscle. The number of fibers in a muscle can be estimated by a "physiological cross section" that cuts as many fibers as possible at right angles. Such a section would be transverse to the long axis of a strap or fusiform muscle but would have to curve around the periphery of a pennate muscle. Because pennate muscles contain many short fibers, they can generate more force per unit of muscle than can muscles with other fiber arrangements. The useful force developed in a pennate muscle is not the force developed along the line of action of the muscle fibers but is the resultant of this force along the axis of the tendon (Fig. 10-6*D*). Force along the other resultant is lost as heat, but the useful force far exceeds that in a parallel-fibered muscle of equal mass.

Pennate muscles are well adapted for forceful isotonic contractions of short extent and for isometric contractions. They often initiate a movement, which is completed by strap or fusiform muscles. Pennate fibers usually do not perform as much work as do parallel-fibered muscles because they do not shorten greatly.

Because they do not shorten much, pennate muscles also do not bulge much during contraction. They sometimes are found where space is limited, such as inside of a lobster's claw, or where body contours need to be maintained, such as on the surface of a fish (Chapter 11). Most muscles develop maximum forces in the order of 3 to 6 kg/cm^2 of cross-sectional surface.

Force and velocity of contraction are inversely related. This can be seen in a force–velocity curve (Fig. 10-7). Note that, as force increases, the velocity of contraction decreases. This has an implication for the total power that a muscle can generate because power is the rate of doing work:

$$\text{Power} = (\text{force}) \times (\text{velocity of contraction})$$

As shown in Figure 10-7, power output can be calculated from a force–velocity curve. When force is low (but above zero) and velocity is high, or force is high and velocity is low, power is low. Maximum power is generated at about 30% of the maximum force. Thus, muscles can move heavy loads (i.e., develop more power) when they contract slowly. Although pennate muscles can develop more force than can others of equal mass, they do not necessarily generate more power because their fibers are short and contraction velocity is low.

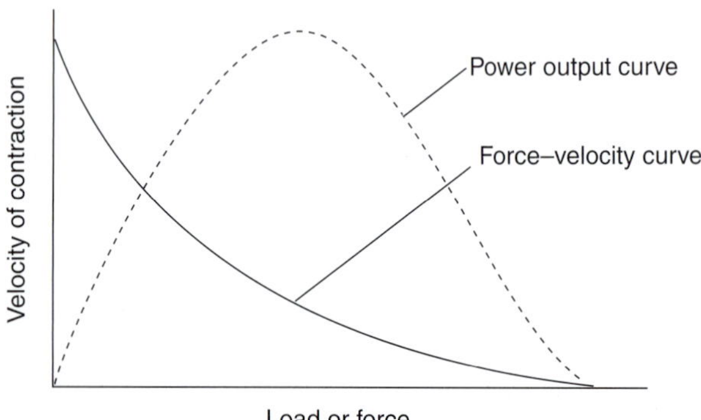

FIGURE 10-7
Force–velocity and power output curves for the contraction of a skeletal muscle.

Muscle Development and Groups

Somatic and Visceral Muscles

A regional grouping of the muscles (e.g., those of the shoulder and arm) or a functional grouping (e.g., protractors and extensors) is useful in dissection or in functional studies, but such groups often include muscles of different evolutionary, or phylogenetic, origins. For example, the shoulder group of muscles of tetrapods includes muscles derived from the appendicular, trunk, and gill arch muscles of fishes (see Fig. 10-21). During the course of evolution, muscles sometimes change their positions in the body and even their attachments. For this reason, the phylogenetic or evolutionary group to which a muscle belongs is most accurately determined not by adult position or attachments (although these can be helpful) but by its embryonic development and nerve supply. The embryonic origin and innervation of muscles are conservative, are the best criteria for homology, and can be recognized in different evolutionary lines.

The small muscles within the iris of the eye develop embryonically from the same ectodermal tissue (the optic cup) that gives rise to the iris and retina (Chapter 12), but all other muscle tissue is derived from mesodermal mesenchyme. Following the useful, but sometimes artificial, division of the body into somatic and visceral parts (Chapter 7), we can classify these muscles as somatic or visceral. Most **somatic muscles** lie in the "outer" tube of the body. As we discussed in Chapter 4, most of these muscles develop from the segmented muscle segments, or **myotomes,** which are derived from the embryonic somites (Fig. 4-16). The ventral extension of each myotome peripheral to the lateral plate mesoderm forms most of the flank musculature (Fig. 10-8), but some flank muscles arise, at least in amniotes, from mesenchyme derived from the outer, somatic layer of the lateral plate. **Visceral muscles** develop in the "inner" tube of the body from the inner, or splanchnic, layer of the lateral plate. Visceral muscles, therefore, are located more deeply in the body than are somatic muscles. They contribute to the walls of most of the digestive tract and other visceral organs and to the walls of the heart and blood vessels. As explained in Chapter 4, blood vessels begin their development in the splanchnic mesoderm but gradually extend to other parts of the body.

The derivation of somatic muscles from the somites and somatic layer of the lateral plate mesoderm, and of visceral muscles from the splanchnic layer of the lateral plate, is quite clear in the trunk. For most of the 20th century, zoologists believed that this was also true in the head. Zoologists followed Goodrich's interpretation (1918, 1930) that the head is segmented in a manner similar to the trunk (Fig. 4-37). Trunk somites continued into the head, although one or more in the otic region are transitory. The somites give rise to the somatic muscles of the head, such as the extrinsic ocular muscles that move the eyeball. A group of muscles, called **branchiomeric muscles** (Gr., *branchia* = gills + *meros* = part), develop in the pharynx wall and attach to the visceral arches. Although techniques available in Goodrich's time did not allow investigators to follow closely the migration of cells during development, it was assumed that the branchiomeric muscles developed from a cranial extension of the splanchnic layer of the lateral plate. This conclusion was reasonable because the branchiomeric muscles lie in the wall of the pharynx. Branchiomeric muscles, therefore, were assumed to be visceral.

Zoologists now realize that segmentation of the head is not as well defined as in the rest of the body, and the embryonic origin of muscles in the head is not as clear as in the trunk. The **paraxial mesoderm,**

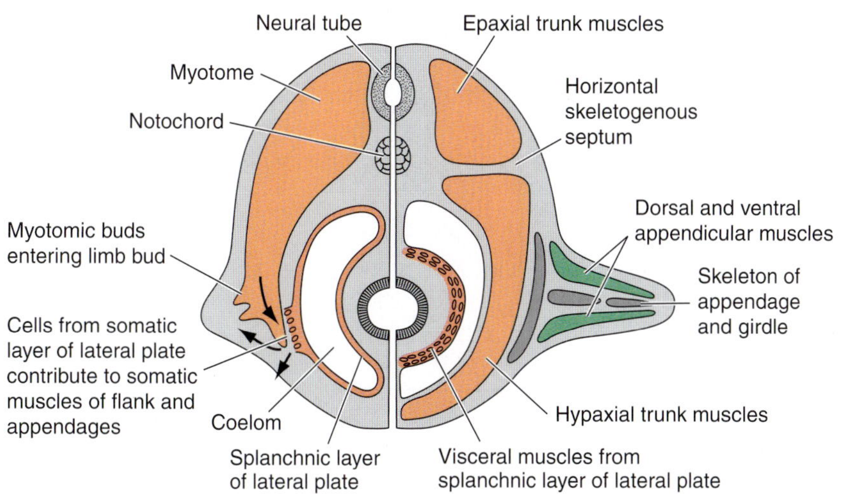

FIGURE 10-8
The difference in the development of somatic and visceral muscles as seen in a cross section through the trunk of a vertebrate at the level of the pectoral girdle. An early developmental stage is shown on the left; a later stage is shown on the right. Axial muscles and visceral muscles are shown in red; appendicular muscles, in green.

which lies beside the neural tube and gives rise to the somites and myotomes in the trunk, does extend into the head. Although the paraxial mesoderm forms somites in the head of chondrichthyans (Fig. 4-37), only the part of the paraxial mesoderm lying caudal to the otic capsule forms distinct somites in amniotes (Figs. 4-39 and 10-9). Four somites and body segments lie in the head caudal to the otic capsule of gnathostomes. The fate of the paraxial mesoderm rostral to the otic capsule only now is being clarified. Studies by Jacobson (1988), Meier (1981, 1982), and Noden (1983, 1991) indicate that this part of the paraxial mesoderm does not become completely segmented in amniotes; rather, it forms a series of somitomeres, or whorls of mesenchymal cells, that are only partially separated from each other by slight indentations. The number of somitomeres varies with taxon, but often seven of them are present. Northcutt (1990) suggests that the first six somitomeres of amniotes can be grouped by twos into units that appear comparable with the first three somites of chondrichthyans, as defined by earlier investigators. The seventh somitomere appears to represent the remains of a fourth somite (Fig. 10-9). If couplets of somitomeres can be equated with well-defined somites, then the head of jawed vertebrates contains eight somites and body segments (four preotic and four postotic ones), as Goodrich believed. Later authors (e.g., Butler and Hodos, 1996) have accepted this interpretation.

To determine the fate of cranial somites and somitomeres, Noden (1983) transplanted groups of Japanese quail head somites and somitomeres into embryos of the domestic chick. He could follow their subsequent development because tissue that differentiated from quail cells carried distinctive nuclear condensations of chromatin. He concluded that the first three somitomeres and the fifth gave rise to the somatic extrinsic ocular muscles that move the eyeball (Fig. 10-10). Similarly, myoblasts from the most caudal head segments and several anterior trunk segments migrate forward beneath the pharynx to form the **hypobranchial muscles** that lie in the floor of the pharynx and enter the tongue of tetrapods. The dorsal parts of these somites form the **epibranchial muscles**, which lie dorsal to the gill region. The origin of ocular, hypobranchial, and epibranchial muscles from somites and somitomeres confirms earlier studies that these are somatic muscles.

The surprising aspect of Noden's study is that the myoblasts that give rise to the branchiomeric muscles also migrate from somitomeres 4, 6, and 7 and from the four postotic somites (Figs. 10-9 and 10-10); they are not derived from a forward extension of the lateral plate. A similar origin for branchiomeric muscles now has been confirmed for most major vertebrate groups. Branchiomeric muscles, therefore, should be considered to be somatic muscles and not visceral ones. In addition to their origin from somites and somitomeres, branchiomeric muscles resemble other somatic muscles in being striated and in being innervated by motor neurons that travel directly to them from the central nervous system. We restrict the term "visceral muscles" to those of the gut wall (except for the branchiomeric ones), other viscera, blood vessels, and heart. All receive their motor innervation through the autonomic nervous system, which, unlike the pattern of innerva-

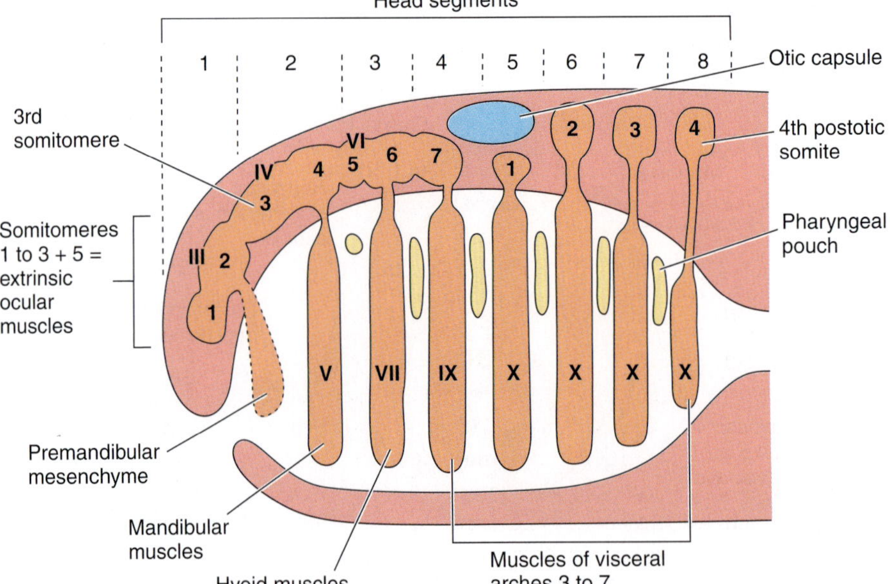

FIGURE 10-9
The segmentation of the head of an amniote and the development of cranial muscles from somites and somitomeres. All cranial muscles are therefore somatic. This diagram is based on studies of a bird embryo by Noden (1983). (*After Northcutt.*)

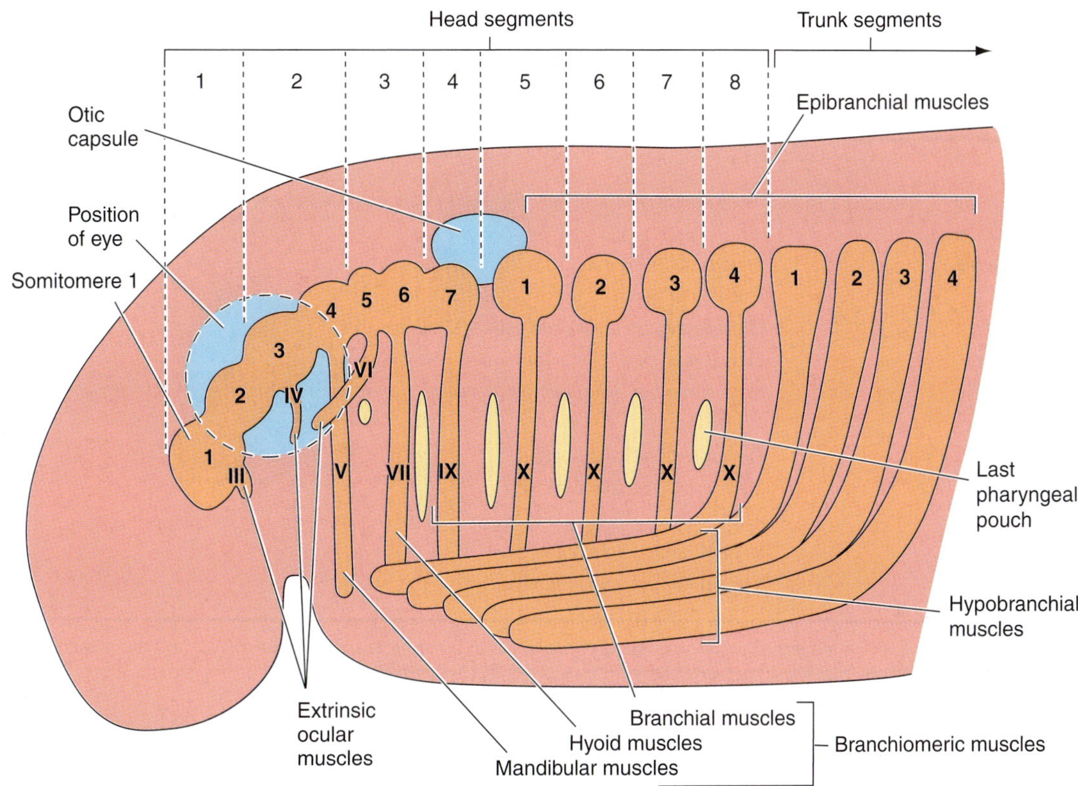

FIGURE 10-10
The continued development of cranial muscles in an amniote. The cranial nerves that innervate the extrinsic ocular muscles and branchiomeric muscles are indicated by Roman numerals. The origin of epibranchial and hypobranchial muscles from postotic and trunk somites also is shown.

tion of somatic muscles, involves a peripheral motor relay (Chapter 13). All visceral muscles, except for those of the heart, are smooth.

It may seem unusual that somatic muscles, normally restricted to the "outer" tube of the body, attach onto the visceral arches located deep in the pharynx wall. However, you will recall (Fig. 4-39*C*) that these arches develop from ectodermal neural crest cells that migrate into the pharynx wall and not from cells in the wall. The connective tissue that permeates the branchiomeric muscles is also of neural crest origin.

Somatic Muscle Groups

Visceral muscles will be discussed with the organs of which they form a part and will not be discussed further in this chapter. Zoologists subclassify the somatic muscles as axial muscles located along the longitudinal axis of the body or appendicular muscles that develop from mesenchyme that migrates from the myotomes (and sometimes from the somatic layer of the lateral plate) into the limb buds (Fig. 10-8 and Table 10-3).

Axial muscles can be further subclassified according to the group of body segments from which they arise. The **extrinsic ocular muscles** develop from the first three and the fifth somitomeres (Fig. 10-10). A specific cranial nerve is associated with each of these somitomeres and supplies the muscle(s) that develop from them (Figs. 4-37 and 10-10): somitomeres 1 and 2 with the **oculomotor nerve** (III), somitomere 3 with the **trochlear nerve** (IV), and somitomere 5 with the **abducens nerve** (VI).

The **branchiomeric muscle** develops from the remaining somitomeres and from the four postotic head somites. A specific cranial nerve supplies the muscles that develop from each (Figs. 4-37, 10-9, and 10-10). **Mandibular muscles** arise from somitomere 4 and are supplied by the **trigeminal nerve** (V). **Hyoid muscles** arise from somitomere 6 and are supplied by the **facial nerve** (VII). The remaining branchiomeric muscles, which are called **branchial muscles,** attach to the third visceral arch (i.e., first branchial arch) and to the remaining four arches. Muscles of the third visceral arch develop from somitomere 7 and are supplied by the **glossopharyngeal nerve** (IX). The remaining

TABLE 10-3 Groups of Somatic Muscles

Muscle Groups	Innervation*
Axial muscles	
Extrinsic ocular muscles	Oculomotor (III), trochlear (IV), and abducens (VI) nerves
Branchiomeric muscles	
Mandibular muscles	Trigeminal (V) nerve
Hyoid muscles	Facial (VII) nerve
Branchial muscles	Glossopharyngeal (IX) and vagus (X) nerves
Epibranchial muscles	Dorsal rami of occipital and anterior spinal nerves
Hypobranchial muscles	Ventral rami of spino-occipital nerves, form hypobranchial nerve
Trunk and tail muscles	
Epaxial muscles	Dorsal rami of spinal nerves
Hypaxial muscles	Ventral rami of spinal nerves
Appendicular muscles	
Dorsal group	Ventral rami of spinal nerves
Ventral group	Ventral rami of spinal nerves

*The innervation shown is for a fish, such as a shark. Some name changes occur in amniotes. Part of the vagus, and the muscles it supplies, becomes the spinal accessory nerve (XI). The hypobranchial nerve becomes the hypoglossal nerve (XII).

branchial muscles arise from the four postotic somites and are supplied by the **vagus nerve** (X). These visceral arches sometimes are given the name of the nerve associated with them: glossopharyngeal arch, vagal arches.

The dorsal parts of the postotic somites and of the first several trunk somites also give rise, at least in fishes, to a small group of **epibranchial muscles** that lie above the gill region. **Hypobranchial muscles** below the gill region develop in all vertebrates from the forward migration of myoblasts from more caudal postotic and the more anterior trunk somites around the back of the gill region (Fig. 10-10). Epibranchial and hypobranchial muscles are innervated in fishes by a group of **occipital nerves,** which leave the central nervous system from the occipital region of the skull, and by several anterior spinal nerves. Branches of the nerves that supply the hypobranchial muscles usually aggregate to form a conspicuous **hypobranchial nerve.** Epibranchial muscles retain their embryonic segmentation in many anamniotes; hypobranchial muscles tend to fuse and form longitudinal bands, but traces of segmentation often remain.

The remaining myotomes form the **muscles of the trunk** and **tail.** These muscles remain segmented in fishes, but parts of many myotomes fuse to form longitudinal bundles and broad sheets in tetrapods. The myotomes become divided in all gnathostomes by the horizontal skeletogenous septum (Chapter 8), which extends from the vertebral column to the body surface at the lateral line (Fig. 10-8). The portions of the myotomes that lie dorsal to the horizontal septum become the **epaxial muscles;** those ventral to it become the **hypaxial muscles.** Branches of the spinal nerves innervate trunk and tail muscles. **Dorsal branches,** or **rami** (L., *ramus* = branch), supply the epaxial group; **ventral rami** supply the hypaxial group.

Appendicular muscles are defined as those that begin their differentiation within the limb bud. The mesenchyme that forms these muscles arises in fishes as buds from the ventrolateral surfaces of a number of adjacent myotomes (Fig. 10-8). Appendicular muscles also are innervated by the ventral rami of spinal nerves, as are the ventrolateral trunk muscles. Appendicular muscles always insert on the girdles or bones of the paired appendages, but the origins of some migrate medially to the trunk. Some trunk and branchiomeric muscles secondarily shift their insertions onto the girdles, but these are not considered part of the appendicular musculature because embryonically they develop from mesenchyme outside of the limb buds. Appendicular muscles can be sorted into dorsal and ventral groups that primitively lie dorsal and ventral to the appendicular skeleton and girdles.

The Evolution of the Axial Muscles

Vertebrates have many individual muscles, and they change greatly as the methods of support, locomotion, feeding, gas exchange, and other activities of vertebrates change during their adaptation to their many habitats and modes of life. It is impossible to give a comprehensive view of muscle evolution in a book of this scope. We focus on the pattern of the muscles in a generalized jawed fish (e.g., a shark), on the major changes that occurred in the adaptation to land, and on additional changes that accompanied the evolution of an active, endothermic terrestrial life. We will consider some additional muscles as we discuss the functional anatomy of particular organ systems. We trace the evolution of muscles by the groups we have defined because these can be recognized in all vertebrates.

Extrinsic Ocular Muscles

The most rostral axial muscles belong to the extrinsic ocular group. These small, strap-shaped muscles arise from the wall of the orbit and insert on the surface of the eyeball. They rotate the eyeball as needed, such as when an animal shifts its field of vision or continues to look at a fixed object while moving. The ability to rotate the eyeball is common to all vertebrates with well-developed eyes, regardless of the habitat in which they live, so these muscles tend to be conservative. They change little during the course of evolution. Nearly all vertebrates have the six muscles shown in Figure 10-11, which are found in a jawed fish. The **ventral oblique, ventral rectus, medial rectus,** and **dorsal rectus** arise embryonically from the first two somitomeres, or first head segment (Fig. 10-10), and so are supplied by the nerve of this segment, the **oculomotor nerve** (III). The **dorsal oblique** arises from the third somitomere (second head segment) and is supplied by the **trochlear nerve** (IV); the **lateral rectus** arises from the fifth somitomere (third head segment) and is supplied by the **abducens nerve** (VI).

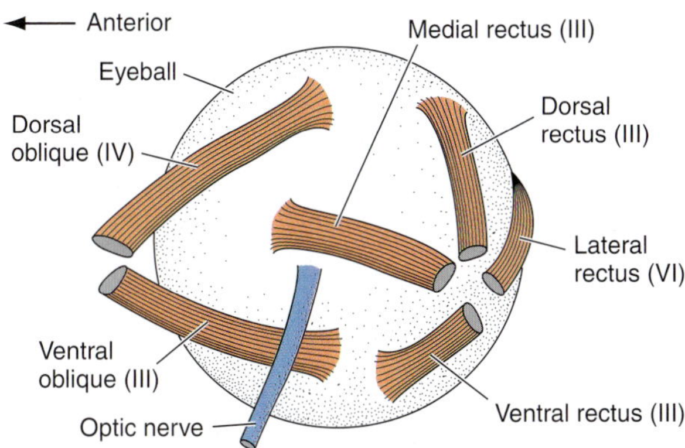

FIGURE 10-11
The extrinsic ocular muscles of a fish. The right eye has been removed from the orbit and is viewed from its back, or medial, side. The cranial nerves that supply the muscles are indicated by Roman numerals.

Most tetrapods have evolved upper and lower eyelids and, often, a transparent nictitating membrane that can move across the surface of the eyeball. These structures, and associated glands, clean and protect the eyeball in an environment where the surface of the eyeball is not bathed in water. Most, but not all, tetrapods have a **retractor bulbi** that can pull the eyeball deeper into the orbit and so facilitates the movement of the nictitating membrane across its surface. Innervation of the retractor bulbi by the abducens nerve indicates its evolution from the third head segment.

Parts of the bird's dorsal and ventral rectus have separated as a **levator palpebrae superioris** and a **depressor palpebrae inferioris** that act on the upper and lower eyelids, respectively. These muscles, as would be expected, are innervated by the oculomotor nerve. Mammals also have a levator palpebrae superioris, and some species have a depressor palpebrae inferioris.

Branchiomeric Muscles

The branchiomeric muscles lie in the lateral wall of the pharynx between the hypobranchial muscles, which lie ventral to the pharynx, and the epibranchial muscles, which lie dorsal to the pharynx (Fig. 10-12). The branchiomeric and hypobranchial muscles act together in breathing movements, capturing food, manipulating food within the mouth and pharynx, and swallowing. In some cases, the epibranchial muscles assist them. The ways in which these activities occur are affected greatly by whether a vertebrate lives in water or on land, so epibranchial, branchiomeric, and hypobranchial muscles change greatly during the course of evolution. We discuss feeding in Chapter 16 and breathing movements in Chapter 18. At this time, we will consider the general pattern of the muscles involved, beginning with the branchiomeric muscles.

Muscles of a Typical Fish Branchiomere Ancestral vertebrates were jawless, so all of their branchiomeric muscles likely acted to move a current of water into the mouth, through the pharynx, and out of the gill slits. Aside from providing for gas exchange, this current brought in small food particles. The branchiomeric muscles associated with each visceral arch were probably similar, and they may have resembled the pattern found in the caudal two thirds of the pharynx in contemporary sharks. A shark's last five visceral arches, or branchial arches, lie adjacent to the pharyngeal lumen, and interbranchial septa extend from all but the last arch outward toward the side of the head. Muscles associated with these branchial arches (*sensu strictu*) collectively are called **branchial muscles.**

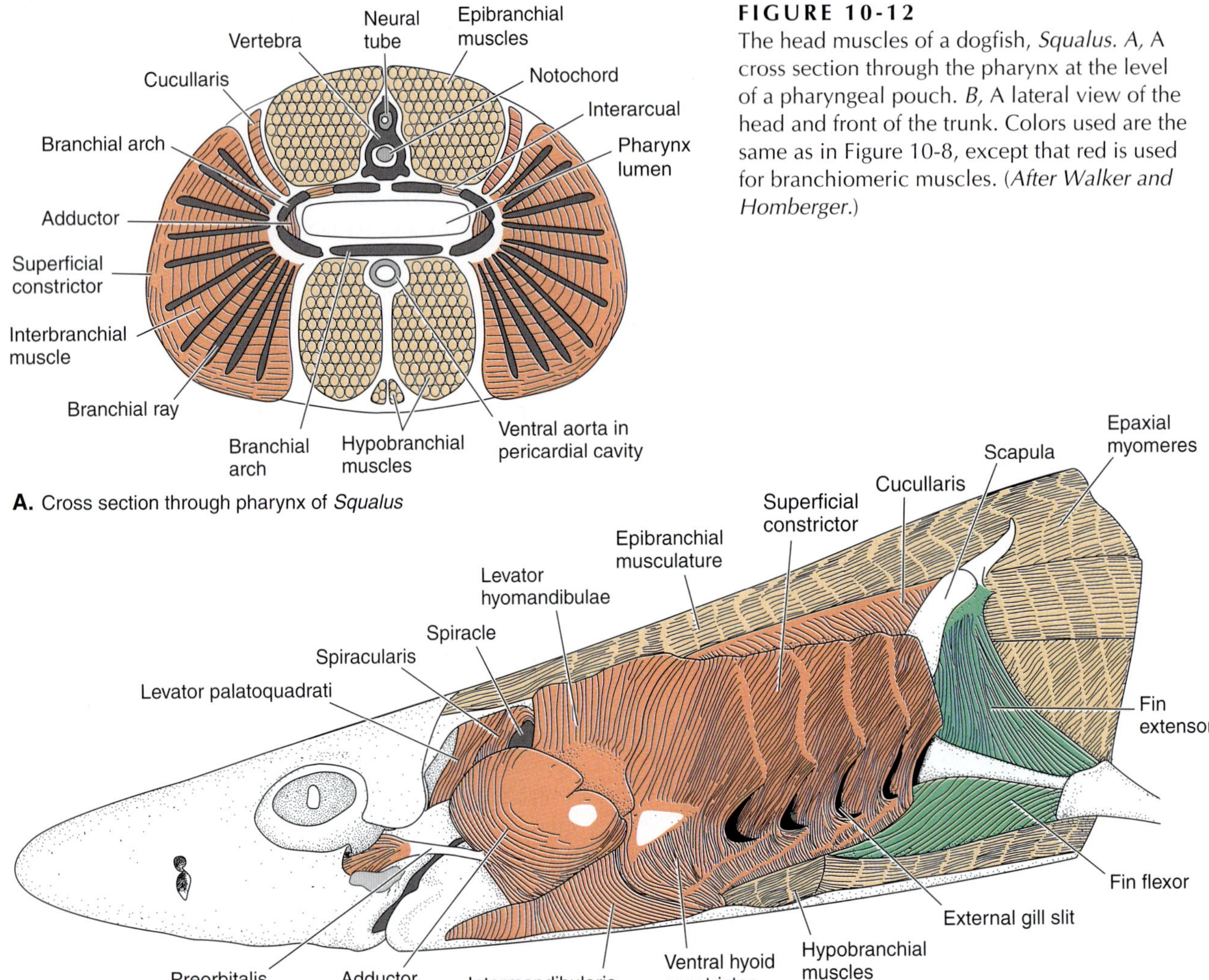

FIGURE 10-12
The head muscles of a dogfish, *Squalus*. *A,* A cross section through the pharynx at the level of a pharyngeal pouch. *B,* A lateral view of the head and front of the trunk. Colors used are the same as in Figure 10-8, except that red is used for branchiomeric muscles. (*After Walker and Homberger.*)

Most of each interbranchial septum is composed of an **interbranchial muscle,** the fibers of which have a circular arrangement (Fig. 10-12*A*). Skeletal branchial rays also extend from the arches into the septa, which they help stiffen. Gills lie on the anterior and posterior surfaces of a representative septum, and pharyngeal pouches lie between successive septa. As the septa approach the body surface, the interbranchial muscle fibers change their direction slightly and form superficial sheets that cover the dorsal and ventral parts of the pharyngeal pouches, leaving only small openings, the external gill slits, to the body surface. These sheets of muscle are the **dorsal** and **ventral superficial constrictor muscles** (Fig. 10-12*B*). Primitively, each branchial arch probably had a **levator muscle** that extended from the fascia overlying the epibranchial muscles to the dorsal part of the arch, and these muscles would have helped expand the pharynx. In contemporary sharks, the levators of all of the branchial arches have united to form a triangular muscle, the **cucullaris** (L., *cucullus* = hood), most of which inserts on the scapular region of the pectoral girdle. This is an example of a branchial muscle that functionally has become associated with the appendicular skeleton, even though part of the muscle still inserts on the last branchial arch. Small **adductor** and **interarcual muscles** interconnect segments of the branchial arches (Fig. 10-12*A*).

Contraction of the branchial muscles (except for the levators) draws the parts of the branchial arches together and compresses the pharynx and pharyngeal pouches. These muscles are responsible for discharging a current of water across the gills and out the external gill slits. The cranium is elevated, the mouth is opened, and the pharynx is expanded, drawing water and food in by the elastic recoil of the branchial arches, assisted by the contraction of the hypobranchial and epibranchial muscles (Chapters 16 and 18).

The Evolution of Mandibular Muscles The muscles of the mandibular and hyoid arches became modified with the evolution of jaws and the incorporation of the hyoid arch as part of the jaw-suspension mechanism. Most of the mandibular musculature of the first arch of sharks forms a powerful **adductor mandibulae,** which closes the jaws (Fig. 10-12*B*). Another part, the **levator palatoquadrati,** helps lift the palatoquadrate cartilage during prey capture. This muscle is serially homologous to the branchial levators. The **spiracularis** of elasmobranchs controls the opening and closing of the spiracle, the **preorbitalis** protracts the jaws during feeding, and an **intermandibularis** helps compress the throat. The intermandibularis is in series with the ventral branchial constrictors. All of the mandibular muscles are innervated by the trigeminal nerve (V; Table 10-4).

Because of their embryonic association with the mandibular arch and their innervation by the trigeminal nerve, the mandibular muscles of other vertebrates can be recognized and are homologous as a group with those of a shark. But during embryonic development, the premuscular mandibular tissue cleaves and differentiates in different patterns among the many vertebrate taxa. Unless the pattern of embryonic differentiation has been studied and compared between groups, it is not always possible to homologize individual mandibular muscles in the adult of one group with those in another group. But often, an approximation can be made based on position, attachments, and other criteria. The same can be said for other groups of branchiomeric muscles and, indeed, for all muscle groups.

Most of the mandibular muscles continue to act on the jaw mechanism during the evolution of tetrapods. The large mandibular adductor complex forms the most conspicuous of the jaw-closing muscles. The division of the adductor mandibulae in the various lines of tetrapod evolution correlates with divergences in their methods of feeding. The pattern seen in *Necturus* is shown in Figure 10-13*A* and Table 10-4. As the jaws

TABLE 10-4 Branchiomeric Muscles*

Group	Shark	Amphibian/Reptile	Mammal
Mandibular muscles (trigeminal nerve)			
	Adductor mandibulae	Adductor mandibulae	Temporalis Masseter Pterygoideus Tensor tympani Tensor veli palati
	Levator palatoquadrati	Levator pterygoidei, Protractor pterygoidei	—
	Spiracularis	—	—
	Preorbitalis	—	—
	Intermandibularis	Intermandibularis	Mylohyoideus Anterior digastric
Hyoid muscles (facial nerve)	Levator hyomandibuli	—	Stapedius
	Dorsal constrictor	Depressor mandibuli Branchiohyoideus	—
	Interhyoideus	Interhyoideus	Stylohyoid Posterior digastric
	Ventral constrictor	Sphincter coli	Platysma Facial muscles
Branchiomeric muscles of remaining arches (glossopharyngeal; vagus; and, in amniotes, spinal accessory nerve)	Cucullaris	Cucullaris Levatores arcuum	Trapezius complex Sternocleidomastoid complex
	Interarcuals, branchial adductors	—	—
	Superficial constrictors and interbranchials	Dilator laryngis Subarcuals Transversi ventralis Depressores arcuum	Intrinsic muscles of the larynx and certain pharyngeal muscles

*Muscles in each group are homologous as a group between taxa, but muscles cleave during development in different ways in each taxon, so it is not always possible to exactly homologize individual muscles between taxa.

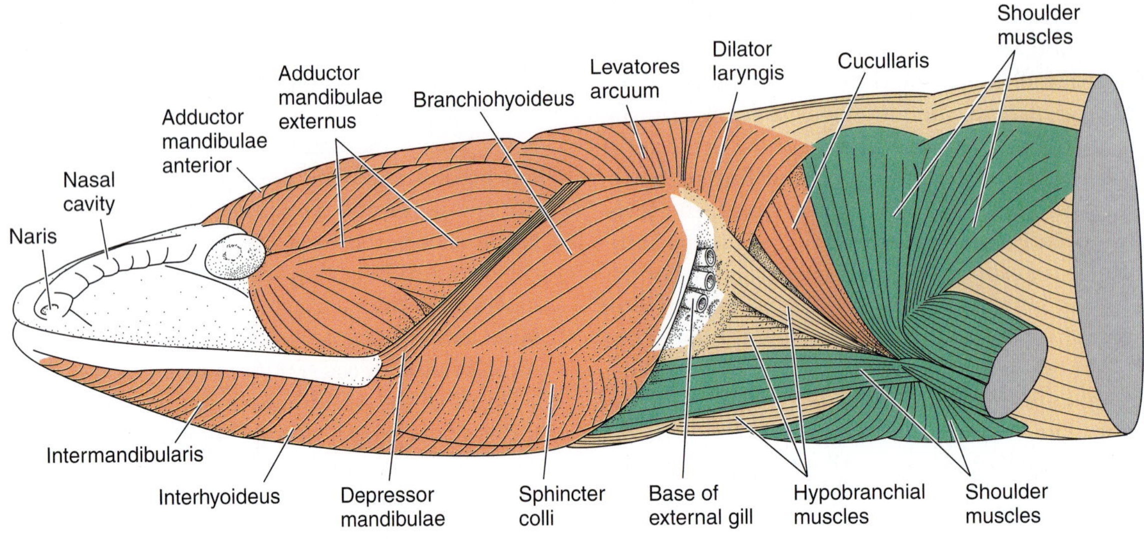

A. Branchiomeric and shoulder muscles of *Necturus*

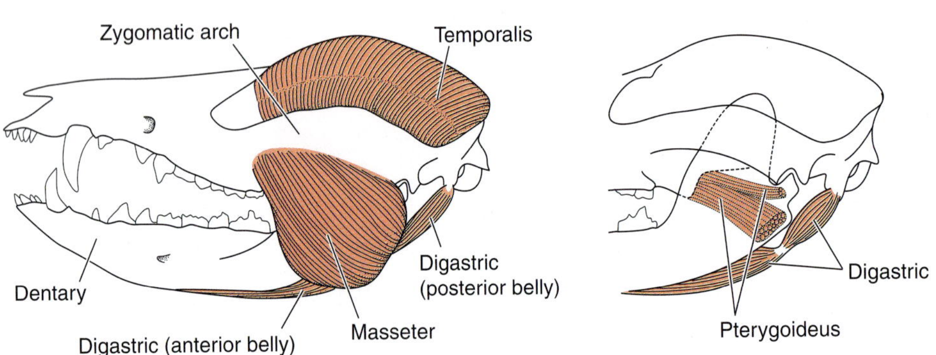

B. Superficial jaw muscle of *Didelphis*

C. Deeper jaw muscle of *Didelphis*

FIGURE 10-13
Evolution of the branchiomeric muscles. *A,* A lateral view of the branchiomeric muscles of *Necturus.* Parts of the adjacent hypobranchial and pectoral muscle are shown. *B* and *C,* Superficial and deeper views of the mandibular muscles of an opossum, *Didelphis.* The posterior belly of the digastric muscle is of hyoid origin. In *C,* the lower jaw is drawn as though it were transparent so you can seen the pterygoideus muscles that attach to its medial surface. (*B and C, After Romer and Parsons.*)

become stronger and their movements more complex in the line of evolution toward mammals, the adductor complex becomes divided into several distinct muscles (Fig. 10-13*B*). A **temporalis** lies in the temporal fossa, one or two **pterygoids** originate from the pterygoid processes on the underside of the skull, and a **masseter** (Gr., *maseter* = one who chews) originates from the zygomatic arch (Fig. 10-13*B*). The evolution of these muscles is closely related to the evolution of temporal fenestration and to the jaw changes that occurred during the evolution of mammals (Chapter 7). In mammals, all of these muscles insert on the dentary bone, but one small part of the original adductor complex remains attached to the mammalian derivative of the mandibular cartilage, that is, the malleus in the middle ear. This muscle is the **tensor tympani.** Mammals also have a small **tensor veli palati,** which lies in the soft palate.

Amphibians, reptiles, and birds with kinetic skulls retain mandibular muscles, such as a **levator pterygoidei** and **protractor pterygoidei** (see Focus 16-1), which attach onto palatal bones and help elevate the cranium when the jaws are opened. Tetrapods with akinetic skulls, including mammals, lack these muscles.

An intermandibularis remains in tetrapods as a compressor of the floor of the mouth (Fig. 10-13*A*). It helps pump air into the lungs of amphibians (Chapter 18). This sheet is represented in mammals by a **mylo-**

hyoid, which extends between the two sides of the lower jaw, and by the anterior part of the **digastric muscle,** the posterior part of which develops from hyoid muscles (Fig. 10-13*B*).

The Evolution of Hyoid Muscle The hyoid muscles of a shark include a **levator hyomandibulae,** a **dorsal** and a **ventral hyoid constrictor,** and an **interhyoideus** (Fig. 10-12*B* and Table 10-4). They help suspend the jaw and compress the pharynx. Hyoid muscles always are supplied by the facial nerve (VII).

Considerable hyoid musculature remains in amphibians. Superficial ventral sheets form an **interhyoideus** and **sphincter colli** (L., *collum* = neck; Fig. 10-13*A*). A **depressor mandibulae** and **branchiohyoideus** lie caudal to the jaw joint. The depressor mandibulae is new to tetrapods. It inserts on a retroarticular process of the lower jaw and, along with some hypobranchial muscles (see subsequent discussion), depresses the lower jaw in nonmammalian tetrapods.

The remodeling of the jaw that occurred during the evolution of mammals has been accompanied by some changes in jaw-opening mechanisms. Some hypobranchial muscles remain important in opening the jaws, but the digastric muscle functionally replaces the depressor mandibulae in placental mammals. The digastric extends between the caudal part of the skull and the ventral border of the dentary bone (Fig. 10-13*B*). In many species of mammals, the digastric is divided into two bellies, as its name implies, by a central tendon that attaches to the hyoid apparatus. The posterior belly develops from the hyoid musculature and is innervated by the facial nerve; the anterior belly develops from mandibular muscles and is supplied by the trigeminal nerve.

A small **stylohyoid** (see Fig. 10-15*B*), which extends from the styloid process of the skull (a derivative of the hyoid arch) to the hyoid, and a **stapedius,** which attaches to the stapes, are the only parts of the hyoid musculature that remain attached to the hyoid arch or its derivatives in mammals. (Recall that the stapes evolved from the hyomandibula.)

The superficial sheets of the hyoid musculature lose their connection to the hyoid in mammals and spread out beneath the skin of the neck and face to form the **platysma** (Gr., *platysma* = flat object) and **facial muscles** (Fig. 10-14). Facial muscles insert into the scalp and the base of the auricle, encircle the eye, and form the fleshy cheeks and lips of mammals. The evolution of fleshy cheeks and lips is coupled with the evolution of nipples on the mammary glands and the ability to suckle. Movements of the lips and face are also

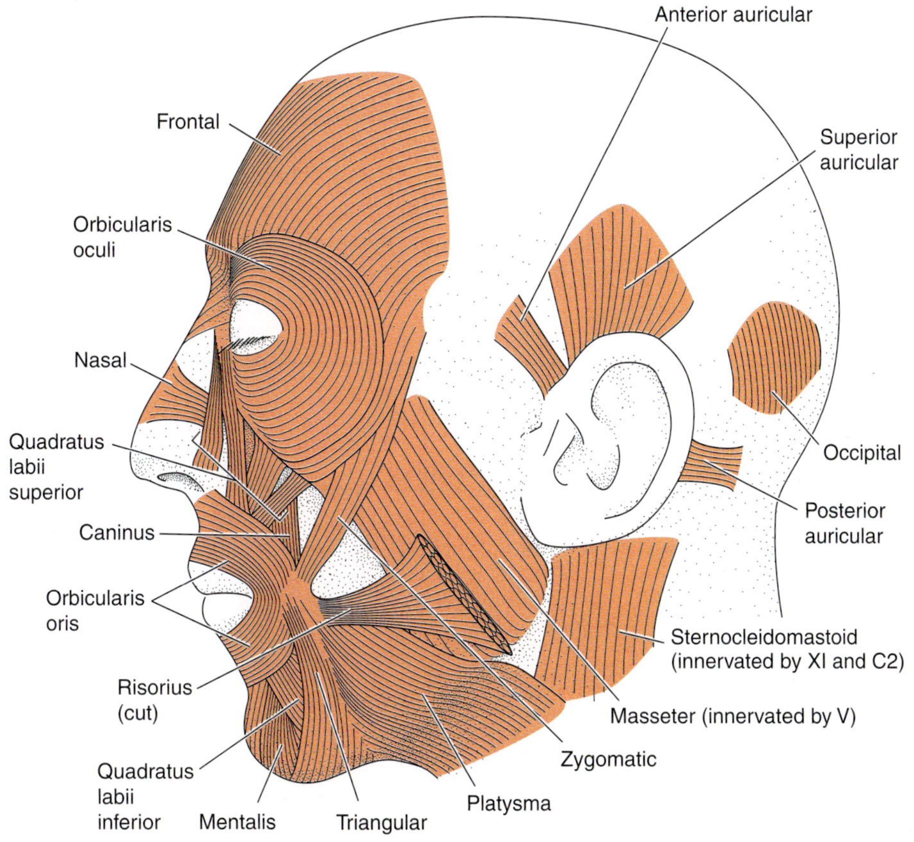

FIGURE 10-14
The platysma, facial muscles, and adjacent muscles of a human. Only the platysma and facial muscles are of hyoid origin and are innervated by the facial nerve (VII). (*After Neal and Rand.*)

an important way of communicating in many mammals. The facial nerve receives it name because it innervates these muscles, but it also supplies the rest of the hyoid musculature.

The Evolution of the Branchial Muscles We described the basic pattern of branchial muscles in a shark earlier. Aquatic salamanders, such as *Necturus,* retain many branchial muscles. However, during the transition from water to land, lungs replaced gills as sites for gas exchange, and many branchial muscles were lost. Those branchial muscles associated with the parts of the visceral skeleton that contributed to the larynx and hyobranchial apparatus remained, as did those that had secondarily shifted their insertion to the pectoral girdle.

In mammals, deep muscles in the wall of the pharynx are of branchial origin. They are supplied by the **glossopharyngeal nerve** (IX). Also, the small **intrinsic muscles of the larynx,** which extend between the laryngeal cartilages, are of branchial origin. It is not surprising that the laryngeal muscles are innervated by a branch of the vagus nerve (X) because these cartilages develop from parts of the fourth, fifth, and sixth visceral arches.

The cucullaris of fishes and amphibians expands considerably in mammals and divides to form a **trapezius group** of muscles on the dorsal surface of the shoulder and a **sternocleidomastoid complex,** which extends from the base of the skull to the sternum and clavicle (see Fig. 10-21*A*). Although branchiomeric in origin, these muscles no longer act on the visceral skeleton but help move the shoulder and head. Their motor innervation in amniotes is from the **spinal accessory nerve** (XI), which evolved from the branch of the vagus that supplied the cucullaris. Branches of cervical spinal nerves also enter these muscles but are believed to carry sensory feedback from muscle activity rather than motor neurons.

Hypobranchial Muscles

The hypobranchial muscles of fishes represent a rostral extension of the hypaxial trunk muscles from the pectoral girdle into the floor of the pharynx. Although not of branchiomeric origin, they have acquired insertions onto the ventral parts of the mandibular arch, hyoid arch, and some of the branchial arches. They help open the jaws and expand the pharynx during feeding and breathing.

Traces of the segmentation that characterize the trunk muscles of fishes may remain, but, for the most part, the hypobranchial muscles form continuous, longitudinal bundles. It is convenient to classify them as **prehyoid muscles,** which lie, or at least insert, rostral to the hyoid arch, and **posthyoid muscles,** which lie caudal to the hyoid arch (Table 10-5). In sharks, the prehyoid group is represented by a **coracomandibularis,** which extends from the mandible toward the coracoid region of the pectoral girdle (Fig. 10-15*A*). A **rectus cervicis** complex, which extends caudad from the hyoid arch to the pectoral girdle, represents the posthyoid group. The rectus cervicis complex of sharks often includes a **coracohyoideus** and **coracoarcuals.** The latter extend deeply to attach onto the branchial arches. In teleosts, a **geniohyoideus** (Gr., *geneion* = chin) separates from the coracomandibularis, and a **sternohyoideus** develops from the rectus cervicis. Hypobranchial muscles of fishes are innervated by the ventral rami of the spino-occipital nerves, which unite to form the **hypobranchial nerve.**

Hypobranchial muscles are quite complex in terrestrial vertebrates because food is not supported and carried by a current of water. The hyobranchial apparatus and a muscular tongue, both controlled by hypobranchial muscles, transport food in the mouth, swallow it, and sometimes even catch it.

TABLE 10-5 Hypobranchial Muscles*

Group	Shark	Amphibian (*Necturus*)	Mammal
Prehyoid muscles	Coracomandibular	Genioglossus	Genioglossus Lingualis
		Geniohyoideus	Geniohyoideus Hyoglossus Styloglossus
Posthyoid muscles	Rectus cervicis (coracohyoideus, coracoarcual, coracobranchials)	Rectus cervicis Omoarcuals Pectoriscapularis	Sternohyoideus Sternothyroideus Thyrohyoideus Omohyoid

*Ventral rami of spino-occipital nerves in anamniotes, hypoglossal nerve and cervical plexus in amniotes.

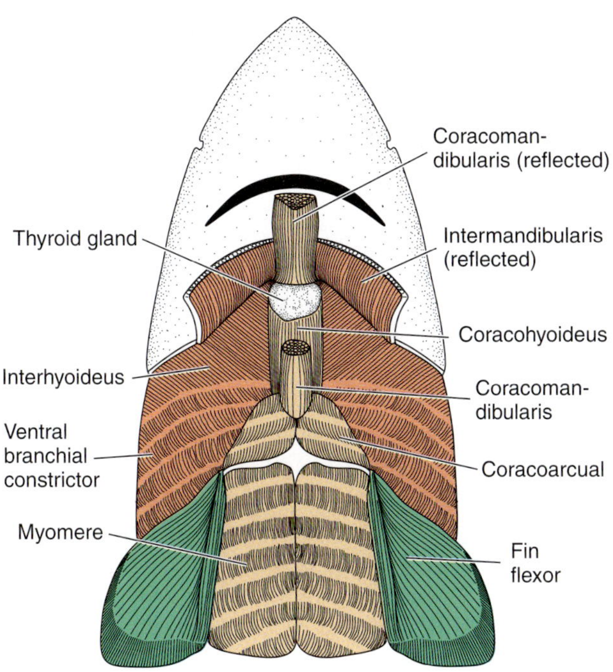

A. Study of the hypobrachial muscles of *Squalus*

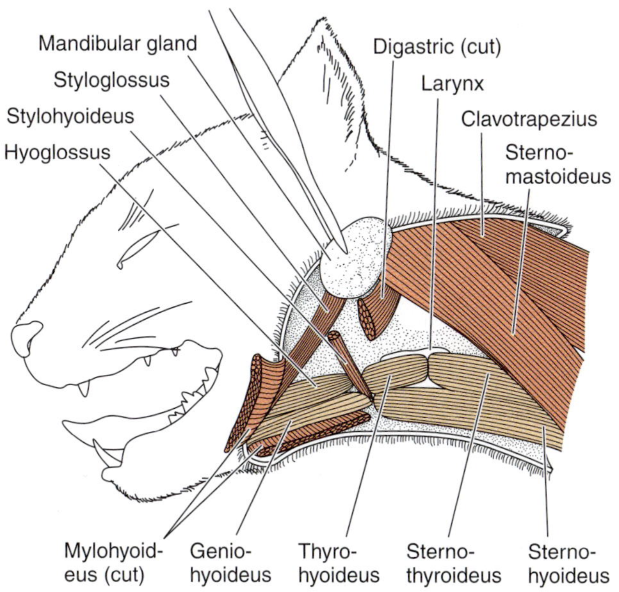

B. Study of the hypobranchial and adjacent muscles of *Felis*

FIGURE 10-15
Evolution of the hypobranchial muscles. Hypobranchial muscles are shown in yellow; adjacent branchiomeric muscles, in red; appendicular muscles, in green. *A,* A ventral view of the hypobranchial muscles of the dogfish, *Squalus. B,* A lateral view of the hypobranchial muscles of a cat, *Felis.*

Hypobranchial muscles lying rostral to the hypobranchial apparatus in amphibians include a **geniohyoideus** and one or more slips that enter the newly evolved, muscular tongue. Often, only a single **genioglossus** (Gr., *glossa* = tongue), which retracts the tongue and moves food back into the gullet, is present. The rectus cervicis of many amphibians often does not differentiate into separated muscles. An amphibian uses fishlike movements of the pharynx floor to move air into the lungs rather than water across the gills (Chapter 18). In *Necturus* and some other species, slips of the rectus cervicis, such as an **omoarcual** (Gr., *omos* = shoulder) and **pectoriscapularis,** attach to the pectoral girdle (Fig. 10-13*A*).

Hypobranchial muscles become more complex in amniotes, especially in mammals that use their tongues to manipulate food between their teeth during mastication (Fig. 10-15*B* and Table 10-5). In addition to the geniohyoideus and genioglossus present in some amphibians, mammals have several additional prehyoid muscles that extend into the tongue: a **hyoglossus,** a **styloglossus,** and a **lingualis.** The lingualis is an intrinsic muscle of the tongue that is confined to the organ and constitutes much of its substance. Muscles extending between the sternum, the thyroid cartilage of the larynx, and the hyoid are a **sternohyoideus,** a **sternothyroideus,** a **thyrohyoideus,** and sometimes an **omohyoid.** The actions of the hypobranchial muscles of mammals are very important in the complex movements of the hyoid apparatus and larynx that occur during jaw opening, food transport, and swallowing (Chapter 16). This group of muscles continues to be innervated by the equivalent of the ventral rami of spino-occipital nerves. The occipital component of this nerve in amniotes emerges from the back of the skull as a distinct cranial nerve, the **hypoglossal nerve** (XII).

Epibranchial Muscles

The rostral part of the epibranchial musculature, which lies dorsal to the gill region, attaches onto the cranium. Epibranchial muscles play a role in some fishes and amphibians in elevating the cranium when the mouth is opened during feeding (Chapter 16). But for the most part, the epibranchial muscles represent a forward extension of the epaxial trunk muscles.

Trunk and Tail Muscles

Fishes The embryonic myotomes of the trunk and tail develop into a series of folded muscle segments, the **myomeres,** of adult fishes (Fig. 10-16). The sequential contraction of the myomeres, acting with the

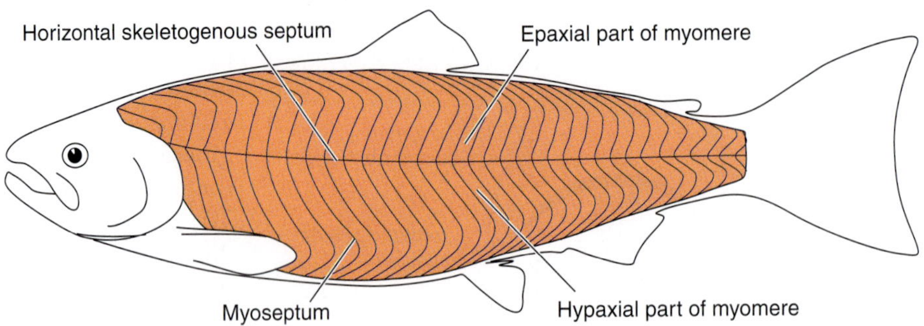

FIGURE 10-16
A lateral view of the trunk muscles of a salmon, *Salmo*. (*After Bond.*)

vertebral column, causes a series of lateral undulations by which fishes swim. Myomere structure is, therefore, correlated closely with the pattern of swimming (Chapter 11). The individual myomeres are separated from each other by connective tissue **myosepta,** which extend inward to the vertebral axis. A horizontal skeletogenous septum does not develop in jawless fishes, but one is present in all jawed fishes and divides the myomeres into **epaxial** and **hypaxial** parts (Fig. 10-8*A* and Table 10-6).

Most parts of the embryonic myotomes transform into the myomeres, but, in addition, small buds that separate from the myotomes become associated with the median fins. The buds differentiate into small mus-

TABLE 10-6 Trunk Muscles

Group	Fish	Amphibian	Mammal	
Epaxial muscles (dorsal rami of spinal nerves)	Epaxial portion of myomeres	Interspinalis	Interspinalis	Transversospinalis
		Dorsalis trunci	Intertransversarii	
			Occipitals	
			Multifidi	
			Spinalis	
			Semispinalis	
			Longissimus dorsi	Longissimus
			Splenius	
			Iliocostalis	Iliocostalis
Hypaxial muscles (ventral rami of spinal nerves)	Hypaxial portion of myomeres	Subvertebralis	Longus colli	Subvertebral
			Psoas minor	
			Quadratus lumborum	
		Levator scapulae	Omotransversarius	Lateral
		Thoraciscapularis	Serratus ventralis (part)	
		External oblique	Serratus ventralis (part)	
			Rhomboideus	
			Serratus dorsalis	
			Scalenus	
			Rectus thoracis	
			External oblique	
			External intercostals	
		Internal oblique	Internal oblique	
			Internal intercostals	
		Transversus	Transversus abdominis	
			Transversus thoracis	
		Rectus abdominis	Diaphragm muscles	Ventral
			Rectus abdominis	

cles that attach to the radial elements in the dorsal and anal fins and control the shapes and movements of the fins. Most of these movements help maintain stability, but some species of teleosts swim slowly by undulations that travel along the fins rather than down the trunk.

Tetrapods Major changes occurred in the trunk and tail muscles of tetrapods because the role of lateral undulations of the trunk and tail decreases in locomotion, whereas the importance of appendicular movements increases. Lateral undulations of the trunk and tail continue to contribute to locomotion in some amphibians (salamanders) and reptiles (squamates), and the myomeres remain segmented. Segmentation of the trunk is greatly reduced in birds and mammals. Although trunk muscles become less important in locomotion, they play an important role in mediating flexion and extension of the spine and in supporting the body against gravity. Trunk muscles also control the movement of the head, which becomes independent of trunk movements, and the actions of some help ventilate the lungs.

Epaxial Muscles of Tetrapods The epaxial muscles are particularly important in supporting the body and moving the vertebral column and head. They are represented in salamanders primarily by a **dorsalis trunci,** which remains segmented (Figs. 10-17*A* and 10-18*B*, Table 10-6). Deep parts form a series of small **interspinalis** muscles that interconnect adjacent vertebrae.

As trunk and head movements became more complex in reptiles and mammals, the epaxial muscles have become more specialized. Typically, three longitudinal bundles extend forward from the sacrum and pelvis to the head (Figs. 10-17*B* and 10-18*C*). The most medial and deepest bundle in reptiles is a **transversospinalis** system, which consists of short fibers extending between adjacent vertebrae, or only across a few body segments. A more lateral **longissimus dorsi** consists of longer fibers, many of which extend from the sacrum to the neck. As the most lateral **iliocostalis** ex-

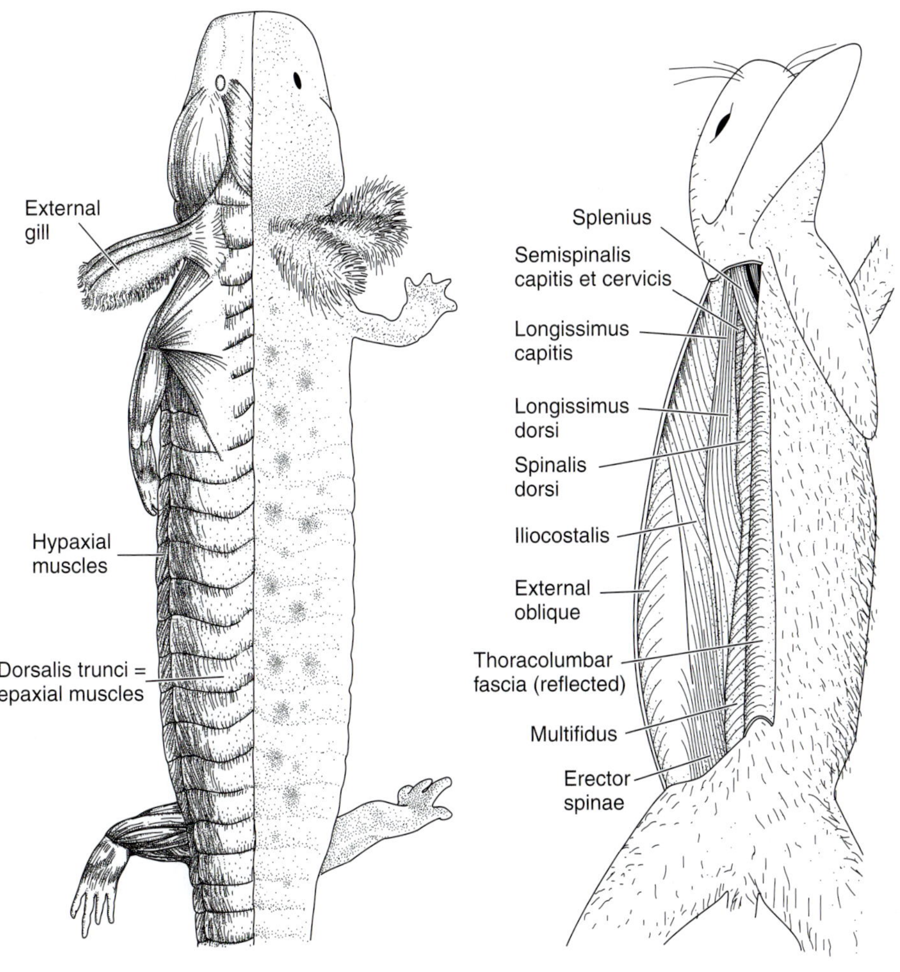

FIGURE 10-17
The evolution of the epaxial muscles. *A*, Dorsal view of the trunk muscles of a salamander, *Necturus*. *B*, Dorsal view of the epaxial muscles of a rabbit, *Oryctolagus*. (*B, after Walker and Homberger.*)

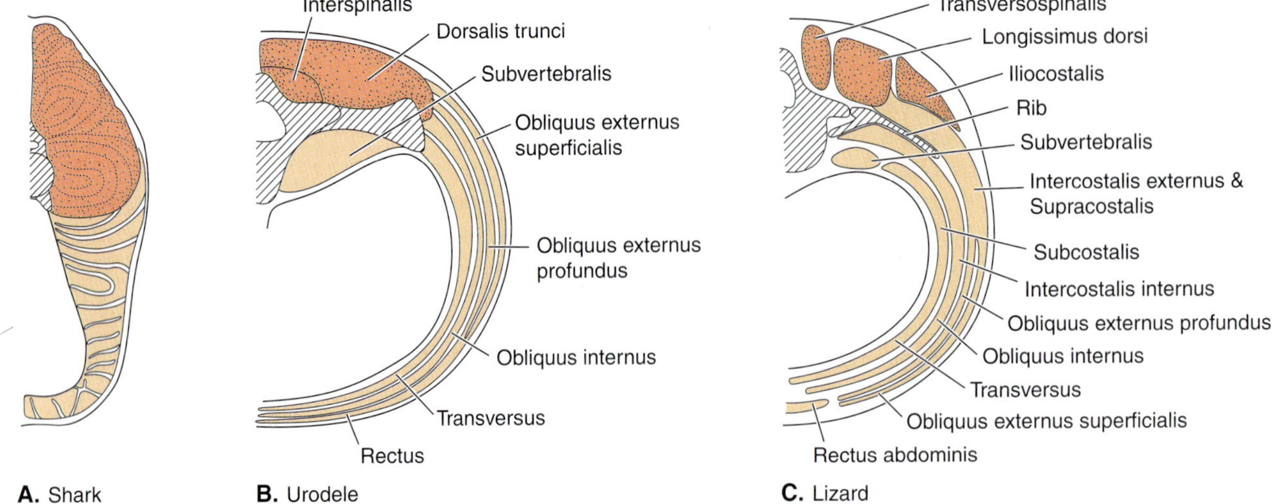

FIGURE 10-18
Evolution of the trunk muscles as seen in cross sections through the trunk. Epaxial muscles are stippled; hypaxial ones are not. *A,* A shark. *B,* A urodele. *C,* A lizard. Lizards have well-developed ribs, and in *C,* a rib is assumed to be present dorsally. The dorsal hypaxial names in *C* are those used in the rib-bearing region; the ventral names are those used in the abdominal region. (*After Romer and Parsons.*)

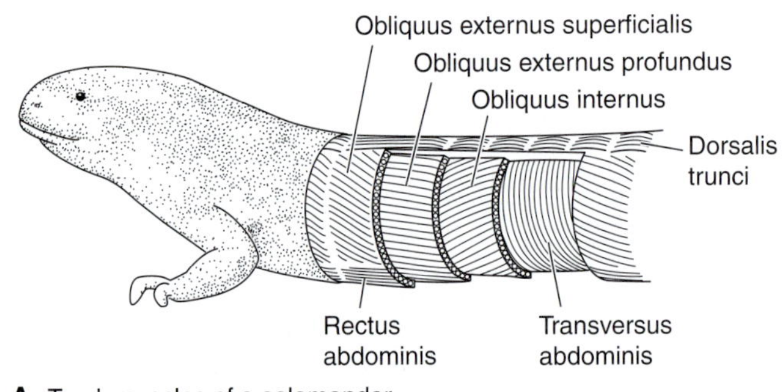

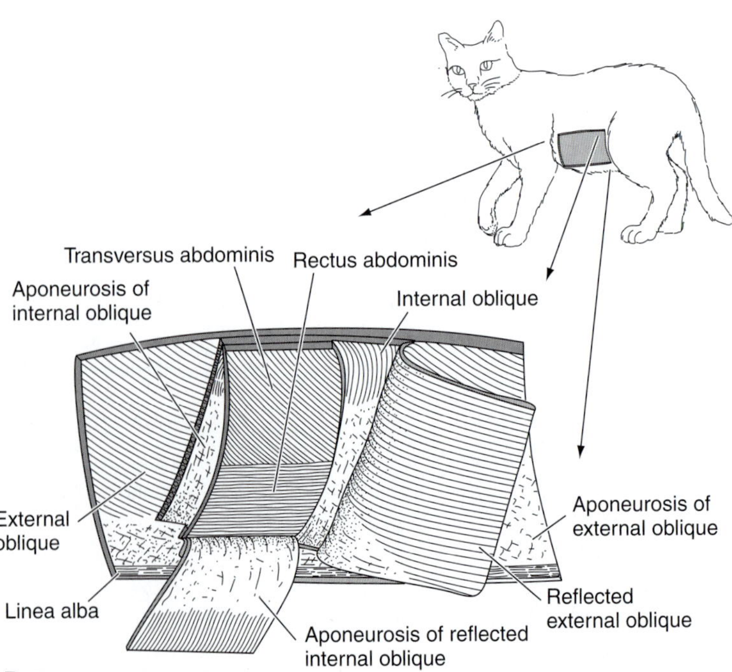

FIGURE 10-19
Evolution of the hypaxial muscles as seen in lateral views. Parts of superficial muscle layers are removed to show deeper layers. *A,* A salamander, *Dicamptodon. B,* A cat. (*A, After Carrier; B, after Walker and Homberger.*)

FOCUS 10-3 *Actions of Trunk Muscles in Tetrapods*

Because all parts of the myomeres of fishes contribute to lateral bending of the trunk and tail, it is logical to assume that this continues in tetrapods, such as salamanders and squamates, in which lateral undulations continue to play a role in locomotion. Frolich and Biewener (1992) demonstrated that activity of the epaxial muscles does correlate with lateral bending in the salamander, *Ambystoma tigrinum.* Does activity of the hypaxial muscles also contribute to lateral bending? Carrier (1993) found that they do participate in lateral bending during swimming of the Pacific giant salamander, *Dicamptodon ensatus.* But during walking, only the two external obliques are active on the flexing side; the internal oblique and transversus are active only on the opposite side. Because the actions of these two sets of muscles are opposing each other, Carrier believes that they could not contribute significantly to lateral flexion. Carrier proposes that the oblique orientation of most of these muscles, together with their different lines of action, enables them to help resist the torsion of the trunk that occurs during walking (see Fig. 8-11*B*).

Ritter (1995, 1996) has studied the locomotor functions of the trunk muscles in the lizards *Iguana iguana* and *Varanus salvator.* He finds that the epaxial iliocostalis and longissimus dorsi are active on the side of the body in which a rear leg is providing support and thrust. Because the front leg on the same side (ipsilateral) of the body tends to be off of the ground and moving forward, the ipsilateral shoulder would tend to fall toward the ground. Ritter proposes that the epaxial muscles act to stabilize the back, that is, to prevent the shoulder from dropping too far. He also finds that lateral bending of the trunk correlates not with epaxial muscle activity, but with the contraction of the hypaxial external oblique and rectus abdominis.

This finding does not completely agree with Carrier's for salamanders. Carrier did find that the action of the external oblique correlated with lateral bending but that muscles on the opposite side of the body opposed it. Carrier did not study the activity of the rectus abdominis. These questions are being studied further. Ritter proposes that with the evolution of amniotes the role of the epaxial muscles shifted from primarily causing lateral bending of the trunk to a supportive and stabilizing role. This correlates with the change in the morphology of the most conspicuous epaxial muscles from being composed of short segmented fibers to the formation of long longitudinal columns.

One or more hypaxial muscles of tetrapods extend from the trunk to the pectoral girdle (see Fig. 11-11). These muscles act on the girdle, often transferring body weight from the trunk to the pectoral girdle and appendage. No direct bony connection exists between the pectoral girdle and the vertebral column, although, in amniotes, the girdle usually has an indirect connection via the sternum and ribs. Among these muscles in mammals are a large serratus ventralis, a rhomboideus, and an omotransversarius (or levator scapulae; Table 10-6).

In addition to their roles in abdominal support, locomotion, and trunk stabilization, hypaxial muscle layers have a major role in respiration in amniotes (Chapter 18). This includes both inspiration and expiration. Other hypaxial muscles may participate in mammals. These include a serratus dorsalis and a scalenus group. The primary inspiratory muscle of mammals lies in the diaphragm, a partition that separates the thoracic and abdominal cavities (Chapter 4). It, too, is of hypaxial origin because it develops embryonically from the ventral part of cervical myotomes. All hypaxial muscles are innervated by the ventral rami of spinal nerves.

It had long been thought that hypaxial muscles have no role in respiration in amphibians, but recent studies have shown that contraction of at least the transversus abdominis is important in expiration (Brainerd et al., 1993; Brainerd and Monroy, 1998).

tends forward from the sacrum and part of the pelvic girdle (the ilium), it gives off slips that insert on the ribs or parts of the cervical vertebrae derived from ribs. All three of these bundles remain partly segmented.

Segmentation of the epaxial musculature is lost in mammals, except in the deeper parts of the transversospinalis system. This band of muscles is represented primarily by the **multifidi** and **spinalis dorsi** (Fig. 10-17*B*). The posterior parts of the mammalian **longissimus dorsi** and **iliocostalis** often partially unite to form a large **erector spinae.** As these bundles approach the head, they become subdivided into many smaller muscles that support and move the head (Table 10-6). All of the epaxial muscles are innervated by dorsal rami of spinal nerves.

Hypaxial Muscles of Tetrapods The hypaxial muscles of tetrapods are subclassified into three major groups (Figs. 10-18*B* and *C* and 10-19):

1. A **subvertebral group** (Table 10-6) lies ventral to the vertebral column, acts on the vertebral axis, and assists the epaxial muscles in supporting the body and bending the vertebral column.
2. A ventral group includes the **rectus abdominis,** which extends longitudinally on either side of the midventral line. It helps support the abdomen, and its longitudinal fibers allow it to contribute to lateral and ventral flexion of the trunk.
3. A **lateral group** lies on the flank and forms three or four layers. These layers are particularly distinct in

the abdomen, where we may find (from the surface inward) an **obliquus externus** (which may be divided into superficial and deep layers in salamanders and squamates), an **obliquus internus,** and a **transversus abdominis** (Figs. 10-18*B* and *C* and 10-19*A* and *B*). Simons and Brainerd (1999) have surveyed this musculature in salamanders and found considerable variation. The obliquus externus may or may not be subdivided, and, in a few species, either the obliquus internus or the transversus abdominis may be absent. Fibers in each layer run at different angles. The layers are thick and continue to be segmented in salamanders, but they form broad, thin sheets in amniotes.

Multiple layers of hypaxial muscles with fibers running in different directions certainly support and strengthen the trunk wall, as multiple layers strengthen plywood. There are three layers of intercostal muscles in the parts of the trunk in which ribs are present, including the mammalian thorax. **External** and **internal intercostals** are well developed in all amniotes (Fig. 10-18*C*). The innermost transverse layer is also well developed in reptiles, where it forms a **subcostalis,** but in mammals, this layer is represented by an incomplete **transversus thoracis**. These muscles also play important roles in locomotion, trunk stabilization, and respiration (Focus 10-3 p. 341).

The Evolution of Appendicular Muscles

The paired appendages of most fishes do not deliver major propulsive thrusts. The fins may provide lift, but they are used primarily in maintaining stability, braking, and maneuvering. Paired fin movements and appendicular muscles usually are quite simple (Fig. 10-20*A*). Often, as in sharks, a single dorsal muscle, called the **extensor,**[1] is located on the dorsal part of the fin. It pulls the fin dorsally. A ventral muscle, the **flexor,** is located on the ventral surface of the fin and pulls the fin ventrally. In species whose fin movements are more complex, the simple flexor and extensor muscles become subdivided and more specialized.

[1]These muscles often are called the abductor (dorsal muscles) and adductor (ventral muscle) in sharks because they respectively pull the fin away from the midventral line of the body, the point of reference for the action of these muscles, or away from it. In actinopterygians, the reference point is the lateral surface of the body, so the dorsal fin muscle is the adductor, and the ventral one, the abductor. Calling them the flexor and extensor avoids this confusion.

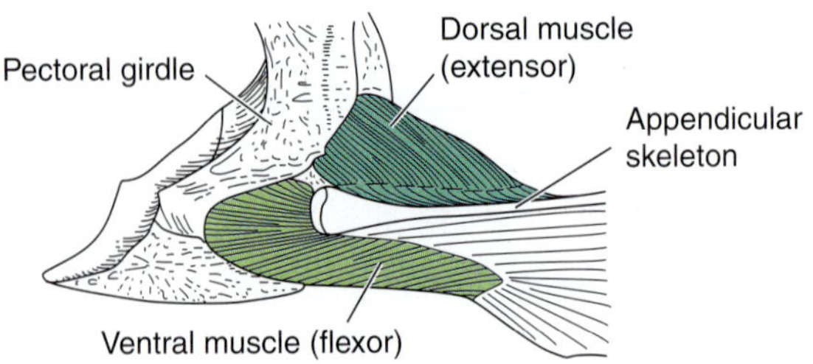

A. Pectoral fin muscles of a sturgeon

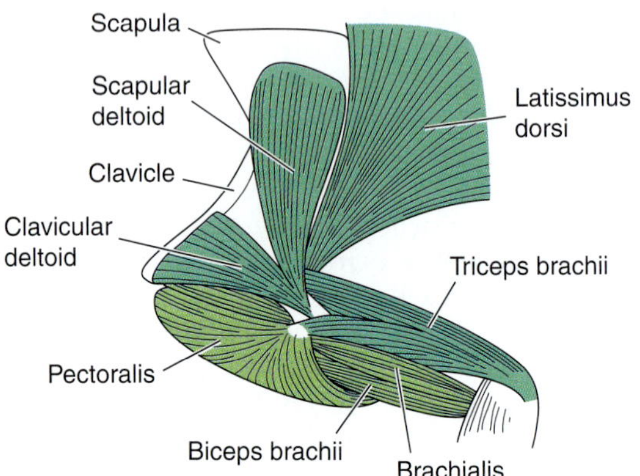

B. Superficial forelimb muscles of a lizard

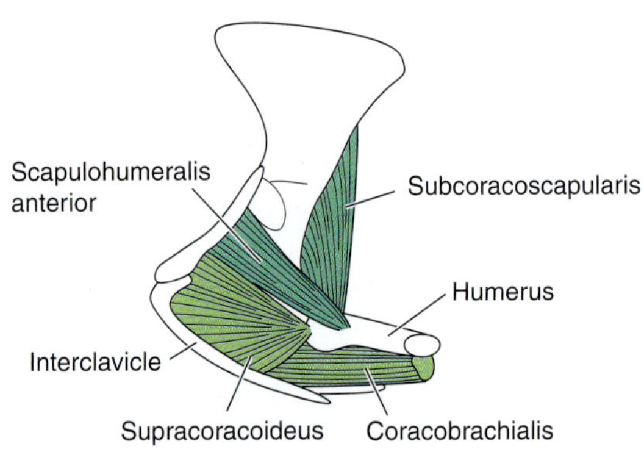

C. Deep forelimb muscles of a lizard

FIGURE 10-20
Evolution of the pectoral appendicular muscles. Dorsal appendicular muscles are shown in dark green; ventral ones, in yellow-green. *A,* Anterolateral view of the muscles in a sturgeon, *Acipenser*; *B* and *C,* superficial and deep lateral views of the muscles in a lizard, *Lacerta*. (*After Romer and Parsons.*)

The structure and movements of the paired appendages of terrestrial vertebrates are far more complex than those of the paired fins of fishes because the limbs support the body and provide the major propulsive thrusts. Indeed, appendicular muscles constitute the bulk of the muscular system in many tetrapods. Despite their complexity, tetrapod appendicular muscles

TABLE 10-7 Appendicular Muscles of the Pectoral Girdle and Appendage

Group	Fish	Reptile (Lizard)	Mammal
Dorsal group	Extensor	Latissimus dorsi	Cutaneous trunci (part) Latissimus dorsi Teres major
		Subcoracoscapularis	Subscapularis
		Deltoid	Deltoid complex
		Scapulohumeralis anterior	Teres minor
		Triceps brachii	Triceps brachii Tensor fasciae antebrachii
		Antebrachial extensors	Antebrachial extensors
Ventral group	Flexor	Pectoralis	Cutaneous trunci (part) Pectoralis complex
		Supracoracoideus	Supraspinatus Infraspinatus
		Biceps brachii	Biceps brachii
		Brachialis	Brachialis
		Coracobrachialis	Coracobrachialis
		Antebrachial flexors	Antebrachial flexors

can be subdivided into dorsal and ventral groups on the basis of their embryonic development from two premuscular masses that lie, respectively, above and below the developing appendicular skeleton. These masses cleave into many components during the later development of a tetrapod. Collectively, the muscles that develop from the dorsal mass are homologous to the fish extensor; those from the ventral mass, to the fish flexor (Table 10-7).

We will discuss the functional anatomy of support and locomotion in Chapter 11. At this time, we can say that the dorsal muscles of both tetrapod pectoral and pelvic limbs abduct and extend the limbs as a whole and their various segments. These actions occur during the swing phase of a step, when each limb in turn is removed from the ground and extended. Ventral muscles do the opposite—they adduct the limbs as a whole and flex their distal segments. These actions occur during the stance phase, when each limb in turn is placed on the ground and develops a propulsive thrust. In addition to these actions, the pectoral and pelvic limbs as a whole are advanced or protracted during the swing phase. A force is developed that could retract or draw the limbs caudad during the stance phase, but because the feet remain on the ground, this force advances the trunk relative to the feet. Both dorsal and ventral muscles participate in these actions, depending on whether their lines of action lie anterior (protraction) or posterior (retraction) to the shoulder and hip joints. Limb muscles also brace the bones across the joints, holding the body up on the legs.

The splayed limb position of amphibians and most reptiles necessitates a powerful ventral musculature that adducts the humerus and femur and flexes the antebrachium and crus so as to raise the body from the ground. Nearly vertical bony columns support the body of many mammals more directly because the limbs have rotated closer to the body axis, and movement of all limb segments is primarily back and forth in a parasagittal plane. Powerful limb adductors no longer are needed. Ventral muscles that act across the shoulder joint become much less important. Ventral parts of the girdle are reduced, as we have seen (Chapter 9), and some ventral muscles are reduced or shifted dorsally. Dorsal muscles are better positioned to protract and retract the mammalian humerus. Some reduction of the ventral muscles that act across the hip joint also has occurred during the evolution of mammals, but reduction of the ventral parts of the pelvic girdle and its muscles is not as pronounced as in the shoulder. Limb movements became more complex as tetrapods evolved into more active and agile animals, and, with this, some subdivision and multiplication of both dorsal and ventral muscles also has occurred.

Appendicular muscles of tetrapods are exceedingly numerous and are best sorted out in the laboratory. We will consider the major muscles of the pectoral girdle and brachium as an example of the evolutionary

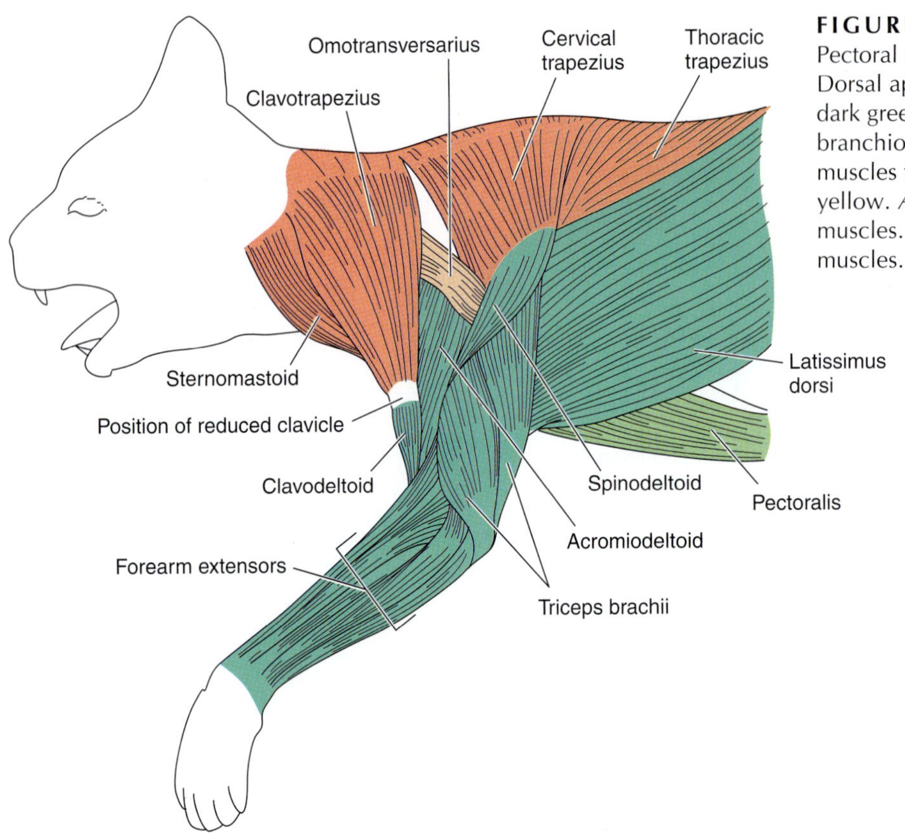

A. Shoulder and forearm muscles of *Felis*

FIGURE 10-21
Pectoral appendicular muscles of a cat. Dorsal appendicular muscles are shown in dark green; ventral ones, in yellow-green; branchiomeric ones, in red; and axial muscles that act on the shoulder, in yellow. *A,* A lateral view of the superficial muscles. *B,* Lateral view of the deeper muscles. *C,* Medial view of the muscles.

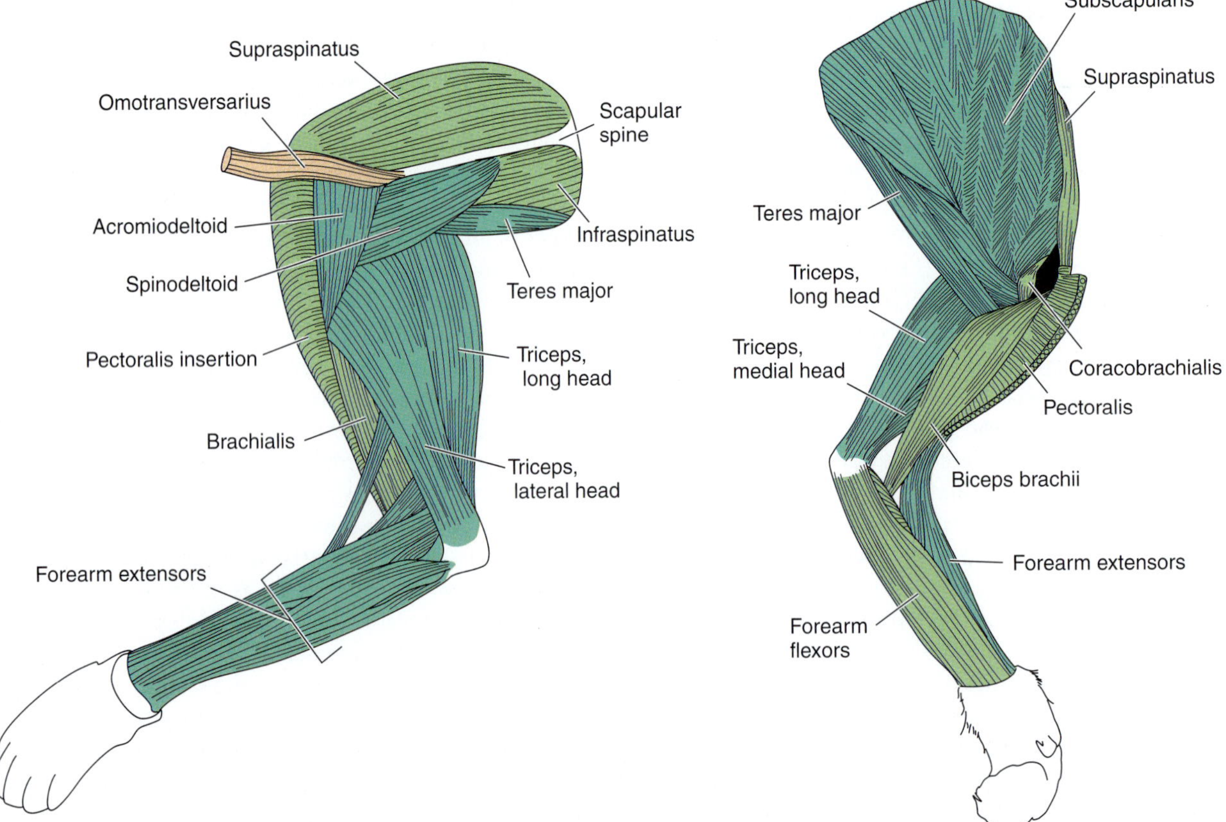

B. Lateral view of deep shoulder and forearm muscles of *Felis*

C. Medial view of shoulder and forearm muscles of *Felis*

changes that occur in appendicular muscles. Major changes can be appreciated by comparing these muscles in a reptile (lizard) and a mammal.

The simple extensor and flexor muscles of fishes cleave into many separate muscles in early tetrapods as limb movements become more complex, but dorsal and ventral muscle groups can be recognized. Dorsal appendicular muscles on the shoulder and brachium of reptiles (Fig. 10-20*B* and *C*, Table 10-7) are a **latissimus dorsi, deltoid,** and **triceps brachii,** and, more deeply, a **scapulohumeralis anterior** and a **subcoracoscapularis.** Notice in particular that the subcoracoscapularis extends from the posterior border of the scapula and coracoid to the proximal end of the humerus. Ventral muscles are large. A **pectoralis, biceps brachii,** and **brachialis** lie superficially, and a **supracoracoideus** and **coracobrachialis** lie more deeply. The supracoracoideus and powerful coracobrachialis arise from the large coracoid plate (which lies deep to the supracoracoideus). The coracobrachialis extends to the distal end of the humerus.

The mammalian shoulder is covered dorsolaterally by muscles that are not true appendicular muscles in the sense that they did not develop from mesenchyme within the limb bud (Fig. 10-21*A*). The **sternomastoid** and the three parts of the **trapezius** are branchiomeric muscles that have secondarily acquired an attachment to the pectoral girdle. They evolved from the fish cucullaris (Fig. 10-12*B*). The **omotransversarius** (also called the levator scapulae) is one of several hypaxial trunk muscles that act on the scapula. Others attach to the medial side of the scapula (Fig. 11-11).

Dorsal appendicular muscles continue to be important, for they help to brace the bones and hold the animal up on its limbs, and they are in a good position to swing the limb fore and aft (Fig. 10-21). The deltoid often is subdivided to a greater extent than in reptiles, and a part of the reptilian latissimus dorsi separates as a **teres major.** The reptilian scapulohumeralis anterior remains as a small **teres minor,** but it is located very deeply and is not shown in the figures. The reptilian subcoracoscapularis enlarges greatly and covers the medial surface of the scapula. It is now called the **subscapularis** because the large coracoid plate of reptiles is reduced in mammals to a small coracoid process on the scapula.

Because the limbs of mammals are no longer splayed but are pulled nearly under the trunk, major changes have occurred in the ventral appendicular muscles. The pectoralis remains an important muscle in adducting the limbs and participating in their protraction and retraction. It is often subdivided into several parts. With the loss of the coracoid plate, the supracoracoideus migrates dorsally during embryonic development onto the lateral surface of the scapula, part passing in front of and part behind the scapular spine. These muscles are now called the **supraspinatus** and **infraspinatus,** respectively, and are in a good position to assist in the fore and aft swing of the appendage. The large coracobrachialis of reptiles remains as a small **coracobrachialis,** which extends from the coracoid process of the scapula only to the proximal end of the humerus.

Many mammals can wiggle their skin in order, for example, to shake off insects. A thin but extensive **cutaneous trunci** fans out from the armpit to insert onto the skin over the trunk. Its origin from the base of the pectoralis and latissimus dorsi indicates its evolutionary origin from these muscles. It is listed in Table 10-6 but not shown in the figures. Some mammals also have a small **tensor fasciae antebrachii,** a derivative of the triceps that inserts into the fascia over the forearm.

Electric Organs

Many tissues, such as muscles, glands, and nerves, generate weak electric currents when they become active. Such currents travel through water because water, even fresh water, contains ions. Many fishes and a few other aquatic vertebrates have evolved **electroreceptors** that detect the weak currents generated by the muscle activity of other animals, and they utilize this information in finding prey (Chapter 12). About 250 species of fishes have evolved specialized **electric organs** that produce stronger electric currents. They use these, coupled with their electroreceptors, for electrocommunication, navigation, and sometimes in prey capture and defense. We will consider electric organs at this time because the tissue in electric organs is usually modified muscle.

Electric organs occur in many different groups, differ in many aspects of their construction, and must have evolved independently many times. Electric organs occur in some tropical teleosts: the gymnotids (relatives of minnows) of Central and South America, the mormyrids (osteoglossomorphs) of Africa, and the electric catfish, *Malapterurus,* of the Nile. The only saltwater fishes with electric organs are several rays, of which *Torpedo* is the most famous, and the stargazer, *Uranoscopus.* The electric organs of mormyrids and most of the gymnotids generate trains of relatively weak electric pulses. Some pulses are used in species and sex recognition; others form an electric field that the fish uses to navigate in the murky waters of its habitat. Objects whose electric conductive properties differ from that of the water distort the field, and the fish detects these distortions. The fish avoids distorting the electric field itself by keeping its trunk and tail straight and swimming slowly by undulating an elongated anal or dorsal fin. A familiar example of a gymnotid is the knifefish (*Apteronotos albifens),* often sold in aquarium stores. It easily moves forward or backward, or stays in position, by undulations of an exceedingly long anal fin, which extends forward to the underside of the head. Other fins are much reduced or absent.

The electric eel, the electric catfish, and the marine species can generate electric currents sufficiently strong to stun, and sometimes kill, prey or predators. *Torpedo* and *Electrophorus* have the most powerful organs. *Torpedo* can generate a current flow as high as 50 amperes at 50 volts. (Amperage is a measure of the amount of current; voltage is the push or force driving the current.) This is a total power of 2500 watts (watts = amperes × volts). *Electrophorus* also can generate a power of 2000 watts or more, but it must do so at a higher voltage, up to 500 volts, because fresh water is a poorer conductor of electricity than is salt water. The amperage in *Electrophorus* is much lower, about 4 amperes. The powerful electric organs likely evolved from weaker ones used for communication and navigation. Once they attained enough power to stun other organisms, natural selection could favor this use. Darwin was hard put to explain the early stages in the evolution of these powerful organs, for fish with weaker organs were unknown in his time.

Different tissues or groups of muscles are utilized in the various fish groups (Fig. 10-22). The electric organ of the electric "eel," *Electrophorus* (a gymnotid), consists of modified hypaxial trunk muscles; that of *Malapterurus* forms a sheath directly beneath the skin of the trunk and tail. It may be specialized glandular tissue. Branchiomeric muscles form the electric organ of *Torpedo,* and extrinsic ocular muscles form that of *Uranoscopus.*

Most electric organs consist of a series of disk-shaped **electroplaques,** each of which represents a modified muscle cell or its motor end plate. In one gymnotid, *Sternarchus,* the electroplaques evolved from the motor neurons themselves rather than from the motor end plates. The structure and activity of an electric organ can be appreciated by examining that of *Electrophorus.* Each electroplaque is a multinucleated, disk-shaped cell that is flat on the innervated side and highly folded on the other side (Fig. 10-23). Most myofibrils have been lost, although traces are sometimes found. At rest the outside of the plasma membrane is positive relative to the inside. The resting potential across the membrane is –84 millivolts (mv), about the same as in a normal muscle cell. When the organ discharges, only the innervated surface becomes depolarized, and its potential rises to +67 mv. This

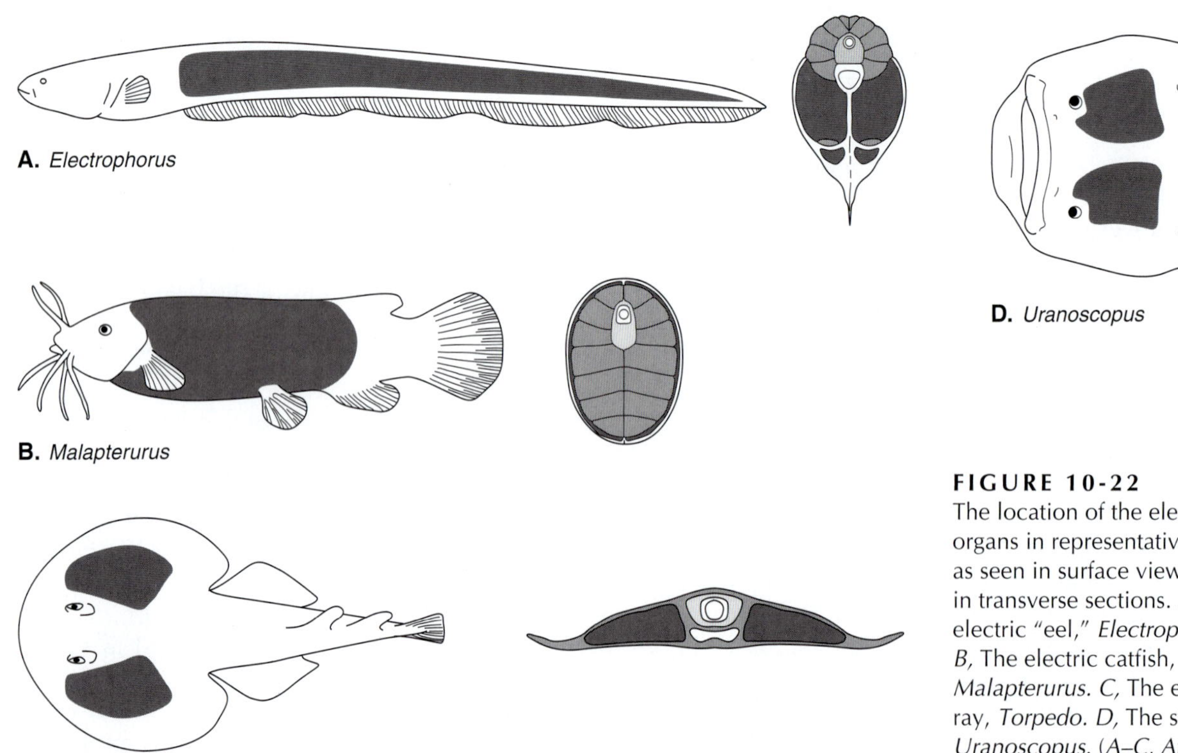

FIGURE 10-22
The location of the electric organs in representative fishes as seen in surface views and in transverse sections. *A,* The electric "eel," *Electrophorus. B,* The electric catfish, *Malapterurus. C,* The electric ray, *Torpedo. D,* The stargazer, *Uranoscopus. (A–C, After Portman; D, after Bond.)*

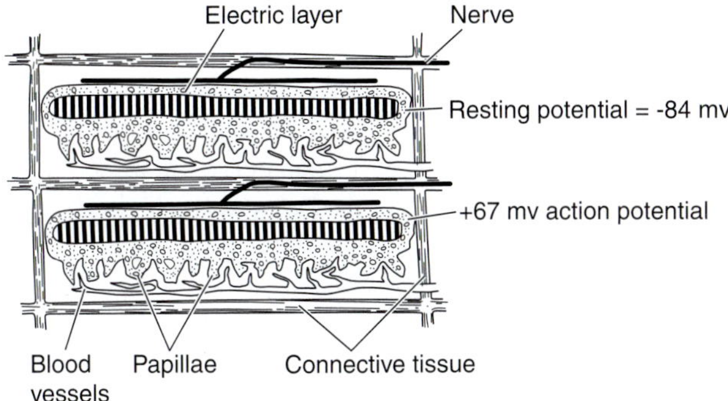

FIGURE 10-23
Two electroplaques from the electric organ of *Electrophorus*. The resting potential is indicated in the top one; the action potential, in the bottom one. Only the innervated surface becomes depolarized. (*After Hoar.*)

gives a total voltage range across the entire electroplaque of 151 mv: +67 mv on the depolarized surface to −84 mv on the other. The exceedingly high voltage of the entire organ in *Electrophorus* derives from the stacking of the electroplaques in long longitudinal columns along the trunk (Fig. 10-22), each containing up to 10,000 electroplaques. About 70 such longitudinal columns are found on each side of the body. Because the electroplaques in each column are in series and discharge simultaneously, the voltage of each one is added to that of the others in the same series. The columns of electroplaques in *Torpedo* extend vertically, so they are much shorter, but as many as 2000 parallel columns are present on each side of the body. The parallel arrangement reduces the resistance to current flow and results in a higher amperage than in *Electrophorus*, but the shorter columns lower the voltage.

SUMMARY

1. Muscles are the major effectors of vertebrates. They move the body and most materials through it, and they help support the body by bracing limb bones across joints.
2. Skeletal muscle tissue consists of long, multinucleated, and striated cells. Contraction is initiated by nerve impulses that reach the cells at the motor end plates of neurons.
3. Cardiac muscle tissue is composed of moderately elongated, striated, and branching cells that are tightly united to each other by intercalated disks. Its myogenic rhythm is modulated by nerve impulses.
4. Smooth muscle tissue is composed of moderately elongated, nonstriated, spindle-shaped cells. Its contraction is slower and more sustained than in other muscle tissue. The control of unitary smooth muscle, which occurs in the wall of most visceral organs, is myogenic but is modulated by nerve impulses. The control of multiunit smooth muscles in the walls of blood vessels is neurogenic.
5. Smooth and cardiac muscle fibers are arranged in sheets and layers within the walls of visceral organs and blood vessels. Skeletal muscle fibers usually attach to parts of the skeleton by the extension of the connective tissue within them to form tendons and aponeuroses.
6. Muscle fibers within a muscle are organized as motor units, each of which consists of a motor neuron and all of the fibers it supplies. Motor units containing few muscle fibers are adapted for delicate movements; those with many fibers are adapted to generate a strong force.
7. Muscle cells or fibers contain protein myofilaments composed primarily of actin and myosin. In striated and cardiac muscle, the myofilaments are grouped into myofibrils, which are barely visible by light microscopy. Muscle contraction results from interactions between actin and myosin myofilaments, which causes the actin myofilaments to be pulled into the array of myosin ones.
8. Muscle may shorten against a fixed load (isotonic contraction), contract when their ends are held in a fixed position (isometric contraction), or lengthen as they generate force (negative work contraction).
9. Muscles contract against an antagonistic muscle or force, which restores the muscle to its resting length. Muscles usually act in synergic groups.
10. Tonic skeletal muscle fibers have multiple motor end plates. The number of motor impulses reaching them grades their contraction. Tonic muscles are usually small and occur where slow, sustained, and carefully graded contractions are needed.
11. Most vertebrate muscles are twitch, or phasic. Twitch fibers have single motor end plates. Their contraction is all or none, but the force of contraction of an entire muscle can be increased by the temporal summation of nerve impulses and by the recruitment of more motor units.
12. The metabolism of twitch fibers may be slow oxidative, fast glycolytic, or intermediate. Slow-oxidative fibers contract slowly and no faster than oxygen can be delivered to them, so they do not fatigue easily. Fast-twitch fibers derive their en-

ergy by anaerobic glycolysis. They contract rapidly but fatigue easily.

13. Many muscles are multigeared and contain populations of different fiber types. The slow fibers are recruited for slow, sustained contractions; the fast ones are recruited when a rapid but brief contraction is needed.

14. The connective tissues surrounding muscle fibers and their tendons are elastic elements that can store elastic energy when they are stretched and release this energy when they recover. Much of the force generated by whole muscles during their cyclic activity derives from elastic energy.

15. Strap-shaped and fusiform muscles containing long, parallel fibers can induce a more extensive movement than pennate muscle containing many short fibers because muscle fibers can contract about one third of their resting length. Strap-shaped and fusiform fibers also contract faster because long fibers contain more sarcomeres in series.

16. Pennate muscles with many short fibers develop more force than do other types of muscle of equal mass because the force that a muscle generates is a function of the number of myosin–actin cross bridges made at one time.

17. Force and velocity of contraction are inversely related, so muscles can generate more force when they contract slowly. Because

$$\text{Power} = (\text{force}) \times (\text{velocity of contraction})$$

maximum power is generated at intermediate forces and velocities of contraction.

18. Somatic muscles, most of which lie in the body wall and appendages, develop from the paraxial somites and somatic layer of the lateral plate mesoderm. Visceral muscles develop from the splanchnic layer of the lateral plate mesoderm and form the wall of most of the gut and visceral organs. Recent studies have shown that the branchiomeric muscles of the pharynx wall also develop from paraxial mesoderm, so they are somatic muscles and not visceral ones. Except for cardiac muscles, which are striated, visceral muscles are smooth, and all are innervated by the autonomic nervous system.

19. Somatic muscles can be classified as axial or appendicular.

20. The most rostral group of axial muscles is the small, extrinsic muscles that move the eyeball. They can be classified into three subgroups innervated by the oculomotor nerve (cranial nerve III), the trochlear nerve (IV), and the abducens nerve (VI). Extrinsic ocular muscles are quite conservative and change only slightly during evolution.

21. The next group of axial muscles is the branchiomeric muscles associated with the visceral skeleton. They can be sorted into the mandibular muscles that attach to the mandibular arch and are innervated by the trigeminal nerve (cranial nerve V), the hyoid muscles that attach onto the hyoid arch and are innervated by the facial nerve (VII), and the branchial muscles that attach onto the remaining branchial arches. Those attaching onto the first branchial arch (visceral arch 3) are supplied by the glossopharyngeal nerve (IX); those attaching to the remaining arches are innervated by the vagus nerve (X).

22. The mandibular muscles primarily act to close the jaws, and, in some species, especially those with kinetic skulls, they help move them back and forth. Jaw movements are quite complex in mammals that masticate their food, and mandibular muscles differentiate into many components. One part of the mandibular musculature, the tensor tympani, remains attached to the malleus of the middle ear, a derivative of the mandibular arch.

23. The hyoid musculature attaches onto the hyoid arch in fishes, so it is often involved in the suspension and movement of the jaws. A part of the hyoid musculature of amphibians forms the depressor mandibulae, the primary jaw opening muscle of nonmammalian tetrapods. The remodeling of the jaw that occurred in the evolution of mammals led to the formation of another jaw-opening muscle, the digastric, which is derived partly from mandibular and partly from hyoid musculature. Most of the hyoid musculature of mammals forms the facial muscles on the head and the superficial platysma on the neck. The small stylohyoid and stapedius muscles, which attach to the stapes in the middle ear, are the only parts of the hyoid musculature that remain attached to derivatives of the hyoid arch.

24. The branchial muscles are quite complex in fishes, for they are the major muscles that compress the pharynx and drive out the respiratory current of water. The superficial levator muscles of the branchial arches, which primitively may have helped to expand the pharynx, aggregate to form a cucullaris, most of which inserts and acts on the pectoral girdle. Most of the branchial muscles are reduced greatly in tetrapods, along with the loss of gills. In mammals, muscles in the wall of the pharynx and larynx represent the branchial muscles. The cucullaris expands in tetrapods and forms in mammals the trapezius and sternocleidomastoid complexes of muscles that extend from the head to the shoulder and sternum.

25. A hypobranchial group of axial muscles lies ventral to the pharyngeal region. Although not of branchiomeric origin, these muscles have acquired attachments to the ventral parts of the mandibular, hyoid, and branchial arches. They expand the pharynx and open the jaws in fishes, drawing in a current of water. Spino-occipital nerves innervate them in fishes. These muscles become divided and more complex in tetrapods, for they supply the newly evolved muscular tongue. In mammals, they participate in food manipulation in the mouth and in swallowing. The occipital component of the spino-occipital nerve of fishes becomes the hypoglossal nerve (i.e., cranial nerve XII).
26. Epibranchial axial muscles elevate the cranium in some fishes and salamanders and thus help open the mouth. They become a part of the dorsal trunk muscles in most tetrapods.
27. The trunk and tail axial musculature of fishes forms a series of segmented myomeres that cause the lateral undulations of the body by which fishes swim. They are divided in all jawed vertebrates into the dorsal epaxial muscles and the ventrolateral hypaxial muscles.
28. Trunk muscles tend to lose their segmental nature in adult tetrapods. The epaxial muscles play a prominent role in supporting the body against the pull of gravity and in moving the head. Most of the hypaxial musculature forms three or four thin sheets that support the body wall, participate in trunk undulations and stabilization in amphibians and reptiles, and participate in breathing movements. The diaphragmatic muscles belong in this group. Hypaxial muscles ventral to the vertebral column assist the epaxial muscles in body support. Several hypaxial muscles connect to the scapula and transfer body weight from the trunk to the pectoral girdle.
29. Appendicular muscles are those that develop from mesenchyme within the limb buds. Some branchial and trunk muscles secondarily attach to the pectoral girdle. The appendicular muscles of sharks are a single dorsal extensor muscle and ventral flexor. Appendicular muscles become large and numerous in tetrapods as the girdles and appendages assume the major role in locomotion. Evolutionary changes in the appendicular muscles parallel the changes in limb position and movements.
30. Muscle tissue of various origins forms the electric organs in most of those species of fishes that have such organs. Electric organs generate weak pulses used in navigation and electrocommunication in some species. This may have been their primitive function. The organs become powerful in other species and can generate currents strong enough to stun prey or predators.

REFERENCES

Bennett, M. V. L., 1969: Electric organs. *In* Hoar, W. S., and Randall, D. J., editors: *Fish Physiology.* New York, Academic Press.

Bond, C. E., 1996: *The Biology of Fishes.* Philadelphia, Saunders College Publishing.

Bourne, G. H., editor, 1972: *The Structure and Function of Muscle.* New York, Academic Press.

Brainerd, E. L., and Monroy, J. A., 1998: Mechanics of lung ventilation in a large aquatic salamander, *Siren lacertina. Journal of Experimental Biology,* 201:673–682.

Brainerd, E. L., Ditelberg, J. S., and Bramble, D. M., 1993: Lung ventilation in salamanders and the evolution of the vertebrate air-breathing mechanism. *Biological Journal of the Linnean Society,* 49:163–183.

Butler, A. B., and Hodos, W., 1996: *Comparative Vertebrate Neuroanatomy: Evolution and Adaptation.* New York, Wiley-Liss.

Carrier, D. C., 1993: Action of the hypaxial muscles during walking and swimming in the salamander *Dicamptodon ensatus. Journal of Experimental Biology,* 180:75–83.

Chen, C. C., 1955: The development of the shoulder region of the opossum, *Didelphis virginiana,* with special reference to the musculature. *Journal of Morphology,* 97:415–472.

Davis, R. L., Weintraub, H., and Lasser, A., 1987: Expression of a single transfected cDNA converts fibroblasts into myoblasts. *Cell,* 51:987–1000.

Dorit, R. L., Walker, W. F., and Barnes, R. D., 1991: *Zoology.* Philadelphia, Saunders College Publishing.

Edgeworth, F. H., 1935: *The Cranial Muscles of Vertebrates.* London, Macmillan.

Fawcett, D. W., 1994: *Bloom and Fawcett: A Textbook of Histology.* New York, Chapman & Hall.

Frolich, L. M., and Biewener, A. A., 1992: Kinematic and electromyographic analysis of the functional role of the body axis during terrestrial and aquatic locomotion in the salamander *Ambystoma tigrinum. Journal of Experimental Biology,* 162:107–130.

Gans, C., and Bock, W., 1965: The functional significance of muscle architecture: A theoretical analysis. *Ergebnisse der Anatomie und Entwicklungsgeschichte,* 38:115–142.

Gilbert, S. C., *Developmental Biology,* 5th edition, 1997. Sunderland, Mass., Sinauer Associates.

Goldspink, G., 1977: Design of muscles in relation to locomotion. *In* Alexander, R. McNeil, and Goldspink, G., editors: *Mechanics and Energetics of Animal Locomotion.* New York, John Wiley & Sons.

Goodrich, E. S., 1930: *Studies on the Structure and Development of Vertebrates.* London, Macmillan.

Goodrich, E. S., 1918: On the development of the segments of the head in *Scyllium. Quarterly Journal of the Microscopical Society,* 63:1–30.

Grundfest, H., 1960: Electric fishes. *Scientific American,* 203:115–124.

Hoar, W. S., 1983: *General and Comparative Physiology.* Englewood Cliffs, NJ, Prentice-Hall.

Huddart, H., 1975: *The Comparative Structure and Function of Muscles.* New York, Pergamon Press.

Jacobson, A. G., 1988: Somitomeres: Mesodermal segments of vertebrate embryos. *Development,* 104:209–220.

Lissmann, H. W., 1958: On the function and evolution of electric organs in fish. *Journal of Experimental Biology,* 35:156–191.

Lou, F., Curtin, N. A., and Woledge, R. C., 1999: Elastic energy storage and release in white muscle from the dogfish, *Scyliorhincus canicula. Journal of Experimental Biology,* 202:135–142.

McMahon, T. A., 1984: *Muscles, Reflexes, and Locomotion.* Princeton, N. J., Princeton University Press.

Meier, S. P., 1981: Development of the chick embryo mesoblast: Morphogenesis of the prechordal plate and cranial segments. *Developmental Biology,* 83:49–61.

Meier, S. P., 1982: The development of segmentation in the cranial region of vertebrate embryos. *Scanning Electron Microscopy,* 3:1269–1282.

Meier, S. P., and Tam, P. P. L., 1982: Metameric pattern development in the embryonic axis of the mouse: I. Differentiation of the cranial segments. *Differentiation,* 21:95–108.

Neal, H. V., and Rand, H. W., 1936: *Comparative Anatomy.* Philadelphia, Blakiston.

Noden, D. M., 1991: Vertebrate craniofacial development: The relation between ontogenetic process and morphological outcome. *Brain, Behavior and Evolution,* 38:190–225.

Noden, D. M., 1984: Craniofacial development: New views on old problems. *Anatomical Record,* 208:1–13.

Noden, D. M., 1983: The embryonic origins of avian cephalic and cervical muscles and associated connective tissues. *American Journal of Anatomy,* 168:257–276.

Northcutt, R. G., 1990: Ontogeny and phylogeny: A reexamination of conceptual relationships and some applications. *Brain, Behavior and Evolution,* 36:116–140.

Portman, A., 1948: *Einführung in der Vergleichenden Morphologie der Wirbeltiere.* Basel, Schwabe.

Ritter, D., 1996: Axial muscle function during lizard locomotion. *Journal of Experimental Biology,* 199:2499–2510.

Ritter, D., 1995: Epaxial muscle function during locomotion in a lizard (*Varanus salvator*) and the proposal of a key innovation in the vertebrate axial musculoskeletal system. *Journal of Experimental Biology,* 198:2477–2490.

Romer, A. S., 1944: The development of tetrapod limb musculature: The shoulder region of *Lacerta. Journal of Morphology,* 74:1–41.

Romer, A. S., and Parsons, T. S., 1977: *The Vertebrate Body,* 6th edition. Philadelphia, Saunders College Publishing.

Simons, R. S., and Brainerd, E. L., 1999: Morphological variation of hypaxial musculature in salamanders (Lissamphibia: Caudata). *Journal of Morphology,* 241:153–164.

Tam, P. P. L., Meier, S. P., and Jacobson, A. G., 1982: Differentiation in the metameric pattern in the embryonic axis of the mouse: II. Somatometric organization of the presomitic mesoderm. *Differentiation,* 21:47–63.

Walker, W. F., and Homberger, D. G., 1992: *Vertebrate Dissection,* 8th edition. Philadelphia, Saunders College Publishing.

Williams, P. L., Bannister, L. H., Berry, M. M., Collins, P., Dyson, M., Dussek, J. E., and Ferguson, M. W. J., editors, 1995: *Gray's Anatomy,* 38th edition. New York, Churchill Livingstone.

11

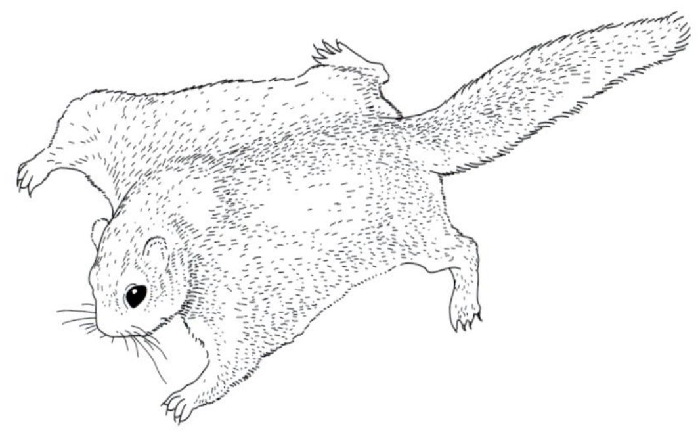

Functional Anatomy of Support and Locomotion

PRECIS

Having studied the structure and evolution of the skeletal and muscular systems, we now can address the functional interrelationships between the individual skeletal units and the muscles that act on them. We will examine the functional anatomy of support and locomotion in this chapter and that of feeding mechanisms in Chapter 16. In a book of this scope, we cannot analyze all of the supporting mechanisms that have evolved among vertebrates nor all of their many patterns of locomotion. The ones we have selected are representative of most vertebrates, and they will illustrate the types of understanding of vertebrate structure that derive from functional analyses.

OUTLINE

Now that you are acquainted with the structure and evolution of the skeletal and muscular systems, we can examine the functional interrelations between them. The individual bones, ligaments, and muscles form interrelated functional units that support the body and are responsible for the locomotion of the animal and the movements of most of its parts. By analyzing these units, we can learn much about the configuration of the musculoskeletal system and the way it works. We will examine the functional anatomy of support and locomotion in this chapter by considering how the problems and solutions differ in the aquatic, terrestrial, and aerial environments. Many aspects of support and locomotion have been reviewed in a symposium organized by Long and Koob (2000).

Support and Locomotion in the Aquatic Environment

Squid and some other aquatic animals swim by ejecting jets of water, but most aquatic craniates swim by undulating or oscillating some parts of the body. Nearly all vertebrates can swim to some extent, and many are excellent swimmers. Swimming, of course, is the sole or primary pattern of locomotion in fishes and larval amphibians. They are **primary swimmers.** Terrestrial vertebrates that have returned to the water are **secondary swimmers.** They have readapted to an aquatic mode of life but still retain at least traces of their terrestrial ancestry. Even whales, which cannot survive on land, surface to breathe air.

Most primary swimmers are also **undulatory swimmers,** for they propel themselves by lateral undulations that travel down the trunk and tail or, in a few cases, along fins that have long attachments to the body (Fig. 11-1). A few secondary swimmers, such as salamanders and crocodiles, also propel themselves with lateral undulations of a flattened tail. But most secondary swimmers, and a few fishes, are **oscillatory swimmers** that propel themselves primarily with oscillations or paddle-like movements of their tail (the flukes of whales) or paired appendages. Undulations of the trunk help in some species. Surfperches and many reef fishes swim by oscillating their pectoral fins, a pattern termed **labriform swimming.** (See Drucker and Jensen, 1996, for

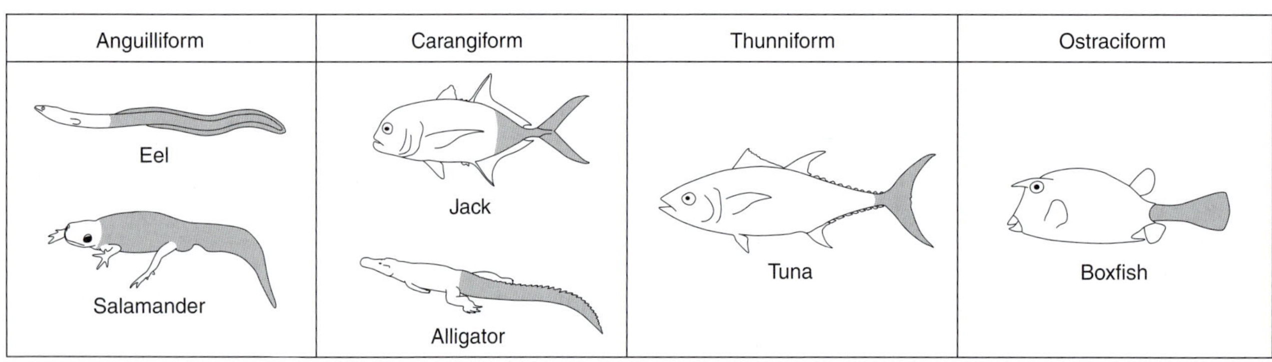

A. Undulatory swimmers that use the trunk and tail

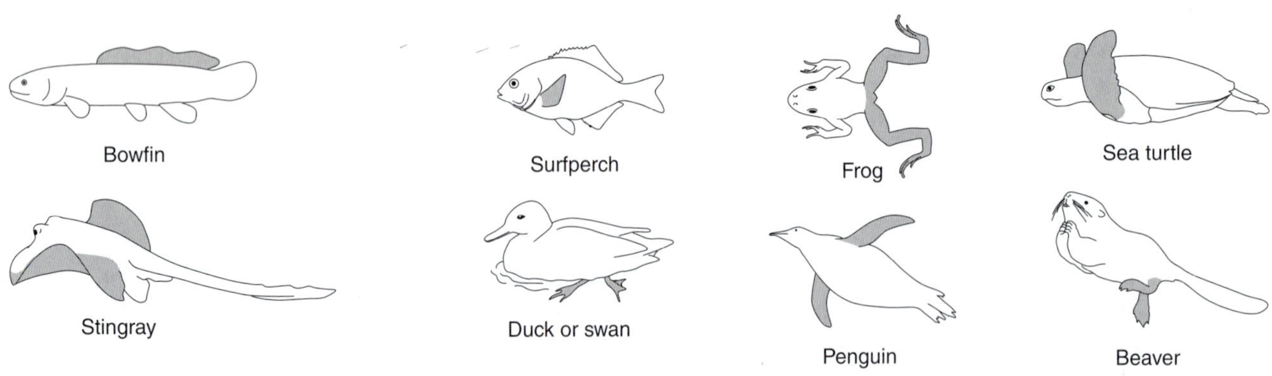

B. Undulatory swimmers that use paired fins **C.** Oscillatory swimmers that use paired fins or limbs

FIGURE 11-1
A–C, Major swimming patterns among vertebrates. The parts of the body that generate the major propulsive thrust are shaded. *(Modified from Webb.)*

an analysis of labriform swimming in a surfperch.) Frogs and ducks swim by oscillatory thrusts of their hind legs and large webbed feet; sea turtles use elongated pectoral flippers; penguins use paddle-like wings; and beavers use large, paddle-like, webbed hind feet. We will analyze undulatory swimming in fishes as an example of an aquatic pattern of locomotion and examine the morphological features associated with this pattern.

Problems of Support and Movement in Water

Swimming animals must first of all support themselves in the water. Support in water or air is governed by a principle discovered by the famous Greek mathematician and physicist, Archimedes (d. 212 BC). According to Archimedes' principle, an object in a fluid (water or air) displaces a weight of fluid equal to its own weight, and the displaced fluid exerts an upward force on the object. Because water is a dense medium compared with air, the upward force exerted on an object in water is substantial and provides considerable buoyancy. Buoyancy is the major force supporting a fish, but flesh is slightly denser than water (density = mass/volume) so the weight of a fish (mass × gravitational acceleration) is slightly greater than the buoyancy force. A swimming fish, therefore, must produce a lift force that overcomes the downward pull of gravity not compensated for by the buoyancy of the water.

Although water offers considerable support for an animal, it also offers more resistance to moving through it than air. The body of a fish must be firm enough to push through the dense water, and a fish must overcome frictional and other forces, collectively called **drag,** that tend to hold it back. It also must be able to maintain stability and to maneuver.

We have seen (Chapter 8) that the design of the vertebral column of fishes helps meet some of these problems. The strong disk-shaped or spool-shaped vertebral centra form a compression strut that resists the compression forces generated as the animal pushes through the water. The vertebral centra also prevent the body from shortening when longitudinal muscle fibers contract and generate the propulsive forces that overcome drag. The vertebral column, or parts of it, must be able to bend laterally during swimming. All or many of the joints between vertebrae allow for lateral bending. Because of the buoyancy of the water, vertical bending forces are negligible and need not be resisted. Only a few fishes with specialized patterns of swimming have zygapophyses.

Drag

Drag forces are substantial, and the design of many aspects of fish structure reduces them. Two major types of drag forces resist the forward motion of a fish: (1) frictional drag and (2) pressure drag. **Frictional drag** (Fig. 11-2*A*) is the force exerted on the surface of a fish due to the viscosity of the water. A thin **boundary layer** of water surrounds a moving fish. The innermost part of this layer is carried along at nearly the same speed as the fish, but peripheral parts of the boundary layer move increasingly slowly, finally equaling the speed of the surrounding water. Shear forces within the boundary layer are responsible for frictional drag. Frictional drag is lowest when the surface area of the fish is minimized relative to its mass, the fish is swimming slowly, and the water flows smoothly across its surface (i.e., the flow is **laminar**). When a fish of the same mass and shape swims more rapidly, the boundary layer increases in thickness, and the increased undulations of the trunk and tail disrupt the smooth flow of water (Fig. 11-2*B*). The boundary layer tends to separate from the surface, especially posteriorly, and this produces eddies. The thicker boundary layer and the turbulent flow and resulting eddies increase a type of drag known as pressure drag (see subsequent discussion). In general, a smooth surface tends to reduce frictional drag, but sometimes a small turbulence in the inner part of the boundary layer, such as those caused by ridges on the scales, prevents the separation of the layer and the formation of larger eddies. The layer of mucus that covers most fishes further reduces frictional drag, for it reduces the viscosity of the water in the boundary layer.

Pressure drag results from differences in water pressure between the front and rear of the fish. Water is displaced as the fish moves forward and then separates behind the fish, forming eddies and a wake (Fig. 11-2*B*). The pressure gradient, with high pressure at the head and a reduced pressure over the posterior part of the body, tends to hold the fish back. Pressure drag depends on body shape. A sphere moving through water would have a very high pressure drag but a low frictional drag because its surface area is small relative to its mass. A long, slender body of the same mass would have a low pressure drag but a high frictional drag because its surface is much larger. An optimal compromise between these shapes is a teardrop-shaped object having a maximum diameter about one third of the way back from the front. Optimal diameter is equal to about one-fifth the length of the object. Fishes that swim rapidly have such a streamlined or **fusiform** shape.

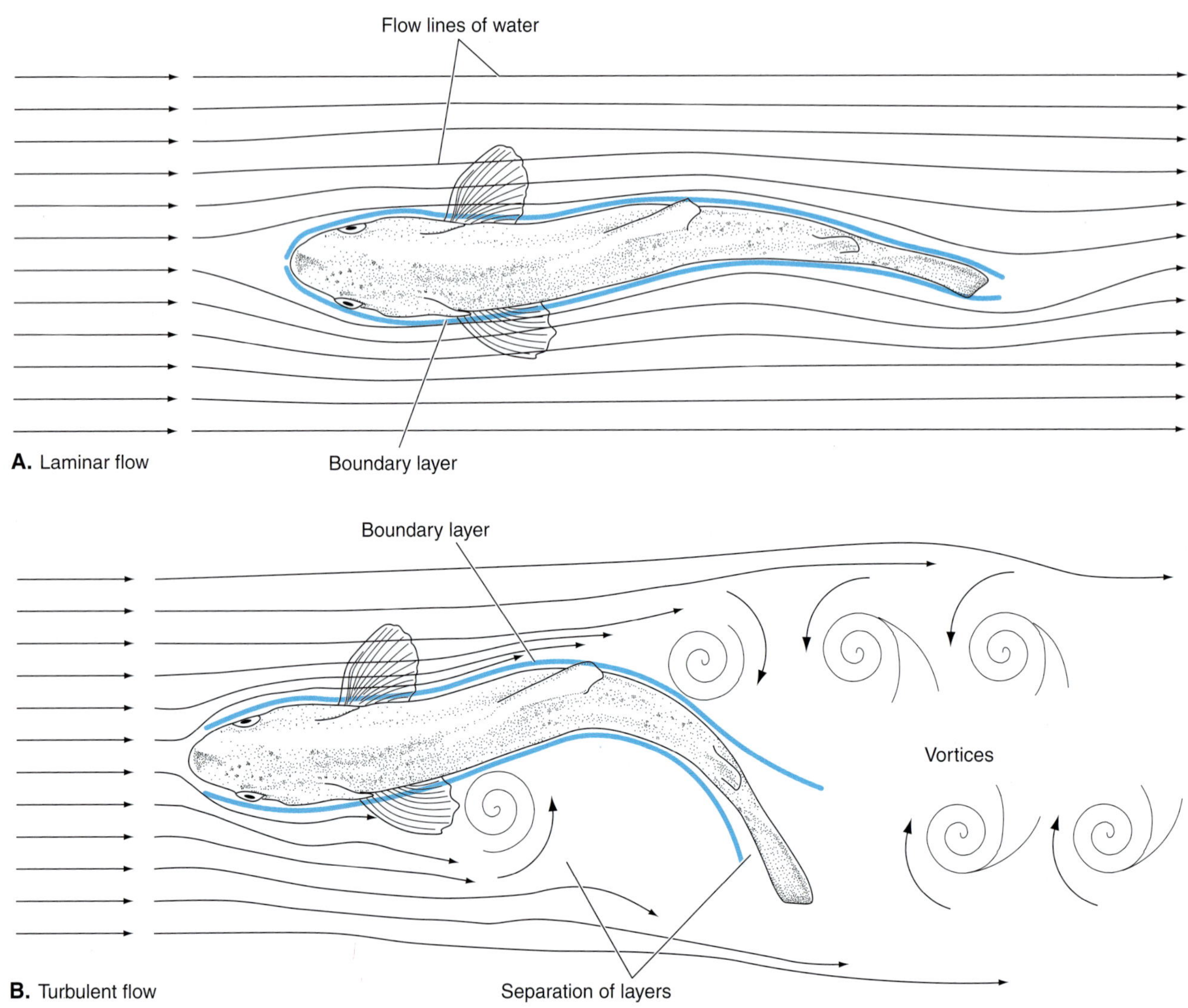

FIGURE 11-2
Forces on a fish. *A,* Frictional drag is low when the flow across the surface of the fish is smooth, or laminar. *B,* Frictional drag increases as the fish swims faster because the boundary layer thickens and begins to separate from the fish. Differential pressure over the body causes pressure drag. (*A, Modified from Triantafyllou and Triantafyllou.*)

Propulsion in Water: Swimming

Most fishes swim by generating propulsive forces by undulations or bends of the body that move caudad along the trunk and tail. As the undulations travel back, they thrust laterally and caudally against the water. Undulatory swimming may be transient or periodic. Many reef fishes, freshwater sunfishes, and bass are examples of **transient swimmers**. They lie quietly in the water much of the time but can accelerate rapidly as they dart forward and turn to escape predators or attack elusive prey. Body shape need not be highly streamlined but must be adapted for acceleration and maneuvering. The body often is short, with a short turning radius, and is quite high from dorsal to ventral. The height presents a large surface area for thrust against the water. Many pelagic fishes, such as tunas and many sharks, are examples of **periodic swimmers**. These fishes can develop sudden bursts of speed but also are adapted for slower cruising, often for long periods of time. Body shape is longer and more streamlined than in transient swimmers.

The amount of body that undulates varies considerably among species of periodic swimmers. Most of the trunk and tail move back and forth in eels and other long-bodied **anguilliform swimmers** (Fig. 11-1). The amount of the trunk and tail that undulates shifts caudad in other fishes in a continuum from undulations of the caudal half of the body, as seen in jacks (**carangiform swimming**), to propulsive movements limited to little more than the tail, as occurs in tunas and some oceanic sharks (**thunniform swimming**).

Boxfishes have a rigid, inflexible trunk and move only their tail (**ostraciform swimming**). Considerable lateral bending occurs throughout the vertebral column of anguilliform and carangiform swimmers, but most bending occurs near the tail base in thunniform swimmers. The vertebrae in this region are diplospondylous (Chapter 8). Centra are more numerous (two per body segment), which increases flexibility. The joints between them are very flexible. Tunas have zygapophyses that restrict movement in more cranial vertebrae.

Thunniform swimmers are the most rapid fishes, and tunas can reach speeds of 75 km/h. These fishes are highly streamlined, and drag is further reduced by limiting movements to the caudal region of the body. The sickle-shaped tail can move back and forth rapidly, and its effectiveness is increased by being very high so it extends above and below the wake of the fish.

In a trout, which has a pattern of swimming intermediate between anguilliform and carangiform, waves of contraction begin near the anterior end of the trunk and sweep down the body, first on one side and then on the other (Fig. 11-3). The push of a bend of the body obliquely backward against the water accelerates a certain mass of water. Because of its inertia, the water generates on the fish an opposite reaction force, the magnitude of which is the product of the mass of water moved and its acceleration. This reaction force can be resolved into forward and lateral components or thrusts. The left and right lateral components cancel each other to some extent but also tend to rotate the fish about its center of gravity; that is, as the tail moves to the right, the front of the body would be expected to swing to the left. This happens only to a limited extent. The gradual tapering of the body toward the caudal end, especially at the narrow tail base, reduces the lateral force. Moreover, the shape of the fish offers great resistance to sideways movements because the heavy front part of the body has more inertia than do the tapering trunk and tail.

The forward components of the reaction forces from all regions of the body combine and propel the fish. The wave of contraction moves caudally along the body at a speed greater than the forward movement of the fish. As a result, the next posterior region of the body to exert a thrust will push on a mass of water already accelerated by the previous region of the body; thus, its effect is magnified. The reaction of the water becomes greater as the waves of contraction travel caudally because of this summation effect and because the waves increase in amplitude. As they increase in amplitude, the waves push more directly in line with the overall direction of movement of the fish, so the forward components of the reaction of the water increase progressively. Because the caudal parts of the body and the tail move from side to side to a greater extent than do the anterior regions, and the tail has a particularly broad, flat surface, the mass of water moved and its acceleration are accordingly greater. We have described a representative pattern of thrust generation, but considerable variation occurs among fishes (Altringham and Ellerby, 1999).

Undulatory movements of the body and tail produce vortices in the water that are shed from the tip of the tail (Fig. 11-2*B*). In a dorsal view of a swimming fish, the vortices appear as a pair of columns of counterclockwise and clockwise whorls, but they are actually rings connected above and below the wake. A jet-like flow of water passes caudad through the vortices. Studies are being made by many investigators to determine the information that the vortices can provide on the energetics of swimming (Triantafyllou and Triantafyllou, 1995; Muller et al., 1997).

Role of Myomeres in Swimming

The segmented myomeres of fishes are well suited to generate undulatory waves of the trunk and tail. Because the vertebral column prevents the body from shortening, contraction of several myomeres on one side pulls the myosepta together and causes a curvature. The zigzag folds of the myomeres and their overlapping, conelike extensions (Fig. 11-4*A* and *C*), which become more pronounced near the tail, presum-

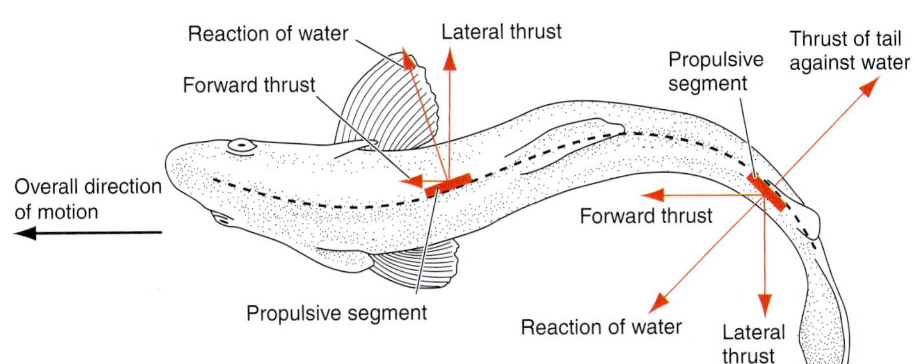

FIGURE 11-3
A dorsal view of a swimming trout showing forces developed by the undulating trunk and tail and their forward and lateral components. (*After Webb.*)

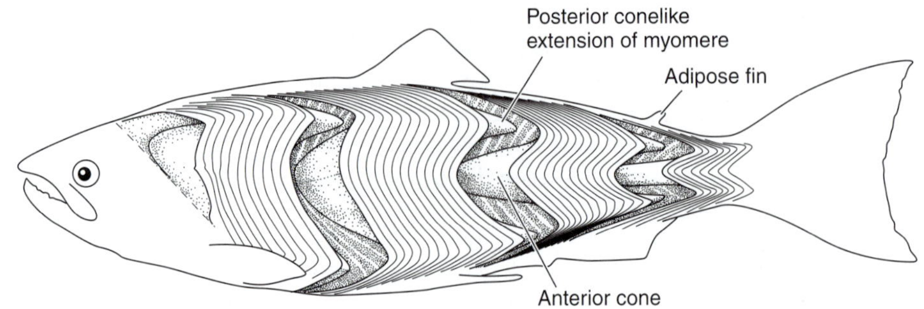

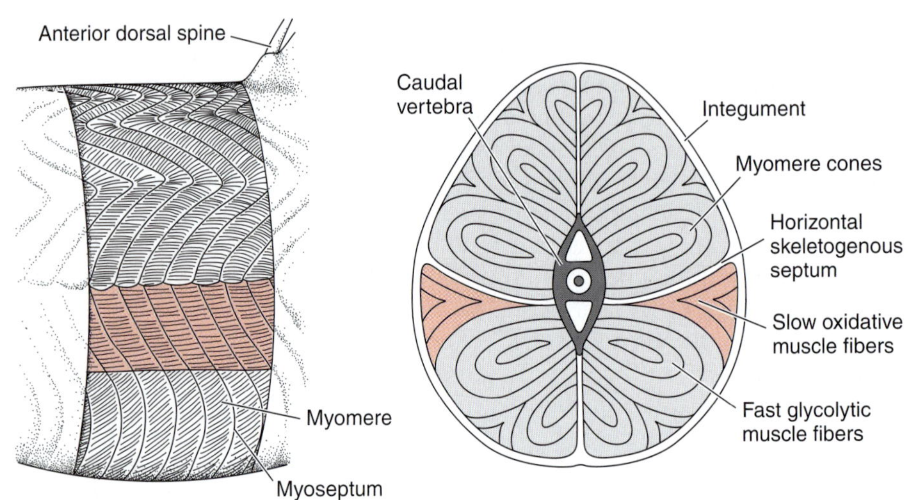

FIGURE 11-4
Myomeres of a fish. *A,* A salmon (*Salmo*) in which myomeres have been removed in several regions to show the conelike extensions of adjacent myomeres. *B,* A surface view of the myomeres of a dogfish (*Squalus*), showing the direction of muscle fibers within the myomeres. *C,* A transverse section through the tail of a dogfish, showing overlapping myomere cones and the distribution of white and red fibers. (*A, From Romer and Parsons, after Greene and Greene; B and C, from Walker and Homberger.*)

ably allow one myomere to exert an influence over a greater body length than would otherwise be the case. The longer folds toward the tail, and tendon-like extensions from the apices of the cones into the tail, cause caudal undulations of increased amplitude and force. The overlap of myomeres ensures a smooth generation of force and flow of undulations. At a given transverse level of the body, several overlapping myomeres may be in different stages of contraction. Contraction of the surface myomere at one level may be reaching maximum force, while the conelike extension to this level of a more anterior myomere is beginning to relax, and the extension to this level of the myomere behind it is just starting to generate force.

In many fishes, the superficial fibers of the myomeres insert into dense connective tissue that forms an **exotendon,** which takes a helical course along the trunk and tail and is bound firmly to the skin. This is why it is difficult to skin a fish without disrupting the myomeres. The exotendon helps maintain a streamlined body shape; transfers forces to the tail; and, like all tendons, can store energy. As a curvature passes down one side of the body, the exotendon on the opposite (convex) side is stretched to a slight extent and stores energy. This energy is released and assists bending when a wave of contraction passes down the previously convex side.

The direction of the muscle fibers is not alike in all parts of a myomere. Those near the center of the side of the body have a longitudinal orientation and are longer than the obliquely oriented fibers near the mid-dorsal and midventral lines (Fig. 11-4*B*). Reasons for this are not well understood but may contribute to a uniform strain distribution within the muscles. Muscle fibers lie at varying distance from the plane of bending, which is the sagittal plane of the body. Those fibers near the skeletogenous horizontal septum on the trunk lie farther from the sagittal plane and may have to shorten more than fibers near the mid-dorsal and midventral lines. The obliquely oriented fibers near the mid-dorsal and midventral lines shorten less and may contract nearly isometrically.

Van Leeuwen (1999) has developed a quantitative model to predict the shape and architecture of myomeres in teleosts. His model also shows that ribs and the support of the skin prevent extensive muscular deformation that do not contribute to bending. Dorsal and ventral (anal) fins are located such that unfavorable mechanical interactions with trunk muscle contractions are avoided.

The physiological properties of muscle fibers in the myomeres are not all alike. The most lateral fibers are slow-oxidative (red) fibers (Fig. 11-4*B* and *C*), which are active during cruising or maintaining position in a current. When a burst of speed is needed, the more dorsal, ventral, and deeper fast-glycolytic (white) fibers are used. Fishes with these two types of muscle fibers have a two-geared system. Some species, including carp, have pink fibers with intermediate properties that are located between the white and red ones; these fishes have three gears.

Many investigators currently are studying the dynamic properties of the swimming musculature. (See Long, 1998; Shadwick et al., 1998; and Wardle et al., 1995, for historical summaries and examples of this type of research in functional anatomy.) Much work remains to be done, but it is becoming clear that swimming is very efficient, partly because of the release and storage of elastic strain energy. As traveling undulatory waves move down the body of a fish, muscle fibers pass through cycles of shortening (on the concave side of a bend) and lengthening (on the convex side of a bend). These cycles are slightly out of phase with the cycles of muscle stimulation because muscle fibers are stimulated and their internal contractile mechanisms initiated while the fibers are still in their lengthening phase, which means that serial elastic elements within or surrounding the muscle fibers are being stretched (strained), and elastic strain energy is stored. This energy is released during the shortening phase of the muscle fibers. Muscle-fiber contraction also stiffens the body. Increased body stiffness increases the speed at which undulatory waves travel down the body and minimize the metabolic cost of bending the body.

Buoyancy

In addition to generating a propulsive thrust sufficient to overcome drag and move forward, a fish must float at an appropriate depth in the water and remain on an even keel as it swims. As we explained earlier in this chapter, flesh is denser than is water, so the pull of gravity, despite the buoyancy effect of the water, causes a fish to sink slowly unless other factors are present to offset the density of the fish. The density of chondrichthyans is reduced by three factors: (1) a skeleton of cartilage is less dense than one of heavily calcified bone, (2) most chondrichthyans store lipid in their liver as an energy reserve (lipid is much less dense than is glycogen, which is the energy store found in the liver of most vertebrates), and (3) the body fluids of marine chondrichthyans contain considerable urea and trimethylamine oxide (TMAO). Although excreted by most vertebrates, chondrichthyans retain considerable urea as part of their unique osmoregulatory mechanism (Chapter 20), and TMAO appears to counter some of the adverse effects of urea on enzyme actions. Both urea and TMAO are less dense than are other solutes, so they also have a substantial buoyancy effect (Withers et al., 1994). The combination of these features helps chondrichthyans approach neutral buoyancy where the forces of gravity and buoyancy are equal. The forces may be equal in some pelagic sharks, but many sharks are still denser than water and can be seen lying on the bottom when not swimming.

Hydrodynamic forces derived from body shape and the actions of certain fins overcome the remaining tendency of sharks to sink and keep them afloat. Sharks, in common with many early fishes, have **heterocercal tails,** in which the upper lobe of the caudal fin is longer than the ventral lobe (Fig. 11-5*A* and *B*). The upper lobe also is stiffened by an extension of the vertebral axis. Undulations of the trunk and tail provide the forward thrust that propels the animal as they do in other fishes, but beyond this, investigators have assumed that the tail also keeps the shark afloat. A long-held classic hypothesis proposed that the lateral movements of the tail during swimming also generate an upward lift force. As the tail moves laterally, the smaller and more flexible lower lobe lags the stiffer upper lobe. The tail becomes inclined at an angle to the horizontal plane, so it pushes downward and laterally against the water (Fig. 11-5*A*, *right side*). The reaction of this force can be resolved into a vertical lift component and a lateral drag component. The lift generated by the tail produces a turning moment about the fish's center of gravity that tends to push the head down, but this is balanced by an opposite turning moment generated by the planing effect of the large pectoral fins placed low on the body. The flat ventral surface of the head contributes to the planing effect. Because the moment arm of tail lift to the center of gravity is longer than that for pectoral fin lift, the lift from the tail must be smaller than that from the pectoral fins and head.

Thomson (1990) and Thomson and Simanek (1977) questioned the classic hypothesis by analyzing slow-motion moving pictures of the swimming movements of many sharks. They found that the lower lobe of the fin often did not lag the upper lobe but often preceded it (Fig. 11-5*B*, *right side*). The tail, therefore, had a negligible lift effect much of the time. They proposed that the shape of the tail, with its stiff upper lobe, would generate a slightly downward and forward thrust through the center of gravity (Fig. 11-5*B*, *line b*). The shark could maintain or change its position in the water easily by using the flexible lower lobe of the fin as a "trimming device." The thrust of the tail could be redirected to pass slightly above the center of gravity (Fig. 11-5*B*, *line a*), in which case the head would be pushed down, or slightly below the center of balance (*line c*), in which case the head would be pushed up.

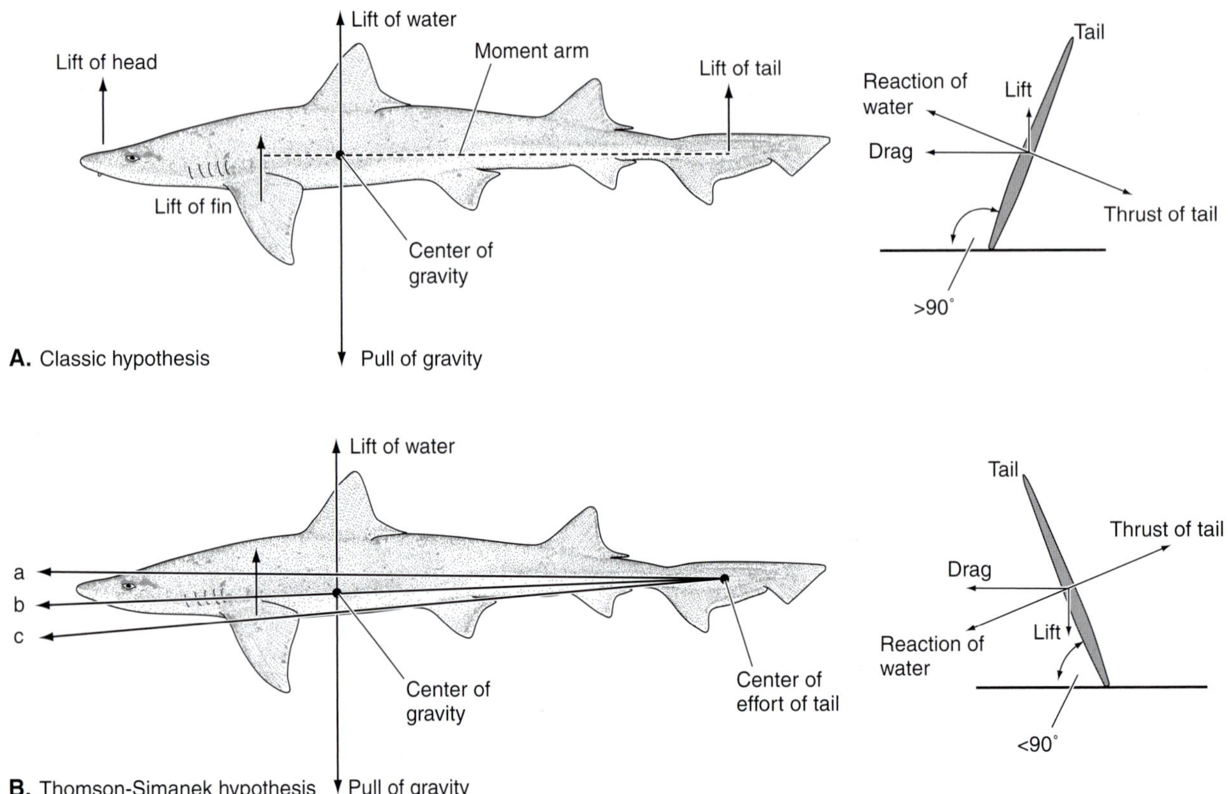

FIGURE 11-5
Possible roles of the heterocercal tail of sharks in providing lift during swimming. *A*, The classic hypothesis. *B*, The Thomson-Simanek hypothesis. Diagrammatic cross sections through the tail on the right side show the resolution of the force developed by the thrust of the tail against the water.

Ferry and Lauder (1996) have reexamined the action of the heterocercal tail, making three-dimensional kinematic analyses of tail actions in leopard sharks (*Triakis semifasciata*). They also used dye-stream visualizations of water movements. Their investigations support the classic hypothesis, at least for leopard sharks. Many sharks have similar shaped tails, but considerable variation exists in the size of the two lobes of the heterocercal tail among sharks.

Sarcopterygians and early actinopterygians have a lunglike sac of air in their body cavity. During the course of actinopterygian evolution, the lungs are transformed into a hydrostatic **swim bladder,** which is located in the body cavity just ventral to the vertebral axis (Fig. 11-6). A sac of air in the body, whether it be lungs or a swim bladder, makes these fishes less dense so they float more easily than did early, heavy-bodied fishes that tended to sink. Teleosts can regulate the amount of gas in the swim bladder (Chapter 18), which gives them the ability to attain neutral buoyancy and float at any level in the water, with little muscular effort.

Fishes with a swim bladder need not have the mechanisms we see in sharks for buoyancy control. Bone replaces much or all of the cartilage in the skeleton, lipids and other weight-reducing molecules need not be accumulated in certain tissues, and the pectoral and caudal fins need not act to generate lift. The pectoral fins of teleosts are smaller and located more dorsally, a position better suited to their role in maneuvering. As we have discussed (Chapter 8), the asymmetrical heterocercal tail becomes symmetrical externally. The caudal fin rays attach to a sharply uptilted tip of the vertebral axis (the urostyle). The axis of rotation of the caudal fin is vertical and is capable of generating a symmetrical thrust. But as Lauder (1989) has pointed out, we can-

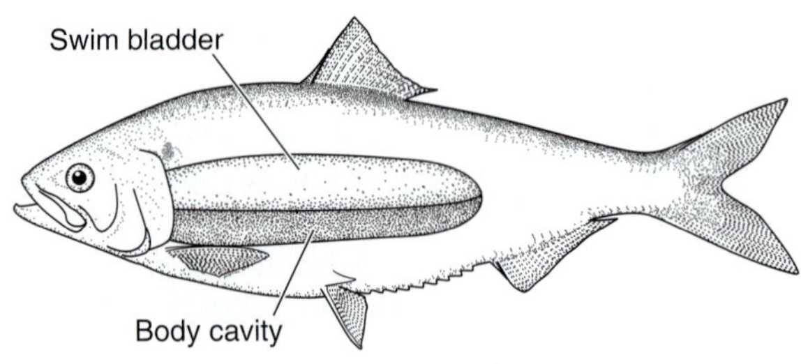

FIGURE 11-6
A lateral view of a teleost showing the location of the swim bladder in the dorsal part of the body cavity.

FOCUS 11-1 *Actions of a Dorsal Fin*

Jayne et al. (1996) examined experimentally the function of the flexible portion of the dorsal fin of the bluegill sunfish (*Lepomis macrochirus*) during steady swimming and during the fast propulsive kick-and-glide and C-start modes (a C-start is one in which the trunk and tail of the fish are strongly flexed into a C-shape just before propulsion). They also examined dorsal fin action during braking. Flexible rays, to the bases of which small **dorsal inclinator muscles** attach, support the flexible portion of the dorsal fin (Fig. A). Jayne et al. implanted electrodes into these muscles and into the adjacent myotomal muscles responsible for trunk undulations to determine when they were active. This information was correlated with trunk and fin movements as seen in high-speed video pictures. During the various types of swimming, the dorsal inclinator muscles contract mainly on the side of the body that is being swept laterally by myotomal action. Myotomes and dorsal inclinator muscles contract on the same side of the body, which is consistent with the hypothesis that the dorsal inclinator muscles stiffen the dorsal fin and oppose its tendency to bend in a direction opposite from that in which the body and fin are moving. The activity of the dorsal inclinator muscles was sufficient to resist dorsal-fin bending during steady swimming but not during the fast, propulsive starts. During braking, the dorsal inclinator muscles are most active and stiffened the fin on the side opposite to myotomal activity, which is consistent with a hypothesis that the dorsal fin contributes to the drag needed to brake. The dorsal fin clearly participates actively during swimming and is not just a passive stabilizer.

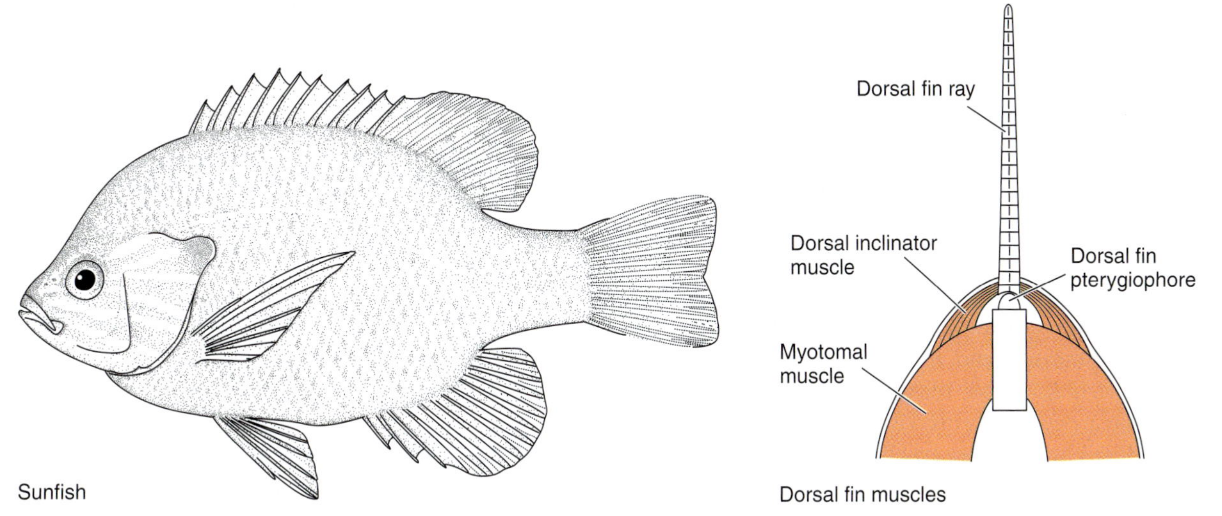

A. A lateral view of a bluegill sunfish, *Lepomis macrochirus* (left), and a vertical section through the flexible part of the dorsal fin (right).

not assume that the homocercal tail always generates a symmetrical thrust. The evolution of the homocercal tail has been accompanied by the evolution of a complex intrinsic caudal fin musculature not present in the heterocercal tail. These muscles probably modulate caudal fin function in connection with different modes of swimming (e.g., slow cruising, fast starts, and changes in depth), but their actions have not been studied.

Stability

A swimming fish is subject to various displacement forces that must be countered if the fish is to remain on an even keel. The action of the tail, as we have seen, tends to cause the head to move from side to side, a motion called **yaw.** The head and median fins counter this. The large and relatively heavy head has considerable inertia and so does not move easily from side to side, and the surface area of the median fins resists lateral movement of the body. Any tendencies for the fish to rotate about its longitudinal axis (**roll**) or for its head to move up and down in the vertical plane (**pitch**) also are countered by the position and movements of the median and paired fins. Apart from these generalized functions of fins, little is known about the ways the fins actually function in fishes with different body shapes and in different types of swimming and maneuvers. These questions now are being investigated (Focus 11-1).

Support and Locomotion in the Terrestrial Environment

Support on Land

Vertebral Support When vertebrates moved from water to land, supporting body weight became a major problem because air is not a dense medium and affords little lift. Amphibians, reptiles, and most mammals rest lying on the ground in a sheltered location (Fig. 11-7*A*), but even under these circumstances, the vertebral column, girdles, and ribs must be capable of preventing body weight from collapsing the lungs and other internal organs. When terrestrial vertebrates walk, they raise their trunks off the ground. Amphibians and most reptiles raise themselves only a short distance (Fig. 10-7*B*) because their humerus and femur project nearly laterally and move back and forth close to the horizontal plane. Mammals carry their trunks well off the ground, for their legs have rotated close to the trunk (Fig. 11-7*C*). Their humerus and femur move close to the vertical plane. The hind legs of bipedal dinosaurs and birds also have rotated under the trunk and move in the vertical plane (Fig. 11-7*D*).

The vertebral column of terrestrial vertebrates is not a compression strut as it is in fishes, but a beam that supports all parts of the body against gravitational forces and transfers weight to the girdles and appendages. Vertical bending of the column must be resisted in amphibians and most reptiles because they carry their trunks close to the ground. But lateral bending must occur because lateral undulations of the trunk and tail still play a role in the locomotion of most species. Vertical bending need not be resisted so much in mammals because they carry their bodies well off the ground. Indeed, vertical bending often participates in the locomotion of many mammals (see Fig. 11-26).

The vertebral column of terrestrial vertebrates is very strong, as we have discussed in Chapter 8. Except for early tetrapods, which spent a great deal of time in the water, and paedomorphic salamanders, the notochord has largely been replaced by solid, well-ossified, and unified centra. Intervertebral disks between successive centra often contain a remnant of the notochord known as the **nucleus pulposus.** Thick, circular layers of connective tissue fibers (Fig. 11-8) surround this. Intervertebral disks allow the vertebral column to bend, act as shock absorbers, and distribute forces evenly over the surface of adjacent centra. If, for example, the vertebral column bent dorsally, the dorsal edge of an intervertebral disk would be compressed, but this force would be distributed through the semifluid disk to all parts of the surface of adjacent centra. The neural

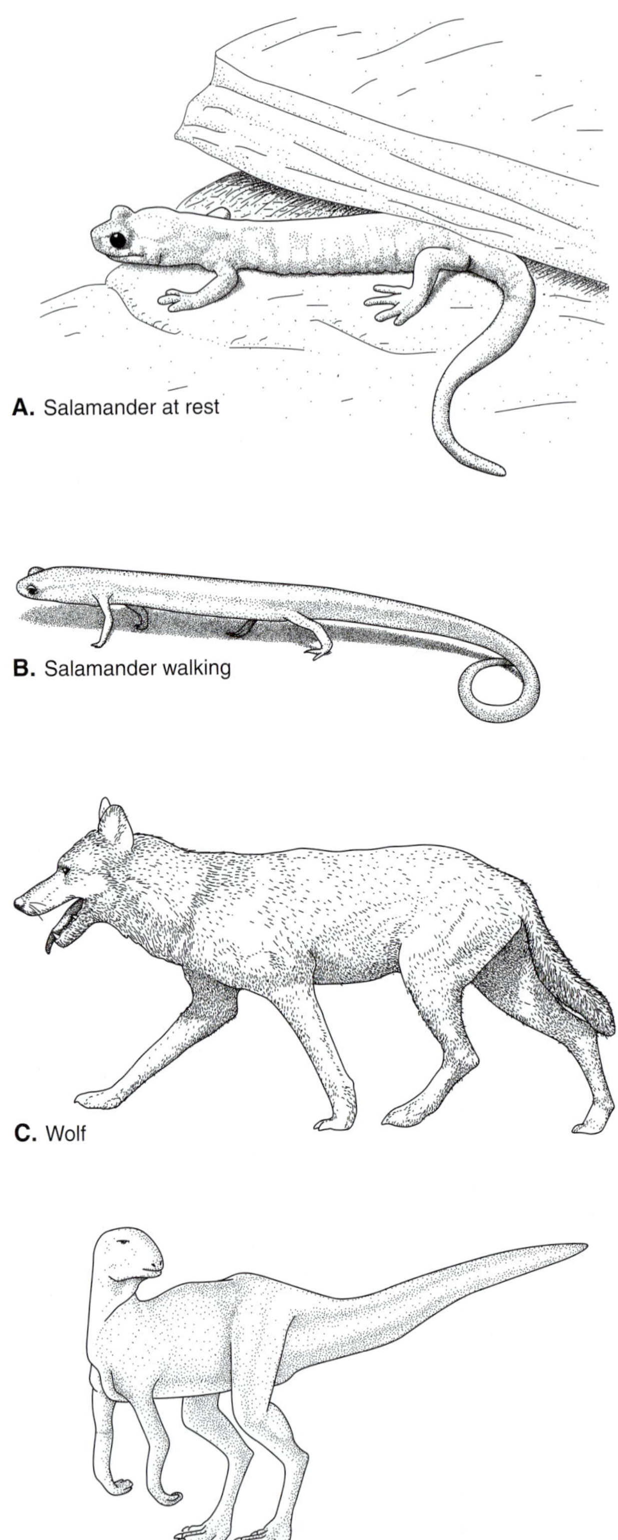

FIGURE 11-7
Limb positions of tetrapods. *A,* A salamander resting on the ground. *B,* A salamander walking with splayed limbs. *C,* A wolf walking with limbs drawn under the body. *D,* A bipedal dinosaur standing with hind limbs under the body.

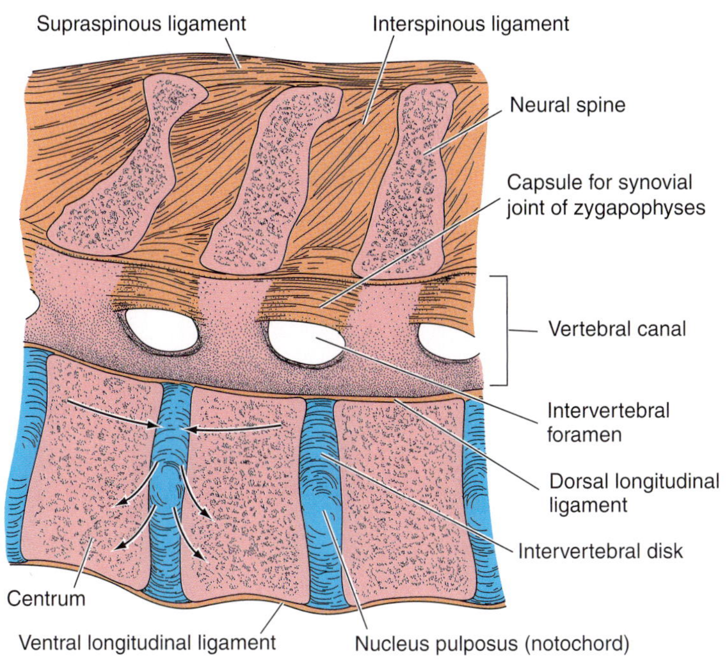

FIGURE 11-8
A sagittal section through several lumbar vertebrae of a mammal showing intervertebral disks and some of the ligaments that unite the vertebrae. Anterior is toward the left. The arrows show the distribution of forces from an extension of the spine. (*After Williams et al.*)

arches are fused to the centra, and those of successive vertebrae are linked by zygapophyses. Depending on the plane of their articular surfaces, zygapophyses restrict bending in some directions (often in the vertical plane) but allow it in other directions.

Strong ligaments (Fig. 11-8) link the vertebrae. Longitudinal ligaments extend across the dorsal and ventral surfaces of the centra, the tops of the neural spines, and through the neural arch. Diagonal interspinal ligaments extend between the neural spines, and the zygapophyses form synovial joints encapsulated by articular capsules. Epaxial muscles link individual vertebrae (interspinalis) or cross several vertebral segments (multifidi, spinalis, erector spinae). The vertebral column and associated ligaments and muscles form a firm but flexible complex.

Many early anatomists tried to analyze the structure of the vertebral axis of terrestrial vertebrates by applying the biomechanical principles of bridge construction. The vertebral axis resembles a bridge in some ways, but the spine is not a static support system. It is a dynamic structure that must support the head and body in many positions, receive the thrust of the legs during locomotion, and often participate during locomotion. A Dutch anatomist, Slijper (1946), compared the trunk skeleton to an elastic bow. The arched trunk skeleton of many mammals resembles an archer's bow (Fig. 11-9*A*). The vertebral centra are comparable to the wood in the bow. They are supporting elements that are under compression. The distribution of the bone trabeculae within them follows the longitudinal stress lines, and the joints between them permit the bow to bend. The bow is flexed by a "bowstring" composed of the sternum and ventral trunk muscles, such as the rectus abdominis. The string is connected to the bow anteriorly by the short, stout anterior ribs, which are stabilized by the scalenus muscle and posteriorly by the pelvic girdle. Subvertebral muscles (psoas and quadratus in mammals) form another partial bowstring. The curvature of the bow can change under the action of various trunk muscles as the animal changes positions and moves, but it cannot sag and permit the trunk to collapse as long as the bowstring is under tension. The trunk skeleton of amphibians and reptiles is flat, as it is in swayback mammals, such as a horse. This arrangement is more comparable to a violin bow, but the principles are the same (Fig. 11-9*B*). The head of mammals is raised, and the cervical vertebrae form another archer's bow that curves in the opposite direction from the trunk bow (Fig. 11-9*A*). Dorsal cervical muscles and nuchal ligaments form the string of the cervical bow.

The neural spines of the vertebrae are lever arms that transmit a force (the pull of muscles on them) to a center of rotation, or fulcrum, located between the centra (Fig. 11-9*A*). Cervical muscles, such as the splenius, acting through the neural spines of the anterior thoracic vertebrae, rotate the neck upward and support the head. Gravity acts to pull the head and neck downward. Here is another example of two sets of turning moments (Chapter 5). The neural spines of the cervical and anterior thoracic vertebrae are the lever arms for the in-force that raises the head, and they are oriented nearly perpendicular to the lines of actions of muscles

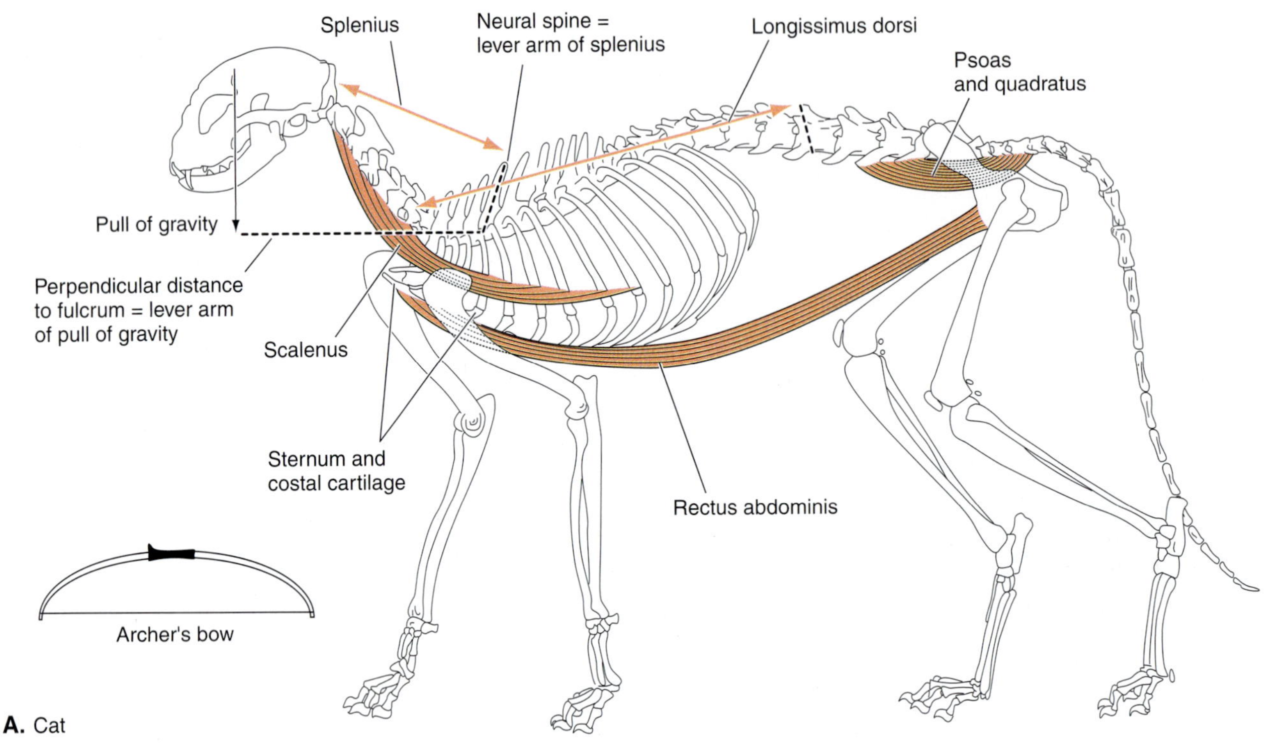

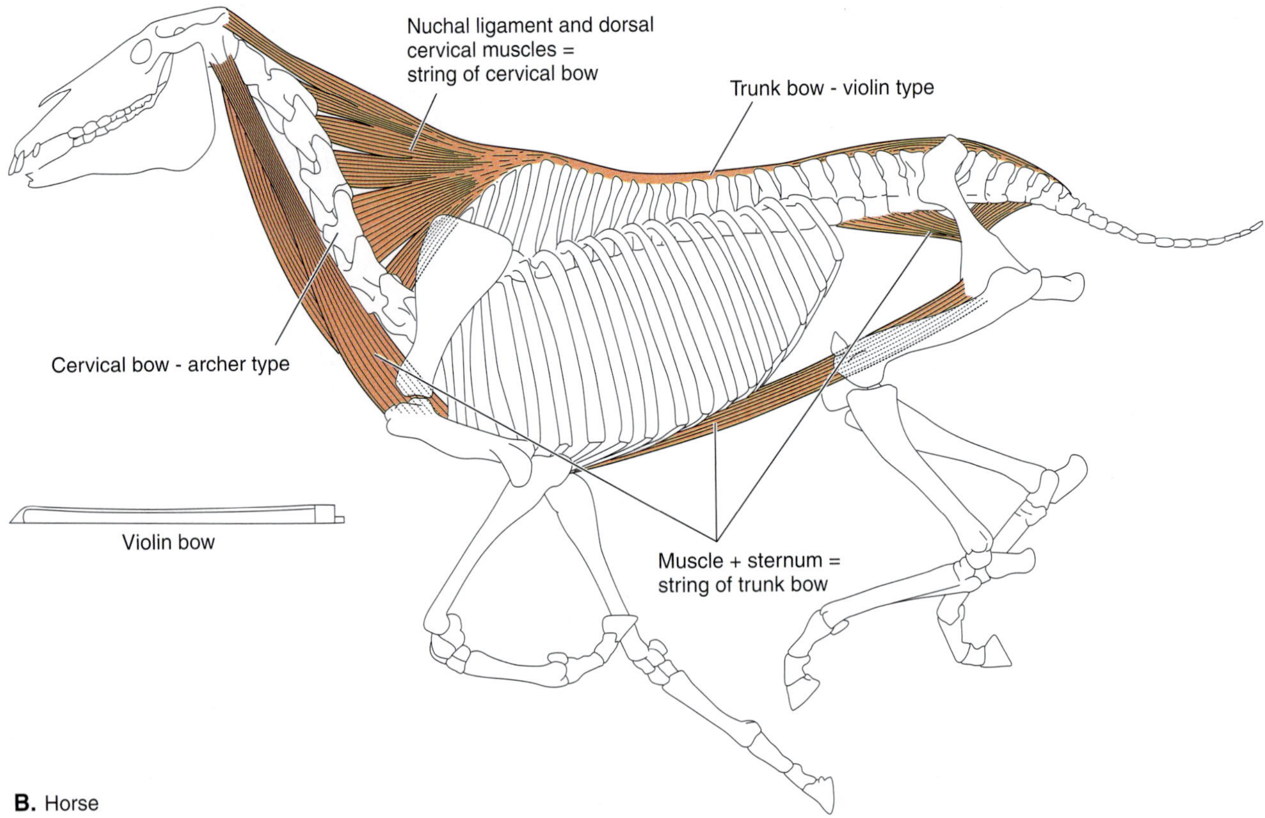

FIGURE 11-9
Diagrams of the biomechanics of the mammalian trunk skeleton. *A,* A cat, in which the trunk vertebrae form a bow analogous to an archer's bow. *B,* A horse, in which the trunk vertebrae form a bow analogous to a violin bow. (*A, After Slijper; B, after Wake.*)

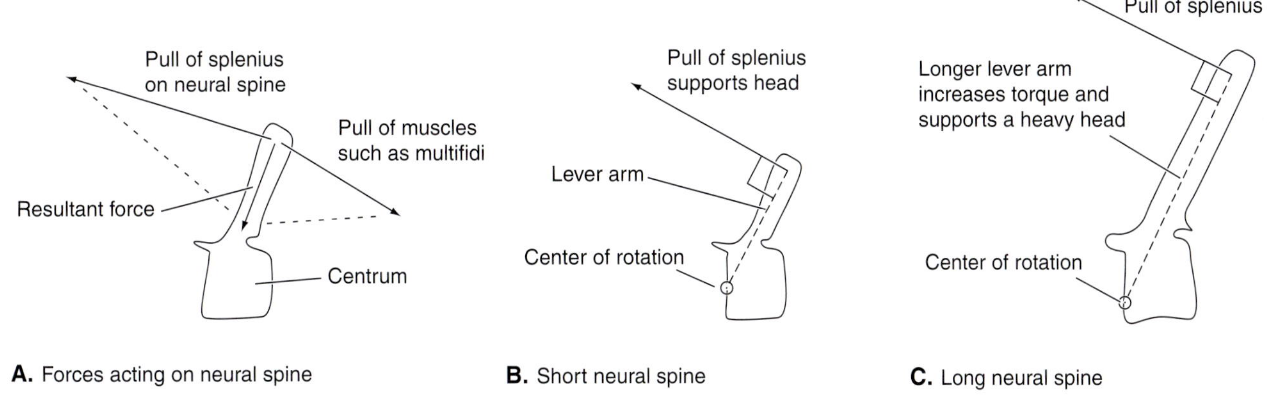

FIGURE 11-10
Biomechanics of the neural spine. *A,* When two or more muscles pull on the neural spine, its inclination is along a line formed by the resultant of the forces. *B* and *C,* The neural spine is a lever arm for muscles attaching to it. An increase in its length when a heavy head is supported by the splenius (*C*) increases the torque generated by the lever system.

acting on them. When several muscles act on the same spines from different directions, the direction of the spines must be the resultant of all of the forces (Fig. 11-10*A*). Because the splenius and similarly oriented muscles are the major ones acting on the anterior thoracic neural spines, the spines tend to incline slightly caudad.

An increase in the length of the neural spines increases the length of the lever arms and increases their mechanical advantage. Recall (Chapter 5) that a turning moment is the product of the muscle force and the length of its lever arm (Fig. 11-10*B* and *C*). Long neural spines of thoracic vertebrae enable the splenius and similar muscles to support a much heavier head. The splenius and its lever arm constitute an in-moment. The out-moment in this case is the product of the out-force (i.e., the downward pull of gravity from the center of mass of the head and neck) and the out-lever arm (i.e., the perpendicular distance from the line of action of gravity to the fulcrum at the base of the neck). No physical lever is present here. Because in-moments and out-moments must balance when the system is in equilibrium, it is not surprising that quadruped mammals with heavy heads have exceptionally long neural spines on their anterior thoracic vertebrae (Fig. 11-9*B*).

Ungulates with large heads that they must lower when they browse or graze also have particularly prominent nuchal ligaments (Fig. 11-9*B*). The ligaments are stretched by the weight of the head when the head is lowered, and the energy that stretched the ligaments is stored as elastic energy. When the animal raises its head, most of the stored elastic energy is released and helps raise the head. Muscle action is still needed, but far less than would otherwise be the case.

Muscle pull on the neural spines is primarily in the sagittal plane or close to it. The situation is analogous to a carpenter's joist, which must resist bending in only one plane and the strength of which is increased by increasing the depth of the joist (Chapter 5). Sagittal forces on the neural spines can be resisted by increasing their width in the sagittal plane. Because lateral forces are not strong, the spines are blade shaped and not thick from side to side.

In addition to the support and movements of the head, many other supporting functions and movements occur along the vertebral column. The complex interaction of many moments appears to explain the regional variation that we find in the length, direction of inclination, and width of the neural spines. For example, the head can be supported by anterior thoracic vertebrae only if they, in turn, are supported by other back muscles, such as the longissimus dorsi, that arise from the lumbar vertebrae (Fig. 11-9*A*). The line of action of the longissimus dorsi in many mammals necessitates that the neural spines of the lumbar vertebrae incline anteriorly. Caudally directed neural spines near the front of the thoracic region and cranially directed spines in the lumbar region characterize many, but not all, quadruped mammals. This condition is referred to as **anticliny.**

Limb Support When a terrestrial vertebrate is standing or moving, not only the vertebral column but also the legs must support body weight. Weight is transferred to the pelvic girdle and limb by way of one or more sacral vertebrae and ribs. The number of vertebrae and ribs involved, and the degree to which they are fused together, correlates with the forces they must transfer. Mammals have more sacral vertebrae than do

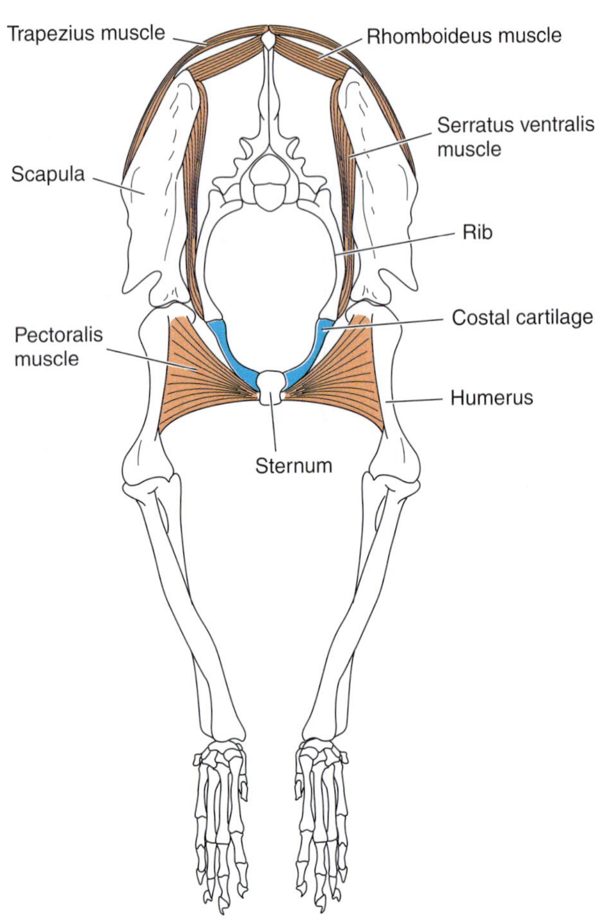

FIGURE 11-11
A cranial view of a transverse section through a cat, showing muscular connections between the axial skeleton and the pectoral girdle and appendage. (*From Walker and Homberger.*)

amphibians and reptiles. Weight is transferred to the pectoral girdle and appendage not by bony connections but by a muscular sling extending between the trunk skeleton and the girdle and appendage (Fig. 11-11). Although the rhomboideus and pectoralis muscles participate to some extent, the major component of the sling in mammals is the serratus ventralis muscle. Because this muscle extends between the distal ends of the bony ribs and the dorsal border of the scapula, the ribs are under compression stress when they are transferring body weight. Bone is well adapted to meet this stress. The costal cartilages, which extend from the ribs to the sternum, need not resist compression, but they must be sufficiently flexible to allow the ribs to move during respiration. Cartilage is the ideal material for this purpose.

In order to support the body, the legs must be drawn under the body by muscle action. Because the legs are not solid pillars but jointed struts, the continued contraction of muscles crossing the joints is needed to stabilize the moments around the joints and prevent the animal from collapsing, even when it is standing still. The amount of muscular energy needed to stabilize joints and hold up an animal is reduced in some mammal species by several factors. To the extent that the segments of the limbs can be aligned vertically, one directly above another, the moments at the joints are reduced. Weight is transferred directly through bones to the ground. Some dinosaurs, elephants, and other heavy terrestrial vertebrates tend to have pillar-like limbs of this type (Fig. 5-25). The tibia of humans also can be brought directly under the femur when standing. The configuration of joint surfaces and ligaments around joints restricts motion in certain directions. The olecranon of the ulna and calcaneus in the ankle form lever arms for muscle attachment, but they also prevent hyperextension at these joints. The cruciate and other ligaments behind the knee joint prevent hyperextension at this joint (Fig. 11-12). Muscle action is not needed to prevent joints of these types from bending backward.

Horses and some other ungulates have one or more passive **stay mechanisms** that hold limb joints in an extended position with minimal muscular effort, thereby enabling them to stand even when asleep. The way in which the knee (stifle) and ankle (hock) joints of a horse are stabilized is an example (Fig. 11-13). The superficial digital flexor muscle has become almost entirely tendinous and forms a strong cord that extends from near the distal end of the femur to the cal-

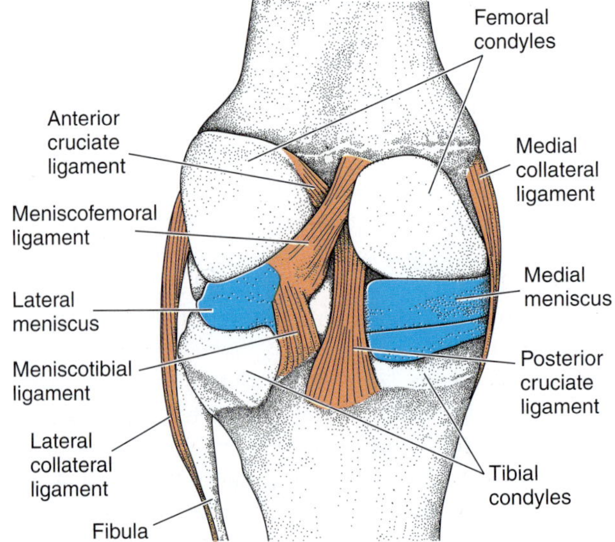

FIGURE 11-12
A posterior view of the left knee (stifle) joint of a horse showing the ligaments that prevent hyperextension of the crus. The articular capsule has been removed. (*After Getty.*)

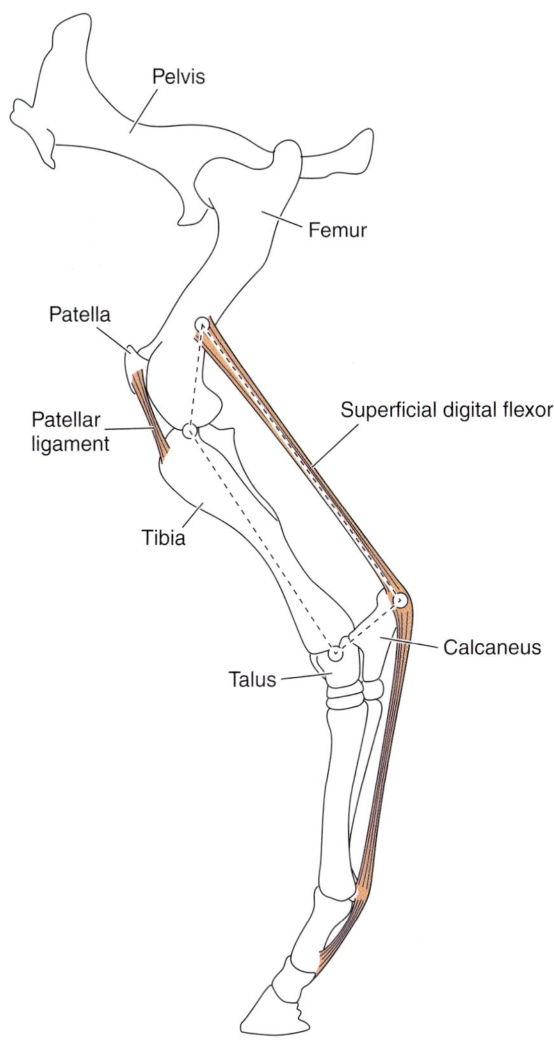

FIGURE 11-13
A lateral view of the pelvis and hind leg of a horse showing the stay mechanism that locks the femur and the tibia in a fixed position when the animal is at rest. Anterior is toward the left. Dashed lines show the parts of the locking parallelogram, and its four angles are shown by small, open circles. (*Modified from Getty.*)

caneus, where it is firmly attached, and then onto the caudal surface of the phalanges. The superficial digital flexor cord forms one side of a parallelogram, the other side of which is formed by the tibia. The distal end of the femur and the talus–calcaneus form the two ends of the parallelogram. As in any parallelogram, none of the angles at the corners can change without all others simultaneously changing. When the animal is resting, a muscle slip can pull the patella above a small crest on the medial side of the femoral trochanter. It is held in this position by strong patellar ligaments extending to the tibial crest. The patella resting above the crest locks the angles of the parallelogram. No change can occur until other muscles pull the patella away from the crest and into the patellar groove.

Walking and Running

Adaptations for Walking Although vertebrates move on the land in many ways (snakes slither, frogs jump, and many species burrow), walking was certainly the ancestral pattern of terrestrial locomotion. Walking probably began in the water before vertebrates ventured onto land (Chapter 9). The choanate fish ancestors of terrestrial vertebrates (Fig. 3-15) swam through the water and probably bottom vegetation by lateral undulations of their trunk and tail, but these fishes also had sturdy, lobate paired fins. These fins probably were used as paddles and occasionally to thrust against the bottom. By advancing and retracting the paired fins, undulations of the trunk and tail would thrust the fins against the bottom. Because the buoyancy of the water supported the fish, the paired fins need not have a supportive function. The number of paired fins in contact with the bottom at any time did not matter.

As fish began to emerge upon the land, the constraints of terrestrial life necessitated considerable changes in their appendicular apparatus. The paired fins evolved into jointed limbs capable of the more complex movements needed to move on the land. Lateral undulations of the trunk and tail continued to help advance and retract the limbs (Fig. 8-11*A*), but a more powerful and complex appendicular musculature became increasingly important in limb movements.

The limbs and girdles strengthened and became capable of supporting the entire weight of the body and of maintaining stability when the animal is walking. Maximum stability is attained when all four feet are placed on the ground, but an animal cannot move unless one or more feet are off the ground and advanced for the next foot placement. We discuss the many sequences of limb movements later in this chapter. Brief periods of instability occur in any sequence when the body is supported by an unbalanced combination of feet upon the ground; for example, one front foot and the contralateral hind foot, or even a single foot. During such periods, the trunk is subjected to torsion forces (Fig. 8-11*B*), which are met by strengthening the vertebral axis (i.e., the evolution of zygapophyses, Chapter 8), and by the contraction of the oblique layers of the abdominal musculature (Chapter 10).

Limb Actions in the Step Cycle of Amphibians and Reptiles A few terms must be understood to analyze walking. A walking vertebrate advances by a succession of **steps**. First, one foot is placed on the ground and develops a thrust that accelerates the body and moves it forward (the **propulsive phase** of a step cycle). Then this foot is removed from the ground and advanced in preparation for the next foot placement (the **swing phase**). The length of a step is the distance that the trunk is advanced during the propulsive phase of a step cycle. Each leg goes through a step cycle, and one cycling of all the legs constitutes one **stride**. That is, the length of a stride is measured from the first placement of one foot on the ground (often the left hind foot is taken as the reference foot) to the next placement of that same foot. A stride for a quadruped is composed of four step cycles, one for each leg. Throughout a stride, the oscillations of the limbs must be carefully coordinated to provide adequate thrust[1] and speed but maintain adequate stability. Moreover, movements and thrusts of the limbs must be accomplished as efficiently as possible so the cost in energy expenditure can be sustained if the animal is traveling far or can be made up soon if the animal must spurt for a short distance.

Of course, not all contemporary amphibians and reptiles walk in the same way. The pattern seen in many lizards is representative and probably is close to that used by ancestral tetrapods. As we have noted, amphibians and most reptiles carry their bodies close to the ground because the limbs are splayed to the side. Brinkman (1981) made a cineradiographic study of hind limb movements in the iguana (*Iguana*), and Jenkins and Goslow (1983) made a similar study of forelimb and girdle movements in the savannah monitor lizard (*Varanus exanthematicus*). The upper limb segment (humerus or femur) moves back and forth close to the horizontal plane (Figs. 11-14 and 11-15). The humerus and femur are carried slightly above the horizontal plane when the upper limb is protracted during the swing phase of the step and slightly below this plane when the limb is retracted during the propulsive phase. The lower limb segment (antebrachium or crus) extends downward at nearly a right angle to the distal end of the humerus or femur and moves close to the vertical plane.

When a foot is placed on the ground at the beginning of the propulsive phase, the upper limb is protracted and the lower limb extended (Figs. 11-14 and 11-15). Extension of the antebrachium is more pronounced than that of the crus. The hand or foot points forward and often slightly laterally. A propulsive thrust is developed by several actions. The actions of appendicular muscles retract the humerus or femur, and, because the foot maintains its position on the ground, this draws the body forward. The humerus or femur also rotates about its longitudinal axis, which rotates the lower leg in the manner of a crank (Fig. 11-16). While the humerus or femur is being retracted and rotated, the

[1]When the foot is first placed on the ground, it has a slight braking effect because it lies anterior to the shoulder or hip joints, but the body is carried forward by other feet. As the body moves over the placed foot, the angle of the limb gives a propulsive thrust. All of these forces for all of the limbs must be coordinated.

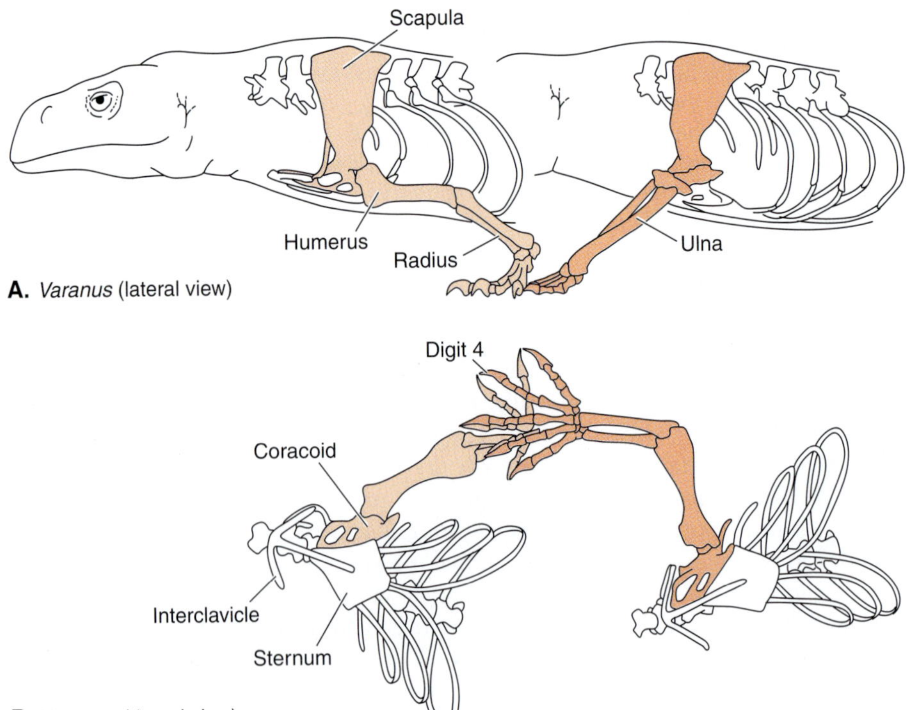

FIGURE 11-14
The movements of the forelimb and pectoral girdle of a lizard, *Varanus exanthematicus*, at the beginning (*right*) and end (*left*) of the propulsive phase of a step. The position of the limb when the foot is first placed on the ground is shown by a darker shade of color. Note that the foot remains in the same position as the body advances. *A*, Lateral view. *B*, Dorsal view. (*After Jenkins and Goslow.*)

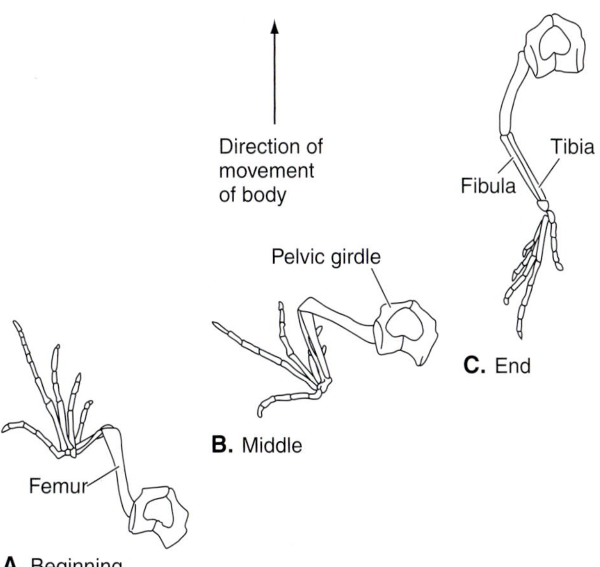

FIGURE 11-15
Dorsal views of the movements of the hind limb and pelvic girdle of the lizard, *Iguana*, at the beginning (*A*), middle (*B*), and end (*C*) of the propulsive phase of a step. The arrow shows the direction of movement. (*After Brinkman.*)

antebrachium or crus flexes as the distal end of the upper limb advances over the point of foot placement, and then the antebrachium or crus extends and helps push the body forward (Figs. 11-14 and 11-15).

The hand continues to point forward throughout the propulsive phase because the radius rotates on the humerus and ulna as the limb is drawn caudad. These are difficult motions to describe, but they can be visualized if you place your arm in the position assumed by an amphibian or reptile when the hand is first placed on the ground, and advance the body while maintaining the position of the hand on the ground. The situation is somewhat different in the hind leg because only a hinge action can occur at the knee joint. Rotation occurs in the ankle, primarily at the mesotarsal joint. The hind foot is directed forward at the beginning of the propulsive phase, but as the femur is drawn posteriorly and the body advances, the foot rotates laterally and points laterally or posteriorly at the end of the propulsive phase (Fig. 11-15). In both the hand and foot, duration of toe contact with the ground is maximized by digits that increase in length from the most medial or first one, to the fourth finger. The length of the fifth digit decreases abruptly. Plantar flexion of the hand or foot adds a final component to the thrust of the limb.

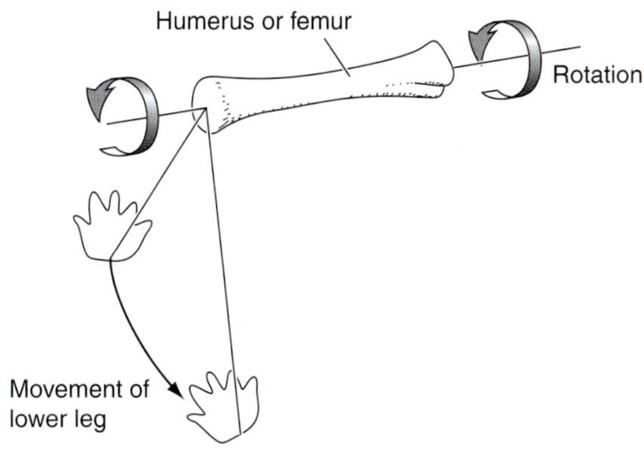

FIGURE 11-16
The effect of rotation of the humerus or femur on the lower leg in early tetrapods.

Swimming-like sinusoidal curvatures of the trunk and tail occur in most amphibians and reptiles (Fig. 8-11*A*). In addition to helping advance and retract the limbs, these motions also rotate the girdles slightly (Figs. 11-14*B* and 11-15), which helps advance and retract the limbs and adds to step length. An extra increment is added to the length of the step of the front leg in lizards by a translational movement of the coracoid plate on the sternum (Fig. 11-14*B*). As the limb is retracted, the girdle also slides caudad on the sternum for a distance equal to about one fourth of the length of the coracoid plate.

The swing phase of a step is a simple recovery. The antebrachium or crus flexes as the hand or foot is removed from the ground. The humerus or femur is protracted, and the forelimb is extended.

All of these movements require a powerful ventral or adductor musculature to draw the humerus or femur ventrally to a slight extent and thus raise and keep the body off the ground. Strong forelimb muscles also are needed to bend the antebrachium or crus and maintain it in a nearly vertical position. As we have seen (Chapter 9), the ventral parts of the girdles are greatly expanded and provide the needed surface for the origins of the adductor musculature. Large tuberosities, trochanters, and muscular ridges on the humerus and femur provide for their insertions. You can appreciate the muscular force needed by crawling with your arms splayed to the side rather than under your body.

Limb Actions in the Step Cycle of Mammals The limbs of mammals are no longer in the primitive splayed position but have rotated far enough under the body so the humerus or femur moves fore and aft closer to the vertical plane. The **stance,** or distance between the placement of left and right feet, is narrower than in amphibians and reptiles, which places the limbs closer to the projection of the body's center of gravity and provides better support with less muscular effort. Limbs situated more or less under the body swing through longer arcs, which adds to step and stride length with no or little extra expenditure of energy.

The degree to which the limbs are beneath the body varies among mammal species. In running, or **cursorial** (L., *cursor* = a runner), species such as ungulates and carnivores, the limbs are well under the body and all parts of the limb move in the same parasagittal plane. But in many mammals, they are not quite so far under. Jenkins (1971) and Jenkins and Weijs (1979) studied limb movements in the opossum (*Didelphis virginiana*) by cineradiography. The elbow joint, which is directed posteriorly, lies slightly lateral to the shoulder joint at the beginning of the propulsive phase (Fig. 11-17*A*). The knee is directed forward and slightly laterally (Fig. 11-18*A*). In all species, both the hand and the foot are directed forward, or nearly so, throughout the propulsive phase. Toe length is more equal. As the propulsive phase continues, the humerus or femur is retracted, drawing the body forward. The forelimb flexes and the feet dorsiflex as the shoulder or hip advances over the point of placement of the feet on the ground. The forelimbs extend and feet plantar flex near the end of the propulsive phase.

The scapula also rotates during the propulsive phase, so at the end of the phase, the scapula is oriented more vertically than at the beginning and the glenoid fossa has moved caudally and ventrally (Fig. 11-17*B*). These movements add to step length. Although the hips are firmly articulated with the vertebral column, a small lateral movement of the vertebral column causes the hips to swivel slightly (Fig. 11-18*A*).

Muscle Recruitment in a Step Cycle Many electromyographic studies have been made during locomotion in humans and quadrupeds. Jenkins and Weijs (1979) studied muscle activity during a step cycle in a mammal (the opossum, *Didelphis virginiana*); Jenkins and Goslow (1983), in a reptile (the lizard *Varanus exanthematicus*); and Dial et al. (1991), during flight in birds (the starling, *Sturnus vulgaris*). Dial et al. also compared the sequence of actions of shoulder and brachial muscles in all three species during a step cycle. The general nature of their conclusions can be seen by examining the activity of a selection of the muscles that they studied (Table 11-1).

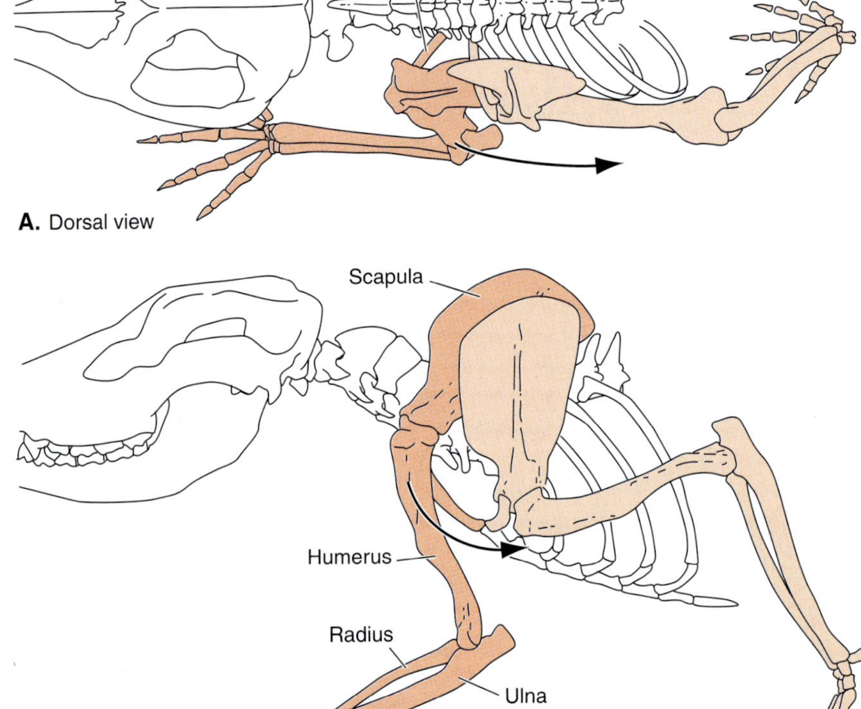

FIGURE 11-17
The movements of the forelimb and pectoral girdle of the Virginia opossum, *Didelphis,* at the beginning (*upper figure*) and end (*lower figure*) of the propulsive phase of a step. *A,* Dorsal view. *B,* Lateral view. The position of the limb when the foot is first placed on the ground is shown in a darker shade of color. The figure shows the limb retraction during the propulsive phase of a step, but in a walking animal, the foot retains its position on the ground and the body advances (e.g., Fig. 11-14). (*After Jenkins and Weijs.*)

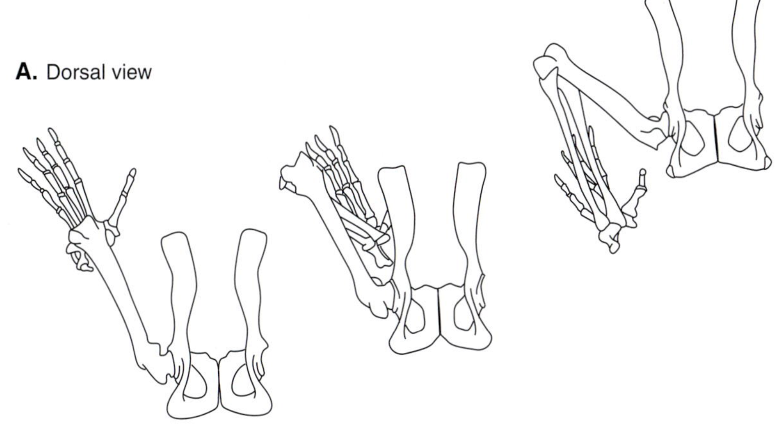

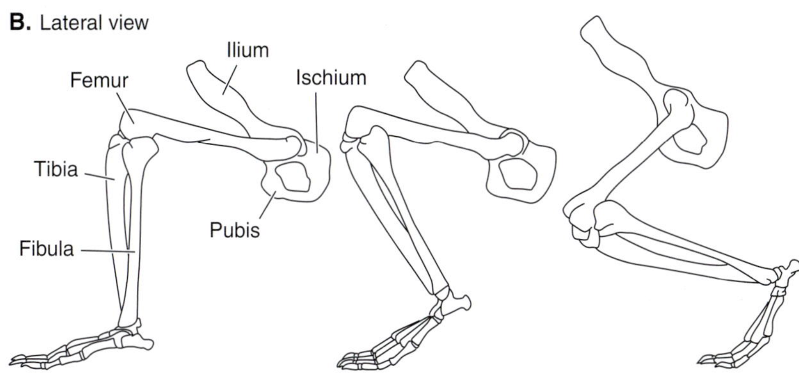

FIGURE 11-18
The positions of the hind limb segments and pelvic girdle of the opossum, *Didelphis,* are shown at three points during the propulsive stage of a step. *A,* Dorsal view. *B,* Lateral view. (*After Jenkins.*)

TABLE 11-1 Muscle Fiber Activity

Examples of the activity of selected shoulder and arm muscles belonging to both dorsal and ventral appendicular groups in a lizard *(Varanus)*, a mammal (*Didelphis,* or dog, *Canis*), and a bird *(Sturnus)* are shown. The periods of electrical activity in the muscles are shown by the placement and length of the horizontal lines. *(After Dial et al.)*

				Propulsion (Downstroke)	Swing (Upstroke)
Shoulder Muscles	Dors.	Deltoideus (scapular)	*Varanus*		
			Didelphis		
			Sturnus		
	Dors.	Subscapularis	*Varanus*		
			Didelphis		
			Sturnus		
	Vent.	Pectoralis	*Varanus*		
			Didelphis		
			Sturnus		
Arm Muscles	Dors.	Triceps (humeral)	*Varanus*		
			Didelphis		
			Sturnus		
	Vent.	Biceps brachii	*Varanus*		
			Canis		
			Sturnus		

FOCUS 11-2 *Muscle Fiber–Type Recruitment in a Step Cycle*

Many of the appendicular muscles of tetrapods are composed of mixtures of two or three physiological types of striated muscle fibers. In mammals, slow-oxidative (red) fibers maintain posture and are active in slower movements that continue over long periods of time. Fast-oxidative glycolytic (pink) fibers join the slow-oxidative fibers as speed increases. Fast-glycolytic (white) fibers become active only at high speeds. Because these are glycolytic fibers, high speed cannot be sustained for long periods of time.

These generalizations apply to ectothermic vertebrates as well, but the muscles of ectotherms must function over a wide range of body temperatures, which poses physiological problems because muscles tend to contract more slowly at lower temperatures, but lizards can run at the same speed at different temperatures. What physiological adaptations do lizards make that allow them to do this? Jayne et al. (1990) have studied this problem in *Varanus exanthematicus* using quantitative electromyography. They selected a thigh muscle to study, the iliofibularis, because red and white fibers are localized in different parts of the muscle. They found that red muscle fibers were active at all speeds studied but that the intensity of red-muscle activity generally increased with speed to a certain maximum. The speed at which maximum red muscle activity was attained was less at 25°C than at 35°C. However, for a given speed, the amplitude of the electromyograms was greater at the lower temperature. The faster white muscle fibers, as one would expect, are recruited only above some threshold speed, and the threshold was less at lower temperatures. Thus, lizards compensate for lower temperature by an increased amplitude of activity in the red fibers and by their ability to recruit more white fibers at lower temperatures.

As might be expected, more muscles are active, and active for longer periods, during the power-generating propulsive phase of a step cycle (or downstroke of a bird's wing) than in the recovery swing phase (or upstroke of a bird's wing).

Limb movements involve integrated actions of many muscles. Although most of the muscles participate in propulsion, the duration of their activity, and the period of the cycle when they are active, differ. The action of the mammalian triceps (*Didelphis*), for example, begins and ends earlier than that of the mammalian biceps (*Canis*).

Homologous muscles tend to be utilized in similar actions across a wide range of taxa: reptile, mammal, and bird. The evolution of appendicular muscles has tended to be conservative, but some changes occur. The subscapularis in reptiles (*Varanus*), although participating in limb movement, is a small muscle and does not play a prominent role. With the changed limb posture of mammals (*Didelphis*), the subscapularis expands considerably and plays a prominent role in propulsion.

Many of the propulsive muscles show electrical activity near the end of the recovery swing phase, before they begin to participate in propulsion. This phenomenon requires further study. Possibly the muscles are simply checking the forward movement of the limb and preparing for propulsion, but some of them likely are doing negative work, that is, beginning to generate a force while still being stretched by their antagonists. This would stretch the elastic components in the muscles and store elastic energy, which would be recovered in the propulsive phase. Muscles involved in limb movements differ not only in periods of activity but also in the mixture of fiber types that they contain (Focus 11-2).

Gaits Walking is a relatively slow gait, but individual tetrapods can increase their speed by increasing stride length, that is, the distance traveled in each stride. A longer stride enables a tetrapod to travel a greater distance in a given period of time. Increasing stride length is accomplished by changes in **gait,** that is, the particular combination of feet that are on and off the ground during a stride. Gaits used by tetrapods depend on many variables, including limb posture and length, degree of stability or maneuverability needed, energetic costs, and speed. Gaits faster than a walk involve increased periods of instability. Because amphibians and most reptiles move close to the ground, they must have gaits that provide good stability so their bellies do not drag.

Turtles usually use a gait that is particularly stable (Fig. 11-19). If the left hind foot is taken as the lead foot, the sequence of limb advancements and foot placements is left hind, left front, right hind, and right front foot. Each foot is on the ground for a relatively long time (about 0.75 of stride duration for a slowly walking turtle). This time, which is the same for each foot, is called the **duty factor.** Limb movement can be expressed in quantitative terms by stating the duty factor and the **relative phase** of the feet (Fig. 11-19*B*). The first foot to be placed (left hind) is given a relative phase of 0.00 because this is the beginning of the stride. The placement of the next (left front), which occurs slightly over one third of the stride interval, has a phase of 0.35. The next one (right hind) is placed at the middle of the stride and has a phase of 0.50. The placement of the last foot (right front) follows the right hind in a lag that is the same as that between the first two feet, so it has a phase of 0.85. This particular gait is called a **lateral-sequence, diagonal-couplet gait.** In such a gait, the placement of one foot (left hind) is followed by the placement of the ipsilateral foot (left front), hence the term "lateral sequence." The contralateral diagonal feet (left front and right hind or right front and left hind) are in closer phase in their movements and placements than any other pair of feet, hence the term "diagonal couplet."

During most of the stride, either four or three feet (Fig. 11-19*A*) support the body, so the center of gravity falls within successive triangles of supporting feet. These triangles are quite wide from left to right because the splayed limb posture gives the animal a wide stance. The gait and wide stance provide the high degree of stability that these low-slung animals need. When the bases of the triangles of support shift between the sides of the body, there are brief periods when the body is supported by four feet (two overlapping triangles) or by two feet (a hind foot and its contralateral front foot). The latter is unstable, and some pitching and rolling occur as the center of gravity crosses the diagonal between supporting feet, but these periods are so brief that the animal does not drag on the ground.

Because mammals carry their limbs more nearly under the body than do amphibians and reptiles, they have less stability. Mammals have a narrower stance and narrower triangles of support, and their center of gravity is higher off the ground. Stability, however, is not as critical as for amphibians and reptiles. Mammals can rise and fall within each stride to a greater extent without dragging on the ground. Only an overall dynamic stability is necessary, as in riding a bicycle. This allows mammals to reduce duty factors more and have fewer feet on the ground at one time, and it gives them more gait options. Decreased stability also increases maneuverability; mammals can change directions more rapidly.

When moving slowly, many mammals use the lateral-sequence, diagonal-couplet walk seen in amphibians and reptiles. Duty factors are lower, however, so the body is supported during part of the stride only by two feet on the same side of the body (Fig. 11-20*A*). This is a potentially unstable situation and could not be used by low-slung amphibians and reptiles. The same gait can be used at a moderate run when the duty factor for each foot falls below 0.50. Periods when the animal is supported by two feet, or even one, increase in length.

Most mammals shift to other gaits when they begin to run. A few long-legged mammals, such as giraffes and camels, change to a **pace** (called a **rack** in horses) when speed increases (Fig. 11-20*B*). Legs on the same side of the body move together in nearly complete phase, and 0.50 out of phase with the legs on the other side of the body. In a running pace, two brief **flight periods** occur in each stride when the body has suffi-

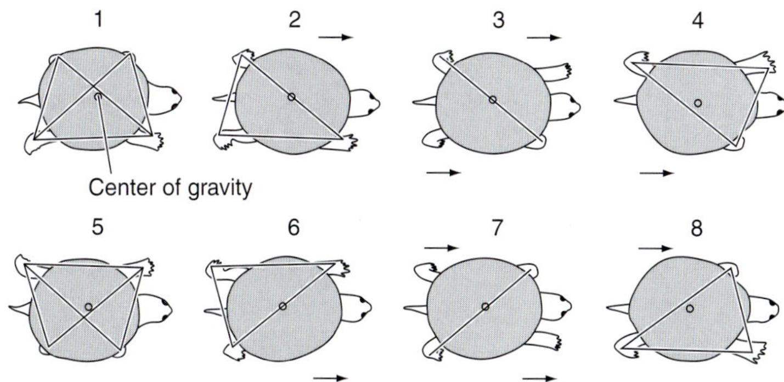

A. Limb movements and foot placements during one stride of a turtle

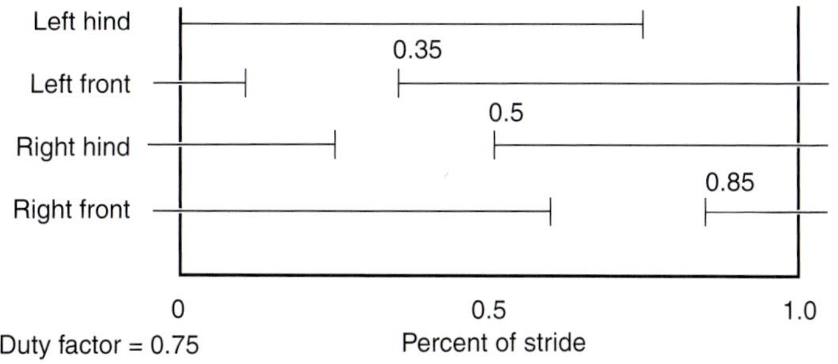

B. Gait diagram of the stride shown in A

FIGURE 11-19
A, Drawings from cinephotographs showing limb movements and foot placements during one stride of a turtle. The stride begins with the placement of the left hind foot. Arrows indicate the feet that are off the ground and moving forward for the next placement. Lines connect feet on the ground. Notice that the animal's center of gravity falls within the triangles of supporting feet during most of the stride. *B,* A gait diagram of the stride shown in *A.* The horizontal lines indicate the period each foot is on the ground. Numbers indicate the lag in foot placement as a percent of the stride. (*After Walker, 1971.*)

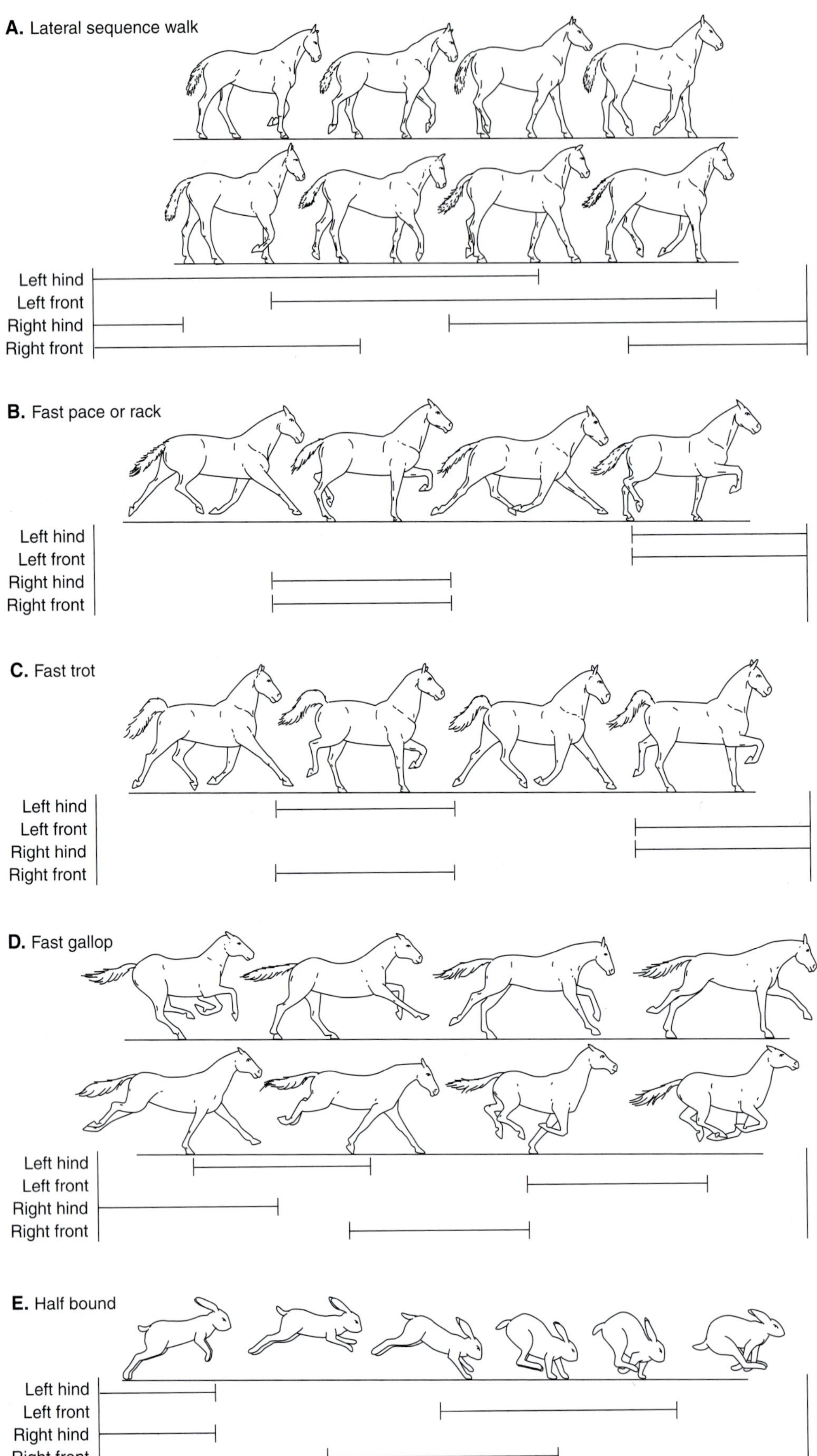

FIGURE 11-20

Some mammalian gaits as seen in a horse (*A–D*) and rabbit (*E*). Horizontal lines below the figures show the duration of placement of each foot. The placement and removal of the feet from the ground also can be seen in the figures above the horizontal lines, but each figure represents just one point in one of the horizontal lines. (*After Gambaryan.*)

cient momentum that it is carried forward with no feet on the ground. These occur when the supporting feet shift from one side of the body to the other.

Horses can be trained to pace, but their natural gait at higher speeds is a **trot** (Fig. 11-20*C*). A contralateral pair of diagonal feet (left front and right hind feet, or right front and left hind feet) move nearly in unison and 0.50 out of phase with the other set. Again, if the trot is fast, two brief flight periods occur within each stride during the shift from one set of supporting feet to the other. Some reptiles can use a walking trot that does not involve any flight periods.

Walks, paces, and trots are described as **symmetrical gaits** because the left and right hind feet, or the left and right front feet, move 0.50 out of phase and are evenly spaced in time. At still higher speeds, most mammals shift to **asymmetrical gaits,** in which the movements of the two hind or two front feet are more nearly in phase. The two hind feet, or the two front feet, tend to thrust at the same time or close to it. A gallop, of which there are several types, is one example. In the **fast gallop** of some ungulates and carnivores (Fig. 11-20*D*), a powerful thrust is initiated by one hind leg (Fig. 11-20*D, first frame*), continued by both, and finally by the other hind leg (Fig. 11-20*D, third frame*). As the hind legs thrust and leave the ground, first one and then the other front leg thrusts and leaves the ground. There are brief periods when both hind feet are off the ground when the animal is supported by a single front leg (Fig. 11-20*D, fifth and seventh frames*). Both front legs complete their thrust and leave the ground before either hind leg is replaced, so there is a brief flight period when all feet are off the ground (Fig. 11-20*D, eighth frame*). The low duty factors and flight period greatly increase stride length in a fast gallop, about 7 m for a racehorse.

The **half-bound** is a gallop-like gait used by rabbits, squirrels, weasels, and some other small mammals (Fig. 11-20*E*). It differs from the fast gallop primarily in that the thrust of the two hind feet is in very close phase and often simultaneous. At the start of the half-bound, the animal elevates the front of the body, bringing the front legs off the ground, and gives a powerful thrust with both hind legs (Fig. 11-20*E, first frame*). There is a brief flight period, when the hind legs leave the ground before either front leg has reached it (Fig. 11-20*E, second frame*). First one and then the other front leg reaches the ground and gives a brief thrust before either hind leg has reached the ground, so there is a second flight period (Fig. 11-20*E, sixth frame*). In some species (e.g., squirrels), the hind legs are so long and the back arches so much that the hind feet are placed on the ground in advance of the front feet, which have just left the ground. In each half-bound, the footprint of the hind foot lies anterior to that of the front feet!

Stride Length and Stride Frequency As we have discussed, increasing **stride length** enables a tetrapod to travel a greater distance in a given period of time. The feet in fast gaits have lower duty factors. They are off the ground for longer periods, during which they are carried forward by the feet on the ground for a greater distance before being placed again. Indeed, as we have seen, all feet may be off the ground briefly during some fast gaits. A second way an individual can increase speed is to increase **stride frequency,** that is, to oscillate the limbs faster, which results in an increased number of strides in a given period. The overall velocity with which a tetrapod travels is the product of stride length and stride frequency. Because stability is so important, low-slung amphibians and reptiles increase speed primarily by increasing stride frequency within a walk, although some can trot at moderate rates. Mammals, for which stability is not so critical, increase speed first by increasing stride frequency but soon shift to gaits that increase stride length. Increasing stride length is less expensive in terms of energy consumed than is increasing stride frequency (Focus 11-3).

Evolutionary Adaptations for Patterns of Limb Action

Limbs as Lever Systems An individual mammal can change speed by changing stride length and stride rate, but mammal limbs also are adapted to the ways in which the limbs are used. The limbs and the muscles that act on them are first-order and second-order lever systems (Chapter 5). In the course of evolution, the length of limb segments, and the length and location of bony processes to which muscles attach, have been subject to selection pressures. In moles and armadillos, which use their front limbs for digging, the limb is adapted to maximize the force delivered at the end of a limb, whereas running ungulates have limbs adapted to maximize the speed or velocity with which the distal end of the limb moves (Fig. 11-21). For purposes of this comparison, we will treat the entire limb from the fulcrum at the shoulder joint to the ground as a single second-order lever that is retracted by muscles (e.g., the latissimus dorsi) that extend from the trunk to the humerus. The out-lever arm is the distance from the shoulder joint to the ground, and the out-force is delivered downward and backward, perpendicular to the out-lever. The in-lever arm is the perpendicular distance from the shoulder joint to the line of action of the retractor muscle. Recall (Chapter 5) that, at equilibrium, the following relationship between lever arms and forces applies:

$$F_i \times L_i = F_o \times L_o$$

FOCUS 11-3 *Energy Costs of Locomotion*

Much energy is stored during locomotion as elastic energy by stretching the tendons and other elastic components of muscles. The contraction of one group of muscles stretches their antagonists (Chapter 10). Sometimes the rise and fall of the body significantly stretches tendons. For example, when the weight of a mammal bears down on a hind foot just placed on the ground, the long calcaneus tendon and others behind the heel and toes are stretched and store elastic energy. The elastic recoil of the tendons and muscles recovers most of this elastic energy and helps raise and advance the animal in the next step cycle (Fig. A).

Although recycling energy is very important, muscles must still contract. The energetic cost of muscle contraction is determined primarily by the energy expended to activate the contractile machinery. Calcium must be pumped into the sarcoplasmic reticulum, and cross bridges between actin and myosin myofilaments must be established and broken. Heglund and Taylor (1988) showed that these costs increase much more rapidly with an increase in stride frequency than with an increase in stride length (change of gaits). They studied the energetic costs of locomotion in 16 mammals ranging in size from a mouse to a horse. To compare animals of such different sizes that travel at very different absolute speeds, they developed the concept of **equivalent speeds.** They compared the preferred trotting speed (as selected by the animal), the speed at the trot–gallop transition, and the preferred galloping speed. They found that the energy cost per gram of animal per stride was not dependent on size; it was the same for a mouse and a horse. But the energy cost did increase as speed increased and the animals changed gaits: 5.0 J/kg per stride at the preferred trotting speed to 7.5 J/kg per stride at the preferred galloping speed. This is a 1.5-fold increase. Small animals, however, with their short legs, had much higher stride frequencies at equivalent speeds. At the trot–gallop transition, a mouse takes six times as many strides as a horse. Because it must cycle its limbs so rapidly, a mouse consumed energy on a per-gram basis at six times the rate of a horse at equivalent speeds.

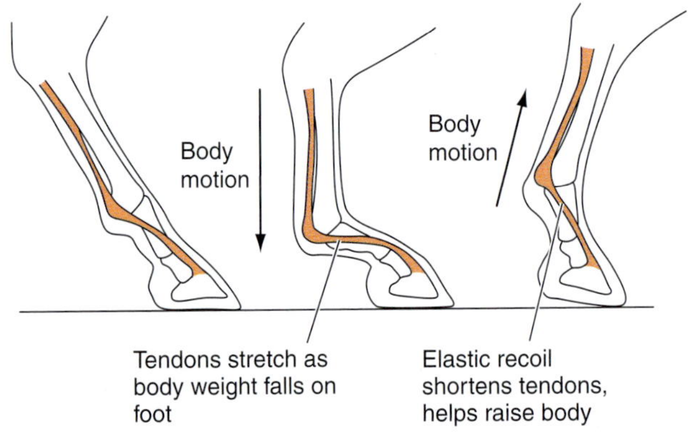

A. Diagrams of the foot of a horse when the weight of the body bears down upon it (left and center) and stretches the calcaneous tendon (orange). The stored elastic energy is released and helps to lift the body as weight on the foot decreases (right).

where F_i is the in-force, L_i is the in-lever, F_o is the out-force, and L_o is the out-lever. By solving this equation for the out-force, it is apparent that an increased value for the out-force can be obtained by increasing the length of the in-lever relative to the out-lever:

$$F_o = (F_i \times L_i)/L_o$$

An increase in the in-lever arm is said to give the muscle acting on it a **mechanical advantage.**

This is what has happened in the evolution of the armadillo. The armadillo has a short out-lever arm relative to most mammals and a distal shift in the point of attachment of the retractor muscle has increased the relative length of the in-lever.

Velocity also is related to the relative length of the lever arms. In a given unit of time, the part of a rotating lever that is located far from the fulcrum will move a greater distance than a point near the fulcrum. Because velocity = distance/time, the velocity of the distant point will be greater than that for the point closer to the fulcrum. This relationship can be expressed as:

$$V_o \times L_i = V_i \times L_o$$

where V_o and V_i represent the velocities at the ends of the out-lever and in-lever arms, respectively (i.e., out-velocity and in-velocity). By solving this equation for the out-velocity, it is apparent that an increased value for the out-velocity can be obtained by lengthening the out-lever arm and by shortening the in-lever arm through a proximal shift in the point of attachment of the retractor muscles:

$$V_o = (V_i \times L_o)/L_i$$

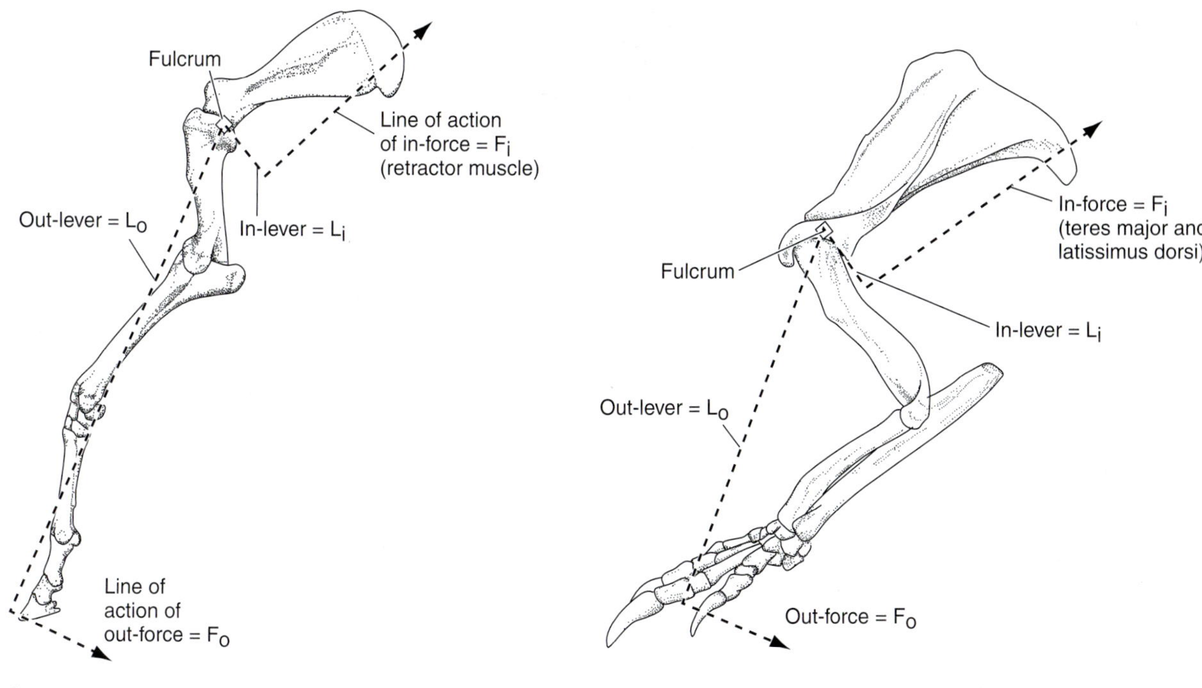

A. Design for speed (*Equus*)

B. Design for power (Armadillo)

FIGURE 11-21
Some limbs are adapted by evolutionary forces to maximize speed or power by changes in the relative length of their in-lever and out-lever arms. A horse's (*Equus*) pectoral girdle and appendage (*A*) are adapted for high velocity; those of an armadillo, *Dasypus* (*B*), are designed for power. The limbs have been drawn to the same size for easy comparison of the lengths of the lever arms. (*After Smith and Savage.*)

This is what has occurred in ungulates and other cursors.

The relationship between force and velocity also can be expressed as a **gear ratio,** which is the length of the out-lever arm divided by the length of the in-lever arm:

$$GR = L_o/L_i$$

When the ratio is low, force is emphasized at the expense of velocity; when the ratio is high, velocity is emphasized at the expense of force. The ratio is low, about 4, for an armadillo, but much higher, about 10, for an ungulate. A low gear in an automobile increases force; a high gear increases velocity.

Other Adaptations for Speed An increase in limb length, which increases stride length, is an obvious adaptation for speed. Some mammals, humans among them, can temporarily increase effective limb length by raising up on the toes when running. Evolutionary changes also have occurred. Early mammals and many others that travel at moderate speeds walk flat footed. They place the soles of their feet flat on the ground, a foot posture termed **plantigrade** (L., *planta* = sole + *gradus* = step; Fig. 11-22). Most primates retain a plantigrade foot, with the first digit articulated in such a way that the hands and feet can grasp branches. In this case the plantigrade foot is an arboreal adaptation. The human foot is also plantigrade but has lost its grasping ability and is specialized for bipedal locomotion. Running species, as we have seen, evolved longer limbs, which increases step and stride length and hence speed. Most carnivores walk on their digits, with the wrist and ankle carried off the ground, a foot posture termed **digitigrade.** Ungulates have very long legs and walk on the tips of those digits that reach the ground, a posture termed **unguligrade** (L., *ungula* = hoof). The terminal phalanx and claw are modified to form a hoof. Digits on each side of the supporting toes are reduced in size or lost. In perissodactyls, the axis of the foot passes through the middle or third digit, which is the largest (Fig. 11-23*A* and *B*). Their foot is described as **mesaxonic.** Rhinoceroses, adapted to run on relatively soft ground, retain three digits, the second to the fourth. This was also the case in ancestral, forest-dwelling horses, but only the third digit remains in contemporary, plains-dwelling species. The foot axis of artiodactyls passes between the third and fourth toes, which are equal in size (Fig. 11-23*C* and *D*). This foot is described as **paraxonic.** Pigs and other species with a primitive foot structure retain small second and fifth digits, but

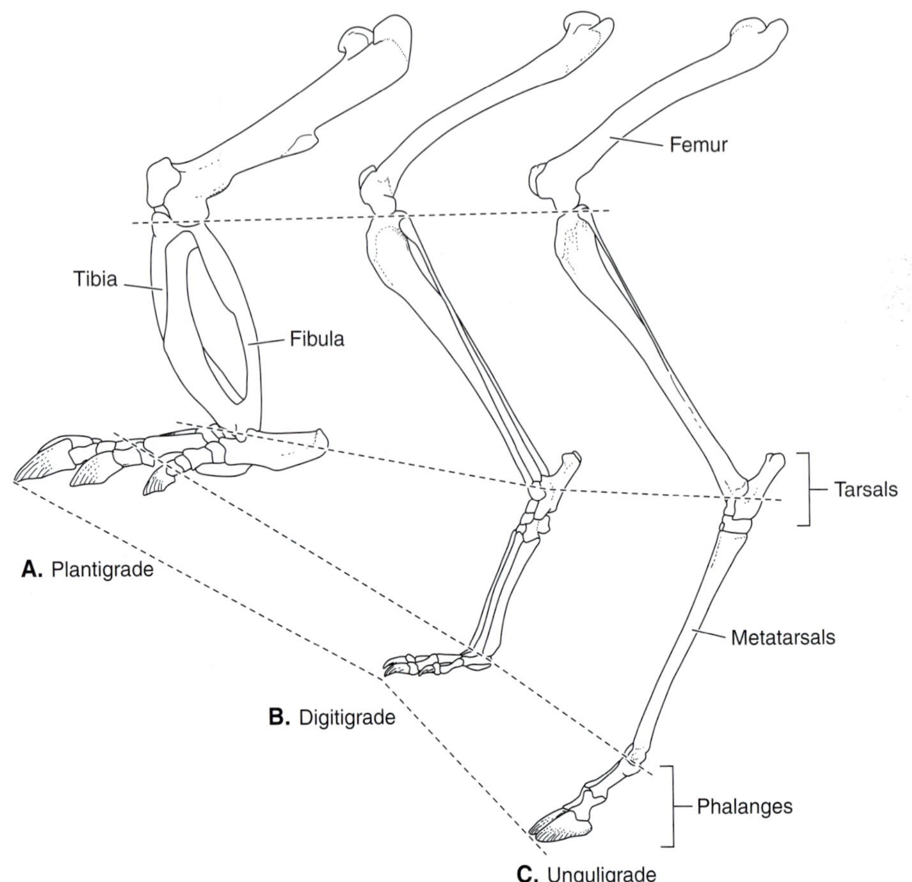

FIGURE 11-22
Limb types and foot postures of mammals. The femur of each species has been drawn to the same size to emphasize changes in proportions of the distal parts of the limbs. The plantigrade foot (*A*) is that of a powerful digger, the armadillo (*Dasypus*); the digitigrade foot (*B*) is that of a good runner, the coyote (*Canis latrans*); the unguligrade foot (*C*) is that of the extremely swift pronghorn (*Antilocapra*). (*After Vaughan.*)

these are lost in more specialized species, including cattle and sheep. Metapodial and podial elements that supported lost toes also are reduced or lost. The two metacarpals and metatarsals that remain in cattle and sheep are fused to form a single **cannon bone** (Gr., *kanon* = measuring rod).

In the evolution of long legs as an adaptation for speed, the foot and distal segment of the limb elongate much more than does the proximal segment (Fig. 11-22). These segments of the limb contain many tendons and less muscle mass than the brachium or thigh. This pattern of elongation minimizes the kinetic energy needed to oscillate the limbs because it keeps most of the mass proximally. The kinetic energy needed to oscillate a limb segment can be expressed by the following formula:

$$\text{Kinetic energy} = (1/2 \text{ mass of limb segment}) \times (\text{velocity of limb segment})^2$$

The proximal part of a limb moves through a shorter arc than does the distal part and therefore has less velocity. Because velocity is squared, it is to a cursor's advantage to have most of the limb muscle mass concentrated here. There is less energetic cost in elongating the distal part of the limb because it contains primarily the long tendons of proximal muscles. Physicists refer to this distribution of mass as minimizing the **moment of inertia.**

A long foot also adds a segment of the limb that oscillates independently of the lower leg and upper limb segment (Fig. 11-24). This too increases speed because the total velocity at the distal end of the limb is the sum of the independent velocities of each segment. The situation is analogous to walking on a moving escalator. The total velocity of the person traveling on the escalator is the sum of the velocity of the escalator plus the velocity of the person's walk.

Other modifications of the appendicular skeleton, in addition to long legs, also occur in many cursorial mammals. The clavicle is essentially a strut that braces the shoulder joint and prevents a displacement of the scapula. This is the primitive condition in tetrapods and mammals. During the propulsive phase of the step cycle, when the front foot is on the ground and the trunk is pulled forward relative to the limb, the clavicle causes the shoulder joint to be deflected laterally (see Fig. 11-17*A*). It acts as a "spoke" that fixes the distance between the sternum and acromion. Terrestrial mammals that retain a clavicle are species in which the legs have not rotated completely beneath the body so that the shoulder and elbow joints are not in the same

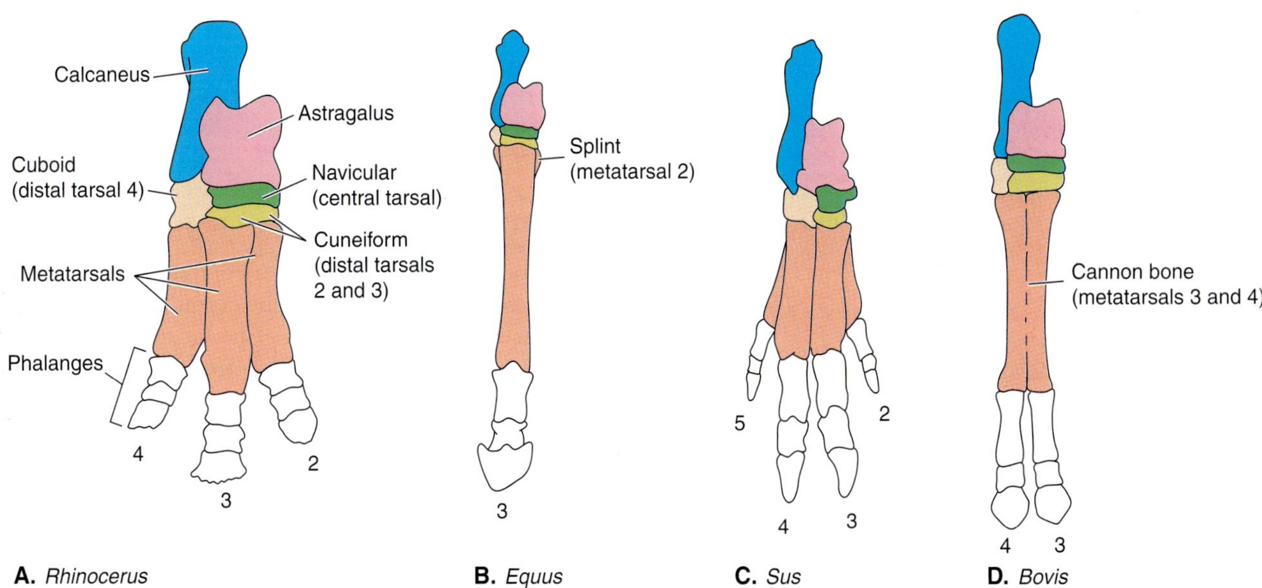

FIGURE 11-23
Front views of the right hind foot of representative ungulates showing the differences in toe reduction between mesaxonic perissodactyls (*A* and *B*) and paraxonic artiodactyls (*C* and *D*). For ease of comparison, homologous tarsal and metatarsal elements are in the same color. *A,* A rhinoceros, *Rhinoceros. B,* A modern horse, *Equus. C,* A pig, *Sus. D,* A cow, *Bovis.*

parasagittal plane. The legs are further under the body in cursorial species, and the shoulder and elbow joints lie in the same parasagittal plane. The clavicle is reduced or lost in such species (Fig. 9-14). The shoulder is not deflected laterally, with a consequent loss of some momentum of the trunk. The loss of the clavicle also permits the scapula to rotate more than in claviculate species. The scapula became, in effect, another limb segment that can add to step length (Fig. 11-24).

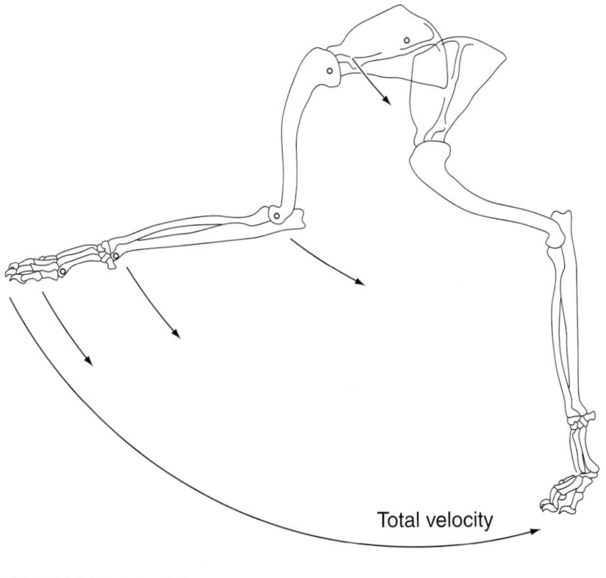

FIGURE 11-24
Rotation of the scapula and pectoral limb segments during locomotion of the cat. The total velocity of the distal end of the limb is the summation of the independent velocities of the individual rotating segments.

The long legs of ungulates are adapted for high speed and strength and perform no other functions. The limbs move fore and aft in a single plane and, unlike the situation in cursorial carnivores, the primitive capacity of the forearm to rotate around its longitudinal axis is lost. The hand is permanently prone. The radius is the primary weight-supporting bone, and the distal end of the ulna frequently is reduced or even lost. The proximal end of the ulna and the olecranon, which are essential for the hinge joint at the elbow and for the insertion of forearm extensor muscles, are retained and may fuse to the radius (Fig. 11-25*A*). The tibia is the main weight-transferring bone in the shank, and the fibula is reduced or lost in many ungulates except for its proximal end, on which muscles attach, and its distal end (the lateral malleolus), which supports the ankle joint (Fig. 11-25*B*). Loss of unused toes, loss of the capacity to rotate the forearm, and loss of associated muscles decrease the mass of the distal part of the limb and reduce the risk for dislocations.

In some running cursors, the vertebral column is adapted to participate significantly. This is the case in many cats, such as the cheetah (Fig. 11-26*A*). A shift in the plane of the articular surfaces of some zygapophyses allows a galloping cheetah to flex and extend its back greatly. It arches (flexes) its back far dorsally during the flight stage in which the legs are gathered under the trunk (left side of figure). By elevating its hips and

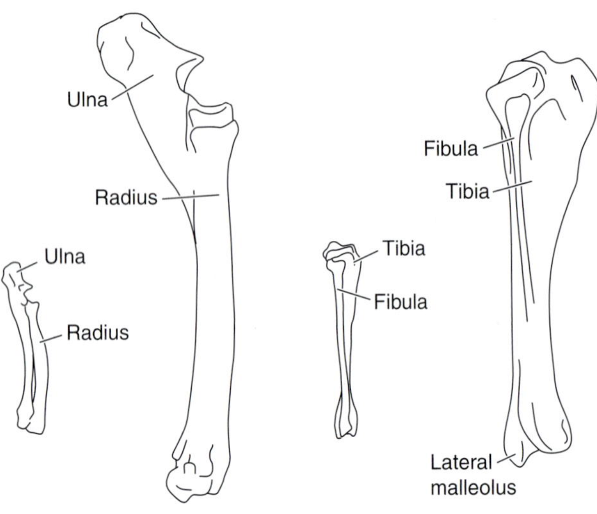

FIGURE 11-25
Lateral views of the bones of the forearm (*A*) and shin (*B*) of an early horse, †*Hyracotherium,* and modern horse, *Equus.*

bending its back far in the opposite direction (extension), it is able to stretch out its fore and hind limbs greatly during the next flight stage (right side of figure). This flexion and extension of the spine greatly increases stride length and speed. Cheetahs can run faster than any other mammal for short periods of time, about 110 km/h for a 60-kg cheetah. Horses are quite heavy cursors, attaining weights of 540 kg or more. Their back flexes and extends only slightly during a gallop (Fig. 11-26*B*). It would be very expensive energetically to raise and lower the posterior part of the body greatly. The differences in galloping between a cheetah and horse correlate with differences in their modes of life. The cheetah is a carnivore that must be able to sprint quickly to capture prey. The horse is an herbivore, adapted for endurance travel over large distances in search of favorable grazing grounds.

Jumping

Most quadruped terrestrial vertebrates can jump or leap to some extent by rapidly extending their hind legs; a few species have become specialized for a jumping or **saltatorial** (L., *saltare, saltatus* = jumping, dancing) mode of locomotion. Among them are frogs and toads, kangaroos, the tarsier (an early, arboreal primate), and some rodents. A saltatorial mode of life imposes certain requirements, and all saltators, regardless of the taxon to which they belong, share many features (Emerson, 1985). This is a good example of convergent evolution. The hind legs, and especially the distal parts, are greatly elongated. The legs are powerful and strongly constructed. The strong thrust of the hind legs during takeoff would twist the body of a jumper if its center of mass is not aligned with the line of action of the propulsive force. Saltators minimize this potential torsion by having the center of mass shifted backward toward the sacrum and by strengthening the vertebral column, which is accomplished by some combination of the following adaptations: shortened trunks; large zygapophyses that firmly unite the vertebrae; a fusion of some cervical vertebrae that reduces head bobbing; reduction in forelimb size; and, sometimes, heavy tails. When not jumping, saltators walk, and modest-sized front legs remain. Frogs and toads do not have a tail, but the tail is long in most mammalian saltators. Kangaroo rats, which also have a tuft of long hair on the end of their tail, use the tail for balance and for guiding the leap. Tendons in the large

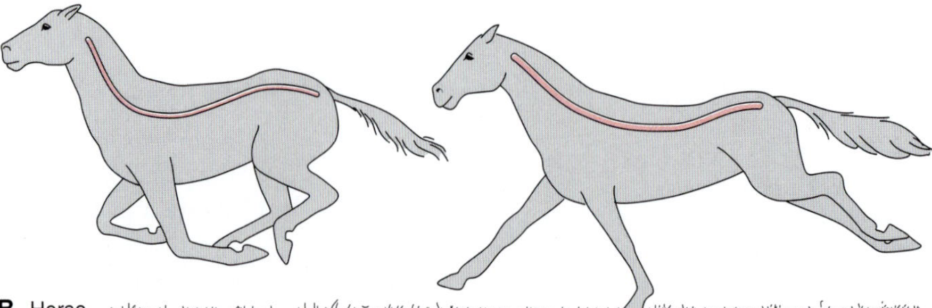

FIGURE 11-26
Dorsoventral flexion and extension of the vertebral column adds greatly to stride length of the cheetah (*A*) but not to that of a horse (*B*). (*After Hildebrand.*)

tails and legs of kangaroos store energy when the animal lands on its feet and tail and release the energy on the next leap. We will examine the specializations of frogs and toads in more detail as an example of saltatorial locomotion.

Frogs and toads use rapid thrusts of their powerful hind legs for not only jumping and hopping but also swimming. The rapid acceleration and distance covered make these modes of locomotion very effective escape mechanisms, as anyone who has tried to catch a frog knows. Jumping likely evolved in this context. Synchronous thrust of the hind legs during swimming appears to have evolved later because very primitive frogs (*Ascaphus* and *Leiopelma*) use their legs alternately when swimming (Abourchid and Green, 1999).

When a frog is at rest on the ground, its limbs are flexed beneath it, and the elongated pelvic girdle bends downward at its joint with the vertebral column (sacroiliac joint). This gives the animal a somewhat humpbacked appearance (Fig. 11-27*A*, *first diagram*

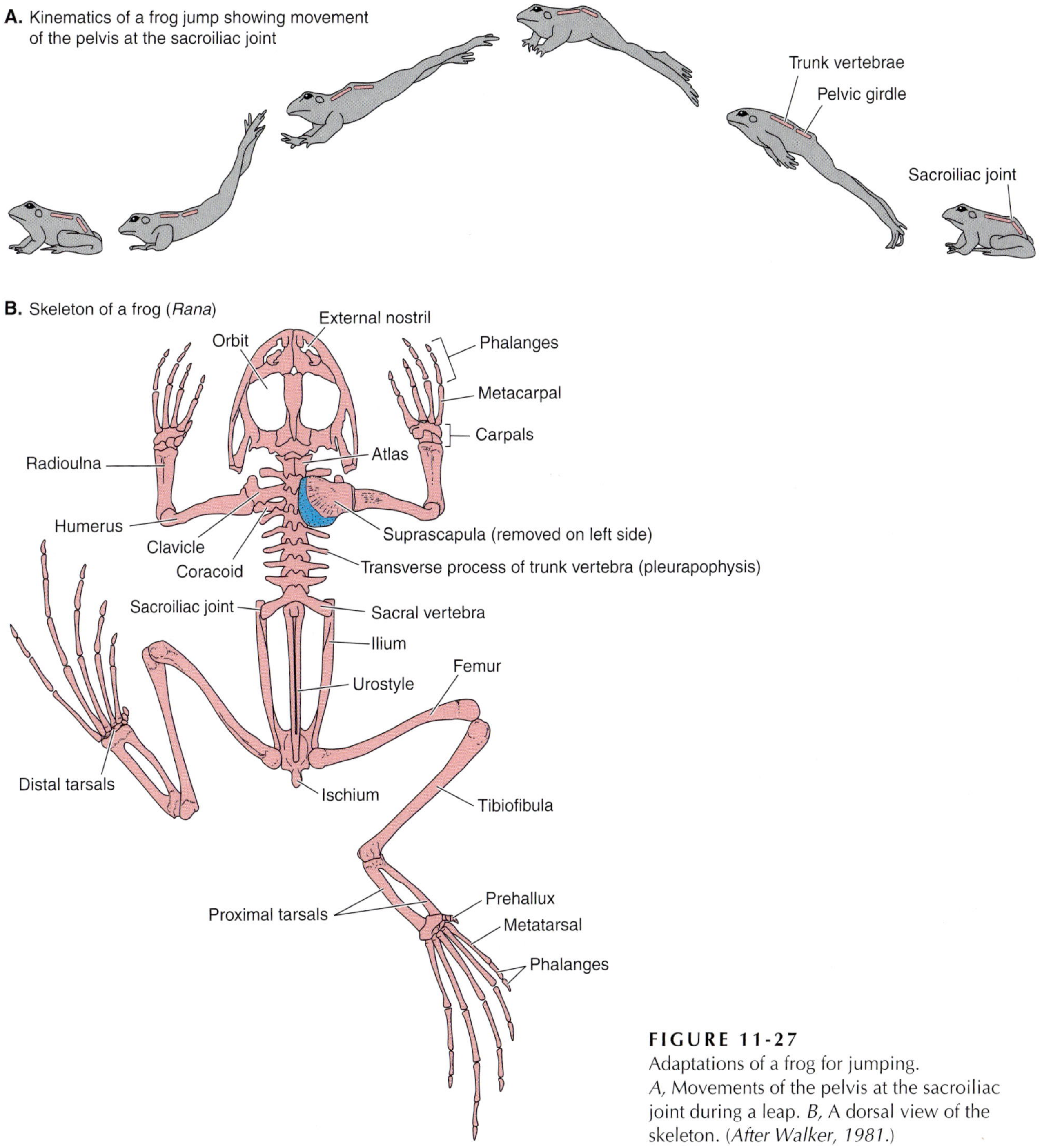

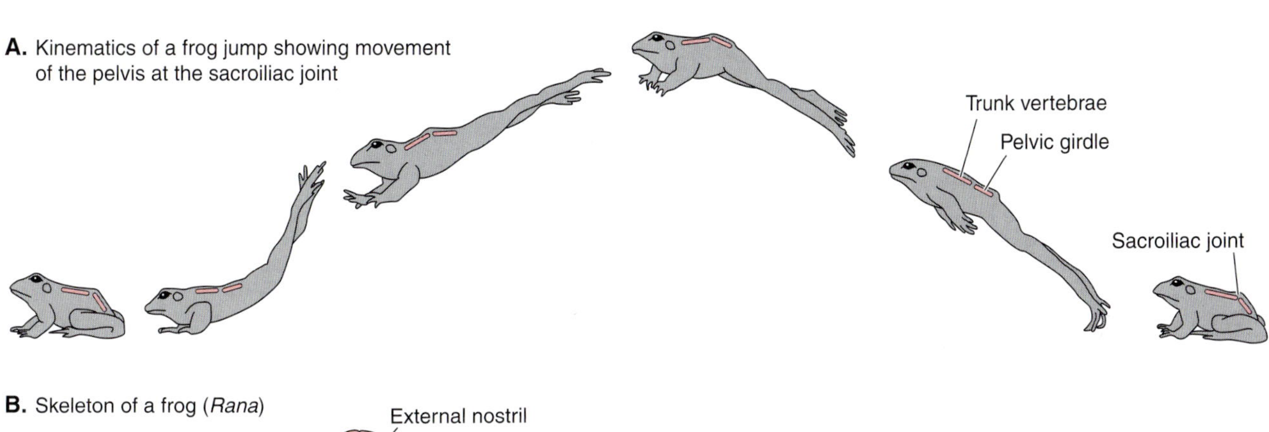

FIGURE 11-27
Adaptations of a frog for jumping. *A,* Movements of the pelvis at the sacroiliac joint during a leap. *B,* A dorsal view of the skeleton. (*After Walker, 1981.*)

FOCUS 11-4 *Metabolic Costs of Amphibian Saltation*

Walton and Anderson (1988) and Anderson et al. (1991) have studied the energetic cost of locomotion in Fowler's toad (*Bufo woodhousii fowleri*) on a treadmill. The gait of these toads combines walking and hopping, with hops becoming more frequent as speed increases. In most terrestrial vertebrates, the energetic cost of locomotion (as measured by oxygen consumption) increases linearly with speeds up to a maximum that can be sustained by aerobic respiration. Speed may increase further, but this speed cannot be sustained. Toads are an exception. Oxygen consumption reaches a maximum at a speed of 0.27 km/h, but toads can sustain speeds nearly twice as fast, up to 0.45 km/h, without additional oxygen consumption. Because the anatomy of toads makes it unlikely that they can store much elastic energy in their muscles and tendons, they must either have the ability to sustain anaerobic metabolism for a prolonged time, or the switch from walking to hopping at higher speeds conserves energy. No evidence shows that toads switch to anaerobic respiration at higher speeds because the amount of phosphocreatine (an energy reserve) in their muscles does not decrease, and lactic acid (a metabolic by-product of anaerobic respiration) does not accumulate. As would be expected, the metabolic cost of a single hop is more than for a single walking stride. However, a toad covers a far greater distance in a single hop than in a single walking stride, and this more than compensates for the greater cost of a hop. The metabolic cost to cover a given distance is 1.9-fold as much for walking as for hopping. By shifting to hopping at higher speeds, toads can go faster with no more energy consumption. This situation is analogous to many mammals changing gaits from a walk, to a trot, to a gallop as speed increases.

from right). During a leap, both hind legs are extended simultaneously. At the same time, the pelvic girdle rotates upward at the sacroiliac joint, providing an extra increment of thrust and bringing the girdle and extending legs in line with the trunk. The animal lands on its front legs (Fig. 11-21*A*, *fifth diagram from right*). When the animal lands, the pelvic girdle rotates downward, bringing the hind legs under the body.

Many specializations accompany this mode of life. The hind legs are very long and strong. The tibia and fibula are fused (Fig. 11-27*B*), and the elongation of the two proximal tarsals and toes greatly lengthens the foot. The foot forms another limb segment that is extended during a leap. Several caudal vertebrae have united to form a long urostyle from which muscles extend to the elongated ilia of the pelvic girdle. This is the mechanism for rotating the pelvic girdle. The vertebral axis is short, the zygapophyses are large, and little rotation occurs around the longitudinal axis of the vertebral column. The front legs act as shock absorbers when the animal lands, and the radius and ulna are fused.

Many studies have been made of body movements and the actions of limb muscles during hopping and swimming. Hopping and swimming are asymmetrical gaits similar in many ways to a gallop or half-bound. After a propulsive phase of about 0.25 of the step cycle, the frog goes into flight (hopping) or an extended glide (swimming), and then the legs flex in preparation for the next thrust. Hopping involves considerable vertical acceleration of body mass against gravity, and swimming must overcome the drag of the water. Acceleration against gravity requires the expenditure of more energy. The electromyographic signals from electrodes implanted in the limb muscles are about twice as large as during swimming. More motor units in the muscles appear to be active (Kamel et al., 1996). Most of the extensor muscles become active 20 ms to 40 ms before muscle shortening begins. This and the fact that many of the extensor muscles connect to relatively long tendons suggest that the muscles are stretched during the initial period of electromyographic activity and store elastic energy that is released later in the jump (Peplowski and Marsh, 1997; Olson and Marsh, 1998). Although individual hops are energetically expensive, hopping is a more efficient way of covering a great distance than is walking (Focus 11-4).

Support and Locomotion in the Aerial Environment

Many lightweight, arboreal vertebrates, including tree frogs and some lizards, can spread their limbs, flatten their bodies, and **parachute** safely to the ground. Parachuting breaks an inadvertent fall and sometimes helps an animal escape a predator. Several lizards and mammals have evolved broad membranes that enable a falling individual to fall more slowly and **glide** a greater horizontal distance. In addition to the benefits of parachuting, glides enable a verte-

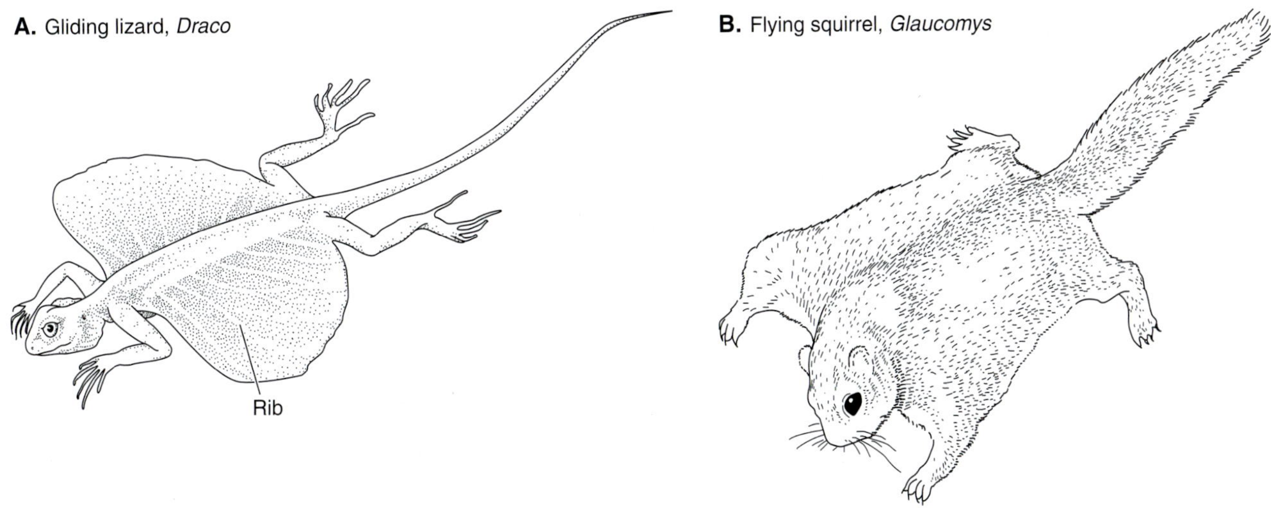

FIGURE 11-28
Two gliding vertebrates. *A,* The gliding lizard, *Draco volans. B,* A flying squirrel, *Glaucomys volans.*

brate to extend its foraging range with little extra expenditure of energy because the animal can move from tree to tree without the need to descend to the ground and climb up again. The gliding lizard, *Draco volans,* has a pair of lateral skinfolds along its trunk that can be extended by six pairs of elongated ribs (Fig. 11-28*A*). Several marsupial mammals (phalangers), the colugo of the Indo-Pacific region, and several rodents (e.g., the "flying" squirrel) glide with lateral membranes that attach to the limbs and are extended when the animal spreads its limbs (Fig. 11-28*B*).

In **flight,** an animal uses wings to sustain itself in the air. As we have seen, wings evolved from pectoral appendages independently in †pterosaurs, birds, and bats. Flight may be passive when an animal simply uses its wings to glide or soar, or it may be powered by the movements of the wings. All types of flight require light weight, and powered flight requires the expenditure of a great deal of energy. Modifications in the digestive, respiratory, and other organ systems make possible the production of the needed energy. Birds and mammals are endothermic. We know little about the metabolism of †pterosaurs, but indications of an integumentary covering of short filoplume-like processes have been found in one fossil. The basic principles of flight are the same in †pterosaurs, birds, and bats, but, because the wings are constructed slightly differently, it is not surprising that certain aspects of flight also differ. We will describe birds as our examples of flying vertebrates. Bird wing structure is described in Chapter 9 and is reviewed in Figure 11-29.

Principles of Flight

The principles that keep a bird aloft are the same in all types of flight. The wing as seen in cross section is an airfoil (Fig. 11-30*A*). The anterior margin of a wing is thicker than its thin trailing margin, so the wing is streamlined, and the air has a smooth laminar flow across it with a minimum of turbulence. Many wings are cambered with a convex upper surface and a concave lower surface. Cambering is particularly pronounced proximally. Because of the wing's shape, air travelling over the dorsal surface moves a slightly longer distance than that over the ventral surface. The airstream necessarily has a greater velocity across the dorsal surface of the wing (V_d) than across its ventral surface (V_v), and this increased speed reduces the pressure on the dorsal surface relative to the ventral surface. This is in accordance with a principle discovered in the 17th century by Bernoulli: in a fluid stream (water or air), pressure is least where the flow is the fastest.

The differential in air pressure above and below the wing also generates a **local air circulation** around the wing (V_l). Air rises at the front of the wing into the reduced pressure area above the wing, moves posteriorly, falls below the wing, and then moves forward on the underside of the wing. The total velocity of air across the top of the wing is $V_d + V_l$ because the airstream and the local circulation are in the same direction. The total velocity across the bottom of the wing is $V_v - V_l$ because the local circulation is in a direction opposite to the airstream. The effect of the local circulation increases

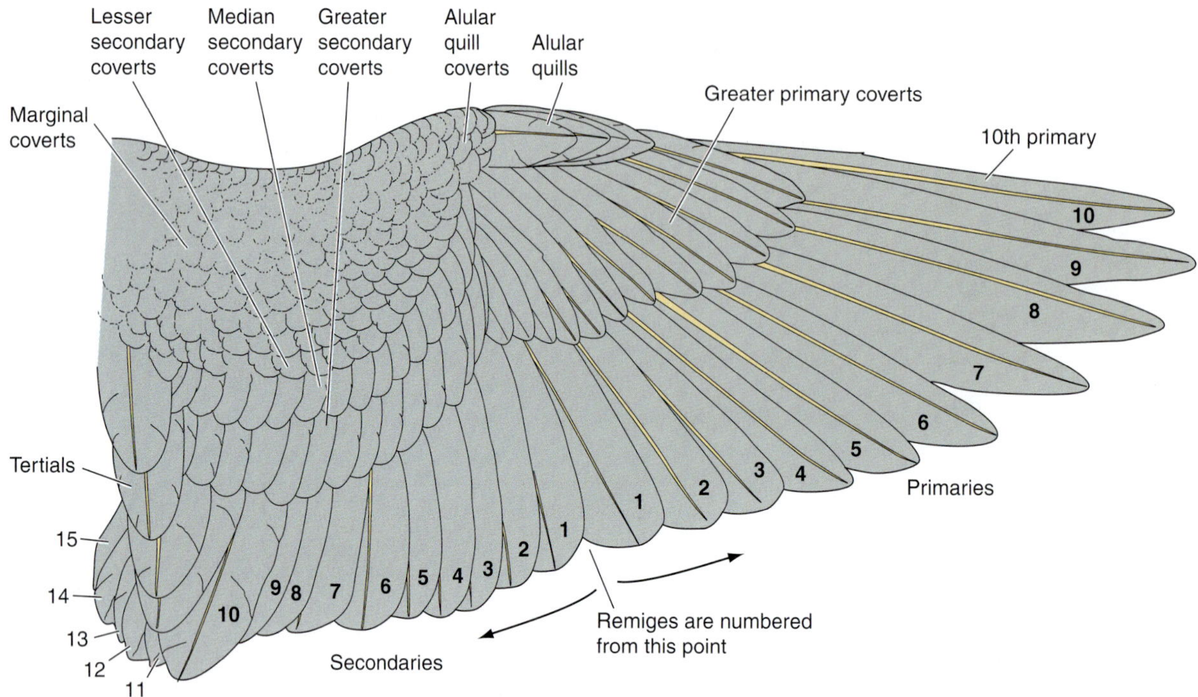

A. Dorsal view of wing surface showing coverts, alular quills and flight feathers (= remiges)

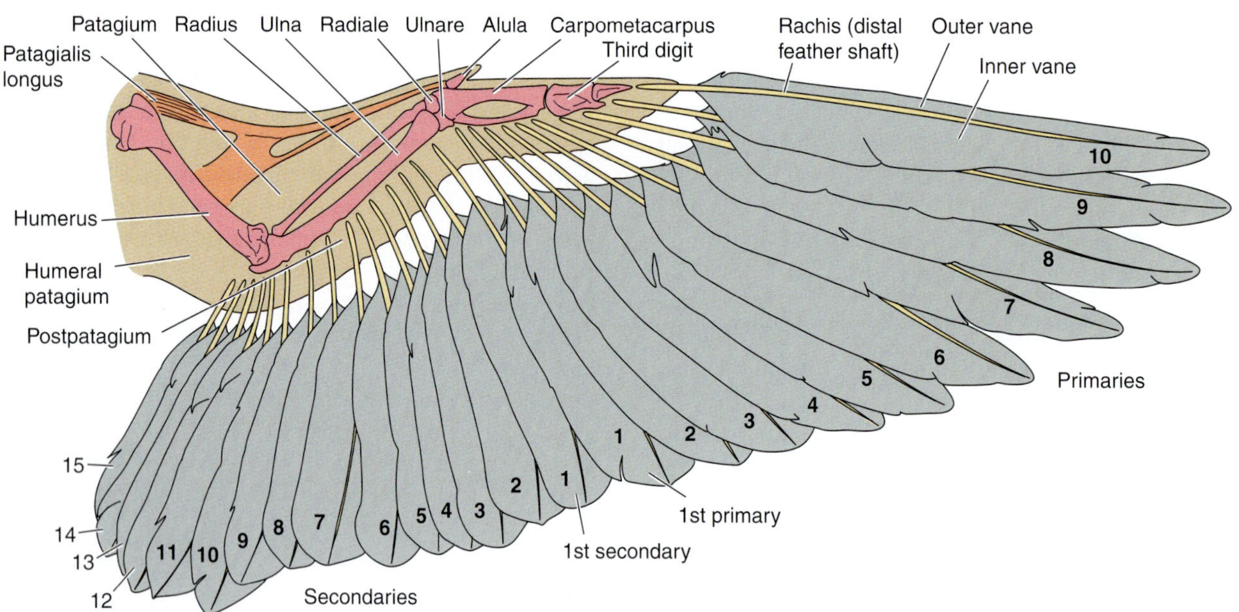

B. Ventral view of wing after removal of covert feathers

FIGURE 11-29

The anatomy of a pigeon's wing. *A*, A view of the feathered dorsal surface. *B*, The skin membranes (patagia) into which remiges insert show after the removal of the coverts, and the locations of skeletal elements within them are indicated. *(After Proctor, Peterson, and Lynch.)*

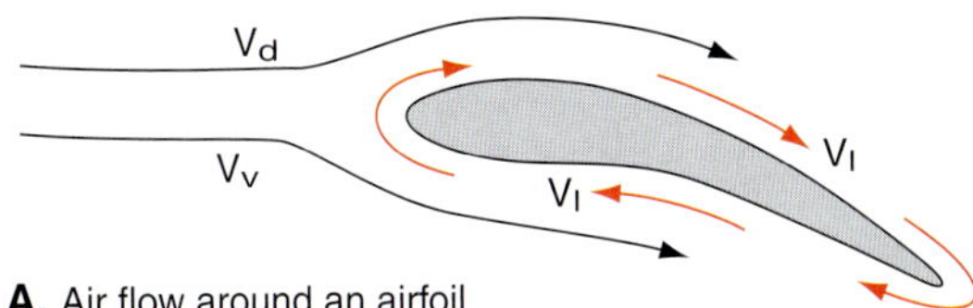

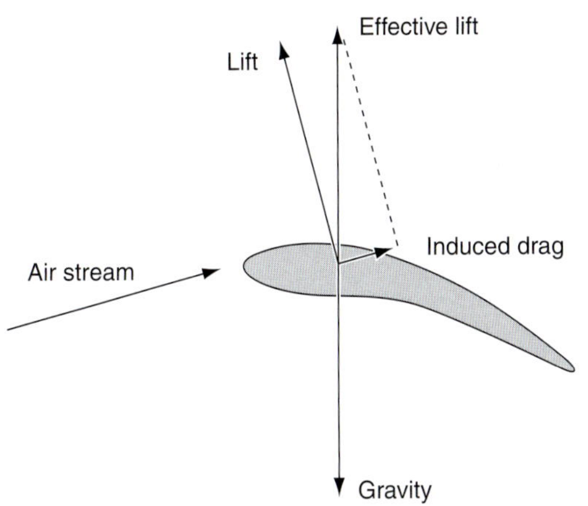

FIGURE 11-30
Aerodynamics of a wing. *A,* The greater velocity of the air flow above the wing (V_d) relative to that beneath the wing (V_v) reduces the air pressure above the wing relative to that beneath the wing. Local air circulation (V_l) around the wing increases this differential. *B,* The local air circulation is shed at the wingtips as tip vortices. *C,* The reduced pressure above the wing generates a lift that is perpendicular to the airstream. This force can be resolved into an effective lift, which opposes the pull of gravity, and into an induced drag, which opposes the forward movement of the bird. Birds also have pressure and frictional drag on the body.

the differential in pressure between the upper and lower surfaces of the wing beyond that produced by the flow of the airstream alone.

Some of the energy of the local circulation around the wing is shed as a series of air currents, which are called **tip vortices,** from the wingtips (Fig. 11-30*B*). It is at the wingtips that air flow equalizes the pressure differences between the two surfaces of the wings. Tip vortices are especially strong in rapidly flying birds because air speed and the pressure difference between the two surfaces of the wing are so great. Tip vortices are less in birds with narrow and pointed wings than in ones with broad and blunt ones because less surface area is present near the wingtip for the air to flow around.

The reduced pressure above the wing relative to that below the wing generates an upward force called **lift.** Lift acts from a center of pressure on the wing normal (perpendicular) to the airstream (Fig. 11-30*C*). If the wing is tilted somewhat, as it usually is, the lift force can be resolved into a vertical **effective lift** component that opposes the pull of gravity, and into an **induced drag** component, which is parallel to the airstream and opposes the forward movement of the bird.

Induced drag is a necessary by-product of the lift force and will vary with the angle at which the airstream hits the wing. Other drag forces are similar to those seen in a fish moving through water. **Frictional drag** results from the friction within the boundary layer of air surrounding a bird. **Pressure drag** results from the pressure difference over the wing, i.e., from air flowing into the reduced pressure area above the wings.

During level flight at a constant speed, lift must equal the weight of the bird and a propulsive force must overcome drag. The formula for lift (L) is:

$$L = 1/2pV^2SC_l$$

where p is the density of the air, V is the air speed, S is the planar surface area of the wing, and C_l is the coefficient of lift.

We can ignore p because it is essentially a constant, but the other variables can change and they have a large effect on lift. Velocity is particularly important because this factor is squared. Birds that soar slowly, such as vultures, need wings with a large surface area to compensate for the low air speed, whereas rapidly flying birds, such as swifts, can and do have shorter and narrower wings. The coefficient of lift depends on several variables, of which the most important for us is the angle of attack of the wing (that is, the extent to which the leading edge of the wing is elevated above the direction of the airstream, (Fig. 11-31*A*). Lift increases as the angle of attack increases, but a trade-off is the increased likelihood of separation of the air flow from the upper wing surface. This increases pressure drag. If the separation occurs far for-

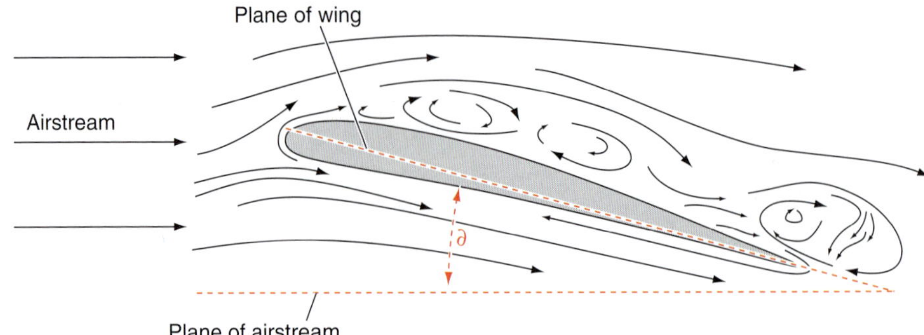

A. A high angle of attack (∂) produces greater lift but also causes lift-reducing turbulence above wing

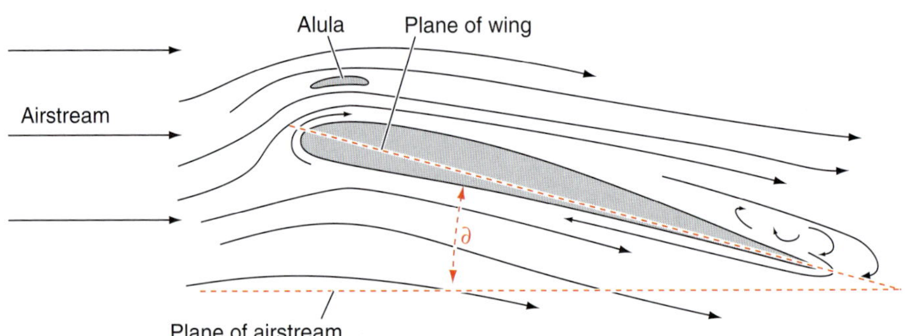

B. Adding a slot at the front of the wing increases air speed above the wing and reduces turbulence

FIGURE 11-31
A, The elevation of the front of the wing above the plane of the airstream (the angle of attack) increases lift but also causes a separation of the air from the upper wing surface, which can cause stall. *B*, The formation of a slot by the alula at the front of the wing increases the velocity of the air flow above the wing and prevents separation.

ward on the wing, lift is suddenly lost and the bird stalls.

Separation can be reduced and lift maintained by creating slots in the wing through which air must flow rapidly (Fig. 11-31*B*). A familiar analogy of the effect of slots is the rapid flow of water between boulders that make "slots" in a river. The river may be flowing slowly, but speed increases if it has to flow between boulders because the volume of water is the same (volume = [cross-sectional area of the river] × velocity). If cross-sectional area decreases, velocity must increase. The alula (Chapter 9) at the front of a bird's wing can be elevated to make one slot. This is particularly helpful during takeoff and landing, when air speed is low. Separating feathers along the trailing edge of the wing and wingtips, where separation of the airflow from the wing is particularly pronounced, forms additional slots.

Powered Flight

Forces in Powered Flight In level powered flight, the propulsive force that overcomes drag derives to a large extent from the downward and forward movement of the wings that occurs during the downstroke (Fig. 11-32*A*). The distal part of the wing, which is formed by the primary flight feathers attached to the hand, rotates so its leading edge is directed slightly downward. As this part of the wing moves through the air, it generates a **local lift** that is perpendicular to the local airflow. Because the wing is inclined, the local lift is also inclined forward. The local lift can be resolved into a vertical component, which helps keep the bird aloft, and a forward propulsive component. This part of the wing acts much like a propeller in generating a forward thrust. Because the hand part of the wing is at a maximum distance from the wing's fulcrum at the shoulder, it is displaced downward more than the proximal part of the wing over the same time period. Its velocity accordingly is greater. This increases the air speed and the forward thrust. In some species, the flight feathers on the distal part of the wing, which are shaped like airfoils (Chapter 6), separate and act as individual propellers. The secondary and tertiary flight feathers attached to the forearm and upper arm function primarily to maintain lift (Fig. 11-29). The upstroke is a simple recovery stroke and generates no lift.

As we have seen, the flow of air leaves the tip of the wings as a series of vortex rings (Fig. 11-30*B*). During steady flight, the vortex rings trail back be-

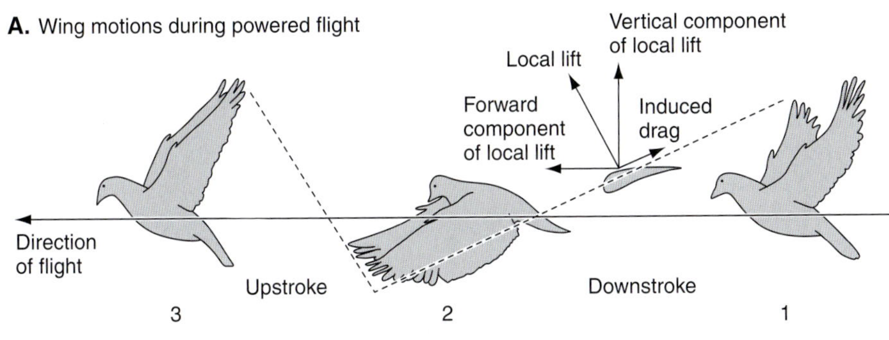

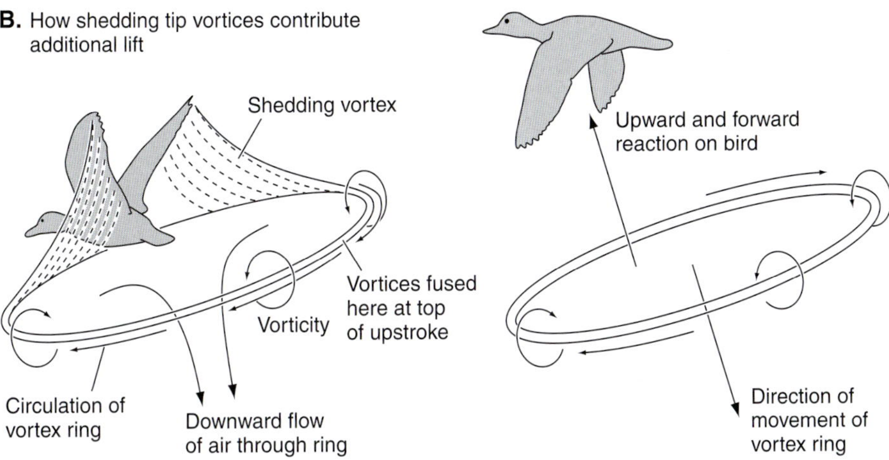

FIGURE 11-32
A, Level powered flight in a bird. *B*, Vorticity of shed vortex rings contributes to lift during takeoff. (*B, After Rayner.*)

hind the bird. But during takeoff and landing, the tips of the left and right wings come close together at the top of the upstroke and bottom of the downstroke. The vortex rings on opposite sides fuse and move downward and backward in the wake of the bird as a series of larger vortex rings (Fig. 11-32*B*). Rayner (1980) and some other investigators believe that the upward and forward reaction of the air to these rings, in accordance with Newton's third law of motion (i.e., for every action there is an opposite and equal reaction), helps push the bird upward and forward.

Wing Actions During Powered Flight Jenkins et al. (1988) studied the movements of the skeletal elements during powered flight by making cineradiographs at 200 frames/s of starlings (*Sturnus vulgaris*) flying in a wind tunnel. During the upstroke (Fig. 11-33*D*), the wing is folded. The humerus lies close to the horizontal plane and is retracted close to the trunk. Forearm and hand bones are flexed at the elbow and wrist. As the upstroke ends and the downstroke begins (Fig. 11-33*A* and *B*), the humerus is protracted and elevated 80° to 90° above the horizontal. Forearm and hand are maximally extended. The humerus is lowered to about 20° below the horizontal near the end of the downstroke (Fig. 11-33*C*); however, the distal part of the wing continues downward because of a slight rotation of the humerus. The dorsal ends of the fused clavicles (the wishbone or furcula) bend laterally during the downstroke and recoil on the upstroke (Fig. 11-33*B* and *C*, *top row, small arrows*). The sternum also moves slightly, ascending and moving posteriorly during the downstroke, and descending and advancing during the upstroke (Fig. 11-33, *bottom row*). Movements of the furcula and sternum may affect the bellows-like air sacs that extend from the lungs and help ventilate the lungs (Chapter 18).

Extension of the forearm at the elbow during the downstroke leads to an automatic extension of the hand. Their flexion during the upstroke also is coupled. Zoologists have long thought that the configuration of the joints at the elbow and wrist caused the radius to slide along the ulna and thereby move the hand in the manner of a set of linked parallel rulers (Fig. 11-34, sketches beneath drawings of wings). Vazquez (1994) has restudied this mechanism. He finds that certain forearm muscles that extend from the distal end of the humerus to the hand, and thereby cross two joints (the elbow and wrist), act as

FIGURE 11-33
Wing, girdle, and sternal movements in a wing-beat cycle of a starling as seen in dorsal (*top row*) and lateral (*bottom row*) views. The figures should be read from right to left. *A*, Upstroke–downstroke transition. *B*, Mid-downstroke. *C*, End of downstroke. *D*, Mid-upstroke. (*After Jenkins et al.*)

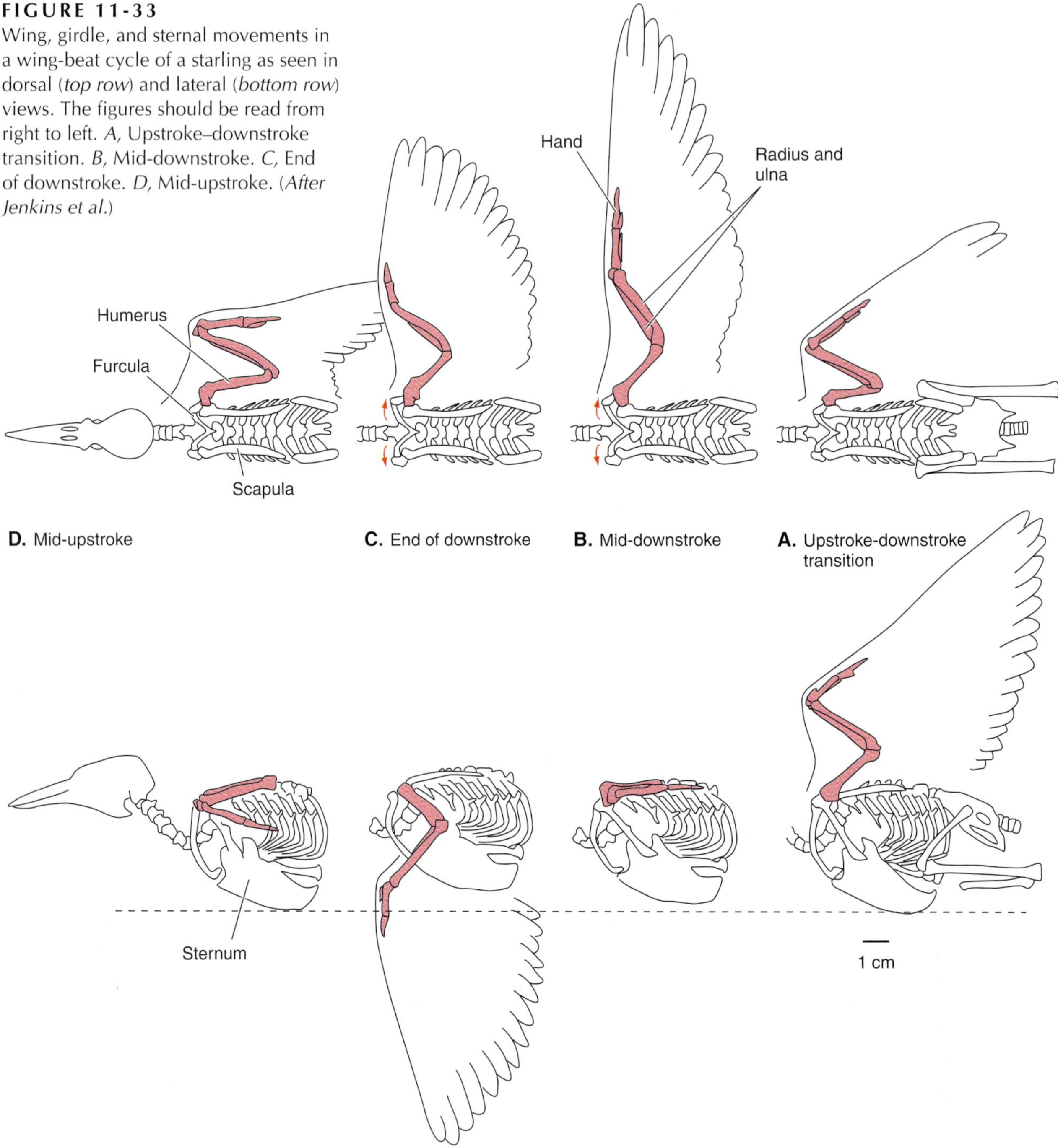

mechanical linkages and are responsible for the automated extension or flexion of the hand. When the triceps muscle contracts and extends the forearm (Fig. 11-34*A*), the extensor metacarpi radialis experiences a strong pull from the humerus. This force is transmitted by the muscle to an extensor process on the carpometacarpus, which causes the hand to extend. When the forearm is flexed by the biceps muscle (Fig. 11-34*B*), the humerus pulls on the origin of the "extensor" carpi ulnaris. This force is transmitted by the muscle to the flexor side of the carpometacarpus, and the hand is flexed. (The "extensor" carpi ulnaris receives its name from its homology to an extensor muscle in other tetrapods, but it acts as a flexor in birds.) Because these muscles are acting as mechanical linkages, their active contraction is not necessary, but their contraction may magnify their passive ability to coordinate movements of the forearm and hand. Automating wing extension and flexion via passive linkages ensures that the various seg-

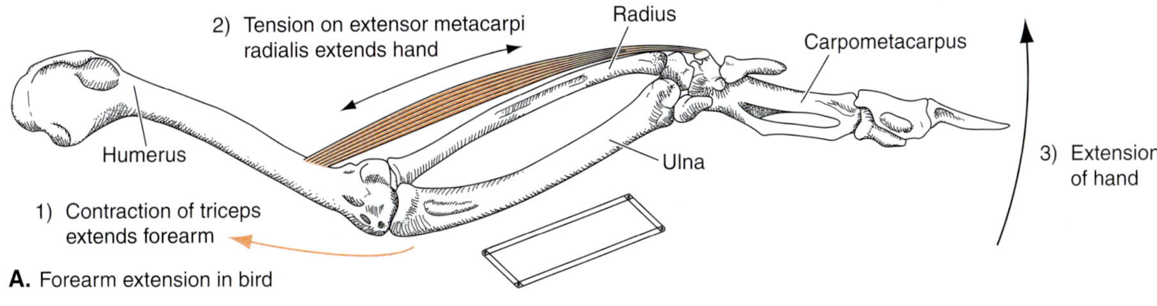

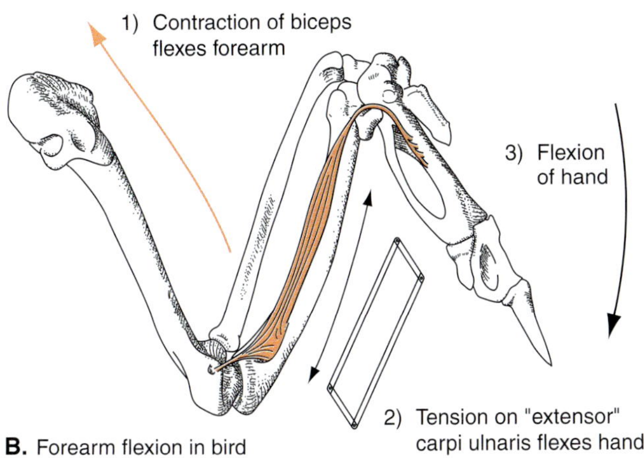

FIGURE 11-34
Automated extension (*A*) and flexion (*B*) of the hand. Forces transmitted passively by certain forearm muscles that cross both the elbow and wrist joints lead to the extension or flexion of the hand when brachial muscles contract and extend or flex the forearm. The diagrams below the wing drawings illustrate an older parallel rule hypothesis for automated movements of the hand. (*After Vazquez.*)

ments of the wing perform appropriately during the various stages of flight. This may be particularly important when the fledglings of tree- or cliff-nesting birds first take flight. The fledglings face serious injury if the wings do not extend and flex at the right times.

Muscle Actions During Powered Flight Many muscles, of course, are involved in the downstroke and upstroke of the wings (Focus 11-5). The two major ones, the pectoralis and supracoracoideus, arise chiefly from the broad, keeled sternum (Fig. 11-35). The pectoralis, a downstroke muscle, inserts on the ventral surface of the humerus, as one would expect. Perhaps unexpectedly, the supracoracoideus, an upstroke muscle, also arises ventrally on the sternum, but its insertion tendon passes through a **foramen triosseum** near the shoulder joint to attach on the dorsal surface of the humerus. The foramen triosseum, as its name implies, is located where three bones (scapula, coracoid, and furcula) come together. The supracoracoideus is a ventral appendicular muscle, but the shift of its tendon of insertion allows it to function as a dorsal muscle. The use of a ventral muscle for the upstroke helps maintain the low center of gravity needed for stability. The supracoracoideus also helps rotate the humerus.

Soaring

Many birds can soar (that is, maintain their position in the air or even rise) by holding their wings extended in the horizontal plane and not flapping them. The wings of birds specialized for soaring provide primarily lift, not power. Adaptations for this are a relatively long arm and forearm, which are the wing parts that generate most of the lift, and a hand that is reduced in size. The arm and forearm of an albatross, a bird that spends most of its life soaring over the southern oceans, form about four fifths of the wing length, and the hand, only one fifth. The arm and forearm of birds that engage primarily in powered flight form about two thirds of wing length; the hand, one third or more. The hand and its primary feathers, as we have seen, act as propellers.

Land birds, such as hawks and vultures, are **static soarers** that gain their power by falling in a rising current of air (Fig. 11-36*A*). They glide downward, but the air around them is rising faster than they are falling, so they stay aloft and may increase their elevation. Rising air currents are formed in many ways. The wind is deflected upward by a cliff, hill, or other obstacle. The ground is heated differentially by the sun. Cities are warmer than the surrounding countryside, open grasslands heat faster than forested land. Differential heating leads to the formation of thermals, large

FOCUS 11-5 *Muscle Activity During a Wing Cycle*

As an illustration of functional anatomy in Chapter 1, we discussed briefly an analysis of electromyographic activity of starling flight muscles made by Dial et al. (1991). We return to this study now in more detail. Important muscles involved are shown in Figure 1-1. The pectoralis is divided into two parts: sternobrachialis (SB) and thoracobrachialis (TB). Figure 1-2 is a diagram of muscle activity during a complete wing cycle, which lasts, on average, 72 ms. The average duration of electromyographic activity of the muscles is shown by the wide portion of the bars for each muscle. Lines extending beyond the wide portion are the standard deviations. The duration bars for downstroke muscles are black in the diagram, upstroke ones are green, and transitional ones are not filled in. Notice that downstroke muscle activity begins in the late upstroke as the upward movement of the wing is slowed. Many muscles contribute to the downstroke. The scapulohumeralis caudalis becomes active during the downstroke–upstroke transition. The supracoracoideus, the major upstroke muscle, becomes active late in the downstroke, decelerating the wing. The pectoralis (on the upstroke) and the supracoracoideus (on the downstroke) are doing negative work for they become active while their antagonists are stretching them. This probably stores elastic energy that is released when they begin to shorten.

Dial (1992) has extended this research to different modes of flight in pigeons (takeoff, ascending, level flight). Gatesy and Dial (1993) also have studied tail-muscle activity, and Dial and Biewener (1993) have calculated the power output of the pectoralis during different modes of flight.

vortices of warm air in which convection currents carry the air upward in the center. As the air rises and cools, it flows downward on the periphery of the thermal. Thermals often separate from the ground and expand as they continue to rise upward into an area of lower air pressure. Some rise for thousands of feet. Static soarers fly slowly and must constantly maneuver and turn to stay in the upward moving air. Because air speed is low, they must have large, broad wings (low aspect ratio) that provide an adequate surface area to generate lift (Fig. 11-36*B*). Wing loading, which is the amount of weight supported by each unit of wing surface, is low. Separating the feathers at the tips of the wing increases air speed across the wing and reduces turbulence and drag (Tucker, 1993).

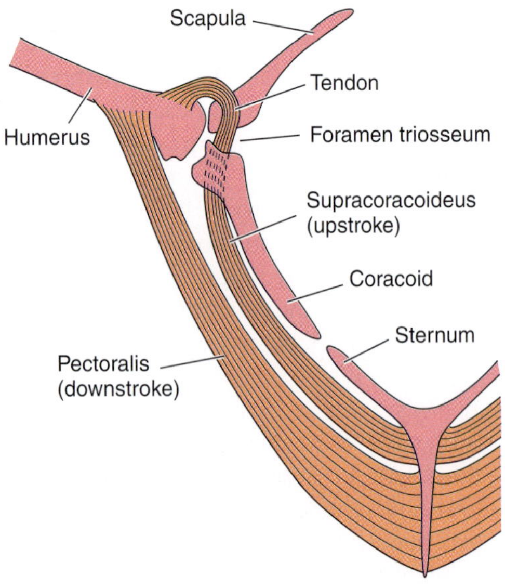

FIGURE 11-35
A transverse section through the shoulder and sternum of a bird, showing the arrangement of two major muscles that move the wing down (the pectoralis) and up (the supracoracoideus). (*After Storer.*)

Although gulls, terns, and other sea birds engage in static soaring when air is deflected upward by a ship, a coastline, or even by a line of waves, they primarily are **dynamic soarers.** As the wind blows steadily across the surface of a large body of water, friction between the air and water causes a gradient in air speed. Speed is lowest at the water surface and increases up to about 15 m, where the speed becomes constant (Fig. 11-36*C*). A dynamic soarer uses this differential. It glides steeply downward with the wind, banks and turns into the wind, and uses the momentum gained from its fall to start to climb. As it ascends, air speed and hence lift increase, which carries the bird upward. (Remember that lift is proportional to the square of the air speed.) Potential energy is converted to kinetic energy as the bird glides downward, and kinetic energy is converted back to potential energy as it soars upward. Dynamic soarers have long, narrow, and pointed wings (high aspect ratio; Fig. 11-36*D*). Wing loading is higher than for static soarers, but increased air speed compensates for the reduced lift generated by the relatively smaller wing area. The wing shape separates the tip vortices and reduces induced drag. Soaring conserves energy relative to powered flight, but some muscle effort is required to hold the wings in a protracted and extended position against the force of air currents, the wind, and the downward pull of gravity on the bird (Meyers, 1993).

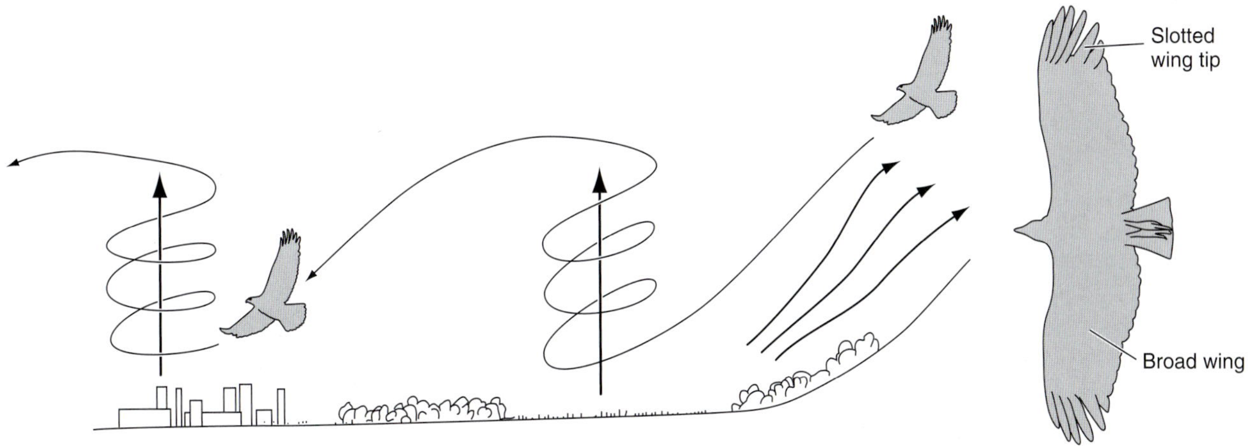

A. Mechanism for static soaring

B. Wings of a static soarer

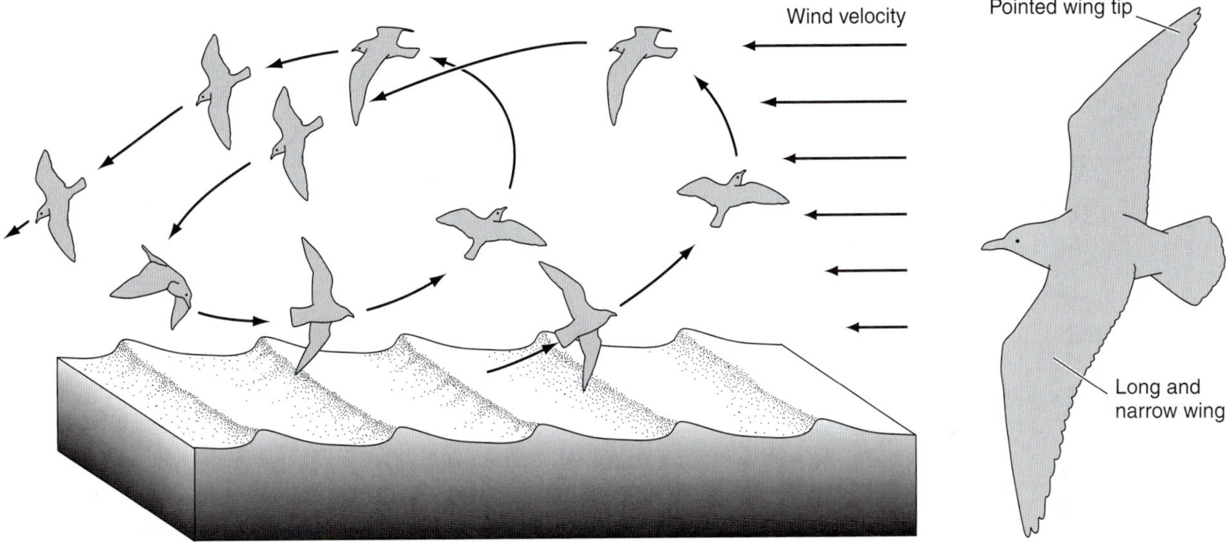

C. Mechanism for dynamic soaring

D. Wings of a dynamic soarer

FIGURE 11-36
Static soarers (*A* and *B*) rely on rising air currents and have relatively short and broad wings (a low aspect ratio). The wing tips are slotted. Dynamic soarers (*C* and *D*) utilize the vertical differential in air speed over large bodies of water and have relatively long and narrow wings (a high aspect ratio). The wing tips are not slotted. (*A and C, After Dorit et al.*)

SUMMARY

1. The vertebral column of fishes need not resist large, vertical bending forces because fishes receive a great deal of support from the water. However, their vertebral column must allow lateral bending and also act as a compression strut that resists shortening of the body when longitudinal muscle fibers in their body wall contract.
2. The teardrop or fusiform shape of fishes reduces pressure drag; the surface area of a fusiform fish, which causes some frictional drag, is not excessive relative to its mass.
3. Most fishes propel themselves by waves of lateral undulations that thrust against the water. Myomere structure is well suited to generate the undulatory waves. Slowly contracting, oxidative (red) fibers in the myomeres are used during normal cruising; fast-contracting, glycolytic (white) fibers are used for short bursts of speed.

4. Although fishes are supported to a large extent by the water they displace, their flesh is denser than water. Other factors also are needed to keep them afloat. The skeleton of cartilage that characterizes chondrichthyans, lipids stored in their livers, and the accumulation of urea and trimethylamine oxide in their tissues reduce their density somewhat. The ventrally flattened head of sharks, the wide pectoral fins set low on the body, and the action of the heterocercal tail can generate additional lift forces. The swim bladder of teleosts enables them to attain neutral buoyancy. Their caudal fin has become symmetrical externally, or homocercal.
5. The median and paired fins of a fish provide stability against roll, pitch, and yaw and participate in maneuvering and braking.
6. The trunk skeleton of terrestrial vertebrates must prevent body weight from collapsing internal organs when the animals rest on the ground. When they walk, the vertebral column resists vertical bending and transfers body weight to the girdles and appendages. Individual vertebrae are strong and firmly linked together.
7. Biomechanically, the vertebral column of mammals resembles an archer's or violin bow. The vertebral column cannot collapse as long as the bowstring (chiefly the sternum and ventral abdominal muscles) is intact.
8. Weight is transferred from the trunk skeleton to the pelvic girdle by way of one or more sacral vertebrae and their ribs and to the pectoral girdle by a muscular sling that attaches to the ribs and scapula.
9. Limbs transfer weight from the girdles to the ground. Because the limbs are jointed struts, muscle action is needed to stabilize the joints and prevent the limbs from collapsing, even when the animal is standing still. The amount of muscle action needed can be reduced by a vertical orientation of the limbs or by passive stay mechanisms that lock the joints in a fixed position.
10. Terrestrial vertebrates walk by a succession of steps. One cycling of all the limbs constitutes one stride.
11. The limbs of amphibians and most reptiles are splayed, so their humerus and femur move back and forth close to the horizontal plane, and their lower limbs move close to the vertical plane. Because a foot remains on the ground during the propulsive phase, the action of retractor and flexor muscles and the cranklike rotation of the humerus and femur advance the trunk. Lateral, sinusoidal curvatures of the trunk assist the limbs in locomotion.
12. The limbs of mammals are more nearly under the body, so all segments of the limbs move close to the vertical plane. The limbs support the body more effectively in this position and swing through longer arcs.
13. Homologous muscles are utilized in similar phases of a forelimb cycle in such diverse species as a lizard, opossum, and starling. Propulsive muscles become active near the end of the swing phase. Similarly, swing-phase muscles become active near the end of propulsion.
14. Many appendicular muscles contain at least two types of muscle fibers. Slow-oxidative (red) fibers maintain posture and are used in normal movement; fast-glycolytic (white) fibers become active only at high speeds.
15. The combination of feet on and off the ground during a stride is known as a gait. Amphibians and reptiles, which carry their bodies close to the ground, utilize very stable gaits in which successive triangles of feet support the body during most of the stride (e.g., a lateral-sequence, diagonal-couplet walk). Mammals also can walk slowly, but they carry their bodies well off the ground and can utilize less stable gaits with fewer feet on the ground when they move faster (pace, trot, gallop, half-bound).
16. The velocity at which a tetrapod travels is the product of the stride frequency and stride length. Tetrapods increase speed first by oscillating their limbs more rapidly, which increases stride frequency. As speed increases, mammals shift to gaits that have fewer feet on the ground at one time, which increases stride length. Increasing stride length is less expensive in terms of energy consumption.
17. In the course of evolution, the limbs of mammals have become adapted for power or speed by changing the relative length of the lever arms for the in-force and out-force.
18. As limbs become adapted for higher speed, limb length increases and foot posture changes from plantigrade through digitigrade to unguligrade. The distal parts of the limbs, which contain less muscle mass, elongate more than do the proximal parts. This keeps most of the muscle mass of the limb proximally and reduces the amount of energy needed to oscillate the limb.
19. Limbs with long feet have, in effect, another limb segment. This additional segment increases velocity because the total velocity of the limb equals the sum of the independent velocities of each segment.
20. The ability to rotate the limb around its longitudinal axis is lost in many cursors. Forelimb bones

fuse, and the distal parts of the ulna and tibia (except for the part supporting the ankle joint laterally) often are lost.

21. Jumping, or saltatorial, tetrapods are characterized by strong, powerful, and long hind legs; shortened trunks; firmly articulated vertebrae; and short front legs. The adaptations of frogs and toads illustrate this pattern of locomotion.

22. Parachuting vertebrates break a fall from a tree by flattening their bodies and spreading their limbs. Gliding vertebrates can spread broad membranes between front and hind limbs and travel a greater horizontal distance. Flying vertebrates have evolved wings and can sustain themselves in the air.

23. The differential in air speed and pressure across a wing generates a lift that acts perpendicularly to the airstream. Lift can be resolved into a vertical effective lift, which opposes gravity, and an induced drag, which acts parallel to the airstream. Induced drag, together with frictional drag and pressure drag, resists a bird's forward motion.

24. The lift is proportional to the area of the wing, the square of the air speed, and the coefficient of lift. Increasing the angle of attack of the wing increases not only the coefficient of lift but also the likelihood of air separation. Slots, which are formed by raising the alula and separating feathers at the wingtip and along the trailing margin, increase lift and reduce air separation. Effective lift must equal the bird's weight in level flight.

25. In steady powered flight, a propulsive force that overcomes drag results from the downward and forward movement of the wings, especially the distal hand portion of the wing that acts as a propeller. During take off, an additional lift and forward thrust comes from the reaction on the bird of vortex rings of air that are shed from the wings and travel downward and backward.

26. Many birds also soar. Static soarers fall in a rising air current. Adequate lift at low air speeds derives from broad wings that are highly slotted at the tip. Oceanic birds are dynamic soarers. They glide downward with the wind, turn into the wind, and use the momentum of their glide to start to ascend. The increasing air speeds they encounter as they rise over the ocean provide additional lift. They have long, narrow, and pointed wings.

REFERENCES

Abourchid, A., and Green, D. M., 1999: Origins of the frog-kick? Alternate-leg swimming in primitive frogs, families Leiopelmatidae and Ascaphidae. *Journal of Herpetology,* 33:657–663.

Alexander, R. McNeil, 1990: Size, speed and buoyancy adaptations in aquatic animals. *American Zoologist,* 30:189–196.

Alexander, R. McNeil, 1984: Elastic energy stores in running vertebrates. *American Zoologist,* 24:85–94.

Alexander, R. McNeil, 1977: Swimming. *In* Alexander, R. McNeil, and Goldspink, G., editors: *Mechanics and Energetics of Animal Locomotion.* New York, John Wiley & Sons.

Alexander, R. McNeil, 1977: Terrestrial locomotion. *In* Alexander, R. McNeil, and Goldspink, G., editors: *Mechanics and Energetics of Animal Locomotion.* New York, John Wiley & Sons.

Altringham, J. D., and Ellerby, D. J., 1999: Fish swimming: Patterns in muscle function. *Journal of Experimental Biology,* 202:3397–3403.

Anderson, B. D., Feder, M. E., and Full, R. J., 1991: Consequence of gait change during locomotion in toads (*Bufo woodhousii fowleri*). *Journal of Experimental Biology,* 158:133–148.

Brinkman, D., 1981: The hind limb step cycle of *Iguana* and primitive reptiles. *Journal of Zoology (London),* 181:91–103.

Dial, K. P., 1992: Activity patterns of the wing muscles of the pigeon (*Columba livia*) during different modes of flight. *Journal of Experimental Zoology,* 262:357–373.

Dial, K. P., and Biewener, A. A., 1993: Pectoralis muscle force and power output during different modes of flight in pigeons (*Columba livia*). *Journal of Experimental Biology,* 176:31–54.

Dial, K. P., Goslow, G. E., Jr., and Jenkins, F. A., Jr., 1991: The functional anatomy of the shoulder in the European starling (*Sturnus vulgaris*). *Journal of Morphology,* 207:227–244.

Drucker, E. G., and Jensen, J. S., 1996: Pectoral fin locomotion in the striped surfperch: I. Kinematic effects of swimming speed and body size. *Journal of Experimental Biology,* 199:2235–2242.

Elder, H. Y., and Trueman, E. R., editors, 1980: *Aspects of Animal Movement.* Society for Experimental Biology, Seminar Series, no. 5. Cambridge, Cambridge University Press.

Emerson, S. E., 1985: Jumping and leaping. *In* Hildebrand, M., Bramble, D. M., Liem, K. F., and Wake, D. B., editors: *Functional Vertebrate Morphology.* Cambridge, Harvard University Press.

Ferry, L. A., and Lauder, G. V., 1996: Heterocercal tail function in leopard sharks: A three-dimensional kinematic analysis of two models. *Journal of Experimental Biology,* 199:2253–2268.

Gambaryan, P. P., 1974: *How Mammals Run*. Translated by H. Hardin. New York, John Wiley & Sons.

Gatesy, S, M., and Dial, K. P., 1993: Tail muscle activity patterns of walking and flying pigeons (*Columba livia*). *Journal of Experimental Biology,* 176:55–76.

Getty, R., 1975: *Sisson and Grossman's The Anatomy of Domestic Animals*, 5th edition. Philadelphia, W. B. Saunders.

Goldspink, G., 1977: Design of muscles in relation to locomotion. *In* Alexander, R. McNeil, and Goldspink, G., editors: *Mechanics and Energetics of Animal Locomotion.* New York, John Wiley & Sons.

Gray, J., 1968: *Animal Locomotion.* London, Weidenfeld and Nicolson.

Heglund, N. C., and Taylor, C. R., 1988: Speed, stride frequency and energy cost per stride: How do they change with body size and gait? *Journal of Experimental Biology,* 138:301–318.

Hildebrand, M., 1985: Walking and running. *In* Hildebrand, M., Bramble, D. M., Liem, K. F., and Wake, D. B., editors: *Functional Vertebrate Morphology.* Cambridge, Harvard University Press.

Hildebrand, M., 1980: The adaptive significance of tetrapod gait selection. *American Zoologist,* 20:255–267.

Hildebrand, M., 1959: Motions of the running cheetah and horse. *Journal of Mammalogy,* 40:481–495.

Jayne, B. C., Bennett, A. F., and Lauder, G. V., 1990: Muscle recruitment during terrestrial locomotion: How speed and temperature affect fiber type use in the lizard. *Journal of Experimental Biology,* 152:101–128.

Jayne, B. C., Lozada, A. F., and Lauder, G. V., 1996: Functions of the dorsal fin in blue gill sunfish: Motor patterns during four distinct locomotor behaviors. *Journal of Morphology,* 228:307–326.

Jenkins, F. A., Jr., 1971: Limb posture and locomotion in the Virginia opossum (*Didelphis virginiana*) and in other non-cursorial mammals. *Journal of Zoology (London),* 165:303–315.

Jenkins, F. A., Jr., and Weijs, W. A., 1979: The functional anatomy of the shoulder of the Virginia opossum (*Didelphis marsupialis*). *Journal of Zoology (London),* 188:379–410.

Jenkins, F. A., Jr., Dial, K. P., and Goslow, G. E., Jr., 1988: A cineradiographic analysis of bird flight: The wishbone in starlings is a spring. *Science,* 241:1495–1498.

Jenkins, F. A., Jr., and Goslow, G. E., Jr., 1983: The functional anatomy of the shoulder of the savannah monitor lizard (*Varanus exanthematicus*). *Journal of Morphology,* 175:195–216.

Kamel, L.T., Peters, S. E., and Bashor, D. P., 1996: Hopping and swimming in the leopard frog, *Rana pipiens:* II. A comparison of muscle activities. *Journal of Morphology,* 230:17–31.

Lauder, G. V., 1989: Caudal fin locomotion in ray-finned fishes: Historical and functional analyses. *American Zoologist,* 29:85–102.

Long, J. H., Jr., 1998: Muscles, elastic energy, and the dynamics of body stiffness in swimming eels. *American Zoologist,* 38:771–792.

Long, J. H., and Koob, T. J., symposium organizers, 2000: Function and evolution of the vertebrate axis. *American Zoologist,* 40:1–155.

Macdonald, D., editor, 1984: *The Encyclopedia of Mammals.* New York, Facts on File Publications.

McMahon, T. A., 1984: *Muscles, Reflexes, and Locomotion.* Princeton, Princeton University Press.

Meyers, R. A., 1993: Gliding flight in the American kestrel (*Falco sparverius*): An electromyographic study. *Journal of Morphology,* 215:213–224.

Muller, U. K., Van den Heuvel, Stamhuis, E. J., and Videler, J. J., 1997: Fish foot prints: Morphology and energetics of the wake behind a continuously swimming mullet (*Chelon labrosus* Risso). *Journal of Experimental Biology,* 200:2893–2906.

Norberg, U. M., 1985: Flying, gliding, and soaring. *In* Hildebrand, M., Bramble, D. M., Liem, K. F., and Wake, D. B., editors: *Functional Vertebrate Morphology.* Cambridge, Harvard University Press.

Nursall, J. R., 1956: The lateral musculature and the swimming of fish. *Proceedings of the Zoological Society of London,* 126:127–143.

Olson, J. M., and Marsh, R. L., 1998: Activation patterns and length changes in hindlimb muscles of the bullfrog *Rana catesbeiana* during jumping. *Journal of Experimental Biology,* 201:2763–2777.

Pelster, B., 1997: Buoyancy. *In* Evans, D. H., editor: *The Physiology of Fishes,* 2nd edition. Boca Raton, CRC Press.

Peplowski, M. M., and Marsh, R. L., 1997: Work and power output in the hind-limb muscles of the Cuban tree frogs *Osteopilus sptentrionalis* during jumping. *Journal of Experimental Biology,* 200:2861–2870.

Peters, S. E., Kamel, L. T., and Bashor, D. P., 1996: Hopping and swimming in the leopard frog, *Rana pipiens:* I. Step cycles and kinematics. *Journal of Morphology,* 230:1–16.

Proctor, N. S., Peterson, R. T., and Lynch, P. J., 1993: *Manual of Ornithology: Avian Structure and Function.* New Haven, Yale University Press.

Rayner, J. M. V., 1980: Vorticity and animal flight. *In* Elder and Trueman, *q.v.*

Shadwick, R. E., Steffensen, J. F., Katz, S. L., and Knower, T., 1998: Muscle dynamics in fishes during steady swimming. *American Zoologist,* 38:755–800.

Simpson, G. G., 1961: *Horses.* New York, Doubleday & Company and the American Museum of Natural History.

Slijper, E. J., 1946: *Comparative Biologic-Anatomical Investigations on the Vertebral Column and Spinal Musculature of Mammals.* Koninklijke Nederlandische Akademie van Wetenschappen, afd. Natuurkunde, Tweede Sectie, Deel XLII, no. 5. Amsterdam, N. V. Noord-Hollandsche Uitgevers Maatschappij.

Smith, J. M., and Savage, R. J. G., 1954: Some locomotory adaptations of mammals. *Zoological Journal of the Linnean Society of London,* 42:603–622.

Snyder, R. C., 1954: The anatomy and function of the pelvic girdle and hindlimb in lizard locomotion. *American Journal of Anatomy,* 95:1–46.

Storer, T. I., 1943: *General Zoology.* New York, McGraw-Hill.

Sukhanov, V. B., 1974: *General System of Symmetrical Locomotion of Terrestrial Vertebrates and Some Features of Movement of Lower Tetrapods.* Translated from Russian.

Published for the Smithsonian Institution and National Science Foundation. New Delhi, Amerind Publishing.

Taylor, C. R., 1978: Why change gaits? *American Zoologist,* 18:153–161.

Thomson, K. S., 1990: The shape of a shark's tail. *American Scientist,* 78:499–501.

Thomson, K. S., and Simanek, D. E., 1977: Body form and locomotion in sharks. *American Zoologist,* 17:343–354.

Triantafyllou, M. S., and Triantafyllou, G. S., 1995: An efficient swimming machine. *Scientific American,* March: 64–70.

Tucker, V. A., 1993: Gliding birds: Reduction of induced drag by wingtip slots between the primary feathers. *Journal of Experimental Biology,* 285–310.

Van Leeuwen, J. L., 1999: A mechanical analysis of myomere shape in fish. *Journal of Experimental Biology,* 202:3405–3414.

Vaughan, T. A., 1978: *Mammalogy,* 2nd edition. Philadelphia, Saunders College Publishing.

Vazquez, R. J., 1994: The automating skeletal and muscular mechanisms of the avian wing. *Zoomorphology,* 114:59–71.

Wake, M. H., 1979: *Hyman's Comparative Anatomy,* 3rd edition. Chicago, University of Chicago Press.

Walker, W. F., 1981: *Dissection of the Frog,* 2nd edition. San Francisco, W. H. Freeman & Co.

Walker, W. F., 1971: A structural and functional analysis of walking in the turtle *Chrysemys picta marginata. Journal of Morphology,* 134:195–214.

Walker, W. F., and Homberger, D. G., 1992: *Vertebrate Dissection,* 8th edition. Forth Worth, Saunders College Publishing.

Walton, M., and Anderson, B. D., 1988: The aerobic cost of saltatory locomotion in Fowler's toad (*Bufo woodhousii fowleri*). *Journal of Experimental Biology,* 136: 273–288.

Wardle, C. S., Videler, J. J., and Altringham, J. D., 1995. Tuning in to fish swimming waves: Body form, swimming mode and muscle function. *Journal of Experimental Biology,* 198:1629–1636.

Webb, P. W., 1997: Swimming. *In* Evans, D. H., editor: *The Physiology of Fishes,* 2nd edition. Boca Raton, CRC Press.

Webb, P. W., 1984: Form and function in fish swimming. *Scientific American,* 251:72–82.

Webb, P. W., and Blake, R. W., 1985: Swimming. *In* Hildebrand, M., Bramble, D. M., Liem, K. F., and Wake, D. B., editors: *Functional Vertebrate Morphology.* Cambridge, Harvard University Press.

Williams, P. L., Bannister, L. H., Berry, M. M., Collins, P., Dyson, M., Dussek, J. E., Ferguson, M. W. J., editors, 1985: *Gray's Anatomy,* 38th edition. New York, Churchill Livingstone.

Withers, P. C., Morrison, G., Hefter, G. T., and Pang, T. S., 1994: Role of urea and methylamines in buoyancy of elasmobranchs. *Journal of Experimental Biology,* 188: 175–189.

part **III**

Integration

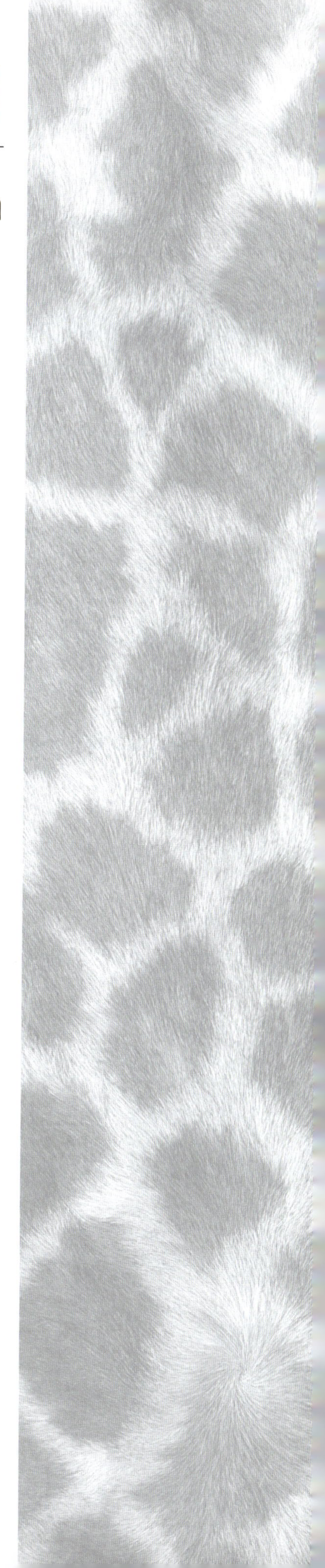

12

The Sense Organs

PRECIS

Vertebrates receive information about changes within their bodies and in the outside world through free nerve endings and a variety of specialized receptor cells located in sense organs. The major sensory systems of vertebrates and some of their functional anatomical changes during evolution will be examined in this chapter.

OUTLINE

To avoid being eaten, to find shelter, food, and mates—in short, to survive in a changing world—vertebrates must detect changes in their external and internal environments and make appropriate behavioral and physiological responses. A study of the sense organs and sensory systems used to do this can reveal much about the biology of a vertebrate. The ability to detect and respond to change is a basic property of life, and, to a degree, all living cells have it, but as animals became more active and complex, certain cells became specialized as **receptors** to monitor the environment for the benefit of the entire organism. Most vertebrate receptors develop embryologically from the neural tube, the neural crest, or the neurogenic placodes (Chapter 4) and connect to other cells of the nervous system. Incoming sensory information is integrated by the nervous system, and responses are generated in the forms of nerve impulses or hormones to activate the appropriate muscles, glands, and other effector organs. Nervous and hormonal integration overlap in many ways, but nervous integration deals with more rapid and discrete types of activity than does hormonal integration. In this and the next two chapters, we will examine receptors and the nervous system; endocrine glands will be described in Chapter 15.

In evaluating the functional anatomy of sensory organs, it is insufficient to demonstrate that a particular type of receptor cell responds physiologically to a certain stimulus because a single type of cell may respond to a variety of imposed stimuli, including pain, temperature, and pressure. Rather, sensory biologists seeking to understand the evolution and functional morphology of a sensory system must demonstrate that an animal actually uses specific types of stimuli to cue behavioral or physiological changes. We will discuss a classic example of this concept later in the section on electroreception (p. 409 and Fig. 12-12).

Receptors

A receptive region of a nerve cell or **neuron** may be stimulated directly by environmental changes within or outside the body and, in response, generate a nerve impulse. This is a common method of reception among "invertebrate" metazoans. Vertebrates retain nerve endings of this type, which are known as **free nerve endings,** but vertebrates also have many specialized types of receptor cells and neurons. **Receptor cells** act as transducers, which are instruments that convert one form of energy into another. They are specialized to detect a minute energy change in a specific environmental signal, such as light, pressure, sound, or taste, and then initiate a nerve impulse in a sensory neuron. Receptor cells have an electrical **resting potential** derived from an unequal distribution of ions across their plasma membrane. The resting potential is maintained in **phasic receptors** until they are stimulated. Upon stimulation of a phasic receptor cell, ion channels in the plasma membrane open, producing an ionic depolarization of the membrane so that the cell develops a **receptor potential.** This response is graded; that is, the receptor potential is proportional to the magnitude of the stimulus. A few receptors are **tonic,** that is, they are always active. Stimulation of tonic receptors increases or decreases the receptor potential. Only when the receptor potential attains a certain magnitude does it initiate a nerve impulse in the sensory neuron. Nerve impulses never vary in magnitude with the magnitude of the receptor potential. A nerve impulse is said to be **all or none;** it either occurs or it does not occur. The magnitude of sensory stimulation above the threshold level needed to initiate a nerve impulse is encoded by the frequency of nerve impulses and sometimes by their pattern. The decoding and perception of particular sensations are functions not of the nerve impulse but of the specificity of the receptors and the connections of their neurons within the nervous system.

Vertebrates evolved receptor mechanisms that detect environmental changes important for their survival; many physiochemical changes that are less critical go undetected. Humans, for example, cannot sense small changes in electric fields, but many aquatic vertebrates, such as chondrichthyans, can. Some receptor types respond to more than one type of incoming stimulus (e.g., certain cutaneous receptors of the skin can detect pressure, which is a mechanical stimulus, and temperature, which is an electromagnetic stimulus). It is convenient to classify the numerous receptors according to the major type of sensory signal to which they are sensitive:

- Chemical receptors
    - Olfactory cells (smell)
    - Taste buds (taste)
- Mechanical, temperature, and electrical receptors
    - Cutaneous receptors
        - Free nerve endings (pain, temperature, and other modalities)
        - Meissner's corpuscles (touch and pressure)
        - Merkel's disks (touch and pressure)
        - Pacinian corpuscles (touch and pressure)
        - Ruffini endings (touch and pressure)
        - Eimer's organs (touch and pressure)
        - Sinus hairs (= whiskers; touch and pressure)
    - Proprioreceptors
        - Tendon and joint receptors (tension)
        - Muscle spindles (degree and rate of contraction)

Lateral line, ear, and electroreceptors
Hair cells (vibrations and gravity)
Ampullary organ cells (electric fields)
Tuberous organ cells (electric fields)
Photoreceptors and specialized thermoreceptors
Rod and cone cells (visible electromagnetic radiation)
Pit organ cells (infrared electromagnetic radiation)

This organization of receptor types is based on the primary sources of incoming stimuli: **chemical, mechanical and electrical,** and **electromagnetic.** It sometimes also is useful to group receptor types, as we did with bones and muscles, according to their locations within the body. Using such a scheme, we can recognize **somatic receptors,** sometimes called **exteroreceptors,** which are located in the outer wall of the body. **Visceral receptors,** which can be termed **enteroreceptors,** lie deeper within the body.

Many free nerve endings and individual receptor cells are scattered through the skin and tissues of the body. Others are aggregated and combined with cells that support and protect them, amplify the environmental stimulus, and help localize the source of the stimulus. We call such aggregations **sense organs.** In this abbreviated treatment, we can examine and compare only a few of the many receptors and sense organs of vertebrates.

Chemoreceptors

All animals detect many chemical changes in their external and internal environments. Vertebrates receive chemosensory signals in four ways. First, internal chemoreceptors continuously monitor certain aspects of the internal environment. Examples include the carotid and aortic bodies, located on major arteries near the heart, which detect changes in the oxygen and carbon dioxide content of the blood. The pH of cerebrospinal fluid, which decreases (i.e., becomes acidic) when carbon dioxide levels in the blood increase, also is monitored by hydrogen ion receptors in the fourth ventricle of the brain. The rates of breathing and blood circulation are reflexively adjusted by information provided by all these receptors. Second, many free nerve endings are present in serous and mucous membranes of the eyes, mouth, and nose and detect noxious chemical stimuli. Lachrymotor agents (i.e., chemicals that make you cry, e.g., the odor of a freshly chopped onion or tear gas) are perceived in part by such endings. Third and fourth, vertebrates have receptors that detect odors and food in the external environment. We have **olfactory receptors** (L., *olfactus* = smell, from *olfacere* = to smell) in our noses and **gustatory receptors** (L., *gustatus* = tasted, from *gustare* = to taste) on our tongues. Although taste and smell have much in common, we think of smell as chemical information carried in the air, and taste as chemical information primarily from food or fluids in contact with parts of the mouth. This distinction becomes blurred for many aquatic craniates, the body surfaces, nose, and mouth of which are bathed in water. In general, olfactory receptors can detect a lower concentration of a substance than can gustatory ones, so they can detect faint traces of distant substances.

Olfactory System

Olfactory receptors are specialized neurons that detect chemical substances called **odorants.** Olfactory neurons are essentially similar in all vertebrates, and they develop embryonically from paired neurogenic placodes that invaginate to form a pair of **nasal sacs** (see Chapter 4 for discussion of neurogenic placodes). The cell body of the neuron, which contains the cell's nucleus, lies in the olfactory epithelium and sends its receptive process, the **dendrite,** to the surface (Fig. 12-1). Nonmotile cilia on the end of the dendrite, which range in number from about 5 to 20, increase the cell's receptive surface several hundred–fold. A long process, or **axon,** extends from the cell body to the main olfactory bulb of the brain. Bundled together, these axons constitute the **olfactory nerve** (cranial nerve I; Chapter 13). By comparison with other sensory systems, olfactory impulses have a particularly short and direct route to the brain.

Odorants in solution in the mucous sheet flowing across the olfactory epithelium bind with **receptor molecules** in the plasma membrane of the olfactory cilia and initiate receptor potentials. Olfactory neurons are extraordinarily sensitive. Dogs can detect a substance known as diacetal in concentrations as low as 1.7×10^{-18} molar. Vertebrates can detect a wide range of odors, possibly as many as 10,000 different odors in the case of humans, but how the distinctions are made is not entirely clear. The distinctive molecular configurations of odorant molecules allow them to bind with and activate specific receptor proteins in the plasma membrane of the cell. The complex genetic control of the synthesis of receptor proteins makes possible the production of different receptor protein families that recognize major differences among odorants. One family, for example, may recognize aromatic compounds such as benzene and its derivatives, another family may recognize aliphatic compounds, and so on. Subtle variations in receptor proteins may en-

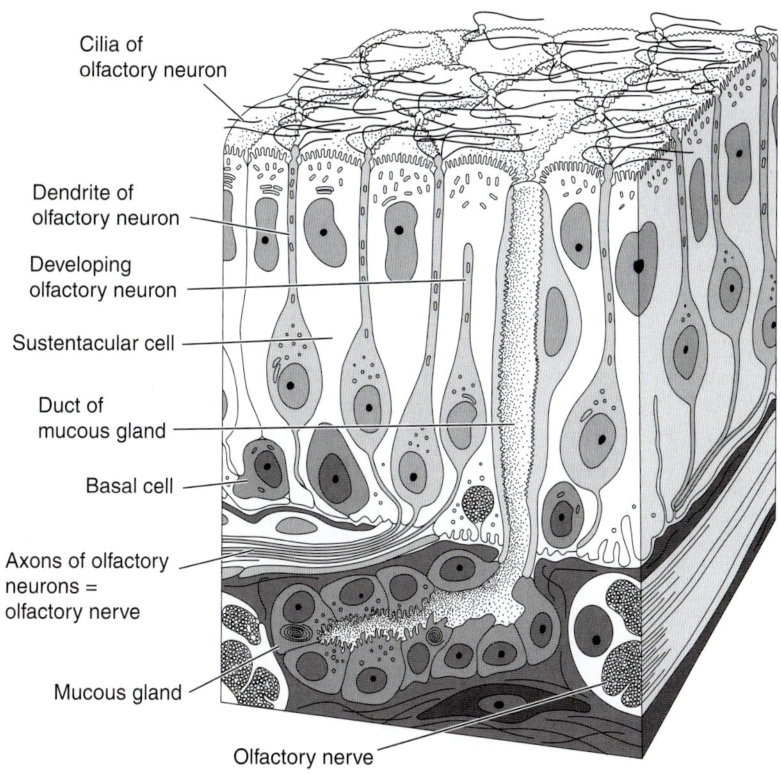

FIGURE 12-1
A portion of the human olfactory epithelium. *(After Williams et al.)*

able the distinction of different benzene derivatives: toluene, xylene, or phenol. But it is still unlikely that there is a different receptor protein for each possible odorant. Each olfactory cell can bind with one or a small group of structurally related odorants. Perception of different odors probably depends on processing in the brain of the combination of information from stimulated olfactory neurons, the intensity of stimulation, and the synaptic pattern that their projections make within the central nervous system. Perhaps an analogy is color vision, in which the processing of information from only three types of photoreceptors (red, green, and blue) enables us to distinguish hundreds of hues.

Olfactory receptor cells show considerable **sensory adaptation.** They are very sensitive to a new odor, but their activity slows down or stops in the continued presence of the same odor. Thus, we notice an odor on first entering the kitchen but fail to detect it after a few minutes. Olfactory cells are interspersed with and supported by **sustentacular cells** (Fig. 12-1). Mucus is secreted by the sustentacular cells and also by simple **goblet cells.** Tetrapods also have large multicellular **mucous glands** in the olfactory epithelium (Fig. 12-1). Cells with motile cilia also occur in the olfactory epithelium of tetrapods; these cilia clear mucus from the surface of the epithelium. **Basal cells** in the epithelium can differentiate into new cells of any type to replace cells that are lost.

Vertebrates move water or air across the olfactory epithelium in different ways. A lamprey has a single nostril on the top of the head that leads to the **olfactory sac** (embryonically paired) and to the **hypophyseal sac** (Fig. 12-2*A*). Respiratory movements of the pharynx vary the pressure on the hypophyseal sac, so that water is alternately sucked into and forced out of the adjacent olfactory sac. In contrast, the paired olfactory sacs of most aquatic gnathostomes have distinct incurrent and excurrent external nostrils, or **nares,** on the surface of the head (Fig. 12-2*B*). Pleats inside the olfactory sacs increase their surface area. Water is drawn into an incurrent naris by ciliary action or by pumping or, in many cases, by suction developed in the oral cavity.

The excurrent openings from olfactory sacs or nasal cavities of choanates[1] are internal nostrils, or **choanae** (Gr., *choane* = funnel), that open in the roof of the mouth (Fig. 12-3). The nasal cavities of choanates are also part of the airways to and from the lungs. The olfactory epithelium of choanates is restricted to the dorsal part of the nasal passage (Fig. 12-4*A*) and is ventilated by the respiratory movements. In lissamphibians and basal sauropsids, which do not ventilate their lungs

[1]Choanata is the group that includes tetrapods as well as fossil taxa such as †Osteolepiformes and †Elpistostegidae; see Chapter 3 and Figure 3-15 in particular.

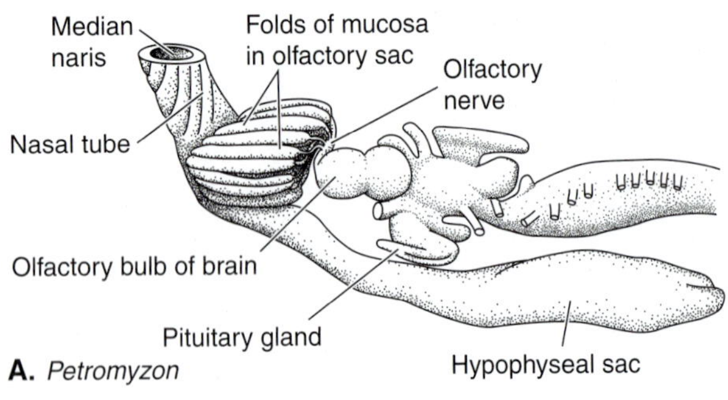

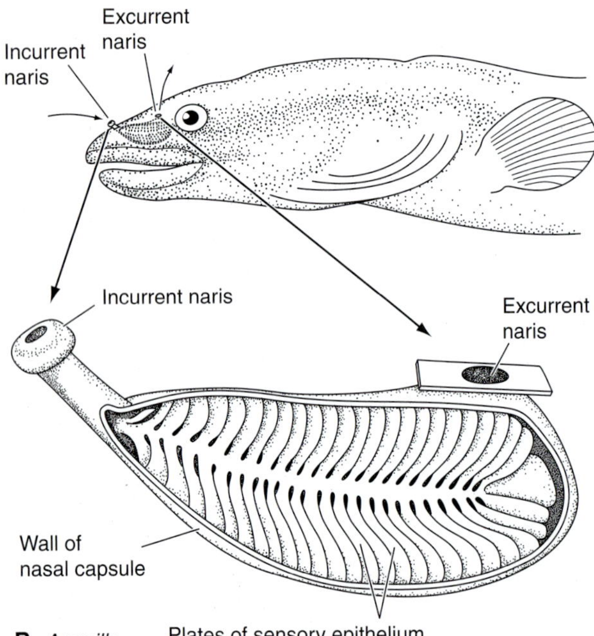

FIGURE 12-2
Olfactory organs of fishes. *A,* Olfactory organ of a lamprey. *B,* Olfactory organ of an eel. *(After Kleerekoper.)*

as frequently as do birds and mammals, these respiratory movements are supplemented by additional pumping movements of the floor of the buccopharyngeal cavity (Chapter 18). The surface area of the olfactory epithelium is increased in amniotes by scrolls of bone, collectively called the **turbinates** (Fig. 12-4*B*). Turbinates are particularly well developed in mammals. They bear the olfactory epithelium and greatly increase the olfactory surface area. Their presence in early mammals, such as †*Morganucodon,* suggests that these animals had an acute olfactory sense that probably reflected a nocturnal mode of life (see Focus 3-5). One of the turbinates lies at the front of the nasal cavity, directly in line with the respiratory current. It helps cleanse, moisten, and warm inspired air, and it cools expired air. As expired air cools, water condenses and is reused to moisten inspired air. Such a mechanism likely evolved in relation to the elevated ventilation rates and endothermy that characterized mammalian evolution.

The sense of smell provides most vertebrates with considerable information about their surroundings. It helps them find food, recognize members of their own species, and avoid enemies. It also aids in homing in some species. For example, young salmon become imprinted to the odors of the stream in which they hatched, and, years later, olfactory clues help them return to the same stream to spawn.

Odor detection also is a part of a communication system based on the secretion into the external environment of species-specific chemical messengers called **pheromones** (Gr., *pherein* = to bear + *hormaein* = to excite). Pheromones have many advantages. Most are

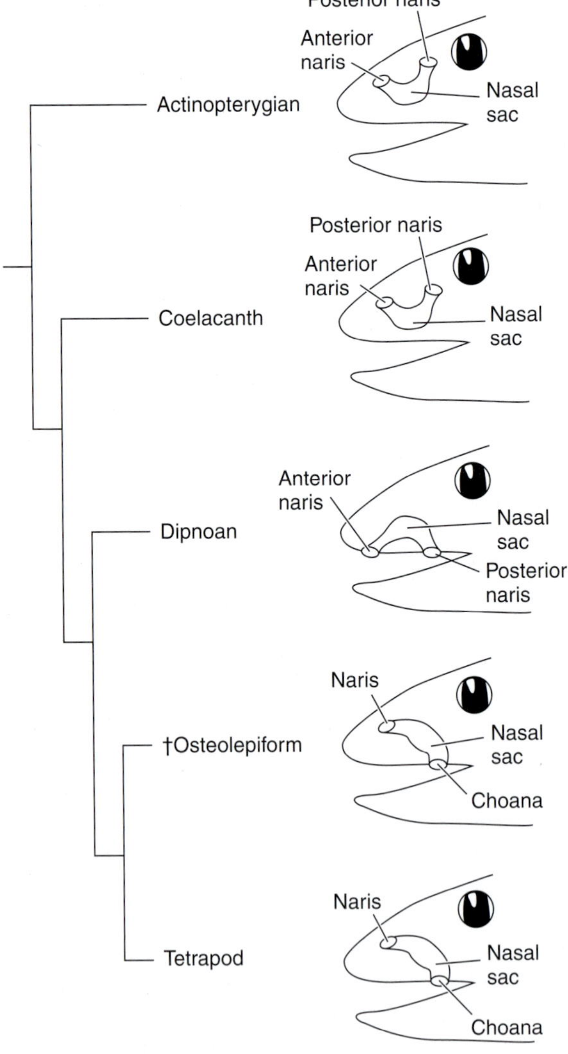

FIGURE 12-3
Evolution of nasal passages and the choana. †Osteolepiforms and tetrapods have a choana, which is a connection of the nasal passages to the oral cavity.

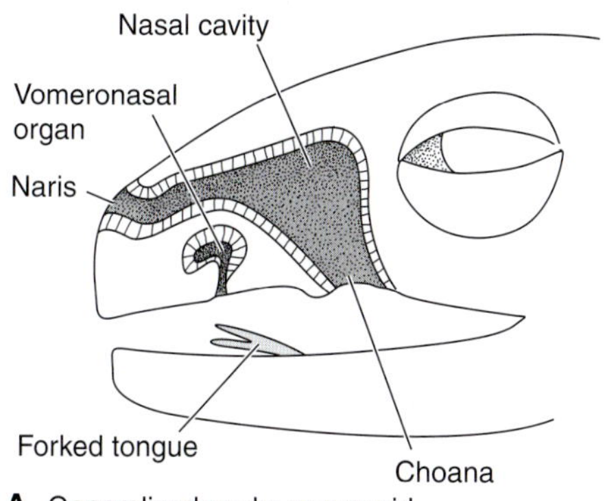

A. Generalized snake or varanid

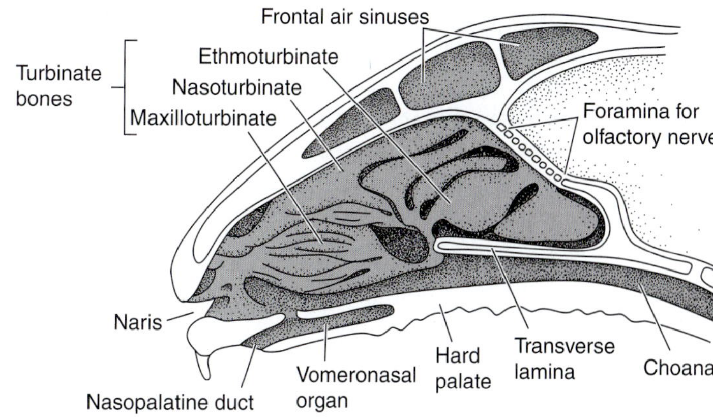

B. Generalized mammalian condition

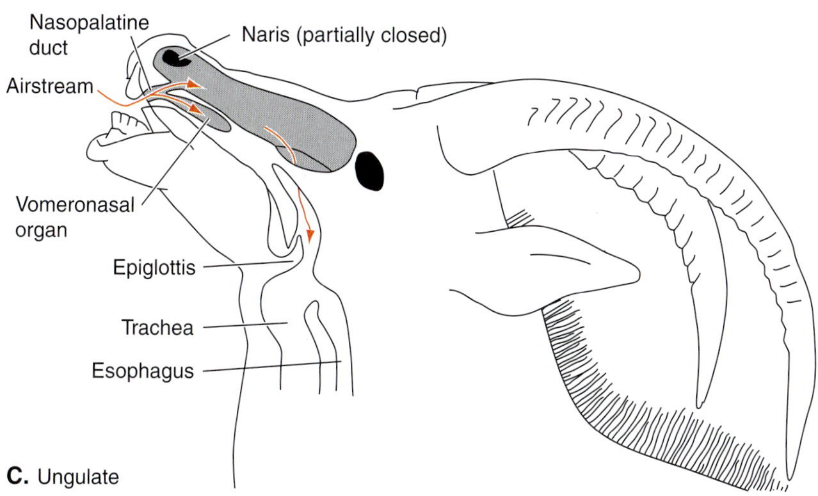

C. Ungulate

FIGURE 12-4
Vomeronasal organ. *A,* Generalized snake or varanid (scleroglossans with forked tongues). *B,* Mammal. *C,* Ungulate. *(A, After Bellairs; B, after Hillenius; C, after Smith.)*

small molecules that are easy to synthesize. They may persist in the environment for hours or days, diffuse around obstacles, and are effective in the dark. The amount of information one pheromone can convey is limited, but some vertebrates secrete several pheromones with different meanings. Pheromones frequently are used to warn conspecifics of danger. For example, injured minnows produce an alarm substance known as *Schreckstoff* that induces others to flee. Other pheromones indicate social status in a hierarchy, are used to establish territories, or signal sexual readiness. For example, female goldfish produce a hormone that promotes oocyte maturation. Some of this hormone also is released as a pheromone into the environment, which sexually arouses mature conspecific males and causes them to liberate sperm (Sorensen et al., 1987).

Odors are less important for arboreal or flying vertebrates than for aquatic and terrestrial ones because olfactory trails do not always cross gaps from tree to tree. Only a few birds, including the nocturnal and ground dwelling kiwi of New Zealand; vultures and other scavengers; and certain marine birds, such as fulmars, have been thought to have a well developed sense of smell. This sense is poorly developed in primates, for they became arboreal animals early in their evolutionary history. Only vestiges of olfactory organs remain in cetaceans, which close their external nostrils under water.

The Vomeronasal Organ

In most tetrapods, the medioventral part of the olfactory epithelium forms a pair of **vomeronasal organs,** or **Jacobson's organ** (Fig. 12-4). These organs are absent from most species of aquatic tetrapods, including larval lissamphibians, most turtles, crocodilians, and aquatic mammals.[2] Birds, most bats, and many primates, including humans, also lack them, although vestiges sometimes appear during embryonic development. The receptive neurons of the vomeronasal organ

[2]Exceptions include adult lissamphibians that are aquatic, including *Xenopus* and *Necturus,* which have vomeronasal systems.

resemble olfactory ones except that the cilia are replaced by microvilli. The axons of the vomeronasal neurons terminate in the **accessory olfactory bulb** of the brain, which has different projections within the brain than does the main olfactory bulb.

The paired vomeronasal organs are not completely separated from the main olfactory chambers in lissamphibians, tuataras, and scleroglossans,[3] and odorants can enter the vomeronasal organ via either the external or internal nostrils. In squamates, the vomeronasal organs usually form a pair of distinct, saclike structures that have their own entrances into the mouth. Derived scleroglossans use them in combination with a forked tongue (Fig. 12-4*A*). Odorants adhere to the tongue as it is darted in and out of the mouth, and the tips of the tongue are brought close to the palatal entrances of the vomeronasal organs. Snakes use this mechanism to follow prey trails and in sexual recognition. In mammals the vomeronasal organs are cul-de-sacs that open into the front of the mouth through the nasopalatine duct, the nasal cavity, or both the nasal cavity and the mouth (Fig. 12-4*B*). Many mammals use the vomero-nasal organ to detect pheromones important to the animal's social and sexual interactions. Males of some species can determine whether a female is in heat by the pheromones she produces. When next you walk the family dog, you may notice it "mouthing" but not swallowing certain objects. This activity, known as the **Flehmen response,** blocks the nostrils and the back of the oral cavity in order to more efficiently suck odorants into the vomeronasal organ (Fig. 12-4*C*).

The Terminal Nerve

All living gnathostomes have a **terminal nerve** that usually is applied closely to the surface of the olfactory nerve.[4] It is tiny in most craniates but large in some species of marine mammals. The distal ends of its neurons terminate in the rostral part of the nasal mucosa, and their cell bodies are scattered along the nerve. The functions of the terminal nerve are not entirely clear, but in jawed fishes, lissamphibians, and mammals, its fibers contain gonadotropin–releasing hormone (GnRH). This hormone, which also is found in many brain cells (notably those of the hypothalamus), helps regulate reproduction (Chapter 21). The presence of GnRH in the terminal nerve suggests that it may be part of a chemosensory system regulating some aspects of reproduction via the nasal detection of pheromones. For example, stimulation of the terminal nerve in male elasmobranchs can cause sperm to be released.

Gustation

Taste is detected by barrel-shaped clusters of 20 to 30 receptor and sustentacular cells of endodermal origin that are called **taste buds** (Fig. 12-5*A*). The surfaces of the taste cells bear microvilli that contain the molecules receptive to chemicals. Taste buds open to the surface by pores. Because taste buds are exposed and subject to wear, mature cells have a life span of only a week or two; undifferentiated cells within the buds continue to divide and transform into replacement cells. The receptive cells are supplied by distinct sensory neurons that return to the brain in cranial nerves from the mouth and pharynx. The facial nerve (VII) carries fibers from taste buds in the oral cavity; the glossopharyngeal (IX) and vagal (X) nerves carry fibers from those in the pharynx (Chapter 13). Taste buds traditionally are classified as visceral sensory organs because most are oriented to the interior of the visceral tube and appear to develop from endoderm.

Although taste buds are used primarily to find and recognize food, they also are important in sexual and other behavioral interactions in many species. Taste buds do not respond to as low a concentration of substances as do olfactory cells, and the substances must be in contact with the buds. Taste buds also respond to a relatively narrower spectrum of chemical substances. Areas of the human tongue are particularly sensitive to salt, sour, sweet, and bitter substances, but you do not need to be a gourmet to detect more than this. As with olfactory cells, different categories of taste buds likely are sensitive to a spectrum of substances. The distinction among tastes probably results from the particular combination of taste buds that are activated and the pattern of their projection in the brain. Our final perception of what we call "flavor" depends on a complex mixture of signals from not only the taste buds but also olfactory cells and tactile receptors. It is not easy to sort out the contributions of each, and the same is probably true for other vertebrates.

Taste buds are distributed throughout the oral cavity and pharynx in fishes and lissamphibians. They also spread onto the skin in many fishes and aquatic lissamphibians. They occur over the entire body surface in catfishes and minnows but are particularly abundant on the barbels around the mouth (Fig. 12-5*B*). Taste buds on the body surface have been thought to arise from endodermal cells that migrate onto the body sur-

[3]Scleroglossa includes several groups, such as gekkos, skinks, varanids, amphisbaenids, and snakes (Chapter 3 and Fig. 3-23 in particular).

[4]The terminal nerve is known as cranial nerve 0 (Chapter 13).

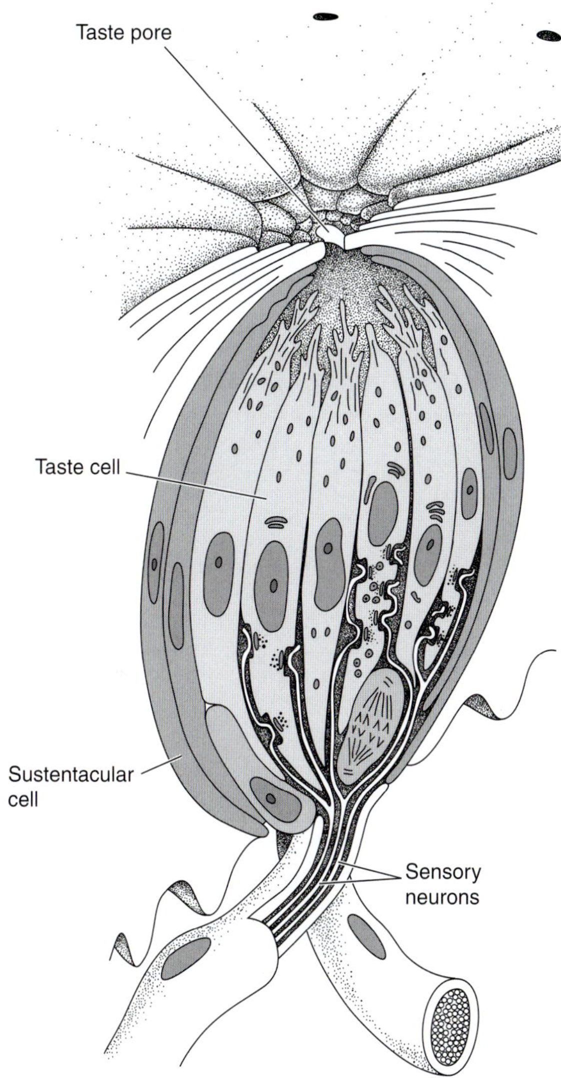

A. Human taste bud

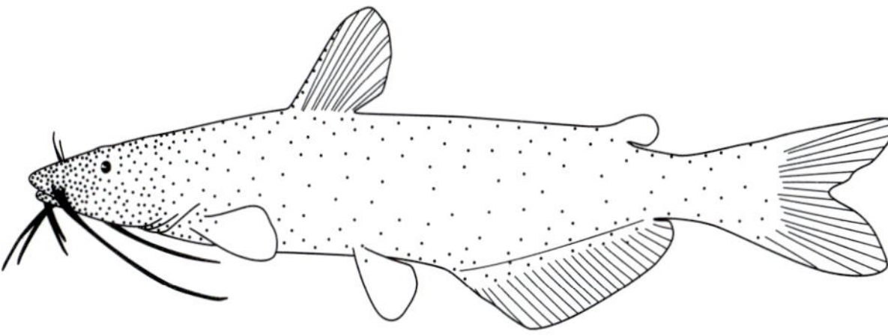

B. Taste buds on catfish skin

FIGURE 12-5
Gustatory organs. *A,* Human taste bud. *B,* The distribution of taste buds on the surface of a catfish. Each dot represents 100 taste buds. Taste buds are too numerous on the barbels to show as individual dots. *(A, After Williams et al.; B, after Atema.)*

face during development.[5] They are supplied by the facial (VII) nerve. In amniotes, taste buds are limited to the oral cavity and pharynx. Many sauropsids (including birds) have taste buds on the back of the tongue and palate. Taste buds are more abundant in mammals. Most are associated with papillae on the tongue, but some are found on the palate, pharynx, and epiglottis.

Cutaneous Receptors

Cutaneous receptors are used for sensing the surfaces of objects or other individuals. For example, many aquatic craniates receive considerable information about their aqueous environment from a special group of cutaneous mechanoreceptors known as the lateral line system, which lies in or just beneath the skin. This system is so intriguing and important that we devote to it a special section (p. 406). All vertebrates have branching free nerve endings in their dermis, and these may penetrate the epidermis. These free nerve endings are activated by vibrations, touch, injuries, abrupt temperature changes, and other external stimuli. Their primary function is to alert the animal to cuts, burns, and other injuries that we perceive as painful. Specialized cutaneous mechanoreceptors also are present. In all cases, a very slight mechanical deformation of some part of the plasma membrane of either a free neuron or a specialized receptor cell initiates the receptor potential. Most cutaneous receptors are confined to the skin and are supplied by spinal nerves. A few spread into the mouth and other mucosae, such as the moist layer of cells covering the eyeball (the cornea; see p. 426). These are supplied by branches of the trigeminal cranial nerve (V), which also supplies cutaneous receptors over the surface of the head. All cutaneous receptors are regarded as somatic receptors.

Mammalian skin contains a bewildering array of touch and pressure receptors. Some are nerve endings layered or laminated with connective tissue fibers. Other nerve endings are not laminated but are typically associated with receptive cells. Probably because they are protected from their surroundings, laminated receptors adapt rapidly to changes in a stimulus and give a response when the stimulus begins or ends. They do not respond to a continuing stimulus. Endings that are

[5]Recent evidence based on experimental embryological studies of axolotls suggests that taste buds also can arise from ectoderm. This finding agrees with the observation that many taste buds on the mammalian tongue occur in areas of epithelium that develop from ectoderm. More experimental evidence on the development of taste buds is needed.

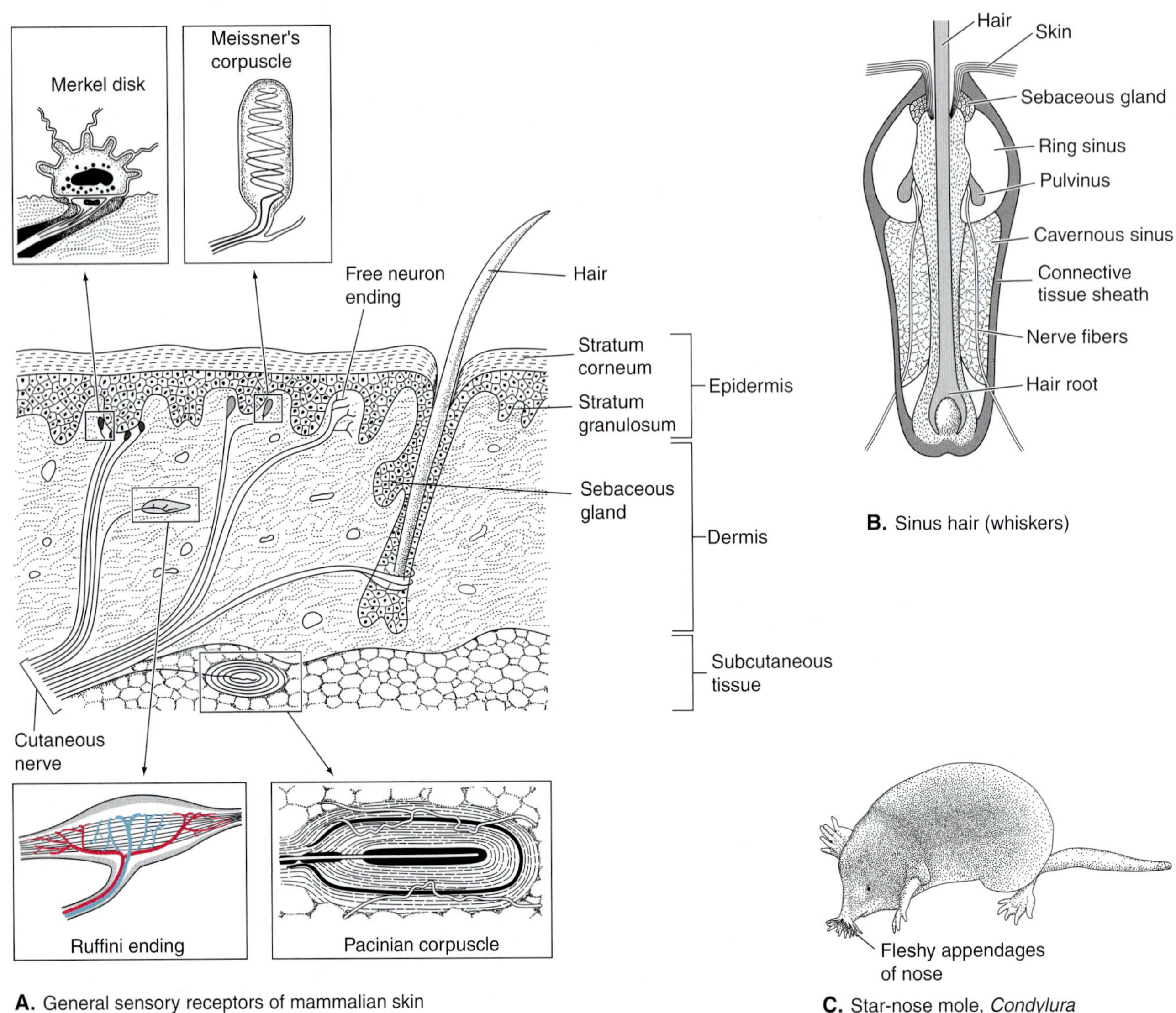

FIGURE 12-6
Cutaneous receptors. *A,* General sensory receptors of mammalian skin as seen in a vertical section. *B,* Sinus hair (whisker). *C,* Star-nose mole, *Condylura.* The nose of this insectivore has 22 fleshy appendages surrounding the nostrils. Each appendage has hundreds of specialized touch receptors known as Eimer's organs. See text for more information. *(A, after Williams et al.)*

not laminated adapt more slowly and remain active for the duration of the stimulus. Those endings, laminated or not, that lie just beneath the epidermis detect stimuli in their immediate vicinity (i.e., their receptive fields are small and well defined). Examples shown in Figure 12-6*A* are **Merkel's disks** (not laminated and slowly adapting) and **Meissner's corpuscles** (laminated and rapidly adapting). Receptors lying deeper in the dermis detect stimuli over a wider and less clearly defined area. Examples are **Ruffini endings** (not laminated and slowly adapting) and **Pacinian corpuscles** (laminated and rapidly adapting).

Mammals also have neurons entwined around hair follicles, so the hairs act as lever arms. Slight movement of the hair tip is magnified at the hair follicle and initiates a nerve impulse. The most specialized of these are the long **vibrissae** (or whiskers) on the snout of many mammals, which are so sensitive that they can respond to air currents as well as to light touch because the base of the hair follicle is suspended within an expanded sinus of tissue fluid that maximizes displacement of any hair movement (Fig. 12-6*B*). Cats, rats, and other nocturnal species use their vibrissae to provide information for moving about in the dark.

Eimer's organs are examples of highly specialized touch receptors. More than 25,000 Eimer's organs are present on the appendages of the nose of the star-nosed mole (Fig. 12-6*C*); this is fivefold as many touch receptors as are found in the human hand! The overwhelming preponderance of one sensory system (in this case, touch) often is correlated with reductions in another sensory system (in this case, moles have very reduced visual systems). Such tradeoffs in the relative development of sensory systems are a common theme across vertebrate evolution.

Less is known about cutaneous mechanoreceptors of nonmammalian tetrapods. Frogs, snakes, and some other species are extraordinarily sensitive to ground (seismic) vibrations and use this information to help detect the presence of predators or prey. Seismic receptors have been found in the skin, although the ear is the more typical route for acquiring such stimuli (e.g., see the discussion of the ear of lissamphibians, p. 419). The lateral margins of the bill of ducks and other water fowl have organs, such as Pacinian corpuscles, that are used to detect food in muddy water. Mechanoreceptors on or near the feather follicles may enable birds to detect an imminent stall and measure airspeed across wings (Brown and Fedde, 1993).

Proprioceptors

Coordinated contraction and relaxation of locomotor and other muscles, in the correct sequence, and with the needed force and velocity, require some sensory feedback from the muscles to centers in the spinal cord and brain that control their activities. A category of mechanoreceptors known as **proprioceptors** (L., *proprius* = one's own + *ceptus* = taken, from *capere* = to take), which are located in muscles, tendons, and joints, continuously provide this information, although we are seldom aware of their activity. Most proprioceptors are associated with somatic muscles, so they are considered somatic receptors.

Tendon organs consist of groups of encapsulated collagen fibers that are entwined by sensory nerve endings (Fig. 12-7*A*). They detect tensions developed by the muscles and provide information on which muscles are active and the magnitude of the forces developed. If forces become dangerously great, then the sensory information reaching the central nervous system will

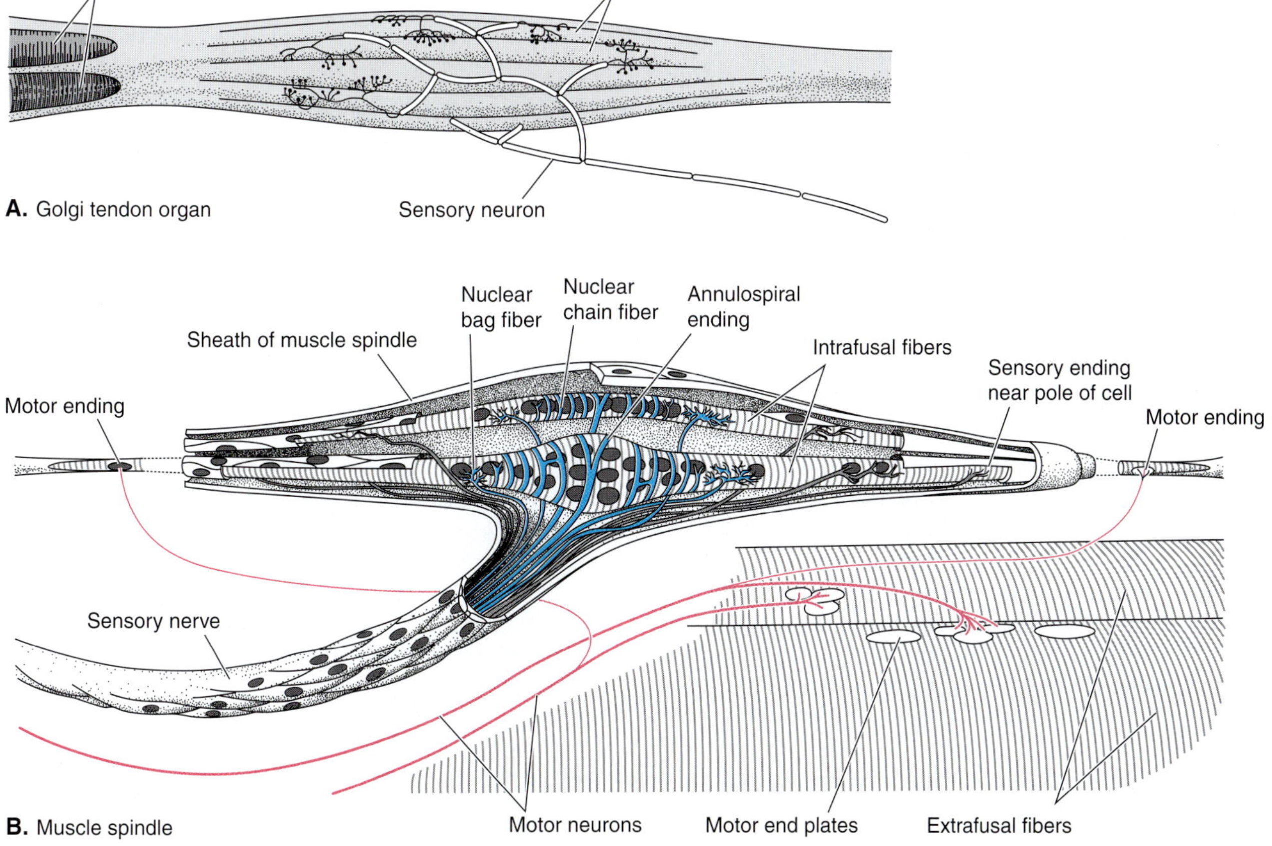

FIGURE 12-7
Proprioreceptors. *A,* Tendon organ. *B,* A muscle spindle. *(After Williams et al.)*

initiate reflexes that reduce the number of motor impulses going to the muscles.

Tetrapods cope with gravitational forces by continuously adjusting their posture. Adjustments are made continuously in the degree and rate of muscle contractions as limb angles and the loads on the muscles change. Tendon organs are more numerous in tetrapods than in nontetrapods, and **muscle spindles** are present within the skeletal muscles of tetrapods. Muscle spindles provide information by which the degree and rate of muscle contraction can be adjusted to meet the changing forces to which the muscles are subjected.

A muscle spindle is a small, fusiform group of specialized muscle fibers inside a sheath (Fig. 12-7*B*). These fibers are called **intrafusal fibers** (L., *fusus* = spindle) because they lie within the sheath. The surrounding, normal muscle fibers are referred to as **extrafusal.** Most of the intrafusal muscle fibers are known as **nuclear bag fibers** because their nuclei are concentrated in the swollen equatorial region of the fibers. The equatorial region lacks contractile myofibrils and is encircled by **annulospiral endings** of sensory neurons. Other, more slender intrafusal fibers, known as **nuclear chain fibers,** receive branching sensory neuron ends nearer their poles. Motor neurons also terminate near the polar ends of each type of intrafusal fiber.

As muscles stretch because of increasing loads on them, the muscle spindles are passively stretched. This is detected by the sensory endings on the nuclear chain fibers. Sensory information returning to the spinal cord initiates nerve reflexes that increase the motor output to the extrafusal fibers in the stretched muscles. The increase causes the extrafusal fibers to contract enough to compensate for their increased load. If, for example, you stand flat-footed and bend your knees, the gastrocnemius and soleus muscles on the back of the shin and the hamstring muscles on the back of the thigh are stretched. If this were not compensated for by an increased contraction of these muscles, you would fall.

During muscle contraction, the polar regions of the nuclear bag fibers also are stimulated and contract. If their rate of contraction matches the rate of contraction of the surrounding extrafusal fibers, then tension will not develop in the nucleated, noncontractile region of the nuclear bag fibers. If the rate of contraction of the extrafusal fibers lags behind that of the nuclear bag fibers, then the nuclear region of the nuclear bag fibers becomes stretched. This is detected by the annulospiral neuron endings. Sensory impulses returning to the spinal cord will initiate an increase in the rate of nerve impulses going to the extrafusal fibers until the rates of contraction of intrafusal and extrafusal fibers are again the same.

Lateral Line System

The lateral line system is one of the most highly variable sensory systems of craniates. In its basic form, it consists of a series of mechanoreceptive organs in the skin organized as a series of lines. It has been lost or modified in many groups and is the source for sense organs specialized to detect other modalities, such as electric fields. In our treatment, we will discuss its basic structure and trace a few of its many evolutionary modifications.

Basic Organization of the Lateral Line System

The lateral line system is present in living hagfishes, lampreys, chondrichthyans, actinopterygians, fishlike sarcopterygians, and larval lissamphibians (including paedomorphic species). This somatic sensory system enables them to detect water disturbances. The lateral

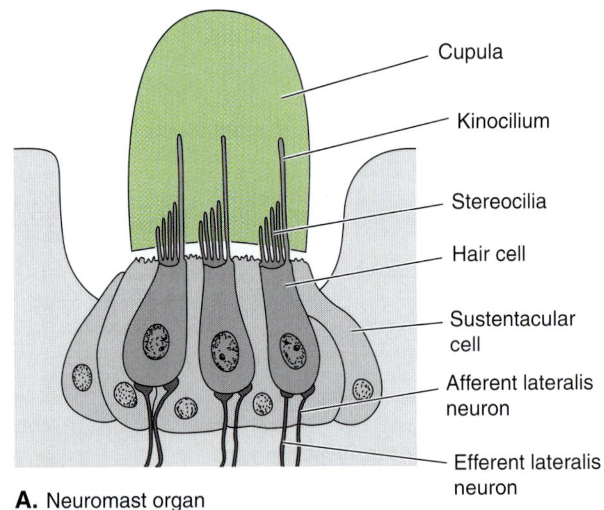

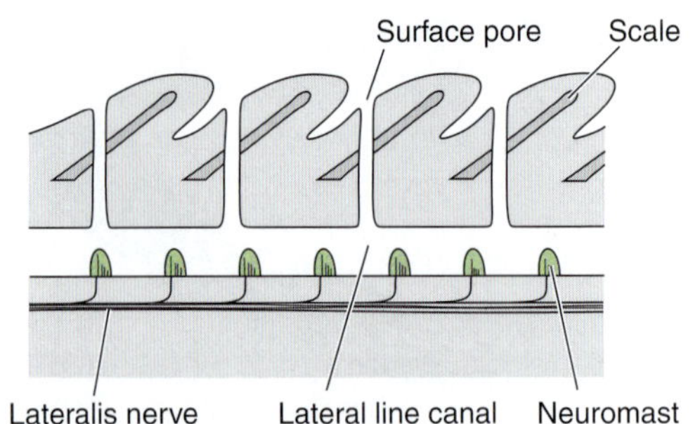

FIGURE 12-8
Neuromast and lateral line canal of a representative teleost. *A,* A neuromast. *B,* A vertical section through the skin and lateral line canal.

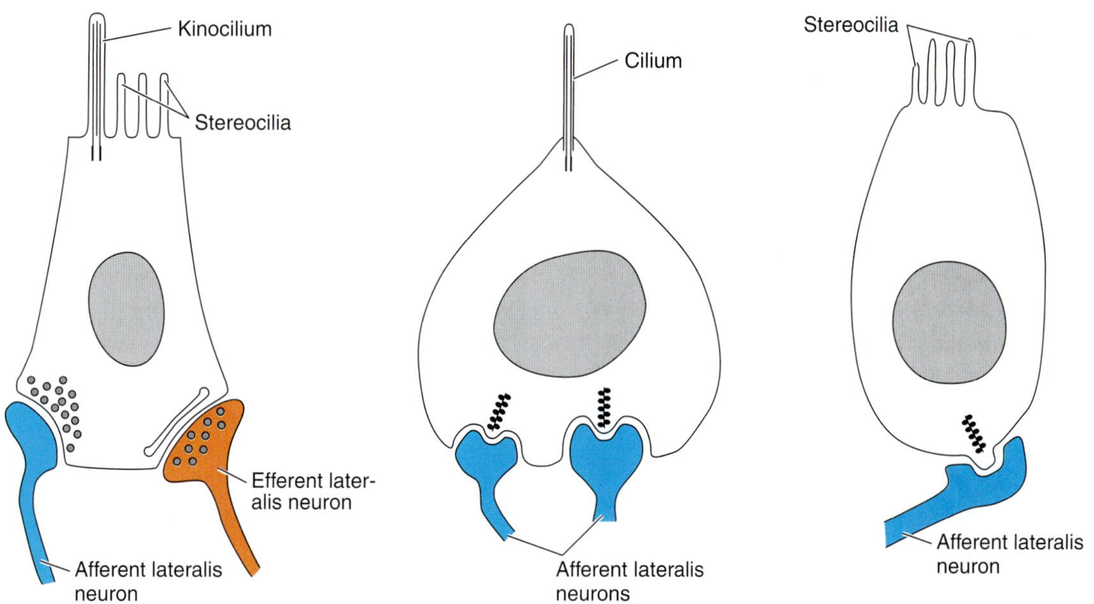

FIGURE 12-9
Neuromast, ampullary organ, and tuberous organ. *A,* Hair cell of a canal neuromast organ typical of chondrichthyans and many other gnathostomes. Note the presence of cilia (a kinocilium) and microvilli (the series of stereocilia) on its apical surface. This cell is a mechanoreceptor and is fundamentally similar to hair cells found in the inner ear of craniates. *B,* Sensory cell of an ampullary organ typical of chondrichthyans and other gnathostomes but not teleosts. Note the single kinocilium and absence of stereocilia. This cell is an electroreceptor. *C,* Sensory cell of an ampullary organ of a teleost, such as a catfish that has secondarily evolved electroreception. Note the absence of a kinocilium and the presence of stereocilia on its apical surface. *(A and B, Modified from Boord and Campbell; C, modified from Northcutt.)*

line system generally is lost at metamorphosis in lissamphibians, although several families of aquatic salamanders retain it as adults. It is entirely absent in all living amniotes, even those such as sea turtles or cetaceans that readapted to an aquatic mode of life. The sense organs within the lateral line system are small clusters of mechanoreceptor and sustentacular cells called **neuromasts** (Fig. 12-8*A*). The individual receptor cells are termed **hair cells** because each bears a single, long **kinocilium** that is followed by a cluster of 15 to 30 **stereocilia** of decreasing length (Fig. 12-9*A*). The kinocilium is a modified cilium containing the characteristic 9 + 2 pattern of microtubules. The stereocilia lack microtubules and are modified microvilli. The kinocilium and stereocilia project into an overlying gelatinous secretion, the **cupula.**

During embryonic development, the cells that will form the hair cells of the neuromasts develop from ectodermal placodes that are adjacent to the one that will give rise to the receptive cells of the inner ear (Chapter 4). Gnathostomes generally have three lateral line placodes rostral to the otic placode and three caudal to it (Figs. 4-27 and 4-30). Some of the neuromasts lie on the skin surface; others lie in **grooves** or **canals;** still others lie in linearly arranged pits, known as **pit organs.**[6] Most neuromasts are on the head. Neuromasts located in the water-filled skin canals or grooves open to the surface by pores (Fig. 12-8*B*). The canals and grooves have a distinct pattern. One canal, known as the **trunk canal,** extends along the flank from head to tail, whereas several others ramify on the head (Fig. 12-10). Particularly noteworthy in the head of gnathostomes are the **supraorbital, infraorbital,** and **preopercular/mandibular canals.** Each canal in the trunk or head is accompanied by a nerve ramus that leads back to the brainstem (Chapter 13). The patterns of the lateral line canals and their innervation across gnathostomes are remarkably conservative and useful for phylogenetic work. **Canal bones** may develop surrounding the neuromasts of the lateral line canals, and these are centers of ossification for many major bones found in the head of Osteichthyes (Chapter 7).

[6]These are not homologous to the thermoreceptive organs of pit vipers; see p. 433.

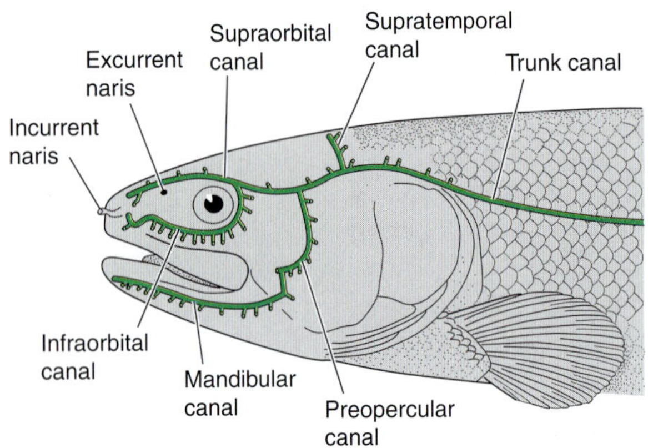

FIGURE 12-10
The distribution of lateral line canals on the head of the bowfin, *Amia*. *(After Jarvik.)*

The hair cells of neuromasts are tonic mechanoreceptors that generate a constant base rate of nerve impulses; bending the cupula alters the rate. Neuromasts enable the animal to detect water movements in different directions because movements that bend the cupula toward the kinocilium increase the rate of nerve impulses that the cell generates, whereas an opposite movement decreases the rate. **Efferent lateralis neurons** that extend from the brain to many hair cells modulate their sensitivity and may suppress background "noise," caused, for example, by the animal's own movements. The sensory neurons that carry impulses from the neuromasts to the medulla of the brain are known as **afferent lateralis neurons.** Traditionally, these neurons have been regarded as components of the facial (VII), vagal (X), and sometimes the glossopharyngeal (IX) cranial nerves, but most neuroanatomists now believe that the lateralis neurons constitute six distinct lateralis nerves (Chapter 13).

The polarity of the neuromasts within a canal is not the same. All of the kinocilia in one set of neuromasts may be located on the caudal edges of the hair cells; those in another set may be rotated 180° and be on the rostral edge. This arrangement, together with the pattern of distribution of the canals, enables a fish to both detect water disturbances and determine their source. "Water flowing from head to tail" may be the message of one set; "water flowing from tail to head" may be the message of the other set. Many sorts of water disturbances can be detected: currents, the movements of nearby animals or of the fish itself, disturbances caused when a fish approaches a rock or other stationary object, and low-frequency vibrations generated by a nearby sound source. The lateral line system has aptly been defined as "distant touch."

You can make some predictions about a gnathostome's sensory world by studying its lateral line system. For instance, placement of neuromasts in canals allows fish to better localize the source of an impinging pressure wave. If, for example, all of the neuromasts were located at the body surface, then they all would be deflected as the animal swims forward. By placing them in canals, however, impinging pressure waves at right angles to the body surface can be detected. If an animal has relatively large pores opening into its canals or if they lie in open grooves, then it is often a relatively slow swimmer; faster swimmers require greater shielding of the canal neuromasts and so have smaller pores.

Electroreceptors

Water conducts electricity well, and **electroreceptors** occur in many groups of aquatic craniates. Most types of electroreceptors develop from placodal tissue adjacent to the tissue that forms typical lateral line mechanoreceptors (Fig. 4-30) and are supplied by lateralis neurons. The similarities and differences between lateral line mechanoreceptors and typical ampullary electroreceptors are summarized in Table 12-1. Like the lateral line sense, electroreception is an ancient sensory system of vertebrates. Living hagfishes do not have any recognizable type of electroreceptors, nor do they have any of the hindbrain nuclei associated with electroreception in other groups of craniates. Also, hagfishes appear unable to physiologically detect even strong electric fields. In contrast, lampreys are electroreceptive, although their electroreceptors are confined to the epidermis and do not extend into the dermis as do the larger organs found in gnathostomes. Thus, we interpret electroreception as a synapomorphy of vertebrates.

The electroreceptors found in many living gnathostomes (e.g., chondrichthyans, basal actinopterygians, coelacanths, lungfishes, and larval lissamphibians) are called **ampullary organs.** In chondrichthyans, groups of ampullary organs (often known as **ampullae of Lorenzini** in this group) are clustered on the head adjacent to the lateral line canals (Fig. 12-11*A* and *B*). Each ampullary organ consists of a subcutaneous tube that lies tangential to the skin's surface. One end of the tube opens by a pore on the body surface, and the other terminates in a slight enlargement, the **ampulla,** which contains modified hair cells (Fig. 12-11*C*). Compared with typical hair cells, the sensory cell of an ampullary organ has a single cilium and no microvilli (e.g., compare Fig. 12-9*A* and *B*). The entire tube of the ampullary organ is filled with a gelatinous, mucopolysaccharide secretion, so a cupula is absent. The jelly has the properties of an electrical capacitor. It has a low electrical resistance and readily holds and con-

TABLE 12-1 Comparison of Chondrichthyan Neuromasts and Electroreceptors

Characteristics	Neuromasts	Electroreceptors
Distribution of receptors	Head, trunk, and tail	Head
Receptor cell specializations	Hair cell (kinocilium and series of stereocilia)	Modified hair cell (cilium; no stereocilia)
Innervation	Anterodorsal, anteroventral, and otic lateral line nerves via ventral roots	Anterodorsal, anteroventral, and otic lateral line nerves via dorsal roots
	Middle, supratemporal, and posterior lateral line nerves via ventral roots	
Peripheral termination	Afferent and efferent	Afferent only
Central termination	Posterior lateral line lobe	Anterior lateral line lobe
Function	Mechanoreception	Passive electroreception
Stimulus	Water movements (particle motion)	DC and low-frequency AC
Role	Orientation and coordination of swimming movements	Electrolocation

Modified from Boord and Campbell (1977).

ducts current. The rate of discharge of the tonic electroreceptive cells in the ampulla is altered by an electric current oriented parallel to the jelly-filled tubes. The electrical resistance of the skin and tube's wall prevents electric currents from reaching the electroreceptive cells except through the jelly. Impulses are sent to the brain on afferent lateralis neurons; there appears not to be any efferent innervation of the sensory cells of an ampullary organ (Fig. 12-9*B*).

Until the middle of the 20th century, no one suspected the electroreceptive role of ampullary organs. Initially regarded as specialized mucous glands, electrophysiological studies of ampullary organs conducted in the 1930s showed that they were exquisitely sensitive to different concentrations of salt water and temperature (indeed, the ampullae of Lorenzini of chondrichthyans are among the most sensitive known temperature receptors, although we have no idea whether any chondrichthyan species uses them to detect temperature changes). For a time, the detection of these stimuli was regarded as their function, even though it seemed doubtful that the acquisition of such information warranted so complex and elaborate a sensory network. Then, behavioral studies convincingly showed that chondrichthyans use ampullary organs for the **passive electrolocation** of prey organisms (Fig. 12-12). A chondrichthyan can detect the weak electric currents generated inadvertently by the contraction of the cardiac and respiratory muscles of prey organisms buried in the sand, even when its other senses have been blocked. The known lower limit of detection is a gradient of 0.001 mV/cm. Sophisticated neurophysiological systems allow an elasmobranch to distinguish external electric fields (signals) from self-generated electric fields (noise) created by their own muscles (see Montgomery and Bodznick, 1999). Different chondrichthyans have different arrangements of the ampullary organs. For example, the size and distribution of the ampullae of Lorenzini in skates correlate with differences in their feeding strategies. The discovery of their electroreceptive role ruled out the old idea that ampullary organs were capable only of temperature or salinity detection. Even though they **are** sensitive to these stimuli, one cannot assume that this is the information that the animal typically uses them to gather. This story shows the importance of investigating not only what a sensory receptor can detect but also what an animal does with the sensory information.

Many organisms, including sharks, can orient themselves in magnetic fields, and some use this ability in navigation and migration. Ampullary organs of chondrichthyans are sensitive enough to detect the minute voltage gradient induced by swimming in the earth's magnetic field, and in some species the use of this sense for navigation has been demonstrated experimentally (Kalmijn, 1978, 1988).

Electroreception based on ampullary organs was lost in the ancestor of neopterygians (Fig. 12-13). This loss eliminated not only the ampullary organs themselves but also their central connections within the brainstem (Chapter 13). Such an evolutionary loss has puzzled

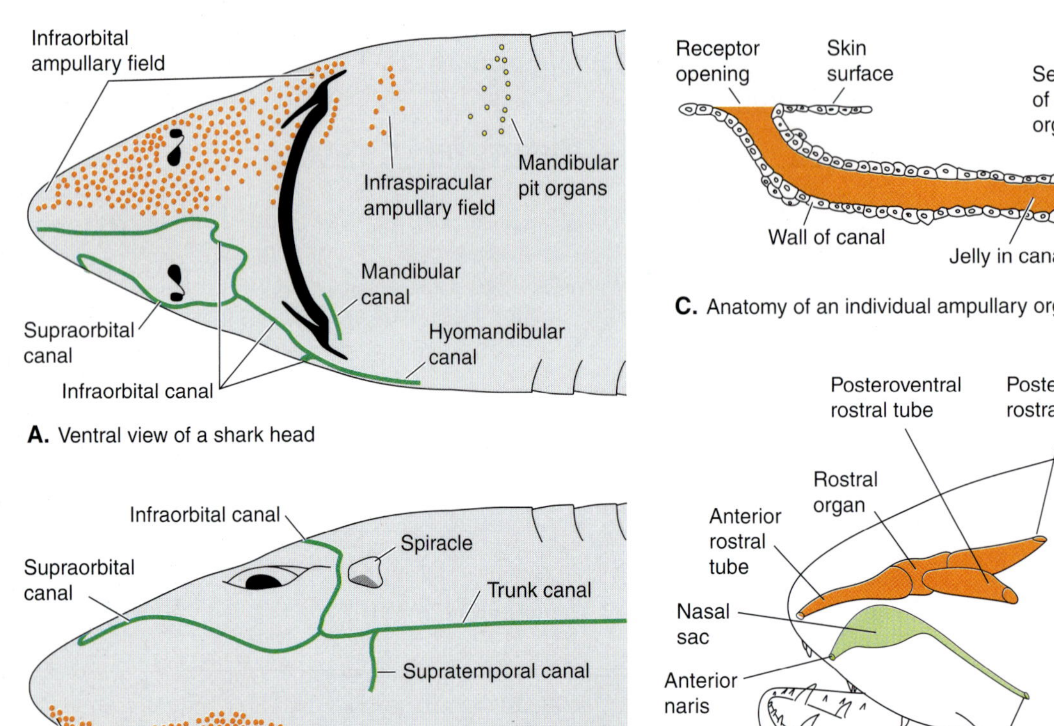

FIGURE 12-11
Electroreceptive organs of sharks and coelacanths. *A,* Ventral view showing the distribution of the lateral line and ampullary organs on the head of a shark. *B,* Dorsal view of the lateral line and ampullary organs on the head of a shark. *C,* Anatomy of an individual ampullary organ. *D,* The rostral organ of *Latimeria. (A and B, After Daniels; C, after Szabo; D, after Bemis and Hetherington.)*

neuroanatomists and behaviorists, and no compelling adaptive explanation exists for the absence of electroreception in groups such as bowfins, gars, and early teleosts. Electroreception was independently lost in frogs and the lineage that led to amniotes (Fig. 12-13). This evolutionary loss is more easily understood because air does not conduct electricity.

Some of the most intriguing examples of convergent derivation of a sensory system are provided by the multiple re-evolution of electroreception, which has occurred at least twice in teleosts [i.e., in the clade including catfishes and South American knifefishes (Gymnotidae) and independently in the clade including elephant snout fishes (Mormyridae) and some allies] and once in amniotes (monotreme mammals, e.g., the platypus).[7] Receptive cells of teleostean electroreceptors lack kinocilia on their apical surfaces (Fig. 12-9*C*). This, together with differences in the way that these organs respond physiologically to electric fields, gives us great confidence that they are not homologous to the typical ampullary electroreceptors found in chondrichthyans. In the case of the platypus, electroreceptors are modified skin glands innervated by the trigeminal (V) cranial nerve.

Catfishes have a single type of electroreceptor that has been termed an ampullary organ because it, too, is

[7]Some neurobiologists suspect that certain shorebirds also are electroreceptive, with putative receptors on the bill.

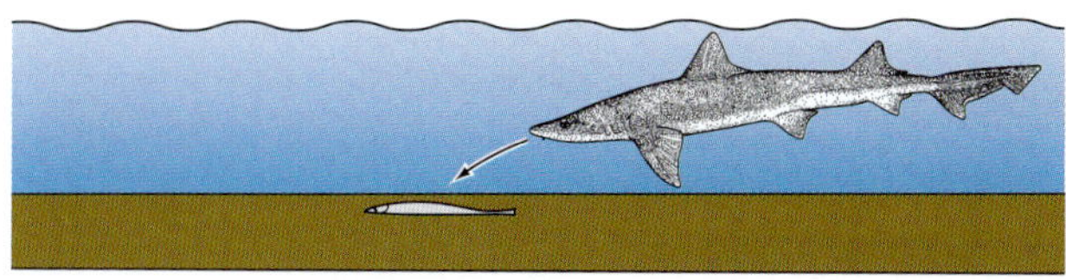

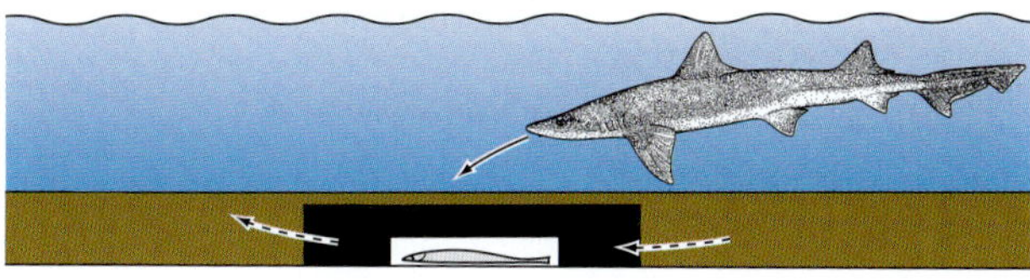

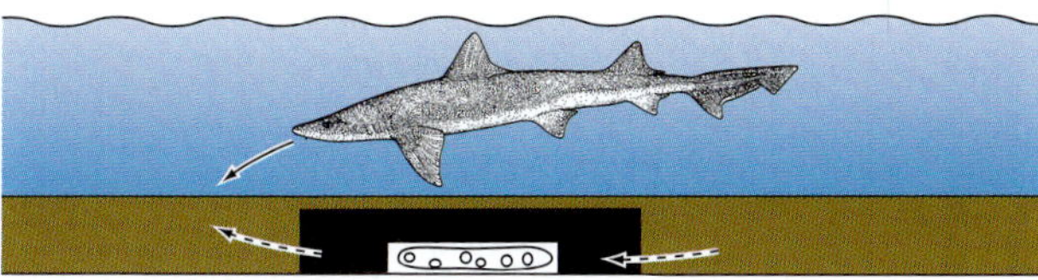

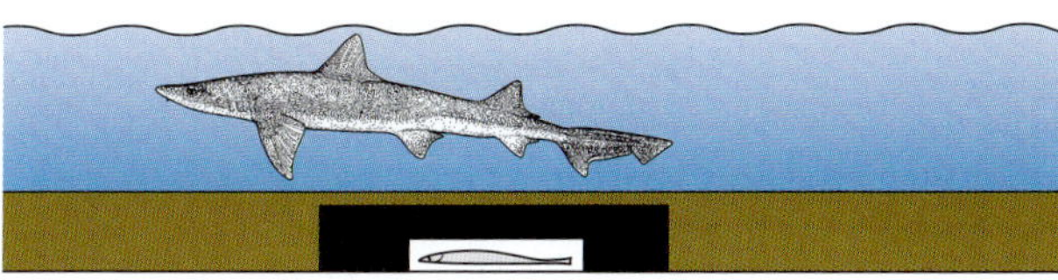

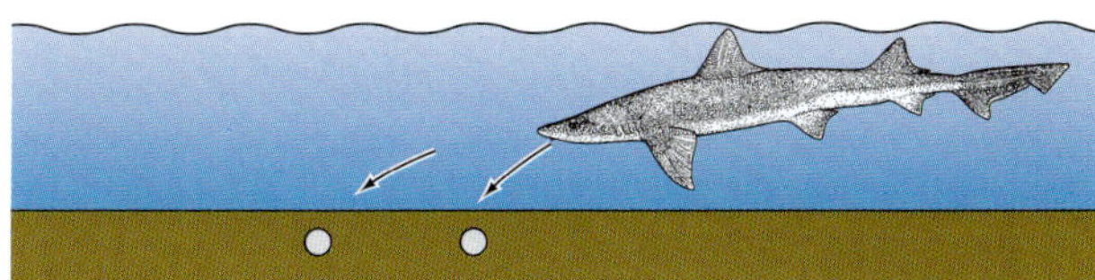

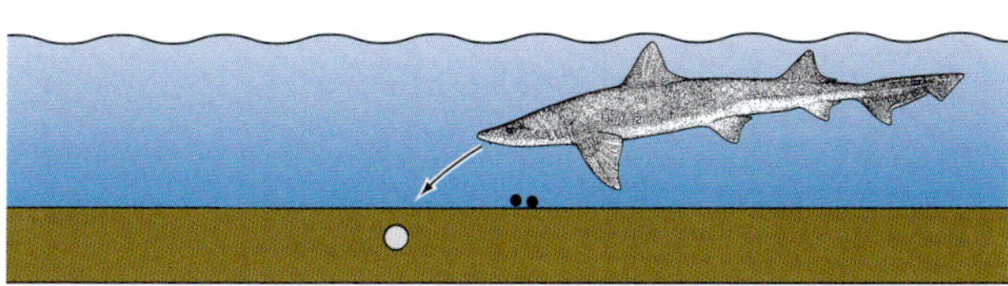

FIGURE 12-12
Experiments on electroreception by sharks. *A,* A live flatfish lying buried in the sand draws an attack by the shark. *B,* A live flatfish in an agar box (transparent to electric fields) through which a current of water is flowing draws an attack directly on itself. *C,* The shark attacks downstream of a dead fish in the agar box where the water stream surfaces. *D,* A live flatfish in the agar box covered by a sheet of plastic, which acts as an electrical insulator and chemical barrier, fails to draw an attack. *E,* Two electrodes buried in the sand draw attacks. *F,* The shark preferentially attacks an active electrode instead of chopped bits of food. *(After Kalmijn.)*

shaped like an ampule or flask.[8] Like the electroreceptors of chondrichthyans, catfish electroreceptors are sensitive to low-frequency (i.e., DC) electric fields. Other electroreceptive teleosts have similar ampullary organs as well as a second type of electroreceptor, known as a **tuberous organ.** Tuberous organs are specialized for detecting rapidly changing (i.e., high-frequency) electric fields and so can be referred to as phasic electroreceptors. Many species of teleosts with tuberous organs also generate high-frequency electric pulses and fields using muscles modified into electrogenic organs (Chapter 10); they can then detect the pulses that they generate using their tuberous organs. Among extant fishes, this capability occurs in the knifefishes of South America and in the elephant snout fishes of Africa and a few related groups. Studies on gymnotids and mormyrids show that they engage in **active electrolocation,** detecting nearby objects in the turbid waters in which they live by the distortions they produce in their own electric fields. They also use electric pulses in **electrocommunication.** Electrocommunication is important in species and sex recognition, in territoriality, and probably in other social interactions.

Among many other interesting variants of ampullary electroreceptors is the **rostral organ** of coelacanths (Fig. 12-11*D*). This large organ, situated in the snout, opens to the outside by three large pores on each side of the head; it is anatomically different from typical ampullary electroreceptors, and its function was unknown for many years. Neuroanatomical and behavioral evidence now confirm that it is electrosensory.

The Ear

For more than 100 years, comparative anatomists have noted the many similarities between the ear and the lateral line system. The receptor cells of both systems are hair cells that are stimulated when liquids or other materials move across their surface and bend their cilia. The two systems develop from adjacent neurogenic placodes (Chapter 4). Neurons from cranial nerve VIII,[9] which carry sensory information from the ear, terminate in the medulla of the brain adjacent to the terminations of lateralis fibers, which carry information

[8]This is a deeply embedded but very unfortunate term in the literature because it suggests that the ampullary organs of catfishes are homologous to the ampullary organs of chondrichthyans; again, we stress that the condition in catfishes appears to have evolved independently.

[9]VIII is commonly called the **statoacoustic nerve** or **octaval nerve**; it also is known as the **vestibulocochlear nerve** in mammals (Chapter 13).

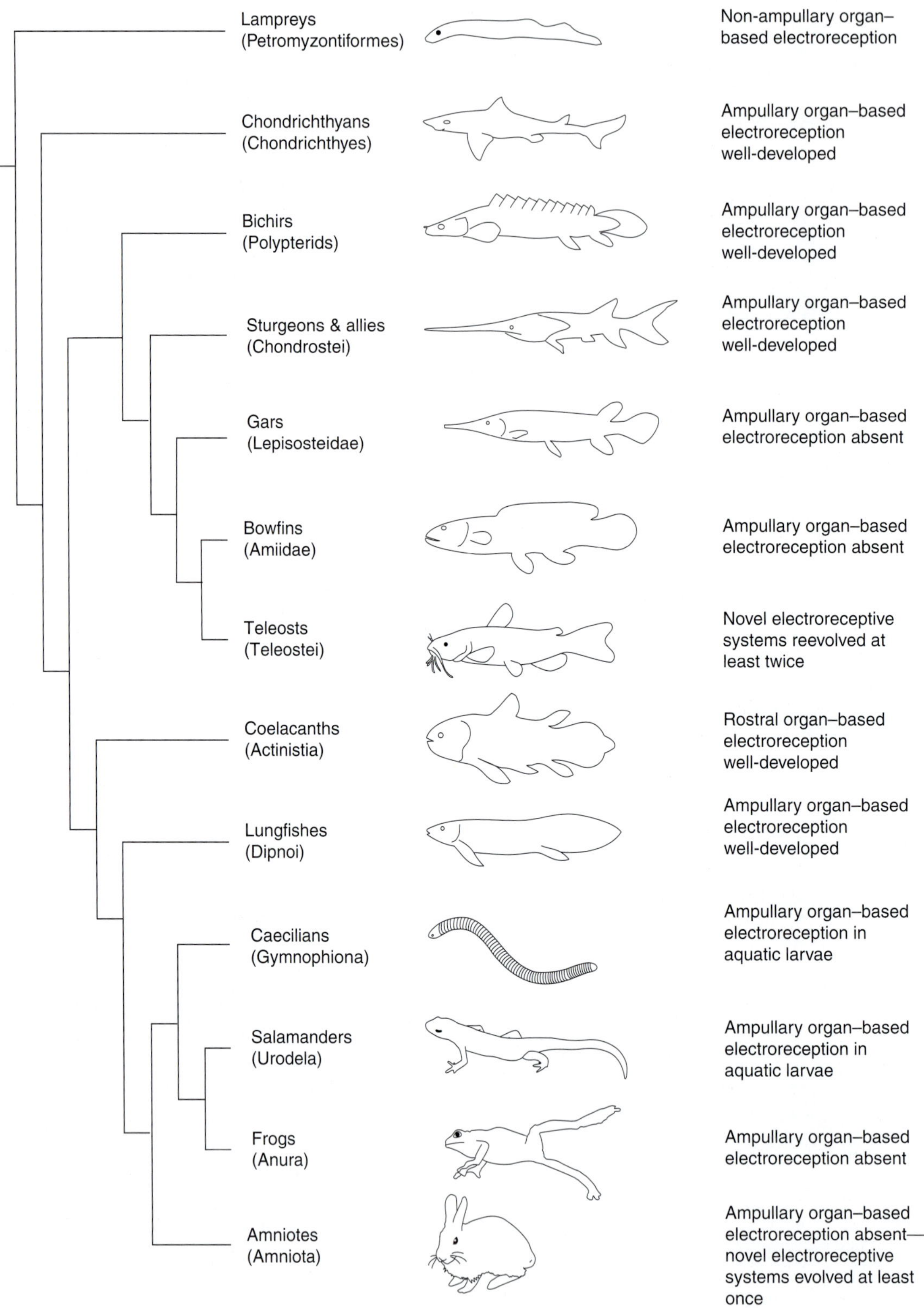

FIGURE 12-13
Cladogram of electroreception. Ampullary electroreception was independently lost in neopterygians (the group of actinopterygian fishes that includes gars, bowfins, and teleosts), frogs, and amniotes. Electroreception re-evolved at least twice in teleosts and at least once in amniotes.

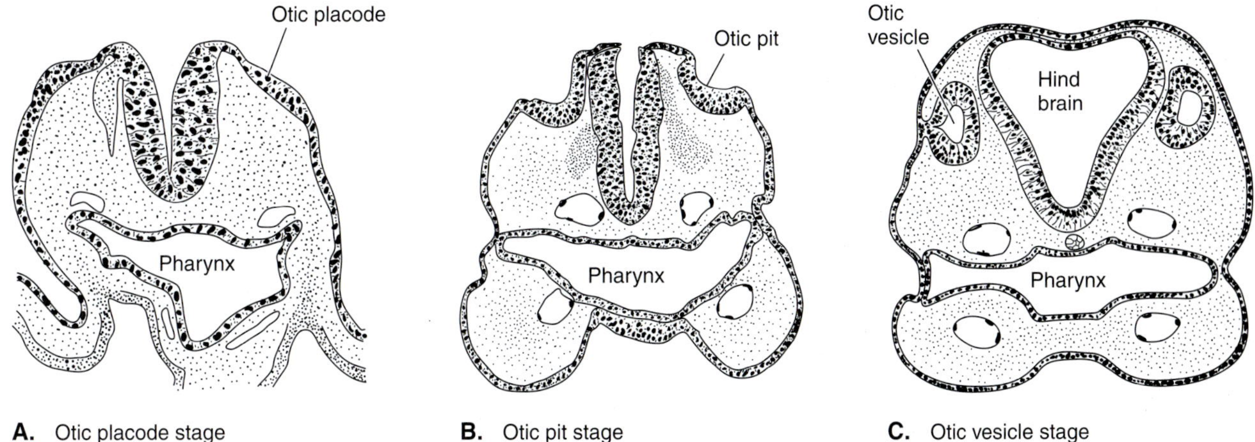

FIGURE 12-14
Three stages in the development of the inner ear. *A*, Otic placode stage. *B*, Otic pit stage. *C*, Otic vesicle stage. *(After Arey.)*

from the lateral line organs and electroreceptors (there are three lateral line nerves anterior to the entry of cranial nerve VIII and three such nerves posterior to its entry; see Chapter 13). Because of these similarities, most investigators regard the lateral line, ampullary electroreceptive system, and ear as components of the **octavolateralis system.** The lateral line part of this system detects water disturbances around aquatic vertebrates, and the ampullary electroreceptors detect electric fields. The part of the ear containing hair cells is invaginated beneath the skin and thus is isolated from external aquatic disturbances. It has, instead, become specialized to detect internal liquid disturbances caused by changes in the orientation and movement of the body (balance and acceleration) and by any external sound waves that are able to reach it.

All vertebrates have an inner ear embedded in the otic capsule of the skull, and this is where the receptive hair cells are located. Living tetrapods also have middle and external ears that are specializations for receiving airborne sound waves and transmitting them to the inner ear.

Inner Ear Structure

The inner ear develops embryonically in all vertebrates as an invagination of the ectodermal **otic placode** to form an **otic vesicle** (Fig. 12-14; also Chapter 4). The vesicle gradually differentiates into a series of membranous sacs and ducts that are filled with a lymphlike fluid called **endolymph.** The entire complex is known as the **membranous labyrinth** (blue in Fig. 12-15). It lies within a system of parallel canals and chambers in the cartilage or bone of the otic capsule of the skull. These spaces are known as the **osseous labyrinth.** The space between the membranous labyrinth and the **osseous labyrinth** is crisscrossed by strands of connective tissue and filled with a liquid called **perilymph** (green in Fig. 12-15).

Many adult anamniotes have an **endolymphatic duct** that opens onto the surface of the head. In other vertebrates, it either is lost or forms a small, deeply

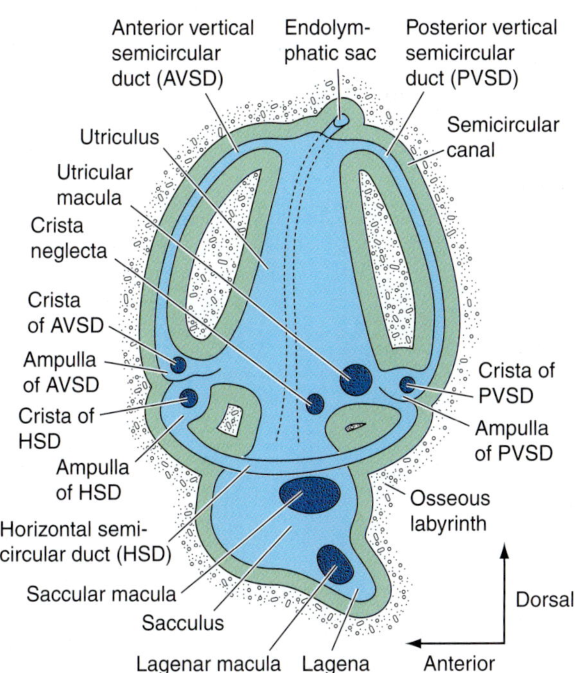

FIGURE 12-15
Inner ear of a typical gnathostome. Blue indicates structures filled with endolymph. Green spaces are filled with perilymph. *(After Kingsley.)*

seated **endolymphatic sac.** In gnathostomes, each membranous labyrinth has **three semicircular ducts** that connect with a chamber known as the **utriculus** (L., *utriculus* = small bag). These ducts are sometimes called canals, but technically the term **semicircular canal** applies to the spaces in the osseous labyrinth in which the semicircular ducts lie. In gnathostomes, two of the semicircular ducts lie in the vertical plane perpendicular to each other; the third lies in the horizontal plane at right angles to the other two. A swelling (termed an **ampulla**; not homologous with the ampulla associated with the electroreceptive organs) is located at one end of each duct. The semicircular ducts are remarkably similar in most gnathostomes (Fig. 12-16). In contrast, lampreys lack the horizontal duct (Fig. 12-16*A*), and hagfishes have only a single semicircular duct (although its orientation allows it to detect the same motions as a lamprey).

Gnathostomes exhibit greater variation in the rest of the membranous labyrinth. For example, cartilaginous fishes have a utriculus that is divided into two parts. In all gnathostomes, the utriculus connects ventrally with a larger sac, called the **sacculus** (L., *sacculus* = small sac), from which a caudoventral evagination of some type arises (Fig. 12-16). In most groups of gnathostomes, the caudoventral evagination of the sacculus forms a small **lagena** (L., *lagena* = flask), and in some diapsids and mammals the lagena develops into a longer duct. The lagena becomes greatly elongated in therians and coils to form the **cochlear duct** (Gr., *kochlias* = snail). The name cochlear duct also is given to the elongate lagena of birds.

Several types of hair-cell groups are located within the membranous labyrinth. A patch of sensory hair cells, termed a **crista** (L., *crista* = crest), occurs in the ampulla of each semicircular duct. Another, called the

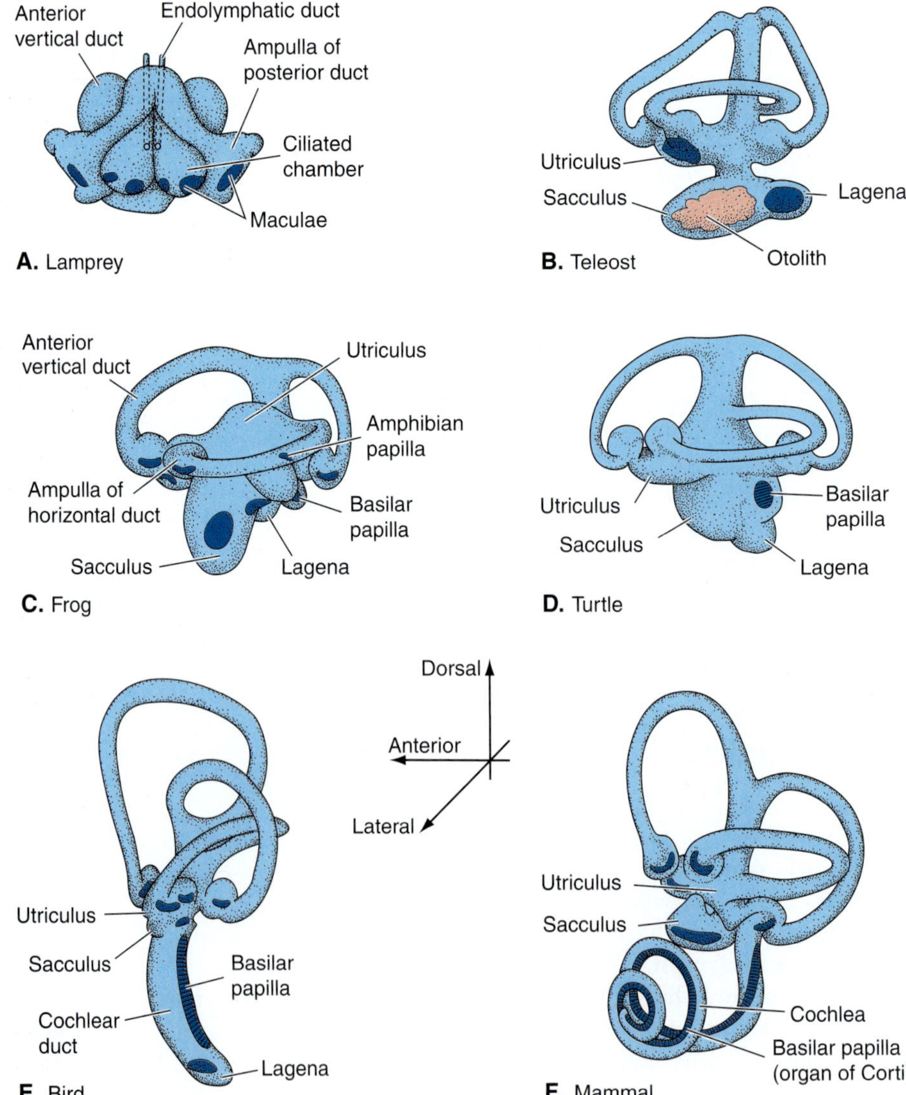

FIGURE 12-16
Lateral views of the labyrinth of representative vertebrates. *A,* A lamprey. *B,* A teleost. *C,* A frog. *D,* A turtle. *E,* A bird. *F,* A mammal. *(A and C, After Retzius.)*

crista neglecta because it was long overlooked, is found in the utriculus (Fig. 12-15). Larger patches of hair cells, termed **maculae** (L., *macula* = spot), occur in the utriculus and sacculus of all vertebrates and in the lagena of many. Maculae often appear as small white spots because each is overlain by small calcareous crystals, termed **statoconia,** which are secreted into a gelatinous membrane. The statoconia are loosely organized in cartilaginous fishes and mixed with sand grains that enter through the endolymphatic duct. Some of the mineral particles are magnetic.[10] Statoconia are consolidated into larger **otoliths** (Gr., *ot* = ear + *lithos* = stone) in many osteichthyans. Because they are so heavily mineralized, otoliths are denser than the rest of the fish. Different groups of teleosts have evolved large otoliths in different chambers within the membranous labyrinth, but typically either the saccular or utricular otolith is enlarged for use in sound detection. Teleostean otoliths often have distinctive shapes in different species. Otoliths of some species grow by the accretion of layers of calcareous crystals, and the number of layers can be counted for use as an indication of the fish's age.

Sensory papillae are patches of hair cells and their associated sustentacular cells that occur only in the sacculus, lagena, or cochlear duct of tetrapods. The cilia of their hair cells impinge on an overlying **tectorial membrane** (L., *tectum* = roof). Movement between the cilia and the membrane activates the receptor cells.

Equilibrium

Hair cells within the ampullae, utriculus, and sacculus of all vertebrates detect changes in position and movement and so provide information that helps an animal maintain its position in space, or its **equilibrium.** Additional information about position and movement comes from sight, proprioceptors, and, in tetrapods, tactile and pressure receptors in the feet. The pull of gravity on the statoconia or otoliths, particularly the large one in the sacculus, registers **static equilibrium,** that is, the present orientation of the body in space. If the animal rolls to one side or pitches forward, then the changed pull of gravity on the statoconia or otoliths in the sacculus and utriculus creates shear forces on the cilia of the hair cells and alters their rate of discharge. Because hair cells in the maculae of the sacculus and utriculus have different polarities, they can detect displacements in different directions. Stops, starts, and other changes in **linear acceleration** also affect the statoconia or otoliths because their inertia causes their movements to lag behind those of the body as a whole.

Angular accelerations, or turning movements of the head, are detected by the cristae in the semicircular ducts. If, for example, a vertebrate turns to the left or to the right, then the crista in the ampulla of the horizontal duct moves at the same rate as the body, but movement of the endolymph in the narrow duct lags slightly and, as a result, bends the cupula overlying the crista, which nearly blocks one entrance into the duct. Similarly, turning movements in the vertical plane will affect some combination of the vertical ducts. Presumably, the large diameter of the two vertical ducts in the ear of a lamprey enables them to detect some horizontal movements of the head, but clearly the gnathostome condition of a dedicated horizontal duct is a better design.

Hearing Mechanisms of Chondrichthyans and Teleosts

To us, hearing is the detection of airborne pressure waves, but to a teleost fish (and to many other aquatic gnathostomes), it is the detection of water disturbances generated by sound sources under water or at the water's surface. Sound waves generated by a vibrating object spread much more rapidly through a dense medium, such as water, than through air. Sound travels at 1500 m/s in water, compared with 330 m/s in air. Sound waves have two components: (1) low-frequency **particle motion** or **displacement waves,** which are somewhat analogous to the ripples produced when a pebble is dropped into the water, and (2) higher frequency **pressure waves,** which result from the alternate compression and rarefaction of molecules in the water. The amplitude of the displacement waves decays very rapidly (proportional to the square of the distance from the sound source) regardless of the sound's frequency. In contrast, the amplitude of a sound pressure wave decays more slowly (linearly with distance) in a frequency-dependent manner (i.e., low-frequency sounds travel much farther than do high-frequency ones). For example, certain very low-frequency sounds generated by instruments in New Zealand can be detected in California, having crossed the entire Pacific Ocean basin. Because of differences in their decay rates, the displacement wave component of sound predominates close to the sound source (that is, in the **near-field**), whereas the pressure wave component is more important at a distance from the sound source (that is, in the **far-field**).

[10]It was proposed that these magnetic particles enabled chondrichthyans to detect the earth's magnetic field, but others believe that this is not true (Hanson et al., 1990).

Because hair cells are stimulated by the displacement of their stereocilia, a displacement wave directly affects the hair cells that it can reach. For example, lateral line organs are well suited to detect and localize the source of low-frequency water particle displacements in the near-field. Maximum sensitivity for such lateral-line detection is for sounds with frequencies of 50 Hz to 150 Hz. Displacement waves may also directly reach the inner ears of chondrichthyans by passing through the skin and jelly-filled parietal fossa in the chondrocranium and affecting the crista neglecta, which lies in part of the utriculus. Skates and galeomorph sharks can hear very well using this system. Teleosts use their otoliths for detecting near-field displacement waves. When a displacement wave impinges on the body, the movements of the dense otoliths lag slightly behind the movements of the rest of the body, which creates a relative movement between the otolith and the hair cells on which it rests, causing the hair cells to bend. Such displacement of the body relative to its otoliths is a way to detect near-field sounds.

Mechanisms whereby teleosts detect higher frequency, far-field pressure wave components are complex. Because the density of a teleost is similar to that of water, sound pressure waves easily pass through and move the body at nearly the same amplitude and frequency as they move the water. In this sense, a teleost's body is transparent to sound. For a sound pressure wave to be detected, the waves must induce a movement over certain hair cells that differs from that of the rest of the body. Neurophysiological evidence suggests that, for many teleosts, the macula in the sacculus is more sensitive to far-field pressure waves than are other parts of the inner ear.

A potentially better mechanism is to use a gas-filled space in the body, such as the swim bladder, as a hydrophone (see Chapter 18 for more on swim bladders). Sound pressure waves passing through the body cannot compress the liquid in the tissues, but they can compress the air in the swim bladder. The changing pressure in the swim bladder vibrates its walls at the same frequency as the impinging sound wave. Carps, minnows, and other ostariophysans have a particularly elegant system. The swim bladder vibrations are transferred to the inner ear by paired sets of small bones, the **Weberian ossicles,** which are derived from ribs (Fig. 12-17). These ossicles extend from the anterior end of the swim bladder to a perilymphatic sac that lies beside the sacculus. The system is analogous to the tympanic membrane and auditory ossicles of mammals (p. 420 and Chapter 22). Ostariophysans have very low auditory thresholds (i.e., they can detect very faint sounds) and can detect some of the highest frequencies

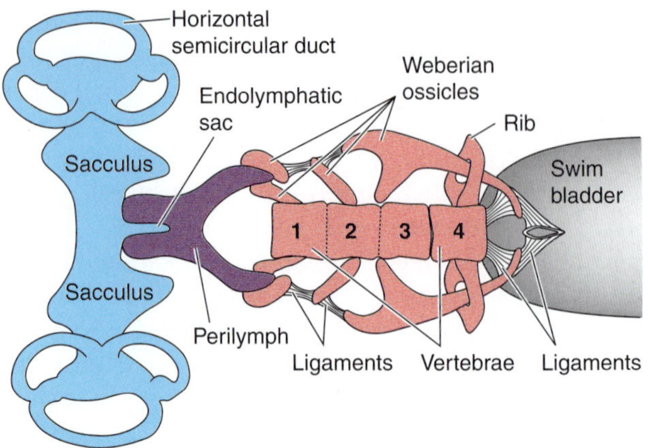

FIGURE 12-17
A dorsal view of the Weberian apparatus of an ostariophysan. *(After Popper.)*

(≤ 5000 Hz) of any teleosts. They are most sensitive to frequencies between 200 Hz and 600 Hz (Fig. 12-18). In contrast, tunas and allies that do not have a swim bladder have far less sensitive ears (Fig. 12-18). Many other groups of teleosts independently evolved swim bladder–ear connections.

Considerable evidence indicates that teleosts, like tetrapods, can detect the direction from which a sound came. Tetrapods detect direction by differences in the time and intensity of the arrival of the pressure waves at the two ears, but it is unlikely that teleosts can do so because pressure waves travel so fast in water, and the distance between the two inner ears is so small. Therefore, directional discrimination in fishes probably depends on differences in the polarity of groups of hair cells in the macula.

The underwater world can be a relatively silent place to a human diver, whose ears are adapted to detect airborne sound waves. But what do chondrichthyans or teleosts hear? Many sorts of noises are detected with underwater listening instruments. Waves, surf, and currents all generate sounds, as do rapid changes in the direction or velocity of swimming animals. Animals feeding on coral, shellfish, and other coarse material make noises. Many teleosts also produce sounds at will by gnashing their teeth or rubbing spines or certain pectoral bones together. Others have muscles that pluck on the swim bladder as one might pluck a bass fiddle, or they drum on the body wall overlying the swim bladder with their pectoral fins. Some species use sounds to call mates, startle predators, or warn conspecifics of danger. Some of the sounds produced by teleosts are very loud. A chorus of marine catfish can generate a noise of 20 decibels—equivalent to a subway train 10 m away.

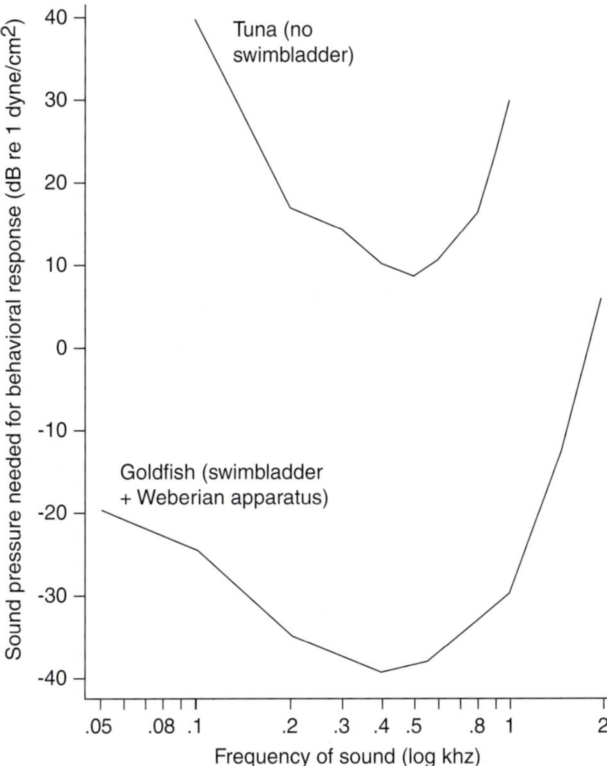

FIGURE 12-18
Audiograms comparing hearing sensitivity of two teleosts: a tuna and a goldfish, which is an ostariophysan. The ear of an ostariophysan is far more sensitive to sound and also can detect a broader range of sound frequencies. *(After Fay and Popper.)*

Clearly, sound can be an important sensory modality for teleosts.

Hearing Systems of Tetrapods

Although a teleost with a swim bladder and a swim bladder–ear transmission system has no difficulty in receiving sound pressure waves, this is not necessarily true for a tetrapod. Low-frequency sound waves (< 1000 Hz) of sufficient intensity may travel through the ground as well as air and can cause a slight displacement of superficial skull bones that can be detected as sound. Caecilians, terrestrial salamanders, some squamates, and all snakes detect low-frequency waves in this way. The most important sounds for a tetrapod, however, are higher in frequency and so travel only as sound pressure waves in air, a medium that is far less dense than water. The force of an airborne sound pressure wave must be amplified considerably to enter the denser liquids of the inner ear to produce a displacement wave that can move certain hair cells relative to overlying structures. Terrestrial vertebrates that detect higher-frequency sound waves have a thin **tympanic membrane** (also called a **tympanum** or **eardrum**) on or near the surface of the head. The term **external ear** may be applied to a tympanum; the term also may refer to accessory structures, such as the **ear pinna,** that help to direct sounds to the tympanum. An air-filled **middle ear,** or **tympanic cavity,** lies on the inside of the tympanic membrane and connects to the pharynx by way of the **auditory tube,** also known as the **Eustachian tube** (Figs. 12-19 and 12-20). The tympanic cavity and auditory tube equalize the air pressure on the two sides of the tympanic membrane, which enables the membrane to respond quickly to high-frequency sounds. The tympanic cavity and auditory tube of an amniote develop from the first embryonic pharyngeal pouch, so they are homologous to the first gill pouch, or spiracle, of a fish. We are uncertain whether this homology strictly applies to the middle ear cavity and auditory tube of lissamphibians, which show certain peculiarities in their development.[11]

The conversion of a sound pressure wave in air to a displacement wave in the perilymph and endolymph of the inner ear is made by the response of the tympanic membrane (or certain skull bones when a membrane is absent). Three other structures are essential (Figs. 12-19 and 12-20):

1. At least one auditory ossicle, known as the **columella** (in lissamphibians and sauropsids) or **stapes** (in mammals), must be present, the lateral end of which connects either to a tympanic membrane or to superficial skull bones on which airborne sound waves impinge. The medial end of the columella or stapes bears a **foot plate** that fits into an opening known as the **oval window** on the lateral surface of the otic capsule. The columella or stapes is homologous to the hyomandibular of fishes (Chapter 7). The large size of the tympanic membrane relative to the size of the foot plate forms a pressure-amplification mechanism. All the energy impinging on the large tympanum is concentrated on the small foot plate. The resulting increase in energy per unit of area on the foot plate enables the high-frequency aerial vibrations to overcome the inertia, or **impedance,** of the liquid in the perilymphatic duct and set up displacement waves in it. The internal and

[11]The "cochlear" nuclei of frogs do not appear to be homologous to the cochlear nuclei of amniotes, which is evidence that the central connections of these sensory systems also evolved independently.

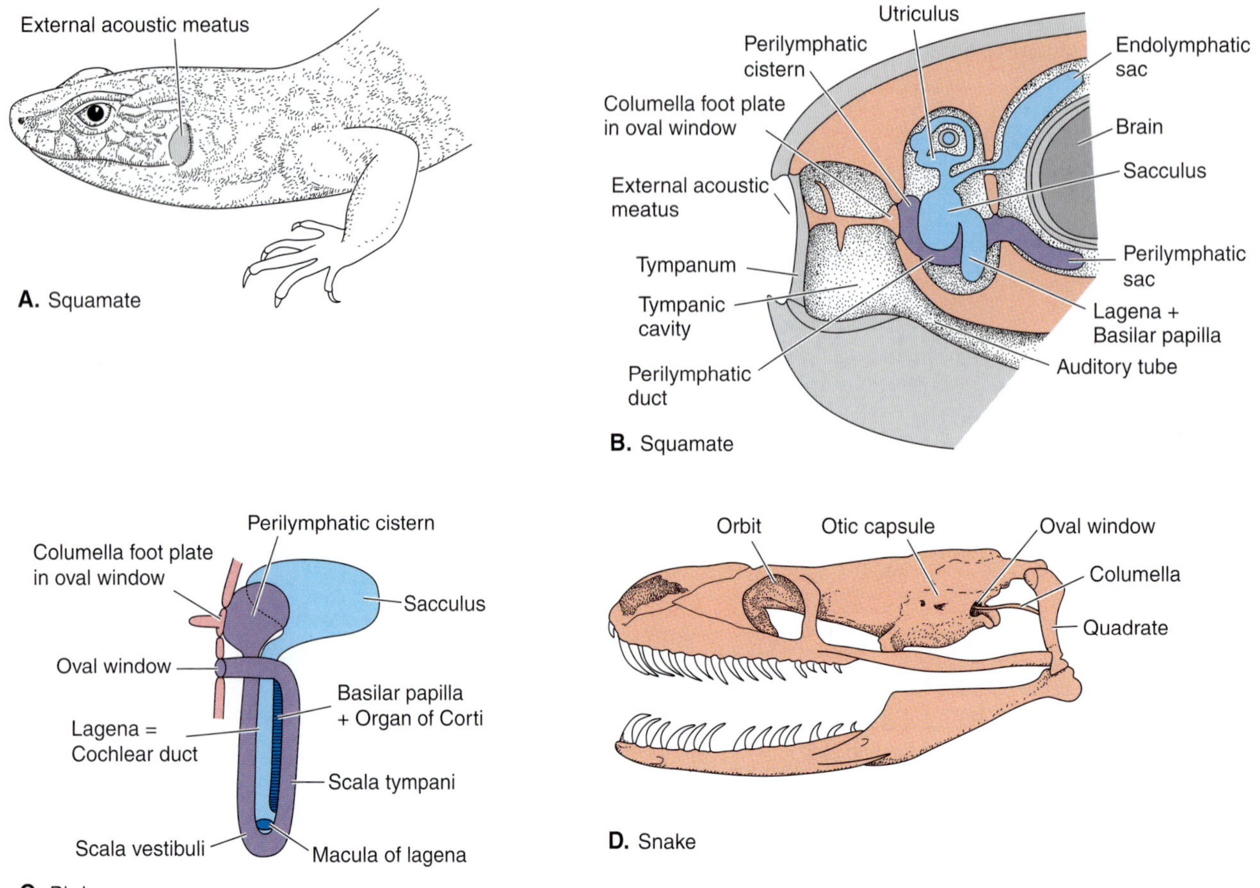

FIGURE 12-19
Ears of squamates and birds. *A,* A surface view of the ear of a squamate. *B,* A transverse section through the ear of a squamate. *C,* The auditory part of the ear of a bird. *D,* The highly modified middle ear of a snake. *(B, After Portmann; C, after Romer and Parsons.)*

middle ears often are described as an impedance-matching device.

2. At least one specialized **perilymphatic duct** must be present. The perilymphatic duct receives displacement waves from the foot plate and carries them past a receptive part of the membranous labyrinth to some point where they can be dissipated. The perilymphatic duct may be expanded to form a **perilymphatic cistern** adjacent to the oval window. The dissipation point is either inside the cranial cavity (e.g., Fig. 12-19*B*) or at the **round window** that opens into the middle ear cavity (e.g., 12-19*C*).
3. Finally, within the membranous labyrinth, at least one sensory area must be specialized to receive the displacement waves from the perilymphatic duct. Tetrapods have one or more unique groups of hair cells known as **papillae,** which are specialized to receive these waves. A papilla consists of hair cells overlain by a **tectorial membrane** rather than by otoliths. Many differences in the location of the papilla or papillae exist among tetrapods.

Different groups of tetrapods detect airborne sound waves in different ways, utilizing some structures that are homologous and some that are not. This suggests considerable independent evolution of auditory mechanisms in tetrapods. Such a pattern of multiple evolutionary origins of auditory specialization parallels that already noted for teleost fishes. In particular, a tympanic membrane was not present in the earliest tetrapods, and perhaps its absence in caecilians and salamanders is a retention of this plesiomorphic condition. An ear utilizing a tympanic membrane evolved independently at least three times in tetrapods: (1) in the lineage that leads to anurans (frogs), (2) in the line of evolution to turtles and diapsids, and (3) in the late synapsid lineage that gave rise to mammals.

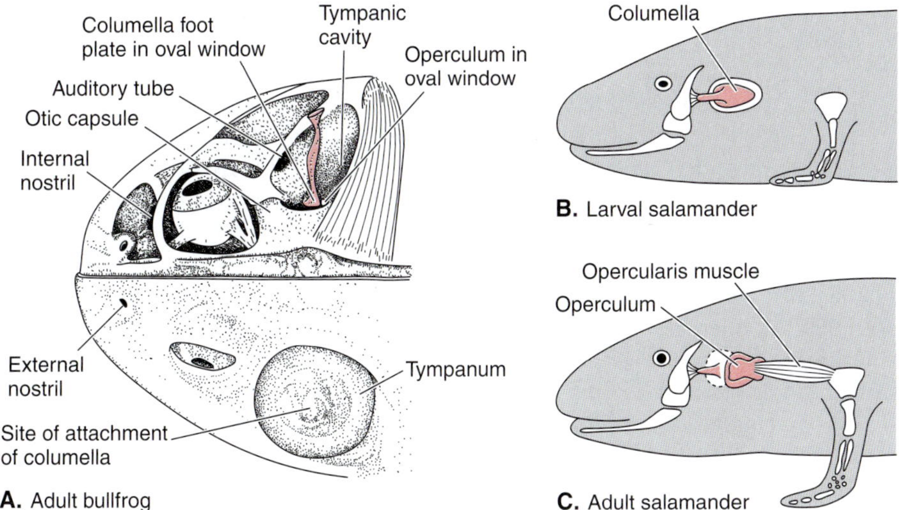

FIGURE 12-20
Lissamphibian ears. *A,* The ear of an adult bullfrog in surface view (*below*) and dissected (*above*). *B,* The auditory mechanisms of a larval salamander. *C,* The auditory mechanisms of an adult salamander. *(A, Modified from Walker and Homberger; B and C, after Kingsbury and Reed.)*

The Ear of Sauropsids

We begin with the ear of sauropsids,[12] in which the tympanic membrane usually is located at the bottom of a recess, the **external acoustic meatus** (Fig. 12-19*A*). The tympanum is described as postquadrate in position because it lies just caudal to the quadrate bone, which usually partly encases the tympanum and middle ear cavity. The columella crosses the tympanic cavity and transfers displacement waves from the tympanum through the oval window to a **perilymphatic cistern** (Fig. 12-19*B*). A **perilymphatic duct** extends from the cistern across the lagena, which contains the receptive **basilar papilla,** to the **perilymphatic sac,** where the waves are dissipated. This sac may lie beside the round window, which opens back into the tympanic cavity, but it often lies in the cranial cavity. This ear is very sensitive and can respond to high-frequency sounds. Some squamates can detect sound waves with a frequency as high as 10 kHz.

The ear of birds is similar. The main differences are that the lagena forms an elongated **cochlear duct,** and the basilar papilla forms an equally long **organ of Corti**[13](Fig. 12-19*C*). Songbirds can detect frequencies as high as 21 kHz.

Many squamates, including all snakes, lack a tympanum, an absence that is regarded as a secondary condition. Instead, the columella connects with the quadrate bone (Fig. 12-19*D*). Such animals detect chiefly ground vibrations, but their ears also are sensitive to low-frequency airborne vibrations. It is probable that the tympanic membrane was secondarily lost in these species as an adaptation to a burrowing mode of life. Those squamates that lack it are, for the most part, burrowers: all amphisbaenians live beneath ground, and we believe that snakes passed through a burrowing phase early in their evolution (most species now live above ground, but basal snakes still burrow).

The Ear of Lissamphibians

Frogs have a large tympanum located high up on the head surface, dorsal to the quadrate bone[14] (Fig. 12-20*A*). The columella of a frog connects to the tympanum and continues across the tympanic cavity to the oval window in the otic capsule. This is referred to as a frog's **tympanic ear system** to distinguish it from a second auditory system, the **opercularis system,** that also is present in lissamphibians. Lissamphibians have an additional auditory ossicle, known as the **operculum,** that also lies in the oval window and fits into a notch on the caudal surface of the footplate of the columella. The operculum develops embryonically from a part of the wall of the otic capsule.[15] The **opercularis muscle** extends from the operculum to the suprascapular element of the pectoral girdle, thereby coupling the inner ear to the girdle, leg, and ground. The opercularis muscle is homologous to the levator scapulae muscle (Chapter 10). The perilymphatic cistern lies just inside the oval window and continues past the sacculus to a perilymphatic sac that extends into the cranial cavity. Two sensory papillae are present in the ear of frogs: a

[12]See Fig. 3-20 for Sauropsida, which includes the living turtles and diapsids.

[13]The organ of Corti of a bird convergently resembles the organ of Corti of a mammal. Both terms are unfortunately embedded in the literature (i.e., one should be changed).

[14]Many Paleozoic amphibians had an otic notch in a similar position that may have held a tympanic membrane.

[15]The operculum of lissamphibians is not homologous to the operculum of osteichthyans or chimeras, which are gill coverings.

basilar papilla near the entrance to the lagena, and a more dorsally located **amphibian papilla,** which is uniquely found in lissamphibians (Fig. 12-16*C*). Larval salamanders have a columella that connects to superficial skull bones (Fig. 12-20*B*), but this is either lost or reduced later in development so that adult salamanders have only the opercular system (Fig. 12-20*C*).

Lissamphibians are secretive creatures that are sensitive to low-frequency seismic vibrations that can alert them to potential danger. Early in the 20th century, Kingsbury and Reed (1909) proposed that the opercularis system detected ground vibrations of this type. Hetherington, Jaslow, and Lombard (1986) corroborated this hypothesis and found that the opercularis is a tonic muscle that is tense and rigid, exactly what would be expected if it was part of a vibration-detecting system. Low-frequency vibrations of this type are detected by the amphibian papilla. The tympanic ear system of frogs is adapted to receive much higher sound frequencies, such as their mating calls. These are detected by the basilar papilla. Salamanders do not have mating calls and, not surprisingly, they lack the tympanic system, including the basilar papilla.

Evolution of the Mammalian Ear

Mammals have a third type of tympanic ear. An external flap, the **auricle** or **pinna,** helps funnel sound waves down the external acoustic meatus to the tympanic membrane (Fig. 12-21). The human auricle is neither very large nor important, but the auricle of many mammals is large and can be moved and directed toward a sound source. A large auricle amplifies sound waves because the waves it gathers are concentrated on the relatively smaller tympanic membrane.

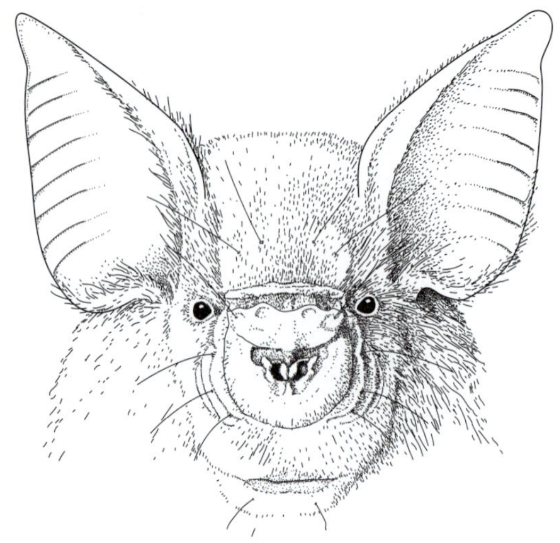

FIGURE 12-21
Face of a giant leaf-nosed bat (Rhinolophidae: *Hipposideros commersoni*) to show enlarged pinna used in ultrasonic hearing. *(After Vaughan et al.)*

Three auditory ossicles, the **malleus, incus,** and **stapes,** extend from the tympanic membrane across the middle ear cavity (Fig. 12-22*A*). The oval window of a mammal is technically known as the **fenestra vestibuli.** The ossicles derive their names from their shapes (L., *malleus* = hammer, L., *incus* = anvil, L., *stapes* = stirrup). As we have seen (Chapter 7), the malleus evolved from the quadrate bone; the incus, from the articular bone and the stapes, from the hyomandibular. The quadrate and articular bones bear the jaw joint in nonmammalian vertebrates. As in other vertebrates with tympanic ears, amplification of a sound wave derives from the large size of the tympanic membrane relative to the foot plate of the stapes. The three auditory ossicles of mammals form a lever system that further increases this amplification about one and a half times, but as in any lever system, the increase in force at one end of the system is accompanied by a reduction in the extent of movement at that end. In humans the total force on the foot plate of the stapes is 22 times that on the tympanic membrane, but the extent of movement of the foot plate is only about one third that of the membrane.

Two small muscles, the **tensor tympani** and **stapedius,** attach to the malleus and stapes, respectively. Because the malleus derives from the mandibular arch, the tensor tympani is a mandibular arch muscle and is innervated by the trigeminal (V) nerve. The stapes is a hyoid arch derivative, and the stapedius muscle is innervated by the facial (VII) nerve. Tension in these muscles dampens oscillations of the ossicles and so protects delicate inner ear structures from excessive movements caused by loud noises, including the sounds that the animal itself may make, such as the ultrasonic sounds of bats and cetaceans (see below), which are emitted at very high energy levels.

The hypothesis that the mammalian ear, with its three ossicles, evolved from a tympanic ear similar to that seen in living sauropsids, in which there is only a single ossicle, presents some serious problems. How could the articular and quadrate, when they became redundant as jaw joint bones, move into the middle ear without disrupting its function? What would be the advantage of adding them to an effective ear? These problems vanish if, as we now believe, mammals evolved from ancestral amniotes that did not have tympanic ears (e.g., animals similar to †captorhinids and early synapsids, such as †sphenacodonts). These animals had a large columella that was connected to the quadrate. In these forms, there was no modification of

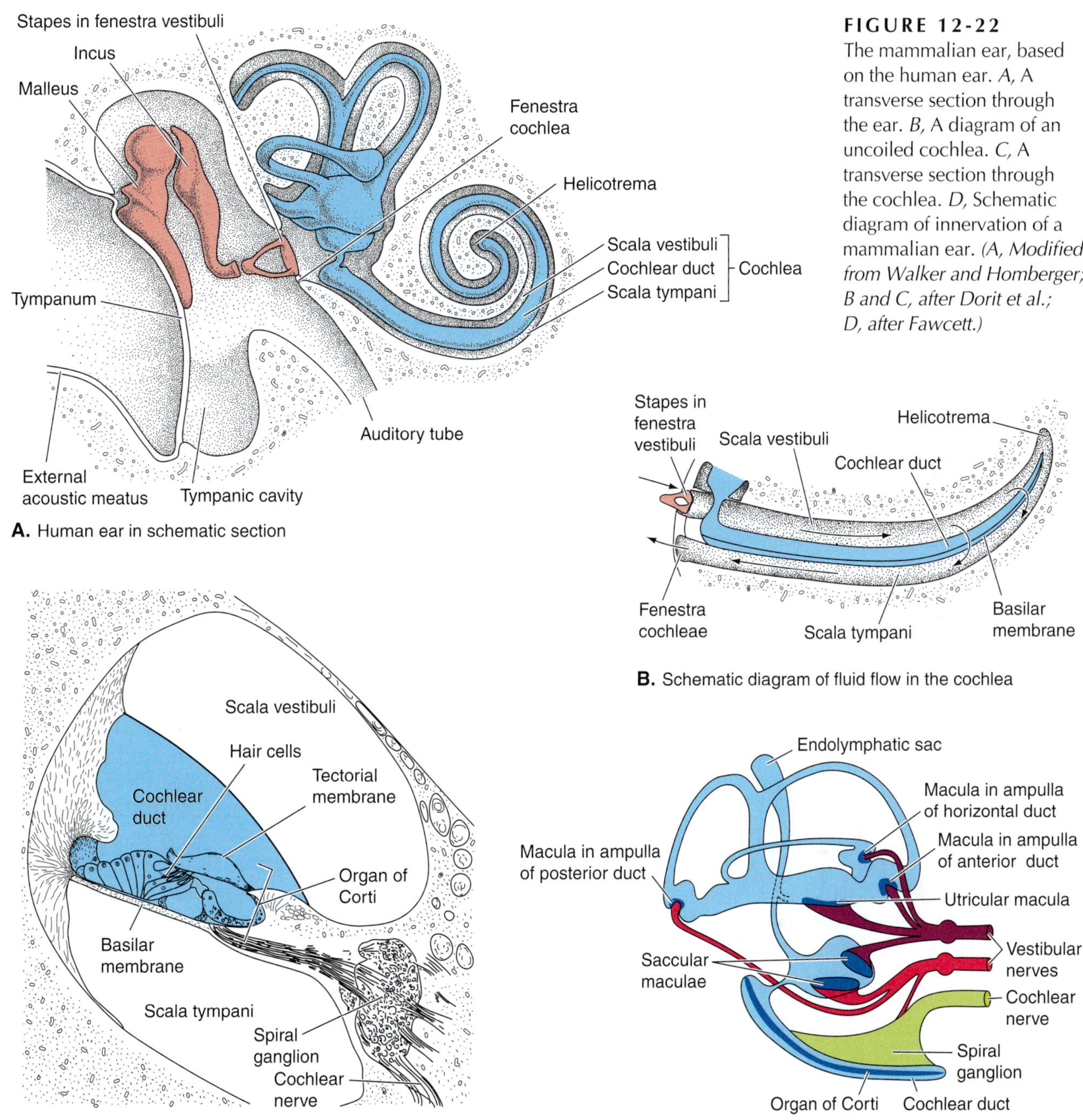

A. Human ear in schematic section

B. Schematic diagram of fluid flow in the cochlea

C. Transverse section through cochlea

D. Schematic diagram of innervation of mammalian ear

FIGURE 12-22
The mammalian ear, based on the human ear. *A,* A transverse section through the ear. *B,* A diagram of an uncoiled cochlea. *C,* A transverse section through the cochlea. *D,* Schematic diagram of innervation of a mammalian ear. *(A, Modified from Walker and Homberger; B and C, after Dorit et al.; D, after Fawcett.)*

the quadrate, as there is in extant sauropsids, that would accommodate a tympanic membrane and middle ear behind it. Presumably, †captorhinids and early synapsids detected ground vibrations and low-frequency airborne vibrations through the skull, including the lower jaw. Vibrations detected by the lower jaw would have been transmitted via the articular, quadrate, and columella to the inner ear.

As the postdentary bones of the jaw and the columella became smaller and lighter (Chapter 7), they became more efficient in transmitting sound. Such improved sound transmission may have been another factor in addition to feeding mechanics in the evolution of changes in jaw mechanisms. More derived synapsids had an unusual flange, the **reflected lamina,** that extended ventrally and caudally from the angular bone (Figs. 7-29 and 7-30). Many investigators believe that the reflected lamina of the angular held a so-called mandibular tympanic membrane that lay adjacent to an air-filled space. Such a space would have further increased the efficiency of the articular, quadrate, and columella in sound transmission. With the final shift of the jaw joint to the dentary and squamosal bone in early mammals, the articular, quadrate, and columella (now modified into the tiny, stirrup-like stapes) became dedicated auditory ossicles. If we are correct in thinking that early mammals, such as †*Morganucodon*, occupied an insectivorous, nocturnal niche, then we may hypothesize that the ability to detect high-frequency sounds may have been advantageous. So, rather than trying awkwardly to fit the articular and quadrate into an already functional ear, we now believe that these bones together with the columella were sound-transmitting bones throughout the early history of tetrapods and especially in the line of evolution to mammals. Changes in jaw mechanics made it possible for these bones to become dedicated and highly efficient auditory ossicles.

The lagena and perilymphatic duct found in ancestral terrestrial vertebrates are elongated in mammals to form the snail-like **cochlea**[16] (Figs. 12-16*F* and 12-22*A*). The cochlea spirals around a core of bone, the **modiolus,** that contains the cochlear branch of cranial nerve VIII, which is called the **vestibulocochlear nerve** in mammals (Fig. 12-22*D*). The cochlea becomes progressively narrow as it approaches its apex. Its spiral ranges from a one-quarter turn in the platypus to four turns in the guinea pig. The human cochlea makes three and a half turns. The presence in early fossil mammals of spaces in the otic capsule for an elongated and partly coiled cochlea is evidence that these mammals had a keen sense of hearing.

The lagena itself forms the **cochlear duct** that contains the elongated **basilar papilla** known as the **organ of Corti** (Fig. 12-22*B* and *C*). The hair cells of the organ of Corti rest on the **basilar membrane,** and their cilia impinge on an overlying tectorial membrane. Displacement waves are received from the stapes at the fenestra vestibuli by a part of the perilymphatic duct called the **scala vestibuli.** This duct passes along one side of the cochlear duct and returns on the other as the **scala tympani.** The scala vestibuli and scala tympani connect at the apex of the modiolus by a small opening known as the **helicotrema.** Displacement waves are released through the scala tympani back into the tympanic cavity at the round window, or **fenestra cochleae.** When displacement waves cross the cochlear duct from the scala vestibuli to the scala tympani, they cause slight movements of the basilar membrane. Shear forces develop between the hair cells and the tectorial membrane, activating the hair cells, which in turn initiate nerve impulses. Because the dimensions and other properties of the cochlea and basilar membrane change from base to apex, traveling waves of different frequencies cause a maximum displacement of the membrane at different levels. High frequencies are detected near the base of the cochlea; low ones register near its apex. Most mammals can detect frequencies as high as 20 kHz.

Bats, particularly microchiropterans, have evolved a very sensitive auditory system that enables them to avoid obstacles in the dark and to find their insect prey by sending out high-frequency sounds and listening to the echoes. Sounds with a frequency greater than 20 kHz are referred to as **ultrasonic** because they are higher than most mammals can detect. A large, ossified larynx allows a high tension to be developed on the vocal cords and, as a result, sounds with frequencies as high as 150 kHz can be generated. The ultrasonic sounds are emitted through the mouth in some species of bats and through the nose in others. Because these high frequencies have very short wavelengths, they have a very high resolution, and such bats can discriminate between objects that are very close together. The ultrasonic sounds are emitted as pulses lasting 1 ms to 4 ms. Pulses can be repeated up to 100 times per second as a bat hones in on an insect. Frequencies drop about an octave during the course of each pulse. Different species of bats listen to different aspects of the echoes: (1) time lag in echo return; (2) decrease in loudness between the emitted sound and its echo; and (3) differences in frequency between the emitted sound and its echo, which are sensed as a "beat note." Considerable independent evolution of these systems has occurred among bats.

Cetaceans have evolved an underwater sonar system that also is based on emitting high-frequency sounds and listening to the reflected echoes. Considerable modification of the ear, originally sensitive to airborne sounds, has occurred so that underwater sounds can be detected. Many cetaceans also have evolved communication systems based on low-frequency "songs" (about 20 Hz) that can travel hundreds and perhaps thousands of kilometers.

Photoreceptors

The ability to detect changes in light is important for nearly all metazoans because their periods of feeding and reproduction and many other aspects of their physiology and behavior are closely attuned to the diurnal cycle and to seasonal changes in day length. The additional ability to detect the source of light can provide information on the approach of predators, location of shelter, and so forth. For those animals with groups of photoreceptive cells numerous enough to discriminate between different light intensities and different wavelengths, and a neuronal circuitry capable of processing these data, light can provide much information about the world. The informational content of a visual

[16]Based on phylogeny, the cochlea of birds is thought to have been derived independently of the cochlea of mammals.

image—its location, size, shape, and movement—can be much greater than that provided by most other senses.

To receive light, an animal must have **photoreceptive cells** containing a **visual pigment** that absorbs quanta of light energy, which initiates chemical changes that generate nerve impulses. The light energy most valuable to vertebrates falls between wavelengths of 380 nm (violet) and 760 nm (red). Shorter wavelengths extending into the ultraviolet quickly are absorbed by water and so are of no value to aquatic animals. Light with such short wavelengths also contains so much energy per quantum that it is potentially destructive to tissues in terrestrial vertebrates. The integument protects deeper tissues from this ultraviolet radiation. In contrast, infrared and longer wavelengths contain too little energy to be biologically useful in most cases, although some vertebrates have special receptors that detect infrared radiation as heat (see section on infrared receptors, known as thermoreceptors, p. 433).

The body surface of most metazoans, including many vertebrates, can detect changes in light, although specific light-sensitive cells have not always been identified (Zimmerman and Heatwolfe, 1990). Image-forming eyes are limited to animals that move about a great deal in well-illuminated environments: many arthropods, cephalopods, and all but a few vertebrates.

Median Eyes

For many vertebrates, the detection of light and the physiological adjustments to changes in light levels and day length are mediated by photoreceptors different from those that form images. In addition to image-forming lateral eyes, anamniotes and many diapsids have one or two light-receptive **median eyes** on top of the head. These develop, as does the retina of image-forming eyes, as embryonic outgrowths from the diencephalic region of the brain.

An adult lamprey has a nonpigmented spot of skin on the top of its head, beneath which lies a **pineal eye** (Fig. 12-23*A*). A second organ, known as the **parietal eye,** lies deep to the pineal eye (Fig. 12-23*A*). A slight right–left asymmetry exists in the position of these eyes. Each is a hollow sphere of cells. The deep cells in both spheres are photoreceptive and partly shielded by pigment. Sensory neurons extend from them to the brain. Experiments on ammocoete larvae of the lamprey show that the median eye complex generates a low level of neuronal activity in dim light that is correlated with the larva's nocturnal activity. Strong light inhibits neuronal activity, and the larva rests during the day. Beyond this, the median eye complex of lampreys is a **neuroendocrine transducer** that translates light signals into

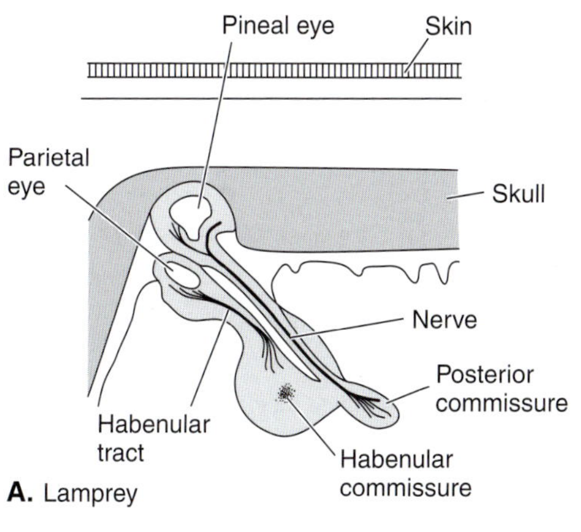

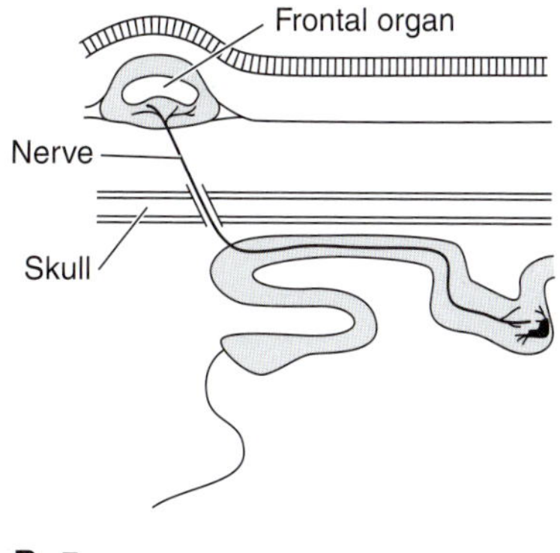

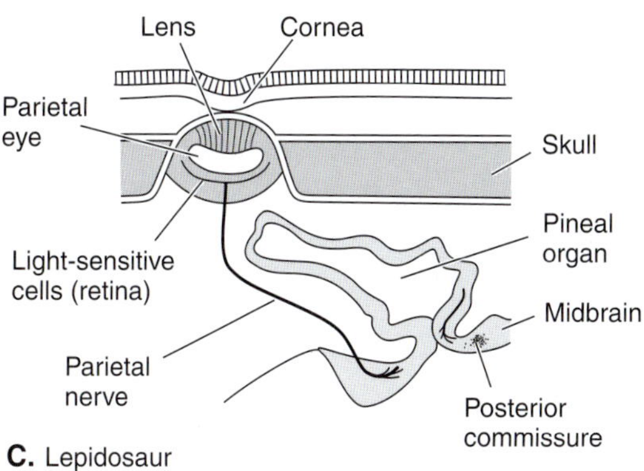

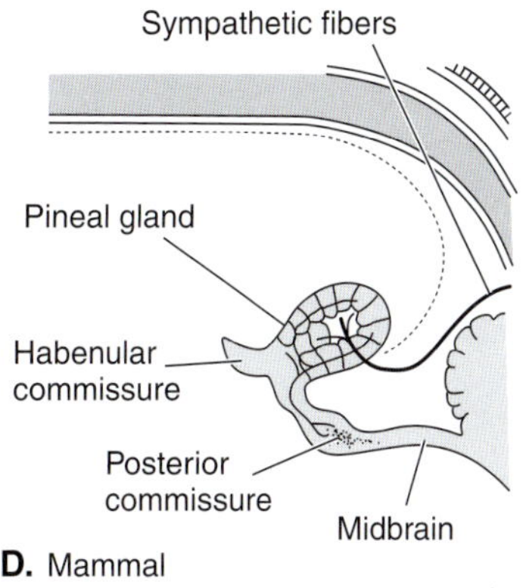

FIGURE 12-23
The median eye complex of representative vertebrates. *A,* Lamprey. *B,* Frog. *C,* Lepidosaur. *D,* Mammal. *(After Studnicka.)*

chemical messages. In the absence of light, the median eye complex produces an enzyme that converts the neurotransmitter serotonin into a hormone **melatonin.** The melatonin is released into the circulation and causes the pigment in the melanophores of the skin to concentrate, and thus the animal blanches in the dark. Light inhibits the formation or activity of this enzyme, pigment becomes dispersed, and the animal is dark during the day. Experiments also show that the median eye complex is essential for metamorphosis and influences the development of sexual maturity in lampreys. All of this indicates that the median eye complex has a role in the daily and seasonal rhythms of lampreys, but the nature of its involvement is not well understood.

Although not universally present, the median eye complex is an ancient and widespread feature of vertebrates. A foramen for it occurs in the skull of many early fossil craniates, such as †osteostracans, †placoderms, early cartilaginous and bony fishes, as well as early tetrapods. Extant species do not always have a foramen for it, but a reduced pineal organ, called the **epiphysis,** is present in chondrichthyans and most actinopterygians. Tadpoles and frogs have a small **frontal organ,** which represents the parietal eye (Fig. 12-23*B*). *Sphenodon* and many squamates have a well-developed parietal eye complete with cornea-like and lenslike structures and a reduced **pineal organ** beneath the skull roof (Fig. 12-23*C*). The parietal eye monitors the level of solar radiation and affects the animal's orientation to the sun and its movements into the sun or shade. The complex is represented in birds and mammals by an endocrine **pineal gland,** the activities of which also are affected by light even though it is located deeply below the skull roof (Fig. 12-23*D* and Chapter 15).

The Structure and Function of Image-Forming Eyes

Except for a few species that live in dark habitats—some deep-sea fishes and many cave-dwelling or burrowing vertebrates, the eyes of which have been secondarily reduced—all vertebrates have a pair of image-forming eyes. The eyes are essentially little biological cameras. They usually are located on the side of the head in such a position that little overlap exists in the left and right visual fields. The eyes are directed more rostrally in some birds and many mammals, so that the two visual fields overlap to a greater extent. A high degree of overlap of the visual fields provides depth perception, or **stereoscopic vision,** in many mammals.

Many important structures are found in the orbit around each eyeball. **Extrinsic ocular muscles** (Fig. 10-11) attach to the eyeball and control its movements. Eyes must be directed at objects in the visual field despite movements of the body and head, and the movements of the two eyeballs usually are synchronized very precisely. **Tear glands** and **movable eyelids** moisten and protect the surface of the eyes in tetrapods (p. 431).

Although the eye varies greatly in adaptive details among vertebrates, its basic structure is the same in all. The human eye is representative of the design typical for a tetrapod (Figs. 12-24 to 12-27). Its wall consists

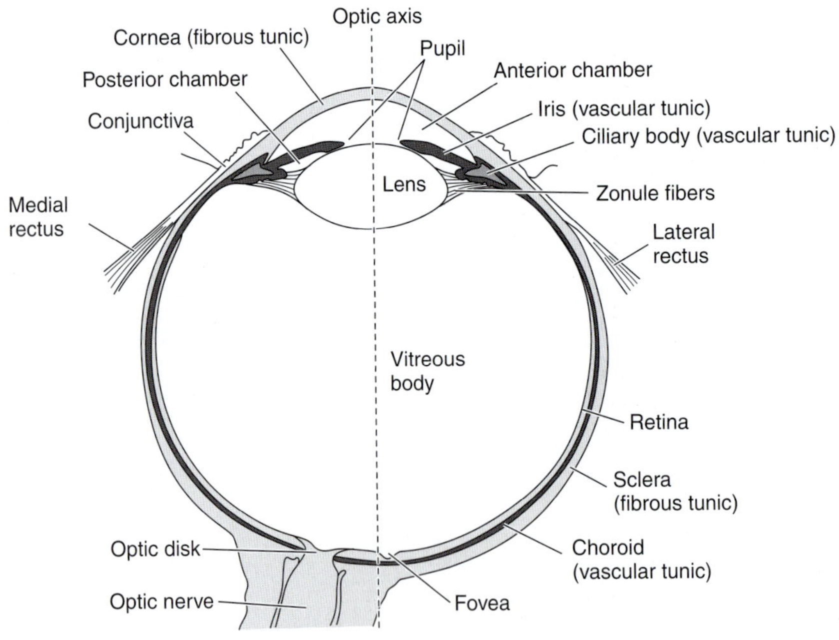

FIGURE 12-24
A mammalian eye.

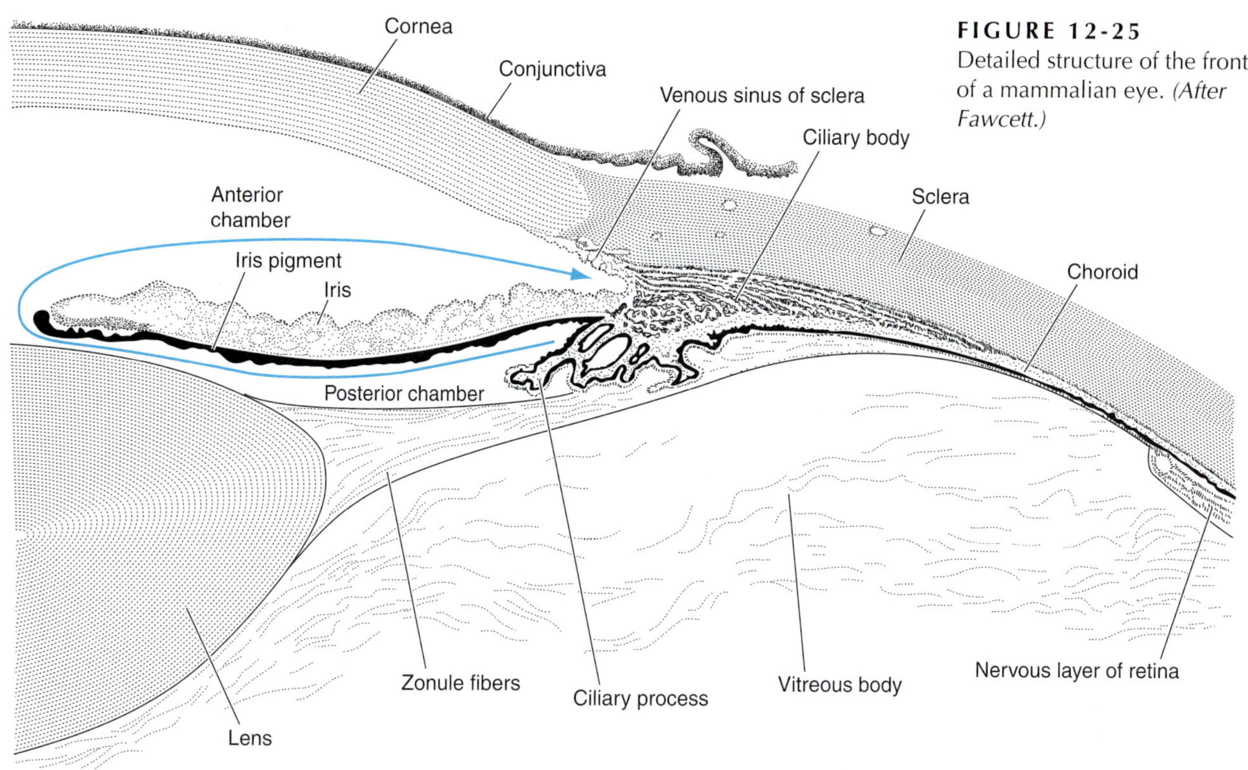

FIGURE 12-25
Detailed structure of the front of a mammalian eye. *(After Fawcett.)*

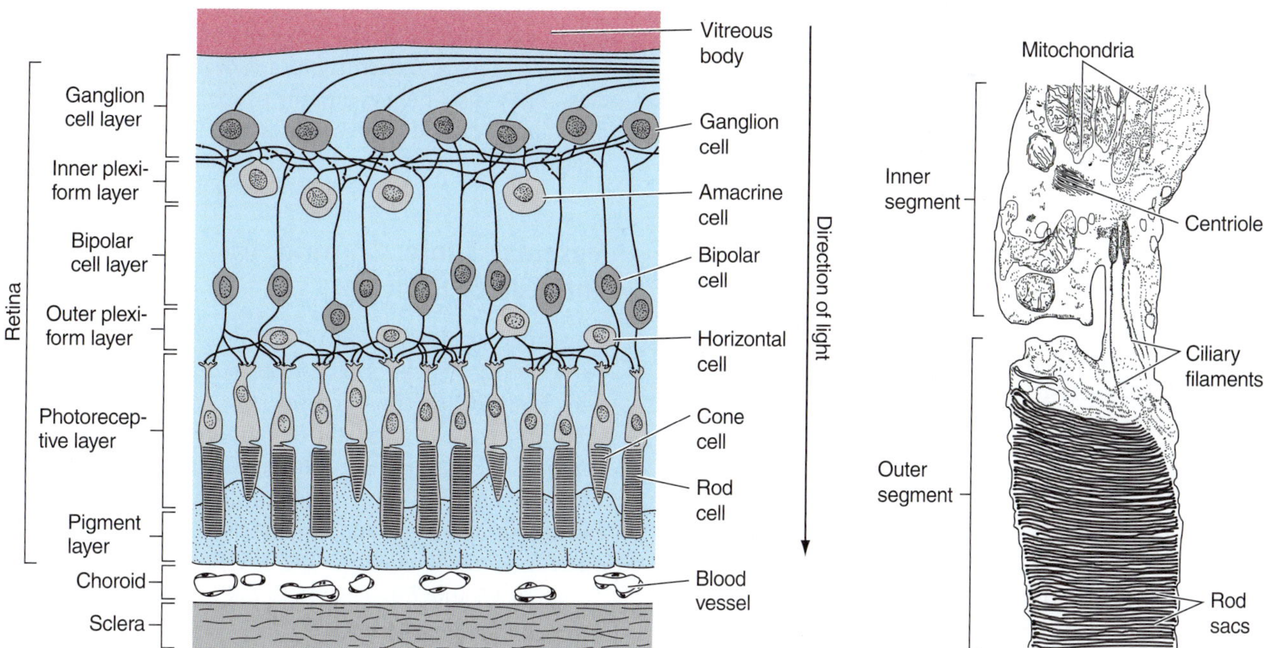

A. Layers of the retina in relation to the sclera, choroid, and vitreous body

B. Ultrastructure of a portion of a rod cell

FIGURE 12-26
Retinal structure. *A,* The layers of the eyeball and the connections among cells within the retina. *B,* An enlargement of the junction between the outer and the inner segment of a rod. *(A, After Dowling and Boycott. B, After Giese.)*

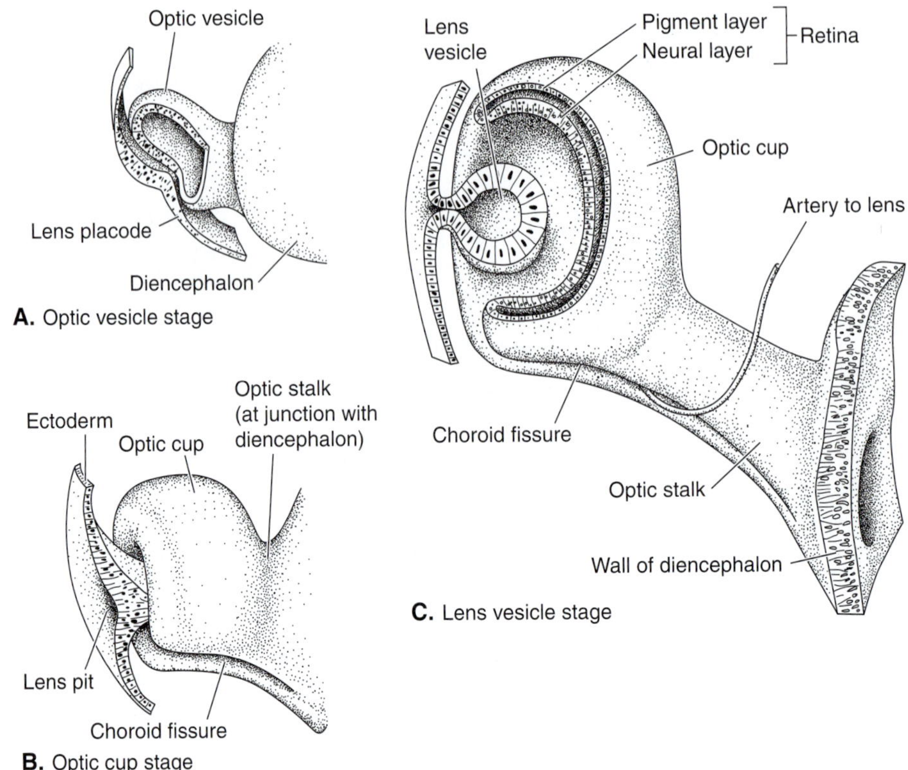

FIGURE 12-27
The development of the eye. *A*, Optic vesicle stage. *B*, Optic cup stage. *C*, Lens vesicle stage. *(After Arey.)*

of three layers of tissue: an outer supportive **fibrous tunic,** a middle nutritive **vascular tunic,** and an inner **retina** containing the photoreceptive cells. Part of the fibrous tunic is modified as a **cornea,** which allows light to enter and which, in tetrapods, plays a significant role in bending the light to focus on the retina. Parts of the vascular tunic and retina are modified to form the **iris,** which controls the amount of light entering the eye, and the **ciliary body,** which supports and focuses the **lens.** A watery **aqueous humor** fills the spaces in the eye in front of the lens, and a more gelatinous **vitreous body** (L., *vitreus* = glassy) lies behind the lens. As light passes through the different parts of the eye, it is bent, or refracted, toward the optic axis and casts an inverted image on the retina.

The Fibrous Tunic

The fibrous tunic is a dense connective tissue that forms the essential supportive framework for the eyeball. Most of it is opaque and white and is known as the **sclera** (Gr., *skleros* = hard). This is the "white" of our eyes. The fibrous tunic also forms the **cornea** (L., *corneus* = horny) at the front of the eye. The cornea is avascular and otherwise modified to facilitate the passage of light. A delicate epithelial layer known as the **conjunctiva** covers the surface of the cornea, turns onto the inner surface of the eyelids, if they are present, and is continuous with the surface layers of the skin. A ring of cartilage or **sclerotic bones** develop in the sclera of many vertebrates. Usually, the sclerotic bones are located at the level of the lens, where they support and strengthen the point of origin of lens muscles. They also help maintain eyeball shape, especially in some birds and other species that have nonspherical eyeballs.

Vascular Tunic: Choroid, Iris, and Ciliary Body

The next layer, the vascular tunic, is richly supplied with blood vessels and also contains some pigment. This layer, called the **choroid** (Gr., *chorioeides* = like a membrane), in the posterior part of the eyeball helps nourish the underlying retina. Choroid folds sometimes extend into the vitreous body. They form the **falciform process** in some teleosts, the **papillary cone** in some squamates, and the **pecten** in birds (Fig. 12-28*A*, *C*, and *D*). The pecten is sometimes elaborate and may contain as many as 30 folds. The function of these choroid folds is not entirely clear, but if the pecten is destroyed, then the oxygen content within the eye decreases greatly. Some investigators also postulate that by casting a shadow on the retina, these choroid processes help the animal detect the movement of an image across the retina. Pigment in the

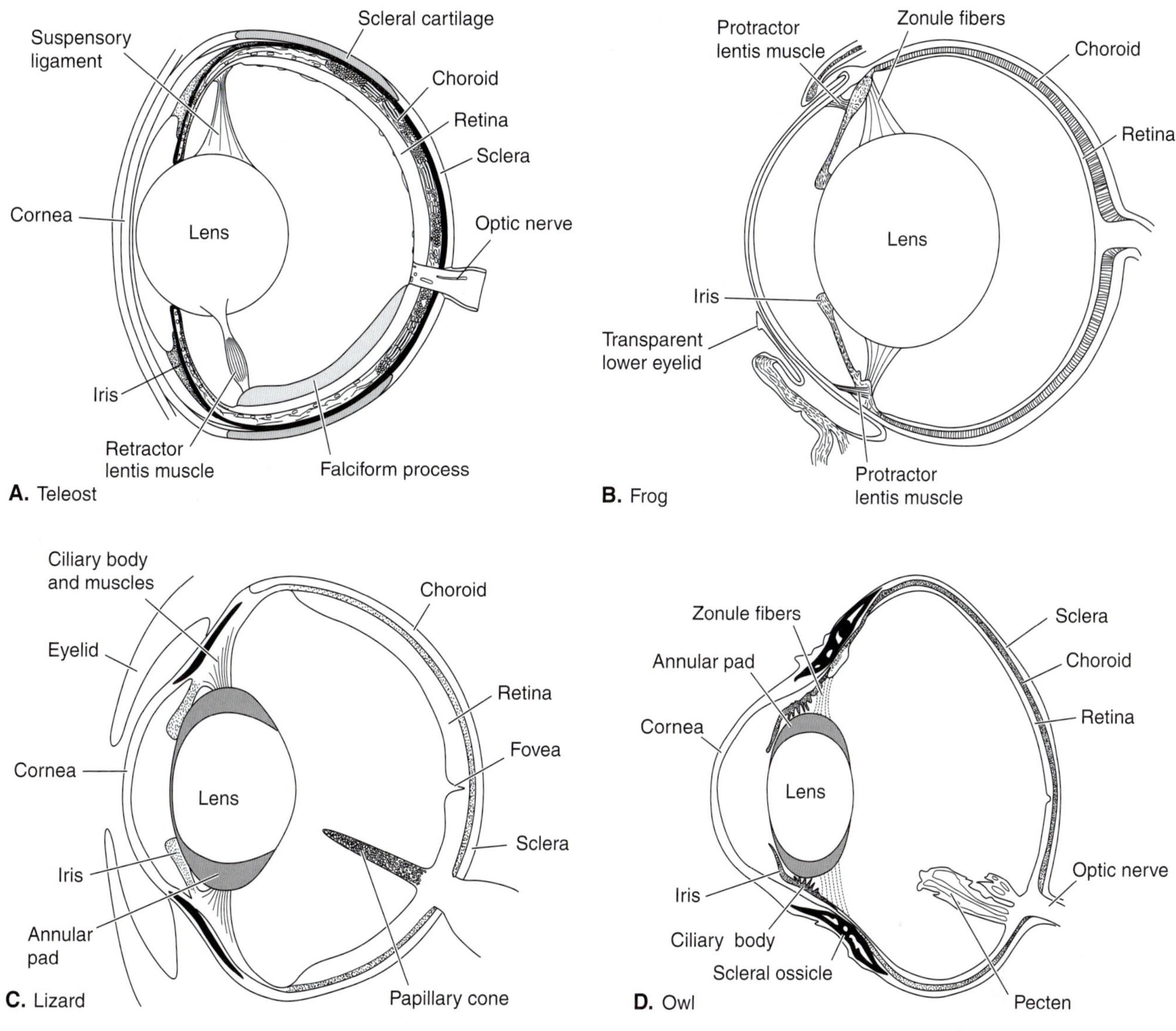

FIGURE 12-28
The eyeballs of representative vertebrates. *A,* A teleost. *B,* A frog. *C,* A lizard. *D,* An owl. *(After Walls.)*

choroid and in the adjacent pigmented layer of the retina prevents a scattering of light and a blurring of the image.

It is advantageous that as much light as possible reaches the photoreceptors of the retina in vertebrates that are active in dim light, such as many chondrichthyans, actinopterygians, crocodilians, and mammals. Many of these animals have a **tapetum lucidum** (L., *tapete* = carpet + *lucidus* = shining) behind part of the retina; it reflects light that has passed through the photoreceptors back onto them. The photoreceptor cells thus are stimulated adequately, but the tradeoff is some blurring of the image. The tapetum may be located in the choroid or in the adjacent part of the retina. The reflective layer can be composed of specially arranged collagen fibers, extracellular plates of guanine, or intracellular purine crystals. The **eyeshine** of nocturnal animals caught in a flashlight or headlights is light that has been reflected by the tapetum.

The vascular tunic and accompanying non-nervous retinal tissue continue beside the lens and turn in front of it to form the **iris** (Gr., *iris* = rainbow). Muscle fibers within the iris regulate the size of its opening, called the **pupil,** and hence the amount of light passing through the lens (Fig. 12-24). These iris muscles, of retinal origin, come from ectoderm and usually are smooth muscle fibers. Both circularly arranged sphincter and radially arranged dilator muscle fibers are present in the iris of most vertebrates. The diameter of the pupil is under autonomic control, meaning that it opens and closes in response to signals from the sympathetic and parasympathetic neurons that innervate it

(Chapter 13). The pupil usually is a round opening, but its shape is slitlike in many vertebrates, such as cats and gekkonids, which are active under a wide range of ambient light levels. The halves of the iris on either side of the slit can be drawn far apart, like curtains, in dim light and almost completely closed in bright light to protect the very sensitive retina. The "bunching up" of contracting tissues around a circular pupil prevents it from being closed as completely.

The vascular tunic and associated non-nervous retinal tissue adjacent to the lens form the **ciliary body.** Delicate **zonule fibers** extend from it to the lens, which they help hold in place (Fig. 12-25). Smooth muscle fibers of retinal origin in the ciliary body focus the lens in different ways in different groups of vertebrates. **Ciliary processes** of the ciliary body also secrete the lymphlike aqueous humor into the eye's **posterior chamber,** the space between the lens and the iris (Fig. 12-25). From here, the aqueous humor flows into the **anterior chamber,** located between the iris and the cornea (Fig. 12-25, *blue arrow*). It finally drains into the bloodstream through the **venous sinus of the sclera,** which encircles the eye at the junction of the iris and cornea. The aqueous humor nourishes the avascular cornea and lens and creates intraocular pressure that helps maintain the shape of the eye.

The Retina

The retina is the third tissue layer of the eyeball. Its most peripheral portion and its **pigment layer** become associated with the choroid, ciliary body, and iris, as we have seen. Its photoreceptive layer and nervous portion are limited to the posterior part of the eyeball and consist of multiple layers of cells (Fig. 12-26*A*). Because of the way the eye evolved and develops embryonically (p. 429), light must pass through all of the neuron layers of the retina and through the bases of the rods and cones to reach the photoreceptive layer. The **rod** and **cone cells** lie deep in the retina, next to the pigment layer. In many nonmammalian vertebrates, processes of the pigment cells extend around the receptive regions of the rod and cone cells. Pigment migrates into the processes during bright illumination, thereby partly shielding the cells, and withdraws when light levels fall. Four neuronal layers lie between the rod and cone cells and the vitreous body. These neural layers are an **outer plexiform layer** composed of horizontal cells, a **bipolar cell layer,** an **inner plexiform layer** formed by amacrine cells, and a **ganglion cell layer** next to the vitreous body. Axons of the ganglion cells course along the surface of the retina and turn inward at the **optic disk** to form cranial nerve II, the **optic nerve** (Fig. 12-24). Because no photoreceptors are present in the optic disk, this is a **blind spot.**

The **outer segment** of a photoreceptive cell is considered to be a modified cilium because it is connected to the rest of the cell by a narrow, cilium-like stalk (Fig. 12-26*B*). The stalk contains the characteristic nine peripheral microtubules but lacks the central two. Membranes within the outer segments of the rods and cones contain photoreceptive **pigments** that absorb light energy and convert it into chemical energy. This in turn changes the cell's membrane potential and affects the amount of neurotransmitter released by the cell (Chapter 13). Photoreceptive pigments in vertebrate eyes consist of a protein, known as an **opsin,** which is combined with an aldehyde of vitamin A_1 or A_2, called **retinene.** The proteins vary in different pigments and determine the spectrum of light energy that is absorbed. The pigment in the rods, known as **rhodopsin,** or **porphyropsin** in some species (or in some life stages of some species that metamorphose from aquatic to terrestrial environments, such as certain lissamphibians), absorbs light over a wide spectral range, but the range varies with the species. Cones contain pigments, known as **iodopsins,** with more restricted bands of spectral absorption. Multiple pigments, sometimes in one cell but more often in different cells, are a prerequisite for color vision. Mammals with color vision have three types of cones, each having an absorption maximum in a different part of the spectrum: 450 nm (blue), 525 nm (green), and 550 nm (red). Retinal photopigments that are sensitive to ultraviolet light as well as to color have been found in many vertebrates, especially teleosts and birds. Studies indicate that many of these animals can detect ultraviolet light, but the adaptive significance of this is not clear (Jacobs, 1992).

Rods are insensitive to different colors, but they are very sensitive to light and have a very high degree of **intraretinal convergence** (i.e., a great many rods synapse with a few bipolar cells, and these synapse with still fewer ganglion cells). The human retina contains approximately 120 million rods and 5 million cones, but there are only about 1 million axons of ganglion cells that form the optic nerve; therefore, a small amount of light falling on many rods in the same convergent pathway can **summate** via bipolar cells to generate nerve impulses in the ganglion cells. The tradeoff, however, is a decrease in resolution or **visual acuity** because a distinction cannot be made between spots of light falling on the same convergent pathway. Rods can function in dim light, but they cannot distinguish colors, and rod vision is slightly blurred. Cones distinguish colors but require brighter illumination. They have far less convergence and so form much sharper images.

The eyes of humans and many other vertebrates contain both rods and cones, but these receptors are not uniformly distributed. A small central area of the retina has a high concentration of cells with little convergence, and the **fovea** within it contains only cones. This is the region of highest visual acuity, and we adjust our eyes so that images fall here when we wish to discriminate fine points of light. The retina is relatively thin in the fovea, so its photoreceptors lie in a little pit. The pit may itself act as a diverging lens and further enhance visual acuity. The number of cones decreases rapidly from the fovea toward the periphery of the retina, whereas the number of rods increases to a maximum about 20° from the fovea.

Many additional types of interconnections occur within the retina. The retinal regions from which ganglion cells receive signals partly overlap. Horizontal connections among rods, cones, and bipolar neurons are made by the horizontal cells in the outer plexiform layer. Horizontal interconnections among bipolar and ganglion cells are made by the amacrine cells of the inner plexiform layer. Some of these interconnections inhibit and others facilitate the transmission of impulses. The significance of these arrangements is not entirely clear, but some sharpen image boundaries and facilitate the detection of motion. Still, it is clear that much signal processing occurs in the retina—not unexpectedly, because the retina is a part of the brain.

The Origin and Development of the Eye

How did an organ as complex as the eye evolve? Why does light pass through so many cell layers before reaching the photoreceptors? Both phylogenetic and developmental clues can be found. For example, an amphioxus has ciliated, photoreceptive cells in the lining layer of the lumen (known as the **ependymal layer**) of its central nervous system. Most lie on the floor or lateral walls of the lumen and they are particularly numerous anteriorly. Although shielded by a pigment spot from light stimuli coming from beneath the animal, they can be activated by light from above because light easily passes through the thin integument and body wall. These photoreceptive cells probably are homologous to the vertebrate rods and cones because the rods and cones also develop embryonically from ciliated **ependymal epithelial** cells that lie in the lining of the neural tube.

As the ancestors of vertebrates became larger and more active animals, we may hypothesize that it was advantageous to concentrate photoreceptive cells near the anterior end of the body, in the lumen of part of the evolving brain. An evagination of this region would bring these cells closer to the body surface. The advantage of having the photoreceptive part of the brain nearer the body surface would be enhanced were adjacent tissues modified to transmit light. The cornea and lens may have evolved in this way.

This scenario is hypothetical, but an essentially similar sequence of events occurs during the embryonic development of the vertebrate eye (also see Chapter 4). The eye initially develops as a single median evagination[17] of the diencephalon that soon bifurcates to form the paired **optic vesicles** (Fig. 12-27). As each optic vesicle grows toward the body surface, its proximal part narrows as the **optic stalk,** and its distal part invaginates to form a two-layered **optic cup.** The **choroid fissure** lies on the ventral surface of the stalk and carries blood vessels for the eye. The outer layer of the optic cup becomes the pigment layer of the retina, whereas the inner layer differentiates into the photoreceptive cells and neuronal layers of the retina. The outer, photoreceptive segments of the rod and cone cells develop from cilia of the ependymal epithelium lining the optic cup. Because of the way the optic cup develops, these cilia are directed toward the lumen. After the lumen has narrowed later in development, the receptive segments of the rods and cones lie next to the pigment layer of the retina, and light must pass through the nervous layers of the retina to reach them.

The optic cup induces the overlying surface ectoderm first to thicken as a **lens placode** and then to invaginate and form a **lens vesicle** that differentiates into the lens. Adjacent mesenchyme encapsulates the lens and optic cup to form the fibrous and vascular tunics. Tissue from the ectodermal optic cup contributes to the iris and ciliary body and forms their muscles. Embryonic blood vessels that cross the vitreous body to supply the developing lens atrophy after birth; otherwise, they would cast shadows on the retina.

Evolutionary Adaptations of the Eye

Eyes and their neuronal projections offer one of the best anatomical predictors of vertebrate behavior and function, so it is important to consider ways in which the eyes have become modified in different groups.

[17]Thus cyclopia (an anomalous condition in which a single median image-forming eye is present) is actually the default developmental condition and will result if bifurcation of the median evagination fails to occur.

The Eyes of Lampreys, Chondrichthyans, and Actinopterygians

Lampreys, chondrichthyans, and actinopterygians live in aquatic environments, and water continuously bathes and cleanses the corneal surface, so that tear glands are unnecessary and never evolved. Most species lack movable eyelids, but a few have stationary skin folds above and below the eye. In some chondrichthyans, a **nictitating membrane** is present in the medial corner of each eye (see Fig. 12-30*A*). It closes to protect the eye during close contact with prey. Many species of teleosts, particularly those that swim at high speeds, such as mackerels (Scombridae), have fared the eyes into the surface of the head using clear **adipose eyelids** (Fig. 12-30*B*). Adipose eyelids help to maintain streamlining of the head by limiting the drag that would be induced were the eye to project out from the head's surface. Light levels change more slowly in most aquatic environments than on land, and the pupillary response of lampreys, chondrichthyans, and actinopterygians is correspondingly slower than that of tetrapods. Dilator fibers are absent in the iris of many species, so that the pupil slowly expands when the sphincter muscles relax.

Because the refractive index of water is close to that of the cornea, light waves pass through the cornea of a lamprey, chondrichthyan, or actinopterygian without being substantially bent or refracted. The lens, therefore, is the primary refractive structure in the eye, and it must be thick—nearly spherical—to provide adequate refraction (Fig. 12-28*A*). The spherical lens is composed of modified layers of epithelial cells. Successive layers have different curvatures and slightly different optical properties, so that light is gradually refracted toward the optic axis as it passes through them. This avoids the image distortion due to spherical aberration that would occur were the lens a homogeneous sphere.

Differences in the shapes of the lens of a teleost and a mammal and their relationship to vision under water are shown in Figure 12-29*A–D*. A teleost's eye cannot

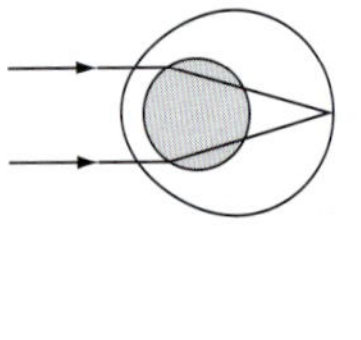

A. Teleost eye in water

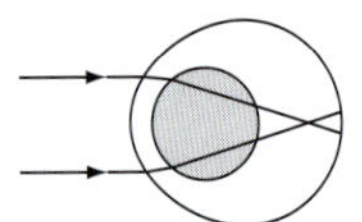

B. Teleost eye in air

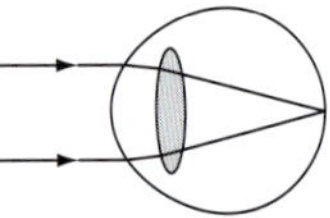

C. Mammal eye in air

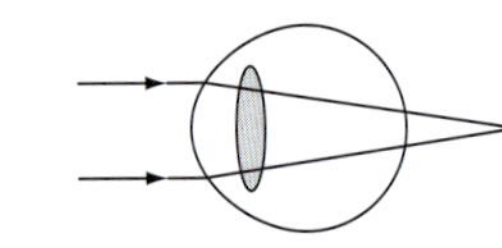

D. Mammal eye in water

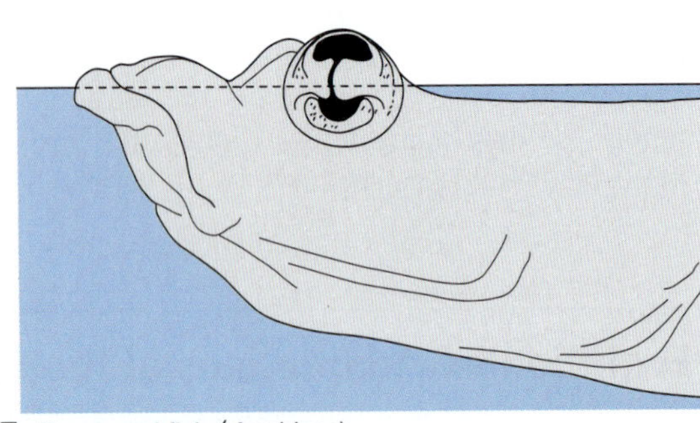

E. Four-eyed fish (*Anableps*)

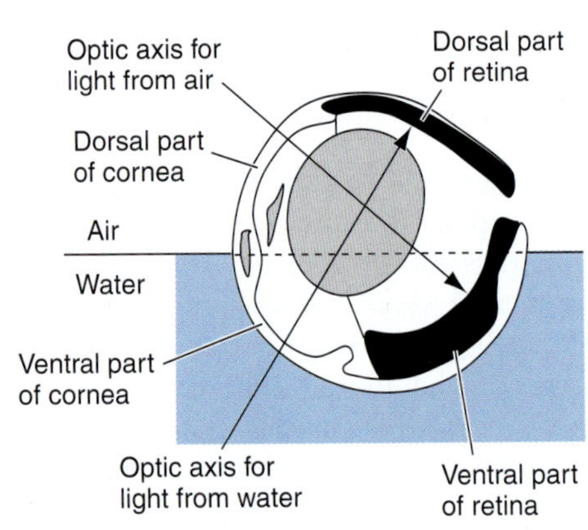

F. Eye of *Anableps*

FIGURE 12-29
Eyes adapted for aquatic versus terrestrial vision. *A,* The eye of a typical teleost in water. *B,* The eye of a typical teleost in air. *C,* The eye of a typical mammal in air. *D,* The eye of a typical mammal in water. *E,* The four-eyed fish (*Anableps*) lives at the water surface and has an eye specialized to see both above and below water. *F,* Detail of the eye of *Anableps*; note the barrel-shaped lens. Light passing into the eye from the air above the fish passes through the short axis of the lens and focuses on the ventral portion of the retina. Light passing into the eye from the water beneath the fish passes through the long axis of the lens and is focused on the dorsal portion of the retina.

focus properly in air because of the additional refraction that occurs when light passes from the air through the cornea. At the opposite extreme, a terrestrial mammal's eye cannot focus properly under water because refraction by the cornea is decreased, which explains why humans need the air space created by a diving mask in front of their eyes to enable them to focus on objects under water.

Four-eyed fishes in the genus *Anableps* provide a particularly clear example of the differences between an eye designed to work in water and an eye designed to work in air (Fig. 12-29*E* and *F*). The eye has two pupils—one directed above the water's surface and the other below. Light passing from the air through the dorsal part of the cornea is refracted and passes through the thin axis of the lens to focus on the ventral part of the retina. Light passes from the water through the ventral part of the cornea, is refracted by the thicker part of the lens, and is focused on the dorsal part of the retina. Thus, a single lens with optical axes of two different lengths is able to focus light from both air and water.

Lampreys, chondrichthyans, and actinopterygians tend to be nearsighted because the lens is close to the cornea and far from the retina. To focus sharply on a distant object, muscles move the lens closer to the retina. A lamprey **accommodates** to (= focuses on) a more distant object by contracting a corneal muscle that pulls the cornea inward and pushes the lens back. Lens-to-retina distance in an aquatic animal is analogous to lens-to-film distance in a camera. (To focus on a near object, the lens-to-film distance is increased by moving the lens away from the film; to focus on a distant object, this distance is decreased). Appropriately for a large predator, the eyes of a shark at rest are focused on somewhat more distant objects. To see a nearer object, an ectodermal **protractor lentis** muscle pulls the lens away from the retina. The eye of an actinopterygian has a small **retractor lentis** muscle (Fig. 12-28*A*) of ectodermal origin that attaches to the lens and pulls it toward the retina. In all three of these cases, intraocular pressures restore the lens to its resting position when the focusing muscle relaxes.

Light intensity is low in many bodies of water, and long wavelengths of light, especially reds and oranges, are readily absorbed. Correlated with this, the retina of many chondrichthyans and actinopterygians consists primarily of rods. The maximum sensitivity for pigments in the photoreceptive cells of teleosts living in marine coastal and fresh water is in the green-yellow range, corresponding to the wavelengths of the maximum available light. Only blue light penetrates far into the water, and species living in deep oceanic waters have a photoreceptive pigment that absorbs maximally in the blue part of the spectrum. Only those species living in brightly illuminated habitats, such as coral reefs or clear, freshwater lakes, have many cones. As would be expected, many chondrichthyans and actinopterygians have a tapetum lucidum that reflects light back onto the retina.

The Eye of Terrestrial Vertebrates

Because they are not constantly bathed in water, the eyes of terrestrial vertebrates must be protected and kept moist in other ways. A tetrapod's eye usually has one or more **eyelids** that can move across its surface and protect and cleanse it. The eye of lissamphibians has a stationary upper eyelid but a movable and transparent lower one (Fig. 12-28*B*). Lissamphibians also can retract the eyeball deeper into the orbit using the retractor bulbi muscle so that the eyelids completely close over it. In amniotes, both upper and lower lids are movable, and a third, transparent eyelid, the **nictitating membrane** (L., *nictare* = to wink), is usually present.[18] The nictitating membrane normally is retracted into the median corner, where upper and lower lids come together, but it can be flicked across the surface of the cornea. In humans, it is reduced to a vestigial **semilunar fold. Eyelashes** are associated with the eyelids of mammals. Tetrapods also have evolved **lacrimal, harderian,** or other **tear glands,** the secretions of which flow across the cornea or are spread across it by the movement of the eyelids (Fig. 12-30). The tears of mammals drain into a **lacrimal duct** that opens into the nasal cavity. Modified sebaceous glands, known as **tarsal glands,** open on the edges of the eyelids (Fig. 12-30*C*); their waxy secretions help prevent the tears from flowing onto the face.

Because its index of refraction is much greater than that of air, the cornea of a terrestrial vertebrate's eye refracts incoming light and plays an essential role in focusing light on the retina. Because of the cornea's role in refraction, a tetrapod's lens performs less refraction than does the lens of a teleost eye, and it is both thinner along the optic axis and less curved (e.g., Figs. 12-29*A* and *C*). For example, in lissamphibians, the lens changes during metamorphosis from a spherical shape in the aquatic larva to a more oval shape in the adult (Fig. 12-28*B*). The lens is even thinner in amniotes (Fig. 12-28*C* and *D*). Distant vision is important for most terrestrial vertebrates, and the lens is positioned relatively close to the retina, so that distant objects are in focus when the eye is at rest.

[18]The nictitating membrane of amniotes is not homologous to the nictitating membrane of sharks.

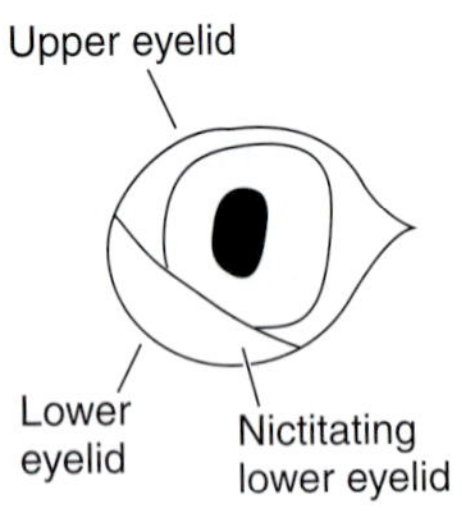

A. Shark eyelids

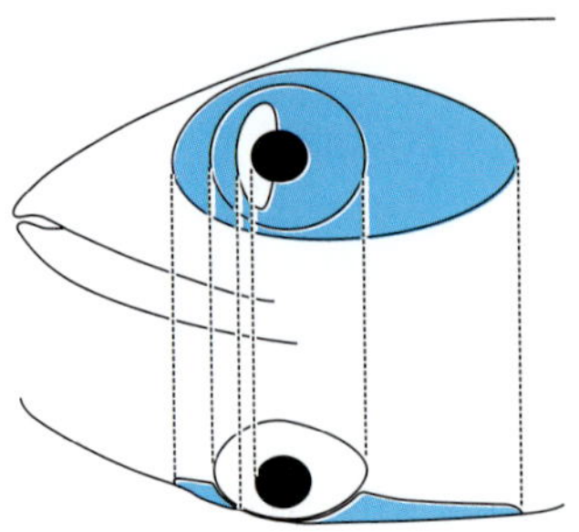

B. Adipose eyelid of a mackerel

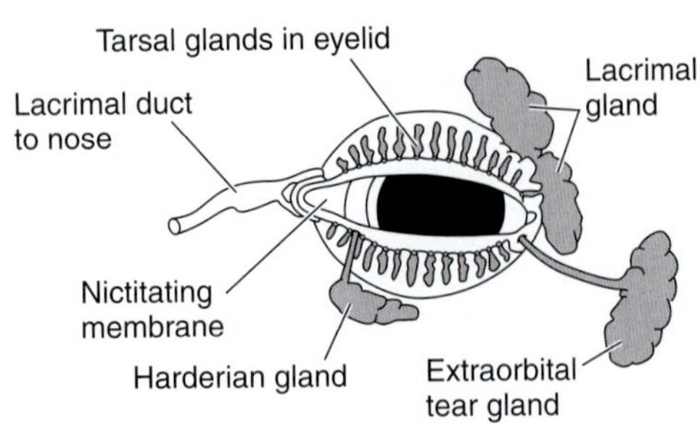

C. Eyelids and glands of mammalian eye

FIGURE 12-30
Eyelids and accessory structures. *A,* The stationary upper and lower eyelids and nictitating membrane of the eye in a shark. *B,* The adipose eyelid of a mackerel is composed of clear tissue. It allows the head to be streamlined without limiting the field of view. (The top view is a lateral view, and the lower figure is a schematic section through the eye and adipose eyelid.) *C,* The lacrimal apparatus of a mammal. *(After Portmann.)*

Different groups of tetrapods have different methods of accommodation. Lissamphibians accommodate for a close object by contracting the **protractor lentis** muscle (Fig. 12-28*B*), which pulls the lens away from the retina and toward the cornea. In contrast, amniotes have elastic lenses and accommodate for near objects by changing lens shape; changing lens shape is analogous to changing lenses on a camera. Except in snakes, the lens of diapsids is encircled by an **annular pad** that lies adjacent to the ciliary body (Fig. 12-28*C*). During accommodation for close objects, a sphincter-like muscle within the ciliary body exerts a force on the annular pad, causing the lens to bulge, its refractive power to increase, and the image to be brought into focus. We think that the eyes of snakes partially degenerated during a period when their ancestors lived as burrowing animals, and then re-evolved as snakes readapted to live on the ground surface. Unlike other squamates, snakes lack ciliary muscles and accommodate for distant objects by the contraction of iris muscles that push the lens deeper into the eyeball.

Flying birds have a keen sense of sight and can quickly adjust to changes in light levels and distance. A bird's eye is extraordinarily large and is strengthened by a ring of **sclerotic bones,** also known as **scleral ossicles** (Fig. 12-29*D*). The eyes of some hawks and owls are larger in absolute size than ours despite their relatively much smaller body size. A bird's iris muscles are striated rather than smooth and respond rapidly to changes in levels of light.[19] Birds retain the annular pad mechanism for accommodation, but their ciliary muscles also act in other ways. One set of muscles pulls the lens forward into a narrowing ring of sclerotic bones, so that the lens bulges; another set pulls on the cornea to slightly change its curvature. The sclerotic bones provide a firm point of origin for these muscles.

In a mammalian eye at rest, intraocular pressure tends to push the wall of the eyeball peripherally, and this force is transferred to the elastic lens by the ciliary body and zonule fibers (Fig. 12-24). The lens is under tension and somewhat flattened, so that distant objects are in focus on the retina. When tension is released, the lens bulges slightly, and nearer objects come into focus. The lens is restored to its resting shape by the relaxation of ciliary muscles and the action of intraocular pressure.

Because levels of illumination can be much higher on land than in water, terrestrial vertebrates frequently have many cones and correspondingly high visual acuity, and many tetrapods have color vision. Most species of lissamphibians, which are secretive and shade-dwelling creatures, have few cones in their retinas, but some have color vision. Cones are abundant in the retinas of squamates and birds, and most species have high visual acuity and well-developed color vision. Diurnal and predacious hawks have exceptionally keen sight. Their eyes are directed forward, so that the two visual fields overlap to a large extent, but not completely. Two foveae—a central one on the optic axis, and one placed more laterally—contain as many as a million cones per square millimeter. Images of objects that are straight ahead of a hawk fall on the central fovea of each eye, and concurrent images of objects to either side fall on the lateral foveae. In contrast to the diurnal birds, early mammals were nocturnal creatures, with few cones but many rods, and many mammals do not have color vision. Color vision evolved independently in primates and a few other primarily diurnal groups of mammals.

[19]This is also true in a few squamates.

Thermoreceptors

Vertebrates can detect thermal changes in their environment. Such an ability is critical for thermal regulation in endothermic birds and mammals, but the mechanism for detecting environmental temperature changes is not well understood. Neurophysiologists have demonstrated small areas of mammalian skin that are sensitive to heat or cold, and certain cells in the mammalian hypothalamus respond to changes in blood temperature. Some investigators propose that specific encapsulated nerve endings in the skin are dedicated hot or cold receptors. Others believe that temperature changes are detected by the combination and pattern of stimulation of encapsulated and free neuron endings.

Specialized infrared detectors are present in some groups of snakes. In boa constrictors and pythons, the infrared receptors are in a series of shallow pits on the scales bordering the mouths. Independently, the rattlesnakes and other pit vipers evolved a pair of deep **pit organs** (hence the name "pit viper") on the face between the eyes and nostrils (Fig. 12-31).[20] Highly branched free neuron endings with many mitochondria lie in a delicate membrane within the pit. These pits are sensitive to slight differences in infrared radiation between the warm-blooded prey of the snake and the prey's surroundings. If the pits are not obstructed, a rattlesnake can strike at a bird or mouse with incredible accuracy, even in the dark.

[20]These infrared-detecting pit organs of pit vipers are not homologous to the pit organs of the lateral line system.

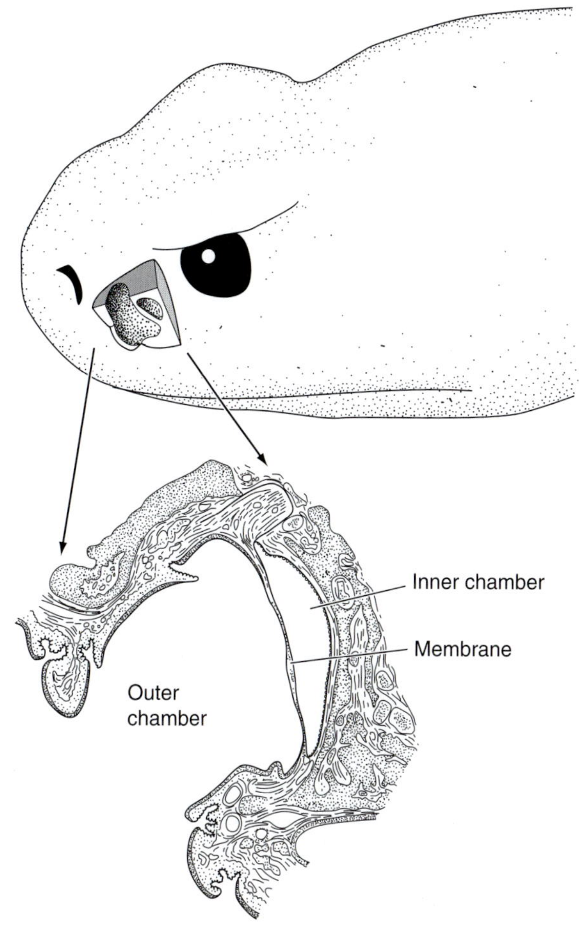

FIGURE 12-31
A section through the pit organ of a rattlesnake. *(After Gamow and Harris.)*

SUMMARY

1. Free nerve endings, receptor cells, and aggregations of receptor cells called sense organs provide vertebrates with essential information about changes in their internal and external environments.
2. The sense of smell, or olfaction, is detected by olfactory neurons. These cells are located in the nasal cavities, and their processes to the brain constitute the olfactory nerve. Vertebrates differ in the ways that they move water or air across these cells.
3. Most terrestrial vertebrates have a distinct group of olfactory cells that form vomeronasal organs (Jacobson's organs). Often, pheromones and other odors of importance in the animal's social interactions are detected by this mechanism. The terminal nerve also may be part of a system for detecting pheromones dealing with reproduction.
4. Gustatory receptors are taste buds of endodermal or ectodermal origin that are usually located within the oral cavity and pharynx. They are supplied by the facial, glossopharyngeal, and vagus nerves. Taste buds may be distributed widely over the body surface in some species of teleosts.
5. The skin of vertebrates contains many free neuron endings that are stimulated by injuries. In addition, mammals have specialized touch and pressure receptors in the skin.
6. Proprioceptors include tendon organs and muscle spindles. They monitor the force, the degree, and the rate of contraction of skeletal muscles. They provide information by which these factors are adjusted during muscle contraction.

7. The lateral line system of anamniotes (including larval lissamphibians) consists of linearly arranged neuromasts usually located within canals in the skin. They provide information on water currents, the animal's own movements, and low-frequency vibrations. Lateralis fibers return to the brain in as many as six lateral line nerves, which usually are associated closely with the facial, glossopharyngeal, and vagus nerves (Chapter 13).
8. The electroreceptors found in chondrichthyans and many osteichthyans are modified parts of the lateral line system. These fishes can sense the electric activity of the muscles of other animals as they search for prey; some others also generate and receive electric pulses in electrolocation and electrocommunication systems.
9. The inner ear is closely related to the lateral line system. It consists of a membranous labyrinth that is surrounded by perilymph and lodged in an osseous labyrinth in the otic capsule. The mechanoreceptive hair cells of the inner ear lie within the membranous labyrinth.
10. Certain groups of hair cells within the membranous labyrinth detect the movements and orientation of an animal in space (equilibrium), and others detect sound waves.
11. A sound source in water generates low-frequency particle displacement waves and higher frequency pressure waves. Many fishes detect low frequencies primarily by the lateral line; they detect higher frequencies by the sacculus. Ostariophysans use the swim bladder and their specialized set of Weberian ossicles as a hydrophone for detecting high-frequency pressure waves.
12. Low-frequency sound waves can be detected by tetrapods through superficial skull bones, but a tympanic membrane beside an air-filled middle ear cavity is needed to convert pressure waves above 1000 Hz into displacement waves. Movements of the tympanic membrane are transmitted by one or more auditory ossicles across the middle ear to the inner ear. Here they set up waves in a special perilymphatic channel that carries them to the receptive cells.
13. A tympanic membrane probably has evolved independently in at least three lines of tetrapods: frogs, sauropsids, and a subgroup of synapsids that includes the mammals.
14. A frog's ear has a tympanic membrane, which connects to the columella, and an operculum, which connects via the opercularis muscle to the pectoral girdle and front leg. The tympanic membrane detects high-frequency mating calls; the opercular system detects low-frequency environmental noises and seismic vibrations. Adult salamanders have only the opercular system.
15. Most sauropsids have an ear with a tympanic membrane and a single auditory ossicle, which is the columella. Amphisbaenians and snakes have secondarily lost the tympanic membrane, and in these forms, the columella abuts on the quadrate bone.
16. Mammals have a tympanic membrane and three auditory ossicles: the malleus, incus, and stapes. The malleus and incus bore the jaw joint in early synapsids, with the malleus being homologous to the articular and the incus homologous to the quadrate.
17. In mammals, the lagena has become the cochlear duct of the inner ear, and perilymphatic channels, known as the scala tympani and scala vestibuli, form the rest of the cochlea.
18. Most early fishes, amphibians, and sauropsids had a median eye complex consisting of a pineal eye, a parietal eye, or both. This complex is retained in many extant groups. The median eye monitors ambient light and appears to initiate physiological adjustments to light levels. The complex is represented in birds and mammals by the pineal gland.
19. The image-forming eyes of vertebrates are essentially similar in all species and consist of three layers of tissue. The fibrous tunic forms the cornea and sclera; the vascular tunic forms the choroid, ciliary body, and iris; and the retina is the deepest layer.
20. Light is received by rod and cone cells. Rods function in dim light, but they cannot distinguish color, and rod vision is slightly blurred. Cones distinguish colors but require brighter illumination. They form much sharper images.
21. Light does not travel far in water. Most lampreys, chondrichthyans, and actinopterygians tend to be nearsighted, with a spherical lens located close to the cornea. They focus on more distant objects by contracting a muscle that retracts the lens toward the retina. Sharks are farsighted and accommodate to close objects by moving the lens toward the cornea.
22. The surface of a tetrapod's eyeball is protected by movable eyelids and bathed by tears secreted by tear glands.
23. The eye of most tetrapods at rest is focused on distant objects. A lissamphibian accommodates for near objects by contracting a muscle that pulls the lens toward the cornea; except for snakes, amniotes focus light by increasing the curvature of the lens.
24. Vertebrates can detect temperature changes, but we are uncertain as to the receptors in many cases. Boa constrictors, pythons, and pit vipers have one or more pairs of thermoreceptive pits on their heads that help them detect warm-blooded prey.

REFERENCES

Allin, E. F., 1975: Evolution of the mammalian middle ear. *Journal of Morphology,* 147:403–438.

Arey, L. B., 1957: *Developmental Anatomy,* 6th edition. Philadelphia, W. B. Saunders.

Atema, J., 1980: Smelling and tasting under water. *Oceanus,* 23:4–18.

Bemis, W. E., and Hetherington, T. E., 1982: The rostral organ of *Latimeria chalumnae:* Morphological evidence of an electroreceptive function. *Copeia,* 1982:467–471.

Bertmar, G., 1981: Evolution of vomeronasal organs in vertebrates. *Evolution,* 35:359–366.

Blaxter, J. H. S., 1987: Structure and development of the lateral line. *Biological Reviews of the Cambridge Philosophical Society,* 62:471–514.

Boord, R. L., 1993: Structure, development, and phylogeny of cranial nerves: Preface. *Acta Anatomica,* 148:69.

Boord, R. L., and Campbell, C. B. G., 1977: Structural and functional organization of the lateral line system of sharks. *American Zoologist,* 17:431–441.

Boord, R. L., and McCormick, C. A., 1984: Central lateral line and auditory pathways: A phylogenetic perspective. *American Zoologist,* 24:765–774.

Brown, R. E., and Fedde, M. R., 1993: Airflow sensors in the avian wing. *Journal of Experimental Biology,* 179:13–30.

Buck, L., and Axel, R., 1991: A novel multigene family may encode odorant receptors: A molecular basis for odor recognition. *Cell,* 65:75–187.

Catania, K. C., 1999. A nose that looks like a hand and acts like an eye: The unusual mechanosensory system of the star-nosed mole. *Journal of Comparative Physiology,* 185:367–372.

Chouchkov, C., 1978: *Cutaneous Receptors.* Berlin, Springer-Verlag.

Cock Buning, T. de, 1983: Thermal sensitivity as a specialization for prey capture and feeding in snakes. *American Zoologist,* 23:363–375.

Coombs, S., Gorner, P., and Munz, H., editors, 1989: *The Mechanosensory Lateral Line, Neurobiology and Evolution.* New York, Springer-Verlag.

Corwin, J. T., 1981: Audition in elasmobranchs. *In* Tavolga, W. N., Popper, A. N., and Fay, R. R., editors: *Hearing and Sound Communication in Fishes.* New York, Springer-Verlag.

Crescitelli, F., 1977: *The Visual System in Vertebrates.* New York, Springer-Verlag.

Crescitelli, F., 1972: The visual cells and visual pigments of the vertebrate eye. *In* Dartnall, H. J. A., editor: *Handbook of Sensory Physiology,* volume 7. Berlin, Springer-Verlag.

Daniels, J. F., 1934: *The Elasmobranch Fishes.* Berkeley, University of California Press.

Dijkgraaf, S., and Kalmijn, A. J., 1966: Function of ampullae of Lorenzini. *Z Vergleichende Physiologie,* 53:187–194.

Dowling, J. E., 1987: *The Retina, an Approachable Part of the Brain.* Cambridge, Harvard University Press.

Dowling, J. E., and Boycott, B. B., 1966: Organization of the primate retina: electron microscopy. *Proceedings of the Royal Society of London, Series B,* 166:80–111.

Eakin, R. M., 1973: *The Third Eye.* Berkeley, University of California Press.

Eddy, J. M. P., 1972: The pineal complex. *In* Hardisty, M. W., and Potter, I. C., editors: *The Biology of Lampreys.* London, Academic Press.

Fay, R. C., and Popper, A. N., 1985: The octavolateralis system. *In* Hildebrand, M., Bramble, D. M., Liem, K. F., and Wake, D. B., editors: *Functional Vertebrate Morphology.* Cambridge, Harvard University Press.

Feng, A. S., and Hall, J. C., 1991: Mechanoreception and phonoreception. *In* Prosser, C. L., editor: *Neural and Integrative Animal Physiology.* New York, Wiley-Liss.

Fritzsch, B., and Wahnschaffe, U., 1983: The electroreceptive ampullary organs of urodeles. *Cell Tissue Research,* 229:483–503.

Gamow, I., and Harris, J. F., 1973: The infrared receptors of snakes. *Scientific American,* 228:94–100.

Giese, A. C., 1979: *Cell Physiology,* 5th edition. Philadelphia, W. B. Saunders.

Gilland, E., and Baker, R., 1993: Conservation of neuroepithelial and mesodermal segments in the embryonic vertebrate head. *Acta Anatomica,* 148:110–123.

Goldsmith, T. H., 1991: Photoreception and vision. *In* Prosser, C. L., editor: *Neural and Integrative Animal Physiology.* New York, Wiley-Liss.

Goodenough, J., McGuire, B., and Wallace, R. A., 1993: *Perspectives on Animal Behavior.* New York, John Wiley & Sons.

Gould, S. J., 1990: An earful of jaw. *Natural History,* 100 (3):12–23.

Hanson, M., Westerberg, H., and Öblad, M., 1990: The role of magnetic statoconia in dogfish (*Squalus acanthias*). *Journal of Experimental Biology,* 151:205–218.

Hartline, P. H., 1971: Physiological basis for detecting sound and vibrations in snakes. *Journal of Experimental Biology,* 54:349–371.

Hetherington, T. E., Jaslow, A. P., and Lombard, R. E., 1986: Comparative morphology of the amphibian opercularis system: I. General design features and functional interpretation. *Journal of Morphology,* 190:42–61.

Hillenius, W. J., 1992: The evolution of nasal turbinates and mammalian endothermy. *Paleobiology* 18:17–29.

Hodgson, E. S., and Mathewson, R. F., editors, 1978: *Sensory Biology of Sharks, Skates, and Rays.* Arlington, Va., Office of Naval Research, Department of the Navy.

Hueter, R. E., symposium editor, 1991: Vision in elasmobranchs. *Journal of Experimental Zoology,* 5(suppl): 1–141.

Jacobs, G. H., 1992: Ultraviolet vision in vertebrates. *American Zoologist,* 32:544–554.

Johansson, R. S., and Vallbo, A. B., 1983: Tactile sensory coding in the globrous skin of the human hand. *Trends in Neurosciences,* 6:27–32.

Kalmijn, A. J., 1988: Electromagnetic orientation: A relativistic approach. *In* O'Conner, M. E., and Lovely, R. H., editors: *Electromagnetic Fields and Neurobehavioral Functions.* New York, John Wiley & Sons.

Kalmijn, A. J., 1978: Experimental evidence of geomagnetic orientation in elasmobranch fishes. *In* Schmidt-Koenig, K., and Keeton, W. T., editors: *Animal Migration, Navigation and Hearing.* New York, Springer-Verlag.

Kalmijn, A. J., 1971: The electric sense of sharks and rays. *Journal of Experimental Biology,* 55:371–383.

Kalmijn, A. J., 1966: Electro-perception in sharks and rays. *Nature,* 212:1232–1233.

Kingsbury, B. G., and Reed, H. D., 1909: The columella auris in Amphibia. *Journal of Morphology,* 20:549–628.

Kleerekoper, H., 1969: *Olfaction in Fishes.* Bloomington, Indiana University Press.

Levine, J. S., 1985: The vertebrate eye. *In* Hildebrand, M., Bramble, D. M., Liem, K. F., and Wake, D. B., editors: *Functional Vertebrate Morphology.* Cambridge, Harvard University Press.

Lombard, R. E., and Bolt, J. R., 1979: Evolution of the tetrapod ear: An analysis and reinterpretation. *Biological Journal of the Linnean Society,* 11:17–76.

Lombard, R. E., and Straughan, I. R., 1974: Functional aspects of anuran middle ear structures. *Journal of Experimental Biology,* 61:71–93.

Meisami, E., 1991: Chemoreception. *In* Prosser, C. L., editor: *Neural and Integrative Animal Physiology.* New York, Wiley-Liss.

Montgomery, J. C., and Bodznick, D., 1999: Signals and noise in the elasmobranch electrosensory system. *Journal of Experimental Biology,* 202:1349–1355.

Northcutt, R. G., 1989: The phylogenetic distribution and innervation of cranial mechanosensory lateral line. *In* Coombs, S., Gorner, P., and Munz, H., editors: *The Mechanosensory Lateral Line, Neurobiology and Evolution.* New York, Springer-Verlag.

Northcutt, R. G., and Davis, R. E., editors, 1983: *Fish Neurobiology.* Ann Arbor, University of Michigan Press.

Parrington, F. R., 1979: The evolution of the mammalian middle and outer ears: A personal review. *Biological Reviews,* 54:369–387.

Parsons, T. S., editor, 1966: The vertebrate ear. *American Zoologist,* 6:368–466.

Parsons, T. S., 1967: Evolution of nasal structure in the lower tetrapods. *American Zoologist,* 7:397–413.

Pettigrew, J. D., 1999: Electroreception in monotremes. *Journal of Experimental Biology,* 202:1447–1454.

Platt, A. C., Popper, A. N., and Fay, R. R., 1989: The ear as part of the octavolateralis system. *In* Coombs, S., Gorner, P., and Munz, H., editors: *The Mechanosensory Lateral Line, Neurobiology and Evolution.* New York, Springer-Verlag.

Popper, A. N., 1980: *Comparative Studies of Hearing in Vertebrates.* New York, Springer-Verlag.

Popper, A. N., and Fay, R. R., 1977: Structure and function of the elasmobranch auditory system. *American Zoologist,* 17:443–452.

Ralph, C. L., Firth, B. T., Gern, W. A., and Owens, D. W., 1979: The pineal complex and thermoregulation. *Biological Reviews,* 54:41–72.

Raschii, W., 1986: A morphological analysis of the ampullae of Lorenzini in selected skates. *Journal of Morphology,* 189:225–247.

Retzius, G., 1881–1884: *Das Gehororgan der Wirbelthiere. Morphologischhistologische Studien,* 2 volumes. Stockholm, Samson & Wallin.

Smith, W. J., 1977: *The Behavior of Communicating: An Ethological Approach.* Cambridge, Harvard University Press.

Sorensen, P. W., Hara, T. J., and Stacey, N. E., 1987: Extreme olfactory sensitivity of goldfish to a steroidal pheromone. *Journal of Comparative Physiology,* 160: 305–314.

Stern, P., and Marx, J., editors, 1999: Making sense of scents. *Science,* 286:707–728.

Studnicka, F. K., 1905: Die Parietalorgane. *In* Oppel, A., editor: *Lehrbuch der vergleichenden mikroskopischen Anatomie der Wirbeltiere,* part 5. Jena, Gustav Fischer.

Szabo, T., 1974: Peripheral and central components in electroreception. *In* Fessard, A., editor: *Handbook of Sensory Physiology,* volume 3, part 3, pp 13–58. Amsterdam, Elsevier.

Tavolga, W. N., Popper, A. N., and Fay, R. C., 1981: *Hearing and Sound Communication in Fishes.* New York, Springer-Verlag.

van Bergeijk, W. A., 1967: The evolution of vertebrate hearing. *In* Neff, W., editor: *Contributions to Sensory Physiology.* New York, Academic Press.

Waldvogel, J. A., 1990: The bird's eye view. *American Scientist,* 78:342–353.

Walker, W. F., Jr., 1981: *Dissection of the Frog,* 2nd edition. San Francisco, W. H. Freeman & Company.

Walls, G. L., 1942: *The Vertebrate Eye and Its Adaptive Radiation.* Bloomfield Hill, Mich., Cranbrook Institute of Science.

Webster, D. B., Fay, R. R., and Popper, A. N., editors, 1992: *The Evolutionary Biology of Hearing.* New York, Springer-Verlag.

Wever, E. G., 1985: *The Amphibian Ear.* Princeton, Princeton University Press.

Wever, E. G., 1978: *The Reptilian Ear: Its Structure and Function.* Princeton, Princeton University Press.

Wurtman, R. J., Axelrod, J., and Kelley, D. E., 1968: *The Pineal.* New York, Academic Press.

Zimmerman, K., and Heatwolfe, H., 1990: Cutaneous photoreception: A new sensory mechanism for reptiles. *Copeia,* 1990:860–862.

13

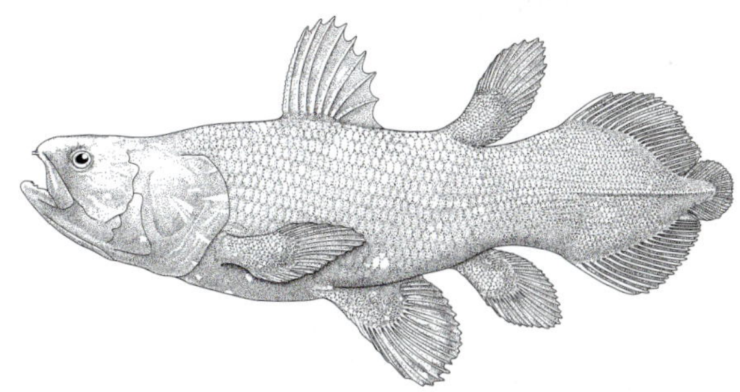

The Nervous System I: Organization, Spinal Cord, and Peripheral Nerves

PRECIS

The nervous system mediates most of a vertebrate's rapid responses to changes in its environment. In this chapter, we examine the structural and functional units of the nervous system, their basic patterns of organization, and the structure and function of the spinal cord and peripheral nerves.

OUTLINE

Most of the rapid responses to changes in a vertebrate's external or internal environments are mediated by the nervous system. Activating the correct combination of effectors so that appropriate responses are made requires a processing of signals within the nervous system that is called **coordination,** or **integration.** The nervous system is constantly bombarded with sensory signals. Many are **filtered** as they pass to higher brain centers so that some are suppressed and others enhanced. Often, sensory inputs from different types of receptors are combined to provide a more complete "picture" of changes taking place. The sensory input may be combined with stored information (memories) about what happened under similar circumstances in the past experience of the individual animal before a response is made. A sheep, for example, senses another animal nearby. If this is another grazing ewe, then the sensory signals may be ignored, but if the approaching animal is a wolf or a ram, then the ewe should run away or perhaps assume a mating stance. Some responses are affected by the activity of endocrine glands. A ewe under the influence of sex hormones in the mating season will behave very differently toward a ram than at other times. The motor responses themselves require a complex pattern of signals. Escaping a predator requires not only the sequential activation of certain muscles and the inhibition of antagonistic ones but also many physiological adjustments. Rates of breathing and circulation must increase, energy reserves must be made available to the limb muscles, and so on. Often, information about the outcome of the response is fed back into the nervous system and reinforces or alters the memories stored there. Conditioned reflex experiments have shown that all groups of vertebrates can be trained to perform simple discrimination tasks. Although much vertebrate behavior is innate and based on inherited neuronal connections, vertebrates learn from experience.

Cellular Components of the Nervous System

The two basic cellular components of the nervous system are neurons, which transmit signals to other neurons, glands, muscles, or receptor cells, and glial cells, which provide for structural and functional needs of neurons. This section briefly describes their structure.

Neurons and Synapses

The primary functional units of the nervous system are the nerve cells, or **neurons** (Gr., *neuron* = nerve). All neurons are ectodermal, for they develop embryonically from the neural tube, from the neural crest, or from ectodermal neurogenic placodes (Chapter 4). They are arrayed in such a way that we can recognize a **central nervous system,** or **CNS,** consisting of the brain and spinal cord, and a **peripheral nervous system,** or **PNS,** consisting of nerves extending between the CNS and the receptors (e.g., sense organs) and effectors (e.g., muscles and glands). Many morphological types of neurons are known, but all contain four essential parts: (1) a cell body, or trophic segment; (2) dendrites, or a receptive segment; (3) an axon, or conductive segment; and (4) a terminal arborization, or transmissive segment (Fig. 13-1*A–C*).

The **cell body** contains the nucleus and metabolic machinery of a neuron. Proteins and neurotransmitters that carry signals between cells are synthesized here. In a **multipolar motor neuron** (Fig. 13-1*A*), the cell body lies very close to the dendrites, and a long axon extends outward from the cell body. The cell body, however, is not always next to the dendrites. It may lie near the middle of an axon, as in a **bipolar cell** from the retina (Fig. 13-1*B*), or it may be set off on one side of the axon, as in a **pseudounipolar sensory neuron** (Fig. 13-1*C*). Pseudounipolar neurons begin their development as bipolar cells in which proximal portions of the dendrite and axon secondarily fuse (Fig. 13-1*D*). Aggregations of cell bodies lying outside the central nervous system, often along the root of a nerve, are known as **ganglia** (see Table 13-1 for definitions of essential terms used in describing CNS and PNS structures). Biologists previously defined dendrites and axons with respect to their location relative to the cell body (as in the multipolar neuron), but they now are defined by their functions. Dendrites receive stimuli and initiate nerve impulses; axons conduct nerve impulses toward other neurons, muscles, or glands. Neurobiologists often use the term **projection** to refer to the target of a group of axons; thus, we may say that a group of axons projects to a particular group of cell bodies in the brainstem.

An **axon** (Gr., *axon* = axle or axis) is a long, slender, cytoplasmic process that often is called a **nerve fiber.** Axons conduct sensory impulses from the dendrites of receptors to the CNS, impulses within the CNS, and motor impulses from the CNS to the effectors. Axons frequently branch. As we have seen (Chapter 10), the axon of a single motor neuron branches and innervates many muscle fibers to form a motor unit. Axons are specialized to rapidly conduct nerve impulses, often over a long distance, without diminution of signal amplitude. A neuron's plasma membrane has an electric **resting potential** of about 60 mV that derives from an unequal distribution of ions across it. As with muscle cells (Chapter 10), a high concentration of extracellular, positively charged sodium ions causes the outside of the membrane to be positive relative to the inside. We say that such a membrane is **polarized.** Further physiological details of a nerve impulse will not be discussed at length, but you need to understand the main events. Briefly, when an axon receives a threshold stimulus, ion channels in the plasma membrane open, membrane permeability changes, and the axon develops an **action potential** (Fig. 13-2). Sodium ions rush into the axon so that membrane polarity is momentarily reversed. Biologists say that such a membrane is **depolarized** because it has lost, at least temporarily, the polarization typical of the resting membrane. In turn, depolarization affects adjacent parts of the membrane, and a **nerve impulse** propagates as a wave of depolarization over the surface of the axon. Depolarization of any segment of the plasma membrane is followed quickly by the pumping out of positively charged ions and the restoration of the resting potential. The action potential of an axon is an **all-or-none response** because it does not vary in amplitude with the strength of the stimulus. Stimulus strength is coded only by whether the axon is active or inactive (a binary on–off system) and by the number and pattern of nerve impulses in a given time.

The branching **terminal arborizations** (also called **telodendria;** Gr., *telos* = end + *dendria* = trees) at the end of an axon make contact with the receptive segments of other neurons at special junc-

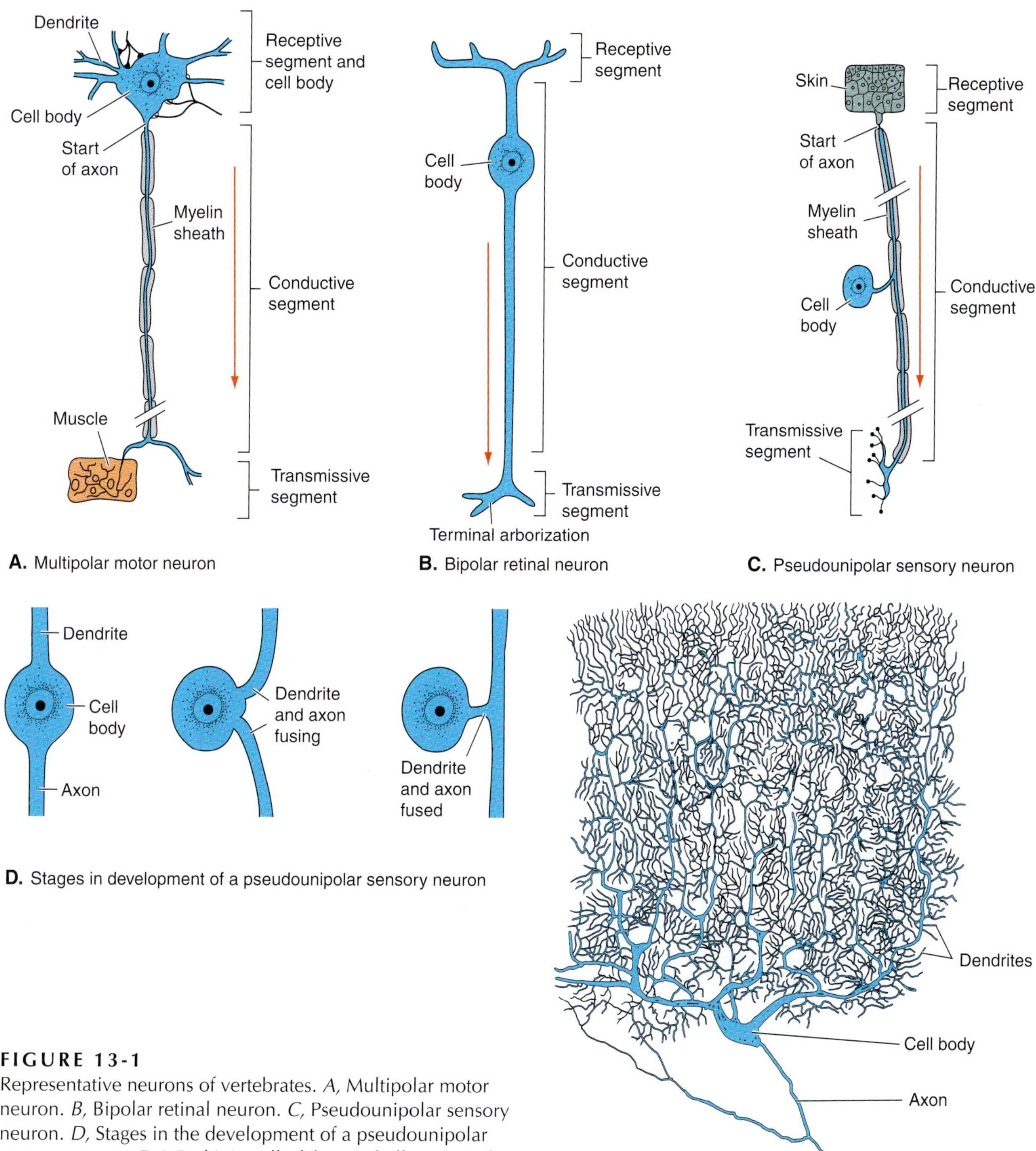

FIGURE 13-1
Representative neurons of vertebrates. *A,* Multipolar motor neuron. *B,* Bipolar retinal neuron. *C,* Pseudounipolar sensory neuron. *D,* Stages in the development of a pseudounipolar sensory neuron. *E,* A Purkinje cell of the cerebellum. *(E, After Ramon y Cajal.)*

tions known as **synapses** (Gr., *synapsis* = union) or with effectors at synapse-like **myoneural** junctions (= junctions with muscle cells) or **axoglandular junctions** (= junctions with glands). The general structure of a synapse is diagrammed in Figure 13-3*A*. A nerve terminal contains many minute **synaptic vesicles,** in which neurotransmitter substance is stored. Many types of **neurotransmitters** are known, among them acetylcholine, noradrenaline (= norepinephrine), serotonin, and dopamine (Table 13-2). As a general rule, any given neuron uses only one type of neurotransmitter at all of its synapses. Thus, neurobiologists speak of neurons that use acetylcholine as **cholinergic,** those that use noradrenaline as **adrener-**

TABLE 13-1 Comparison of Terms for Structures and Cells in the Peripheral and Central Nervous Systems

Structure	Name of the Structure in the Peripheral Nervous System	Name of the Structure in the Central Nervous System
Aggregation of cell bodies	A **ganglion**, often associated with a spinal or cranial nerve or lying as a chain of ganglia adjacent to the spinal cord	A **nucleus** within the gray matter or a **cortex** within gray matter in the roof of certain brain regions (sometimes also called a ganglion, a locus, an area, or other terms)
A bundle of myelinated axons	A **nerve**	A **tract** (or, rarely, a **nerve,** as for the optic nerve)
Glial cells responsible for myelination	**Schwann cells**	**Oligodendrocytes**
Other types of glial cells		**Astrocytes, microglia**

gic, and so forth. Synapses are an important site of action for pharmacological agents, and many drug-based therapies rely on our ability to target a specific class of synapses and neurons based on the particular neurotransmitter that they use. When a nerve impulse reaches the terminal, some of the neurotransmitter is released from synaptic vesicles into the extracellular spaces beyond its tip, where it crosses the narrow **synaptic cleft,** binds with **receptor sites** on the **postsynaptic cell,** and excites or inhibits this cell (Fig. 13-3*A*). An axon can be stimulated experimentally at its center, and a nerve impulse will travel away from the point of stimulation in both directions. The impulse cannot reverse itself, however, because it cannot go back over previously stimulated portions of the neuron's membrane until the membrane's resting potential has been restored. The polarity of communication within the nervous system thus is determined by synapses because nerve impulses normally cross them in only one direction: from the **presynaptic neuron** containing the synaptic vesicles to the **postsynaptic** cell containing receptor binding sites for the neurotransmitters.

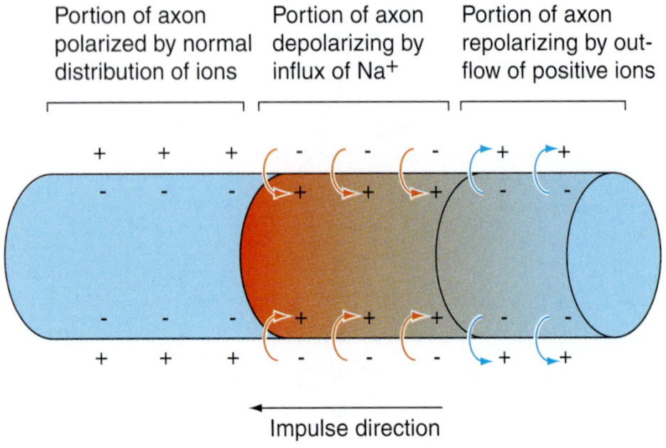

The receptive segment of a neuron includes the branching **dendrites** (Gr., *dendria* = trees) and often the cell body. Dendrites of some neurons form extensive, branching trees (Fig. 13-1*D*). A receptive segment can be stimulated directly in the case of free nerve endings, by synapses with receptor cells, or by the terminals of other neurons. Thousands of synapses of many different neurons may occur on the dendrites and cell body of a single neuron, and synapses between terminals also may be present (Fig. 13-3*B*). The degree of polarization of the plasma membrane of the dendrites and cell body may be in constant flux as different neurotransmitters impinge on it. Some neurotransmitters are **excitatory** because they partially depolarize the postsynaptic membrane; others, released by different neurons, hyperpolarize the membrane and thus have an **inhibitory** effect. A threshold level of depolarization must be reached for a nerve impulse to be initiated in the postsynaptic cell's axon. Signals thus may be aborted, delayed, accelerated, or processed in other ways at the receptive region of a neuron. Whether a nerve impulse is generated in the postsynaptic cell's axon depends on the number of excitatory and inhibitory impulses impinging on the cell's receptive region, on their frequency, and on the nature of the neurotransmitter receptor sites.

FIGURE 13-2
Movements of ions during propagation of an action potential.

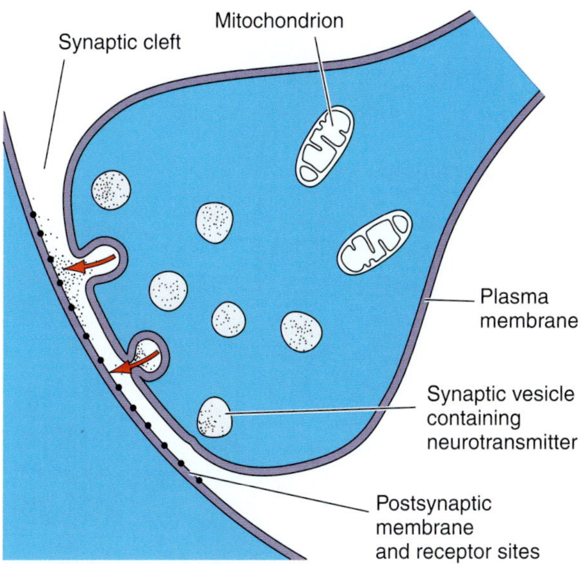

A. Structure of a single synapse

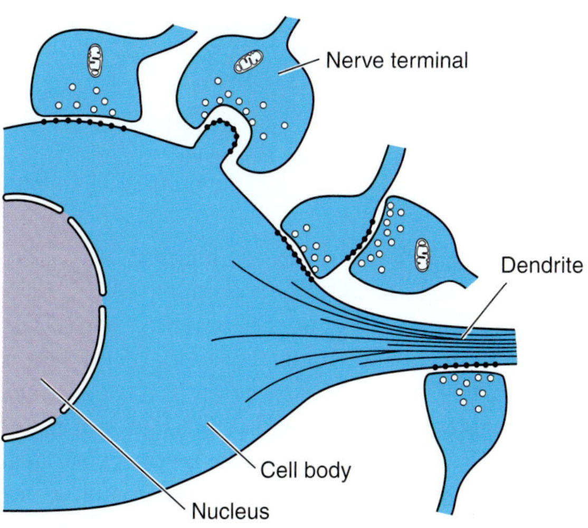

B. Multiple synapses on a neuronal cell body

FIGURE 13-3

Synapses. *A,* Structure of a single synapse. The nerve terminal is separated from the postsynaptic cell by a narrow synaptic cleft. *B,* Multiple synapses on a cell body and dendrite. Synapses between nerve terminals also are possible. *(A, After Noback; B, after Williams et al.)*

Schwann Cells and Neuroglia

Glia is the general term for the non-neuronal cells of the nervous system (Gr., *glia* = glue). All axons in the peripheral nerves of gnathostomes are surrounded during embryonic development by glial components known as **Schwann cells,** which develop from the neural crest and neurogenic placodes. In gnathostomes, an inner, tonguelike process of these cells grows around a segment of the axon many times so that many layers of the Schwann cell's plasma membrane are laid down to form a **myelin sheath** (Gr., *myelos* = marrow) that is visible by light microscopy (Fig. 13-4*A*). Because the plasma membrane of cells consists of a bilayer of lipids, a myelin sheath is fatty. Some axons are surrounded by as many as 100 layers of plasma membrane. The myelin sheath is interrupted between successive Schwann cells by **nodes of Ranvier** (Fig. 13-4*B*). The Schwann cells associated with some of the peripheral axons, especially those of many autonomic nerves, envelop the axons but do not continue to grow to form a sheath around them. Such axons are described as **nonmyelinated** because the layers of Schwann cell plasma membrane are too thin to be detected by light microscopy. Because the nodes of Ranvier are the only regions where the plasma membrane of a myelinated axon is close to the aqueous-based extracellular fluid, the wave of depolarization that constitutes the nerve impulse jumps from node to node. As a result, impulse transmission is much more

TABLE 13-2 Examples of Neurotransmitters and the Chemical Groups to Which They Belong

Chemical Group	Neurotransmitter	Function
Cholinergic	Acetylcholine	Excitatory or inhibitory, depending on the type of receptor
Biogenic amines	Norepinephrine, epinephrine, dopamine, serotonin, histamine	Excitatory or inhibitory, depending on the type of receptor
Amino acids	Glutamate, aspartate	Excitatory
	GABA (gamma-aminobutyric acid), glycine	Inhibitory

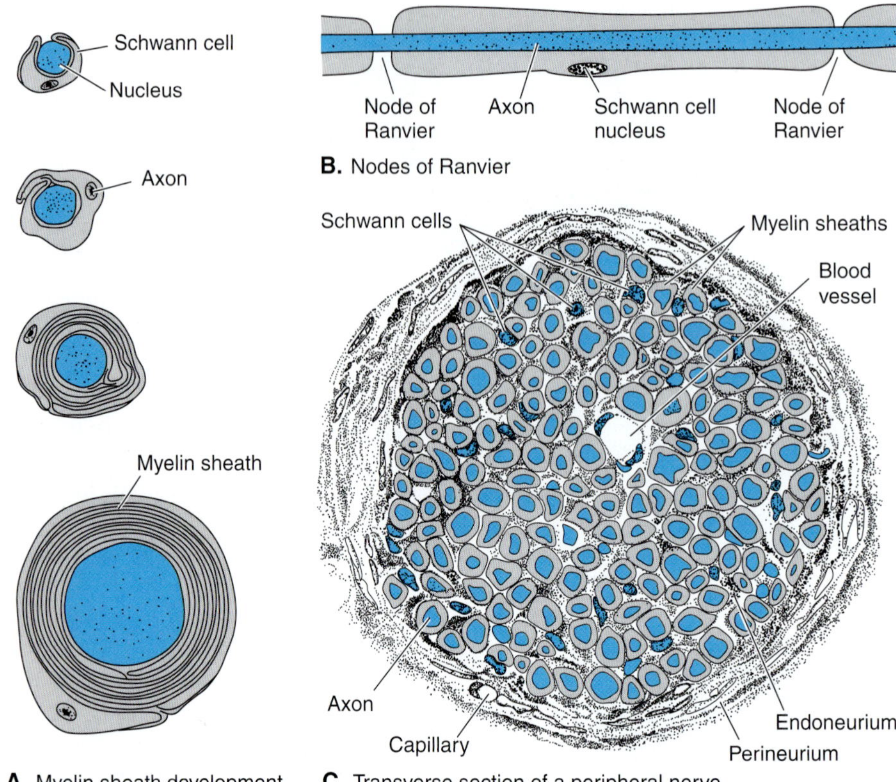

FIGURE 13-4
Axons and nerves. *A,* The development of myelination of an axon (transverse sections). *B,* Nodes of Ranvier along a myelinated axon. *C,* Transverse section of a peripheral nerve, showing the connective tissue sheaths that invest individual neurons (endoneurium) and the entire bundle of axons (perineurium). The blood supply of the nerve also is shown. *(A, After Williams et al.; C, after Fawcett.)*

rapid in myelinated than in nonmyelinated axons. (Transmission velocity also is affected by temperature and by axon diameter, being faster at higher temperatures and in neurons of large diameter.) **Mesodermal connective tissue** envelops individual neurons and groups of peripheral neurons to form peripheral **nerves** (Fig. 13-4*C*). Nutritive blood vessels follow the connective tissue into the larger nerves. The general terminology used for the branching patterns of an individual anatomical nerve is reasonably well standardized. A large nerve or nerve **trunk** divides to form two or more **rami,** which in turn give off two or more **ramules.** Additional hierarchical terms are available to describe the more distal portions of a complex branching pattern.[1]

Many other types of glial cells, collectively called the **neuroglia,** occur in the CNS, where they are 30 to 50 times more numerous than are the neurons. **Oligodendrocytes,** which develop from the neural plate rather than neural crest, send out processes that wrap around and myelinate axons within the CNS (Fig. 13-5). Each oligodendrocyte on average myelinates 15 axons. This is a slow process, and myelination is not completed in the human brain until many months after birth. Regions of the CNS containing myelinated axons are whitish in unstained sections of the spinal cord and brain and consequently are called **white matter.** Most of the white matter consists of bundles of axons extending between parts of the CNS. Such bundles of axons within the CNS are called **tracts** (see Table 13-1 for definitions and comparisons of structures in the CNS and PNS). Regions containing nonmyelinated axons, dendrites, and cell bodies are termed **gray matter.** Dense aggregations of cells bodies within the gray mater are known as **nuclei** (Table 13-1). Nuclei are important landmarks within the CNS because they indicate areas involved in processing sensory input or motor outflow. In other regions of the CNS, cell bodies may be densely concentrated in a sheet, as in the **cortices** found in the roof of certain brain regions, in which the sheets of cell bodies are laminated (like plywood) between sheets of fibers.

Star-shaped glial cells known as **astrocytes** are interposed among the neuronal cell bodies (Fig. 13-5). In addition to their role in supporting adjacent cells, astrocytes perform many essential physiological functions, such as regulation of ionic composition and electrical balance of extracellular fluids. In particular, they take up excess potassium ions that flow out of very active neurons and extrude these potassium ions through their end feet to the surfaces that they contact: capillaries, the pia mater on the brain surface, and the special-

[1]No single, standard set of terms exists for the distal branching pattern of complex nerves; see discussion in Northcutt and Bemis (1993) for more information.

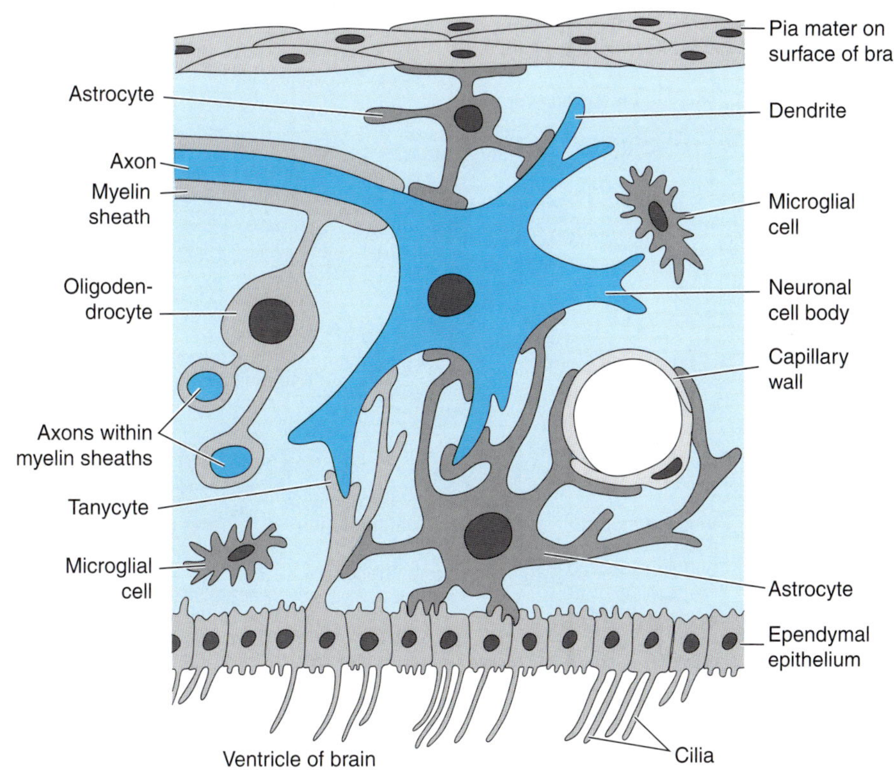

FIGURE 13-5
An idealized transverse section of the wall of the brain to show neuroglial cells; other non-nervous cells; and their relationships to neurons, the ventricular system, and blood vessels. *(Modified from Williams et al.)*

ized ependymal epithelium lining the nervous system. Astrocytes also degrade metabolites, including neurotransmitters released into synaptic clefts. Because they contact both capillaries and neuron cell bodies, astrocytes also may have a nutritive role.

Oligodendrocytes and astrocytes develop from the neural tube ectoderm and collectively are sometimes called **macroglia.** Smaller **microglia** develop from macrophages of mesodermal origin (Fig. 13-5). They have a phagocytic role, removing foreign and degenerative products. **Tanycytes** in the ependymal epithelium also extend processes inward to many neuronal cell bodies and presumably participate in the metabolism of these cells.

Organization of the Nervous System

The basic structure and plan of organization of the nervous system is similar in all vertebrates. Because of the way they develop (Chapter 4), the spinal cord and brain of a vertebrate are hollow and lined by a non-nervous and partly ciliated **ependymal epithelium** (Fig. 13-5). The adult spinal cord has a small **central canal** (Fig. 13-6), which enlarges within parts of the brain to form a series of interconnected **ventricles** (Chapter 14). **Cerebrospinal fluid,** or **CSF,** slowly circulates within these spaces. Bilateral features of the nervous system often are referred to by the contrasting terms **ipsilateral** (L., *ipsi* = the same + *lateral* = side) and **contralateral** (L., *contra* = the opposite + *lateral* = side). These terms are especially important in describing the targets of particular tracts of nerve cells. Typical **spinal nerves** attach to the cord at segmental intervals by **dorsal** and **ventral roots** (Fig. 13-6). In most vertebrates (lampreys being a prominent exception that is discussed in detail later), the dorsal and ventral roots of a body segment unite slightly peripheral to the spinal cord to form a spinal nerve from which **dorsal, ventral,** and **communicating rami** extend to different parts of the body. Most of the cranial nerves appear to be serially homologous to either a ventral or a dorsal root of a typical spinal nerve.

The Three Connectional Categories of Neurons and Their Distribution

Regardless of differences in their morphologies, all neurons belong to one of three categories: (1) **primary sensory,** or **afferent, neurons** (L., *ad* = toward + *ferens* = carrying, from *ferre* = to carry) carry impulses from free nerve endings or specialized receptor cells into the CNS; (2) **motor,** or **efferent, neurons** (L., *ex* = from + *ferens* = carrying, from *ferre* = to carry) carry signals from the CNS to effectors, such as

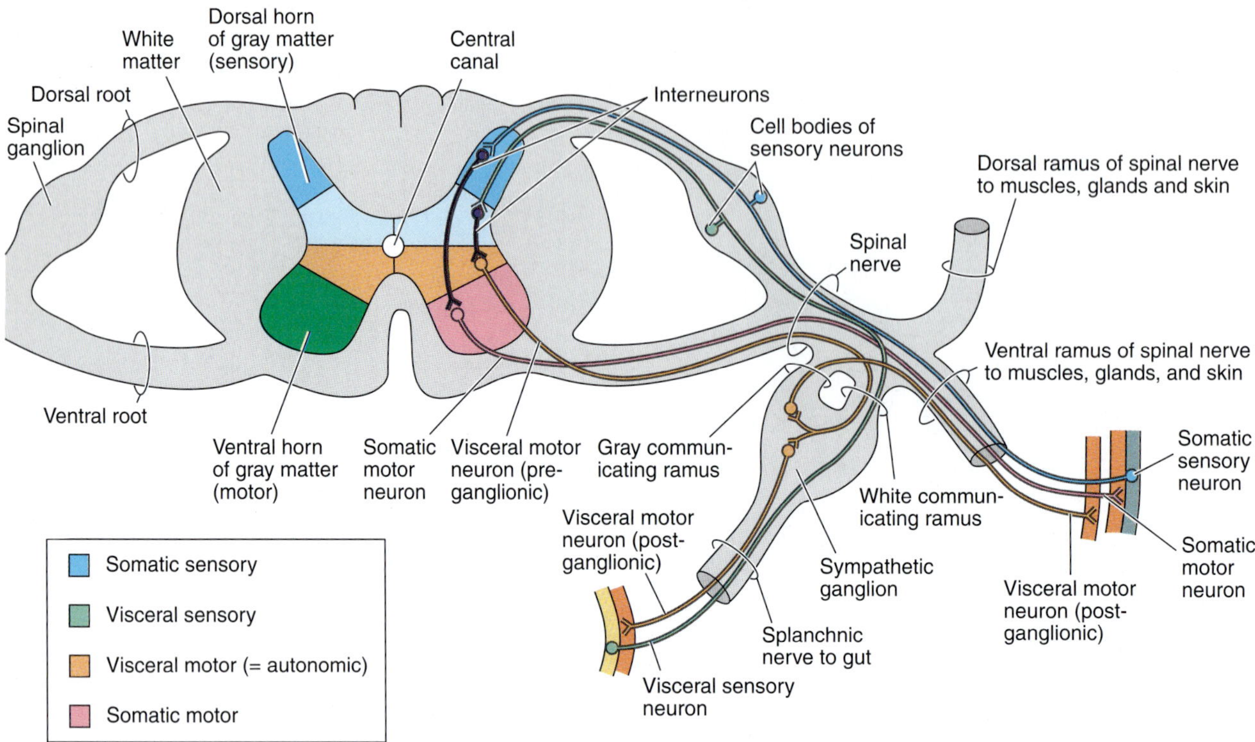

FIGURE 13-6
A schematic transverse section through the spinal cord of a mammal to show the four functional groups of neurons that constitute the nervous system and their interrelationships. On the left side, the gray matter is color coded to demarcate its four general regions; note that the dorsal half of the gray matter functions in sensory input, whereas the ventral half of the gray matter functions in motor outflow. On the right side, the sensory and motor neurons are color coded to show the pathways that they follow in the peripheral nerves.

muscles or glands[2]; and (3) **interneurons** (L., *inter* = between), which are confined to the CNS and receive signals from the primary sensory neurons, integrate this information through their often exceedingly complex connections with other interneurons, and ultimately send signals out to the periphery by motor neurons. Sensory and motor neurons are mixed in the spinal nerves, but they segregate beside the spinal cord (Fig. 13-6). Sensory neurons always enter the cord through the dorsal roots of the spinal nerves, and in amniotes the motor neurons leave through the ventral roots.

Sensory neurons of spinal nerves develop from neural crest cells. The cell bodies of these neurons remain peripheral to the CNS in small swellings called **spinal ganglia,** which are located on the dorsal roots (Fig. 13-6).[3] Inside the gray matter of the spinal cord, the sensory neurons form synapses with the dendrites and cell bodies of interneurons or directly with motor neurons. Interneurons and most motor neurons develop from ectodermal neural tube cells, and their cell bodies remain in the gray matter that primitively surrounds the central canal and ventricles.

In a transverse section of the spinal cord, the gray matter has the appearance of a butterfly (Fig. 13-6). A pair of **dorsal horns** and a pair of **ventral horns** interconnect across the midline by the **gray commissure.** The dorsal horns contain the dendrites and cell bodies of interneurons that receive many of the terminations of the primary sensory neurons. These interneurons relay information to other parts of the CNS. The ventral horns contain the dendrites and cell bodies of the motor neurons. Motor neurons receive impulses from some primary sensory neurons and from many in-

[2]Note that the terms "afferent" and "efferent" here refer to the direction of information flow relative to the CNS. For information traveling within the CNS, we may also speak about neurons that are afferent or efferent with respect to a particular nucleus or cortical area.

[3]A common synonym for the term "spinal ganglion" is "dorsal root ganglion."

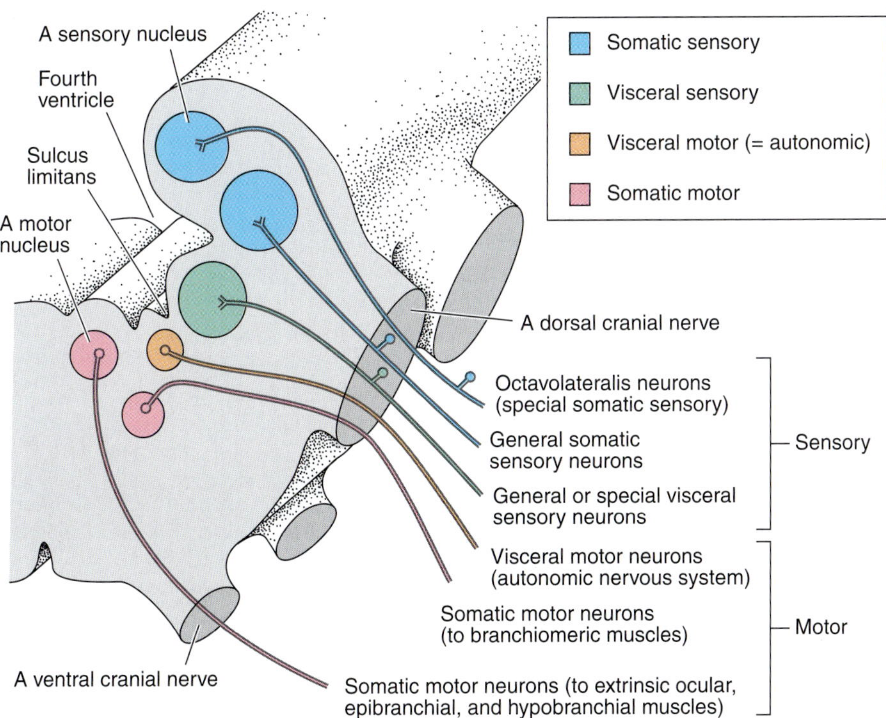

FIGURE 13-7
The locations of the nuclei of the cranial nerves in the brainstem in a composite transverse section. The four functional groups of neurons and the brainstem nuclei from which they originate or to which they project are color coded as in Figure 13-6.

terneurons that originate in the dorsal horns or in higher brain centers. **Dorsal horns thus are sensory in function, whereas ventral horns are motor in function.** Surrounding the butterfly-shaped gray matter are white matter columns formed by axonal tracts and divided into three major groups: the **dorsal, lateral,** and **ventral funiculi.** The dorsal funiculus contains only sensory tracts that carry information to the brain; the lateral and ventral funiculi carry both sensory information to the brain and motor outflow from higher centers (see Fig. 13-11*B*).

Dorsal sensory and ventral motor regions are particularly distinct in the spinal cord of an embryo because a longitudinal groove, the **limiting sulcus** (= sulcus limitans), is present between them on the lateral wall of the central canal and in many of the ventricles of the brain. The limiting sulcus also can be seen in the caudal parts of the adult brain of many vertebrates (Fig. 13-7).

Functional Groups of Sensory and Motor Neurons

Sensory and motor neurons and the regions where they terminate or originate can be subclassified into functional groups associated with the somatic and visceral parts of the body (Table 13-3). **Somatic sensory neurons** from somatic receptors enter the dorsal horn of the spinal cord by the dorsal root. Many of these neurons immediately synapse with interneurons in the dorsal horn (Fig. 13-6). **Visceral sensory neurons** from visceral receptors terminate in the **visceral sensory cell column. Visceral motor neurons** to visceral muscles in the gut, heart, and other internal organs and to exocrine glands arise in the **visceral motor cell column,** and **somatic motor neurons** to somatic muscles arise in the **somatic motor cell column,** located in the most ventral part of the ventral horn.

Anatomists used to classify the visceral motor neurons as autonomic ones supplying glands and visceral organs or branchiomeric ones going to the branchiomeric muscles, which are located only in the cranial region. As we have discussed (Chapter 10), evidence indicates that the branchiomeric muscles are somatic, and thus the neurons supplying them are also somatic (see additional discussion in the later section on cranial nerves). Thus visceral motor neurons supply only glands and the muscles in the walls of visceral organs. These neurons form the **autonomic nervous system.** We will return to the details of the autonomic nervous system later in this chapter, but at this time you should note that visceral motor neurons differ from somatic motor ones in not extending directly from the CNS to the organs they supply. They always relay in some peripheral **autonomic ganglion,** such as the **sympathetic ganglion** shown in Figure 13-6. **Preganglionic visceral motor neurons** extend from the spinal cord (or brain) to the peripheral ganglion,

TABLE 13-3 Four Functional Groups of Neurons

Functional Group	Subgroup	Information Carried	Comments
Somatic sensory	Special somatic sensory	Sensory afferents from ear, eye, lateral line, and electroreceptive organs	Restricted to certain cranial nerves
	General somatic sensory	Sensory afferents from pain, temperature, and touch receptors in skin	Found in spinal and cranial nerves
Visceral sensory	Special visceral sensory	Sensory afferents from taste buds	Restricted to certain cranial nerves
	General visceral sensory	Sensory afferents from wall of visceral tube	Found in spinal and cranial nerves
Visceral motor (= autonomic)	—	Motor to glands and muscles of the visceral tube	Found in spinal and cranial nerves
Somatic motor	—	Motor to skeletal muscles of the body wall, including branchiomeric muscles	Found in spinal and cranial nerves

where they synapse with a **postganglionic visceral motor neuron,** which continues to the effectors. The postganglionic neurons always develop from neural crest cells. In mammals, the axons of the preganglionic neurons leave the spinal nerve to reach their sympathetic ganglion in a **white communicating ramus,** so called because these fibers are myelinated and thus white in color. Some postganglionic axons return to the spinal nerve to be distributed with the dorsal and ventral rami of the spinal nerve; these nonmyelinated fibers reach the spinal nerve through a **gray communicating ramus** (Fig. 13-6).

The generalizations about the spinal cord and neurons apply with some modifications to the cranial nerves and to the brainstem to which they attach. The **brainstem** is the "stalk" of the brain—that is, the brain without the cerebellum and forebrain (which, in turn, consists of the cerebrum and diencephalon; see Chapter 14). As the cell columns in the spinal cord continue into the brain, they form discrete islands of cell bodies within the gray matter known as **nuclei** (Figs. 13-7 and 13-8). A nucleus within the CNS is essentially a collection of nerve cell bodies and is analogous to the ganglia located in the PNS (Table 13-1). The brainstem nuclei have the same relationship to each other as do the cellular columns of gray matter in the spinal cord, (i.e., sensory nuclei are located dorsally, and motor nuclei are located ventrally).

As we discuss later in this chapter, most of the cranial nerves are serially homologous to the type of spinal nerve we find in lampreys, in which the dorsal and ventral roots form separate dorsal and ventral nerves that do not unite. Like spinal nerves, some cranial nerves contain the axons of somatic sensory neurons from the skin and visceral sensory neurons from receptors in the gut wall and other visceral organs. In cranial nerves, these are called, respectively, **general somatic sensory** and **general visceral sensory** neurons to differentiate them from **special sensory neurons** that travel with certain cranial nerves but that never are found with spinal nerves (Table 13-3). The special sensory fibers found in the head are **octavolateralis neurons** from the ear, lateral line, and electrosensory organs (**special somatic sensory neurons;** Table 13-3), and **gustatory neurons** from taste buds (**special visceral sensory neurons;** Table 13-3). Octavolateralis and gustatory neurons develop embryonically from ectodermal neurogenic placodes, not from neural crest cells as do most other sensory neurons.[4] As you would expect from observations mentioned earlier about the organization of the spinal cord, all sensory neurons regardless of their functional group enter the brain through dorsal nerves. There, they terminate in **dorsal sensory nuclei,** which have characteristic locations (Figs. 13-7 and 13-8). Special somatic sensory neurons (= octavolateralis neurons) terminate in the dorsal-most nuclei, and general somatic sensory ones, in a nucleus ventral to it (but still in a relatively dorsal position). As

[4]Among other examples of sensory neurons that do not arise from the neural crest cells are the olfactory neurons, which develop from the nasal placodes, and retinal sensory cells, which develop from the ependymal epithelium of the neural tube; see subsequent discussion and Chapters 4 and 12.

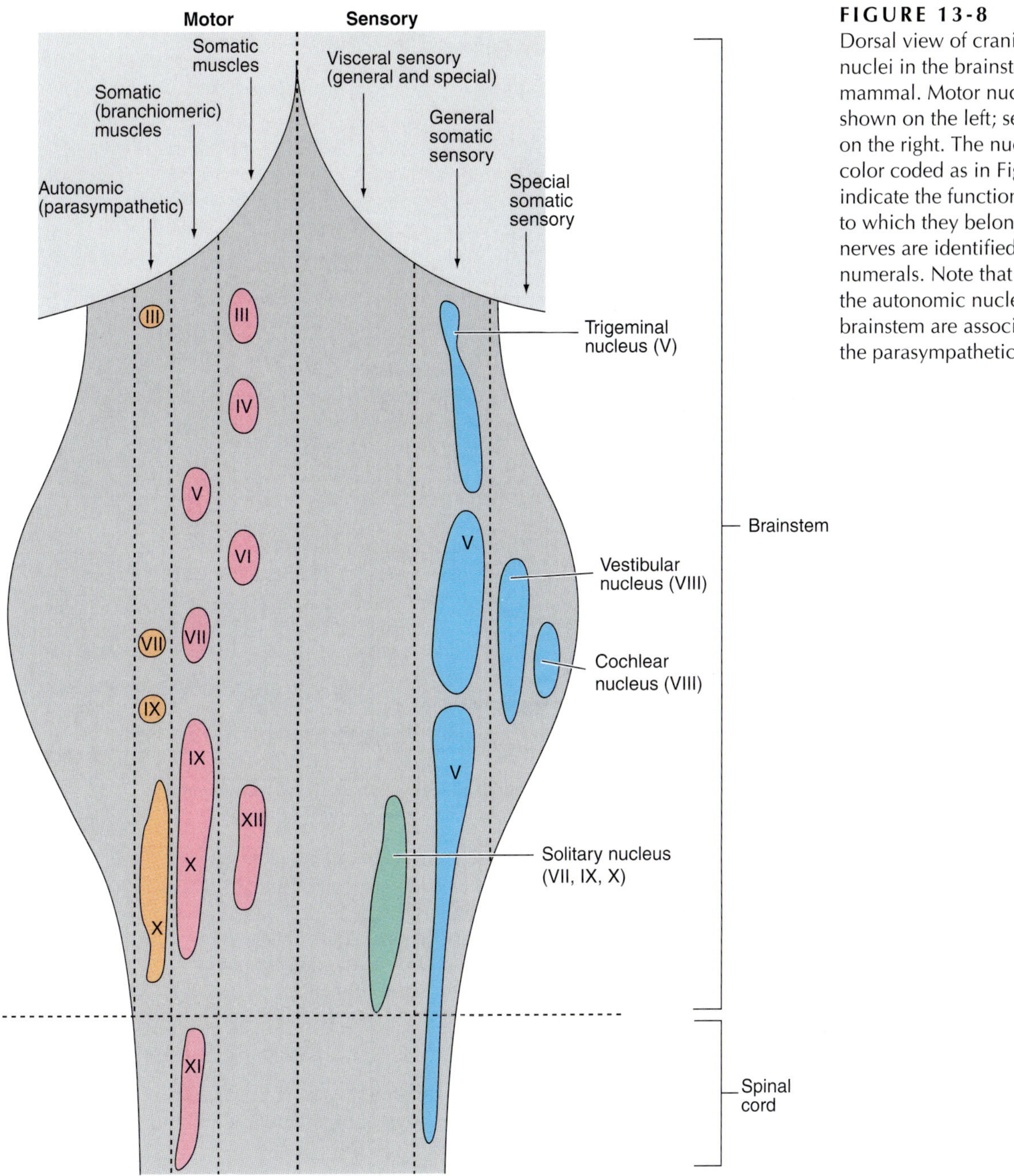

FIGURE 13-8
Dorsal view of cranial nerve nuclei in the brainstem of a mammal. Motor nuclei are shown on the left; sensory ones on the right. The nuclei are color coded as in Figure 13-6 to indicate the functional group to which they belong. Cranial nerves are identified by Roman numerals. Note that all of the autonomic nuclei in the brainstem are associated with the parasympathetic division.

shown in Figures 13-7 and 13-8, both general and special visceral sensory neurons terminate in the solitary nucleus ventral to the somatic sensory ones, but the special visceral sensory neurons (= gustatory neurons) terminate in the rostral part of this nucleus, and the general visceral sensory neurons terminate in the caudal part. As in the spinal cord of a lamprey, motor neurons of cranial nerves originate in ventral nuclei but can exit the brainstem through either a dorsal or ventral root, depending on which cranial nerve the fibers belong to.

Development of the Spinal Cord

Some features of spinal cord development are shown in Figure 13-9. Soon after closure of the neural folds (Chapter 4), the wall of the neural tube forms a pseudostratified columnar epithelium known as the **germinal** or **matrix zone** (Fig. 13-9*A*). The term "pseudostratified" refers to the fact that each cell within the matrix zone extends from the central canal to the outermost edge of the neural tube, even though a casual

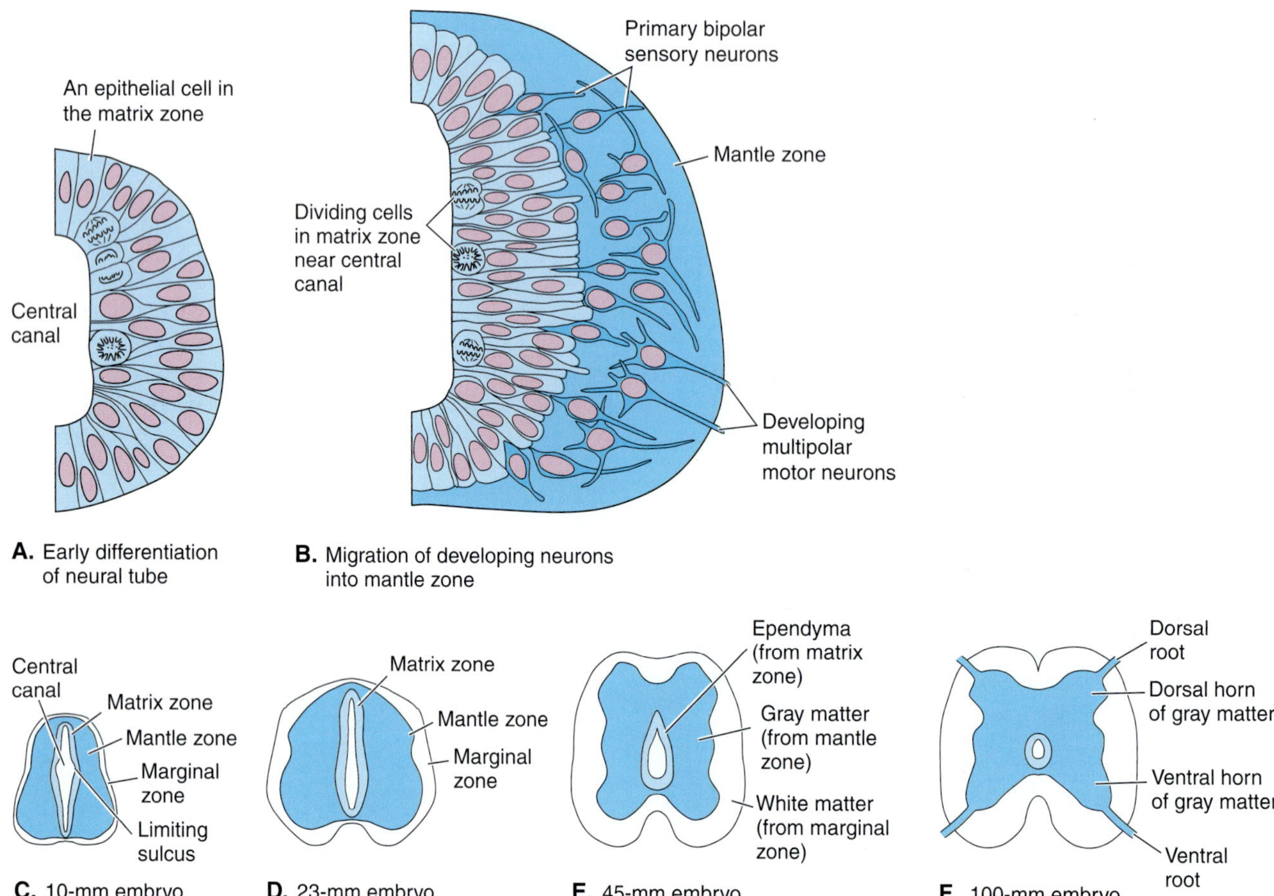

FIGURE 13-9
Development of the spinal cord. *A,* Early differentiation of the neural tube, showing proliferation of cells in the pseudostratified epithelium of the matrix zone. *B,* Migration of developing neurons into the mantle zone. Primary bipolar sensory neurons and multipolar motor neurons can be seen already. *C,* Transverse section through the spinal cord of a 10-mm pig embryo. A very thin marginal zone, which is fated to form the white matter, now can be seen. *D,* Schematic transverse section through spinal cord of a 23-mm pig embryo. *E,* Transverse section through spinal cord of a 45-mm pig embryo. *F,* Transverse section through spinal cord of a 100-mm pig embryo. (*Modified from Balinsky.)*

inspection might suggest that two or three layers of cells are present. Mitosis occurs very rapidly in the matrix zone, but these cell divisions occur only near its lumenal surface. As more and more presumptive neurons are generated, some begin to migrate outward from the matrix zone into the **mantle zone** (Fig. 13-9*B*). Such migrating neurons are initially bipolar; those in the ventral portion of the mantle that are destined to become motor neurons soon develop their typical multipolar appearance.

Axons of the nerves developing in the mantle zone next begin to grow outward into the **marginal zone** of the spinal cord (Fig. 13-9*C*). As diagrammed in Figure 13-9*D* and *E,* the marginal zone will give rise to the tracts of white matter on the outside of the neural tube, whereas the mantle forms the columns of gray matter. In adults, the matrix layer that was the source of all of these neurons is restricted to the very thin ependymal layer immediately surrounding the central canal.

One of the more striking aspects of spinal cord development is the tremendous overproduction of neurons and subsequent cell death that occurs. The explanation for these phenomena derives from the way in which neurons connect with their targets. For example, a group of multipolar motor neurons fated to supply muscles in the arm start their outgrowth before the arm has even formed. They extend their processes outward into the periphery and eventually reach the general site where the relevant target muscles are developing. Those neurons that establish contacts with cells within the muscle in turn receive trophic factor from the muscle cells and survive. Those neurons that do not establish appropriate contacts die. Such neuronal cell death eliminates more than 50% of the outgrowing neurons in some portions of the spinal cord.

Basic Neuronal Circuitry

Activities of the nervous system are controlled by many types of circuits among sensory neurons, interneurons, and motor neurons. Neurobiologists are only beginning to understand the complex pathways that mediate the integrative functions of the brain, including learning and memory. Many of the simpler circuits are well known, and we will consider some of them in the next chapter (Chapter 14) on the brain. Here, we examine three basic types of neuronal pathways common to all vertebrates: (1) **reflexes,** (2) **ascending pathways** (from lower centers to higher brain centers), and (3) **descending pathways** (from the brain to lower centers).

Much activity is controlled by **reflexes,** mediated by cells located in the spinal cord (Fig. 13-10). If you touch a hot stove, for example, impulses travel on primary sensory neurons and synapse with interneurons in the dorsal horn of the spinal cord. These interneurons stimulate and inhibit the appropriate combination of motor neurons, and you jerk your hand away. An integrated movement such as this requires that certain muscles contract with the appropriate force and at the correct time while the tension in antagonistic muscles is reduced in a comparable way. This particular reflex is termed a **three-neuron reflex arc,** for it includes three types of neurons: (1) a sensory neuron, (2) an interneuron, and (3) a motor neuron. In contrast, the familiar knee-jerk reflex requires only two types of neurons because the sensory neurons synapse directly with the motor ones. Reflexes such as the two-neuron and three-neuron reflex arcs involve neurons on only one side of the body and one body segment. Other reflexes include neurons that cross the body's midline and affect the other side of the body. Neuronal tracts that cross the midline are called **commissures,** or **decussations.**[5] Reflexes also may involve neurons that ascend or descend along the body axis and affect many body segments. These are described as **intersegmental reflexes.** For example, the coordinated movements of swimming and walking are integrated reflexively by pools of intersegmental and decussating neurons.

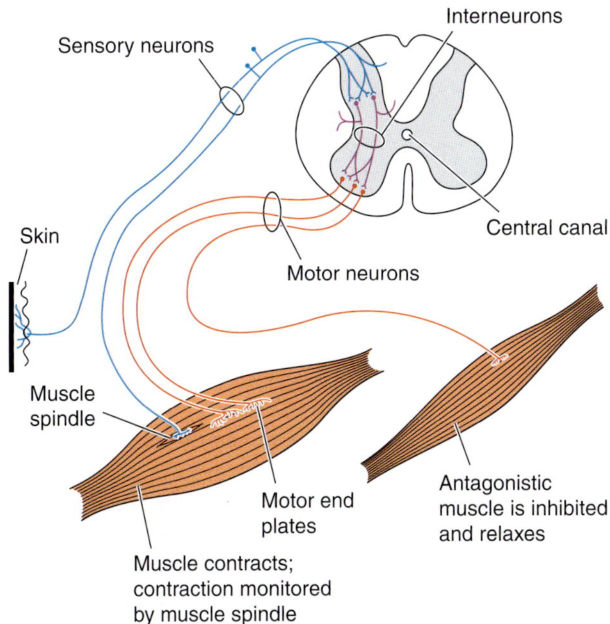

FIGURE 13-10
A simple reflex arc.

Swallowing, the rate of breathing, the rate of heartbeat, and many other vital processes are controlled reflexively by centers in the brainstem. Many reflexes in the cord and brainstem are the products of a long evolutionary history. They are **innate reflexes** and are the same in all individuals of a species. Others, called **conditioned reflexes,** develop during the lifetime of an individual as a result of an animal's repetitive experiences. When learning to drive a car, you must think about each action, but many of these actions become conditioned with experience. Conditioned reflexes are not inherited directly, although the capacity for them to develop is.

Many sensory impulses ascend from the spinal cord or brainstem to higher centers in the brain (Fig. 13-11*A*). Sensory impulses usually ascend on interneurons, but impulses from many proprioceptors and some touch receptors that enter the cord may ascend on the primary sensory neurons as far as the caudal end of the brainstem. Groups of axons with similar functions terminate in the same part of the brain and ascend together in common tracts that are described by their points of origin and termination—for example, the **spinocerebellar tracts** from the spinal cord to the cerebellum, the **spinothalamic tract** to the thalamus in the forebrain, and so on (Fig. 13-11*B*). These tracts are indistinguishable at the gross anatomical level in the spinal cord except for their general location within the three major funiculi of the white matter. Instead, they must be studied by tract-tracing methods (see Focus 14-1). Most sensory impulses decussate on their way to higher centers so that impulses originating on

[5]In neuroanatomy, the term "commissure" refers to a neuronal tract that projects to a group of cell bodies in the same basic location in the nervous system but on the opposite side of the body (i.e., the projection is to the group of cells that is the mirror image of the source group). In contrast, "decussation" refers to a neuronal tract that projects to a group of cell bodies in a different location in the nervous system and on the opposite side of the body.

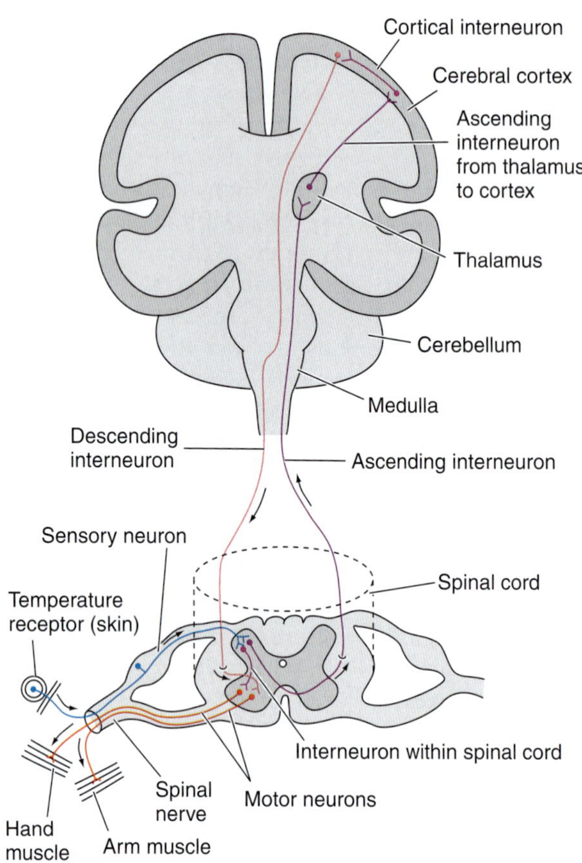

A. Pathways for temperature reception

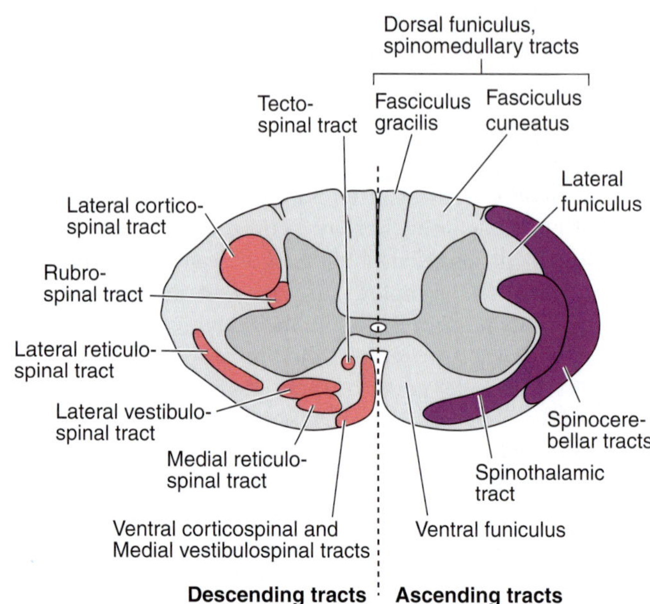

B. Descending and ascending tracts in the spinal cord

FIGURE 13-11
Pathways to and from the brain. *A*, The ascending pathway, taken by an impulse from a temperature receptor, and the descending pathway to an effector, such as a muscle, are shown. *B*, The location in the spinal cord of major descending and ascending tracts. Descending tracts are shown on the left; ascending ones on the right. *(A, After Dorit et al.; B, after Williams et al.)*

the left side of the body terminate in the right side of the brain, but the decussation is not always complete. The reasons for decussation are not well understood, other than the obvious need for the two halves of the body to function as an integrated unit and not as isolated left and right sides.

Many ascending interneurons of mammals terminate in nuclei in the **thalamus** (Fig. 13-11*A*). This important region of the diencephalon is considered in more detail in Chapter 14. Briefly, it serves as a site from which additional ascending neurons project to their targets in other brain regions. Sensory signals may be filtered in the thalamus, and certain ones continue by other interneurons to specific parts of the gray matter of the cerebrum, which is a cortex composed of alternating sheets of the cell bodies of interneurons and fiber tracts located on the outside surface of the forebrain (Chapter 14). When such signals reach specific parts of the **cerebral cortex** in humans, we become aware of the sensation. You have already jerked your hand away from that hot stove you touched, but now you sense that you burned your fingers. You may decide to do something, perhaps turn down the stove and fetch the first aid kit. Impulses descend through the CNS on other interneurons, which decussate at some level of the brain and continue down the cord in specific tracts, such as the **corticospinal tracts** (Fig. 13-11*B*), to terminate in neuronal pools that will activate and inhibit the appropriate muscles.

The Spinal Cord and Spinal Nerves

The spinal cord exhibits interesting and functionally important variation among vertebrates, some aspects of which are detailed in this section.

Spinal Cord

The spinal cord lies between the brain and the spinal nerves, but the cord is far more than a simple passageway for nerve impulses between receptors and effectors and the brain. Some interneurons of the spinal cord span from one segment of the trunk or tail to one or more other segments and thus are known as **intersegmental neurons.** Their presence makes possible considerable sensory-motor integration within the spinal cord. The rhythmic and coordinated movements of swimming and walking, for example, are controlled and integrated reflexively by pools of intersegmental and decussating spinal neurons, and such spinal circuits act as **central pattern generators.** If the spinal cord of a cat has been transected behind the skull, the cat can still walk on a moving treadmill with a nearly normal pattern of limb extension and flexion, although the animal must be supported. The explanation for this is that passive movements of the limbs caused by the treadmill activate cutaneous receptors in the skin and proprioreceptors in the muscles and tendons, and this is sufficient to start the central pattern generators. Continued sensory input is not necessary to maintain rhythmic patterns, but the synergy among muscles is less well coordinated without it. Thus, the details of locomotor integration are controlled in the spinal cord, but centers in the brain normally are responsible for initiating activity of the central pattern generators; maintaining balance; modulating speed and force of muscle contraction; and, of course, integrating goal-directed movements.

The complexity of the spinal cord, and the degree to which the brain exerts control over spinal activity, increases during the course of vertebrate evolution. Hagfishes and lampreys have relatively simple spinal cords. Because none of the axons are myelinated, a sharp distinction cannot be made between white and gray matter in the spinal cord of a lamprey (Fig. 13-12*A*). The cell bodies of the motor neurons are located, as one would expect, in the ventral part of the spinal cord, but their dendrites ramify into the peripheral part of the cord, where synapses occur with sensory neurons and interneurons. Although the cell bodies of a few large sensory neurons lie dorsally in the peripheral part of the cord, the cell bodies of most sensory neurons are located in the spinal ganglia. A few interneurons decussate and ascend and descend within the cord, but tracts are not well developed. The best-defined one is composed of the axons of giant **Müller cells.** The axons of the Müller cells descend from cell bodies located near the caudal end of the brain and synapse on the dendrites of motor neurons. Their function is to initiate rapid escape reactions. Although Müller axons have a diameter of 50 μm, they are nonmyelinated, and their conduction velocity is only 5 m/s. This speed is faster than in other lamprey neurons, but far slower than in the smaller but myelinated fibers found in gnathostomes. The spinal cord of a lamprey is relatively thin and avascular. Its flattened shape facilitates the diffusion of gases, nutrients, and other materials.

The spinal cord is larger, well vascularized, and rounder in gnathostomes (Fig. 13-12*B*). The evolution of a highly vascular nervous system was an important factor in vertebrate evolution, for it allowed larger and more complex nervous systems to evolve. As in lampreys, the cell bodies of interneurons and motor neurons lie in the gray matter, but many of their dendrites and the terminations of sensory neurons ramify more highly in the peripheral part of the cord, where many synapses among them are made. As more ascending and descending fiber tracts evolved, the white matter increased and bulged outward except near the midline, where the **dorsomedian sulcus** and **ventromedian fissure** remain (Fig. 13-12*B*).

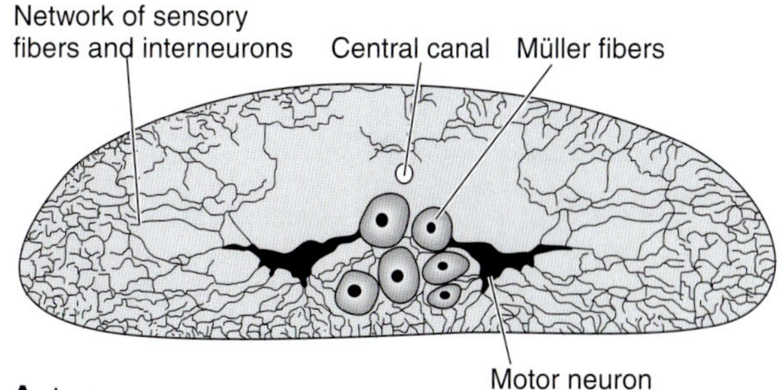

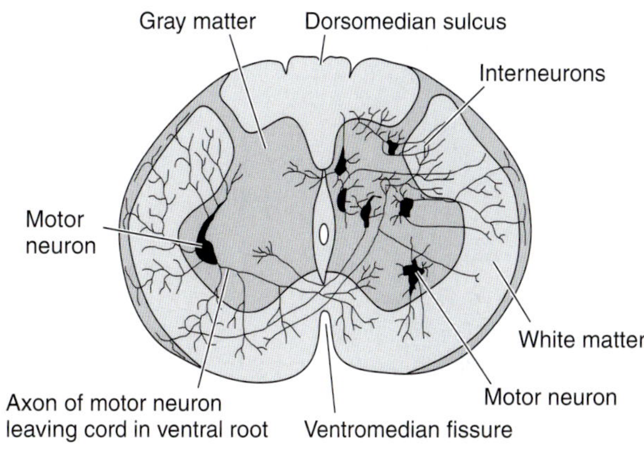

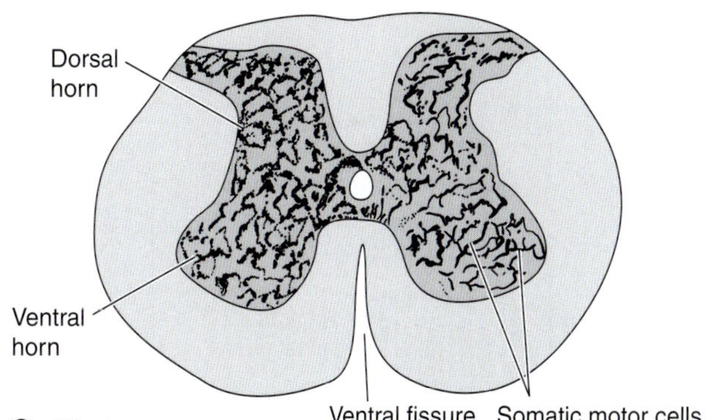

FIGURE 13-12
Transverse sections of the spinal cord of representative vertebrates. *A*, A lamprey. *B*, A frog. *C*, An alligator. *(A, After Tretjakoff; B, after Gaupp.)*

A. Dissection of the spinal cord of a human being

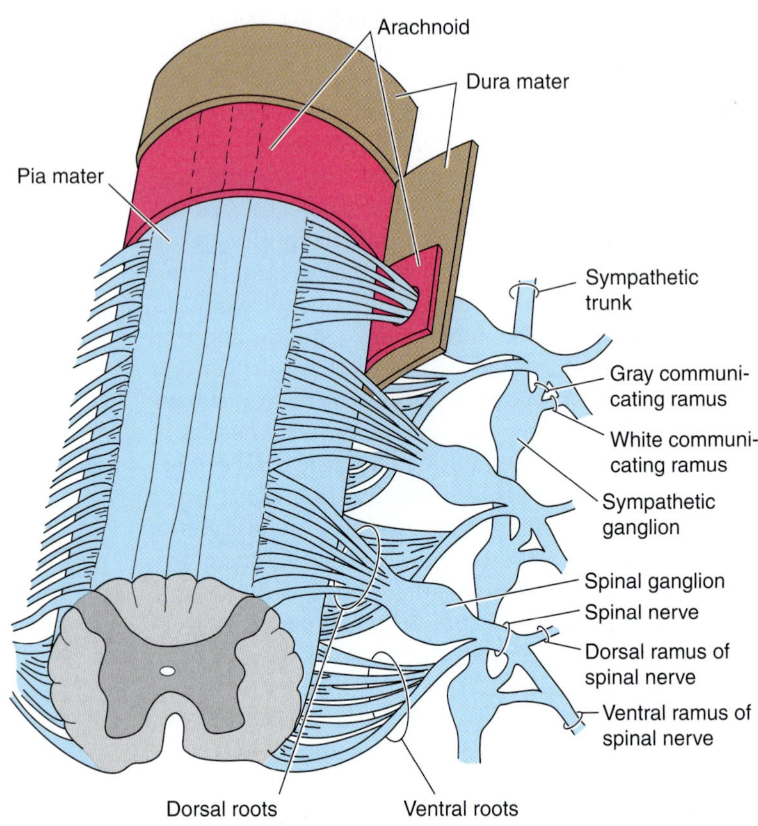

B. Organization of spinal roots, ganglia, and nerves in trunk

FIGURE 13-13

A, A dorsal view of the spinal cord and nerves of a human. *B,* The organization of spinal roots, ganglia, and nerves in the trunk of a mammal. The dura mater, arachnoid, and pia mater are colored to show their relationships to the spinal cord. In mammals, the dorsal and ventral roots are divided into rootlets. Also see Figures 13-6 and 14-3. *(A, Modified from Woodburne; B, modified from Panchen.)*

The gray matter in the spinal cord of an amniote has the characteristic butterfly shape when seen in transverse section (Fig. 13-12*C*). Synapses among neurons of amniotes are confined to the gray matter; white matter contains only ascending and descending fiber tracts. The brains of birds and mammals exert more control over body activity than do the brains of other vertebrates, and tracts to and from the brain are correspondingly more numerous. The degree of vascularization of the cord increases, and the number of neuroglial cells that service the neurons also is greater in birds and mammals.

The spinal cord is nearly as long as the vertebral column in most vertebrates, although in mammals, frogs, and a few teleosts, the cord is shorter because it grows more slowly than does the rest of the body (Fig. 13-13*A*). As a consequence, the more caudal spinal nerves form a bundle, known as the **cauda equina,** as they extend from their attachment to the cord caudally through the vertebral canal to the intervertebral foramina through which they leave. A thin **terminal filament,** composed of the connective tissues surrounding the spinal cord, extends from the end of the cord to the caudal end of the vertebral canal. Vertebrates with well-developed limbs have large numbers of neurons supplying them, and the diameter of the spinal cord is correspondingly greater in the cervical and lumbar regions, where most nerves to the limbs originate. The spinal cord of mammals is protected by the thin **pia mater,** which overlies the neural tissue; the highly vascularized **arachnoid layer** outside of it; and the tough **dura mater,** which ensheaths the entire cord and roots of the spinal nerves (Fig. 13-13*B*). These three layers also surround the brain and are discussed in greater detail in Chapter 14.

Spinal Nerves

As we have seen, the spinal nerves of vertebrates usually attach by roots to the spinal cord, but in lampreys, the homologues of the dorsal and ventral roots form distinct **dorsal** and **ventral spinal nerves.** The ventral nerves of lampreys are segmentally spaced and lead directly into the myomeres. They carry only somatic motor neurons (Fig. 13-14*A* and *B*). The dorsal nerves of lampreys are intersegmental and pass between the myomeres to the body surface and gut regions. They contain both somatic and visceral sensory neurons and

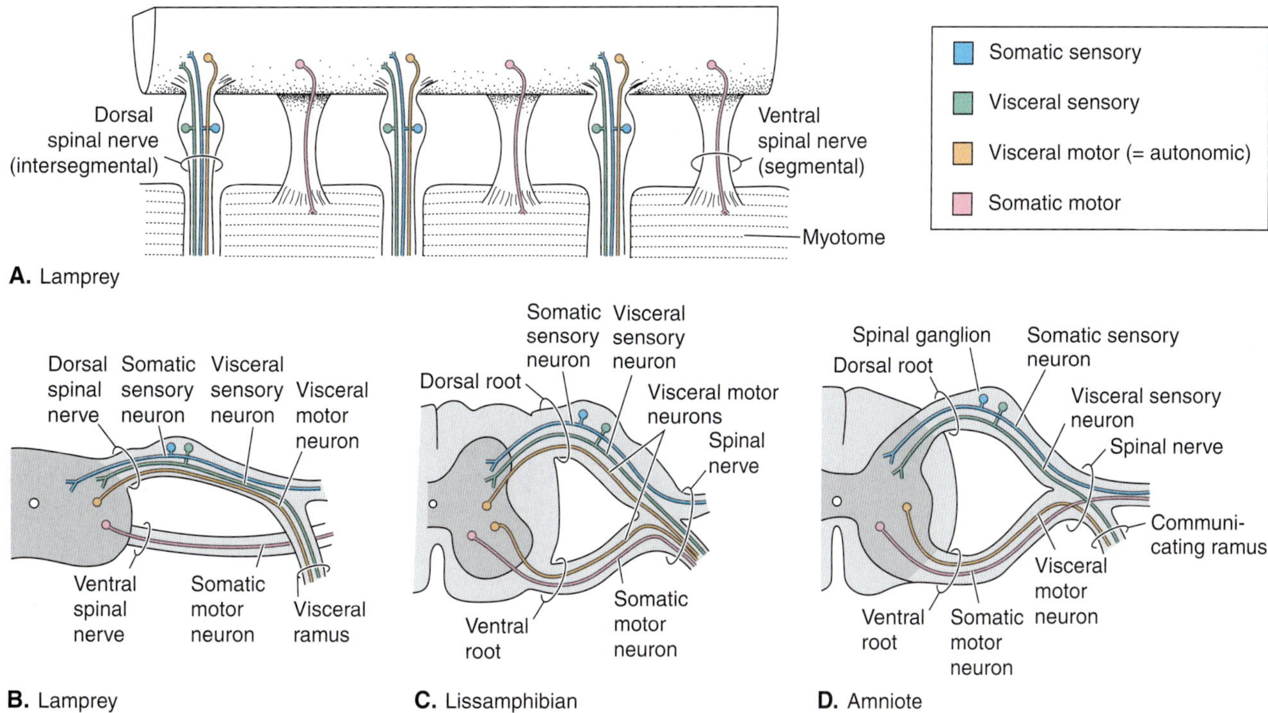

FIGURE 13-14
Evolution of spinal nerves. *A,* A dorsal view of the spinal cord and nerves of a lamprey. *B,* Transverse section through the spinal cord and nerves of a lamprey. *C,* Transverse section through the spinal cord of a lissamphibian. *D,* Transverse section through the spinal cord of an amniote. The neurons are color coded to indicate the functional groups to which they belong. *(Modified from Romer and Parsons.)*

the small number of visceral motor neurons that leave the CNS through the spinal nerves. Most visceral motor (= autonomic) neurons reach the viscera of lampreys through a cranial nerve, the vagus (described in the later section on the autonomic nervous system). The presence of distinct dorsal and ventral spinal nerves, each with its own types of neurons, is believed to be a plesiomorphic character of vertebrates.

In all other craniates, including hagfishes, the distinct dorsal and ventral spinal nerves of a trunk segment unite to form a definitive spinal nerve (Fig. 13-14*C* and *D*). The originally separate dorsal and ventral nerves now are referred to as **dorsal** and **ventral roots** of the spinal nerve of that segment. All sensory fibers continue to enter through the dorsal root, and somatic motor fibers continue to leave through the ventral root. In anamniotes, many visceral motor fibers leave through the dorsal roots, but some also leave through the ventral root (Fig. 13-14*C*). In amniotes, all visceral motor fibers leave through the ventral roots along with the somatic motor neurons (Fig. 13-14*D*).

The segmental pattern of development and distribution of the spinal nerves in elasmobranchs is shown in Figure 13-15. In an early embryo, before the nerves have extended very far from the neural tube, a series of myotomal muscle buds extends into the prospective pectoral, pelvic, and two dorsal fins (Fig. 13-15*A*). As diagrammed in the hypothetical intermediate developmental stage (Fig. 13-15*B*), nerves that will supply the pectoral fin arise from spinal segments 1 to 13; those supplying the pelvic fin from the spinal segments 25 to 35. Thus, the fins and their muscles and nerves have a discrete and predictable segmental organization. In the adult, the nerves that extend into the pectoral and pelvic fins bend toward the bases of their fins (Fig. 13-15*C*).

The number of spinal nerves varies with the number of segments in the trunk and tail. A frog, specialized to thrust powerfully with its hind legs as it swims or jumps, has a short trunk, a reduced tail, and only ten pairs of spinal nerves. A snake, which moves by lateral undulations of its long trunk and tail, may have several hundred pairs of spinal nerves. Spinal nerves are named only when body regions are well defined, as they are in mammals, which have **cervical spinal nerves, thoracic spinal nerves, lumbar spinal nerves, sacral spinal nerves,** and **caudal spinal nerves.** The spinal

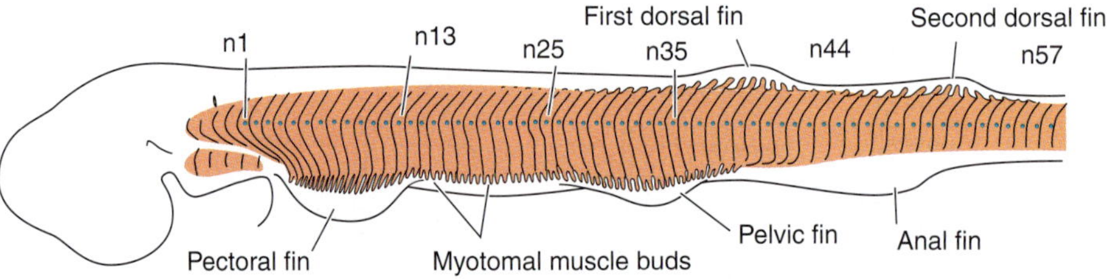

A. Spinal nerves, somites, and associated structures of a 19-mm embryo of *Squalus*

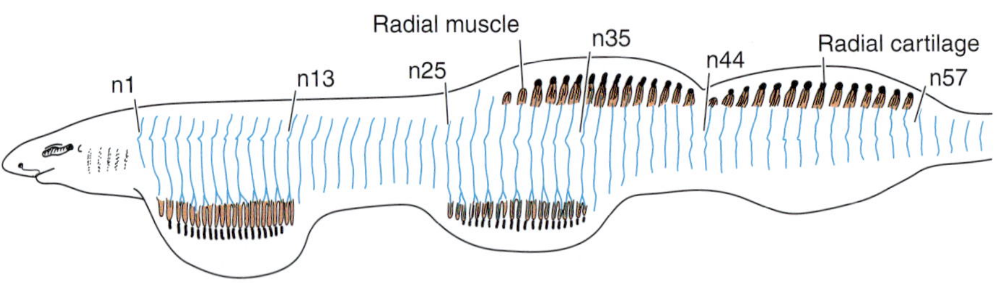

B. Hypothetical intermediate developmental stage

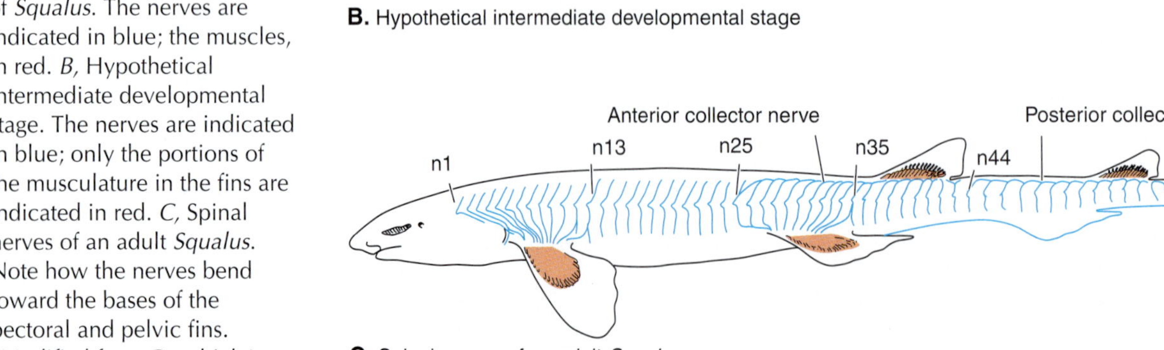

C. Spinal nerves of an adult *Squalus*

FIGURE 13-15 Nerves to fins of an elasmobranch. *A,* Spinal nerves, somites, and associated structures of a 19-mm embryo of *Squalus.* The nerves are indicated in blue; the muscles, in red. *B,* Hypothetical intermediate developmental stage. The nerves are indicated in blue; only the portions of the musculature in the fins are indicated in red. *C,* Spinal nerves of an adult *Squalus.* Note how the nerves bend toward the bases of the pectoral and pelvic fins. *(Modified from Goodrich.)*

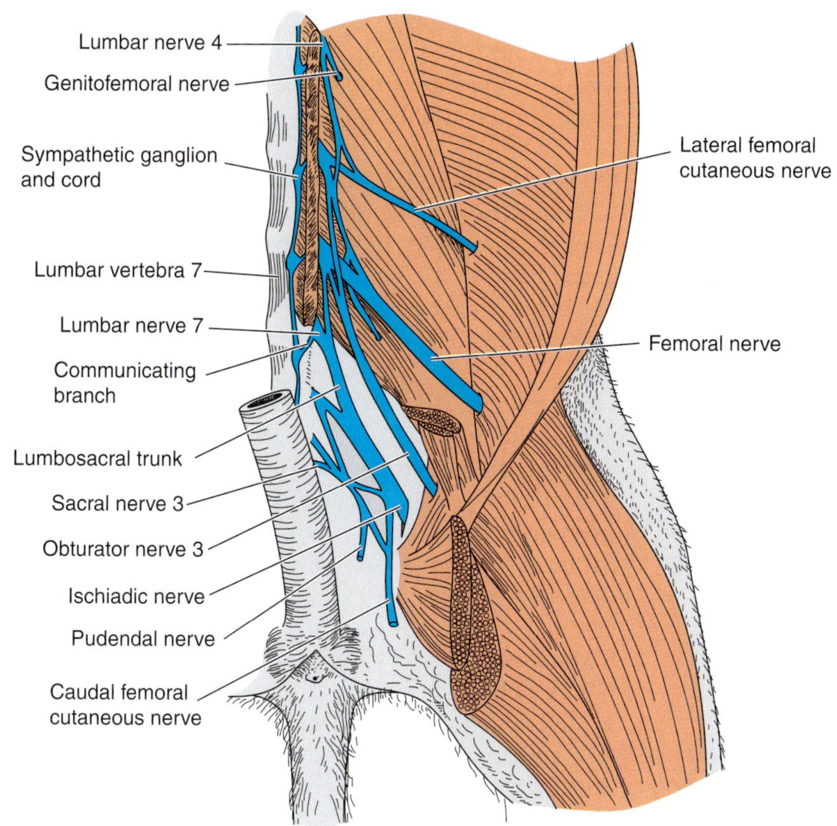

FIGURE 13-16
The lumbosacral plexus of a cat. *(After Walker and Homberger.)*

nerves within each of these regions are assigned simple abbreviations, such as C1, C2, T1, and T2.

Epaxial muscles are segmented in most vertebrates, and the **dorsal rami** of spinal nerves supplying these muscles retain a segmental pattern (Figs. 13-13*A*). Even in mammals, in which many of the epaxial muscles form longitudinal muscle bundles, the deeper epaxial muscles remain segmented. Correlated with the complexity of embryonic development of the hypaxial muscles, and especially of the appendicular muscles, many **ventral rami** unite to form complex networks or **plexuses** before nerves are distributed to the muscles (Fig. 13-16). The innervation of muscles is phylogenetically conservative. Motor neurons establish connections very early in embryonic development with developing muscle fibers in the myotomes, and they tend to follow their target tissue even if it migrates or unites with tissue from other myotomes. One muscle in a limb, for example, may have developed from muscle tissue that came embryonically from myotomes 8 to 10; another muscle, from tissue derived from myotomes 9 to 12. Another striking example of the segmental organization of the limbs comes from the pattern of innervation of the skin (Fig. 13-17). In

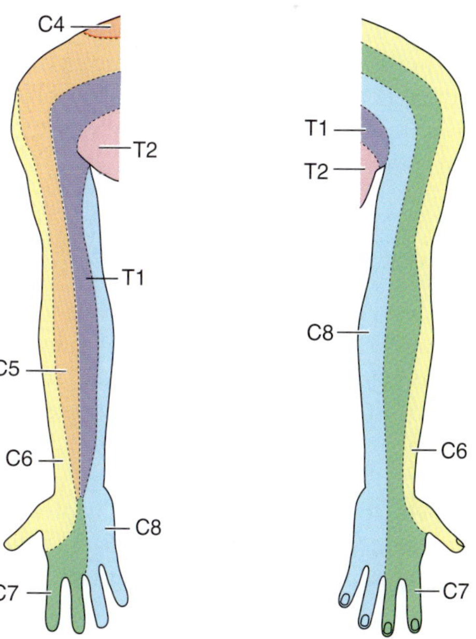

FIGURE 13-17
Dermatomes of the human arm. *A,* Ventral surface. *B,* Dorsal surface. *(Modified from Panchen.)*

humans, cervical segments C4 to C8 and thoracic segments T1 and T2 innervate the **dermatomes** (= dermal segments) of the arm.[6] To reach the muscles, dermatomes, and other structures of the limbs, some of the neurons in the spinal nerves come together in the plexus at the base of the limb. Such plexuses occur in all gnathostomes, and they reach their highest complexity among mammals and birds in which the **cervical plexus** supplies many ventral neck muscles, the **brachial plexus** supplies the pectoral appendage, a **lumbosacral plexus** supplies the pelvic appendage, and a **coccygeal plexus** supplies some of the pelvic muscles.

Cranial Nerves of Gnathostomes as Represented by Chondrichthyans

An understanding of the structure and function of the cranial nerves can provide remarkable insight into the anatomical organization of the entire cranial region.

The simplified distribution pattern of the cranial nerves of an idealized gnathostome based on a shark is shown in Figure 13-18, and the functional components of these nerves are listed in Table 13-4. The nerves have names descriptive of some aspect of their structure or function, and most also are expressed as Roman numerals that correspond to their sequence in humans. Study of the cranial nerves can be simplified somewhat if we interpret some as unique sensory nerves of the head, others as serially homologous to the ventral spinal nerves of lampreys and probably ancestral vertebrates, and still others as serially homologous to dorsal spinal nerves of lampreys.

Sensory Nerves Unique to the Head

Sensory nerves unique to the head include four sensory nerves and two general categories of nerves found only in the head and composed only of special somatic sensory fibers or special visceral sensory fibers (Fig. 13-18*A*, blue lines).

Terminal Nerve (0) As we have seen (Chapter 12), the special somatic sensory terminal nerve may be a part of a chemosensory system regulating certain aspects of reproduction in response to olfactory pheromones. It is given the designation 0 because it was discovered after the conventional numbering system had been established.

Olfactory Nerve (I) The olfactory nerve is a special somatic sensory nerve that returns chemosensory fibers from the nasal sac. It is a short nerve in most vertebrates because the nasal sac and the olfactory bulb of the brain, to which its axons go, usually lie next to each other.

Optic Nerve (II) Because the retina develops as a part of the brain (Chapters 4 and 12), the optic nerve is technically a brain tract and not a nerve. Its axons are those of the ganglion cells of the retina. These are interneurons that receive impulses from the bipolar retinal neurons (Fig. 12-26) and carry them to the optic lobes of the brain.

Statoacoustic, Vestibulocochlear, or Octaval Nerve (VIII) This nerve is known by different names in different groups of vertebrates. Because their ears lack a cochlea, nerve VIII of an anamniote is often called the statoacoustic nerve instead of the vestibulocochlear nerve, which is the term used for mammals. A name that avoids confusing references to its different functions in different vertebrates is the "octaval nerve," which is preferred by many comparative neuroanatomists. Regardless of its name, nerve VIII can be characterized as a special somatic sensory nerve that returns fibers from all parts of the inner ear (i.e., parts related to balance and parts related to hearing).

Lateral Line Nerves (not numbered) The vast majority of anamniotes have lateral line and electroreceptive organs (Chapter 12) innervated by a series of six lateral line nerves that develop from six discrete neurogenic placodes on the surface of the embryonic head (Chapter 4). Our system for numbering cranial nerves is based on the condition in humans. Because humans and other amniotes lack lateral line nerves, the lateral line nerves have never been numbered. Conventionally, comparative anatomists have regarded lateral line nerves as components of cranial nerves VII, IX, and X, and this interpretation still is presented in many textbooks. More accurately, the anterodorsal (= **ADLLN**), anteroventral (= **AVLLN**) and otic (= **OLLN**) lateral line nerves form their own discrete ganglia near the trigeminal and facial ganglia, and their roots enter the brainstem adjacent to those of nerve VII (Figs. 13-18 and 13-19). The middle (= **MLLN**), supratemporal (= **STLLN**), and posterior (= **PLLN**) lateral line nerves form ganglia and enter the brainstem near the roots of nerves IX and X.

[6]Like the skeleton and muscles of the arm, dermal structures of the arm also arise from specific trunk segments during embryogenesis and retain this segmental pattern into adulthood; the innervation pattern of the dermatomes provides clear evidence of this.

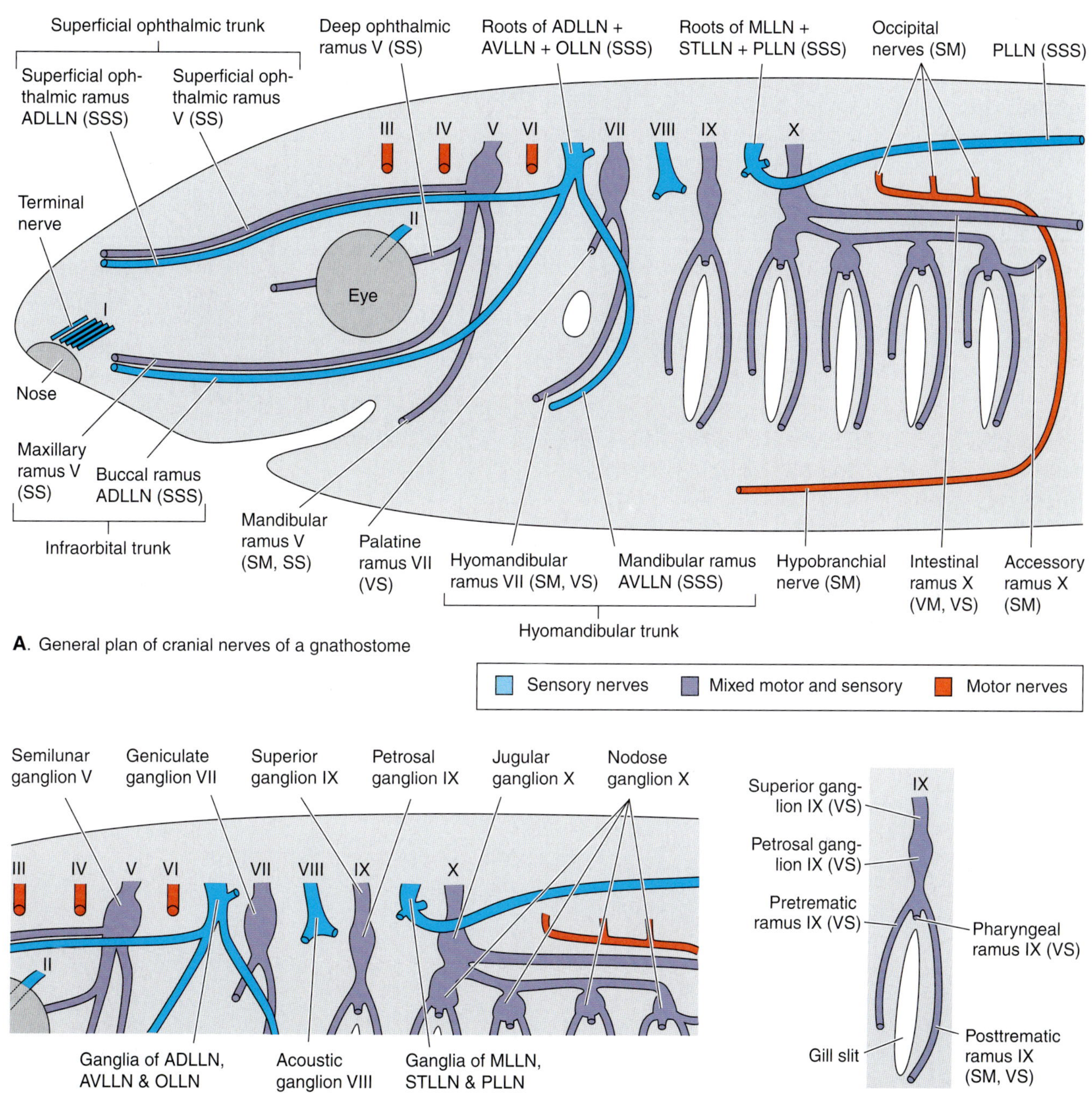

FIGURE 13-18

The cranial nerves and their major branches in a typical gnathostome, based on their arrangement in spiny dogfish (*Squalus*) and coelacanths (*Latimeria*). This diagram exaggerates certain features and omits others in order to emphasize general patterns. Nerves are identified by Roman numerals or abbreviations; see Table 13-4. *A,* General plan of cranial nerves of a gnathostome. Many texts still treat the six lateral line nerves (ADLLN, AVLLN, OLLN, MLLN, STLLN, and PLLN) as components of other cranial nerves, but we have adopted the research findings of Northcutt (1990) that these are ontogenetically and phylogenetically distinct cranial nerves. *B,* General terminology for cranial nerve ganglia of a gnathostome. *C,* Detail of the major rami of nerve IX, the glossopharyngeal nerve. The arrangement of the pharyngeal, pretrematic, and posttrematic rami is indicated. SM = somatic motor, SS = somatic sensory, SSS = special somatic sensory, VM = visceral motor (autonomic), VS = visceral sensory. *(See Northcutt and Bemis, 1993.)*

TABLE 13-4 Cranial Nerves and Their Major Functional Components

Nerves	Chondrichthyans						Condition in Amniotes
	Somatic Sensory		Visceral Sensory		Visceral Motor	Somatic Motor	
	General and Proprioceptive	*Special*	*General*	*Special*			
0 (Terminal)		X					No change
I (Olfactory)		X					Vomeronasal component usually added
II (Optic)		X					Less decussation of fibers
III (Oculomotor)					X	X	No change
IV (Trochlear)						X	No change
ADLLN Anterodorsal lateral line nerve		X					Lost
AVLLN Anteroventral lateral line nerve		X					Lost
OLLN Otic lateral line nerve		X					Lost
V (Trigeminal)	X					X	No change
VI (Abducens)						X	No change
VII (Facial)	X		X	X		X	Autonomic fibers to lacrimal glands and rostral salivary glands added
VIII (Statoacoustic or vestibulocochlear)		X					No change
MLLN Middle lateral line nerve		X					Lost
STLLN Supratemporal lateral line nerve		X					Lost
PLLN Posterior lateral line nerve		X					Lost
IX (Glossopharyngeal)	X		X	X		X	Autonomic fibers to parotid salivary glands added
X (Vagus)	X		X	X	X	X	Reduction of somatic motor fibers; accessory branch becomes cranial nerve XI
Occipitals	X					X	Becomes cranial nerve XII

Gustatory Nerves (not numbered) The specialized visceral sensory fibers that carry taste sensations develop from a series of neurogenic placodes originally located dorsal to the branchiomeres of an embryo and thus known as the **epibranchial placodes.** The gustatory fibers become associated with cranial nerves VII, IX, and X. Their status as separate and discrete cranial nerves is discussed in the later section on the origin of special sensory neurons.

Ventral Cranial Nerves

Those cranial nerves interpreted as ventral cranial nerves resemble the ventral spinal nerves of lampreys because they contain primarily somatic motor neurons going to many of the somatic muscles in the head (compare Figs. 13-7 and 13-14*A* and *B*). This group consists of those cranial nerves that supply the extrinsic ocular muscles and the epibranchial and hypobranchial muscles. Some

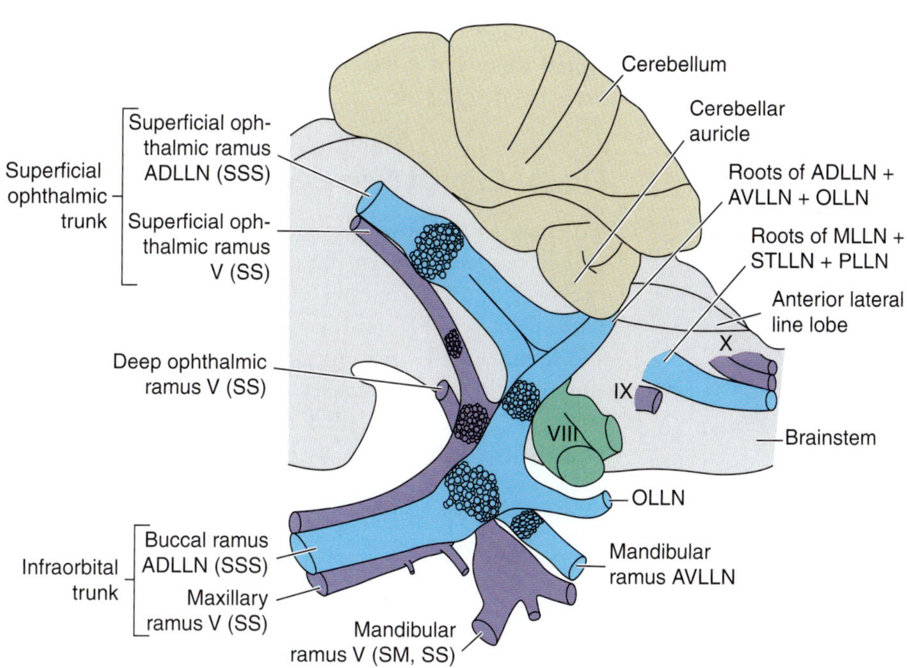

FIGURE 13-19
Dissection of the lateral line nerves of smooth dogfish (*Mustelus*) near their entrance to the brainstem. Compare this figure with Figure 13-18. SS = somatic sensory, SSS = special somatic sensory. *(Modified from Boord and Campbell.)*

also carry somatic sensory proprioceptive neurons, which return information about the degree of contraction of the extrinsic ocular muscles. The hypothesized patterns of derivation of these muscles from somitomeres and somites in the head are shown in Figure 13-20. For an overview of the ventral cranial nerves, see the nerves indicated with red lines in Figure 13-18*A*.

Oculomotor Nerve (III) The oculomotor nerve supplies the four extrinsic ocular muscles that develop from the first two somitomeres: dorsal rectus, medial rectus, ventral rectus, and inferior oblique. Anatomists are unsure about the location of proprioceptive neurons returning from these muscles in fishes. In humans, the proprioreceptive neurons travel in the trigeminal nerve. The oculomotor nerve has a small **ciliary branch,** which carries autonomic neurons (visceral motor) into the eyeball to supply the ciliary and iris muscles.

Trochlear Nerve (IV) The trochlear nerve (L., *trochlea* = pulley), so named because the muscle it supplies passes through a connective tissue pulley in

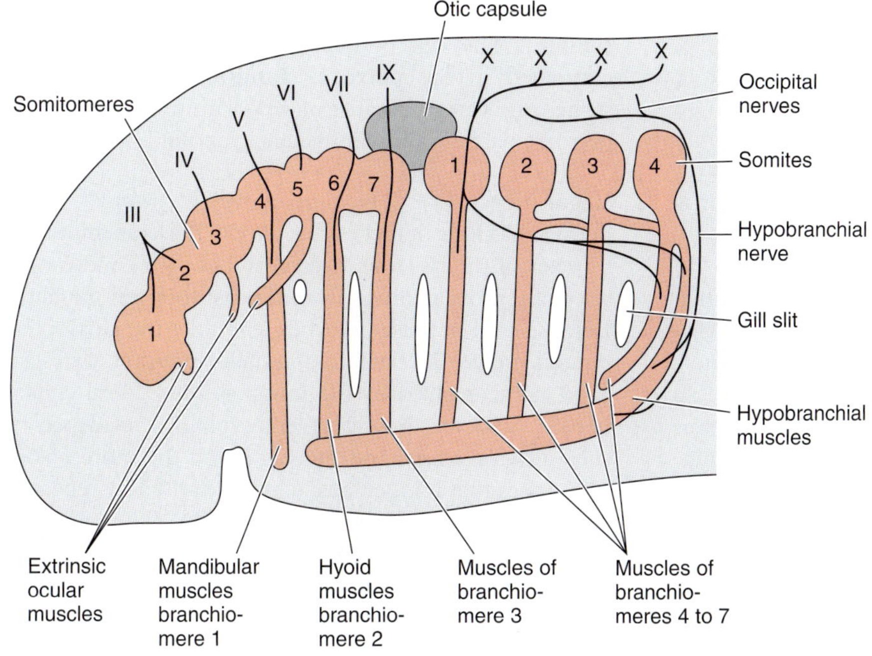

FIGURE 13-20
Relationship between cranial nerves, somitomeres, somites, and visceral arches during the development of the head of an early vertebrate. Nerves are identified by Roman numerals, and somitomeres and somites, by Arabic numerals. *(Modified from Northcutt.)*

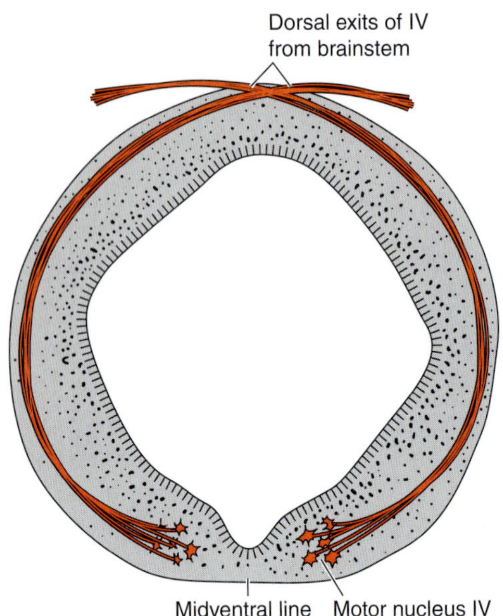

FIGURE 13-21
Exit of the roots of the trochlear nerve. *(Modified from Balinsky.)*

mammals, supplies the superior oblique muscle, the one extrinsic ocular muscle that develops from the third somitomere. Although its fibers actually leave the brainstem from a dorsal position, they clearly originate near the midventral line of the brain (Fig. 13-21).

Abducens Nerve (VI) The abducens nerve (L., *abducens* = leading away) arises from the ventral surface of the brain and supplies the lateral rectus muscle, the only muscle to develop from the fifth somitomere in many living vertebrates. In some other living vertebrates, the abducens nerve supplies an additional cranial muscle derived from somitomere 5. For example, in the coelacanth (*Latimeria*), it supplies the large basicranial muscle (Fig. 13-22), which is believed to be homologous to the retractor bulbi muscle of some tetrapods (a good example is a frog; watch how a frog retracts its eyes when swallowing, a movement that is caused by the retractor bulbi).

Occipital Nerves Epibranchial and hypobranchial muscles of fishes, which are located in the caudal region of the head and beneath the pharynx, are innervated by three or four occipital nerves and by a variable number of anterior spinal nerves. The occipital nerves are called occipital rather than cranial or spinal because the location of the caudal border of the skull varies phylogenetically. In some species, these nerves emerge from the brainstem and spinal cord within the skull; in other species, they emerge just behind the skull. Their dorsal rami supply the epibranchial muscles, and their ventral rami unite to form the **hypobranchial nerve,** which follows the hypobranchial musculature into the floor of the pharynx.

Dorsal Cranial Nerves

The next four cranial nerves (V, VII, IX, and X) appear to be comparable with the dorsal spinal nerves of lampreys. They are mixed nerves that contain some combination of sensory and motor neurons (Figs. 13-7, 13-8, and 13-14). All supply somatic motor neurons to the branchiomeric muscles of the visceral arches, so they sometimes are called branchiomeric nerves. Beyond this, some carry visceral motor (= autonomic neurons) and visceral sensory neurons, and all have somatic sensory neurons. These nerves are shown in purple in Figure 13-18.

Glossopharyngeal Nerve (IX) We start our treatment of the dorsal cranial nerves out of numerical sequence because nerve IX offers a relatively simple prototype for understanding the basic branching pattern of a dorsal cranial nerve (Fig. 13-18*C*). Its primary branch is a **posttrematic branch** (Gr., *tremat* = hole or cleft), which passes caudal to the first complete gill pouch, carries somatic motor neurons to the muscles of the third visceral arch, and returns visceral sensory neurons from the adjacent part of the gill pouch. Other branches are its **pretrematic branch,** which goes in front of the first gill pouch, and a small **pharyngeal branch** to the roof of the pharynx, both of which carry only visceral sensory neurons. In mammals, the free nerve endings and specialized receptors that are the source of the sensory information lie on the back of the tongue and adjacent parts of the pharynx; hence the name for this nerve (Gr., *glossa* = tongue). The cell bodies of its sensory neurons are contained in the **petrosal ganglion** (Gr., *petros* = stone), so called because it lies within the stony, or petrous, part of the mammalian temporal bone.

Vagal Nerve (X) The vagal nerve (L., *vagus* = wandering) has four branches that supply the branchiomeric muscles of the last four visceral arches (4–7) and return visceral sensory neurons from the last four gill pouches. Because each of these branches resembles the glossopharyngeal nerve, investigators usually interpret that the vagal nerve evolved by the union of four dorsal cranial nerves; however, no direct embryological evidence of such a union exists, and the vagus and the more posterior arches that it innervates may have arisen by repeated genesis of the "last arch" (Gilland, 1994). The vagus has a short **accessory branch,** which carries somatic motor neurons to the cucullaris muscle, and a long **intestinal branch,** which carries autonomic (= visceral motor)

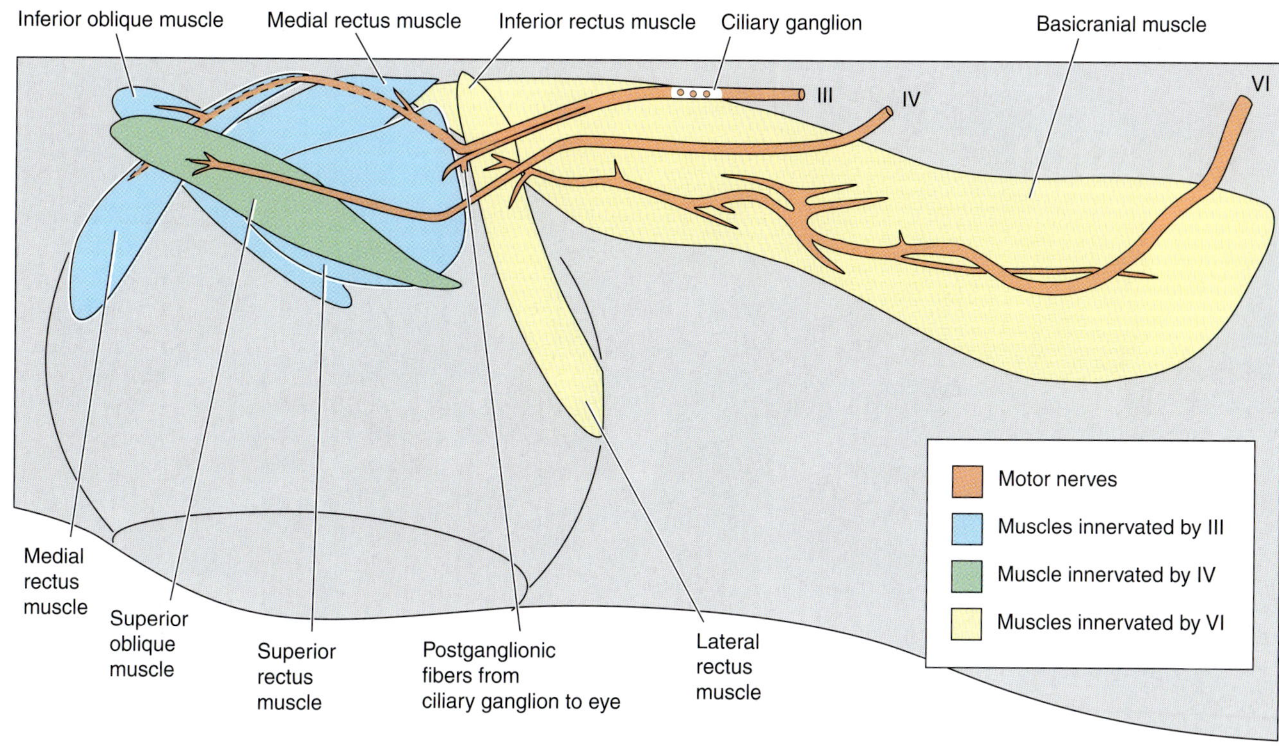

A. Basicranial and extrinsic eye muscles of *Latimeria* in dorsal view showing innervation by III, IV, and VI

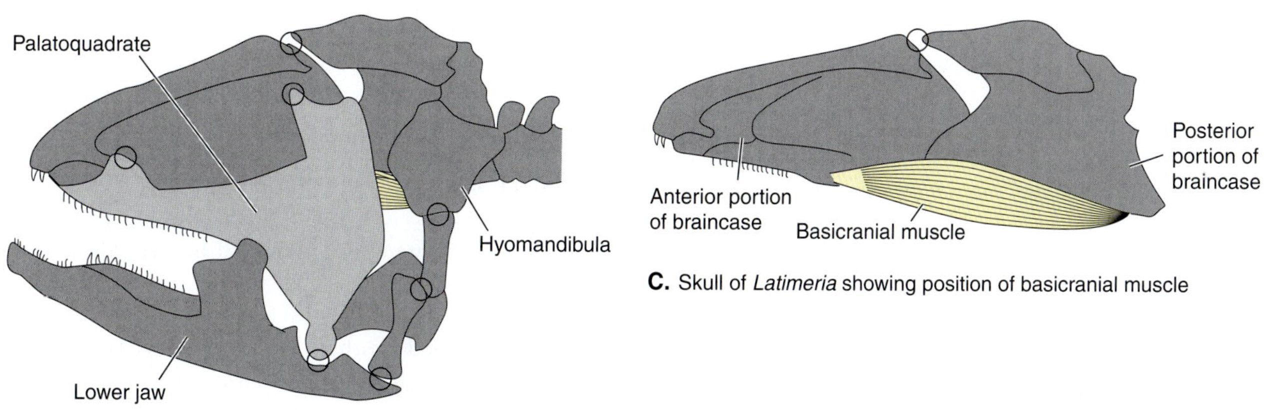

B. Skull and lower jaw of *Latimeria* in lateral view

C. Skull of *Latimeria* showing position of basicranial muscle

FIGURE 13-22
Nerves to the basicranial and extrinsic ocular muscles of the coelacanth, *Latimeria*. *A,* Basicranial and extrinsic eye muscles of *Latimeria* in dorsal view, showing innervation by nerves III, IV, and VI. *B,* Skull and lower jaw of *Latimeria* in lateral view. *C,* Skull of *Latimeria,* showing position of basicranial muscle. *(A, Modified from Northcutt and Bemis.)*

neurons to the abdominal viscera and returns visceral sensory neurons. The cell bodies of the sensory neurons in this nerve lie in the **jugular ganglion** and in the series of **nodose ganglia** (L., *nodus* = knob).

Facial Nerve (VII) and Trigeminal Nerve (V) Gill pouches are reduced rostral to the glossopharyngeal nerve, so it is not surprising that the more rostral dorsal cranial nerves lack a pretrematic branch. Like the glossopharyngeal nerve, the facial nerve has a pharyngeal branch, which is large and known as the **palatine ramus.** In contrast, the trigeminal lacks a pharyngeal branch, although whether this is because it was secondarily lost or because it was never present is unknown. The posttrematic branch of the facial nerve is its **hyomandibular ramus,** and that of the trigeminal nerve

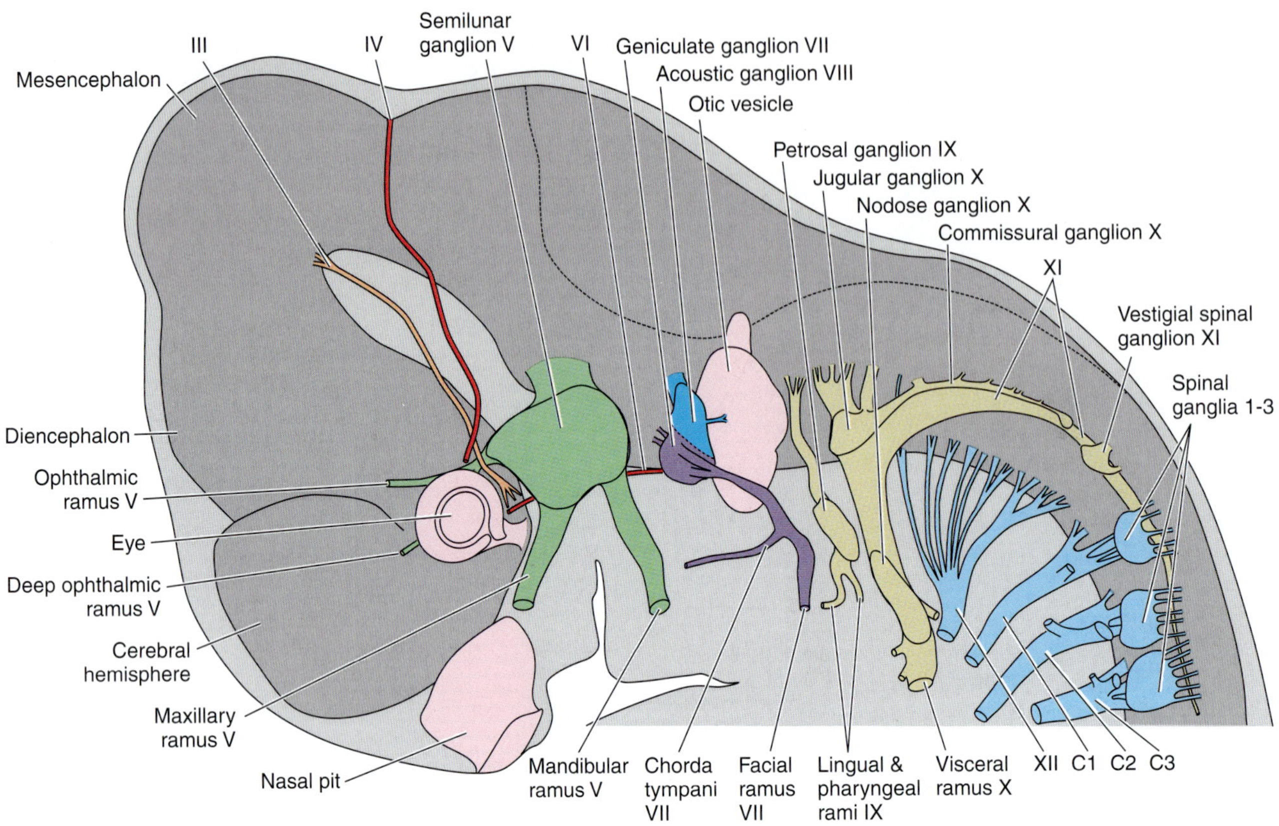

FIGURE 13-23
Developing cranial nerves and ganglia of a 10-mm pig embryo. *(Modified from Balinsky.)*

is the **mandibular ramus.** These nerves carry somatic motor neurons to the branchiomeric muscles of the hyoid and mandibular arches, respectively. The hyomandibular nerve also carries visceral sensory neurons, and the mandibular ramus carries general somatic sensory neurons from the skin over the lower jaw. The dorsal branches of the trigeminal nerve bifurcate and pass dorsal and ventral to the orbit. In anamniotes, branches of the anterodorsal lateral line nerve (its **superficial ophthalmic** and **buccal rami**) travel with these branches of the trigeminal nerve and return special somatic sensory lateral line neurons from most of the head. The dorsal branches of the trigeminal nerve (its **superficial ophthalmic** and **maxillary rami**) return general somatic sensory neurons from the skin over the head. The cell bodies of the sensory neurons in the facial nerve lie in the **geniculate ganglion** (L., *geniculum* = little knee); those of the trigeminal nerve lie in the **semilunar ganglion** (Fig. 13-23).

One of the most intriguing portions of the trigeminal nerve complex is the **deep ophthalmic ramus,** also known as the **profundal nerve.** Its fibers develop from a separate neurogenic placode in many anamniotes, and it has its own discrete ganglion, the **profundal ganglion,** which lies adjacent to the semilunar ganglion.

Evolution of the Cranial Nerves

We have interpreted certain cranial nerves as the serial homologues of ventral spinal nerves, and others as the serial homologues of dorsal cranial nerves. This assumption is based on the interpretation that the vertebrate head, or at least a substantial part of it, evolved from a series of segments comparable to the trunk segments. Segmentation of the vertebrate trunk is fundamentally a segmentation of the locomotor myomeres, but this imposes segmentation on the axial skeleton to which the myomeres attach and on the spinal nerves and blood vessels that supply them. The head also may have been segmented in ancestral vertebrates, as is the front of the body in amphioxus. As activity increased and a high degree of cephalization evolved in early vertebrates, the head acquired new sensory, integrative, feeding, and respiratory functions. Because of these functions, and because cranial structures develop embryonically under the influence of other segmental patterns imposed by neural crest cells and ectodermal neurogenic placodes, we should not be surprised to see differences among segments in the head and trunk. However, if the head is segmented, we should be able to account for a ventral and a dorsal nerve for each of the head segments. We have discussed (Chapter 10 and Fig. 10-7) the segmental nature of the cranial paraxial mesoderm. Four distinct somites lie caudal to the otic capsule, and seven less clearly demarcated somitomeres lie rostral to the otic capsule (Fig. 13-20). Northcutt (1990) proposed that two somitomeres are the equivalent of a head segment. If one somitomere in the otic region has been lost (and evidence from chondrichthyan embryos shows the loss of premuscular tissue in this region), then vertebrates have four preotic head segments and four postotic segments (Table 13-5). Do some of the cranial nerves relate to these segments in the same way that separate dorsal and ventral spinal nerves of lampreys (and probably ancestral vertebrates) relate to trunk segments? We cannot be certain at this time, but Goodrich argued for serial homology in his classic synthesis (1930), and Northcutt (1990) tentatively supported this view.

The cranial nerves we have interpreted as ventral cranial nerves certainly resemble the ventral spinal nerves of lampreys. They contain only (or primarily) somatic motor neurons, and they lack sensory neurons to the skin or viscera. They enter the brainstem ventrally, with the exception of the trochlear nerve, which leaves the brainstem from a dorsal position (e.g., Figs. 13-21 and 13-23). Despite its point of exit from the brain, the nucleus of the trochlear nerve is located ventrally just as we would expect. As you can see in Figure 13-20 and Table 13-5, the oculomotor (III), trochlear

TABLE 13-5 Segmentation of Cranial Muscles and Nerves

Head Segment	Origin	Visceral Arch	Muscles	Ventral Nerves (somatic motor)	Dorsal Nerves (somatic motor, general somatic, and visceral sensory)	Octavolateralis Nerves	Chemosensory Nerves
							0 (Terminal) I (Olfactory)
1	Somitomere 1 Somitomere 2		Most extrinsic ocular muscles	III (Oculomotor)	Deep ophthalmic branch of V? (Profundus)		
2	Somitomere 3 Somitomere 4	1	Mandibular arch and dorsal oblique	IV (Trochlear)	V (Trigeminal)		
3	Somitomere 5 Somitomere 6	2	Hyoid arch and lateral rectus	VI (Abducens)	VII (Facial)	ADLLN, AVLLN, OLLN	X (Gustatory)
						VII (Stato-acoustic)	
4	Somitomere 7 Lost?	3	First branchial arch		IX (Glossopharyngeal)	MLLN	X (Gustatory)
5	Somite 1	4	Second branchial arch		X (Vagus)	STLLN	X (Gustatory)
6	Somite 2	5	Third branchial arch and hypobranchial muscles	Occipital	X (Vagus)	PLLN	X (Gustatory)
7	Somite 3	6	Fourth branchial arch and hypobranchial muscles	Occipital	X (Vagus)	PLLN	X (Gustatory)
8	Somite 4	7	Fifth branchial arch and hypobranchial muscles	Occipital	X (Vagus)	PLLN	X (Gustatory)

(IV), and abducens (VI) nerves appear to be the ventral nerves of the first three head segments (somitomeres 1–6). Segments 4 and 5 lack ventral nerves, probably because the evolutionary expansion of the otic capsule in this region eliminated one somitomere and possibly part of a somite. The segmented series of ventral cranial nerves resumes with the occipital nerves. These appear to be the ventral nerves of cranial segments 6, 7, and 8 (postotic segments 1–4).

The trigeminal nerve (V), facial nerve (VII), glossopharyngeal nerve (IX), and vagal nerve (X) appear to be serially homologous to the dorsal spinal nerves because they have most of the characteristics seen in dorsal spinal nerves. In particular, they attach more dorsally on the brainstem than do the ventral cranial nerves, they contain general somatic sensory and general visceral sensory neurons, and they carry motor neurons to the striated branchiomeric muscles associated with the visceral arches (Fig. 13-20). As we have discussed (Chapter 10), the branchiomeric muscles have the characteristics of somatic muscles in the only group of vertebrates (birds) in which the origin of these muscles is clear (i.e., they arise from paraxial mesoderm and are striated). Thus, we have provisionally interpreted the branchiomeric muscles as somatic muscles in all craniates. Motor neurons supplying the branchiomeric muscles have traditionally been called special visceral motor neurons, but, because we consider that they supply somatic muscles, they should be called somatic motor neurons (Northcutt, 1990). They resemble other somatic motor neurons (and differ from the visceral motor neurons of the autonomic nervous system) in traveling all the way to the muscles they supply without relaying in a peripheral ganglion. Also, the nuclei of these neurons are located far ventrally in the brainstem (Figs. 13-7 and 13-8), a location consistent with the view that they are somatic motor neurons. The hypothesis that the dorsal cranial nerves are serially homologous to dorsal spinal nerves requires that somatic motor neurons in ancestral vertebrates left the CNS through both dorsal and ventral nerves. If this were also true in spinal nerves, then such fibers subsequently must have been lost from dorsal spinal nerves. Some dorsal nerves of amphioxus contain somatic motor neurons, which supports the interpretation that somatic motor fibers were present originally in all dorsal and ventral nerves.

As you can see in Figure 13-20 and Table 13-5, the trigeminal nerve (V) is the dorsal nerve of the second head segment (somitomeres 3–4); the facial nerve (VII), that of the third segment (somitomeres 5–6); the glossopharyngeal nerve (IX), that of the fourth segment (somitomere 7); and the four branchiomeric branches of the vagus (X), the dorsal nerves of the last four head segments. Some investigators interpret the deep ophthalmic ramus of the trigeminal nerve complex, which is a distinct nerve in lampreys and many anamniotes and which originates developmentally from a neurogenic placode, as the missing dorsal nerve of the first head segment.

We leave the topic of cranial nerve evolution with a caveat: much information about the nerves and their embryology still does not fit neatly into any of the many models concerning head segmentation and evolution. For example, we still do not understand the embryological or evolutionary significance of the multiple origins of the trigeminal complex from neurogenic placodes and neural crest. Broad phylogenetic studies of the trigeminal complex and its development likely will prove essential to modeling head segmentation. Similar problems exist for each of the cranial nerves, and this area of research certainly will remain fruitful for the foreseeable future.

The Origin of Special Sensory Neurons

Some of the cranial nerves carry special sensory neurons for sense organs, such as electroreceptors and taste buds, that do not develop in the trunk and therefore never are represented in spinal nerves.[7] As we have seen (Chapter 12), neurons from the ear and lateral line (including electroreceptors) have much in common: they receive stimuli from hair cells and have similar connections within the brain. Biologists often call them the **octavolateralis system.** Many of the special sensory neurons now travel with other sensory neurons in dorsal cranial nerves, but were they always a part of these nerves? Cole (1896) and Landacre (1910) pointed out many years ago that the octavolateralis and chemosensory neurons differ from other sensory neurons in important ways. Octavolateralis and chemosensory neurons develop from neurogenic placodes and not from neural crest cells. Moreover, the cell bodies of octavolateralis and chemosensory neurons tend to lie more distally in ganglia than do those of other sensory neurons; in the case of the octavolateralis neurons, the cell bodies lie in distinct ganglia. Histologically, octavolateralis neurons are very distinct from other sensory neurons, having larger cell bodies, for example. The octavolateralis and chemosensory systems also project to different areas within the brain than do other sensory neurons (e.g., see Figs. 13-7 and 13-8).

Northcutt (1989 and 1990) proposed that the octavolateralis and chemosensory neurons were not originally components of the dorsal nerves but formed dis-

[7]Where such sense organs do occur on the trunk, as is true for taste buds in certain teleosts, their presence is always secondary (i.e., they reach their final sites in the back by migrating out from their origins on the head).

tinct nerves. In the case of the octavolateralis system, Northcutt's hypothesis seems increasingly well justified. The inner ear and octaval neurons develop embryonically from one neurogenic placode, but the lateralis neurons develop from as many as three preotic and three postotic neurogenic placodes (Chapter 4). Therefore, the earliest vertebrates likely had as many as six distinct lateral line nerves as well as nerve VIII, and this is the interpretation that we adopt for this book. Such a pattern is apparent in many living vertebrates, such as coelacanths (*Latimeria*; Northcutt and Bemis, 1993).

However, evidence that the gustatory nerves ever constituted a separate series of independent cranial nerves is less compelling. Although the gustatory fibers develop from at least three epibranchial neurogenic placodes, no vertebrate has separate gustatory nerves. Instead, the epibranchial neurogenic placodes give rise to ganglia and nerves that also relay pain and temperature information from the endodermal lining of the pharynx. Thus, for now, we maintain the traditional interpretation that the gustatory fibers are components of nerves VII, IX, and X.

Changes in the Cranial Nerves of Amniotes

The pattern of the cranial nerves we have seen in chondrichthyans persists with predictable modifications in amniotes (Table 13-4, Figs. 13-23 and 13-24). A few evolutionary changes have occurred in the nerves returning from the special sensory organs. Among these changes are the following:

1. Most terrestrial vertebrates have a vomeronasal organ, and they acquire with it a vomeronasal component to the olfactory nerve (see Chapter 12 for evolution of the vomeronasal system).
2. In most nonmammalian vertebrates, most of the fibers of the optic nerve decussate in the **optic chiasm,** located at the point of attachment of the optic nerves to the brain. In many mammals, only half of the fibers decussate, and the others remain on the same side of the body (= ipsilateral). This change may correlate with overlapping visual fields and the development of stereoscopic vision.
3. A cochlea is well developed in some sauropsids, especially birds, as well as in mammals, so nerve VIII in amniotes is often termed the vestibulocochlear nerve.
4. The lateral line system is entirely lost in amniotes, so the six lateral line nerves are absent.

Few changes have occurred in the ventral cranial nerves. The oculomotor, trochlear, and abducens nerves change little except for the additional extrinsic ocular muscles, known as the palpebral muscles, that they supply (Chapter 10). The basicranial muscle of *Latimeria,* which also was present in the rhipidistians ancestral to tetrapods, became modified in amphibians to form the retractor bulbi muscle. The level of the head–trunk transition is stabilized in amniotes, and the occipital nerves attach to the back of the brain and emerge through a skull foramen as the **hypoglossal nerve (XII).** The hypoglossal nerve extends beneath the tongue (Gr., *hypo* = under + *glossa* = tongue) to supply the muscles of the tongue and other hypobranchial muscles.

Because more evolutionary changes have occurred in those structures that are supplied by the dorsal cranial nerves, these nerves in amniotes have changed more than have the ventral cranial nerves.

The trigeminal nerve of amniotes consists of ophthalmic, maxillary, and mandibular branches. Its ophthalmic branch incorporates both the superficial ophthalmic and the deep ophthalmic branches seen in

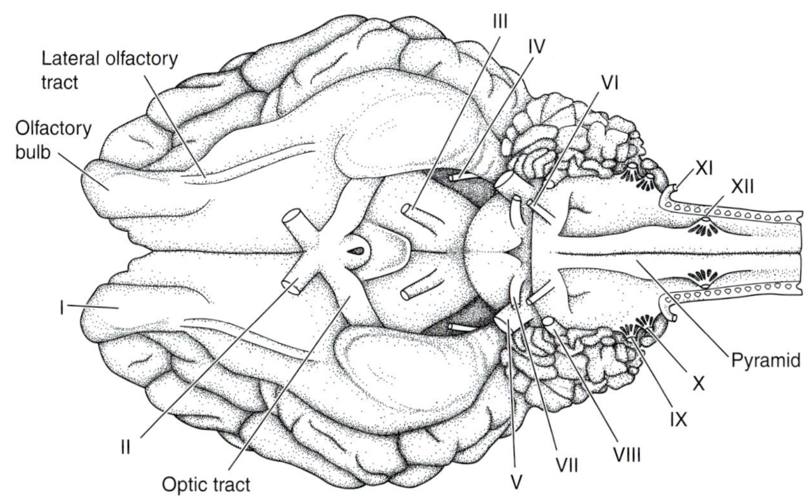

FIGURE 13-24
A ventral view of a sheep brain, showing the locations of the cranial nerves. *(After Noback et al.)*

chondrichthyans. The other branches of nerve V in amniotes are essentially the same as in chondrichthyans. All the branches carry general somatic sensory fibers from receptors in the skin of the head, in the teeth, in the mouth, and on the tongue. Those on the tongue are from general receptors and not from taste buds. Some investigators believe that most, if not all, of the proprioceptive fibers from the extrinsic ocular muscles of amniotes also return in the trigeminal nerve because the cell bodies of some of these neurons have been found in the semilunar ganglion of nerve V. The motor fibers of the trigeminal nerve are confined to its mandibular branch. They are somatic motor fibers to the mandibular muscles and to the tensor tympani muscle. (Recall from Chapter 7 that the malleus, to which the tensor tympani attaches, is a mandibular arch derivative.) The trigeminal nerve supplies some uniquely evolved receptors in amniotes. These include the infrared-receptive pit organs of snakes (Chapter 12) and the unique mechanosensory and electrosensory organs in the bill of the platypus, *Ornithorhynchus.*

The facial nerve of amniotes continues to carry somatic motor fibers to the hyoid musculature. Most of this musculature forms the platysma and facial muscles in mammals, but the stapedius, stylohyoid, and posterior belly of the digastric remain attached to skeletal elements derived embryonically from the hyoid arch. Many new cranial glands evolve in tetrapods, and the facial nerve carries new visceral motor (= autonomic) neurons that supply the lacrimal glands, the mandibular and sublingual salivary glands, and glands in the palatal and nasal mucosa. The major sensory fibers in the facial nerve of amniotes are special visceral sensory ones from taste buds on the palate and anterior two thirds of the tongue. Those on the tongue leave with somatic sensory fibers of the trigeminal in its lingual branch but then separate from the lingual, cross the tympanic membrane in the **chorda tympani,** and enter the brainstem together with the facial nerve.

The glossopharyngeal nerve of amniotes continues to supply visceral receptors, primarily taste buds on the caudal parts of the pharynx and tongue, and a few muscles in the wall of the pharynx. It has acquired visceral motor (autonomic) neurons to the parotid salivary gland.

The major branch of the vagus nerve of amniotes is homologous to the intestinal branch of fishes. This branch follows the trachea and esophagus carrying visceral motor (= autonomic) and general visceral sensory neurons to the heart, lungs, and abdominal viscera. It has lost most of its taste neurons and somatic motor ones, apart from those to the larynx.

The accessory branch of the vagus observed in chondrichthyans is distinct from the vagus in amniotes and forms its own cranial nerve, known as the **accessory nerve (XI),** which carries somatic motor fibers to derivatives of the chondrichthyan cucullaris muscles, including the trapezius and sternocleidomastoid complexes found in mammals (Chapter 10). Proprioceptive fibers from these muscles are believed to return in the spinal nerves that also innervate them.

The Autonomic Nervous System

Skeletal and branchiomeric muscles of the body are supplied by somatic motor neurons that travel directly from the CNS to the effectors. As noted earlier in this chapter, exocrine glands and the muscles of the gut wall, heart, and other internal organs are innervated by visceral motor neurons (= autonomic[8]) that undergo a pe-

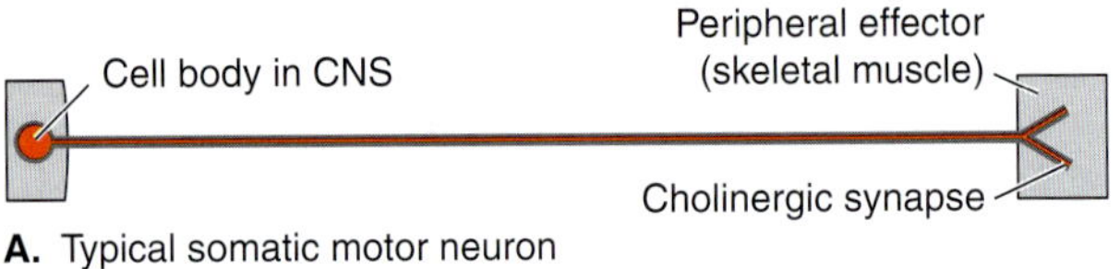

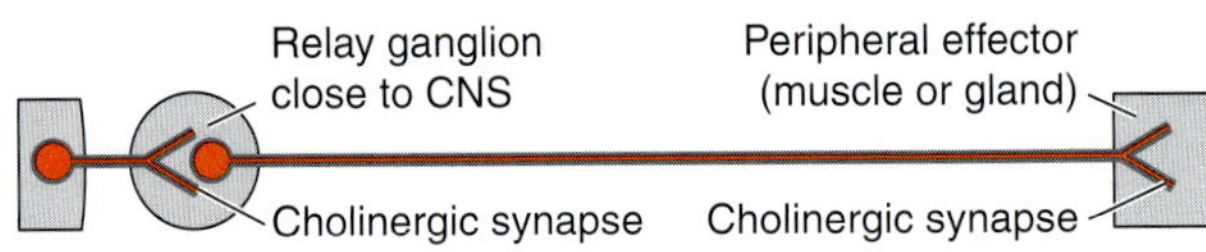

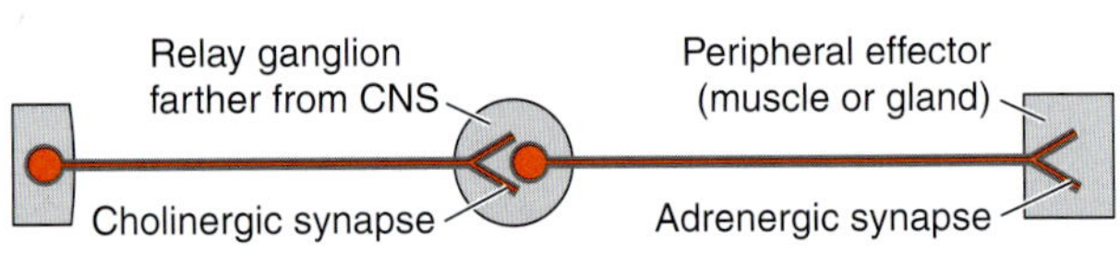

FIGURE 13-25
Comparison of the organization of somatic motor and visceral motor (= autonomic) nerves. *A,* Typical somatic motor neuron. A single nerve cell extends from the CNS to a peripheral effector, such as a skeletal muscle. *B,* Typical parasympathetic motor neurons form a two-neuron chain in which the relay ganglion is relatively close to the CNS. *C,* Typical sympathetic motor neurons form a two-neuron chain in which the relay ganglion is relatively far from the CNS. See also Table 13-8.

[8]These neurons long were known as general visceral motor neurons. This terminology stems from the long-held belief that branchiomeric muscles were visceral components, making it necessary to distinguish the neurons supplying them as special visceral motor neurons. Because we now regard the branchiomeric muscles as somatic, and their neurons as somatic motor, this distinction no longer needs to be made. Accordingly, the only visceral motor neurons are those of the autonomic nervous system.

ripheral relay. This group of neurons constitutes the autonomic nervous system. Lightly myelinated **preganglionic neurons** leave the CNS and synapse in peripheral ganglia with the dendrites and cell bodies of **postganglionic neurons** (Fig. 13-25). The axons of the postganglionic neurons, which are not myelinated, continue to the effectors. Considerable divergence and convergence of information occur within the autonomic ganglia. A single preganglionic fiber usually synapses with many postganglionic neurons. Conversely, a single postganglionic neuron may receive many preganglionic fibers from different sources. The principal neurotransmitter secreted by the preganglionic neurons is acetylcholine, but a variety of peptide neurotransmitters also may be released. These neurotransmitters appear to modulate the activity of acetylcholine. Thus the autonomic ganglia are not simply relay stations but points where divergence and integration of impulses can occur. General visceral sensory fibers, which return from the viscera to the CNS in the same nerves that carry the motor fibers, provide essential feedback for the central regulation of autonomic activity. Nevertheless, these sensory fibers are not considered to be part of the autonomic system, which is defined in functional terms as a visceral motor system that regulates involuntary visceral activity and helps the body maintain homeostasis.

The Autonomic Nervous System of Mammals

The mammalian autonomic nervous system, which regulates many involuntary functions and which is relatively well known, consists of three divisions: the (1) **sympathetic** (Gr., *syn* = with + *pathos* = feeling), (2) **parasympathetic** (Gr., *para* = beside or accompanying), and (3) **enteric** (Gr., *enteros* = intestine) **divisions.** These divisions differ in function, in the locations of their preganglionic neurons, and in the organization and activity of their postganglionic neurons (Tables 13-6 and 13-7).

The enteric division is restricted to the gut wall, but most visceral organs, including the digestive tract, receive both sympathetic and parasympathetic

TABLE 13-6 Examples of the Effects of Sympathetic and Parasympathetic Stimulation in Mammals

Organs	Sympathetic Stimulation	Parasympathetic Stimulation
Endocrine gland		
Adrenal medulla	Secretes	Innervation not present
Cardiovascular system		
Heart rate	Increases	Decreases
Force of ventricular contraction	Increases	—
Coronary arteries	Dilate	Constrict
Arteries in skeletal muscles	Active ones dilate	—
Lungs		
Muscles in bronchioles	Relax	Contract
Digestive organs		
Salivary glands	Secretion of mucus	Secretion of enzymes
Gastric glands	—	Secretion
Pancreas	—	Secretion
Liver	Releases sugar into blood	Bile flows
Visceral blood vessels	Constrict	Dilate
Intestinal muscles	Relax	Contract
Anal sphincter	Contracts	Relaxes
Urogenital organs		
Bladder sphincter	Contracts	Relaxes
Arteries of external genitalia	Constrict	Dilate
Skin		
Hair muscles	Contract	—
Sweat glands	Secretion	—
Cutaneous arteries	Constrict	—
Eye		
Iris sphincter	—	Contracts
Iris dilator	Contracts	—
Ciliary muscles	Relax	Contract

TABLE 13-7 Comparison of Characteristics of the Parasympathetic and Sympathetic Divisions of the Autonomic Nervous System

Characteristic	Parasympathetic	Sympathetic
Location of cell body within CNS and outflow point for preganglionic neurons	Hindbrain and sacral regions of spinal cord (= craniosacral)	Thoracic and lumbar regions of spinal cord (= thoracolumbar)
Location of ganglia	Close to target organ	Close to spinal cord
Length of preganglionic axons	Long	Short
Length of postganglionic axons	Short	Long
Preganglionic neurotransmitter	Acetylcholine	Acetylcholine
Postganglionic neurotransmitter	Acetylcholine	Norepinephrine

CNS = central nervous system.

innervation. Because their postganglionic fibers release different transmitters at the neuron–effector junctions, the sympathetic and parasympathetic divisions have different effects: one system stimulates, and the other inhibits, the activity of different organs (Table 13-6).

Postganglionic parasympathetic fibers secrete acetylcholine and so can be described as **cholinergic fibers.** Postganglionic sympathetic fibers secrete noradrenaline (also called norepinephrine) and so are **adrenergic fibers** (Fig. 13-25). Noradrenaline is chemically similar to the adrenaline secreted by the medullary cells of the adrenal gland (Chapter 15). Indeed, sympathetic stimulation and adrenal medulla secretion supplement each other in enabling a vertebrate to adjust to stresses, such as escaping a predator or chasing prey. This is often called the **fight-or-flight** reaction. In essence, this reaction provides more energy to the skeletal muscles of the body. The rate and force of cardiac muscle contraction increase, blood pressure increases, bronchioles in the lungs dilate, more blood is sent to the active skeletal muscles and less to the skin and viscera, and blood sugar levels increase. Hairs also are elevated, sweating increases, and the pupils dilate. Gut motility, digestive enzyme secretion, and sexual activity are inhibited. Parasympathetic stimulation has the opposite effects and may be called **rest and digest.** Energy is conserved and restored. Heart rate and force of contraction decrease, blood pressure decreases, bronchioles constrict, more blood is diverted to the digestive tract, gut motility and digestive enzyme secretion increase, bile flows, sugars are stored, and defecation and urination are promoted. The pupils contract, and sexual activity may be stimulated.

Morphological distinctions between the sympathetic and parasympathetic divisions are not always as clear as their pharmacological and physiological differences. Morphologists describe the mammalian sympathetic system as being a **thoracolumbar outflow** from the CNS and the parasympathetic system as a **craniosacral outflow.** Preganglionic sympathetic fibers in humans leave through the thoracic and first three or four lumbar spinal nerves, whereas preganglionic parasympathetic fibers leave through the oculomotor, facial, glossopharyngeal, vagus, and sacral nerves 2 to 4 (Fig. 13-26). But exceptions exist, and a few sympathetic fibers occur in the vagus and sacral nerves of some mammalian species.

Preganglionic sympathetic fibers that send impulses to cervical, cranial, and cutaneous organs (including the iris of the eye and its ciliary muscles as well as hair muscles, sweat glands, and cutaneous blood vessels) enter the thoracolumbar portion of the paired **sympathetic cords** and relay in **sympathetic ganglia** along the cord (these ganglia are sometimes known as sympathetic chain ganglia). The cords extend through the neck to the base of the head and bear cervical ganglia in which the relays to cervical and cranial organs occur, but these cervical extensions of the sympathetic cords do not receive any additional fibers from the cervical spinal nerves. Most of the preganglionic sympathetic fibers to abdominal viscera pass directly through the sympathetic cords and continue as **splanchnic nerves** to their relay sites in ganglionic masses that are located at the bases of major abdominal arteries. These ganglia take their names from the arteries with which they are associated: **coeliac, superior mesenteric,** and **inferior mesenteric ganglia.** Postganglionic sympathetic fibers to all cutaneous structures and to the blood vessels in skeletal muscles re-enter the spinal nerves by a **gray ramus communicans** (described earlier and diagrammed in Fig. 13-7). Sympathetic fibers going to the skin of the head travel with the trigeminal nerve. Postganglionic fibers to other organs in the head, thorax, and abdomen follow along the arteries.

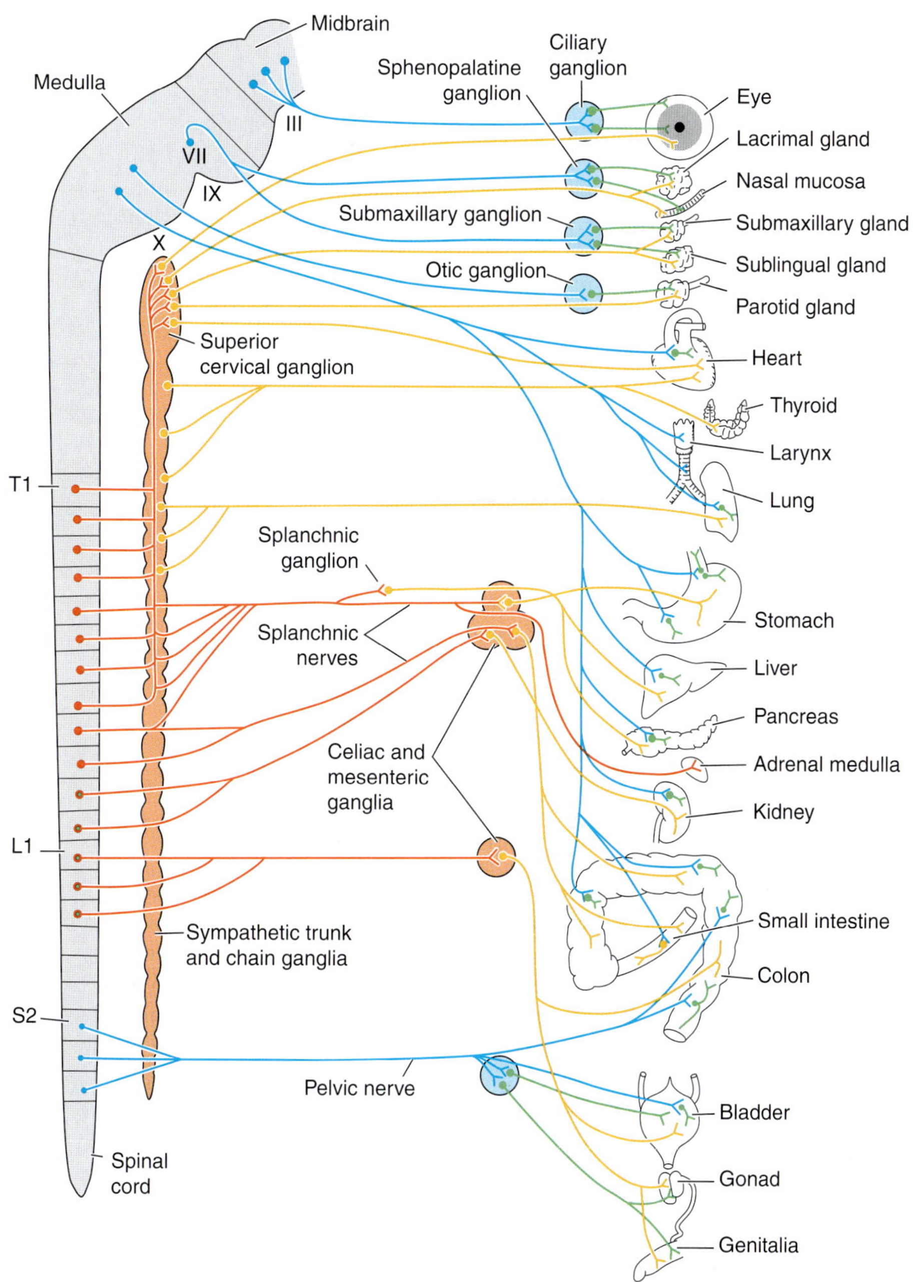

FIGURE 13-26
The mammalian autonomic nervous system. Sympathetic neurons are shown in red (first neuron) and yellow (second neuron in a two-neuron chain); parasympathetic ones in blue (first neuron) and green (second neuron in a two-neuron chain). Sympathetic fibers to the skin and blood vessels are not shown.

Preganglionic parasympathetic fibers leaving through the oculomotor nerve (III) carry impulses destined for the ciliary body and iris of the eye; these fibers have their relay point in the ciliary ganglion, which is located along the course of nerve III (Table 13-8). Parasympathetic fibers in the facial and glossopharyngeal nerves carry information destined for the lacrimal and salivary glands and to other glands in the mucous membranes of the nose, mouth, and pharynx. The parasympathetic ganglia for these pathways are the submandibular, sphenopalatine, and ethmoidal ganglia of nerve VII and the otic ganglion of nerve IX (Table 13-8). The vagus carries the largest outflow of parasympathetic fibers, which extend from their relay sites to the thoracic and abdominal viscera. The sacral outflow of parasympathetic information is carried by the sacral nerves, which unite to form a small **pelvic nerve,** which extends to reach the pelvic viscera. Preganglionic parasympathetic fibers are much longer than those of the sympathetic system because they extend nearly all the way to the effectors, whereas the postganglionic fibers are very short. The peripheral relays often occur in minute ganglia in the walls of the organs they supply. No parasympathetic fibers go to the body wall or extend into the limbs, so cutaneous organs and smooth muscles in the walls of most blood vessels do not have a parasympathetic supply.

TABLE 13-8 Ganglia Associated with Cranial Nerves of an Amniote

Nerve	Sensory Ganglia: *Proximal*	Sensory Ganglia: *Distal*	Parasympathetic Ganglia
0 (Terminal)			
I (Olfactory)			
II (Optic)			
III (Oculomotor)			Ciliary ganglion
IV (Trochlear)			
V (Trigeminal)	Semilunar ganglion		
VI (Abducens)			
VII (Facial)	Geniculate ganglion (dual origin from neural crest and neurogenic epibranchial placode)		Submandibular ganglion, sphenopalatine ganglion, ethmoidal ganglion
VIII (Statoacoustic or vestibulocochlear)	Acoustic ganglion		
IX (Glossopharyngeal)	Superior ganglion	Petrosal ganglion	Otic
X (Vagus)	Jugular ganglion	Nodose ganglion	
XI (Accessory)			
XII (Hypoglossal)			

The degree of divergence also differs between these two divisions. Each preganglionic sympathetic fiber synapses with about ten postganglionic neurons, so that divergence is extensive and sympathetic effects are diffuse. Parasympathetic stimulation is more narrowly targeted because each preganglionic fiber synapses with only about three postganglionic neurons.

The enteric division is composed of two interconnected neuronal networks: (1) a **myenteric plexus** lies between the longitudinal and circular muscle layers of the gut wall, and (2) a **submucosal plexus** lies just beneath the mucosa. Sensory neurons that are restricted to the gut wall detect changes in gut muscle tension and in the chemical environment in the gut lumen. Interneurons and motor neurons that are confined to the gut wall control gut motility, blood vessel tone, and the secretion of gastric and intestinal glands. The enteric system can function autonomously, but a second level of control is exercised by sympathetic and parasympathetic neurons that terminate in the plexuses. Much gut activity also is influenced by locally produced gastrointestinal hormones (Chapter 15).

Evolution of the Autonomic Nervous System

The autonomic system of other amniotes is similar to that of mammals. Anamniotes, too, have ways of regulating visceral activity, but only parts of the autonomic system of selected species have been studied. In general, the distinction between sympathetic and parasympathetic divisions of anamniotes is not as clear as it is in amniotes, and few organs have a dual innervation.

The autonomic nervous system of hagfishes and lampreys is rudimentary. Fibers believed to be sympathetic supply some blood vessels, but sympathetic ganglia are absent. The vagus contains parasympathetic fibers, but no parasympathetic fibers travel to the eyeball in the oculomotor nerve.

Chondrichthyans do not have a well-organized sympathetic cord but have segmental, sympathetic ganglia in the trunk, which are interconnected by a few plexuslike, longitudinal fibers (Fig. 13-27). Nearly all of the peripheral relays occur in these ganglia, and postganglionic fibers follow arteries to the viscera. A number of sympathetic fibers also travel in the vagus, and these fibers relay close to the organs they supply. In chondrichthyans (the skin of which, of course, does not have the thermoregulatory role that it has in birds and mammals), no extension of sympathetic fibers into the head or to the cutaneous chromatophores and blood vessels occurs. Parasympathetic fibers to the eye travel to it by the oculomotor nerve; those going to the heart and stomach travel in the vagus nerve. Parasympathetic neurons are not present in other cranial nerves because the large glands that evolved in conjunction with feeding systems of tetrapods (Chapter 15) are not present

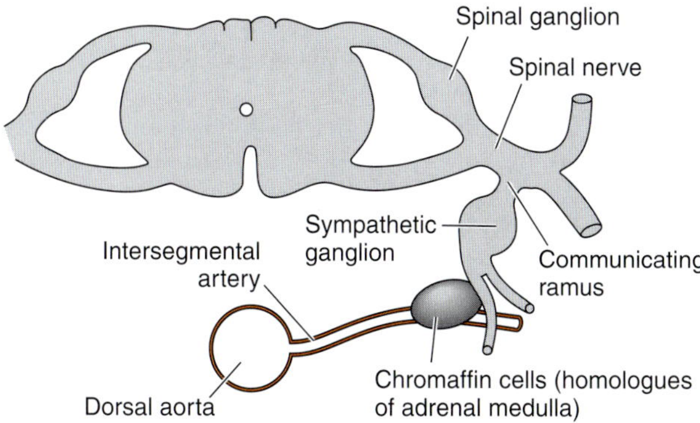

FIGURE 13-27
A sympathetic ganglion of a dogfish and its connections. The chromaffin cells constitute part of an endocrine gland (Chapter 15).

in the head of chondrichthyans. Parasympathetic neurons have an inhibitory effect on the rate of heartbeat, as in amniotes. Only the stomach receives both a parasympathetic and a sympathetic supply. The primary excitatory fibers that help cause gastric secretion are the sympathetic ones—just the opposite of the effect of sympathetic stimulation on the stomachs of amniotes. The function of the parasympathetic fibers to the stomach is less clear.

The autonomic system is more clearly organized in teleosts and amphibians. The sympathetic ganglia lie along a pair of sympathetic cords that extend to the base of the head. Sympathetic fibers re-enter the spinal nerves of teleosts, innervate the chromatophores, and cause the pigment in these cells to concentrate. A pituitary hormone has the opposite effect. The heart receives a sympathetic innervation in amphibians, and its effect is opposite that of the parasympathetic fibers in the vagus. Indeed, the antagonistic roles of sympathetic and parasympathetic innervation were first discovered in studies of frog hearts. Lissamphibians have tear and salivary glands, and a few parasympathetic fibers travel to them in the facial and glossopharyngeal nerves. A few frogs have a parasympathetic outflow through their sacral nerves, but no outflow occurs in this region in teleosts or salamanders.

SUMMARY

1. The nervous system integrates sensory information and initiates most of a vertebrate's rapid responses to changes in its environment.
2. Neurons are the structural and functional units of the nervous system. Many types exist, but each consists of a trophic segment (the cell body), a receptive segment (dendrites), a conducting segment (axon), and a transmissive segment (terminal arborization).
3. Nerve impulses usually are carried in one direction across synapses between neurons, and across myoneural (= neuromuscular) and axoglandular (= termination on a gland) junctions, by transmitter substances. These substances may have excitatory or inhibitory effects.
4. Most peripheral neurons become myelinated during development by Schwann cells.
5. Several types of non-nervous neuroglia are abundant in the CNS. Oligodendrocyctes myelinate central axons. Astrocytes regulate ionic balances in extracellular fluid, degrade metabolites, and may have a nutritive role. Microglia are phagocytes.
6. Myelinated axons form the white matter of the CNS; nonmyelinated axons and neuron cell bodies lie in gray matter.
7. The central cavity of the nervous system expands into ventricles within the brain. These contain cerebrospinal fluid.
8. The nervous system consists of only three broad categories of neurons: (1) sensory, or afferent; (2) motor, or efferent; and (3) interneurons. Interneurons are confined to the CNS. The cell bodies of most sensory neurons are located in peripheral ganglia. The cell bodies of the motor neurons are located in central gray matter and in autonomic ganglia.
9. Sensory and motor neurons can be sorted into functional groups: general somatic sensory neurons from cutaneous receptors and proprioceptors, special somatic sensory neurons from special sense organs, general visceral sensory neurons from most visceral receptors, special visceral sensory neurons from taste buds, visceral motor (autonomic) neurons to gut and heart muscles and to exocrine glands, and somatic motor neurons to somatic muscles. All have characteristic terminations or origins within the CNS.
10. Neurons form reflex arcs within the spinal cord and brainstem that are responsible for much locomotor and visceral activity. Neurons also form pathways that ascend to and descend from the brain. Most pathways decussate on the way to and from the brain. Ascending pathways relay in the thalamus before projecting to the cerebrum.
11. During the course of vertebrate evolution, the spinal cord became more vascular and enlarged as

more neuronal tracts developed between the spinal cord and brain.

12. The dorsal and ventral roots of spinal nerves form separate nerves in lampreys, but these unite in all gnathostomes. Sensory neurons always enter through dorsal roots of spinal nerves, and somatic motor ones always leave through the ventral roots. Visceral motor neurons leave through the dorsal roots of lampreys but have shifted during evolution, so all lie in the ventral roots of amniotes.
13. The cranial nerves of gnathostomes, such as sharks, can be sorted into three groups, each with distinctive characteristics: (1) the terminal, olfactory, and statoacoustic nerves and lateral line nerves are unique to the head and contain only special somatic sensory neurons; special visceral sensory (= gustatory) fibers from taste buds may constitute what were phylogenetically separate cranial nerves but that now travel with other nerves; (2) the oculomotor, trochlear, abducens, and occipital nerves supply somatic motor neurons to the extrinsic ocular, epibranchial, and hypobranchial muscles; they rarely contain other types of neurons and appear to be serially homologous to the ventral spinal nerves of lampreys; and (3) the trigeminal, facial, glossopharyngeal, and vagus nerves supply somatic motor neurons to the branchiomeric muscles, and they also contain a variety of sensory neurons; they appear to be serially homologous to dorsal spinal nerves of lampreys.
14. The heads of ancestral vertebrates probably contained eight segments. The nerves we have interpreted as serially homologous to ventral and dorsal cranial nerves can be related to these segments.
15. The optic nerve does not properly belong to any of the groups of cranial nerves because it is a brain tract.
16. During the course of gnathostome evolution, few changes occurred in the ventral cranial nerves other than the emergence of the occipital nerves from within the skull as the hypoglossal nerve (XII). Major changes occurred in other cranial nerves. For example, amniotes lost the lateral line nerves. The facial and glossopharyngeal nerves evolved visceral motor neurons (autonomic) to tear, salivary, and other glands of the head. The vagus nerve lost most of its somatic motor neurons apart from those to the larynx. The accessory branch of the vagus nerve also became separated as a distinct nerve, known as the accessory nerve (XI).
17. The autonomic nervous system is the motor system for gut, heart, and other visceral muscles and for exocrine glands. It is characterized by motor relays located in peripheral ganglia, where some divergence and integration (i.e., processing) of impulses occur.
18. The autonomic nervous system of amniotes consists of separate sympathetic and parasympathetic divisions and an enteric division, which is confined to the gut wall. Most visceral organs, including the enteric system, receive both sympathetic and parasympathetic innervation. In general, sympathetic innervation prepares an animal for fight or flight, whereas parasympathetic innervation prepares an animal to rest and digest. The sympathetic and parasympathetic divisions exit from different parts of the CNS and relay at different sites. The autonomic nervous system is poorly developed in hagfishes, lampreys, and chondrichthyans, but it is more evident in bony fishes and amphibians.

REFERENCES

All references on the nervous system are collected at the end of Chapter 14.

14

The Nervous System II: The Brain

PRECIS

After examining the development of the brain and its meninges, we will consider the structure and organization of a fish brain as an example of a craniate brain that displays the basic organizational features of all craniates. We will then summarize the major evolutionary changes seen in amphibians and amniotes before considering some important mammalian pathways.

OUTLINE

Craniates differ from early chordates in having a markedly enlarged brain; however, a brain is not a completely new craniate feature. We find a well-developed brain in many protostomes, including the octopus and arthropods. Although cephalochordates, the sister group of craniates, do not have a grossly visible brain, cell groups have been identified in the cranial end of their neural tube that appear to be homologous to groups of cells found in the craniate forebrain, midbrain, and hindbrain. Moreover, the embryonic development of these cell groups in cephalochordates and craniates is regulated by homologous genes. Indeed, homologues of these same genes regulate brain development in protostomes (see Butler, 2000, for a review). Although an incipient brain was present in the early chordate ancestors of craniates, it does enlarge greatly during craniate evolution—a process called **encephalization.**

We may never be certain as to all of the factors that led to this enlargement, but increasing activity certainly played a major role. The front of the body of an active animal (and we believe that ancestral craniates were far more active than their early chordate ancestors) is the first part of the body to encounter environmental changes.

Ancestral craniates probably evolved or increased the concentration of olfactory receptors, gustatory receptors, photoreceptors, neuromasts of the lateral line and ear, and probably electroreceptors on their heads. Processing the input from these receptors and directing the activity of segmental and branchial structures would require a larger aggregation of neurons and the enlargement of the rostral end of the neural tube.

Embryonic Development and Regions of the Brain

A brief consideration of the embryonic development of the mammalian brain will acquaint you with the regions of the brain found in all craniates and with the major structures associated with each region. As we have explained (Fig. 4-20), the brain develops by an enlargement of the cranial end of the neural tube. First, three brain expansions can be recognized: (1) a **forebrain,** or **prosencephalon** (Gr., *pros* = before + *enkephalos* = brain); (2) a **midbrain,** or **mesencephalon** (Gr., *mesos* = middle); and (3) a **hindbrain,** or **rhombencephalon** (Gr., *rhombos* = lozenge).

The prosencephalon first gives rise to the **optic vesicles,** which will form the retina of the eye (Fig. 4-20), and to a midventral **infundibulum.** The pituitary gland, or **hypophysis,** attaches to the infundibulum (Fig. 14-1*A*); and its posterior portion, the **neurohypophysis,** develops as an outgrowth of the infundibulum. Another brain region, the **telencephalon** (Gr., *telos* = end), develops soon after as a pair of rostrolateral extensions from the prosencephalon (Fig. 14-1*B*). These differentiate into the paired **cerebral hemispheres,** which together constitute the **cerebrum,** and into the paired **olfactory bulbs** (Fig. 14-1*C*). Paired chambers known as the **lateral ventricles** lie in the cerebral hemispheres and may extend into the olfactory bulbs (Fig. 14-2). Brain ventricles are lined by the **ependymal epithelium,** as are all cavities within the central nervous system.

The rest of the prosencephalon remains in the midline and differentiates as the **diencephalon** (Gr., *dia* = between). Its cavity, the **third ventricle,** remains connected to the lateral ventricles by the paired **interventricular foramina of Monro** (Fig. 14-2). The lateral walls of the diencephalon thicken to form the **thalamus** (Gr., *thalamos* = inner chamber); the floor becomes the **hypothalamus;** and the roof, most of which remains very thin, is the **epithalamus.** As they enter the brain, the optic nerves partially decussate, forming an **optic chiasm**[1] just rostral to the hypothalamus. The median eye complex, which is present in some craniates, develops from a part of the epithalamus called the **epiphysis.**

The mesencephalon remains undivided. A narrow **cerebral aqueduct of Sylvius** passes through it and connects the third ventricle with the **fourth ventricle** in the rhombencephalon. The dorsal part of the mesencephalon, known as the **tectum** (L., *tectum* = roof), enlarges as a pair of **optic lobes** in nonmammalian vertebrates (Fig. 14-2) and a more deeply situated and more caudal pair of auditory lobes called the **tori semicirculares.** The tectum often contains an extension of the cerebral aqueduct called the **optic ventricle.** The mesencephalic tectum of mammals similarly differentiates into the paired **superior colliculi** (visual) and **inferior colliculi** (auditory; Fig. 14-1*C*). The ventral part of the mesencephalon is known as its **tegmentum** (L., *tegmentum* = covering).

The rostral one third of the roof of the rhombencephalon differentiates into a **metencephalon** (Gr., *meta* = after), of which the most conspicuous part is the **cerebellum.** During the embryonic development of birds and mammals, neuroblasts migrate from the cerebellum into the ventral part of the rhombencephalon and differentiate into pontine and other nuclei, which relay information between the cerebrum and cerebellum, and a conspicuous band of transverse fibers. This region is known as the **pons** (L., *pons* = bridge; Fig. 14-1*C*). A pons does not differentiate in reptiles and anamniotes, so all of the rhombencephalon, except for the cerebellum, forms the fifth brain region, the **myelencephalon,** which leads to the spinal cord (Gr., *myelos* = marrow, or spinal cord). The myelencephalon forms the adult **medulla oblongata** and contains the fourth ventricle.

The major divisions of the brain lie along the same horizontal plane in most adult vertebrates (Fig. 14-2). Brain regions become folded on one another in birds and mammals because the head extends forward from an upward-curving neck. **Mesencephalic, cervical,** and **pontine flexures** develop sequentially in the positions shown in Figure 14-1*B*.

The Meninges and Cerebrospinal Fluid

The brain and spinal cord become covered by one or more layers of connective tissue that develop partly from mesoderm and partly from neural crest cells. These are the **meninges** (sing., meninx; Gr., *meninx* = membrane). Fishes have a single layer, the **primitive meninx,** which closely invests the central nervous system. Strands of connective tissue extend from it to a layer of connective tissue that lines the cranial cavity and vertebral canal. A gelatinous material fills the space between the primitive meninx and the surrounding cartilage and bone. Cerebrospinal fluid circulates

[1]The Latin term for this structure is *chiasma opticus.* This often is partly anglicized as "optic chiasma"; "optic chiasm" is the full anglicization of the term.

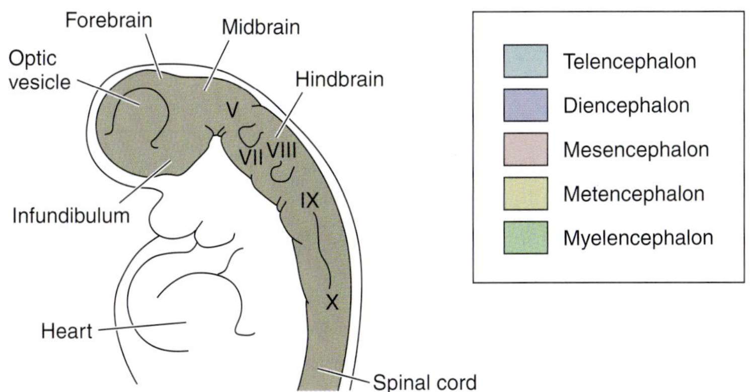

A. Human brain at 3 mm long

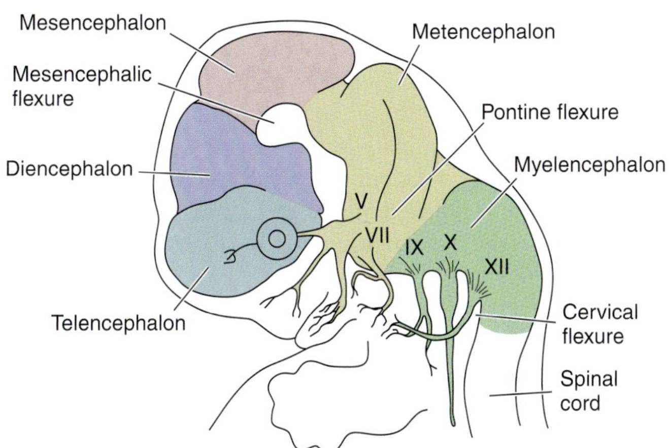

B. Human brain and cranial nerves at 7 weeks' gestation

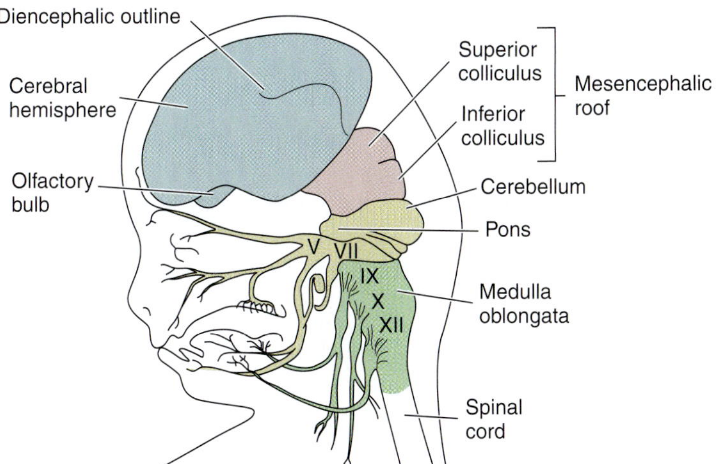

C. Human brain and cranial nerves at 3 months' gestation

FIGURE 14-1
A–C, Diagrams in lateral view of three stages in the development of the human brain, showing the differentiation of the principal brain regions. (*After Corliss.*)

in the cavities within the central nervous system, and some is secreted into spaces around it. In amphibians and reptiles the primitive meninx is divided into a dense **dura mater** (L., *dura mater* = hard mother), which unites with the connective tissue lining the cranial cavity, and a more delicate and vascular **secondary meninx,** which covers the brain and spinal cord.

Mammals and birds have three meninges because the secondary meninx has differentiated into two layers (Fig. 14-3*A*). A **pia mater** (L., *pia mater* = tender mother) closely invests the surface of the central ner-

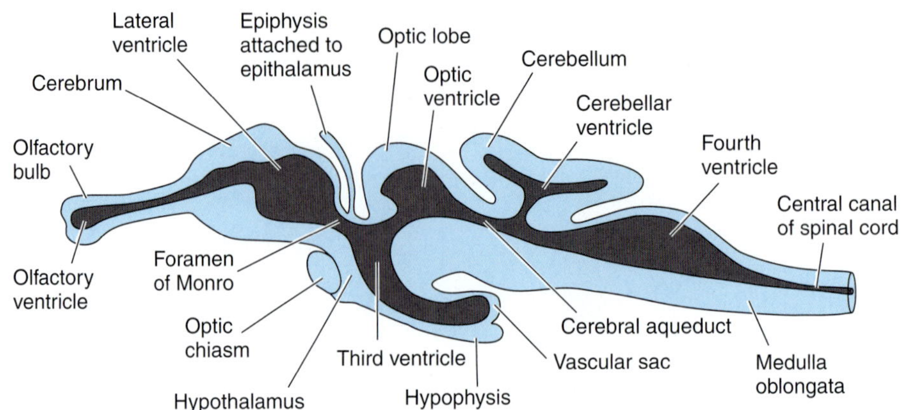

FIGURE 14-2
A sagittal section of the brain of a dogfish, showing the major parts of the brain and its ventricular system. (*After Walker and Homberger.*)

vous system and follows all of the convolutions of the brain. An **arachnoid** (Gr., *arachne* = spider + *eidos* = form) lies peripheral to the pia mater and crosses many of the crevasses on the brain surface. Strands of connective tissue extend like spider webs from the arachnoid across a **subarachnoid space** and attach to the pia mater. The cerebrospinal fluid that circulates around the central nervous system of birds and mammals is confined to the subarachnoid space and cavities of the brain. Many small blood vessels lie in the strands of the arachnoid and over the pia mater and follow the pia into the convolutions of the brain. The dura mater lies peripheral to the arachnoid. It is fused with the endosteum lining the cranial cavity. In mammals, a longitudinal, sickle-shaped septum of dura, known as the **falx cerebri** (Fig. 14-3*A*), extends between the cerebral hemispheres, and a transverse dural septum, the **tentorium,** extends between the cerebrum and the cerebellum (Fig. 14-3*B*). The latter is ossified in many species. These septa help stabilize the brain and hold it in place, especially during rapid rotations of the head.

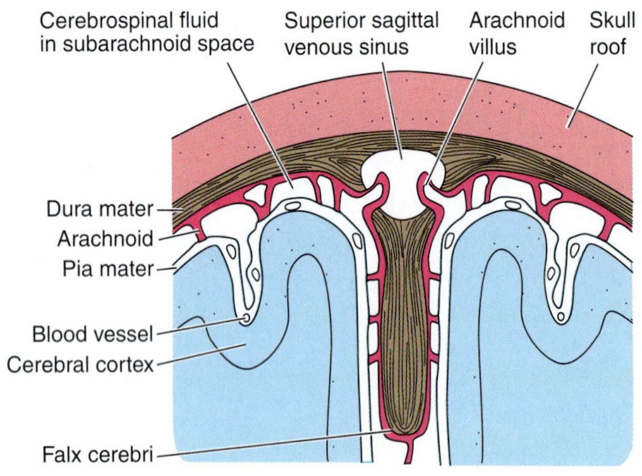

A. Transverse section through skull roof, meninges, and cerebrum

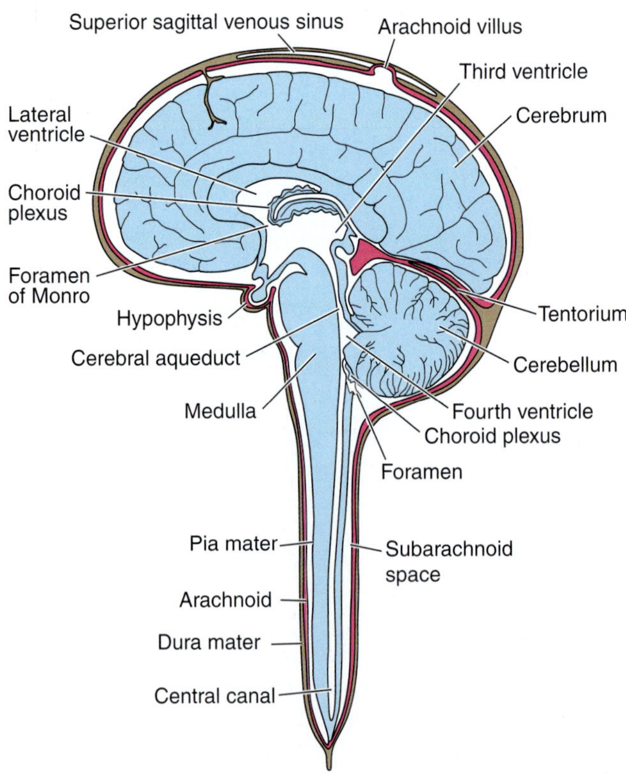

B. Sagittal section of central nervous system

FIGURE 14-3
The meninges and cerebrospinal fluid in the mammalian central nervous system. *A,* The meninges as seen in a transverse section through the center of the dorsal part of the cerebrum. *B,* A sagittal section of the central nervous system showing the choroid plexuses, the meninges, and the circulation of the cerebrospinal fluid. (*From Walker and Homberger.*)

Part of the wall of many ventricles is very thin, consisting only of the ependymal epithelium and the adjacent vascular meninx. An area of this type is known as a **tela choroidea.** Vascular tufts extending from the tela choroidea form **choroid plexuses,** which secrete the cerebrospinal fluid. The choroid plexuses extend into the ventricles in mammals, but in anamniotes and reptiles, some balloon into the space around the brain. Mammals have choroid plexuses in the floor of the lateral ventricles and in the roofs of the third and fourth ventricles (Fig. 14-3*B*). Cerebrospinal fluid flows toward the fourth ventricle, partly by secretion pressure and partly by the action of cilia on the ependymal cells. Cerebrospinal fluid escapes through several pores in the thin roof of the fourth ventricle and circulates slowly in the subarachnoid space. It eventually returns to the blood, primarily by diffusion through **arachnoid villi,** which project into large venous sinuses in the dural septa.

The cerebrospinal fluid forms a liquid cushion around the brain and spinal cord, helping support these delicate structures and buffering them from blows to the head or body. Because the subarachnoid spaces follow blood vessels deep into the brain, the cerebrospinal fluid also enters the brain and becomes continuous with the extracellular fluid. The extracellular fluid bathing brain neurons and glial cells thus has two origins: (1) capillaries within the brain and (2) the choroid plexuses.

Blood–brain barriers exist between the extracellular fluids and the blood in the choroid plexuses and in the capillaries of the brain. The endothelial cells of capillaries in the choroid plexuses and brain have tight junctions between them and are less permeable than are other capillaries, and the capillaries in the brain often are surrounded by feet of astrocyte cells (Fig. 13-5). The choroid plexuses and cerebral capillaries act as selective barriers and active transport sites for many substances, thereby carefully regulating the composition of cerebral extracellular fluid. Lipid-soluble molecules, such as oxygen, carbon dioxide, and some drugs (e.g., heroin), easily cross these barriers; but proteins, many hormones, many enzymes, and waste products in the blood and other substances are held back. Glucose, amino acids, and other nutrients are actively transported across. Active transport is bidirectional, and many waste products and metabolites are removed in this way.

Because the brain has no lymphatic system, the cerebrospinal fluid also acts as a subsidiary drainage system for the brain, carrying off excess water, carbon dioxide, and metabolites, which are discharged into the blood at the arachnoid villi. The cerebrospinal fluid makes an important contribution to brain nutrition and drainage.

The Fish Brain: A Prototype for Vertebrates

Neurons are the functional and structural units of the nervous system, so their organization within the brain and their interconnections are of considerable importance. Many techniques to trace neurons and to determine functions of parts of the brain are now available (Focus 14-1). Although less is known about the organization and functions of the brains of fishes than of the brains of mammals, neuroanatomists have learned a great deal in recent years. We begin by discussing the structure and organization of the brains of fishes. Fish brains are far less complex than are those of amniotes, so you can become acquainted with the basic organization of the brain that is common to all craniates. This will serve as a point of departure for tracing major trends in the evolution of the brain.

The Hindbrain

The Medulla Oblongata and Reticular Formation

The hindbrain of fishes consists of the medulla oblongata and the cerebellum (Fig. 14-4). The medulla oblongata is essentially a forward extension of the spinal cord. It contains nuclei that receive the sensory input of cranial nerves attaching to it and gives rise to the axons of motor neurons leaving via these nerves (the trigeminal nerve, V, through the vagus nerve, X; Fig. 13-8).

The sensory nuclei of the hindbrain receive fibers from general cutaneous receptors, taste buds, the lateral line receptors (including electroreceptors), vestibular receptors (equilibrium and acceleration), and sound receptors. In carp, minnows, and catfish, which have extensive taste receptors on the body surface, the sensory nuclei that receive this information are enormous and form large **facial** and **vagal lobes** on the medulla oblongata (Fig. 14-4*D*). The **octavolateralis nuclei,** which receive sensory input from the lateral line organs and ear, also may form a pair of bumps on the dorsal surface of the medulla near the entrance of the statoacoustic or octaval nerve (cranial nerve VIII) and lateralis nerves. Most of the somatic motor nuclei supply the branchiomeric muscles of the jaws and branchial arches, but some supply epibranchial and hypobranchial muscles. Visceral motor nuclei of fishes send preganglionic parasympathetic fibers through the oculomotor nerve to the eyeball and through the vagus nerve to postcranial organs.

Although some of the integration of medullary functions occurs through direct connections between the sensory and motor nuclei, most occurs through the

FOCUS 14-1 *Determining Central Nervous System Pathways and Functions*

The spinal cord and brain are vital and complex organs. Investigators can see gross changes that have occurred during evolution by dissecting the CNS of representative vertebrates, but these changes tell us little about functions. To begin to understand function, it is first necessary to make a "road map," or "wiring diagram," of neuronal organization within the CNS. This will show us the pathways neurons take and give us some indication of which parts of the CNS may influence other parts. Many techniques have been used. Particularly important ones involve the injection into parts of the spinal cord or brain of isotopically labeled amino acids (e.g., tritiated thymidine), horseradish peroxidase, or other materials. These are taken up by neurons at synapses or other sites such as the cell body. Depending on the nature of the material, or proteins incorporating the labeled amino acids, the injected material will travel forward along an axon to its termination (**anterograde tracing**) or backward into the cell body (**retrograde tracing**), or both. Some substances will cross one or more synapses and label an entire pathway. At a later time the experimental animal is sacrificed, microscopic sections are made through parts of the spinal cord and brain, and neuronal tracts or nuclei containing the markers can be identified.

One of the most exciting developments in neuroanatomy in recent decades has been the introduction of dyes sometimes called intercalating dyes because they intercalate into the cell membrane of neurons. Examples include the materials known as DiI and DiO, which are fluorescent and thus can be detected using fluorescence microscopy even when they are present in very small concentrations. The methodology for tract tracing with DiI is very straightforward and can be used on preserved specimens. A small crystal of the dye is placed adjacent to the neurons to be traced at either their terminal arborizations or their cell bodies. The specimen is left for some period of time during which the dye intercalates into the neuronal cell membranes and becomes distributed throughout the length of the neurons. At a later time, sections of any part of the pathway can be studied using fluorescent microscopy to visualize the neurons carrying the dye. This method has made it possible to investigate neuronal tracts in preserved specimens, or those species that cannot be easily maintained for laboratory study. This opens up the possibility of collecting neuroanatomical information for a much wider selection of vertebrates.

Neurobiologists can refine the "wiring diagrams" and gain some indications of function by electrophysiological experiments. In one type of experiment, a stimulating electrode activates a nerve or brain center. A recording electrode is used to penetrate the brain in regions likely to be activated, stopping at approximately 100-μm intervals to determine whether neurons in that region can be driven by the stimulus, that is, to detect **evoked potentials**. In another type of experiment, a stimulating electrode is rigidly mounted to the skull and is advanced in small steps into the region being investigated, a stimulus is given, and the response of the animal is monitored.

Neurologists first learned a great deal about human brain function by observing the changes in behavior

reticular formation. The reticular formation is essentially a rostral extension into the medulla oblongata and mesencephalic tegmentum of the deeper layers of the dorsal horn of the spinal cord (Fig. 14-5). It performs the same functions in the brainstem that these layers do in the spinal cord, namely receiving sensory inputs and serving to integrate complex intersegmental reflexes.

For descriptive purposes, the reticular formation often is divided into ascending and descending parts. The ascending reticular system receives sensory or afferent impulses from the lower parts of the body via **spinoreticular tracts** and sends them forward to higher brain centers, including the cerebellum, optic tectum, and hypothalamus (Fig. 14-5*A*). Other parts of the ascending reticular formation, at least in mammals, reach the cerebrum and affect processes such as sleep, arousal, and attention. Attention is the ability to focus on a particular task by eliminating extraneous signals.

Descending or efferent impulses reach the reticular system from the striatum, tectum, and cerebellum (Fig. 14-5*B*). Descending pathways from the reticular system are the **reticulobulbar tract** to motor neurons in the brainstem and the **reticulospinal tract** to motor centers in the spinal cord. Through these tracts, the reticular formation coordinates the activity of neuron pools, which are called **central pattern generators,** in the brainstem and spinal cord that control rhythmic motor patterns. These patterns involve eye movements, feeding and swallowing, breathing, heart rate, swimming, and generating electric discharges in electric fishes. Swimming and eye movements, for example, can be directed toward food. In fishes and many nonmammalian vertebrates, the reticulospinal tract is the major pathway by which the brain can affect lower parts of the body because no direct pathways from the cerebrum exist.

The descending reticular formation of lampreys, teleosts, and many amphibians includes a single pair of

caused by trauma or pathological lesions and then locating the damaged area in postmortem examination. The French surgeon Pierre Paul Broca, as an example, performed an autopsy in 1861 on a patient with a severe speech defect. The patient could speak isolated words but could not speak grammatically or in full sentences. A lesion was found in a cerebral area that is now known as Broca's area (p. 493). Neurosurgeons have determined other functions during surgical procedures, which often are done with the patient under local anesthesia because the brain contains no pain receptors. As a surgeon probes for a tumor, he or she can observe a twitch of a muscle or other changes in behavior, and the patient can comment on sensations.

Many additional brain functions have been determined by neurobiologists in experimental animals by observing the changes in behavior that result from the microsurgical destruction of areas by cautery, and the local administration with micropipets of neurotransmitters and other drugs. Immunocytological procedures also are being used to localize neurons containing particular neurotransmitters.

Many procedures for neuronal tracing and for studying functions are of necessity invasive and lead to the sacrifice of experimental animals. As we discussed above, tract tracing can now be done on preserved material.

Recently, advancements in imaging techniques have allowed neurologists to explore the structure and function of a living brain without using disruptive invasive procedures. A particularly interesting example is **positron emission tomography** (PET), which produces computer-generated images of a series of planes or sections through a brain. Unlike X-ray computerized tomography (CT scans), which identifies different densities in the brain, positron emission tomography shows the distribution of isotopes that emit radiation in the form of positrons. A positron-emitting isotope can be bound to an important biological compound and given to a subject, and various biological processes can be explored. For example, an analogue of glucose, which can be taken up by cells but not leave them, can be labeled with a radioactive isotope and given to a subject. Because particularly active cells take up the analogue of glucose, investigators can compare the sites and degree of glucose utilization of a person at rest and then performing various tasks. In this way, investigators have determined which parts of the brain become active when the subject is viewing an object, listening to a sound, memorizing something, or doing other things. Neurologists are learning a great deal about the internal organization of the central nervous system and its functions, but our knowledge is far from complete.

As humans, we have benefited tremendously from the insights that tract tracing and studies of CNS functions have brought to the study of human health and disease; as biologists, we benefit by understanding more completely the nervous system and biology of other vertebrates. Studies of tract tracing and CNS functions promise to continue to reveal important new insights about comparative vertebrate anatomy.

giant **Mauthner cells,** the large, myelinated axons of which decussate and descend the full length of the spinal cord, giving off numerous branches to the motor neurons. Because the speed of a nerve impulse is proportional to axon diameter, transmission on Mauthner axons is very rapid. The system mediates rapid escape reactions, such as rapid flexion of the trunk in aquatic species or rapid extension of the hind legs in frogs.

The Cerebellum Hagfishes have no trace of a cerebellum, and that of lampreys is rudimentary at best, consisting primarily of part of the octavolateralis nucleus. It probably is comparable to part of the auricles of the cerebellum of sharks (see subsequent discussion). The cerebellum of other fishes is quite large and can be divided into a median **body of the cerebellum** and a pair of lateral **auricles of the cerebellum** (Fig. 14-4*B*). Actinopterygian fishes have in addition an expanded part of the body of the cerebellum called the **valvula cerebelli** (Fig. 14-4*C*). The auricles receive input primarily from the vestibular parts of the ear and the lateral line; the body of the cerebellum receives inputs from all sensory systems throughout the body, with the exception of olfactory and possibly gustatory receptors. Proprioceptive input is extensive. The body of the cerebellum also has connections with other parts of the brain, including visual and auditory centers. Considerable integration occurs among the three layers of cells that form the cerebellar surface. Its efferent fibers go first to deep neurons within the cerebellum (Fig. 14-5*B*). Efferents from these cells connect with the reticular formation, which, in turn, affects motor centers in the brainstem and spinal cord.

The connections of the cerebellum allow it to monitor the position and movements of the body; muscle activity; and the visual, auditory, spatial, and electric "landscape" about the fish. It plays an important role in coordinating muscle activity, including eye movements, ensuring that the degree, duration, and timing

FIGURE 14-4
Representative fish brains.
A, A lamprey (lateral view).
B, A dogfish (lateral view).
C, A mormyrid (dorsal view).
D, A carp (lateral view).
(*C, After Hopkins;*
D, after Bond.)

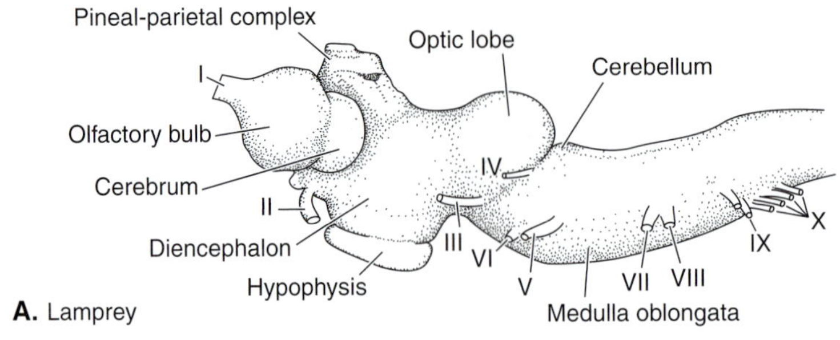

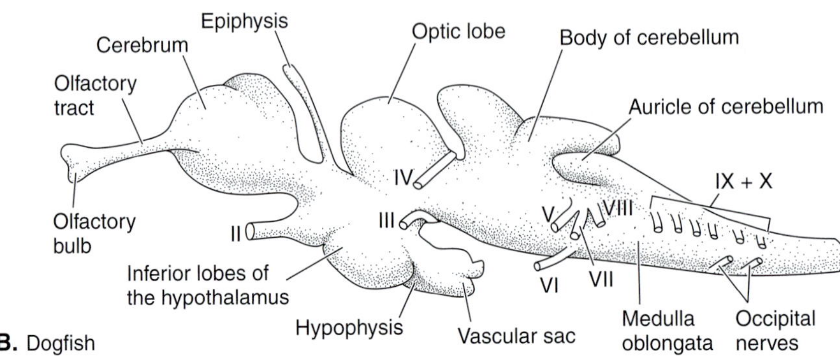

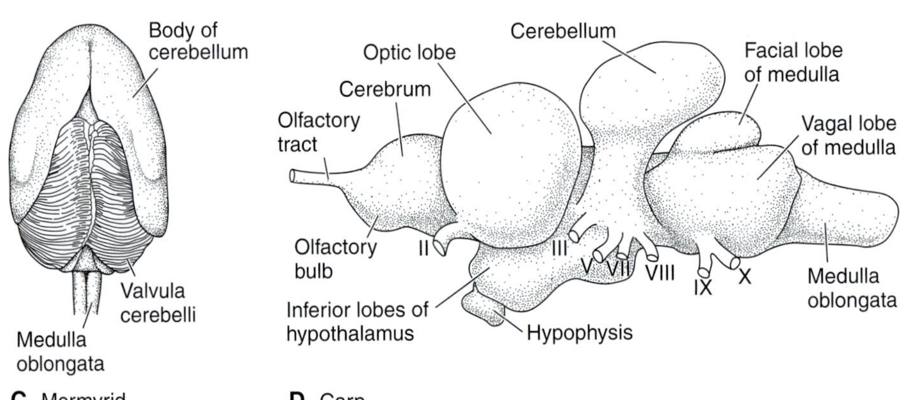

FIGURE 14-5
Lateral views of important neuronal pathways in the brain of fishes, based on a shark. In this and similar diagrams, nuclei and other areas of gray matter are shaded. *A,* Afferent pathways from sensory centers and the reticular formation to higher brain centers. *B,* Efferent pathways from higher brain centers to the reticular formation and motor centers.

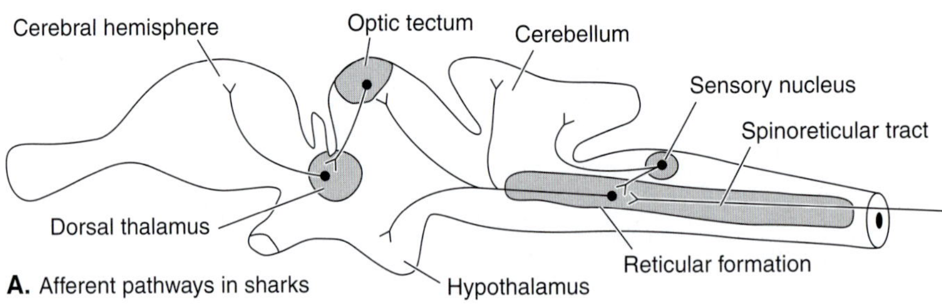

of muscle contraction occur with reference to the position and movements of the fish. The cerebellum also participates in learned motor behaviors.

The cerebellum is enormous in mormyrid fishes, which have highly developed electroreceptive and electrolocation systems. The cerebellum is so large that in a dorsal view, it covers the entire brain except for a portion of the medulla oblongata (Fig. 14-4*C*). Its valvula cerebelli is highly folded, providing a large surface for an increased number of neurons. If unfolded, it would be over ten times the length of the fish.

The Midbrain

The Tegmentum Because the floor or **tegmentum** of the midbrain is a part of the brainstem, much of its structure and many of its functions are similar to those of the medulla oblongata. The tegmentum contains the midbrain part of the reticular formation, part of the nucleus of the trigeminal nerve, and the nuclei of the oculomotor and trochlear nerves (see Fig. 13-8).

Beyond this, the tegmentum includes nuclei in lampreys and all jawed fishes that act as a **midbrain locomotor region,** or **command generator.** These initiate coordinated, purposeful sequences of locomotor movements, particularly of the paired fins, or limbs in tetrapods, that are sent to central pattern generators in the spinal cord. Although lacking in lampreys, the tegmentum of some fishes also contains a **red nucleus,** which receives projections from the cerebellum. The red nucleus connects with the spinal cord by a **rubrospinal tract** and is another way by which the cerebellum affects locomotor movements. The red nucleus and rubrospinal tract are better developed in tetrapods (see Fig. 14-12).

The Midbrain Roof Most of the midbrain roof of fishes forms a conspicuous pair of bulges called the **optic tectum** or **optic lobes** (Fig. 14-4*B* and *D*). Caudally, the roof contains a second pair of deeply seated centers, each called a **torus semicircularis** (L., *torus* = knot, bulge), which comprise the **auditory tectum.** The torus semicircularis receives ascending auditory, lateral line, and electroreceptive fibers and relays impulses forward (Fig. 14-6). The size of the torus semicircularis correlates closely with the importance of auditory and both the mechanoreceptive and electroreceptive components of the lateral line system in the lifestyle of the fish.

The optic tectum is a very important integrating center in fishes, comparable in some respects to the cerebrum of amniotes. Excepting galeomorph sharks and rays, which have a large cerebrum, the optic lobes are usually the largest part of the brain. Its neurons are arranged in layers. Most of the fibers from the retina terminate here in the most superficial layers of neurons (Fig. 14-6); a few end in the dorsal thalamus. The inputs form a precise (although inverted) spatial map of the visual field. Although the primary input is visual, the optic tectum is multisensory and also receives fibers in deeper layers from the auditory, somatosensory, and electroreceptive systems. These too form spatial maps, which are in register with the visual map. That is, a location of the visual map would correspond to one in the auditory map. Descending fibers lead to tegmental motor centers and to the reticular formation (Fig. 14-5*B*). The connections of the tectum allow a fish to localize stimuli in space, a falling food particle, for example, and direct eye and locomotor movements toward it.

The Diencephalon

The Epithalamus The diencephalon is the caudal part of the forebrain, and it can be subdivided into an **epithalamus** dorsally, a **thalamus** laterally, and a **hypothalamus** ventrally (Fig. 14-2). The epithalamus contains a small center, the **habenula,** which has connections to the telencephalic limbic system (see subsequent discussion). The habenula appears to serve as an interface for limbic and motor system pathways. The **pineal–parietal complex** consists of the photoreceptive median eyes discussed earlier (Chapter 12), or an epiphyseal stalk, or the pineal endocrine gland (Chapter 15). The major function of this complex is the regulation of cyclic behaviors in reference to the diurnal and seasonal cycles of day length.

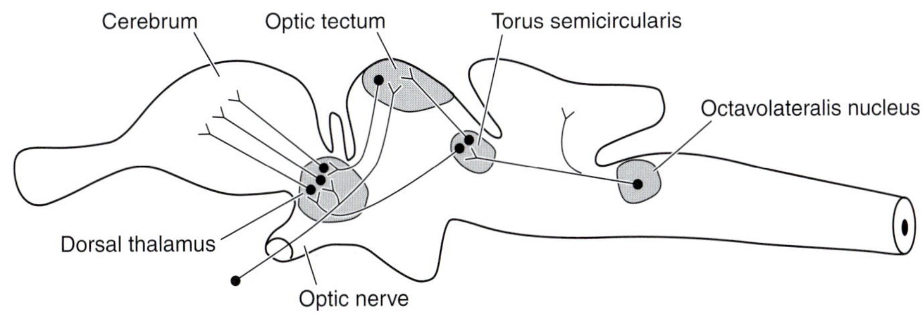

FIGURE 14-6
Lateral view of major optic and auditory pathways in a shark.

The Thalamus The thalamus is the wall of the brain lateral to the third ventricle. Most of it forms a group of **dorsal thalamic nuclei,** but a small ventral thalamus, or **subthalamus,** is closely associated with a cerebral motor center called the striatum (see subsequent discussion). Dorsal thalamic nuclei relay ascending somatosensory, visual, and auditory and lateral line information to the telencephalon, and they also have other reciprocal interconnections with the telencephalon (Figs. 14-5*A* and 14-6). The evolution of the dorsal thalamus is closely linked to that of the telencephalon. Dorsal thalamic nuclei occur in all vertebrates, but they are not as large and complex in fishes and amphibians as in amniotes. Uniquely, fishes have additional diencephalic nuclei that relay ascending sensory information to the telencephalon, including ventral parts of the thalamus, the hypothalamus, and a more caudal region called the posterior tuberculum.

The Hypothalamus Heart rate, respiratory rate, and some other visceral activities are coordinated in the reticular formation, but the hypothalamus is the major center for visceral integration. It is particularly important for activity mediated through the autonomic nervous system and through many of the endocrine glands. The structure and function of the hypothalamus are best known from studies on amniotes, but many of these features and activities have been identified in fishes. Although the hypothalamus has connections that enable it to receive most types of sensory information, its primary sensory inputs are gustatory and other visceral sensory inputs relayed from the reticular formation (Fig. 14-5*A*), and special somatic sensory olfactory signals relayed from parts of the telencephalon. Some of its cells also respond directly to the levels of glucose, water, salts, hormones, and other factors in the blood. Interacting with the limbic system (see subsequent discussion), the hypothalamus of all vertebrates affects behaviors essential for the survival of the individual and the species. Individual survival includes finding and obtaining the appropriate amount of food and water and regulating gut movements and digestion, blood sugar levels, and water and salt balance. The hypothalamus also regulates periods of rest and activity. Species survival includes many reproductive behaviors, such as courtship, mating, and parental care. Some parts of the hypothalamus have a stimulating, and others an inhibitory, effect. It controls many functions directly via efferent neurons to motor centers in the thalamus, reticular formation, and spinal cord. Other functions are controlled indirectly via its influence over the secretion of the hypophysis (Chapter 15).

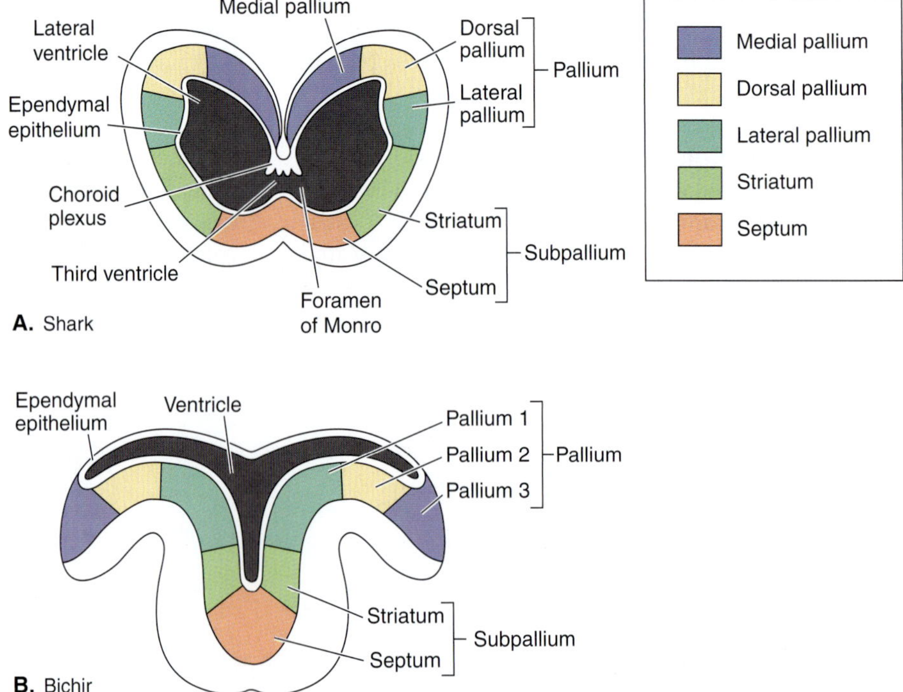

FIGURE 14-7
Transverse sections through the cerebral hemispheres showing the two types of hemispheres found in vertebrates. Dorsal to top. *A,* The evaginated hemispheres as seen in a cartilaginous fish with a laminar distribution of gray matter. *B,* The everted hemispheres as seen in an actinopterygian fish (Bichir) with a laminar distribution of gray matter. Different regions of gray matter are shown in color.

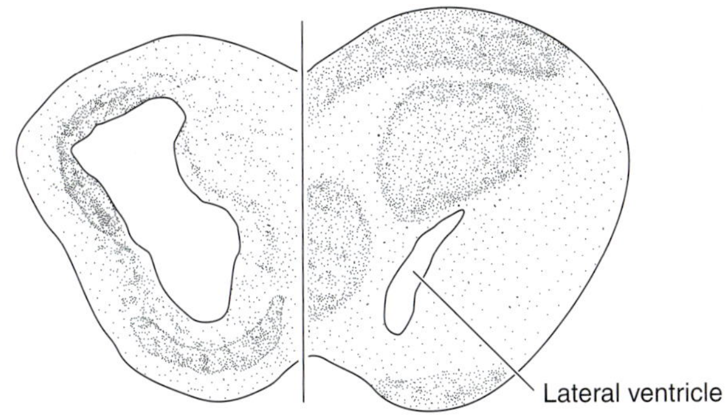

FIGURE 14-8
Transverse sections through the cerebral hemispheres of two types of sharks, showing the relative abundance and distribution of neuron cell bodies. *A (left),* In the laminar, or group I, hemisphere of an early shark (a squaliform shark), cell bodies have migrated only a short distance from their embryonic position close to the lateral ventricle. *B,* In the elaborated, or group II, cerebrum of a galeomorph shark, cell bodies are more numerous and are dispersed throughout the hemisphere. (*Modified after Butler and Hodos.*)

The Telencephalon

Organization of the Telencephalon As we pointed out earlier, the cerebral hemispheres of the telencephalon arise as outgrowths from the rostral end of the neural tube. In all vertebrates, except for actinopterygian fishes, the cerebral hemispheres **evaginate** as a pair of outgrowths, each containing a lateral ventricle (Fig. 14-7*A*). The telencephalon of actinopterygians develops as a pair of **eversions** that grow laterally and curve ventrally. The central canal of the neural tube enlarges to form a single median ventricle covered dorsally only by the ependymal epithelium and primitive menix (Fig. 14-7*B*).

Within each of the major craniate radiations (i.e., jawless fishes, chondrichthyans, actinopterygians, and sarcopterygians) early species in the radiation retain cerebral hemispheres in which the cell bodies of the neurons have not migrated far from their embryonic position close to the ventricle (Fig. 14-8*A* and Table 14-1). This pattern is called **laminar,** or **Group I,** by Northcutt (1978) and Butler and Hodos (1996). The number of neurons has increased greatly in derived members of each radiation, and the cell bodies have migrated extensively through the wall of the cerebral hemispheres, forming a more complex **elaborated,** or **Group II,** pattern (Fig. 14-8*B*).

Cerebral Regions and Functions The gray matter of the cerebrum can be divided into a dorsal portion, called the **pallium** (L., *pallium* = cloak), and a ventral **subpallium** (Fig. 14-7 and Table 14-2). Three pallial regions can be recognized in most craniates, but their presence in hagfishes is questionable. In craniates with evaginated cerebral hemispheres, these regions are described by their relative positions: **medial pallium, dorsal pallium,** and **lateral pallium.** In actinopterygians with everted hemispheres, the medial pallium lies most laterally, but, because homologies are unclear, pallial regions are simply numbered (Fig. 14-7*B*). Subpallial regions are a medial **septum** and a more lateral **striatum** (Fig. 14-7).

The lateral pallium is the **primary olfactory center.** It receives fibers directly from the olfactory bulbs via olfactory tracts and sends this information to other pallial areas. At one time, zoologists believed that the cerebral hemispheres of fishes were nearly exclusively olfactory centers. Olfactory input is extensive, and the cerebral hemispheres are the primary centers for olfactory integration and motor responses to olfactory stimuli. However, it is now recognized that the cerebrum of fishes has other functions as well.

TABLE 14-1 Types of Cerebral Hemispheres in Vertebrates

	Laminar Cerebrum Group I	Elaborated Cerebrum Group II
Jawless fishes	Lamprey	Hagfishes
Chondrichthyans	Squalomorph sharks, chimaeras	Galeomorph sharks, skates, rays
Actinopterygians	Bichirs, sturgeons, gars, bowfin*	Teleosts
Sarcopterygians	Lungfishes, amphibians	Reptiles, birds, mammals

*All show a limited migration of cells away from the ventricle.

TABLE 14-2 Terminology for Regions of Gray Matter in Evaginated Cerebral Hemispheres

Cerebral Gray Matter	Current Terms	Mammalian Structures	Former Terms
Pallium	Medial pallium	Hippocampus	Archipallium
	Dorsal pallium	Isocortex	Neopallium
	Lateral pallium	Piriform lobe	Paleopallium
Subpallium	Striatum	Striatum	Striatum
	Septum	Septum	Septum

The medial pallium, which is called the **hippocampus** in mammals, is a major part of the **limbic system.** Other telencephalic parts of the limbic system are nuclei in the septum and the **amygdala,** which is a deep part of the lateral pallium and striatum (Gr., *amygdale* = almond). The limbic system also includes parts of the hypothalamus and the habenula within the diencephalon. The limbic system has an extensive sensory input from all of the sensory systems. Olfactory impulses come directly from the olfactory bulbs and also from the lateral pallium; most of the other sensory inputs are relayed from the dorsal thalamus or other pallial regions. Its major efferent pathways go back to other pallial regions and to the septum, habenula, and hypothalamus. Acting with the hypothalamus, the limbic system of mammals influences many aspects of behavior, especially motivational and emotional behaviors related to self-preservation and species preservation. These behaviors include fighting, fleeing, sexual behavior, and care of the young. It also has a role in memory and learning. Although the limbic system of anamniotes has not been studied as extensively, homologues of all of the major components are present, and studies that have been made suggest that it has the same functions as in mammals.

Receiving somatosensory, visual, auditory, and other sensory input from the dorsal thalamus, and not receiving direct projections from the olfactory bulbs, are defining features of the enlarged amniote dorsal pallium. Whether the relatively small pallial area tentatively identified as the dorsal pallium in fishes and other anamniotes is homologous to the amniote dorsal pallium is unresolved. Supporting the hypothesis of a homology is the fact that the dorsal pallium receives some visual, auditory, lateral line, and electrosensory projections from the dorsal thalamus, and olfactory input, when present, is not extensive. In contrast to amniotes, most ascending sensory information in anamniotes goes to the medial pallium. Integration of some sensory information occurs in the dorsal pallium, and efferents project through the striatum to the brainstem (Fig. 14-5*B*). The striatum of all vertebrates is involved with the regulation of body movements.

Major Trends in Tetrapod Brain Evolution

The brains of fishes and tetrapods vary enormously in size and in complexity of organization (Focus 14-2). But the brain regions of tetrapods, the structures they contain, and their basic organizational features are the same as in fishes. In discussing the evolution of the tetrapod brain, we need only consider the major differences between them and fishes. The cerebrum, thalamus, mesencephalic tectum, and cerebellum were the regions most affected as vertebrates adapted to terrestrial life and became more active and inquisitive creatures with a wide range of behavioral responses.

The Amphibian Brain

The amphibian brain is little different from the brain of many fishes (Fig. 14-9*A*). The major sensory input into the cerebrum continues to be olfactory, although some somatosensory, visual, auditory, lateral line, and probably other sensory stimuli are projected to it from dorsal thalamic nuclei. A few sensory impulses from caudal parts of the body reach the thalamus directly on **spinothalamic tracts,** but most continue to be relayed to the dorsal thalamus through the reticular formation. The small dorsal pallium projects to the lateral and medium pallium from where cerebral efferents extend to the hypothalamus, tectum, and reticular formation. None go directly to the spinal cord. The cerebrum of amphibians is Group I, or laminar, so its organization is simpler than in many fishes with a Group II, or elaborate, cerebrum. The amphibian cerebrum certainly mediates responses to olfactory stimuli, but we know little about its other functions. A decere-

FOCUS 14-2 *Brain Size*

Large vertebrates tend to have larger sensory and motor surfaces on and in the body, so they naturally have more neurons and larger brains than do smaller vertebrates. Brain size alone is a meaningless figure, but comparing brain size as expressed by weight with body weight gives a useful figure. Because body weight of vertebrates ranges from 1 g or less to many tons, weights are expressed on a logarithmic scale (Fig. A). Data points for species in each major group of vertebrates cluster, and the clusters have been outlined in the figure by polygrams. Notice the tremendous variation even within a single vertebrate group, such as mammals, and the considerable overlap of groups. The most significant comparisons are among species with the same body weight. Any vertical line connects species with the same body size, and the line shown in the figure comes close to intersecting all groups that are shown.

Brain size is largely a function of the number of neurons it contains. As the number of neurons increases, so does the potential for more complex repertoires of behaviors. One might expect, therefore, that derived species and groups would have larger brains than early groups and their species. To some extent, this is true. Hagfishes and lampreys have the smallest brains, and birds and mammals, which have evolved the most complex behaviors, have the largest brains relative to their body size. Although many birds and mammals of the same size have similar-sized brains, some mammals have larger brains than do birds of the same size. Cartilaginous fishes, which we usually regard as an early group, have larger brains than do the ray-finned fishes of the same size. Moreover, the brain size of some cartilaginous fishes overlaps that of birds and mammals of the same size. Turtles and diapsid reptiles have brains that are no larger than those of ray-finned fishes of the same size.

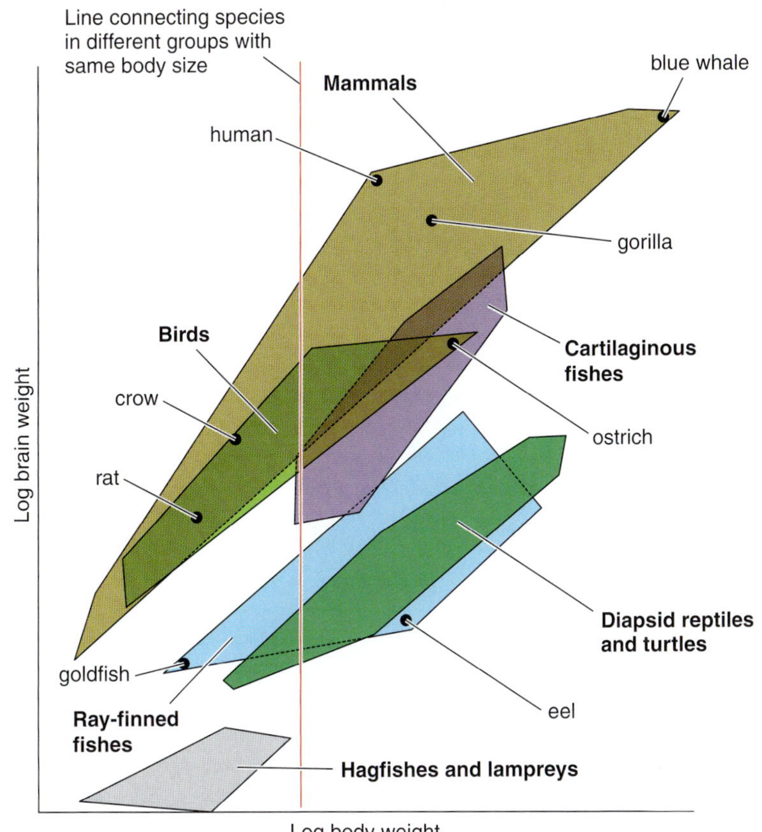

A. The logarithm of brain weight plotted as a function of body weight in approximately 200 species representing major craniate clades. (Modified from Butler and Hodos, after Northcutt.)

brated frog does not seek shelter or feed spontaneously.

The optic tectum remains the primary integration center of the body. It continues to project to the reticular formation, but **tectospinal tracts** have evolved that extend directly to motor columns in the rostral part of the spinal cord and are involved with neck movements. As in fishes, electrical stimulation of parts of the tectum elicits many coordinated locomotor and feeding movements. Stimulation of certain parts of the

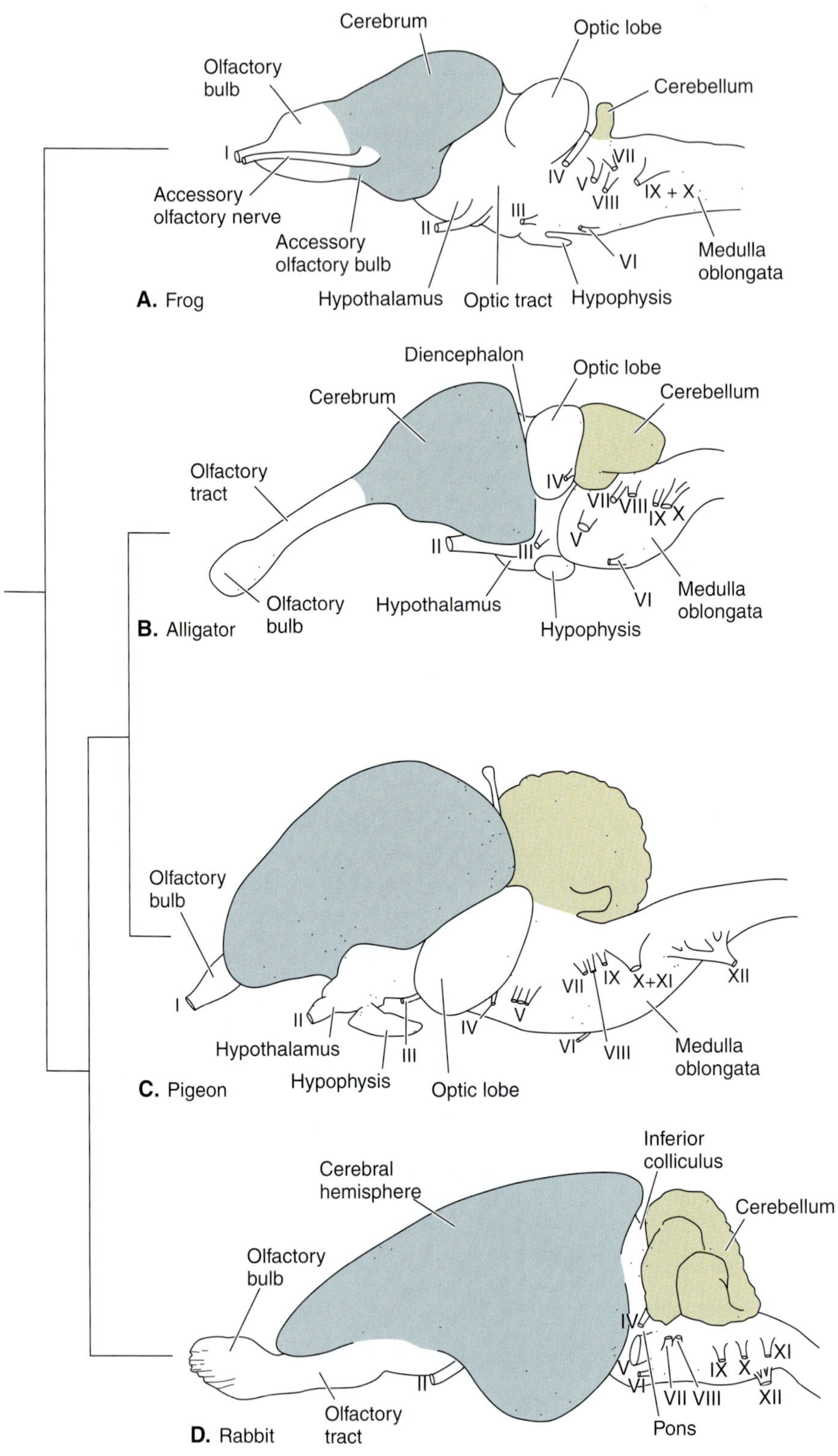

FIGURE 14-9
The evolution of the brain in tetrapods, as seen in lateral view, emphasizing the changes in size of the cerebrum, optic lobes, and cerebellum. Four stages are shown: amphibian (frog, *A*); diapsid reptile (alligator, *B*); bird (pigeon, *C*); and mammal (rabbit, *D*). (*A, After Gaupp; B, after Weidersheim; C, after Pettingill; D, after Barone et al.*)

tectum also inhibits certain activities, including the clasping reflex in a male frog.

Locomotor movements in amphibians are not as complex as in many fishes, and the cerebellum is smaller than in most fishes. Its connections are essentially the same as in fishes. Adult amphibians that spend most of their adult lives on land have lost the lateral line system and, with it, brain centers that process this information. Interestingly, ambystomid and some other salamanders redevelop the lateral line system when reentering ponds to reproduce. Paedomorphic species, such as the salamander, *Necturus,* and highly aquatic frogs, such as *Xenopus*, retain the system.

The Brain of Amniotes

The Cerebrum, Dorsal Thalamus, and Optic Tectum We will consider major sensory and motor pathways in the mammalian brain in the next section. At this time, we present an overview of the evolution of the brain in amniotes. The evolution of the cerebrum, dorsal pallium, and tectum are closely linked. The cerebrum of all amniotes is greatly enlarged compared with that of amphibians, for it has the elaborated, or Group II, pattern of neurons. The expanded cerebral hemispheres have grown caudally, partly (reptiles and birds) or nearly completely (mammals) covering the dorsal and lateral surfaces of the diencephalon (Fig. 14-9*B–D*). This enlargement results primarily from the growth and expansion of the dorsal pallium. The medial (limbic) pallium is "pushed" medially and, in mammals, rolls into the medial side of the lateral ventricle to form the **hippocampus** (Fig. 14-10*D*). The lateral pallium is "pushed" laterally and ventrally. In mammals, it forms the **piriform lobe** (primary olfactory cortex), which is separated from the dorsal pallium by a longitudinal **rhinal sulcus** (Gr., *rhin* = nose + L., *sulcus* = furrow; Figs. 4-10*D* and 14-11*B*). These regions of the brain have essentially the same functions in amniotes as they have in anamniotes.

The enlargement of the dorsal pallium in amniotes is closely linked to an expansion of the dorsal thalamus, which relays more somatosensory, optic, and auditory impulses forward to parts of the dorsal pallium. The evolution of the dorsal thalamus and dorsal pallium among different groups of amniotes is controversial. We follow a hypothesis put forward by Butler (1994) and elaborated by Butler and Hodos (1996). Butler recognizes two distinct groups of nuclei in the dorsal thalamus and two recipient regions in the dorsal pallium. A **collothalamus** receives visual, auditory, and some somatosensory information from the midbrain tectum (which forms the colliculi in mammals) and

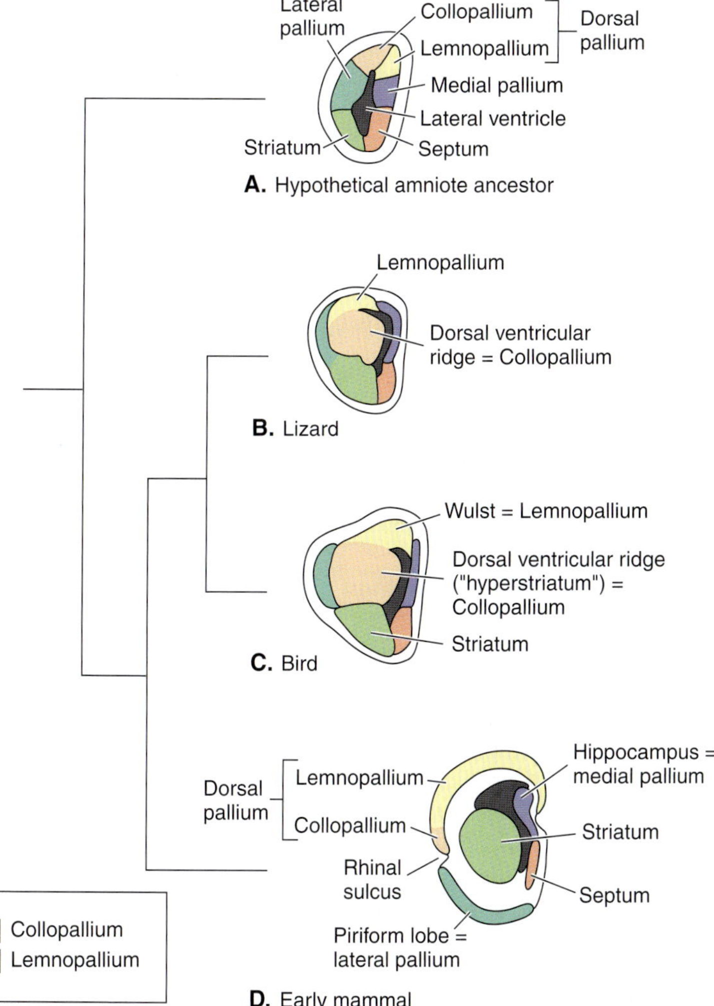

FIGURE 14-10
A–D, The Butler-Hodos hypothesis of the evolution of the cerebrum in amniotes as seen in transverse sections through the left cerebral hemisphere. Lateral is toward the left. Colors are as in Figure 14-7, with the addition of colors for the collopallium and lemnopallium. (*After Butler and Hodos.*)

sends it forward to a **collopallium** (Fig. 14-10). A **lemnothalamus** predominantly receives somatosensory and visual fibers more directly from primary sensory centers. These fibers bypass the midbrain tectum by travelling along ribbon-like fiber tracts, such as the **lateral lemnisci** (Gr., *lemniscus* = ribbon) and part of the **optic tract,** which go directly to the lemnothalamus. The lemnothalamus relays this information to the **lemnopallium.**

Butler further hypothesizes that the dorsal pallium of ancestral amniotes already had differentiated into a lateral collopallium and a medial lemnopallium (Fig. 14-10*A*). In the line of evolution to turtles, diapsids, and birds, an enlarged **dorsal ventricular ridge** evolved that pushed into the lateral ventricle (Fig. 14-10*B* and *C*). This ridge lies above the striatum and often is named the **hyperstriatum, neostriatum,** and **ectostriatum** in birds in the mistaken belief that it was a part of the striatum. Neurobiologists generally agree that the dorsal ventricular ridge evolved from the pallium. Butler believes that it represents an expanded collopallium; others argue that it arose from the lateral pallium. The lemnopallium also expands, but not as much. In birds, it forms a conspicuous dorsal bulge on the medial side of a cerebral hemisphere that is called the **Wulst** (German, *Wulst* = bulge, hump). In the line of evolution to mammals, it is the lemnopallium that expands the most (Fig. 14-10*D*).

The dorsal pallium of amniotes becomes increasingly important as a sensory integration center. In turtles, diapsids, and birds, the greatly expanded dorsal ventricular ridge (collopallium) interacts closely with the optic tectum, which in birds forms large optic lobes that protrude from beneath the cerebral hemisphere (Fig. 14-9*C*). For example, most fibers of the optic nerve terminate in the optic lobes, from which impulses are relayed through the collothalamus to the dorsal ventricular ridge (collopallium). Other fibers go directly to the lemnothalamus, from which they are relayed to the lemnopallium (Wulst). Vision is very important in birds and other amniotes, all of which have at least two interacting systems.

In the line of evolution to mammals (Fig. 14-10*D*), most fibers in the optic nerve, for example, go directly to the lemnothalamus and on to an optic region of the lemnopallium (i.e., **primary visual cortex**). Other optic fibers still go to the optic lobes, which form a pair of **superior colliculi** (L., *colliculus* = little hill; Fig. 14-11*C*). These remain centers for optic reflexes, such as directing eyeball movements toward stimuli, and in spatial orientation. In most mammals, the optic tectum also localizes an object in the visual field, and the visual cortex identifies the object. Most projections of somatosensory impulses also go to the lemnothalamus by the medial lemniscus and then on to the somatosensory cortex within the lemnopallium. Auditory impulses related to auditory reflexes (e.g., turning your head toward a sound) are mediated in the torus semicircularis, which in mammals becomes a pair of small **inferior colliculi** on the surface of the mesencephalic roof. Most auditory impulses are relayed to the collothalamus and on to the auditory part of the collopallium (**primary auditory cortex**).

The lateral pallium and medial pallium are characterized by containing three neuronal layers. Such a pattern is called an **allocortex.** The more complex mammalian dorsal pallium forms a six-layered **isocortex,** or **neocortex,** on the surface of the cerebral hemispheres. The surface of the isocortex remains relatively smooth in many mammals, including the duckbilled platypus, opossum, rabbit (Fig. 14-9*D*), and many rodents. A smooth surface probably was the ancestral condition of the isocortex. As mammals increased in size in many evolutionary lineages, the isocortex became highly convoluted, forming surface folds called **gyri** (Gr., *gyrus* = circle), which are separated by grooves, the **sulci** (Fig. 14-11*A*). A convoluted surface is found in such diverse species as kangaroos, the spiny anteater, most primates, the larger carnivores and herbivores, and cetaceans. A convoluted surface accommodates the increased number of neuron cell bodies needed to process the increased sensory input and motor output of these generally larger mammals. A simple increase in thickness of the cortical layer would disrupt its intricate, six-layered structure.

Efferents from the dorsal pallium of reptiles pass through the striatum and go on to the reticular formation before going to motor centers. None go directly to motor centers in the brainstem or spinal cord. Birds and mammals retain these ancestral pathways. In addition, some dorsal pallial efferents of birds go directly to motor centers in the brainstem and spinal cord. Such a direct pathway from the dorsal pallium (isocortex) of mammals is well developed and forms the **pyramidal system.** We will return to this topic when we consider motor pathways in the mammalian brain.

The Cerebellum The duckbilled platypus among mammals has a unique electroreceptive system in its bill, which it uses to find food, but all amniotes have lost the lateral line and electroreceptive systems. The cerebellum no longer receives input from these systems, but it still has enlarged greatly relative to the cerebellum of amphibians and many fishes (Fig. 14-9*B–D*). Amniotes have evolved an extraordinarily wide range of movements of their head, trunk, tail, limbs, and digits as they have adapted to the more difficult problems of support, locomotion, and feeding on land and to the variety of terrestrial, arboreal, and aerial modes of life open to them.

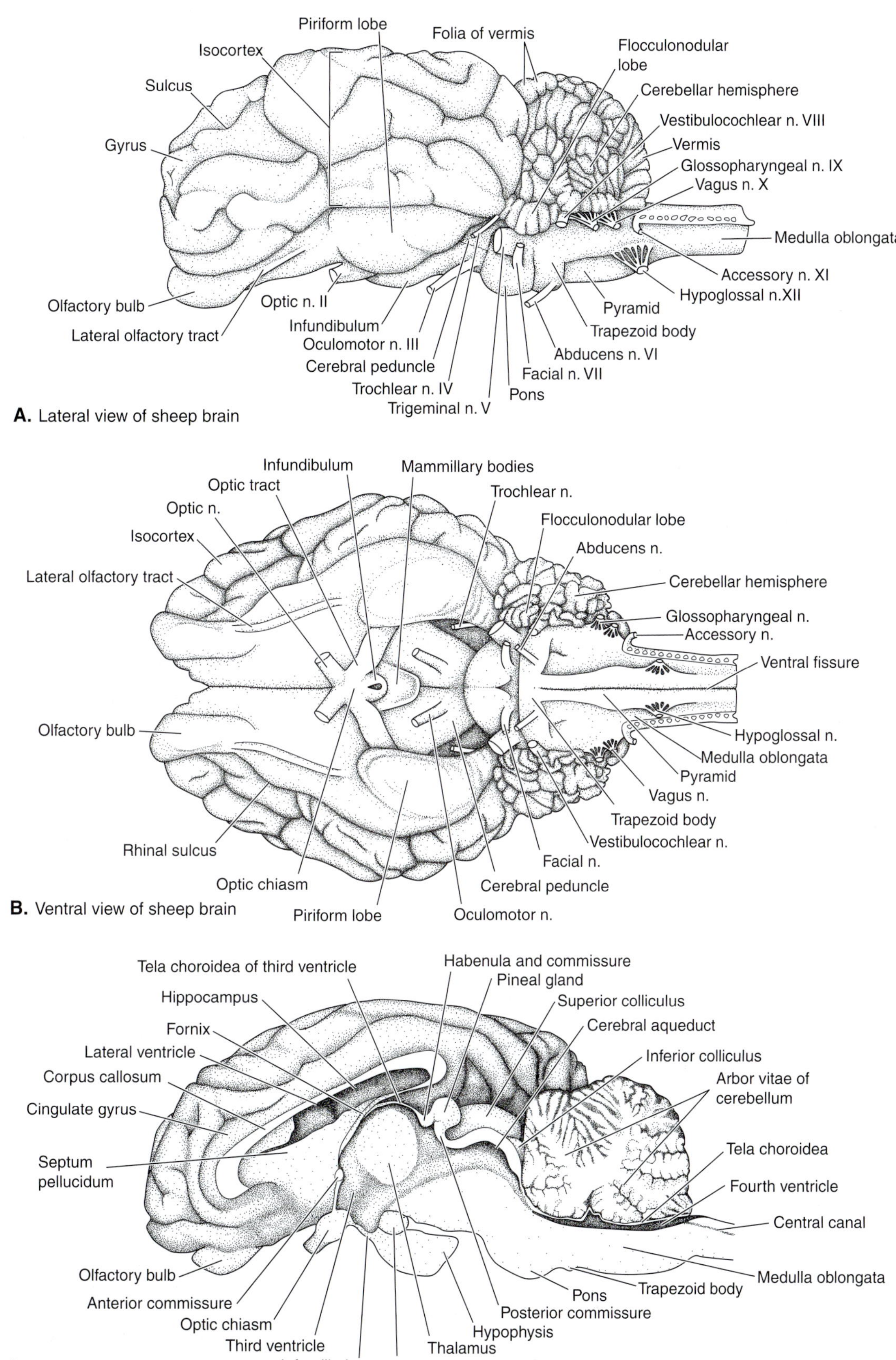

FIGURE 14-11
The sheep brain, a representative mammalian brain, in lateral (*A*), ventral (*B*), and sagittal (*C*) views. (*From Walker and Homberger.*)

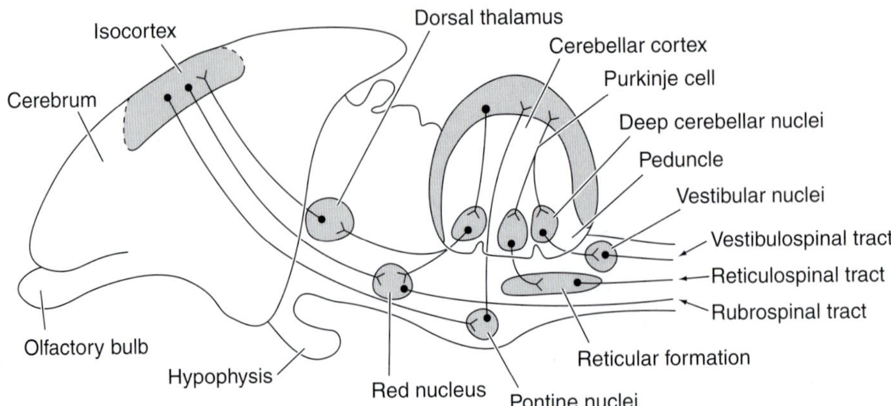

FIGURE 14-12
Lateral view of the major efferent connections of the mammalian cerebellum and its interconnections with the cerebrum.

The cerebellum is particularly large in birds and mammals, and its surface area is increased by the formation of many, tightly packed, leaflike folds, or **folia** (L., *folium* = leaf). The auricles of the cerebellum remain vestibular centers and form **flocculonodular lobes** (L., *flocculus* = a tuft of wool) in mammals (Fig. 14-11*A*). The body of the cerebellum, which is represented in mammals by most of the **vermis** (L., *vermis* = worm), continues to receive an extensive input from most sensory modalities. Proprioceptive input is particularly important because receptors in this system monitor the state and degree of muscle contraction. A prominent feature of the avian and mammalian cerebellum is a pair of lateral expansions of the cerebellar body, which form the **cerebellar hemispheres.** Among their many interconnections in mammals are ones with the isocortex (Fig. 14-12). Fibers going into and out of the cerebellum pass through three stalklike **cerebellar peduncles,** or **brachi,** which attach to the brainstem.

The extensive sensory input and motor output of the cerebellar cortex pass through a deep layer of white matter, which has a branching, treelike configuration. It is called the **arbor vitae** (L., *arbor vitae* = tree of life; Fig. 14-11*C*). All efferent impulses leave on axons of large Purkinje cells, which occupy the middle of the three layers of cells in the cortex. Cells in the other two layers make many interconnections with each other and with the extensive dendritic tree of each Purkinje cell (Fig. 13-1*E*). As a result of the excitation and inhibition that occur among these interconnections, the output of the Purkinje cells can be fine-tuned to meet changing conditions. The targets of the Purkinje cells are the **deep cerebellar nuclei** present in all tetrapods (Fig. 14-12). The number of these nuclei varies, but mammals have four of them.

Efferent fibers from the deep cerebellar nuclei go to brainstem motor nuclei, especially the ones that control eye movements (not shown in Fig. 14-12), to vestibular centers, and forward to the red nucleus and dorsal thalamic nuclei. Projections from the thalamus affect the motor activity of the isocortex, which in turn affects major motor pathways. Fibers from the isocortex that go to the red nucleus in the tegmentum, together with cerebellar fibers that go directly to this nucleus, influence the output of the rubrospinal pathway (Fig. 14-12). Other fibers from the isocortex travel to **pontine nuclei** in the newly evolved **pons,** which, as explained earlier, develops by the migration of cells from the cerebellum into the portion of the hindbrain lying ventral to the cerebellum. In a sense, the cerebellum can monitor motor directives initiated in the isocortex and modify them as needed. In short, the cerebellum is in a position to influence the motor output of nearly all descending motor pathways.

An important overall effect of the cerebellum is to influence the extent, duration, and timing of muscle contraction. These factors are essential for the smoothness of muscle contraction, for maintaining balance, and for muscle coordination. Beyond this, evidence shows that the cerebellum plays a role in motor learning, that is, the ability to execute precise, carefully timed, and complex motor movements.

Important Mammalian Pathways

Major Sensory Pathways in the Mammalian Brain

Ascending Sensory Pathways Many sensory impulses continue to ascend to higher brain centers by relays through the reticular formation, but in mammals, to a greater extent than in other vertebrates, ascending systems have evolved that carry sensory impulses more directly and rapidly to the dorsal thalamus and isocortex. Although different sensory stimuli have different pathways, those coming from pain, temperature, pressure, touch, and taste receptors have several common features. Impulses ascend on three neuron chains. The

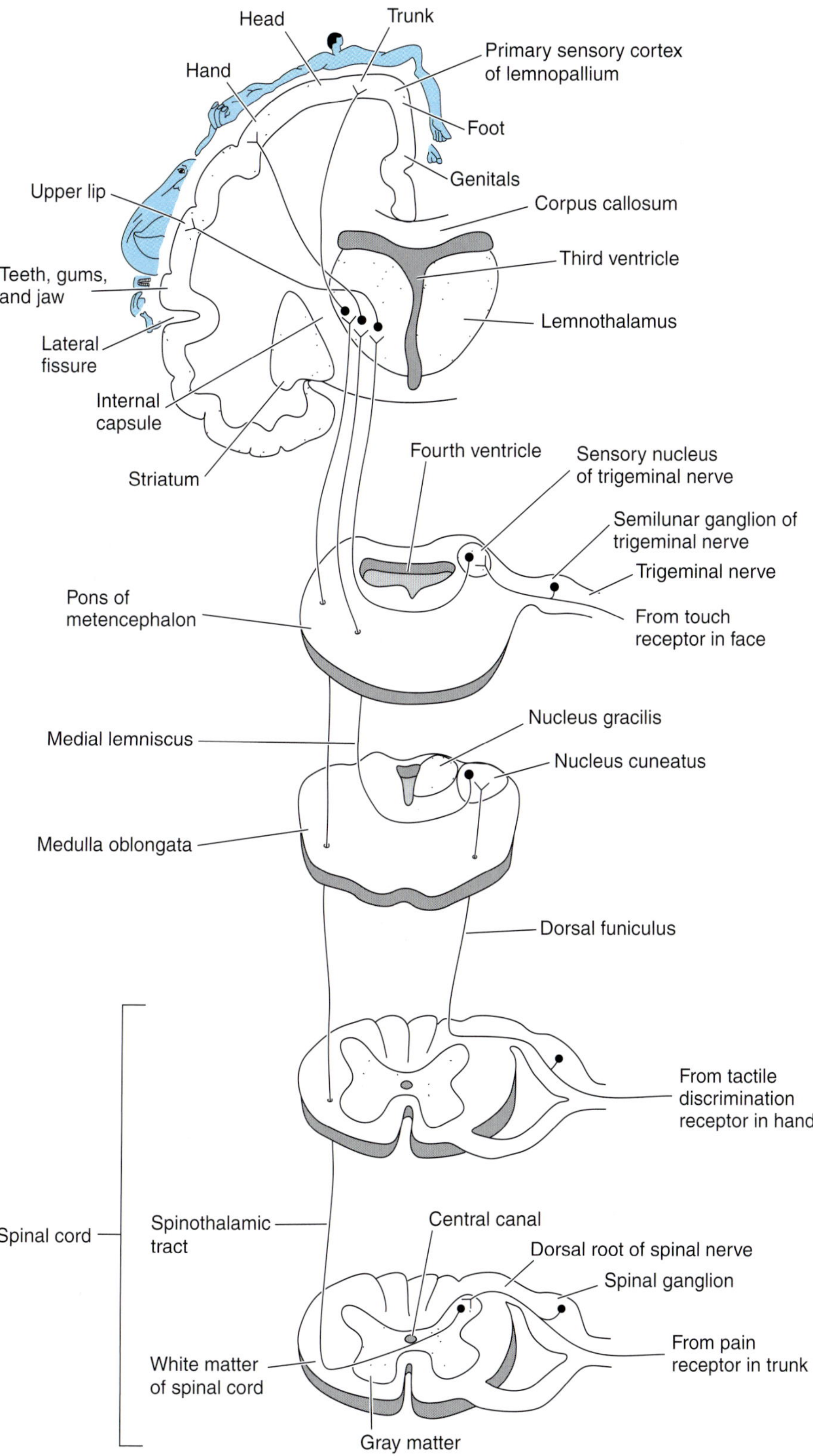

FIGURE 14-13
Transverse sections through the spinal cord and parts of the brain of a mammal, showing the ascending somatosensory pathways from receptors to the primary sensory cortex. Projections from parts of the body to the sensory cortex are indicated. The extent of the cortical areas receiving impulses is illustrated by the size of the figures beside the cortex.

primary sensory neurons from the receptors usually terminate on interneurons in the dorsal horn of the spinal cord, or sensory nuclei of cranial nerves, at the level at which they enter the central nervous system (Fig. 14-13). The sensory neurons from receptors for tactile discrimination on the trunk and limbs, and from proprioceptors, are an exception to this generalization because they ascend in the dorsal funiculi of the spinal cord all the way to the **cuneate** and **gracile nuclei** in the medulla. Tactile discrimination pathways are par-

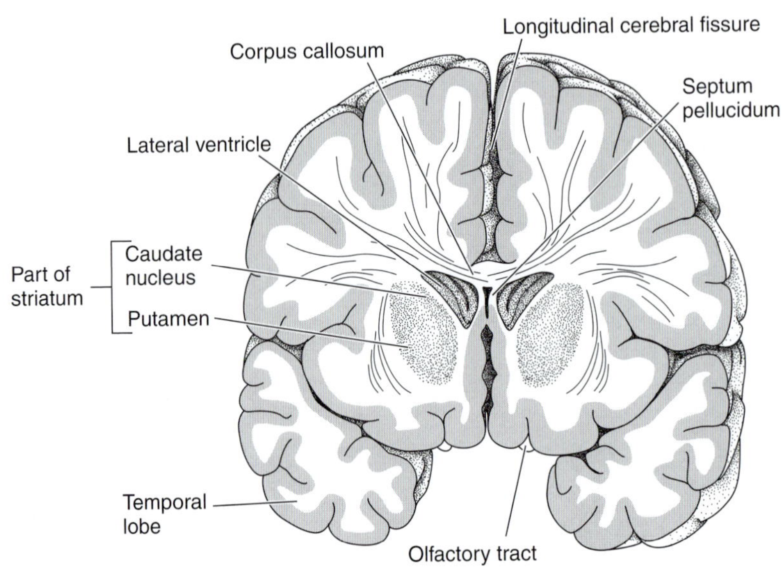

A. Transverse section through cerebrum and corpus striatum of human

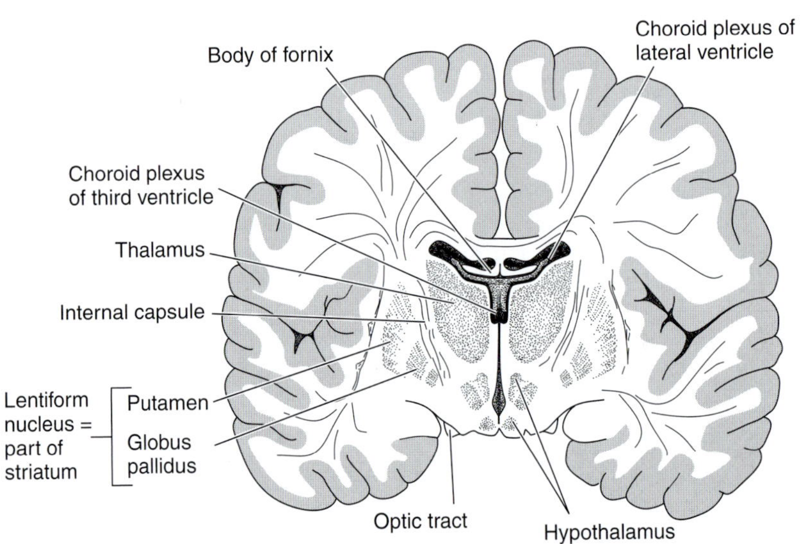

B. Transverse section through cerebrum and thalamus of human

FIGURE 14-14
Two transverse sections through the human brain, showing the positional relationships to each other of the cerebral hemispheres, striatum, internal capsule, thalamus, and hypothalamus. *A,* A section through the cerebrum and corpus striatum. *B,* A section slightly posterior to *A,* passing through parts of the cerebrum and thalamus. (*After Noback et al.*)

ticularly well developed in primates and other groups that use their hands to handle objects. **Second-order neurons** continue from the dorsal horns or brainstem nuclei, usually decussate, and ascend on lemniscal tracts to specific dorsal thalamic nuclei on the opposite, or contralateral, side of the body. A few fibers do not cross and remain ipsilateral. **Third-order neurons** extend from the dorsal thalamic nuclei through the **internal capsule** (Fig. 14-14*B*), a bundle of fibers passing between the striatum and thalamus, to terminate in very specific parts of the **primary sensory cortex.** In general, sensory information from different parts of the body is projected topographically upside down in the primary sensory cortex. The amount of this cortex that receives signals from different parts of the body is related not to the absolute size of the part of the body but to the density of receptors. Impulses from the fingers, lips, and tongue of humans, for example, occupy a disproportionately large share of the primary sensory cortex (Fig. 14-13). The primary sensory cortex is located just caudal to the **central sulcus of Rolando** (Fig. 14-15).

Impulses are processed in each of the nuclei through which they pass. Among other things, extraneous information ("noise"') tends to be suppressed, and the important information ("signal") is enhanced. Except for pain and certain combinations of signals that are interpreted as pleasurable, which are centered in the thalamus, consciousness of stimuli is attained when the impulses reach the cerebral cortex.

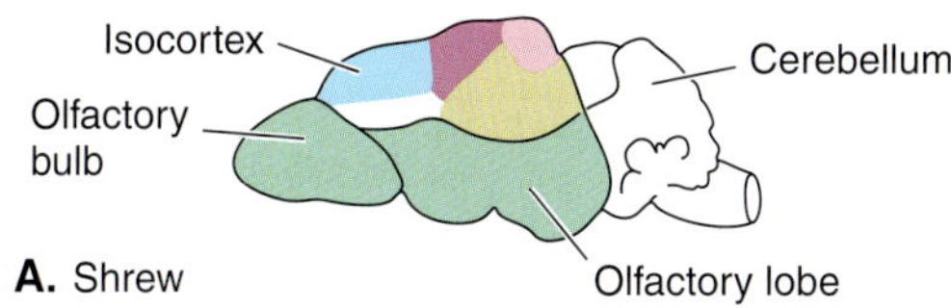

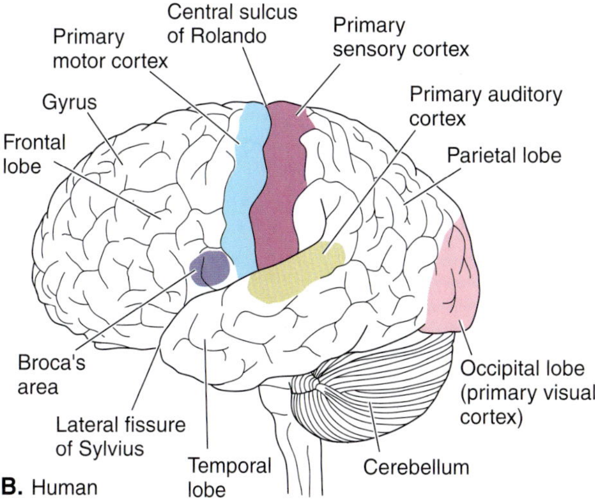

FIGURE 14-15
The primary sensory and primary motor areas of the isocortex of a shrew (*A*) and a human (*B*). Much of the human cortex is occupied by association areas (the blank spaces). One of them, Broca's area that deals with speech, is shown. (*After Romer and Parsons.*)

The Optic System Optic, auditory, and olfactory pathways are different from those of other senses. The fibers in the optic nerve are those of retinal **ganglion cells** (Chapter 12). Because the retina is a part of the brain, the ganglion cells are not primary sensory neurons but interneurons of a higher order. Many of the optic fibers of a mammal remain ipsilateral, unlike the optic fibers in other vertebrates, all or most of which decussate in the optic chiasm. Cetaceans are one exception; all of their optic fibers decussate. The degree of decussation correlates with the degree to which the visual fields of the two eyes overlap. Herbivores and other species subject to predation tend to have laterally placed eyes, which give them wide fields of vision that allow them to see approaching predators. The fields of vision of the two eyes overlap only slightly, and most of their optic fibers decussate. In carnivores, and especially in advanced primates, the eyes have rotated forward. The visual fields of the two eyes overlap extensively, and many optic fibers are ipsilateral (Fig. 14-16). As many as 50% of the optic fibers project ipsilaterally in some primates; the others continue to decussate. Objects in the center of the visual field are seen by each eye but from slightly different angles. The image in each eye projects to both sides of the brain, and the processing of these images permits stereoscopic vision and depth perception. This is very important for a carnivore, which must judge prey distance, and for a primate that scampers through the trees.

As in other vertebrates, the mammalian retina projects to many brain areas: the tectum, a small pretectal area lying just rostral to the tectum; tegmentum; hypothalamus; and thalamus. Mammals differ from anamniotes and other amniotes in that most of their optic fibers go to a dorsolateral lemnothalamic nucleus known as the **lateral geniculate body,** rather than to the optic tectum. Fibers beginning in the lateral geniculate body pass through the **optic radiation** of the internal capsule (Fig. 14-16) and terminate in the **primary visual,** or **striate, cortex,** which is located in the occipital lobe of the dorsal pallium (Fig. 14-15*B*). This is the primary visual pathway. Processing of the visual signals occurs at all levels: retina, lateral geniculate body, and primary visual cortex. This processing analyzes pattern, color, and depth of the visual image and leads to perception. The processing of the visual maps is hierarchically organized, with the neurons at each level "seeing" more and more. The primary visual cortex projects to the **extrastriate cortex** (visual association area) lying rostral to the striate cortex. This area is essential for conceptualizing visual relationships (position) of objects and for learning to identify objects by their appearance.

In addition to receiving fibers from the optic tract, the superior colliculi also receive input from the extrastriate part of the cortex. Following processing of the retinal and cortical inputs, the superior colliculi send fibers to a collothalamic nucleus, the **pulvinar,** from which impulses are sent directly to the extrastriate cortex. Extrastriate cortices project to multiple visual cortical areas, including the inferotemporal area in the temporal lobe. Various visual cortical areas are involved in maintaining visual attention and analyzing motion.

The superior colliculi also participate in visually related reflexes, such as movements of the head and eye in the direction of an unexpected stimulus, and localizing an object in the visual field. The superior colliculi, together with the pretectal area, also mediate congruent movements of the eyes to follow a moving object, and eye movements that keep the eyes fixed on an object when the head moves. Eye movements involve inputs to the superior colliculi from the retina, visual cortex, and vestibular system. Output is to the motor nuclei of the extrinsic ocular muscles (Fig. 14-16). The superior colliculi and pretectal area also mediate pupillary and accommodation reflexes through their effect on autonomic nuclei. A few tectal fibers go to a small nucleus that sends efferent fibers back to the retina through the optic nerve. These fibers, which also are present in some other verte-

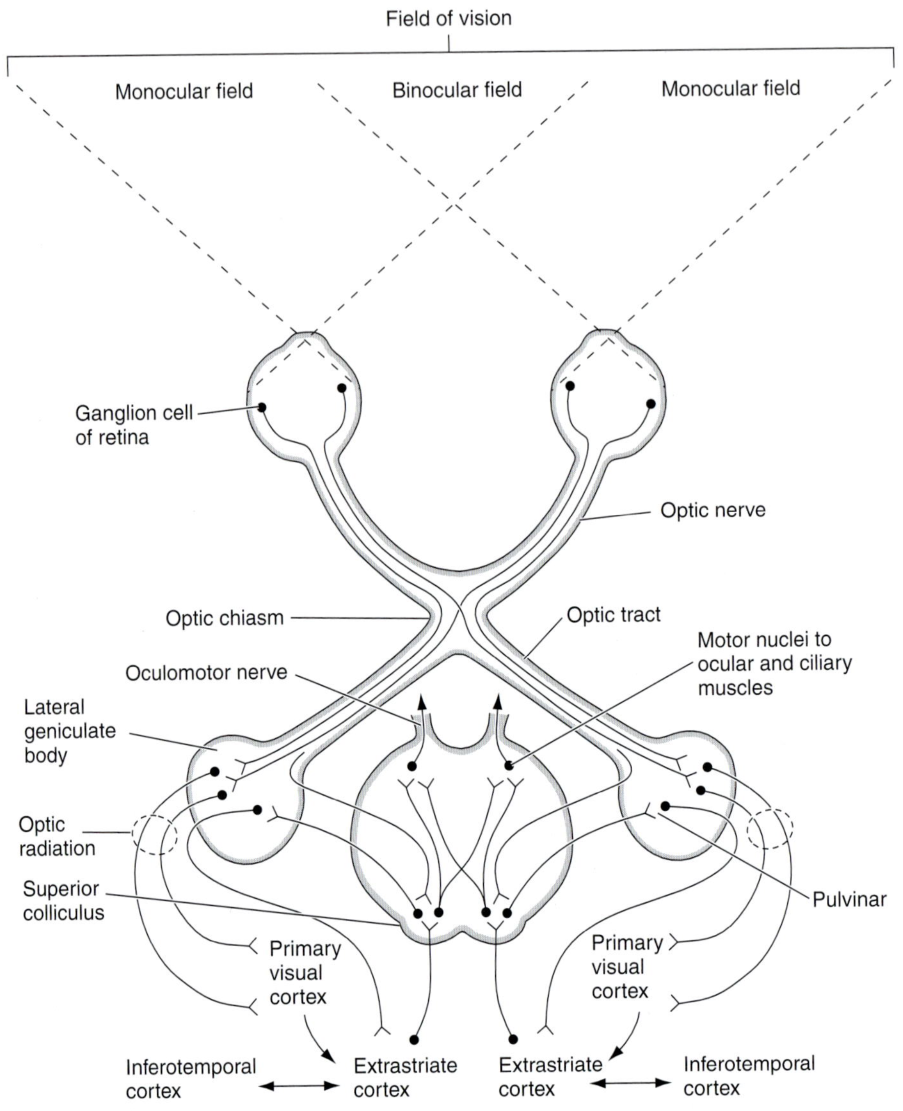

FIGURE 14-16
Dorsal view of the major mammalian optic pathways.

brates, may modulate retinal sensitivity and may affect image processing in the retina.

The Auditory System The primary sensory neurons of the auditory system begin in the organ of Corti of the cochlea (Chapter 12), travel in the vestibulocochlear nerve, and terminate in **cochlear nuclei** in the medulla (Fig. 14-17). Many second-order neurons from the cochlear nuclei decussate in the **trapezoid body,** located on the ventral surface of the medulla (Fig. 14-11*B*), and terminate in the **superior olivary complex of nuclei** in the medulla. Some second-order neurons remain ipsilateral. Third-order neurons that originate in the superior olivary complex, with second-order fibers from the cochlear nuclei that bypass the olivary complex, continue on a tract called the **lateral lemniscus** to the **inferior colliculi** of the midbrain tectum. Some fibers project from here to the reticular formation and motor nuclei of the brainstem. The tectum mediates reflexes that move the head and sometimes the auricles of the ear toward the sound sources. The inferior colliculus projects to the **medial geniculate body** of the collothalamus. Thalamic fibers travel through the internal capsule to the **primary auditory cortex,** which is located in the temporal lobe of the dorsal pallium (Fig. 14-15). As is the case with the optic system, processing of signals occurs at all levels. Two points should be emphasized. First, the cochlear nuclei and other ascending centers are **tonotopically** organized, that is, different sounds are "dissected" into their separate components, and these are finally synthesized in the primary auditory cortex and adjacent areas where the awareness and interpretation of sounds occur. Second, auditory signals from each ear ascend on each side of the brain, which enables the brain to localize the source of sounds by

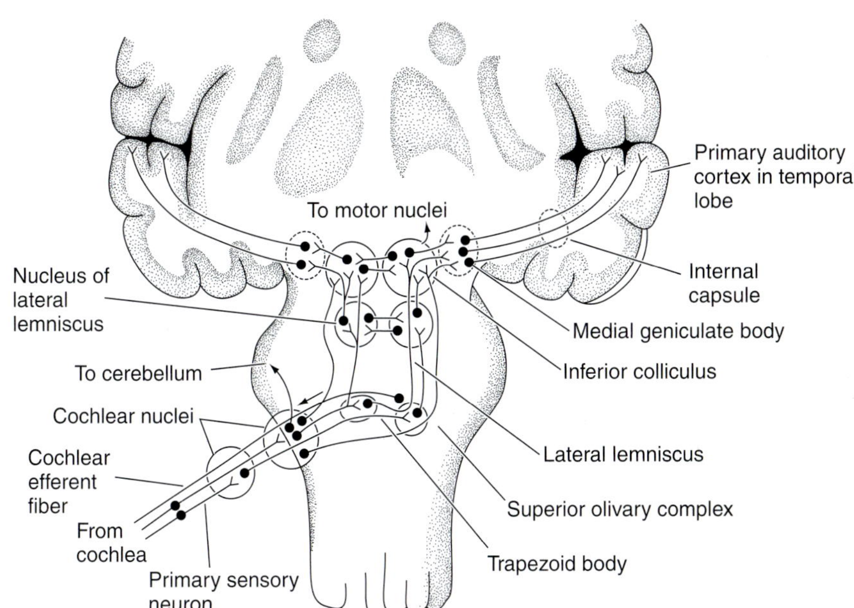

FIGURE 14-17
The major mammalian auditory pathways projected onto a dorsal view of the brain. (*After Noback et al.*)

comparing differences in the timing and intensity of sounds arising in each ear.

A few efferent fibers return from the superior olivary nuclei through the vestibulocochlear (octaval) nerve to the organ of Corti. These fibers modulate the activity of the organ of Corti by enhancing the meaningful signal and suppressing extraneous "noise."

The Olfactory and Limbic Systems The olfactory and limbic systems are closely integrated because the sense of smell is so important to many limbic functions. These systems include the olfactory bulbs, lateral pallium (piriform lobe of the cerebrum), medial pallium (hippocampus), amygdala, and septum. Although some of these centers have other functions, they may have evolved in early vertebrates as centers for processing olfactory impulses.

Olfactory neurons from the nasal cavities extend the short distance to the **main olfactory bulb,** where they terminate in hundreds of complex tangles of neuronal processes called **glomeruli** (Fig. 14-18*A*). Among other connections in the glomeruli, the olfactory neurons synapse with large **mitral cells,** the axons of which are the primary ones leaving the olfactory bulb. Each glomerulus may receive as many as 25,000 (in rabbits) olfactory neurons and the terminations of nearly 100 other neurons. Many of these are from short neurons within the glomerulus, collaterals of mitral cells that feed back to the glomerulus, and neurons coming from the contralateral olfactory bulb. Obviously a great deal of signal processing, some known to be inhibitory, occurs here. Considerable convergence also occurs because far fewer mitral axons leave the glomeruli than olfactory neurons enter.

The axons of mitral cells form the **olfactory tract,** which leads to the cerebrum (Fig. 14-11*B*). Some mitral cells, or their collaterals, enter a small **anterior olfactory nucleus,** where they synapse with neurons that pass through the anterior part of the **anterior commissure** to the contralateral olfactory bulb (Fig. 14-18*A*). The main target of the mitral cells is the piriform lobe (L., *pirum* = pear + *forma* = shape), which is located on the lateroventral surface of the cerebrum (Fig. 14-11*B*). The piriform lobe is homologous to the lateral pallium of nonmammalian vertebrates, and it continues to be the primary olfactory area. Olfactory discrimination and awareness probably occur here, and conditioned olfactory reflexes are lost when this area is destroyed. Olfactory learning also involves extensive parts of the forebrain. The piriform lobe projects to dorsal thalamic nuclei which, in turn, project to the isocortex immediately dorsal to the rhinal sulcus. It also projects to the habenula in the epithalamus, from where efferents lead to the reticular formation and motor nuclei.

Other mitral fibers from the main olfactory bulb project to the septal region, a small rostral part of the hippocampus, and to the amygdala (Fig. 14-18*A*). A different part of the amygdala receives fibers from the vomeronasal organ via the accessory olfactory bulb. The amygdala is located deeply in the ventral part of the cerebrum lateral to the optic chiasm. It evolved from part of the lateral pallium and striatum but is now a part of the limbic system. The amygdala receives other fibers from the thalamus, cerebral cortex, and reticular formation. It influences many aspects of behavior, primarily through efferent fibers that go to the hypothalamus. Among its functions are mediating

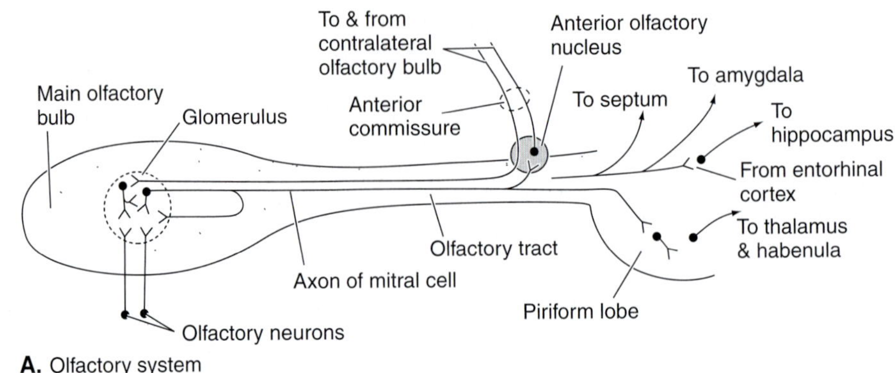

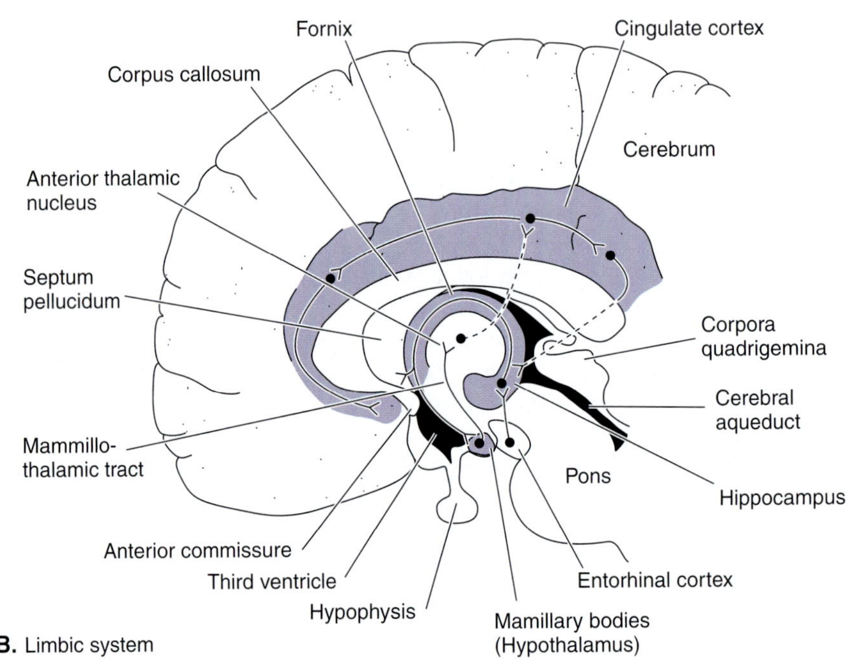

FIGURE 14-18 The major components of the mammalian olfactory and limbic systems. *A,* The major olfactory pathways in a ventrolateral view. The size of the glomerulus is greatly exaggerated; hundreds are present in an olfactory bulb. *B,* The limbic system as projected onto a dissected sagittal section of the brain. Two major parts of the limbic system are shown in purple. (*Modified after Noback et al.*)

sniffing, licking, and other olfactory-based reflexes. It also sends fibers to parts of the thalamus and dorsal pallium.

The **hippocampus** is an important part of the limbic system. As we have seen, it represents the medial pallium, which has been shifted medially by the expansion of the dorsal pallium. It receives olfactory fibers from the **entorhinal cortex,** which lies near the caudal part of the piriform lobe (Fig. 14-18*A*). The hippocampus also receives gustatory and other visceral sensory fibers from the reticular formation; fibers from the **cingulate cortex** on the medial side of a cerebral hemisphere (Fig. 14-11*C*); and auditory, visual, and general somatic sensory projections, which are relayed into the hippocampus via the entorhinal cortex. It retains all of the visceral integrating functions that it has in fishes. Beyond this, the hypothalamus is an important center in both birds and mammals for controlling body temperature and the high level of metabolism that characterize these endothermic vertebrates. Efferents from the hippocampus aggregate to form a prominent tract known as the **fornix** (L., *fornix* = arch or vault), which makes an arc in the base of the cerebrum and then turns ventrally to the **mammillary bodies** of the hypothalamus (Figs. 14-18 and 14-11*B*). A small section of the fornix goes to septal nuclei, which in turn project to the hypothalamus. A prominent **mammillothalamic tract** from the mammillary bodies returns to the thalamus. Relays in anterior thalamic nuclei and the cingulate cortex return impulses to the hippocampus. The complex pathway from hippocampus to mammillary bodies, to anterior thalamic nuclei, to the cingulate cortex, and back to the hippocampus is called the **Papez circuit** (Fig. 14-18*B*). This circuit integrates the hypothalamus into the limbic system.

Because of their extensive connections, the hippocampus, amygdala, hypothalamus, and other parts of the limbic system influence many aspects of behavior, especially motivational and emotional behaviors related to self-preservation and species preservation. These behaviors include feeding, drinking, fighting, fleeing, reproduction, and care of the young. The limbic system exerts much of its influence by inhibiting the hypothalamus and tegmental part of the reticular formation. For example, electrical stimulation of parts of the limbic system causes a mammal to stop an activity in which it is engaged. Conversely, destruction of parts of the limbic system releases the mammal from normally inhibitory stimuli and leads to an overreaction to stimuli. Destruction also causes some behaviors to become repetitive, and complex sequences are not completed in an orderly fashion. A female rat may continuously pick up and drop a newborn infant, apparently not sure what to do next.

The limbic system also has been implicated in the formation of short-term memories. Humans with lesions in the limbic system can remember events of times long past, but because they are unable to bring together and reinforce the signals needed to establish new memories, they cannot recall events that occurred a few minutes ago. The formation of some memories may require an emotional input, especially of the type that is essential to survival.

Motor Control and Pathways

The motor systems allow vertebrates to maintain posture and balance and to move the body and its parts in response to sensory clues and (in many vertebrates) the desires of the animal. Pools of neurons in the brainstem and spinal cord control many reflexes, the rhythmic, stereotyped movement of body parts during locomotion, and other movements.

Pathways descending from the brain also initiate or modulate motor activity. As we have described, tracts descend from the optic tectum (tectospinal tract), red nucleus (rubrospinal tract), reticular formation (reticulospinal tract), and vestibular nuclei (vestibulospinal tract). Beyond these, amniotes retain the motor pathway seen in other vertebrates that descends from the cerebrum through the striatum to the reticular formation and motor centers (Fig. 14-5*B*). Mammals also have a direct **corticospinal tract** that descends directly from the pallium to motor centers. Now we examine these two pathways from the cerebrum.

The Striatum The mammalian striatum is quite large. Its most conspicuous nuclei are the **caudate nucleus,** lying in the floor of the lateral ventricle (Fig. 14-14*A*), and the **lentiform nucleus** (= **putamen** + **globus pallidus**), lying lateral to the internal capsule (Fig. 14-14*B*). Additional, deeper nuclei also exist. The striatum receives a major input from the isocortex, and its major outflow is to part of the dorsal thalamus. It is also highly interconnected with the subthalamus (ventral thalamus) and a group of tegmental nuclei (Fig. 14-19), including a heavily pigmented nucleus called the **substantia nigra.** Neurons of the striatum and substantia nigra produce many neurotransmitter and neuromodulator substances, including acetylcholine, noradrenalin, dopamine (the pigment in the substantia nigra is associated with dopamine production), GABA (gamma aminobutyric acid), and serotonin. These substances allow for many excitatory and inhibitory reactions as the striatum processes information reaching it.

Via its many interconnections, the mammalian striatum regulates motor output from the isocortex to tegmental nuclei and the reticular formation, from which impulses are sent to motor nuclei in the brainstem and the spinal cord. In general, the striatum tends to smooth out what might otherwise be jerky muscle contractions by inhibiting undesirable movements. The importance of the striatum and substantia nigra in

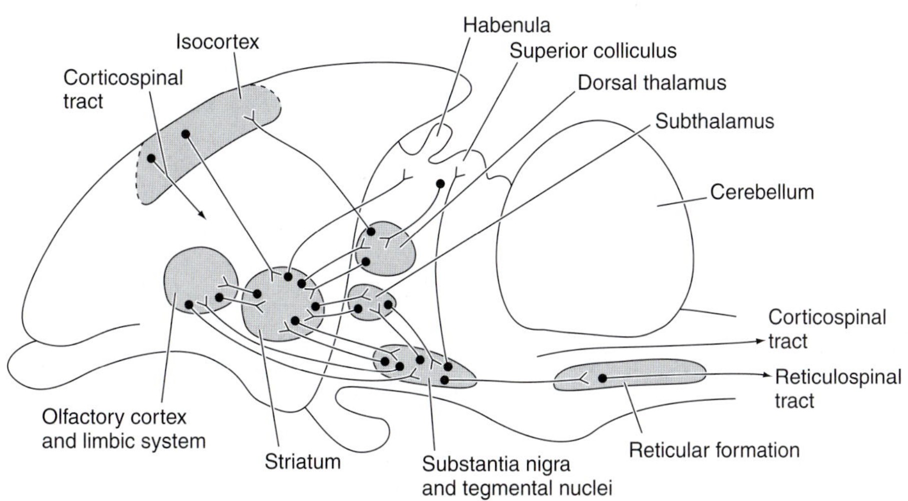

FIGURE 14-19 Lateral view of the major connections of the mammalian striatum and associated substantia nigra in the tegmentum.

motor control is evident when disease or lesions destroy parts of them or alter their complex "cocktail" of neurosubstances, for example, in Parkinson's disease or Huntington's chorea. These diseases cause changes in muscle tone, tremors in limb and body movements, and disturbances in gait and eye movements. In some cases, patients can think about and plan to make body movements, such as lifting a leg or speaking, but are unable to execute them effectively. In other cases, uncontrollable hyperactivity may occur, for example, constantly turning the head or moving a limb.

The Pyramidal System A new and important cortical level of motor control evolved along with the isocortex of mammals (Fig. 14-20). It is called the **pyramidal system** and consists of direct pathways from the isocortex to motor centers in the brainstem and spinal cord. These pathways are the **corticobulbar tract** and the **corticospinal tract.** The pyramidal system is the highest of the hierarchically arranged levels of motor control. The pyramidal system allows mammals to execute complex, voluntary motor activities with precision. For humans, this includes speech, manipulating tools with their hands, playing a piano, or batting a ball. These activities are purposeful and may be initiated at will. They also are learned to a large extent, so performance improves with practice.

The fibers of the pyramidal system going to motor centers begin with neuronal cell bodies located in the precentral gyrus, or **primary motor cortex,** of the dorsal pallium (Fig. 14-15). Some of these are large, pyramid-shaped cells. The amount of cortex occupied by cells supplying different parts of the body is proportional not to the size of the part but to the number of motor units it contains (Fig. 14-20). Neurons going to motor centers supplying the lips, tongue, and hands occupy a disproportionately large share of the motor cortex in humans. As with the primary sensory cortex, the body is represented more or less upside down in the primary motor cortex. Axons of these motor cells form the corticobulbar and corticospinal tracts that descend through the internal capsule and contribute to conspicuous bulges, the **cerebral peduncles,** that can be seen on the lateroventral part of the mesencephalon (Fig. 14-11*B*). The corticobulbar fibers decussate and terminate in the motor nuclei of cranial nerves, but the corticospinal fibers form the **pyramids** (which can be seen on the ventral surface of the medulla), decussate, and continue to the motor horns of the spinal cord. These motor cells send collateral axons into the **pontine nuclei** in the ventral portion of the metencephalon from where impulses are sent to the cerebellum. Additional collateral axons enter the red nucleus, which also receives impulses back from the cerebellum.

Cortical Integration

The isocortex of mammals is divided into **frontal, parietal, temporal,** and **occipital lobes,** which, in humans, underlie skull bones bearing the same names (Fig. 14-15*B*). The boundaries between the lobes are not well defined in many mammals, but a prominent **lateral fissure of Sylvius** separates the human temporal lobe from the others, and a **central sulcus of Rolando** separates the frontal and parietal lobes. The **primary sensory cortex** lies in the parietal lobe just posterior to the central sulcus; the **primary motor cortex** lies in the frontal lobe just rostral to the central sulcus; the **primary auditory cortex** lies in the temporal lobe; and the **primary visual cortex** is in the posterior part of the occipital lobe. Cortical regions between these sensory and motor cortices are called **association areas.** Apart from differences in their afferent and efferent connections, the cortical areas differ in cell density, thickness of their six layers, types of interconnections, and other cytological details. On the basis of these differences, Brodmann, an early 20th-century German anatomist, subdivided the cortex into nearly 50 areas. His numbering system is still used in detailed analyses.

The different areas of the cortex have many interconnections. **Short association fibers** pass from one gyrus to adjacent ones, **long association fibers** interconnect more distant lobes within one hemisphere, and **commissural fibers** pass between the two hemispheres (Fig. 14-20). The phylogenetically older olfactory and limbic parts of the two hemispheres are interconnected by the **anterior commissure** (Fig. 14-18*A*) and by the **commissure of the fornix.** Major parts of the temporal lobes also are interconnected by the anterior commissure. The rest of the isocortex of the two hemispheres is interconnected in eutherian mammals by a very large commissure called the **corpus callosum** (L., *corpus callosum* = hard body; Figs. 14-11*C* and 14-20). Monotremes and marsupials lack the corpus callosum, but marsupials have an exceptionally large anterior commissure that includes dorsal pallial commissural fibers. In this and many other ways, marsupials are not inferior to eutherians but have evolved similar capabilities in different ways.

The size of the isocortex varies greatly among mammals. In insectivores and other early eutherians, it constitutes about 20% of the telencephalon. Association areas are small, and the primary sensory and motor areas (Fig. 14-15*A*) occupy most of the cortex. In more derived eutherians, the isocortex is much larger, and in humans it forms about 80% of the telencephalon (Fig. 14-15*B*).

Some of the association areas supplement the activity of the adjacent primary sensory or motor areas. The

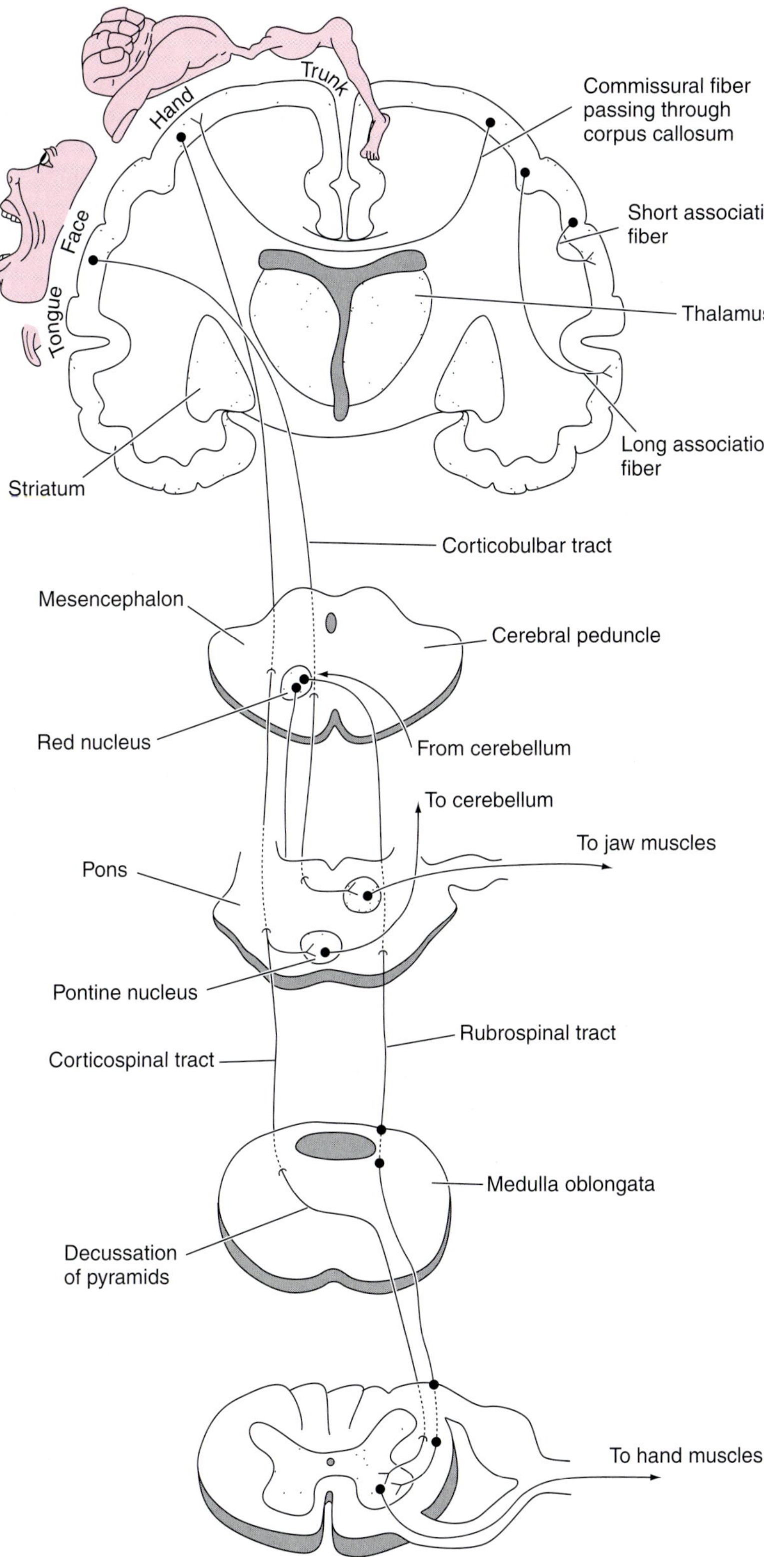

FIGURE 14-20
Transverse section through the spinal cord and parts of the brain in a mammal showing the pyramidal system, or corticobulbar and corticospinal pathways, which extend directly from the primary motor cortex to motor centers in the brainstem and spinal cord. The location in the motor cortex of motor regions for parts of the body is shown, and the approximate area devoted to each part is indicated by the size of the figures beside the cortex. Collateral fibers from the corticobulbar and corticospinal tracts enter the red nucleus and pontine nuclei, thereby establishing connections from and to the cerebellum, respectively.

region preceding the primary visual cortex, for example, is necessary for conceptualizing visual relationships and learning to recognize objects. Positron emission tomography studies have shown that thoughts and anticipation of movements begin in the association area rostral to the primary motor cortex, which is called the **premotor cortex.** The premotor cortex also interacts with the primary motor cortex in such a way that a complex motor activity, such as playing a piano, is not a random series of movements but forms an integrated pattern. The final motor impulses to motor centers leave from the primary motor cortex. The most rostral part of the frontal lobe, sometimes called the **prefrontal lobe,** has extensive connections with other parts of the isocortex and with the limbic system. If these connections are severed, as neurosurgeons formerly did in treating certain cases of severe depression, by an operation known as a prefrontal lobotomy, a person's motivation, ability to formulate goals and plan for the future, and ability to concentrate decrease.

Association areas also integrate separate visual, tactile, olfactory, and other sensory signals to provide an overall perception of the environment and of changes taking place. Association areas are essential for numerous mental processes, such as learning, memory, communication, reasoning, and conceptual and symbolic thought. The two cerebral hemispheres of humans perform somewhat different functions. In right-handed people and most left-handed people, the left hemisphere has a dominant role in communication skills, such as speech, reading, and writing; the right hemisphere dominates in spatial recognition and other artistic skills. In some left-handed people, these functions are reversed. **Broca's area**, a speech center, is located in the ventral part of the left frontal lobe (Fig. 14-15*B*).

The human brain probably contains 10 billion or more neurons, and it processes millions of signals simultaneously. Neurobiologists have learned a great deal about brain structure and function by a process of reducing the brain to its elementary pathways and localizing many functions. But the brain is far more than sets of pathways and functions. It is now necessary to put the units together and learn how the brain functions as an integrated whole. This will require more complex methods of analysis. Studies have begun, but comparative neuroanatomy will continue to be a challenging and fruitful field for research long into the future.

SUMMARY

1. The brain develops from three embryonic enlargements of the neural tube, which later differentiate into five regions. A forebrain differentiates into a telencephalon and diencephalon. The midbrain, or mesencephalon, remains undivided. The hindbrain divides into the metencephalon and myelencephalon. Cavities within the brain enlarge to form a series of interconnected ventricles.
2. The brain is covered by connective tissue meninges, the innermost of which contributes to the choroid plexuses that secrete most of the cerebrospinal fluid. The cerebrospinal fluid helps support, protect, and nourish the brain.
3. The hindbrain forms the medulla oblongata and cerebellum. The medulla oblongata contains the nuclei of many cranial nerves and a network of neurons known as the reticular formation. The reticular formation integrates sensory information received in this region and relays impulses forward to higher brain centers and backward to the spinal cord. Many eye movements, feeding and swallowing movements, respiratory movements, cardiac rate, and blood pressure are controlled here.
4. The cerebellum of fishes consists of paired auricles, which receive vestibular and lateral line impulses, and a central body, which receives tactile, proprioceptive, and electroreceptive information. The cerebellum interacts with other centers in motor coordination.
5. The tegmentum, or midbrain floor, contains an extension of the reticular formation and also a red nucleus with connections to the cerebellum and spinal cord.
6. The midbrain roof, known as the tectum, is the major integration center of the fish brain, and it usually is the largest part of the brain. It comprises the optic tectum, or optic lobes, and a nucleus called the torus semicircularis, which has auditory, lateral line, and electroreceptor connections.
7. The diencephalon is divided into an epithalamus dorsally, a thalamus laterally, and the hypothalamus ventrally.
8. The epithalamus contains the habenula and the pineal–parietal complex.
9. Most of the thalamus contains nuclei that relay impulses to the telencephalon. It is not large in fishes.
10. The hypothalamus is a major center for visceral integration. Acting with the limbic system, it affects many activities related to individual and species survival. Its output is primarily through the autonomic nervous system and by its control of the hypophysis, an endocrine gland.
11. The telencephalon of nearly all vertebrate groups develops as a pair of evaginations from the rostral end of the neural tube. In actinopterygians, it develops as a pair of eversions. Early members of the major vertebrate radiations have a laminar pattern for the distribution of neurons within the telencephalon. More advanced, derived species have an elaborated pattern.
12. Several regions can be recognized in the gray matter of a cerebral hemisphere. The lateral part of its

roof forms a lateral pallium, which is the primary olfactory area. The medial pallium (hippocampus of mammals) is part of the limbic system. Acting with the hypothalamus the limbic system regulates emotional and motivational behavior related to self-preservation and species preservation. The dorsal pallium between the other two pallial regions receives ascending sensory information from the dorsal thalamus. This is not extensive in most fishes. The floor of the cerebrum includes the striatum and septum.

13. The amphibian brain is little different from that of some fishes with laminar brains. In most species, the cerebellum has no lateral line or electroreceptive input.

14. All amniotes have an expanded cerebrum with an elaborate pattern of neuron distribution. This expansion results primarily from the great expansion of the dorsal pallium, which pushes the lateral pallium laterally and ventrally as the piriform lobe and the medial pallium medially as the hippocampus.

15. The enlargement of the dorsal pallium is closely related to the expansion of the dorsal thalamus. Some investigators recognize two parts of the dorsal pallium: (1) a lateral collopallium and (2) a medial lemnopallium. The collopallium receives sensory projections from the part of the dorsal thalamus that, in turn, receives projections predominantly from the tectum (the colliculi of mammals). The lemnopallium receives projections from the part of the dorsal thalamus that predominantly receives sensory information more directly on lemniscal tracts.

16. Both parts of the dorsal pallium expand in amniotes, but in the line of evolution to diapsids and birds, the collopallium expands the most. It rolls into the floor of the lateral ventricle, forming a large dorsoventricular ridge. In the line of evolution to mammals, the lemnopallium expands the most and moves to the surface of the cerebrum.

17. The midbrain tectum continues to be important in diapsids and birds and interacts with the collopallium in integrative processes. Most functions of the optic tectum in mammals are shifted to the visual part of the lemnopallium. The optic lobes (superior colliculi) remain as reflex centers for certain optic reflexes. The torus semicircularis forms a pair of surface bulges (the inferior colliculi) that mediates certain auditory reflexes and relays ascending auditory input to the dorsal thalamus.

18. A cortex of three cell layers characterizes the lateral and medial palliums. The isocortex of the dorsal pallium has six cell layers. The surface area of the isocortex is increased in large mammalian species by forming a series of folds, the gyri, separated by grooves, the sulci.

19. The cerebellum also has enlarged as head, body, and limb movements have become more complex. Its surface area is increased in birds and mammals by the formation of folds, the folia. The cerebellar auricles form the flocculonodular lobes in mammals, the body forms most of the vermis, and lateral expansions form the cerebellar hemispheres. The cerebellum retains its earlier connections and has evolved new ones with the isocortex via the pons. The cerebellum influences the extent, duration, and timing of muscle contractions, so it plays an essential role in motor coordination.

20. Many ascending sensory pathways have evolved. All those from lower centers go to dorsal thalamic nuclei, where some processing of signals occurs before they are sent to the primary sensory cortices of the isocortex.

21. Most of the optic fibers of mammals project via the lateral geniculate bodies of the dorsal thalamus to the primary visual cortex of the isocortex. Processing of the visual image occurs in the retina, thalamus, and primary visual cortex. The primary visual cortex, in turn, projects to the extrastriate cortex, an area essential for learning to recognize objects. Some optic fibers continue to go to the optic lobes (superior colliculi), which are linked by pathways to the extrastriate cortices. This system appears to be involved in maintaining visual attention. The superior colliculi also participate in congruent eye movements, pupillary reflexes, and accommodation.

22. Auditory fibers go first to the cochlear nuclei in the medulla. From there, they are projected on several pathways toward the inferior colliculi of the tectum. The inferior colliculi project to the medial geniculate body of the dorsal thalamus, which in turn projects to the primary and other auditory cortical areas where the awareness and interpretation of sounds occur.

23. Olfactory neurons terminate in clusters of neuronal processes in the olfactory bulbs, where considerable processing of the impulses occurs. The primary target of neurons leaving the olfactory bulbs is the lateral pallium, or piriform lobe. This is the primary olfactory center, and olfactory discrimination, awareness, and learning probably occur here.

24. The medial pallium, or hippocampus, still receives some olfactory signals. The hippocampus is a central part of the limbic system, which integrates the olfactory centers, hypothalamus, thalamus, and part of the cerebrum into a system that is important in motivational and emotional behaviors related to survival. The limbic system has also been implicated in short-term memory formation.

25. Mammals retain motor pathways that descend from the isocortex through the striatum, and through tegmental nuclei and the reticular formation to lower motor centers. The striatum of mammals has connections with many parts of the brain. The striatum has an important role in regulating motor output from the isocortex to motor centers. Disruptions of its structure or balance of neuroactive chemicals causes changes in muscle tone, tremors in limb and body movements, disturbances in gaits, eye movements, and other motor activity.

26. Mammals also have evolved new corticobulbar and corticospinal tracts that lead directly from the isocortex to motor centers in the brainstem and spinal cord. These pathways allow mammals to execute complex, voluntary motor activities.

27. Areas within the isocortex can be localized for specific sensory and motor functions. The rest of the isocortex consists of association areas, where considerable processing and integration of sensory and motor signals occur and where complex neuronal processes for learning, memory, and conceptual thought are believed to occur. In early mammals, these association areas were small, but in advanced primates, they occupy most of the cerebrum. Numerous interconnections occur between areas in one cerebral hemisphere, and commissural tracts allow for coordination of activity between the two cerebral hemispheres.

REFERENCES FOR CHAPTERS 13 AND 14

Adelman, G., editor, 1987: *Encyclopedia of Neurosciences.* Boston, Birkhauser.

Barone, R., Pavaux, C., Blin, P. C., and Cuq, P., 1973: *Atlas d'Anatomie du Lapin.* Paris, Masson & Cie.

Benzo, C. A., 1986: The brain. *In* Sturkie, P. D., editor: *Avian Physiology*, 4th edition. New York, Springer-Verlag.

Butler, A. B., 2000: Chordate evolution and the origin of craniates: An old brain in a new head. *Anatomical Record,* 261:111–125.

Butler, A. B., 1994: The evolution of the dorsal thalamus in jawed vertebrates, including mammals: Cladistic analysis and a new hypothesis. *Brain Research Reviews,* 19:29–65.

Butler, A. B., 1994: The evolution of the dorsal pallium of amniotes: Cladistic analysis and a new hypothesis. *Brain Research Reviews,* 19:66–101.

Butler, A. B., and Hodos, W., 1996: *Comparative Vertebrate Neuroanatomy: Evolution and Adaptation.* New York, Wiley-Liss.

Cole, F. J., 1896: On the cranial nerves of *Chimaera monstrosa* (Linn.); with a discussion of the lateral line system and the morphology of the chorda tympani. *Transactions of the Royal Society of Edinburgh,* 38:630–680.

Corliss, C. E., 1976: *Patten's Human Embryology.* New York, McGraw-Hill.

Crosby, E. C., and Schnitzlein, H. N., 1982: *Comparative, Correlative Neuroanatomy of the Vertebrate Telencephalon.* New York, Macmillan.

Davis, R. E., and Kassel, J., 1983: Behavioral functions of the teleostean telencephalon. *In* Davis, R. E., and Northcutt R. G., editors: *Fish Neurobiology*, volume 2, *Higher Brain Areas and Functions.* Ann Arbor, Mich., The University of Michigan Press.

Davis, R. E., and Northcutt, R. G., editors, 1983: *Fish Neurobiology,* volume 2, *Higher Brain Areas and Functions.* Ann Arbor, Mich., The University of Michigan Press.

Demski, L. S., 1984: Evolution of the neural systems in the vertebrates: Functional-anatomical approaches. *American Zoologist,* 24:689–833.

Dorit, R. L., Walker, W. F., Jr., and Barnes, R. D., 1991: *Zoology.* Philadelphia, Saunders College Publishing.

Ebbesson, S. O. E., and Northcutt, R. G., 1976: Neurology of anamniotic vertebrates. *In* Masterton, R. B., Campbell, C. B. G., Bitterman, M. E., and Hotton, N., editors: *Evolution of Brain and Behavior*, volume 1. Hillsdale, N. J., Lawrence Erlbaum & Associates.

Gaupp, E., 1891–1904: *Anatomie des Frosches.* Braunschweig, Friedrich Vieweg und Sohn.

Goodrich, E. S., 1930: *Studies on the Structure and Development of Vertebrates.* London, Macmillan. (Reprinted 1996, Chicago, University of Chicago Press.)

Goodrich, E. S., 1918: On the development of the segments of the head in *Scyllium*. *Quarterly Journal of Microscopical Sciences,* 63:1–30.

Goslow, G. E., Jr., 1985: The neural control of locomotion. *In* Hildebrand, M., Bramble, D. M., Liem, K. F., and Wake, D. B., editors: *Functional Vertebrate Morphology.* Cambridge, Harvard University Press.

Grillner, S., and Wallen, P., 1985: Central pattern generators for locomotion, with special reference to vertebrates. *Annual Review of Neuroscience,* 8:233–261.

Hopkins, C. D., 1983: Functions and Mechanisms in Electroreception. *In* Northcutt, R. G., and Davis, R. E., *q.v.*

Kandel, E. C., Schwartz, J. H., and Jessell, T. M., editors, 1991: *Principles of Neural Science*, 3rd edition. New York, Elsevier.

Kappers, C. U. A., Huber, G. C., and Crosby, E., 1936: *The Comparative Anatomy of the Nervous System of Vertebrates, Including Man.* New York, Macmillan.

Karten, H. J., 1991: Homology and the evolution of the "neocortex." *Brain, Behavior and Evolution,* 38:264–272.

Kimelberg, H. K., and Norenberg, M. D., 1989. Astrocytes. *Scientific American,* 260:66–76.

Landacre, F. L., 1910: The origin of the cranial ganglia in *Ameiurus*. *Journal of Comparative Neurology,* 20:309–411.

Larsell, O., 1967–1972: *The Comparative Anatomy and Histology of the Cerebellum.* Minneapolis, University of Minnesota Press.

Nieuwenhuys, R., 1982: An overview of the organization of the brain of actinopterygian fishes. *American Zoologist,* 22:287–310.

Noback, C. R., Strominger, N. L., and Demarset, R. J., 1991: *The Human Nervous System: Introduction and Review,* 4th edition. Philadelphia, Lea & Febiger.

Noden, D. M., 1991: Vertebrate craniofacial development: The relation between ontogenetic process and morphological outcome. *Brain, Behavior and Evolution,* 38:190–225.

Norris, H. W., and Hughes, S. P., 1920: The cranial, occipital and anterior spinal nerves of the dogfish. *Journal of Comparative Neurology,* 31:293–395.

Northcutt, R. G., 1990: Ontogeny and Phylogeny: A reevaluation of conceptual relationships and some applications. *Brain, Behavior and Evolution,* 36:116–140.

Northcutt, R. G., 1989: The phylogenetic distribution and innervation of the mechanosensory lateral line. *In* Coombs, S., Gorner, P., and Munz, H., editors: *The Mechanosensory Lateral Line, Neurobiology and Evolution.* New York, Springer-Verlag.

Northcutt, R. G., 1987: Evolution of the vertebrate brain. *In* Adelman, G., editor: *Encyclopedia of Neurosciences.* Boston, Birkhauser.

Northcutt, R. G., 1985: Brain phylogeny, speculations on pattern and cause. *In* Cohen, M. J., and Strumwasser, J., editors: *Comparative Neurobiology, Modes of Communication in the Nervous System.* New York, John Wiley & Sons.

Northcutt, R. G., 1985: Evolution of the vertebrate central nervous system. *American Zoologist,* 24:701–716.

Northcutt, R. G., 1978: Brain organization in the cartilaginous fishes. *In* Hodgson, E. S., and Mathewson, R. F., editors: *Sensory Biology of Sharks, Skates and Rays,* volume 1. Arlington, Va., Department of the Navy.

Northcutt, R. G., and Davis, R. E., editors, 1983: *Fish Neurobiology,* volume 1, *Brain Stem and Sense Organs.* Ann Arbor, Mich., The University of Michigan Press.

Penfield, W., and Rasmussen, T., 1950: *The Cerebral Cortex of Man.* New York, Macmillan.

Pettingill, O. S., Jr., 1985: *Ornithology in Laboratory and Field,* 5th edition. Orlando, Academic Press.

Ramon y Cajal, S., 1909: Histologie du Systeme Nerveux de l'Homme & des Vertebres. Translated by L. Azoulay. Paris, Maloine. (Reprinted, 1952, Madrid, Instituto Ramon y Cajal.)

Romer, A. S., and Parsons, T. S., 1986: *The Vertebrate Body,* 6th edition. Philadelphia, Saunders College Publishing.

Sawro, Von W. A., 1990: Vergleichende Mikrostruktur der ventralen Ruckenmarkshorner einiger Wirbeltiere. *Zoologische Jahrbuch Abteil Anatomie Ontogonie Tiere,* 120:143–162.

Spector, R., and Johanson, C. E., 1989. The mammalian choroid plexus. *Scientific American,* 261:68–74.

Stensio, E. A., 1963. The brain and cranial nerves in fossil, lower craniate vertebrates. *Skrifter Norske Videnskaps-Akademi 1 Oslo I. Mat.-Naturv. Klasse. Ny Serie,* 13:1–120.

Streit, W. J., Kincaid-Colton, C. A., 1995: The brain's immune system. *Scientific American,* 273:54–61.

Tretjakoff, D., 1927: Das periphere Nervensystem des Flussneunauges. *Zeitschrift Wissenschaftliche Zoologie,* 129:359–952.

Ulinski, P. S., 1990: Nodal events in forebrain evolution. *Netherlands Journal of Zoology,* 40:215–240.

Vanegas, H., 1983: Organization and physiology of the teleostean optic tectum. *In* Davis, R. E., and Northcutt, R. G., editors: *Fish Neurobiology,* volume 2. Ann Arbor, Mich., The University of Michigan Press.

Walker, W. F., and Homberger, D. G., 1992: *Vertebrate Dissection,* 8th edition. Fort Worth, Saunders College Publishing.

Wiedersheim, R., 1906: *Vergleichende Anatomie der Wirbeltiere.* Jena, Gustav Fischer.

Young, J. Z., 1933: The autonomic nervous system of selachians. *Quarterly Journal of Microscopical Science,* 75:571–624.

15

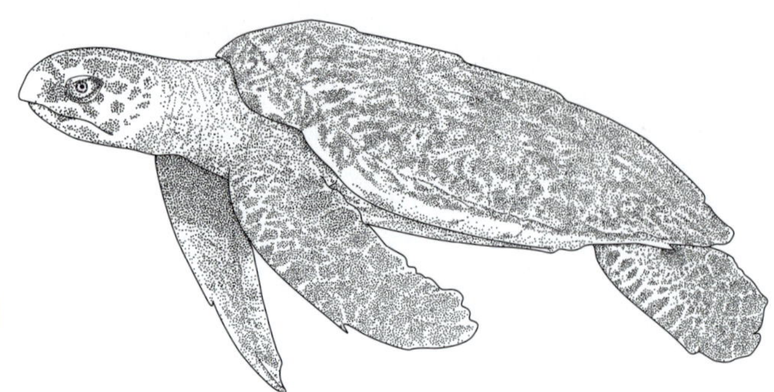

Endocrine Integration

PRECIS

After examining the general nature of chemical controls and integration, we will focus on the endocrine system. We will first explore the similarities and differences between nervous and endocrine control and their interrelationship, go on to consider the nature of hormones and how they act, and then discuss the major endocrine glands. Those that integrate reproduction will be considered later with the reproductive system.

OUTLINE

Much activity of the body is regulated by the nervous system, but many functions are controlled by approximately two dozen ductless endocrine glands that are widely scattered throughout the body. **Endocrine glands** (Gr., *endon* = within + *krino* = to separate) differ from exocrine glands by discharging their secretions, known as **hormones** (Gr., *hormaein* = to excite), into body fluids rather than onto epithelial surfaces. This fluid usually is the blood, but some hormones enter the lymph or cerebrospinal fluid.

Endocrine regulation is a special case of chemical regulation, which is ubiquitous and as old as life itself. Individual cells have chemical control systems within them that regulate many processes. An example of such an **intracrine** (L., *intra-* = within) control mechanism is the activation within a cell of a second messenger by an external first messenger. The second messenger, in turn, activates a series of enzymes that lead to some biological effect, perhaps the synthesis and release by the cell of some secretory product. At the other extreme, organisms may release **semiochemicals** (Gr., *semeion* = a sign or signal) into their environment that affect the behavior of other organisms. **Pheromones** (Gr., *pherein* = to carry) are semiochemicals that trigger a response among members of the same species, perhaps enabling opposite sexes to find each other. **Allomones** (Gr., *allos* = other, different) trigger a response in a different species. The allomones released by a threatened skunk discourage potential enemies.

Between intracrines and semiochemicals, many levels of chemical control occur among the cells within an organism. **Paracrines** are signaling molecules that are released into the interstitial fluid between cells and send information through the extracellular matrix from one cell to nearby cells (Fig. 4-2*B*). They include embryonic inducing substances, immune regu-

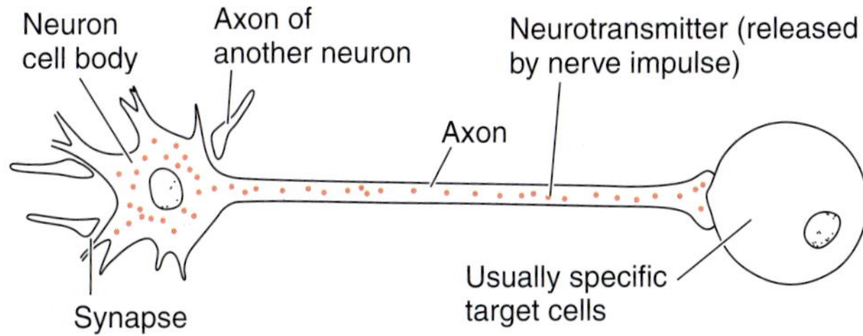

A. Myoneural or axoglandular junction

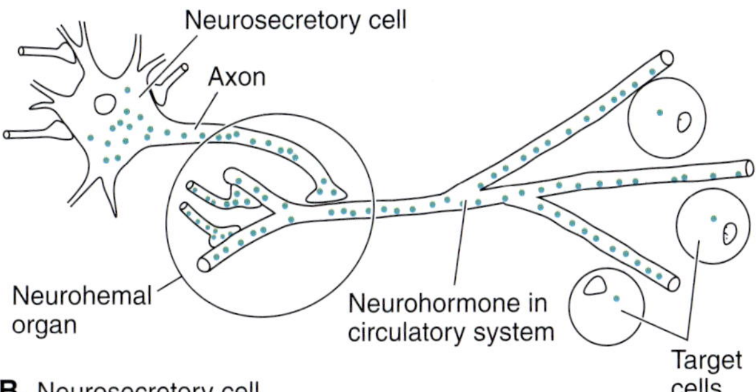

B. Neurosecretory cell

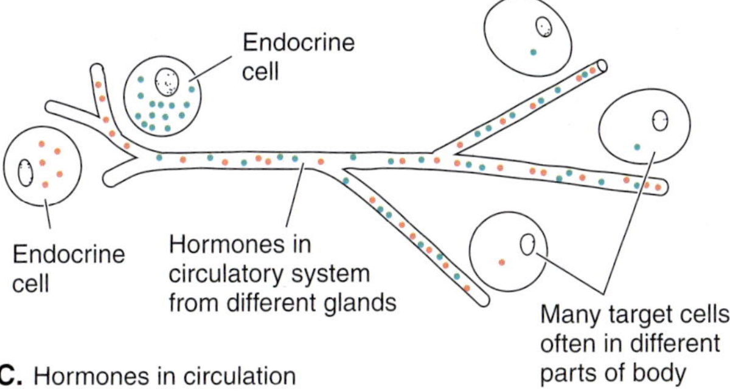

C. Hormones in circulation

FIGURE 15-1
A comparison of neural and endocrine control systems. *A,* The action of a neuron on its target cell. *B,* Release by a secretory neuron of neurohormones into the circulatory system and their transport to the target cells. *C,* Release of two hormones by different endocrine cells into the circulatory system and their transport to target cells. Only target cells with the appropriate receptor respond.

lators, growth factors, and others. **Neurotransmitters,** as we have seen, are released into synaptic clefts by a presynaptic neuron and activate, modulate, or inhibit a postsynaptic neuron (Fig. 15-1*A*). **Neurohormones** (also called neurosecretions) are synthesized by neurons and released into the blood or cerebrospinal fluid, which carries them to their target cells. Often, the neuron terminals, which temporarily accumulate the neurohormones, and associated blood vessels are organized into a **neurohemal organ** (Fig. 15-1*B*). Finally, **endocrine glands** release hormones into the blood, which carries them to target organs a considerable distance away (Fig. 15-1*C*).

In this chapter, we will focus on endocrine glands and neurohormones. After considering similarities and differences between nervous and endocrine control, we explore the nature of hormones and their modes of action and then discuss most of the endocrine glands. A discussion of the hormones that regulate reproduction is deferred until we consider the reproductive system (Chapter 21). We know far more about mammalian hormones and neurohormones than about those of other vertebrates, so, for each endocrine gland, we will consider the mammalian condition before making comparisons with nonmammalian vertebrates. An overview of the location of the major endocrine glands of mammals is given in Figure 15-2, and a list of the major vertebrate hormones and their primary effects is given in Table 15-1.

A Comparison of Endocrine and Neural Integration

Most multicellular animals have a nervous system and a set of endocrine glands. Both of these control systems resemble each other by exerting their effects by stimu-

Text continues on page 508

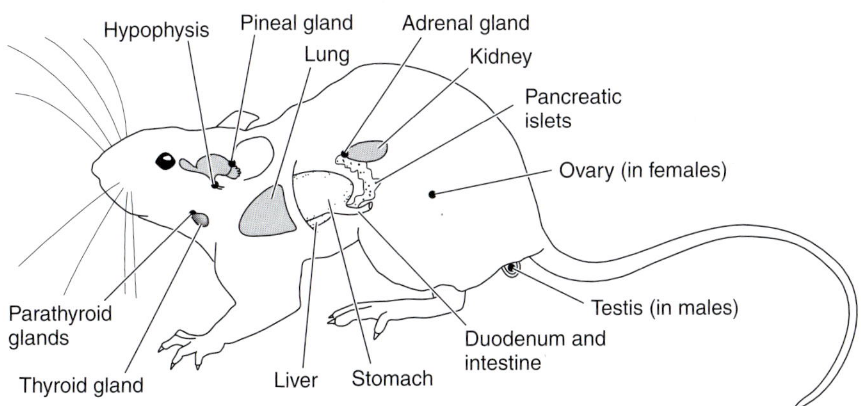

FIGURE 15-2
The location of the major endocrine glands in a rat. Testes and ovaries are not present in the same individual.

TABLE 15-1 Major Mammalian Hormones and their Primary Effects

Gland and Hormone	Major Actions
Adenohypophysis	
Thyrotropin (TSH)	Secretion of thyroid hormones
Follicle-stimulating hormone (FSH)	Development of ovarian follicles
	Development of seminiferous tubules of testis
	Development of Sertoli cells in testis
Luteinizing hormone (LH)	Transformation of follicle into corpus luteum
	Production of testosterone by interstitial cells
Growth hormone (GH)	Protein synthesis and growth
	Release of somatomedin from liver
Prolactin (PRL)	Synthesis of milk
	Affects growth, reproduction, water and electrolyte balances
Corticotropin (ACTH)	Release of glucocorticoids and androgens from adrenal cortex
Melanotropin (MSH)	Dispersal of pigment in amphibian melanophores
	Synthesis of melanin
Hypothalamic hormones affecting adenohypophysis	
Thyrotropin-releasing hormone (TRH)	Release of thyrotropin
Gonadotropin-releasing hormone (GnRH)	Releases follicle-stimulating hormone and luteinizing hormone
Growth hormone release-inhibiting hormone (GHRIH)	Blocks release of growth hormone
Somatocrinin	Releases growth hormone in absence of GHRIH
Prolactin-releasing hormone (PRH)	Releases prolactin
Prolactin release-inhibiting hormone (PRIH)	Inhibits release of prolactin
Corticotropin-releasing hormone (CRH)	Releases corticotropin
Melanotropin-releasing hormone (MRH)	Releases melanotropin
Melanotropin release-inhibiting hormone (MRIH)	Inhibits release of melanotropin
Hypothalamic neurohormones released from neurohypophysis	
Vasopressin (VP) or antidiuretic hormone	Prevents diuresis
Oxytocin (OXY)	Contraction of smooth muscle, especially in uterus and mammary glands
Pineal gland	
Melatonin	Adjusts physiological rhythms to diurnal and seasonal cycles
	Inhibiting effect on gonad development
Adrenal medula	
Norepinephrine in young mammals, epinephrine in adults	Reinforces sympathetic nervous system in adjusting body to short-term stress
Adrenal cortex	
Glucocorticoids: Cortisol, corticosterone	Energy metabolism, adjusts body to long-term stress
Mineralocorticoid: Aldosterone	Sodium reabsorption and potassium secretion by kidney tubules
	Raises blood pressure
Cortical androgen	Protein synthesis, muscle growth
Thyroid gland	
Triiodothyronine (T_3), tetraiodothyronine (T_4), (thyroxine)	Accelerate oxidative metabolism
	Growth, development, and other effects through its permissive role on other cells
Skin, liver, kidneys	
Production of 1,25-dihydroxycholecalciferol (1,25-DHC) from cholesterol and vitamin D in skin	Carrier protein for absorption of calcium

TABLE 15-1 *(Continued)*

Gland and Hormone	Major Actions
Parathyroid glands	
Parathormone (PTH)	Removal of calcium from bone
	Calcium reabsorption and potassium excretion by kidney tubules
C cells in thyroid gland	
Calcitonin (CT)	Inhibits action of parathormone on bone
Liver	
Somatomedin	Body growth
Kidneys	
Renin	Converts angiotensinogens in blood to angiotensin I
Lungs	
Angiotensin I converted to angiotensin II and III	Release of aldosterone from kidney cortex
Pancreatic islets	
Insulin (I)	Lowers blood sugar
Glucagon (G)	Raises blood sugar
Gastrointestinal tract	
Gastrin	Raises stomach acid, which promotes release of pepsinogen
Somatostatin (SS)	Inhibits acid release
Secretin	Release of water and bicarbonate from pancreas
Cholecystokinin (CCK)	Release of pancreatic enzymes
	Release of bile
Peptide YY (PYY)	Inhibits gastric acid secretion
Glucose-dependent insulinotropic peptide (GDIP)	Insulin release from pancreatic islets
Motilin	Increases motility of gastrointestinal tract
Testis	
Testosterone	Growth of male reproductive organs
	Development of male secondary sex characters
	Controls male reproductive behavior
Ovary	
Estradiol	Growth of female reproductive organs
	Development of female secondary sex characters
	Proliferation of uterine lining
	Affects LH surge and ovulation
	Inhibits release of GnRH
Progesterone	Maintains uterine lining
Relaxin	Relaxes pelvic ligaments before birth
	Promotes growth of uterine muscles
	Stimulates development of mammary glands
Placenta	
Chorionic gonadotropin (CG)	Maintains corpus luteum until placenta produces gonadal steroids
Estradiol	Same as ovarian estradiol
Progesterone	Same as ovarian progesterone
Chorionic somatomammotropin (placental lactogen)	Milk synthesis
Relaxin	Same as ovarian relaxin

lating or inhibiting target cells by means of chemical messengers, neurotransmitters, and hormones, respectively. In a few cases, hormones and neurotransmitters are the same products. A major difference between nervous and hormonal integration, as stated earlier, is the way the chemical messengers are transmitted, whether along neurons to the target cells or through the circulatory system (Fig. 15-1). The nervous and endocrine control systems also differ in other important ways. Neurons are activated by receptor cells or by other neurons. A few endocrine glands also produce and release their hormones in response to a nerve impulse, but most are activated differently. Some endocrine glands respond to changes in the level of a substrate. For example, the entrance of food into the duodenum from the stomach promotes the release of secretin from certain duodenal cells, and this hormone stimulates a copious secretion of pancreatic juice (Chapter 17). Most endocrine glands respond to changes in the blood level of their own hormone or to that of some other endocrine gland. Often, complex feedback mechanisms between endocrine glands control the hormone levels in the blood.

Once activated, neurons and endocrine glands send their signals to the target cells, but hormones do not go directly to their target cells, as nerve impulses do. Hormones are carried passively in the blood, either in solution or loosely bound to carrier plasma proteins. They must pass into and out of capillaries, so transit time is relatively long. The blood carries the entire spectrum of hormones throughout the body, but only target cells with receptor molecules that closely match the molecular configuration of a particular hormone will respond to this hormone. A "lock-and-key" mechanism operates. A given target cell may have thousands of receptors for a particular hormone, and thousands more for other hormones to which it can respond. Nearly all of the cells of the body respond to some hormones, such as thyroxin, which is produced by the thyroid gland. Some cells respond to a narrower range of hormones; for example, only the cells responsible for producing the aqueous and alkaline portion of the pancreatic juice respond to secretin. Another hormone promotes the synthesis and release of the pancreatic enzymes.

The effect of a nerve impulse on the target cell is momentary because enzymes that enter the synaptic cleft or neuron-effector junctions rapidly metabolize and degrade the neurotransmitter substances. Additional nerve impulses are needed for a continued response. Hormones, in contrast, have much longer effects because they are relatively stable compounds; however, they must be continuously synthesized because they are lost in various ways. They are taken out of circulation when they bind with receptors, and the hormone–receptor complex is eventually metabolized. Many hormones also are lost through the excretory system.

Because of these differences, endocrine integration is slower and has longer-lasting effects than does neuronal integration. Hormones are particularly well suited to regulate metabolism, growth, metamorphosis, slow skin-color changes, water and mineral balances, sexual development, and reproduction. They play a major role in the maintenance of homeostasis. Hormones are less effective than are nerve impulses in rapidly and briefly stimulating specific organs. Although endocrine and neuronal integration are distinct control systems, they interact with each other in many ways.

The Nature of Hormones and Hormonal Action

The secretory cells of endocrine glands develop embryonically from epithelial cells or from neural crest cells that have migrated into the epithelia. Despite the great diversity of hormones, all are derivatives of either amino acids or steroids. Hormones produced by cells of ectodermal or endodermal origin are synthesized from amino acids, or modified amino acids. The hypothalamic hormones, which control the release of hormones from the adenohypophysis (a part of the pituitary gland), are small neuropeptides secreted by neurons and consisting of 3 to 44 amino acids. Insulin, the first hormone the structure of which was determined, is a small protein consisting of 51 amino acids. Most of the amino acid–derived hormones produced by the adenohypophysis are large proteins with molecular weights of 25,000 to 30,000. Norepinephrine and epinephrine, which are produced by the medulla of the adrenal gland, are catecholamine derivatives of the amino acid tyrosine. Hormones produced by the thyroid gland combine tyrosine with iodine. Hormones produced by cells of mesodermal origin, which are those of the adrenal cortex, testis, ovary, and placenta, are all steroids derived from cholesterol.

Peptide and protein hormones are water soluble, so they go into solution in the blood, but most become loosely bound to carrier plasma proteins, which facilitate their transport and prevent circulating proteases from breaking them down. Steroids are not soluble in water, so they must be bound to carriers.

Because they are water soluble, peptide and protein hormones (the **first messengers**) cannot pass through the lipid plasma membranes of their target cells; rather, they bind with specific membrane protein receptors

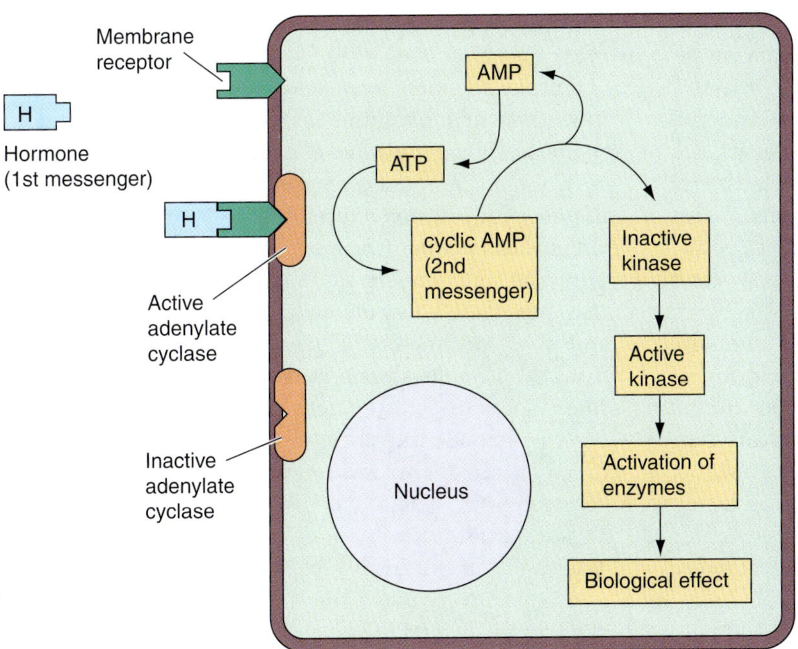

A. Hormone binds with membrane receptor leading to enzyme activation.

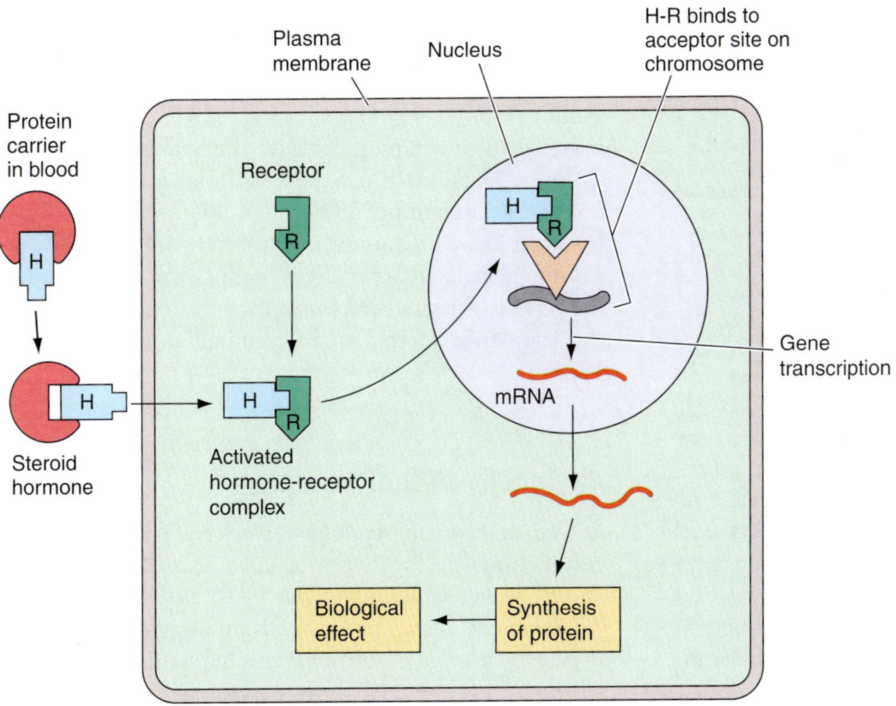

B. Hormone passes through membrane, binds to receptor, then enters the nucleus and stimulates gene transcription.

FIGURE 15-3
Methods of hormone action. *A,* A peptide or protein hormone (the first messenger) binds with a receptor in the plasma membrane of the cell. By way of intermediate G proteins (not shown), this union activates a second messenger on the inside of the membrane, and its activation sets off a series of reactions that lead to the biological effect of the cell. *B,* A steroid hormone easily passes through the plasma membrane and binds with a cytoplasmic receptor. The receptor hormone complex enters the nucleus and acts on a gene. In some cases, the receptor is in the nucleus.

(Fig. 15-3*A*). The way this binding leads to the activation of a cell was first elucidated by research on how epinephrine from the medulla of the adrenal gland promotes the conversion of glycogen to glucose in liver cells. By way of intermediate G proteins (not shown in Fig. 15-3*A*), this union activates an enzyme, in this case, adenylate cyclase, on the inner side of the plasma membrane. The activated adenylate cyclase then leads to the production of a **second messenger** (cyclic adenosine monophosphate, or cAMP), which, in turn, leads to the activation of a series of enzymes (kinases) and ultimately to the biological effect of the cell.

Because steroid and thyroid hormones are lipid soluble, they easily pass through the lipid plasma membranes of cells, where they unite with receptor molecules in either the cytoplasm or nucleus of the cell (Fig. 15-3*B*). If the receptor is in the cytoplasm, the hormone–receptor complex enters the nucleus and binds to an acceptor site on a chromosome. This binding activates a gene and leads to the synthesis of proteins (enzymes) that produce the biological effect of the hormone.

Hormones can be effective in very small amounts. Many are present in the blood at concentrations of only 10^{-12} molar. Considerable amplification occurs during their action. A peptide hormone–receptor complex may activate only a few molecules of an enzyme, but these, in turn, activate more and more at each step in an enzyme cascade. A steroid hormone–receptor complex activates a gene, but the gene can synthesize many protein molecules.

The Hypothalamo-Hypophyseal Axis

The pituitary gland, or **hypophysis cerebri** (Gr., *hypo* = under + *physis* = growth + *cerebri* = of the brain), and the **hypothalamus** of the brain are very closely associated morphologically and functionally. Together, they form a complex known as the **hypothalamo–hypophyseal axis.**

Development and Structure of the Hypophysis

The hypophysis attaches to the ventral surface of the hypothalamus. It develops embryonically in all vertebrates from two ectodermal evaginations that meet and unite (Fig. 15-4*A*). An **infundibulum** grows ventrally from the diencephalon of the brain, and **Rathke's pouch** extends dorsally from the roof of the developing mouth, or **stomodaeum.**

The infundibulum remains connected to the floor of the diencephalon, which becomes the hypothalamus, and gives rise to the part of the gland known as the **neurohypophysis** (Fig. 15-4*B*). In mammals, the neurohypophysis consists of a **median eminence,** next to the brain, and a narrow **infundibular stalk,** which leads to an expanded **pars nervosa** (Fig. 15-4*B*). The neurohypophysis contains no secretory cells, but axons of neurosecretory neurons in the hypothalamus carry neurohormones to the median eminence and pars nervosa. Hormones are stored and released from the pars nervosa into networks of blood vessels. Much of the neurohypophysis is, therefore, a neurohemal organ (Fig. 15-4*C*).

Rathke's pouch loses its connection with the stomodaeum in most adult vertebrates and gives rise to the rest of the gland, the **adenohypophysis** (Gr., *aden* = gland). The adenohypophysis forms a thick **pars distalis** and, in most vertebrates, a **pars intermedia,** which lies between the pars distalis and the neurohypophysis (Fig. 15-4*B*). The lumen of Rathke's pouch may persist as a narrow cleft between the pars distalis and pars intermedia. A **pars tuberalis** may extend from the pars distalis along the infundibular stalk. Distinct types of secretory cells characterize the different parts of the adenohypophysis. The neurohypophysis is called the **posterior lobe of the pituitary** in the older literature; the adenohypophysis, the **anterior lobe.** These terms accurately describe their positions in mammals but not in all vertebrates.

Few neurons enter the adenohypophysis in mammals, but it is functionally connected to the hypothalamus by a unique vascular supply known as the **hypophyseal portal system** (Fig. 15-4*C*). Hypothalamic neurohormones are discharged by axons into capillaries of the median eminence, and these are carried by small portal veins to capillaries in the adenohypophysis, where they control the synthesis and release of its hormones. (Portal veins lie between two capillary beds rather than returning directly to the heart.) The adenohypophysis of mammals is described as being under neurovascular control. The adenohypophysis also can be affected by hormones in the cerebrospinal fluid because specialized ependymal epithelial cells, called **tanycytes,** can transfer hormones from this fluid into the capillaries of the median eminence (see also Fig. 13-5).

Phylogeny of the Hypophysis

A well-developed hypophyseal system with functional connections to the hypothalamus is unique to craniates. The neural gland of ascidians and Hatschek's pit of amphioxus may be partly homologous to the hypophysis. They are epithelial organs that arise from or close to the oral cavity and are closely associated with the nervous system. At least one hypophyseal hormone (**luteinizing hormone** [LH]) has been identified by immunological tests in Hatschek's pit.

Considerable variation occurs in the details of hypophyseal structure among craniates (Fig. 15-5). An important variable is the control of secretions of the

Text continues on page 513

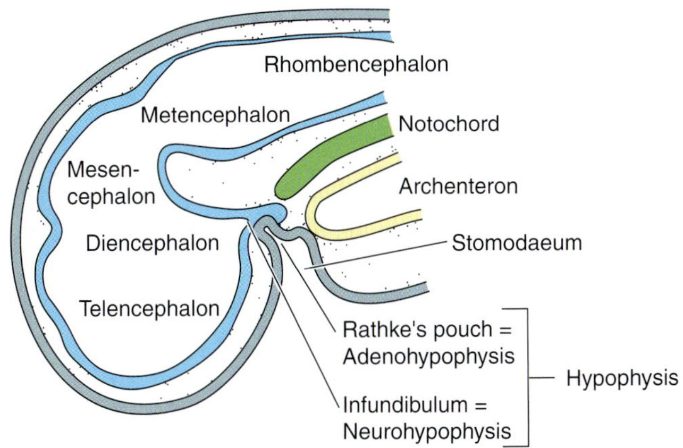

A. Embryonic development

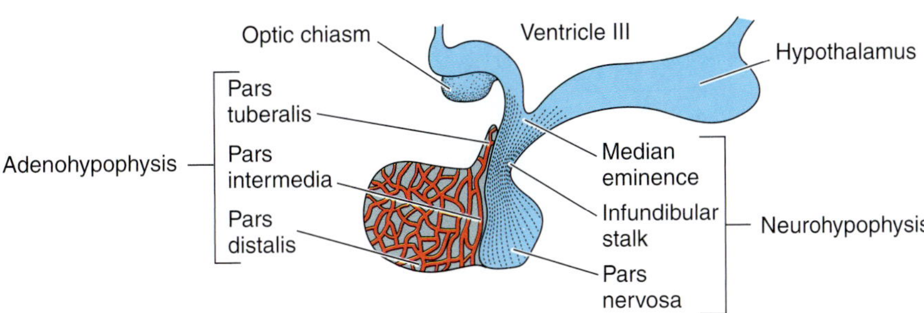

B. Parts of the hypophysis of an adult mammal

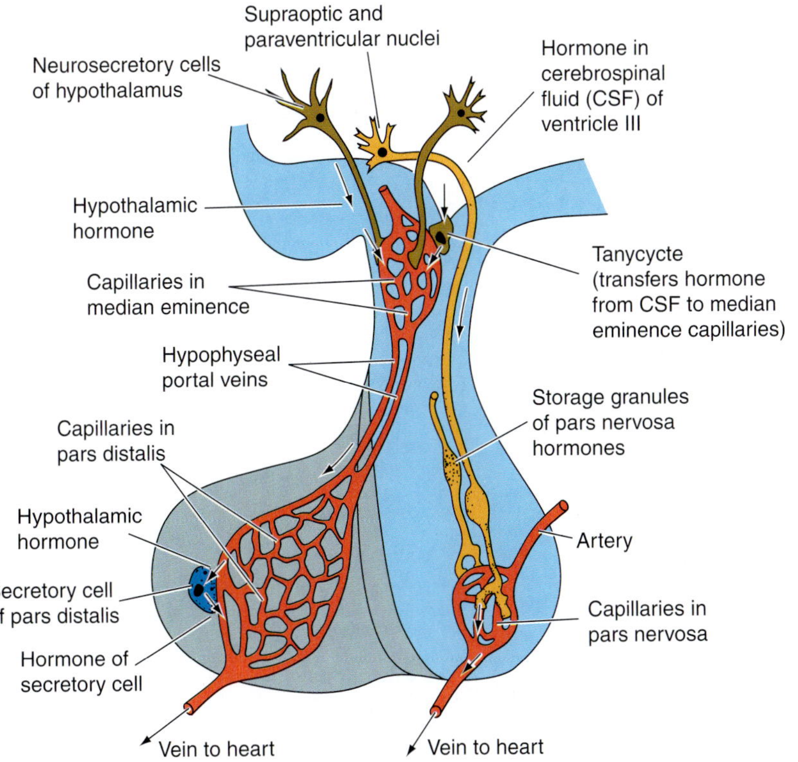

C. Neuronal and vascular connections between the hypothalamus and hypophysis

FIGURE 15-4
The mammalian hypothalamo-hypophyseal axis. *A,* The embryonic development of the hypophysis from evaginations of the diencephalon and stomodaeum. *B,* Parts of the adult hypophysis. *C,* Neuronal and vascular connections between the hypothalamus and hypophysis.

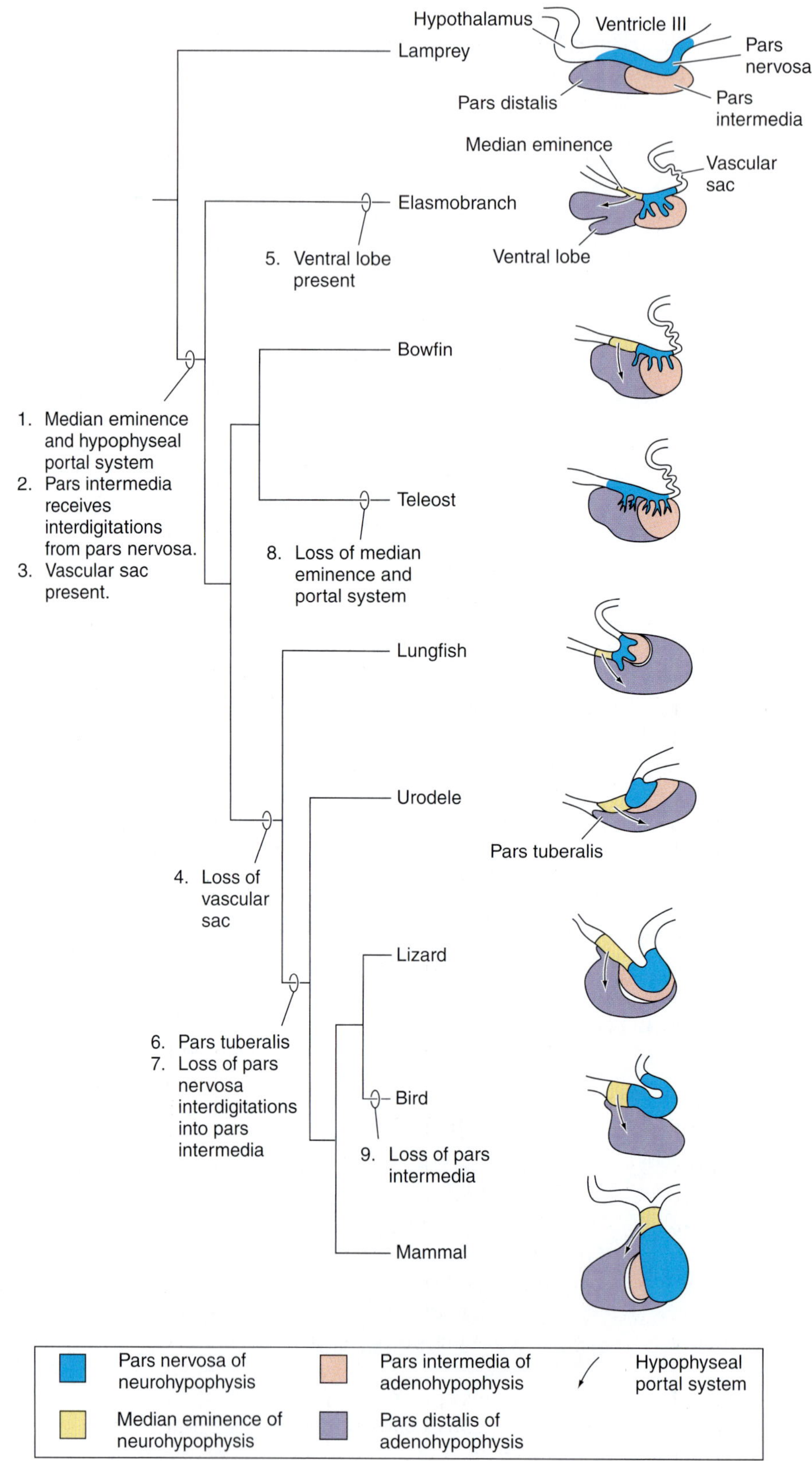

FIGURE 15-5
A cladogram showing the evolution of the hypophysis in representative vertebrates.

adenohypophysis. Mammals and other tetrapods have a **neurovascular control** with neurohormones from hypothalamic neurons passing from the median eminence of the neurohypophysis into a vascular portal system that carries them to all parts of the adenohypophysis. Except for teleosts and some jawless fishes, fishes also have a medial eminence and a portal system extending into the pars distalis of the adenohypophysis, so this part of the gland has some neurovascular control. A unique feature of most fishes is the extension of finger-like projections from the pars nervosa of the neurohypophysis into the pars intermedia of the adenohypophysis. These processes carry hypothalamic neurons into the pars intermedia and presumably bring it under direct **neuroglandular control.** Similar processes extend into the pars distalis in many fishes, so this region may have a dual control: neurovascular and neuroglandular. The control is exclusively neuroglandular in teleosts because they lack a median eminence and portal system. Some jawless fishes have a median eminence, others do not, and a portal system is not always present. Evidence shows, however, that material diffuses from the hypothalamus into the adenohypophysis.

Another variable is the presence in all jawed fishes, except lungfishes, of a **vascular sac** extending from the caudal part of the hypothalamus dorsal to the hypophysis (Fig. 15-5). Its significance is unknown, and it may not have an endocrine function. Elasmobranch fishes differ from all other vertebrates in having a **ventral lobe** attached to the underside of the adenohypophysis (Fig. 15-5). A **pars tuberalis** is a consistent feature of the adenohypophysis of all tetrapods (Fig. 15-5). The pars intermedia is absent in birds and some mammalian species, including such diverse species as whales, elephants, beavers, and adult humans.

Hormones of the Adenohypophysis

Seven major hormones are produced by the adenohypophysis in nearly all vertebrates (Table 15-1). Most act by controlling the activity of other endocrine glands. They can be grouped into three categories based on similarities in their chemical structure (amino acid sequences) and biological activity. Hormones in each category likely had a common evolutionary origin.

Category I hormones are glycoproteins. Each is composed of two polypeptide chains containing specific carbohydrates. One chain is the same in each hormone. **Thyroid-stimulating hormone (TSH),** also known as **thyrotropin** (L., *trophe* = nourishment; a tropin "nourishes," or activates, an endocrine gland), increases the rate of synthesis and release of thyroid hormones from the thyroid gland, which increase the rate of oxidative metabolism in endothermic birds and mammals.

Two additional category I hormones are **follicle-stimulating hormone (FSH)** and **luteinizing hormone (LH).** Together, these hormones are called **gonadotropins.** These two hormones were first discovered in female mammals, where they promote the development of the ovarian follicles and their transformation after ovulation into corpora lutea. The ovarian follicles and corpora lutea, in turn, produce estradiol and progesterone, hormones essential for the development of female secondary sex characters and the female reproductive cycles (Chapter 21). Later discoveries have shown that these hormones also are present in male mammals. FSH promotes the development of the seminiferous tubules in the testis, which produce sperm. LH (sometimes called interstitial cell–stimulating hormone in males) promotes the synthesis of the male sex hormone, testosterone, by the interstitial cells of the testis.

Category II hormones are **growth hormone (GH)** and **prolactin (PRL).** Both are large, folded polypeptide chains with considerable structural and functional overlap. GH does not have a single target but affects many tissues. It channels amino acids into protein synthesis and has other effects that promote body growth. Some of the effects of GH result from its stimulating the release of other growth factors, such as **somatomedins** from liver cells. A deficiency of GH during human childhood leads to certain types of dwarfism, and an excess, to giantism. GH and somatomedins have been found in all tetrapod groups and some fishes.

Prolactin was named from the discovery that it stimulates the synthesis (but not the release) of milk by the mammary glands. PRL occurs in all groups of craniates, with the possible exception of hagfishes and lampreys, and it appears to have the widest range of actions of any of the adenohypophyseal hormones. Interacting with other hormones, it affects growth, reproduction, the development of integumentary features, and water and electrolyte balances, and it may modulate the immune response. For example, along with GH, PRL stimulates the secretion of somatomedins by liver cells. It is essential in many ways for reproduction. PRL helps sustain the production of testosterone by the testis and progesterone by the corpus luteum. It stimulates reproductive migrations in many animals, including the movement of certain salamanders to water. It promotes nest building and maternal care in some teleosts and many birds. Indeed, roosters that have been given sufficient amounts of PRL will brood the chicks. In some birds, PRL interacts with other hormones to produce brood patches—ventral skin regions that lose feathers and transfer maternal body heat more efficiently to the eggs. In pigeons, it stimulates the desquamation of cells in the crop and the formation of a crop milk that is fed to newly hatched young. The effects of PRL are not limited to reproduction. In some fishes, it helps control water and salt balances and is thus essential for certain anadromous species to enter fresh water during their spawning runs. Similar effects have been found in nonmammalian tetrapods, but the role of PRL in mammalian osmoregulation is not yet clear.

The two **category III hormones,** (1) **adrenocorticotropic hormone (ACTH),** also known as **corticotropin,** and (2) **melanocyte-stimulating hormone (MSH),** also known as **melanotropin,** are the smallest of the adenohypophyseal hormones. Both share a common prohormone that later differentiates.

Corticotropin stimulates the synthesis and release of most of the hormones produced in the cortex of the adrenal gland, an organ located near the kidney. Adrenocortical hormones, in turn, affect a wide range of metabolic processes (Table 15-1 and p. 518).

Melanotropin was named for its action in amphibians and reptiles, where it causes the dispersal of pigment granules in the melanophores of the skin, and the animal becomes dark. In the absence of the hormone, the pigment granules are concentrated in the cell, and the animal is lighter in color. These **physiological color changes** help the animal adapt to changes in the tone of its background. Temperature and emotional state in some reptiles also affect the release of MSH. Although produced in pigment-producing cells, skin pigment in birds and mammals is deposited outside of these cells, so these animals do not have physiological color changes. However, MSH controls the synthesis of melanin, which is seasonal in some species. The coat of the short-tailed weasel is white in the winter, but melanin production causes the summer coat to be brown.

Melanotropin has been found in all vertebrates tested, but its physiological role in fishes is unknown. Pigment changes in fishes occur rapidly under sympathetic–parasympathetic control. MSH is produced only in the pars intermedia of the adenohypophysis in those vertebrates with a pars intermedia. When the pars intermedia is absent, as in birds and some mammals, it is produced in the pars distalis. All the other adenohypophyseal hormones are produced only in the pars distalis in tetrapods, but in fishes, some also are produced in the pars intermedia.

Neurohormones of the Hypothalamus Affecting the Adenohypophysis

The synthesis and release of adenohypophyseal hormones are regulated primarily by many small **releasing hormones** composed of peptides. These are synthesized by neuron cell bodies in hypothalamic nuclei. Axons of these cells carry the releasing hormones to the median eminence, where they enter the hypophyseal portal system to be carried in the blood to the adenohypophysis (Table 15-1). Here, they activate cells with the appropriate receptors. In some cases, autonomic innervation may have an additional stimulatory or inhibiting effect. A great deal is known about factors that affect the output of the adenohypophysis, but much of this information is beyond the scope of this book. In general, each hormone produced by the adenohypophysis requires the presence of its particular releasing hormone, except that the release of both FSH and LH is controlled by a single **gonadotropin-releasing hormone (GnRH).**

As the level of a particular adenohypophyseal hormone increases in the blood, its continued synthesis and release are reduced in various ways. In most cases, a **negative-feedback mechanism** operates in which the increasing level of the adenohypophyseal hormone either reduces the sensitivity of the cells producing the hormone to its releasing hormone, reduces the sensitivity of the cells producing the releasing hormone, or both. Negative feedback brings the level of the hypophyseal hormone down to normal. In other cases, negative feedback operates more indirectly by promoting the production of a **release-inhibiting hormone,** which then acts on the cells producing the releasing hormone (Table 15-1). Sometimes the release of the inhibition by various means is the primary factor in promoting secretion.

Neurohormones of the Hypothalamus in the Neurohypophysis

Two chemically related peptide hormones are synthesized in the **paraventricular** and **supraoptic nuclei** in the hypothalamus of mammals (Fig. 15-4*C*). Axon tracts carry them to the pars nervosa of the hypophysis, where they are stored and from which they are released in response to neural stimulation. **Vasopressin (VP)** was first recognized and named for its effect in raising blood pressure (a pressor effect). Later, scientists discovered that the pressor effect occurred only at high concentrations. Its primary biological effect at lower concentrations is to prevent **diuresis,** the excretion of a copious and dilute urine, by promoting the reabsorption of water from parts of the kidney tubules (Chapter

20). Its alternate name, **antidiuretic hormone,** derives from this. **Oxytocin (OXY)** stimulates the contraction of certain smooth muscles. It is important in mammals in initiating uterine contractions at birth and releasing milk from the mammary glands in response to the suckling stimulus of the infant (Chapter 21).

Homologous hormones, ones with a few amino acid substitutions, occur in all vertebrates. In other tetrapods, they also promote water reabsorption from kidney tubules, the urinary bladder, or skin (in amphibians), and in birds, the contraction of oviduct muscles during egg laying. Their biological functions in fishes are not well known.

We saw (Chapter 14) that the hypothalamus is critical for the neuronal integration of visceral functions. Because of its neurosecretory relationships with the hypophysis, the hypothalamus also is the major control center for the hypophysis, which in turn is a central part of the endocrine system because its hormones affect so many other glands and physiological processes. The hypothalamus responds to the blood level of many hormones, but it is also a major **neuroendocrine transducer,** for it converts nerve impulses to hormonal signals and thereby places much of the endocrine system under some degree of nervous control. Endocrine secretion can be affected by emotional states, by changes in many physiological processes, and by light and other environmental changes detected by sense organs, which then transmit impulses through the nervous system. Hormonal levels influence much nervous activity, in turn. A female dog, for example, comes into heat under the influence of reproductive hormones and behaves differently toward male dogs than when she is not in heat. The two control systems of the body, nervous and endocrine, thus are integrated closely.

The Pineal Gland

A **pineal gland** (L., *pineus* = pertaining to pine, so called because of its shape) is found in many reptiles, birds, and mammals. It is absent in cetaceans. The pineal gland (Fig. 15-6) is a small, median, cone-shaped structure attached to the roof of the diencephalon, and its cellular components, known as **pinealocytes,** produce the hormone **melatonin.** Melatonin is synthesized from the amino acid tryptophan by several intermediate products, one of which is serotonin. Light inhibits the activity of certain enzymes in this pathway, so melatonin is produced at night. In some species of birds, light may affect the activity of the pineal gland directly, but in most species and in mammals, the pinealocytes are under neural control. The pinealocytes, therefore, are neuroendocrine transducers because nervous signals control the synthesis of enzymes producing melatonin.

The melatonin level in the blood has a distinct **circadian rhythm** (L., *circa* = approximately + *dies* = day), being higher at night. This rhythm is built in, or **endogenous,** to a large extent because it persists for many days when an experimental mammal is kept in constant darkness. The controlling mechanism is the **suprachiasmatic nucleus** in the hypothalamus of the brain, which acts as a "biological clock" (Fig. 15-6). The genetic mechanisms of the "biological clock" of vertebrates and many other organisms rely on genetic oscillators that produce certain proteins, the accumulation of which, in turn, reduces the activity of the genes that produced them. (See Gekakis et al., 1998, for a review).

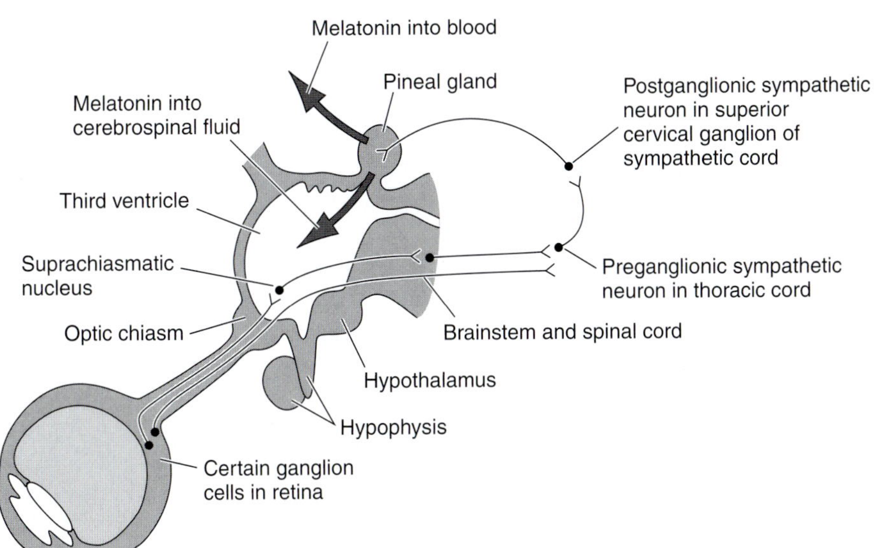

FIGURE 15-6
Sagittal section through the mammalian eye, diencephalon, and pineal gland, showing the neuronal pathway by which light inhibits the synthesis of melatonin by the pineal gland. Melatonin is synthesized during dark periods. Most enters the blood, but some enters the cerebrospinal fluid.

The activity of the suprachiasmatic nucleus is "set," or **entrained,** by information received in the retina on changes in day length. The receptive pigment is not one of those in the rods and cones because mice in which the rods and cones have been destroyed still can respond to changes in day length. The pigment appears to be a **cryptochrome** in other retinal cells. (See Barinaga, 1999, for a brief review.) Information from the retina about day length is carried by the axons of some ganglion cells that leave the optic pathway and enter the suprachiasmatic nucleus (Fig. 15-6). A neuronal pathway from the suprachiasmatic nucleus leads to the brainstem and spinal cord, and from the spinal cord to the pineal gland by way of the sympathetic nervous system. Neural signals promote the activity of the enzymes producing melatonin.

Melatonin is released into the blood, which carries it throughout the body. It also is released into the cerebrospinal fluid of the third ventricle, from which it is transferred to the median eminence of the neurohypophysis by tanycytes. Thus, it may influence the activity of the adenohypophysis more directly. Because of its cyclic production and wide distribution, the primary action of melatonin is to adjust many endogenous physiological rhythms, such as sleep, hormone levels, and appetite, to diurnal and seasonal cycles. The reduction of thyroid secretion at night in many mammals appears to result from the increased level of melatonin. Scientists even have proposed that "jet lag" may result from the melatonin rhythm getting out of phase with the light–dark cycle (Fevre-Montage et al., 1978).

Studies on rodents have shown that melatonin has an inhibiting effect on gonad development in mammals but that an increase in light levels reduces this effect. Through its effects on the median eminence and adenohypophysis, the pineal gland may play a critical role in the control of seasonal reproductive cycles. Many species reproduce in the spring, when day length increases. Clinical studies in humans suggest that a decrease of about 75% in the circulating level of melatonin between the ages of 7 and 12 years is a factor in promoting the onset of puberty.

The pineal gland has evolved from a part of the epiphyseal complex of anamniotes, which includes a median light-receptive pineal eye, parietal eye, or both (Fig. 12-23). The photoreceptive cells in these eyes, as well as those of the lateral image–forming eyes, respond directly to the absence of light and produce some melatonin. Melatonin acts in opposition to MSH and causes the pigment in the melanophores of some fishes and amphibians to concentrate (hence the name for this hormone). Its higher level at night causes these animals to blanch. It also appears to regulate other diurnal and seasonal cycles. The parietal eye of lizards controls the extent to which they expose themselves to sunlight. The pinealocytes of amniotes are modified photoreceptive cells.

Gern et al. (1986) proposed that melatonin in ancestral vertebrates was not a hormone but a paracrine because it acted only locally in the eyes. Elevated levels of melatonin at night caused pigment in the retina to concentrate and thus facilitated the stimulation of the photoreceptive cells by dim light. However, some melatonin spilled over into the blood, and, because a regular nightly surge of melatonin occurred, it could be co-opted to time physiological processes that would be most advantageous if they occurred at regular intervals. Melatonin became a hormone that acted on more distant cells and tissues, which are involved in diurnal color changes, degree of exposure to sunlight, and other rhythms.

The Urophysis

Teleosts have a neurohemal organ known as the **urophysis** (Gr., *oura* = tail + *physis* = growth) located beneath the caudal end of their spinal cord. Neurosecretory neurons in the caudal part of the spinal cord send axons into the urophysis, where they come close to blood vessels. Peptide hormones synthesized by these cells have been called **urotensins** because they can increase blood pressure in teleosts. The urophysis resembles a neuroendocrine transducer. It appears to play a role in osmoregulation, but more research is needed to clarify its biological function. Chondrichthyans have a comparable group of neurosecretory cells, but their axons do not aggregate to form such a well-defined organ.

The Adrenal Glands

The paired **adrenal glands** (L., *ad* = toward + *ren* = kidney), also called **suprarenal glands,** receive their name from their location beside or above the kidneys in mammals (Fig. 15-7). Each mammalian gland consists of two distinct parts, the cortex and the medulla, which have different origins, structures, and functions.

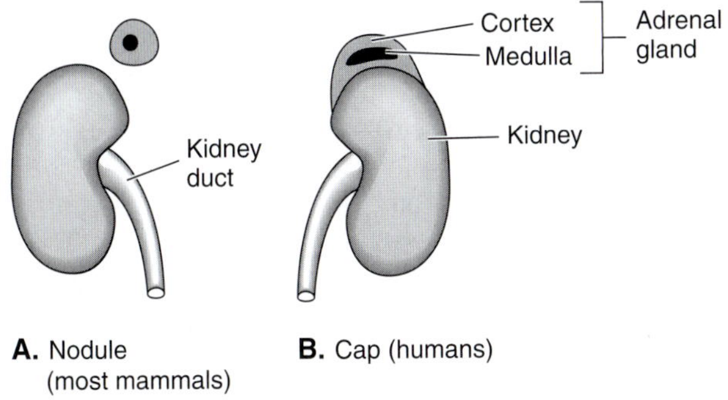

FIGURE 15-7
Structure of the mammalian adrenal gland. *A,* It is a nodule near the superior end of the kidney in most mammals. *B,* It forms a cap over the kidney in humans.

The Adrenal Medulla

The **adrenal medulla** of mammals is composed of **chromaffin cells,** which are arrayed around the periphery of venous sinuses, into which they discharge their secretions (Fig. 15-8). Investigators have identified two populations of chromaffin cells. One population produces **norepinephrine,** and the second population of cells contains enzymes that methylate norepinephrine to **epinephrine.** These catecholamine hormones also could be called neurohormones because norepinephrine is the same product as the neurotransmitter released by postganglionic sympathetic neurons at their junction with effectors. Norepinephrine production predominates in young mammals, but epinephrine production predominates in adult mammals. These hormones are released when the chromaffin cells are stimulated by preganglionic sympathetic neurons. Because of the identity of norepinephrine to the neurotransmitter of the postganglionic sympathetic neurons, and the close chemical relationship of norepinephrine to epinephrine, these medullary hormones reinforce the action of the sympathetic nervous system. Both sympathetic stimulation and the action of the medullary hormones mobilize resources needed to increase metabolism and adjust the body to short-term stress (i.e., the flight-or-fight reaction; Chapter 13). Because the level of norepinephrine production is low in the adrenal medulla of adult mammals, medullary norepinephrine has a secondary role to norepinephrine produced by postganglionic sympathetic neurons. In addition to supplementing the effects of norepinephrine, epinephrine also can stimulate the conversion of stored muscle glycogen into glucose and energy. Norepinephrine mobilizes glycogen stored only in liver cells.

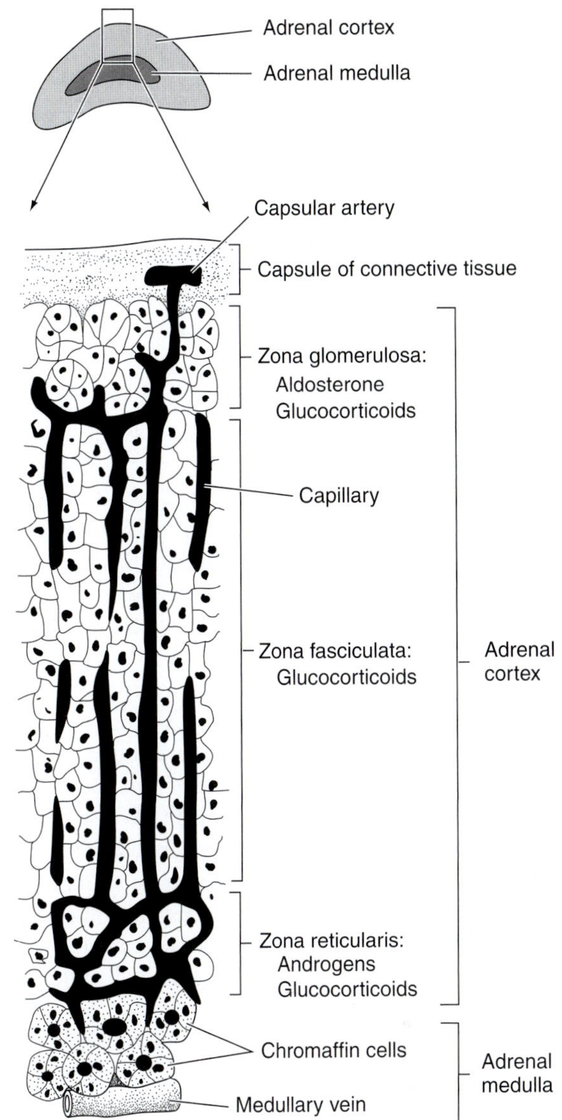

FIGURE 15-8
The histological structure of the mammalian adrenal gland.

The similarity of their products points to an affinity between chromaffin cells and postganglionic sympathetic neurons. This notion is supported by the innervation of both by preganglionic sympathetic neurons and by embryonic studies showing that both are derived from neural crest cells (Chapter 4). At one stage during their migration to the adrenal medulla, the precursors of the chromaffin cells are associated with the sympathetic ganglia that contain the cell bodies of developing postganglionic sympathetic neurons. A few chromaffin cells remain in the sympathetic ganglia, or in small, adjacent clusters called **paraganglia.** Chromaffin cells seem to be modified postganglionic sympathetic neurons.

Other chromaffin-like cells occur in the lining of the gastrointestinal and respiratory tracts and in the connective tissue of the gut, liver, and pancreas. All perform certain common biochemical steps (decarboxylation of tyrosine) in the synthesis of norepinephrine, epinephrine, and related hormones. Some investigators believe that these dispersed chromaffin and chromaffin-like cells are a **diffuse neuroendocrine system** that supplements or amplifies the actions of the sympathetic nervous system and adrenal medulla.

The Adrenal Cortex

The **adrenal cortex** develops by the proliferation of coelomic epithelial cells into the mesenchyme adjacent to the developing kidneys and gonads. The cortex produces many steroid hormones that chemically are similar to those produced by the ovary (estrogen) and testis (testosterone), which also develop partly from a proliferation of coelomic epithelium adjacent to the embryonic kidney (Chapter 21). The primate adrenal cortex is surprisingly large during fetal life because it produces substances that are precursors for the production of estrogen by the placenta, part of which is derived from fetal cells. Failure of the fetal zone to produce these precursors results in the termination of gestation. The fetal zone of the cortex regresses after birth. The adult mammalian adrenal cortex consists of irregular clusters and cords of epithelial cells interspersed with vascular spaces. Three zones are recognized: **zona glomerulosa, zona fasciculata,** and **zona reticularis** (Fig. 15-8). The functions of the cortical hormones overlap to some extent, but they can be sorted into three groups: glucocorticoids, mineralocorticoids, and cortical androgens.

The secretion of **glucocorticoids,** of which **cortisol** and **corticosterone** are the most important, is stimulated by the secretion of ACTH by the adenohypophysis. Most glucocorticoids are produced in the zona fasciculata, and they have a strong effect on energy metabolism. They inhibit blood glucose utilization by many tissues (especially skeletal muscle) and thus give nervous tissues, such as the brain, a preferential source of glucose. They stimulate amino acid uptake by liver cells and its conversion to glucose and storage as glycogen. They mobilize fat stores. These reactions help an animal adjust to certain aspects of chronic stress. High levels of glucocorticoids also suppress many of the body's normal inflammatory and immunological responses. They inhibit capillary dilation, decrease capillary permeability, and inhibit the mobilization of lymphocytes. These reactions, too, help an animal handle certain aspects of stress. Mammals that are subject to chronic stress develop enlarged adrenal cortices.

Aldosterone, the most important **mineralocorticoid,** is produced in the zona glomerulosa. It affects electrolyte balances and blood pressure. It promotes the reabsorption of sodium ions by the kidney tubules and the excretion of potassium and hydrogen ions. Because of the osmotic effect of sodium reabsorption, the amount of water in the interstitial fluid and blood increases, which leads to an increase in blood pressure. The secretion of aldosterone is not regulated by ACTH but by a complex series of reactions involving the kidney, liver, and lungs (Focus 15-1).

Cortical androgens are produced in the zona reticularis. They resemble the male sex hormone, testosterone, and at high blood levels they have a masculinizing effect on females. At normal levels, they promote protein synthesis and muscle growth in both sexes.

Phylogeny of the Adrenal Glands

All craniates have groups of cells homologous to the mammalian adrenocortical and chromaffin tissues, but they are scattered in and near the kidneys in fishes (Fig. 15-9). Hagfishes and lampreys have clusters or islets of chromaffin cells located dorsal to the coelom along the posterior cardinal veins and anterior end of the kidneys. Although cortical hormones have been found in hagfish and lamprey blood, their source is not known. The islets of chromaffin tissue of elasmobranchs lie between the kidneys beside the sympathetic ganglia in a position similar to the mammalian paraganglia. Most of the chromaffin tissue of elasmobranchs forms one or more **interrenal bodies** between the kidneys. In other fishes, chromaffin and cortical tissue form scattered islets in various parts of the kidneys.

The cortical and chromaffin tissues come together to form adrenal glands in tetrapods (Fig. 15-9). Urodeles have a series of small adrenal glands along the ventral surface of the kidneys. Amniotes have a single pair of adrenal glands. The cortical tissue lies peripheral to the chromaffin tissue, but only mammals have a continuous renal cortex. Norepinephrine is the major secretion of the chromaffin cells in nonmammalian vertebrates as well as in embryonic mammals. The methylation of norepinephrine to epinephrine in adult mammals requires cortisol, which is produced in the cortex, to synthesize a medullary enzyme. The close association of the cortex and medulla in mammals allows cortisol to reach the medulla directly because some blood flows through the cortex to the medulla.

FOCUS 15-1 *The Renin–Angiotensin–Aldosterone Mechanism*

The control of aldosterone release is a good example of the complexity of hormone interactions. We begin with the release of the hormone **renin** from the kidneys in response to a decrease in blood pressure within the kidneys or low sodium levels in the blood (Fig. A). The sensory signals are received by, and renin is produced in a special cluster of cells called, the **juxtaglomerular apparatus,** which is associated with the kidney tubules. (The kidney tubules remove waste products from the blood and selectively remove or reabsorb water, ions, and many other substances according to circumstances; Chapter 20). Once in the blood, renin reacts with circulating **angiotensinogen** or renin substrate, which is produced in the liver, to form **angiotensin I.** On reaching the lungs, angiotensin I is converted to **angiotensin II** and **III** by **angiotensin-converting enzyme,** which is produced in the lungs. Both angiotensin II and III stimulate the production and release of aldosterone by certain adrenal cortex cells. Angiotensin II is the more abundant in humans; angiotensin III, in rats. As we mentioned, aldosterone then promotes sodium reabsorption and potassium secretion by the kidney tubules. The osmotic effect of this increases fluid volume in the blood and blood pressure, but angiotensin II and III are themselves potent vasoconstrictors and increase blood pressure.

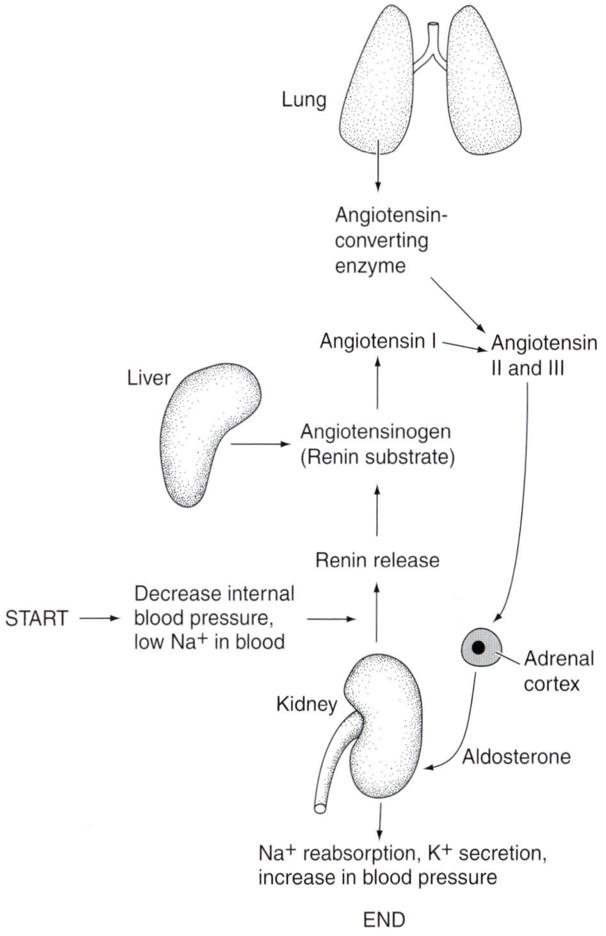

Lung
Angiotensin-converting enzyme
Angiotensin I
Angiotensin II and III
Liver
Angiotensinogen (Renin substrate)
Renin release
START
Decrease internal blood pressure, low Na^+ in blood
Adrenal cortex
Kidney
Aldosterone
Na^+ reabsorption, K^+ secretion, increase in blood pressure
END

A. A diagram of the regulation of blood pressure by the renin-angiotensin-aldosterone mechanism.

The Hypothalamo-Hypophyseal–Thyroid Axis

The **thyroid gland** arises embryonically as a midventral outgrowth from the pharynx floor between the first and second pairs of pharyngeal pouches (Figs. 4-39*G* and 15-10). The thyroid primordium soon loses its connection with the pharynx and differentiates into many small, spherical epithelial follicles, which migrate a variable distance caudad. In chondrichthyans, the follicles remain near their site of origin as a discrete median thyroid gland in the floor of the pharynx, but in the hagfish, lamprey, and most teleosts, the thyroid follicles become widely scattered in small groups throughout the pharynx. Some even enter the anterior part of the kidney (head kidney, Chapter 20), liver, and heart in teleosts. In an amphibian, the follicles form a pair of thyroid glands in the pharynx floor. The thyroid gland of sauropsids migrates far caudally, near the posterior end of the trachea, and may remain as a median organ or become paired. In mammals, the thyroid gland is a bilobed structure that lies near the anterior end of the trachea or, in humans, ventral to the thyroid cartilage (Gr., *thyreoeides* = resembling an oblong shield) of the larynx, hence its name.

The thyroid follicles are embedded in a highly vascularized connective tissue (Fig. 15-11). As is common in endocrine glands, many of the capillaries surrounding them are **fenestrated.** That is, exceptionally thin, window-like areas in their walls facilitate the passage of larger molecules. The single layer of epithelial cells that forms the follicle wall synthesizes a proteinaceous fluid called **colloid** that is secreted into the lumen of the follicle and stored there. The colloid contains **thyroglobulin,** a protein rich in the amino acid tyrosine, and iodine. Tyrosine is united with iodine within the follicle, and then the iodinated tyrosines are coupled to form

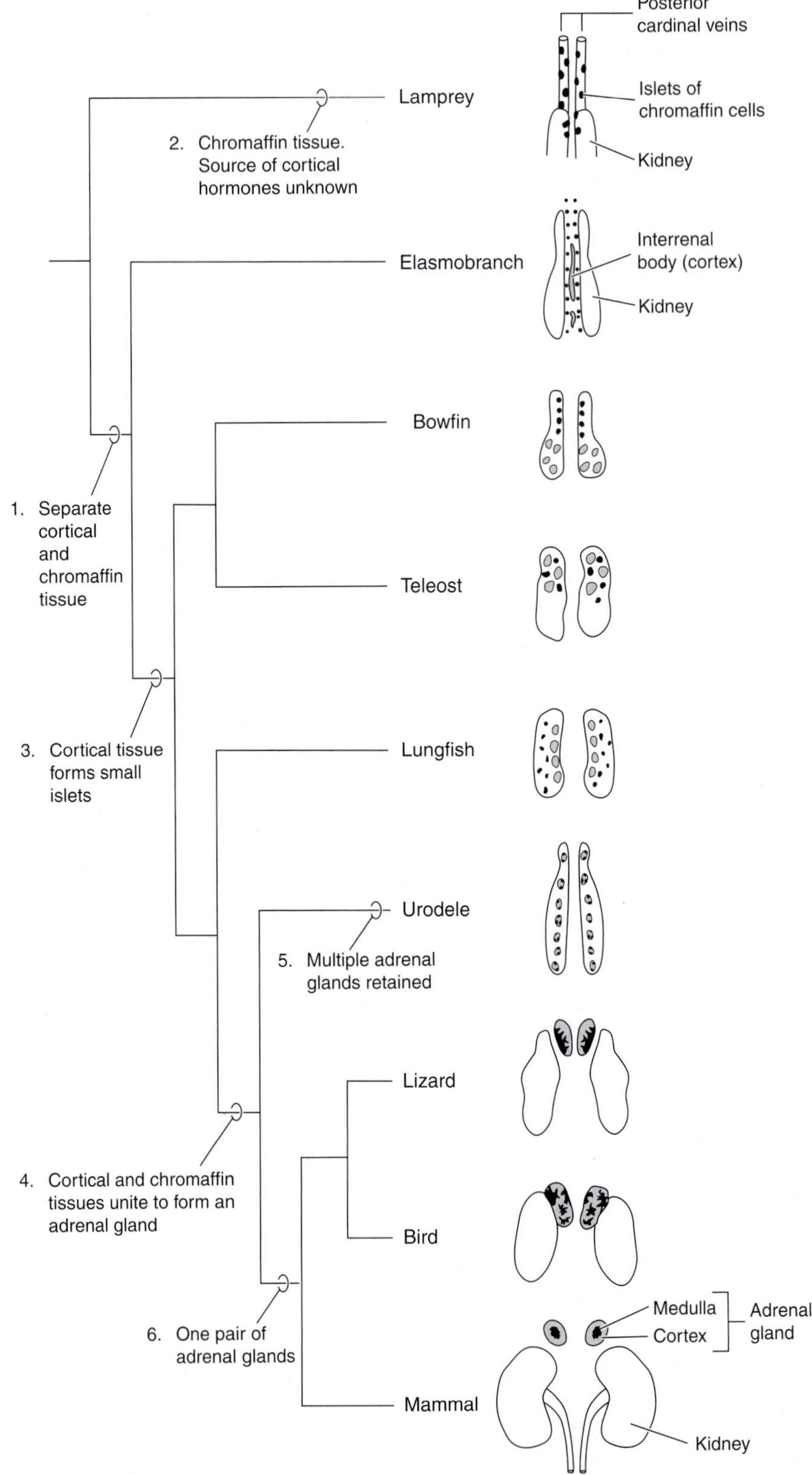

FIGURE 15-9
A cladogram suggesting the pattern of the evolution of the adrenal gland in representative vertebrates.

the thyroid hormones: **triiodothyronine (T3)** and **tetraiodothyronine (T4),** also called **thyroxine.** Sufficient dietary iodine is essential for the synthesis of the thyroid hormones. People living in parts of the world where the soil or water is deficient in iodine have low levels of thyroid hormones, and a hypothyroid enlargement of the gland results in a **goiter.** An enlargement occurs because increasing amounts of TSH are produced in an "attempt" to compensate for the low levels of thyroid hormones. This was a problem in sections of

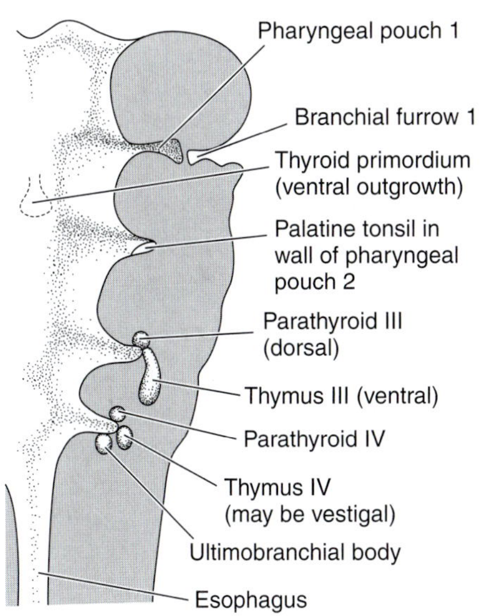

A. Early stage of pharynx development

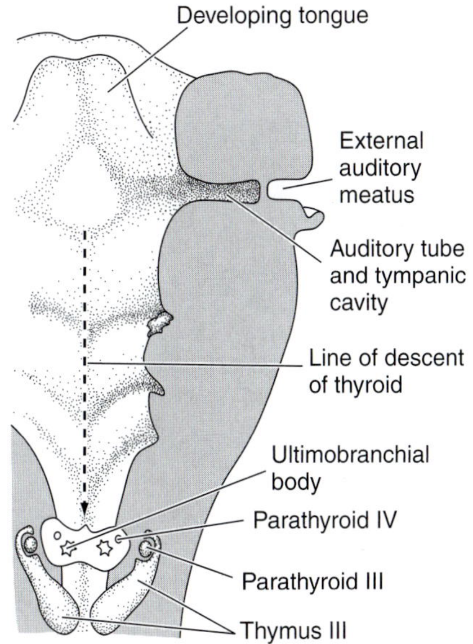

B. Later stage of pharynx development

FIGURE 15-10
Frontal sections of the embryonic development of the mammalian pharynx and its derivatives. An early stage is shown on the left, a later one, on the right. The thymic primordia are invaded by lymphocytes and become a part of the body's immune system. (*After Corliss.*)

the United States (i.e., parts of the Great Lake region, Appalachia, and the Rocky Mountains) until the introduction of iodized salt into the diet. It still is a problem in developing countries. Goiters also may result from an overactive thyroid gland, but this is less common.

A reduction in the amounts of thyroid hormones in the blood stimulates the release of TSH-releasing hormone (TRH) by the hypothalamus, and, on reaching the adenohypophysis, TRH promotes the synthesis and release of TSH. TSH causes the microvilli on the lumen side of the epithelial cells of the follicle to elongate and envelop bits of colloid. The colloid is partly digested within the cells, and the hormones are released from the colloid and discharged from the opposite surface. Blood levels of thyroid hormones are restored to normal. The thyroid gland is unique among endocrine glands in being able to store its hormones.

Thyroid hormones easily pass through the lipid plasma and nuclear membranes of cells and exert their influence by binding with **thyroid hormone receptors** on the chromosomes. This binding, in turn, activates (or, in a few cases, represses) the transcription of certain genes into proteins. Because so many cells are affected, the functions of the thyroid hormones in mammals are

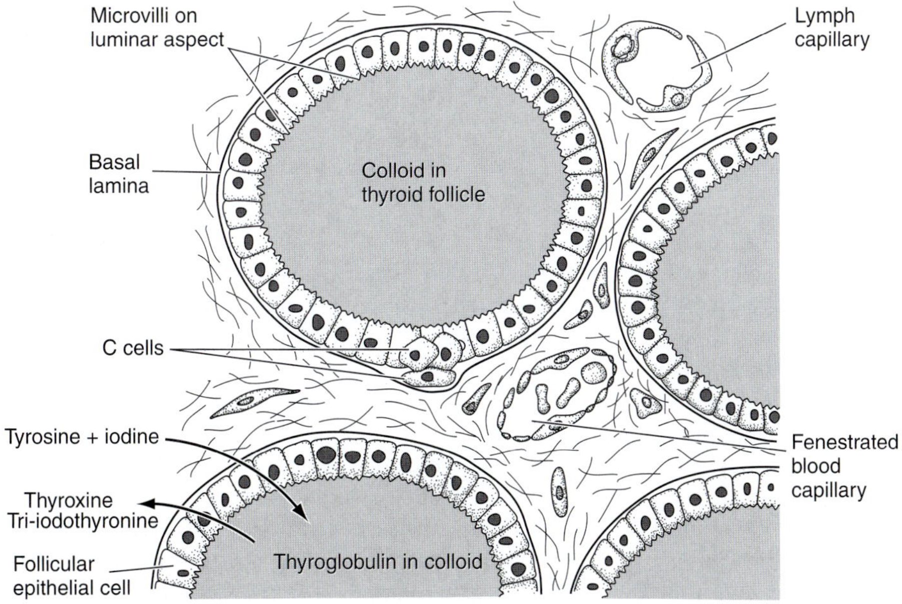

FIGURE 15-11
The histological structure of the mammalian thyroid gland. (*After Williams et al.*)

difficult to generalize. Metabolism, growth, development, and reproduction are among the many processes that are affected. Many of the effects are **"permissive"** in the sense that the thyroid hormones maintain body cells in a state of readiness to respond to other hormones. Thyroid hormones in mammals and birds accelerate oxidative metabolism throughout the body, and they are needed to maintain the high level of metabolism that characterizes endotherms. The production of these hormones decreases greatly in hibernating mammals. Production peaks when these mammals begin to emerge from hibernation and generate considerable body heat. Thyroid hormones are essential for normal growth and development, possibly by their permissive action in making cells more responsive to GH. Nervous tissues are particularly sensitive to a lack of thyroid hormones during development. A low level of these hormones in human infants causes a severe retardation of mental development known as **cretinism.** The maturation of the ovary and testis also requires thyroid hormones, which apparently affect the release of gonadotropins by their influence on the hypothalamus and its production of GnRH. Hair molting in some mammals involves both thyroid hormones and cortisol. A low thyroid level and high cortisol level causes a molt; the reverse induces hair growth.

Every adult craniate has a thyroid gland, but the capacity to iodinate tyrosine and synthesize iodinated proteins is not confined to craniates. Many invertebrate groups can do so, and the iodinated proteins usually are a component of their skeletons, pharyngeal teeth, and other hard parts. The significance of this ability is not understood, for no clear evidence shows that invertebrates make use of these proteins. Tunicates and amphioxus synthesize these proteins in certain cells of their endostyle. They are released into the pharynx and digested and absorbed as iodinated tyrosines in the intestine. This pattern of production, release, and absorption implies a function, but its nature is unclear. The endostyle-like **subpharyngeal gland** of the ammocoetes larva of the lamprey also produces and releases iodinated proteins into the pharynx. Part of this gland transforms into the thyroid gland in adults. We see in this developmental sequence the transformation of an exocrine gland into an endocrine gland. The craniate thyroid gland seems to have evolved as a means of storing iodinated proteins and releasing them directly into the blood.

The functions of the thyroid hormones of anamniotes and reptiles have not been thoroughly studied. Adult hagfishes and lampreys produce thyroid hormones, but their functions are not yet known. As expected, the thyroid hormones of ectothermic fishes, amphibians, and reptiles do not increase metabolic rate as they do in endothermic birds and mammals, but they appear to have many of the other effects seen in birds and mammals. They promote growth (except in amphibians, see subsequent discussion), molting, gonad maturation, and some aspects of reproductive behavior (migration, spawning).

Thyroid hormones have a negative effect on amphibian growth, but they do induce metamorphosis, a period in which growth stops, and profound changes occur in the morphology and functions of most organ systems as the animal changes from an aquatic tadpole to a terrestrial adult. The dramatic effect of thyroid hormones on amphibian metamorphosis was observed many years ago, when scientists found that early metamorphosis could be induced by providing frog tadpoles with thyroxine or, in some cases, extra iodine. Removal of the tadpole's thyroid gland inhibited metamorphosis. Many cases of paedomorphosis in salamanders result from low levels of either TSH or thyroid hormones.

Since these early observations, many studies have been made on the roles of thyroid hormone and thyroid-hormone receptors at different stages of amphibian development. The thyroid-hormone receptors of responding tissues appear to be particularly important (Wolffe and Shi, 1999). Patterns of gene expression leading to metamorphosis vary with the distribution and abundance of thyroid-hormone receptors and the degree of their binding to the thyroid hormones. Gene expression patterns are very complex. For example, the rapid growth of the hind legs of a frog tadpole during metamorphosis involves the activation of 14 genes and the decrease in activity of 5 others. To complicate matters further, investigators have discovered that thyroid hormones interact with others, in particular corticosterone and PRL, which also increase rapidly during metamorphosis.

Endocrine Regulation of Calcium and Phosphate Metabolism

Calcium ions (Ca^{2+}) and phosphate ions (HPO_4^{2-}) are essential for the formation of bone and teeth and for a multitude of biochemical processes. They are taken in with the diet, stored in bones and teeth, and maintained in the blood in a dynamic equilibrium. In mammals, they are regulated in such a way that their product equals some constant:

$$[Ca^{2+}][HPO_4^{2-}] = k$$

This constant changes with the physiological state of the animal, being greater, for example, in a growing mammal than in an adult. The relationship between these ions means that an increase in Ca^{2+} will lead to a drop in HPO_4^{2-}, and vice versa. We can focus, therefore, primarily on factors affecting calcium homeostasis.

Calcium homeostasis in mammals is dependent on the interaction of three hormones from different sources: **parathyroid hormone (PTH**; also known as **parathormone**) and **calcitonin (CT),** both of which are polypeptides, and the steroid **1,25-dihydroxycholecalciferol (1,25-DHC).** Parathormone is produced by the **parathyroid glands.** Most tetrapods have two pairs of parathyroid glands that develop as endodermal epithelial buds from the third and fourth pharyngeal pouches, but lissamphibians often have a third pair that develops from the second pharyngeal pouch (Figs. 15-10 and 15-12). The parathyroid epithelial buds usually form on the ventral surface of the pouches, but those of mammals develop dorsally. The epithelial buds separate from the pharyngeal pouches and come to lie near the thyroid gland or, as in many mammals, become embedded in its dorsal surface.

Calcitonin is produced by **parafollicular,** or **C, cells.** The C cells arise from the neural crest but usually become lodged in **ultimobranchial bodies** that develop in all vertebrates from the ventral or posterior surface of the last pair of pharyngeal pouches (Fig. 15-12). The ultimobranchial bodies are vestigial in most mammals, but the C cells lodge within the thyroid gland (Fig. 15-11).

1,25-dihydroxycholecalciferol is a complex hormone, the synthesis of which begins in the skin by the action of ultraviolet light, which converts cholesterol to **cholecalciferol,** or vitamin D. Cholecalciferol is converted in the liver to **25-hydrocholecalciferol (25-HC)** and in the kidney to 1,25-DHC (Fig. 15-13).

Calcium and phosphate ions enter the body through the intestine (Fig. 15-13). Calcium absorption needs a calcium-binding protein the synthesis of which requires 1,25-DHC. Phosphate ions passively follow the calcium ions. Once in the blood, the level of calcium ions is maintained primarily by the action of PTH. PTH is released from the parathyroid glands when calcium levels in the blood decrease, and its release is inhibited when calcium levels increase. PTH causes the removal of calcium ions from bones, partly by activating osteoclasts, and the reabsorption of calcium ions by the kidney tubules. 1,25-DHC facilitates both of these actions. Calcium ions are lost in the urine only when their blood levels are high. PTH has the opposite effects on phosphate ions, promoting its excretion by the kidney, for example. Calcitonin (CT) inhibits the action of PTH on bone, hence it promotes the storage of calcium ions in bone. Calcium homeostasis is essential for nerve and muscle metabolism, and prolonged low levels cause muscle tetany and death from suffocation.

The regulation of calcium and phosphate levels in the blood in other tetrapods is similar to that in mam-

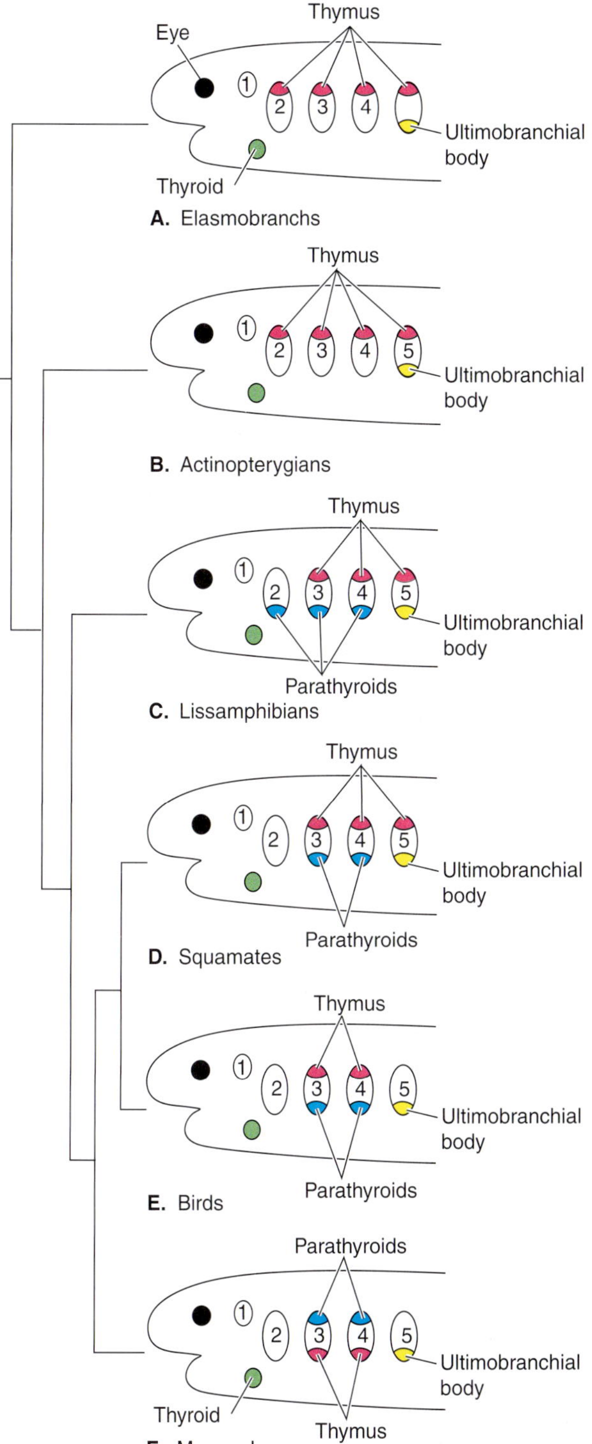

FIGURE 15-12

A–F, The embryonic development of the pharyngeal glands and thymus in the major vertebrate groups. The embryonic pharyngeal pouches are shown as ovals and are numbered. The thymus, the buds of which generally arise from the surface of pharyngeal pouches opposite to the surface forming the parathyroid primordia, is a part of the immune system (Chapter 19). The thyroid gland develops as an outgrowth of the floor of the pharynx.

FIGURE 15-13
The primary factors involved in calcium homeostasis in mammals. The pathway for the formation of 1,25-dihydroxycholecalciferol (1,25-DHC) from cholesterol and vitamin D in the skin is shown at the top right side. 1,25-DHC promotes the absorption of calcium ions from the intestine. Phosphate ions follow the absorption of calcium ions (Ca^{2+}) from the intestine, and parathormone (PTH) has an action on phosphate opposite to its action on calcium.

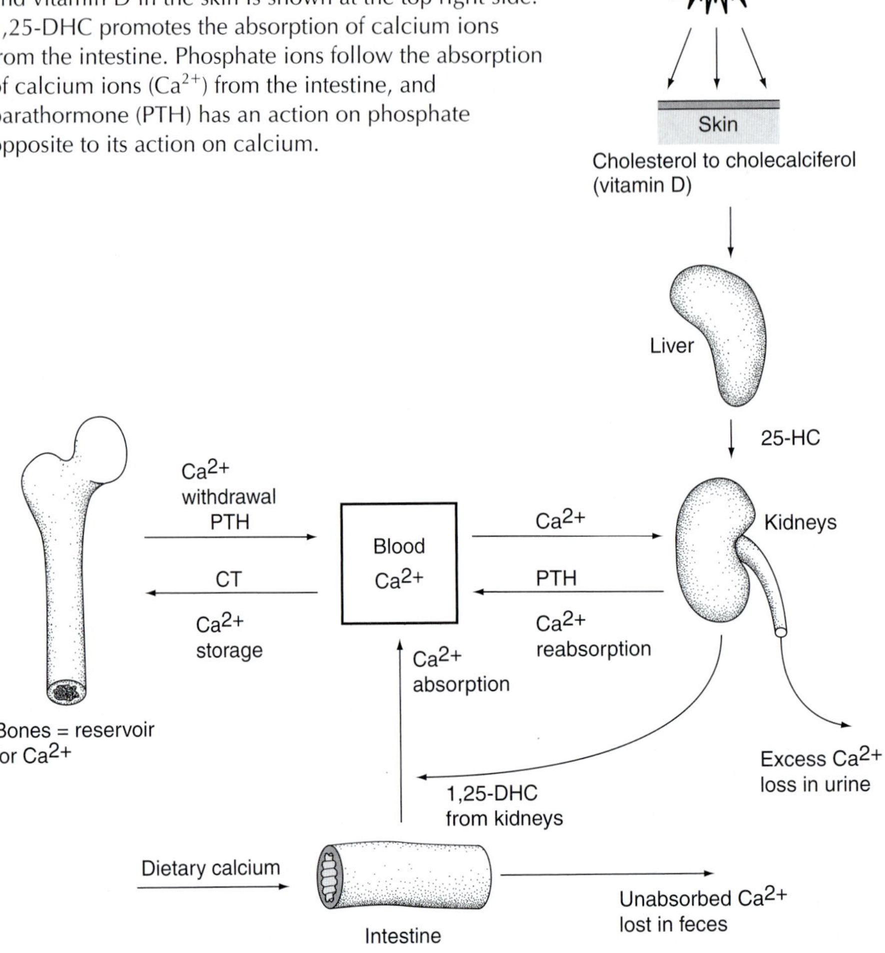

mals. All tetrapods have parathyroid glands and ultimobranchial bodies, and the vitamin D complex appears to be active. Calcium-binding protein has been found in the intestinal mucosa of amphibians. Fishes are quite different: All lack parathyroid glands, but parathormone, which increases calcium levels in the blood, has been found in the hypophysis. Calcium-lowering factors appear to come from the ultimobranchial bodies and small **corpuscles of Stannius,** which are embedded in the kidneys.

Endocrine Regulation of Metabolism

Animals derive their energy by the intracellular breakdown of carbohydrates, primarily the simple sugar glucose. Excess glucose can be stored as a more complex carbohydrate (glycogen), as lipids, and as protein. All of these storage products can be reconverted to glucose when needed. Hormones produced by clumps of endocrine cells, which are known as the **pancreatic islets (islets of Langerhans),** are critical in the utilization of glucose and in these metabolic conversions.

The pancreatic islets of most tetrapods are small clusters of endocrine cells that are scattered in the pancreas (Fig. 15-14). (The pancreas is a large exocrine gland the secretory lobules [the acini] of which secrete many digestive enzymes that are discharged through ducts that usually join the bile duct from the liver.) Although pancreatic exocrine tissue is present, many fishes do not have a distinct pancreas (hagfishes and lampreys). Even when a pancreas is present in fishes, the islet cells do not form discrete islets within it; rather, they tend to be scattered in the intestinal wall, along abdominal blood vessels,

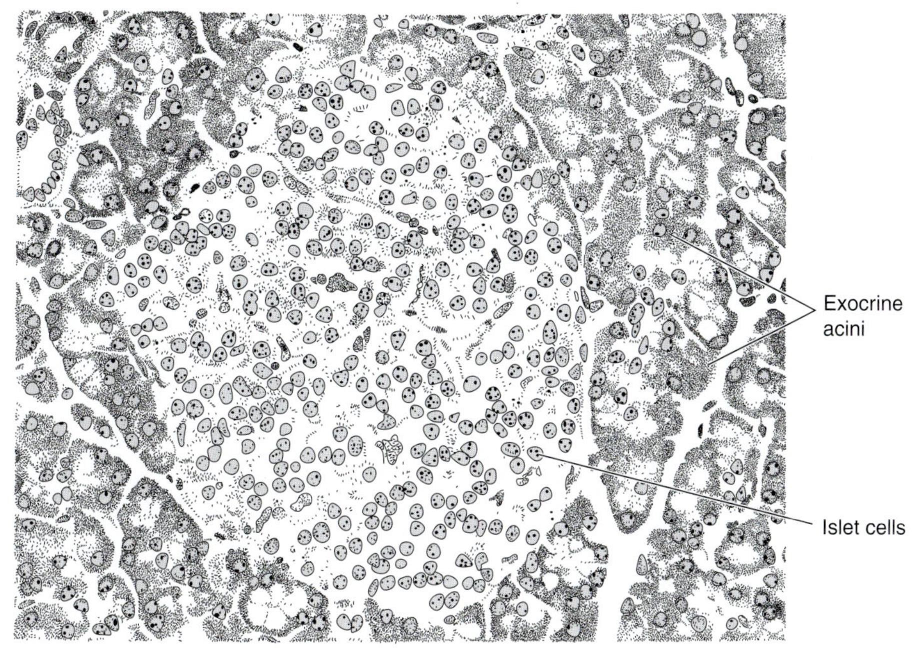

FIGURE 15-14
A drawing based on a photomicrograph of an endocrine pancreatic islet in the pancreas of a mammal. Most of the pancreas is composed of exocrine cells, which secrete the digestive enzymes. These form small secretory clumps (acini), some of which show around the islet.

within the liver, and along the bile duct, and sometimes they form layers around the smaller pancreatic ducts. Lungfishes do have a compact pancreas within the intestinal wall, which does contain islets. The islet cells of most teleosts are aggregated to form masses of tissues called **Brockmann bodies.** Some species have many Brockmann bodies; in others (Atlantic cod, bullheads), most of the islets cells are contained in a single one.

Several distinct cell types, which produce different hormones, have been identified in mammalian pancreatic islets. **Insulin (I),** so named because it came from the pancreatic islets (L., *insula* = island), was the first pancreatic hormone to be discovered. An increase in blood sugar levels, following a meal, for example, directly stimulates the insulin-producing cells to produce insulin. Insulin decreases blood sugar levels by facilitating the transport of glucose into cells, particularly into muscle and fat cells. Insulin also increases the uptake of amino acids and fatty acids by target cells. In addition, insulin enhances the action of certain enzymes involved in the utilization of glucose. As blood sugar levels decrease to normal, insulin secretion decreases. Autonomic neurons also terminate on many islet cells and appear to modulate their response to blood sugar. Parasympathetic stimulation promotes insulin release, and sympathetic stimulation blocks it. An insufficient production of insulin or an unresponsiveness of the target cells to insulin leads to **diabetes mellitus,** a disease in which glucose is not adequately utilized. Glucose levels increase in the blood, glucose is excreted in the urine along with a great deal of water, and fats and proteins are used as a source of energy to compensate for the inability to utilize glucose.

A second islet hormone, **glucagon,** is a simple polypeptide. Its secretion is promoted by a decrease of blood sugar. Its primary action is to mobilize stored carbohydrate (glycogen) in liver cells and release it as glucose into the blood. It also promotes the breakdown of stored fat. Glucagon is one of several agents (e.g., GH, epinephrine) that increase blood sugar levels. In mammals, glucagon is particularly important in fasting or poorly fed individuals. Other islet cells produce **pancreatic polypeptide** and **pancreatic somatostatin.** Little is known about their actions. Somatostatin is believed to be a paracrine and to act locally within the islets as an inhibitor of the release of insulin and glucagon.

Insulin and glucagon are conservative molecules, and only a few changes in their amino acid composition have occurred during vertebrate evolution. The effects of insulin and glucagon are generally the same in other vertebrates as in mammals. Lizards and birds, however, are far less sensitive to insulin, and glucagon appears to be their primary regulator of carbohydrate metabolism. Glucagon-like molecules have been found in amphioxus, tunicates, and crabs. Insulin is a very ancient molecule, and insulin-like molecules occur in protozoans, bacteria, and fungi!

Endocrine Regulation of the Gastrointestinal Tract

The steplike process of digesting foods as they pass through the digestive tract necessitates a precise integration between food movements and the release of digestive enzymes from the stomach and small intestine. This control in mammals is effected partly by the autonomic nervous system and partly by peptide hormones secreted by scattered cells in the epithelial lining of the stomach and first parts of the small intestine (duodenum and jejunum). In general, parasympathetic stimulation by the vagus nerve promotes gut movements and secretion, and sympathetic stimulation inhibits these actions. Many types of gastrointestinal, peptide-producing cells have been discovered. Because these cells show an affinity to chromaffin cells and to postganglionic parasympathetic neurons, both of which are derived from the neural crest, some investigators believe that these cells also are derived from the neural crest. All of the cells lie next to the gut lumen so they can respond directly to the presence of digestive products in the gut.

Vagal stimulation and the presence of food in the stomach promote the secretion of **gastrin** by cells in the distal or pyloric region of the stomach (Fig. 15-15

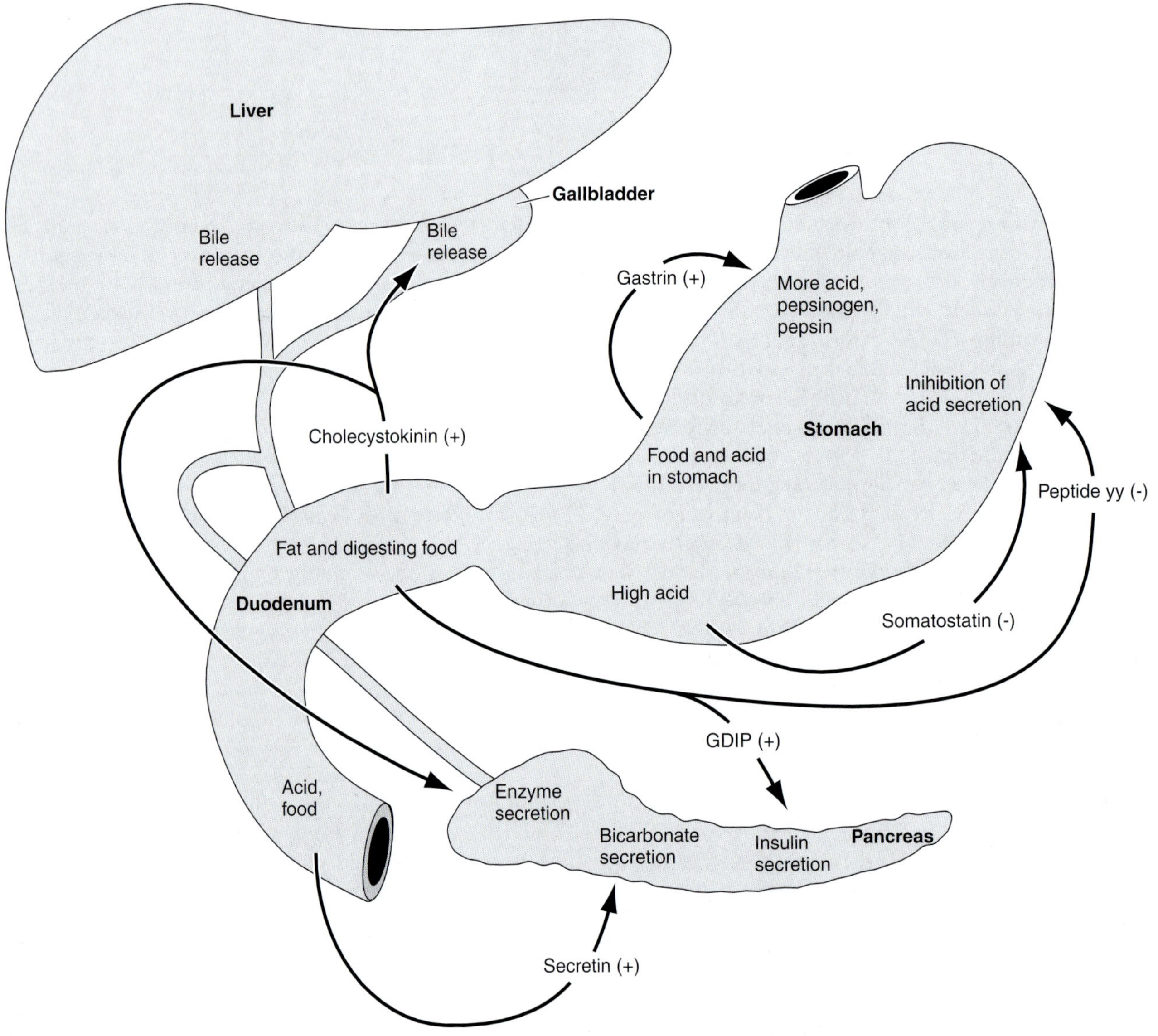

FIGURE 15-15
The major hormones that affect gastric and pancreatic secretion and bile release.

TABLE 15-2 Major Controls on Digestive Functions

Stimulus	Target	Outcome
Vagus impulse	Parietal cells of gastric glands	Acid release
Acid and food in pyloric region	Pyloric cells	Gastrin release
Gastrin	Parietal cells of gastric glands	More acid release
Acid via local neurone reflex	Chief cells of gastric glands	Pepsinogen release
Acid	Pepsinogen	Pepsin
High acidity	Pyloric cells	Somatostatin release
Somatostatin	Parietal cells	Inhibition of acid release
Entry of acid food into duodenum	Duodenal cells	Secretin release
Secretin	Exocrine pancreas	Bicarbonate solution release
Fat and digesting food	Duodenal cells	Cholecystokinin release
Cholecystokinin	Exocrine pancreas	Enzyme release
	Gallbladder	Bile release
Digesting food	Duodenal cells	Peptide YY release
Peptide YY	Parietal cells	Further inhibition of acid release
Digesting food	Duodenal cells	Glucose-dependent insulinotropic hormone release
GI-dep insulinotropic hormone	Endocrine pancreas	Insulin release if glucose rising in blood
Digesting food	Duodenal cells	Motilin release
Motilin	Intestine	Increased gut motility

and Table 15-2). Gastrin then causes the secretion of hydrochloric acid by the parietal cells of the gastric glands. The resulting increase in acid levels in the stomach evokes a local parasympathetic reflex that causes the release of the proenzyme pepsinogen from the chief cells of the gastric glands. Pepsinogen is changed to active pepsin by the acids in the stomach lumen. As acid levels increase in the pyloric region of the stomach, certain cells there release **somatostatin,** which inhibits acid release.

The discharge of the partly digested acid contents of the stomach stimulates certain duodenal cells to produce **secretin.** This hormone, discovered by Bayliss and Starling in 1902, was the first hormone the effects of which could be clearly demonstrated. It was believed at first that secretin alone stimulated all of the pancreatic secretion. Scientists now recognize that secretin promotes only the release of water, salts, and bicarbonate ions, which neutralize the acids coming from the stomach. The release of the pancreatic enzymes requires another hormone originally called **pancreozymin,** the secretion of which is stimulated by the presence of fat and peptides. Parasympathetic stimulation supplements hormonal enzyme release. Yet another hormone, **cholecystokinin,** was believed to promote the contraction of the gallbladder and the release of bile from the liver. Although bile contains no digestive enzymes, it has a digestive role because it emulsifies fats and makes them more susceptible to the action of fat-splitting enzymes. Subsequent research showed that pancreozymin and cholecystokinin are the same peptide. This peptide was first identified by using both names, cholecystokinin–pancreozymin, but most endocrinologists now simply call it cholecystokinin, which is the older name. In addition to its other functions, cholecystokinin also slows down the rate of emptying of the stomach.

Another peptide, originally called enterogastrone, is produced by the intestinal mucosa in response to the presence of digesting food. Scientists now recognize that enterogastrone is three different peptides: (1) **peptide YY,** which, together with somatostatin, inhibits gastric acid secretion; (2) **glucose-dependent insulinotropic peptide (GDIP),** which stimulates insulin release from the pancreatic islets provided that blood glucose levels also are elevated; and (3) **motilin,** which stimulates motility of the gastrointestinal tract. Yet other peptides have been found.

By using immunoreactive studies, investigators have found that many nonmammalian vertebrates have a number of the gastrointestinal peptides found in mammals, but few studies have been made on their biological function. Many fundamental differences occur between the digestive pattern of mammals and nonmammalian vertebrates (Chapter 17). Some fishes lack stomachs, for example.

SUMMARY

1. Endocrine regulation of body activities tends to be slower, more diffuse, and longer lasting than nervous regulation because hormones must enter and leave the blood, cells with receptors for specific hormones often are scattered widely in the body, and hormones have a longer life than do neurotransmitters. Endocrine integration is particularly well suited for controlling continuing processes, such as metabolism, growth, slow body color changes, metamorphosis, mineral and water balances, sexual development, and reproduction.
2. Despite their diversity, hormones belong to only a few categories of chemical compounds: amino acids or modified amino acids, peptides, proteins, and steroids. Peptide and protein hormones, which are soluble in the blood, do not diffuse through the lipid plasma membrane of cells; rather, they bind to receptor molecules in the membrane. This activates a second messenger within the cell, which then is responsible for the effect of the hormone.
3. Lipid-soluble steroid hormones and thyroid hormones easily pass through the plasma membrane and ultimately exert their effects by influencing genes. Either method involves considerable amplification of the original signal, so hormones are effective in very small amounts.
4. The hypothalamus of the brain and the hypophysis (pituitary gland) are closely integrated structurally and functionally to form the hypothalamo–hypophyseal axis. Some cells in hypothalamic nuclei secrete two neurohormones that are stored and released from the neurohypophysis. VP promotes water reabsorption, and OXY stimulates smooth-muscle contraction.
5. Other hypothalamic nuclei produce other neurohormones that are carried by small portal veins to the adenohypophysis, where they control the synthesis and release of its numerous hormones. These hormones affect many other endocrine glands and body functions: TSH (thyroid gland), gonadotropin (ovary and testis), ACTH (adrenal cortex), GH, PRL, and MSH.
6. Because of the wide range of activities affected by hypophyseal hormones and the close integration of the hypophysis with the hypothalamus, the nervous system exerts considerable control over the endocrine system. Conversely, some hormones (reproductive hormones) affect the nervous system and the behavior of animals.
7. The pineal gland evolved from the photoreceptive epiphyseal complex (pineal and parietal eyes) of anamniotes and reptiles. It is the source of melatonin, a hormone the synthesis of which is inhibited by light. Melatonin production, therefore, has a distinct circadian rhythm, which is controlled by a "biological clock" in the hypothalamus.
8. The adrenal glands of mammals consist of quite different cortical and medullary cells. Chromaffin cells of the medulla are derived from the neural crest and appear to be modified postganglionic sympathetic neurons. Chromaffin cells secrete norepinephrine and (primarily in adult mammals) epinephrine, a methylated norepinephrine. In addition to supplementing the action of epinephrine and the sympathetic nervous system in adjusting the body to short-term stress, epinephrine stimulates the conversion of stored muscle glycogen to glucose.
9. The adrenal cortex produces three groups of steroid hormones: (1) glucocorticoids (cortisol and corticosterone) have a strong effect on energy metabolism, tending to promote the uptake of amino acids and their conversion to glucose and storage as glycogen; (2) mineralocorticoids (aldosterone) affect electrolyte balances and blood pressure; and (3) cortical androgens promote protein synthesis.
10. Chromaffin cells and adrenocortical tissue are distinct in fishes, forming small patches on or between the kidneys. These two tissues come together in tetrapods, but only in mammals does a complete cortex surround the medulla. This association appears to facilitate the methylation of norepinephrine to epinephrine.
11. All vertebrates have a thyroid gland that develops as a ventral outgrowth from the pharynx. Tyrosine is united with iodine within thyroid follicles to form the thyroid hormones: T3 and T4 (thyroxine). These hormones bind with receptors within the nucleus and have wide-ranging effects on metabolism, growth, development, and reproduction. Many of the effects are said to be "permissive" because the thyroid hormones maintain body cells in a state of readiness to respond to other hormones. Thyroid hormones are essential for endothermic birds and mammals to maintain their high levels of metabolism. Thyroid hormones are essential for amphibian metamorphosis.
12. Calcium and phosphate homeostasis depends on the interaction of three hormones: (1) 1,25-DHC, a derivative of vitamin D, promotes the active absorption of calcium ions (phosphate ions passively follow); (2) PTH, which is produced by

the parathyroid glands of tetrapods, promotes the removal of calcium ions from bone when their blood level decreases and their excretion when levels increase; and (3) calcitonin, produced by neural crest cells that become lodged in the ultimobranchial bodies or in the thyroid gland of mammals as the C cells, inhibits the action of PTH on bone, hence it promotes the storage of calcium. Fishes lack parathyroid glands, but PTH has been found in their hypophysis.

13. Insulin, which is produced by certain cells in the pancreatic islets, is essential for the utilization of glucose and interconversions between glucose, other carbohydrates, proteins, and fats. Insulin decreases blood sugar by facilitating the transport of glucose into cells and its use by the cells. Glucagon, a second islet hormone, mobilizes carbohydrate stored as glycogen in liver cells and releases it into the blood. Autonomic nervous stimulation also is important in the release and storage of glucose.

14. The sequential movements of the gut and secretion of digestive enzymes as food passes through the digestive tract are regulated by the autonomic nervous system and by many gastrointestinal peptides secreted by cells next to the gut lumen.

REFERENCES

Barinaga, M., 1999: The clock plot thickens. *Science,* 284:421–422.

Barrington, E. J. W., 1975: *An Introduction to Comparative and General Endocrinology*, 2nd edition. New York, Oxford University Press.

Bentley, P. J., 1976: *Comparative Vertebrate Endocrinology.* Cambridge, Cambridge University Press.

Bern, H. A., 1985: The elusive urophysis: Twenty-five years in pursuit of caudal neurohormones. *American Zoologist,* 25:763–769.

Carmichael, S. W., and Winkler, H., 1985: The adrenal chromaffin cell. *Scientific American,* 235:40–49.

Clark, N. B., Norris, D. O., and Peter, R. E., symposium organizers, 1983: Evolution of endocrine systems in lower vertebrates: A symposium honoring Professor Aubrey Gorbman. *American Zoologist,* 23:593–748.

Corliss, C. E., 1976: *Patten's Human Embryology.* New York, McGraw-Hill.

DeGroot, L. J., editor, 1989: *Endocrinology*, 2nd edition. Philadelphia, W. B. Saunders.

Ellis, L. C., symposium organizer, 1976: Endocrine role of the pineal gland. *American Zoologist,* 16:3–101.

Epple, A., and Brinn, J. E., 1987: *The Comparative Physiology of the Pancreatic Islets.* Berlin, Springer-Verlag.

Fevre-Montage, M., Von Cauter, E., Refetoff, S., et al., 1978: Effects of "jet lag" on hormonal patterns: II. Adaptation of melatonin circadian periodicity. *Journal of Clinical Endocrinology and Metabolism,* 52:642–649.

Fritzsch, B. M., 1990: The evolution of amphibian metamorphosis. *Journal of Neurobiology,* 21:1011–1021.

Gekakis, N., Staknis, D., et al., 1998: Role of the clock protein in the mammalian circadian mechanism. *Science,* 280:1564–1569.

Gern, W. A., Duvall, D., and Nervina, J. M., 1986: Melatonin: A discussion of its evolution and actions in vertebrates. *American Zoologist,* 26:985–996.

Gorbman, A., Dickhoff, W. W., Vigna, S. R., Clark, N. B., and Ralph, C. L., 1983: *Comparative Endocrinology.* New York, John Wiley & Sons.

McIntosh, C. H. S., 1995: Control of gastric acid secretion and the endocrine pancreas by gastrointestinal regulatory peptides. *American Zoologist,* 35:455–465.

Norris, D. O., 1996: *Vertebrate Endocrinology,* 3rd edition. San Diego, Academic Press.

Orci, L., Vassalli, J. D., Perrelet, A., 1988: The insulin factory. *Scientific American,* 259:85–94.

Pang, P. K. T., and Epple, A., editors, 1980: *Evolution of Vertebrate Endocrine Systems.* Lubbock, Tex., Texas Tech Press.

Reiter, R. J., editor, 1981: *The Pineal Gland.* Boca Raton, CRC Press.

Sherwood, N. M., and Parker, D., 1990: Neuropeptide families: An evolutionary perspective. *Journal of Experimental Biology,* Supplement 4:63–71.

Tamarkin, L., Curtis, J. B., and Almeida, O. F. X., 1985: Melatonin: A coordinating signal for mammalian reproduction? *Science,* 277:714–719.

Turek, F. W., 1985: Circadian neural rhythms in mammals. *Annual Reviews of Physiology,* 47:49–64.

Uvnas-Moberg, K., 1989: The gastrointestinal tract in growth and reproduction. *Scientific American,* July: 78–83.

Williams, P. L., Bannister, L. H., Berry, M. M., Collins, P., Dyson, M., Dussek, J. E., Ferguson, M. W. J., editors, 1995: *Gray's Anatomy*, 38th edition. New York, Churchill Livingstone.

Wolffe, A. P., and Shi, Yun Bo, 1999: A hypothesis for the transcriptional control of amphibian metamorphosis by the thyroid hormone receptor. *American Zoologist,* 39:807–817.

part **IV**

Metabolism and Reproduction

16

The Digestive System: Oral Cavity and Feeding Mechanisms

PRECIS

We examine how animals obtain the raw materials that they need to sustain life and utilize them in their metabolism in Part IV. Acquiring food for energy to drive metabolism and other physiological processes is central to the survival of the individual. Competition for food resources is often intense in the animal's natural surroundings. Increased efficiency in acquiring and processing of food is at a premium in natural selection. Many adaptations in the feeding mechanisms of vertebrates are correlated with their relative efficiency in competitive interactions. Many animals are predacious. Other factors, such as elusiveness and numerous defense strategies of the prey, also may have played a role in the evolution of various adaptive features in the vertebrate feeding apparatus.

OUTLINE

For most vertebrates, metabolic rate and all life processes depend on regular ingestion of food items, which in many cases must be mechanically broken down in the oral cavity prior to chemical digestion. Thus, food is taken into the mouth (**ingestion**), processed (**mastication**), and then swallowed (**deglutition**). Chewing, or mastication, serves two functions: (1) food or prey items are mechanically broken down to a condition suitable for swallowing and (2) the resulting increase in surface area of the masticated food facilitates the penetration of digestive enzymes and so increases the rate of chemical breakdown. Foods with resistant cell walls, such as leaves, stems, and grasses, require extensive mechanical maceration for the digestive enzymes to be effective. Vertebrates have evolved many different designs with which a vast array of food resources is exploited. In this chapter, we provide a brief review of the functional design of the feeding apparatus of aquatic and terrestrial feeding within an evolutionary perspective.

The Development of the Digestive Tract

Most of the digestive tract develops from the embryonic **archenteron** (Gr., *arche* = beginning + *enteron* = gut), which is continuous in many vertebrates with a yolk sac (Fig. 16-1*A*). The endoderm of the archenteron forms only the lining of the digestive tract and its various derivatives. These derivatives are the lining of the respiratory passages, which evaginate near the front of the archenteron, and the secretory cells of the liver, pancreas, and other glandular outgrowths of the archenteron. Connective tissues and muscles in the walls of those organs located in the body cavity, and the coelomic

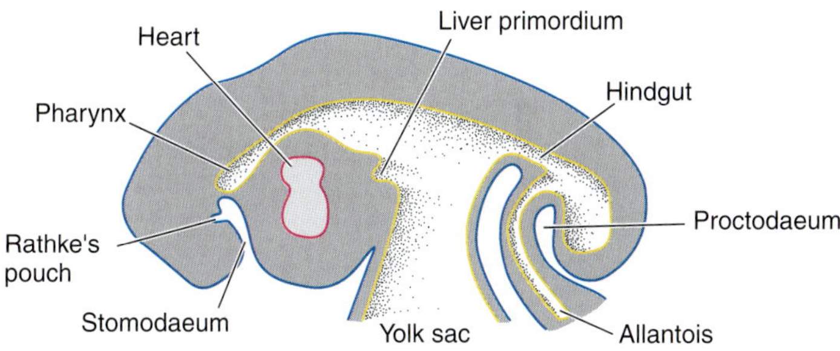

A. Early stage in development of digestive tract

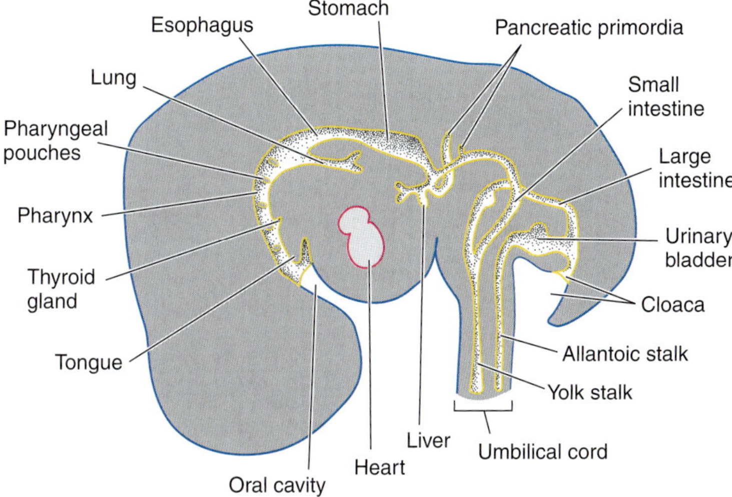

B. Later stage in development of digestive tract

FIGURE 16-1
Lateral views of the development of the digestive tract of a terrestrial vertebrate. *A,* An early stage in which the archenteron has differentiated into a foregut and hindgut, but a yolk sac is still present. *B,* A later stage in which gut regions are differentiating. Blue = ectoderm; yellow = endoderm.

epithelium covering them, develop from the adjacent splanchnic layer of the lateral plate mesoderm. In the head and branchial region, the muscles are of somitic and somitomeric origin, and the connective tissue comes from neural crest cells, as do the visceral arches (Chapter 4). At first, the archenteron has a broad connection with the yolk sac, but the formation of head and tail folds and lateral body folds gradually separates the embryo from the yolk sac. The archenteron becomes differentiated into a **foregut,** extending toward the head; a **hindgut,** extending toward the tail; and a **midgut,** which remains connected to the yolk sac.

As the embryo continues to take shape and separate from the underlying yolk mass, the digestive tract becomes tubular and lengthens (Fig. 16-1*B*). The pharynx differentiates from the anterior part of the archenteron. A series of lateral diverticula from the pharynx become the pharyngeal pouches. The first of these form the spiracle (fishes) or the auditory tube and middle ear cavity of a tetrapod that has these structures (Chapter 4). Epithelial buds from certain pouches form the parathyroid gland and thymus. The thyroid gland is an evagination from the rostral part of the pharynx floor; the lungs and respiratory passages are evaginations from the caudal part of the floor. We discussed these glandular derivatives in Chapter 15. We return to the respiratory derivatives in Chapter 18. An esophagus connects the pharynx with the developing stomach, and the intestine follows the stomach. The liver develops as a ventral evagination of the archenteron just caudal to the stomach and heart, and the pancreas develops from one or more outgrowths with or near the liver evagination. A urinary bladder grows out from the floor of the hindgut in most tetrapods. In amniote embryos, this diverticulum enlarges as the allantois, one of the extraembryonic membranes (Chapter 4).

An ectodermal pocket, the **stomodaeum** (Gr., *stoma* = mouth + *hodaion* = way), invaginates at the front of the embryo and extends toward the archenteron. It forms the oral, or buccal, cavity. When the oral plate between the stomodaeum and the archenteron breaks down, the oral cavity and pharynx become continuous. A similar ectodermal **proctodaeum** (Gr., *proktos* = anus) invaginates at the caudal end of the embryo. It is separated from the archenteron for a short period by the cloacal membrane. When this breaks down, the proctodaeum and caudal end of the archenteron form a chamber called the **cloaca** (L., *cloaca* = sewer), which receives the terminations of the digestive, urinary, and reproductive tracts.

The Mouth and Oral Cavity

The mouth opening and the **oral,** or **buccal, cavity** are variable parts of the digestive tract because vertebrates gather and ingest many kinds of food in numerous ways. Even the position of the mouth opening is not the same (Fig. 16-2). The mouth opening of a jawless hagfish or lamprey develops near the caudal end of the stomodaeum. In most jawed fishes, the mouth opening develops at a level between the hypophyseal sac and the pair of embryonic nasal placodes. Both the entrance and the exit from each nasal sac thus come to lie rostral to the mouth opening. In choanate fishes and tetrapods, the mouth opening is at the level of the nasal placodes. An external nostril leading into each nasal sac lies rostral to the mouth, and an internal nostril from each sac lies within the oral cavity.

Basic Modes of Feeding

A brief discussion of feeding methods will provide perspective because oral structures are adapted to the way food is gathered. (Additional aspects of feeding will be examined with the teeth, jaws, and tongue.) The urochordates, the cephalochordates, and the ammocoetes larva of the lamprey are jawless **suspension-,** or **filter-, feeders.** Some combination of ciliary currents, the movement of paired velar flaps at the entrance to the pharynx, and an alternate expansion and contraction of the buccopharyngeal cavity moves a current of water through the mouth and pharynx. Suspended food particles in the water are entrapped by mucus and carried back into the esophagus. Methods of entrapment vary. In the ammocoetes larva mucus is secreted by goblet cells in the laterally placed pharyngeal pouches, which contain the gills. Tributary strands of mucus from each pouch move into the pharynx lumen and unite to form a longitudinal mucous cord to which food particles adhere. This cord is carried caudally by ciliary action. The endostyle in the floor of the pharynx is not the source of the mucus, as formerly believed, but may contribute digestive enzymes. The first vertebrates lacked jaws, but their increased activity in comparison with urochordates and cephalochordates, and their better developed sensory apparatus, may have allowed them to be predators of small, soft-bodied animals. Some early jawless fishes had bony plates bearing denticles on the borders of their mouths. These denticles may represent the evolutionary beginning of teeth.

The evolution of teeth and jaws enabled vertebrates to feed in other ways, grasping and chewing larger prey. Most fishes are **suction-feeders** and capture prey by drawing the water with the prey into the mouth cavity. However, some fishes are **ram-feeders** that simply overtake the prey with their mouths wide open. Many primitive terrestrial vertebrates transport food within the mouth by rapidly advancing the head relative to the food, the inertia of which keeps the food stationary. Gans (1969) has called this **inertial-feeding,** for the inertia of the food carries it back during a succession of rapid forward darts of the head. Other terrestrial vertebrates use their tongues to transport food to the back of the mouth. Most vertebrates swallow their food whole or in large pieces, but most mammals and some reptiles chew, or **masticate,** their food.

Although vertebrates feed in many ways, suspension-feeding is particularly efficient in habitats where large quantities of relatively small food organisms are available. Many jawed vertebrates from fishes to mammals have re-evolved adaptations for this mode of life. Many convergences occur. All suspension-feeders are aquatic, or at least feed in the water. They must filter a large volume of water to obtain enough food. Suction-feeding or ram-feeding usually ingests the food. Teeth are reduced or lost, and some type of filtering device traps food. Among the jawed suspension-feeders are the basking sharks, manta rays, herring, tadpoles, ducks, flamingos, and the enormous baleen whales. (See Sanderson and Wassersug, 1990, for a review of this topic.)

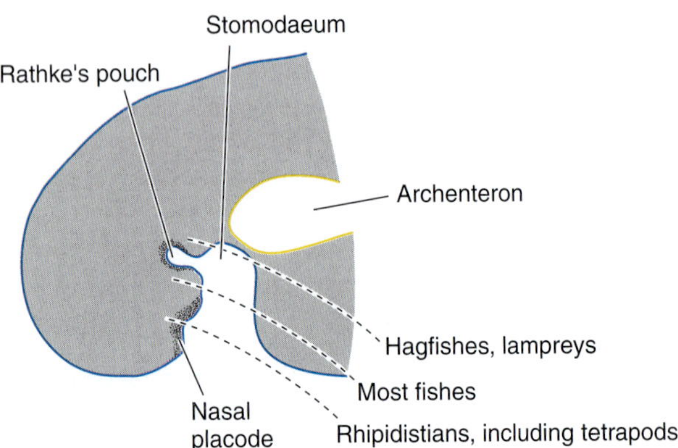

FIGURE 16-2
The stomodaeum and associated structures in a craniate embryo, indicating the level at which the mouth opening develops in different groups. Colors as in Figure 16-1.

Teeth

Living jawless vertebrates lack bony teeth, but horny (keratinous) cones called teeth are associated with the mouth and tongue of adult hagfishes and lampreys. Enamel-like proteins have been demonstrated in the horny teeth of hagfishes. Horny teeth are part of the lamprey's highly specialized method of feeding, in which the animal clings to its prey, rasps its flesh, and sucks the blood and other body fluids.

True teeth evolved along with jaws and are present in some part of the oral cavity of gnathostomes, unless they have been secondarily lost. A representative adult tooth is composed of bonelike dentine covered on the exposed surface by a layer of hard enamel (Fig. 16-3*A*). The structure of these materials is similar to that in bony scales (Chapter 6). The blood vessels and nerves that maintain the tooth enter its base and lie in the **pulp cavity.** The entrance into the pulp cavity is wide during tooth development and growth but becomes relatively narrow in a mature tooth. The part of the tooth above the gum and subject to wear is its **crown;** that embedded in the gum and sometimes in the jaw is its **root.**

The development of the dentine and enamel of teeth is the same as their development in bony scales. Indeed, teeth most likely evolved from the denticular parts of scales (Fig. 6-6*A*) as jaws began to form. The first indication of tooth formation in the embryo is the growth into the gum of a longitudinal ridge of epithelial cells called the **dental lamina** (Fig. 16-3*B*). Underlying **dental papillae** push into the dental lamina and interact with the epidermal cells to form a series of **tooth germs.** The mesenchyme within these papillae is of neural crest origin. The dental lamina soon atrophies, but tooth germs remain in the jaws throughout the life of most vertebrates and during the embryonic period of mammals. The individual teeth develop from **tooth buds,** which are derived secondarily from the tooth germs (Fig. 16-3*C*). The portion of the tooth bud derived from the epidermal dental lamina forms a cap-shaped **enamel organ** that overlies the dental papilla. The central cells of the enamel organ secrete glycoproteins into the intercellular spaces and separate to form a diffuse reticulum of star-shaped cells, but the surface of the enamel organ remains epithelial in nature. The inner layer of the epithelium differentiates as column-shaped **ameloblasts,** which lay down enamel. The adjacent cells of the dental papilla form columnar **odontoblasts,** which produce the dentine.

In fishes, teeth may be distributed throughout the oral cavity and pharynx—on the jaws, palate, tongue,

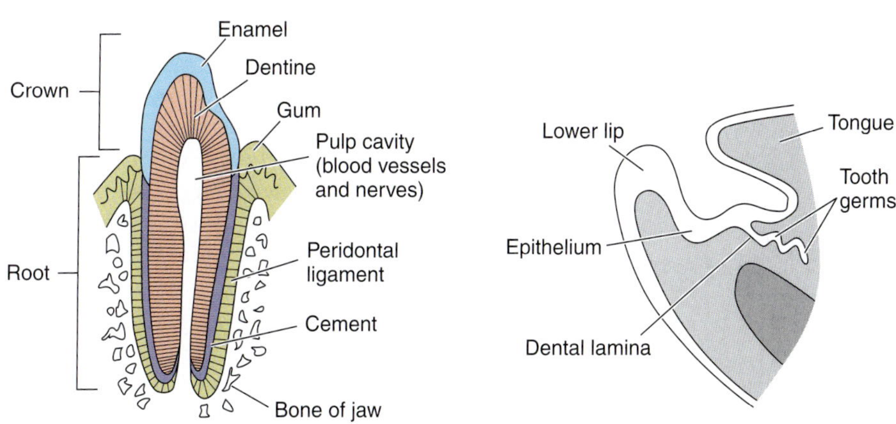

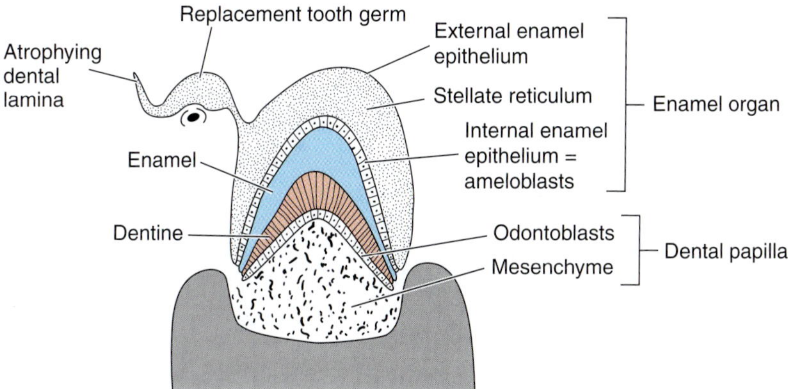

FIGURE 16-3
Tooth structure and development. *A,* A mature mammalian tooth. *B,* An early stage in tooth development. *C,* A later stage, in which the tooth bud is producing enamel and dentine. Blue = enamel; red = dentine; purple = cement; yellow = gum.

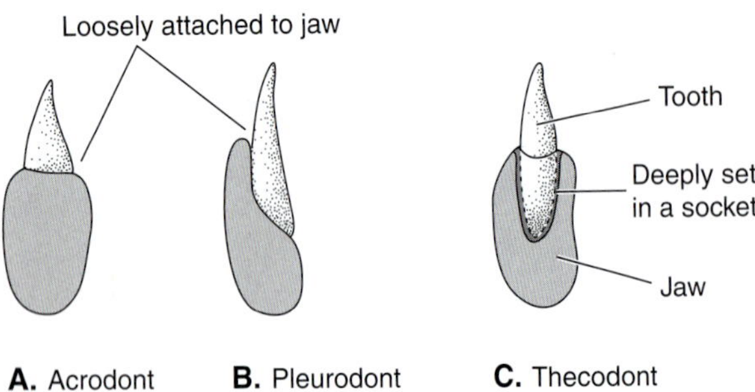

FIGURE 16-4
Vertical sections through the teeth and lower jaw of representative vertebrates, showing methods of tooth placement and attachment to the jaw.

and some of the branchial arches. Tetrapod teeth usually are limited to the jaw margins and sometimes the palate. They are attached to underlying structures by connective tissue fibers that form a **periodontal ligament** and sometimes also by cement. **Cement** is an acellular and avascular type of bone. As can be seen in Figure 16-4, the root is frequently attached loosely to the top or side of the jaw margin. A superficial attachment to the jaw is called **acrodont** or **pleurodont.** Acrodont teeth differ from pleurodont teeth primarily in being attached to the top or inside edge of the jaw rather than to the outside edge and in having a more limited replacement. Teeth also may be set within deep sockets in the jaw, an attachment termed **thecodont** (Gr., *theke* = case), which enables the teeth to withstand strong forces. The thecodont teeth of the therapsid ancestors of mammals and of mammals, together with many other changes in oral structures, are associated with the evolution of increased activity. More food must be ingested, and it must be processed more rapidly. Food is therefore masticated to increase the surface area available for the action of the digestive enzymes.

In most vertebrates, new teeth can develop from the retained tooth germs throughout the animal's life as older teeth fall out or wear out. Such vertebrates are described as **polyphyodont** (Gr., *polyphyes* = manifold + *odont* = tooth). A new tooth begins to form before an old one is lost. As the new tooth matures, the root of the old one is reabsorbed. The tooth loosens and eventually falls out. Loss and replacement are not random processes but follow a complex cycle. Waves of activation inducing new tooth formation travel slowly along the jaws from anterior to posterior. Successive waves of replacement, and sometimes overlapping waves, follow each other closely. Thus, at any one time, areas exist where old teeth have been lost, where newly formed ones are just coming in, and where fully formed teeth are present. Empty sockets tend to be flanked by fully formed or newly erupted teeth. The mechanism ensures that half or more of the teeth are always functional. Not all vertebrates are polyphyodont. Most mammals are **diphyodont** and have only two sets of teeth: milk (or deciduous) and permanent. The toothed whales are **monophyodont** and have only a single set.

Although tooth size may vary, all the teeth of most fishes and most amphibians and reptiles have a similar shape, a condition termed **homodont.** The shape depends on how the teeth are used (Fig. 16-5). In most fishes and early tetrapods, they are simple cones, for they function primarily to prevent caught prey from escaping. Shark teeth are often triangular, with sharp and sometimes serrated edges. Sharks pierce and cut up the organisms on which they prey into large chunks. Piecing and cutting compliant materials, such as flesh, pose unique physical problems different from those that apply to cutting a noncompliant material, such as a piece of wood (Frazzetta, 1988). Frazzetta finds that smooth teeth are particularly well suited for piercing and that serrated teeth have a greater cutting efficiency. Fishes and few tetrapods that feed on shellfish and plants have flattened teeth that may fuse to form crushing tooth plates. Snakes that prey on small mammals that are capable of biting back have evolved methods of prey immobilization. One method utilizes large grooved or hollow teeth—**fangs**—to inject the victim with a poison secreted by modified salivary glands.

Teeth have been reduced or lost where they have no value to a species' method of feeding. Frog tadpoles feed primarily on plant material. Instead of teeth, a tadpole has horny papillae, called **labial teeth,** around the mouth opening that help it cling to plants. Small bits of plant food are scraped up by horny beaks at the front of the jaws and filtered from the respiratory current within the buccopharyngeal cavity. Carnivorous larval salamanders, in contrast, are suction-feeders and have small teeth to hold their prey. Many adult terrestrial frogs and salamanders lack true teeth on the jaw margins. Insect prey is caught by a flick of the tongue and held in the mouth by palatal teeth. A horny sheath on the jaw margins replaces teeth in turtles and present-day birds.

The ingestion and mastication of food by mammals have been accompanied by the evolution of teeth specialized for different functions. Such a dentition with teeth that differ in shape is said to be **heterodont** (Fig. 16-6). Primitive placental mammals have three small **incisor teeth** (abbreviated I) at the rostral end of each side of the upper and lower jaws. These are followed by a single **canine** (C), four **premolars** (P), and three **molars** (M). The number of each type of tooth can be expressed as a dental formula in which the numerator indicates the number on each side of the upper jaw and

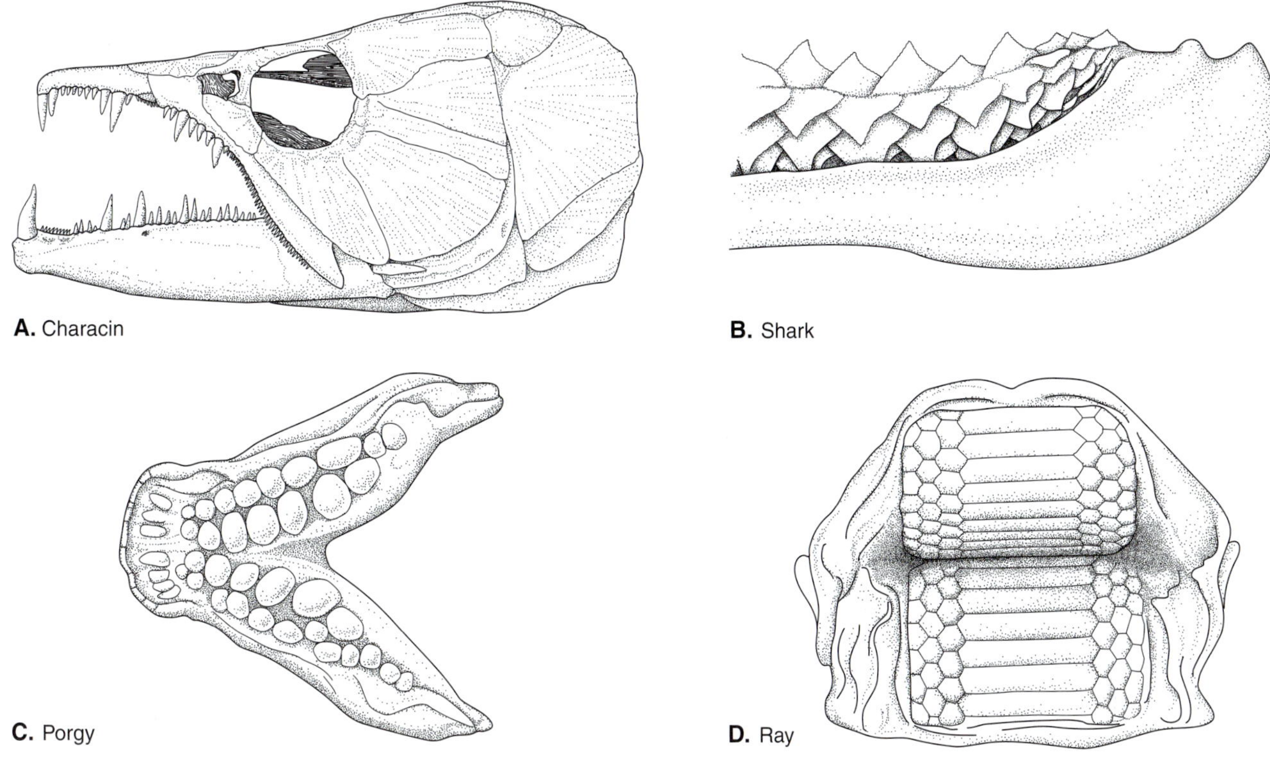

FIGURE 16-5
Examples of different types of teeth in several fishes. *A,* Conical teeth of a pikelike characin (an order of South American and African fishes). *B,* Triangular functional and replacement teeth of a shark. *C,* Peglike teeth of a porgy (a perciform fish). *D,* Crushing tooth plates of a ray. (*A, After Gregory; C and D, after Lagler et al.*)

the denominator, the number in a lower jaw: I 3/3, C 1/1, P 4/4, M 3/3. The number tends to be reduced in more advanced placentals. The dental formula for humans is I 2/2, C 1/1, P 2/2, M 3/3. Marsupials and extinct groups of early mammals are characterized by different dental formulas. For example, the maximum number of teeth that a marsupial possesses is I 5/4, C 1/1, P 2/2, M 4/4.

In most mammals, teeth begin to emerge as the suckling period ends. As the jaw grows larger, deciduous incisors, canines, and premolars emerge. As the jaws continue to grow, these teeth are replaced by larger, permanent ones. The good occlusion that is needed for mastication is maintained. Later in life, the molar teeth erupt in sequence, in the human case at approximately 6 (M_1), 12 (M_2), and 18 (M_3, the "wisdom teeth") years of age. The molars are not replaced, for final jaw length is attained by the time the last one erupts. Precise occlusion and limited tooth replacement are among the diagnostic features or synapomorphies for mammals.

The configurations of particular teeth are adapted to their use. Incisors are typically small, spade-shaped teeth used for cutting, cropping, and picking up food. Rodents have only two pairs (in contrast to the human four pairs) of incisors that are greatly enlarged and specialized for gnawing plant food. The pulp cavity of each one remains open, and the tooth grows throughout life at the same rate that it wears away at the tip. Enamel is limited to the front surface and forms a sharp cutting edge as the softer dentine behind it wears away more rapidly. The tusks of elephants and spiraled tusks of narwhals are examples of highly modified incisors.

Canines usually are large, conical teeth used in seizing, piercing, and killing prey. They are particularly large in carnivores (Fig. 16-6). The canines of some male pigs are modified as defensive tusks. The crowns of human canine teeth resemble and function as incisors, but the roots are canine-like and much larger than those of incisors.

Primitively, the premolars are puncturing teeth, and the molars have a combined cutting and crushing action. These teeth, collectively called the **cheek teeth,** are the most complex and variable. Except for the first premolar, each has two or more roots, and the crown bears a complex pattern of conical cusps connected by sharp crests. The molars of the earliest known mammals

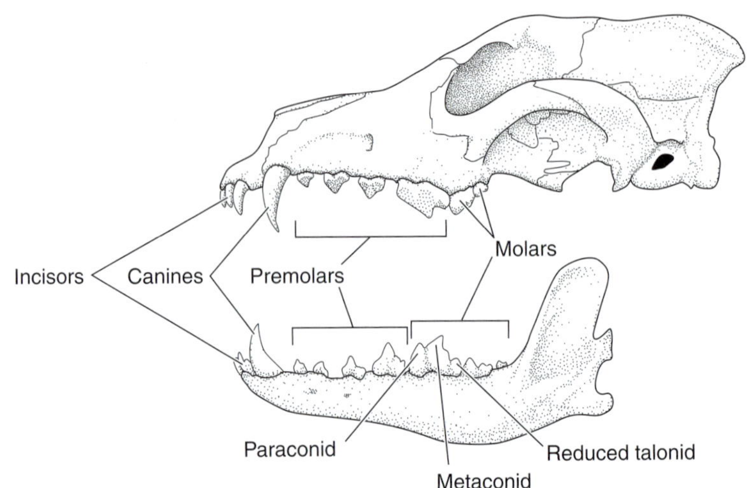

FIGURE 16-6
A lateral view of the skull and lower jaw of the African hunting dog, *Lycaon,* to illustrate the types and number of teeth in eutherian mammals. This dog has lost the last upper molar that is present in more primitive eutherians. The last upper premolar and first lower molar of this species and in most recent terrestrial carnivores are specialized as shearing carnassial teeth. (*After Vaughan.*)

(the late Triassic †morgonucodontians) had one large, central cusp flanked by a pair of smaller ones (Fig 16-7*A*). At first, these cusps were in a linear sequence, but in the more derived early mammals close to the ancestry of therians (the †symmetrodonts), the cusps had shifted to form a triangle with the primary cusp at the apex (Fig. 16-7*B*). The apex of the upper molar was directed lingually; that of the lower molar turned labially. The teeth were spaced such that a lower molar fit between two upper ones, and their crests formed good shearing surfaces as the teeth slid past each other (Fig. 16-7*C*). The triangular configuration, called a **trigon** in the upper molar and a **trigonid** in the lower one, increased the number of shearing facets compared with the more primitive, linear arrangement and thus made for more efficient mastication.

An additional cone, the **protocone,** evolved on the lingual apex of the upper molar in the ancestors of marsupials and placental mammals. Two of the original three cusps remained near the base of the trigon as the **paracone** (rostrally) and **metacone** (caudally; Fig. 16-7*D*). Small accessory cones may be present. The original three cusps of the lower molar remained as the **protoconid, paraconid,** and **metaconid;** a low heel, known as the **talonid,** evolved from a ridge on the caudal border of the trigonid. The central part of the talonid formed a shallow basin that was flanked by small cusps: an **entoconid** on the lingual side, an **ectoconid** on the labial side. During tooth occlusion, the protocone of the upper molar fell on the talonid basin, thereby providing some crushing action that supplemented the shearing action as the crests of the trigon and trigonid slid past each other. These types of molars, which are called **tribosphenic,** occur in all marsupial and placental mammals. Tribosphenic molars in their original form still occur in primitive insectivorous placentals but become modified in many derived taxa as diets changed.

The shearing action of the cheek teeth is accentuated in carnivores (Fig. 16-6). In recent species of the Carnivora (e.g., cats and dogs), the fourth upper premolar and the first lower molar are specialized as cutting **carnassials** (L., *carnis* = flesh). The cusps of the fourth upper premolar are linearly arranged and form a good shearing face on the lingual side of the tooth. This upper shearing face crosses the shearing ridge on the labial side of the lower molar, which is formed by the paraconid and metaconid. The protoconid and talonid are reduced in most carnivores and have been lost in cats. Postcarnassial teeth may be reduced or lost.

The crushing action of the molars is greater in omnivores and herbivores. A **hypocone** evolves on the caudolingual corner of the upper molar beside the protocone. This converts the trigonid into a square-shaped tooth (Fig. 16-8, left side). The lower molar also becomes squared by the loss of the paraconid and the elevation of the talonid to the height of the rest of the trigonid (Fig. 16-8, right side). The cusps of both upper and lower molars become round hillocks rather than sharp cones. Crushing and grinding surfaces are thus formed. Humans, pigs, and primitive herbivores have **bunodont molars** of this type. The overall height of the tooth does not increase, and the tooth is described as **low crowned** (Fig. 16-9*C*).

The four cusps of the bunodont molar, originally separate, and certain accessory cusps fuse to form ridges, or lophs, in more specialized herbivores. **Lophodont molars** of this type occur among perissodactyls, elephants, and rodents (Fig. 16-9*A*). In artiodactyls, the original four cusps remain independent but become elongated rostrocaudally to become crescent shaped. This is a **selenodont molar** (Fig. 16-9*B*). Lophodont and selenodont teeth are functionally identical. Frequently, many of the premolars

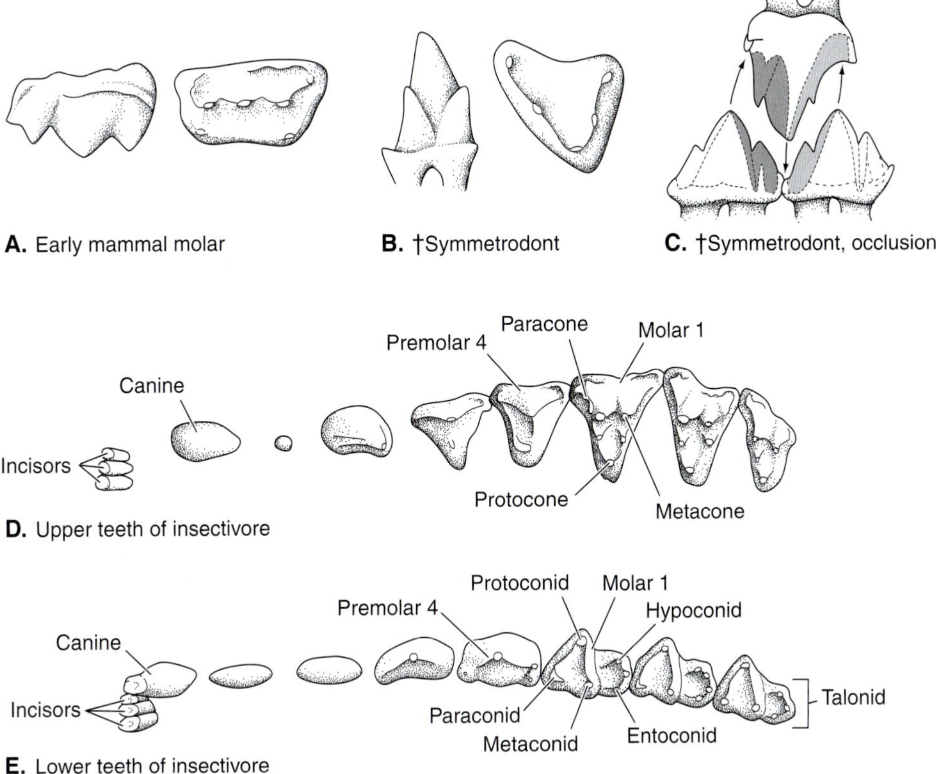

FIGURE 16-7
The evolution of mammalian molar teeth. *A,* The upper molar of an early mammal (a †morganucodontian) in the lateral (*left*) and occlusal (*right*) views. *B,* Teeth of a †symmetrodont close to the origin of therian mammals: lateral view of the lower left molar (*left*); occlusal view of the upper right molar (*right*). *C,* Lateral view of the occlusion of the upper and lower molars of a symmetrodont. *D* and *E,* The dentition of an insectivore with tribosphenic molars. By showing the upper right teeth in the top row (*D*) and the lower left teeth in the bottom row (*E*), one can visualize how the teeth occlude. Notice that the protoconids of the lower molars fit between the protocones of the upper molars when the teeth come together.

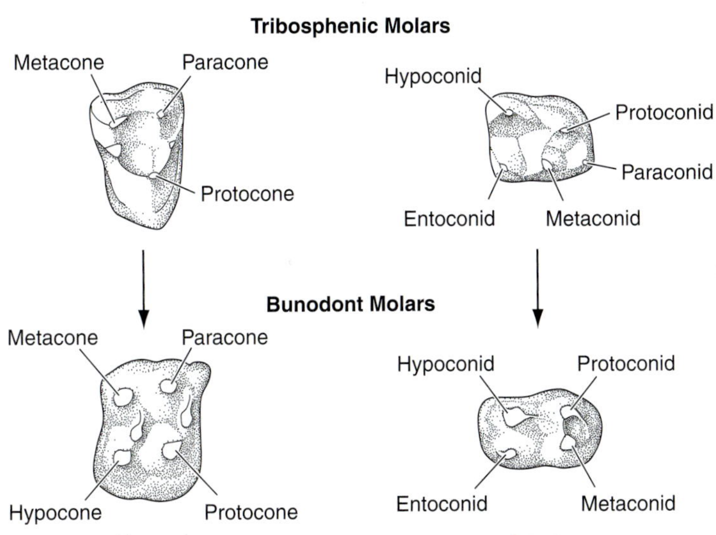

FIGURE 16-8
The evolution of molars. Primitive tribosphenic molars are shown in the top row, and bunodont molars, in the bottom row. Upper right molars are shown in the left column, and lower left molars, in the right column. (*After Romer and Parsons.*)

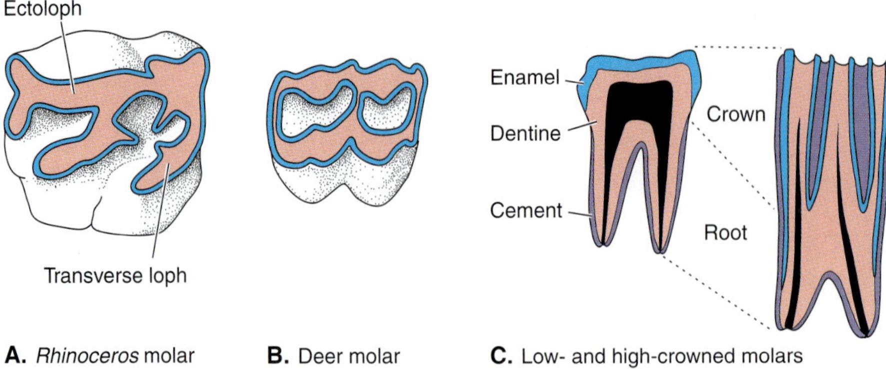

FIGURE 16-9
Some adaptations of molar teeth for grinding. *A,* Occlusal view of the lophodont molar of a rhinoceros. *B,* Occlusal view of the selenodont molar of a deer. *C,* Vertical sections through a low-crowned molar (*left*) and a high-crowned, or hypsodont, molar (*right*). Colors as in Figure 16-3. (*After Vaughan.*)

in these groups of mammals become molarized and assume the structure of molars. Sometimes premolars and molars can be distinguished only by their pattern of embryonic development. (Recall that deciduous premolars are replaced by permanent ones; molars are not replaced.) The cheek teeth of species that feed on gritty grasses or other abrasive plant food are subject to considerable wear. Adaptations for this are an increase in the height of the ridges or crescent-shaped cusps and a migration of cement from the root over the surface of the tooth and into the valleys between the ridges and cusps. This is a **high-crowned,** or **hypsodont, molar** (Fig. 16-9*C*). The tooth becomes more resistant to wear, and more tooth is available to wear away. As the ridges wear away, alternating layers of cement, enamel, and dentine are exposed. Because these materials have different degrees of hardness, enamel being the hardest and dentine the softest, differential wear occurs. Sharp ridges of enamel become flanked on one side by cement and on the other by dentine.

The Feeding Mechanisms of Vertebrates

The study of feeding mechanisms in vertebrates has undergone great progress because of the introduction of high-speed video to determine bone movements, simultaneous electromyographic recording of muscle activity, and the use of strain gauges to permit accurate determination of deformation. These new experimental approaches have yielded a more comprehensive understanding of not only the precise functions of structural systems but also the transformations of the skull and associated muscles during vertebrate evolution. Here we will discuss the vertebrate head from a functional perspective.

Feeding in the Aquatic Medium: Fishes and Amphibians

The starting point of the evolution of the feeding mechanism in ray-finned fishes (Actinopterygii) is the pattern found in the primitive "†paleoniscoid" fishes (Fig. 16-10) in which the mouth is opened by epaxial muscles that lift the head (Fig. 16-10*D*). At the same time ventral body muscles (hypaxial) and the interhyoideus (sternohyoideus) muscle, which runs from the pectoral girdle to the hyoid, pull down the lower jaw. These actions produce a large gape but no increase in volume of the mouth cavity. It is hypothesized that these fish captured their prey by overtaking it with their widely opened mouth and a sudden forward movement. Once the prey was overtaken, the mouth was closed by a substantial set of jaw-closing muscles, the adductor mandibulae (Fig. 16-10*C*). Because the predator overtakes the prey by a sudden forward movement, this mode of feeding is called ram-feeding.

Sharks also open their jaws by lifting the head with epaxial muscles while the mandible is pulled down by the ventral muscles, which run between the hyoid and pectoral girdle (rectus cervicis or coracohyoideus or sternohyoideus) and between the mandible and pectoral girdle (the coracomandibularis; Fig. 16-11*A*). Some sharks can produce an enormous gape while overtaking their prey. Most also will protrude their upper jaws while simultaneously enlarging the volume of the mouth and pharynx. Such movements produce suction, during which the prey is drawn into the mouth cavity. Most sharks capture their prey by a combination of ram-feeding and suction-feeding. Adduction of the jaw results in a formidable bite due to the very large adductor mandibulae and preorbitalis muscles (Fig. 16-11*B*). During the recovery phase, the volume of the mouth and pharynx is restored by the levator palatoquadrati, interhyoideus, and intermandibularis muscles, which pull the hyoid and palatoquadrate forward and upward. The insertion site of the adductor mandibulae

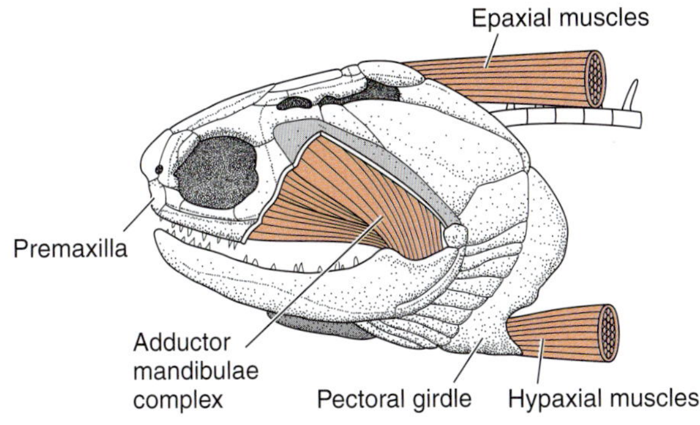

A. Muscles used in feeding

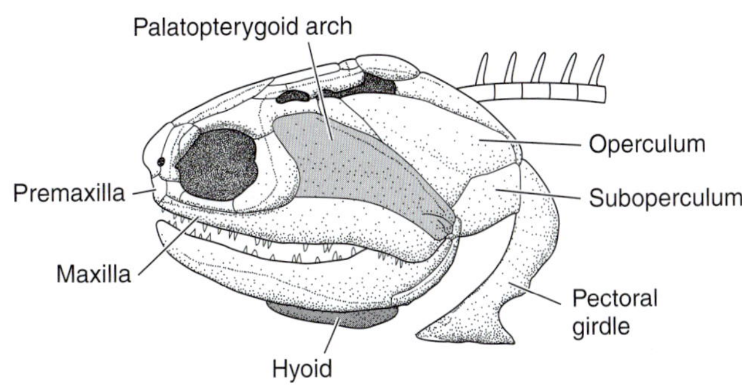

B. Skeletal components

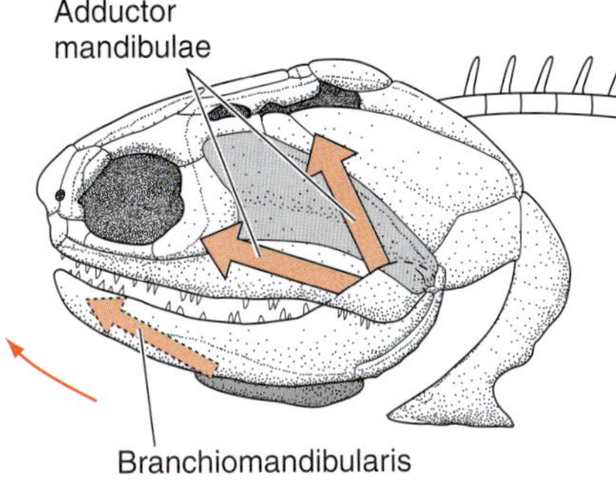

C. Jaw closing muscles

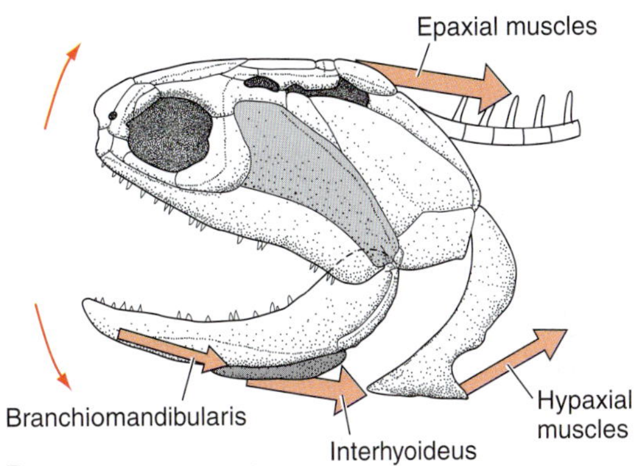

D. Jaw opening muscles

FIGURE 16-10
Lateral views of the head of a fossil primitive actinopterygian (a "†paleoniscoid"), showing the feeding mechanism. *A,* The head with a reconstruction of the muscles used in feeding. *B,* The major skeletal components of the head. *C,* The lines of actions of muscles (*wide orange arrows*) that close the jaw. *D,* The lines of actions of muscles that open the jaw (*wide orange arrows*). (*Modified from Lauder.*)

in relation to the jaw joint is such that it produces a great torque around the jaw joint and therefore an optimized out-force at the teeth. The dogfish (*Squalus acanthias*) and some other sharks (especially the predacious great white shark, which is the largest living marine carnivorous fish in the world) can produce very large gapes and biting forces with sharp, serrated teeth. Wilga and Motta (1998) have made a comprehensive electromyographic and kinematic analysis of the feeding mechanism of a squaliform shark.

A new skull design and the dermatocranium evolved in the actinopterygian fishes and is characterized by a progressively greater number and mobility of bony elements (Fig. 16-12). During the evolution of the ray-finned fishes, the following structural changes occurred:

1. In the upper jaw, the premaxilla becomes greatly enlarged, highly mobile, and the only toothed element, while the highly mobile maxilla loses its teeth and can make forward and backward swinging movements (Fig. 16-12).
2. The palatopterygoid arch, which forms the jaw suspension, changes from an oblique (Fig. 16-10*B*) to a more vertical position so that the volume of the mouth cavity is substantially increased.
3. Concomitant with the vertical orientation of the palatopterygoid arch, the jaw joint was shifted from a position posteroventral to the orbit (Fig. 16-10*B*) to an anteroventral location (Fig. 16-12*B*).
4. The gill cover (operculum), which was originally rigidly attached to the skull (Fig. 16-10*B*), is now separate from the skull and articulates with a ball-and-socket joint to the palatopterygoid arch (Fig. 16-12*A*). It can rotate around its joint and flare outward. Its movements are transferred to the other two bones of the gill cover (i.e., the suboperculum and interoperculum) and by a ligament to the posterior corner of the mandible.
5. The hyoid becomes connected to the gill cover by a ligament, so that movement and forces of the hyoid are also indirectly transferred to the lower jaw. In this advanced design, the mouth is opened by lifting

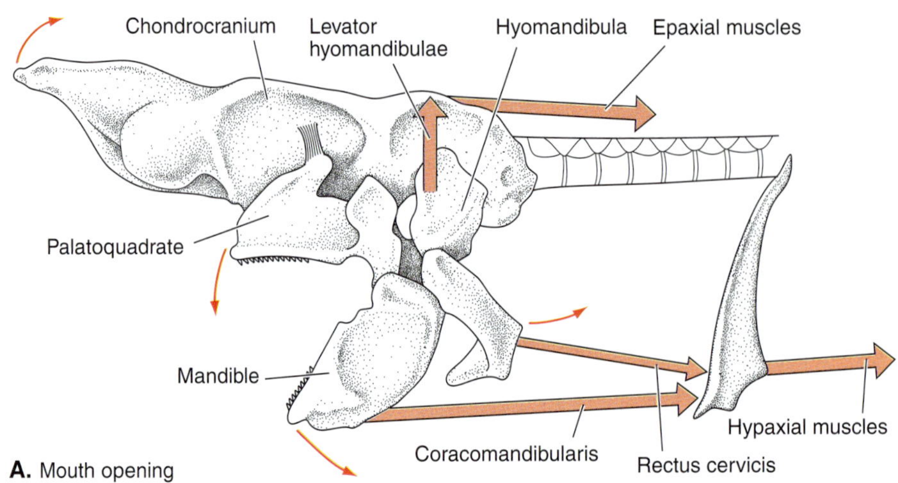

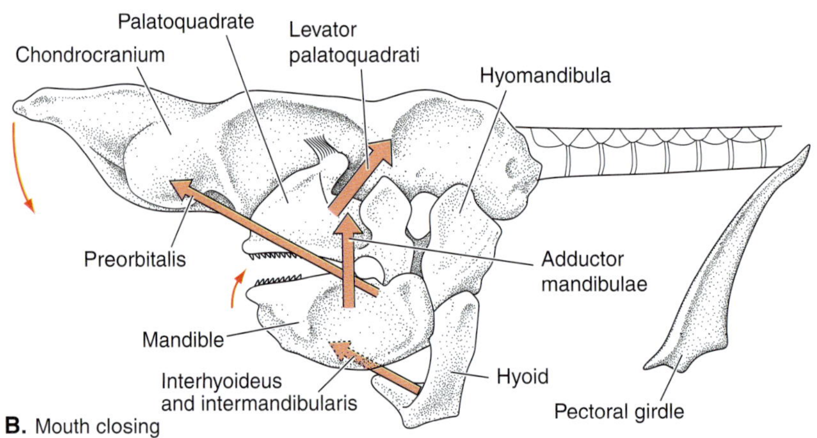

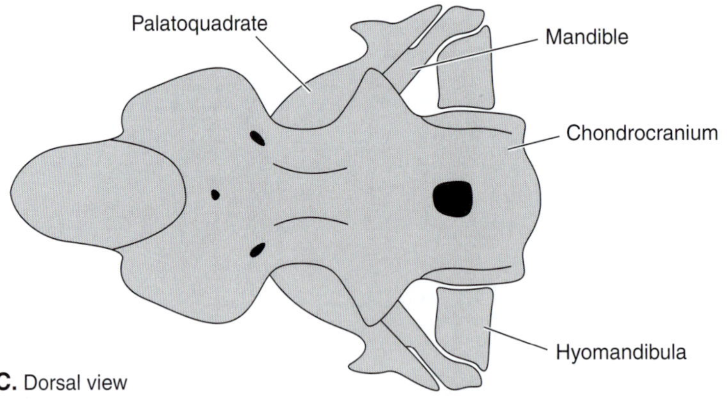

FIGURE 16-11
Lateral views of the head of a shark, showing the feeding mechanism. Wide orange arrows show the lines of actions of muscles, with the width of the arrows depicting the relative level of muscle activity. Narrow arrows show the direction of movement of skeletal components. *A,* Mouth opening. *B,* Mouth closing. *C,* Dorsal view.

of the head by action of the epaxial muscles, while simultaneous action of the hypaxial and sternohyoideus muscles pulls the hyoid backward and downward (Fig. 16-12*B*). This backward and downward movement of the hyoid is transmitted by a ligament to the interoperculum, which pulls by a ligament on the lower, posterior corner of the mandible, causing it to drop.

In advanced teleost fishes, a third mechanism to open the mouth evolved. It involves the rotation of the gill cover by the action of a muscle that lifts the operculum (levator operculi). This rotation of the gill cover is transmitted by the ligament to the posterior corner of the mandible, causing it to drop (Fig. 16-12*B*). While the lower jaw is lowered, the premaxilla slides forward. Such a movement is

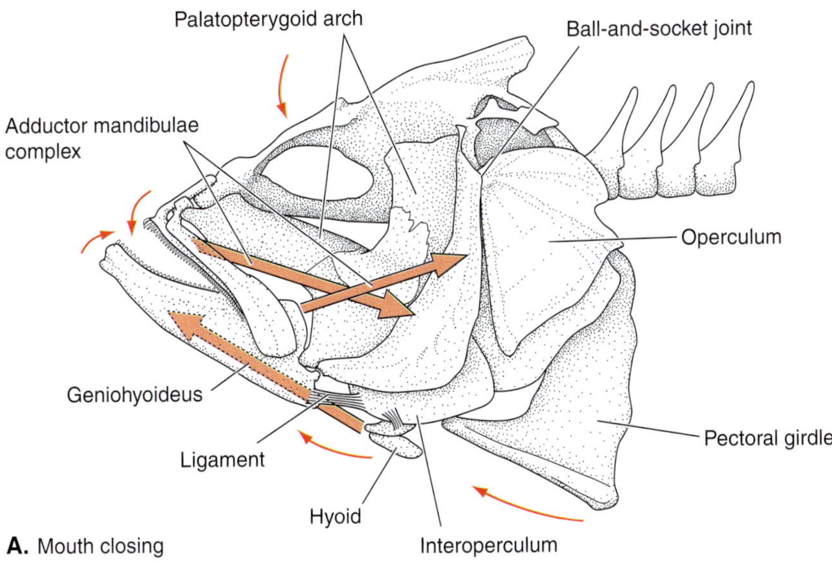

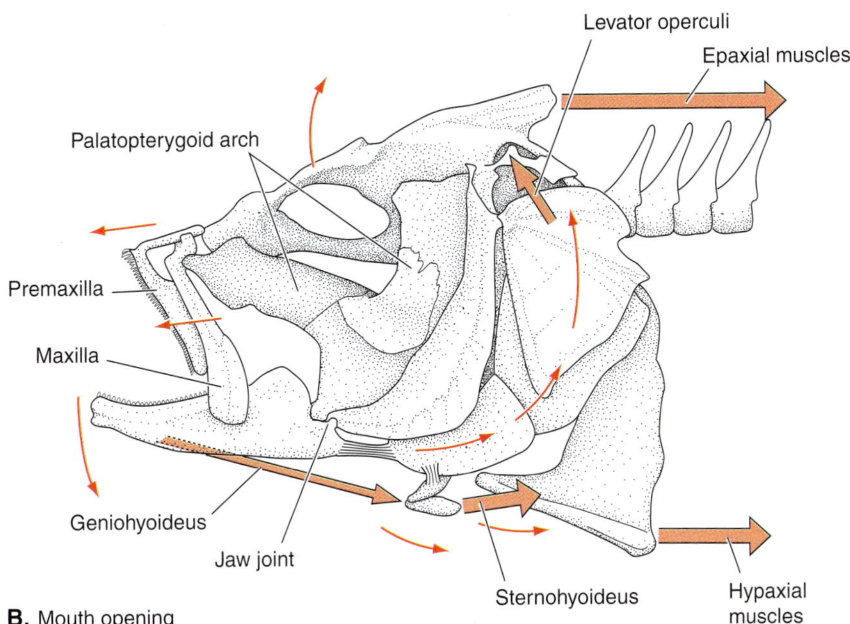

FIGURE 16-12
Lateral views of the head of an advanced teleost fish, showing the feeding mechanism. Wide orange arrows show the lines of actions of muscles, with the width of the arrows depicting the relative level of muscle activity. Narrow arrows show the direction of movement of skeletal components. *A,* Mouth closing. *B,* Mouth opening.

called jaw protrusion. At the same time, the volume of the buccal cavity increases significantly because the palatopterygoid arch, which forms the sidewalls of the mouth cavity, flares out while the floor of the buccopharynx, formed by the hyoid, drops. Such a sudden increase in volume of the buccopharyngeal cavity when the mouth is opened draws water into the mouth. In this way, prey is sucked into the mouth. Once the prey has entered the mouth cavity, the jaws are closed by actions of the adductor mandibulae muscles (Fig. 16-12*A*), which is followed by a recovery phase in which the hyoid is drawn forward and up to its original position by the geniohyoideus (which is the coracomandibularis in the shark) muscle. The original volume of the mouth is restored and water escapes from under the gill cover while the prey is retained.

This basic functional design with highly mobile components in the skull operated by a multitude of muscles is retained throughout the adaptive radiation of the over 25,000 species of ray-finned fishes, even though numerous variations have evolved. The unsurpassed efficiency and versatility of the combination of suction-feeding and ram-feeding in water have enabled these fishes to exploit any conceivable food resource from the 9000-m depths of the abyss to the 3000-m heights of the mountain streams. The gapes of bottom-feeders are directed ventrally, whereas the mouths of surface-feeders are directed upward. Fish eaters have long jaws, whereas algae scrapers have short jaws. The vast majority of fishes have evolved a nearly circular gape, which is the most efficient design for suction, analogous to a pipette.

Aquatic salamanders also feed by suction. Suction-feeding patterns in salamanders resemble those of actinopterygian fishes. The skull is lifted by the epaxial muscles coincident with the lowering of the lower jaw by the depressor mandibulae muscle, which runs from the quadrate to insert on the articular bone of the mandible, caudal to the jaw joint (Fig. 16-13). A depressor mandibulae is a new muscle in tetrapods, not present in fishes. Mouth opening is further assisted by contractions of the hypaxial and rectus cervicis (sternohyoideus) muscles that greatly enlarge the volume of the mouth and pharyngeal cavities (Fig. 16-13*B*). The sudden increase in mouth and pharyngeal volume draws the water and prey into the mouth. The mouth then is closed by the adductor mandibulae muscle complex, which pulls up the lower jaw. As in fishes, mouth closure is followed by a recovery phase, during which the position of the hyoid is restored by the geniohyoideus muscle (Fig. 16-13*A*) and the water is forced out of the mouth.

Feeding in the Aquatic Medium: Turtles

Ancestral turtles, like all amniotes, were terrestrial. But many groups have readapted to an aquatic habitat and evolved prey-capture mechanisms that resemble those of fishes and amphibians in many, but not all, ways. Aquatic-feeding turtles resemble fishes and amphibians in using large-amplitude hyoid depression to expand the buccopharyngeal cavity and draw water in. But the prey organisms are not sucked in; rather, they are captured by a simultaneous rapid extension of the neck and head (ram-feeding). The suction created by hyoid depression appears to hold the prey in place and prevents the rapid forward movement of the head from pushing it away. Summers et al. (1998) call this **compensatory suction.**

Swallowing in Aquatic Anamniotes

Once prey or food has been captured, it must be transported from the jaws to the pharynx and esophagus. Aquatic anamniotes use a hydraulic transport. In both fishes and aquatic amphibians, water currents within the mouth and pharynx transport prey. These currents are created by repeated cyclical movements of the jaws and hyoid that closely resemble the mouth opening, closing, and recovery phases during prey capture.

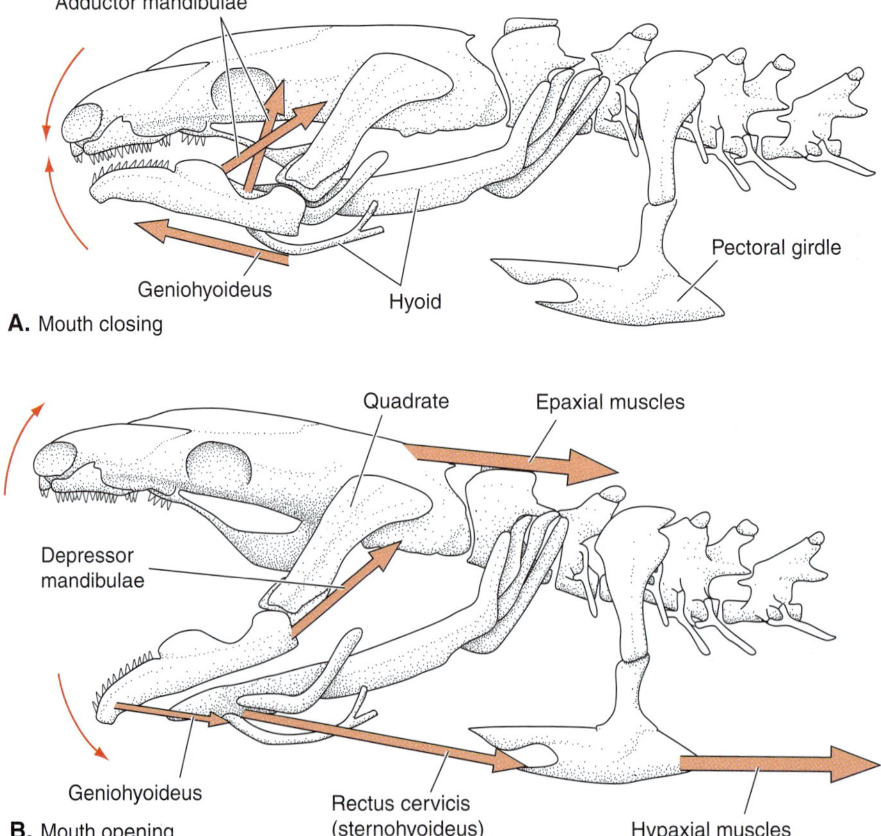

FIGURE 16-13
Lateral views of the head of an aquatic salamander, showing the feeding mechanism. Wide orange arrows show the lines of actions of muscles, with the width of the arrows depicting the relative level of muscle activity. Narrow arrows show the direction of movement of skeletal components. *A*, Mouth closing. *B*, Mouth opening. (*Modified from Reilly and Lauder.*)

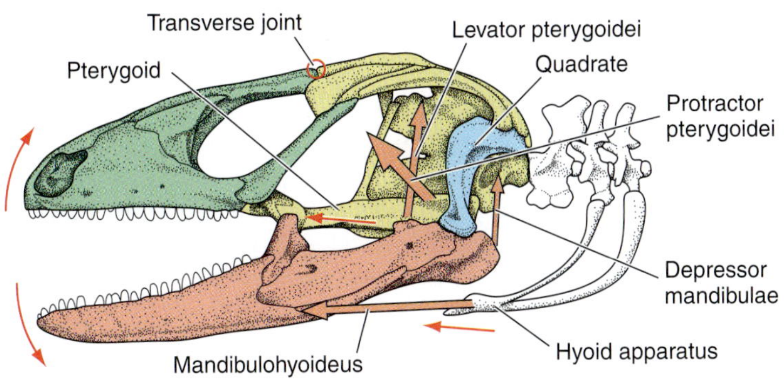

A. Slow opening

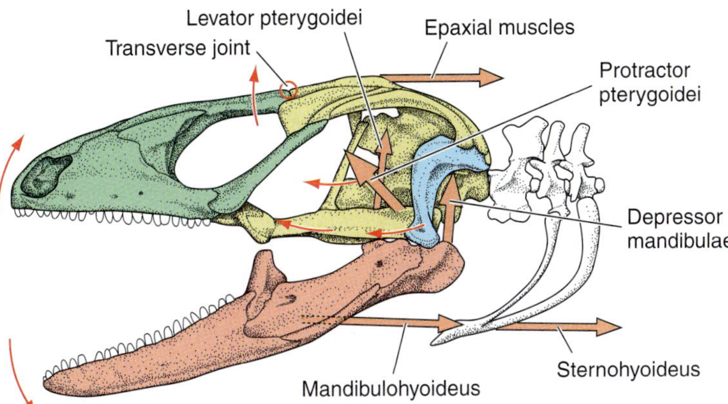

B. Fast opening

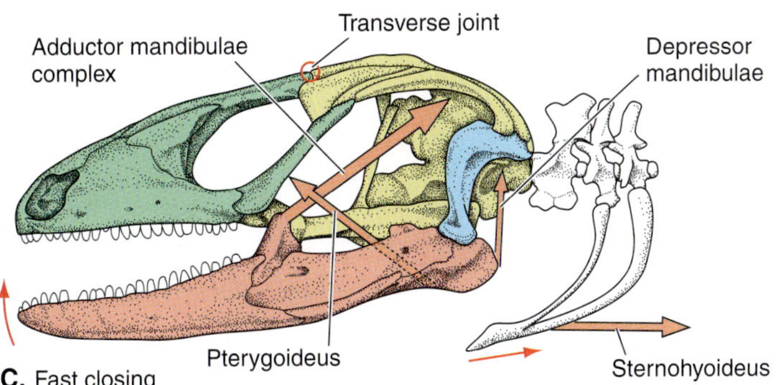

C. Fast closing

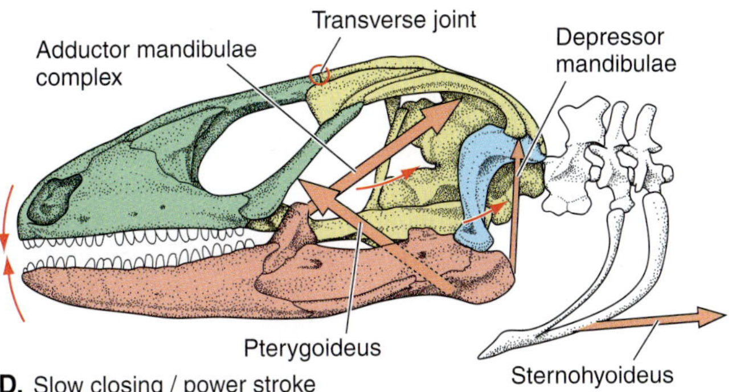

D. Slow closing / power stroke

FIGURE 16-14

Lateral views of the head of a lizard, showing the feeding mechanism. Wide orange arrows show the lines of actions of muscles, with the width of the arrows depicting the relative level of muscle activity. Narrow arrows show the direction of movement of skeletal components. *A*, Slow opening. *B*, Fast opening. *C*, Fast closing. *D*, Slow closing/power stroke.

FOCUS 16-1 *Cranial Kinesis or Intracranial Mobility*

As we have seen, "cranial kinesis" is the term for the mechanism of relative motion between adjacent parts of the skull (e.g., between braincase and muzzle; Fig. 16-14*B*). Cranial kinesis is widespread among a great many vertebrates but is lost in mammals. It is a primitive feature of tetrapods, and it has undergone different elaborations in many evolutionary lineages. In vipers and pit vipers (Figs. A and B), cranial kinesis is most extremely expressed. The braincase itself is solid, but the loss of the temporal roof frees the squamosal bone to swing laterally and rostrally. The quadrate articulates movably with the squamosal. A ligament that permits them to separate connects the rostral ends of the two mandibles. Contraction of the protractor pterygoidei and levator pterygoidei muscles pulls the palatopterygoid forward and up. The highly specialized fang-bearing maxilla is pushed forward and rotates, thereby erecting the fang. At the same time, the squamosal moves upward and the quadrate swings anteriorly and laterally while the mandible is lowered by the depressor mandibulae, producing an enormous gape.

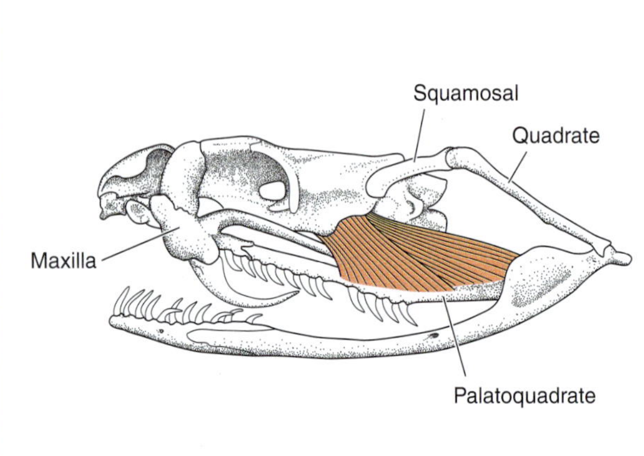

A. Rattlesnake with fangs folded

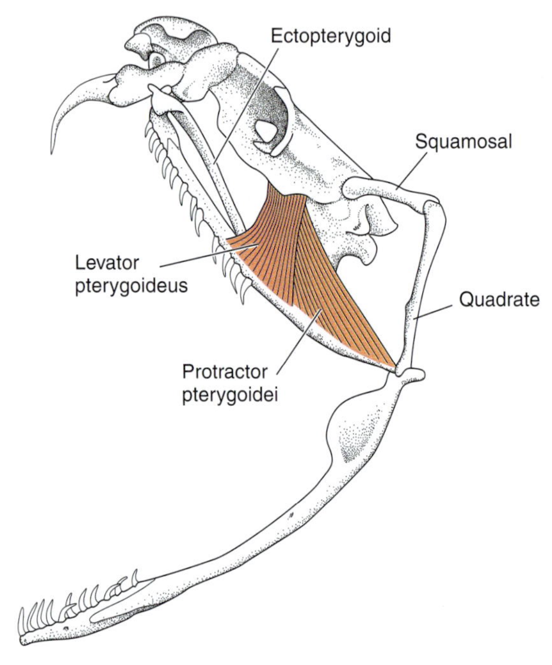

B. Rattlesnake with fangs erected for striking

In teleosts the dorsal and most posterior ventral elements of the gill arches are specialized to form toothed structures, the pharyngeal jaws, with retractor and protractor muscles. These pharyngeal jaws can rake prey and food into the gullet (esophagus), working in conjunction with hydraulic transport. In many fishes, the pharyngeal jaws and muscles become specialized so they can function not only for transport of prey and food, but also for mastication prior to swallowing.

Terrestrial Feeding: Amniotes

One of the major features to change during the aquatic-to-terrestrial transition in vertebrate evolution is the feeding mechanism. Significantly different designs of the feeding apparatus are required because air is less dense than is water, so that prey or food in air cannot be sucked in. As we have seen, suction is almost universal in aquatic vertebrates, but it has no role in the prey capture mechanism of terrestrial tetrapods. The essential difference between aquatic and terrestrial feeding is that, in land-dwelling vertebrates, coordinated movements of the mandible and tongue replace water flow.

The kinematics in a lizard serve as a model of a generalized terrestrial vertebrate feeding cycle (Fig. 16-14). In all terrestrial vertebrates, including amphibians; reptiles and birds with kinetic skulls; and mammals, which lack cranial kinesis, the feeding cycle consists of four stages: (1) slow opening, (2) fast opening, (3) fast closing, and (4) slow closing/power stroke. The cycle begins with **slow opening** of the mandible. In lizards and other amniotes with kinetic

Closing of the mouth results in depression and a retraction of the palatopterygoid arch. During swallowing, the palatopterygoid arch is alternately protracted and retracted. The arch in effect acts as a ratchet, and each side of the arch can be protracted and retracted independently. Together, they pull the large prey into the esophagus. Cranial kinesis enables snakes to swallow whole prey of enormous size after it has been immobilized by constriction or by injecting poison.

Kinetic skulls also reach a high degree of development during the spectacular adaptive radiation of all modern birds. Functionally, the avian skull consists of four units: (1) the braincase, (2) the bony palate and the quadrate, (3) the upper jaw, and (4) the mandible (Figs. C and D). The quadrate allows for anteroposterior movements. The bony palate also slides back and forth. A hinge joint is present at the frontonasal junction. The protractor pterygoidei activates kinesis by imparting a forward push to the base of the upper jaw, which causes the bill to rotate upward about the transverse hinge joint at the frontonasal junction. The depressor mandibulae lowers the mandible. The pterygoideus and adductor mandibulae muscles effect closing of the mandible and lowering of the upper jaw.

Scientists hypothesize that cranial kinesis in birds enhances the manipulatory repertoire of the jaws. A versatile and precise grip on food is especially valuable in birds for opening seeds and shucking out the edible parts and for gripping other materials in nest building, preening, and other activities. Cranial kinesis also offers shock absorption in birds such as woodpeckers, and allows the bird to maintain a steady line of sight while feeding.

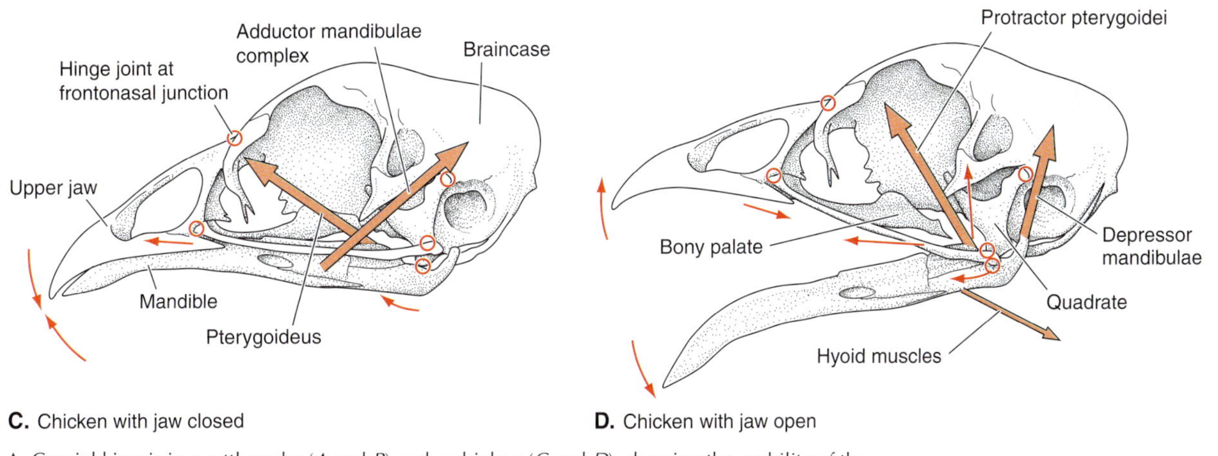

C. Chicken with jaw closed

D. Chicken with jaw open

A. Cranial kinesis in a rattlesnake (*A* and *B*) and a chicken (*C* and *D*), showing the mobility of the palatopterygoid and quadrate complex and the hinge joint at the frontonasal joint.

skulls, the snout lifts up in relation to the braincase. This movement of one part of the skull in relation to the braincase is called **cranial kinesis,** which is made possible by the presence of a transverse hinge across the skull roof. In lizards this transverse joint is located just above the posterior margin of the orbit (Fig. 16-14*A*). Kinesis is effected by actions of the protractor pterygoidei and levator pterygoidei muscles, which draw the pterygoid bones forward and upward relative to the braincase. Because the pterygoids are firmly articulated to the anterior components of the skull, the snout is lifted. The mandible is slightly lowered by moderate contraction of the depressor mandibulae, while the hyoid is pulled forward by the mandibulohyoideus (geniohyoideus) muscle. This stage is followed by **fast opening,** during which a sudden and rapid opening of the mouth to maximum gape occurs (Fig. 16-14*B*). Fast opening is caused by strong contractions of the depressor mandibulae, sternohyoideus, and mandibulohyoideus muscles. After fast opening, the **fast-closing** mechanism is activated by strong contractions of the adductor mandibulae and pterygoideus muscles (Fig. 16-14*C*). The mandible is lifted and the gape decreases rapidly. The final stage is **slow closing/power stroke,** during which the snout is depressed by the strong action of the pterygoideus muscle, which pulls the pterygoid bones caudally and downward (Fig. 16-14*D*). At the same time continuous strong contraction of the adductor mandibulae complex causes a strong bite against the prey or food. In some reptiles the power stroke includes cutting, slicing, and crushing of the prey. Cranial kinesis in snakes and birds is discussed in Focus 16-1.

The four stages in the feeding cycle also are found in all mammals studied, including humans. **Slow opening** begins with actions of both the anterior and posterior bellies of the digastric muscle (Fig. 16-15*A*). The role of the digastric in jaw opening is a new feature characteristic of mammals. It replaced the older system of jaw opening in reptiles and amphibians in which the depressor mandibulae muscle pulls on a process of the articular bone of the mandible. In mammals the articular bone is transformed into an auditory ossicle. Slow opening is primarily effected by the digastric. The hyoid moves forward because of contraction of the geniohyoideus. As in reptiles, slow opening is followed by **fast opening,** in which the gape increases rapidly and the sternohyoideus and geniohyoideus muscles become active, pulling down the mandible over a large distance (Fig. 16-15*B*). This action is made possible because the hyoid is stabilized by simultaneous contractions of the anterior and posterior bellies of the digastric. In the next stage (**fast closing**), the mandible is lifted rapidly over a great distance by the contractions of the masseter and pterygoideus muscles (Fig. 16-15*C*). The final stage, **slow closing/power stroke,** is characterized by strong contractions of the masseter, pterygoideus, and temporalis muscles, while the hyoid moves backward (Fig. 16-15*D*). During this stage, food is cut into smaller pieces.

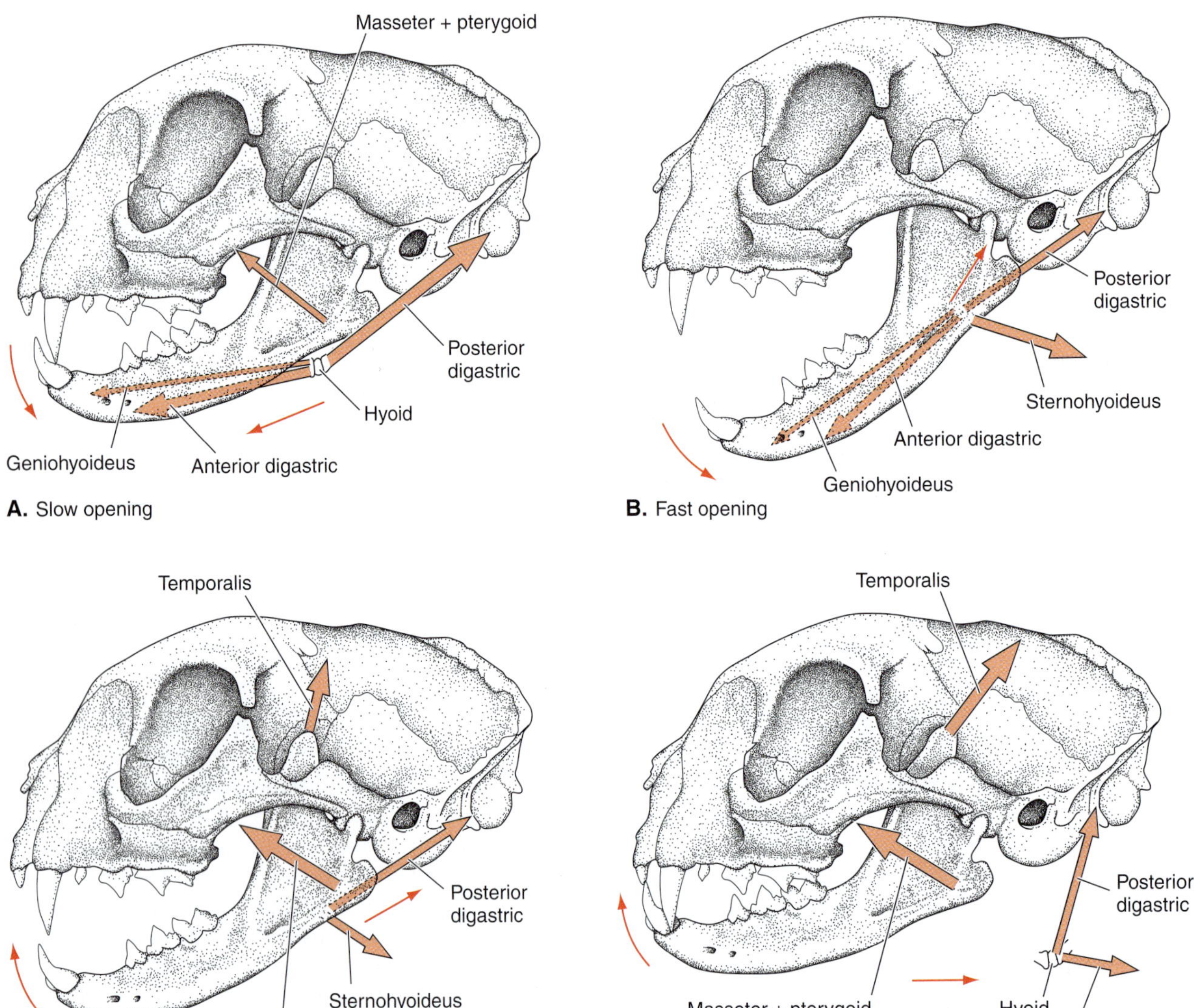

FIGURE 16-15
Lateral views of the head of a cat, showing the feeding mechanism. Wide orange arrows show the lines of actions of muscles, with the width of the arrows depicting the relative level of muscle activity. Narrow arrows show the direction of movement of skeletal components. *A*, Slow opening. *B*, Fast opening. *C*, Fast closing. *D*, Slow closing/power stroke.

Mastication requires a precise positioning of the lower teeth relative to the upper ones, and the positions may change as food is gathered and processed. A forward movement of the lower jaw enables rodents to engage the incisor teeth when they gnaw food; then, a backward movement of the lower jaw enables them to engage the cheek teeth and grind the food. Incisors and cheek teeth are not engaged at the same time. Rodents and elephants grind their food by fore and aft movements of the lower jaw, whereas most other herbivores chew by side-to-side movements. A carnivore moves its lower jaw somewhat laterally to engage the carnassial teeth first on one side and then on the other before masticating. A mammal usually chews on only one side of its mouth at a time.

Teeth, the shape and size of the jaws, the shape and position of the jaw joint, and the size and arrangement of the jaw muscles are adapted to the type of food eaten. A comparison of the jaw mechanics of a representative carnivore and herbivore is a good example (Fig. 16-16). The jaw joint of the carnivore is on the same horizontal plane as that of the teeth, which is effective for cutting food because it brings the cheek teeth together like scissors blades. The carnassial teeth, which are nearest the joint, engage first. The mandibular joint is a hinge with a cylindrical mandibular condyle that fits within a transverse, groove-shaped mandibular fossa. The configuration of the joint restricts fore and aft movements of the lower jaw but allows it to move a little from side to side as well as up and down. The jaw joint of a herbivore, in contrast, lies on a different plane than do the teeth. This is well above the plane of the teeth in rabbits, but sometimes it is below the tooth plane, as in some herbivorous dinosaurs. As a consequence, all the cheek teeth on one side engage concurrently, like the two surfaces of a nutcracker. The mandibular condyle and fossa form relatively flat surfaces, which allows the freedom of movement of the lower jaw needed in grinding food.

The temporal muscle is large in carnivores and constitutes more than half the adductor mass. It arises in a large temporal fossa and inserts on a long coronoid process of the mandible. This configuration gives the temporalis a long lever arm and hence a good mechanical advantage. The lever arm is the perpendicular distance between the muscle's line of action and the fulcrum (jaw joint) about which it acts (Fig. 16-16*A*, *C*, and *E*). The vertical component of the muscle's line of action provides a powerful bite force, and the horizontal component pulls the mandible backward into the mandibular fossa and maintains the integrity of the joint. This backward pull is particularly important for a carnivore, for it compensates for the large forward force on the front of the jaw as the animal pulls on its prey with its canine teeth. The more modest masseter muscle on the lateral surface of the jaw and the pterygoid muscle on the medial surface contribute to the bite force and form a muscular sling that moves the lower jaw from side to side to engage the carnassial teeth.

In sharp contrast, the herbivore's masseter and pterygoid muscles form most of the adductor mass. The high position of the jaw joint and the enlarged angular process of the mandible increase the length of the lever arm and mechanical advantage of these muscles (Fig. 16-16*B*, *D*, and *F*). The masseter usually is composed of two layers of fibers with lines of action nearly perpendicular to each other. These lines of action provide strong fore and aft as well as vertical forces. The masseter and pterygoid position the teeth and provide most of the complex forces needed to grind the food. No strong forward force is directed at the front of the jaws. The temporal muscle, temporal fossa, and coronoid process are small.

The Palate

The roof of the oral cavity is called the **palate**. In aquatic vertebrates, the palate is flat and smooth, whereas in terrestrial vertebrates, it is vaulted with relief. Many fishes and early tetrapods have palatal teeth. A pair of internal nostrils, the **choanae,** open near the rostral end of the palate in choanate fishes, amphibians, reptiles, birds, and mammals (Fig. 16-17). The choanae of choanate fishes are thought to be the outlets for the olfactory currents of water circulating through their nasal cavities. Those of tetrapods are a part of the air passages to and from the lungs. Most tetrapods have **vomeronasal organs,** and the entrances to them are associated with the choanae or perforate the palate independently (Chapter 12). In many reptiles and birds, small **palatal folds** lie lateral and caudal to the choanae and form grooves that continue the air passages further caudad (Fig. 16-17*B*). A bony **hard palate,** which further separated the oral cavity and nasal passages, evolved from similar folds in crocodiles, the therapsid ancestors of mammals, and mammals (Chapter 7). A fleshy **soft palate** continues caudad from the mammalian hard palate and divides the rostral part of the pharynx into nasal and oral parts. Together, the hard and soft palates constitute the **secondary palate** (Fig. 16-17*C*). The hard palates of many mammals bear cornified, transverse palatal rugae that may help hold the food as it is chewed. The cornified baleen plates used in suspension-feeding by the toothless whales develop from similar ridges.

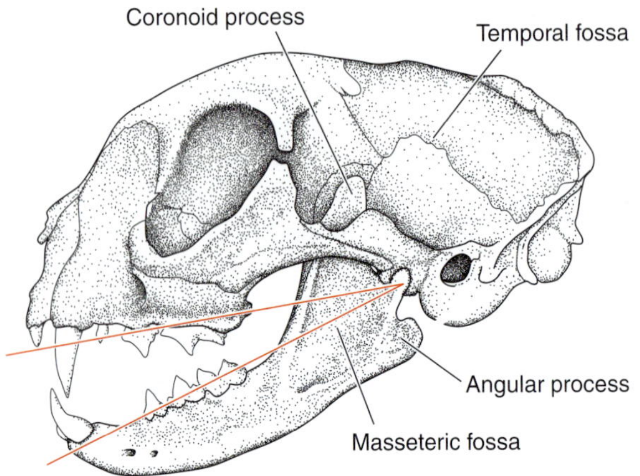

A. Cat skull

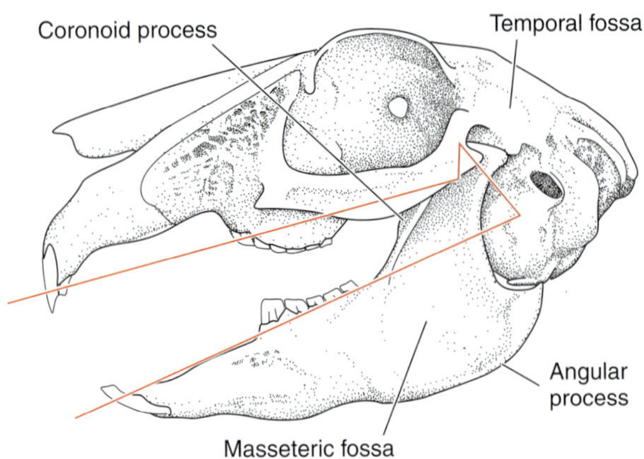

B. Rabbit skull

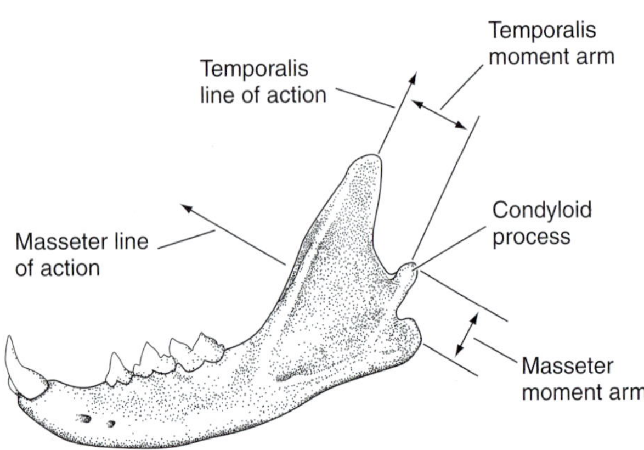

C. Mandible of a cat

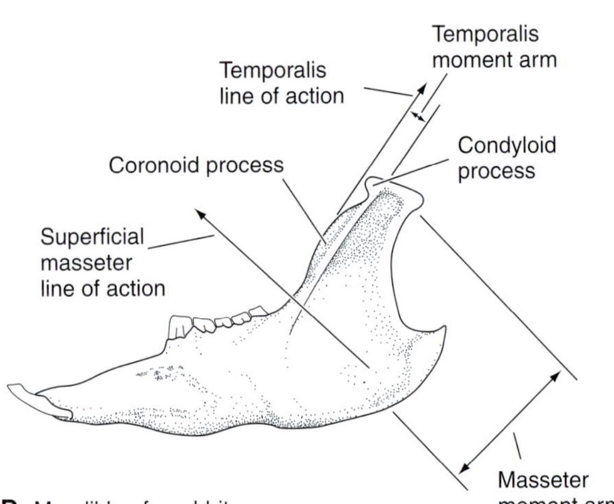

D. Mandible of a rabbit

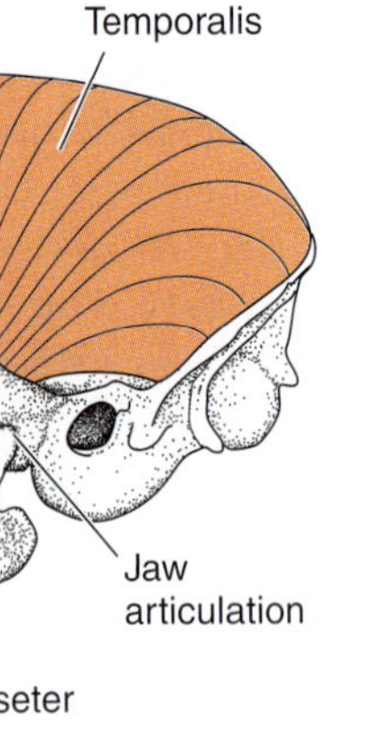

E. Jaw-closing muscles of a cat

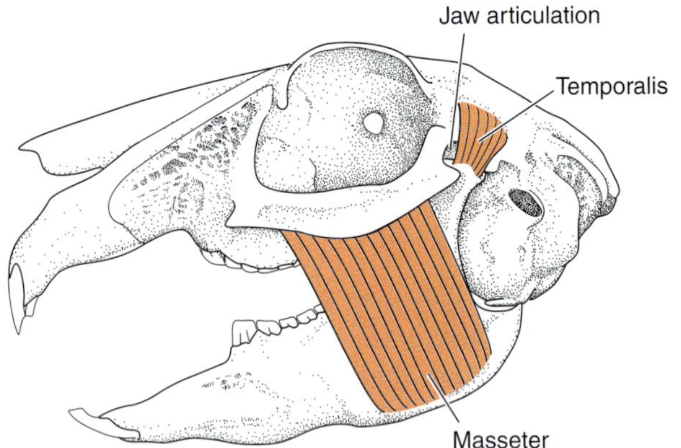

F. Jaw-closing muscles of a rabbit

FIGURE 16-16
The jaw mechanics of a carnivore, based on a cat (*left column*), and a herbivore, based on a rabbit (*right column*). Lines through the teeth indicate the way that the upper and lower teeth come together as the jaws close: as in a pair of scissors in a carnivore (*A*) or as in a nutcracker in a herbivore (*B*). In a carnivore (*C*), the longer moment arm of the temporalis enhances the muscle's function; whereas the herbivore (*D*) has a more powerful masseter because of its longer moment arm. Location of jaw articulations and the relative sizes of the masseter and temporalis muscles differ significantly between carnivores and herbivores (*E* and *F*). (*A–D, After Walker and Homberger.*)

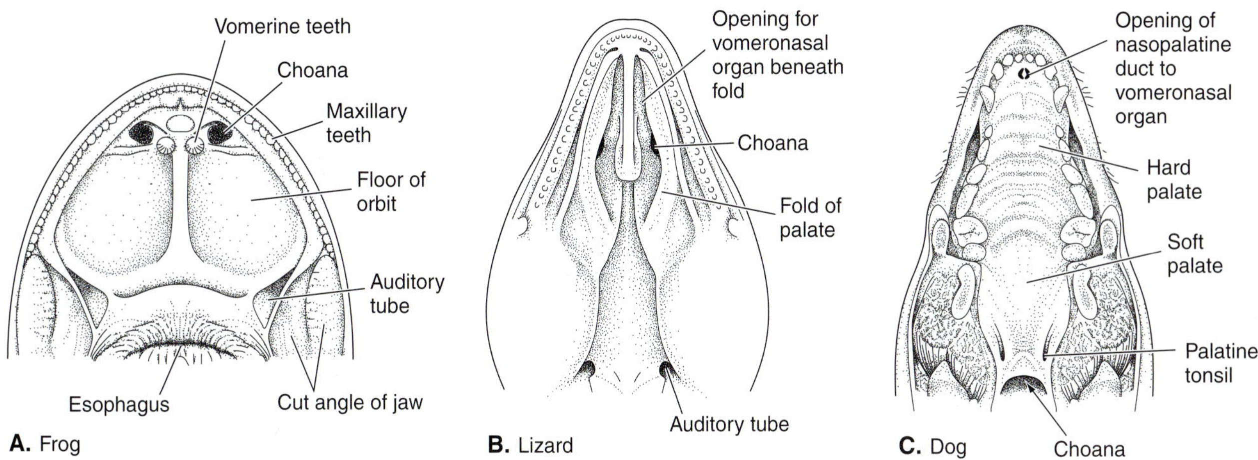

FIGURE 16-17
Ventral views of the palate of representative tetrapods. *A,* A frog. *B,* A lizard. *C,* A dog. (*A, After Walker; B and C, after Romer and Parsons.*)

Tongue, Cheeks, and Lips

The primary difference between aquatic and terrestrial feeding is the mechanism of food transport. In aquatic feeding, water currents created by movements of the bones of the skull and hyoid transport prey within the mouth and pharyngeal cavity. In contrast, terrestrial vertebrates rely on movements of the tongue. A tongue develops in the floor of the buccopharyngeal cavity in many vertebrates. A lamprey has a protrusible "tongue" that bears horny teeth and is used to rasp its prey's flesh. In most fishes, the basihyal, a median ventral element of the hyoid, forms a tonguelike structure called the **primary tongue,** which is not muscular. In some species, it bears teeth. The fish's primary tongue does not play a role in transporting or manipulating the food.

Tetrapods have evolved mobile muscular tongues that are supported by the hyoid and anterior branchial arches and invaded by hypobranchial musculature (Chapter 10). Only the most caudal part of the tetrapod tongue is homologous to the fish's primary tongue. The rest of the tongue of an amphibian is called the **gland field,** and it develops as a ventromedial elevation between the hyoid and the mandibular arches. In amniotes, a pair of more rostral **lateral lingual swellings** are added to the gland field, now called the **tuberculum impar,** and primary tongue (Fig. 16-18*A*). The muscular tongue of

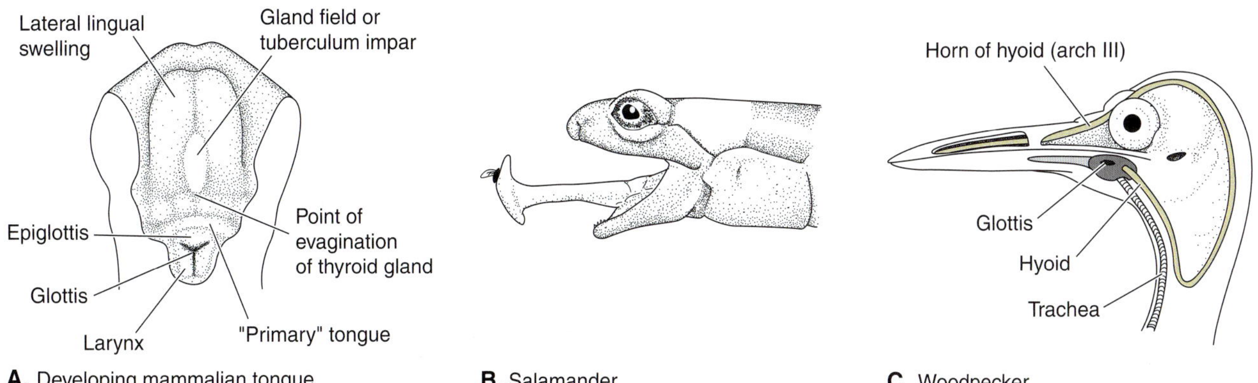

FIGURE 16-18
Tongue development and structure. *A,* A dorsal view of the development of the mammalian tongue. *B,* The tongue of a plethodontid salamander catching an insect. *C,* The tongue and supporting hyoid apparatus of a woodpecker. (*A, After Corliss; B, modified from Noble.*)

all tetrapods plays an important role in transporting and swallowing the food. In mammals, it also positions the food between the teeth of upper and lower jaws.

Tongue movements have been studied by X-ray motion pictures of tongues implanted with radio-opaque markers (Fig. 16-19). In both reptiles and mammals, tongue movements are correlated with movements of the hyoid and jaws in the four stages we have discussed. During slow opening, the lower jaw opens slightly, and the tongue and hyoid move forward and upward. Contractions and expansions also occur within different parts of the muscular tongue (Fig. 16-19). First, the anterior region expands and moves forward, under the food. It then contracts back under the food, and the tongue, as a whole, moves forward, holding the food against the palate. At the end of slow opening, the tongue and hyoid are in their most anterior position, and the tongue is under the food (Fig. 16-19*B*). During fast opening, the gape increases rapidly while the tongue and hyoid are drawn posteriorly, carrying the food back in the mouth cavity (Fig. 16-19*C*). Fast closing is characterized by rapid adduction of the jaw while the tongue and hyoid continue to move backward. The cycle concludes with slow closing without a strong force being applied by the adductor muscles. The tongue and hyoid reverse their directions and move forward (Fig. 16-19*D*). A varying number of transport cycles will carry food through the oral cavity. Transport is regularly interrupted by biting cycles, during which the tongue moves forward or sideways during fast opening to place the food between the teeth for tooth contact in fast closing.

In mammals, the movements of the tongue are also the result of movements of the mandible and hyoid and movement produced within the body of the tongue itself. Tongue movements in cats are associated with characteristic profiles of jaw and hyoid movements, as described in the four stages of the transport cycle in reptiles (Fig. 16-19), with the exception of the long duration (>45% of cycle duration) of the slow-opening stage. The tongue moves forward during the slow-opening stage and backward during most of the rest of the transport cycle. During this extended slow-opening stage, the cat's tongue expands and protrudes under the food. The extension of the tongue is about 60% longer than the retracted tongue. The forward-moving tongue with the food on it is elevated to the palatal rugae. Such contact with the rugae would hinder the forward movement of the food on the advancing tongue and would leave it more posteriorly placed. Retraction and shortening of the tongue in the other steps of the cycle take place with the tongue away from the palate. Protraction and exten-

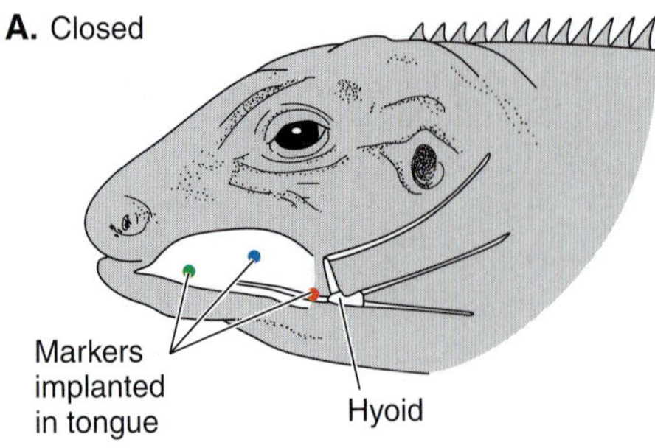

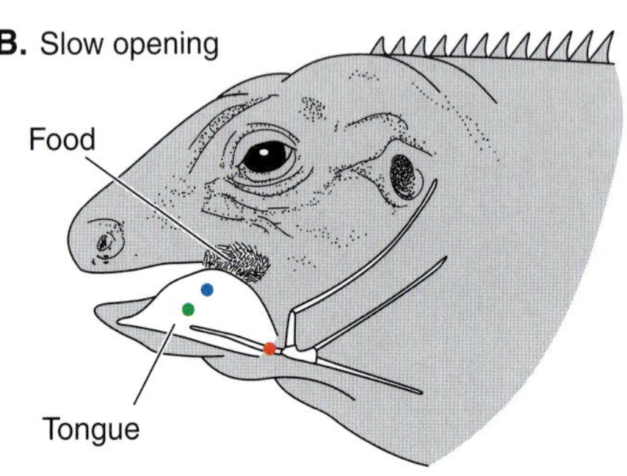

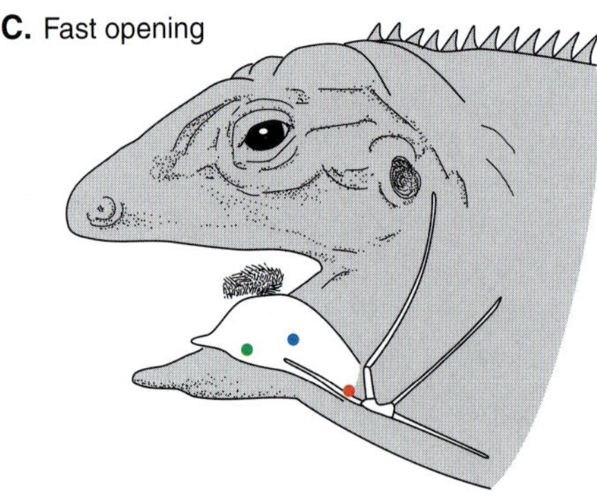

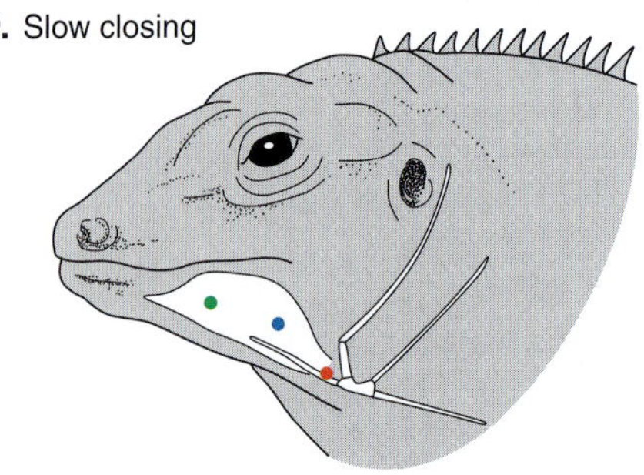

FIGURE 16-19
A–D, Drawings of four stages of food transport in the lizard, *Ctenosaurus,* made from cineradiographic film. Small dots in the tongue region are tongue and hyoid markers. (*From Smith.*)

sion of the tongue against the palatal rugae followed by tongue retraction and shortening away from the palate form the basis of the food-transport mechanism in the oral cavity of mammals (Thexton and McGarrick, 1989).

The tongue often is used in food gathering. Most frogs and salamanders and some lizards, such as Old World chameleons, can rapidly project part of their tongues from the mouth and capture an insect (Fig. 16-18*B*). The tip of the tongue is covered by a sticky mucus. A woodpecker has a long and spiny tongue that is supported by an elaborate hyobranchial apparatus, part of which is housed in a bony tube that curves around the back of the skull (Fig. 16-18*C*). After a woodpecker has chiseled a hole in a tree with its bill, the long tongue is protruded to shuck out an insect or grub. Ant-eating and termite-eating mammals also gather their food with long, sticky tongues. Certain of the tongue muscles of the giant anteater extend as far caudad as the posterior end of the sternum. Spiny papillae on the tongues of cats and other carnivores help the animals rasp flesh from a bone.

The tongue also is used in many other ways. It bears taste buds and other receptors so it is an important sense organ. Snakes and some lizards have protrusible, forked tongues that are used in association with the vomeronasal organ to detect certain olfactory clues (Chapter 12). Evaporation of water from the surface of the large, moist tongues of dogs and other panting mammals helps regulate body temperature. We, of course, use our tongues in speech as well as the manipulating and swallowing of food.

The fleshy lips and cheeks that characterize marsupial and placental mammals assist the muscular tongue in manipulating food within the mouth, and they are used by newborns in sucking milk from their mothers' nipples or teats. Monotremes have horny beaks and cannot suckle; their young lap up milk that is secreted onto hairs rather than discharged through nipples. In many species of rodents, part of each cheek evaginates beneath the facial muscles to form a pair of large cheek pouches in which food can be temporarily stored.

Oral Glands

Most fishes lack oral glands, apart from scattered mucus-secreting cells. Lampreys are a notable exception. They have a pair of large glands that secrete the anticoagulant needed to keep their prey's blood flowing freely as they feed on it. Oral glands, including multicellular salivary glands, are well developed in terrestrial vertebrates. Their mucous and serous secretions lubricate the food and facilitate food manipulation and swallowing in the terrestrial environment. Modified salivary glands of some snakes produce the hemolytic or neurotoxic poisons that are injected by fangs to immobilize their prey. The secretion of digestive enzymes is not an important function of the salivary glands of most terrestrial vertebrates, although a few amphibians and mammals synthesize a small amount of salivary amylase, a starch-splitting enzyme.

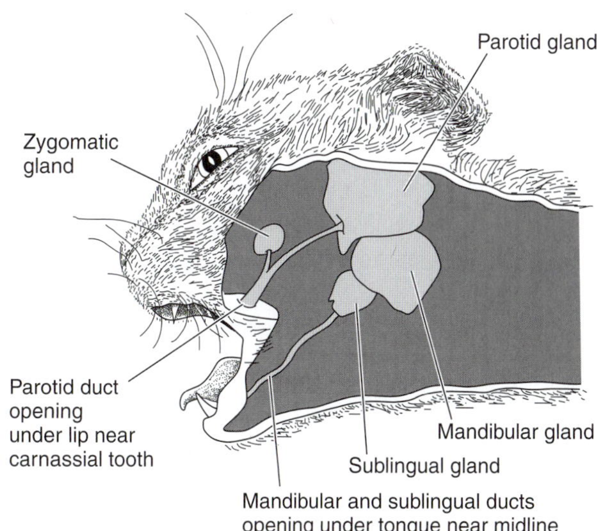

FIGURE 16-20
The salivary glands of a cat. (*After Walker and Homberger.*)

The oral cavities of mammals are particularly well equipped with scattered secretory cells and large salivary glands (Fig. 16-20). Most mammals have well-developed **parotid, mandibular,** and **sublingual glands.** Some species also have a **zygomatic gland** in the floor of the orbit, and **buccal** and **molar glands** beneath the mucous membrane of the lips. Saliva is composed primarily of serous and mucous secretions that are important for lubricating the food during mastication and swallowing. Humans among the few mammals that secrete salivary amylase. Starch digestion is initiated during mastication and continues in the stomach until the high acidity of the gastric secretions stops the action of amylase. The swallowed saliva of a cow provides a medium in a chamber of the stomach, the rumen, in which bacteria multiply and cellulose is broken down (Chapter 17). A cow may produce more than 100 L/day of saliva. Toxins are produced by the salivary glands of some insectivores, and an anticoagulant is secreted by the salivary glands of vampire bats.

FIGURE 16-21
The evolution of feeding patterns of vertebrates. (*Modified from Reilly and Lauder.*)

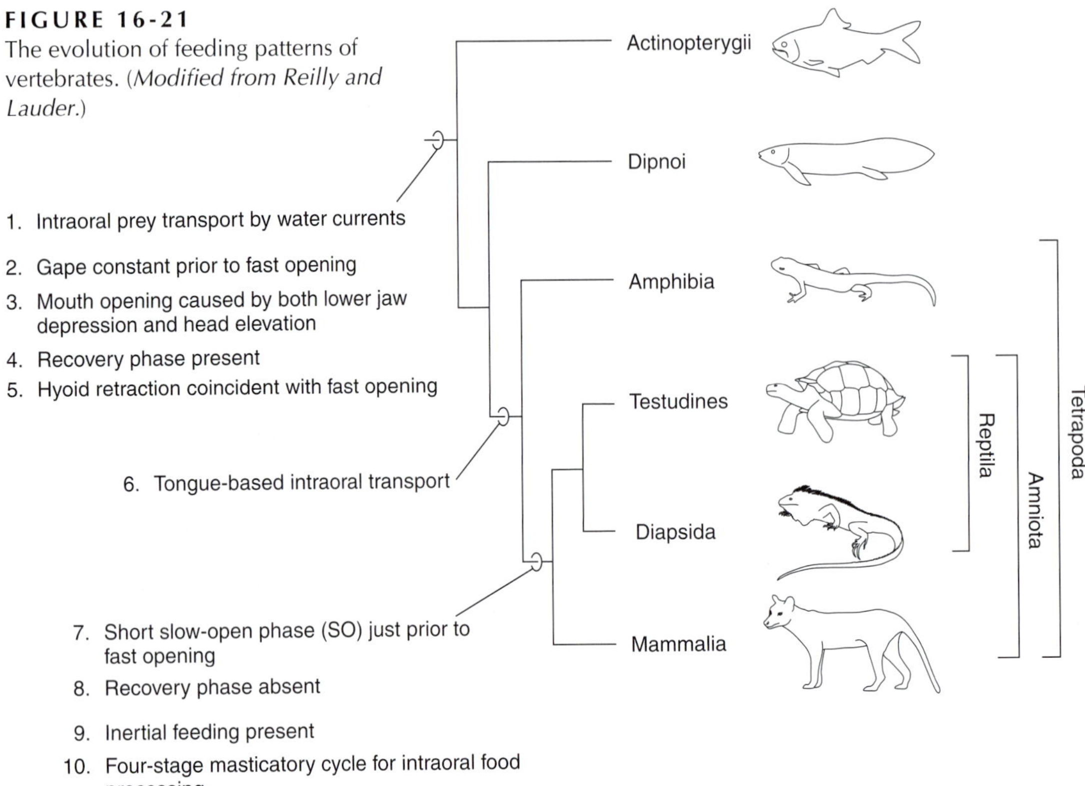

Evolution of the Vertebrate Feeding Mechanism

Major changes in the feeding apparatus and behavior have emerged during vertebrate evolution (Fig. 16-21). Reilly and Lauder (1990) have identified the following five features as primitive:

1. Prey transport is effected by water currents within the oral cavity.
2. The gape remains constant prior to fast opening.
3. Mouth opening is produced by both pronounced lower jaw depression (sternohyoideus muscle) and head elevation (epaxial muscles).
4. A recovery phase follows prey ingestion.
5. Hyoid retraction by the sternohyoideus is coincident with the fast-opening stage.

These characteristics are present in all bony fishes and sharks. The only feature that remains constant during the evolution of all vertebrates is the fifth feature. In tetrapods, a muscular tongue first evolved for prey transport within the oral cavity. This feature becomes further elaborated and important in the amniotes. Three more features become modified in the amniotes: (1) the chewing and prey transport cycle begins with a slow-opening stage prior to fast opening, (2) the gape is produced by lowering the mandible rather than by head lifting, and (3) the recovery phase is no longer distinguishable. Two evolutionary innovations emerged in amniotes: inertial-feeding and a four-stage masticatory cycle for food processing in the mouth. In many reptiles and birds, especially those with reduced or specialized tongues, inertial-feeding takes place. Once the prey is captured, the head is pulled rapidly backward, imparting a posteriorly directed motion to the prey (Smith, 1986). The jaws are then opened very fast and the head thrusted forward to surround the prey. These inertial thrusts are repeated until the prey reaches the esophagus. Most birds, snakes, and carnivorous lizards employ inertial-feeding, in which, within a few tenths of a millisecond, complete acceleration–deceleration of the skull and cervical complex occurs.

With the emergence of the complex, four-stage, masticatory cycle, amniotes do not use their jaws and tongues simply to transport whole prey or large items of food to the esophagus, but they bite, align, masticate, pack, and transport the food to the esophagus. This highly efficient processing of food enhances the rate of digestion of a wide variety of foods.

SUMMARY

1. Vertebrates obtain the food and other nutrients needed to sustain their metabolism through the digestive system. Food must be ingested, sometimes broken down mechanically, transported through the mouth and swallowed, and chemically digested.
2. The lining and glandular derivatives of the digestive tract develop from the endodermal archenteron; the rest of the wall develops from mesenchyme, most of which comes from splanchnic mesoderm. An ectodermal stomodaeum forms the oral cavity, and an ectodermal proctodaeum contributes to the cloaca.
3. Urochordates, cephalochordates, and larval lampreys are filter-feeders. Many types of feeding mechanisms have evolved. Most fishes and aquatic tetrapods are suction-feeders. Many fishes are ram-feeders. In general, many primitive terrestrial vertebrates are inertial-feeders, and mammals masticate their food.
4. The enamel and dentine of teeth are similar to these products on bony scales. Enamel is produced by ameloblasts of the enamel organ; dentine, by odontoblasts.
5. In most vertebrates, the teeth are loosely bound to the jaws or to underlying skeletal elements (acrodont and pleurodont), but where forces acting on the teeth are large, they are set in sockets (thecodont).
6. Successive waves of new-tooth formation replace teeth throughout the lives of most vertebrates, but in most mammals, replacement is limited to a deciduous and a permanent set.
7. All the teeth have a similar shape in most vertebrates (i.e., homodont), and the shape is related to the type of food eaten and to the method of obtaining it. In late therapsids and in most mammals, teeth differentiate and serve different functions (i.e., heterodont).
8. Teeth begin to emerge in most mammals as the suckling period ends, and their pattern of eruption and replacement ensures the good occlusion needed for mastication.
9. Incisors typically are small teeth used for cutting, cropping, and picking up food. The canines usually are large, conical teeth used in seizing, piercing, and killing prey. Primitively, the premolars are puncturing teeth, and the molars have a combined cutting and crushing action.
10. The molars of ancestral mammals were triconodont, with three cusps in a linear sequence. A shift of the cusps to a triangular configuration in †symmetrodonts increased the number of shearing facets. Ancestral marsupials and placentals had tribosphenic molars, in which a low heel was added to the lower molars as a crushing function was added to the original shearing action.
11. Teeth, especially the premolars and molars, became modified as mammals adapted to different diets. Contemporary carnivores have a set of shearing carnassials that evolved from the last upper premolar and first lower molar. Omnivores and primitive herbivores have low-crowned, bunodont molars that typically bear four rounded cusps. The cusps form ridges (lophodont molars) or crescents (selenodont molars) in more specialized herbivores. Herbivores that feed on highly abrasive food have evolved high-crowned molars.
12. Primitive ray-finned fishes ("†paleoniscoids") are ram-feeders. Raising the head and lowering the lower jaw produces a large gape, with which the fish overtakes its prey. The jaws then are closed.
13. Most sharks capture prey by a combination of ram-feeding and suction-feeding. They can form a large gape by raising the skull and lowering the lower jaw.
14. Ray-finned fishes have become suction-feeders. Many parts of the skull are highly mobile, the size of the buccopharyngeal cavity is rapidly increased, and food is sucked in along with a current of water. Evolutionary modifications in the head for suction-feeding include the following: (1) the enlargement and increased mobility of the premaxilla, accompanied by changes in the maxilla that enable it to swing backward and forward; (2) an increasing vertical orientation of the palatopterygoid arch; (3) a forward shift of the jaw joint; (4) separation of the operculum from the skull roof and the evolution of a mobile articulation between the operculum and palatopterygoid arch; and (5) the evolution of mechanisms that transmit the movement and forces developed on the hyoid to the lower jaw.
15. Aquatic salamanders are also suction-feeders. Patterns of feeding resemble those of actinopterygian fishes, except that a depressor mandibulae muscle lowers the lower jaw.
16. Food is hydraulically transported through the buccopharyngeal cavity of aquatic anamniotes by a current of water. Repeated movements of the jaws and hyoid that resemble the movements used in prey capture produce the current.
17. Coordinated movements of the mandible and a muscular tongue replace water flow through the mouth of terrestrial vertebrates. The feeding cycle

always involves (1) slow opening, (2) fast opening, (3) slow closing, and (4) fast closing with the development of sufficient force to firmly grasp, cut, crunch, or masticate the food. These movements are accompanied by cranial kinesis in most reptiles and birds but not in mammals.

18. The fleshy cheeks and lips of marsupials and eutherians assist in food manipulation and are used by newborns in suckling.

19. Apart from scattered mucus-producing cells, most fishes lack oral glands, but these glands are well developed in tetrapods. The salivary glands of mammals include the parotid, mandibular, sublingual, zygomatic, molar, and buccal glands. The glands of some mammals produce salivary amylase.

20. Choanae open onto the primary palate in choanate fishes, amphibians, reptiles, and birds. In mammals, a hard palate carries the choanae to the pharynx, and a soft palate divides the rostral part of the pharynx into nasal and oral parts.

REFERENCES

References for this chapter are combined with others on more caudal parts of the digestive system at the end of Chapter 17.

17

The Digestive System: Pharynx, Stomach, and Intestine

PRECIS

Vertebrates must chemically break down their food to absorb the contained nutrients that are needed to sustain their metabolism. The structure and evolution of the digestive system reflect adaptations to the type of food eaten, to the animal's level of metabolism, and to body size.

OUTLINE

Food that has been transported through the oral cavity and masticated in some species next enters the pharynx. From there, it passes through the esophagus, stomach, and intestine. Digestion occurs primarily in the stomach and intestine. Many digestive enzymes acting in sequence are needed to break down the food into small molecules. These molecules must then be absorbed along with minerals and other nutrients. The large quantities of water that are released into the digestive tract as a component of the digestive secretions must be reabsorbed. Finally, the undigested residues, cells sloughed off the digestive tract lining during the passage of the food, and the bacteria that have multiplied in parts of the digestive tract must be eliminated, or **egested.** During these processes, the lining of the digestive tract is subjected to wide changes in pH, to the action of powerful proteolytic enzymes, and to considerable abrasion. It must be protected and repaired.

The structure of the digestive tract and its evolution are affected by many factors, including the type of food eaten, the level of activity and metabolism of the animal, and the size of the animal. Because no craniate can synthesize the enzyme **cellulase,** needed to digest the woody cellulose in plant food, herbivorous craniates must have colonies of microorganisms in some part of their digestive tract that can synthesize cellulase. These microorganisms may be lodged in special compartments of the stomach or intestine or in an exceptionally long intestine. Levels of activity and metabolism are coupled with the rate at which food must be digested and absorbed. An active, endothermic animal needs a large digestive and absorptive surface area to process an increased amount of food. Regardless of other factors, small animals also need more digestive and absorptive surface area per unit of body mass than do large ones because their metabolism is relatively higher.

The Pharynx and Its Derivatives

Development

The **pharynx** arises from the anterior part of the archenteron and is the part of the digestive tract from which the paired endodermal **pharyngeal pouches** arise (Figs. 15-10 and 16-1). Jawed fishes have six pairs of pouches that extend laterally, meet ectodermal furrows, and open to the surface. The first pouch is reduced to a narrow passage, called the **spiracle.** The remaining ones develop into the branchial chambers that contain the gills. Because of its importance in respiration (Chapter 18), the pharynx of fishes is a large chamber. As food passes through the pharynx of a fish, it is prevented from entering the branchial chambers by a filter of **gill rakers** attached to the bases of the branchial arches. The lungs of tetrapods develop as ventral outgrowths from the floor of the caudal part of the pharynx. In tetrapods, the pharynx is quite short—in adult tetrapods, it is little more than a connecting segment between the oral cavity and the esophagus. Although pharyngeal pouches develop in embryonic tetrapods, the caudal ones are crowded together and reduced in number. Embryonic mammals have only four pairs of pouches, but the last one incorporates the fifth pouch. The pharyngeal pouches of tetrapods do not open to the body surface, except in larval amphibians. Even the middle ear cavity of tetrapods, a modified second pharyngeal pouch, is closed off from the outside world by a membrane (Chapter 12). Food passing through the pharynx is prevented from entering the air passages to and from the lungs by the closure of a slitlike **glottis** in the floor of the pharynx. In addition, mammals have a flaplike **epiglottis,** which lies rostral to the glottis (Fig. 18-21) and deflects food around the glottis. As we have discussed (Chapter 15), several endocrine glands and glandlike structures also develop from the pharyngeal epithelium.

The Thymus and Tonsils

A **thymus** develops in all vertebrates from the endodermal epithelium of certain pharyngeal pouches and from the adjacent ectodermal epithelium. In fishes, all the pouches, or the first four, contribute to thymus formation, but in tetrapods, the number is more restricted. In mammals, only the third and fourth are involved, and the contribution of the third is by far the greater (Fig. 15-12). Thymic epithelium is derived from the dorsal surface of the pouches in question except in mammals, in which it comes from the ventral surface. After separating from the pouches, the epithelial thymic buds receive stem cells from the spleen and bone marrow by its blood supply. These cells differentiate into **T lymphocytes** (often simply called **T cells**), which are released, circulate in the blood, and populate lymph nodes and other lymphoid organs, where they continue to multiply and differentiate under the influence of at least one thymic hormone (**thymopoietin**). T cells, of which several types exist, play a key role in certain of the body's immune responses (Chapter 19). Many T cells participate in a **cell-mediated response,** in which the cells directly attack virus-infected cells, fungi, and other foreign cells. Other T cells assist in the proliferation and activation of other types of lymphocytes (B lymphocytes, Chapter 15), and others have yet other functions. In newborn humans, the thymus is relatively large, but it continues to grow at a slower rate until puberty. Thereafter, it is invaded by fat cells and begins to regress, but some thymic tissue persists and remains active into old age.

Tonsils are also lymphoid organs that participate in the tetrapod body's immune responses and appear in amphibians in the dorsal epithelium of the last pouch. They develop around the wall of the anterior part of the pharynx, a position that exposes their cells to invading antigens. The paired palatine tonsils of mammals begin their development as epithelial derivatives of the second pair of pharyngeal pouches (Fig. 15-10), and they, too, are invaded by lymphocytes.

Gut Tube Structure

The postpharyngeal digestive tract is a tubular structure, parts of which often are enlarged to form saccular chambers. The lining of the gut tube is a stratified squamous epithelium in those parts subject to considerable abrasion: the oral cavity, pharynx, esophagus, and cloaca. It is a simple columnar epithelium in all or most of the stomach and in the intestine. A loose, fibrous connective tissue and muscle form the rest of the wall. The structure and arrangement of these layers in the mammalian intestine are representative (Fig. 17-1).

The innermost layer, called the **mucous membrane,** or **mucosa,** is the most complex. It consists of the simple columnar epithelium lining the intestine and a fibrous connective tissue containing many blood and lymphatic vessels that receive absorbed materials. Aggregations of white blood cells in the connective tissue may form **lymphocyte nodules.** They are part of the

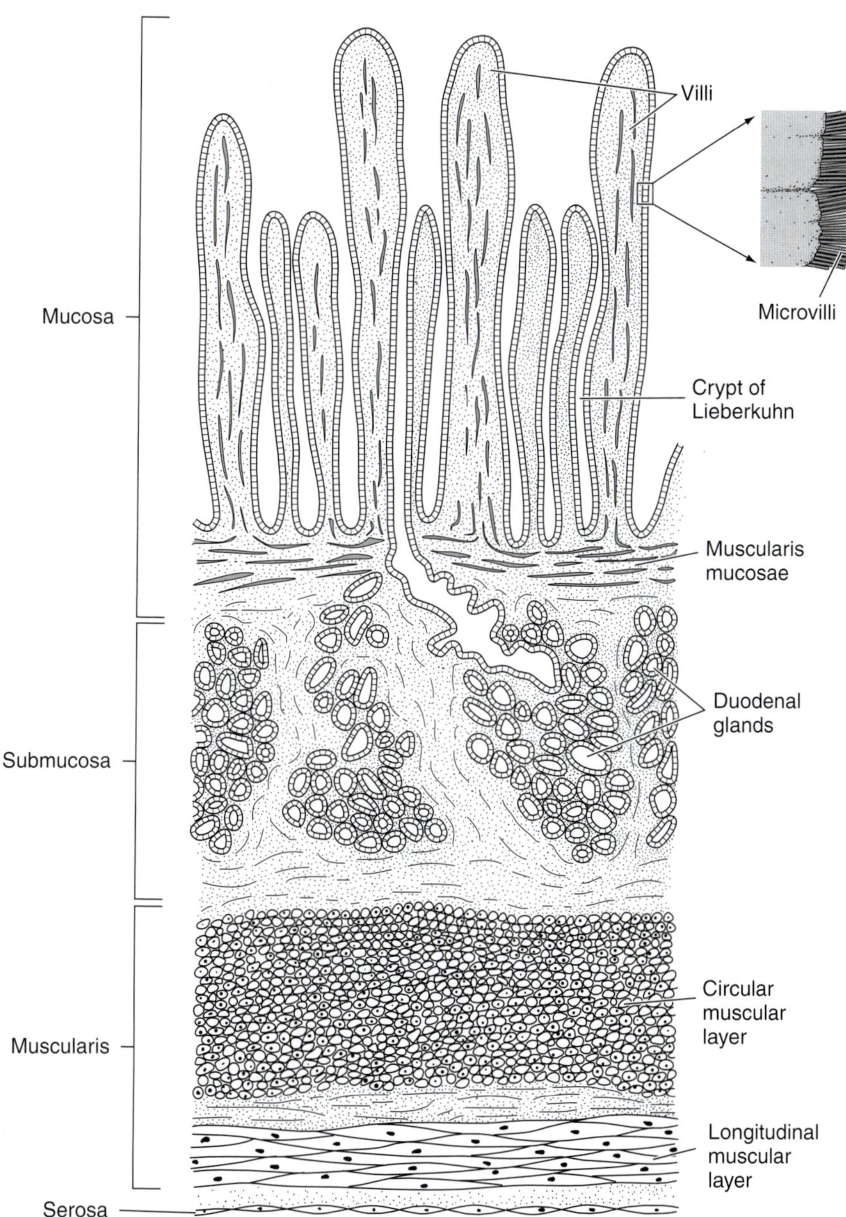

FIGURE 17-1
A longitudinal section through the duodenum of a mammal. (*After Williams et al.*)

body's immune system. Electron microscopy has shown that the surface of the epithelial cells bears numerous microvilli that greatly increase the digestive and absorptive surface area. The mucosal surface is further increased by larger folds, the nature of which varies greatly among vertebrates and between gut regions. The mammalian small intestine has many circular folds, called the **valves of Kerckring,** and the lining is studded with numerous minute, finger-like projections, called **villi**. Many of the epithelial cells, known as **goblet cells,** secrete mucus that lubricates the lining and protects it against autodigestion. Other lining cells produce the intestinal enzymes. These enzymes are not free within the lumen but appear to act among the microvilli at the cell surface. Digestive products thus are concentrated at the cell surface; this facilitates absorption. Many simple tubular glands, the **crypts of Lieberkuhn,** invaginate between the bases of the villi into deeper parts of the mucosa. Undifferentiated stem cells multiply in the crypts and spread over the intestinal lining, replacing cells that are sloughed off. Zymogenic cells deep in the crypts may secrete lysozyme, an antibacterial substance. Scattered **argentaffin cells** in the crypts secrete peptide hormones that help regulate digestive activities (Chapter 15). A thin layer of smooth muscle, the **muscularis mucosae,** forms the outer border of the mammalian mucosa. Some of the muscle cells enter the villi. The muscularis mucosae mediates those movements of the villi and mucosa that are independent of the movement of the gut as a whole.

A second layer, the **submucosa,** which is composed of a very vascular, fibrous connective tissue, lies deep to the mucosa. Compound glands invaginate into the submucosa of the mammalian duodenum from the mucosa (Fig. 17-1). A layer known as the **muscularis** lies peripheral to the submucosa. It is composed of two layers of muscle that spiral around the intestine and are antagonistic to each other. The outer, **longitudinal layer** forms such an open spiral that its fibers are seen primarily in lateral view in a longitudinal section; the inner, **circular layer** forms such a tight spiral that its fibers are seen in cross section. These muscles are smooth. Their activity is integrated by the enteric part of the autonomic nervous system, which consists of a myenteric plexus, located between the two layers of the muscularis, and a submucosal plexus, located between the muscularis and submucosa (Chapter 13). Not all of the smooth muscle cells receive a direct innervation from the enteric system, but action potentials spread slowly from cell to cell. Peristaltic contraction of the muscle layers propels the food mass through the digestive tract and churns it, thereby helping break it up mechanically and mix it with digestive secretions. Churning is particularly pronounced in the stomach. An increase in thickness of the circular muscle layer also forms sphincters, such as the **pyloric sphincter** at the caudal end of the stomach, that retain the food for a certain time. Connective tissue covered by coelomic epithelium forms a thin layer known as the **serosa** on the surface of the wall of those parts of the digestive tract lying within the body cavity.

The Esophagus

The "esophagus" of the lamprey represents the dorsal part of the larval pharynx, the ventral part of which forms the respiratory tube (Fig. 17-2*A*). In other vertebrates, the **esophagus** (Gr., *oisophagos* = gullet) is a connecting segment between the pharynx and the stomach or, in those vertebrates that lack stomachs, between the pharynx and the intestine. The esophagus primarily transports food but performs additional functions in some species. It is specialized to crush eggs in egg-eating snakes, and in some reptiles and birds, it serves for the temporary storage of partly swallowed prey. Grain-eating and seed-eating birds and a few other vertebrates have a **crop,** a sac that develops from the caudal portion of the esophagus. Food can be gathered quickly and stored here, and seeds may be softened by water and some bacterial fermentation. Under the influence of prolactin secreted by the adenohypophysis (Chapter 15), the crops of pigeons and doves of both sexes produce a milky secretion that is fed to the young.

The crop and lower esophagus of the hoatzin, a South American bird, are specialized as a fermentation chamber, analogous to the chambered stomach of cows and other ruminants (Grajal et al., 1989). This bird feeds on soft, young leaves, which are ground up by cornified epithelial ridges in the highly muscular and very large crop. A bacterial colony in the crop produces cellulase, which breaks down cellulose in the leaves into short-chain organic acids, which are absorbed from the crop or more distally along the digestive tract. As the bacteria multiply, they produce microbial protein that the bird digests in its stomach and intestine. A large fermentation chamber is an unusual specialization for a bird because most birds conserve weight as an adaptation for flight, but the hoatzin is a poor flier.

The esophagus is lined by a stratified squamous epithelium that is slightly cornified in some turtles, most birds, herbivorous mammals, and other species that swallow coarse food. Mucus-secreting and serous-secreting cells usually are present, but larger glands are seldom found. The lining of the esophagus frequently bears papillae or longitudinal folds. Unlike more caudal parts of the gut tube, the musculature in its wall often is striated. Normally, its lumen is collapsed, but it can be distended as food is swallowed. The esophagus is short in fishes and amphibians but much longer in amniotes because not only has the heart moved caudad, but other viscera have also shifted posteriorly, and a well-defined neck is present (Figs. 17-2 to 17-4).

The Stomach

Tunicates, amphioxus, ammocoetes, larvae, and adult hagfishes and lampreys lack stomachs (Fig. 17-2*A*). This may have been the primitive craniate condition. The small food particles on which these animals fed—probably more or less continuously—could have been processed directly by the intestine. Presumably, a stomach evolved as vertebrates began to feed on larger organisms that were caught at less frequent intervals. The stomach is, first of all, a chamber for the storage of food. Some investigators have postulated that hydrochloric acid production by the gastric glands evolved in the context of killing bacteria and helping preserve the food. The synthesis of pepsin, an endopeptidase that cleaves long protein chains into shorter ones, may have evolved later.

A stomach is not essential for digestion because endopeptidases also are produced by the pancreas. If an

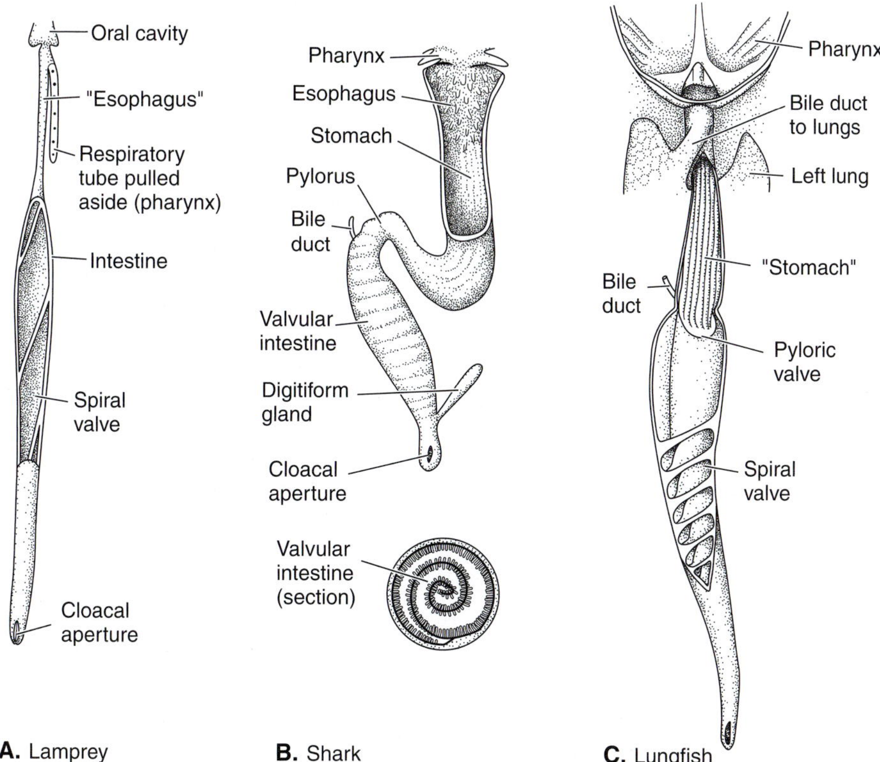

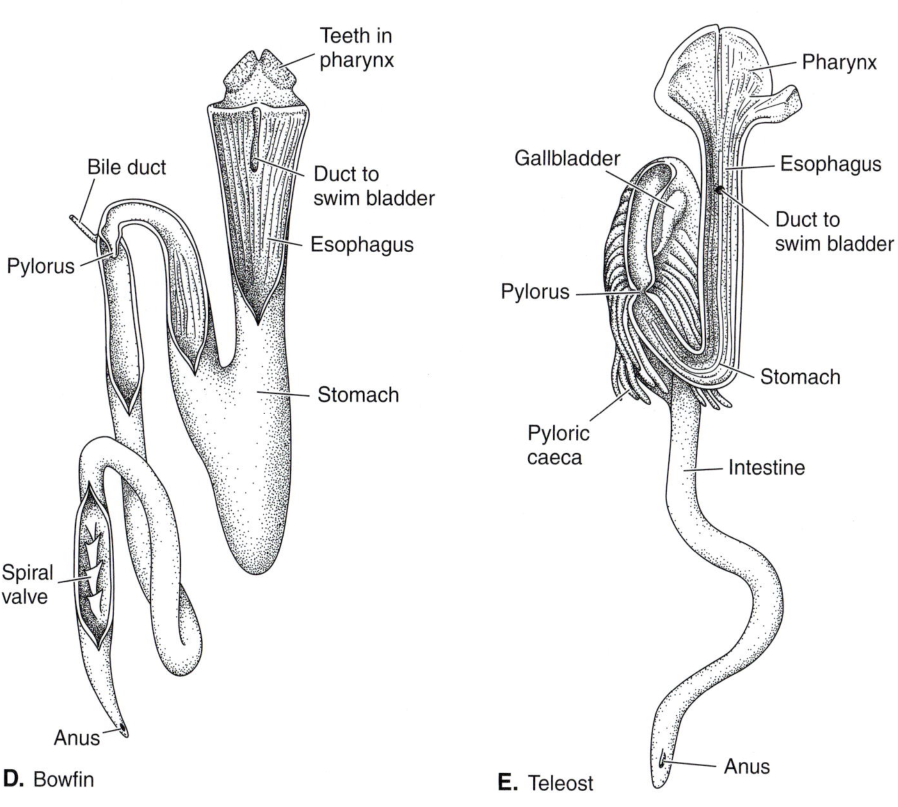

FIGURE 17-2
Ventral views of the digestive tract of representative fishes. The drawings extend from the caudal end of the pharynx to the cloaca or anus. The liver and pancreas have not been included, but the point of entrance of the bile duct is shown. *A,* A Lamprey. *B,* A shark; the insert shows the spiral valve in a cross section through the valvular intestine. *C,* A lungfish, *Protopterus.* *D,* A bowfin, *Amia. E,* A teleost, *Salmo.* (*Modified from Bolk et al.*)

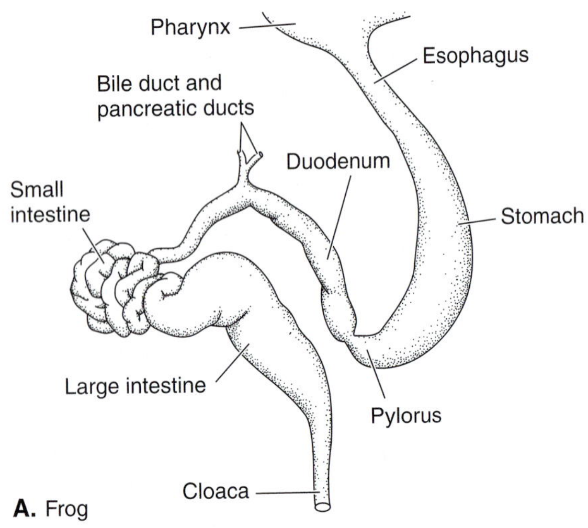

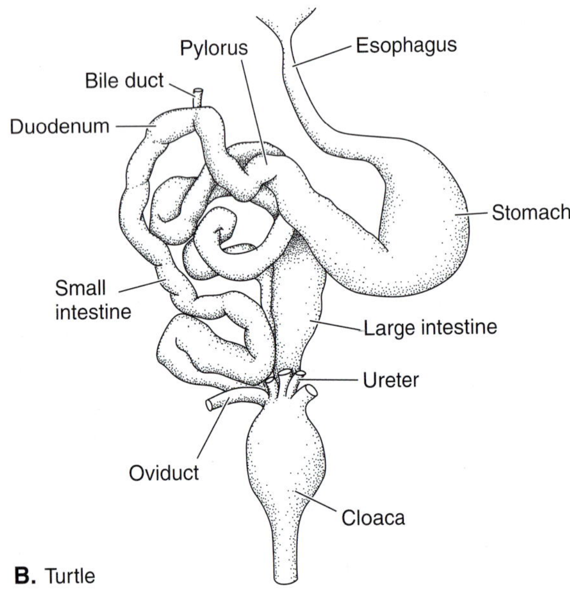

FIGURE 17-3
Ventral views of the digestive tracts of primitive tetrapods. The liver and pancreas have not been included, but the point of entrance of the bile duct is shown. *A*, A frog, *Rana*. *B*, A turtle, *Emys*. (*A, after Walker; B, after Bolk et al.*)

animal feeds on very small particles, or if the food is finely ground in its mouth or pharynx, a stomach may be secondarily lost. This loss presumably occurred in chimeras, lungfishes (Fig. 17-2*C*), and all carplike fishes and minnows, which lack stomachs.

The stomach usually is a J-shaped sac (Figs. 17-2*B* and *D* to 17-4), although it is straight in some slender, long-bodied vertebrates: eels, many salamanders, snakes, and lizards. A pyloric sphincter, or **pylorus** (Gr., *pyloros* = gatekeeper), at its posterior end normally is contracted, so food is retained in the stomach until it has been broken down—mechanically by churning movements and chemically by pepsin—into a consistency that can be processed by the intestine. A **cardiac sphincter** sometimes is present at the esophageal end. Topographically, the stomach can be divided into a pyloric region, adjacent to the pylorus, and a body, or corpus. In mammals, it also has a cardiac region, nearest the heart, and a dome-shaped bulge, the fundus, adjacent to the esophageal entrance (Fig. 17-4).

A stratified, squamous, esophageal type of epithelium may extend far into the stomach in some turtles, rodents, ruminants, and other vertebrates that feed on coarse food, but usually the stomach is lined by a simple columnar epithelium. Goblet cells are abundant throughout the lining and, together with branched tubular glands in the pyloric and cardiac regions, secrete a copious amount of mucus, which helps protect the stomach from autodigestion. Simple tubular **gastric glands,** which are found primarily in the body and fundus, secrete **hydrochloric acid** and **pepsinogen.** Part of the pepsinogen molecule is cleaved from the rest of the molecule by hydrochloric acid in the stomach lumen. This cleavage forms the smaller but active enzyme, **pepsin.** Both acid and pepsinogen are synthesized by the same cells in most vertebrates, but a division of labor has evolved in mammals. **Parietal cells** produce the acid; **chief cells** produce the pepsinogen.

Other enzymes also may be synthesized in the stomachs of some vertebrates. Many amphibians, reptiles,

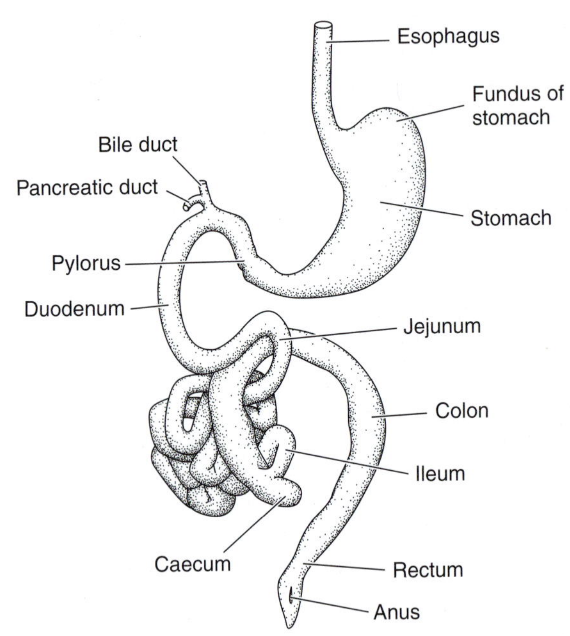

FIGURE 17-4
Ventral view of the digestive system of a cat. The liver and pancreas have not been included, but the points of entrance of the bile and pancreatic ducts are shown.

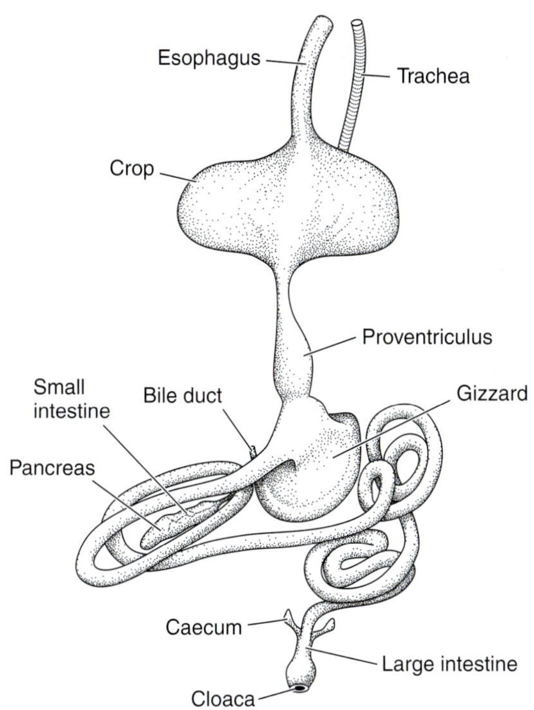

FIGURE 17-5
Ventral view of the digestive tract of a pigeon. The liver and pancreas have not been included, but the point of entrance of the bile duct is shown.

birds, and some mammals produce **chitinase,** an enzyme that hydrolyzes the insect cuticle. Mammals, especially young ones, secrete **renin,** an enzyme that curdles milk and causes milk protein to remain in the stomach long enough to be acted on by pepsin.

Some vertebrates have unusual specializations of the stomach. In puffer and porcupine fishes, the stomach, or a diverticulum from it, can be inflated with water. When threatened or disturbed, the fish increases in size and assumes a globular shape with erect spines, becoming much less appealing to a predator. A **gizzard,** which can grind up food, has evolved from all or much of the stomach in gizzard shad and certain other fishes; some reptiles, including crocodiles; and all birds (Fig. 17-5). The gizzard has a thick muscular wall and a tough proteinaceous lining, and it usually contains small stones that have been swallowed. It is important for birds because they have lost their teeth and cannot otherwise mechanically break down their food. The bird's gizzard develops from the posterior part of the stomach; the anterior part, known as the **proventriculus,** produces gastric secretions that mix with the food as it is ground up.

A number of mammals have evolved complex, chambered stomachs that enable them to process plant food. Plant food is abundant and easier to gather than catching prey, but its energy and protein content are relatively low. A large volume of plant food must be processed to overcome these disadvantages. This processing requires that some part of the gut, such as the stomach, have a large capacity and that passage time of the digesting food mass be slow, which provides space and allows time for colonies of microorganisms to ferment and break down cellulose. Hippopotamuses, camels, giraffes, and ruminants (i.e., sheep and cattle) have large, multichambered stomachs. Enlarged stomachs are found in artiodactyls and some other plant-eating mammals, including kangaroos, sloths, and some monkeys even though they do not remasticate and reswallow the food. An enlarged and somewhat divided stomach also is found in many whales, but its functions are unclear. Fermentation of chitin in the skeletons of invertebrate food may occur in the whale's stomach.

Details of the morphology of chambered stomachs vary, but the stomach of a cow is representative of a cud chewer, i.e., it remasticates and reswallows the food. A cow's stomach consists of four chambers arranged functionally in linear sequence: the (1) **rumen,** (2) **reticulum,** (3) **omasum,** and (4) **abomasum** (Fig. 17-6). Only the abomasum is lined by a simple columnar epithelium and contains gastric glands. The other chambers are lined with a stratified, esophageal type of epithelium. The rumen is the largest chamber and has an open connection with the reticulum, and the two function more or less as a unit. The reticulorumen of a large cow may contain 300 L of material. Plant food and saliva accumulate in the reticulorumen, and the food is acted on by colonies of anaerobic bacteria and ciliated protozoa that synthesize the enzyme **cellulase.**

Cellulose is fermented and broken down into acetic, butyric, and other short-chain organic acids and carbon dioxide and methane. The saliva, which is chiefly a weak sodium bicarbonate solution, buffers the acids, serves as a medium in which the microorganisms grow and multiply, and helps detoxify alkaloids and other toxins that may be in the food. Simple nitrogenous compounds in the food are converted to ammonia, from which the microorganisms synthesize new protein. Additional nitrogen for protein synthesis is derived from swallowed air and by diffusion of nitrogen and urea from the blood. Urea is a particularly important nitrogen source for bacterial protein in camels, goats, and other species that feed on a low-nitrogen diet. The bacteria also synthesize most B-complex vitamins. Periodically, a ruminant belches, releasing some of the gas and regurgitating some of the food. The animal remasticates and reswallows the food, that is, it ruminates, or chews, its cud (L., *ruminare* = to chew the cud). This process is repeated several times. Most of the organic acids are absorbed directly from the

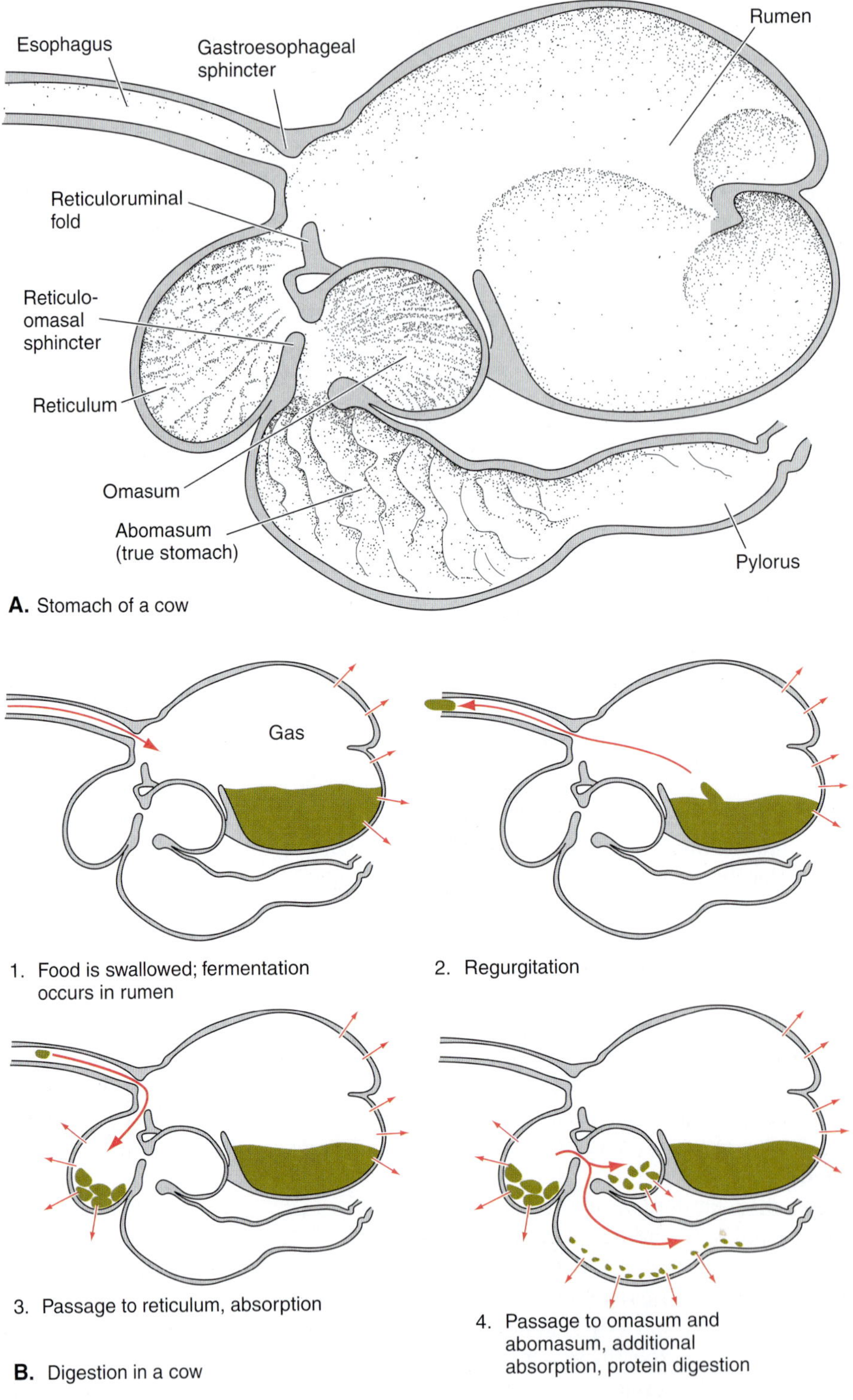

FIGURE 17-6
A, The four chambers of the ruminant "stomach" as represented by a cow. Only the fourth chamber, the abomasum, is comparable with the stomach of other mammals. *B,* Rumination in a cow. Swallowed food accumulates in the rumen; is regurgitated through the esophagus and remasticated in the mouth, possibly several times; and finally passes through the reticulum and omasum into the abomasum. (*Modified after Nickel et al., and Langer.*)

rumen, the internal surface of which is increased greatly by numerous small, leaf-shaped folds and through the reticulated surface of the reticulum (chefs know the wall of this chamber as "tripe"). Up to 70% of a cow's caloric requirements are met by the direct absorption of organic acids from the reticulorumen. Only food that is broken down into very fine particles, liquid, and microorganisms can pass between the deep, leaflike folds of the omasum and go into the abomasum, where normal protein digestion begins. Nearly all of a ruminant's amino acids are derived from microbial proteins harvested from the reticulorumen.

The Intestine and Cloaca

Apart from initial starch digestion in the mouth of some vertebrates and initial protein digestion in the stomach, the intestine is the primary site for both digestion and absorption in all but a few vertebrates, notably the ruminants. Secretions of both the liver and pancreas are added to the intestinal secretions and play an important role in digestion. Although bile released by the liver contains no enzymes, it has a twofold digestive role. Its alkalinity neutralizes the acidity of the gastric contents as they are discharged into the intestine and produces a pH favorable for the action of pancreatic and intestinal enzymes. Its bile salts emulsify fats, breaking them down into small globules that collectively have a large surface area on which pancreatic and intestinal lipase can act.

Both bile and pancreatic secretions are released at a continuous, slow rate in many vertebrates that feed much of the time. Bile and pancreatic secretions are released only when the gastric contents enter the intestine in intermittent feeders. This release is controlled by a combination of nerve reflexes and hormonal action (Chapter 15). Bile may accumulate in a **gallbladder** in some vertebrates. A gallbladder is more likely to be found in intermittent feeders than in continuous feeders, but some herbivores (e.g., some ruminants and rabbits) have one. Carnivores, the diets of which include considerable fat, have a conspicuous gallbladder, where bile is concentrated between meals by the reabsorption of water. Little bile concentration occurs in those herbivores that have a gallbladder.

Both the pancreas and intestine produce **lipase,** which hydrolyzes the small fat globules formed by the action of bile into fatty acids and glycerol. **Pancreatic amylase** continues the breakdown of starch (initiated in some mammals by salivary amylase) into double sugars or disaccharides. Several intestinal **disaccharidases** degrade different disaccharides into single sugars or monosaccharides. The intestine produces **enterokinase,** an enzyme that activates trypsin and chymotrypsin secreted by the pancreas. **Trypsin** and **chymotrypsin** are endopeptidases that resemble gastric pepsin in cleaving long protein chains into shorter fragments, but they act at different biochemical sites. Several **exopeptidases** produced by both the pancreas and intestine degrade polypeptides into the individual amino acids. **Nucleases** from the pancreas hydrolyze nucleic acids.

Digested organic materials, mineral ions, water (much of which was released in the digestive secretions), and other small molecules are absorbed from the intestinal lumen. Fatty acids and fat-soluble vitamins simply **diffuse** through the lipid plasma membrane of intestinal cells, following their concentration gradients. Water and water-soluble materials, such as glycerol, cannot pass directly through the plasma membrane but diffuse through small, protein-lined pores in the membrane. Some minerals, including calcium and iron, bind to a carrier molecule that moves across the membrane in a process called **facilitated diffusion.** Sodium ions bind to a different carrier molecule that moves across the cell as sodium is actively pumped out of the opposite surface of the cell. Most of the amino acids and monosaccharides also bind to the same carrier and are actively transported across the plasma membrane with the sodium in a process called **cotransport.** Undigested residues, together with bacteria that multiply in the intestine, are egested as feces.

The surface area of the intestine available for these activities, like the sites at which they occur, varies considerably among craniates according to the type of food eaten and the animal's level of activity and metabolism. Hagfishes, lampreys, chondrichthyan fishes, and some bony fishes have short, nearly straight intestines that extend from the stomach to the end of the trunk (Fig. 17-2*A–C*). The intestine usually opens into a **cloaca,** a chamber that also receives the excretory and reproductive ducts. The cloaca opens to the body surface by a **cloacal aperture.** In hagfishes and lampreys, the surface area of the intestine is increased slightly by a longitudinal fold that makes a low-pitched spiral along the intestine. In cartilaginous and primitive bony fishes, a tightly pitched fold, the **spiral valve,** increases it much more. The spiral valve considerably increases both transit time and the absorptive surface compared with a straight intestine. This type of intestine shows little regional differentiation, and the entire intestine is called the **valvular intestine.** The terminal part, which in cartilaginous fishes receives the contents of a salt-excreting **digitiform,** or **rectal, gland,** (Chapter 20), is sometimes called the rectum even though it is not homologous to the mammalian rectum.

In more derived bony fishes, such as *Amia*, the spiral valve is reduced or lost, and the intestine lengthens and coils (Fig. 17-2*D*). Herbivorous teleosts have particularly long intestines, increasing transit time of the food. In some species, the intestine lodges a bacterial colony that digests cellulose. Internal longitudinal and transverse folds and, in some species (e.g., mullets), small, finger-like villi further increase the surface area.

In addition to a moderately long and partly coiled intestine, most teleosts have **pyloric caeca** that evaginate from the intestine just distal to the pylorus (Fig. 17-2*E*). Caeca range in number from three to several hundred, and they further increase digestive and absorptive surface area. Sardines and anchovies, which feed on waxy copepods, have caeca that secrete a special **wax lipase.** A cloaca is absent from most teleosts,

FOCUS 17-1 *Gut Structure and Anuran Metamorphosis*

The mode of life of frog tadpoles changes markedly during metamorphosis as they undergo an environmental shift from water to land. All organ systems are affected, but changes in the digestive system are particularly striking. In common ranid frogs, the shift involves a change from a microphagous and herbivorous diet to that of a carnivorous predator. Ranid tadpoles are suspension feeders, ingesting more or less continuously detritus and other small particles, some of which they scrape from the surface of aquatic plants and other objects with their horny beaks. They also reingest much of their own fecal material, in which some microbial digestion of cellulose probably occurs. Minute food particles are trapped by sheets of mucus in the buccopharyngeal cavity, and these sheets are carried caudally by ciliary action into the intestine. Water escapes across the gills into the gill chamber and finally leaves through the spiracle (Fig. A, *top*).

A stomach has barely differentiated in tadpoles, so the food enters a very long intestine, which is approximately 10 times body length in ranid frogs. It is accommodated in the body cavity by forming a tight double coil (Fig. A, *bottom*). A long intestine increases passage time and allows for the processing of a large volume of low-energy food. Food is partially broken down on its first passage through the gut, and more nutrients are extracted on the second passage. The gut does not contain a large microbial colony for cellulose digestion, as formerly believed, but the recycled feces do contain some bacteria. Extracellular digestion is supplemented by some intracellular digestion of phagocytosed food particles.

At the metamorphic climax, when changes are particularly rapid, tadpoles cease feeding. Programmed cell death leads to the loss and reabsorption of much of the digestive tract, and an extensive remodeling occurs. The buccopharyngeal filtering mechanism is completely lost, and a tongue, with which the frog catches and ingests prey, develops. A well-developed stomach, complete with gastric glands, forms, and the intestine shortens to one third or less of its larval length. The gut becomes increasingly muscularized, and all digestion is extracellular. The frog becomes a discontinuous carnivorous predator.

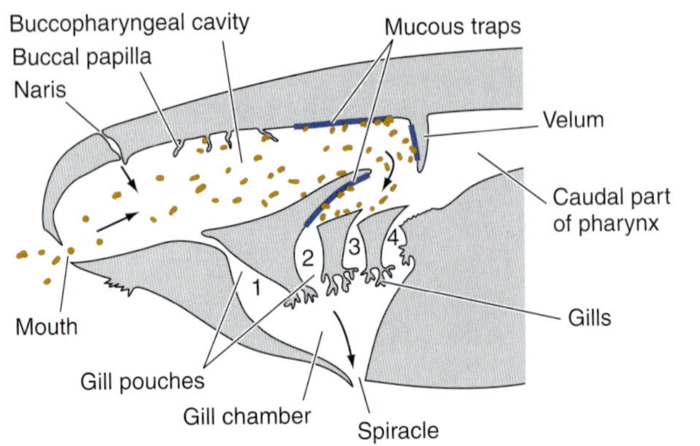

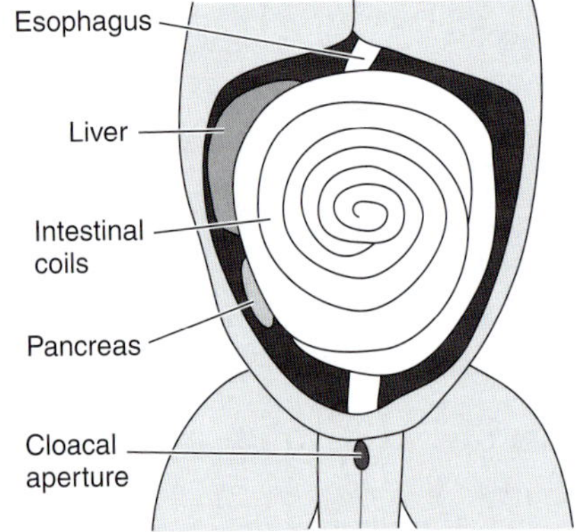

A. Structure of the digestive tract of tadpoles. *Top,* A sagittal section through the buccopharyngeal cavity, showing the filtering mechanism. *Bottom,* A ventral view of the intestinal coils. (*Top, After Hourdry et al.; bottom, after Pretty et al.*)

and the intestine terminates in an **anus** on the body surface.

Lampreys, salmon, and some other fishes stop feeding as they move toward their breeding grounds. Their intestines undergo autodigestion and become greatly reduced.

The intestines are moderately long in most amphibians and reptiles (Fig. 17-3) and very long in herbivorous species, in which bacterial fermentation of cellulose occurs. Particularly striking changes occur during the metamorphosis of frogs as they change from a microphagous and herbivorous diet to a carnivorous one (Focus 17-1). Intestinal surface area also is increased in amphibians and reptiles by internal folds and occasionally by a few villi. The intestine can be divided into a **small intestine** and a slightly wider **large intestine.** The large intestine is short in amphibians and primarily reabsorbs water and temporarily stores feces. Herbivorous lizards and turtles have a large and often sacculated large intestine. A short **intestinal caecum** may be present at the proximal end of the large intestine. The large intestine lodges a bacterial colony that ferments plant food in the same way as occurs in the chambered stomach of some mammals. The resulting organic acids are absorbed directly. A cloaca is present.

As would be expected, the intestines of endothermic birds and mammals are longer and have much more internal surface area than those of amphibians and rep-

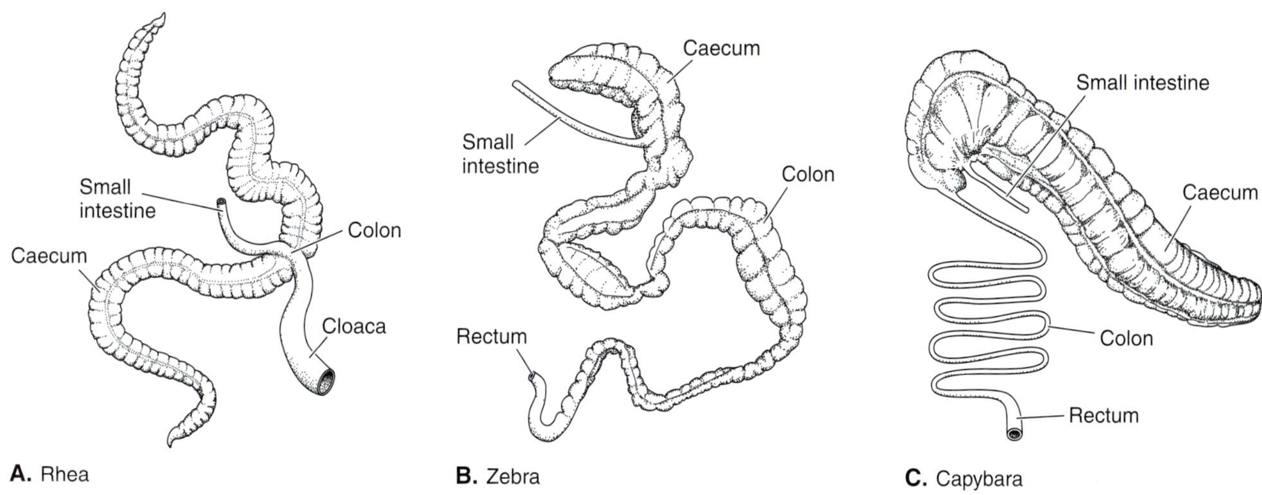

FIGURE 17-7
Modifications in the caecum and colon in hindgut fermenters. *A,* The *Rhea,* a flightless bird. *B,* The zebra. *C,* The capybara, a large South American rodent. (*After Stevens and Hume.*)

tiles (Figs. 17-4 and 17-5). The length of the intestine as a whole is several body lengths in insectivorous and carnivorous species of birds and mammals and much longer in herbivorous and carnivorous species. The length of the intestine of some herbivorous mammals that feed on bulky, low-caloric foods may be 25 body lengths. Most birds do not feed on such low-caloric foods and do not have such long intestines, the weight of which would impair flight. Grouse, which fly only short distances, and rheas, which do not fly at all, feed on plant food. They have a pair of exceptionally long **intestinal caeca** that lodge bacterial colonies, and much digestion and absorption of plant food occurs here (Fig. 17-7*A*).

The internal surface area of the small intestines of birds and mammals is increased greatly by numerous finger-shaped or leaf-shaped intestinal villi (Fig. 17-1). Villi continue into the large intestines of birds but not of mammals. The small intestine can be differentiated into a **duodenum, jejunum,** and **ileum** (Fig. 17-4). The basis for these regions is primarily the degree of vascularity and the number and types of glands. Functions of these regions overlap, but, in general, more digestive activity occurs in the duodenum and jejunum, and more absorption occurs in the ileum. These regions are more clearly defined in mammals than in birds.

A **caecum** of variable size is present at the point where the small intestine of mammals enters the large intestine, or **colon** (Fig. 17-4). The colon continues to discharge into a **cloaca** in birds and monotreme mammals (Fig. 17-8). A copulatory organ is located in the ventral cloacal wall of the cloaca, or tail base, in male reptiles and a few birds. Birds have a dorsal cloacal diverticulum, the **bursa of Fabricius,** which is the site for the maturation of B lymphocytes. These cells are involved in certain immune reactions. In marsupials and eutherians, the embryonic cloaca becomes divided during development. The dorsal part contributes to the **rectum,** which opens at the anus on the body surface; the ventral part contributes to the urogenital passages (Chapter 20).

The caecum and colon are of moderate length and size in carnivorous and omnivorous mammals, such as cats and humans (Fig. 17-4). A **vermiform appendix** may lie at the end of the caecum of some mammalian species, including humans. Its walls contain many nodules of lymphocytes, and it probably is a part of the body's immune system. The mammalian colon contains bacteria that produce cellulase, and the resulting organic acids are absorbed from the colon. Mammals

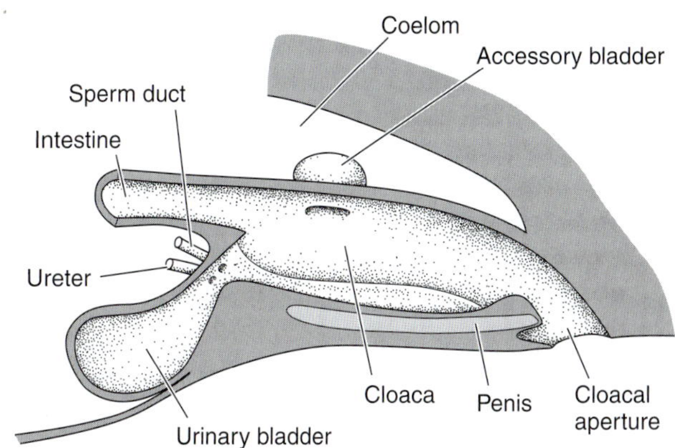

FIGURE 17-8
A longitudinal section through the cloaca of a male turtle.

FOCUS 17-2 *The Functional Anatomy of Large Mammalian Herbivores*

Zoologists recognize two groups of hoofed mammals, or ungulates: (1) the odd-toed perissodactyls (i.e., tapir, rhinoceros, horse) and (2) the even-toed artiodactyls (i.e., pigs, deer, camels, sheep, antelope, and cattle). The perissodactyls evolved in the Paleocene, when widespread tropical forest provided a wide choice of plant food, including high-quality, low-fiber herbage. Perissodactyls were at first predominant, but the increasing spread of grasslands in the Miocene, in which the forage had a high fiber content, favored the artiodactyls.

The foregut-fermenting artiodactyls need only partly chew their food as they gather it. They then regurgitate and thoroughly masticate the food, often when and where they can keep alert to predators. A relatively long time is needed for food to be broken down mechanically and chemically in the reticulorumen to the very fine particles that can pass through the omasum to the abomasum. Transit time through the digestive tract of a cow averages about 80 hours. A disadvantage is that not all sugars and proteins released from the macerated cells can be utilized by the animal. Some are degraded by the fermentation process. An advantage is that some of the nitrogen released as ammonia during fermentation in the reticulorumen is utilized by the microorganisms to synthesize their own protein, which is nearly fully utilized as the food passes through the abomasum and intestine. Plant alkaloids and other toxins also can be detoxified as the food ferments in the reticulorumen.

Hindgut-fermenting perissodactyls masticate their food as they eat. Plant toxins are absorbed as the food mass passes through the stomach and intestine and must be detoxified in the liver. Digestion of cellulose occurs in the caecum, colon, or both but is only 70% as efficient as in foregut fermenters. (You may have noticed the relatively high fiber content of horse feces compared with cow dung.) Little of the microbial protein synthesized in the caecum and colon can be utilized because large hindgut fermenters do not eat their fecal material. It would seem that foregut fermenters have all of the advantages, but hindgut fermenters have some. Soluble nutrients (sugars and proteins) released from macerated cells are not degraded by fermentation because they are processed before the food reaches the hindgut fermentation chamber. Hindgut fermenters also have the advantage of a relatively fast transit time, averaging 48 hours in a horse. When forage is abundant, a hindgut fermenter can obtain a great deal of energy relatively quickly by processing a large volume of plant food, but the food must contain the essential nutrients and calories. Because of the slower passage time for food, foregut fermenters cannot do this.

Janis (1976), who first made comparisons of the advantages and disadvantages between foregut and hindgut fermenters, argued that horses do better than ruminants on low-quality diets because they can compensate for the low quality (i.e., high fiber) by processing more food rapidly. Janis assumed that where foregut fermenters (antelope) and hindgut fermenters (zebras) share a habitat, the zebra eats the poorer food. They do eat the grasses, but observations by McNaughton (1985) show that they select grasses with high-quality seeds rich in carbohydrates and protein. Eating stems may be incidental to harvesting the seeds. Alexander (1993) has developed a quantitative model that compares the relative merits of foregut and hindgut fermenters. His model predicts that the optimum gut for a mammal feeding on poor-quality (i.e., high-fiber) food has a large foregut fermentation chamber.

Foregut fermenters appear to have the overall advantage, even when food quality is low. Their geologic history and present-day distribution support this. Artiodactyls replaced perissodactyls as grassland became more widespread in the Miocene. It is artiodactyls and not perissodactyls that can thrive in habits with low-quality (i.e., high-fiber) foods. Examples are mountain sheep and goats in alpine areas and camels in semiarid or desert areas. The two groups of ungulates, however, have coexisted successfully since the late Eocene by engaging in resource partitioning and utilizing different plants or parts of plants.

Many species of ruminating antelopes also coexist on the eastern African plains. Whether these species coexist by partitioning food resources based on their body size or on the quality of food that they choose is unclear. Gordon and Illius (1996) have reexamined the evidence and concluded that, in fluctuating resource environments, selection would favor the ability to use poor-quality food and hence favor large body size, but some small species would survive by out-competing the larger species for the limited high-quality food items available.

with relatively short colons may utilize a moderate amount of cellulose in this fashion. Some vitamins (including vitamin K, needed for blood clotting) also are synthesized by colonic bacteria and absorbed from the colon. But the colon of most mammals is primarily a site for water reabsorption and for the consolidation of the feces.

In perissodactyls (rhinoceros, horse), rodents, and lagomorphs (rabbits), the caecum, colon, or both are very long, large, and sacculated (Fig. 17-7*B* and *C*). These mammals do not have a chambered stomach. They are often called **hindgut fermenters,** in contrast to the **foregut fermenters** with a chambered stomach. The caecum and colon lodge the bacterial flora that ferments plant food, and absorption of the breakdown products occurs here. Bacteria multiply in the caecum and colon, utilizing as nitrogen sources residual nitrogen in the food and nitrogen and urea that diffuse in from the blood. However, most hindgut fermenters harvest less microbial protein than do foregut fermenters because the hindgut lies posterior to the primary digestive and absorptive surfaces of the gut. Foregut and hindgut fermenters represent different solutions among mammals to the effective utilization of plant food. Each method has certain advantages and disadvantages (Focus 17-2).

Many small mammalian hindgut fermenters (e.g., rodents and lagomorphs) are able to obtain more calories and nutrients from their food than can larger species (e.g., perissodactyls and artiodactyls) by eating 25% to 60% of their feces. This practice is known as **coprophagy** (Gr., *kopros* = dung + *phagein* = to eat). Coprophagy is important for small mammalian species because they must sustain relatively high metabolic rates. Metabolic rates for the large mammalian herbivores are much lower and, because of their size, they are able to consume and process a volume of food sufficiently large to meet their metabolic requirements. Coprophagous mammals produce two types of feces: hard pellets during the part of the day that they are active (usually nighttime), and a softer, more watery feces when they are resting or sheltering from predators. It is the soft feces that they eat. Alexander (1993) has developed a quantitative model of coprophagy. His model predicts that coprophagy is particularly advantageous for small hindgut fermenters.

The specializations of the digestive tracts of birds and mammals enable them to quickly digest and absorb the large amount of food that they need to sustain their high level of metabolism. They grind their food in the mouth or gizzard, digestive glands are numerous, a high body temperature accelerates enzyme action, their intestinal surface area is large, the walls of the intestine are very vascular, gut muscles are well developed, and gut mobility is high. Transit time through the gut is far faster in birds and mammals than in other vertebrates. A snake may require a week or more to digest a mouse, but an owl or fox can digest a mouse in a day.

The Liver and Pancreas

We have seen that the liver and pancreas have important digestive roles. These organs also have additional vital functions. The liver develops as a ventral evagination from the archenteron at the level of the transverse septum and just caudal to the stomach and heart. It remains connected to the intestine by the **common bile duct** (Figs. 16-1 and 17-9). The pancreas develops from one or more primordia. Usually, a **ventral pancreatic bud** arises from the liver primordium, and one or sometimes two **dorsal pancreatic buds** arise as separate archenteron evaginations. All the pancreatic primordia usually extend into the dorsal mesentery and unite. Either the stalk of the ventral pancreas bud, that of the dorsal pancreatic bud, or both may persist in the adult as pancreatic ducts.

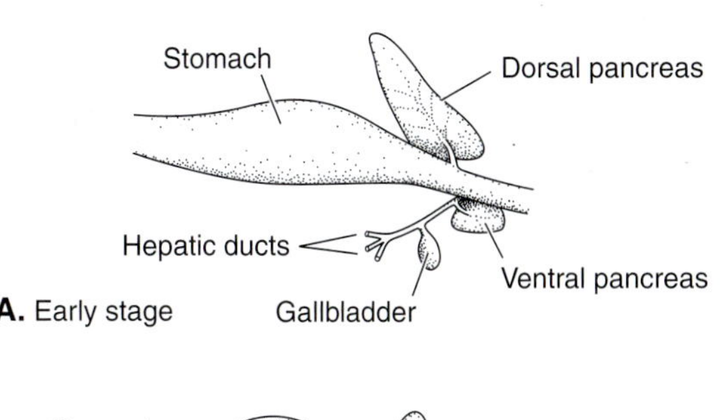

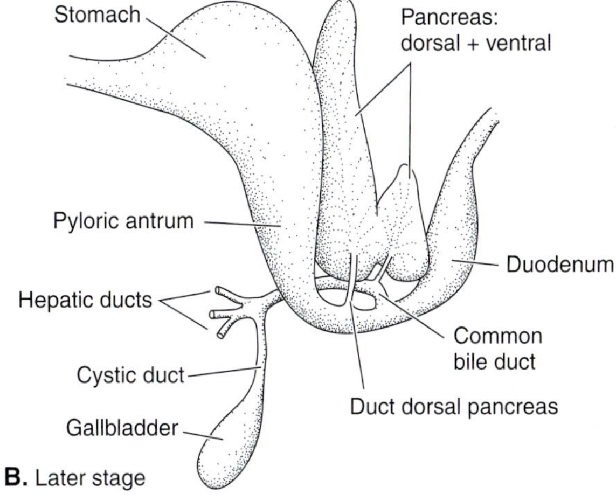

FIGURE 17-9
Two stages in the development of the mammalian pancreas. *A,* An early stage, when the organs have appeared. *B,* A later stage, when the organs are differentiating.

The Liver

In the adults of all vertebrates, the liver is the largest organ in the body cavity, and it occupies most of the space in the abdomen. Microscopically, the liver consists of many groups, or **lobules,** of **hepatic plates** (Gr., *hepat-* = liver) composed of hepatic cells (Fig. 17-10). The plates of each lobule usually are organized around a **central vein,** which is a tributary of the hepatic vein (Chapter 19) and so leads directly to the heart or into a vein that enters the heart. Branches of the **hepatic artery,** containing oxygenated blood, and branches of the **hepatic portal vein,** which has drained

FIGURE 17-10
The mammalian liver. *A,* A microscopic view of a cross section of parts of several lobules showing their structure and relationship to associated blood vessels and ducts. *B,* A stereodiagram of a portion of a liver lobule. (*A, After Hammersen; B, after Elias.*)

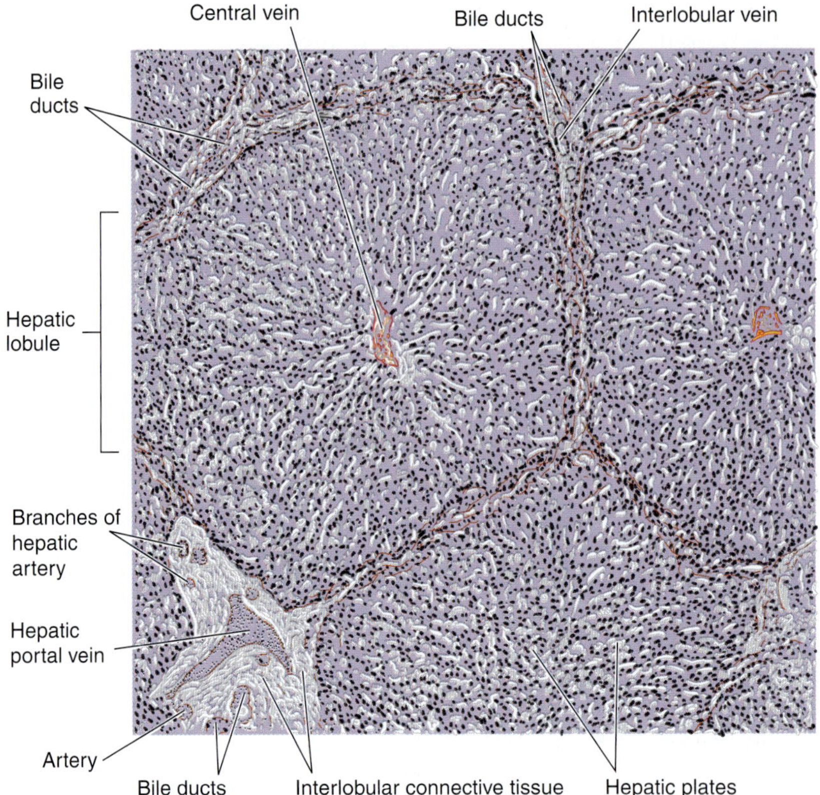

A. Microscopic section of mammalian liver

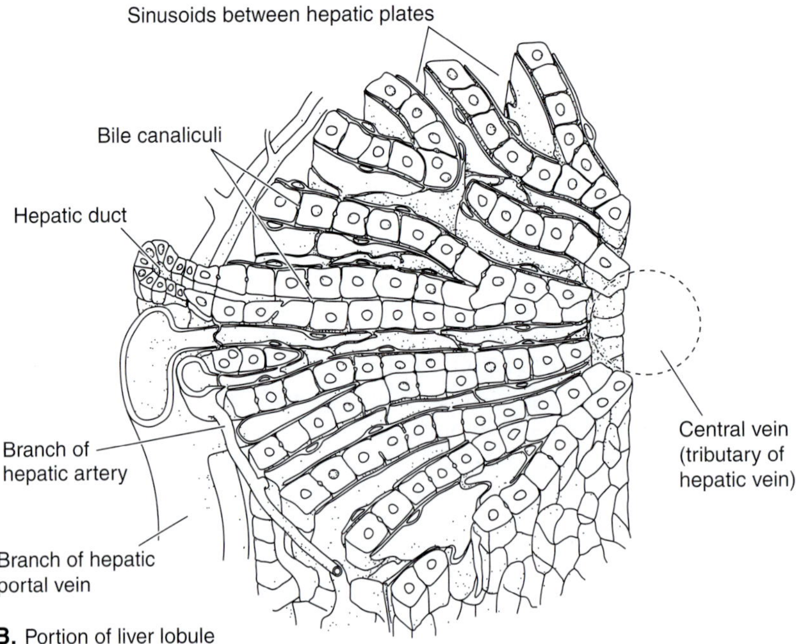

B. Portion of liver lobule

the stomach, intestine, and spleen, lie peripheral to each hepatic lobule. **Hepatic sinusoids** pass between and through the hepatic plates as they carry blood from the hepatic artery and hepatic portal vein to the central vein. Although the hepatic sinusoids are the size of capillaries, they lack the complete endothelial lining of capillaries, so blood in them comes into direct contact with many hepatic cells and with phagocytic cells within the sinusoids. Minute **bile canaliculi** lie between hepatic cells and drain bile, the secretion of the hepatic cells, into **hepatic ducts** that leave the liver and unite to form the **common bile duct** (Fig. 17-9*B*).

Often, a **gallbladder** evaginates from the common bile duct and remains connected to it by the **cystic duct** (Fig. 17-9*B*). A sphincter at the intestinal entrance of the common bile duct normally is contracted so that bile backs up into the gallbladder, where it is stored and sometimes concentrated by water reabsorption. When the gastric contents are discharged into the intestine, this sphincter relaxes, the gallbladder contracts, and bile flows into the intestine.

Liver structure and functions are related closely to both the digestive and the circulatory systems. Bile is a very alkaline solution containing **bile pigments,** which are excretory products derived from the breakdown of hemoglobin in the liver, and **bile salts,** which, as we have discussed, emulsify fats.

Other liver functions derive from its intimate circulatory connections. Worn-out red blood cells are destroyed in the spleen, but the hemoglobin that is released is carried to the liver in the hepatic portal system and taken up by hepatic cells. The iron part of the molecule is salvaged, but the rest of the molecule is converted to bile pigments and excreted in the bile. New blood cells develop within the connective tissues of the liver in the embryos of all vertebrates and in the adults of anamniotes. All products that are absorbed from the intestine (except for much of the fat, which enters the lymphatic system) are carried to the liver by the hepatic portal vein and pass through the hepatic sinusoids before being distributed to the body. Many harmful materials are phagocytosed and detoxified. Numerous interconversions of glucose, amino acids, and fats occur. Many food molecules in excess of the animal's immediate needs are converted to glycogen or, in many fishes, lipids and are stored in the hepatic cells. When amino acids undergo conversion into other materials, the amino group is removed first and converted within the hepatic cells into ammonia, urea, or other waste products, which are carried by the blood to the kidneys or other excretory surfaces and excreted. Between meals, when blood sugar levels decrease, food stored in the hepatic cells is reconverted to glucose and released into the circulatory system. Most proteins in the blood plasma are synthesized in the liver, as is the yolk that is transferred by female vertebrates to their eggs. Under the influence of pituitary growth-stimulating hormone, liver cells also synthesize the hormone **somatomedin,** which stimulates growth.

The Pancreas

Every craniate has pancreatic tissue. In lampreys, lungfishes, and teleost fishes, however, the pancreatic tissue is embedded in the wall of the intestine or scattered in the mesentery and is not a grossly visible pancreas. In other vertebrates, the pancreas is a complex gland, containing both exocrine and endocrine portions. We considered the endocrine portion in Chapter 15. The exocrine portion, which constitutes most of the organ, is a compound acinar gland (Fig. 15-14). Numerous minute, roundish acinar units, each of which is made up of several types of cells, synthesize the many digestive enzymes that are released into the intestine in an alkaline solution when the gastric contents are discharged.

SUMMARY

1. The pharynx of fishes is a large and important part of the respiratory system. In terrestrial vertebrates, it is a relatively short food passage.
2. The thymus and palatine tonsils begin their development as epithelial buds from certain pharyngeal pouches. They soon are invaded by lymphocytes and become part of the body's immune system.
3. The gut tube is composed of a mucosa that contains mucus-producing, enzyme-producing, and hormone-producing cells; a connective tissue submucosa; two layers of smooth muscles; and a connective tissue and epithelial serosa.
4. The esophagus is particularly long in terrestrial vertebrates, for the pharynx is short and a neck has developed. It is modified as a crop in grain-eating birds, but normally it is a simple food-conducting passage.
5. Stomachs were probably absent from the first vertebrates. They later evolved as storage chambers when vertebrates began to feed intermittently on larger food. Gastric glands produce hydrochloric acid and the endopeptidase pepsinogen. They also produce chitinase in some tetrapods and renin in young mammals.
6. The stomach is modified as a gizzard that grinds food in some fishes, some reptiles, and all birds. The stomach of ruminants and some other mammals is a multichambered organ in which cellulose is digested by cellulase, which is produced by a colony of bacteria and other microorganisms. Considerable absorption occurs in the reticuloru-

men, and much of a ruminant's amino acids are derived from harvesting bacteria.

7. Secretions of the liver and pancreas are added to intestinal secretions and play an important role in digestion. Bile from the liver neutralizes the acid gastric contents when they enter the intestine, and the liver's bile salts emulsify fats. Many pancreatic and intestinal enzymes break down all categories of food particles into small molecules that can then be absorbed along with water, minerals, and other needed materials. Undigested residue and a considerable volume of bacteria are eliminated as feces.
8. The configuration and divisions of the intestine vary greatly among vertebrates. They are closely related to the animal's type of food, body size, and level of activity. In general, ectotherms have shorter and simpler intestines with less surface area than do endotherms. The intestine and urogenital ducts of most vertebrate groups terminate in a common chamber, the cloaca. In therian mammals, the cloaca becomes divided.
9. Intestinal bacteria, like those in the chambered stomach of some mammals, can ferment cellulose, and the resulting organic acids can be absorbed from the intestine. This process is particularly pronounced in herbivorous tetrapods, and their intestinal caecum, parts of the colon, or both are greatly enlarged to accommodate the bulky plant food and bacterial colony. Fermentation in the foregut (stomach) or hindgut has certain advantages and disadvantages.
10. The liver produces bile, which emulsifies fat and facilitates its absorption. Blood returning from the digestive tract passes through liver sinusoids. Hepatic cells detoxify many noxious substances, interconvert many absorbed food products, store excess food as glycogen or lipid, release food into the blood as needed, synthesize urea and many plasma proteins, assist in iron metabolism, and produce the hormone somatomedin.
11. The pancreas develops as outgrowths of the liver diverticulum and adjacent parts of the intestine. Pancreatic tissue remains embedded in the intestinal wall in lampreys, lungfishes, and teleosts but forms a distinct organ in other vertebrates. Many of the digestive enzymes that act in the intestine are synthesized in the pancreas.

REFERENCES FOR CHAPTERS 16 AND 17

Aerts, P., Osse, J. W. M., and Verraes, W., 1987: Model of jaw depression during feeding in *Astatoltilapia elegans* (Teleostei: Cichlidae): Mechanisms of energy storage and triggering. *Journal of Morphology,* 194:85–109.

Alexander, R. McNeil, 1993: The energetics of coprophagy: A theoretical analysis. *Journal of Zoology (London),* 230:629–637.

Alexander, R. McNeil, 1993: The relative merits of foregut and hindgut fermentation. *Journal of Zoology (London),* 231:391–401.

Alexander, R. McNeil, 1970: Mechanics of feeding action of various teleost fishes. *Journal of Zoology (London),* 162:145–156.

Barel, C. D. N., 1983: Towards a constructional morphology of cichlid fishes (Teleostei, Perciformes). *Netherlands Journal of Zoology,* 33:357–424.

Barrington, E. J. W., 1945: The supposed pancreatic organs of *Petromyzon fluviatilis* and *Myxine glutinosa. Quarterly Journal of Microscopical Science,* 85:391–417.

Bels, V. L., and Baltus, I., 1989. First analysis of feeding in *Anolis* lizards. *Progress in Zoology,* 35:141–145.

Bolk, L., 1931: *Handbuch der Vergleichenden Anatomie der Wirbeltiere,* 6 volumes. Reprinted Amsterdam, A. Asher and Co., 1967.

Bramble, D. M., and Wake, D. B., 1985: Feeding mechanisms of lower tetrapods. *In* Hildebrand, M., Bramble, D. M., Liem, K. F., and Wake, D. B., editors: *Functional Vertebrate Morphology.* Cambridge, Harvard University Press.

Butler, P. M., and Joysey, K. A., editors, 1978: *Development, Function and Evolution of Teeth.* New York, Academic Press.

Carrier, D. R., 1995: Mechanism of lung ventilation in the caecilian *Dermophis mexicanus. Journal of Morphology,* 226:289–295.

Chivers, D. J., and Hladik, C. M., 1980: Morphology of the gastrointestinal tract in primates: Comparisons with other mammals in relation to diet. *Journal of Morphology,* 166:337–386.

Chivers, D. J., and Langer, P., editors, 1994: *The Digestive System in Mammals: Food, Form and Function.* Cambridge, Cambridge University Press.

Clark, R. B., 1964: *Dynamics of Metazoan Evolution: The Origin of the Coelom and Its Segments.* Oxford, Oxford University Press.

Crompton, A. W., 1971: The origin of the tribosphenic molar. *In* Kermack, D. M., and Kermack, K. A., editors: *Early Mammals.* London, Linnean Society.

Crompton, A. W., 1995: Masticatory function in nonmammalian cynodonts and early mammals. *In* Thomason, J. J., editor: *Functional Morphology in Vertebrate Evolution.* Cambridge, Cambridge University Press.

Crompton, A. W., Thexton, A. J., Hiiemae, K. M., and Cook, P., 1975: The movement of the hyoid apparatus during chewing. *Nature,* 258:69–70.

Dahlberg, A. A., editor, 1971: *Dental Morphology and Evolution.* Chicago, University of Chicago Press.

Denison, R. H., 1961: Feeding mechanisms of Agnatha and early gnathostomes. *American Zoologist,* 1:177–181.

Doran, G. A., 1975: Review of the evolution and phylogeny of the mammalian tongue. *Acta Anatomica,* 91:118–129.

Elias, H., 1949: A re-examination of the structure of the mammalian liver: II. The hepatic lobule and its relation to the vascular and biliary systems. *American Journal of Anatomy,* 85:379–456.

Elias, H., and Sherrick, J. C., 1969: *Morphology of the Liver.* New York, Academic Press.

Frazzetta, T. H., 1988: The mechanics and form of shark teeth (*Chondrichthyes, Elasmobranchii*). *Zoomorphology,* 108:93–107.

Frazzetta, T. H., 1962: A functional consideration of cranial kinesis in lizards. *Journal of Morphology,* 111:287–320.

Gans, C., 1969: Comments on inertial feeding. *Copeia,* 1969:855–857.

Gordon, I. J., and Illius, A. W., 1996: The nutritional ecology of African ruminants: A reinterpretation. *Journal of Animal Ecology,* 65:18–28.

Grajal, A., Strahl, S. D., Parra, R., Dominguez, M. G., and Neher, A., 1989: Foregut fermentation in the hoatzin: A neotropical leaf-eating bird. *Science,* 245:1236–1238.

Hammersen, F., 1980: *Histology: A Color Atlas of Cytology, Histology, and Microscopic Anatomy.* Baltimore, Urban & Schwarzenberg.

Herring, S. W., 1993: Functional morphology of mammalian mastication. *American Zoology,* 33:289–299.

Hiiemae, K. M., and Crompton, A. W., 1985: Mastication, food transport, and swallowing. *In* Hildebrand, M., Bramble, D. M., Liem, K. F., and Wake, D. B., editors: *Functional Vertebrate Morphology.* Cambridge, Harvard University Press.

Hourdry, J., L'Hermite, A., and Ferrand, R., 1996: Changes in the digestive tract and feeding behavior of anuran amphibians during metamorphosis. *Physiological Zoology,* 69:219–251.

Janis, C., 1976: Evolutionary strategy of the Equidae and origins of the rumen and cecal digestion. *Evolution,* 30:757–774.

Janis, C. M., and Fortelius, M., 1988: On the means whereby mammals achieve increased functional durability of their dentitions, with special reference to limiting factors. *Biological Reviews,* 63:197–230.

Jennings, J. B., 1972: *Feeding, Digestion and Assimilation in Animals,* 2nd edition. London, Macmillan, St. Martin's Press.

Kardong, K. V., 1980: Evolutionary patterns in advanced snakes. *American Zoologist,* 20:269–282.

Lagler, K. F., Bardach, J. E., and Miller, R. E., 1962: *Ichthyology.* New York, John Wiley & Sons.

Langer, P., 1988. *The Mammalian Herbivore Stomach: Comparative Anatomy, Function and Evolution.* New York, G. Fischer.

Lauder, G. V., 1982. Patterns of evolution in the feeding mechanism of actinopterygian fishes. *American Zoologist,* 22:275–285.

Lauder, G. V., 1985: Aquatic feeding in lower vertebrates. *In* Hildebrand, M., Bramble, D. M., Liem, K. F., and Wake, D. B., editors: *Functional Vertebrate Morphology.* Cambridge, Harvard University Press.

Lauder, G. V., 1979: Feeding mechanisms in primitive teleosts and the halecomorph fish *Amia calva. Journal of Zoology (London),* 187:543–578.

Lauder, G. V., and Prendergast, T., 1992: Kinematics of prey capture in the snapping turtle, *Chelydra serpintena. Journal of Experimental Biology,* 164:55–78.

Leeuwen, J. L. van, and Muller, M., 1984: Optimum sucking techniques for predatory fish. *Transactions of the Zoological Society of London,* 37:137–169.

Liem, K. F., 1990: Aquatic versus terrestrial feeding modes: Possible impact on the trophic ecology of vertebrates. *American Zoologist,* 30:209–221.

Liem, K. F., 1980: Adaptive significance of intra- and interspecific differences in the feeding repertoires of cichlid fishes. *American Zoologist,* 20:295–314.

Liem, K. F., 1978: Modulatory multiplicity in the functional repertoire of the feeding mechanism in cichlid fishes: I. Piscivores. *Journal of Morphology,* 158:323–360.

Lombard, E., and Wake, D. B., 1976: Tongue evolution in the lungless salamanders, family Plethodontidae: I. Introduction, theory, and a general model of dynamics. *Journal of Morphology,* 148:265–286.

McNaughton, S. K., 1985: Ecology of a grazing ecosystem: The Serengeti. *Ecological Monographs,* 55:259–294.

Mitchell, P. C., 1916: Further observations of the intestinal tracts of mammals. *Proceedings of the Zoological Society of London,* 1916:183–251.

Mitchell, P. C., 1901: On the intestinal tract of birds. *Transactions of the Linnean Society of London,* series 2, 8:173–275.

Moss, M., 1970: Enamel and bone in shark teeth, with a note on fibrous enamel in fishes. *Acta Anatomica,* 77:161–187.

Motta, P. J., and Wilga, C. A., 2000: Advances in the study of feeding behaviors, mechanisms, and mechanics of sharks. *Environmental Biology of Fishes,* 20:1–26.

Motta, P. J., and Wilga, C.A., 1999: Anatomy of the feeding apparatus of the nurse shark, *Ginglymostoma cirratum. Journal of Morphology,* 241:33–60.

Motta, P. J., and Wilga, C. A., 1995: Anatomy of the feeding apparatus of the lemon shark, *Negaprion brevirostris. Journal of Morphology,* 226:309–329.

Nickel, R., Schummer, A., and Seiferle, E., 1979: *The Viscera of the Domestic Mammals,* 2nd edition. New York, Springer-Verlag.

Noble, G. K., 1931: *The Biology of the Amphibia.* New York, McGraw-Hill.

Oguri, M., 1964: Rectal glands of marine and fresh water sharks: Comparative histology. *Science,* 144:1151–1152.

Osborn, J. W., and Crompton, A. W., 1973: The evolution of mammalian from reptilian dentitions. *Brevoria,* 399:1–18.

Osse, J. W. M., 1969: Functional morphology of the head of the perch (*Perca fluviatilis L.*): An electromyographic study. *Netherlands Journal of Zoology,* 19:289–392.

Peyer, B., 1968: *Comparative Odontology.* Translated by R. Zangerl. Chicago, University of Chicago Press.

Pough, F. H., 1983: Adaptive radiation within a highly specialized system: The diversity of feeding mechanisms in snakes. *American Zoologist,* 23:337–460.

Pretty, R., Naitoh, T., and Wassersug, R. J., 1995: Metamorphic shortening of the alimentary tract in anuran larvae (*Rana catesbeiana*). *Anatomical Record,* 242:417–423.

Preuss, F., and Fricke, W., 1979: Comparative schemata on the histology of the liver with consequences in terminology. *Journal of Morphology,* 162:211–220.

Reilly, S. M., and Lauder G. V., 1990. The evolution of tetrapod feeding behavior: Kinematic homologies in prey transport. *Evolution,* 44:1542–1557.

Robbins, C. T., Spalinger, D. E., and van Hoven, W., 1995: Adaptation of ruminants to browse and grass diets: Are anatomical-based browser-grazer interpretations valid? *Oecologia,* 103:208–213.

Roger, T., Cabanie, P., and Ferre, J. P., 1991: Microscopic and functional anatomy of the ileal papilla and caecocolic valve in the rat. *Acta Anatomica,* 142:299–305.

Rogers, T. A., 1958: The metabolism of ruminants. *Scientific American,* 198:34–38.

Ruckebusch, Y., and Thivend, P., editors, 1980: *Digestive Physiology and Metabolism of Ruminants.* Westport, Conn., AVI.

Sanderson, S. L., and Wassersug, R., 1990: Suspension-feeding vertebrates. *Scientific American,* 262:96–101.

Schaeffer, B., and Rosen, D. E., 1969: Major adaptive levels in the evolution of the actinopterygian feeding mechanism. *American Zoologist,* 1:187–204.

Schwenk, K., 1986: Morphology of the tongue in the tuatara, *Sphenodon punctatus* (Reptilia: Lepidosauria), with comments on function and phylogeny. *Journal of Morphology,* 188:129–156.

Schwenk, K., and Throckmorton G. S., 1989: Functional and evolutionary morphology of lingual feeding in squamate reptiles: Phylogenetics and kinematics. *Journal of Zoology (London),* 219:153–175.

Schwenk, K. (editor), 2000: *Feeding: Form, Function, and Evolution in Tetrapod Vertebrates.* New York, Academic Press.

Skoczylas, R., 1978: Physiology of the digestive tract. *In* Gans, C., et al., editors: *Biology of the Reptilia*, volume 8. New York, Academic Press.

Smith, K. K., 1992: The evolution of the mammalian pharynx. *Zoological Journal of the Linnean Society,* 104:313–349.

Smith, K. K., 1984: The use of the tongue and hyoid apparatus during feeding in lizards (*Ctenosaura similis* and *Tupinambis nigropunctatus*). *Journal of Zoology (London),* 202:115–143.

Stevens, C. E., and Hume, I. D., 1995: *Comparative Physiology of the Vertebrate Digestive System,* 2nd edition. Cambridge, Cambridge University Press.

Summers, A. P., Darouian, K. F., Richmond, A. M., and Brainerd, E. L., 1998: Kinematics of aquatic and terrestrial prey capture in *Terrapene carolina,* with implications for the evolution of feeding in cryptodire turtles. *Journal of Experimental Zoology,* 281:280–287.

Thexton, A. J., and McGarrick J. D., 1989. Tongue movement in the cat during the intake of solid food. *Archives of Oral Biology,* 34:239–248.

Walker, W. F., Jr., 1998: *Dissection of the Frog*, 2nd edition, corrected reprint. New York, W. H. Freeman & Co.

Wells, L. J., 1954: Development of the human diaphragm and pleural sacs. *Carnegie Contributions to Embryology,* 35:107–134.

Wilga, C. D., 1998: Conservation and variation in the feeding mechanisms of the spiny dogfish, *Squalus acanthias. Journal of Experimental Biology,* 201:1345–1358.

Wilga, C. D., and Motta, P. J., 1998: Feeding mechanism of the Atlantic guitarfish *Rhinobatos lentiginosus:* Modulation of kinematic and motor activity. *Journal of Experimental Biology,* 201:3167–3184.

Wilga, C. D., and Motta, P. J., 2000: Durophagy in sharks: Feeding mechanics of the hammerhead shark, *Sphyrna tiburo. Journal of Experimental Biology,* 203:2781–2796.

Wilga, C. D., Wainwright, P. C., and Motta, P. J., 2000: Evolution of jaw mechanics in vertebrates: Insights from Chondrichthyes. *Biological Journal of Linnean Society,* 71.

Williams, P. L., Warwick, R., Dyson, M., and Bannister, L. H., editors, 1989: *Gray's Anatomy*, 37th edition. Edinburgh, Churchill Livingstone.

18

The Respiratory System

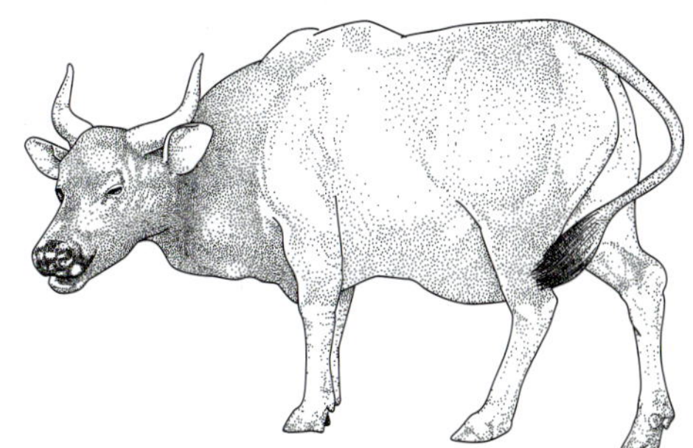

PRECIS

The metabolic energy used by animals is derived from the cellular oxidation of food molecules. The oxygen that is needed, like the carbon dioxide produced as a by-product, enters and leaves the body by diffusion at a respiratory membrane. Water or air, which contains oxygen and carbon dioxide, must be moved across the respiratory membrane by a ventilation process. Many evolutionary changes occurred as vertebrates moved from water to land and eventually became endothermic animals.

OUTLINE

Most of the body's energy needs are supplied by the cellular oxidation of absorbed food products. In very small animals, the oxygen needed for this, and the carbon dioxide produced as a by-product, can be carried between the external environment and the cells by diffusion, but diffusion is a slow process, and it is efficient only for short distances. Diffusion plays a role in vertebrates in the movement of gases over short distances, but systems also are needed for the **bulk flow,** or movement by a muscular pump, of the medium containing the gases over longer distances. First, some sort of a pump must move the medium containing the gases—water or air—across a thin, moist, and vascular membrane through which the gases can diffuse and be exchanged with those in the blood. A membrane of this type is called a **respiratory membrane.** Second, another pump, the heart, must move the blood through vessels that extend between the respiratory membrane and the vicinity of the cells, where diffusion of gases again takes place. We will consider the first set of problems, that is, the nature of the respiratory membrane and the bulk flow of the external medium across it, a process called **ventilation,** in this chapter.

An efficient respiratory system optimizes the diffusion and exchange of gases between the body and the external environment, but many factors affect the nature of the respiratory membrane and the way it is ventilated. The rate of diffusion, R, follows the general equation for diffusion:

$$R = D \times A \times (\Delta p/d)$$

D is a diffusion constant, the value of which depends on the properties of the medium (density, temperature) through which diffusion takes place. A

is the surface area across which diffusion is occurring (i.e., the respiratory membrane). Δp is the difference in partial pressures of the gas on either side of the respiratory membrane, and d is the distance (i.e., the thickness of the respiratory membrane) along which diffusion occurs. High diffusion rates require a large surface area (A) in the respiratory membrane and a short diffusion distance (d) between the external medium and the blood. Flow rates of the medium across the respiratory membrane, and of blood through it, must be slow enough to juxtapose the medium and blood long enough for diffusion to occur. Flow rates must also be fast enough to maintain a difference in the concentration of gases in the medium and blood, that is, to maintain a diffusion gradient (Δp). Ventilation rates or the amount of respiratory membrane in use at any particular time, or both, should be adjustable so that gas exchange matches the needs of the animal as they vary with the animal's level of activity. To provide more respiratory surface or ventilation than an animal needs during its greatest activity would be uneconomical. The quantity of gas in the water or air also affects the size of the membrane and its rate of ventilation. The density and viscosity of the medium affect the amount of energy required to move it across the respiratory membrane. The movement of other diffusible molecules, such as water, salts, and nitrogenous wastes, across the membrane may need to be reduced or enhanced. Body size is an important factor because of its effect on surface–volume relationships. These considerations show that respiratory membranes and the methods of ventilating them differ greatly among vertebrates.

The Respiratory System of Fishes

A fish's respiratory system must be adapted to two major constraints of life in water. First, the amount of oxygen dissolved in water is much less than the concentration of oxygen in air. The quantity dissolved in water depends on the partial pressure of the oxygen in the air, which is about 160 mm Hg at sea level, because this is the force that drives oxygen into the water. The amount also depends on solubility, and oxygen is not very soluble in water. Although 20°C air at sea level contains 210 mL/L of oxygen, fresh water under the same conditions contains only 6.6 mL oxygen/L. Salt water contains somewhat less, 5.3 mL/L. Solubility of oxygen increases slightly in cold water, so fresh water at 0°C holds 10.3 mL/L. The second problem for a fish is that the water in which the oxygen is dissolved is much denser and more viscous than air. Because of these two constraints, a fish must have a rather large respiratory surface and move a large volume of water across it at considerable energetic expense. Any design features of the respiratory system that reduce these energetic costs would be to a fish's advantage.

Unloading carbon dioxide is less of a problem because it is highly soluble in water. It does, however, combine with water to form carbonic acid (H_2CO_3). The pH of the medium may decrease, but most fishes live in sufficiently large bodies of near neutral water (pH 7) that this is of no consequence.

Several other problems derive from the close proximity of blood and water across the respiratory membrane. Heat is quickly exchanged, so most fishes cannot maintain an overall body temperature different from the water in which they live (Focus 3-4). Water, valuable salts, and nitrogenous wastes also will diffuse through the membrane.

Gills

Urochordates and cephalochordates are sufficiently small and inactive animals for the body or the pharynx wall to serve as a respiratory membrane. A feeding and respiratory current of water flows through the mouth, into the pharynx, and out numerous pharyngeal slits; gills are not present. Ancestral craniates became larger and more active animals. Gills, which provide a large surface area for the respiratory membrane, evolved where a current of water could ventilate them.

The larvae of many fish species have **external gills,** which are highly vascularized filamentous processes with large surface areas attached to the lateral surface of the head between certain gill slits. Adult fishes, on the other hand, have **internal gills.** These consist of a great many vascularized plates, the **primary gill lamellae,** attached to the walls of the gill or branchial pouches (Gr., *branchia* = gills) or to the gill arches. Minute and tightly packed **secondary gill lamellae,** where gas exchange occurs, attach perpendicularly and transversely to the primary lamellae (see Fig. 18-3).

The Structure and Development of Internal Gills

The pharyngeal pouches develop embryonically as lateral, endodermal evaginations from the portion of the archenteron that will become the pharynx (Fig. 18-1*A*). Ectodermal furrows push inward from the body surface to meet the endodermal pouches, and the intervening tissue soon breaks down. In most fishes, the epithelium covering the gills appears to be of ectodermal origin. The plates of tissue between successive pouches, to which the primary gill lamellae attach, are

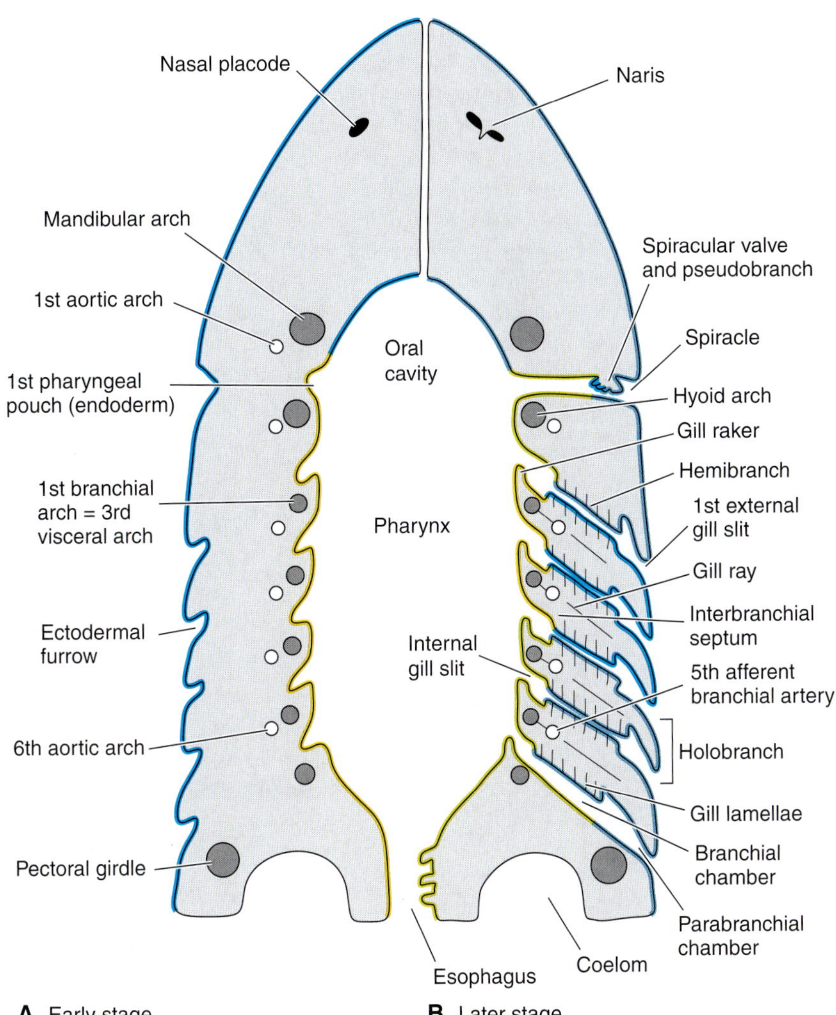

FIGURE 18-1
A frontal section through the pharynx of an elasmobranch. An early developmental stage is shown in *A;* the adult condition, in *B.*

called **interbranchial septa** (Fig. 18-1*B*). A skeletal visceral arch lies in each interbranchial septum. The first visceral arch (the mandibular arch of gnathostomes) lies rostral to the first embryonic pouch, and the last one (the seventh visceral arch) lies caudal to the last pouch. Skeletal, supporting **gill rays** usually extend peripherally from the visceral arches into the interbranchial septa or primary gill lamellae. Muscles and nerves are associated with the skeletal elements, and each of the first six interbranchial septa of a jawed fish contains an embryonic artery, known as the **aortic arch,** that will supply the gills.

Gill pouches were numerous in primitive craniates. Some extinct jawless fishes had 15 pairs. One recent species of hagfish has 14 pairs of pouches, but the lamprey has only 7 (Fig. 18-2*B*). Most chondrichthyan fishes and early bony fishes have six pairs of pouches, the first of which is reduced to a pair of **spiracles** (Figs. 18-1*B* and 18-2*C*). Teleosts have lost the spiracles and so have only five pairs of pouches (Fig. 18-2*D*). All the gill pouches are lined by gill lamellae in living lampreys and hagfishes, and we believe that this resembled the primitive ancestral craniate condition. The distribution of lamellae is more limited in jawed fishes. Interbranchial septa that bear gill lamellae on both surfaces constitute a complete gill, or **holobranch.** Jawed fishes usually have four of these. In addition, some sharks, some lungfishes, and some chondrosteans have a half-gill, or **hemibranch,** on the posterior surface of the hyoid septum (Fig. 18-1*B*). Gill lamellae seldom are present on the posterior surface of the last gill pouch, for no aortic arch exists here to supply them. The African lungfish, *Protopterus,* is an exception, for it has a hemibranch in this position that is supplied by a branch of the sixth aortic arch.

Only a small, gill-like structure lies in the spiracle. Because it receives oxygenated blood from other gills (Chapter 19), it is called a **pseudobranch** (Fig. 18-1*B*). Its function in elasmobranchs is unclear. In teleosts, most of which have retained a pseudobranch even though they have lost the spiracle, it may function as a sense or salt-regulatory organ.

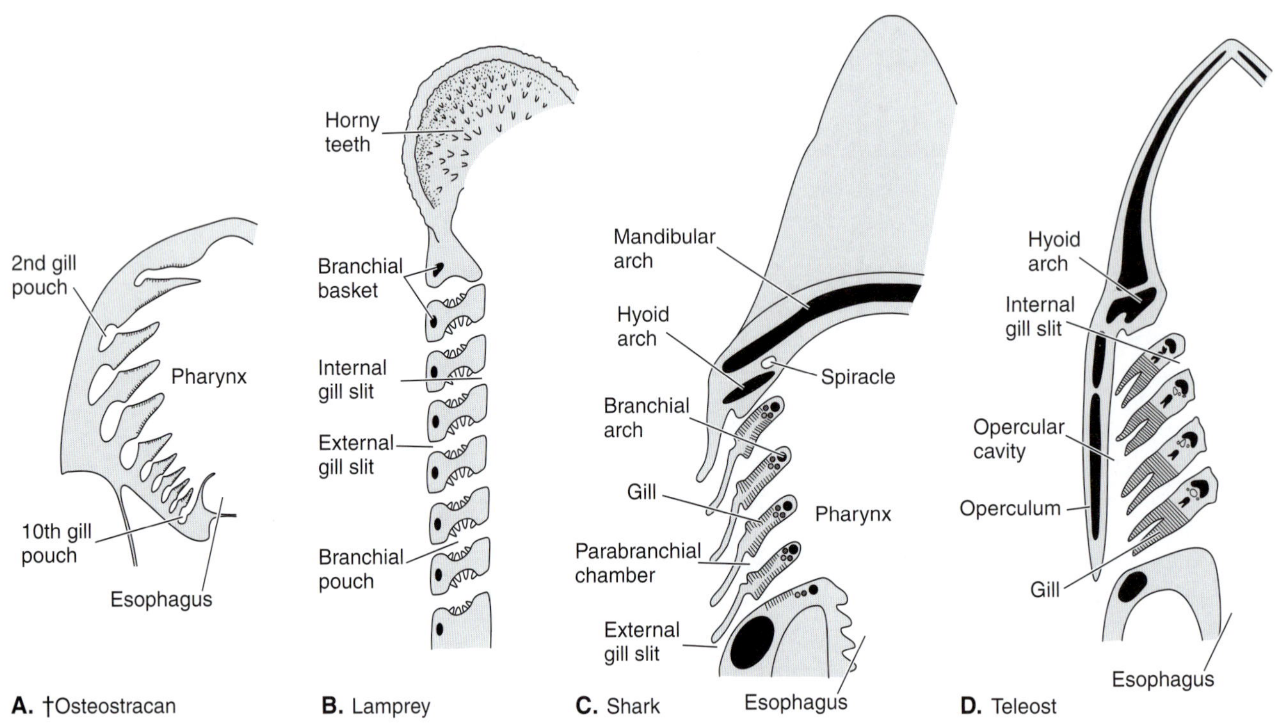

FIGURE 18-2
Frontal sections through four fishes, showing the configuration and numbers of branchial pouches and the types of gills. *A,* Pouched gills of an †osteotrachan. *B,* Pouched gill of a lamprey. *C,* Septal gills of a shark. *D,* Aseptal gills of a teleost.

Gills of Lampreys

Internal gills are arranged in several different ways among fishes, and different patterns of ventilation exist. Jawless fishes have large, saccular branchial pouches that are lined with the primary gill lamellae (Fig. 18-2*A* and *B*). These are called **pouched gills.** Water normally is drawn into the pharynx through the mouth, enters the branchial pouches through pore-shaped **internal gill slits,** and leaves the pouches through pore-shaped **external gill slits.** One of the specializations of the lamprey for its sucking mode of feeding is a longitudinal division of the pharynx into a dorsal food passage, the "esophagus," and a ventral, blind **respiratory tube** from which the internal gill slits arise (Fig. 17-2*A*). When the lamprey is feeding on another fish, water must be pumped in and out of the branchial pouches through the external gill slits.

Gills of Elasmobranchs

The branchial pouches of chondrichthyan fishes are narrow chambers, and the gill lamellae are borne on the interbranchial septa, which continue to the body surface (Figs. 18-1*B* and 18-2*C*). These are **septal gills.** A vertically elongated internal gill slit leads from the pharynx into each **branchial chamber. Gill rakers** at the bases of the interbranchial septa keep food in the pharynx. The structure of the gill rakers correlates with the type of food eaten. They are short processes in predacious sharks but form numerous, long, thin filaments in the plankton-feeding basking and whale sharks. **Parabranchial chambers** lie between the branchial chambers and the small, slit-shaped external gill slits. The distal tips of the interbranchial septa act as valves that can close the external gill slits.

The embryonic aortic arches give rise to arteries that supply and drain the gills. Branches of **afferent branchial arteries** (L., *ad* = toward + *ferent* = carrying) carry blood low in oxygen content into all of the primary gill lamellae. Tributaries of **pretrematic** and **posttrematic arteries** collect aerated blood from the gills. The pretrematic and posttrematic arteries lead to an **efferent branchial artery** (L., *ex* = out + *ferent* = carrying), which carries blood to the dorsal aorta that distributes the blood to the body. Vascular beds where gas exchange occurs lie in the secondary lamellae between the afferent and the pretrematic and posttrematic arteries (Fig. 18-3*A* and *B*). Blood enters small vessels in the interbranchial septum from the afferent branchial artery and then flows through vascular spaces in the secondary lamellae. It is collected by small

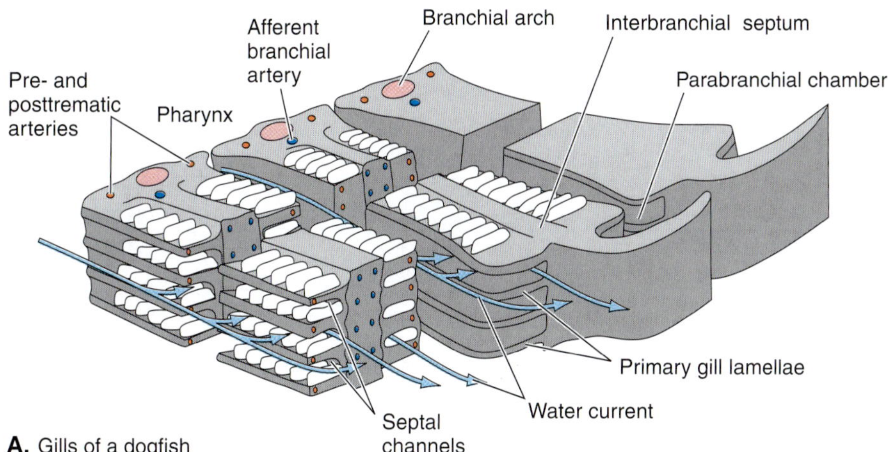

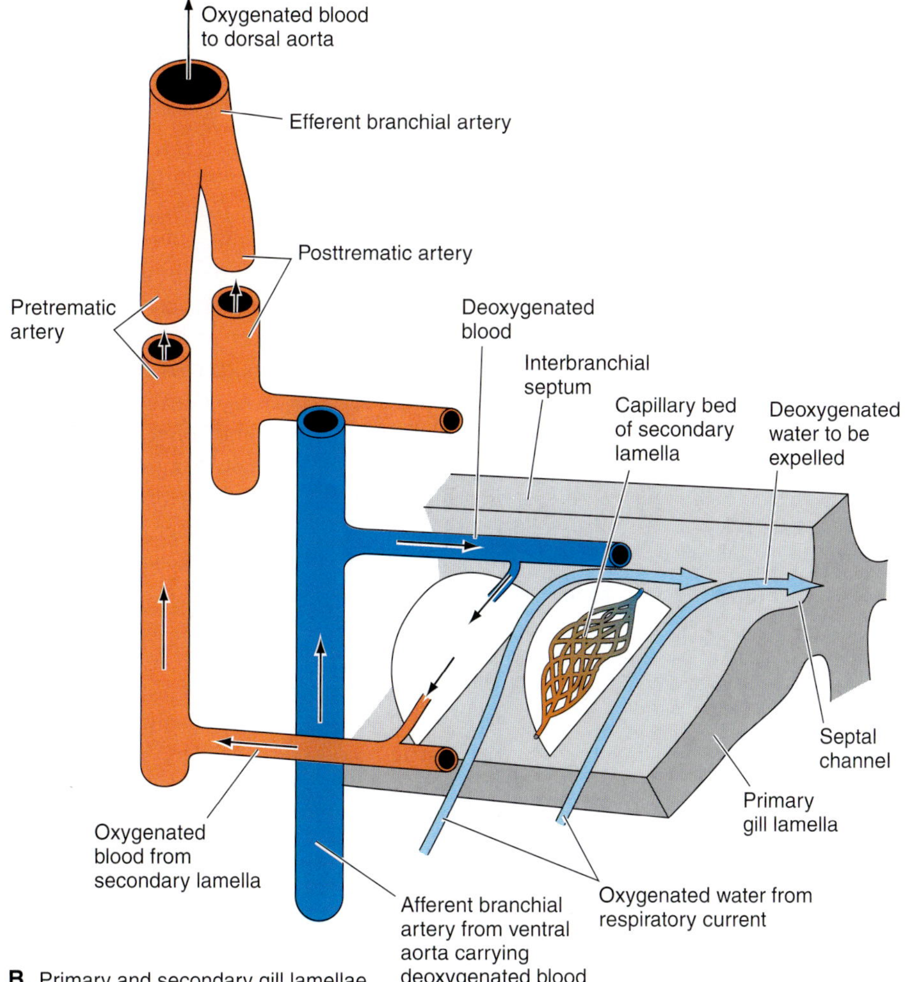

FIGURE 18-3
A, A stereodiagram of portions of the gills of a dogfish. Water (*arrows*) flows between the secondary lamellae to septal channels located beside the interbranchial septum, whence it is discharged. Blood flow (not shown) is in the opposite direction through the secondary lamellae. *B,* Enlarged portion of primary and secondary gill lamellae with associated blood flow (*black arrows*) and water flow (*blue arrows*). *(A, After Hughes.)*

vessels and carried to the pretrematic and posttrematic arteries at the gill base. As it flows laterally from the internal to the external gill slits, most of the water passes between the secondary lamellae into **septal channels** beside the interbranchial septum from which it is discharged. Blood and water flow in opposite directions through and across the secondary lamellae (Fig. 18-3*B*). This countercurrent flow affords a considerably more efficient gas exchange than blood and water moving in the same direction, that is, concurrently (Fig. 18-4). In concurrent flow, oxygen would diffuse from the water and enter the blood until an equilib-

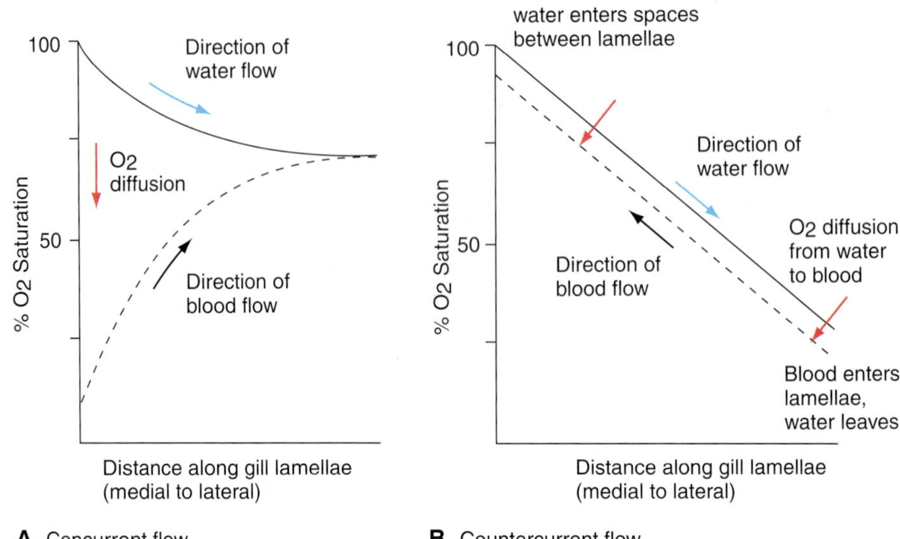

FIGURE 18-4
Graphs showing the changes in oxygen content of blood and water during concurrent flow (*left*) and countercurrent flow (*right*).

rium was reached. Because of the presence of hemoglobin that binds with oxygen, the blood would finally contain more oxygen than the water, but considerable oxygen would remain in the water. In countercurrent flow, well-aerated blood leaving the secondary lamellae encounters water that has not yet crossed the secondary lamellae and so contains more oxygen than the blood. The opposite conditions prevail at the septal-channel end of the secondary lamellae. There, water that has lost most of its oxygen is beside blood containing even less oxygen. Thus, a gradient for the diffusion of oxygen from the water to the blood exists along the entire length of the secondary lamellae, and most of the oxygen (80–95%) in the water enters the blood.

The pathway described is a **respiratory pathway** because it aerates the blood in the gills and delivers oxygen-rich blood to the tissues. In addition, elasmobranchs and most other fishes have a **nonrespiratory pathway,** in which some blood can be diverted away from the gill lamellae by directly entering the efferent part of the system or the venous drainage of the gills. This appears to be a mechanism whereby no more blood is aerated than necessary to meet the fish's current level of metabolism because blood passing through the gills in most fishes also may lose or gain water from the environment (Chapter 20). However, the role of the nonrespiratory system in elasmobranchs is less clear because their blood is iso-osmotic to seawater.

During inspiration in sharks, the mouth and spiracles open and valves that close the gills shut. The pharynx expands by contractions of the coracomandibularis and rectus cervicis (sternohyoideus) muscles (Fig. 18-5*A* and *B*), which reduces the pressure within the pharynx relative to that in the surrounding water. Pressure is reduced even more in the parabranchial chambers by the outward bowing of the valves closing the external gill slits. The pharynx and parabranchial chambers act together as a **suction pump,** creating a pressure gradient that draws water into the pharynx through the mouth and spiracles, across the gills, and into the parabranchial chambers. During expiration (Fig. 18-5*C* and *D*), the mouth and spiracles close, the pharynx and branchial chambers are compressed by the adductor mandibulae and most of the branchial muscles, the external gill slits open, and water is driven out. The pharynx and branchial chambers act together as a **force pump.** Branchial muscles are most active during expiration, compressing the pharynx and branchial chambers. The visceral skeleton also is compressed, and considerable energy is stored in the bent cartilages. Expansion of the system during inspiration occurs partly by the elastic recoil of the visceral skeleton and partly by the contraction of somatic hypobranchial muscles that pull the pharynx floor ventrally. All of this activity minimizes energy consumption. The shark conserves still more energy by moving water fairly steadily in only one direction, rather than in and out external gill slits. The mass of the water need not be alternately accelerated and decelerated. Most sharks rely on an equal balance of the force and suction pumps in their respiratory cycle. Some fast-swimming sharks, and also some teleosts, open their mouths after reaching a certain speed and use the forward motion that is generated by their trunk and tail muscles to drive water across the gills. This is called **ram ventilation.** Sharks using ram ventilation have reduced or lost their spiracles. Skates and rays, which are bottom-dwelling fish

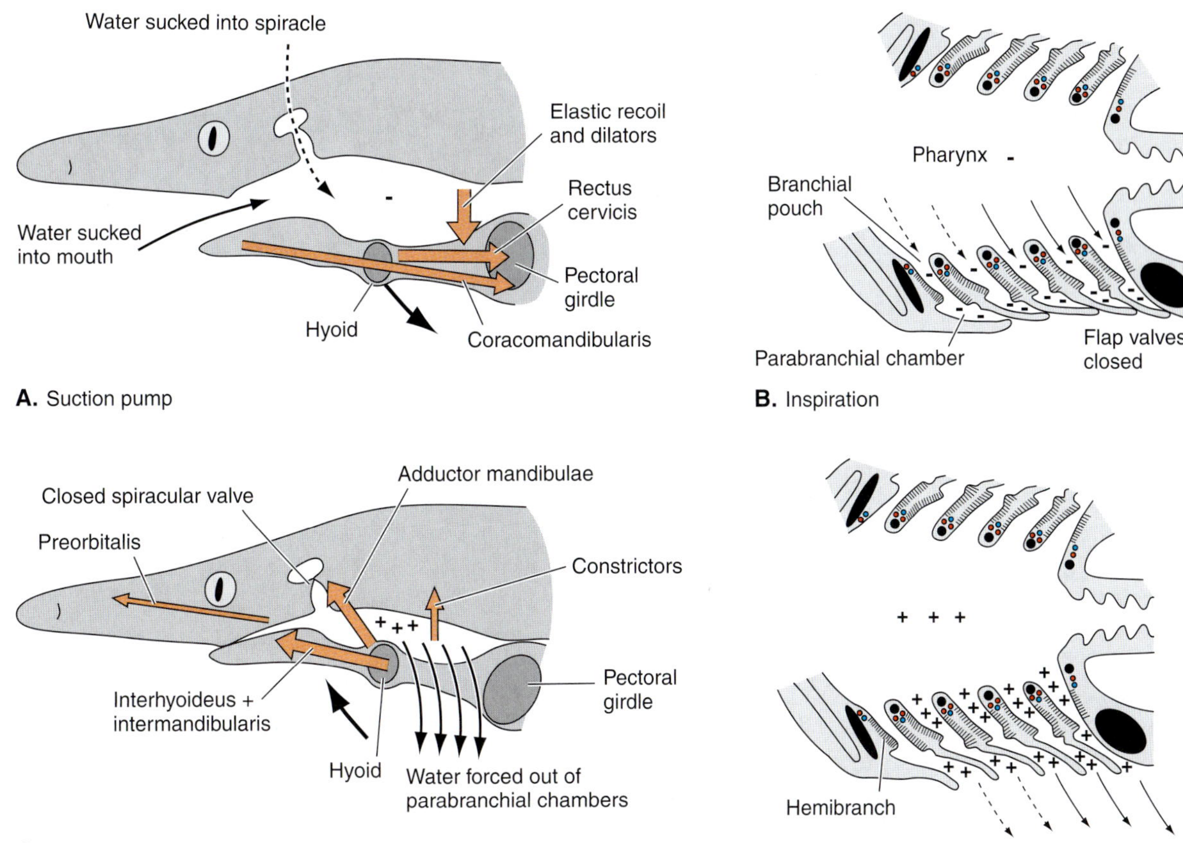

FIGURE 18-5
The mechanics of gill ventilation in the dogfish as seen in lateral views (*left*) and frontal sections (*right*) of the pharynx. Relative pressures are indicated by + and −. Narrow black arrows indicate the flow of water that entered the pharynx through the mouth; dashed black arrows, that of water that entered through the spiracle. Orange arrows indicate muscle activity, with the width expressing the relative level of activity. Wide black arrows depict movements of the hyoid. *A* and *B* show inspiration; *C* and *D* show expiration. *(B and D, After Hughes.)*

with ventrally placed mouths, have exceptionally large spiracles through which most of the water enters the pharynx. Skates and rays ventilate their gills primarily by the suction pump, whereas the force pump plays only a minor role.

Gills of Bony Fishes

Bony fishes evolved a branchial apparatus that is somewhat different from that of elasmobranchs. A flap of body wall supported by bones, known as the **operculum,** extends from the hyoid arch region of the head laterally and caudally over the gills. There is a large, common **opercular cavity** for all the gills, and one valved, external gill slit, rather than a series of small parabranchial chambers, each with its own external gill slit (Figs. 18-2*D* and 18-6). The interbranchial septa are reduced, greatly so in teleosts, so the primary gill lamellae extend freely into the opercular cavity. The gills are described as **aseptal.** With the emergence of the opercular system, teleosts have a respiratory cycle that produces a continuous flow of water in one direction from the oropharynx to the opercular cavity over the gills. The cycle consists of two stages. The suction pump stage is activated by an expansion of the mouth cavity and pharynx by the sternohyoideus muscle and branchial levators (Fig. 18-6*A* and *C*). Because the volume of the oropharynx is increased, a pressure lower than the ambient surrounding pressure is created, and water flows into the mouth. At the same time, the opercular cavity is greatly enlarged by dilator muscles, which result in an even lower pressure, causing the water from the oropharynx to be drawn over the gills into the opercular cavity (Fig. 18-6*B* and *C*). In the second stage, the force pump is activated by the adduc-

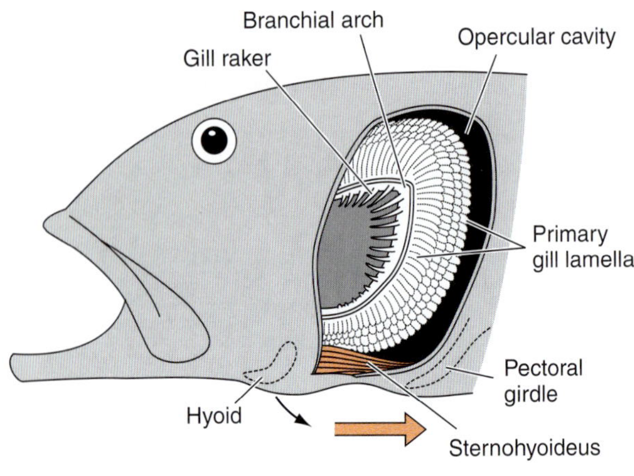

A. Lateral view

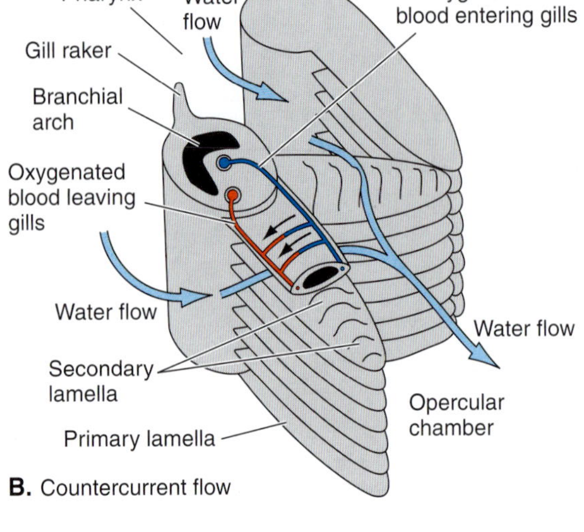

B. Countercurrent flow

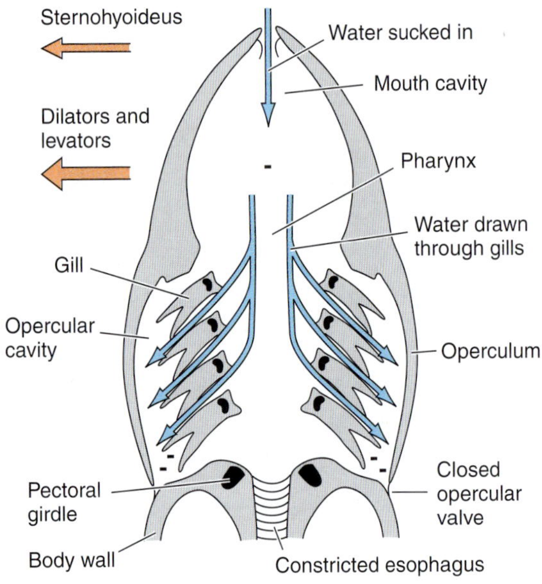

C. Inspiration (expansion stage)

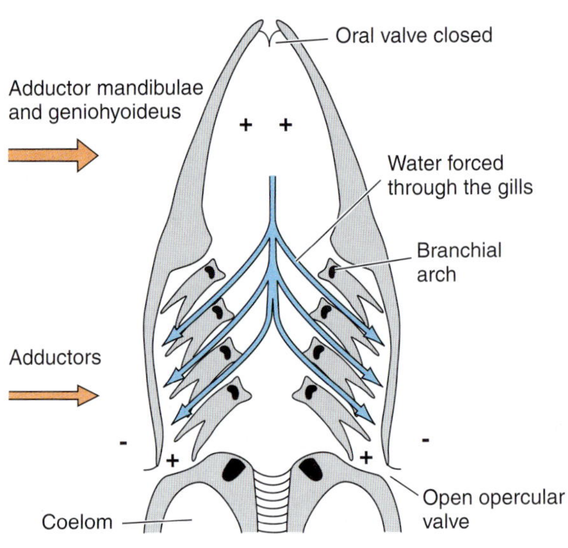

D. Inspiration (compression stage)

FIGURE 18-6
The gills in a teleost. *A,* A lateral view of the opened opercular chamber. The black arrows indicate the movements of the hyoid. *B,* The countercurrent flow of water and blood in a gill. Blue arrows indicate the flow of water across the gills; black arrows indicate the flow of blood within the gills. *C,* Mechanics of inspiration. Blue arrows show the direction of water movement. *D,* Mechanics of inspiration. *(B, After Dorit et al.)*

tors and geniohyoideus muscles (Fig. 18-6*D*). The oropharynx is compressed, creating a positive pressure that is even more than the increasing pressure in the opercular cavity. As a result, water continues to flow from the oropharynx through the gills into the opercular cavity and thence to the outside through the opened opercular cavity. The efficiency of the teleostean respiratory system is based on the maintenance of a differential pressure between the oropharynx and opercular cavity, which are separated by a curtain of gills (Fig. 18-6*B*). It produces an uninterrupted unidirectional flow of respiratory water that runs in the opposite direction of the blood flow for an optimized countercurrent system.

The structure of the enormous number of secondary lamellae provides a large surface area (*A* in the diffusion equation) and an exceptionally short diffusion distance (*d* in the diffusion equation) between water and blood. The surface epithelium on each side of a secondary lamella often is only one cell thick (Fig.

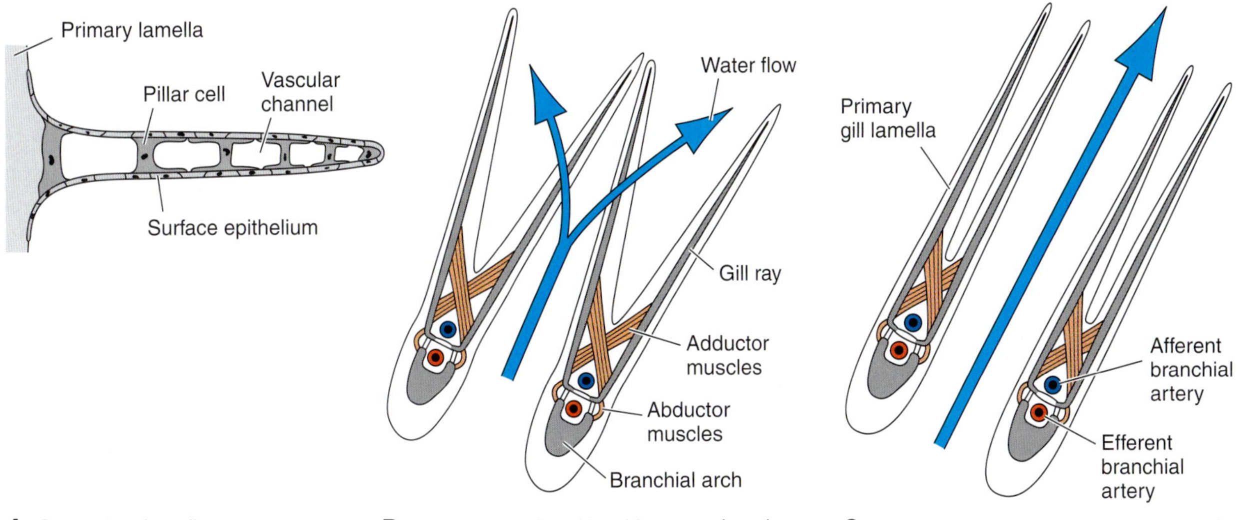

FIGURE 18-7
Gill structure of a teleost. *A,* A transverse section through a secondary lamella, showing the close proximity of the vascular channels through which blood flows and the surface across which water flows. *B,* Primary gill lamellae of adjacent gills meet when adductors relax, so water must cross the secondary lamellae. *C,* Primary gill lamellae of adjacent gills separate when the adductor muscles constrict, and much water leaves the opercular chamber without crossing the secondary lamellae. *(A, After Hughes; B and C, after Bijtel.)*

18-7*A*). **Pillar cells** bearing thin cytoplasmic flanges that spread out beneath them hold these epithelial layers apart. Blood flows in the narrow spaces among the pillar cells rather than through typical capillary beds.

The total cross-sectional area of all the vascular channels within the secondary lamellae is considerably greater than the sum of the cross-sectional area of the branchial arteries supplying them. Similarly, the cross-sectional area of all the spaces for water passage between the large number of secondary lamellae is greater than the cross-sectional area of the pharynx. Because the velocity of a liquid moving from a narrow area into a larger area decreases, as does water flowing from a stream into a pond, the rate of blood flow through the lamellae and of water flow across them is reduced. Adequate time for diffusion is available.

The combined surface area of the secondary lamellae varies greatly among phylogenetically unrelated species of teleosts and probably is adapted to their modes of life. Hughes (1984) calculated that a sluggish, bottom-dwelling toadfish has 151 mm^2 of gill surface area per gram of body weight, compared with 1241 mm^2/g for an active, pelagic menhaden. Diffusion distance across the gill lamellae is also less in active fishes, and the hemoglobin content of their blood is higher.

Gas exchange across the gills is accompanied by the diffusion of other small molecules. Considerable ammonia in most fishes and urea in some teleosts are excreted through the gills. Chondrichthyan fishes are an exception because their gills are impervious to the diffusion of urea, and a large quantity of this molecule accumulates in their bodily fluids and tissues. Water and salts also are lost or gained through the gills, depending on their relative concentrations in the external environment and bodily fluids. Water, for example, is gained by osmosis through the gills of freshwater fishes but lost in most marine species, the environment of which is saltier than are their bodily fluids (Chapter 20).

Because of the attendant osmotic problems, it is to a bony fish's advantage not to ventilate its gills, or perfuse blood through them, more rapidly than needed to meet its oxygen requirements at any particular time. Several mechanisms help match the rate of gas exchange with a fish's metabolic requirements:

1. Ventilation rate can be regulated.
2. The primary lamellae of adjacent gills can be brought together or moved farther apart (Fig. 18-7*B* and *C*). When adductor muscles within the gills are relaxed, the elasticity of the skeletal gill rays, assisted by the contraction of small abductor muscles, spreads the primary lamellae apart. The tips of adjacent gills meet, a damlike mechanism is formed, and most of the water passing across the gills must pass through the spaces between the secondary

lamellae. When the adductor muscles contract, the tips of adjacent gills are pulled apart and much of the water leaves the branchial chambers without crossing the secondary lamellae.

3. The nonrespiratory pathway also can control the amount of blood perfusing the secondary lamellae. Vascular shunts divert some of the blood directly from afferent to efferent branchial arteries or into the venous drainage of the gills without going through the secondary lamellae. In many other species, the number of secondary lamellae being perfused is subject to control.

Accessory Air Respiratory Organs

The gills of bony fishes are such efficient respiratory organs that 80% to 95% of the limited oxygen available in the water is taken up by the blood. But in some situations, the gills alone cannot provide for a fish's needs. Oxygen levels can be very low in shallow, warm pools or in swamps where considerable decay occurs, and in other bodies of stagnant water. A fish could not live in these habitats without an accessory air respiratory organ that enables it to use the oxygen in the air. Gills are ineffective for this purpose because the delicate gill lamellae cannot be supported in air; they collapse and clump together, greatly reducing the respiratory surface.

Many teleost fishes can supplement aquatic gill breathing with accessory air respiratory organs. These fishes mostly, but not exclusively, live in waters low in oxygen, and they become bimodal breathers, breathing both water and air in various proportions, depending on environmental conditions. Some teleosts even become obligate air breathers. Virtually all bimodal breathers retain gills and simultaneously evolve specialized air-breathing organs. Some use a vascular skin as a respiratory organ, such as freshwater eels, which often migrate over land. However, other bimodal breathers possess modifications of many parts of the gut: the linings of the mouth, pharynx, esophagus, intestine, and rectum. Mud or rice eels, which are eel-shaped, perch-like Synbranchiformes, have a highly vascularized mouth, pharynx, and esophagus as an air-breathing organ. They gulp air, which they hold for 30 minutes to 3 hours, taking up oxygen. These fishes exhibit amphibious life and can remain on land for up to six months. Gouramis, the climbing perch (Fig. 18-8*A*), the Siamese fighting fish, snakeheads, and the walking catfish, among others (Fig. 18-8*B*), develop a dorsal outpocketing above the gills to form a **suprabranchial air chamber,** which can be filled with air. One of the gill arches develops an elaborate, folded, highly vascularized structure called the **labyrinth organ,** or **arborescent organ,** which protrudes into the air chamber and functions as a "lung," enabling the fish to live in oxygen-poor waters or even to spend considerable time on land. Some armored catfishes and loaches have parts of their intestines lined with folded epithelia, into which blood capillaries grow, transforming these intestinal segments into accessory air-breathing organs. When in oxygen-poor waters, these fishes supplement gill-breathing with air-breathing by swallowing air into their respiratory intestines and temporarily suspending digestive functions.

Teleosts ventilate their suprabranchial air chambers by passing the spent air low in oxygen into the expanding buccal cavity. From here, it is expelled through an opened mouth at the surface of the water. Fresh, oxygen-rich air is sucked in, and the mouth closes when the buccal cavity is filled. Compression of the buccal cavity will force the fresh air into the chamber. Thus, the chamber is emptied by suction and filled by

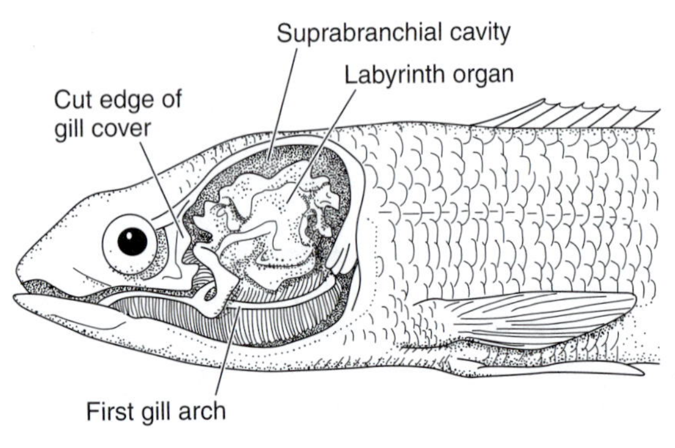

FIGURE 18-8
Air-breathing adaptations in teleosts. *A,* Climbing perch. *B,* Walking catfish.

compression of the buccal cavity, which acts as a pump, specifically as a **pulse pump,** by alternating suction with compression. Similarly, an equivalent pulse pump ventilates air in the intestine.

Lungs

Lungs are but one of many aerial respiratory organs to have evolved in fishes. Many bony fishes have either lungs or swim bladders, except many bottom-dwelling species, in which they have been secondarily lost, so these organs are a common character of the group. The organ is most lunglike in the primitive members of the group, so lungs appear to have arisen early in evolution. Lungs may have evolved in certain †placoderms and primitive bony fishes that were living in stagnant freshwater habitats subject to periodic droughts during the late Silurian and Devonian periods (Chapter 3). Geological evidence suggests that the earth at this time was subject to pronounced fluctuations in rainfall. Dry and wet periods alternated as they do in some tropical habitats today. During the hot, dry seasons, many bodies of fresh water would have become smaller, and water temperatures would have increased. Oxygen levels would have decreased. Some ponds probably became stagnant swamps or dried up completely. Chondrichthyan fishes and †placoderms, many of which were marine, would not have been affected by these climatic fluctuations, but early freshwater bony fishes, including the rhipidistian ancestors of tetrapods, would have been unable to survive unless they had accessory air respiratory organs, such as lungs.

Among actinopterygians with respiratory lungs are the gar (*Lepisosteus*), the bichirs and reed fish (*Polypterus* and *Erpetoichthys,* Focus 18-1), the bowfin (*Amia*), and the pirarucu (*Arapaima*). Actinopterygian fishes ventilate their lungs with a characteristic four-stroke buccal pump: The buccal cavity expands so that "spent" oxygen-poor air from the lung can pass into it. Once filled, it compresses to expel the spent air. The empty buccal cavity then expands again to take in fresh air and finally compresses to force the new air into the lungs. The bichirs and lungfishes have retained their lungs throughout their evolutionary history in their tropical freshwater environments (Fig. 18-9).

Lungs develop embryonically in bichirs, lungfishes, and amphibians as a ventral evagination from the floor of the digestive tract (pharynx or, in some cases, esophagus) just caudal to the last pair of pharyngeal pouches. The primordia of the lungs resemble a pair of displaced pharyngeal pouches in some amphibian larvae, and this has led some investigators to suggest a homology between actinopterygian, lungfish, and amphibian lungs. This would imply that the lungs of tetrapods, commonly thought to be an adaptation for life on land, are in fact a primitive feature of all bony fishes and their descendants. In the bichir (*Polypterus*), the African lungfish (*Protopterus*), and amphibians, the bilobed lungs extend into the pleuroperitoneal cavity and grow ventrally, one on each side of the digestive tract (Fig. 18-9). In other lungfishes, the single lung primordium extends caudally and dorsally on one side of the digestive tract and may subsequently become bilobed. The vascular connections of the lungs of lungfishes and those of *Polypterus* and amphibians are similar. In lungfishes and primitive tetrapods, a slit-shaped opening, the **glottis,** on the floor of the pharynx or esophagus leads to the lungs.

The lungs of lungfish, *Polypterus,* and indeed all vertebrates with lungs contain specialized cells that secrete a surface-film of lipoprotein known as a **surfactant.** However, some variation in chemical composition of the surfactant has occurred throughout vertebrate evolution. As we shall see (p. 599), mammalian lungs are subdivided into many small chambers (alveoli) in which gas exchange occurs. The composition of their surfactant greatly reduces surface tension in the alveoli and hence the resistance to lung expansion. Less energy is needed to expand them. The lungs of fishes and early tetrapods have far less compartmentalization. The need to decrease surface tension is less. The composition of their surfactant is such that it acts more as an "antiglue," preventing adhesion of adjacent epithelial surfaces at low lung volumes. This may have been the primitive function of a surfactant (Daniels et al., 1995).

When breathing air, a lungfish comes to the surface and first begins to inhale fresh air. It does so by expanding its buccal cavity and sucking in fresh air (Fig. 18-10). During this process, the glottis opens and spent air is transferred from the lungs to the buccal cavity by elastic recoil of the pressurized lungs and contractions of smooth muscles that surround the lungs. As a result, the spent air mixes with the just inhaled fresh air in the buccal cavity. Excess mixed air escapes from the open mouth. The fish closes its mouth and compresses its buccal cavity, which acts as a force pump, driving the mixed air into the lungs. The glottis closes, and air is held in the lungs under pressure. Thus, the lungfish's ventilation pattern is typically a two-stroke pulse pump. It expands and compresses the buccal cavity only once during the respiratory cycle. Because mixed air is delivered to the lung, this pattern also is called the **mixed-air buccal pump system.** This mode of ventilation is characteristic of sarcopterygian fishes except coelacanths. High-speed X-ray cinematography and direct mea-

FOCUS 18-1 *The First Aspiration Pump in the Evolution of Vertebrates*

Virtually all air-breathing fishes and amphibians ventilate their lungs by using a pulse pump. In pulse-pump systems, air forces the lung to expand, whereas in aspiration systems, air is sucked into the already expanding lung. Air-breathing fishes and amphibians begin to fill the lung only after the mouth is closed and the compressive muscles act on the buccal force pump. A remarkable exception is found in the polypterid fishes *Polypterus* (bichir) and *Erpetoichthys* (reedfish). The first appearance of lung ventilation by aspiration in the evolutionary history of vertebrates is exhibited by the pattern of ventilation in these primitive actinopterygians. Polypterid fishes are encased in a stiff scale jacket analogous to the dermal armor of early tetrapods (Fig. A, Part I). Cineradiographic analysis, pressure recordings, and strain-gauge recordings sensing deformations of the "dermal armor" by Brainerd et al. (1989) have elucidated an evolutionarily early aspiration pump. The fish exhales first by contraction of smooth muscles around the lung (Part II). Spent air is forced into the pharynx and exits from the opercular slit (Part III). The reduction in size of the lungs will cause a decrease in volume of the space occupied by organs within the peritoneal cavity, causing the ventral dermal armor to buckle inward (Part III). Such buckling is made possible by the nature of the peg and socket articulations between the thick scales. These articulations have the capacity to undergo recoil after the initial deformation. As the dermal armor passively recoils to its original resting position, it creates a subatmospheric pressure in the pleuroperitoneal cavity, causing the lung to expand (Parts IV and V). The resulting subatmospheric pressure in the expanding lung will draw air in through the widely opened mouth very rapidly. The use of recoil aspiration by polypterid fishes demonstrates that a stiff body wall capable of storing energy can contribute to inhalation in vertebrates.

The ribs of polypterid fishes are confined to the dorsal part of the body and are not directly involved in the ventilation process even though they provide stiffness of the body, thus allowing for deformation of the belly. The discovery of recoil aspiration in polypterid fishes has important implications for our attempts to understand the ventilation mechanics of early Paleozoic amphibians. Early Paleozoic amphibians retained ventral, bony scales from their air-breathing fish ancestors. These V-shaped scale rows are very similar to the rhomboid scale body armor of polypterid fishes. The configuration of their ribs suggests that they confer stiffness on the body wall. The presence of recoil aspiration in polypterid fishes encased in a thick dermal scale jacket suggests that the ventral dermal armor of early amphibians may well have played an important role in aspiration breathing.

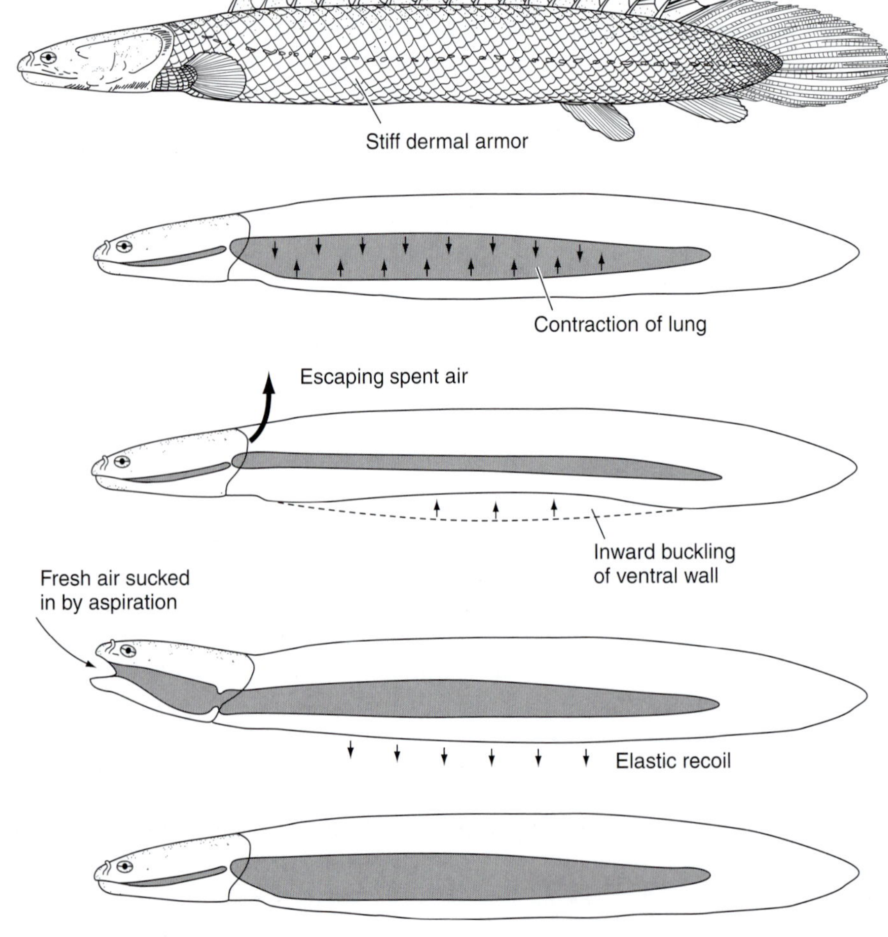

A. A diagrammatic summary of the recoil aspiration model for polypterid ventilation. (*After Brainerd et al.*)

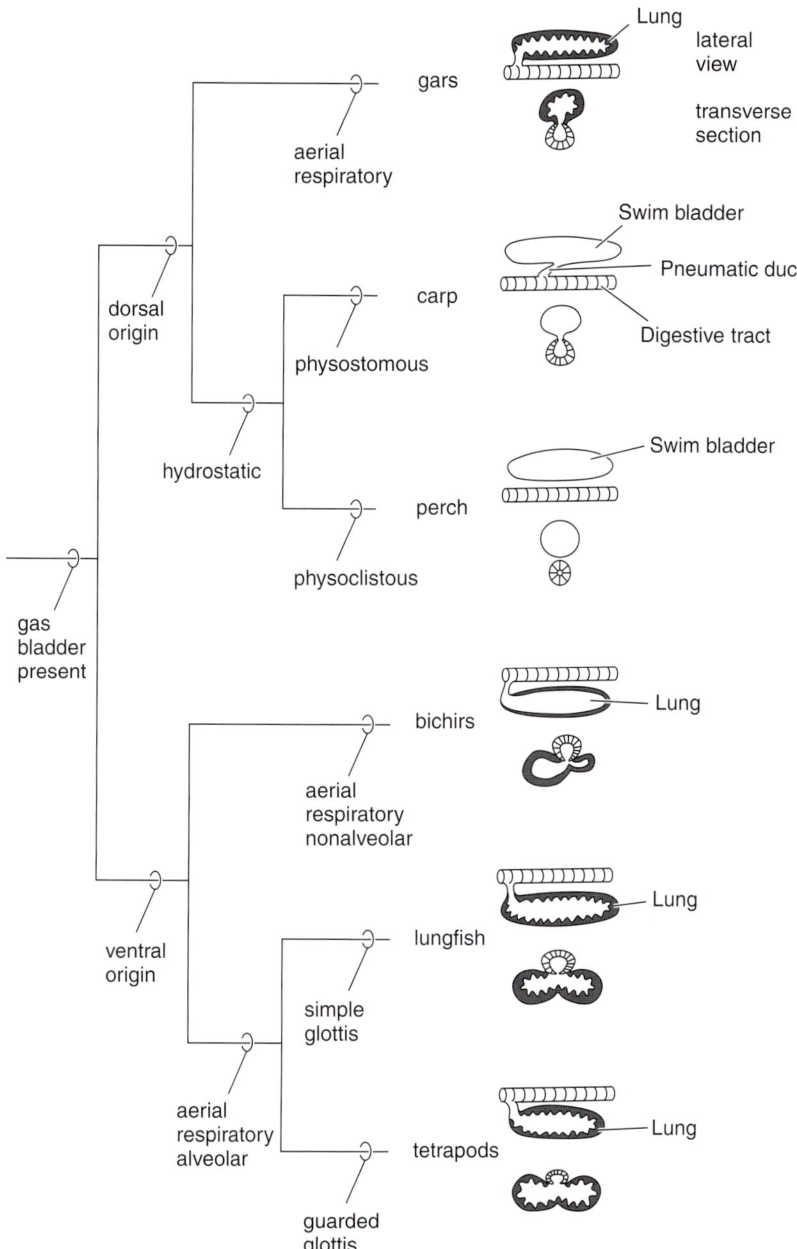

FIGURE 18-9
A cladogram showing the evolution of lungs and swim bladders.

surements of the gas mixture have verified this (Brainerd et al., 1993). Lungfishes hold air in the lung for a considerable time as the oxygen slowly is used. Breathing is not continuous. Long periods of breath holding, or **apnea,** alternate with short periods of lung ventilation.

Some compartmentalization of the lung by internal septa provides a large surface area for gas exchange (Fig. 18-9). The Australian lungfish, *Neoceratodus,* uses its lung to supplement the gills in oxygen-poor water. It need not use them when oxygen supplies are normal. Internal lung septa are more numerous in *Protopterus,* which has lost its first two gills and is an obligate air breather. It will drown if prevented from reaching the surface to gulp air. The South American lungfish, *Lepidosiren,* is also an obligate air breather.

Carbon dioxide accumulates in the lungs during the periods of apnea. The hemoglobin of lungfishes, like that of terrestrial vertebrates, must be adapted to bind with oxygen in the presence of more carbon dioxide than normally is present in water. Accumulated carbon dioxide is released during expiration, but the frequency of lung ventilation is too low to unload all the carbon dioxide. Most carbon dioxide continues to diffuse from the gills because it is very soluble in water.

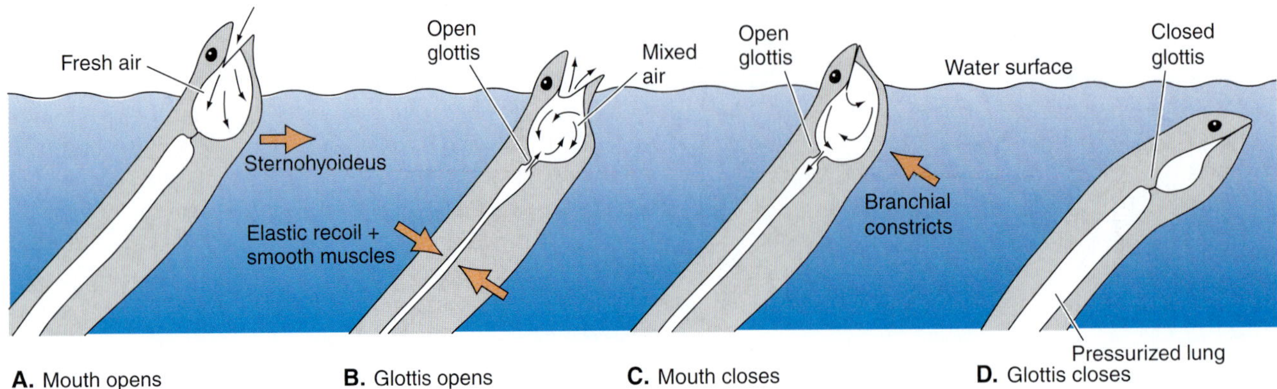

FIGURE 18-10
Ventilation of the lung by a buccal pulse pump in the South American (*Lepidosiren*) and African (*Protopterus*) lungfishes. Black arrows indicate air flow; orange arrows indicate muscle action. *A,* A lungfish at the surface opens its mouth and takes fresh air into its oropharyngeal cavity. *B,* The glottis opens, and spent air expelled from lungs mixes with fresh air. Excess air escapes through the mouth. *C,* The mouth closes, and mixed air is forced into the lungs. *D,* The glottis closes, and air is held in the lungs.

The Swim Bladder

The lungs of primitive actinopterygian fishes have transformed into **swim bladders** in most neopterygian species and coelacanths. Neopterygian fishes live in oxygen-rich waters, and selection would favor the conversion of lungs into effective hydrostatic organs, which would endow these fishes with neutral buoyancy and therefore with highly efficient locomotory capacities. In a primitive actinopterygian, the swim bladder arises as a dorsal outgrowth from the caudal end of the esophagus and remains connected to the esophagus by a pneumatic duct (Fig. 18-9). The pneumatic duct remains in **physostomous** species (Gr., *physa* = bladder + *stoma* = mouth), but is lost in more advanced teleosts, known as **physoclistous species** (Gr., *kleiein* = to close).

The swim bladder of primitive actinopterygians continues to be an important site for the uptake of oxygen, but it also has acquired a hydrostatic function. In a teleost, it is primarily a hydrostatic organ that can be regulated, enabling the fish to attain neutral buoyancy and so maintain its position in the water with minimal muscular effort. Neutral buoyancy requires that the density of the fish equal that of the water. Density equals mass or weight divided by volume. One milliliter of pure water weighs 1 g and so has a density of 1. But 1 mL of flesh weighs slightly more than 1 g. Because flesh alone has a density slightly greater than that of water, a fish without a swim bladder, such as many bottom-dwelling fishes and elasmobranchs, tends to sink. A sac of air in the fish decreases weight and can make its overall density equal to that of water. Problems arise, however, when a fish changes depth. When it goes deeper, water pressure increases, the swim bladder is compressed and becomes smaller, and the fish's density increases. The fish would sink faster except for the addition of enough gas to the swim bladder to maintain it at the appropriate size. When the fish rises, surrounding water pressure decreases, the swim bladder tends to expand, and gas must be removed from the bladder to maintain neutral buoyancy.

The gas in the swim bladder of most species is about 80% oxygen, and it is secreted into the bladder by a **gas gland** on one surface of the organ (Fig. 18-11). Often, gas must be secreted against a considerable concentration gradient, for water pressure increases rapidly with depth. At a depth of 500 m, for example, the fish and the gases in its swim bladder are subjected to a pressure of 50 atm, and the partial pressure of oxygen in the swim bladder is 40 atm. The partial pressure, or tension, of oxygen in the water and in the fish's blood, however, is 0.2 atm. ("Tension" is the term applied to the partial pressure of a gas when in solution.) But a fish can secrete oxygen into the swim bladder against a pressure gradient and keep it there. Oxygen is kept in the bladder by a **rete mirabile** (L., *rete* = net + *mirabile* = wonderful), a set of relatively long, parallel capillaries located just before the gas gland (Fig. 18-11). Together, the rete mirabile and gas gland are called the **red body.** The tension of the oxygen in the venous capillaries just leaving the swim bladder is high because it is in equilibrium with the partial pressure of oxygen in the swim bladder. Oxygen starts to be carried off, but the venous capillaries are surrounded by arterial capillaries carrying blood with a lower oxygen tension in the opposite direction. A countercurrent exchange occurs: oxygen diffuses from the venous blood

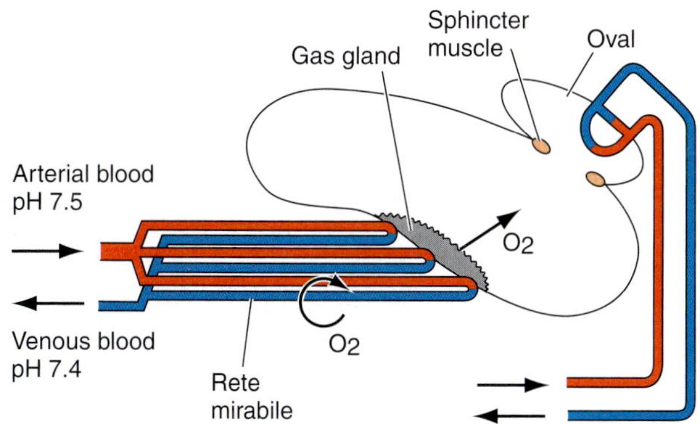

FIGURE 18-11
The operation of the swim bladder in a physoclistous teleost. *(Modified from Alexander.)*

into the arterial blood and is carried back to the swim bladder. The rete mirabile acts as an oxygen gate, and the longer the capillaries in this organ, the less oxygen that escapes.

Secretion of oxygen into the swim bladder requires a local increase in the acidity of the blood. As the blood becomes more acidic, hemoglobin releases bound oxygen. Acid has two sources. Carbon dioxide entering the blood combines with water to form carbonic acid. This reaction is accelerated greatly in the presence of the enzyme carbonic anhydrase in the red blood cells. In addition, considerable lactic acid is secreted by the gas gland. The release of oxygen by hemoglobin is a far faster reaction than its recombination, so considerable unbound oxygen accumulates in the blood in and near the gas gland, and the rete mirabile prevents most of it from escaping. Additional oxygen is continually being brought into the system by the arterial blood, and it, too, is released. Eventually, the tension of oxygen in the blood in the gas gland exceeds its partial pressure in the swim bladder, and oxygen diffuses into the swim bladder. The rete mirabile and gas gland together form a **countercurrent multiplier system.**

Removal of oxygen from the swim bladder as a fish rises in the water is a far simpler process. Oxygen is released through the pneumatic duct in physostomes, but it is reabsorbed from a special, vascularized compartment of the swim bladder known as the **oval** in physoclists. A sphincter that normally separates the oval from the main part of the swim bladder relaxes; oxygen enters the oval and, because of its high partial pressure, rapidly diffuses into the blood. Oxygen does not diffuse back into the blood through other parts of the swim bladder because its wall contains a diffusion barrier of guanine plates. The swim bladder, with its energy-efficient regulatory capacity to maintain the size under different pressure regimes, is considered a biological marvel, a true evolutionary innovation that has enabled teleosts to penetrate and occupy countless aquatic habitats in the world.

Although the swim bladder is primarily a hydrostatic organ, it has additional functions in certain species. It contains a store of oxygen that can be used in cellular respiration if needed. It is part of the hearing mechanism in most ostariophysan fishes because the bladder can act as a resonator and amplifier (Chapter 12). Drums and some other species can pluck on the swim-bladder wall with specialized muscles to produce sounds.

Respiration in Early Tetrapods

Because dry, 20°C air at sea level contains 210 mL of oxygen per liter, adult terrestrial vertebrates have far more oxygen available to them than do fishes, but they need a moist respiratory membrane because the oxygen must be in solution to diffuse into the blood. Major problems for terrestrial vertebrates are how to expose and ventilate the respiratory membrane without an excessive loss of body water, and how to prevent collapse of the respiratory organ in air, which is less dense than is water. Lungs, which terrestrial vertebrates inherited from fishes, are well adapted to meet these problems. They usually contain internal septa or divisions of the airways that increase the surface area of the respiratory membrane and provide the needed support. Air within the lungs contains a great deal of water vapor; in mammals, it is normally saturated. The respiratory surface is kept moist without a great deal of water loss because the rate of ventilation is low, the air is conditioned by mucous glands in the airways prior to entering the lungs, and not all the air in the lungs is exchanged in each breathing cycle. The low rates of ventilation and exchange are made possible by the high oxygen content of the air in the lungs. Water vapor, which is a gas, reduces the partial pressure of oxygen in the lungs to about 100 mm Hg, compared with 160 mm Hg for dry air at sea level, but this is still ample.

Amphibian Respiratory Organs

The larvae of two of the three genera of contemporary lungfishes have external gills attached to the surface of their heads, and the larvae of rhipidistian ancestors of tetrapods probably had them, too. Amphibian larvae are aquatic and retain external gills. Fossil evidence indicates that the larvae of some early neotetrapods had external gills. Salamanders retain external gills throughout their larval period (Fig. 18-12), and

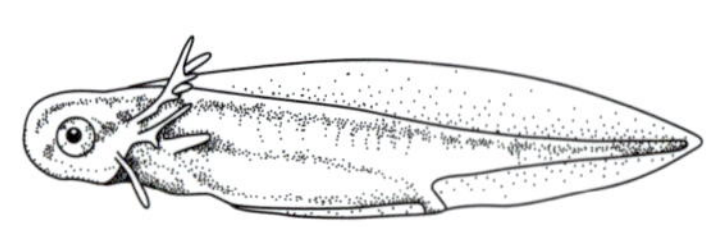
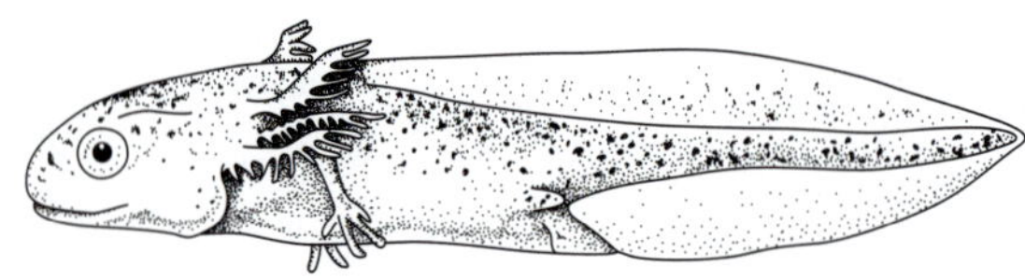

FIGURE 18-12
The development of the external gills in the larva of a salamander. *(After Glaesner.)*

neotenic species, such as the mudpuppy, *Necturus*, have them throughout life. The external gills of young frog tadpoles become covered by an **opercular fold** that opens on the body surface through a single pore called the **spiracle** (this term is unfortunate because the pore is neither homologous nor analogous to the spiracle of a fish). The operculum is extensive and even envelops the developing forelimbs. Older tadpoles lose the external gills as they develop deeper ones attached to the branchial arches. We are uncertain of whether these deeper gills are homologous to the internal gills of fishes. Lungs develop and begin to function late in larval life. Gills are lost at metamorphosis, the forelimbs push through the operculum, and the remains of the operculum fuse with the body wall.

The frog respiratory system (Fig. 18-13) traditionally has been selected as a representative of adult amphibians even though new studies on salamanders have provided new functional and evolutionary perspectives. In frogs, valved, external nostrils lead to short nasal cavities that open through choanae at the front of the palate. Air passes through the buccopharyngeal cavity and enters the glottis, which is supported by small, **lateral laryngeal cartilages** derived from the sixth visceral arch. The glottis leads to a small, triangular **laryngotracheal chamber,** from which the lungs emerge. All these passages are lined by cilia and by mucous and serous secretory cells. These secretions keep air moist and trap dirt particles and carry them away from the lungs. The lungs of frogs are sac-shaped organs; in salamanders, they are elongated. The respiratory surface of frog lungs is increased by primary septa, and sometimes also by secondary ones, which form a series of pockets. The interior of the lungs is open.

Ventilation of the lungs of a frog resembles that of lungfishes, except that air enters and leaves through the nares and nasal cavities rather than through the mouth opening (Fig. 18-14). In a representative breathing cycle, a lowering of the pharynx floor draws fresh air through the nasal cavities and into the buccopharyngeal cavity. The glottis then opens, and positive pressure in the lungs drives stale air out. Expiration is assisted by the elastic recoil of the lungs, and by the contraction of flank (hypaxial) muscles (DeJongh and Gans, 1969) and, specifically, the transversus abdominis, a newly evolved hypaxial muscle of tetrapods (Brainerd, 1999). The expired air is mainly expelled over the fresh air, which is sequestered to some degree in the floor of the buccopharyngeal cavity, but some mixing of air does occur. The nostrils are closed, the pharynx floor is raised, air is forced into the lungs, and the glottis closes. This pattern resembles the two-stroke cycle of lungfishes. The pumping action of the pharynx may be repeated until the lungs are well inflated. Modifications of this sequence have been described (Vitalis and Shelton, 1990).

In addition to buccal-pump breathing for inhalation, salamanders use the transversus abdominis muscle to increase body-cavity pressure and force air out of the lungs. Thus, salamanders exhale actively by means

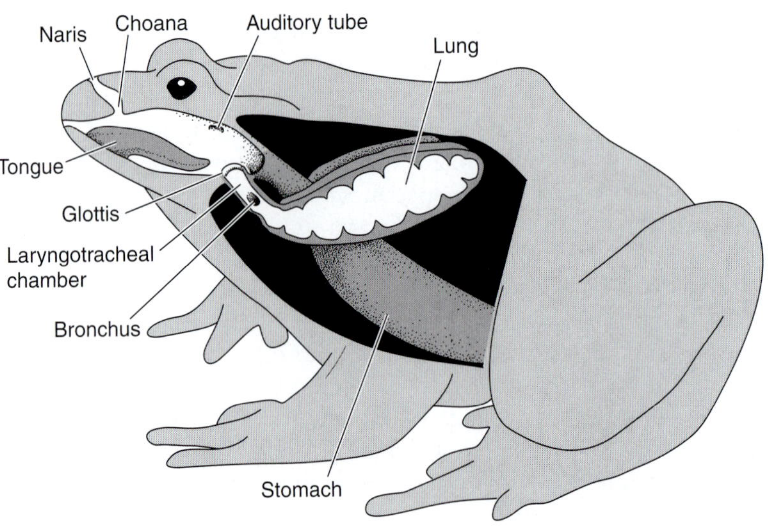

FIGURE 18-13
A schematic dissection of the respiratory system of the frog. *(After Dorit et al.)*

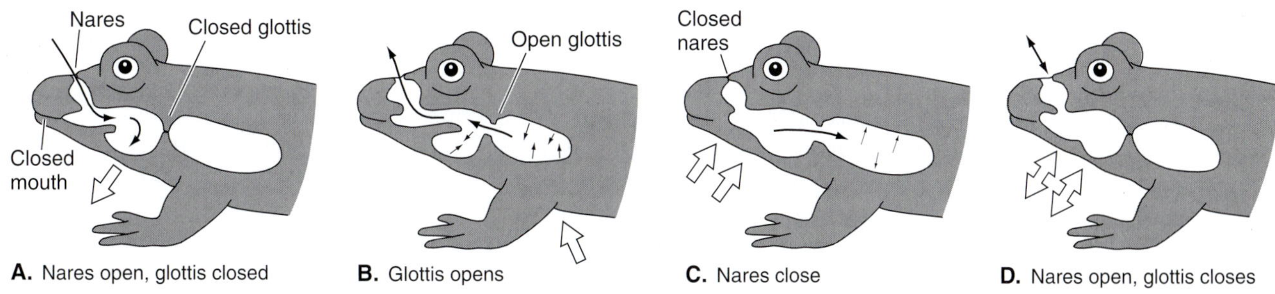

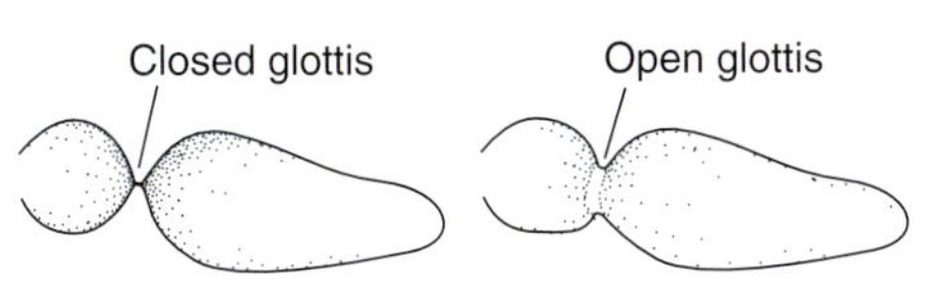

FIGURE 18-14
Lung ventilation in amphibians. *A–D,* The exchange of air in a frog. Long arrows show air movements; short arrows, expansion and contraction of the lungs and buccal cavity; open arrows, the movements of the hyoid and floor of the buccopharyngeal cavity and pleuroperitoneal cavity. *E,* Drawings from videographs of a larval salamander showing closing and opening of the glottis. *(A–D, After DeJongh; E, after Brainerd.)*

of the transversus abdominis muscle while relying on the buccal pump to inhale. This pattern of inhalation by buccal pump and exhalation by the action of the transversus abdominis muscle is widespread among salamanders and also is found in frogs that have been studied. As in lungfish, amphibians can tolerate prolonged periods of apnea alternating with brief periods of ventilation. Because their metabolism and oxygen needs are low, a lung full of air can provide an amphibian with oxygen for many minutes. Because ventilation rates are low, eliminating carbon dioxide is more problematic than is obtaining oxygen.

In addition to lungs, the thin, moist, vascular skin of most extant amphibians acts as a respiratory membrane. Considerable cutaneous gas exchange occurs in most extant species, but much variation exists. Some arboreal, tropical frogs that live in dry habitats have very little cutaneous gas exchange. On the other hand, one large family of salamanders (the plethodontids) have lost their lungs and depend exclusively on cutaneous gas exchange. Loss of lungs is an unexpected adaptation for a terrestrial vertebrate, and various scenarios have been proposed as to how this may have come about (Reagan and Verrell, 1991). Whatever the causes, reliance on cutaneous gas exchange restricts plethodontids to habitats where temperatures are cool and the oxygen supply is dependable. Some live on the bottom of cool mountain streams; others live beneath rocks and logs in damp woods. Plethodontids are also small animals with a surface–mass ratio favorable for gas exchange through the skin, and they are not very active.

In species in which both the lungs and skin are used as respiratory membranes, the lungs are usually more important for oxygen uptake than carbon dioxide loss, and the skin is more important for carbon dioxide loss than oxygen uptake (Shoemaker et al., 1992). But the role of the lungs and skin varies with temperature and activity. As temperature rises and/or activity increases, the lungs assume a greater role in the exchange of both gases. Conversely, as temperature or activity decreases, the skin becomes more important. At very low temperatures, the lungs may not be used at all.

Cutaneous gas exchange also exposes amphibians to considerable water loss. Amphibians mitigate this loss by living in wet or moist habitats where they can make up for the lost water, and, in some species (i.e., toads), by being most active during the early morning and evening, when humidity is high.

As part of their reproductive behavior, frogs have evolved a method of vocalization. **Vocal cords** are located in the laryngotracheal chamber and can be vibrated by passing air across them. Although both sexes have vocal cords, only the males have resonating **vocal sacs.** These are evaginations from the floor or lateral walls of the pharynx that can be filled with air. The laryngotracheal chamber of males also is significantly larger than is that of females of comparable size in species in which this feature has been examined. It is the distinctive call of male frogs in the spring that attracts conspecific females.

The Reptile Respiratory System

Gas exchange in the embryos of reptiles, birds, and egg-laying mammals is through the allantois, a vascularized extraembryonic membrane that extends from the hindgut to unite with the chorion just beneath the eggshell (Chapter 4). The chorioallantoic membrane contributes to the placenta in placentals and a few marsupials.

Neck length increases in reptiles, and the primitive laryngotracheal chamber becomes divided into a **lar-**

ynx (Gr., *larynx* = gullet) and **trachea** (Gr., *tracheia* = rough artery). The larynx wall is supported by a pair of **arytenoid cartilages** that flank the glottis and by a ring-shaped **cricoid cartilage.** These cartilages are derivatives from the lateral laryngeal cartilages of amphibians. Vocal cords are absent, but some reptiles can squeak, hiss, or bellow by rapidly expelling air. Cartilaginous rings, or partial rings, in the tracheal wall keep the trachea open for the free flow of air.

The horny skin of most reptiles precludes effective cutaneous gas exchange, so the lungs of adults are the primary or only site for gas exchange. Certain aquatic turtles exchange some gases through their skin and also the buccopharyngeal cavity and cloaca, in and out of which they can pump water. Some species of turtles have accessory, vascular bladders that evaginate from the cloaca for this purpose (Fig. 17-8). *Sphenodon* retains amphibian-like lungs with only a slight increase in internal surface area, but most reptiles have lungs that are relatively more compartmentalized and larger than are those of amphibians (Fig. 18-15). A wide central **bronchus** (Gr., *bronchos* = windpipe) leads from the trachea into each lung, secondary bronchi branch from it, and **alveolar sacs** of varying size bud off them. Lung structure varies considerably among species. The anterior part of the lung is more compartmentalized in lizards and snakes than is the posterior part, which is frequently a simple sac with poorly vascularized walls. The posterior region appears to serve for air storage or acts as a bellows to help ventilate the rest of the lung. This region parallels the development of air sacs in birds. One lung usually is lost in amphisbaenians and long-bodied squamates, including snakes. As one would expect, large species have more compartmentalization of the lungs than do smaller ones. This maintains an adequate ratio between respiratory surface and body mass.

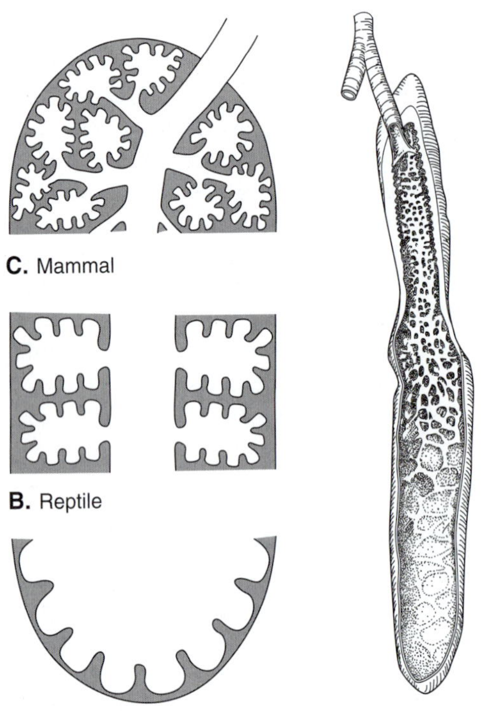

FIGURE 18-15
A–C, The degree of compartmentalization of the lungs of different tetrapods: *A,* an amphibian; *B,* a reptile; *C,* a mammal. *D,* The lung of a lizard, *Ophiosaurus. (After Portmann.)*

Of particular significance is the reptilian method of ventilating the lungs. Reptiles use an **aspiration pump.** During inspiration, contraction of intercostal muscles simultaneously on both sides of the body enlarges the pleuroperitoneal cavity in which the lungs lie by abducting the ribs and expanding the ribcage, pressure within the pleuroperitoneal cavity decreases to below atmospheric pressure, the lungs expand, and air is sucked into them, hence the term "aspiration pump." The glottis is closed, and air is held in the lungs until the next breathing cycle. Prolonged periods of apnea occur, as in lungfishes and amphibians. Expiration results from opposite rib movements and from the contraction of smooth muscle fibers in the lung wall. Aspiration by a suction-pump mechanism is more efficient than a gular (buccal) force pump because air can be transferred into the lungs in one movement. Thus, in general, reptiles breathe by aspiration breathing, but some lizards also use a gular (buccal) pump to assist lung ventilation, especially during or just after moderate to rapid locomotion. Lateral undulations that accompany locomotion appear to reduce the effectiveness of costal respiratory movements because intercostal muscles are being used to help bend the body from side to side. They contract first on one side and then on the other and not simultaneously as in costal respiration (Owerkowicz et al., 1999). The first aspiration pump in vertebrate evolution has been documented for a primitive actinopterygian fish, *Polypterus* (Focus 18-1), even though it evolved here as an analogous system that does not use a costal pump.

Crocodilians evolved an unusual method of ventilation. Their lungs lie in separate pleural cavities anterior to the liver. The contraction of a unique **diaphragmatic muscle,** which extends from the liver to the pelvic girdle, pulls the liver caudad and enlarges the pleural cavities (Fig. 18-16), causing fresh air to be sucked into the lungs. (This muscle is not homologous to the mammalian diaphragm.) Expiration occurs by the contraction of abdominal flank muscles, which increases intra-abdominal pressure and pushes the liver forward. Thus, crocodilians ventilate their lungs with a hepatic **piston pump** in addition to the primitive **costal** aspiration pump. Because turtles are encased by a dorsal carapace and a ventral plastron, they cannot

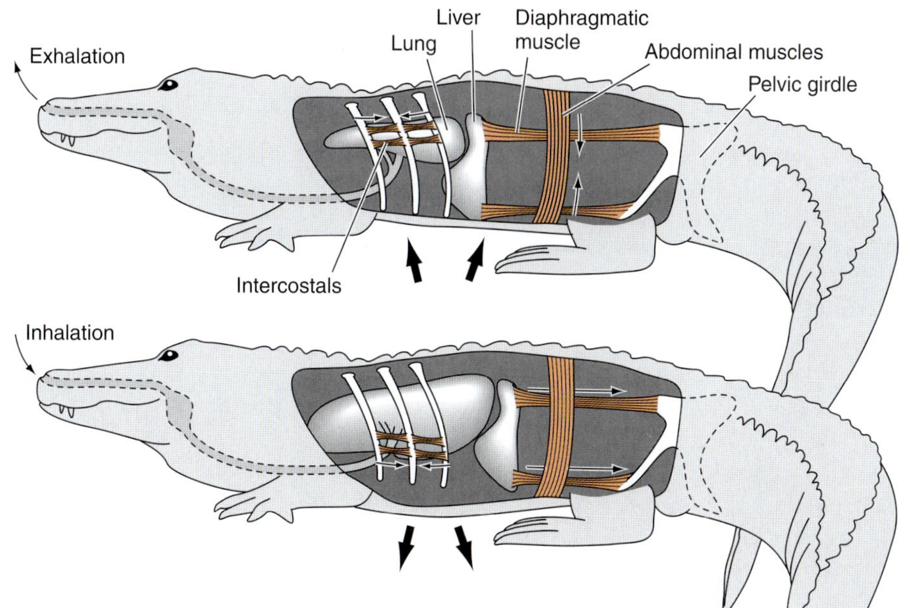

FIGURE 18-16 Ventilation of the lungs in a caiman. Small arrows within the body show contraction of different muscles; large arrows beneath the animal, the movements of the abdominal wall. *(Modified from Gans and Clark.)*

use ribs for aspiration pumping. At first it was believed that they can change the pressure in the lungs by in-and-out movements of their limbs, but recent studies have shown that these movements contribute insufficiently. Instead, they possess a transversus abdominis and a "diaphragmaticus" muscle to lift the ventral plastron toward the carapace. This decreases the volume of the body cavity, causing the lungs to exhale. Inhalation occurs when the obliquus abdominis muscle increases the body cavity volume.

Respiration in Birds

Because they are endothermic and flying animals, birds need an exceptionally efficient and compact respiratory system that will sustain a high level of metabolism and not add greatly to their weight. Bird lungs are relatively small organs that adhere to the dorsal wall of the pleural cavities and do not change appreciably in volume (Fig. 18-17*A*). The lungs themselves are only half the size of those of a mammal of comparable size, but they connect to a system of **air sacs** that pass among the viscera and even extend into many of the bones. Together, lungs and air sacs have a volume two to three times that of the lungs of a comparable mammal. The air-sac walls are not heavily vascularized, so they do not contribute to the gas-exchange surface. Rather, the air sacs, in combination with a unique pattern of airways within the lungs, make possible a unidirectional flow of air through the lungs. In other pulmonate vertebrates, in which the air moves in and out of the lungs through the same passages, a certain amount of stale air always mixes with inspired air. But in birds, the one-way flow carries relatively fresh air, with more oxygen and less carbon dioxide, across the respiratory surfaces during both inspiration and expiration.

The morphology of the avian respiratory system is complex. A number of investigators have studied the system using various techniques, including inserting minute probes into the airways to determine the composition of the air within them and the direction of its flow. The numerous air sacs can be grouped functionally into an anterior and a posterior set (Fig. 18-17*A*). The trachea bifurcates into a pair of **primary bronchi,** each of which passes rather directly through the center of a lung in a **mesobronchus,** which connects with the posterior air sacs (Fig. 18-18*A*). Air in the posterior sacs returns to a lung through **mediodorsal bronchi,** which connect with thousands of small **parabronchi.** Innumerable short **air capillaries** bud off the parabronchi and branch and anastomose with adjacent air capillaries (Fig. 18-17*B*). The air capillaries are interwoven by dense vascular capillary beds, forming a **respiratory labyrinth** in which gas exchange occurs. Distances are short, and diffusion appears to keep the composition of the air in the air capillaries nearly in equilibrium with that flowing unidirectionally through the parabronchi. The parabronchi lead to medioventral bronchi, which in turn connect with the anterior air sacs and primary bronchus (Fig. 18-18*B*). The parabronchi are parallel to each other throughout the lung in early birds and in most of the lung in other species. This part of the lung is called the **paleopulmo.** A small **neopulmo,** in which the parabronchi form a network, also is incorporated into the lungs of many birds.

The lungs are ventilated by rocking movements of the sternum, which alternately expand and compress

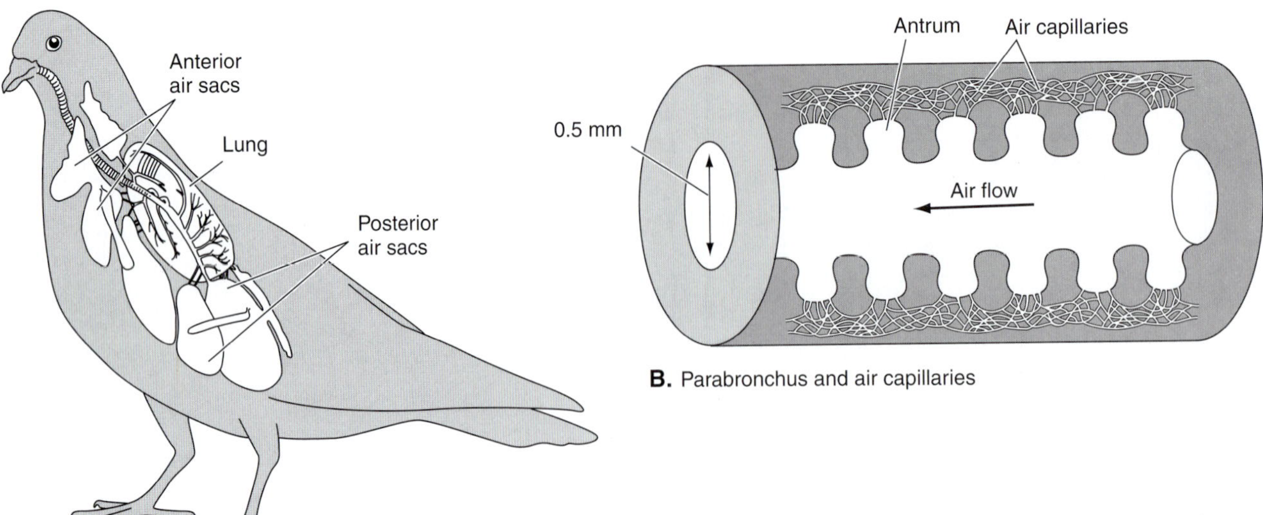

A. Lungs and air sacs

B. Parabronchus and air capillaries

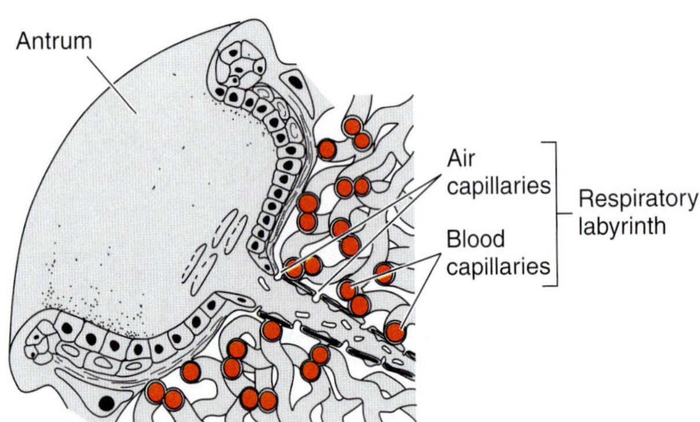

C. Antrum and respiratory labyrinth

FIGURE 18-17
The anatomy of the respiratory system of a bird. *A,* A lateral view of the lungs and major air sacs. *B,* A lateral view of a parabronchus and air capillaries. Arrows show the unidirectional flow of air across the respiratory surfaces. *C,* A small portion of the wall. *(A, After Salt; B and C, modified from Smith et al.)*

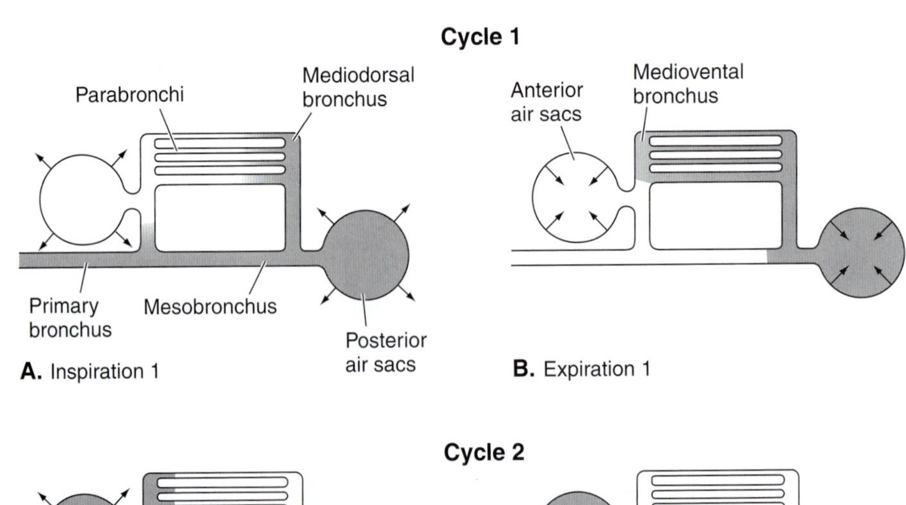

A. Inspiration 1

B. Expiration 1

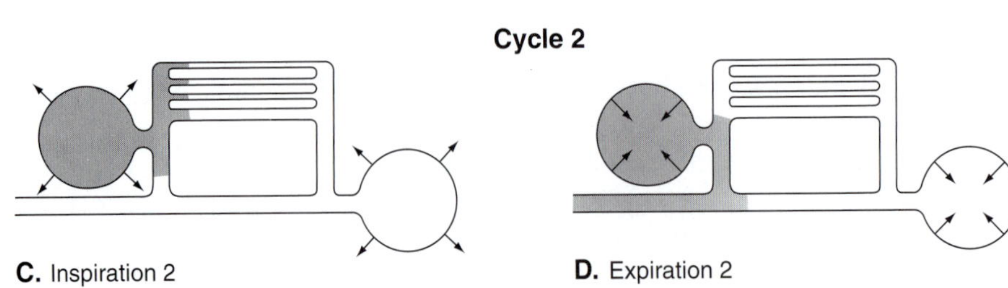

C. Inspiration 2

D. Expiration 2

FIGURE 18-18
The movement of a volume of air (*shaded*) through the lungs and air sacs of a bird. Two cycles of inspiration and expiration are needed to move a specific mass of air through the system. *(After Bretz and Schmidt-Nielsen.)*

FOCUS 18-2 *Aerodynamic Valves in the Avian Lung*

In the absence of flaps or other mechanical valves in the avian airways, how the unidirectional flow of air is maintained has been a puzzle. The diameters of the openings into the various bronchi and air sacs, and the direction in which they face, certainly play an important role in determining the path of least resistance. But many investigators have postulated that **aerodynamic valves,** the operation of which depends on subtle factors, such as air pressure and composition, also operate. An **inspiratory valve** of this type has been found in geese by Wang et al. (1992). The caudal region of the primary bronchus has considerable circularly arranged smooth muscle in its wall. This section, which they call the **accelerating segment,** contracts during inspiration (Fig. A). As a consequence, air passing through it accelerates so that it bypasses the openings of the medioventral bronchi and continues into the mesobronchus and onto the posterior air sacs and mediodorsal bronchi. The stimulus for changes in dimensions of the accelerating segment appears to be changing carbon dioxide (CO_2) levels. The smooth muscle fibers relax, and the segment dilates in the presence of increased level of CO_2, which would occur during expiration.

Whether an aerodynamic **expiratory valve** exists that would help air leaving the posterior air sacs to bypass the mesobronchus and enter the mediodorsal bronchi and parabronchi is unknown. Brown et al. (1995) believe that they may have found one in geese. They present evidence that the mesobronchus does narrow during expiration, which, of course, would help direct air into the mediodorsal bronchi. The mechanism for this is not entirely clear but appears to involve a "dynamic compensation," in which a reduced air pressure in the mesobronchus during expiration of air from the anterior air sacs causes its wall to partially collapse. Valve efficacy was positively correlated with the rate of expiratory air flow. It was 95% efficient at flow rates assumed to occur during exercise, when dynamic compensation would be the greatest, and less efficient at lower flow rates.

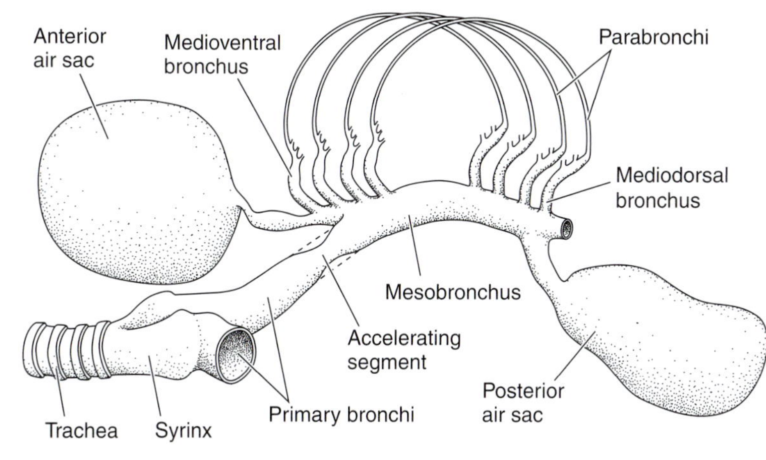

A. The principal airways in the lung of a goose, showing the contracting segment during inspiration. Its diameters during expiration are shown by the dashed lines. (*Modified from Wang et al.*)

the bellows-like air sacs. Hinges between the dorsal and the ventral parts of the ribs allow the sternum to move (Fig. 8-17*A*). The uncinate processes on the dorsal parts of the ribs act as lever arms on which certain respiratory muscles attach. Because flight muscles also cause movements of the sternum, clavicle, and ribs, ventilation and the flight movements are coupled and occur in synchrony (Jenkins et al., 1988). The glottis is held open most of the time because bird lungs are ventilated continuously, and no prolonged periods of apnea occur. Two inspirations and expirations are needed to move a unit of air through the system (Fig. 18-18). During the first inspiration, fresh air is drawn through the primary bronchus and mesobronchus into the posterior air sacs (Part A). This air moves through the mediodorsal bronchi and parabronchi during the first expiration (Part B). It is drawn into the anterior air sacs during the second inspiration (Part C) and finally is expelled from the system during the second expiration (Part D). Air, of course, enters and leaves the lungs and air sacs during each cycle of inspiration and expiration, but it is a different volume of air. During inspiration, one volume of air passes through the lungs to the posterior air sacs (Part A), and a volume of air inhaled previously is drawn out of the lungs into the anterior air sacs (Part C). During expiration, air in the posterior sacs (Part B) is driven into the lungs, and the stale air is expelled from the anterior air sacs through the main bronchi and trachea to the outside (Part D).

Air flows through the system along the line of least resistance, and this is determined partly by the diameters and directions of the openings of the various bronchi and partly by "aerodynamic valves" (Focus 18-2).

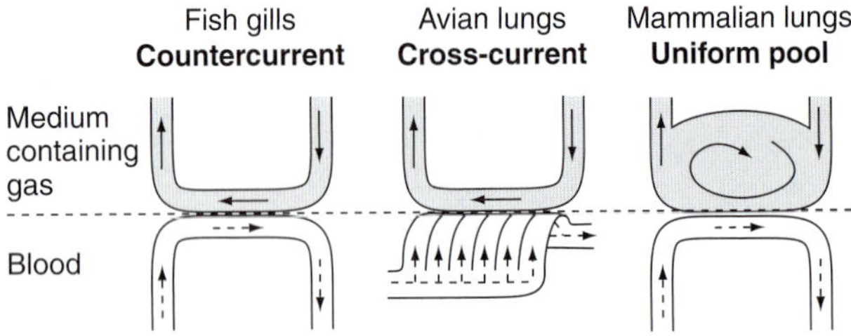

FIGURE 18-19
A comparison of the movement of blood and the external medium (water or air) across the respiratory surface of a fish, bird, and mammal. *(After Piiper and Scheid.)*

Blood in the vascular capillaries flows transversely to the air flow in the parabronchi and air capillaries in a pattern called **cross-current flow** (Fig. 18-19). As air flows from right to left in the model, it loses oxygen to the blood; blood flowing across the air gains oxygen. Because air and blood flow across each other, a gradient for the transfer of oxygen exists at each crossing of parabronchi and vascular capillaries, but the gradient is different at each intersection. The net result of this system, as in the countercurrent flow of blood and water in a fish's gills, is that the blood leaving the system has removed most of the oxygen from the air. In the lungs of other vertebrates, the oxygen content of the blood and air reach an equilibrium. The unidirectional flow of air through a bird's lungs, the cross-current flow of air and blood, and the very large gas-exchange surface combine to produce an extraordinarily efficient respiratory system. Birds extract a greater proportion of oxygen from the air than do mammals, and they can live and fly at high altitudes where mammals cannot survive.

Birds have also evolved a vocal apparatus that is associated with the airways. The calls and songs of birds are used in species recognition, for sounding alarms, in establishing territories, and in reproductive behavior. The vocal box is not the larynx, as in frogs and mammals, which is situated at the top of the trachea, but a **syrinx** (Gr., = panpipe) that is located at the bifurcation of the trachea into the main bronchi (Fig. 18-20*A*). Syrinx structure varies among species, but usually one or more tympanum-like membranes lie between cartilaginous rings in its wall, and they are vibrated by air moving across them. Air vibrations are transformed into meaningful sounds by changes in tension on these tympani, by the configuration of the trachea and buccopharyngeal cavity, and by tongue movements. Trumpeter swans and whooping cranes have exceedingly long tracheae, half or more of which loops within the sternal keel (Fig. 18-20*B*). This gives their calls their deep, resonating, and somewhat trombone-like quality.

Respiration in Mammals

The synapsid line of evolution culminating in mammals and the sauropsid line to reptiles and birds diverged millions of years ago, near the time of the origin of amniotes. Although mammals, too, evolved an efficient respiratory system, needed by endothermic animals, the mammalian system differs in many ways from that of birds because it evolved its complex design from the generalized amniote condition independently. The evolution of the secondary palate in mammals and their therapsid ancestors made possible a separation of food and respiratory passages (Chapter 7). The paired **nasal cavities** of mammals, which lie dorsal to the hard palate, are relatively larger than those of other vertebrates. Each contains three folds or scrolls of bones, known as the **turbinates,** or **conchae,** which greatly increase the surface area (Fig. 18-21). The mucous membrane that covers the turbinates and lines the other respiratory passages as they continue into the lungs is very vascular and ciliated and contains

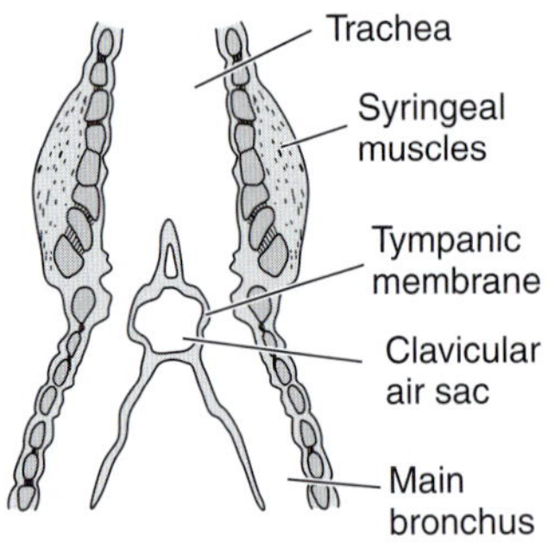

A. Syrinx of a blackbird

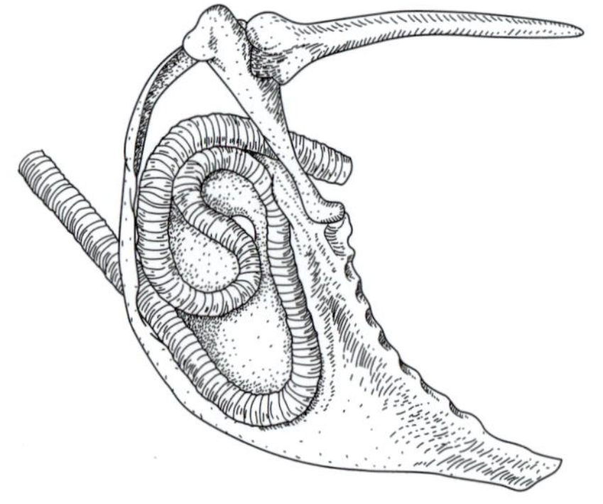

B. Sternum and trachea of a whooping crane

FIGURE 18-20
Vocalization in birds. *A,* A frontal section through the syrinx of a male blackbird, *Turdus merula. B,* A lateral view of a dissection of the sternum and trachea of a whooping crane. *(A, After Pettingill; B, after Portmann.)*

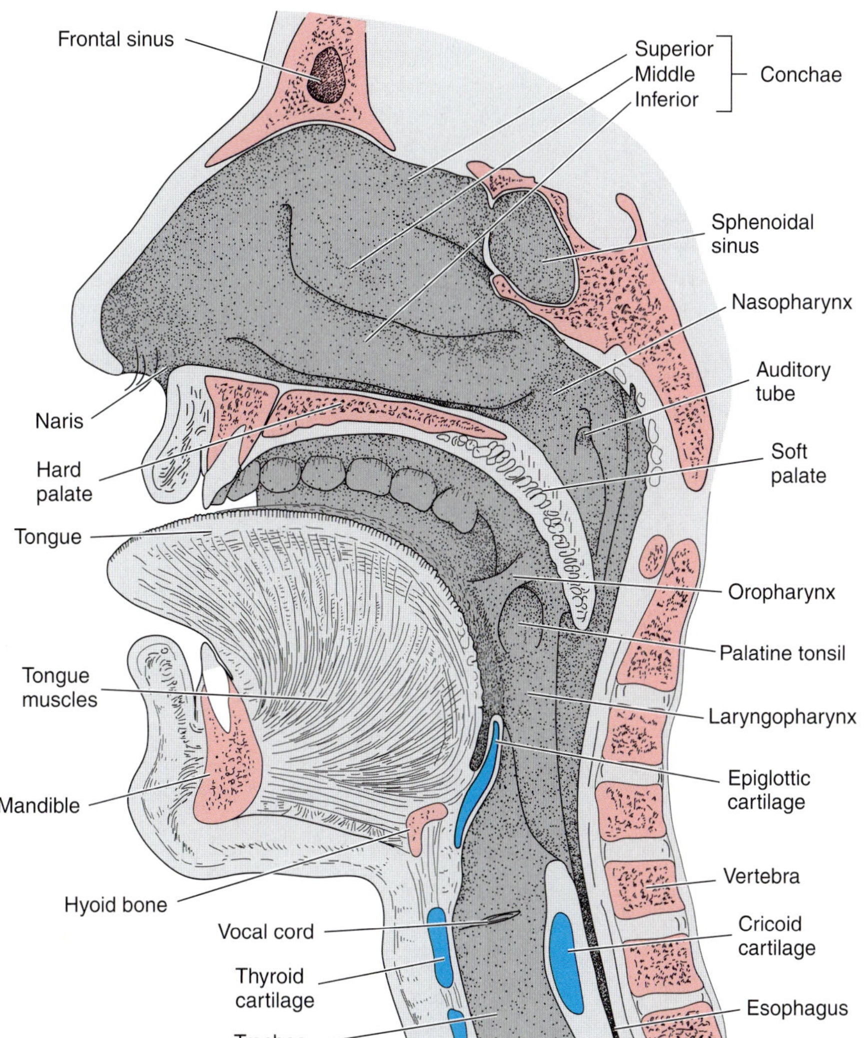

FIGURE 18-21
A sagittal section through a part of the head of a human showing the nasal and mouth cavity, pharynx, esophagus, and trachea.

cells that produce mucous and serous secretions. Considerable conditioning of the air thus occurs as it passes across these surfaces to the lungs. The air is warmed and moistened, and dirt trapped in the mucus is carried by ciliary action to the throat to be swallowed or expectorated.

The nasal cavities connect through the **choanae** with the **nasopharynx,** which is separated from the oropharynx by the soft palate. Food and air passages cross in the laryngopharynx, but food normally does not enter the respiratory passages because the larynx and hyoid apparatus are pulled forward against the base of the tongue during swallowing. A troughlike fold, the **epiglottis,** flips back over the glottis or deflects food around it into the esophagus. A newly evolved cartilage that appears to have no homologue in other species supports the epiglottis. Except when food is swallowed, the **glottis,** which is bounded laterally by the **vocal cords,** is held open because mammalian lungs are ventilated continuously.

The mammalian **larynx** is more complex than that of other tetrapods (Fig. 7-26). It continues to be supported caudally by the ring-shaped **cricoid cartilage,** and the paired arytenoid cartilages, which also extend into the vocal cords, support its rostrodorsal wall. A new **thyroid cartilage** supports its lateroventral wall. The thyroid cartilage evolved from the fourth and fifth visceral arches as they became dissociated from the hyobranchial apparatus. The intrinsic muscles of the larynx that shape it and control the vocal cords are muscles of the fourth, fifth, and sixth visceral arches; as would be expected, they are innervated by a branch of the vagus nerve. Vibrations of air produced by the vocal cords are "shaped" by movements of the pharynx, soft palate, tongue, and lips.

The trachea continues down the neck and gives rise within the thorax to primary bronchi, which enter the lobes of the lungs. The internal passages within mammalian lungs are more finely compartmentalized than are those of amphibians or reptiles, so a great increase

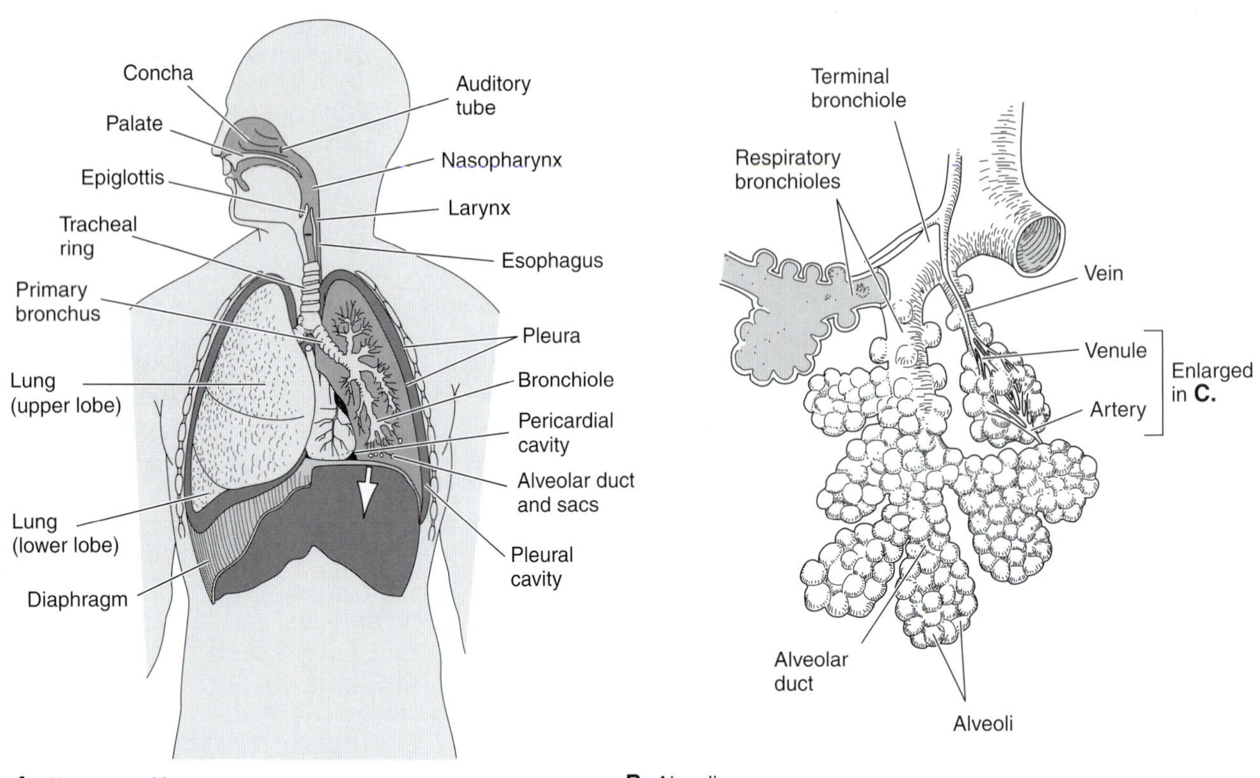

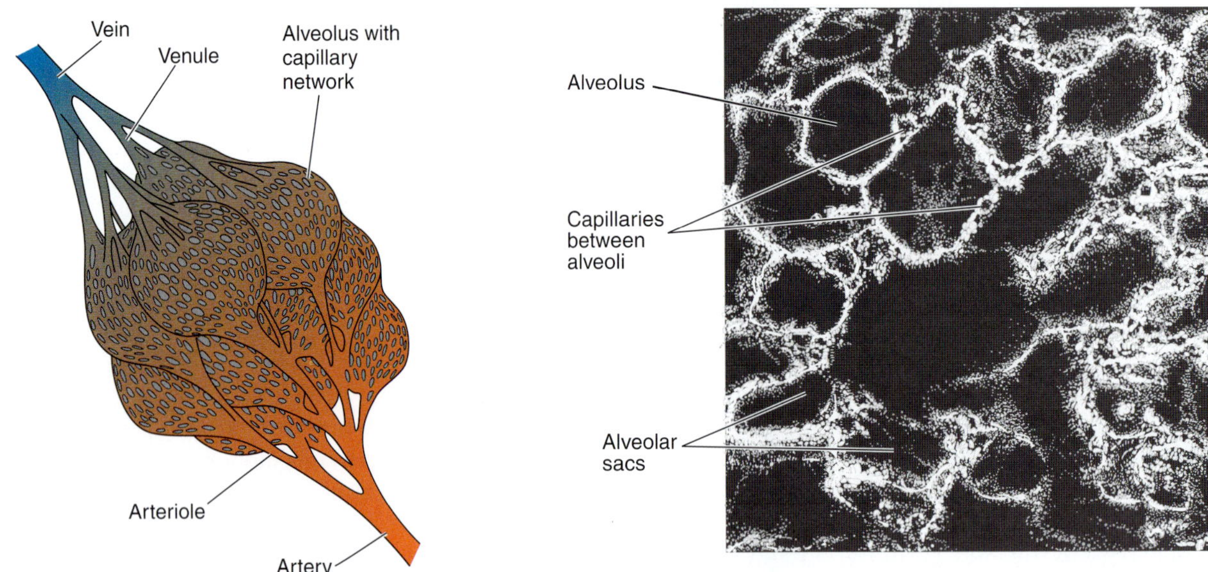

FIGURE 18-22
Diagrams of the mammalian respiratory system. *A,* An overview of the location and structure of the airways and lungs. *B,* An enlargement of the alveoli. *C,* The capillary bed that covers the alveoli. *D,* A drawing based on an electron micrograph of a portion of a mammalian lung. Notice that most of the capillaries have alveoli on two of their surfaces. *(A–C, After Dorit et al.; D, after Kessel and Kardon.)*

in internal surface area has occurred (Figs. 18-15*C* and 18-22). Whereas a frog has about 20 cm^2 of lung surface area per cubic centimeter of lung tissue, a human has 300 cm^2 per cubic centimeter. The airways within the lungs branch and rebranch at least 20 times, forming a **respiratory tree** that terminates in **bronchioles, alveolar sacs,** and individual **alveoli.** The walls of the airways become progressively thinner along the course of the respiratory tree. Supporting cartilages, smooth muscle, and secretory and ciliated cells gradually disappear. The walls of the alveoli, through which gas exchange occurs, consist only of an exceedingly thin squamous epithelium that is nearly completely covered by vascular capillaries. Diffusion distance is about 0.2 μm. Only in birds is the distance shorter, on the order of 0.1 μm. The alveoli are tightly packed, and the capillaries lie between them. This arrangement allows for diffusion of gases to occur across two or more surfaces of the capillaries (Fig. 18-22*C*).

A costal aspiration pump ventilates mammalian lungs. Air is moved in and out by changes in the size of the pleural cavities, alternately increasing and decreasing the pressure within the lungs relative to atmospheric pressure. Pleural cavity size is increased primarily by the contraction and caudal movement of the muscular **diaphragm** (Fig. 18-22*A*). During stronger inspiration, the external intercostal and other respiratory muscles pull the ribs rostrally and increase the dimensions of the thorax. Pressure within the lungs must be reduced enough during inspiration to overcome the frictional resistance of the numerous airways to the inflow of air and also to overcome the surface tension of the walls of the many thousands of alveoli. This surface tension tends to collapse the alveoli. The resistance of the airways is reduced by a pattern of branching that maximizes the diameters of the major airways, by cartilaginous supports in their walls, and by the secretion of surfactants (p. 585). Surfactants are particularly important in mammals because of the small size and large surface area of their alveoli. Expiration in quadrupeds is largely a passive process that depends on the elastic recoil of the lungs, ribs, and abdominal viscera, which are pushed caudally by the contraction of the diaphragm. Internal intercostal, abdominal, and other respiratory muscles become active during forced expiration. Although certain intercostal muscles become active during strong inspiration and expiration, the primary function of these muscles is to maintain the integrity of the chest wall so that the intercostal spaces do not bulge outward or cave in as pressure within the thorax changes. In mice, the ribs are used in lung ventilation when the animal is at rest, whereas the diaphragm becomes the major ventilatory component during activity.

Evolution of Respiratory Patterns in Air-Breathing Vertebrates

Traditionally, lung ventilation has been viewed as a clear dichotomy: fishes and amphibians breathe with a buccal pulse pump, whereas amniotes breathe with an aspiration pump generated by body muscles. There is no doubt that the primitive pattern of vertebrate air-breathing is the buccal pulse pump found in actinopterygian fishes. This primitive buccal pump pattern is characterized by four-stroke breathing; the buccal cavity expands during exhalation to draw "spent" air out of the lung or accessory air-breathing organ and then compresses fully to expel all of the exhaled gas (Fig. 18-23). The buccal cavity then expands a second time to take fresh air in and compresses to force fresh air into the lungs or other accessory air-breathing organs. In sarcopterygian fishes and salamanders the two-stroke breathing cycle evolved. Here, the buccal cavity expands to draw in fresh air, while spent air also flows

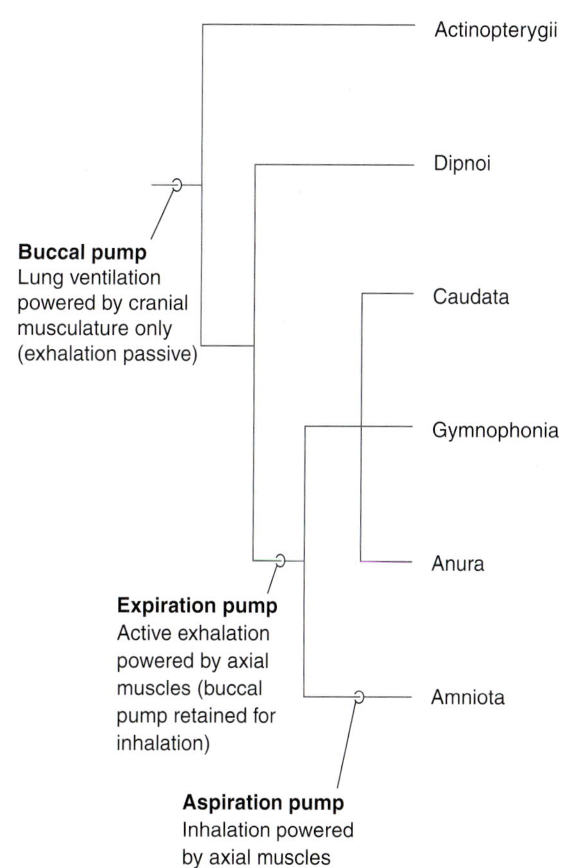

FIGURE 18-23
A cladogram showing major stages in the evolution of lung ventilation. *(After Brainerd.)*

in to mix with the fresh air. Some of this mixture is subsequently pumped into the lungs, and the rest exits by the mouth and gill slits.

Brainerd (1999) has experimentally established that salamanders exhibit a primitive intermediate evolutionary step toward aspiration breathing (Fig. 18-23). In addition to the two-stroke buccal pump to inhale, salamanders exhale by using the transversus abdominis muscle. Amniotes employ full aspiration breathing, in which body muscles are used for not only exhalation but also inhalation. In all amniotes, except for turtles, intercostal muscles and ribs are recruited for inhalation by costal aspiration. The costal aspiration mechanism is primitive in amniote evolution. The patterns seen in turtles, crocodilians, and birds are evolutionary specializations. In mammals, a new structure, the diaphragm, is interposed between the pleural and peritoneal cavities. Contractions of the diaphragm assume the primary function in perfecting the mammalian aspiration pump.

SUMMARY

1. Gases diffuse between the external medium (water or air) and the blood in the body through a respiratory membrane. Vertebrates require a means of ventilating this membrane by the bulk flow of the medium across it.
2. The respiratory system of a fish must be adapted to the limited supply of oxygen available in water and to the high density of this medium. Unloading carbon dioxide is not a problem for a fish because this gas is very soluble in water. Heat, salts, and water are also exchanged across the respiratory membrane.
3. Many larval fishes have external gills, but gas exchange in adults occurs through internal gills.
4. The gill pouches of fishes develop from endodermal pharyngeal pouches that meet ectodermal furrows extending inward from the body surface. The first pouch is reduced to a spiracle or has been lost.
5. A complete gill, or holobranch, bears primary gill lamellae on each surface. These support numerous secondary gill lamellae through which gas exchange occurs. A skeletal branchial arch, gill rays, vascular derivatives of an embryonic aortic arch, muscles, and nerves lie within the holobranch, often in an interbranchial septum.
6. The pouched gills of agnathans line spherical gill pouches. The pouches are numerous.
7. Elasmobranchs have septal gills borne on interbranchial septa and facing narrow branchial chambers. Each chamber opens into a parabranchial chamber and to the surface through an external gill slit.
8. A countercurrent flow—whereby water flows across the secondary lamellae in the opposite direction that blood flows through them—increases the efficiency of gas exchange.
9. The valvular action of the mouth and external gill slits allows the pharynx and parabranchial chambers, by their concurrent expansion and contraction, to act together first as a suction pump and then as a force pump in ventilating the gills.
10. Teleosts have aseptal gills, which extend into a common opercular cavity that opens to the surface through a single gill slit. The absence of septa increases the efficiency of the countercurrent exchanges of gases between blood and water. The pharynx and opercular cavity act alternately and in sequence as suction and force pumps. The combination of suction and force is described as a pulse pump.
11. The structure of the secondary lamellae provides a large surface area and a short diffusion distance between water and blood.
12. Because of the attendant osmotic problems, a fish does not ventilate its gills more than necessary to meet its oxygen needs. Various mechanisms regulate the amount of ventilation.
13. Lungs probably evolved in ancestral bony fishes as accessory respiratory organs that permitted them to live in unstable freshwater environments.
14. Lungfishes ventilate their lungs with a pulse pump. Some mixing of fresh and spent air occurs, and this mixed air is held in the lungs under positive pressure for a prolonged period of apnea. Some carbon dioxide accumulates in the lungs, but most is eliminated through the gills.
15. The lungs of the first bony fishes have transformed into hydrostatic swim bladders in most contemporary species.
16. Fishes attain neutral buoyancy by regulating the amount of air in the bladder. In most species, oxygen is secreted into the bladder by a gas gland and held in the bladder by a countercurrent rete mirabile. Oxygen can be reabsorbed through the oval.
17. Terrestrial vertebrates must avoid an excess loss of body water through their lungs. The large amount

of oxygen in air and a low metabolic rate allow amphibians to avoid water loss by ventilating their lungs at a low rate.

18. Contemporary adult amphibians ventilate their lungs by a buccopharyngeal pump and retain air in the lungs under positive pressure for a prolonged period of apnea. Oxygen also is gained and carbon dioxide is eliminated by cutaneous respiration.

19. Plethodontid salamanders have lost their lungs and rely on cutaneous respiration for the exchange of gases.

20. Lung surface area is relatively larger in reptiles than in amphibians. Reptiles use rib movements and (in crocodiles) an unusual muscle that pulls the liver caudad to ventilate the lungs. Reptiles continue to have prolonged periods of apnea. Reptiles have a costal aspiratory pump.

21. Birds have small lungs that are connected to an extensive system of air sacs. Their lungs are ventilated continuously by rib and sternal movements and a bellows-like action of the air sacs.

22. The airways of birds are arranged in such a way that a unidirectional flow of air passes through numerous parabronchi and across air capillaries. Little stale air is retained in the lungs. A cross-current flow of blood over the parabronchi and of air within them thoroughly aerates the blood.

23. The airways of mammals end in numerous, thin-walled alveoli within the lungs that are covered by dense capillary networks. A large surface area is available for gas exchange, but new air entering the lungs always mixes with some stale air. The lungs are ventilated continuously by a well-developed muscular diaphragm.

REFERENCES

Alexander, R. McNeil, 1966: Physical aspects of swim bladder function. *Biological Reviews,* 41:141–176.

Bijtel, H. J., 1949: The structure and the mechanism of the movement of the gill filaments in the Teleostei. *Archives Neerlandaises de Zoologie,* 8:1–22.

Brainerd, E. L., 1999: New perspectives on the evolution of lung ventilation mechanisms in vertebrates. *Experimental Biology Online,* 4:11–18.

Brainerd, E. L., 1994: The evolution of lung-gill bimodal breathing and the homology of vertebrate respiratory pumps. *American Zoologist,* 34:289–299.

Brainerd, E. L., Liem, K., and Samper, C. T., 1989: Air ventilation and recoil aspiration in Polypterid fishes. *Science,* 246:1593–1595.

Brainerd, E. L., and Monroy, J. A., 1998: Mechanics of lung ventilation in a large aquatic salamander, *Siren lacertina. Journal of Experimental Biology,* 201:673–682.

Brainerd, E. L., Ditelberg, J. S., and Bramble, D. M., 1993: Lung ventilation in salamanders and the evolution of vertebrate air-breathing mechanisms. *Biological Journal of the Linnean Society,* 49:163–183.

Bramble, D. M., and Carrier, D. R., 1983: Running and breathing in mammals. *Science,* 291:251–256.

Bretz, W. L., and Schmidt-Nielsen, K., 1971: Bird respiration: Flow patterns in the duck lung. *Journal of Experimental Biology,* 54:103–118.

Brown, R. E., Kovacs, C. E., Butler, J. P., Wang, N., Lehr, J., and Banzett, R. B., 1995: The avian lung: Is there an expiratory valve? *Journal of Experimental Biology,* 198: 2349–2357.

Burggren, W. W., and Bemis, W. E., 1992: Metabolism and ram ventilation in juvenile paddlefish, *Polyodon spathula* (Chondrostei: Polyodontidae). *Physiological Zoology,* 65: 515–539.

Burggren, W. W., and Roberts, J., 1991: Respiration and metabolism. *In* Prosser, C. L., editor: *Comparative Animal Physiology,* 4th edition. New York, Wiley-Liss.

Clements, J. A., Nellenbogen, J., and Trahan, H. J., 1970: Pulmonary surfactant and evolution of the lungs. *Science,* 169:603–604.

Daniels, C. B., Orgeig, S., and Smits, A. W., 1995: The evolution of the vertebrate pulmonary surfactant system. *Physiological Zoology,* 68:539–566.

DeJongh, H. J., and Gans, C., 1969: On the mechanism of respiration in the bullfrog, *Rana catesbeiana:* A reassessment. *Journal of Morphology,* 127:259–290.

Dejours, P., 1981: *Principles of Comparative Respiratory Physiology,* 2nd edition. Amsterdam, North Holland Co.

Dorit, R. L., Walker, W. F., Jr., and Barnes, R. D., 1991: *Zoology.* Philadelphia, Saunders College Publishing.

Farmer, C., 1997: Did lungs and the intracardiac shunt evolve to oxygenate the heart in vertebrates? *Paleobiology,* 23:358–372.

Farmer, C. G., and Jackson, D. C., 1998: Air-breathing during activity in the fishes *Amia calva* and *Lepisosteus oculeatus. Journal of Experimental Biology,* 201:943–948.

Feder, M. E., and Burggren, W. W., 1985: Skin breathing in vertebrates. *Scientific American,* 235:126–142.

Gans, C., 1970: Strategy and sequence in the evolution of the external gas exchangers in ectothermal vertebrates. *Forma et Functio,* 3:61–104.

Gans, C., 1970: Respiration in early tetrapods: The frog is a red herring. *Evolution,* 24:723–734.

Gans, C., and Hughes, G. M., 1967: The mechanism of lung ventilation in the tortoise *Testudo graeca* Linnaeus. *Journal of Experimental Biology,* 47:1–20.

Gatten, R. E., Jr., 1985: The use of anaerobiosis by amphibians and reptiles. *American Zoologist,* 25:945–954.

Gaunt, A. S., editor, 1973: Vertebrate sound production. *American Zoologist,* 13:1139–1255.

Gaunt, A. S., and Gans, C., 1969: Mechanics of respiration in the snapping turtle *Chelydra serpentina. Journal of Morphology,* 128:195–228.

Graham, J. B., 1997: *Air-Breathing Fishes: Evolution, Diversity and Adaptation.* San Diego, Academic Press.

Hughes, G. M., 1984: General anatomy of the gills. *In* Hoar, W. S., and Randall, D. J., editors: *Fish Physiology,* volume 10. New York, Academic Press.

Hughes, G. M., editor, 1974: *Respiration of Amphibious Vertebrates,* 2nd edition. London, Heinemann.

Hughes, G. M., 1963: *Comparative Physiology of Vertebrate Respiration.* Cambridge, Harvard University Press.

Hughes, G. M., and Morgan, M., 1973: The structure of fish gills in relation to their respiratory function. *Biological Reviews,* 48:419–475.

Jenkins, F. A., Jr, Dial, K. P., and Goslow, G. E., 1988: A cineradiographic analysis of bird flight: The wishbone in starlings is a spring. *Science,* 241:1495–1498.

Johansen, K., 1971: Comparative physiology: Gas exchange and circulation in fishes. *Annual Review of Physiology,* 33:569–599.

Johansen, K., 1968: Air-breathing fishes. *Scientific American,* 219:102–111.

Jones, F. R. H., and Marshall, N. B., 1953: The structure and function of the teleostean swimbladder. *Biological Reviews,* 28:16–83.

Jones, J. D., 1972: *Comparative Physiology of Respiration.* London, Arnold Publishing Co.

Kessel, R. G., and Kardon, R. H., 1979: *Tissues and Organs: A Text-Atlas of Scanning Electron Microscopy.* San Francisco, W. H. Freeman & Co.

Lagler, K. F., Bardach, J. E., Miller, R. E., 1962: *Ichthyology.* New York, John Wiley & Sons.

Liem, K. F., 1988: Form and function of lungs: The evolution of air breathing mechanisms. *American Zoologist,* 28:739–759.

Liem, K. F., 1987: Functional design of the air ventilation apparatus and overland excursions by teleosts. *Fieldiana,* 37:1–29.

Liem, K. F., 1985: Ventilation. *In* Hildebrand, M., Bramble, D. M., Liem, K. F., and Wake, D. B., editors: *Functional Vertebrate Morphology.* Cambridge, Harvard University Press.

McClelland, B. E., and Wilczynski, W., 1989: Sexual dimorphic laryngeal morphology in *Rana pipiens. Journal of Morphology,* 201:293–299.

McMahon, B. R., 1969: A functional analysis of the aquatic and aerial respiratory movements of the African lungfish, *Protopterus aethiopicus,* with reference to the evolution of lung ventilation mechanisms in vertebrates. *Journal of Experimental Biology,* 51:407–430.

Metcalfe, J. D., and Butler, P. J., 1986: The functional anatomy of the gills of the dogfish (*Scyliorhinus canicula*). *Journal of Zoology London,* 208:519–530.

Munshi, J. S. D., Olson, K. R., Ojha, J., and Ghosh, T. K., 1986: Morphology and vascular anatomy of the accessory respiratory organs of the climbing perch, *Anabas testudineus. American Journal of Anatomy,* 176:321–331.

Owerkowicz, T., Farmer, C. G., Hicks, J. W., and Brainerd, E. L., 1999: Contribution of gular pumping in lung ventilation in monitor lizards. *Science,* 284:1661–1663.

Pettingill, O. S., Jr., 1985: *Ornithology in Laboratory and Field,* 5th edition. Orlando, Academic Press.

Piiper, J., editor, 1978: *Respiratory Function in Birds, Adult and Embryonic.* Berlin, Springer-Verlag.

Piiper, J., and Scheid, P., 1977: Comparative physiology of respiration: Functional analysis of gas exchange organs in vertebrates. *International Review of Physiology,* 14: 219–253.

Randall, D. J., Burggren, W. W., Farrell, A. P., and Haswell, M. S., 1981: *The Evolution of Air Breathing in Vertebrates.* Cambridge, Cambridge University Press.

Reagan, N. L., and Verrell, P. A., 1991: The evolution of plethodontid salamanders: Did terrestrial mating facilitate lunglessness? *American Naturalist,* 138:1307–1313.

Scheid, P., and Piiper, J., 1972: Cross-current gas exchange in avian lungs: Effects of reversed parabranchial air flow in ducks. *Respiration Physiology,* 16:304–312.

Schmidt-Nielsen, K., 1971: How birds breathe. *Scientific American,* 225:72–79.

Shoemaker, V. H., Hillman, S. S., Hillyard, S. D., Jackson, D. C., McClanahan, L. L., Withers, P. C., and Wygoda, M. L., 1992: Exchange of respiratory gases, ions, and water in amphibious and aquatic amphibians. *In* Feder, M. E., and Burggren, W. W., editors: *Environmental Physiology of the Amphibians.* Chicago, University of Chicago Press.

Simons, R. S., 1996: Lung morphology of cursorial and noncursorial mammals: Lagomorphs as a case study for a pneumatic stabilization hypothesis. *Journal of Morphology,* 230:299–316.

Smith, J. H., Meier, J. L., Lamke, C., Neill, P. J. G., and Box, E. D., 1986: Microscopic and submicroscopic anatomy of the parabronchi, air sacs, and respiratory spaces in the budgerigar (*Melopsittacus undulatus*). *American Journal of Anatomy,* 177:221–242.

Vitalis, T. Z., and Shelton, G., 1990: Breathing in *Rana pipiens:* The mechanism of ventilation. *Journal of Experimental Biology,* 154:537–556.

Wang, N., Banzett, R. E., Nations, C. S., and Jenkins, F. A., Jr., 1992: An aerodynamic valve in the avian primary bronchus. *Journal of Experimental Zoology,* 262:441–445.

Weibel, E. R., and Taylor, C. R., 1981: Design of the mammalian respiratory system. *Respiration Physiology,* 44:1–164.

19

The Circulatory System

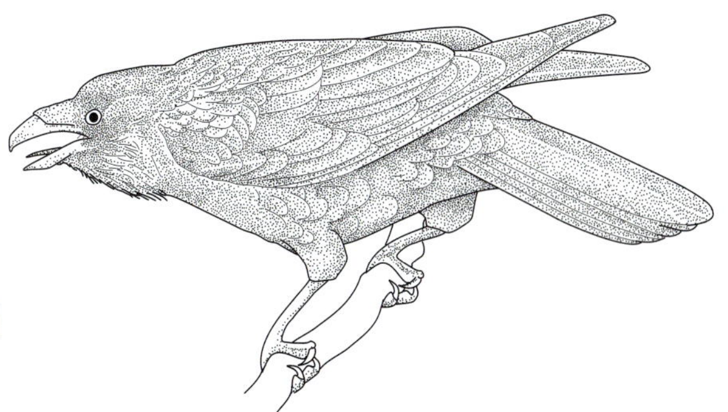

PRECIS

The circulatory system is essential for the transport of materials throughout the body and for the maintenance of homeostasis. Blood performs the biochemical and physiological functions of the system, the heart propels the blood, and the vessels transport it. The heart and vessels changed greatly as vertebrates moved from water to land during their evolution and became endothermic animals.

OUTLINE

The circulatory system is first of all the transport system of the body. It picks up oxygen, nutrients, and other needed materials at sites of intake, transports them to the tissues, and then returns carbon dioxide, nitrogenous wastes, and other excess substances in the interstitial fluids to sites of removal from the body. Materials enter and leave the blood by diffusion, supplemented for some molecules by active transport. In essence, the circulatory system shortens diffusion distance by providing for the bulk flow of materials between various parts of the body. This is essential for animals that grow more than a few millimeters in diameter and lead active lives. Beyond this, the circulatory system distributes heat between parts of the body, carries hormones, and generally plays a vital role in maintaining a constant internal environment, which is called **homeostasis** (Gr., *homoios* = alike + *stasis* = standing). Many components of the blood also are needed to prevent an excess loss of blood after an injury, heal wounds, and defend the body against disease organisms.

Components of the Circulatory System

The Spleen and Other Hemopoietic Tissues

The biochemical and physiological functions of the circulatory system are performed by blood, which is composed of a liquid that carries cells of several types. In humans, all blood cells are relatively short-lived, with life spans ranging from 1 or 2 days for some leukocytes to 120 days for erythrocytes. **Hemopoietic tissues** (Gr., *haima* = blood + *poietikos*

= producing) constantly produce new blood cells. Hemopoietic tissues are delicate, reticular connective tissues containing stem blood cells that multiply and differentiate. Hemopoietic tissues are widespread in embryos but become restricted in the adults, especially of the amniotes. The spleen, kidneys, and liver are important sites for hemopoiesis in adult anamniotes. Most of the blood cells of adult mammals develop in the red bone marrow.

The **spleen** is a conspicuous organ attached by a mesentery to the left side of the stomach in most vertebrates. Hagfishes, lampreys, and lungfishes lack definitive spleens, but splenic tissue is embedded in the wall of part of their digestive tracts. Arteries enter the spleen through a network of dense, connective tissue trabeculae that extend into the organ from its connective tissue capsule (Fig. 19-1). Many of the small arterioles into which the arteries break up are surrounded by masses of lymphocytes that constitute the **white pulp.** Blood eventually is collected into venous sinusoids that lead to the veins draining the organ. A great deal of blood and many blood cells are present throughout the interstices of a delicate **reticulum** that constitutes the **red pulp.** How blood enters and leaves these spaces is unclear. An "open theory" holds that the arterioles open directly into the interstices and drain into the venous sinusoids. Proponents of a "closed theory" believe that the arterioles or their terminal capillaries lead directly to the venous sinusoids. Some blood passes directly through the sinusoids, and some passes back and forth through the walls of the venous sinusoids to the interstices of the red pulp. Both pathways likely operate because isotopically labeled blood shows that most blood moves through the spleen rapidly (the "closed pathway"), but a smaller amount has a much slower transit time (the "open pathway"). The spleen performs many functions. Lymphocytes multiply in the white pulp, and many enter the red pulp to participate in the body's immune responses. Numerous macrophages line the interstices of the red pulp and remove damaged or aging erythrocytes and other debris. Breakdown products of erythrocyte destruction go to the liver, where iron is salvaged and other materials are excreted as bilirubin in the bile. Many hemopoietic cells occur in the red pulp of nonmammalian vertebrates and embryonic mammals. Finally, numerous erythrocytes are stored in the red pulp and venous sinusoids. They are released by the contraction of the spleen when a sudden need for an increase in the blood's oxygen-carrying capacity occurs, as during vigorous exercise or a hemorrhage.

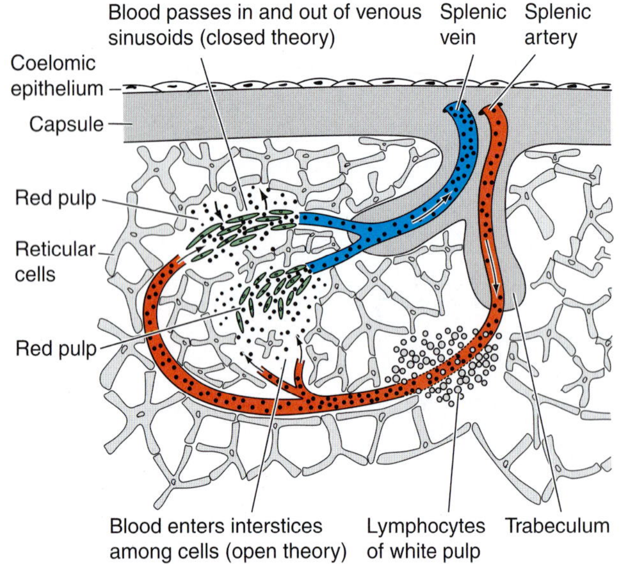

FIGURE 19-1
The structure of the spleen. Both open and closed hypotheses of the circulation through the organ are shown. Evidence suggests that both patterns of circulation occur. *(Modified from Williams et al.)*

The Heart and Blood Vessels

The circulatory system of vertebrates is closed, for the blood is confined within blood vessels. By definition, **arteries** are vessels that carry blood from the heart to capillary networks in the tissues; **veins** are vessels that return blood to the heart. Usually, arteries carry blood high in oxygen content; veins carry blood low in oxygen. But this is not always true, for the type of blood in vessels depends on the site of gas exchange. The pulmonary artery of a mammal, for example, carries blood low in oxygen content from the heart to the lungs.

The **heart** is a pump that receives blood low in hydrostatic pressure and increases the pressure sufficiently to drive the blood through the system. It is lined by a simple squamous epithelium that is called the **endothelium** and is covered by a coelomic epithelium, the **pericardium.** The **myocardium** between these epithelial layers is composed of a connective tissue "skeleton" and cardiac muscle. Cardiac muscle resembles skeletal muscle in having striated fibers that contract rapidly with considerable force (Fig. 10-1*B*); it resembles smooth muscle in its control. Cardiac muscle has an inherent rhythm of contraction that can be modified by autonomic nerve fibers, most of which terminate on a pacemaker. The pacemaker, also called the **sinoatrial node** in mammals, is composed of modified cardiac muscle fibers and is located in the heart wall near where veins enter the heart (Fig. 19-2). Action potentials generated at the pacemaker spread from fiber to fiber, passing through specialized junctions. Birds and mammals also have a specialized conducting system in the ventricle composed of modified cardiac muscle known as **Purkinje fibers.** Once an action po-

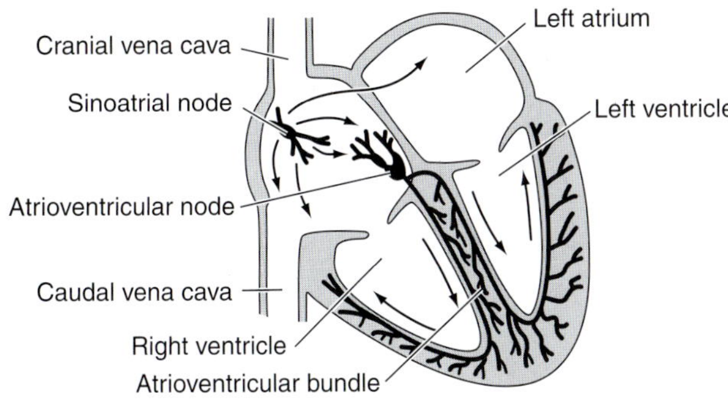

FIGURE 19-2
Ventral view of the electrical conducting system of the human heart. *(After Dorit et al.)*

tential traveling through the atrial muscles reaches the **atrioventricular node,** located at the beginning of the conducting system, a wave of excitation is transmitted rapidly through Purkinje fibers to all parts of the ventricle.

The heart not only is a device to pump blood but also functions as an endocrine organ. The walls of one of the chambers, the atrium, produce **atrial natriuretic peptide,** a hormone that acts on the kidneys to increase the excretion of salt and water.

Endothelium lines all the blood vessels. The rest of the wall of the vessels is composed of varying amounts of elastic fibers, smooth muscles, and collagen fibers (Fig. 19-3). The forces acting on the vessels and their functions determine the amount of each. Vessels are often described as having three layers of tissue. A **tunica intima** includes the endothelium and elastin fibers, if present; a **tunica media** is composed primarily of smooth muscles; and the **tunica externa** contains primarily collagen fibers. The larger arteries, and especially those near the heart, have a relatively large amount of elastic fibers. They absorb kinetic energy and stretch as blood is pumped into them from the heart. When the heart relaxes, the elastic recoil of these fibers keeps the blood flowing. The small arterioles

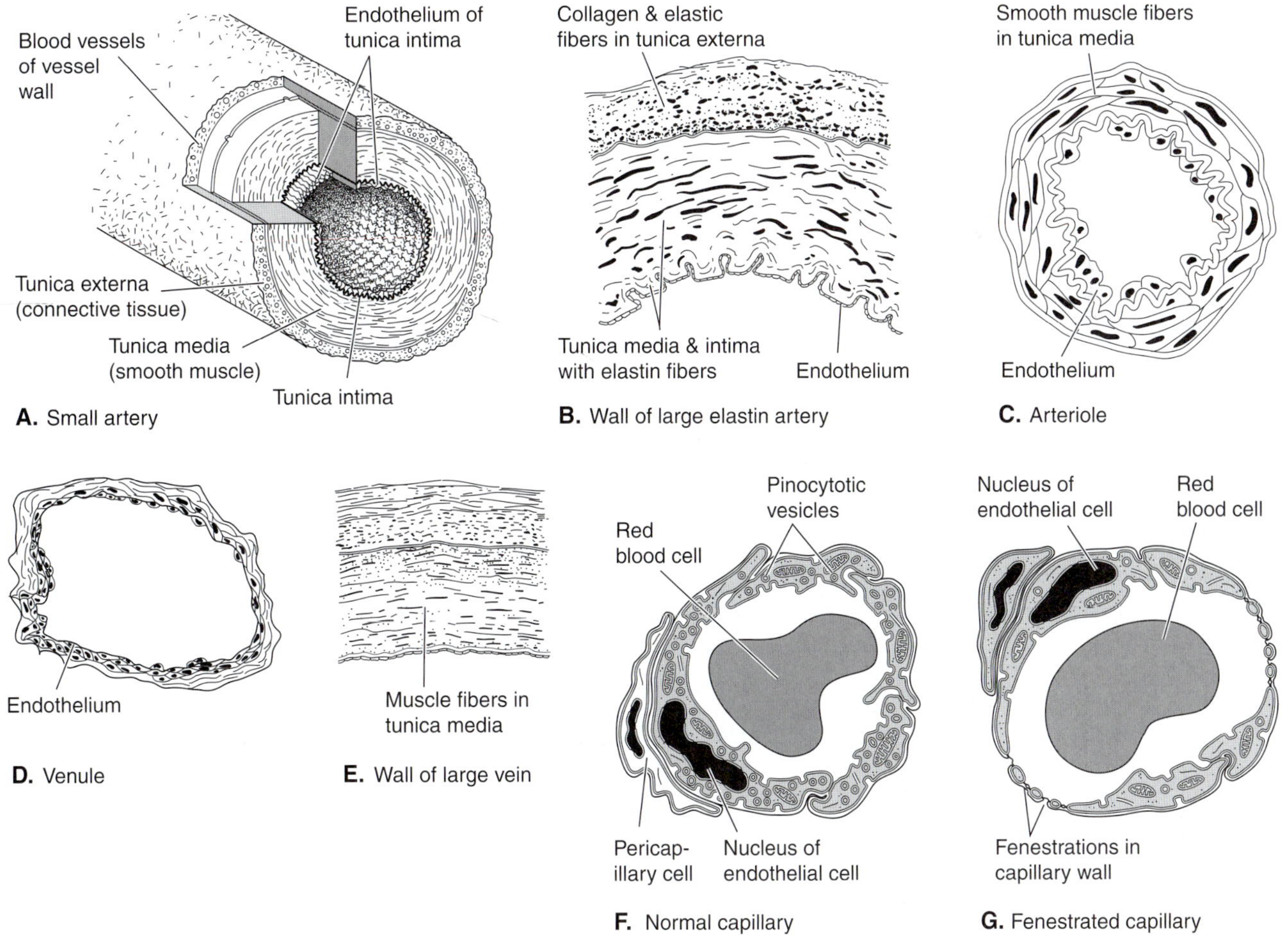

FIGURE 19-3
A–G, The structure of representative blood vessels. The structure of the capillaries is based on electron micrographs, and capillaries are drawn to a much larger scale than are the other vessels. Capillaries have a diameter not much larger than that of the erythrocytes within them. *(Modified from Williams et al.)*

preceding the capillaries have a relatively large amount of smooth muscle. Their degree of contraction regulates peripheral resistance and the amount of blood flowing through particular capillary beds.

The walls of **capillaries** consist only of endothelium. Here, exchanges occur between the blood and the interstitial fluid bathing the cells. Diffusion is the primary factor in the exchanges, but **pinocytotic vesicles** are known to form on one surface of the endothelial cells, move across the cells, and discharge on the other side. In some organs, such as many endocrine glands and the kidneys of amniotes, the capillaries are **fenestrated:** small slits are present in the endothelial cells, so blood and interstitial fluid are separated only by the delicate basal lamina of the endothelial cells.

The **venules** that receive blood from the capillaries, and the larger veins to which they lead, have much thinner walls than those of arteries of comparable size. Collagen and some smooth muscles are the walls' major components. Veins are also much larger in diameter than are the arteries they accompany. In dogs, they contain nearly 70% of total blood volume. Veins are, therefore, an important reservoir of blood. A slight contraction of the veins and a decrease in their diameter can add greatly to the volume of blood flowing through the heart at any given time.

As arteries branch and lead to capillary beds in the tissues, the diameter of the vessels decreases. Because the number of vessels has increased, however, their combined cross-sectional area and circumference increase (Table 19-1). By the time the capillaries reach the intestine, for example, their combined cross-sectional area is about 770 times the cross-sectional area of the aorta that supplies them, and their combined circumference is nearly 1 million times the circumference of the aorta. As capillaries lead to veins and the veins coalesce into fewer and larger vessels, total cross-sectional area and circumference decrease. These changes have important functional consequences. A huge surface area (the circumference) is available in the capillary beds where exchanges occur. Also, because the total blood flow through the system at any one time is a constant, the velocity of flow changes greatly. It decreases as blood flows to the capillaries, and the total cross-sectional area increases. More time is available for the diffusion of materials between the blood and the interstitial fluid. Flow rate increases as the blood collects in the veins and total cross-sectional area decreases. An analogy is the decreasing velocity of a stream as it widens into a pond and the increasing velocity as the pond flows into a narrowing outlet. Blood pressure, on the other hand, continues to decrease throughout the system because it is reduced continuously by friction. It decreases particularly rapidly in capillary beds because of the very small diameter of the capillaries and their combined large surface area. It is very low in veins. Veins usually contain valves that prevent a backflow of blood. Such valves allow the surrounding skeletal muscles to squeeze the blood in the direction of the heart. Valves in veins assist the unidirectional flow of blood in both aquatic and terrestrial vertebrates. In Chapter 6, we saw how the regulation of blood capillary volume and flow plays an important role in the control of not only body temperature but also blood pressure.

The heart and all the vessels so far described constitute the **cardiovascular system.** Gnathostomes also have a **lymphatic system** of vessels. Blind lymphatic capillaries in most of the body tissues lead to progressively larger lymphatic vessels, which eventually discharge into the veins, usually near the heart, where venous pressure is very low. The lymphatic system is a supplementary drainage system that is needed in vertebrates with relatively high blood pressures to return excess liquid and plasma proteins, some of which escape from the blood. The return of the plasma proteins is essential to maintain the balance between the blood's osmotic and hydrostatic pressures that is needed for water exchange.

TABLE 19-1 Blood Vessels in the Mesenteric Vascular Bed of an Adult Dog[a]

	Diameter (mm)	Number of Vessels	Total Cross-sectional Area (mm^2)	Circumference of Individual Vessels (mm)	Total Circumference (mm)
Aorta	10.0	1	78.5	31.4	31.4
Capillaries	0.008	1 billion	60,319.0	0.025	30,159,289.0
Vena cava	12.5	1	122.7	39.3	39.3

[a]Dimensions and number of vessels from A. C. Burton.

The control and structure of the blood, the heart, and all the vessels enhance the flow of blood and maintain a relatively constant composition of the interstitial fluid, despite changes in the level of metabolism of particular tissues or changes in the external environment of animals.

Embryonic Development of the Blood Vessels and Heart

Blood vessels develop within the mesenchyme of the embryo. Groups of mesenchymal cells first aggregate as small clusters called **blood islands** (Fig. 19-4*A*). Within the islands, the mesenchyme differentiates into blood cells; while on the periphery of the islands, it forms the vessel walls. The islands gradually coalesce to form small vessels. The first vessels develop in the splanchnic layer of the lateral plate mesoderm adjacent to the yolk-laden archenteron, or yolk sac. Some of these vessels unite to form a pair of **vitelline,** or **subintestinal, veins** that extend forward beneath the developing pharynx. The vitelline veins unite and expand beneath the pharynx to form a tubular heart (Figs. 19-4*B* and 19-5*A*). A **ventral aorta** extends forward from the heart and gives rise to a series of paired **aortic arches** that extend dorsally through the branchiomeres to the **dorsal aortae.** Six pairs of aortic arches eventually differentiate from anterior to posterior adjacent to the first six visceral arches, but the most anterior ones are lost or highly modified before the more posterior ones form. The dorsal aortae are paired above the pharynx but unite more posteriorly to form a single vessel that continues caudally, lying ventral to the developing vertebral column. The first branches to develop from the dorsal aorta are **vitelline arteries,** which transport blood to the intestine or yolk sac. A visceral vitelline circulation, whereby blood from the yolk sac runs into the heart, then to the first aortic arches, and returns to the yolk sac, is now established, enabling the embryo to utilize energy reserves stored in the yolk (Fig. 19-4*B*).

As the liver develops from a ventral outgrowth of the archenteron just behind the heart, it expands around

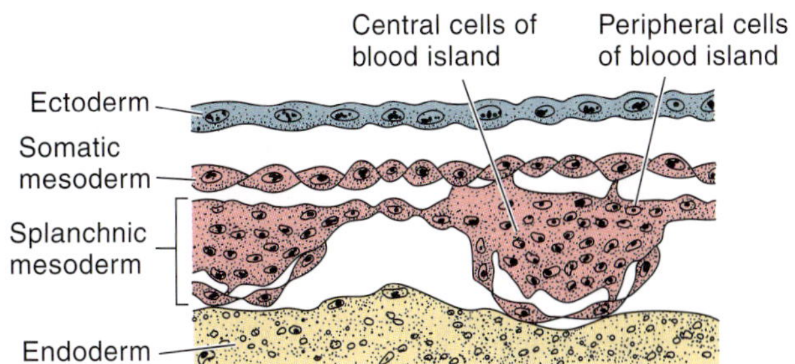

A. Early stage in formation of blood vessels and cells

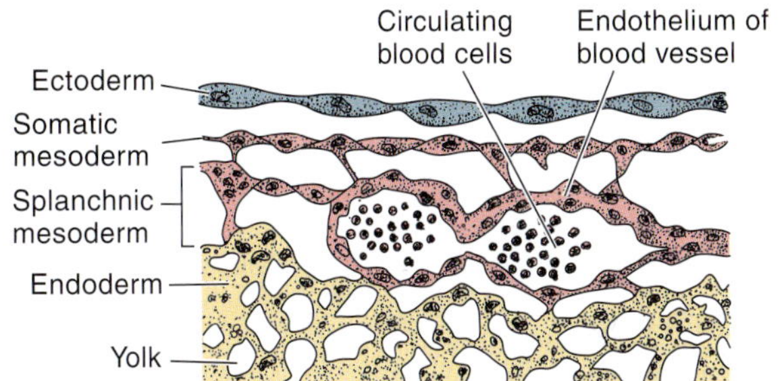

B. Later stage in formation of blood vessels and cells

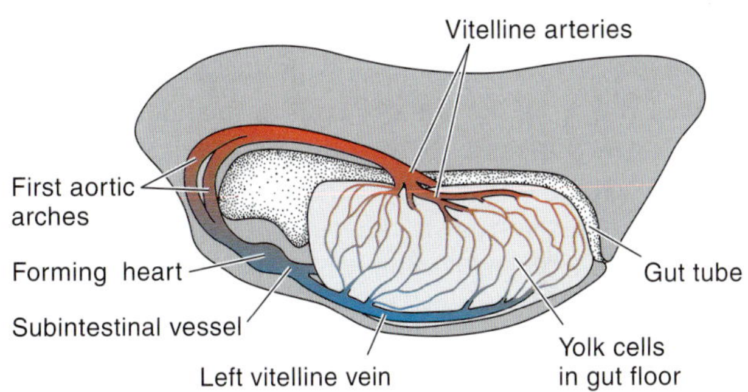

C. Early circulation of an amphibian embryo

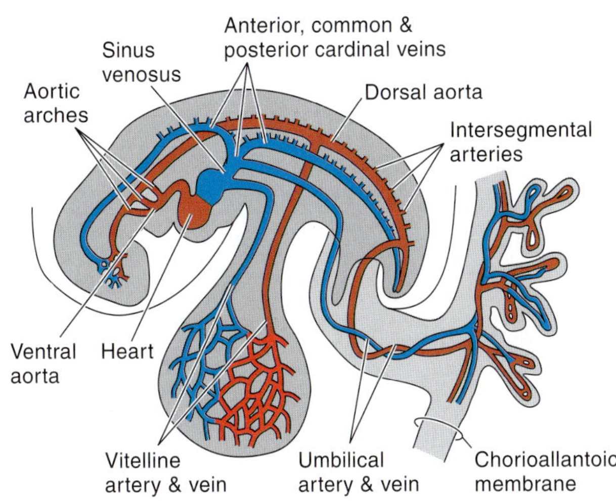

D. Early circulation of a human embryo

FIGURE 19-4
The development of the circulatory system. *A* and *B,* Two stages in the formation of blood cells and blood vessels from blood islands in the splanchnic mesoderm. *C,* A lateral view of the vitelline circulation of an early amphibian embryo. *D,* A lateral view of the major blood vessels in a human embryo at 26 days' gestation. *(A, After Corliss; B, after Romer and Parsons; C, after Moore.)*

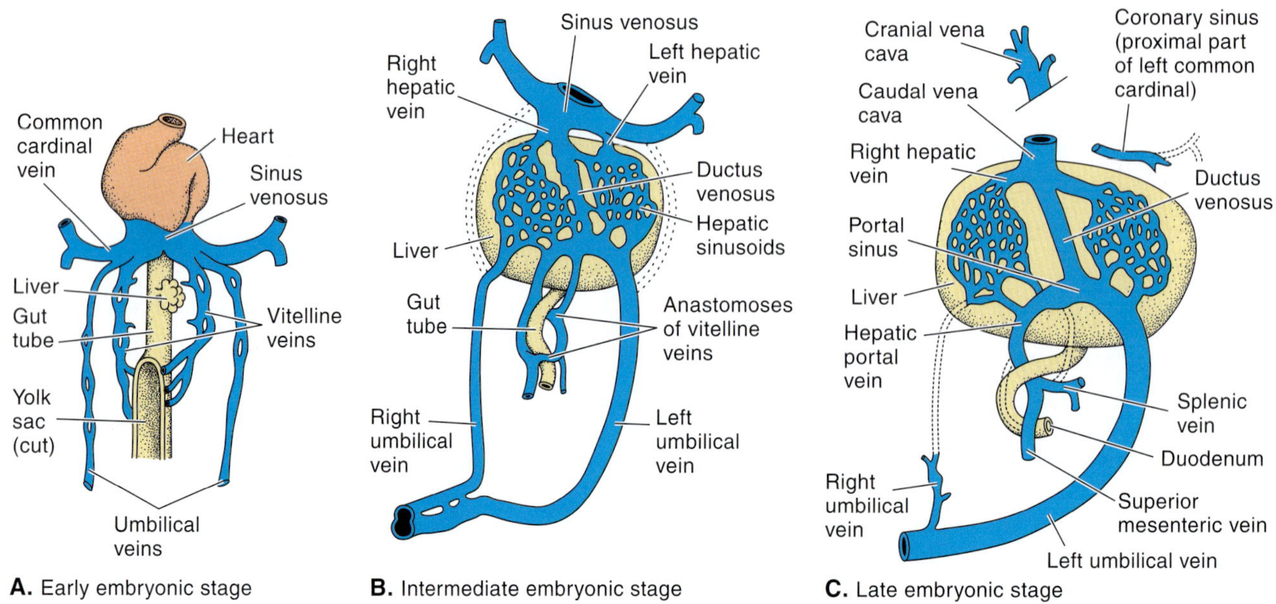

A. Early embryonic stage **B.** Intermediate embryonic stage **C.** Late embryonic stage

FIGURE 19-5
A–C, Ventral views of three stages in the development of the hepatic portal system and related veins in a mammalian embryo. *(After Corliss.)*

the vitelline veins, which break up within the liver into capillary-sized **hepatic sinusoids** (Fig. 19-5*A* and *B*). The sinusoids drain into a pair of **hepatic veins,** which differentiate from the anterior ends of the vitelline veins and lead to the heart. Caudal to the liver, the vitelline veins anastomose and parts atrophy as they form the **hepatic portal system** leading to the hepatic sinusoids (Fig. 19-5*C*). Portal systems are defined as groups of veins that drain one set of organs and lead to capillaries or sinusoids in another organ. The hepatic portal system brings blood from the intestinal region into close contact with the hepatic cells of the liver (Chapter 17).

As these changes are occurring, a somatic circulation begins to develop. Paired arteries leave the aorta to supply the body wall, appendages, kidneys, and developing gonads (Fig. 19-4*C*). A pair of **anterior cardinal veins** returns blood from the head, and a pair of **posterior cardinal veins** returns blood from most of the posterior parts of the body. Anterior and posterior cardinals on each side unite to form a **common cardinal,** which extends ventrally to the heart, passing through the dorsal part of the transverse septum that it helped form (Chapter 14). The appendages are drained primitively by a pair of **lateral abdominal veins,** which enter the common cardinals (see Fig. 19-7*A*).

The pattern of vessels now established resembles the pattern in adult fishes and amphibians. In the embryos of amniotes, some of these primitive channels are lost and replaced by others. Many of the differences are correlated with differences in the sites of nutrient, gas, and excretory exchanges. In the embryos of eutherian mammals, a pair of **allantoic** or **umbilical arteries** develops in the wall of the allantois and lead from the dorsal aorta to the placenta. A pair of **allantoic** or **umbilical veins** returns blood to the embryo (Fig. 19-4*D*). The umbilical veins develop from the primitive lateral abdominal veins and at first enter the common cardinals, but they soon establish a new passage through the liver, the **ductus venosus,** which leads into the right hepatic vein (Fig. 19-5*B*).

The Circulatory System of Early Fishes

The structure of the heart and circulatory patterns changed considerably in the course of vertebrate evolution. Sites of gas, ion, and temperature exchange between the body and the external environment evolved as vertebrates became active, endothermic animals. A particularly important variable is the site for gas exchange, so the evolution of the circulatory system is closely coupled to the evolution of the respiratory system (e.g., gills and lungs; see Chapter 18).

The Heart

In primitive fishes without lungs (e.g., elasmobranchs, Fig. 19-6) the heart and the pericardial cavity in which it lies are located beneath the posterior end of the

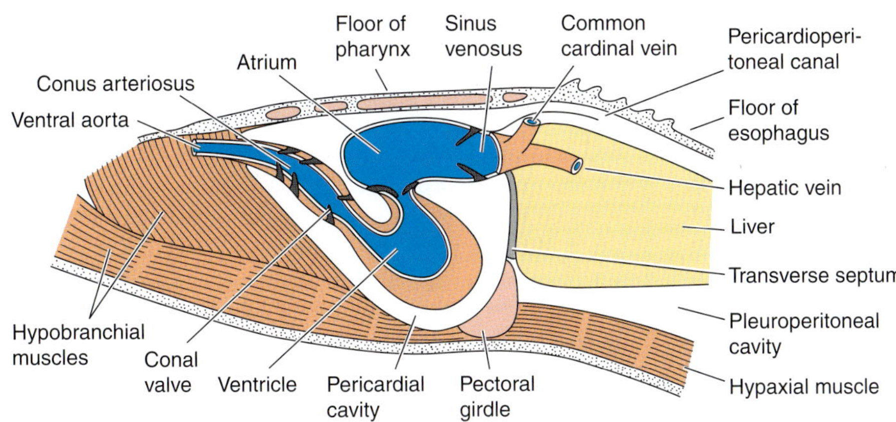

FIGURE 19-6
A sagittal section through the heart of a dogfish and the structures that surround it in the pharynx floor.

pharynx floor. This position is close to the gills through which the heart must drive the blood before it is distributed to the body. Because the tubular embryonic heart grows in length more than does the pericardial cavity, the adult heart has folded upon itself and forms an S-shaped loop that lies in the vertical plane (Fig. 19-6). The fish heart is not divided into left and right sides, as is a mammalian heart, and it receives and pumps only blood low in oxygen content. In most fishes, the heart has differentiated into four chambers, lying in linear sequence. Valves between the chambers prevent backflow of blood. The most caudal and dorsal chamber is a thin-walled **sinus venosus,** which continuously receives blood very low in pressure from the common cardinal and hepatic veins. The sinus leads to the somewhat thicker-walled **atrium,** a large chamber with valves at each end that accumulates blood. When it contracts, blood is pumped into the thicker-walled **ventricle** with sufficient force to slightly stretch the ventricular muscles. This process is advantageous because stretching the muscle maximizes the force with which it can contract. Contraction of the ventricle increases blood pressure sufficiently to drive the blood into the last heart chamber, the **conus arteriosus,** which is a new feature of gnathostomes, then through the capillary-like vascular beds in the gills, and on to the body. The conus arteriosus, which also has thick, muscular walls and several tiers of valves, acts as a buffer. It absorbs the abrupt increase in pressure during ventricular systole (Gr., *systole* = drawing together) and maintains pressure during ventricular diastole (Gr., *diastole* = drawing apart). It tends to even out and prolong blood flow to the gills. Teleost fishes have lost the muscular conus arteriosus. Instead, they have developed an elastic chamber, the **bulbus arteriosus,** that lies in the pericardial cavity. The elastic properties of the bulbus arteriosus reduce the pulsations generated by the ventricle. Even so, blood flow through the gills is not perfectly even, but heartbeat and respiratory movements are so synchronized that the maximum flow of blood through the gills coincides with the maximum flow of water across them.

The pressure of blood in the ventral aorta of a dogfish ranges from 28 (diastole) to 38 (systole) mm Hg (Satchell, 1999). Pressure drops considerably as blood passes through the gills to a range of 24 to 32 mm Hg in the dorsal aorta. This is adequate to drive the blood through the tissues because a fish is supported by water and blood pressure need not also overcome high gravitational forces, as in terrestrial vertebrates. Additional pressure decreases occur as blood passes through tissue capillaries and the renal and hepatic portal systems that supply and drain, respectively, the kidney and liver. Pressures in the large venous sinuses preceding the heart are very low, ranging from slightly less than ambient pressure to slightly more (-0.4 to $+0.1$ mm Hg).

Stiff tissues (Fig. 19-6) surround the pericardial cavity in which the heart lies. A large basibranchial cartilage lies dorsal to the pericardial cavity, the pectoral girdle and liver lie posterior to it, and the rest of the cavity is surrounded by the thick hypobranchial musculature. Because stiff walls surround the pericardial cavity, it does not collapse when the ventricle contracts, the volume within the pericardial cavity increases slightly, and pressure within the cavity decreases. Concurrent contraction of the respiratory hypobranchial muscles, many of which extend between the visceral arches and the wall of the pericardial cavity, helps reduce pressure more. Reduced pressure in the pericardial cavity allows the thin-walled sinus venosus to expand and suck blood from the large venous sinuses. The heart acts first as a suction and then as a force pump. To work effectively, the system requires that secreted serous liquid not be permitted to accumulate in the pericardial cavity. Elasmobranchs and many primitive bony fishes have a **pericardioperitoneal canal** that passes through the transverse septum and continuously drains the pericardial cavity.

The Arterial System

The six embryonic aortic arches become interrupted in fishes by the capillary-like networks in the gills. The ventral part of the first aortic arch is lost in all living vertebrates, but the ventral parts of the other five form the **afferent branchial arteries,** which lead into each of the gills (Fig. 19-7). The dorsal parts of most of the arches form **collector loops** around the internal gill slits. Utilization of the entire inner surface of the gill pouches is made possible by establishing complete loops around each gill pouch. The anterior and posterior limbs of the loop are called, respectively, the **pretrematic** and **posttrematic arteries** (Fig. 18-3*B*). The collector loops receive oxygenated blood from the gills and lead to **efferent branchial arteries** that continue to the dorsal aorta. The first embryonic gill pouch is reduced to a spiracle in many fishes or has been lost entirely. When a spiracle is present, it commonly contains a rudimentary, gill-like structure, the **pseudobranch** (Chapter 18). The dorsal part of the first aortic arch contributes to a **spiracular artery,** which carries aer-

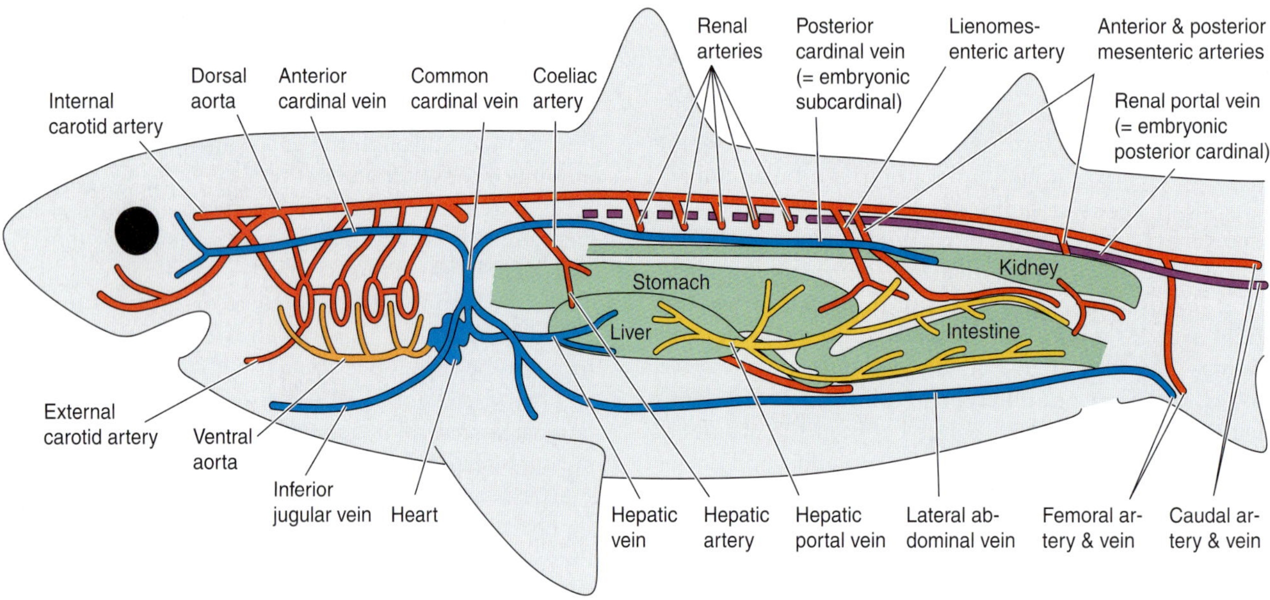

Major arteries and veins of an elasmobranch

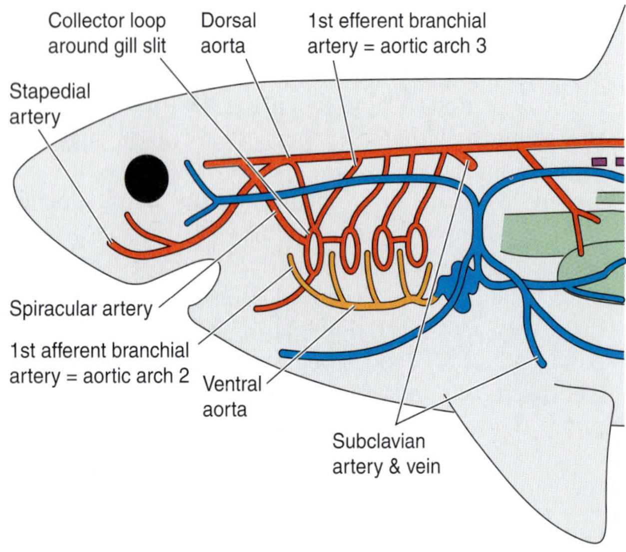

Details of vessels in the head and pharyngeal region

FIGURE 19-7
Lateral view of the major arteries and veins of an elasmobranch. Intersegmental vessels are not shown except for the subclavian and iliac vessels, which supply the paired fins. The dashed blue line above the kidney indicates the portion of the embryonic posterior cardinal that atrophies. This portion is replaced functionally by a subcardinal vein, located ventral to the kidney, which becomes a part of the adult posterior cardinal vein.

ated blood from the first collector loop through the pseudobranch to the internal carotid artery.

The paired **internal carotid arteries,** which supply blood to most of the head, are rostral extensions of the embryonically paired dorsal aortae. One branch of this vessel enters the skull to supply the brain. Another branch, the **stapedial artery** (so called because it runs through the stapes of embryonic and primitive mammals), supplies the eye and most of the outside of the head. The external carotid artery of fishes extends forward from the ventral part of the first collector loop and supplies only tissues near the lower jaw.

The **dorsal aorta** continues caudally beneath the vertebral column to the tail, where it is known as the **caudal artery.** Several median visceral arteries, called **coeliac** and **mesenteric arteries,** extend through the mesenteries to most of the abdominal viscera. The gonads and kidneys are supplied by a series of small, paired **gonadal** and **renal arteries.** Larger paired **intersegmental arteries** extend between the myomeres of the trunk and tail. In gnathostomes two pairs of enlarged intersegmental arteries, the **subclavian** and **iliac arteries,** lead to the **brachial** and **femoral arteries,** which enter the pectoral and pelvic fins.

The Venous System

Blood from the abdominal viscera, apart from the kidneys, drains into tributaries of the **hepatic portal vein.** After flowing through the capillary-like hepatic sinusoids in the liver, this blood is collected by a pair of **hepatic veins,** which pass through the transverse septum to the sinus venosus. All of these vessels develop from the embryonic vitelline veins (Figs. 19-5 and 19-7). Arterial blood reaches the hepatic sinusoids by a **hepatic artery,** a branch of the celiac artery.

The **anterior cardinal veins,** also known as the **lateral head veins,** drain most of the head. After receiving superficial tributaries from the orbit and adjacent areas, and three deeper cerebral veins from within the skull, each anterior cardinal extends caudally to the **common cardinal.** The ventral surface of the head is drained by **inferior jugular veins,** which enter the common cardinal veins independently.

The tail and trunk are drained by the posterior cardinal system. In jawless fishes and in the early embryos of other fishes, the **caudal vein** bifurcates at the posterior end of the trunk into a pair of **posterior cardinal veins,** which extend forward above the kidneys to the common cardinal. During the subsequent development of jawed fishes, new **subcardinal veins** form ventral to the kidneys and tap into the anterior part of the posterior cardinals. The midsections of the posterior cardinals, shown by the dashed line in Figure 19-7, then atrophy. The posterior parts of the early embryonic posterior cardinals have now been converted into **renal portal veins,** for they carry the blood they receive from the tail to capillary beds on the kidney tubules. The kidneys are drained by the subcardinals, which, with the anterior part of the embryonic posterior cardinals, now form the adult posterior cardinals.

The significance of the renal portal system, which is lacking in mammals, is not entirely clear. The kidney tubules of vertebrates are supplied by two capillary beds. A knot of capillaries, known as the **glomerulus,** protrudes into the beginning of each tubule, and an artery from it leads to a second capillary bed lying over the rest of the tubule (Fig. 20-3). The renal arteries lead first to the glomeruli, where a filtrate of the blood passes into the kidney tubules (Chapter 20); in contrast, the renal portal system leads directly to the tubular capillary beds, where materials may be added to or reabsorbed from the tubules. Because the renal arteries are very small in fishes, a renal portal system ensures the passage of a large volume of blood from the tail through the kidneys. Although pressure is often low in the caudal vein, Satchell (1971) showed that the pressure can build up and equal that in the dorsal aorta by the operation of a "caudal pump." The caudal artery and vein are completely encased by the hemal arches of the caudal vertebrae, which protect them from the waves of muscle contraction that sweep down the tail as the fish swims (Fig. 19-8). Muscle contractions do, however, squeeze the tributaries of the caudal vein and drive blood into the caudal vein with considerable force. Valves at the junctions of the tributary veins and caudal vein prevent backflow. Valves, which normally are found only in veins, are present at the origins of the intersegmental arteries to the tail muscles and prevent

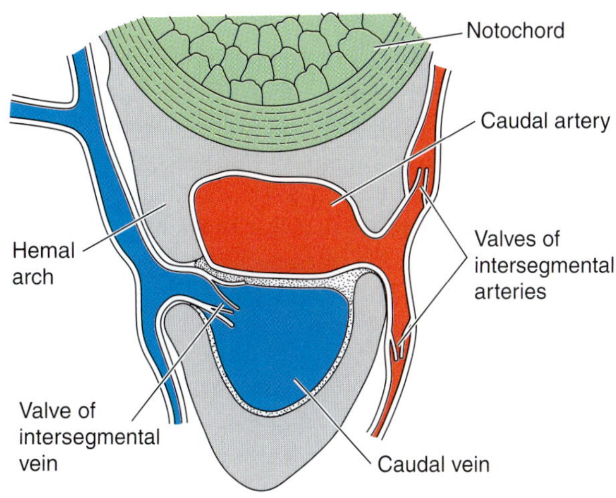

FIGURE 19-8
A transverse section through the hemal arch and enclosed caudal artery and caudal vein of a herring. *(After Satchell.)*

arterial blood from being forced back into the caudal artery. Arterial blood enters the tail muscles during periods of muscle relaxation.

In jawed fishes, the paired fins are drained by veins that enter the lateral abdominal system (Fig. 19-7). The **lateral abdominal veins** are absent from teleosts. The **femoral veins** from the pelvic fins of teleosts enter the renal portal veins, and the **subclavian veins** from the pectoral fins enter the common cardinals.

Evolution of the Heart and Arteries

Few changes of evolutionary significance occur in the branching pattern of the dorsal aorta. All vertebrates have several **median visceral arteries** that supply the digestive tract and its derivatives; paired **lateral visceral arteries** to the gonads and kidneys; and paired **intersegmental arteries** to the trunk, tail, and paired appendages. Terrestrial vertebrates usually have fewer paired vessels than do fishes—often only a single pair of **gonadal arteries** and another pair of **renal arteries.** Longitudinal anastomoses between the peripheral branches of the intersegmental arteries enable one artery to supply a more extensive area than in fishes. This is accompanied by a reduction in the number of points of origin of the intersegmental arteries from the dorsal aorta.

In contrast, major changes occur in the heart and aortic arches during the evolution of the vertebrates. These are correlated with the change from using gills to lungs for primary gas exchange that accompanied the shift from water to land (Focus 19-1).

Lungfishes

Branchial circulation is a very efficient means of gas exchange in water-breathing fishes. The countercurrent flow of water and blood in the gills enables a fish to take up most of the limited amount of oxygen in the water and to unload carbon dioxide, which is very soluble in water. But fishes that live in water subject to a reduced oxygen tension or to periodic drought cannot rely solely on gills. They have evolved accessory respiratory organs, of which lungs are of particular interest because they also were present in the choanate fishes ancestral to terrestrial vertebrates. Unfortunately, the nature of gas exchange in these fishes is unknown, but we infer that it was similar to that of contemporary lungfishes. The tropical habitat of living lungfishes is subject to water stagnation and seasonal drought, not unlike the environment of the choanate fishes ancestral to tetrapods. Lungfishes take up oxygen from the air in the lungs by means of a **pulmonary circulation,** but they have not dispensed with a **branchial circulation.**

Gills continue to play an important role in oxygen uptake and the discharge of carbon dioxide. The gills are fully functional when oxygen tension in the water is greater than that in the blood. But when lungfishes are in oxygen-poor water, blood is diverted to the lungs and away from the gills, from which oxygen could be lost. Experimental and cineradiographic studies have revealed that the lungfish vascular system permits considerable flexibility. By constricting certain segments of the aortic arches, a lungfish can direct the flow of blood to certain areas and restrict or prevent flow to other areas. These arterial segments that can either constrict or dilate function in an analogous way to valves in industrial plumbing. Lungfishes have evolved a heart and a pattern of aortic arches that allows for both branchial and pulmonary circulation and maintains a high degree of separation between oxygen-rich and oxygen-poor bloodstreams (Fig. 19-9). A new vessel, the **pulmonary vein,** has evolved and carries blood from the lung to a partially divided atrium (Fig. 19-9*A*). We will now review two extreme modes of the flexible circulation in the lungfish, *Protopterus.*

During the aquatic breathing mode, oxygen-poor blood from the body and head is collected in the sinus venosus (Fig. 19-9*A*). The deoxygenated blood is pumped sequentially into the atrium, ventricle, conus arteriosus, and the aortic arches. The oxygen-poor blood is directed to gills on the second, fifth, and sixth arches, where the blood is oxygenated. It is then collected by the efferent branchial arteries and carried to the dorsal aorta. This pattern resembles the branchial circulation in water-breathing fishes (Fig. 19-7). In the aquatic breathing mode, a constriction of the pulmonary artery prevents blood already oxygenated in the gill of the sixth arch from entering the lung (Fig. 19-9*A*). With this segment constricted ("valve" closed), oxygenated blood passes through the remaining segment of the sixth efferent branchial artery, called the **ductus arteriosus,** into the dorsal aorta. Oxygen-poor blood is prevented from passing through the third and fourth aortic arches, which have no gills, by constrictions ("valves" closed) of their bases. In this way, all oxygen-poor blood in the anterior arches passes through the gills on the second arch for oxygenation (Fig. 19-9*A*).

During the aerial-breathing mode, the pattern changes drastically and foreshadows the circulation in tetrapods. Oxygen-poor blood from the sinus venosus enters the right side of the atrium and is propelled by the single ventricle through the conus arteriosus, which has a **spiral valve** (Fig. 19-9*B*). This valve directs the oxygen-poor blood into the sixth aortic arch. This oxygen-poor blood is shunted to bypass the gills

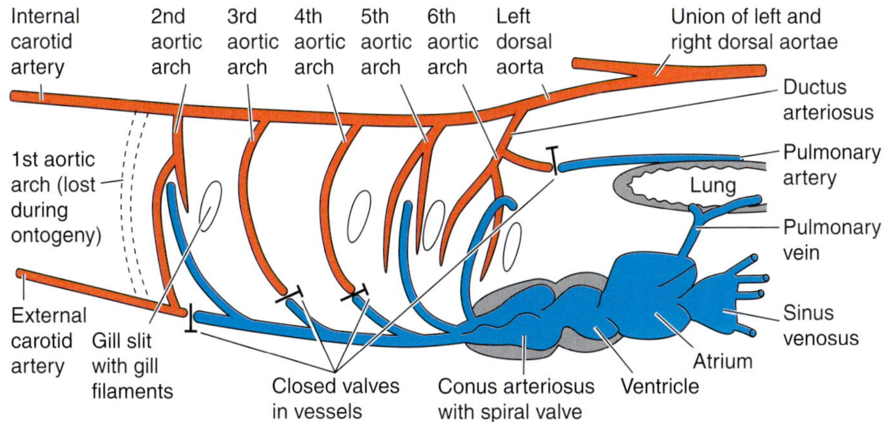

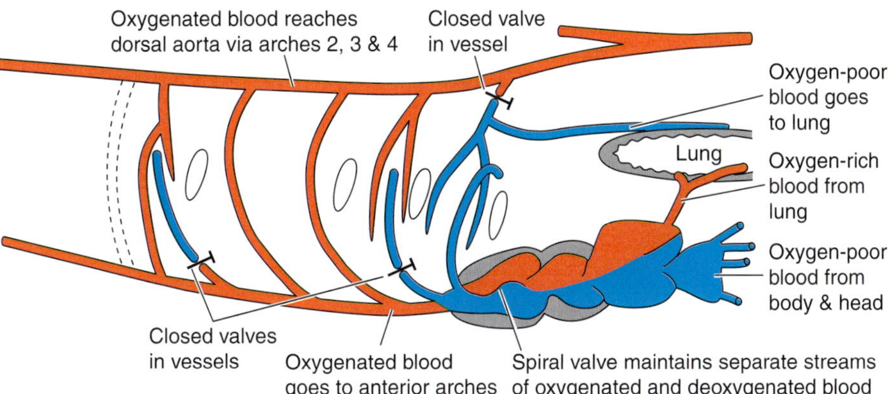

FIGURE 19-9
Lateral views of the aortic arches of the African lungfish, *Protopterus*. The attached heart is diagrammed in a ventral view, with its chambers shown in linear sequence. *A,* The pattern of circulation when the fish relies on gill irrigation in highly oxygenated water. *B,* The pattern when the fish uses its lungs exclusively for air breathing. Derivatives of the embryonic aortic arches are indicated by Arabic numerals.

on the sixth arch. It enters the lung by the pulmonary artery because the ductus arteriosus is constricted ("valve" closed) (Fig. 19-9*B*). Oxygenated blood from the lung is returned to the left compartment of the atrium by the pulmonary vein. Radiographic studies have shown that the oxygen-rich and oxygen-poor bloodstreams remain separated to a high degree as they pass through the partly divided atrium, ventricle, and conus arteriosus. Oxygen-rich blood is deflected to the anterior arches 2 to 4. (The first embryonic arch has been lost.) Arches 3 and 4 do not bear gills, so they form direct passages to the dorsal aorta. Oxygen-rich blood is distributed to the head and body by tributaries of the dorsal and ventral aorta. The pattern of circulation of lungfishes during aerial respiration foreshadows tetrapod circulation. For example, an open and subsequently closed ductus arteriosus forms the basis of fetal and adult circulation in placental mammals (p. 622).

The lungfish system is of interest for two reasons. It probably is close to the ancestral condition from which tetrapod patterns evolved. It also illustrates the high degree of functional flexibility needed by vertebrates living in an environment where the availability of oxygen changes. The two bloodstreams are separated to a large extent, but different mixings and volumes of blood may be sent to different parts of the system according to environmental conditions and the extent to which gills or lungs are being used.

Lungs are accessory respiratory organs in the Australian lungfish, *Neoceratodus,* which retains gills on all the arches, but they are essential in *Protopterus* and *Lepidosiren,* the gills of which are reduced.

Amphibians

The hearts of amphibians and most nonavian sauropsids remain incompletely divided, and central shunts in or next to the heart permit some mixing or directed flows of oxygen-rich and oxygen-poor blood. This appeared to be a functional inefficiency to many earlier biologists, who believed that the incomplete divisions of the heart were necessary transitional stages in the evolution of the double circulation of birds and mammals, with a completely divided heart and separation of the two bloodstreams. Recently, it was experimentally demonstrated that the central shunts are very effective adaptations to the various modes of life of amphibians and reptiles.

Amphibians and reptiles resemble lungfishes in having long periods of apnea interspersed with brief periods of lung ventilation. They do not ventilate their lungs continuously as do birds and mammals. As oxygen in the lungs is consumed, it would be energetically inefficient to continue to pump a large volume of blood through them. The shunts between the left and the right sides of the heart permit the volumes of blood flowing through the lungs and body to be unequal and to be altered as circumstances and the demand for oxygen change. In birds and mammals, in contrast, the volume of blood passing through the two sides of the completely divided heart is equal at all times.

Larval amphibians retain gills and acquire a pulmonary circulation (Fig. 19-10*A*). At metamorphosis gills are lost, the heart migrates caudally closer to the lungs, and the animals rely on their lungs and pulmonary circulation for most of their oxygen uptake. Amphibians have a relatively low level of metabolism, so their oxygen needs are not great. Some oxygen also is taken in through the skin, especially when the animals are under water.

The aortic arches of adult amphibians are reduced to a greater extent than in lungfish (Fig. 19-10*B* and *C*). The second embryonic arch is lost as well as the first, so the external carotid artery originates from the base of the third arch. The third arch itself is now a part of the internal carotid artery. The ventral aorta between the third and the fourth arches becomes the **common carotid artery.** The small enlargement at the branching of the external and internal carotid arteries is called the **carotid body.** Its function in amphibians is unclear. Some evidence suggests that it is a receptor that monitors the oxygen content of the blood; other evidence, that it detects pressure changes; but other evidence, that it secretes epinephrine. The segment of the dorsal aorta between the third and the fourth arches, which is called the **carotid duct,** has been lost in most adult amphibians, so the common carotid artery supplies both the external and the internal carotids. Adult salamanders usually retain both the

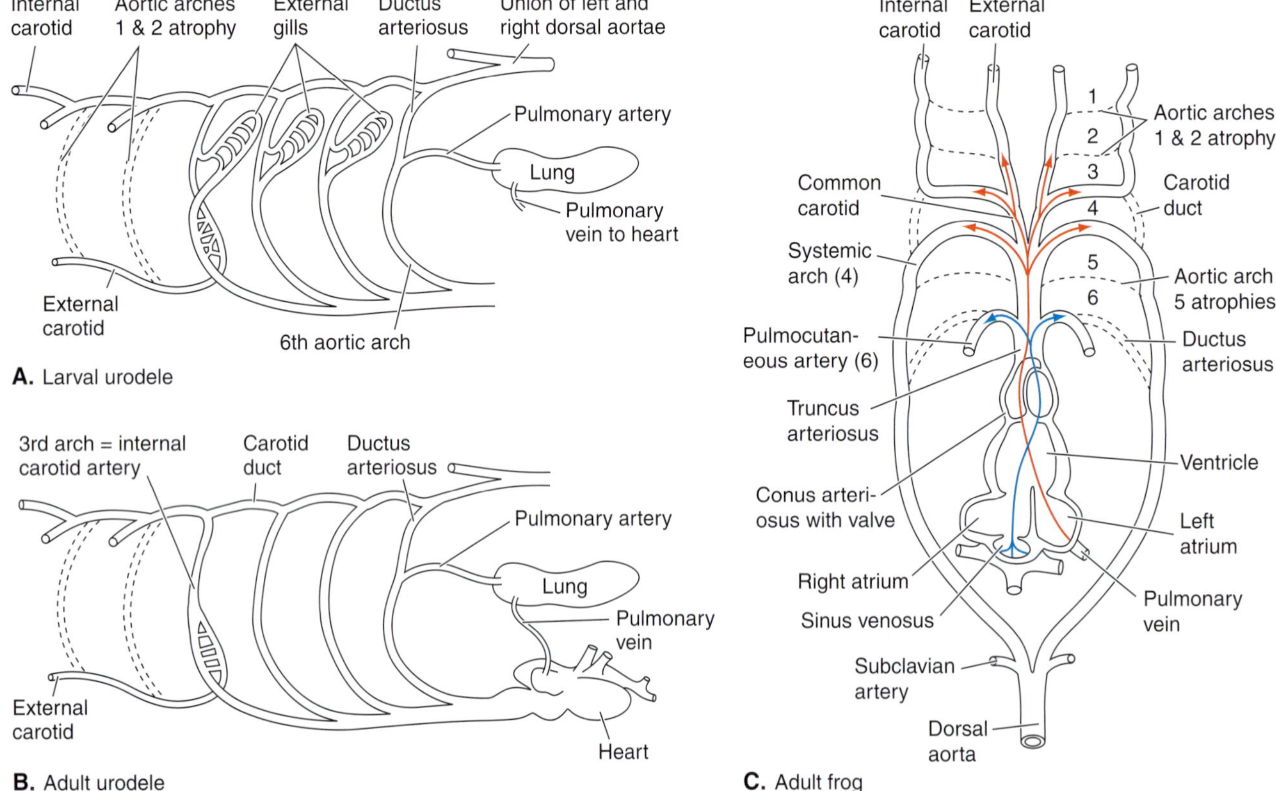

FIGURE 19-10
The aortic arches of amphibians. *A* and *B,* Lateral views of the aortic arches in a larval and an adult urodele, respectively. *C,* Ventral view of the aortic arches and heart of an adult frog. The heart chambers are diagrammed in linear sequence. *(Modified from Goodrich.)*

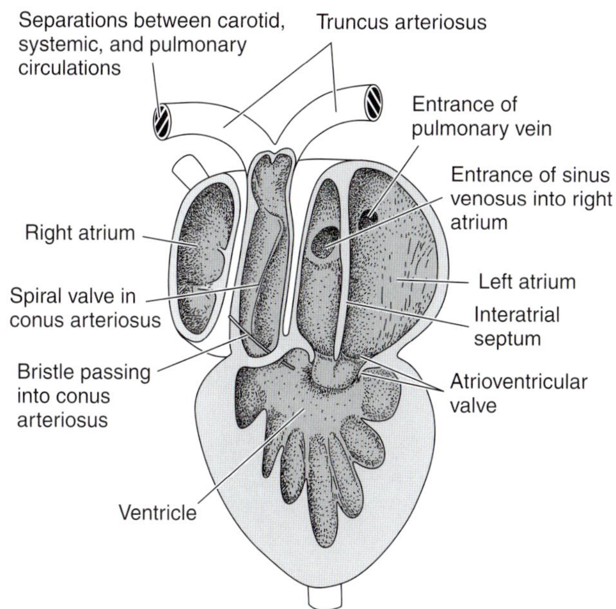

FIGURE 19-11
A ventral dissection of a frog heart, *Rana*. *(After Kerr.)*

fourth and the fifth arches as **systemic arches** leading to the dorsal aorta and trunk. The fifth arch commonly is smaller, and it has been lost in frogs. The sixth arch continues into the **pulmocutaneous artery** to the lungs and skin. Its embryonic connection to the dorsal aorta, known as the **ductus arteriosus,** usually disappears in adults.

The frog heart has been studied extensively and is representative for most amphibians (Fig. 19-11). Radiographic studies and analysis of the amount of oxygen and blood pressures in various vessels indicate that the two bloodstreams entering the heart from the body and lungs remain separated within the heart to a considerable degree when the animals are on land and the lungs are functioning. The atrium is completely divided by an **interatrial septum.** Although the ventricle is undivided, the oxygen-rich and oxygen-poor bloodstreams appear to be separated to a large extent by slight differences in the time they enter and leave and by the spongy wall of the ventricle. The streams finally are sorted out by a complex spiral valve in the conus arteriosus and by divisions within the bifurcated ventral aorta, now called the **truncus arteriosus.** Oxygen-rich blood returns from the lungs to the left atrium, and most of it is directed to the carotid and systemic arches. Blood entering the sinus venosus and right atrium has come from both the body and the skin, where carbon dioxide is removed and some oxygen is taken up. Although this blood does not contain as much oxygen as does blood from the lungs, it contains more than oxygen-poor blood that enters the sinus venosus of a fish. Most of this blood is directed to the pulmocutaneous artery. This pattern of circulation through the heart sends blood richest in oxygen to the head, a somewhat mixed blood to the rest of the body, and blood lowest in oxygen to the lungs and skin.

The volume of blood flowing to the body, skin, and lungs probably can be altered to meet different circumstances. Evidence shows that pulmonary resistance increases as oxygen in the lungs is consumed, especially when the animal is under water, and blood is diverted from the lungs to the skin and body. A muscular sphincter in the pulmonary artery just distal to the origin of the cutaneous artery regulates the amount of flow to the skin and lungs.

Chelonia and Lepidosauria

Embryonic chelonians and lepidosaurians never have a branchial circulation; gas exchange is through the chorioallantoic membrane. Aortic arches appear during development and are transformed into a pattern similar to that in adult frogs (Fig. 19-12). The third arches contribute to the internal carotid arteries, only the fourth arches form the systemic or aortic arches to the body because the fifth arches are lost, and the sixth arches are a part of the pulmonary arteries to the lungs. No branches exist from the pulmonary arteries to the skin because reptiles have a method of ventilating the

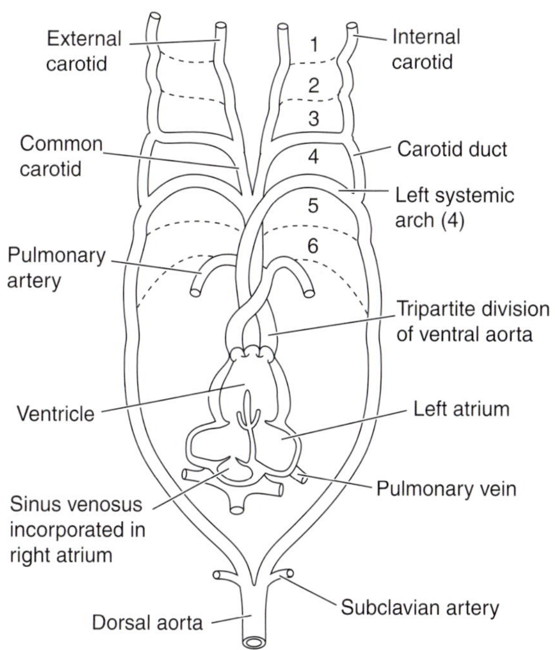

FIGURE 19-12
A ventral view of the heart and aortic arches of a reptile. The heart structure applies to most reptiles but not to crocodiles (see Fig. 19-13*B*). *(After Goodrich.)*

lungs that is adequate for both oxygen uptake and carbon dioxide removal. The ductus arteriosus, present in embryos, disappears in adults, but the carotid duct often persists. The various features of contemporary chelonians and lepidosaurians are the incorporation during embryonic development of the conus arteriosus into the ventricle, and an unusual tripartite division of the ventral aorta that carries the origins of the pulmonary artery, left systemic arch, and right systemic arch from the ventricle. The right systemic arch gives rise to the entire carotid circulation to the head and then curves caudally to join the left systemic arch to form the dorsal aorta.

The atria are completely separated, as in most amphibians. Pulmonary veins return to the left atrium, and systemic veins, to the right one. The sinus venosus is much reduced in most reptiles. The ventricle is completely divided in crocodiles and partly divided in other reptiles. The condition in lizards is representative for most nonavian sauropsids. The interventricular septum is complete caudally near the apex of the ventricle (Fig. 19-13*A*). More anteriorly, slightly caudal to the entrances from the atria, the interventricular septum is represented by a muscular ridge that extends into the ventricle from its ventral surface, curves to the right, and partly divides the ventricle into three functional compartments (Fig. 19-13*A*). Dorsally, a **cavum arteriosum** receives oxygen-rich blood from the left atrium, and a **cavum venosum** receives oxygen-poor blood from the right atrium. A wide **interventricular canal** connects these two cava, but the opening of the atrioventricular valves as the ventricle fills (Fig.

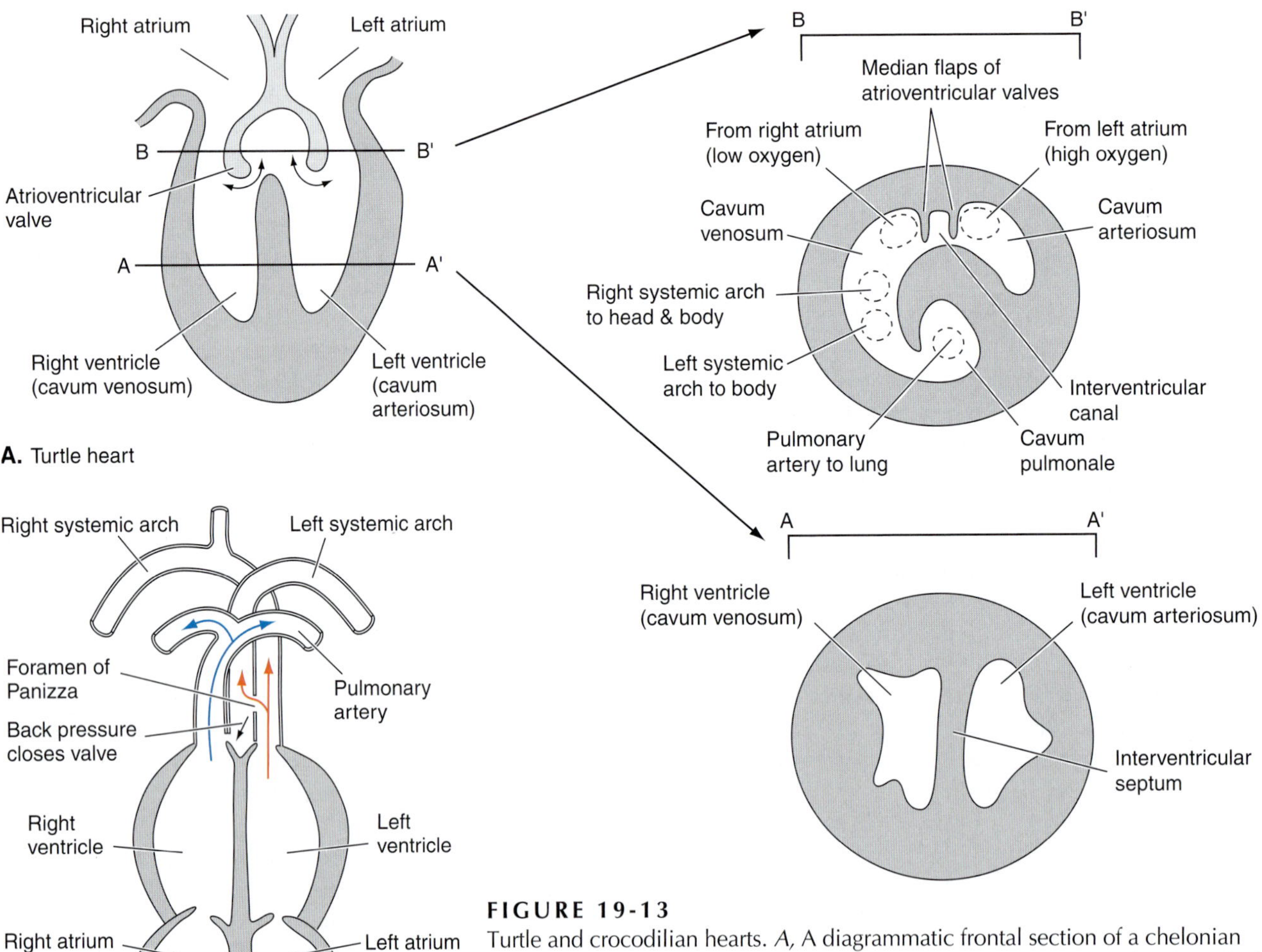

FIGURE 19-13
Turtle and crocodilian hearts. *A*, A diagrammatic frontal section of a chelonian (turtle) heart (lizard hearts are similar). Transverse lines indicate the level at which the transverse sections shown to the right were cut. On the top right, a transverse section through the ventricle just caudal to the atria and the point from which the arteries leave the ventricle is shown. The location of the vessels that enter the atria and exit the ventricle is indicated by the dotted circles. On the bottom right, a transverse section through the ventricle is shown. *B*, The crocodile heart. The chambers have been diagrammed in linear sequence. *(A, After White.)*

19-13*B*) closes it off. A third compartment, the **cavum pulmonale,** lies ventral to the curvature of the muscular ridge and also receives oxygen-poor blood from its broad connection with the cavum venosum. The pulmonary artery to the lungs leaves from the cavum pulmonale, and both systemic arches leave from the cavum venosum. When the ventricle begins to contract, most of the oxygen-poor blood in the cavum venosum moves into the cavum pulmonale. This blood is sent out the pulmonary artery rather than the systemic arches because the lungs offer less peripheral resistance than the body. Concurrently, the atrioventricular valves close. This action opens the interventricular canal, and oxygen-rich blood begins to move from the cavum arteriosum, where it has been sequestered, into the cavum venosum. Continued contraction of the ventricle brings its ventral wall up against the crest of the muscular ridge and prevents most of the oxygen-rich blood from entering the cavum pulmonale. Oxygen-rich blood, now in the cavum venosum, leaves through the systemic arches. In many cases, both systemic arches receive oxygen-rich blood, but their exits from the cavum venosum are positioned in such a way that when some mixing of oxygen-rich and oxygen-poor blood occurs, the distribution of the blood to the two systemic arches is not the same. The right systemic arch, which leaves nearer the cavum arteriosum, receives more oxygen-rich blood, whereas the left systemic arch, the origin of which is closer to the cavum pulmonale, receives more oxygen-poor blood.

The system is very labile in lepidosaurs and turtles because the distribution of the two types of blood depends so much on the relative peripheral resistance in the lungs and body. When the lungs are being ventilated and pulmonary resistance is low, as much as 60% of cardiac output goes to the lungs. Even some of the blood that has been through the lungs is recycled to the lungs. A **left-to-right shunt** occurs, and the blood becomes thoroughly saturated with oxygen. When the lungs are not being ventilated, oxygen in the lungs is consumed, pulmonary arterioles constrict, and pulmonary resistance increases. The same thing occurs when the animals dive. Under these circumstances, less blood is sent to the lungs, and some oxygen-poor blood is recycled to the body; a **right-to-left shunt** occurs. The admixture of some oxygen-poor blood with oxygen-rich blood going to the body does not necessarily mean that less oxygen is delivered to the tissues. The increased carbon dioxide content of this mixed blood causes hemoglobin to unload a greater amount of its bound oxygen than it normally would (the Bohr effect). The hypothesis is that shunts optimize the delivery of oxygen to the tissue under different circumstances and optimize the removal of carbon dioxide from them. To gain a better understanding of this effect, the volume of blood passing through the various vessels, the degree to which the lungs are ventilated, and the oxygen content of the blood must be measured simultaneously and various valves visualized. Studies of this sort are difficult but have been conducted (Hicks and Wang, 1996; Wang and Hicks, 1996; Axelsson et al., 1996; Franklin and Axelsson, 2000).

Crocodilians

The crocodilian heart differs from that of other sauropsids in having a complete interventricular septum (Fig. 19-13*B*). The right systemic arch arises from the *left* ventricle, and the left arch arises from the *right* ventricle together with the pulmonary artery. The **foramen of Panizza,** which is a narrow channel, connects the two systemic arches as they depart from the ventricle (Fig. 19-13*B*). Physiological studies of the crocodilian hearts have shown that when the ventricles contract, blood flows to the vessels of least resistance. In a resting crocodilian that is breathing air, oxygen-rich blood is propelled at a high pressure from the *left* ventricle to the *right* systemic arch, but because its pressure is higher than that of the left systemic arch, the oxygen-rich blood flows through the foramen of Panizza into the left systemic arch. High pressure in the left systemic arch keeps the valves at its base closed, leaving the pulmonary trunk as the only exit of the right ventricle. As a result, both the left and right systemic arches carry oxygen-rich blood to the systemic tissues, whereas the pulmonary trunk carries oxygen-poor blood to the lungs.

However, when the crocodile dives, the pressure in the right ventricle rises substantially because of the high resistance of the pulmonary trunk, which undergoes vasoconstriction together with the lung vessels. Consequently, the pressure in the *left* systemic arch, which carries oxygen-poor blood from the *right* ventricle, exceeds that of the *right* systemic arch. As a result, oxygen-poor blood flows from the left systemic to the right systemic arch, which has oxygen-rich blood. This is called the *right-to-left shunt* through the foramen of Panizza. (Note that the terms "right" and "left" refer to the origin of the blood from the right and left ventricles, respectively, and not to the systemic arches that have crossed in crocodilians.) The right-to-left shunt increases the efficiency of the delivery of blood to the body and provides a lung bypass in the absence of air breathing. The shunt across the foramen of Panizza allows shunting not only from right to left but also from left to right, depending on pressure differences in the left and right systemic arches, the pulmonary artery, and positions of the associated valves. It is a versatile device that can adjust quickly to oxygen requirements necessitated by various environmental conditions.

The vascular arrangements in chelonians, lepidosaurians, and crocodilians optimize the use of energy by matching the degree of blood flow through the lungs with the extent to which the lungs are being used. They permit a greater muscularization of the ventricles and the development of a blood pressure significantly higher than in fishes and amphibians. Increased pressure and an adjustable separation of the bloodstreams enable chelonians, lepidosaurians, and crocodilians to be considerably more active and penetrate a greater range of habitats.

Birds

Birds are closely related to dinosaurs and crocodiles and placed with these vertebrates in the taxon Archosauria (Fig. 3-25), but they have become very active endothermic vertebrates. Possibly, the ancestors of birds had a heart and pattern of aortic arches similar to that of crocodiles. Because birds evolved endothermy and lungs that are continuously ventilated, shunts bypassing the lungs, as found in crocodiles, would have had no adaptive value. Equal volumes of blood are sent to the lungs and body at all times, which appears to have been simply the result of the loss of the left systemic arch (Fig. 19-14*A*). Birds have a single systemic or aortic arch, the right one; as in reptiles, it gives rise to the carotid circulation before turning caudad to the rest of the body. As might be expected, the subclavian arteries supplying the large flight and wing muscles are very well developed.

Adult Mammals

The synapsid line of evolution to mammals and the sauropsid line to reptiles and birds diverged in the Carboniferous more than 250 million years ago, so these groups have had a long, independent evolution. As mammals became active, endothermic vertebrates with lungs that were ventilated continuously, they too evolved completely divided hearts as in birds and crocodiles. Oxygen-poor blood is received directly by the right atrium and pumped by the right ventricle to the lungs (Fig. 19-14*B*) because the very reduced sinus venosus has been incorporated into the wall of the atrium to become the sinoatrial node, or pacemaker. Oxygen-rich blood returns to the left atrium and is pumped by the left ventricle to the body. Superficially, the mammalian heart resembles those of birds and crocodiles, but the interventricular septum evolved independently and develops embryonically in a slightly different way, so it is not homologous to the interventricular septum of these vertebrates.

Six embryonic aortic arches appear during the embryonic development of a mammal, but as in most other terrestrial vertebrates, only three persist (Figs. 19-14*B*

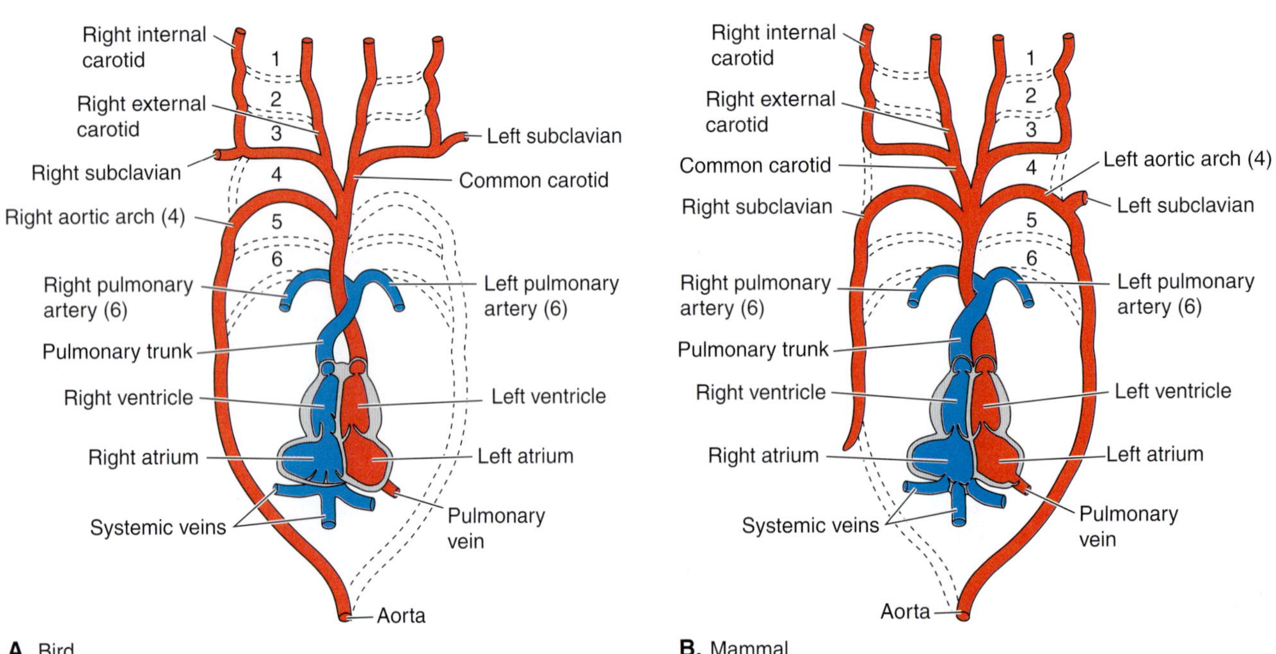

FIGURE 19-14
A ventral view of the heart and aortic arches of a bird (*A*) and a mammal (*B*). *(After Goodrich.)*

and 19-15). The third arches contribute to the internal carotid arteries. The left fourth aortic arch alone forms the arch of the aorta. The right one is present but supplies only the subclavian artery on that side of the body. Its connection to the dorsal aorta is lost. The sixth arches contribute to the pulmonary arteries. The dorsal part of the left sixth arch, the ductus arteriosus, is present in the embryo. The ventral aorta and conus arteriosus divide during development, so the arch of the aorta arises directly from the left ventricle to carry oxygenated blood to the head and body; the pulmonary arteries arise by a common pulmonary trunk from the right ventricle to bring oxygen-poor blood to the lungs.

Embryonic and primitive mammals have a carotid circulation in which the stapedial branch of the internal carotid artery supplies most of the outside of the head, as in other vertebrates, but in most adult eutherian mammals, the external carotid taps into the distal branches of the stapedial (Fig. 19-15). It takes over these branches when the proximal part of the stapedial artery (the segment passing through the stapes) atrophies. Thus, the adult external carotid artery supplies the entire outside of the head. The term "external carotid," although used in all vertebrates, derives from the distribution of the artery in eutherian mammals. The internal carotid artery carries blood primarily to the cerebral hemispheres. The rest of the blood supply to the brain comes from derivatives of the intersegmental arteries that ascend with the spinal cord and from anastomoses with branches of the external carotid. These anastomoses usually are very small. Groups of specialized receptor cells are found at the bifurcation of the common carotid artery. These aggregations of receptors are called **carotid bodies.** Their function is to sense blood carbon dioxide levels.

The complete division of the ventricle allows for different degrees of muscularization of its wall and the development of different systemic and pulmonary pressures. The left ventricular musculature is far more massive than is that of the right side. The left ventricle develops a mean pressure of about 100 mm Hg—far higher than that developed by the comparable part of the reptilian ventricle, which is between 30 and 40 mm Hg. A high systemic pressure, an efficient system for conducting action potentials through the ventricular musculature, and a rapid heartbeat combine to circulate blood quickly and efficiently through the body. Blood moves as rapidly through the lungs, but it would be un-

Text continues on page 622

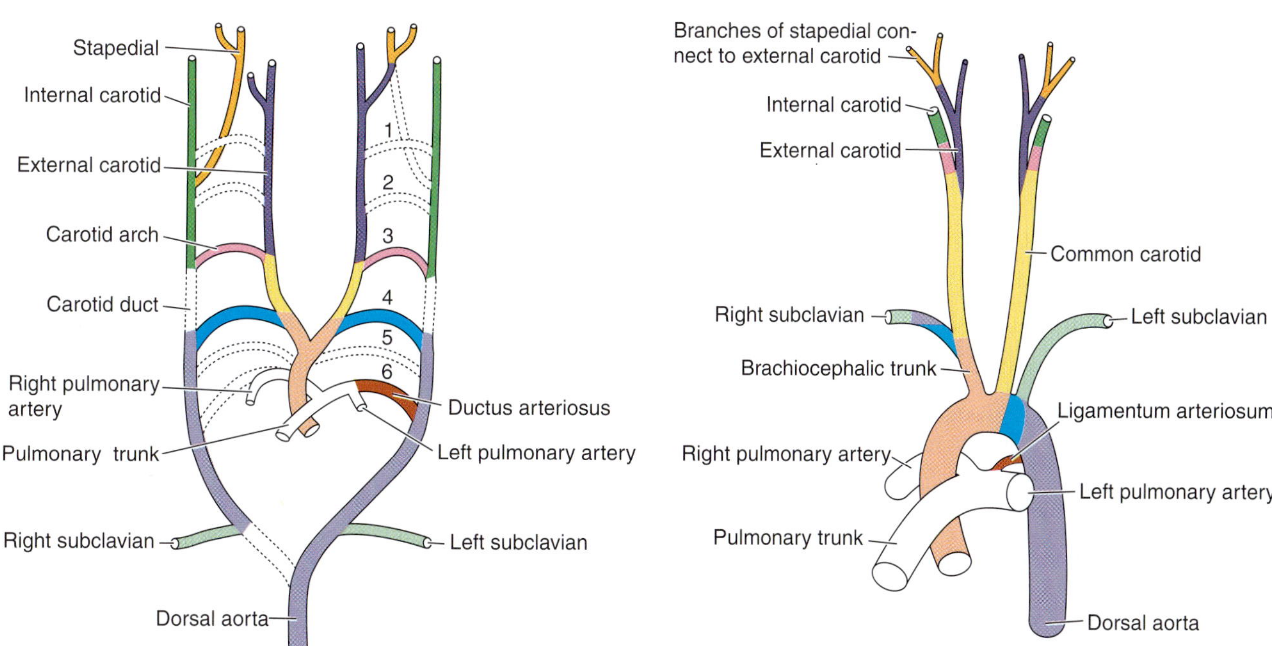

FIGURE 19-15
Ventral view of the aortic arches and carotid circulation in a mammal. *A,* Two developmental stages are expressed; the earlier is on the left, and the later is on the right. The external carotid artery taps into the stapedial (a branch of the internal carotid, left side) and takes over its peripheral branches (right side). After the atrophy of the basal part of the stapedial, the external carotid artery supplies the entire outside of the head in most adult mammals. *B,* The pattern in an adult human. *(After Barry.)*

FOCUS 19-1 *Evolution of the Aortic Arches*

One of the hallmarks in comparative anatomy is the discovery and broadly based explanation of the evolutionary pattern of the heart and great vessels of vertebrates. It still has profound and pervasive implications in comparative biology. A brief summary is shown (Fig. A). When vertebrates first appeared, they must have possessed a ventral and dorsal aorta with aortic arches between them (1). The ventral aorta supplies oxygen-poor blood to the afferent parts of the aortic arches, which carry the blood to the capillary beds of the gills for oxygenation. Oxygen-rich blood flows from the efferent parts of the aortic arches to the dorsal aorta for distribution to the body. A distinct pressure decrease occurs between ventral and dorsal aortae, caused by the intervening capillary beds. Lampreys and hagfishes have eight or more arches. This high number is a specialization for the living jawless fishes (2). In the gnathostomes, the number of aortic arches is limited to six, which is maintained in the embryos of all of its members (3). Within this lineage, the Chondrichthyes, the first aortic arch is virtually lost, and becomes a spiracular artery, whereas the remaining five arches all give rise to specialized collecting loops with pretrematic and posttrematic branches (4). This pattern deviates significantly from the sister clade, the Osteichthyes, in which both the first and second aortic arches are lost (5). This large clade includes the Actinopterygii, Dipnoi, Amphibia, Chelonia, Lepidosauria, Crocodilia, Aves, and Mammalia. The Actinopterygii retain gill circulation with four distinct afferent and efferent aortic arches without collecting loops (6). It reflects a fundamental adaptation to maximize gas exchange with the aquatic medium. On the other hand, in the sister clade of the actinopterygians, the sarcopterygians, the gill circulation is supplemented with lung ventilation. As a result, the pulmonary artery and vein and a functional ductus arteriosus (7) arose as a major evolutionary innovation from the sixth arch, giving the organism a flexible shunt to balance blood supply to and from gills and lungs according to environmental conditions. In the living Dipnoi, except *Neoceratodus,* the third and fourth aortic arches form a complete connection between the dorsal and ventral aortae because the respective arches bear no gills and lack capillary beds (8). Consequently, the characteristic decrease in blood pressure from ventral to dorsal aorta is eliminated for the first time in vertebrate evolution. The tetrapod clade develops a complete atrial septum (9) and loses the fifth aortic arch altogether. Adult amphibians lose a short section of the dorsal aorta between the third and fourth aortic arches, the carotid duct (10), which is maintained in other tetrapods. In salamanders, the internal carotid is formed by a forward extension of the dorsal aorta (11), whereas frogs' blood circulation evolves a characteristically well-developed pulmocutaneous artery (12), which reflects the importance of skin respiration that plays a major role in humid environments of most frogs. In adult amniotes, the ductus arteriosus, which is functional in embryos, becomes ligamentous and forms the ligamentum arteriosum (13), whereas the carotid arteries arise generally from the third aortic arch and parts of the ventral and dorsal aortae (15). Chelonia and Lepidosauria have an incompletely divided ventricle with the characteristic interventricular canal as a shunt (14), whereas crocodiles develop a complete interventricular septum and thus a four-chambered heart. However, crocodiles maintain a shunt, but not within the ventricle as is true in turtles and lepidosaurians. Instead, the foramen of Panizza (16) at the base of the systemic arches functions as a most effective shunt, which allows for right-to-left shunts during dives and rest. In birds, the right systemic arch is functional, whereas the left never develops (17). In contrast, the left systemic arch is the only one differentiated in mammals because the right systemic arch becomes the right subclavian artery (18).

This phylogenetic pattern shows effects of developmental canalization in the retention of the segmental aortic arches in either adult or embryonic vertebrates. Functionally, the aortic arches in their primitive configuration maximize gas exchange in aquatic respiration but undergo transformations when air respiration evolved as an accessory mechanism in lungfishes. The development of the pulmonary artery and the ductus arteriosus, first seen in lungfishes, is an evolutionary innovation with which vertebrates made their first explorations of terrestrial habitats. In tetrapods, separate systemic and pulmonary circulations are formed by the subdivision of the atrium and ventricle, whereas the conus arteriosus produces the bases of the trunks of the pulmonary and left and right aortic trunks. The importance of hemodynamic functional efficiency is reflected by the convergent evolution of the single systemic arch in birds and mammals, the right and left, respectively. Both evolved independently, with the identical mechanisms functionally adapted for elevated metabolism and active life. The embryonic ductus arteriosus, which has made its appearance in lungfishes, shunts blood directly from the ventral to the dorsal aorta, bypassing the lung. This capacity to shunt blood away from the fetal lung has made possible the development of mammal embryos within the uterus (see p. 622). It is the best documented example of how humans' distant lungfish heritage has made it functionally and structurally possible for humans to develop as a placental mammal within the uterus.

A. The phylogenetic pattern of major evolutionary transformations of the great vessels and heart of the living vertebrates includes:

1. The presence of ventral and dorsal aortae in adults, embryos, or both
2. Eight or more aortic arches in lampreys and hagfishes
3. The presence of six aortic arches during development
4. Collector arterial loops with pretrematic and posttrematic vessels in Chondrichthyes
5. Loss or modification of first and second aortic arches
6. Four afferent and efferent branchial arches accompanied by a pressure decrease in actinopterygians
7. Differentiation of the pulmonary artery and vein and an embryonic ductus arteriosus
8. Third and fourth aortic arches not interrupted by gill capillary beds
9. Full development of interatrial septum
10. Loss of carotid duct in adult amphibians

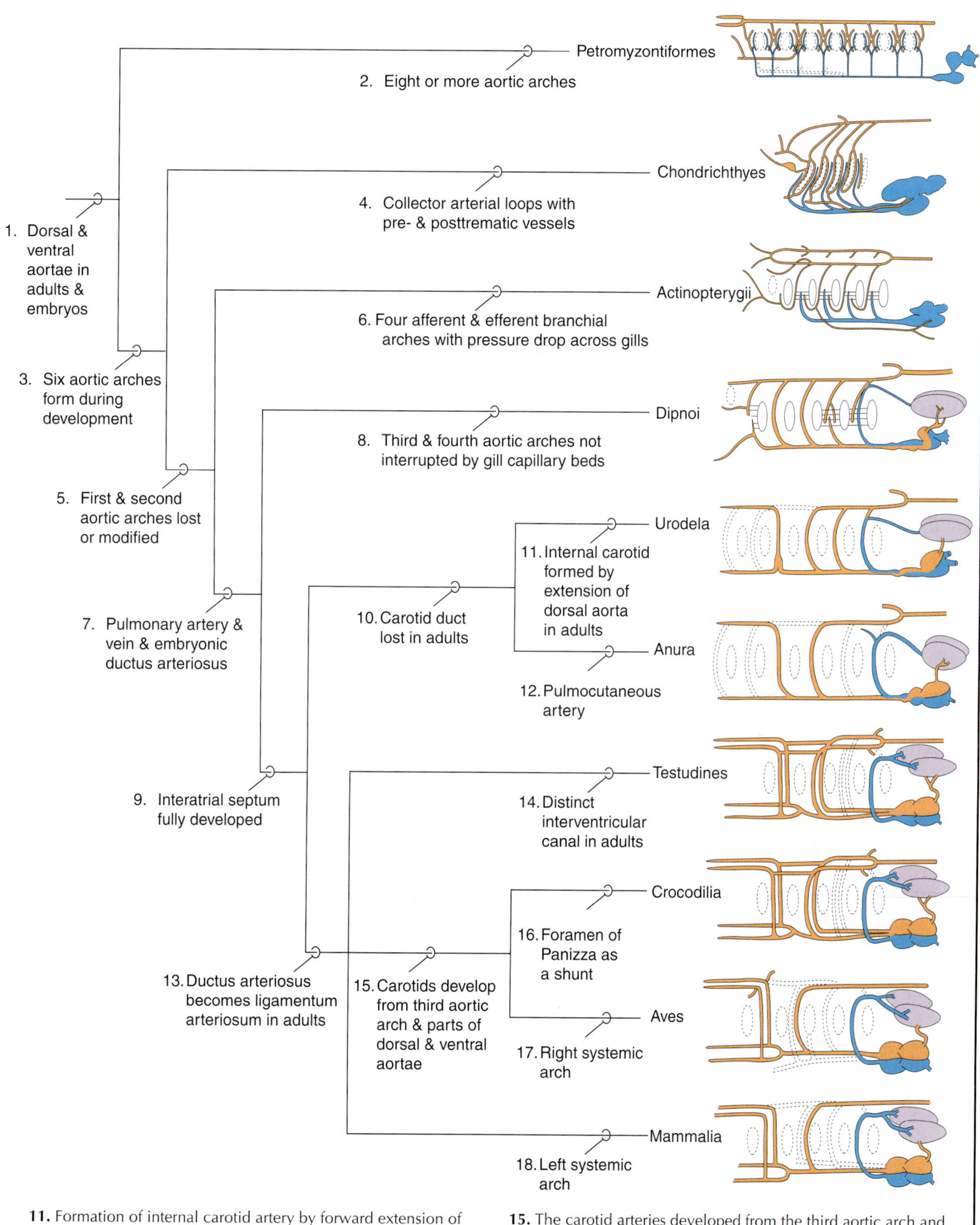

11. Formation of internal carotid artery by forward extension of the dorsal aorta in adult urodeles
12. The pulmocutaneous artery in anurans
13. Change of the ductus arteriosus into a ligamentum arteriosum in adults
14. A distinct interventricular canal in the adult reptilian heart
15. The carotid arteries developed from the third aortic arch and from parts of the dorsal and ventral aorta in crocodiles and birds
16. A foramen of Panizza as a shunt in crocodiles
17. A right systemic arch in birds
18. A left systemic arch in mammals

desirable for them to have such a high pressure. High pressure would cause an excessive filtration of water from the pulmonary capillaries into the alveoli. Pulmonary resistance is much lower, and pulmonary pressure is therefore only about one fifth the systemic pressure, on the order of 15 to 20 mm Hg, which is comparable to the pulmonary pressures in reptiles.

The relatively massive and compact cardiac muscles of mammals necessitate a well-developed **coronary circulation** to supply their metabolic needs. A pair of coronary arteries leaves the very base of the arch of the aorta. Coronary veins return to a coronary sinus, which empties into the right atrium beside the entrances of the systemic veins. The coronary system is present but less well developed in amphibians and reptiles because their heart walls are thin and spongy enough to be supplied to a large extent by the blood flowing through the heart. Elasmobranch and teleost fishes also have coronary arteries that derive from the efferent branchial circulation and bring oxygenated blood to the heart, which is needed because only oxygen-poor blood flows through their hearts. We summarize the remarkable history of the aortic arches in Focus 19-1.

Fetal and Neonatal Mammals

Although an adult mammal has no shunts between the systemic and pulmonary circuits, a fetal mammal does because the placenta, and not the lungs, is the site for gas exchange (Fig. 19-16*A*). In eutherians, blood rich in oxygen returns to the fetus from the placenta in the umbilical vein. Most of this blood enters a direct passageway through the liver, the **ductus venosus,** and joins venous blood in the **caudal or inferior vena cava** returning from the posterior parts of the fetus. The entrance of the caudal vena cava into the right atrium is situated in such a way that most of this blood passes through a valved opening, the **foramen ovale,** in the interatrial septum and enters the left atrium. This blood, which has the highest oxygen content of blood in the fetus, has bypassed the lungs. The left ventricle pumps it into the arch of the aorta, and much of it leaves through the first branches of the aorta to the head and shoulders. Oxygen-poor blood returns from the front of the body in the **cranial or superior vena cava**. Its entrance into the right atrium is positioned in such a way that this blood, together with a small admixture of blood from the caudal vena cava, enters the right ventricle to be pumped toward the lungs. Fetal lungs are collapsed and offer a higher resistance to blood passage than does the systemic circuit, so most of this blood flows through the open **ductus arteriosus** to join the aorta caudal to its branches to the head and shoulders. The ductus arteriosus is a part of the left sixth aortic arch. A mixed blood is distributed to the rest of the body and, by the umbilical arteries, to the placenta.

In addition to diverting most of the blood from the lungs, these shunts provide both ventricles with a sufficient volume of blood to pump so that their musculature can develop normally. The ductus arteriosus sometimes is described as the "exercise channel" of the right ventricle because it permits the ventricle to pump considerable blood even though little of it can go through the lungs. Similarly, the foramen ovale is the "exercise channel" of the left ventricle, for it provides the left ventricle with a reasonable volume of blood and compensates for the small amount of blood that returns from the lungs.

Pressures within the circulatory system change abruptly at birth. The lungs inflate, and pulmonary resistance decreases to less than systemic resistance. Blood from the right ventricle now goes to the lungs rather than through the constricting ductus arteriosus to the body (Fig. 19-16*B*). The return of a large volume of aerated blood from the lungs raises pressure in the left atrium sufficiently to close the valve in the foramen ovale. Accordingly, all of the blood in the right atrium now flows to the right ventricle and lungs. Soon the interatrial valve grows onto the interatrial septum, but a depression, the **fossa ovalis,** remains here throughout life. The ductus arteriosus remains open for a few hours or days after birth and continues to act as a shunt, but because pulmonary resistance is so low, it recirculates some aerated blood to the lungs rather than diverting blood from them. This is of considerable physiological importance during the neonatal period, when a changeover from fetal to adult types of hemoglobin occurs. Fetal hemoglobin is adapted to take up oxygen at relatively low tensions because the fetus must be able to take oxygen away from the mother's blood, but fetal hemoglobin also does not unload oxygen in the tissues as easily as does adult hemoglobin. As long as fetal hemoglobin still circulates in the newborn, the neonatal blood must become as thoroughly saturated with oxygen as possible to compensate for the greater tendency of fetal hemoglobin to hang on to its oxygen. A recirculation of some aerated blood through the lungs accomplishes this task. After the hemoglobin changeover has occurred, the ductus arteriosus contracts fully, and the adult circulatory pattern is established (Fig. 19-16*C*). Eventually, the lumen of the ductus arteriosus is filled with connective tissue, and the duct becomes the adult **ligamentum arteriosum.**

Marsupial fetuses also have cardiac shunts that bypass the lungs during intrauterine life. Because marsupials are born at an earlier stage of development and complete their development in the marsupium, it is not surprising that their shunts are not identical to those of

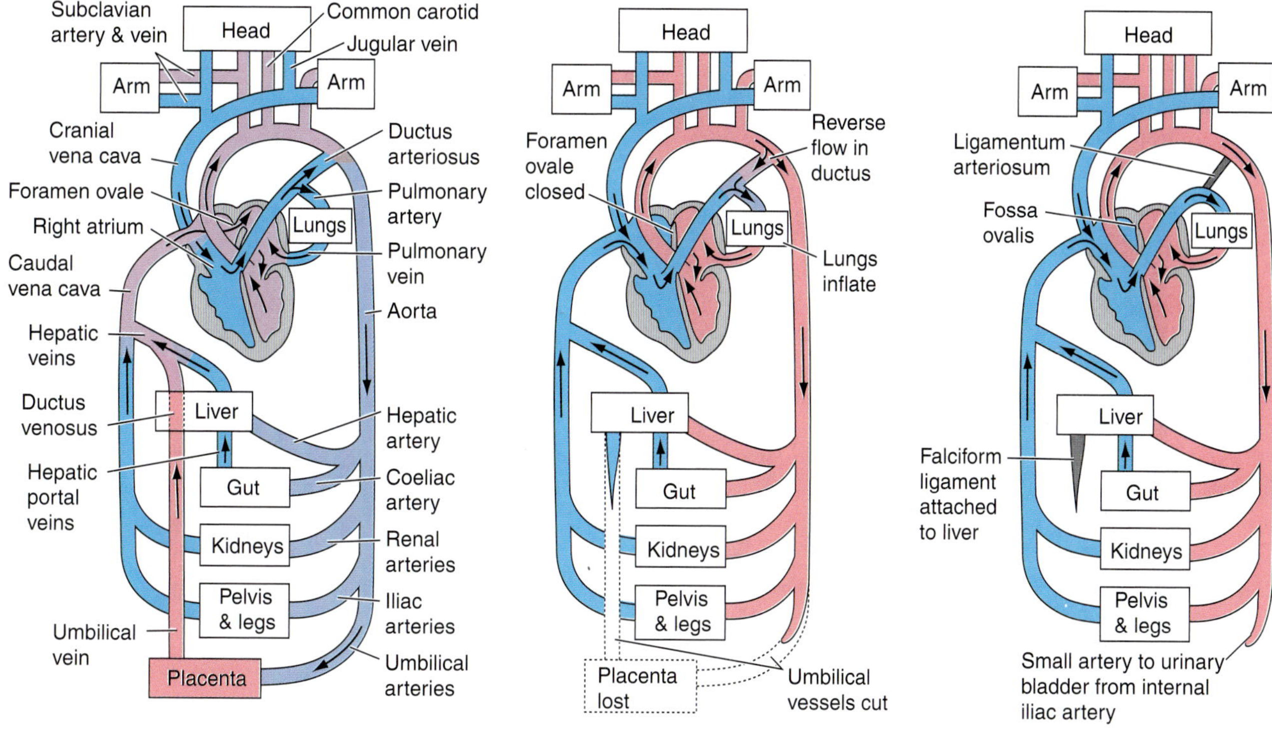

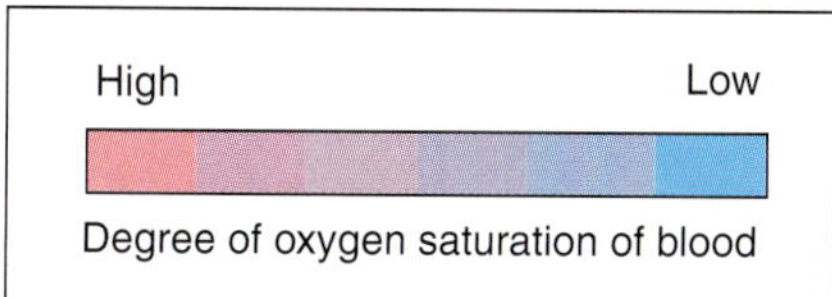

FIGURE 19-16
A–C, Ventral views of the circulation in a fetal, neonatal, and adult mammal. Relative oxygen content of the blood in the vessels is shown by the color gradation. *(Modified from Dorit et al.)*

eutherians. A marsupial fetus has a ductus arteriosus and an interatrial connection, but the septum that partly closes the foramen ovale in eutherians is fenestrated in marsupials. It does not fill in by tissue proliferation for days after birth. In addition, the interatrial septum is incomplete in some marsupials at the time of birth. As in other organ systems, marsupials have evolved somewhat different, but no less effective, strategies than have eutherians (Runciman et al., 1995).

Shunts in fetal and neonatal mammals perform analogous functions to shunts in reptiles. They divert blood from the lungs when pulmonary resistance is high. They recirculate some blood through the lungs when pulmonary resistance is low and a more thorough oxygenation of the blood is needed.

Evolution of the Venous System

The **hepatic portal system** is conservative, and few changes occur from the primitive condition seen in fishes. All vertebrates have hepatic portal systems, but the pattern of tributaries varies slightly with the configuration of the visceral organs that they drain. The liver sinusoids of amphibians and reptiles also receive some of the drainage from the hind legs because the veins draining their hind legs connect either with the hepatic portal system or directly with the liver (Fig. 19-17*C*). These vascular connections to the liver enable this important organ to process a large volume of blood.

Pulmonary veins appear in lungfishes and terrestrial vertebrates (p. 612) and carry oxygenated blood from the lungs to the heart.

The remaining changes in the venous system occur in the drainage of the appendages; in the renal portal and posterior cardinal systems, which are replaced by a **caudal vena cava;** and in the anterior cardinal, which is replaced by jugular veins and a **cranial vena cava.**

Drainage of the Appendages

Jawless fishes have no paired appendages to be drained (Fig. 19-17*A*), but a pair of **lateral abdominal veins** drain the appendages of primitive jawed fishes, such as elasmobranchs (Fig. 19-17*B*). In am-

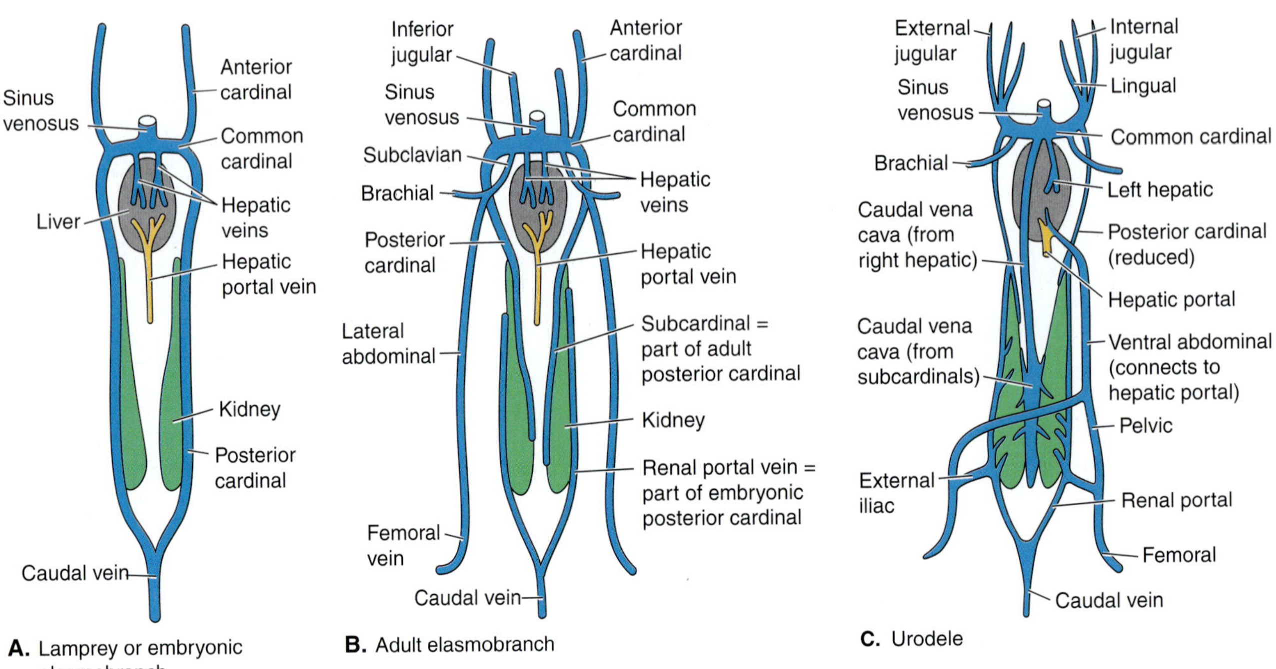

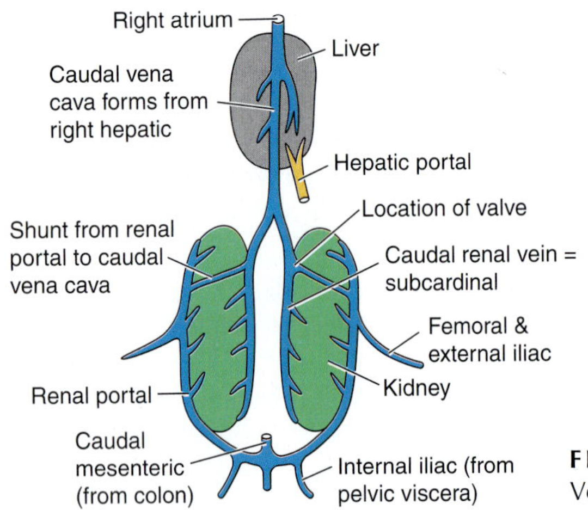

FIGURE 19-17
Ventral views showing important stages in the evolution of the venous system. *A,* A lamprey. *B,* An elasmobranch. *C,* A urodele. *D,* A bird. The reptile system is similar to that of a bird.

phibians, the lateral abdominal veins establish a connection with the liver. The part of the lateral abdominal vein cranial to this connection then disappears in the adults, and the **brachial** and subclavian veins from the pectoral appendage enter the common cardinal directly. The part of the lateral abdominal system caudal to the liver remains paired in reptiles but fuses to form a single, median, **ventral abdominal vein** in amphibians (Fig. 19-17*C*). The **femoral veins** from the pelvic appendages of amphibians and reptiles acquire a new connection, the **external iliac veins,** with the renal portal system. Blood from the hind legs is now diverted either to the liver by the lateral abdominal system or to the kidneys by the renal portal system. These important organs receive more blood than they do in fishes. The external iliac connection to the renal portal system finally becomes dominant, and the lateral abdominal vein itself disappears in adult birds and mammals (Figs. 19-17*D* and 19-18). In embryonic birds and mammals, however, part of the primitive lateral abdominal system persists as the **allantoic** or **umbilical veins,** which return blood from the allantois in birds and from the placenta in mammals (Fig. 19-4*C*).

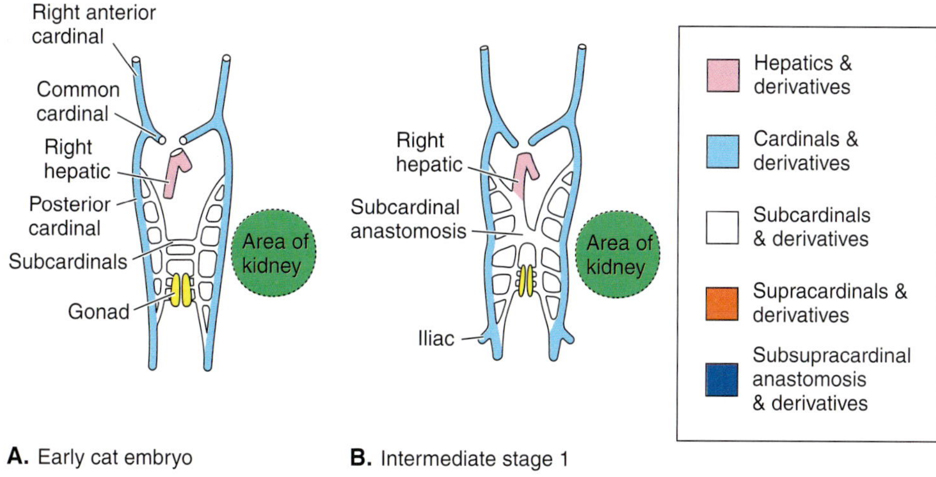

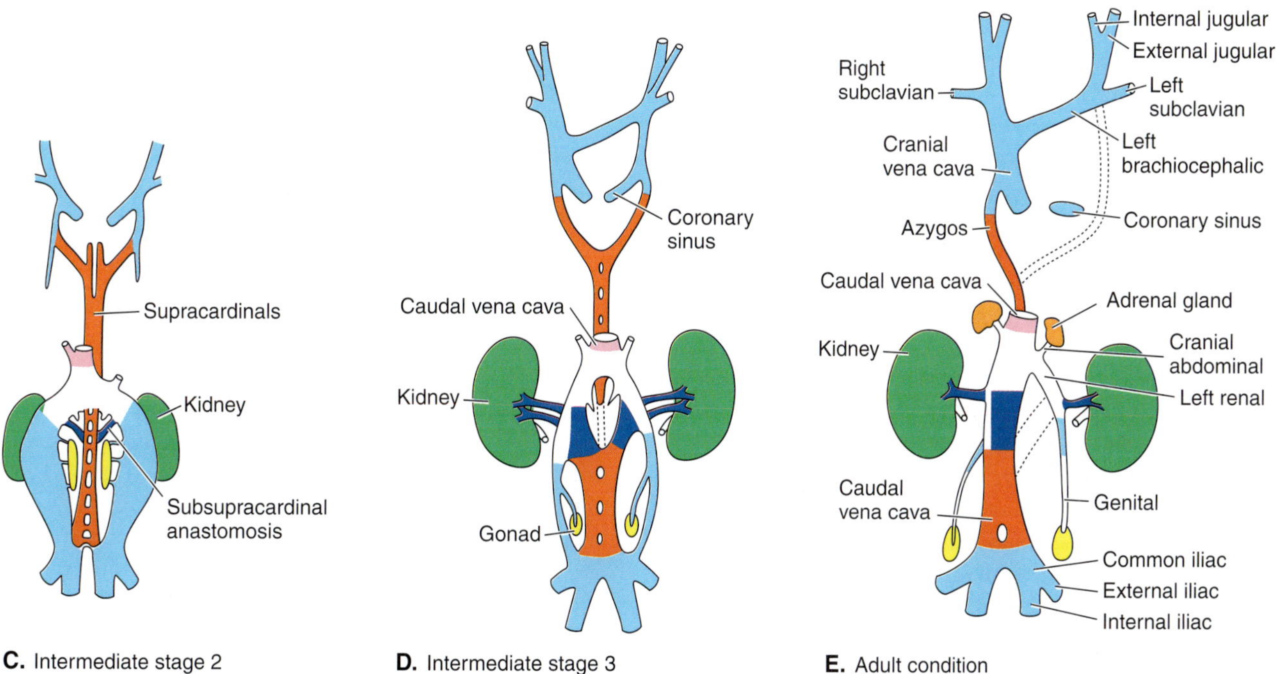

FIGURE 19-18
Ventral view of the development of the veins of a cat. The primitive cardinal system in an early embryo (*A*) is transformed through intermediate stages (*B–D*) into a cranial and caudal vena cava (*E*). Only the stump of the segment of the caudal vena cava that develops from the right hepatic vein is shown. *(After Huntington and McClure.)*

The Caudal Vena Cava

In most fishes, the renal portal and posterior cardinal systems (Fig. 19-17*A* and *B*) drain the tail, kidneys, and trunk. The beginning of an alternative course for the blood, the **caudal** or **inferior vena cava,** first appears in lungfishes and amphibians. It is a large vein that extends caudally from the sinus venosus and incorporates the now-fused embryonic subcardinal veins (Fig. 19-17*C*). (Recall that the subcardinals form much of the posterior cardinals in the adult jawed fishes.) The anterior part of the caudal vena cava develops in lungfishes by an enlargement of the anterior portion of the embryonic right posterior cardinal, but in all terrestrial vertebrates, it forms as a caudal extension of the right hepatic vein from the liver (Fig. 19-17*C*).

The caudal vena cava of urodeles drains the kidneys and also receives most of the drainage from the tail, for most of this blood enters the kidneys through the renal portal system. Some blood from the tail continues forward in small posterior cardinal veins, which retain their connection with the renal portal system. (Recall that the renal portal system develops embryonically from the caudal portion of the posterior cardinals.) This connection has been lost in frogs, reptiles, and birds (Fig. 19-17*D*), so the caudal vena cava receives all the drainage of the kidney and tail, as well as most of the blood from the hind legs that enters the renal portal system. Not all of this blood goes through the capillary beds over the kidney tubules, for reptiles and birds have one or more direct shunts from the renal portal system to the caudal vena cava. Valves within the shunts regulate the amount of blood going through the renal peritubular capillaries or directly to the caudal vena cava. When the valves are open, blood is returned more rapidly to the heart, but the factors that control the operation of these valves are not well known.

In the evolution of mammals, the caudal vena cava has extended caudally to unite with the most caudal section of the renal portal system, the part receiving the external iliac and caudal veins (Fig. 19-18*A*). The section of the renal portal system anterior to this connection is then lost, so blood from the tail and hind legs goes directly into the caudal vena cava rather than through the kidneys. How this occurs is best seen in the embryonic development of the veins. The pattern of veins in an early mammalian embryo is similar to that of an elasmobranch (Fig. 19-18*A*). A bit later, a caudal vena cava develops from the right hepatic and taps into the subcardinals. The embryo's veins now resemble those of a urodele (Fig. 19-18*B*). The caudal extension of the caudal vena cava involves the development of a pair of **supracardinal veins** located dorsal to the kidneys. These tap into both the anterior and the posterior ends of the embryonic posterior cardinals (Fig. 19-18*C*). They also acquire a connection, known as the **subsupracardinal anastomosis,** with the partly fused subcardinals. Shortly after this, the cranial and caudal sections of the supracardinals become disconnected from each other (Fig. 19-18*D*). Blood from the tail and hind legs now has two routes forward. The early tetrapod route includes the caudal segment of the posterior cardinal system (the adult renal portals), passages through the kidney, fused subcardinals, and the caudal extension of the right hepatic vein. The other route, which will become the adult mammal's caudal vena cava, includes the caudal segment of the posterior cardinal system, the caudal half of the supracardinals (the one on the right side usually becomes dominant), the right subsupracardinal anastomosis, the subcardinal (primarily the right one), and the caudal extension of the right hepatic (Fig. 19-18*D*). Other parts of the embryonic vessels form the renal, gonadal, and cranial abdominal veins.

Most of the thoracic wall of a mammal is drained by intercostal veins that lead into an **azygos vein** (Gr., *a* = without + *zygon* = yoke), which develops from the cranial half of the embryonic supracardinal plus a small segment of the embryonic posterior cardinal (Fig. 19-18*E*). Usually, the azygos vein receives the intercostal veins from both sides of the body, but in some species, those of the left side enter a distinct **hemiazygos vein,** which enters the coronary sinus (the left common cardinal).

Why the renal portal system and much of the posterior cardinal system are replaced by the caudal vena cava is not entirely clear. Replacement may be related to an increase in arterial systemic blood pressure. The renal arteries provide the kidney tubules with a sufficient volume of blood to be cleared of nitrogenous and other wastes, so an additional venous contribution is not needed. The caudal vena cava is also a more direct passageway for blood from the caudal parts of the body and speeds up venous return to the heart.

The Cranial Vena Cava

The head of a jawed fish is drained by the anterior cardinals (also called the lateral head vein) and inferior jugulars, both of which enter the common cardinals (Figs. 19-17*B* and 19-19*A*). In tetrapods, the deeper tributaries of the anterior cardinal, which are within the skull, coalesce to form an **internal jugular vein,** which emerges through the jugular foramen (Figs. 19-17*C* and 19-19*B*). In mammals, this foramen is located near the back of the skull. The glossopharyngeal, vagus, and accessory nerves accompany the internal jugular vein. Part of the internal jugular is homologous to the fish's posterior cerebral vein, which also emerges through a foramen accompanied by comparable nerves. The superficial tributaries of the anterior cardinal coalesce to form the **external jugular vein.** This vein is located in the head and neck more superficially than the anterior cardinal. In some species, the external jugular also receives blood from within the skull by anastomoses in the orbital region with the internal jugular. The inferior jugular, now known as the **lingual vein,** acquires a new connection with the external jugular.

The evolution of the external and internal jugular veins is accompanied by the loss of the primitive anterior cardinal, except for its most caudal portion, which receives the jugulars and usually the subclavian vein as well. After the jugulars and subclavian veins unite, the base of the anterior cardinal and the common cardinal

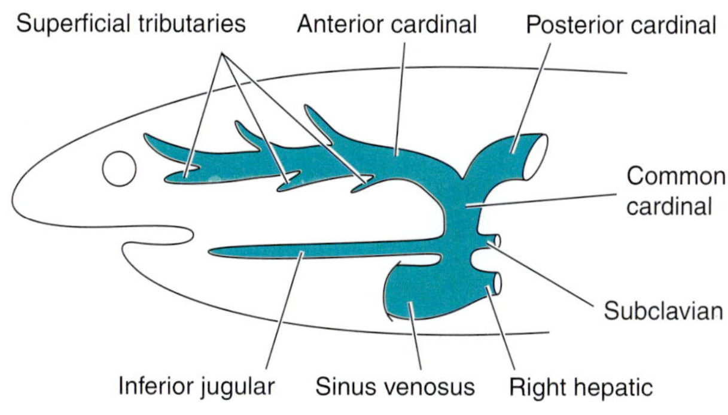

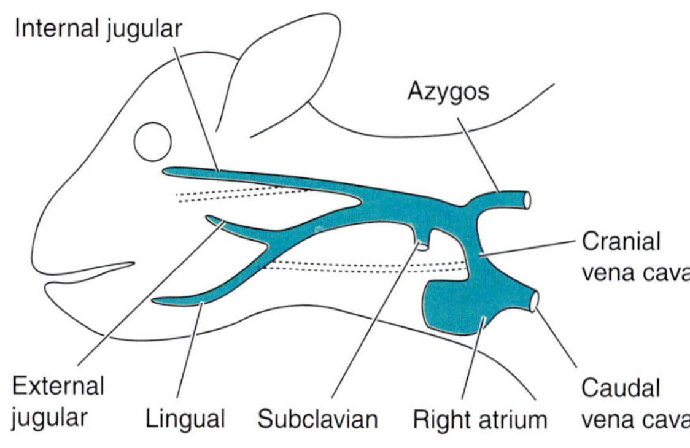

FIGURE 19-19
Lateral view of the veins of the head to illustrate the development of the internal and external jugular veins from the anterior cardinal and inferior jugular veins. *A,* A fish. *B,* A mammal. The broken lines in *B* represent atrophied vessels.

into which it leads are known as the **cranial,** or **superior, vena cava.** Many adult mammals retain the embryonic condition, in which both a left and a right cranial vena cava enter the right atrium (Fig. 19-18*D*). To reach the right atrium, the left cranial vena cava crosses the dorsal surface of the heart. As it does, it receives the coronary veins draining the heart, so this segment of the left cranial vena cava is also a **coronary sinus** (Fig. 19-18*D*). In many other mammals, including humans and cats, a cross connection called the **left brachiocephalic vein** extends from the left jugulars and subclavian to join the right cranial vena cava (Fig. 19-18*E*). After this connection has been established, the base of the left anterior cardinal and the common cardinal become very small or disappear, except for the segment of the left common cardinal serving as the coronary sinus.

Evolution of the Lymphatic System

Hagfishes, lampreys, chondrichthyans, and teleosts have networks of small, blind vessels that accompany the veins and help the cardiovascular capillaries drain the tissues. These vessels develop from the veins and empty into them at frequent intervals. Although similar to lymphatic vessels, they sometimes contain a few erythrocytes and may represent an independently evolved system with lymphatic-like functions. The condition in lungfishes is not known.

Tetrapods have evolved distinct **lymphatic systems,** in which **lymphatic capillaries** (p. 606) help drain most of the tissues of the body. (Exceptions include the eye, inner ear, and central nervous system, which have their own drainage patterns: humors of the eye, endolymph, and cerebrospinal fluid, respectively). Bone marrow, cartilage, and the deep parts of the spleen and liver also lack lymphatic drainage.) The pattern of the large vessels into which the lymphatic capillaries enter varies considerably. Amphibians typically have **subcutaneous ducts** draining the skin and superficial muscles, **subvertebral ducts** draining deeper tissues, and **visceral plexuses** in the major internal organs (Fig. 19-20*A*). Large lymph sinuses are associated with many of these channels. Lymph is propelled by the contraction of surrounding body muscles and pulsating sections of the vessels, the **lymph hearts,** which keep it moving more or less in one direction. Valves are not present. The lymphatic vessels usually discharge into the anterior cardinal and renal portal veins.

Mammals have a single, large lymph sinus, the **cisterna chyli** (Gr., *chylos* = juice), located dorsal to the aorta just behind the diaphragm (Fig. 19-20*B*). It receives all of the drainage caudal to the diaphragm. Valves are numerous in mammalian lymphatic vessels. Vessels coming from the intestinal region, the **lacteals** (L., *lac* = milk), often are prominent and have a whitish appearance, for they are the primary route for the passage of fat molecules absorbed in the intestine. A large subvertebral duct, which is called the **thoracic duct** in mammals, continues forward from the cisterna chyli, receives lymphatics from the thoracic wall, and joins the left subclavian vein near its union with the jugulars. Vessels draining the left side of the head, neck, and the left shoulder and arm join the thoracic duct near its entrance into the subclavian. Those draining comparable parts of the right side of the body form a short **right lymphatic duct,** which enters the right subclavian vein. Lymph hearts are absent, and lymph is moved by the contraction of surrounding body muscles. Numerous valves prevent a backflow.

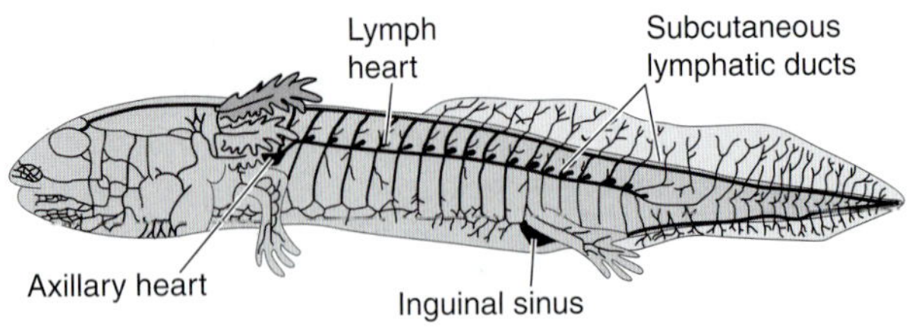

A. Larval salamander

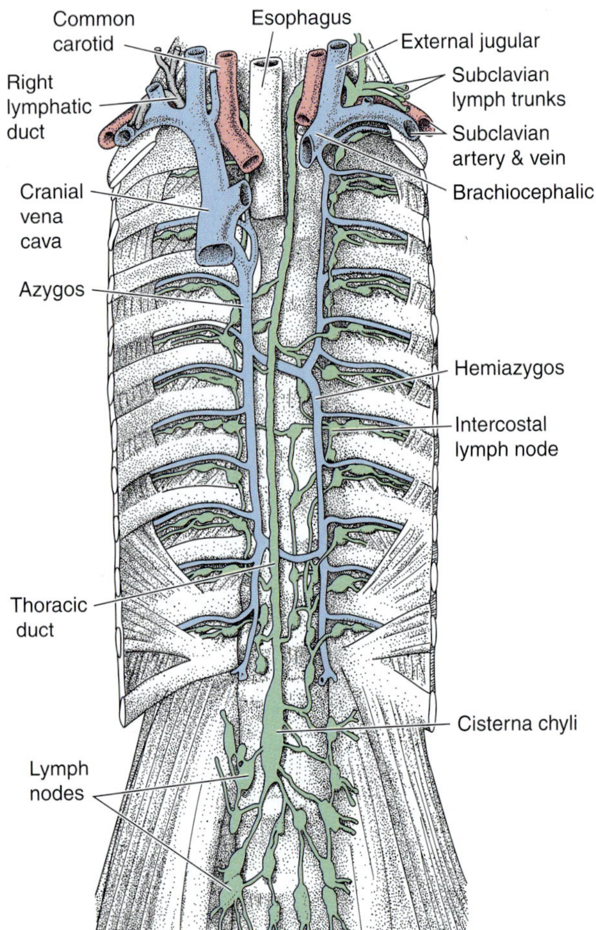

B. Human

FIGURE 19-20
The lymphatic system. *A,* The superficial lymphatic vessels of a salamander larva. *B,* Deep lymphatic vessels and nodules of a human. *(A, After Hoyer and Udziela; B, after Williams et al.)*

Lymph nodes occur in a few aquatic birds and are abundant in mammals. Many are found in the neck, armpits, groin, and base of the mesentery, where many lymphatic vessels converge. Each node consists of many groups of lymphocytes, called **lymph follicles,** enmeshed in a network of reticular fibers (Fig. 19-21). The follicles are separated by dense connective tissue trabeculae that penetrate the node from its outer capsule. Several afferent lymphatic vessels enter the periphery of a node, and the lymph seeps through sinuses of reticular fibers around, between, and through the lymph follicles. Many macrophages line the sinuses and phagocytize foreign material in the lymph. The lymphocytes often initiate immune reactions to invading antigens. Lymph leaves through a single efferent vessel. Small arteries and veins also supply the nodes.

Additional aggregations of lymphatic tissue occur in other organs. Many lymph nodules are embedded in the mucosa and submucosa of the digestive tract, just beneath the lining epithelium. Among these are the **lingual, pharyngeal,** and **palatine tonsils,** which form a ring of lymphatic tissue around the mammalian pharynx at the level of the embryonic second pharyngeal pouch. Another large group, **Peyer's patches,** occur in the small intestine of young mammals; many lymph nodules occur in the wall of the vermiform appendix. Young birds have a distinct patch of lymph nodules in the **bursa of Fabricius,** a dorsal pouch of the cloaca. It plays a vital role in the development of the avian immune system.

The thymus, located in the base of the neck, is a lymphatic organ of particular importance. It develops in all vertebrates as epithelial buds from certain pharyngeal pouches (Chapter 17), but the epithelial tissue is invaded by and dominated by lymphocytes. Lymphocytes enter the organ from other hemopoietic tissues, but they proliferate here. A population of lymphocytes known as the **T lymphocytes** matures in the thymus

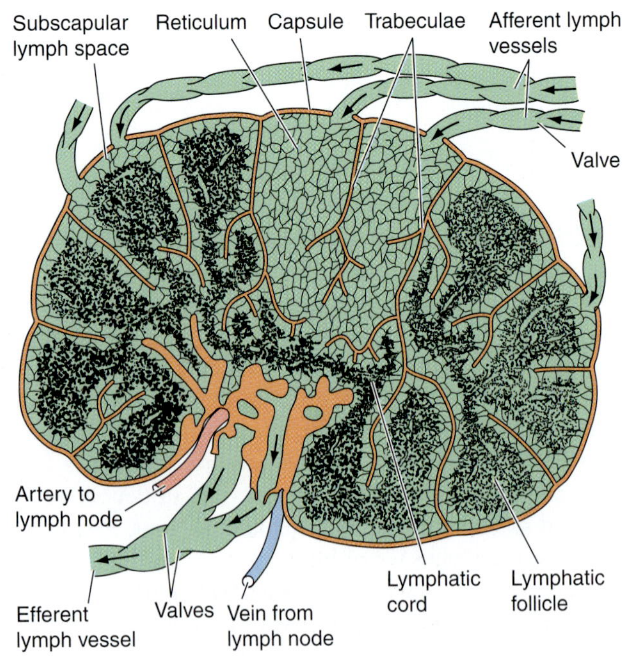

FIGURE 19-21
The structure of a mammalian lymph node. *(After Fawcett.)*

before spreading to the lymph nodes and other lymphatic tissues. The remaining lymphocytes of the body are known as **B lymphocytes.** In birds, they mature in the bursa of Fabricius, and in mammals, possibly in some of the intestinal lymphoid nodules. On appropriate antigenic stimulation, B lymphocytes transform into cells that synthesize and release circulating antibodies in a humeral immune response.

SUMMARY

1. The circulatory system transports materials throughout the body and helps maintain homeostasis.
2. The biochemical and physiological functions of the system are performed by blood, which consists of a liquid plasma and several types of cells.
3. The spleen functions as a hemopoietic organ, a site for the destruction of old erythrocytes, a site for phagocytosis, and a storage organ for erythrocytes.
4. The heart is the pump that creates the hydrostatic pressure that drives the blood through the vessels.
5. Cardiac muscle has an inherent rhythm that in advanced vertebrates is initiated at the sinoatrial node and modulated by autonomic nerves terminating on the node and heart. Birds and mammals have specialized systems for conducting impulses in the ventricles.
6. Arteries are defined as vessels that lead away from the heart; veins, as vessels leading to the heart. Either may contain blood high or low in oxygen content. Vessels are lined by endothelium and have a wall containing elastic fibers, smooth muscle, and connective tissue. Wall structure correlates with the forces within the vessels and their function.
7. Capillaries are very small vessels the walls of which consist only of endothelium. Exchanges between the blood and the interstitial fluid occur here.
8. Because of changes in the combined cross-sectional area and circumference of vessels in different parts of the system, the rate of blood flow decreases as blood moves toward the capillaries and increases as blood flows from the capillaries into the veins. Pressure falls throughout the system. The fall is especially pronounced in the capillaries.
9. Blood vessels and cells begin their development as blood islands in the splanchnic mesoderm. An early vitelline circulation to and from the yolk is established before the somatic circulation. The pattern of the embryonic vessels in a mammal is similar to their pattern in an adult fish.
10. An early fish, such as an elasmobranch, has a four-chambered heart: sinus venosus, atrium, ventricle, and conus arteriosus. The heart receives low-pressure, oxygen-poor blood and increases the pressure sufficiently to drive it through the gills, where aeration occurs, and on to the body.
11. The location of the heart in a pericardial cavity surrounded by stiff tissues that do not collapse enables the heart to act alternately as a suction and force pump.
12. The six embryonic aortic arches are transformed into a branchial circulation that consists of afferent branchial arteries that lead to the gills, collector loops that receive aerated blood, and efferent branchial arteries that lead to the dorsal aorta, from which blood is distributed to the body.
13. The branches of the aorta consist of median visceral (coeliac and mesenteric) arteries that supply the digestive tract and its derivatives, lateral visceral (gonadal and renal) arteries to the gonads and kidneys, and intersegmental arteries to the body wall and appendages.
14. A hepatic portal system carries blood from the abdominal viscera to the sinusoids in the liver. Two hepatic veins carry blood from the liver to the sinus venosus.
15. In jawed fishes, a renal portal system that carries blood from the tail to capillaries over the renal tubules has evolved. A caudal pump provides the necessary increase in pressure.
16. The kidneys and trunk are drained by a pair of posterior cardinals; the appendages, by lateral abdominal veins; and the head, by the paired anterior cardinals and inferior jugulars. All these vessels enter the paired common cardinals that lead to the sinus venosus.
17. Lungfishes have evolved a pulmonary circulation that carries blood to the lungs and back to the heart as well as retaining most of the branchial circulation.
18. Amphibians lose the branchial circulation during ontogeny; it is never present in reptiles. The number of aortic arches is reduced, but the third pair remains as part of the internal carotids to the head, the fourth and sometimes the fifth are systemic arches to the dorsal aorta and body, and the sixth pair provides pulmonary arteries to the lungs.

19. Blood returning to the heart from the body is separated from that returning from the lungs in a partially divided atrium (lungfishes) or a completely divided atrium (amphibians and reptiles). The division of the ventricle, conus arteriosus, and ventral aorta varies morphologically in these groups, but in general, oxygen-poor and oxygen-rich blood are physiologically separated to a high degree when the lungs are functioning. Blood richest in oxygen goes to the head and body; that lowest in oxygen, to the gills and lungs.

20. The system is very flexible, and when the lungs are not functioning, the incomplete division of the heart permits blood to be diverted from the lungs. The volume of the pulmonary and systemic flow need not be equal.

21. Birds and mammals have independently evolved completely divided hearts. The right side receives oxygen-poor blood from the body and sends it to the lungs, and the left side receives oxygen-rich blood from the lungs and sends it to the body. Blood flow through the pulmonary and systemic circuits is equal. The complete division of the heart permits the evolution of high pressures in the systemic circuit and the retention of modest pressures in the pulmonary circuit.

22. Both birds and mammals retain the paired third aortic arch as part of the internal carotids and the paired sixth arches as part of the pulmonary arteries. In mammals, the left fourth arch is the systemic arch to the body; in birds, the right fourth arch is the systemic arch.

23. Fetal mammals retain bypasses in the heart that divert blood from the lungs, and one of these is used temporarily in neonatal eutherian mammals to divert more blood to the lungs. Pulmonary and systemic volumes are not the same.

24. The hepatic portal system and pulmonary veins have changed little during vertebrate evolution.

25. The pectoral appendages of tetrapods drain into the common or anterior cardinals. A lateral (or ventral) abdominal vein continues to receive the drainage of the pelvic appendages in amphibians and reptiles, but it carries blood to the hepatic or renal portal system rather than to the common cardinals. The liver and kidney process a great deal of the blood returning from the hind legs.

26. In lungfishes, amphibians, reptiles, and birds, a caudal vena cava evolves from part of the right hepatic vein and the subcardinals. It receives most or all of the drainage from the kidneys, posterior part of the trunk, and hind legs.

27. The caudal vena cava is extended caudal to the kidneys in mammals and directly receives the drainage of the pelvic region and hind legs. The renal portal system has been lost. The extension of the caudal vena cava results from the evolution of a supracardinal system of veins and a subsupracardinal anastomosis. Higher blood pressures in mammals deliver a large volume of blood to the kidneys by way of the renal arteries, and the loss of the renal portal system enables blood from the caudal parts of the body to return more rapidly to the heart.

28. By a fusion of its deep and superficial tributaries, the anterior cardinal system is converted to the external and internal jugulars and the cranial vena cava.

29. The lymphatic system in tetrapods is a subsidiary drainage system of the tissues that returns excess water and plasma proteins to the veins. It is absent in fishes.

30. Lymph nodes that are associated with the lymphatic system have evolved in mammals. With lymph nodules in the wall of the digestive system, they are part of the body's immune system.

REFERENCES

Adolph, E. F., 1967: The heart's pacemaker. *Scientific American,* 216:3:32–37.

Altman, P. L., and Dittmer, D. S., editors, 1971: *Biological Handbooks: Respiration and Circulation.* Bethesda, MD, Federation of American Societies for Experimental Biology.

Axelsson, M., Franklin, C. E., Lofman, C. O., Nilsson, S., and Grigg, G. C., 1996: Dynamic anatomical study of cardiac shunting in crocodiles using high-resolution angioscopy. *Journal of Experimental Biology,* 199:359–365.

Baker, M. A., 1979: A brain-cooling system in mammals. *Scientific American,* 240:5:130–139.

Barnett, C. H., Harrison, R. J., and Tomlinson, J. D. W., 1958: Variations in the venous system of mammals. *Biological Reviews,* 33:442–487.

Barry, A., 1951: The aortic arch derivatives in the human adult. *Anatomical Record,* 111:221–238.

Bourne, G. H., editor, 1980: *Hearts and Heart-Like Organs.* New York, Academic Press.

Burggren, W. W., and Johansen, K., 1987: Circulation and respiration in lungfishes (Dipnoi). *In* Bemis, W. E., Burggren, W. W., and Kemp, N. E., editors: *The Biology and Evolution of Lungfishes.* New York, Alan R. Liss.

Burggren, W. W., and Johansen, K., 1982: Ventricular hemodynamics in the monitor lizard, *Varanus exanthematicus:* Pulmonary and systemic separation. *Journal of Experimental Biology,* 96:343–354.

Burton, A. C., 1972: *Physiology and Biophysics and the Circulation,* 2nd edition. Chicago, Year Book Medical Publications.

Butler, P. J., and Jones, D. R., 1982: The comparative physiology of diving in vertebrates. *Advances in Comparative Physiology and Biochemistry,* 8:179–364.

Carey, F. G., 1973: Fishes with warm bodies. *Scientific American,* 228:36–44.

Caro, C. G., Pedley, T. J., Schroter, R. C., and Seed, W. A., 1978: *The Mechanics of Circulation.* Oxford, Oxford University Press.

DeLong, K. T., 1962: Quantitative analysis of blood circulation through the frog heart. *Science,* 138:693–694.

Farrell, A. P., 1991: Circulation of body fluids. *In* Prosser, C. L., editor: *Environmental and Metabolic Animal Physiology.* New York, Wiley-Liss.

Folkow, B., and Neil, E., 1971: *Circulation.* New York, Oxford University Press.

Foxon, G. E. H., 1955: Problems of the double circulation of vertebrates. *Biological Reviews,* 30:196–228.

Franklin, C. E., and Axelsson, M., 2000: An actively controlled heart valve. *Nature,* 406:847–848.

Groom, A. C., and Schmidt, E. E., 1990: Microcirculatory blood flow through the spleen. *In* Browdler, A. J., editor: *The Spleen. Structure, Function and Significance.* London, Chapman & Hall Medical.

Hicks, J. W., 1998: Cardiac shunting in reptiles: Mechanisms, regulation, and physiological functions. *In* Gans, C., and Gaunt, A. B., editors: *The Biology of the Reptilia,* volume 19. *Morphology G., Visceral Organs.* Chicago, University of Chicago Press.

Hicks, J. W., and Wang, T., 1996: Functional role of cardiac shunts in reptiles. *Journal of Experimental Zoology,* 275: 204–216.

Holmes, E. B., 1975: A reconsideration of the phylogeny of the tetrapod heart. *Journal of Morphology,* 147:209–228.

Huntington, G. S., and McClure, C. F. W., 1920: The development of the veins in the domestic cat. *Anatomical Record,* 20:1–31.

Ishimatsu, A., Hicks, J. W., and Heisler, N., 1996: Analysis of cardiac shunting in the turtle *Trachemys (Pseudemys) scripta:* Application of the three outflow vessel model. *Journal of Experimental Biology,* 199:2667–2677.

Johansen, K., and Burggren, W., editors, 1985: *Cardiovascular Shunts.* Copenhagen, Munksgaard.

Johansen, K., and Burggren, W., 1980: Cardiovascular function in the lower vertebrates. *In* Bourne, G. H., editor: *Hearts and Heartlike Organs: Comparative Anatomy and Development,* volume 1. New York, Academic Press.

Johansen, K., and Hanson, D., 1968: Functional anatomy of the hearts of lungfishes and amphibians. *American Zoologist,* 8:191–210.

Jones, D. R., and Shelton, G., 1993: The physiology of the alligator heart: Left aortic flow patterns and right-to-left shunts. *Journal of Experimental Biology,* 176:247–269.

Kampmeier, O. F., 1969: *Evolution and Comparative Morphology of the Lymphatic System.* Springfield, IL, Charles C Thomas.

Lillywhite, H. B., 1988: Snakes, blood circulation and gravity. *Scientific American,* 259:92–98.

O'Donoghue, C. H., and Abbott, E., 1928: The blood–vascular system of the spiny dogfish, *Squalus acanthias* Linne and *Squalus sucklii* Gill. *Transactions of the Royal Society of Edinburgh,* 55:823–890.

Randall, D. J., 1968: Functional morphology of the heart of fishes. *American Zoologist,* 8:179–189.

Robinson, T. F., Factor, S. M., and Sonnenblick, E. H., 1986: The heart as a suction pump. *Scientific American,* 254:84–91.

Runciman, S. I. C., Gannon, B. J., and Baudinette, R. V., 1995: Central cardiovascular shunts in the perinatal marsupial. *Anatomical Record,* 243:71–83.

Satchell, G. H., 1999: Circulatory System: Distinctive Attributes of the Circulation of Elasmobranch Fish. *In* Hamlett, W. C., 1999: *Sharks, Skates, and Rays: The Biology of Elasmobranch Fishes.* Baltimore, Johns Hopkins Press.

Satchell, G. H., 1991: *Physiology and Form of Fish Circulation.* Cambridge, Cambridge University Press.

Satchell, G. H., 1971: *Circulation in Fishes.* Cambridge, Cambridge University Press.

Shelton, G., and Burggren, W. W., 1976: Cardiovascular dynamics of the chelonia during apnoea and lung ventilation. *Journal of Experimental Biology,* 64:323–343.

Szidon, J. P., Lahiti, S., Lev, M., and Fishman, A. P., 1969: Heart and circulation of the African lungfish. *Circulation Research,* 25:23–38.

Wang, T., and Hicks, J. W., 1996: The interaction of pulmonary ventilation and the right–left shunt on arterial oxygen levels. *Journal of Experimental Biology,* 199:2121–2129.

White, F. N., 1969: Redistribution of cardiac output in the diving alligator. *Copeia,* 567–570.

White, F. N., 1968: Functional anatomy of the heart of reptiles. *American Zoologist,* 8:211–219.

Young, B. A., Lillywhite, H. B., and Wassersug, R. J., 1993: On the structure of the aortic valves in snakes (Reptilia: Serpentes). *Journal of Morphology,* 216:141–159.

20

The Excretory System and Osmoregulation

PRECIS

The excretory system includes the kidneys and their ducts. With the skin, gills, lungs, and special salt-excreting or salt-absorbing structures, it maintains the constancy of the internal environment by eliminating the nitrogenous wastes of metabolism and helping regulate the salt and water balances of the body. Its structure and evolution must be considered in the context of the environments in which vertebrates live and their level of metabolism.

OUTLINE

The composition of the interstitial fluid that bathes the cells of vertebrates must be maintained within narrow limits. Cellular metabolism, however, produces carbon dioxide, water, and nitrogenous wastes that diffuse into the interstitial fluid. Other metabolic processes add organic acids, phosphates, sulfate ions, and other materials. The interstitial fluid may gain or lose water or salts depending on the vertebrate's environment: fresh water, salt water, or land. Despite these destabilizing factors, the composition of the interstitial fluid is held relatively constant by exchanges between it and the blood, and the composition of the blood, in turn, is maintained by carefully controlled exchanges between the animal and its external environment. These exchanges occur through many organs, including the skin, liver, gills, lungs, kidneys, and special structures that excrete or absorb salt. The liver eliminates bile pigments, and the respiratory system, assisted sometimes by the skin, removes carbon dioxide. Here we will examine the kidneys and other organs that remove the nitrogenous and other wastes of metabolism and concurrently maintain the water and salt balances of the body. The removal of waste products of metabolism is called **excretion.** Excretion should not be confused with defecation, which is the removal primarily of undigested residues and bacteria from the digestive tract. Only the bile pigments in the feces are by-products of cellular metabolism.

Morphologically, the kidneys and their ducts are associated intimately with the reproductive system. Kidneys and gonads develop from adjacent tissues, and after the excretory or urinary ducts have developed, the reproductive system usually taps into them or their derivatives. These two systems often are treated together as the **urogenital system,** but we will first examine the excretory system and relate the reproductive system to it in the next chapter.

The Renal Tubules

Renal Tubule Structure and Function

The kidneys of vertebrates are of mesodermal origin. They develop from the pair of embryonic **nephric ridges** (Gr., *nephros* = kidney) or **mesomeres** that are located between the somites (epimeres) and the lateral plate (hypomere) (Figs. 4-16 and 20-1) and extend the length of the coelom. The segmentation of the somites extends into the nephric ridges of primitive vertebrates and divides them into a series of segmented **nephrotomes** (Gr., *tome* = cutting), each of which differentiates into a **nephron** or **renal tubule** (L., *renes* = kidneys). The segmentation may be limited to the anterior part of the nephric ridges. The renal tubules are the structural and functional units of the kidneys. The proximal end of a representative tubule forms a two-layered, cup-shaped capsule of simple squamous epithelium that is known as **Bowman's capsule** or a

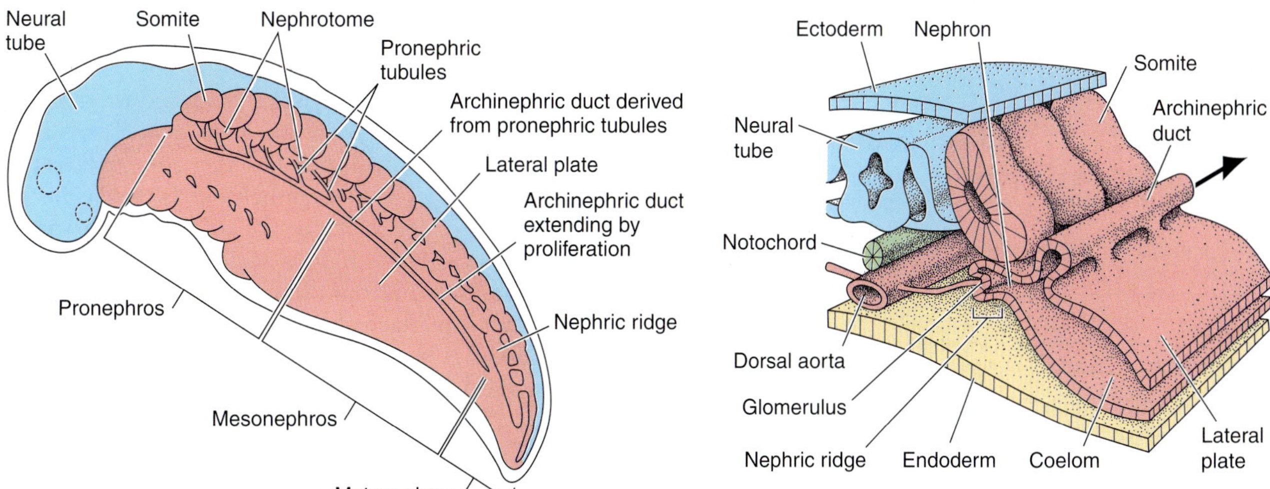

FIGURE 20-1
The embryonic development of the nephrons. *A,* A lateral view of an amniote embryo, showing the sequential differentiation (from anterior to posterior) of nephrons in the nephric ridge. The nephric ridge lies between the somites and the lateral plate mesoderm. *B,* A stereodiagram of the nephric ridge and adjacent structures. *(A, After Pough et al.; B, after Williams et al.)*

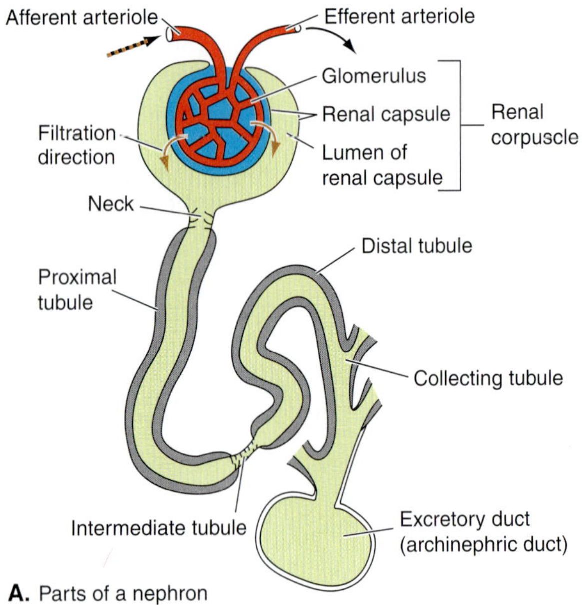

A. Parts of a nephron

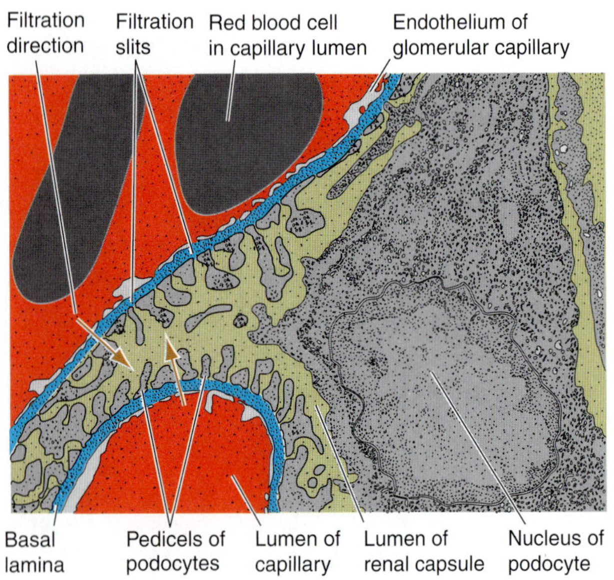

B. Glomerular capillary/renal capsule interface

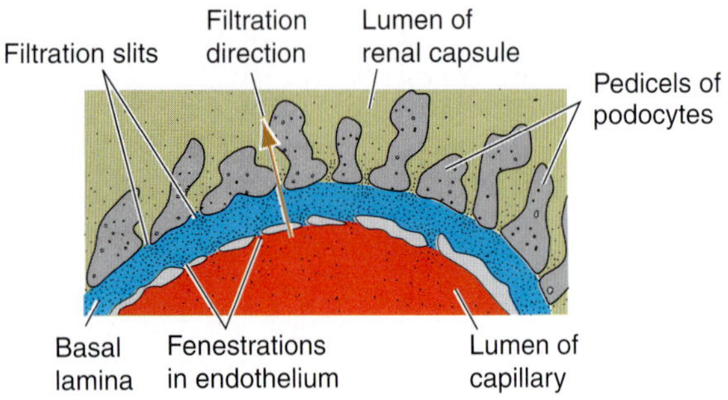

C. Enlargement of the interface

renal capsule (Fig. 20-2*A*). The capsule envelops a tangled knot of capillaries known as a **glomerulus** (L., *glomus* = ball). The glomerulus receives an **afferent arteriole** from the renal artery and is drained by an **efferent arteriole.** Together, the renal capsule and glomerulus form a **renal corpuscle.** The rest of each tubule is composed of a simple epithelium, but the nature of the epithelial cells and the length and pattern of the nephron vary considerably among vertebrates according to the excretory and osmoregulatory problems that each group faces. Often, especially in anamniotes, a short **neck** of ciliated cells connects the renal corpuscle with a thicker **proximal tubule** of secretory or absorptive cells. A thin **intermediate tubule,** which varies in length and sometimes contains ciliated cells, may lie between the proximal and **distal tubules.** In amniotes, and some anamniotes, **collecting tubules** receive a number of renal tubules and, in turn, enter the excretory duct.

The renal corpuscles are **filtration mechanisms** from which a filtrate of the blood plasma enters the renal tubule. Two mechanisms increase the amount of filtrate formed in the renal corpuscles compared with capillary beds in most other organs. First, the capillaries of the glomerulus lie between two arterioles (Fig. 20-2*A*). The diameter of the efferent arteriole is smaller than that of the afferent arteriole, and its contraction under some conditions increases the hydrostatic pressure of the blood in the glomerular capillaries. This increases filtration pressure, which is the difference between the hydrostatic and the colloidal osmotic pressures of the blood. Second, diffusion distance is short because the walls of the capillaries and renal capsule are exceptionally thin. Details have not been studied in all vertebrates, but electron-microscope studies of mammalian renal corpuscles have shown that the endothelial cells that form the walls of the capillaries are perforated by small slits, i.e., they are **fenestrated** (Fig. 20-2*B* and *C*). The epithelial cells that form the adjacent wall of the renal capsule are called **podocytes.** Podocytes do not lie tightly against the basal lamina that separates them from the capillary endothelial cells; rather, they are attached only by footlike pedicels with **filtration slits** between them. A similar condition was seen in the "solenocytes" of amphioxus (Chapter 2).

FIGURE 20-2
The basic structure of the vertebrate nephron. *A,* The parts of the nephron of an amphibian. *B,* A drawing of an electron micrograph of the interface between the renal capsule and two adjacent capillaries. *C,* An enlargement of the interface. Basal lamina (blue) is not shown in *A* because it is too small. *(A, From Walker and Homberger; B and C, after Kessel and Kardon.)*

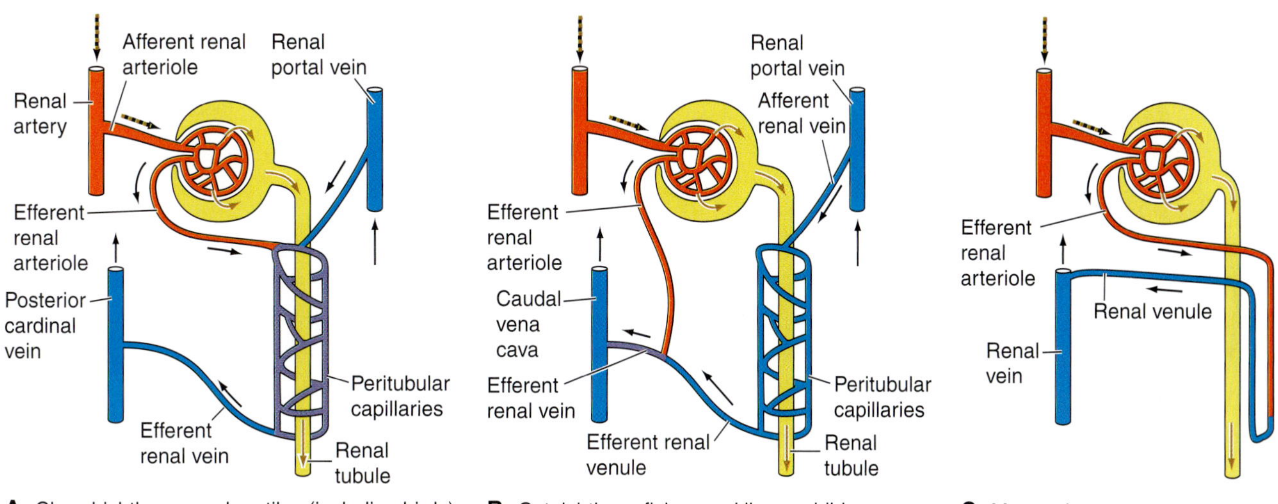

FIGURE 20-3
The blood supply to the nephrons of various vertebrates. *A,* Chondrichthyans, reptiles and birds. *B,* Osteichthyan fishes and lissamphibians. *C,* Mammals. *(From Walker and Homberger.)*

The cavities of the glomerular capillaries and renal capsule are separated in many places only by the very thin basal lamina of the epithelial cells. This lamina is composed of delicate collagen fibers and glycoproteins that we believe are secreted by both the endothelial cells and the podocytes. During filtration, most of the large plasma proteins are held back, but filtration is nonselective for smaller molecules, which occur on the filtrate in the same proportion as in the blood. The filtrate contains nitrogenous wastes and other materials that must be eliminated, as well as glucose, amino acids, and other molecules that should be saved. Whether water or salts need to be conserved or eliminated depends on the animal's environment.

The renal tubules and collecting tubules are surrounded by **peritubular capillaries** that receive their blood in different ways among the different groups of vertebrates (Fig. 20-3). The peritubular capillaries of chondrichthyan fishes, reptiles, and birds receive blood from both the efferent renal arteriole and afferent renal vein leaving, respectively, the glomeruli and renal portal vein (Fig. 20-3*A*). Those of osteichthyan fishes and amphibians receive blood only from the renal portal vein because the efferent renal arteriole enters the renal vein directly (Fig. 20-3*B*). The renal portal system is lost in mammals, so their peritubular capillaries receive blood only from the efferent renal arterioles (Fig. 20-3*C*). In all cases, blood leaving the peritubular capillaries drains into renal veins. As the filtrate passes down the tubules, substances that must be saved are selectively **reabsorbed** and enter the capillaries around the tubules. Reabsorption of some products is active and requires the tubular cells to do metabolic work; for other products, reabsorption is by passive diffusion. In some species, the amount of waste products in the filtrate is augmented by **selective secretion** by the tubular cells. Waste products are concentrated in this way as the filtrate passes down the tubules to be discharged as **urine.**

Renal Tubule Evolution

Most vertebrates have renal tubules of the type described, in which the renal tubules do not connect with the coelom and the glomeruli are surrounded by renal capsules. These are known as **internal glomeruli** (Fig. 20-4*D*). The first few tubules that develop at the anterior end of the nephric ridges of ammocoetes and some larval amphibians open into coelomic recesses through a ciliated **nephrostome,** and the glomeruli bulge into this recess. These are **external glomeruli** (Fig. 20-4*B*). In larval vertebrates possessing external glomeruli, the filtrate is discharged into the coelomic recesses and is drawn by ciliary action into the nephrostomes. The glomeruli are internal in adult vertebrates, but many of the tubules of adult elasmobranchs, primitive actinopterygians, and many amphibians retain nephrostomes and the renal capsule connects to the coelom through a narrow coelomic funnel (Fig. 20-4*C*). Nephrostomes are absent from other adult vertebrates.

This variation in nephron types and their pattern of distribution suggest an evolutionary sequence. Ancestral craniates probably had an external glomerulus and

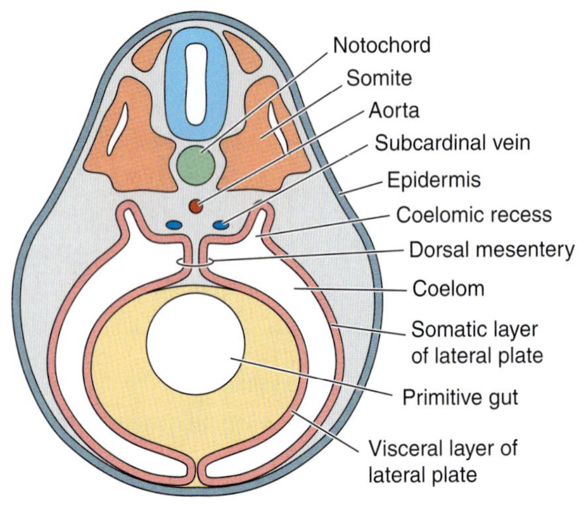

A. Section of vertebrate embryo

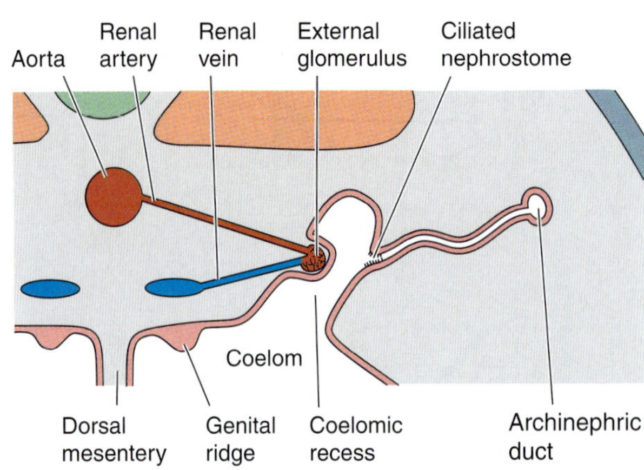

B. Ammocoetes and larval lissamphibians

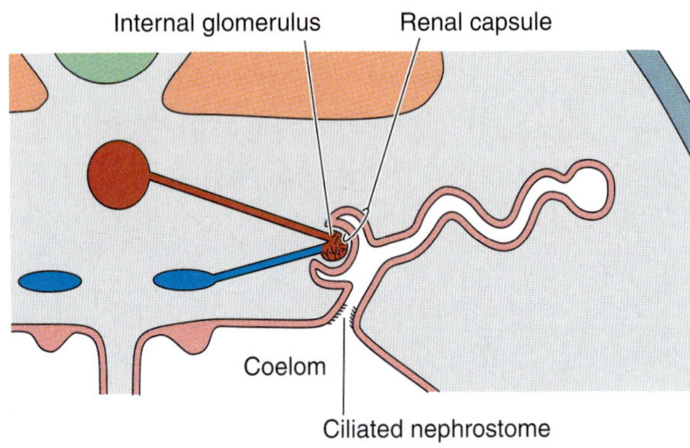

C. Elasmobranchs and some actinopterygians

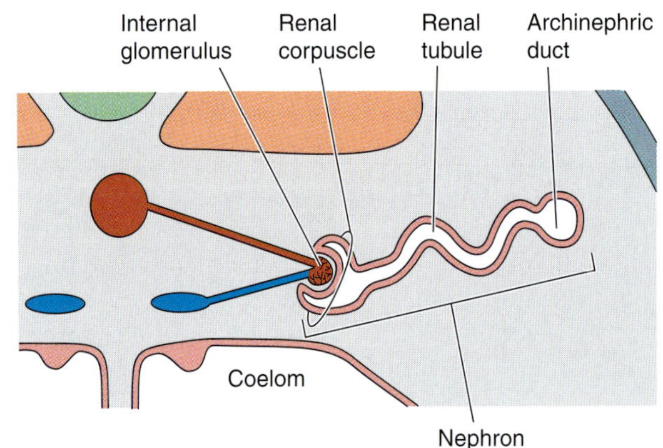

D. Most other vertebrates

FIGURE 20-4
A, Cross section of a primitive vertebrate embryo, showing the coelomic recess that contributes to the formation of the renal capsule. *B–D,* The probable sequence in the evolutionary development of the renal capsule. *B,* External glomeruli as seen in the ammocoete larva of the lamprey and some larval lissamphibians. *C,* The internal glomeruli of elasmobranchs and some actinopterygians retain a connection to the coelom. *D,* The internal glomeruli of most other vertebrates. *(From Walker and Homberger.)*

nephrostomes, as do the first few to develop in very primitive craniates. The vascular filtration surface was overlain with podocytes, which allowed the escape of many small molecules, including water, into the coelom. This filtrate was drawn off through coelomic connections (nephrostomes) into tubules where a selective reabsorption occurred to form urine. This type of excretory organ was once thought to be unique to craniates, but it is now known to be widespread among invertebrates, including urochordates and cephalochordates, and especially in their larvae (Ruppert, 1994). The mechanism would become more efficient as the coelomic recess into which each glomerulus discharged became a part of the tubule, that is, grew around the glomerulus as a renal capsule. The glomerulus becomes internal. The nephrostomes were lost during subsequent evolution, leaving the type of renal tubule found in most vertebrates.

Kidney Development and Evolution

The Holonephros

Like somites and many other embryonic structures, nephrons differentiate sequentially from anterior to posterior during embryonic development along most of the nephric ridge, but not all of them become func-

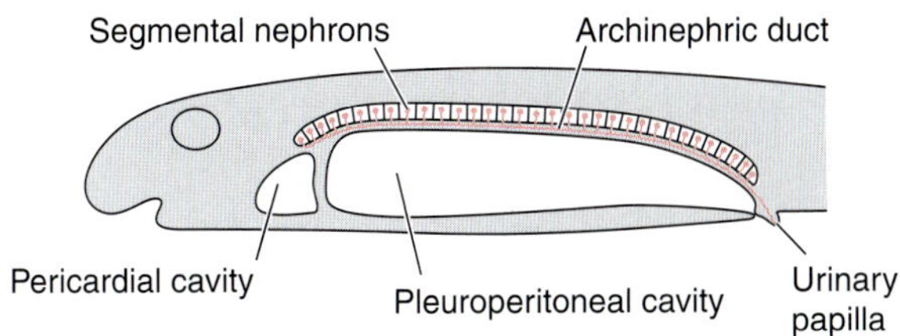

A. Theoretical holonephros

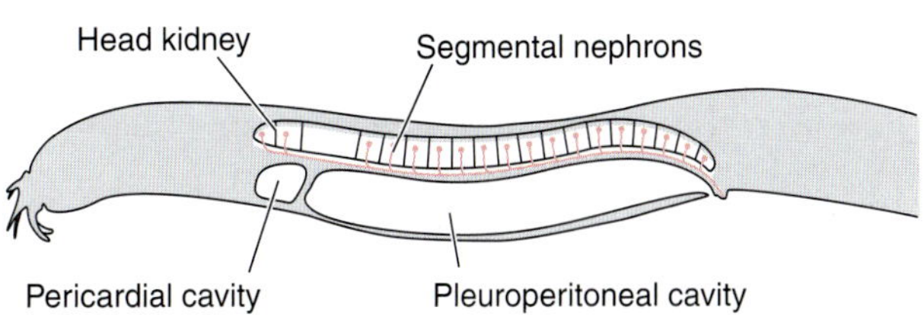

B. Primitive opisthonephros

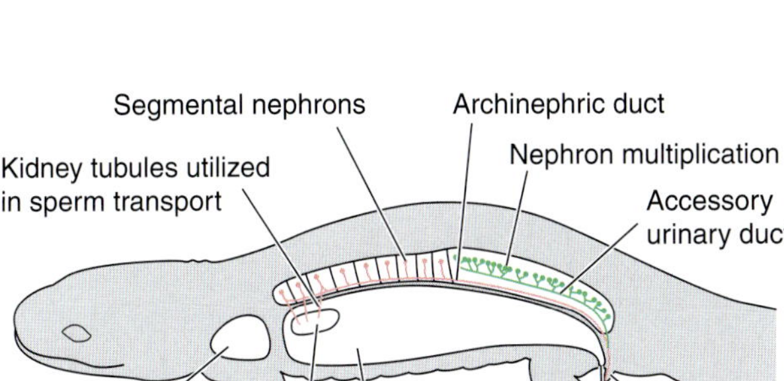

C. Advanced opisthonephros

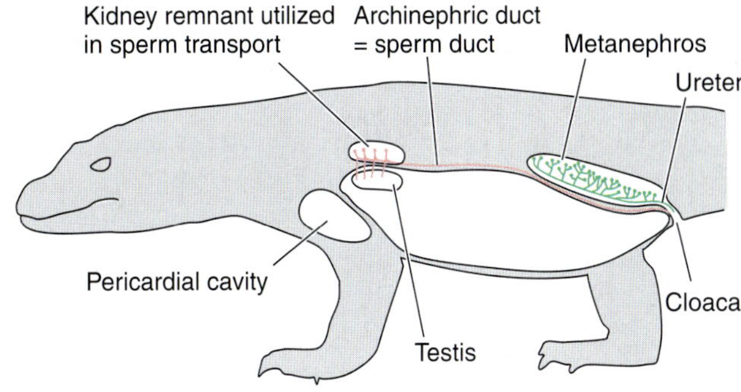

D. Metanephros

FIGURE 20-5
Lateral views of the evolution of the kidney and its ducts in adult craniates. *A,* The theoretical holonephros. *B,* The primitive opisthonephros found in a hagfish. *C,* The advanced opisthonephros characteristic of most fishes and amphibians. *D,* The metanephros of amniotes.

tional in most vertebrates (Fig. 20-1*A*). Analysis of the variation in the tubules that do become functional in vertebrate larvae and adults has led anatomists to postulate that in ancestral vertebrates, segmental tubules, all of which were functional, developed from the entire nephric ridge and drained into an archinephric duct. This hypothetical ancestral kidney is called an **archinephros,** or **holonephros** (Fig. 20-5*A*). Larval hagfishes and caecilians have kidneys that resemble the holonephros. Segmental tubules develop from the entire nephric ridge, but not all are functional concurrently.

The Pronephros

In all vertebrate embryos, the kidney begins with the differentiation of a few renal tubules from the anterior end of the nephric ridge overlying the pericardial cavity. The distal ends of these tubules unite to form an archinephric duct that rapidly grows caudally to the cloaca. This early-developing embryonic kidney is called the **pronephros** (Figs. 20-1*A* and 20-6*A*). A dozen or more pronephric tubules appear in some anamniotes, but most species have only four or five, and even some of these soon regress. Only one to three pronephric tubules develop in amniotes. In anamniotes, pronephric tubules are segmental and usually have coelomic funnels connecting to the coelom. Many of these tubules form a functional pronephric kidney in the embryos and larvae of hagfishes and lampreys, many bony fishes, and amphibians. But the pronephros is not functional in the embryos of chondrichthyan fishes or amniotes. Its only role in these vertebrates is to initiate the formation of the archinephric duct. Adult hagfishes and a few teleosts retain a pronephros, but the extent to which it continues to function is unclear. It is called a **head kidney** because it is located far forward over the pericardial cavity and is separated from the more caudal and functional kidney by a gap in the nephric ridge (Fig. 20-5*B*). The pronephros has been lost in the adults of other vertebrates.

The Mesonephros

Kidney tubules fail to differentiate in a few body segments caudal to the pronephros, so in all vertebrates, a gap is present between the pronephros and more caudal renal tubules. As the pronephros regresses, the archinephric duct induces the sequential differentiation of tubules in the more caudal parts of the nephric ridge. These tubules tap into the archinephric duct and become functional (Fig. 20-6*B*). The new tubules at first are segmental, and those of anamniotes retain

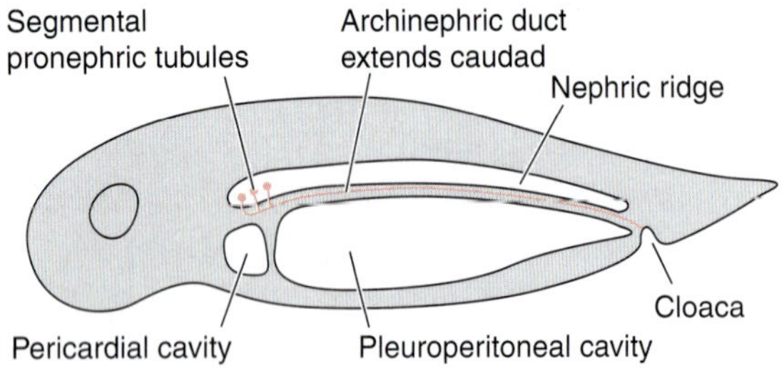

A. Pronephros in early embryo

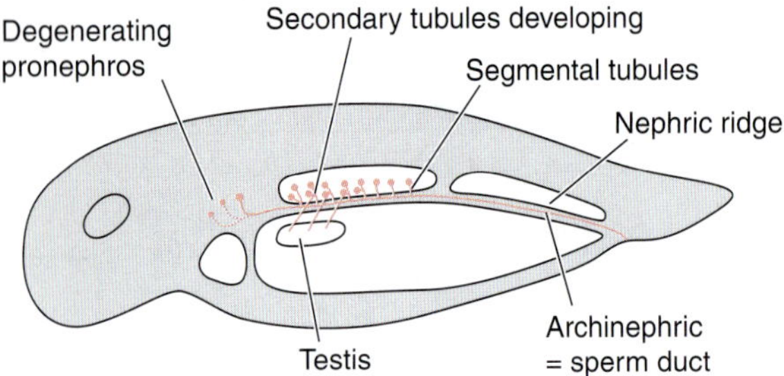

B. Mesonephros in intermediate embryo

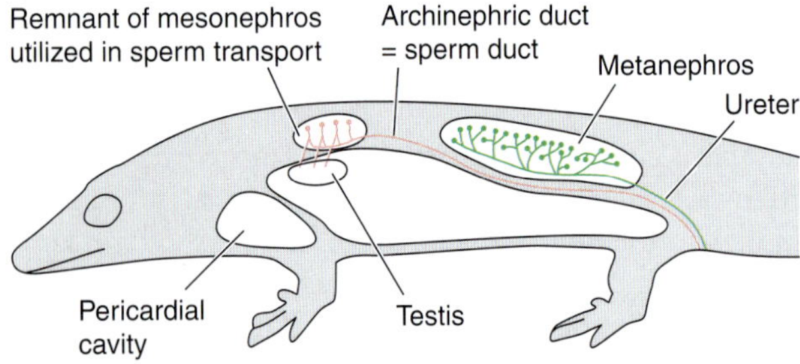

C. Metanephros in late embryo and adult

FIGURE 20-6
A–C, Lateral views of the sequence of kidneys that occurs during the embryonic development of an amniote.

coelomic funnels. As development continues, additional tubules usually grow out from the base of each primary one, and the segmental nature of the kidney becomes obscure. Secondary and tertiary tubules lack coelomic funnels. Tubules having a common origin unite to form a **collecting duct** before entering the archinephric duct.

Tubules that differentiate in the middle part of the nephric ridge form a kidney called the **mesonephros** (Figs. 20-1*A* and 20-6*B*). This kidney functions in the embryos and larvae of all vertebrates, but the degree to which it functions in mammalian embryos is inversely related to the amount of excretion that occurs through the placenta. The mesonephric tubules are distinct from tubules that develop in adjacent parts of the nephric ridge in the embryos of amniotes. The cranial mesonephric tubules acquire a connection with the developing gonad in most groups of vertebrates. Teleosts are a notable exception.

The Opisthonephros

In amniote embryos, a gap is present between the mesonephros and tubules that differentiate in the caudal part of the nephric ridge, so the mesonephros is recognizable as a distinct entity (Fig. 20-6*B*). In contrast, in anamniotes, the sequential differentiation of kidney tubules continues without interruption throughout the nephric ridge. New tubules function as soon as they tap into the archinephric duct. Adult hagfishes have functional, segmental renal tubules with coelomic funnels throughout most of the nephric ridge (Fig. 20-4*B*). This kidney in adult hagfishes approaches in its structure the holonephros postulated for ancestral vertebrates, except for the specialization of the pronephric region and the gap in tubule sequence just caudal to the pronephros. Because this kidney includes the embryonic mesonephros and also tubules that develop in the caudal part of the nephric ridge, it is called an **opisthonephros** (Gr., *opisthen* = behind, at the back). More specifically, it is a **primitive opisthonephros** because of the primitive, segmental nature of the tubules (Fig. 20-5*B*).

Other fishes and amphibians have an **advanced opisthonephros** (Fig. 20-5*C*). Renal tubules multiply, especially in the caudal portion of the kidney, so they are no longer segmentally arranged. As a consequence, the caudal part of the kidney usually is enlarged, and most urine production occurs here. The cranial part of the organ, which is derived from the embryonic mesonephros, is slender in chondrichthyan fishes and urodeles, produces little or no urine, and in males receives sperm. The opisthonephric kidneys of teleosts are extremely variable. Some are long and slender and the two kidneys are partly united; some are divided into more or less separate anterior and posterior parts; but others are short, compact organs confined to the caudal part of the trunk. Frogs have short trunks, and their kidneys also are short, compact organs. One or more **accessory urinary ducts** may bud off the archinephric duct in elasmobranchs and urodeles and enter the thickened, urinary portion of the opisthonephros (Fig. 20-5*C*). As accessory urinary ducts develop, they may separate completely from the archinephric duct and enter the cloaca independently. Accessory urinary ducts are more likely to be present in males than in females. When present, most of the urine is transported in them, and the archinephric duct transports primarily sperm.

The Metanephros

Although the embryonic mesonephros contributes to the adult opisthonephros in most anamniotes, in amniotes it is a transitional kidney and functions only in their embryos (Fig. 20-6*B*). Later in the development of amniotes, a **ureteric bud** extends from the caudal end of the archinephric duct, grows into the caudal end of the nephric ridge, and branches extensively (Fig. 20-7). The branching ureteric bud induces the differentiation of many renal tubules that tap into it. The ureteric bud itself forms the **collecting tubules** and the **ureter** that drain the adult kidney. This type of kidney, called a **metanephros,** occurs in all adult amniotes (Figs. 20-5*D* and 20-6*C*). It is homologous only to the posterior portion of the opisthonephros. A metanephros is always drained exclusively by the ureter, which, in reptiles and birds, separates from the archinephric duct and enters the cloaca independently. In therian mammals, it enters the **urinary bladder**.

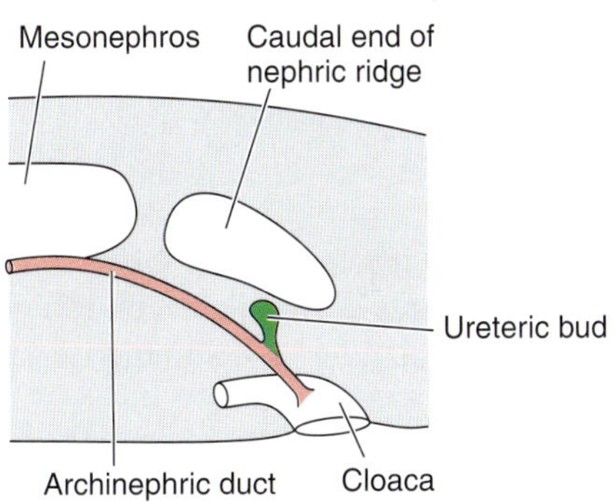

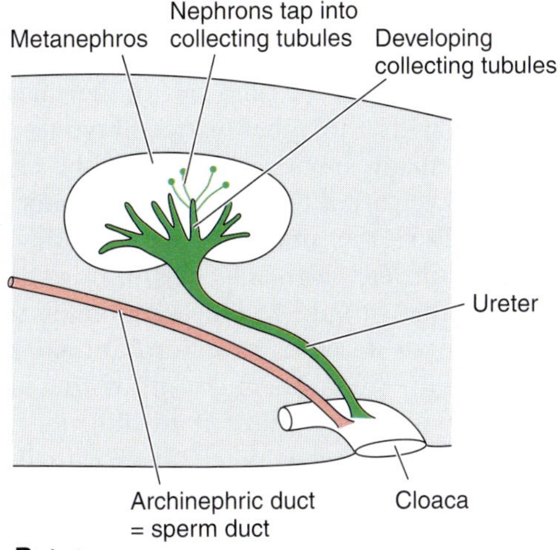

FIGURE 20-7
A and *B*, Lateral views showing how a ureteric bud induces the formation of nephrons in the metanephric region of the nephric ridge.

The ureter develops in the same way as an accessory urinary duct and is its homologue. As the metanephros and ureter develop and become functional, the amniote mesonephros and archinephric duct regress, except for those parts in males that are connected to the testis. The cranial mesonephric tubules and archinephric duct become part of the system of ducts that carry sperm (Chapter 21).

The kidneys develop dorsal to the coelom, but they commonly bulge into it as the tubules multiply during development. The dorsal surfaces of the kidneys lie against back muscles and are not covered by coelomic epithelium. This position is described as **retroperitoneal** (L., *retro* = backward). Although the metanephric kidneys develop from the caudal part of each nephric ridge, they migrate anteriorly during development and come to lie retroperitoneally just caudal to the liver. Much of this migration results from differential growth of parts of the body in this area.

In mammals, the kidneys are usually compact, bean-shaped organs and tend to be subdivided into many lobes in large species, including many ungulates, bears, seals, and cetaceans. The subdivisions shorten the length of the collecting tubules and thus may facilitate the flow of urine.

During the evolution of the kidney, urinary functions tend to shift and become concentrated in the posterior part of the nephric ridge (Fig. 20-5). An ancestral holonephros has been replaced by an opisthonephros as the functional adult kidney of most adult anamniotes. Urinary functions frequently are concentrated in the caudal part of the opisthonephros, and one or more accessory urinary ducts may drain this region. A metanephros replaces the opisthonephros as the adult kidney in amniotes. A posterior shift also occurs in urinary functions during the embryonic development of amniotes (Fig. 20-6). A very short and transitory pronephros forms the archinephric duct. The mesonephros uses this duct and becomes the functional embryonic kidney. It is replaced in adults by a metanephros.

The Urinary Bladder and Cloaca

The caudal ends of either the archinephric ducts or the accessory urinary ducts are slightly enlarged in many fishes, and these areas are sometimes called **urinary bladders,** or a **urogenital sinus** (Fig. 20-8*A*). In some species, the caudal ends of the left and right ducts are conjoined, so the bladders are partly united. All of these sacs are small, and their functional significance unknown. A continuous urine discharge occurs

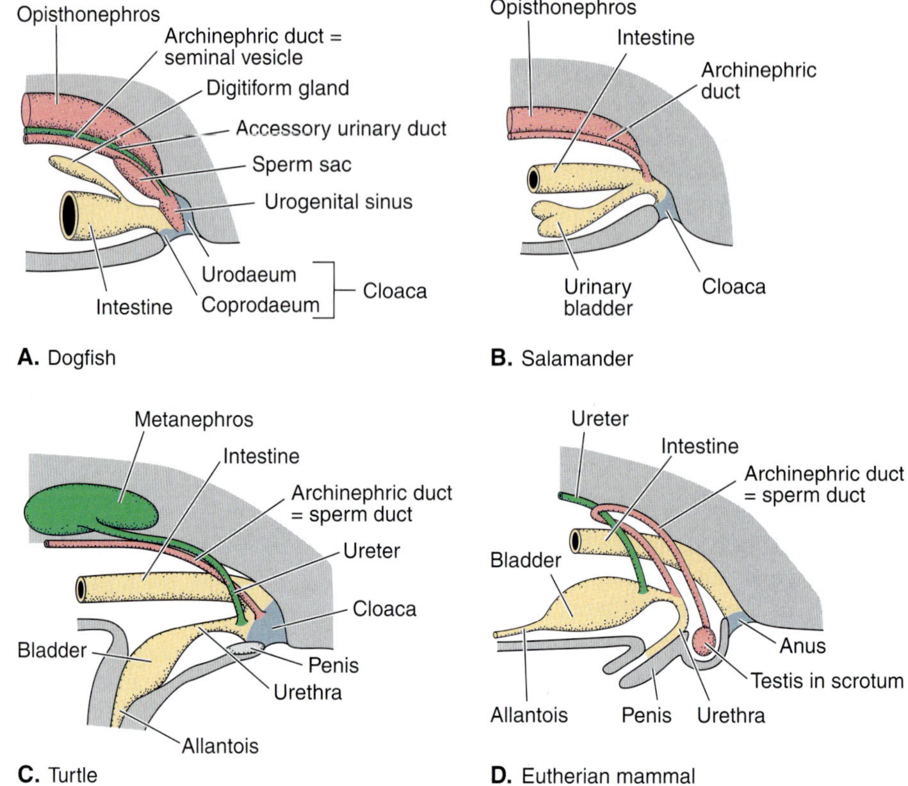

FIGURE 20-8
Lateral views of the cloacal region of representative male vertebrates, showing the relationship to each other of terminations of the intestine, urinary ducts, and associated structures. The way the cloaca becomes divided in eutherian mammals so as to continue the opening of the intestine and urinary ducts to the surface is considered in the next chapter. *A,* A dogfish. *B,* A salamander. *C,* A turtle. *D,* A eutherian mammal.

in the freshwater environment. In elasmobranchs, the ducts and sinus open into the cloaca dorsal and caudal to the opening of the digestive tract. In many fish species, the cloaca is partially divided by folds on its lateral walls into a dorsal **urodaeum,** receiving urine and genital products, and a ventral **coprodaeum** (Gr., *kopros* = dung + *hodaion* = way), receiving feces.

Amphibians have large, commonly bilobed urinary bladders that evaginate from the midventral part of the cloaca into the body cavity (Fig. 20-8*B*). The epithelium of the urinary bladder is of endodermal origin, but the wall becomes muscularized and vascularized by mesoderm. The epithelium is an unusual type known as transitional, which allows for changes in the lining of structures that expand and contract. **Transitional epithelium** is composed of two layers of cells that are connected to the basement membrane and a third layer wedged in between the tops of the cells of the basal layer. The shapes of the cells and their distributional configuration give the epithelium the appearance of being stratified when the bladder is empty. When the bladder is fully distended, the cells are spread apart, resulting in two layers of flattened cells: one basal layer of triangular cells alternating with inverted triangles and a second layer of very flattened cells wedged in between the basal cells. The archinephric ducts open into the dorsal part of the cloaca, and urine flows into the bladder by gravity and the contraction of the cloacal wall. Urinary discharge is not continuous in the terrestrial environment. Urine accumulates temporarily in the amphibian urinary bladder, and some water is reabsorbed from here.

A homologous cloacal evagination develops in the embryos of amniotes and expands beyond the confines of the body into the extraembryonic coelom (Fig. 20-8*C*). This outgrowth is the **allantois,** and with the embryonic chorion, it forms an important embryonic excretory and respiratory organ (Chapter 4). At hatching or birth, the extraembryonic part of the allantois is lost, together with the other extraembryonic membranes, but in many species, the part of the allantois within the body enlarges to become the **urinary bladder** and its duct, the **urethra.** Some frogs, a few teleosts, most turtles, *Sphenodon,* and lizards have urinary bladders that, in addition to temporarily storing urine, are a site for water reabsorption. Certain aquatic turtles can pump water in and out of the bladder through the cloaca and utilize it in aquatic gas exchange. Indeed, some turtle species have paired **accessory urinary bladders** that evaginate from the cloaca for this purpose (Fig. 17-8). The urinary bladder has been lost in the adults of other reptiles and most birds, which excrete nitrogenous waste products as a semisolid paste. Mammals produce a liquid urine that is stored temporarily in a urinary bladder, although no water is reabsorbed.

The ureters of most reptiles, birds, and monotremes continue to open in the dorsolateral wall of the cloaca, and urine must cross the cloaca to enter the bladder in those species having one. The division of the cloaca in marsupials and eutherian mammals (Chapter 21) sorts out urinary and reproductive products and undigested residues. The ureters join the bladder, and the bladder is drained by a urethra that continues to the body surface or to a urogenital canal (Fig. 20-8*D*).

Excretion and Osmoregulation

Nitrogen Excretion

Although some nitrogenous wastes derive from the metabolism of nucleic acids, most come from the deamination of amino acids. This process occurs primarily in the liver. Each removed amino group (NH_2) picks up a hydrogen ion and transforms into an ammonia molecule (NH_3; Fig. 20-9). Ammonia is toxic, so it must be either removed quickly from the tissues and body or converted to a less toxic substance. Because ammonia is very soluble in water, it can be flushed out of the tissues by a rapid water turnover, that is, by a rapid entry and removal of water from the body. This occurs in freshwater teleosts. If the requisite enzymes are available, ammonia can be converted to other materials. Urea, which contains two amino groups, is far less toxic than is ammonia and need not be removed so rapidly. Alternatively, more amino groups can be combined to form the larger uric acid molecule. Uric acid has a very low solubility in water, so it precipitates from solution as minute crystals. It is chemically inert and nontoxic. It can be stored in tissues, but it is discharged by adult vertebrates as a paste-like mass with little or no water being required to carry it off. Other forms of nitrogenous wastes occur in small amounts, but ammonia, urea, and uric acid are the primary ones. Because they require different amounts of water for their elimination, the molecular form in which a vertebrate eliminates nitrogen is closely linked to the availability of water and so to the problem of osmoregulation.

The Environment of Ancestral Craniates

Because excretion and osmoregulation are coupled, the structure of the renal tubules of primitive living craniates may help us understand the environment of ancestral craniates and the circumstances in which the renal tubules evolved. If craniates evolved from some earlier chordate group (tunicates or cephalochordates), as we believe, their remote ancestors must have resembled the early chordates in being marine. The early chordates, like other marine invertebrates, have an inorganic salt content in their bodily fluids that is similar to the salt water in which they live, so their bodily fluids have the same osmotic pressure as does sea water. Their bodily fluids are **iso-osmotic** to their environment.

Possibly, the earliest craniates also were marine. Evidence is suggested by the contemporary but primitive hagfish, which (1) appears to be more primitive than all known living vertebrates, (2) is marine, and (3) the bodily fluids of which are nearly iso-osmotic with sea water. The hagfish is the only craniate to have an inorganic salt content in its plasma equivalent to that of sea water (Table 20-1). It differs from the early chordates and resembles many vertebrates in having renal tubules with exceptionally large renal corpuscles that filter a great deal of water from the blood (Fig. 20-10). This water need not be reabsorbed because water easily reenters the iso-osmotic bodily fluids. Why an iso-osmotic animal needs to eliminate so much water through the kidneys when water diffuses so easily through the gills is not entirely clear. Kidneys do, however, supplement the gills in flushing out ammonia. Limited paleontological evidence also indicates that the earliest vertebrates may have been marine. Whether a fossil bed was formed in marine or freshwater deposits usually can be determined by the presence or absence of certain types of invertebrates. Echinoderms, for example, have been a marine group throughout their history, and their fossils are found only in marine deposits. The earliest known fragments of the armor of early jawless vertebrates come from deposits that have been interpreted in this way as marine or brackish. But remember that the remains of freshwater organisms can be carried by rivers into marine deposits.

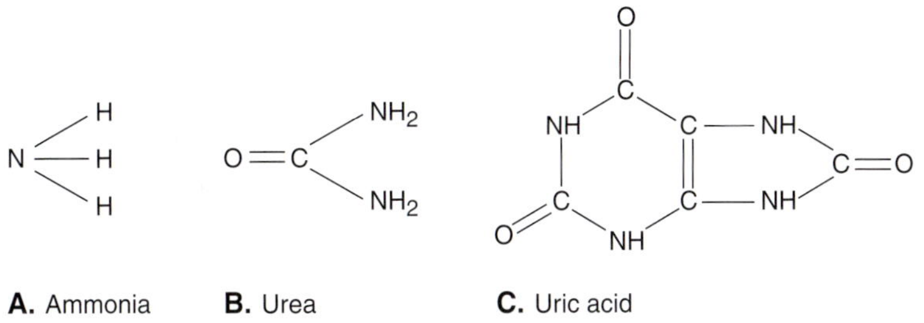

N—H, H, H

A. Ammonia

O=C(NH2)(NH2)

B. Urea

O=C, NH, C—NH, C=O, C—NH, NH, O

C. Uric acid

FIGURE 20-9
The structural formulas of the three most common types of nitrogenous excretory products.

TABLE 20-1 Osmotic Concentration of Inorganic Salts and Urea in Plasma*

	Habitat	Osmotic Concentration (mOsm/L)	Urea (mOsm/L)
Sea water		≈1000	
Hagfish *(Myxine)*	Marine	1152	
Lamprey *(Petromyzon)*	Marine	317	9
Dogfish *(Squalus)*	Marine	1000	354
Freshwater ray *(Potamotrygon)*	Fresh water	308	1+
Goldfish *(Carassius)*	Fresh water	259	
Toadfish *(Opsanus)*	Marine	392	
Eel *(Anguilla)*	Marine	371	
Eel *(Anguilla)*	Fresh water	323	
Coelacanth *(Latimeria)*	Marine	1181	355
Frog *(Rana)*	Fresh water	200	1+

*The amount of urea included in the osmotic concentration is shown separately if it is 1 milliosmol per liter or greater and therefore osmotically significant. (Data from K. Schmidt-Nielsen.)

Whether the first craniates were marine or fresh water, it is clear that they adapted to a freshwater environment early in their evolutionary history. Except for marine cartilaginous fishes and a few other species (discussed later), all vertebrates have an inorganic salt content in their plasma that is substantially less than that of salt water (Table 20-1). Their bodily fluids are **hypo-osmotic** to sea water. This means that vertebrates have evolved physiological mechanisms to become ionically independent from their environment, whether this is salt water, fresh water, or land. Maintaining ionic independence (osmoregulation) requires the expenditure of considerable energy to eliminate or conserve water and salts. It is unlikely that ionic independence, and the adaptation of cellular enzymes to function at lower levels of inorganic salts than those

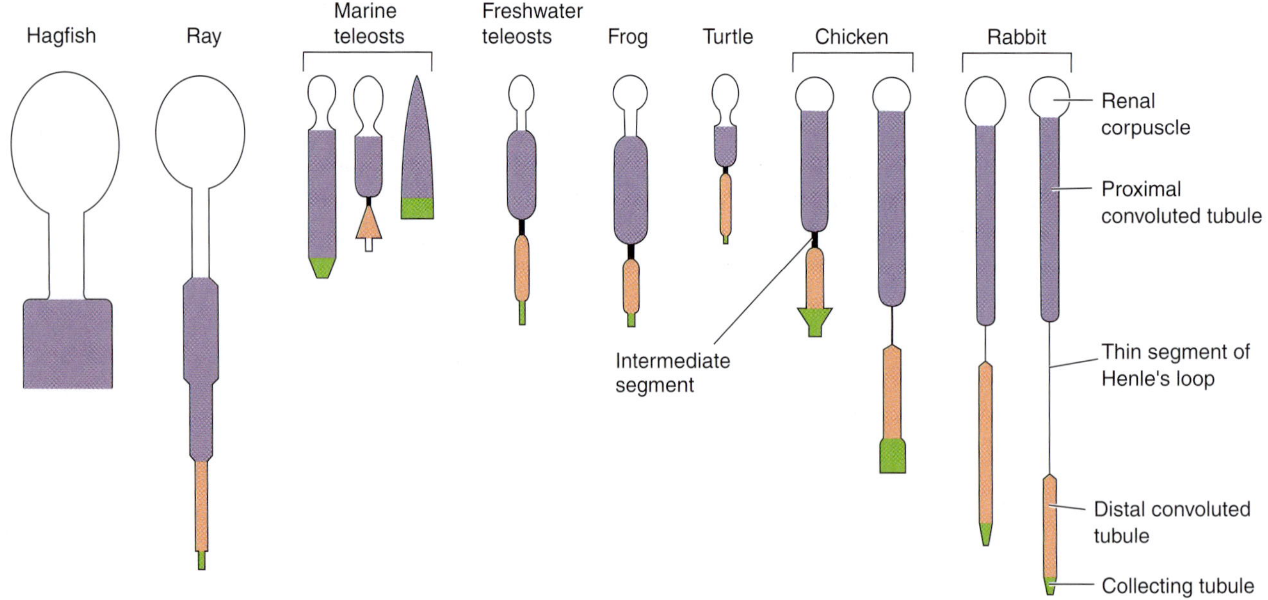

FIGURE 20-10
Nephron structure in representative craniates. The thin and intermediate segments are ciliated. *(After Prosser.)*

present in sea water, would have evolved except as an adaptation to life in fresh water.

Renal corpuscles, which occur in nearly all craniates, are well suited to remove the excess water that floods the tissues in a freshwater fish. The presence of renal corpuscles in hagfish suggests that renal corpuscles may have evolved first in the marine environment as a means of ensuring a high water turnover for the elimination of ammonia. Their presence in the earliest craniates would have facilitated the entry of craniates into fresh water, where much water would diffuse into the body and have to be pumped out. The fossil record also supports the hypothesis that early craniates adapted to fresh water: most of the primitive jawless craniates and other early fish fossils are found in mostly fresh water and only some marine deposits.

Freshwater Fishes

About 90% of the nitrogenous wastes of freshwater fishes is excreted as ammonia; most of the rest, as urea (Table 20-2). The gills are the primary excretory organs, for about six times as much nitrogen is lost by diffusion through them as through the kidneys. The kidneys supplement the gills in nitrogen excretion and are essential in osmoregulation, especially in removing water. The bodily fluids of a freshwater fish have a salt content far greater than that of the environment in which it lives, so water enters the body by osmosis through any diffusible surface, such as the gills, gut lining, and sometimes the skin. A freshwater fish must produce a copious, dilute urine that is hypo-osmotic to its bodily fluids. The renal tubules, with their large renal corpuscles, are admirably suited for water removal because they produce a large filtrate volume. The tubule contains two ciliated segments: (1) the narrow part immediately following the capsule and (2) the intermediate segment (Fig. 20-10). The cilia work like water pumps to move the filtrate at a high velocity.

Although a freshwater fish produces a copious and dilute urine that flushes out water and helps eliminate ammonia, water entry into the body must be limited so that more does not enter than can be removed (Fig. 20-11*A*). Mucus secreted by the skin reduces osmosis through the body surface. Although some water is unavoidably swallowed during feeding, the fish minimizes water uptake through the gut by not drinking extra water.

Bodily salts also are lost by a freshwater fish by diffusion through the gills. Additional salts are lost through the filtrate, but some salts are actively reabsorbed by distal parts of the renal tubules. Lost salts are regained in the food and by the active uptake of the limited amount of salt available in fresh water by special cells on the gills, called **ionocytes.** A freshwater fish, however, cannot obtain enough salt to maintain a salt level as high as that in sea water. Its cells must be able to function at lower salt levels. Also, a salt level that is too high is detrimental in fresh water.

Saltwater Teleosts

When early vertebrates reentered the sea from the freshwater habitat to which they had become adapted, they were confronted with a different set of problems. Because the salt content of their bodily fluids was less than that of salt water, water would leave their bodies by osmosis. An additional problem would be taking in too much salt. Actinopterygians and cartilaginous fishes have evolved different solutions to these problems. Marine teleosts conserve some water by a reduction in the size of their renal corpuscles (Fig. 20-10), but because they continue to eliminate most of their nitrogen as ammonia, they need a high water turnover. Their hypo-osmotic urine is not as copious as is that of freshwater fishes (Fig. 20-11*B*). The marine fish tubule has lost the ciliated neck and intermediate segments (Fig. 20-10). Consequently, the filtrate moves much

TABLE 20-2 Major Form of Nitrogen Excretion

Animal	Environment	Type of Nitrogen Excretion
Teleost	Fresh water	Ammonia, some urea
Teleost	Marine	Ammonia, some urea
Elasmobranch	Marine	Urea
Larval amphibian	Fresh water	Ammonia, some urea
Adult amphibian	Terrestrial	Urea, some ammonia
Reptile	Terrestrial	Uric acid
Bird	Terrestrial	Uric acid, urea in some species
Mammal	Terrestrial	Urea, small amounts of ammonia, and sometimes uric acid

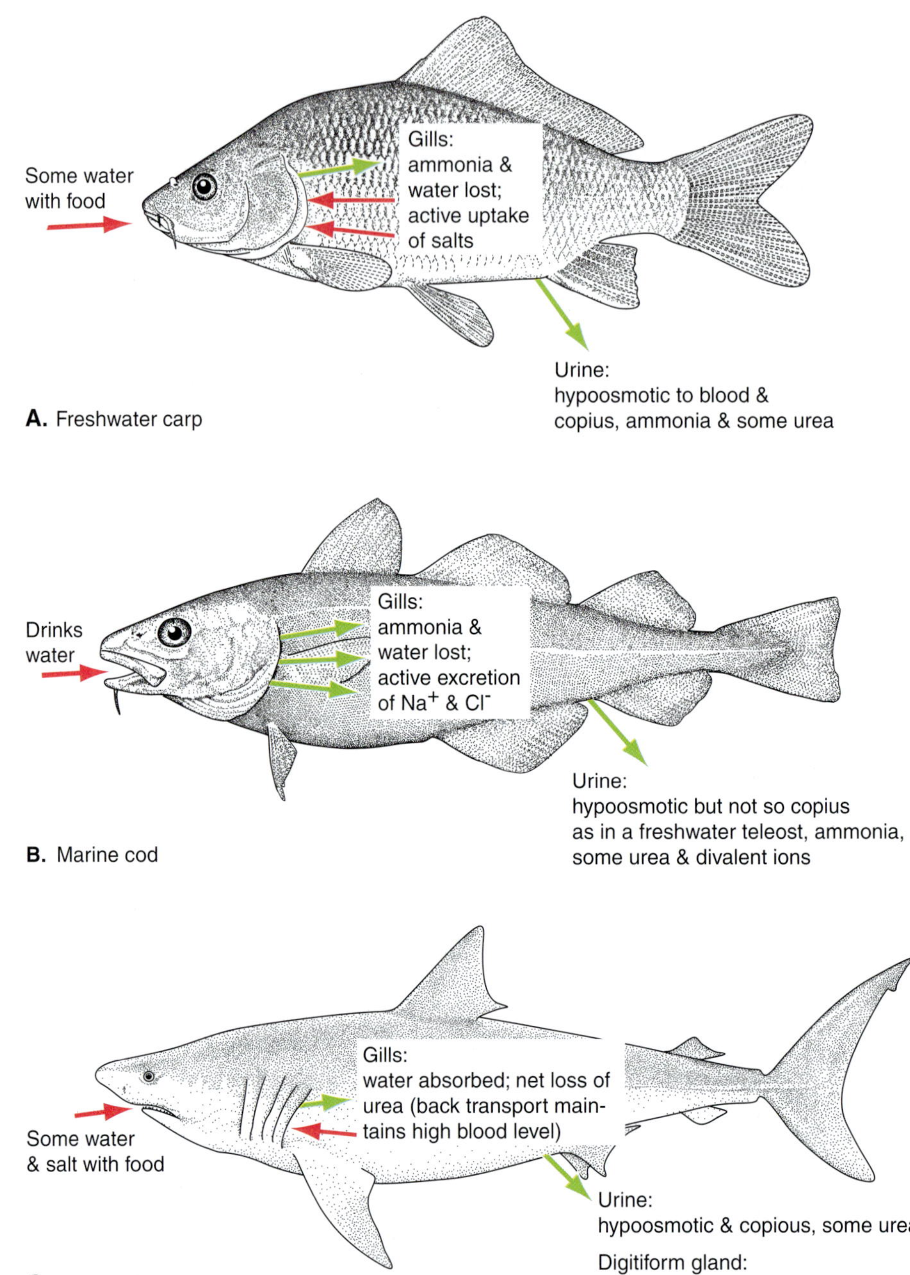

FIGURE 20-11
Osmoregulation and excretion in fishes. The major sites for the exchanges of water, salts, and nitrogenous wastes are shown. *A,* A freshwater teleost (carp). *B,* A saltwater teleost (cod). *C,* A saltwater shark.

slower because it is not acted on by cilia. In this way, the volume of urine is reduced and water loss is minimized. Saltwater teleosts make up for water lost in the urine and in osmosis by drinking sea water and then actively excreting the excess salts. Little sodium and chloride are lost in the urine, but ionocytes in their gills can actively excrete these ions. Note that ionocytes can move salts in either direction: they take up salts in the freshwater environment and excrete them in the marine environment. The divalent magnesium and sulfate ions taken in with sea water cannot be eliminated through the gills, but they are excreted actively by the renal tubules.

Some teleosts, such as eels and salmon, move between salt and fresh water during their life cycle. Their physiology and behavior resemble those of freshwater fishes in fresh water and marine fishes in salt water. Salmon are born in fresh water; migrate to salt water, where they mature; and then return to fresh water to spawn. They are called **anadromous.** In the presence of the pituitary hormone prolactin, the kidney functions as that of a freshwater fish, producing copious amounts of dilute urine. When prolactin is absent, the renal tubule eliminates excess ions actively, and water is moved slowly because the cilia are inactivated. In contrast, eels are **catadromous,** because they hatch in the sea, migrate to fresh water to mature, and return to the sea to spawn. Eels have a slightly higher salt content in their bodily fluids when in the ocean than when in fresh water (Table 20-1). When in fresh water, prolactin makes the kidney perform as that of a freshwater fish. As soon as the eel is in sea water, prolactin production diminishes, and the kidney assumes the function of a saltwater fish. In the course of their evolution, many marine fishes have reentered fresh water and readapted to this habitat. Indeed, the fossil record and analyses of the distribution of lungs and swim bladders indicate that contemporary freshwater teleosts had a marine phase in their ancestry.

A few marine teleosts have lost their renal corpuscles (Fig. 20-10). This certainly lessens the problem of water loss through the kidneys, for it eliminates filtration. All of the blood reaching the peritubular capillaries now comes from the renal portal system (Fig. 20-3*B*). Nitrogen can be removed by renal tubule secretion and by diffusion through the gills. Most deep-sea fishes have lost the renal corpuscles, resulting in considerable water conservation. The blood of some Antarctic fish species contains many relatively small glycoprotein molecules that act as an antifreeze by lowering the blood's freezing point. Investigators hypothesize that these molecules are retained in the bodily fluids because of the loss of renal corpuscles.

Chondrichthyan Fishes

Marine chondrichthyans took a different evolutionary pathway. They convert ammonia to the less toxic urea and eliminate most of their nitrogenous wastes in this form (Table 20-2). They also retain sufficient urea in their bodily fluids to raise their internal osmotic pressure to a level comparable to that of sea water (Table 20-1). Their tissues have become adapted to function at urea levels that most other vertebrates could not tolerate for long. Investigators have long believed that urea is retained because the gills are structurally impervious to its diffusion. But most nitrogen is lost as urea, and little urea escapes through the kidneys because it is reabsorbed by the kidney tubules. Urea is lost through the gills, but a back transport mechanism that returns considerable urea to the blood seems to be present (Wood et al., 1995).

Urea retention gives marine cartilaginous fishes the same osmotic problem as freshwater fishes (Fig. 20-11*C*). Water diffuses into their bodies by osmosis and must be eliminated by the production of a copious and dilute urine. Their renal corpuscles are large, even larger than those of freshwater teleosts, most of which evolved from marine ancestors (Fig. 20-10). Although cartilaginous fishes do not drink sea water, some salt water enters the gut during feeding, and additional salts enter the body through the gills. Excess salts are excreted by a special **digitiform** or **rectal gland** that discharges into the caudal end of the intestine. Very few cartilaginous fishes are able to reenter fresh water for extended periods because the high urea level of their bodily fluids would draw in more water than could be eliminated. The few species that have adapted to fresh water retain less urea than do marine species and have reduced digitiform glands (Table 20-1).

The retention of urea by adult cartilaginous fishes is an interesting solution to the problem of osmoregulation in the marine environment, but the synthesis of urea and the habituation of the tissues to high urea levels may have evolved first as embryonic adaptations. The young of cartilaginous fishes develop either in egg cases deposited in the sea or in the reproductive tracts of their mothers (Chapter 21). In either environment, water turnover in the embryos is relatively low, and they could not survive unless ammonia were converted to the less toxic urea.

From Water to Land: Lungfishes and Amphibians

Unfortunately, nothing is known about excretion and osmoregulation in the early freshwater choanate fishes that were the ancestors of terrestrial vertebrates. The related coelacanth, *Latimeria,* which has survived, tells us nothing; it is a marine genus, gives birth to living young as do most cartilaginous fishes, and solves its osmotic problem by urea retention (Table 20-1)—an interesting example of convergence with chondrichthyans. Both contemporary lungfishes, some of which live in bodies of fresh water that dry up, and amphibians inhabit environments like those of the ancestors of terrestrial vertebrates. Ancestral terrestrial vertebrates must have faced similar problems and probably excreted and osmoregulated in similar ways.

Living lungfishes and larval amphibians live in fresh water and have the same problems as do freshwater fishes. Excess water enters their bodies by osmosis and must be eliminated by a copious urine that is hypoosmotic to their bodily fluids. Although some salts are reabsorbed in the renal tubules, the net loss must be made up by obtaining salts from food or directly from the environment. Given a high water turnover, most nitrogen is eliminated as ammonia, but some urea is produced (Table 20-2).

When water is in short supply, lungfishes and amphibians conserve it by excreting most of their nitrogen as urea. The African lungfish, *Protopterus,* can estivate in a dried mud cocoon when its pond dries up. Because metabolism is very low during this period, the lungfish produces less waste nitrogen than usual, and all of it is converted into urea. The urea is not eliminated but accumulates in the tissues, often for several years, until water returns.

Most adult amphibians spend considerable time on land and must conserve water. They, too, convert most of their waste nitrogen to urea. A great deal of water and urea are filtered through the modest-sized renal corpuscles, and additional urea is added by tubular secretion (Fig. 20-10). Amphibian renal tubules resemble those of their freshwater fish ancestors. They have the same two ciliated segments (the narrow neck after the capsule and the intermediate segment) to move large amounts of filtrate rapidly. The urine remains hypo-osmotic to the bodily fluids, so a great deal of water is lost this way, although some is reabsorbed from the urinary bladder (Fig. 20-12*A*). Additional water is lost by evaporation through the thin and vascular skin, which functions as a respiratory membrane (Chapter 18). Evaporation through the skin of many amphibians occurs at the same rate as from a free water surface under similar conditions of temperature and humidity. Amphibians minimize this route of water loss by living in cool, moist habitats or being active during the cooler times of day, when the humidity is higher. Despite these water-saving mechanisms, amphibians lose a great deal, and most return to water periodically to regain water by osmosis through the skin (Fig. 20-12*B*). Their skin cells also can actively take up salts from the water.

Several anurans are able to live in environments where very little water is available by retaining a great deal of urea in their tissues. This solution is used by the crab-eating frog, *Rana cancrivora*, of Southeast Asia, which lives in saltwater mangrove swamps; by several toads; and by some salamanders that live in dry soil for several months during the year. Another frog, *Cyclorana* (*Chirolepsis*), lives in the deserts of central Aus-

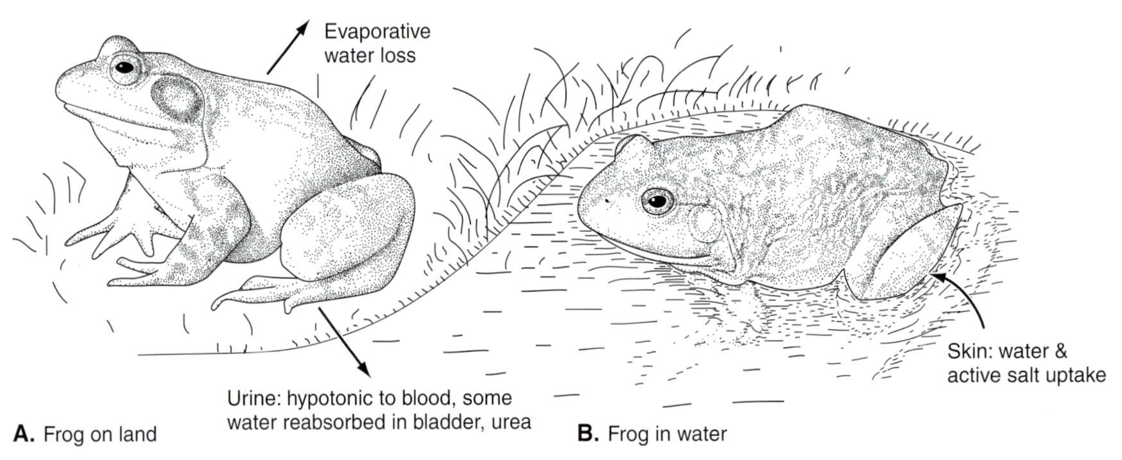

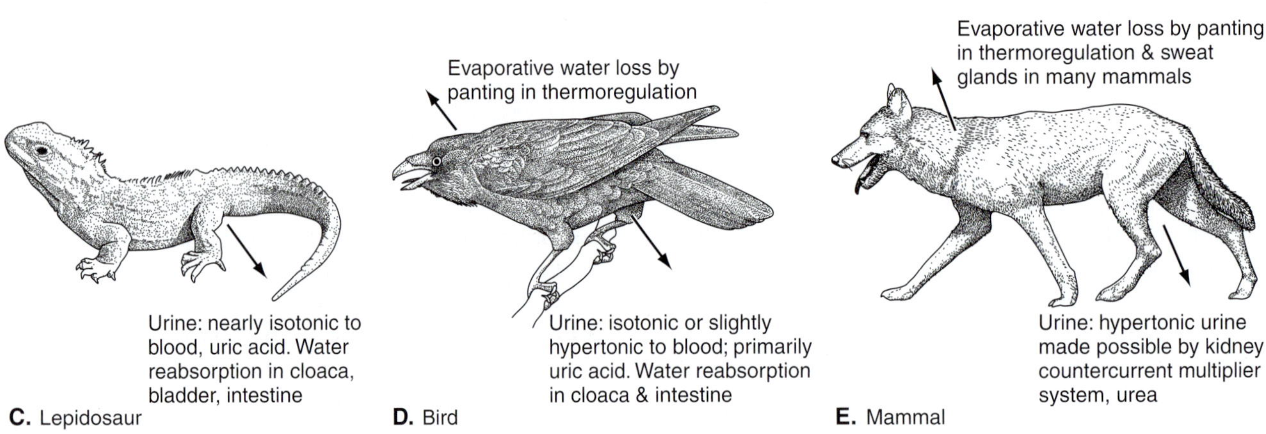

FIGURE 20-12
Osmoregulation and excretion in terrestrial vertebrates. The major sites for the exchanges of water, salts, and nitrogenous wastes are shown. *A,* A frog on land. *B,* A frog in the water. *C,* The tuatara (*Sphenodon*). *D,* A bird. *E,* A dog.

tralia. It soaks up water during the rainy periods and produces a copious and very dilute urine that is stored in the urinary bladder. As much as one third of the animal's weight is represented by this urine. Additional water is stored in the bloated subcutaneous lymph sacs. During the dry periods, which may last two years or more, the frog estivates within a cocoon in deep burrows and gradually uses up the stored water. A few arboreal frogs, including *Chiromantis* and *Phyllomedusa*, conserve water, as do most reptiles, by excreting their nitrogen primarily as uric acid. Uric acid accumulating in the urinary bladder becomes a semisolid material. Cells in the neck of the bladder of these animals are ciliated and the cilia help move the uric acid into the cloaca (Bolton and Beuchat, 1991). Amphibians and semiaquatic reptiles (most turtles and crocodiles) that produce urea as their primary excretory product do not have these cilia.

Adaptations for Terrestrial Life: Reptiles

Vertebrates could not live and be active under truly terrestrial conditions unless they evolved mechanisms that enabled them to conserve water more successfully than do most amphibians and to eliminate nitrogenous wastes at the same time. Water loss by respiration is reduced in all amniotes (Fig. 20-12*C*). They do not use cutaneous gas exchange or ventilate their lungs more than necessary (Chapter 18). Cornification of the skin of reptiles, birds, and mammals also reduces evaporative water loss. A desert tortoise, for example, loses as little as 3 mg of water by cutaneous evaporation per square centimeter of body surface per hour, but reptiles that have readapted to an amphibious environment, such as crocodiles and some turtles, lose much more. The more terrestrial amniotes also have excretory adaptations that conserve considerable water.

In addition to being able to convert waste nitrogen into urea, many amniotes also can synthesize uric acid, which is chemically inert and requires little water for its removal. The enzymes needed for uric acid synthesis in several frogs may have first evolved in amniotes as an embryonic adaptation. Adaptation to land by vertebrates involved the evolution of a cleidoic (amniotic) egg that can be laid on land or retained in the body of the mother. The egg provides all of the metabolic needs of the embryo, a free-living aquatic larval stage is suppressed, and the young hatch or are born as miniature adults. Waste nitrogen must be stored within the allantois as uric acid because water turnover between the embryo and its environment is insufficient to eliminate it as ammonia or urea.

Adult tortoises, lizards, snakes, and other terrestrial reptiles continue to eliminate most of their waste nitrogen as uric acid (Fig. 20-12*C*). Their renal corpuscles are exceptionally small (Fig. 20-10) and contain fewer glomerular capillaries than in most other vertebrates. Indeed, many tubules are aglomerular in some species of reptiles. Tubular filtration is reduced greatly, and with it, the loss of water. Some uric acid is filtered, but most enters the tubules by active tubular secretion, which is possible because the renal portal system provides the peritubular capillaries with a large volume of blood that is independent of the blood supply to the reduced glomeruli. The urine is iso-osmotic or slightly hypo-osmotic to the bodily fluids. Reptiles are unable to produce a urine that is more concentrated in salts than their bodily fluids. Considerable water is reabsorbed in the cloaca and the urinary bladder, if one is present. Material in the cloaca can back up into the intestine, and water also is reabsorbed here. Uric acid leaves the body with the feces as a whitish, pastelike material.

Although all reptiles synthesize uric acid, the more aquatic species have a higher water turnover, produce a more dilute urine, and eliminate some of their nitrogen in other ways. Crocodiles and sea snakes remove a great deal of their waste nitrogen as ammonia; aquatic turtles eliminate it as urea. Reptiles that spend much or all of their lives in the sea also must have salt-excreting glands that eliminate excess salts. The salt-excreting glands of sea turtles discharge into the orbits; those of the marine iguana, into the nasal cavities; those of sea snakes, into the oral cavity.

High Metabolism and Endothermy: Birds and Mammals

The excretory and osmoregulatory problems of birds and mammals have been compounded by the evolution of a high level of metabolism and endothermy. Birds and mammals must cope with an increased volume of nitrogenous wastes and also conserve body water and salts. Water conservation is particularly important because birds and mammals unavoidably lose some water through the respiratory passages and skin in their thermoregulation. Mammals conserve water by passing air over the respiratory turbinates in their nasal cavities. Birds and mammals have an increased number of renal tubules. Although some very primitive vertebrates have only one pair per body segment, and most ectothermic vertebrates have only a few hundred or thousand, estimates of the number of renal tubules in a human range from 2 to 4 million. Filtration pressures also are higher in birds and mammals than in lower vertebrates because the complete division of the heart makes possible the development of a high systemic blood pressure in conjunction with a lower pulmonary pressure. The increased number of renal tubules and high filtration pressures enable birds and mammals to clear nitrogenous wastes from a large volume of plasma. Chickens filter a volume of plasma four times that of their bodily fluids in 24 hours; dogs can filter eight times the volume of their bodily fluids. In reptiles and anamniotes, filtration volume in a 24-hour period is less than bodily fluid volume.

Birds conserve water partly in the same way as do reptiles. Renal corpuscles are small, and uric acid is the primary excretory product (Figs. 20-9 and 20-12*D*). Some uric acid is filtered, but much is added by tubular secretion from the blood that reaches the peritubular capillaries through the renal portal system. Birds, too, reabsorb water in the cloaca and from material that backs up into the intestine from the cloaca. In addition, some species of birds have evolved long, narrow loops of the renal tubules that dip into the medullary region of the kidney. This type of tubule enables the reabsorption of considerable water and the production of a very concentrated urine that is hyperosmotic to the bodily fluids; that is, it contains less water than do the bodily fluids. Urea is the primary nitrogenous waste filtered in birds with many tubules of this type; precipitation of much uric acid would clog the narrow medullary loops.

The mechanism for the production of a hyperosmotic urine is understood most clearly in mammals, in which a similar type of tubule evolved independently. The mechanism depends on the configuration of the renal tubules and their arrangement within the kidney. A gross section of the kidney shows distinct **cortical** and **medullary regions** (Fig. 20-13*A*). The renal corpuscles and both the proximal and the distal convoluted tubules lie in the cortex (Fig. 20-13*B, left*). A highly specialized, new mammalian segment of the tubule forms a thin, hairpin loop, known as the **medullary loop of Henle,** which dips into the medulla. Most mammals have two types of tubules: some with short loops of Henle and others with long loops that dip far into the medulla. Long **collecting tubules** begin in the cortex, where they receive the distal convoluted tubules of many nephrons and extend through the medulla to open into the **renal calices,** which, in turn, lead to the **renal pelvis.** The renal

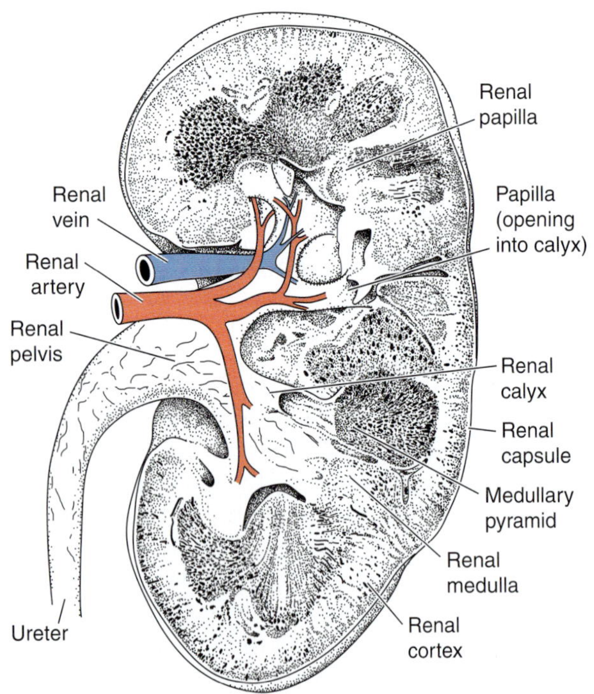

A. Section through a mammalian kidney

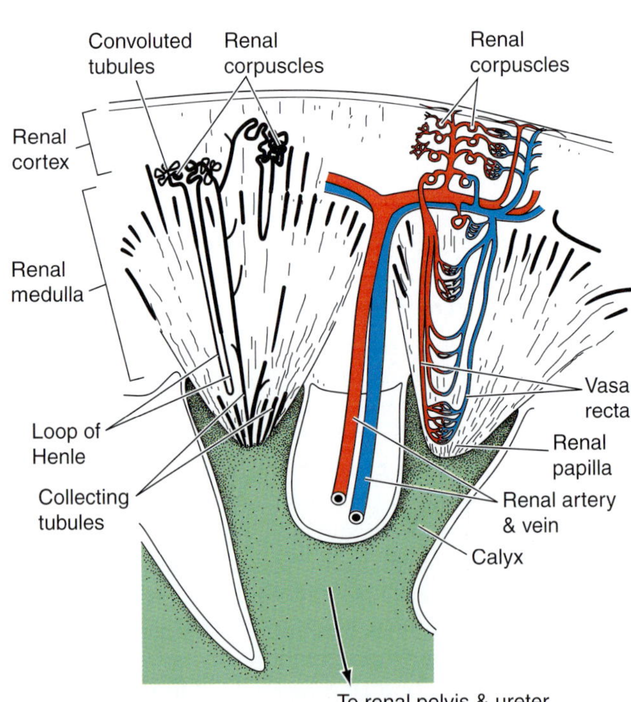

B. Mammalian kidney tubules in relationship to circulation

FIGURE 20-13
Mammalian kidney structure. *A,* A longitudinal section through a human kidney, showing the major regions and their relationship to renal blood vessels and to the ureter. *B,* An enlargement of two renal pyramids, showing kidney tubules and their relationship to circulation. The location of the parts of a nephron is shown in the left pyramid, and the vascular architecture, in the right one. The blood vessels and nephrons overlap in the kidney.

pelvis is an expansion of the ureter with the kidney. The long loops of Henle and the collecting tubules are aggregated in the medulla to form one or more **renal pyramids,** the apices of which form **renal papillae,** which protrude into the renal calices. Thin vascular loops, known as the **vasa recta,** follow the loops of Henle into the medulla (Fig. 20-13*B, right*).

The filtrate produced in the renal corpuscles contains a great deal of water and all of the same types of smaller solute molecules present in the blood. As the filtrate passes through the tubule, sugars, amino acids, vitamins, and other solutes needed by the body are actively reabsorbed from the filtrate by the cells of the proximal and distal convoluted tubules, as they are in other vertebrates. Excesses of any of these substances are allowed to stay in the urine. The ability to reabsorb water and concentrate the urine derives from the unique properties and topography of the loops of Henle, collecting tubules, and vasa recta (Fig. 20-14). Although the wall of most of the loop of Henle is composed of thin epithelial cells, cuboidal cells in the upper half of the ascending limb of the loop have the

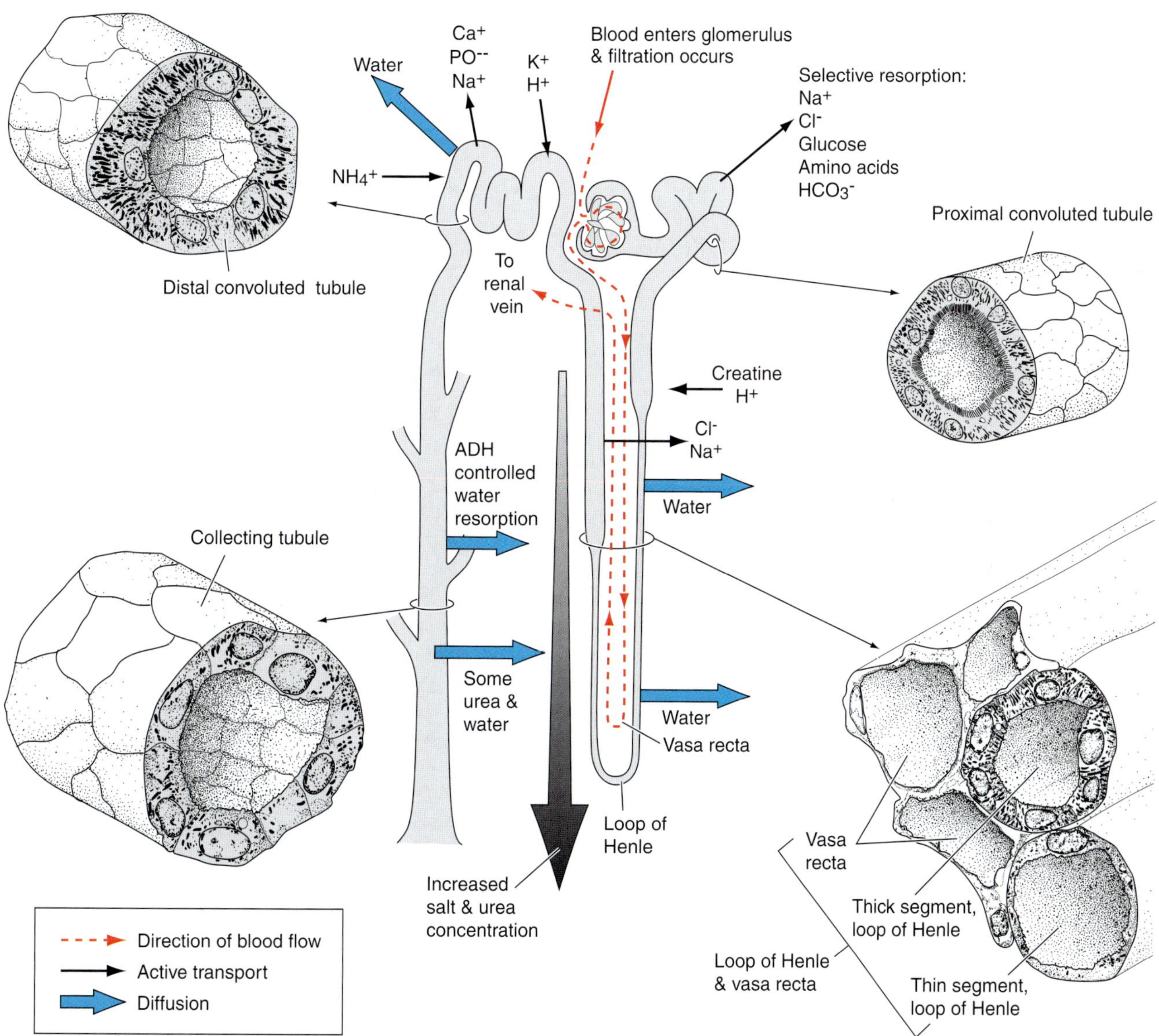

FIGURE 20-14
A mammalian nephron. Dashed red arrows represent blood flow. The regions where materials are exchanged by active transport (*narrow black arrows*) or by passive diffusion (*wide blue arrows*) are shown. The combined result of kidney action is the production of a hypertonic urine. *(Modified from Williams et al.)*

metabolic machinery to pump salt out of the tubular contents and into the interstitial fluid surrounding the loop. In particular, negatively charged chloride ions are actively pumped out, and positively charged sodium ions follow the chlorine. As more salts enter the filtrate from the blood, only to be pumped out again in the ascending limb of the loop of Henle, salts accumulate in the medullary interstitial fluid. Few are carried off by the medullary capillaries because the blood flow is sluggish and the descending arterial and ascending venous limbs of the vasa recta, into which the capillaries lead, lie adjacent to each other. These two limbs of the vasa recta form a countercurrent mechanism. Venous and arterial blood run in opposite directions in the vasa recta, so salts that start to be carried off by the ascending venous limb pass into the descending arterial limb. As a result, the salt gradient in the medullary interstitial tissue is maintained. The venous component of the vasa recta returns blood to the renal vein with recaptured water and blood that has been cleared of waste products. The countercurrent flow of blood in the vasa recta and the pumping out of salts in the ascending limbs of the loops of Henle combine to form a very efficient **countercurrent multiplier system** that establishes a salt gradient, which becomes increasingly salty deeper and deeper in the medulla. The amount of salt that can be concentrated depends on the lengths of the loops of Henle and vasa recta.

This salt gradient enables water to diffuse passively out of the tubular contents. As the dilute tubular contents in the descending limbs of the loops of Henle enter an ever-saltier environment, water moves out by osmosis. The water does not reenter the ascending limbs of the loops of Henle because their cells are impervious to its passage. As salts are pumped out of the ascending limbs of the loops of Henle, and as other solutes are actively reabsorbed in the distal convoluted tubules, the tubular contents again become so dilute that water passively diffuses from the distal convoluted tubules. As the tubular contents reenter the medulla in the collecting tubules, the contents again pass through a progressively saltier environment, and more water diffuses out. Water that enters the interstitial fluid diffuses into the peritubular capillaries and vasa recta and is carried off. Water is not affected by the countercurrent flow in the vasa recta because the plasma proteins stay in the blood and exert sufficient osmotic pressure to retain the water. As the valuable solutes are actively reabsorbed from the tubular contents and water diffuses out, the urea in the tubules becomes increasingly concentrated. Most of the urea remains in the tubules because the cells of their walls are impervious to its diffusion, but the cells of the collecting tubules permit some urea to diffuse out. The presence of this urea in the interstitial fluid of the medulla intensifies the salt gradient that promotes the reabsorption of water.

The amount of water that must be conserved depends on the mammal's environment. Most or all of the tubules of desert rodents have very long loops of Henle, and some of these rodents can produce an extremely concentrated urine that has 25 times the salt content of their blood. These mammals need drink no water, for they lose very little, and they meet their cellular needs with the water produced as a by-product of metabolism. All of the kidney tubules of a beaver, in contrast, have short medullary loops. Beavers produce a urine that is only twice as concentrated as are their bodily fluids. About one third of human renal tubules have long loops of Henle, and our urine is about four times as concentrated as is our blood.

The lengths of the loops of Henle are correlated with the capability to form concentrated urine. The longer the loops, the more concentrated the urine that can be produced. Mammals inhabiting dry environments, as well as those that live on diets rich in salts and other solutes, possess very long loops of Henle and collecting tubules. The extraordinary lengths of the collecting tubules are accommodated in very long papillae. The length and mass of the renal papillae are correlated with the power of the kidney to produce concentrated urine.

Urine concentration varies within limits in a particular species of mammal, depending on water intake. A major controlling factor is the level of **antidiuretic hormone** (Gr., *dia* = through + *ouron* = urine) in the blood. This hormone is synthesized by certain hypothalamic cells and released from the neural lobe of the pituitary gland (Chapter 15). A high level of antidiuretic hormone increases the permeability of the cells of the collecting tubules to water and so promotes a high concentration of urine. If water levels in the blood increase too much, antidiuretic hormone synthesis and release are decreased, water permeability of the collecting ducts is decreased, and the urine becomes more dilute.

Birds and mammals that spend much of their lives in marine environments have the additional problem of salt balance. Salt intake can be problematic, especially if they feed on marine invertebrates and plants that are iso-osmotic with salt water. Fish-eating marine species have less of a problem, for they are ingesting food with a substantially lower salt content. (Recall that all craniates, except for the hagfishes and those vertebrates that retain urea in their tissues, are hypo-osmotic to sea water.) However, some excess salt is taken in. Most birds have small **salt-excreting glands** that excrete

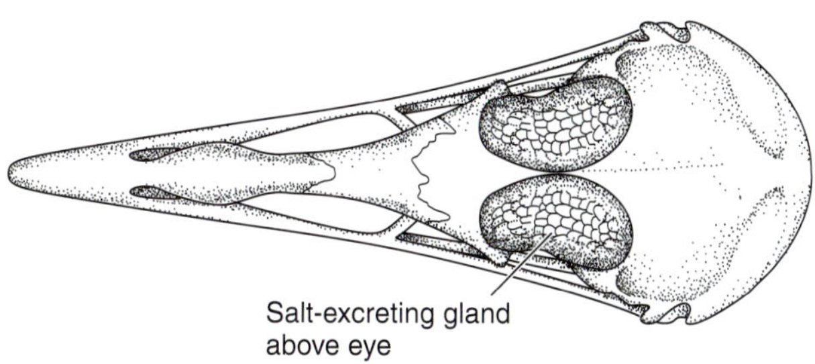

FIGURE 20-15
Dorsal view of the skull of a seagull, showing the location of the salt gland. *(After Schmidt-Nielsen.)*

some salt. The glands are located in the orbit but discharge into the nasal cavities. These glands are large in marine species and can excrete a very salty solution into the nasal cavities when the bird is under salt stress (Fig. 20-15). The kidneys of most mammals cannot excrete excess amounts of salt because they cannot produce a urine that is more concentrated than is sea water, although they produce one more concentrated than is their blood. If a person drinks sea water, 1350 mL of urine is produced to remove the excess salts in every liter of salt water, so he or she would dehydrate rapidly. The kidneys of many marine species, however, can produce urine with a higher salt concentration than that of sea water. If a whale consumes 1 L of sea water, only 650 mL of urine are produced to carry off the salts. There is a net gain of water. Marine mammals excrete excess salt in this way rather than by specialized salt-excreting glands.

Developmental Canalization, Convergent and Reversed Evolution, and Evolutionary Innovation

The sequence from anterior to posterior of the functional pronephros, mesonephros, and metanephros, as exhibited in the embryology of derived mammals, is one of the best examples of developmental canalization among vertebrates. The succession is maintained even in more primitive vertebrates that have an earlier endpoint (the opisthonephros, which combines attributes of the mesonephros and metanephros). Vertebrates share fundamental developmental patterns in kidney differentiation, and when they are excreting urea or uric acid as adults, they faithfully recapitulate the evolutionary sequence in their ontogeny, excreting, in succession, ammonia, urea, and uric acid. Thus, it is not surprising that some striking examples of convergent mechanisms have occurred in the renal biology during vertebrate evolution. When water conservation is at a premium, in fishes the nitrogenous excretory product urea, which is less toxic than is ammonia, is produced and sequestered in the body muscles to increase the osmotic pressure in various fishes to match or approach the exterior. This remarkable mechanism has evolved independently in the marine Chondrichthyes, estivating lungfish, and the eel-shaped mud eel, a derived tropical teleost fish belonging to the Synbranchiformes that burrows in mud during droughts. To solve the problem of water loss, squamates and birds convert nitrogenous wastes into the almost insoluble and wholly nontoxic uric acid. Excretion of uric acid requires no water loss because the acid is voided as a white solid, often with the feces. Primitively nitrogenous wastes are eliminated in solution as ammonia that is readily dissolved in water as in teleosts. However, this primitive mechanism reappeared in some aquatic turtles and crocodilians because they occupy aquatic habitats in which ammonia can be eliminated promptly from the body.

The mammalian nephron deviates to a greater extent from that of all vertebrates. Its unique structure is considered an evolutionary innovation because it enables mammals to penetrate and exploit a great diversity of habitats, including deserts, seas, swamps, and high-altitude habitats. The middle part of the urinary tubule has become greatly elongated and narrow and undergone a sharp bend to form a hairpin loop. This mammalian specialization is named Henle's loop and is adapted specifically for water recovery by means of the highly efficient complex countercurrent multiplier mechanism. As a result, mammals have decoupled or separated the original function of reabsorption of transported materials in the filtrate from the new function to recover water and produce hypertonic urine. The longer the Henle's loop, the more hypertonic the urine.

Developmental canalization of the kidney is profound, as reflected by the maintenance of the archinephric or Wolffian duct, the glomeruli, and the nephric tubules. These parts remain functional in relatively primitive form in the developing embryo. But adults require more elaborate and efficient systems to serve their more demanding body functions. As a result, part of the kidney remains functional in relatively primitive form while more elaborate new systems are added for the adult. The strong selective pressures favoring developmentally functional excretory organs, coincident with the increasing functional selection for greater adult efficiency, must have canalized the successive shifts from one type of kidney to another during the evolution of the vertebrates.

SUMMARY

1. With the skin, lungs, and special salt-excreting or salt-absorbing cells, the excretory system maintains the stability of the internal environment by eliminating nitrogenous wastes and by helping regulate the salt and water balances of the body.
2. The structural and functional units of the excretory system are the kidney tubules, or nephrons, which develop from the mesodermal nephric ridge.
3. A filtration process occurs in the renal corpuscles; selective reabsorption and secretion occur in other parts of the tubules.
4. Ancestral kidney tubules may have had external glomeruli that filtered materials into the coelom; wastes are then drained into the tubules through coelomic recesses. Glomeruli are internal in most tubules and filter directly into the tubules, but traces of the coelomic recesses remain in some primitive tubules as coelomic funnels.
5. An ancestral vertebrate may have had a holonephric kidney, in which segmental tubules developed along the entire nephric ridge. A larval hagfish has a kidney that approaches this.
6. The pronephros is an embryonic kidney that develops over the pericardial cavity and forms the archinephric duct. It is functional in the larvae or embryos of hagfishes and lampreys and also in many bony fishes and amphibians but not in other groups.
7. The mesonephros is an embryonic kidney that differentiates from the portion of the nephric ridge that overlies the anterior half of the pleuroperitoneal cavity. Its tubules tap into the archinephric duct and are functional in the larvae or embryos of most vertebrates. The ducts from the testis acquire a connection with some of the cranial mesonephric tubules in most male vertebrates.
8. The adult kidney of hagfishes is a primitive opisthonephros. It includes the embryonic mesonephros and segmental tubules that develop from the caudal part of the nephric ridge.
9. Most adult fishes and amphibians have an advanced opisthonephros, which differs from the primitive opisthonephros by the multiplication of tubules, especially in the caudal part of the kidney. One or more accessory urinary ducts may drain the caudal part.
10. An adult amniote has a metanephros that develops only from the caudal part of the nephric ridge and is drained by a ureter. In males, part of the embryonic mesonephros and the archinephric duct remain and transport sperm.
11. Urinary and genital ducts lead to a cloaca in most vertebrate groups. Most tetrapods, except for birds, have a urinary bladder where urine accumulates and from which water is sometimes reabsorbed. The cloaca becomes divided in therian mammals, and the digestive and urogenital passages then lead directly to the body surface.
12. Nitrogen is eliminated primarily as ammonia if the animal has a high water turnover or as less toxic urea and uric acid if water must be conserved.
13. Although the bodily fluids of the marine hagfish are iso-osmotic with sea water, the animal has large renal corpuscles and a high water turnover, which may help flush out ammonia.
14. Ancestral craniates probably also were marine and iso-osmotic with their environment, as are urochordates and cephalochordates. But vertebrates must have adapted early in their evolution to fresh water because all contemporary species are ionically independent of their environment.
15. Freshwater fishes have an internal salt concentration greater than that of their environment. They must produce a copious urine that is hypo-osmotic to their bodily fluids to eliminate excess water that enters their bodies by osmosis. Salts must be conserved and replaced through salt-absorbing ionocytes on their gills.
16. Saltwater teleosts have a lower internal salt concentration than that of their environment, so they must conserve water and eliminate salts. Their urine, although hypo-osmotic to their bodily fluids, is less copious. Their renal tubules have very small glomeruli or are aglomerular. They drink sea water and excrete the excess salts through ionocytes on their gills and through their kidney tubules.
17. Because chondrichthyans retain considerable urea in their tissues, they have a salt concentration greater than the marine environment where most species live. They must eliminate excess water and salts.
18. Adult amphibians continue to produce a hypo-osmotic urine but conserve water by staying in moist microhabitats and reabsorbing some from their urinary bladders. They must return to water periodically to regain lost water and salts, partly by absorption through their skin.
19. The cornified skin of reptiles reduces water loss, their renal corpuscles are very small, and much ni-

trogen is excreted as uric acid. Most of the small amount of water used to remove uric acid from the kidneys is reabsorbed in the cloaca, urinary bladder, or caudal end of the intestine.

20. Because they are endothermic, birds and mammals produce a large volume of nitrogenous wastes, and they also lose some water in their thermoregulation. Most birds conserve water by excreting nitrogen as uric acid and reabsorbing water in the cloaca and caudal end of the intestine.

21. Mammals excrete nitrogen as urea. Water is conserved by the production of a hyperosmotic urine. This is made possible by the evolution of a countercurrent multiplying system formed by the loops of Henle of the renal tubules and the accompanying vasa recta.

REFERENCES

Beeuwkes, R., 1982: The two solute model of the mammalian nephron. *In* Taylor, C. R., Johansen, K., and Bolis, L., editors: *A Companion to Animal Physiology.* Cambridge, Cambridge University Press.

Bentley, P. J., 1976: Osmoregulation. *In* Gans, C., and Dawson, W. R., editors: *Biology of the Reptilia,* volume 5. New York, Academic Press.

Bentley, P. J., 1971: *Endocrines and Osmoregulation: A Comparative Account of the Regulation of Water and Salt in Vertebrates.* New York, Springer-Verlag.

Bolton, P. M., and Beuchat, C. A., 1991: Cilia in the urinary bladder of reptiles and amphibians: A correlate of uric acid production? *Copeia,* 1991:711–717.

Cambaryan, S. P., 1994: Microdissectional investigation of the nephrons in some fishes, amphibians, and reptiles inhabiting different environments. *Journal of Morphology,* 219:319–339.

Chase, S. W., 1923: The mesonephros and urogenital ducts of *Necturus maculosus* Rafinesque. *Journal of Morphology,* 37:457–532.

Dantzler, W. H., and Braun, E. J., 1980: Comparative nephron function in reptiles, birds, and mammals. *American Journal of Physiology,* 239:R197–R213.

Denison, R. H., 1956: A review of the habitat of the earliest vertebrates. *Fieldiana, Geology,* 11:359–457.

Dobbs, G. H. III, Lin, Y., and DeVries, A. L., 1974: Aglomerulism in Antarctic fish. *Science,* 185:793–794.

Foskett, J. K., and Scheffey, C., 1982: The chloride cell: Definitive identification of the salt-secretory cell of teleosts. *Science,* 215:164–166.

Fox, H., 1977: The urogenital system of reptiles. *In* Gans, C., and Parsons, T. S., editors: *The Biology of the Reptilia,* volume 6. New York, Academic Press.

Fox, H., 1963: The amphibian pronephros. *Quarterly Review of Biology,* 38:1–25.

Franklin, C. E., and Grigg, G. C., 1993: Increased vascularity of the lingual salt glands of the estuarine crocodile, *Crocodylus porosus,* kept in hyperosmotic salinity. *Journal of Morphology,* 218:143–151.

Fraser, E. A., 1950: The development of the vertebrate excretory system. *Biological Reviews,* 25:159–187.

Gordon, M. S., Schmidt-Nielsen, K., and Kelly, H. M., 1961: Osmotic regulation in the crab-eating frog (*Rana cancrivora*). *Journal of Experimental Biology,* 38:659–678.

Hickman, C. P., and Trump, B. F., 1969: The kidney. *In* Hoar, W. S., and Randall, D. J., editors: *Fish Physiology,* volume 1. New York, Academic Press.

Holmes, W. N., and Phillips, J. G., 1985: The avian salt gland. *Biological Reviews,* 60:213–256.

Kessel, R. G., and Kardon, R. H., 1979: *Tissues and Organs: A Text-atlas of Scanning Electron Microscopy.* San Francisco, W. H. Freeman & Co.

Kirschner, L. B., 1991: Water and ions. *In* Prosser, C. L., editor: *Environmental and Metabolic Animal Physiology.* New York, John Wiley & Sons.

Lacy, E. R., and Reale, E., 1991: Fine structure of the elasmobranch renal tubule: Intermediate, distal, and collecting duct segments of the little skate. *American Journal of Anatomy,* 192:478–497.

Moffat, D. B., 1975: *The Mammalian Kidney.* Cambridge, Cambridge University Press.

Murrish, D. E., and Schmidt-Nielsen, K., 1970: Water transport in the cloaca of lizards: Active or passive? *Science,* 170:324–326.

Nicolson, S. W., and Lutz, P. L., 1989: Salt gland function in the green sea turtle, *Chelonia mydas. Journal of Comparative Biology,* 144:171–184.

Pang, P. K. T., Griffith, R. W., and Atz, J. W., 1977: Osmoregulation in elasmobranchs. *American Scientist,* 17:365–377.

Peaker, M., and Linzell, J. L., 1975: *Salt Glands in Reptiles and Birds.* Cambridge, Cambridge University Press.

Pough, F. H., Heiser, J. B., and McFarland, W. N., 1996: *Vertebrate Life,* 4th edition. Upper Saddle River, N.J., Prentice-Hall.

Prosser, C. L., 1973: *Comparative Animal Physiology,* 3rd edition. Philadelphia, W. B. Saunders.

Rankin, J. C., and Davenport, J., 1981: *Animal Osmoregulation.* New York, John Wiley & Sons.

Richter, S., 1995: The opisthonephros of *Rana esculenta* (Anura): I. Nephron development. *Journal of Morphology,* 226:173–187.

Riegel, J. A., 1972: *Comparative Physiology of Renal Excretion.* Edinburgh, Oliver & Boyd.

Rupert, E. E., 1994: Evolutionary origin of the vertebrate nephron. *American Zoologist,* 34:542–553.

Schmidt-Nielsen, K., 1979: *Desert Animals: Physiological Problems of Heat and Water.* Reprinted 1964. New York, Dover Publications.

Schoemaker, V. H., and Nagy, K. A., 1977: Osmoregulation in amphibians and reptiles. *Annual Reviews of Physiology,* 39:449–471.

Sever, D. M., 1991: Comparative anatomy and phylogeny of the cloacae of salamanders (Amphibia: Caudata): I. Evolution at the family level. *Herpetologica,* 47:165–193.

Skadhauge, E., 1981: *Osmoregulation in Birds.* Berlin, Springer-Verlag.

Smith, H. W., 1932: Water regulation and its evolution in fishes. *Quarterly Review of Biology,* 7:1–26.

Wood, C. M, Pärt, P., and Wright, P. A., 1995: Ammonia and urea metabolism in relation to gill function and acid–base balance in a marine elasmobranch, the spiny dogfish, *Squalus acanthias. Journal of Experimental Biology,* 198:1545–1558.

21

The Reproductive System and Reproduction

PRECIS

Vertebrates reproduce and perpetuate their species by sexual reproduction. The method of sex determination, the pattern of reproduction, and the structure of the gonads and reproductive passages vary greatly among vertebrates and relate to their modes of life and the environments in which they live. Reproduction is integrated by gonadal hormones that interact with the hypothalamus and hypophyseal hormones.

OUTLINE

Tunicates can reproduce asexually by budding, as well as sexually, but the reproduction of cephalochordates and craniates is exclusively sexual. Sexual reproduction is thought to ensure the recombination of genetic material necessary for the maintenance of genetic variation and evolutionary flexibility in a population. However, some vertebrate species and populations within a species are entirely female. The eggs are activated in ways other than by normal fertilization, a chromosome doubling restores the diploid number, and the eggs develop **parthenogenetically** (Gr., *parthenos* = virgin + *genesis* = descent or birth). This would seem to reduce genetic recombination, but the known parthenogenetic populations have evolved by interspecific hybridization, so considerable genetic diversity is maintained by the way parental chromosomes become distributed during meiosis and subsequent chromosome doubling. Examples are seen in the whip-tailed lizards, *Cnemidophorus,* in some populations of the salamander *Ambystoma* (which are triploid), and in a tropical teleost species, the Amazon molly, *Poecilia.*

Sex Determination

Genetic recombination through sexual reproduction requires that some individuals in a population be female, and others, male. Determining sex is far more complicated than generally realized. Sex is determined in birds and mammals genetically at the time of fertilization. This pattern is known as **genotypic sex determination.** In mammals, the expression of a gene known as *SRY* directs the differentiation of the embryonic gonad in the male direction (Haqq et al., 1994). *SRY* promotes the synthesis of the male hormone (**testosterone**) and the production of

Müllerian-inhibiting substance, which inhibits the differentiation of female structures. (The embryonic oviduct is called the Müllerian duct.) Usually the genetic factors are carried on distinctive **sex chromosomes.** Male mammals are the heterogametic sex because their sperm-forming cells have X and Y chromosomes. The *SRY* gene is located on the Y chromosome. The sex chromosomes segregate during meiosis to form two kinds of sperm, X and Y. Female mammals are homogametic (XX) and produce only one kind of egg, X. Although determined genetically at the time of fertilization, sex may be altered during development by hormonal influences. Genotypic sex determination also determines sex in birds, but the females are the heterogametic sex. To avoid confusion, different symbols are used in birds: females are WZ, and males, ZZ.

Genotypic sex determination also occurs in snakes and some lizards. Whether the male or the female is the heterogametic sex varies considerably. In turtles, many lizards, and crocodiles, sex is determined by the incubation temperature, a phenomenon known as **temperature-dependent sex determination.** The critical period during development occurs before the developing gonad begins to express the distinctive characteristics of a testis or an ovary. Turtle eggs incubated at temperatures above 30°C produce a preponderance of females; those incubated below 25°C hatch primarily as males. Although the first studies were done in the laboratory, field studies of the green sea turtle have confirmed these results. Mostly females emerge from nests on open beaches, whereas mostly males emerge from shaded nests. The position of an egg within a clutch can be a determinant when incubation temperatures are close to the thresholds. Eggs near the center of a clutch are exposed to more metabolic heat from their siblings than peripheral eggs, and they hatch as females. Most of the peripheral eggs hatch as males. Less is known about sex determination in anamniotes.

Just how temperature determines sex is not clear. If the female sex hormone, **estrogen,** is administered to eggs during the critical period, it overrides the effects of lower temperatures (which produce males) and the eggs hatch as females (Wibbels et al., 1994). Factors that determine sex in cases of temperature-dependent sex determination are now being studied at several laboratories.

As far as we know, temperature-dependent sex determination is found only among egg-laying reptiles. Reptiles that retain the young in the body of the female until hatching have genotypic sex determination. This appears to be necessary because the body temperature of reptiles, even though they are ectotherms, tends to be maintained at a high level during their diurnal period of activity. Reliance on temperature-dependent sex determination would produce so many individuals of the same sex that the continuation of the species would be endangered. The evolution of endothermy by birds and mammals also may have been a factor in the evolution of their sex chromosomes.

When sex is not rigorously determined by genetic factors, hermaphroditism may evolve. **Hermaphroditism** (Gr. mythology, the son of Hermes and Aphrodite became united in one body with a nymph) is the presence in one individual of both testes and ovaries. Many tunicates are **synchronous hermaphrodites:** both testes and ovaries function concurrently. A few teleosts, such as certain sea bass, are also synchronous hermaphrodites and have a combined **ovotestis** (Fig. 21-1). Self-fertilization, possible in these cases, ensures reproduction in the absence of the other sex. This could be an advantage to sessile tunicates, but cross fertilization normally occurs because of slight differences in the timing of gamete release or because spawning behavior is sometimes necessary to stimulate gamete release.

Many species in five orders of teleost fishes are **sequential hermaphrodites,** for the gonad changes function during the life of an individual. Sea basses, most wrasses, and parrot fishes are protogynous. **Protogynous** (Gr., *gyne* = woman) species are first female and then male. In protogynous species, males usually are territorial and mate selectively with large females. **Protandrous** (Gr., *andr* = man) species begin as functional males and change into females later in life. Protandrous species are typically nonterritorial and mate randomly. Many sex reversals occur at some species-specific age when gonadal tissues and ducts are capable of responding to a shift in the balance of male and female hormones. But other factors, including social interactions, may cause a reversal. For example, in some species of coral reef fishes, in which a group of females breeds with a single male, the death of the

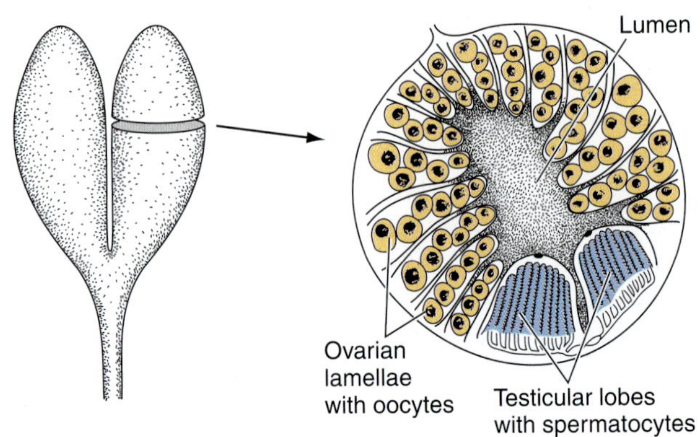

FIGURE 21-1
A section through the ovotestis of a sea bass (Serranidae). *(After D'Ancona.)*

male (or his experimental removal from the population) stimulates the dominant female to transform into a male. An advantage of sex reversals is that every individual is a potential female, and the population can increase rapidly if some misadventure occurs.

Reproductive Patterns

We considered the structure of mature gametes, fertilization, and early development in Chapter 4. Here, we will examine other aspects of the reproductive system and the major patterns of reproduction. The reproductive system consists of the **gonads** (Gr., *gone* = seed), testes or ovaries, which produce the **gametes** (Gr., *gamet* = spouse), sperm or eggs, and usually a set of ducts that carry them from the body. Sperm are small, motile cells surrounded only by their plasma membrane. Eggs are much larger cells that contain food reserves. They are surrounded by their plasma membrane, the **plasmolemma,** and by an acellular **primary envelope** that is deposited while they are in the ovary.

Most male vertebrates use certain excretory passages for the discharge of sperm. Parts of these ducts or associated glands produce secretions that help carry the sperm and sometimes assist in their nutrition and maturation. If fertilization is internal, the male usually has a **penis** or other type of copulatory organ by which sperm are transferred to the female.

Most female vertebrates have evolved a separate oviduct for the transport of eggs. Parts of it usually secrete **secondary envelopes** around the eggs—jelly, albumenous layers, or shells that help nourish and protect the embryos. The structure and function of the oviduct are closely related to the pattern of reproduction of vertebrates.

Most fishes, amphibians, reptiles, and all birds are **oviparous** (L., *ovum* = egg + *parere* = to bring forth). Oviparous fishes and amphibians lay macrolecithal or mesolecithal eggs containing sufficient yolk for the embryos to develop into free-swimming larvae that can provide for their own needs. Courtship behavior at the onset of the reproductive season ensures that the males and females are together, and sperm is discharged over the eggs as they are laid. Large numbers of eggs and sperm must be produced by these oviparous species, for not all eggs are fertilized, and the mortality rate of eggs and larvae is high. Some teleosts, such as the cod, may lay many millions of eggs in a season. Some toads of the genus *Bufo* lay as many as 30,000 eggs in a season. Salamanders and frogs that brood their eggs, however, lay only a dozen or two. Most species of fishes and amphibians are repeat reproducers, or **iteroparous** (L., *iterare* = to repeat). They spawn many times in their lifetime. In contrast, most lampreys and some teleosts (e.g., the salmon and eel) are **semelparous** (L., *semel* = once), meaning that they reproduce only once in a lifetime. A semelparous strategy in anamniotes is adopted in those species with stable, predictable, well-circumscribed spawning grounds.

The probability of successful fertilization and embryonic development is greater in oviparous reptiles and birds. Fertilization is internal; they lay cleidoic eggs with extraembryonic membranes and enough yolk to provide for the needs of the embryos until they hatch as miniature adults. The eggs are encased in shells and are deposited in hidden or protected nests. Egg numbers are reduced. Turtles may lay nearly 100, but lizard and snake clutch sizes range from 1 to about 40. Reduction in egg numbers is correlated with the emergence of some degree of parental care.

Other vertebrates, including chondrichthyans, some teleosts, probably all caecilians, a very few frogs and salamanders, about 20% of squamates, and all living mammals, have internal fertilization and retain their embryos in an expanded part of the oviduct, called the **uterus**, for all of their embryonic development. These are described as live-bearing, or **viviparous,** species (L., *vivus* = *living* + *parere* = to bring forth) because their embryos are born as miniature adults. The source of inorganic and organic nutrients for the developing embryos ranges along a continuum from a complete dependence on material stored in the egg while it is in the ovary to a complete dependence on the transfer of materials in the uterus from the mother. The first extreme has been called **ovoviviparity,** whereas the term "viviparity" has sometimes been limited to the last extreme. No fully satisfactory terms exist for intermediate conditions, and in many cases, the degree of dependence on the mother is unknown. For these reasons, many investigators prefer to use the terms **placental viviparity** for cases in which a fully functional placenta is present, and **aplacental viviparity** for cases of less dependence on the mother but in which the young are born as miniature adults. Fertilization is internal in all viviparous species and usually occurs in the upper part of the oviduct. Fewer sperm and eggs need be produced than in oviparous species because the likelihood of individual survival is much greater, but the eggs must have a great deal of yolk if few nutrients are transferred to them from the mother while they are in the uterus. In a sense, viviparous species are mobile incubators.

In addition to the primary sex organs that produce and carry gametes, many vertebrates develop **secondary sex characteristics** that are important in sex recognition, as signals of sexual maturity and receptivity, in courtship behavior, and sometimes in combat among males for access to females (Fig. 21-2). Secondary sex characteristics include body colors, special ornamentation, antlers, beards, and so on. Another is body size. Male mammals commonly are larger than

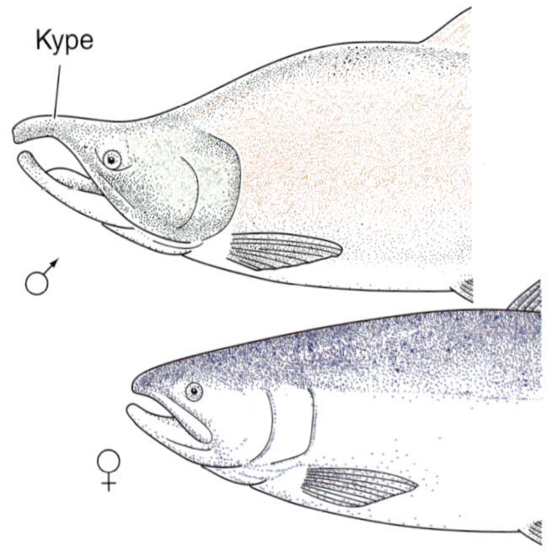

A. *Oncorhynchus* (breeding male above, female below)

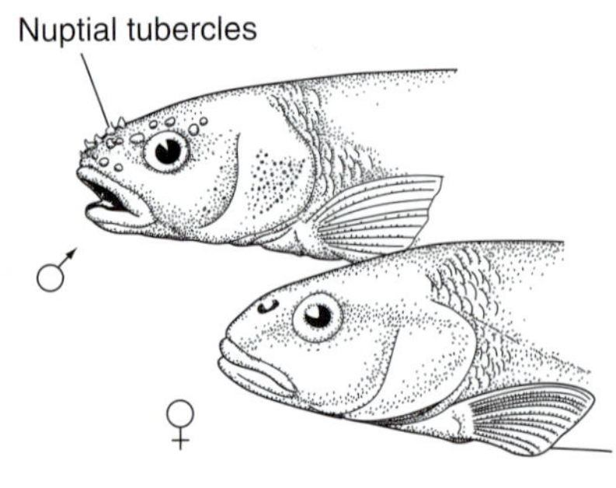

B. *Semotilus* (male and female)

C. *Odocoileus* (male)

D. *Homo* (male)

FIGURE 21-2 Examples of secondary sex characteristics. *A,* The shape of the snout and body of a sexually mature male salmon (*Oncorhynchus*) differs from those of a female. *B,* The male creek chub (*Semotilus*) has nuptial tubercles on its head. These are not present in females. *C,* Male mule deers have antlers. *D,* A bearded man. *(A and B, After Lagler et al.; C, after Vaughn; D, after Austin and Short.)*

are females, but the reverse is true in many other vertebrates. For instance, the females of many raptorial birds (hawks, kestrels) are larger than the males. Secondary sex characteristics develop at maturity under the influence of sex hormones. In cases in which sex reversal occurs, the secondary sex characteristics often change too, as exemplified by the large parrot fishes. Behavior also is affected strongly by the sex hormones. The biological purpose of life is reproduction, and much of an animal's structure and activity is devoted to this end. The number of viable offspring produced commonly measures biological fitness.

Development of the Reproductive System

The early development of the gonads and reproductive ducts is the same in both sexes. The embryo passes through a **sexually indifferent stage,** in which the primordia of both male and female structures are present. Potentially, the embryo may become either a male or a female. If subsequent differentiation is toward a male, male structures continue to develop, and female ones atrophy. The opposite occurs during the differentiation of a female. Because of their similarity in early development, many male and female reproductive organs may be homologized. Often, vestiges of female organs remain in males, and vice versa. The development of the reproductive system is best known in mammals but is similar in most other vertebrates. Teleosts are a notable exception and are considered separately.

Development of the Gonads

In all embryos, gonad development begins as a thickening of the coelomic epithelium and mesenchyme adjacent to the medial border of the mesonephric kidneys. As mesenchyme accumulates, a **genital ridge** appears on each side of the body (Fig. 21-3*A*). Only

FIGURE 21-3
Transverse section through the mesonephros and developing gonad. *A* and *B*, The development of the sexually indifferent gonad. *C*, An early stage in the differentiation of an ovary (*left*) and testis (*right*). *D*, A later stage in the differentiation of the ovary and testis. *(After Williams et al.)*

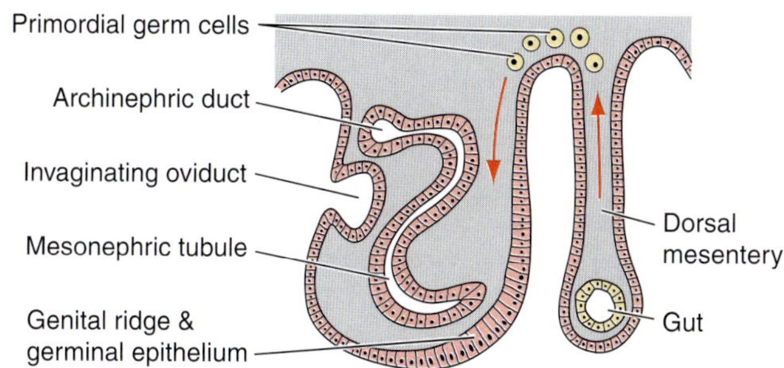

A. Early development of indifferent gonad

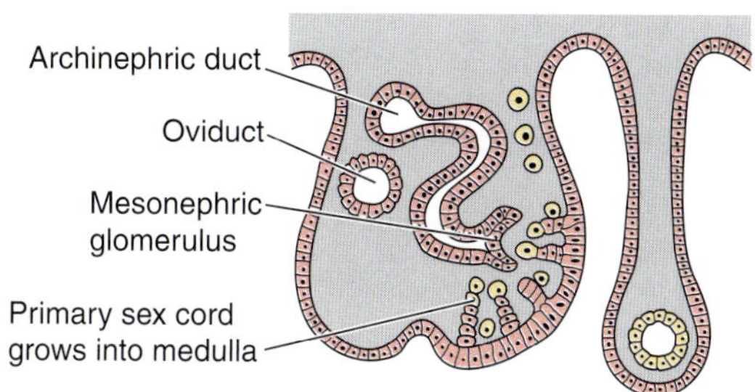

B. Later development of indifferent gonad

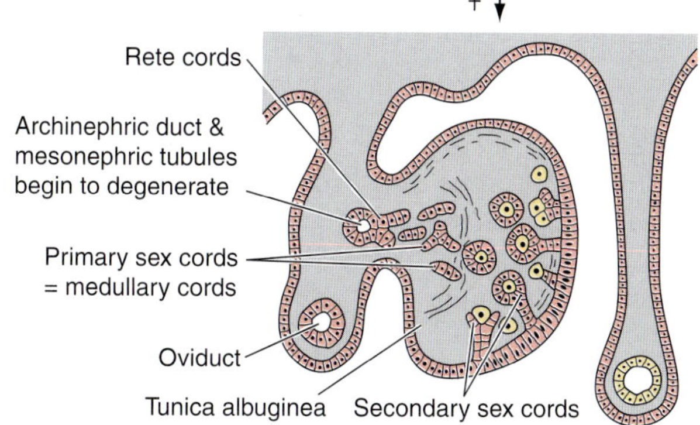

C. Early development of female gonad

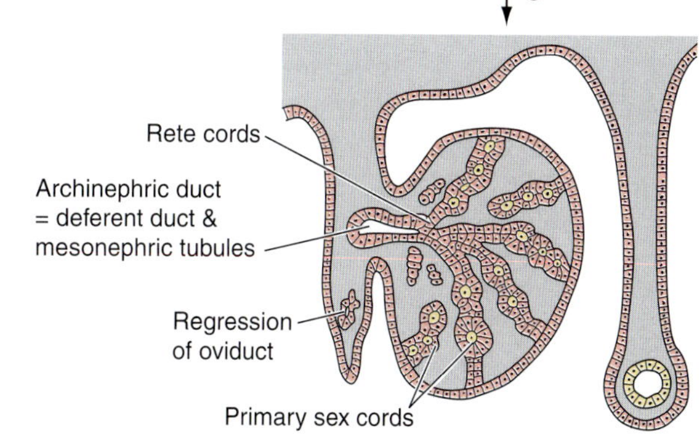

D. Early development of male gonad

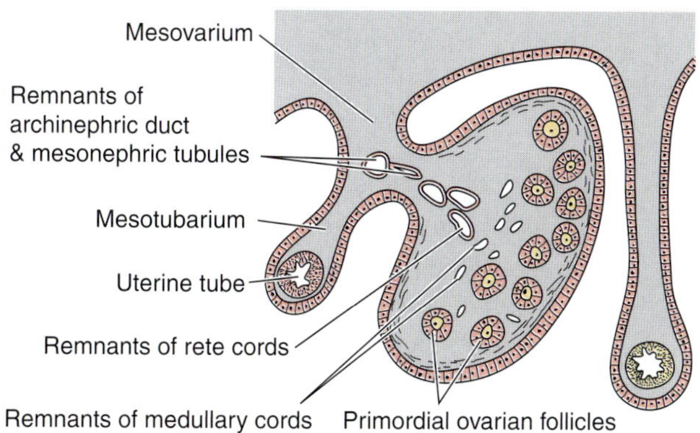

E. Later development of female gonad

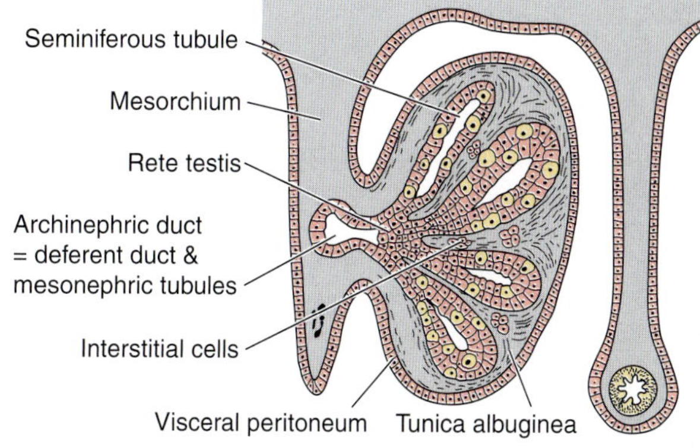

F. Later development of male gonad

the central part of the genital ridge forms the functional gonad in most vertebrates; more cranial and caudal parts regress. The coelomic epithelium overlying the developing gonad thickens as a **germinal epithelium,** and **primordial germ cells,** which are the progenitors of the gametes, invade it. The primordial germ cells can be recognized first as large, distinctive cells in the wall of the **yolk sac** or gut next to the endoderm. The primordial germ cells migrate, primarily by amoeboid motion, through the dorsal mesentery, and into the germinal epithelium. Although few in number at first (20–30 in a human embryo), they multiply en route and in the germinal epithelium. By this stage, the indifferent gonad has a well-defined **medulla** of mesenchymal origin and a **cortex** of epithelial origin. Cords of germinal epithelium then grow into the medullary tissue to form the **primary sex cords** (Fig. 21-3*B*).

If the embryo becomes a male, the primary sex cords, probably under the influence of medullary tissue, multiply, enlarge, and differentiate into the sperm-producing **seminiferous tubules** of the testis (Fig. 21-3*C* and *D, right column*). All the primordial sex cells become incorporated into the primary sex cords in male amniotes, but some sperm-forming cells remain in other parts of the testis in anamniotes. Next, a peripheral extension of mesenchyme separates the cords from the cortex and, in mammals, thickens to form a dense, connective tissue, the **tunica albuginea,** at the surface of the testis (Fig. 21-3*F, right column*). The testis is primarily an organ that develops in the medulla of the indifferent gonad. The cortex regresses, leaving only the visceral peritoneum that covers the surface of the testis and the tunica albuginea. Some other parts of the medulla, and possibly some epithelial cells that do not contribute to the seminiferous tubules, form the **interstitial cells** of the testis that produce the male sex hormone. As the testis enlarges, it separates from the mesonephros or its derivative, to varying extents in different groups of vertebrates, but it remains connected to the kidney by a mesentery known as the **mesorchium** (Gr., *mesos* = middle + *orchis* = testis).

A few primary sex cords also invade the medulla of the gonad during the differentiation of the female. These are known as the **medullary cords,** and a rudimentary tunica albuginea separates them from the cortex (Fig. 21-3*C* and *D, left column*). Other primary cords remain in the cortical region, and they are joined in mammals and some other vertebrates by a proliferation of **secondary sex cords.** In mammals, the remaining primordial germ cells in the germinal epithelium are carried into the ovary by the secondary sex cords, but some sex cells remain in the germinal epithelium in many female vertebrates. As the cortex is thickening, blood vessels and connective tissue invade it from the medulla, and the medulla itself regresses, although traces of the medullary cords may remain. The secondary sex cords in the cortex break up into spherical clusters of cells, each of which contains a developing egg surrounded by a layer of epithelial cells. These groups of cells are the **primordial follicles.** The ovary, unlike the testis, develops primarily in the cortex of the indifferent gonad. As the ovary enlarges, it, too, separates from the mesonephros, but it remains connected to the body wall by a mesentery known as the **mesovarium.**

Development of the Reproductive Ducts

Lampreys and hagfishes, as we shall see, lack reproductive ducts, but they are present in other vertebrates. As the testis or ovary is developing, a network of potential tubules begins to form deep in the medullary region of the indifferent gonad between the primary sex cords and the anterior mesonephric tubules. They are called **rete cords,** or **cords of the urogenital union,** for they have the potential to connect the gonad and cranial mesonephric tubules (Fig. 21-3*C* and *D*). The rete cords canalize during the development of most male vertebrates, and the route through the mesonephros and its archinephric duct becomes the definitive sperm passage. The mesonephros and archinephric duct are also parts of the excretory system in male anamniotes but not in amniotes that evolve a metanephros and ureter (Chapter 20). The rete cords regress in females or, at most, leave functionless vestiges. Female anamniotes also retain mesonephric tubules and the archinephric duct as parts of their excretory system, but in female amniotes, these, too, regress or become vestigial.

Both sexes also begin to develop an **oviduct** (often called the **Müllerian duct** in embryos). The oviduct of chondrichthyans and some amphibians arises by a longitudinal splitting of the archinephric duct (Fig. 21-4). Its anterior opening into the coelom, known as the **ostium tubae** (L., *ostium* = door), appears to represent the coelomic funnels of one or more cranial renal tubules. This may be the way an oviduct evolved, but in other vertebrates, it simply arises as a longitudinal, groovelike invagination of the coelomic epithelium on the lateral surface of the mesonephros (Fig. 21-3*A* to *D, left*). As the groove extends caudad, it deepens, folds on itself, and forms the oviduct, which continues to grow posteriorly to the cloaca. The oviduct regresses in males, sometimes leaving traces, but in females, it continues to develop and becomes the passage for the removal of eggs. As the oviduct enlarges, it

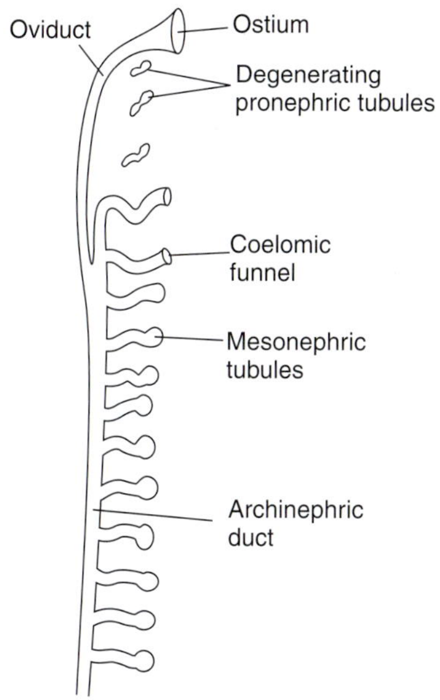

FIGURE 21-4
A ventral view of the developing oviduct of a chondrichthyan in which the oviduct forms by a longitudinal splitting of the archinephric duct. *(Modified from van den Broek.)*

separates from the mesonephros but remains connected to the body wall by a mesentery known as the **mesotubarium**.

The Mature Gonads

Although their structure varies to some extent among vertebrates, the gonads are relatively conservative organs, and those of mammals are representative. Important differences in other groups will be indicated.

Testes

Amniotes. The testes of amniotes are compact organs composed of long, highly convoluted **seminiferous tubules,** which lead to a system of ducts (cords of the urogenital union or rete testis) through which sperm leave. Many tubules end blindly, but others branch and anastomose or form loops, both ends of which connect to the duct system of the testis. Humans may have as many as 600 seminiferous tubules in a testis, some of which are 80 cm long. The tubules are surrounded by a loose connective tissue that contains the **interstitial cells of Leydig,** which secrete the male sex hormone, testosterone (Fig. 21-5*A*). Tubules often are grouped into lobules by dense connective tissue septa.

The tubule wall consists of sperm-forming cells and **Sertoli cells** (Fig. 21-5*B*). Sertoli cells rest on the basal lamina on the periphery of a seminiferous tubule and extend inward to the tubule lumen. They are large and irregularly shaped, with many recesses that surround groups of developing sperm cells, and probably contribute to their nutrition and maturation. Processes of Sertoli cells branch extensively in the wall of the tubule and contribute to its supporting framework.

The stem sperm-forming cells are diploid **spermatogonia,** which lie in the periphery of the tubule (Fig. 21-5*B*). Spermatogonia divide mitotically near the beginning of a reproductive season or, in humans and other species without reproductive seasons, throughout reproductive life. Some cells always remain as spermatogonia; others continue to grow to form **primary spermatocytes** (Fig. 21-6). The primary spermatocytes then divide once meiotically to form **secondary spermatocytes,** and these undergo a second meiotic division to form haploid **spermatids.** During the process of their development, the sperm-forming cells move from basal compartments in the Sertoli cells to compartments next to the lumen. Spermatids do not divide again but undergo a complex transformation to become the motile **spermatozoa.** A flagellate tail and acrosome form, and much of the cytoplasm is lost (Chapter 4). Sometimes the separation of daughter cells is incomplete and double-headed sperm are formed. The whole process of sperm formation, or **spermatogenesis,** requires several weeks (64 days in humans). In species with extended reproductive periods, successive waves of spermatogenesis travel along the seminiferous tubules, so different regions of the tubules are in different stages of spermatogenesis.

The testes lie within the body cavity close to their site of embryonic development in all vertebrates except for many mammals, in which the testes descend into an external pouch known as the **scrotum** (L., = pouch). In a few mammals, including monotremes, edentates, elephants, sirenians, and cetaceans, the testes remain in the abdomen. The testes of bats, most rodents and lagomorphs, and some carnivores and ungulates descend into the scrotum during the breeding season and are withdrawn into the body cavity after reproduction (Focus 21-1).

The scrotum is a pair of sacs formed by the evagination of all the layers of the body wall. Early during development, a cord of tissue known as the **gubernaculum** (L., = rudder) extends from the caudal end of each developing testis, through the incompletely

FIGURE 21-5
Seminiferous tubules and spermatogenesis in mammals. *A,* A transverse section of a part of the testis, showing seminiferous tubules and interstitial cells. *B,* Detail of a portion of a seminiferous tubule, showing Sertoli cells and some of the stages in spermatogenesis. Many of the stages lie in recesses of the Sertoli cells. *(After Fawcett.)*

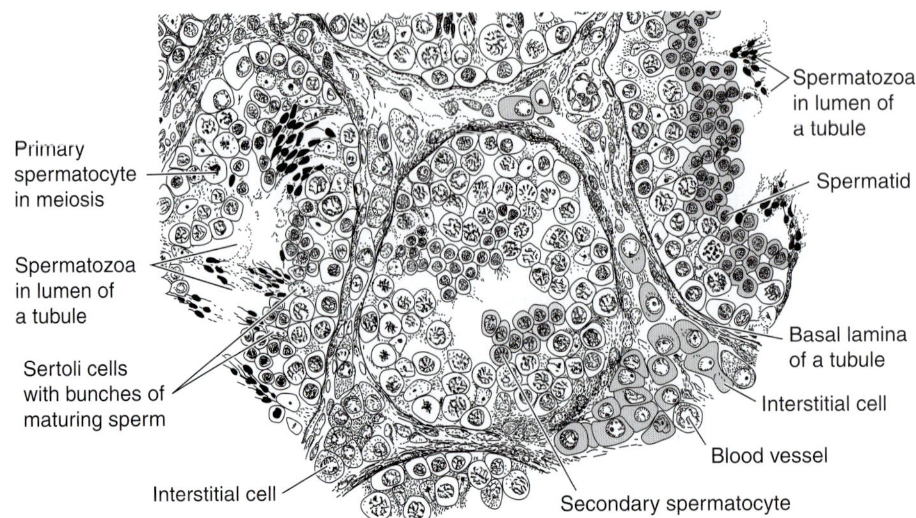

A. Transverse section of testis through several seminiferous tubules

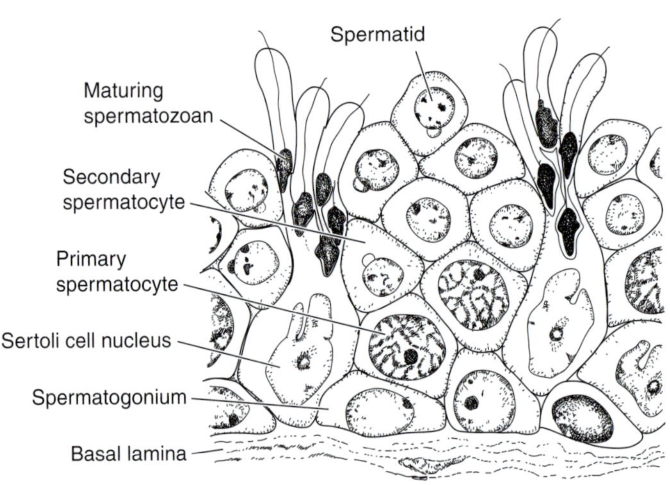

B. Detail of seminiferous tubule

formed body wall in the region of the groin, and into the skin of the future scrotum (Fig. 21-7*A* and *B*). Each testis descends in a retroperitoneal position soon afterward. As a testis approaches the groin, its descent is accompanied by an evagination of the muscular and connective tissue layers of the body wall and by a coelomic sac, the **vaginal sac** (or processus vaginalis; L., *vagina* = sheath), which enfolds the testis with its sperm duct, nerve, and blood vessels. These layers and the skin constitute the wall of the scrotum, but the muscular and connective tissue layers form a well-defined sac that travels a short distance beneath the skin of the groin before entering the scrotal skin (Fig. 21-7*C*). It sometimes is convenient to refer to this sac as the **cremasteric pouch,** for the muscular tissue within its wall is the **cremasteric muscle** (Gr., *kremastos* = hung). The area of the body wall from which the cremasteric pouch arises is the **inguinal canal.** The cord of blood vessels, nerve, and sperm duct that travels to and from the testis within the vaginal sac is the **spermatic cord.** The vaginal sac remains open in most mammals, but its proximal part normally atrophies in humans, leaving only a distal part surrounding the testis.

The mechanisms for the descent of the testis are not well known. The gubernaculum, being composed of soft tissue, may provide a route of low resistance. The failure of the gubernaculum to elongate after its formation and the continued growth of surrounding regions of the body may, in effect, pull the testis caudad. In species in which the testis migrates, intra-abdominal pressure contributes to its descent, and the contraction

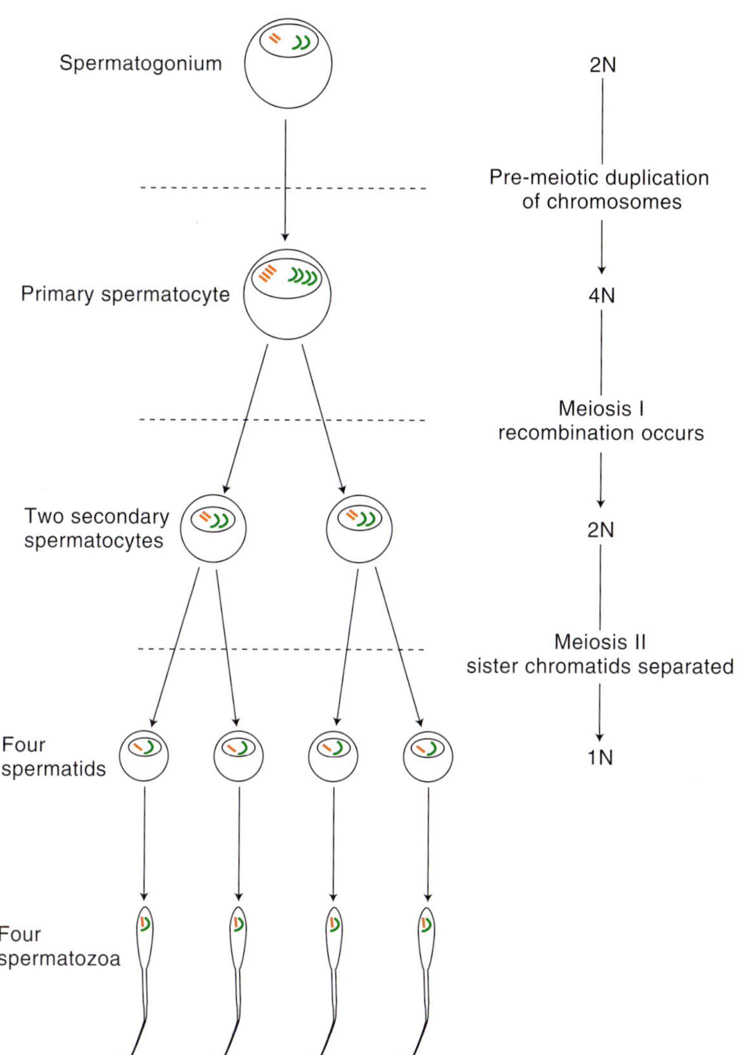

FIGURE 21-6
Mammalian spermatogenesis. *(After Fawcett.)*

of the cremasteric muscles shortens the pouch and pulls the testis inward.

Anamniotes. Fertilization is external in most anamniotes, so enormous amounts of sperm must be produced to ensure fertilization. The testes of most fishes are large and often fill the body cavity prior to spawning. The paired testes unite to form a single organ in lampreys and many teleosts. The testes of frogs are composed of seminiferous tubules, but those of most anamniotes consist of **seminiferous ampullae,** or lobules, of developing sperm cells. Cells in the walls of the ampullae appear to function as Sertoli cells. The ampullae are completely evacuated when sperm are discharged, and the ampullae themselves may regress. Before the next reproductive season, stem sperm-producing cells move into old and newly formed ampullae from a permanent population of spermatogonia that reside in other parts of the testis. Hormone-secreting interstitial cells are not present in the testes of all anamniotes, but they have been found in those of elasmobranchs and many amphibians. Branching, finger-shaped **fat bodies** are attached to the gonads of many amphibians and contain reserves of energy probably utilized in sperm and egg production (Fig. 21-8). In temperate species, gamete production begins during the late winter period of dormancy, when the animals are not feeding.

Developmental stages are found in some sexually immature anamniotes, in which the anterior part of the gonad resembles an ovary while the posterior part more closely resembles a testis. This is not hermaphroditism but a reflection of the potential of the embryonic gonad to develop into either a testis or an ovary. Only one part of the gonad normally functions, but the potential for sex reversal exists. Male toads of the family Bufonidae have a **Bidder's organ** at the front of the testis that resembles an arrested ovary (Fig.

FOCUS 21-1 *The Descent of the Testes*

The evolutionary reasons why the testes descend into an external scrotum in many mammals are not entirely clear. Temperatures are several degrees lower in the scrotum than intra-abdominal temperatures, 3°C in humans. Moore (1926) proposed that the lower scrotal temperatures are critical for certain stages in spermatogenesis, especially for the transformation of the spermatids into spermatozoa. If the testes fail to descend in young boys, normal spermatogenesis does not occur at puberty. Conversely, if scrotal temperatures are elevated in experimental animals, sterility results. Several factors reduce scrotal temperatures. The scrotum is exposed, its skin is thin and often sparsely haired, sweat glands are abundant, and subcutaneous fat is absent. Moreover, each testicular artery and vein is closely entwined to form a **pampiniform plexus** (L., *pampinus* = tendrile + *forma* = shape), which acts as a heat exchanger (Fig. A). Much of the heat in the warm blood going to the testis in the artery is transferred to the cooler venous blood returning from the testis before the arterial blood reaches the testis. The cremasteric muscle and smooth muscle fibers in the dartos tunic, which insert into the scrotal skin, fine-tune the mechanism. If temperatures become too cool, contraction of these muscles pulls the testes closer to the warmth of the body wall. Bedford (1977) has proposed that the critical thermosensitive stage is not a part of spermatogenesis in the testis but rather the storage of spermatozoa in the adjacent epididymis.

Experiments confirm that in many species, sperm cannot develop normally or survive at the higher body core temperatures. Several additional lines of evidence support the hypothesis that lower temperatures are needed for sperm development or storage. The testes are permanently descended in many species, they migrate into the scrotum during the reproductive season in other species, and some species that have a high core body temperature but lack a scrotum have alternate cooling mechanisms. Cetaceans appear to have a countercurrent cooling mechanism in which branches of veins bringing cool blood from the dorsal fins and flukes entwine with an arterial plexus that supplies the testes (Rommel et al., 1992). Birds, which do not have scrota, also have high core temperatures, although air sacs may lower the temperature in the immediate vicinity of their testes.

Frey (1991) proposed an alternate hypothesis that he believes explains why some mammals have descended testes and others do not. He points out that the last stage in spermatogenesis (when spermatids are transformed into spermatozoa) is highly sensitive to fluctuations in pressure. Oogenesis lacks this last pressure-sensitive stage. He further points out that intra-abdominal pressures fluctuate widely in mammals that gallop or jump or the vertebral column of which is strongly flexed and extended during locomotion. The testes of mammals with these patterns of locomotion must be shielded from the damaging fluctuations in intra-abdominal pressures. Frey reasons that a scrotum evolved as a protection against pressure fluctuations as many mammals adopted new locomotor strategies. Mammals without these specialized patterns of locomotion had no need for a scrotum. Once the testes were descended, they may have become secondarily adapted to the cooler scrotal temperature. Frey also believes that the pampiniform plexus is more a peripheral pump than a heat exchange device. Pulsations of the testicular artery help drive venous blood back into the caudal vena cava and thus prevent its peripheral accumulation, which could increase pressures in the testis. Not all investigators agree with Frey. Indian elephants, in which the testes are intra-abdominal, are used to move logs and do other heavy work, and this activity probably raises intra-abdominal pressures. Clearly more studies are needed to unravel mysteries of descended testes.

21-8). Seasonal changes in its size suggest an endocrine function, but when the testis ceases to function in old males, a Bidder's organ may become a functional ovary.

Ovaries

The eggs undergo their maturation, or **oogenesis,** within the ovaries. The ovaries of most vertebrates are paired organs, but in lampreys and many teleosts, they unite to form a single organ. In some species, either the left or the right ovary atrophies. Many viviparous sharks retain only the right one. The ovaries are paired in a few birds, including hawks, but most birds retain only the left one. Monotremes have both ovaries, but eggs mature only in the left one. The adaptive significance of the loss or reduction in function of an ovary is unclear.

Mammalian Ovaries. The ovaries of mammals are relatively smaller organs than in other vertebrates. Few eggs are produced in a reproductive period, and they contain less yolk than do the eggs of reptiles or birds. Therian embryos undergo at least part of their development in a uterus, but monotremes are ovoviviparous. The eggs of monotremes are retained for a while and laid at a relatively later stage, at about 4 mm in diameter, when they leave the ovary; a human egg has a

Werdelin and Nilsonne (1999) have offered a new phylogenetic perspective of the scrotum and testicular descent in mammals. They propose that the presence of a scrotum and descended testicles is primitive for living mammals and that evolution has generally proceeded from a scrotal condition to progressively more ascrotal and to the complete loss of testicular descent. The most derived nondescended testicles occur in monotremes, elephants, sirenians, elephant shrews, and hyracoids. True seals (Phocidae) have retained testicular descent even though they lack a scrotum and have subcutaneous testicles, whereas whales (Cetacea) have their testicles near their kidneys but retain the pampiniform plexus, an indication of the former presence of a scrotum. They conclude that evolution has proceeded from scrotal to ascrotal, but never the reverse, and that the loss of testicular descent is rare. Werdelin and Nilssone conclude that the descent of the testes into the scrotum may have evolved earlier together with endothermy as an adaptation to a constant high core temperature. Subsequent loss of the scrotum and absence of testicular descent are thought to be restricted to those mammals that have evolved special mechanisms to cool the intraabdominal testes (Pabst, et al., 1995). Thus pressure fluctuations and high intra-abdominal temperatures have been implicated as causal factors in the descent of the testes and evolution of the scrotum in mammals.

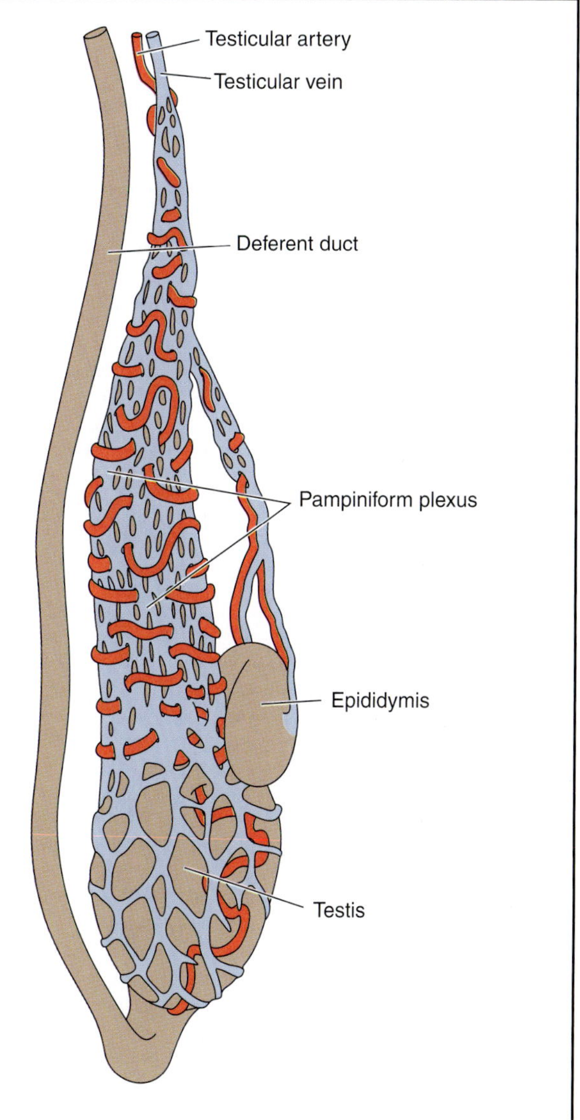

A. The pampiniform plexus in a human. (After Frey.)

diameter of about 140 μm when it leaves the ovary. Mammalian ovaries, like the testes, become connected by a gubernaculum to the body wall during their development, but they undergo only a partial descent. The gubernaculum often persists in an adult female as a **round ligament of the ovary** extending between the ovary and the groin.

Mature mammalian ovaries are dense organs consisting of a vascularized, connective tissue stroma in which the follicles are embedded (Fig. 21-9). Unlike spermatogonia, the oogonia cease multiplying during embryonic life. Eggs in the **primordial follicles,** which are located around the periphery of the ovary, have reached the **primary oocyte** stage of oogenesis. A human ovary contains approximately 2 million of these follicles at birth, but most atrophy, leaving only about 40,000 at puberty. Most of these also atrophy, and only 400 or so will ovulate. As the remaining eggs and follicles continue to develop after puberty, the follicular cells multiply and form many layers of **granulosa cells** around each egg. Surrounding layers of connective tissue contribute a sheath, or **theca,** to the wall of the follicle. The follicle protects and transfers nutrients to the egg. It is also an endocrine gland that secretes female sex hormones. As a follicle continues to develop, a **follicular liquor** accumulates in spaces among the granulosa cells. These spaces gradually enlarge, coalesce, and form a large central cavity, making

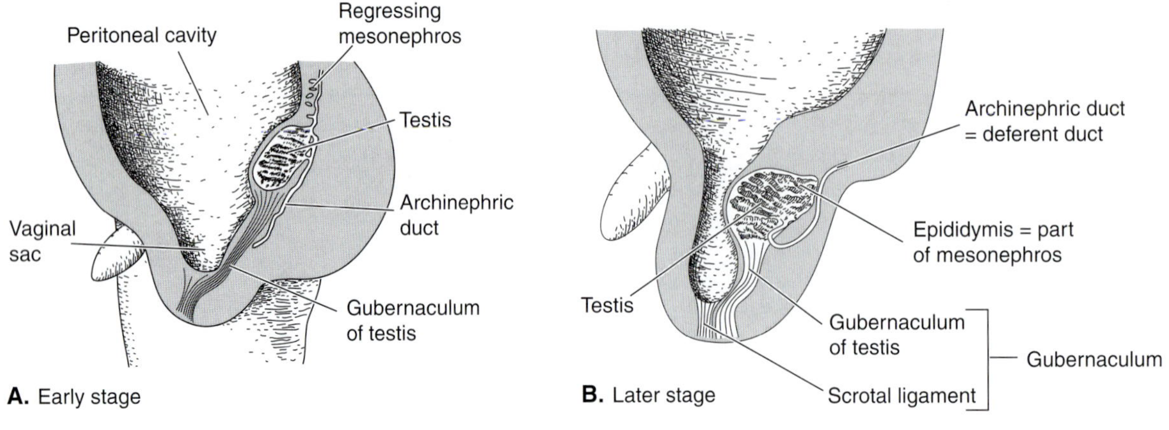

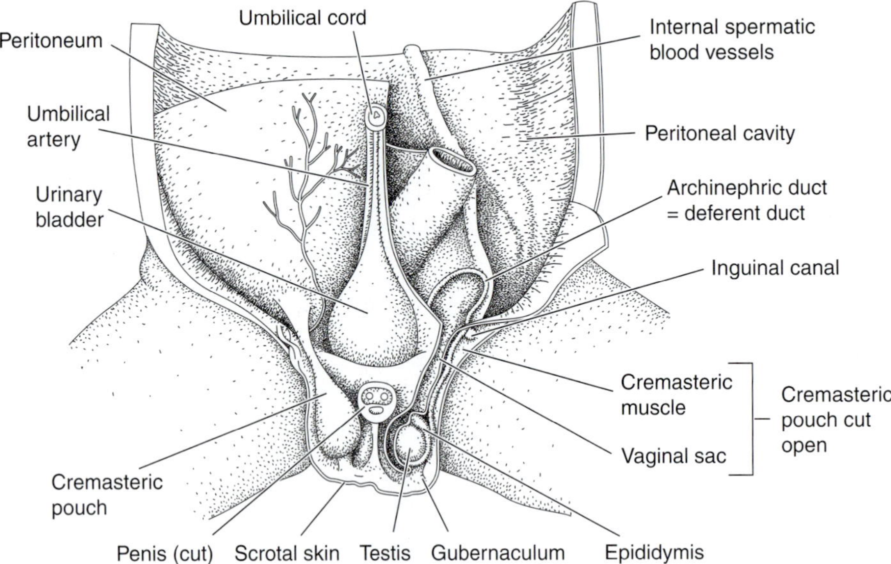

FIGURE 21-7
The descent of the testis based on a male human. *A* and *B,* An early and a later stage in a lateral view. *C,* A ventral dissection of a nine-month fetus to show the descended testis and its relationship to surrounding structures. *(After Corliss.)*

the egg eccentric in position. A **mature** or **Graafian follicle** may be 10 mm or more in diameter, and it bulges outward on the ovary surface. During these processes, the primary oocyte enlarges and undergoes its first meiotic division to form a **secondary oocyte** and a minute, **first polar body.** Food reserves are kept within one functional cell. The second meiotic division, which will give rise to the mature **ootid** and a **second polar body,** begins shortly before ovulation but normally requires the stimulus of fertilization to be completed. As an egg cell matures, the follicular cells secrete around it the primary envelope, known in mammals as the **zona pellucida.** The number of follicles that mature during a reproductive period depends on whether the species normally gives birth to single or multiple young.

Each mature follicle ruptures at the ovary surface at **ovulation,** and the eggs, still surrounded by some granulosa cells, are discharged into the coelom to be picked up by the reproductive tract. A ruptured follicle is converted to a hard, yellowish body, the **corpus luteum,** which also secretes female sex hormones, estrogen, and progesterone. Corpora lutea begin to atrophy before the next reproductive cycle.

Nonmammalian Ovaries. The ovaries of other vertebrates are relatively much larger than those of mammals. So many eggs are produced by oviparous

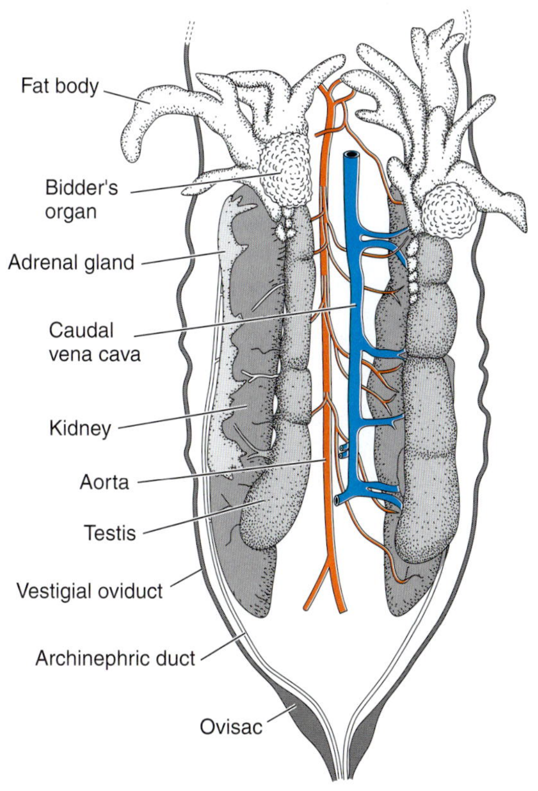

FIGURE 21-8
A ventral view of the urogenital system of a male toad. *(After Turner and Bagnara.)*

anamniotes, and the eggs of oviparous amniotes contain so much yolk, that the ovaries fill all the available space in the body cavity prior to egg laying; they regress to a small size after the reproductive season. The ovaries do not have so extensive a stroma as in mammals (Fig. 21-10). Often they are hollow and contain a lymph space. Oogenesis occurs within follicles and passes through the same stages as in mammals. Oogonial divisions are completed during embryonic life in birds, some reptiles, and chondrichthyans, but the germinal epithelium remains active in most other vertebrates, and they produce many eggs during their reproductive lives. The follicles of most vertebrates have very thin walls and consist of only one or a few layers of granulosa cells. As an egg accumulates yolk, it enlarges and fills the follicle. A follicular cavity is not present. Usually, several generations of developing follicles are present at one time. The remains of the follicles usually are reabsorbed after ovulation. Functional corpora lutea may develop from the follicles in some species, but chiefly those that have some degree of viviparity: most sharks, several amphibians, and some lizards and snakes. Birds lack corpora lutea.

Evolution of the Reproductive Passages

Hagfishes and lampreys lack reproductive ducts. Both sperm and eggs are discharged into the coelom at the time of spawning. Gametes leave the caudal end of the coelom through a pair of **genital pores** that enter the most caudal part of the archinephric ducts, where these ducts conjoin to form a **urogenital papilla** that protrudes into the cloaca (Fig. 21-11). The genital pores open under the influence of sex hormones only a few weeks before spawning. Genital pores may be the ancestral vertebrate method of gamete discharge. This certainly is the simplest way morphologically, and the discharge of gametes into the coelom also occurs in many invertebrate groups. Chondrichthyan fishes have a pair of **abdominal pores** of uncertain function that lead from the coelom to the cloaca. Some investigators regard them as vestiges of a primitive method of gamete discharge, but all gnathostomes have evolved reproductive ducts that remove the gametes.

Male Reproductive Passages

Amphibians. The simplest morphologically, and probably the ancestral pattern of male reproductive ducts, is found in oviparous amphibians. The duct system of the testis, which develops embryonically from the rete testis, converges to a **testis canal,** located in the center or along one margin of the testis (Fig. 21-12*A*). A variable number of **connecting tubules,** also called **efferent ductules** in amphibians (L., *ex* = out + *ferent* = carrying), carry sperm from the testis canal, through the mesorchium, to the anterior end of the opisthonephros, where they connect to a **lateral kidney canal** or directly to kidney tubules. The lateral kidney canal, when present, develops from anterior mesonephric tubules. The connecting tubules appear to develop from the testicular duct system (rete testis), but contributions from mesonephric tubules cannot be ruled out. Sperm then pass through a few **anterior opisthonephric tubules** to the **archinephric duct,** through which they continue to the cloaca. In caecilians and some salamanders, sperm-carrying kidney tubules retain glomeruli, but in other groups, the tubules lose them and their urinary function. In a few salamanders, the adjacent part of the archinephric duct becomes highly coiled and stores sperm. This part of the duct sometimes is given the name of its mammalian homologue, the **duct of the epididymis.** The archinephric duct of most amphibians carries both sperm and urine, although not at the same time (Fig. 21-8). A few salamanders have evolved one or more

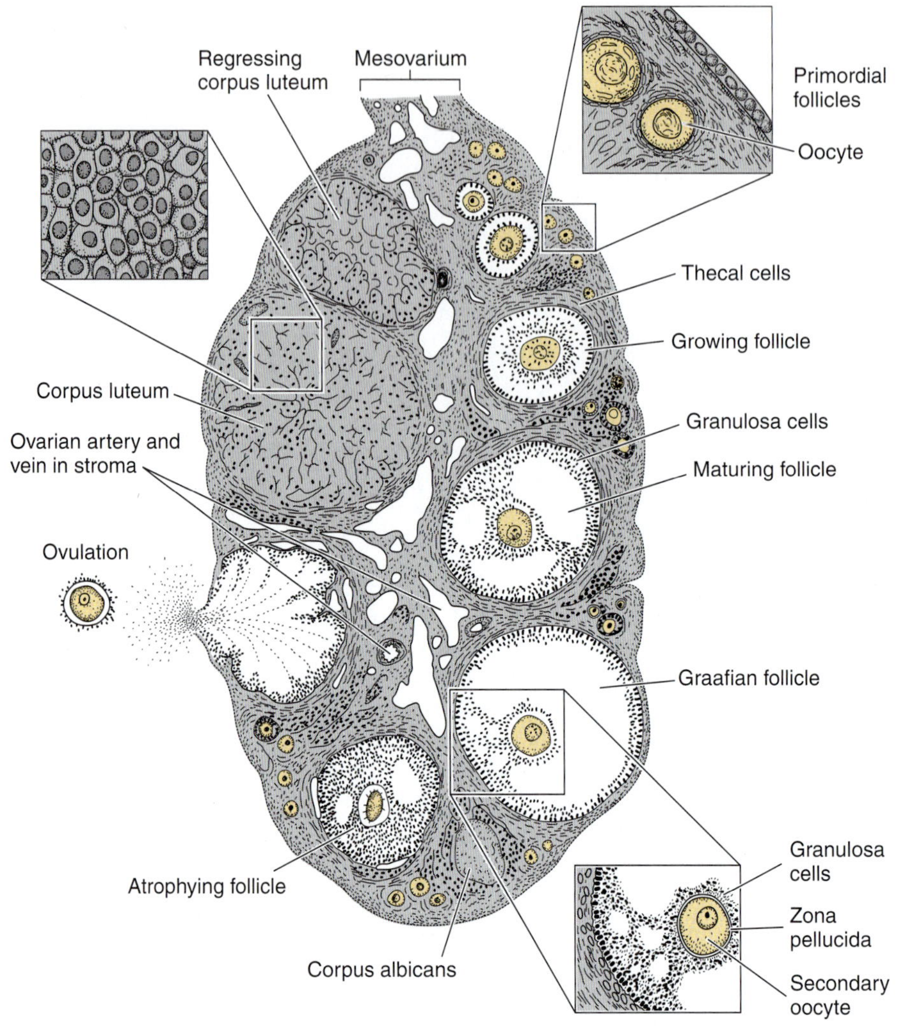

FIGURE 21-9
A longitudinal section of a mammalian ovary showing developing follicles, ovulation, and the development and regression of the corpus luteum. All of these stages would not be seen at the same time. *(After Turner and Bagnara).*

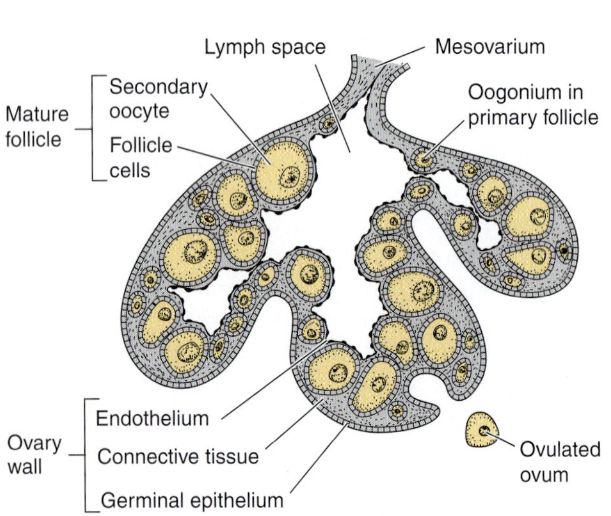

FIGURE 21-10
A section through an amphibian ovary.

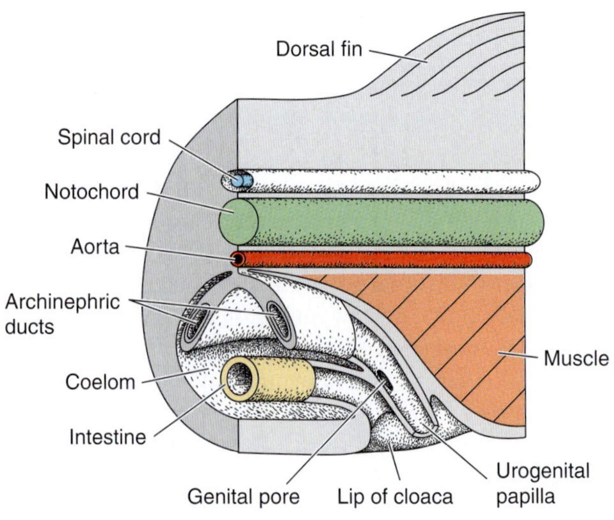

FIGURE 21-11
A stereodiagram of the cloacal region of an adult lamprey, showing the genital pore. *(After Knowles.)*

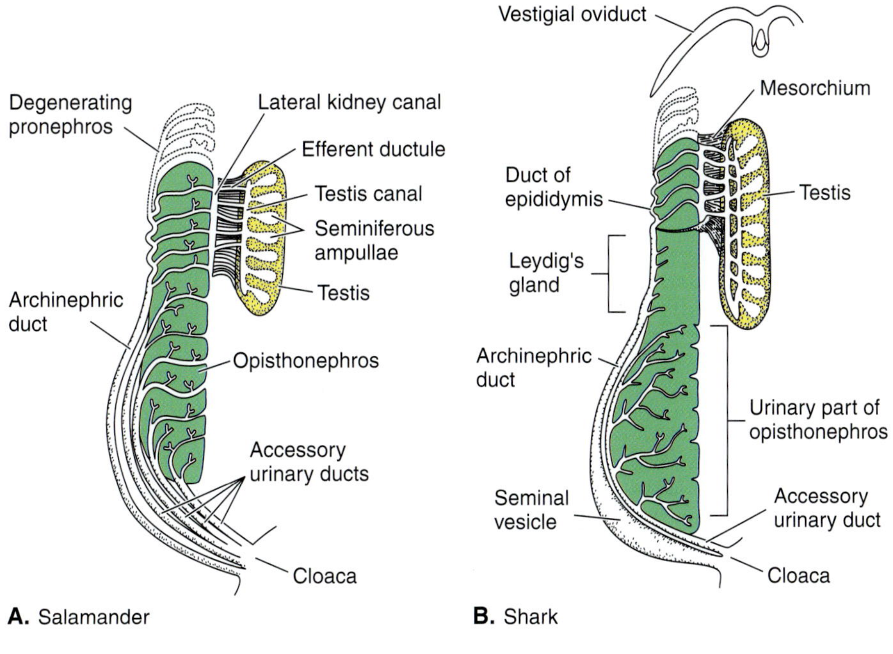

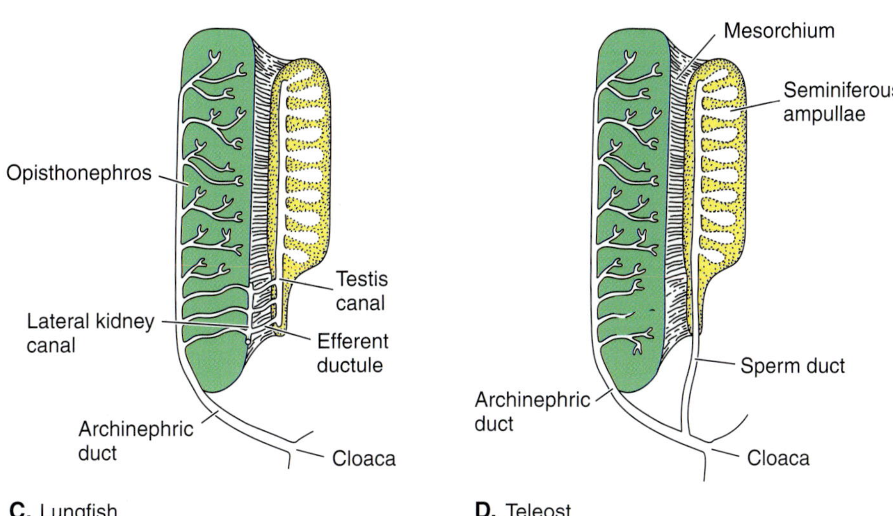

FIGURE 21-12
Ventral views of the male reproductive ducts and their relationship to excretory ducts in representative male anamniotes. *A,* A salamander. *B,* A shark. *C,* The South American lungfish, *Lepidosiren.* *D,* A teleost. *(After Portmann.)*

accessory urinary ducts that transport most of the urine (Fig. 21-12*A*). The embryonic Müllerian ducts often remain as vestigial oviducts in males (Fig. 21-8).

Chondrichthyan Fishes. The reproductive passages of male cartilaginous fishes represent a modification of the pattern, seen in amphibians, that evolved with internal fertilization. One or more accessory urinary ducts drain the caudal urinary portion of the kidney, and the rest of the kidney and its archinephric duct are taken over by the reproductive system (Fig. 21-12*B*). Most of the opisthonephric tubules between those receiving sperm and those carrying urine are modified as **Leydig's gland** and secrete a **seminal fluid.** Seminal fluid transports the sperm to the female and probably has some of the nutritive and activating functions that it has in mammals (p. 670). The archinephric duct is wide as it crosses the urinary part of the kidney and forms a **seminal vesicle,** which contributes to the seminal fluid and helps store sperm. Part of each pelvic fin is modified as a copulatory **clasper** that transfers sperm to the female. A long **siphon** with muscular walls is associated with each clasper. Water taken into the siphon plus its own secretions help propel sperm into the female. Vestiges of the oviducts commonly occur in males.

Bony Fishes. Lungfishes and early actinopterygians retain the primitive gnathostome pattern of male reproductive passages, except that the testis connects

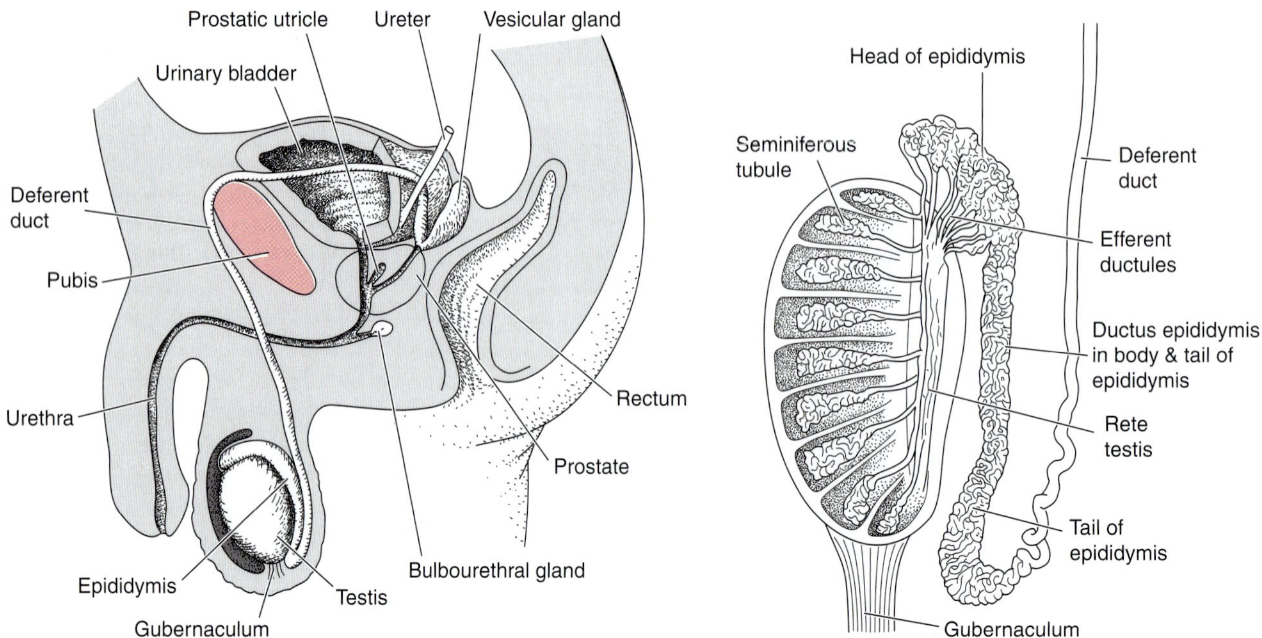

FIGURE 21-13
A and *B,* The reproductive system of a male amniote, based on a human. The pattern of the accessory sex glands varies among mammals. In male reptiles and birds, the testes do not descend, the ducts open into a cloaca, and the penis is embedded in the ventral cloacal wall.

with caudal rather than cranial renal tubules in some lungfishes (Fig. 21-12*C*). Teleosts have evolved a unique **sperm duct,** which develops as an extension of the testis canal, completely bypasses the kidneys, and continues caudad to open on the outside of the body adjacent to the excretory openings. It joins the caudal end of the archinephric duct in a few species (Fig. 21-12*D*).

Amniotes. Except for the location of the testis within a scrotum, the pattern of reproductive ducts in all male amniotes is similar to that in mammals (Fig. 21-13). The kidney, now a metanephros, is drained by a ureter. The archinephric duct and the part of the embryonic mesonephros connecting to the testis function only as reproductive passages. The developing testis does not separate far from the mesonephros, so the **rete testis** leads directly into modified mesonephric tubules, now called the **efferent ductules.** The efferent ductules of amniotes develop from mesonephric tubules; they are not homologous to the connecting tubules (sometimes called efferent ductules) of anamniotes, which develop from the rete testis and pass through a rather wide mesorchium to the testis. The distal ends of the efferent ductules are highly coiled and become the head of a band of tissue, called the **epididymis** (Gr., *epi* = upon + *didymos* = testis), that lies on the surface of the testis. The body and tail of the epididymis are composed of an extensively convoluted part of the archinephric duct called the **duct of the epididymis.** The amniote epididymis is homologous to the cranial end of the anamniote opisthonephros and adjacent parts of the archinephric duct. The continuation of the archinephric duct, now called the **deferent duct** (L., *de* = from + *ferent* = carrying), extends caudally to the cloaca or to the part of the mammalian urethra that is derived from the cloaca (p. 682). In mammals with a scrotum, each deferent duct travels in the spermatic cord and enters the body cavity through the inguinal canal. Vestiges of caudal mesonephric tubules (the **paradidymis**) and an embryonic Müllerian duct (**appendix testis, prostatic utricle**) sometimes persist in adult males (Fig. 21-14 and Table 21-1).

Spermatozoa are stored in the epididymis and deferent duct, where secretions nourish them and appear to play a role in their maturation. Sperm removed from the epididymis before they have stayed there for a few hours are incapable of becoming motile and fertilizing eggs.

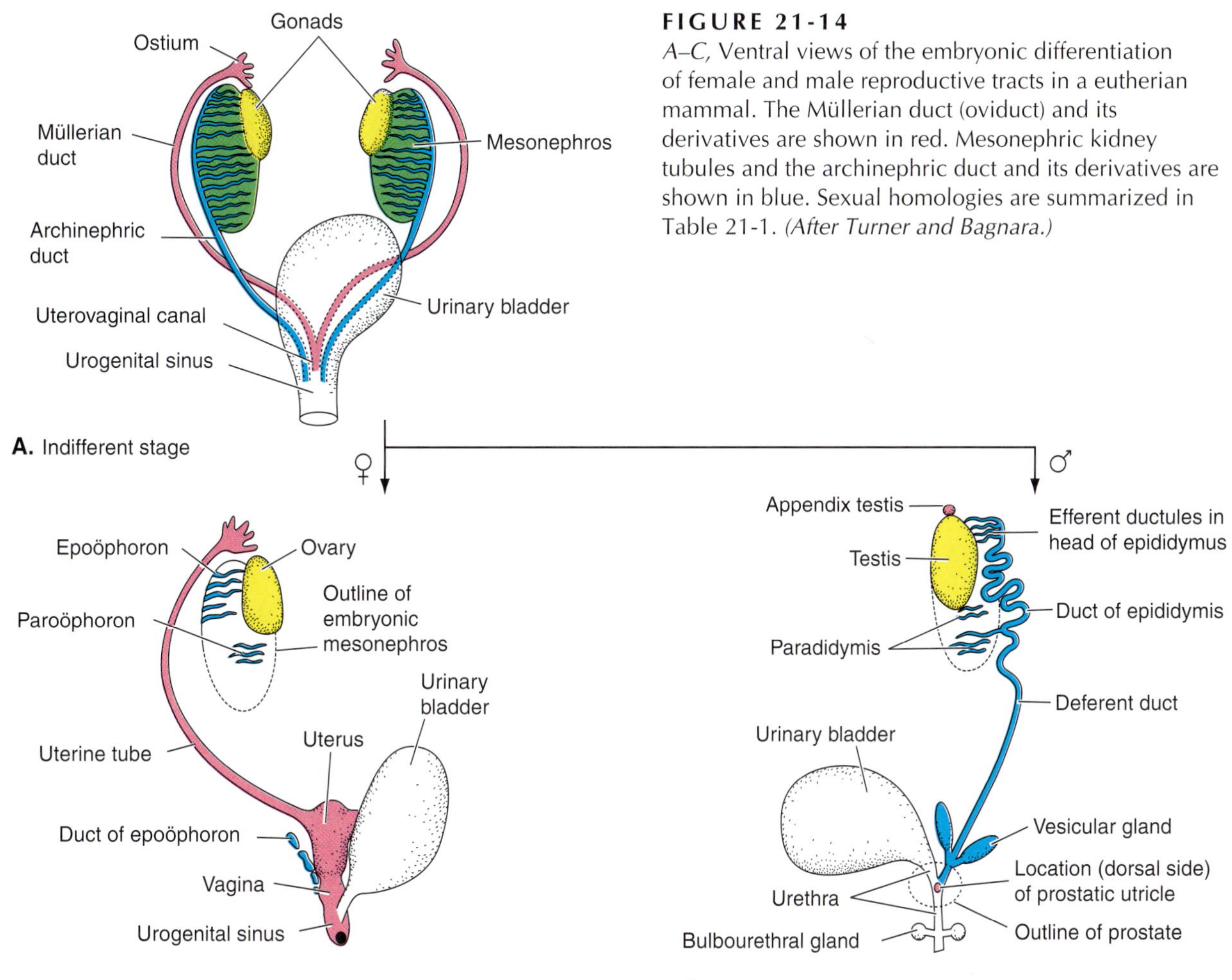

FIGURE 21-14
A–C, Ventral views of the embryonic differentiation of female and male reproductive tracts in a eutherian mammal. The Müllerian duct (oviduct) and its derivatives are shown in red. Mesonephric kidney tubules and the archinephric duct and its derivatives are shown in blue. Sexual homologies are summarized in Table 21-1. *(After Turner and Bagnara.)*

Several accessory sex glands that secrete the seminal fluid develop along the sperm passages. They are particularly well developed in mammals and may include the **prostate gland, vesicular gland,** and **bulbourethral gland** (Figs. 21-13*A* and 21-14*C*). In addition to transporting sperm, the seminal fluid of mammals nourishes the sperm, helps make them motile, and neutralizes acids and other materials in the female reproductive tract that would interfere with their survival.

Most male amniotes have evolved a single or bifid intromittent organ with which sperm are transferred to the female. This organ, known in mammals as a **penis,** develops in the ventral wall of the cloacal region. It lies in the tail base in reptiles. When not engorged with blood and erect, it usually is withdrawn into a sheath. Most male birds transfer sperm to the females simply by a brief cloacal apposition, but an intromittent organ is present in phylogenetically more primitive birds, such as the ratites, ducks, geese, and a few other species.

Female Reproductive Passages and Reproduction in Nonmammalian Vertebrates

Embryonic mesonephric tubules and the archinephric ducts are never used in gamete transport by females; rather, the eggs take an independent route through the oviducts. The mesonephric tubules in female anamniotes are incorporated in the opisthonephros, which is drained by the archinephric duct. These structures are vestigial in female amniotes (**paroöphoron, epoöphoron**), for the metanephros and ureter are the functional excretory organs (Fig. 21-14*B* and Table 21-1).

Amphibians. The reproductive passages of female oviparous amphibians, like those of the male, are closer to the ancestral vertebrate pattern than are the more specialized passages of most fishes (Fig. 21-15*A*). A pair of long, convoluted **oviducts** extend along the dorsal body wall beneath the kidneys from the anterior end of the body cavity to the cloaca. The anterior end of each

TABLE 21-1 Embryology and Sexual Homology of Mammalian Reproductive Organs*

Embryo	Adult Male	Adult Female
Primary sex cords	Seminiferous tubules	(Medullary cords of ovary)
Secondary sex cords	—	Ovarian follicles
Gubernaculum	Gubernaculum	Round ligament of ovary
Rete testis	Rete testis	—
Mesonephros		
Cranial tubules	Efferent ductules in head of epididymis	(Epoöphoron)
Caudal tubules	(Paradidymis)	(Paroöphoron)
Archinephric duct	Duct of epididymis	(Duct of epoöphoron)
	Deferent duct	(Duct of epoöphoron)
Oviduct (Müllerian duct)		
Cranial end	(Appendix testis)	Uterine tube
		Uterus
Caudal end	(Prostatic utricle)	Vagina
Allantois	Urinary bladder	Urinary bladder
	Part of urethra	Urethra
Urogenital sinus	Part of urethra	Vaginal vestibule
Genital tubercle	Much of penis	Clitoris
Genital groove	Part of urethra	Vaginal vestibule
Genital folds	Ventral part of penis	Labia minora
Scrotal swellings	Cutaneous part of scrotum	Labia majora

*Parentheses identify vestigial structures; a dash indicates that the structure has been lost.

oviduct is expanded slightly to form a funnel-shaped **infundibulum** (L., *infundibulum* = funnel), which opens into the coelom through an **ostium tubae.** Its caudal end sometimes expands to form an **ovisac,** in which eggs accumulate for a brief period before they are laid. Reproductive hormones stimulate the development of cilia on the peritoneum, so as the eggs are ovulated, they are swept toward the ostium. Ciliary action carries them into the ostium, and ciliary action within the oviduct and muscular contractions of its wall carry them caudad. As the eggs pass down the oviduct, they become coated with secondary envelopes secreted by oviductal cells. The secondary envelopes of amphibians are jelly-like layers that imbibe water and swell after the eggs have been laid and fertilized. These layers help protect and insulate the eggs.

Fertilization is external. Usually, sperm are sprayed over the eggs as they are laid. Before egg laying, many male salamanders release their sperm in mucous clumps called **spermatophores** (Gr., *phoros* = bearing), which the female collects with her cloacal lips and stores for a short time in cloacal recesses. Only a few species build nests. Most eggs are simply deposited in the water, but they often clump to each other and to surrounding objects. A few salamanders that lay in damp locations on land wrap their bodies around their eggs and brood them. This helps maintain a high level of moisture.

Some degree of viviparity has evolved in most or all caecilians, several frogs, and some salamanders of the genus *Salamandra* that retain eggs in the oviducts. A number of tropical frogs have evolved other ways to protect and brood their tadpoles, including carrying them on their backs, in vocal sacs, or even in their stomachs.

Chondrichthyan Fishes. Primitive cartilaginous fishes are oviparous, but the male has a copulatory organ, and fertilization is internal. Although only a few eggs are produced, these contain enough yolk to provide for the young until they hatch. Eggs are large—5 cm or more in diameter. Both oviducts curve ventrally anterior to the liver and unite. They share a common ostium tubae in the falciform ligament (Fig. 21-15*B*). The ventral position and large size of the ostium tubae are specializations for receiving the exceptionally large eggs. Secondary envelopes, which take the form of a horny, proteinaceous **egg case,** are secreted by a **nidamental gland** (L., *nidamentum* = nesting material). The egg cases of oviparous species bear tentacles that become entangled in seaweed or other material, and the young pass through a larval stage in the egg cases (Fig. 21-16). Embryos of vivipa-

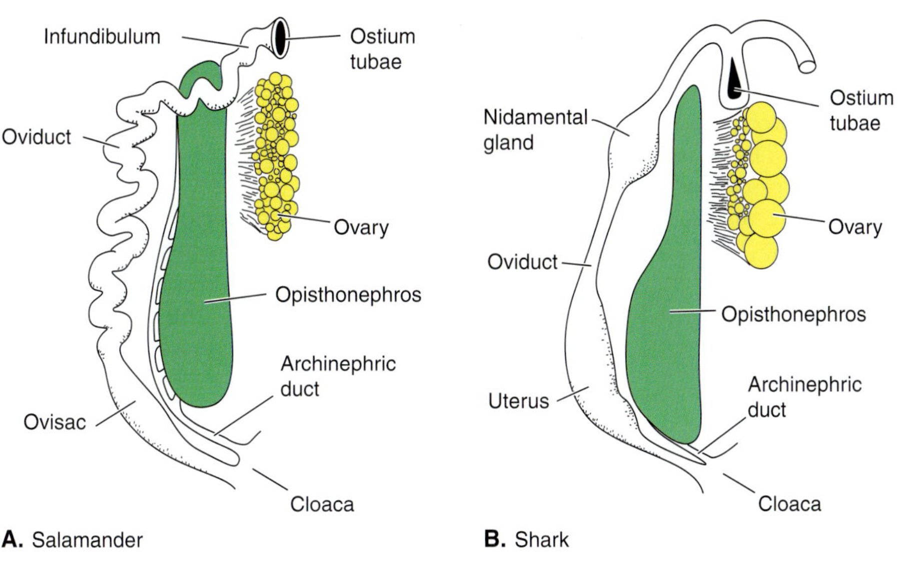

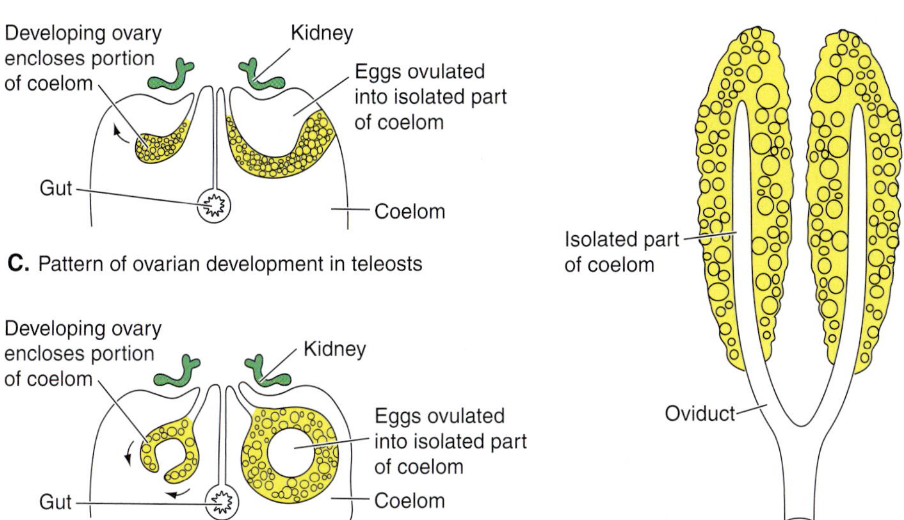

FIGURE 21-15
The reproductive tracts of female anamniotes. *A,* A salamander. *B,* A shark. *C* and *D,* Transverse sections through the developing ovary of teleosts, illustrating two ways the hollow ovary develops. *E,* A ventral view of the ovaries and oviducts of a teleost. *(A and B, Modified from Portmann.)*

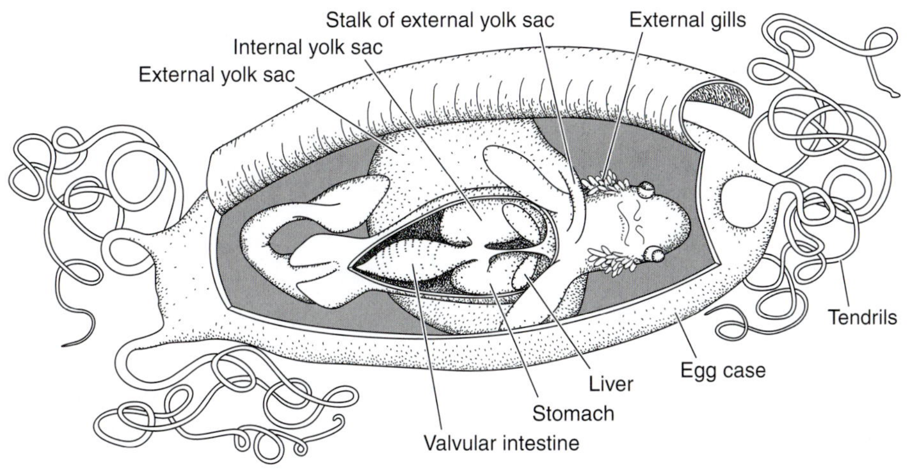

FIGURE 21-16
A section of the egg case of an oviparous shark. Some of the yolk stored in the external yolk sac is transferred to an internal yolk sac later in development. At birth, the external yolk sac is very small, but food reserves remain in the internal yolk sac. *(After Hamlett et al., 1993.)*

rous species are lodged in an expanded **uterus** that develops from the caudal part of the oviduct. Only thin-walled egg cases are secreted around the eggs in these species, and they are later reabsorbed. The degree of vascularity and the complexity of folding within the uterus correlate with the degree to which the embryos depend on the mother for nutrients, from very little to full placental viviparity (reviewed by Hamlett and Koob, 1999). The placenta, when present, is formed by the apposition of the embryo's vascular yolk sac to the uterine lining, so it is called a **yolk sac placenta.**

Bony Fishes. Lungfishes and nearly all actinopterygians are oviparous. The oviducts of primitive species receive eggs that are ovulated into the general body cavity, but most teleosts have evolved a unique ovary and oviduct. The ovary contains a large central cavity representing a part of the coelom that has become separated from the main body cavity by a folding of the ovary against the body wall or on itself (Fig. 21-15*C* and *D*). Eggs are discharged into this cavity and are not free in the general coelom. The oviduct usually is a simple tubelike extension of the ovary, so it is not homologous to the oviducts of other vertebrates (Fig. 21-15*E*). Both oviducts fuse caudally and open through a genital papilla located near the anus and excretory openings. These modifications are correlated with the large number of eggs, sometimes numbering in the millions, that teleosts release in a season. Fertilization is external in oviparous species. Some teleosts build nests, and a few brood their young. One or both parents protect them from predators and help circulate water across them. The embryos of the few viviparous species develop in the oviduct, in the coelomic space within the ovary, or in the follicles themselves.

Reptiles and Birds. Reptiles, such as turtles, have a pair of long, convoluted oviducts (Fig. 21-17*A*), but the right one is reduced or lost in most birds, together with the right ovary (Fig. 21-17*B*). The eggs of reptiles enter the oviduct from the coelom, but in birds, they directly enter an exceptionally large ostium tubae of the infundibulum, which partially invests the ovary and mature follicles. The eggs are fertilized in the upper region of the oviduct, and the glandular oviduct wall secretes the secondary membranes.

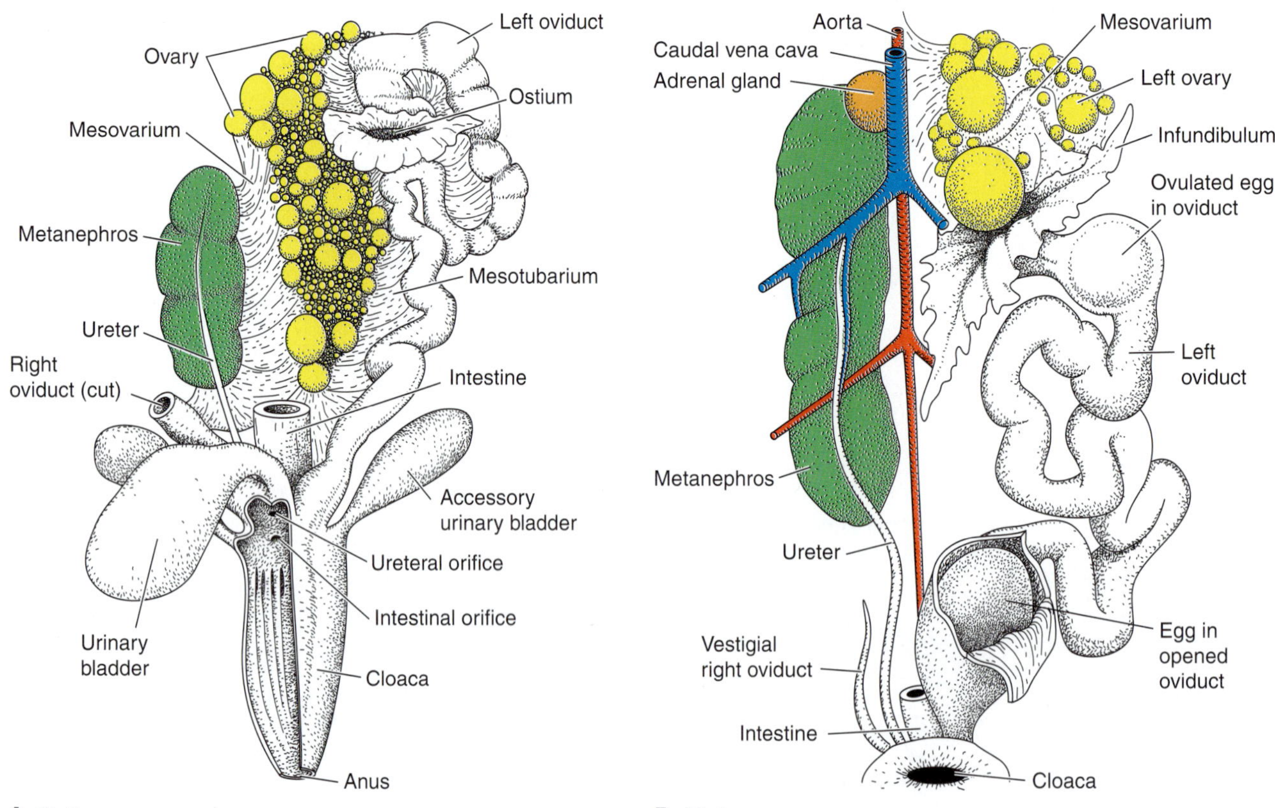

FIGURE 21-17
Ventral views of the urogenital systems of female reptiles and birds. *A,* A turtle. *B,* A bird. Only the left ovary is shown for the turtle, but a right one is present; most birds have only the one shown. *(A, After Turner and Bagnara; B, After Portmann.)*

Most reptiles and all birds are oviparous, but the evolution of a cleidoic egg permits them to bypass the aquatic larval stage and reproduce in a terrestrial environment (Chapter 4). Large amounts of yolk are stored in the eggs, and other materials are provided by the albumen that is secreted around the eggs by the oviduct. The egg is completely surrounded by a shell membrane and a shell, which are secreted by the lower end of the oviduct. The shell is composed of a meshwork of proteinaceous fibers impregnated with calcium salts. In turtles and crocodilians, the eggshell is more calcified than in lepidosaurs, in which the eggshell is thin and parchment-like; the bird eggshell is more heavily calcified. Gas exchange and, depending on the external environment, some loss or gain of water vapor occurs through it. The need for gas exchange precludes cleidoic eggs from being laid in the water. Most reptiles bury their eggs in nests on land, where they will be incubated by environmental temperatures, but a few reptiles and all birds brood their eggs. As the embryo develops, its extraembryonic chorion and amnion help protect it, and its allantois allows gas exchange and stores excretory products. The eggs of reptiles and birds are large because they contain so much yolk. Some turtles and snakes may lay nearly 100 eggs, but most reptiles lay fewer, and birds lay only 1 to 20 in a season. Viviparity has evolved independently in many lizards and snakes, and many degrees of dependence on the mother are known (Focus 21-2).

Passages and Reproduction in Mammals

Monotremes. In monotremes, the oviducts have become differentiated into many of the distinctive regions that characterize other mammals (Fig. 21-18*A*). Each oviduct has an infundibulum that partly enfolds the ovary, and a narrow **uterine tube,** or **Fallopian tube,** that continues to an expanded **uterus.** The two uteri independently enter a **urogenital sinus** that, after also receiving the excretory products, continues to the cloaca. As the fertilized eggs pass down the uterine tube and enter the uterus, the secondary envelopes that characterize cleidoic eggs are secreted around them, but the shell is very thin, permeable, and expansile. The egg accumulates additional food from uterine secretions while in the uterus and grows from its ovulation diameter of 4 mm to 15 mm by the time it is laid. The shell thickens after an egg has reached its final size. Embryonic development begins while the egg is in the female reproductive tract. The embryos of monotremes and some lizards and snakes first depend on food stored in the egg while it is in the egg and later become dependent on food provided by uterine secretions. This shift appears to be an intermediate stage in the evolution of placental viviparity. An advantage of this intermediate condition is less waste of maternal resources if the egg does not become fertilized or fails to develop. Eggs are brooded by the female after they are laid; echidnas develop a temporary skin pouch that folds over them. Hatched young lap up milk secreted by the mammary glands. Nipples are not present. The evolution of the female's ability to secrete milk after the hatching or birth of the young, and the maternal care associated with nursing, are particularly important features that have contributed greatly to the success of mammals.

Marsupials. All therian mammals are placentally viviparous, but metatherians and eutherians have evolved different patterns. In marsupial females, the caudal end of each oviduct becomes specialized as a **vagina,** and the vaginae enter the **urogenital sinus** (Fig. 21-18*B* and *C*). The urogenital sinus opens directly to the outside because the cloaca becomes divided in all therians (p. 680). The anterior ends of the vaginae unite to form a **vaginal sinus,** into which the uteri enter. Each uterus has its own neck, or **cervix,** so it is independent, a condition described as **duplex.** A narrow uterine tube continues from each uterus to the infundibulum. The penis of many male marsupials is bifid, and sperm travel up both vaginae. Shortly before birth in primitive marsupials, such as the opossum, the vaginal sinus grows caudad and connects directly with the urogenital sinus to form a shorter and more direct birth canal. This pathway, known as the **pseudovaginal canal,** is a permanent structure in most marsupials following the first parturition (e.g., in the kangaroo).

Fertilized marsupial eggs receive a mucoid coat and a shell membrane as they travel down the uterine tubes to the uteri, but no shell is secreted. The embryos develop within the shell membrane during the first two thirds of their gestation period. Gestation periods are very brief, ranging from ten days in the opossum to about a month in some kangaroos. Nutrients secreted by the uterus are absorbed through the shell membrane, and gas exchange also occurs across it. After the reabsorption of the shell membrane, an expanded and vascular yolk sac reaches the chorionic ectoderm (Fig. 21-19*A*) and forms a **yolk sac,** or **choriovitelline placenta,** that is applied loosely to the uterine lining. The early expansion of the yolk sac prevents the allantois from reaching the chorion in most species, but the allantois does reach it in a few marsupials, including the koala, and forms an additional **chorioallantoic placenta** (Fig. 21-19*B*). These placentas do not penetrate the uterine lining in most species, but in the bandicoot, the chorioallantoic placenta is invasive.

FOCUS 21-2 *Squamate Placentation*

Of the approximately 20% of squamates that are viviparous, most are aplacental and derive their inorganic and organic nutrients primarily from material stored in the yolk. Blackburn (1993a) proposed that these species, together with oviparous ones, be called **lecithotrophic** (Gr., *lekithos* = egg yolk + *trophe* = nourishment). Only a few species exhibit true placental viviparity and derive an important quantity of nutrients by exchanges with the mother. They are **placentotrophic.**

Viviparous squamates have evolved many types of placentas, differing in the fetal extraembryonic membranes (Fig. 4-17) that make contact with the uterine lining. (A placenta, by definition, is an intimate apposition or fusion of fetal and maternal tissues for physiological exchange.) This subject has been studied by many investigators, and different aspects have been reviewed by Blackburn (1993) and Stewart (1993). The yolk sac contributes to several types of placentas, and its unique morphology needs a brief description. The yolk sac of squamates is unusual in that the splanchnic layer of mesoderm invades the yolk mass and forms a **yolk cleft** that separates the distal part of the yolk mass (the **isolated yolk mass**) from the main **yolk sac** (Fig. A). The isolated yolk mass is surrounded only by ectoderm and endoderm. This layer is called the **bilaminar omphalopleure** (Gr., *omphalos* = navel). The most distal pole of the yolk is not covered by extraembryonic membranes and forms a **yolk navel.**

Early in development, only the wall of the yolk sac, which becomes vascularized by splanchnic mesoderm, contacts the uterine lining, forming a yolk sac, or **choriovitelline placenta** (L., *vitellus* = egg yolk) This is a transitory placenta in squamates. After the isolated yolk mass forms, its avascular wall (the bilaminar omphalopleure) contacts the uterine lining, forming an **omphaloplacenta.**

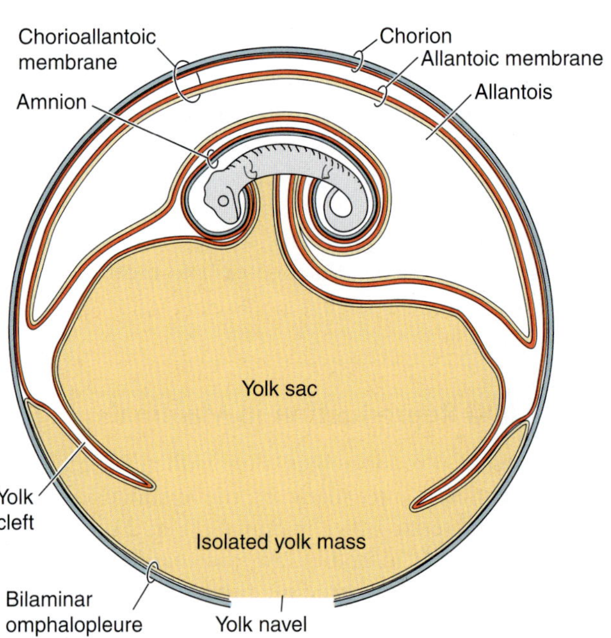

A. The extraembryonic membranes of a squamate. The yolk becomes separated by a yolk cleft into the yolk sac and an isolated yolk mass. Part of the chorioallantoic membrane contacts the uterine lining to form the allantoplacenta. Part of the bilaminar omphalopleure contacts the uterine lining to form the omphaloplacenta. *(After Blackburn.)*

The placental relationship of marsupials lasts only a short time before the young are born. The young complete their development attached to the nipples of the mammary glands, which often are located within a skin pouch, the **marsupium** (L., *marsupium* = pouch). A marsupium is particularly well developed in hopping and arboreal species. The reasons for an early birth are not entirely understood. Because half of the embryo's genes are paternal, the embryo is partly foreign tissue to the mother. It is hypothesized that the shell membrane, which is maternal tissue, prevents foreign embryonic antigens from reaching the mother and stimulating the synthesis of antibodies. Antibodies are synthesized, however, when the shell membrane is reabsorbed and a placenta is established because the chorionic ectoderm, or **trophoblast,** does not function as an immunological barrier, as it appears to do in eutherian mammals (p. 144). Thus, the early birth is partly an immunological rejection.

In some species, the expanding allantois later invades the yolk cleft and contributes to the omphaloplacenta, forming an **omphaloallantoic placenta.** In all squamates, the allantois expands as development continues and contacts the chorion throughout the dorsal hemisphere of the egg, forming a chorioallantoic membrane**.** On contacting the uterine lining, the chorioallantoic membrane forms a type of chorioallantoic placenta, often called the **allantoplacenta** in squamates. The allantoplacenta is similar in many ways to the chorioallantoic placenta of eutherian mammals.

In all of these placentas, a thin shell membrane at first lies between the fetal extraembryonic membranes and uterine lining but apparently does not interfere with transfers between embryo and mother. It degenerates later in development in most cases. Most oxygen and carbon dioxide exchanges are believed to occur through the allantoplacenta. The chorioallantoic membrane is the only vascularized extraembryonic membrane available to accomplish this task during most of gestation. The transitory choriovitelline placenta could participate earlier in development, and the omphaloallantoic placenta could participate late in gestation in species in which the yolk mass regresses sufficiently to bring the vascular surface of this placenta close to the chorion. Water easily passes into the egg and could be picked up by any of the placental types.

Because most squamates have a large supply of yolk and are highly lecithotrophic, their inorganic and organic nutrients are derived from the yolk. Some nutrients could be derived from the mother and enter the egg through the omphaloplacenta and omphaloallantoic placentas. The embryonic component of these placentas have hypertrophied cells and other specializations, as do the adjacent uterine cells, that imply a role in nutritive transfer. This has yet to be confirmed experimentally. In the few squamates that have a reduced yolk supply and are highly placentotrophic, nutritive transfers must occur through the allantoplacenta. This placenta is the most complex found in squamates (Fig. B). The uterine lining adjacent to the uterine artery and vein forms a series of deep infoldings that interdigitate with comparable foldings of the chorioallantoic membrane. Both the cells lining the uterus and the adjacent cells on the surface of the chorion show characteristic structural features of absorptive epithelial cells.

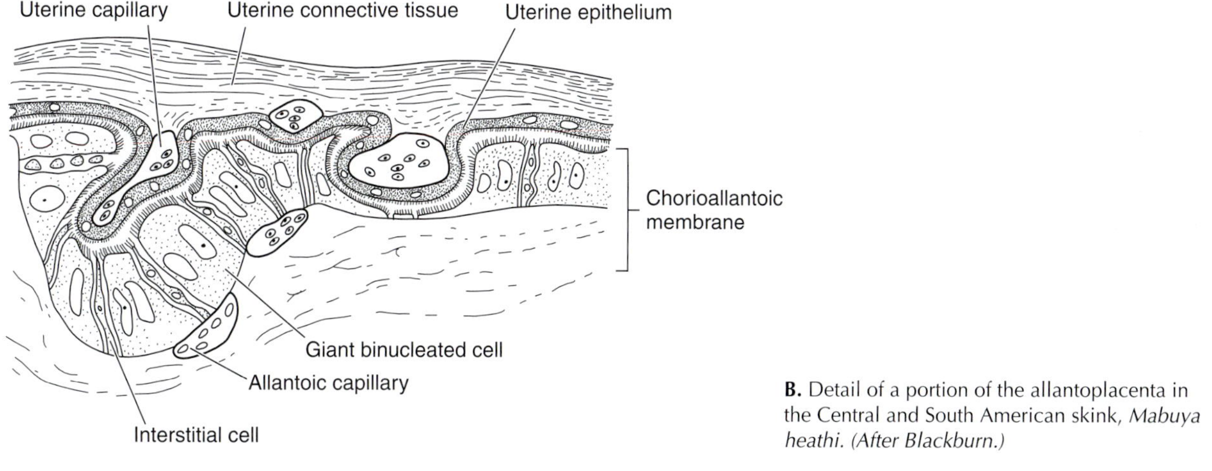

B. Detail of a portion of the allantoplacenta in the Central and South American skink, *Mabuya heathi. (After Blackburn.)*

The pattern of marsupial reproduction should not be regarded as a step toward the eutherian chorioallantoic placenta and relatively long intrauterine life. Marsupials and eutherians evolved independently from an ancestral therian group. The reproductive success of marsupials is at least as high as in eutherians. Marsupial reproduction is particularly well adapted to the stressful and unpredictable environments where most recent species live. Food is plentiful on the Australian continent after a rainfall, but it becomes scarce and patchy during droughts, some of which last for years. Few maternal resources are invested in the young during their short intrauterine life, and young in a pouch can easily be aborted when environmental resources fail. Marsupial mothers live to reproduce again, but eutherian mothers, drawing on their own stored reserves to support the embryos when food is in short supply, may weaken themselves and jeopardize their lives. Moreover, marsupials can reproduce very rapidly during favorable periods because pregnancy does not interrupt

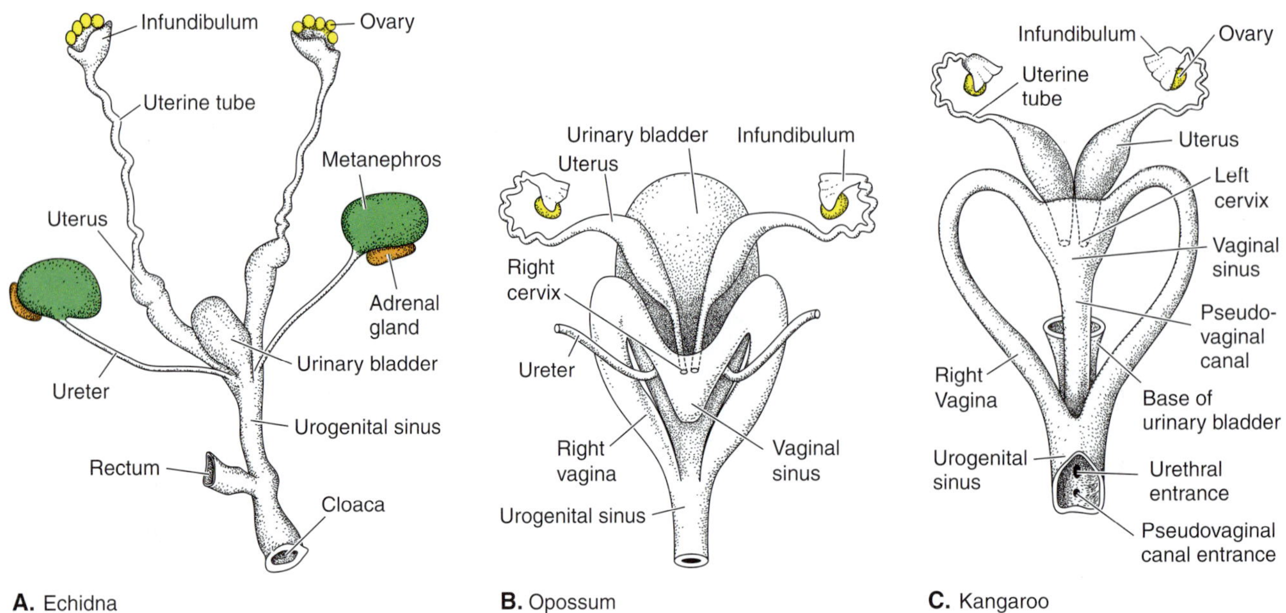

FIGURE 21-18
The reproductive system of female monotremes and marsupials. *A,* A ventral view of an echidna. *B,* A dorsal view of an opossum. *C,* A dorsal view of a kangaroo. *(A, After Griffiths; B and C, after Portmann.)*

the cyclical production of eggs, as it does in eutherians. A kangaroo may have one joey hopping about at heel, which occasionally crawls into the marsupium for milk; a younger offspring attached to a nipple and being supplied with a different type of milk; and a third in utero. The development of the intrauterine embryo is arrested in an early stage because the lactation for its older siblings prevents further development and additional pregnancies. When they are weaned, development of the intrauterine embryo is resumed and additional pregnancies may occur.

Eutherians. An infundibulum, uterine tube, uterus, and vagina also differentiate along the oviducts of eutherian mammals, but the vaginal portions unite to form a single, median organ that leads to the urogenital sinus, or **vaginal vestibule** (Fig. 21-20*A*). The uteri are **duplex** in lagomorphs, many rodents, and elephants, for each enters the vagina independently and has its own **cervix.** The caudal ends of the uteri have fused in most eutherians to form a small **uterine body** from which a pair of long, convoluted **uterine horns** extend. **Uterine tubes** extend from the horns. Development occurs within the horns. A slight partition remains within the uterine body of the **bipartite uterus** of most carnivores and many ungulates, but not in the **bicornuate uterus** of other ungulates, many carnivores, and whales (Fig. 21-20*B* and *C*). A complete fusion of the uteri into a single, large body occurs in armadillos and most primates, including humans (Fig. 21-20*D*). Although called a **simplex uterus,** this type represents the most advanced degree of uterine fusion.

Secondary envelopes are not secreted around the fertilized eggs of eutherians during their passage down the uterine tubes to the uterus. Rather, each embryo quickly develops a surface **trophoblast** and establishes a placental relationship with the mother soon after reaching the uterus. The placenta is **chorioallantoic,** for the allantois and not the yolk sac vascularizes the trophoblast, which is the mammalian homologue of the chorionic ectoderm, very early in development (Fig. 21-19*C*). The degree of union between the embryo's chorioallantoic membrane and the mother's uterine lining, or **endometrium** (Gr., *metra* = womb), varies considerably among groups. The fetal component of the placenta does not penetrate the endometrium in a few ungulates, including pigs, so the uterine lining is not disrupted and is not shed at birth. This is described as a **nondeciduous placenta** (L., *decidere* = to fall down or off). In most eutherians, however, the early embryo becomes partly or completely embedded within the endometrium. Much of the uterine lining is cast off at birth with these **deciduous placentas.** The degree to which the endometrium is eroded by the fetal trophoblast also varies; the terms used describe the maternal and fetal tissues that are in contact. No disruption of the endometrium is seen in

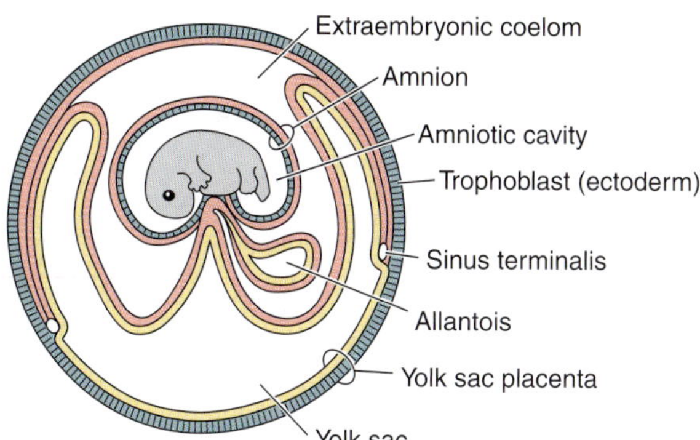

A. Yolk sac placenta typical of most marsupials

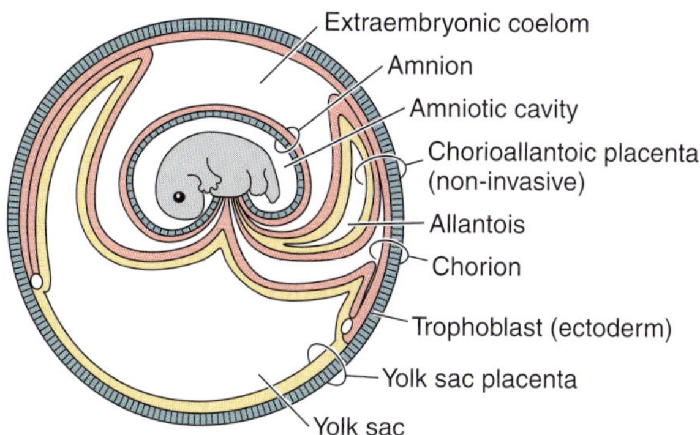

B. Yolk sac and chorioallantoic placentas of a koala

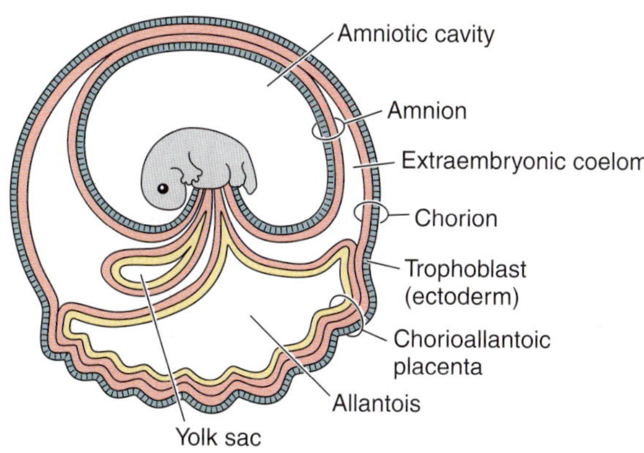

C. Chorioallantoic placenta of a typical eutherian

FIGURE 21-19
The placentas of therian mammals. *A,* The yolk sac placenta characteristic of most marsupials. *B,* The combined yolk sac and chorioallantoic placenta of the koala. *C,* The chorioallantoic placenta of a eutherian. *(After Dawson.)*

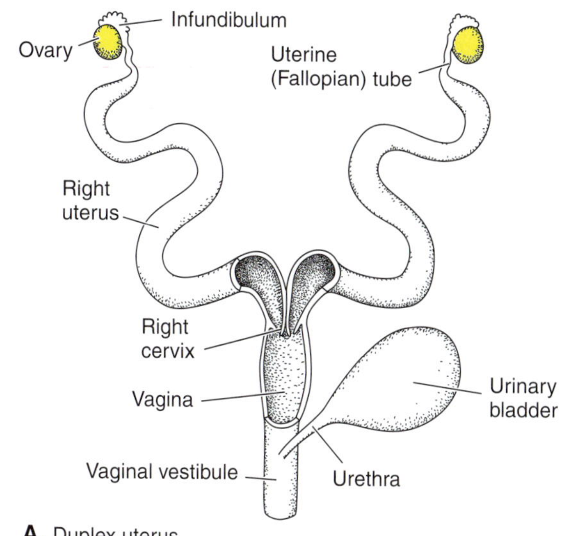

A. Duplex uterus

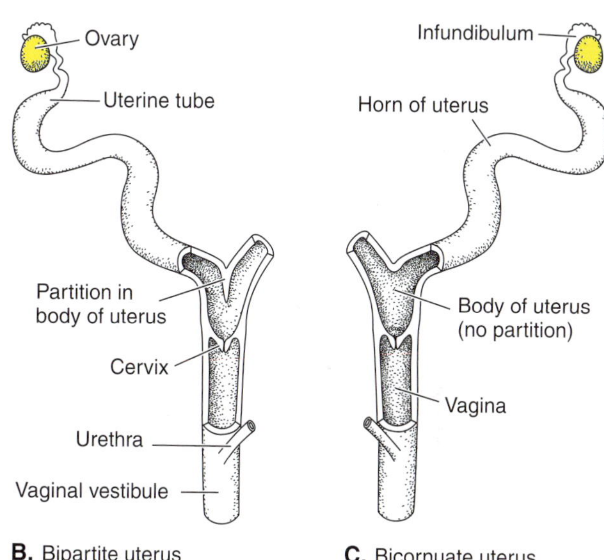

B. Bipartite uterus

C. Bicornuate uterus

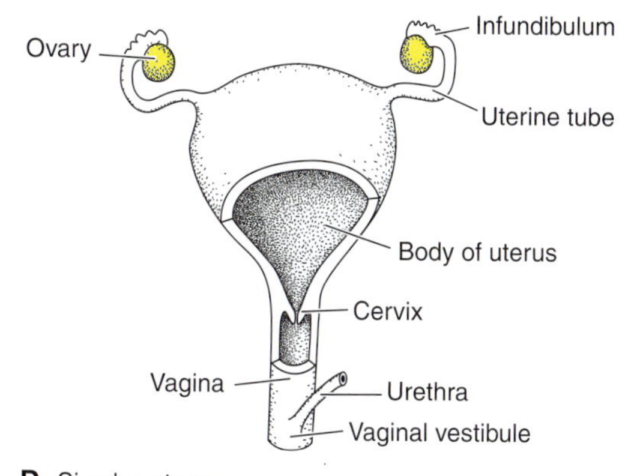

D. Simplex uterus

FIGURE 21-20
A–D, Diagrams of the reproductive tracts of eutherian mammals showing types of uteri.

the **epitheliochorial placenta** of pigs (Fig. 21-21*A*). With a **hemochorial placenta,** found in insectivores, bats, many rodents, and anthropoid primates, maternal blood vessel walls break down so that villi on the surface of the fetal chorion are bathed by maternal blood (Fig. 21-21*B*). Lagomorphs and some rodents have a

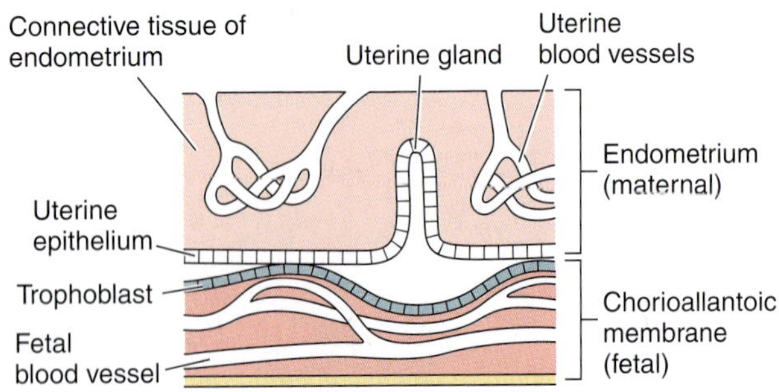

A. Epitheliochorial placenta

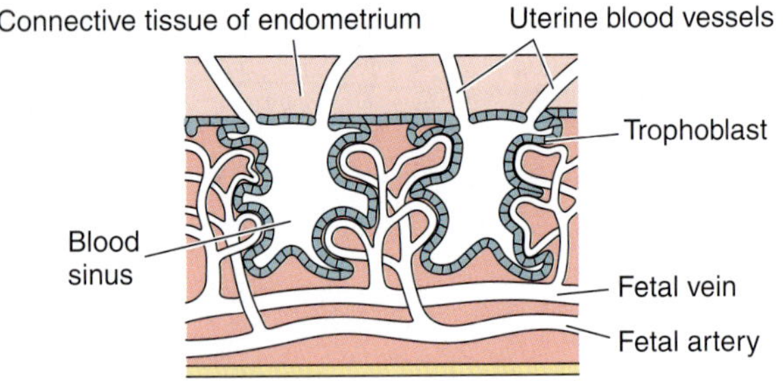

B. Hemochorial placenta

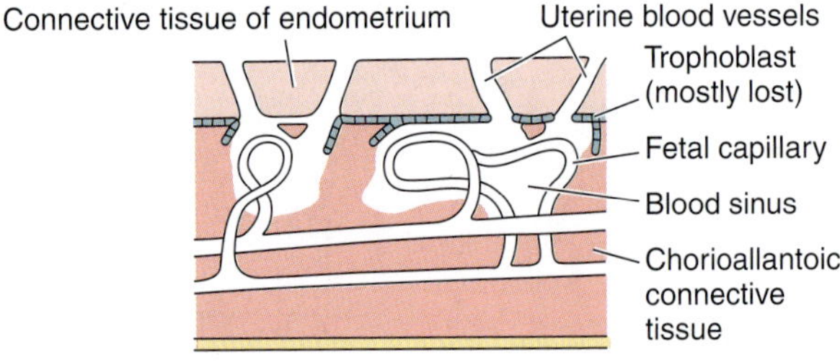

C. Hemoendothelial placenta

FIGURE 21-21
Representative chorioallantoic placentas, showing the degree of union between maternal and fetal tissues. *A,* Epitheliochorial placenta. *B,* Hemochorial placenta. *C,* Hemoendothelial placenta. *(A and B, Modified from Witschi.)*

hemoendothelial placenta, for most of the trophoblast on the surface of the chorion also is lost, and maternal blood flows across the endothelial walls of fetal capillaries (Fig. 21-21*C*).

Division of the Cloaca: The External Genitalia

Most vertebrates have a common chamber, the **cloaca,** that receives the intestine and the excretory and reproductive ducts (Fig. 20-7). Most actinopterygian fishes have lost the cloaca, and their excretory and reproductive ducts open on the body surface, often through a

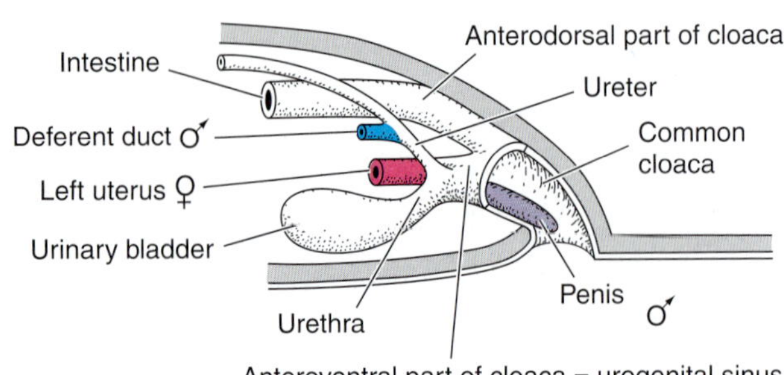

FIGURE 21-22
Lateral view of the cloaca of a monotreme. The point of entry into the cloaca of both male and female organs is shown, but an individual would have only one or the other. *(After Grant.)*

common papilla, just posterior to the anus. The urinary bladder of amphibians, and those amniotes that have one, is a ventral outgrowth from the cloaca. Monotremes retain the cloaca, but it has become partly divided into a dorsal portion, which receives the intestine, and a ventral **urogenital sinus,** which receives the urinary bladder, ureters, and female uteri (Fig. 21-22). The points of entrance of the ureters have shifted more ventrally than in reptiles, and they enter the urogenital sinus beside the neck of the bladder, or **urethra.** The deferent ducts of the male join the distal end of each ureter. The union of a ureter and deferent duct reflects the embryonic condition, in which the ureter is an outgrowth from the caudal end of the archinephric duct (now the deferent duct). The intestinal and urogenital parts of the cloaca enter a common chamber posteriorly.

The cloaca of an adult therian mammal is completely divided, so the intestine and urogenital passages have separate openings on the body surface. The way the division comes about can be seen clearly during embryonic development. An early, sexually indifferent embryo has a cloaca very similar to that of monotremes (Fig. 21-23*A*). A **urorectal fold** grows caudally and gradually divides the cloaca into a dorsal part, which becomes the **rectum,** and a ventral **urogenital sinus.** The allantois, the base of which will form the urinary bladder, connects with the anteroventral part of the urogenital sinus, and the still conjoined archinephric ducts and ureters attach close to the entrance of the allantois. The oviducts also enter the urogenital sinus close by. A **genital tubercle** begins to form on the anteroventral wall of the cloaca. In a later, sexually indifferent stage, the ureters separate from the archinephric ducts and shift to the developing bladder (Fig. 21-23*B*). The neck of the bladder begins to narrow to form the urethra, which joins the urogenital sinus. A **genital groove** flanked by **genital folds** develops along the cloacal side of the enlarging genital tubercle.

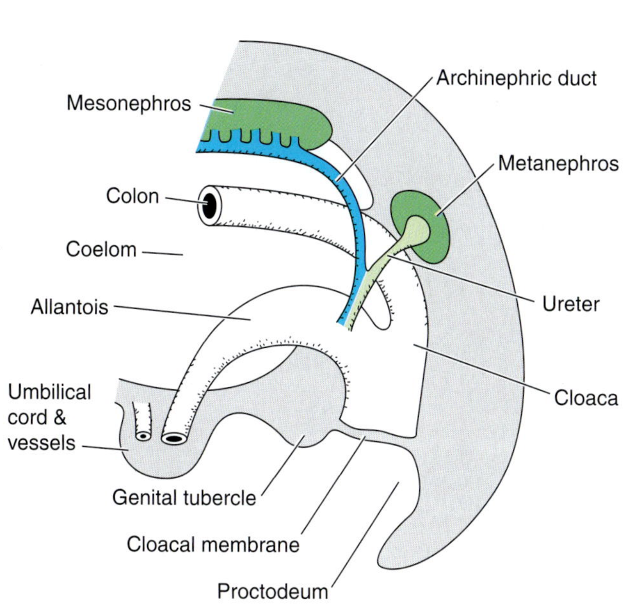

A. Early sexually indifferent stage

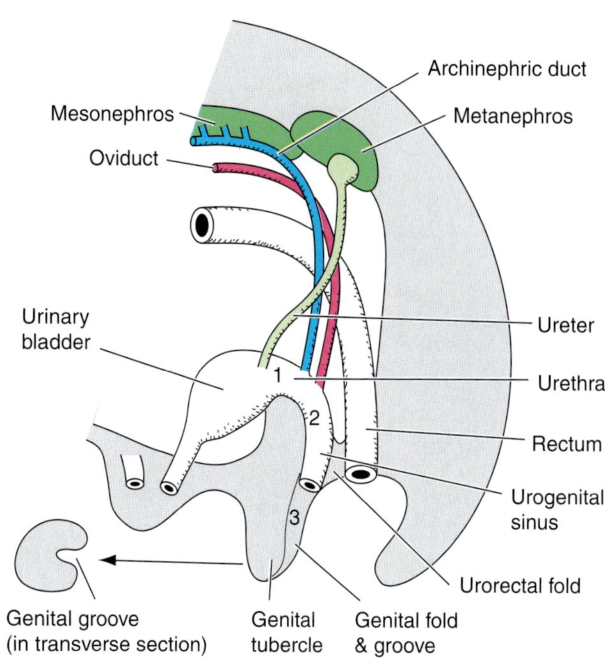

B. Later sexually indifferent stage

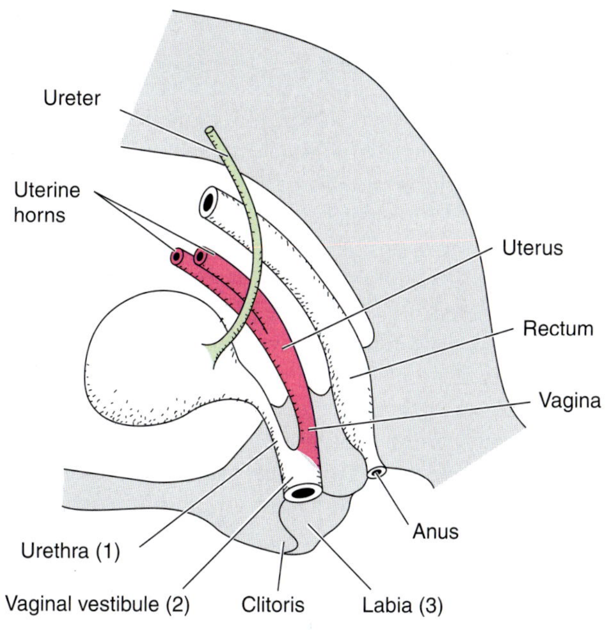

C. Differentiation of female

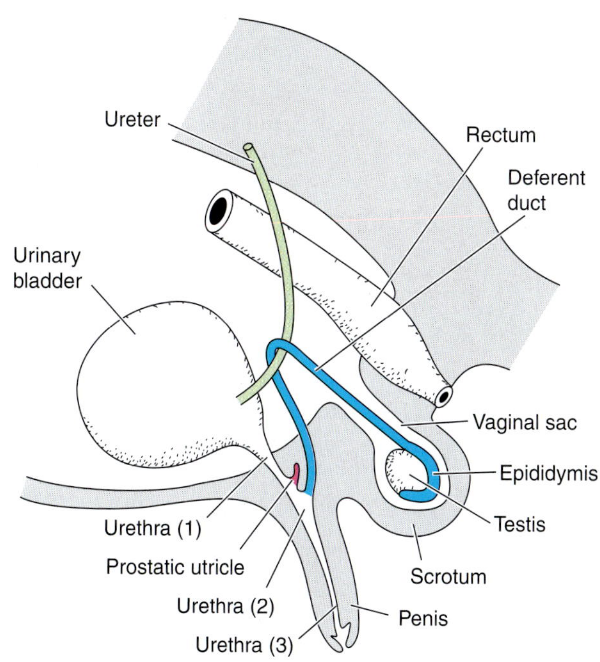

D. Differentiation of male

FIGURE 21-23
Lateral diagrams of the division of the cloaca in a eutherian. *A* and *B,* Early and later sexually indifferent stages. *C,* Differentiation in a male. *D,* Differentiation in the female. Portions of the urogenital passages leading from the urinary bladder are identified by the same numbers: 1, portion developing from the neck of the bladder (allantois); 2, portion developing from the urogenital sinus; 3, portion or area developing from the genital groove. *(After Walker and Homberger.)*

If the embryo differentiates into a male (Fig. 21-23*D*), the oviducts regress, but their distal ends often remain as a **prostatic utricle** within the prostate (Fig. 21-13). The urogenital sinus has narrowed and continues the urethra through the prostate and pelvic canal. The genital tubercle enlarges further to form the **penis,** and the genital folds unite to enclose the genital groove within it as the penile part of the urethra. The male urethra is composed of three segments that develop from the neck of the urinary bladder, the urogenital sinus, and the genital groove. The scrotum first appears as a pair of scrotal swellings. As the testes descend into them, the **scrotal swellings** unite around the base of the penis. Several columns of spongy and vascular **erectile tissue** develop within the penis. The erectile tissue becomes engorged with blood during an erection. A **corpus spongiosum** forms around the penile urethra, and its distal end expands as the **glans penis.** A pair of **corpora cavernosa** forms on the opposite side of the penis (Fig. 21-24). The glans penis is encased in a pocket of skin called the **prepuce.**

If the embryo differentiates as a female (Fig. 21-23*C*), the archinephric ducts are lost, and in eutherians, the lower ends of the oviducts unite to form a median vagina. The vagina and urethra enter the urogenital sinus, which becomes the **vaginal vestibule** in an adult female. Separate openings of the urethra and vagina are only seen in primates and a few other eutherian species. The female urethra is comparable to only the first segment of the male urethra, that is, to the part that develops from the neck of the bladder. The vaginal vestibule is a long, tubular passage in most female mammals, but it is a shallow depression in some, including humans, that is flanked by liplike skinfolds. The **labia minora** develop from the genital folds, so they are the sexual homologues of the part of the male penis surrounding the urethra. The **labia majora** are homologous to the scrotal swellings. A **glans clitoridis** lies at the anterior junction of the labia minora and is comparable to the male glans penis. Small columns of erectile tissue form the **body of the clitoris** and are comparable to the male corpora cavernosa.

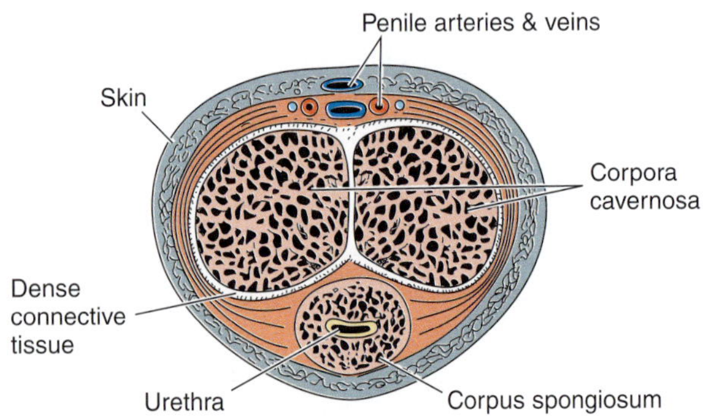

FIGURE 21-24
A cross section of a human penis. *(After Fawcett.)*

Because the male and female reproductive systems develop from common primordia present in a sexually indifferent embryo, they share many features. The pattern of development of the entire reproductive system of both male and female amniotes is shown in Figure 21-14, and a comparison of adult structures is given in Table 21-1.

Reproductive Hormones

The Hormones and Their Effects

Besides producing eggs and sperm, the gonads also are endocrine glands that secrete sex hormones under the influence of the adenohypophyseal gonadotropins: **follicle-stimulating hormone** (FSH) and **luteinizing hormone** (LH). FSH and LH have important roles in the production and release of the sex hormones, as we shall see. The major male sex hormones, known as androgens, are **cortical androgen,** secreted by the adrenal cortex (Chapter 15), and **testosterone,** secreted by the interstitial cells of the testis. The major female sex hormones are estrogens (chiefly **estradiol**) and **progesterone,** which are produced by the ovary. Androgens, estrogens, and progesterone have much in common, for they are steroid derivatives of cholesterol, and the biochemical pathways by which they are synthesized overlap considerably. Enzymes needed for their synthesis occur in both the ovary and the testis, as well as in some other tissues, so it is not surprising that some androgens are synthesized by females, and some estrogens and progesterone, by males. Males and females differ not in the types of hormones they produce but in their quantity. Mature male humans have only about one tenth of the blood level of estrogens found in females at ovulation, and mature females have only one fifth of the amount of androgens as males.

The balance of these hormones is an important factor in controlling the direction of differentiation from the sexually indifferent stage of the embryo. Androgens promote protein synthesis and growth in both sexes. High levels of androgen are necessary for the maturation of the male gonads and reproductive passages and for the development of male secondary sex characteristics. In many species, abnormally high androgen levels during development can cause genetic females to differentiate in a male direction and become intersexes, that is, have morphological features that are intermediate between the two sexes. Unlike true hermaphrodites, intersexes usually either are sterile or function only as one sex. Normal differentiation in the female direction results from low androgen levels, supplemented in many species by an increase in the level of estrogens. Estrogens are needed for females to mature and develop their secondary sex characteristics.

Reproductive periods also are regulated largely by gonadotropins and sex hormones. A few vertebrates, including humans, rats, and domestic chickens, reproduce throughout the year. Mature males of these species produce sperm continuously, but egg production is cyclical. During a mammalian **ovarian cycle,** the follicles and eggs enlarge, the eggs mature and ovulate, and the follicles transform into corpora lutea, which finally regress. Cyclical changes in the size of the female reproductive passages, in the amount of their glandular secretions, and often in behavior accompany the ovarian cycle. Many female mammals come into heat, or **estrus** (Gr., *oistros* = gadfly; frenzy), near the time of ovulation. Their behavior advertises their sexual condition, and they will permit copulation only at this time. This behavior increases the probability of fertilization. The length of an ovarian cycle varies greatly. It is 28 days in humans but only 4 to 5 days in rats. Species with repetitive cycles throughout the year are called **polyestrous.**

Reproduction is seasonal in most vertebrates. Birds and most ectotherms of the temperate region are spring breeders, but deer and many other mammals copulate in the autumn and give birth in the spring. Dogs and cats typically have two or three reproductive periods in a year. Seasonal breeders have one or more ovarian cycles during their reproductive season, and then they enter a long period of **anestrus:** testes and ovaries become small and nonfunctional, and reproductive passages become smaller and lose their secretory state.

Integration of Reproduction

These seasonal and cyclical changes are controlled and integrated by the hypothalamus, which acts by the cyclical production of **gonadotropin–releasing hormone** (Chapter 15). This releasing hormone travels through the hypophyseal portal system to the adenohypophysis and stimulates the synthesis and release of FSH and LH. These gonadotropins, in turn, stimulate the secretion of sex hormones by the gonads. The cyclical activity of the hypothalamus is controlled to a large extent by negative feedback from the fluctuating levels of the sex hormones, but it also is affected by nerve stimuli in many cases. The hypothalamus of some species may have an inherent rhythm, or biological clock, that is modulated by the sex hormones and nerve stimuli. Control mechanisms have been studied most thoroughly in birds and mammals, but considerable evidence indicates that these mechanisms operate in many other species as well. Much more research in this area is underway.

Seasonal breeders clearly respond to one or more seasonal environmental changes that affect the hypothalamus. Many environmental changes have been shown to act on the hypothalamus in some species: food quantity, day length, temperature, and even social interactions in colonial seabirds. Day length is particularly important. As the hours of light increase in late winter, spring breeders' gonadotropin and sex hormone production increases, and the gonads and reproductive passages enlarge. In fall breeders, a decrease in day length triggers these events. Light, which can penetrate the skull and brain in small species, may stimulate the hypothalamus directly, but stimuli usually reach the hypothalamus from the image-forming eyes or, in some species, by the median eye or pineal gland.

Females. Under the influence of gonadotropin-releasing hormone produced by the hypothalamus, the adenohypophysis of mature female mammals produces some FSH and LH throughout the reproductive period. But the amounts of these gonadotropins fluctuate. Increasing levels of both LH and FSH are needed for the growth and enlargement of the follicles and the synthesis by the follicles of estrogen (Fig. 21-25). The thecal cells of the follicles have receptors only for LH. Under the influence of LH, they produce androgen, the male sex hormone that is the biochemical precursor for estrogen. The granulosa cells of the follicles at first have receptors only for FSH, and FSH activates the enzymes that convert androgen to estrogen. Estrogen acting with LH and FSH promotes the rapid growth of one or more follicles.

In some species of mammal, estrogen also causes the uterine lining to thicken in preparation for the reception of an embryo. Estrogen may also bring the female into heat. Estrogen causes the granulosa cells to develop receptors for LH. The positive feedback to the hypothalamus of increased and sustained levels of estrogen produced by one or more follicles nearly ready to ovulate promotes a surge in the secretion of LH. Because all of the follicle cells can now respond to LH, the high levels of LH also cause them to release a proteolytic enzyme that breaks down the follicle wall. Ovulation occurs. Most mammals are **spontaneous** or **cyclic ovulators** because ovulation occurs regardless of whether males are present. The shrew, mink, rabbit, cat, and a few other species are **induced ovulators**—the hypothalamus does not release the hormone needed for an LH surge unless nerve stimuli from copulation reach it. A single copulation induces ovulation

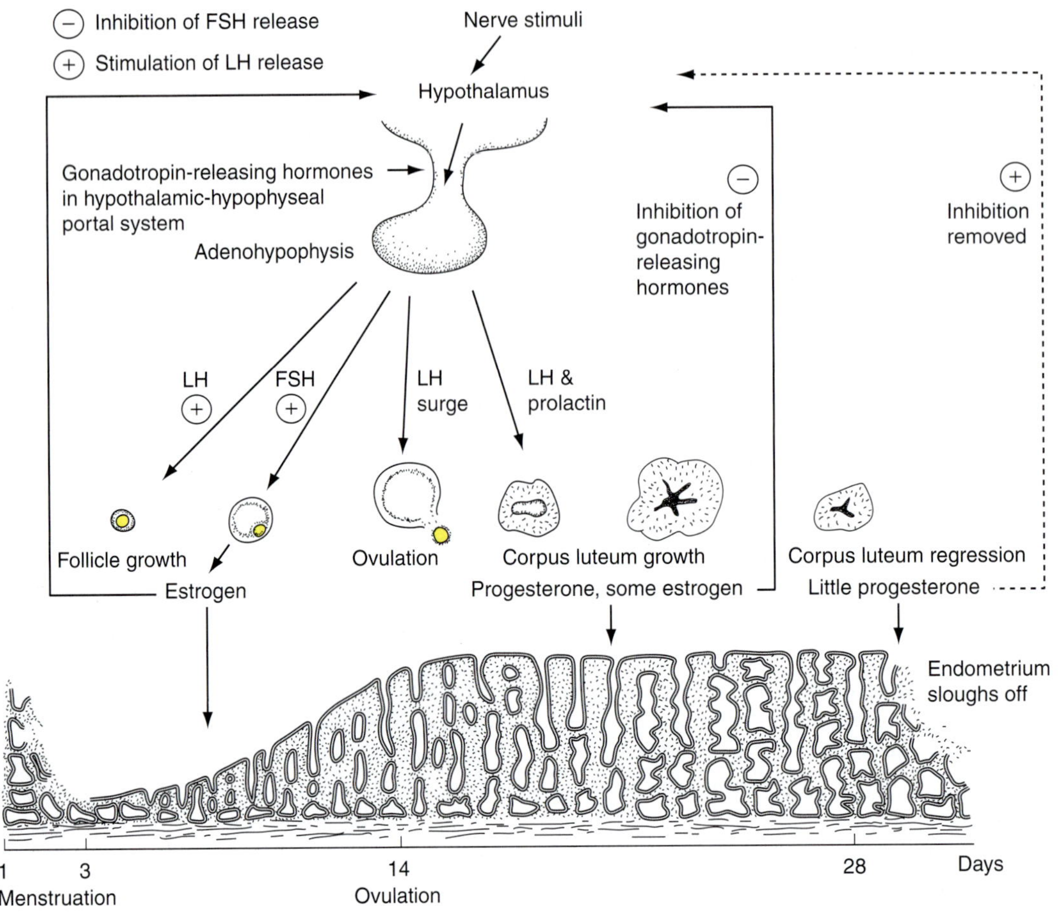

FIGURE 21-25
Cyclical changes during an ovarian cycle in the ovary and endometrium of a eutherian mammal. Day lengths and the endometrial cycle are based on the human cycle.

in rabbits, but repetitive and frequent copulations are necessary in many induced ovulators.

Continued production of LH causes the ovulated follicles to transform into corpora lutea. **Prolactin,** which also is secreted by the adenohypophysis at this time, has luteinizing effects in rats and mice but not in all mammals. It does help maintain the corpora lutea. The corpora lutea secrete some estrogens, but their primary product is progesterone. Progesterone increases the vascularity and glandular secretion of the endometrium. It also feeds back to the hypothalamus and inhibits the release of gonadotropin-releasing hormone in eutherian mammals. Follicle development stops when this releasing hormone decreases to low levels. If pregnancy does not occur, the corpora lutea soon regress, progesterone secretion stops, and the inhibition of the hypothalamus on gonadotropin-releasing hormone production no longer exists. Under the influence of this gonadotropin, FSH and LH production increase, follicles begin to develop, and the ovarian cycle starts anew, unless the animal goes into a period of anestrus. The uterine lining cannot be maintained in its receptive condition in the absence of progesterone. It sloughs off abruptly in most primates in a period known as **menstruation.**

If a eutherian becomes pregnant, the embryonic trophoblast of many species soon secretes **chorionic gonadotropin.** This hormone promotes the growth of the corpora lutea and the continued secretion of estrogens and progesterone, which are needed to maintain the uterine lining and form the placenta. Human pregnancy tests are based on the detection of this gonadotropin. A continued secretion of prolactin by the adenohypophysis, rather than chorionic gonadotropin, maintains the corpora lutea in other species. Corpora lutea persist until late in pregnancy in many mammals, but the placenta itself synthesizes estrogens and progesterone in humans and some other species. When this placental activity begins, chorionic gonadotropin secretion falls off and the corpora lutea regress. Because the continued high levels of progesterone block the ovarian cycles in eutherians, they cannot become pregnant again. This blockage does not occur in marsupials.

As pregnancy continues, the ovaries and placentas of many mammals secrete **relaxin.** This hormone peaks just before birth and relaxes the pelvic symphysis and other pelvic ligaments. The increased flexibility of the pelvis facilitates the passage of the fetus through the pelvic canal. Many other hormonal changes occur at birth. Estrogen and progesterone secretion decrease, and prolactin production increases. The release of **oxytocin** from the neural lobe of the hypophysis helps bring about the uterine contractions that expel the fetus.

Estrogens, progesterone, and **placental lactogen** stimulate the proliferation of the ducts and alveoli of the mammary glands during pregnancy, but not the secretion of milk. Milk secretion is, in fact, blocked by progesterone. It does not begin until after birth, when progesterone secretion abruptly decreases and the adenohypophysis increases its output of prolactin. Prolactin is essential for milk production, and it also promotes maternal behavior in many species. The ejection of milk is caused by a combined neuronal and hormonal reflex (Fig. 21-26). When nerve impulses resulting from the tactile stimulus of the infant's sucking reach the hypothalamus, oxytocin is released from the neural lobe of the hypophysis. Oxytocin reaches the mammary glands through the circulatory system and causes smooth muscle contraction and milk ejection. Mammary gland growth, milk secretion, and milk ejection are separate processes mediated by different hormones. Prolactin secretion continues as long as the infant is suckling, and its presence inhibits the release of gonadotropic hormones and new ovarian cycles in both eutherians and marsupials. Most mammals cannot become pregnant during lactation. When nursing stops, prolactin secretion falls off and ovarian cycles begin again.

Males. The control of male reproduction is diagrammed in Figure 21-27. FSH and LH also are produced in males under the influence of hypothalamic releasing hormone. Their production is cyclical in seasonal breeders, though it is not as strongly cyclical

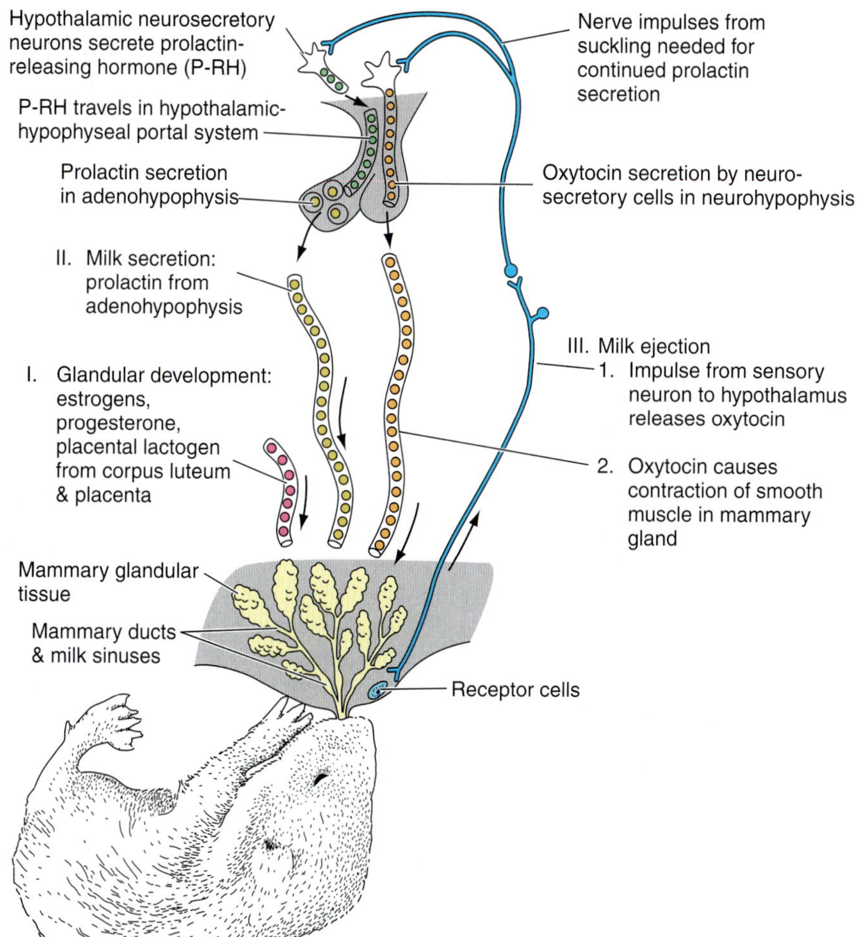

FIGURE 21-26
The hormones and nerve impulses needed for mammary gland development, milk secretion, and milk ejection.

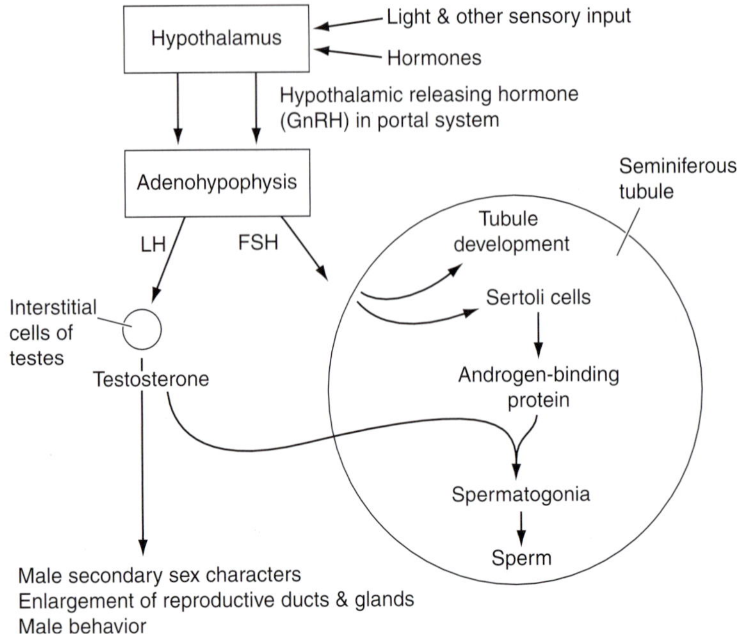

FIGURE 21-27
The factors that control male reproduction.

as in polyestrous females. FSH is necessary for the development of the seminiferous tubules and for the enlargement of the testes at maturity and during the reproductive season in seasonal breeders. It also promotes the synthesis by the Sertoli cells of an androgen-binding protein. LH acts on the interstitial cells and causes them to secrete testosterone. Some testosterone enters the seminiferous tubules, unites with **androgen-binding protein,** enters the germinal epithelium, and promotes the development and maturation of the sperm. Testosterone traveling throughout the body causes the enlargement and secretion of the reproductive passages and accessory sex glands; the development of the secondary sex characteristics; and usually, by its effects on the nervous system, courtship and other behavioral changes.

Major Features of Urogenital Evolution

The reproductive and excretory apparati are functionally very disparate, but developmentally they are so similarly canalized that it is logical to review their evolutionary transformations in unison. As shown in Figure 21-28, in all craniates, the archinephric duct develops in embryogeny (1). It is only in the adult female of the amniotes that the archinephric duct is much reduced or lost. Hagfishes develop a single gonad (5) and possess functional genital pores through which gametes are extruded in the absence of reproductive ducts. The archinephric duct is used to drain urine. Hagfishes also have a functional pronephros in the larva and an adult kidney that approaches the holonephros in its morphology (4).

All other craniates, the vertebrates, have differentiated a new kidney, which in anamniotes is an opisthonephros (2). Lampreys also have a single gonad (7), but the resemblance to that of hagfishes is only superficial because the single gonad develops ontogenetically as a fusion of two gonads. In addition to the presence of an opisthonephros, primitive vertebrates have seminiferous ampullae (3) that make up their testes.

All gnathostomous vertebrates have reproductive ducts (6) that transport gametes from the gonads to the outside. Internal fertilization first appears among the chondrichthyans, the males of which possess intromittent organs, the claspers (9). Male cartilaginous fishes also have the anterior part of their opisthonephros taken over by the reproductive system to become Leydig's gland (10), which secretes seminal fluid used during internal fertilization. The posterior part of the opisthonephros maintains renal functions. Female chondrichthyan fishes develop an oviduct, also called the Müllerian duct, by a longitudinal splitting of the archinephric duct (8).

Among actinopterygian fishes, teleosts and *Amia* have lost the cloaca in both males and females, and they have unique reproductive ducts. The teleostean sperm duct is a direct extension of the testis canal, whereas the oviduct is a simple tubelike extension of the central canal of the ovary (12).

In all remaining vertebrates (i.e., coelacanths, lungfishes, amphibians, reptiles, birds, and mammals), the

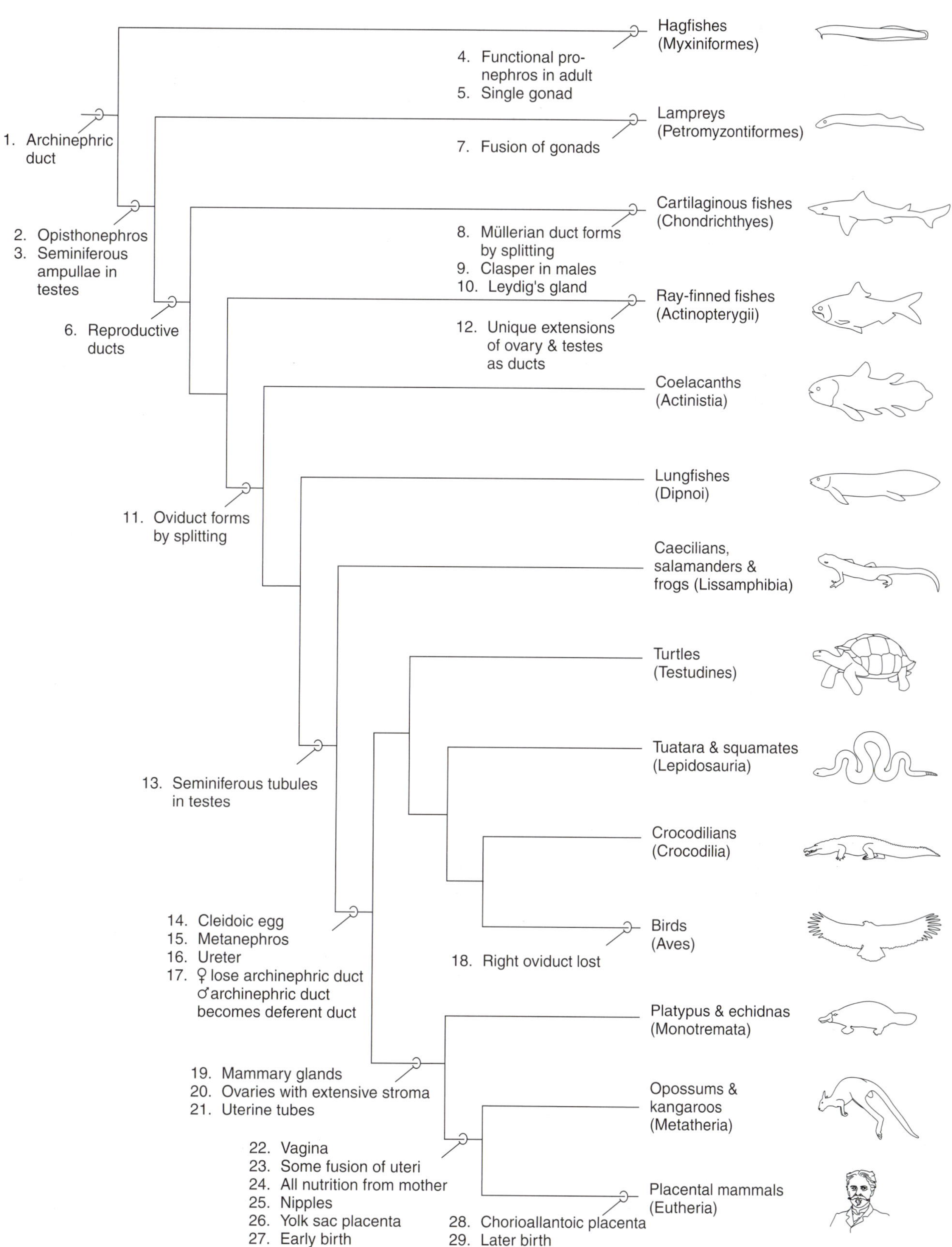

FIGURE 21-28
Major features of urogenital evolution in craniates.

oviduct arises in ontogeny as a longitudinal, groove-like invagination of the coelomic epithelium on the lateral surface of the mesonephros (11). Coelacanths and lungfishes resemble each other and primitive fishes in the retention of the cloaca and drainage of urine by the posterior archinephric duct. Frogs among amphibians and the amniotes have males with testes that are composed of seminiferous tubules (13), which differ from ampullae in being long, highly convoluted ductules.

Amniotes share several urogenital specializations (14–17). All possess a cleidoic egg with its extraembryonic membranes that makes reproduction and development on land possible. The metanephros becomes the functional kidney and it is drained by the new ureter. All male amniotes have an epididymis to store spermatozoa, and the archinephric duct, which in females becomes reduced or lost, becomes the deferent duct (17). Birds are unique in having only the left oviduct with an exceptionally large ostium tubae. Their right oviduct is lost (18).

All female mammals have evolved mammary glands (19) and an ovary with an extensive connective tissue stroma not seen before (20). Each oviduct has a new narrow cranial portion, the uterine or fallopian tube (21), which widens posteriorly to form the uterine horns and uterine body. Monotremes have separate uteri and the mammary glands. The distal end of the oviducts differentiates as a vagina in Metatheria and Eutheria (22). The vaginae and usually parts of the uteri fuse in metatherians and eutherians (23). Nipples (25) are associated with the mammary glands in both metatherians and eutherians. Although a placental relationship between mother and embryo has evolved independently in a few members of many groups, all metatherians and eutherians are distinguished from all other vertebrates in being completely dependent on the mother for supplying their nutritional needs (24). This is accomplished early in development by uterine secretions and later by a very well-developed placenta. The Metatheria possess the yolk sac placenta (26) and are born relatively early (27); Eutheria develop a chorioallantoic placenta (28), and they are born later in development (29).

SUMMARY

1. Vertebrates reproduce sexually. The eggs of a few hybrid populations of fishes, amphibians, and reptiles can develop parthenogenetically.
2. Mammals, birds, and a few other vertebrates have genotypic sex determination using heteromorphic sex chromosomes. The environmental condition in which the eggs develop is an important determinant of sex in many other vertebrates. Temperature-dependent sex determination is widespread among egg-laying reptiles.
3. A few fishes are synchronously hermaphroditic, but cross fertilization normally occurs. Many fishes are sequential hermaphrodites.
4. Primitive members of each craniate group are oviparous. Viviparity has evolved independently in some species of every craniate group except for hagfishes, lampreys, and birds.
5. In viviparous species, the embryos depend on maternal food, which the mother transfers either to the eggs while the eggs are in the ovary (ovoviviparity or aplacental viviparity) or to the embryos while they are in the uterus (placental viviparity). Many intermediate conditions between these extremes occur.
6. The primary sex cords that invade the medullary region of the gonad become the seminiferous tubules or ampullae of the testis. They regress in females, and secondary sex cords that invade the gonad cortex become the ovarian follicles.
7. Rete cords develop in the gonad between the primary sex cords and mesonephric tubules. They canalize in males and become part of the passageway for sperm; they regress in females.
8. An oviduct begins to develop in both sexes, usually by a folding of the coelomic epithelium adjacent to the mesonephros.
9. The testes of amniotes contain seminiferous tubules, in which spermatogenesis occurs, and interstitial cells, which secrete the male sex hormone. The testes of most anamniotes are composed of seminiferous ampullae. Spermatogonia multiply throughout sexual maturity.
10. The testes of most mammals lie permanently in a scrotum or descend to the scrotum during the reproductive season, but the testes of some mammals remain in the abdominal cavity. The reasons for the evolution of a scrotum are not entirely clear. Hypotheses have been proposed that the lower scrotal temperature is necessary for the production or storage of sperm or that a scrotum removes the testes from the inhibiting effects on spermatogenesis of fluctuations in intra-abdominal pressure resulting from certain patterns of locomotion. Evolution has proceeded from scrotal to

ascrotal, but never the reverse, and the loss of testicular descent is rare and restricted to mammals that evolved special cooling mechanisms for the testes.

11. Oogonia cease dividing in mammalian ovaries prior to birth. The eggs mature within the ovarian follicles and are discharged into the coelom at ovulation. In mammals, each follicle transforms into a corpus luteum after ovulation.

12. Mammalian ovaries are relatively small, but those of other vertebrates are large during the reproductive season because many eggs are produced, and in oviparous species, they are laden with yolk. Oogonial divisions continue throughout sexual maturity in the ovaries of most vertebrates except for mammals, birds, a few reptiles, and chondrichthyan fishes. Corpora lutea develop from ovulated follicles primarily in species that have some degree of viviparity.

13. Hagfishes and lampreys lack reproductive ducts. Sperm as well as eggs are shed into the coelom. They leave through a genital pore that opens into the caudal end of the archinephric duct.

14. The sperm of amphibians, and probably ancestral jawed vertebrates, are carried from the testis by connecting tubules to cranial opisthonephric tubules, and thence down the archinephric duct to the cloaca. Urine also is carried by the archinephric duct in some species but by accessory urinary ducts in others.

15. Fertilization is internal in chondrichthyans, and cranial parts of the kidney and much of the archinephric duct are specialized in males to produce a seminal fluid.

16. Teleosts have a unique sperm duct that leads directly from the testis to the outside, or to the caudal end, of the archinephric duct. Kidney tubules have no role in sperm transport.

17. The passageway for sperm is the same in amniotes as in amphibians, but the terminology is different. Sperm leave the testis through the rete testis (connecting tubules); pass through efferent ductules (mesonephric tubules) to the duct of the epididymis (part of the archinephric duct); enter the deferent duct (the rest of the archinephric duct); and continue to the cloaca or, in therians, to the urethra. The efferent ductules and duct of the epididymis form the epididymis.

18. Specializations for internal fertilization in male amniotes include the evolution of accessory sex glands that produce the seminal fluid and, usually, a penis. Birds lack a penis. The accessory sex glands of mammals include the prostate, vesicular, and bulbourethral glands.

19. A pair of oviducts carries the eggs of amphibians and primitive jawed vertebrates from the coelom to the cloaca. Secondary envelopes usually are secreted around the eggs as they travel down the oviducts.

20. The upper part of the oviduct of chondrichthyans is specialized to secrete a shell around the fertilized eggs, and the lower part forms a uterus in viviparous species.

21. Most cartilaginous fishes are viviparous. The dependence of the embryos on maternal food transferred in the uterus ranges from a little to a lot. A yolk sac placenta is present in some species.

22. Teleosts discharge their eggs into a hollow ovary, part of which forms a unique oviduct that leads directly to the outside.

23. Reptiles and birds have oviducts (usually only the left one in birds) that secrete the secondary envelopes that characterize the cleidoic egg: albumen, a shell membrane, and a shell.

24. Each oviduct of monotremes has differentiated into a uterine tube and uterus. A shell membrane and shell are secreted around the eggs. The eggs continue to accumulate food from maternal secretions and begin to develop while in the uterus. Eggs are brooded after they are laid, and newly hatched young are fed milk.

25. The caudal end of each marsupial oviduct differentiates as a vagina, and these open independently into the urogenital sinus.

26. A shell membrane, but no shell, is secreted by marsupial mothers around their embryos. The embryos are nourished early in gestation by uterine secretions, which are absorbed through the shell membrane. When the membrane is reabsorbed, a brief placental relationship is established between the uterine lining and, usually, the embryonic yolk sac. Birth occurs soon after the placenta forms, apparently because the fetal trophoblast is not an immunological barrier. The embryos attach to nipples, which often are located in a marsupium, and complete their development outside of the uterus.

27. Eutherians have only a single vagina, and the uterine portions of the oviducts usually unite to some degree.

28. Secondary envelopes are not secreted around eutherian eggs, and a chorioallantoic placenta is established very early. The trophoblast appears to be an effective immunological barrier, so the embryos are not rejected.

29. Details of placental structure, including the intimacy of the union between maternal and fetal tissues, vary considerably among eutherian groups.

30. The cloaca becomes divided in therians, so the intestine and urogenital passages continue directly to the surface of the body.

31. The male sex hormones are cortical androgen, produced by the adrenal cortex, and testosterone, produced by the interstitial cells of the testis; the female hormones are estrogen and progesterone, secreted by the ovarian follicles and corpora lutea. All of the sex hormones are chemically related steroids. Males have some female hormones; females have some male ones.
32. Androgens promote protein synthesis and growth in both sexes. High levels of cortical androgens and testosterone allow males to mature and develop their secondary sex characteristics. Low androgen levels and high estrogen levels permit females to mature and develop their secondary sex characteristics.
33. Reproductive periods, ovarian cycles, the production of sperm, and many aspects of reproductive behavior are integrated by complex interactions among a hypothalamic releasing hormone, gonadotropins secreted by the adenohypophysis, and the gonadal hormones.

REFERENCES

Andrews, R. M., 1999: Evolution of viviparity in squamate reptiles (*Sceloporus* spp.): A variant of the cold-climate model. *Journal of Zoology, London,* 250:243–253.

Andrews, R. M., Mathies, T., Qualls, C. P., and Qualls, F. J., 1999: Rates of embryonic development of sceloporus lizards: Do cold climates favor rapid development? *Copeia,* 1999:691–699.

Andrews, R. M., Qualls, C. P., and Rose, B. R., 1997: Effects of low temperature on embryonic development of *Sceloporus* lizards. *Copeia,* 1997:827–833.

Asdell, S. A., 1964: *Patterns of Mammalian Reproduction,* 2nd edition. Ithaca, N.Y., Cornell University Press.

Austin, C. R., and Short, R. V., editors, 1976: *Reproduction in Mammals,* book 6: *The Evolution of Reproduction.* Cambridge, Cambridge University Press.

Bedford, J. M., 1978: Anatomical evidence for the epididymis as the prime mover in the evolution of the scrotum. *American Journal of Anatomy,* 152:483–508.

Bedford, J. M., 1977: Evolution of the scrotum, The epididymis as prime mover? *In* Calaby, J. H., and Tyndale-Biscoe, C. H., editors: *Reproduction and Evolution.* Australian Academy of Science.

Blackburn, D. G., 1998. Structure, function, and evolution of the oviducts of squamate reptiles, with special reference to viviparity and placentation. *Journal of Experimental Zoology,* 282:560–617.

Blackburn, D. G., 1993: Chorioallantoic placentation in squamate reptiles: Structure, function, and evolution. *Journal of Experimental Zoology,* 266:414–430.

Blackburn, D. G., 1993: Standardized criteria for the recognition of reproductive modes in squamate reptiles. *Herpetologica,* 49:118–132.

Bull, J. J., 1980: Sex determination in reptiles. *Quarterly Review of Biology,* 55:3–21.

Cowles, R. B., 1958: The evolutionary significance of the scrotum. *Evolution,* 12:417–418.

Duvall, D., Guillette, L. J., Jr., and Jones, R. E., 1982: Environmental control of reptilian reproductive cycles. *In* Gans, C., and Pough, F. H., editors: *Biology of the Reptilia,* volume 13. New York, Academic Press.

Freeman, S., 1990: The evolution of the scrotum: a new hypothesis. *Journal of Theoretical Biology,* 145:429–445.

Frey, von R., 1991: Zur Ursache des Hodenabstiegs (*Descensus testiculorum*) bei Saugetieren. *Zeitschrift Zoologische Systematische Evolution Forschung,* 29:40–65.

Grant, T., 1984: *The Platypus.* Kensington, Australia, New South Wales University Press.

Greven, H., 1998. Survey of the oviduct of salamandrids with special reference to the viviparous species. *Journal of Experimental Zoology,* 282:507–525.

Griffiths, M., 1978: *The Biology of Monotremes.* New York, Academic Press.

Hamlett, W. C., and Hysell, M. K., 1998: Uterine specializations in elasmobranchs. *Journal of Experimental Zoology,* 282:438–459.

Hamlett, W. C., and Koob, T. J., 1999. Female reproductive system. *In* Hamlett, W. C., editor: *Sharks, Skates and Rays: The Biology of Elasmobranch Fishes.* Baltimore, Johns Hopkins.

Hamlett, W. C., Eulitt, R. L., Jarrell, R. L., and Kelly, M. A., 1993: Uterogestation and placentation in elasmobranchs. *Journal of Experimental Zoology,* 266:347–367.

Hamlett, W. C., Knight, D. P., Koob, T. J., Jezior, M., Luong, T., Rozycki, T., Brunette, N., and Hysell, M. K., 1998: Survey of oviductal gland structure and function in elasmobranchs. *Journal of Experimental Zoology,* 282:399–420.

Haqq, C. M., King, C. Y., Ukiyama, E., Falsafi, S., Haqq, T. N., Donahoe, P. K., and Weiss, M. A., 1994: Molecular basis of mammalian sexual differentiation: Activation of Müllerian inhibiting gene expression by SRY. *Science,* 266:1494–1500.

Jameson, E. W., Jr., 1988: *Vertebrate Reproduction.* New York, John Wiley and Sons.

Jones, R. E., editor, 1978: *The Vertebrate Ovary: Comparative Biology and Evolution.* New York, Plenum Press.

Lagler, K. F., Bardach, J. E., and Miller, R. E., 1962: *Ichthyology.* New York, John Wiley and Sons.

Lofts, B., 1974: Reproduction. *In* Lofts, B., editor: *Physiology of the Amphibia.* New York, Academic Press.

Moore, C. R., 1926: The biology of the mammalian testis and scrotum. *Quarterly Review of Biology,* 1:4–50.

Packard, G. C., 1977: The physiological ecology of reptilian eggs and embryos, and the evolution of viviparity within the class Reptilia. *Biological Reviews,* 52:71–105.

Pabst, D. A., Rommel, S. A., McLellan, W. A., Williams, T. M., and Rowles, T. K., 1995: Thermoregulation of the intra-abdominal testes of the bottlenose dolphin (*Tursiops truncatus*) during exercise. *Journal of Experimental Biology,* 198:221–226.

Parkes, A. S., editor, 1952–1966: *Marshall's Physiology of Reproduction.* London, Longmans Green.

Perry, J. S., 1972: *The Ovarian Cycle of Mammals.* New York, Nafner Publishing.

Potts, G. W., and Wootton, R. J., editors, 1984: *Fish Reproduction: Strategies and Tactics.* Orlando, FL, Academic Press.

Rommel, S. A., Pabst, D. A., McLellan, W. A., Mead, J. G., and Potter, C. W., 1995: Anatomical evidence for a countercurrent heat exchanger associated with the dolphin testis. *Anatomical Record,* 232:150–156.

Sadleir, R. M. F. S., 1973: *The Reproduction of Vertebrates.* New York, Academic Press.

Sever, D. M., and Brizzi, R., 1998. Comparative biology of sperm storage in female salamanders. *Journal of Experimental Zoology,* 282:460–476.

Shine, R., 1995: A new hypothesis for the evolution of viviparity in reptiles. *American Nature,* 145:809–823.

Standora, E. A., and Spotila, J. R., 1985: Temperature dependent sex determination in sea turtles. *Copeia,* 1985: 711–722.

Stewart, J. R., 1993: Yolk placentation in reptiles: Structural innovation in a fundamental vertebrate nutritional system. *Journal of Experimental Zoology,* 266:431–449.

Stewart, J. R., and Thompson, M. B., 1998. Placental ontogeny of the Australian scincid lizards *Niveoscincus coventryi* and *Pseudemoia spenceri. Journal of Experimental Zoology,* 282:535–559.

Taylor, D. H., and Guttman, S. I., editors, 1977: *The Reproductive Biology of Amphibians.* New York, Plenum Press.

Turner, C. D., and Bagnara, J. T., 1976: *General Endocrinology,* 6th edition. Philadelphia, W. B. Saunders.

Tyndale-Briscoe, C. H., and Renfree, M., 1987: *Reproductive Physiology of Marsupials.* Cambridge, Cambridge University Press.

Viets, B. E., Ewert, M. A., Talent, L. G., and Nelson, C. E., 1994: Sex-determining mechanisms in squamate reptiles. *Journal of Experimental Zoology,* 270:45–56.

Wake, M. H., 1993: Evolution of oviductal gestation in amphibians. *Journal of Experimental Zoology,* 266:394–413.

Wake, M. H., 1998: Oviduct structure and function and reproductive modes in amphibians. *Journal of Experimental Zoology,* 282:477–506.

Werdelin, L., and Nilsonne, A., 1999: The evolution of the scrotum and testicular descent in mammals: A phylogenetic view. *Journal of Theoretical Biology,* 196:61–72.

Wibbels, T., Bull, J. J., and Crews, D., 1994: Temperature-dependent sex determination: A mechanistic approach. *Journal of Experimental Zoology,* 270:71–78.

Witschi, E., 1956: *Development of Vertebrates.* Philadelphia, W. B. Saunders.

part **V**

Conclusion

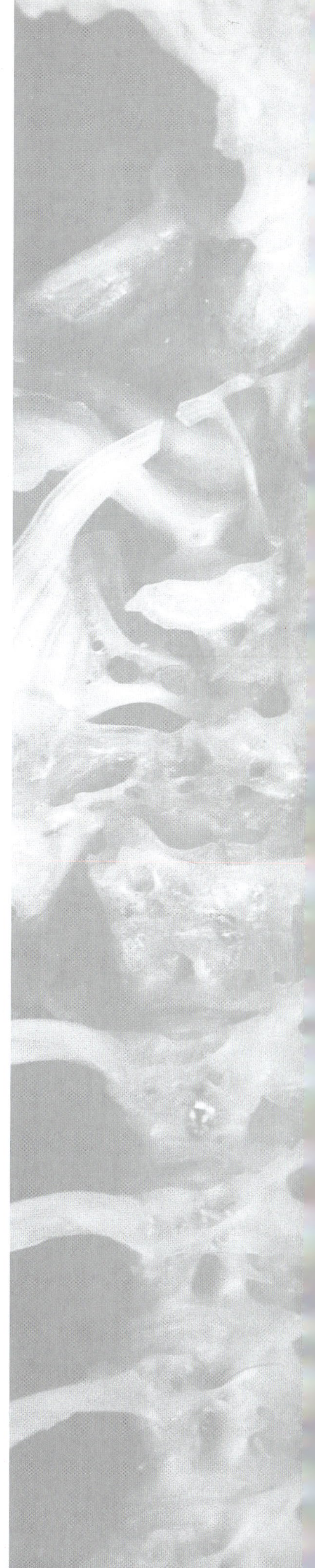

22

Epilogue

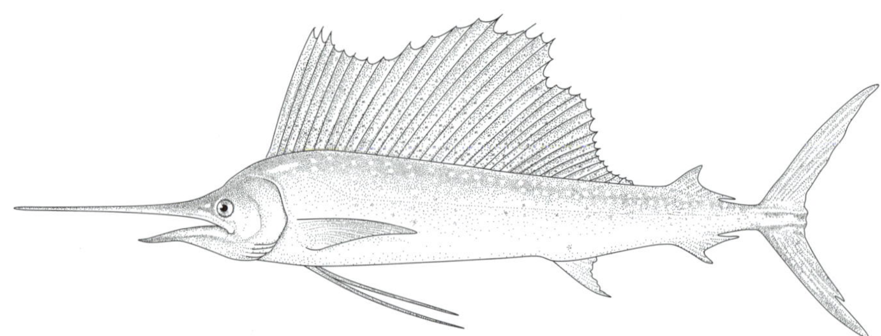

OUTLINE

The study of animal form has spanned more than two centuries. It underwent several periods of extensive productivity, punctuated by virtual stasis, declines, and revivals. This textbook has explored briefly the vast structural basis of comparative anatomy (e.g., the vertebrate cheiropterygium, Chapter 9), new functional insights that emerged with the integration and advent of new experimental methods (e.g., bird flight, Chapter 11; the vertebrate feeding apparatus, Chapter 16), and theories of comparative methods (e.g., the evolution of vertebrate ventilation, Chapter 18). Despite this long research activity, fewer than 6% of vertebrate species are anatomically well known. This paucity of structural knowledge proves to be a distinct hindrance in ongoing and future research that tries to explain the nature and causal factors underlying biodiversity.

Rather than trying to furnish you an overall but necessarily superficial and abstract conclusion of this vast subject, we discuss a summary based on one model in concrete terms. This model faithfully reflects in depth the goals and achievements of comparative anatomy and illustrates the direction of future studies in this field. We have selected the vertebrate hearing apparatus, especially with respect to the middle ear, as an example.

Comparative Anatomy and Homology

The formation of the hypothesis that the hyomandibula in fishes is the homologue of the amphibian columella revolutionized comparative anatomy by illustrating the morphological transition from aquatic to terrestrial life and explaining the origin of the mammalian auditory system. Even though the fish hyomandibula is strictly a suspensory element for the jaw apparatus (Fig. 22-1*A*), it has a distinct synovial joint with the cranium and a distal articulation with the quadrate. Both of these joints foreshadow the future configuration of the hyomandibula, which occurred by a slight shift of its proximal cranial connection to a window in the otic capsule and by a caudodorsal shift of its distal end from the quadrate to the tympanum, or eardrum (Fig. 22-1*B*). The early ontogeny of the amphibian columella is virtually identical to that of the fish hyomandibula. The originally vertical dorsoventral position of the hyomandibula evidently changed to a much more horizontal lateromedial position across the new tympanic or middle ear cavity (Chapter 12 and Fig. 22-1*A* and *B*) within the tetrapods. The middle ear cavity develops from the first embryonic pharyngeal pouch, as does the spiracle of a fish, indicating that these regions are homologous. This supports the homology of the fish hyomandibula and amphibian columella. Additional anatomical evidence also shows positional similarity between the amphibian and sauropsid columella. The prominent sauropsid columella also traverses the middle ear from the otic capsule proximally to the superficial eardrum or tympanum distally, when such a membrane is present (Fig. 22-1*C*). The tympanum lies near the quadrate in sauropsids. Additional structural analysis within a phylogenetic context has led to the remarkable discovery that in synapsids, thought to be ancestral to mammals, both the quadrate and articular have become reduced and less firmly articulated with their surrounding bones, reducing their jaw-joint–bearing role (Chapter 7). This trend culminated with the incorporation of the quadrate, the columella (which remains articulated with the quadrate), and the articular into the expanded middle ear in mammals (Fig. 22-1*D*). These bones are very small in mammals, and their names have been changed to reflect their mammalian shapes: articular = malleus (hammer), quadrate = incus (anvil), and columella = stapes (stirrup). Developmental studies of the jaw joint and ear region of the opossum clearly show how the angular bone of the lower jaw develops a reflected lamina that becomes the tympanic bone that holds the tympanum and partly surrounds the middle ear (Fig. 22-2). Such studies also

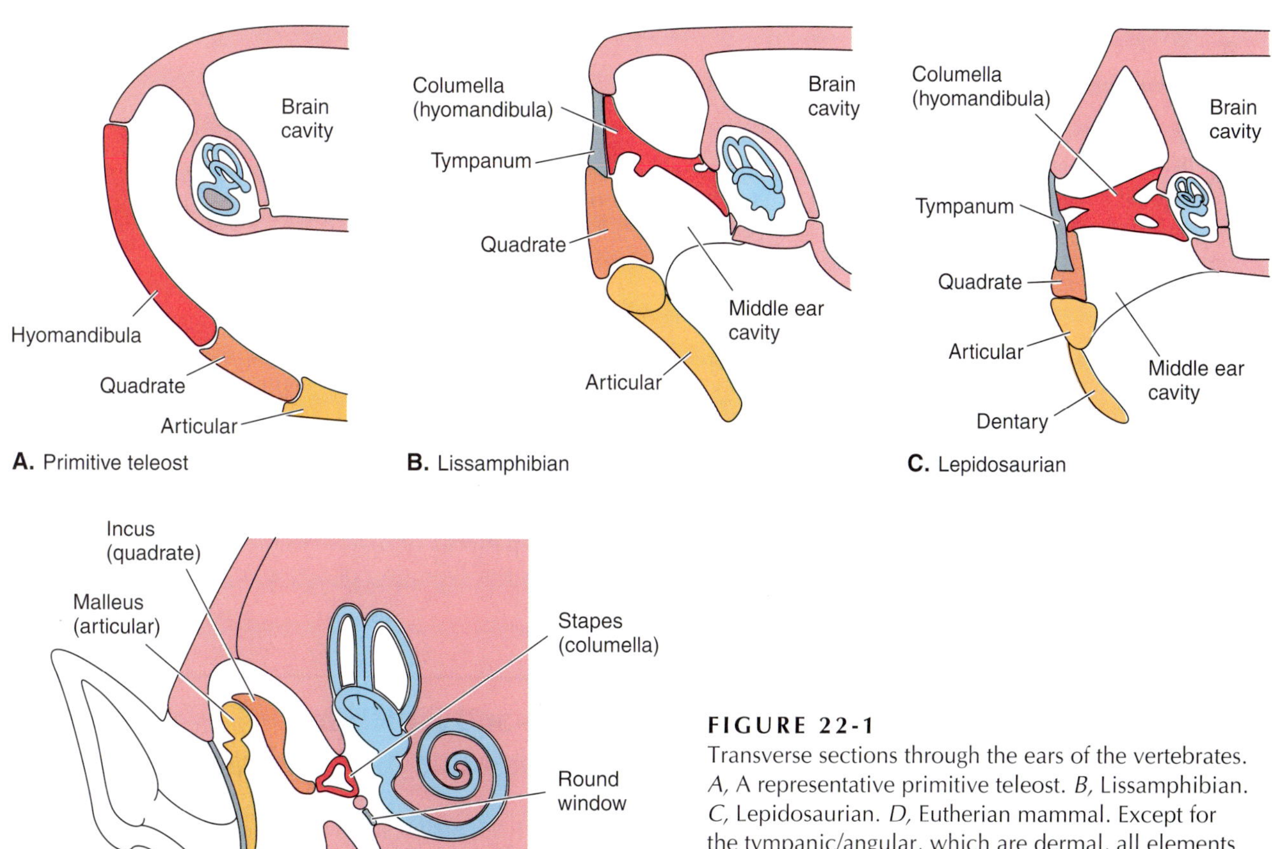

FIGURE 22-1
Transverse sections through the ears of the vertebrates. *A,* A representative primitive teleost. *B,* Lissamphibian. *C,* Lepidosaurian. *D,* Eutherian mammal. Except for the tympanic/angular, which are dermal, all elements differentiate from the mandibular arch and are cartilage replacement bone. Blue = undifferentiated cartilage of the mandibular or Meckel's cartilage. *(Modified from Romer.)*

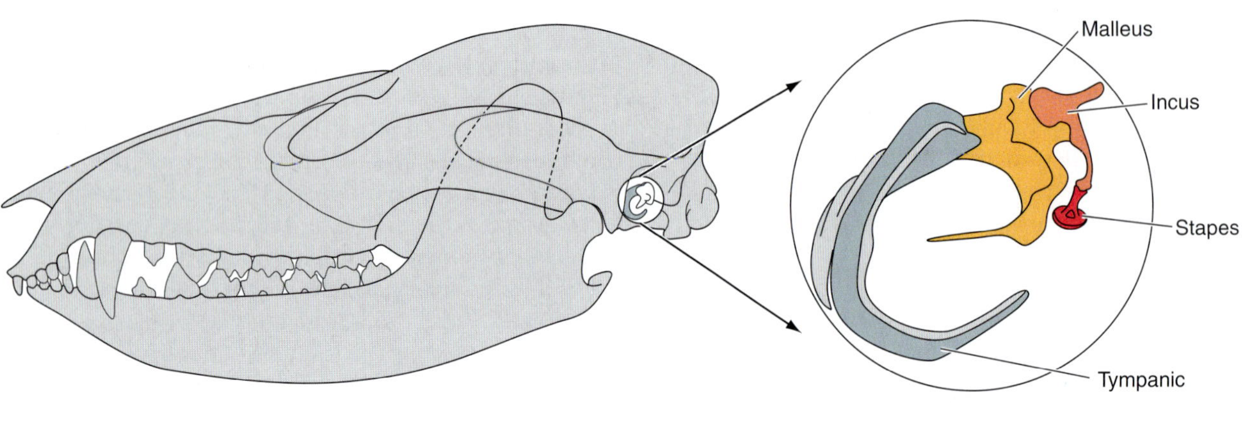

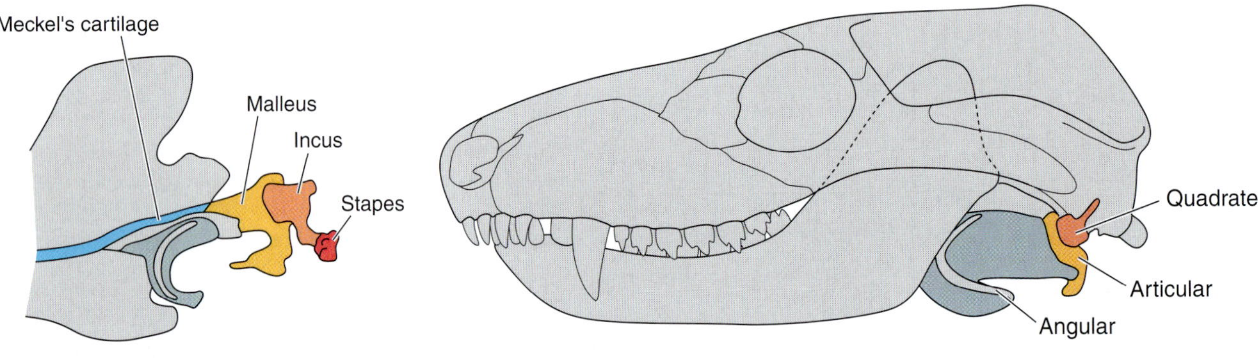

FIGURE 22-2
A, A skull of the adult opossum, *Didelphis virginiana,* showing the mammalian jaw joint. *B,* Adult ear ossicles. *C,* The middle ear elements in early development of the pouch young opossum. *D,* Skull of a cynodont *Thrinaxodon,* to show the primitive jaw joint. Except for the tympanic/angular, which are dermal, all elements differentiate from the mandibular and hyoid arches and are cartilage replacement bones. *(Modified from Hopson).*

show how the malleus and incus differentiate from the originally modified jaw-joint–bearing bones to become the mammalian ear ossicles. The derivation of the malleus from the articular is further supported by the fact that it is acted on by the tensor tympani muscle and innervated by the trigeminal (V) nerve of the mandibular arch (Chapter 13). In contrast, the stapes, a modified hyomandibula, is a hyoid derivative, and its stapedius muscle is innervated by the facial (VII) nerve.

This homology of the mammalian ear ossicles and dissimilar-looking elements in other vertebrates is concurrent with many other significant hypotheses of homology. Among these are the homology of the mammalian sperm duct with the anamniote archinephric duct (Chapter 21) and the homology of components of the mammalian larynx with elements of certain gill arches (Chapter 7). General congruence among the numerous homologies that have been hypothesized in comparative anatomy represents the strongest type of support for Darwin's theory of descent with modification. When the study of form is combined with developmental biology within a comparative evolutionary framework, it becomes especially powerful in elucidating the patterns of diversity among vertebrates. It also explains the evolution of major organ systems.

Form and Function: Renaissance in Vertebrate Anatomy

As the field progressed, many anatomists asked questions beyond structure and homology; they shifted their focus from the description of pattern to explaining causally the differences in vertebrate form. As we shall see from the studies of the hearing apparatus, their approach is essentially functional.

Teleost Hearing without Weberian Ossicles

The body density of fishes is about the same as that of water, so impinging sounds move the body at approximately the same amplitude and frequency as they do the water. As a consequence, fishes should be unable to perceive sound. However, fishes have two structures with very different densities than water, so they move very differently from the rest of the fish body: the swim bladder (Chapter 18) and otoliths (Chapter 12). When the swim bladder is set in motion by sound waves, it functions as a hydrophone and acts as a secondary radiator, setting up an acoustic wave that has sufficient energy to stimulate the inner ear. Because the heavier and denser otoliths are not as easily set in motion as is the surrounding tissue, their phase and amplitude of movement differ from those of the rest of the body, which causes a shearing motion of the cilia of the hair cells and consequently the detection of sound. Fishes with a swim bladder can detect sound waves well, but fishes without a swim bladder are not only less sensitive to sounds in general but also much more limited in frequency-range sensitivity than are fishes with a swim bladder (Fig. 22-3). This shows the importance of the hearing function of the swim bladder. Removal of the swim bladder invariably results in a marked loss of sensitivity to pressure signals above 160 Hz.

The application of modern empirical tools for data collection (e.g., sound chambers and neurophysiological tools) is now being used by functional morphologists interested in analyzing the reaction of an animal to its environment. Neurobiologists can experimentally record the greatest mechanical efficiency that the structure of the organ under study permits.

Teleost Hearing with Weberian Ossicles

A functional acoustic coupling between the inner ear and swim bladder can change the physical relationship between the two. In the Otophysi (a subgroup of the Ostariophysi containing most of its species), the swim bladder and inner ear are mechanically coupled by the Weberian ossicles (Fig. 22-4). As is discussed in Chapter 12, the Weberian ossicles consist of three or four small ossicles derived from vertebral elements. The ossicles articulate movably with elastic cartilage to transmit vibrations from the swim bladder to the inner ear (Fig. 22-4). Movement of the swim-bladder walls during acoustic stimulation causes anteroposterior movements of the Weberian ossicles. The swim bladder has a double wall, a collagenous tough tunica externa, and

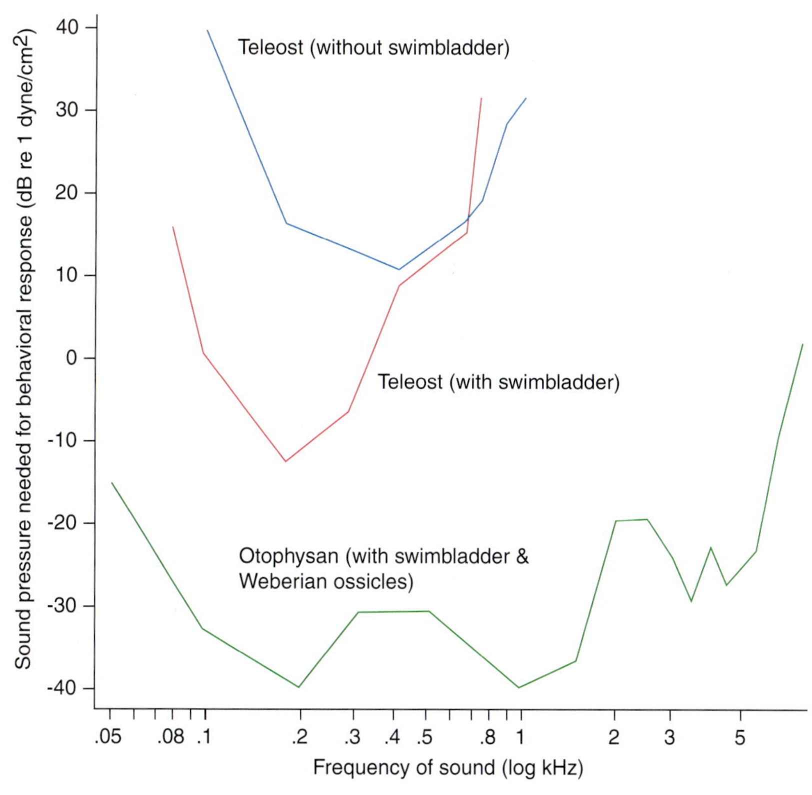

FIGURE 22-3
Comparative audiograms showing the greatest sensitivity of the otophysan fish ears. Notice that the least sensitive ears are those of a fish lacking a swim bladder, whereas the most sensitive ear is one with Weberian ossicles. The fish with a swim bladder but without Weberian ossicles occupies an intermediate position. *(Modified from Fay and Popper.)*

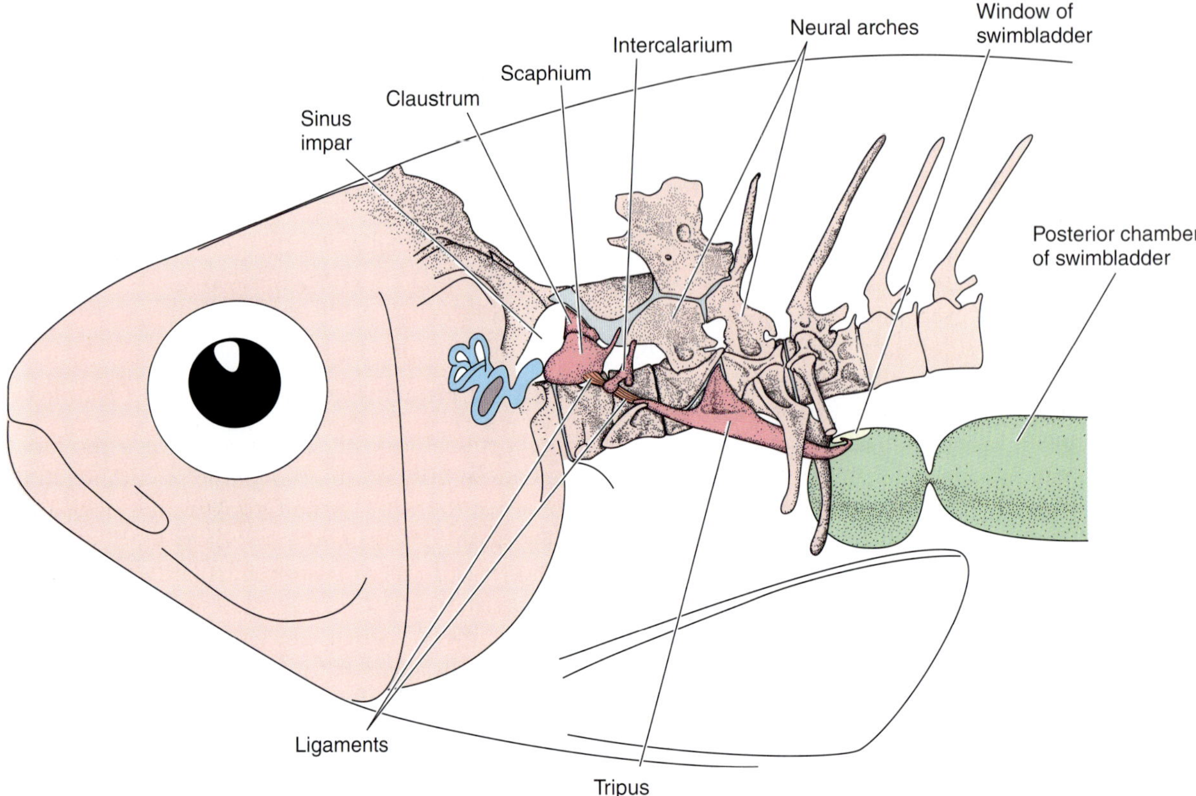

A. Lateral view of Weberian apparatus

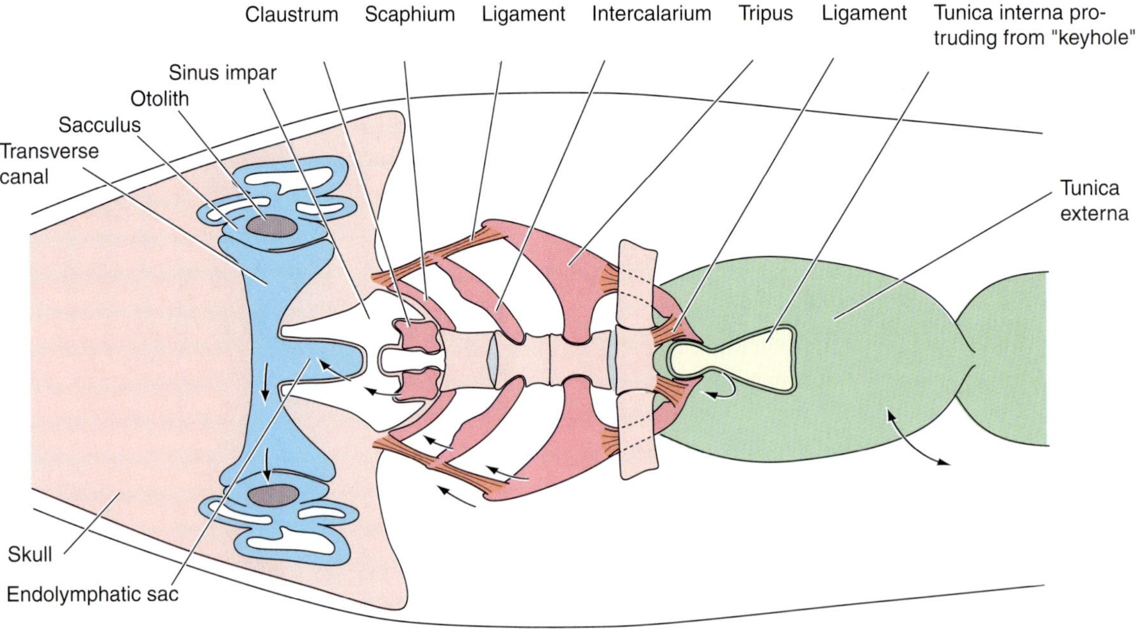

B. Dorsal view of Weberian apparatus

FIGURE 22-4
A, Lateral view of an otophysan fish with the Weberian ossicles and swim bladder. *B,* Detail of the dorsal view of the swim bladder and Weberian ossicles. Vibrations, indicated by an arrow, set up in the swim bladder are transmitted to the tripus, and other ossicles that are connected by ligaments to the sinus impar and transverse canals hence to the sacculus, where hearing is perceived. *(B, Modified from Alexander).*

an elastic inner tunica interna. Dorsally, the tunica externa has a gap in the shape of a keyhole (Fig. 22-4) exposing the elastic tunica interna. During vibration of the gas within the swim bladder, the elastic tunica interna pops out and in from the "keyhole," causing the edges of the keyhole to deform. Because the posterior arm of the most posterior Weberian ossicle, the tripus, is attached to the edge of the keyhole, it undergoes anteroposterior movements, which are transmitted by ligaments to the succeeding ossicles, the intercalarium and scaphium (Fig. 22-4), and finally to the claustrum. The moving scaphium and claustrum cause fluid movements to the sinus impar, which is a specialized perilymphatic chamber. The sinus impar, in turn, transfers the movements to the transverse canal and the sacculi, causing the saccular otoliths to "rock." A lever action occurs in the ossicles, especially with respect to the tripus (Fig. 22-4). The posterior arm of the tripus is farther from the fulcrum (which is the articulation to the third vertebra) than is the anterior arm, which is attached to the intercalarium. Thus, the ossicles not only transfer but also amplify the pressure waves from the swim bladder to the inner ear. The architecture of the tunica externa with the dorsal keyhole through which the tunica interna can move, together with the amplifying lever design of the tripus, explains the exceptional efficiency of hearing in the Otophysi. Functionally, the Otophysi have superior hearing capabilities in terms of frequency and sensitivity (Fig. 22-3).

Hearing in Tetrapods

Primitively, in tetrapods (e.g., in sauropsids), sound travels from the eardrum, or tympanum (on the surface of the head), to the endolymph of the inner ear by one auditory, rodlike ossicle that traverses the middle ear cavity, the columella. The columella attaches laterally to the tympanum and medially with a special

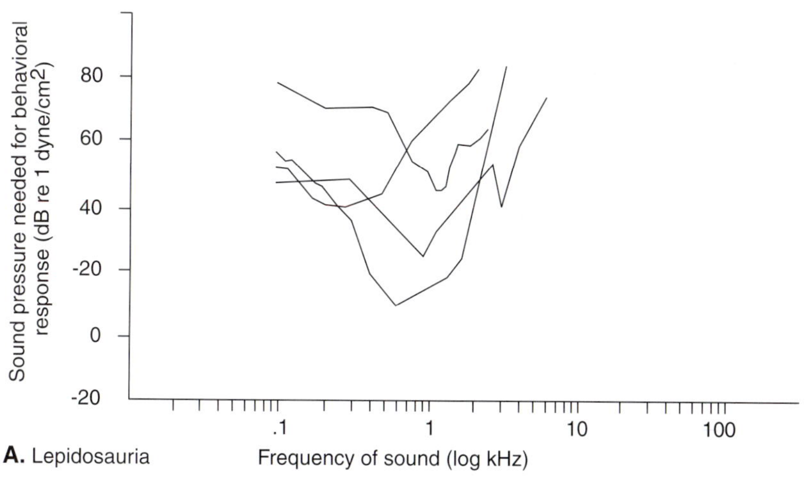

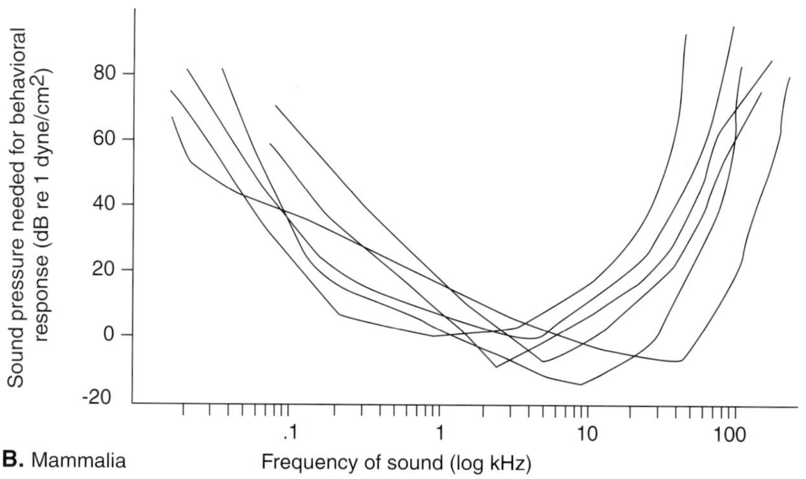

FIGURE 22-5
Comparative audiograms showing differences in the sensitivities and the frequency ranges of the ears of lepidosaurians (*A*) and mammals (*B*).

footplate to the oval window of the otic capsule (Chapter 12). A correlation exists between the ratio of tympanic membrane and the size of the footplate of the columella and hearing sensitivity. The larger the ratio, the more sensitive the hearing. All of the energy impinging on a large tympanum is concentrated on the small footplate. The resulting increase in energy per unit area on the footplate enables the vibrations to overcome the inertia, or impedance, of the liquid of the inner ear. Certain sauropsid ears (those of lepidosaurs) can respond to frequencies as high as 10 kHz (Fig. 22-5*A*).

In mammals, two more auditory ossicles, the incus and malleus, have been added to the columella, now the stapes. Together, the three ossicles form a lever system that increases the amplification about 1.5-fold. Additional amplification results from the area of the tympanic membrane, which is 22 times the area of the footplate of the stapes. As a result, mammals have very sensitive auditory systems (Fig. 22-5*B*).

Thus, functionally, the swim bladder with the Weberian ossicles in the Otophysi is very similar to the middle ear cavity with the three auditory ossicles in mammals. In both groups, amplification is enhanced by maximizing the lever system of the ossicles, the architecture of the swim bladder in otophysans, and the ratio of the tympanum and footplate of the stapes in mammals. Consequently, both mammals and otophysans have more sensitive ears that respond to a greater range of frequencies (Figs. 22-3 and 22-5) than do their primitive sister groups lacking such devices. Because the functional performance is so similar, and the phylogenetic relationships so distant between Mammalia and Otophysi, the Weberian ossicles and the three mammalian auditory ossicles are one of the best known examples of convergent evolution in vertebrate biology. The evolutionary origins of the two systems were independent of each other, but the same functional specialization was achieved in both cases by comparable amplifying lever systems of the ossicles.

This study and comparison of two very efficient and independently evolved hearing systems is a triumph in comparative functional anatomy achieved by merging anatomy, physiology, biomechanics, and developmental biology within a phylogenetic context. Another example of such landmark contributions is the functional evolution of vertebrate respiration (Chapter 18). Modern comparative anatomy demonstrates that major advances are made by being able to integrate previously divided methods, concepts, and theories. The merger of previously disparate disciplines is especially evident in comparative anatomy. Vitality is correlated with pluralism in research and education, while history has shown that stasis invariably occurs when a discipline is subdivided into separate sharply demarcated subdisciplines.

Evolutionary Innovation and Diversity

The emergence of the three auditory ossicles within the middle ear cavity in mammals is considered by many biologists to be an evolutionary innovation, as is the appearance of Weberian ossicles connecting the swim bladder with the inner ear in otophysan fishes. Even though clearly convergent, the two systems furnish each group unique hearing abilities. Ecologically and behaviorally, mammals are known to use their hearing acuity and sensitivity in mate recognition, territoriality, reproduction, aggressive and alarm communication, predation and foraging, and parental care. Indirect evidence shows selective advantage for the insectivorous primitive mammals in their nocturnal habitat to have the improved ability to detect high-frequency sounds. This innovation could have been a major factor for the eventual great diversity of Mammalia.

However, one of the challenges in comparative anatomy goes beyond a simple assertion that the possession of a particular innovation underlies the diversification of a group. Lauder (1981) has proposed the use of phylogenetic patterns to evaluate the relative role of an innovation in generating diversification. The problem here is to examine whether the mammalian ear ossicles have been partially responsible for the eventual great diversity of Mammalia. When we compare Mammalia, which have the three auditory ossicles, with their close and distant outgroups within the Synapsida (Figs. 3-31 and 22-6) lacking the innovation, it is apparent that the latter have not become very diverse as far as can be told by the fossil record (Fig. 22-6). Furthermore, the limited diversity within the synapsid outgroups seems correlated with the lack of penetration into the very habitats for which enhanced hearing abilities is beneficial. To further evaluate the causal relationship between the mammalian innovation of hearing ossicles and diversity, we examine a convergent aspect of an innovation, which produces enhanced hearing and its probable effect on diversity in the otophysan fishes. As in mammals, superior hearing capabilities made possible by the Weberian ossicles may have contributed to the ability of the Otophysi to penetrate innumerable habitats. Hearing, especially in unclear freshwater habitats, is used in reproductive encounters, aggression, homing, prey location, species recognition, migration, and orientation. Otophysi has undergone even more divergent evolution (with more

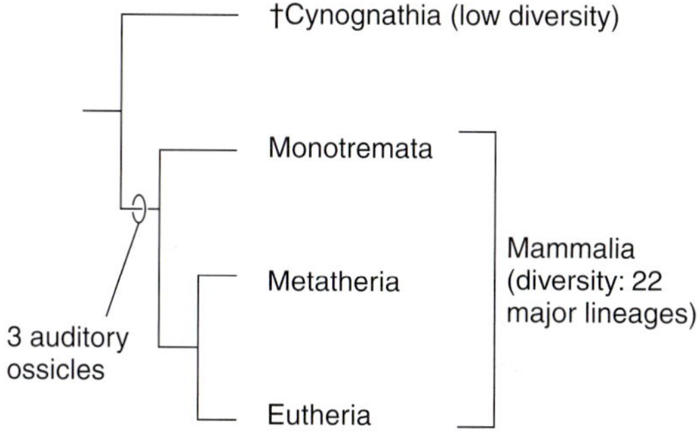

A. Auditory ossicles and probable effect on diversity in Mammalia and in its outgroup lacking the innovation

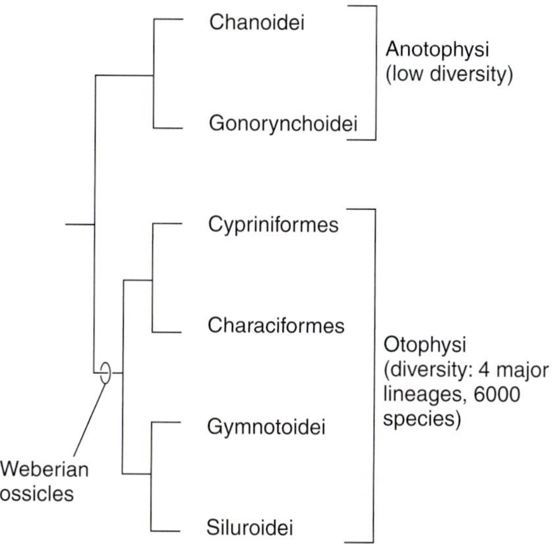

B. Weberian ossicles and probable effect on diversity in Otophysi and in its outgroup lacking the innovation

FIGURE 22-6
A, A phylogenetic comparison of the effect of the possible presence of the three auditory ossicles on the divergence of mammals with 17 ordinal lineages and the absence of divergence of the primitive synapsid outgroup, the cynodonts, lacking the three ossicles. *B,* A phylogenetic comparison of the effect of the presence of the Weberian apparatus on divergent evolution of the Otophysi, and the lack of divergence in its primitive sister group, the Anotophysi, without the Weberian ossicles.

than 6000 species) than mammals (Fig. 22-6). In sharp contrast the primitive outgroup of the otophysans within the Ostariophysi, the Anotophysi, lacking such an innovation, have not undergone diversification (Fig. 22-6). Thus, these contrasting phylogenetic patterns between mammals and otophysan fishes and their respective outgroups support the notion that differentiation of innovation of hearing in Mammalia and Otophysi is causally correlated with divergent evolution. Only after a comparison within a phylogenetic context can we conclude that functional design may well have a role in influencing ecological, behavioral, and evolutionary patterns.

The Contemporary Role of Comparative Anatomy

Because comparative anatomy has become truly multidisciplinary in its approach, it can now be regarded an integral part of many biological fields, ranging from phylogeny to evolutionary cell and molecular biology. Moreover, it is spearheading the very integration to which so many biology departments and societies aspire. It is the new morphology that plays an exemplary role beyond giving just lip service to integrative biology. Although the modern theory of the phylogeny of the vertebrates is based mostly on comparative anatomy (Chapter 3), new notions on the nature and fate of segmentation of head muscles, branchial skeleton, and cranial nerves have resulted from the integration of comparative anatomy, systematics, developmental biology, and molecular genetics. Major advances in explaining fish swimming and vertebrate flight have been made as a result of the application and incorporation, respectively, of hydrodynamic and aeronautical engineering methods and theories in comparative anatomy and behavior (Chapter 11). New integrative approaches have yielded an entirely new explanation of vertebrate feeding, substituting the previous speculative notions based on the derivation of function from form, with theories substantiated by experimental data. These data have been obtained with the selective use of highly sensitive, miniature pressure transducers in fishes; electromyography; high-speed light and radiographic cinematography; strain gauges to measure stress, strain, and bone deformation; and miniaturized cinecameras within the animal's mouth. These powerful empirical tools have been adopted from medicine, physiology, and the nuclear power plant industry and integrated with neurobiology and comparative anatomy. The results are briefly summarized in Chapter 16, with an emphasis on suction and ram-feeding fishes, and pattern generators and stereotypical mastication in amniotes, and in Chapters 18 and 19, where the mechanics of lung ventilation and the separation of blood flows without a physical barrier are discussed.

Comparative anatomy not only serves as a foundation for other biological disciplines as it has done so successfully for more than a century but also, in its revitalized form, has become a forerunner in integrative biology. As such, it is a path-breaking, pioneering discipline in identifying and solving new problems and generating novel theories crossing the traditional interdisciplinary barriers of phylogenetics, developmental biology, experimental zoology, behavior, ecology, evolution, and engineering science. Revitalized, integrative comparative anatomy generates new questions of general biological interest and tries to explain causally the evolutionary effects of functional design on biological diversity. Thus, comparative anatomy has entered a new research landscape with a very bright horizon.

REFERENCES

The references listed below represent a brief cross section of current trends in the science of form. They emphasize how comparative anatomy is spearheading the emerging comparative integrative mission of biology.

Aerts, P., 1990: Mathematical biomechanics and the what, how and why in functional morphology. *Netherlands Journal of Zoology,* 40:153–172.

Aerts, P., and deVree, F., 1993: Feeding performance and muscular constraints in fish. *Journal of Experimental Biology,* 177:129–147.

Aerts, P., 1998: Vertical jumping in *Galago senegalensis*: The quest for an obligate mechanical power amplifier. *Philosophical Transactions of the Royal Society of London Series B: Biological Sciences,* 353:1607–1620.

Alexander, R. McNeil, 1985: The ideal and the feasible: Phylogenetic constraints on evolution. *Biological Journal Linnean Society,* 26:345–358.

Alexander, R. McNeil, 1981: Factors of safety in the structure of animals. *Scientific Progress, Oxford,* 67:1099–1130.

Biewener, A. A., 1998: Muscle function in *vivo*: A comparison of muscles used for elastic energy savings *versus* muscles used to generate mechanical power. *American Zoologist,* 38:703–717.

Bramble, D. M., and Jenkins, F. A., 1994: Mammalian locomotor-respiratory integration: Implications for diaphragmatic and pulmonary design. *Science,* 262:235–240.

Burke, A. C., Nelson, C. E., Morgan, B. A., and Tabin, C., 1995: *Hox* genes and the evolution of vertebrate axial morphology. *Development,* 121:333–346.

Carrier, D., 1996: Function of the intercostal muscles in trotting dogs: ventilation or locomotion? *Journal of Experimental Biology,* 199:1455–1465.

Carrier, D., 1991: Conflict in the hypaxial musculo-skeletal system: Documenting an evolutionary constraint. *American Zoologist,* 31:644–654.

Chardon, M., and Vandewalle, P., 1997: Evolutionary trends and possible origin of the Weberian apparatus. *Netherlands Journal of Zoology,* 47:383–403.

Emerson, S. B., and Koehl, M. A. R., 1990: The interaction of behavioral and morphological change in the evolution of a novel locomotor type: "Flying frogs." *Evolution,* 44:1931–1946.

Farmer, C. G., and Carrier, D. R., 2000: Pelvic aspiration in the American alligator (*Alligator mississippiensis*). *Journal of Experimental Biology,* 203:1679–1687.

Ferry-Graham, L. A., and Lauder, G. V., 2000. Aquatic prey capture in fishes: A century of progress and new directions. *Journal of Morphology,* in press.

Fink, W., and Fink, S., 1981: Interrelationships of the ostariophysan teleost fishes. *Journal of the Linnean Society London,* 72:297–353.

Fish, F. E., 1993: Influence of hydrodynamic design and propulsive mode on mammalian swimming energetics. *Australian Journal of Zoology,* 42:79–101.

Galis, F., 1996: The application of functional morphology to evolutionary studies. *Trends in Ecology and Evolution,* 11: 124–129.

Gatesy, S. M., and Dial, K. P., 1996: From frond to fan: Archaeopteryx and the evolution of short-tailed birds. *Evolution,* 50:2027–2048.

Gatesy, S. M., and Dial, K. P., 1996: Locomotor modules and the evolution of avian flight. *Evolution,* 50:331–340.

Gould, S. J., and Lewontin, R. C., 1979: The spandrels of San Marco and the Panglossian paradigm: A critique of the adaptationist programme. *Proceedings Royal Society of London, B.* 205:581–598.

Goslow, G. E., Dial, K. P., and Jenkins, F. A., 1989: The avian shoulder, an experimental approach. *American Zoologist,* 29:287–301.

Greene, H. W., and Cundall, D., 2000: Evolutionary biology: Limbless tetrapods and snakes with legs. *Science,* 287:1939–1941.

Hall, B. K., 1996: Bauplane, phylotypic stages and constraint: Why are there so few types of animals? *In* Hecht, M. K., McIntyre, R. J., and Clegg, M. T., editors: *Evolutionary Biology,* volume 29. New York, Plenum Press.

Herrel, A., Aerts, P., Fret, J., and deVree, F., 1999: Morphology of the feeding system in agamid lizards: Ecological correlates. *Anatomical Record,* 254:496–507.

Holland, P. W. H., and Garcia-Fernandez, J., 1996: *Hox* genes and chordate evolution. *Developmental Biology,* 173:382–395.

Holland, P. W. H., Garcia-Fernandez, J., Williams, N. A., and Sidow, A., 1994: Gene duplications and the origins of vertebrate development. *Development,* (Supplement): 125–133.

Lauder, K. F., 2000: Function of the caudal fin during locomotion in fishes: Kinematics, flow visualization, and evolutionary patterns. *American Zoologist,* 40:101–122.

Lauder, G. V., 1981: Form and function: Structural analysis in evolutionary morphology. *Paleobiology,* 7:430–442.

Liem, K. F., 1991: Toward a new morphology: Pluralism in research and education. *American Zoologist,* 31:759–767.

Liem, K. F., and Summers, A. P., 2000: Integration of versatile functional design, population ecology, ontogeny and phylogeny. *Netherlands Journal of Zoology,* 245–259.

Maynard Smith, J., Burian, R., Kauffman, S., Alberch, P., Campbell, J., Goodwin, B., Lande, R., Raup, D., and

Wolpert, L., 1985: Developmental constraints and evolution. *Quarterly Review Biology,* 60:267–287.

Mueller, G. B., 1990: Developmental mechanisms as the origin of morphological novelty: A side hypothesis. *In* Nitecki, M. H., editor: *Evolutionary Innovations.* Chicago, University of Chicago Press.

Muller, M., 1993: The angles of femoral and tibial axes with respect to the cruciate ligament four-bar system in the knee joint. *Journal of Theoretical Biology,* 161:221–230.

Osse, J. W. M., van den Boogaart, J. G. M., 1997: Size of flatfish larvae at transformation, functional demands and historical constraints. *Journal of Sea Research,* 37: 229–239.

Pabst, D. A., 2000: To bend a dolphin: convergence of force transmission designs in cetaceans and scombrid fishes. *American Zoologist,* 40:146–155.

Popper, A. N., 1983: Organization of the inner ear and auditory processing. *In* Northcutt, R. G., and Davis, R. E., editors: *Fish Neurobiology.* Ann Arbor, The University of Michigan Press.

Rayner, J. M. V., 1991: Complexity and a coupled system: Flight, echolocation and evolution in bats. *In* Schmidt-Kittler, N., and Vogel, K., editors: *Constructional Morphology and Evolution.* New York, Springer-Verlag.

Rosowsky, J. J., 1992: Hearing in transitional mammals: Predictions from the middle-ear anatomy and hearing capabilities of extant mammals. *In* Webster, D. B., Fay, R. R., and Popper, A. N., editors: *The Evolutionary Biology of Hearing.* New York, Springer-Verlag.

Sanderson, S. L. and Cheer, A. Y., 1993. Fish as filters: an empirical and mathematical analysis. *Contemporary Mathematics,* 141:135–160.

Schaefer, S. A., and Lauder, G. V., 1986: Testing historical hypotheses of morphological change: Biomechanical decoupling in loricarioid catfishes. *Evolution,* 50: 1661–1675.

Schmidt-Kittler, N., and Vogel, K., 1991: *Constructional Morphology and Evolution.* Springer-Verlag, New York.

Van Leeuwen, J. L., 1999: A mechanical analysis of myomere shape in fish. *Journal of Experimental Biology,* 202: 3405–3414.

Van Leeuwen, J. L., and Spoor, C. W., 1992: Modelling mechanically stable muscle architectures. *Philosophical Transactions of the Royal Society of London, Series B: Biological Science,* 336:275–292.

Vermeij, G., 1974: Adaptation, versatility, and evolution. *Systematic Zoology,* 22:466–477.

Wagner, G. P., and Schwenk, K., 2000: Evolutionary stable configurations: functional integration and the evolution of phenotypic stability. *Evolutionary Biology,* 31:155–217.

Wainwright, P. C., 1996: Ecological explanation through functional morphology: the feeding biology of sunfishes. *Ecology,* 77:1336–1343.

Wake, D. B., 1991: Homoplasy: The result of natural selection, or evidence of design limitations? *American Naturalist,* 138:543–567.

Wake, D. B., and Roth, G., 1989: *Complex Organismal Function: Integration and Evolution in Vertebrates.* New York, John Wiley & Sons.

Wake, M. H., 1997: Amphibian locomotion in evolutionary time. *Zoological Analysis of Complex Systems,* 100: 141–151.

Glossary

This Glossary includes many of the technical terms that you will encounter in this book. Within each definition all **boldface** terms are defined elsewhere in this glossary. The phonetic spelling in brackets, when given, indicates the pronunciation of a term. The principal stressed syllables are indicated by a prime (′); other syllables are separated by hyphens (-). The macron (¯) is used for long vowels and the breve (˘) is used for short vowels. The schwa (ə) represents an unstressed neutral vowel. Its sound can vary according to the vowel it represents or the sounds surrounding it (about, item). ä = the vowel sound in "car"; î that in "pier"; ô that in "paw"; o͝o that in "took"; o͞o that in "boot"; û that in "urge"; zh = the "s" in "vision"; th = the "th" in "them".

For most terms, the classic derivation is given in parentheses because the derivation is descriptive of some aspect of the term and will help you learn and remember it. Because many word roots are repeated in different combinations in other terms, you will soon become familiar with the more common roots. The derivations typically include three components: (1) the language of the original word (Gr. = Greek, L. = Latin); (2) the original word in *italics;* and (3) the meaning of the original word. Usually only the nominative is given for Greek and Latin nouns, but the genitive (gen.) is included when this is necessary to recognize the root. For Greek and Latin verbs, the original word usually is shown in the first person, singular, present tense because this form of the word is closer to the root term than is the infinitive. The past participle (pp.) or present participle (pres.p.) also is given when this is necessary to recognize the root. The English meaning is given in the infinitive. In some cases, a noun or verb is given in the form it takes when used in combination with other words. This is indicated by a hyphen before or after the word (e.g., *odonto-* = tooth, as in **odontoblast**).

When the English and classic terms are identical, only the classic meaning is given:

Abducens nerve (L., *abducens,* pres.p. = leading away)

When two or more successive terms use the same root, the derivation is given only for the first one:

Archicerebellum (Gr., *arche* = origin, beginning + L., *cerebellum* = small brain)

Archinephric duct (Gr., *nephros* = kidney)

The origin of many repetitive terms is given under the first entry of the term. For example, *ligamentum arteriosum* is defined the way this combination of words is used, but the derivation of "ligamentum" and "arteriosum" will be found under the terms "ligament" and "artery."

The pronunciations, derivations, and definitions of additional terms can be found in *Dorland's* or *Stedman's* medical dictionaries or in unabridged dictionaries.

A

Abdomen [ăb′də-mən] (L., *abdomen* = from + *abdo* = to conceal). The part of the body cavity containing the **viscera.**

Abducens nerve [ăb-dū′sənz] (L., *abducens,* pres.p. = leading away). The nerve that innervates the lateral rectus **muscle** of the eyeball; cranial nerve 6.

Abductor [ăb′dŭk′tər] (L., *ab-* = prefix meaning away from + *duco,* pp. *ductus,* = to lead). Describes a muscle that abducts or moves a structure away from the midventral line of the body or some other point of reference, e.g., the abductor femoris.

†Acanthodians [ăk-ăn-thō′dē-ənz] (Gr., *akanthodes* = spiny). An extinct group of early **bony fishes** containing the earliest known jawed **vertebrates.** Although commonly called the spiny sharks, they may be more closely related to **bony fishes (Osteichthyes)** than to sharks **(Chondrichthyes).**

Acceleration. The rate of increase in speed, often expressed as meters per second per second.

Accessory nerve. The eleventh cranial nerve of **amniotes,** which innervates the sternocleidomastoid and trapezius complex of muscles.

Acetabulum [ăs′ĭ-tăb′yo͞o-lŭm] (L., = vinegar cup). The socket in the **pelvic girdle** that receives the head of the **femur.**

Acinar [ăs′ĭ-när] (L., *acinus* = berry). A berry-shaped group of glandular cells.

Acoelous vertebra [ā-sē′ləs] (Gr., *a* = without + *koiloma* = hollow). A vertebral body that lacks cavities and is flat on each surface.

Acousticolateralis system [ŭ-ko͞os′tĭ-kō-lăt-ər-ă′lĭs] (Gr., *akoustikos* = related to hearing). The ear and **lateral line system** of **fishes** and larval **amphibians;** also called the octavolateralis system.

Acrodont tooth [ăk'rō-dŏnt] (Gr., *akron* = tip). A tooth that is loosely attached to the crest or inner edge of the jaw.

Acromion [ă-krō'mē-ən] (Gr., *omos* = shoulder). The process on the **scapula** to which the **clavicle** articulates in **species** with a well-developed **clavicle.**

Acrosome [ăk'rō-sōm] (Gr., *soma* = body). The cap at the apex of a **sperm** head that contains enzymes needed for the **sperm** to penetrate the **egg.**

Actinopterygians [ăk'tĭn-ŏp-tə-rĭj'ē-ənz] (Gr., *aktin* = ray + *pteryg-* = fin or wing). A group (Actinopterygii) containing about half of all living **vertebrate species,** or 95% of living **fish species;** the **ray-finned fishes.** See **fish.**

Action potential. The electrical changes that occur across the plasma membrane of **muscle** and **nerve** cells when they become active.

Adaptation. A feature suited for a particular environment or mode of life. Also used to refer to the **evolutionary process** by which organisms became fitted to their environment. Identified retrospectively, this term is often used ambiguously.

Adaptive radiation. A term for a retrospectively identified hypothetical process in which a lineage **speciates** and evolves to occupy different habitats and modes of life.

Adductor [ă-dŭk'tər] (L. *adducere,* pp. = lead to). Describes a **muscle** that adducts, or moves a structure toward the midventral line of the body or other point of reference, such as the adductor mandibulae.

Adenohypophysis [ăd'ə-nō-hī-pŏf'ə-sĭs] (Gr., *aden* = gland + *hypophysis* = under growth). The secretory portion of the **pituitary gland,** or **hypophysis,** that develops as an outgrowth from the **stomodaeum;** secretes **hormones** that regulate pigment production, growth, and the activity of the **thyroid gland, adrenal gland,** and **gonads.**

Adrenal gland [ə-drē'nəl] (L., *ad* = toward + *rene* = kidney). An **endocrine gland** next to the **kidney,** consisting of distinct cortical and **medullary** parts. Major **hormones** of the **cortex** are **cortisol, aldosterone,** and cortical androgen; the major **hormone** of the **medulla** is **epinephrine.**

Adrenaline. See **epinephrine.**

Adrenergic fibers [ăd-rə-nûr'jĭk] (L., *ad* = toward + *rene* = kidney). **Postganglionic sympathetic neurons** that release **noradrenaline (norepinephrine)** at the neuroeffector junctions.

Adrenocorticotropic hormone [ə-drē-nō-kôr'tĭ-kō-trō'pĭk] (Gr., *trophe* = nurture). A **hormone** produced by the **adenohypophysis** that promotes the synthesis and release of adrenocortical **hormones.**

Aestivation. See **estivation.**

Afferent [ăf'ər-ənt] (L., *ad* = toward + *fero* = to carry). Describes structures that travel toward a point of reference, such as **neurons** toward the **central nervous system** or **arteries** toward the **gills.**

Agnatha [ăg-nā'thə] (Gr., *a* = without + *gnathos* = jaw). A **nonmonophyletic** group including the jawless **vertebrates** (e.g., **ostracoderms** and **cyclostomes**) with a fossil record extending back over 500 million years and represented today by the living hagfishes and lampreys.

Aldosterone [ăl-dŏs'tər-ōn]. A **hormone** of the adrenal cortex that helps regulate mineral metabolism.

Allantois [ă-lăn'tō-ĭs] (Gr., *allas* = sausage + *eidos* = form). The extraembryonic membrane that develops as an outgrowth of the hind gut. It serves for respiration and **excretion** in **reptile** and bird **embryos,** contributes to the **placenta** in **eutherians,** and forms the **urinary bladder** and part of the **urethra** in adult **amniotes.**

Allocortex [ă-lŏ-kôr'tĕks] (Gr., *allos* = other + L. *cortex* = bark). The part of the **mammalian brain cortex** characterized by three neuronal layers, the **lateral** and **medial pallium.**

Allometry [ă-lŏm'ə-trē] (Gr., *metron* = measure). The study of relative growth in which the proportions of a part of an animal change at a different rate than another part.

Alula [ăl'yōō-lə] (L., *ala* = wing + *-ule* = suffix denoting diminutive). The tuft of **feathers** borne by the first digit of a bird's wing.

Alveolus [ăl-vē'ō-ləs] (L., = small pit). A small pit or cavity, such as a tooth socket or a small saclike structure in a **lung,** where gas exchange occurs.

Ameloblasts [ă-mĕl'ō-blăsts] (Middle English, *amel* = enamel + Gr., *blastos* = bud). Cells that secrete **enamel.**

Ammocoete [ăm'ə-sēt] (Gr., *ammocoetes* = something bedded in sand). The **larva** of the lamprey.

Ammonia. The first breakdown product of nitrogen metabolism; very toxic and soluble in water and requires a high water turnover for its elimination; often converted to the less toxic **urea,** and the nontoxic and stable **uric acid.**

Amnion [ăm'nē-ŏn] (Gr., *amnion* = fetal membrane). The innermost of the extraembryonic membranes that surround the **embryo** and encase it in amniotic fluid.

Amniote [ăm'nē-ōt]. A **vertebrate** the **embryo** of which has an **amnion; tetrapods** other than **amphibians.** See **anamniote.**

Amphibians [ăm-fĭb'ē-ənz] (Gr., *amphi* = both, double + *bios* = life). A group of **vertebrates** including frogs, salamanders, and numerous fossil forms. Amphibians usually have aquatic **larvae** and terrestrial adults. See **lissamphibians.**

Amphicoelous vertebra [ăm-fə-sē'ləs] (Gr., *koilma* = hollow). A **vertebral** body that is concave on each surface.

Amphistylic suspension [ăm-fĭ-stī'lĭk] (Gr., *stylos* = pillar). A type of jaw suspension in **fishes** in which the upper jaw is supported by connections to both the **chondrocranium** and **hyoid** arch.

Ampulla [ăm-pōōl'ə] (L., = flask), *pl.* **-pullae** [-pōōl'ī]. A small, membranous vesicle, such as that on the end of a **semicircular duct.**

Ampullary organ. **Ampulla**-shaped electroreceptors in the skin of some **fishes** and **amphibians,** such as the **ampullae** of Lorenzini of sharks.

Amygdala [ə-mĭg'də-lə] (Gr., *amygdale* = almond). A deep **nucleus** of **gray matter** in the **cerebral hemisphere;** part of the **lateral pallium** and **limbic system.**

Anagenesis. Evolutionary change that does not involve branching (as opposed to **cladogenesis**). Usually used to refer to gradual changes within a **species** or a lineage. See **cladogenesis, speciation.**

Analogy [ăn-ăl'ə-jē] (Gr., *analogia* = correspondence). A functional similarity among nonhomologous organs. For example, the wing of a butterfly can be considered analogous (but not **homologous**) to the wing of a bird. A type of **homoplasy.**

Anamniote [ăn-ăm'nē-ōt] (Gr., *an* = without + *amnion* = fetal membrane). A **vertebrate** without an **amnion,** such as a **fish** or **amphibian.** Clearly a **nonmonophyletic group** (as in **fish**). See **amniote.**

†Anapsid [ăn-ăp'sĭd] (Gr., *a, an* = without + *apsid* = loop or bar). Without arches; a **vertebrate** skull with a complete roof of **bone** in the temporal region, or a **taxonomic group** of **reptilian species** (†Anapsida) with such a skull type.

Anastomosis [ăn-ăs'tə-mō'sis] (Gr., = opening, outlet). A peripheral union between blood vessels or other structures.

Androgen [ăn'drō-jən] (Gr., *aner,* gen. *andros* = male). A **hormone** that promotes the development of male characteristics.

Anestrus [ăn-ĕs'trəs] (Gr., *oistros* = gadfly, frenzy). The nonbreeding period of sexually mature animals.

Angle of attack. The angle at which the leading edge of a bird's wing is elevated above the horizontal; an increase in angle of attack increases lift up to the stalling point.

Antagonist [ăn-tăg'ə-nĭst] (Gr., *anti* = against + *agona* = contest). A structure, usually a **muscle,** that opposes or resists the action of another.

Antebrachium [ăn-tə-brā'kē-əm] (L., *ante* = before + *brachium* = arm). The forearm.

Anterior chamber. The space within the eyeball located between the **iris** and **cornea.**

Anterior commissure. An **olfactory commissure** within the **cerebrum** located just rostral to the columns of the fornix. See **commissure.**

Antidiuretic hormone [ăn'tē-dī-yo͞o-rĕt'ĭk] (Gr., *anti* = against + *dia* = through + *ouresis* = urination). A **hormone** produced in the **hypothalamus** and stored and released from the **neurohypophysis;** promotes water reabsorption from part of the **kidney** tubule and so concentrates the urine.

Antler [ănt'lər] (L., *ante* = before + *oculus* = eye). One of the **bony,** branching, and **deciduous** horns of members of the deer family; usually restricted to males.

Anura [ăn-yo͝or'ə] (Gr., *a* = without + *oura* = tail). The **amphibian** taxon to which frogs belong; also called Salientia. Note: "Anura" is a taxon; "anurans" (a vernacular) refers to a group.

Anus [ā'nŭs] (L., = anus). The **caudal** opening of the digestive tract.

Aorta [ā-ôr'tə] (Gr., *aorte* = great artery). A large artery; if unspecified, the dorsal aorta that carries **blood** from the **heart** to the body.

Aortic arches. Embryonic arteries that pass between the pharyngeal pouches as they carry **blood** from the ventral to the dorsal **aorta.**

Apnea [ăp'nē-ə] (Gr., *a* = without + *pnoia* = breathing). The cessation of breathing, during which the breath is held and the **lungs** are not ventilated.

Apomorphic character. A **character** hypothesized to be uniquely derived for (i.e., diagnostic of) a particular **monophyletic taxon.** In evolutionary terms, an **apomorphy** would be a peculiar feature shared by the members of a **monophyletic taxon** that was inherited by each of those members from a hypothetical common ancestor. Some authors divide the term "apomorphy" into two subcategories: "autapomorphy" for terminal **taxa** (distal-most branches of a given **cladogram**), and "synapomorphy" for **monophyletic groups** of **taxa** (branch points of a given **cladogram**). **Synapomorphy** (shared derived **character**) is the most commonly used form. **Synapomorphy** is equivalent to **phylogenetic homology** or derived **character.** Examples: the presence of **feathers** for **Aves** (birds), the presence of jaws for **Gnathostomata.** See **cladistics, derived character, homology.**

Apomorphy. [ăp'ō-môr-fē]. See **apomorphic character.**

Aponeurosis [ăp'ō-nyo͝or-ō'sis] (Gr., *apo* = away from + *neuron* = nerve, sinew). A sheetlike **tendon** of a **muscle.**

Appendix (L., *appendere* = to hang upon). A dangling extension of another organ, such as the vermiform appendix on the **caecum.**

Aqueduct of Sylvius (*Franciscus Sylvius,* 1614–1672, Dutch anatomist). See **cerebral aqueduct.**

Aqueous humor [ā'kwē-ŭs hyo͞o'mer] (L., *aqua* = water + *humor* = liquid). The **lymph**like liquid filling the anterior and posterior chambers of the eye.

Arachnoid [ə-răk'noid] (Gr., *arachne* = spider + *eidos* = form). Tissue surrounding the **central nervous system,** lying peripheral to the **pia mater** and underlying the **dura mater.**

Arbor vitae [är'bər vīt'ē] (L., *arbor* = tree + *vita* = life). The treelike configuration of white fibers entering and leaving the **mammalian cerebellum.**

†Archaeornithes [är'kē-ôr'nə-thēz] (Gr., *arche* = beginning + *oris* = bird). A primitive group of birds, which includes †*Archaeopteryx.*

Archenteron [ärk-ĕn'tər-ŏn] (Gr., *arche* = origin, beginning + *enteron* = intestine, gut). The **embryonic** gut cavity, lined with **endoderm.**

Archicerebellum [är'kē-sĕr-bəl'əm] (Gr., *cerebellum* = small brain). The part of the **cerebellum** that receives vestibular impulses from the ear and impulses from the **lateral line system;** the flocculonodular lobes in **mammals.**

Archinephric duct [är'kə-nĕf'rĭk] (Gr., *nephros* = kidney). The first-formed **kidney duct,** which drains the **kidney** of most **anamniotes** and becomes the ductus deferens of male **amniotes.**

Archinephros. See **holonephros.**

Archipallium. See **medial pallium.**

Archipterygium [är'kē-tə-rĭj'ē-əm] (Gr., *ptery-* = fin or wing). The paired fins of lungfishes in which radials extend from each side of a central **axis;** once believed to be the ancestral morphotype of paired fins.

Archosaurs [är'kō-sôrz] (Gr., *archon* = ruler + *sauros* = lizard). The **reptilian** group (Archosauria) that includes the two extinct orders of dinosaurs, the extinct **†pterosaurs** (†flying **reptiles**), the contemporary crocodiles, and birds.

Arcualia [är'kyōō-ā-lē-ə] (L., *arcus* = bow, arch). Small arches of **cartilage** or **bone** that often contribute to the formation of a **vertebra.**

Artery [är'tər-ē] (L., *arteria* = artery). A vessel that carries **blood** away from the **heart.** The **blood** may be high or low in oxygen content.

Articular [är-tĭk'yə-lər] (L., *articulus* = joint). Pertaining to a joint.

Artiodactyls [är'tē-ō-dăk'təlz] (Gr., *artios* = even + *daktylos* = finger or toe). The **mammalian** group (Artiodactyla) that includes **ungulates** with an even number of toes: pigs, deer, cattle.

Arytenoid cartilage [ăr-ə-tē'noid] (Gr., *arytainoeides* = ladle-shaped). The ladle-shaped **cartilage** of the **mammalian larynx** that attaches to and modifies the tension of the **vocal cords.**

Aspiration pumping. A method of **lung** ventilation in which air is sucked into the **lungs;** occurs primarily in **amniotes.**

Astrocytes [ăs'trō-sītz] (Gr., *astron* = star + *kytos* = hollow vessel or cell). Star-shaped nutritive and supportive **glial** cells of the **central nervous system.**

Atlas [ăt'ləs] (Gr. mythology, a god supporting the Earth upon his shoulders). The first **cervical vertebra** of terrestrial **vertebrates,** which articulates with the **skull;** nodding movements of the head occur between the atlas and **skull.**

Atrium [ā'trē-əm] (L., = entrance hall). A chamber, such as the atrium of the **heart,** that receives **blood** from the **sinus venosus** or **veins.**

Atrophy [ăt'rō-fē] (Gr., *a* = without + *trophe* = nourishment). The decrease in size and sometimes loss of a structure.

Auditory [ô'dĭ-tôr-ē] (L., *audio* = to hear). Pertaining to the ear.

Auditory tube. A tube that extends between the **tympanic cavity (middle ear)** and **pharynx** of most **tetrapods** and equalizes the air pressure on both sides of the **tympanic membrane; homologous** to the **spiracle** of **fishes.** Sometimes called the eustachian tube.

Auricle [ô'rĭ-kəl] (L., *auricula* = external ear). The external flap of the mammalian ear. Also, an ear-shaped appendage on the atrium of mammals.

Autonomic nervous system [ô'tə-nŏm'ĭk] (Gr., *autos* = self + *nomos* = law). The part of the nervous system carrying **visceral** motor fibers to the **viscera** and **glands.**

Autostylic suspension [ô'tō-stī'lĭk] (Gr., *stylos* = pillar). A type of jaw suspension in which the upper jaw is attached to the rest of the **skull** by its own processes.

Aves [ā'vēz] (L., = birds). The **vertebrate taxon** that contains the birds.

Axillary [ăk'sə-lĕr-ē] (L., *axilla* = armpit). Pertaining to the armpit: axillary **artery.**

Axis [ăk'sis] (L., = axle, axis). The second **cervical vertebra** of **mammals;** rotary movements of the head occur between the axis and **atlas.**

Axon [ăk'sŏn'] (Gr., = axle, axis). The long, slender process of a **neuron** specialized for the transmission of **nerve** impulses.

Azygos vein [ā-zī'gəs] (Gr., *a* = without + *zygon* = yoke). An unpaired **vein** that drains most of the intercostal spaces on both sides of the **mammalian thorax.**

B

Basal lamina. The thin layer, or **lamina,** of **matrix** that underlies epithelial surfaces; composed primarily of collagen fibrils that do not organize as fibers; formerly called the basement membrane.

Basal nuclei. A group of **nuclei** in the **striatum** of the **mammalian cerebrum.**

Basapophysis [bā'zə-pŏf'ə-sĭs] (Gr., *basis* = base + *apo* = away from + *phys* = growth). A transverse process low on a **vertebral** body to which a subperitoneal rib of a **fish** attaches; serially **homologous** to a **hemal arch.**

Biceps [bī'sĕps] (L., *bi* = two + *ceps* = head). A structure with two heads, such as the biceps **muscle.**

Bicornuate [bī-kôr'nyōō-āt] (L., *cornu* = horn). A structure with two **horns,** such as a bicornuate **uterus.**

Bile [bīl] (L., *bilis* = bile). The secretion of the **liver,** containing bile pigments and fat-emulsifying bile salts.

Biped [bī'pĕd] (L., *bi* = two + *pes,* gen. *pedis* = foot). A **tetrapod** that stands upright, such as a human.

Bladder. A membranous sac filled with air or liquid.

Blastocoele [blăs'tō-sēl] (Gr., *blastos* = bud + *koilos* = hollow). A cavity of the **blastula** that becomes obliterated during **gastrulation** and **mesoderm** formation.

Blastocyst [blăs'tō-sĭst] (Gr., *kystis* = bladder). The modified **blastula** of a **eutherian mammal.**

Blastodisk [blăs'tō-dĭsk] (Gr., *diskos* = disk). The disk of cells formed during **cleavage** that lies on the top of the yolk of large-yolked eggs of **fishes** and **reptiles** (including birds) and on the top of the **yolk sac** of **mammals.**

Blastomere [blăs'tō-mîr] (Gr., *meros* = part). One of the cells of the **blastula.**

Blastopore [blăs'tō-pôr] (Gr., *poros* = pore). The opening into the **archenteron** that is formed during **gastrulation.**

Blastula [blăs'tyū-lə] (L., diminutive of Gr. *blastos* = bud). The ball of cells formed during **cleavage,** usually containing a **blastocoele.**

Blood. The liquid circulating in the **arteries, capillaries,** and **veins,** consisting of a liquid plasma and cellular elements.

Blood–brain barrier. The structural and physiological barriers that regulate the exchange of materials between the **blood, brain tissue,** and **cerebrospinal fluid.**

Bone. The hard, skeletal material of **vertebrates** that consists of **collagen** fibers to which calcium phosphate crystals are bound, usually arranged in alternating layers of **matrix** and **bone**-forming cells.

Bony fishes. See **osteichthyes.**

Boundary layer. The layer of water or air surrounding a moving aquatic or flying animal in which **shear** forces occur; causes frictional **drag.**

Bowman's capsule. (*Sir William Bowman,* British anatomist, 1816–1892). See **renal capsule.**

Brachial [brā'kē-əl] (L., *brachium* = upper arm). Pertaining to the upper arm; armlike.

Brachium conjunctivum [kŏn-jŭngk-tī'vəm] (L., *conjungo,* pp. *conjunctus* = to join together). The most cranial **cerebellar** peduncle; an armlike **neuronal** tract of **mammals** through which impulses enter and leave the **cerebellum.**

Brachium pontis [pŏn'tĭs] (L., *pons,* gen. *pontis,* = bridge). The middle **cerebellar** peduncle of mammals carrying impulses into the **cerebellum** from the **pons.**

Brain. The enlarged **cranial** portion of the **central nervous system** enclosed by the **cranium;** the major integrative center of the **central nervous system.**

Braincase. The **cartilages** and **bones** that encase the **brain.**

Brainstem. The **brain** exclusive of the **cerebellum** and **forebrain** (**diencephalon** and **cerebrum**).

Branchial [brăng'kē-əl] (Gr., *branchia* = gills). Pertaining to the **gills.**

Branchial arches. Those **visceral arches** (numbers 3–7) that support the **gills** in **fishes.**

Branchiomeres [brang'kē-ō-mērz] (Gr., *meros* = part). The embryonic branchial segments lying between the pharyngeal pouches, including a visceral arch and associated muscle and nerves. Most also contain an **aortic arch.**

Branchiomeric. Pertaining to **muscles** and other structures associated with the **visceral arches.**

Bronchus, pl. **bronchi** [brŏng'kŭs] (Gr., *bronchos* = windpipe). A branch of the **trachea** that enters the **lungs.**

Buccal [bŭk'al] (L., *bucca* = cheek). Pertaining to the mouth, as in *buccal cavity.*

Bulbourethral glands [bŭl'bō-yōō-rē'thrəl] (L., *bulbus* = a bulbous root + Gr., *ourethra* = urethra). Accessory sex **glands** of male **mammals** that are located near the base of the **penis** and discharge into the **urethra.**

Bulla [bōōl'ə] (L., = bubble). A bubble-like expansion of some structure, such as the tympanic bulla on the temporal **bone.**

Bunodont [bōō'nō-dŏnt] (Gr., *bounos* = mound + *odont-* = tooth). **Molar** teeth with low, rounded cusps.

Bursa [bûr'sə] (L., = purse). A saclike cavity.

Bursa of Fabricius (*Giralamo Fabricius,* Italian anatomist and embryologist, 1533–1619). A dorsal **cloacal** diverticulum of birds, site of the maturation of B lymphocytes.

C

Caecilians [sē-sĭl'ē-ənz] (L., *caecilia* = blindworm). Tropical wormlike burrowing **amphibians** of the **taxon Gymnophiona.**

Caecum [sē'kəm] (L., *caecus* = blind). A blind-ending pouch attached to part of the **intestine,** such as the one at the beginning of the **mammalian** large **intestine.**

Calcaneus [kăl-kā'nē-əs] (L., = heel). The large proximal **tarsal** bone that forms the "heel bone" of **mammals.**

Calcitonin [kăl-sĭ-tō'nĭn] (L., *calx* = lime + Gr., *tonos* = tension). A **hormone** produced by the C cells of the **ultimobranchial bodies** or **thyroid gland;** its actions oppose those of **parathormone,** for it promotes the deposition of calcium in **bone** and reduces its level in the **blood.**

Calyx, pl. **calyces** [kā'lĭks, kăl'ĭ-sēz] (Gr., *kalyx* = cup). A cuplike compartment, such as the **renal** calyces or subdivisions of the **renal** pelvis.

Canaliculi [kăn'ə-lĭk'yōō-lī] (L., *canaliculi* = small channels). Small canals in **bone matrix** that contain the processes of the **osteocytes.**

Canine [kā'nīn] (L., *canis* = dog). The **mammalian** tooth behind the **incisors,** usually longer than other teeth.

Cantilever. A projecting beam or other structure that is supported at only one end.

Capillary [kăp'ə-lĕr'ē] (L., *capillus* = hair). One of the minute **blood** vessels between **arteries** and **veins** through which exchanges between the **blood** and **tissue** fluids occur.

Capitulum [kə-pĭch'yə-ləm] (L., = small head). A small, articulating knob on the end of a **bone,** such as a rib.

†Captorhinida [kăp'tō-rī'nĭd-ə] (L., *capus* = capture + Gr., *rhis* = nose). An early, extinct group of primitive **Sauropsida.**

Carapace [kâr'ə-pās] (Spanish, *carapacho* = covering). The dorsal shell of a turtle; the chitinous covering of a crustacean.

Cardiac [kär'dē-ăk] (Gr., *kardia* = heart). Pertaining to the **heart.**

Cardinal vein [kär'dən-əl] (L., *cardinalis* = principal). One of the principal **veins** of **embryonic vertebrates** and adult **anamniotes.**

Carnassials [kär-năs'ē-əlz] (L., *caro,* gen. *carnis* = flesh). The specialized shearing teeth of **carnivores;** the fourth upper **premolar** and first lower **molar.**

Carnivore [kär'nə-vôrz] (L., *-vorous* = devouring). An animal that feeds on other animals.

Carotid [kə-rŏt'ĭd] (Gr., *karotides* = large neck artery, from *karoo* = to put to sleep, because compressing the artery causes unconsciousness). Pertaining to a large **artery** in the neck or to nearby structures.

Carpal [kär'pəl] (Gr., *karpos* = wrist). One of the small **bones** of the wrist.

Cartilage [kär'təl-əj] (Gr., *cartilago* = cartilage). A firm but elastic skeletal **tissue** the **matrix** of which contains proteoglycan molecules that bind with water. Occurs in all **embryos,** in adult **cartilaginous fishes,** and in parts of the skeleton of other **vertebrates** providing firmness as well as flexibility.

Cartilage-replacement bone. Bone that develops within and around the **embryonic endoskeleton.**

Cartilaginous fish. See **chondrichthyes.**

Caudal [kôd'l] (L., *cauda* = tail). Pertaining to the tail.

Caudata [kô'dä-tə]. The **amphibian** group that includes the salamanders.

Cecum. See **caecum.**

Cenozoic. The era of geologic time ranging from about 65 million years before the present to the present.

Center of buoyancy. The point in the body of an aquatic **vertebrate** through which the resultant force of buoyancy acts.

Center of gravity. The point in the body of an animal through which the resultant force of gravity acts.

Central nervous system. That part of the nervous system located in the longitudinal **axis** of the body; consists of the **brain** and **spinal cord.**

Central pattern generator. Groups of **neurons** in the **spinal cord** and **brain** the activity of which is responsible for innate, cyclical movements of body parts, as occur in swimming and walking.

Centrum [sĕn'trəm] (Gr., *kentron* = center). The central part of the **vertebral** complex, lying ventral to the **vertebral** or **neural** arch.

Cephalic [sĕ-făl'ĭk] (Gr., *kephale* = head). Pertaining to the head.

Cephalization [sĕf'ə-lĭ-zā'shən]. The development of a well-defined head.

Cephalochordata [sĕf'ə-lō-kôr-dä'tə] (L., *chordata* = string). The **taxon** of **chordates** that includes amphioxus.

Ceratotrichia [sĕr'ə-tō-trĭk'ē-ə] (Gr., *kerat-* = horn + *trich-* = hair). The horny fin rays of **Chondrichthyes.**

Cerebellum [sĕr'ə-bĕl'əm] (L., = small brain). The dorsal part of the **metencephalon,** which is a center for motor coordination.

Cerebral aqueduct [sĕ'rə-brəl] (L., *cerebrum* = brain). The narrow passage within the **brain** that extends between the third and fourth **ventricles;** also called the **aqueduct of Sylvius.**

Cerebral hemispheres (Gr., *hemi* = half + *sphaira* = globe, ball). The pair of hemispheres that form most of the **telencephalon.** They are the major integrating centers of the **brain** in **mammals.**

Cerebrospinal fluid. A **lymph**like fluid that circulates within and around the **central nervous system,** which it helps protect and nourish.

Cerebrum. The two **cerebral hemispheres** of the **brain** in **vertebrates.**

Cervical [sûr'vĭ-kəl] (L., *cervix* = neck). Pertaining to the neck.

Cervix [sûr'vĭks]. The necklike portion of an organ, such as the neck of the **uterus.**

Character. Any feature that is an observable part or attribute of an organism. Congruent **characters** that diagnose groups of organisms are conjectures of **homology,** or **synapomorphies.** See **congruence, homology, synapomorphy.**

Cheek teeth. A collective term for **premolar** and **molar** teeth of **mammals.**

Cheiropterygium [kī-rō-tə-rĭj'ē-əm] (Gr., *chiro-* = hand + *pteryg-* = fin or wing). The paired appendage of a terrestrial **vertebrate.**

Chelonia [kə-lō'nē-ə] (Gr., *chelone* = tortoise). The **taxon** to which turtles belong.

Chimaera [kī-mîr'ə] (Gr., *chimaira* = monster). A **cartilaginous fish** belonging to **Holocephali.**

Choana [kō-ăn'ə] (Gr., *choane* = funnel). One of the paired openings from the **nasal** cavities into the **pharynx;** an internal nostril.

Chondrichthyes [kŏn-drĭk'thĭ-ēz] (Gr., *chondros* = cartilage + *ichthyos* = fish). The **cartilaginous fishes,** including sharks, skates, rays, and **chimaeras.**

Chondroblast. An early embryonic cell that is destined to produce **cartilage.**

Chondrocranium [kŏn'drō-krā'nē-əm] (Gr., *chondros* = cartilage + *kranion* = skull). **Cartilages** that encase the **brain** and major **sense organs** in **embryos** and the adults of some **vertebrates,** also called the neurocranium.

Chondrocyte [kŏn'drō-sīt] (Gr., *kytos* = hollow vessel or cell). A mature **cartilage** cell; develops from a **chondroblast.**

Chondrosteans [kŏn-drôs'tē-ənz] (Gr., *osteon* = bone). A primitive group of **actinopterygians,** including the contemporary sturgeons and paddlefishes.

Chordamesoderm [kôr'də-mĕz'ō-dûrm]. The longitudinal, mid-dorsal group of **mesodermal** cells that moves into the roof of the **archenteron** during **gastrulation** and gives rise to the **notochord.**

Chordates [kôr'dāts]. The group to which **tunicates, amphioxus,** and **craniates** belong; characterized by having a **notochord** at least at some stage of their life cycle.

Chorion [kô'rē-ŏn] (Gr., *chorion* = skinlike membrane enclosing the fetus). The outermost extraembryonic membrane of **amniotes.**

Choroid [kôr'oid] (Gr., *chorioeides* = like a membrane). The highly vascularized middle **tunic** of the eyeball that lies between the **fibrous tunic** and the **retina.**

Choroid plexus [plĕk'səs] (L., *plexus* = network). The vascular network of the **telachoroidea** that secretes the **cerebrospinal fluid;** it may invaginate into certain ventricles or evaginate into the space around the brain.

Chromaffin cells [krō'mə-fĭn] (Gr., *chromo-* = color + L., *affinis* = affinity). Cells in the **medulla** of the **adrenal gland** of **neural crest** origin that secrete **norepinephrine** and **epinephrine** and have an affinity for chromic stains.

Chromatophore [krō-măt'ə-fôr] (Gr., *phoros* = bearing, from *pherein* = to bear). A **vertebrate** cell of **neural crest** origin that carries pigment or reflective granules.

Cilia [sĭl'ē-ə] (L., *cilia* = hairs). Minute, movable processes of some **epithelial** cells that contain a **characteristic** pattern of nine peripheral and two central **microtubules.**

Ciliary body. A part of the **vascular tunic** of the eyeball that secretes the **aqueous humor** and contains **muscle** fibers used in focusing the eye.

Circadian rhythm [sər-kā'dē-ən] (L., *circa* = about + *dies* = day). A metabolic or behavioral pattern with a cycle of approximately 24 hours.

Cisterna chyli [sĭ-stûr'nə kīl'ē] (L., *cisterna* = an underground reservoir, cistern + Gr., *chylos* = juice). The sac that receives **lymph** from the **viscera** and **caudal** parts of the body.

Clade. A **monophyletic group** (as monophyletic is used here).

Cladistics [klə-dĭs'tĭks] (Gr., *clados* = branch). A method of investigating evolutionary relationships. Cladistics does not generally try to identify specific ancestors but rather attempts to interpret the relative interrelationships among **taxa** by calculating the most efficient (i.e., most **"parsimonious"**) **hierarchical** arrangement of an **empirical** data set. The data consist of **characters** thought to be uniquely derived for particular groups (e.g., the presence of **vertebrae** in **vertebrates,** the presence of jaws in **gnathostomes,** the presence of **feathers** in birds). **Congruent** data (**homologies** or **synapomorphies**) and **parsimony** are used to make **cladograms** and **phylogenetic** trees. Cladistics is currently the most widely used method of investigating evolutionary relationships. See **character, congruence, homology, parsimony, synapomorphy.**

Cladogenesis. A process theory of lineage multiplication involving branching (i.e., splitting) and divergence between **species** over time. See **anagenesis, speciation.**

Cladogram. A type of branching diagram that uses a **hierarchical** organization of data to construct a putative **phylogeny** of **taxa.** The **characters** on a cladogram that exhibit **congruence** are thought to be indicative of evolutionary relationship. These **characters** are termed **apomorphies** or **synapomorphies.** For example, **feathers** are thought to be uniquely derived for birds (and their presence in all birds is hypothesized to be due to inheritance from a common ancestor). See **apomorphy, cladistics, hierarchy, monophyletic group, parsimony.**

Clasper. The modified part of the **pelvic** fin of male **chondrichthyan fishes** used to transfer **sperm** to the female.

Clavicle [klăv'ĭ-kəl] (L., *clavicula* = small key, nail). A **dermal bone** of the **pectoral girdle** extending medially from the **scapula** to the **interclavicle** or **sternum.**

Cleavage [klē'vĭj]. The mitotic divisions by which the single-celled **zygote** is converted to a multicellular **blastula** of the same size.

Cleidoic egg [klī-dō'ĭk] (Gr., *kleid-* = clavicle, key). The self-contained eggs of **amniotes** in which a free **larval** stage is bypassed; modified in **viviparous species.**

Cleithrum [klī'thrəm] (Gr., *kleithron* = bar). A bar-shaped **dermal** element of the **pectoral girdle** of some **fishes** and early **tetrapods;** located dorsal to the **clavicle.**

Clitoris [klĭt'ər-ĭs] (Gr., *kleitoris* = hill). The small erectile organ of a female **mammal** that corresponds to the male **glans penis** and **corpora cavernosa penis.**

Cloaca [klō-ā'kə] (L., *cloaca* = sewer). The posterior chamber of most **fishes,** nonmammalian **tetrapods,** and **monotreme mammals** into which the digestive tract and **urogenital** passages discharge.

Coccyx [kŏk'sĭks] (Gr., *kokkyx* = cuckoo). Several fused **caudal vertebrae** of humans; does not reach the body surface but serves for the attachment of certain **muscles.**

Cochlea [kŏk'lē-ə] (L., = snail shell). The snail-shaped part of the **mammalian inner ear,** consisting of the **cochlear duct** and the **scala vestibuli** and **scala tympani.**

Cochlear duct. The **duct** within the **cochlea** that is a part of the **membranous labyrinth** and contains the receptive cells for sound.

Coelacanths [sē'lə-kănths] (Gr., *koilos* = hollow + *akantha* = spine). A group of **sarcopterygian fishes** with a long diverse fossil record going back several hundred million years but represented today by only a single living species.

Coelom [sē'ləm] (Gr., *koiloma* = a hollow). A body cavity that is completely lined by an **epithelium** of **mesodermal** derivation.

Collagen [kŏl'ə-jən] (Gr., *kolla* = glue + *genos* = descent). A protein produced by **fibroblasts;** forms most of the extracellular fibers of **connective tissues** and skeletal **tissues.** It is composed of ultramicroscopic fibrils that usually are organized into fibers that differ in size among the types of collagen.

Collecting ducts. The small tubules that receive material from the **kidney** tubules and lead to the renal pelvis or urinary duct.

Colliculus, pl. **colliculi** [kə-lĭk'yōō-ləs] (L., = little hill). One of the small elevations on the dorsal surface of the **mesencephalon** of **mammals** that is a center for certain optic (superior colliculus) or auditory (inferior colliculus) **reflexes.**

Colon [kō'lən] (Gr., *kolon* = colon). The large **intestine** of **tetrapods** exclusive of the **caecum** and **rectum.**

Columella [kŏl'yə-mĕl'ə] (L., = small column). The single, rod-shaped auditory **ossicle** of nonmammalian **tetrapods** that transmits vibrations from the **tympanic membrane** to the **inner ear;** called the **stapes** in **mammals.**

Commissure [kŏm'ə-shōōr] (L., *commissura* = seam). A band of nervous **tissue,** or a sensory canal, that crosses the midline of the body. Neuronal commissures interconnect comparable structures of the two sides of the **central nervous system.** See **decussation.**

Common bile duct. The principal **duct** carrying **bile** to the **intestine,** formed by the confluence of **hepatic** ducts from the **liver** and, when present, the **cystic duct** from the **gallbladder.**

Compression. A stress that results when two parallel **forces** move toward each other.

Concha [kŏng'kə] (Gr., *konkhe* = seashell). One of several folds within the **mammalian nasal** cavities that increase their surface area; also called a **turbinate bone.**

Condyle [kŏn'dīl] (Gr., *kondylos* = knuckle). Any convexly rounded articular surface, such as the occipital condyles of most **vertebrates** or mandibular condyles of **mammals.**

Congruent. Nonconflicting. Congruent data are those data that fit together **hierarchically,** with no conflict. The degree of congruence is dependent on the percentage of total **characters** in a data set that are congruent with each other. See **hierarchy.**

Conjunctiva [kŏn-jŭngk-tī'və] (L., *conjunctus* = joined together). The **epithelial** layer that lines the eyelids and reflects over the **cornea.**

Connective tissue. A widespread body **tissue** characterized by an extensive extracellular **matrix** of fibers. It connects other **tissues** and supports the body; includes fibrous **tissue,** fat, **cartilage,** and **bone.**

Contralateral. Descriptive of a structure that is located on the opposite side of the body from the point of reference.

Conus arteriosus [kō'nəs är-tîr'ē-ō'səs]. The fourth chamber of the **heart** of most **fishes** that extends between the **ventricle** and the **ventral aorta.**

Convergence. For phylogenetic context, see **homoplasy.** In neuroanatomy, this term refers to multiple neurons or receptor cells projecting to a smaller number of target cells.

Convergent evolution. See **homoplasy.**

Coprodaeum [kŏp-rō-dē'əm] (Gr., *kopros* = dung + *hodaion* = way). The portion of the **cloaca** that receives the feces.

Coprophagy [kŏ-prŏf'ə-jē] (Gr., *phagein* = to eat). The reingestion of feces; characteristic behavior of many rodents and lagomorphs.

Coracoid [kôr'ə-koid] (Gr., *korax* = crow + *eidos* = form). A **cartilage-replacement bone** that forms the posteroventral part of the **pectoral girdle,** reduced to a small process shaped like a crow's beak in **therians.**

Cornea [kôr'nē-ə] (L., *corneus* = horny). The transparent part of the **fibrous tunic** at the front of the eyeball.

Cornua [kôr'nōō-ə] (L., = horns). Hornlike processes of a structure, such as the cornua of the **hyoid bone.**

Corpora quadrigemina [kôr'pər-ə kwŏd'rə-jĕm'ə-nə] (L., *corpus,* pl., *corpora,* = body + *quadrigeminus* = fourfold). A collective term for the paired superior and inferior **colliculi** on the roof of the **mesencephalon** of **mammals.**

Corpus callosum [kôr'pəs kə-lō'səm] (L., *callosus* = hard). The large **commissure** interconnecting the two **cerebral hemispheres.**

Corpus cavernosum penis [kăv'ər-nō'səm pē'nəs] (L., *caverna* = hollow place). One of a pair of columns of **erectile tissue** that forms much of the **penis.**

Corpus luteum [lōō-tē'əm] (L., *luteus* = yellow). The hard, yellowish body that develops from an **ovulated** follicle and acts as an **endocrine gland.**

Corpus spongiosum penis. A column of **erectile tissue** that surrounds the penile portion of the **urethra.**

Corpus striatum. See **striatum.**

Cortex [kôr'tĕks] (L., = bark). A layer of distinctive **tissue** on the surface of an organ, such as the adrenal cortex or the cerebral cortex.

Cortisol [kôr'tĭ-sôl]. A **hormone** produced by the adrenal cortex that helps regulate carbohydrate metabolism.

Cosmine [kŏz'mēn] (Gr., *kosmios* = well ordered). A type of **dentine** found in certain **bony** scales in which there are **dentine** tubules grouped into radiating tufts.

Cosmoid scale. A thick, **bony** scale with a conspicuous layer of **cosmine,** characteristic of early **sarcopterygians.**

Costal [kŏs'təl] (L., *costa* = rib). Pertaining to the ribs.

Cowper's gland (*William Cowper,* British anatomist, 1666–1709). See **bulbourethral gland.**

Cranial kinesis [kĭ-nē'sĭs] (Gr., *kinesis* = movement). Movement of parts of the **skull,** exclusive of the lower jaw, relative to each other, occurs during feeding in many nonmammalian **vertebrates.**

Craniate. [krā'nē-āte] (Gr., *kranien* = skull or braincase). The subgroup of **chordates** in which the **brain** is encased in a **cranium;** includes the hagfishes and **vertebrates.**

Cranium [krā'nē-ŭm]. The **skull,** especially the part encasing the **brain.**

Cremasteric pouch [krē'mə-stĕr'ĭk] (Gr., *kremaster* = suspender). Layers of the body wall that suspend the **testis;** the **scrotal** wall apart from the **skin.**

Cribriform plate [krĭb'rə-fôrm] (L., *cribrum* = sieve + *forma* = shape). The perforated portion of the sphenoid **bone** through which groups of **olfactory neurons** pass.

Cricoid cartilage [krī'koid] (Gr., *krikos* = ring + *eidos* = form). Ring-shaped **cartilage** of the **mammalian larynx.**

Crop. The distal part of the **esophagus** of certain birds, especially grain-eating **species,** that stores food.

Crossopterygians [krôs'ŏp-tə-rĭj'ē-ănz] (Gr., *krossoi* = tassels + *pteryg-* = fin or wing). A collective name sometimes used for three groups of **sarcopterygians: coelacanths, rhipidistians,** and **tetrapods.** Some earlier authors use this term in a **nonmonophyletic** sense by excluding **tetrapods.**

Crus, pl. **crura** [krōōs, krōōr'ə] (L., = leg). The lower leg, shank, or shin of a **tetrapod.**

Crypt of Lieberkühn (*Johann N. Lieberkühn,* German anatomist, 1711–1756). Glandlike **invaginations** from the small **intestine** of **mammals; epithelial** cells multiply here, spread over the **intestinal** lining to replace worn-out cells, and some release digestive enzymes.

Ctenoid scale [tĕn'oid] (Gr., *ktenoeides* = like a cock's comb). A thin, **bony** scale having comblike processes on its posterior margin.

Cupula [kyōō'pyə-lə] (L., a small tub). A cup-shaped, jellylike secretion that caps the group of **hair cells** in a **neuromast.**

Cursorial [kûr-sôr'ē-əl] (L., *cursor* = runner). Pertaining to a **vertebrate** specialized for running.

Cutaneous [kyōō-tā'nē-əs] (L., *cutis* = skin). Pertaining to the **skin.**

Cycloid scale [sī'kloid] (Gr., *kyklos* = circle). A thin, **bony** scale having a smooth surface and rounded margins.

Cyclostome [sī'klō-stōm] (Gr., *stoma* = mouth). A **nonmonophyletic group** of convenience used to refer to **Petromyzontiformes** (lampreys) and **Myxiniformes** (hagfishes). Thought to be **nonmonophyletic** by most systematic ichthyologists.

Cystic duct [sĭs'tĭk] (Gr., *kystis* = bladder). The **duct** of the **gallbladder.**

D

Deciduous [dĭ-sĭd'yōō-əs] (L., *deciduus* = falling off). Teeth (or other elements) that are shed, e.g., the first set of teeth of a **mammal,** which are replaced by the **permanent teeth.**

Decussation [dĕk'ə-sā'shən] (L., *decusso,* pp. *decussatus* = to divide crosswise in an X). The crossing of **neuronal** tracts in the midline of the **central nervous system.** Decussations generally do not connect comparable parts of the two sides of the central nervous system. See **commissure.**

Defecation [dĕf'ĭ-kā'shən] (L., *defaeco,* pp. *defaecatus* = to remove the dregs). The elimination of undigested residue and bacteria from the digestive tract.

Deferent duct [dĕf'ər-ənt] (L., *defero,* pres.p. *deferens* = to carry away). The **sperm duct** of **amniotes, homologous** to the **archinephric duct** of **anamniotes.**

Delamination [dē-lăm'ə-nā'shən] (L., *de* = from + *lamina* = small plate). The splitting off of cells to form a new layer.

Dendrite [dĕn'drīt] (Gr., *dendrites* = relating to a tree). Branching **neuronal** processes that receive **nerve** impulses.

Density. The **mass** or **weight** of a body divided by its volume.

Dentine [dĕn-tēn'] (L., *dens,* gen. *dentis* = tooth). **Bone**-like material that forms the substance of a tooth deep to the superficial **enamel.**

Derived character. A **character** thought to have been uniquely evolved for a particular group (e.g., **feathers** for birds and closely related **species; mammary glands** for **mammals**); also called an **apomorphic** or **synapomorphic character.** See **character, primitive character.**

Dermal bones [dûr'məl] (Gr., *derma* = skin). Superficial **bones** that lie in or just beneath the **skin** and develop from the direct deposition of **bone** in **connective tissue;** also called **membrane bones.**

Dermal denticle [dĕn'tĭ-kəl] (L., *denticulus* = small tooth). A small, toothlike scale often found in the **skin** of **cartilaginous fishes;** also called a placoid scale.

Dermatocranium [dûr-mə-tō-krā'nē-əm] (Gr., *kranion* = skull). The portion of the **skull** composed of **dermal bones.**

Dermatome [dûr'mə-tōm] (Gr., *tone* = a cutting). The lateral portion of a **somite,** which will form the **dermis** of the **skin.**

Dermis [dûr'mĭs]. The dense **connective tissue** layer of the **skin** deep to the **epidermis.**

Deuterostome [do͞o'tə-rō-stōm] (Gr., *deuteros* = second + *stoma* = mouth). The group of **coelomate** animals in which the **stomodaeum** rather than the **blastopore** forms the adult mouth; includes echinoderms and **chordates.**

Diagnose (with regard to **taxa**). To define a group so it can be distinguished from all other groups within a particular **taxon.**

Diaphragm [dī'ə-frăm] (Gr., *dia* = through, across + *phragma* = a partition wall). The membranous and **muscular** partition between the thoracic and the abdominal cavities in **mammals.**

Diaphysis [dī-ăf'ĭ-sĭs] (Gr., *physis* = growth). The shaft of a limb **bone.**

Diapophysis [dī-ə-pŏf'ĭ-sĭs] (Gr., *apo* = away from + *physis* = growth). A transverse process that extends from the **vertebral arch** and receives the tuberculum of a rib.

Diapsid [dī-ăp'sĭd] (Gr., *di-* = two + *apsis* = arch). Pertaining to a **reptilian skull** in which two temporal **fenestrae** and two arches of **bone** are present, or to a **reptile** with such a **skull.**

Diarthrosis [dī'är-thrō'sĭs] (Gr., *arthron* = joint). A joint allowing considerable movement between the elements, including a hinge action, sliding, and rotation.

Diastole [dī-ăs'tə-lē] (Gr., = dilation). The period during which the **ventricle** of the **heart** relaxes and fills with **blood.**

Diencephalon [dī'ĕn-sĕf'ə-lŏn] (Gr., *dia* = through, across + *enkephalos* = brain). The region of the **brain** between the **telencephalon** and **mesencephalon,** consisting of the **epithalamus, thalamus,** and **hypothalamus.**

Digit [dĭj'ĭt] (L., *digitus* = digit). A finger or toe.

Digitigrade [dĭj'ĭ-tĭ-grād] (L., *gradus* = step). Walking with the heel and ankle raised off the ground so only the **digits** bear the body weight.

Diphycercal tail [dĭf'ĭ'sûr'kəl] (Gr., *diphyes* = twofold + *kerkos* = tail). A **caudal** fin in which the **vertebral axis** is straight and divides the fin margin into roughly symmetrical upper and lower lobes.

Diphyodont [dī-fī'ə-dŏnt] (Gr., *di-* = two + *phyo* = to produce + *odont-* = tooth). Pertaining to **mammals** with two sets of teeth, **deciduous** and **permanent.**

Diplospondyly [dĭp'lō-spŏn'də-lē] (Gr., *diploos* = double + *spondylos* = vertebra). A condition in which two **vertebral** centra per body segment are present; found in some early **tetrapods** and the **caudal** region of some **fishes.**

Dipnoan [dĭp'nō-ən] (Gr., *di-* + *pnoe* = breath). Lungfish.

Divergence. In neuroanatomy, a neuronal pathway that projects to many targets.

Dorsal pallium [păl'ē-əm] (L., *pallium* = cloak). The dorsalmost part of the **pallium,** forms the **isocortex** in **mammals.**

Drag. The resistance to the movement of an animal through the water or air in which it lives.

Duct [dŭkt] (L., *ductus* = conveyance, channel). A small, tubular passage.

Duct of Cuvier (*Baron Georges Cuvier,* 18th-century French scientist). The common **cardinal vein.**

Ductus arteriosus [dŭk'təs är-tîr'ē-ō'səs]. The dorsal part of the sixth **aortic arch,** may serve as a bypass of the **lungs** in **larval** or fetal stages.

Ductus venosus [və'nō'səs]. An **embryonic** connection between the umbilical **vein** and the **caudal vena cava;** bypasses the hepatic **sinusoids.**

Duodenum [do͞o'ō-dē'nəm] (L., *duodeni* = 12 each). The first portion of the **tetrapod** small **intestine,** which is 12 fingerbreadths long in humans.

Dura mater [do͝or'ə mā'tər] (L., = hard mother). The dense outer **meninx** surrounding the **mammalian central nervous system.**

E

Ear [ēr] (Anglo-Saxon, *eare* = ear). The organ of hearing.

Ectoderm [ĕk'tō-dûrm] (Gr., *ektos* = outside + *derma* = skin). The outermost of the three **embryonic germ layers;** forms the **epidermis, nervous system,** and **neural crest.**

Ectothermy [ĕk'tō-thûrm-ē] (Gr., *thermos* = heat). A condition in which an animal derives its body heat primarily from the external environment, so its body temperature is about the same as the ambient temperature; also known as **poikilothermy.**

Effector [ĭ-fĕk'tər] (L., = producer). Any organ or cell that responds in some way to a stimulus.

Efferent [ĕf'ər-ənt] (L., *ex* = out + *fero,* pres.p. *ferens* = to carry). Pertaining to structures that carry something away from a point of reference, such as efferent **neurons** leading from the **central nervous system.**

Efferent ductules. Minute, **sperm**-transporting **ducts;** the cords of the **urogenital** union in **anamniotes** and

mesonephric tubules in the head of the **epididymis** in **amniotes.**

Egest [ē-jĕst′] (L., *egestus* = taken out). The elimination of material from the **caudal** end of the digestive tract; also called **defecation.**

Elasmobranchs [ē-lăz′mō-brangks] (Gr., *elasmos* = thin plate + *branchia* = gills). The **taxonomic group** of **cartilaginous fishes** that includes sharks, skates, and rays.

Electric organ. An organ composed of modified **muscle** or glandular **tissue** that produces electric currents. Electric organs are used for electrolocation, defense, and communication; found primarily in certain fishes.

Electroplaque [ē-lĕk′trō-plăk] (Gr., *electron* = amber, from which electricity can be produced by friction + French, *plaque* = plate). The plates of modified **muscular tissue** that form the **electric organs** of some **fishes.**

Embryo [ĕm′brē-ō] (Gr., *embryon* = ingrowing). An early stage in the development of an organism that is dependent for energy and nutrients on materials stored within itself or obtained from a mother; embryos are not free living.

Empirical. Based on experimental and/or descriptive data.

Enamel [ē-năm′əl] (Middle English, *amel* = enamel). The very hard material on the surface of teeth and some **bony** scales; consists almost entirely of crystals of hydroxyapatite.

Endocrine glands [ĕn′dō-krĭn] (Gr., *endo* = within + *krino* = to separate). Ductless **glands** that discharge their secretions **(hormones)** into the **blood.**

Endoderm [ĕn′dō-dərm] (Gr., *derma* = skin). The innermost of the three **germ layers;** forms the lining of most of the digestive and respiratory tracts and **glandular** cells derived from these structures.

Endolymph [ĕn′dō-lĭmf] (L., *lympha* = liquid). The liquid within the **membranous labyrinth.**

Endometrium [ĕn′dō-mē′trē-əm] (Gr., *metra* = womb). The mucous membrane lining the **uterus.**

Endoskeleton [ĕn′dō-skĕl′ĭ-tn]. The part of the skeleton that lies deep within the body wall, appendages, and **pharynx;** composed of **cartilage** or **cartilage-replacement bone.**

Endostyle [ĕn′dō-stīl] (Gr., *stylos* = pillar). An elongated, **ciliated** groove in the **pharynx** floor of **tunicates** and **amphioxus.**

Endothelium [ĕn′dō-thē′lē-əm] (Gr., *thele* = delicate skin). Delicate **epithelium** lining **blood** vessels and the **heart.**

Endothermy [ĕn′dō-thûr′mē] (Gr., *therme* = heat). A condition in which an animal derives its body heat from internal metabolic processes, so it maintains a high and relatively constant body temperature despite variations in ambient temperature; also known as **homiothermic.**

Enterocoele [ĕn′tə-rō-sēl′] (Gr., *enteron* = gut + *koilos* = hollow). A **coelom** that develops primitively as buds from the gut cavity.

Epaxial [əp-ăk′sē-əl] (Gr., *epi* = upon + *axon* = axle, axis). Pertaining to structures that lie above or beside the **vertebral axis.**

Ependymal epithelium [ĭ-pen′də-məl] (Gr., *ependyma* = garment). The **epithelial** layer that lines the **central nervous system.**

Epiboly [ē-pĭb′ə-lē] (Gr., *epibole* = act of throwing on). The spreading of animal hemisphere cells over vegetal hemisphere cells during the **gastrulation** of some **vertebrates.**

Epidermis [ĕp-ĭ-dûr′mĭs] (Gr., *epi* = upon + *derma* = skin). The **epithelial** layer that forms the surface of the **skin.**

Epididymis [ĕp′ĭ-dĭd′ə-mĭs] (Gr., *didymoi* = testes). A band of **tissue** on the **amniote testis** that is **homologous** to the **cranial** part of the **opisthonephros** and part of the **archinephric duct** of **anamniotes.**

Epiglottis [ĕp′ĭ-glŏt′ĭs] (Gr., *glottis* = entrance to the windpipe). The flap of **fibrocartilage** that deflects food around the entrance of the **mammalian larynx.**

Epimere. See **somite.**

Epinephrine [ep′ə-nef′rin] (Gr., *epi-* = upon + *nephros* = kidney). The **hormone** produced by the adrenal **medulla;** it resembles **norepinephrine** produced by the postganglionic sympathetic **neurons** and it helps the body adjust to stress. Also called **adrenaline.**

Epiphysis [ĭ-pĭf′ĭ-sĭs] (Gr., *physis* = growth). The end of a **mammalian** long **bone;** a threadlike outgrowth from the roof of the **diencephalon** of **cartilaginous fishes.**

Epithalamus [ĕp′ĭ-thăl′ə-mŭs] (Gr., *thalamos* = chamber, bedroom). The roof of the **diencephalon** lying above the **thalamus;** part of it is an **olfactory** center.

Epithelial. See **epithelium.**

Epithelium [ĕp′ĭ-thē′lē-əm] (Gr., *thele* = delicate skin). The delicate cellular **tissue** that covers surfaces and lines cavities. **Epithelial,** *adj.*

Epoöphoron [ĕp′ō-ŏf′ə-rən] (Gr., *oon* = egg + *phero* = to bear). A **vestigial** organ near the **ovary** of **amniotes** that is **homologous** to the male **epididymis.**

Erectile tissue. A **tissue** containing cavernous vascular spaces that swell when they become filled with **blood.**

Esophagus [ĭ-sŏf′ə-gəs] (Gr., *oisophagos* = gullet). The part of the digestive tract between the **pharynx** and **stomach,** or between the pharynx and **intestine** if a **stomach** is absent.

Estivation (also aestivation) [ĕs′tə-vā′shən] (L., *aestivus* = summer). A period of inactivity and dormancy during periods of hot, dry weather.

Estradiol [ĕs′trə-dī′ôl]. The primary **hormone** produced by the **ovarian follicle;** promotes the development of female secondary sex **characteristics** and the development of the **uterine** lining during an **ovarian** cycle. Its feedback to the **hypothalamus** promotes the **luteinizing hormone** surge needed for **ovulation** in many **mammals.**

Estrus [ĕs′trəs] (Gr., *oistros* = gadfly, frenzy). A period in some female **mammals** of increased sexual excitement about the time of **ovulation** during which copulation may occur.

Euryapsid [yŏŏr′ē-ăp′sĭd] (Gr., *eurys* = wide + *apsis* = arch). Pertaining to a **reptilian** skull in which a single temporal **fenestra** is present high on the skull and a wide arch of **bone** beneath it; a **taxonomic group** (Euryapsida) of **reptilian species** with such a **skull** type.

Eustachian tube (*Bartolomeo Eustachio,* a 16th-century Italian anatomist). See **auditory tube.**

Eutherians [yōō-thîr'ē-ənz] (Gr., *eu* = true, good + *therion* = wild beast). The group of **therian mammals** with a relatively long **gestation period;** the **placental mammals.**

Evagination [ĭ-văj'ə-nā'shən] (L., *e* = out of + *vagina* = sheath). An outgrowth from another structure, or the process that gives rise to the outgrowth.

Evolution. Unidirectional (noncyclic) change.

Evolutionary homology. Fundamentally similar parts in different organisms that have evolved from a common precursor in an ancestral **species;** they may or may not resemble each other superficially or functionally.

Evolutionary process. A historical causal explanation for the apparent orderliness and systematic patterns of the biological world.

Excretion [ĕk-skrē'shən] (L., *ex* = out + *cretus* = separated). The elimination of nitrogenous wastes.

Exocrine glands [ĕk'sō-krĭn] (Gr., *ex* = out + *krino* = to separate). **Glands** the secretions of which are discharged through a **duct** onto some surface or into a cavity.

Extension [ĭk-stĕn'shən] (L., *tendere* = to stretch). A movement that carries a distal limb segment away from the next proximal segment, retracts a limb at the shoulder or hip, or moves the head or a part of the trunk toward the mid-dorsal line.

External acoustic meatus [ə-kōō'stĭk mē-ā'təs]. The external **ear** canal of **amniotes** extending from the body surface to the **tympanic membrane.**

External nostrils. See **nares.**

Extrinsic [ĭk-strĭn'sĭk] (L., *extrinsicus* = from without). Acting from outside the organ in question; applied to **muscles** that are not within or a part of the organ to which they attach.

Extrinsic ocular muscles [ŏk'yə-lər]. The group of small **muscles** that extend from the wall of the **orbit** to the eyeball and control the movements of the eyeball.

F

Facial [fā'shəl] (L., *facies* = face). Pertaining to the face; applied to **muscles,** the seventh **cranial nerve,** and other structures.

Facial nerve. The seventh **cranial nerve;** innervates facial and other **muscles** associated with the second **visceral arch,** some **salivary glands,** and taste receptors on the front of the **tongue.**

Fallopian tube. (*Gabriele Fallopio,* 16th-century Italian anatomist). See **uterine tube.**

Falx cerebri [fălks sĕr'ə-brē] (L., *falx* = sickle + *cerebrum* = brain). The sickle-shaped fold of **dura mater** that projects between the **cerebral hemispheres.**

Fascia [făsh'ē-ə] (L., = band, bandage). Sheets of **connective tissue** that lie beneath the **skin** (superficial fascia) or ensheathe groups of **muscles** (deep fascia or perimysium).

Fasciculus [fă-sĭk'yə-ləs] (L., = small bundle). A small bundle of **muscle** or **nerve** fibers.

Feathers [fĕ*th*'ərz] (Old English, *fether* = feather). **Skin** derivatives, characteristic of birds, that consist primarily of keratinized **epidermal** cells, provide insulation, and form the flying surfaces of the wing and tail.

Femur [fē'mər] (L., = thigh). The thigh or the **bone** within the thigh.

Fenestra [fə-nĕs'trə] (L., = window). A relatively large opening, such as a temporal fenestra in the **skull.**

Fenestra cochleae [kŏk'lē-ē] (L., *cochlea* = snail shell). The opening in the wall of the **otic capsule** through which pressure waves are released from the **cochlea** to the **tympanic cavity;** also called the **round window.**

Fenestra vestibuli [vĕ-stĭb'yə-lē] (L., *vestibulum* = antechamber). The opening in the wall of the **otic capsule** through which vibrations of the **auditory ossicles** establish pressure waves in the **cochlea;** also called the **oval window.**

Fibroblast [fī'brō-blăst] (L., *fibra* = fiber + Gr., *blastos* = bud). An irregularly shaped **connective tissue** cell that produces the extracellular **matrix,** including **collagen** fibers.

Fibrocartilage. A variety of **cartilage** composed mainly of fibers similar to **connective tissue** fibers that sometimes is found in **bone** grooves and articulations.

Fibrous tunic [tōō'nĭk] (L., *tunica* = coat). The dense **connective tissue** forming the outer layer of the eyeball; divided into the transparent **cornea** and opaque **sclera.**

Fibula [fĭb'yə-lə] (L., = buckle). The slender **bone** on the lateral side of the shin of **tetrapods.**

Filtration [fĭl-trā'shən]. The nonselective passage of molecules in the **blood,** other than plasma proteins, from the **glomerulus** into the **renal tubule.**

Fish. A **nonmonophyletic group** name of convenience commonly used for **vertebrates** other than **tetrapods.** Among living **vertebrates,** this includes **lungfishes,** the **coelacanth,** cartilaginous fishes (e.g., sharks, batoids, and **holocephalians**), and **actinopterygians (ray-finned fishes).** For a single **species** the plural form is fish; for more than one **species,** the plural form is fishes.

Fissure [fĭsh'ər] (L., *fissura* = cleft). A deep groove or cleft in certain organs, such as the **brain** and **skull.**

Flexion [flĕk'shən] (L., *flexus* = bending). A movement that brings a distal limb segment toward the next proximal segment, advances a limb at the shoulder or hip, or bends the head or a part of the trunk toward the midventral line.

Follicle-stimulating hormone. A **hormone** of the **adenohypophysis** that promotes the development of the **ovarian follicles.**

Foramen, pl. **foramina** [fə-rā'mən, -răm'ə-nə] (L., = opening). A perforation of an organ, usually a small opening.

Foramen magnum [măg'nəm] (L., *magnus* = large). The large opening in the **skull** for passage of the **spinal cord.**

Foramen of Monro (*Alexander Monro Secundus,* 1759–1808, Scottish anatomist). See **interventricular foramen.**

Foramen of Panizza. An opening between the bases of the left and right systemic arches in crocodilians; shunts **blood.**

Foramen ovale [ō-văl'ē]. A valved opening in the interatrial septum of **fetal mammals** that allows some **blood** to pass from the right to the left **atrium,** thereby bypassing the **lungs;** becomes the adult **fossa ovalis.**

Force. The product of **mass** and **acceleration.**

Forebrain. See **prosencephalon.**

Fornix [fôr'nĭks] (L., = vault, arch). An arch-shaped **neuronal** tract deep in the **cerebrum** that carries impulses from the **hippocampus** to the **hypothalamus.**

Fossa [fŏs'ə] (L., = ditch). A groove or depression in an organ.

Fossa ovalis [ō-vä'ləs]. A depression in the interatrial septum that represents the **fetal foramen ovale.**

Fossorial [fŏ-sôr'ē-əl] (L., *fossorius* = adapted for digging). Descriptive of an animal adapted for digging, such as a mole.

Fovea [fō'vē-ə] (L., = a pit). A small depression, such as the fovea in the **retina** that contains a concentration of cones.

Friction. The resistance to motion of an object resulting from its contact with the surface on which it is moving or the medium through which it is moving.

Frontal [frŭn'tl] (L., *frons,* gen. *frontis* = forehead). Pertaining to the forehead, such as the frontal **bone.**

Fulcrum [fŏŏl'krem] (L., = bedpost). The point of rotation or pivot in a **lever** system.

Funiculus [fyŏŏ-nĭk'yə-ləs] (L., = slender cord). A bundle or column of **white matter** in the **spinal cord.**

Furcula [fûr'kyə-lə] (L., = small fork). The united **clavicles** or wishbone of a bird.

Fusiform [fyōō'zə-fôrm] (L., *fusus* = spindle + *forma* = shape). A spindle-shaped or streamlined object.

G

Gait. The repetitive sequence for moving and placing the feet on the ground during locomotion of **tetrapods.**

Gallbladder [gôl'blăd-ər] (Old English, *galla* = bile). A small sac attached to the **liver** in which **bile** accumulates before its discharge into the **intestine.**

Gamete [găm'ēt] (Gr., *gamet-* = spouse). The haploid **germ cell:** mature **sperm** or egg.

Ganglion [găng'glē-ŏn] (Gr., = little tumor, swelling). A group of **neuron** cell bodies that lie peripheral to the **central nervous system** in **craniates.**

Ganoid scale [găn'oid] (Gr., *ganos* = sheen). A **bony** scale with a thick layer of surface **ganoine,** characteristic of the scales of early **actinopterygians.**

Ganoine [găn'ō-ən]. Enamel or enamel-like material deposited in layers on the surface of some **bony** scales.

Gastralia [găs-trā'lē-ə] (Gr., *gaster* = stomach). Riblike structures in the ventral abdominal wall of some **reptiles.**

Gastric [găs-trĭk] (Gr., *gaster* = stomach). Pertaining to or resembling the **stomach**.

Gastrulation [găs-trōō-lā'shən] (Gr., *gastrula* = little stomach). The process by which a single-layered **blastula** is converted into a two-layered **gastrula** with an **archenteron; mesoderm** formation often accompanies gastrulation.

Gear ratio. An expression of the relationship between **force** and **velocity;** determined by dividing the **length** of the **out-lever** by the length of the **in-lever.**

Genus name [je'nəs] (L., = race). The **taxon** that comprises very closely related **species,** and the first term in the binomial name for a **species.**

Germ layers (L., *germen* = bud). The three **epithelial tissue** layers **(ectoderm, mesoderm, endoderm)** in an early **embryo** from which all organs will arise.

Gestation period. The period in which the young are carried in the **uterus** before birth; from conception to birth.

Gills. The respiratory organs of aquatic **vertebrates,** consisting of platelike or filamentous outgrowths from a surface across which water flows.

Girdles. The skeletal elements in the body wall that support the **pectoral** and **pelvic** appendages.

Gizzard [gĭz'ərd] (Old French, *gezier* = gizzard). A **muscular** compartment of the **stomach** that usually contains swallowed stones with which food is ground up.

Gland [glănd] (L., *glans* = acorn). A group of **secretory** cells.

Glans clitoridis [glănz klĭ-tôr'ĭ-dĭs] (Gr., *kleitoris* = hill). The small mass of **erectile tissue** at the distal end of the **clitoris** of a female **mammal.**

Glans penis [pē'nəs] (L., *penis* = tail, penis). The bulbous distal end of the **penis** of a **mammal.**

Glenoid fossa [glĕn'oid] (Gr., *glene* = socket + *eidos* = form). The socket in the **pectoral girdle** of **tetrapods** that receives the head of the **humerus.**

Glia. See **neuroglia.**

Glide. A controlled descent at a low angle to the horizontal.

Glomerulus [glō-mĕr'yə-ləs] (L., *glomus* = ball). A ball-like network of **capillaries** that is surrounded by the **renal capsule** at the proximal end of a **renal tubule.** Also refers to clusters of short **neurons** and neuron processes among which the **olfactory** neurons terminate.

Glossal [glô'səl] (Gr., *glossa* = tongue). Pertaining to the **tongue;** also used to describe certain **muscles,** such as the genioglossus.

Glossopharyngeal nerve [glô'sō-fə-rĭn-jē-əl] (Gr., *pharynx* = throat). The ninth **cranial nerve,** which innervates **muscles** of the third **visceral arch** and returns sensory fibers from the part of the **pharynx** near the base of a **tongue.**

Glottis [glŏt'ĭs] (Gr., = opening of the windpipe). The opening near the base of the **tongue** that leads from the **pharynx** to the **larynx.**

Glucagon [glōō'kə-gŏn] (Gr., *glykys* = sweet). A **hormone** produced by the **pancreatic islet** cells of the **pancreas** that promotes the breakdown of glycogen and the release of sugar from the **liver;** increases **blood** sugar level.

Gnathostomes [năth'ə-stōmz] (Gr., *gnathos* = jaw + *stoma* = mouth). A collective term for all the jawed **vertebrates** (Gnathostomata).

Gonads [gō'nădz] (Gr., *gone* = seed). The **gamete**-producing reproductive organs, the **ovaries** and **testes.**

Graafian follicle [grä'fē-ən] (*Rijnier de Graaf,* Dutch anatomist, 1641–1673). See **ovarian follicle.**

Grade. An artificial (i.e., **nonmonophyletic**) **taxon.** The term is usually used to indicate a group defined by convergent or primitive characters (e.g., **"mammal**-like **reptiles"** or **"Pisces"** in their traditional usage). Such groups are often defined by "niche adaptation" rather than by geneology. This term is more relevant to paleoecological studies than to evolutionary studies. See **clade.**

Graviportal [grav'i-pōr'tăl] (L., *gravitas* = weight + *portare* = to carry). Pertaining to **appendicular** and other adaptations that support great weight, as in elephants.

Gray matter. Tissue in the **central nervous system** consisting of **neuron** cell bodies and unmyelinated **nerve** fibers.

Gubernaculum [gōō'bər-năk'yə-ləm] (L., = small rudder). A cord of **tissue** that extends between the **embryonic testis** of **therian mammals** and the developing **scrotum** and guides the descent of the **testis.**

Gustatory [gŭs'tə-tôr'ē] (L., *gusto,* pp. *gustatus* = to taste). Pertaining to the sense of taste.

Gymnophiona [jĭm'nō-fē'ən-ə] (Gr., *gymnos* = naked + *ophidion* = snake). A **taxon** of tropical **amphibians** that includes the wormlike, burrowing **caecilians.**

Gyrus [jī'rəs] (Gr., *gyros* = circle). One of the folds on the surface of the **cerebrum.**

H

Habenula [hə-bĕn'yōō-lə] (L., = small strip). A small, **epithalamic nucleus** with **olfactory** connection.

Hagfishes. See **myxiniformes.**

Hair. A filamentous **skin** derivative of **mammals** that consists primarily of **keratinized epidermal** cells; helps provide insulation. See **epidermis.**

Hair cells. The receptive cells of the **ear** and **lateral line system,** so called because they bear superficial cytoplasmic processes, most of which are modified **microvilli.**

Halecomorphi [hăl'ē-kō-môr'fī] (L., *halec* = herring + Gr. *morphe* = form). The division of the **neopterygians** that includes the living *Amia calva,* and numerous fossil **taxa.**

Hallux [hăl'əks] (Gr., = big toe). The first or most medial **digit** of the foot.

Hard palate. A shelf of **bone** in **mammals** that separates the **oral cavity** from the **nasal** cavities; together with the **soft palate** it forms the **secondary palate.**

Harderian gland (*Johann Harder,* a 17th-century Swiss anatomist). A tear **gland** present in certain **mammals** and located rostral to and beneath the eyeball; also called the **gland** of the **nictitating membrane.**

Haversian system. See **osteon.**

Head kidney. A group of **pronephric** renal tubules that persists in the adults of hagfishes and some **teleosts.**

Heart. A hollow-chambered **muscular** organ that pumps **blood** through the body.

Heliothermy [hē'lē-ō-thûrm'ē] (Gr., *helios* = sun + *thermos* = heat). The maintenance of a high body temperature by regulation of the body's exposure to the sun; characteristic of many **reptiles.**

Hemal arch [hē'məl] (Gr., *haima* = blood). A skeletal arch on the ventral surface of a **caudal vertebra** that forms a canal around the **caudal artery** and **vein.**

Hemibranch [hĕm'ē-brăngk] (Gr., *hemi* = half + *branchia* = gills). A **gill** of **fishes** with **gill** filaments or **lamellae** present on only one surface of the interbranchial septum, often the first **gill.**

Hemichordate [hĕm'ē-kôr'dāt] (Gr., *chorde* = string). A group (Hemichordata) of marine invertebrates showing some affinity to the **chordates.** Contains the acorn worms.

Hemopoietic tissue [hē'mō-poi-ĕt'ĭk] (Gr., *haima* = blood + *poietikos* = producing). A **tissue** in which **blood** cells are formed.

Hepatic [hĭ-păt'ĭk] (Gr., *hepar,* gen. *hepatikos* = liver). Pertaining to **blood** vessels, **ducts,** or other structures associated with the **liver.**

Hepatic portal system. A system of **veins** that drain the abdominal digestive organs and lead to **sinusoids** within the **liver.**

Hepatic vein. One of the **veins** that receives **blood** from the **hepatic sinusoids** and leads to the **heart** or **caudal vena cava.**

Herbivore [hûr'bə-vôr] (L., *herba* = herb + *-vorous* = devouring). Animal specialized to feed on plant material.

Hermaphrodite [hər-măf'rə-dīt] (Gr. mythology, the son of Hermes and Aphrodite who became united in one body with a nymph). An animal with both male and female reproductive organs.

Heterocercal tail [hĕ'tə-rō-sûr'kəl] (Gr., *heteros* = other + *kerkos* = tail). A **caudal fin** of **fishes** in which the **vertebral axis** turns upward into an enlarged dorsal lobe.

Heterochrony [hĕ'tər-ō-krō-nē] (Gr., *chronos* = time). A genetic shift in the timing of the development of a body part or process relative to the ancestral condition.

Heterodont [hĕ'tər-ō-dŏnt'] (Gr., *odous, odont-* = tooth). Pertaining to dentition in which the teeth are differentiated and perform different functions, as in **mammals.**

Hibernation. (L., *hibernus* = wintery). The period of torpor in which some **vertebrates** pass the winter.

Hierarchy. An organization of things arranged in order of rank; a pattern of subsets nested within larger sets.

Hindbrain. See **rhombencephalon.**

Hippocampus [hĭp'ə-kăm'pəs] (Gr., = seahorse). The **medial pallium** of **mammals,** which has shifted medially and protrudes into the lateral **ventricle;** part of the **limbic system.**

Holoblastic cleavage [hŏl'ō-blăs'tĭk] (Gr., *holos* = whole + *glastos* = bud). A pattern of **cleavage** in which the **cleavage** furrows pass through the entire egg.

Holocephalians [hŏl'ō-sĭf-ā'lē-ənz] (Gr., *holos* = whole + *kephale* = head). The group of **cartilaginous fishes** that includes the chimaeras.

Holonephros [hŏl'ō-nĕf'rəs] (Gr., *nephros* = kidney). The hypothetical ancestral **vertebrate kidney** consisting of segmented renal tubules that develop along the full length of the **nephric ridge;** also called an **archinephros.**

Homeostasis [hō'mē-ō-stā'sĭs] (Gr., *homoios* = alike + *stasis* = standing). The condition in which a constant internal environment is maintained despite factors that tend to destabilize it.

Homeothermic [hō'mē-ō-thûr'mĭk] (Gr., *homios* = like + *therme* = heat). Pertains to **vertebrates** in which the body temperature remains relatively constant despite variations in ambient temperature; **endothermic.** See **endothermy.**

Homocercal tail [hō'mō-sûr'kəl] (Gr., *homos* = same + *kerkos* = tail). A **caudal fin** that is superficially symmetrical but retains a slight uptilt in the skeleton of the **vertebral axis; characteristic** of **teleosts.**

Homodont [hō'mō-dŏnt] (Gr., *odous, odont-* = tooth). Pertaining to dentition in which all the teeth are essentially alike, differing only in size.

Homologous. To show **homology.**

Homology [hō-mŏl'ə-jē] (Gr., *homologia* = agreement). The use of this term today is variable, and thus confusing, but it generally refers to some aspect of "sameness" (e.g., structures that match each other in origin, position, shape, or composition). Several types of homology have been defined. With regard to systematic **evolutionary** studies, the most commonly used and relevant type is **"phylogenetic** homology." **Phylogenetic** homology is the fundamental similarity among organs in different organisms thought to be the result of their **evolution** from a precursor organ in a common ancestor. Empirically defined, **phylogenetic** homology at its most **taxonomically** inclusive level is the equivalent of **synapomorphy.** Also see **serial homology, sexual homology.**

Homoplasy [hō'mō-plā'zē] (Gr., *plastos* = molded). **Morphological** resemblance among organs that are not phylogenetically **homologous.** In an **evolutionary** context, these are similarities that have originated independently with different lineages. A number of authors have attempted to distinguish among different kinds of homoplasy (e.g., parallelism and convergence) based on presumed recency of common ancestry, **adaptation,** and other hypothetical factors, but the authors believe that such a distinction cannot be unambiguously made based on **empirical** grounds. See **analogy, homology.**

Hormones [hôr'mōnz] (Gr., *hormono,* pres.p. *hormon* = to rouse or set in motion). The secretions of the **endocrine glands.**

Horn (Anglo-Saxon, = horn). A **bony** projection from the **skull** of many **ruminants** that is covered by layers of **keratinized epidermis** and is not shed; usually occurs in both sexes.

***Hox* genes.** Short sections of DNA that occur in clusters called homeoboxes; nearly identical sequences have been found in many invertebrate and **vertebrate** groups and regulate the expression of genes that determine the features **characteristic** of each body segment.

Humerus [hyōō'mər-əs] (L., = upper arm). The **bone** of the upper arm.

Hyaline cartilage [hī'ə-lĭn] (Gr., *hyalos* = glass). **Cartilage** with a clear, translucent **matrix.**

Hyobranchial apparatus. The group of **visceral arches** that support the **tongue** and **larynx** of **tetrapods;** includes the **hyoid** arch and one or more other arches.

Hyoid [hī'oid] (Gr., *hyoeides* = shaped like the letter *ypsilon* = Y). Pertaining to structures associated with the second **visceral arch,** known as the hyoid arch.

Hyoid apparatus. See **hyobranchial apparatus.**

Hyomandibula [hī'ō-măn-dĭb'yōō-lər] (L., *mandibula* = jaw). The dorsal element of the **hyoid** arch of **fishes** that extends from the **otic capsule** to the posterior end of the upper jaw.

Hyostylic suspension [hı'ō-stī'lĭk] (Gr., *stylos* = pillar). A type of jaw suspension in **fishes** in which the upper jaw is attached to the **skull** by the hyomandibula.

Hypaxial [hī-păk'sē-əl] (Gr., *hypo* = under + *axon* = axle, axis). Pertaining to structures that lie ventral to the **vertebral axis.**

Hyperosmotic [hī'pər-ŏz-mŏt'ik] (Gr., *hyper* = above + *osmos* = action of pushing). A condition in which the concentration of osmotically active solutes in the liquid in question is greater than in the comparison liquid.

Hypobranchial [hī'pō-brăng'kē-əl] (Gr., *hypo* = under + *branchia* = gills). Pertaining to **muscles** or other structures located ventral to the **gills.**

Hypocercal tail [hī'pō-sûr'kəl] (Gr., *kerkos* = tail). A **caudal fin** in which the **vertebral axis** turns into an enlarged ventral lobe.

Hypoglossal nerve [hī'pō-glŏs'əl] (Gr., *glossa* = tongue). The 12th **cranial nerve** of **amniotes,** which innervates **muscles** in the **tongue; homologous** to the **hypobranchial nerve** of **anamniotes.**

Hypomere. See **lateral plate.**

Hypo-osmotic [hī'pō-ŏz-mŏt'ik] (Gr., *osmos* = action of pushing). A condition in which the concentration of osmotically active solute in the liquid in question is less than that in the comparison liquid.

Hypophysis [hī-pŏf'ĭ-sĭs] (Gr., *physis* = growth). The pituitary gland. An endocrine organ, consisting of two lobes, which is located at the base of the skull.

Hypothalamus [hī'pō-thăl'ə-məs] (Gr., *thalamos* = chamber, bedroom). The ventral part of the **diencephalon** that lies beneath the **thalamus;** an important center for **visceral** integration.

Hyposodont [hĭp'sō-dŏnt] (Gr., *hypsos* = height + *odont-* = tooth). A high-crowned tooth.

I

Ileum [ĭl'ē-əm] (L., = small intestine; from Gr., *eileo* = roll up, twist). The caudal portion of the small **intestine** of **tetrapods.**

Iliac [ĭl'ē-ăk]. Pertains to structures near or supplying the **ilium,** such as the iliac artery.

Ilium [ĭl'ē-əm] (L., = groin, flank). The dorsal **bone** of the **tetrapod pelvic girdle** that attaches onto the **sacrum.**

Incisor [ĭn-sī'zər] (L., = the cutter; from *incido* = to cut into). One of the front teeth of **mammals** lying rostral to the **canine;** used for cutting or cropping food.

Incongruence. Homoplasy, or characters that do not show congruence with other **character** data. See **congruence, homoplasy.**

Incus [ĭng'kəs] (L., = anvil). The anvil-shaped middle **auditory ossicle** of **mammals, homologous** to the quadrate **bone.**

Induction [ĭn-dŭk'shən] (L., *inductus* = led in). An **embryonic** process whereby a **tissue** causes an adjacent tissue to differentiate in a characteristic way.

Inertia [ĭ-nûr'shə] (L., *iners* = sluggish). The tendency of a body at rest to remain at rest, or of one in motion to remain in motion.

Infundibulum [ĭn'fŭn-dĭb'yə-ləm] (L., = little funnel). A funnel-shaped structure, such as the expansion of the **oviduct** that contains its coelomic entrance; also a ventral **evagination** of the **hypothalamus** that forms the **neurohypophysis.**

Ingest [ĭn-jĕst'] (L., *ingestus* = taken in). To take material into the mouth.

Ingroup. A relative term referring to all **species** within the particular **taxon** of reference (e.g., all **taxa** within the **taxon** being analyzed **phylogenetically**). See **outgroup.**

Inguinal [ĭng'gwə-nəl] (L., *inguen,* gen. *inguinis* = groin). A term used to describe structures in or near the groin.

Inguinal canal. A passage through the body wall of **mammals** that leads from the **abdominal** cavity into the **vaginal** cavity of the **scrotum;** the ductus deferens as well as the **blood** vessels and **nerves** supplying the **testis** pass through it.

In-lever. The **lever** arm through which a **force** is delivered into a **lever** system; it is the perpendicular distance from the line of action of the in-**force** to the **axis** of rotation of the **lever** system.

Innate behavior [ĭ-nāt'] (L., *innatus* = inborn). Those aspects of behavior that are inherited or instinctive and not learned.

Inner ear. That portion of the **ear** that lies within the **otic capsule** of the **skull** and contains the receptive cells for equilibrium and hearing.

Insectivore [ĭn-sĕk'tĭ-vôr] (L., *insectum* = insect + *-vorous* = devouring). An insect-eating animal, specifically the group of **eutherian mammals** that includes the shrews and moles.

Insertion [ĭn-sûr'shən] (L., *insertio* = a planting). That point of attachment of a **muscle** that moves the most when the **muscle** shortens; it is the most distal end of limb **muscles.**

Insulin [ĭn'sŭ-lĭn] (L., *insula* = island). The **hormone** produced by the **pancreatic islets** that decreases **blood** sugar by promoting the uptake of glucose by cells and its conversion into glycogen in **liver** and **muscle** cells.

Integument [ĭn-tĕg'yŏŏ-mənt] (L., *integumentum* = covering). The **skin.**

Integumentary skeleton. Hard structures such as plates of dermal **bone, bony** scales, and teeth that develop in or just beneath the skin. (See also **dermal bone.**)

Intercentrum [ĭn'tər-sĕn'trŭm] (L., *inter* = between + Gr., *kentron* = center). The **ventral** body that lies between the **pleurocentra.**

Interclavicle [ĭn'tər-klăv'ĭ-kəl] (L., *clavicula* = small key). The ventromedian element of the **pectoral girdle** that lies between the **clavicles.**

Internal capsule. A sheet of white fibers passing through the **striatum,** which carries most impulses to and from the **cerebral** cortex.

Internal nostrils. See **choanae.**

Interneurons [ĭn'tər-nŏŏr'ŏnz] (L., *inter* = between + Gr., *neuron* = nerve, sinew). **Neurons** within the **central nervous system** that lie between the motor and sensory **neurons.** Their connections are responsible for most of the integrative activity of the **central nervous system.**

Interstitial cells [ĭn'tər-stĭsh'əl] (L., *interstitium* = space between). Cells of the **testis** between the **seminiferous tubules** that produce **testosterone.**

Interstitial fluid. A **lymph**like fluid that lies in the minute spaces between the cells of the body.

Interventricular foramen [ĭn'tər-vĕn-trĭk'ū-lər] (L., *ventriculus* = belly + *foramen* = hole). The opening between the lateral **ventricles** and third **ventricle** of the **brain;** also called the **foramen of Monro.**

Intervertebral disk [ĭn'tər-vər'tē-brəl] (L., *vertebratus* = jointed). Disks of **fibrocartilage** that lie between the **vertebral bodies (centra)** of **mammals** and some other **vertebrates.**

Intervertebral foramen. An opening between successive **vertebral arches** through which a spinal **nerve** passes.

Intestine [ĭn-tĕs'tĭn] (L., *intestinus* = the intestine). The portion of the digestive tract between the **stomach** and **cloaca** or **anus;** site of most digestion and absorption.

Intrinsic [ĭn-trĭn'sĭk] (L., *intrinsicus* = on the inside). A structure that is an inherent part of an organ, such as the ciliary **muscles** of the eyeball.

Invagination [ĭn-vă'jə-nā'shən] (L., *in* = into + *vagina* = sheath). An ingrowth or the process that gives rise to an ingrowth.

Involution [ĭn'vō-lŏŏ'shən] (L., *involutus* = rolled up). A process that occurs during **gastrulation** of some **vertebrates** by which surface cells roll over the lip of the **blastopore** and move into the **archenteron.**

Ipsilateral [ĭp'sĭ-lăt'ər-əl] (L., *ipse* = the same + *latus* = side). Pertaining to structures on the same side of the body.

Iris [ī'rĭs] (Gr., *iris* = rainbow). The part of the **vascular tunic** of the eyeball that lies in front of the **lens,** with the **pupil** in its center.

Ischium [ĭs'kē-əm] (Gr., *ischion* = hip). The ventral and posterior element of the **pelvic girdle.**

Islets of Langerhans (*Paul Langerhans,* 19th-century German physician). See **pancreatic islets.**

Isocortex [ī'sō-kôr'tĕks] (Gr., *isos* = equal + L., *cortex* = bark). The expanded **dorsal pallium** of **mammals;** characterized by six **neuronal** layers. Sometimes called the neopallium.

Isometric contraction [ī'sō-mĕt'rĭk] (Gr., *metron* = measure). A **muscle** contraction in which **force** is developed but the **muscle** does not shorten.

Iso-osmotic [ī'sō-ŏs-mŏt'ĭk] (Gr., *osmos* = action of pushing). A condition in which the concentration of osmotically active solutes in the liquid in question is the same as in the comparison liquid.

Isotonic contraction [ī'sō-tŏn'ĭk] (Gr., *tonos* = tension). A **muscle** contraction in which the tension developed remains the same and the muscle shortens.

J

Jacobson's organ (*Ludwig L., Jacobson,* 19th-century Danish surgeon and anatomist). See **vomeronasal organ.**

Jejunum [jə-jōō'nəm] (L., *jejunus* = empty). Approximately the first half of the **mammalian** postduodenal small **intestine;** usually found to be empty at autopsies.

Jugular veins [jŭg'yū-lər] (L., *jugulum* = throat). Major **veins** in the neck of **mammals** that drain the head.

K

Keratin [kĕr'ə-tĭn] (Gr., *keras* = horn). A **horny** protein synthesized by the **epidermal** cells of many vertebrates.

Kidney [kĭd'nē]. The organ that removes waste products, especially nitrogenous wastes, from the **blood** and produces urine.

Kinetic skull [kĭ-nĕt'ĭk] (Gr., *kinein* = to move). A **skull** in which the upper jaw and palate can move relative to other parts, found in many **fishes,** squamates, and birds.

L

Labia [lā'bē-ə] (L., = lips). Liplike structures.

Labyrinth [lăb'-ə-rĭnth] (Gr., *labyrinthos* = labyrinth). An intricate system of connecting pathways, such as the **membranous labyrinth** of the **inner ear.**

Lacrimal apparatus [lăk'rĭ-məl] (L., *lacrima* = tear). Pertaining to **glands** and associated structures that produce and transport the tears.

Lactation [lăk-tā'shən] (L., *lac* = milk). The production and release of milk.

Lacuna [lə-kōō'nə] (L., = pit). A small cavity, such as one in **bone** that contains an **osteocyte.**

Lagena [lə-jē'nə] (Gr., *lagenos* = flask). A posteroventral **evagination** of the **sacculus;** homologous to the **cochlear duct.**

Lamella [lə-mĕl'ə] (L., = small plate). A thin plate or layer of **tissue,** such as the lamellae in **fish gills** where gas exchange occurs.

Laminar flow. The smooth, nonturbulent flow of water or air across the surface of the body.

Lamprey. See **Petromyzontiformes.**

Larva [lär'və] (L., = mask). A free-living developmental stage that is markedly different from the adult.

Larynx [lăr'ĭngks] (Gr., = larynx). A chamber at the entrance to the **trachea;** contains the **vocal cords** in many **tetrapods** other than birds.

Lateral line nerves. A group of six special somatic sensory **cranial nerves** (not numbered) that are found in aquatic anamniotes. These nerves return impulses from the **lateral line system.** They are sometimes considered to be parts of the facial, glossopharyngeal, and vagus nerves, but they have an independent phylogenetic origin and an embryonic origin from neurogenic **placodes.**

Lateral line system. A sensory system of **fishes** and **larval amphibians** that detects low-frequency water disturbances; parts are sometimes modified as **electroreceptors.**

Lateral pallium [păl'ē-əm] (L., *pallium* = cloak). The lateral-most part of the pallium; includes the **amygdala** and primary olfactory cortex (piriform lobe) of **mammals.** Sometimes called the paleopallium.

Lateral plate. The most lateral or ventral portion of the **mesoderm** that contains the **coelomic cavity;** also called the hypomere.

Lemniscus [lem-nis'kŭs] (Gr., *lemniskos* = ribbon). A ribbon-shaped **neuronal** tract ascending from sensory relay **nuclei** to parts of the **thalamus.**

Lens [lĕnz] (L., = lentil). The part of the eyeball that focuses light on the **retina.**

Lepidosaurs [lĕp'ĭ-dō-sôrz] (Gr., *lepsis* = scale + *sauros* = lizard). A group of **diapsid reptiles,** including *Sphenodon,* lizards, snakes, and amphisbaenians.

Lepidotrichia [lĕp'ĭ-dō-trĭk'ē-ə] (Gr., *trich-* = hair). **Bony fin rays** of **fishes** composed of rows of small, tube-shaped segments, thought to be modified scales.

Levers. Rodlike mechanical devices that exert a force by turning about a pivot or fulcrum.

Leydig cells. (*Franz von Leydig,* German anatomist, 1821–1908). See **interstitial cells.**

Lift. An upward force generated by a stream of water or air flowing across a fin or wing. The line of action of the lift **force** is perpendicular to the stream.

Ligament [lĭg'ə-mənt] (L., *ligamentum* = band, bandage). Strong **connective tissue** band that extends between structures, usually skeletal elements; also describes certain **mesenteries.**

Ligamentum arteriosum [lĭg'ə-mĕn'təm är'tîr-ē-ō'səm]. The **connective tissue** band extending between the **pulmonary** artery and **aorta;** a remnant of the **embryonic ductus arteriosus.**

Limbic system [lĭm'bĭk] (L., *limbus* = border). A **brain** region that encircles the **diencephalon** and leads to the **hypothalamus;** includes the **amygdala, hippocampus, fornix,** cingulate **cortex,** and part of the **hypothalamus.** Important in behaviors related to survival of the species, such as feeding and sexual activity.

Lingual [lĭng'gwəl] (L., *lingua* = tongue). Pertaining to the **tongue,** such as lingual **muscles.**

Linnean classification system. A convention of **hierarchical** ranking that allows organization of **taxa** as sets and subsets. Major categories are (in descending order of rank) kingdom, phylum, class, order, family, **genus,** and **species. Species** names are binomial (two-word) names, and all higher categories are one-word names. Additional rank categories between the seven major ranks are available through the use of prefixes, such as "sub-," "super-," "supra-," and "infra-." The rank of tribe is occasionally used as a suprageneric rank, and the word "section" has been used at several different intermediate levels.

Lissamphibians (Gr., *lissos* = smooth + *amphibianz*). A group containing contemporary **amphibians.** See **amphibians.**

Liver [lĭv'ər] (Anglo-Saxon, *lifer* = liver). A large gland that develops from the floor of the **archenteron** just behind the **stomach;** secretes **bile** and processes **blood** brought to it in the **hepatic portal system.**

Loop of Henle (*Friedrich G. J. Henle*, German anatomist, 1809–1885). See **medullary loop.**

Lophodont [lŏf'ə-dŏnt] (Gr., *lophos* = crest + *odont-* = tooth). A **cheek tooth** the cusps of which have united to form ridges.

Lumbar [lŭm'bər, -bär] (L., *lumbus* = loin). Descriptive of structures in the back between the **thorax** and **pelvis,** such as lumbar **vertebrae.**

Lung [lŭng]. One of a pair of respiratory organs of terrestrial **vertebrates** that develops as an outgrowth from the floor of the **pharynx.** Lungs are present in some **fishes** but are sometimes single and develop **caudal** to the **pharynx.**

Luteinizing hormone [lo͞o'tē-ə-nī'zĭng] (L., *luteus* = yellow). A **hormone** produced by the **adenohypophysis** that promotes maturation of **ovarian follicles, ovulation,** and the growth of the **corpus luteum.**

Lymph [lĭmf] (L., *lympha* = clear water). A clear liquid derived from **interstitial fluid** that flows through the lymphatic vessels.

Lymph heart. Muscular sections of lymphatic vessels of some **amphibians** and **reptiles** the contractions of which help propel the **lymph.**

Lymph node. Nodules of lymphatic **tissue** along the course of the lymphatic vessels; the contained lymphocytes respond to invading antigens and initiate immune responses.

M

Macroevolution. (Gr., *makros* = large). Generally used to refer to large-scale **evolutionary** processes or patterns at the species level and above. See **microevolution.**

Macrolecithal [măk'rō-lĕs'ə-thəl] (Gr., *lekithos* = yolk). An egg with a large amount of yolk, found in many **fishes, reptiles,** and birds.

Macrophage [măk'rō-fāj] (Gr., *phagein* = to eat). Large cells that **phagocytose,** or **ingest,** foreign material. See **phagocytosis, ingest.**

Macula [măk'yū-lə] (L., = spot). Spot or patch, specifically clusters of **hair cells** in the **sacculus** and **utriculus** of the **inner ear.**

Malleus [măl'ē-ŭs] (L., = hammer). The outermost of the three **mammalian auditory ossicles; homologous** to the **articular bone.**

Mammalia. See **mammals.**

Mammal-like reptiles. A **nonmonophyletic group** name of convenience, commonly used in the past, that contains many groups of extinct **amniotes.** The name is misleading because these **vertebrates** are not **reptiles** at all, but a series of basal lineages more closely related to **mammals** than to crocodiles, lizards, snakes, dinosaurs, and birds.

Mammals [măm'əlz] (L., *mamma* = breast). The **vertebrate** group (Mammalia) characterized by **mammary glands** and **hair.**

Mammary glands. Cutaneous glands that secrete milk. Unique to **mammals.**

Mandibular arch [măn-dĭb'yū-lər] (L., *mandibula* = lower jaw). The first **visceral arch** of jawed **vertebrates.**

Mandibular cartilage. The ventral part of the **mandibular arch;** forms the lower jaw of **cartilaginous fishes.** Sometimes called Meckel's cartilage.

Mandibular gland. A **mammalian salivary gland** that is located near the **caudal** end of the mandible, or lower jaw.

Manus [mā'nəs] (L., = hand). The hand.

Marsupials [mär-so͞o'pē-əlz] (L., *marsupium* = pouch). The pouched **mammals.** See also **metatheria.**

Marsupium [mär-sū'pē-ŭm]. The pouch of a **marsupial** in which the young are carried.

Mass. The quantity of material an object contains, usually measured by **weight.**

Matrix [mā'trĭks] (L., = womb, a female set aside for breeding). **1.** The medium in which a substance is embedded, specifically the extracellular material in **connective tissues. 2.** A tabular illustration of **empirical data** arranged in columns and rows.

Meatus [mē-ā'təs] (L., = passage). A passage such as the **external acoustic meatus,** which leads to the **tympanic membrane.**

Meckel's cartilage. (*Johann F. Meckel*, 18th-century German anatomist). See **mandibular cartilage.**

Medial pallium [păl'ē-əm] (L., *pallium* = cloak). The medial-most part of the **pallium;** becomes the **hippocampus** of **mammals.** Sometimes called the archipallium.

Mediastinum [mē'dē-ə-stī'nəm] (L., *mediastinus* = medial, from *medius* = middle). The area between the two **pleural cavities** of **mammals** that contains the **pericardial cavity, thymus,** and other structures.

Medulla [mĭ-dŭl'ə] (L., = core, marrow). The central part of an organ, often as opposed to its periphery or **cortex.**

Medulla oblongata. The posterior region of the **brain** that is continuous with the **spinal cord.**

Medullary loop. Portion of the **renal tubule** of **mammals** and some birds and **reptiles** that loops into the **medulla** of the **kidney,** essential in establishing the interstitial salt gradient needed for the production of a concentrated urine. Sometimes called the loop of Henle.

Melanophore [mĕl'ə-nō-fôr'] (Gr., *melas* = black + *-phore* = bearing). A cell of **neural crest** origin in the **skin** that produces and carries the black pigment melanin.

Melanophore-stimulating hormone. See **Melanotropin.**

Melanotropin (L., *trophe* = nourishment). **Hormone** produced by the intermediate part of the **adenohypophysis;** causes the dispersal of melanin granules in some animals. Sometimes called melanophore-stimulating hormone.

Melatonin [mĕl'ə-tō'nĭn] (Gr., *tonos* = stain). A **hormone** produced by the **pineal gland** in inverse proportion to the amount of light received; may be important in regulating sexual development and biorhythms.

Membrane bone. See **dermal bone.**

Membranous labyrinth [mĕm'brə-nŭs]. The sacs and **ducts** of the **inner ear** that are filled with **endolymph** and contain the receptive cells for balance and hearing.

Meninges [mə-nĭn'jēz] (Gr., *meninx*, pl., *meninges* = membrane). **Connective tissue** membranes that surround the central nervous system, namely, the **dura mater, arachnoid,** and **pia mater.**

Meniscus [mə-nĭs′kəs] (Gr., *meniskos* = crescent). A crescent-shaped disk of **fibrocartilage** found in some joints, including the knee joint.

Meroblastic cleavage [mĕr′ō-blăs′tĭk] (Gr., *meros* = part + *blastos* = bud). The partial **cleavage** of **macrolecithal** eggs.

Mesectoderm [mĕz-ĕk′tō-dûrm] (Gr., *mesos* = middle + *ektos* = outside). **Mesoderm**-like **tissue** in the head of **vertebrates** that arises from **neural crest** cells.

Mesencephalon [mĕz′ĕn-sĕf′ə-lŏn] (Gr., *mesos* = middle + *enkephalos* = brain). The **midbrain,** which dorsally forms the **optic lobes** or **corpora quadrigemina.**

Mesenchyme [mĕz′ən-kĭm] (Gr., *enchein* = to pour in). An **embryonic tissue** that consists of star-shaped, wandering cells and gives rise to most adult **tissues,** except for **epithelium.**

Mesentery [mĕz′ən-tĕr′ē] (Gr., *enteron* = intestine). Any fold of **coelomic epithelium** that suspends **visceral** organs or extends between them, carrying **blood** vessels and **nerves;** in a limited sense, the membrane that suspends the small **intestine.**

Meso- [mĕz′ō] (Gr., *mesos* = middle). A term that, when combined with the name of a **visceral** organ, denotes a **mesentery** suspending that organ, such as the mesocolon suspending the **colon.**

Mesoderm [mĕz′ō-dûrm] (Gr., *derma* = skin). The central **germ layer** of an early **embryo;** gives rise to most of the **connective tissue, muscles,** and **blood.**

Mesolecithal [mĕz′ō-lĕs-ə-thəl] (Gr., *lekithos* = yolk). An egg, such as that of an **amphibian,** with a moderate amount of yolk.

Mesomere. See **nephric ridge.**

Mesonephric duct. See **archinephric duct.**

Mesonephros [mĕz′ō-nĕf′rəs] (Gr., *nephros* = kidney). An **embryonic kidney** that develops in the central part of the **nephric ridge;** contributes to the adult kidney of **anamniotes** and the **epididymis** of male **amniotes.**

Mesozoic. The era of geologic time ranging from about 230 to 65 million years before the present.

Metacarpal [mĕt′ə-kär′pəl] (Gr., *meta* = after + *karpos* = wrist). One of the skeletal elements in the palm of the hand.

Metamerism [mĕ-tăm′ər-ĭz-əm] (Gr., *meros* = part). The condition in which the body is divided into similar segments.

Metamorphosis [mĕt′ə-môr′fə-sĭs] (Gr., = transformation). The rapid change in form from a **larva** to an adult.

Metanephros [mĕt′ə-nĕf′rōs] (Gr., *nephros* = kidney). The adult **kidney** of **amniotes,** which develops from the **caudal** part of the **nephric ridge.**

Metatarsal [mĕt′ə-tär′səl] (Gr., *tarsos* = sole of the foot). One of the skeletal elements of the sole of the foot.

Metatheria [mĕt′ə-thîr′ē-ə] (Gr., *therion* = wild beast). The group of **therian mammals** that includes the **marsupials.**

Metencephalon [mĕt′ĕn-sĕf′ə-lŏn] (Gr., *enkephalos* = brain). The **brain** region that includes the **cerebellum** and, in birds and **mammals,** the **pons.**

Microevolution. Generally used to refer to small-scale processes of change working at the level of interbreeding groups of organisms (populations). Such processes are thought by evolutionists to represent the mechanisms responsible for **speciation.** See **macroevolution.**

Microglia [mī-krŏg′lē-ə] (Gr., *micros* = small + *glia* = glue). Small **neuroglial** cells of mesodermal origin, some of which are phagocytic. See **phagocytosis.**

Microlecithal [mī′krō-lĕs′ĭ-thəl]. An egg with a small amount of yolk.

Microvilli [mī′krō-vĭl′ī] (L., *villus* = shaggy hair). Minute, nonmotile cytoplasmic processes on the surface of many **epithelial** cells; they greatly increase surface area.

Midbrain. See **mesencephalon.**

Middle ear. That portion of the **ear** of **tetrapods** that usually contains the **tympanic cavity** and one or more **auditory ossicles** that transmits vibrations from the body surface (usually from a **tympanic membrane**) to the **inner ear.**

Middle ear cavity. See **tympanic cavity.**

Modulus of elasticity. A measure of the elastic properties of a material; equals **stress** divided by **strain.** Structures with a low modulus of elasticity are more elastic than ones with a high modulus.

Molar [mō′lər] (L., *mola* = millstone). One of the teeth in the most posterior group of **mammalian** teeth, usually adapted for crushing or grinding.

Moment. The product of a **force** times the perpendicular distance from the line of action of the **force** to an **axis** of rotation; also called a **torque.**

Monophyletic [mŏn′ō-fī-lĕt′ĭk] (Gr., *monos* = single + *phyle* = tribe). See **monophyletic group.**

Monophyletic group. A **taxon** or group of organisms that includes all known descendants of a hypothetical ancestor and no other members. Putatively monophyletic groups are identified by **hierarchies** of special similarities (also referred to as **characters, phylogenetic homologies,** or **synapomorphies**), such as **hair** and **mammary glands** for **mammals, feathers** for birds, jaws for **gnathostomes,** and the presence of **bone** for **osteichthyans.** See **characters, homology.**

Monophyletic taxa. See **monophyletic group** (monophyletic taxa = monophyletic groups).

Monotremes [mŏn′ō-trēmz] (Gr., *monos* = single + *trema* = hole). A group including the extant **prototherians;** includes the platypus and spiny anteater.

Morphogenesis [môr′fō-jĕn′ə-sĭs] (Gr. *morphe* = form + *genesis* = production). The development of form.

Morphological data. Descriptive data based on the anatomy of organisms.

Morphology [môr-fŏl′ō-jē] (Gr., *morphe* = form + *logos* = discourse). The study of structure.

Motor unit. A motor **neuron** and the **muscle** fibers it supplies.

Mucosa [mū-kō′sə] (L., *mucosus* = mucous, slimy). The lining of the gut or other **visceral** organs, consisting of **epithelium** and associated **connective tissue.**

Mucus [mū′kəs]. (L., = slime). A slimy material produced by some **epithelial** cells that is rich in the glycoprotein mucin. The adjective is mucous.

Muscle [mŭs′əl] (L., *musculus* = muscle). A contractile **tissue** primarily responsible for the movement of an animal or its parts; discrete groups of **muscle** cells with a common origin and insertion.

Myelencephalon [mī′ə-lĕn-sĕf′ə-lŏn] (Gr., *myelos* = core, marrow + *enkephalos* = brain). The most **caudal** region of the **brain;** consists of the **medulla oblongata** and leads to the **spinal cord.**

Myelin sheath [mī′ə-lĭn]. A sheath around most **axons,** composed of lipid materials.

Myocardium [mī′ō-kär′dē-əm] (Gr., *my-* = muscle + *kardia* = heart). The **muscular** layer of the **heart.**

Myoepithelial cells [mī′ō-ĕp-ə-thē′lē-əl]. Elongated **epithelial** cells with contractile properties, such as those associated with **sweat glands.**

Myofilaments [mī′ō-fĭl′ə-mənts]. Ultramicroscopic filaments of actin and myosin that form the contractile mechanism of **muscle** cells.

Myoglobin [mī′ō-glō′bĭn] (L., *globus* = globe). A hemoglobin-like molecule in red **muscle.**

Myomere [mī′ō-mîr] (Gr., *meros* = part). A **muscle** segment, usually applied to adult segments.

Myometrium [mī′ō-mē′trē-əm] (Gr., *metra* = uterus). The **muscular** layer of the **uterus.**

Myoseptum [mī′ō-sĕp′təm]. A **connective tissue** septum between **myomeres.**

Myotome [mī′ō-tōm] (Gr., *tome* = cutting). A muscle segment, usually applied to **embryonic** segments.

Myxiniformes [mĭx′ĭn-ə-fôr′mēz] (Gr., *myxa* = slime + L., *forma* = form). A group of jawless **fishes** including the hagfishes.

N

Nares [nĕr′ēs] (L., *naris,* pl., *nares* = nostrils). The paired openings from the outside into the **nasal** cavities; **external nostrils.**

Nasal [nā′zəl] (L., *nasus* = nose). Pertaining to the nose, as in nasal bone.

Neocerebellum [nē′ō-sĕr-ə-bĕl′əm] (Gr., *neos* = new + L., *cerebellum* = small brain). The portion of the **mammalian cerebellum** that has connections with the **cerebrum;** includes the **cerebellar hemispheres** and part of the **vermis.**

Neocortex. See **isocortex.**

Neognathous birds [nē′ō-nâth-əs] (Gr., *gnathos* = jaw). The group of birds (Neognathae) with a relatively advanced, nonreptilian type of palate; includes most orders of birds. See **paleognathous birds.**

Neonatal [nē′ō-nā′təl] (L., *natus* = born). Newborn.

Neopallium. See **isocortex.**

Neopterygians [nē′ŏp-tə-rĭj′ē-ənz] (Gr., *neos* = new + *pteryg-* = fin or wing). The group of **actinopterygian fishes** that includes gars, bowfins, and **teleosts.**

Neornithes [nē-ôr′nə-thēz] (Gr., *neos* = new + *ornis* = bird). The group of birds that has lost many of the primitive features of the **Archaeornithes** (including the long tail); essentially modern birds.

Neoteny [nē-ŏt′ən-ē] (Gr., *teinein* = to extend). **Paedomorphosis** that results from the slowing down of **somatic** development relative to reproductive development; it occurs in many salamanders.

Nephric ridge [nĕf′rĭk] (Gr., *nephros* = kidney). The region of the **mesoderm** between the **somite** and **lateral plate** that gives rise to the **kidneys** and **gonads;** also called nephrogenic ridge and mesomere.

Nephron [nĕf′rŏn]. A **renal tubule,** the structural and functional unit of the **kidneys.**

Nerves [nûrvz] (L., *nervus* = nerve). A cordlike group of **axons** and associated **connective tissue** that lies outside the **brain** and **spinal cord; nerves** connect the **central nervous system** with other organs of the body.

Neural arch [nŏŏr′əl]. See **vertebral arch.**

Neural crest (Gr., *neuron* = nerve, sinew). A pair of ridges of **ectodermal** cells that develop along the top of the **neural tube** as the neural folds close; this **derived character** of **craniates** gives rise to many of their distinctive features, including the **visceral** skeleton, pigment cells, sensory and postganglionic **neurons,** the **dentine**-producing cells of teeth, and certain **bony scales.**

Neural tube. The tube formed in the **embryo** by the joining of the pair of **neural** folds; the precursor of the **brain** and **spinal cord.**

Neurilemma [nŏŏr′ə-lĕm′ə] (Gr., *lemma* = husk). The thin sheath formed by cells of **neural crest** origin that surrounds an unmyelinated **axon,** or, after having myelinated an **axon,** lies on the surface of the **myelin sheath.**

Neurocranium. See **chondrocranium.**

Neuroectoderm (Gr., *ektos* = outside + *derma* = skin). That portion of the **ectoderm** that gives rise to the **neural tube** and **neural crest.**

Neurogenic placode. See **placode.**

Neuroglia [nŏŏ-rŏg′lē-ə] (Gr., *glia* = glue). Cells in the **central nervous system** that help support, protect, and maintain the **neurons;** they include **astrocytes, oligodendrocytes,** and **microglia.**

Neurohemal organ [nŏŏr′ō-hē′məl] (Gr., *haima* = blood). An organ, such as the **neurohypophysis,** formed by the termination of a group of neurosecretory **neurons** and the **blood** vessels into which they discharge their products.

Neurohypophysis [nŏŏr′ō-hī-pŏf′ə-sĭs] (Gr., *hypo* = under + *physis* = growth). The posterior part of the **hypophysis** that develops from the **infundibulum** of the **brain;** its **hormones** promote the reabsorption of water and smooth **muscle** contraction.

Neuromast [nŏŏr′ō-măst] (Gr., *mastos* = knoll, breast). An aggregation of sensory **hair cells** and supporting cells in the **lateral line system** that is overlain by a gelatinous **cupula.**

Neuron [nŏŏr′ŏn]. A **nerve** cell, the structural and functional unit of the nervous system.

Neurosecretory cells [nŏŏr′ō-sĭ-krē′-tə-rē]. **Neurons** that secrete **hormones.**

Neurotransmitters [nŏŏr′ō-trăns′mĭt-ərz]. Substances released by **neurons** at **synapses** and neuroeffector junctions that activate or inhibit the target cells.

Nictitating membrane [nĭk′tĭ-tā′tĭng] (L., *nicto,* pp. *nictatus* = to wink). A third eyelid of many **amniotes** that helps protect and cleanse the surface of the eyeball.

Nidamental gland [nī′də-mĕn′təl] (L., *nidamentum* = nesting material). An aggregation of **glands** in the **oviduct** that secrete coverings for the eggs.

Nipple [nĭp′əl] (Old English, *neb* = small nose). A **papilla** that bears the openings of the **ducts** from the **mammary glands.**

Nonmonophyletic. See **nonmonophyletic group.**

Nonmonophyletic group. A group that is not **monophyletic** (see **monophyletic group**). Some authors attempt to divide nonmonophyletic groups into two types (**paraphyletic** and **polyphyletic**), but these subcategories often are defined inconsistently. This distinction is thought here to be relatively unimportant.

Noradrenaline. See **norepinephrine.**

Norepinephrine. [nôr-ep′ə-nef′rin] (L., *nor* = short for normal + Gr., *epi-* = upon + *nephros* = kidney.) The **hormone** produced by **postganglionic** sympathetic **fibers** and by **chromaffin cells** of the **adrenal medulla.**

Notochord [nō′tō-kôrd] (L., *notos* = back + *chorda* = string, cord). A rod of vacuolated cells encased by a firm sheath that lies ventral to the **neural tube** in **vertebrate embryos** and some adults.

Nucleus [nōō′klē-əs] (L., = kernel). An organelle within a cell that contains the genetic material; a group of **neuron** cell bodies within the **brain.**

O

Obturator foramen [ŏb′tə-rā′tər] (L., *obturo,* pp. *obturatus* = to stop up). A **foramen** in the **pubis** of **reptiles,** or an opening between the pubis and **ischium** in **mammals;** the obturator **muscles** arise from the periphery of the obturator foramen and close it.

Occipital nerves [ŏk-sĭp′ĭ-təl] (L., *occiput* = back of the head). **Nerves** that emerge from the occipital region of the **skull,** or just behind it, in **fishes** and some **amphibians;** they become the hypoglossal nerve of **amniotes.**

Occlusion [ō-klōō′zhən] (L., *occludo,* pp. *occlusus* = to shut up). The closing of a passage; the coming together of the surfaces of the teeth of upper and lower jaws.

Octavolateralis system [ŏk-tā′vō-lăt-ə-rā′lĭs] (L., *octavus* = the eighth + *latus* = side, flank). The combined vestibuloauditory and **lateral line systems** of **fishes** and **amphibians;** fibers from the **ear** return in the eighth **nerve,** and those from the **lateral line system** return in the adjacent **lateral line nerves.** Fibers of the lateral line nerves are closely associated with the seventh, ninth, and tenth **nerves.**

Oculomotor nerve [ŏk′yōō-lō-mō′tər] (L., *oculus* = eye + *motorius* = moving). The third cranial **nerve,** which innervates most of the **extrinsic muscles** of the eyeball and carries autonomic fibers into the eyeball.

Odontoblast [ō-dŏn′tō-blăst] (Gr., *odont-* = tooth + *blastos* = bud). Cell of **neural crest** origin that produces the **dentine** of teeth or certain **bony scales.**

Olecranon [ō-lĕk′rə-nŏn] (Gr., *olene* = elbow + *kranion* = head). A process on the proximal end of the **ulna** to which the triceps **muscle** attaches.

Olfactory [ōl-făk′tə-rē] (L., *olfacio,* pp. *olfactus* = to smell). Pertaining to the nose.

Olfactory bulb. A rostral enlargement of the **brain** in which the **olfactory nerve** terminates.

Olfactory nerve. The first cranial **nerve,** consisting of **neurons** returning from the nose to the **olfactory bulb.**

Oligodendrocytes [ŏl′ĭ-gō-dĕn′drə-sīts] (Gr., *oligos* = few + *dendron* = tree + *kytos* = hollow vessel or cell). **Neuroglial** cells of **ectodermal** origin that myelinate **axons** in the **central nervous system.**

Omentum [ō-mĕn′təm] (L., = fatty membrane). The **peritoneal** fold, sometimes containing a great deal of fat, which extends between the body wall and **stomach** (greater omentum), or between the **stomach** and **liver** and **duodenum** (lesser omentum).

Omnivore [ŏm′nə-vôr] (L., *omnis* = all + *-vorous* = devouring). An animal that eats a wide variety of food, both plant and animal.

Ontogeny [ŏn-tŏj′ə-nē] (Gr., *on* = being + *genesis* = birth or descent). The development of an individual.

Oogenesis [ō-ō-jĕn′ə-sĭs] (Gr., *oon* = egg). The development and maturation of an egg.

Operculum [ō-pûr′kyə-ləm] (L., = covering). The **gill** covering of **fishes** and some **amphibian larvae** consisting of **bone** and soft **tissue** (the **bone** by itself is referred to as the opercle); also an **auditory ossicle** in contemporary **amphibians.**

Ophthalmic nerve [ŏf-thăl′mĭk] (Gr., *ophthalmos* = eye). One of the main branches of the **trigeminal nerve;** passes through the orbit.

Opisthocoelous vertebra [ō-pĭs′thō-sē′lŭs] (Gr., *opisthen* = behind + *kolima* = hollow). A **vertebral body** that is concave on the posterior or **caudal** surface and convex on the anterior surface.

Opisthonephros [ō-pĭs′thō-nĕf′rəs] (Gr., *nephros* = kidney). The adult **kidney** of most **anamniotes; kidney** tubules are concentrated caudally.

Optic [ŏp′tĭk] (Gr., *optikos* = pertaining to the eyes). Pertaining to the eyes.

Optic chiasm [kī-ăz′mə] (Gr., *chiasma* = cross, from the Greek letter *chi* = X). The complete or partial **decussation** of the **optic nerves** on the floor of the **diencephalon.**

Optic lobes. A pair of enlargements of the roof of the **mesencephalon** that are important integration centers for sight and other senses in **anamniotes.**

Optic nerve. The second cranial **nerve,** which carries impulses from the **retina.**

Oral cavity [ôr′əl] (L., *os,* gen. *oris* = mouth). The mouth cavity, also called the **buccal** cavity.

Orbit [ôr′bĭt] (L., *orbis* = circle, eye). The cavity in the **skull** for the eyeball.

Organ of Corti (*Marquis Alfonso Corti,* Italian anatomist, 1822–1888). The sound receptive organ in the **mammalian cochlea.**

Origin [ôr'ĭ-jĭn] (L., *origio* = beginning). The starting point of a structure; that end of a **muscle** that attaches to the more fixed part of the skeleton, which is the proximal end in limb **muscles.**

Osmosis [ŏs-mō'sĭs] (Gr., *osmos* = action of pushing). The movement of water through a semipermeable membrane, through which solute molecules do not pass, from an area of high water concentration to one with a lower water concentration.

Osmotic pressure. The pressure that results from the movement of water by **osmosis** into a solution surrounded by a semipermeable membrane.

Ossicle [ŏs'ĭ-kəl] (L., *ossiculum* = small bone). Any small bone, such as one of the **auditory** ossicles.

Osteichthyes [ŏs'tē-ĭk'thēz] (Gr., *osteon* = bone + *ichthyes* = fishes). The group of animals in which all or part of the **endoskeleton** ossifies; includes most **vertebrates.** This term is sometimes also used in a **nonmonophyletic** way excluding **tetrapods.**

Osteoblast [ŏs'tē-ō-blăst] (Gr., *blastos* = bud). A cell that produces the **bone matrix.**

Osteoclast [ŏs'tē-ō-klăst] (Gr., *klastos* = broken). A cell that removes **bone** and calcified **cartilage** during the process of **bone** remodeling and growth.

Osteocyte [ŏs'tē-ō-sīt] (Gr., *kytos* = hollow vessel or cell). A mature **osteoblast** that is surrounded by the **matrix** it has produced.

Osteoderm [ŏs'tē-ō-dûrm] (Gr., *derma* = skin). A small **bone** embedded in the **skin** of some **vertebrates.**

Osteon [ŏs'tē-ən]. A cylindrical unit of **bone** consisting of concentric layers that have developed around a central cavity containing **blood** vessels; also called a haversian system.

Ostium [ŏs'tē-əm] (L., = entrance, mouth). The entrance to an organ, such as the **oviduct.**

Ostracoderms [ŏs-trā'kō-dûrmz] (Gr., *ostrakon* = shell + *derma* = skin). A **nonmonophyletic group** name of convenience applied to several orders of **Paleozoic** jawless fishes that are characterized by the extensive development of **bone** in the **skin.**

Otic capsule [ō'tĭk] (Gr., *otikos* = pertaining to the ear). The portion of the **chondrocranium** that houses the **inner ear.**

Otolith [ō'tō-lĭth] (Gr., *oto-* = ear + *lithos* = stone). A calcareous structure found in the **sacculus** and **utriculus** of **vertebrates;** its movement with respect to gravity stimulates underlying **hair cells** and allows an animal to detect its position and movement.

Outgroup. Any **taxon** outside the **taxon** of reference or study. The closest outgroup is the **sister group.**

Out-lever. The **lever** arm through which a **force** is delivered out of a **lever** system to its point of application; it is the perpendicular distance from the line of action of the outforce to the **axis** of rotation of the **lever** system.

Oval window. See **fenestra vestibuli.**

Ovarian follicles [ō-vĕr'ē-ən] (L., *ovarium* = ovary). Groups of **epithelial** and **connective tissue** cells in the **ovary** that invest and nourish maturing eggs. The mature **follicle** is sometimes calld a **graafian follicle.**

Ovary [ō'və-rē] (L., *ovarium* = ovary). One of a pair of female reproductive organs containing the **ovarian follicles** and eggs.

Oviduct [ō'və-dŭkt] (L., *ovum* = egg + *ducere,* pp. *ductus* = to lead). The tube that carries eggs from the **coelomic** cavity to the outside.

Oviparous [ō-vĭp'ər-əs] (L., *pario* = to bear). A pattern of reproduction in which eggs are laid and then develop outside the body of the mother.

Ovoviviparous [ō'vō-vī-vĭp'ər-əs] (L., *viviparus* = bringing forth alive). A pattern of reproduction in which the eggs are retained within the **uterus** and the **embryos** are born as miniature adults. The term is often limited to aplacental **viviparity,** for all or most of the needed nutrients and energy are contained within the egg.

Ovulation [ŏv'yo͞o-lā'shən]. The rupture of the **ovarian follicle** and the discharge of the eggs from the **ovary** into the **ceolomic** cavity, or sometimes directly into the surrounding infundibulum.

Ovum [ō'vəm] (L., = egg). The mature egg cell.

Oxytocin [ŏk'sĭ-tō'sĭn] (Gr., *okytokos* = swift birth). A **hormone** produced by the **neurohypophysis** that promotes the contraction of **uterine muscles** at birth and the release of milk during **lactation.**

P

Paedomorphosis [pē'dō-môr'fə-səs] (Gr., *paid-* from *pais* = child + *morphe* = shape). The retention of juvenile **characters** into the adult stage.

Palate [păl'ĭt] (L., *palatum* = palate). The roof of the mouth. See **hard palate, soft palate.**

Palatoquadrate cartilage [păl'ə-tō-kwŏd'rāt] (L., *quadratus* = square). The dorsal part of the **mandibular arch.**

Paleocerebellum [pā'lē-ō-sĕr'ə-bĕl'əm] (Gr., *palaios* = ancient + *cerebellum* = small brain). The part of the **cerebellum** that receives proprioceptive impulses; the flocculonodular lobes in **mammals.**

Paleognathous birds [pā'lē-ō-năth'əs] (Gr., *gnathos* = jaw). Birds that retain a primitive **reptile**-like palate; the kiwi, emu, ostrich, and similar birds, most of which are flightless.

Paleopallium. See **lateral pallium.**

Paleozoic. An era of geologic time ranging from about 600 to 230 million years before present.

Pallium [păl'ē-əm] (L., *pallium* = cloak). The dorsal portion of the **cerebral gray matter;** most forms a surface **cortex** in **mammals.** See **dorsal pallium, lateral pallium, medial pallium.**

Pampiniform plexus [păm-pĭn'ĭ-fôrm] (L., *pampinus* = tendril + *forma* = shape). A convoluted network of veins in **mammals** that surrounds the spermatic **artery.**

Pancreas [păn'krē-əs] (Gr., *pan* = all + *kreas* = flesh). A large glandular outgrowth of the **duodenum** that secretes many digestive enzymes; also contains the **pancreatic islets.**

Pancreatic islets. Small clusters of endocrine cells in the **pancreas** that produce **hormones** that regulate sugar metabolism; also called the **islets of Langerhans.**

Papilla [pă-pĭl'ə] (L., = nipple). A small, conical protuberance.

Paracrines [para'a-krinz] (Gr., *para* = alongside of + *krino* = to separate). Signalling molecules released by cells that transmit information between cells that are close together, unlike **endocrines,** which transmit messages through the blood between more distant cells.

Paradidymis [pĕr'ə-dĭd'ə-məs] (Gr., *para* = beside + *didymoi* = testes). A small group of **vestigial** mesonephric tubules in mammals located beside the **epididymis** and **testis.**

Paraganglia [pĕr'ə-găng'glē-ə] (Gr., *ganglion* = little tumor). Small groups of **chromaffin cells** that lie beside the sympathetic **ganglia.**

Parallel evolution. See **convergent evolution.**

Parallelism. See **homoplasy.**

Paraphyletic. See **nonmonophyletic group.**

Parapophysis [pĕr'ə-pŏf'ə-sĭs] (Gr., *apo* = away from + *physis* = growth). A transverse process on a **vertebral body** to which the head of a rib attaches, or the facet on a **vertebral body** for such an attachment.

Parasympathetic nervous system [pĕr'ə-sĭm-pə-thĕt'ĭk] (Gr., *syn* = with + *pathos* = feeling). The portion of the **autonomic nervous system** that, in mammals, leaves the **central nervous system** through certain **cranial** and **sacral nerves;** promotes metabolic processes that produce and store energy.

Parathormone [pĕr'ə-thôr'mōn] (Gr., *horma,* pres.p. *hormon* = to rouse or set in motion). The **hormone** of the **parathyroid gland;** helps regulate mineral metabolism.

Parathyroid glands [pĕr'ə-thī'roid] (Gr., *thyreos* = oblong-shaped shield + *eidos* = form). **Endocrine glands** of **tetrapods** located dorsal to or near the **thyroid gland;** their **hormone** regulates calcium and phosphate metabolism.

Paraxial mesoderm. That portion of the **mesoderm** that lies just lateral to the **neural tube,** differentiates into **somites** in the trunk and **caudal** part of the head and into **somitomeres** more rostrally.

Parietal [pə-rī'ə-təl] (L., *paries* = wall). Pertaining to the wall of some structure, such as the parietal **bone** or parietal **peritoneum.**

Parietal eye. A median, photoreceptive eye of some **fishes** and **reptiles;** lies between the **parietal bones.**

Parotid gland [pə-rŏt'ĭd] (Gr., *para* = beside + *otikos* = pertaining to the ear). A **mammalian salivary gland** located **caudal** to the **ear.**

Parsimony. In systematics, this term refers to the maximum amount of congruence among data. The most parsimonious hypothesis is the one that requires the fewest assumptions (e.g., steps of **character** transformation) about a data set. Parsimony is a necessary methodological tool to empirically select the preferred hypothesis of relationship from a potentially infinite number of possible phylogenetic hypotheses on the basis of **character congruence.** See **congruence.**

Parthenogenesis [pär'thə-nō-jĕn'ĭ-sĭs] (Gr., *parthenos* = virgin + *genesis* = descent or birth). Activation and development of an egg without fertilization.

Patella [pə-tĕl'ə] (L., = small plate). The kneecap.

Pectoral [pĕk'tōr-əl] (L., *pectoralis,* pertaining to the breast; from *pectus* = breastbone). Pertaining to the chest, as in pectoral appendage, pectoral **muscles.**

Pectoral girdle. A series of **bones** or **cartilages** for the attachment of the **pectoral fins** or limbs.

Pelvic [pĕl'vĭk] (L., *pelvis* = basin). Pertaining to basin-shaped structures, such as the human **pelvic girdle,** or to structures near the **pelvic girdle.**

Pelvic girdle. A series of **bones** or **cartilages** for the attachment of the **pelvic fins** or limbs.

†Pelycosaurs [pĕl'ĭ-kō-sôrz] (Gr., *pelyx,* gen. *pelykos* = bowl, axe + *sauros* = lizard). An early group of **synapsids** of questionable **monophyly,** most of which have narrow, deep, axe-shaped skulls.

Penis [pē'nĭs] (L., = tail, penis). The male copulatory organ.

Pericardial cavity [pĕr'ĭ-kär'dē-əl] (Gr., *peri* = around + *kardia* = heart). The portion of the **coelom** that surrounds the **heart.**

Perichondrium [pĕr'ĭ-kŏn'drē-əm] (Gr., *chondros* = cartilage). The **connective tissue** covering of a **cartilage.**

Perilymph [pĕr'ə-lĭmf] (L., *lympha* = a clear liquid). The **lymph**like fluid that surrounds the **membranous labyrinth** of the **inner ear.**

Periosteum [pĕr'ē-ŏs'tē-əm] (Gr., *osteon* = bone). The **connective tissue** covering of a **bone.**

Peripheral nervous system. The portion of the nervous system lying peripheral to the **brain** and spinal cord; the cranial and spinal **nerves.**

Perissodactyls [pə-rĭs'ō-dăk'təlz] (Gr., *perissos* = odd + *daktylos* = finger or toe). The **mammalian** group that includes those **ungulates** with an odd number of **digits** (three or one): the rhinoceros, tapirs, horses.

Peritoneal. Pertaining to the **peritoneum.**

Peritoneal cavity [pĕr'ĭ-tə-nē'əl] (Gr., *peritonaion* = to stretch over). The part of the **mammalian coelom** that surrounds the **viscera.**

Peritoneum [pĕr'ĭ-tə-nē'əm]. The **connective tissue** and **epithelial** layer that lines the **peritoneal cavity,** forms **mesenteries,** and covers the **viscera.**

Permanent teeth. The teeth of **mammals** that replace the milk, or **deciduous,** teeth.

Pes [pĕz] (L., = foot). Foot.

Petromyzontiformes [pĕ'trō-mī'zŏn-tĭ-fôr'mēz] (Gr., *petros* = stone + *myzo* = to suck in + L., *forma* = form). The lampreys.

Phagocytosis [făg'ō-sī-tō'sĭs] (Gr., *phagein* = to eat + *kytos* = hollow vessel or cell). The **ingestion** and breaking down of foreign particles by a cell.

Phalanges [fə-lăn'jēz] (Gr., *phalanx,* pl. *phalanges* = battle line of soldiers). **Bones** of the **digits** that extend beyond the palm or sole.

Pharynx [fĕr′ĭngks] (Gr., = throat). The portion of the digestive tract from which the pharyngeal pouches develop in an **embryo;** lies between the **oral cavity** and **esophagus;** the crossing place of digestive and respiratory tracts.

Pheromones [fĕr′ə-mōnz] (Gr., *pherein* = to bear + *horma*, pres.p. *hormon* = to rouse or set in motion). Chemical secretions that act as signals for another individual of the same **species.**

Phylogenetic homology. See **homology.**

Phylogenetic hypothesis. See **phylogeny.**

Phylogeny [fī-lŏj′ə-nē] (Gr., *phylon* = race + *genesis* = birth or descent). A hypothesis of **evolutionary** relationships among the members of a **monophyletic group.** A phylogeny may be fully resolved (normally containing only dichotomous branching) or remain partly unresolved (containing polychotomous branch points or nodes). Also often referred to as an **evolutionary** tree.

Physoclistous [fī′sō-klī′stəs] (Gr., *physa* = bladder + *kleien* = to close). Pertaining to **fishes** in which the **swim bladder** is not connected to the digestive tract.

Physostomous [fī′sō-stō′məs] (Gr., *stoma* = mouth). Pertaining to the **fishes** in which the **swim bladder** remains connected to the digestive tract by a **pneumatic duct.**

Pia mater [pī′ə mā′tər] (L., = tender mother). The delicate vascular membrane that invests the **brain** and **spinal cord;** the innermost of the three **mammalian meninges.**

Pineal eye [pĭn′ē-əl] (L., *pineus* = relating to pine; from *pinus* = pine tree). A dorsal outgrowth of the **diencephalon** that forms a light-sensitive eye in some **fishes** and **amphibians** and becomes the **pineal gland** in **mammals.**

Pineal gland. An **endocrine gland** that produces **melatonin,** especially in the dark. **Melatonin** is believed to adjust many endogenous physiological processes to diurnal and seasonal cycles.

Pisces [pī′sēz] (Gr., = fishes). A **nonmonophyletic group** name of convenience for all **vertebrates** excluding **tetrapods** (i.e., **fishes**). See **monophyletic group.**

Pitch. The vertical rotation of a swimming or flying **vertebrate** about its longitudinal **axis.**

Pituitary gland [pĭ-too′ĭ-tĕr′ē]. See **hypophysis.**

Placenta [plə-sĕn′tə] (L., = flat cake). The apposition or union of parts of the **uterine** lining and fetal extraembryonic membranes through which exchanges between mother and **embryo** occur.

Placental mammals. See **eutherians.**

Placode [plăk′ōd] (Gr., *placodes* from *plax* = plate + *eidos* = like). A thickened disk of **ectoderm** that gives rise to certain **sense organs** and **nerves. Neurogenic placodes** give rise to some receptor cells and sensory **neurons** in the head.

†Placoderms [plăk′ō-dûrmz] (Gr., *derma* = skin). A group of **Paleozoic** jawed **fishes** characterized by the extensive development of **bone** in the head and **thorax.**

Placoid scale. See **dermal denticle.**

Plantigrade [plăn′tĭ-grād] (L., *planta* = sole of the foot + *gradus* = step). Walking with the sole of the foot on the ground.

Plastron [plăs′trən] (French, = breastplate). The ventral shell of a turtle.

Plesiomorphic character [plē′sē-ō-môr′fĭk] (Gr., *plesios* = near + *morphe* = shape). A primitive or ancestral **character.** Meaningful only in a relative sense (e.g., the presence of **mammary glands** is primitive when considering only apes, but it is derived and **apomorphic** when considering all of **Mammalia**).

Pleura [ploor′ə] (Gr., = side, rib). The **coelomic epithelium** in the **pleural cavities.**

Pleural cavities. The **coelomic** spaces that enclose the lungs of **mammals.**

Pleurapophysis [ploor′ə-pŏf′ĭ-sĭs] (Gr., *apo* = away + *physis* = growth). A **vertebral** transverse process that incorporates a rib.

Pleurocentrum pl. **Pleurocentral** [ploor′ō-sĕn′trəm] (L., *centrum* = center). A dorsolateral element of the **vertebral** body of **vertebrates** that becomes the main **vertebral** body of **amniotes.**

Pleurodont tooth [ploor′ō-dŏnt] (Gr., *odont-* = tooth). A tooth that is loosely attached to the outside edge of the jaw.

Pleuroperitoneal cavity [ploor′ō-pĕr′ĭ-tən-ē′əl]. The **peritoneal cavity** and potential **pleural cavities** of **anamniotes** and some **reptiles;** contains the **viscera** and **lungs** (if present).

Plexus [plĕk′səs] (L., = a braid). A network of nerves or **blood** vessels.

Pneumatic duct [noo-măt′ĭk] (Gr., *pneuma* = air). The **duct** that connects the **swim bladder** with the **pharynx** in **physostomous fishes.**

Poikilothermic [poi′kĭ-lō-thûr′mĭk] (Gr., *poikilos* = varied + *thermos* = heat). Pertains to **vertebrates** in which the body temperature varies with the ambient temperature; **ectothermic.**

Pollex [pŏl′ĕks] (Gr., = thumb). The thumb.

Polyphyletic. See **nonmonophyletic group.**

Polyphyodont [pŏl′ē-fī′ō-dŏnt] (Gr., *polyphyes* = manifold + *odont-* = tooth). Pertaining to many successive sets of teeth.

Pons [pŏnz] (L., = bridge). The ventral part of the **metencephalon** of birds and **mammals;** has a conspicuous, superficial band of transverse fibers.

Portal veins [pôr′təl] (L., *porta* = gate). **Veins** that drain one **capillary** bed and lead to another one in a different organ, such as the **hepatic portal** and **hypophyseal** portal systems.

Posterior chamber. The cavity within the eyeball located between the **iris** and the **ciliary body.**

Postganglionic fiber [pōst′găng-glē-ŏn′ĭk] (Gr., *ganglion* = small tumor). A **neuron** of the **autonomic nervous system** with its cell body in a peripheral **ganglion** and its **axon** extending to the **effector** organ.

Posttrematic [pōst′trē-măt′ĭk] (Gr., *trema* = hole). Pertaining to **blood** vessels or **nerves** that lie **caudal** to a **branchial** pouch.

Power. The rate of doing **work.**

Preadaptation. The evolution of a feature that enables an animal to exploit a new environment, such as the **evolution** of **lungs** in certain **fishes.**

Preganglionic fibers [prē'găng-glē-ŏn'ĭk] (Gr., *ganglion* = small tumor). A **neuron** of the **autonomic nervous system** with its cell body in the **brain** or **spinal cord** and its **axon** extending to a peripheral **ganglion.**

Premolars [prē-mō'lərz] (L., *molaris* = millstone). **Cheek teeth** that lie rostral to the **molars** and may be specialized for cutting or grinding.

Pressure. Force per unit area, such as grams per square centimeter.

Pretrematic [prē'trē-măt'ĭk] (Gr., *trema* = hole). Pertaining to **blood** vessels or **nerves** that lie rostral to a **branchial** pouch.

Primates [prī'māts] (L., *primus* = one of the first). The **eutherian** group that includes lemurs, monkeys, apes, and humans.

Primitive character. A **character** state that is the ancestral condition for a group; also called a **plesiomorphic character.** For example, the presence of bone is primitive for birds, but, conversely, the presence of **feathers** is derived for birds. See **character, derived character.**

Primitive streak. A longitudinal thickening of cells on the **blastoderm** of large-yolked eggs, through which prospective **chordamesoderm** and **mesoderm** cells move inward; **homologous** to the **blastopore.**

Primordium [prī-môr'dē-əm] (L., = beginning). The first indication of the formation of a structure in an **embryo.**

Processus vaginalis [prō-sĕs'əs vă'jī'năl'ĭs] (L., = process + *vagina* = sheath). A sac that contains the **mammalian testis** and its sperm **duct** and **blood** vessels, as well as the **coelomic vaginal** cavity; located in the **scrotum,** also called **vaginal** sac.

Procoelous [prō-sē'ləs] (Gr., *koilios* = hollow). A **vertebral body** with a concavity on its cranial surface.

Proctodaeum [prŏk'tō-dē'əm] (Gr., *proktos* = anus + *hodaion* = way). An **ectodermal invagination** near the **caudal** end of the **embryo** that contributes to the **cloaca.**

Progenesis [prō-jĕn'ĭ-sĭs] (Gr., *pro* = before + *genesis* = origin). **Paedomorphosis** that, in theory, results from the acceleration of reproductive maturity relative to **somatic** development.

Progesterone [prō-jĕs'tə-rōn] (L., *gesto,* pp. *gestatus* = to bear). A **hormone** produced by the **corpus luteum** and later by the **placenta;** prepares the **uterus** for the reception of a fertilized egg and maintains the **uterine** lining during pregnancy.

Prolactin [prō-lăk'tĭn] (L., *lac, lact-* = relating to milk). A **hormone** produced by the **adenohypophysis** that promotes maternal behavior and milk production.

Pronephros [prō-nĕf'rəs] (Gr., *pro* = before + *nephros* = kidney). The first formed **kidney** of a **vertebrate embryo,** which lies dorsal to the **pericardial cavity** and forms the **archinephric duct** before it **atrophies.**

Proprioceptor [prō'prē-ō-sĕp'tər] (L., *proprius* = one's own + *capio,* pp. *ceptus* = to take). A receptor in **muscles, tendons,** and joints that monitors the activity of **muscles.**

Prosencephalon [prŏs'ĕn-sĕf'ə-lŏn] (Gr., *pro* = before + *enkephalos* = brain). The **embryonic forebrain,** which gives rise to the **telencephalon** and **diencephalon.**

Prostate [prŏs'tāt] (Gr., *prostates* = one who stands before). An accessary sex **gland** of male **mammals** that surrounds the **urethra** just before the **urinary bladder.**

Protandry [prō-tăn'drē] (Gr., *protos* = first + *andr-* = man). Sequential **hermaphroditism** in which the **gonad** functions first as a **testis** before it acts as an **ovary.**

Protochordates [prō'tō-kôr'dāts] (L., *chorda* = string, cord). A **nonmonophyletic group** of convenience for the noncraniate **chordates:** the **tunicates** and **cephalochodates.**

Protogyny [prō-tŏj'ə-nē] (Gr., *gyne* = woman). Sequential **hermaphroditism** in which the **gonad** functions first as an **ovary** before it acts as a **testis.** See **hermaphrodite.**

Protostomes [prō'tō-stōmz] (Gr., *stoma* = mouth). The group of **coelomate** animals in which the **blastopore** forms or contributes to the mouth; includes mollusks, annelids, and arthropods.

Prototherians [prō'tō-thîr'ē-ənz] (Gr., *therion* = wild beast). A primitive or basal branch of **Mammalia (mammals);** includes the contemporary, egg-laying **monotremes.**

Protraction [prō-trăk'shən] (L., *pro* = before + *traho,* pp. *tractus* = to pull). Muscle action that moves the entire appendage of a quadruped forward.

Proventriculus [prō'vĕn-trĭk'ū-lŭs] (L., *ventriculus* = small-belly). The anterior, glandular portion of the **stomach** of birds.

Pseudobranch [sū'dō-brănk] (Gr., *pseudes* = false + *branchia* = gills). A small first gill of some **fishes,** without a respiratory function.

†Pterosaur [tĕr'ə-sôr] (Gr., *pteryg-* = fin or wing + *sauros* = lizard). An extinct order of flying **reptiles.**

Pterygiophores [tə-rĭj'ē-ō-fôrz] (Gr., *phoros* = bearing). The supporting **cartilages** or **bones** of the **fin rays.** Also called radials or basals.

Pubis [pyū'bĭs] (L., *pubes* = genital hair). The cranioventral **bone** of the **pelvis** of **tetrapods.**

Pulmonary [pŏŏl'mə-nĕr'-ē] (L., *pulmo* = lung). Pertaining to the **lungs,** as the pulmonary **artery.**

Pupil [pyū'pəl] (L., *pupilla* = pupil). The central opening through the **iris** of the eye.

Pygostyle [pī'gō-stīl] (Gr., *pyge* = rump + *stylos* = pillar). The fused, **caudal vertebrae** of a bird, which support the tail **feathers.**

Pylorus [pī-lôr'əs] (Gr., *pyloros* = gatekeeper). The **caudal** end of the **stomach,** which contains a **sphincter muscle.**

Pyramidal system [pĭ-răm'ĭ-dəl] (Gr., *pyramis* = pyramid). The direct motor pathway in **mammals** from the **cerebrum** to the motor **nuclei** and columns.

R

Radius [rā'dē-əs] (L., = ray). A **bone** of the forearm of **tetrapods** that rotates around the **ulna;** located on the thumb side when the hand is supine.

Ramus [rā′məs] (L., = branch). A branch such as those of a spinal nerve.

Rathke's pouch (*Martin H. Rathke,* German anatomist, 1793–1860). A dorsal **evagination** of the **stomadaeum** that forms the **adenophypophysis.**

Ray-finned fishes. See **actinopterygians.**

Receptor [rē-sĕp-tər] (L., = receiver). A specialized cell or **neuron** ending that responds to a specific stimulus and initiates a **nerve** impulse.

Rectum [rĕk′təm] (L., *rectus* = straight). The terminal segment of the **intestine** that leads to the **anus.**

Reflex [rē′flĕks] (L., *reflecto,* pp. *reflexus* = to bend backward). An innate reaction in response to a peripheral stimulus.

Releasing hormones. Hormones produced by the **hypothalamus** that travel in the **hypophyseal** portal system and promote the release of specific adenohypophyseal **hormones.** In several cases, inhibiting **hormones** are also known.

Renal [rē′nəl] (L., *ren* = kidney). Pertaining to the **kidneys.**

Renal capsule. The dilated end of a **kidney** tubule that surrounds a knot of **capillaries.**

Renal portal system. A system of **veins** that drains the tail and hind legs of most nonmammalian **vertebrates** and leads to the peritubular **capillaries** of the **kidneys.**

Renal tubule. A **kidney** tubule or **nephron.**

Reptiles [rĕp′tīlz] (L., *reptilis* = creeping). A **nonmonophyletic group** name of convenience. This term has most often referred to a group of **amniotes** including turtles, lizards, snakes, and crocodiles (among living **vertebrates**). Phylogenetically based classifications recognize that crocodilians and dinosaurs are more closely related to birds than to lizards and snakes, and that the precise relationships of turtles are yet unclear.

Resultant of force. A **vector** that expresses the interaction between two or more **vectors.**

Rete cords [rē′tē] (L., *rete* = net). Minute cords in the **embryo** that interconnect the primary **sex cords** and the cranial mesonephric tubules; they contribute to the **sperm** passages in males and regress in females.

Rete mirabile [mĭ-rä′bĭ-lə] (L., = wonderful net). A network of small **arteries** or **capillaries,** such as those associated with the gas **gland** of the **swim bladder.**

Reticular formation [rĭ-tĭk′yə-lər] (L., *reticulum* = small net). A network of short **interneurons** in the **brainstem** that forms a primitive integrating system. In **mammals,** it also projects to the **cerebrum** and helps maintain the level of arousal.

Reticulate speciation. A process theory involving the origin of a new **species** through hybridization of two different **species.**

Retina [rĕt′-n-ə]. The innermost layer of the eyeball; contains pigment cells, photoreceptive cells, and **neurons.**

Retraction [rĭ-trăk′shən] (L., *retractio* = a drawing back). **Muscle** action that moves the entire appendage of a quadruped backward.

Retroperitoneal [rĕ′trō-pĕr-ĭ-tən-ē′əl] (L., *retro* = backward + Gr., *peritonaion* = to stretch over). Pertaining to structures, such as the **kidneys,** that lie dorsal to the **peritoneal cavity.**

Rhinal [rī′nəl] (Gr., *rhin-* = nose). Pertaining to the nose.

Rhipidistians [rĭ′pĭ-dĭs′tē-ənz] (Gr., *rhipis* = fan). Often used as **nonmonophyletic group** name of convenience for certain **sarcopterygian fishes,** including the presumed ancestors of **tetrapods.** We use it here in a **monophyletic** sense. See **sarcopterygians.**

Rhombencephalon [rŏm′bĕn-sĕf′ə-lŏn] (Gr., *rhombos* = lozenge-shaped + *enkephalos* = brain). The hindbrain, the most posterior of the three primary divisions of the developing **brain;** subdivides into the **metencephalon** and **myelencephalon.**

Roll. Rotation of a swimming or flying **vertebrate** around its longitudinal **axis.**

Round window. See **fenestra cochleae.**

Rudiment [ro͞o′dĭ-mənt] (L., *rudimentum* = first attempt). An early stage in the development of an organ; a **primordium.**

Rumen [ro͞o′mən] (L., *rumen* = gullet). The first and largest chamber of the **ruminant stomach.**

Ruminants [ro͞o′mĭ-nənts] (L., *rumino* = to chew the cud). Those **artiodactyls** with chambered **stomachs,** including deer, sheep, and cattle.

S

Sacculus [săk′yū-ləs] (L., = small sac). The most ventral chamber of the **membranous labyrinth.**

Sacral. Pertaining to the **sacrum.**

Sacrum [sā′krəl, sā′krəm] (L., *sacrum* = sacred). The **vertebrae,** or the union of two or more **vertebrae** and their ribs, by which the **pelvis articulates** with the **vertebral** column.

Salientia. See **anurans.**

Salivary gland [săl′ĭ-vĕr′ē] (L., *saliva* = saliva). A **gland** that produces the saliva; the major ones in **mammals** are the **parotid, mandibular,** and sublingual glands.

Salt gland. A **gland** or secretory cells that secrete excess salt; found near the nose and eye in certain marine **reptiles** and birds and on the **gill** of certain marine **fishes.**

Saltatorial [săl′tə-tôr′e-əl] (L., *saltatio* = to dance). Adapted for leaping.

Sarcopterygians [sär′kŏp-tə-rĭj′ē-ənz] (Gr., *sarkodes* = fleshy + *pteryg-* = fin or wing). The group of **Osteichthyes** with fleshy ("lobed"), paired **fins,** including **coelacanths, rhipidistians,** lungfishes, and **tetrapods.** Traditionally, this group included only **fishes** (i.e., excluded **tetrapods**), but sarcopterygians are more closely related to **tetrapods** than they are to **ray-finned fishes** or sharks.

Sauropsida. A group (as used here) containing **reptiles** and birds.

Scala tympani [skā′lə tĭm′pă-nē] (L., *scala* = ladder + *tympanum* = drum). The **perilymphatic duct** through which pressure waves pass from the **cochlea** to the **tympanic cavity.**

Scala vestibuli [vĕs-tĭb′yū-lē] (L., *vestibulum* = antechamber). The **perilymphatic duct** through which pressure waves enter the **cochlea** from the **auditory ossicle.**

Scales. Hard, platelike structures on the surface of the skin in many vertebrates.

Scaling. Analyzing the relationship between the size of a structure, or level of activity of a process, and body size.

Scapula [skăp′yū-lə] (L., = shoulder blade). The dorsal element of the **pectoral girdle** that ossifies from **cartilage.**

Schizocoele [skĭz′ə-sēl] (Gr., *schizo* = to cleave + *koilos* = hollow). A **coelom** formed by cavitation of the **mesoderm** rather than by **enterocoelic** pouches, characteristic of **protostomes.**

Schwann cells (*Theodor Schwann,* German histologist, 1810–1882). See **neurilema.**

Sclera [sklîr′ə] (Gr., *skleros* = hard). The opaque, "white" portion of the **fibrous tunic** of the eyeball.

Sclerotic bones [sklə-rŏt′ĭk] (Gr., *oto-* = ear). A ring of **bones** that develops in the **sclera** of some **vertebrates** and strengthens the eyeball wall.

Sclerotome [sklîr′ə-tōm] (Gr., *tomos* = a cutting). The medial portion of a **somite** that forms the **vertebrae** and the **caudal** part of the **chondrocranium.**

Scrotum [skrō′təm] (L., = pouch). The sac that encases the **mammalian testes;** it includes all of the layers of the body wall.

Sebaceous glands [sĭ-bā′shē-ŭs] (L., *sebum* = tallow). **Mammalian cutaneous glands** that secrete oily and waxy materials.

Secondary palate. A **palate** that separates the food and air passages; in **mammals,** it consists of a **bony hard palate** that separates the **oral** and **nasal cavities** and a fleshy **soft palate** that separates the **oral pharynx** from the **nasal pharynx.**

Secretin [sĭ-krēt′ĭn] (L., *secerno,* pp. *secretus* = to secrete). A **hormone** produced by the **duodenal mucosa** that promotes the secretion of the aqueous portion of the **pancreatic** juice.

Segmentation [sĕg′mən-tā′shən]. Refers to the division of the body into a longitudinal series of segments.

Selachian [sĭ-lā′kē-ən] (Gr., *selachios* = resembling a shark). A **taxonomic** subdivision of sharks.

Selenodont [sĭ-lē′nō-dŏnt] (Gr., *selene* = crescent + *odont-* = tooth). **Mammalian cheek teeth** with crescent-shaped cusps.

Semicircular duct [sĕm′ĭ-sûr′kyə-lər]. One of the **ducts,** shaped like a half-circle, of the **membranous labyrinth;** semicircular ducts are located within a set of semicircular canals in the **otic capsule** of the **skull.**

Seminal fluid [sĕm′ə-nəl] (L., *semen* = seed). The fluid secreted by male reproductive **ducts** and accessory sex **glands** that carries the **sperm.**

Seminal vesicle. See **vesicular gland.**

Seminiferous tubules [sĕm′ə-nĭf′ər-əs]. The tubules within the **testis** that produce the **sperm.**

Sense organ. An aggregation of receptive cells and associated cells that support them and may amplify a stimulus.

Septum. A partition between two structures. Also, a group of small nuclei within the rostral ventromedial wall of the **subpallium.**

Serial homology. A type of **homology** referring to similarity between different parts of a series of structures within a single organism (e.g., different leaves on a branch, different segments of a worm, different limbs of a **tetrapod**). See **homology.**

Sertoli cells (*Enrico Sertoli,* Italian histologist, 1842–1910). **Epithelial** cells of the **seminiferous tubules** that play a role in the maturation of the **sperm.**

Sesamoid bone [sĕs′ə-moid] (Gr., *sesamon* = sesame seed + *eidos* = form). A **bone** that develops in the **tendon** of a **muscle** near its insertion and facilitates the movement of a **muscle** across a joint, acts as a **lever** arm, or alters its direction of pull; the **patella** and pisiform are examples.

Sessile [sĕs′əl] (L., *sessilus* = fit for sitting). Describes an animal that lives attached to its substratum.

Sex cords. **Embronic** cords of **epithelium** and **primordial** germ cells that give rise to the **seminiferous tubules** or **ovarian follicles.**

Sexual homology. Parts in different sexes of the same **species** that develop from the same type of **primordium.**

Shear. A **stress** that results from two parallel but not directly opposite **forces** that are moving toward each other.

Sinus [sī′nŭs] (L., = a cavity). A cavity or space within an organ.

Sinusoids [sī′nə-soidz] (Gr., *eidos* = form). **Capillary**-sized **blood** spaces in the **liver** or other organs that are not completely lined by **endothelial** cells.

Sinus venosus [vē-nō′səs]. The most **caudal** chamber of the **heart** of **anamniotes** and some **reptiles;** receives the systemic **veins.**

Sister group. The closest **monophyletic group** outside the **ingroup.** See **ingroup, outgroup, monophyletic group.**

Skin. See **integument.**

Skull [skŭl] (Old English, *skulle* = bowl). The group of **bones** and **cartilages** that encase the **brain** and major **sense organs** and form the jaws; the lower jaw sometimes is not considered to be a part of the **skull.**

Soaring. A type of flight in which the wings are held stationary and the animal remains aloft by utilizing upward air currents (static soaring) or differential air speeds at different elevations (dynamic soaring).

Soft palate. A fleshy **palate** in **mammals** that separates the **nasal** and **oral pharynx;** part of the **secondary palate.**

Somatic [sō-măt′ĭk] (Gr., *somatikos* = bodily). Refers to structures that develop in the body wall or appendages as opposed to those in the gut tube, such as the somatic **muscles,** somatic skeleton.

Somite [sō′mīt]. One of the series of dorsal segments, or divisions of the **paraxial mesoderm,** in the trunk and **caudal** part of the head in a developing **embryo;** also called an epimere.

Somitomere [sō′mə-tō-mîr] (Gr., *meros* = part). One of the partial divisions of the **paraxial mesoderm** in the rostral part of the head of a developing **embryo.**

Specialization. Presumed **adaptations** to a particular habitat and mode of life.

Speciation. The process leading to the origin of new **species** through time. See **anagenesis, cladogenesis, reticulate speciation.**

Species [spē'shēz] (L., = particular kind). Several different definitions of the term "species" exist, some of which conflict. In general, most of these definitions specify reproductive coherence due to genetic and behavioral compatibility of the sexes (in the case of sexually reproducing organisms), uniqueness of **evolutionary** role (due to genetic isolation from other species), an origin (time of speciation), and an end (extinction or **cladogenesis,** either past or predicted for the future). Some systematists believe the species are neither more nor less real than higher **taxa** and that they should be defined as the smallest discernible **monophyletic group.** See **cladogenesis, monophyletic group, speciation.**

Sperm (Gr., *sperma* = seed). The mature male **gametes,** also called spermatozoa.

Spermatogenesis [spûr-măt'ō-jĕn'ĭ-sĭs] (Gr., *genesis* = birth, descent). The formation and maturation of the **sperm.**

Spermatophore [spûr-măt'ō-fôr] (Gr., *phoros* = bearing). A clump of **sperm** encapsulated in mucoid material; deposited by some male salamanders.

Sphincter [sfĭngk'tər] (Gr., *sphinkter* = band, lace). A circular **muscle** that closes the opening of an organ or surrounds another structure, e.g., the pyloric sphincter, sphincter colli muscle.

Spinal column [spī'nəl] (L., *spina* = thorn, backbone). The **vertebral** column.

Spinal cord. The **central nervous system caudal** to the **brain.**

Spiracle [spĭr'ə-kəl] (L., *spiraculum* = air hole). The reduced first **gill** pouch of some **fishes** through which water may enter the **pharynx;** also, the opening from the **gill** chamber of frog tadpoles.

Spiral valve (L., *spira* = coil). A helical coil in the **intestine** of early **fishes;** also a fold within the **conus arteriosus** and **ventral aorta** of lungfishes and some **amphibians** and **reptiles** that helps separate pulmonary and systemic bloodstreams.

Splanchnic [splăngk'nĭk] (Gr., *splanchnon* = gut, viscus). Descriptive of structures that supply the gut, such as the splanchnic **nerves.**

Splanchnocranium [splăngk'nō-krā'nē-əm] (Gr., *kranion* = skull). The portion of the cranial skeleton composed of the **visceral arches.**

Spleen [splēn] (Gr., *splen* = spleen). A vascular organ near the **stomach** in which **blood** cells may be produced, stored, and eliminated.

Squamates [skwā'mātz] (L., *squama* = scale). The **reptilian** division that includes the lizards, amphisbaenians, and snakes.

Stall. Sudden loss of lift by the wings.

Stapes [stā'pēz] (L., = stirrup). The most medial of the three **auditory ossicles** of **mammals; homologous** to the **hyomandibula** of **fishes** and **columella** of nonmammalian **tetrapods.**

Step. The distance a **tetrapod** moves forward by the action of one leg and foot.

Sternum [stûr'nəm] (Gr., *sternon* = chest). The breastbone of **tetrapods.**

Stomach [stûm'ək] (Gr., *stomacos* = stomach). The part of the digestive tract where food is stored temporarily and where digestion usually is initiated.

Stomodaeum [stō'mə-dē'əm] (Gr., *stoma* = mouth + *hodaion* = on the way). An **ectodermal invagination** at the front of the **embryo** that forms the **oral cavity.**

Strain. The deformation in a material that results from **stress.**

Stratum [străt'əm] (L., = layer). A layer of **tissue,** such as the stratum corneum on the **skin** surface.

Stress. The **force** per unit area that is applied to a material.

Striatum [strī-ā'tŭm] (L., *striatus* = striped). A group of **nuclei** in the base of the **cerebrum** through which white fibers pass; part of the **subpallium.**

Stride. The distance a **tetrapod** moves forward from the placement of one foot on the ground to the next placement of the same foot; equivalent to four **steps** in a quadruped.

Subpallium [sŭb-păl'ē-əm] (L., *sub* = under + *pallium* = cloak). **Gray matter** of the **cerebrum** lying ventral to the **pallium;** includes the **striatum** and **septum.**

Sulcus [sŭl'kəs] (L., = groove). A groove on the surface of an organ, such as the sulci on the **cerebrum** of a **mammal.**

Sulcus limitans [lĭm'ĭ-təns]. A groove in the central canal of the nervous system that delineates the dorsal sensory areas of **gray matter** from the ventral motor ones.

Summation [sŭm-ā'shən]. The addition of successive events that come in rapid sequence to produce a response, or a response of greater magnitude.

Suprarenal gland. See **adrenal gland.**

Suprasegmental control. A level of integration by parts of the **brain** that is superimposed on the basic pattern of activity of lower centers.

Surfactant [sûr-făk'tənt] (L., *superficius* = superficial + *actio,* pp. *actus* = to do). A surface tension depressant found on the lining of the **lungs.**

Suture [sū'chər] (L., = seam). An immovable joint (and type of **synarthrosis**) in which the **bones** are separated by a septum of **connective tissue,** such as those between **dermal bones** of the **skull.**

Sweat glands. Mammalian cutaneous glands that secrete a watery solution (eccrine sweat glands) or odoriferous materials (apocrine sweat glands).

Swim bladder. A sac of gas, located dorsally in the body cavity of most **actinopterygians,** that has a hydrostatic function.

Sympathetic nervous system [sĭm'pə-thĕt'ĭk] (Gr., *sym* = with + *pathos* = feeling). The part of the **autonomic nervous system** that, in **mammals,** leaves the **central nervous system** from parts of the **spinal cord;** its activity helps an animal adjust to stress by promoting physiological processes that increase the energy available to **tissues.**

Symphysis [sĭm'fĭ-sĭs] (Gr., *physis* = growth). A joint (and type of **synarthrosis**) between bones that permits limited movement by the deformation of the **fibrocartilage** between them, as the **pelvic** symphysis; usually occurs in the midline of the body.

Synapomorphy. A shared derived **character** or **character** state at its most **taxonomically** inclusive level. A **character diagnosing** a **monophyletic group** (e.g., the presence of jaws for Gnathostomata). See **apomorphy, character.** Equivalent to a **phylogenetic homology.** See **homology.**

Synapse [sĭn'ăps] (Gr., *synapsis* = union). The junction at which an impulse passes from one **neuron** to another.

Synapsid [sĭ-năp'sĭd] (Gr., *apsid* = loop or bar). A **skull** with a single laterally placed temporal **fenestra,** or a group of **vertebrates** with such a **skull,** such as the **Synapsida.**

Synapsida. A group containing **Mammalia** plus a number of closely related extinct lineages or **taxa.**

Synarthrosis [sĭn'är-thrō'sĭs] (Gr., *arthron* = joint). A joint with fibrous or **cartilaginous** material between the adjacent elements; growth can occur here but no or only limited movement.

Synchondrosis [sĭn'kŏn-drō'sĭs] (Gr., *chondros* = cartilage). A joint (or type of **synarthrosis**) in which **cartilage** separates two **bony** elements, found between **bones** that ossify in the **chondrocranium;** growth can occur but only limited movement.

Synergy [sĭn'ər-jē] (Gr., *ergon* = work). Pertaining to different **muscles** or other organs that interact to produce a common effect.

Synovial fluid [sĭ-nō'vē-əl] (L., *synovia* = joint oil). A clear fluid that serves as a lubricant in movable joints.

Synovial joint. See **diarthrosis.**

Synsacrum [sĭn-sā'krəm] (L., *sacrum* = sacred). The group of fused **vertebrae** and their ribs in birds that **articulates** with the **pelvis.**

Syrinx [sîr'ĭngks] (Gr., = panpipe). The voice box of birds, located at the distal end of the **trachea.**

Systemic circulation [sĭ-stĕm'ĭk] (Gr., *systema* = a whole composed of several parts). The circulation through the body as a whole, exclusive of the circulation through the respiratory organs (**branchial** or pulmonary circulation) or **heart** (coronary circulation).

Systole [sĭs'tə-lē] (Gr., = a drawing together). The period during which the **ventricle** of the **heart** contracts and expels blood.

T

Tapetum lucidum [tə-pē'təm lū'sĭd-əm] (L., *tapete* = carpet + *lucidus* = clear, shining). A layer within or behind the **retina** of some **vertebrates** that reflects light back onto the photoreceptive cells.

Tarsal [tär'səl] (Gr., *tarsos* = sole of the foot). One of the **bones** in the ankle.

Taxon [tăk'sŏn], pl. **taxa** (Gr., *taxis* = arrangement). A group of organisms given a proper name for the sake of classification. A taxon can be (and should be, in the authors' opinion) **monophyletic** (e.g., the **genus** *Clupea,* which includes the true herrings, or the family Acipenseridae, which includes the sturgeons). Occasionally taxon is also used to refer to **nonmonophyletic groups** such as the traditional **"Reptilia"** (a group that excludes some of its putative descendants, i.e., birds). Taxa are defined through **characters** that are discovered through empirical investigation. See **character, Linnean classification system, monophyletic group.**

Taxonomic group. See **taxon.** (Taxonomic group = a taxon.)

Tectum [tĕk'təm] (L., = roof). A roof, specifically the roof of the **mesencephalon.**

Tegmentum [tĕg-mĕn'təm] (L., *tegmen* = covering). The floor of the **mesencephalon** or **metencephalon.**

Telachoroidea [tē'lə-kə-roi'dē-ə] (L., *tela* = web + Gr., *chorion* = membrane enclosing the fetus + *eidos* = form). A thin membrane composed of the **ependymal epithelium** and the vascular **meninx** that forms the roof or wall of some **ventricles.**

Telencephalon [tĕl'ĕn-sĕf'ə-lŏn] (Gr., *telos* = end + *enkephalos* = brain). The rostral part of the **forebrain** from which the **olfactory bulbs** and **cerebral hemispheres** develop.

Teleosts [tē'lē-ôsts] (Gr., *osteon* = bone). An extremely speciose group of **vertebrates** including all living **neopterygian fishes** other than gars and bowfins.

Telodendria [têl'ō-dĕn'drē-ə] (Gr., *dendria* = trees). The terminal branches of an **axon.** See **terminal arborization.**

Tendon [tĕn'dən] (L., *tendo* = to stretch). A cord of dense **connective tissue** that extends between a **muscle** and its attachment.

Tendon organ. A **proprioceptor** in **tendons** that is stimulated by **tension** developed by **muscle** contraction.

Tension. The **stress** that results from two parallel **forces** pulling directly away from each other.

Tentorium [tĕn-tôr'ē-əm] (L., = tent). The septum of **dura mater,** ossified in some **species,** that extends between the **cerebrum** and **cerebellum.**

Terminal arborization. The terminal branching of a **neuron.**

Terminal nerve. A small **nerve** present beside the **olfactory nerve** in most **vertebrates;** its function is unclear, but it may have a role in detecting **pheromones** and regulating reproductive functions.

Testis [tĕs'tĭs] (L., = witness, originally an adult male, testis). The male reproductive organ, which produces **sperm** and male sex **hormones.**

Testosterone [tĕs-tŏs'tə-rōn]. The male sex **hormone** produced by the **testis;** promotes the development of male secondary sex characteristics and of **sperm.**

Tetrapods [tĕt'rə-pŏdz] (Gr., *tetra* = four + *pous, podos* = foot). A common name for terrestrial **vertebrates;** they have four feet unless some have been secondarily lost or converted to other uses.

Thalamus [thăl'ə-məs] (Gr., *thalamos* = chamber, bedroom). The lateral walls of the **diencephalon;** an important center between the **cerebrum** and other parts of the **brain.**

Thecodont teeth [thē'kō-dŏnt] (Gr., *theke* = case + *odous, odont-* = tooth). Teeth that are set in sockets.

Therapsid [thə-răp'sĭd] (Gr., *therion* = wild beast + *apsis* = arch). A group of **synapsids** very closely related to **mammals.**

Therians [thîr'ē-ənz]. The group of **mammals** that includes the **marsupials** and **eutherians.**

Thoracic duct [thə-răs'ĭk] (Gr., *thorax* = chest). The large lymphatic **duct** of **mammals** that passes through the **thorax** and enters the large **veins** near the **heart.**

Thorax [thôr'ăks]. The region of the **mammalian** body encased by the ribs.

Thymus [thī'məs] (Gr., *thymos* = thyme, thymus; so called because it resembles a bunch of thyme). The lymphoid organ that develops from certain pharyngeal pouches, necessary as the site where certain T lymphocytes mature.

Thyroid gland [thī'roid] (Gr., *thyroides* = resembling an oblong shield). An **endocrine gland** that develops from the floor of the **pharynx** and in humans is located adjacent to the thyroid **cartilage** of the **larynx;** its **hormones** increase the rate of metabolism.

Thyroid-stimulating hormone. A **hormone** produced by the **adenohypophysis** that promotes the secretion of the **thyroid gland.**

Thyroxine [thī-rŏk'sēn] (Gr., *oxo-* = oxygen). One of the **hormones** released by the **thyroid gland.**

Tibia [tĭb'ē-ə] (L., = the large shinbone). The **bone** on the medial side of the lower leg, in line with the first **digit.**

Tissue [tĭsh'ū] (Old French, *tissu* = woven). An aggregation of cells that together perform a similar function.

Tongue [tŭng] (Old English, *tunge*). A **muscular** mobile organ on the floor of the **oral cavity** of **tetrapods** that often helps gather food and manipulates it within the mouth.

Tonsil [tŏn'səl] (L., *tonsilla* = tonsil). One of the lymphoid organs that develops in the wall of the **pharynx** near the level of the second pharyngeal pouch.

Torque. A turning **force** equal to the product of the **force** and the perpendicular distance between the line of action of the **force** and the **fulcrum** about which it acts; also called a **moment.**

Trabeculae [trə-bĕk'yū-lē] (L., = little beams). Small, rodlike skeletal structures, such as the trabeculae within **bones.**

Trachea [trā'kē-ə] (L., *tracheia arteria* = rough artery, windpipe). The respiratory tube between the **larynx** and the **bronchi.**

Tract [trăkt] (L., *tractus* = a drawing out). A group of **axons** of similar function traveling together in the **central nervous system.**

Transverse septum. The partition of **epithelium** that separates the pericardial from the **pleuroperitoneal cavity.**

Trigeminal nerve [trī-jĕm'ə-nəl] (L., *trigeminus* = threefold). The fifth cranial nerve, which has three branches in **mammals;** it innervates the **muscles** of the **mandibular arch** and returns sensory fibers from **cutaneous receptors** over most of the head.

Triiodothyronine [trī-ī'ō-dō-thī'rə-nēn] (Gr., *tri* = three + *iodo* = violet-like or iodine + *thyroides* = resembling an oblong shield). A **hormone** produced by the **thyroid gland.**

Trochanter [trō-kăn'tər] (Gr., = a runner). One of the processes on the proximal end of the **femur** to which thigh **muscles** attach.

Trochlear nerve [trŏk'lē-ər] (L., *trochlea* = pulley). The fourth cranial **nerve,** which innervates the superior oblique **muscle;** the **mammalian muscle** passes through a **connective tissue** pulley before inserting on the eyeball.

Trophoblast [trō'fō-blăst] (Gr., *trophe* = nourishment + *blastos* = bud). The outer layer of the **mammalian blastocyst;** initiates **placenta** formation; **homologous** to the **chorionic ectoderm.**

Tunic [tū'nĭk] (L., *tunica* = coating). Describes a layer of an organ, such as the layers of the eyeball.

Tunicates [tū'nĭ-kĭts]. The group of **chordates** that includes the sea squirts and their allies; also called urochordates.

Turbinate bones [tûr'bə-nāt] (L., *turbinatus* = top-shaped, whirlwind). Scroll-shaped **bones** in the **nasal** cavities of **mammals** that increase the surface area of the cavities; also called **conchae.**

Turbulent flow. A disrupted flow of fluid along the surface of a swimming or flying **vertebrate.**

Tympanic cavity (L., *tympanum* = drum). The **middle ear** cavity, which lies between the **tympanic membrane** and the **otic capsule** containing the **inner ear.**

Tympanic membrane. The eardrum.

U

Ulna [ŭl'nə] (L., = elbow). The bone of the **antebrachium** of **tetrapods** that extends behind the elbow, lying on the side adjacent to the fifth finger when the hand is supine.

Ultimobranchial bodies [ŭl'tə-mō-brăng'kē-əl] (L., *ultimus* = farthest + Gr. *branchia* = gills). Derivatives of the **caudal** surface of the last **branchial** pouch; in **fishes,** they contain the C cells, the **hormone** of which, calcitonin, helps regulate mineral metabolism.

Ungulates [ŭng'gyə-lĭts] (L., *ungula* = hoof). A collective term for the hoofed **mammals: artiodactyls** and **perissodactyls.**

Unguligrade [ŭng'gyə-lĭ-grād] (L., *gradus* = step). Walking on the toe tips.

Urea [yŏŏ-rē'ə] (Gr., *ouron* = urine). A breakdown product of nitrogen metabolism; occurs in **elasmobranchs,** some **amphibians,** and **mammals.**

Ureter [yŏŏ-rē'tər] (Gr., *oureter* = ureter, from *ouron* = urine). The **duct** of **amniotes** that carries **urine** from a **metanephric kidney** to the **urinary bladder.**

Urethra [yŏŏ-rē'thrə] (Gr., *ourethra* = urethra). The **duct** in **amniotes** that carries urine from the **urinary bladder** to the **cloaca** or outside; part of it also carries **sperm** in males.

Uric acid [yŏŏr'ĭk] (Gr., *ouron* = urine). A breakdown product of nitrogen metabolism; occurs chiefly in **reptiles** and birds, requires that little water be removed from the body.

Urinary bladder [yŏŏr'ə-nĕr-ē]. A saccular organ in which urine accumulates before discharge from the body.

Urodeles [yŏŏr'ō-dēlz] (Gr., *oura* = tail + *delos* = visible). The **amphibian** subgroup that includes the salamanders; also called **Caudata.**

Urogenital [yŏŏr'ō-jĕn'ĭ-tl] (Gr., *ouron* = urine + L., *genitalis* = genital). Pertains to structures that are common to the urinary and genital systems, such as certain urogenital **ducts.**

Urophysis [yŏŏr'ō-fī'sĭs] (Gr., *oura* = tail + *physis* = growth). A neurosecretory organ on the **caudal** end of the **spinal cord** in **elasmobranchs** and **teleosts.**

Uropygeal gland [yŏŏr'ō-pī'jē-əl] (Gr., *pyge* = rump). An oil-secreting **gland** of birds, located dorsal to the tail base.

Urostyle [yŏŏr'ō-stīl] (Gr., *stylos* = pillar). An elongated **bone** of **anurans** that represents fused **caudal vertebrae.**

Uterine tube [yū'tər-ĭn] (L., *uterus* = womb). The portion of the **mammalian oviduct** that carries eggs from the **coelom** to the **uterus;** also called the **fallopian tube;** site of fertilization.

Uterus [yū'tər-əs]. The portion of an **oviduct** in which **embryos** develop in live-bearing **species.**

Utriculus [yū-trĭk'yə-ləs] (L., = small sac). The upper chamber of the **membranous labyrinth** to which the **semicircular ducts** attach.

V

Vagina [və-jī'nə] (L., = sheath). The passage in female **therians** that leads from the **uterus** to the **vaginal vestibule.**

Vaginal vestibule [vĕs'tə-byŏŏl] (L., *vestibulum* = antechamber). The passage or space in female **therians** that receives the **vagina** and **urethra;** also called the **urogenital sinus.**

Vagus nerve [vā'gəs] (L., = wandering). The tenth cranial **nerve;** carries motor fibers to the **muscles** of the last four **visceral arches,** autonomic fibers to the **heart** and **viscera,** returns sensory fibers from these areas, and supplies the **lateral line canal** in **fishes** and **larval amphibians.**

Vas deferens [văs]. See **deferent duct.**

Vasa efferentia [vā'sə]. See **efferent ductules.**

Vascular tunic [văs'kyə-lər] (L., *vasculum* = small vessel). The middle layer of the eyeball; it forms the **choroid, ciliary body,** and **iris.**

Vasopressin. See **antidiuretic hormone.**

Vector [vĕk'tər] (L., *vector* = bearer). A quantity, such as a **force,** that has both a magnitude and a direction.

Vein [vān] (L., *vena* = vein). A vessel that conveys **blood** toward the **heart;** most veins contain blood low in oxygen content, but pulmonary veins from the **lungs** are rich in oxygen.

Velocity [və-lŏs'ĭ-tē]. Distance traveled divided by the time unit.

Vena cava [vē'nə cā'və] (L., = hollow vein). One of the major **veins** of lungfishes, **amphibians,** and **amniotes;** leads directly to the **heart.**

Ventral aorta [ā-ôr'tə]. An artery that leads from the **heart** to the **aortic arches** and their derivatives; contributes to the arch of the **aorta** and base of the **pulmonary artery** in **mammals.**

Ventricle [vĕn'trĭ-kəl] (L., *ventriculus* = small belly). The chamber of the **heart** that greatly increases **blood** pressure and sends **blood** to the **arteries;** also a chamber within the **brain.**

Vermis [vûr'mĭs] (L., = worm). The "segmented" medial portion of the **amniote cerebellum.**

Vertebra [vûr'tə-brə] (L., = joint, vertebra). One of the skeletal units that make up the spinal column.

Vertebral arch. The arch of a **vertebra** that surrounds the **spinal cord;** also called a **neural arch.**

Vertebral body. The main supporting component of a **vertebra,** lying ventral to the **vertebral arch;** also called the **centrum.**

Vertebrates [vûr'tə-brāts]. The subgroup of **craniates** that contains **species** with at least an incipient **vertebral** column; excluding hagfishes, all **craniates** are **vertebrates.**

Vesicular gland [vĕ-sĭk'ə-lər] (L., *vesicus* = small bladder). One of the accessory sex **glands** of males that contributes to the **seminal fluid.**

Vestibular apparatus [vĕ-stĭb'yə-lər] (L., *vestibulum* = entrance). The portion of the **inner ear** that detects changes in position and **acceleration.**

Vestibulocochlear nerve [vĕs-tĭb'yə-lō-kŏk'lē-ər] (L., *cochlea* = snail shell). The eighth cranial **nerve,** which returns fibers from the parts of the **inner ear** related to equilibrium and sound detection; often called the statoacoustic **nerve** in **anamniotes.**

Vestige [vĕs'tĭj] (L., *vestigium* = trace). A remnant in one organism of a structure that is well developed in another organism and has no function or a different function from that of its well-developed **homologue.**

Vestigial. See **vestige.**

Villi [vĭl'ī] (L., = shaggy hairs). Multicellular but minute, often finger-shaped projections of an organ that increase its surface area, such as the **intestinal** villi.

Viscera [vĭs'ər-ə] (L., = internal organs). A collective term for the internal organs.

Visceral arches. The skeletal arches that develop in the wall of the **pharynx;** include the **mandibular, hyoid,** and **brancial arches.**

Vitelline [vĭ-tĕl'ĭn] (L., *vitellus* = yolk). Pertains to structures associated with the **embryonic yolk sac,** such as the vitelline **arteries** and **veins.**

Vitreous body [vĭ'trē-əs] (L., *vitreus* = glassy). The clear, viscous material in the eyeball between the **lens** and **retina.**

Viviparity [vĭv'ə-pĕr'ĭ-tē] (L., *vivus* = living + *pario* = to bring forth). A pattern of reproduction in which the **embryos** are born as miniature adults. The term is often limited to **placental** viviparity, in which the **embryos** are completely dependent on materials transferred from the mother.

Vocal cords [vō'kəl] (L., *vocalis* = pertaining to the voice). Folds of mucous membrane within the **larynx** of many **tetrapods** the vibrations of which produce sound.

Vomeronasal organ [vō'mər-ō-nā'zəl]. An accessory **olfactory** organ located between the palate and **nasal** cavities of most **tetrapods,** important in feeding and sexual behavior. See also **Jacobson's organ.**

Vulva [vŭl'və] (L., = covering). The external genitalia of a female.

W

Weberian ossicles [ŏs'ĭ-kəlz] (*E. H. Weber,* German anatomist, 1795–1878 + L., *ossiculum* = small bone). A set of small **bones** that transmit sound waves from the **swim bladder** to the **inner ear** in some **teleosts.**

Weight. The mass of a structure times the acceleration of gravity.

White matter. Tissue in the **central nervous system** that consists primarily of myelinated **axons.**

Wing loading. The weight of a bird divided by the area of its wings.

Wolffian duct. (*K. F. Wolff,* 18th-century German embryologist). A term often applied to the **embryonic archinephric duct.**

Work. The product of a **force** and the distance through which it acts.

Y

Yaw. The tendency for the head of swimming or flying **vertebrates** to move from left to right about its longitudinal **axis.**

Yolk sac [yōk] (Anglo-Saxon, *geula* = yellow). The yolk-containing sac attached to the ventral surface of **embryos** that develops from **macrolecithal** eggs.

Z

Zygapophysis [zī'gə-pŏf'ĭ-sĭs] (Gr., *zygon* = yoke + *apo* = away from + *physis* = growth). A process of a **vertebral arch** that articulates with a comparable process on an adjacent arch; also called **articular process.**

Zygomatic arch [zī'gō-măt'ĭk] (Gr., *zygoma* = bar, yoke). The arch of **bone** in a **mammalian skull** that lies beneath the **orbit** and connects the **facial** and cranial regions of the **skull.**

Zygote [zī'gōt] (Gr., *zygotes* = yoked together). The cell formed by the union of a **sperm** and an egg.

Index

(Pages followed by "f" indicate figures and by "t" indicate tables. Dagger symbols "†" denote extinct taxa.)

C

G

H

P

Q

R

S